W9-BKT-684

Principles of Medicinal Chemistry

Principles of Medicinal Chemistry
Fourth Edition

William O. Foye, Ph.D., D.Sc. (hon.)
Sawyer Professor of Pharmaceutical Sciences Emeritus
Massachusetts College of Pharmacy and Allied Health Sciences
Boston, Massachusetts

Thomas L. Lemke, Ph.D.
Professor of Medicinal Chemistry
College of Pharmacy
University of Houston
Houston, Texas

David A. Williams, Ph.D.
Professor of Chemistry
Massachusetts College of Pharmacy and Allied Health Sciences
Boston, Massachusetts

A Lea & Febiger Book

Williams & Wilkins

BALTIMORE • PHILADELPHIA • HONG KONG
LONDON • MUNICH • SYDNEY • TOKYO

A WAVERLY COMPANY

1995

Executive Editor: Donna Balado
Developmental Editor: Lisa Stead
Production Coordinator: Marette D. Magargle
Project Editor: Rebecca Krumm

Cover art created by Dr. James Henkel

Williams & Wilkins
Rose Tree Corporate Center
1400 North Providence Road
Building II, Suite 5025
Media, PA 19063-2043 USA

Accurate indications, adverse reactions, and dosage schedules for drugs are provided in this book, but it is possible they may change. The reader is urged to review the package information data of the manufacturers of the medications mentioned.

Printed in the United States of America

First Edition 1974

Library of Congress Cataloging in Publication Data

Principles of medicinal chemistry / [edited by] William O. Foye, Thomas L. Lemke, David A. Williams.—4th ed.
p. cm.
Includes bibliographical references and index.
ISBN 0-683-03323-9
1. Pharmaceutical chemistry. I. Foye, William O. II. Lemke, Thomas L. III. Williams, David A., 1938– .
[DNLM: 1. Chemistry, Pharmaceutical. QV 744 P957 1995]
RS403.P75 1995
615′.7—dc20
DNLM/DLC
for Library of Congress 94-29481
CIP

95 96 97 98
1 2 3 4 5 6 7 8 9 10

The authors and publishers have made every effort to trace the copyright holders for borrowed material. If they have inadvertently overlooked any, they will be pleased to make the necessary arrangements at the first opportunity.

PREFACE

At the time of the first edition of this book, the role of the underlying biochemical interactions in the activity of drugs was beginning to be understood. The concept of receptor sites for drugs on proteins, enzymes, and other macromolecules was developing rapidly, but the specific chemical groups and functions responsible for interaction at a receptor site were still largely undetermined. As knowledge about both chemistry and biology advanced, and the role of the computer in presenting drug-receptor interactions was developed, a greater degree of complexity unfolded in many cases than simple models had predicted. A variety of disciplines has attempted to describe and make clear these interactions, including pharmacology, molecular biology, biochemistry, and medicinal chemistry. The mechanisms of many drug substances can now be explained at the chemical level, but aspects from all of these disciplines are often interwoven to present a relatively complete picture of the various interactions and transformations involved in the course of events leading to drug activity.

The task of presenting a coherent and clear account of the variety of drugs and the means by which they exert their physiologic effects, along with relatively complete coverage of the various categories of drugs now in use, is daunting for authors of a book of one volume. The attempt to do this, which was formulated in the first edition, involves the discussion of activity of a few major drugs in each category, along with a listing in tables of other drugs of interest. This approach is still followed for many drug categories. Some subjects, however, which include a fairly large and diverse group of chemical structures, deviate from this organization and require a larger discussion of activities of leading substances. A uniform approach to explaining drug activities is thus defeated by the growing number of diverse structures found for some drug classes.

One result of this complex picture of drug interactions and activities, necessary for the researcher and drug industry, has been an equally increased difficulty for the student to comprehend these advances. Whereas the first editions of this text generally evoked student comments that this book "could be read," we now hear complaints that the book "is difficult to understand." There is probably no simple way to present a complex subject, without sacrifice of essential details, but the fact that students are still able to use the book, as well as others in the field, and pass their courses in medicinal chemistry, however presented, attests to the skill and ability of the authors of the various chapters.

This edition follows the same classification of drugs as used in the previous editions. New chapters address areas in which recent advances have been substantial or are likely to be made, including molecular modeling, pharmaceutical biotechnology, biopharmaceutical properties of drug substances, and approaches to anti-AIDS agents. Other chapters have been substantially broadened, and with the inclusion of a significant number of new authors, many chapters have been reorganized.

With the growing knowledge of the complexity of drug action as a result of astonishing advances in biology, biochemistry, and medicinal chemistry, the task of presenting the material in a manner understood by the beginning student becomes equally more difficult. With a broad and voluminous literature to follow, often from several different fields, the author requires a considerable range of knowledge. Writing a clear and readily grasped account of drug activities also requires some ability, and the authors who have made their complex subjects clear deserve our admiration and our thanks. The necessary contraction of a subject in the selection of topics for a textbook chapter is always a difficult task, and bound to be unappreciated by some readers. It also serves as a means of consolidating a science that large, comprehensive treatments are often unable to accomplish. We hope that such treatment of the subject will continue to be of value to those in the related areas of biology, chemistry, and pharmacology.

For the fourth edition of this text, I acknowledge the assistance of two new editors, Thomas L. Lemke and David A. Williams. With their fresh viewpoints and extensive teaching experience, the book should continue to serve an audience, now world wide, that has found it useful. Grateful acknowledgment is also made to all who have assisted in various ways with the completion of the book: colleagues who have read manuscripts, those using the book who have made suggestions regarding content, students who have found errors, and professors in various locations who have pointed to the importance of specific topics in their countries, and always the engaging assistance of the secretaries who put the chapters in final form.

Boston, Massachusetts W. O. Foye

CONTRIBUTORS

Nitya Anand, Ph.D.
Senior Scientist
Central Drug Research Institute
Chattar Manzil
Lucknow, India

Alireza Banijamali, Ph.D.
Research Chemist
Uniroyal Chemical Company
Naugatuck, Connecticut

Eric M. Billings, Ph.D.
School of Pharmacy
University of Connecticut
Storrs, Connecticut

Raymond G. Booth, Ph.D.
Assistant Professor of Medicinal Chemistry
School of Pharmacy
University of North Carolina
Chapel Hill, North Carolina

Ronald F. Borne, Ph.D.
Professor of Medicinal Chemistry
School of Pharmacy
University of Mississippi
University, Mississippi

Daniela Braghiroli, Ph.D.
Department of Pharmaceutical Science
University of Modena
Modena, Italy

Robert W. Brueggemeier, Ph.D.
Professor of Medicinal Chemistry
College of Pharmacy
The Ohio State University
Columbus, Ohio

Raymond E. Counsell, Ph.D.
Professor of Pharmacology and Medicinal Chemistry
Department of Pharmacology
The University of Michigan Medical School
Ann Arbor, Michigan

Robert W. Curley, Jr., Ph.D.
Associate Professor of Medicinal Chemistry
College of Pharmacy
The Ohio State University
Columbus, Ohio

Francesco Dall'Acqua, Ph.D.
Chairman, Department of Pharmaceutical Sciences
University of Padua
Padua, Italy

Maria Di Bella, Ph.D.
Department of Pharmaceutical Science
University of Modena
Modena, Italy

Raymond W. Doskotch, Ph.D.
Professor Emeritus of Medicinal Chemistry and Pharmacognosy
College of Pharmacy
The Ohio State University
Columbus, Ohio

Dennis R. Feller, Ph.D.
Professor of Pharmacology
College of Pharmacy
The Ohio State University
Columbus, Ohio

William O. Foye, Ph.D., D.Sc. (hon.)
Sawyer Professor of Pharmaceutical Sciences Emeritus
Department of Chemistry
Massachusetts College of Pharmacy and Allied Health Sciences
Boston, Massachusetts

David S. Fries, Ph.D.
Professor of Medicinal Chemistry
School of Pharmacy
University of the Pacific
Stockton, California

Gerald Gianutsos, Ph.D.
Associate Professor of Pharmacology
School of Pharmacy
University of Connecticut
Storrs, Connecticut

Robert K. Griffith, Ph.D.
Professor of Medicinal Chemistry
School of Pharmacy
West Virginia University Health Sciences Center
Morgantown, West Virginia

Larry M. Hagerman, Ph.D.
Senior Research Scientist
Adria Laboratories
Columbus, Ohio

Richard H. Hammer, Ph.D.
Professor of Medicinal Chemistry
College of Pharmacy
J. Hillis Miller Health Center
University of Florida
Gainesville, Florida

Edmund J. Hengesh, Ph.D.
Professor of Pharmaceutical Chemistry
School of Pharmacy
Ferris State College
Big Rapids, Michigan

James G. Henkel, Ph.D.
Associate Professor of Medicinal Chemistry
School of Pharmacy
University of Connecticut
Storrs, Connecticut

Ronald L. Jacoby, Ph.D.
Professor of Pharmaceutical Chemistry
School of Pharmacy
Ferris State College
Big Rapids, Michigan

Sunil S. Jambhekar, Ph.D.
Associate Professor of Industrial Pharmacy
Massachusetts College of Pharmacy and Allied Health Sciences
Boston, Massachusetts

Elizabeth A. Johnson, Ph.D.
Assistant Professor of Behavioral Medicine and Neurology
School of Medicine
West Virginia University
Health Sciences Center
Morgantown, West Virginia

V. Craig Jordan, Ph.D., D.Sc.
Professor of Cancer Pharmacology
Robert H. Lurie Cancer Center
Northwestern University Medical School
Chicago, Illinois

Giulio Jori, Ph.D.
Professor of Molecular Biology
University of Padua
Padua, Italy

Lemont B. Kier, Ph.D.
Professor of Pharmaceutical Chemistry
School of Pharmacy
Medical College of Virginia Campus
Richmond, Virginia

Roy L. Kisliuk, Ph.D.
Professor of Biochemistry
Department of Biochemistry and Pharmacology
Tufts University School of Medicine
Boston, Massachusetts

Danny L. Lattin, Ph.D.
Professor of Medicinal Chemistry
College of Pharmacy
University of Arkansas for Medical Sciences
Little Rock, Arkansas

Thomas L. Lemke, Ph.D.
Professor of Medicinal Chemistry
College of Pharmacy
University of Houston
Houston, Texas

Matthias C. Lu, Ph.D.
Associate Professor of Medicinal Chemistry
College of Pharmacy
University of Illinois at Chicago
Chicago, Illinois

Timothy J. Maher, Ph.D.
Professor of Pharmacology
Massachusetts College of Pharmacy and Allied Health Sciences
Boston, Massachusetts

Ian W. Mathison, D.Sc., Ph.D.
Professor of Pharmaceutical Chemistry and Dean
School of Pharmacy
Ferris State College
Big Rapids, Michigan

Ahmed S. Mehanna, Ph.D.
Assistant Professor of Chemistry
Massachusetts College of Pharmacy and Allied Health Sciences
Boston, Massachusetts

Duane D. Miller, Ph.D.
Professor of Medicinal Chemistry
College of Pharmacy
University of Tennessee
Memphis, Tennessee

Lester A. Mitscher, Ph.D.
Professor of Medicinal Chemistry
School of Pharmacy
University of Kansas
Lawrence, Kansas

Prem Mohan, Ph.D.
Assistant Professor of Medicinal Chemistry
College of Pharmacy
University of Illinois at Chicago
Chicago, Illinois

John L. Neumeyer, Ph.D.
President, Research Biochemicals, Inc.
Natick, Massachusetts

Howard A. I. Newman, Ph.D.
Professor of Pathology
Ohio State University Medical Center
Columbus, Ohio

Karl A. Nieforth, Ph.D.
Professor of Medicinal Chemistry
School of Pharmacy
University of Connecticut
Storrs, Connecticut

Gary O. Rankin, Ph.D.
Professor of Pharmacology
School of Medicine
Marshall University
Huntington, West Virginia

Sisir K. Sengupta, Ph.D.
Research Professor in Biochemistry
Boston University School of Medicine
Boston, Massachusetts

Manohar L. Sethi, Ph.D.
Associate Professor of Medicinal Chemistry
The College of Pharmacy and Pharmacal Sciences
Howard University
Washington, D.C.

Robert D. Sindelar, Ph.D.
Associate Professor of Medicinal Chemistry
School of Pharmacy
University of Mississippi
University, Mississippi

William E. Solomons, Ph.D.
Professor of Chemistry
University of Tennessee at Martin
Martin, Tennessee

Richard R. Tidwell, Ph.D.
Associate Professor of Pathology
School of Medicine
University of North Carolina
Chapel Hill, North Carolina

Michael S. Tute, Ph.D.
Canterbury, Kent
England, U.K.

Julius A. Vida, Ph.D.
International Pharmaceutical Consultants
Greenwich, Connecticut

Jamey P. Weichert, Ph.D.
Research Investigator
Department of Radiology
The University of Michigan Medical School
Ann Arbor, Michigan

Eugene D. Weinberg, Ph.D.
Professor of Microbiology
Biology Department and Medical Sciences Program
Indiana University
Bloomington, Indiana

David A. Williams, Ph.D.
Professor of Chemistry
Massachusetts College of Pharmacy and Allied Health Sciences
Boston, Massachusetts

Donald T. Witiak, Ph.D.
Professor of Medicinal Chemistry and Dean
School of Pharmacy
University of Wisconsin
Madison, Wisconsin

CONTENTS

Chapter 1

INTRODUCTION AND HISTORY

William O. Foye

Medicinal chemistry, according to Burger, "tries to be based on the ever-increasing hope that biochemical rationales for drug discovery may be found."[1] In contrast, he described pharmaceutical chemistry as being concerned primarily with modification of structures having known physiologic or pharmacologic effects and with analysis of drugs. Medicinal chemistry as practiced encompasses both definitions, but finding the biochemical pathways through which drugs exert their beneficial effects has become a dominating activity of the medicinal chemist. This activity has branched into two main directions: one essentially biologic and the other essentially physicochemical. The biologic direction has added the roles of enzymologist and molecular biologist to the group of research scientists working under a medicinal chemical designation. The physicochemical direction has required that a quantum mechanician, spectroscopist, and biopharmacist be included. Attempts to correlate or reconcile the results of biochemical measurements with physicochemical calculations also occupy the attention of medicinal chemists.

A logical approach to the study of drugs and their activities is the recognition of the basic principles behind the biochemical events leading to drug actions. Biochemical pathways of action for drugs are gradually being elaborated, and a rapidly increasing amount of biochemical information about drug action is now found in the literature. An amazing amount of insight into the behavior of drugs at the macromolecular level has been developed, and there is much direct and indirect evidence supporting these biochemical postulations of drug action. A review of the historical development of our knowledge of enzymes and related aspects of interest to medicinal chemists, including enzyme activities and structure, and the effect of drugs on these activities, might provide a logical introduction to this volume. The effect of drugs on enzyme systems has occupied the greatest share of attention the medicinal chemist has devoted to interaction of drugs with cellular macromolecules. A brief survey of important events is given in Table 1–1.

A chronologic survey regarding important discoveries of drugs and other biologically important molecules can be found in Burger's *Medicinal Chemistry;* no attempt, therefore, has been made here to reproduce this information, but historical surveys can be found in most chapters. The primary function of the medicinal chemist is still to discover new drugs, but a knowledge of the underlying principles of biochemical action should be of immense value for the design of new drug molecules. Molecular orbital or other calculations designed to elucidate electronic and conformational aspects of molecules are now used to attempt to predict optimal structures for selective biologic activity, based on certain physical and biochemical properties.

In the so-called prescientific period, natural products having a history as folk remedies were in use, but little of the drug therapy of today is based on these remedies. Some of the natural products currently used, either themselves or as derivatives, were often used originally for other purposes, such as arrow poisons, part of religious or other rituals, or even cosmetics. Examples of such products include opium, belladonna, cinchona bark, ergot, curare, nutmeg, calabar bean, foxglove, and squill. Many drugs originally used as folk remedies, on the other hand, have been abandoned.

Development of drug therapy could not progress until knowledge of anatomy and physiology had reached the status of science. The empiric observations of Harvey and Sydenham were of great importance to this development in the seventeenth century. The work of Magendie (1783–1855), an instructor of anatomy in Paris, probably represents the first application of the experimental method to medicine. He administered a Javanese arrow poison (nux vomica) to animals by various routes and described the resulting convulsions and asphyxia. This was probably the first experiment in drug absorption. By removing or sectioning the spinal cord, he concluded that this was the site of action of the active component. This was subsequently isolated and named strychnine. Magendie and his students studied a number of other drugs and physiologic problems, and they isolated several other alkaloids. He eventually published a formulary based on pure compounds.

Following the French Revolution, the sciences broke with their previous dependence on logic rather than observation and became more empiric, a development necessary for real advancement. The study and classification of diseases made considerable progress during the first half of the nineteenth century, and a new spirit of inquiry developed. Ineffective remedies were recognized as such and discarded. Although the German university system was well established by 1850, and definite programs of research were instituted, much of the drug discovery in the nineteenth century resulted from the investigations of either amateurs or investigators in the chemical industry, then mainly concerned with dyes. It was not until well

Table 1–1. Important Events Concerning Enzymes and Coenzymes

Year	Event
1811	Kirchhoff observed that a glutinous component of wheat can convert starch to sugar and dextrin.
1825	Schwann described pepsin.
1830	Robiquet and Boutron discovered hydrolysis of amygdalin by bitter almonds. Liebig and Wohler (1837) and Robiquet (1838) named the enzyme "emulsin."
1831	Leuchs described the diastatic action of ptyalin.
1833	Payen and Persoz separated active amylase from malt.
1837	Berzelius described fermentation as a catalytic process.
1856	Corvisart described trypsin.
1858	Pasteur noted that a mold ferments *dextrorotatory* but not *levorotatory* tartaric acid.
1862	Danielewski separated pancreatic amylase from trypsin by adsorption.
1870	Liebig developed a purely chemical theory of enzyme action.
1878	Kuhne designated *unorganized ferments,* such as pepsin and diastase, as *enzymes.*
1894	Emil Fischer began investigations leading to present ideas of enzyme specificity.
1897	Buchner discovered that a cell-free yeast extract can cause alcoholic fermentation.
1897	Bertrand observed that certain enzymes require dialyzable substances to exert catalytic activity. These substances were called *coenzymes*.
1898	Croft-Hill performed the first synthesis catalyzed by an enzyme, that of isomaltose.
1902	Henri's work on invertase led to development of enzyme kinetics.
1903	Henri proposed that an enzyme and its substrate combine to form a complex.
1909	Sörensen pointed out the dependence of enzyme activity on pH.
1910	Hudson and Michaelis gave the first theoretic explanation of enzyme activity-pH curves.
1913	Michaelis and Menten treated the concept of an enzyme-substrate complex according to ideas of chemical equilibria.
1923	Barger and Stedman found physostigmine to be an inhibitor of cholinesterase
1923	Ribonuclease, a phosphodiesterase enzyme, was discovered by Jones and Perkins.
1923	Hartridge and Roughton designed a rapid mixing device for measurement of rapid reactions and transient states.
1924	Kuhn recognized that the action of β-amylase on starch involves an inversion of configuration.
1925	Keilin used spectrophotometry to characterize hemoproteins.
1925	Briggs and Haldane showed that a steady-state treatment could be applied to enzyme kinetics.
1925–1935	Protein nature of enzymes was demonstrated in several laboratories by work on flavins.
1926	Jansen and Donath isolated thiamine.
1926	Sumner prepared crystalline urease.
1930	Northrop crystallized pepsin.
1930	Lohmann showed that transfer of phosphate from adenosine triphosphate (ATP) to a phosphate receptor, in the hexokinase reaction, requires magnesium.
1930	Catalytic amines were used as carboxylase models by Langenbeck.
1931	Uridylic acid was first isolated as a constituent of nucleic acids by Levene and Bass.
1931	Aeschlimann showed that neostigmine inhibits cholinesterase.
1931–1936	Isolation and identification of the pyridine nucleotide coenzymes by Warburg and Christian and by von Euler, Albers, and Schlenk.
1932	Waugh and King showed that ascorbic acid undergoes reversible oxidation-reduction to dehydroascorbic acid.
1932	Warburg showed that "old yellow enzyme" contains riboflavin.
1932	Acetylcholinesterase was discovered in blood by Stedman and co-workers.
1933	The hypothesis that free thiol groups are essential for the activity of some enzymes was developed by Hellerman and his associates and by Bersin and Logemann.
1937	Hellerman proposed a metal bridge complex in arginase.
1938	Crystalline pyridoxine was isolated from natural sources by Gyorgy, Kuhn, and Wendt, and by Lepkovsky.
1938	Flavin adenine dinucleotide was isolated as the coenzyme of D-amino acid oxidase by Warburg and Christian.
1938	Use of electrophoresis was made by Tiselius to purify pepsin.
1940	Keilin and Mann found that carbonic anhydrase is a zinc-containing enzyme.

Table 1–1. *(Continued)*

1940	Link found that dicumarol is a vitamin K antagonist.
1940	Mann and Keilin reported the inhibition of carbonic anhydrase by sulfanilamide.
1940	Fildes theorized that substances structurally related to essential metabolites could be chemotherapeutic by a competitive antagonism.
1941	Folic acid was isolated from natural sources by Mitchell, Snell, and Williams.
1943	Nachmansohn and Machado discovered that acetylation of choline by rabbit brain extracts does not proceed unless ATP is present as a source of energy.
1945	Lipmann showed that biologic acetylations require not only ATP but also another cofactor, which he called *coenzyme A*.
1945	Snell demonstrated coenzyme functions of pyridoxal phosphate.
1946	Diisopropyl phosphorofluoridate was found to be an inhibitor of cholinesterase by McCombie and Saunders.
1946	The structure of folic acid was determined at Lederle Laboratories.
1946	Use of an antimetabolite, methotrexate, for the treatment of leukemia was made by Farber and associates.
1948	O'Kane and Gunsalus showed that a factor later named *lipoic acid* is essential for oxidation of pyruvic acid.
1948	Enzymic incorporation of inorganic pyrophosphate into an organic molecule was noted by Kornberg on formation of ATP and nicotinamide mononucleotide from diphosphopyridine nucleotide (DPN).
1950	Michaelis and Wollman demonstrated free radical formation from α-tocopherol.
1950	Anionic and esteratic sites in acetylcholinesterase were recognized by Adams and Whittaker and by Wilson and Bergmann.
1950	An imidazole group was suggested as being at the active site of acetylcholinesterase by Wilson and Bergmann.
1951	Functions of glutathione in enzyme reactions and cell respiration were elaborated by Barron.
1951	Pyrithiamine was shown to be a thiamine antagonist by Cerecedo and coworkers.
1951	Amino acid sequence of insulin was established by Sanger and Tuppy.
1952	An azomethine chelation mechanism of action for pyridoxal was established by Metzler and Snell and others.
1954	Pullman, San Pietro, and Colowick established that the pyridine ring in the coenzyme DPN is reduced to a 1,4-dihydropyridine.
1954	Oxythiamine was shown to be a competitive inhibitor of thiamine by Naber and his researchers.
1954	The amino acid composition of crystalline carboxypeptidase was determined by Smith and Stockell.
1954	Vallee and Neurath showed that pancreatic carboxypeptidase contains 1 gram atom of zinc per mole of protein.
1954	Eigen and co-workers developed relaxation methods permitting the measurement of reaction rates with time constants as short as 10^{-10} sec.
1955	Kosower concluded that reduction of the pyridine nucleotide coenzyme DPN involves the charge transfer type of complexing (Milliken, 1952).
1955	Kennedy and Weiss first demonstrated a cytidine nucleotide as an enzymic cofactor.
1956	Pyridine-2-aldoxime methiodide was found by Wilson to reactivate alkyl phosphate-inhibited acetylcholinesterase.
1956	Sutherland detected cyclic adenosine monophosphate (3′, 5′-AMP) in animal tissues.
1957	Cunningham and Westheimer postulated the concerted action of a serine and a histidine residue in the active center of chymotrypsin.
1958	Koshland pointed out that enzyme proteins undergo conformational changes on binding of substrates to enzymes.
1958	Smith proposed that a high energy thiol ester bond is present in papain and is essential for its activity.
1958	Friden observed metal ion-induced aggregation with glutamic dehydrogenase.
1958	Kendrew and co-workers determined the structure of myoglobin to 2Å resolution by x-ray crystallography.
1959	Metals associated with flavoproteins were found in several laboratories to be capable of oxidation-reduction during enzyme catalysis.
1960	Malmstrom used electron paramagnetic resonance (EPR) to determine the nature of ligands in metalloenzymes.
1960	Fine structure of a genetically modified enzyme, dihydropteroic acid synthetase, was determined by Hotchkiss and Evans.

(continued)

Table 1–1. *(Continued)*

1961	For enzyme studies, Baker, Shaw, and Singer developed active site-directed reagents.
1962	Baker showed that 4-iodoacetamidosalicylic acid is an active site-directed inhibitor of lactic dehydrogenase.
1962	Shaw showed that the chloromethyl ketone of tosylphenylalanine is an active site-directed inhibitor of chymotrypsin.
1962	Inhibition of dihydropteroic acid synthetase was established as mode of action of the sulfonamide drugs by Woods.
1963	Merrifield developed a method of solid-phase peptide synthesis used to prepare insulin and ribonuclease.
1965	Strominger and Tipper found that transpeptidase is selectively acylated by the β-lactam ring of penicillin and cephalosporin antibiotics.
1965	Species differences among dihydrofolate reductases were recognized as a basis for chemotherapy by Hitchings.
1966	Incorporation of nitroxide radicals in proteins as environmental probes for electron spin resonance studies was described by McConnell.
1968	Interaction of rifamycin with bacterial RNA polymerase was found by Wehrli et al.
1969	Bactericidal effect of rifampicin was recognized as due to inhibition of ribonucleic acid nucleotidyltransferase by Lancini et al.
1969	Knowledge of the enzymes required for the biosynthesis of dihydropteroic acid established by Richey and Brown.
1970	An altered dihydrofolate reductase was found to be associated with drug resistance in plasmodia by Ferone et al.

into the twentieth century that the search for new drug entities or classes took place in university laboratories.

The first use of synthetic organic chemicals for interference with life processes was probably when ether and chloroform were introduced for anesthesia during the first half of the nineteenth century. In consequence, early efforts to find synthetic drugs were concentrated on anesthetics and hypnotics and eventually analgesics. Chloral hydrate appeared in 1869 and paraldehyde in 1882, and the sulfone hypnotics were discovered by accident in 1888. The local anesthetic properties of ethyl p-aminobenzoate were known in 1890 and led to the development of procaine hydrochloride (Novocain), the structure of which is based on some features of the cocaine molecule. Cocaine was introduced as a local anesthetic in 1884.

Phenacetin also appeared during this period, and its discovery resulted from observations of the hydroxylation and conjugation of aniline in the animal body. This was probably the first drug to be designed as a result of knowledge of a biochemical transformation. Aspirin was introduced in 1899, and it resulted from an attempt to reduce the nausea caused by the salicylates, which had been used as antipyretics. Antipyrine was discovered from investigations of the chemistry of quinine, at this time, and the urethane hypnotics also resulted from the study of compounds produced by the chemical industry.

The next period in the development of medicinal agents was dominated by Paul Ehrlich. He was appointed Director of the Institute for Experimental Therapy in Frankfurt in 1899, at the age of 45. By this time, synthetic analgesics, anesthetics, and antipyretics were being manufactured by the German chemical industry. It occurred to him that because these molecules differentiate between cells in man, other molecules might differentiate between the cells of man and his parasites. This belief was strengthened by his previous experience in studying the selective staining of various tissues of the mammalian body by dyes, as well as by his studies of the selectivity of an antibody for the corresponding antigen.

Ehrlich was responsible for the discovery of a number of biologically active compounds, although few of them were put to use. Perhaps his greatest contributions to the advance of medicinal chemistry were the original ideas he had on the modes of drug action. He proposed that receptors exist in mammalian cells and that both antigens and chemotherapeutic agents (a term he coined) possess haptophoric (anchorer) and toxophilic (poisoner) groups. Chemotherapeutic agents, he considered, combine with the receptor areas of the cells by ordinary chemical reactions, a concept still valid, although modified to include more types of bond formation. He also stated that "the union between the alkaloid and the chemoreceptor is labile and reversible and not firmly bound"[2]—a view still held.

His early work on tissue staining techniques led to the discovery of the antimalarial activity of methylene blue in 1891. Further work with dyes resulted in the discovery of the trypanocidal action of trypan red and later of Trypaflavin. A direct result of the latter discovery was the finding of the antibacterial acriflavine. His work with arsenicals led to the introduction in 1910 of Salvarsan for treatment of syphilis, only 5 years after the causative organism of syphilis had been identified. He established the structure of Atoxyl, previously found by Breinl and Thomas to have a favorable effect on human trypanosomiasis.

Drug resistance was also discovered in Ehrlich's laboratories and supported his hypothesis of drug action through chemical combination with cellular receptors. Ehrlich concluded that resistance developed when the drug was no longer absorbed by the parasite (trypanosomes), and in the case of the arsenicals, the amino group present was the haptophore and the arsenoxide the toxophile. His ideas were thus supported by experimental facts, but they were slow to gain acceptance.

Several discoveries were made in the ensuing period by modification of compounds found to be active by Ehrlich and his school. The synthetic antimalarials Plasmoquine (1926) and Atabrine (1932) were based on the structure of methylene blue, which was found to have antimalarial activity by Ehrlich in 1891. Two of Ehrlich's arsenicals were found to have clinical application for diseases other than trypanosomiasis: carbarsone for amebiasis and oxophenarsine for syphilis. A method for growing *Entamoebae* in vitro was not developed until 1928, which handicapped the discovery of amebicides before that time.

This period following Ehrlich's discoveries and his postulation of a theory of drug action and prior to the discovery of the sulfonamides and antibiotics was characterized by increasing knowledge of the chemistry of natural substances, particularly enzymes. This was greatly assisted by new physical, chemical, and biologic methods, perhaps the most important being microchemical techniques for determination of structure.

Optical techniques, such as x-ray analysis and ultraviolet spectroscopy, were also introduced during this period and became of great value. The x-ray investigations of Astbury and others, along with increasingly sensitive analytic methods, showed protein molecules to be ordered chains of amino acids joined by polypeptide linkages. It was perceived that the properties of protein molecules were largely determined by spatial orientations of the constituent amino acids, but it was not learned until recently how a particular polypeptide orientation determines a particular biologic effect. This has come about through a knowledge of amino acid sequence, active or binding groups, cofactors, and three-dimensional structure.

The dynamic states of metabolic or enzymatic processes were also revealed by the findings of Schönheimer, Rittenberg, and others. Knowledge of nucleic acids and carbohydrates advanced, and much progress was made in the discovery of enzymes and the reactions they catalyze. The relationship between coenzymes and vitamins was recognized, and metabolic processes, such as carbohydrate breakdown, were beginning to be understood. Perhaps a statement by F. Gowland Hopkins can characterize the state of knowledge of life processes in that era:

> Chemical systems exist, and have existed ever since living organisms appeared in the world, in which the exact coordination in time and space of a multitude of reactions is an essential and indispensable condition—if chemistry is to provide the help that biology so urgently needs, it must approach living organisms not as chemical stores but as seats of chemical events.[3]

The emerging science of enzymology advanced rapidly during this post-Ehrlich period. The fermentative capacity of cell-free yeast juice had been first observed by Buchner in 1897, but by 1926, Sumner had crystallized an enzyme (urease). Other enzymes were later crystallized, and the amino acid compositions were determined.

Accurate methods for molecular weight determination were developed during the 1930s and 1940s. It was not until the 1950s, however, that chemical methods were available that permitted determination of amino acid sequences in proteins. Three-dimensional structure of proteins was not elaborated until Kendrew and co-workers determined the structure of myoglobin in 1958 by x-ray diffraction studies. By 1970, at least 15 other protein structures had been determined to near-atomic resolution, and electron density maps in two dimensions had been calculated for more than a dozen other proteins.

The first direct evidence of enzymatic involvement by a drug was found by Loewi and Navratil, who showed in 1926 that physostigmine inhibits the enzymatic hydrolysis of acetylcholine. Other drug-enzyme interactions were found in the following years, but it was not until the sulfonamides had been introduced and their mode of action studied that it became evident that drugs might in general interfere with normal chemical processes in the cells.

By 1940, when Fildes put forward his thesis that chemotherapeutic substances might be designed to compete with essential metabolites or growth factors, thereby destroying an invading organism by competitive antagonism of the function of the essential metabolite, the stage was prepared for the development of present-day medicinal chemistry. There was some knowledge of active compounds for many of the branches of chemotherapy and pharmacodynamics, methods for the determination of structure of enzymes and other biologic macromolecules were being elaborated, and the importance of interaction of drug molecules with biologically important macromolecules was realized. Accounts of the historical development of the various classes of medicinally active compounds may be found in the following chapters.

Techniques for the study of drug-macromolecule interactions have come into use in the modern, or post-1940, period. The development of theory and technique for measurement of steady-state kinetics, following the initial work of Michaelis and others, took place largely in this period. Since then, methods for measuring rapid reactions and transient states, having time constants approaching the range of those of molecular vibrations ($\sim 10^{-13}$ sec), have been elaborated. The study of stereospecificity of enzyme reactions through the use of optically pure chemical isomers has also been developed to a high degree, and steric specificity is of fundamental importance for some classes of drug molecules.

Other methods of importance for some classes of medicinals include those for studying proton abstractions and electron transfers, formation of chemical bonds, such as those involved in Schiff base formations, and degree of binding to metal constituents of enzyme systems. Nuclear magnetic resonance, electron spin resonance, and x-ray crystallography have also been used in studying drug-macromolecule interactions.

Chemical modification of drug molecules to locate the member of a series having optimal effect has long been carried out, is still widely used, and will probably continue to be a factor necessary to drug discovery. Chemical modification of enzymes by reagents directed to the active site, developed principally by Baker and Shaw, is now used and is important in design of drug structures. Although this procedure has been of great value in delineating active sites of enzymes and focusing attention on the reactive chemical groups involved at the active site, it was not until Koshland showed in 1958 that enzymes undergo conformational changes on substrate interaction that the importance of the whole protein molecule in enzyme action was realized. His concept of induced fit at the active site, with resultant thermodynamic changes, has explained enzyme—small molecule interactions that could not be elucidated on the basis of reaction with a specific chemical group, and it has broadened the concepts of molecular requirements for enzyme inhibitions. Other events in the development of our knowledge of enzymatic reactions of present and potential value to medicinal chemists are listed in Table 1–1.

The concept of the drug "receptor," the portion of a macromolecule that undergoes reaction or "coupling" with a drug molecule to produce a physiologic change or pharmacologic response, has undergone much modification from the 1960s to the 1990s. In contrast to the static "lock and key" view of Ehrlich, receptors are now considered flexible and capable of some variation in different locations (analogous to enzymes) as well as of lateral movement within the framework in which they are embedded (often in the plasma membrane or cell organelle membrane). Recognized receptors are lipoproteins, proteins (as in enzymes), occasionally lipids, and nucleic acids. Many antibiotics and anticancer agents interfere with DNA replication or transcription, and steroid hormones react with DNA. The work of many investigators has contributed to the development of the receptor concept, but the contributions of Belleau, Ariens, Mautner and Bloom, among medicinal chemists, and Changeux, Colquhoun, Cuatracasas, Seeman, Snyder, Triggle, and Yamamura, among molecular pharmacologists, stand out. Much development in the receptor theory is taking place, and some theories will ultimately be discarded. The use of computer graphics to portray drug-receptor interaction has also been a notable development during the past decade.

The recognition of the numerous involvements of adenylate cyclase and guanylate cyclase in drug mechanisms has also been a major advance of the 1970s and 1980s. Activation of adenylate cyclase produces cyclic AMP, a widely distributed "second messenger" that is involved with transmission of hormones, drugs, and neurotransmitters to effector organs. Other systems use guanylate cyclase as a messenger, producing cyclic GMP, which has effects mostly opposite to those of cyclic AMP. The complexity of drug and hormone interactions is becoming appreciated, and the hope of a simple drug-effector response is vanishing.

The approach to the practice of medicinal chemistry has developed from an empiric one involving organic synthesis of new compounds, based largely on modification of structures of known activity, to a more logical and less intuitive approach. This is mostly because of the findings in molecular biology, pharmacology, and enzymology, and the mechanics of drug-receptor interactions.

Medicinal chemistry has therefore become a collaborative effort between a variety of chemists, biologists, spectroscopists, geneticists, and biotechnologists. As knowledge of receptor structure, function, and stereochemistry has become available, the use of molecular graphics with computer programs illustrating drug-receptor interactions has become important in drug design. The problems involved in predicting drug activity are still formidable, however, but the synthesis of analogs of "lead" compounds has become less laborious or extensive with the knowledge of the biochemical make-up of receptors. The term receptor is now meant to include membranes, ion channels, and enzymes, as well as other protein, lipid, or carbohydrate surfaces.

Manfred Wolff[4] refers to the present development of medicinal chemistry as a renaissance, stating that: "Underlying this new age is a foundation that includes the explosive development of molecular biology since 1960, the advances in physical chemistry and physical organic chemistry made possible by high-speed computers, and new powerful analytic methods. . . ." Perhaps this exciting situation will continue for some time before mathematics and physical measurements convert medicinal chemistry to a more exact, but much less intriguing, science.

REFERENCES

1. A. Burger, *Medicinal Chemistry,* 3rd ed. New York, Wiley-Interscience, 1970, p. 2.
2. P. Ehrlich, *Chem. Ber., 42,* 17(1909).
3. F.G. Hopkins, Gluckstein Memorial Lecture, Institute of Chemistry, 1932.
4. M.E. Wolff, *Burger's Medicinal Chemistry,* 4th ed., Part 1, The Basis of Medicinal Chemistry. New York, Wiley-Interscience, 1980, p. vi.

SUGGESTED READINGS

A. Albert, *Selective Toxicity, The Physico-chemical Basis of Therapy,* 7th ed., New York, Methuen, 1984.

C. Hansch, P.G. Sammes, and J.B. Taylor, *Comprehensive Medicinal Chemistry,* Vols. 1–6, Oxford, Pergamon Press, 1990.

J. Nogrady, *Medicinal Chemistry, A Biochemical Approach,* New York, Oxford University Press, 1985.

B.G. Reuben and H.A. Wittcoff, *Pharmaceutical Chemicals in Perspective,* New York, John Wiley & Sons, 1989.

S.M. Roberts and B.J. Price, *Medicinal Chemistry, The Role of Organic Chemistry in Drug Research,* New York, Academic Press, 1985.

W. Sneador, *Drug Discovery: The Evolution of Modern Medicines,* New York, John Wiley & Sons, 1985.

Chapter 2

MEDICINALS OF PLANT ORIGIN: HISTORICAL ASPECTS

William O. Foye

The use of plants or plant extracts for medicinal purposes has been going on for thousands of years, and herbalism and folk medicine, both ancient and modern, have been the source of much useful therapy. Some of the plant products currently used, either in their natural form or as derivatives, were often used originally for other purposes, such as arrow poisons, as part of religious or other rituals, and even as cosmetics. Examples of such products include opium, belladonna, cinchona bark, ergot, curare, nutmeg, calabar bean, foxglove, and squill. Many drugs originally used as folk remedies have been abandoned. The intent, in this brief review, is to present some of the more important drugs, of both the past and present, derived from herbs and nonwoody plants. This excludes the antibiotics, obtained from micro-organisms; fungal metabolites such as ergot; and drugs from woody plants, such as quinine and quinidine from cinchona bark, curare from chondrodendon bark, physotigmine from the seeds of a woody vine, and pilocarpine from a Brazilian shrub. All of these are treated in other chapters.

Perhaps the earliest recorded use of a plant medicinal is that of the herb called "Ma Huang," a species of Ephedra used medicinally in China for over 5000 years. The important constituent of this plant is ephedrine, which has been used successfully in the treatment of bronchial asthma, hay fever, and other allergic conditions. Probably the major use of ephedrine and other ephedra alkaloids today is a antitussives and oral decongestants for the common cold. Commercial ephedra consists of the dried young branches of various species, including Ephedra sin-

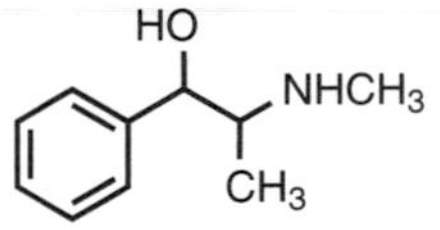

Ephedrine

ica and E. equisetina, natives of China, and E. gerardiana and E. nebrodensis, indigenous to India. The latter are still used as sources of the drug. Ephedrine was first isolated from the herb in 1887 and was introduced into Western medicine about 1925 by K. K. Chen of the Eli Lilly Company.

Ephedra occupies a peculiar position in the vegetable kingdom because it belongs to a very small class of Gymnospermae, common to the conifers, instead of in a seed vessel, the Gnetales. In the gymnosperms, the seeds are situated on an open scale, as in the true flowering plants, or angiosperms. The Gnetales number about 40 living species, 35 belonging to Ephedra. They occur in temperate and subtropical regions of Asia, America, and the Mediterranean; most of the American species, however, are devoid of the alkaloidal drugs. They have marked resemblances to the angiosperms and are considered by some to be relatives of the transitional plant types that existed during the Cretaceous Period before the advent of the true flowers.

Related alkaloids (nitrogen-containing basic organic molecules) having structures similar to that of ephedrine and of medicinal importance today are found in many plants. For example, β-phenethylamine, a pressor amine (which increases blood pressure), occurs in mistletoe (Viscum album L.). Hordenine occurs in barley (Hordeum vulgare L.) and in the cactus Anhalonium fissuratum Engelm. Dopamine, another relative of β-phenethylamine, is found in the banana. The latter compound, under slightly alkaline conditions, undergoes atmospheric oxidation with ease, giving rise to a black pigment.

Mescaline, another related pressor amine, is present in peyotl (Lophophora williamsi [Lemaire] Coulter), and is responsible for the hallucinations caused by this cactus. Peyotl has long been used in religious ceremonies by the Indians of Mexico and the southwestern United States, and its use is now apparently spreading. Mescaline is

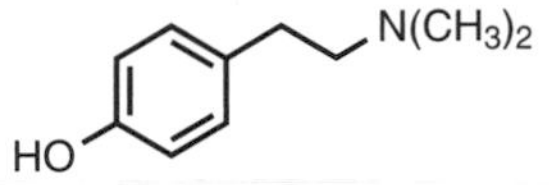

Hordenine

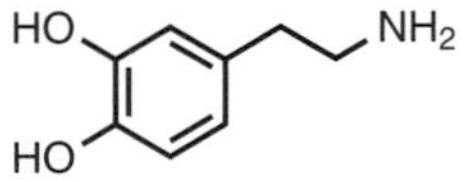

Dopamine

probably the most widely known psychotomimetic drug, i.e., one that produces changes in thought perception and mood without causing major disturbances of the autonomic nervous system. These drugs are used experimentally to produce model psychoses, which can be studied in the same way as true mental disorders.

An alkaloidal drug having a much more complicated chemical structure, but part of which is a β-phenylethylamine moiety, is morphine, a major constituent of opium. Morphine was first isolated from opium by Sertürner in 1805 and was the first plant alkaloid to be isolated in relatively pure form. Opium consists of the dried latex from the unripe fruit of Papaver somniferum, a species of poppy. Opium has been produced commercially in Asia Minor, Turkey, Bulgaria, Yugoslavia, Iran, India, and China for extraction of morphine; the clandestine production, which is considerable, is probably more widespread.

Morphine is used mainly to relieve pain, but it has hypnotic (sleep-producing) and cough-suppressing properties as well. Codeine and noscapine, also present in P. somniferum, are used extensively today as antitussives in anticold preparations. The early Egyptians were aware of the sleep-producing properties of opium. The habit of opium eating for narcotic purposes was established in Eastern Europe in the seventeenth century. DeQuincy described the sensations that precede the sleep caused

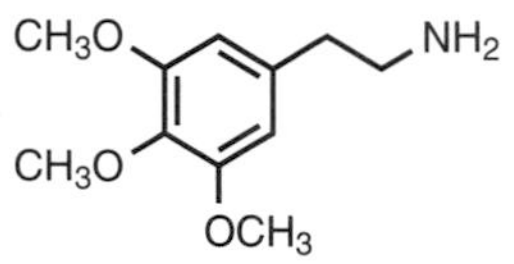

CH_3O NH_2 CH_3O OCH_3

Mescaline

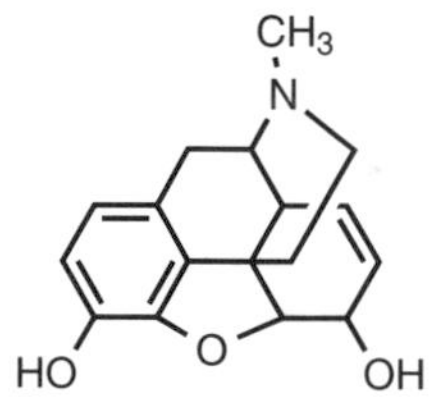

CH_3 N HO O OH

Morphine

by opium in his *Confessions of an Opium Eater.* Despite its ultimately harmful tendency, morphine is still one of the most important drugs today. Sydenham, the noted seventeenth century physician and founder of the clinical method, remarked that "without opium I would not care to practice medicine." Actually, 25 alkaloids have been isolated from opium, but with the exception of morphine and codeine (morphine methyl ether), the other alkaloids are of limited medicinal importance.

A number of species of the family Solanaceae contain some therapeutically useful alkaloids, including belladonna, henbane, thornapple, and mandrake. All of these plants contain alkaloids composed of a common base, tropine, combined with various organic acids. Cocaine, found in the leaves of the coca plant, is also a tropine derivative, containing two organic acids. The solanaceous alkaloids produce hallucinations and were the sorcerer's drugs of the Middle Ages.

Both the leaf and the root of the deadly nightshade (Atropa belladonna) have been used as sources of belladonna. This has been employed by physicians for treating colic, asthmatic and intestinal spasms, and whooping cough and is presently found in oral decongestant preparations. It also relieved the night sweats of tuberculosis and has been used to diminish the activity of the salivary

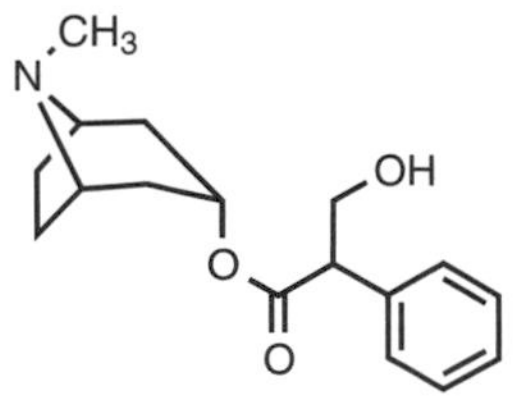

CH_3 N OH O O

(+/-) Atropine
(-) Hyoscyamine

and gastric glands. The name belladonna (beautiful lady) arose through the use of the juice of the plant by women in ancient times as a cosmetic to dilate the pupil of the eye. One of the major uses today of the constituent alkaloid atropine is in ophthalmology for dilation of the pupil. A single drop of solution containing 1 part of the atropine in 40,000 parts of water is sufficient to dilate the pupil of a cat. Solanaceous plants grow commonly in central and southern Europe and have been cultivated extensively, particularly in Hungary. Although atropine can be synthesized, most of it is still obtained by extraction of the Duboisa species of the Solanaceae found in Australia.

Hyoscyamus niger, or henbane, is a common herb in Europe and is found as far east as India. It has been cultivated for medicinal purposes both in England and in Europe, but the major plant source has been Hyoscyamus muticus, found in Egypt. Hyoscyamus has a smaller amount of total alkaloids than belladonna.

The thornapple, Datura stramonium, grows wild in southern Europe and has been cultivated in England, France, and Germany. I found some growing within the walls of one of the castles of the French kings in the Loire valley. The main use of the isolated alkaloids has been

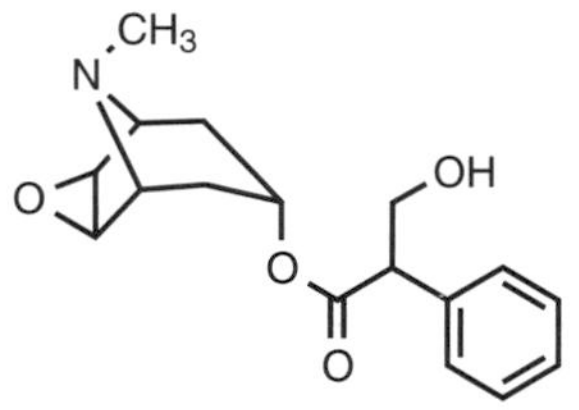

CH_3 N O OH O O

(-) Hyoscine

the relief of spasmodic asthma and with some success in the treatment of postencephalitic parkinsonism. Datura metel, belonging to the same genus as stramonium and indigenous to India, has been the principal source of the alkaloid hyoscine, or scopolamine. The base, scopoline, is a close relative of tropine. Hyoscine has been used for its hypnotic and sedative properties. Combined with morphine, hyoscine has also been used for presurgical anesthesia and to allay the pain of childbirth.

Alkaloids of the same class also occur in mandrake (Mandragora officinarum L.), a solanaceous plant of the Mediterranean region, especially Greece. Mandrake is one of the oldest drugs known in Western medicine and at one time was regarded as an almost universal cure for bodily ills. Taken with wine, mandrake was used to relieve toothache; it was also considered a charm. It is used little, if at all, today because more powerful synthetic analgesics and local anesthetics are available.

The working model on which the organic structures of today's local anesthetics are based is cocaine, an alkaloid found in the leaves of the coca plant. Coca, "the divine plant of the Incas," has been cultivated in Peru and Bolivia for ages. The Indians of South America chew the dried leaves of the shrub after mixing them with slaked lime or plant ash to allay the onset of hunger and fatigue. The leaves from three varieties of the plant have been used commercially: Erythroxylum truxillense, or Peruvian coca, E. coca, or Bolivian coca, and E. spruceanum, Burck, grown in Java. Although cocaine of plant origin is used to a limited extent today as a local anesthetic, its

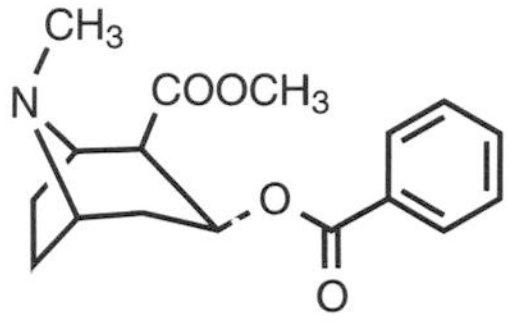

Cocaine

use by addicts is considerable. Cocaine addiction causes physical ill health and the wildest hallucinations. Cocaine addiction became one of the common forms of narcotic addiction during the trench warfare of World War I. Chewing the leaves, however, does not lead to addiction or ill effects. Cocaine addicts generally sniff a few milligrams of the ground leaf powder up one nostril.

Cocaine was not used in medicine until the latter half of the nineteenth century, and plantations of coca were established in Ceylon and Java. In 1882, Koller and Freud, in Vienna, discovered the anesthetic effect of cocaine on the eye, which made eye surgery possible. Cocaine was also used for operations on the throat and larynx and for extraction of teeth. Cocaine and other local anesthetics are frequently injected with adrenaline, a β-phenylethylamine derivative, which contracts the blood vessels, localizes the anesthetic action, and also reduces the toxic effects on the central nervous system. Cocaine is still considered superior to the synthetic local anesthetics for operations involving the nasal septum.

The root of Cephaëlis ipecacuanha Rich., a small plant found in Brazil, belonging to the family Rubiaceae, contains several alkaloids of medicinal value, the most important being emetine. Ipecacuanha preparations made from the whole drug have been useful as expectorants in treating bronchitis and whooping cough and as an emetic and diaphoretic. Emetine, a constituent alkaloid having two cyclic β-phenylethylamine moieties in its structure, has been an important treatment for amebic dysentery. This is a common tropical disease, caused by the protozoan Entamoeba histolytica. It causes painful ulcers in the intestinal mucosa and should not be confused with bacillary dysentery. Although emetine is effective in removing the amebae, it has cumulative toxic effects, so there has been a long search for better antiamebics.

An alkaloid of complex structure, reserpine, was isolated from the root of Rauwolfia serpentina. Benth. ex Kurz, a plant indigenous to India, at the Ciba research laboratories in Basel in 1952. Its introduction into Western medicine revolutionized the treatment of hypertension because it has the double effect of lowering high blood pressure and acting as a tranquilizer. Its use is now considerable, especially for patients whose hypertension is aggravated by anxiety or emotional disturbances. The root of R. serpentina has been used for centuries in India to treat anxiety, insanity, and snakebite. An embargo imposed on the export of the plant by the Indian government led to the search for the alkaloid in other plants, and it was found in many species of Rauwolfia in various tropical countries and in other genera of the Apocynaceae.

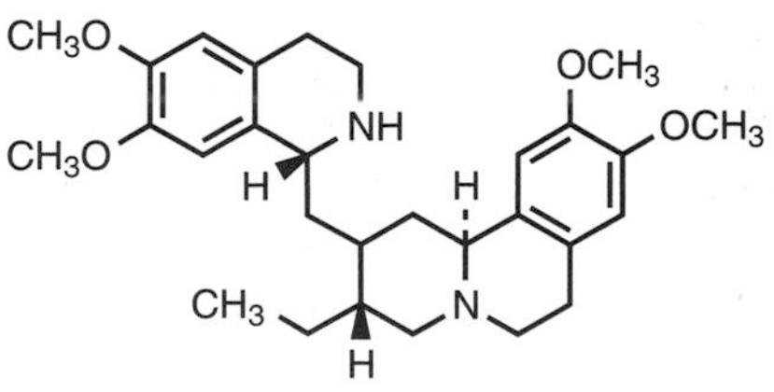

Emetine

Other plant drugs acting on the cardiovascular system include the so-called cardiac glycosides, which are nonalkaloid compounds having a steroid nucleus attached to a sugar derivative (see Chapter 19). This group of plant drugs is comprised of digitalis, strophanthus, and squill, and their constituents act as heart stimulants. Digitalis consists of the leaves of the purple foxglove, Digitalis purpurea, a plant widely distributed throughout Europe.

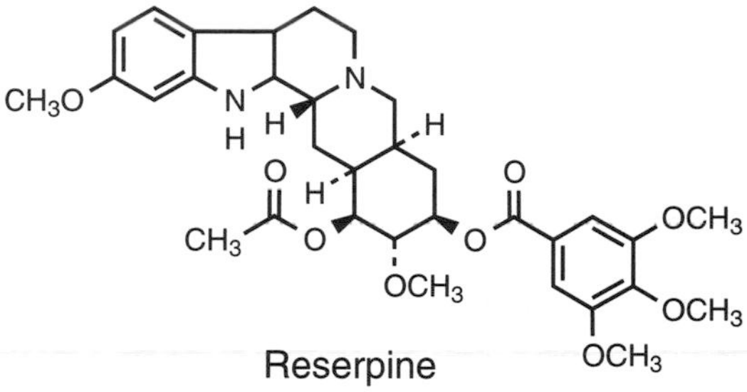

Reserpine

More frequently used today is digoxin, a related glycoside obtained from Digitalis lanata, which is three to four times more potent as a heart stimulant than digitalis. These glycosides also have a diuretic action that increases their utility in treating congestive heart failure.

Strophanthus consists of the ripe seeds of Strophanthus kombé, a climbing plant of considerable size found in tropical eastern Africa. Extracts of the seeds have been used as arrow poisons. The active constituent is the glycoside strophanthin, of which the aglycone is strophanthidin. It acts similarly to digitalis but is more powerful and requires greater caution in its use. Squill, or white squill, is the bulb of Urginea maritima, a plant abundant in the Mediterranean countries; Indian squill is obtained from U. indica. The action of squill on the heart is much more powerful than that of digitalis, and it is no longer used for this purpose. Red squill contains similar active principles, but also some that are toxic for rats, and it has therefore been used as a rat poison.

An alkaloidal drug currently used as a coronary vasodilator is papaverine, originally obtained from Papaver somniferum along with opium. It is now produced synthetically.

The autumn crocus or meadow saffron, Colchicum autumnale, has been used as remedy for gout and as an antiarthritic agent. The plant is found in the limestone regions of Central Europe and is common in southwest England. The active principle, colchicine, was first isolated in 1820 (see Chapter 25). It is toxic and has the property of bringing cell division to a halt at a particular stage. It was thought at one time that this property would make the compound valuable in treating various tumors, but this promise has not been fulfilled. It has been useful in cytologic studies, however, and its use has led to the discovery of methods for the artificial production of polyploid varieties of many plants of economic importance.

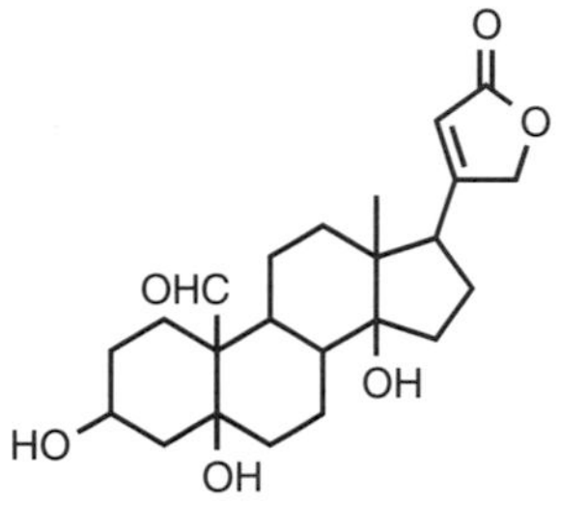

Strophanthidin

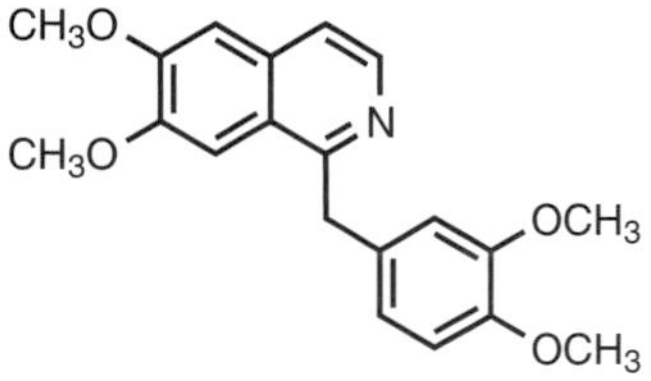

Papaverine

Several compounds of plant origin have been investigated for anticancer activity in clinical trials. The most important of these are the alkaloids derived from the Madagascar periwinkle, Catharanthus roseus (Vinca rosea) (see Chapter 37). These alkaloids are called leurocristine and vincaleukoblastine. More than 500 species of plants, many of them folklore remedies, have been shown to have antitumor activity, but practically all of those investigated have lacked clinical utility.

Probably the most widely used alkaloid in the world is caffeine, noted for its central nervous system stimulating effects. Two alkaloids closely related in structure, theophylline and theobromine, present also in coffee and tea, are presently used to a small extent as bronchodilators (see Chapters 15 and 21). Their commercial source is Theobroma cacao, or coca. Caffeine and theobromine have also had past use as diuretics, but synthetic diuretics of much greater potency are now available.

Other drugs of plant origin, excluding the antibiotics, which are derived largely from microorganisms, fungal metabolites such as ergot, and compounds such as quinine that are obtained from woody plants, many of which are folklore remedies, still enjoy considerable use. Many laxatives and purgatives are of vegetable origin, such as senna (from Cassia species), aloes (from Aloe species, Liliaceae), rhubarb (from Rheum species), podophyllum (from Podophyllum peltatum, the May apple), and psyllium seed husks (from Plantago species). Licorice powder probably owes most of its medicinal value to the powdered senna leaf it contains.

The active principle of podophyllum is podophyllotoxin. The juice expressed from the root of the May apple was used by American Indians as a purgative and anthelmintic, and the dried resin was included in the first edition of the *United States Pharmacopeia* and later in the *British Pharmacopeia* as a purgative. It caused severe irritation of the intestinal mucosa, however, and its use as a purgative was discontinued. Its irritant effects led to its use as an escharotic and to remove superficial tumors. Its recognition as a mitotic poison led to the testing of various derivatives as antitumor agents, and in 1970, Sandoz Laboratories reported the antitumor activity of etoposide, now in clinical use for non–small cell lung cancer and some types of testicular cancer.

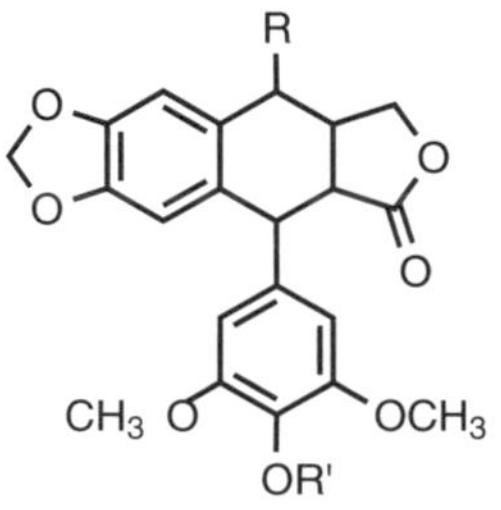

Podophyllotoxin, R = OH, R' = CH_3

Etoposide, R' = H, R = O–(sugar)

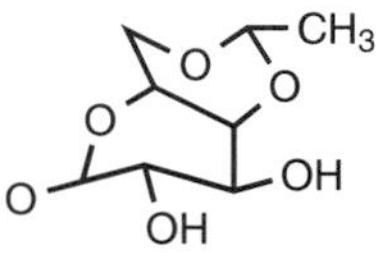

A Chinese herb drug obtained from the rhizome of Ligusticum chaunxiong Hort. has been in use in China as a "blood activating" agent for over 2000 years. It has recently been found to inhibit hyperaggregation of the blood platelets and in so doing alleviates one of the deleterious effects of high doses of ionizing radiation.[1] One of the active constituents of this herbal drug preparation is a derivative of the Harmala alkaloids, having the following structure.

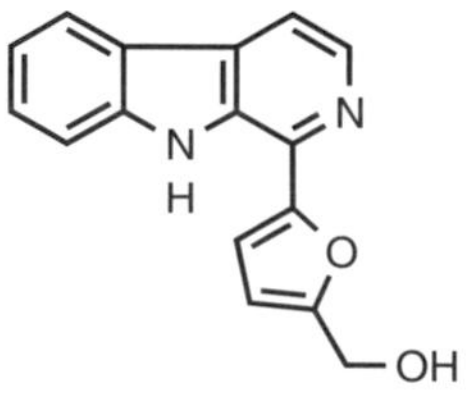

1-(5-Hydroxymethyl-2-furyl)-9H-pyrido(3,4-b)indole

Previously, drugs of plant origin provided many of the diuretics, emmenagogues, carminatives, rubefacients and dermatologic remedies, expectorants, anthelmintics, miotics, and antidiarrheal and hemorrhoidal preparations and have been either omitted or briefly discussed here. Although drugs of plant origin are still employed for some of these uses, synthetic drugs now constitute the major part of the products used. It may be interesting to note that figures from more than 1 billion prescriptions dispensed from pharmacies in the United States during 1967 show that about 243 million, or about 23%, of all prescriptions contained one or more products of plant origin.[2] This survey excluded products of microbial origin, and the products included represented about 50 pure compounds and 40 crude or semipurified types of plant extracts. About 50 genera of plants were represented. This percentage of plant drugs had not changed appreciably from that seen in a survey in 1962, and it is probably much the same today. Plant drugs, therefore, still constitute an important part of the medicinals used today.

With the increasing loss of much of the world's forests, particularly in the tropics, the potentially remarkable properties of plant constituents not yet discovered and threatened with extinction could be forever lost. If this occurred, many future drugs and other useful plant products would remain undiscovered, and the often surprising chemical structures produced by the genetic diversity of plants might not be envisioned by future chemists.

REFERENCES

1. H.-F. Wang, et al., *J. Ethnopharmacology, 34,* 215(1991).
2. N. R. Farnsworth, in *Phytochemistry,* Vol. III, L.P. Miller, Ed., New York, Van Nostrand Reinhold Company, 1973.

SUGGESTED READINGS

N. L. Allport, *The Chemistry and Pharmacy of Vegetable Drugs,* London, George Newnes, Ltd., 1943.

R. H. F. Manske and H. L. Holmes, Eds. *The Alkaloids,* Vols. 1–8, New York, Academic Press, 1950–1960.

G. A. Swan, *An Introduction to the Alkaloids,* New York, John Wiley & Sons, Inc., 1967.

V. E. Tyler, et al., *Pharmacognosy,* 7th ed., Philadelphia, Lea & Febiger, 1976.

H. W. Youngken, *Textbook of Pharmacognosy,* 6th ed., Philadelphia, The Blakiston Company, 1950.

N. R. Farnsworth, *Plants and modern medicine: where science and folklore meet, Traditional Medicine,* World Health Forum, *6,* 76(1985).

W. H. Lewis and M. P. E. Lewis, *Medical Botany,* New York, Wiley-Interscience, 1977.

Chapter 3

BIOPHARMACEUTICAL PROPERTIES OF DRUG SUBSTANCES

Sunil S. Jambhekar

Throughout its history, the pharmacy profession has been concerned primarily with the manner in which drugs produce their pharmacologic effects and the dosage forms through which drugs are administered. Since the early twentieth century, efforts have been directed to determining, understanding, and providing rational explanation of drug effects on biologic systems, limited only by our ability to correlate the observed physiologic events with a reasonable hypothesis or concept. Pharmacists, at one time, were closely involved in formulating a prescription written by a physician for a patient. Today, most of the formulating is done by the pharmaceutical manufacturer.

Early descriptions of drug action were confined to their reference as tonic or toxic effects. This approach was followed by the concept of receptor theory, which for decades remained primarily an operational concept that was useful for discussing the new actions of drugs on a molecular level.[1] Research in receptor theories, however, has provided evidence that the drug receptors do exist as distinct entities, and a limited success has been attained in the characterization of receptors.[2,3]

An extension of the receptor theory of drug action is an increased emphasis on the importance of physicochemical properties of the drug and the relationship of such properties to pharmacologic responses. Because these properties play an important role in determining biologic action of pharmaceuticals, it is appropriate to refer to these properties as biopharmaceutical properties of drug substances. Examples of such properties include solubility, partition coefficients, diffusivity, degree of ionization, and polymorphism, which, in turn, are determined by the chemical structure and stereochemistry of drug substances.

A consideration of these biopharmaceutical properties is fundamental to discussing several important aspects of the overall effects. For a given chemical entity (drug), there will often be a difference in physiologic availability and, presumably, in clinical responses. This difference is primarily attributable to the fact that drug molecules must cross various biologic membranes and interact with intercellular and intracellular fluids before reaching the elusive region called the "site of action." Under these conditions, the biopharmaceutical properties of the drug must contribute favorably to facilitate absorption and distribution processes to augment the drug concentration at various active sites. Furthermore, equally important is the fact that these biopharmaceutical properties of a drug must ensure a specific orientation on the receptor surface so that a sequence of events is initiated that leads to the observed pharmacologic effects. Drug molecules deficient in the required biopharmaceutical properties may display marginal pharmacologic action or be totally ineffective.

Biopharmaceutics may be defined as the study of the influence of formulation factors on the therapeutic activity of a drug product or dosage forms. It involves the study of the relationship between some of the physicochemical properties of a drug and the biologic effects observed after the administration of a drug in various dosage forms or drug delivery systems. Almost any alteration in a drug delivery system is likely to alter the drug delivery rate and the amount of the drug delivered to the desired place in the body. Factors with such an effect include the chemical nature of the drug (ester, salts, complexes, etc.), the particle size and surface area of the drug, the type of dosage forms (solution, suspension, capsule, tablet), and the excipients and processes used in the manufacturing of the drug delivery systems.

Drugs, via drug delivery systems, are most often administered to human subjects by the oral route. Compared to other routes of drug administration, especially intravenous access, the oral route is unusually complex with respects to the physicochemical conditions existing at the absorption site. Therefore, before we discuss how the biopharmaceutical properties of a drug in a dosage form may affect the availability and the action of a drug, it is prudent for the reader to review the gastrointestinal physiology.

GASTROINTESTINAL PHYSIOLOGY

Figure 3–1 schematically represents the gastrointestinal tract and some of the problems encountered in consideration of drug absorption from the site following administration of a drug via dosage forms. The stomach may be divided into two main parts: the body of the stomach and the pylorus. Histologically, these parts correspond to the pepsin and HCl-secreting area and the mucus-secreting area of the gastric mucosa, respectively. In the human, the stomach contents are usually in the pH range of 1 to 3.5, with the most common being pH 1 to 2.5. Furthermore, humans have a diurnal cycle of

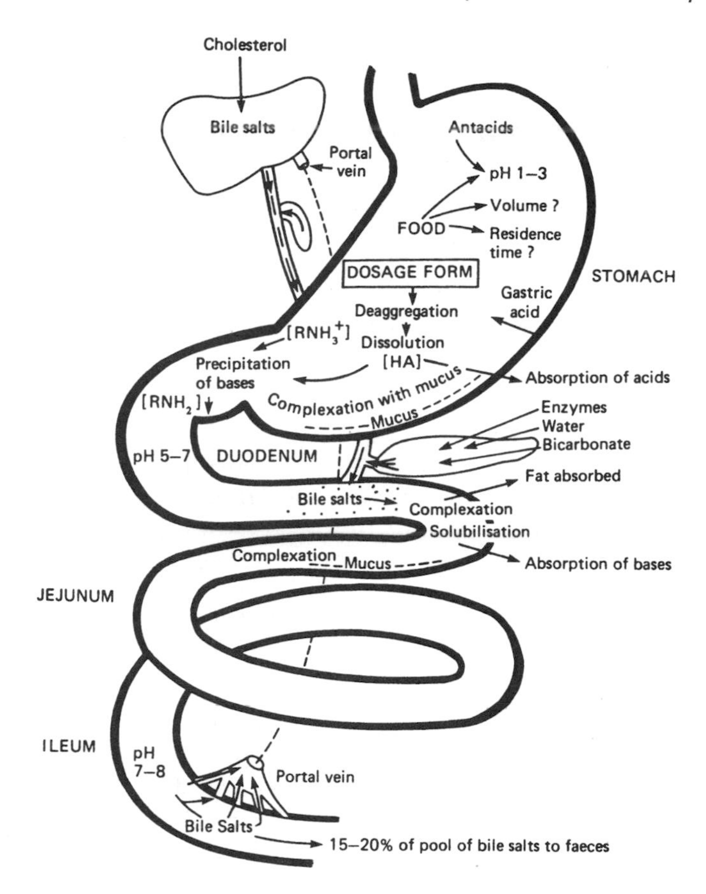

Fig. 3–1. Processes occurring along with drug absorption in the gastrointestinal tract, and the factors that affect drug absorption. (From A. T. Florence and D. Attwood, *Physicochemical Principles of Pharmacy,* 2nd ed., New York, Chapman and Hall, 1988.)

gastric acidity. During the night, stomach contents are usually more acidic, i.e., pH is about 1.3, whereas during the day, because of food consumption, the pH is less acidic. The recovery of stomach acidity occurs rapidly, however. The presence of protein, being amphoteric, acts as an excellent buffer, and as digestion proceeds, the liberated amino acids increase the neutralizing capacity enormously.

The small intestine is divided anatomically into three sections: duodenum, jejunum, and ileum. All three areas are involved in the digestion and absorption of food. The available absorbing area is increased by surface folds in the intestinal lining. The surface of these folds possess villi and microvilli (Fig. 3–2). The duodenal contents in the human are usually in the pH range of 5 to 7. Acidity gradually decreases along the length of the gastrointestinal tract, with the ultimate pH being 7 to 8 in the lower ileum. It has been estimated that approximately 8 liters of fluid enter the upper intestine per day, approximately 7 liters of which arise from digestive juices and fluids, and about 1 liter from oral intake.

Over the entire length of the large and small intestines and the stomach is the brush border, consisting of a uniform coating (3 mm thick) of mucopolysaccharide. This coating layer acts as a mechanical barrier to bacteria or food particles.

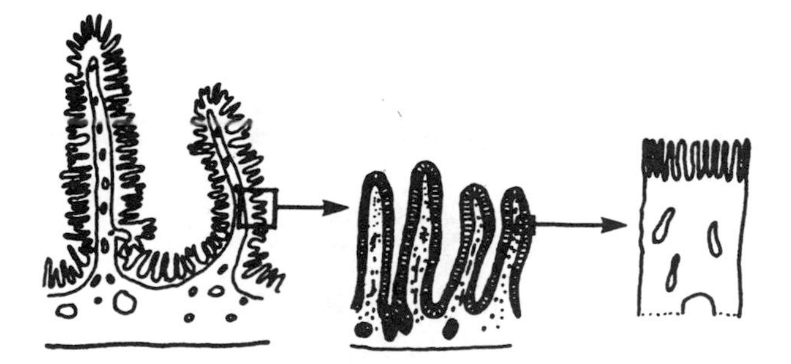

Fig. 3–2. The epithelium of the small intestine at different levels of magnification. From left to right: the intestinal villi and microvilli that constitute the brush border.

When a dosage form containing a drug or drug molecules moves from the stomach through the pylorus into the duodenum, the dosage form encounters a rapidly changing environment with respect to pH. Furthermore, digestive juices secreted into the small bowel contain many enzymes not found in the gastric juices. Digestion and absorption of foodstuff occur simultaneously in the small intestine. Intestinal digestion is the terminal phase of preparing foodstuff for absorption and consists of two processes: (1) completion of the hydrolysis of large molecules to smaller ones that can be absorbed, and (2) bringing the finished product of hydrolysis into an aqueous solution or emulsion.

Drug absorption, whether from the gastrointestinal tract or other sites, requires the passage of the drug in a molecular form across the barrier membrane. Most drugs are presented to the body as solid or semisolid dosage forms, and the drug particles must first be released from that form. These drug particles must first dissolve, and if the drug possesses the desirable biopharmaceutical properties, it will pass from a region of high concentration to a region of low concentration across the membrane into the blood or general circulation (Fig. 3–3).

Knowledge of biologic membrane structure and its general properties is pivotal in understanding absorption processes and the role of the biopharmaceutical properties of drug substances.

Biologic Membrane

The prevalent view is that the gastrointestinal membrane consists of a bimolecular lipoid layer covered on each side by protein with the lipid molecule oriented

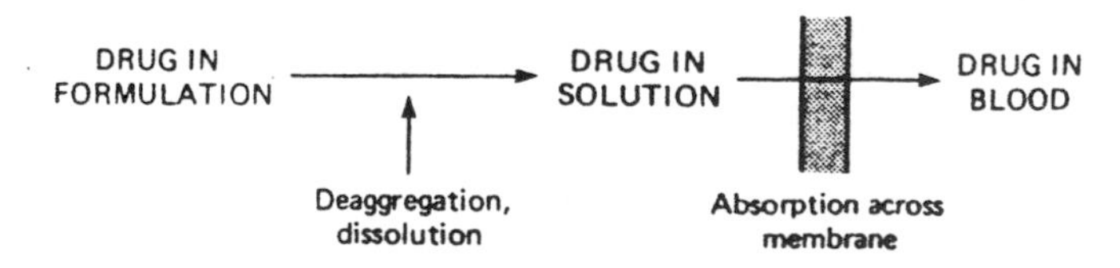

Fig. 3–3. Sequence of events in drug absorption from formulations.

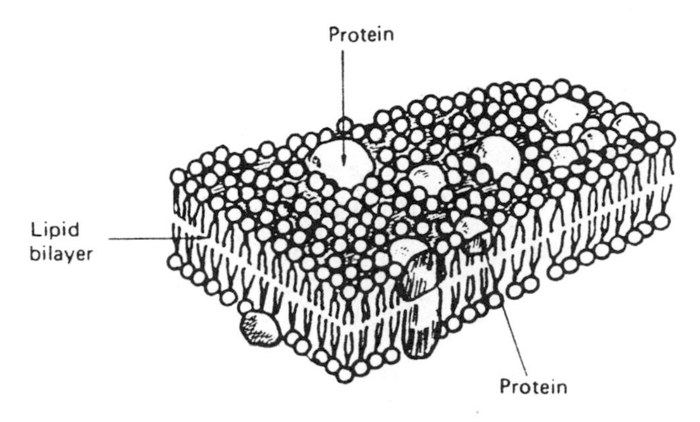

Fig. 3–4. The fluid mosaic model. (From A. T. Florence and D. Attwood, *Physicochemical Principles of Pharmacy,* 2nd ed., New York, Chapman and Hall, 1988.)

perpendicular to the cell surface (Fig. 3–4). The lipid layer is interrupted by small water-filled pores that are approximately 4 angstrom in radius, and a molecule with a radius of 4 angstrom or less may pass through the water-filled pores. Thus, membranes have a specialized transport system to assist the passage of water-soluble material and ions through the lipid interior, a process sometimes referred to as "convective absorption." The rate of permeation of such small molecules through the pore is affected not only by the relative sizes of the holes and the molecules but also by the interaction between permeating molecules and the membrane. When permeation through the membrane occurs, the permeating substance is considered to have transferred from solution in the luminal aqueous phase to the lipid membrane phase, then to the aqueous phase on the other side of the membrane. Biologic membranes differ from a polymeric membrane in that they are composed of small amphipathic molecules, phospholipid and cholesterol. The protein layer associated with membranes is hydrophobic. Therefore, biologic membranes have a hydrophilic exterior and hydrophobic interior. Cholesterol is a major component of most mammalian biologic membrane, and its removal renders the membrane highly permeable. Cholesterol complexes with phospholipids and its presence reduces the permeability of the membrane to water, cations, glycerides, and glucose. The shape of the cholesterol molecule allows it to fit closely with the hydrocarbon chains of unsaturated fatty acids in the bilayer. It is the general opinion that the cholesterol makes the membrane more rigid. The flexibility of the biologic membrane to reform and adapt to a changed environment is its important feature. The details of membrane structure are still widely debated, and a more recent membrane model is shown in Figure 3–4.

In addition to biopharmaceutical factors, several physiologic factors may also affect the rate and extent of gastrointestinal absorption. These factors are as follows: properties of epithelial cells, segmental activity of the bowel, degree of vascularity, effective absorbing surface area per unit length of gut, the surface and interfacial tensions, the electrolyte content and their concentration in luminal fluid, the enzymatic activity in the luminal contents, and gastric emptying rate of the drug from stomach.

MECHANISMS OF DRUG ABSORPTION

Drug transfer is often viewed as the movement of a drug molecule across a series of membranes and spaces, which, in aggregate, serve as a macroscopic membrane. The cells and interstitial spaces lying between the gastric lumen and the capillary blood or structure between sinusoidal space and the bile canaliculi are examples. The cellular membrane and space may impede drug transport to varying degrees and, therefore, any one of them can be a rate-limiting step to the overall process of drug transport. This complexity of structure makes quantitative prediction of drug transport difficult. A qualitative description of the processes of drug transport across functional membranes are as follows.

Passive Diffusion

The transfer of most drugs across biologic membrane occurs by passive diffusion, a natural tendency for molecules to move from higher concentration to one of lower concentration. This movement of drug molecules is caused by the kinetic energy of the molecules. The rate of diffusion depends on the magnitude of the concentration gradient (dC) across the membrane and can be represented by the following equation:

$$-\frac{dC}{dt} = K^*dC = K(C_{abs} - C_b)$$

(Equation 1)

in which $-dC/dt$ is the rate of diffusion across a membrane, K is a complex proportionality constant that includes the area of membrane, the thickness of the membrane, the partition coefficient of the drug molecule between the lipophilic membrane and the aqueous phase on each side of the membrane, and the diffusion coefficient of the drug.

The gastrointestinal absorption of a drug from an aqueous solution requires transfer from the lumen to the gut wall followed by penetration of the epithelial membrane by a drug molecule to the capillaries of the systemic circulation. On entering the blood, the drug distributes itself rapidly. Because of the volume differences at absorption and distribution sites, the drug concentration in blood (C_b) is much lower than the concentration at the absorption site (C_{abs}). This concentration gradient is maintained throughout the absorption process, i.e., (C_b − C_{abs}). As a result, the concentration gradient, (dC in equation 1), is approximately equal to C_{abs} and, hence, equation 1 can be written as

$$-\frac{dC}{dt} = K^*C_1$$

(Equation 2)

Because absorption by passive diffusion is a first-order

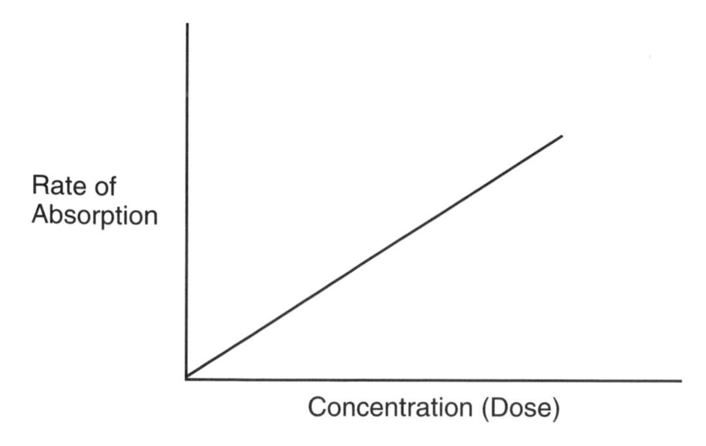

Fig. 3–5. Effect of drug concentration on the rate of absorption owing to passive diffusion.

process, the rate of absorption (dC/dt in equation 2) is directly proportional to the concentration at the site of absorption (C_1). The greater the concentration of drug at the absorption site, the faster the rate of absorption (Fig. 3–5). The percent of dose absorbed at any time, however, remains unchanged.

A major source of variation is membrane permeability, which depends on the lipophilicity of the drug molecule. This feature is often characterized by its partition between oil and water. The lipid solubility of a drug, therefore, is an important physicochemical property governing the rate of transfer through a variety of biologic membrane barriers. Figure 3–6 illustrates the role of partition

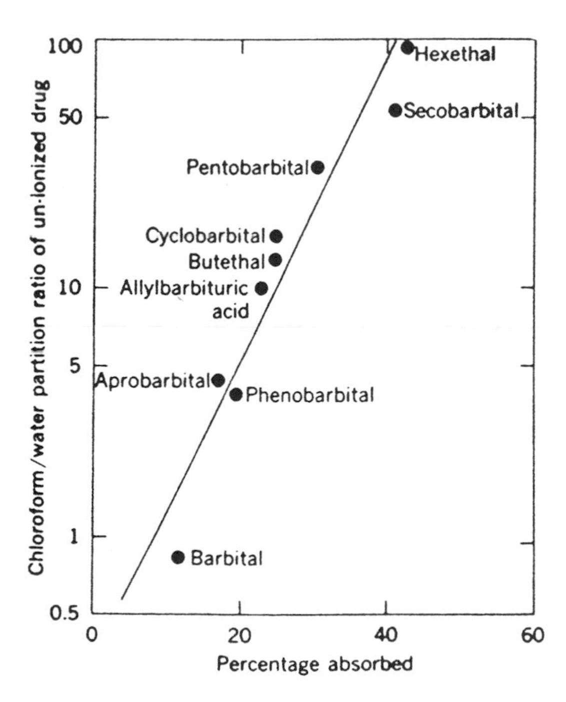

Fig. 3–6. Comparison between colonic absorption of barbiturates in the rat and lipid-to-water partition coefficient of the un-ionized form of the barbiturates. (From L. S. Schanker, *J. Pharmacol. Exp. Ther., 126,* 283(1959).

coefficients in the drug absorption process. It is clear that a good correlation exists between the percentage of drug absorption and the partition coefficient of an un-ionized drug.

Carrier Mediated or Active Transport

Although most drugs are absorbed by passive diffusion, some drugs of therapeutic interest and some chemicals of nutritional value are absorbed by a carrier-mediated or active transport mechanism. In this type of transport, membranes have a specialized role. The usual requirement for active transport is structural similarities between the drug and the substrate normally transported across the membrane. Active transport differs from passive diffusion in the following ways: (1) the transport of the drug occurs against a concentration gradient, (2) the transport mechanism can become saturated at high drug concentration, and (3) a specificity for a certain molecular structure may promote competition in the presence of similarly structured compound. This competition, in turn, may decrease the absorption of a drug.

Active or facilitated absorption of a drug is usually explained by assuming that carriers in membranes are responsible for shuttling these solutes in mucosal or serosal direction. The number of apparent carriers in membranes, however, is limited. Therefore, the rate of transfer may be described by the equation:

$$\text{Absorption rate} = \frac{dC}{dt} = \frac{V_{max}*C}{K_m + C}$$

(Equation 3)

in which C is the solute concentration at the absorption site and V_{max} (the maximum theoretical transfer rate) and K_m (the concentration of drug at ½V_{max}) are constants. In low doses or concentration, when $K_m \gg C$, equation 3 reduces to

$$\frac{dC}{dt} = \frac{V_{max}}{K_m}*C = K*C$$

(Equation 4)

Equation 4 indicates that the apparent first-order kinetics is observed. Under these conditions, a sufficient number of carriers are available so that a constant proportion of solute molecules presented to the membrane is transported. As the solute concentration increases, the number of free carriers is reduced and the proportion of solute molecules transferred across the membrane is reduced until a maximum absolute number is reached. When $C \gg K_m$, then

$$\text{Absorption rate} = \frac{dC}{dt} = V_{max}$$

(Equation 5)

Equation 5 indicates that a further increase in solute con-

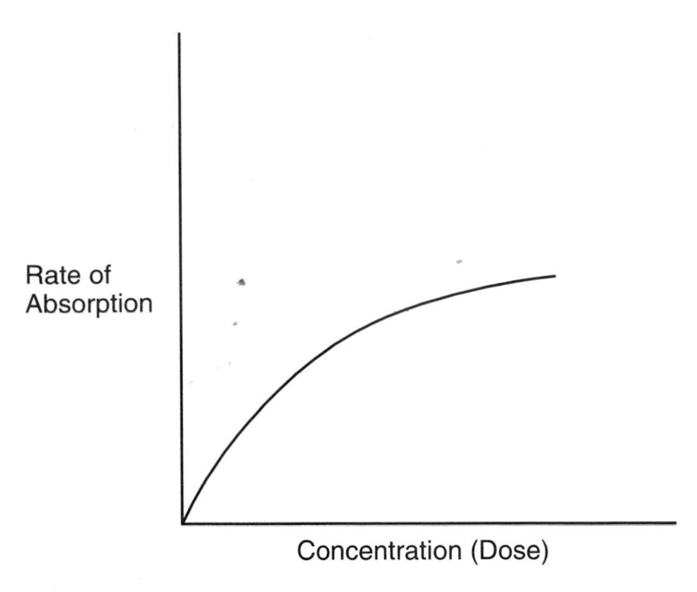

Fig. 3–7. The relationship between drug concentration and rate of absorption owing to active transport.

centration will not result in any further increase in the rate of absorption (Fig. 3–7).

The capacity-limited characteristics of carrier-mediated processes suggest that the bioavailability of drugs absorbed in this manner should decrease with increasing doses. Therefore, the use of a large single oral dose of these drugs is irrational and, if larger daily doses are necessary, divided doses are preferable. Examples of substances that are actively transported include amino acids, methyldopa, 5-fluorouracil, penicillamine, and levodopa.

Convective Absorption

The absorption of small molecules, with molecular radii less than about 4 angstroms, through water-filled pores of biologic membrane is referred to as convective absorption. The rate of such absorption is equated to the product of a sieving coefficient, the rate of fluid or water absorption and the concentration of solute in the luminal content. The sieving coefficient is indirectly related to the relative sizes of the pores and the molecules.

Ion Pair Absorption

In 1967, Higuchi suggested that highly ionized compounds, such as quaternary ammonium compounds, may be absorbed by ion pair mechanism.[6] In vitro, a relatively large organic anion can combine with a relatively large cation to form an ion pair that crosses a water-organic solvent interface and transfers to an organic phase.

PHYSICOCHEMICAL FACTORS VERSUS DRUG ABSORPTION

The pH-Partition Hypothesis

Drug absorption is influenced by many physiologic factors. Additionally, it also depends on many physicochemical properties of the drug itself. Shore, Brodie, Schanker, and colleagues[7–12] concluded from their research that most drugs are absorbed from the gastrointestinal tract by a process of passive diffusion of the un-ionized moiety across a lipid membrane. Furthermore, the dissociation constant, lipid solubility, and pH of the fluid at the absorption site determine the extent of absorption from a solution. The interrelationship among these parameters is known as the pH-partition theory. This theory provides a basic framework for the understanding of drug absorption from the gastrointestinal tract and drug transport across biologic membrane. The principal points of this theory are as follows:

1. The gastrointestinal and other biologic membranes act like lipid barriers.
2. The un-ionized form of the acidic or basic drug is preferentially absorbed.
3. Most drugs are absorbed by passive diffusion.
4. The rate of drug absorption and the amount of drug absorbed are related to its oil-water partition coefficient; the more lipophilic the drug, the faster its absorption.
5. Weak acidic and neutral drugs may be absorbed from the stomach, but basic drugs are not.

When a drug is administered intravenously, it is immediately available to body fluids for distribution to the "site of action." All extravascular routes, however, can influence the overall therapeutic activity of the drug, primarily because of its dissolution rate: a step necessary for a drug to be available in a solution form. When a drug is administered orally in a dosage form such as a tablet, capsule, or suspension or intravenously, the rate of absorption across the biologic membrane frequently is controlled by the slowest step in the following sequence:

$$\text{Dosage form} \xrightarrow{\text{dissolution}} \text{Drug in solution form} \xrightarrow{\text{absorption}} \text{Drug in general circulation}$$

In many instances, the slowest or rate-limiting step in the sequence is the dissolution of the drug. When dissolution is the controlling step, any factors that affect the rate of dissolution must also influence the rate of absorption. This rate, in turn, affects the extent and duration of action. Several factors can influence the dissolution rate of drug from solid dosage forms and, therefore, the therapeutic activity. These factors include solubility of a drug, particle size and surface area of drug particles, crystalline and salt form of a drug, and the rate of distintegration.

The absorption rate of drugs can also be affected by interaction or formation of complexes in the gastrointestinal tract. Generally, such complex formation reduces the concentration of free drug at the absorption site. Because the complexed drug is absorbed either slowly or not at all, the net effect is the reduction of concentration of drug at the absorption site and a slower rate of absorption.

Ionization and pH at Absorption Site

The fraction of the drug existing in its un-ionized form in a solution is a function of both the dissociation constant of a drug and the pH of the solution at the absorption site. The dissociation constant, for both weak acids and bases, is often expressed as pK_a (the negative logarithm of a dissociation constant, K_a). The Henderson-Hasselbach equation for the ionization of a weak acid, HA, can be derived from the equation:

$$HA + H_2O \rightleftharpoons A^- + H_3O^+$$

(Equation 6)

We may express the equilibrium constant as follows:

$$K_a = \frac{{}_aH_3O^+ * {}_aA^-}{{}_aHA}$$

(Equation 7)

in which K_a is the equilibrium or dissociation constant and a is the activity coefficient. Assuming the activity coefficients approach unity in dilute solutions, the activity coefficient may be replaced by concentration terms, and equation 7 becomes:

$$K_a = \frac{[H_3O^+][A^-]}{[HA]}$$

(Equation 8)

The negative logarithm of K_a is referred to as pK_a. Thus,

$$pK_a = -\log K_a$$

Taking the logarithm of the expression for the dissociation constant of a weak acid in equation 8 yields:

$$-\log K_a = -\log[H_3O] - \log \frac{[A^-]}{[HA]}$$

(Equation 9)

in which A^- is the ionized form of a weak acid and HA is the un-ionized form.

$$pH - pK_a = \log \frac{[\text{Ionized}]}{[\text{Un-ionized}]}$$

(Equation 10)

Assuming α is the fraction of ionized species and $1 - \alpha$ is the fraction remaining as the un-ionized form, equation 10 can be written as:

$$pH - pK_a = \log \frac{\alpha}{1 - \alpha}$$

(Equation 11)

or

$$\frac{\alpha}{1 - \alpha} = \text{Antilog } (pH - pK_a)$$

(Equation 12)

From equation 12, the fraction or percentage absorbable and nonabsorbable form of a weak acid can be calculated, provided the pH condition at the site of administration is known. Analogously, the dissociation or basicity constant for a weak base is derived as follows:

$$B + H_2O \leftrightarrow BH^+ + OH^-$$

The dissociation constant, K_b

$$K_b = \frac{{}_aOH * {}_aBH}{{}_aB} = \frac{[OH^-][BH]}{[B]}$$

$$\text{and } pK_b = -\log K_b$$

(Equation 13)

The pK_a and pK_b values provide a convenient means of comparing the strength of weak acids and bases. The lower the pK_a, the stronger the acid; the lower the pK_b, the stronger the base. pK_a and pK_b values of conjugate acid-base pairs are linked by the expression:

$$pK_a + pK_b = pK_w$$

(Equation 14)

in which pK_w is the negative logarithm of the dissociation constant of water. Taking logarithm of equation 13 and rearranging yields:

$$-\log K_b = -\log[OH^-] - \log \frac{[BH^+]}{[B]}$$

(Equation 15)

Although the dissociation constant of a weak base is described by the term K_b, it is conventionally expressed in terms of K_a because of the relationship expressed in equation 14.

Equation 15 can be written as:

$$pH = pK_w - pK_b - \log \frac{[BH]}{[B^-]}$$

(Equation 16)

Because $pK_w - pK_b = pK_a$, equation 16 takes the following form for a weak base. BH^+ is the un-ionized form and B^- is the ionized form:

$$pK_a - pH = \log \frac{[\text{Ionized}]}{[\text{Un-ionized}]}$$

(Equation 17)

Again, assuming α is the fraction of ionized species and $1 - \alpha$ is the fraction of un-ionized species, equation 17 becomes:

$$pK_a - pH = \log \frac{\alpha}{1 - \alpha}$$

or

$$\frac{\alpha}{1 - \alpha} = \text{Antilog } (pK_a - pH)$$

(Equation 18)

From equation 18, one can calculate the fraction or percent of absorbable and nonabsorbable form of a weak base, given the pH condition at the site of drug absorption. Figure 3–8 shows the pK_a values of several drugs and the relative acid or base strength of these compounds.

The relationship between pH and pK_a and the extent of ionization is given by equations 12 and 18 for weak acids and weak bases, respectively. Accordingly, most weak acidic drugs are predominantly in the un-ionized form at lower pH of the gastric fluid and may, therefore, be absorbed from the stomach as well as from the intestine. Some very weak acidic drugs, such as phenytoin and many barbiturates, with pK_a values are greater than 7.0, are essentially un-ionized at all pH values. Therefore, for these weak acidic drugs, transport is more rapid and independent of pH, provided the un-ionized form is lipophilic or nonpolar. Furthermore, it is important to note that the fraction un-ionized changes dramatically only for weak acids, with pK_a values between 3 to 7. Therefore, for these compounds a change in the rate of transport with pH is expected (Fig. 3–9).

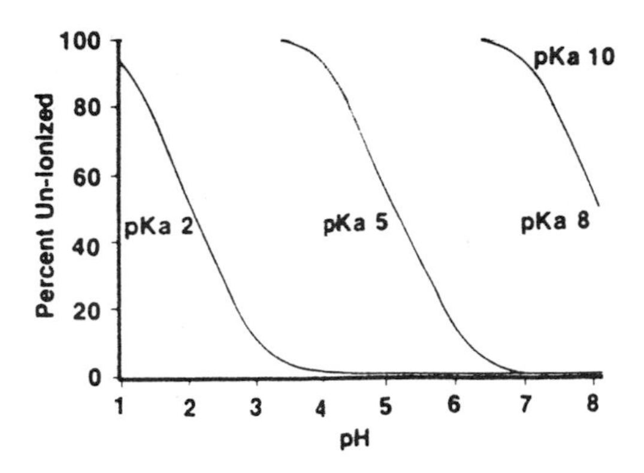

Fig. 3–9. For very weak acids, pK_a values greater than 8.0 are predominantly un-ionized at all pH values between 1.0 and 8.0. Profound changes in the fraction un-ionized occur with pH for an acid with a pK_a value that lies within the range of 2.0 to 8.0. Although the fraction un-ionized of even strong acids increases with hydrogen ion concentration, the absolute value remains low at most pH values shown. (From M. Rowland and T. Tozer, *Clinical Pharmacokinetics: Concepts and Application,* 2nd ed., Philadelphia, Lea & Febiger, 1989.)

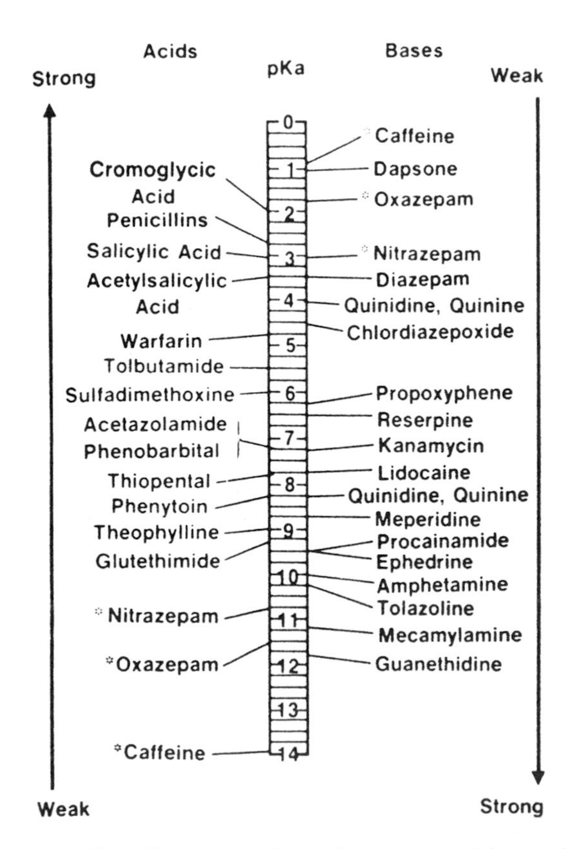

Fig. 3–8. The pK_a values of certain acidic and basic drugs. Drugs denoted with an asterisk are amphoteric. (From M. Rowland and T. Tozer, *Clinical Pharmacokinetics: Concepts and Application,* 2nd ed., Philadelphia, Lea & Febiger, 1989.)

Although the transport of weak acids with pK_a values of less than 3.0 should depend on pH, the fraction un-ionized is so low that transport across the gut membrane may be slow, even under the most acidic conditions.

Most weak bases are poorly absorbed, if at all, in the stomach because they are present largely in the ionized form at low pH. Codeine, a weak base with pK_a of about 8.0, has about one in every million molecules in its un-ionized form at gastric pH of 1.0. Weakly basic drugs such as dapsone, diazepam, and chlordiazepoxide are essentially un-ionized through the intestine. Strong bases, those with pK_a values between 5 and 11, show pH-dependent absorption. Stronger bases such as guanethidine ($pK_a > 11$) are ionized throughout the gastrointestinal tract and tend to be poorly absorbed.

The evidence of the importance of dissociation in drug absorption is found in the result of studies in which pH at the absorption site is changed (Tables 3–1 and 3–2). Table 3–1 clearly shows the decreased absorption of a weak acid at pH 8.0, compared to pH 1.0. On the other hand, an increase to pH 8.0 promotes the absorption of a weak base with practically nothing absorbed at pH 1.0. The data in Table 3–2 permits a comparison of intestinal absorption of acidic and basic drugs from buffered solutions ranging from pH 4.0 to 8.0. These results are in agreement with pH-partition hypothesis.

Lipid Solubility

Some drugs may be poorly absorbed after oral administration even though they are available predominantly in an un-ionized form in the gastrointestinal tract. This factor is attributed to the low lipid solubility of the un-ionized molecule. A guide to the lipid solubility or lipophilic

Table 3–1. Comparison of Gastric Absorption of Acids and Bases at pH 1 and 8 in the Rat[14]

Acids	pK_a	Percent Absorbed at pH 1	Percent Absorbed at pH 8
5-Sulfosalicyclic acid	<2.0	0	0
5-Nitrosalicyclic acid	2.3	52	16
Salicyclic acid	3.0	61	13
Thiopental	7.6	46	34
Bases			
Aniline	4.6	6	56
p-Toluidine	5.3	0	47
Quinine	8.4	0	18
Dextromethorphan	9.2	0	16

nature of a drug is provided by a property called partition coefficient. This parameter, therefore, influences the transport and absorption processes of drugs, and is one of the most widely used properties in quantitative structure-activity relationships.

The movement of molecules from one phase to another is called partitioning. Drugs partition themselves between the aqueous phase and lipophilic membrane; preservative emulsions partition between the water and oil phases; antibiotics partition from body fluids to microorganisms; and drug and other adjutants can partition into the plastic and rubber stoppers of containers. Therefore, it is important that this process is understood.

If two immiscible phases are placed adjacent to each other, one containing a solute soluble in both phases, the solute will distribute itself into two immiscible phases until equilibrium is attained and, therefore, no further transfer of solute occurs. At equilibrium, the chemical potential of the solute in one phase is equal to its chemical potential in the other phase. If we consider an aqueous (w) and an organic (o) phase:

Table 3–2. Comparison of Intestinal Absorption of Acids and Bases in the Rat at Several pH Values[15]

Acids	pK_a	Percent Absorbed from Rat Intestine			
		pH 4	pH 5	pH 7	pH 8
5-Nitrosalicyclic acid	2.3	40	27	0	0
Salicyclic acid	3.0	64	35	30	10
Acetylsalicyclic acid	3.5	41	27	—	—
Benzoic acid	4.2	62	36	35	5
Bases					
Aniline	4.6	40	48	58	61
Amidopyrine	5.0	21	35	48	52
p-Toluidine	5.3	30	42	65	64
Quinine	8.4	9	11	41	54

$$\mu_w^\theta + RT \ln a_o = \mu_o^\theta + RT \ln a_o$$

(Equation 19)

in which a represents the activity coefficient of a solute. Rearranging this equation yields:

$$\frac{\mu_w^\theta + \mu_o^\theta}{RT} = \ln \frac{a_w}{a_o}$$

in which the term on the left side is a constant at a given temperature and pressure. Therefore,

$$\frac{a_w}{a_o} = \text{constant or } \frac{a_o}{a_w} = \text{constant}$$

(Equation 20)

These constants are the partition or distribution coefficients, P. If the solute under consideration forms an ideal solution in both phases or solvent, the activity coefficient can be replaced by the concentration term, and equation 20 becomes:

$$P = \frac{C_o}{C_w}$$

(Equation 21)

which is used conventionally to calculate the partition coefficient of a drug. In equation 21, C_0, the concentration of drug in the organic or oil phase, is divided by the concentration in the aqueous phase. The greater the value of P, the higher the lipid solubility of the solute. It has been demonstrated for several systems that the partition coefficient can be approximated by the solubility of the solute in the organic phase divided by the solubility in the aqueous phase. Therefore, the partition coefficient is a measure of the relative affinities of the solute for an aqueous or nonaqueous or oil phase. The effect of lipid solubility and, hence, partition coefficient on the absorption of a series of barbituric acid derivatives is shown in Table 3–3 and Figure 3–6.

Table 3–3. Comparison of Barbiturate Absorption in Rat Colon and Partition Coefficient (Chloroform/Water) of Undissociated Drug[9]

Barbiturate	Partition Coefficient	Percent Absorbed
Barbital	0.7	12
Apobarbital	4.9	17
Phenobarbital	4.8	20
Allylbarbital	10.5	23
Butethal	11.7	24
Cyclobarbital	13.9	24
Pentobarbital	28.0	30
Secobarbital	50.7	40
Hexethal	>100	44

Although drugs with greater lipophilicity and, therefore, partition coefficient are better absorbed, it is imperative that drugs exhibit some degree of aqueous solubility. This feature is essential because the availability of drug molecule in solution form is a prerequisite for drug absorption, and the biologic fluids at the site of absorption are aqueous in nature. Therefore, from a practical viewpoint, drugs must exhibit a balance between hydrophilicity and lipophilicity. This factor is always taken into account when a chemical modification is considered as a way of improving the efficacy of a therapeutic agent.

The critical role of lipid solubility in drug absorption is a major guiding principle in the process of drug discovery and development. Polar or hydrophilic molecules such as gentamicin, ceftrixine, and streptokinase are poorly absorbed after oral administration and must therefore be administered parenterally. Lipid-soluble drugs with favorable partition coefficients are generally well absorbed after oral administration. Often, the selection of a compound with higher partition coefficient from a series of research compounds provides improved pharmacologic activity. Occasionally, the structure of an existing drug is modified to develop a similar pharmacologic activity with improved absorption. Chlortetracycline, which differs from tetracycline by the substitution of a chlorine at C-7, substitution of an n-hexyl (Hexethal) for a phenyl ring in phenobarbital, or replacement of the 2-carbonyl of pentobarbital with a 2-thio group (thiopental) are examples of agents with enhanced lipophilicity (Fig. 3–10).

It is important to note that even a minor molecular modification of a drug may promote the risk of also altering the efficacy and safety profile of a drug. For this reason, medicinal chemists prefer the development of a lipid-soluble prodrug of a drug with poor oral absorption characteristics.

Prodrugs are designed in some cases to improve permeability and oral absorption of the parent drug. They are more lipid soluble than the parent drug, and should be rapidly converted to the parent compound during absorption from the gut wall or the liver. Pivampicillin, a prodrug of ampicillin, is an ester and is more lipid soluble. Therefore, it is absorbed more efficiently than the parent compound ampicillin.[16]

The pH partition theory provides a basic framework for the understanding of drug absorption and is, sometimes, an oversimplification of a more complex process. For example, experimentally observed pH-absorption curves are less steep (Fig. 3–11) than those expected

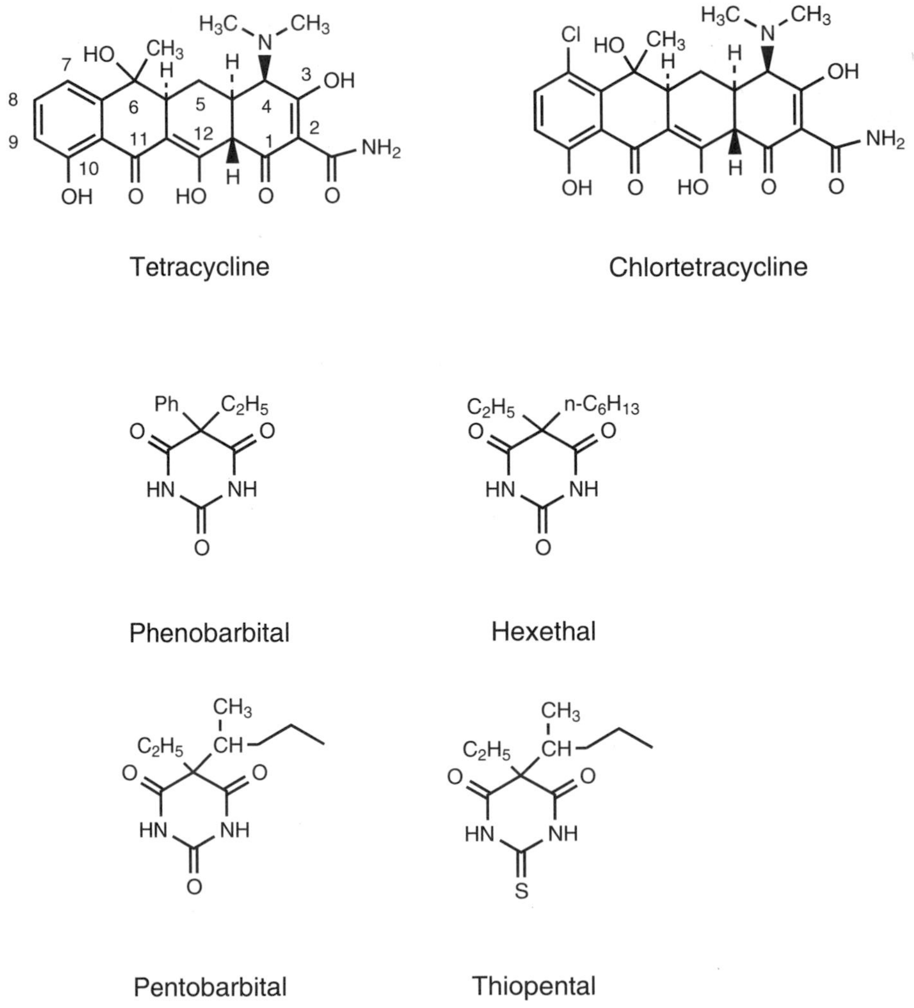

Tetracycline

Chlortetracycline

Phenobarbital

Hexethal

Pentobarbital

Thiopental

Fig. 3–10. Drug pairs in which chemical modification enhances lipophilicity.

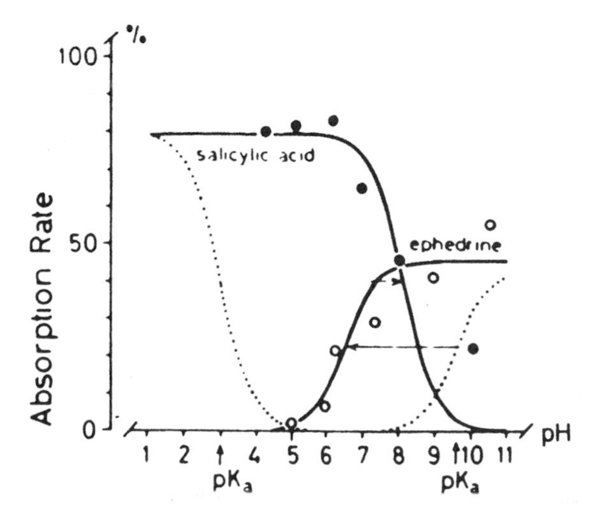

Fig. 3–11. Relationship between absorption rates of salicylic acid and ephedrine and bulk phase pH in the rat small intestine in vivo. *Dashed lines,* curves predicted by the pH-partition theory in the absence of an unstirred layer. (From D. Winne, The influence of unstirred layers on intestinal absorption, in Intestinal Permeation Workshop Conference Hoechest, Vol. 4, M. Kramer, and F. Lauterbach, Eds., Amsterdam-Oxford, Excerpta Medica International Congress series #391, 1977, pp. 58–64.)

theoretically and are shifted to higher pH values for acids and lower pH values for bases.

This deviation, observed experimentally, is attributed by several investigators to factors such as limited absorption of ionized species of drugs, the presence of an unstirred diffusion layer adjacent to the cell membrane, and a difference between lumenal pH and cell membrane surface pH.

DRUG DISSOLUTION VERSUS DRUG ABSORPTION

Solid Dosage Forms and Suspension

When a drug is administered orally, in tablet, capsule, or suspension form, the rate of absorption is often controlled by how fast the drug particles dissolve in the fluid at the site of administration. Hence, dissolution rate is often the rate-limiting (slowest) step in the process (see page 16). When dissolution of the drug controls the rate of absorption, it is said to be dissolution rate limited. Figure 3–12 illustrates the absorption of aspirin from solution and two different types of tablets. It is clear that aspirin absorption is more rapid from solution than from tablet formulations. This rapid absorption is an indication of the absorption being dissolution rate limited. A general relationship describing the dissolution of a drug was first reported by Noyes and Whitney.[19] The equation derived by Noyes and Whitney states that:

$$\frac{dC}{dt} = KS\,(C_s - C)$$

(Equation 22)

in which dC/dt is the dissolution rate, K is a constant, S is the surface area of the dissolution solid, C_s is the equilibrium solubility of drug in the solvent, and C is the concentration of drug in the solvent at time t.

The constant K in this equation has been shown to be equal to D/h, in which D is the coefficient of the dissolving material of the drug and h is the thickness of the diffusion layer surrounding the dissolving solid particles. This diffusion layer is a thin stationary film of a solution adjacent to the surface of a solid particle (Fig. 3–13) and is saturated with drug, i.e., drug concentration in the diffusion layer is equal to C_s, the equilibrium solubility. The term $(C_s - C)$ in equation 22 represents the concentration gradient for the drug between the diffusion layer and the bulk solution. If dissolution is the rate-limiting step in the absorption process, the term C in this equation is negligible compared to C_s. Under this condition, equation 22 is reduced to:

$$\frac{dC}{dt} = \frac{DSC_s}{h}$$

(Equation 23)

which describes a diffusion-controlled dissolution process. When solid drug particles are introduced to the fluids at the absorption sites, the drug promptly saturates the diffusion layer (see Fig. 3–13). This step is followed by the diffusion of drug molecules from the diffusion layer into the bulk solution, which are instantly replaced in the diffusion layer by molecules from the solid crystal or particle. This process is continuous. Even though it oversimplifies the dynamics of the dissolution process, equation 23 is a qualitatively useful equation and clearly indicates the effects of some important factors on the dissolution, and therefore, the absorption rate of drugs. When dissolution is the rate-limiting factor in the absorption, bioavailability is affected. These factors are listed in Table 3–4.

The Noyes-Whitney equations (equations 22 and 23) demonstrate that equilibrium solubility (C_s) is one of the

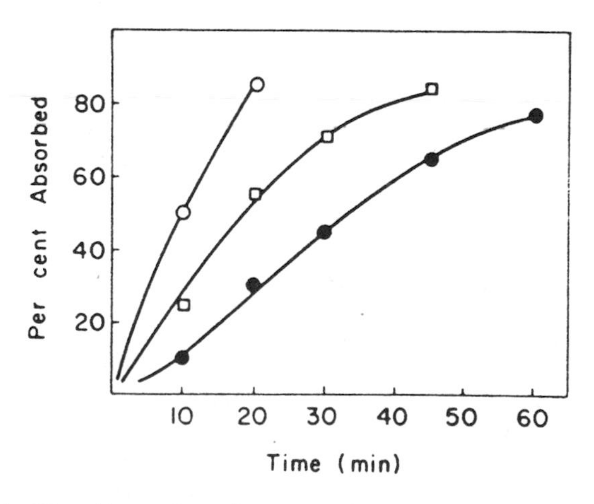

Fig. 3–12. Absorption of aspirin after oral administration of a 650-mg dose in solution (○), in buffered tablets (□), or in regular tablets (●). (Data from G. Levy, et al., *J. Pharm. Sci., 54,* 1719 (1965).

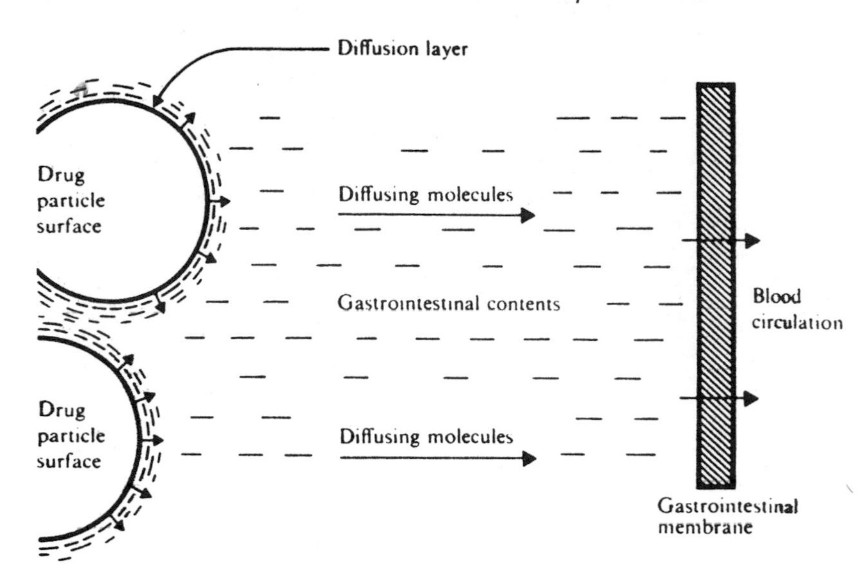

Fig. 3–13. Dissolution from a solid surface. (From A. T. Florence and D. Attwood, *Physicochemical Principles of Pharmacy,* 2nd ed., New York, Chapman and Hall, 1988.)

major factors determining the rate of dissolution. Changes in the characteristics of solvents, such as pH, that affect the solubility of the drug, affect its dissolution rate. Similarly, the use of a different salt or other physicochemical form of a drug that has a solubility different from the parent drug usually affects the dissolution rate. Increasing the surface area of a drug exposed to the dissolution medium, by reducing the particle size, usually increases the dissolution rate. In the discussion to follow, some of the more important factors affecting dissolution, and therefore absorption, are presented in greater detail.

pH

The solubility of weak acids and bases is a function of the pH of the medium. Therefore, differences in the dissolution rate are expected to occur in different regions of the gastrointestinal tract. The solubility of weak acid is obtained by:

$$C_s = [HA] + [A^-]$$

(Equation 24)

in which [HA] is the intrinsic solubility of the un-ionized acid, i.e., C_o, and $[A^-]$ is the concentration of its anion, which can be expressed in terms of its dissociation constant, K_a, and C_o:

$$C_s = C_o + \frac{K_a C_o}{[H^+]}$$

(Equation 25)

Analogously, the solubility of a weak base is obtained by:

$$C_s = C_o + \frac{C_o[H^+]}{K_a}$$

(Equation 26)

By substituting equations 25 and 26 into equation 23 for the term C_s, the following dissolution rate equations are obtained:

$$\text{For weak acids: } \frac{dC}{dt} = \frac{K'(C_o + K_a C_o)}{[H^+]}$$

(Equation 27)

Table 3–4. How to Change Parameters of Dissolution Equation to Increase (+) or Decrease (−) Rate of Solution

Equation Parameter	Comments	Effect on Rate of Solution
D (diffusion coefficient of drug)	May be decreased in presence of substances that increase viscosity of the medium	(−)
A (area exposed to solvent)	Increased by micronization and in "amorphous" drugs	(+)
δ (thickness of diffusion layer)	Decreased by increased agitation in gut or flask	(+)
c_s (solubility in diffusion layer)	That of weak electrolytes altered by change in pH, by use of appropriate drug salt or buffer ingredient	(−)(+)
c (concentration in bulk)	Decreased by intake of fluid in stomach, by removal of drug by partition or absorption	(+)

From A.T. Florence and D. Attwood: Physicochemical Principles of Pharmacy, 2nd ed., New York, Chapman and Hall, 1988.

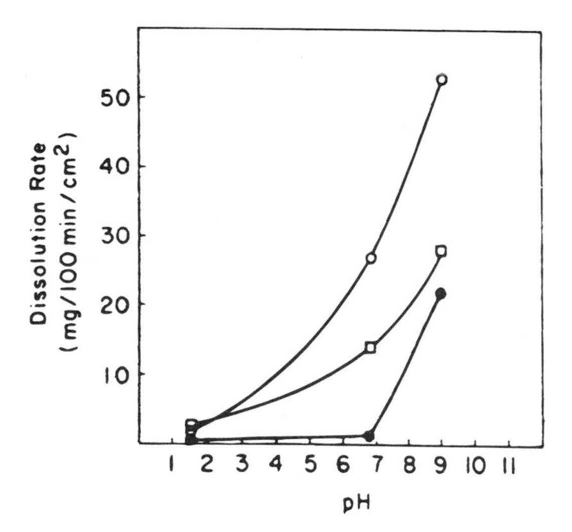

Fig. 3–14. pH-Dependent dissolution of salicylic acid (○), benzoic acid (□), and phenobarbital (●). (From M. Gibaldi, *Biopharmaceutics and Clinical Pharmacokinetics,* 3rd ed., Philadelphia, Lea & Febiger, 1984.)

or

$$\frac{dC}{dt} = \frac{K'C_o(1 + K_a)}{[H^+]}$$

(Equation 28)

$$\text{For a weak base: } \frac{dC}{dt} = \frac{K'C_o(1 + [H^+])}{K_a}$$

(Equation 29)

In equations 27 through 29, K' is equal to DS/h. Equations 28 and 29 clearly suggest that the dissolution rate of weak bases decreases with increasing pH. Hence, the dissolution rate of weak bases is maximum in gastric fluid, but for weak acids, it is at a minimum. Furthermore, the dissolution rate of weak acids increases as the solid drug particles move to the more alkaline regions of the gastrointestinal tract. Figure 3–14 illustrates the dissolution rates of weak acids as a function of pH.

The absorption of a salt of weak acid or base can be explained by using the following:

$$\text{Salt} \xrightarrow{\text{dissolution}} \text{Ionized form} \underset{K_2}{\overset{K_1}{\rightleftharpoons}} \text{Un-ionized form}$$

in which K_1 and K_2 represent the rate constants associated with the formation of un-ionized and ionized species of a compound, respectively. The ratio of these two rate constants represents the dissociation constant of a compound. The absorption of the un-ionized species of a molecule disturbs the equilibrium of the process. To regain the equilibrium, some of the ionized species are converted into un-ionized species, which are then absorbed through the membrane. This process, being continuous, permits the absorption of the un-ionized species. Therefore, a drug molecule eventually is absorbed.

The relatively poor dissolution of weak acids at the pH of gastric fluid diminishes further the importance of the stomach as a drug absorption site. Although gastric absorption of weak acids may occur from solution, it is unlikely that much of the drug dissolves and is absorbed during the short residence time as solid dosage form in the stomach. Ogata and co-workers[21] proposed that the critical value of solubility that separates acid drugs from the absorption sites (stomach or intestine) is about 30 mg/ml in 0.1 N HCl when 1 g of drug is administered orally. The authors found that if the solubility of a drug is less than 3 mg/ml, practically no absorption occurs in the stomach. Changes in the gastric pH also alter the solubility of certain drugs and may affect the dissolution and absorption rates. A patient with achlorhydria has a higher gastric pH and absorbs aspirin more rapidly than a normal subject. On the other hand, similar differences were not observed with respect to the absorption rates of acetaminophen, a weaker acid, the solubility of which would be unaffected by changes in pH.[22]

The relationship between dissolution rate and hydrogen ion concentration, described in equations 28 and 29, are approximations and tend to overpredict the dissolution rate of weak acids in the small intestine and weak bases in the stomach. In reality, the hydrogen ion concentration $[H^+]$ of the bulk is not equal to the hydrogen ion concentration $[H^+]$ of the diffusion layer.

Salts

The dissolution rate of a particular salt is usually different from that of a parent compound. Sodium or potassium salts of weak acids dissolve more rapidly than the free acid. The same is true with HCl or other salts of weak bases. Table 3–5 illustrates the dissolution rate differences between some weak acids and their sodium salts.

The differences in the dissolution rates of salt and parent compound can be explained by considering the pH

Table 3–5. Dissolution Rate of Weak Acids and Their Sodium Salts*

		Dissolution Rate (mg/100 min/cm²)		
Compound	pK_a	0.1 N HCl pH 1.5	0.1 M Phosphate pH 6.8	0.1 M Borate pH 9.0
Benzoic acid	4.2	2.1	14	28
Sodium salt		980	1770	1600
Phenobarbital	7.4	0.24	1.2	22
Sodium salt		~200	820	1430
Salicylic acid	3.0	1.7	27	53
Sodium salt		1870	2500	2420
Sulfathiazole	7.3	<0.1	~0.5	8.5
Sodium salt		550	810	1300

* Data from E. Nelson, *J. Am. Pharm. Assoc. (Sci. Ed.), 47,* 297 (1958).

of the diffusion layer. At a given pH, regardless of salt or free acids/bases, a drug has a fixed solubility. The classic dissolution equation predicts a slower dissolution of a salt of a drug, and the concept of diffusion layer becomes useful.

For sodium or potassium salts of weak acids, the pH of the solution in a diffusion layer is greater than the pH of the diffusion layer for the corresponding weak acid. On the other hand, the pH of the solution in the diffusion layer for hydrochloride salts of weak bases is always less than that of the diffusion layer of the corresponding free base. Therefore, effective solubility and the dissolution rate of soluble salts are greater than those of its corresponding free acid or base. Many examples of the effects of soluble salts on drug absorption are available. The potassium salt of penicillin V yields a higher peak plasma concentration of antibiotic than the corresponding free acid.[24] Sodium salts of barbiturates are reported by Anderson[25] to provide a rapid onset of sedation. Some salts have a lower solubility and dissolution rate than their parent compounds. Examples include aluminum salts of weak acids and palmoate salts of weak bases. In these particular examples, insoluble films of either weak acids or palmoic acid appear to form in the dissolving solids, which further retards the dissolution rate.

Surface Area and Particle Size

A drug dissolves more rapidly when its surface area is increased. This change is usually accomplished by reducing the particle size of a drug. Therefore, many poorly soluble and slowly dissolving drugs are currently available in micronized or microcrystalline form. The problems of low water solubility and particle size were not fully appreciated, but the result has been a reduction in the therapeutic dose of some drugs without sacrificing therapeutic efficacy. The formulation of spironolactone, for example, has been reduced from 500 mg to 25 mg owing to micronization. A similar result has been obtained for griseofulvin.

SUMMARY

At one time, it was common to assume that the biologic response to a drug was simply a function of the intrinsic pharmacologic activity of the drug molecule. Today, while assessing the potency of most drugs, more consideration is given to plasma drug concentration-response than to dose-response relationships. The concentration of a drug in the plasma depends on the rate and extent of absorption, which in turn is influenced by the physicochemical properties of drug substances. Drug absorption may profoundly affect the onset and intensity of a biologic response of a drug. Clinically significant differences in the absorption of closely related drugs such as lincomycin and clindamycin, penicillin and pivampicillin, or secobarbital and sodium secobarbital are invariably attributable to significant differences in their physicochemical properties.

Dissolution is simply a process by which a solid substance goes into solution. The determination of dissolution rates of pharmaceutical substances from dosage forms does not predict their bioavailability and/or their in vivo performance; rather, it indicates the potential availability of drug substance for absorption. Therefore, it is essential for pharmacists and pharmaceutical scientists to know and understand the importance of dissolution and its potential influence on the rate and extent of absorption and availability for drugs.

Factors affecting the dissolution rate of a drug from a dosage form can be related to the physicochemical properties of a drug, the formulation of a dosage form, and the dissolution apparatus and test parameters.

REFERENCES

1. E. J. Ariens, *J. Cardiovasc. Pharmacol., 5,* 58(1983).
2. J. C. Venter and C. M. Fraser, *Mechanism of action and regulation,* in *Neurotransmitter Receptors,* S. Kito, et al., Eds., New York, Plenum Press, 1984.
3. R. J. Lefkowitz, et al., *Annu. Rev. Biochem., 52,* 159(1983).
4. A. T. Florence and D. Attwood, *Physicochemical Principles of Pharmacy,* 2nd ed., New York, Chapman and Hall, 1988.
5. L. S. Schanker, *J. Pharmacol. Exp. Ther., 126,* 283(1959).
6. J. G. Wagner, *Biopharmaceutics and Relevant Pharmacokinetics,* 1st ed., Hamilton, The Hamilton Press, 1988, p. 31.
7. P. A. Shore, et al., *J. Pharmacol. Exp. Ther., 119,* 361(1957).
8. C. A. M. Hogben, et al., *J. Pharmacol. Exp. Ther., 125,* 275(1959).
9. L. S. Schanker, *J. Med. Pharm. Chem., 2,* 343(1960).
10. L. S. Schanker, *Annu. Rev. Pharmacol. Toxicol., 1,* 29(1961).
11. L. S. Schanker, *Passage of drugs across the gastrointestinal epithelium in drugs and membrane,* C. A. M. Hogben, Ed., Proceedings of the First International Pharmacology Meeting, Vol. 4, New York, The Macmillan Company, 1963.
12. L. S. Schanker, *Physiological transport of drug,* in *Advances in Drug Research,* N. J. Harper and A. B. Simons, Eds., London, Academic Press, 1966.
13. M. Rowland and T. Tozer, *Clinical Pharmacokinetics: Concepts and Application,* 2nd ed., Philadelphia, Lea & Febiger, 1989.
14. L. S. Schanker, *J. Pharmacol. Exp. Ther., 128,* 81(1958).
15. J. J. Zimmerman and S. Feldman, *Physical-chemical properties and biologic activity,* in *Principles of Medicinal Chemistry,* 3rd ed., W. O. Foye, Ed., Philadelphia, Lea & Febiger, 1988.
16. E. L. Foltz, et al., *Antimicrob. Agents Chemother.,* 442(1970).
17. D. Winne, *The influence of unstirred layers on intestinal absorption,* in *Intestinal Permeation Workshop Conference Hoechest,* Vol. 4, M. Kramer and F. Lauterbach, Eds., Amsterdam-Oxford, Excerpta Medica International Congress series #391, 1977, pp. 58–64.
18. G. Levy, et al., *J. Pharm. Sci., 54,* 1719(1965).
19. N. A. Noyes and W. R. Whitney, *J. Am. Chem. Soc., 19,* 930(1987).
20. M. Gibaldi, *Biopharmaceutics and Clinical Pharmacokinetics,* 4th ed., Philadelphia, Leb & Febiger, 1991.
21. Ogata, H., et al., *Chem. Pharm. Bull., 27,* 1281(1979).
22. A. Pottage, et al., *J. Pharm. Pharmacol., 26,* 144(1974).
23. E. Nelson, *J. Am. Pharm. Assoc. (Sci. Ed.), 47,* 297(1958).
24. H. Juncher and F. Raaschou, *Antibiotic Med., 4,* 497(1957).
25. K. W. Anderson, *Arch. Int. Pharmacodyn. Ther., 147,* 171(1964).

Chapter 4

STRUCTURAL FEATURES AND PHARMACOLOGIC ACTIVITY

Ian W. Mathison, William E. Solomons, and Richard R. Tidwell

A drug is subjected to many complex processes from the time it is administered to the time the biologic response is effected. Processes such as passage of the drug through biologic membranes and penetration to sites of action are major phenomena to be considered in a drug's action and depend to a large extent on the physical properties of the molecule. The stereochemistry of a molecule, that is, the relative spatial arrangement of the atoms, or three-dimensional structure of the molecule, also plays a major role in the pharmacologic properties, because many of these processes are stereospecific.

Although stereochemistry does play a major role in a drug's biologic action, factors such as the lipid: water distribution function, the pK value, or perhaps the rate of hydrolysis or metabolism may differ between isomeric pairs and account for the observed differences in pharmacologic activity. Care, therefore, must be taken when considering structure-activity relationships to determine if major differences in physical properties exist before one makes firm correlations with the steric arrangement of the molecules. In many reported examples of steric relationships, these factors are ignored.

When stereochemistry alone is responsible for differences in the degree of pharmacologic activity between isomers, conclusions may be made regarding the steric requirements of the drug receptor site. On the other hand, if isomers differ in effectiveness and in other parameters, such as partition coefficient or pK values, the ability of one isomer to preferentially reach the active site or to be metabolized more readily is a real possibility. In these cases, therefore, steric factors are indirectly responsible for the observed differences in action.

The influence of steric factors on pharmacologic activity is considered under three major headings: (1) Optical and Geometric Isomerism and Pharmacologic Activity; (2) Conformational Isomerism and Pharmacologic Activity; (3) Isoterism and Pharmacologic Activity.

OPTICAL AND GEOMETRIC ISOMERISM AND PHARMACOLOGIC ACTIVITY

Optical Isomerism

The fact that enantiomorphic pairs, that is, optical isomers, may exhibit different biologic activities is by no means a recent discovery. Enantiomorphs were separated and their different activities observed as early as 1858 by Louis Pasteur.[1] He separated the isomers of tartaric acid by manually picking out the different crystals under magnification, and observed that one of the isomers of ammonium tartrate inhibits the growth of the mold Penicillium glaucum, while the other isomer has no effect on the growth of the mold.

Early in the twentieth century, information relating biologic activity with optical isomerism slowly began to appear with the work of investigators such as Cushny[2] and Easson and Stedman,[3] who provided important initial studies. Today, the studies are extensive and of primary importance in drug design.

Optical isomers may be defined simply as compounds that differ only in their ability to rotate the plane of polarized light. The (+), or dextrorotatory (d), isomer rotates light to the right (clockwise) and the (−), or levorotatory (l), isomer rotates light to the left (counterclockwise). Enantiomorphs (also called optical antipodes or enantiomers) may be defined as optical isomers in which the atoms or groups about an asymmetric center are arranged in such a manner that the two molecules differ only as does the right hand from the left. In other words, they are nonsuperimposable mirror images. Enantiomorphs rotate the plane of polarized light in equal amounts but in opposite directions. Because there is no difference in physical properties between enantiomorphs, their difference in biologic activity must be due to their spatial arrangement, or stereochemistry.

The earliest nomenclature of asymmetric centers is the (D) and (L) designations. These terms were chosen arbitrarily to coincide with the (+) and (−) isomers of glyceraldehyde. The (D) configuration was assigned to the (+) isomer of glyceraldehyde, and other isomers that rotated light in the same direction were also designated (D). From this initial concept, it can be said that (D) and (L) represent configuration, whereas (+) and (−) denote the rotation of the plane of polarized light by the compound. X-ray crystallography has proved that many of the absolute configurations are in fact not related to the rotation, as previously thought, and it is possible, therefore, to have a D(−) or an L(+) isomer. The terms D and L are used only when referring to the absolute configuration and should not be confused with d and l, which denote rotation.

The more recent approach to nomenclature of asymmetric centers, widely reported in today's scientific literature, is the Cahn, Ingold, and Prelog system. This system

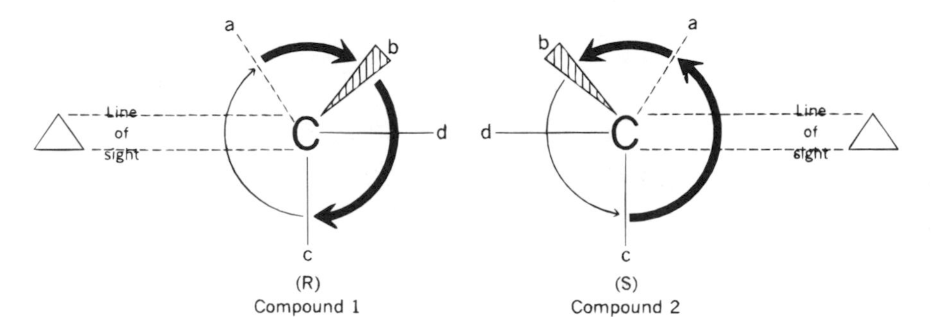

Fig. 4–1. Pair of enantiomorphs in which groups a, b, c, and d are in order of decreasing atomic numbers, i.e., a > b > c > d.

is more convenient in most cases because it allows one to include the actual arrangements of the groups at a center in the standard nomenclature for a particular compound. An oversimplified explanation of this system is as follows: (1) the atoms or groups surrounding an asymmetric center are given priorities according to atomic number and various sequence considerations (for the purposes of this discussion, the atoms with highest atomic number will be given highest priority); (2) the molecule is rotated so that the group with lowest priority is away from the viewer; and (3) following descending priorities, one forms either clockwise or counterclockwise motion about the center. If the motion is clockwise, the compound is termed (R); if the motion is counterclockwise, the notation is (S). Figure 4–1 shows a pair of enantiomorphs with groups a, b, c, and d, the order of atomic number being a > b > c > d. The molecule is sighted to the side opposite the smallest group, in this case d. In Figure 4–1, therefore, compound 1 has the R configuration, and compound 2 has the S configuration.

In Figure 4–2, the optical isomers of tartaric acid are drawn and their (+) and (−) rotations are given along with the (R) and (S) nomenclature. Compounds 3 and 4 are enantiomorphs or nonsuperimposable mirror images, and compound 5 is a diastereoisomer of compounds 3 and 4, i.e., an isomer (with two or more asymmetric centers) that is not a mirror image of any of the others. Diastereoisomers may exhibit significant differences in physical properties, such as solubility, partition coefficient, and melting point. Differences in biologic activity between diastereoisomers may, therefore, be due to differences in physical properties. Compound 5 shows no optical activity, because it possesses a plane of symmetry (dotted line), and the (+) and (−) rotation around the two asymmetric centers is equal and opposite with a resultant cancellation of optical rotatory properties.

Influence on Pharmacologic Activity

The differences in biologic activity between optical isomers depend on their ability to react selectively at an asymmetric center in the biologic system. The reason for differences in activity at asymmetric centers is shown simply in Figure 4–3. In the diagram, A, B, C, and D on the molecule represent functional groups that either bind or have a place of fit on the asymmetric surface. The corresponding letters on the asymmetric surface are the individual sites of binding, or fit. It is easily seen from this diagram that of the two enantiomorphs, only one (compound 6) has the correct orientation for all three groups to fit at their respective sites.

In the case of optical isomers, the observed differences in biologic activity may be due to a difference in the distribution of the isomers or to a difference in the properties of the drug-receptor combination if less than the optimal number of binding groups is suitably located for binding. Differences in distribution occur as optical isomers are selected by some other asymmetric center in the biologic system before the isomer reaches the specific receptor, possibly due to optically active processes such as selective penetration of membranes, selective metabolism, or selective absorption at sites of loss. Figure 4–4 shows the selective phases that optical isomers may be subjected to prior to the biologic response. An optically active drug may not be subjected to all of them, but these processes may contribute to superiority of biologic effect of one isomer.

The difference in reactivity of enantiomorphs at the receptor site has been elegantly demonstrated in the receptor site and drug interaction hypothesis proposed by Beckett.[4] Figure 4–5 shows the interaction of the optical isomers of epinephrine at the proposed receptor site. Only the (−) isomer has the OH group in the correct

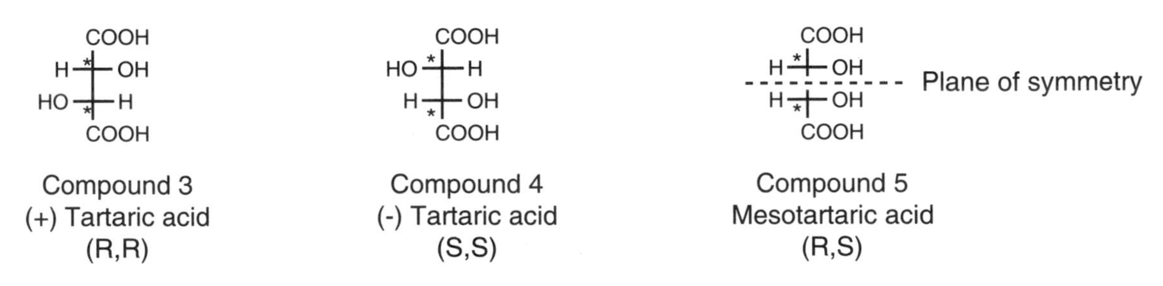

Fig. 4–2. Optical isomers of tartaric acid. *, center of asymmetry.

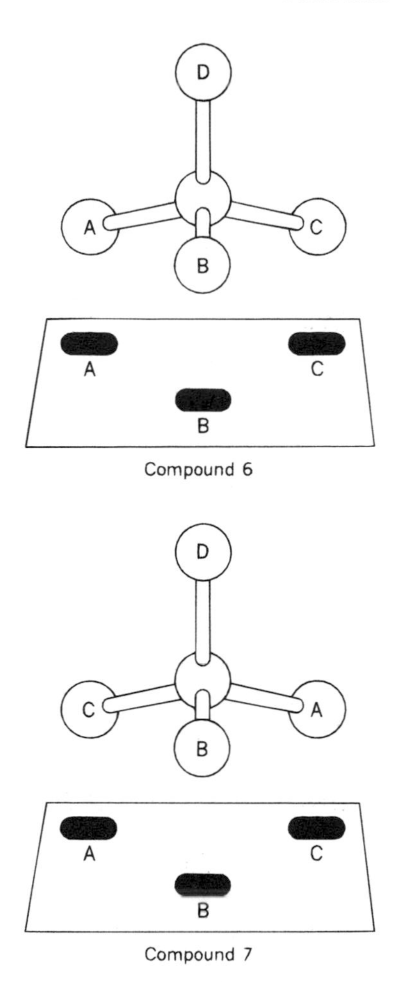

Fig. 4–3. Optical isomers. Only in compound 6 do the functional groups A, B, and C align with the corresponding sites of binding on the asymmetric surface.

orientation to allow perfect binding by all groups. This explanation has been proposed as the reason for the high pressor activity of (−) epinephrine, whereas the (+) isomer shows only minimal activity.

One of the drawbacks of stating that the difference in biologic activity of enantiomorphs is due to the difference in reactivity at the receptor site is that often isomers do not reach the receptor site in the same concentrations, as has already been demonstrated in discussing Figure

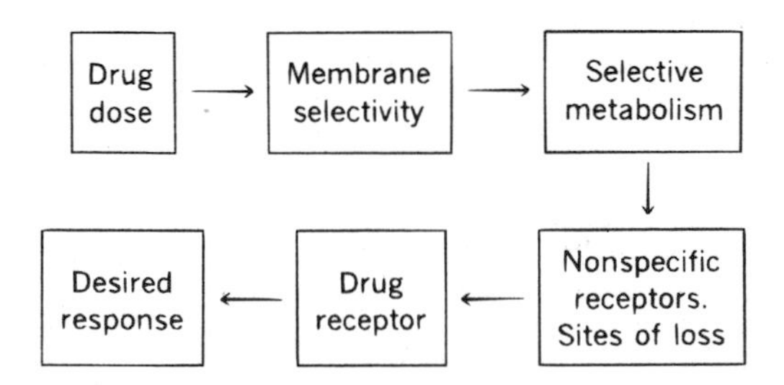

Fig. 4–4. Selective phases to which optical isomers may be subjected prior to biologic response.

4–4, and thus would not be expected to possess quantitatively equal activities. This point is especially true of in vivo systems, i.e., the simple intravenous injection of a drug, in which the factors in Figure 4–4 become significant in elicitation of the drug response. A brief discussion of documented examples of the possible points of stereoselectivity therefore seems appropriate.

The quantitative differences in activity shown by the optical isomers of muscarine are thought to be good examples of selectivity at the receptor site. Because muscarine-type molecules are not susceptible to enzyme hydrolysis by enzymes such as cholinesterase, their muscarinic potencies on isolated tissue preparations may be interpreted as being a direct measure of the interaction at the receptor site.[5] The (+) muscarine isomer (5S, 4R, 2S), shown in Figure 4–6, is 200 to 800 times more active than the (−) isomer with respect to its ability to contract various types of muscle tissue, such as guinea pig ileum.[6]

It has long been recognized that stereochemical configuration can play a role in the metabolism of optically active drug molecules. The metabolizing enzymes (optically active) in binding with a racemic drug (substrate) will clearly produce diastereoisomeric complexes possessing differing physical and chemical properties, thus providing the possibility for metabolic reactions to occur at differing rates, i.e., stereoselective drug metabolism. A metabolized drug resulting from this stereoselective process may have decreased or increased activity, depending on whether the drug or the metabolite elicits the response. An example of how stereoselective metabolism may affect the potency of a drug is the action of the optical isomers of hexobarbital. Two major oxidative metabolic pathways exist for hexobarbital: (1) allylic oxidation yielding the 3-hydroxy and 3-ketohexobarbitals,[7] and (2) epoxidation, which leads to the formation of 1,5-dimethylbarbituric acid. The L-hexobarbital has a longer latency of action and longer half-life and achieves higher blood levels than does the D-hexobarbital when injected into male rats. It has been proposed that this difference in activity is a result of the D enantiomer being more rapidly metabolized than the L species.[8]

Selective metabolism may act through several stereoselective metabolic enzyme systems, e.g., amino acid oxidases, decarboxylases, hydrolytic enzymes, dehydrases, and dehydrogenases, all of which in specific cases are stereoselective. As an example, cytochrome P450, a widely distributed oxidase enzyme system involved in the oxidation of a variety of drug molecules, demonstrates metabolic stereoselectivity in the oxidation of benzo (a) pyrene. Cytochrome P450 epoxidation of this molecule[9] initially yields the 4S5R, 7R8S and 9S10R epoxides. Of these, only the 7R8S epoxide is a precursor in the bioactivation of benzopyrene as a carcinogen. The stereoselective metabolic activation of this agent is summarized in Figure 4–7.

Erythrocyte acetylcholinesterase, a hydrolytic enzyme that hydrolyzes lactoylcholine to its constituent lactic acid and choline, also demonstrates steric selectivity. The L(+) lactoylcholine undergoes hydrolysis by the enzyme more readily than does the D(−) isomer. A hypothetic interaction between the enzyme and lactoylcholine has

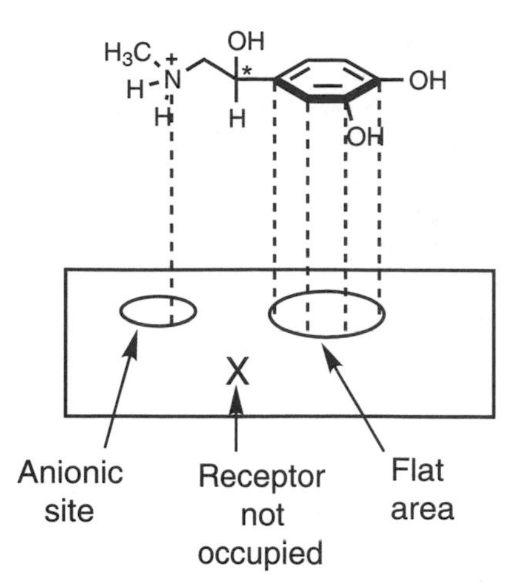

(+) Epinephrine—less active

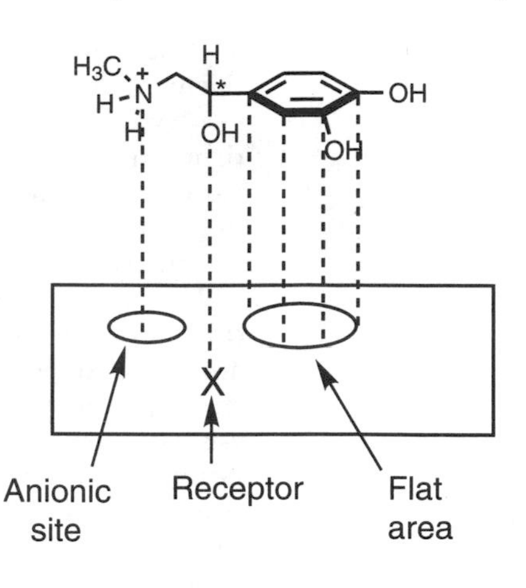

(-) Epinephrine—more active

Fig. 4–5. Interaction of optical isomers of epinephrine at the proposed receptor site; *, center of asymmetry.

been proposed,[10] and Figure 4–8 shows how the spatial arrangement about the asymmetric center is important in lining up the other sites on the enzyme. The alternative arrangement about the asymmetric carbon atom would obviously not allow binding of either the OH grouping or the CH_3 grouping.

A rarely observed but interesting stereoselective metabolic phenomenon is inversion of configuration during metabolism. In man, R(−) ibuprofen (lesser active isomer) is excreted predominantly as the S(+) enantiomer (or its hydroxy or carboxy derivative) as a result of an isomerase enzyme thought to be present in the gut wall[11] (Table 4–1).

Selectivity of passage of a drug through a membrane may occur as a result of asymmetric centers associated with the membrane, and thus may play a major role in determining the relative potencies of pairs of isomers. The asymmetric centers on the optically active drug molecule may bind selectively to the centers on the membrane, contributing to differences in penetration. If a drug must cross a membrane to reach a receptor, selectivity at the membrane is important in biologic response.

Adsorption of only the (+) isomer of α-naphthylglycolic acid (a dye) by wool is a simple example of this type of surface adsorption. If a racemic mixture—equal amounts of both (+) and (−) isomers—of the dye is placed in a solution with wool, the wool preferentially adsorbs the (+) dye, leaving a significantly higher percentage of the (−) dye in the solution when the wool is removed.[12]

Transportation of molecules across a membrane by

H3C * CH3N(CH3)3 * O * HO

(+) Muscarine

Fig. 4–6. Muscarine isomer. *, center of asymmetry.

transporting enzymes known as permeases may also be a selective process. The ability of only the L isomers of the amino acids valine, leucine, and isoleucine to penetrate the cell walls of bacteria such as Escherichia coli demonstrates this type of stereospecific transportation. In general, however, stereoselectivity in drug transport is not observed. Similarly, plasma and tissue binding of drugs generally shows little stereoselectivity, with the notable exceptions of oxazepam,[13] probably other benzodiazepines, and tryptophan.[14]

The enantiomers of a drug molecule may also show differences in activity because of stereoselective reactions at nonspecific receptors, or sites of loss, i.e., one isomer reacts with a nonspecific receptor that is sterically similar to the specific receptor for the required response. Reaction of the isomer at this second receptor may cause no response and thus an overall loss of drug activity, or it may show a response different from the intended one.

It is possible, therefore, that the potentially more potent isomer may effect the lower response because it has reacted at the nonspecific site. In these cases, the activity may be enhanced by tying up the nonspecific receptor by reaction with another molecule that has an asymmetric center similar to that of the drug. As an illustration, a certain dosage of quinine fails to produce a complete response against chick malaria. When the same dosage is administered in combination with derivatives of quinine having the same optical asymmetry as quinine, the antimalarial response is enhanced. Optically dissimilar derivatives in combination with quinine do not enhance the response.[15] Displacement of the quinine from the nonspecific receptor by sterically related molecules allows the quinine to react at the specific site and effect good antimalarial activity.

The most important and most likely area for stereoselectivity is the specific receptor site. As seen from the previously mentioned studies, however, one must be careful in drawing such conclusions. In a review article by Patil and co-workers,[16] the various sites for selectivity are presented for the adrenergic drugs. At adrenergic synapses, stereoselectivity has been observed for: (1) all biosyn-

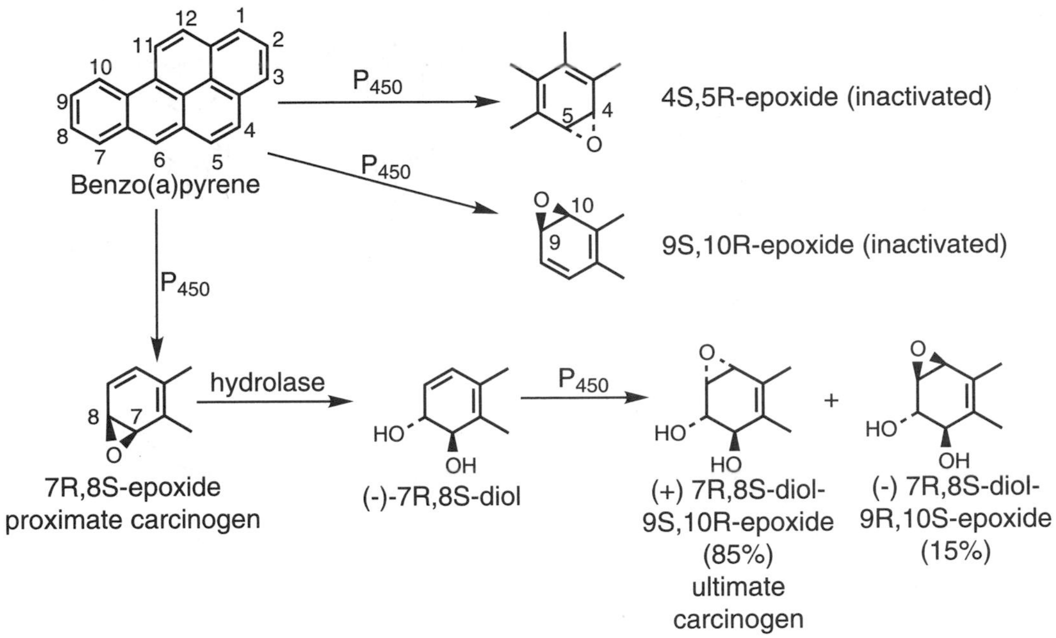

Fig. 4–7. Stereoselective metabolic activation of benzo(a)pyrene.

thetic pathways; (2) transport in the adrenergic neuron; (3) binding and retention of the adrenergic drug in the granule; (4) action of the enzyme monoamine oxidase on the drug; and (5) the α-adrenergic and β-adrenergic receptors. The total action of these drugs therefore involves complex series of stereoselective events.

Regardless of where the selectivity takes place, the general conclusion is that many pairs of optically isomeric drugs exhibit quantitatively different responses.

The classic example is ephedrine. Of the optical isomers of ephedrine and its diastereoisomer pseudoephedrine, only the D(−) ephedrine significantly blocks the β-adrenergic receptor,[17] thereby lowering blood pressure. This high degree of specificity is seen from the structures of the four molecules in Figure 4–9. The Cahn, Ingold, and Prelog designation for each asymmetric carbon is given in parentheses to the right of each asymmetric atom. For these molecules, the configuration around both asymmetric centers is particularly important, and as can be seen from Figure 4–9, the β center must be (R) and the α center (S) for maximal activity.

Isomeric potency differences of 100-fold are common in many of the phenylthylamines, whereas in the isomeric imidazolines synthesized to date,[18] adrenergic potency differences of only 5- to 10-fold have been observed. This disparity indicates that the adrenergic receptors have stringent steric requirements for the former group as compared to the imidazolines.

Another drug in which the optical isomers have been

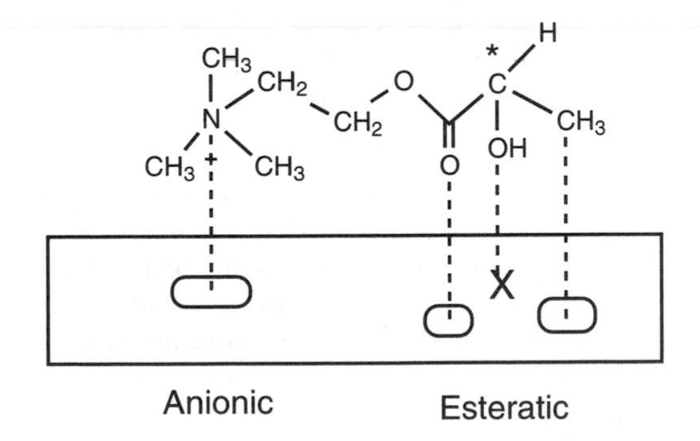

Fig. 4–8. Hypothetic interaction between L(+) lactoylcholine and erythrocyte acetylcholinesterase. *, center of asymmetry.

Table 4–1. Inversion of Configuration of Ibuprofen in Humans

Enantiomer	Administered Ibuprofen S/R Isomer Ratio	Excreted S/R Ratio
II	95:5	95:5
III	6:94	80:20
I	50:50	70:30

Ibuprofen

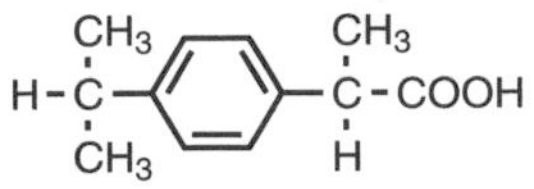

I: racemate
II: (S) (+)
III: (R) (−)

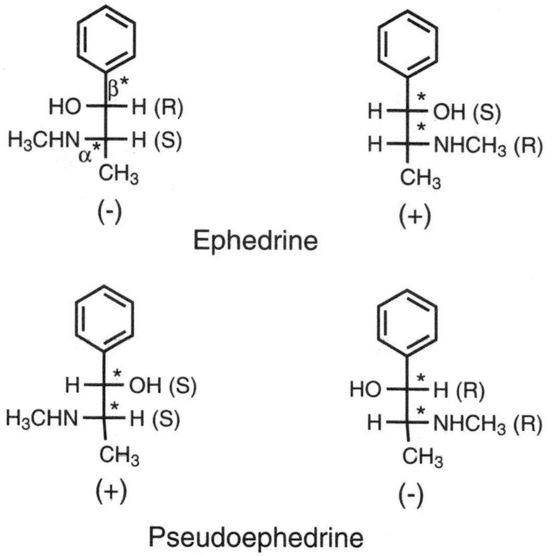

Fig. 4–9. For maximal activity of these molecules, the β center must be (R) and the α center (S). *, center of asymmetry.

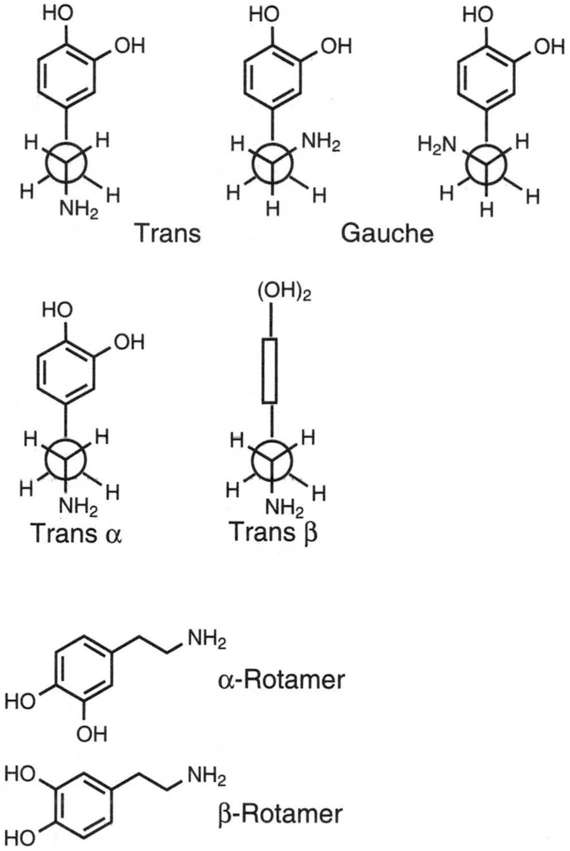

Fig. 4–11. Some dopamine conformations.

isolated and the biologic activities studied is 1-(4-nitrophenyl)-2-isopropylaminoethanol (INPEA). This drug was tested in rats for its ability to antagonize several β-adrenergic receptor responses, namely: (1) the positive chronotropic response to epinephrine; (2) the calorigenic action produced by epinephrine; and (3) the arterial depressor response to isoproterenol. In each case, the L(+) INPEA had little or no effect, whereas the (D) − INPEA (Fig. 4–10) had a high degree of activity at the same dose levels.[19]

Optical stereoselectivity is shown in both dopaminergic and antidopaminergic agents, even though dopamine itself lacks asymmetry. Dopamine is a flexible molecule, however, and can thus assume a conformation to enable a "fit" with dopamine receptors. It is generally thought that of the possible dopamine conformations (Fig. 4–11), the *trans* β-dopamine is preferred by the dopamine receptor.[20] This preferred conformation, in addition, has α and β rotamers. Rigid rotameric analogs have been studied to determine the more desirable rotamer, with varying conclusions.[20] Optically active dopamine agonists possess specific orientations, i.e., optical stereoselectivity, to elicit activity. For example, it has been shown that 6aS (+) apomorphine[21] is devoid of dopaminergic activity, whereas the 6aR (−) isomer possesses significant activity.

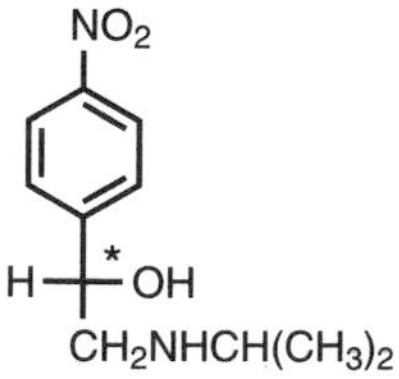

Fig. 4–10. D(−)INPEA. *, center of asymmetry.

In the dopamine antagonists, the neuroleptic (+) butaclamol is able to block the agonist effects of apomorphine whereas (−) butaclamol is an inactive antagonist in this screening protocol.[22]

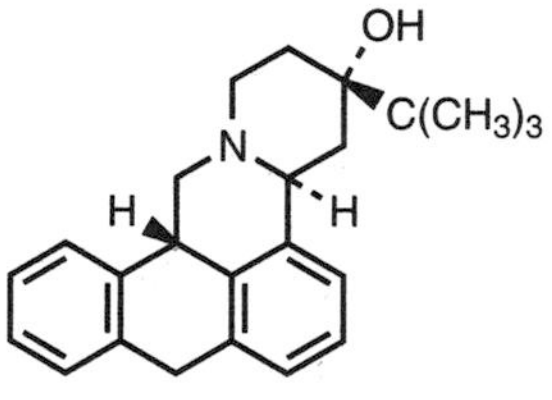

An article by Portoghese and Williams[23] clearly demonstrates the complexities involved in correlating the biologic activity with the optical isomeric configurations of drugs. When isomethadol was tested for analgesic activity, it was discovered that of the four possible isomers (Fig. 4–12), only one, the 3S, 5S, β(+) isomer, possessed high analgesic potency.

Upon acetylation of the OH grouping to give acetylisomethadol, the isomers possessing greatest activities are the α(+) and the β(−) isomers. The change in activity is clearly shown by the values for the median effective dose (ED_{50}) for the isomers of both isomethadol and acetylisomethadol (Table 4–2). The most potent iso-

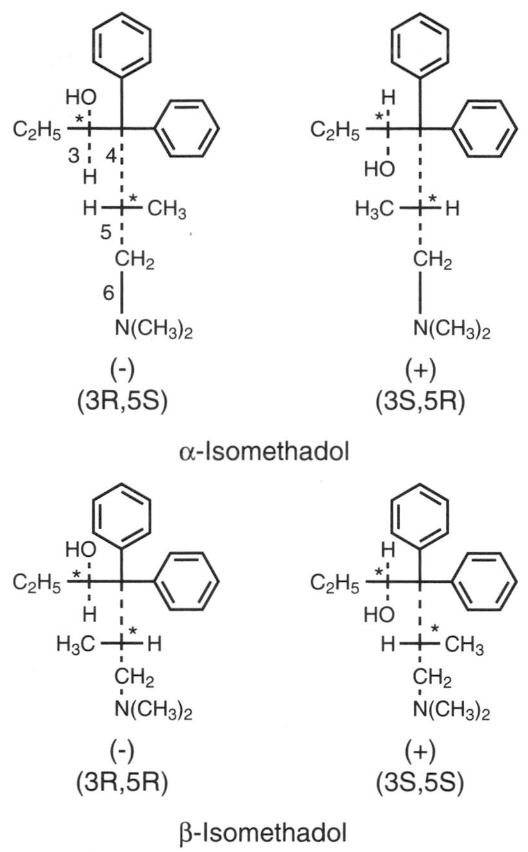

(-) (3R,5S) (+) (3S,5R)

α-Isomethadol

(-) (3R,5R) (+) (3S,5S)

β-Isomethadol

Fig. 4–12. Of these four isomers, only the 3S, 5S, [β(+)] one has high analgesic potency. *, center of asymmetry.

methadol isomer becomes the least potent on acetylation and vice versa.

Two possible explanations may account for this difference in activity. The first incorporates the suggestions that two sites on the same receptor but located in dissimilar topographic environments, one of which is a proton acceptor capable of binding by hydrogen binding to the proton-donating OH group of isomethadol, and the second is a proton donor dipole capable of binding with the proton-accepting ester molecule (acetylisomethadol). Each site is stereoselective. The second explanation is that there are two types of analgesic receptors having different stereoselectivities: one binds isomethadol and the other binds acetylisomethadol.

Table 4–2. Analgesic Potency of Isomers of Isomethadol and Acetylisomethadol

Isomethadol	ED_{50} (mg/kg)
α(+)	60.7
β(−)	58.7
α(−)	91.7
β(+)	6.2
Acetylisomethadol	ED_{50} (mg/kg)
α(+)	2.7
β(−)	10.9
α(−)	62.7
β(+)	70.6

Other studies have demonstrated optical stereoselectivity in the calcium blockers: (−) verapamil, (−) nimodipine, and (−) D600 are the more potent enantiomers in regard to negative inotropic activity.[24]

R(+) etomidate is a potent short-acting nonbarbiturate hypnotic; its S(−) enantiomer is devoid of hypnotic activity.[25]

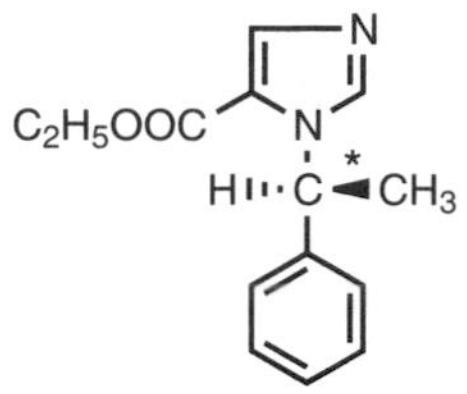

Etomidate

Indacrinone, a potent diuretic of long duration with only transient uricosuric activity, is an optically active indanone derivative that is used clinically as the racemate. Its diuretic activity has been shown to reside predominantly with the (−) isomer. The (+) enantiomer, which possesses minimal diuretic activity, has a relatively high uricosuric/diuretic ratio. Enrichment of the isomeric ratio favoring the (+) enantiomer enables a moderate lowering of plasma uric acid, i.e., improvement of therapeutic efficacy by improving the uricosuric activity of the racemic drug by enantiomeric manipulation.[26]

Emetine is a naturally occurring levorotatory alkaloid with four asymmetric centers. Its activity, resulting from its effect on protein synthesis, has been shown to be highly stereoselective. The synthesis of other than the naturally occurring stereoisomer has yielded products with lesser antiamebic activity.[27] Configurational comparisons of (−) emetine with (−) cycloheximide (a protein synthesis inhibitor) show distinct similarities (Fig. 4–13) from which a steric pharmacophore for amebicidal activity and protein synthesis inhibition has been proposed.[28]

Stereoselectivity has also been observed in the biologically active peptides. Naturally occurring peptides are generally composed of L amino acids with only a few exceptions in the animal kingdom. For example, dermorphin, an opioid peptide found in amphibian skin, contains D-alanine. In plant and bacterial peptides, however, D configurations are found more frequently. Bacterial cell walls, for example, contain D-alanine, and D-penicillamine is found in penicillin. Substitution of the D configurational analog for the L amino acids produces conformational changes and induces resistance to enzymatic cleavage of the peptide bonds. This latter effect results in an increased duration of action of the peptide, thereby providing a potential approach to the production of long-acting peptide drugs. In some instances, however, substitution of the enantiomeric amino acid produces an analog devoid of activity. For example, the naturally occurring opioid pentapeptide enkephalins possess L-tyrosine at the critical position 1 of the chain. Substitution of D-tyrosine for this amino acid inactivates this molecule in

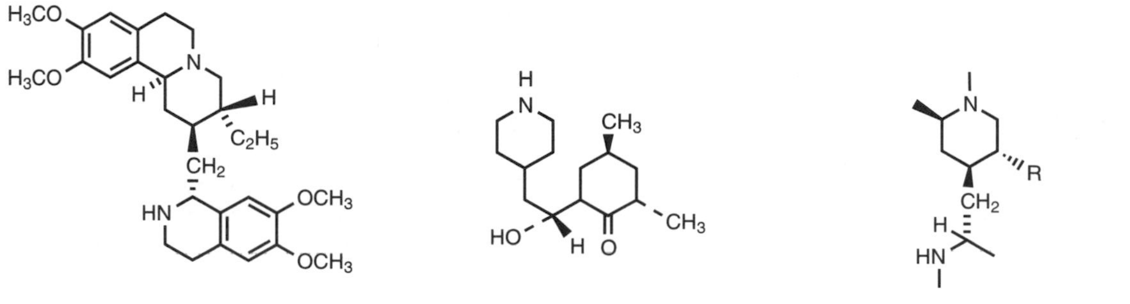

Emetine

Cycloheximide

Proposed pharmacophore for protein synthesis inhibition

Fig. 4–13. Stereochemistry of some protein synthesis inhibitors.

regard to morphine-like actions of the parent peptide. Replacement of other amino acids in the chain by the enantiomeric D amino acids reduces potency considerably, e.g., D-phenylalanine at position 4 and D-leucine at position 5. Assuming that the biologic activity of peptides is primarily a function of the nature of the side chain interactions with receptors, an all-D analog of an all-L peptide is expected to possess overall similarity in shape and similar orientation of the side chains which would impart similar pharmacologic activity. The synthesis of all-D analogs (retroanalogs) of active L peptides, e.g., D-bradykinin and D-oxytocin, however, yielded inactive compounds.[29] The backbone of the peptide chain is thus implicated in the biologic response of these compounds.

Other examples of stereoselectivity in peptides have been reviewed for the ACTH neuropeptides, the luteinizing hormone-releasing hormone (LH-RH) peptide, and the arginine vasopressin (AVP) peptide.[30]

Geometric Isomerism

Geometric isomerism is another steric feature important in the action of many drug molecules. The term geometric isomerism (or *cis-trans* isomerism) indicates a type of diastereoisomer that occurs as a result of restricted rotation around a bond, as in olefinic compounds. The designation *cis* is used when identified groups are on the same side of the plane of the molecule and *trans* when the groups are on the opposite sides of the plane.

Geometric isomerism does not necessarily impart optical isomerism to the compound. If the structure is asymmetric (or dissymmetric), however, geometric isomers may exhibit optical activity. In Figure 4–14, an example is given of geometric isomerism for both an olefin (2-butene) and a cyclic alkane (1,2-dimethylcyclohexane). It can be seen that *trans*-1,2-dimethylcyclohexane can exist as an enantiomorphic pair.

Influence on Pharmacologic Activity

When biologic differences among geometric isomers are discussed, one must remember that the distribution of the compound in the biologic system varies because of great differences in physical properties of the isomers. Unlike enantiomers (i.e., D or L isomers), it may be difficult to correlate the differences in biologic activity solely with the difference in the spatial arrangement (stereochemistry) of the isomers. For example, one isomer may be more highly ionized at physiologic pH, resulting in a difference in adsorption at surfaces and penetration of membranes, resulting in significant differences in biologic activity.

Differences in reactivity of geometric isomers at the receptor site may be demonstrated by the proposed receptor-drug interactions shown in Figure 4–15. In compound 1, if the three substituents (A, B, and C) on the cyclopentane ring are required for binding to the receptor surface, only the *cis* arrangement would allow all three groups to bind and thus elicit the expected biologic response. A similar situation is shown for an olefinic, compound 2.

The observed differences in biologic activity of geometric isomers may be due in part to differences in the interatomic distance of the groups essential for the elicitation of the response. The classic example is diethylstilbestrol, a drug synthesized to mimic the natural estrogenic hormone estradiol. The *trans* isomer of diethylstilbestrol has 14 times the estrogenic activity of the *cis* compound.[31] Figure 4–16 reveals that the interatomic distance between the OH groups of the *trans*-diethylstil-

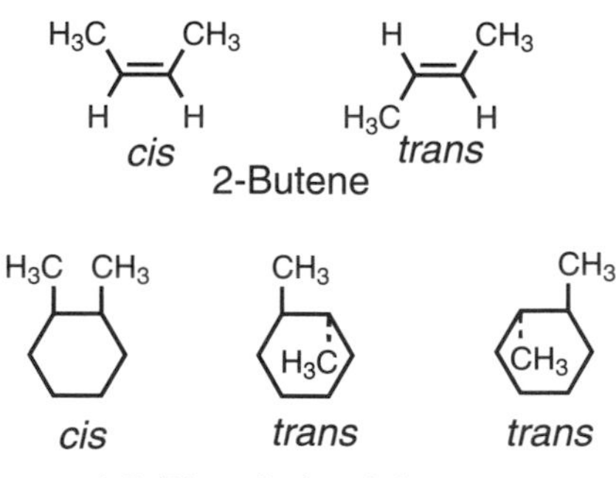

Fig. 4–14. Geometric isomerism. The *trans*-1,2-dimethylcyclohexane can exist as an enantiomorphic pair.

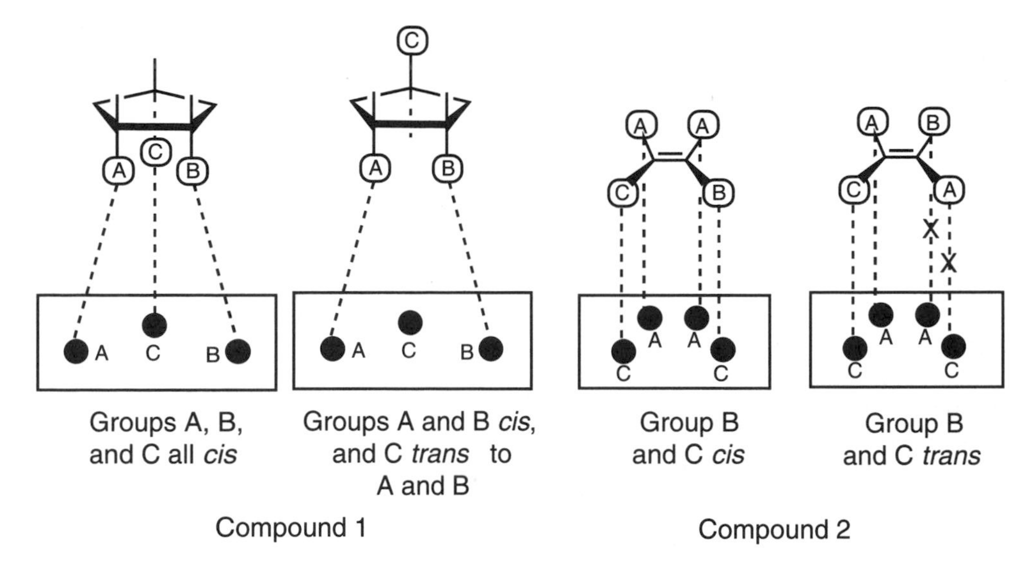

Fig. 4–15. Only the *cis* arrangements allow binding of substituents A, B, and C, thereby eliciting the expected biologic response.

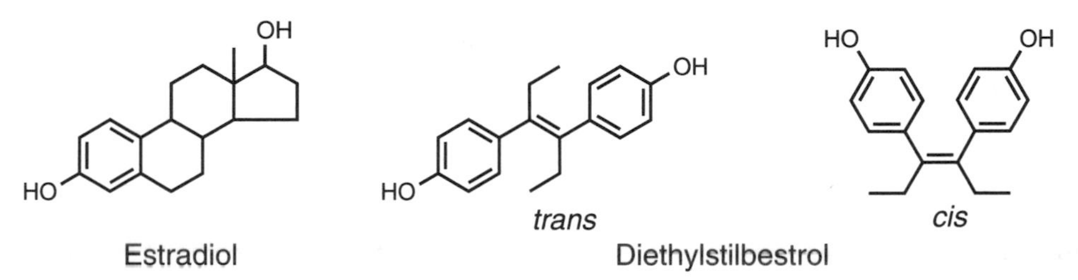

Fig. 4–16. The interatomic distances between the OH groups in *trans*-diethylstilbestrol and in estradiol are similar, accounting for the greater estrogenic activity of the *trans* isomer.

bestrol closely approximates the distance between the OH groups in estradiol.

The correlation between geometric isomerism and biologic activity has received little attention because of the sometimes vast differences in physical properties of the geometric isomers. Casy and Parulkar,[32] however, working on the antihistaminic activity of some 2-butenes possessing the general structure shown in compound 1 of Figure 4–17, demonstrated that the *cis* (Ar/H) arrangement is necessary for optimal antihistaminic activity. The antihistamine pyrrobutamine (Co-Pyronil), compound 2 in Figure 4–17, was developed on the basis of these findings. Bombykol, a 10-*trans*-12-*cis*-hexadecadienol insect pheromone, has been shown to be 10^9 to 10^{13} times more potent as an insect sex attractant than other possible geometric isomers.[33]

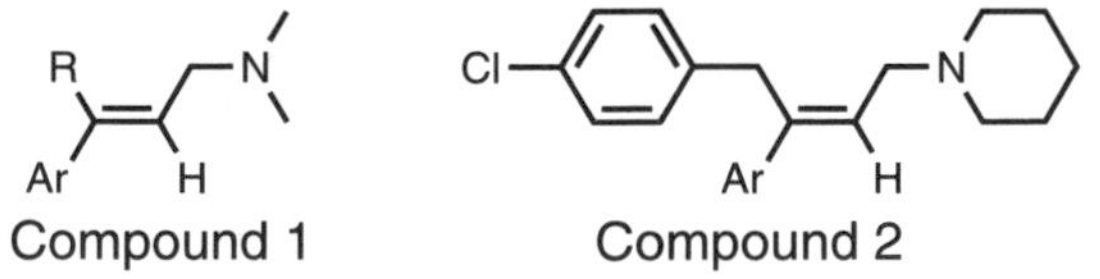

Compound 1 Compound 2

Fig. 4–17. The discovery that the *cis* (Ar/H) arrangement of compound 1 (a 2-butene) is necessary for optimal antihistaminic activity led to the development of compound 2 (pyrrobutamine).

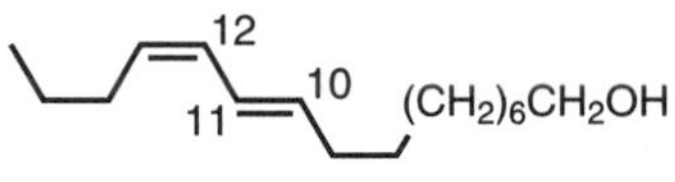

$(CH_2)_6CH_2OH$

Bombykol

CONFORMATIONAL ISOMERISM AND PHARMACOLOGIC ACTIVITY

The second important steric phenomenon in the action of drug molecules is the effect of conformational isomerism. This topic continues to receive attention in the search for improved medicinal agents.

Conformational isomerism is defined by Eliel et al.[34] as the nonidentical spatial arrangement of atoms in a molecule, resulting from rotation about one or more single bonds. Different conformations result only from rotation about bonds between atoms having at least one other substituent. A molecule such as water, H—O—H, would not exist in different conformations, because rotation of the O—H bonds produces no distinctive arrangements. Hydrogen peroxide, H—O—O—H, would result in distinctive conformations with rotation of the O—O bond (Fig. 4–18).

Although these simple molecules demonstrate the concept of conformational isomerism, drug molecules are more complex structures, and a brief discussion of con-

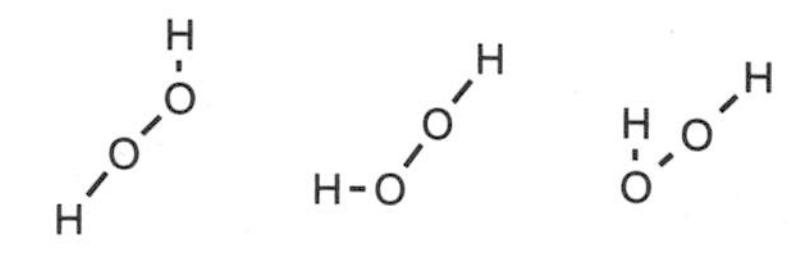

Fig. 4–18. Different conformations of H_2O_2. Notice the changing relationship of the hydrogen atoms as the O-O bond is rotated.

formational isomerism as it relates to acyclic and cyclic structures common in drug molecules is appropriate.

The concept that free rotation existed about single bonds in acyclic compounds was disproved in 1936. Kemp and Pitzer[35] observed that the experimentally determined thermodynamic values of ethane differed consistently from the calculated values, and the only explanation was the existence of a barrier of approximately 3.0 kcal/mol to rotation about the carbon-carbon bond. Subsequent investigations verified that a barrier of 2.8 kcal/mol does exist in ethane. This barrier to rotation is due to the decreasing distance between the hydrogen atoms on adjacent carbon atoms as the carbon-carbon bond is rotated.

Figure 4–19 illustrates both the sawhorse representation and the Newman projection formula, which presents a view of the molecule along the carbon-carbon bond axis. The eclipsed conformation is the one in which the hydrogen atoms are in closest proximity and is the least stable of all possible conformations. The staggered conformation allows the greatest separation of the hydrogens and, consequently, is the most stable conformation. An infinite number of conformations is possible as the carbon-carbon bond is rotated. The eclipsed and staggered conformations represent extremes—the highest and lowest energy states, respectively.

With substitution of groups other than hydrogen on ethane and with longer chain molecules, similar relationships are observed. For example, rotation of the bond between carbon atoms 2 and 3 in butane results in the conformations depicted by the Newman formulae shown in Figure 4–20.

The skew, or gauche, conformations are less stable by 0.8 kcal/mol than the completely staggered form because of the nearness of the methyl groups in the skew form. In general, the most stable conformer of a given molecule

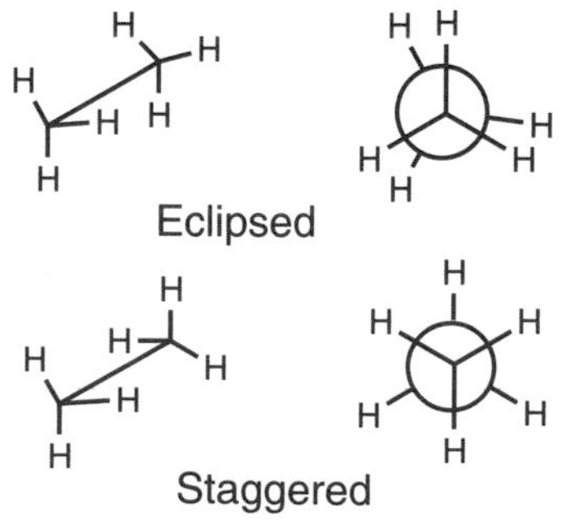

Eclipsed

Staggered

Fig. 4–19. The eclipsed and staggered conformation of ethane.

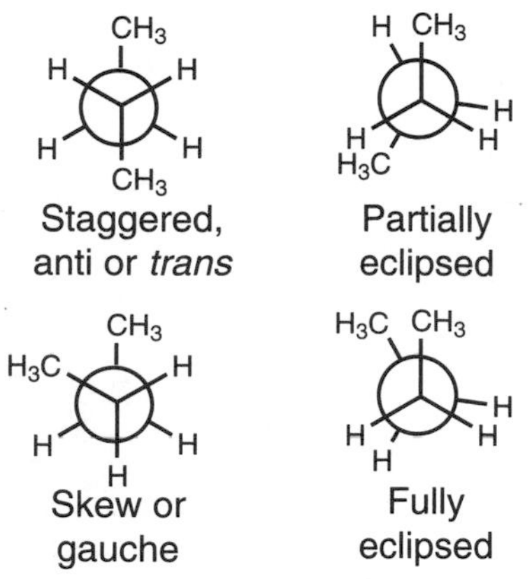

Staggered, anti or *trans*

Partially eclipsed

Skew or gauche

Fully eclipsed

Fig. 4–20. Rotational (or conformational) isomers of butane.

possesses the smallest number of butane skew interactions, regardless of whether the adjacent interacting groups be methyl or some other group of similar or larger size. The favored conformation of acylic molecules usually prevails when the larger groups are staggered and separated from one another by as great a distance as possible. The only exception is when forces of attraction rather than repulsion occur between the groups or atoms. An example of this type of situation is a molecule capable of intramolecular hydrogen bonding in which skew, or gauche, relationship is favored between the interacting groups (Fig. 4–21).

The conformation of cyclic molecules is probably of most interest to medicinal chemists. Several ring systems containing three to seven carbons are commonly found in all classes of medicinal agents. Perhaps even more common are heterocyclic ring systems in which atoms such as nitrogen, oxygen, sulfur, and sometimes phosphorus are included along with carbon. The conformation of these heterocyclic rings usually behaves similarly to that of carbocyclic rings.

The cyclopropane ring (present in the general anesthetic cyclopropyl methyl ether) is a planar system; cyclobutane (a four-membered heterocyclic ring present in penicillin and cephalosporin) and cyclopentane (present in cyclopentamine, a nasal decongestant) deviate from a planar system. The conformations of cyclopentane, for example, are shown in Figure 4–22. Ring systems larger than six carbons are usually puckered and are not discussed in this chapter.

The cyclic molecule of most interest in medicinal chemistry is cyclohexane. This ring system is present in

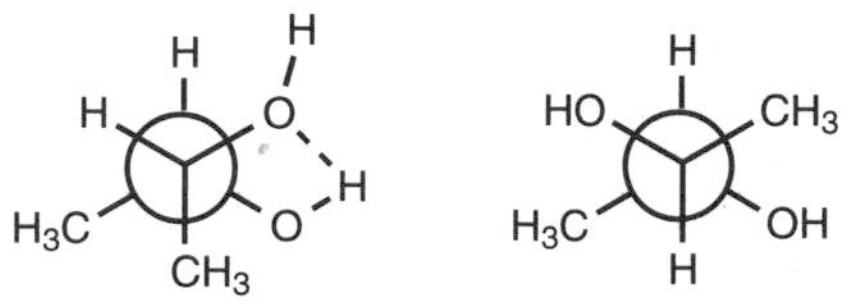

Fig. 4–21. Intramolecular hydrogen bonding causes the skew conformation to be more favorable than the staggered one.

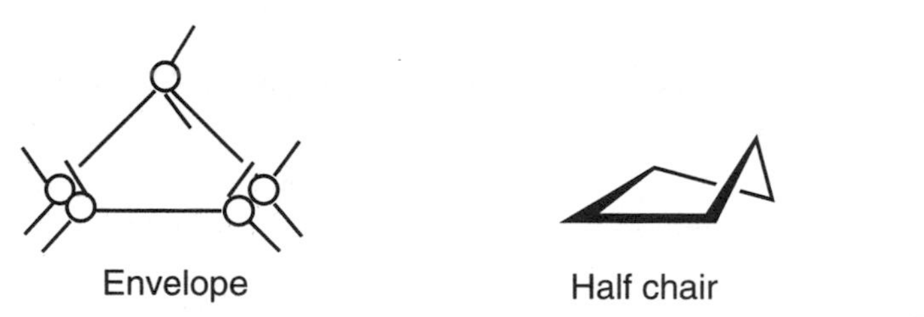

Fig. 4–22. The favored conformation of cyclopentane.

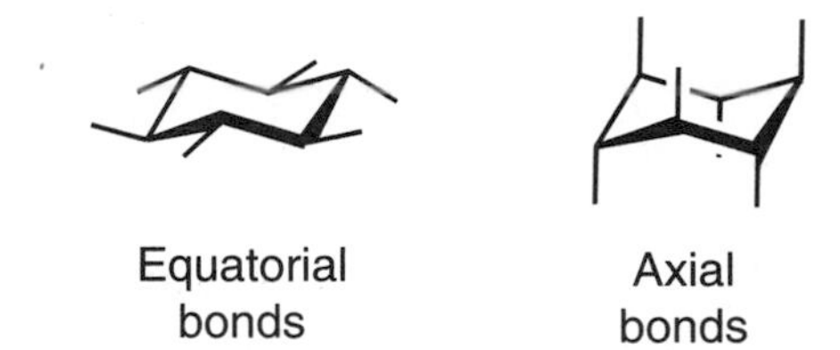

Fig. 4–25. The equatorial and axial bonds of cyclohexane.

a large number of drugs and naturally occurring bio-organic molecules. Cyclohexane can exist in two conformations in which each carbon atom has tetrahedral bond angles of 109° 28′. The two conformations are the chair form and the boat, or flexible, form (Fig. 4–23).

The chair form is favored over the boat form because all bonds are staggered in the chair form, whereas two eclipsed bond interactions and a van der Waals repulsive interaction between the two hydrogens that are directed toward each other (often called the flagpole—bowsprit interaction) are present in the boat form. This combination of factors renders the chair form more stable than the boat form by about 6.9 kcal/mol. The boat conformer may relieve the flagpole—bowsprit interaction and to some extent reduce the eclipsed bond interactions by converting to a conformation halfway between two boat forms. This conformation is known as the skew-boat, twist, or stretch form (Fig. 4–24) and is only 5.3 kcal/mol less stable than the chair form. Unless unusual circumstances prevail, cyclohexane adopts the chair conformation. Polycyclic compounds, which contain more than one cyclohexane ring (for example, steroids), assume the conformation that allows the largest number of chair forms. This also applies to six-membered heterocyclic ring systems.

In the chair conformation of cyclohexane (Fig. 4–25), six hydrogens extend radially outward from an axis passing through the center of the molecule, and these are termed equatorial (e) bonds. The other six hydrogens are parallel with the axis of the ring, and three of the

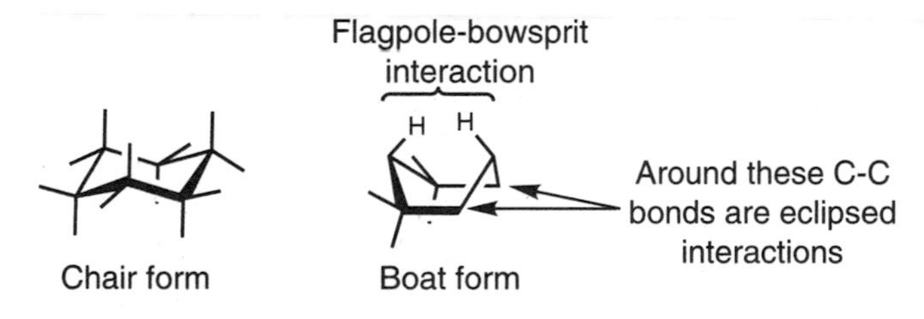

Fig. 4–23. The chair and boat conformations of cyclohexane.

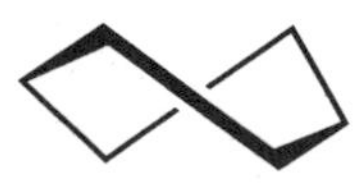

Fig. 4–24. Skew-boat, twist, or stretch conformation of cyclohexane.

bonds are above and three below the plane of the molecule. These are called axial (a) bonds.

By expending energy of approximately 10 kcal/mol, cyclohexane can flip from one chair form into another, in which case the bonds that were axial become equatorial and those that were equatorial become axial (Fig. 4–26). This action is known as chair-chair flipping, or chair inversion. At room temperature, the flipping, occurs rapidly, and it slows to the extent that the axial and equatorial hydrogens can be differentiated by nuclear magnetic resonance only after being cooled to −90 to −100°C. An indication of the rapidity of the flipping is the calculation that at −66.7°C, the rate is 105 flips/sec.[36]

Cyclohexane rings can exist with substituents at either axial or equatorial positions. Polysubstitution of cyclohexane results in both *cis-trans* (geometric) isomerism and optical isomerism when the ring is dissymmetrically substituted (Fig. 4–27).

In summation, the following generalizations may be made regarding the conformation of substituted cyclohexanes.

1. The cyclohexane ring prefers the chair over the skew-boat conformation by 1000:1 at room temperature.
2. In substituted cyclohexanes, the chair conformation with the greatest number of equatorial groups is usually most stable. If one of the substituents is bulky (for example, a tertiary butyl group), the conformation that allows this group to be equatorial is preferred. Exceptions include *trans*-1,2-dihalocyclohexanes in which the large, highly polar halogen atoms repel one another in the diequatorial conformation and assume the diaxial conformation in an attempt to relieve the repulsive interaction. (See *trans*-1,2-dichlorocyclohexane in Fig. 4–27.) Also, chair conformations with axial substituents are favored if the substituents can undergo attractive interactions, such as intramolecular hydrogen bonding in the axial positions. For example, the diaxial conformation is favored in *cis*-1,3-dihydroxycyclohexane (Fig. 4–28).

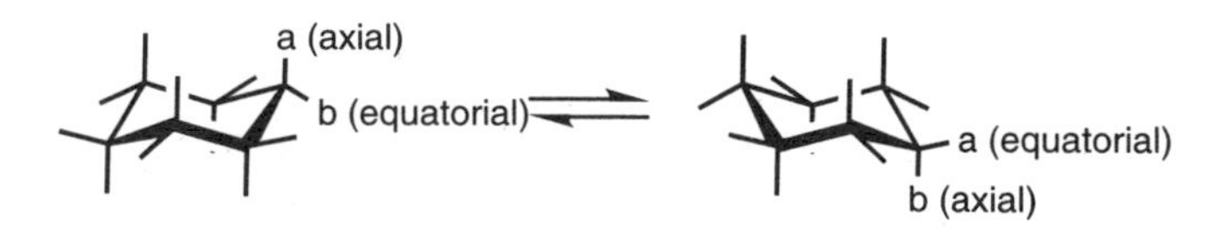

Fig. 4–26. Chair inversion of cyclohexane.

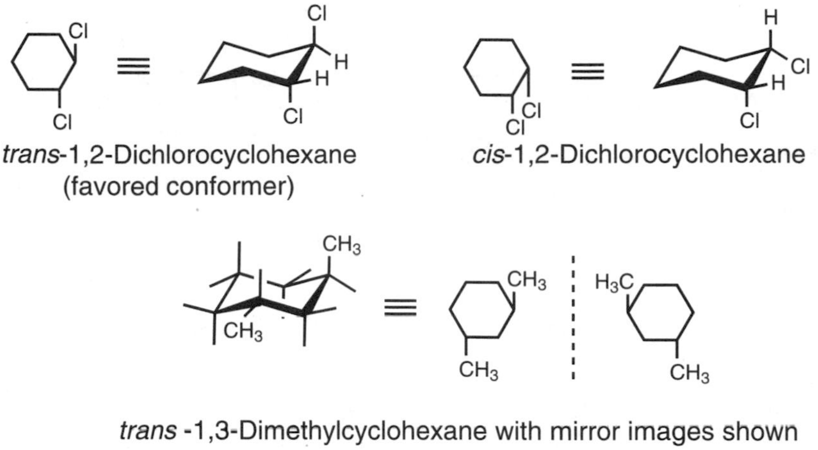

trans-1,2-Dichlorocyclohexane (favored conformer)

cis-1,2-Dichlorocyclohexane

trans-1,3-Dimethylcyclohexane with mirror images shown

Fig. 4–27. Examples of *cis-trans* isomerism *(above)* and optical isomerism *(below)*.

3. The foregoing generalizations are applicable to six-membered heterocyclic systems.

Conformation of Rigid and Semirigid Molecules

Some cyclic compounds are rigid or semirigid in the sense that rotation about single bonds or chair inversion is prohibited or made more difficult by some inherent structural feature of the molecule. This fact is well demonstrated in norbornane, in which the six-membered cyclic portion of the molecule is held in the boat form by a methylene bridge (Fig. 4–29).

Rigidity is less obvious, but nevertheless present, in the *trans*-decalin series. This molecule contains two fused cyclohexane rings that cannot undergo chair-chair flipping because of the *trans* ring juncture (Fig. 4–30).

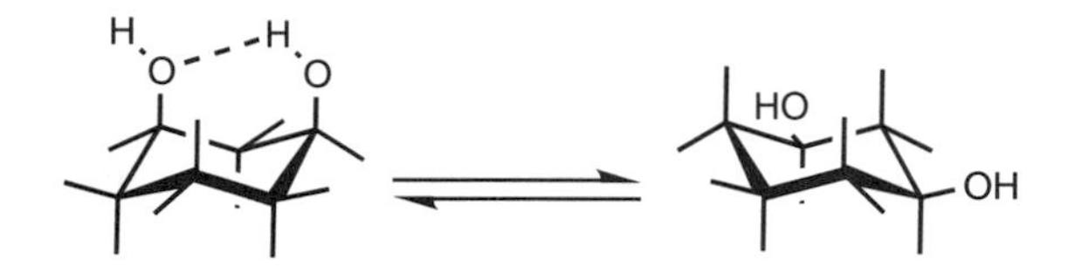

Fig. 4–28. The diaxial conformation of *cis*-dihydroxycyclohexane is favored because of intramolecular hydrogen bonding.

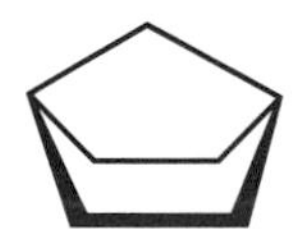

Fig. 4–29. Norbornane contains a cyclohexane ring held in the boat conformation.

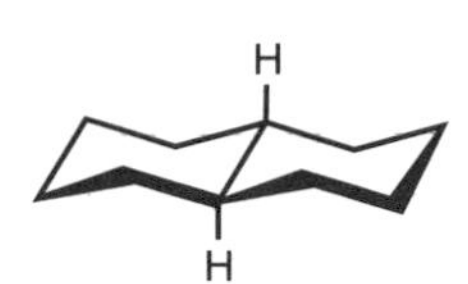

Fig. 4–30. The rigid *trans*-decalin ring system.

Each chair form can convert to the less stable flexible twist form; the double chair conformation, however, is greatly favored.

Another class of medicinally important compounds that are in most instances conformationally rigid is the steroids. These compounds can exist in either the 5α or the 5β form (Fig. 4–31). Both forms are rigid.

Substituted ring systems of the foregoing types can exhibit configurational isomerism because of the restricted rotation of single bonds and the inability to undergo chair–chair flipping. For example, the *trans*-decahydroisoquinoline ring system similarly substituted in one case axially and in another case equatorially results in diastereomers (Fig. 4–32). Each diastereomer has a nonsuperimposable mirror image, which means that each will consist of an enantiomorphic pair (dl pair). In a strict sense, these configurational isomers do not fulfill the requirements for conformational isomerism, but they are generally considered fixed, or frozen, conformations.

Because axial substituents are in an entirely different environment from equatorial substituents, differences in their respective chemical reactivities and physical proper-

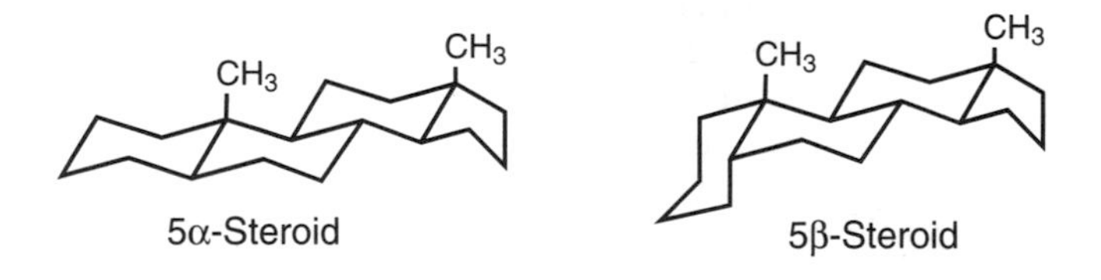

Fig. 4–31. The 5α and 5β conformations of steroid molecules.

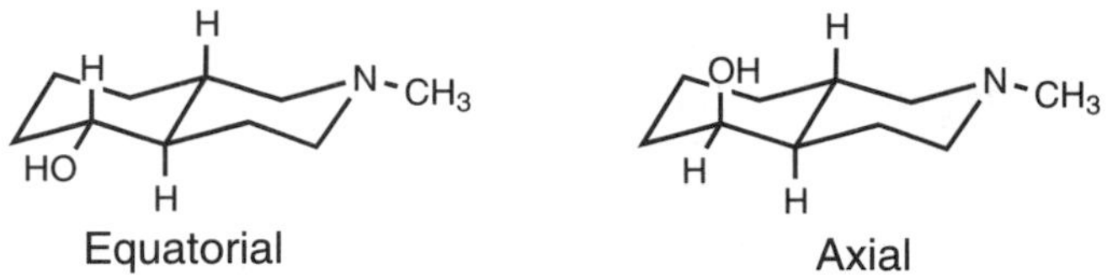

Equatorial

Axial

Fig. 4–32. Diasteromers of 5-hydroxy-2-methyl-*trans*-decahydroisoquinoline.

ties are expected and observed. The difference in chemical reactivity of the two positions has been thoroughly studied in the rigid steroid ring system.

Physical Methods of Conformational Analysis

Conformation may be determined by several physical methods. Perhaps the most commonly used is interpretation of the nuclear magnetic resonance (NMR) spectrum of a molecule. This method is used to measure the magnetic moments of hydrogen atoms in organic molecules.

The magnitude of the magnetic moment of a hydrogen atom depends on the atoms surrounding it in the molecule (i.e., its environment). If a molecule exists as two separate conformers in slow equilibrium, the NMR spectrum may show the spectrum of each conformer or, if the equilibrium is fast, the average of the individual conformers.

The average spectrum can usually be separated into the spectrum of the individual conformers by lowering the temperature at which the spectrum is determined, thereby slowing down the equilibrium. For instance, the interconversion of the chair forms of cyclohexane is slowed sufficiently at $-66.7°C$ so that the average peak representing the axial and equatorial hydrogens just begins to split, and splitting continues with further cooling. If two peaks representing the same proton in different conformations can be separated and identified at a certain temperature, the ratio of the areas of the two peaks represents the ratio of the two conformers present in equilibrium at this temperature. This information is only a portion of what can be gained from an NMR study of conformation.

Infrared spectroscopy is also useful in conformational analysis. Characteristic stretching frequencies in the infrared spectrum may indicate that the molecule has an intramolecular hydrogen bond that would be possible in only one conformer. Different bands in the spectrum may originate from the same group in different conformations of the molecule; by obtaining the spectrum at different temperatures, the equilibrium constant can be determined for the interconversion of the conformers. It has been found that an equatorial group attached to a cyclohexane ring system has a stretching vibration at higher frequency than does the corresponding axial group.

Other physical methods used in the determination of the conformation of molecules include dipole moment measurements, optical rotatory properties, pK values, and mass spectrometry studies. These methods usually provide only fragments of information concerning the conformation of a molecule, and only when the combined data are obtained can a complete conformational analysis be realized.

Some physical methods yield the complete details of the structure of conformation of a molecule. These methods include x-ray crystallography, electron diffraction, and microwave spectroscopy. The compound must be in the gaseous state for analysis by electron diffraction or microwave spectroscopy and in the crystalline state for x-ray crystallography. These requirements are drawbacks to these methods, because the preferred conformations of a molecule in the crystalline or gaseous state are not necessarily the same as those preferred in solution.

This limitation is accentuated in assigning conformation to drug molecules by these methods and then attempting to relate the conformation to biologic activity, because most drugs in the body are in an aqueous or lipid solution. The active conformation of a drug molecule at physiologic temperature and in a physiologic medium, therefore, may be entirely different from the preferred conformation in the crystalline or gaseous state.

Theoretic methods for determining the preferred conformations of flexible drug molecules by using molecular orbital theory have been developed by Kier,[37] Neely,[38] and others. Calculations of the lowest energy conformation for several different biologically important molecules have given results that compare closely with experimentally determined conformations. Again, caution must be exercised in attempts to correlate the calculated most stable conformation with biologic activity, because a less stable conformation may be the species that reacts in vivo with the drug receptor to produce the biologic response.

Effect of Conformational Isomerism on Biologic Activity of Drugs

The knowledge that active sites of enzymes and certain drug receptor sites are stereoselective or stereospecific justifies study of the conformation of the drug molecules that may interact at these sites. Although little is known about the chemical constitution of receptors in biologic tissue, it is generally accepted that the receptor is protein or lipoprotein and is part of the membrane of the cell.

Studies have shown that on interaction with a substrate molecule, enzymes undergo conformational changes.[39] Because the active sites of enzymes and drug receptor sites in tissue are considered similar, theories have been proposed that on interaction with a drug molecule, the receptor undergoes a conformational change, which is ultimately observed as the pharmacologic response.[40]

A receptor site may bind only one of many conformations of a flexible drug molecule. This pharmacophoric conformation has the correct spatial arrangement of all the binding groups of the drug molecule for alignment with the corresponding binding sites on the receptor, as shown in Figure 4–5. Molecules that can adopt the conformation required for binding may act as agonists or antagonists to the action of the receptor. An antagonist molecule may be bound to a receptor site and therefore not trigger the pharmacologic response because of the absence of some key functional groups in the molecule. Alternatively, in the bound state, the molecule may be conformationally constrained in such a way that the functional group that effects the response cannot interact with the receptor to elicit the reaction.

For example, in Figure 4–33*a*, groups A and B are essential for binding, and group C triggers the response. A molecule possessing groups A and B and not possessing group C (Fig. 4–33*b*) will probably bind to the receptor surface but the receptor will not elicit a response because of the absence of group C's effect on it. In Figure 4–33*c*, the optical isomer of the original molecule (Fig. 4–33*a*)

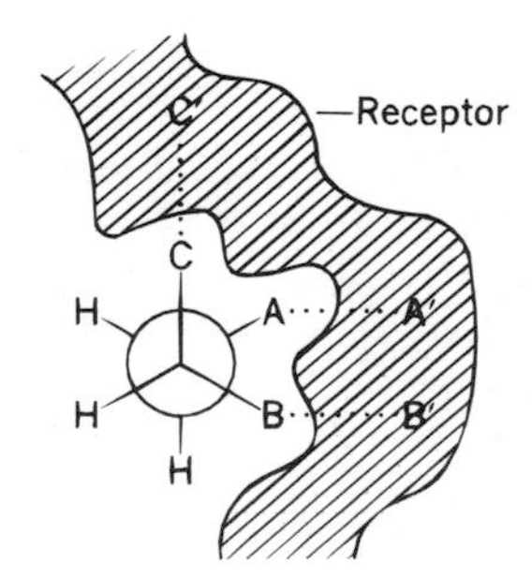

a. Agonist molecule with essential groups for binding and eliciting response

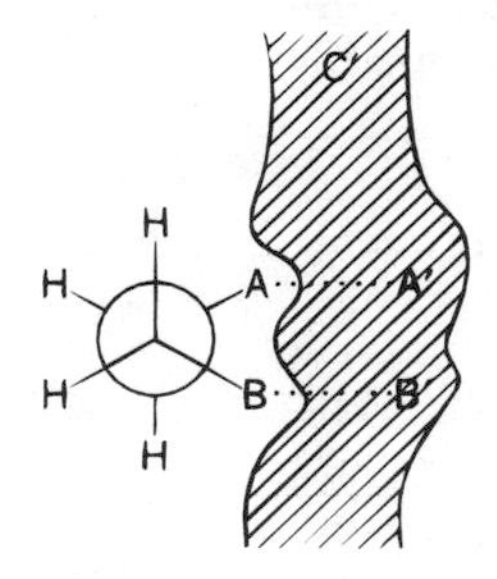

b. Antagonist molecule with essential binding groups but lacking the group necessary for eliciting response.

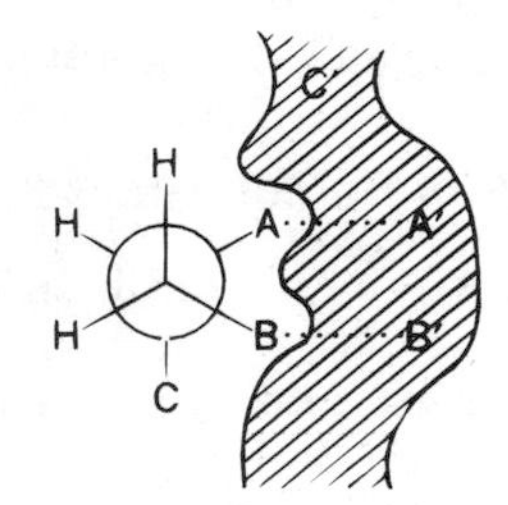

c. Antagonist and optical isomer of molecule in *a*, which can bind but cannot elicit response.

Fig. 4–33. The conformation of these hypothetic molecules determines the ability to bind to the receptor surface and to elicit the response (A′, B′, and C′ are complementary binding sites for A, B, and C).

would possibly be an antagonist, because it can adopt a conformation that permits binding to the receptor but cannot adopt the conformation that permits both binding and correct alignment of group C, which elicits the response.

The pharmacophoric conformation of a molecule that allows binding to the receptor and eliciting of the response is not necessarily the preferred conformation in the crystalline state or in solution; it may be a thermodynamically unstable conformation. In some cases, the energy of binding to a receptor site may overcome the barrier to the formation of an unstable conformer.

Belleau and Chevalier[41] have shown that the thermodynamically unstable conformation of a substrate (a 2,2′-bridged biphenyl analog of benzoylphenylalanine methyl ester) is bound by the enzyme chymotrypsin. A drug-receptor interaction, therefore, may involve the favored conformer or an unstable conformer. The lack of knowledge about the conformational changes that occur in the drug and receptor during the drug-receptor interaction is the primary difficulty in efforts to determine stereostructure-activity correlations through conformational analysis.

Attempts to overcome this difficulty have resulted in a unique approach by which one is able to determine with a degree of certainty the pharmacophoric conformation of drug molecules. This method involves the synthesis and testing of conformationally rigid analogs of the flexible drug molecule. Conformationally rigid analogs are advantageous, because the key functional groups are constrained in one position, and the configuration of the pharmacophoric conformation can be determined.

One drawback to this approach is that to produce a rigid analog of a flexible molecule, new atoms and bonds must be added to the original molecule. This imparts different chemical and physical properties, which must

a. The "cisoid" conformation of acetylcholine.

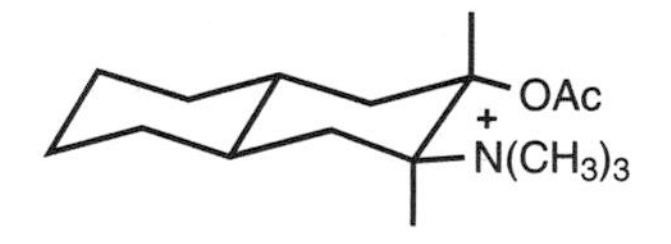

b. The rigid *trans*-decalin analog of the "cisoid" conformation of acetylcholine.

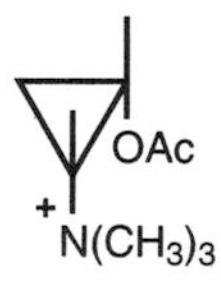

c. The rigid cyclopropyl analog of "cisoid" acetylcholine.

d. The "transoid" conformation of acetylcholine.

e. The rigid *trans*-decalin analog of the "transoid" conformation of acetylcholine.

f. The rigid cyclopropyl analog of "transoid" acetylcholine.

Fig. 4–34. Examples of rigid analogs of acetylcholine.

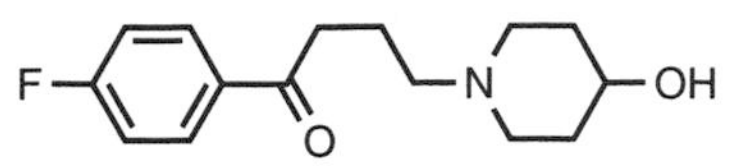

Fig. 4–35. The tranquilizing drug 4-(4-hydroxypiperidino)-4′-fluorobutyrophenone.

be taken into consideration when evaluating biological data, because there is no way of predicting the effect of these changes on activity. Greatest success with this approach usually is achieved when the rigid analog and the drug molecule are as similar in size or molecular weight as possible.

The flexible cholinergic molecule, acetylcholine, has been the subject of many studies involving this approach. Acetylcholine acts at two receptors, the muscarinic and the nicotinic, producing appropriate responses, and it is hydrolyzed by the enzyme acetylcholinesterase. Several studies involving rigid analogs of acetylcholine have indicated that a "cisoid" conformation may be required by the nicotinic receptor, whereas a "transoid" conformation of acetylcholine seems to be required by the muscarinic receptor and by the hydrolytic enzyme acetylcholinesterase.

Figure 4–34 shows the "cisoid" and "transoid" conformations of acetylcholine and two examples of rigid analogs of acetylcholine. The *trans*-decalin analogs (Fig. 4–34*b* and *e*) represent a large addition of atoms in order to make the acetylcholine moiety rigid. There is therefore a great change in the chemical and physical properties of these analogs, which results in a major loss in acetylcholine type of activity.[42]

The cyclopropyl analogs (Fig. 4–34*c* and *f*), however, have minimal changes in structure and properties, because only one methylene (—CH_2—) group has been added. The result is that little loss in activity is observed.[43] The pharmacologic activities of each of these series of rigid analogs of acetylcholine support the postulate that the "transoid" conformation is responsible for muscarinic activity and is hydrolyzed more readily by the enzyme acetylcholinesterase.

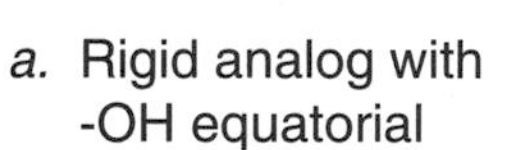

b. Rigid analog with -OH axial

c. Rigid analog with an oxygen function in both the axial and the equatorial position

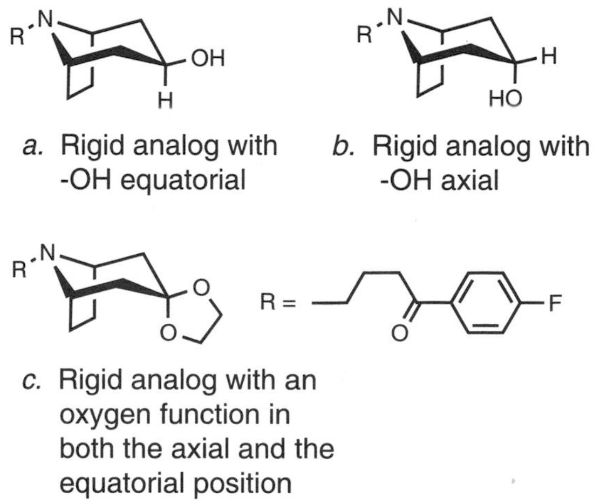

Fig. 4–37. Three rigid analogs of the tranquilizer 4-(4-hydroxypiperidino)-4′-fluorobutyrophenone, using the nortropine ring system to introduce rigidity.

A thorough study using rigid conformers was conducted by Li and Biel[44] on the tranquilizing drug 4-(4-hydroxypiperidino)-4′-fluorobutyrophenone (Fig. 4–35). The objective was to determine possible steric requirements of the hydroxyl group. Of the possible conformations of the substituted piperidine ring, the chair forms in Figure 4–36*a* and *b* were considered the only two that might be active at the receptor. The twist forms in Figure 4–36*c* and *d* were excluded because their high energies would make them unstable to the degree that their involvement in the drug-receptor interaction would be highly unlikely.

Rigid analogs of the chair forms in Figure 4–36*a* and *b* were synthesized, and their structures are shown in Figure 4–37. Analog *a* mimics the conformer in Figure 4–36 with the hydroxyl group in the equatorial position. Analog *b* mimics the conformer with an axial hydroxyl group, and analog *c* mimics both conformers because it has an oxygen function in both the axial and the equatorial position. These analogs, when subjected to muscle relaxation tests, exhibited the following relative order of activity or potency: $b > c > a$. These results indicate that, for maxi-

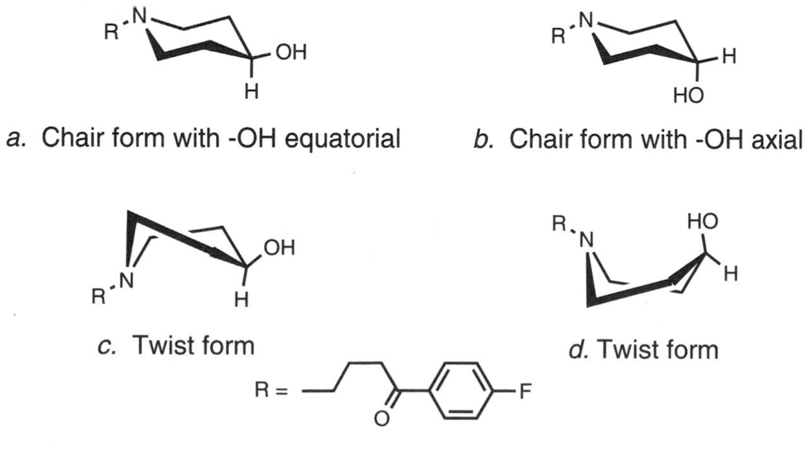

a. Chair form with -OH equatorial

b. Chair form with -OH axial

c. Twist form

d. Twist form

Fig. 4–36. The possible conformations for 4-(4-hydroxypiperidino)-4′-fluorobutyrophenone.

Threo-α,β-dimethyl-acetylcholine

Erythro-α,β-dimethyl-acetylcholine

Fig. 4–38. The threo-α,β-dimethylaminoacetylcholine and erythro-α,β-dimethylacetylcholine diasteromers. (Only one enantiomer of each diasteromer is shown).

mal potency, the hydroxyl groups or oxygen function should preferably be in the axial position. The probable pharmacophoric conformation of 4-(4-hydroxypiperidino)-4′-fluorobutyrophenone would be *b* in Figure 4–36, the conformer with the axial hydroxyl group.

The preceding discussion included only two examples of many published studies in which rigid analogs of flexible drug molecules were used to assess stereostructure—activity relationships. These reports include further studies and reviews on cholinergic drugs,[45–48] analgesic drugs,[49–51] and adrenergic drugs.[52–55]

Conformational analysis may also be used to explain differences in biologic activity of diastereoisomeric drugs. In many instances, the slight differences in chemical and physical properties seem insufficient to explain the great differences in drug potency. The assumption is that one diastereomer may more readily adopt a conformation that fits the receptor and elicits a response.

An example of this approach is the interpretation of the differences in the muscarinic activity of (+)-threo-α-β-dimethylacetylcholine and (+)-erythro-α-β-dimethylacetylcholine in Figure 4–38. The erythro racemate (d, l mixture) is more active than the threo racemate. As noted previously, several studies have established that the muscarinic receptor requires that the more active drugs have the conformation in which the acetoxy group and the trimethylammonium group are in a "transoid" position. The ease with which a muscarinic agent can adopt this pharmacophoric conformation therefore determines its potency.

An examination of the conformation of each diastereomer of the α, β-dimethylacetylcholines in which the acetoxy and trimethylammonium groups are *trans* (Fig. 4–39) shows that the threo conformer would be less favored than the more stable erythro conformer. The "transoid" conformation of the erythro isomer has greater stability because the methyl groups are farther apart than in the "transoid" threo isomer. If the predictions of conformational preference are accurate, one is able to correlate the low potency of the threo isomer with the fact that rotation to a less favorable conformation must occur to attain the correct orientation for interaction with the receptor.

Threo conformer

Erythro conformer

Fig. 4–39. The "transoid" conformation of (+)-threo-α,β-diethylacetylcholine and (+)-erythro-α,β-dimethylacetylcholine in which the acetoxy and trimethylammonium groups are *trans.*

ISOSTERISM AND PHARMACOLOGIC ACTIVITY

Chemical Isosterism

The concept of chemical isosterism is credited to Langmuir,[56,57] who in a series of papers in 1919 elaborated on the similarities in physicochemical properties of atoms, groups, radicals, and molecules with similar electronic structures. Similarities occurred most often in atoms that were in the same vertical columns of the periodic table, where the outer shells of electrons are identical or almost identical, and in atoms that are not so far separated that their variation in size and mass is not great.

To a lesser degree, the same trend is recognized for contiguous atoms in a horizontal row. For example, the chemical properties of chlorine and bromine are more similar than those of carbon and chlorine or chlorine and iodine. Although chlorine and iodine are in the same vertical group and their outer electron shells are identical, they differ in size, as indicated by the van der Waals' radii and atomic weights (chlorine 1.80 Å, 35.46; iodine 2.15 Å, 126.91). The crux of Langmuir's concept was that the number and arrangement of electrons in the outermost shell must be similar.

Langmuir's treatises were concerned almost entirely with the elements, inorganic molecules, ions, and small organic molecules, such as diazomethane and ketene. Table 4–3 is a comparison of the physical properties of N_2O and CO_2 and illustrates some of the data compiled by Langmuir; similar relationships have been shown for N_2 and CO.

In 1925, Grimm[59] formulated a set of hydride displacement rules. Vertical columns of isosteric groups were formed by displacing one place to the right successively the elements of a horizontal row and adding a hydrogen atom (or hydride ion) and continuing this process as long

Table 4–3. Comparison of Physical Properties of N_2O and CO_2

Property	N_2O	CO_2
Viscosity at 20°C	148×10^{-6}	148×10^{-6}
Density of liquid at +10°C	0.856	0.858
Refractive index of liquid, D line 16°C	1.193	1.190
Dielectric constant of liquid at 0°C	1.593	1.582
Solubility in alcohol at 15°C	3.250	3.130

Table 4–4. Grimm's Concept for Hydride Displacement

C	N	O	F	Ne
	CH	NH	OH	HF
		CH_2	NH_2	OH_2
			CH_3	NH_3
				CH_4

as possible. Table 4–4 is an example. Each vertical column represented a group of isosteres. Table 4–5 shows how the process is continued with the next horizontal row of elements.

Included in these two tables are many elements contained in naturally occurring organic compounds, with the exception of silicon, fluorine, and the inert gases. From a more modern viewpoint, such empiric rules do not include parameters such as basicity, acidity, electronegativity, polarizability, bond angles, size, shape of molecular orbitals, electron density, and partition coefficients, which make important contributions to the overall physiochemical and, therefore, biologic properties of a molecule.

Erlenmeyer and co-workers[60–65] broadened the concept of isosterism and at the same time added restrictions that could not be followed. They felt that the peripheral layers of electrons must be almost identical in shape, size, and polarity, and that the compounds should be isomorphic or cocrystallizable; the molecules of an isosteric pair should fit into the same crystalline lattice. Although some such sets of compounds were observed, it was too stringent a condition to follow; examples of compounds capable of mixed crystal formation are found in the pairs shown in Table 4–6.

Table 4–5. Grimm's Concept for Hydride Displacement (Cont.)

Si	P	S	Cl	Ar
	SiH	PH	SH	HCl
		SiH_2	PH_2	SH_2
			SiH_3	PH_3
				SiH_4

Table 4–6. Compounds Capable of Mixed Crystal Formation

$C_6H_5-N{=}N-C_6H_5$ $C_6H_5-CH{=}CH \cdot C_6H_5$	$C_6H_5-O-CH_2 \cdot C_6H_5$ $C_6H_5-NH-CH_2 \cdot C_6H_5$
$C_6H_5-O-CH_2 \cdot C_6H_5$	$C_6H_5-NH-NH-C_6H_5$
$C_6H_5-CH_2 \cdot CH_2 \cdot C_6H_5$	$C_6H_5-CH_2 \cdot CH_2 \cdot C_6H_5$
$CH_3-C_6H_4-C_6H_4-CH_3$ $Cl-C_6H_4-C_6H_4-Cl$	

Hinsberg[66] first postulated the isosteric replacement of CH═CH by S and perhaps initiated recognition of the interchanging of the various aromatic rings, such as thiophene, benzene, pyridine, pyrrole, and furan, as isosteric group replacements.

Bioisosterism

Langmuir's work was not directly related to biologically active molecules; however, the groundwork was laid and the stage was set for a carry-over into the biologic sciences.

If a molecularly modified drug is to react at the same sites as the original drug to bring about the desired action, the molecular modification should not be too drastic. Isosterism, therefore, naturally fits into the scheme of drug design and molecular modification in the creation of new and improved therapeutic agents.

In the transition from pure chemistry to medicinal chemistry, Freidman[58] in 1951 aptly coined the term bioisosterism, and since then the meaning of the term has gradually broadened.

Subsequently, Burger[67] classified and subdivided bioisosterism along its evolutionary path into the following categories:

- I. Classical bioisosteres
  - A. Monovalent atoms and groups
  - B. Divalent atoms and groups
  - C. Trivalent atoms and groups
  - D. Tetrasubstituted atoms
  - E. Ring equivalents
- II. Nonclassical bioisosteres
  - A. Exchangeable groups
  - B. Rings versus noncyclic structures

The classic monovalent bioisosteres included the halogens and the groups—XH_n, in which X is C, N, O, and S. The divalent atoms and groups comprise R—O—R′, R—NH—R′, R—CH_2R′, and R—Si—R′. The trivalent bioisosteres are limited to C and N in the formation of trivalent groups, such as R—N═R′ and R—CH═R′. The tetra-substituted atoms comprise only the three elements ═C═, ═N═, and ═P═. The manner in which these three functions are drawn is not indicative of three-dimensional characteristics, which in most instances would approximate a tetrahedron. The group relating to ring equivalents involves the interchange of —CH═CH—, —S—, —O—, —NH, and —CH_2—. The nonclassic bioisosteres do not rigidly fit the steric and electronic rules of the classic bioisosteres.

Recent Bioisosteric Applications

An excellent example of exchangeable groups is the sulfonamido isosteres of the catecholamines.[68] An alkylsulfonamido group may be substituted for the phenolic hydroxy group of certain catecholamines. Some of the resulting compounds have agonist activity; others are antagonists. Analogies may be drawn between the catecholamines and the alkylsulfonamidophenethanolamines, shown in Figure 4–40, via the pK_a value of the alkylsulfonamido compound (pK_a = 9.1) and of phenylephrine (pK_a = 9.6). It is believed that the alkylsulfonamido

Table 4–7. Isosteric Replacement of C for O and O for X

Parent Compound	Bioisostere	Activity of Parent Compound	Reference
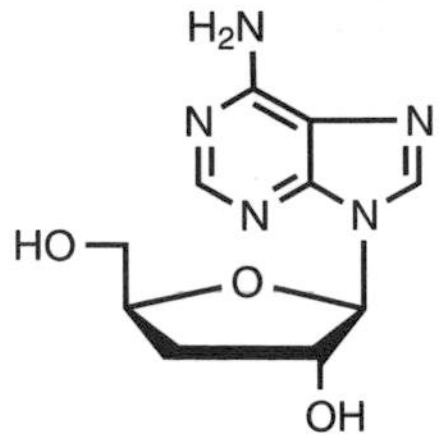 1'-(9-Adenyl)-β-D-3'-deoxyribose	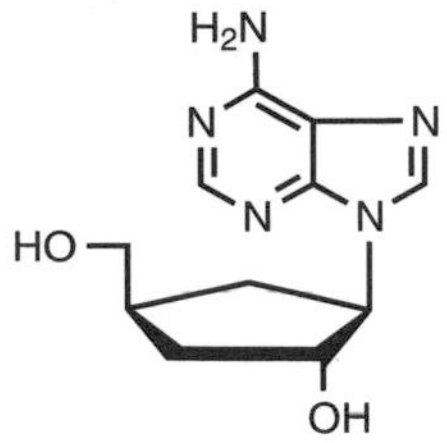 1-(9-Adenyl)-2-*trans* -hydroxy-4-cis-hydroxymethylcyclopentane	Adenosine deaminase (cleavage of adenosine to the purine and carbohydrate moiety) (−)*	73
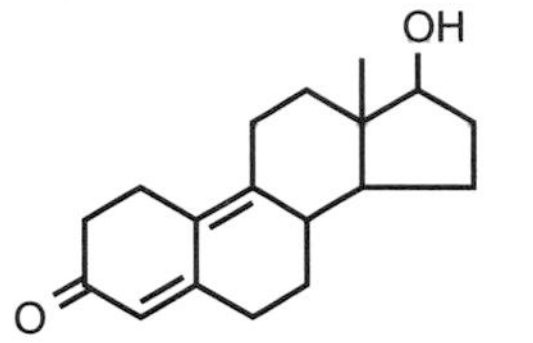 17β-hydroxyestra-4,9(10)-dien-3-one	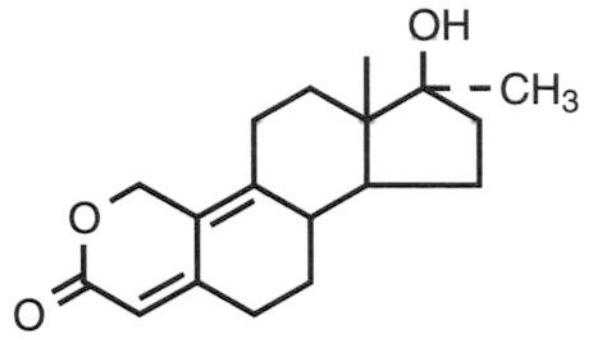 17α-Methyl-17β-hydroxy-2-oxaestra-4,9(10)-dien-3-one	Myotropic, androgenic (+)*	74
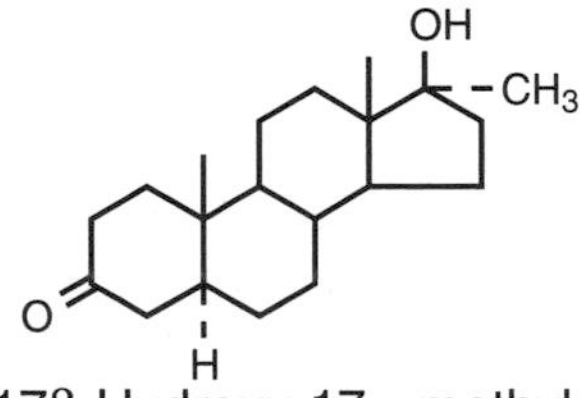 17β-Hydroxy-17α-methyl-4,5-dihydrotestosterone	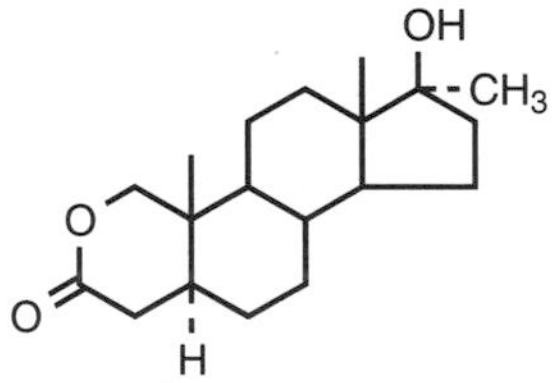 Oxandrolone	Myotropic, androgenic (+)*	75
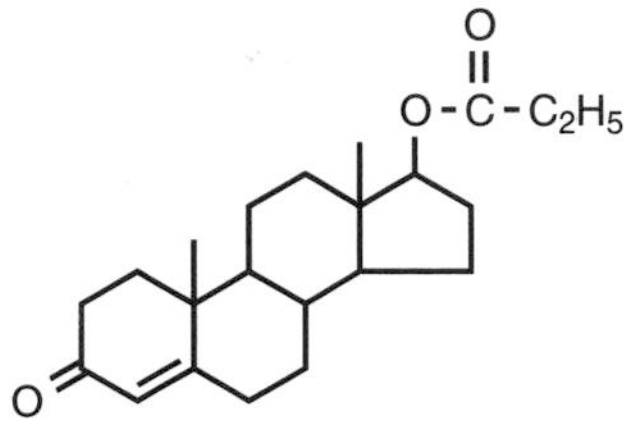 Testosterone propionate	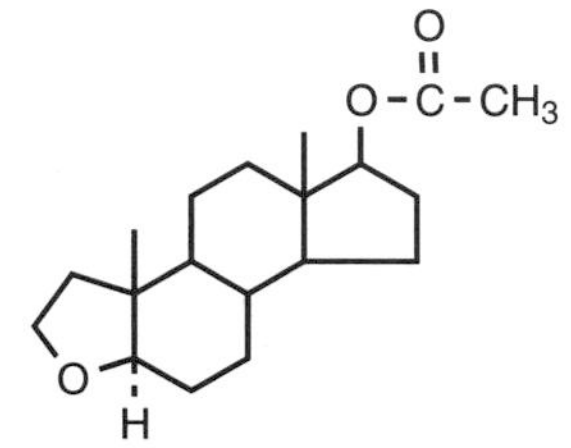 3-Oxa-5α-A-norandrostane-17β-ol acetate	Androgenic (−)*	76
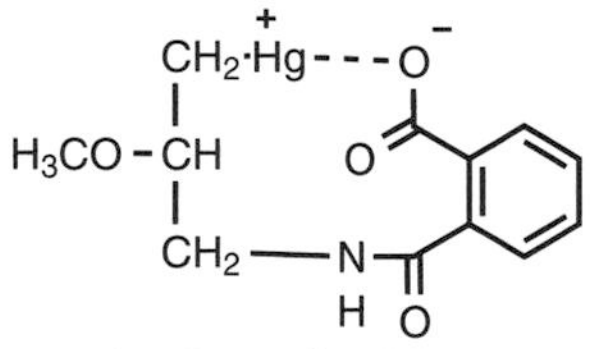 Anhydro-3-hydroxy-mercuri-2-methoxy-propyl hydrogen phthalamide	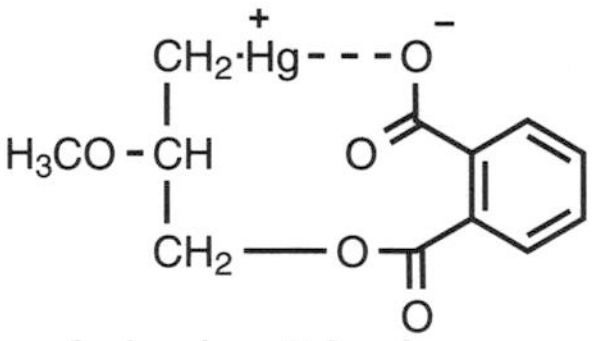 Anhydro-3-hydroxy-mercuri-2-methoxy-propyl hydrogen phthalate	Diuretic (+)*	77
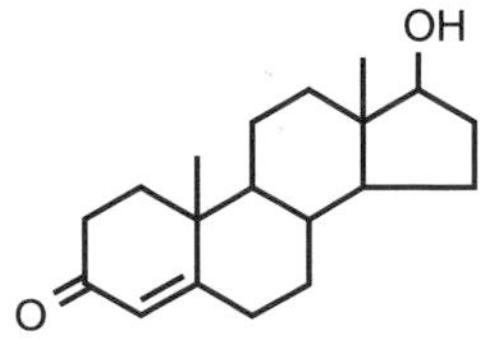 Testosterone	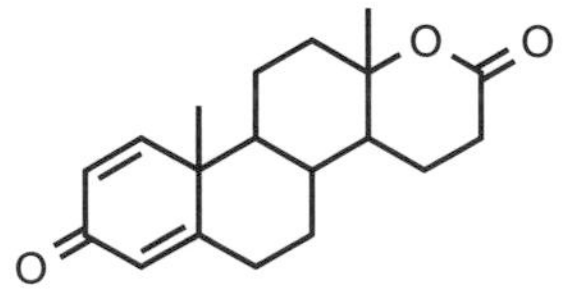 17α-Oxa-D-homo-1,4-androstadiene-3,17-dione (or Δ^1 testolactone)	Mammary gland antineoplastic (+)* Bioisostere has greatly decreased androgenic effects	78

Table 4–7. *(Continued)*

Parent Compound	Bioisostere	Activity of Parent Compound	Reference
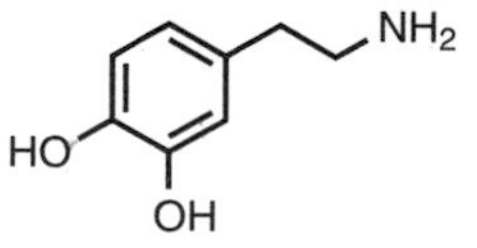 Dopamine	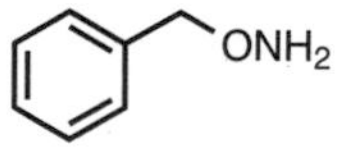 Benzyloxyamine 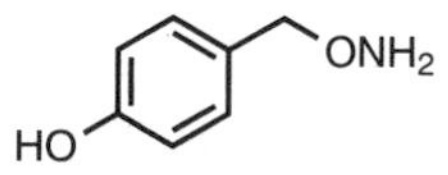 p-Hydroxybenzyloxyamine	Dopamine converted to norepinephrine via dopamine β-oxidase (−)* Addition of p-hydroxyl grouping increases inhibition	79

(+)*—activity of bioisostere similar to parent compound.
(−)*—activity of bioisostere not similar to parent compound.

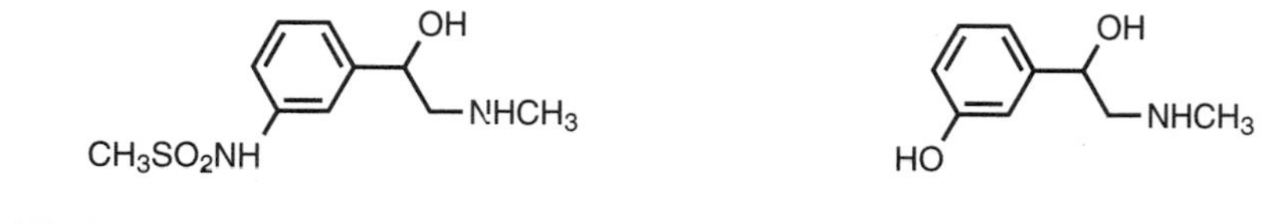

Alkylsulfonamidophenethanolamine   Phenylephrine

Fig. 4–40. Comparison of alkylsulfonamidophenethanolamine and phenylephrine.

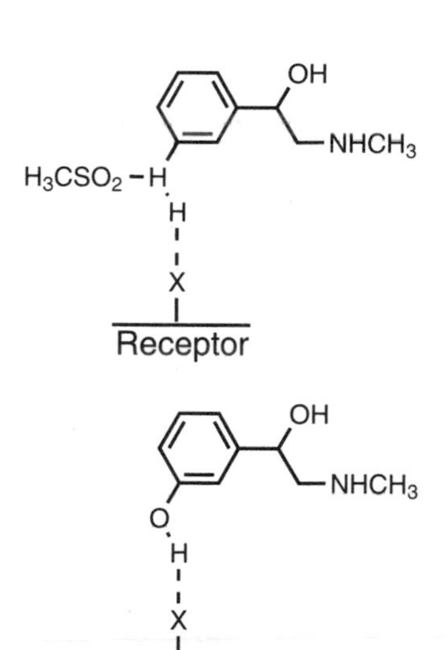

Fig. 4–41. Alkylsulfonamidophenethanolamine *(upper)* and phenylephrine *(lower)* at the receptor site.

group and the phenolic hydroxyl group are capable of aligning themselves similarly at the receptor site. Both compounds thereby elicit a response at the adrenergic receptor, as shown in Figure 4–41.

Both compounds have almost the same activity in the thiopental-barbital anesthetized dog; when given intravenously, 0.004 mg/kg of the alkylsulfonamido compound and 0.002 mg/kg of phenylephrine cause a 20% increase in blood pressure.

The classic example of cyclic versus noncyclic bioiso-

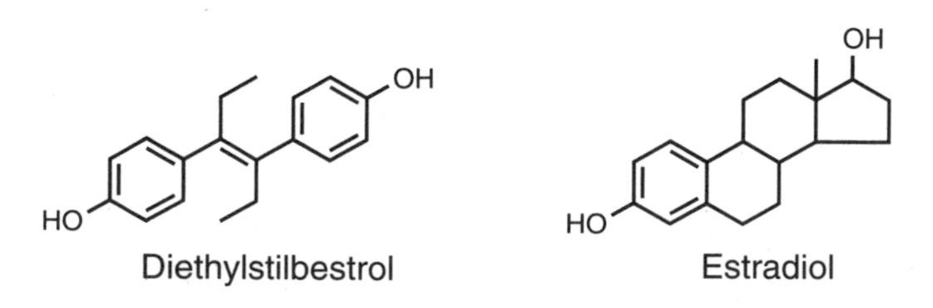

Fig. 4–42. Cyclic versus noncyclic bioisoterism.

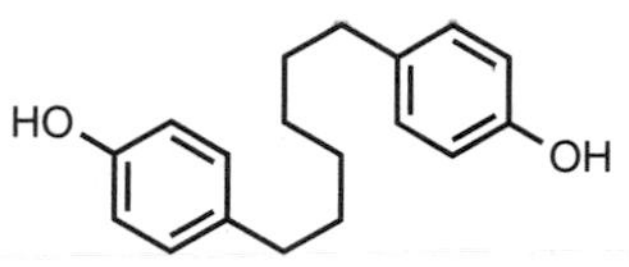

1,6-bis-(p-hydroxyphenyl)hexane

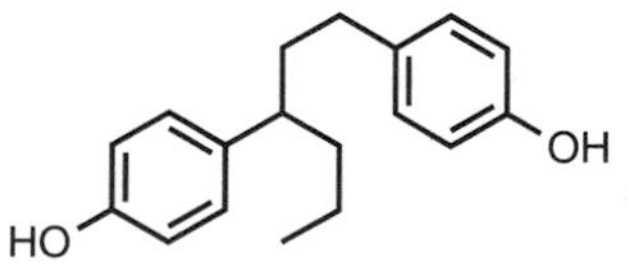

1,3-bis-(p-hydroxyphenyl)hexane

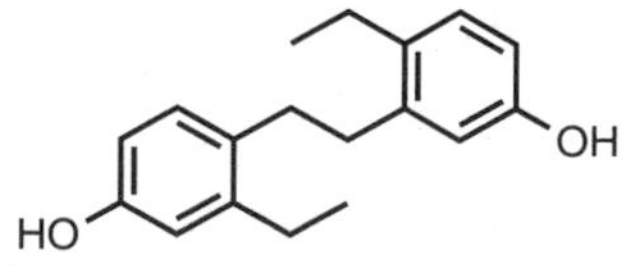

1,3-bis-(2-ethyl-4-hydroxyphenyl)ethane

Fig. 4–43. Noncyclic analogs of estradiol.

Table 4–8. Isosteric Replacement of N for X

Parent Compound	Bioisostere	Activity of Parent Compound	Reference
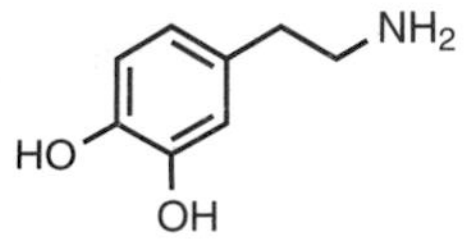 Dopamine	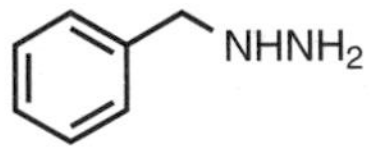 Benzylhydrazine; 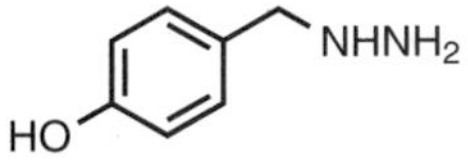 ρ-Hydroxybenzylhydrazine	Dopamine converted to norepinephrine via dopamine β-oxidase (−)* Addition of p-hydroxyl grouping in bioisostere increases inhibition	79
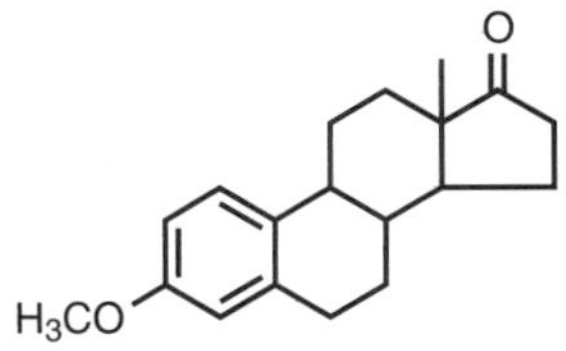 Estrone-3-methyl ether	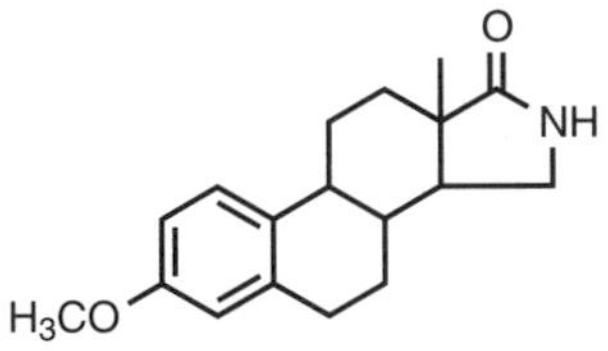 16-Azaestrone-3-methyl ether	Decrease in testes and prostate weight in rat (+)*	80
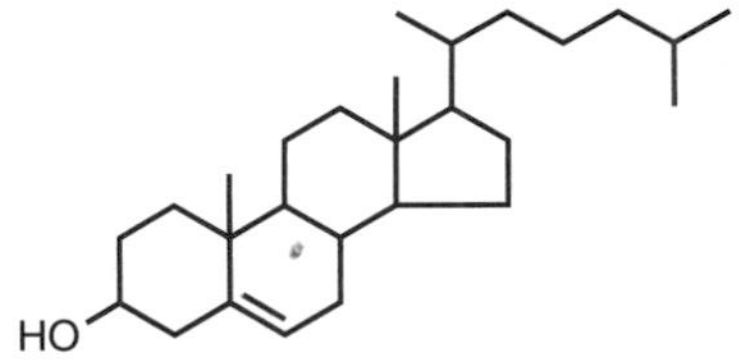 Cholesterol	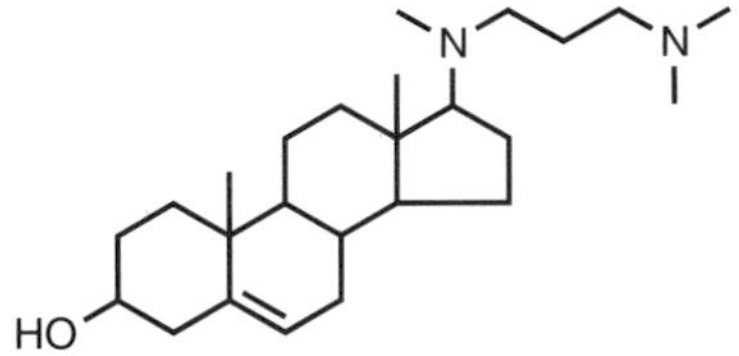 20,25-Diazacholesterol	Important biologic metabolite (−)*	81
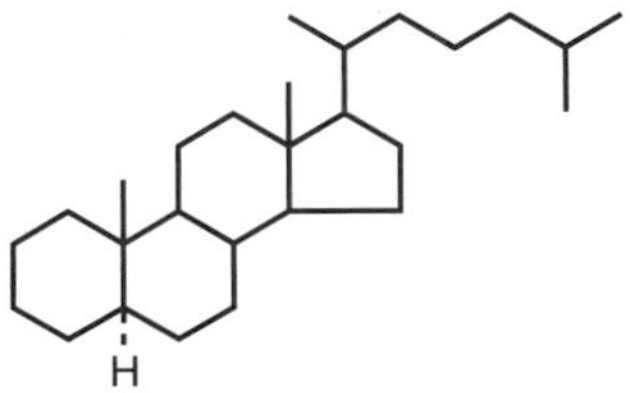 5α-Cholestane	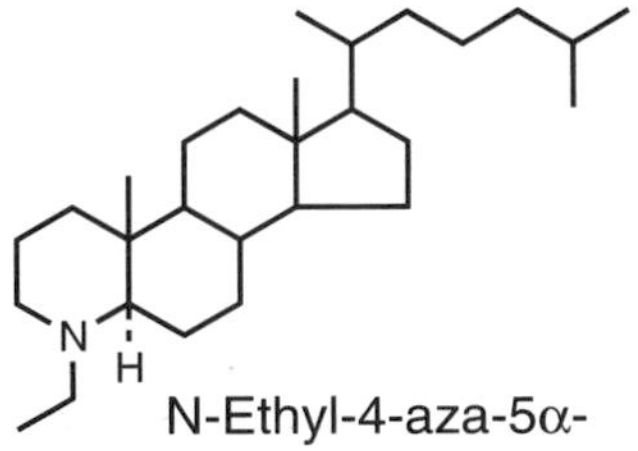 N-Ethyl-4-aza-5α-cholestane	Inhibition of reduced nicotinamide adenine dinucleotide oxidation with membrane fragments from *Bacillus subtilis* (activity of the isostere)	82
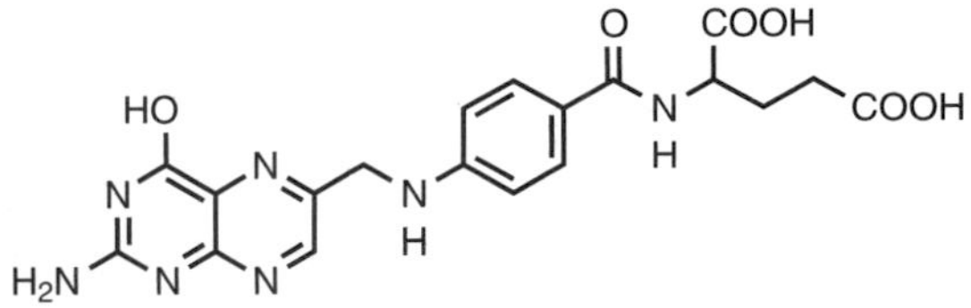 Folic acid	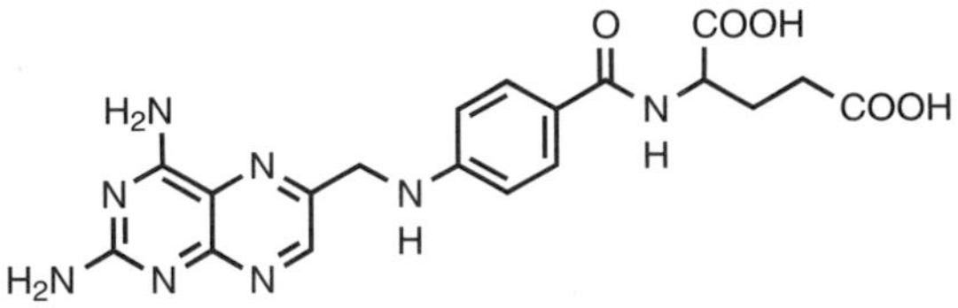 Aminopterin	Metabolite, substrate of folic acid reductase (−)*	83

(+)*—activity of bioisostere similar to parent compound.
(−)*—activity of bioisostere not similar to parent compound.

Table 4–9. Isosteric Replacement of S for X

Parent Compound	Bioisostere	Activity of Parent Compound	Reference
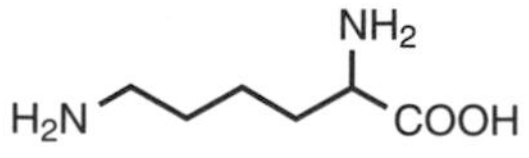 Lysine	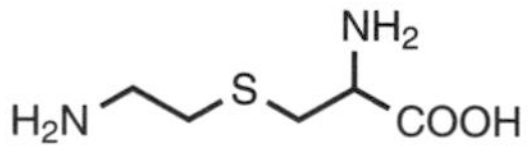 Thialysine	Nutrient for certain bacteria incorporated into rat bone marrow (−)*	84
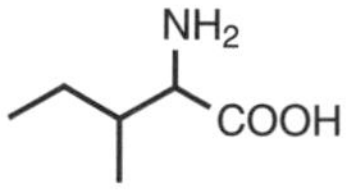 Isoleucine	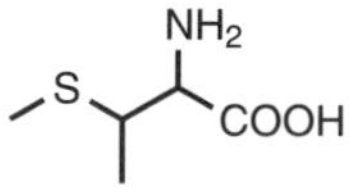 Thiaisoleucine	Nutrient for growth of K-12 strain of Escherichia coli; may be incorporated into t-RNA; is transaminated with α-ketoglutarate as amino acceptor (−)* and (+)*	85
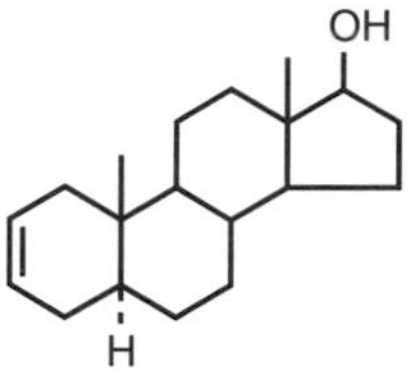 17β-Hydroxy-5α-androst-2-ene	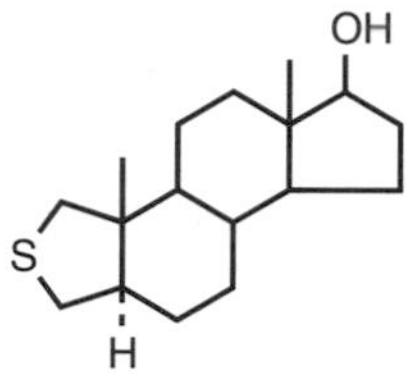 2-Thia-A-nor-5α-androst-17β-ol	High order of androgenic and myotropic activity (+)*	86
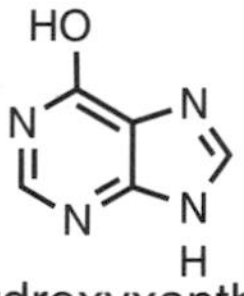 Hydroxyxanthine	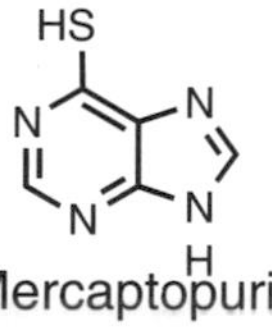 6-Mercaptopurine	A common purine metabolite (−)* and (+)* Bioisostere: a neoplastic suppressant, active in certain enzymic reactions, inactive in others	87

(+)*—activity of bioisostere similar to parent compound.
(−)*—activity of bioisostere not similar to parent compound.

Table 4–10. Isosteric Replacement of Si for C

Parent Compound	Bioisostere	Activity of Parent Compound	Reference
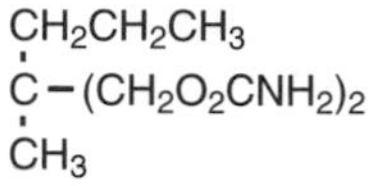 $CH_3CH_2CH_2$–C(–CH_3)–$(CH_2O_2CNH_2)_2$ Meprobamate	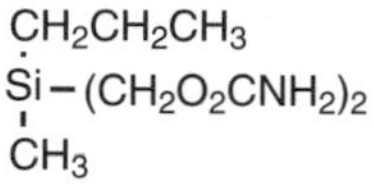 $CH_3CH_2CH_2$–Si(–CH_3)–$(CH_2O_2CNH_2)_2$ Silameprobamate	Tranquilizer (+)* by intraperitoneal injection	88
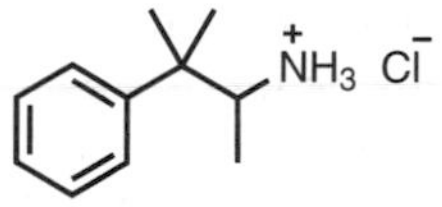 α-Methyl-β-dimethyl phenethylamine hydrochloride	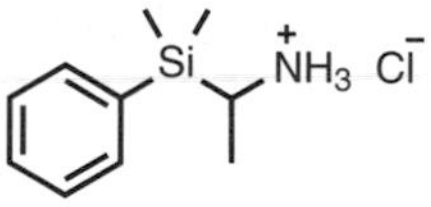 (α-Aminomethyl)-dimethyl phenylsilane hydrochloride	Increased rate and depth of respiration (+)*	89
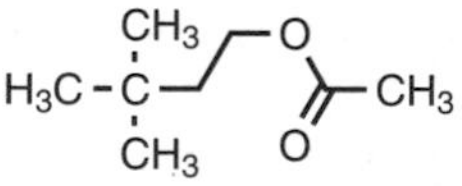 Acetyl carbocholine	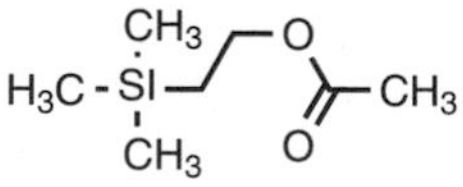 Acetylsilicholine	Indirect acting cholinergic agents (+)*	90
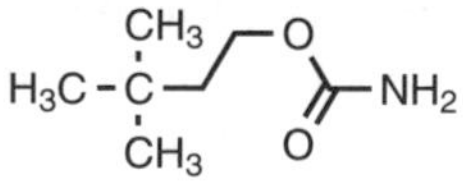 3,3-Dimethylbutane-1-carbamate	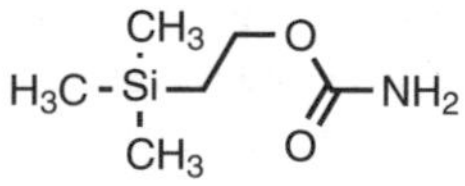 3,3-Dimethyl-2-sila-butane-1-carbamate	Antimuscarinic (+)*	91

(+)*—activity of bioisostere similar to parent compound.

Table 4–11. Isosteric Replacement Involving Cyclic versus Noncyclic Analogs

Parent Compound	Bioisostere	Activity of Parent Compound	Reference
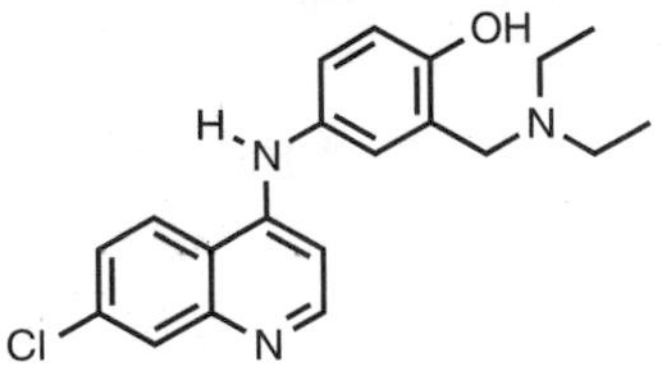 Amodiaquine	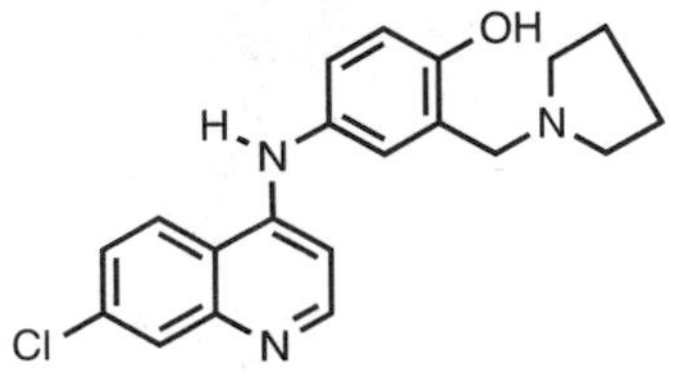 Amopyroquine	Antimalarial (+)*	92
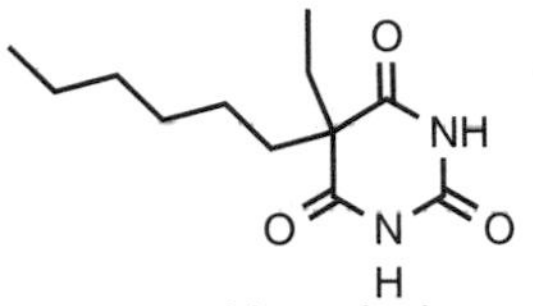 Hexethal	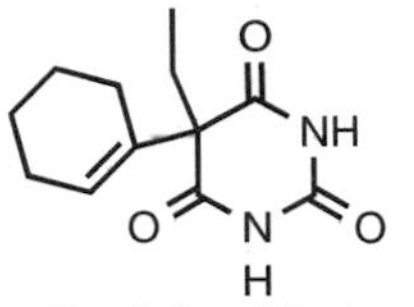 Cyclobarbital	Sedative and hypnotic (+)*	93
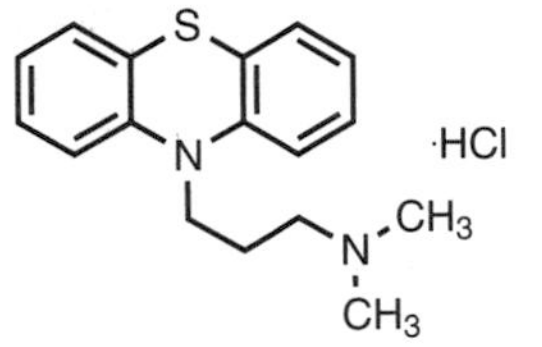 Promazine hydrochloride	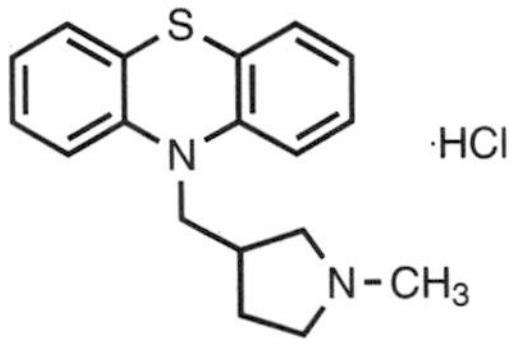 Methdilazine hydrochloride	Central nervous system depressant (+)*	93
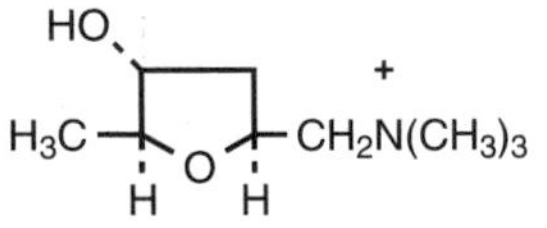 L-(+)-Muscarine	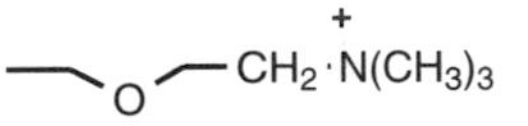 Choline ethyl ether	Muscarinic activity (+)*	94

(+)*—activity of bioisostere similar to parent compound.

sterism is diethylstilbestrol and estradiol (Fig. 4–42). Diethylstilbestrol has about the same potency as the naturally occurring estradiol. Many noncyclic analogs related to estradiol have been synthesized, some of which have estrogenic activity. The analogs shown in Figure 4–43, which are partial replicas of estradiol, have little if any activity.[69,70]

The central double bond of diethylstilbestrol is highly important for the correct orientation of phenolic and ethyl groups at the receptor site.[71] The *cis* isomer is only about $\frac{1}{14}$ as active as the *trans* isomer,[72] as noted previously.

Physicochemical properties, such as electronegativity, polarizability, bond angles, van der Waals' radii, number of substituents, charge, and acidity or basicity of the atom can greatly influence the physicochemical characteristics of the molecule. Drug molecules in turn exert their effect by influencing receptor sites in living systems through their physicochemical properties. It follows, therefore, that alteration of an atom or group in a molecule will change the physicochemical properties of the molecule and thereby the biologic response to it.

Tables 4–7 through 4–12 contain a variety of bioisosteres, including classic and nonclassic examples, incorporating marketed drugs as well as interesting experimental compounds taken from the recent literature.

SUMMARY

The primary role of the medicinal chemist has become one of improving existing drugs by increasing their potency and duration of action and by decreasing toxic side effects, as well as creating new drugs from naturally occurring biochemical substances by molecular modification. This goal is achieved through organic synthesis (occasionally through enzymic reactions of microorganisms) concomitant with pharmacologic testing.

The incorporation of steric considerations in the development of new drug molecules is desirable if advances are to be made in the elucidation of receptors for specific pharmacologic responses at the molecular level. The elucidation of the steric features of known pharmacologic agents, or so-called pharmacophoric conformations, holds much promise for progress in drug design, enabling medicinal chemists to synthesize new drugs with not only the essential functional groupings but also the capability of adopting the optimal steric arrangements.

Table 4–12. Exchangeable Groups

Parent Compound	Bioisostere	Activity of Parent Compound	Reference
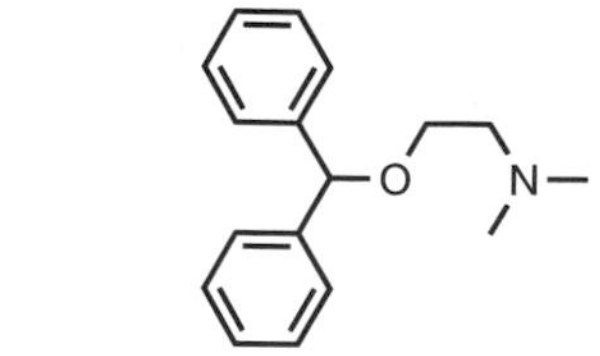 Diphenhydramine (Benadryl)	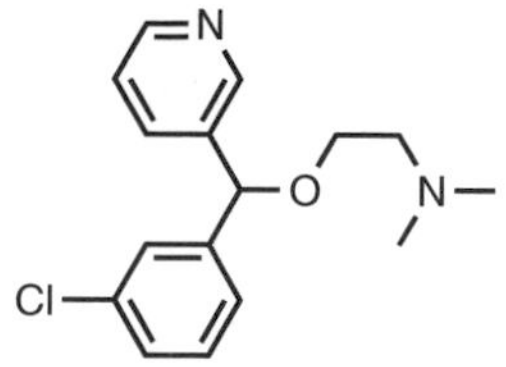 d-Carbinoxamine (Clistin)	Antihistamine (+)*	95
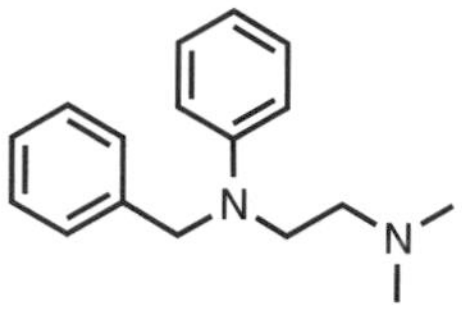 Antergan	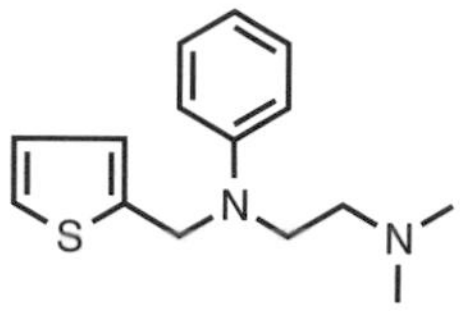 Diatrin	Antihistamine (+)*	95
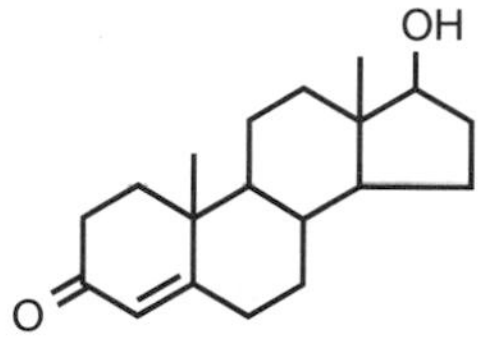 Testosterone	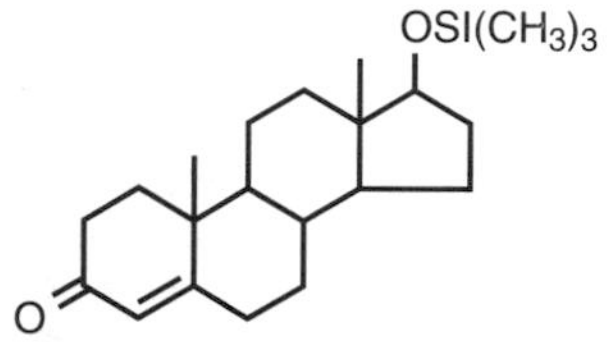 Testosteroxytrimethyl silane	Androgenic (+)*	96
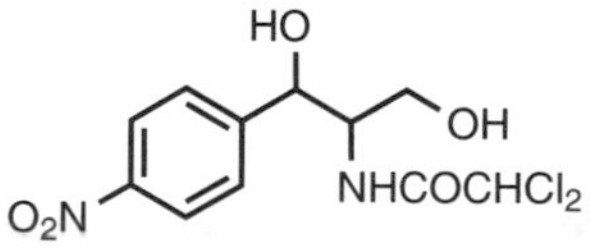 Chloramphenicol	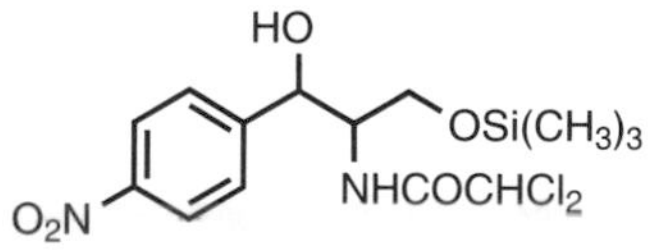 2-Dichloroacetamido-1-ρ-nitro-phenyl-3-trimethylsilyloxy-1-hydroxy propane	Antibacterial (+)*	97
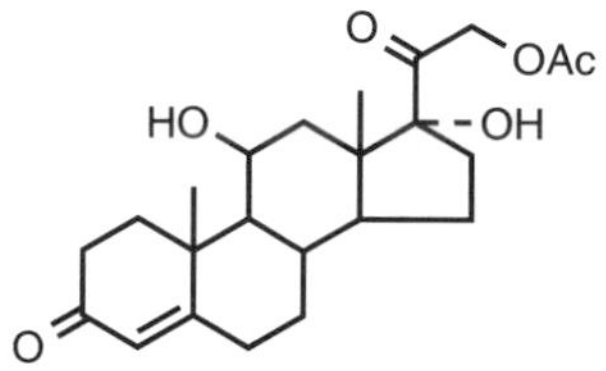 Hydroxycortisone acetate	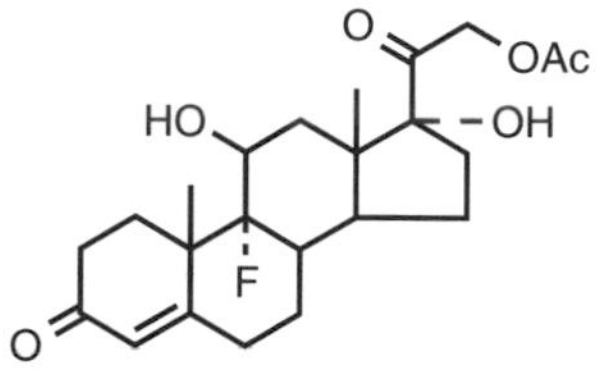 9α-Fluorohydrocortisone acetate	Anti-inflammatory (+)*	98
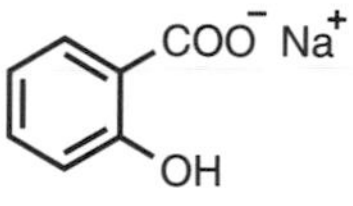 Sodium salicylate	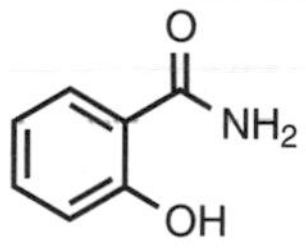 Salicylamide	Analgesic, antipyretic (+)*	99
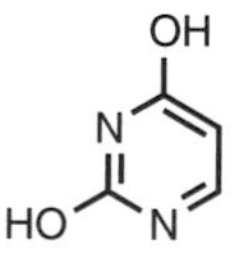 Uracil	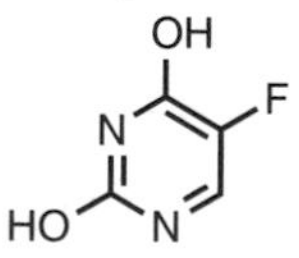 5-Fluorouracil	Bioisostere: neoplastic suppressant, active in certain enzymic, reactions, inhibitory in others (−)* and (+)*	100

(+)*—activity of bioisostere similar to parent compound.
(−)*—activity of bioisostere not similar to parent compound.

REFERENCES

1. L. Pasteur, *Compt. Rend., 46,* 615(1858).
2. A. R. Cushny, *Biological Relations of Optically Isomeric Substances,* London, Balliere, Tindall and Cox, 1926.
3. L. H. Easson and E. Stedman, *Biochem. J., 27,* 1257(1933).
4. A. H. Beckett, *Prog. Drug. Res., 1,* 445(1959).
5. A. H. Beckett, et al., *J. Pharm. Pharmacol., 15,* 362(1963).
6. L. Gyemch and K. R. Unna, *Proc. Soc. Exp. Biol. Med., 98,* 882(1958).
7. M. T. Bush and W. L. Weller, *Drug Metab. Rev. I,* 249(1972).
8. R. L. Furner, et al., *J. Pharmacol. Exp. Ther., 169,* 153(1969).
9. D. M. Jerina, et al., Fifth International Symposium on Microsomes and Drug Oxidations, R. Sato and K. Kato, Eds. Japan Scientific Societies Press, p. 195, (1982).
10. J. V. Auditore and B. V. Rama Sastry, *Arch. Biochem., 105,* 506(1964).
11. D. G. Kaiser, et al., *J. Pharm. Sci., 65,* 269(1976).
12. C. W. Porter and C. T. Hirst, *J. Am. Chem. Soc., 41,* 1264(1919).
13. W. E. Muller and V. Wollert, *Mol. Pharmacol., 11,* 52(1975) and *Res. Commun. Chem. Pathol. Pharmacol. 9,* 413(1974).
14. T. P. King and M. Spencer, *J. Biol Chem., 245,* 6134(1970).
15. H. Veldstro, *Pharmacol. Rev., 8,* 339(1956).
16. P. N. Patil, et al., *J. Pharm. Sci., 59,* 1205(1970).
17. P. N. Patil, *J. Pharmacol. Exp. Ther., 160,* 308(1968).
18. D. D. Miller, et al., *J. Med. Chem., 19,* 1382(1976); *J. Med. Chem., 23,* 1232(1980); *J. Pharmacol. Exp. Ther., 217,* 1(1981).
19. L. Almirante and W. Murmann, *J. Med. Chem., 9,* 650(1966).
20. A. S. Horn and J. R. Rodgers, *J. Pharm. Pharmacol., 32,* 521(1980); B. Costall, et al., *J. Pharm. Pharmacol., 34,* 246(1982).
21. W. S. Saari, et al., *J. Med. Chem., 16,* 171(1973).
22. M. G. Bogaert and W. A. Buylaert, in *Stereochemistry and Biological Activity of Drugs,* A. J. Ariens, et al. Eds., Oxford, Blackwell Scientific, 1983, pp. 147–148.
23. P. S. Portoghese and D. A. Williams, *J. Med. Chem., 13,* 626(1970).
24. P. B. M. W. M. Timmermans, in *Stereochemistry and Biological Activity of Drugs,* A. J. Ariens et al., Eds., Oxford, Blackwell Scientific, 1983, pp. 166–169.
25. J. J. P. Heykants, et al., *Arch. Int. Pharmacodyn. Ther., 216,* 113(1975).
26. J. A. Tobert, et al., *Clin. Pharmacol. Ther., 29,* 344(1981).
27. M. Barash, et al., *J. Chem. Soc., 3530*(1959); 2157(1959). H. T. Openshaw, et al., *J. Chem. Soc.,* 101(1969).
28. A. P. Grollman, *Science, 157,* 84(1967).
29. O. Hechter, et al., *Proc. Natl. Acad. Sci. U.S.A., 72,* 563(1975).
30. A. Witter, in *Stereochemistry and Biological Activity of Drugs,* A. J. Ariens et al., Eds., Oxford, Blackwell Scientific, 1983, pp. 154–159.
31. E. Walton and G. Browle, *Nature, 151,* 305(1943).
32. A. F. Casy and A. P. Parulkar, *Can. J. Chem., 47,* 423(1969).
33. A. Butenandt and E. Hecker, *Angew. Chem., 73,* 350(1961).
34. E. L. Eliel, et al., *Conformational Analysis.* New York, Interscience, 1967, p. 1.
35. J. D. Kemp and K. S. Pitzer, *J. Chem. Phys., 4,* 749(1936).
36. E. L. Eliel, et al., *Conformational Analysis.* New York, Interscience, 1967, p. 41.
37. L. B. Kier, in *Fundamental Concepts in Drug-Receptor Interactions,* J. F. Danielli, J. F. Moran, and D. J. Triggle, Eds., London, Academic Press, 1970, p. 15.
38. W. B. Neely, *J. Med. Chem., 12,* 16(1969).
39. G. M. Edelman and W. O. McClure, *Accts. Chem. Res., 1,* 65(1968).
40. B. Belleau, *J. Med. Chem., 7,* 776(1964).
41. B. Belleau and A. Chevalier, *J. Am. Chem. Soc., 90,* 6864(1968).
42. E. E. Smissman, et al., *J. Med. Chem., 9,* 458(1966).
43. C. Y. Chiou, et al., *J. Pharmacol. Exp. Ther., 166,* 243(1969).
44. J. P. Li and J. H. Biel, *J. Med. Chem., 12,* 917(1969).
45. F. W. Schueler, *J. Am. Pharm. Assn., 45,* 197(1956).
46. S. Archer, et al., *J. Med. Pharm. Chem., 5,* 423(1962).
47. E. E. Smissman and G. S. Chappell, *J. Med. Chem., 12,* 432(1969).
48. M. Martin-Smith, et al., *J. Pharm. Pharmacol., 19,* 561(1967).
49. E. E. Smissman and M. Steinman. *J. Med. Chem., 10,* 1054(1967).
50. M. R. Boots and S. G. Boots, *J. Pharm. Sci., 58,* 553(1969).
51. P. S. Portoghese, et al., *J. Med. Chem., 11,* 219(1968).
52. E. E. Smissman and W. H. Gastrock, *J. Med. Chem., 11,* 860(1968).
53. F. Meyer, et al., *Pharmazie, 20,* 333(1965).
54. P. N. Patil, et al., *J. Pharm. Sci., 59,* 1205(1970).
55. P. S. Portoghese, *Annu. Rev. Pharmacol Toxicol., 10,* 51(1970).
56. I. Langmuir, *J. Am. Chem. Soc., 41,* 868(1919).
57. I. Langmuir, *J. Am. Chem. Soc., 41,* 1543(1919).
58. H. L. Freidman, *Influence of Isosteric Replacements Upon Biological Activity,* National Academy of Sciences—National Research Council Publication No. 206, Washington, D.C., U.S. Government Printing Office, 1951, p. 295.
59. H. G. Grimm, *Z. Elektrochemie, 31,* 474(1925).
60. H. Erlenmeyer and E. Berger, *Biochem. Z., 252,* 22(1932); *255,* 429(1932); *262,* 196(1933).
61. H. Erlenmeyer, et al., *Helv. Chim. Acta., 16,* 733(1933).
62. H. Erlenmeyer, et al., *Helv. Chim. Acta., 29,* 1960(1946).
63. H. Erlenmeyer, and M. Leo, *Helv. Chim. Acta., 15,* 1171(1932); *16,* 897(1933); *16,* 1381(1933).
64. H. Erlenmeyer and H. Rey Bellet, *Helv. Chim. Acta., 37,* 234 (1954).
65. H. Erlenmeyer and E. Willi, *Helv. Chim. Acta., 18,* 740(1935).
66. O. Hinsberg, *J. Prakt. Chem., 93,* 302(1916).
67. A. Burger, in *Medicinal Chemistry,* 3rd ed., A. Burger, Ed., New York, Wiley-Interscience, 1970, p. 74.
68. A. A. Larsen and P. M. Lish, *Nature, 203,* 1283(1964).
69. E. W. Blanchard, et al., *Endocrinology, 32,* 307(1943).
70. B. R. Baker, *J. Am. Chem. Soc., 65,* 1572(1943).
71. E. C. Dodds, et al., *Nature, 141,* 247(1938).
72. E. Walton and G. Brownlee, *Nature, 151,* 305(1943).
73. H. J. Schaeffer, et al., *J. Pharm. Sci., 53*(11), 1368(1964).
74. E. F. Nutting and D. W. Calhoun, *Endocrinology, 84,* 441(1969).
75. H. D. Lennon and F. J. Saunders, *Steroids, 4,* 689(1964).
76. L. J. Lerner, et al., *Steroids, 6,* 223(1965).
77. W. R. Jones, et al., *Arch. Int. Pharmacodyn. Ther., 138,* 175(1962).
78. *1973 Physicians' Desk Reference.* Oradell, N.J., Medical Economics, 1973, p. 1412.
79. C. R. Creveling, et al., *Biochem. Biophys. Res. Commun., 8,* 215(1962).
80. A. Boris, *Steroids, 11,* 681(1968).
81. R. E. Counsell, et al., *J. Med. Chem., 8,* 45(1965).
82. F. Varrichio, *Appl. Environ. Microbiol., 15,* 206(1967).
83. M. S. Silver, *J. Am. Chem. Soc., 88,* 4247(1966).
84. E. Work, *Biochem. Biophys. Acta., 62,* 173(1962).
85. A. Szentirmai and H.E. Umbarger, *J. Bacteriol., 95,* 1666(1968).

86. M. E. Wolff and G. Zanati, *J. Med. Chem., 12,* 629(1969).
87. L. L. Bennett, et al., *Cancer Res., 23,* 1574(1963).
88. R. J. Fessenden and M. D. Coon, *J. Med. Chem., 8,* 604(1965).
89. R. J. Fessenden and M. D. Coon, *J. Med. Chem., 7,* 561(1964).
90. P. T. Henderson, et al., *J. Pharm. Pharmacol., 20,* 26(1968).
91. R. J. Fessenden and R. Rittenhouse, *J. Med. Chem., 11,* 1070(1968).
92. A. I. White, in *Textbook of Organic Medicinal Chemistry,* C. O. Wilson, et al., Eds., Philadelphia, J. B. Lippincott, 1966, p. 308.
93. T. C. Daniels and E. C. Jorgensen, in *Textbook of Organic Medicinal Chemistry,* C.O. Wilson, et al., Eds., Philadelphia, J. B. Lippincott, 1971, pp. 407, 428.
94. O. Gisvold, in *Textbook of Organic Medicinal Chemistry,* C. O. Wilson, et al., Eds., Philadelphia, J. B. Lippincott, 1971, pp. 498, 500.
95. R. F. Doerge, in *Textbook of Organic Medicinal Chemistry,* C. O. Wilson, et al., Eds., Philadelphia, J. B. Lippincott, 1971, pp. 682, 683.
96. E. Chang and V. K. Jain, *J. Med. Chem., 9,* 433(1966).
97. Upjohn Company British Patent 1,129,240 (1968): Chemical Abstracts 70: 29050X "Chloramphenicol Trialkylsilyl Ethers."
98. O. Gisvold, in *Textbook of Organic Medicinal Chemistry,* C. O. Wilson, et al., Eds., Philadelphia, J. B. Lippincott, 1971, pp. 811–812.
99. T. O. Soine and R. E. Willette, in *Textbook of Organic Medicinal Chemistry,* C. O. Wilson, et al., Eds., Philadelphia, J. B. Lippincott, 1971, pp. 736, 738.
100. R. Duschinsky, et al., *J. Am. Chem. Soc., 79,* 4559(1957).

SUGGESTED READING

E. J. Ariens, W. Soudijn, and P. B. M. W. M. Timmermans, Eds., *Stereochemistry and Biological Activity of Drugs,* Oxford, Blackwell Scientific, 1983.

Chapter 5

THEORETIC ASPECTS OF DRUG DESIGN

Michael S. Tute and Lemont B. Kier

The development of a useful structure-activity relationship (SAR) from a body of physical, chemical, and biologic experimental work is an intellectual exercise. No experiments are performed. The data are assembled in several ways, examined, manipulated, correlated, and reexamined many times in an effort to find relating characteristics. All of this is done in the mind of the investigator, or on the computer as an extension of the mind. Perceived relationships are frequently verified by further synthesis and testing, but the search for an SAR is a nonexperimental part of drug design and study. It is the theoretic aspect of the drug design process.

HISTORICAL REVIEW

Some 400 years ago, the astronomer and physicist Galileo suggested that in order to understand the universe, man must pay attention to the quantitative aspects of his surroundings and discover the mathematic relationships that exist between them. This suggestion from the founding father of modern science leads us into our discussion.

Drug design requires a certain knowledge of biologic systems and how they are modulated, and an appreciation of the physicochemical properties and structure of molecules. Given this knowledge, suggestions for drug design may be founded on a biochemical rationale or on the trial-and-error testing of analogs of natural agonists or known drug molecules.

Trial and error approaches, or what has often been called Edisonian research, in which a large number of experiments are performed in the search for some objective, often have not taken full advantage of what has been learned and recorded along the way. These approaches are giving way to one in which information derived from testing is used to build up an SAR. When "activity" is expressed quantitatively, e.g., as the concentration of a substance required to elicit a particular response, and when physicochemical property or structure is also described by numbers, the SAR may be cast in the form of a mathematic equation, or QSAR—quantitative structure-activity relationship.

Mathematics has been essential for developing our concepts of atomic structure and chemical reactivity, from Mendeleev's periodic table through the Rutherford-Bohr model of the atom in 1913, and thence to modern quantum mechanics. It has been the same for developing concepts of molecular structure and biologic "reactivity." The first quantitative correlations were made between activity and property, rather than structure. Before 1900, it was shown that the narcotic effect of a homologous series of alcohols varied in proportion to molecular weight, and that the toxicities of many organic compounds were inversely proportional to solubility in water, or proportional to partition coefficients between water and oil. But before 1900, the concept of structure was ill-defined and one spoke more of constitution, i.e., the relative number of atoms of various kinds making up the molecule. It was not until 1929 that belief in a symmetric benzene ring was confirmed by x-ray studies!

Contemporary x-ray crystallography, not only of small organic molecules but also of proteins, including the enzymes that are the targets for many of our drugs and even of enzyme-drug complexes, has led us to an understanding of structure and some aspects of interaction between molecules. Protein crystallography has proved invaluable not only for developing our concepts of intermolecular interactions (the diagnosis of drug-receptor geometry), but also for confirming the nature of interactions suggested by QSAR.[1]

Studies of structure and function of enzymes have taught us how to quantify the activity of drugs acting as enzyme inhibitors. Yet other drugs have been developed that act directly on cell surface receptors to mimic (as agonists) or block (as antagonists) the action of neurotransmitters such as acetylcholine or noradrenaline. Pharmacologic studies of neurotransmitters and physiologic studies of cell structure and function have given us our understanding of how drugs act at the cellular level, and have provided us with relevant data on the activity of agonist or antagonist drugs that can be used to develop a QSAR.

A QSAR can be expressed in its most general form by the following equation:

$$\text{Biologic activity} = f(\text{physicochemical and/or structural parameters}) \quad \text{(Equation 1)}$$

The overall objective is to find parameters from experiment or theory that, when substituted into one of the many forms of the equation along with biologic activity for a series of molecules, give a statistically significant correlation. A good correlation can lead directly to an explanation of perceived biologic changes in a series, expressed in physicochemical or structural terms. The equation can be used to diagnose or support a mechanism. If a good equation, or model, is found, it may be used to predict other molecules with greater activity. This application is the most exciting aspect of theoretic methods.

Because many (Q)SAR methods have been proposed, it is necessary to limit our consideration to those that have been applied in many cases to drug activity data, giving mechanistic insight and allowing useful predictions to be made. We apologize to those investigators whose methods are in an embryonic state and hence are not included in this chapter. As their general value emerges, they will be described in subsequent revisions.

THEORETIC APPROACHES

Property-Activity Relationships

A widely used theoretic method in the drug industry today is that based on the work of Hansch and collaborators.[2] This method is popular because it seeks relationships between biologic activity and physicochemical properties such as partition coefficient (hydrophobicity), degree of ionization, or molecular size, and because these properties can be perceived as of likely importance through the derivation of a model equation (the Hansch equation). These properties may be measured (e.g., a pK_a) but more typically are derived by extrapolation or calculation. The commonly used hydrophobicity descriptor, log P (octanol:water partition coefficient) is often calculated by summation of increments of log P (known as fragment, f − values) for substructures and of empirically derived correction factors.[3] Hansch analysis is that method that seeks to fit physicochemical parameters to equations of the general form of equation 1, but of a specific form that is a model of the drug-receptor binding and/or drug transport process.[4]

In a general model for drug action, three steps are necessary to elicit a biologic response.[5] (1) The drug reaches the neighborhood of some specific receptor by passive transport, that is, it is applied to some external region and then by a "random walk" process, an effective number of molecules manage to cross several membranes to reach the target area. (2) A drug molecule-receptor complex forms. (3) The complex may undergo some chemical reaction or conformational change.

In any particular case, not all of these steps are necessarily critical or limiting. Thus, for most pharmacodynamic agents, no chemical change, such as covalent bond formation, occurs at the receptor and step 2 will lead to a biologic response. On the other hand, for many chemotherapeutic agents, a step 3 is necessary; thus, penicillin must acylate its target, the transpeptidase enzyme of the bacteria. Again, in comparing a series of penicillin derivatives for their relative antibacterial activity, although step 3 is necessary, it may not be limiting, because all derivatives may be equal in their ability to acylate the receptor, but may differ in their ability to cross the bacterial cell wall to reach the target area.

The activity of a drug can be related to the probability, *p*, of a drug molecule going through steps 1 to 3, and is thus related to the probability of its going through the individual steps by:

$$p = p_a p_b p_c$$

If k is the proportionality constant and C is the applied molecular concentration,

$$\text{Activity} = k \cdot C \cdot p = k \cdot C \cdot p_a \cdot p_b \cdot pc$$

It is usual in structure-activity work to measure, if possible, the dose necessary to give a constant biologic response, for example, the ID_{50}, which is the dose that causes a 50% inhibition in enzyme activity or in physiologic response. Considering activity to be a constant, we may rewrite the preceding equation and take logarithms to derive

$$\log \frac{1}{C} = \log p_a + \log p_b + \log p_c + k'$$

A pharmacodynamic agent can be supposed to act by reversible inhibition of an enzyme. It may do this by forming a complex, in step 2, with a receptor site on the enzyme, thus stabilizing some inactive conformation of that enzyme. Using K to represent the equilibrium constant between drug and receptor, our model is represented by the scheme:

$$\text{Dose, C} \xrightarrow[\text{random walk}]{} | \overset{K}{\rightleftharpoons} | \text{ receptor} \longrightarrow \text{response}$$

It is intuitive to replace p_b, the probability of forming a drug-receptor complex, by the equilibrium constant, K. Similarly, if a chemical reaction occurred and was rate (response) limiting, we could replace p_c by a rate constant. We can now write

$$\log \frac{1}{C} = k_1 \log p_a + k_2 \log K + k_3$$

(Equation 2)

and have expressed potency (the log 1/C term) as a linear combination of terms relating to drug transport (log p_a) and intrinsic activity (log K), with k_1, k_2, and k_3 being proportionality constants.

Intrinsic Activity and Linear Free-Energy Relationships

If a drug is directly applied to the neighborhood of a receptor (e.g., an in vitro enzyme assay), then $p_a = 1$, $\log p_a = 0$, and as a model, the equation becomes

$$\log \frac{1}{C} = k_1 \log K + k_2$$

(Equation 3)

If K is the equilibrium constant for a derivative, for example, a substituted benzoic acid, and K_0 is the equilibrium constant for the parent, we can define a parameter by writing

$$\log \frac{K}{K_0} = \rho\sigma$$

in which ρ is a constant for the reaction or equilibrium being studied, and σ is a parameter characterizing the electronic influence, for example, electron release or withdrawal, of the substituent on the reaction center. This can be rewritten as

$$\log K = \rho\sigma + k$$

or in general as

$$\log K = k_1\sigma + k_2$$

(Equation 4)

in which σ is a Hammett parameter characterizing electronic effects of the substituent on the reaction center. The parameters σ are universally applicable and are found from tables of physicochemical data. Equation 4 is called a linear free energy equation (LFE). When this equation is used to substitute for log K in equation 3, we get an expression of the form

$$\log \frac{1}{C} = k_1\sigma + k_2$$

(Equation 5)

Given results (C) for a series of N compounds, it is possible to assign σ values and to submit N simultaneous equations to linear multiple regression analysis to find the best values of k_1 and k_2. Now k_1 represents the sensitivity of the biologic reaction to electronic effects. Such an equation is predictive in that from a knowledge of σ parameters one can calculate the potency, and diagnostic in that any compound obeying the relationship is likely to be acting by the same mechanism.

Antibacterial Effect of Sulfonanilides

The effect of the substituent R on the sulfonamide group was parameterized by σ. An excellent relationship was found between σ and potency.[6]

$$\log \frac{1}{C} = 1.05\sigma - 1.28$$

$$(r = 0.97, N = 17)$$

The conclusion was that within this set, R does not bind to the receptor but only influences potency insofar as it affects ionization. These results agree well with the accepted mode of action of sulfonamides as antagonists of the para-aminobenzoic acid anion. The critical reaction is ionization in the bulk phase.

In few studies does an electronic parameter alone account for all variance in the data. When adsorption of drug onto enzyme, penetration of lipid membranes, or serum protein binding is critical, a linear free energy parameter related to lipophilicity (hydrophobicity) change is required. By analogy with Hammett, Hansch proposed:[7]

$$\log \frac{P_X}{P_H} = \rho\pi$$

defining ρ as 1 for the model system of partitioning between octanol and water, in which P_X and P_H are partition coefficients of derivative and parent. We can rewrite this equation as

$$\pi = \log P_X - \log P_H$$

Now, either π, a substituent parameter from tables, or log P, which may be measured or estimated by adding π values to log P_H, can be used as a hydrophobicity parameter. Expanding equation 5 to include hydrophobic bonding, a generalized model equation in vitro is

$$\log \frac{1}{C} = k_1\pi + k_2\sigma + k_3$$

Drug Transport

In vivo, drug transport to the receptor neighborhood competes with drug excretion (fast for water-soluble, hydrophilic compounds), localization of drug in fatty tissue (extensive for very lipophilic compounds), and drug metabolism (fast for lipophilic drugs). Intuitively, drug molecules should possess some idealized lipophilicity according to the application for which they are intended. If drugs have to pass through several lipid membranes to reach the receptor, then a drug molecule that is not lipophilic will not pass easily into a membrane; conversely, one that is too lipophilic will not leave the membrane easily to continue its journey to the receptor. We can well imagine the existence of some optimum lipophilicity for a set of molecules. This idea is usually expressed by the model of a normal distribution between the probability of access to the receptor and the lipophilicity, or, using logarithms and the log P parameter, by the parabolic relationship of the equation:

$$\log p_a = k_1 \log P - k_2(\log P)^2 + k_3$$

and the optimum lipophilicity can be found from the regression coefficients as:

$$\log P_0 = \frac{k_1}{2k_2}$$

In a series of molecules with a common skeleton and varying substituent, π could replace log P in the equation. Substituting appropriate expressions for log p_a and log K into equation 2 results in the following equation as a general mathematic model for drug potency as a function of transport and receptor binding:

$$\log \frac{1}{C} = k_1 \pi - k_2 \pi^2 + k_3 \sigma + k_4$$

(Equation 6)

We note the following features of this equation.

1. Not all of the terms may be necessary in any particular case.
2. If hydrophobicity plays a part in drug transport, the equation does not allow us to distinguish the sensitivity of the receptor to hydrophobic binding.
3. Although equation 6 is the most used model, in general, a plot of potency against log *P* is not a parabola, but is more aptly described as bilinear, with ascending and descending parts of different slopes, and a parabolic part in the region of the optimum. This shape can be predicted by a rigorous mathematic model in which partitioning through a series of membranes separating lipid and aqueous compartments of different volumes is considered.[8]
4. Even in vitro, with no drug transport to consider, parabolic or bilinear plots of potency against log *P* or π are found if compounds of sufficiently wide range of lipophilicity are examined. This result may be attributed to a hydrophobic binding site of restricted size or to limiting solubility or micelle formation with higher homologs.

Structure-Activity Relationships

Since the introduction of Hansch analysis, there have been many variations in the development of physical models for drug action, using essentially property-based parameters reflecting changes in hydrophobic, electronic, or steric character. Property-based parameters can be criticized, however. It is not always easy to measure or calculate such parameters, in particular ones related to steric effects, and one rarely knows which properties are likely to be relevant. Furthermore, property values often are redundant, that is, two or more structures can have the same log P or same pK_a, and, after all, we are interested in designing drug *structures.*

In sharp contrast to a QSAR model built up from a relationship between activity and a physical property is a model in which the structure of a molecule is encoded. Basically, the structure is a count of some primitive components or features (atoms, atom types, atom connections) or is a statement of the probability of a location of such components.

Consider the narcotic potency of a series of normal alcohols (Fig. 5–1). We may measure a physical property, say the boiling point. A plot is made of this data point and a QSAR is calculated. If a quality relationship is found, we have a basis for an interpretation. That interpretation is mechanistic, that is, we can conclude that a rate-limiting physicochemical event governs the measured potency and that event is related in some way to the boiling point. Everything that we know and understand about the boiling point can now be harvested to illuminate our understanding of the mechanism and potency variation measured in this series. Note that our understanding based on this equation is bounded by potency measurements and by the physical property measured. We have no structure information directly related to this model. Moreover, there is a danger in assigning causality to boiling point as a determinant of narcosis, and faulty ideas of mechanism may arise.

In Figure 5–1*B*, we plot a relationship between potency and a purely structural attribute, the carbon count. Again, a QSAR model is derived. This model is purely structural in its description. No physicochemical phenomenon is encoded, no mechanistic interpretation is inherent, no direct relationship to property is inferred. We can draw conclusions about the structural influence on potency, but only to the extent that the carbon count is our metric.

These examples are two different paradigms. We can-

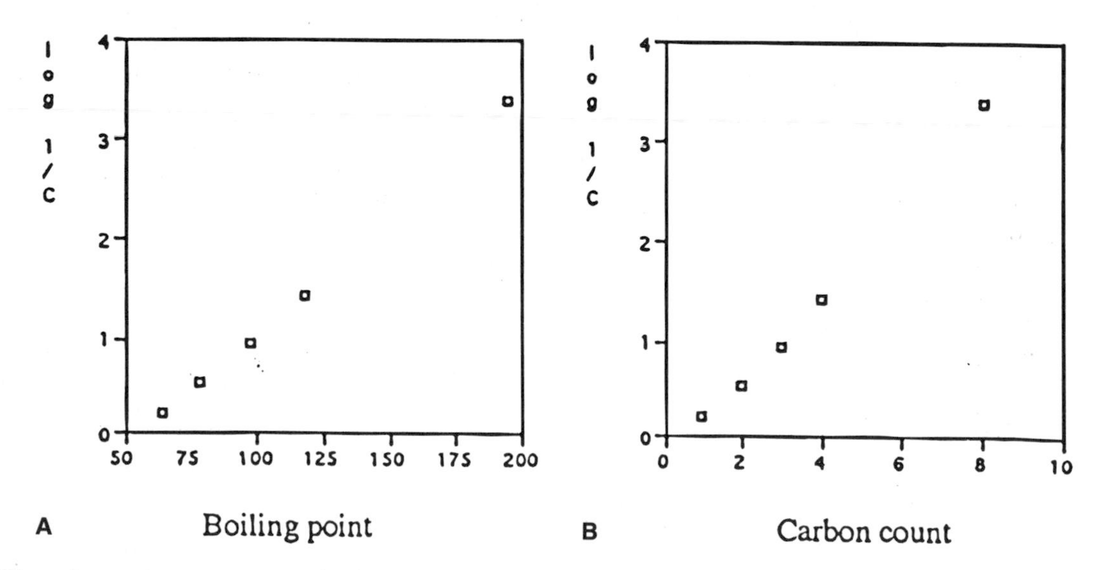

Fig. 5–1. Plots of narcotic potency versus boiling point *(A)* and narcotic potency versus carbon count *(B)*. Narcotic potency expressed as log 1/C after Leo et al.[9]

not interpret Figure 5–1*A* in terms of molecular structure, but must perform an additional analysis (relating boiling point to structure) to transform this model into the information needed by the synthetic chemist to predict the activity of new candidate molecules. With Figure 5–1*B*, we are blind as to any insight into mechanism. We can, however, speak about certain structural features that may enhance potency.

Theoretic methods have therefore been developed in which the actual structure or structural characteristics more fundamental than property have been examined and quantified. At the simplest level, this method consists of comparing changes in biologic activity within a series of related molecules to the presence or absence of defined substituents, or molecular features, at any part of a common molecular skeleton.[10] A set of simultaneous equations is written in the form

$$\log A = \Sigma G_i X_i + k$$

and solved by computer to give "best fit" values for all Gs and for k. Each G_i represents a feature, and X = 1 or X = 0 according to its presence or absence from the particular structure. A limitation of this approach is the assumption that the contribution of a substituent will be constant and additive, independent of its environment.

At an intermediate level, Kier and Hall developed the method of molecular connectivity,[11] which has its origins in topology. This method leads to a series of numeric indices reflecting the presence in molecules of structural features such as branching, unsaturation, cyclization, and heteroatom position. These indices are unique, depend on intramolecular environment, and can be correlated with physicochemical or biologic properties using standard statistical methods.

At the ultimate level of sophistication and complexity is a quantum mechanical description of structure. Quantum mechanics attempts to describe the nature of matter at the atomic and molecular level. Because of its complexity, it is limited to simple systems, and for practical use, the approximation known as molecular orbital theory (MO) has been developed. The MO calculations on drug molecules give numbers that reflect electronic aspects of structure in terms of electron location and energy. A summation of energy for a particular conformation allows the assessment of the probability of occurrence of that conformation, and so MO theory has been used widely for conformational analysis.[12] Apart from the need for massive computing power, the principal disadvantage of MO methods lies in their (usual) neglect of solvent effects. Most calculations are performed in vacuo.

QUANTITATIVE MODEL CONSTRUCTION

Property-Based Models

To develop a property-based model, the investigator must determine the dataset, the properties, and the most expedient method of statistical analysis. Although the choices may largely be determined by availability of data and resources, the investigator must also bear in mind the objective of the investigation. Is this model mechanistic, i.e., is an equation sought in order to support or suggest a mechanism of action, or is the objective to predict useful directions for analog synthesis? All choices are interdependent.

In choosing the dataset, one generally wants as many compounds as possible, but it is best to omit uniquely insoluble compounds or perhaps the one acid in a series of neutral substances, for inclusion of data on such irregular compounds may be misleading. For subsequent analysis by use of multiple regression techniques, the aim should be to include at least five compounds for every descriptor considered.

In choosing the (property) descriptors, the investigator should bear in mind the model being tested. If in vivo data are being analyzed, addition of a squared term in log P or π will be appropriate because an optimum value of hydrophobicity is likely. Electronic properties can be represented by a measured (pK_a, NMR shift), tabulated (Hammett σ), or calculated (MO charge) descriptor. Steric properties can be represented by a measured distance or volume using a computer graphics tool, or otherwise calculated as seems appropriate to the model being tested or developed.

The method of data analysis depends on the quantity and quality of data and on the desired outcome. If merely classification is required, e.g., into actives/inactives or agonists/antagonists, then discriminant analysis[13] or other pattern recognition techniques may be chosen. Typically, if limited but precise data are available and mechanistic information is sought, multiple correlation analysis is used.

Some recently developed theoretic methods are capable of generating a huge number of descriptors. For example, one can construct a grid of points in space around each molecule in a dataset, and compute the hydrophobic, electrostatic, and steric components of interaction at each point on the grid with some hypothetic receptor, in a method known as Comparative Molecular Field Analysis (CoMFA).[14] To handle such data requires an advanced statistical tool known as partial least squares (PLS)[15] and the interpretative skills of a statistician.

Structure-Based Models

We define molecular structure as a summation of attributes encoding the presence, form, and location of atoms, including their adjacency relationships or topology. At the most complex level, we can use quantum mechanical methods to calculate the most probable location of electrons.

To start at the simplest level, the chemical graph is a familiar and commonly used illustration of a molecular structure. In this book, the depiction of structures is through the use of some variant of the chemical graph. Much information is conveyed by these portrayals of a molecule, but only recently have numeric indices encoding quantitative information been derived from them.

Kier and Hall developed a general procedure for the quantitative description of certain attributes of molecular structure based on a branching index proposed by Randic.[16] This procedure leads to numeric parameters called

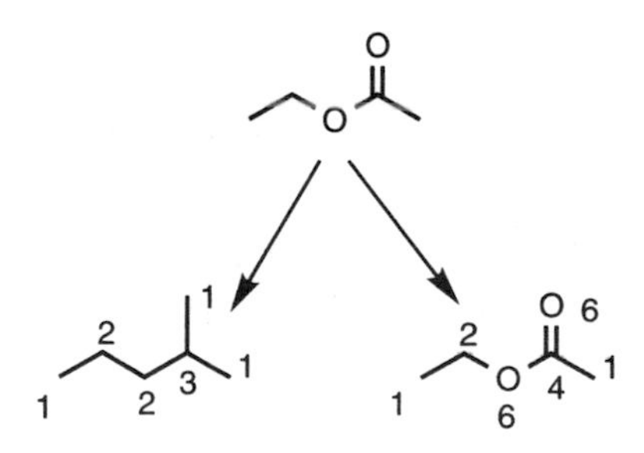

Fig. 5–2. Dissection of ethyl acetate into delta values.

molecular connectivity indices.[11] In practical terms, the method extends the elementary example of structure description such as the count of atoms in Figure 5–1*B* and extends it to encode attributes of structure such as atom, bond, and fragment counts and types, unsaturation, and cycles. In general, topology and valence electron counts are a part of the information within the various indices calculated by this method.

The molecular connectivity indices are calculated from the hydrogen suppressed chemical graph commonly referred to as the skeleton structure of a molecule. Using ethyl acetate as an example, we can identify the nonhydrogen atoms by two numbers. The first number is the adjacency or the count of sigma electrons contributed by an atom to other, nonhydrogen atoms, called the simple delta index. The second number is the count of all valence electrons on an atom other than those bonding hydrogen atoms, called the valence delta (Fig. 5–2).

These two sets of atom descriptions form the bases for parallel calculation of molecular connectivity indices. The calculation begins with the dissection of the molecule into various fragments composed of one or more adjacent bonds called paths. A first-order dissection is one in which all bonds are separated and identified with their delta or valence delta indices.

A value for each fragment is calculated as the reciprocal square root of the product of the delta values defining each fragment. These fragment values are then summed to give a whole molecule index called chi of the first order ($^1\chi$):

$$^1\chi = \Sigma(\delta_i\delta_j)^{-0.5}$$

If the valence delta values are used, the chi value is the valence chi index, $^1\chi^v$. Higher orders of indices are calculated by dissecting the molecule into fragments of larger numbers of adjacent paths.

An extensive family of structural indices is thus available for development of QSAR models. The information in the indices is purely structural and leads to interpretations with direct meaning to the synthetic chemist for the design of new compounds.

Building on this general approach, Kier and Hall developed additional structural indices derived from the chemical graph. These indices include those describing molecular shape and flexibility, known as the kappa (κ) indices[17] and, more recently, a set describing the electronic and the topologic states of atoms within molecules, known as the electrotopologic state indices[18] (S_E).

Quantum Mechanics

Molecular connectivity is limited to a description of atom type, location, and topology. For a complete and fundamental description of matter, including the electron, we must turn to quantum mechanics or to the approximation known as molecular orbital (MO) theory.

Quantum mechanics sets down a number of equations that encode the most probable positions and energies of the electrons in atoms and molecules. These equations are complex and defy exact solution except for the simplest molecules. For practical use, as with drug molecules, MO theory is invoked: an assumption is made that electrons are associated with orbitals embracing the entire molecular skeleton, and that each orbital can hold up to two paired electrons. Each MO is represented as a linear combination of atomic orbitals (LCAO) centered on each atom. Thus, each molecular orbital, ψ, is expressed as a combination of atomic orbitals $\psi_{a...n}$ over a molecule of n atoms.

$$\psi_i = c_{ia}\psi_a + c_{ib}\psi_b + c_{ic}\psi_c + \cdots\cdots c_{in}\psi_n$$

in which the c's are coefficients representing the contributions of the atomic orbitals to the molecular orbital.

In practice, we solve for the numeric value of the coefficients using, according to the size of the problem and available resources, either an ab initio or an appropriate semiempiric technique. In ab initio methods, all electrons are explicitly included in the calculation. In semiempiric methods, only valence electrons are explicitly included and the effect of core electrons is parameterized. Within both the ab initio and the semiempirical framework, a hierarchy of methods is available.[12]

Properties Calculated by MO Theory

Having obtained the orbital coefficients, *c*, we can obtain a useful picture of electron location, and hence both charge and electrostatic potential or dipole moment. The value of c^2 is a measure of the probability of location at a region of space in the molecular orbital. There are two electrons per orbital, so charge density q_i around atom i is obtained by

$$q_i = \Sigma\, 2c_i^2$$

Having computed charge densities at regions in space (these may optionally be at skeleton atoms), it is but a short step to compute electrostatic potential or dipole moment. Both these properties have been used in property-based QSAR, but they were derived by structure-based calculation rather than by measurement.

Another important property calculated by MO methods is relative energy. Electrons in molecular orbitals possess energies depending on the orbital and its relationship to the nuclei of the atoms within the molecule. This energy is approximately that necessary to remove the electron from the orbital, and is known as the ionization energy, E, which is related to the wave function ψ of the MO by the Schrödinger equation:

$$H\psi = E\psi$$

The H is the Hamiltonian operator, which includes the kinetic and potential energies of the electron. Solution for E for each MO in the molecule gives a series of energy levels that reflect the susceptibility of the molecule to both donate and receive electrons to and from other molecules—characteristics of fundamental importance to chemical reactivity and to spectrophotometric events, sometimes also to some events taking place when drugs combine with their receptors.[19]

Molecular Conformation

The most active area of theoretic research using MO theory has been in the prediction of the preferred conformation of molecules. The shape (size plus conformation) that a molecular prefers is of great importance in determining whether the interaction of that molecule with a receptor will bring the appropriate atoms near each other. The proximity of atoms of drug and receptor is postulated to account for biologic events following the interaction of an agonist with its target receptor.

The preferred conformation of a molecule is a structural characteristic that arises largely from the interactions of atoms within the molecule. It is also influenced by the environment, i.e., differences in conformation are often apparent when going from gas phase to solid (crystal), to solution, or to a possibly hydrophobic environment in the proximity of a receptor.

A molecule assumes its conformation as a response to attractions and repulsions present throughout the molecule. Minimization of these energies is the driving force.

The minimization energy is a function of bond angles, bond lengths, and torsion angles. By varying these parameters in a systematic way and calculating the total energy as a sum of orbital energies, one can find a minimum energy structure. The first such study reported was on acetylcholine,[20] and it has been followed by a large number of predictions using many MO procedures (see reference 12 for a review).

Computer Graphics Simulations

Computer graphics[21] is now widely used to "build" the electronic equivalent of a Dreiding molecular model and to display both the model and some properties, e.g., surfaces represented as dots, or electrostatic potentials represented as contours, color coded and in stereo, on a high-resolution video display unit. Computer graphics can be used as a tool to check the meaning of a model-based QSAR. For example, QSARs have been generated for series of ligands interacting with dihydrofolate reductase, carbonic anhydrase, trypsin, and papain, enzymes with structures known from x-ray work. Models of the ligands were built, refined by molecular mechanics calculations, and "docked" into the receptor constructed from its x-ray coordinates.[1] By color coding solvent-accessible surfaces of the enzyme according to property (polar, hydrophobic, or charged), it became possible to "see" the very interactions suggested by the QSAR. Moreover, new binding sites could be postulated. The technique has both diagnostic and predictive potential.

Computer graphics has helped us to appreciate and visualize structure and to measure structural attributes. Molecules are "built" by linking together fragments from a database of x-ray structures or by using standard bond lengths and angles. Subsequently, such structures may be refined and conformations selected by molecular mechanics,[22] optimizing structure by reference to a "force-field" of standard geometry and a listing of the energy penalties for distortion of bond lengths, angles, and nonbonded interactions.

Goodford[23] introduced a quantitative technique for probing the environment of a protein to identify putative binding sites for drugs. A charge is assigned to each atom of the protein. Then, using molecular mechanics principles, the interaction of the protein with a probe group is computed at points on a regular grid encompassing the structure. Using an oxygen anion as probe on dihydrofolate reductase, a binding site adjacent to an arginine residue was detected. This finding explains why the derivative (I, R = $(CH_2)_5CO_2H$), suggested for synthesis after examining this binding site,[24] has 50 times the affinity of the parent antibacterial drug trimethoprim (I, R = CH_3).

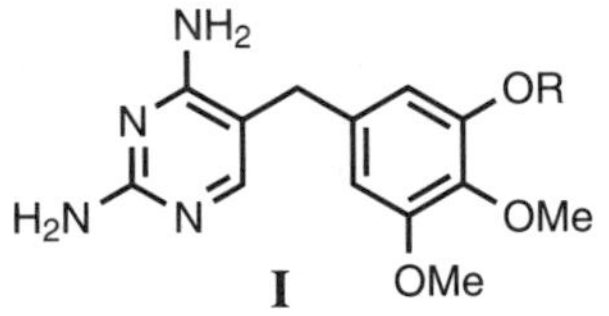

I

The probe technique of Goodford, when used on a series of drug molecules or ligands, spawned the CoMFA technique, which was initially described by Cramer and coworkers[14] in 1988 and is now used in conjunction with computer graphics displays.

SUMMARY AND PROJECTION

The increased availability of computers for database storage, computation of property, optimization of structure, and the rapid analysis and correlation of these data have made theoretic drug design a practical procedure. Pharmaceutical companies now employ individuals with specialist knowledge of theoretic methods whose responsibility it is to understand and use the relevant technique and associated technology for any particular problem. We anticipate greater use of theoretic methods and the development of novel and more effective drugs as a result within the next decade.

REFERENCES

1. C. Hansch and T. E. Klein, *Acc. Chem. Res., 19,* 392(1986).
2. C. Hansch, *Drug Dev. Res., 1,* 267(1981).
3. C. Hansch and A. J. Leo, *Substituent Constants for Correlation Analysis in Chemistry and Biology,* New York, John Wiley & Sons, 1979.
4. C. Hansch, *Acc. Chem. Res., 2,* 232(1969).
5. J. W. McFarland, *J. Med. Chem., 13,* 1192(1970).

6. C. Hansch, *Drug Design,* Vol. 1, E. J. Ariens, Ed., New York, Academic Press, 1971, p. 271.
7. T. Fujita, et al., *J. Am. Chem. Soc., 86,* 5175(1964).
8. H. Kubinyi, *J. Med. Chem., 20,* 625(1977).
9. A. Leo, et al., *J. Med. Chem., 12,* 766(1969).
10. S. M. Free, Jr. and J. W. Wilson, *J. Med. Chem., 7,* 395(1964).
11. L. B. Kier and L. H. Hall, *Molecular Connectivity in Chemistry and Drug Research,* New York, Academic Press, 1976.
12. G. H. Loew and S. K. Burt, *Quantum Mechanics and the Modeling of Drug Properties,* in *Comprehensive Medicinal Chemistry,* C. Hansch, et al., Eds., New York, Pergamon Press, 1990.
13. Y. C. Martin, et al., *J. Med. Chem., 17,* 409(1974).
14. R. D. Cramer, III, et al., *J. Am. Chem. Soc., 110,* 5959(1988).
15. L. Stahle and S. Wold, Multivariate Data Analysis and Experimental Design in Biomedical Research, in *Progress in Medicinal Chemistry,* Vol. 25, G. P. Ellis and G. B. West, Eds., Amsterdam, Elsevier, 1988.
16. M. Randic, *J. Am. Chem. Soc., 97,* 6609(1975).
17. L. H. Hall and L. B. Kier, *The Molecular Connectivity Chi Indexes and Kappa Shape Indexes in Structure-Property Modeling,* in *Reviews of Computational Chemistry,* D. B. Boyd and K. Lipkowitz, Eds., New York, VCH, 1991.
18. L. H. Hall, et al., *Quant. Struct.-Act. Relat., 10,* 43(1991).
19. A. K. Debnath, et al., *J. Med. Chem., 34,* 786(1991).
20. L. B. Kier, *Mol. Pharmacol., 3,* 497(1967).
21. *Molecular Graphics and Drug Design,* in *Topics in Molecular Pharmacology,* Vol. 3, A. S. V. Burgen, et al., Eds., Amsterdam, Elsevier, 1986.
22. E. Osawa and H. Musso, *Top. Stereochem., 13,* 117(1982).
23. P. J. Goodford, *J. Med. Chem., 28,* 849(1985).
24. L. F. Kuyper, et al., *J. Med. Chem., 25,* 1120(1982).

SUGGESTED READINGS

P. R. Andrews, et al., *J. Med. Chem., 27,* 1648(1984).

R. Franke, *Theoretical Drug Design Methods,* Amsterdam, Elsevier, 1984.

L. B. Kier and L. H. Hall, *Molecular Connectivity in Structure Activity Analysis,* Letchworth, England, Research Studies Press Ltd., 1986.

L. B. Kier and L. H. Hall, *Adv. Drug Res., 22,* 1(1992).

Y. C. Martin, *Quantitative Drug Design,* New York, Marcel Dekker, 1978.

W. G. Richards, *Quantum Pharmacology,* 2nd ed., London, Butterworths, 1983.

J. G. Topliss (Ed.), *Quantitative Structure-Activity Relationships of Drugs,* New York, Academic Press, 1983.

S. H. Yalkowsky, et al., (Eds.), *Physical Chemical Properties of Drugs,* New York, Marcel Dekker, 1980.

Chapter 6

*MOLECULAR MODELING**

James G. Henkel and Eric M. Billings

During the 1980s, molecular modeling grew from a basic research activity in a few university laboratories to a critical component of rational drug design, with application across the spectrum of the medical chemistry enterprise.[1] The explosive growth of this discipline is the result of a fortunate combination of factors. One factor is the tremendous decrease in the cost of computing capability, thus making it more available to many scientists. Another is the development of powerful graphics display hardware, which allows the manipulation of even highly complex structures with little or no visual delay when redrawing the altered image. The result is a highly visual interface between the computer (and its data) and the scientist. Rather than dealing with pages of numbers on a computer printout, the medicinal chemist can now visualize the results of a computation or simulation as a picture or series of pictures. Within this visual framework, the resulting numbers (and the properties represented by them) are easier to assimilate. A third factor in the growth is the development of powerful and robust modeling software with ergonomic interfaces, which have been developed from both academic and commercial sources. This software permits the use of powerful computational algorithms by those who are more interested in the results of the computation than its details.

The field of molecular modeling is commonly thought of as being composed of several interlinked activities, including molecular graphics, computational chemistry, statistical modeling, and, to some degree, molecular data and information management. The molecular graphics aspect represents the drug molecules and their associated molecular properties in a visual way, so that one may gain greater insight into their pharmacologic behavior. The computational chemistry component is concerned with the simulation of atomic and molecular properties of compounds of medicinal interest through equations, and with the numeric methods used to solve these equations on the computer. Statistical modeling encompasses the search for quantitative relationships between the structures or properties of a series of compounds and their resultant biologic activities. This aspect of the medicinal chemistry enterprise, called quantitative structure-activity relationships (QSAR), is discussed in Chapter 5. The chemical data/information management part of the enterprise actually has several components. One component organizes the properties of thousands of compounds into an extensive database, capable of being searched for highly promising compounds with the right combinations of properties to make them candidates for pharmacologic evaluation. Another component aids the chemist in the synthesis of new drugs by providing strategies and choices of ways to accomplish the organic synthesis of a series of drug candidates. Yet a third component may help to organize the attendant molecular properties of a series of compounds to make them easier to subject to statistical analysis, such as QSAR.

The common component of all of these activities is the computer. Thus, when all of the aforementioned activities are taken together, they constitute the field of medicinal chemistry known as computer-assisted molecular design (CAMD) or computer-assisted drug design (CADD). The other common name, computer-assisted molecular modeling (CAMM), is redundant, because by our definition, molecular modeling is computer assisted.

RATIONAL DRUG DESIGN

The rational drug design process has changed the way in which potential new drugs are discovered. In the past, many drugs were developed through a search of natural sources, most notably plants and microorganisms. Drugs such as morphine, penicillin G, and digitalis are examples of efforts of this type. Subsequent synthetic modifications of these natural products extended the formulary by improving on the therapeutic profiles of the natural products. The whole family of semisynthetic penicillins and digitalis derivatives are examples of these modifications. The increasing cost of screening compounds and the decreasing yield of new and unique lead compounds from natural sources has made this approach less favored in recent years. One estimate holds that only one compound in every 20,000 screened randomly will make it into the clinic,[2] and this already low yield is expected to decrease even further in the future.

The cost of taking a compound with potential therapeutic value from the laboratory bench to the pharmacy shelf has become almost prohibitive in recent years. Commonly used pharmaceutical industry figures indicate that in the late 1980s and early 1990s, bringing a drug to market required approximately $120 to $150 million and about 8 to 10 years. This cost, plus profits, then must be

* The images contained in this chapter were generated with Insight II (version 2.2.0), Discover (version 2.8), and Delphi (version 2.4) from Biosym Technologies, Inc., using a Silicon Graphics 4D/70GT workstation running the IRIX 4.0.5 operating system.

made up through the sales of the drug over the 7 to 10 years of its expected useful product life.

These costs are the result of the thoroughness and caution prescribed by the drug product development process. Development includes several steps, including (1) the compound's synthesis and its initial screening for pharmacologic activity, (2) the requisite preclinical animal studies for both short-term and long-term toxicity, (3) phase I clinical trials in healthy volunteers, (4) phase II clinical trials in a limited cohort of patients with the target disease, and (5) phase III clinical trials in a broad population of affected patients. Only after these steps have occurred and stringent criteria have been met at each step can FDA approval take place so the drug can be marketed. Another significant cost built into the drug product development process is the expense of synthesizing and testing all of the unsuccessful drug candidates investigated along the way toward approval of the successful drug candidate. These costs can be substantial. Literally thousands of compounds are synthesized in step 1 and subjected to in vitro or in vivo pharmacologic screening in the search for the best candidate for the preclinical animal studies that occur in step 2. Dozens of compounds might be subjected to preclinical animal studies before one or two compounds emerge that are suitable for clinical trials. Of those compounds in clinical trials, several are usually eliminated from further consideration when unforseen side effects occur. These side effects can occur in any phase of the trial regimen, and the further along a compound is in development when the side effects are discovered, the more expensive the effort becomes.

Although some progress has been made in recent years toward reducing the time and cost of drug product development by modifying the approval and clinical trial process, a successful drug product will never be either cheap or easy to produce. Certainly one way to make the drug product development process more cost effective is to improve the yield of promising drug candidates. This improvement can be accomplished by "working smarter" during the drug design process, applying both our knowledge of the mechanistic basis of a target disease and our knowledge of the molecular characteristics of the compounds to have an effect on the disease state, thereby reducing the number of unsuccessful drug candidates. This approach to therapeutic development is called the rational drug design approach. Molecular modeling in all of its forms is intimately linked to our ability to carry out rational drug design.

The rational drug design process starts with an understanding of the fundamental physiologic and biochemical aspects of the disease or target, rather than a random screening process. As a first step, instead of just testing existing or newly synthesized compounds against a particular disease, rational drug design prescribes an investigation of the fundamental characteristics of the disease of interest. If the disease could be treated by the inhibition of a particular enzyme, then a study of the structure and detailed mechanism of action of that enzyme would be undertaken. If the crystal structure of the enzyme could be obtained, then we would have an exact model of the drug design target that we could approach directly. More often than not, we cannot obtain a model for a direct design process, so we must use the so-called indirect methods, which rely on structure-activity relationships and modeling of hypothetical active sites and target receptors.[3]

The next step in the rational design process is the identification of a lead compound, that is, a compound to serve as the starting point for the drug design process. This lead compound could be the natural substrate for the current example enzyme or some other compound that has a side effect indicative of potential therapeutic activity. After development of one or more lead compounds, a congener set (a series of carefully designed analogs of the lead compound) is synthesized and tested for activity in an appropriate biologic test system. The resulting biologic activity is then analyzed with statistical techniques and graphic and computational techniques are used to develop a quantitative model of its structure-activity relationships. Based on this model, a new congener set is devised, synthesized, and tested. The process is repeated until either a final clinical candidate compound is reached or no further progress is made toward a clinically useful drug.

We now consider in more detail both the molecular graphics and the computational chemistry components of molecular modeling, followed by some examples of recent applications of the techniques to drug design. The statistical aspects of molecular modeling are covered in Chapter 5. The chemical information management aspects of modeling have been reviewed,[4] and further information concerning this activity is found there and in the references cited therein.

PRINCIPLES OF MOLECULAR GRAPHICS

Rendering images of molecules on a computer screen and manipulating them with the aid of software is fairly simple in concept but rather complex in practice because of the necessary inclusion of the numerous conditions and options that make the software powerful enough to be of general use in the drug design process. In the simplest terms, the image of the molecule is drawn by converting each atomic coordinate into a corresponding screen coordinate, then connecting the appropriate atoms by lines representing bonds. Depth cueing (conveying the perception that the molecule's third dimension goes into the screen) is accomplished on the more advanced systems by dimming the images of the atoms that are more distant from the viewer and by rendering the image in perspective rather than in orthographic projection. Translation of the image (moving it on the screen) is accomplished by adding or subtracting an appropriate distance along one or more of the coordinate axes on the screen, then redrawing it. Rotation is accomplished by applying the standard trigonometic equations to the coordinates, then redrawing the transformed image on the screen. For example, we can rotate a molecule about the Z-axis by an angle Θ, causing it to move like the second hand of a clock (Fig. 6–1). Each atom at point (X_1, Y_1, Z_1) is moved to point (X_2, Y_2, Z_2) by transforming it according to the equations

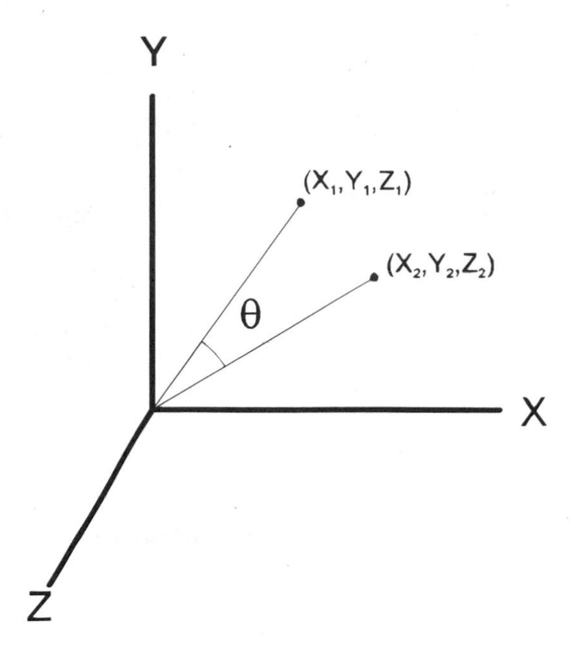

Fig. 6–1. Rotation of a point about the Z-axis by an angle Θ. The starting point is at coordinates (X_1, Y_1, Z_1). The final point is at coordinates (X_2, Y_2, Z_2). The Z-coordinate does not change in this rotation, i.e., $Z_1 = Z_2$.

$$X_2 = X_1(\cos\Theta) + Y_1(\sin\Theta)$$
$$Y_2 = Y_1(\cos\Theta) - X_1(\sin\Theta)$$
$$Z_2 = Z_1$$

The Z-axis values do not change in this particular transformation. In practice, these rotation and translation operations, along with others, are usually done all at once as matrix operations, which is the way the computer works best. The computer must be able to perform these calculations rapidly and the graphics display system must be able to redraw the resulting images on the screen quickly. Under these conditions, the rapidly redrawn image appears to move about the screen in real time, giving the impression of realistic motion in space, usually under the control of a mouse or other pointing device.

To gain the most insight from a series of molecular graphics images, several different types of molecular image renderings are needed. Some common types of viewing formats of small molecules include the standard stick view, the ball-and-stick view, and the space-filled or CPK (Corey-Pauling-Koltun) view. Figure 6–2*A* to *C* shows the morphine molecule in each of these respective viewing formats. The stick view of a molecule (Fig. 6–2*A*) is the most uncluttered, and will best show the structural relationships among the atoms. The CPK view (Fig. 6–2*B*) gives a better rendering of the actual shape of the molecule, but at the expense of the structural and stereochemical detail. The overlapping spheres representing the atoms in a CPK view are usually drawn with radii proportional to the van der Waals radii. The ball-and-stick display (Fig. 6–2*C*) offers a compromise between these two viewing modes. An additional viewing format is the dot surface display (Fig. 6–2*D*), in which a "cage" of dots is placed around the stick figure to denote the surface formed from the van der Waals radii or some other property of the molecule, such as the solvent accessible surface.[5] This view shows both the stereochemical detail of the stick figure and the shape of the molecule's particular surface of interest.

For the display of macromolecules such as proteins and nucleic acids, these display techniques must be augmented by some additional ones because of the complexities of the molecules involved.[6] One of the most useful utilities for protein structure display is the topology diagram, sometimes called the ribbon display. This type of structural diagram traces the amino acid backbone throughout the structure as a flat or rounded ribbon, thereby simplifying the view of the secondary and tertiary structure of the protein. Without such a simplification, the display is usually too cluttered to provide much useful information. The ribbon view permits easy identification of structural features such as α-helices, β-strands and sheets, and connecting loops. Figure 6–3 shows three different renderings of the crystal structure of dihydrofolate reductase,[7] generated from coordinates[8] obtained from the Protein Data Bank.[9] Dihydrofolate reductase (DHFR) is a crucial enzyme for the biosynthesis of the one-carbon units that are used in the biosynthesis of the nucleic acid thymidylate. Figure 6–3*A* shows the complete enzyme in stick view. Clearly, little information is available from this view because of its complexity. Too much information in the image has rendered it incomprehensible. Figure 6–3*B* shows the same enzyme in a CPK view, which provides some insight into the shape of the enzyme but no insight into its internal organization. Finally, Figure 6–3*C* shows the enzyme backbone in ribbon view, from which we can identify a number of substructures, including 4 helices, a β-sheet with 8 strands, and 15 turns. Although they are not labeled in this view, two sites have specific functions. One is composed of residues binding an NADPH cofactor, and the other interacts with the natural substrate (dihydrofolate) or its congeners. These sites can be displayed separately if desired.

The power of the molecular graphics techniques is best

Morphine

Methotrexate

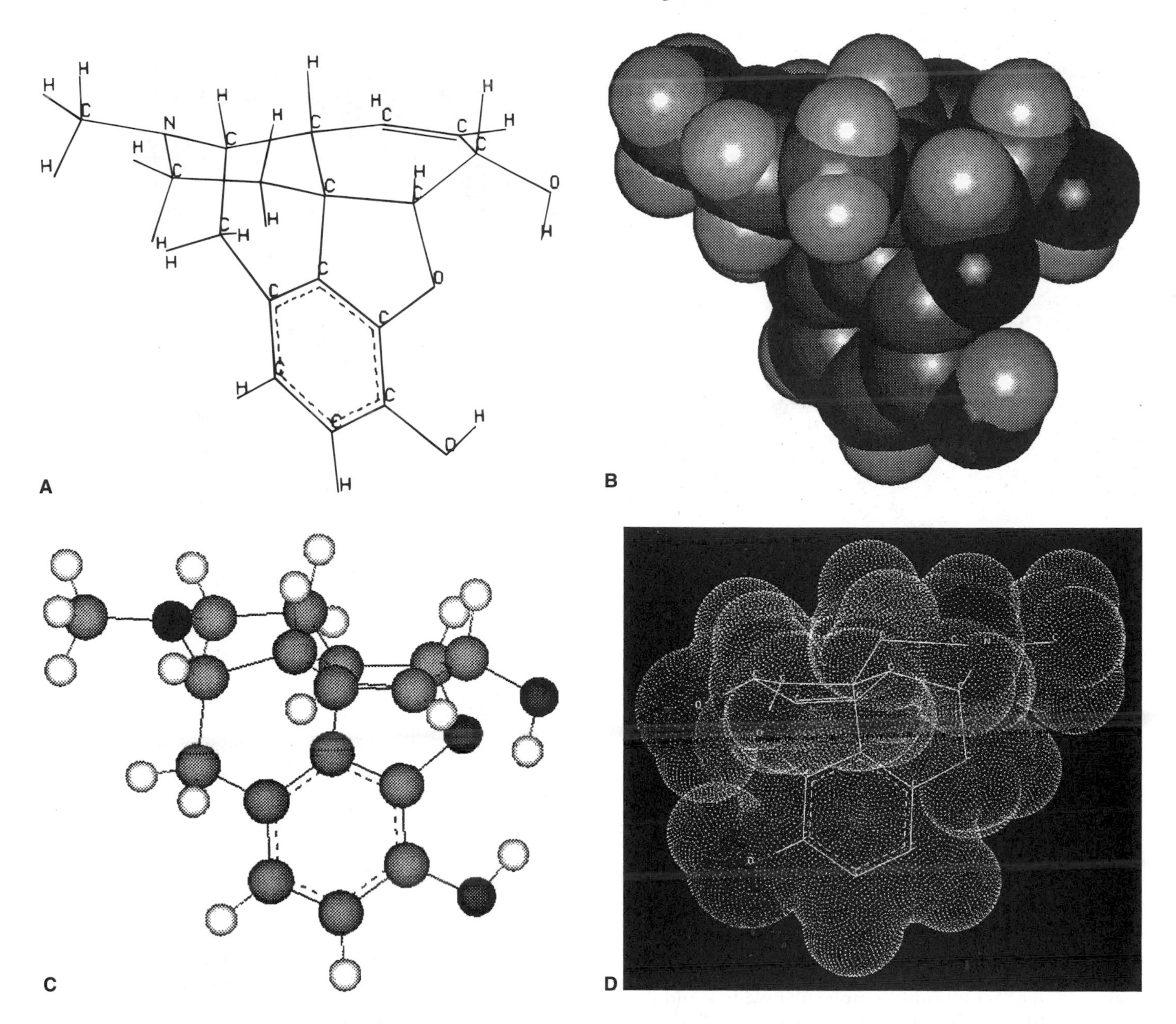

Fig. 6–2. Four displays of the morphine molecule: a stick display *(A)*; a CPK (space-filled) display *(B)*, a ball-and-stick display *(C)*, showing both the atoms as distinct entities and the bonds as lines between them; and a dot surface display *(D)*, which shows both the stereochemistry of the atoms and the realistic actual shape of the molecule.

shown when several of the techniques are combined to provide some real insight into the action of a drug. Continuing with the (DHFR) example, the substrate analog methotrexate is a powerful inhibitor of the DHFR enzyme. It finds wide use treating many types of cancer, as well as other disease states. This drug binds tightly to the active site of DHFR, and its binding process has been modeled and well studied.[10] Figure 6–4 shows a graphic representation of methotrexate in the active site of DHFR. The active site of the enzyme is rendered in a CPK view, as is the drug. The remaining backbone of the protein is displayed in the ribbon format. The pteridine portion of the molecule (the polycyclic aromatic ring) is embedded in the active site pocket, with the p-aminobenzoylglutamate portion binding to the more polar surface of the enzyme. Using the power of molecular graphics, real insight into the action of a drug, and thus ideas on ways to improve its action, can be gained.

PRINCIPLES OF COMPUTATIONAL CHEMISTRY

The computational aspects of molecular modeling simulate particular behaviors and properties of molecules. To the extent that these properties can be calculated and validated, the molecular factors that contribute to a compound's action as a drug can be predicted or explained. Some of the molecular properties commonly computed include a compound's potential energy in a

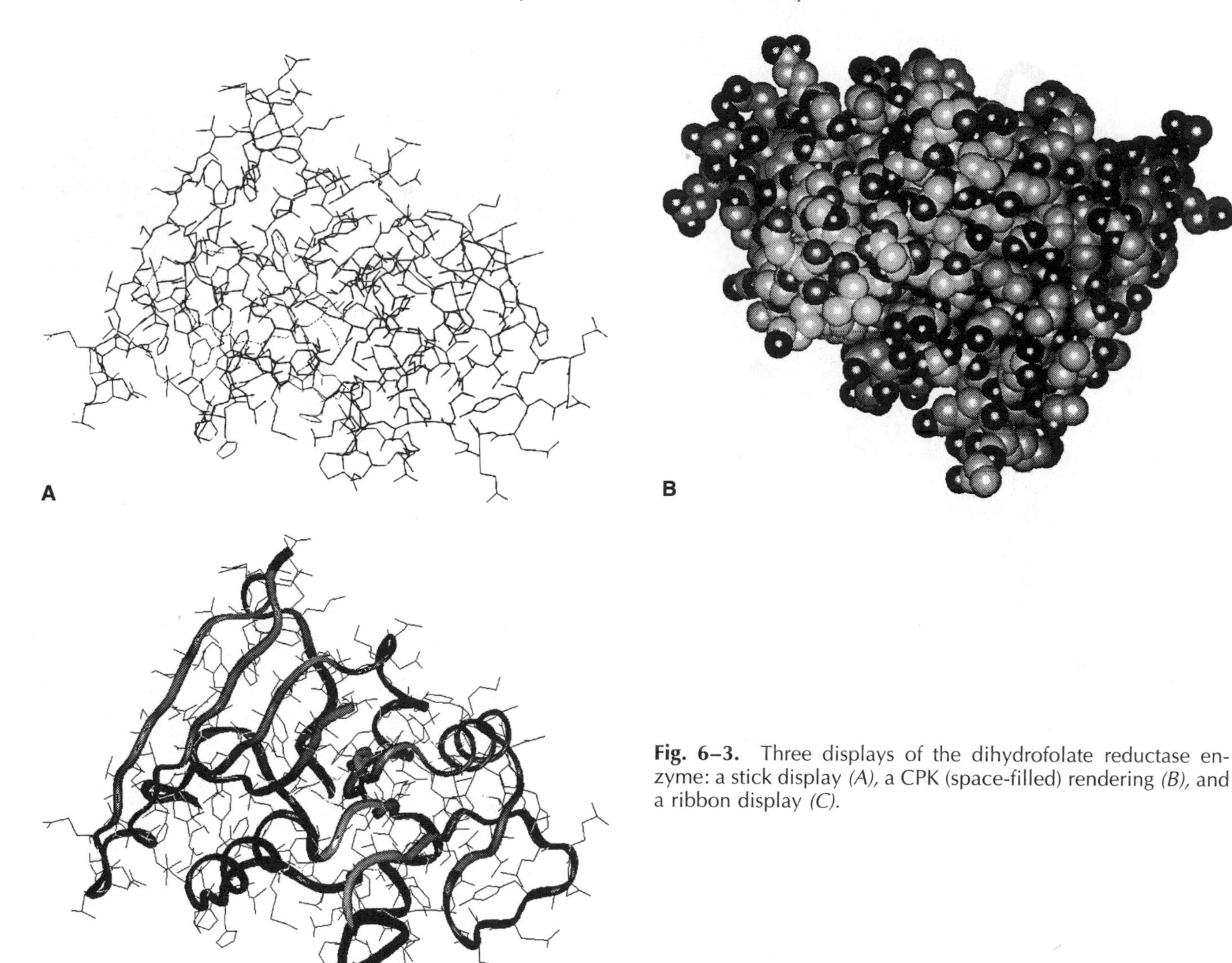

Fig. 6–3. Three displays of the dihydrofolate reductase enzyme: a stick display *(A)*, a CPK (space-filled) rendering *(B)*, and a ribbon display *(C)*.

particular conformation, the electron density at each atom, and the compound's molecular volume and shape. The advances in computational chemistry are what have given molecular modeling its great utility.

Several computational approaches are used in molecular modeling, ranging from the computationally expedient to the highly rigorous. In the following paragraphs, some of these approaches are considered briefly. For more detailed discussion of these complex topics, the reader should consult the cited references.

Molecular Mechanics

The most computationally expedient approach to modeling is that based on the so-called molecular mechanics formalism.[11] In this approach, each atom is treated as a mass proportional to its atomic mass and each of its bonds is treated as an analog of a mechanical spring, which has a force constant associated with it. Each bond has a restoring force to preserve its optimal geometry, including one for the stretching or compressing of the bond, one for bending of a bond angle away from the ideal angle, and one for torsional (twisting) distortions of a bond. Nonbonded interactions are also considered, including van der Waals steric interactions and electrostatic or charge interactions. These latter interactions are called coulombic interactions, and they result from the attraction or repulsion between two fully or partially charged atoms.

The molecular mechanics approach assumes that the total potential energy of a molecule is made up of contributions from each of the terms for each of the atoms and bonds in the molecule. The equation expressing this condition is formalized as

$$E_{total} = \Sigma E_{stretch} + \Sigma E_{bend} + \Sigma E_{torsion} + \Sigma E_{vdW} + \Sigma E_{Coulomb}$$

in which E represents the energy attributable to that particular bond distortion. Some additional terms (e.g., H-bonding) are often included in the equation if the molecules under study need them. The calculation sums the energy resulting from all interactions among all atoms,

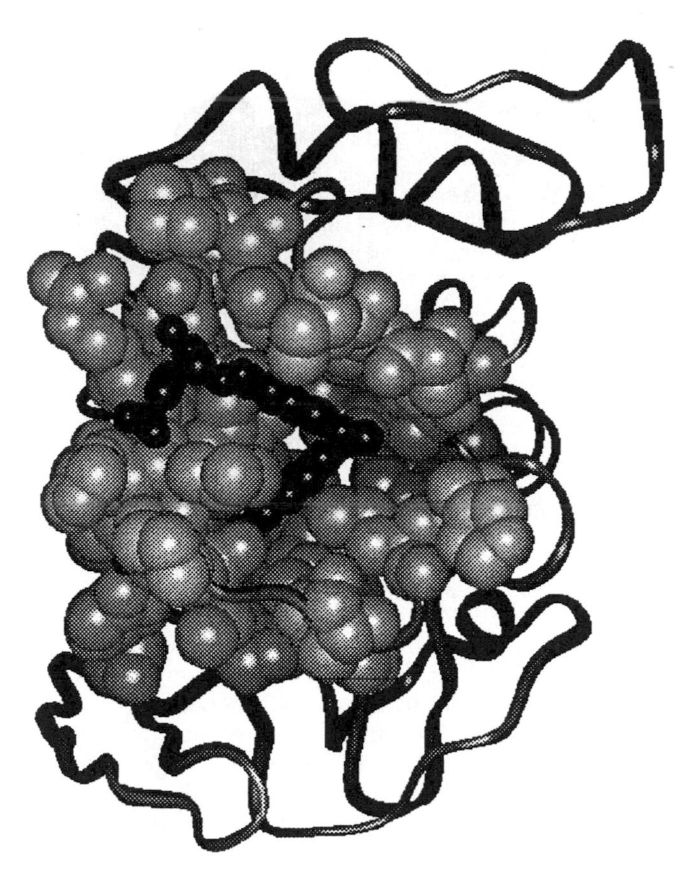

Fig. 6–4. A ribbon view of the methotrexate inhibitor bound to the active site of the dihydrofolate reductase enzyme. The drug and the active site of the enzyme are rendered in CPK format; the remaining portion of the enzyme is drawn in ribbon format.

both bonded and nonbonded, as appropriate. Some terms are more influential than others in their contributions to the overall energy of the molecule's orientation. For example, the $E_{stretch}$ term, although it exists only for bonded atoms, is a dominant interaction term. It can be computed to a reasonable approximation by the Hooke's law equation for harmonic motion:

$$E_{stretch} = k(r - r_0)^2$$

This equation says that causing a bond to deviate from its natural or ideal length by stretching or compressing it requires energy. The magnitude of this increase in energy is a function of a force constant k, the bond distance r, and the ideal bond distance r_0. The values of k and r_0 are specific to the atom types (e.g., $-C-D-$, $-C{=}C-$, and $-C-N-$). Actually, we know that the energy needed to compress a bond is greater than that needed to stretch a bond, i.e., the equation is not truly symmetric as the equation implies. As long as the bond length r is not too far from the ideal r_0, however, the Hooke's law equation works reasonably well. To be more accurate, a Morse function, which better approximates the energy of bond stretch and compression, could be used.[12] Similarly, distortion of bond angles and dihedral angles from their equilibrium or minimum energy positions adds to the overall energy of the molecular system.

The nonbonding interactions are the van der Waals term (E_{vdW}) and the coulombic term ($E_{Coulomb}$), and are present among all of the atoms in the molecule. The interactions of this type between bonded atoms are usually not considered because the other terms describe these interactions (e.g., bond stretching). The importance of the nonbonded terms in the energy expression is strongly dependent on the distance between the interaction atoms. Close interatomic distances produce highly significant contributions, whereas the interactions that occur at several angstroms are less important. For electrostatic coulombic interactions, the force of attraction or repulsion depends on the magnitude of the charge, the distance between the atoms, and the dielectric constant of the medium (D). The equation for the interaction between two point charges,

$$E_{Coulomb} = \frac{q_i q_j}{D r_{ij}}$$

is relatively simple, in which q is the charge on atom i or j, D is the dielectric constant of the medium between the two charges, and r_{ij} is the distance between them.

Van der Waals nonbonded interactions represent even more complex behavior. At distances close enough for atomic overlap, the interaction term is strongly repulsive, because two atoms cannot occupy the same space at the same time. At greater distances, the interaction force is weakly attractive due to the so-called London dispersion forces. A function based on the Lennard-Jones potential,

$$E_{vdW} = \epsilon \left[\left(\frac{r_{min}}{r} \right)^{12} - 2 \left(\frac{r_{min}}{r} \right)^{6} \right]$$

in which r_{min} is the distance between the atoms at minimum energy, ϵ is the minimum energy at r_{min}, and r is the actual distance between the two atoms, has been found to approximate this behavior.

Because these equations are based on classic models, the computer time necessary to solve them is relatively short, at least compared to the quantum chemical methods. Using these molecular mechanics approaches, energies of small molecules (less than 100 atoms) may be computed on ordinary desktop personal computers, often in seconds. These methods are especially useful for investigating conformational energies and for studies on macromolecules, including proteins and polymers.

The advantages of molecular mechanics approaches lie in their ability to simulate large systems (more than 10,000 atoms) and in their relative speed. The simulations based on these approaches can be corrected for temperature and pressure, so reliable atomic trajectories (i.e., the molecule's path of motion in space) may be computed. These in turn allow molecular mechanics to provide information about binding modes of ligands to receptors, vibrational modes, and the importance of steric pharmacophores. More advanced studies can determine the free energy of binding for a ligand and even transport properties such as diffusion coefficients. Some disadvantages of the molecular mechanics approaches include the fact that they are not useful for computing electronic properties such as electron density, and that the force constants for each of the energy terms must be well characterized and appropriate for the molecule under study. To overcome the latter disadvantage, different sets of force field parameters have been developed, including the original MM2,[11] AMBER,[13] CVFF,[14] and CHARMm.[15] The choice of which force field to use is one of the first decisions the medicinal chemist must make before undertaking a modeling study.

Because the molecular mechanics methods are concerned only with the nuclei of the molecule and not with the properties and distribution of the molecule's electrons, they are not as useful for computing the characteristics of the molecule that depend on the movement of its electrons. These characteristics, which include the electron density at various atoms of the molecule and the energies of the highest occupied and lowest unoccupied molecular orbitals (the HOMO and LUMO energies), must be computed using one of the quantum chemical methods.

Quantum Chemical Methods

A more rigorous approach to molecular modeling lies in the realm of quantum chemical calculations.[16] Several properties of molecules are based on the arrangement of their electrons and on the interaction of those electrons with the electrons and nuclei of other molecules. To consider these electronic properties, we must consider the quantum chemical nature of matter. Several computational methods have been developed to explicitly include the molecule's electrons as well as nuclei, which necessarily makes use of quantum chemistry. These calculations are more complex than those used in molecular mechanics, and require considerably more computer resources, on the order of 100 to 100,000 times more. Despite this limitation (and it is a real one for larger molecules), the results offer insight into the quantized, wavelike behavior of electrons and nuclei, something an empiric force field calculation cannot do adequately, if at all.

The quantum chemical methods are based on finding solutions to the Schrödinger wave equation

$$H\Psi = E\Psi$$

in molecular orbital theory. Unfortunately, solving the Schrödinger equation has proven difficult, and it may only have exact, analytic solutions for the simplest of systems—such as the hydrogen atom. A look at the approach used to solve the equation for the hydrogen atom reveals how complex a solution would be for a larger system.

The total energy of the hydrogen atom (E) can be described as the sum of the kinetic energy (K) and potential energy (U) of its two components, in this case a proton and an electron. This approach is analogous to the approach taken in the development of the empiric force field in molecular mechanics. The Schrödinger equation for this relationship can be written as

$$H\Psi = (K + U)\Psi = E\Psi$$

in which Ψ is the molecular wave function (the unknown term) and H is called the Hamiltonian operator. It represents the functions necessary to determine the component energies of the system. H "operates" and Ψ to compute the component energies. The presence of Ψ on both sides of the equation is a reflection of the wave nature of the electron.

The time-independent, nonrelativistic form of the equation is suitable for much of quantum chemistry:

$$\left(-\frac{1}{2}\nabla^2 + \frac{1}{R}\right)\Psi = E\Psi$$

in which $-\frac{1}{2}\nabla^2$ is an operator giving the kinetic energy of the system and $1/R$ represents the attractive coulombic potential in atomic units. This simplification is due to the well-known Born-Oppenheimer approximation, which considers the nuclei to be stationary, because the electrons are roughly 1800 times lighter than protons and neutrons, and therefore adapt faster to their environment. Solutions to this equation are in the form of discrete atomic orbitals. These solutions are the familiar *s, p, d,* and *f* orbitals discussed in beginning chemistry courses.

The Hamiltonian operator for the hydrogen molecule (two bonded hydrogen atoms) serves as a good example of how the computation for a larger molecule is actually performed. This molecule has four particles—two electrons at positions r_1 and r_2, and two proton nuclei at positions R_1 and R_2. All of their interactions enter into the total energy of the molecule, and a term for the interactions of each pair of particles is included in the Hamiltonian operator:

$$H = -\frac{1}{2}\nabla_1^2 - \frac{1}{2}\nabla_2^2 + \frac{1}{R_1R_2} - \frac{1}{R_1r_1} - \frac{1}{R_1r_2} - \frac{1}{R_2r_1} - \frac{1}{R_2r_2} + \frac{1}{r_1r_2}$$

The first two terms represent the kinetic energies of the two electrons. The third term represents the coulombic interaction between the two proton nuclei. The next four terms represent the electron-proton interactions, and the last term represents interaction between the two electrons. As complex as this operator is for only four particles, imagine how complex it would be for even a small drug molecule of, say, 30 atoms and 180 electrons!

The computation of the electron-electron interactions is costly in terms of computer resources, and an additional approximation is often made to simplify the computation process. This second approximation (developed by Hartree and Fock) is lumping an electron's interactions with other electrons into an effective field, V. This results in a simplified form of the Hamiltonian operator for each electron *i:*

$$H_i = -\frac{1}{2}\nabla_i^2 - \frac{1}{R_1 r_i} - \frac{1}{R_2 r_i} + V$$

Even this simplification (with only four particles) has no direct, analytic solution. In fact, we must use other computer techniques to approximate solutions for all molecules larger than three particles—which means everything of interest! These approximate solutions are found through the use of numeric methods. A numeric method is a computerized, iterative process that starts with a trial solution to the equation—a guess at the answer, if you will. It then tests the solution for accuracy, and attempts to improve the trial solution as necessary. The refinement of the trial solution is repeated until it converges to a solution that is within acceptable limits. The solution to the simplified equation is determined by making an initial guess for each electron's molecular orbital. The initial guesses can be simple, linear combinations of the hydrogen atomic orbitals. Each electron's orbital is determined, and its contribution to the effective potential is calculated. As each of the electrons is evaluated in turn, the overall solution improves steadily. When the individual orbitals do not change significantly from one iteration to the next, a self-consistent field (V) solution has been found.

The numeric methods are designed for computer-based operation. They offer several levels of approximation, from the overly simplistic to the interminably rigorous, but they generally fall into two categories. The most theoretically rigorous are the ab initio (literally, "from the beginning") methods, which treat each electron of the molecule explicitly and keep track of its interaction with every other electron and proton in the molecule. Ab initio methods (e.g., the GAUSSIAN series of programs[17]) are the most rigorous and reliable of the quantum chemical methods, but they are also by far the most costly, computationally speaking. Because of the computational demands of the ab initio approach, it is not suitable for studies of large molecules, even on the fastest supercomputer. It is also not suitable for investigations of problems of drug-receptor interactions for this same reason.

One way to simplify a quantum chemical calculation so it can be applied to problems of interest in drug design is to approximate or simplify many of the atomic orbital interactions that would have to be computed if an ab initio method were used. The methods that use these approximations are collectively referred to as semiempiric methods, because certain interactions can be estimated from experimental data. Because of the large number of interactions to compute, varying degrees of simplification can be introduced to reduce computing time. Depending on the simplifications made, the semiempiric methods are roughly 1000 times faster than ab initio methods, yet they are nearly as accurate for most applications, even though they do not have as rigorous a theoretic basis. Because of this great decrease in the need for computing resources, most quantum chemical calculations are performed using one of the semiempiric approaches.

There are many different flavors of semiempiric methods, each with its strengths and weaknesses. The most commonly used approaches in drug design include the modified neglect of differential overlap (MNDO) approach and the perturbative configuration interaction using localized orbitals (PCILO) approach. These methods neglect interactions among nonvalence orbitals, which form the inner core of an atom's electrons. Some also neglect interactions between electrons in different molecular orbitals, such as those that do not interact to form a bond or those that are not centered on adjacent atoms. An even greater simplifying strategy is to treat some of the electronic interactions between molecular orbitals as constants, rather than to compute either all or some of the more significant interactions individually. If care is used, the results of such approximations do not contain intolerable errors, but one has to know when the approximations can be made and when they introduce an unacceptable level of inaccuracy.

Collectively, the quantum chemical modeling methods have accounted for the greatest computer use, certainly more than the molecular mechanics methods. From the drug design perspective, several properties can be computed using quantum methods, including the following:

1. **Examining molecular orbitals.** The molecular orbitals for a given molecule combine to determine the electron density of the molecule. The two specific molecular orbitals of particular interest in drug design are the highest occupied molecular orbital (HOMO) and the lowest unoccupied molecular orbital (LUMO). The HOMO is the highest energy orbital containing an electron. An electron in this orbital is the most likely electron to undergo a reaction. The other electrons are at lower, more stable energies. A review of the geometry of the HOMO reveals the location of the most reactive electron in the molecule. In an electron-rich nucleophile, the location of the HOMO can give insight into a possible orientation for the transition state during nucleophilic attack, e.g., in an enzymatic reaction. The LUMO is vacant, as the name implies. It is the orbital most likely to be occupied if one additional electron is added to the molecule. In an electron-deficient (electrophilic) molecule, the LUMO shape and location may give some insight into a possible reaction geometry. This information can, for example, lead to an understanding of the details of an enzymatic process, which in turn can lead to a series of better inhibitors or substrates for the enzyme.
2. **Computing electric and magnetic properties.** The density of the electron cloud surrounding the nuclei will shift under the influence of other atoms and charges in the vicinity. As a result, a neutral molecule may contain local regions bearing partial positive or negative charges, thereby producing a dipole moment. The quantum chemical methods can compute this electric dipole from the electron densities and the positions of the nuclei. The magnitudes of the electrostatic fields surrounding the molecule also can be computed from the electron density and nuclei positions. Knowledge of the position and

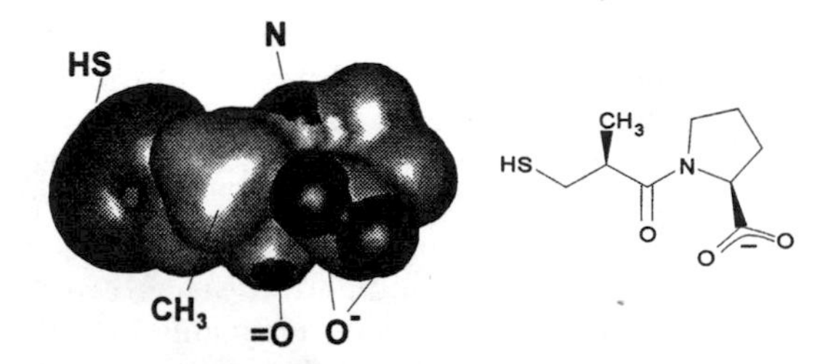

Fig. 6–5. The calculated molecular orbitals of captopril, with the chemical structure alongside (see text for discussion).

magnitudes of these electrostatic potential fields is especially useful in designing ligands for receptors and enzyme active sites.

Figure 6–5 shows the molecular orbitals of captopril, drawn as contour plots, computed using the semiempiric MNDO-type approach implemented in the MOPAC program.[18] Its structure is shown next to the contour for reference. Captopril is an inhibitor of angiotensin converting enzyme (ACE), which is one of the blood pressure regulatory enzymes.[19] This drug binds reversibly to the active site of ACE through the formation of two ionic bridges, one at the carboxylate group and the other at the sulfur atom.

The sulfur is known to bind to an electron-deficient Zn^{++} atom in the active site of the ACE. The basis for the formation of the two ionic bonds can be seen by considering the total electron density, the HOMO, and the LUMO of the molecule. The darker areas on the surface represent areas of greater electron density. As expected, these include the oxygen atoms, the nitrogen atom, and the sulfur atom. The most reactive of the molecular orbitals is the HOMO. This is the site that would react most easily with an electron-deficient moiety. It can be seen surrounding the carboxylate group that binds to a positively charged active site residue.

The location of the LUMO, was calculated to surround the thiol group, which indicates that the sulfur atom is the atom that can most easily accommodate an additional electron. One way of generating an additional electron is by the removal of a proton. Because the extra electron is easily accommodated, the attached proton becomes more labile or acidic. Given that the sulfur must bind as the deprotonated species, the binding is facilitated by this easy loss of the proton.

3. **Tracing reaction pathways.** As we know, the quantum chemical methods are based on a Hamiltonian operator that determines the energy of the system for a specific configuration of the atoms of a molecule. A single-point calculation can determine the energy (and molecular orbitals) of the particular configuration being investigated. Because a chemical reaction is nothing more than a series of changes in the configurations of two or more reactants as they are converted to products, it is possible to slowly change the configurations of the atoms involved in the reaction from those of the reactants to those of the products. If this adjustment is done in small increments and small changes are made in the positions of the nonreacting atoms to minimize the energy of the system, the overall energy and path of the chemical reaction can be traced and studied. For example, the irreversible binding of an enzyme inhibitor to the enzyme active site can be modeled and studied, with a view toward examining the structural requirements for a better inhibitor. Many studies of this type exist in the literature.
4. **Computing thermodynamic data.** The set of molecular orbitals resulting from a quantum chemical calculation can be used to determine several thermodynamic values. The heat of formation can be calculated, representing the energy required to bring the particles to the current configuration. This value determines the relative energy of a given conformation. Obtaining these data is an advantage over the empiric (molecular mechanics) force fields, which compute excess conformational energy and do not include the energy of formation of the bonds. An absolute comparison of the two molecules' energies is possible, whereas empiric force fields allow comparisons of conformations of the same molecule only. The entropy of a molecular system (ΔS) can be approximated by calculating vibrational modes near the equilibrium configuration. Also, the free energy of solvation can be approximated by a semiempiric method.

APPROACHES TO GEOMETRY OPTIMIZATION

One of the more powerful features of modern molecular modeling is the ability to determine a minimum energy conformation of a molecule. Again, although the concept of minimization is fairly simple, in practice it involves considerable complexity. Optimization of a structure is most often performed using molecular mechanics calculations, but the quantum chemical methods can also be used at a cost of increased computer time. In the simplest terms, a minimization calculation begins with an assigned starting geometry and computes the steric or potential energy of the molecule in that geometry. The positions of the atoms of the molecule are then adjusted in a systematic way to lower the energy of each atom, and then the energy of the entire molecule is recomputed. If the energy of the new geometry is less than the starting energy, the new geometry is adopted as the revised starting geometry. Further adjustments of the positions of the molecule's atoms are made and the process is repeated until no further reduction in energy occurs.

One of the disadvantages of all standard geometry optimization approaches is that the calculation always finds the nearest minimum energy orientation to the starting geometry, which is not necessarily the global minimum energy orientation, i.e., the lowest energy orientation the molecule can assume. An analogy that can be used to describe the geometry optimization process is that of pouring water on a hilly terrain. The terrain represents the conformational energy surface of the molecule, with the higher altitude points having more potential energy than the lower altitude points. The water represents the various conformations of the molecule under investigation. Somewhere on the surface is the lowest altitude point (the global energy minimum), but it cannot be detected using this "water technique" unless the water is poured on a hill adjacent to the global minimum. If the water is poured at a random point on the surface of the terrain, it will flow downhill until it comes to the lowest spot at the bottom of the nearest hill. It will not, however, climb up another hill to see if there is an even lower spot on the other side. These hills and ridges represent conformational barriers of varying sizes that relate to conformational changes of a flexible molecule.

One method of finding the minimum energy conformation for a molecule is to rotate each bond systematically by a fixed amount, followed by a geometry minimization for each new conformation. Continuing the water flow analogy, this technique is like pouring water in several selected places on the terrain and measuring the altitudes of all the ponds that result from the downhill flows. In the limit, if enough starting points are chosen to cover the whole terrain, the global minimum, i.e., the deepest point, will be found. This approach is both valid and effective for small drug molecules with only a few rotatable bonds. If a compound has three rotatable bonds and the rotation increment is 30° (a common value), then 12 minimization operations would be needed for each bond (360/30). To search the whole energy surface and cover all of the possible orientations, we would need 12^3 ($12 \times 12 \times 12$) = 1728 minimization operations. This computation is done easily on modern workstations. If each minimization takes 1 minute of computer time, the global minimum could be found in just over 24 hours. The problem becomes more difficult (and even virtually impossible) when a compound has more rotatable bonds. A molecule with 10 rotatable bonds (a fairly common occurrence) would need 12^{10} = 61,917,364,224 (almost 62 billion) minimizations. Even at 1 minute each (an unrealistic time, because each minimization takes longer because of the greater number of atoms involved), it would take 117,803 years to complete the calculation! Obviously, such a task is impossible with present-day computing resources. One can follow certain strategies to cut down this time, but even using these techniques, the task is still daunting.

Molecular Dynamics

An alternative approach to determining conformational energies, as well as evaluating interactions between macromolecules and potential drug molecules, is through molecular dynamics. In this approach, the computer is used to solve Newton's equations of motion for one or more molecules over time. The result is a series of predicted coordinates that, when plotted, trace the movement of atoms within the molecule. On most high-end systems, the atoms' trajectories can be displayed as a movie by rapidly displaying the sequence of individual frames. Associated with each frame is the total energy of the molecule in that particular orientation. By inspecting a graph of the energy versus time, a family of low energy orientations can be found and investigated further.

The motion of the nuclei of molecules follows the laws of classic physics. Because they are more massive than the electrons, their quantum behavior can be neglected. They vibrate at frequencies determined by the masses of the atoms they comprise and by the force constants of the bonds connecting them. For each atom, the force exerted by the atom as it moves is expressed by Newton's classic equation $F_i = m_i a_i$, which states that the force F on atom *i* at any moment is the product of the mass of that atom times its acceleration. To relate this fundamental property to the motion of the molecule over time, the expression must be recast into one that is based on the coordinates of the atom (r_i) and the potential energy of the atom (V). The differential equation that meets these criteria is as follows:

$$-\frac{\delta V}{\delta r_i} = m_i \frac{\delta^2 r_i}{\delta t_i^2}$$

Unfortunately, there is no exact solution to this equation for more than two atoms, but a solution can be approximated using a Taylor series expansion, resulting in the equation

$$r_{(t+\Delta t)} = r_{(t)} + \frac{\delta r}{\delta t}\Delta t + \frac{\delta^2 r}{\delta t^2}\frac{\Delta t^2}{2} + \cdots$$

This complex looking expression simply says that we can compute the position of the atom after a short time interval Δt if we know $r(t)$ (the coordinates of the atom at time t), $\delta r/\delta t$ (the velocity of the atom, i.e., its first derivative with respect to time), and $\delta^2 r/\delta t^2$ (the acceleration of the atom, i.e., its second derivative with respect to time), along with an approximation of the higher order terms of the series. The behavior of each atom is integrated over time and over all atoms of the molecule by evaluating the average velocity of each atom during the time interval Δt, then updating the coordinates of the atom:

$$r_{(t+\Delta t)} = r_{(t)} + v_{avg}\Delta t$$

This equation requires some intermediate calculations that are not discussed here in the interest of simplicity. It also makes some assumptions that limit the method somewhat. The most important assumption the equation makes is that the time interval Δt must be small enough to simulate the motions of all the atoms, that is, it must be able to evaluate the motion of the fastest vibrating atoms. If it cannot do this, the molecule's calculated internal energies will become unstable and the calculation will fail. The fastest moving atoms in any molecule are the hydrogen atoms, which have a C-H stretching periodicity of about 10^{-14} sec. To capture the motion of these atoms accurately, we need to take about 10 samples for each vibration, which requires a time interval (Δt) of 10^{-15} sec or 1 femtosecond (fs). Because the motion of the heavy atoms (C, N, O, etc.) occurs on a picosecond (ps) time scale (10^{-12} sec or 1000 fs), a molecular dynamics run must cover at least a 10,000-fs (10-ps) time interval, and a 100,000-fs (100-ps) interval would be better yet.

The strategy for finding the minimum energy orientation using molecular dynamics typically includes the following steps:

1. Place the molecule in a local energy minimum by running a molecular mechanics geometry optimization.
2. Run a number of steps of dynamics at 1-fs time intervals, the number depending on the computer time available. Record the coordinates and energy of each time point.
3. Save the conformation representing every 1000-fs (1-ps) steps. During this time, the heavy atoms will have moved significantly.
4. Perform a geometry optimization on each of these saved conformations.
5. Examine the minimized conformations, looking for the one or more with the lowest potential energy. These low energy conformations may give insight into one or more potential active orientations of the drug or the macromolecule.

To recall the running water analogy used previously, molecular dynamics has allowed it to rain on the terrain. The "water" will collect in the low spots (lowest energy conformations) of the potential energy surface. We can then look for the lowest altitude pond on the terrain surface, which will be the global minimum.

Table 6–1 provides a summary of the features, uses, and requirements for each of the different molecular modeling methods, along with some of the programs that perform the calculations.

HARDWARE CONSIDERATIONS

As mentioned previously, two factors that have made the development of modern molecular modeling possible over the last decade are the great increases in the speed and capacity (along with a simultaneous decrease in cost) of computing hardware and the development of hardware-based graphics display systems with enough speed to manipulate complex images in real time, i.e., depicting the motion of images without waiting for the system to redraw them.

Graphics Hardware

Traditionally, the two types of graphics display hardware used for molecular graphics have been the vector display and the raster display. The vector display technology produces its images by drawing lines on the face of a special color cathode ray tube. The electron beam that produces the lines is directed only at the points on the screen that are necessary to draw the image specified by the software. The electron beam's movements are minimal, thus the refresh time (the time it takes to redraw the image) for smaller molecules can be higher without producing a flicker of the image. The operation of the raster display is analogous to a television set, with a series of closely spaced lines constantly being painted across the screen of the color cathode ray tube. The display hardware must repeatedly sweep over the entire screen, hitting every point on the screen, independent of the

Table 6–1. A Summary of Molecular Modeling Methodologies and Their Uses

Use	Empiric Forcefield	Quantum Semiempiric	Quantum Ab Initio
Geometry optimization	Yes	Yes	Yes
Molecular dynamics	Yes	Limited	No
Vibrational spectra	Yes	Yes	Yes
Thermodynamic values			
Heat of formation ΔH	No	Yes	Yes
Relative ΔG	Yes	No	No
Absolute ΔS	No	Yes	Yes
Chemical reactions			
Topology	Fixed	Variable	Variable
Transition states	No	Yes	Yes
Reaction coordinate	No	Yes	Yes
Force constants	No	Yes	Yes
Practical differences			
Treatable number of atoms	~10,000	<100	<50
Relative CPU time/iteration	1	1000	100,000
Disk space requirement	10 MB	20 MB	>100 MB
Memory requirement	4–16 MB	8–32 MB	16–512 MB
Popular programs	AMBER CHARMm Discover MM2/MM3 SYBIL	MOPAC AMPAC PCILO	GAUSSIAN GAMESS HONDO

image being displayed. It turns each picture element (pixel) on or off in a particular color, as specified by the graphics software, to produce the image. The advantages of a vector display include the sharpness of the image and the speed with which the images can be redrawn and moved about the screen, which are possible because the only points on the screen the electron beams hit are those where lines are located. The advantage of the raster display is that it can produce more complex images. Until the development of a graphics display technology that puts the computational load of many of the graphic transformation operations into the graphics hardware itself, raster technology was used mostly for producing static images. Most modern molecular graphics systems are raster based, and it is commonplace to manipulate even the most complex images effortlessly.

The molecular graphics hardware systems of choice in recent years at the higher end of the spectrum are the graphics workstations produced by companies such as Silicon Graphics and Evans and Sutherland. At the lower end of the spectrum are the personal computer-based systems, including both MS-DOS and Macintosh systems. The great increase in cost-effective computing power has equipped even the low-end systems with enough power to draw and manipulate molecules of moderate size, although the graphics display technology found in most desktop computers has not yet come close to matching the power of the higher end graphic workstations. This gap will probably narrow somewhat in coming years as technology improves and costs decrease further.

Computational Capabilities

Many factors contribute to the computing capacity of a computer, including the speed and internal architecture of its central processing unit (CPU), the effective speed of its memory-accessing circuitry, the speed and capacity of its disk-based memory, and several other engineering considerations. Phenomenal advances in all of these areas have been made over the past 15 to 20 years. One common index of the capacity of a given computer is the speed of its CPU, both in terms of millions of instructions it can carry out per second (MIPS) and the millions of floating point operations it can perform per second (MFLOPS) in a standardized test suite of software. The MIPS rating is commonly considered more appropriate for nonscientific applications, whereas the MFLOPS rating is accepted as more appropriate for scientific computations. The MIPS number actually measures how quickly the processor executes instructions, not how much actual work it can do per second, and so many people consider the MIPS rating an overly simplistic measure of performance. Nevertheless, it can serve as a more or less one-dimensional measure of how far computing has come over the last 15 to 20 years.

To illustrate the progress made over the last few years, Table 6–2 details some of the characteristics of some common processors and computing systems. The numbers in this table should not be taken as absolutes, because they were derived from a number of different sources and many of them are rough estimates subject to interpretation. In the microcomputer domain, computing power has increased approximately 340-fold (34,000%) since the personal computer was introduced in 1981, as measured by MIPS values. This growth has brought the desktop personal computer (PC) well beyond the speed of the 1970s mainframe computers. This is not to say that the capacities of mainframe computers have been exceeded by the PCs; the mainframes were (and are) capa-

Table 6–2. Performance Characteristics of Typical Computer Processors or Systems

Processor/System	Clock Speed (MHz)	Instructions/Second $\times 10^6$ (MIPS)	Floating Point Operations/Second $\times 10^6$ (MFLOPS)
PC/Macintosh Systems			
Intel 8086/8087/8088 (IBM PC1)	5	0.33	0.007
Intel 80286/80287 (IBM PC/AT)	8	1.2	0.012
Motorola 68000 (Original Macintosh)	8	1.3	0.004
Intel 80386/80387	20	7.5	0.17
Motorola 68040 (Macintosh Quadra 700)	25	22	1.4
Intel 80486DX2 (Gateway 2000)	33/66	30	2.4
Intel Pentium (Next generation PC)	66/132	112	8.5
Workstations (RISC)			
Sun SPARCstation 1	n/a	12.5	1.4
MIPS R3000 (SGI 4D/35)	33	30	4.3
MIPS R4000 (SGI Indigo)	50/100	85	15
MIPS R4400 (Next generation SGI)	75/150	128	23
IBM RS/6000 (Model 540)	30	44	19
Mainframes			
IBM 370/158	6.7	0.80	0.23
IBM ES/9000-520	111	n/a	38
Cray Y-MP C90	238	n/a	387

ble of multiprocessing, i.e., performing more than one task at a time, whereas PCs are only now beginning to have this capability. The current-generation graphics workstation (e.g., a Silicon Graphics Indigo or equivalent) was unknown 10 years ago, for the enabling technology did not exist at the time. Today, the most powerful multiprocessor-equipped workstations are beginning to approach the performance of the supercomputer class of systems for certain types of computing jobs. Mainfames also have improved dramatically, gaining over 6000-fold in speed (as expressed by MIPS). The fastest of the current supercomputers (e.g., the Cray Y-MP) are truly marvels of speed and capacity, although they are still not fast enough to perform the calculations necessary to solve many of the problems facing medicinal chemists today, especially those concerning proteins and other macromolecules. If the rate of increase in computing speed and capacity continues at the dramatic pace of the last two decades, the day will come when computing power will not be the limiting factor in modeling and drug design. At that point, humankind's creativity will certainly have been enhanced.

SOFTWARE CONSIDERATIONS

The computer software that implements molecular modeling is the crucial link between the computer and the medicinal chemist. Essentially all of the leading molecular modeling software packages on the market provide powerful computational and molecular graphics capabilities to the medicinal chemist. Included in any suite of software programs will be an interactive graphics display and manipulation capability, as well as a computational package for performing geometry optimization. In addition to these minimum capabilities, many of the packages include special modules integrated into the user interface that perform small molecule modeling and computation, macromolecular modeling, modeling of drug-receptor interactions, and drug design assistance.[20] Other more specialized capabilities include the following, among others:

1. **Molecule sketching and structure generation.** Central to any modeling effort is the ability to obtain structures of the molecules of interest. For small molecules, it is convenient to be able to sketch the structure much as a chemist would write it on a piece of paper, then translate it into a full three-dimensional coordinate file of the molecule of interest. Another useful feature is the ability to build up a structure from molecular fragments stored on disk. This capability is especially valuable when creating new oligopeptides and small proteins; each new amino acid in the chain can be added by simply pointing to a residue with the mouse or other pointing device.
2. **Protein modeling by homology.** The analysis and prediction of protein structure is one of the most challenging tasks facing science today. As drug design moves to become a more exact science, greater numbers of drugs will be designed by direct methods, i.e., by fitting the drug candidates to a well-defined target molecule. These target molecules will typically be proteins, the structures of which will have been determined by x-ray crystallography or by indirect spectroscopic techniques, such as nuclear magnetic resonance.[21] With rapid developments in biotechnology, it has become straightforward to determine the amino acid sequence of a protein by reading its genetic code and to make pure quantities of the protein.[22] It is more difficult to determine the secondary and tertiary structure of the protein, i.e., to predict how it folds to produce a functional enzyme or receptor system. The structural and sequence similarities among various proteins is well known. Current-generation protein modeling software provides the capability to search a large database of known protein structures, looking for sequence similarity between the members of the database and the new protein of unknown structure. Quite often, protein fragments with similar amino acid sequences will be found to fold in similar ways. At least one package[23] assists in the prediction of an unknown protein structure by comparing its sequence to those of known structure, thereby extending our knowledge to new macromolecular targets.
3. **Quantum chemical computational capability.** Almost all modeling packages have at least semiempiric quantum chemical computational capabilities, typically an interface to MOPAC[18] or another MNDO-based package. The more complete packages also have access to ab initio techniques and to more advanced capabilities, such as the ability to compute electrostatic fields in the presence of solvent, rather than only in a vacuum.[24]
4. **Drug design using artificial intelligence techniques.** One of the newest capabilities of high-end drug design software involves the use of the computer for the de novo design of potential drug molecules. These compounds are often designed as ligands for target protein active sites (the structures of which may be either known or unknown), but they also can be created for any target about which some activity features or requirements are known.

One of the most exciting approaches in this area is found in the Ludi program.[25] Ludi uses fragment libraries and sets of search rules to fit together combinations of fragments from an extensive fragment library. These fragment combinations are chosen to match the stereochemical requirements of a particular drug design target. If the detailed structure of the target macromolecule is known, this program analyzes the geometry of the active or target site, then suggests combinations of fragments that will fill the active site and interact with hydrogen donor or acceptor centers as well as aliphatic and aromatic hydrophobic areas, as appropriate. These families of suggested fragments may then be connected by the medicinal chemist using the modeling features of the software to produce a collection of potential lead compounds suitable for further investigation.

If the active site of the target molecule is not known, this approach can help the medicinal chemist to construct a hypothetic active site, which is generated by analyzing the structural features of compounds known to bind to that site, looking for the hydrogen bonding sites and hydrophobic sites they have in common. This computer-based application of the active analog approach was first elucidated by Marshall.[26] Once the model of the hypothetic active site is established, the drug design process occurs almost as if the structure of the active site were known. The risk of some unsuitable potential lead compounds coming out of a less precise analysis such as this may be somewhat greater than the situation in which the target site is well characterized, but useful insights into new leads are still obtainable and valuable.

A different but complementary approach to semiautomated drug design is through the use of an expert system, one that involves the techniques of artificial intelligence to analyze a set of compounds and predict biologic activity and reliability indices for each compound based on a set of rules and a proposed pharmacophore. Apex[27] is one program of this type. This approach uses the structures of both a series of active compounds (minimally five or six compounds) and a collection of inactive compounds. It applies a set of selection rules to identify those features present in the active compounds and absent in the inactive compounds. In doing so, it may compute hydrophobicity indices, locate hydrogen bonding sites, and calculate atomic charges and electron density indices. It then subjects these characteristics to statistical analysis and constructs a pharmacophore for the activity of interest in the form of a set of rules. The medicinal chemist can then apply those rules to a new set of compounds that the program rates with a suitability or reliability index and a predicted activity. One or more of the highly ranked compounds may then be made and tested. The outcome of these tests will either further validate the model or cause the programs to modify the rules to refine the model.

Clearly, approaches such as these are going to assume greater prominence in the future as the problems of designing a highly selective drug become more complex.

5. **Statistical analysis (QSAR) in drug design.** Some commercial modeling packages have the capability to analyze and correlate the results of biologic testing of a series of compounds with their structural and physicochemical parameters, and then to display the results of the analysis graphically. Such fully integrated capabilities provide an extremely powerful modeling tool for the medicinal chemist. Besides the standard Hansch-type analysis discussed in Chapter 5, more advanced modules are available. One of these, Comparative Molecular Field Analysis (CoMFA),[28] elegantly and powerfully extends the QSAR capabilities of the medicinal chemist from two dimensions into three dimensions.

The CoMFA is based on modeling the forces resulting from both steric and electrostatic interactions within a three-dimensional lattice. The program computes values for both van der Waals and coulombic energies of interaction between a probe "pseudoatom" and each molecule of interest by constructing a lattice into which each of the molecules in the series is placed. Intersections of the lattice are spaced a finite distance apart (typically 2 Å). Each of the computed energy of interaction values is placed in a column of a large table, which contains one row for each compound in the series, along with the biologic activity of the compound. Obviously, many more columns than rows results from this process. A powerful statistical technique know as partial least squares analysis[29] is then used to derive a series of relationships among the molecular field values and the biologic activities. These results are subjected to rigorous cross-validation to be sure that the resulting relationships are not artifactual. Because the molecular field values are based on three-dimensional shape relationships, the graphic representations resulting from the analysis indicate the regions in space in which favorable and unfavorable interactions occur in the series. Moreover, the equations coming from the analysis are predictive of the affinity the compounds of interest will have in the test system.

The power of the CoMFA technique lies in the fact that one needs no direct knowledge of the geometries of the target molecules to reach useful drug design conclusions. Techniques such as CoMFA support a bright future for the computer-based rational drug design process.

AN EXAMPLE OF COMPUTER-ASSISTED MOLECULAR DESIGN IN ACTION: A STUDY OF THE β-LACTAMASE ENZYME ACTIVE SITE

Over the past decade, countless reports of the use of modeling in the drug design enterprise have been pub-

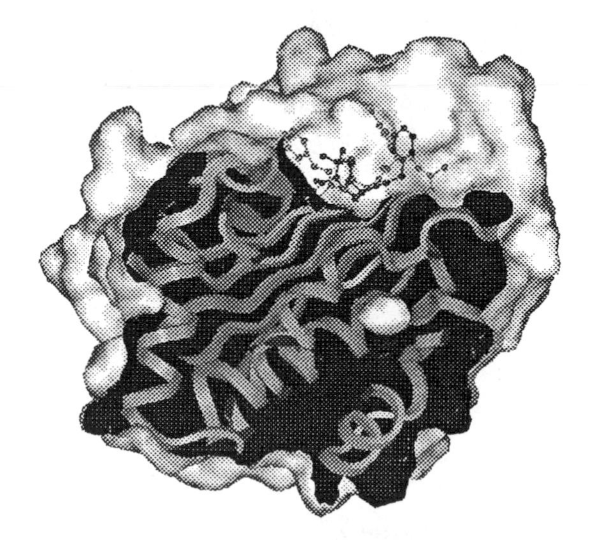

Fig. 6–6. A stylized cross-sectional view of the β-lactamase enzyme from B. licheniformis 749/C.

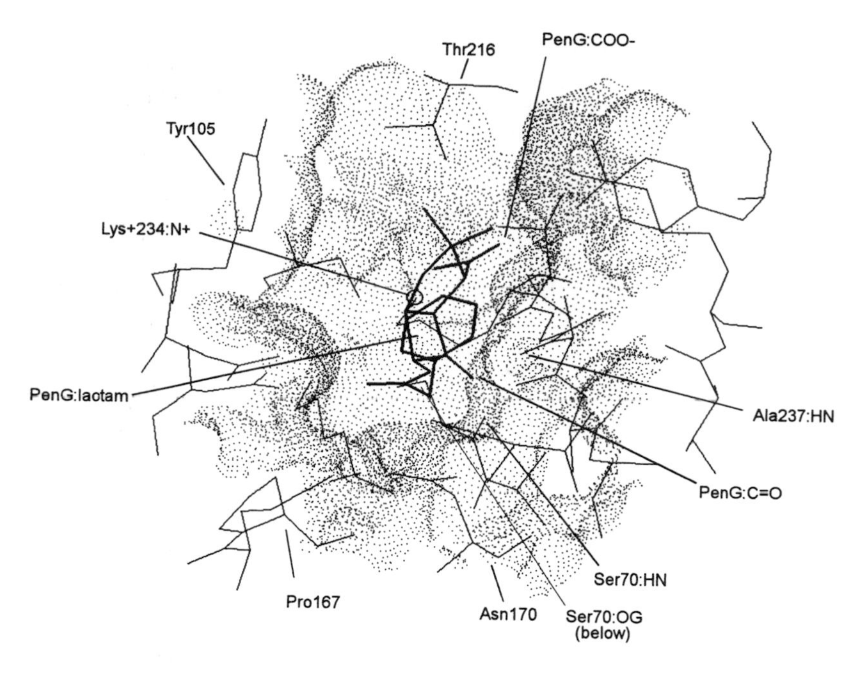

A

Penicillin-G

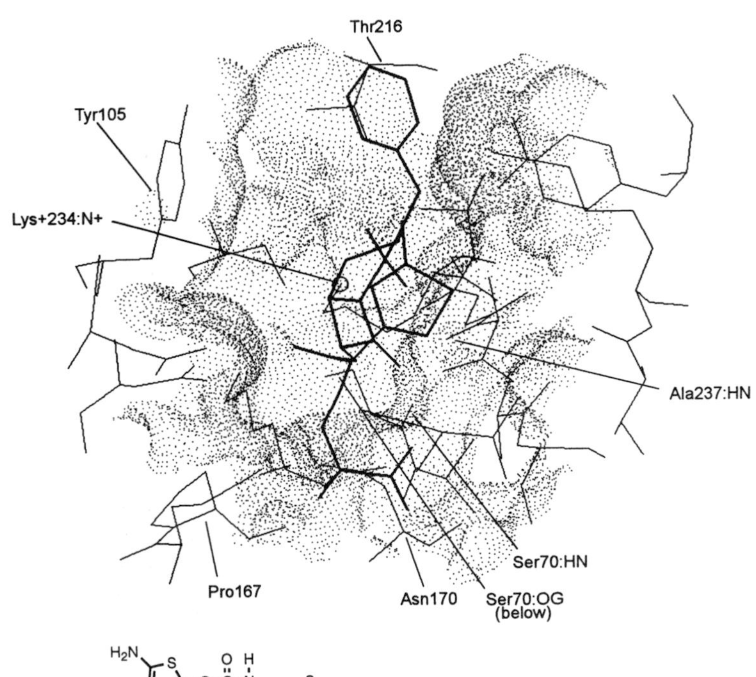

B

Ceftazidime

Fig. 6–7. Two views of the active site of the B. licheniformis *β*-lactamase enzyme, each containing an antibiotic molecule. *A,* Binding of the first-generation antibiotic penicillin G. *B,* Less-effectively bound third-generation cephalosporin ceftazidime (see text for discussion).

lished. The following example illustrates the use of some of these techniques.

The β-lactamases constitute a family of enzymes produced by a number of penicillin-resistant microorganisms. They have evolved over the years, apparently in response to the continued use of β-lactam antibiotics (penicillins, cephalosporins, and monobactams) for the treatment of bacterial infections. These enzymes protect the organism from the lethal actions of the β-lactam antibiotics by hydrolyzing the β-lactam ring of the antibiotic before it can inhibit the cell wall synthesizing enzymes. This enzyme has been isolated and studied in a number of organisms,[30] and recently was isolated and crystallized from Bacillus licheniformis 749/C.[31] Figure 6–6 shows a graphic representation of the enzyme, based on its crystal structure coordinates,[32] and illustrates the usefulness of modern display methods and modeling in helping to understand the molecular basis of the enzyme's action and ways to overcome the effects of this enzyme. Figure 6–6 shows a cross-sectional view through the enzyme's active site. The internal secondary structural elements (helices, loops, and sheets) are shown with a ribbon diagram. The surface of the enzyme is covered to denote the solvent-accessible surface. The active site of the enzyme is located at the top of the molecule as it is displayed, in a "saddle point" or cleft on the molecule, into which the antibiotic molecule can fit. The portion of the enzyme closest to the reader has been clipped away to obtain a better view of this active site. Two antibiotics, a first-generation β-lactam (penicillin G, the darker structure) and a third-generation cephalosporin (ceftazidime, the lighter structure) are positioned in the active site. The substrates must (and do) conform to the steric requirements of the enzyme's folds in its active site, although the penicillin binds more efficiently than the cephalosporin. Because the enzyme is a bacterial detoxification enzyme, the better substrate the drug is for the enzyme, the more effectively the drug will be hydrolyzed by the enzyme, generally speaking. The converse of this fact is that the drug that does not fit as well into the active site will not be destroyed as quickly, which will allow it to be more effective against the bacteria.

A more detailed view of the geometry of the active site is shown in Figure 6–7, although the quality of this view is somewhat less than those seen on the workstation screen itself, because it is compromised by the lack of color and depth cueing. In Figure 6–7*A*, some of the relevant features of the active site and the substrate are labeled for clarity. The view is down into the active site cleft. The penicillin G molecule is oriented so that the β-lactam part of the molecule is near the bottom of the cleft, with the "V" envelope fold at the bottom. The carboxylate group of the antibiotic is oriented downward, strongly bound to the protonated side-chain nitrogen of lysine-234, although the binding does not show well in this view. A fairly steep wall forms this cleft. The "north face" of the wall is formed by the residues near threonine-216; the "south face" is made up of the residues from proline-167 to aparagine-170. Surprisingly, there is substantial room on the "west side" of the cleft, because the drug binds to the residues on the "east face" of the cleft. The hydrolytic "machinery" includes several binding points on the east face, along with a series of residues that deliver a water molecule to accomplish the hydrolysis. These crucial points include a strong hydrogen bond between the β-lactam carbonyl oxygen and two amide hydrogens, one on alanine-237 and the other on serine-70. The activated nucleophilic oxygen, which attacks the activated lactam carbonyl carbon, is the γ-hydroxyl of serine-70. As the model shows, penicillin G fits nicely into the active site and as one would expect, it is hydrolyzed efficiently by this enzyme.

The third-generation antibiotic ceftazidime binds less effectively to the β-lactamase enzyme (and is therefore less subject to inactivation by this route) because it has additional substituents that sterically "bump" into the cleft of the active site, as shown in Figure 6–7*B*. Specifically, the larger groups on the bicyclic ring structure interact with the asparagine-170 and the proline-167. For this drug to bind to the enzyme active site, steric distortions must be introduced into the side chain. Because this action requires energy, the drug is not as good a substrate for the enzyme and is thus not hydrolyzed as quickly.

How could a better antibiotic come out of a study such as this one? One could compare the geometry of this active site with the geometry of the active site of the cell wall synthesizing enzyme. By analyzing the structural requirements of both target enzymes, one might come up with a lead compound that could have inhibitory activity at *both* active sites, thereby shutting off the β-lactamase protection enzyme and killing the organism. Subsequent synthesis and testing of such compounds would either validate the hypothesis or yield more facts on which a revised hypothesis would be based, following the principles of rational drug design discussed previously.

SUMMARY

Molecular modeling has become a valuable and essential tool to the medicinal chemist working in the drug design enterprise. The advent of powerful computers and the evolution of high-quality software has resulted in a flexible array of methods that facilitate the rational drug design process. These modeling methods can speed the refinement of a lead compound from the laboratory bench to the pharmacy shelves.

Molecular modeling is based on several computational methods, ranging from quantum chemistry to simple parameterization of experimental results. The techniques extend from simply visualizing a small drug molecule on a personal computer to computing the intricacies of drug-target interactions on a supercomputer. Although the complexities and difficulties of the methods vary, all of the molecular modeling operations have a common element—they are attempts to rationalize the behavior and activity of bioactive agents. This common element may, in fact, be the most important, because it provides a path for humankind's constant search for an understanding of nature and for the relief of human and animal diseases.

REFERENCES

1. Volume 4, *Quantitative Drug Design*, in *Comprehensive Medicinal Chemistry*, C. Hansch, P. G. Sammes, and J. B. Taylor, Eds., Oxford, Pergamon Press, 1990.

2. M. Murcko, Vertex Pharmaceutical Corp., personal communication, 1993.
3. G. H. Loew, et al., *Pharm. Res., 10,* 475(1993) and references cited therein.
4. C. Humblet and J. B. Dunbar, Jr., *Ann. Rep. Med. Chem., 28,* 275(1993).
5. M. L. Connolly, *Science, 221,* 709(1983).
6. C. Branden and J. Tooze, *Introduction to Protein Structure,* New York, Garland Publishing, Inc., 1991, pp. 1–302.
7. J. T. Bolin, et al., *J. Biol. Chem., 257,* 13,650(1982).
8. Entry 3DFR, version of June 1982.
9. F. C. Bernstein, et al., *J. Mol. Biol., 112,* 535(1977); E. Abola, et al., *Protein Data Bank,* in *Crystallographic Databases—Information Content, Software Systems, Scientific Applications,* F. H. Allen, G. Bergerhoff, and R. Sievers, Eds., Data Commission of the International Union of Crystallography, Bonn/Cambridge/Chester, 1987, pp. 107–132.
10. D. J. Filman, et al., *J. Biol. Chem., 257,* 13,663(1982).
11. U. Burkert and N. L. Allinger, *Molecular Mechanics,* Washington D.C., American Chemical Society, 1982, pp. 1–339.
12. A. T. Hagler, et al., *J. Am.. Chem. Soc., 101,* 813(1979).
13. S. J. Weiner, et al., *J. Comput. Chem., 7,* 230(1986).
14. U. Dinur and A. T. Hagler, *New Approaches to Empirical Force Fields,* in *Reviews in Computational Chemistry,* Vol. 2, K. B. Lipkowitz and D. B. Boyd, Eds., Weinheim, VCH, 1991.
15. B. R. Brooks, et al., *J. Comput. Chem., 4,* 187(1983).
16. T. Clark, *A Handbook of Computational Chemistry,* New York, Wiley-Interscience, 1985, pp. 1–332.
17. W. J. Hehre, et al., *Ab Initio Molecular Orbital Theory,* New York, John Wiley & Sons, 1986.
18. J. J. P. Stewart, *J. Comput. Aided Mol. Des., 4,* 1(1990).
19. D. McAreavey and J. I. S. Robertson, *Drugs, 40,* 326(1990).
20. N. C. Cohen, et al., *J. Med. Chem., 33,* 883(1990).
21. J. W. Erickson and S. W. Fesik, *Ann. Rep. Med. Chem., 27,* 271(1992).
22. S. W. Zito, Ed., *Pharmaceutical Biotechnology, a Programmed Text,* Chap. 1, Lancaster, PA, Technomic Publishing Co., 1992.
23. *Homology,* San Diego, CA, Biosym Technologies, Inc.
24. *DelPhi,* San Diego, CA, Biosym Technologies, Inc.; M. Gilson, K. Sharp, and B. Honig, *J. Comput. Chem., 9,* 327(1988).
25. *Ludi,* San Diego, CA, Biosym Technologies, Inc.; H.-J. Bohm, *J. Comput. Aided Mol. Des., 6,* 61(1992).
26. G. R. Marshall, et al., in *Computer-Aided Drug Design,* E. C. Olsen and R. E. Christofferson, Eds., Washington, D.C., American Chemical Society Symposium Series No. 112, 1979, p. 205.
27. *Apex,* San Diego, CA, Biosym Technologies, Inc.; V. E. Golender and A. B. Rozenblit, *Logical and Combinatorial Algorithms for Drug Design,* Research Studies Press, New York, Wiley and Sons, 1983.
28. *CoMFA,* St. Louis, MO, Tripos Associates; R. D. Cramer, III, D. E. Patterson, and J. D. Bunce, *J. Am. Chem. Soc., 110,* 5,959(1988).
29. W. Lindberg, et al., *Anal. Chem., 55,* 643(1983).
30. A. Coulson, *Biotechnol. Gen. Eng. Rev., 3,* 219 (1985); J.-M. Frère and B. Joris, *Crit. Rev. Microbiol., 11,* 299(1985).
31. J. R. Knox and P. C. Moews, *J. Mol. Biol., 220,* 435(1991).
32. J. R. Knox, personal communication, Entry 4BLM of the Brookhaven Protein Data Bank, 1994.

Chapter 7

RECEPTORS AND DRUG ACTION

Timothy J. Maher

The human body is an example of an exquisitely designed, extremely complex machine that functions day-in and day-out to allow for the survival of the organism in response to a never-ending onslaught of external challenges. When one considers the enormous variety of environmental stressors to which the body is continually subjected, the existence of a multitude of checks and balances associated with its physiologic and biochemical systems is not surprising. These systems, including endocrinal, nervous, and enzymatic, typically function in concert to adapt to changing environmental conditions. Whereas some systems are designed to respond quickly, i.e., within milliseconds, and for a short time, others are designed to act more slowly, but usually have significantly longer durations, i.e., months to years. Together, these systems support the survival of the organism. Misfunctioning of the control of such systems, however, often leads to disease and potentially the eventual demise of the individual.

The use of specific chemical compounds to treat disease dates back to early humans. Many primitive cultures used plants and other raw materials in an attempt to mitigate the influences of evil spirits and other factors rooted in superstition, which were believed to be the foundations of such illnesses. Over the centuries, serendipitous observations involving the ability of largely botanical preparations to alter disease processes laid the foundation for the modern day, more systematic approach to the discovery of medicinals for therapeutic use. The collaboration of chemical and biologic scientists continue this quest for the "magic bullet" to treat those diseases that challenge the individual's well being.

HISTORICAL PERSPECTIVES

In the middle 1800s, Bernard was first to demonstrate that some drugs are capable of producing their effects by acting at specific sites within the body with his classic experiments involving curare.[1] He showed that this neuromuscular blocking agent, which was used as an arrow poison by South American natives, prevented skeletal muscle contraction following nerve stimulation, but was without effect when the muscle was stimulated directly. This work demonstrated a localized site of action for a drug and, most importantly, suggested that a gap or synapse existed between the nerve and the muscle. From these findings, he also postulated that some substance normally communicated the information between the nerve and the target tissue. These findings established the foundations for what is known today as chemical neurotransmission, a process frequently disrupted by diseases, and likewise the target of many therapeutic agents.

Early work by Langley[2] around the turn of the century established the initial foundations for the interaction of drugs with specific cellular components, later to be identified and termed receptors. Before this time, many leading experts believed that most drugs acted nonselectively on virtually all of the cells in the body to produce their responses, a response resulting from their general physical characteristics, e.g., lipid solubility, and not related to specific structural features of the compound. Langley noted that compounds like pilocarpine, which mimic the parasympathetic division of the autonomic nervous system, were selective and extremely potent. Additionally, a compound like atropine was capable of blocking, in rather selective fashion, the effects of pilocarpine and parasympathetic nervous system stimulation. Importantly, he concluded that the two compounds interacted with the same component of the cell.

Paul Ehrlich,[3] a noted microbiologist during the late nineteenth and early twentieth centuries, is credited with coining the term *receptive substance* or *receptor*. His observations that various organic compounds appeared to produce their antimicrobial effects with a high degree of selectivity led him to speculate that drugs produced their effects via binding to such a receptive substance. The interaction of the drug with the receptor was analogous to a lock and a key. Thus, certain organic compounds would fit properly into the receptor and activate it, leading to a high degree of specificity. Although such a situation might be considered ideal for drug therapy, in actuality, few drugs interact only with their intended receptors. The frequency of side effects not associated with a simple extension of their desired actions indicates that drug molecules can also combine with other receptors or nonreceptor entities on or within cells to produce a host of other, and often undesirable, effects.

Some drugs produce their desired effects without interaction with a specific receptor. For instance, osmotic diuretics produce their pharmacologic effects simply by creating an osmotic gradient in the renal tubules, thereby fostering the elimination of water in the urine. This action is purely the result of a physical characteristic of the drug. Similarly, antacids produce their beneficial effects by chemically neutralizing the hydrochloric acid found in the gastrointestinal tract. No absorption of the drug is even required for their effects to be observed.

More sophisticated mechanisms can also be involved in the nonreceptor actions of therapeutic agents. For instance, the antineoplastic agent mechlorethamine, a nitrogen mustard, produces its beneficial (pharmacologic) and adverse (toxicologic) effects via interaction with many cellular components, in both cancerous and normal cells. By its conversion to a highly reactive electrophilic ethylenimmonium ion intermediate, this agent reacts with neucleophilic cellular components, such as hydroxyl, sulfhydryl, phosphate, carboxyl, and imidazole groups. In particular, by alkylating the number 7 nitrogen of guanine in DNA, this agent produces miscoding (cytosine normally base pairs with guanine in DNA, although thymine now substitutes for cytosine) and the eventual death of the cell results.[4] When one realizes that all replicating cells contain a number 7 nitrogen of guanine in their DNA, it is easy to see why mechlorethamine, and most other antineoplastic agents, produce nonselective destruction of cells throughout the body. Thus, no specific receptor is involved in the actions of this class of pharmacologic agent.

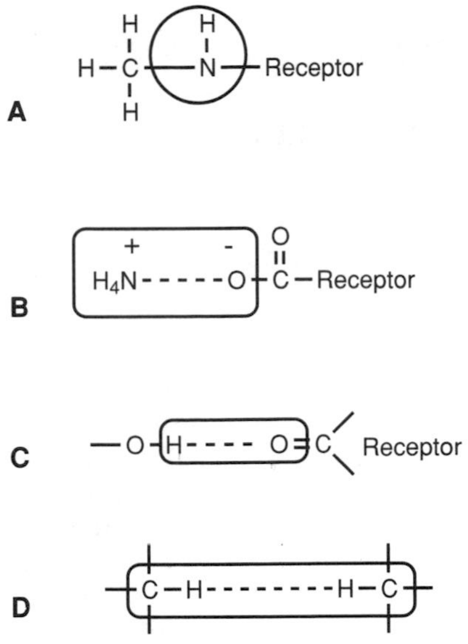

Fig. 7–1. Various drug-receptor bonds. *A*, covalent; *B*, ionic; *C*, hydrogen; *D*, hydrophobic.

AFFINITY—THE ROLE OF CHEMICAL BONDING

In the early 1990s, Hill used nicotine and curare in isolated muscle preparations and noted the effects of temperature.[1] He concluded that the ability of a drug to produce an effect must result from specific chemical interactions between the drug and specific sites. He also noted that the effects of many drugs were reversible, because washing the isolated tissue often restored the sensitivity of the tissue to nerve stimulation. These studies set the foundation for our understanding of the chemical interactions between drugs and receptors.

When a drug interacts with a receptor, several chemical attractive forces are responsible for the initial interaction. Compounds that are attracted to a receptor macromolecule are said to have affinity for that receptor, and may be classified as agonists or antagonists. Additionally, compounds with affinity are also referred to as ligands. Agonists are those compounds that have affinity for the receptor and are also capable of producing a biologic response as a result of its interaction with the receptor.[5] As noted subsequently, the ability to produce a response is referred to as efficacy or intrinsic activity. Drugs that are capable of interacting with the receptor but not activating it to produce a response are termed antagonists. This class of drug is said to have affinity, but lacks intrinsic activity. The basis for the extent of the affinity of a compound for a receptor depends on its three-dimensional characteristics from a physical and electrochemical standpoint.

Assuming a compound has been distributed properly to the general area of the receptor in the body, a function simply of its physical characteristics, the ability of that compound to diffuse in close enough proximity to the receptor to interact with it, depends initially on the types of chemical bonds that can be established between the drug and the receptor. The overall strengths of these bonds vary (Fig. 7–1), and determine the degree of affinity between the drug and the receptor.

The strongest of bonds involved in drug-receptor interactions is the covalent bond, in which two atoms share a pair of electrons yielding atoms that satisfy the octet rule, i.e., results in a stable electron configuration. Double and triple bonds also can form as a result of two or three pairs of electrons, respectively, sharing in the formation of a covalent bond. The extent of sharing of electrons in a covalent bond may not be completely equal between the two atoms, thus establishing slight polarity whereby the atom with the greater electron attraction, i.e., electronegativity, develops a relative negative charge with respect to its slightly positively charged partner. Because of the significant strength of such a bond, 50 to 150 kcal/mol, covalent bonding often results in a situation in which the receptor is irreversibly bound by the ligand, leading to its eventual destruction via endocytosis and phagocytosis. Full recovery of cellular function therefore requires the synthesis of new receptors.

An example of an irreversible covalent bond formation between drug and receptor involves the longlasting blockade of α-adrenoceptors by phenoxybenzamine. Once converted to a highly reactive carbonium ion intermediate, this haloalkylamine can covalently link, via alkylation, the amino, sulfhydryl, and carboxyl groups in the α-adrenoceptor. The receptor is thus rendered irreversibly nonfunctional and is eventually destroyed. New receptor synthesis requires a number of days, thus accounting for the extremely prolonged duration of the block associated with this agent. As will be discussed, this property of phenoxybenzamine, to irreversibly bind the α-adrenoceptor, was critical for the demonstration of spare receptors.[6,7] Because other receptors and cellular components also contain molecular groups that are likewise capable of interacting with the activated phenoxybenzamine intermediate, it is not surprising that receptors that mediate the actions of other neurotransmitters (e.g., acetylcholine, serotonin, and histamine) are also subject to alkylation and blockade, demonstrating the lack of selectivity of this compound.

When two ions of opposite charge are attracted to each other through electrostatic forces, an ion bond is formed. The strength of this type of bond varies between 5 and 10 kcal/mol, and decreases proportionally to the square of the distance between the two atoms. The ability of a drug to bind to a receptor via ionic interactions therefore increases significantly as the drug molecule diffuses closer to the receptor. Additionally, the strength associated with the ionic bond is strong enough to support a transient interaction between the receptor and the drug, but, unlike the covalent bond, it is not so strong as to prevent dissociation of the complex.

The tendency of an atom to participate in ionic bonding is determined by its degree of electronegativity. Hydrogen, as a standard, has an electronegativity value of 2.1 (Linus Pauling units). Atoms or functional groups with greater attraction for electrons include: fluorine, chlorine, hydroxyl, sulfhydryl, and carboxyl groups. On the other hand, chemical groups that have a weaker tendency than hydrogen to attract electrons include alkyl groups.

Hydrogen that is linked via covalent bonding to a strongly electronegative atom, such as oxygen, chlorine, or fluorine, develops a relative positive charge and will be attracted to another atom possessing a relative negative charge by what is termed hydrogen bonding. Water molecules, which have an electronic dipolar nature (the hydrogens are relatively positive because of the attraction of electrons by the oxygen), can bond easily to other water molecules. At 2 to 5 kcal/mol, a single hydrogen bond is relatively weak and would not be expected to support a drug-receptor interaction alone. When multiple hydrogen bonds are formed between drugs and receptors, however, as is typically the case, a significant amount of stability is conferred on the drug-receptor interaction. Thus, hydrogen bonding is most likely an essential requirement for many drug-receptor interactions.

Hydrophobic interactions between nonpolar organic molecules also can contribute to the binding forces that attract a ligand to its receptor. The strength of these bonds typically are weak, at 1 to 4 kcal/mol.

Van der Waals forces, sometimes referred to as London's forces, are those attractive forces that exist between two neutral, nonpolar molecules as they come in close proximity to one another. Theorists suggest that for these forces to operate, a momentary dipolar structure has to exist to allow for such association. This dipolar structure may occur as a result of a temporary imbalance of charge distribution within molecules. These forces are weak (0.5 to 1 kcal/mol) and decrease proportionally to the seventh power of the interatomic distance.

Most therapeutically useful drugs bind only transiently to their intended receptor. The combination of a variety of bonds including ionic, hydrogen, and van der Waals attractive forces can contribute to the initial binding of a drug to the receptor. Once the drug has bound, a biologic response may result (if an agonist). Following binding of the receptor, a conformational change in the receptor may occur that initiates the activation of the biologic response and also changes the attractive environment between the drug and the receptor, thereby allowing for the dissociation of the drug-receptor complex.

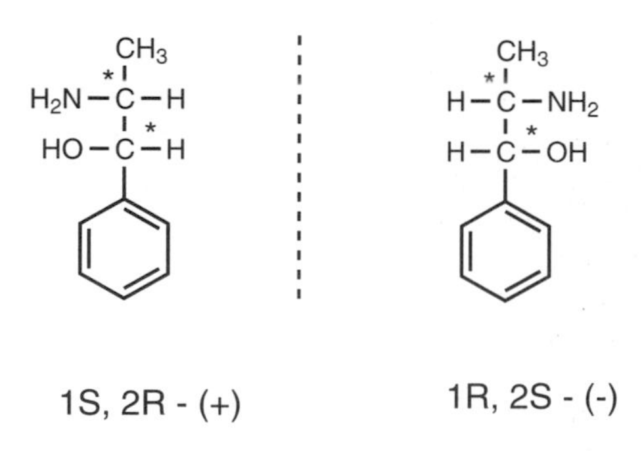

Fig. 7–2. Projection formulae of 2-amino-3-phenyl-1-propanol stereoisomers. (From T. J. Maher and D. A. Johnson, *Drug Dev. Res., 24,* 149(1991).)

Specific three-dimensional requirements must be satisfied for a compound to act effectively as an agonist. The specificity of a drug for its receptor is demonstrated elegantly by the unique three-dimensional characteristics of chiral compounds. As early as 1901, Pasteur[8] noted the significance of assymetric compounds in biologic systems. Since that time, much has been learned from chiral compounds regarding three-dimensional binding requirements of receptors. For instance, although the individual enantiomers (nonsuperimposible mirror images) of norephedrine (2-amino-3-phenyl-1-propanol; Fig. 7–2) have identical molecular weights, melting points, lipid solubility, and empiric formulae, these compounds have significantly different α-adrenoceptor agonistic activities. The 1R,2S enantiomer (levo) is about 100 times more potent than the 1S,2R enantiomer (dextro) in vivo and in vitro.[9,10] (Because there are two chiral centers, there is also another set of stereoisomers, 1S,2S and 1R,2R, that have an even different pharmacologic profile.) Thus, the greater efficacy of the 1R,2S enantiomer most likely depends on its ability to bind and activate the receptor as a result of its preferential fit into the receptor. Many synthetically prepared therapeutic agents are a mixture of two enantiomers (racemic mixtures), with one enantiomer largely responsible for the desired pharmacologic effect.[11] The other may be inactive or may even contribute more significantly to the toxicity of the therapeutic agent. In the future, knowledge of chirality should play a significant role in the advances gained in receptor theory as well as in therapeutics.

DOSE-RESPONSE RELATIONSHIPS

A. J. Clark is generally given credit for being the first to apply the law of mass action principles to the concept of drug-receptor interactions, thus providing further evidence for dose-effect phenomenon.[12] More recently, theorists have doubted the principles of the law of mass action in receptor-drug interactions. The law of mass action applies to compounds dissolved in fluids that are allowed to diffuse freely. Now that much is known about the anchoring of most receptors to, or within, cell membranes where the drug interactions are thought to occur, this

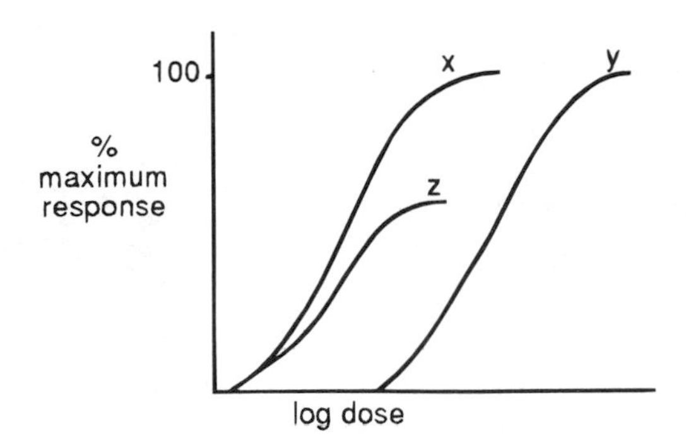

Fig. 7–3. Dose-response relationship.

environment would actually constitute a solid-liquid interface, and thus the law of mass action might not be completely applicable.

The following equation illustrates the interaction of a drug [D] with a receptor [R], which results in a drug-receptor complex [DR] and a biologic response. The interaction between the drug and the receptor is generally reversible.

$$[D] + [R] \rightleftharpoons [DR] \rightarrow \text{Biological Response}$$

Drug-receptor interaction

After administration of a drug, one can monitor the biologic responses produced. As illustrated in Fig. 7–3, the effect produced is typically plotted versus the log of the dose administered. This relationship yields a sigmoidal function that possesses a relatively linear portion of the curve about its central point. Quantitative and qualitative indications of potency and efficacy can be obtained from such a representation. Potency is inversely related to the dose required to produce a given response (typically half-maximum), whereas efficacy is the ability of a drug to produce a full response (100% maximum). In Figure 7–3, drug X is equally efficacious as drug Y, and drug X is more potent than drug Y. Additionally, drug Z is more potent than drug Y, and drug Z is equipotent to drug X.

RECEPTOR LOCATIONS

When an action potential arrives down the axon of the nerve cell, a depolarization-induced exocytosis of neurotransmitter from its storage sites in the presynaptic terminal continues the flow of information to the target site, typically the postsynaptic cell. The neurotransmitter is believed to diffuse across the extracellular fluid-filled space known as the synapse and then interact with postsynaptic receptors; however, the released neurotransmitter may also be capable of interacting with presynaptic receptors located on the neurons that just released the neurotransmitter. The function of these receptors involves the regulation of nerve transmission, and they are referred to as autoreceptors because the neurotransmitters that activate them control their own release.

An exquisite example of such autoreceptor control of neurotransmission is observed in norepinephrine-containing postganglionic neurons of the sympathetic nervous system[13] (Fig. 7–4). Norepinephrine, which is capable of stimulating both α- and β-adrenoceptors, initially is released and is present in low concentration in the synapse. Low concentrations of this agent are capable of preferentially stimulating β-adrenoceptors located presynaptically, which function to increase the release of more neurotransmitter and thereby magnify the intended response. Additionally, the epinephrine released from the adrenal medulla during sympathetic stimulation might also play an important role in facilitating neuronal norepinephrine release. This situation is an example of a positive feedback system, which allows for a rapid rise in the concentration of the neurotransmitter and thus the intended signal. After this initial period of robust norepinephrine release, norepinephrine concentrations in the synaptic cleft are high, which then leads to stimulation of other presynaptic autoreceptors, this time that terminate the additional release of neurotransmitter. This negative-feedback system allows the signal to be terminated quickly.

Together, the presynaptic facilitatory β-adrenoceptor-mediated mechanism and the presynaptic inhibitory α-adrenoceptor-mediated mechanism allow delivery of a

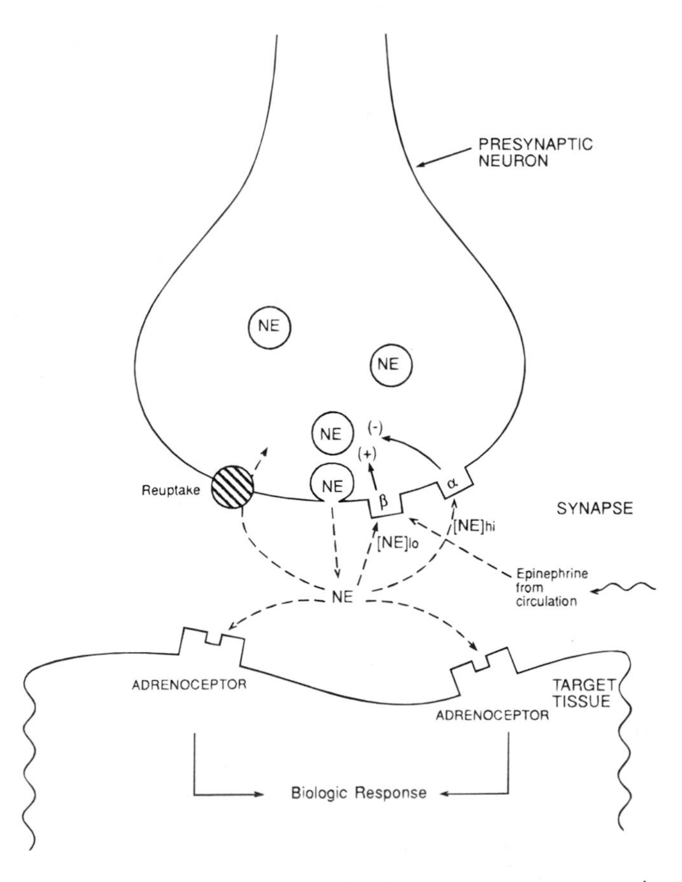

Fig. 7–4. Autoreceptor control of neurotransmission as observed in norepinephrine-containing postganglionic neurons of the sympathetic nervous system. NE, norepinephrine; NE[lo], low norepinephrine concentration; NE[hi], high norepinephrine concentration, α, alpha adrenoceptor; β, beta adrenoceptor; (−), inhibit; (+), stimulate.

rapid, robust, and well-controlled signal. If one were to design a system that was to respond quickly to stressors, such as the sympathetic nervous system is designed to do, a system that turns on rapidly and can be terminated quickly would be ideal, and presumably of an evolutionary advantage. Additionally, many other neurotransmitter autoreceptors have been identified, for instance, in the serotoninergic, dopaminergic, and histaminergic transmitter systems. Even in some examples, a neurotransmitter can interact with a presynaptic receptor to influence the release of a different neurotransmitter. For instance, norepinephrine released from neurons in the gastrointestinal tract can decrease acetylcholine release.

RECEPTORS AND THE BIOLOGIC RESPONSE

When attached to an agonist, receptors typically are capable of interacting with an effector, which functions to transduce the intended signal via amplification into a biologic response. One family of receptors studied in great detail is the G-protein-coupled receptor[14] (Fig. 7–5). These receptors begin with their amino terminus on the extracellular surface of the membrane and span back and forth across the membrane seven times, terminating with their carboxy terminal intracellularly. Binding of an agonist to the extracellular domains is believed to cause a conformational change on the intracellular side that is linked to a GTP binding protein complex[15] (Fig. 7–6). This change then stimulates the binding of GTP, which activates the G-protein and, in turn, regulates the activity of a number of effectors such as adenylyl cyclase, phospholipase C and A_2, and ion channels for Ca^{++}, Na^+, and K^+.

Different G-proteins exist that may modify the activity of adenylyl cyclase. The G_s form activates whereas the G_i form inhibits. The levels of cyclic AMP available in the cell as a result of the level of activity of adenylyl cyclase control the amount of phosphorylation of protein kinase, which in turn open or close calcium channels that mediate the observed biologic responses. Additionally, phospholipase C, which is activated by G_s and possibly by other

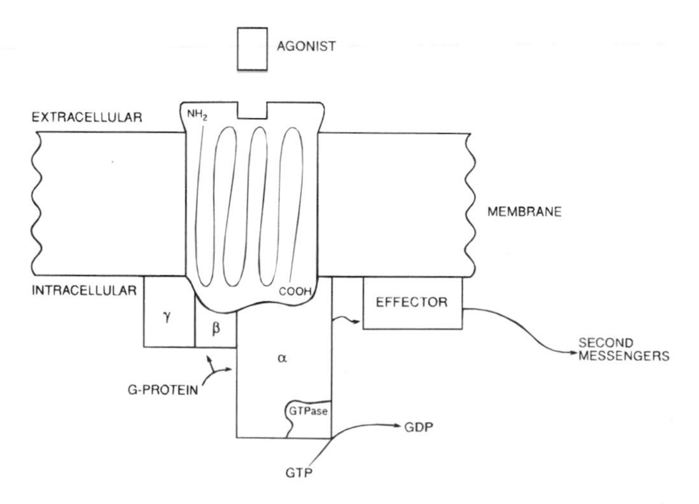

Fig. 7–5. G-protein coupled receptor. α, β, γ correspond to G-protein subunits.

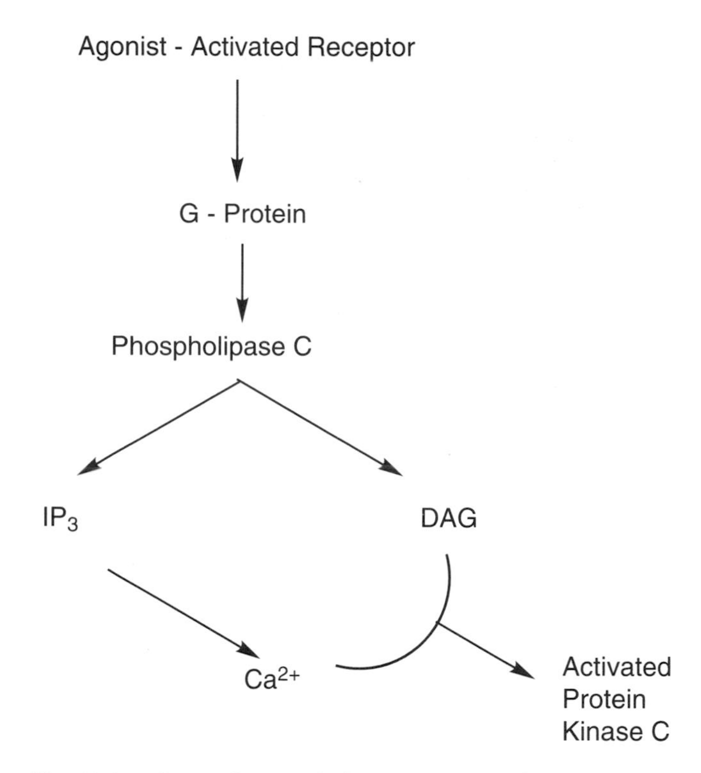

Fig. 7–6. G-protein coupled receptor transducing systems. DAG, diacylglycerol; IP_3, inositol-1,4,5-triphosphate.

less well-defined G-proteins, termed $G_?$, increases the hydrolysis of phosphoinositol bisphosphate to inositol triphosphate (IP_3), which then enhances intracellular calcium release.[16] The increased intracellular calcium, in addition to the other by-product of phospholipase C activity, diacylglycerol (DAG), activate protein kinase C, which regulates the activity of a number of important intracellular enzymes (see Fig. 7–6).

Another family of receptors are catalytic receptors, which function as protein kinases. A single globular protein most likely spans the membrane such that activation by an agonist will increase the activity of the protein kinase inside the cell, resulting in a biologic response. A guanylyl cyclase-linked receptor functions in a similar fashion.[17]

Some receptors, by virtue of their location through the membrane, form ion channels that allow for changes in the conductance of ions, and result in depolarization or hyperpolarization of target cells.[18] These receptors have extracellular agonist recognition sites that, when activated, open to allow a given ion to flow down its electric or chemical gradient. Examples of such receptors include the nicotinic cholinergic receptor, the excitatory amino acid receptors (e.g., N-methyl-D-aspartate, kainate), the γ-aminobutyrate-A (GABA-A) receptor, and the glycine-A (strychnine-sensitive) receptor.[19]

When the nicotinic cholinergic receptor is bound by two molecules of acetylcholine, the ion channel opens and sodium flows into the cell down its electrochemical gradient to cause depolarization. On the other hand, when the GABA-A receptor is activated, chloride ions flow

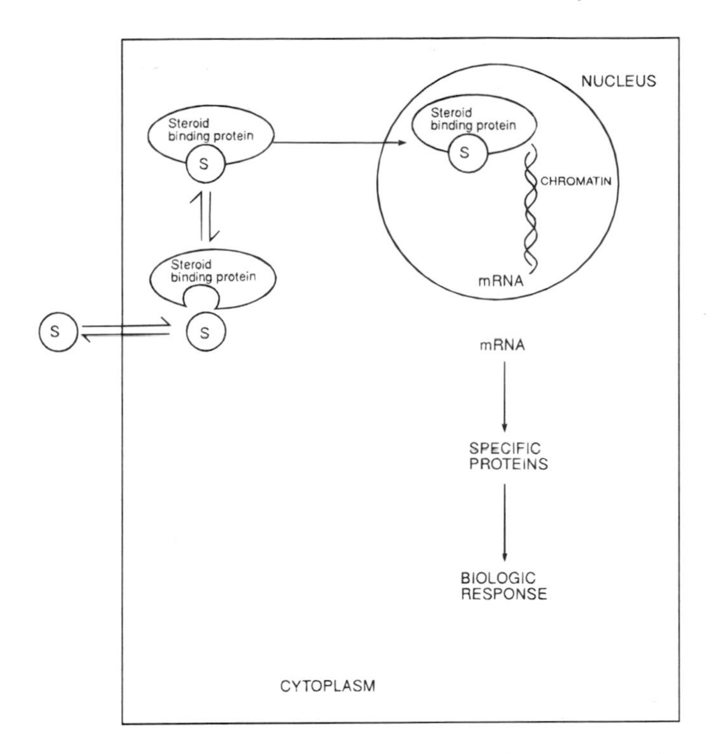

Fig. 7–7. Mechanism of action of steroid drugs. S, steroid.

into the cell, resulting in hyperpolarization. The selectivity of drugs for these receptors depends on their ability to bind to the recognition sites on the extracellular surface; which ion will then flow is a unique characteristic of the receptor-formed ion channel.

Although most receptors are on the surface of the cell, some important receptors are located within the cytoplasm of the cell. The best studied example is the receptor for the steroid hormones.[20] To be active, steroids must enter the cell and combine with a receptor that happens to be a soluble DNA-binding protein that regulates the transcription of specific genes (Fig. 7–7). The binding proteins appear to have three major domains: a carboxy terminus region that recognizes and binds the steroid, a central portion that recognizes the site on the nuclear DNA, and an amino terminus region, the function of which is less well understood. When the binding protein receptor is not occupied by a steroid, it has a negative regulatory function to prevent the gene from being transcribed. When occupied, however, the negative regulatory function is prevented and gene transcription is allowed to proceed.

RECEPTOR SUBTYPES

Careful examination of the effects of a series of sympathomimetics by Ahlquist[21] led him to postulate the existence of at least two types of adrenoceptors, which he termed α and β. Realizing that adrenoceptor agonists were capable of causing either relaxation or contraction of isolated smooth muscles, he noted that while a compound like norepinephrine had potent excitatory actions but weak inhibitory actions, another catecholamine, isoproterenol, had potent inhibitory actions but weak excitatory actions. When a series of related compounds are tested for potency in various tissues, one can demonstrate that for the α-adrenoceptor, epinephrine $\geq$ norepinephrine $>>>$ isoproterenol, and for the β-adrenoceptor, isoproterenol $>$ epinephrine $\geq$ norepinephrine. Following the findings of Ahlquist, others used specific antagonists that had become available to lend further support to this designation of receptor subtypes. Additionally, with the development of highly selective antagonists, the classification of receptors into subtypes has expanded at a tremendous rate.[22] For example, the α-adrenoceptor just noted can be subclassed as α1-A, 1-B, 1-C, 1-D and α2-A, 2-B, and 2-C on the basis of cloning experiments. The therapeutic significance of such distinctions is not known, however, because of a lack of selective agonist and/or antagonists. Some therapeutic distinction can be made between α1 and α2 in a general way in that α1 vasoconstriction antagonism by prazosin, or central nervous system α2-adrenoceptor stimulation by clonidine, are useful in the treatment of hypertension. Similarly, β-adrenoceptor antagonists are available that antagonize β1-adrenoceptors with some selectivity. Thus, the use of metoprolol, a a selective β1-antagonist, is effective and relatively safe in hypertensive patients with compromised airway function, whereas the use of a nonselective (β1 and β2) antagonist such as propranolol would be contraindicated in such a patient.

A summary of the most important receptor subtypes from a therapeutic standpoint is presented in Table 7–1. The reader should realize that this summary is a simplification of what is currently known about the various receptor subtypes. For instance, within the general category of serotonin receptors, at least 13 subtypes can be identified from cloning experiments. The lack of selective agonists and antagonists to characterize the pharmacology of each of these subtypes, however, has hindered our understanding of their individual functions and importance from a therapeutic standpoint. Our ability to eventually develop drugs that manipulate selectively such receptor subtypes has enormous therapeutic implications.

SPARE RECEPTORS

Biologic systems often have built-in safety factors to enhance the efficiency of receptor-stimulus coupling and thereby assure the desired neurotransmission. In many tissues containing α-adrenoceptors, only a small percentage of the available receptors need to be occupied to produce a maximum response, depending on the particular tissue studied and the agonist used. Occupancy of 100% of the available receptors is therefore not always required because spare receptors or receptor reserve are present. Studies with phenoxybenzamine, which alkylates the α-adrenoceptor and therefore irreversibly inactivates the receptor, indicate that only 5 to 10% of the available receptors need be activated to elicit a maximum response to a full or strong agonist such as norepinephrine or phenylephrine.[6,7,23] To obtain a maximum response to a partial agonist like ephedrine, however, nearly 100% of the receptors must be occupied. The explanation behind

Table 7–1. A Survey of Receptor Subtypes

Receptor Class	Subtype	Selective Agonist	Selective Antagonist	Effector	Cloned
Adrenoceptor	α_1	Phenylephrine	Prazosin	IP_3/DAG	Yes
	α_2	Clonidine	Yohimbine	↓ cAMP	Yes
	β_1	Dobutamine	Atenolol	↑ cAMP	Yes
	β_2	Terbutaline	Butoxamine	↑ cAMP	Yes
Dopamine receptors	D_1	Fenoldopam	Dihydroxidine	↑ cAMP	Yes
	D_2	Bromocriptine	(−) Sulpiride	↓ cAMP	Yes
Excitatory amino acid receptor	NMDA	NMDA	D-AP5	↑ Na^+/Ca^{++}	Yes
	AMPA	AMPA	CNQX	↑ Na^+	Yes
	Kainate	Kainate	?	↑ Na^+/K^+	Yes
GABA receptors	GABA-A	Muscimol	Bicuculline	↑ Cl^-	Yes
	GABA-B	Baclofen	Saclofen	↓ cAMP	No
Histamine receptor	H_1	2-(m-Fluorophenyl) Histamine	Mepyramine	IP_3/DAG	Yes
	H_2	Dimapril	Ranitidine	↑ cAMP	Yes
Muscarinic receptors	M_1	Oxotremorine	Pirenzepine	IP_3/DAG	Yes
	M_2	Oxotremorine	AF-DX116	↓ cAMP	Yes
Nicotinic receptor	N_{Muscle}	?	Decamethonium	↑ Na^+/Ca^{++}	Yes
	$N_{Neuronal}$	?	Hexamethonium	↑ Na^+/Ca^{++}	Yes
Opioid receptor	Mu	Sufentanil	CTAP	↓ cAMP	Yes
	Delta	[DAla^2]deltorphin	Naltrindole	↓ cAMP	Yes
	Kappa	Dynorphin	Nor-binaltorphimine	↓ cAMP	Yes
Serotonin receptor (5-HT)	$5\text{-}HT_{1A}$	8-OH-DPAT	Spiperone	↓ cAMP	Yes
	$5\text{-}HT_{2A}$	α-Methyl-5-HT	Ketanserin	IP_3/DAG	Yes

?, no known selective compounds available; cloned, receptor subtype has been cloned and the amino acid structure is known. Chemical abbreviations used: AF-DX116, 11-([2-{(diethylamino)methyl}-1-piperidinyl]acetyl)-5-11-dihydro-6H-pyrido[2,3-b] [1,4]benzodiazepine-6-one; AMPA, D,L-α-amino-3-hydroxy-5-methyl-4-isoxalone propionic acid; cAMP, cyclic adenosine 3′,5′-monophosphate; CNQX, 6-cyano-7-nitroquinoxaline-2,3-dione; CTAP, DPhe-Cys-Tyr-DTrp-Arg-Thr-Pen-Thr-NH2; DAG, diacyl glycerol; D-AP5, D-amino-5-phosphonopentanoate; GABA, γ-aminobutyric acid; 5-HT, 5-hydroxytryptamine, serotonin; IP-3, inositol 1,4,5-triphosphate; NMDA, N-methyl-D-aspartate; 8-OH-DPAT, 8-hydroxy-2-(di-n-propylamino)tetralin.

this difference may involve a less-than-ideal receptor-drug interaction for partial agonists. A partial agonist may function as an antagonist if it interferes with the ability of a full agonist to bind to its receptor and produce a response. In the absence of a full agonist, however, the partial agonist only displays agonistic activity.

THE DYNAMIC NATURE OF RECEPTORS

As is characteristic of most of the individual components of living systems, receptors are constantly in a state of dynamic adaptation. One could envision these protein molecules floating within the fluid mosaic of the biologic membrane awaiting interaction with normal physiologic signals. The function of such receptors, once stimulated, involves attempts at correcting perturbations of the normal physiology of the cell or organism. This role in maintaining homeostasis within the organism requires constant adaptations to the changing environment. One approach in responding to such unpredictable challenges to homeostasis that appears to have developed in just about every species studied involves the ability of receptors to change in response to such external assaults. Receptors are known to be able to decrease in actual number or in their affinity for an agonist when stimulated at higher than normal frequencies. This alteration in the availability or functional capacity of a given receptor most likely constitutes an adaptive mechanism whereby the cell or organism is protected from agonist overload. For example, the chronic administration of a β-adrenoceptor agonist such as isoproterenol is known to produce a desensitization of the β-adrenoceptors in the heart.[24] During this period of overstimulation, the cell somehow recognizes the abnormal intensity of stimulation and initiates an adaptive change in the cell to protect its homeostasis. This adjustment is accomplished by a process of down-regulation of the receptor. As a general principle, the body attempts to maintain homeostasis, whether perturbed by environmental challenges, disease processes, or even the administration of drugs. Actually, the body sees the administration of drugs as a perturbation of homeostasis and usually attempts to overcome the effects of the drug by invoking receptor adaptations. With appropriate dosing schedules, however, drugs often can be used with little observed receptor adaptation such that the desired pharmacologic effect is maintained.

In a similar fashion to the example just described, chronic administration of the β-adrenoceptor antagonist propranolol leads to a state of receptor supersensitivity or up-regulation. The cells within the tissue, such as the heart, sense an alteration in the normal rate of basal

β-adrenoceptor stimulation, and thus respond by increasing either the number or the affinity of the receptors for their natural agonists, norepinephrine and epinephrine. Additionally, enhanced efficiency of the interaction between the receptor and its transducing systems may also account for a portion of the observed supersensitivity. The knowledge that such a receptor adaptation occurs has paramount practical therapeutic implications. Abrupt withdrawal of this class of agents may precipitate acute myocardial infarction, and thus this practice should be scrupulously avoided.

Some pathophysiologic states are characterized by such perturbations in receptor dynamics. Prinzmetal's angina is thought to be characterized by an imbalance between vasodilatory $\beta 2$-adrenoceptor function and vasoconstrictor $\alpha 1$-adrenoceptor function. In this disease state, the excessive α vasoconstriction of coronary arteries leads to myocardial ischemia and pain. The inadvertent use of a β-adrenoceptor antagonist, which is used safely in typical angina pectoris, by preventing β-adrenoceptor vasodilation may leave unopposed the α vasoconstrictor influences and actually precipitate anginal pain. Thus, an understanding of the role receptors play in physiology, pathophysiology, and pharmacology is essential for optimal therapeutic interventions.

FUTURE DIRECTIONS

Our understanding of the nature and role of receptors has increased tremendously since the work of Langley and Ehrlich. With advances made in the field of molecular biology, it is now possible to clone individual receptor subtypes (see Table 7–1) and determine their function in cell culture. By modifying the amino acid structure at those sites believed to be involved in agonist binding, we can appreciate the interaction of drugs currently available and the rational design of those awaiting discovery. Additionally, as we begin to determine the structure of receptor subtypes through cloning techniques, we hopefully will better understand those disease processes that result from, or lead to, receptor adaptations or dysfunction.

REFERENCES

1. C. D. Leake. *A Historical Account of Pharmacology to the 20th Century,* Springfield, Charles C Thomas 1975.
2. J. N. Langley. *J. Physiol., 33,* 374(1905).
3. P. Ehrlich. in *Collected Papers of Paul Ehrlich,* Vol III, F. Himmelweit, Ed., London, Pergamon, 1957.
4. C. C. Price, *Chemistry of alkylation,* in *Antineoplastic and Immunosuppressive Agents,* Part II, *Handbuch der Experimentellen Pharmakologie,* Vol. 38, A. C. Sartorelli and D. G. Johns, Eds., Berlin, Springer, 1975, pp. 1–5.
5. M. Nickerson, *Nature, 178,* 697(1956).
6. K. P. Minneman and P. W. Abel, *Naunyn Schmiedebergs Arch. Pharmacol., 327,* 238(1984).
7. J. C. Besse and R. F. Furchgott, *J. Pharmacol. Exp. Ther., 197,* 66(1976).
8. L. Pasteur, *On the asymmetry of naturally occurring organic compounds, the foundations of stereochemistry,* in *Memoirs by Pasteur, Van't Hoff Le Bel and Wislicenus,* G. M. Richardson, Ed., Stuttgart, Birkhauser, 1901, pp. 103–142.
9. F. A. Moya-Huff and T. J. Maher, *J. Pharm. Pharmacol., 40,* 876(1988).
10. D. A. Johnson and T. J. Maher, *Drug Dev. Res., 23,* 159(1991).
11. T. J. Maher and D. A. Johnson, *Drug Dev. Res., 24,* 149(1991).
12. J. Parascandola, *Trends Pharmacol. Sci., 4,* 421(1982).
13. K. Starke, *Rev. Physiol. Biochem. Pharmacol., 107,* 73(1987).
14. B. K. Kobilka, *Annu. Rev. Neurosci., 15,* 87(1992).
15. A. G. Gilman, *Annu. Rev. Biochem., 56,* 615(1987).
16. M. J. Berridge, *Annu. Rev. Biochem., 56,* 159(1987).
17. M. Chinkers, et al., *Nature, 338,* 78(1989).
18. A. Brisson and P. N. T. Unwin, *Nature, 315,* 474(1985).
19. J. A. Kemp and P. D. Leeson, *Trends. Pharmacol. Sci., 14,* 20(1993).
20. R. M. Evans, *Science, 240,* 889(1988).
21. R. P. Ahlquist, *Am. J. Physiol., 153,* 586(1948).
22. K. P. Minneman and T. A. Esbenshade, *Annu. Rev. Pharmacol. Toxicol., 34,* 117(1994).
23. R. F. Furchgott, *The use of beta-haloalkylamines in the differentiation of receptors and in the determination of dissociation constants of receptor-agonist complexes,* in *Advances in Drug Research,* Vol. 3, N. J. Harper and A. B. Simmonds, Eds., New York, Academic Press, 1966, pp. 21–55.
24. A. E. Tattersfield, *Bull. Eur. Physiopathol. Respir., 21,* 1s(1985).

SUGGESTED READINGS

E. J. Ariens, *Arch. Int. Pharmacodyn. Ther., 99,* 32(1954).

J. W. Black and P. Leff, *Proc. R. Soc. Lond. [Biol.], 220,* 141(1983).

D. B. Bylund, *FASEB J., 6,* 832(1992).

J. W. Kebabian and J. L. Neumeyer, *The RBI Handbook of Receptor Classification,* Natick, MA, Research Biochemicals International, 1994.

T. Kenakin, *Pharmacologic Analysis of Drug-Receptor Interaction,* 2nd ed., New York, Raven, 1993.

R. J. Tallarida and L. S. Jacobs, *The Dose Response Relation in Pharmacology.* New York, Springer, 1979.

Trends in Pharmcolgoical Sciences: 1994 Receptor & Ion Channel Nomenclature Supplement, Oxford, Elsevier Science Ltd., 1994.

Chapter 8

DRUG METABOLISM

David A. Williams

Humans are exposed throughout their lifetime to a large variety of drugs and nonessential exogenous (foreign) compounds (collectively referred to as xenobiotics) that may pose as a health hazard. Drugs taken for therapeutic purposes as well as occupational or private exposure to the vapors of volatile chemicals or solvent pose possible health risks; smoking and drinking involve the absorption of large amounts of substances with potential adverse health effects. Furthermore, the ingestion of natural toxins in vegetables and fruits, pesticide residues in food, as well as carcinogenic pyrolysis products from fats and protein formed during the charbroiling of meat have to be considered. Most of these xenobiotics undergo enzymatic biotransformations by xenobiotic-metabolizing enzymes in the liver and extrahepatic tissues, and are eliminated by excretion as hydrophilic metabolites. In some cases, especially during oxidative metabolism, numerous chemical procarcinogens form reactive metabolites capable of binding covalently to proteins or nucleic acids—a critical step to mutagenicity, cytotoxicity, and carcinogenicity. Therefore, insight into the biotransformation and bioactivation of xenobiotics becomes an undisputable prerequisite for the assessment of drug safety and risk estimation of chemicals and drugs.

Detoxication and toxic effects of drugs and other xenobiotics have been studied extensively in various mammalian species. Frequently, differences in sensitivity to these toxic effects were observed and can now be attributed to a difference between species in the isoenzyme/isoforms* of cytochrome P450 monooxygenases (CYP450). The level of expression of the CYP450 enzymes is regulated by a variety of endogenous factors such as hormones, sex, age, disease, and the presence of environmental factors such as inducing agents.

Drugs can no longer be regarded as chemically stable entities that elicit the desired pharmacologic response and then are excreted from the body. Drugs undergo a variety of chemical changes in the animal organism by enzymes of the liver, intestine, kidney, lung, and other tissues, with consequent alterations in the nature of their pharmacologic activity, duration of activity, and toxicity. Thus, the pharmacologic and toxicologic activity of a drug (or xenobiotic) is in many ways the consequence of its metabolism.

Drug therapy is becoming oriented more to controlling metabolic, genetic, and environmental illnesses (such as cardiovascular disease, mental illness, cancer, and diabetes) than to short-term therapy associated with infectious diseases. In most cases, drug therapy lasts for months or even years and the problems of drug toxicity from long-term therapy become more serious. Therefore, a greater knowledge of drug metabolism is becoming essential.

The practice of prescribing several drugs simultaneously is also increasing. An awareness of possible drug interactions is essential to avoid catastrophic synergistic effects and chemical, enzymic, and pharmacokinetic interactions that may produce toxic side effects.

The study of xenobiotic metabolism has developed rapidly during the past few decades.[1–3] These studies have been fundamental in the assessment of drug efficacy, safety, and the design of dosage regimens; in the development of food additives and the assessment of potential hazards of contaminants; in the evaluation of toxic chemicals; and in the development of pesticides and herbicides and their metabolic fate in insects, other animals, and plants. The metabolism of drugs and other xenobiotics is fundamental to many toxic processes such as carcinogenesis, teratogenesis, and tissue necrosis. Often the same enzymes involved in drug metabolism also carry out the regulation and metabolism of endogenous substances. The inhibition and induction of these enzymes by drugs and xenobiotics may consequently have a profound effect on the normal processes of intermediary metabolism, such as tissue growth and development, hemopoiesis, calcification, and lipid metabolism.

The increased knowledge of drug metabolism, fed by the need for greater safety evaluation of drugs and chemicals, has resulted in a proliferation of publications[2] and a series of monographs that present the current state of knowledge of foreign compound metabolism from biochemical and pharmacologic viewpoints.[3]

PATHWAYS OF METABOLISM

Drugs, plant toxins, food additives, environmental chemicals, insecticides, and other chemicals foreign to the body undergo enzymic transformations that usually result in the loss of pharmacologic activity. The term detoxication describes the result of such metabolic changes.

* Isoform and isoenzyme (isozyme) are often used interchangeably. An isoenzyme is a different enzyme with a different amino acid sequence, which acts on the same substrate(s) to produce the same product(s). Different CYP450s act on different substrates (although overlap does exist) or different CYP450s bind the same substrate but make different products. Hence, they are not true isoenzymes by the classic definition, and the term isoform has come into favor.

Although drug metabolism usually leads to detoxication, the processes of oxidation, reduction, and other enzyme-catalyzed reactions may lead to the formation of a metabolite having therapeutic or toxic effects. This process is often referred to as bioactivation.

The pathways of xenobiotic metabolism have been divided into two major categories. Phase 1 reactions (biotransformations) include oxidation, hydroxylation, reduction, and hydrolysis. In these enzymatic reactions, a new functional group is introduced into the substrate molecule, an existing functional group is modified, an acceptor site for phase 2 transfer moieties is exposed, thus making the xenobiotic more polar and therefore more readily excreted. Phase 2 reactions (conjugation) are enzymatic syntheses whereby a functional group is masked by the addition of a new group, for example, acetyl, sulfate, glucuronic acid, or certain amino acids, which further increases the polarity of the drug or xenobiotic. Most substances undergo both phase 1 and phase 2 reactions, sequentially. Those xenobiotics that are resistant to metabolizing enzymes or are already hydrophilic are excreted largely unchanged. Generally, most of these reactions occur at the more reactive centers of the molecule, such as hydroxyl or amino groups, or aromatic rings. This basic pattern of xenobiotic metabolism is common to all animal species, including man, but species may differ in details of the reaction and enzyme control.

Several factors influencing xenobiotic metabolism include:

1. **Genetic factors.** Species differences are seen in the biotransformation and conjugation of xenobiotics. Individual variations may also result from genetic differences in the expression of the metabolizing enzymes (see section on genetic polymorphism).
2. **Physiologic factors.** Age is a factor because the very young and the old have impaired metabolism. Hormones (including those induced by stress), sex differences, pregnancy, changes in the intestinal microflora, diseases (especially those involving the liver), and nutritional status can also influence xenobiotic metabolism.
3. **Pharmacodynamic factors.** Dose, frequency, and route of administration, plus tissue distribution and protein binding of the drug, affect its metabolism.
4. **Environmental factors.** Competition with other drugs and xenobiotics for the metabolizing enzymes and poisoning of enzymes by toxic chemicals, such as carbon monoxide or pesticide synergists, alter metabolism. Induction of enzyme expression (the number of enzyme molecules increased but activity is constant) by other drugs and xenobiotics is another consideration.

Such factors may change not only the kinetics of an enzyme reaction, but also the whole pattern of metabolism, thereby altering the pharmacologic activity or the toxicity of a xenobiotic. Species differences in response to xenobiotics must be considered in the extrapolation of pharmacologic and toxicologic data from experiments in animals to humans. The primary factors in these differences are probably the rate and pattern of drug and xenobiotic metabolism in the various species.

One of the earliest examples of bioactivation was the reduction of Prontosil to the antibacterial agent sulfanilamide. Other examples of drug metabolism leading to therapeutically active drugs include the hydroxylation of acetanilid to acetaminophen, and the N-demethylation of the antidepressant impiramine to desipramine and the anxiolytic diazepam to desmethyldiazepam. The insecticide parathion is desulfurized by both insects and mammals to paraoxon.

Most drugs and other xenobiotics are metabolized by enzymes normally associated with the metabolism of endogenous constituents, e.g., steroids and biogenic amines. The liver is the major site of drug metabolism, although other xenobiotic-metabolizing enzymes are found in nervous tissue, kidney, lung, plasma, and the gastrointestinal tract (digestive secretions, bacterial flora, and the intestinal wall).

Because the liver is the principal site for xenobiotic and drug metabolism, liver disease should have important effects on the metabolism of drugs and on the duration of drug action.[4,5] Several factors identified as major determinants of the metabolism of a drug in the diseased liver are the nature and extent of liver damage, the drug involved, the dosage regimen, and the degree of participation of the liver in the pharmacologic activity of the drug. Liver disease affects the elimination half-life of some drugs but not of others, although all undergo hepatic biotransformation (Table 8–1). Some results have shown that the capacity for drug metabolism is impaired in chronic liver disease, which could lead to drug overdosage. Consequently, because of the unpredictability of drug effects in the presence of liver disorders, drug therapy under these circumstances is complex, and more than usual caution is needed.[5]

Table 8–1. The Effect of Liver Disease in Humans on the Elimination Half-life of Various Drugs

Difference Reported	No Difference Reported
Acetaminophen (±)*	Chlorpromazine
Amylbarbital	Dicoumarol
Carbenicillin	Phenytoin
Chloramphenicol	Phenylbutazone
Clindamycin	Salicylic acid
Diazepam	Tolbutamide
Hexobarbital	
Isoniazid	
Lidocaine	
Meperidine	
Meprobamate	
Pentobarbital	
Phenobarbital	
Prednisone	
Rifamycin	
Tolbutamide	
Theophylline	

* ±, drugs for which clearance is disputable but may be increased.

Table 8–2. Examples of Drugs Exhibiting First Pass Metabolism

Acetaminophen	Methyltestosterone
Albuterol	Metoprolol
Alprenolol	Nifedipine and other dihydropyridines
Aspirin	Nortriptyline
Cyclosporin	Organic nitrates
Desmethylimipramine	Oxprenolol
Fluorouracil	Pentazocine
Hydrocortisone	Propoxyphene
Imipramine	Propranolol
Isoproterenol	Salicylamide
Lidocaine	Terbutaline
Meperidine	Verapamil

Many reactions occurring in the gastrointestinal tract are associated with the bacterial and other microflora of the tract (see section on extrahepatic metabolism).[6] When drugs are administered orally, or when there is considerable biliary excretion of a drug or its metabolites into the gastrointestinal tract, e.g., with a parentally administered drug, the bacterial flora can be involved in drug metabolism. The activity or orally administered conjugated estrogens, e.g., Premarin, involves the hydrolysis of the sulfate conjugates by sulfatases, releasing estrogens to be reabsorbed from the intestine into the portal circulation. Other ways in which bacterial flora can affect metabolism are the following: (1) production of toxic metabolites, (2) formation of carcinogens from inactive precursors, (3) detoxication, (4) exhibition of species differences in drug metabolism, (5) exhibition of individual differences in drug metabolism, (6) production of pharmacologically active metabolites from inactive precursors, and (7) production of metabolites not formed by animal tissues.

Substances influencing drug and xenobiotic metabolism (other than enzyme inducers) include lipids, proteins, vitamins, and metals. Dietary lipid and protein deficiencies diminish microsomal drug metabolizing activity. Protein deficiency leads to a reduction in hepatic microsomal protein and lipid deficiency; oxidative metabolism is decreased because of an alteration in endoplasmic reticulum membrane permeability affecting electron transfer. In terms of toxicity, protein deficiency would increase the toxicity of drugs and xenobiotics by reducing their oxidative microsomal metabolism and clearance from the body.

FIRST PASS METABOLISM

Although hepatic metabolism continues to be the most important route of metabolism for xenobiotics and drugs, other biotransformation pathways play a significant role in the metabolism of these substances. Among the more active extrahepatic tissues capable of metabolizing drugs are the intestinal mucosa, kidney, and lung (see the subsequent section on extrahepatic metabolism). The ability of the liver and extrahepatic tissues to metabolize substances to either pharmacologically inactive or bioactive metabolites before reaching systemic blood levels is called the presystemic first pass effect.

The intestinal mucosa is particularly important for orally administered drugs prone to microsomal oxidation and glucuronidation and sulfation conjugation pathways. Because the intestinal mucosa is enriched with glucuronosyl transferases and sulfotransferases, presystemic first pass metabolism of susceptible drugs results in their low systemic availability. Sulfation and glucuronidation are presystemic intestinal first pass metabolism major pathways in humans for terbutaline, albuterol, fenoterol, and isoproterenol. Low oral bioavailability for a given drug may be the result of either presynaptic intestinal metabolism (e.g., cyclosporin, dihydropyridine calcium antagonists, estrogens, and progestins) or hepatic first pass metabolism with cytochrome P450 monooxygenases or conjugation reactions. Other routes of administration for susceptible drugs have been investigated in an attempt to overcome the pronounced presystemic metabolism, (e.g., subcutaneous, intravenous, inhalation, and nasal).

Several orally administered drugs are known to undergo liver first pass metabolism during their transport to the systemic circulation from the gastrointestinal tract. Thus, the liver can remove substances from the blood after their absorption from the gastrointestinal tract, thereby preventing distribution to other parts of the body. This effect can seriously impair the bioavailability of an orally administered drug, reducing the amount of the drug that reaches the systemic circulation and ultimately its receptor to produce its pharmacologic effect. Drugs subject to first pass metabolism are listed in Table 8–2.

Studies are being performed to determine the effect of presynaptic and hepatic first pass metabolism on the toxicity and carcinogenicity of xenobiotics. For a nontherapeutic toxic substance, the existence of a first pass effect is desirable because the liver can bioinactivate it, preventing its distribution to other parts of the body. On the other hand, first pass metabolism may increase its toxicity by biotransforming the toxicant to a more toxic metabolite, which can re-enter the blood and exert its toxic effect. First pass metabolism may also occur by the enzymes in the mucosa of the gastrointestinal tract before reaching the systemic circulation. The extent of first pass metabolism depends on the drug delivery system, because a formulation may increase or decrease the rate of dissolution, the residence time of a drug in the gastrointestinal tract, and the dose. The more prolonged the residence time, the greater the efficiency of first pass metabolism. The drug form and delivery system should yield optimal bioavailability and pharmacokinetic profiles resulting in a reproducible clinical response.

ELIMINATION PATHWAYS

Most drugs and xenobiotics are lipid soluble and are altered chemically by the metabolizing enzymes, usually into less toxic and more water-soluble substances, before being excreted into the urine (or bile in some cases). The formation of conjugates with sulfate, amino acids, and glucuronic acid is particularly effective in increasing the polarity of drug molecules. The principal route of excretion of drug and their metabolites is in the urine. If drugs and other compounds foreign to the body were not metabolized in this manner, substances with a high lipid-water partition coefficient could be reabsorbed readily from the urine through the renal tubular membranes and into the plasma. Such substances would therefore continue to be recirculated and their pharmacologic or toxic effects would be prolonged. Very polar or highly ionized drug molecules are often excreted in the urine unchanged.

Tubular reabsorption is greatly reduced by conversion of a drug into a more polar substance with a lower partition coefficient. In general, the more resistance a drug has to the metabolizing enzymes, the greater its therapeutic action and the smaller the dose needed to achieve a particular therapeutic goal.

Urine is not the only route for excreting drugs and their metabolites from the animal body. Other routes include bile, saliva, lungs, sweat, and milk. The bile has been recognized as a major route of excretion for many endogenous and exogenous compounds. Thiocyanate, the detoxication product of cyanide, is excreted principally in the saliva. Some drugs may be excreted into the milk and affect the breast-fed baby.

Enterohepatic Cycling of Drugs

Steroid hormones, bile acids, drugs and their respective conjugated metabolites, when eliminated in the bile, are available for reabsorption from the duodenal-intestinal tract into the portal circulation, undergoing the process of enterohepatic cycling (EHC).[7] Nearly all drugs are excreted in the bile, but only a few are concentrated in the bile. For example, the bile salts are so efficiently concentrated in the bile and reabsorbed from the gastrointestinal tract that the entire body pool recycles several times per day. Therefore, EHC is responsible for the conservation of bile acids, steroid hormones, thyroid hormones, and other endogenous substances. In humans, compounds excreted into the bile usually have a molecular weight greater than 500, whereas with rats, the molecular weight is greater than 325. Compounds with a molecular weight between 300 and 500 are excreted in both urine and bile. Some compounds would not be expected to be excreted in the bile because of a molecular weight of less than 300 and a relatively nonpolar structure. Consequently, biliary excretion is more common in the rat than in humans. Compounds excreted into bile are usually strongly polar substances that are charged (anionic) (e.g., dyes) or uncharged (e.g., cardiac glycosides and steroid hormones). Biotransformation of this type of compound by means of phase 1 and phase 2 reactions would produce a conjugated metabolite, which is usually anionic, more polar, and of a molecular weight greater than that of the parent compound. They are most often present at their glucuronide conjugates, because glucuronidation adds 176 to the molecular weight of the parent compound. Unchanged drug in the bile is excreted with the feces, metabolized by the bacterial flora in the intestinal tract, or reabsorbed into the portal circulation.

Not unexpectedly, the bacterial intestinal flora plays a direct involvement in EHC and the recycling of drugs through the portal circulation (see subsequent discussions on extrahepatic metabolism). Conjugated drug and metabolites excreted in the bile may be hydrolyzed by enzymes of the bacterial flora releasing the parent drug, or its phase 1 metabolite, for reabsorption into the portal circulation.[6] Among the numerous compounds metabolized in the enterohepatic circulation are the estrogenic and progestational steroids, digitoxin, indomethacin, diazepam, pentaerythritol tetranitrate, mercurials, arsenicals, and morphine. The oral ingestion of drugs or other xenobiotics inhibiting the gut flora (i.e., nonabsorbable antibiotics) can affect the pharmacokinetics of the initial drug.

The impact of EHC on the pharmacokinetics and pharmacodynamics of a drug depends on the importance of biliary excretion of the drug relative to renal clearance and on the efficiency of gastrointestinal absorption. Enterohepatic cycling becomes dominant when biliary excretion is the major clearance mechanism for the drug. Because the majority of bile is stored in the gallbladder and released with the ingestion of food, intermittent spikes in the plasma drug concentration are observed following re-entry of the drug from the bile via EHC. From a pharmacodynamic point of view, the net effect of EHC is to increase the duration of a drug in the body and to prolong its pharmacologic action.

Chronic treatment with the enzyme inducer phenobarbital enhances the biliary excretion of some drug molecules and their metabolites by increasing liver size, bile flow, and more efficient transport into the bile. This behavior is not shared by all inducers of the cytochrome P450 monooxygenases. The route of administration may also influence excretion pathways. Direct administration into the portal circulation might be expected to result in more biliary excretion than could be expected via the systemic route.

DRUG METABOLISM AND AGE

By the year 2000, approximately 20% of the population will be 65 years of age or older and will be responsible for 40% of the national drug expenditures. People in this age range represent a significant portion of the population, are the most medicated, and receive 25% of all prescription drugs dispensed. The average elderly patient in a health care facility receives 10 medications daily, which results in the potential for a greater incidence of adverse drug reactions. The widespread use of medications in the elderly also results in a large incidence of drug-related interactions. Not unexpectedly, these interactions are related to changes in drug metabolism and clearance from the body (Table 8–3). The interpretation of the age-related alteration in drug response must consider the contributions of absorption, distribution, metabolism, and excretion.[8] Drug therapy in the elderly will become one of the more significant problems in clinical medicine. It

Table 8–3. Effect of Age on the Clearance of Some Drugs

No Change	Decrease
Acetaminophen (±)*	Acetaminophen (±)
Aspirin	Alprazolam
Diclofenac	Amitriptylene
Digitoxin	Carbenoxolone
Diphenhydramine	Chlordiazepoxide
Ethanol	Chlormethiazole
Flunitrazepam	Clobazam
Heparin	Desmethyldiazepam
Lormetazepam	Diazepam
Midazolam	Labetalol
Nitrazepam	Lidocaine
Oxazepam	Lorazepam
Phenytoin (±)	Morphine
Prazosin	Meperidine
Propylthiouracil	Nifedipine and other dihydropyridines
Temazepam	Norepinephrine
Thiopental (±)	Nortriptyline
Tolbutamide (±)	Phenytoin
Warfarin	Piroxicam
	Propranolol
	Quinidine
	Quinine
	Theophylline
	Verapamil

* ±, drugs for which clearance is disputable but may be increased.

is well documented that the metabolism of many drugs and their elimination is impaired in the elderly.

The decline in drug metabolism because of advanced age is associated with physiologic changes that have pharmacokinetic implications affecting the steady-state plasma concentrations and renal clearance for the parent drug and its metabolites.[9] Those changes relevant to the bioavailability of drugs in the elderly are decreases in hepatic blood flow, glomerular filtration rate, hepatic microsomal enzyme activity, plasma protein binding, and body mass. Because for any drug the rate of elimination from the blood through hepatic metabolism is determined by hepatic blood flow, protein binding, and intrinsic clearance, a reduction in hepatic blood flow can lead to an increase in drug bioavailability and decreased clearance, with symptoms of drug overdose and toxicity as the outcome. Drugs for which elimination depends on hepatic blood flow have a high extraction ratio and undergo extensive first pass metabolism when administered orally. Available evidence suggests that age is associated with a reduction in first pass metabolism of some but certainly not all drugs. Those drugs exhibiting such a reduction in the elderly include the dihydropyridine calcium antagonists, chlormethiazole, diazepam, lorazepam, chlordiazepoxide, alprazolam, propranolol, verapamil, labetalol, theophylline, morphine, amitriptyline, and nortriptyline when administered orally. The bioavailability of drugs with low extraction ratios depends on the percent of drug-protein binding and not on first pass hepatic metabolism. In as much as drug binding to plasma proteins is an important factor in the rate of drug metabolism, it appears not to be a significant factor in the elderly.

Age-related changes in drug metabolism are a complicated interplay between age-related physiologic changes, genetics, environmental influences (diet and nutritional status, smoking, and enzyme induction), concomitant disease states, and drug intake. In most studies, the elderly appear as responsive to drug-metabolizing enzyme activity (phase 1) as young individuals. All of the common phase 2 pathways of drug conjugation, including glucuronidation, sulfation, and glycine conjugation, are variably affected by aging. Given the number of factors that determine the rate of drug metabolism, it is not surprising that the effects of aging on drug elimination by metabolism have led to variable results even for the same drug. Therefore, the bioavailability of a drug in the elderly and the potential for drug toxicity depend upon its extraction ratio and mode of administration. Given the possibility that drug elimination is altered in old age, initial doses of metabolized drugs should be reduced in older patients and then modified according to the clinical response.[9] A decrease in hepatic drug metabolism coupled with age-related alterations in clearance, volume of distribution, and receptor sensitivity can lead to prolonged plasma half-life and increased drug toxicity (see Table 8–3).

The ability of the human fetus and placenta to metabolize drugs and xenobiotics is well established. A 1973 clinical study reported that women ingest an average of 10 drugs during pregnancy, not including anesthetics, intravenous fluids, vitamins, iron, nicotine, cosmetic products, artificial sweeteners, or exposure to environmental contaminants. The majority of these substances readily cross the placenta, thus exposing the fetus to a large number of xenobiotic agents. The knowledge of the effects of prenatal exposure to drugs, environmental pollutants (e.g., smoking), and other xenobiotics (e.g., ethanol) on the fetus has led to a decrease in the exposure to these substances during pregnancy. The human fetus is at risk from these substances because of the presence of only the cytochrome P450 monooxygenase 3A subfamily, which is capable of metabolizing xenobiotics during the first part of gestation. Placentas of tobacco smokers have shown a significant increase in the rate of placental cytochrome P450 monooxygenase activity (CYP1A subfamily). Concern for this type of enzyme activity is increasing because this enzyme system is known to catalyze the formation of reactive metabolites that are capable of binding covalently to macromolecules producing permanent effects (e.g., teratogenic, hepatotoxic, and/or carcinogenic) in the fetus and newborn. A more disturbing fact is that the other conjugation enzymes (i.e., glucuronosyl transferases, epoxide hydrolase, glutathione transferase, and sulfotransferase), which are important for the formation of phase 2 conjugates of these reactive metabolites, are found in low to negligible levels, increasing the exposure of the fetus to these potentially toxic metabolites.

Fetal drug metabolism functions either as a protective mechanism against environmental xenobiotics to transform active molecules into inactive molecules, or as a toxifying system when transforming innocuous substances into reactive molecules. The placenta is not a barrier that protects the fetus from xenobiotics, because almost every drug present in the maternal circulation crosses the placenta and reaches the fetus. Depending on the pharmacologic activity of the parent substance and/or its metabolites, both fetal and adult maternal drug metabolism may be viewed as complementary yet contradictory. Because metabolites are generally more water soluble than the parent substance, drug metabolites formed in the fetus may be trapped and accumulate on the fetal side of the placenta. Such accumulation can result in drug-induced toxicities or developmental defects. The difference between fetal and adult metabolism can be used advantageously, however, and constitutes the rationale for transplacental therapy; the administration of betamethasone several days before delivery can increase the production of surfactant in the fetal lung and prevent respiratory distress syndrome in the neonate.

The activity of CYP3A isoforms in the human fetal liver is similar to that seen in adult liver microsomes. The fetal activity for CYP3A7 isoform is unusual as most other fetal isoforms of cytochrome P450 exhibit 5 to 40% of the adult isoforms. Fetal and neonatal drug metabolizing enzyme activities may differ from those in the adult.

From birth, the neonate is exposed to drugs and other foreign compounds persisting from pregnancy as well as those transferred in breast milk. Fortunately, many of the drug-metabolizing enzymes operative in the neonate developed during the fetal period. The routine use of therapeutic agents during labor and delivery and during pregnancy is widespread, and the fact that potentially harmful metabolites can be generated by the fetus and newborn warrants consideration. Consequently, the use of drugs that are capable of forming reactive metabolic intermediates should be avoided during pregnancy, delivery, and the neonatal period. Activity of drug-metabolizing enzymes in liver microsomes of neonates is also increased after treatment during pregnancy with known enzyme inducers (e.g., phenobarbital). Important factors in any maternal-fetal exposure—drug, environmental, and/or chemical—that should be considered include the dose, frequency of administration, and timing during gestation, as well as genetic and fetal susceptibility.

DRUG BIOTRANSFORMATION PATHWAY (PHASE 1)

Human Hepatic Cytochrome P450 Enzyme System

Oxidation is probably the most common reaction in xenobiotic metabolism catalyzed by a group of membrane-bound mixed-function oxidases in the smooth endoplasmic reticulum of the liver and other extrahepatic tissues, called the cytochrome P450 monooxygenase enzyme system (the abbreviation CYP450 will be used for this enzyme system).[10] CYP450 is also known as mixed-function oxidase (MFO) or microsomal hydroxylase. The subcellular fraction containing the smooth endoplasmic reticulum is called the microsomal fraction. CYP450 functions as a multicomponent electron-transport system responsible for the oxidative metabolism of a variety of endogenous substrates (such as the steroids, fatty acids, prostaglandins, and bile acids) and exogenous substances (xenobiotics), including drugs, carcinogens, insecticides, plant toxins, environmental pollutants, and many other foreign chemicals. Central to the functioning of this unique superfamily of heme proteins is an iron-protoporphyrin coordinated to the sulfur of cysteine residue of the apoprotein. The ability of CYP450 to form a biologically inactive ferrous carbonyl complex with carbon monoxide, with its major absorption band at 450 nm, led to its discovery and name. CYP450 has an absolute requirement for NADPH (reduced for of nicotinamide adenine dinucleotide phosphate) and molecular oxygen (dioxygen). The rate at which various compounds are metabolized by this system depends on the species, strain, nutritional status, tissue, age, and pretreatment of the animals. The variety of reactions catalyzed by CYP450 (Table 8–4) include the oxidation of alkanes and aromatic compounds; the epoxidation of alkenes, polycyclic hydrocarbons, and halogenated benzenes; the dealkylation of secondary and tertiary amines and ethers; the deamination of amines; the conversion of amines to N-oxides, hydroxylamine, and nitroso derivatives; and the dehalogenation of halogenated hydrocarbons. It also catalyzes the oxidative cleavage of organic thiophosphate esters, the sulfoxidation of some thioethers, the conversion of phosphothionates to the phosphate derivatives, and the reduction of azo and nitro compounds to primary aromatic amines.

The most important function of CYP450 is its ability to "activate" molecular oxygen (dioxygen), permitting the incorporation of one atom of oxygen into an organic substrate molecule concomitant with the reduction of the other atom of oxygen to water (see Table 8–4). The introduction of a hydroxyl group into the hydrophobic substrate molecule provides a site for subsequent conjugation with hydrophilic compounds (phase 2), thereby increasing the aqueous solubility of the product for its transport and excretion from the organism. This enzyme system catalyzes xenobiotic transformations in ways that usually lead to detoxication, but in some cases, they lead to products with greater cytotoxic, mutagenic, or carcinogenic properties. Another enzyme, a nonheme, microsomal flavoprotein monooxygenase, is responsible for the oxidation of certain nitrogen and sulfur-containing organic compounds (see the section on flavin monooxygenase).

CYP450 consists of at least two protein components: a heme protein called cytochrome P450 and a flavoprotein called NADPH-cytochrome P450 reductase containing both FMN and FAD. Cytochrome P450 is the substrate- and oxygen-binding site of the enzyme system, whereas the reductase serves as an electron carrier, shuttling electrons one at a time from NADPH to cytochrome P450. A third component essential for electron transport from NADPH to cytochrome P450 is a lipid, phosphatidylcholine. The phospholipid does not function in the system as an electron carrier, but may facilitate the transfer of electrons from NADPH-cytochrome P450 reductase to cytochrome P450. For some isoforms, cytochrome b5 increases the rate of catalysis because of its ability to shuttle in the second electron.

Of the three components involved in microsomal oxi-

Table 8–4. Hydroxylation Mechanisms Catalyzed by CYP450

Aromatic hydroxylation	$CH_3CO-N(H)-C_6H_5 \xrightarrow{[OH]} CH_3CO-N(H)-C_6H_4\text{-}OH$
Aliphatic hydroxylation	$R-CH_3 \xrightarrow{[OH]} R-CH_2\text{-}OH$
Deamination	$R-CH(NH_2)\text{-}CH_3 \xrightarrow{[OH]} R-C(OH)(NH_2)\text{-}CH_3 \longrightarrow R-CO\text{-}CH_3 + NH_4^+$
O-Dealkylation	$R-O\text{-}CH_3 \xrightarrow{[OH]} R-O\text{-}CH_2\text{-}OH \longrightarrow R-OH + CH_2O$
N-Dealkylation	$R-N(CH_3)_2 \xrightarrow{[OH]} [R-N(CH_2OH)(CH_3)] \longrightarrow R-NH\text{-}CH_3 + CH_2O$
	$R-NH\text{-}CH_3 \xrightarrow{[OH]} [R-NH\text{-}CH_2OH] \longrightarrow R-NH_2 + CH_2O$
N-Oxidation	$(CH_3)_3\text{-}N \xrightarrow{[OH]} [(CH_3)_3-NOH] \longrightarrow (CH_3)_3-NO + H^+$
Sulfoxidation	$R-S-R' \xrightarrow{[OH]} [R-S(OH)-R'] \longrightarrow R-S(=O)-R' + H^+$

dative xenobiotic metabolism, cytochrome P450 is important because of its vital role in oxygen activation and substrate binding. The active site of cytochrome P450 consists of a hydrophobic substrate-binding domain in which is embedded an iron protoporphyrin (heme) prosthetic group exactly like that of hemoglobin, peroxidase, and the b-type cytochromes. X-ray studies reveal that in the ferric state, the two nonporphyrin ligands are water and cysteine thiolate (Fig. 8–1). Four of the coordination positions of iron are occupied by the tetradentate porphyrin ring. The fifth (proximal) ligand is cysteine thiolate in all states of the enzyme and is absolutely essential for the formation of the reactive oxenoid intermediate. The sixth (distal) coordination position is occupied by an easily exchangeable ligand, most likely water, which is labile and exchanged easily for stronger ligands such as CN-, amines, imidazoles, and pyridines. The ferrous form loses the water ligand completely, leaving the sixth position open for binding ligands such as O_2 and CO. The resting substrate-free protein is a low-spin hexacoordinate octahedral ferric complex. Substrate binding is coupled to the displacement of the weakly bound water and the formation of a substrate-bound pentacoordinate (bipyramidal) high-spin ferric complex, with a lower reduction

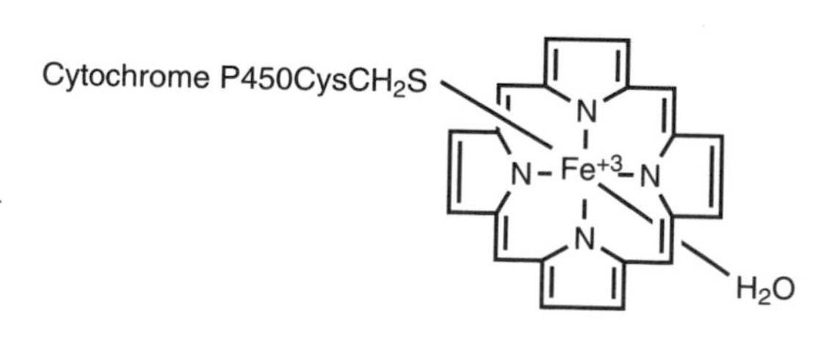

Fig. 8–1. Ferric heme thiolate catalytic center of cytochrome P450. The porphyrin side chains are deleted for clarity.

potential facilitating the reduction of Fe^{III} by the flavoprotein NADHP-cytochrome P450 reductase. Oxygen binding to reduced CYP450 is rapid.

The vast array of xenobiotics presents a unique challenge to the human body to metabolize these lipophilic foreign compounds, and makes it impractical to have one enzyme for each compound or each class of compounds. Therefore, whereas most cellular functions are usually very specific, xenobiotic oxidation requires CYP450s with diverse substrate specificities and regioselectivities (multiple sites of oxidation). Several types of CYP450 molecules can be found in a single species of animal. For example, the rat has more than 40 different CYP450 genes, each coding for a different version of the enzyme (isoform) that can metabolize almost any lipophilic compound to which they are exposed.

Nebert classified the cytochrome P450 supergene family on the basis of their structural (evolutionary) relationships.[11] The CYP450 monooxygenases resulting from this supergene family have been subdivided into families with greater than 40% amino acid homology and subfamilies with greater than 55% homology.[11] CYP450s are named using the root symbol CYP (CYtochrome P450), followed by an Arabic numeral designating the family member, a letter denoting the subfamily, and another Arabic numeral representing the individual gene. Thus, CYP1A1 is P450 form 1 in the A subfamily of family 1. Names of genes are written in italics. The nomenclature system is based solely on sequence similarity among the CYP450s and does not indicate the properties or function of individual P450s. The mammalian superfamily consists of six families (CYP7, CYP11, CYP17, CYP19, CYP21, and CYP27) involved in steroid and bile acid metabolism, and four families (CYP1, CYP2, CYP3, and CYP4) are almost exclusively responsible for xenobiotic metabolism (Table 8–5).[12] CYP450s probably evolved initially for the regulation of endogenous substances, such as metabolizing cholesterol to maintain membrane integrity and for steroid biosynthesis and metabolism, rather than for metabolizing foreign compounds. The CYP450s are either involved in highly specific steroid hydroxylations located in the inner mitochondrial membrane or bound to the endoplasmic reticulum of the cell having broad substrate specificity. In evolutionary terms, CYP450s evolved from a common ancestor and only more recently during the last 100 million years have CYP450 genes taken on the role of producing enzymes for metabolizing a vast array of lipophilic foreign compounds. The emergence of the xenobiotic-CYP450 genes probably evolved from the steroidogenic CYP450s for enhancing animal survival by synthesizing new CYP450s for metabolizing plant toxins in the food chain. It is not surprising that animals and humans possess a large array of diverse CYP450 enzymes capable of handling a multitude of xenobiotics. Interindividual variation in the expression of xenobiotic-CYP450 genes (genetic polymorphism) or their inducibility may be associated with differences, for example, in individual susceptibility to cigarette smoke carcinogenesis. Certain CYP450 isoforms that clearly exhibit genetic polymorphism are known to metabolize and generally inactivate therapeutic agents. The extent of CYP450 polymorphisms in humans is being investigated to determine the risk or protection against cancer. Food mutagens are typically carcinogens in tissue, but they are activated by CYP1A2 in the liver and CYP3A. Specific forms of CYP450 in hepatic microsomes are regulated by hormones (e.g., CYP3A subfamily), and are induced or inhibited by drugs, food toxins, and other environmental xenobiotics (see the section on induction and inhibition of cytochrome P450 isoforms). Identification of a specific CYP450 isoform as the major form responsible for metabolism of a drug in humans permits reconciliation of its toxicity or other pharmacologic effects.

No evidence exists that the active oxygenating species differs between CYP450s, suggesting the substrate specificities, substrate affinity, regioselectivity, and rates of reaction are probably a consequence of topographic features of the active site of apoproteins.[13,14] Because a primary function of these enzymes is the metabolism of hydrophobic substrates, it is likely that hydrophobic forces are important in the binding of many substrates to the apoproteins. Nonspecific binding is consistent with the multiple substrate orientations in the active site necessary for the broad regioselectivities observed. A specific binding requirement would decrease the diversity of substrates. Some CYP450 isoforms have constrained binding sites and thus metabolize small molecules (e.g., CYP2E1), others have planar binding sites and only metabolize planar compounds (i.e., CYP1A2 metabolizes polycyclic aromatic hydrocarbons [PAH]), another subfamily of CYP450s have high affinities with specific apoprotein interactions (hydrogen bonds, ion-pair formation) for specific substrates (i.e., CYP2D6 metabolizes a variety of lipophilic amines), and yet another isoform has broader affinity for large lipophilic substrates (e.g., CYP3A4). If the CYP450 isoforms are tightly membrane bound, substrate access to the active site would be limited to compounds that can diffuse through the membrane, whereas a different CYP450 isoform may be bound less tightly and will metabolize hydrophilic compounds.

In the past, the CYP450s were often referred to as having broad and overlapping specificities, but it became apparent that the broad substrate specificity is attributed to multiple isoenzymic forms of CYP450. The phenotype of an individual with respect to the forms and amounts of the individual CYP450s expressed in the liver can determine the rate and pathway of the metabolic clearance of a compound (see discussion on genetic polymorphism). Significant differences exist between humans and animal species with respect to the catalytic activities and regulation of the expression of the hepatic drug metabolizing CYP450s. These differences often make it difficult to extrapolate to humans the results of CYP450-mediated metabolism studies performed experimentally in animal species. Caution is warranted in the extrapolation of rodent data to humans, because some isoforms are similar between species (e.g., CYP1A and CYP3A subfamilies), whereas other subfamilies are different (e.g., CYP 2A, 2B, 2C, and 2D).

The unique and diverse characteristics of the CYP450 ensure that predicting the metabolism of xenobiotics will be difficult. To date, no crystal structure for a mammalian

Table 8–5. cDNA Expressed Human Cytochrome P450 Isoforms

P450	Synonyms	Tissue	Reaction Type	Substrates
CYP1A1	P_1-450, MC-5, P450c	Many		Text
CYP1A2	P450$_{PA}$, P_3, P450d, HLd, LM_4, MC-1	Liver	Table 8–6	Table 8–6
CYP2A6	IIA3, P450(1)	Liver Lung	Table 8–6	Table 8–6
CYP2B6	LM2, IIB1, PB-4, hIIB, P450$_{PB}$	Liver		
CYP2C8	mp-12, pB8, IIC2	Liver	Table 8–8	Table 8–8
CYP2C9	mp-1, P450$_{MP}$, mp-4, IIC1	Liver	Table 8–8	Table 8–8
CYP2C10	P450$_{TB}$, mp, mp-8	Liver	Table 8–8	Table 8–8
CYP2C18		Liver	Table 8–8	Table 8–8
CYP2C19		Liver		
CYP2D6	P450$_{DB}$, db1	Liver Kidney	Table 8–9	Table 8–9
CYP2E1	P450$_j$, j, HLj, LM3a	Liver	Table 8–10	Table 8–10
CYP2F1	IIF1	Lung		
CYP3A3	HLp	Liver Intestine		
CYP3A4	P450$_{NF}$, nf-25, hPCN1	Liver Extrahepatic	Table 8–12	Table 8–12
CYP3A5	hpCN3, HLp2, LM3c	Liver		
CYP3A7	HFLa	Fetal liver		
CYP4B1		Lung	ω- and ω-1 hydroxylation of fatty acids + prostaglandin	Prostaglandins Medium-long chain fatty acids
CYP7	P450$_{7\alpha}$	Liver	7α-Hydroxylation of cholesterol to bile acids	Cholesterol
CYP11A1	P450$_{scc}$	Adrenal cortex Mitochondria	Side chain oxidation of cholesterol to pregnenolone	Cholesterol
CYP11B1	P450$_{11\beta}$	Adrenal cortex Mitochondria	Steroid 11β-hydroxylation	11-Desoxycorticosterone
CYP17	P450$_{17\alpha}$	Liver	Steroid 17α-hydroxylation	Progesterone Pregnenolone
CYP19	P450$_{arom}$	Liver, placenta Ovaries	Aromatization of androgens to estrogens	Testosterone Androstenedione
CYP21A2	P450$_{C21}$	Adrenal Microsomes Liver	Steroid 21-hydroxylation	Progesterone 17α-Hydroxy-progesterone
CYP27		Liver	25-Hydroxylation of cholesterol	Cholesterol

membrane-bound CYP450 isoform has been described. The active oxygen is thought to be an iron-oxene species, and because most oxidations occur by one-electron process, the oxygen is likely to have a triplet state. If the catalytic cycle is initiated in the absence of substrate or if the substrate leaves the active site, uncoupling pathways are available to prevent the oxidation of active site amino acids and the self-destruction of the CYP450s. Uncoupling pathways include the destruction of superoxide and peroxide.

Cytochrome P450 Isoforms[15]

Family 1

The CYP1A subfamily plays an integral role in the metabolism of two important classes of environmental carcinogens, polycyclic aromatic hydrocarbons (PAH) and aryl amines (Table 8–6).[16] The PAH are commonly present in the environment as a result of industrial combustion processes and in tobacco products. Several potent carcinogenic aryl amines result from the pyrolysis of

Table 8–6. Substrates and Reaction Type for Human Subfamilies CYP1A and 2A

CYP1A2	CYP2A6
Phenacetin (O-deethylation)	Coumarin (7-hydroxylation)
Acetaminophen (benzoquinone imine)	Aflatoxin B1
Theophylline (N^1- and N^3-demethylation)	
Caffeine (N^1- and N^3-demethylation)	
Imipramine (N-demethylation)	
Estradiol (2- and 4-hydroxylation)	
Antipyrine (N-demethylation)	
Fluoroquinolones (3′-hydroxylation of piperazine ring)	
Fluvoxamine	

amino acids in cooked meats and can cause colon cancer in rats. Environmental and genetic factors can alter the expression of this subfamily of these enzymes. CYP1A1 (also called aromatic hydrocarbon hydroxylase, AHH) is expressed only in the liver, small intestine, placenta, skin, and lung in response to the presence of inducers such as PAH (i.e., in cigarette smoke, and the carcinogen 3-methylcholanthrene), β-naphthoflavone (a noncarcinogenic inducer related to dietary flavones), and indole-3-carbinol (found in Brussels sprouts and related vegetables). CYP1A1 metabolizes a range of PAH, including a large number of procarcinogens and promutagens. Diethylstilbestrol and catecholestrogens (2- and 4-hydroxyestradiol) are oxidized by CYP1A1 to their quinone analogs, which are normally reduced to inactive metabolites.[17] In the absence of a detoxifying reduction step, however, the quinones may accumulate and initiate carcinogenic processes or cell death by covalently damaging DNA or cellular proteins. Interindividual variation in the inducible expression of CYP1A1 might be related to a genetic difference in Ah receptor expression, which could explain differences in individual susceptibility to cigarette smoke-induced lung cancer. Therefore, genetic factors appear to be important in the expression of the CYP1A1 gene in humans and its involvement in human carcinogenesis. Women who smoke are at greater risk than men of developing lung cancer (adenocarcinoma) and chronic obstructive pulmonary diseases. Interestingly, CYP1A1 activity was present in the small intestinal mucosa of untreated rats, suggesting the presence of low levels of chemical inducers in the rat chow. The mechanism for the induction of the CYP1A1 gene begins with binding of the inducing agent(s) to a cytosolic receptor protein, the Ah receptor, which is translocated to the nucleus and binds to the DNA of the CYP1A1 gene, thus enhancing its rate of transcription. The presence of the Ah receptor in hepatic and intestinal tissues may have implications beyond xenobiotic metabolism and may play a role in the induction of other genes for the control of cellular growth and differentiation. On the other hand, CYP1A1 may metabolize procarcinogens to hydroxylated inactive compounds that are not mutagenic. The question of how the bowel protects itself from ingested compounds known to be activated by CYP1A1 (i.e., PAH) remains unanswered.[16]

CYP1A2 (also known as phenacetin O-deethylase, caffeine demethylase, or antipyrine N-demethylase) metabolizes aryl amines, nitrosoamines, and aromatic hydrocarbons, including the bioactivation of promutagens and procarcinogens such as aflatoxin B1 and the aryl amines, caffeine, and other substances (Tables 8–6 and 8–7). It is expressed in the liver, intestine, and stomach, and is induced by smoking, PAH, and isosafrole (a noncarcinogenic dietary compound). CYP1A2 is primarily responsible for the activation of the carcinogen aflatoxin B1 under ordinary conditions of human exposure. The pneumotoxin ipomeanol is activated by CYP1A2 in the liver, not as originally thought by the primary lung CYP450s CYP2F1 and CYP4B1. Evidence for polymorphism of this isoform has been reported, and it is likely that low CYP1A2 activity will be associated with altered susceptibility to the bioactivation of procarcinogens, promutagens, and other xenobiotics known to be substrates for this enzyme. The expression of the CYP1A2 gene in the stomach becomes an important issue for gastric carcinogenesis induced by smoking and the metabolic activation of the procarcinogens, aryl amines, to mutagens.[16] Clinical studies have suggested that the N-demethylation of imipramine is greater in smokers than in nonsmokers.

Family 2

CYP2A6 carries out the 7-hydroxylation of coumarin (coumarin 7-hydroxylase) and aflatoxin B1 (Table 8–6), as well as the bioactivation of nitrosoamines, and procarcinogens. The CYP2A subfamily hydroxylates testosterone at its 7α- and 15α-position in the rat and mouse but not in humans.

The transcriptional controlled expression of the human CYP2B6 gene into an isoform has not been positively demonstrated, even though phenobarbital appears to induce the formation of CYP2B6 isoform. The role of CYP2B6 subfamily in human drug metabolism is questionable and the induction of the CYP3A subfamily by phenobarbital in humans may ultimately be responsible for many of the well-documented interactions between barbiturates and other drugs.[15] Rat CYP2B1 ($CYP450_{PB}$) is 76% similar to human CYP2B6. The human CYP2C subfamily consists of CYP2C8, CYP2C9, CYP2C10, CYP2C18, and CYP2C19, metabolizing a range of drugs and other compounds including (S)-warfarin, (S)-mephenytoin, and tolbutamide (Table 8–8). Polymorphism for

Table 8–7. Some Procarcinogens and Other Toxins Activated by Human Cytochrome P450s

CYP1A1	CYP1A2	CYP2E1	CYP3A4
Benzo[a]pyrene and other polycyclic aromatic hydrocarbons	4-Aminobiphenyl	Benzene	Aflatoxin B1
	2-Naphthylamine	Styrene	Aflatoxin G1
	2-Aminofluorene	Acrylonitrile	Estradiol
	2-Acetylaminofluorene	Vinylbromide	6-Aminochrysene
	2-Aminoanthracene	Trichloroethylene	Polycyclic hydrocarbon dihydrodiols
	Heteropolycyclic amines (2-aminoquinolines)	Carbon tetrachloride	
	Aflatoxin B1	Chloroform	
	Ipomeanol	Methylene chloride	
		N-nitrosodimethylamine	
		1,2-Dichloropropane	
		Ethyl carbamate	

the metabolism of (S)-mephenytoin and tolbutamide has been detected (see the section concerning genetic polymorphism). CYP2C9 and CYP2C10 are involved in tolbutamide methyl hydroxylation, and CYP2C9 is a factor in the 4′-hydroxylation of phenytoin and (R)-mephenytoin. CYP2C18 ((S)-mephenytoin hydroxylase) may be the isoform associated with the 4′-hydroxylation of (S)-mephenytoin. The CYP2C subfamily apparently is not inducible in humans.

CYP2D6 is one of the two most important isoforms with a distinct substrate specificity for metabolizing drugs and does not appear to be inducible. It metabolizes debrisoquine and at least 30 other drugs (Table 8–9). Debrisoquine metabolism is one of the best studied examples of metabolic polymorphism, with its molecular basis of defective metabolism well understood (see the discussion on genetic polymorphism). The isoform metabolizes a wide variety of lipophilic amines and is probably the only CYP450 for which a charged or ion-pair interaction is important for CYP2D6 substrate binding. It also appears to preferentially catalyze the hydroxylation of one enantiomer (stereoselectivity). Quinidine is an inhibitor of CYP2D6 and concurrent administration with CYP2D6 substrates results in increased blood levels and toxicity for these substrates. If the pharmacologic action of the CYP2D6 substrate depends on the formation of active metabolites, quinidine inhibition results in a lack of a therapeutic response. The interaction of two substrates for CYP2D6 can prompt a number of clinical responses. The first pass hepatic metabolism of one CYP2D6 substrate may be inhibited or its rate of elimination may be prolonged, resulting in higher plasma concentration, depending on which substrate has the better affinity for the isoform.

CYP2E1 plays an integral role in the metabolism of numerous halogenated hydrocarbons (including volatile general anesthetics) and a range of low-molecular weight organic compounds, including dimethyformamide, ace-

Table 8–8. Some Substrates and Reaction Type for Human Subfamily CYP2C

CYP2C8	CYP2C9,10	CYP2C18
Warfarin (7-hydroxylation)	Warfarin (7-hydroxylation)	(S)-mephenytoin (4′-hydroxylation)
Tolbutamide (methyl hydroxylation)	Tolbutamide (methyl hydroxylation)	Mephobarbital (side chain hydroxylation)
Retinoic acid	Retinoic acid	Omeprazole (hydroxylation)
Retinol	Retinol	Propranolol (side chain hydroxylation)
	Tolbutamide (methyl hydroxylation)	Diazepam (N-demethylation)
	Naproxen (O-demethylation)	Imipramine (N-demethylation)
	Ibuprofen (i-butyl hydroxylation)	Proguanil (cyclization)
	Diclofenac (4′-hydroxylation)	(R)-mephenytoin (N-demethylation)
	Tienilic acid (thiophene ring hydroxylation)	
	Delta-1 THC (7-hydroxylation)	
	Phenytoin (4-hydroxylation)	
	Hexobarbital (3′-hydroxylation)	
	(R)-mephenytoin (4′-hydroxylation)	
	Testosterone (16α-hydroxylation)	
	Phenylbutazone (4-hydroxylation)	
	Sulfinpyrazone	
	Chloramphenicol	

Table 8–9. Some Substrates and Reaction Type for Human CYP2D6 Isoform

Cardiovascular Drugs

Debrisoquine (4-hydroxylation)
Quinidine (hydroxylation)
Flecainide (O-dealkylation)
Propafenone (4-hydroxylation)
Mexiletine (4-hydroxylation and methyl hydroxylation)
Guanoxan (6- and 7-hydroxylation)
Indoramin (6-hydroxylation)
Lidocaine (3-hydroxylation)
Encainide (N-demethylation, O-demethylation)
Captopril

Beta-Adrenergic Blockers

Propranolol (4′-hydroxylation)
Bifuralol (1′-hydroxylation)
Metoprolol (O-demethylation)
Timolol (O-dealkylation)
Alprenolol (4-hydroxylation)

Antidepressants

Amitriptyline (10-hydroxylation)
Clomipramine (hydroxylation)
Nortriptyline (10-hydroxylation)
Imipramine (2-hydroxylation)

Other Psychotropic Drugs

Haloperidol (oxidation of reduced metabolite)
Clozapine (aromatic hydroxylation)
Methoxyphenamine (4-hydroxylation, N-demethylation)
Thioridazine (aromatic hydroxylation)
Perphenazine
Methoxyamphetamine (O-demethylation and B-oxidation)

Opioids

Codeine (O-demethylation)
Dextromethorphan (O-demethylation)

Miscellaneous Drugs

Perhexiline (4′-hydroxylation)
Phenformin (4-hydroxylation)
Sparteine (N-oxidation)

tonitrile, acetone, ethanol, benzene, as well as in the activation of acetaminophen to the reactive metabolite N-acetyl-p-benzoquinoneimine and of other substances (Table 8–10).[18,19] This isoform is expressed in the liver, kidney, intestine, and lung, and is inducible by ethanol, isoniazid, and other chemicals (Table 8–11). It is also known as microsomal ethanol oxidizing system (MEOS), benzene hydroxylase, or aniline hydroxylase. The rat appears to be an excellent model for human CYP2E1 expression and function, but the rabbit is not. Most of the same compounds that induce CYP2E1 are also substrates for the enzyme. The induction of this enzyme in humans can cause enhanced susceptibility to the toxicity and carcinogenesis of CYP2E1 substrates. Some evidence shows interindividual variation in the in vitro liver expression of this isoform. Diabetes and dietary alterations (i.e., fasting, obesity) result in the induction of CYP2E1. Ketogenic diets (increased serum ketone levels), including those deficient in carbohydrates or high in fat, are known to enhance the metabolism of halogenated hydrocarbons in rats.[19] The mechanism of induction appears to be a combination of an increase in CYP2E1 transcription, m-RNA translation efficiency, and stabilization of CYP2E1 against proteolytic degradation. The induction of CYP2E1 resulting from ketosis (i.e., starvation, a high-fat diet, uncontrolled diabetes, obesity) or exposure to alcoholic beverages or other xenobiotics may be detrimental to individuals simultaneously exposed to halogenated hydrocarbons (increased hepatotoxicity from halothane, chloroform). Chronic alcohol intake is known to enhance the hepatotoxicity of halogenated hydrocarbons. Testosterone appears to regulate CYP2E1 levels in the kidney and pituitary growth hormone for regulating hepatic levels of CYP2E1. Kidney damage from halocarbons was greater for male rats but not for female rats.

Table 8–10. Some Substrates and Reaction Type for Human CYP2E1 Isoform

Acetaminophen (p-benzoquinone imine)
Chlorzoxazone (6-hydroxylation)
Styrene (epoxidation)
N-nitrosodimethylamine
Theophylline (C-8 oxidation)
Disulfiram
Methoxypsoralen

Halogenated Hydrocarbons

Chloroform (dehalogenation)
Carbon tetrachloride (dehalogenation)
Trichloroethylene
Methylene chloride
Vinyl chloride/bromide
Ethylene dibromide/dichloride (dehalogenation)

General Anesthetics

Enflurane
Halothane
Methoxyflurane
Sevoflurane
Desflurane

Miscellaneous Organic Solvents

Ethanol (to acetaldehyde)
Glycerin
Dimethylformamide (N-demethylation)
Acetone
Carbon disulfide
Diethylether
Benzene (hydroxylation)
Aniline (hydroxylation)
Acetonitrile (hydroxylation to cyanohydrin)
Pyridine (hydroxylation)

Table 8–11. Inhibitors and Inducers of CYP450 Subfamilies

Inducers	Inhibitors
CYP1A	**CYP3A**
3-Methylcholanthrene and PAH	Imidazole ("conazoles") antifungals (ketoconazole, fluconazole, clotrimazole, miconazole)
Tetrachlorodioxin	Erythromycin and analogs
Smoking	17α-Ethynylestradiol
β-Naphthoflavone	Triacetyloleandomycin
Indole-3-carbinol (from cruciferous plants)	Gestodene
CYP2E	Quinidine
Ethanol	**Other**
Dimethyl sulfoxide	Chloramphenicol (CYP2B, CYP2C)
Acetone	21-Halosteroids (CYP2C, CYP21)
Isoniazid	Quinidine (CYP2D6)
Pyrazoles	Cimetidine (nonselective)
CYP3A	Fluoroquinolones (CYP1A2)
Phenobarbital (and CYP2B, CYP2C)	Sulfinpyrazone (CYP2C9/10)
Phenytoin	Disulfiram (CYP2E1)
Carbamazepine	Methoxypsoralen (CYP2E1)
Dexamethasone (glucocorticoids)	Fluvoxamine (CYP1A2)
Rifampicin, rifabutin, and analogs	SKF525A (CYP2B, CYP2C, CYP3A)
Erythromycin, its analogs, and other macrolides	Spironolactone
Phenylbutazone	
"Conazoles" antifungals	
CYP4A	
Clofibrate and hypolipidemics	

This finding may have implications for sexual differences in the nephrotoxicity of CYP2E1 substrates in humans.

Family 3

The CYP3A subfamily includes the most abundantly expressed CYP450s in the human liver and intestine (see the section on extrahepatic metabolism), but only two forms have been characterized, CYP3A4 and CYP3A5. Approximately two thirds of the CYP450 in the liver is the CYP3A subfamily. CYP3A5 has been detected more in adolescents than in adults, is polymorphically expressed, and does not appear to be inducible, but CYP3A4 is glucocorticoid inducible. Its metabolic capabilities are limited and slower as compared to CYP3A4. CYP3A7 is expressed only in fetal livers (approximately 50% of total fetal CYP450 enzymes), and little is known about its substrate specificity, except of the 16-hydroxylation of dehydroepiandrosterone-3-sulfate, and its ability to hydroxylate allylic and benzylic carbons. The CYP3A subfamily metabolizes a range of clinically important drugs (Table 8–12) and is inhibited by a number of xenobiotics, including erythromycin (see Table 8–11). It also appears to activate aflatoxin B1 and possibly benzo[a]pyrene metabolism. The interindividual differences reported for the metabolism of nifedipine, cyclosporin, triazolam, and midazolam are probably related to changes in induction and not to polymorphism. CYP3A binding is predominantly lipophilic.[14] Drugs known to be substrates for CYP3A have a low and variable oral bioavailability that may be explained by prehepatic metabolism by the intestinal CYP3A subfamily. Therefore, it is the expression and function of the CYP3A subfamily that governs the rate and extent of metabolism of the substrates for the CYP3A subfamily.

Despite the explosion of information on the human CYP450s, 12 human xenobiotic-metabolizing CYP450 isoforms are characterized to varying degrees, but much is yet to be discovered. The vast data base for the CYP450s suggests that more human CYP450s exist and need to be characterized.[15] Because of the inconsistent supplies of human livers, model systems of human xenobiotic metabolism need to be developed and refined. It is clear that no one or combination of animal models reflects the metabolic capabilities of humans. By having a complete understanding of the factors (such as inducers, inhibitors, and effect of disease state) that alter the expression and activity of the enzyme responsible for the metabolism of a compound, it may be possible to predict drug interactions by the in vitro determination of the responsible isoforms and the phenotyping of patients, and eventually the metabolic clearance of the compound.

Catalytic Cycle of Cytochrome P450[20]

At the most fundamental level, the prosthetic groups of all enzymes and proteins that directly interact with molecular oxygen have a common characteristic, the ability to provide either low-energy d-molecular orbitals (i.e., metal ions such as iron and copper) for stabilizing unpaired electrons or extensively delocalized molecular orbitals (i.e., organic cofactors such as flavin, pterin, or porphyrin).

The unreactivity of elemental oxygen (dioxygen) is explained by its ground-state diradical (unpaired triplet) electronic configuration, represented as •O=O• or ↑O=O↑ to indicate the unpaired electron spins. Alternatively, singlet dioxygen, with paired electron spins (↑O=O↓) is too reactive. Free oxygen atoms (oxenes) are highly reactive and not known to occur in any biochemical process, although various oxenoid species have been invoked as reactive intermediates in the mechanisms of monooxygenases. Direct splitting of dioxygen into singlet (paired electron configuration ↑O↓) or triplet (unpaired electron ↑O↑ configuration) oxygen atoms is not a feasible way to circumvent the unreactivity of ordinary dioxygen. The inescapable conclusion is that the dioxygen molecule must be reduced to one of its more reactive relatives such as peroxide, hydroxyl radical, or oxygen atom:

$$O_2 + 2\,e^- + 2\,H^+ \longrightarrow H_2O_2 \qquad \text{peroxide}$$

$$O_2 + 2\,e^- + 2\,H^+ \longrightarrow 2\,HO\cdot + H_2O \qquad \text{hydroxide radical}$$

$$O_2 + 2\,e^- + 2\,H^+ \longrightarrow O + H_2O \qquad \text{oxygen atom}$$

In each case, two reduction equivalents are required. Any of the three reactive oxygen species shown in these equations could oxidize an organic substrate with the net insertion of an oxygen atom. Unlike oxygen transport proteins (e.g., hemoglobin), a monooxygenase can efficiently couple the reductive generation of a reactive oxygen intermediate to substrate oxidation as follows:

$$O_2 + 2\,e^- + 2\,H^+ \longrightarrow O + H_2O$$

$$O + RH \longrightarrow R\text{-}OH$$

$$\overline{O_2 + 2\,e^- + 2\,H^+ + RH \longrightarrow R\text{-}OH + H_2O}$$

The investment of two reduction equivalents from NADPH is the unavoidable cost of introducing an oxygen functionality at a relatively unreactive position in an organic substrate. The generation of a carbon-centered radical and a hydroxyl radical with triplet oxygen atom has been found to be relevant to a number of enzymatic and chemical reactions, involving oxenoids (oxygen rebound mechanism).[21]

The reason that reduced states of oxygen must occur during enzymatic processing lies partly in the unreactivity and triplet character of normal dioxygen. For example, during the initial reaction of triplet dioxygen with a singlet organic molecule such as hexane, a high energy, triplet transition state occurs. This transition state cannot collapse to a stable, bonded molecular product (i.e., hexyl hydroperoxide) until one of the triplet spins inverts itself, allowing spin pairing (i.e., chemical bonding) to occur. The high energy transition state never exists long enough for spin inversion to take place. Instead, the transitory molecular complex reverts to reactant molecules. This same situation exists when the organic molecule being attacked produces an unstabilized, high energy radical. On the other hand, if the resulting organic radical is stabilized, as for instance by α-heteroatoms or by extensive delocalization, the transition state complex has a sufficient lifetime for spin inversion to occur and for the addition of dioxygen to proceed. The transition metals, iron and copper, promote rapid spin inversion for the formation of covalent bonds to dioxygen. Thus, dioxygen is a thermodynamically poor one-electron oxidant.

The function of CYP450 monooxygenases is usually the hydroxylation of a substrate. A reactive radical-like iron oxenoid intermediate is generated, reactive enough to split aliphatic C-H bonds, add to π-bonds, or remove single electrons from heteroatoms. The mechanisms of cytochrome P450 are not fully understood and the reactive oxygen intermediate has not been isolated or even spectroscopically observed. The many variant isoforms isolated show a remarkable uniformity of the catalytic mechanism.

The current view illustrating the cyclic pattern of the reduction and oxygenation of cytochrome P450 as it interacts with substrate molecules, electron donors, and oxygen is shown in Figure 8–2 and can be summarized as follows:[20]

A. The ferric cytochrome P450 binds reversibly with a molecule of the substrate (RH) resulting in a complex (Fe^{III}*RH) analogous to an enzyme-substrate complex. The binding of the substrate facilitates the first one-electron reduction step.
B. The substrate complex of ferric-cytochrome P450 undergoes reduction to a ferrous-cytochrome P450 substrate complex (Fe^{II}*RH) by an electron originating from NADPH and transferred by the flavoprotein, NADPH-cytochrome P450 reductase, from the $FNMH_2$/FADH complex.

C. The reduced cytochrome P450 complex readily binds dioxygen as the ferrous iron sixth ligand to form oxycytochrome P450 ($Fe^{II}*O_2*RH$) complex.
D. Oxycytochrome P450($Fe^{II}*O_2*RH$) undergoes auto-oxidation to a superoxide anion ($Fe^{III}*O_2^{-1}*RH$)
E. The ferric superoxide anion ($Fe^{III}*O_2^{-1}*RH$) undergoes further reduction by accepting a second electron from the flavoprotein (or possibly cytochrome b5) to form the equivalent of a two-electron reduced complex, peroxycytochrome P450 ($Fe^{III}*O_2^{-2}*RH$). If the second electron is not delivered, the superoxide radial anion ($\bullet O_2^{-1}$ can disproportionate to hydrogen peroxide and dioxygen with regeneration of the ferric heme protein-substrate complex. This abortive "uncoupling" of the oxycytochrome P450 complex diverts the ternary complex of oxygen, heme protein, and substrate from its role in oxygen activation and subsequent substrate hydroxylation.

$$Fe^{II}*O_2 \longleftrightarrow Fe^{III}*O_2^{-1} \longrightarrow Fe^{III} + \bullet O_2^{-1}$$

$$2\ \bullet O_2^{-1} + 2\,H^+ \longrightarrow H_2O_2 + O_2$$

F. The ferric peroxycytochrome P450 complex undergoes heterolytic cleavage of peroxide anion to water and to a highly electrophilic perferryl oxenoid intermediate $(Fe^V{=}O)^{+3}$ or a perferryl oxygen-cysteine-porphyrin resonance-stabilized complex. This perferryl oxygen species represents the catalytically active oxygenation species.
G. Depending on the substrate, hydroxylation occurs in two steps involving the abstraction of a hydrogen atom to form a carbon-centered radical–perferric hydroxide pair $[R\bullet(Fe^{IV}\text{-}OH)^{+3}]$, radical addition of $(Fe^V{=}O)^{+3}$ to a π-bond, or electron abstraction from a heteroatom to form a heteroatom-centered radical-cation perferryl intermediate $[R\bullet(Fe^{IV}{=}O)^{+3}]$. Subsequent radical recombination (oxygen rebound) or electron-transfer (deprotonation (yields the hydroxylated product and the regeneration of the ferric cytochrome P450 enzyme complex.

Table 8–12. Substrates and Reaction Type for Human CYP3A4 Isoform

Nifedipine, nicardipine, felodipine, etc. (aromatization)
Cyclosporin (N-demethylation and methyl oxidation)
Erythromycin (N-demethylation) and analogs
Midazolam (methyl hydroxylation)
Dihydroergotamine (proline hydroxylation)
Codeine (N-demethylation)
Diazepam (C-7-hydroxylation)
Dextromethorphan (N-demethylation)
Quinidine (N-oxidation and C-3 hydroxylation)
Lidocaine (N-deethylation)
Diltiazem (N-deethylation)
Tamoxifen (N-demethylation)
Lovastatin (6-hydroxylation)
Verapamil (N-demethylation)
Delta-1 THC (6β-hydroxylation)
Amiodarone (N-deethylation)
Cocaine (N-demethylation)
Dapsone (N-oxidation)
Terfenadine (N-dealkylation, methyl hydroxylation)
Trazolam
Imipramine (N-demethylation)
Clotrimazole
"Conazole" antifungals
Rifampin, rifabutin, and related compounds
Glibenclamide
Valproic acid (hydroxylation and dehydrogenation)
Theophylline (C-8 oxidation)
Carbamazepine
Steroids
Testosterone (6β-hydroxylation)
Progesterone (6β-hydroxylation)
Estradiol (2- and 4-hydroxylation)
17α-Ethynylestradiol (2- and 4-hydroxylation)
Norethisterone (2-hydroxylation)
Hydrocortisone (6β-hydroxylation)
Prednisone (6β-hydroxylation)
Prednisolone (6β-hydroxylation)
Dexamethasone

At step F, it has been suggested that the electron-rich thiolate coordinated *trans* to the peroxide polarizes the peroxide O—O bond to promote heterolysis (asymmetric bond splitting). Heterolytic cleavage of the coordinated peroxide results in the displacement of H_2O, and the remaining iron-oxenoid complex would retain both of the oxidation equivalents of the original peroxide at the formal Fe^V oxidation state, the active oxygenation species. The chemical reactivity and steric requirements of $(Fe^V{=}O)^{+3}$ would control the attack on the substrate. Stereochemical experiments with the enzymes support the heterolytic mechanism. Currently, most evidence indicates that the reactive oxygen intermediate is the perferryl iron oxenoid complex, $(Fe^V{=}O)^{+3}$.

The expected chemical reactivity for $(Fe^V{=}O)^{+3}$ can be assessed by considering the various resonance forms possible when the ligated thiolate and porphyrin are included. The thiolate carries one negative charge and the porphyrin two, with a net zero electrical charge for the structures shown below:

$[\,PP\text{-}S\text{-}Fe^{III}\text{-}O\text{:}\,]^0 \longleftrightarrow [\,PP\text{-}S\text{-}Fe^{IV}\text{-}O\cdot\,]^0$

$\longleftrightarrow [\,PP\text{-}S\text{-}Fe^{V}\text{=}O\cdot\,]^0$

$[\,PP\text{-}S\cdot\ Fe^{V}\text{=}O\,]^0 \longleftrightarrow [\,PP\text{-}S\cdot\ Fe^{III}\text{-}O\cdot\,]^0$

$\longleftrightarrow [\,PP^{+}\text{-}S\cdot\ Fe^{III}\text{=}O\,]^0$

$[\,PP^{+}\text{-}S\text{-}Fe^{IV}\text{=}O\,]^0 \longleftrightarrow [\,PP^{+}\text{-}S\text{-}Fe^{III}\text{-}O\cdot\,]^0$ (PP = protoporphyrin IX)

The first structure is a useful representation of the intermediate as an oxygen atom coordinated to ferric iron. The oxygen atom contains both oxidation equivalents. Subsequent resonance forms move one or both of the oxidation equivalents to the iron, the sulfur, or the porphyrin π-system. This extensive delocalization stabilizes the high energy intermediate and puts radical character on virtually all atoms in the complex. The radical character of the oxygen atom explains how it is able to attack the substrate, although its presence in the porphyrin ring allows some substrate-derived radicals to covalently attach to the porphyrin ring through N-alkylation of a pyrrole nitrogen rather than to recombine with $(Fe^{IV}\text{-}OH)^{+3}$. This deviation from the normal course of reaction explains the suicide inhibition exhibited by some compounds, e.g., the oral contraceptives.[22]

Up to step G in the mechanism, the hydroxylatable substrate has mostly been an inactive spectator in the chemical events of oxygen activation. None of the preceding oxygenated intermediates has been sufficiently reactive to abstract hydrogen from the substrate, which has certainly been available in the active site. The perferryl iron oxenoid complex, however, is a competent hydrogen abstractor, even for relatively inert terminal methyl groups on hydrocarbon chains. Evidence shows that the oxidant is selective in its choice of hydrogen atoms, balancing stability of the resulting carbon radical with stereochemical constraints. The process of removal of the hy-

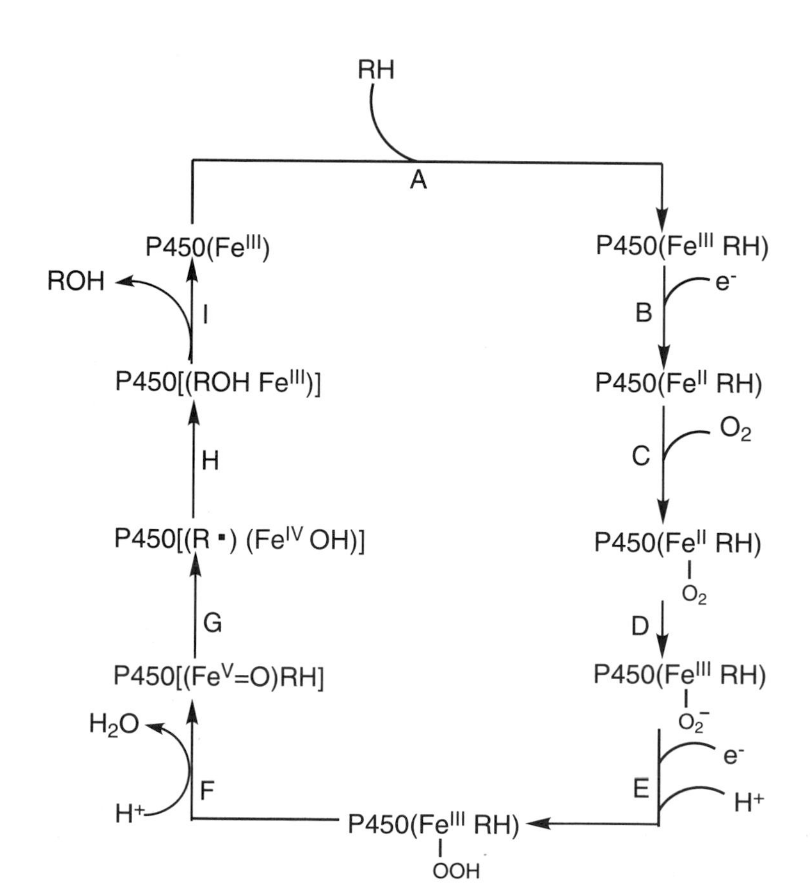

Fig. 8–2. The cyclic mechanism for cytochrome P450. The substrate is RH and the valence state of the heme iron in cytochrome P450 is indicated.

drogen atom and formation of a discrete carbon-centered radical has been documented experimentally. Because the inert aliphatic region of the substrate has been converted to a highly reactive radical, the process is described as substrate activation. Indeed, the transient existence of a substrate-derived radical opens the possibility for alternative pathways leading to abnormal products during P450-mediated oxidation. Following the abstraction of a hydrogen atom from the substrate, a transient diradical configuration exists (see Fig. 8–2). During this period, the carbon-centered substrate radical reorients and recombines with the oxenoid radical still coordinated to iron to form a covalent bond yielding the oxidized product ROH. The recombination event quenches all radicals in the active site. The freedom of movement afforded any particular substrate-derived radical is a function of its size and its overall complementarity to the enzyme active site shape. The resulting product, whether it be an alcohol, a secondary amine, or an N-oxide, diffuses from the enzyme (see Fig. 8–2, step J). A new cycle can be initiated by the binding of fresh substrate.

One problem faced by the radical-generating CYP450 monooxygenases is not in making the oxygen reactive but in controlling the reactive radicals produced. With some CYP450 isoforms (e.g., CYP3A isoforms), the steric constraints are weak, a consequence of their broad substrate selectivity and active site conformation. Therefore, it is common with these isoforms to observe several isomeric hydroxylated products and even multiple hydroxylations on a single substrate molecule. Substrate-derived radicals that are able to migrate from the point of their generation, however, could lead to unproductive and even highly destructive reactions. This problem has not yet been fully solved by nature. The radical mechanism requires a two-step reaction, implying a reactive intermediate. This intermediate radical is at the active sites of CYP450, and is the source of the toxic consequences associated with some of the substrates biotransformed. Thus, the ultimate price paid for the ability to hydroxylate aliphatic positions in a variety of xenobiotics includes not only the investment of NADPH and the extra trouble of providing associated electron-transfer proteins but also some loss of control of the outcome of the oxidative events.[23]

A unified theory of substrate activation reactions by cytochrome P450 proposes initial one-electron oxidation by $(Fe^{V}=O)^{+3}$ in all cases.[23] Various studies have shown that the hydroxylation proceeds not by a direct one-step insertion of the oxygen atom, but by a two-step, two-electron process involving radical and/or cationic substrate intermediates with subsequent radical recombination (oxygen rebound).[21] Depending on the substrate, the activation event would be hydrogen atom abstraction to a carbon-centered radical, radical addition of $(Fe^{V}=O)^{+3}$ to a π-bond, or electron removal from a heteroatom (N, O, P, S, Cl, Br, or I) to a heteroatom-centered radical-cation intermediate. The regio- and stereoselectivity of substrate hydroxylation are determined by the proximity of the oxidizable site to the perferryl oxygen and its reactivity potential and the structural features specific to each CYP450.[23]

Heteroatom-containing substrates usually undergo hydroxylation adjacent (α) to the heteroatom, as compared to other positions. Reactions of this type include N-, O-, S-dealkylation, dehydrohalogenations, and oxidative deamination reactions. Two mechanisms have been suggested. One is the abstraction of a hydrogen atom from the carbon adjacent to the heteroatom, and the resultant carbon radical is stabilized by the heteroatom. Alternatively, abstraction of an electron from the heteroatom to form a heteroatom radical cation subsequently transfers a hydrogen atom from the more labile α-carbon to generate a carbon radical. Collapse of the carbon radical-perferric hydroxide radical pair hydroxylates the carbon adjacent to the heteroatom, generating an unstable geminal hydroxy heteroatom-substituted intermediate (e.g., carbinolamine, halohydrin, hemiacetal, hemiketal, or hemithioketal) that breaks down, releasing the heteroatom and forming a carbonyl compound.[23]

Xenobiotics containing heteroatoms (N, S, P, and halogens) frequently are metabolized by heteroatom oxidation to its corresponding heteroatom oxide (tertiary amine to its N-oxide, sulfides to sulfoxides, phosphines to phosphine oxides). Heteroatom oxidation can also be attributed to a microsomal flavin-containing monooxygenase. As in the case for heteroatom α-hydroxylation, one electron oxidation of the heteroatom occurs as the first step to form the heteroatom cation perferric hydroxide radical intermediate, which collapses to generate the heteroatom oxide. This reaction is favored by the absence of α-hydrogens and stability of the heteroatom radical-cation.[23]

All of the known oxidative reactions catalyzed by CYP450 monooxygenase can be described in the context of a mechanistic scheme involving the ability of a high valent iron oxenoid species to bring about the stepwise one-electron oxidation through the abstraction of hydrogen atoms, abstraction of electrons from heteroatoms, or the addition of π-bonds. A series of radical recombination reactions completes the oxidation process.

Induction and Inhibition of Cytochrome P450 Isoforms

Induction

Many drugs, environmental chemicals, and other xenobiotics enhance the metabolism of themselves or of other co-ingested/inhaled compounds, thereby altering their pharmacologic and toxicologic effects.[24,25] Prolonged administration of a drug or xenobiotic can lead to enhanced metabolism of a wide variety of other compounds. Enzyme induction is a dose-dependent phenomenon.

Drugs and xenobiotics exert this effect by inducing transcription of CYP450 mRNA and synthesis of xenobiotic-metabolizing enzymes in the smooth endoplasmic reticulum of the liver and other extrahepatic tissues.[24] This phenomenon is called enzyme induction, and the term has been used to describe the process by which the rate of synthesis of an enzyme is increased relative to the rate of synthesis in the uninduced organism. In many older studies of mammalian systems, the term induction was inferred from the increase in enzyme activity, but the amount of enzyme protein had not been determined.

Enzyme induction is important in interpreting the results of chronic toxicities, mutagenicities, or carcinogenesis and explaining certain unexpected drug interactions in patients.

Many drugs and xenobiotics stimulate the activity of the CYP450 isoforms, as shown in Table 8–11. These stimulators have nothing in common as far as their pharmacologic activity or chemical structures are concerned, but they are all metabolized by one or more of the CYP450 isoforms. Most are lipid soluble at physiologic pH. Polycyclic aromatic hydrocarbons (PAH) in cigarette smoke, xanthines and flavones in foods, halogenated hydrocarbons in insecticides, polychlorinated biphenyls, and food additives are but a few of the environmental chemicals that alter the activity of CYP450 enzymes.[26]

Enzyme induction is important in evaluating the pharmacologic, toxicologic, and clinical implications for the therapeutic actions of a drug. As a result of induction, a drug may be either metabolized more rapidly to metabolites that are more potent or toxic than the parent drug or to less active metabolites. Induction can also enhance the activation of procarcinogens or promutagens. Not all inducing agents enhance their own metabolism; e.g., phenytoin induces CYP3A, but is hydroxylated by CYP2C9/10, which is constitutive. Some of the more common enzyme inducers of CYP450 subfamilies, which may also be substrates for the same CYP450 isoform, include phenobarbital (CYP2B, CYP2C, and CYP3A), phenytoin (CYP3A), rifampicin (CYP3A), and cigarette smoke (CYP1A) (see Table 8–11). The broad range of drugs metabolized by these CYP450 subfamilies (see Tables 8–6 to 8–12) and that are also affected by these enzyme inducers raises the issue of clinically significant drug interactions and their clinical implications. An example of a clinical CYP450 drug interaction involves rifampin and oral contraceptives. Rifampin induces the formation of CYP3A4, thereby reducing the serum levels of the oral contraceptive because of increased oxidative metabolism to less active metabolites, and increasing the risk for pregnancy. Drugs poorly metabolized by CYP450 enzymes are probably little affected by enzyme induction. Inducers of CYP450 isoforms also stimulate the oxidative metabolism or synthesis of endogenous substances, such as the hydroxylation of androgens, estrogens, progestational steroids (synthetic oral contraceptives), glucocorticoids, vitamin D, and bilirubin, decreasing their biologic activity. These enzyme inducers might also be implicated in deficiencies associated with these steroids and vitamin D. For example, the induction of C-2 hydroxylation of estradiol and synthetic estrogens by phenobarbital, dexamethasone, or cigarette smoking in women results in the increased formation of the principal and less active metabolite of these estrogenic substances, reducing their effectiveness.[27] Thus, cigarette smoking in premenopausal women could result in an estrogen deficiency, increasing the risk of osteoporosis and early menopause. Postmenopausal women who smoke and take estrogen replacement therapy may lose the effectiveness of the estrogen.

In addition to enhancing metabolism of other drugs, many compounds, when administered for a long time, stimulate their own metabolism, thereby decreasing their therapeutic activity and producing a state of apparent tolerance. This self-induction may explain some of the change in drug toxicity observed in prolonged treatment. The sedative action of phenobarbital, for example, becomes shorter with repeated doses and can be explained in part on the basis of increased metabolism.

The time course of induction varies with different inducing agents and different isoforms, except that CYP1A induction involves the Ah receptor. Increased transcription of CYP450 mRNA has been detected as early as 1 hour after the administration of phenobarbital, with maximum induction after 48 to 72 hours. After the administration of PAH, such as 3-methylcholanthrene and benzo-[a]pyrene, maximum induction of CPA1A subfamily is reached within 24 hours. Less potent inducers of hepatic drug metabolism may take as long as 6 to 10 days to reach maximum induction.[21]

Exposure to a variety of xenobiotics may preferentially increase the hepatic content of specific forms of CYP450.[24,25] Therefore, the process of enzyme induction involves the adaptive increase in the content of specific enzymes in response to the enzyme-inducing agent. The proposed mechanism of induction involves binding of the particular RNA polymerase to the promoter segment of the gene causing expression of the respective CYP450 structural gene with increased transcription of mRNA, resulting in increased CYP450 isoform synthesis. On the other hand, PAH interact with a specific cytoplasmic receptor (Ah receptor) that may enter the nucleus, analogous to steroid receptors. The PAH-receptor complex interacts with a regulatory site on the CYP1A gene, resulting in increased transcription of mRNA. Proliferation of the smooth endoplasmic reticulum may be regarded as a morphologic expression of enzyme induction. The induction of CYP450 is also associated with an increase in levels of NADPH-oxidase and NADPH-cytochrome P450 reductase in the endoplasmic reticulum. Other inducible metabolizing enzymes include UDP-glucuronosyl transferase and glutathione transferase.

Phenobarbital and rifampin are probably the enzyme inducers studied most extensively, and decrease the effects of many concurrently administered drugs listed in Tables 8–8 (CYP2C) and 8–12 (CYP3A).

Cigarette smoke has been shown to increase the hydrocarbon-inducible isoforms CYP1A1 and CYP1A2 in the lungs, liver, small intestine, and placenta of cigarette smokers. A decrease in the pharmacologic action and/or stimulation of the metabolism of several drugs is the end result. Cigarette smoking has been reported to lower the blood levels for theophylline, imipramine, estradiol, pentazocine, and propoxyphene; decrease urinary excretion of nicotine; and decrease drowsiness from chlorpromazine, diazepam, and chlordiazepoxide, although the plasma levels, half-life, or total clearance for diazepam are unchanged.

A diet containing Brussels sprouts, cabbage, and cauliflower stimulated monooxygenase activity in rat intestine.[29] It was subsequently determined that indole derivatives (indole-3-carbinol) were responsible for the enzyme induction. Other examples of chemicals found naturally in foods that enhance metabolism in animals are flavones, safrole, eucalyptol, xanthines, β-ionone, and organic peroxides. Volatile oils in soft woods (e.g., cedar) have been shown to be enzyme inducers.

Sober alcoholics show an increase in microsomal CYP2E1 enzyme activity, leading to more rapid clearance of drugs and xenobiotics that are substrates for this isoform from the body. As discussed previously, hepatic CYP2E1 oxidizes ethanol, and chronic ethanol intake increases the activity of CYP2E1 through enzyme induction.[18,19] When intoxicated, alcoholics are more susceptible to the action of various drugs because of inhibition of drug metabolism by the excessive quantity of alcohol in the liver and an additive or synergistic effect in the central nervous system. The basis for this inhibition is unknown. Furthermore, moderate ethanol consumption reduces the clearance of some drugs, presumably because of competition between ethanol and the other drugs for hepatic biotransformation. The changes in drug metabolism in alcoholics can also be attributed to other factors, such as malnutrition, other drugs, and the trace chemicals that determine the flavor and odor of alcoholic beverages. Heavy drinkers metabolize phenobarbital, tolbutamide, and phenytoin more rapidly than nonalcoholics, which may be clinically important because of problems in adjusting drug therapy in alcoholics.

Certain inducers of liver microsomes have been used therapeutically for hyperbilirubinemia in children with jaundice and for Cushing's syndrome. These observations suggest the possibility of treating some genetic diseases with suitable enzyme inducers.

Inhibition

Another method of altering the in vivo effects of drugs or other xenobiotics metabolized by CYP450s is through the use of inhibitors (see Table 8–11). The CYP450 inhibitors can be divided into three categories according to their mechanism of action: reversible inhibition, metabolite intermediate complexation of CYP450, or mechanism-based inactivation of CYP450.[25,28] The polysubstrate nature of CYP450 is responsible for the large number of documented interactions associated with the inhibition of drug oxidation and drug biotransformation.

Reversible inhibition of CYP450 is the result of reversible interactions at the heme-iron active center of CYP450, the lipophilic sites on the apoprotein, or both. The interaction occurs before the oxidation steps of the catalytic cycle, and their effects dissipate quickly when the inhibitor is discontinued. The most effective reversible inhibitors are those that interact strongly with both the apoprotein and the heme-iron. It is widely accepted that inhibition has an important impact on the oxidative metabolism and pharmacokinetics of drugs whose metabolism cosegregates with that of an inhibitor (see Tables 8–6 to 8–12).[28] Drugs interacting reversibly with CYP450 include the fluoroquinolone antimicrobials, cimetidine, the "conazole" antifungals (ketoconazole, fluconazole, and others), quinidine (specific for CYP2D isoforms), and diltiazem. Cimetidine is the only H-2 antagonist that inhibits CYP450 by interacting directly with the CYP450 heme-iron through one of its imidazole ring nitrogen atoms. Cimetidine is not a universal inhibitor of CYP450 oxidative metabolism, but it does bind differentially to several CYP450 isoforms. Cimetidine inhibits the oxidation of theophylline (CYP1A), chlordiazepoxide (CYP2C), diazepam (CYP2C), propranolol (CYP2C and CYP2D), warfarin (CYP2C), and antipyrine (CYP1A), but not that of ibuprofen (CYP2C), tolbutamide (CYP2C), mexilitine (CYP2D), 6-hydroxylation of steroids (CYP3A), and carbamazepine.[25] The imidazole-based conazole antifungals are potent inhibitors of CYP3A and of the CYP450-mediated biosynthesis of endogenous steroid hormones. The conazole antifungals exert their fungiostatic effects through inhibition of fungal P450, inhibiting the oxidative biosynthesis of lanosterol to ergosterol, thereby affecting the integrity and permeability of the fungal membranes.

Noninhibitory alkylamine drugs have the ability to undergo CYP450-mediated oxidation to nitrosoalkane metabolites (Fig. 8–3), which have a high affinity for forming a stable complex with the reduced (ferrous) heme intermediate of CYP450. This process is called metabolite intermediate complexation.[28] Thus, CYP450 is unavailable for further oxidation and synthesis of new enzyme is required to restore CYP450 activity. The process relies on at least one cycle of the CYP450 catalytic cycle to generate the required heme intermediate. The macrolide antibiotics, trioleandomycin, erythromycin, clarithromycin, and their analogs, are selective inhibitors of CYP3A4, and are capable of inducing the expression of hepatic and extrahepatic CYP3A4 mRNA and induction of their own biotransformation into nitrosoalkane metabolites. The clinical significance of this inhibition with CYP3A4 is the long-lived impairment of the metabolism of a large number of co-administered substrates for this isoform and the associated changes in pharmacokinetics for these drugs (see Table 8–12). For the macrolides to be so metabolized, they must possess an unhindered dimethylamino sugar and the whole compound must be lipophilic. Other alkylamine-based drugs demonstrating this type of inhibition include orphenadrine (antiparkinson drug) and SKF525A (the original CYP450 inhibitor), which generates complexes with CYP2B, CYP2C, and CYP3A subfamilies. Methylenedioxyphenyl compounds (i.e., the insecticide synergist piperonyl butoxide and the flavoring agent isosafrole) generate me-

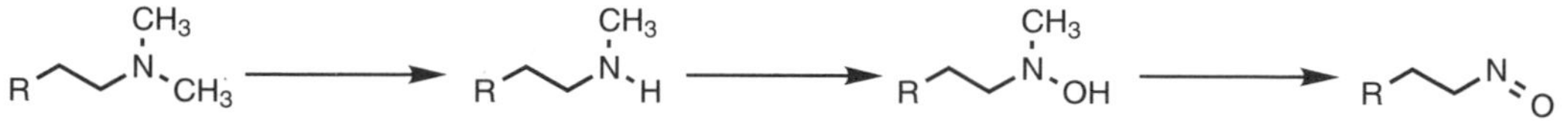

Fig. 8–3. The sequence of oxidation of dialkylamine to nitroso metabolite intermediate.

Table 8–13. Examples of Phase 1 Xenobiotic Oxidative Biotransformation Catalyzed by CYP450

Substrate	Product(s)
1. Side Chain Oxidation	
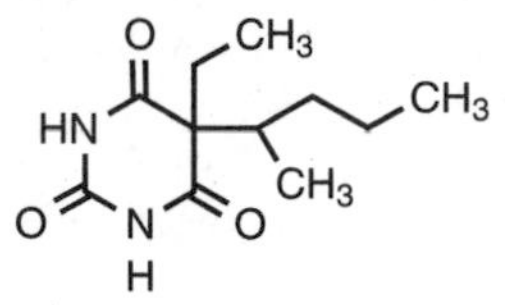 Pentobarbital	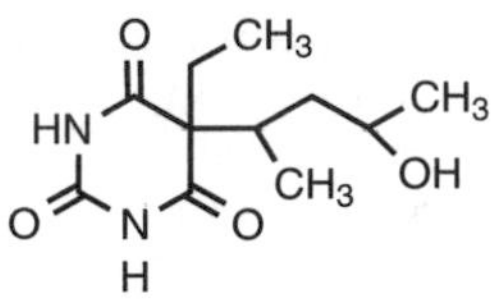
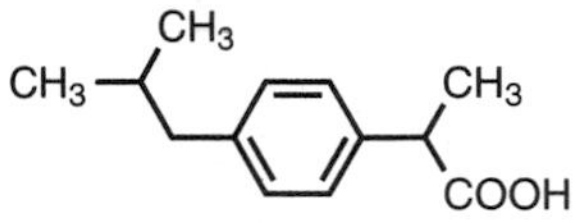 Ibuprofen	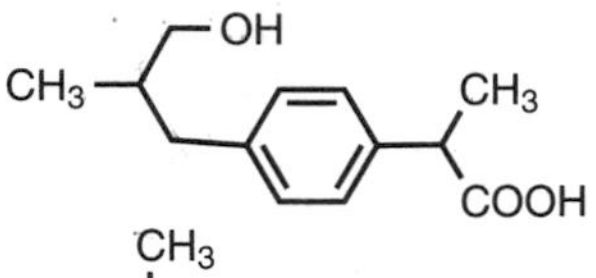 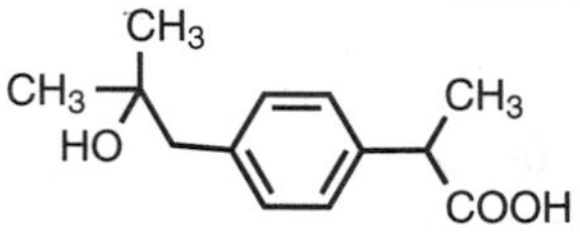
2. Aromatic Ring Oxidation	
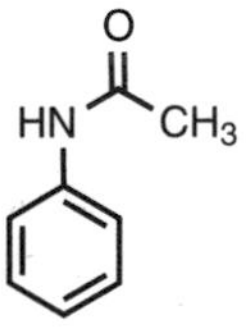 Acetanilide	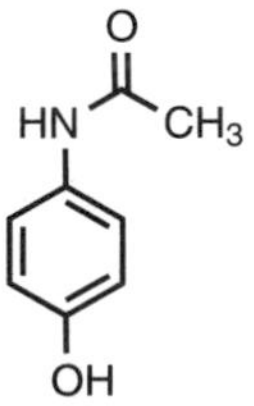 Acetaminophen
3. Methyl Oxidation	
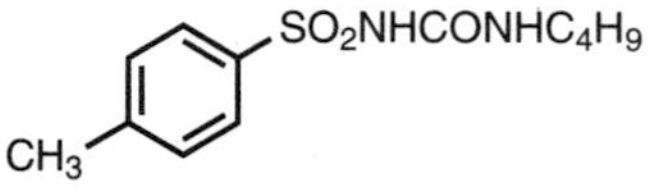 Tolbutamide (CH_3–C_6H_4–$SO_2NHCONHC_4H_9$)	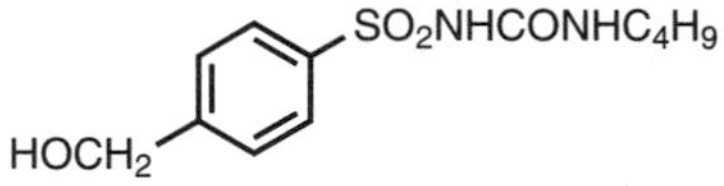 $HOCH_2$–C_6H_4–$SO_2NHCONHC_4H_9$
4. Heterocyclic Ring Oxidation	
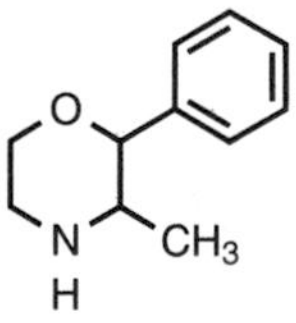 Phenmetrazine	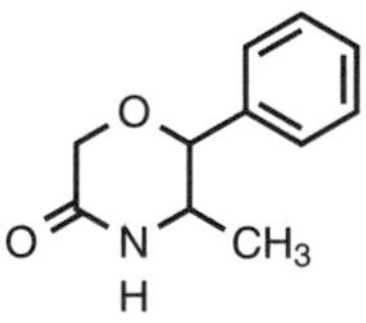
5. N-Dealkylation	
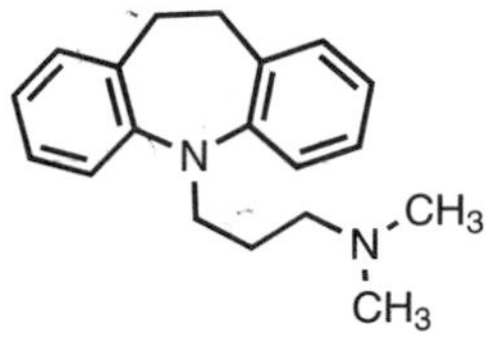 Imipramine	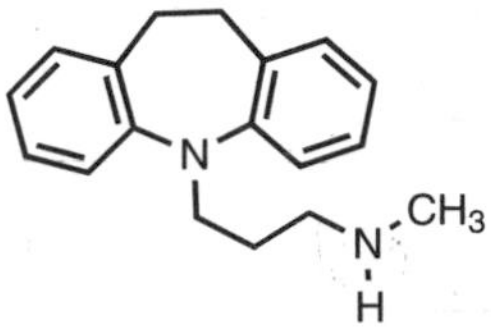 Desimipramine

Table 8–13. *(Continued)*

Substrate	Product(s)
6. O-Dealkylation	
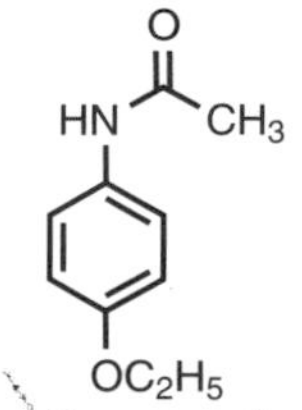 Phenacetin	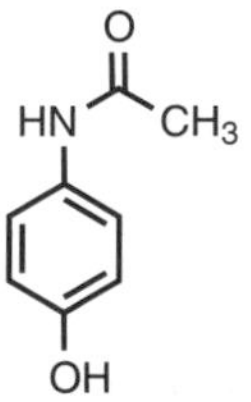 Acetaminophen
7. S-Dealkylation	
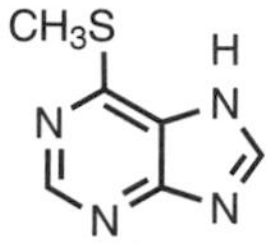 6-Methylmercaptopurine	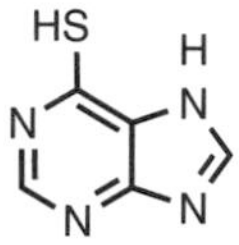 6-Mercaptopurine
8. Deamination	
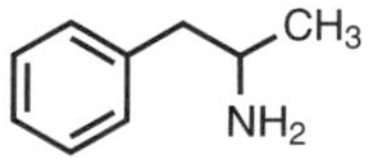 Amphetamine	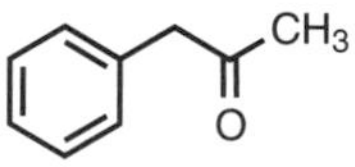 + NH_4^+
9. N-Oxidation	
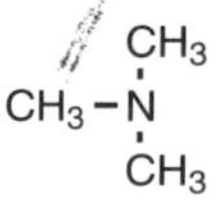 Trimethylamine	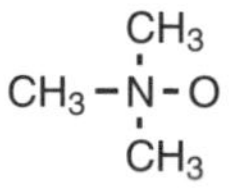 Trimethylamine oxide
10. Sulfoxidation	
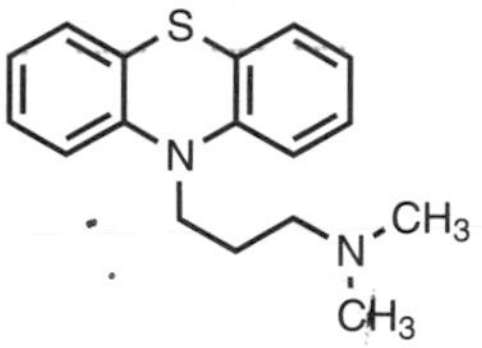 Chlorpromazine	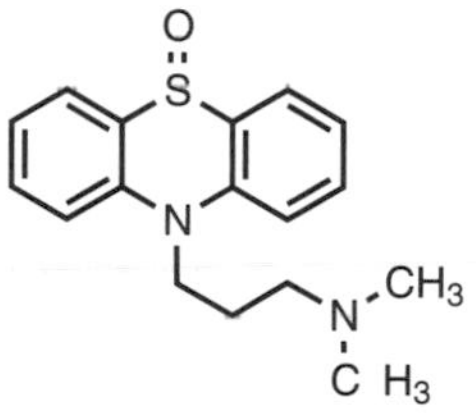 Chlorpromazine sulfoxide
11. Azoreduction	
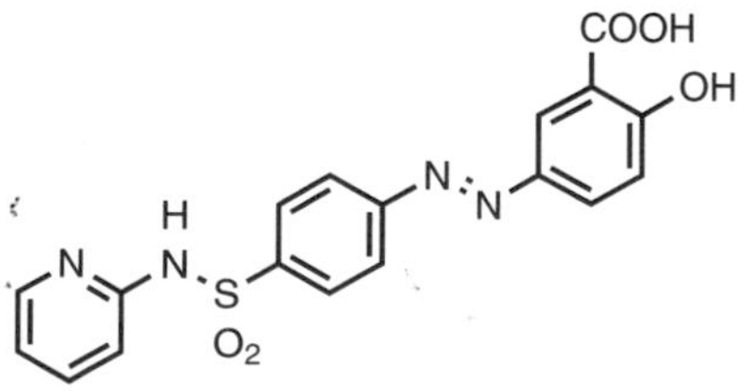 Azulfidine	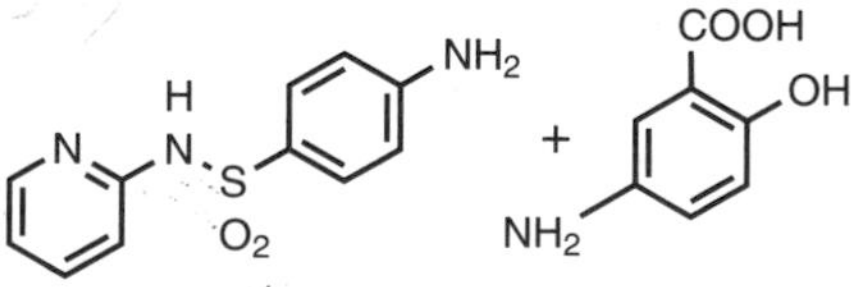 Sulfapyridine + p-Aminosalicylic acid

tabolite intermediates that form stable complexes with both the ferric and ferrous state of CYP450.

Certain drugs that are noninhibitory of CYP450 contain functional groups that, when oxidized by CYP450, generate metabolites that bind irreversibly to the enzyme. This process is called mechanism-based inhibition ("suicide inhibition") and requires at least one catalytic CYP450 cycle during or subsequent to the oxygen transfer step when the drug is activated to the inhibitory species. Alkenes and alkynes were the first functionalities found to inactivate CYP450 by generation of a radical intermediate that alkylates the heme structure (see the section on alkene and alkyne hydroxylation).[22,28] Iron is lost from the heme and abnormal N-alkylated porphyrins are produced. Drugs that are mechanism-based deactivators of CYP450 include the 17α-acetylenic estrogen, 17α-ethynylestradiol, the 17α-acetylenic progestin, norethindrone (norethisterone), and their radical intermediate that N-alkylates heme; chloramphenicol and its oxidative dechlorination to an acyl moiety that alkylates CYP450 apoprotein; cyclophosphamide and its generation of acrolein and phosphoramide mustard; spironolactone and its 7-thio metabolite that alkylates heme; 21-halosteroids; halocarbons; and secobarbital. The selectivity of CYP450 isoform destruction by several of these inhibitors indicates the involvement of this isoform in its bioactivation of such drugs.

Oxidations Catalyzed by Cytochrome P450 Isoforms

Aliphatic and Alicyclic Hydroxylations

The accepted mechanism for the hydroxylation of alkane carbon-hydrogen bonds proceeds by the stepwise abstraction of a hydrogen atom by the perferryl oxygen intermediate ([Fe^{V}=O]) to generate a carbon radical. This process is followed by recombination of the radical with the perferryl hydroxide intermediate ([Fe^{IV}-OH]) terminating the carbon radical (oxygen rebound), and generating the hydroxylated product (Fig. 8–4). The regioselectivity observed can be attributed to factors associated with substrate binding to the active site of the enzyme or with the reactivity of the carbon being hydroxylated.

The principal metabolic pathway for the methyl group is oxidation to the hydroxymethyl derivative followed by its nonmicrosomal oxidation to the carboxylic acid (e.g., tolbutamide, Table 8–13). On the other hand, some methyl groups are oxidized only to the hydroxymethyl derivative, without further oxidation to the acid. Where there are several equivalent methyl groups, usually only one is oxidized. For aromatic methyl groups, the para methyl is the most vulnerable.

Alkyl side chains are often hydroxylated on the terminal or the penultimate carbon atom (e.g., pentobarbital, Table 8–13). The isopropyl group is an interesting side chain showing hydroxylation at the tertiary carbon and at one of the equivalent methyl groups (e.g., ibuprofen, Table 8–13). Hydroxylation of alkyl side chains attached to an aromatic ring does not follow the general rules for alkyl side chains, because the aromatic ring influences the position of hydroxylation. Generally, oxidation occurs preferentially on the benzylic methylene group and to a lesser extent at other positions on the side chain.

The methylene groups of an alicycle are readily hydroxylated, generally at the least hindered position, and/or at an activated position, e.g., α to a carbonyl (cyclohexanone); α to a double bond (cyclohexene); α to a phenyl ring (tetralin). The products of hydroxylation often show stereoisomerism. Nonaromatic heterocycles generally undergo oxidation at the carbon adjacent to the heteroatom (e.g., phenmetrazine, Table 8–13).

The toxicity of alkylnitriles (isobutyronitrile > propionitrile > acetonitrile) is thought to result from cyanide release during oxidative metabolism by CYP2E1, which presumably occurs at the electron-deficient α-carbon to form an unstable cyanohydrin that decomposes and releases cyanide and an aldehyde (see reference 44).

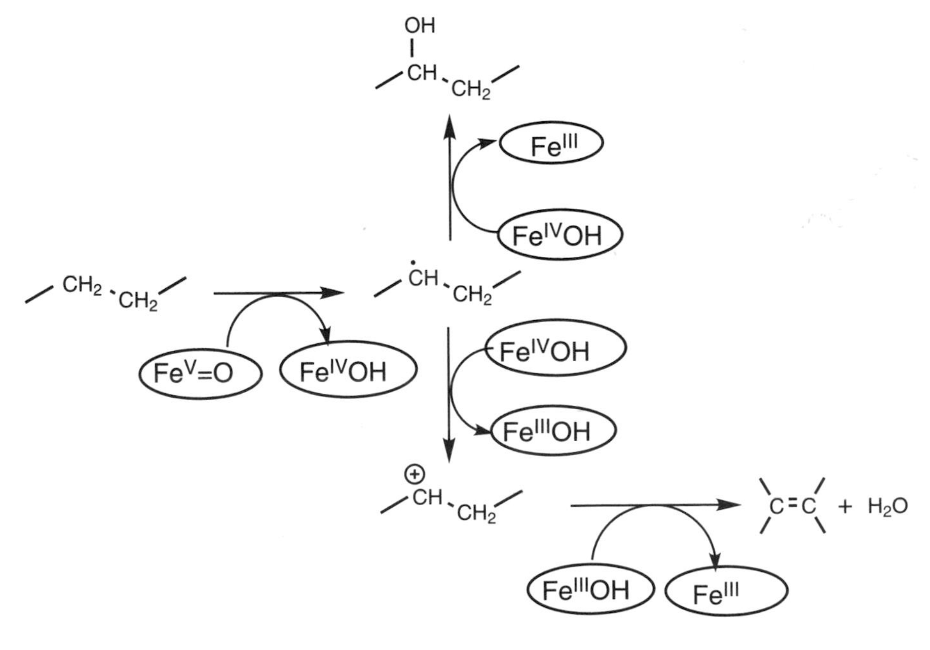

Fig. 8–4. Proposed mechanisms for the hydroxylation and dehydrogenation of alkanes.

In addition to hydroxylation reactions, CYP450s can catalyze the dehydrogenation of an alkane to an alkene (olefin) via an electron transfer between the carbon radical and perferryl complex. It is proposed that the carbon radical transfers an electron to the perferryl complex, generating a carbocation, which deprotonates to a dehydrogenated product(s) (alkene[s]) rather than a hydroxylated product (see Fig. 8–4).[23] Some evidence exists for the direct loss of the hydrogen radical to give an alkene. An example of the ability of CYP450 to function both as a dehydrogenase and monooxygenase has been demonstrated with the anticonvulsant valproic acid (VPA). At least four metabolic routes for VPA have been identified in humans:[30] β-oxidation and acylgluronidation are the major pathways, whereas the minor pathways include dehydrogenation to (E)2-ene, 4-ene, and (E)2,4-diene; hydroxylation to the 3-, 4-, and 5-hydroxyl products; and oxidation to 3-oxo VPA. Presumably, the CYP3A subfamily catalyzes these reactions. The (E)2-ene-VPA metabolite is reported to have anticonvulsant activity comparable to VPA, but only the 4-ene-VPA metabolite has been linked to hepatotoxicity, teratogenicity, and destruction of CYP450 (heme N-alkylation) via the carbon radical intermediate.[30] The factors determining whether CYP450 catalyzes hydroxylation (oxygen rebound/recombination) or dehydrogenation (electron transfer) remains unknown, but hydroxylation is generally favored. In some instances, dehydrogenation may be the primary product, i.e., 6,7-dehydrogenation of testosterone.

Alkene and Alkyne Hydroxylation

The oxidation of alkenes yields primarily epoxides and a series of products derived from 1,2-migration and N-alkylation of the heme-porphyrin ring (Fig. 8–5). Despite considerable experimental evidence, the proposed mechanism and intermediates of monooxygenation of alkenes and alkynes (acetylenes) remains controversial.[13,23] The proposed mechanism for the oxidation of π-bonds involves a stepwise sequence of one-electron transfer between the radical complex and the perferryl oxygen intermediate ($[Fe^V{=}O]$) (Fig. 8–5). Following the initial formation of an alkene-P450 π-complex, the one-electron transfer yields either a radical cation (pathway 1) or a radical σ-complex (pathway 2). The radical cation transfers an electron to the perferryl complex (α-deprotonation) and subsequent recombination with the perferryl hydroxide intermediate ($[Fe^{IV}\text{-}OH]$) terminates the radical generating a hydroxylated product. The radical σ-complex can either collapse to the epoxide or transfer an electron to generate a σ-complex, which can proceed to an epoxide, N-alkylation of the heme-porphyrin ring, or 1,2-group migration to a carbonyl product. The stereochemical configuration of the alkene is retained during epoxidation.

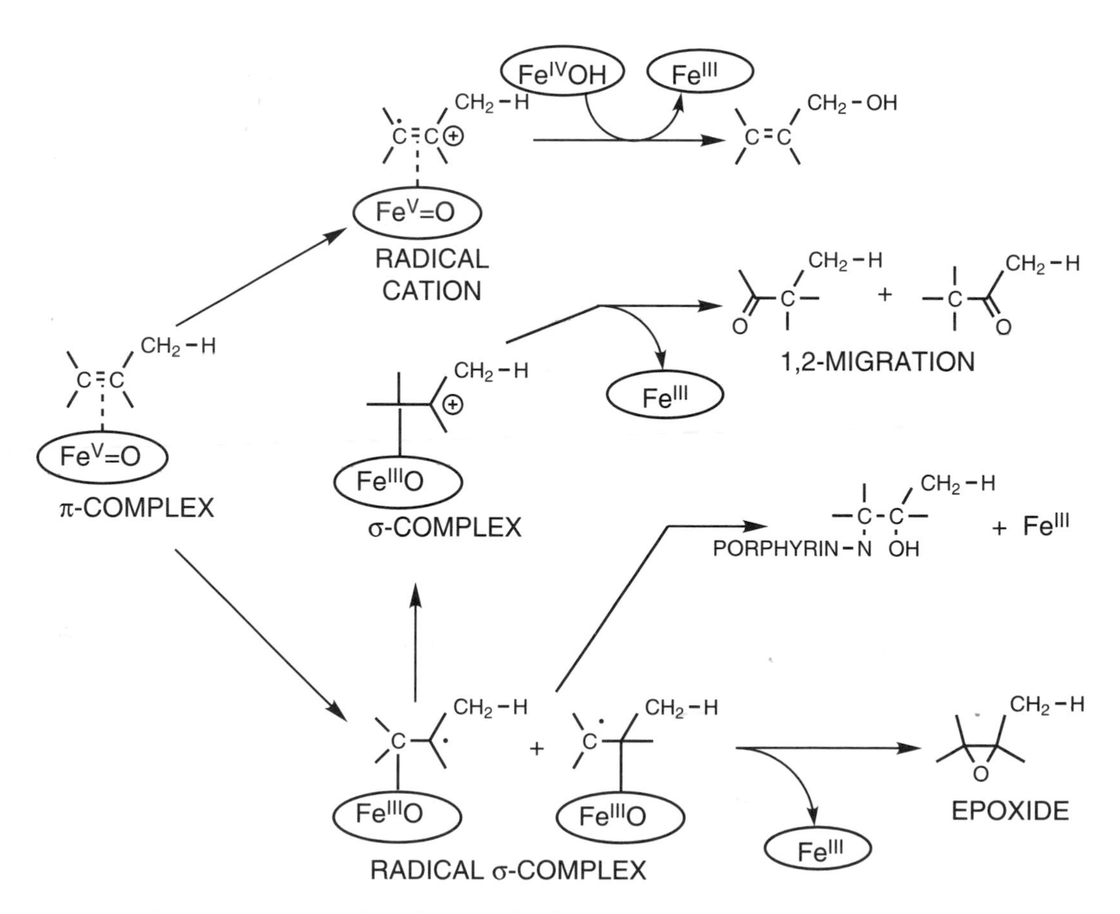

Fig. 8–5. Postulated mechanisms for the epoxidation and oxidation of alkenes.

The epoxides can differ in reactivity. Those that are highly reactive either undergo pH catalyzed hydrolysis to excretable vicinal dihydrodoils or react covalently (alkylate) with macromolecules such as proteins or nucleic acids leading to tissue necrosis or carcinogenicity. Moreover, the ubiquitous epoxide hydrolase can catalyze the rapid hydrolysis of epoxides to nontoxic vicinal dihydrodiols. Several drugs (carbamazepine, cyproheptadine, and protriptyline), however, were found to form stable epoxides at the 10,11-position during biotransformation. The fact that these epoxides could be detected in the urine indicates these oxides are not particularly reactive and should not readily react covalently with macromolecules.

The epoxidation of terminal alkenes is accompanied by the mechanism-based ("suicide") N-alkylation of the heme-porphyrin ring. If the σ-complex attaches to the alkene at the internal carbon, the terminal carbon of the double bond can N-alkylate the pyrrole nitrogen of the porphyrin ring.[22] The heme adduct formation is mostly observed with monosubstituted, unconjugated alkenes (i.e., 17α-ethylenic steroids and 4-ene metabolite of valproic acid).

In addition to the formation of epoxides, heme adducts, and hydroxylated products, carbonyl products are also created. These latter products result from the migration of atoms to adjacent carbons, 1,2-group migration. For example, during the CYP450 catalyzed oxidation of trichloroethylene ($Cl_2C = CHCl$), a 1,2-shift of chloride occurred to yield chloral (Cl_3CHO).

Like the alkenes, alkynes (acetylenes) are readily oxidized but usually faster. Depending on which of the two alkyne carbons are attacked, different products are obtained.[22] If attachment occurs on the terminal alkyne carbon, a hydrogen atom migrates, forming a ketene intermediate (R-CH=C=O) that readily hydrolyzes with water to form an acid or that can alkylate nucleophilic protein side chains (i.e., lysinyl or cysteinyl) to form a protein adduct (Fig. 8–6). The effect of attaching the perferryl oxygen at the internal alkynyl carbon is N-alkylation of a pyrrole nitrogen in the porphyrin ring by the terminal acetylene carbon, with the formation of a keto heme adduct (Fig. 8–6). The latter mechanism has been proposed for the irreversible inactivation of CYP3A4 with 17α-alkynyl steroids (i.e., 17α-ethynylestradiol).

Aromatic Hydroxylation

The metabolic oxidation of aromatic carbon atoms by CYP450 depends on the isoform catalyzing the oxidation and the oxidation potential of the aromatic compound. The products usually are phenolic products and the position of hydroxylation can be influenced by the type of substituents on the ring according to the theories of aromatic electrophilic substitution. For example, electron-donating substituents enhance p- and o-hydroxylation, whereas electron-withdrawing substituents reduce or prevent m-hydroxylation. Moreover, steric factors must also be considered, because oxidation usually occurs at the least hindered position. For monosubstituted benzene compounds, parahydroxylation usually predominates, with some ortho product being formed (e.g., acetanilid, see Table 8–13). When there is more than one phenyl ring, usually only one is hydroxylated (e.g., phenytoin).

The hydroxylation of aromatic compounds by CYP450 has traditionally been considered to be mediated by an arene oxide (epoxide) intermediate. This view has been favored since the discovery that the migration of groups from the para to the meta position of aromatic compounds occurs during para hydroxylation of aromatic compounds, the "NIH shift." The accepted mechanism of cytochrome P450-mediated enzymatic oxidation of aromatic compounds is similar to that for the alkenes, a stepwise one-electron transfer between the radical and/or cationic substrate intermediates and the perferryl oxygen intermediate with the formation of the arene oxide and other phenolic products (Fig. 8–7).[23] Following the initial formation of an arene-P450 π-complex, one-electron transfer yields either a σ-complex (pathway 1) or a radical σ-complex (pathway 2). The radical σ-complex can collapse to the arene oxide or transfer an electron

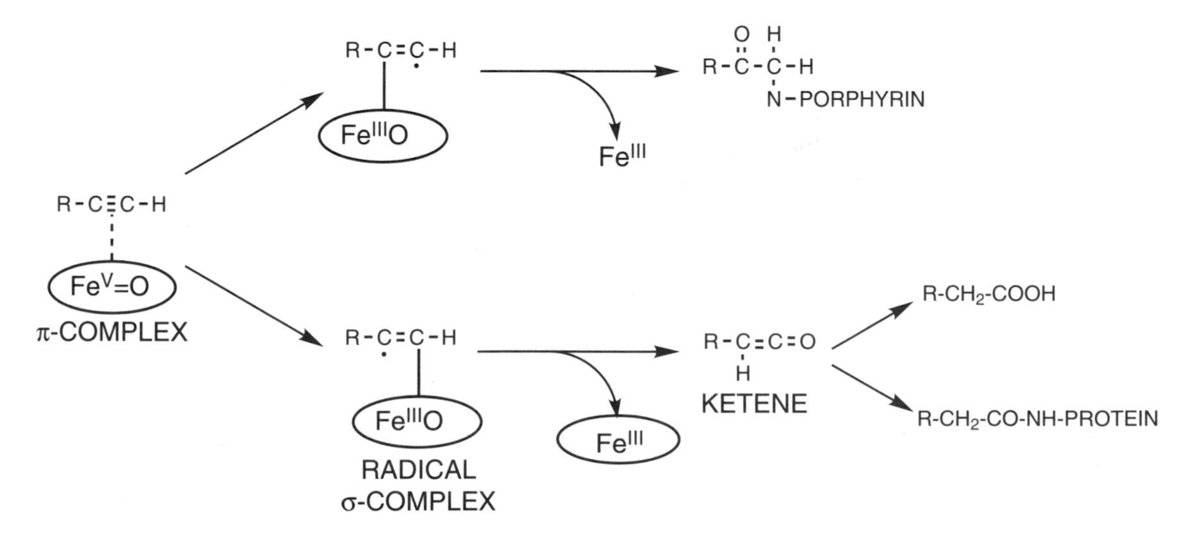

Fig. 8–6. Postulated mechanisms for the oxidation of alkynes.

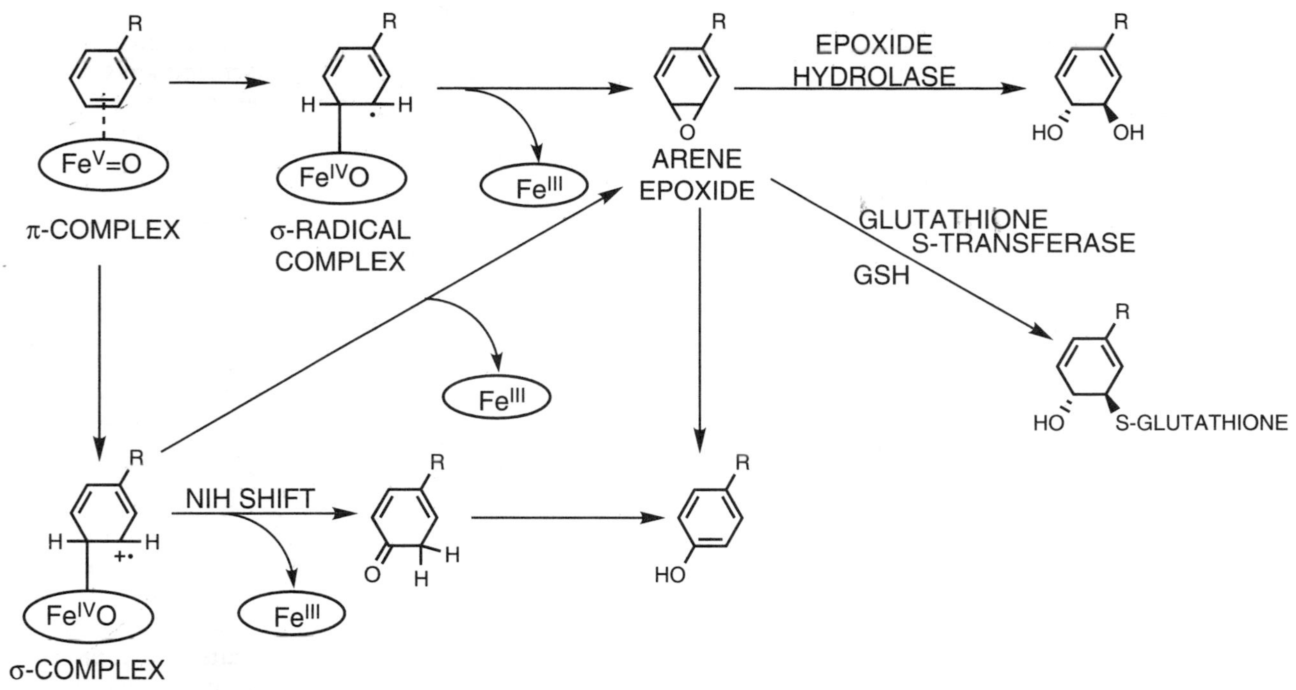

Fig. 8–7. Proposed mechanisms for the oxidation of aromatic compounds (arenes).

to generate a σ-complex, which can proceed to an arene oxide or NIH shift (1,2-group migration) to a phenolic product. Arene oxides are highly unstable entities and rearrange (NIH shift) nonenzymatically to phenols or hydrolyzed enzymatically with epoxide hydrolase to 1,2-dihydrodiols (*trans* configuration), which are dehydrogenated to 1,2-diphenols. The oxidation of aromatic compounds can be highly specific to individual CYP450 isoforms, suggesting that substrate binding and orientation in the active site may dominate the mechanism of oxidative catalysis.

Alternatively, the substituent attached to the aromatic ring (halogen, other heteroatoms, or alkyl) could undergo one-electron oxidation of the substituent to a radical cation intermediate. It would collapse to the meta-substituted cationic intermediate that can close to an arene oxide, analogous to that from the aromatic electrophilic substitution pathway.

The formation of phenols and the isolation of urinary dihydrodiols, catechols, and glutathione conjugates (mercapturic acid derivatives) implicates arene oxides as intermediates in the metabolism of benzene and substituted benzenes in mammalian systems. The arene oxides are also susceptible to conjugation with glutathione to form premercapturic acids (see the section on glutathione conjugation).

The relative importance of these three competitive pathways is influenced by the electronic properties (i.e., stability) of the arene oxide. Electron-withdrawing substituents stabilize arene oxides, whereas electron-donating substituents have a destabilizing effect. For example, benzene yields phenol as the major metabolite (minor amounts of catechol and mercapturic acid), suggesting an unstable epoxide that rearranges spontaneously, whereas halobenzenes and PAH afford 1,2-dihydrodiols and glutathione conjugates, suggesting a stable epoxide capable of being metabolized by epoxide hydrolase and glutathione S-transferase (see Fig. 8–7). It should be pointed out, however, that where other competitive pathways of biotransformation exist, the importance of arene oxide formation can be diminished. More vulnerable substituents will be metabolized preferentially, thus facilitating excretion. For example, toluene is metabolized mainly to benzoic acid.

The CYP1A2 and CYP3A subfamilies are important contributors to 2- and 4-hydroxylation of estradiol, and CYP3A4 for the 2-hydroxylation of the synthetic estrogens, for example, 17α-ethynylestradiol.[27] The principal metabolite (as much as 50%) for estradiol is 2-hydroxyestradiol, with 4-hydroxy and 16α-hydroxyestradiol as the minor metabolites. The 2-hydroxy metabolite of both estradiol and ethynylestradiol have limited or no estrogenic activity, whereas the C-4 and C-16α-hydroxy metabolites have a potency similar to estradiol. In humans, 16α-hydroxyestradiol is the major estrogen metabolite in pregnancy and in breast cancer. The metabolites 16α-hydroxyestrone and 4-hydroxyestrone may be carcinogenic in specific cells because they are capable of damaging cellular proteins and DNA after their further activation of quinone intermediates.

Xenobiotic-metabolizing enzymes not only detoxify xenobiotics but also cause the formation of active intermediates (bioactivation), which in certain circumstances may elicit a diversity of toxicities, including mutagenesis, carcinogenesis, and hepatic necrosis.[26] Some nucleophiles, in addition to glutathione, such as other sulfhydryl compounds (most effective), alcohols, and phosphates, can react with arene oxides. Many of these nucleophiles are found in proteins and nucleic acids. The covalent binding of these bioactive epoxides to intracellular macromolecules provides a molecular basis for these toxic effects (see the discussion on toxicity from oxidative metabolism).

N-Dealkylation and Oxidation

The dealkylation of secondary and tertiary amines to yield primary and secondary amines, respectively, is one of the most important and frequently encountered reactions in drug metabolism. The proposed mechanism for oxidative N-dealkylation involves abstraction of an electron from the nitrogen by the perferryl oxygen intermediate (Fig. 8–8), forming an aminium radical cation (1). The aminium radical cation transfers an electron from the adjacent carbon to form an α-amino carbon radical (2) (deprotonation) that collapses to an imminium ion (3) via another electron transfer. The imine subsequently hydrolyzes to a carbinolamine (4) and then to the dealkylated amine and a carbonyl product.[23] Evidence for a α-amino carbon radical has been provided by the isolation of ring expanded metabolites of 1-phenylcyclobutylamine and the irreversible inhibition of CYP450 (heme N-alkylation) by cyclopropylamines. Reportedly, the cyclopropane ring opens to a methylene radical that in turn N-alkylates the heme-porphyrin ring (i.e., mechanism-based inactivation).

Some of the N-substituents removed by oxidative dealkylation are methyl, ethyl, n-propyl, isopropyl, n-butyl, allyl, benzyl, and others having an α-hydrogen. Usually, dealkylation occurs with the smaller alkyl group initially. Substituents that are more resistant to dealkylation include the tertbutyl (no α-hydrogen) and the cyclopropylmethyl. In general, tertiary amines are dealkylated to secondary amines faster than secondary amines are dealkylated to primary amines. This difference in rate has been correlated with lipid solubility. Appreciable amounts of secondary and primary amines therefore accumulate as metabolites that are more polar than the parent amine, thus slowing their rates of diffusion across membranes and reducing their accessibility to receptors.

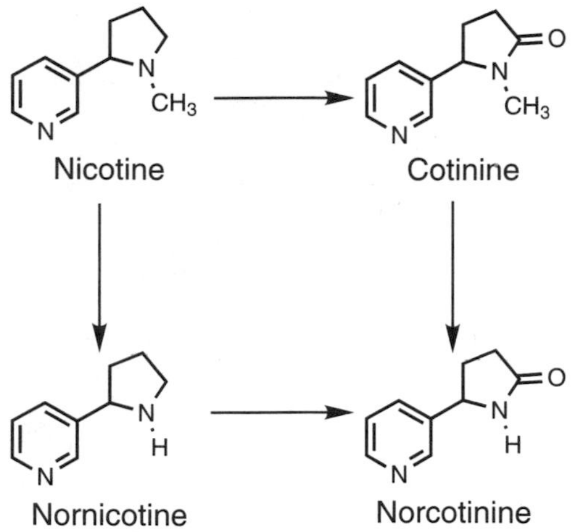

Fig. 8–9. The oxidative metabolism of nicotine to cotinine and their N-demethylated products.

Frequently, these amine metabolites contribute to the pharmacologic activity of the parent substance (e.g., imipramine, see Table 8–13) or produce unwanted side effects, such as hypertension, resulting from the N-dealkylation of N-isopropylmethoxamine to methoxamine. The design of an analogous drug without these unwanted drug metabolites can be achieved by proper choice of replacement substituents, for example, substituting the N-isopropyl group in N-isopropylmethoxamine with a tert-butyl (N-tert-butylmethoxamine or butoxamine). N-dealkylation of substituted amides and aromatic amines occurs in a similar manner. N-substituted nonaromatic nitrogen heterocycles undergo oxidation on the α-carbon as well as N-dealkylation (e.g., nicotine to cotinine, Fig. 8–9). The resultant carbinolamine can undergo further oxidation to a lactim (e.g., cotinine, Fig. 8–9).

In general, N-oxygenation of amines to stable N-oxides occurs with tertiary amines and less frequently with primary and secondary amines, and to hydroxylamines when no α-protons are available (e.g., arylamines) or the amin-

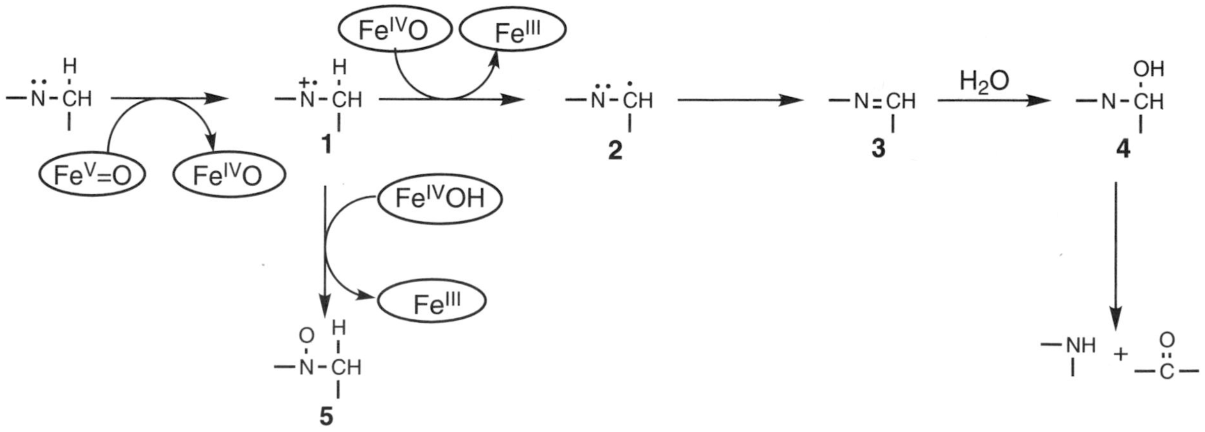

Fig. 8–8. Proposed mechanisms for CYP450 catalyzed N-dealkylation reactions.

ium radical cation is stabilized by neighboring electron donation. The proposed mechanism for the generation of N-oxides (*5*, Fig. 8–8) involves the transfer of an electron from the aminium radical cation to the perferryl hydroxide complex regenerating CYP450 (see subsequent section on N-oxidation with flavin monooxygenase). The dealkylation of tertiary amines via the N-oxide occurs through rearrangement of the N-oxide to the *α*-carbon generating a carbinolamine, which in turn collapses to the secondary amine. The tertiary amine must be capable of forming an N-oxide as well as having a hydrogen on the adjacent carbon for this mechanism to be operative. The amine metabolites can be N-conjugated, increasing their excretion.

O- and S-Dealkylation

Oxidative O-dealkylation of ethers is a common metabolic reaction with a mechanism of dealkylation analogous to that of N-dealkylation; oxidation of the *α*-carbon and subsequent decomposition of the unstable hemiacetal to an alcohol (or phenol) and a carbonyl product.[23] The substituent alkyl group leaves as a carbonyl derivative. Thioethers are also dealkylated by the same mechanism to hemithioacetals.

$$AR\text{-}OCH_3 \longrightarrow AR\text{-}\overset{\bullet +}{O}CH_2 \longrightarrow AR\text{-}OCH_2OH \longrightarrow AR\text{-}OH + CH_2O$$

The majority of ether groups in drug molecules are aromatic ethers, e.g., codeine, prazocin, and verapamil. For example, codeine is O-demethylated to morphine (see Table 8–13). The rate of O-dealkylation is a function of chain length, i.e., increasing chain length or branching reduces the rate of dealkylation. Steric factors and ring substituents influence the rate of dealkylation, but are complicated by electronic effects. Some drug molecules contain more than one ether group, in which case, usually only one ether is dealkylated. The methylenedioxy group undergoes variable rates of dealkylation to the 1,2-diphenolic metabolite, as well as being capable of forming a stable complex with and inhibiting CYP450.

Aliphatic and aromatic methyl thioethers undergo S-dealkylation to thiols and carbonyl compounds. For example, 6-methyl-thiopurine is demethylated to give the active anticancer drug, 6-mercaptopurine (see Table 8–13). Other thioethers are oxidized to sulfoxides, e.g., chlorpromazine (see Table 8–13).

Deamination

The mechanism of oxidative deamination follows a pathway similar to that of N-dealkylation. Initially, oxidation to the imminium ion occurs, followed by decomposition to the carbonyl metabolite and ammonia. Oxidative deamination can occur with *α*-substituted amines, exemplified by amphetamine (see Table 8–13). Disubstitution of the *α*-carbon inhibits deamination (e.g., phentermine). Some secondary and tertiary amines, and amines substituted with bulky groups, can undergo deamination directly, without N-dealkylation, for example, fenfluramine. Apparently, this behavior is associated with increased lipid solubility.

Dehalogenation

Many halogenated hydrocarbons, such as insecticides, pesticides, general anesthetics, plasticizers, flame retardants, and commercial solvents, undergo many different dehalogenation biotransformations.[18,19] Because of our potential exposure to these halogenated compounds as drugs and environmental pollutants in air, soil, water, or food, it is important to understand the interactions between metabolism and toxicity. Some form glutathione or mercapturic acid conjugates, whereas others undergo dehydrohalogenation and reductive dehalogenation catalyzed by CYP2E1. In many cases, reactive intermediates are produced that may react with a variety of tissue molecules.

Halogenated hydrocarbons differ in their chemical reactivity as a result of the electron-withdrawing properties of the halogens on adjacent carbon atoms, resulting in the *α*-carbon developing electrophilic character. The halogen atoms also have the ability to stabilize *α*-carbon cations, free radicals, carbanions, and carbenes.

Oxidative dehydrohalogenation is a common metabolic pathway for many halogenated hydrocarbons.[23] Cytochrome P450-catalyzed oxidation generates the transient *gem*-halohydrin (analogous to alkane hydroxylation) that can eliminate the hydrohalic acid to form carbonyl derivatives (aldehydes, ketones, acyl halides, and carbonyl halides). This reaction requires the presence of at least one halogen and one *α*-hydrogen. *gem*-Trihalogenated hydrocarbons (Fig. 8–10) are more readily oxi-

$$CHCl_3 \xrightarrow{P450} HO\text{-}CCl_3 \longrightarrow COCl_2 + Cl^-$$

$$COCl_2 + NH_2\text{-}PROTEIN \longrightarrow PROTEIN\text{-}NH\text{-}CO\text{-}NH\text{-}PROTEIN$$

$$COCl_2 + H_2O \longrightarrow CO_2 + 2HCl$$

Fig. 8–10. The CYP2E1-catalyzed oxidation of chloroform to phosgene.

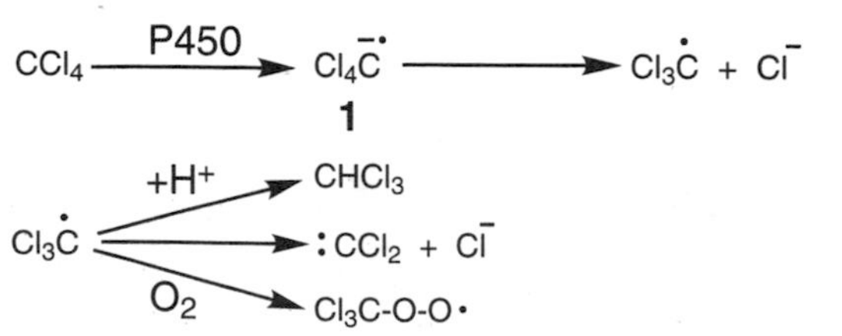

$$CCl_4 \xrightarrow{P450} Cl_4\dot{C}^{-} \ (\mathbf{1}) \longrightarrow Cl_3\dot{C} + Cl^-$$

$$Cl_3\dot{C} \xrightarrow{+H^+} CHCl_3$$
$$Cl_3\dot{C} \longrightarrow :CCl_2 + Cl^-$$
$$Cl_3\dot{C} \xrightarrow{O_2} Cl_3C\text{-}O\text{-}O\cdot$$

Fig. 8–11. The CYP2E1-catalyzed oxidation of hepatotoxic carbon tetrachloride.

dized than are the *gem*-dihalogenated and monohalogenated compounds. The acyl halides and the carbonyl halides are the more reactive intermediates that can react either with water to form less toxic carboxylic acids and halide ions or nonenzymatically with tissue molecules, eliciting the toxicity of the parent molecule. Chloramphenicol ($RNHCOCHCl_2$) is biotransformed into an acyl halide ($RNHCOCOCl$) that selectively acylates the apoprotein of CYP450.[29]

Reductive metabolism of polyhalogenated hydrocarbons can undergo one-electron reduction to a transient radical anion intermediate (1) that can eliminate halide ion to form a carbon radical (2). This radical can either abstract a hydrogen from an unsaturated lipid in the endoplasmic reticulum membrane, initiating lipid peroxidative changes in the membrane to form a reduced polyhalogenated hydrocarbon; accept a second electron to generate a carbanion that can degrade into either a carbene or an olefin and a halide ion; or react with oxygen to yield a peroxyradical. Carbon tetrachloride is an example of this pathway (Fig. 8–11). The reductive dehalogenation of DDT (dichlorodiphenyltrichloroethane) to DDE (dichlorodiphenyldichloroethene) and DDD (dichlorodiphenyldichloroethane) occurs through the radical intermediate. DDE accumulates in fatty tissue and appears to be relatively stable to further metabolic degradation. DDD is dehalogenated to dichlorophenylacetic acid. Apparently, insects resistant to DDT have evolved a glutathione transferase system that is capable of bioinactivating DDT.

Monohalogenated, *gem*-dihalogenated, and vicinal dihalogenated alkanes undergo glutathione transferase-catalyzed conjugation reactions to produce S-substituted glutathione derivatives that are metabolically transformed into the more stable and less toxic mercapturic acids. This common route of metabolism occurs through nucleophilic displacement of a halide ion by the thiolate anion of glutathione (see the discussion on glutathione conjugation). The mutagenicity of the 1,2-dihaloethanes has been attributed to the formation of the S-(2-haloethyl) glutathione, which becomes a reactive episulfonium ion, a DNA alkylating intermediate. Many of the halogenated hydrocarbons exhibiting nephrotoxicity undergo the formation of similar S-substituted cysteine derivatives.

The hepato- and nephrotoxicity of the fluorinated inhalation anesthetics halothane, CF_3-CHClBr and the "fluranes" (enflurane, CHF_2-O-CF_2-CHClF; isoflurane, CHF_2-O-CHCl-CF_3; desflurane, CHF_2-O-CHF-CF_3; and methoxyflurane, CH_3O-CF_2-$CHCl_2$) are intimately related to their metabolism catalyzed by CYP450.[31,32] The fluranes have been suspected of causing liver damage and cross-reactivity with halothane because of similar patterns in their metabolism.[32] Halothane has received the most attention because of its ability to cause "halothane-associated" hepatitis, involving an immunologic reaction after its repeated exposure in surgical patients. Because of this hepatic damage, halothane has been replaced with one of the "less toxic" flurane anesthetics (Fig. 8–12).

The toxicity of halothane and the fluranes is related to their metabolism to either an acid chloride (or fluo-

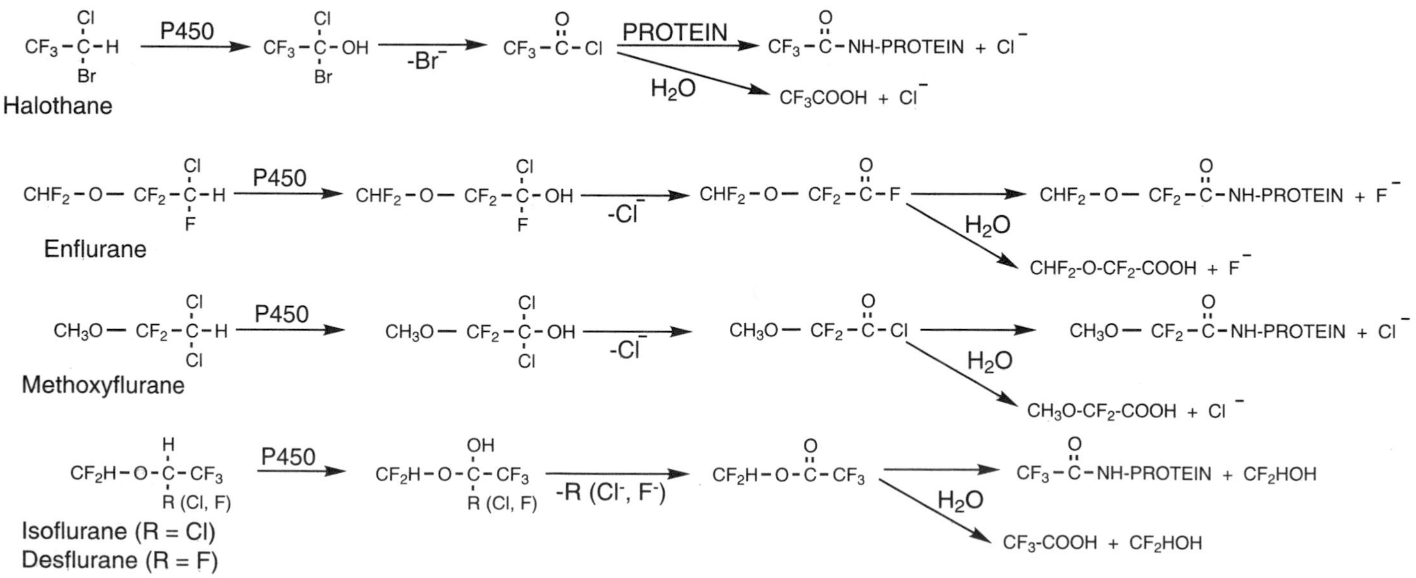

Halothane:
$$CF_3\text{-}CHClBr \xrightarrow{P450} CF_3\text{-}CClBr\text{-}OH \xrightarrow{-Br^-} CF_3\text{-}C(=O)\text{-}Cl \xrightarrow{PROTEIN} CF_3\text{-}C(=O)\text{-}NH\text{-}PROTEIN + Cl^-$$
$$CF_3\text{-}C(=O)\text{-}Cl \xrightarrow{H_2O} CF_3COOH + Cl^-$$

Enflurane:
$$CHF_2\text{-}O\text{-}CF_2\text{-}CClFH \xrightarrow{P450} CHF_2\text{-}O\text{-}CF_2\text{-}CClF\text{-}OH \xrightarrow{-Cl^-} CHF_2\text{-}O\text{-}CF_2\text{-}C(=O)\text{-}F \longrightarrow CHF_2\text{-}O\text{-}CF_2\text{-}C(=O)\text{-}NH\text{-}PROTEIN + F^-$$
$$CHF_2\text{-}O\text{-}CF_2\text{-}C(=O)\text{-}F \xrightarrow{H_2O} CHF_2\text{-}O\text{-}CF_2\text{-}COOH + F^-$$

Methoxyflurane:
$$CH_3O\text{-}CF_2\text{-}CCl_2H \xrightarrow{P450} CH_3O\text{-}CF_2\text{-}CCl_2\text{-}OH \xrightarrow{-Cl^-} CH_3O\text{-}CF_2\text{-}C(=O)\text{-}Cl \longrightarrow CH_3O\text{-}CF_2\text{-}C(=O)\text{-}NH\text{-}PROTEIN + Cl^-$$
$$CH_3O\text{-}CF_2\text{-}C(=O)\text{-}Cl \xrightarrow{H_2O} CH_3O\text{-}CF_2\text{-}COOH + Cl^-$$

Isoflurane (R = Cl), Desflurane (R = F):
$$CF_2H\text{-}O\text{-}CH(R)\text{-}CF_3 \ (R = Cl, F) \xrightarrow{P450} CF_2H\text{-}O\text{-}C(OH)(R)\text{-}CF_3 \xrightarrow{-R\ (Cl^-, F^-)} CF_2H\text{-}O\text{-}C(=O)\text{-}CF_3 \longrightarrow CF_3\text{-}C(=O)\text{-}NH\text{-}PROTEIN + CF_2HOH$$
$$CF_2H\text{-}O\text{-}C(=O)\text{-}CF_3 \xrightarrow{H_2O} CF_3\text{-}COOH + CF_2HOH$$

Fig. 8–12. The CYP2E1-catalyzed metabolism of the fluorinated volatile anesthetics halothane, enflurane, isoflurane, desflurane, and methoxyflurane.

ride) or a trifluoroacetate intermediate (see Fig. 8–12). The CYP2E1 has been identified as the isoform catalyzing the biotransformation of the fluranes.[19,31] The hydroxylated intermediate decomposes spontaneously to reactive intermediates, an acid chloride (or fluoride) or trifluoroacetate, which can either react with water to form Cl^-, Br^-, and/or F^- and a fluorinated carboxylic acid, or bind covalently to tissue proteins producing an acylated protein, CF_3CO-NH-protein or trifluoroacetylate (TFA)-protein. The acylated protein becomes a "hapten" stimulating an immune response and a hypersensitivity reaction. The patient is sensitized to future exposures of the volatile anesthetic. After subsequent exposure to a fluorinated anesthetic, the antigenic TFA protein stimulates an immune response, producing halothane-like hepatitis. Because of the common metabolic pathway involving CYP2E1 for enflurane, isoflurane, desflurane, and methoxyflurane, halothane-exposed patients who had halothane hepatitis can show cross-sensitization to one of the other fluranes, triggering an idiosyncratic hepatic necrosis. The formation of antigenic TFA protein is related to the amount of CYP2E1 catalyzed metabolism for each agent; halothane (20 to 40%) > enflurane (2 to 8%) > isoflurane (0.2 to 1%) > desflurane (< 0.1%). Enough fluoride ion is generated from oxidative dehalogenation during flurane anesthesia to produce subclinical nephrotoxicity. Interestingly, female rats metabolize halothane more slowly than do males and are less susceptible than males to hepatotoxicity. For patients with pre-existing liver dysfunction, isoflurane or desflurane may be a better anesthetic choice.

In today's environment, most humans have been exposed to many CYP2E1 enzyme-inducing agents (including alcohol, recreational, and industrial or agricultural chemicals), having an unknown effect on hepatic toxicity from volatile anesthetics. Enhanced activity for CYP2E1 has been observed in obesity, isoniazid therapy, ketogenic diets, and alcoholism.

Azo and Nitro Reduction

In addition to the oxidative systems, liver microsomes also contain enzyme systems that catalyze the reduction of azo and nitro compounds to primary amines. A number of azo compounds, such as Prontosil and sulfasalazine (see Table 8–13), are converted to aromatic primary amines by azoreductase, an NADPH-dependent enzyme system in the liver microsomes. Evidence exists for participation of cytochrome P450 in some reductions. Nitro compounds, for example, chloramphenicol and nitrobenzene, are reduced to aromatic primary amines by a nitroreductase, presumably through nitrosoamine and hydroxylamine intermediates. These reductases are not solely responsible for the reduction of azo and nitro compounds, probably because of reduction by the bacterial flora in the anaerobic environment of the intestine.

N- and S-Oxidations Catalyzed by Flavin Monooxygenase

The major hepatic monooxygenase systems responsible for the oxidation of many drugs, carcinogens, pesticides, aromatic polycyclic hydrocarbons, and other xenobiotics containing nitrogen, sulfur, or phosphorus are cytochrome P450 monooxygenase and microsomal flavin-containing monooxygenase, FMO.[33] The FMO exhibits broader substrate specificities than CYP450 monooxygenases and has a mechanism distinctly different from the CYP450 monooxygenase. Because oxygen activation occurs before substrate addition, any compound binding to 4*α*-hydroperoxyflavin, the enzyme-bound monooxygenating FMO intermediate, is a potential substrate. Typically, FMO catalyzes oxygenation of the N and S heteroatoms ("soft nucleophiles") (Table 8–14) but not heteroatom dealkylation reactions. The products formed from FMO catalyzed oxidation are consistent with a direct two-electron oxidation of the heteroatom. Thus, FMO constitutes an alternative biotransformation pathway for N- and S-containing lipophilic xenobiotics. Flavin monooxygenase is not normally inducible by phenobarbital nor affected by cytochrome P450 inhibitors. With few exceptions, however, xenobiotic substrates for FMO are also substrates for the isoforms of CYP450 producing similar oxidation products. Which monooxygenase is responsible for the oxidation can be readily determined because FMO is thermally labile in the absence of NADPH, whereas CYP450 is stable.

Of the many nitrogen functional groups in drugs and xenobiotics, only secondary and tertiary acyclic, cyclic, and arylamines, hydroxylamines, and hydrazines are oxidized by FMO and excreted in the urine (see Table 8–14). The tertiary amines from stable amine oxides, and secondary amines are sequentially oxidized to hydroxylamines, nitrones, and a complex mixture of products. Secondary *N*-alkylarylamines can be N-oxygenated to reactive N-hydroxylated metabolites that are responsible for the toxic, mutagenic, and carcinogenic activity of these aromatic amines. For example, the chemically unstable hydroxylamine intermediates of aromatic amines degrade into bladder carcinogens (see the discussion of mechanism under glucuronic acid conjugation), and the hydroxamic acid intermediates of N-arylacetamides are bioactivated into liver carcinogens. Hepatic FMO, however, will not catalyze the oxidation of primary alkyl- or arylamines, except for the carcinogenic N-hydroxylated derivatives of 2-aminofluorene, 2-aminoanthracene, and other amino polycyclic aromatic hydrocarbons. S-Oxidation occurs almost exclusively by FMO (see Table 8–14). Sulfides are oxidized to sulfoxides and to sulfones, thiols to disulfides, and thiocarbamates, mercaptopyrimidines, and mercaptoimidazoles (i.e., the antithyroid drug methimazole) via sulfenates to sulfinates, all of which are eliminated in the urine. Flavin monooxygenase does not catalyze epoxidation reactions or hydroxylation at unactivated carbon atoms of xenobiotics. Primary aromatic amines and amides, aromatic heterocyclic amines and imines, and the aliphatic primary amine phentermine are N-oxidized by CYP450 to hydroxylamines. CYP450 oxidizes carbon disulfide to carbon dioxide and hydrogen sulfide (through a thioperoxy anion, HOS-) and the antipsychotic phenothiazines to sulfoxides.

The major steps in the catalytic cycle for FMO are

Table 8–14. Examples of Substrates and Reaction Mechanisms for Flavin Monooxygenase

Nitrogen Compounds

Tert-acyclic and cyclic amines to N-oxides

$$R_3\text{—N} + \text{Enz-FAD-OOH} \rightarrow R_3\text{—N} \rightarrow \text{O} + \text{Enz-FAD-OH}$$

Chlorpromazine	Imipramine	Amitriptyline
Fluphenazine	Nicotine	Diphenhydramine
Atropine	Dimethylaniline	

Sec-acyclic and cyclic amines to N-hydroxides and nitrones

$$R_2\text{—NH} + \text{Enz-FAD-OOH} \rightarrow R_2\text{—NOH} + \text{Enz-FAD-OH}$$

$$R_2\text{—NOH} + \text{Enz-FAD-OOH} \rightarrow R_2\text{—}\underset{\underset{\text{O}}{|}}{\text{N}}\text{=O} + \text{Enz-FAD-OH}$$

Desmethylimipramine	N-methylbenzylamine
Desmethyltrifluperazine	N-methylaniline

N-alkyl and N,N-dialkylaryl amines to N-hydroxides

$$\text{R—NH-Ar} + \text{Enz-FAD-OOH} \rightarrow \text{R—}\underset{\underset{\text{Ar}}{|}}{\text{N}}\text{OH} + \text{Enz-FAD-OH}$$

1,1-Hydrazines

$$\text{R—}\underset{\underset{\text{R}}{|}}{\text{N}}\text{H-NH}_2 + \text{Enz-FAD-OOH} \rightarrow \text{R—}\overset{\overset{\text{O}}{|}}{\underset{\underset{\text{R}}{|}}{\text{N}}}\text{NH}_2 + \text{Enz-FAD-OH}$$

Sulfur Compounds

Thiols $\text{R—SH} \rightarrow \text{R—SS—R}$

Disulfides $\text{R—SS—R} \rightarrow \text{R—}\underset{\underset{\text{O}}{|}}{\text{SS}}\text{—R}$

Sulfides $\text{R—S—R} \rightarrow \text{R—}\underset{\underset{\text{O}}{|}}{\text{S}}\text{—R} \rightarrow \text{R—}\overset{\overset{\text{O}}{|}}{\underset{\underset{\text{O}}{|}}{\text{S}}}\text{—R}$

Cimetidine	Ranitidine
Sulindac (as sulfide)	Thioridazine

Thioamides

Mercaptopurines, mercaptoimidazoles, mercaptopyrimidines

Methimazole

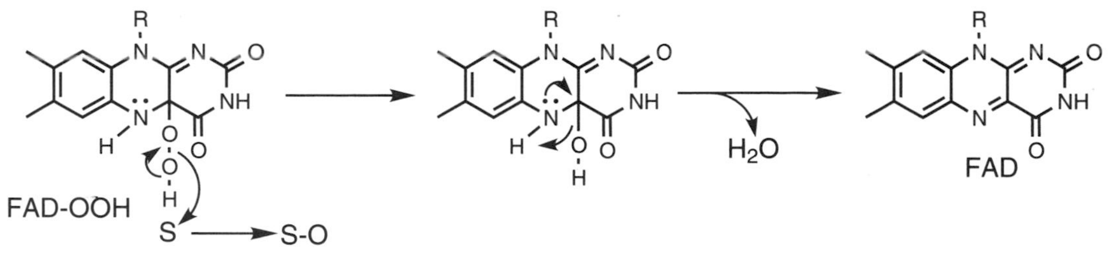

Fig. 8–13. The mechanism of FMO-catalyzed oxidation of a substrate.

shown in Figures 8–13 and 8–14.[34] Like most of the other monooxygenases, FMO requires NADPH and oxygen as cosubstrates to catalyze the oxidation of the xenobiotic substrate. But unlike CYP450, the xenobiotic being oxidized does not need to be bound to the 4α-hydroperoxyflavin intermediate (FAD-OOH) for oxygen activation to occur. Apparently, FMO is present within the cell in its enzyme-bound activated-hydroperoxide (Enz-FAD-OOH) state ready to oxidize any suitable lipophilic substrate that binds to it. Flavin monooxygenase uses a nonradical, nucleophilic displacement type of mechanism binding dioxygen with a reduced flavin. The reactive oxygen intermediate is a reactive derivative of hydrogen peroxide, flavin-4α-hydroperoxide, which is reactive enough to successfully attack a lone electron pair on a heteroatom such as nitrogen or sulfur but not reactive enough to attack a typical C-H bond. These studies suggest the xenobiotic substrate (S) interacts with the 4α-hydroperoxyflavin form of FMO and is oxidized by oxygen transfer from Enz-FAD-OOH to form the oxidized product (SO).

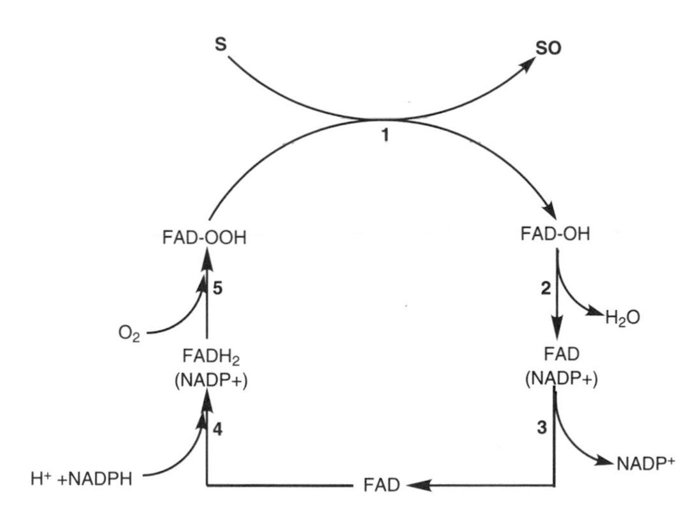

Fig. 8–14. The FMO catalytic cycle. Oxygenated product (SO) is formed by nucleophilic attack of a substrate (S) on the terminal oxygen of the enzyme-bound hydroperoxyflavin (FAD-OOH) followed by heterolytic cleavage of the peroxide (1). The release of H_2O or of NADP+ (3) is rate limiting for reactions catalyzed by liver FMO. Reduction of flavin (FAD) by NADPH (4) and the addition of oxygen (5) complete the catalytic cycle by regenerating the oxygenated FAD-OOH.[34]

Neither S nor its product SO is essential for any other steps in the cycle. Steps 2 to 5 simply regenerate the oxygenating agent Enz-FAD-OOH from Enz-FAD-OH, NADPH, oxygen, and a proton. Any compound readily crossing cell membranes by passive diffusion and penetrating to the FMO-bound hydroperoxyflavin intermediate is a potential substrate, thus explaining the broad substrate specificity exhibited for FMO. The fact that the xenobiotic substrate is not required for activation of the FMO-hydroperoxyflavin state distinguishes FMO from CYP450 monooxygenases, in which substrate binding initiates the P450 catalytic cycle and activation of oxygen to the perferryl oxygenating agent. It is not unusual for FMO oxidation products to undergo reduction to the parent xenobiotic, which can enter into repeated redox reactions called metabolic cycling.

Results of substrate specificity studies suggest that the number of ionic groups on endogenous substrate is an important factor enabling FMO to distinguish between xenobiotic and endogenous substrates, preventing the indiscriminate oxidation of physiologically important amine and sulfur compounds.[34] Without exception, FMO readily catalyzes the oxidation of uncharged amines or sulfur compounds (in equilibrium with its respective monocation or monoanion; sulfur compounds, the charge is on sulfur atom), but will not catalyze the oxidation of dianions (e.g., thiamine pyrophosphate), dications (e.g., polyamines), dipolar ions (e.g., amino acids and peptides), or other polyionic compounds with one or more anionic groups (i.e., COO-) distal to the heteroatom (e.g., coenzyme A).

Unlike the CYP450 system, only three isoforms of hepatic FMO have been characterized in the adult human liver:[35] minor form I (or FMO 1A1) but the major form in fetal tissue; major form II (FMO 1D1); and form III, of which little is currently known. The substrate specificities for these isoforms have not been reported. The availability of different forms of FMO may be of clinical importance in the pharmacologic and toxicologic properties of FMO-dependent drug oxidations.

Peroxidases and Other Monooxygenases

Peroxidases are hemoproteins and perhaps are the most closely related enzymes to CYP450 monooxygenase.[36] The normal course of peroxide catalyzed oxidation is:

$$ROOH + Fe^{III} \longrightarrow [FeO]^{+3} + ROH$$
$$[FeO]^{+3} + AH_2 \longrightarrow [FeO]^{+2} + AH^{\cdot}$$
$$[FeO]^{+2} + AH_2 \longrightarrow Fe^{III} + AH^{\cdot}$$
$$2\,AH^{\cdot} \longrightarrow A + AH_2$$
or
$$[FeO]^{+2} + AH + H^{+} \longrightarrow Fe^{III} + A + H_2O$$

The $[FeO]^{+3}$ intermediate is analogous to the perferryl complex in cytochrome P450 and can perform heteroatom oxygenation (N-dealkylation) and aromatization of 1,4-dihydropyridines.

Other monooxygenases catalyzing oxidation reactions similar to CYP450 include dopamine β-monooxygenase, a mammalian copper containing enzyme catalyzing carbon hydroxylation, epoxidation, S-oxygenation, and N-dealkylation reactions, and non-heme iron-containing enzymes from bacteria and plants.

Nonmicrosomal Oxidations

In addition to the microsomal oxidases, other oxidases and dehydrogenases that catalyze oxidation reactions are present in the mitochondrial and soluble fractions of tissue homogenates.

Oxidation of Alcohols

Alcohol dehydrogenase is an NAD-specific enzyme located in the soluble fraction of tissue homogenates. It exhibits a broad specificity for alcohols; most primary alcohols are readily oxidized to their corresponding aldehydes. Some secondary alcohols are oxidized to the ketones, whereas other secondary and tertiary alcohols are excreted unchanged or as a conjugate metabolite (e.g., cyclohexanol). Some secondary alcohols show mixed activity because of steric factors and lack of affinity of the substrate for the enzyme.

$$C_2H_5-OH \longrightarrow H_3C-CHO$$

$$H_3C-CH(OH)-CH_3 \longrightarrow H_3C-C(=O)-CH_3$$

Oxidation by alcohol dehydrogenase is the principal pathway for ethanol metabolism, but the microsomal isoform CYP2E1 also plays a significant role in ethanol metabolism and tolerance. Apparently, two thirds of ingested ethanol is oxidized by alcohol dehydrogenase and the remainder by CYP2E1; during intoxication, however, ethanol induces the expression of CYP2E1. The induction of CYP2E1 contributes to the activation of xenobiotics, increasing the vulnerability of the heavy drinker to anesthetic drugs, over-the-counter analgesics, prescription drugs, and chemical carcinogens. In turn, the excessive amounts of acetaldehyde generated causes hepatotoxicity, lipid peroxidation of membranes, formation of protein adducts, and other cellular changes.

Although the toxicity of methanol in humans has long been recognized, frequent reports of such toxicity are not surprising given the number of consumer products containing methanol. Methanol (wood alcohol, methyl alcohol) is commonly used as a solvent in organic synthetic procedures, and is available to consumers in a variety of products ranging from solid fuels (Sterno), paint removers, solvent for "ditto" copy machines, motor fuels, antifreeze, and alcoholic beverages. Oral methanol toxicity in humans is characterized by its rapid absorption from the gut followed by a latent period of many hours before metabolic acidosis (lowered blood pH and bicarbonate levels) and ocular toxicity is evident. The metabolic acidosis and blindness result from the excessive accumulation of formic acid and the inability of the hepatic tetrahydrofolate pathway to oxidize formate to carbon dioxide. The rate of elimination of methanol from the blood is relatively slow compared to ethanol, accounting for its long latency period. Its half-life ranges from 2 to 3 hours at low blood concentration to 27 hours at high blood concentration. Evidence supports the singular role of liver alcohol dehydrogenase in the metabolism of methanol to formaldehyde, although it is oxidized slowly by alcohol dehydrogenase (about one sixth the rate of ethanol). The demonstration that methanol is a substrate for alcohol dehydrogenase provides a rational basis for the use of ethanol in the treatment of methanol toxicity. Ethanol depresses the rate of methanol oxidation by acting as a competitive substrate for alcohol dehydrogenase, reducing the formation of formaldehyde. On the other hand, formaldehyde is not usually detected in the blood because of its rapid metabolism by aldehyde dehydrogenase to formate. Although human exposure to methanol vapor is less prevalent, methanol is rapidly absorbed through the skin or by inhalation and can result in methanol poisoning, depending on severity of exposure.

Alcohol dehydrogenase also functions as a reductase when it catalyzes the reduction of an aldehyde or ketone to an alcohol. In addition, other NADP- or NAD-dependent dehydrogenases in the cytosol are capable of reducing a variety of ketones. Ketones are stable to further oxidation and consequently yield reduction products as major metabolites. Examples of reduction are chloral hydrate to trichloroethanol, naltrexone to 6β-hydroxynaltrexol, and methadone to α-methadol. These alcohol metabolites are all pharmacologically active.

Oxidation of Aldehydes

Xanthine oxidase, aldehyde oxidase, and an NAD-specific aldehyde dehydrogenase catalyze the oxidation of endogenous aldehydes, such as those produced by the oxidation of primary alcohols or the deamination of biogenic amines, and of exogenous aldehydes to the corresponding carboxylic acids. Aldehyde oxidase and xanthine oxidase are metalloflavoprotein enzymes. Aldehyde dehydrogenase shows polymorphism in Orientals.

$$H_3C-CHO \longrightarrow H_3C-COOH$$

Oxidative Deamination of Amines

Monoamine oxidase (MAO) and diamine oxidase (DAO) catalyze oxidative deamination of amines to the aldehydes in the presence of oxygen. The aldehyde products can be metabolized further to the corresponding alcohol or acid by aldehyde oxidase or dehydrogenase. Monoamine oxidase is a mitochondrial membrane flavin-containing enzyme that catalyzes the oxidative deamination of monoamines according to the following equation:

$$R\text{-}CH_2-NH_2 + O_2 \longrightarrow [R\text{-}CH=NH] \longrightarrow R-CHO + NH_4^+$$

Substrates for this enzyme include several monoamines, secondary and tertiary amines in which the amine substrates are methyl groups. The amine must be attached to an unsubstituted methylene group, and compounds having substitution at the α-carbon atom are poor substrates for MAO, e.g., aniline, amphetamine, and ephedrine, but are oxidized by the microsomal CYP450 enzymes rather than by MAO (see Table 8–16). For secondary and tertiary amines, alkyl groups larger than a methyl and branched alkyl groups, i.e., isopropyl, t-butyl, or β-phenylisopropyl, inhibit MAO oxidation, but such substrates may function as reversible inhibitors of MAO. Nonselective irreversible inhibitors of MAO include hydrazides (phenelzine) and tranylcypromine, and the MAO-B selective inhibitors pargyline and selegiline. Monoaminic oxidase is important in regulating the metabolic degradation of catecholamines and serotonin (5-HT) in neural tissues, and hepatic MAO has a crucial defensive role in inactivating circulating monoamines or those that originated in the gastrointestinal tract and were absorbed into the systemic circulation (e.g., tyramine). Two types of MAO isolated are MAO-A and MAO-B. They show dissimilar substrate preferences and different sensitivities to inhibitors. MAO-A is found mainly in peripheral adrenergic nerve terminals and shows substrate preference for 5-hydroxytryptamine, norepinephrine, and epinephrine; MAO-B is found principally in platelets and shows selectivity for nonphenolic, lipophilic β-phenethylamines. Common substrates to both MAO-A and MAO-B are dopamine, tyramine, and other monophenolic phenylethylamines.

MPTP (1-methyl-4-phenyl-1,2,3,6-tetrahydropyridine), a contaminant in the synthesis of reversed esters of meperidine, was discovered to be a highly selective neurotoxin for dopaminergic cells, producing parkinsonism.[37] The neurotoxicity of MPTP is associated with cellular destruction in the substantia nigra along with severe reductions in the concentration of dopamine, norepinephrine, and serotonin. The remarkable neurotoxic action for MPTP involves a sequence of events beginning with the metabolic activation of MPTP to the toxic metabolite MPP^+ (1-methyl-4-phenylpyridinium ion) by MAO-B, specific uptake and accumulation of MPP^+ in the nigrostriatal dopaminergic neurons, and ending with the inhibition of oxidative phosphorylation (of NADH dehydrogenase in complex I). This inhibition results in mitochondrial injury depriving the sensitive nigrostriatal cells of oxidative phosphorylation with their eventual cell

N-CH3
MPTP

N+-CH3
MPP+

death and exhibition of the neurotoxic actions of MPP^+. MAO-B inhibitors (e.g., deprenyl) blocked this biotransformation.

Diamines, such as $H_2N\text{-}(CH_2)_n\text{-}NH_2$, in which n is less than six, are not attacked and show little affinity for MAO. If the intermolecular distance between the amine groups is increased, the rate of oxidation by MAO increases. Evidently, the second amine group interferes with attachment of the amine to the enzyme.

Diamine oxidase attacks both diamines and histamine in much the same way MAO attacks monoamines, forming aldehydes. This enzyme is inhibited by carbonyl-blocking reagents and produces hydrogen peroxide, supporting the role of pyridoxal phosphate and the flavin prosthetic groups in the catalytic action of the enzyme. Diamine oxidase is recovered in the supernatant after centrifugation and removal of particulate matter. It is present in kidneys, intestines, liver, lung, and nervous tissue. It limits the biologic effects of histamine and the

$H_2N-(CH_2)_4-NH_2$
Putrescine

$H_2N-(CH_2)_5-NH_2$
Cadaverine

polymethylene amines putrescine and cadaverine. It also attacks monoamines, but at a higher substrate concentration.

Plasma amine oxidases are in blood plasma of mammals and include spermine oxidase, which deaminates spermine and other polyamines.

Purine Oxidation

Many purine derivatives undergo oxidation to uric acid derivatives, presumably by xanthine oxidase, a metalloflavoprotein that normally catalyzes the oxidation of hypoxanthine and xanthine to uric acid. The compound 6-mercaptopurine undergoes oxidation to 6-mercapturic acid. Allpurinol is an inhibitor of xanthine oxidase, and it inhibits the formation of uric acid in patients with gout.

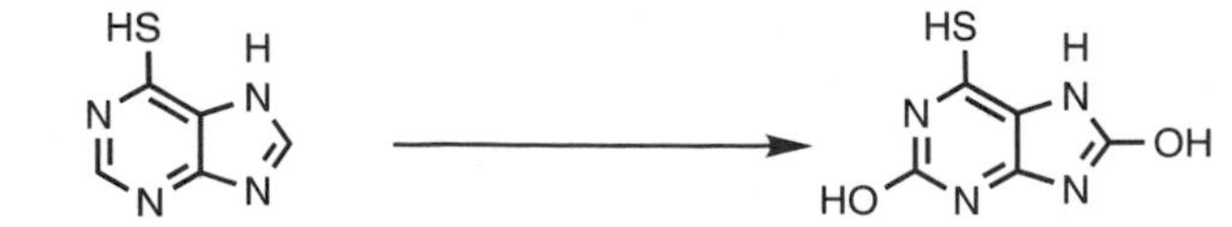

6-Mercaptopurine → 6-Mercapturic acid

Miscellaneous Reductions

Disulfides (e.g., disulfiram), sulfoxides (e.g., dimethylsulfoxide), N-oxides, double bonds such as those in progestational steroids, and dehydroxylation of aromatic and aliphatic hydroxyl derivatives are examples of reductions occurring in microsomal and/or nonmicrosomal (usually cytosol enzymes) fractions.

$(C_2H_5)_2-NCSS\text{-}SSCN\text{-}(C_2H_5)_2 \longrightarrow (C_2H_5)_2-NCSS-H$

Disulfiram → Diethyldithiocarbamic acid

$CH_3SOCH_3 \longrightarrow CH_3SCH_3$

Dimethylsulfoxide → Dimethylsulfide

Various studies on the biotransformation of xenobiotic ketones have established that ketone reduction is an important metabolic pathway in mammalian tissue. Because carbonyl compounds are lipophilic and may be retained in tissues, their reduction to the hydrophilic alcohols and subsequent conjugation are critical to their elimination. Although ketone reductases may be closely related to the alcohol dehydrogenases, they have distinctly different properties and use NADPH as the cofactor. The metabolism of xenobiotic ketones to free alcohols or conjugated alcohols has been demonstrated for aromatic, aliphatic, alicyclic, and unsaturated ketones (e.g., naltrexone, naloxone, hydromorphone, and daunorubicin). The carbonyl reductases are distinguished by the stereospecificity of their alcohol metabolites.

β-Oxidation

Alkyl carboxylic acids are metabolized by β-oxidation as their coenzyme A (CoA) thioesters, e.g., valproic acid. This pathway involves the oxidative cleavage of two carbon units at a time (as acetate), beginning at the carboxyl terminus and continuing until no more acetate units can be removed. The reaction is terminated when a branch or aromatic group is encountered. The metabolism of even and odd phenylalkyl acids can serve as an example:

Hydrolysis

In general, esters and amides are hydrolyzed by enzymes in the blood, liver microsomes, kidneys, and other tissues. Esters and certain amides are hydrolyzed rapidly by a group of enzymes called carboxylesterases. The more lipophilic the amide, the more favorable it is as a substrate for this enzyme. In most cases, the hydrolysis of an ester or amide bond in a toxic substance results in bioinactivation to hydrophilic metabolites that are readily excreted. Some of these metabolites may yield conjugated metabolites (i.e., glucuronides). Carboxylesterases include cholinesterase (pseudocholinesterase), arylcarboxyesterases, liver microsomal carboxylesterases, and other unclassified liver carboxylesterases. Cholinesterase hydrolyzes choline-like esters (succinylcholine) and procaine, as well as acetylsalicylic acid. Genetic variant forms of cholinesterase have been identified in human serum (e.g., succinylcholine toxicity when administered as ganglionic blocker for muscle relaxation). Meperidine is hydrolyzed only by liver microsomal carboxylesterases (Fig. 8–15). Diphenoxylate is hydrolyzed to its active metabolite, diphenoxylic acid, within 1 hour (Fig. 8–15). Presumably, the peripheral pharmacologic action of diphenoxylate is attributed to zwitterionic diphenoxylic acid, which is readily eliminated in the urine.

Esters that are sterically hindered are hydrolyzed more slowly and may appear unchanged in the urine. For example, approximately 50% of a dose of atropine appears unchanged in the urine of humans. The remainder appears to be unhydrolyzed biotransformed products.

As a rule, amides are more stable to hydrolysis than are esters, and it is not surprising to find amides excreted largely unchanged. This fact has been exploited in developing the antiarrhythmic drug procainamide. Procaine is not useful because of its rapid hydrolysis, but 60% of a dose of procainamide was recovered unchanged from the urine of humans, with the remainder mostly N-acetylprocainamide. On the other hand, the deacylated metabolite of indomethacin (a tertiary amide) is one of the major metabolites detected in human urine. Amide hydrolysis of phthalylsulfathiazole and succinylsulfathiazole by bacterial enzymes in the colon releases the antibacterial agent sulfathiazole.

In summary, phase 1 metabolic transformations introduce new and polar functional groups into the molecule, which may produce one or more of the following changes:

1. Decreased pharmacologic activity-deactivation
2. Increased pharmacologic activity-activation
3. Increased toxicity-carcinogenesis, mutagenesis, cytotoxicity
4. Altered pharmacologic activity

$$C_6H_5-(CH_2)_{10}\text{-}COOH \longrightarrow C_6H_5-(CH_2)_{10}\text{-}CO\text{-}CoA \longrightarrow C_6H_5-COOH$$

$$C_6H_5-(CH_2)_{11}\text{-}COOH \longrightarrow C_6H_5-(CH_2)_{11}\text{-}CO\text{-}CoA \longrightarrow C_6H_5-CH_2\text{-}COOH$$

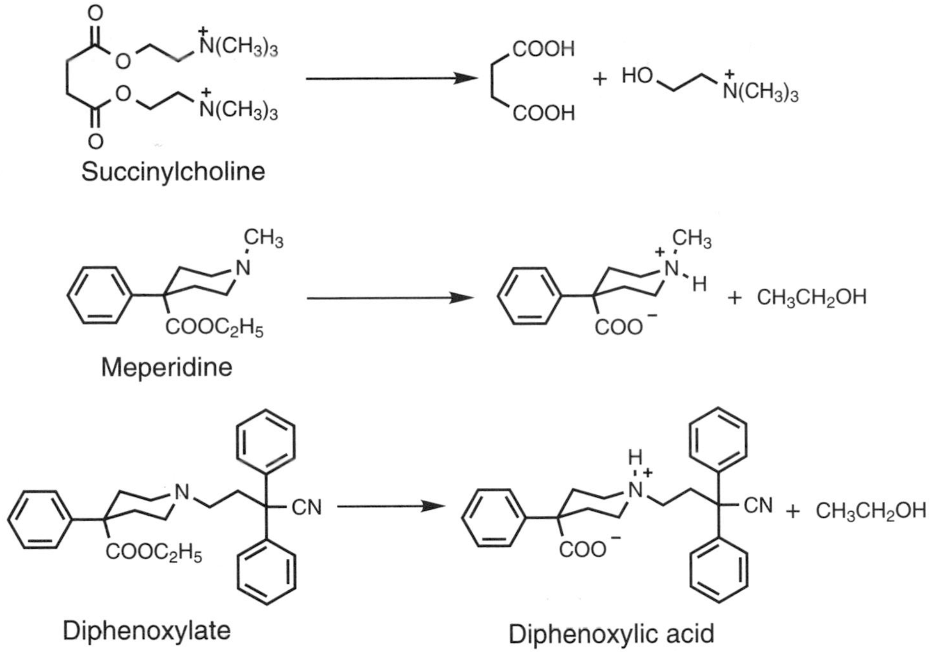

Succinylcholine

Meperidine

Diphenoxylate

Diphenoxylic acid

Fig. 8–15. Examples of hydrolysis reactions.

Drugs exhibiting increased activity or activity different from the parent drug generally undergo further metabolism and conjugation, resulting in deactivation and excretion of the inactive conjugates.

DRUG CONJUGATION PATHWAYS (Phase 2)

Conjugation reactions represent probably the most important xenobiotic biotransformation reaction.[38] Xenobiotics are usually lipophilic, well absorbed in blood, but slowly excreted in the urine. Only after conjugation (phase 2) reactions have added an ionic hydrophilic moiety, such as glucuronic acid, sulfate, or glycine to the xenobiotic, is water solubility increased and lipid solubility decreased enough to make urinary elimination possible. The major proportion of the administered drug dose is excreted as conjugates into the urine and/or bile. Conjugation reactions may be preceded by phase 1 reactions. For xenobiotics with a functional group available for conjugation, conjugation alone may be its fate.

Traditionally, the major conjugation reactions (glucuronidation and sulfation) were thought to terminate pharmacologic activity by transforming the parent drug or phase 1 metabolites into readily excreted ionic polar products. Moreover, these terminal metabolites would have no significant pharmacologic activity, i.e., poor cellular diffusion and affinity for the active drug's receptor. This long-established view changed, however, with the discovery that morphine 6-glucuronide has more analgesic activity than morphine in humans and minoxidil sulfate is the active metabolite for the antihypertensive minoxidil. For most drugs and xenobiotics, conjugation is a detoxification mechanism. Some compounds, however, form reactive intermediates that have been implicated in carcinogenesis, allergic reaction, and tissue damage.

Sequential conjugation for the same substance gives rise to multiple conjugated products (see p-aminosalicylic acid in Fig. 8–16). The xenobiotic can be a substrate for more than one metabolizing enzyme. For example, different conjugation pathways could compete for the same functional group. The outcome is an array of metabolites excreted in the urine or feces. The factors determining the outcome of this interplay include availability of cosubstrates, enzyme kinetics (V_{max}), substrate affinity (km) for the metabolizing enzyme, and tissues. When a cosubstrate is low or depleted, the competing reactions can take over. The reactivity of the functional group determines all subsequent events. For example, major conjugation reactions at the phenolic hydroxyl groups are sulfation, ether glucuronidation, and methylation; for amine, acetylation, sulfation, glucuronidation; for carboxyl groups amino acid conjugation, ester glucuronidation.

Conjugation enzymes may show stereospecificity toward enantiomers when a racemic drug is administered. The metabolite pattern of the same drug administered orally and intravenously may be different because of presystemic intestinal conjugation.

A current and in-depth review of the different phase 2 conjugations is available in reference 38.

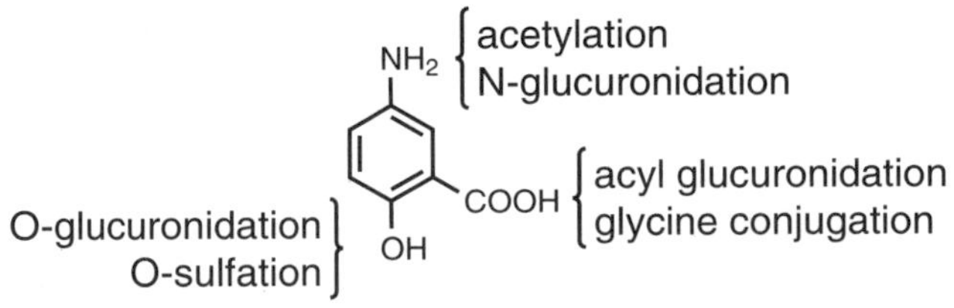

acetylation
N-glucuronidation
acyl glucuronidation
glycine conjugation
O-glucuronidation
O-sulfation

Fig. 8–16. Sequential conjugation pathways for p-aminosalicylic acid.

Glucuronic Acid Conjugation

Glucuronide formation is one of the most common routes for drug and xenobiotic metabolism to water-soluble metabolites, and accounts for a major share of the conjugated metabolites.[38] Its significance lies in the readily available supply of glucuronic acid in liver and in the many functional groups forming glucuronide conjugates, e.g., phenols, alcohols, carboxylic acids, and amines. The reaction involves the direct condensation of the drug or xenobiotic (or its phase 1 product) with the activated form of glucuronic acid, uridine diphosphate glucuronic acid (UDPGA). The overall scheme of reactions is shown in Figure 8–17.

The reaction between UDPGA and the acceptor compound is catalyzed by UDP-glucuronosyl transferases, a group of ubiquitous microsomal enzymes present in liver and a variety of extrahepatic tissues. Their location in the endoplasmic reticulum has important physiologic effects in the neutralization of reactive intermediates generated by the CYP450 system and in controlling the levels of reactive metabolites present in these tissues. The resultant glucuronide has the β-configuration about carbon 1 of glucuronic acid. With the attachment of the hydrophilic carbohydrate moiety containing an easily ionizable carboxyl group (pK_a 3–4), a lipid-soluble substance is converted into a conjugate that is poorly reabsorbed by the renal tubules from the urine and is excreted more readily into the urine or, in some cases, into the bile. Some UDP-glucuronosyl transferases are inducible by barbiturates or cigarette smoking.

The drugs and other xenobiotics forming glucuronides with alcohols and phenols are ether glucuronides; aromatic and some aliphatic carboxylic acids form ester (acyl) glucuronides; aromatic amines form N-glucuronides, and sulfhydryl compounds form S-glucuronides, both of which are more labile to acid than are the O-glucuronides (see Fig. 8–17). Some tertiary amines (e.g., tripelennamine) have been reported to form quaternary ammonium N-glucuronides. Substances containing a 1,3-dicarbonyl structure, e.g., phenylbutazone, can undergo formation of C-glucuronides by direct conjugation without prior metabolism. The acidity of the methylene carbon of the 1,3-dicarbonyl group determines the degree of C-glucuronide formation.

Drug-acyl glucuronides are reactive conjugates at physiologic pH. The acyl group of the C^1-acyl glucuronide can migrate via transesterification from the original C-1 position of the glucuronic acid to the C-2, C-3, or C-4 positions. The resulting positional isomers are not hydrolyzable by β-glucuronidase, giving the appearance of a new unknown conjugate(s). Under physiologic or weakly alkaline conditions, however, the C^1-acyl glucuronide can hy-

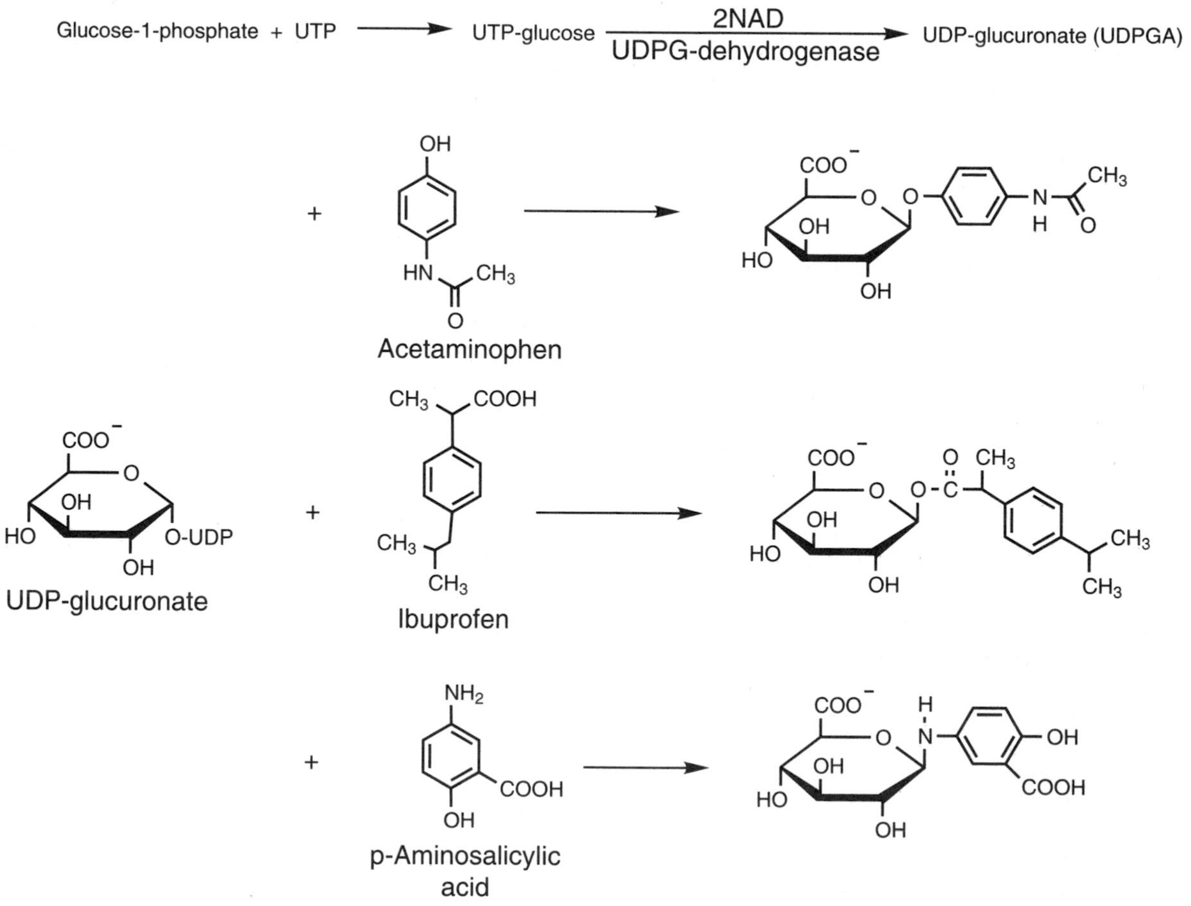

Glucose-1-phosphate + UTP → UTP-glucose —2NAD / UDPG-dehydrogenase→ UDP-glucuronate (UDPGA)

Fig. 8–17. Glucuronidation pathways.

drolyze in the urine to the parent substance (aglycone) or undergo acyl migration to an acceptor macromolecule. The pH catalyzed migration of the acyl group from the drug C^1-O-acyl glucuronide to a protein or other cellular constituent occurs with the formation of a covalent bond to the protein.[39] The acylated protein becomes a "hapten" and could stimulate an immune response against the drug, resulting in the expression of hypersensitivity reaction, or other forms of immunotoxicity. A high incidence of anaphylactic reactions have been reported for several nonsteroidal anti-inflammatory drugs (NSAID) (benoxaprofen, zomepirac, indoprofen, alclofenac, ticrynafen, and ibufenac) that have been removed from the market. All of these NSAID are metabolized by humans to acyl glucuronides. Similar reactions have been reported for other NSAID including tolmetin, sulindac, ibuprofen, ketoprofen, and acetylsalicylic acid. The frequency of the immunotoxic response may be related to the stability of the acyl glucuronide, the chemical rate kinetics for the migration of the acyl group, and the concentration and stability/half-life of the antigenic protein. When the acyl glucuronide is the primary metabolite, patients with decreased renal function (i.e., elderly individuals) or when probenecid is coadministered, renal cycling of the unconjugated (aglycone) parent drug or metabolite is likely to occur, resulting in the plasma accumulation of the aglycone. The reduced elimination of the acyl glucuronide increases its hydrolysis back to the aglycone or the migration of the C^1-O-acyl group to an acceptor macromolecule.

Endogenous substances conjugated with glucuronic acid include steroids, bilirubin, and thyroxine. Not all glucuronides are excreted by the kidneys. Some are excreted into the intestinal tract with bile (see the discussion about enterohepatic cycling), where β-glucuronidase present in the intestinal flora hydrolyzes the C^1-O-glucuronide back to the aglycone (drug or xenobiotic or their metabolites) for reabsorption into the portal circulation.

Generally, glucuronides are biologically and chemically less reactive than their parent molecules, and are readily eliminated without interaction with intracellular substances. Some glucuronide conjugates, however, are more active than the parent drug.[40] Morphine, for example, forms the 3-O- and 6-O-glucuronides in the intestine and in the liver. The 3-O-glucuronide is the primary glucuronide metabolite of morphine with a blood concentration 20 times morphine. Pharmacologically, it is an opiate antagonist. On the other hand, 6-O-glucuronide, with a blood concentration twice that of morphine, is a more potent μ-receptor agonist and analgesic than morphine in humans, whether administered orally or parentally. Thus, the analgesic effects of morphine is the result of a complex interaction of the drug and its two metabolites with the opiate receptor. Apparently, the 6-O-glucuronide can pass into the brain more easily than anticipated.

Glucuronidation is also capable of promoting cellular injury (e.g., hepatotoxicity, carcinogenesis) by facilitating the formation of reactive electrophilic (electron-deficient) intermediates and their transport into target tissues.[26] The induction of bladder carcinogenesis by aromatic amines may occur as the result of the glucuronidation of N-hydroxylarylamine. These O-glucuronides become concentrated in the urine, where they are readily hydrolyzed by the acid pH of the urine back to the N-hydroxylarylamines, eliminating water under these conditions to the electrophilic arylnitrenium species. This reactive species can bind covalently with endogenous cellular constituents, e.g., nucleic acids and proteins, initiating carcinogenesis.

Sulfation and glucuronidation occur side by side, often competing for the same substrate (most commonly phenols, i.e., acetaminophen), and the balance between sulfation and glucuronidation is influenced by such factors as species, doses, availability of cosubstrates, and inhibition and induction of the respective transferases.

Sulfate Conjugation

Sulfation is the primary route of metabolism for catecholamine neurotransmitters, steroid hormones, thyroxine, bile acids, phenolic drugs, and other foreign compounds.[38] Moreover, the tyrosinyl group of peptides and proteins is the site of sulfation, resulting in a possible change in their properties. A drug of xenobiotic is sulfated by transfer of an active sulfate from 3′-phosphoadenosine-5′-phosphosulfate (PAPS) to the acceptor molecule, a cytosolic reaction catalyzed by sulfotransferases (Fig. 8–18); PAPS is formed enzymatically from ATP and inorganic sulfate. The availability of PAPS and its precursor inorganic sulfate determines the reaction rate. Evidence suggests the presence of sulfotransferases specific for steroids and phenols. In humans, sulfotransferases are found in the liver, small intestine, brain, kidneys, and platelets. A "thermolabile" (TL) form of human phenol sulfotransferase catalyzes the sulfation of dopamine and other monoamines, whereas a "thermostable" (TS) form catalyzes the sulfation of a variety of phenolic substances. The latter is controlled by genetic polymorphism. For example, phenol is sulfated by a sulfotransferase in the liver, kidneys, and intestines, whereas steroids are sulfated only in the liver.

The total pool of sulfate is usually limited and can be readily exhausted. With increasing doses of a drug, conjugation with sulfate becomes a less predominant pathway. At high doses for a competing substrate (i.e., acetaminophen), glucuronidation usually predominates over that of sulfation, which prevails at low doses. Other precursors for sulfate include L-methionine and L-cysteine. When PAPS, inorganic sulfate, or the sulfur amino acids are low or depleted or when a substrate for sulfation is given in high doses, competing reactions with glucuronidation can take over. O-methylation is a competing reaction for catecholamine. As with glucuronidation, sulfation is a detoxication reaction, although sulfate conjugates have been reported to be pharmacologically active (e.g., minoxidil sulfate, dehydroepiandrosterone sulfate, and morphine 6-sulfate) or are converted into unstable sulfate conjugates that form reactive intermediates implicated in carcinogenesis and tissue damage. Sulfation of an alcohol generates a good leaving group and can be

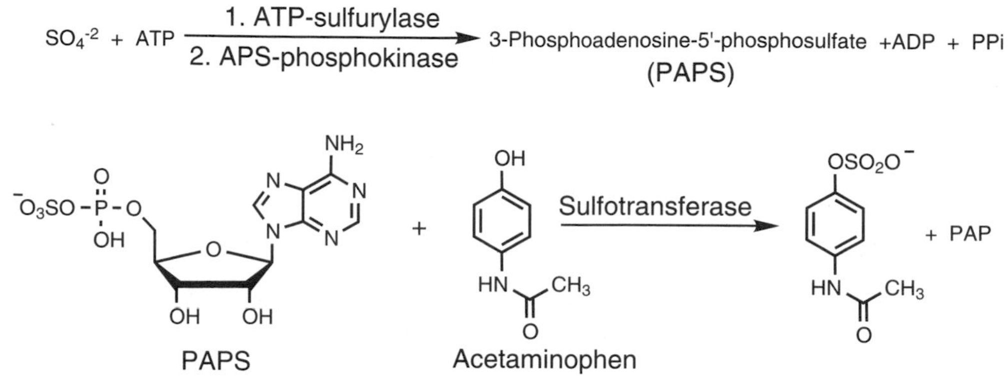

SO_4^{-2} + ATP —1. ATP-sulfurylase / 2. APS-phosphokinase→ 3-Phosphoadenosine-5'-phosphosulfate +ADP + PPi (PAPS)

PAPS + Acetaminophen —Sulfotransferase→ + PAP

Fig. 8–18. Sulfation pathways.

an activation process for alcohols to produce a reactive electrophilic species.[26]

Sulfate conjugation is principally a reaction of phenols, and to a lesser extent of alcohols, to form highly ionic and polar sulfates (R-O-SO_2H). N-sulfates, like the N-glucuronides, are capable of promoting cytotoxicity by facilitating the formation of reactive electrophilic intermediates. Sulfation of N-oxygenated aromatic amines is an activation process for some arylamines that can eliminate the sulfate to an electrophilic species capable of reacting with proteins or DNA, e.g., 2-acetylaminofluorene. N-sulfation of arylamines to arylsulfamic acids, R-$NHSO_3$H, is a minor pathway.

Sulfate conjugates are excreted mostly in the urine, but biliary elimination is common for steroids. On hydrolysis of biliary sulfate conjugates in the intestine by sulfatases, the parent drug (or xenobiotic) or its metabolites may be reabsorbed into the portal circulation for eventual elimination in the urine as sulfate conjugate (see the section on enterohepatic cycling). The rate of sulfation appears to be age dependent, decreasing with age.

An important site of sulfation, especially after oral administration, is the intestine. The result is a presystemic first pass effect, decreasing drug bioavailability for several drugs for which the primary route of conjugation is sulfation. Drugs such as isoproterenol, albuterol, steroid hormones, α-methyldopa, acetaminophen, fenoldopam are affected. Competition for intestinal sulfation between coadministered substrates may influence their bioavailability with either an enhancement of or decrease in therapeutic effects. An example would be coadministration of acetaminophen and the oral contraceptive ethynylestradiol.

Conjugation with Amino Acids

Conjugation with amino acids is an important route in the conjugation of drug and xenobiotic carboxylic acids for elimination.[38] Glycine, the most common amino acid, forms water-soluble ionic conjugates with aromatic, arylaliphatic, and heterocyclic carboxylic acids. These amino acid conjugates are usually less toxic than their precursor acids and are excreted readily into the urine and bile. These reactions involve the formation of an amide or peptide bond between the xenobiotic carboxylic acid and the amino group of an amino acid, usually glycine. The xenobiotic must first be activated to its coenzyme A (CoA) thioester before reacting with the amino group (Fig. 8–19). The formation of the xenobiotic acyl CoA thioester is of critical importance in intermediary metabolism of lipids as well as intermediate and long chain fatty acids.

The major metabolic biotransformations for xenobiotic carboxylic acids include conjugation with either glucuronic acid or glycine. The metabolic fate of these carboxylic acids depends on the size and type of substituents adjacent to the carboxyl group. Most unbranched aliphatic acids are completely oxidized and do not usually form conjugates, although branched aliphatic and arylaliphatic acids are resistant to β-oxidation and form glycine and/or glucuronide conjugates. Interestingly, substitution of the α-carbon favors glucuronidation rather than glycine conjugation. Benzoic and heterocyclic aromatic acids are principally conjugated with glycine. Glycine conjugation is preferred for xenobiotic carboxylic acids at low doses, and glucuronidation is preferred at high doses with broad substrate selectivity. In humans and some species of monkeys, glutamine forms a conjugate with phenylacetic acids and related arylacetic acids. Bile acids form conjugates with glycine and taurine by the action of enzymes in the microsomal fraction rather than in the mitochondria.

2-Arylpropionic acids ("profens") are a major group of nonsteroidal anti-inflammatory drugs (NSAID) that exist in two enantiomeric forms.[41] The anti-inflammatory activity (inhibition of cyclooxygenase) for the NSAID resides with the S(+) enantiomer. The intriguing aspect for the metabolism of the NSAID is their unidirectional chiral inversion from the R(−) to the S(+) enantiomer (Fig. 8–20). The NSAID acyl CoA thioester is the critical intermediate for this chiral inversion of the 2-arylpropionic acids and the formation of the thioester is stereospecific for the pharmacologically inactive R-enantiomer.[42] Racemic ibuprofen and related anti-inflammatory 2-arylpropionic acids (e.g., benoxaprofen, carprofen, cicliprofen, clidanac, fenoprofen, indoprofen, ketoprofen, loxoprofen, and naproxen) undergo in vivo metabolic inversion to the more active S-enantiomer via the formation, epimerization, and hydrolysis of their respective acyl CoA thioesters.[41] The unidirectional R- to S-inversion of ibuprofen is attributed to the stereoselective thioester formation of (R)-ibuprofen CoA, and not to the

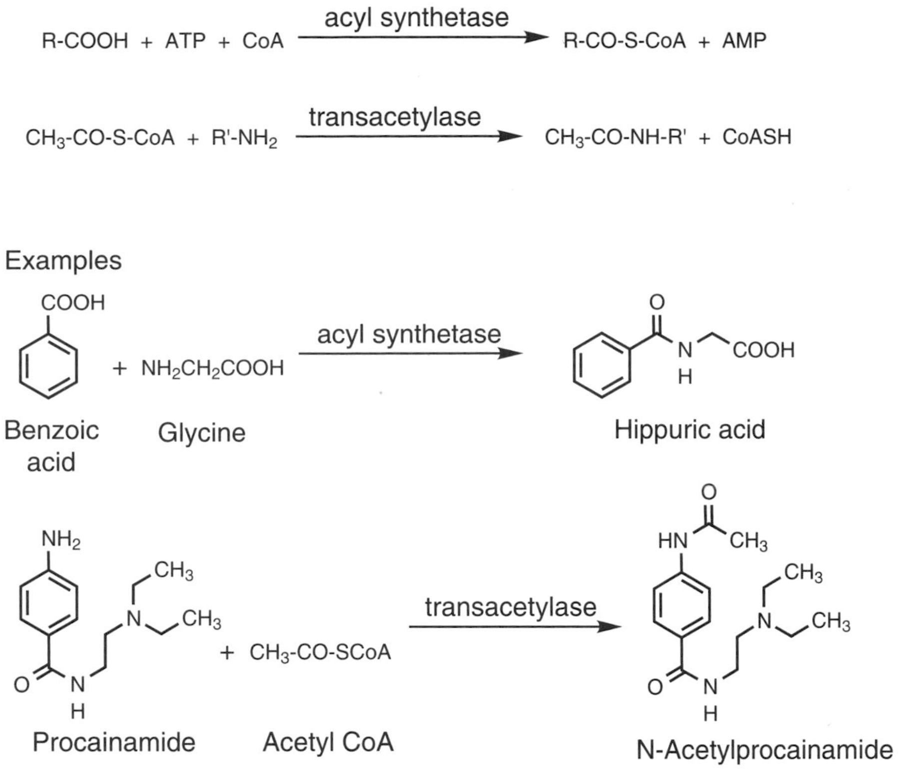

$$R\text{-}COOH + ATP + CoA \xrightarrow{\text{acyl synthetase}} R\text{-}CO\text{-}S\text{-}CoA + AMP$$

$$CH_3\text{-}CO\text{-}S\text{-}CoA + R'\text{-}NH_2 \xrightarrow{\text{transacetylase}} CH_3\text{-}CO\text{-}NH\text{-}R' + CoASH$$

Examples

Benzoic acid + Glycine (NH_2CH_2COOH) —acyl synthetase→ Hippuric acid

Procainamide + Acetyl CoA ($CH_3\text{-}CO\text{-}SCoA$) —transacetylase→ N-Acetylprocainamide

Fig. 8–19. Amino acid conjugation pathways of carboxylic acids with glycine and acetylation pathways.

stereoselectivity of either the epimerization or hydrolysis steps.[42] S(+)-ibuprofen does not form its CoA thioester in vivo. Because the formation of 2-arylpropionyl CoA thioester is analogous to the activation and metabolism of medium and long chain fatty acids, it seems possible that conditions either elevating (e.g., diabetes or fasting) or depleting CoA may alter CoA thioester formation of the 2-arylpropionic acids and their in vivo metabolic inversion. Amino acid conjugation (i.e., CoA activation) is more sensitive to steric hindrance than is glucuronidation (e.g., arylacetic acids).

In contrast to the enhanced reactivity and toxicity of the various glucuronide, sulfate, acetyl and glutathione conjugates, amino acid conjugates have not proven toxic. It has been proposed that amino acid conjugation is a detoxication pathway for reactive acyl CoA thioesters.

Acetylation

Acetylation is principally a reaction of amino groups involving the transfer of acetyl CoA to an aromatic primary or aliphatic amine, amino acid, hydrazine, or sulfonamide group.[38] The liver is the primary site of acetylation, although extrahepatic sites have been identified. Sulfon-

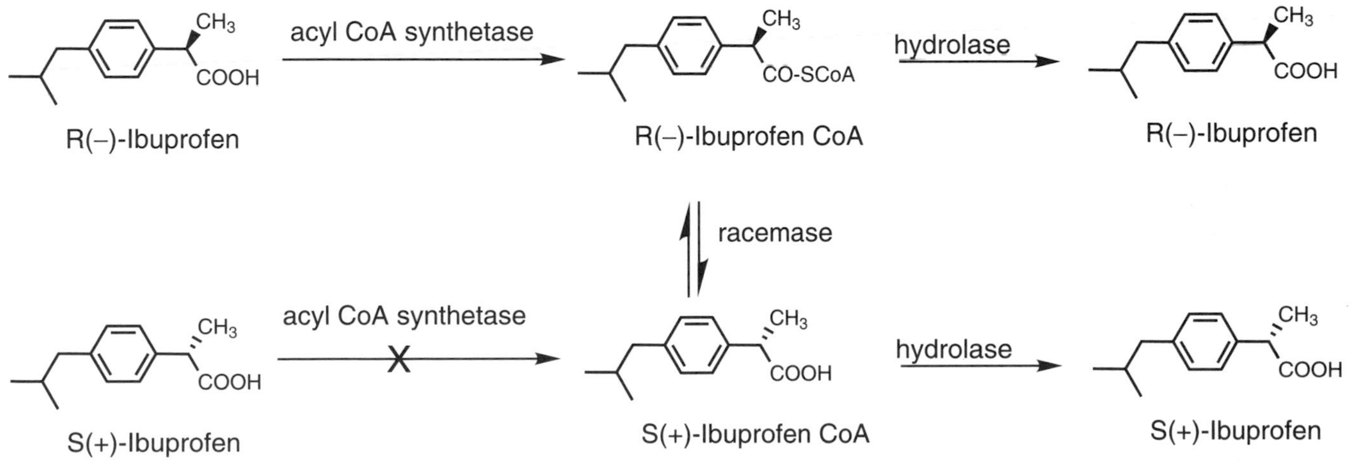

R(−)-Ibuprofen —acyl CoA synthetase→ R(−)-Ibuprofen CoA —hydrolase→ R(−)-Ibuprofen

racemase

S(+)-Ibuprofen —acyl CoA synthetase (X)→ S(+)-Ibuprofen CoA —hydrolase→ S(+)-Ibuprofen

Fig. 8–20. Stereospecific inversion of R(−) to S(+)-ibuprofen.

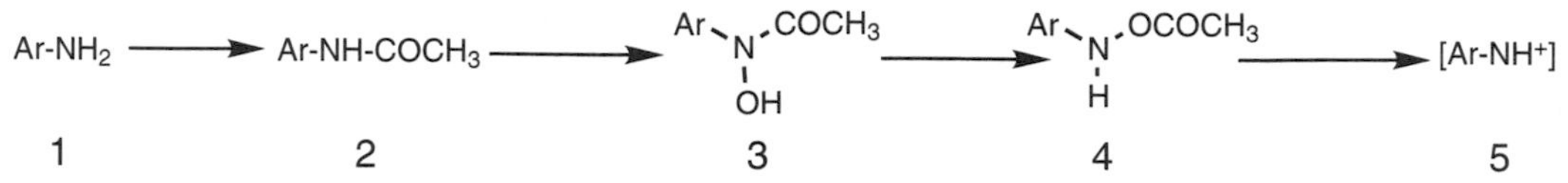

$$Ar\text{-}NH_2 \longrightarrow Ar\text{-}NH\text{-}COCH_3 \longrightarrow Ar\text{-}N(OH)\text{-}COCH_3 \longrightarrow Ar\text{-}N(H)\text{-}OCOCH_3 \longrightarrow [Ar\text{-}NH^+]$$

1 2 3 4 5

Fig. 8–21. Bioactivation of acetylated arylamines.

amides, being difunctional, can form either N^1 or N^4 acetyl derivatives, and, in some instances, the diacetylated derivative has been identified. Secondary amines are not acetylated. Acetylation may produce conjugates that retain the pharmacologic activity of the parent drug, e.g., N-acetylprocainamide (see Fig. 8–19).

The existence of genetic polymorphism in the rate of acetylation has important consequences in drug therapy and tumorigenicity of xenobiotics. Acetylation polymorphism has been associated with differences in human drug toxicity between the two acetylator phenotypes, slow and fast acetylators. Slow acetylators are more prone to drug-induced toxicities and accumulate higher blood concentrations of the unacetylated drug (e.g., hydralazine and procainamide-induced lupus erythematous, isoniazid-induced peripheral nerve damage, sulfasalazine-induced hematologic disorders) than rapid acetylators. Fast acetylators eliminate the drug more rapidly by conversion to its relatively nontoxic N-acetyl metabolite. For some drug substances, however, fast acetylators may pose at greater risk of liver toxicity than slow acetylators because they produce toxic metabolite(s) more rapidly, e.g., isoniazid forms the hepatotoxic monoacetylhydrazine metabolite. The difference in acetylator phenotype could also influence adverse drug reactions, e.g., isoniazid and rifampin and isoniazid.

The possibility arises that genetic differences in acetylating capacity may confer differences in susceptibility to chemical carcinogenicity from arylamines. The tumorigenic activity of arylamines (1) may be the result of a complex series of sequential metabolic reactions commencing with N-acetylation (2), oxidation to arylhydroxamic acids (3), metabolic transformation to acetoxyarylamines (4) by N, O-acyltransferase, and loss of the acetoxy group to form an arylnitrenium ion (5) that is capable of covalently binding to nucleic acids and proteins (see the section on glucuronidation) (Fig. 8–21).[43] The rapid acetylator phenotype is expected to form the acetoxyarylamine at a greater rate than the slow acetylator, and thereby presents a greater risk for the development of bladder and liver tumors than the slow acetylator.

Glutathione Conjugation and Mercapturic Acid Synthesis

Mercapturic acid are S-derivatives of N-acetylcysteine synthesized from glutathione (GSH).[38] It is generally accepted that most compounds metabolized to mercapturic acids first undergo conjugation with glutathione, catalyzed by an enzyme called glutathione S-transferase, found in the soluble supernatant liver fractions. The reaction is depicted in Figure 8–22.

Glutathione transferase apparently increases the ionization of the thiol group of GSH, increasing its nucleophilicity toward electrophiles and conjugating with these potentially harmful electrophiles, thereby protecting other vital nucleophilic centers in the cell such as nucleic acids and proteins. Glutathione is also capable of reacting nonenzymatically with nucleophilic sites on neighboring macromolecules. Once conjugated with GSH, the electrophiles are usually excreted in the bile and urine.

A wide range of functional groups yield thioether conjugates of GSH, as well as products other than thioethers. The reaction may be viewed as the result of nucleophilic attack by GSH on electrophilic carbons with leaving groups such as halogen (alkyl, alkenyl, aryl, or aralkyl halides), sulfate (alkylmethanesulfonates), and nitro (alkyl nitrates) groups, ring opening of small ring ethers (epoxides, β-lactones, such as β-propiolactone), and the addition to the activated β-carbon of an α,β-unsaturated carbonyl compound. Organic nitrate esters (e.g., the coronary vasodilator nitroglycerin) undergo a dismutation reaction that results in the oxidation of GSH to GSSG (through formation of the labile S-nitrate conjugation product), and reduction of the nitrate ester to an alcohol and inorganic nitrite. The lack of substrate specificity gives argument to the fact that glutathione transferase has undergone adaptive changes to accommodate the variety of xenobiotics to which it is exposed. Usually, the conjugation of an electrophilic compound with GSH is a reaction of detoxication, but some carcinogens have been activated through conjugation with GSH.[23]

The enzymatic conjugation of GSH with epoxides provides a mechanism of protecting the liver from injury caused by certain bioactivated intermediates (see the subsequent section on toxicity from oxidative metabolism). Not all epoxides are substrates for this enzyme, but the more chemically reactive epoxides appear to be better substrates. Important among the epoxides that are substrates for this enzyme are those produced from halobenzenes, PAH, and others by the action of CYP450 monooxygenase. Epoxide formation exemplifies bioactivation because the epoxides are reactive and potentially toxic, whereas their GSH conjugates are inactive. Conjugation of GSH with the epoxides of aryl hydrocarbons eventually results in the formation of hydroxymercapturic acids (premercapturic acids), which undergo acid-catalyzed dehydration to the mercapturic acids. The halobenzenes are usually conjugated in the para-position. The mutagenicity of 1,2-dihaloalkanes (e.g., ethylene dibromide) results from reactions with GSH displacing bromide and

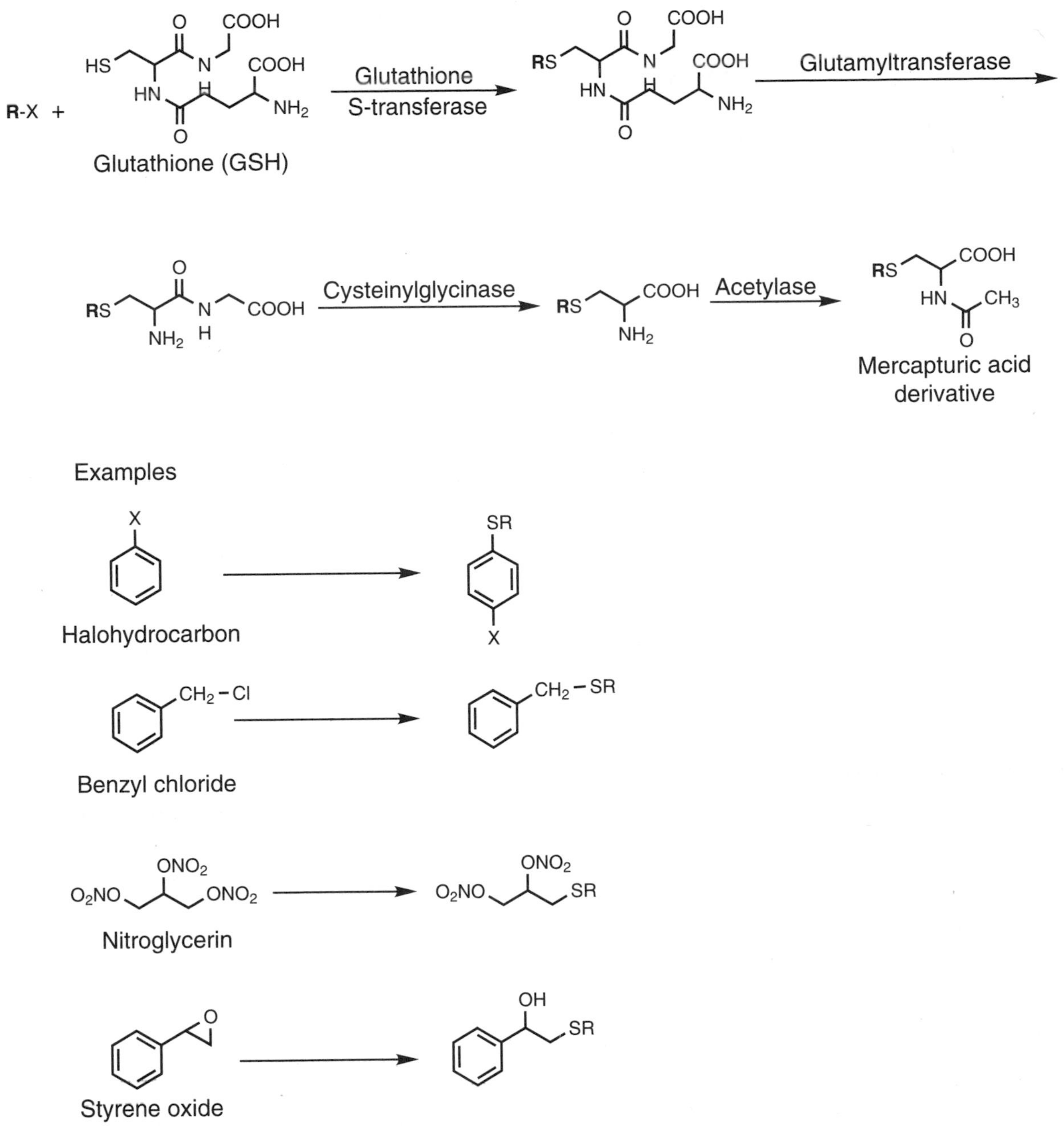

Fig. 8–22. Glutathione and mercapturic acid conjugation pathways.

rearrangement to an episulfonium electrophile that can in turn react with DNA.

The mercapturic acid pathway appears to have evolved as a protective mechanism against xenobiotic-induced hepatotoxicity or carcinogenicity, serving to detoxify a large number of noxious substances that we inhale or ingest (e.g., drugs, xenobiotics, and plant toxins in food) or that are produced daily in the human body. A correlation exists between the hepatotoxicity of acetaminophen and levels of GSH in the liver. The probable mechanism of toxicity that has emerged from animal studies is that acetaminophen is CYP1A2 and CYP2E1 oxidized to the N-acetyl-p-benzoquinoneimine intermediate that conjugates with and depletes hepatic GSH levels (Fig. 8–23). This action allows the benzoquinoneimine to bind covalently to tissue macromolecules. The mercapturic acid derivative of acetaminophen represents approximately 2% of the administered dose of acetaminophen. Thus, the possibility exists that those toxic metabolites usually are detoxified by conjugating with GSH exhibit their hepatotoxicity (or perhaps carcinogenicity) because the liver has been depleted of GSH and is incapable of inactivating them. Pretreatment of animals with phenobarbital often hastens the depletion of GSH by increasing the formation of epoxides or other reactive intermediates.

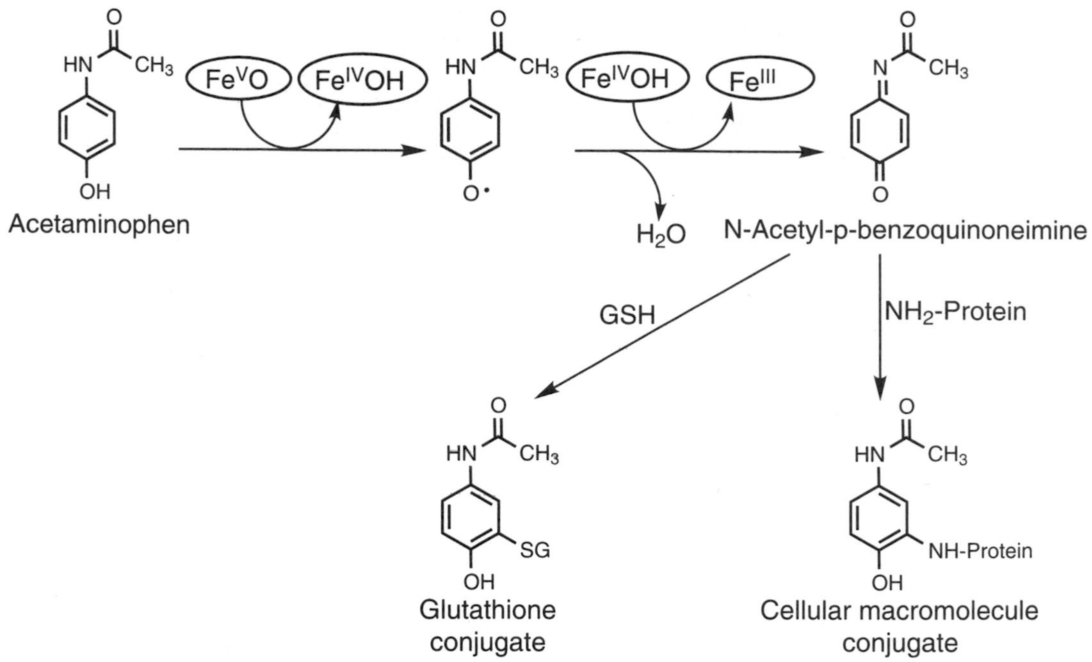

Fig. 8–23. Proposed mechanism for the CYP450-catalyzed oxidation of acetaminophen to its N-acetyl-p-benzoquinoneimine intermediate, which can further react with either glutathione (GSH) or cellular macromolecules (NH_2-PROTEIN).

Methylation

O- and N-methylation is a common biochemical reaction but appears to be of greater significance in the metabolism of endogenous compounds than for drugs and other foreign compounds.[38] Methylation differs from other conjugation processes in that the O-methyl metabolites formed may in some cases have as great or greater pharmacologic activity and lipophilicity than the parent molecule (e.g., the conversion of norepinephrine to epinephrine). Methionine is involved in the methylation of endogenous and exogenous substrates because it transfers its methyl group via the activated intermediate S-adenosylmethionine (SAM) to the substrate under the influence of methyl transferases (Fig. 8–24). Methylation results principally in the formation of O-methylated, N-methylated, and S-methylated products.

The process of O-methylation is catalyzed by the magnesium-dependent enzyme catechol-O-methyltransferase (COMT) transferring a methyl group to the meta- or less frequently to the paraphenolic-OH (regioselectivity) of catecholamines (e.g., norepinephrine), as well as their deaminated metabolites. It does not methylate monohydric or other dihydric phenols. The meta/para product ratio depends greatly on the type of substituent attached to the catechol ring. Substrates specific for COMT include catechol-like structures: the catecholamines, norepinephrine, epinephrine, and dopamine; amino acids, L-DOPA, and α-methyl-DOPA; and 2- and 4-hydroxyestradiol metabolites of estradiol. The enzyme is thought to function in the biologic inactivation of the adrenergic neurotransmitter norepinephrine as well as other endogenous and exogenous catechol-like substances. It is found in liver, kidneys, nervous tissue, and other tissues.

Hydroxyindole-O-methyltransferase, which O-methylates N-acetylserotonin, serotonin, and other hydroxyindoles, is also found in the pineal gland for the formation of melatonin. This enzyme differs from COMT in that it does not methylate catecholamines and has no requirement for magnesium iron.

The N-methylation of various amines is among several conjugate pathways for metabolizing amines. Specific N-methyltransferases catalyze the transfer of active methyl groups from SAM to the acceptor substance. Phenylethanolamine-N-methyltransferase (PNMT) methylates a number of endogenous and exogenous phenylethanolamines (e.g., normetanephrine, norepinephrine, and norephedrine) but does not methylate phenylethylamines. Histamine-N-methyl transferase specifically methylates histamine, producing the inactive metabolite N^1-methyl-histamine. Amine-N-methyltransferase will N-methylate a variety of primary and secondary amines from a number of sources, including endogenous biogenic amines (serotonin, tryptamine, tyramine, and dopamine) and drugs (desmethylimipramine, amphetamine, and normorphine). Amine-N-methyl transferases seem to have a role in recycling N-demethylated drugs.

Thiols are generally considered toxic, and the role of thiol S-methyl transferases is among other detoxication pathways for the oxidative metabolism of these compounds (see the discussion of flavin-containing monooxygenases). The S-methylation of sulfhydryl compounds also involves a microsomal enzyme requiring SAM. Although a wide range of exogenous sulfhydryl compounds is S-methylated by this microsomal enzyme, none of the endogenous sulfhydryl compounds, e.g., cysteine and GSH, can function as substrates. S-methylation clearly

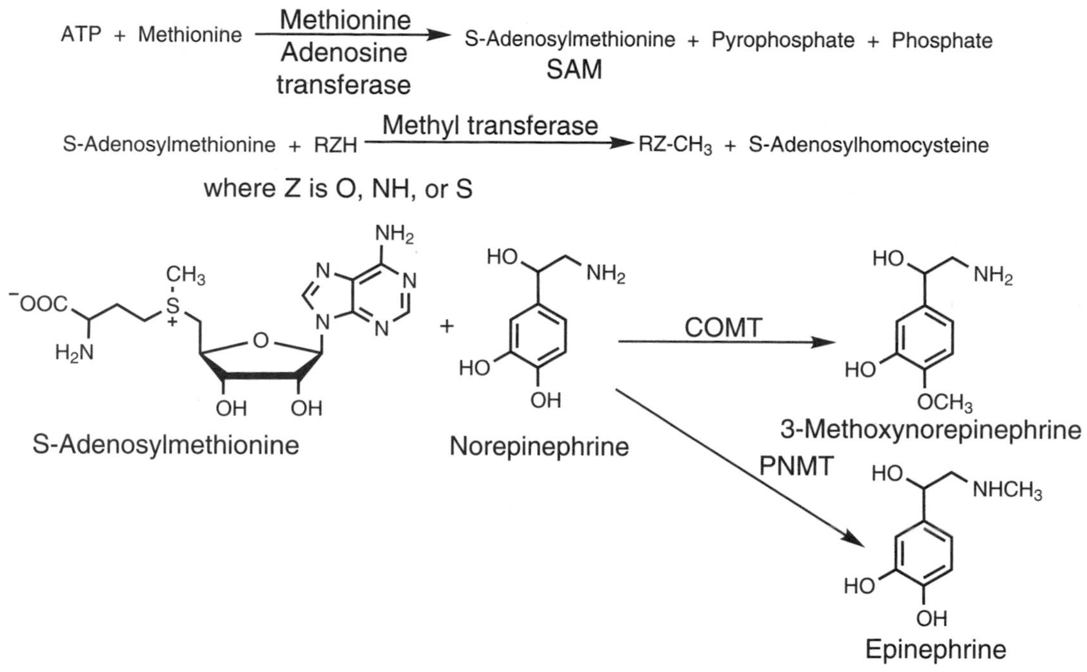

$$\text{ATP} + \text{Methionine} \xrightarrow[\text{Adenosine transferase}]{\text{Methionine}} \underset{\text{SAM}}{\text{S-Adenosylmethionine}} + \text{Pyrophosphate} + \text{Phosphate}$$

$$\text{S-Adenosylmethionine} + \text{RZH} \xrightarrow{\text{Methyl transferase}} \text{RZ-CH}_3 + \text{S-Adenosylhomocysteine}$$

where Z is O, NH, or S

S-Adenosylmethionine + Norepinephrine → (COMT) 3-Methoxynorepinephrine

→ (PNMT) Epinephrine

Fig. 8–24. Methylation pathways.

represents a detoxication step for thiols. Dialkyldithiocarbamates (e.g., disulfiram) and the antithyroid drugs (e.g., 6-propyl-2-thiouracil), mercaptans, and hydrogen sulfide (from thioglycosides as natural constituents of foods, mineral sulfides in water, fermented beverages, and bacterial digestion) are S-methylated. Other drugs undergoing S-methylation include captopril, thiopurine, penicillamine, and 6-mercaptopurine.

Conjugation of Cyanide

The toxicity of hydrogen cyanide is the result of its ionization to cyanide ion in biologic tissues. It is a powerful metabolic inhibitor that arrests cellular respiration by inactivating cytochrome enzymes fundamental to the respiratory process, as well as combining with hemoglobin to form cyanomethemoglobin, which is incapable of transporting oxygen to tissues. With the wide prevalence of cyanoglycosides in plant materials, the ability to detoxity cyanide is a vital function of the liver, erythrocytes, and other tissues. Rhodanese, a mitochondrial enzyme in liver and other tissues, catalyzes the formation of thiocyanate from cyanide rapidly in the presence of thiosulfate and colloidal sulfur, but cysteine and GSH are poor sulfur donors. The detoxification of cyanide depends on the availability of a physiologic pool of thiosulfate, the origin of which is not known. A possible source for thiosulfate is the transamination of cysteine to β-mercaptopyruvate and transfer of the mercapto group by a sulfur transferase to sulfite producing thiosulfate. Depletion of this pool increases cyanide toxicity. In the presence of excess cyanide, however, minor pathways for cyanide metabolism may occur, including oxidation to cyanate (NCO^-), reaction with cobalamin (vitamin B12) to form cyanocobalamin, and the formation of 2-iminothiazolidine-4-carboxylic acid from the nonenzymatic reaction between cystine and cyanide.[44]

$$S_2O_3^{2-} + CN^- \longrightarrow CNS^- + SO_3^{2-}$$

GENETIC POLYMORPHISM

The variability between individuals in the therapeutic response to the same dose of a given drug may be the result of an alteration in the metabolism, absorption, distribution, or elimination of the drug. A decrease in the rate of drug metabolism can lead to an increase in drug blood levels, increasing the likelihood of drug interactions or toxicity. On the other hand, increasing the rate of drug metabolism may result in a reduced therapeutic response. The wide range of interindividual rates of CYP450-dependent drug metabolism cannot be explained solely in terms of genetics, but could include the induction state of CYP450 as well as liver function and perfusion, disease, sex, age, exercise, occupational exposures, drugs, circadian and seasonal variation, dietary factors, renal function, stress, starvation, alcohol intake, and smoking. Polymorphisms have now been detected in many xenobiotic-metabolizing enzymes, and many xenobiotics may be bioactivated or detoxicated by polymorphic enzymes.[45–47] Polymorphism of drug-metabolizing enzymes gives rise to subgroups in the population differing in their ability to perform certain biotransformation reactions with obvious clinical ramifications.[47] Metabolic polymorphism may have several consequences, e.g., when enzymes detoxicate drugs that are used either therapeutically or socially, adverse drug reactions may occur in individuals lacking this enzyme. This interindividual variation in the metabolism of drugs and other xenobiotics is the result of genetic differences in the natural expression of a CYP450 isoform. The discovery of genetic polymorphism resulted from the observation of unusual or exaggerated drug reactions after normal doses of drugs to patients.

Cytochrome P450 isoforms (CYP1–3 families) represent the best studied group of xenobiotic-metabolizing enzymes. it is widely accepted that genetic factors have an important impact on the oxidative metabolism and pharmacokinetics of drugs whose metabolism cosegregates with that of debrisoquine (see Table 8–9). The debrisoquine type of genetic polymorphism has been studied extensively with the discovery that poor metabolizers of debrisoquine also have an impaired metabolism of over 20 other drugs, including β-adrenergic blockers, antiarrythmics, antidepressants, opiods, and many other clinically used drugs.[45,46] The debrisoquine/sparteine-type polymorphism is a clinically important genetic variation of drug metabolism characterized by two phenotypes, the extensive metabolizer (EM) and the poor metabolizer (PM). The PM phenotype is inherited as an autosomal recessive trait with a defective CYP2D6 gene leading to zero expression or the expression of a nonfunctional protein. The majority of compounds listed in Table 8–9 are either cardiovascular or psychoactive drugs, and PM are associated with the risk of serious adverse effects if any of these drugs have a narrow therapeutic index or when two or more drugs are given simultaneously. Codeine is ineffective in PM because it cannot be O-demethylated to morphine. Individuals with PM phenotype are also characterized by loss of CYP2D6 stereoselectivity in hydroxylation reactions. The biggest difference occurs with hydroxylation reactions and less with dealkylation reactions.

Clinical studies have demonstrated that the PM phenotype of debrisoquine polymorphism represents a high-risk population group (approximately 5 to 10% of the Caucasian population, about 2% Oriental, and 1% Arabic) with a propensity to develop adverse drug effects, or even cancer.

It can be anticipated that large differences in steady-state concentration for CYP2D6 substrates will occur between individuals of the PM and EM phenotypes when they receive the same dose. The end result is wide interindividual variations in the intensity and duration of drug effects and side effects. For some drugs (see Table 8–9) genetic factors controlling drug metabolism are a major source of variation in concentration and hence therapeutic response and side effects. Depending on the drug and reaction type, a 10- to 30-fold difference in blood concentrations be observed in the PM phenotype debrisoquine polymorphism.[47]

The activities of the cytochrome isoforms vary among individuals and environmental influences in the bioactivation of chemical carcinogens (i.e., CYP1A1). Such variations can have a major influence in determining drug toxicity and the risk of cancer. The rate of carcinogen activation may underlay strong interindividual variations and the influence of hereditary factors on the effect of inducing compounds. The presence or absence of a particular CYP450 isoform may be a rationale predicting the individual risk from exposure to carcinogenic compounds. A suggestion has been made that EM phenotypes may be more prone than PM phenotypes to develop cancers because they are better able to activate procarcinogens into active carcinogens.

The large interindividual variability observed in the therapeutic response to the anticonvulsant drug mephenytoin can now be attributed to polymorphism.[47] The defect is with the human isoform (CYP2C18 or CYP2CMP) catalyzing the p-hydroxylation of the (S)-mephenytoin. The R-enantiomer is N-demethylated by CYP2C8 and its metabolism is similar for both PM and EM. A polymorphism in the metabolism of (S)-mephenytoin has been detected in 2 to 6% of the Caucasians and 15 to 20% of Orientals unable to p-hydroxylate (S)-mephenytoin. The metabolism of a number of therapeutically important drugs that cosegregate with (S)-mephenytoin are listed in Table 8–8. Polymorphic isoforms of CYP2C are expressed in human livers and catalyze polymorphic hydroxylation of the clinically important mephenytoin; polymorphism for tolbutamide hydroxylation is catalyzed by CYP2C9/10, but is less active with CYP2C8.

Acetylation, a nonmicrosomal form of metabolism, also exhibits polymorphisms and was first demonstrated in the acetylation metabolism of isoniazid (see the section on acetylation). Slow acetylators are autosomal recessive and fast acetylators are either heterozygotes or homozygotes. Several forms of acetyl transferase occur in humans. Some clinically used drugs undergoing polymorphic acetylation include isoniazid, procainamide, hydralazine, phenelzine, dapsone, caffeine, some benzodiazepines, and possibly the carcinogenic secondary N-alkylaryl amines (2-aminofluorene, benzidine, and 4-aminobiphenyl). Intestinal acetyl transferase appears not to be polymorphic (i.e., 5-aminosalicylic acid). The proportion of fast acetylation phenotype in Caucasians is about 30 to 45%, in the Oriental population, 80 to 90%, and in Canadian Eskimos, 100%. The incidence of drug-induced systemic lupus erythematosus from chronic procainamide therapy is more likely to appear with slow acetylators.

The only FMO pathway exhibiting polymorphism is the genetic disease trimethylaminuria, in which individuals excrete diet-derived free trimethylamine in the urine. Usually, trimethylamine undergoes extensive FMO N-oxidation.

Polymorphism has been associated with serum cholinesterases, alcohol dehydrogenases, aldehyde dehydrogenases, epoxide hydrolase, and xanthine oxidase.[47] Approximately 50% of the Oriental population lack aldehyde dehydrogenase, resulting in high levels of acetaldehyde following ethanol ingestion and causing the symptoms of nausea and flushing. People with genetic variants of cholinesterase respond abnormally to succinylcholine, procaine, and other related choline esters. The clinical consequence of reduced enzymic activity for cholinesterase is succinylcholine and procaine are not hydrolized in the blood, experiencing prolongation of their respective pharmacologic activities.

It is clear that genetic polymorphism has contributed a great deal to our understanding of interindividual variation in the metabolism of drugs, and how to change dose accordingly to minimize drug toxicity and to improve therapeutic efficacy. In humans, drugs not subject to polymorphic metabolism also exhibit substantial interindividual variation in their disposition, attributed to a great extent to environmental factors such as inducing agents, smoking, and alcohol ingestion.

EXTRAHEPATIC METABOLISM

Because the liver is the primary tissue for xenobiotic metabolism, it is not surprising that our understanding of mammalian CYP450 monooxygenase is based chiefly on hepatic studies. Although the tissue content of CYP450s is highest in the liver, CYP450 enzymes are ubiquitous and their role in extrahepatic tissues remains unclear. The CYP450 pattern in these tissues differs considerably from that in the human liver.[48] In addition to liver tissue, CYP450 enzymes are found in lung, nasal epithelium, intestinal tract, kidney and adrenal tissues, and brain. It is possible that the expression and induction of the isoforms in the extrahepatic tissues by environmental xenobiotics may affect the expression and activity of the isoforms of CYP450 in the biosynthesis and metabolism of endogenous steroids. Therefore, characterization of CYP450 in extrahepatic tissues is important to our overall understanding of the biologic importance of this family of isoforms to improved drug therapy, design of new drugs and dosages forms, toxicity, and carcinogenicity.

The mucosal surfaces of the gastrointestinal tract, the nasal passages, and lungs are major portals of entry for xenobiotics into the body and as such are continuously exposed to a variety of orally ingested or inhaled airborne xenobiotics including drugs, plant toxins, environmental pollutants, and other chemical substances. As a consequence of this exposure, these tissues represent a major target for necrosis, tumorigenesis, and other chemically induced toxicities. Many of these toxins and chemical carcinogens are relatively inert substances that must be bioactivated in order to exert their cytotoxicity and/or tumorigenicity. The epithelial cells of these tissues are capable of metabolizing a wide variety of exogenous and endogenous substances, and provide the principal and initial source of biotransformation for these xenobiotics during the absorptive phase. The consequences of such presystemic biotransformation is either a decrease in the amount of xenobiotic available for systemic absorption by facilitating the elimination of polar metabolites, or toxicification by activation to carcinogens, which may be one determinant of tissue susceptibility for the development of intestinal cancer. The risk of colon cancer may depend on dietary constituents that contain either procarcinogens or compounds modulating the response to carcinogens.

The intestinal mucosa is particularly important for orally administered drugs prone to microsomal oxidation[49] and glucuronidation and sulfation conjugation pathways.[38] Mounting evidence shows that many of the clinically relevant aspects of CYP450 may in fact occur at the level of the intestinal mucosa and differences among patients in dosing requirements. The highest concentrations of CYP450s occur in the duodenum with a gradual tapering into the ileum. CYP2E, CYP3A, CYPC8-19, and CYP2D6 have been identified in the human intestine. Therefore, intestinal CYP450 isoforms provide potential presystemic first pass metabolism of ingested xenobiotics affecting their oral bioavailability (e.g., hydroxylation of naloxone) or bioactivation of carcinogens or mutagens.

The CYP3A subfamily plays a significant role in the intestinal prehepatic metabolism of its numerous drug substrates (see Table 8–12).[48–50] The concentration of functional CYP3A in the gut is the major determinant of a drug's oral bioavailability, which determines to a great extent blood levels and therapeutic response, and is influenced by genetic disposition, induction, and inhibition. Xenobiotics ingested orally can modify the activity of intestinal CYP3A enzymes by induction, inhibition, and stimulation. By modulation of the isoform pattern in the intestine, a drug or xenobiotic could alter its own metabolism and that of others in a time- and dose-dependent manner. Its concentration in the intestine is comparable to that of the liver. Drugs known to be substrates for CYP3A usually have a low and variable oral bioavailability that may be explained by prehepatic metabolism by the small intestine CYP450 monooxygenases. A comparison of cyclosporin pharmacokinetics after oral and intravenous application has raised the issue of possible prehepatic metabolism of cyclosporin. These studies have shown the poor bioavailability of cyclosporin is attributed to intestinal metabolism of cyclosporin by CYP3A4.[50] The oral administration of dexamethasone induces the formation of CYP3A and erythromycin inhibits it. The glucocorticoid inducibility of CYP3A4 may also be a factor in differences of metabolism between males and females. Recent studies[27] have suggested that intestinal CYP3A4 C-2 hydroxylation of estradiol contributed to the oxidative metabolism of endogenous estrogens circulating with the enterohepatic recycling pool. Norethisterone has a low oral bioavailability of 42% because of oxidative first pass metabolism (CYP3A), but levonorgestrel is completely available in women with no conjugated metabolites.

Several clinically relevant drug interactions between orally co-administered drugs and CYP3A subfamily can be explained by a modification of drug metabolism at the CYP450 level. Drugs that are known inhibitors, inducers, and substrates for CYP3A can potentially interact with the metabolism of a coadministered drug.[29] Inducers can reduce absorption and oral bioavailability, whereas these same factors are increased by inhibitors. For example, erythromycin can enhance the oral absorption of another drug by inhibiting its metabolism in the small intestine by a CYP3A isoform. Because CYP3A is a major metabolic pathway for the metabolism of synthetic glucocorticoids, prednisone, prednisolone, and methylprednisolone (but not dexamethasone) have been found to be competitive inhibitors of CYP3A, by virtue of being competitive substrates for CYP3A. In addition to co-administered drugs,

Table 8–15. Conjugation Reactions Occurring in Intestinal Wall

Reaction	Substrate
Glucuronidation	
Ester (acyl)	Retinoic acid Deoxycholic acid
Ether	Buprenorphine Ethynylestradiol Beta-adrenergic agonist Naloxone Morphine (3-O-) Acetaminophen Salicylamide Danthron
Sulfation	Ethynylestradiol (major) Danthron Beta-adrenergic agonists Acetaminophen Salicylamide
Glutathione conjugation (mercapturates)	Arene oxides Ethacrynic acid Isorbide dinitrate
Acetylation	Sulfonamides 5-Aminosalicylic acid N-hydroxyaminoflurene

metabolism interactions with exogenous CYP3A substrates secreted in the bile are possible.

It is not surprising that dietary factors can affect the intestinal CYP450 isoforms. For example, a two-day dietary exposure to cooked Brussels sprouts significantly decreased the 2α-hydroxylation of testosterone, yet induced CYP1A2 activity for PAH.

Because the intestinal mucosa is enriched with glucuronosyltransferases, sulfotransferases, and glutathione transferases (Table 8–15), presystemic first pass metabolism for orally administered drugs susceptible to these conjugation reactions results in their low oral bioavailability.[7] Presystemic metabolism often exceeds liver metabolism for some drugs. Other routes of administration (e.g., subcutaneous, intravenous, inhalation and nasal) for these susceptible drugs have been investigated in an attempt to overcome the pronounced presystemic metabolism. For example, more than 80% of intravenously administered albuterol is excreted unchanged in urine with the balance as glucuronide conjugates, whereas when it is administered orally, less than 5% is systemically absorbed because of intestinal sulfation and glucuronidation. Presystemic intestinal first pass metabolism is a major pathway in humans for most β-adrenergic agonists, e.g., glucuronides and/or sulfates for terbutaline, fenoterol, albuterol, and isoproterenol), morphine (3-O-glucuronide), acetaminophen (O-sulfate), and estradiol (3-O-sulfate). The bioavailability of orally administered estradiol or ethynylestradiol in females is about 50%, whereas mestranol (3-methoxyethynylestradiol) has greater bioavailability because it is not significantly conjugated. Levodopa has a low oral bioavailability because of its metabolism by intestinal L-aromatic amino acid decarboxylase. The activity of this enzyme depends on the percent bound of its cofactor pyridoxine (vitamin B6). Tyramine, which occurs in fermented foods such as cheeses and red wines, ripe bananas, and yeast extracts, is metabolized by both MAO-A and MAO-B in the gut wall.

The extensive first pass sulfation of phenolic drugs, for example, can lead to increased bioavailability of other drugs by competing for the available sulfate pool, resulting in the possibility of drug toxicity.[38] Concurrent oral administration of acetaminophen with ethynylestradiol resulted in a 48% increase in ethynylestradiol blood levels. Ascorbic acid, which is sulfated, also increases the bioavailability of concurrently administered ethynylestradiol. Sulfation and glucuronidation occur side by side, often competing for the same substrate, and the balance between sulfation and glucuronidation is influenced by several factors such as species, doses, availability of cosubstrates, inhibition, and induction of the respective transferases.

The occurrence of intestinal CYP450 enzymes and bacterial enzymes in the microflora allows the metabolism of relatively stable environmental pollutants and food-derived xenobiotics (i.e., plants contain a variety of protoxins, promutagens, and procarcinogens) into mutagens and carcinogens.[7] For example, cruciferous vegetables (Brussels sprouts, cabbage, broccoli, cauliflower, and spinach) are all rich in indole compounds (e.g., indole3-carbinol), which with regular and chronic ingestion are capable of inducing some intestinal CYP450s (CYP1A subfamily) and inhibiting others (CYP3A subfamily). It is likely these vegetables would also alter the metabolism of food-derived mutagens (e.g., heterocyclic amines produced during charbroiling of meat are CYP450 N-hydroxylated and become carcinogenic in a manner similar to aryl amines) and carcinogens.

The extent of a drug's metabolism in the small bowel and its role in clinically relevant drug interaction remains to be evaluated and must be taken into account in oral pharmacokinetics analysis of future drug interaction studies. Clinically significant interaction will not always occur when a drug is combined with other isoform subfamily substrates. Oral co-administration of a drug with drugs interacting with its metabolism need not be avoided, but the blood concentration of the drug must be monitored closely and the dose should be adapted.

Potential substrates for biotransformation by the gut microflora include orally administered drugs, xenobiotics, and other environmental substances. Moreover, the intestinal microflora may play an important role in the presystemic first pass metabolism of compounds that are poorly or incompletely absorbed by the gut mucosa (e.g., sulfasalazine). In contrast to the predominantly hepatic oxidative and conjugative metabolism of the liver, gut microflora is largely degradative, hydrolytic, and reductive, with a potential for both metabolic activation and detoxication of xenobiotics.

The intestinal microflora plays an important role in the enterohepatic recirculation of xenobiotics via their conjugated metabolites (e.g., digoxin, oral contraceptives norethisterone and ethynylestradiol, and chloramphenicol) and endogenous substances (steroid hormones, bile acids, folic acid and cholesterol), which re-enter the gut via the bile.[7] Compounds eliminated in the bile are conjugated with glucuronic acid, glycine, sulfate, glutathione, and once secreted into the small intestine, the bacterial β-glucuronidase, sulfatase, nitroreductases, and various glycosidases catalyze the hydrolysis of the conjugates. The clinical use of oral antibiotics (e.g., erythromycin, penicillin, clindamycin, and aminoglycosides) has a profound effect on the gut microflora and the enzymes responsible for the hydrolysis of drug conjugates undergoing EHC. Bacterial reduction include nitro reduction of nitroimidazole, azo reduction of sulfasalazine (a prodrug) to 5-aminosalicylic acid (the active moiety) and sulfapyridine, and reduction of the sulfoxide of sulfinpyrazone to its sulfide. The sulfoxide of sulindac is reduced by both gut microflora and hepatic CYP450s.

Most of the hepatic xenobiotic biotransformation pathways are also operative in the lung.[51] Because of the differences in organ sizes, the total content of the pulmonary xenobiotic-metabolizing enzyme systems is generally lower than in the liver, creating the impression for a minor role of the lung in xenobiotic elimination. The specific activities of the microsomal CYP450 and FMO, epoxide hydrolase, and the phase 2 conjugation pathways, however, are comparable to those in the liver. Thus, the lungs may play a significant role in the metabolic elimination or activation of drugs and xenobiotics. When drugs are injected intravenously, intramuscularly, or subcutaneously, or after skin absorption, the drug initially enters the pulmonary circulation after which the lung becomes the organ of first pass metabolism for the drug. The blood levels and therapeutic response of the drug are influenced by genetic disposition, induction, and inhibition of the pulmonary metabolizing enzymes. By modulation of the CYP450 isoform pattern in the lung, a drug or xenobiotic could alter its own metabolism and that of others in a time- and dose-dependent manner. Because of its position in the circulation, the lung provides a second pass metabolism for xenobiotics and their metabolites exiting from the liver, but it is also susceptible to the cytotoxicity or carcinogenicity of hepatic activated metabolites. Antihistamines, beta blockers, opioids, and tricyclic antidepressants are among the basic amines known to accumulate in the lungs as a result of their binding to surfactant phospholipids in lung tissue. The significance of this relationship to potential pneumotoxicity remains to be seen.

The nasal mucosa is recognized as a first line of defense for the lung against airborne xenobiotics because it is constantly exposed to the external environment.[52,53] Drug metabolism in the nasal mucosa is an important consideration not only in drug delivery, but also for toxicologic implications because of xenobiotic metabolism of inhaled environmental pollutants or other volatile chemicals. CYP450 enzymes in the nasal epithelial cells can convert some of the airborne chemicals to reactive metabolites, increasing the risk of carcinogenesis in the nasopharynx and lung (e.g., nitrosamines in cigarette smoke). The most striking feature of the nasal epithelium is that CYP450 catalytic activity is higher than in any other extrahepatic tissue, including the liver. Nasal decongestants, essences, anesthetics, alcohols, nicotine, and cocaine have been shown to be metabolized in vitro by CYP450 enzymes from the nasal epithelium. The fact that the CYP450s in the nasal mucosa are active, first pass metabolism should be considered when delivering susceptible drugs to the nasal tissues. Flavin monooxygenases, carboxylesterases, aldehyde dehydrogenase, and other conjugation (phase 2) enzymes are also active in the nasal epithelium.

The isoforms of CYP450s and their regulation in the brain are of interest in defining the possible involvement of CYP450s in central nervous system toxicity and carcinogenicity. The CYP450s present in the kidney and adrenal tissues include isoforms primarily involved in the hydroxylation of steroids, arachidonic acid, and 25-hydroxycholcalciferol.

STEREOCHEMICAL ASPECTS OF DRUG METABOLISM

In addition to the physicochemical factors that affect drug metabolism, stereochemical factors play an important role in the biotransformation of drugs and other xenobiotics. This involvement is not unexpected because the drug-metabolizing enzymes are also the enzymes that metabolize certain endogenous substrates, which for the most part are chiral molecules. Most of the enzymes show stereoselectivity, i.e., one stereoisomer enters into biotransformation pathways preferentially, but not exclusively. On the other hand, stereospecificity indicates complete stereoselectivity. Metabolic stereochemical reactions can be categorized as follows: substrate stereoselectivity, when two enantiomers of a chiral substrate are metabolized at different rates; product stereoselectivity, in which a new chiral center is created in a symmetric molecule and one enantiomer is metabolized preferentially; substrate-product stereoselectivity, in which a new chiral center of a chiral molecule is metabolized preferentially to one of two possible diastereomers.[54] An example of substrate stereoselectivity is the preferred decarboxylation of S-α-methyldopa to S-α-methyldopamine, with almost no reaction for R-α-methyldopa. The reduction of ketones to stereoisomeric alcohols and the hydroxylation of enantiotropic protons or phenyl rings by monooxygenases are examples of product stereoselectivity. For example, phenytoin undergoes aromatic hydroxylation of only one of its two phenyl rings to create a chiral center at C-5 of the hydantoin ring, methadone is reduced preferentially to its α-diastereometric alcohol, and naltrexone is reduced to its 6-β-alcohol. An example of substrate-product stereoselectivity is the reduction of the enantiomers of warfarin and the β-hydroxylation of S-α-methyldopamine to (1R:2S)-α-methylnorepinephrine, whereas R-α-methyldopamine is hydroxylated to only a negligible extent. In vivo studies of this type can often be confused by the further biotransformation of one stereoisomer, giving the false impression that only one stereoisomer was formed preferentially. Moreover, some compounds show stereoselective absorption, distribution, and excretion, which proves the importance of also performing in vitro studies. Although studies on the stereoselective biotransformation of drug molecules are not yet extensive, those that have been done indicate that stereochemical factors play an important role in drug metabolism and, in some cases, could account for the differences in pharmacologic activity and duration of action between enantiomers (see the discussion of chiral inversion of the NSAID).

TOXICITY FROM OXIDATIVE METABOLISM

The mechanism of toxicity for xenobiotics, drugs, and plant and chemical toxins is generally accepted as resulting from bioactivation to reactive intermediates, before their toxicity manifests. Many of these metabolites have pharmacologic and toxicologic effects, some of which may differ from those of the parent drug. Many carcinogens elicit their biologic property through a covalent linkage to DNA. This process can lead to mutations and potentially to cancer. Most chemical carcinogens of concern are relatively inert and require activation by the xenobiotic-metabolizing enzymes before they can undergo reaction with nucleophilic groups on DNA and proteins (cytotoxicity). There are many ways to bioactivate procarcinogens, promutagens, plant toxins, drugs, and other xenobiotics.[26] Oxidative bioactivation reactions are by far the most studied and common. Conjugation reactions (phase 2), however, are also capable of activating these xenobiotics to electrophiles, using the conjugating derivative as a good leaving group. These reactive intermediates are mostly electrophiles (electron-deficient substances), such as epoxides, quinones, or free radicals formed by the CYP450 monooxygenases and FMO. Reactive intermediates tend to be oxygenated in sterically hindered positions, making them nonacceptable substrates for subsequent detoxicating enzymes, such as epoxide hydrolase and glutathione S-transferases. Therefore, their principal fate is covalent linkage with intracellular macromolecules, including enzyme proteins and DNA. Experimental studies indicate that the CYP1A subfamily can oxygenate C-substrates (e.g., PAH) in sterically hindered positions, and activation by N-hydroxylation (e.g., aryl acetamides) appears to depend either on FMO or CYP450 isoforms. The formation of chemically reactive metabolites is important because they frequently cause a number of different toxicities, including tumorigenesis, mutagenesis, tissue necrosis, and hypersensitivity reactions.

Some toxic chemicals exert their toxic action by lethal injury or biologic autooxidation (lipid peroxidation). Lethal injury involves the disruption of cellular energy metabolism by inhibition of oxidative phosphorylation (e.g., MPTP) or ATPase, resulting in disruption of subcellular organelles, cell death, and tissue necrosis. Because the early stages of lethal injury are reversible, complete recovery may occur. Autooxidation is the process whereby cellular components are irreversibly oxidized and damaged by free radicals or free-radical generating systems, resulting in the oxidation and depletion of glutathione and various thiol enzymes, to lipid peroxidation and the disruption of cellular membranes, and to cell death, tissue necrosis, and the death of the organism. When cell death does not occur, nonlethal changes such as mutations and malignant transformations are likely.

Our understanding of these reactions was advanced by the studies of Gillette and co-workers.[55] They proposed that the proportion of the dose that binds covalently to critical macromolecules could depend on the quantity of the reactive intermediate that forms.

A scheme illustrating the complexities of metabolically induced toxicity is shown in Figure 8–25. The figure is for chemical carcinogen metabolism, but it can be applied to other toxicities. Reactions that proceed via the hatched arrows eventually lead to neoplasia. Some carcinogens may form the ultimate carcinogen directly through CYP450 isoform metabolism; others, like the PAH (e.g., benzo[a]pyrene), appear to involve a minimum three-step reaction sequence forming an epoxide, then the diol, and perhaps a second epoxide group on another part of the molecule. Others form the N-hydroxy intermediate that requires transferase-catalyzed conjugation (e.g., O-glucuronide, O-sulfate) to form the ultimate carcinogen. The quantity of ultimate carcinogen formed should relate directly to the proportion of the dose that binds or alkylates DNA.

The solid-arrow reaction sequences are intended to show detoxification mechanisms, which involve several steps. First, the original chemical may form less active products (phenols, diols, mercapturic acids, and other conjugates). Second, the ultimate carcinogen may rearrange so as to be diverted from its reaction with DNA (or whatever is the critical macromolecule). Third, the covalently bound DNA may be repaired. Fourth, immunologic removal of the tumor cells may occur. Several mechanisms within this scheme could regulate the quantity of covalently bound carcinogen: (1) the activity of the rate-limiting enzyme, such as epoxide hydrolase, monooxygenase, or one of the transferases, could be involved; (2) the availability of co-substrates, e.g., glutathione, UDP-glucuronic acid, or PAPS, may be rate limiting; (3) specific monooxygenase activities for detoxification and activation must be considered; (4) availability of alternate reaction sites for the ultimate carcinogen, e.g., RNA and protein may be involved; (5) possible specific transport mechanisms that deliver either the procarcinogen or its ultimate carcinogen to selected molecular or subcellular sites.

Most lipid-soluble exogenous aromatic compounds and most compounds with olefinic unsaturation are metabolized in humans and in animals through epoxide formation. The importance of metabolically produced epoxides in mediating adverse biologic effects has aroused concern about clinically used drugs known to be metabolized to epoxides. Drugs possessing structural features prone to metabolic epoxidations are abundant. Metabolically produced epoxides have been reported for allobarbital, secobarbital, protriptyline, carbamazepine, cyproheptadine, and are implicated with diethylstilbesterol, phenytoin, phensuximide, phenobarbital, mephobarbital, lorazepam, and imipramine.[56] The alarming biologic effects of some epoxides, however, do not imply that all epoxides have similar effects. Epoxides vary greatly in mo-

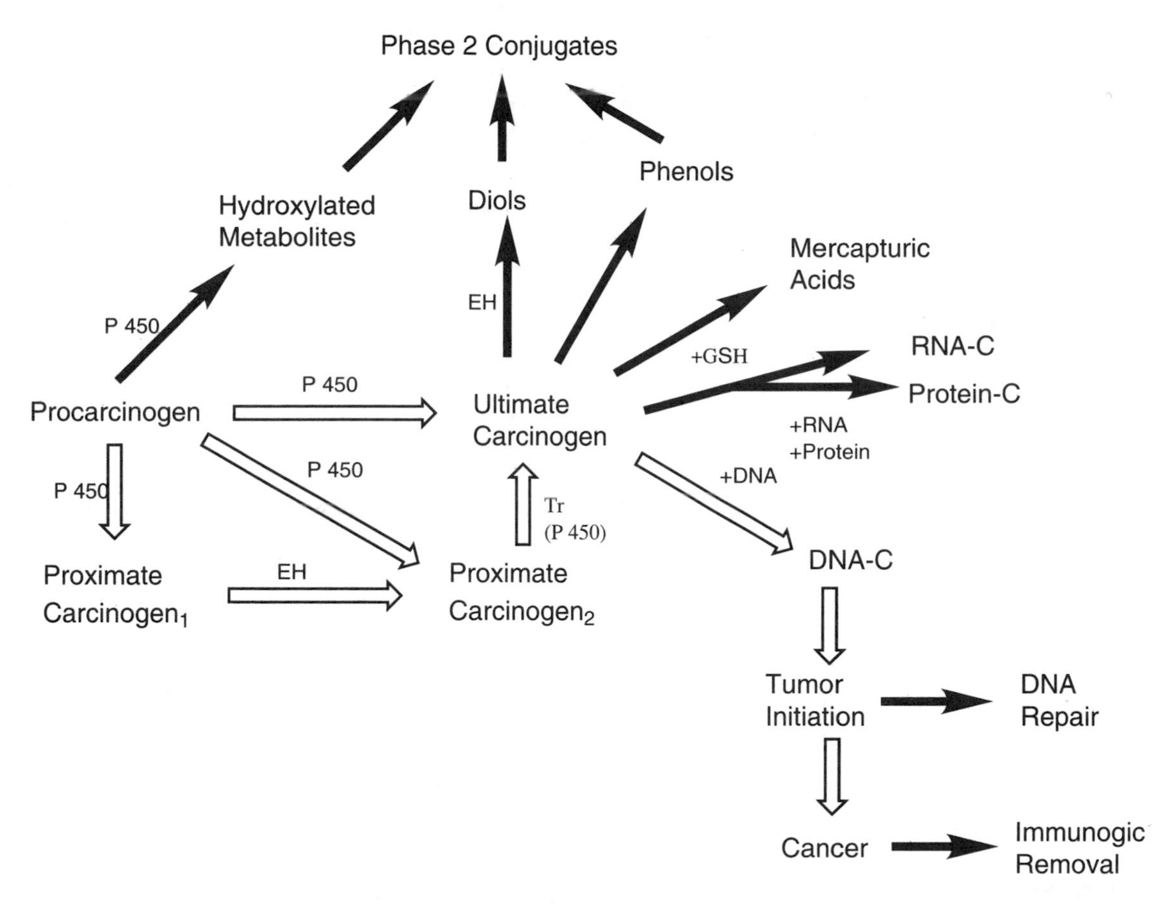

Fig. 8–25. Metabolism of chemical carcinogens.

lecular geometry, stability, electrophilic reactivity, and relative activity as substrates for epoxide-transforming enzymes (e.g., epoxide hydrolase, glutathione S-transferase, and others). Firm predictions of biologic effects of a given epoxide based on any of these factors are as yet not possible. When epoxidation was accepted as a route of metabolism for aromatic compounds, it was believed that K-region epoxides (the 9,10-oxide of phenanthrene is an example of a K-region epoxide) were the "reactive epoxides" sought as intermediates in the metabolism of PAH. It was found, however, that K-region epoxides of carcinogenic hydrocarbons were less carcinogenic than the parent compounds, and some authors propose that the reactive epoxides are dihydrodiol epoxides. The metabolic activation of benzo[a]pyrene illustrates this point (Fig. 8–26). Substantial evidence suggests that of the two racemic benzo[a]pyrene-7,8-dihydrodiol-9,10-epoxides, the dihydrodiol epoxide-2 is the ultimate carcinogenic metabolite of benzo[a]pyrene in mammalian tissue. Hydrogenation of the 9,10 double bond in the 7,8-dihydrodiol affords a noncarcinogenic molecule. When the dihydrodiol epoxide reacted with polyguanylic acid or calf thymus DNA, the respective adducts were isolated linked to the 2-amino group of 2′-deoxyguanosine.[57,58] The covalent DNA adducts are formed by attack of the N^2-amino group of guanosine residues on DNA at the C-10 of the diastereomeric diol epoxides resulting in the formation of *trans* adducts (see Fig. 8–26). These and other studies have shown that the "reactive epoxides" are capable of reacting with DNA, proteins, and other macromolecules in vitro and in vivo.[58]

Other studies have revealed toxic metabolites for isoniazid, iproniazid, acetaminophen, phenacetin, furosemide, and cephaloridine.[59] Isoniazid is acetylated to its major metabolite acetylisoniazid, which is hydrolyzed to acetylhydrazine and isonicotinic acid. Acetylhydrazine is metabolized by the CYP450 monooxygenase to an N-hydroxy intermediate that dehydrates into an acylating intermediate, which can then initiate the process that leads to hepatic necrosis. Iproniazid is hydrolyzed to isopropylhydrazine, which undergoes oxidation to a reactive alkylating intermediate, which has been trapped as S-isopropylglutathione. Acetaminophen is among the safest of all minor analgesics when taken in normal therapeutic doses, but it has been known to cause acute hepatic necrosis, sometimes fatal in both adults and children, when taken in excessive doses or after prolonged or repeated administration. The oxidation of acetaminophen to the chemically reactive N-acetyl-p-benzoquinoneimine, catalyzed by the isoforms CYP1A2 and 2E1, can either bind covalently with glutathione to form an inactive product, or with cellular macromolecules, initiating the processes leading to hepatic necrosis (see Fig. 8–23). Furosemide, a frequently used diuretic drug, is reportedly a teratogen. When administered in large doses, it produces hepatic necrosis in mice. The hepatic toxicity apparently results

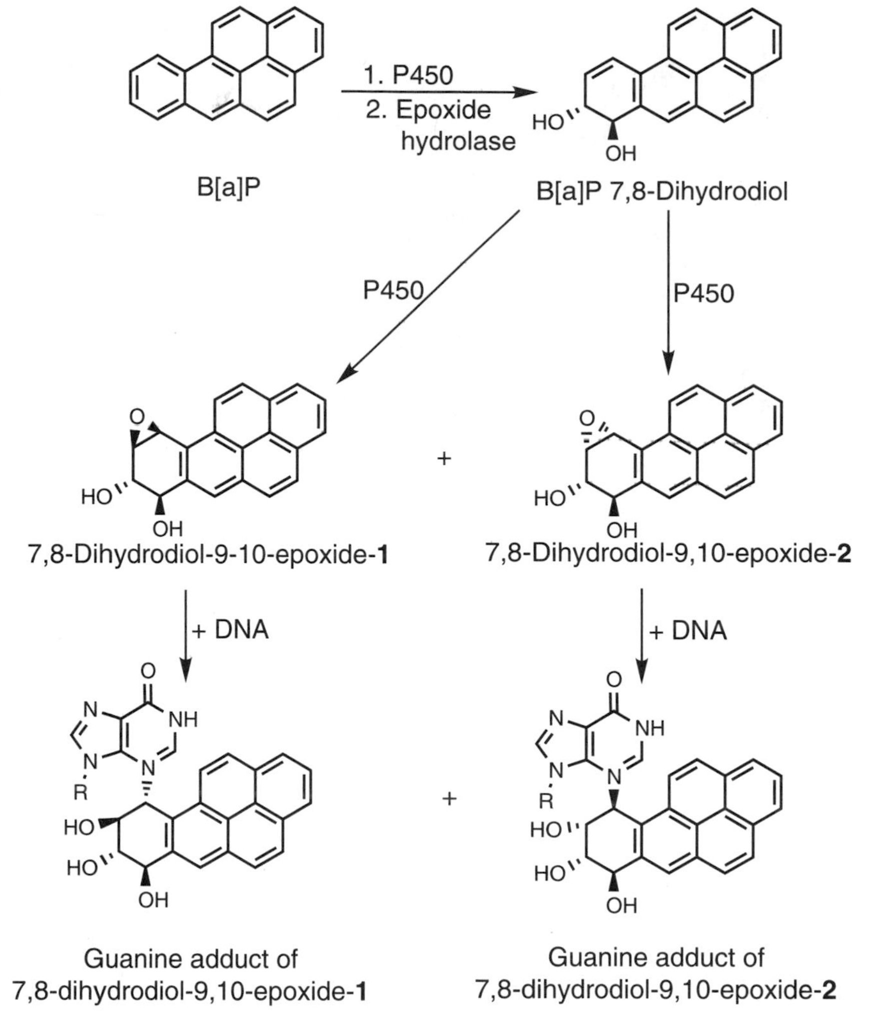

B[a]P

1. P450
2. Epoxide hydrolase

B[a]P 7,8-Dihydrodiol

P450

P450

7,8-Dihydrodiol-9-10-epoxide-**1**

+

7,8-Dihydrodiol-9,10-epoxide-**2**

+ DNA

+ DNA

Guanine adduct of 7,8-dihydrodiol-9,10-epoxide-**1**

+

Guanine adduct of 7,8-dihydrodiol-9,10-epoxide-**2**

Fig. 8–26. P450 metabolic activation of benzo[a]pyrine to benzo[a]pyrene diol epoxides in mouse skin. Covalent binding with DNA-guanosine groups occurs at position 10 of the diastereomeric diol epoxides.

from metabolic activation of the furan ring, possibly through an epoxide. Cephaloridine produces renal necrosis via metabolic activation of the thiophene ring in a manner similar to that described for furosemide. The concept that the pattern of metabolism can change with the dose accounts for the fact that some drugs may not be toxic unless a certain threshold dose is exceeded. For example, with acetaminophen and furosemide, no necrosis or covalent binding occurs until a dose of 100 mg/kg in animals is exceeded. Enzyme induction may enhance the formation of reactive intermediates and the severity of the tissue injury. Glutathione plays a protective role in the hepatic tissue injury produced by acetaminophen (see the section on glutathione conjugation), but not by furosemide. One method for preventing liver injury in the event of acetaminophen overdosage is the use of acetylcysteine, which traps the reactive metabolite similar to that with GSH.

The fact that numerous organic compounds that are essentially nontoxic as long as their structure is preserved can be converted into cytotoxic, teratogenic, mutagenic, or carcinogenic compounds by normal biotransformation pathways in animals and in humans is now well established.[26] The "reactive intermediate" (an electrophile) then reacts with cellular constituents, forming either nontoxic products or binding covalently with essential macromolecules, and initiating processes that eventually lead to the toxic effect. A better understanding of the mechanisms underlying these reactions may lead to more rational approaches to the development of nontoxic therapeutic drugs. For the present, it seems that new advances in drug therapy cannot occur without some risk of causing structural tissue lesions. Special attention to risk factors is required for drugs that will be used for long periods in the same patient.

N-nitrosoamines ($R_2N\text{-}N=O$) have been implicated as mutagenic, carcinogenic, and teratogenic substances in animals, and have been suspected in some human cancers. Although they occur naturally in trace amounts, they can be formed by reaction of nitrite with secondary or tertiary amines. Among the amines to which people are exposed are drugs and agricultural chemicals. Some of these amines (e.g., methapyrilene, an antihistamine in over-the-counter sleeping aids) have been shown to form N-nitroso compounds by reaction with nitrite and to be mutagenic with bacteria. These reactions can occur in food preserved with nitrite or in the stomach from dietary nitrites. Nitrosoamines have to be activated metabolically before they can exert their carcinogenic or other toxic effects. One of the proposed mechanisms is hydroxylation of the α-carbon by CYP450 (similar to N-dealkylation) with subsequent N-dealkylation and spontaneous rearrangement to possible alkyl carbonium ions, which, in turn, can covalently bind to cellular macromolecules.[23] Studies with dimethylnitrosamine have shown methylation to occur at the N-7 position in guanine as well as S-methylation of cysteine. The molecular basis of nitrosoamine-induced carcinogenesis or mutagenesis is not clearly understood.

AMES TEST FOR CARCINOGENICITY

The toxicity of many chemicals is the result of their metabolic conversion to reactive electrophiles that interact irreversibly with critical nucleophilic sites on cellular macromolecules. In laboratory animals and in humans, the CYP450 system has been shown to be the major enzyme system involved in the activation of most xenobiotics. For the majority of carcinogens, the initiation stage of carcinogenesis involves a critical activation step linking the formation of an electrophile with an inheritable alteration in the cellular DNA (mutation). Of all the tests used for determining the potential mutagenicity of a chemical, the Ames test has received the greatest attention. This test measures the ability of a xenobiotic following microsomal activation to revert histidine-requiring strains of Salmonella typhimurium back to the wild type, regaining the ability to grow in a histidine-deficient environment. The test can determine whether a chemical causes genetic damage following metabolism and predicts the potential carcinogenicity of chemicals. It has not always been possible, however, to show a direct relationship between mutagenicity and carcinogenicity. The Ames Salmonella test is carried out by combining the test compound with the Salmonella strain, a microsomal formation of CYP450 enzymes, and agar. The mixture is poured into an agar plate and incubated, and growth is observed. Because of the different solubilities and ionic properties of test molecules, a mutagenic response is not always observed, so variations in preparing the media are carried out to ensure solubility of the test compound in the agar media. The metabolic activation of the chemical in the microsomal fraction and only the electrophile escaping the detoxication mechanisms will induce bacterial mutagenesis. A negative result should be further investigated before concluding the chemical is nonmutagenic in the Ames test.

DRUG INTERACTIONS

Many drugs can alter the metabolic detoxification of other drugs and thereby change pharmacologic response. In these interactions, the drug vital to the patient's therapy may be adversely affected, leading perhaps to a synergistic or antagonistic effect and producing and undesirable or toxic side effects. One reason for the increased incidence of drug interactions is the practice of simultaneously prescribing several potent drugs. Moreover, it is common for patients to ingest nonprescription products concurrently with potent prescribed medications.

The ability of drugs and other foreign substances to stimulate metabolism of other drugs has already been discussed. Phenobarbital, for example, stimulates metabolism of a variety of drugs, e.g., phenytoin and coumarin anticoagulants. Stimulation of bishydroxycoumarin metabolism can create a problem in a patient undergoing anticoagulant therapy. If phenobarbital administration is stopped, the rate of metabolism of the anticoagulant decreases, resulting in enhanced anticoagulant activity and the increased possibility of hemorrhage. Serious side effects have resulted from this type of interaction.

These observations point out that combined therapy of a potent drug, e.g., bishydroxycoumarin, and a stimulator of drug metabolism, e.g., phenobarbital, can create a hazardous situation if the enzyme stimulator is withdrawn and therapy with the potent drug is continued without an appropriate decrease in dose.

Some drugs are competitive inhibitors of nonmicrosomal metabolic pathways. Serious reactions have been reported in patients treated with an MAO inhibitor, such as trancypromine or iproniazid, because they usually are sensitive to a subsequent dose of a sympathomimetic amine, e.g., amphetamine, or a tricyclic antidepressant, e.g., amitriptyline, which is metabolized by MAO.

Besides interactions with other drugs, interactions with normal body constituents or naturally occurring compounds, such as food, may be important. Severe hypertensive reactions have occurred when patients treated with MAO inhibitors have ingested cheeses and other foods rich in the biogenic amine tyramine. Allopurinol, a xanthine oxidase inhibitor used for the treatment of gout, inhibits metabolism of 6-mercaptopurine and other drugs metabolized by this enzyme. A serious drug interaction results from the concurrent use of allopurinol for gout and 6-mercaptopurine to block the immune response from a tissue transplant or as antimetabolite in neoplastic diseases. In some cases, however, allopurinol is used in conjunction with 6-mercaptopurine to control the increase in uric acid elimination from 6-mercaptopurine metabolism. The patient should be supervised closely, because allopurinol, an inhibitor of purine metabolism, given in large doses, may have serious effects on bone marrow.

SEX DIFFERENCES IN DRUG METABOLISM

The role of sex as a contributor to variability in drug and xenobiotic metabolism is not clear, but increasing numbers of reports show differences in metabolism between men and women, raising the intriguing possibility that endogenous sex hormones, or hydrocortisone, or their synthetic equivalents may influence the activity of inducible CYP3A. For example, N-demethylation of erythromycin was significantly higher in females than males. Nevertheless, the N-demethylation was persistent throughout adulthood. In contrast, males exhibited unchanged N-demethylation values.

Sex-dependent differences of metabolic rates have been detected for some drugs. Side chain oxidation of

Table 8–16. Metabolic Pathways of Common Drugs

Drug	Pathway
Amphetamines 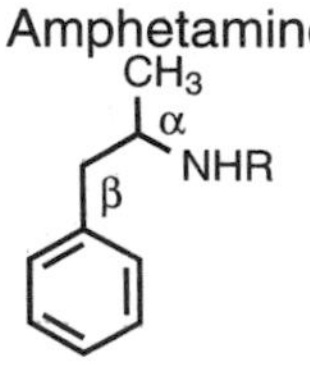	Deamination (followed by oxidation and reduction of the ketone formed) N-oxidation N-dealkylation Hydroxylation of the aromatic ring Hydroxylation of the β-carbon atom Conjugation with glucuronic acid of the acid and alcohol products from the ketone formed by deamination
Barbiturates 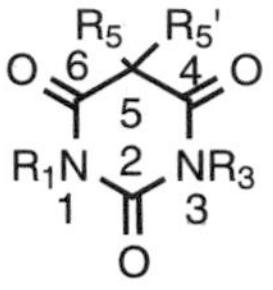	Oxidation and complete removal of substituents at carbon 5 N-dealkylation at N^1 and N^3 Desulfuration at carbon 2 (thiobarbiturates) Scission of the barbiturate ring at the 1:6 bond to give substituted malonylureas
Phenothiazines 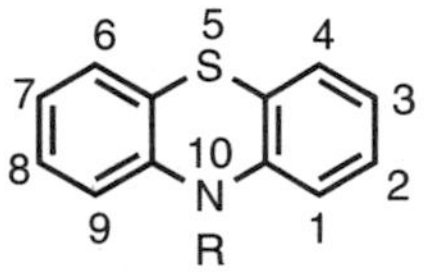	N-dealkylation in the N^{10} side chain N-oxidation in the N^{10} side chain Oxidation of the heterocyclic S atom to sulfoxide or sulfone Hydroxylation of one or both aromatic rings Conjugation of phenolic metabolites with glucuronic acid or sulfate Scission of the N^{10} side chain
Sulfonamides 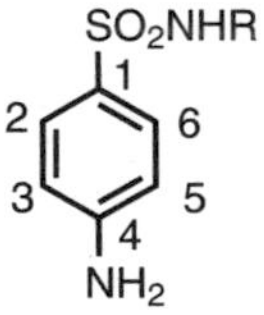	Acetylation at the N^4 amino group Conjugation with glucuronic acid or sulfate at the N^4 amino group Acetylation or conjugation with glucuronic acid at the N^1 amino group Hydroxylation and conjugation in the heterocyclic ring, R
Phenytoin 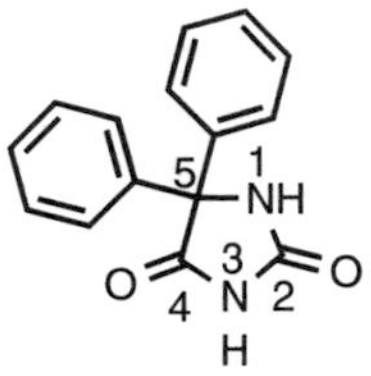	Hydroxylation of one aromatic ring Conjugation of phenolic products with glucuronic acid or sulfate Hydrolytic scission of the hydantoin ring at the bond between carbons 3 and 4 to give 5,5-diphenylhydantoic acid
Meperidine 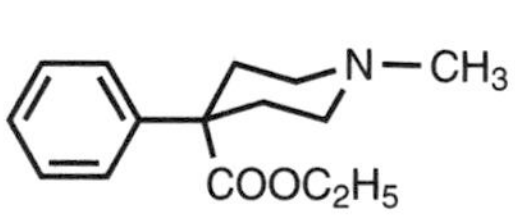	Hydrolysis of ester to acid N-dealkylation Hydroxylation of aromatic ring N-oxidation Both N-dealkylation and hydrolysis Conjugation of phenolic products
Pentazocine 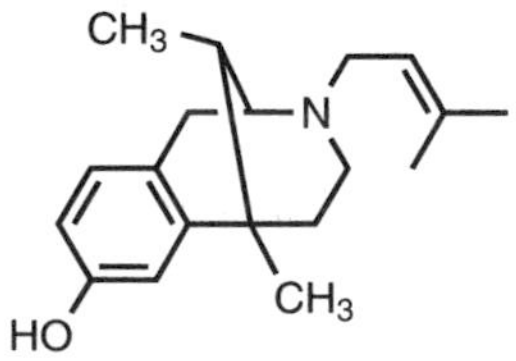	Hydroxylation of terminal methyl groups of the alkenyl side chain to give *cis* and *trans* (major) alcohols Oxidation of hydroxymethyl product of the alkenyl side chain to carboxylic acids Reduction of alkenyl side chain and oxidation of terminal methyl group

Table 8–16. *(Continued)*

Drug	Pathway
Cocaine 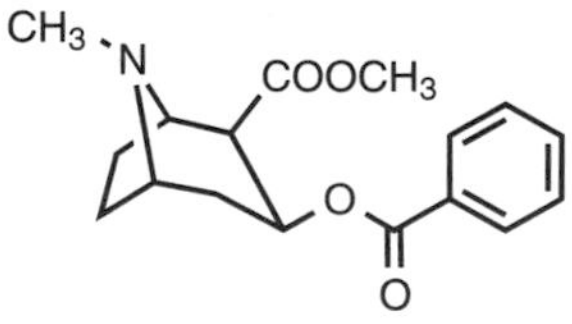	Hydrolysis of methyl ester Hydrolysis of benzoate ester N-dealkylation Both hydrolysis and N-dealkylation
Phenmetrazine 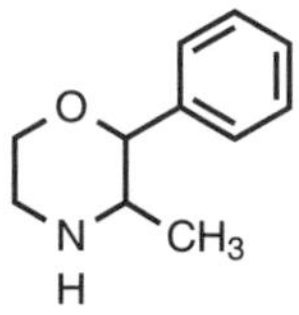	Oxidation to lactam Aromatic hydroxylation N-oxidation Conjugation of phenolic products
Ephedrine 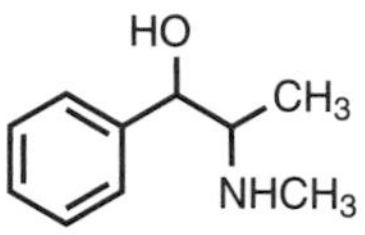	N-dealkylation Oxidative deamination Oxidation of deaminated product to benzoic acid Reduction of deaminated product to 1,2-diol
Propranolol 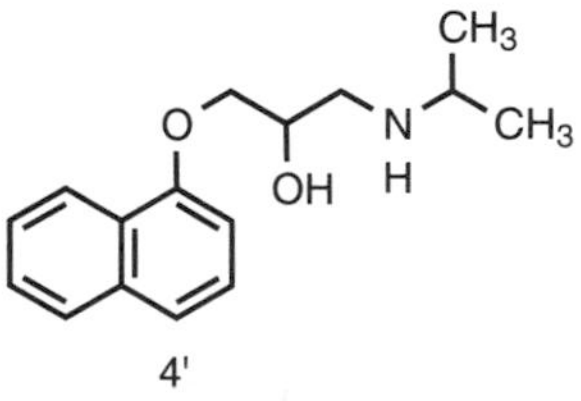	Aromatic hydroxylation at C—4′ N-dealkylation Oxidative deamination Oxidation of deaminated product to naphthoxylactic acid Conjugation with glucuronic acid O-dealkylation
Indomethacln 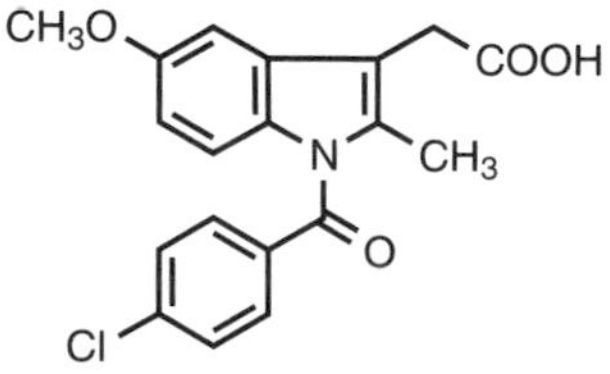	O-demethylation N-deacylation of p-chlorobenzoyl group Both O-dealkylation and N-deacylation Conjugation of phenolic products with glucuronic acid Other conjugation products
Diphenoxylate 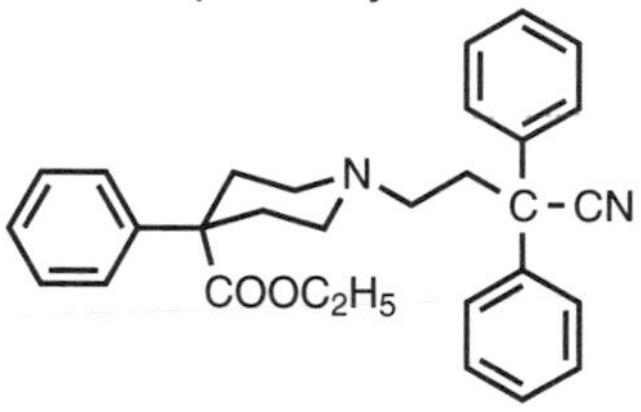	Hydrolysis of ester to acid Hydroxylation of one aromatic ring attached to the N-alkyl side chain
Diazepam 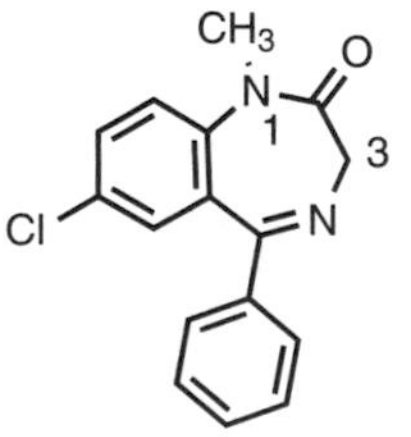	N-dealkylation at N^1 Hydroxylation at carbon 3 Conjugation of phenolic products with glucuronic acid Both N-dealkylation of N^1 and hydroxylation at carbon 3

(continued)

Table 8–16. *(Continued)*

Drug	Pathway
Prostaglandins 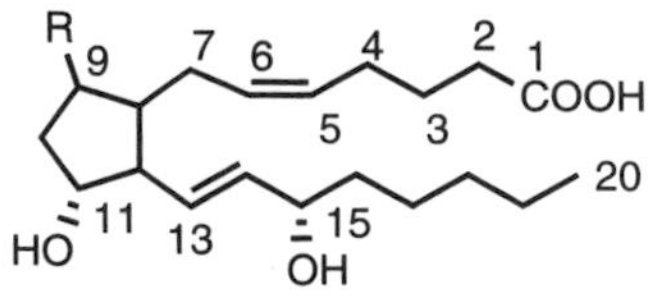	Reduction of double bonds at carbons 5 and 6, and 13 and 14 Oxidation of 15-hydroxyl to ketone β-Oxidation of carbons 1, 2, 3 and 4 ω-Oxidation of carbon 20 to acid
Cyproheptadine 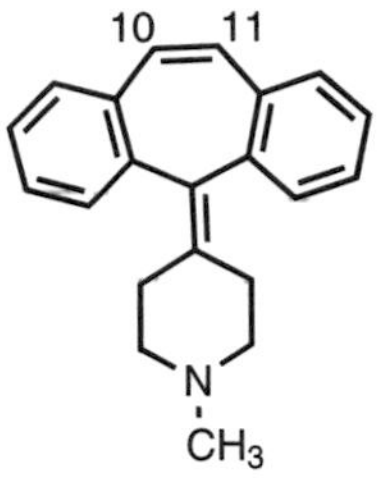	N-dealkylation 10,11-Epoxide formation Both N-dealkylation and 10,11-epoxidation
Hydralazine 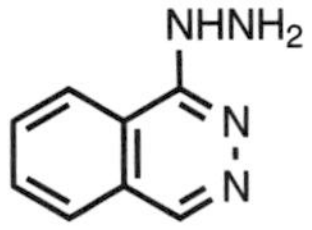	N-acetylation with cyclization to a methyl-s-triazolophthalazine N-formylation with cyclization to an s-triazolophthalazine Aromatic hydroxylation of benzene ring Oxidative loss of hydrazinyl group to 1-hydroxy Hydroxylation of methyl of methyl-s-triazolophthalazine Conjugation with glucuronic acid
Methadone 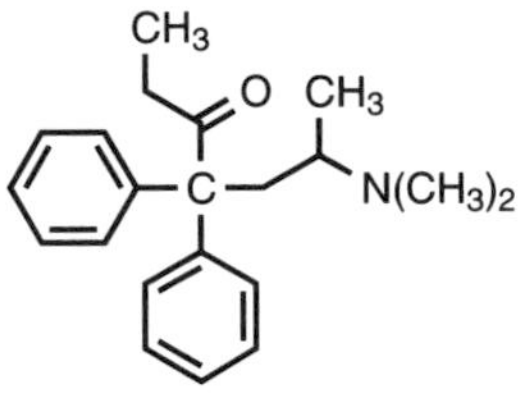	Reduction of ketone to hydroxyl Aromatic hydroxylation of one aromatic ring N-dealkylation of alcohol product N-dealkylation with cyclization to pyrrolidine
Lidocaine 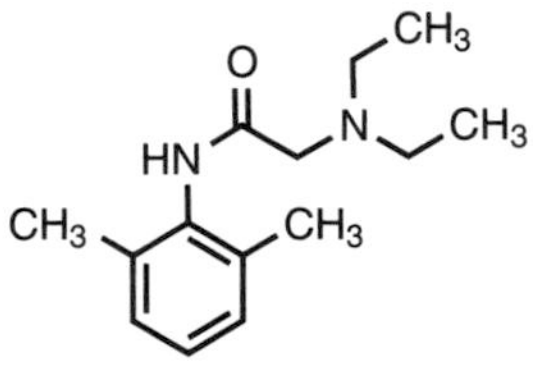	N-dealkylation Oxidative cyclization to a 4-imidazolidone N-oxidation of amide N Aromatic hydroxylation ortho to methyl Hydrolysis of amide
Imipramine 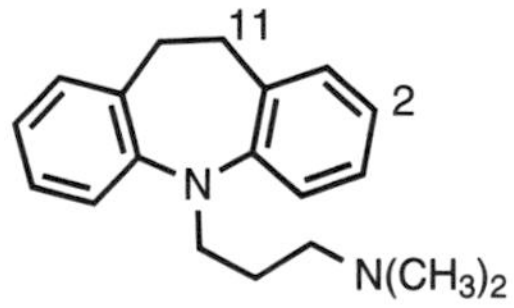	N-dealkylation Hydroxylation at C-11 Aromatic hydroxylation (C-2) N-oxidation Both N-dealkylation and hydroxylation
Cimetidine 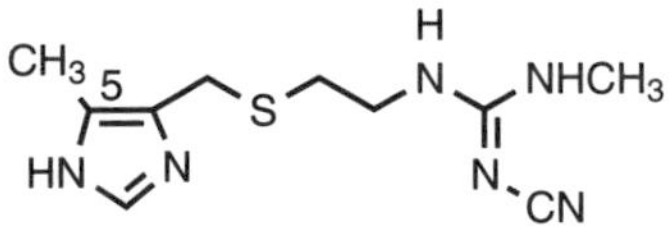	S-oxidation Hydroxylation of 5-methyl

Table 8–16. *(Continued)*

Drug	Pathway
Terfenadine 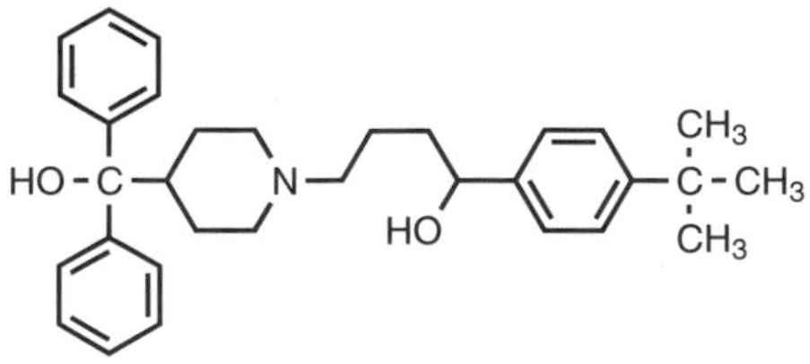	N-demethylation Methyl hydroxylation to CH_2OH CH_2OH oxidation to COOH
Valproic Acid 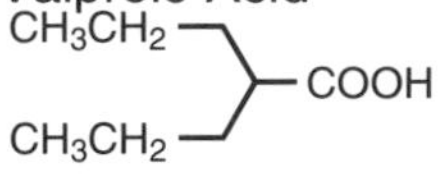	CoA thioester Dehydrogenation to (E) 2-ene Dehydrogenation to (E) 2,4diene Dehydrogenation to 4-ene 3-Hydroxylation
Piroxicam 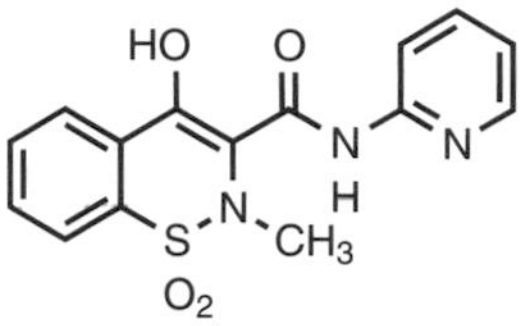	Pyridine 3′-hydroxylation Hydrolysis of amide Decarboxylation
Caffeine 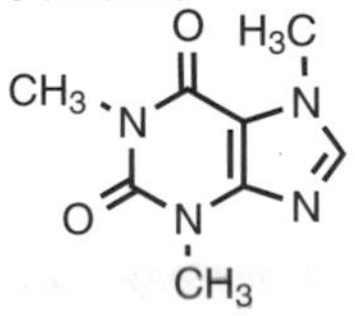	N^3-demethylation N^1-demethylation N^7-demethylation to theophylline C-8 oxidation to uric acids Imidazole ring opened
Theophylline 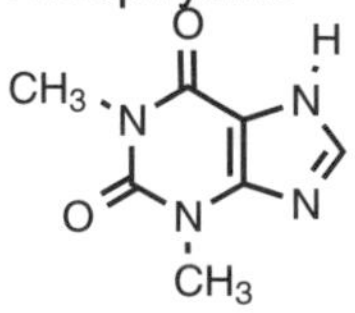	N^3-demethylation N^1-demethylation C-8 oxidation to uric acids 1-Me xanthine to 1-Me uric acid with xanthine oxidase Imidazole ring opened
Nicotine 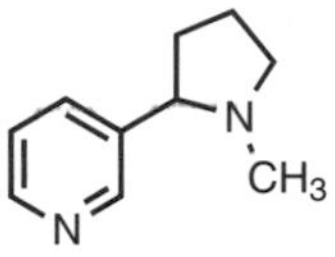	Pyrrolidine 5′-hydroxylation to cotinine Pyrrolidine N-oxidation (FMO) N-demethylation (nornicotine and norcotinine) Pyridine N-methylation 3′-Hydroxylation of cotinine
Ibuprofen 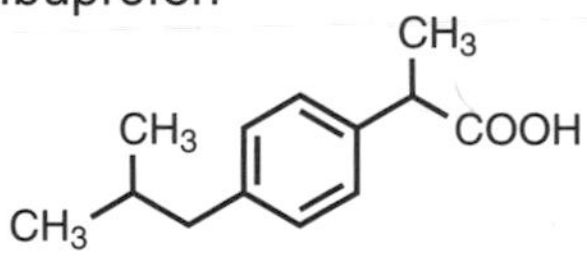	CoA thioester and epimerization of R− to S+ enantiomer Methyl hydroxylation to CH_2OH CH_2OH to COOH Acylglucuronide
Tamoxifen 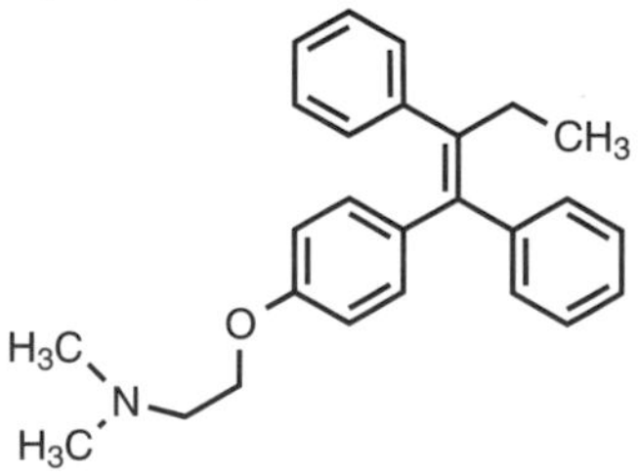	N-demethylation 4′-Hydroxylation N-oxidation (FMO) 4-O-sulfate 4-O-glucuronide

(continued)

Table 8–16. *(Continued)*

Drug	Pathway
Lovastatin 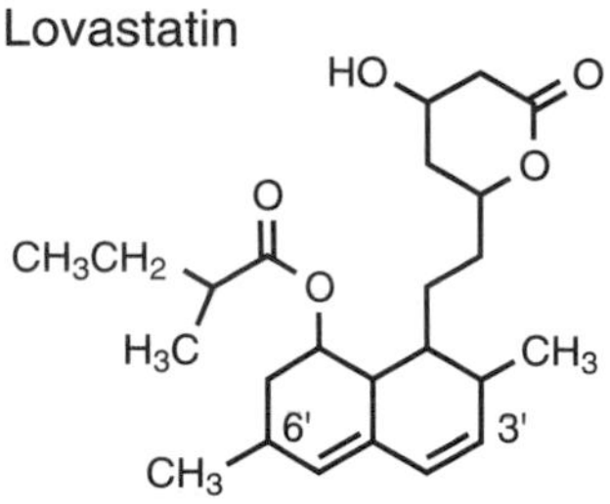	6′-Hydroxylation 3′-Side chain hydroxylation 3′-Hydroxylation β-oxidation of lactone O-glucuronides
Ciprofloxacin 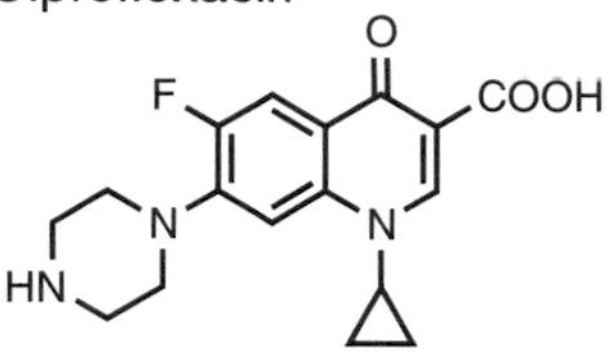	Piperazine 3′-hydroxylation N-sulfation
Labetalol 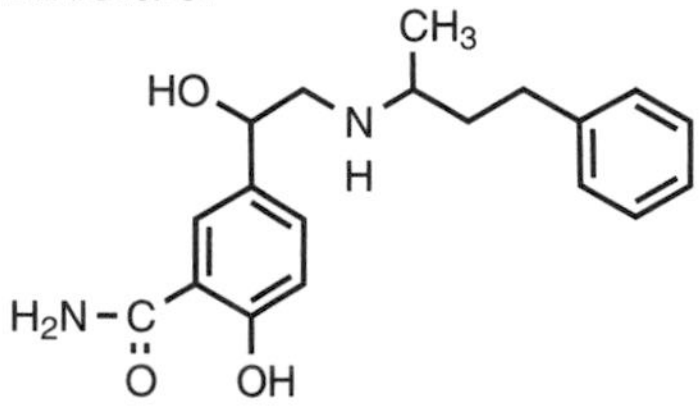	O-sulfate (major) O-glucuronide
Acetaminophen 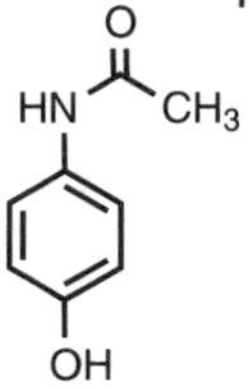	O-glucuronide O-sulfate Oxidation to N-acetyl-p-benzoquinoneimine Conjugation of N-acetyl-p-benzoquinoneimine with glutathione
Tripelennamine 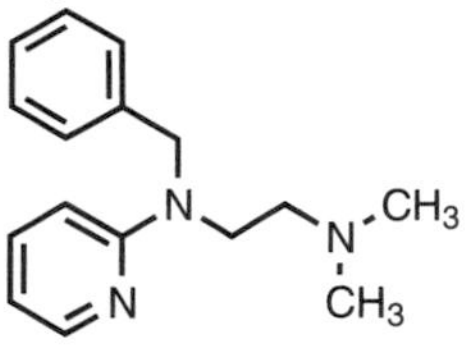	p-Hydroxylation Benzylic C-hydroxylation N-depyridinylation N-debenzylation
Felodipine 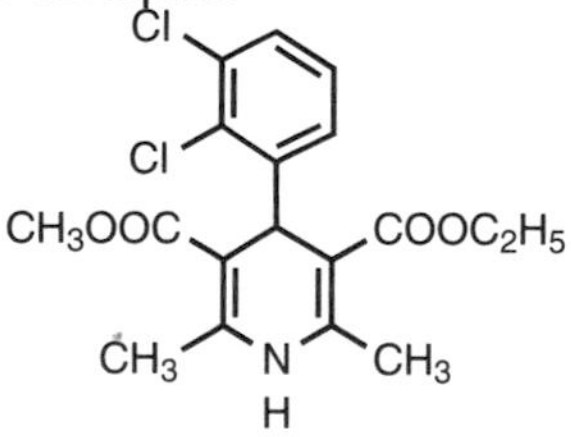	Aromatization Ester hydrolysis Methyl hydroxylation

propranolol was 50% faster in males than in females, but no differences between sexes were noted in aromatic ring hydroxylation. N-demethylation of meperidine was depressed during pregnancy and for women taking oral contraceptives. Other examples of drugs cleared by oxidative drug metabolism more rapidly in men than in women included chlordiazepoxide and lidocaine. Diazepam, prednisolone, caffeine, and acetaminophen are metabolized slightly faster by women than by men. No sex differences have been observed in the clearance of phenytoin, nitrazepam, and trazodone, which interestingly are not substrates for the CYP3A subfamily. Sex differences in the rate of glucuronidation have been noted.

More investigation is warranted, and future pharmacokinetic studies examining the alteration in drug metabolism in one sex needs to be re-examined with respect to the other sex. Even in postmenopausal women, CYP3A function may be altered and influenced by the lack of estrogen or the presence of androgens.

MAJOR PATHWAYS OF METABOLISM

Many drugs undergo both phase 1 and phase 2 reactions. Some of the metabolic pathways for common drugs are listed in Table 8–16.

REFERENCES

1. R. T. Williams, *Detoxication Mechanisms,* 2nd ed., New York, John Wiley and Sons, 1959.
2. *Drug Metabolism Reviews,* New York, Marcel Dekker; *Drug Metabolism and Disposition,* Baltimore, Williams & Wilkins; *Xenobiotica,* London, Taylor and Francis.
3. M. Anders, Ed., *Bioactivation of Foreign Compounds,* New York, Academic Press, 1985; J. Caldwell and W. Jakoby, Eds., *Biological Basis of Detoxication,* New York, Academic Press, 1983; W. Jakoby, Ed., *Enzymatic Basis of Detoxication,* Vols. I. and II, New York, Academic Press, 1980; W. Jakoby, J. R. Bend, and J. Caldwell, Eds., *Metabolic Basis of Detoxication-Metabolism of Functional Groups,* New York, Academic Press, 1982; G. J. Mulder, Ed., *Conjugation Reactions in Drug Metabolism: An Integrated Approach,* London, Taylor and Francis, 1990; P. R. Ortiz de Montellano, Ed., *Cytochrome P450 Structure, Mechanism, and Biochemistry,* New York, Plenum Press, 1986; B. Testa and P. Jenner, *Drug Metabolism: Chemical and Biochemical Aspects,* New York, Marcel Dekker, 1976; B. Testa and P. Jenner, *Concepts in Drug Metabolism,* New York, Marcel Dekker, 1981.
4. G. R. Wilkinson and S. Schenker, *Drug Metab. Rev., 4,* 139(1975).
5. A. J. McLean and D. J. Morgan, *Clin. Pharmacokinet., 21,* 42(1991).
6. K. F. Ilett, et al. *Pharmacol. Ther., 46,* 67(1990).
7. M. R. Dobrinska, *J. Clin. Pharmacol., 29,* 577(1989).
8. D. L. Schmucker, *Pharmacol. Rev., 37,* 133(1985).
9. C. Durnas, et al. *Clin. Pharmacokinet., 19,* 359(1990); K. Woodhouse and H. A. Wynne, *Drugs & Aging, 2,* 243(1992).
10. S. D. Black. *FASEB J. 6,* 680(1992).
11. F. J. Gonzalez and H. V. Gelboin, *Environ. Health Perp., 98,* 81,085(1992); D. W. Nebert, et al., *DNA Cell Biol., 10,* 1(1991).
12. F. J. Gonzalez, *Trends Pharmacol. Sci., 13,* 346(1992).
13. F. P. Guengerich, *Crit. Rev. Biochem. Mol. Biol., 25,* 97(1990).
14. D. S. Smith and B. C. Jones, *Biochem. Pharmacol., 44,* 2089(1992).
15. S. A. Wighton and J. C. Stevens, *Crit. Rev. Toxicol., 22,* 1(1992).
16. P. G. Traber, et al., *Biochim. Biophys. Acta, 1171,* 167(1992).
17. D. Roy, et al., *Arch. Biochem. Biophys., 296,* 450(1992).
18. D. R. Koop, *FASEB J. 6,* 724(1992).
19. J. L. Raucy, et al. *Crit. Rev. Toxicol., 23,* 1(1993).
20. R. E. White and M. J. Coon, *Annu. Rev. Biochem., 49,* 315(1980); R. E. White, *Pharmacol. Ther., 49,* 21(1991).
21. J. T. Groves, *J. Chem. Educ., 62,* 928(1985).
22. P. R. Ortiz de Montellano and N. O. Reich, *Inhibition of Cytochrome P450 Enzymes,* Chap. 8, in *Cytochrome P450 Structure, Mechanism, and Biochemistry,* P. R. Ortiz de Montellano, Ed., New York, Plenum Press, 1986; P. R. Ortiz de Montellano, *Alkenes and Alkynes,* in *Bioactivation of Foreign Compounds,* M. Anders, Ed., New York, Academic Press, 1985, pp. 121–155.
23. F. P. Guengerich and T. L. MacDonald, *Accts. Chem. Res., 17,* 9(1984); F. P. Guengerich and T. L. MacDonald, *FASEB J., 4,* 2453(1990); P. R. Ortiz de Montellano, *Trends Pharmacol. Sci., 10,* 354(1989).
24. A. B. Okey, *Pharmacol. Ther., 45,* 241(1990).
25. M. Barry and J. Feely, *Pharmacol. Ther., 54,* 17(1992).
26. F. P. Guengerich, *Pharmacol. Ther., 54,* 17(1992).
27. C. P. Martucci and J. Fishman, *Pharmacol. Ther., 57,* 237(1993).
28. M. Murray and G. F. Reidy, *Pharmacol. Rev., 42,* 85(1990); M. Murray, *Clin. Pharmacokinet., 23,* 132(1992).
29. A. H. Conney, et al., *Clin. Pharmacol. Ther., 22,* 707(1977).
30. T. A. Ballie, *Pharm. Weekbl. [Sci.], 14,* 122(1992).
31. K. E. Thummel, et al., *Drug Metab. Dispos., 21,* 350(1993).
32. R. H. Elliot and L. Strunun, *Br. J. Anaesth., 70,* 339(1993).
33. D. M. Zeigler, *Drug Metab. Rev., 19,* 1(1988).
34. D. M. Zeigler, *Trends Pharmacol. Sci., 11,* 321(1990).
35. C. Dolphin, et al., *J. Biol. Chem., 266,* 12,379(1991); J. R. Cashman, et al., *Drug Metab. Dispos., 21,* 492(1993).
36. P. F. Hollenberg, *FASEB J., 6,* 686(1992); L. J. Marnett, et al., *Comparison of the Peroxidase Activity of Hemeproteins and Cytochrome P450,* Chap. 2, in *Cytochrome P450 Structure, Mechanism, and Biochemistry,* P. R. Ortiz de Montellano, Ed., New York, Plenum Press, 1986.
37. T. P. Singer and R. R. Ramsay, *FEBS Lett., 274,* 1(1990).
38. G. J. Mulder, Ed., *Conjugation Reactions in Drug Metabolism: An Integrated Approach,* New York, Taylor and Francis, 1990.
39. H. Spahn-Langguth and L. Z. Benet, *Drug Metab. Rev., 24,* 5(1992).
40. G. J. Mulder, *Trends Pharmacol. Sci., 13,* 302(1992).
41. J. Caldwell, et al., *Biochem. Pharmacol., 37,* 105(1988).
42. T. S. Tracy, et al., *Drug Metab. Dispos., 21,* 114(1993).
43. S. D. Nelson, *Arylamines and Arylamide: Oxidation Mechanisms,* in *Bioactivation of Foreign Compounds,* M. Anders, Ed., New York, Academic Press, 1985, pp. 349–375.
44. A. E. Ahmed, et al., *Nitriles,* in *Bioactivation of Foreign Compounds,* M. Anders, Ed., New York, Academic Press, 1985, pp. 485–489.
45. U. A. Myer, et al., *Pharmacol. Ther., 46,* 297(1990).
46. M. Eichelbaum and A. S. Gross, *Pharmacol. Ther., 46,* 377(1990).
47. A. K. Daly, et al., *Pharmacol. Ther., 57,* 129(1993).
48. P. B. Watkins, *Semin. Liver Dis., 10,* 235(1990).
49. L. S. Kaminsky, and M. J. Fasco, *Crit. Rev. Toxicol., 21,* 407(1991).
50. U. Christians and K.-F. Sewing, *Pharmacol. Ther., 57,* 291 (1993).
51. R. A. Roth and A. Vinegar, *Pharmacol. Ther., 48,* 143(1990).
52. C. J. Reed, *Drug Metab. Rev., 25,* 173(1993).
53. M. A. Sarkar, *Pharmacol. Res., 9,* 1(1992).
54. B. Testa and P. Jenner, *Drug Metabolism: Chemical and Biological Aspects,* New York, M. Dekker, 1976.
55. J. R. Gillette, et al., *Annu. Rev. Pharmacol. Toxicol., 14,* 271(1974); D. J. Reed, *Cellular Defense Mechanisms Against Reactive Metabolites,* in *Bioactivation of Foreign Compounds,* M. Anders, Ed., New York, Academic Press, 1985, pp. 71–108.
56. F. Oesch, *Biochem. Pharmacol., 25,* 1935(1976).
57. M. Koreeda, et al., *Science, 199,* 778(1978).
58. D. Thakker, et al., *Polycyclic Aromatic Hydrocarbons: Metabolic Activation of Ultimate Carcinogens,* in *Bioactivation of Foreign Compounds,* M. Anders, Ed., New York, Academic Press, 1985, pp. 177–242.
59. S. D. Nelson, et al., *Role of Metabolic Activation in Chemical-Induced Tissue Injury,* in *Drug Metabolism Concepts,* D. M. Jerina, Ed., ACS Symposium Series 44, Washington, D.C. American Chemical Society, 1977, pp. 155–185.

SUGGESTED READINGS

M. Anders, Ed., *Bioactivation of Foreign Compounds,* New York, Academic Press, 1985.

J. Caldwell and W. Jakoby, Eds., *Biological Basis of Detoxication,* New York, Academic Press, 1983.

W. Jakoby, Ed., *Enzymatic Basis of Detoxication,* Vols. I and II, New York, Academic Press, 1980.

W. Jakoby, J. R. Bend, and J. Caldwell, Eds., *Metabolic Basis of Detoxication-Metabolism of Functional Groups,* New York, Academic Press, 1982.

G. J. Mulder, Ed., *Conjugation Reactions in Drug Metabolism: An Integrated Approach,* London, Taylor and Francis, 1990.

P. R. Ortiz de Montellano, Ed., *Cytochrome P450 Structure, Mechanism, and Biochemistry,* New York, Plenum Press, 1986.

B. Testa and P. Jenner, *Drug Metabolism: Chemical and Biochemical Aspects,* New York, Marcel Dekker, 1976.

B. Testa and P. Jenner, *Concepts in Drug Metabolism,* New York, Marcel Dekker, 1981.

R. T. Williams, *Detoxication Mechanisms,* 2nd Ed., New York, John Wiley and Sons, 1959.

Chapter 9

VOLATILE ANESTHETICS

Ronald L. Jacoby and Karl A. Nieforth

General anesthetics are depressant drugs that produce partial or total loss of the sense of pain, which may be accompanied by loss of consciousness. This state of insensibility is known as anesthesia. Before the development of effective anesthetic agents, individuals were prepared for surgery by several less than humane methods, such as physical restraint, shock, or large doses of alcohol or opium. These procedures not only discouraged patients from seeking the assistance of surgeons, but also restricted development of surgical techniques.

The history of inhalation of intoxicating agents is colorful and suggests that general anesthesia should have been in practice well before its eventual discovery. Only when chemists prepared synthetic agents that were serendipitously found to possess anesthetic activity did the dream of painless surgery become a reality.

Early experiments with diethyl ether and nitrous oxide as dental and surgical anesthetics failed to receive adequate publication, and it remained for Horace Wells, a Hartford dentist, to introduce surgical anesthesia in 1844. While attending a public demonstration of laughing gas (nitrous oxide), Wells first observed the anesthetic effects of the gas. One of the volunteers who inhaled the gas, Samuel Cooley, a local pharmacy clerk, injured his leg during the demonstration but experienced no pain while under the influence of the gas. Wells questioned Cooley about the incident and, on the following day, began using nitrous oxide during dental procedures, thereby introducing painless dentistry. This was shortly followed by William Morton's demonstration of ether anesthesia in Boston and James Simpson's use of chloroform in Edinburgh.

Since then, numerous compounds have been introduced as anesthetics, and several have received extensive clinical use. One might question the continuing search for new anesthetics were it not for several shortcomings of those currently in use.

SIGNS AND STAGES OF ANESTHESIA

As evidenced by alterations in physiologic responses, general anesthetics act by depressing nervous function. As higher concentrations of anesthetic agents in nervous tissue are obtained, depth of anesthesia increases. Anesthesia with diethyl ether was divided by Guedel into four separate stages, which were subsequently subdivided by Gillespie, as shown in Figure 9–1.

Stage 1. Analgesia
A mild depression of higher cortical centers, suitable for surgical procedures not requiring extensive muscular relaxation.

Stage 2. Delirium
Excitement as a result of depression of cortical motor centers resulting in possible extensive involuntary activity.

Stage 3. Surgical anesthesia
Subdivided into four planes representing progressive increases in depth of anesthesia as characterized by: first, loss of spinal reflexes; second, decreased muscle reflexes; third, paralysis of intercostal muscles; and fourth, disappearance of muscle tone.

Stage 4. Respiratory paralysis
So designated by Guedel because of his use of diethyl ether as a model, but more correctly referred to as toxic or overdose stage involving respiratory and vasomotor cessation.

Stage 1, analgesia, may be related to depression of higher levels of the cortex and thalamic centers responsible for transmittal of painful stimuli. All anesthetics do not produce analgesia. They do bring about insensitivity to pain, but this result is related to loss of consciousness rather than to specific analgesic response.

Stage 2 extends from the loss of consciousness to the beginning of surgical anesthesia and may be accompanied by uncoordinated muscular movements associated with delirium, urinary incontinence, and related responses. This excitement and involuntary activity need not always occur during this stage or may be prolonged in conditions such as alcoholism. Depression of cerebral cortex and reticular formation may cause excitement, which in turn increases heart rate, blood pressure, and respiration. Irregularities in respiration are more marked with odoriferous anesthetics because of vapor-induced irritation. The first two stages are termed the induction period, which is ideally short and free of excitement, involuntary activity, and irritation.

Most surgical procedures are carried out during stage 3, which is characterized by regular, involuntary breathing, roving eyeball movements, and the absence of certain reflexes. As surgical anesthesia deepens, these physical signs progressively change, differentiating the four planes of surgical anesthesia. With diethyl ether, muscle tone decreases, eye reflexes disappear, and thoracic respiration gradually disappears.

Cessation of respiration signifies stage 4, and breathing must be assisted until anesthetic effect decreases. Clinical symptoms of toxicity (stage 4) are not frequently mani-

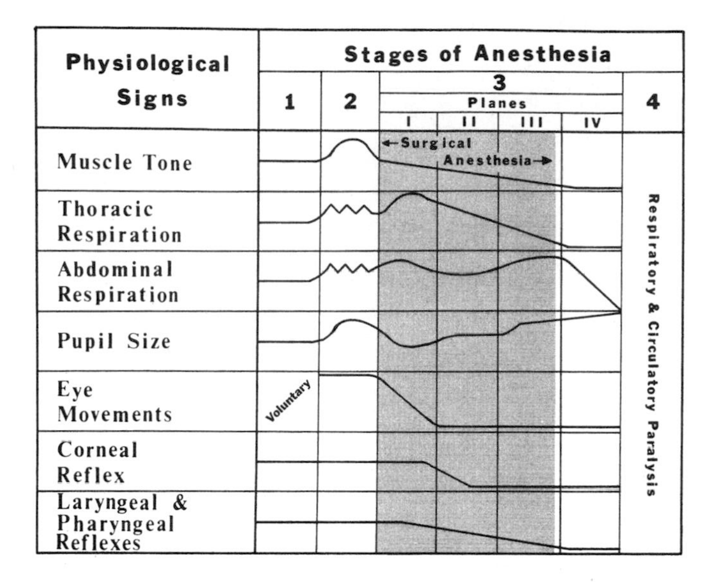

Fig. 9–1. Stages and planes of diethyl ether anesthesia related to various physiologic changes. (Adapted from D. R. Laurence, *Clinical Pharmacology,* 3rd ed., London, J. & A. Churchill, 1966, p. 187.)

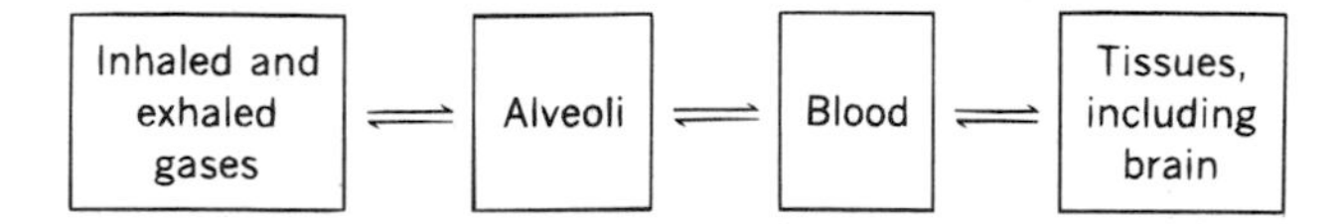

Fig. 9–2. The complex equilibria existing during the state of anesthesia.

fested by the volatile anesthetics, probably because of their short-term use in controlled situations. All of Guedel's predictable signs for ether anesthesia are not observed with all other anesthetics. Each agent has its own set of clinical signs of anesthesia, and many modern anesthetics produce such rapid induction that Guedel's signs cannot be clearly observed. Also, anesthetic practice involves the use of barbiturates for induction, muscle relaxants, and other central nervous system (CNS) depressants that obscure the agent's signs of anesthesia.

UPTAKE, DISTRIBUTION, AND ELIMINATION OF ANESTHETIC AGENTS

Anesthetic agents may be divided into three classes: gases, volatile liquids, and intravenous or fixed anesthetics. The last class, ultrashort-acting barbiturates, is considered in another chapter. This discussion is restricted to gas and volatile liquid anesthetics. Various techniques and equipment have been devised to administer anesthetic gas or vapor. The open drop method consists of dropping the liquid anesthetic on gauze or other absorbent material supported over the patient's nose and mouth by a wire frame, forming a mask. As the patient inhales, anesthetic vapor is drawn into the lungs.

Although this method is simple and inexpensive, it provides the anesthesiologist with little information of the concentration of anesthetic inhaled and also allows potentially toxic vapors to infiltrate the surrounding area. These and other complications of the open drop method have been overcome with the development of anesthetic machines. These devices allow the anesthesiologist to control oxygen, carbon dioxide, and anesthetic concentrations by means of flowmeters, absorbers for expired carbon dioxide and water, and vaporizers. The reader is referred to other sources for a detailed description of inhalation appliances.[1]

During administration of an anesthetic, its concentration in the blood rapidly increases toward that in the inspired gas. As the anesthetic enters tissues, this concentration approaches that of the arterial blood supply. Organs transfused by a large amount of blood, such as the brain, rapidly acquire high concentrations of anesthetic. Depth of anesthesia is a function of the amount of anesthetic in the brain, and the time of induction and recovery are determined by the rate of change of concentration in the brain. A simplified diagram of this system is shown in Figure 9–2.

Factors that determine the amount of anesthetic in the brain are alveoli ventilation, rate of blood flow through the lungs, solubility of the agent in blood and tissues, and rate of blood flow through the brain. Induction and

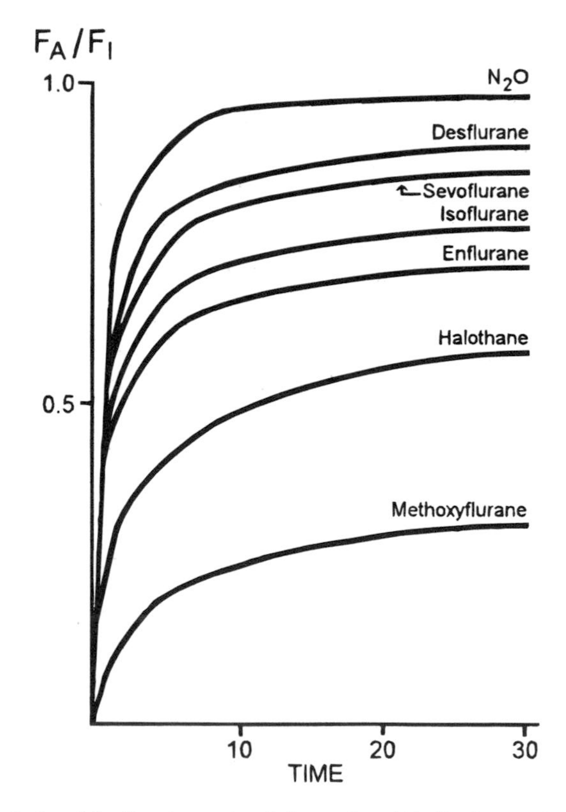

Fig. 9–3. Idealized graph of the ratio of F_A/F_I shows the rate at which the end-tidal (exhaled) anesthetic (F_A) approaches the concentration of anesthetic being breathed (F_I). (Adapted from E. I. Eger II, *Isoflurane (Forane): A Compendium and Reference,* Madison WI, Ohio Medical Products, 1981; N. Yasuda et al., *Anesthesiology, 74,* 489[1991].)

Table 9–1. The Partition Coefficients, MACs, and Metabolism of Volatile Anesthetics

Anesthetic	Partition Coefficient at 37°C*		MAC (% of 1 Atm.)†		% Metabolized‡
	Oil/Gas	Blood/Gas	Without N_2O	With N_2O (%)	
Methoxyflurane	970	12	0.16	0.07(56)	50
Halothane	224	2.3	0.77	0.29(66)	20
Enflurane	98.5	1.19	1.7	0.60(70)	2.4
Isoflurane	90.8	1.4	1.15	0.50(70)	0.17
Sevoflurane	53.4	0.60	1.71	0.66(64)	4–6§
Desflurane	18.7	0.42	6.0	2.83(60)	0.02§
Nitrous oxide	1.4	0.47	104	—	0.004

*Partition coefficients are from reference 1.
† MAC is the minimum alveolar concentration, expressed as volume percent, that is required to produce immobility in 50% of middle-aged humans undergoing surgical incision. MAC values are from reference 1.
‡ The percent of absorbed anesthetic recovered as metabolites is from reference 2.
§ The data for sevoflurane and desflurane are from reference 3.

recovery in particular are determined by the solubility of the agent in blood.

Based on their blood/gas partition coefficients in Table 9–1, anesthetics can be classified as highly blood soluble (methoxyflurane), of medium solubility (halothane, enflurane, and isoflurane), and of low solubility (sevoflurane, desflurane, and nitrous oxide). The highly blood-soluble methoxyflurane has slow induction and recovery time. Before appreciable amounts of methoxyflurane diffuses into brain tissue, the blood reservoir must be nearly saturated. Blood-soluble agents require several passages of the blood through the lungs. On the other hand, blood concentrations of insoluble agents like nitrous oxide quickly approach saturation, resulting in rapid transfer to brain tissues. This relationship between blood solubility of anesthetic and the rate at which it saturates blood (measured through exhaled gas) is shown in Figure 9–3.

Although small amounts of gas diffuse through the skin or are metabolized, the main route of elimination is the lungs. When administration of gas is discontinued, elimination of the agent begins and the factors that controlled uptake operate in reverse. Agents of high blood solubility are released slowly from the blood, leading to slow recovery. Conversely, blood-insoluble agents, like nitrous oxide, leave the blood and are exhaled rapidly, resulting in rapid recovery. Interest in agents with low blood solubility, like sevoflurane and desflurane, is renewed because they are potentially useful in outpatient surgery for which rapid recovery is essential.

THEORIES OF ANESTHESIA

Since the development of clinical anesthesia over a century ago, there has been a vast amount of research and speculation concerning the mechanism of action of anesthetic agents. Despite these efforts, the exact mechanism remains unknown. Many theories of narcosis do not explain how unconsciousness is produced at a molecular level, but instead relate some physicochemical property of anesthetic agents to their anesthetic potency or describe biochemical or physiologic events that occur during anesthesia. Nonetheless, most of the evidence discussed subsequently supports the view that anesthetics produce their effect by inhibiting transmission in the CNS. The human brain contains an estimated 10^{12} neurons, many of which are interconnected to produce complex networks capable of processing sensory inputs, producing coordinated muscle movements, and storing information in memory. The brain cells are connected one to another through synapses, and it is this synaptic transmission that is inhibited to produce anesthesia. In the subsequent discussion, we examine the structure and function of a simplified neuronal network of the cerebral cortex (Fig. 9–4) before exploring the mechanisms by which the many different anesthetics produce their effects.[6]

Cerebral Cortex Structure and Function

Cerebral cortex neurons are of two main types, pyramidal and stellate cells. The pyramidal cells are the main excitatory neurons of the cortex. Their axons are the major output projects of the cortex, extending to other areas of the brain and the spinal cord. Pyramidal cells are excitatory neurons and the amino acid, glutamic acid, is the excitatory neurotransmitter of these cells. A single pyramidal cell may receive hundreds, even thousands, of inputs from other pyramidal cells, stellate cells, and specific afferents from the thalamus. It is the summation of all these inputs that determines whether an impulse is generated by the pyramidal cell. The second type of cortex neuron is the stellate cell, which relays inputs to the pyramidal cells. These cells do not have long axons that project out of the cerebral cortex, but they instead connect one neuron to another in a local area of the cortex. Thus, stellate cells are known as interneurons. They are either excitatory or inhibitory neurons. Recent research concerning the mechanism of action of anesthetics has focused on the inhibitory interneurons that contain γ-aminobutyric acid (GABA), the major inhibitory neurotransmitter of the brain. In the center of Figure 9–4 is a

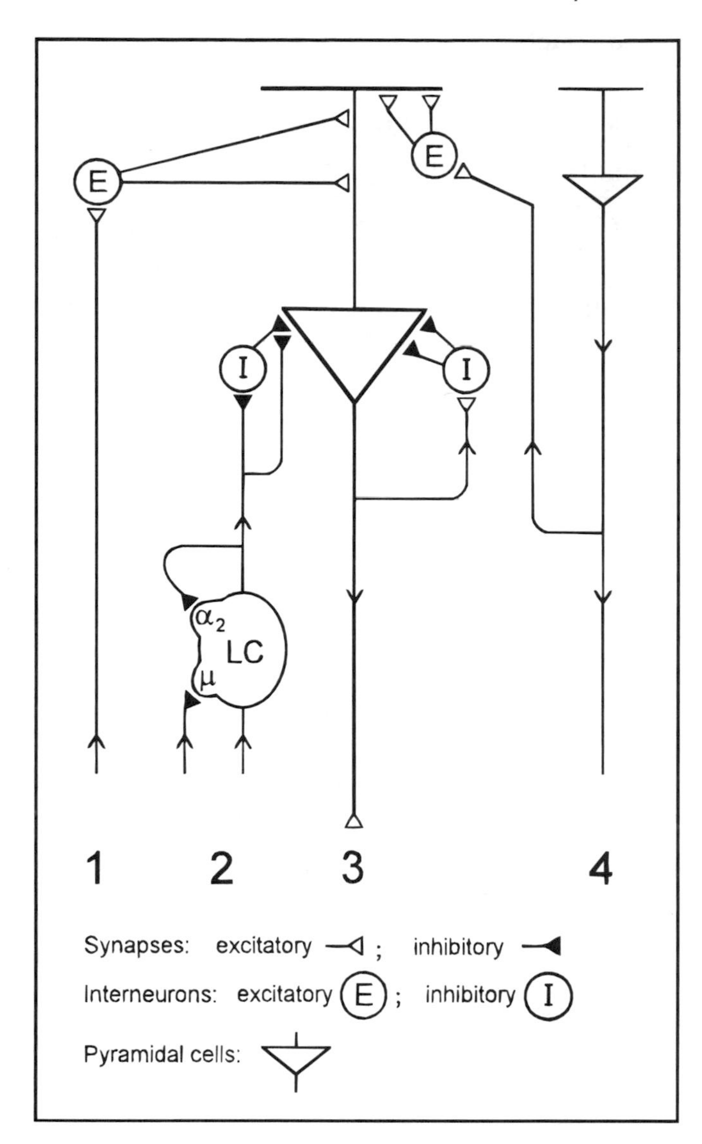

Fig. 9–4. Schematic diagram of the principal neuron types in the cerebral cortex. Consciousness is associated with excitatory output of the pyramidal cells (3). Pyramidal cells receive multiple synaptic contacts from specific afferents of the thalamus (1), other brain areas as locus coeruleus (LC), and other pyramidal cell collaterals (4). These inputs are relayed to pyramidal cells through inhibitory and excitatory interneurons (stellate cells).

pyramidal cell (labeled 3) with excitatory inputs from the thalamus, labeled 1 (specific afferents), and from other pyramidal cells, labeled 4. Notice that these inputs are relayed to the pyramidal cell through stellate cells that are shown as excitatory, but they could also be inhibitory interneurons. The pyramidal cell, 3, is also shown to receive input from an inhibitory interneuron that is modulated by noradrenergic fibers from the locus coeruleus (LC), which are themselves inhibitory. Thus, activation of LC neurons attenuates inhibitory GABA-ergic interneurons to ultimately facilitate CNS activation. Pyramidal cells have collateral fibers that synapse with GABA-containing inhibitory interneurons. Firing of the pyramidal cell to produce output also activates the inhibitory interneuron, which then opposes additional firing of the pyramidal cell. This self-inhibition is shown on the right side of cell 3 in Figure 9–4.

Most of the individual synaptic connections shown in Figure 9–4 have been proven, but their complete assembly, as shown, is simplified and hypothetic. The figure is shown for two purposes. First, it illustrates the complexity of cerebral neuronal networks. Normal consciousness involves a sensitive balance between excitatory (glutamate) and inhibitory (GABA) neurotransmitters. Other neurotransmitters are involved, such as norepinephrine, serotonin, acetylcholine, and histamine, which control the general level of activity and excitability of the CNS. For example, the LC shown in Figure 9–4 is a tiny nucleus located in the upper brain stem. This nucleus has been linked with various responses, such as attention and vigilance. High levels of noradrenergic input from the LC to the cerebral cortex facilitate the processing of signals from other sensory centers, but the LC does not produce high levels of cerebral activity and output by itself.[7] Second, Figure 9–4 illustrates most of the synaptic connections postulated as being involved in the mechanism of action of the many different anesthetics.

Glutamic acid, the excitatory neurotransmitter elaborated by pyramidal cells, binds to postsynaptic N-methyl-D-aspartate (NMDA) receptors to produce synaptic transmission (Fig. 9–5).[8] When stimulated, this NMDA recep-

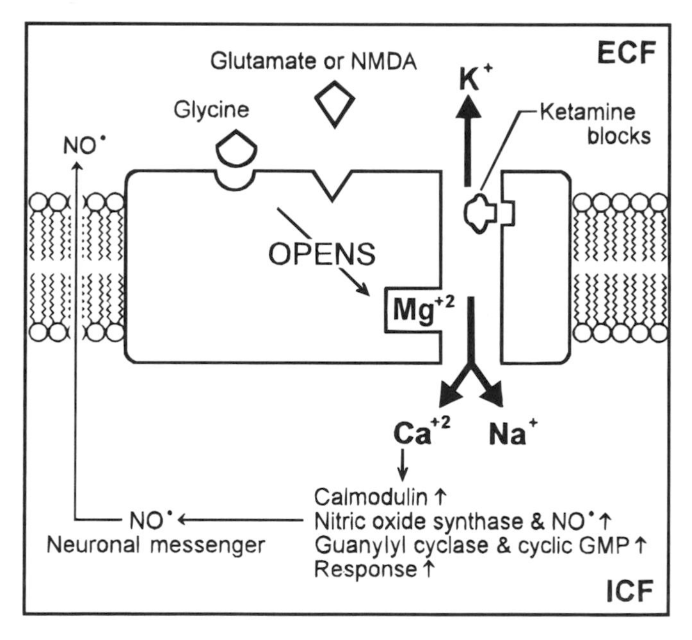

Fig. 9–5. Glutamate or NMDA receptors in the central nervous system. Binding of agonists (glutamate or NMDA) opens channel allowing potassium ions to flow outward to extracellular fluid (ECF) and sodium and calcium ions to flow into the nerve cells. Increased intracellular (ICF) calcium ion concentration triggers a cascade that produces a response and liberates the neuronal messenger nitric oxide. Ketamine may produce anesthesia by blocking these NMDA-controlled channels, which are located at excitatory synapses on pyramidal cells (3) in Figure 9–4. (Adapted from L. B. Wingard, et al., *Human Pharmacology,* St. Louis, Mosby-Year Book, 1991, pp. 20–22.)

tor opens channels, allowing entry of Na^+ and CA^{+2} ions into the postsynaptic cell and exit of K^+ ions. This increased intracellular Ca^{+2} binds to calmodulin, which activates nitric oxide synthase (NOS), converting arginine to citrulline and nitric oxide (NO). Nitric oxide activates soluble guanylyl cyclase to form cyclic guanosine 5′-monophosphate (cGMP), which can induce a variety of functions including protein phosphorylation, gating of ion channels, and phosphodiesterase activity. These effects are not restricted to the postsynaptic neuron, because NO can freely diffuse to exert its effects in surrounding neurons. It can diffuse back into the presynaptic neuron to reinforce synaptic transmission or into adjacent neurons to alert them of local synaptic action. Nitric oxide may even diffuse into local capillaries, dilating them and thus providing enhanced blood flow to support increased neuronal activity. These observed effects have led to the hypothesis that NO is an important mediator of consciousness.[9] It is believed that the injectable anesthetic ketamine produces anesthesia by blocking the NMDA-controlled channels, as shown in Figure 9–5. Also, the NOS inhibitor, nitro^G-L-arginine methyl ester (L-NAME) has been shown to reversibly block the production of NO and thus reduce the threshold for anesthesia.[10] In these examples, ketamine and L-NAME produce CNS depression by blocking excitatory synaptic transmission and its attendant events.

In a second case, anesthesia is produced by enhancing inhibitory neuronal activity. Figure 9–6 illustrates the $GABA_A$ receptor/chloride channel complex, which mediates the events associated with the inhibitory transmitter GABA.[11] When bound to its site, GABA causes the chloride-conducting channel to open with the resulting influx of chloride ions, leading to hyperpolarization of the postsynaptic membrane and inhibition of neuronal firing. Barbiturates, benzodiazepines, and steroid anesthetics have distinct binding sites on the $GABA_A$ receptor/chloride channel complex that enhance GABA binding and/or chloride conductance. Other agents as propofol, volatile anesthetics, and alcohol also enhance GABA action, but it is not known if unique binding sites exist for these agents. All of these agents promote GABA inhibitory transmission.

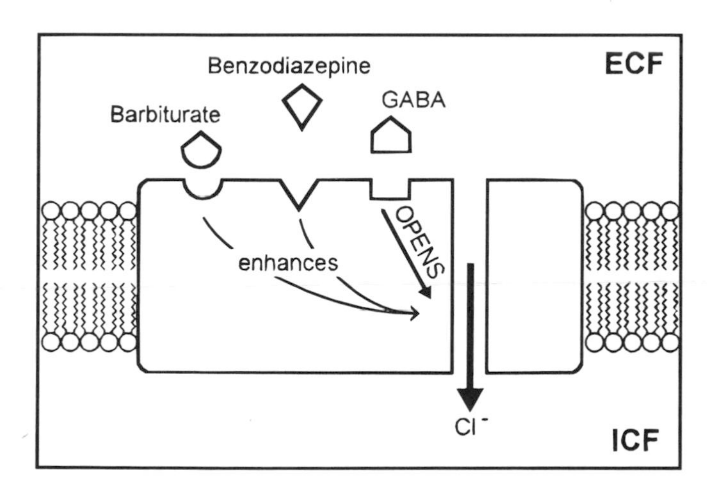

Fig. 9–6. GABA type A receptor controls chloride ion channel. GABA binds to its receptor, opening the chloride ion channel, resulting in hyperpolarization of the neuron. Benzodiazepines and barbiturates may produce anesthesia by allosterically enhancing GABA opening of chloride channels, which are located at inhibitory synapses on pyramidal cells (3) in Figure 9–4. ECF = extracellular fluid; ICF = intracellular fluid. (Adapted from L. B. Wingard, et al., *Human Pharmacology,* St. Louis, Mosby-Year Book, 1991, pp. 20–22.)

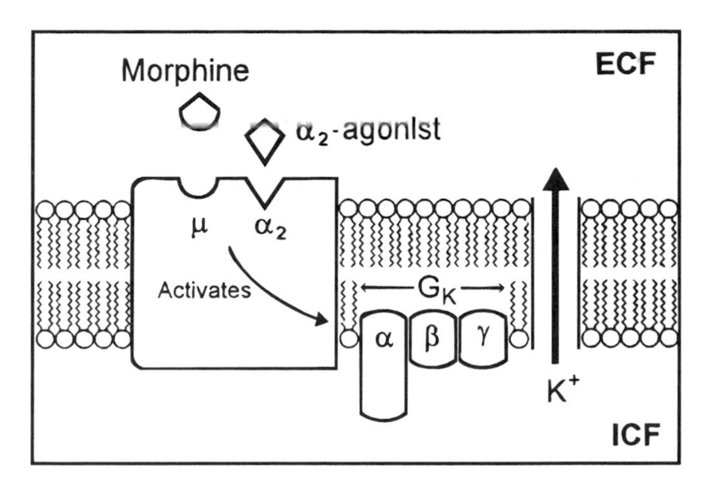

Fig. 9–7. Morphine and α_2-agonists activate their respective G-proteins, which hyperpolarize neurons by lowering intracellular fluid (ICF) potassium ion concentration. Anesthesia may be produced by this disinhibition of locus coeruleus inhibitory effects, (2) in Figure 9–4. (Adapted from L. B. Wingard, et al., *Human Pharmacology,* St. Louis, Mosby-Year Book, 1991, pp. 20–22.)

Lastly, Figure 9–7 shows how clonidine and dexmedetomidine, α_2-adrenoceptor agonists, or morphine may produce their hypnotic-anesthetic effects through interactions in the LC of the brain stem.[12] It is believed that both of these agents bind to LC receptors that are coupled to regulatory G-proteins that control inward-directed potassium channels.

H_3C H CH_3 CH_3 N N H

Dexmedetomidine

Inhibition of these channels reduces intraneuronal K^+ leading to hyperpolarization and silencing of the LC neurons. Locus coeruleus axons project into several areas of the brain and cerebral cortex where they extensively ramify, contacting most, if not all, cerebral neurons. Thus, reduction of LC suppression of inhibitory interneurons by α_2-agonists has a profound effect on cerebral function. The shift in balance toward cortical GABAergic interneurons results in generalized depression.

To this point, discussion has focused on the actions of injectable anesthetics because much is known about their biochemical mechanisms of action. The picture that has developed is one of complex neuronal connections in the brain where anesthesia may be produced in a variety of ways. Agents that inhibit excitatory pathways or those that promote inhibitory neurons produce anesthesia. Anesthetics produce their effects in many parts of the brain in addition to the cerebral cortex. For example, the brain stem reticular activating system, which is implicated in control of wakefulness, is affected by a variety of agents. Consciousness involves the coordinated actions of many parts of the brain and consequently, no exclusive region is the site of anesthesia.

Volatile Anesthetics

As is the case with the anesthetics discussed previously, volatile anesthetics produce their depressant effects by blocking synaptic transmission. With their simple structures however, these agents do not interact with receptors, but instead inhibit synaptic transmission through nonspecific interactions with synaptic membranes and/or their associated proteins. For instance, during induction of anesthesia, inhibitory neurons are suppressed first, leading to stage 2 delirium (see Fig. 9–1). This occurrence is consistent with the idea that excitatory neurons typically fire unless they are restrained by inhibitory controls. Low concentrations of anesthetic remove the inhibitory controls allowing for brief, but often significant, hyperactivity.

Most authorities agree that the interactions between volatile anesthetics and membranes are hydrophobic in nature, although it has not been determined if the site of hydrophobic interaction is with membrane lipids or with hydrophobic regions of membrane proteins or ion channels.

Lipid Theory

The lipid theory, independently developed by Meyer and Overton, states that a direct relationship exists between lipid solubility of an agent and its anesthetic potency. This theory does not purport a mechanism of action for anesthetics, but rather describes an excellent correlation between anesthetic potency of a compound and its oil/water or oil/gas partition coefficient, as shown in Figure 9–8. From this theory, one would predict compounds of high lipid solubility to be potent anesthetic agents. Yet, this is not always the case, because many fat-soluble substances, such as aromatic compounds, do not produce anesthesia. The Meyer-Overton theory does not offer an explanation of anesthesia at the molecular level, but it does strongly suggest that the site of action of anesthetics is hydrophobic.

Anesthetic-Membrane Interactions

It is known that anesthetics dissolve in cell membranes and cause disordering or increased fluidity of the constituent lipids. This disordering of membrane lipids causes expansion of the membrane, which results in compression of the membrane components. Ion channels or other excitable proteins are no longer able to function.

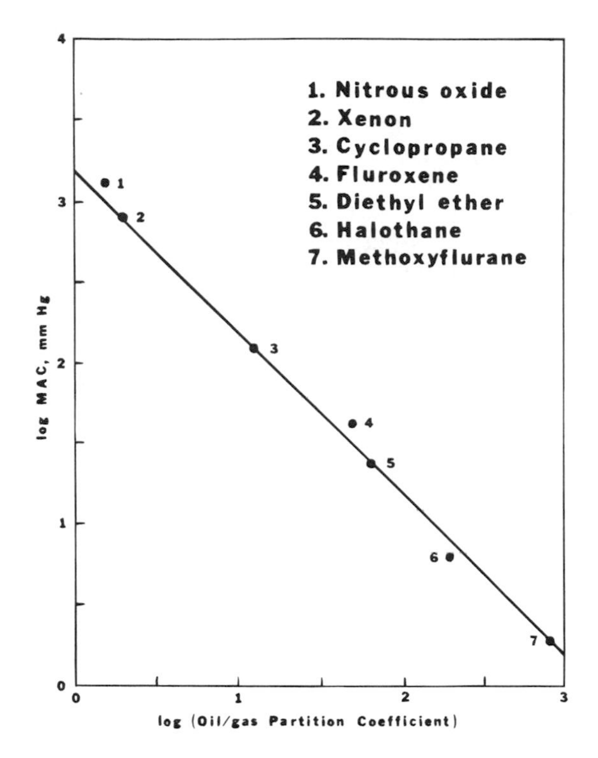

Fig. 9–8. Plot of the logarithm of the minimum alveolar concentration (MAC) versus the logarithm of the oil/gas partition coefficient of various volatile anesthetics at 37°C in the dog. (Data from E. I. Eger et al., *Anesthesiology, 26,* 771 [1965]; reproduced from E. R. Larsen, et al., in *Drugs Affecting the Central Nervous System,* Vol. 2, A. Burger, Ed., New York, Marcel Dekker, 1968, p. 8.)

For instance, the $GABA_A$ receptor/chloride channel complex discussed previously could be deformed in such a way that it would remain open, resulting in Cl^- ion influx leading to synaptic hyperpolarization with the resultant inhibition of neuronal firing. The membrane expansion hypothesis is supported by the finding that application of pressures above 100 atmospheres restores consciousness in animals anesthetized at 1 Atm. This information suggests that the elevated pressure counteracts the expansion and/or disordering of the "anesthetized" membrane, thus restoring consciousness.

Alternatively, agents could produce anesthesia by directly interacting with membrane proteins. Anesthetics could bind to hydrophobic protein regions in ion channels, thus preventing ion flux. Other suggested targets for anesthetic binding and inhibition are G-proteins and their related intracellular secondary messenger systems. Considerable research focused on signal transduction systems as related to neuronal activity, memory, learning, and consciousness is expected to provide new insight into the mechanisms of brain activity and anesthesia. Two books contain detailed accounts of the effects of anesthetics on membranes and of the mechanisms of anesthesia.[13,14]

Anesthetic Agents

Two of the earliest anesthetic agents, diethyl ether and chloroform, may serve as the starting points, because to-

Table 9–2. Structures and Properties of Volatile Anesthetics

Generic Name (Trade Name)	Structure	Boiling Point (°C)	Chemically Stable*
Desflurane (Suprane)	$CHF_2—O—CHF—CF_3$	23.5	Yes
Enflurane (Ethrane)	$CHF_2—O—CF_2—CHFCl$	56.5	Yes
Halothane (Fluothane)	$CF_3—CHBrCl$	50.2	No
Isoflurane (Forane)	$CHF_2—O—CHCl—CF_3$	48.5	Yes
Methoxyflurane (Penthrane)	$CH_3—O—CF_2—CHCl_2$	104.7	No
Nitrous oxide	N_2O	−88.0	Yes
Sevoflurane	$(CF_3)_2CH·O—CH_2F$	58.5	No

* Indicates stability to soda lime, UV light, and common metals.

gether they exhibit most of the undesirable properties of general anesthetics.

Diethyl ether and chloroform are potent anesthetics that produce excellent analgesia and skeletal muscle relaxation. Unfortunately, ether is flammable, and mixtures of it with air, oxygen, or nitrous oxide are explosive. Induction with ether is slow and unpleasant. Recovery is prolonged and often accompanied by nausea and vomiting. Ether is irritating to the respiratory tract, resulting in excessive salivation and secretion of bronchial mucus.

The use of chloroform as an anesthetic is severely restricted because of its narrow margin of safety, toxicity to liver and kidneys, and undesirable circulatory effects, i.e., cardiac arrest, arrhythmia, and hypotension.

Obviously, ether and chloroform left much to be desired, and their use has ceased. Flammability was effectively reduced through the addition of chlorine with a concomitant increase in toxicity. Advances in fluorine chemistry associated with nuclear weapon research permitted the rapid introduction of several new, efficacious agents.

In general, an ideal anesthetic would have the following characteristics: (1) be nonflammable; (2) be inexpensive; (3) require uncomplicated equipment for administration; (4) provide adequate muscular relaxation; (5) produce rapid and uncomplicated induction and emergence; (6) have no effect on myocardium or respiration at anesthetic doses; (7) be chemically and metabolically stable; (8) be sufficiently potent to permit adequate oxygen supply; (9) have a wide margin of safety.

Most of the problems associated with diethyl ether and chloroform anesthesia have been overcome. The names and structures of several general anesthetics are given in Table 9–2. Methoxyflurane is seldom used because of its potential to cause renal toxicity, and N_2O is usually not used as the sole anesthetic because of its low potency. Two new agents, desflurane and sevoflurane, are as yet not fully evaluated, leaving the agents enflurane, isoflurane, and halothane, which are commonly used as general anesthetics. Table 9–3 provides a comparative assessment of their properties. Of those listed, isoflurane appears to be the agent most closely approaching ideality.[2,17]

A new agent, desflurane, was introduced in the United States in 1992.[18] Clinical studies suggest that its pharmacologic properties are much like those of isoflurane; however, its low blood and tissue solubilites provide for rapid recovery, which makes it a candidate for use in outpatient surgery. Also, desflurane is metabolized to the least extent of all halogenated anesthetics, which suggests that it may have low toxicity.

Sevoflurane, used in Japan for several years but not yet available in the United States, appears in many ways similar to desflurane.[2] This agent, however, decomposes to some extent, when exposed to soda lime, to potentially toxic products. It is also metabolized to a greater extent (4 to 6%) than other new anesthetics. High levels of me-

Table 9–3. Comparative Assessment of Enflurane (E), Halothane (H), and Isoflurane (I)

Property	Superior		Intermediate	Inferior
Stability	I	=	E	H
Blood solubility	I		E	H
Pungency	H		I	E
Respiratory depression	H		I	E
Circulatory depression	I		H	E
Induction of arrhthymias	I		E	H
Muscle relaxation	I	=	E	H
Increased intracranial pressure/cerebral blood flow	I		E	H
Seizure activity	H	=	I	E
Metabolism	I		E	H
Toxicity	I		E	H

Adapted with permission from J. G. Wade and W. C. Stevens, *Anesth. Analg., 60,* 679, 1981; see also reference 2.

Table 9–4. Relative Flammability of "Nonflammable" Anesthetics*

	Halothane (%)	Enflurane (%)	Isoflurane (%)
Minimum flammable concentration (MFC) of agent in 30% O_2 with remaining atmosphere N_2O	4.75	5.75	7.0
Minimum effective alveolar concentration (MAC) of agent given in above atmosphere	0.28	0.65	0.46
MAC in man in absence of N_2O	0.75	1.68	1.15
MFC/MAC in N_2O	17	8.9	15.2

* Data taken from references 15 and 16.

tabolism can lead to toxicity. Further studies and experience are needed with both desflurane and sevoflurane to determine the place these agents will take in modern anesthesia.

The occurrence of fires in operating theaters is of great concern to all participants in the surgical procedure. Although the introduction of "nonflammable" agents such as halothane, enflurane, and isoflurane has substantially decreased the hazard, such fires still occur. Three essential ingredients are required for any combustion: (1) an ignition source such as a laser, (2) a combustible material (gauze, drapes, rubber tubes, etc.), and (3) an oxidizing agent (O_2, N_2O). Many substances are flammable in pure oxygen, N_2O, or mixtures, but not air. Certain substances are flammable in N_2O at concentrations that are too low to permit ignition in pure oxygen.[15] The concentrations required for combustion as indicated in Table 9–4 are higher than those generally encountered, except possibly during induction.

Nonhalogenated Hydrocarbons

Alkanes, alkenes, and alkynes exhibit anesthetic activity that increases in each homologous series as chain length increases. Nevertheless, high toxicity, particularly to the cardiovascular system, precludes the use of most hydrocarbons as general anesthetics. Cyclopropane is the only hydrocarbon that was extensively used in clinical practice, but because of its explosive properties, its use has been discontinued.

Ethers

Low-molecular weight hydrocarbon ethers have anesthetic activity that increases along with toxicity as chain length increases. Alkane, alkene, alkyne, and alicyclic ethers have been investigated as potential anesthetic agents, but only ethyl-substituted and vinyl-substituted ethers have been found clinically useful. Introduction of unsaturation into an aliphatic ether increases potency and also shortens induction and emergence. It also reduces chemical stability and often increases toxicity. All ethers are flammable and form explosive mixtures with anesthetic gases. For this reason, they are no longer used.

Halogenated Anesthetic Agents

The addition of halogen atoms to ethers or hydrocarbons decreases or eliminates flammability and often increases anesthetic potency. Unfortunately, many halogenated compounds produce an unacceptably high incidence of arrhythmias and renal or hepatic damage, which precludes their use as anesthetics.

Chloroform was first used in this country in 1847 shortly after Simpson's demonstration in Edinburgh. It was to claim its first victim within 2 months.

Halogenated compounds containing only bromine are not useful as anesthetics because of low volatility and low stability. Brominated compounds containing other halogens, chlorine and fluorine, however, like halothane ($CF_3CHBrCl$), may be useful. Chlorinated hydrocarbons have had limited use. These compounds, including chloroform, ethyl chloride, and trichloroethylene, are far from ideal anesthetics because of high hepatotoxicity and cardiac arrhythmias.

More recently introduced anesthetics have been fluorinated ethers or hydrocarbons. Halothane, the first fluorinated agent introduced, was followed by fluoxene and then methoxyflurane. Newer agents such as desflurane and sevoflurane routinely appear as the search for the perfect anesthetic continues.

Fluorination of ethers or hydrocarbons confers various desirable anesthetic properties. Addition of fluorine atoms decreases flammability. Fluorination of hydrocarbons decreases boiling points to permit certain hydrocarbon skeletons, unsuitable in the unhalogenated form, to be used as anesthetics. Catechol-induced arrhythmias decrease in frequency as fluorine atoms replace other halogens, which appear to induce arrhythmias in proportion to molecular size. Interesting structure-activity relationships among the fluorinated compounds await unraveling as more information about mechanisms of action becomes available.

Fluorinated hydrocarbons and ethers appear to have a convulsant and anticonvulsant component, which may be accentuated or attenuated by structurally related compounds. The drug bis-(2,2,2-trifluoroethyl) ether (Flurothyl, Indoklon) is a convulsant that has been suggested as a substitute for electroshock therapy. The convulsant response demonstrates a dose ceiling, above which anesthesia occurs, in reasonably good agreement with that predicted by lipid solubility.[19] Differential distribution within membrane subregions that house gating mechanisms for GABA, glutamate, or acetylcholine receptors has been proposed to explain convulsant or anesthetic predominance.[20]

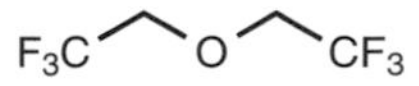

Fluroethyl (Indoklon)

The drug 1,1,1,3,3,3,-hexafluoroisopropyl methylether (Isoindoklon) is a weak but good anesthetic, which can inhibit the convulsive actions of Indoklon.

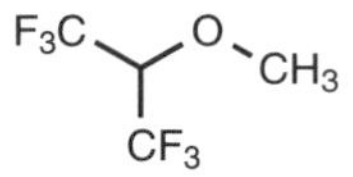

Isoindoklon

Identification of specific receptor sites may be questioned because of the inability of halothane and chloroform to inhibit Indoklon-induced convulsions.

Nitrous Oxide

Nitrous oxide (N_2O) is the least potent anesthetic in use today, and for that reason it is not used alone to produce surgical anesthesia. The MAC value for N_2O has been estimated to be 105 to 120%, so hyperbaric conditions are required to use it as a sole anesthetic agent.[21] Therefore, it is used primarily in combination with other more potent anesthetics, which permits use a lower concentration of the more potent agent and reduces the likelihood of toxic reactions. Anesthetics are often administered in a gas mixture of 70% N_2O and 30% oxygen, which reduces the requirement of the more potent agent by one half to two thirds, as is shown in Table 9–1. Nitrous oxide is an excellent analgesic in subanesthetic concentrations accounting for its use during dental procedures and labor. Recent studies and analysis of the clinical pharmacology of the gas support the continued use of N_2O, even though some criticism has been published.[22] New interest in N_2O toxicity has resulted from the finding that N_2O inactivates the vitamin B_{12} component of methionine synthetase, which can lead to impairment of DNA synthesis. Fortunately, this toxic effect is a potential problem only with chronic exposure.

Ketamine Hydrochloride

The compound 2-(o-chlorophenyl)-2-methylaminocyclohexanone hydrochloride (Ketalar) is a potent, rapid-acting injectable anesthetic with a short duration of action. It is used for short (10 to 25 minute) surgical procedures that do not require skeletal muscle relaxation.

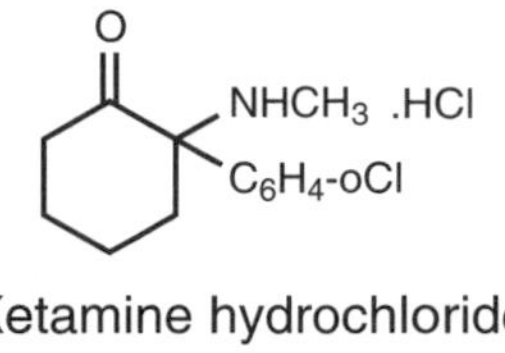

Ketamine hydrochloride

Ketamine is believed to produce anesthesia by blocking the effects of glutamate, one of the primary, excitatory neurotransmitters in the brain. Ketamine binds in the NMDA receptor-controlled channels as shown in Figure 9–5. This channel blockade could then prevent excitatory synaptic transmission and also attenuate the increase in intracellular Ca^{+2} levels usually evoked by NMDA or glutamate.[23] The structurally related "street drug" phencyclidine (PCP, HOG, THC, Angel Dust) also binds to the same channel-binding site. Perhaps action at this site accounts for the psychotomimetic side effects of ketamine.

Peak plasma levels are reached within 1 minute of intravenous injection. Redistribution of ketamine from brain to other tissues is responsible for termination of its anesthetic action. Ultimately, ketamine is metabolized in the liver and then excreted in the urine as water-soluble metabolites or glucuronide conjugates.

Ketamine and other dissociative anesthetics are characterized by dreams or hallucinations during emergence. These psychic disturbances occur more frequently in adults than in children under age 15 years. Benzodiazepines such as diazepam are most effective in preventing these unpleasant emergence reactions.

N-demethylation of ketamine is a major metabolic route in man. Although the metabolite does possess depressant effects, animal studies suggest that a contribution to the hypnotic effect of ketamine by its metabolite is unlikely. Ketamine is also metabolized by hydroxylation of the cyclohexane ring in the 4,5 or 6 position. Ketamine has one asymmetric center. The isomers have been separated and the S(+) isomer was found to be three to five times more potent than its enantiomer. Cost considerations, however, will probably prevent the marketing of this pure, more active isomer. The status of ketamine in anesthesiology has been reviewed in detail.[23]

Propofol

Propofol is an intravenous agent unrelated to the barbiturates or other intravenous anesthetics.[24] Chemically, propofol is 2,6-diisopropylphenol (Dipravan), and it exists as an oil at room temperature with low water solubility.

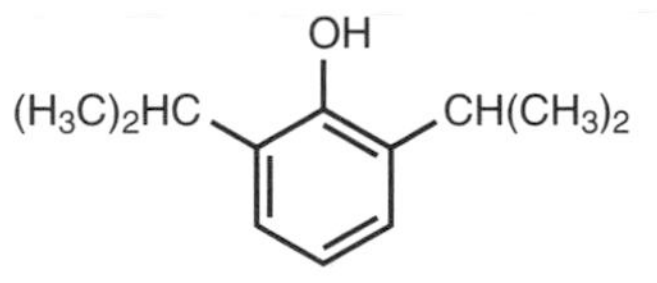

Propofol

It is formulated as an emulsion containing 1 or 2% propofol, 10% soybean oil, 1.2% egg lecithin, and 2.25% glycerol. Propofol appears to produce its hypnotic effect by enhancing GABA-ergic effects in the brain. Its action is similar to that of the benzodiazepines, but it does not bind directly to the benzodiazepine receptors.[25] Intravenous injection of a therapeutic dose produces hypnosis within 1 minute from the start of injection. A single bolus dose of 2 to 2.5 mg/kg will produce anesthesia for about 5 minutes. Additional propofol may be administered to extend the time of anesthesia or a volatile anesthetic may be used for extended surgical procedures. The hypnotic effect of propofol is terminated by tissue redistribution, and the agent is ultimately eliminated in the urine as glucuronide and sulfate conjugates. The agent is used for both induction and maintenance anesthesia with the advantages of a low incidence of emetic episodes and rapid recovery, making it well suited for outpatient surgical procedures.

Ultrashort-acting Barbiturates

The ultrashort-acting barbiturates are used to produce both surgical and basal anesthesia. When injected intravenously, they rapidly produce unconsciousness, and emergence is rapid. Ultrashort-acting barbiturates are frequently used in combination with a volatile anesthetic, such as N_2O. The barbiturate produces rapid, pleasant anesthesia, which is then maintained with the volatile anesthetic. The ultrashort-acting barbiturates are discussed in detail elsewhere in this book.

TOXICITY

Clinical symptoms of toxicity are not frequently manifested with volatile anesthetics, probably because of their infrequent and short-term use. Hepatotoxicity, however, occurs in 1 in 10,000 to 30,000 patients exposed to halothane, and nephrotoxicity has been observed after exposure to methoxyflurane. Both of these toxic reactions are thought to be caused by chemically reactive intermediate metabolites of the anesthetics.[26]

Hepatic Toxicity

Halothane hepatitis is believed to be mediated by covalent bonding of a reactive metabolite to hepatic tissue, which is then recognized as ''foreign'' and stimulates the formation of antibodies. Subsequent exposure to the anesthetic causes a hypersensitivity reaction to develop that is manifested as hepatotoxicity.[27,28]

Under hypoxic conditions, metabolic reduction of halothane occurs (Table 9–5), which results in free radical elaboration. This highly reactive species is potentially hepatotoxic, but recent investigations shed doubt on its involvement in halothane hepatotoxicity.

An alternative hypothesis for the genesis of halothane hepatitis points to the oxidative metabolic pathway of halothane, which produces trifluoroacetyl chloride as an intermediate (Table 9–5). Some patients with halothane hepatitis have been found to have specific antibodies that react with liver microsomal proteins altered by reaction with trifluoroacetyl chloride. These antibodies, found only in halothane hepatitis patients, have been shown to participate in antibody-dependent cell-mediated cytotoxic response to halothane-altered hepatocytes.

The similarity in metabolism proposed for desflurane, enflurane, halothane, isoflurane, and methoxyflurane leaves the possibility of cross-sensitivity. It is known that metabolism of enflurane and isoflurane produces intermediates that covalently alter hepatic protein in a fashion similar to that observed with halothane. Also, enflurane-altered hepatic tissue is known to cross-react with halothane-induced antibodies. Thus, evidence continues to mount in support of the view that acetylation of tissue and subsequent development of hypersensitivity accounts for halothane hepatitis, and it is also consistent with the observation that anesthetics like enflurane and isoflurane, which are metabolized to a lesser extent, infrequently produce this severe hepatotoxicity.

Nephrotoxicity

Renal toxicity of fluorinated anesthetics is related to the metabolic elaboration of inorganic fluoride. Polyuric failure has been identified as a consequence of methoxyflurane anesthesia and has been related to serum fluoride levels. The usually reversible nephropathy is observed when levels of serum fluoride ion exceed 50 μmol/L. The relative rates of metabolic defluorination of fluorinated anesthetics are methoxyflurane $>$ enflurane $=$ sevoflurane $>$ isoflurane $>$ halothane and desflurane. Only metabolism of methoxyflurane easily yields nephrotoxic levels of fluoride.

Malignant Hyperthermia

Malignant hyperthermia is a rare but potentially fatal complication of anesthesia that occurs in 1 per 14,000 anesthetic uses. All fluorinated anesthetics can trigger this complication in genetically susceptible individuals. It is characterized by intense hypermetabolic reaction in skeletal muscle. Uncontrolled muscle contraction occurs with large amounts of heat generated. Untreated, malignant hyperthermia can rapidly progress to death. Treatment for the syndrome includes intensive supportive therapy and administration of the muscle relaxant dantrolene.

Chronic Toxicity

Concern has been raised about adverse reactions that may occur in personnel who are chronically exposed to low levels of waste anesthetic gas in the operating room and dental operatory. Several epidemiologic studies have uncovered the presence of an occupational hazard, but the magnitude of the risk is unknown. Increased incidence of spontaneous abortion, congenital malformations in offspring, and an increase of cancer in operating room personnel have been reported. In response to these concerns, scavenging systems to reduce trace concentrations of anesthetic gases in operating room atmosphere have been installed. Besides waste anesthetic, operating room personnel are exposed to high levels of stress and also radiation, which must be evaluated as alternate explanations for these adverse reactions.

METABOLISM

Contrary to earlier belief, volatile anesthetics are at least partially metabolized. Metabolic studies have revealed that halogenated anesthetics are metabolized

Table 9–5. Proposed Metabolites of Fluorinated Anesthetics*

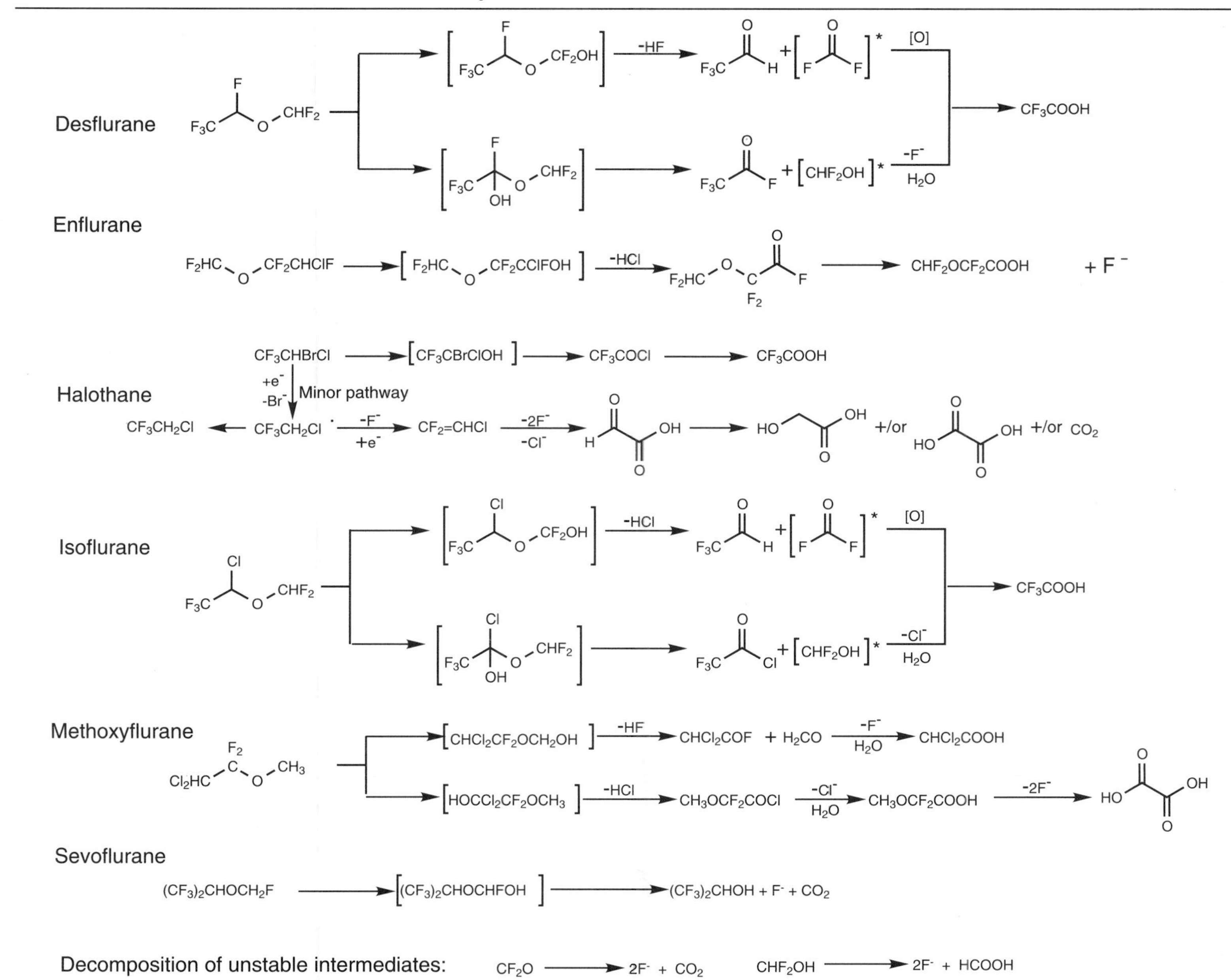

* Data from references 27 and 28.

from 0.004 (desflurane) to 50% (methoxyflurane). Halogenation usually hinders metabolism, and the rate of metabolism is inversely related to the bond energy of carbon-halogen bonds. Thus, the rate of metabolism of halogen substituted agents is: Br > Cl > F. In fact, perfluoro hydrocarbons are not known to be metabolized.

Most metabolism occurs after administration of anesthetic is discontinued, so agents that accumulate in the body have the potential for lengthy and often extensive metabolism. The agents in Table 9–1 are arranged in order from highest to lowest lipid solubility, which is paralleled by a decreasing extent of metabolism. Methoxyflurane with its high lipid solubility is extensively metabolized, whereas desflurane is nearly inert to metabolic degradation.

Metabolism of halogenated anesthetics is catalyzed by liver cytochrome P450 enzymes. The initial step in the oxidative metabolism involves hydroxylation at a terminal carbon atom or at a carbon adjacent to an ether oxygen. The intermediate alcohols are unstable and subsequently decompose, releasing inorganic halogens ions and small organic metabolites that may be further oxidized or hydrolyzed. Reductive metabolism, also catalyzed by cytochrome P450, occurs only with halothane. This reduction is thought to occur under hypoxic conditions and yields the reactive and toxic free radical, $CF_3\,CHCl\cdot$, as shown in Table 9–5.

Metabolism of all fluorinated anesthetics results in liberation or inorganic fluoride, and blood and/or urine levels of fluoride are often determined as an indirect measure of the extent to which an anesthetic is metabolized. Serum fluoride levels are also of concern because of the kidney damage caused by high levels of that ion. Another metabolite often found is trifluoroacetic acid or a similar α-substituted acetic acid derivative. Levels of these metabolites are also measured to quantify anesthetic metabolism. These substituted acetic acids may also be related to anesthetic toxicity, as was discussed in the hepatotoxicity section.

ADJUNCTS TO GENERAL ANESTHESIA

Suitable combinations of narcotic analgetics, sedatives, anticholinergics, skeletal muscle relaxants, and antianxiety agents are used for preanesthetic medication. These drugs decrease anxiety, reduce the amount of anesthetic required, and minimize some undesirable side effects of anesthetic agents, such as salivation, bradycardia, and postanesthetic vomiting.

Narcotic analgetics, such as morphine or meperidine, are used to reduce pain, anxiety, and the amount of anesthetic needed. Unfortunately, analgetics delay awakening from general anesthesia and frequently cause hypotension and gastrointestinal disturbances.

The benzodiazepines are now preferred over other opiates and barbiturates as preanesthetic medications. The benzodiazepines provide sedation and anterograde amnesia, and reduce anxiety without producing adverse CNS effects, hypotension, or postoperative nausea and vomiting. Other antianxiety agents and neuroleptics in use for premedication are hydroxyzine and droperidol.

Anticholinergics, such as atropine and scopolamine, are used to inhibit excessive respiratory secretions seen during administration of many volatile anesthetics. These agents also prevent parasympathetically induced cardiovascular effects, such as hypotension and bradycardia.

Muscle relaxants are used to provide skeletal muscle relaxation to facilitate intubation of the trachea and to provide optimal surgical working conditions. Skeletal muscle relaxants are available with different durations of action. An agent with rapid onset and brief duration of action like succinylcholine or mivacurium is useful when intubation of the trachea is the reason for giving a muscle relaxant. Long-acting agents like doxacurium, pancurium, and vecuronium provide for a longer duration of skeletal muscle relaxation, making them useful for reducing muscle tone with the surgical field and for facilitating mechanical respiration of patients.

Clonidine, an α_2-adrenergic agonist used for the treatment of hypertension, is known to produce sedative side effects and to reduce the dose requirements of analgesics and anesthetics. These effects have been exploited for the premedication of patients for general anesthesia. Now, dexmedetomidine, a highly selective α_2-agonist, is being evaluated for use as a preanesthetic medication. It effectively reduces the amount of analgesic and anesthetic required without respiratory depression, and provides for improved hemodynamic stability in the patient.[12] It is proposed that dexmedetomidine produces its effects by stimulating α_2-receptors in the locus coeruleus, as was described in the section on mechanism of action.

DRUG INTERACTIONS

Although anesthetics are given for relatively short periods of time and leave the body rapidly once administration is discontinued, several drug interactions have been observed with these agents.

The potential for drug interactions with volatile anesthetics occurs in two different ways. First, interactions can occur between anesthetics and drugs that the patient has been taking chronically or, second, drug interactions may occur between the agents administered during anesthesia.

Antihypertensive agents have long been of concern to the anesthetist. It was feared that these agents would alter cardiovascular function, leading to difficult-to-manage hypotension and myocardial depression. Some anesthetists have recommended that antihypertensive therapy be discontinued prior to anesthesia, but that notion is no longer universally accepted. In fact, most believe it is desirable to continue chronic, antihypertensive therapy up to the time of anesthesia, because this practice will give the patient a stable cardiovascular system to help withstand the stress of anesthesia and surgery.

Centrally active agents present a similar situation. The administration of levodopa, phenothiazine tranquilizers, and tricyclic antidepressants is usually continued up to the time of anesthesia. The one exception is with monoamine oxidase inhibitors (MAOI). It is recommended that this group of agents be discontinued 2 weeks prior to anesthesia because of possible interactions with anesthetics and the adjuncts to anesthesia.

The potential is great for drug interactions to occur between the agents that the anesthetist administers preparatory to anesthesia, during induction, and finally through surgical anesthesia. Many of these agents, such as barbiturates, benzodiazepines, phenothiazines, and opioids, are CNS, circulatory, an respiratory depressants, so their potentiation when used in combination is not surprising. Potentiation or additive effects of neuromuscular blocking agents by volatile anesthetics is common and of major importance. Care is always taken to properly adjust doses of these agents when they are used in combination. Lastly, certain anesthetics may sensitize the heart to catecholamine-induced arrhythmias. Because catecholamines are used routinely to combat hypotension during anesthesia, precautions must be taken to avoid ventricular fibrillation when these agents are used. Interactions among these agents are so common that they are always considered during anesthetic procedures. A complete discussion of these and other interactions is available.[29]

REFERENCES

1. R. K. Stoelting and R. D. Miller, *Basics of Anesthesia,* 2nd ed., New York, Churchill Livingstone, 1989.
2. W. C. Stevens and H. G. G. Kingston, *Inhalation Anesthesia,* in *Clinical Anesthesia,* 2nd ed., P. G. Barash, B. F. Cullen, and R. K. Stoelting, Eds., Philadelphia, J. B. Lippincott, 1992.
3. A. Devcic and D. C. Warltier, *Anesthesia Today, 3,* 1(1993).
4. E. I. Eger II, *Isoflurane (Forane): A compendium and Reference,* Madison, WI, Ohio Medical Products, 1981.
5. N. Yasuda et al., *Anesthesiology, 74,* 489(1991).
6. J. H. Martin, *Cortical Neurons, the EEG, and the "Mechanisms of Epilepsy,"* in *Principles of Neural Science,* 2nd ed., E. R Kandel and J. H. Schwartz, Eds., New York, Elsevier, 1985.
7. M. Scheinin and D. A. Schwinn, *Anesthesiology, 76,* 873(1992).
8. M. L. Mayer, et al., *Ann. N.Y. Acad. Sci., 648,* 194(1992).
9. T. M. Dawson, et al., *Ann. Neurol., 32,* 297(1992).
10. R. A. Johns, et al. *Anesthesiology, 77,* 779(1992).
11. D. L. Tanelian, et al., *Anesthesiology, 78,* 757(1993).
12. M. Maze and W. Tranquilli, *Anesthesiology, 74,* 581(1991).
13. H. R. Roth and K. W. Miller, Eds., *Molecular and Cellular Mechanisms of Anesthetics,* New York, Plenum, 1986.
14. R. C. Aloia, et al., Eds., *Drug and Anesthetic Effects on Membrane Structure and Function,* New York, Wiley-Less, 1991.
15. L. B. Perry, et al., *Anesth. Analg., 54,* 152(1975).
16. E. I. Eger II, *Anesthetic Uptake and Action,* Baltimore, Williams & Wilkins, 1974.
17. J. G. Wade and W. C. Stevens, *Anesth. Analg., 60,* 666(1981).
18. J. H. Tinker, Ed., *Anesth. Analg., 75,* S1(1992).
19. D. Koblin, et al., *Anesth. Analg., 60,* 464(1981).
20. E. M. Landau, et al., *J. Med. Chem., 22,* 325(1979).
21. G. B. Russell, et al., *Anesth. Analg., 70,* (1990).
22. E. I. Eger II, et al., *Anesth. Analg., 71,* 575(1990).
23. J. Church and D. Lodge, *N-Methyl-D-aspartate (NMDA) antagonism is central to the actions of ketamine and other phencyclidine receptor ligands,* in *Status of Ketamine in Anesthesiology,* E. F. Domino, Ed., Ann Arbor, NPP Books, 1990.
24. P. S. Sebel and J. D. Lowdon, *Anesthesiology, 71,* 260(1989).
25. V. A. Peduto, et al., *Anesthesiology, 75,* 1000(1991).
26. M. T. Baker and R. A. VanDyke, *Biochemical and toxicological aspects of the volatile anesthetics,* in *Clinical Anesthesia,* 2nd ed., D. G. Barash, B. F. Cullen, and R. K. Stoelting, Eds., Philadelphia, J. B. Lippincott, 1992.
27. D. D. Koblin, *Anesth. Analg., 75,* S10(1992).
28. D. D. Christ, et al., *Anesthesiology, 69,* 833(1988).
29. J. L. Neigh, *Inhalation anesthetic agents.* In *Drug Interactions in Anesthesia,* 2nd ed., N. T. Smith and A. N. Corbascio, Eds., Philadelphia, Lea & Febiger, 1986.

SUGGESTED READINGS

J. R. Cooper, et al., *The Biochemical Basis of Neuropharmacology,* 6th ed., New York, Oxford University Press, 1991.

A. G. Gilman, et al., *The Pharmacological Basis of Therapeutics,* 8th ed., New York, Pergamon Press, 1990.

W. C. Stevens and H. G. G. Kingston, *Inhalation anesthesia,* in *Clinical Anesthesia,* 2nd ed., P. G. Barash, B. F. Cullen, and R. K. Stoelting, Eds., Philadelphia, J. B. Lippincott, 1992.

R. K. Stoelting and R. D. Miller, *Basics of Anesthesia,* 2nd ed., New York, Churchill Livingstone, 1989.

Chapter 10

CENTRAL NERVOUS SYSTEM DEPRESSANTS: SEDATIVE-HYPNOTICS

Julius A. Vida

THE STRUCTURE OF THE NERVE CELL

The fundamental unit of the nervous system is the neuron, or nerve cell, the most important property of which is its excitability. A neuronal unit (nerve cell) is enclosed in a matrix of cellular tissues surrounded, in turn, by a nutrient fluid derived from blood. Every neuron is an anatomically discrete cellular entity. Each neuron consists of a cell body (soma) that contains the nucleus. From the soma stems an extensive network of branches, the axon (long process) and the dendrites (short process). The neuron, including the axon and the dendrites, is enclosed within a surface membrane. The membrane consists of a semipermeable double (bimolecular) layer of phospholipids. Large particles such as organic molecules (amino acids, sugars, etc.) and charged particles (ions) cannot pass across it. There are proteins embedded within the bimolecular lipid layer, however, that provide the membrane with its permeability to many substances (organic compounds and ions) and are also responsible for the functional activity of the living membrane.

Two types of proteins are enclosed in the cell membrane: (1) rod-like α-helix proteins that are receptors for extracellular messengers such as hormones, neurotransmitters, and antibodies; and (2) globular proteins that act as channels through which ions can cross the membrane. Ion channels are important in the function of nerve cells and may be opened or closed by voltage or chemical changes or may not be controlled. Hence, the classification of voltage-gated, chemically gated, or nongated ion channels.

The nerve cell body is made up of a gel-like colloidal solution of lipoproteins, containing a high proportion of water in which various substances (glucose, salts, oxygen, carbon dioxide, etc.) are dissolved. The composition of the salt solution inside the membrane is usually different from that on the outside, because of differences in the permeability of membranes to the various ions (K^+, Na^+, Ca^{2+}, Cl^-, HCO_3^-, etc.). In the resting state, the neuronal membrane is polarized, the inside being normally more negative than the outside. This is due to a relatively high concentration of Na^+ outside the cell and a relatively high concentration of K^+ inside. In the resting cell, the concentration differences are maintained by relatively low ion permeabilities across the membrane and by a metabolic pump that is responsible for driving the sodium ions and potassium ions in opposite directions across the membrane, requiring metabolic expenditure of energy. The pump's energy is supplied by ATP. The pump serves as a functional link between the electrical properties of the membrane and the metabolic system of the cell. The individual neuron possessing a resting potential can respond to an exogenous stimulus by changes in electrical activity. As a result, an action potential is generated. The excess potential may be discharged in the form of nerve impulses. Most signalling in the nervous system involves changes in the membrane potential of neurons.

Synaptic Transmission

Neurons communicate with each other either by passage of electrical current (neural conduction) produced by movement of ions (an electrical process) or by the release of chemical transmitters (a chemical process). The term synapse is used to describe the transfer of neuronal activity between neurons (interneuronal transmission) through the axons. The process of synaptic transmission involves the transfer of energy from the terminal branches of the axon of one neuron to another nerve cell to which they are not connected. Transmission of the nerve impulse across synaptic junctions is a chemical process.

Synaptic transmission takes place between the presynaptic neuron and the postsynaptic neuron across a synaptic space of about 250 Å. At the point where the axon approaches a postsynaptic cell, it breaks up into numerous fine branches (teledendrites), which are 1 μm in diameter or less. These branches have knob-like endings (terminal boutons), which are 4 μm in diameter. A continuous membrane covers the surface of each terminal bouton. As many as 50 terminal boutons may be in contact with a postsynaptic cell soma, which is also covered with a continuous surface membrane.

There is no anatomic continuity between the presynaptic and postsynaptic nerve cells, because both are within the boundaries of continuous membranes that are separated by the fluid-filled synaptic space. As the wave of depolarization sweeps down the axon, it causes a discrete amount (quantum) of chemical transmitter to be secreted by exocytosis from the presynaptic nerve ending into the synaptic cleft. The transmitter diffuses across the synaptic space and binds to the receptors of the postsy-

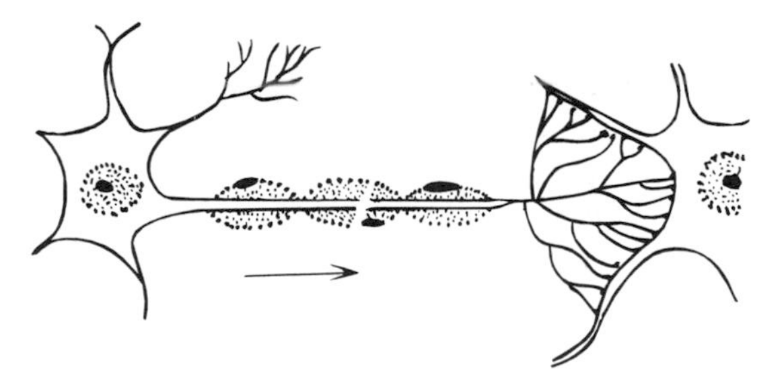

Fig. 10–1. Representation of an exosomatic synaptic junction.

naptic membrane. The binding causes conformational changes in the molecules of the membrane, which in turn produces changes in ion permeabilities or affects the metabolism of the cells. The ion fluxes alter the membrane potential of the postsynaptic neuron producing either excitation or inhibition. Once the transmitter has interacted with the postsynaptic cell, it is either removed by active uptake processes in the neuronal or surrounding glial cells or destroyed by metabolic deactivation. A schematic representation is shown in Figure 10–1.

Synaptic Excitation

If the chemical transmitter promotes the activity of the postsynaptic cell by increasing its excitability, synaptic transmission is facilitated.

Presynaptic depolarization leads to activation of voltage-gated calcium channels and the influx of calcium. The influx of calcium in turn leads to transmitter release. A quantum is the smallest amount of transmitter that is normally released, and the quantal release has been demonstrated at all chemically transmitting synapses. If the transmitter is excitatory, the electrical charge of the postsynaptic membrane decreases. In this case, the postsynaptic response to a transmitter substance consists of a net inward movement of positive charge (synaptic depolarization). After an interval of 0.3 to 0.5 msec (the synaptic delay), there is rapid increase (1.0 to 1.5 msec) to a peak to produce the excitatory postsynaptic potential (EPSP).

Because the decay of this potential is slow (5 msec), the potential increases as the number of activated synapses increases. The sum of depolarizations produced by each synapse may reach a critical value (threshold for impulse generation) when the resting potential of the postsynaptic cell reverses to produce an action potential. Thus, synaptic depolarization may ultimately evoke the discharge of an impulse that sweeps down its axon.

The classic view of chemical transmission was that only one specific chemical transmitter is released from any presynaptic nerve cell, and that the effect of the transmitter may be excitatory or inhibitory depending on the nature of postsynaptic cells. For example, acetylcholine released at motoneuron nerve terminals to skeletal muscles has an excitatory action at the neuromuscular junction leading to contraction of the muscle, whereas acetylcholine released from vagal nerve terminals has an inhibitory action in the heart, slowing the heart beat. Thus, an important corollary to Dale's law is that the action of a transmitter substance is determined by specific mechanisms at the particular postsynaptic sites.

It is now known, however, that many nerve cells release more than one transmitter and that they may have different actions, although any presynaptic neuron always releases the same neurotransmitter or the same combination of neurotransmitters from the membrane-bound synaptic vesicles.

Presynaptic Inhibition

This type of inhibition is exerted by a chemical transmitter on the excitatory presynaptic terminals rather than on the postsynaptic membrane. Presynaptic inhibitors decrease the excitatory synaptic action by decreasing the output of excitatory transmitter substances in the presynaptic terminals.

Postsynaptic Inhibition

The result of postsynaptic inhibition is the opposite of synaptic excitation. Postsynaptic inhibition is exerted on the postsynaptic membrane by decreasing the excitability of the postsynaptic neuron. Inhibitory synapses cause an increase in the postsynaptic membrane potential, producing the inhibitory postsynaptic potential (IPSP).

In synaptic excitation, progressive depolarization of the postsynaptic neurons occurs. In postsynaptic inhibition, the postsynaptic neuron is in the state of hyperpolarization, during which its excitability is decreased. Like synaptic excitation, postsynaptic inhibition is associated with changes in the permeability of the postsynaptic membrane. In synaptic excitation, nonspecific transmembranal migration of several ions takes place. In synaptic inhibition, the permeability of the postsynaptic membrane is limited to K^+ and Cl^- ions, which when hydrated, are smaller than 4 Å. In this case, there is an increased outward movement of positive charge (K^+) and an increased inward movement of negative charge (Cl^-) resulting in a net outward movement of positive charge. These ion flows produce a membrane potential that is more polarized (hyperpolarized). Synapses of this type prevent the membrane from depolarizing and hence work against the initiation of impulses.

It has been suggested that inhibitory transmission is mediated by adrenergic substances as well as by amino acids related to GABA. A schematic representation is shown in Figure 10–2.

Postsynaptic Potentials

Most neurons fire spikes as a result of the activities of neurotransmitters that exert their effects on receptor-coupled ion conductances, which are voltage independent; that is, the receptor alters the coupled ion channel regardless of the membrane potential at the moment.

There are, however, some neurons that do fire spikes unconventionally by substituting influx of Ca^{++} ions (a voltage-sensitive calcium conductance) for the conventional influx of Na^+ ions. This activity may be caused by nonclassic transmitters that operate on receptors coupled to voltage-sensitive conductances. Instead of producing

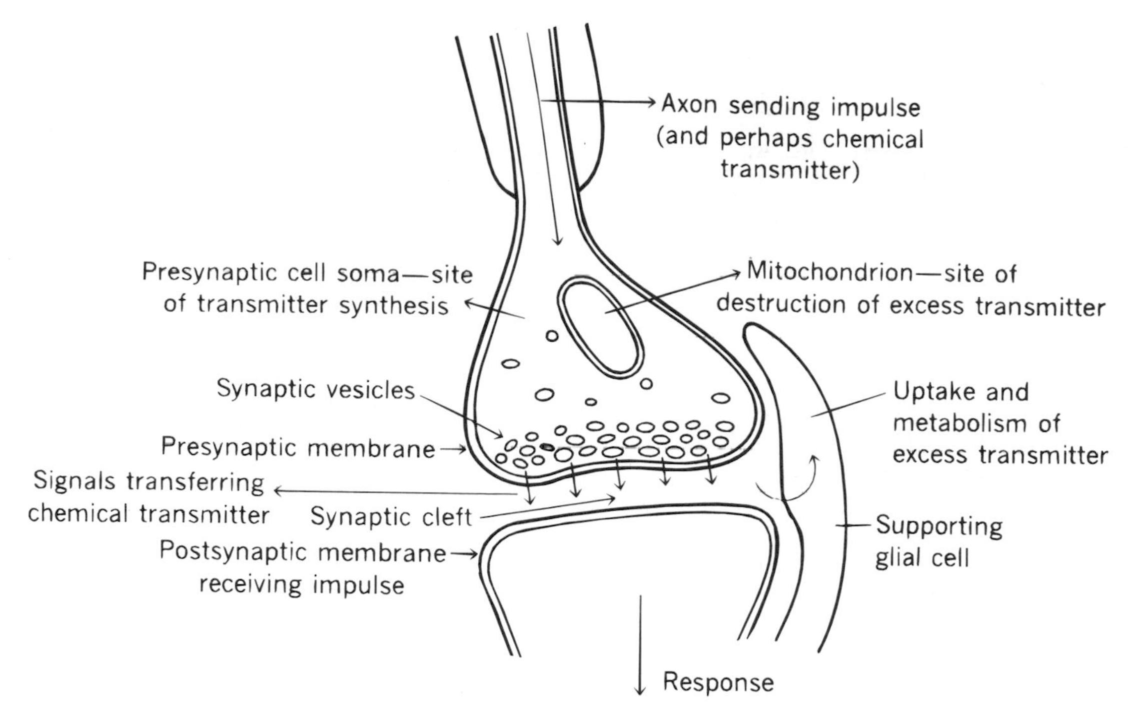

Fig. 10–2. Representation of a synapse.

a depolarizing voltage shift by direct transmitter action, these nonclassic transmitters exert a "modulatory" action on ionic conductances. These modulatory actions may be produced, for example, by activation of cyclic nucleotide synthesis, or by an increase in the secretion of phospholipase C, which is located in cell membranes and catalyzes the breakdown of phosphatidylinositol.

Chemical Transmitters

Some evidence shows that a number of chemical mediators are involved in synaptic transmission. To identify a chemical transmitter, three criteria have to be fulfilled: (1) The substances must be released at the presynaptic terminal on stimulation. (The substance may be synthesized at or transported to the active site.) (2) The substance applied exogenously must have the effect of nerve stimulation at the postsynaptic terminal identical to that produced by synaptic excitation. (3) The substance must be removed or inactivated at the site of action, and specific antagonists must block its synaptic action.

Considerable progress has been made in identifying chemical transmitters, in large part because of the microelectrode technique, and immunoassay, immunohistochemical staining, and radiochemical techniques, as well as electron microscopy, spectrophotofluorometry, and high-performance liquid chromatography techniques. In addition, the development of the monoclonal antibody technique provides a particularly powerful technique for the study of the neurotransmitters and their receptors.[1] These transmitters are either biogenic amines (dopamine, epinephrine, norepinephrine, 5-hydroxytryptamine, and histamine) or amino acids (glycine, glutamate, γ-aminobutyric acid and most probably, aspartate). One transmitter, acetylcholine, does not contain a primary amine group.

The important neurotransmitters of the central nervous system (CNS) are shown in Table 10–1.

Biogenic Amines

Under normal conditions, many amines found in the brain have been implicated in neuronal functions; these include amines devoid of an OH group. The most important of these, histamine, is widely distributed in the CNS. Histamine is formed from the amino acid histidine by the enzyme histidine decarboxylase. In the CNS, histamine has a less important role than in mast cells, basophils, platelets, bronchial smooth muscle, or gastric mucosa.

Histidine —(histidine decarboxylase)→ Histamine

Some phenolic amines and catecholamines that are important neurotransmitters include dopamine, tyramine, octopamine, norepinephrine, epinephrine, and serotonin. (Their biogenesis and metabolism are shown in Figs. 10–3 and 10–4).

The most efficient inhibitory of the central dopadecarboxylase, m-hydroxybenzylhydrazine, causes only a slight loss of the central monoamines. The decarboxylase is even less inhibited by α-methyldopa, which is a substrate

Table 10–1. Putative Neurotransmitters in Mammalian CNS

Structure	Name	Marker Enzyme	Predominant Activity
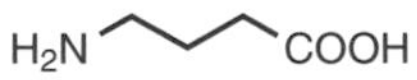 H_2N COOH	γ-Aminobutyric acid GABA	Glutamic acid decarboxylase (GAD)	Inhibitory
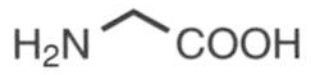 H_2N COOH	Glycine Amino acetic acid	Serine hydroxymethylase	Inhibitory
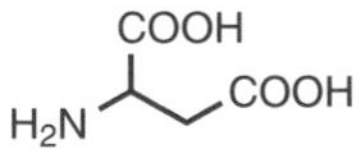 COOH H_2N COOH	Aspartic acid Aspartate	Aspartic acid transaminase	Excitatory
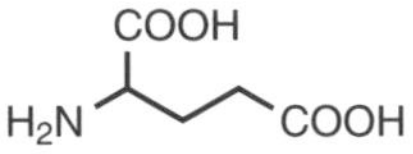 COOH H_2N COOH	Glutamic acid Glutamate	Glutamic acid-keto glutaric acid transaminase	Excitatory
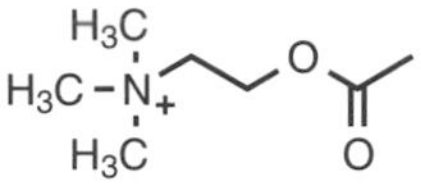 H_3C H_3C-N^+ H_3C O O	Acetylcholine	Choline acetyl transferase	Inhibitory
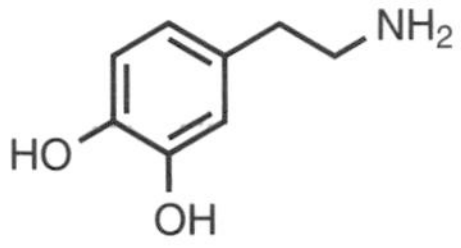 NH_2 HO OH	Dopamine	Tyrosine hydroxylase	Inhibitory
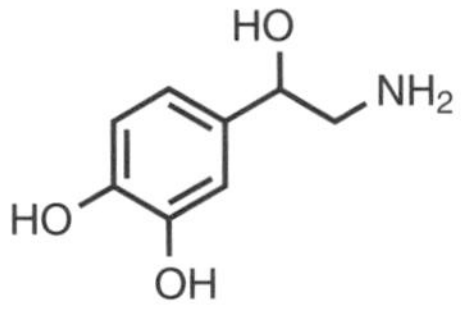 HO NH_2 HO OH	Norepinephrine	Dopamine-β-hydroxylase	Inhibitory
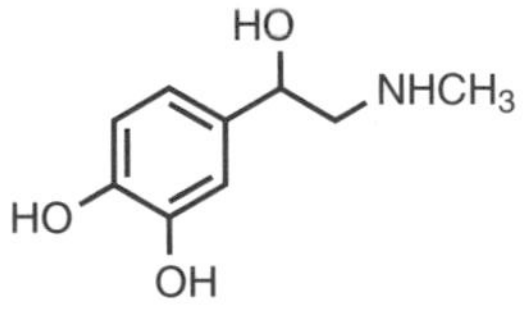 HO $NHCH_3$ HO OH	Epinephrine	Phenethanolamine-N-methyl transferase	Complex
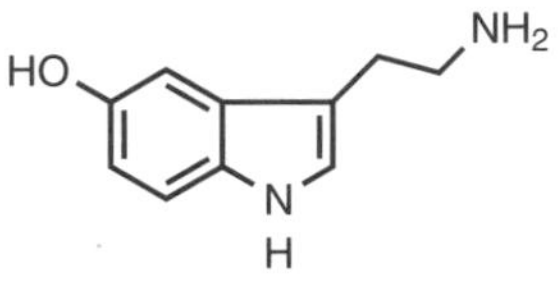 HO NH_2 N H	5-Hydroxytryptamine	Tryptophan hydroxylase	Inhibitory
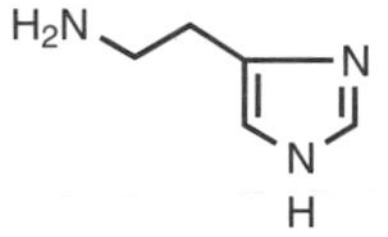 H_2N N N H	Histamine	Histidine decarboxylase	Complex

and is transformed to α-methyldopamine and further to α-methylnorepinephrine. The peripheral decarboxylase inhibitors benserazide (RO4–4602) and carbidopa (MK 486) inhibit all the decarboxylase except that in the central monoamine nerves. These two compounds potentiate the central effects of dopa and 5-hydroxytryptophan (5-HTP) by preventing decarboxylation of these amino acids outside the CNS. Dopamine β-hydroxylase is inhibited by compounds chelating its copper, such as disulfiram, and its metabolite, diethyldithiocarbamate, as well as FLA–63 and fusaric acid.

The role of acetylcholine at cholinergic synapses has been well established (see Chapter 18).

Amino Acids

Glycine is an inhibitory transmitter. During synaptic inhibition, changes in membrane polarization induce depolarization of the neurons. The postsynaptic actions of the electrophoretically administered glycine and GABA are also due to depolarization.

The inhibitory and depolarizing action of glycine is reversibly abolished by strychnine, a compound that blocks spinal postsynaptic inhibition. Antagonism by strychnine makes glycine a likely inhibitory transmitter[2] at postsynaptic sites.

GABA (γ-aminobutyric acid) is present in all portions of the brain and inhibits virtually all CNS neurons by

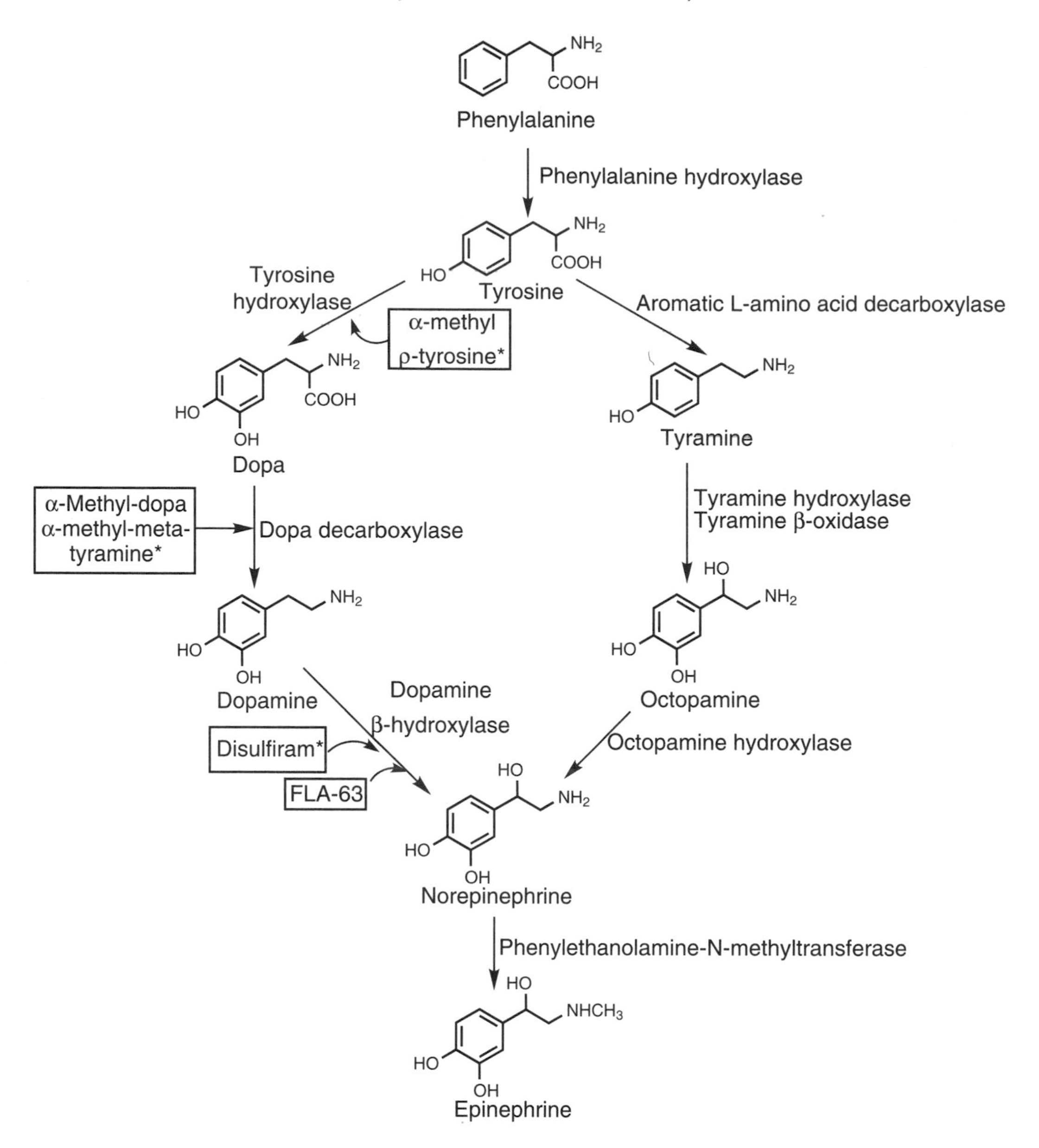

Fig. 10–3. Biogenesis and metabolism of dopamine, tyramine, octamine, norepinephrine, and epinephrine. *, Drugs that inhibit synthesis or catabolism are set in boxes.

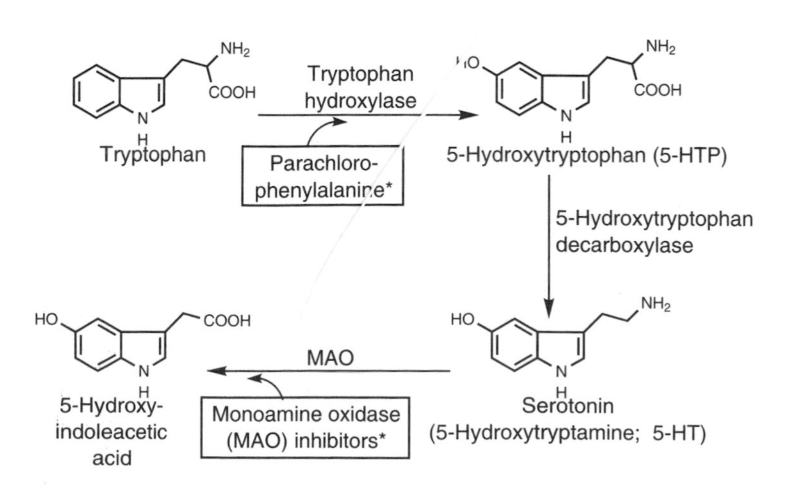

Fig. 10–4. Biogenesis and metabolism of serotonin. *, Drugs that inhibit synthesis or catabolism are set in boxes.

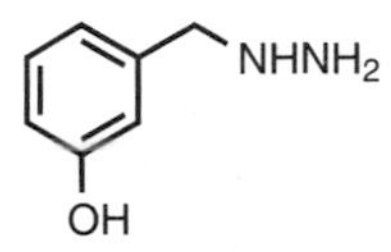

m-Hydroxybenzylhydrazine

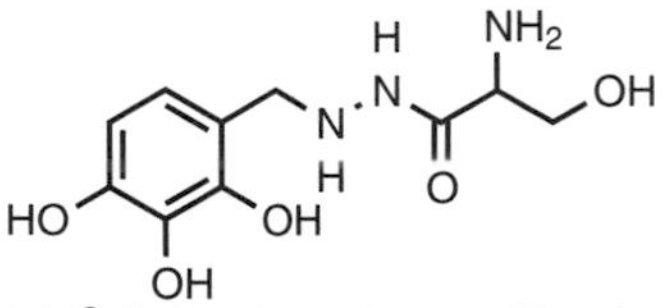

N^1-Seryl-N^2-(trihydroxybenzyl)hydrazine
(R04-4602)

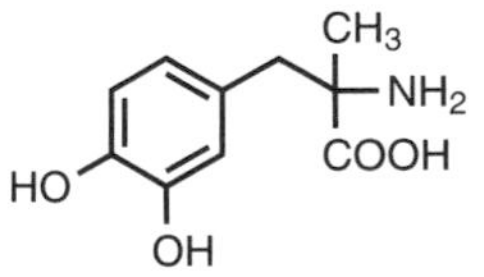

α-Methyldopa

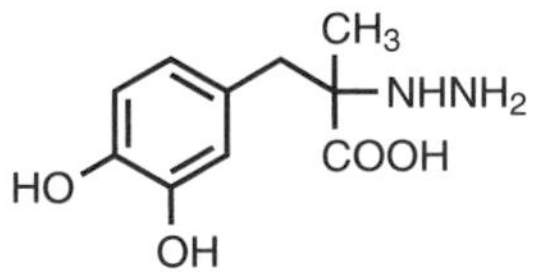

α-Methyldopahydrazine
(MK 486)

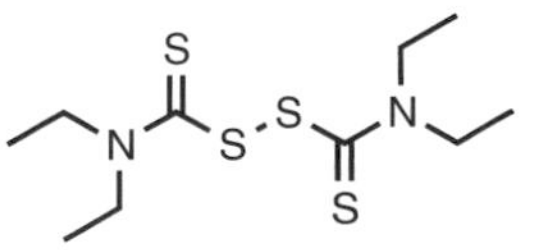

Disulfiram

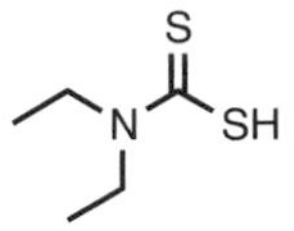

Diethyldithiocarbamic acid

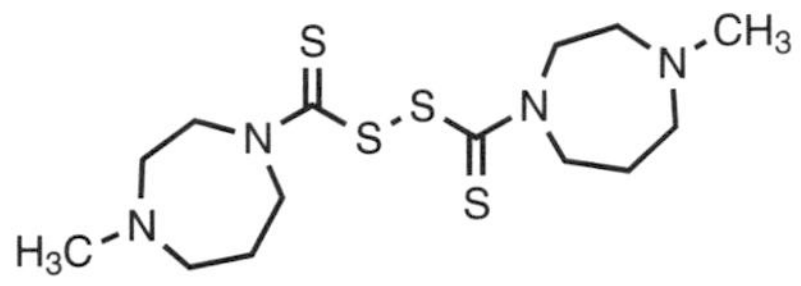

Bis(4-methyl-1-homopiperazinyl-thiocarbonyl)-disulfide
(FLA-63)

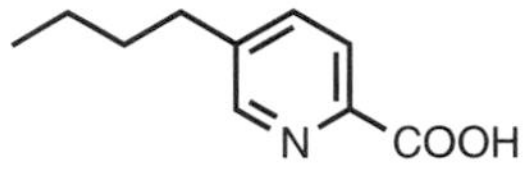

Fusaric acid

increasing cell membrane permeability to chloride ions, thereby stabilizing resting membrane potential to remain in a depolarized state near the resting potential. Most

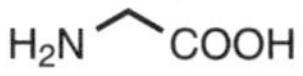

Glycine

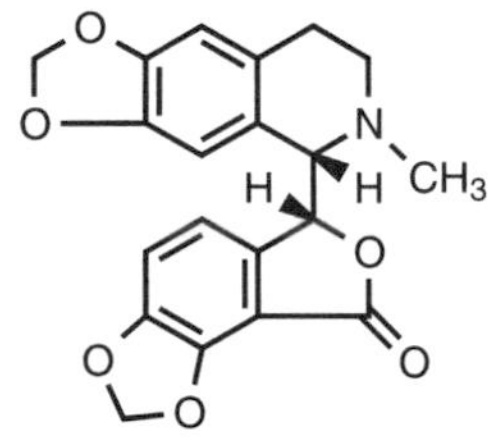

Bicuculline

GABA responses are mediated by the postsynaptic $GABA_A$ receptors that are activated by the selective agonists isoguvacine and muscimol and are blocked by the selective antagonists bicuculline and picrotoxin. By contrast, the presynaptic $GABA_B$ receptors are insensitive to bicuculline. Baclofen (β-p-chlorophenyl GABA) acts as an agonist at $GABA_B$ receptors, whereas Saclofen and 2-hydroxysaclofen are selective $GABA_B$ antagonists.[3]

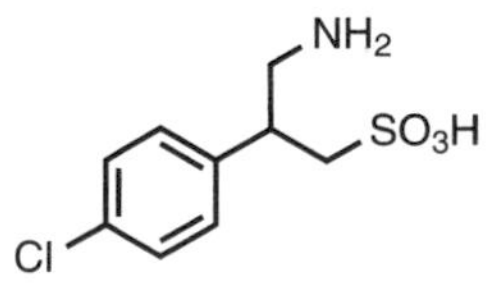

Saclofen

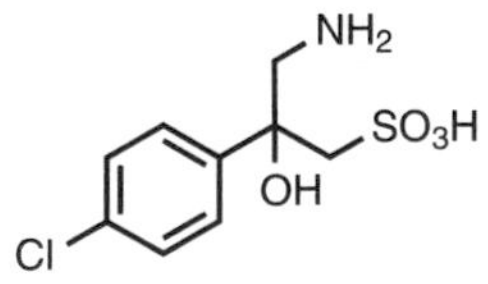

2-OH-Saclofen

Both L-glutamate and L-asparate are reported to be excitatory transmitters[4] because of their depolarizing effects. The antagonists for glutamic acid and aspartic acid, however, are not particularly potent species. The biogenesis and metabolism of L-glutamic acid, GABA, and L-aspartic acid are shown in Tables 10–2 and 10–3.

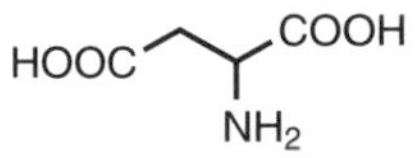

Aspartic acid

The role of glutamate and GABA in the brain has been summarized. Taurine, the depressant effects of which are blocked by strychnine, has been suggested as an inhibitory transmitter.[5]

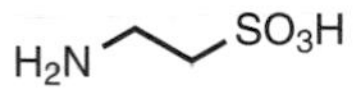

Taurine

Cyclic Nucleotides

The key role of these compounds in almost every tissue in the body, coupled with the high concentration of cyclic nucleotide synthesizing and catabolizing enzymes in the CNS, suggests that cyclic nucleotides may also act as neurotransmitters.[6] Electrophysiologic, neurochemical, and immunocytochemical evidence has been obtained that the synaptic action of norepinephrine (and perhaps that of dopamine as well) is mediated by an interaction through the adenylate cyclase in certain cells of the cerebral cortex, indicating that cyclic AMP is a synaptic sec-

Table 10–2. Biogenesis and Metabolism of L-Glutamic Acid and GABA

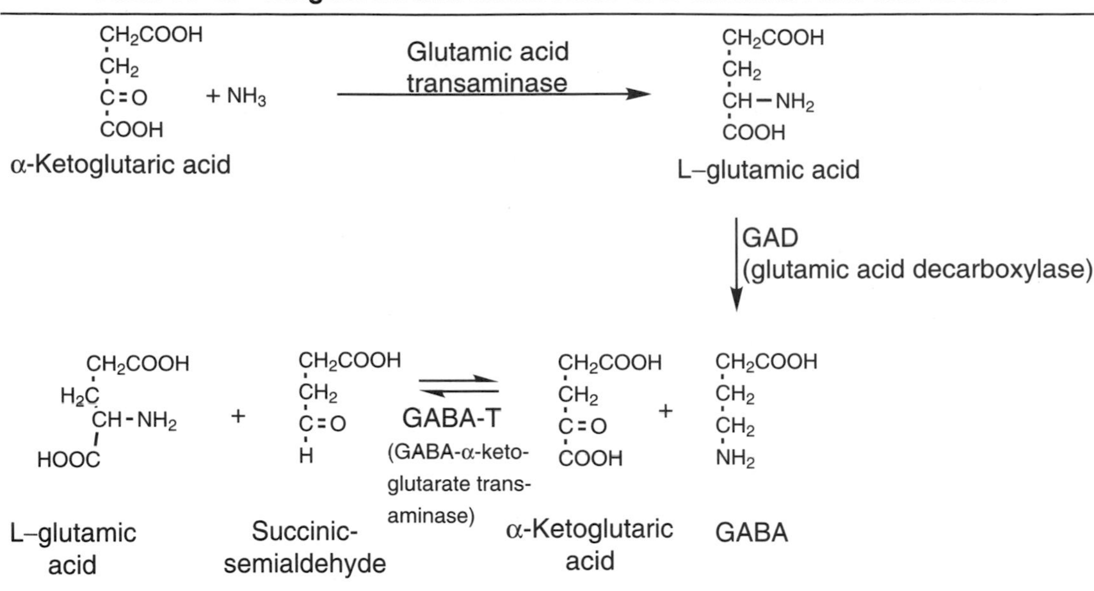

ond messenger. Similarly, it has been suggested that cyclic GMP may be a mediator at the muscarinic cholinergic receptors. These findings suggest the following events in synaptic transmission. Neuronal stimulation evokes release of neurotransmitters that reach the postsynaptic cells, where they activate adenylate cyclase to increase the rate of production of the cyclic nucleotide. The cyclic nucleotide, in turn, activates a specific protein kinase, the role of which is to phosphorylate through ATP some protein substrate, an important step in protein synthesis. The phosphorylated protein may become incorporated into the membrane and/or may change the ion permeability properties of membranes with a resulting depolarization or hyperpolarization, depending on the equilibrium potential of the various ions (K^+, Na^+, Ca^{++}, Cl^-, etc.) in the cells. The phosphate transfer is terminated by a phosphoprotein phosphatase that removes the phosphate group from the protein. The cyclic nucleotides are catabolized by a 3′,5′-phosphodiesterase that produces 5′-AMP (Fig. 10–5).

Neuropeptides

Many of the peptide hormones of the endocrine and neuroendocrine systems are also present in the CNS and can function as neurohormones or as neuromodulators/neuroregulators. Table 10–4 lists the 40 neuropeptides described to date in neurons and nerve terminals within the mammalian CNS. It has been found that, like the neurotransmitter-releasing neurons, the peptide-releasing neurons are also Ca^{++} dependent in their neuropeptide (transmitter) release and some of the postsynaptic effects may be mediated by changes in membrane ion flux or in cyclic nucleotide synthesis.

A significant overlap exists in the distribution of the classic neurotransmitters and neuropeptides, and both may be present in the same nerve terminals. This finding suggests that more than one transmitter can coexist within one neuron and they may be released together in response to nerve stimulation. Also suggested is the possibility of a co-transmission with the neurotransmitter and the neuropeptide regulating the functional role of each other. Such examples include dopamine along with cholecystokinin, 5-hydroxytryptamine along with substance P, and acetylcholine along with vasoactive intestinal polypeptide.

Other Substances

Prostaglandins[7] probably have an action on the release of transmitter from adrenergic terminals and may also

Table 10–3. Biogenesis and Metabolism of L-Aspartic Acid

CH₂COOH / C=O / COOH	+	CH₂COOH / CH₂ / CH—NH₂ / COOH	Aspartic acid transaminase ⇌ Glutamic acid α-ketoglutaric acid transaminase	CH₂COOH / CH—NH₂ / COOH	+	CH₂COOH / CH₂ / C=O / COOH
L–glutamic acid		L–glutamic acid		Aspartic acid		α-Ketoglutaric acid

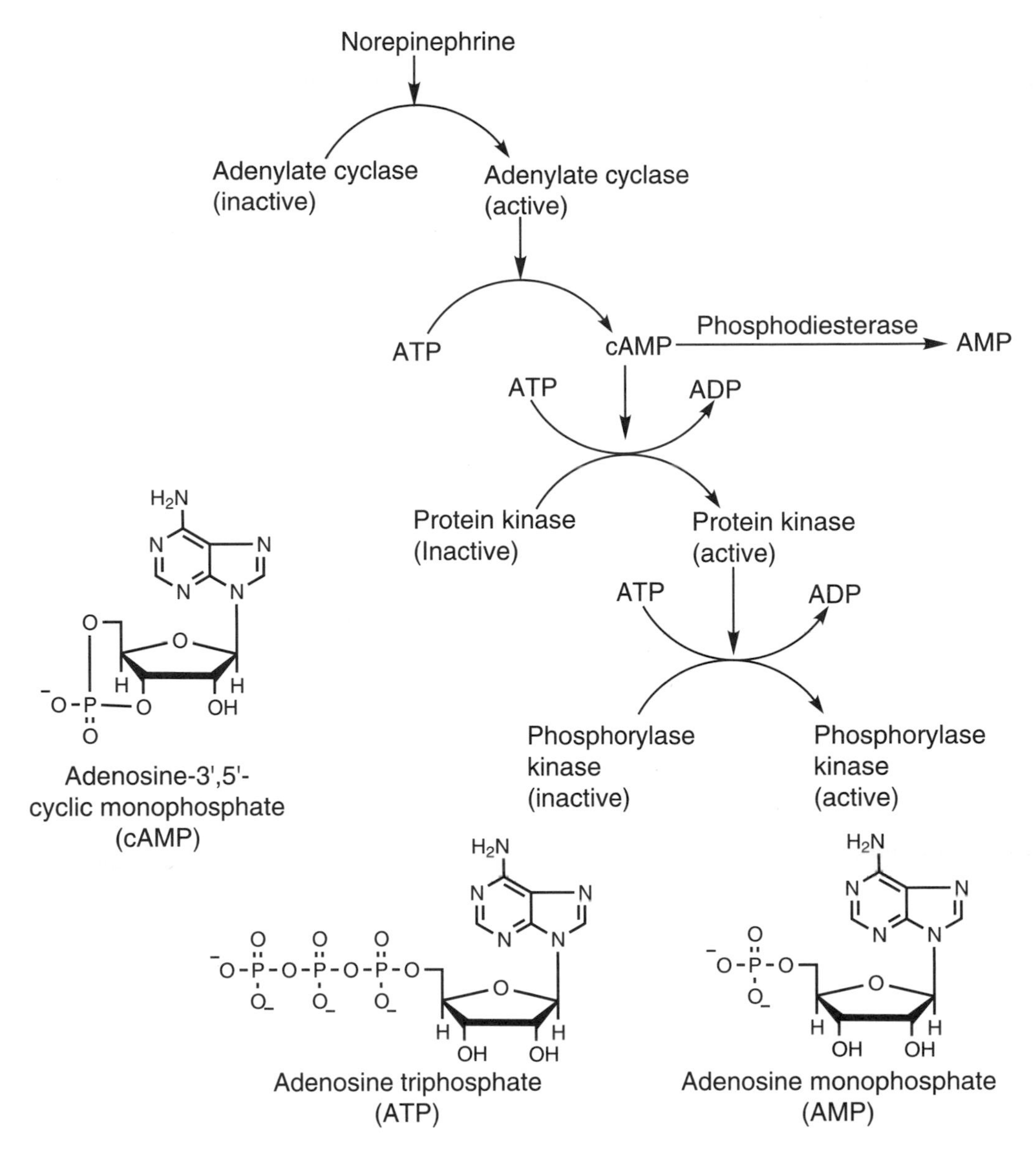

Fig. 10–5. Cascade of reactions mediated by cAMP.

influence the postsynaptic response evoked by norepinephrine stimulation. Specifically, PGE_1 and PGE_2 may inhibit the postsynaptic response to norepinephrine stimulation.

Chemical Transmission

When a transmitter substance reacts with the postsynaptic receptor, it may produce either excitation or inhibition. The action of the transmitters results in selective increases or decreases of ionic conductance or permeability of the membrane for ions. Accordingly, a neuron will increase or decrease its rate of discharge or generate an action potential in the postsynaptic cell. Excitation is usually caused by increased permeability to sodium ions resulting in marked influx of sodium into the neurons. Inhibition is usually caused by permeability changes to potassium or chloride or both. In excitation, an electrochemical equilibrium is reached by the influx of sodium, which depolarizes the membrane potential, making it easier to fire off an action potential. In inhibition, the influx of potassium and chloride produces an electrochemical equilibrium that is near the resting potential, thereby preventing the firing of the neuron.

Experimentally, electrical stimulation of postsynaptic cells has no effect on the presynaptic cells, showing that the synapse allows current to pass along in one direction only, that is, from the presynaptic to the postsynaptic cells, but not in the reverse direction. To generate a current, the presynaptic cell must be more positive than the postsynaptic cell, which can be achieved by depolarizing the presynaptic axon (i.e., making its interior less negative).

Chemical transmitters in the neurons are in a state of flux, being continuously synthesized, released, and metabolized, producing profound changes in the activity of

Table 10–4. Neuropeptides

Pituitary peptides
- Corticotrophin (ACTH)
- Growth hormone (GH)
- β-Lipotropin
- α-Melanocyte-stimulating hormone (α-MSH)
- Oxytocin
- Thyroid-stimulating hormone (TSH)
- Vasopressin

Circulating hormones
- Angiotensin
- Calcitonin
- Glucagon
- Gastric inhibitory peptide
- Insulin

Gut hormones

Avian pancreatic polypeptide
- Bombesin
- Calcitonin gene-related peptide (CGRP)
- Cholecystokinin (CCK)
- Gastrin
- Physalaemin
- Eleidosin
- Cerulein
- Motilin
- Neurotensin
- Neuropeptide Y
- Neurokinin A and B
- Pancreatic polypeptide (PP)
- Secretin
- Substance P
- Vasoactive intestinal polypeptide (VIP)

Opioid peptides
- Dynorphin
- β-Endorphin
- Met-enkephalin
- Leu-enkephalin
- Kyotorphin

Hypothalamic releasing hormones
- Bradykinin
- Corticotropin-releasing factor
- Gonadotropin-releasing hormone (GnRH)
- Growth hormone-releasing factor (GHRF)
- Luteinizing hormone releasing hormone (LHRH)
- Somatostatin
- Thyrotropin releasing hormone (TRH)

Miscellaneous peptides
- Bombesin
- Bradykinin
- Camosine
- Neuropeptide γ
- Neurotensin
- Proctolin

These 40 peptides have been described in neurons and nerve terminals within mammalian CNS, other than those related to endocrine or neuroendocrine functions.

nerves. Yet the amount of catecholamine transmitters (and other transmitters) in tissues remains constant, attributable to the various autoregulatory processes that take place as a result of feedback mechanisms.

When sympathetic nerves are stimulated, the tyrosine hydroxylase activity is increased, converting increased amounts of tyrosine to dopa, which in turn gives rise to increased amounts of dopamine and norepinephrine. It is known, however, that catecholamines interfere with the pteridine cofactor necessary to the function of tyrosine hydroxylase. Thus, the negative feedback effect observed is overcome by an increase in nerve firing brought about by stress using up more catecholamines and thereby reducing the amounts of catecholamines found in the nerve terminals. This change reduces the negative feedback effect of catecholamines and increases tyrosine hydroxylase activity even further. As a result, more tyrosine is converted to dopa and in turn to catecholamines. Conversely, when nerve activity is decreased, the level of catecholamines increases, decreasing the activity of tyrosine hydroxylase and, in turn, the level of dopa and catecholamines in the nerves.

CLASSIFICATION OF CENTRAL NERVOUS SYSTEM DRUGS

Drugs affecting the CNS display multiple activities. Based on their therapeutic usefulness, CNS drugs may be divided into three categories.

1. *General CNS Depressants.* These drugs exert their effect by nonselective depression of the synaptic structures, including presynaptic and postsynaptic tissues. The drugs stabilize neuronal membranes by depressing the postsynaptic structures, accompanied by a decrease in the amount of chemical transmitter released by the presynaptic neuron.

Drugs may exert their effect on the postsynaptic nerve cell by interfering with the generation of postjunctional potential. This action is usually accomplished by the depolarization of the postsynaptic cell, which in turn is caused by an increase in the membrane permeability of the postsynaptic nerve cell. As soon as the postsynaptic neuron is depolarized, it is in a refractory state, i.e., inexcitable by presynaptic impulses, which are normally excitatory. Alternatively, these drugs may interfere with synaptic transmission by raising the threshold of the postsynaptic nerve cell. As a result, release of transmitter substances will not bring about depolarization. The higher postsynaptic threshold may be caused by a decrease in the membrane permeability of the postsynaptic cell.

2. *General CNS Stimulants.* These drugs exert their action nonselectively by either of the following mechanisms: blockade of postsynaptic inhibition (strychnine) or direct neuronal excitation (picrotoxin). Direct neuronal excitation may be achieved by depolarization of the presynaptic cell, increase in presynaptic release of transmitter, attenuated transmitter action, labilization of the neuronal membrane, or decrease in synaptic recovery time.

3. *Selective CNS Drugs.* These may be either depressants or stimulants, acting by diverse mechanisms and including anticonvulsants, centrally acting muscle relaxants, analgesics, and psychopharmacologic agents.

The mechanism of action of many drugs is still poorly understood. Included are CNS depressants such as anesthetics, anticonvulsants, analgesics, and sedatives.

SEDATIVE-HYPNOTICS

These agents are general depressants. A sedative produces mild depression and calms anxiety and excitation without causing drowsiness or impaired performance. A hypnotic compels the user to sleep, a stronger form of depression. An anesthetic produces very strong depression (surgical anesthesia). Drugs that exert a general depressing effect on the CNS can be used—in principle—to elicit any degree of depression from mild sedation to anesthesia, depending on the dose and route of administration.

The Physiology of Sleep

At one time, sleep was considered a passive process. The revolutionary discovery of the ascending reticular activating system by Moruzzi and Magoun changed this theory. Today sleep is considered an active process,[8] regulated by the reticular activating system. Increase in the activity brings about wakefulness, and diminution of the activity brings about the two states of sleep. This theory was substantiated by the finding that sleeping animals can be awakened through stimulation of electrodes implanted in the midbrain reticular formation.

States of Sleep

Three life states can be distinguished: wakefulness, slow-wave sleep (SWS) (nonrapid eye movement —NREM sleep), and paradoxic sleep (PS) (rapid eye movement—REM sleep).[9,10]

In the wakeful state, one is fully aware of oneself and the environment; external stimulus brings about full response. On the other hand, sleep is a period of inertia and low responsiveness.[11] It can be terminated by insistent external stimuli.

The discovery of the electroencephalogram[12] provided evidence that the brain displays continuous electrical activity. Electroencephalographic (EEG) studies by Aserinsky and Kleitman led to the discovery of the two kinds of sleep.[13] Dement and Kleitman subdivided slow-wave sleep into four stages, during which rapid eye movements are absent.[14] The fifth stage is the paradoxic sleep characterized by rapid eye movement. Electroencephalographic studies established a cyclic pattern of states of sleep.

Each state of sleep is characterized by different EEG patterns. Wakefulness is characterized by low-voltage fast activity of the EEG, high muscle activity, and numerous rapid eye movements indicating intensive interaction with the environment. In wakefulness, the electromyographic (EMG) recording shows a high level of muscular activity in the head and neck areas. The electrooculogram (EOG) shows frequent eye movements. When sleep overtakes wakefulness, the transition is gradual. At first, drowsiness (stage 1 sleep) sets in, characterized by 8 to 12 cps α-rhythm in the EEG and by muscular tone relaxation and slow eyeball oscillation in the EOG. After a few minutes, stage 2 is reached, which involves definite sleep characterized by a 12- to 15-cps EEG pattern. From 45 to 50% of the total sleep period is spent in stage 2 sleep. As stage 3 sleep approaches, the EEG voltage continues to increase and the frequency continues to decrease. A gradual change into stage 4 sleep characterized by delta waves (high amplitude, slow wave) is observed. After the NREM sleep has been completed (stage 1 to stage 4 combined), a sudden burst of activities begins as REM sleep takes over, characterized by rapid eye movements and a mixed-frequency EEG pattern.

The Sleep Cycle

Under normal circumstances, normal young adults at night display a regular pattern of sleep, 20 to 25% of which is REM sleep and 75 to 80% of which is NREM sleep. NREM always precedes REM sleep. After about 90 minutes of NREM sleep, the first REM sleep occurs, with a mean duration of about 20 minutes. Afterwards, REM periods occur cyclically at intervals of about 90 minutes. Throughout a night, four or five periods of dreaming (REM sleep) take place, accounting for about 20% of the total sleep time.

The effect of dream deprivation was extensively studied by Dement.[15] If the subject was awakened every time the EOG and EEG indicated that dreaming had just begun, he or she became selectively deprived of REM sleep; because every time he or she fell asleep again, the sleep cycle began with NREM sleep.

As a result of selective REM sleep deprivation, REM sleep appears at shorter and shorter intervals, and the subject has to be awakened more often. As a result, a pressure for REM sleep builds up.

This REM sleep deprivation is also manifested by the so-called rebound effect. When the subject previously deprived of REM sleep is allowed to sleep, the REM sleep period is longer than normal. Most hypnotics and addicting drugs suppress REM sleep, and after their withdrawal, a rebound excess of REM sleep occurs.

The need for NREM sleep has also been demonstrated. Subjects deprived of all sleep spend more time in NREM sleep.

Many autonomic, physiologic, and biochemical changes are associated with wakefulness and slow-wave and paradoxic sleep. The reader should consult references 16 to 19 for details.

The Effects of Drugs on the Sleep Cycle

Most compounds that act on the CNS decrease REM sleep after one dose. Repeated administration leads to tolerance and the REM sleep returns to occupying the normal 20 to 25% of the night.

Although sleep patterns appear relatively normal in the addicted state, abnormalities do exist, as observed on withdrawal. Discontinuation of the drug causes a rebound increase in REM exceeding normal levels, and the onset of REM sleep takes place sooner than normally.

Despite the apparent normal patterns in the addicted state, a chronic deficit of REM sleep develops. This deficit may be caused by the reduced activity of vividness of dreams and the reduced profusion of eye movements in the addicted state. Despite claims to the contrary, most studies demonstrate that all hypnotics reduce REM sleep.[20] A possible exception is chloral hydrate (500 mg). Benzodiazepines reduce REM sleep, but sometimes only after two or three nights' administration. They also reduce the slow-wave cycle of the NREM sleep. Benzodiazepines used as minor tranquilizers have the same effects on the sleep pattern as benzodiazepines used as hypnotic drugs.

Because most hypnotics available on the market suppress REM sleep, prolonged administration of them must be considered bad practice. Use for an extended period is detrimental, because these drugs do not induce natural sleep; tolerance develops, and there is a danger of dependence. In addition, many hypnotics cause a hangover effect, indicating that psychologic impairment and electrophysiologic changes are inevitable.

Sleep-Inducing and Sleep-Facilitating Factors

While sleep-inducing (hypnogenic) factors are directly involved in inducing sleep, the role of sleep-facilitating factors is indicated by the fact that REM sleep duration in rodents can be substantially increased by keeping the animals in a warm environment. The characteristics of these two different factors are outlined in Table 10–5.

Table 10–5. Sleep-Inducing and Sleep-Facilitating Factors

Sleep-Promoting Factors (Hypnogenic Factors, Sleep Inducers)

—Should be endogenous, in the brain, CSF, or blood
—Should increase during instrumental deprivation of sleep
—Act directly on executive mechanisms of sleep
—Are probably different for NREM and REM sleep, but may have the same ancestor molecule
—Their administration at the receptor level should trigger and increase NREM and/or REM sleep *provided that permissive systems are blocked*
—Inactivation of these factors should result in either total insomnia or selective suppression of NREM or REM sleep; such suppression could be obtained for a long time and should not be followed by any secondary rebound.

Sleep-Facilitating Factors

—May be exogenous (warm temperature) or endogenous
—Do not obligatorily increase during sleep deprivation
—Do not act directly on executive mechanisms of sleep, but control the permissive mechanisms that impair sleep onset
—Their administration should increase either NREM or REM sleep or both
—In case of inactivation of sleep-promoting factors, the administration of sleep-facilitating factors may induce sedation, drowsiness, but not physiologic sleep
—Inactivating sleep-facilitating factors may delay sleep onset temporarily, but a secondary rebound of sleep should occur later

From M. Jouvet in *Sleep Mechanisms,* A. Borbely and J. L. Valatx, Eds., Berlin, Springer, 1984, P. 82.

Biogenic Amines and Sleep

NREM Sleep (SWS). This controversial and rapidly changing area of research is further complicated by the fact that major differences are found between humans and animals. In the cat, reserpine suppresses NREM sleep for 12 hours. Reserpine depletes nonspecifically the serotonin, dopamine, and norepinephrine concentrations in the brain. If administration of reserpine in the cat is followed by injection of 5-HTP, a precursor of serotonin (5-HT), which restores brain serotonin levels to normal, NREM sleep is restored to normal. It is postulated, therefore, that depletion of serotonin levels in the brain is responsible for the suppression of NREM sleep. On the other hand, if administration of reserpine is followed by injection of dihydroxyphenylalanine (dopa), which restores brain catecholamine levels to normal, REM sleep is also restored to normal. Accordingly, it is suggested that catecholamines in the brain are responsible for inducing REM sleep.[21]

These effects were found to be specific in the cat. When the same experiments were repeated in the rat and rabbit, the opposite results were obtained, e.g., reserpine increased REM sleep time. Reserpine also increased REM sleep time in humans.[22] It was also found that NREM sleep was decreased in humans by the administration of 5-HPT with or without the simultaneous administration of a peripheral decarboxylase inhibitor, which blocks the conversion of 5-HTP to 5-HT.[23]

A direct relationship exists in animals between the cerebral serotonin level and the amount of NREM sleep.[24] An increase of cerebral serotonin increases NREM sleep and vice versa. The serotonin level of the brain is raised by administration of either 5-HTP, a precursor in the biogenesis of serotonin, or MAO inhibitors. Short-term administration of parachlorophenylalanine (PCPA), which

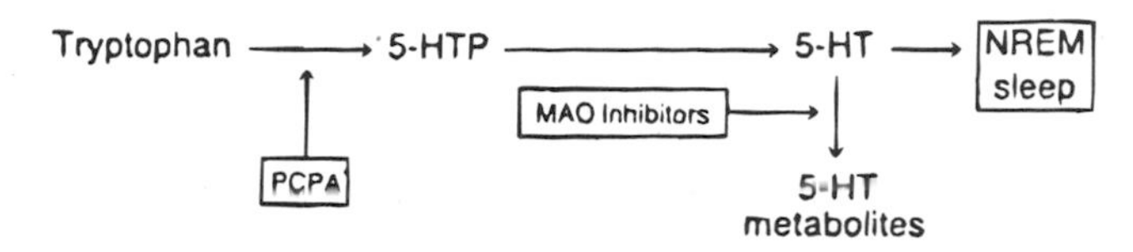

Fig. 10–6. The effect of serotonin on NREM sleep in animals. An increased 5-HT level, caused by HTP or MAO inhibitors, increases NREM sleep. A decreased 5-HT level, brought about by PCPA, decreases NREM sleep. If PCPA is followed by 5-HTP, NREM sleep rebounds.

inhibits the action of the tryptophan hydroxylase enzyme, selectively decreases brain serotonin (5-HT) and in turn NREM sleep. If, in animals, administration of PCPA is followed by injection of 5-HTP, the normal serotonin level of the brain is restored, causing an NREM sleep rebound (Fig. 10–6). These findings support the hypothesis that in animals, serotonin is necessary to trigger or sustain NREM sleep.

Contradictory results were obtained, however, as a result of long-term administration of PCPA. It was found that sleep returned to normal levels in animals in about 1 week. Furthermore, it was also found that in humans PCPA administration did not affect NREM sleep. These findings indicate that in humans no correlation exists between serotonin concentration and NREM sleep. The serotonergic theory, implying that serotonin acts as a hypnogenic neurotransmitter, is also subject to criticism for the following reason: the serotonergic neurons are most active and serotonin release is most abundant during arousal, whereas the activity of these neurons is decreased during NREM and even more during REM sleep. This finding is inconsistent with the postulation that serotonin is a sleep neurotransmitter. Further studies are needed to clarify the role of serotonin in NREM sleep.

REM Sleep (Paradoxic Sleep). The REM stage of sleep commences only after a certain amount of NREM sleep has occurred. Because 5-HT is implicated in NREM sleep, a correlation between brain 5-HT levels and REM sleep is likely—5-HT may prime the norepinephrine mechanism, which in turn may trigger REM sleep. A certain minimal threshold concentration of 5-HT is necessary to trigger REM sleep.

Two conditions must be met for the onset of REM sleep:

1. Some REM sleep-promoting (hypnogenic) factor must be present. Depletion of 5-HT suppresses the synthesis or release of REM sleep factor, leading to REM-sleep suppression.
2. The activation of permissive systems must be suppressed. If the permissive systems are activated, REM sleep is suppressed. Most drugs that suppress REM sleep (tricyclic antidepressants, MAO inhibitors, amphetamines) activate the permissive systems.

Administration of 5-HTP not only increases the serotonin level, and NREM sleep, but also suppresses REM sleep, indicating that the brain serotonin level influences not only NREM sleep (in a direct relationship) but, secondarily, also REM sleep. It appears that over a threshold value, an indirect relationship exists between brain serotonin levels and REM sleep. (The greater the serotonin level, the less the amount of REM sleep.) This relationship is also supported by the fact that MAO-inhibiting drugs that increase brain monoamine levels nonspecifically (such as serotonin, norepinephrine, and dopamine) also suppress REM sleep and, at the same time, increase NREM sleep in the cat. In addition, it was reported that serotonin reuptake inhibitors (zimelidine and alaproclate) that increase postsynaptic serotonin levels increase NREM sleep but decrease REM sleep latency and REM sleep in the cat.

Monoamine oxidase, therefore, appears to have a role in promoting the transition from NREM to REM sleep. Because MAO-inhibiting drugs suppress REM sleep, it is possible that deaminated metabolites of serotonin effect the transition from NREM to REM sleep. Paroxetine, a new antidepressant drug with serotonin reuptake inhibitory properties, was found to reduce REM sleep dose dependently after all paroxetine doses. This provided further evidence for the inverse relationship of brain serotonin levels and REM sleep.

It appears that a minimum norepinephrine (NE) threshold concentration is necessary for the onset of REM sleep. This is supported by the following findings:

1. Administration of the presynaptic and postsynaptic α_2-adrenoreceptor agonist clonidine reduces REM sleep. This effect can be ascribed to clonidine's selective NE activity, manifested by its inhibitory activity on the firing of NE neurons in the brain, probably by stimulating α_2-adrenoreceptors present in the cell body. As a result, the concentration of NE is below the threshold value required for REM sleep.
2. Administration of phenoxybenzamine, thymoxamine, piperoxane, or yohimbine, all α_2-adrenoreceptor blockers, produces a facilitation of REM sleep. Their α_2-adrenoreceptor blocking activity produces enhanced norepinephrine release with the NE concentration exceeding the threshold value required for REM stimulation.
3. If administration of piperoxane is preceded by administration of α-methylparatyrosine (AMPT), an inhibitor of catecholamine synthesis, the overall result is REM suppression attributed to the absence of NE in presynaptic terminals.
4. Chlorpromazine (CPZ) has a biphasic effect. Small doses of CPZ facilitate REM sleep, probably because of its α_2-antagonistic effect, which enhances release of NE to reach threshold NE concentration. Large doses of CPZ, on the other hand, produce a decrease in REM sleep because of an increased level of NE, to be discussed later.
5. If administration of a small dose of CPZ is preceded by administration of AMPT, an inhibitor of catecholamine synthesis, the overall result is REM sleep suppression. This may be due to the absence of NE in presynaptic terminals.

On the other hand, there is also evidence that above a minimum threshold value that is necessary to induce

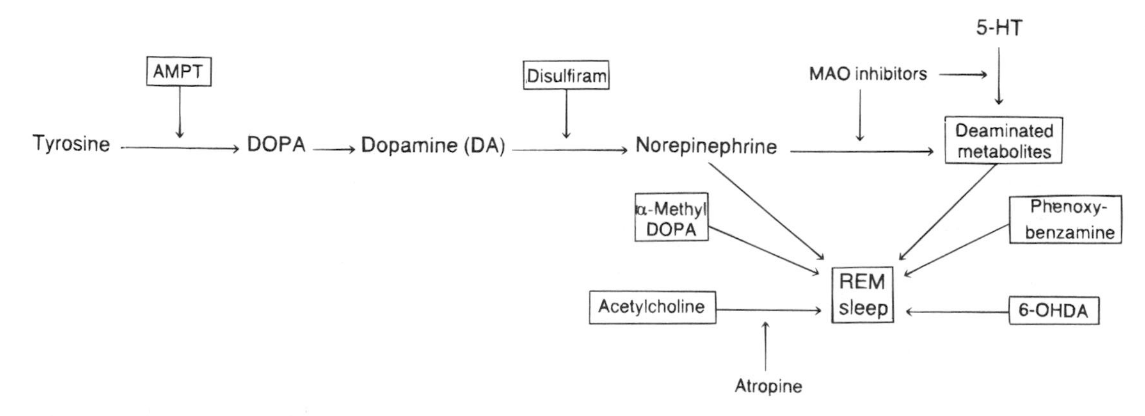

Fig. 10–7. The effect of drugs on REM sleep. (Drugs that facilitate REM are set in boxes.) An extremely high norepinephrine level caused by DOPA induces wakefulness. Depletion of norepinephrine caused by α-methyparatyrosine, disulfiram, or α-methyldopa increases REM sleep. Depletion of deaminated metabolites of 5-HT or norepinephrine suppresses REM sleep. Acetylcholine triggers REM sleep. Anticholinergic drugs (atropine) suppress REM sleep. The α-adrenergic blocker phenoxybenzamine or thymoxamine increases REM sleep. The catecholamine depletor 6-hydroxydopamine increases REM sleep.

REM sleep, an inverse relationship exists between the concentration of catecholamines and REM sleep.

Norepinephrine does not cross the blood-brain barrier. Administration of large doses of dopa, a precursor of both dopamine (DA) and NE, increases wakefulness. In small to moderate doses, L-dopa produces an effect on REM sleep but no effect on NREM sleep. The direction of change in REM sleep was not consistent. It appears that L-dopa may produce a transient increase in REM sleep. These results have to be confirmed. Norepinephrine is involved in the transition from NREM to REM sleep because this transition can be influenced at several points in the biogenesis of NE, as shown in Figure 10–3. Drugs such as α-methylparatyrosine (AMPT) or disulfiram, which inhibit the formation of norepinephrine, as well as α-methyldopa (a false transmitter), may facilitate the transition (Fig. 10–7). Above a minimal NE threshold concentration, an inverse relationship exists between brain NE level and REM sleep. This conclusion is supported by the following facts:

1. Administration of AMPT in both animals and man reduces brain levels of dopamine and NE and in turn increases REM sleep.
2. Administration of the dopamine-β-hydroxylase inhibitors fusaric acid or disulfiram in the rat produced an increase in REM sleep 8 to 24 hours after administration. Because fusaric acid decreases NE levels selectively, it shows that NE and not DA is responsible for the inverse relationship.
3. An inverse relationship exists between 3-methoxy-4-hydroxyphenylglycol (MHPG) excretion and REM sleep. That MHPG is a metabolite of NE but not a direct metabolite of DA suggests the selective involvement of NE in the inverse relationship of catecholamines and REM sleep, because the higher the level of MHPG, the lower the level of NE.
4. Direct intraventricular administration of 300 μg NE but not DA in the rat increases waking and decreases REM sleep.
5. Administration of 6-hydroxydopamine, which destroys catecholamine-containing neurons, increases REM sleep.
6. Monoamine oxidase inhibitors (MAOI) suppress REM sleep. This effect may be ascribed to the increased levels at postsynaptic receptors caused by the MAOI. It is possible, however, that deaminated metabolites derived from serotonin (5-HT) and catecholamines (DA and NE) may be responsible for inducing REM sleep, because the absence of these metabolites caused by the presence of MAOI suppresses the transition from NREM to REM sleep.
7. Administration of antidepressant drugs (imipramine and desipramine) suppresses REM sleep because of increased NE levels caused by antidepressant drugs.

A cholinergic mechanism may be implicated in triggering REM sleep. Injection of acetylcholine and cholinergic agonists (arecholine, bethanechol) helps, whereas administration of anticholinergic drugs (atropine, scopolamine, and hemicholinium-3) hinders the transition to REM sleep. Reports indicate an increase in the rate of release of acetylcholine from the cerebral cortex during REM sleep, as in arousal and alertness.[25]

PHARMACOLOGIC TESTING

In pharmacologic terms, sedation represents depression or calmness. It is characterized by consciousness without any loss of the righting reflex. Sedation is a rapidly reversible state; sensory stimulation quickly brings about full arousal and righting.

On the other hand, hypnosis is characterized by temporary loss of arousability and the righting reflex. The depression is so deep that animals are unable to right themselves.

Tests for Hypnotics

Pharmacologic screening for sedative-hypnotic effects involves abolition of the righting reflex in experimental animals. A compound is hypnotic if it induces sleep in animals. Sleeping time is defined as the period during which the animal has lost its righting reflex, that is, when it does not have the ability to touch a resting surface with all its paws and can be placed on its side or back without rapidly righting itself. The end of the sleeping period is the point at which the animal no longer lies on its side or back but voluntarily returns to its normal upright position. Induction time and sleeping time are both recorded.

The observed pharmacologic effects of many drugs are dose related; small doses cause sedation, larger doses cause hypnosis, and still larger doses bring about surgical anesthesia. The hypnotic dose (HD) is the amount that abolishes the righting reflex in the animal.

In larger animals, electrophysiologic and electroencephalographic studies are often helpful in gaining information about the site of action of CNS depressants as well as about induced sleep patterns.

Tests for Sedatives

Drugs causing sedation exert effects on the CNS and should be screened by other methods as well.

A method of measuring the degree of sedation by scoring reduction of spontaneous activity changes in posture and extent of eye closure has been suggested.[26] Another method involves an activity cage, in which each movement of the animal is electrically recorded. The difference in movement of a group of sedated mice compared to that of control mice is a measure of sedation. Yet another method involves cages in which the animals are isolated from external stimuli. If the drug decreases the time required to bring about sleep, it possesses a sedative effect. The degree of sedation is measured by the time required to bring about sleep.

Another method is based on the fact that CNS depressants potentiate a subhypnotic dose of hexobarbital to a hypnotic one.[27] This method is a means of detecting mild sedatives, but it also gives positive results with other CNS depressants, such as anticonvulsants, analgesics, and anxiolytics.

Experimental evaluation of sedative-hypnotics has been reviewed.[28]

Human sleep laboratory studies have been used increasingly in the evaluation of the effectiveness of hypnotic drugs. These studies have also been reviewed.[29]

STRUCTURE OF HYPNOTICS

The sedative-hypnotic drugs are not characterized by common structural features. Instead, a wide variety of chemical compounds have been used in clinical therapy. An arbitrary classification is as follows:

- Barbiturates
- Aldehydes
- Acetylene derivatives
- Acyclic hypnotics containing nitrogen
- Piperidinediones
- Benzodiazepines
- Imidazopyridines
- Cyclopyrrolones
- Others (including endogenous substances and hypnotics originating from plants)

Barbiturates

The barbiturates exert a depressant effect on the cerebrospinal axis. These drugs depress neuronal activity, as well as skeletal muscle, smooth muscle, and cardiac muscle activity. Depending on the compound, dose, and route of administration, the barbiturates can produce different degrees of depression: sedation, hypnosis, or anesthesia.

Mode of Action

Barbiturates modify the mechanism of synaptic transmission rather than intraneuronal conduction (Fig. 10–8). These drugs in sufficient concentrations reduce the excitability of the postsynaptic cell by altering the permeability of the cell membrane. In general, excitatory synaptic transmission is depressed by barbiturates, whereas inhibitory synaptic transmission is usually unaffected or enhanced.

Barbiturates exert their action on the central synaptic transmission process of the reticular activating system, and the cerebral cortex becomes deactivated. Barbiturates are antidepolarizing blocking agents because they prevent the generation of excitatory postsynaptic potential. These drugs raise the threshold and extend the refractory period of the postsynaptic cell.

The depressant activity of barbiturates can be followed very well on the electroencephalogram. The cerebral electrical activity of normal individuals increases with emotional tension, anxiety, or consumption of a CNS stimulant (such as amphetamine, lysergic acid diethylamide, or caffeine). The increased cerebral electrical activity is caused by intensified reticular activation of the cortex. Administration of barbiturates in sufficient doses has a general calming effect, which is observed clearly on the

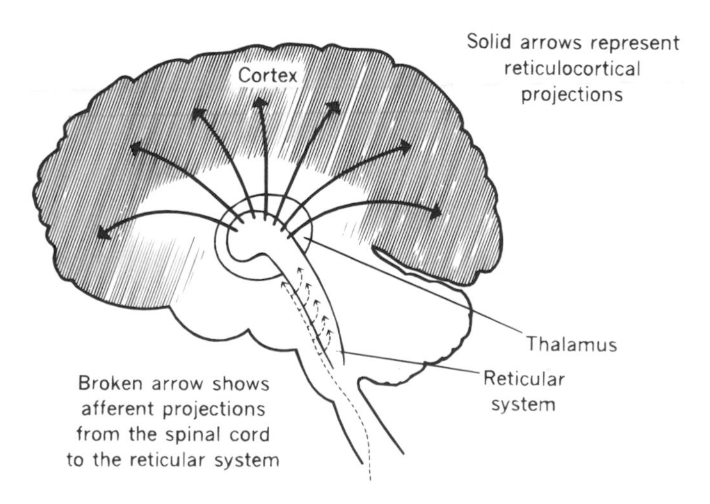

Fig. 10–8. The mode of action of barbiturates. These drugs depress synaptic transmission in the reticular activating system of the brain. Moderation of cerebral electrical activity ensues.

electroencephalogram. Cerebral electrical activity returns to normal as the result of the suppressing influence of barbiturates on reticular activation.

A paradoxic effect of barbiturates also occurs. Small doses bring about hyperexcitation and agitation instead of sedation, because the barbiturate concentration is not great enough to depress the reticular activating system. On the other hand, the concentration of barbiturate is great enough to impede the inhibitory synapses normally present within the cortex. Axodendritic excitability in the cortex is therefore increased.

In addition to the effect on the reticular activating system, barbiturates act on the limbic, hypothalamic, and thalamic synaptic systems.[30]

Biochemical Effects

The effect of barbiturates is marked by a decrease in oxidative metabolic processes and functional activities in the brain. Many barbiturates uncouple oxidative phosphorylation and inhibit the electron transport system. Barbiturates potentiate and/or prolong the action of GABA. The primary action of barbiturates does not involve alterations in the release, reuptake, or metabolism of GABA. Instead, the barbiturates exert their effect at the GABA receptor-ionophore complex by influencing conductance at the chloride channel (ionophore) associated with the GABA receptor. Barbiturates potentiate the GABA-mediated chloride ion conductance and might also interact at the picrotoxin binding site, which is also related to the chloride ionophore. In sufficient concentrations, the GABA-mediated pre- and postsynaptic inhibition is enhanced. In high concentrations, barbiturates activate GABA receptors, thereby altering transmission action that, in turn, produces a marked shift in the balance between excitatory and inhibitory synapses. Whether barbiturates decrease or increase norepinephrine, dopamine, or serotonin levels and, in turn, their synaptic activities is controversial. Barbiturates reduce acetylcholinergic and glutaminergic transmission, while they augment GABAergic activity. These drugs affect the transport of sugars and inhibit cerebral carbonic anhydrase activity. Barbiturates are noted for the induction of liver microsomal enzymes. The biochemical effects of these drugs have been summarized.[31,32]

Effect on Sleep and Dependence

Barbiturates do not induce natural sleep. A reduction in REM sleep (dream time) and eye movement frequency has been observed. A decrease of dream recall has also been reported. Barbiturate addiction at first suppresses REM sleep. As the subject becomes accustomed to the drugs, REM sleep becomes normal but the intensity of eye movements does not. When the barbiturates are withdrawn, excess REM sleep occurs.

Oswald et al. proposed that "drugs capable of causing dependence are drugs which both suppress paradoxical (REM) sleep and provoke a rebound enhancement of paradoxical sleep when withdrawn."[33]

Physical dependence on barbiturates may be induced by various doses of the drugs.[34] As a rule, drug dependence is followed by tolerance, in which increasing doses are required to obtain the same pharmacologic effect.

The decreased effect and shorter duration of activity are brought about by faster metabolism, which is caused by the barbiturates. These drugs speed up their own oxidative breakdown by enzyme induction in the liver. This action accelerates the conversion of lipid-soluble drugs to water-soluble metabolites and their removal from the cells to the blood.[35]

Because barbiturates fail to induce natural sleep and cause tolerance and often dependence, their use is rarely justified.

Chemical Structure and Biologic Activity

Thousands of barbiturates have been synthesized on a trial-and-error basis.[36] Although many structural features required for hypnotic activity were recorded, no clear correlation between structure and activity emerged. In 1951, Sandberg[37] made his fundamental postulation that, to possess good hypnotic activity, a barbituric acid must be a weak acid and must have a lipid/water partition coefficient between certain limits.

On the basis of the first criterion, barbituric acid derivatives are divided into two classes: hypnotics and inactive compounds.

Hypnotic Class

5,5,-Disubstituted barbituric acids
5,5,-Disubstituted thiobarbituric acids
1,5,5-Trisubstituted barbituric acids

Inactive Class

1-Substituted barbituric acids
5-Substituted barbituric acids
1,3-Disubstituted barbituric acids
1,5-Disubstituted barbituric acids
1,3,5,5-Tetrasubstituted barbituric acids

The 1,3,5,5-tetrasubstituted barbituric acids, which are not acidic, are inactive. Their metabolism, however, produces 1,5,5-trisubstituted barbituric acids, which are active.

In addition to their depressant effects (hypnotic activity or anesthesia), many active compounds also produce convulsions, which prevents their clinical use. Whether a particular compound within the active class displays activity or not is determined by the second criterion, the partition coefficient. The following structural features are required for hypnotic activity:

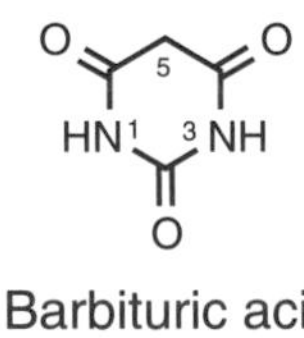

Barbituric acid

(1) For optimal hypnotic activity, at carbon 5, the sum of carbon atoms of both substituents should be between 6 and 10. (2) Within the same series, the branched chain isomer has greater activity and shorter duration. The greater the branching, the more potent the drug (pentobarbital > amobarbital). (3) Stereoisomers have approxi-

mately the same potencies. (4) Within the same series, the unsaturated allyl, alkenyl, and cycloalkenyl analogs have greater potency than the saturated analogs with the same number of carbon atoms. (5) Alicyclic and aromatic substituents impart greater potency than the aliphatic substituent with the same number of carbon atoms. (6) Introduction of a halogen atom into the 5-alkylsubstituent increases potency. (7) Introduction of polar groups (OH, NH_2, RNH, CO, COOH, and SO_3H) into the 5-alkyl substituent destroys potency. (8) Methylation of one of the imide hydrogens (the transition from 5,5-disubstituted barbituric acid to 1,5,5-trisubstituted barbituric acid) increases onset and decreases duration of action. (9) Replacement of sulfur for the carbonyl oxygen at carbon 2 results in thiobarbiturates with quick onset and short duration of action. (10) Introduction of more sulfur atoms (2,4-dithio) decreases potency. The 2,4,6-trithio, 2-imino, 4-imino, 2,4-diimino, and 2,4,6-triimino barbituric acids lack activity.

The effects of these structural features on the hypnotic activity can be rationalized in terms of the partition coefficient.

Modifications at the Fifth Carbon. As the number of carbon atoms at the fifth carbon position increases, the lipophilic character of the substituted barbituric acids also increases. Branching, unsaturation, replacement of alicyclic or aromatic substituents for alkyl substituents, and introduction of halogen into the alkyl substituents all increase the lipid solubility of the barbituric acid derivatives. A limit is reached, however, because as the lipophilic character increases, the hydrophilic character decreases.

Although lipophilic character determines the ability of compounds to cross the blood-brain barrier, hydrophilic character is also important because it determines solubility in biologic fluids and ensures that the compound reaches the blood-brain barrier. Introduction of polar groups into the alkyl substituent decreases lipid solubility below desirable levels.

Modification at the Nitrogen. Substitution of one imide hydrogen by alkyl groups increases lipid solubility. The result is a quicker onset and a shorter duration of activity. As the size of the alkyl substituent increases (methyl → ethyl → propyl), the lipid solubility increases and the hydrophilic character decreases beyond limits. Furthermore, attachment of large alkyl groups (starting with the ethyl group) to the nitrogen imparts convulsant properties to barbiturates. Attachment of alkyl substituents to both nitrogens renders the drugs nonacidic, making them inactive.

Modification at the Oxygen. Replacement of oxygen by sulfur increases lipid solubility. Because maximal thiobarbiturate brain levels are quickly reached, onset of activity is rapid. As a result, these drugs are used as intravenous anesthetics. Replacement of a sulfur or an imine group for more than one carbonyl oxygen decreases hydrophilic properties beyond required limits, thereby destroying activity.

Acidity in Terms of Tautomeric Forms. All four hydrogens of barbituric acid may be exchanged for deuterium.[38] The acidic nature of hydrogens is ascribed to lac-

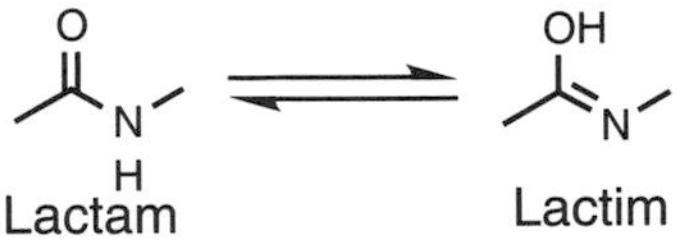

Lactam Lactim

tam-lactim tautomerism. Because barbituric acid contains three lactam groups, in principle, one, two, or all three groups may take up the structure of the lactim group as shown:

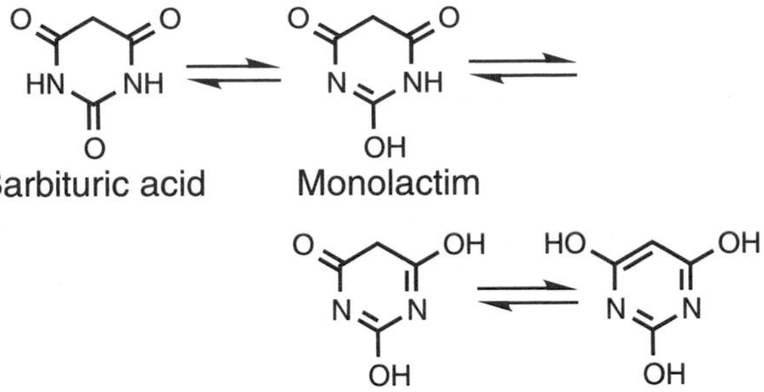

Barbituric acid Monolactim

Dilactim Trilactim

In the crystalline state, barbituric acid exists as the trioxo tautomer, as shown by x-ray analysis.[39,40] The trihydroxy tautomer (trilactim) was ruled out in aqueous solution, because no ultraviolet bands, which are characteristic of a trihydroxyhexahydropyrimidine structure, were found.[41] Ultraviolet spectroscopy also revealed that barbituric acid, 1-methyl barbituric acid, and 1,3-dimethylbarbituric acid display similar spectra, thereby providing further evidence against the existence of the trilactim.[42] Barbituric acid in aqueous solution, therefore, may be characterized by an equilibrium among barbituric acid, monolactim, and dilactim.

Ultraviolet spectroscopic studies were also conducted with monosubstituted and 5,5-disubstituted barbituric acids.[43–46] It was shown that in aqueous solutions, these compounds predominate either in the dioxo tautomeric form (in alkaline medium) or in the trioxo tautomeric form (in acid medium).

The acidity of barbiturates in aqueous solution depends on the number of substituents attached to barbituric acid. The dissociation constant (pK) of unsubstituted barbituric acid is 4.12; the pK value of 5,5-disubstituted barbituric acids ranges from 7.1 to 8.1.[47] Unsubstituted, 1-substituted, 5-substituted, 1,3-disubstituted, and 1,5-disubstituted barbituric acids are strongly acidic because these compounds exist in the tautomeric form. The dissociation of one hydrogen (at C-5) takes place readily; salts of barbituric acids are easily formed by treatment with bases.

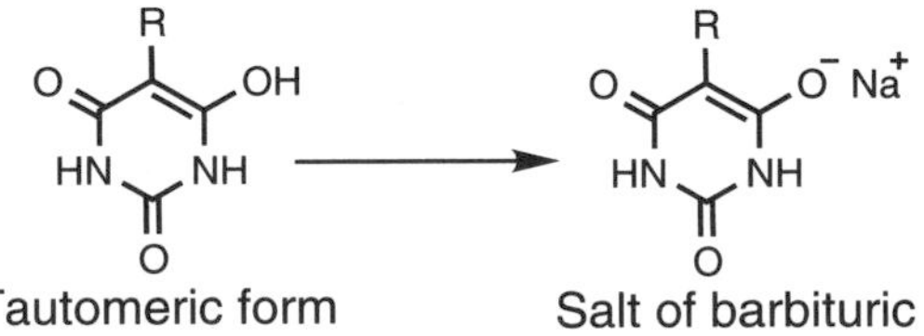

Tautomeric form of barbituric acid → Salt of barbituric acid

The 5,5-disubstituted barbituric acids, 5,5-disubstituted thiobarbituric acids, and 1,5,5-trisubstituted barbituric acids are weakly acidic because these compounds exist predominantly in the trioxo tautomeric form. Although these substances are relatively weak acids, as shown by their dissociation constants in the range of 7.1 to 8.1, salts of these barbiturates are easily formed by treatment with bases.

It has also been reported that 5,5-disubstituted barbituric acids can undergo a second ionization as well.[48] The pK values are in the range of 11.7 to 12.7. It appears reasonable, therefore, to assume that dialkali metal salts of 5,5-disubstituted barbituric acids could be prepared, provided a strong enough base was used. Indeed, the preparation of a dialkali salt of phenobarbital has recently been reported, as shown at the top of p. 160.[49]

Both the monoalkali and the dialkali salts lead to N-substitution rather than O-substitution on reaction with electrophiles.

Metabolic Inactivation

Barbiturates lose their activities through metabolic transformations, which take place principally in the liver. In the course of metabolism, the lipophilic character of barbiturates decreases. The loss of lipid solubility diminishes the concentration of barbiturates in the cerebral tissue with concomitant loss in depressant activity. The loss of lipophilic properties is the result of the following metabolic changes:

1. Oxidation of substituents at carbon 5 occurs. The products are alcohols, phenols, ketones, or carboxylic acids, which because of their hydrophilic character are insoluble in cell lipids.
2. Hydrolytic cleavage of the barbituric ring leads to formation of acetamide and acetylurea derivatives, which are lipophobic (or hydrophilic).
3. Desulfuration of 2-thiobarbiturates takes place readily to yield the more hydrophilic barbiturates.
4. Dealkylation at the nitrogen does not proceed rapidly. The dealkylated products, however, may be excreted more slowly and therefore accumulate in the course of therapy with N-alkylated barbiturates. A definite blood level of phenobarbital has been established in the course of mephobarbital therapy. The N-dealkylation decreases lipophilic character.

Synthesis

Condensation reactions are usually used in the preparation of barbituric acid derivatives. These reactions may take place in acidic, neutral, or basic media.

Condensation reactions in an alkaline medium involve malonic esters, cyanoacetic esters, and malonic amides

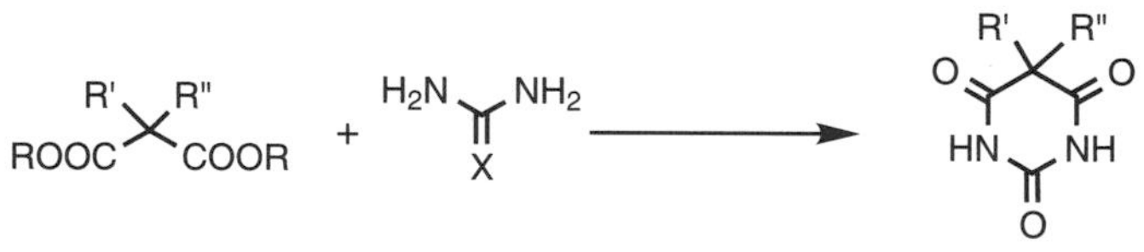

R' R'' ROOC COOR + H_2N NH_2 X → R' R'' O O HN NH O

on one hand, and urea or thiourea on the other hand, in which X = O or S.

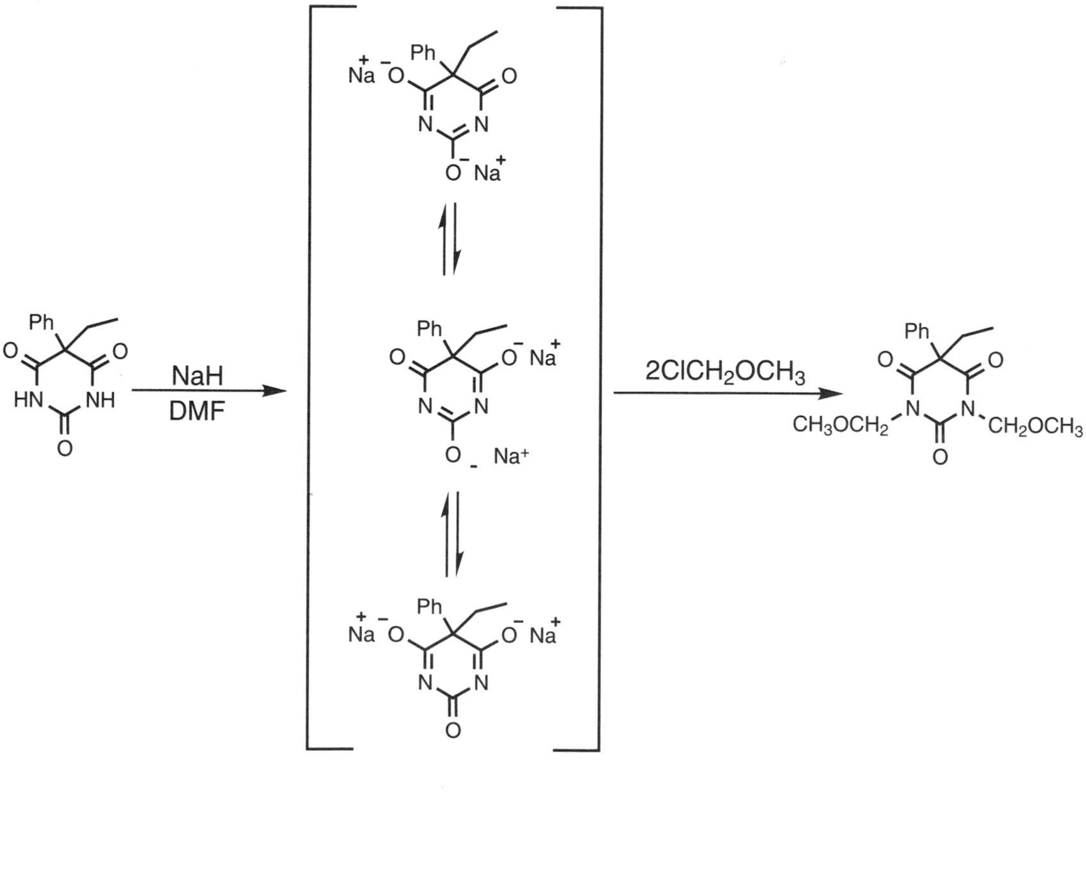

Ph O O HN NH O —NaH / DMF→ [Ph Na⁺ ⁻O O N N ⁻O Na⁺ ⇌ Ph O ⁻O Na⁺ N N O⁻ Na⁺ ⇌ Ph Na⁺ ⁻O ⁻O Na⁺ N N O] —$2ClCH_2OCH_3$→ Ph O O CH_3OCH_2 N N CH_2OCH_3 O

Cyclization of N-substituted ureas in an alkaline medium also produces barbiturates.

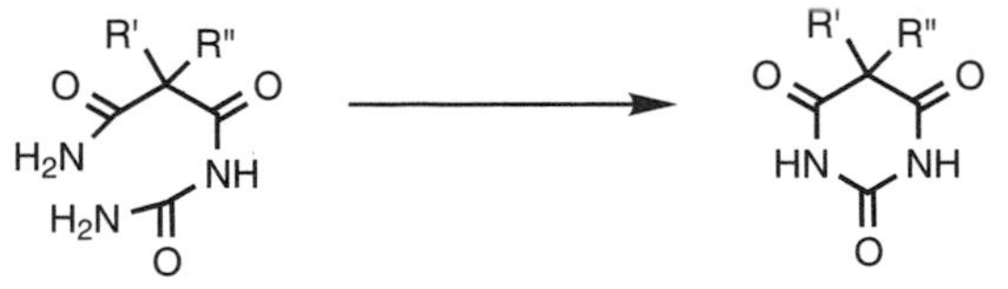

Condensation reactions in a neutral or acidic medium take place readily between malonyl chlorides or malonic acids and urea or thiourea, in which Z = OH or Cl and X = O or S.

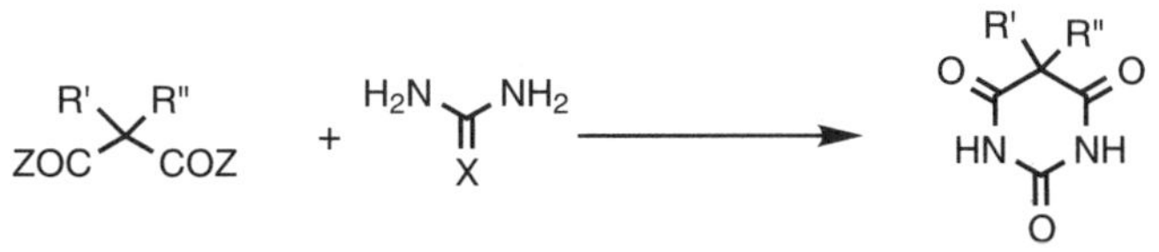

A review of barbituric acids has been published.[50]

Sedative-Hypnotic Barbiturates in Clinical Use

The currently available barbiturates are listed in Table 10–6.

Chloral Derivatives

Because chloral is an aldehyde, it undergoes the usual reactions of an aldehyde. Its behavior with water and ethanol, however, is unique. When chloral, an oily liquid, is treated with water or ethanol, a crystalline solid, chloral hydrate (melting point: 57°C) or chloral alcoholate (melting point: 46°C) is formed. Because of the electron-withdrawing effect of the CCl_3 group, these compounds are stable; the water and ethanol can be removed only by treatment with concentrated sulfuric acid.

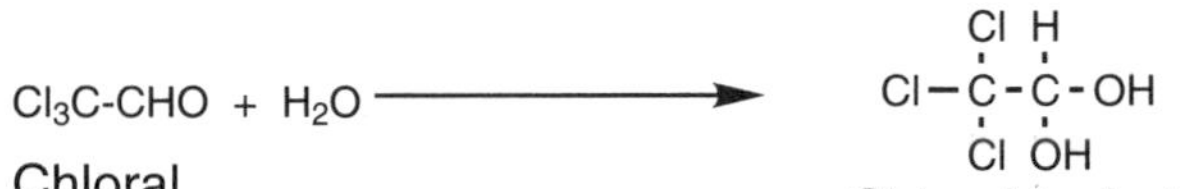

$Cl_3C\text{-}CHO + H_2O \longrightarrow Cl_3C\text{-}CH(OH)_2$

Chloral — Chloral hydrate

Chlordal hydrate has received considerable attention, because it does not significantly suppress REM sleep in doses of 500 and 1000 mg.[22] In another definitive study, 500 mg of chloral hydrate increased total sleep time, decreased waking, and decreased sleep latency.[51] REM and stage 3–4 sleep were unaffected.

Chloral hydrate is readily absorbed from the gastrointestinal tract following oral or rectal doses of 500 mg to 2 g. Sleep occurs within 1 hour and lasts 4 to 8 hours. Chloral hydrate is quickly reduced to trichloroethanol in the liver and other tissues. This metabolic transformation is so quick that it is difficult to detect appreciable chloral hydrate blood levels.[52]

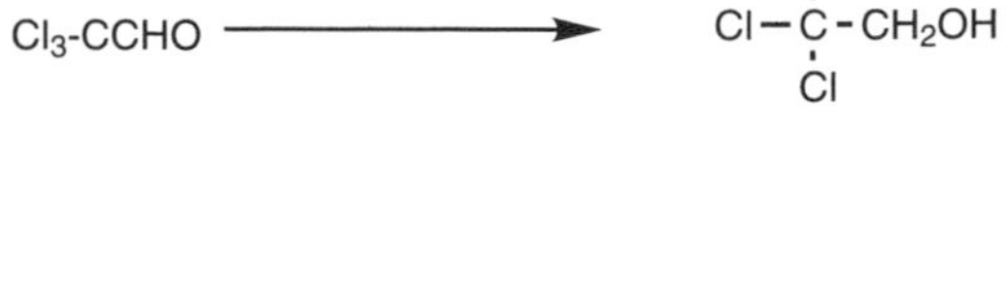

$Cl_3\text{-}CCHO \longrightarrow Cl_3C\text{-}CH_2OH$

It is believed that the initial hypnotic effect of chloral hydrate is exerted by the drug, but the more prolonged effect is caused by trichloroethanol.

Trichloroethanol and chloral hydrate are equipotent, and both are detoxified by metabolic oxidation to the inactive metabolite, trichloroacetic acid.

The site of chloral hydrate action is the rostral reticular activating system. The deactivation is observed on the electroencephalogram. Chloral hydrate has no analgesic or tranquilizing effect and is devoid of adverse respiratory effects. In addition to the unpleasant taste and odor, it often irritates the gastrointestinal tract, causing nausea and vomiting.

Many chloral derivatives have been marketed and supposedly lack the undesirable side effects of chloral hydrate. These compounds produce chloral hydrate or trichloroethanol in the gastrointestinal tract. These derivatives include the following:

$Cl_3\text{-}C\,CH(OH)_2 \cdot (CH_3)_3\overset{+}{N}CH_2CO\overset{-}{O}$

Chloral betaine
an adduct of betaine and chloral

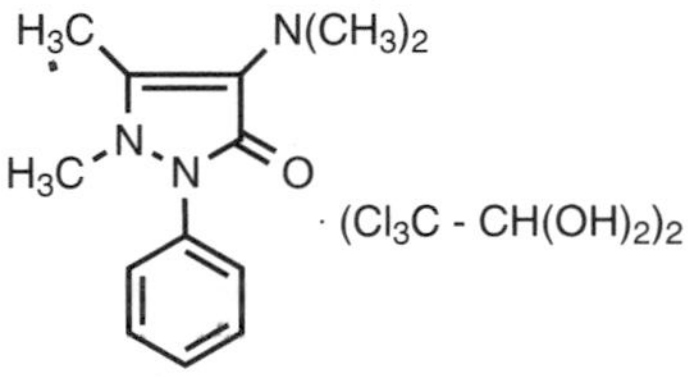

$\cdot\,(Cl_3C\text{-}CH(OH)_2)_2$

Dichloralphenazone

Acetylene Derivatives

Ethchlorvynol (1-chloro-3-ethyl-1-penten-4-yn-3-ol)

This drug is a mild hypnotic with a quick onset and short duration of activity. It resembles the short-acting barbiturates, because it produces similar changes in the electroencephalogram. On the other hand, it produces less initial arousal than barbiturates. Ethchlorvynol also has anticonvulsant and muscle-relaxing properties. It may induce habituation and tolerance. The sedative dose is 100 to 200 mg; the hypnotic dose is 500 mg.

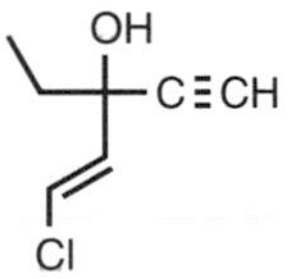

Ethchlorvynol

Acyclic Hypnotics Containing Nitrogen

Amides

Several amides have been marketed, many of them as tranquilizers and muscle relaxants. The following amides possess good sedative properties:

Table 10–6. Sedative-Hypnotic Barbiturates in Clinical Use

Barbituric Acid	Generic Name	Trade Name	R^5	R^5	Other Modifications	Duration of Action (hr)	Onset of Action (hr)	Average Adult Hypnotic Dose (grams)
5-Ethyl-5-isopentyl	Amobarbital	Amytal	C_2H_5	$(CH_3)_2CHCH_2CH_2$		2–8	0.25–0.5	0.1–0.3
5-Allyl-5-isopropyl	Aprobarbital	Alurate, Numal	$CH_2{=}CHCH_2$	$(CH_3)_2CH$		2–8	0.25–0.5	0.065–0.13
5-sec-Butyl-5-ethyl	Butabarbital	Butisol	C_2H_5	$CH_3CH_2CH(CH_3)$		2–4	0.5	0.1
5-Allyl-5-isobutyl	Butalbital	Sandoptal	$CH_2{=}CHCH_2$	$(CH_3)_2CHCH_2$		2–4	0.5	0.2–0.4
5-Ethyl-1-methyl-5-phenyl	Mephobarbital	Prominal	C_2H_5	C_6H_5	1-CH_3	1–4	0.25	0.1–0.2
5-Allyl-5-(1-methyl-2-pentynyl)	Methohexital	Brevital	$CH_2{=}CHCH_2$	$CH_3CH_2C{\equiv}CCH(CH_3)$	1-CH_3	1–4	0.25	0.05–0.1
5-Ethyl-5-(1-methylbutyl)	Pentobarbital	Nembutal	C_2H_5	$CH_3(CH_2)_2CH(CH_3)$		2–4	0.5	0.1–0.2
5-Ethyl-5-phenyl	Phenobarbital	Luminal	C_2H_5	C_6H_5		4–12	0.5–1	0.1–0.2
5-Isopropyl-5-ethyl	Probarbital	Ipral	C_2H_5	$(CH_3)_2CH$		4–12	0.5–1	0.12–0.25
5-Allyl-5-(1-methylbutyl)	Secobarbital	Seconal	$CH_2{=}CHCH_2$	$CH_3(CH_2)_2CH(CH_3)$		1–4	0.25	0.1–0.2

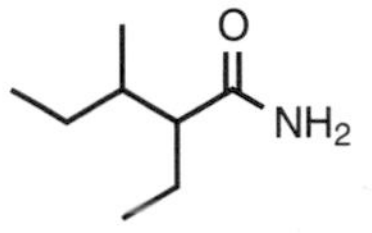

Valnoctamide
(2-ethyl-3-methylvaleramide)

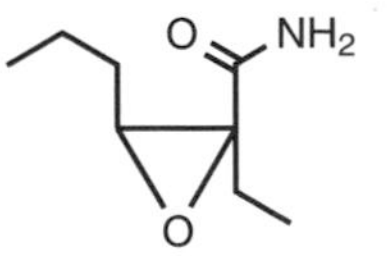

Oxanamide
(2,3-epoxy-2-ethylhexanamide)

Valnoctamide and oxanamide are also potent muscle relaxants. It has been reported that the therapeutic index of simple amides (lethal dose/effective dose) increases rapidly with increasing size in the alkyl substituent. (Allyldiethylacetamide compared to triethylacetamide gives a therapeutic index ratio of 7:1). Because the lipophilic property also increases rapidly, however, allyldiethylacetamide appears to have the optimal structure.

Ureides

Acylation of urea yields ureides. For example, acetylchloride and urea form acetylurea.

$$CH_3COCl + NH_2CONH_2 \longrightarrow CH_3CONHCONH_2$$

Dicarboxylic acids and their derivatives react with urea to form cyclic ureides. For example, oxalic acid and urea from parabamic acid (oxalyurea).

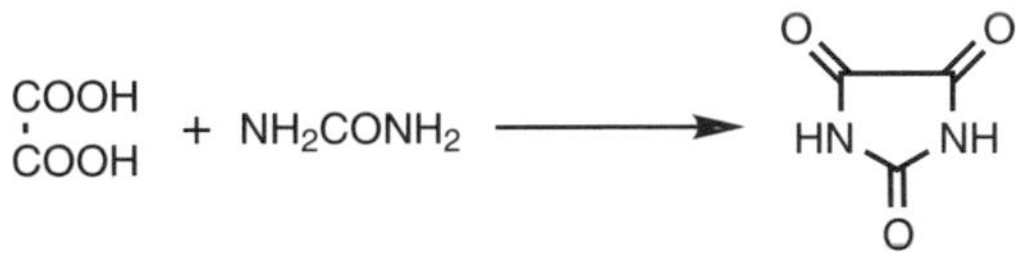

$$\begin{matrix} COOH \\ COOH \end{matrix} + NH_2CONH_2 \longrightarrow$$

A similar reaction involves malonic acid and urea, which form barbituric acid, a cyclic urea. Many of the ureides are useful hypnotics, as summarized in Table 10–7.

The usual oral dose of these compounds is 250 to 900 mg; 250 to 500 mg for sedation and 600 to 900 mg for hypnotic action. Bromisovalum and carbromal have become popular primarily because of their safety. They have been used mainly as daytime sedatives in the treatment of simple anxiety and nervous tension. Because bromide ion is released in vivo, prolonged use of these compounds is not recommended because of the possible development of bromism in patients.

Piperidinediones

Encouraged by the success of barbiturates in sedative-hypnotic therapy, researchers prepared and tested for hypnotic activity many heterocyclic compounds containing the lactam function. Because of the outstanding hypnotic potency observed in experimental animals, methyprylon and glutethimide were introduced into clinical therapy.

Methyprylon (3,3-diethyl-5-methyl-2,4-piperidinedione)

This compound was introduced into clinical therapy in 1955. It was practically a neutral compound when compared to the barbiturates.

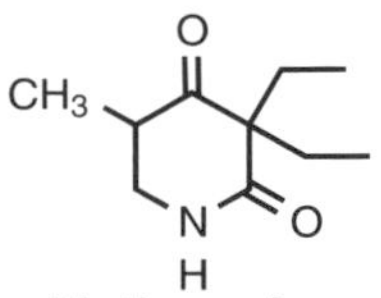

Methyprylon

Methyprylon has been compared to amobarbital in potency and duration of action. Like barbiturates, it suppresses REM sleep in a dose of 300 mg.[54] Methyprylon has not been used as an anesthetic because of its weak potency. It also lacks analgesic, tranquilizing, and muscle-relaxant properties. It has no adverse respiratory or gastrointestinal effects.

The mechanism of action resembles that of barbiturates, and the side effects and toxicity of methyprylon are similar to those of barbiturates. Methyprylon also causes dependence.[55,56]

Glutethimide (2-ethyl-2-phenylglutarimide)

Glutethimide was introduced in 1954 as a sedative-hypnotic drug. It is among the most active nonbarbiturate hypnotics. Glutethimide has been compared to pentobarbital; 500 mg of glutethimide is as potent as 100 mg of pentobarbital.[57] Glutethimide lacks anesthetic, analgesic, tranquilizing, and muscle-relaxant properties. Because it does not affect respiration or blood pressure over a period of several hours, glutethimide is considered safer than phenobarbital.[57]

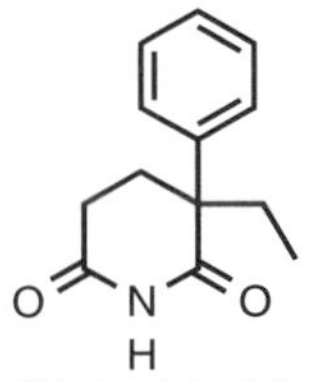

Glutethimide

The deactivating electroencephalographic pattern produced by glutethimide is similar to that caused by secobarbital.[58] Like barbiturates, it is a rostral reticular depressant. It exhibits potent anticholinergic activity,[59] suggesting that it might be a nondepolarizing blocking agent.

Glutethimide significantly suppresses REM sleep at the hypnotic dose of 500 mg.[60] Withdrawal causes a rebound increase in REM sleep.[61–64] Several reports of addiction have been published.[65,66]

Metabolic studies indicate that glutethimide is hydroxy-

Table 10–7. Ureide Hypnotics

General Structure

Generic Name	Chemical Name	R^1	R^2
Acecarbromal	(1-acetyl-3-(2-bromo-2-ethylbutyryl)urea	CH_3CO-	$(CH_3CH_2)_2-C(Br)-$
Bromisovalum	(2-bromo-3-methyl butyryl)urea	H	$(CH_3)_2CH\cdot CH(Br)-$
Carbromal	(2-bromo-2-ethyl-butyryl)urea	H	$(CH_3CH_2)_2-C(Br)-$

lated at various positions, including the glutarimide and benzene rings as well as the ethyl group The product of metabolic detoxification is excreted after conjugation with glucuronic acid at the hydroxyl group. Glutethimide is now regarded as a general CNS depressant with no advantages over barbiturates. A warning has been issued that glutethimide poisoning is more dangerous than barbiturate poisoning.[66]

Benzodiazepines

Many compounds of the 1,4-benzodiazepine series display tranquilizing, muscle-relaxant, anticonvulsant, and sedative effects. Today, many benzodiazepines are widely used as daytime sedatives, tranquilizers, sleep inducers, anesthetics, anticonvulsants, and muscle relaxants.[67]

Close examination of the basic pharmacologic properties of benzodiazepines reveals that, based on the data, no 1,4-benzodiazepine marketed to date can be selected exclusively as a hypnotic agent in preference to other benzodiazepines.[68] Despite this conclusion, many benzodiazepines are specifically promoted as sleep inducers. This rationalization is based on two pharmacokinetic properties of benzodiazepines: speed of onset of action and duration of action.

The speed of onset of action depends on the mode of administration. The onset of sedative activity of intravenously administered benzodiazepines is rapid, with a range of 15 to 30 seconds (a single circulation time) to a few minutes (depending on the patient's' sensitivity, pharmacologic response, and size of the dose). The rapid onset of action of intravenously administered diazepam coupled with its anticonvulsant activity explains its preferred use in stopping repeated epileptic fits (status epilepticus).

The speed of onset of action of orally administered benzodiazepines depends on the speed of availability of the drug from the dosage form, the speed of absorption, and the speed of penetration from the blood into the brain. Several benzodiazepines are specifically promoted as sleep inducers, but this practice does not preclude the use of other anxiolytic benzodiazepines for this indication. Thus, two orally administered anxiolytic benzodiazepines, diazepam and clorazepate, are rapidly absorbed, enter the brain quickly, and are therefore also used as sleep inducers.

On the other hand, among the benzodiazepines specifically indicated for the treatment of insomnia, flurazepam is absorbed most rapidly, triazolam has an intermediate rate of absorption, and temazepam is absorbed slowly and may be given 1 to 2 hours before bedtime.

The durations of action of various benzodiazepines differ considerably. Duration of action is determined by the distribution and elimination (including biotransformation) of the benzodiazepines. The "anxiolytic" benzodiazepine diazepam has a long elimination half-life because of its metabolic transformation to N-desmethyl diazepam (nordiazepam). Nordiazepam has antianxiety and sedative activities on its own and has a long plasma half-life (over 100 hours). Nordiazepam is also the major metabolite of clorazepate. Given the long half-life of nordiazepam, the duration of action after a single dose of diazepam is determined by the distribution. On the other hand, the duration of action of the "sleep-inducing" benzodiazepines triazolam and midazolam with an ultrashort half-life (less than 4 hours) is determined by their elimination half-life.

Flurazepam has a similar short half-life, but it has an active metabolite, desalkylflurazepam, which has a long half-life.[69] Because the rate of elimination of desalkylflurazepam is slow (40 to 150 hours), flurazepam pharmacokinetically is indistinguishable from many other benzodiazepines with slow elimination rates that are used primarily as anxiolytics. Nitrazepam, a hypnotic drug used in Europe, has a moderately long elimination half-life of about 30 hours and may cause "hangover" effects as well as accumulation on repeated use.

The presence of active agent or active metabolite in blood and brain tissue is responsible for the residual effects of hypnotics in the daytime. These effects may consist of "hangover" effects and oversedation, and may be so severe, especially in the elderly, as to cause tremors, ataxia, and confusion. Benzodiazepines that are rapidly eliminated by biotransformation into an inactive metabolite (temazepam) have gained popularity as sleep inducers, especially in elderly individuals.

Benzodiazepines versus Barbiturates

The effects of benzodiazepines on the sleep pattern are similar to those of barbiturates. Both benzodiazepines and barbiturates decrease body movements, number of awakenings, sleep latency (the time required to fall asleep), total REM sleep, and number of shifts in sleep stages. Both increase total sleep, stage 2 sleep, and fast activity (β-activity) in the EEG.[70] Benzodiazepines may decrease stage 3 and 4 sleep more than barbiturates, but reports of this are not yet confirmed. A possible advantage of some benzodiazepines, especially flurazepam, appears to be that their hypnotic efficacy persists during long-term administration.

Benzodiazepines, when compared to barbiturates and other hypnotics in short-term studies, are equally effective and also equally likely to produce hangover and persistent psychomotor impairment.[71] A major advantage of benzodiazepines is their relative safety. Fatalities resulting from benzodiazepine overdosage alone are rare. When it is taken together with other drugs, intoxication probably depends largely on the type and quantity of nonbenzodiazepine.[72] Benzodiazepines do not cause significant enzyme induction in man, and therefore their tendency to interact with other drugs is smaller than that of barbiturates.

The disadvantage of benzodiazepines appears to be the slowness of their elimination. Drug-induced abnormalities in the electroencephalogram, hangover effects, impairment of psychomotor performance, and liability to fall asleep persist longer with nitrazepam than with sodium amobarbital.[73]

Both benzodiazepines and barbiturates produce rebound abnormalities on withdrawal, with the patients spending a disproportionate amount of time in REM sleep with increased intensity. It must be concluded that a prescription for long-term hypnotic use is rarely indicated. The short-term "situational stress" variety of insomnia is most appropriate for drug treatment.

Mode of Action

The identification of specific, high-affinity binding sites for the benzodiazepines[74,75] in the CNS was achieved by using radiolabeled benzodiazepines. In the CNS, the benzodiazepines interact with a macromolecular membrane complex that also has recognition sites for GABA (γ-aminobutyric acid) and a chloride ionophore. Consequently, several subtypes of benzodiazepine receptors have been identified.[76–78]

It is possible, but not proven, that each of these receptor subclasses may mediate independently the anxiolytic, sedative, muscle relaxant, anticonvulsant, and similar properties, respectively, associated with benzodiazepines.

Because enkephalins and endorphins are endogenous ligands for the opiate receptors, many researchers have attempted to isolate an endogenous ligand for benzodiazepine receptors. So far, the search has not produced a substance with the properties of an endogenous ligand, although several factors have been proposed (purines,[79] peptides,[80] β-carbolines,[81] prostaglandins,[82] d-thyroxine,[83] melatonin,[84] and thromboxane A_2[85]). The discovery of the benzodiazepine receptor, however, has resulted in the identification of several classes of nonbenzodiazepine anxiolytic agents, some of which are more anxioselective than the benzodiazepines.[86]

Benzodiazepines have a potent interaction with the GABA inhibitory neurotransmitter system. It has been shown that benzodiazepines synergistically increase the effects of ionotophoretically applied GABA and, similarly, elevation of brain GABA levels by GABA transaminase inhibitors increases the electrophysiologic effects of benzodiazepine. Conversely, inhibition of GABA synthesis can abolish these benzodiazepine-related effects.

Barbiturates, like benzodiazepines, also interact with the GABA system by increasing the postsynaptic responses to GABA. Barbiturates can also enhance benzodiazepine binding to the benzodiazepine receptor through an increase in receptor affinity. It should be remembered, however, that although barbiturates and benzodiazepines display cross-tolerance, and both classes of compounds interact with the benzodiazepine receptor complex, the barbiturates are weak anxiolytics and muscle relaxants. This fact may indicate that the mechanism of sedative action of the benzodiazepine is different from that of the anxiolytic action of benzodiazepine. As a result, reports on the discovery of selective anxiolytic agents are not surprising. It is likely that selective muscle relaxants, anticonvulsants, and sedative-hypnotics, respectively, of the benzodiazepine and nonbenzodiazepine class will also be discovered.

Increased benzodiazepine binding caused by GABA also involves chloride ions that act synergistically with GABA. It is postulated that binding of drugs to either the GABA receptor or chloride ionophore can allosterically modify the benzodiazepine receptor that leads to enhanced affinity of radiolabeled benzodiazepine to the receptor. This theory leads to the prediction that drugs (barbiturates, β-carbolines, purines, etc.) may elicit their pharmacologic actions by either direct occupation of the benzodiazepine receptor site or by allosteric interaction with another component of the benzodiazepine/GABA/chloride ionophore receptor complex.

Sleep-Inducing Benzodiazepines

Eighteen benzodiazepines are marketed either exclusively as hypnotics or used extensively as sleep inducers in addition to their use as antianxiety agents. The structures of these benzodiazepines are shown as follows. Their effects on sleep have been investigated in great detail. Definitive reports are available on 0.25 and 0.5 mg brotizolam;[87,88] on 1 and 2 mg delorazepam;[89] 5 mg[90]

Flunitrazepam

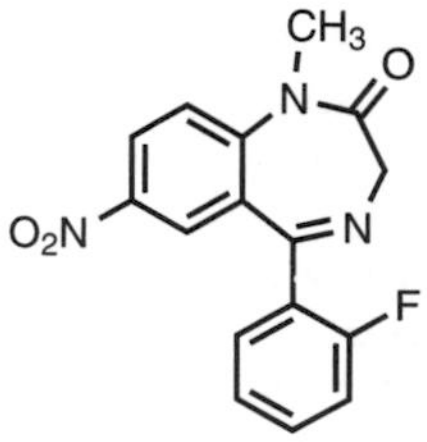

Triazolo-benzodiazepines
Estrazolam: X = H, Y = H
Triazolam: X = CH_3, Y = Cl

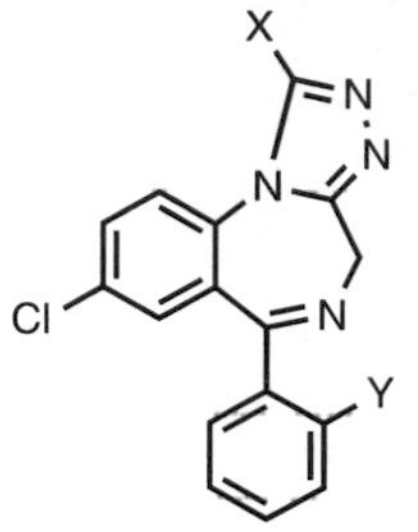

Bromazepam

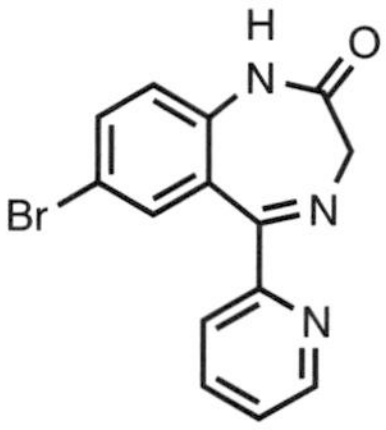

Doxefazepam

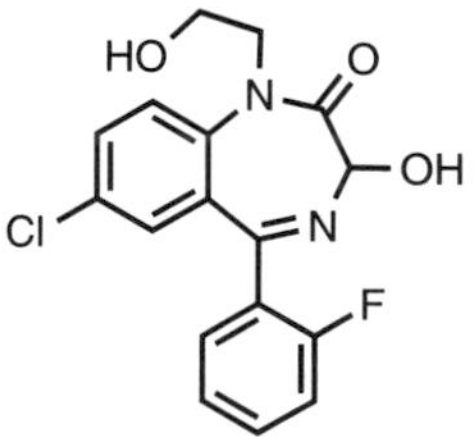

Lorazepam

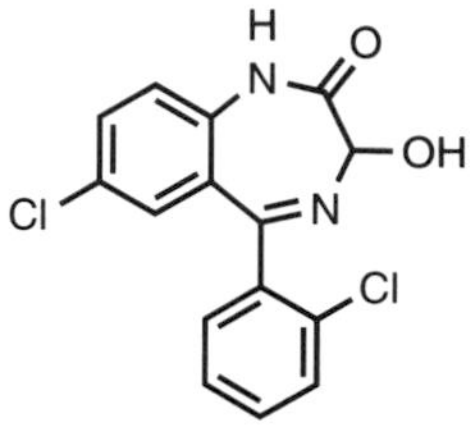

Temazepam

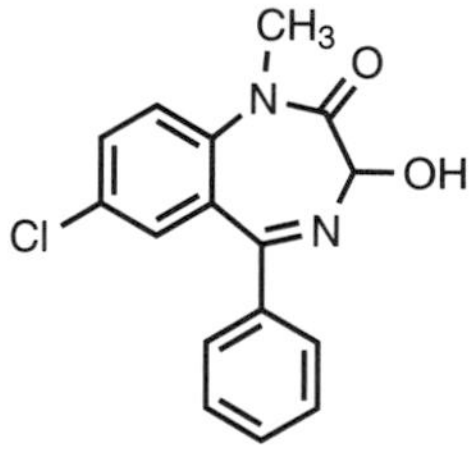

Nimetazepam

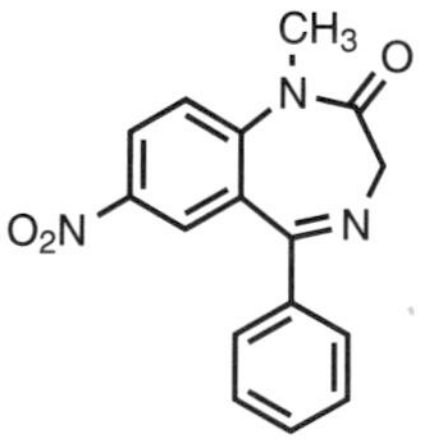

Brotizolam

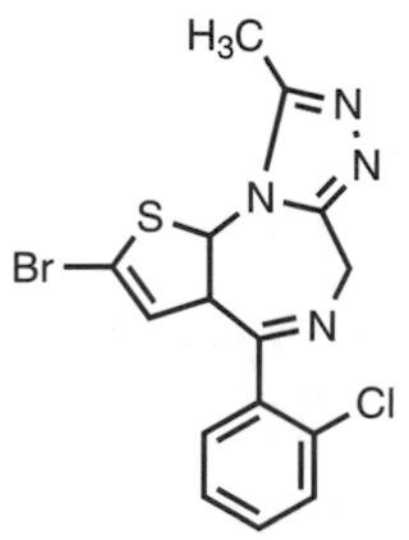

Nitrazepam

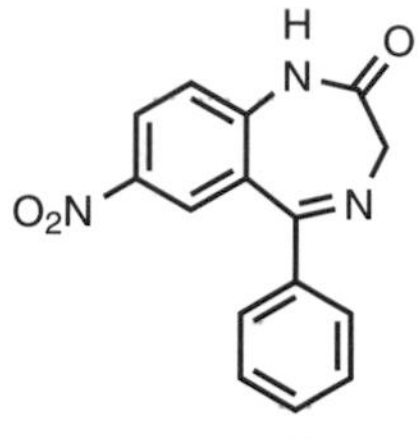

Nordazepam

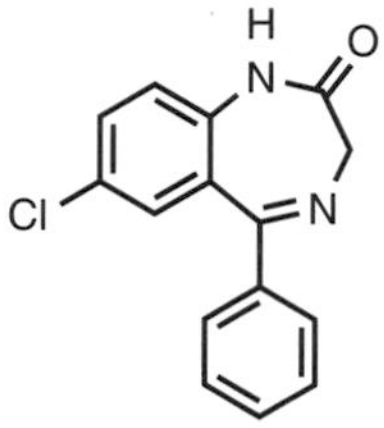

Loprazolam

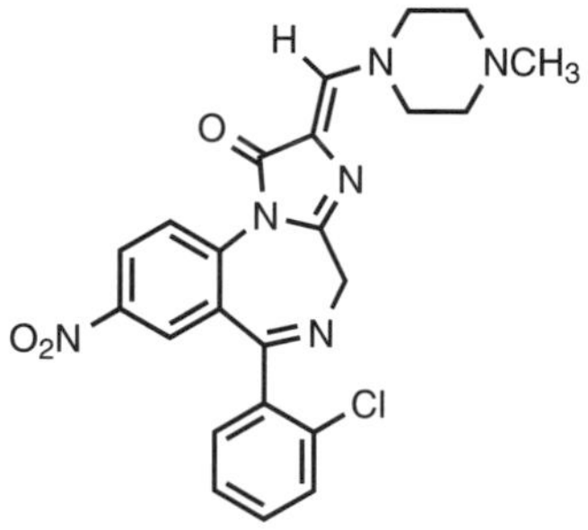

Potassium Clorazepate

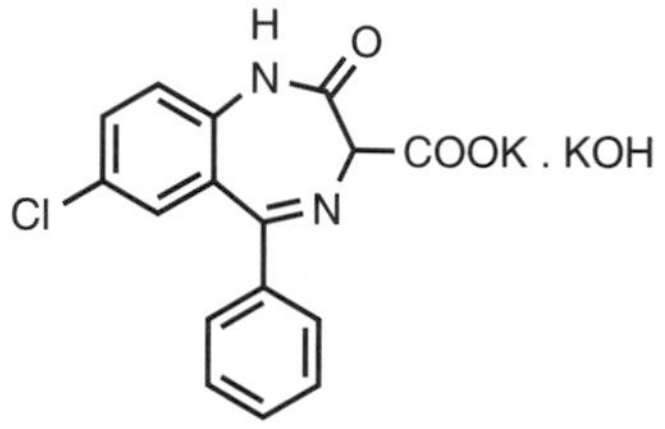

Quazepam

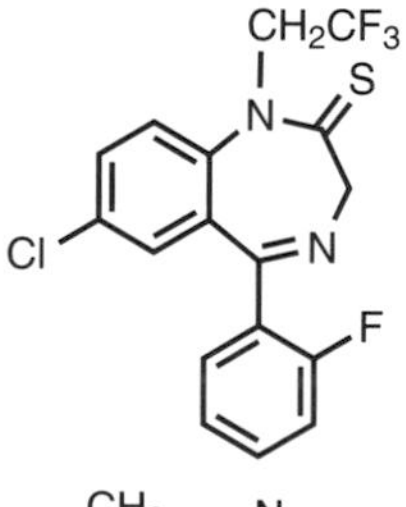

Midazolam

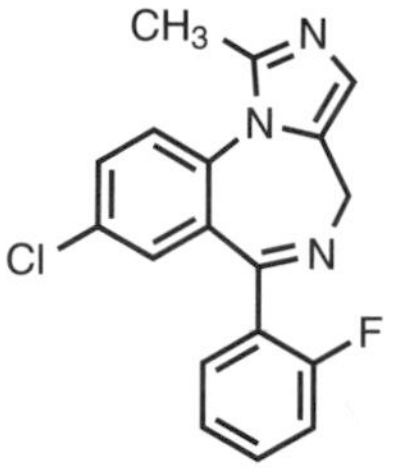

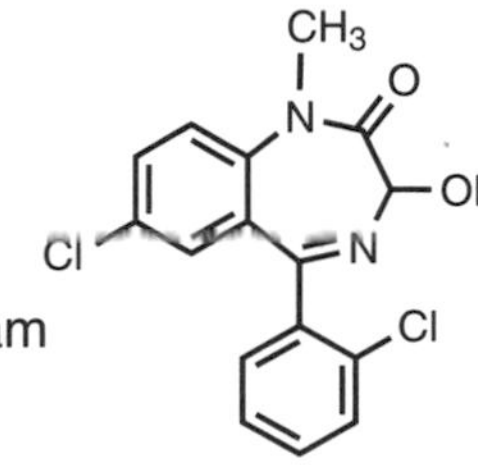

Lormetazepam

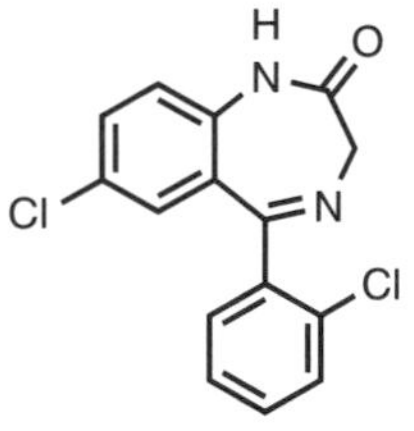

Chlordesmethyldiazepam

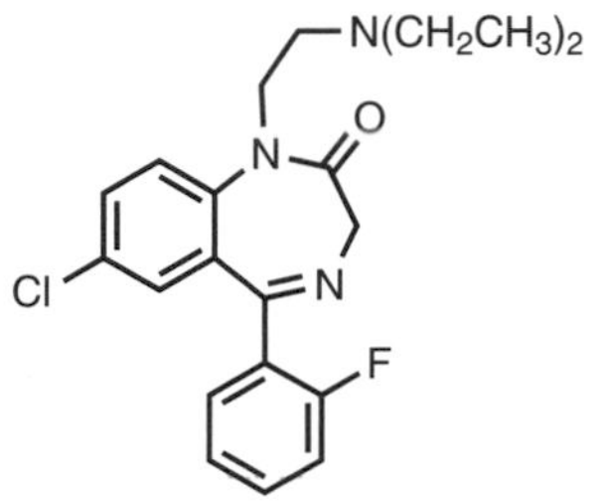

Flurazepam

and 10 to 40 mg[91] doxefazepam; 15 ms[92,93] and 30 mg flurazepam;[94–98] 2 mg intravenously administered,[99] 1 and 2 mg orally administered,[100,101] 2 mg orally administered,[102] and 2 and 4 mg orally administered flunitrazepam;[103] 2.5 to 5 mg[104] and 2 to 4 mg lorazepam;[105] 60 mg,[106] 60 and 80 mg,[107] and 100 mg[108] fosazepam; 5 mg[109] and 10 mg nitrazepam;[110] 5 and 10 mg nordiazepam;[111] 15 mg chlorazepate dipotassium;[111] 10, 25, and 50 mg quazepam;[112] 5 mg doxefazepam;[90] 15, 20, and 30 mg,[113] 10 and 20 mg[114] and 10 to 30 mg temazepam;[115] 2 and 4[116] and 2, 3, and 4 mg of estazolam,[117] 0.5 mg triazolam,[118–120] 0.5, 1.0, and 2 mg,[121] 1 and 2 mg,[122] and 1 and 2.5 mg[123] lormetazapam; 0.5 and 1 mg[124] and 2 mg[125] loprazolam; and 15 mg,[126] and 15 and 20 mg[127] midazolam.

In general, nocturnal wakefulness, body movements, number of awakenings, sleep latency, and stages 3 and 4 sleep are decreased, whereas total sleep and EEG fast activity are increased. Most controversial is the effect of benzodiazepines on REM sleep. Contrary to initial claims, the majority of the studies demonstrate that benzodiazepines decrease total REM sleep. In addition, on withdrawal of benzodiazepines, REM sleep rebound develops. Benzodiazepines are, therefore, similar to barbiturates in their effects on sleep.

Synthesis

1,4-Benzodiazepine-4-oxides. These compounds (chlordiazepoxide and cyprazepam) may be prepared by the following sequence of reactions.[128]

Structure-activity studies have revealed that chlordiazepoxide is the most potent member of this family. The following points are of interest:

1. Increasing the size of the substituent at carbon 2 decreases the potency.
2. Replacement of the phenyl group at carbon 5 by other substituents decreases the potency.

1,4-Benzodiazepine-2-ones. Various methods are known for the synthesis of these compounds. The two methods most often used are outlined.

The transformation of 1,4-benzodiazepine-4-oxides to 3-hydroxy-1,4-benzodiazepines may be achieved by treatment first with acetic anhydride followed by treatment with water.

The following conclusions were drawn from extensive structure-activity investigations: (1) Electron-withdrawing substituents (chloro, bromo, nitro, trifluoromethyl, and cyano groups) at carbon 7 confer high activity. Larger groups or electron donators decrease potency. Any substitution at carbons 6, 8, and 9 decreases potency. (2) Substitution at nitrogen 1 (methyl) and carbon 3 (hydroxy) is allowed. Replacement of sulfur for oxygen at carbon 2 decreases potency. Hydrogenation of the double bond at the fourth and fifth positions reduces the activity. (3) Replacement of the phenyl group at carbon 5 by other

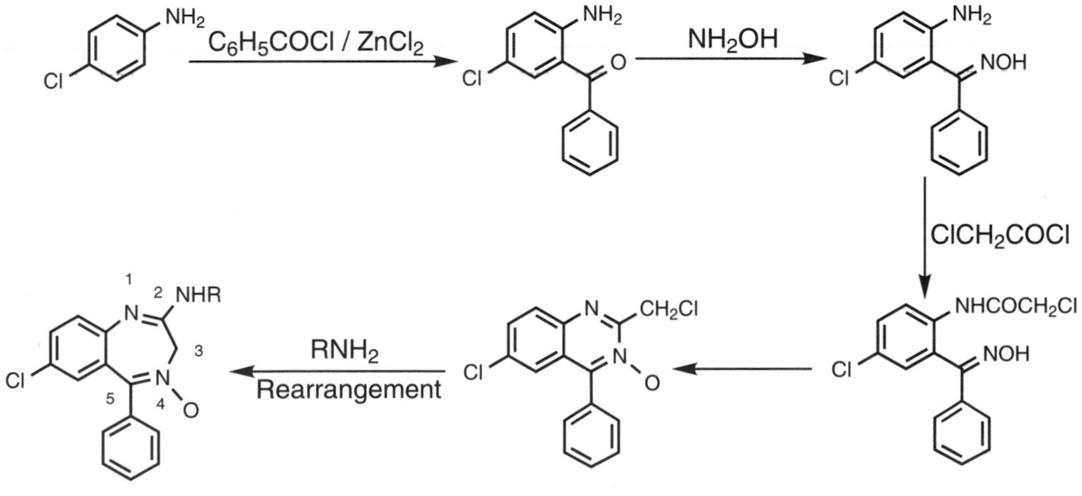

C_6H_5COCl / $ZnCl_2$ → NH_2OH → $ClCH_2COCl$ → → RNH_2 Rearrangement

substituents decreases the potency. Chloro, bromo, or fluoro substituents at carbon 2′ may enhance potency. Other substitutions at carbon 2′, 3′, and 4′ decrease potency.

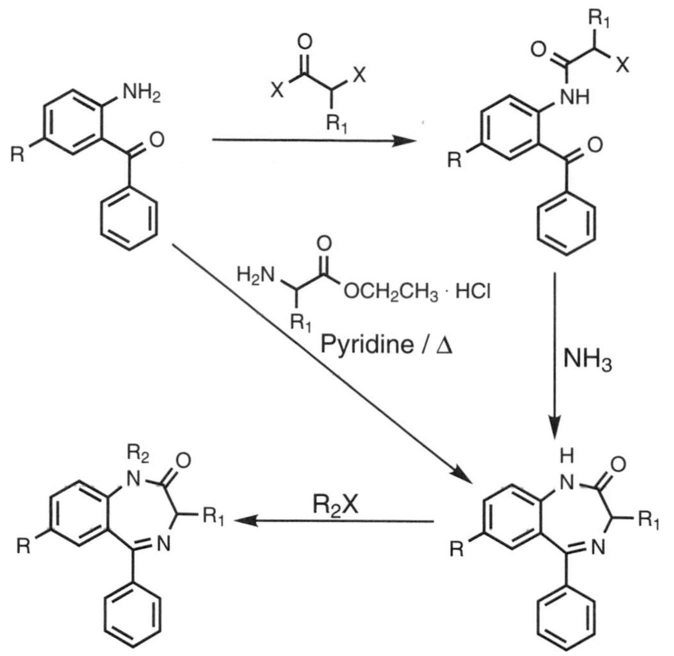

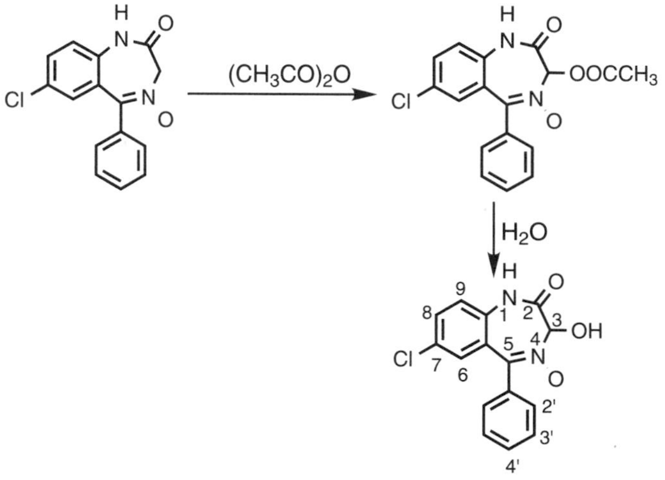

Oxazepam

Imidazopyridines

Zolpidem is the first compound of the imidazopyridines to reach the market.[129] Although a nonbenzodiazepine hypnotic agent, it acts at the high-affinity benzodiazepine receptor subtype in the brain. Zolpidem reduces the time to onset and increases the duration of sleep in patients with insomnia. Zolpidem has no major effects on sleep stages. In addition to the rapid onset of action, it also has a short elimination half-life and no rebound effects on withdrawal of the drug, which are attractive features of a sleep-inducing drug. Unlike benzodiazepines, zolpidem has no muscle relaxant or anticonvulsant effects and its antianxiety effects appear to be minor. Another agent of the same class, alpidem, displays primarily antianxiety and not sedative effects.

Zolpidem is usually given in 5- to 20-mg dosages nightly. Adverse effects are primarily CNS and gastrointestinal. It appears that the dependence liability of zolpidem is minimal.

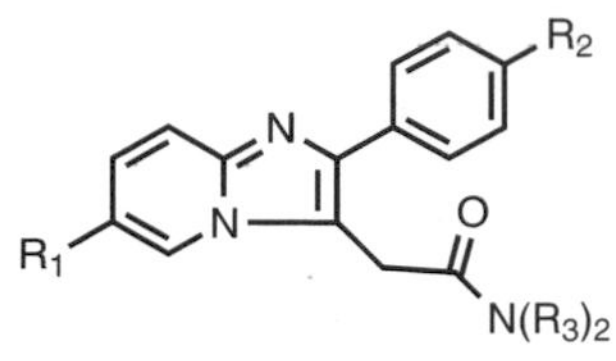

Zolpidem: $R_1 = CH_3$, $R_2 = CH_3$, $R_3 = CH_3$
Alpidem: $R_1 = Cl$, $R_2 = Cl$, $R_3 = CH_2CH_2CH_3$

Cyclopyrrolones

Zopiclone is a new hypnotic agent belonging to the cyclopyrrolone chemical class.[130] Zopiclone, a nonbenzodiazepine hypnotic agent, acts at the high affinity benzodiazepine receptor subtype in the brain.[131] At a 7.5-mg dose, zopiclone decreases sleep latency, increases total sleep duration, reduces the number of awakenings, and increases the sleep efficiency.[132] It effects on sleep stages differ from those observed with benzodiazepines, because REM sleep is substantially unaffected by zopiclone and slow-wave sleep is either unaffected or increased.

The day after bedtime administration of zopiclone, no residual effects or residual impairment of cognitive functions have been observed. Discontinuation of zopiclone is not accompanied by rebound effect and withdrawal symptoms. Its elimination half-life is 5 to 6 hours and no accumulation exists on repeated administration. Zopiclone is devoid of abuse and dependence potential. Adverse effects consist of bitter taste and dry mouth with a minimal incidence of CNS depressant effects.

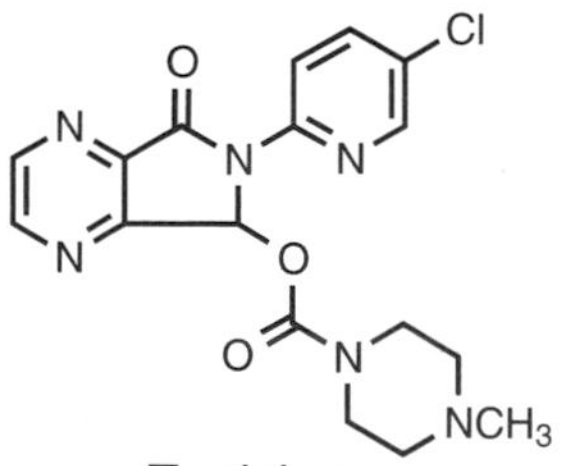

Zopiclone

Other Sedative-Hypnotics

Fenadiazole is used as a sedative mainly for treatment of nervous tension states and frequent awakenings.[132]

Fenadiazole

Etomidate is used as a short-acting hypnotic in anesthesiologic practice. Intravenously administered etomidate induces sleep that is deep and long enough to allow the normal induction and maintenance of anesthesia.[133]

Etomidate

Clomethiazole has been used in the treatment of alcohol withdrawal, delirium tremens, and agitated states. As a hypnotic, it produces a favorable effect on sleep latency but only after long-term, regular therapy.[134]

Clomethiazole

Endogenous Sleep Factors

A dialysate obtained from the cerebral blood of rabbits that were kept asleep by electrical stimulation of the thalamus could induce sleep in normal rabbits. The factor responsible for inducing sleep was later identified as a nonapeptide and was called delta-sleep-inducing peptide (DSIP). Subsequently, DSIP was synthesized and the synthetic DSIP was found to be identical to the natural peptide in its sleep-inducing properties.[135] The amino acid sequence is:

Trp-Ala-Gly-Gly-Asp-Ala-Ser-Gly-Glu.

Sleep-inducing factors were also isolated from goat[136] and rat brain.[137]

Sedatives of Plant Origin

Pharmacologic compounds have been extracted from plants, including Radix valerianae and Glandulae lupuli. Some reportedly have sedative effects.[138]

Antihistamines and Anticholinergic Compounds

Some of the useful antihistamines and centrally acting anticholinergic drugs also exhibit sedative-hypnotic properties.

The antihistamines diphenyhydramine and doxylamine have often been used as sedatives.[139]

Diphenhydramine

Doxylamine

Miscellaneous Compounds

A variety of other compounds has been reported to have sedative-hypnotic effects.[140]

REFERENCES

1. C. A. K. Borrebaeck and J. W. Larrick, Eds., *Therapeutic Monoclonal Antibodies,* New York, Freeman, 1992.
2. A. B. Young and S. H. Snyder, In *Neuroregul. Psychiatr. Disord. Proc. Conf.,* E. Usdin, D. A. Hamburg, and J. D. Barchas, Eds., New York, Oxford Univ. Press, 1977. pp. 515–525.
3. N. G. Bowery and G. D Pratt, *Arzneimittelforschung, 42,* 215(1992).
4. P. L. Herrling, *Arzneimittelforschung, 42,* 202(1992).
5. P. Mandel and H. Pasante-Morales, *Adv. Biochem. Psychopharmcol., 15,* 141(1976).
6. J. Nathanson, *Physiol. Rev., 57,* 157(1977).
7. K. Hillier, *Drugs of Today, 13,* 418(1977).
8. M. W. Johns, *Drugs, 8,* 448(1975).
9. E. L. Hartmann, *The Functions of Sleep,* New Haven, Yale University Press, 1973.
10. W. C. Dement, *Some Must Watch While Some Must Sleep,* San Francisco, W. H. Freeman and Company, 1972, 1974.
11. B. Wyke, *Principles of General Neurology,* Amsterdam, Elsevier, 1969, pp. 207–212.
12. H. Berger, *Arch. Psychiat. Nervenk, 87,* 527(1927).
13. I. Oswald, *Annu. Rev. Pharmacol., 13,* 243(1973).
14. I. Oswald, *Postgrad. Med. J., 52,* 15(1976).
15. G. W. Fenton, *Postgrad Med. J., 52,* 5(1976).
16. A. A. Borbely, et al. *Psychiatr. Clin. Neurosci., 241,* 13(1991).
17. D. J. Greenblatt and R. I. Shader, *Drugs Exp. Clin. Res., 1,* 417(1977).
18. J. Raese, in *Psychopharmacology, from Theory to Practice,* J. D. Barchas, et al., Eds. New York, Oxford University Press, 1977, pp. 306–317.
19. W. B. Mendelson, in *Psychopharmacology, The Third Generation of Progress,* H. Y. Meltzer, Ed., New York, Raven Press, 1987, pp. 1305–1311.
20. I. Oswald, *Br. J. Psychiatry,* Spec. Publ. No. *9,* 272(1975).
21. M. Touret et al., *Exp. Brain Res., 86,* 117(1991).
22. A. Carlsson, in *Psychopharmacology, a Generation of Progress,* M. A. Lipton, et al., Eds. New York, Raven Press, 1978, pp. 1057–1070.
23. R. J. Wyatt and J. C. Gillin, in *Pharmacology of Sleep,* R. L. Williams and I. Karacan, Eds. New York, John Wiley, 1976, pp. 239–274.
24. M. Jouvet, in *Sleep, Physiology and Pathology,* A. Kales, Ed. Philadelphia, J. B. Lippincott, 1969, pp. 89–100.
25. H. Kametani and H. Kawamura, *Life Sci., 47,* 421(1990).
26. R. K. S. Lim, in *Pharmacologic Techniques in Drug Evaluation,*

J. H. Nodine and P. E. Siegler, Eds. Chicago, Year Book Medical Publishers, 1964, pp. 291–297.
27. J. F. Reinhard and J. V. Scudi, *Proc. Soc. Exp. Biol. Med., 100,* 381(1959).
28. R. N. Straw, in *Hypnotics, Methods of Development and Evaluation,* F. Kagan, et al., Eds. New York, Spectrum Publications, Inc. 1975, pp. 65–85.
29. A. Kales, et al., in *Hypnotics, Methods of Development and Evaluation,* F. Kagan, et al., Eds. New York, Spectrum Publications, Inc., 1975, pp. 109–126.
30. J. A. Richter and J. R. Holtman, *Prog. Neurobiol. 18,* 275(1982).
31. L. Decsi, *Prog. Drug Res., 8,* 53(1965).
32. R. Nicoll, in *Psychopharmacology, a Generation of Progress,* M. A. Lipton, et al., Eds. New York, Raven Press, 1978, pp. 1337–1348.
33. I. Oswald, et al., in *Scientific Basis of Drug Dependence,* H. Steinberg, Ed. New York, Grune and Stratton, 1969, pp. 243–247.
34. D. R. Wesson and D. E. Smith, *Barbiturates, Their Use, Misuse and Abuse,* New York, Human Sciences Press, 1977.
35. H. Remmer, in *Scientific Basis of Drug Dependence,* H. Steinberg, Ed. New York, Grune and Stratton, 1969, pp. 111–128.
36. W. J. Doran, *Barbituric Acid Hypnotics,* in *Medicinal Chemistry,* Vol. 4. F. F. Blicke and R. H. Cox, Eds., New York, John Wiley and Sons, 1959.
37. F. Sandberg, *Acta Physiol. Scand., 24,* 7(1951).
38. H. Erlenmeyer, et al., *Helv. Chim. Acta, 19,* 354(1936).
39. G. A. Jeffrey, et al., *Acta Crystallogr.* (Kopenhavn), *14,* 88(1961).
40. W. Bolton, *Acta Crystallogr.* (Kopenhavn), *16,* 166(1963).
41. A. I. Scott, *Ultraviolet Spectra of Natural Products,* New York, Pergammon Press, 1964, pp. 194–195.
42. R. E. Stuckey, *Q. J. Pharm. Pharmacol., 13,* 312(1940).
43. R. E. Stuckey, *Q. J. Pharm. Pharmacol., 15,* 377(1942).
44. O. Rosen and F. Sandberg, *Acta Chem. Scand., 4,* 666(1950).
45. J. J. Fox and D. Shaugar, *Bull. Soc. Chim. Belg., 61,* 44(1952).
46. A. J. Petro, et al., *J. Am. Chem. Soc., 78,* 3040(1956).
47. V. Havlioek, *Int. J. Neuropharmacol., 6,* 8(1967).
48. T. C. Butler, et al. *J. Am. Chem. Soc., 77,* 1488(1955).
49. C. M. Samour, et al. *J. Med. Chem., 14,* 187(1971).
50. C. C. Cheng and B. Roth, in *Progress in Medicinal Chemistry,* G. P. Ellis and G. B. West, Eds., New York, Appleton-Century-Croft, 1971, pp. 66–81.
51. J. I. Evans and O. Ogunremi, *Br. Med. J., 3,* 310(1970).
52. E. K. Marshall and A. H. Owens, *Bull. Hopkins Hosp., 95,* 1(1954).
53. A. Kales, et al., *Biol. Psychiatry, 1,* 235(1969).
54. A. Kales, et al., *Arch. Gen. Psychiatry, 23,* 211(1970).
55. K. Rickels and H. Bass, *J. Med. Sci., 245,* 142(1963).
56. H. Berger, *JAMA, 177,* 63(1961).
57. B. Isaacs and M. B. Glasg, *Lancet, 1,* 558(1957).
58. F. Gross, et al., *Verh. Naturforsch. Ges. Basel, 67,* 479(1956).
59. H. A. Ladwig, A. M. A., *Arch. Neurol., 74,* 351(1955).
60. H. Turrian and F. Gross, *Helv. Physiol. Pharmacol. Acta, 16,* 208(1958).
61. C. Allen, et al., *Psychonom. Sci., 12,* 329(1968).
62. L. Goldstein, et al., *J. Clin. Pharmacol., 110,* 258(1970).
63. A. Kales, et al., *Arch. Gen. Pshchiatry, 23,* 226(1970).
64. A. Kales, et al., *JAMA, 227,* 513(1974).
65. H. Cohen, *N. Y. State J. Med., 60,* 280(1960).
66. J. Holland, et al., *N. Y. State J. Med., 75,* 2343(1975).
67. L. O. Randall, et al., in *Psychopharmacological Agents,* Vol. 3, M. Gordon, Ed., New York, Academic Press, 1974, pp. 175–281.
68. D. J. Greenblatt and R. I. Shader, *Benzodiazepines in Clinical Practice,* New York, Raven Press, 1974, pp. 183–196.
69. L. O. Randal and B. Kappell, in *The Benzodiazepines,* S. Garattini, et al., Eds., New York, Raven Press, 1973, pp. 27–51.
70. D. C. Kay, et al., in *Pharmacology of Sleep,* R. L. Williams and I. Karacan, Eds., New York, John Wiley, 1976. pp. 83–211.
71. J. Arnold, *J. Clin. Psychiatry, 52,* 11(1991).
72. D. J. Greenblatt, et al., *Clin. Pharmacol. Ther., 21,* 497(1977).
73. I. Oswald, et al., in *The Benzodiazepines,* S. Garattini et al., Eds., New York, Raven Press, 1973, pp. 613–625.
74. R. F. Squires and C. Braestrup, *Nature, 266,* 722(1977).
75. H. Mohler and T. Okada, *Science, 198,* 848(1977).
76. C. A. Klepner, et al., *Pharmacol. Biochem. Behav., 11,* 457(1979).
77. R. W. Olsen, *Annu. Rev. Pharmacol. Toxicol., 22,* 245(1982).
78. A. C. Bowling and R. J. Delorenzo, *Science, 216,* 1247(1978).
79. P. Skolnick, et al., *Life Sci., 23,* 1473(1978).
80. A. Guidotti, et al., *Nature, 275,* 553(1978).
81. M. Nielsen, et al., *Life Sci., 25,* 679(1979).
82. T. Asano and N. Ogasawara, *Eur. J. Pharmacol., 80,* 271(1982).
83. S. Nagy and A. Lajtha, *J. Neurochem., 40,* 414(1983).
84. P. J. Marangos, et al., *Life Sci., 29,* 259(1981).
85. A. I. Ally, et al., *Neurosci. Lett., 7,* 31(1978).
86. M. Williams, *Prog. Neuropsychopharmacol. Biol. Psychiatry, 8,* 209(1984).
87. H. Lohmann, et al., *Br. J. Clin. Pharmacol., 16,* 403S(1983).
88. E. Goetzke, et al., *Br. J. Clin. Pharmacol., 16,* 407S(1983).
89. C. Zimmermann-Transella, et al., *J. Clin. Pharmacol., 16,* 481(1976).
90. M. Babbini, et al., *Arzniemittelforsch., 25,* 1294(1975).
91. G. Rodriguez, et al., *Neurophysiology, 11,* 133(1984).
92. M. W. Johns and J. P. Masterton, *Pharmacology, 11,* 358(1974).
93. W. C. Dement, et al., in *The Benzodiazepines,* S. Garattini, E. Mussini, and L. O. Randall, Eds., New York, Raven Press, 1973, pp. 599–611.
94. E. Hartmann, *Psychopharmacologia, 12,* 346(1968).
95. J. Kales, et al., *Clin. Pharmacol. Ther., 12,* 691(1971).
96. T. M. Itil., et al., *Pharmakopsychiatr. Neuropsychopharmakol., 7,* 265(1974).
97. A. Kales, et al., *Clin. Pharmacol. Ther., 19,* 576(1976).
98. A. Kales, et al., *J. Clin. Pharmacol., 17,* 207(1977).
99. I. Freuchen, et al., *Curr. Ther. Res., 20,* 361(1976).
100. V. Samec, *Wien Med. Wochenschr., 126,* 23(1976).
101. E. O. Bixler, et al., *J. Clin. Pharmacol., 17,* 569(1977).
102. U. J. Jovanovic, *J. Int. Med. Res., 5,* 77(1977).
103. G. Cerone, et al., *Eur. Neurol., 11,* 172(1974).
104. J. W. Dundee, et al., *Br. J. Clin. Pharmacol., 4,* 706(1977).
105. R. I. H. Wang, et al., *Clin. Pharmacol. Ther., 19,* 191(1976).
106. S. Allen and I. Oswald, *Br. J. Clin. Pharmacol., 3,* 165(1976).
107. A. N. Nicholson, et al., *Br. J. Clin. Pharmacol., 3,* 533(1976).
108. A. M. Risberg, et al. *Eur. J. Clin. Pharmacol., 12,* 105(1977).
109. K. Adam, et al., *Br. Med. J., 1,* 1558(1976).
110. I. Haider and I. Oswald, *Br. Med. J., 2,* 318(1970).
111. A. N. Nicholson, et al., *Br. J. Clin. Pharmacol., 3,* 429(1976).
112. A. Kales, *Pharmacotherapy, 10,* 1(1990).
113. I. Hindmarch, *Arzneimittelforschung, 25,* 1836(1975).
114. A. N. Nicholson and B. M. Stone, *Br. J. Clin. Pharmacol., 3,* 543(1976).
115. L. K. Fowler, *J. Int. Med. Res., 5,* 295(1977).
116. H. Isozaki, et al., *Curr. Ther. Res., 20,* 493(1976).
117. T. Momose, et al., *Curr. Ther. Res., 19,* 277(1975).
118. L. F. Fabre, et al., *J. Int. Med. Res., 4,* 247(1976).

119. A. Kales, et al., *J. Clin. Pharmacol., 16,* 399(1976).
120. A. Kales, *Hosp. Pract. 25 suppl 3,* 7(1990).
121. A. N. Nicholson and B. M. Stone, *Br. J. Clin. Pharmacol., 13,* 433(1982).
122. R. C. Hill and T. V. A. Harry, *J. Int. Med. Res., 11,* 325(1983).
123. K. Adam and I. Oswald, *Br. J. Clin. Pharmacol., 17,* 531(1984).
124. K. Adam, et al., *Psychopharmacology, 82,* 389(1984).
125. J. Krieger, et al., *Drug Dev. Res., 3,* 143(1983).
126. K. S. Lachnit, et al., *Br. J. Clin. Pharmacol., 16,* 173S(1983).
127. G. W. Vogel and G. Vogel, *Br. J. Clin. Pharmacol., 16,* 103S(1983).
128. L. H. Sternbach, et al., in *Drugs Affecting the Central Nervous System,* Vol. 2, A. Burger, Ed., New York, Marcel Dekker, 1968, pp. 237–264.
129. H. D. Langtry and P. Benfield, *Drugs, 40,* 219(1990).
130. B. Musch and F. Maillard, *Int. Clin. Psychopharmacol., 5, suppl. 2,* 147(1990).
131. J. C. Blanchard, et al., *Life Sci., 24,* 2417(1979).
132. P. J. Chaudoir, et al., *J. Int. Med. Res., 11,* 333(1983).
133. C. E. Famewo and C. O. Odugbesan, *Can. Anaesth. Soc. J., 24,* 35(1977).
134. F. Roeth, et al., *Therapiewoche, 27,* 1462(1977).
135. G. A. Schoenenberger and M. Monnier, *Proc. Natl. Acad. Sci. U.S.A., 74,* 1282(1977).
136. J. Pappenheimer, et al., *J. Neurophys., 38,* 1299(1975).
137. H. Nagasaki, et al., *Proc. Japan Acad., 50,* 241(1974).
138. H. Braun, *Arzneipflanzen Lexikon,* Stuttgart, Gustav Fischer, 1979.
139. R. M. Russo, et al., *J. Clin. Pharmacol., 16,* 284(1976).
140. F. H. Clarke, Ed., *Annu. Rep. Med. Chem.,* Vol. 12, New York, Academic Press, 1977.

SUGGESTED READINGS

A. Borbely and J. L. Valatx, *Sleep Mechanisms,* Berlin, Springer, 1984.
P. B. Bradley, *Introduction To Neuropharmacology,* London, Wright, 1989.
A. G. Brown, *Nerve Cells and Nervous Systems,* London, Springer, 1991.
J. R. Cooper, et al., *The Biochemical Basis of Neuropharmacology,* 6th ed., New York, Oxford University Press, 1991.
Z. L. Kruk and C. J. Pycock, *Neurotransmitters and Drugs,* 3rd ed., London, Chapman & Hall, 1991.

Chapter 11

ANTICONVULSANTS

Julius A. Vida

As early as 2000 B.C., it was recognized that some people suffered from convulsive seizures. The term epilepsy, based on the Greek word epilambanein (meaning "to seize"), is first mentioned by Hippocrates. In the world's first scientific monograph on epilepsy, entitled *On the Sacred Disease* (ca. 400 B.C.), Hippocrates disputed the myth that the cause of epilepsy is supernatural and the cure magic. He described epilepsy as a disease of the brain, which should be treated by diet. At the same time, Hippocrates provided the first classification of epilepsy, which is still used. He distinguished true, or idiopathic, epilepsy (a disorder for which the cause is unknown) from symptomatic, or organic, epilepsy (a disorder resulting from a physiologic abnormality, such as brain injury, tumor, infection, intoxication, or metabolic disturbances).

Two opinions were put forward as to the causes of epilepsy. One was that epilepsy is a single disease entity, and all forms of it have a common cause. On the other hand, it was proposed that different types of epilepsy result from different chemical, anatomic, or functional disorders. At the Symposium on Evaluation of Drug Therapy in Neurologic and Sensory Disease, the general opinion was that "epilepsy is a symptom complex characterized by recurrent paroxysmal aberrations of brain functions, usually brief and self-limited."[1]

All forms of epilepsy originate in the brain. It is believed to be the result of changes in neuronal activity. These changes, such as an excessive neuronal discharge, may in turn be brought about by a disturbance of physicochemical function and electrical activity of the brain. The cause of this abnormality, however, is not clearly understood.

The most important property of the nerve cell is its excitability. It responds to excitation by generating an action potential, which may lead to repeated discharges. All normal neurons may become epileptic if subjected to excessive excitation. DeRobertis et al. list two possible mechanisms for convulsive disorders: a loss of the normal inhibitory control mechanism, and a chemical supersensitivity that increases excitability of neuronal elements.[2]

The origin of the seizures was established as early as the nineteenth century by Jackson.[3] According to him, an intense discharge of gray matter in various regions of the brain initiates the seizures. As a result, it is only a normal reaction of the brain to initiate convulsive seizures. The discharge of excessive electrical (nervous) energy has indeed been substantiated by brain-wave studies made possible by electroencephalography.

Attempts to classify epileptic seizures have been only partially successful, primarily because of limited knowledge of the pathologic processes of the brain. At the turn of the century, a classification of seizures had been published.[4] Even more attempts appeared in the last two decades.[5–9] In 1981, the Commission on Classification and Terminology of the International League Against Epilepsy put forward a new proposal.[10] The classification outlined in Table 11–1 is a short version of this proposal, which is based on clinical seizure type, ictal (seizure induced) electroencephalographic (EEG) expression, and interictal (occurring between attacks or paroxysms) EEG expression.

PARTIAL CONVULSIVE SEIZURES

In partial seizures, the initial neuronal discharge originates from a specific, limited cortical area. The EEG seizure patterns are restricted to one region of the brain, at least at the onset. Depending on the anatomic or functional system involved, several types of partial seizures can occur.

When the discharge does not propagate to circuits involved in consciousness, awareness and/or responsiveness are retained. When consciousness is not impaired, the seizure is classified as a simple partial seizure, usually with unilateral hemispheric involvement. When consciousness is impaired, the seizure is classified as a complex partial seizure, frequently with bilateral hemispheric involvement. The ictal electroencephalogram displays a burst of unusual rhythmic discharge of spikes and slow waves, more or less localized over one, sometimes both, hemispheres. The interictal electroencephalogram shows intermittent local discharges, generally over one hemisphere only. The seizures are related to a variety of local brain lesions, such as post-traumatic scar, tumor, or infection.

These types of seizures are possible at all ages but are most frequent in the elderly. Different types of seizures usually occur independently in different patients. All types of partial seizures, however, may develop into generalized seizures. The diagnosis of partial seizures is therefore difficult. The most frequently occurring seizures are the focal motor, jacksonian, autonomic, and psychomotor types. Computerized tomography (CT) scans and magnetic resonance imaging (MRI) of the head are used in virtually all patients with suspected epilepsy to identify the seizure type.

Focal motor attacks most commonly start in one hand,

Table 11–1. Classification of Epileptic Seizures

I. Partial (local, focal) seizures
  A. Simple (consciousness not impaired)
  B. Complex partial seizures (psychomotor seizures)
    1. Beginning as simple partial seizures, progressing to complex seizures
    2. With impairment of consciousness at onset
  C. Partial seizures evolving to secondarily generalized tonic-clonic convulsions
II. Generalized seizures (convulsive or nonconvulsive)
  A. Absence seizures
    Typical (petit mal)
    Atypical
  B. Myoclonic
  C. Clonic
  D. Tonic
  E. Tonic-clonic (grand mal)
  F. Atonic
III. Unclassified epileptic seizures (includes some neonatal seizures)

one foot, or one side of the face. Often, however, the onset of focal seizures is not specific. The tonic contraction of the involved muscles is followed by clonic movement.

In focal motor seizures without march, the clonic twitching and numbness are confined to a single limb or muscle group, usually in one side of the body. If the focal motor seizures spread to contiguous cortical areas, there may be an orderly sequence of repeated events (movement of hands, face, and legs) known as epileptic march. This type of seizure is called the jacksonian seizure. Consciousness is usually retained. If the unilateral movements characteristic of focal motor or jacksonian seizures steadily spread to the other half of the body, generalized seizures may follow.

Complex partial seizures usually originate from the temporal lobe as a localized discharge but quickly spread to both hemispheres of the brain. These seizures are common in adults and are often preceded by an aura, which is in reality a simple partial seizure. Most complex seizures are accompanied by automatism.

Seizures with automatism describe involuntary motor activity during the state of clouding of consciousness, either in the course of or after an epileptic seizure and usually followed by amnesia for the event.

Partial seizures respond fairly well to anticonvulsant drugs.

GENERALIZED CONVULSIVE SEIZURES

These convulsive disorders are characterized by seizures that are generalized from the start, regardless of symptoms and causes. It is not possible to single out one anatomic or functional system localized in one hemisphere of the brain that is responsible for the clinical symptoms. The first clinical changes indicate that both hemispheres are involved from the outset. Consciousness is usually impaired from the start. The seizures are accompanied by motor changes, which are usually generalized or at least bilateral and symmetric.

The seizures are often accompanied by sudden autonomic discharge. The initial neuronal discharge spreads quickly into the entire, or at least the greater part of, the gray matter. The EEG pattern consists of bilateral, essentially synchronous and symmetric discharges from the start and indicate the widespread nature of neuronal discharge. The cause is rarely known, but it is usually attributed to diffuse lesions, toxic and metabolic disturbances, or constitutional genetic factors. People of all ages are affected by generalized convulsions.

Absences

Both typical and atypical absences bring about brief loss of consciousness (usually less than 20 seconds). During the attack, the subject appears to stare, and any ongoing motor activity is interrupted. The subject is unresponsive to stimuli. After the brief interruption in consciousness, the activity that was in progress before the seizure is resumed. The individual usually has no memory of the seizure and no postictal confusion. Simple absences (petit mal or typical absences) usually occur in children. The electroencephalogram shows a rhythmic spike and wave pattern of 3 cps. If this pattern is replaced by irregular spike-and-slow wave complexes, fast activity, or other paroxysmal activity, the condition is classified as atypical absence (or petit mal variant).

Complex absences bring about other phenomena, too, such as mild clonic jerks (absences with mild clonic components); increase of postural tone (retropulsive absences with tonic components), diminution or abolition of postural tone (absences with atonic components), and automatism (absences with automatism or with autonomic components). Petit mal status is characterized by frequent absences. Approximately one third of the patients with absence seizures also have generalized tonic-clonic seizures. The occurrence of generalized tonic-clonic seizures in patients with absence attacks requires modification of the therapeutic regimen, which is usually different from that of patients who have only absence seizures.

Absences respond fairly well to anticonvulsant drugs. Oxazolidinediones (trimethadione and paramethadione), succinimides (methsuximide, phensuximide, ethosuximide), and sodium valproate are particularly effective.

Myoclonic Seizures

Myoclonic seizures consist of sudden involuntary muscle contractions that may be irregular or rhythmic, generalized, or limited to certain muscle segments. These seizures cause a prolonged loss of consciousness. Myoclonic jerks affect people of all ages and are electroencephalographically characterized by a polyspike and wave pattern of sharp and slow waves. Sudden flexor contractions of muscles in the arms may result in dropping objects held in the hand, and falling to the ground may occur when the legs are involved. Clonazepam or valproic acid are used most often to treat myoclonic seizures.

Generalized Clonic, Tonic, and Tonic-Clonic Seizures

Clonic seizures consist of more prolonged successive myoclonic contractions that are fast. The attacks may last up to 60 seconds. Clonic seizures, which occur especially in children, are characterized by a mixture of fast (10 or more cps) and slow waves with occasional spike and wave patterns (discharge). The loss of consciousness and postural tone is followed by bilateral clonic contractions. Tonic seizures also occur mainly in children and are characterized by a low-voltage fast activity or a fast rhythm (10 or more cps). Tonic seizures consist of extension of the extremities and rigid stretching of the body (axial tonic seizure). Tonic leg flexion may result in falling: tonic extension of lower extremities may keep the subject erect. Vocalization may occur as a result of contraction of thoracic muscles forcing air past the larynx. Autonomic changes may accompany these seizures.

The tonic-clonic seizures (grand mal) are less frequent in children than are the other forms of generalized seizures. Generalized tonic-clonic (GTC) seizures represent a maximal epileptic response of the brain. Like the others, the tonic-clonic seizures are best characterized by the great ease and speed with which the initial discharge spreads into other parts of the brain. The ictal electroencephalogram consists of massive, fast spiking rhythms at 10 or more cps, decreasing in frequency and increasing in amplitude during the tonic phase, interrupted by slow waves during the clonic phase. These seizures are characterized by tonic contraction of all muscle groups causing the patient to fall. The initial contraction may be flexor and is rapidly followed by prolonged extension. During the transition, rapid vibratory tremors may appear that change to violent flexor contractions and relaxation before reaching the clonic phase. The clonic contractions spread through the body, then slow until the seizure ends. Tonic-clonic seizures respond well to barbiturates, hydantoins, and carbamazepine. On the other hand, oxazolidinediones and succinimides are ineffective.

In atonic seizures, a sudden decrease in muscle tone occurs, leading to a head drop, drooping of a limb, or loss of all muscle tone. The person may slump to the ground and may suffer injury from projecting objects. When these attacks are brief, they are known as drop attacks. Electroencephalograms show polyspike and wave or flattening or low-voltage fast activity.

PHARMACOLOGIC TESTING

In 1975, the Epilepsy Branch of the National Institute of Neurological and Communicative Disorders and Stroke (NINCDS) established an anticonvulsant drug screening project that is still ongoing. Since 1975, several thousand chemical compounds have been tested for anticonvulsant activity and neurotoxicity.[11,12] During this program, a six-phase anticonvulsant drug-screening flow chart has been developed[13] (Table 11–2). This strategy of testing is widely accepted and supersedes previously used methodology.[14]

The strategy of testing involves determination of the anticonvulsant activity by a battery of tests. These tests detect, on one hand, the anticonvulsant activity resulting from the prevention of seizure spread (MES test and strychnine test) and, on the other hand, the anticonvulsant activity related to the elevation of seizure threshold (s.c. Met test, bicuculline test and picrotoxin test). In addition, a battery of toxicity tests is performed, designed to determine the minimal median neurotoxic dose, the 24-hour median lethal dose, and the profile of the overt toxic manifestations.

The sequence of testing consists of determination of the anticonvulsant activity in laboratory animals in electrically induced seizures and chemically induced seizures followed by determination of the acute toxicity.

Electrically Induced Seizures

Maximal Electroshock Seizure (MES) Test

Animals are stimulated at the previously determined time of peak effect (TPE), through corneal electrodes by a 60-cycle alternating current applied for 0.2 or 0.3 second.[15,16] In mice, a 50-mA and in rats a 150-mA current is usually used. The characteristics of electroshock seizures are a tonic limb flexion of 1 to 2 seconds followed by a tonic limb extension of roughly 10 to 12 seconds, and finally generalized clonic movements for 12 seconds. The total duration of the seizure is approximately 25 seconds. Only abolition of the hindlimb tonic-extensor spasm is recorded as the measure of anticonvulsant potency. The tonic component is considered abolished if the hindleg extension does not exceed a 90° angle with the plane of the body.

Following drug administration, the animals are challenged at intervals of ½, 1, 2, and 4 hours. The intervals at which the greatest percentage of animals are protected by the approximate ED_{50} dose are recorded as the time of peak effect (TPE). Abolition of the tonic-extensor component of the electroshock seizures indicates that the compound has the ability to prevent seizure spread and helps to select drugs likely to be effective in generalized seizures (except absences) and complex partial seizures.

Time of Peak Effect (TPE). During the phase-1 evaluation of a compound, the approximate dose levels for no anticonvulsant activity, some anticonvulsant activity, no neurotoxicity, and some neurotoxicity are detected.

The approximate ED_{50} (the dose of the compound that elicits an anticonvulsant response in 50% of the animals) is determined by giving four groups of animals (two animals in each group) the four dose levels of the compound just described and subjecting the animals to the appropriate anticonvulsant or neurotoxicity test(s) at ½, 1, 2, and 4 hours, respectively. The data obtained are plotted against time on linear graph paper. The TPE is obtained by visual inspection of the graph.

Median Effective Dose (ED_{50}). Using one of the anticonvulsant procedures, eight animals are injected with the dose used in the determination of the TPE. The percent of animals protected is plotted against the dose. This procedure is repeated with a different dose until at least three points have been established between the dose level that protects 0% of the animals and the dose level that protects 100% of the animals.

These data are then subjected to statistical analysis and

Table 11–2. Flow Chart for Anticonvulsant Drug Screening*

- Phase 1. Anticonvulsant identification
  - a. Mice I.P.
    - Dose range: 30, 100, and 300 mg/kg
    - Tests: MES, scMet, rotorod, and general behavior
    - Time of test: $\frac{1}{2}$ and 4 hr
  - b. Rats P.O.
    - Dose: 50 mg/kg
    - Tests: MES or scMet and minimal neurotoxicity
    - Time of test: $\frac{1}{4}$, $\frac{1}{2}$, 1, 2, and 4 hr
- Phase 2. Anticonvulsant quantification
  - a. Mice I.P.
    - TPE: MES, scMet, rotorod
    - ED_{50}: MES, scMet
    - TD_{50}: rotorod
  - b. Mice P.O.
    - TPE: MES, scMet, rotorod
    - ED_{50}: MES, scMet
    - TD_{50}: rotorod
  - c. Rats P.O.
    - TPE: MES, scMet, minimal neurotoxicity
    - ED_{50}: MES, scMet
    - TD_{50}: minimal neurotoxicity
- Phase 3. Anticonvulsant drug differentiation
  - a. Mice I.P.
    - ED_{50}: scBic
    - ED_{50}: scPic
    - ED_{50}: scStr
  - b. Mouse whole brain (in vitro)
    - BDZ receptor binding
    - GABA receptor binding
    - Adenosine uptake
  - c. Mice I.P.
    - Timed i.v. infusion of Metrazol
  - d. Rats P.O.
    - TPE: EST (kindled)
    - ED_{50}: EST (kindled)
- Phase 4. Toxicity profile, mice I.P.
  - Behavior induced by $1TD_{50}$, $2TD_{50}$s, and $4TD_{50}$s
    - TPE: loss of righting reflex
    - HD_{50}: loss of righting reflex
- Phase 5. Subchronic administration and overt tolerance studies
- Phase 6. Pharmacodynamic interactions (with prototype drugs), mice I.P.
  - TPE: MES, scMet, rotorod
  - ED_{50}; MES, scMet
  - TD_{50}; rotorod

* The various neurotoxicity, anticonvulsant, and liver studies listed have been integrated into an anticonvulsant drug development procedure designed to identify, quantify, and evaluate the anticonvulsant potential and metabolic interactions of candidate chemical substances. MES, maximal electroshock seizure test; scMet, subcutaneous pentylenetetrazol (Metrazol) seizure threshold; I.P., intraperitoneally; P.O., per os (orally); TPE, time of peak effect; ED_{50}, median effective dose; TD_{50}, median toxic dose; scBic, subcutaneous bicuculline test; scPic, subcutaneous picrotoxin test; scStr, subcutaneous strychnine test. From E.A. Swinyard, et al., *General Principles, Experimental Selection, Quantification and Evaluation of Anticonvulsants* in *Antiepileptic Drugs,* Third edition, R. Levy, et al., Eds., New York, Raven Press, 1989, p. 88.

the ED_{50}, 95% confidence interval, and slope of the regression line are recorded in the experimental protocol.

The median toxic dose (TD_{50}), median hypnotic dose (HD_{50}), and 24-hour median lethal dose (LI_{50}) are determined in a similar fashion.

Chemically Induced Seizures

Subcutaneous Strychnine Seizure Pattern Test

The CD_{97} of strychnine (1.20 mg/kg) is injected subcutaneously in a volume of 0.01 ml/g body weight into each of a number of mice at the previously determined TPE.[17] The mice are placed in isolation cages and observed for 30 minutes for the presence or absence of the hindleg tonic extensor component of the seizure. Abolition of the hindleg tonic extensor component is taken as the endpoint for this test and indicates that the test substance has the ability to prevent seizure spread.

Subcutaneous Metrazole Seizure Threshold (sc Met) Test

A pentylenetetrazole (Metrazole) dose of 85 mg/kg subcutaneously in mice causes seizures in more than 97% of animals.[15] This is called the convulsive dose 97, or CD_{97}. In rats, the CD_{97} is 70 mg/kg. This test measures the ability of anticonvulsant drugs to provide complete protection against seizures induced by subcutaneous injection of a CD_{97} of pentylenetetrazole. A threshold convulsion is defined as one episode of clonic spasms that persists for at least 5 seconds. The ability of the compound to elevate the pentylenetetrazole seizure threshold is determined by the absence of clonic seizures during a 30-minute period. In this test, succinimides and oxazolidinediones are particularly effective. Because these drugs are effective in the treatment of absences, the assumption is that scMet tests indicate drugs likely to be active against absences.

Subcutaneous Bicuculline Seizure Threshold Test

The CD_{97} of bicuculline (2.70 mg/kg) is injected subcutaneously in a volume of 0.01 ml/g body weight into each of several mice at the previously determined TPE. The mice are placed in isolation cages and observed for the next 30 minutes for the presence or absence of a threshold convulsion. Absence of a threshold convulsion indicates that the test substance has the ability to elevate the bicuculline seizure threshold.

Subcutaneous Picrotoxin Seizure Threshold Test

The CD_{97} of picrotoxin (3.15 mg/kg) is injected subcutaneously in a volume of 0.01 ml/g body weight into each of several mice at the previously determined TPE.[18] The mice are placed in isolated cages and observed for the next 45 minutes for the presence or absence of a threshold convulsion. Absence of a threshold convulsion is taken as the endpoint and indicates that the test substance has the ability to elevate the picrotoxin seizure threshold.

Determination of Acute Toxicity

Acute toxicity induced by a chemical compound in laboratory animals is usually characterized by some type of neurologic abnormality. In mice, the endpoint for the toxicity is measured by the rotorod test and the righting reflex. In rats, the endpoint for the toxicity is measured by the rotorod test, positional sense test and, gait and stance test. If the rat fails to perform normally on at least two of the three tests, a neurologic deficit exists.

Rotorod Test. A normal mouse maintains its equilibrium for a long time when placed on a rod that rotates at a speed of 6 rpm.[19] Neurologic deficit is shown by the mouse if it fails to maintain its equilibrium for 1 minute on this rotation rod in each of three trials.

Positional Sense Test. If the hind leg of a normal mouse or rat is gently lowered over the edge of a table, the animal quickly lifts its leg back to a normal position. Neurologic deficit is displayed if the animal fails to rapidly correct such an abnormal position of the limb.

Gait and Stance Test. Neurologic deficit is indicated by a circular or zigzag gait, ataxia, abnormal spread of the legs, abnormal body posture, tremor, hyperactivity, lack of exploratory behavior, somnolence, stupor, or catalepsy, among other signs.

Muscle Tone Test. Normal animals have a certain amount of skeletal muscle tone that is apparent to the experienced handler. Neurologic deficit is indicated by the loss of skeletal muscle tone characterized by flaccidity or hypotonia.

Righting Test. If a mouse or rat is placed on its back, the animal will quickly right itself and assume a normal posture. Neurologic deficit is indicated by the animal's inability to rapidly correct for the abnormal body posture. This test is also used as a measure of the median hypnotic dose (HD_{50}).

Receptor Binding Studies

Crude synaptic membranes are prepared by the method described by Enna and Snyder.[20] Tritiated flunitrazepam binding studies as measure of benzodiazepine binding follow Braestrup and Squires' method[21] and GABA receptor binding studies are performed by a method described by Zukin et al.[22]

Adenosine Uptake Studies

Crude synaptic membranes are prepared by the method described by Cotman,[23] and tritiated adenosine uptake is determined by the method of Phillis et al.[24]

Timed Intravenous Infusion of Metrazol

The minimal seizure threshold in each animal is determined by recording the time in seconds from the start of the infusion of a convulsant solution (0.5% Metrazol in 0.9% sodium chloride containing 10 USP units/ml of heparin sodium) to the appearance of the first twitch and onset of clonus). The infusion into the tail vein is set at a constant rate of 0.37 ml/min. The mean doses and standard errors (first twitch and clonus) are obtained by converting the values to mg/kg of Metrazol for each experimental and control animal.

Kindled Rats

Drugs that increase minimal seizure threshold are evaluated in electroshock-kindled rats. At least 24 hours after the rats have been kindled (2 sec stimulation, 8 mA, 60 Hz, corneal electrodes, twice daily for 5 to 7 days) to evoke seizures as described by Racine,[25] the test substance is administered orally. At the previously determined TPE, each animal subjected to the indicated electrical stimulation will display the presence or absence of forelimb clonus. Abolition of the forelimb clonus is taken as the endpoint, which indicates that the test substance has the ability to elevate seizure threshold in the kindled animal.

Acute Toxicity Profile

Six mice are randomly divided into three groups of two mice each, and each group is given a dose equivalent to either 1 TD_{50}, TD_{50}s, or 4 TD_{50}s of the test substance.[25] A comprehensive assessment of the symptoms of toxicity is made 10, 20, and 30 minutes and 1, 2, 4, 6, 8, and 24 hours after administration of the test substance.

This test provides information on the overt signs and symptoms of acute toxicity induced by a chemical compound and also provides the approximate level of the median effective hypnotic (HD_{50}) and median lethal (LD_{50}) doses of the test substance.

Correlation of Laboratory Test Results with Clinical Use

As shown in Table 11–3, all three clinically useful drugs (phenytoin, carbamazepine, and valproate) in generalized tonic-clonic seizures and complex partial seizures prevent seizure spread (anti-MES activity) and may (valproate) or may not (phenytoin and carbamazepine) increase Metrazol seizure threshold. On the other hand, the three drugs effective in generalized absence seizure (ethosuximide, valproate, and clonazepam) increase Metrazol seizure threshold and may (valproate) or may not (ethosuximide and clonazepam) have an effect on seizure spread (anti-MES activity). The two drugs useful in myoclonic seizures (valproate and clonazepam) are effective in all three threshold tests (scMet, scBic, and scPic). The broad anticonvulsant profile of valproate, being the only one of five drugs effective by all five tests, correlates well with its wide spectrum of clinical efficacy.

STRUCTURES OF ANTICONVULSANT DRUGS

The primary use of anticonvulsant drugs is in the prevention and control of epileptic seizures. According to Toman,[26] the ideal antiepileptic drug, among other things, should completely suppress seizures in doses that do not cause sedation or other undesired central nervous system toxicity. It should be well tolerated and highly effective against various types of seizures, and devoid of undesirable side effects on vital organs and functions. Its onset of action should be rapid after parenteral injection for control of status epilepticus, and it should have a long duration of effect after oral administration for prevention of recurrent seizures.

The first effective remedy, potassium bromide, was introduced by Locock[26] in 1857. This agent was largely replaced by phenobarbital in 1912 when Hauptmann[28] tried this sedative in epilepsy. Its great value was recognized at once, and it is still one of the best drugs available.

The usefulness of both bromide and phenobarbital in convulsive disorders was discovered by chance, but phenytoin was developed in 1937 as the result of a study of potential anticonvulsant drugs in animals by Putnam and Merritt.[29,30] It is highly effective in man and is nonsedative.

Treatment of convulsive disorders by using bromide, phenobarbital, and phenytoin constitutes an important advance in clinical therapy. Since 1937, many other drugs have been introduced with the hope that they might be anticonvulsant, nonsedative, and nontoxic (Table 11–4). Most of the anticonvulsant drugs contain the ureide structure (Table 11–5). In addition, phenacemide (phenyl-

Table 11–3. Correlation between the Profile of Anticonvulsant Activity in Laboratory Animals* and the Clinical Use in Man

Antiepileptic Drug	Seizure Spread		Seizure Threshold			Clinical Use†
	MES	scStr	scMet	scBic	scPic	
Phenytoin	+	±	–	–	–	GTC, CP
Carbamazepine	+	±	–	–	+	GTC, CP
Valproate	+	+	+	+	+	GA, GTC, CP myoclonic
Ethosuximide	–	±	+	±	+	GA
Clonazepam	–	–	+	+	+	GA, myoclonic

* Mice; drugs administered I.P.

† Clinical spectrum: GTC, generalized tonic-clonic seizures; CP, complex partial seizures; GA = generalized absence seizures. MES, maximal electroshock seizure test; scStr, subcutaneous strychnine test; scMet, subcutaneous pentylenetetrazol (Metrazol) seizure threshold test; scBic, subcutaneous bicuculline test; scPic, subcutaneous picrotoxin test, +, effective in nontoxic doses; ±, effective in minimal doses; –, ineffective. From E.A. Swinyard, J.H. Woodhead, H.S. White, and M.R. Franklin, *General Principles, Experimental Selection, Quantification and Evaluation of Anticonvulsants in Antiepileptic Drugs,* Third edition, R. Levy, R. Mattson, B. Meldrum, J.K. Penry, and F.E. Dreifuss, Eds., New York, Raven Press, 1989, p. 100.

Table 11–4. Antiepileptic Drugs Marketed in the United States

Year Introduced	International Nonproprietary Name	U.S. Trade Name	Company
1912	Phenobarbital	Luminal	Winthrop
1935	Mephobarbital	Mebaral	Winthrop
1938	Phenytoin	Dilantin	Parke-Davis
1946	Trimethadione	Tridione	Abbott
1947	Mephenytoin	Mesantoin	Sandoz
1949	Paramethadione	Paradione	Abbott
1950	Phenthenylate*	Thiantoin	Lilly
1951	Phenacemide	Phenurone	Abbott
1952	Metharbital	Gemonil	Abbott
1952	Benzchlorpropamide†	Hibicon	Lederle
1953	Phensuximide	Milontin	Parke-Davis
1954	Primidone	Mysoline	Ayerst
1957	Methsuximide	Celontin	Parke-Davis
1957	Ethotoin	Peganone	Abbott
1960	Aminoglutethimide‡	Elipten	Ciba
1960	Ethosuximide	Zarontin	Parke-Davis
1968	Diazepam§	Valium	Roche
1974	Carbamazepine	Tegretol	Geigy
1975	Clonazepam	Clonopin	Roche
1978	Valproate	Depakene	Abbott
1981	Clorazepate§	Tranxene	Abbott
1993	Felbamate‖	Felbatol	Wallace
1994	Gabapentin	Neurontin	Parke-Davis

* Withdrawn from the market in 1952.
† Withdrawn from the market in 1955.
‡ Withdrawn from the market in 1966.
§ Approved by FDA as an adjunct.
‖ FDA may request withdrawal because of several cases of aplastic anemia caused by felbamate.

Table 11–5. Structure of Anticonvulsant Drugs Containing the Ureide Structure

R_1 R_2, O, N, O, X

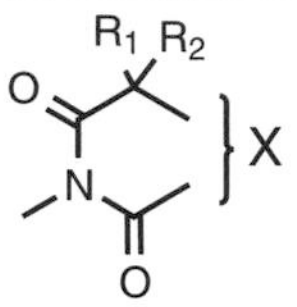

Ureide Structure

Class of Compounds	X
Barbiturates	O, NH
Hydantoins	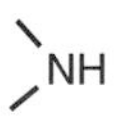 NH
Glutarimides	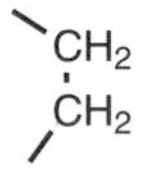 CH_2–CH_2
Oxazolidinediones	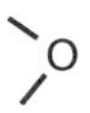 O
Succinimides	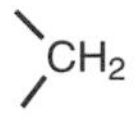 CH_2

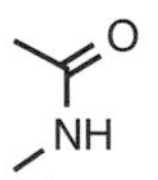

acetylurea), an open-chain compound lacking a heterocyclic ring common to the other anticonvulsants already listed, also possesses the ureide structure.

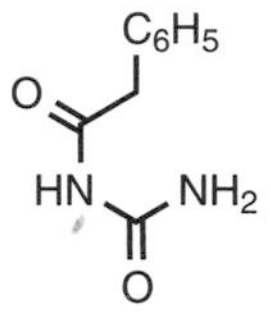

C_6H_5, O, HN, NH_2, O

Phenacemide

Other anticonvulsant drugs not possessing the ureide structure are primidone, valproic acid, benzodiazepines, and carbamazepine.

The application of the ureide drugs in various kinds of epilepsies is shown in Figure 11–1. This illustration is based on clinical therapy.

Barbiturates

Although many barbiturates display sedative-hypnotic activity, only a few have anticonvulsant properties. Paradoxically, many barbiturates cause convulsions at larger doses. Phenobarbital occupies a special place among the anticonvulsant barbiturates. Mephobarbital and metharbital are also anticonvulsants in clinical use.

Phenobarbital (5-ethyl-5-phenylbarbituric acid) is the drug used most commonly for convulsive disorders and

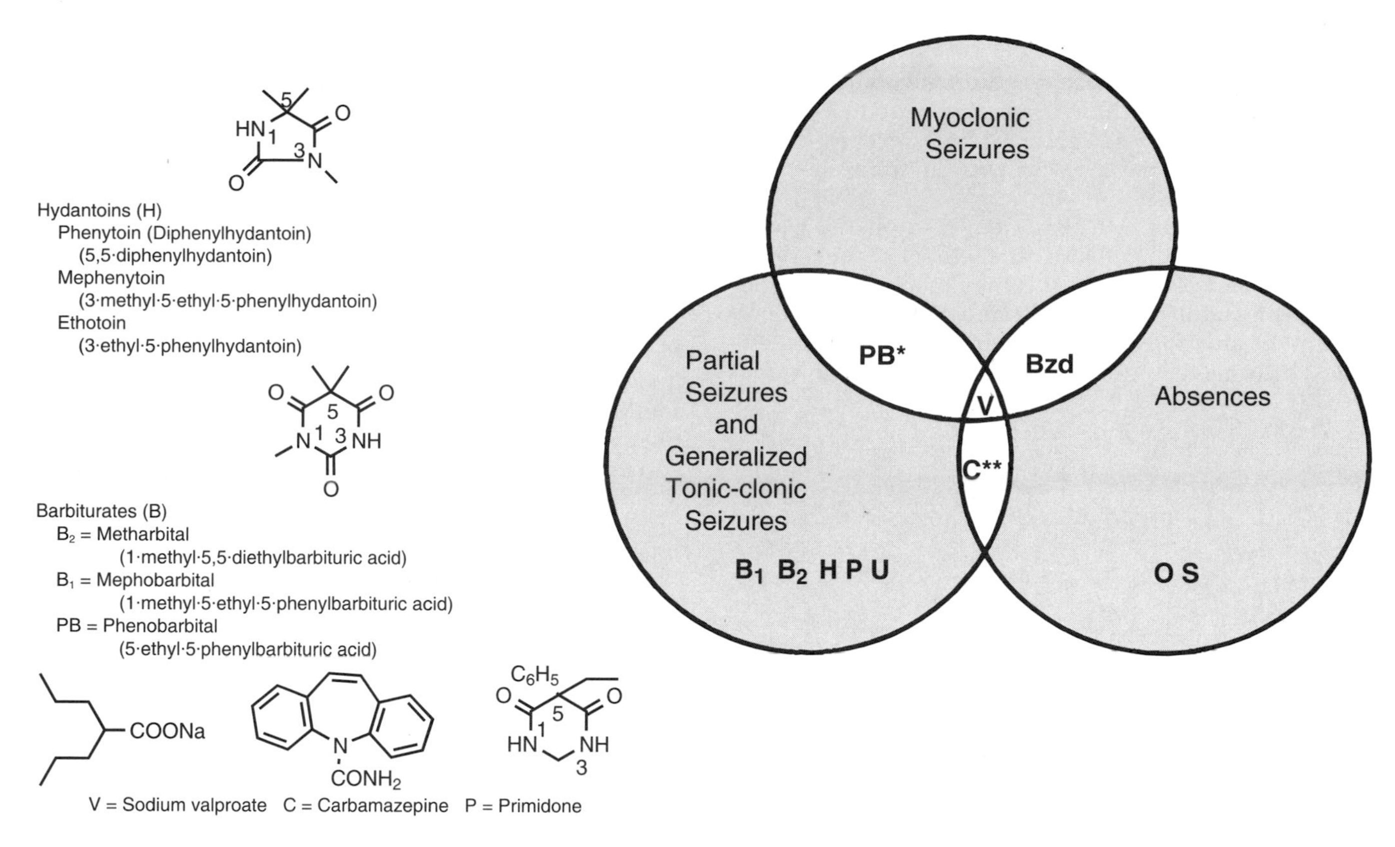

Fig. 11–1. Drug applications in various types of epilepsies. *, J. K. Penry, *Epilepsy, Diagnosis Management Quality of Life,* New York, Raven Press, 1986, p. 17; **, For treatment of absences, carbamazepine is used only in combination with ethosuximide. From R. J. Porter in *New Anticonvulsant Drugs,* B. S. Meldrum and R. J. Porter, Eds., London, John Libbey, 1986, p. 10.

is the drug of choice for infants and young children. The usual dose is 1 to 5 mg/kg daily, and the therapeutic plasma concentration range is 10 to 40 μg/ml with a peak time of 1 to 3 hours. Drowsiness is the most common side effect. The sedative-hypnotic effect of phenobarbital limits its use in older children and adults. Phenobarbital, like the other anticonvulsant barbiturates, exhibits anticonvulsant effects only in doses that usually cause some sedation. Tolerance to the sedative effects often develops, however. The anticonvulsant effects are nonselective. In laboratory animals, phenobarbital is effective by several tests in nontoxic doses. It is active against electrically induced seizures (MES), and it elevates the threshold for chemical (sc Met) stimulation.

Phenobarbital is used clinically, mainly in most generalized and partial seizures, but it is also used in absences, as indicated in Figure 11–1. This drug displays protective indices (the ratio of TD_{50} and ED_{50}) of 5 to 8 by the various test methods. Phenobarbital is distinguished by its long plasma half-life, 3 to 6 days, yielding an extremely stable plasma concentration. A good correlation exists between phenobarbital doses and serum concentrations. There are wide individual differences, however, between serum concentrations and anticonvulsant effectiveness on one hand and observed toxicity on the other hand. It is possible to control seizures with barbiturate serum concentrations below 10 μg/ml, at which levels usually no sedative toxicity is observed. Symptoms in most patients are controlled with barbiturate serum levels of 10 to 30 μg/ml, at which concentrations low levels of sedative are observed. As concentrations exceed 30 μg/ml, sedation is increasingly encountered.

Phenobarbital is metabolized primarily by hydroxylation to 5-p-hydroxyphenyl-5-ethyl-barbituric acid, an inactive metabolic product. That is conjugated with glucuronic acid and sulfuric acid and excreted in the urine as glucuronide or sulfate. Phenobarbital is a potent liver enzyme-inducing agent and increases the ability of the liver to metabolize all drugs normally metabolized in the microsomal fraction. Phenobarbital, therefore, increases the biotransformation of many other drugs taken concurrently.

Mephobarbital (1-methyl-5-ethyl-5-phenyl-barbituric acid) is used in the treatment of generalized and partial seizures. The first step of its metabolism involves loss of the N-methyl group, and it is converted to phenobarbital. No evidence exists that it is more effective or less sedating than phenobarbital in equivalent doses.

Metharbital (1-methyl-5,5-diethylbarbituric acid) is largely demethylated to barbital (5,5-diethylbarbituric acid) in the body, and its effects are probably attributable to barbital. Because metharbital is less effective and more sedating than phenobarbital, it is used as a reserved drug in generalized seizures.

The mode of action of barbiturates is described in Chapter 10.

Hydantoins

The hydantoins were tested as anticonvulsants by Merritt and Putnam.[31,32] These drugs suppressed electrically induced convulsions in animals but were ineffective against convulsions induced by pentylenetetrazole, picrotoxin, or bicuculline.

Phenytoin (diphenylhydantoin, 5,5-diphenyl-2,4-imidazolidinedione), the most active member of the series, is an effective anticonvulsant. It is the first choice for older children and adults and is being challenged by carbamazepine as the alternative to phenobarbital for younger children. The usual daily dose is 3 to 10 mg/kg, administered in one or two doses. Clinically, it is most effective in controlling tonic-clonic (grand mal), clonic, and tonic seizures. Phenytoin is also widely used in partial (psychomotor) seizures. Phenytoin is not used in absence seizures because it may increase their frequency.

The absorption of phenytoin is limited by the extremely low solubility of the un-ionized form in the gastrointestinal fluids and in the plasma. Absorption is a nonlinear saturation process that depends, on one hand, on the passage of phenytoin from the stomach to the duodenum, where a large portion of the drug is present in the ionized form, and on the other hand on the rate at which it can enter the blood. The ionized phenytoin is highly soluble in the intestinal fluid. After entering the blood, phenytoin is removed from the blood by storage in fat and tissues, where it is bound to proteins and phospholipids. Some of the phenytoin is also bound to the plasma proteins, thereby decreasing the amount of free phenytoin in the blood that could enter the tissues.

As with phenobarbital, the efficacy of phenytoin may be encountered at various plasma levels. Toxicity is usually not observed below 10 μg/ml plasma levels, but it increases in severity with increasing serum concentrations that may be necessary to control the seizure. Because phenytoin does not possess sedative properties, it was initially considered less toxic than phenobarbital. The wide range of toxic effects (gastric distress, nausea, vertigo, ataxia, nystagmus, neuropathy, gingival hyperplasia, and occasionally megaloblastic anemia) associated with higher plasma levels of phenytoin (therapeutic plasma concentration ranges from 10 to 20 μg/ml) has resulted in increased uses of carbamazepine and valproic acid in place of phenytoin. The average plasma half-life of phenytoin after oral administration is 24 hours, with a peak time of 4 to 8 hours. It is rapidly detoxified in the liver, mainly to 5-(p-hydroxyphenyl)-5-phenylhydantoin, which is conjugated with glucuronic acid and excreted in the urine. Because this metabolite does not possess anticonvulsant activity, it is believed that phenytoin is the active anticonvulsant agent in the body. Other metabolic products are also formed by hydroxylation of one or both phenyl groups in meta and para positions.

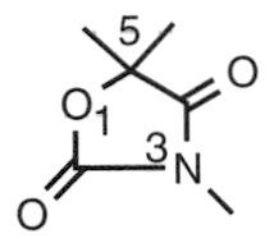

Oxazolidinediones (O)

O_1 = Trimethadione (3,5,5-trimethyloxazolidine-2,4-dione)

O_2 = Paramethadione (3,5-dimethyl-5-ethyloxazolidine-2,4-dione)

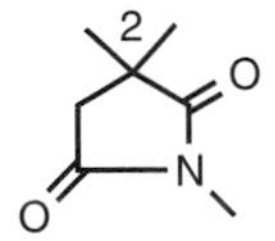

Succinimides (S)

S_1 = Methsuximide ($N_1$2-dimethyl-2-phenylsuccinimide)

S_2 = Phensuximide (N-methyl-2-phenylsuccinimide)

S_3 = Ethosuximide (2-ethyl-2-methylsuccinimide)

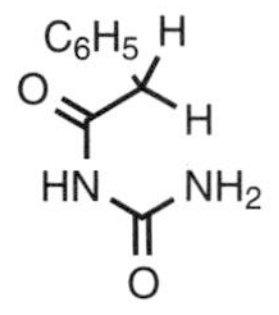

U = Phenacemide

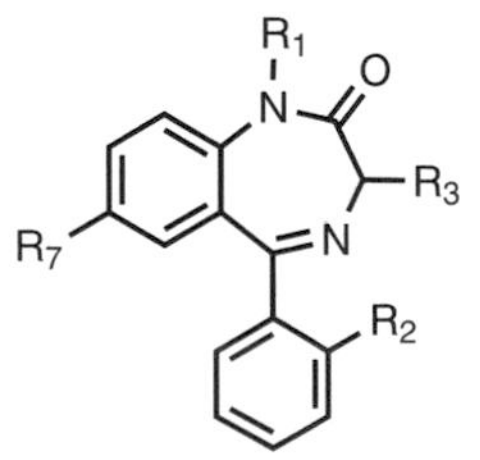

Benzodiazepines (Bzd)

	R_1	R_2	R_3	R_7
Bzd_1=Clonazepam	H	Cl	H	NO_2
Bzd_2=Clorazepate	H	H	COOK	Cl
$BzZD_3$=Diazepam	CH_3	H	H	Cl

Mephenytoin (3-methyl-5-ethyl-5-phenyl-hydantoin) is rapidly metabolized by N-demethylation to 5-ethyl-5-phenylhydantoin, a relatively toxic substance that may produce hepatitis, pancytopenia, exfoliative dermatitis, and aplastic anemia. As a result, mephenytoin has a reserve drug status used only when phenytoin has failed. The combined use of mephenytoin and phenytoin in lower doses of each, however, has lead to improvement in seizure control without side effects.[33]

Ethotoin (3-ethyl-5-phenylhydantoin) is less toxic but produces greater sedation than phenytoin. Although the other side effects of ethotoin are less severe than those of either phenytoin or mephenytoin, the clinical impression is that ethotoin is minimally efficacious and must be given in large doses, a minimum of four times a day. Therefore, ethotoin is not considered a primary anticonvulsant agent.

Mode of Action

Unlike trimethadione and phenobarbital, hydantoins do not suppress activity at the primary focus or protect against pentylenetetrazole-induced seizures. The predominant effect of hydantoins is to reduce the spread of seizure discharge from the focus to neurons initially not involved.[34]

The neurophysiologic effects of phenytoin include elevation of membrane potential, elevation of threshold, reduction of conduction velocity, and reduction or abolition of repetitive firing. Post-tetanic potentiation in synaptic transmission (caused by rapid repetitive presynaptic stimulation) is strongly inhibited by phenytoin.

In excitable tissues (e.g., peripheral nerve) phenytoin exerts a membrane-stabilizing effect[35] because it prevents or interrupts repetitive electrical activity. This feature is attributable to a decreased sodium influx caused by phenytoin,[36] which blocks the sodium channels in a manner similar to that of local anesthetics. Phenytoin also decreases intracellular calcium[37] by limiting membrane permeability to calcium, which in turn blocks the release of excitatory transmitters from presynaptic vesicles. In parallel to its sodium transport inhibitory effect, phenytoin also inhibits protein synthesis[38] and neurotransmitter synthesis.[39]

Phenytoin also exerts a GABA-like effect on presynaptic and postsynaptic membrane to enhance presynaptic and postsynaptic inhibition.[40]

In nonexcitable tissues (e.g., epithelial tissues of the intestinal mucosa, choroid plexus, glia, salivary gland, etc.), however, phenytoin increases the permeability of the basal cell membrane to sodium, which leads to an increase in the activity of the sodium pump.[41] In high doses, phenytoin is excitatory; this may be ascribed to increased transmitter release that, in turn, is caused by inhibition of the uptake of intracellular calcium by mitochondria with the net result of increased intracellular calcium concentration.[37] In addition, a net accumulation of the excitatory neurotransmitters acetylcholine and glutamic acid may take place, depending on the proper combination of factors. An increase in acetylcholine and glutamic acid levels may account for some of the excitatory effects of phenytoin (e.g., potentiation of pentylenetetrazole, strychnine, and picrotoxin-induced convulsions.)

Like phenobarbital, primidone and carbamazepine, agents that block MES seizures and are effective in generalized tonic-clonic seizures, phenytoin inhibits both cyclic AMP and cyclic GMP accumulation produced by veratridine interfering with their depolarizing (excitatory) activity.[42]

In conclusion, phenytoin displays both excitatory and inhibitory activities. The net result is either excitation (at high doses of phenytoin) if excitatory pathways predominate over inhibitory ones or inhibition (anticonvulsant effect) at therapeutic doses, if inhibitory pathways predominate over excitatory pathways.

The possible mechanisms of action of phenytoin have been reviewed.[43]

Chemistry

The Bucherer reaction involves use of a ketone (or aldehyde), ammonium carbonate, and potassium cyanide (usually in a mole ratio of 1:3:2).

$$RR'C{=}O \xrightarrow[(NH_4)_2CO_3]{KCN} [R'-C(R)(CONH_2)-NHCOONH_4] \rightarrow \text{hydantoin}$$

Cyanoacetamides can be oxidized with sodium hypobromite to produce hydantoins through the intermediate isocyanates.

$$R'-C(R)(CONH_2)-CN \rightarrow [R'-C(R)(CONH_2)-CONH_2] \rightarrow [R'-C(R)(CONH_2)-NCO] \rightarrow \text{hydantoin}$$

intermediate isocyanates

Hydroxy acids, on condensation with urea, produce hydantoins.

$$R'-C(R)(OH)-COOH + NH_2CONH_2 \rightarrow \text{hydantoin}$$

Aminonitriles, which are readily obtained from aldehydes and primary amines, are condensed with cyanates to yield 1-substituted hydantoins.

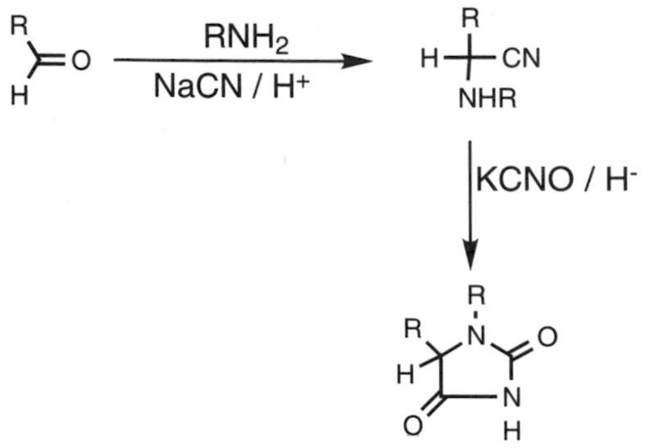

Oxazolidinediones

These compounds were introduced into anticonvulsant therapy between 1946 and 1948. At that time, no effective drugs were available to control absences (petit mal disorders). Acceptance of trimethadione in 1946 and paramethadione in 1948, therefore, was rapid.

Trimethadione (3,5,5-trimethyloxazolidine-2,4-dione) has been widely used to control absences.[44–49] It is ineffective against other generalized major (grand mal) and partial seizures. In animal tests, it is characterized by its effectiveness in the pentylenetetrazole test and its limited value in the electroshock seizure test (MES). Trimethadione can cause nephrosis, aplastic anemia, and bone marrow depression. Monthly urinalysis for evidence of nephrosis and frequent blood counts therefore are essential. Other adverse effects include photophobia, hemeralopia (light blindness), and drowsiness. Because of the many possible side effects, trimethadione is now used only in patients who do not respond to or cannot tolerate a succinimide drug or valproic acid.

Succinimides

Because oxazolidinediones are toxic, an extensive search was undertaken to replace them with less toxic drugs. The succinimides were introduced in 1951 and were widely accepted for the treatment of the absence condition. Phensuximide (N-methyl-2-phenylsuccinimide), methsuximide (N,2-di-methyl-2-phenylsuccinimide), and ethosuximide (2-ethyl-2-methylsuccinimide) were introduced between 1951 and 1958.

Ethosuximide is the most effective and least toxic in the treatment of absence seizures, but it can cause drowsiness, ataxia, rashes, rarely, and hepatic or renal dysfunction. In addition to being less toxic than trimethadione, ethosuximide offers a wider range of protection against different kinds of absences. It is also used in combination with other drugs without adversely affecting other generalized seizures in patients with mixed seizure patterns. Unlike ethosuximide, trimethadione may exacerbate some generalized seizures. Methsuximide is used in the control of absences and some partial seizures.

Succinimides are readily absorbed from the gastrointestinal tract and reach maximal blood concentration in 3 to 7 hours. Ethosuximide is administered at a starting dose of 20 mg/kg, which may be increased to a 35 mg/kg level. A maintenance therapy, symptoms in most patients are controlled with ethosuximide serum levels of 40 to 100 μg/ml. The drugs are evenly distributed through tissues and body fluids and are metabolized via N-demethylation, aromatic and aliphatic hydroxylation, and ring cleavage. The carboxylic acid formed during metabolism is excreted in the urine.

When the succinimides were subjected to the battery of animal tests used in evaluation of anticonvulsants, it was found that they are particularly effective in the scMet seizure threshold test. The precise mechanism of action of succinimides is unknown. It has been postulated that ethosuximide enhances inhibitory processes in the brain, perhaps by some effect on specific inhibitory neurotransmitter systems.[50]

Chemistry

Condensation of an aldehyde or ketone with ethylcyanoacetate yields an unsaturated cyanoester. Michael addition of hydrogen cyanide and hydrolysis of the dicyanide to a dicarboxylic acid is followed by formation of the succinimide ring.

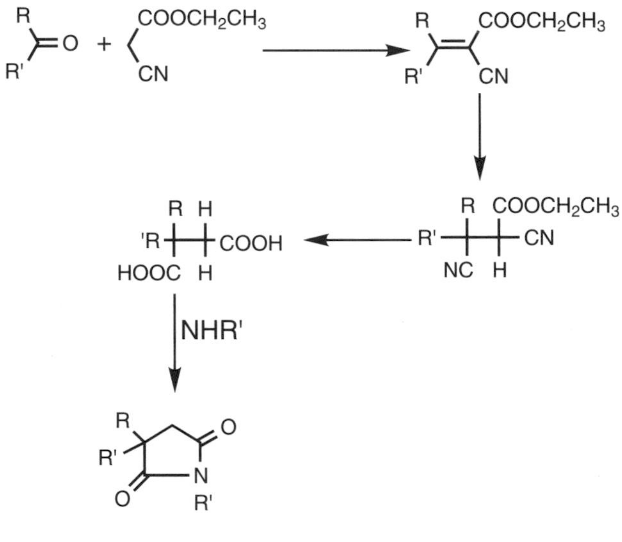

Acylureas

In the search for potent anticonvulsants, acylureas were investigated. These compounds are derived from barbiturates by cleavage of the ring. The most active member of this series, phenacemide, however, causes serious adverse effects, such as severe liver injury, psychoses, and agranulocytosis; as a result, the compound is not used often.[51]

Primidone

This compound is the 2-desoxy analog of phenobarbital. It has been demonstrated that primidone displays independent anticonvulsant activity, and the spectrum of activity and toxicity is different from that of phenobarbital. In laboratory animals, primidone is more effective in the MES and less effective in the scMet test than phenobarbital. Most importantly, primidone is less toxic than phenobarbital.

Primidone is extensively metabolized to phenylethylmalondiamide[52,53] (PEMA) and phenobarbital PEMA has a half-life of 40 hours in epileptic patients and possesses weak anticonvulsant activity in experimental animals. PEMA, however, potentiates the anticonvulsant activity of phenobarbital in concentrations that have no other demonstrable activity.[54] The metabolic conversion of primidone to phenobarbital by hepatic microsomal oxidase enzyme functions is relatively slow but can be induced by other anticonvulsants (e.g., phenytoin and carbamazepine). The plasma half-life of primidone is about 14 hours in patients receiving primidone alone and about 6½ hours in patients also receiving other anticonvulsants. If primidone is given alone, the steady-state ratio or primidone to phenobarbital is about 1:1. If primidone is administered in conjunction with diphenylhydantoin or carbamazepine, the steady-state ratio of primidone to phenobarbital is about 1:3. The daily dose is 8 to 20 mg/kg and the time to peak plasma level is 2 to 4 hours. Therapeutic steady-state levels are 5 to 15 μg/ml, reached within 2 to 4 days. In addition, a therapeutic steady-state level of 15 to 35 μg/ml of phenobarbital, a major metabolite of primidone, is obtained in 15 to 20 days after the therapeutic dose has been attained.

The presence or absence of phenobarbital in the plasma can be correlated with the toxic reactions associated with primidone. When primidone is administered to patients who have not been receiving phenobarbital, acute toxic symptoms consisting of dizziness, nausea, drowsiness, vomiting, and double vision are observed. Tolerance to this syndrome develops when treatment continues and the dose of primidone is increased. This syndrome is not seen in patients previously exposed to phenobarbital. It has been postulated that the metabolism of primidone to phenobarbital is essential for the induction of tolerance.

Primidone is used primarily to control generalized seizures, but it is also used to control partial seizures (psychomotor type). The anticonvulsant potency is lower than that of phenobarbital. Its effectiveness as an anticonvulsant is also limited by its sedative properties.

On the basis of animal and clinical studies, it may be concluded that primidone is an active anticonvulsant drug the mechanism of action of which is most similar to those of phenytoin and carbamazepine. The precise mechanism of action for primidone has not yet been established.

Chemistry

The only method of practical significance consists of reductive desulfurization of 2-thiophenobarbital, which in turn is obtained by condensation of diethyl ethylphenylmalonate and thiourea.

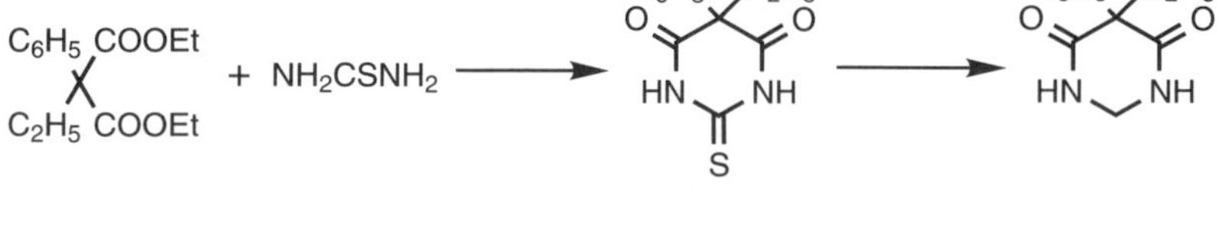

Benzodiazepines

These drugs have been widely used as antianxiety agents and sedative-hypnotics. In laboratory animals, benzodiazepines display outstanding anticonvulsant properties against seizures induced by maximal electroshock (MES) and pentylenetetrazole (Met).

Intravenously administered diazepam is the drug of choice for status epilepticus.[55] For infants, 1 to 2 mg, and for children and adults, 5 to 30 mg are usually administered intravenously and slowly. Diazepam often provides rapid control of status epilepticus seizures. Because of its high lipid solubility, intravenously administered diazepam enters the central nervous system rapidly. The initially high brain concentration, however, is reduced quickly because of redistribution of the drug, and status epilepticus may return.[56] Concomitant intravenous injection of diazepam and phenobarbital has been suggested to overcome this difficulty.[57]

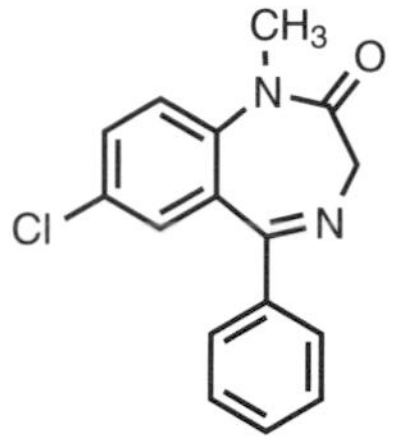

Diazepam

Orally administered diazepam is less effective because tolerance to the anticonvulsant effects of diazepam develops within a short period. On the other hand, when diazepam is effective in seizure control, side effects, particularly sedation, are a serious limiting factor.

A single dose of 10 mg diazepam is known to produce 200 to 500 ng/ml plasma concentration, which declines rapidly in 4 hours. Diazepam is slowly excreted, mainly as metabolites, which have been found in the urine plasma 14 days after a 10-mg dose.

Clonazepam was approved in 1975 for the treatment of akinetic, myoclonic, and absence variant seizures. Clonazepam was also found to be effective in controlling absences, but because of the high incidence of side effects, it is rated second to ethosuximide. It may be useful, however, in absences when succinimide therapy has failed. Used intravenously, clonazepam suppresses various types of status epilepticus seizures, but because of its greater cardiorespiratory depressant effect and difficulty in finding the dose, it is rated second to diazepam.

Although clonazepam displays a wide spectrum of anti-

C_6H_5 COOEt / C_2H_5 COOEt + NH_2CSNH_2 → (2-thiophenobarbital: C_6H_5, C_2H_5; HN, NH; S) → (primidone: C_6H_5, C_2H_5; HN, NH)

convulsant activities, the degree of effectiveness is not as high as that of standard agents. Side effects such as drowsiness, ataxia, somnolence, lethargy, and fatigue are frequent during clonazepam therapy. The anticonvulsant effects of clonazepam may be transient because of development of tolerance. The incidence of tolerance, however, is generally considered less than that observed with other benzodiazepines.

The initial dosage is 0.01 and 0.03 mg/kg given in divided doses; this amount is gradually increased to a 0.1 to 0.3 mg/kg maintenance dose. The half-life for elimination of clonazepam from plasma is 1 to 1½ days. Plasma levels in patients receiving 6 mg clonazepam for up to 12 months vary between 30 and 80 ng/ml.

In addition to their anxiolytic and sedative-hypnotic properties, several other benzodiazepines also display anticonvulsant activity. Although many of the marketed benzodiazepines have not been approved for anticonvulsant use, they are occasionally used in the treatment of epileptic patients. In addition, several already-marketed benzodiazepines are in clinical trials to also obtain the anticonvulsant indication (clobazam, lorazepam, and nitrazepam).

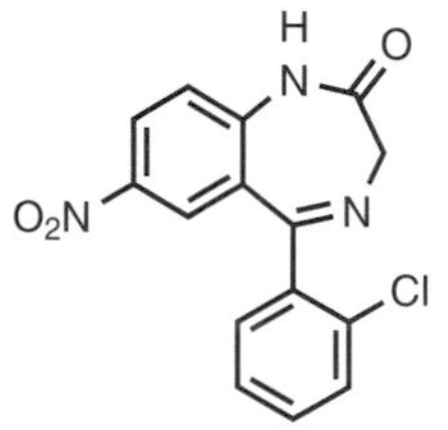

Clonazepam

Clorazepate dipotassium was approved for use as an antiepileptic drug by the U.S. Food and Drug Administration in 1981 as adjunctive therapy for the management of partial seizures. Clorazepate is decarboxylated in the acid of the stomach to desmethyldiazepam, which is also the major active metabolite of diazepam. Clorazepate, however, appears to be less sedating than diazepam. Clorazepate, as an add-on drug may be more useful in children than in adults with partial seizures.

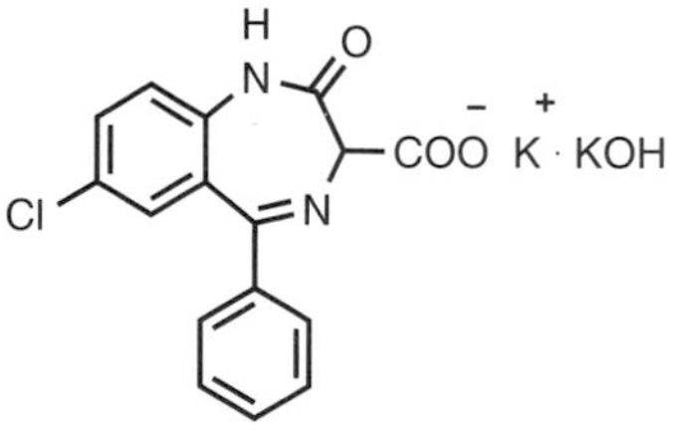

Clorazepate dipotassium

The mode of action of benzodiazepines is described in Chapter 10.

Valproic Acid (Sodium Valproate)

The anticonvulsant properties of valproic acid (dipropylacetic acid) were discovered serendipitously when sparingly soluble compounds of different chemical classes dissolved in dipropylacetic acid were effective in animals.[58] Follow-up experiments revealed that sodium valproate (sodium dipropylacetic acid) is moderately effective against electrically and chemically induced seizures in animals and possesses a satisfactory margin of safety.[59] It was postulated that valproic acid exerts anticonvulsant activity by elevating brain levels of the inhibitory neurotransmitter GABA.[60,61] The doses of valproic acid required to produce an elevated brain GABA level in animal model systems is, however, higher than the doses required to produce an anticonvulsant effect in humans (10 to 60 mg/kg). In favor of the GABA hypothesis, selective augmentation of postsynaptic GABA-mediated inhibition has been demonstrated in mouse spinal cord preparations,[62] but the results must be confirmed at lower concentrations and in other mammalian systems. Valproic acid also displays inhibitory effects on membrane conductance and permeability.[63] Possible direct membrane effects of valproic acid may therefore contribute to the mechanism of action of valproic acid.

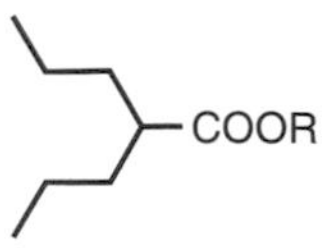

Valproic acid: R = H
Sodium valproate: R = Na

In new patients with typical absence seizures, sodium valproate may become the drug of first choice. In a comparative trial, sodium valproate and ethosuximide were equally effective when either drug was given alone or in combination with other anticonvulsant drugs in children with typical absence seizures. In absence variant seizures (Lennox-Gastaut syndrome), sodium valproate is more effective, whereas in myoclonic seizures, it is less effective than clonazepam. In addition, sodium valproate may be useful in the treatment of generalized (grand mal) seizures. It is less effective in the partial epilepsies.[64,65] Initial dosage of sodium valproate consists of 10 mg/day given in divided doses, with increases up to 60 mg/kg daily for maintenance. The plasma half-life of sodium valproate is 8 to 12 hours. The correlation between dose and blood levels is poor with a range of 40 to 100 μg/ml serum level and the time to peak plasma level is ½ to 2 hours.

Valproic acid is relatively free of side effects; gastrointestinal disturbances are observed most frequently. Tolerance to the anticonvulsant effects of valproic acid has not been observed. Valproic acid was approved by the U.S. Food and Drug Administration in 1978 for the treatment of seizure disorders.

Carbamazepine

This drug was approved by the U.S. Food and Drug Administration in 1974 for use in psychomotor and grand mal epilepsy (generalized and partial seizures) in adults who have not responded to other agents. In animals, the profile of anticonvulsant properties of carbamazepine is similar to that of phenytoin. Carbamazepine is effective in the MES test. It is ineffective against pentylenetetrazole-induced seizures (scMet test). Because many studies on the mechanism of action of carbamazepine have been carried out at drug concentrations far above its therapeutic range, it is difficult to draw definite conclusions concerning the mechanism of action of carbamazepine. Like phenytoin, carbamazepine prevents the spread of seizures produced by post-tetanic potentiation in the brain. Carbamazepine depresses sodium and potassium conductances and depresses synaptic transmission in the reticular activating system, thalamus, and limbic structures.

Carbamazepine increases acetylcholine and decreases choline brain levels, and inhibits ouabain-induced increases in cAMP levels. Carbamazepine-induced alteration in cyclic nucleotide levels may contribute to the mechanism of action of carbamazepine.

A review of the mechanism of action of carbamazepine has been published.[66]

In a double-blind crossover study in patients whose seizures were not controlled completely by combinations of anticonvulsant drugs, carbamazepine was equal in efficacy to phenobarbital of phenytoin in controlling seizure frequency, and side effects were minimal.

Several additional studies have demonstrated that carbamazepine is effective in reducing the seizure frequency of generalized and partial seizures. The drug is not effective for absences.

Toxicity with carbamazepine in most cases is relatively minor and less pronounced than that observed with phenytoin. As a result, carbamazepine is gaining in popularity as compared to phenytoin. The most frequent side effects involve mild drowsiness, diplopia, and gastric irritation. In a few patients, however, carbamazepine is known to produce bone marrow depression and hematologic injury. Therefore, patients receiving carbamazepine must have periodic blood count determinations. The usual dose range is 4 to 30 mg/kg, and serum concentrations fall in a range of 2 to 15 μg/ml. The plasma half-life of carbamazepine in chronically treated patients is 8 to 20 hours, with a peak time of 2 to 6 hours.

Like phenytoin, carbamazepine is highly protein bound (70 to 80%).

N

$CONH_2$

Carbamazepine

Anticonvulsants in Clinical Trial

The most promising anticonvulsants in clinical trials may be categorized by their chemical functional groups into the following chemical classes:

- Amino acids
  - Gabapentin
  - Vigabatrin
- Amides
  - Felbamate
  - Oxcarbazepine
  - Progabide
- Amines
  - Flunarizine
  - Lamotrigine
- Benzodiazepines
  - Clobazam
- Sulfonamides
  - Acetazolamide
  - Zonisamide

Gabapentin

Although gabapentin, 1-aminomethyl-cyclohexaneacetic acid is structurally related to the neurotransmitter GABA, a direct GABA-mimetic action as the mechanism of action can be excluded. Gabapentin is active not only in seizures evoked by impairment of inhibition, but also in a wide variety of other models. Clinically, gabapentin has been tried in patients with refractory epilepsies of various seizures types. Gabapentin appears to be as active in partial and generalized tonic-clonic seizures as conventional antiepileptic drugs, and the antiepileptic effect appears to be maintained over 12 months. Gabapentin is well tolerated. Both preclinical (toxicology, carcinogenecity) and clinical studies are in progress.

Vigabatrin

Vigabatrin (γ-vinyl GABA), a synthetic analog of GABA, was designed to enhance inhibitory neurotransmission. It is a selective, irreversible inhibitor of GABA-transaminase that is responsible for GABA catabolism. Vigabatrin increases brain GABA concentrations in laboratory animals, and this effect is presumed to be responsible for its antiepileptic action. Vigabatrin is an effective anticonvulsant in a variety of experimental models of epilepsy. It has been marketed from 1990 to 1992 in a number of European countries including the United Kingdom and Sweden for the treatment of epileptic patients who are refractory to standard antiepileptic drugs. Only the S(+)-enantiomer possesses pharmacologic activity; however, the racemic mixture has been marketed.

The drug is most useful for partial seizures, but it is also of use for primarily generalized seizures. Based on limited data, vigabatrin is not useful in absences or myoclonic seizures.

Felbamate

Felbamate is structurally related to meprobamate, but it is more potent than meprobamate in experimental models of epilepsy. It is effective in the maximal electroshock, Metrazol, and picrotoxin tests, but ineffective in the bicuculline and strychnine tests. In phase 3 clinical trials, felbamate appears to be most useful as an add-on drug in the treatment of patients with complex partial seizures who do not respond to currently available anticonvulsant drugs.

Oxcarbazepine

Oxcarbazepine is chemically related to carbamazepine and has a similar therapeutic profile, but it is better tolerated than carbamazepine. When used in 50% higher doses, the efficacy of oxcarbazepine is comparable to that of carbamazepine. It is useful in the treatment of generalized tonic-clonic convulsions and partial seizures with or without generalization.

Progabide

Progabide, a derivative of γ-aminobutyramide, produces GABA-mimetic effects by direct stimulation of the GABA receptors. It is highly protein bound and undergoes rapid metabolic degradation. It is less active than phenytoin in the maximal electroshock and strychnine tests, but unlike phenytoin, it also has similar activities in the bicuculline, picrotoxin, and Metrazol tests to those of valproate. In the clinic, progabide has shown modest efficacy in primary generalized, secondary generalized, and partial epilepsy.

Flunarizine

Because it was postulated that calcium currents may play an important role in the genesis of epilepsies, several calcium channel blockers were investigated for their anticonvulsant activities. In experimental models, flunarizine displayed activities with a profile similar to those of phenytoin and carbamazepine. In the clinic, about one third of therapy-resistant epileptic patients with partial complex seizures (with or without secondary generalization) had reduced seizure frequency as a result of adding on flunarizine. Efficacy was maintained without the development of tolerance to the drug. The more general value of flunarizine as an antiepileptic drug in patients who are not resistant to conventional anticonvulsant drugs can be assessed only when monotherapy trials are performed.

Lamotrigine

On the basis of the observation that treatment with several major antiepileptic drugs may lead to a disturbance of folate metabolism, many antifolate compounds were tested for anticonvulsant activity. Lamotrigine was found to be a potent compound, although its activity is not related to its antifolate effect. Lamotrigine probably acts by reducing the release of excitatory amino acids (glutamate) from presynaptic terminals. Lamotrigine has been marketed in the United Kingdom since December 1991 for the adjunctive treatment of partial and secondarily generalized tonic-clonic seizures that are not satisfactorily controlled by other anticonvulsants. The benefit appears to be modest (about a 25 to 30% reduction in mean seizure frequency). More studies are required to better define the role of lamotrigine in treating epilepsy.

Clobazam

Clobazam, a 1,5-benzodiazepine, has displayed anticonvulsant action in a variety of experimental models. Compared to 1,4-benzodiazepines, clobazam appears to have a lower incidence of side effects and greater antiepileptic effects, but no double-blind controlled clinical studies of clobazam with other antiepileptic drugs have been undertaken. Clobazam is effective against generalized tonic-clonic and partial seizures. A major limitation to its long-term use is the development of tolerance. Clobazam is used as adjunctive treatment with other standard anticonvulsant drugs in patients with intractable seizures. Clobazam has been marketed primarily as an anxiolytic agent, but it has also been approved for antiepileptic indication by several health authorities, including those in the United Kingdom and Germany.

Acetazolamide

This compound (5-acetamido-1,3,4-thiadiazole-2-sulfonamide) inhibits carbonic anhydrase activity in the

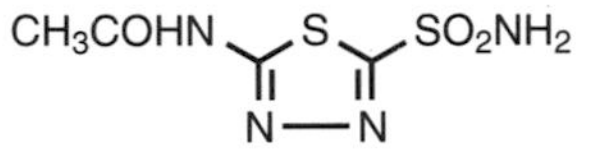

CH_3COHN–(1,3,4-thiadiazole: S, N—N)–SO_2NH_2

brain. The carbonic anhydrase enzyme is responsible for the intracellular reversible hydration of carbon dioxide to carbonic acid.

$$CO_2 + H_2O \rightleftharpoons H_2CO_3$$

By inhibiting carbonic anhydrase activity, acetazolamide causes CO_2 accumulation in the brain and the spinal cord. The anticonvulsant effect is related directly to this inhibition. The ability of a drug to inhibit this activity in the brain has been used to test anticonvulsant potency.

Tolerance to acetazolamide develops after a few weeks of continuous treatment, which limits its use. Acetazolamide is useful, however, for longer periods when administered as adjunct to primary anticonvulsant drugs (e.g., ethosuximide, phenytoin).

Acetazolamide, like CO_2, increases GABA levels in the brain, which may account for some of its anticonvulsant effects.

Zonisamide

This benzisoxazole derivative is active in the prevention of electrically induced seizures, but is not active in the Metrazol seizure test. Unlike acetazolamide, which also contains a sulfonamide group, zonisamide does not inhibit carbonic anhydrase activity. Clinical studies indicate that zonisamide will find its usefulness as an adjunct in partial and generalized tonic-clonic seizures when added to treatment regimens of other antiepileptic drugs.

NH_2 COOH (cyclohexane)

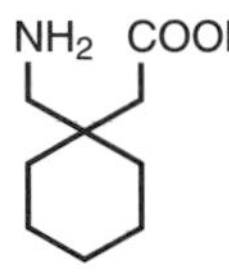

Gabapentin

Flunarizine

Vigabatrin

Lamotrigine

Felbamate

Clobazam

Oxcarbazepine

Acetazolamide

Progabide

Zonisamide

REFERENCES

1. F. M. Forster, Ed., *Report of the Panel on Epilepsy.* Madison, WI, University of Wisconsin Press, 1961, p. 91.
2. E. DeRobertis, G. R. DeLores-Arnaiz, and M. Alberici, *Ultrastructural neurochemistry,* in *Basic Mechanisms of the Epilepsies,* H. H. Jasper, et al., Eds., Boston, Little, Brown and Co., 1969, pp. 137–158.
3. J. H. Jackson, in *Selected Writings of John Hughlings Jackson,* Vol. 1, J. Taylor, Ed., London, Hodder and Stoughton, 1931.
4. W. R. Gowers, *Epilepsy and Other Chronic Convulsive Diseases: Their Causes, Symptoms and Treatment,* 2nd ed., London, J. and A. Church, 1901.
5. H. Gastaut, *Epilepsia (Amst.), 5,* 297(1964).
6. W. G. Lennox, *Epilepsy and Related Disorders.* Boston, Little, Brown and Co., 1960.
7. W. Penfield and H. H. Jasper. *Epilepsy and the Functional Anatomy of the Human Brain,* Boston, Little, Brown and Co., 1954, p. 20.
8. H. Gastaut, *Epilepsia (Amst.), 10,* S2(1969).
9. R. L. Masland, *Epilepsia (Amst.), 10,* S22(1969).
10. Proposal for Revised Clinical and Electroencephalographic Classification of Epileptic Seizures, *Epilepsia,, 22,* 489(1981).
11. R. L. Krall, et al., *Epilepsia, 19,* 393(1978).
12. R. L. Krall, et al., *Epilepsia, 19,* 409(1978).
13. E. A. Swinyard, et al., *General Principles, Experimental Selection, Quantification and Evaluation of Anticonvulsants* in *Antiepileptic Drugs,* Third edition, R. Levy, et al., Eds., New York, Raven Press, 1989, p. 88.
14. J. A. Vida, in *Principles of Medicinal Chemistry,* Second edition, W. O. Foye, Ed., Philadelphia, Lea & Febiger, 1981.
15. E. A. Swinyard, et al., *J. Pharmacol. Exp. Ther., 106,* 319(1952).
16. L. A. Woodbury and V. D. Davenport, *Arch. Int. Pharmacodyn., 92,* 97(1952).
17. A. Vernadakis and D. M. Woodbury, *Epilepsia (Amst.), 10,* 163(1969).
18. O. Barrada and S. I. Oftedal, *Electroencephalogr. Clin. Neurophysiol., 29,* 220(1970).
19. N. W. Dunham and T. A. Miya, *J. Am. Pharm. Assoc. Sci. Ed., 46,* 208(1957).
20. S. J. Enna and S. H. Snyder, *Mol. Pharmacol., 13,* 442(1977).
21. C. Braestrup and R. F. Squires, *Proc. Natl. Acad. Sci. USA, 74,* 3805(1977).
22. S. R. Zukin, et al., *Proc. Natl. Acad. Sci. USA, 71,* 4802(1974).
23. C. W. Cotman, *Methods Enzymol., 31A,* 445(1974).
24. J. W. Phillis, et al., *Gen. Pharmacol., 12,* 67(1981).
25. R. J. Racine, *Electroencephalogr. Clin. Neurophysiol., 32,* 281(1972).
25. S. Irwin, *Psychopharmacologia, 13,* 222(1968).
26. J. E. P. Toman, in *The Pharmacological Basis of Therapeutics,* 3rd ed., L. S. Goodman and A. Gilman, Eds., New York, Macmillan, 1965, p. 217.
27. C. Locock, *Lancet, 1,* 528(1857).
28. A. Hauptmann, *Munch. Med. Wochenschr., 59,* 1907(1912).
29. T. J. Putnam and H. H. Merritt, *Science, 85,* 525(1937).
30. T. J. Putnam and H. H. Merritt, *JAMA, 111,* 1068(1938).

31. H. H. Merritt and T. J. Putnam, *Arch. Neurol. Psychiat., 39,* 1003(1938).
32. H. H. Merritt and T. J. Putnam, *Epilepsia (Amst.), 3,* 51(1945).
33. A. S. Troupin, et al., *Epilepsia, 17,* 403(1976).
34. F. Morrell, et al., *Neurology, 9,* 492(1959).
35. R. S. Neumann and G. B. Frank, *Can. J. Physiol. Pharmacol., 55,* 42(1977).
36. P. De Weer, *Adv. Neurol., 27,* 353(1980).
37. J. H. Pincus, et al., *Adv. Neurol., 27,* 363(1980).
38. A. V. Delgado-Escueta and M. P. Horan, *Adv. Neurol., 27,* 377(1980).
39. R. J. De Lorenzo, *Adv. Neurol., 27,* 399(1980).
40. D. M. Woodbury, *Adv. Neurol., 27,* 447(1980).
41. T. G. Riddle, et al., *Eur. J. Pharmacol., 33,* 189(1975).
42. J. A. Ferrendelli and D. A. Kinscherf, *Ann. Neurol., 5,* 533(1979).
43. D. M. Woodbury in *Antiepileptic Drugs,* D. M. Woodbury, J. K. Penry, and C. E. Pippenger, Eds., New York, Raven Press, 1982, pp. 269–281.
44. J. E. P. Toman, *Pharmacol. Rev., 4,* 168(1952).
45. D. W. Esplin and E. M. Curto, *J. Pharmacol. Exp. Ther., 201,* 320(1957).
46. J. T. Miyahara, et al., *J. Pharmacol. Exp. Ther., 154,* 118(1966).
47. D. M. Woodbury, *Adv. Neurol., 27,* 249(1980).
48. T. C. Pellmar and W. A. Wilson, *Science, 197,* 912(1977).
49. P. M. Diaz, *Neuropharmacology, 13,* 615(1974).
50. J. A. Ferrendelli and W. E. Klunk in *Antiepileptic Drugs,* D. M. Woodbury, et al., Eds., New York, Raven Press, 1982, pp. 655–661.
51. F. A. Gibbs, et al., *Dis. Nerv. Syst., 10,* 245(1949).
52. J. Y. Bogue and H. C. Carrington, *J. Pharmacol. Exp. Ther., 108,* 428(1952).
53. T. C. Butler and W. J. Waddell, *Proc. Soc. Exp. Biol. Med., 93,* 544(1956).
54. B. B. Gallagher and R. I. Baumel, in *Antiepileptic Drugs,* D. M. Woodbury, et al., Eds., New York, Raven Press, 1972, pp. 361–371.
55. H. Gastaut, et al., *Epilepsia (Amst.), 6,* 167(1965).
56. R. H. Mattson, in *Antiepileptic Drugs,* D. M. Woodbury, et al., Eds., New York, Raven Press, 1972, pp. 497–518.
57. B. B. Gallagher, in *Anticonvulsants,* J. A. Vida, Ed., New York, Academic Press, 1977, p. 49.
58. G. Meunier, et al., *Therapie, 18,* 435(1963).
59. G. Carraz, et al., *Therapie, 19,* 917(1964).
60. S. Simler, et al., *J. Physiol. (Paris), 60,* 547(1968).
61. Y. Godin, et al., *J. Neurochem., 16,* 869(1969).
62. R. L. MacDonald and C. K. Bergey, *Brain Res., 170,* 558(1979).
63. G. E. Slater and D. Johnston, *Epilepsia, 19,* 379(1978).
64. D. Simon and J. K. Penry, *Epilepsia (Amst.), 16,* 549(1975).
65. R. M. Pinder, et al., *Drugs, 13,* 81(1977).
66. R. M. Julien, in *Antiepileptic Drugs,* D. M. Woodbury, et al., Eds., New York, Raven Press, 1982, pp. 543–547.

SUGGESTED READINGS

R. H. Levy, et al., *Antiepileptic Drugs,* 3rd ed., New York, Raven Press, 1989.

B. S. Meldrum and R. J. Porter, *New Anticonvulsant Drugs,* London, John Libbey, 1986.

T. A. Pedley and B. S. Meldrum, *Recent Advances in Epilepsy,* Number 5, Edinburgh, Churchill Livingstone, 1992.

J. A. Vida, Ed., *Anticonvulsants, Medicinal Chemistry,* A. Series of Monographs, Vol. 15, New York, Academic Press, 1977.

Chapter 12

NEUROLEPTICS AND ANXIOLYTIC AGENTS

John L. Neumeyer and Raymond G. Booth

Macbeth: How does your patient, doctor?
Doctor: Not so sick, my lord, as she is troubled with thick-coming fantasies, that keep her from her rest.
Macbeth: Cure her of that: Canst thou not minister to a mind diseas'd; pluck from the memory a rooted sorrow; raze out the written troubles of the brain; and with some sweet oblivious antidote cleanse the stuff'd bosom of that perilous stuff which weighs upon the heart?
Doctor: Therein the patient must minister to himself.
—William Shakespeare, *Macbeth*

CLASSIFICATION OF PSYCHOACTIVE DRUGS

One out of every five prescriptions dispensed in the United States is for drugs intended to affect mental processes: to stimulate, sedate, or otherwise modify feelings or behavior. The drugs primarily affecting mental processes can be divided into three major categories depending on their clinical usefulness: neuroleptics, antidepressants and mood-stabilizers, and anxiolytics. Each of these drug categories contains entirely unrelated chemical structures that affect specific anatomic structures and biochemical reactions and produce the desired clinical effects. The neuroleptics, called antipsychotic agents (formerly ataractics or major tranquilizers), are used in the treatment of psychoses. Phenothiazines, thioxanthenes, and butyrophenones are the most important representatives in this group. Clinically, these agents counteract or minimize hallucinations and delusions, alleviate psychomotor excitement, and facilitate social readjustment. Pharmacologically the essential effect of these agents is to reduce dopaminergic activity in the brain.

Mood-altering agents (discussed in Chapter 15), also called thymoleptics, include tricyclic amines and other heterocyclic antidepressants, the monamine oxidase (MAO) inhibitors (also sometimes called psychic energizers), lithium salts, and applications of agents originally developed as anticonvulsants (especially carbamazepine and valproic acid). These drugs, except for lithium carbonate, whose action is still unclear, may act by increasing the activity of norepinephrine and serotonin at their postsynaptic receptor sites in the brain and other tissues. These effects can be initiated in several ways. One is by inhibition of reuptake of biogenic amines at the nerve terminals; another is by inhibition of MAO enzyme systems by drugs such as tranylcypromine and phenelzine; both effects lead to complete neuronal responses that remain incompletely understood.

The anxiolytics (also called antianxiety agents, minor tranquilizers, relaxants, and antineurotic agents) are used in the treatment of anxiety associated with psychoneurotic and psychosomatic conditions. The benzodiazepines, such as chlordiazepoxide, are the most important members of this class of drugs, and older agents, such as the propanediol carbamates, are now rarely used. The anxiolytic action is probably manifested in the limbic system and may involve a potentiation of the neuronal inhibitory effects of γ-aminobutyric acid (GABA), possibly including reduction in turnover of norepinephrine or other monoamine neurotransmitters.

Problems in classification and terminology of psychoactive drugs have arisen as a result of the many therapeutic uses of such drugs other than for their psychotropic effects. For example, some of the benzodiazepines and propanediol carbamates are also used as muscle relaxants (Chapter 13). The antihistaminic sedatives are discussed briefly in this chapter and elsewhere (Chapter 22). The barbiturate sedatives are discussed in Chapter 10, yet a discussion of sedatives would be incomplete without mention of the benzodiazepines. Some agents improve mood and hence are classified as stimulants or euphoriants, but in higher doses they may cause agitation or psychosis. The amphetamines, for example, can be considered both stimulant euphoriants and "psychotogenic" agents.

A classification of psychoactive drugs according to chemical structure with examples in each category is given in Table 12–1.

MECHANISM OF NERVE TRANSMISSION

To provide a background for a discussion of nerve transmission, and in particular how neurotransmitter receptors may be linked to the action of drugs that affect behavior, a review of current concepts of the mechanisms by which nerve cells communicate chemical messages and translate these within the cell into physiologic actions is appropriate.[1]

The delivery of a neurotransmitter message to the interior of a receptive cell appears to be achieved by complex multistepped mechanisms that remain incompletely understood. First, transmitters attach to specialized receptor sites on the surface of the cell, and second, they influence membrane intracellular biochemical processes from the outside. The receptors embedded in the cell membrane are selective in their ability to bind a specific transmitter, presumably because the molecular configuration of the receptor allows the transmitter molecule to "fit" the receptor or to align itself precisely in a thermo-

Table 12–1. Psychoactive Drugs

Hypnotics and sedatives
- Barbiturates (phenobarbital, pentobarbital)
- Benzodiazepines (chlordiazepoxide, oxazepam)
- Carbamates (ethinamate, meprobamate)
- Chloral derivatives (chloral hydrate)
- Acetylenic alcohols (ethchlorvynol)
- Paraldehyde
- Piperidinediones (glutethimide, methyprylon)

Neuroleptics
- Phenothiazines (chlorpromazine, thioridazine, perphenazine, fluphenazine.)
- Thioxanthenes (chlorprothixene, thiothixene, flupenthixol)
- Butyrophenones (haloperidol, duperidol)
- Dibenzazepines (loxapine, clozapine)
- Dihydroindolones (molindone, oxypertine)
- Diphenylbutylpiperidines (pimozide, penfluridol)
- Benzamides (sulpiride, clebopride)
- Benzoquinolizines (tetrabenazine)
- Rauwolfia alkaloids (reserpine)

Anxiolytics
- Benzodiazepines (chlordiazepoxide, diazepam)
- Azaspirodecanediones (buspirone)
- Antihistaminic-diphenylmethanes (hydroxyzine)
- Propanediol carbamates (meprobamate, tybamate)

Antidepressants
- Tricyclic antidepressants (imipramine, amitriptyline)
- Monoamine oxidase inhibitors (tranylcypromine, phenelzine)
- Miscellaneous heterocyclics (trazodone, bupropion)

Mood-stabilizers
- Lithium salts
- Carbamazepine
- Valproic acid

Stimulants, hallucinogens, psychotomimetics, psychotogens
- Lysergic acid diethylamide (LSD)
- Marijuana
- Mescaline
- Amphetamines

dynamically favorable location. The extent to which a transmitter influences its target cell depends on the concentration of the transmitter in the fluid outside the cell as well as its affinity for the membrane receptor. How then does the transmitter, once bound to a receptor, transmit its message to the interior of the cell?

One answer to this question arises largely from the work of Sutherland and collaborators,[2] who found that the membranes of many cells contain an enzyme, adenylyl cyclase, that converts adenosine triphosphate (ATP) to 3,5-(cyclic) adenosine monophosphate (cyclic AMP) (Fig. 12–1). It was reasoned that there is a functional link between the binding of a transmitter to a receptor on the outside of the membrane and the activation of adenylyl cyclase on the inside. Because, however, there are a wide variety of neurotransmitters (e.g., dopamine, norepinephrine, acetylcholine) as well as hormones such as cortisone and the sex hormones, for which different kinds of cells possess differing receptors, any given neurotransmitter or hormone alters the level of cyclic AMP in its target cells but not in other cells. It is currently believed that the cyclic AMP generated by adenylyl cyclase in response to the binding of a neurotransmitter to the receptor acts as a "second messenger" to relay the message of the neurotransmitter (the first messenger) from the membrane to the cell interior.

The activation or inhibition of adenylyl cyclase to alter cyclic AMP levels in response to a neurotransmitter occurs through a class of guanosine 5′-triphosphate (GTP) binding transmembrane glycoproteins known as G-proteins. Some members of the G-protein family stimulate (G_s) adenylyl cyclase, while others inhibit (G_i) the enzyme. G-proteins act as signal transducers such that when a neurotransmitter stimulates its G-protein coupled receptor, causing release of a molecule of guanosine 5′-diphosphate (GDP) and binding of a molecule of GTP, the G-protein can dissociate into subunits to activate its effector (adenylyl cyclase) to modulate second messenger (cyclic AMP) production. The second messenger may then affect intracellular (intraneuronal) processes, thus linking stimulation of the neurotransmitter receptor to physiological effects (Fig. 12–2). Other neurotransmitter receptors linked to G-proteins include β-adrenergic and muscarinic cholinergic types, the latter operating through polyphosphoinositide phosphodiesterase, a phospholipase C (PLC) protein, to affect intracellular phosphoinositol metabolism, producing the intracellular second messengers inositol 1,4,5-triphosphate and diacylglycerol (Fig. 12–2). Thus, how a particular chemical stimulus may affect a neurotransmitter-sensitive second messenger system is important to understand how a neurotransmitter receptor is linked to the action of drugs affecting behavior.

If a drug has a molecular configuration analogous to that of an endogenous neurotransmitter, it may be able to bind to the membrane receptor for that neurotransmitter and mimic the neurotransmitter's action. Such drugs are known as receptor agonists. If a drug has a configuration similar to that of a neurotransmitter but not quite as similar as that of a receptor agonist, it may be able to bind to the same receptor without activating it or to another functionally related receptor with an opposite or regulatory effect. In that case, the drug prevents the activation of the agonist receptor by a neurotransmitter. Drugs of this category ar called receptor antagonists (Fig. 12–3). Drugs that influence behavior could be either agonists or antagonists. Some may have a mixed action by binding to the neurotransmitter receptor and causing partial activation or partial inactivation; some may affect the receptor indirectly by altering the quantity and degree of binding of a neurotransmitter.

Drugs that act as dopamine agonists, such as apomorphine, stimulate the dopamine receptors in the basal ganglia. Dopamine agonists, by supplementing the dopamine that is missing because of the degeneration of

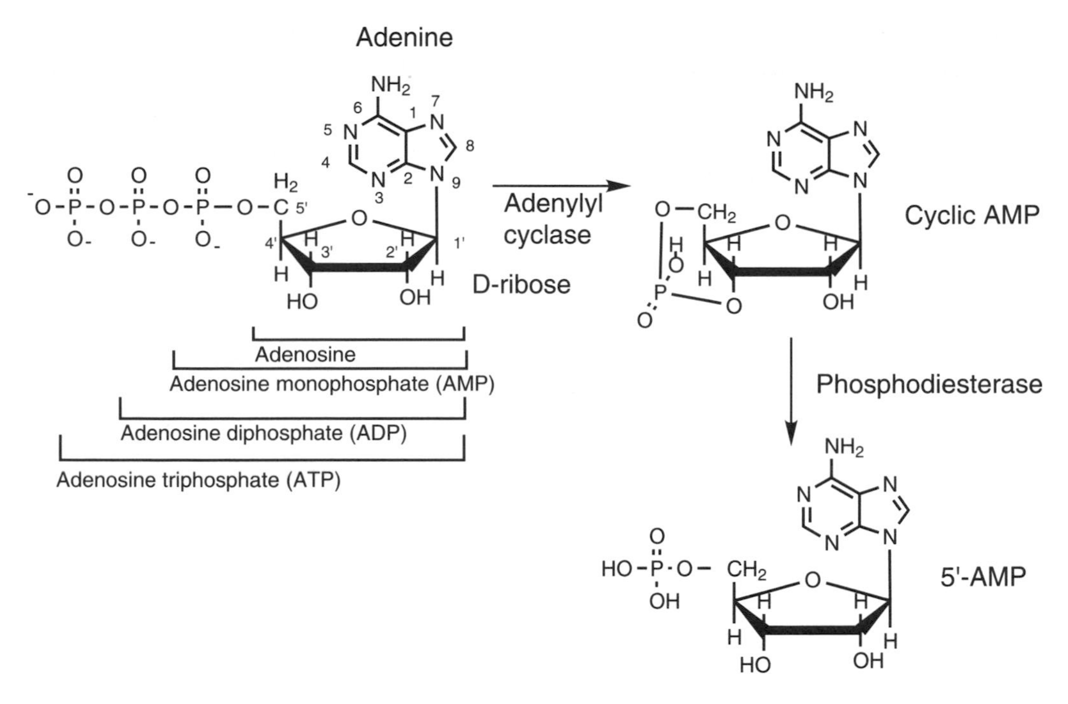

Fig. 12–1. The structure of adenosine triphosphate (ATP), cyclic adenosine monophosphate, and 5′-adenosine monophosphate and the enzymes associated with their synthesis and degradation. Adenylyl cyclase converts the energy-carrier ATP into cyclic AMP by removing two phosphate groups from the molecule and forming a ring. Phosphodiesterase inactivates cyclic AMP by opening the phosphate ring, converting the molecule into an inert form of AMP.

neurons in this region of the brain, are useful in the treatment of Parkinson's disease (Chapter 13). Drugs that act as antagonists of the dopamine receptor, such as the phenothiazine chlorpromazine and butyrophenones such as haloperidol, are useful therapeutic agents in the treatment of psychoses and certain neurologic disorders that may involve excess dopaminergic function, such as Huntington's and Tourette's diseases. One of the current theories, the dopamine hypothesis of psychosis, suggests that dopamine systems in the brain are overactive in psychosis and may contribute to some psychotic symptoms. It is clear that antipsychotic agents interfere with the actions

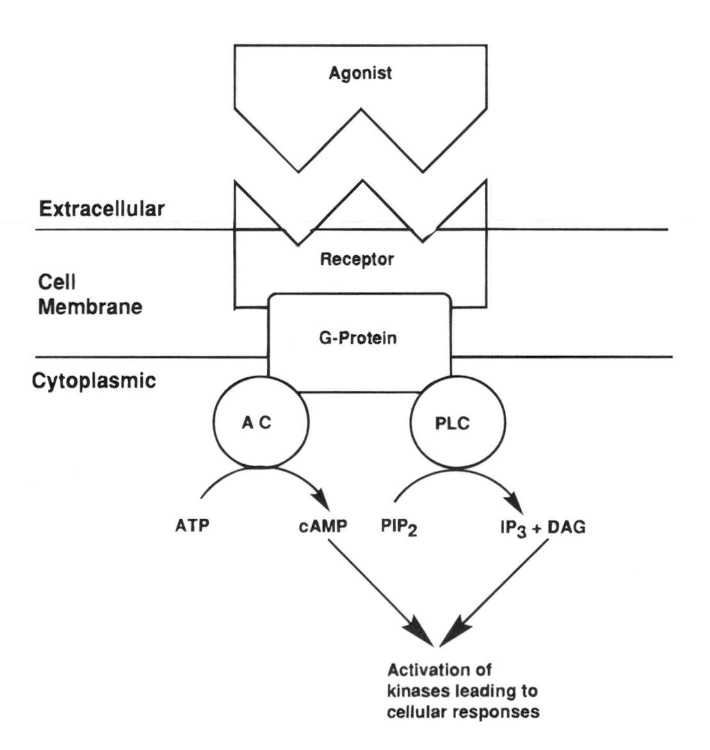

Fig. 12–2. Interaction of an agonist ligand with its G-protein coupled receptor activates effector molecules adenylyl cyclase (AC) or phospholipase C (PLC) to catalyze formation of intracellular second messengers cyclic AMP (cAMP) from ATP or inositol 1,4,5-triphosphate (IP_3) and diacylglycerol (DAG) from phosphatidylinositol 4,5-diphosphate (PIP_2). The second messengers activate target kinases or may have direct effects (e.g., IP_3 releases Ca^{++} from intracellular stores), ultimately leading to cellular responses.

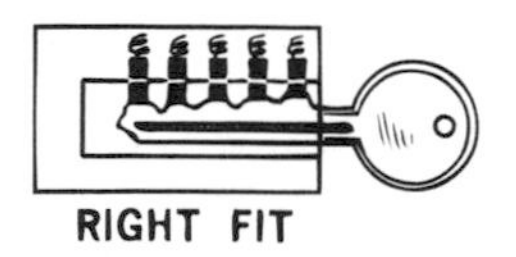

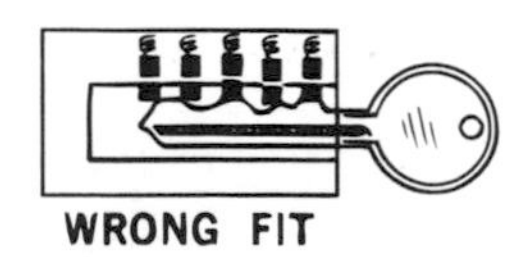

Fig. 12–3. If a particular drug (the key) has a molecular configuration analogous to that of an endogenous neurotransmitter, and complementary to its receptor, it may be able to bind to the membrane receptor (the lock) for that neurotransmitter and mimic the action of the neurotransmitter (i.e., the key has the right fit and can turn the barrel of the lock). Such drugs are known as *receptor agonists.* If a drug has a configuration similar to that of a neurotransmitter but not quite as similar as that of a receptor agonist, it may be able to bind to the receptor without activating it (i.e., the key fits into the lock but has the wrong fit to turn the barrel of the lock). Drugs of this category are called *receptor antagonists.*

of dopamine as synaptic neurotransmitters. This hypothesis is discussed in greater detail below.

DOPAMINE HYPOTHESIS IN PSYCHOSIS OR MANIA

Almost 100 years ago, William James noted in his classic work, *Principles of Psychology,* that "chemical action must, of course, accompany brain activity, but little is known of its exact nature." Since that time, the neurosciences have generated a wealth of information about brain physiology and chemistry.

Dopamine was not identified as a neurotransmitter until 1958 (before that time it was believed to be only a precursor of norepinephrine). Information about the role of dopamine has increased markedly in the past decade through the study of dopamine receptors,[3] which are located throughout the body and particularly in different areas of the brain. The highest concentrations of dopamine are located in the basal ganglia (especially the putamen in humans) (Fig. 12–4). Dopaminergic activity within the basal ganglia is essential for normal posture, muscle tone, and coordinated motion. Dopamine is also a major physiologic modulator of the pituitary hormone, prolactin, and may be the "prolactin inhibitory factor."

In 1963, Carlsson and colleagues produced evidence suggesting that the principal mechanism of action of neuroleptics is by blocking dopamine receptors.[4] Following the inhibition of MAO, the catecholamine metabolite, 3-methoxytyramine, accumulated in mouse brain (more than normetanephrine) (see Chapter 13, Fig. 13–5) more rapidly following small doses of chlorpromazine and haloperidol.

The antihistaminic phenothiazine, promethazine (see Fig. 12–10, VII), a drug that is not effective in the treatment of psychosis, did not alter the concentrations of these methoxylated metabolites. Haloperidol, a butyrophenone derivative also possessing antipsychotic actions similar to those of chlorpromazine but at smaller doses, was correspondingly more potent in elevating the concentrations of these metabolites. Carlsson rationalized that the phenothiazines block catecholamine receptor sites, whereupon a message is conveyed by means of a neuronal feedback to the neuronal cell body: "We receptors are not receiving enough transmitter; send more catecholamines." The catecholamine-generating neurons accordingly fire more rapidly and are required thereby to increase their synthesis of catecholamine metabolites. Indeed, clinically effective phenothiazines and butyrophenones all stimulate dopamine synthesis. The

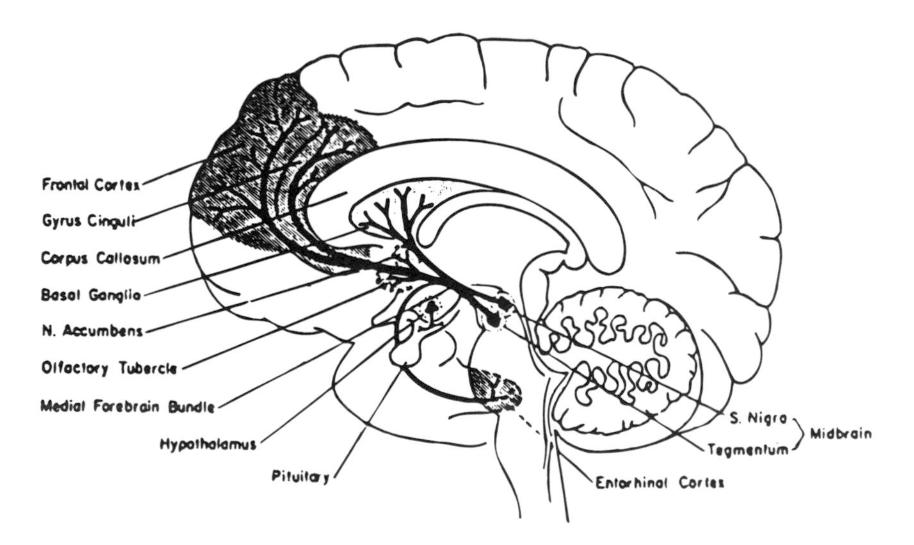

Fig. 12–4. Dopamine-containing neurons in the human brain. The major systems involving dopamine are the nigrostriatal pathway from the zona compacta of the midbrain substantia nigra to the basal ganglia (caudate-putamen); mesolimbic projections from midbrain tegmentum through the lateral hypothalamus to limbic structures, including the septal nuclei (for example, nucleus accumbens septi) and olfactory tubercle; and related mesocortical projections, also arising in the midbrain tegmentum, and projection particularly to mesial-prefrontal and temporal areas of the cerebral cortex. There is also a tuberoinfundibular dopamine-containing (TIDA) system within the hypothalamus that provides dopamine, by secretion, to the pituitary gland by way of the hypophysioportral circulation. (Adapted from R. J. Baldessarini, *Chemotherapy in Psychiatry: Principles and Practice,* Cambridge, Harvard University Press, 1985.)

influence of these drugs on dopamine synthesis correlates much better with clinical effects than does their action on norepinephrine synthesis, although antiadrenergic contributions to their clinical effects have not been entirely ruled out. Many neuroleptics can antagonize the effects of norepinephrine at certain anatomic sites. Additional findings have accumulated over the years so that today the proposal that brain dopamine is involved in psychosis strongly supports the theory that antipsychotic agents interfere with actions of dopamine as a synaptic neurotransmitter in the brain.[3,5,6] They do not prove, however, that antidopamine effects are either necessary or sufficient for antipsychotic efficacy. They do strongly suggest that some of the extrapyramidal neurologic effects and neuroendocrine actions (notably, prolactin release) of this class of agents may be produced by antagonism of dopamine receptors in the basal ganglia and pituitary.

Because dopamine is prominent in the hypothalamus and in the nigroneostriatal tract, which links the midbrain and basal ganglia, these areas of the brain are strongly implicated in the sites of action of antipsychotics. There are also dopamine-containing projections from midbrain nuclei to forebrain regions associated with the limbic system as well as temporal and medial prefrontal cerebral cortical areas closely linked with the limbic system (Fig. 12–4). In these regions of the brain, dopamine serves as a neurotransmitter at nerve terminals that contain the metabolic apparatus also found in sympathetic adrenergic nerve terminals, with the exception that the enzyme dopamine-β-hydroxylase, which converts dopamine to norpinephrine, is missing (Fig. 12–5).

CLASSIFICATION OF DOPAMINE RECEPTORS

Neuroleptic drugs may act at dopamine receptors in different brain regions.[3] Within each of these dopamine-containing systems, dopamine receptors are located postsynaptically on either the cell bodies or dendrites of other neurons or the terminals of afferent fiber projections; others may occur presynaptically on either dopamine neuronal cell bodies and dendrites or dopamine nerve terminals (see Fig. 12–5). In addition to the postsynaptic dopamine receptors thought to lie on cell bodies of cholinergic and probably GABAergic interneurons, there is evidence for the existence of dopamine receptors on presynaptic dopamine terminals within the striatum, and these dopamine autoreceptors act to modulate dopamine synthesis and release negatively. It is currently believed that low concentrations of certain dopamine agonists (such as N-n-propylnorapomorphine) interact with autoreceptors at low doses and so decrease dopaminergic action.

Neuroleptic drug action is further complicated by the abundant evidence in favor of more than one type of brain dopamine receptor in most brain regions. Until recently, compilation of histochemical, electrophysiologic, biochemical, and behavioral evidence suggested only two subtypes of dopamine receptors.[7] Modern molecular biologic methods involving recombinant DNA techniques, however, have resulted in the cloning of multiple receptor subtypes for several neurotransmitters, including dopamine. Currently five different dopamine receptors (Table 12–2) have been cloned and identified using molecular biologic techniques. The amino acid sequence of all these receptors, as deduced from their es-

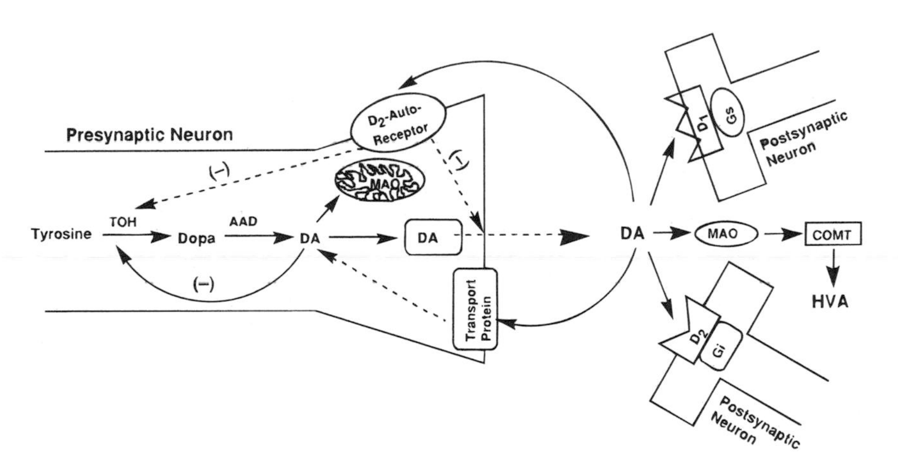

Fig. 12–5. Tyrosine is hydroxylated in a rate-limiting step by tyrosine hydroxylase (TOH) to form dihydroxylphenylalanine (DOPA), which is decarboxylated by L-aromatic amino acid decarboxylase (AAD) to form dopamine (DA). Newly synthesized DA is stored in vesicles, from which release occurs into the synaptic cleft by depolarization of the presynaptic neuron in the presence of Ca^{++}. DA released into the synaptic cleft may go on to stimulate postsynaptic D_1- and D_2-type receptors or presynaptic D_2-type autoreceptors that negatively modulate DA synthesis (via inhibition of TOH) and release. The action of synaptic DA is inactivated largely via reaccumulation into the presynaptic neuron by high-affinity DA neurotransport proteins located on the nerve terminal membrane. Free cytoplasmic DA negatively modulates DA synthesis via end-product (feedback) inhibition of TOH by competition with biopterin cofactor. Cytoplasmic pools of DA may undergo metabolic deamination by monoamine oxidase (MAO), an enzyme bound to the outer membrane of mitochondria, to form dihydroxyphenylacetaldehyde, which oxides to dihydroxyphenylacetate (DOPAC). DA or DOPAC may undergo methylation by catechol-O-methyltransferase (COMT) ultimately to form homovanillic acid (HVA), a metabolite excreted in urine.

Table 12–2. Cloned Dopamine Receptors

Criteria	D_1	D_2	D_3	D_4	D_5
No. of amino acids	446	415–444	446	387	477
Effect of agonist on adenylyl cyclase	+	−	−	−	+
Affinity (nM) for:					
Agonists					
Dopamine	10^3	10^3	10^1	10^2	10^2
(±) SKF 38393	10^2	10^4	10^4	10^4	10^2
Quinpirole	$>10^4$	10^3	10^1	10^1	$>10^4$
Antagonists					
(+) SCH-23390	10^{-1}	10^3	10^3	10^3	10^{-1}
(−)Sulpiride	$>10^4$	10^1	10^1	10^1	$>10^4$
Clozapine	10^2	10^2	10^2	1	10^2

Modified from Civelli et al.[8] and Sokoloff et al.[9]

tablished nucleotide sequence, shows that they belong to the G-protein coupled superfamily of receptors that are structurally characterized by a seven transmembrane spanning region.[8,9] A paucity of selective ligands has hampered pharmacologic characterization of each dopamine receptor subtype; thus, it is still convenient to distinguish them as either D_1-like (D_1, D_5) or D_2-like ($D_{2\ short}$, $D_{2\ long}$, D_3, D_4), based on effects on adenylyl cyclase and affinity for known dopamine receptor ligands (see Table 12–2).

The selective benzazepine derivative, SKF 38393, whose selectivity resides almost exclusively in the R(+) isomer, is used as a selective D_1 *partial* agonist.[10] The structurally related D_1 receptor antagonist (SCH 23390) is also available. The rigid benzophenanthridine derivative, dehydrexidine, has been introduced as a full efficacy agonist (produces stimulation of adenylyl cyclase equivalent to dopamine itself) at D_1-type receptors.[11] Selective D_2 agonists, such as the pyrazole derivative quinpirole (LY 141865), and antagonists such as sulpiride are also now available. The dibenzodiazepine, clozapine, shows relatively greater affinity for the D_4 subtype and is proposed to have a superior antipsychotic clinical profile with low incidence of extrapyramidal side effects (Fig. 12–6). Tritiated versions of several D_1 (e.g., [^{3}H]-SCH 23390) and D_2 (e.g., [^{3}H]-(-)sulpiride) selective ligands are used in radioreceptor assays to determine the affinity of new compounds for dopamine receptors.

In addition to D_1-type receptors, located for the most part postsynaptically, and D_2-type receptors, which may be located both presynaptically and postsynaptically, there are dopamine-binding proteins located at dopaminergic nerve terminals that act to transport dopamine that has been released into the synaptic cleft back into the presynaptic neuron (Fig. 12–5). Several neurotransmitter transport proteins have been cloned and expressed, including those present in nerve terminal membranes of GABAergic, adrenergic, serotoninergic, and

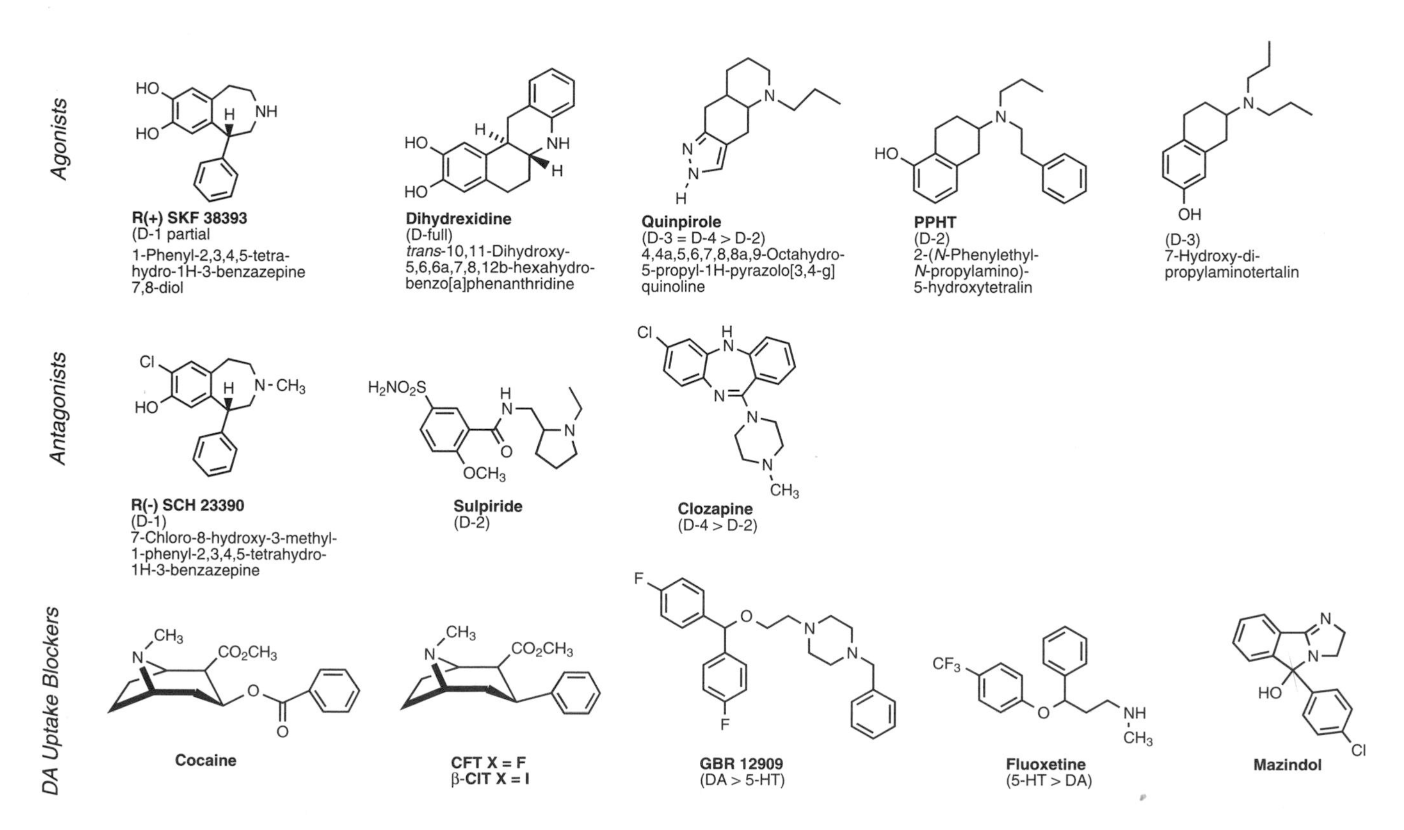

Fig. 12–6. Selective dopamine receptor agonists and antagonists and DA reuptake blockers.

dopaminergic neurons.[12] These neurotransporters are thought to be responsible for terminating neurotransmitter action by reuptake and appear to be the site of action for many antidepressant and psychostimulant drugs. The dopamine transporter is believed to be the principal brain receptor for the behavioral reinforcing and psychomotor stimulant properties of cocaine[13] and also accumulates the neurotoxin 1-methyl-4-phenylpyridinium (MMP^+) into dopaminergic nerve terminals, where it inhibits mitochondrial respiration (see Chapter 13). In addition to binding to the dopamine transporter to inhibit neuronal reuptake of dopamine, cocaine also inhibits the reuptake of norepinephrine and serotonin. Several tropane-based derivatives of cocaine (CFT, CIT; see Fig. 12–6) have been developed that show binding characteristics similar to cocaine and greater affinity at cocaine recognition sites.[14,15] [^{123}I] β-CIT (2β-carbomethoxy-3-β-(4-iodophenyl)tropane) labels dopamine transporters. SPECT (single-photon emission computed tomography) imaging with [^{123}I] β-CIT was demonstrated to be a useful marker for the loss of striatal DA terminals in patients with Parkinson's disease.[16] The nontropane compound mazindol inhibits neuronal uptake of both dopamine and norepinephrine, whereas GBR-12909 and fluoxetine are selective dopamine and serotonin uptake inhibitors (see Fig. 12–6). The norepinephrine and serotonin transporters, but not the dopamine carrier, share high-affinity recognition for antidepressant drugs (see Chapter 15).

BLOOD–BRAIN BARRIER

The concept of a blood–brain barrier (BBB) arose when it was found that some substances present in the blood failed to penetrate the brain but reached other body tissues without difficulty. The substances found not to penetrate included physiologic metabolites such as epinephrine and dopamine, pathologic metabolites such as are found in bile, and antibiotics such as streptomycin and penicillin. The limited ability of certain drugs to penetrate the brain with respect to other organs is the so-called BBB. BBB signifies restricted distribution from blood to brain and appears to result from unique characteristics of brain capillaries. For a drug to penetrate the capillary wall, it must leave the plasma, enter the capillary endothelial cell membrane, leave the membrane, enter the cell cytoplasm, and then again penetrate the outer cell membrane to enter the surrounding extracellular fluid in the brain. The sum of these transitions, which determines the selective permeability of the brain capillary wall, is referred to as the BBB and is illustrated in Fig. 12–7.

It can now be appreciated that a drug molecule, to be effective in the brain, must possess certain physical and chemical characteristics: (1) It must have an appreciable affinity for lipids relative to water. (2) It must be acid-resistant to survive the stomach passage and must resist hydrolysis and degradation by other intestinal contents. (3) It should not be too firmly bound to proteins in the blood. Because the passage through the BBB requires a high degree of lipid solubility (usually expressed as a partition coefficient between oil and water), ionized molecules do not readily penetrate the BBB. Small ions, such as Li^+, Ca^{++}, Na^+, and K^+, can pass the BBB because of their size.

Functional groups such as —OH, COO^-, and —$\overset{+}{N}H_3$ are hydrophilic, and long hydrocarbon chains or aromatic compounds make the molecule lipophilic. Functional groups that are capable of ionizing, such as amino or carboxy moieties to —R_2NH^+ and —CO_2^-, when dissolved in water or physiologic fluids, greatly increase their water solubility and decrease their lipid solubility.

The degree of ionization depends on both the strength as base or acid (pKa) and the pH of the environment (7.4 for blood). The Henderson-Hasselbalch equation,

$$pH = pKa + \log \frac{[A^-]}{[HA]}$$

in which HA is a weak acid (with a dissociation constant pK_a), can be used to calculate the extent to which a drug is charged or ionized at a particular pH.

When this concept is applied to drugs such as amphetamine (pK_a 9.8), one can determine that at the physiologic pH of 7.4, 99.6% of the drug will be in the charged form (see appendix).

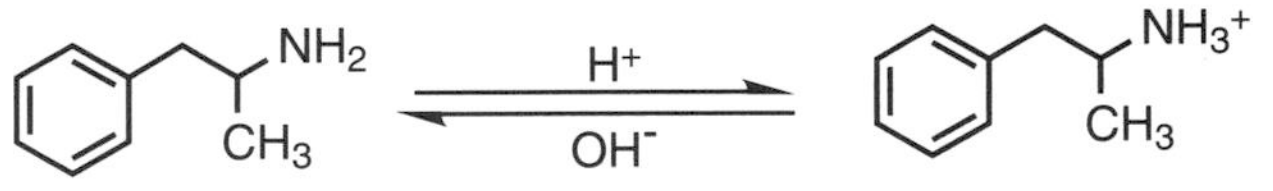

NH_2 / CH_3 $\underset{OH^-}{\overset{H^+}{\rightleftharpoons}}$ NH_3^+ / CH_3

Amphetamine

We know that amphetamine produces cerebral effects, so that even though only 0.4% of the drug is in its lipid-soluble, uncharged form, this seems to be sufficient because of the oil:water partitioning to allow the drug to penetrate the BBB. Addition of a single hydroxy group as in tyramine markedly decreases its penetration into the brain. Addition of another hydroxy group on the benzene ring and on the alkyl side chain, as in norepinephrine, diminishes lipid solubility to such an extent that these substances are no longer active by mouth and do not penetrate the BBB. Strong first-pass metabolism also contributes to its poor oral bioavailability. It should be noted that the hydroxy groups on the benzene ring are weak acids and at physiologic pH are ionized less than 1%.

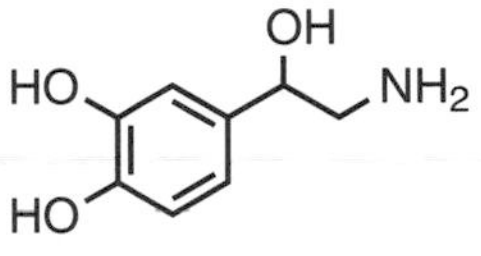

HO, HO, OH, NH_2

Norepinephrine

When one considers a more complex molecule such as chlorpromazine, which contains two benzene rings, it can be predicted that it will be sufficiently lipid-soluble to penetrate the brain, as indeed it does. In contrast to the simpler aromatic amines, chlorpromazine is a weaker base (pK_a 9.2), and at pH 7.4, 98.4% is in the charged

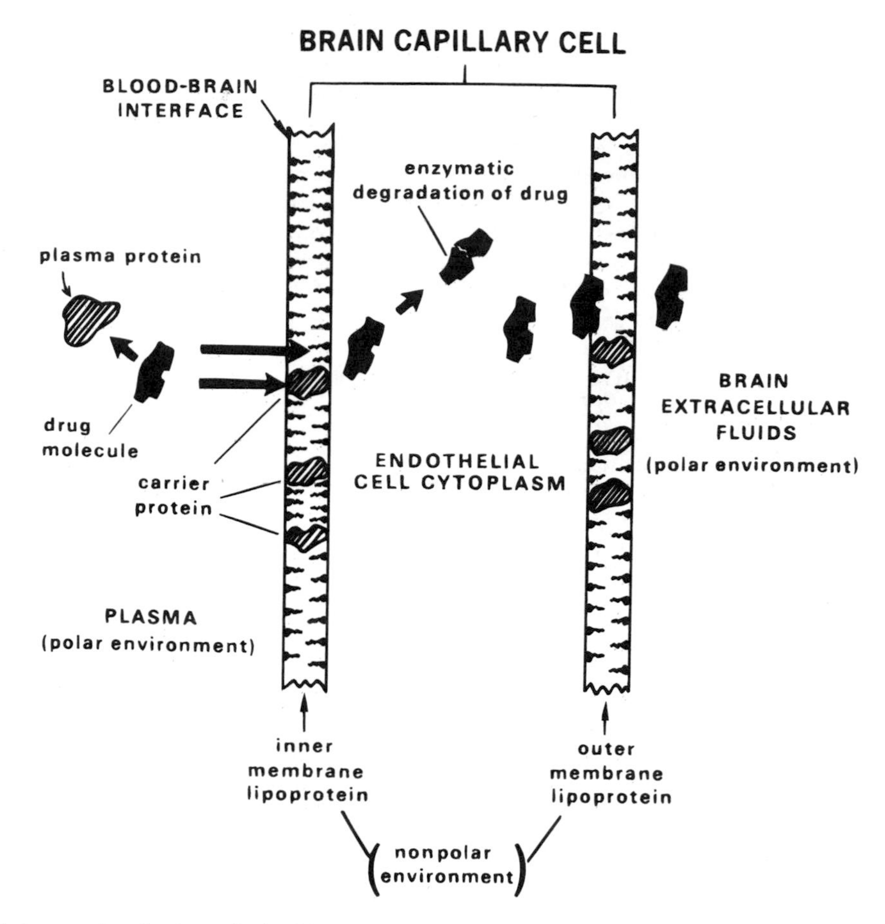

Fig. 12–7. Passage of drug molecule through the blood–brain barrier. The ability of a drug to penetrate the BBB is dependent on its relative affinity to (1) plasma water, (2) plasma protein, (3) membrane lipid, and (4) membrane carrier protein. During its transcellular passage, a drug is exposed to cytoplasmic enzymes and may be degraded before getting to the brain. (Adapted from Olendorf and Dewhurst.[17])

form. The high lipid-solubility of the uncharged form of the drug, however (the chlorine substituent on the benzene ring further enhances lipid solubility), causes this drug to partition to a greater extent in lipids than in aqueous media.

Strategies for Drug Delivery

A pharmacologic strategy that has been developed to deliver drugs to the brain is drug latentiation, or the con-

S N Cl

$CH_2CH_2CH_2N(CH_3)_2$

Chlorpromazine

version of hydrophilic drugs into lipid-soluble drugs. Latentiation to make drugs more lipid-soluble typically involves masking of three primary functional groups: hydroxyl, carboxyl, and primary amine. Hydroxyl groups, for example, must be blocked because BBB permeability to a drug decreases by about 10-fold for each hydroxyl. A classic example of drug latentiation is the conversion of the highly polar morphine to diacetyl morphine (heroin), which is less water-soluble and diffuses through the BBB approximately 100-fold faster than morphine.[18] Once within the brain, pericapillary pseudocholinesterase allows for deacetylation and reconversion back to the parent morphine.

An interesting approach to the delivery of drugs across the BBB is the work of Bodor and colleagues.[19] Their concept, called chemical delivery systems, is based on a dihydropyridine-pyridinium salt redox system, which yields a positively charged drug-pyridinium salt within the brain. After diffusion of the dihydropyridine drug precursor into the brain, the redox product is charged and trapped in the brain, where it slowly releases the drug through enzymatic hydrolysis (Fig. 12–8). This technique has been applied experimentally to the delivery of phenethylamine and dopamine to the brain.[19]

Additional strategies for delivery of drugs to the brain arose from an understanding of the physiology of transport processes at the BBB for nutrients and peptides. For instance, soluble nutrients cross the BBB through carrier

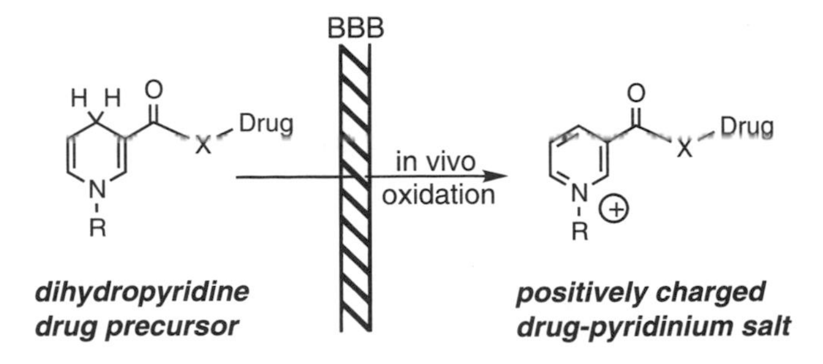

Fig. 12–8. Chemical delivery system based on a dihydropyridine-pyridinium salt redox system.[19]

mediation, and some circulating peptides may cross the BBB via receptor-mediated transcytosis of the peptide through the endothelial cytoplasm.[18] Neutral amino acids in blood are transported in brain interstitium through a specific neutral amino acid carrier system localized in both the lumenal and the antilumenal membranes of the BBB. This neutral amino acid transport system mediates the bidirectional movement of neutral amino acid between blood and brain. Neutral amino acid drugs such as α-methyldopa, L-dopa, α-methyltyrosine, L-tryptophan, and phenylalanine mustard or melphalan gain access to brain interstitium despite being water-soluble because they are transported through the BBB by the neutral amino acid carrier. All of these drugs are probably also transported into peripheral tissues by carrier-mediated systems.

DRUGS USED IN THE TREATMENT OF PSYCHOSES

Several classes of drugs are effective in the treatment of psychoses. They are mostly used in the therapy of schizophrenia, organic psychoses, the manic phase of manic-depressive illness, and other acute or chronic idiopathic psychotic illnesses. They are also indicated in major depression with psychotic features. Currently the most important chemical classes of drugs having antipsychotic or neuroleptic activity are the phenothiazines, thioxanthenes, and butyrophenones. Other chemical classes discussed are the dibenzazepines, benzoquinolizines, benzamides, diphenylbutylpiperidines, reduced indolones, and the rauwolfia alkaloids (Fig. 12–9).

Mechanism of Action

The mechanisms of action of the neuroleptic drugs are only partially known. Although the neuroleptic drugs represent a wide variety of chemical structures (Fig. 12–9), their pharmacologic and clinical activities are remarkably similar. The theory that the antagonism of dopamine-mediated synaptic neurotransmission is an important action of these drugs has been discussed in the previous section.

The neuroleptic agents in current use in the United States all produce a variety of disorders of the control of posture, muscle tone, and movement, presumably as a result of dysfunction of the extrapyramidal nervous system. Several effective but atypical neuroleptic drugs now available, however, only weakly induce acute neurologic reactions (dystonia, parkinsonism, and akathisia or restlessness), indicating that such side effects need not be produced by all neuroleptic agents. This includes thioridazine, sulpiride, molindone, and most notably clozapine (a dibenzodiazepine; see Table 12–5). Because the pharmacologic basis of selectivity of the so-called atypical neuroleptic agents is not fully understood, the antipsychotic activity of an agent such as clozapine is not adequately predicted by current theory or preclinical laboratory tests. Some of the atypical neuroleptic agents such as clozapine and thioridazine are strongly anticholinergic, but molindone and sulpiride are not. Most of the atypical agents have weak antidopamine effects in vitro and have little ability to induce supersensitivity to dopamine agonists.

The neuroleptic drugs differ from most other central nervous system depressants in several ways:

1. In general, neuroleptic agents have limited ability to induce the sedative or hypnotic effects common to sedatives and other central nervous system depressants.
2. The neuroleptic drugs have a remarkably low potential for lethality as a result of coma or respiratory depression on acute overdose.
3. Neuroleptic agents lack euphoriant effects and have a low potential for abuse and dependence.
4. Routine or early tolerance to the main effects of the neuroleptic drugs is not evident, although tolerance is found with some side effects of neuroleptics (including sedation, hypotension, anticholinergic effects, acute dystonic reactions, and possibly parkinsonism) but not others (akathisia), and tolerance to antipsychotic effects may occur after sustained use of potent agents for several years in a minority of patients.
5. Neuroleptics are more capable of diminishing conditioned behavioral responses than depressing unconditioned responses.
6. Neuroleptic drugs have an inhibitory effect on autonomic and motoric expressions of arousal in animals. These effects are presumably mediated in the limbic forebrain and hypothalamus.

The dopamine receptor-antagonism theory discussed previously probably accounts for some of the neurologic side effects of neuroleptic drugs. These effects also contribute to the useful antidyskinetic actions as in Huntington's disease and Gilles de la Tourette's syndrome and explains why most neuroleptic agents elevate levels of prolactin (prolactin release is inhibited by dopamine and dopamine agonists at postsynaptic D_2-type receptors in the anterior pituitary).

The chemoreceptor trigger zone (CTZ) or emetic center located in the brain stem is not within the BBB. The antidopamine effects at these sites contribute to the antinausea and antiemetic effects of most typical neuroleptics.

In general, the less potent neuroleptics are less selective in blocking dopamine receptors and interact with many other systems as well. Some of the side effects associ-

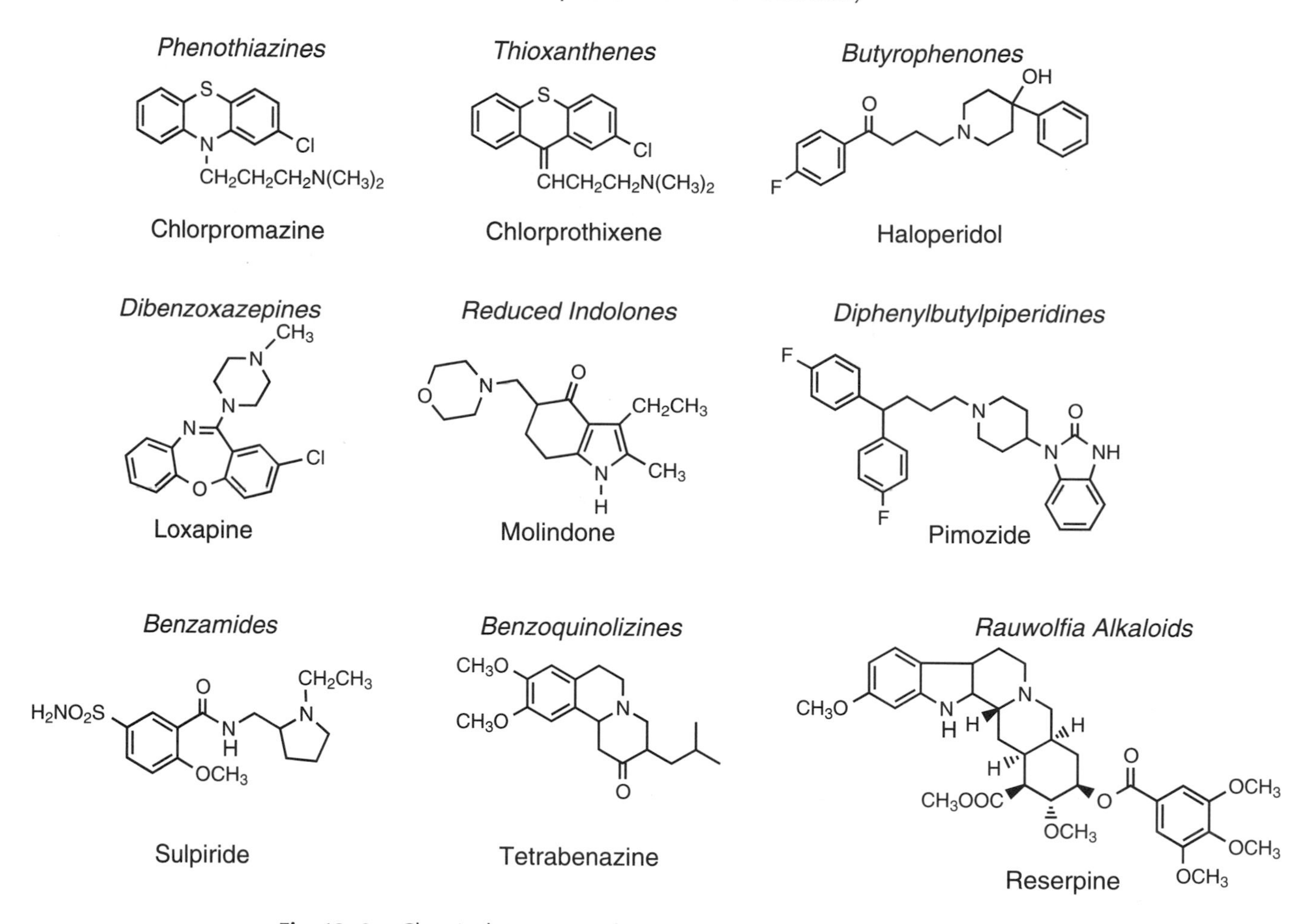

Fig. 12–9. Chemical structures of major chemical classes of antipsychotic drugs.

ated with these drugs, such as sedation, hypotension, sexual dysfunction, and other autonomic effects, may reflect their ability to block adrenergic and histamine receptors.

The anticholinergic actions of the neuroleptics on cardiac, ophthalmic, gastrointestinal, bladder, and genital tissue have been ascribed to the effect of these agents on parasympathetic muscarinic acetylcholine (AChm) receptors. Such antimuscarinic actions are characteristic of the less potent agents and are especially prominent with clozapine and thioridazine.[20] With the phenothiazines and butyrophenones, and virtually all other classes of neuroleptic drugs, extrapyramidal side effects vary directly with the ratio of their anti-D_2:anti-AChm potency.[21]

It has also been suggested that some neuroleptics interact with serotonin ($5\text{-}HT_2$) receptors, particularly those in forebrain regions other than the extrapyramidal motor system. Such interaction with serotonin receptors is characteristic of butyrophenones. The antipsychotic effects of reserpine and other Rauwolfia alkaloids and tetrabenazine are due primarily to their ability to deplete amine stores from cells containing catecholamines and indoleamines in the brain. Their limited efficacy, side effects (hypotension, diarrhea), and tendency to induce excessive sedation have led to their virtual abandonment for the treatment of psychosis.[5]

Several clinically effective neuroleptic drugs, including haloperidol, bind with high affinity to putative sigma (σ) receptors in brain. The σ binding site previously was proposed to be an opiate receptor subtype; however, it subsequently was shown to display reversed stereoselectivity for opiate receptor ligands and is not sensitive to the opiate antagonist naloxone. Although the biologic function of the σ receptor remains unclear and no endogenous ligand has been identified, it has been proposed as a locus of neuroleptic drug action.[22]

The antidopaminergic effects of these drugs in the limbic system, cerebral cortex, and reticular formation correlate with their behavioral and clinical effects. The anti-D_2 effects on the hypothalamus, pituitary, and basal ganglia seem to account for many of the autonomic, neuroendocrine, and neurologic side effects of these agents. The basis of the antipsychotic effects of atypical neuroleptic drugs such as clozapine is unknown and may involve blockade of D_1-type receptors, M_1 muscarinic antagonist activity, effects on serotonin ($5\text{-}HT_{1C}$) receptors, or inhibition of mesolimbic DA release.[23] Such agents encour-

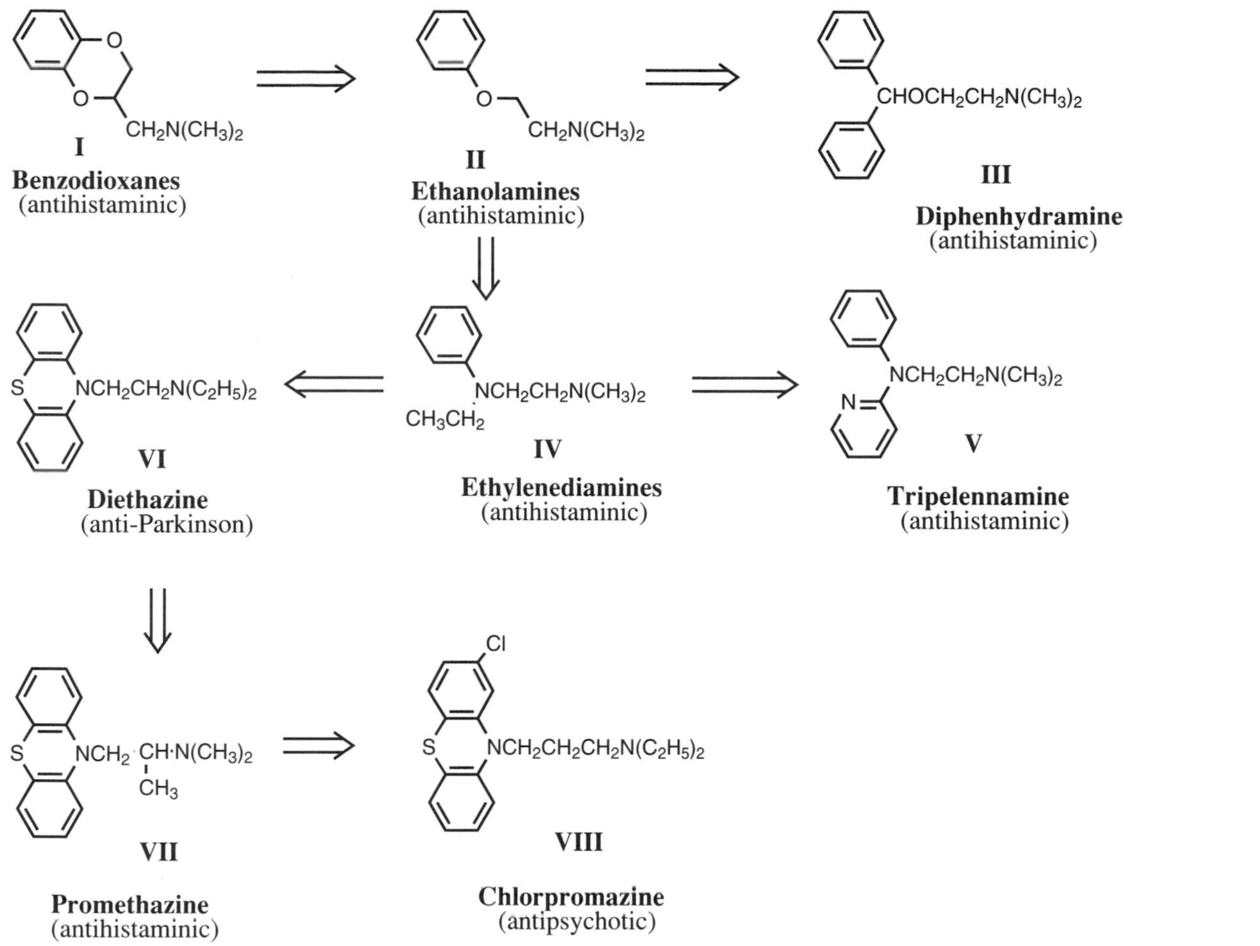

I
Benzodioxanes
(antihistaminic)

II
Ethanolamines
(antihistaminic)

III
Diphenhydramine
(antihistaminic)

VI
Diethazine
(anti-Parkinson)

IV
Ethylenediamines
(antihistaminic)

V
Tripelennamine
(antihistaminic)

VII
Promethazine
(antihistaminic)

VIII
Chlorpromazine
(antipsychotic)

Fig. 12–10. Evolution of the phenothiazine antipsychotic agents.

age the hope that more selective neuroleptic agents with less serious neurologic side effects can be developed.

Phenothiazines and Thioxanthenes

The phenothiazines as a class, and especially chlorpromazine, are the most widely used neuroleptics and may be divided into subclasses—aliphatic, piperidine, piperazine—as can the thioxanthenes, based on differences in their chemical structures, pharmacologic actions, and potency (Table 12–3).

Discovery of Phenothiazines and Neuroleptics

Although phenothiazine was synthesized in 1883 and has been used as an anthelmintic for many years, it has no antipsychotic activity. The discovery of this class of antipsychotic drugs provides an outstanding case history of modern drug development and also points out the unpredictability of biologic activity from structural modification of a prototype drug molecule. The basic structural type from which the phenothiazine antipsychotic drugs trace their origins is the antihistamines of the benzodioxane type I (Fig. 12–10). Bovet hypothesized in 1937 that specific substances antagonizing histamine ought to exist, tried various compounds known to act on the autonomic nervous system, and was the first to recognize antihistaminic activity.[24] With the benzodioxanes as a starting point, many molecular modifications were carried out in various laboratories in a search for other types of antihistamines. The benzodioxanes led to ethers of ethanolamine of type II, which after further modifications led to the benzhydryl ethers (III) characterized by the clinically useful antihistamine, diphenhydramine, or to the ethylenediamine type IV, which led to drugs such as tripelennamine (V). Further modification of the ethylenediamine type of antihistamine resulted in the incorporation of one of the nitrogen atoms into a phenothiazine ring system, which produced the phenothiazine (VI), a compound that was found to have antihistaminic properties and, similar to many other antihistaminic drugs, a strong sedative effect. Diethazine (VI) is more useful in the treatment of Parkinson's disease

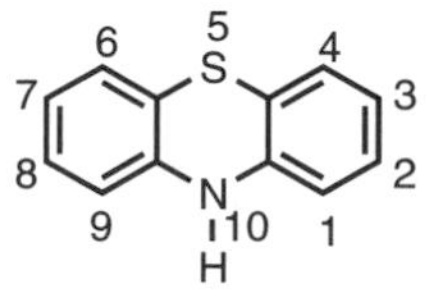

Phenothiazine

Table 12–3. Phenothiazine and Thioxanthene Derivatives Used as Neuroleptics*

Generic Name (Proprietary Name)	R_{10} (phenothiazine: S, N–R_{10}, R_2)	R_2	Adult Antipsychotic Oral Dose Range (mg/day)	Side Effects†			Other Effects
				Sedative Effects	Extrapyramidal Effects	Hypotensive Effects	
Phenothiazines	Propyldialkylamino side chain						
Chloropromazine hydrochloride (Thorazine)	$(CH_2)_3N(CH_3)_2{\cdot}HCl$	Cl	300–800	+ + +	+ +	Oral + + IM + + +	Antiemetic dose 10–25 mg every 4–6 hrs
Triflupromazine hydrochloride (Vesprin)	$(CH_2)_3N(CH_3)_2{\cdot}HCl$	CF_3	100–150	+ +	+ + +	+ +	Antiemetic dose 5–15 mg every 4–6 hrs
Promazine hydrochloride (Sparine)	$(CH_2)_3N(CH_3)_2{\cdot}HCl$	H	25–50 mg/4–6 hrs				Antiemetic dose 5–40 mg every 4–6 hrs
	Alkylpiperidyl side chain						
Thioridazine hydrochloride (Mellaril)	$(CH_2)_2$–(ring N–CH_3) ·HCl	SCH_3	200–600	+ + +	+	+ +	
Mesoridaine besylate (Serentil)	$(CH_2)_2$–(ring N–CH_3) $\cdot C_6H_5SO_3H$	$S(=O)CH_3$	75–300	+ + +	+	+ +	
Piperacetazine (Quide)	$(CH_2)_3N$(ring)–CH_2CH_2OH	$COCH_3$	20–160	+ +	+ +	+	

	Alkylpiperazine side chain						
Perphenazine (Trilafon)	$(CH_2)_3N$-piperidine-CH_2CH_2OH	Cl	8–32	+ +	+ + +	+	
Prochlorperazine edisylate maleate (Compazine)	$(CH_2)_3N$-piperazine-NCH_3	Cl	75–100	+ +	+ + +	+	Antiemetic dose 5–10 mg every 4–6 hrs
Fluphenazine hydrochloride (Permitil, Prolixin)	$(CH_2)_3N$-piperazine-$N-CH_2CH_2OH$ ·2HCl	CF_3	1–20	+	+ + +	+	
Trifluoperazine hydrochloride (Stelazine)	$(CH_2)_3N$-piperazine-$N-CH_3$ ·2HCl	CF_3	6–20	+	+ + +	+	
Acetophenazine maleate (Tindal)	$(CH_2)_3N$-piperidine-CH_2CH_2OH	$COCH_3$	60–120	+ +	+ +	+	
Thiethylperazine maleate (Torecan)	$(CH_2)_3N$-piperidine-CH_2CH_2OH	SCH_2CH_3					Antiemetic dose 10–30 mg daily
	Thioxanthene ring (S, =CHR, R_2)						
Thioxanthenes	***R***	***R_2***					
Chlorprothixene (Taractan)	$(CH_2)_2N(CH_3)_2$	Cl	50–400	+ + +	+ +	+ +	
Thiothixene hydrochloride (Navane)	$(CH_2)_2N$-piperazine-NCH_3 ·HCl	$SO_2N(CH_3)_2$	6–30	+ +	+ +	+ +	

* The phenothiazine derivatives that are effective in the treatment of nausea and vomiting are included in this listing.
† + + +, high; + +, medium; +, low.

(owing to its potent antimuscarinic action) than in allergies, whereas promethazine (VII) is clinically used as an antihistaminic. The ability of promethazine to prolong barbiturate-induced sleep in rodents was discovered and the drug was introduced into clinical anesthesia as a potentiating agent.[25]

In an effort to enhance the sedative effects of such phenothiazines, Charpentier, directing the chemistry, and Courvoisier, the pharmacology, evaluated many modifications of promethazine. This research effort eventually led to the synthesis of chlorpromazine (VIII) in 1950[26] at the Rhône-Poulenc Laboratories. Soon thereafter, the French surgeon Laborit and his co-workers[25] described the ability of this compound to potentiate anesthetics and produce artificial hibernation. They noted that chlorpromazine by itself did not cause a loss of consciousness but produced only a tendency to sleep and a marked disinterest in the surroundings. The first attempts to treat mental illness with chlorpromazine alone were made in Paris in 1951 and early 1952 by Paraire and Sigwald. In 1952, Delay and Deniker began their important work with chlorpromazine.[27] They were convinced that chlorpromazine achieved more than symptomatic relief of agitation or anxiety and that this drug had an ameliorative effect on psychosis. Thus, what initially involved minor molecular modifications of an antihistamine that produced sedative side effects resulted in the development of a major class of drugs that initiated a new era in the drug therapy of the mentally ill. More than anything else in the history of psychiatry, the phenothiazines and related drugs have positively influenced the lives of schizophrenic patients. They have enabled many patients, relegated in earlier days to a lifetime in mental institutions, to assume a greatly improved role in society.

About 24 phenothiazine and thioxanthene derivatives are used in medicine, most of them for psychiatric conditions. The structures, generic and proprietary names, dose, and side effects of phenothiazines and thioxanthenes currently used as neuroleptics are listed in Table 12–3.

Interactions of Phenothiazines and Thioxanthenes with Dopamine Receptors

How do the phenothiazines, which are complex, multiringed structures, interact with the receptor for dopamine or norepinephrine? Examination of the x-ray structures of chlorpromazine and dopamine in what is assumed to be the preferred conformation shows that these two structures can be partly superimposed (Fig. 12–11).[28] Such studies provide clues as to the molecular mechanism and site of action of the phenothiazines. In the preferred conformation of chlorpromazine, its side chain tilts away from the midline toward the chlorine-substituted ring. When thioxanthene derivatives that contain an olefinic double bond between the tricyclic ring and the side chain are examined, it can be seen that such structures can exist in either the *cis* or *trans* isomeric configuration. The *cis* isomer of the neuroleptic chlorprothixene is several times more active than both the *trans* isomer and the compound obtained from saturation of the double bond. Structure D in Figure 12–11 shows the nonsuperimposability with dopamine of a possible conformer of chlorpromazine that would be predicted to be inactive.

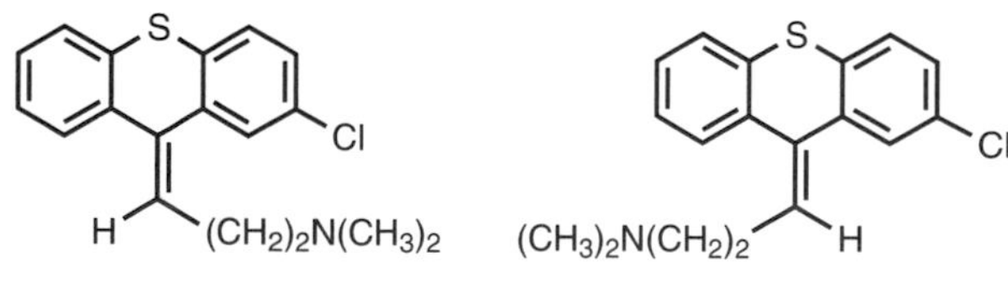

S Cl H $(CH_2)_2N(CH_3)_2$

Cis isomer
Chlorprothixene

S Cl $(CH_3)_2N(CH_2)_2$ H

Trans isomer

The chlorine atom on ring *a* is responsible for imparting asymmetry to this molecule, and the tilt of the side chain toward the ring containing the chlorine atom indicates an important structural feature of such molecules. Compounds lacking a chlorine atom are, in most cases, inactive as neuroleptic drugs. In addition to the ring *a* substituent, another major requirement for therapeutic efficacy is that the side chain amine of phenothiazines contains three carbons separating the two nitrogen atoms (Fig. 12–11). Phenothiazines with two carbon atoms separating the two nitrogen atoms lack antipsychotic efficacy. Compounds such as promethazine (VII) (see Fig. 12–10) are primarily antihistaminic and are less likely to assume the preferred conformation. It should be borne in mind that the side chain of dopamine possesses unlimited flexibility and unrestricted rotation about the β-carbon-phenyl bond. Thus, information concerning the conformational requirements of both dopamine and the dopamine receptor can be obtained from such drugs. (For a more detailed discussion of the conformational requirements of dopamine, see Fig. 13–8, Chapter 13).

Long-Acting Neuroleptics

The duration of action of many of the neuroleptics with a free OH can be considerably prolonged by the preparation of long-chain fatty acid esters. Thus, fluphenazine decanoate and fluphenazine enanthate were the first of these esters to appear in clinical use and are longer acting with fewer side effects than the unesterified precursor. The ability to treat patients with a single intramuscular injection every 1 to 2 weeks with the enanthate or every 2 to 3 weeks with the decanoate ester means that problems associated with patient compliance to the drug regimens and with drug malabsorption can be reduced. Long-acting forms of other phenothiazine, thioxanthene, and butyrophenone derivatives are also available. Only the fluphenazine esters are available in the United States, but other long-acting phenothiazines and thioxanthenes are marketed in other countries. Table 12–4 lists such long-acting neuroleptics.[29]

Pharmacologic Profile of Phenothiazines and Thioxanthenes

Chlorpromazine is commonly taken as the prototype for the many neuroleptic drugs in clinical use today.

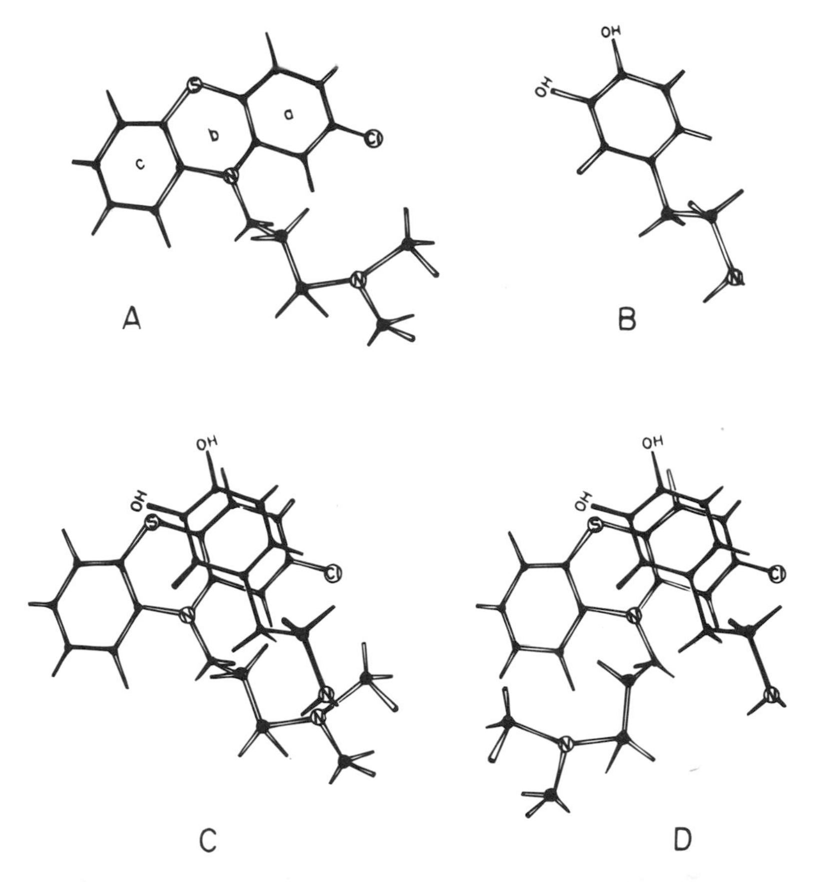

Fig. 12–11. Conformations of chlorpromazine (A), dopamine (B), and their superposition (C), determined by x-ray crystallographic analysis. The a, b, and c in (A) designate rings. D shows another conformation in which the alkyl side chain of chlorpromazine is in the *trans* conformation (ring a and amino side chain), which is not superimposable on dopamine. (Adapted from A. S. Horn and S. H. Snyder, *Proc. Natl. Acad. Sci. U.S.A., 68,* 2325[1971].)

Many of these drugs, especially the low-potency agents such as chlorpromazine, have sedative effects. Neuroleptic drugs also have some antianxiety effects, but this class of agents is not generally used for such purposes. The term neuroleptic, which was introduced to characterize the effects of chlorpromazine and reserpine on psychiatric patients, contrasts the effects of these agents with those of the classic central nervous system depressants (i.e., anesthetics, sedatives, hypnotics, and analgesics). Neuroleptics suppress spontaneous movements and complex behavior, having spinal reflexes and unconditioned nociceptive avoidance behaviors intact. In humans, the neuroleptic drugs reduce initiative and interest in the environment, and they reduce displays of emotion. There may also be some slowness in response to external stimuli and drowsiness. Psychotic patients become less agitated and restless. Aggressive and impulsive behavior diminishes and psychotic symptoms of hallucinations, delusions and disorganized or incoherent thinking tend to disappear. The effects of chlorpromazine and related phenothiazines and most neuroleptics available in the United States are neurologic, including bradykinesia, mild rigidity, some tremor, and subjective restlessness (akathisia) that resemble the effects of Parkinson's disease. For a detailed discussion of the pharmacologic properties of these agents, several reviews should be consulted.[5,30,31]

General Psychophysiologic and Behavioral Effects

Most of the neuroleptics available in clinical practice can diminish spontaneous motor activity in every species of animal studied, including humans. One of the effects of chlorpromazine referred to earlier, akathisia, is manifested paradoxically by an increase in restless activity. The cataleptic immobility of animals treated with phenothiazines resembles the catatonia seen in some psychotic patients. Also, as discussed previously, phenothiazines and many other antipsychotic drugs often produce parkinsonism and other extrapyramidal effects.

Effect on Sleep. The effect of antipsychotic drugs on sleep patterns is not consistent, but the drugs tend to normalize sleep disturbances characteristic of many psychoses.

Effects on Conditioned Avoidance Responses. In animals, chlorpromazine impairs the ability to make a conditioned avoidance response. This procedure has become the basis for screening for neuroleptic agents, although the validity of this test to provide new neuroleptic agents is questionable.

Effects on Complex Behavior. Neuroleptic drugs impair vigilance in human subjects performing a variety of tasks.

Effects on Specific Areas of the Nervous System. The effects of neuroleptic drugs are apparent at all levels in

Table 12–4. Long-acting Neuroleptics

(General structure: phenothiazine ring with N-R at position 10 and R_2 at position 2)

Generic Name	R	R_2	Route of Administration	Dosage Range (mg)	Mean Duration of Action (weeks)
Phenothiazines					
Fluphenazine enanthate	$(CH_2)_3N$(piperazine)$N-CH_2CH_2O-C(=O)-(CH_2)_5CH_3$	CF_3	IM	25–100	1–2
Fluphenazine decanoate	$(CH_2)_3N$(piperazine)$N-CH_2CH_2O-C(=O)-(CH_2)_8CH_3$	CF_3	IM	25–200	2–3
Perphenazine enanthate	$(CH_2)_3N$(piperazine)$N-CH_2CH_2O-C(=O)-(CH_2)_5CH_3$	Cl	IM	25–100	1–2
Pipotiazine undecylenate	$(CH_2)_3N$(piperidine)$-CH_2CH_2O-C(=O)-(CH_2)_8CH=CH_2$	$SO_2N(CH_3)_2$	IM	100–450	1–2
Pipotiazine palmitate	$(CH_2)_3N$(piperidine)$-CH_2CH_2O-C(=O)-(CH_2)_{14}CH_3$	$SO_2N(CH_3)_2$	IM	50–600	4
Thioxanthenes					
Flupenthixol decanoate	Thioxanthene ring with R_2; $=C(H)-(CH_2)_2-N$(piperazine)$N-(CH_2)_2-OR$; $R = -C(=O)-(CH_2)_8CH_3$	CF_3	IM	20–100	1–2

Adapted from Simpson and Lee[27] and Baldessarini.[5]

the nervous system and, as previously discussed, antagonize the actions of dopamine in the basal ganglia and limbic portions of the forebrain. In addition, the effects of these agents on the hypothalamus or pituitary produce endocrine changes, the most prominent being an increase in the secretion of prolactin.

Effect on Chemoreceptor Trigger Zone. Most of the neuroleptic agents have a protective action against the nausea and emesis–inducing effect of apomorphine and certain of the ergot alkaloids, which are dopamine agonists. Several of the neuroleptic (and nonneuroleptic) phenothiazine derivatives (see Table 12–3) have found use in the treatment of nausea and vomiting.

Metabolic Pathways for Phenothiazines and Thioxanthenes

There is increasing evidence that the metabolism of neuroleptic drugs is of major significance in the effects of these drugs.[32] Although considerable information about the metabolism of the extensively studied chlorpromazine is available, information about many of the other drugs administered for prolonged periods is scant. It is becoming more important to understand the metabolic fate of neuroleptic drugs and to measure levels of both parent drug and active metabolites.

The metabolic pathways for neuroleptic drugs are similar to those for many other drugs (see Chapter 8). Some metabolic pathways for chlorpromazine are shown in Fig. 12–12. It should be kept in mind that, during metabolism, several processes can and do occur for the same molecule. For example, chlorpromazine can be demethylated, sulfoxidized, hydroxylated, and glucuronidated to yield 7-O-glu-nor-CPZ-SO. The combination of such processes leads to more than 100 identified metabolites. Metabolic pathways are significantly altered, both quantitatively and qualitatively, by a number of factors, including species, age, sex, interaction with other drugs, and route of administration.

There is evidence (by radioimmunoassay) that the 7-hydroxylated derivatives and possibly other hydroxylated derivatives as well as the mono- and di-desmethylated products (nor_1-CPZ, nor_2-CPZ), are active in vivo and at D_2 receptors, whereas the sulfoxide (CPZ-SO) is inactive.

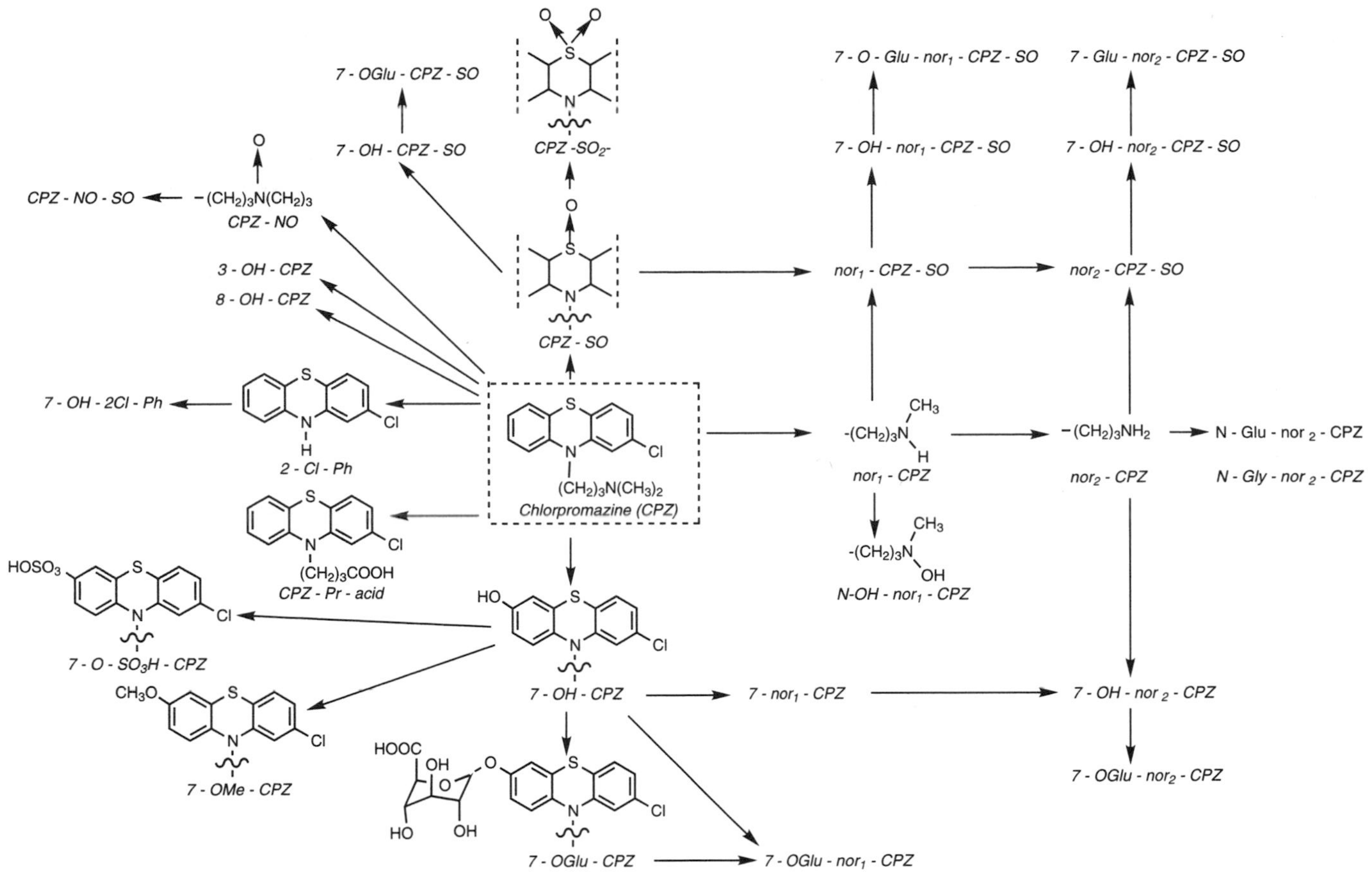

Fig. 12–12. Some metabolic pathways for chlorpromazine. Abbreviations: CPZ, chlorpromazine; NO, N-oxide; SO, sulfoxide; SO_2, sulfone; O-Glu, O-glucuronide; Ph, phenothiazine; Pr-acid, propionic acid; O-SO_3H, sulfate.

Although the thioxanthenes are closely related to the phenothiazines in their pharmacologic effects, there seems to be at least one major difference in metabolism: Most of the thioxanthenes do not form ring-hydroxylated derivatives.

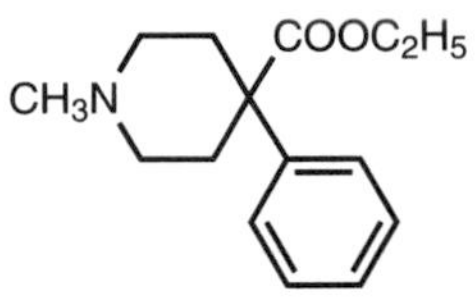

Butyrophenones

A series of butyrophenones of high neuroleptic potency was developed by Janssen and co-workers.[33] To increase the analgesic potency of meperidine, a number of normeperidine analogs were prepared, including the propiophenone and butyrophenone analogs. The propiophenone analog has 200 times the analgesic potency of meperidine, but the butyrophenone analog (in addition to a morphine type of activity) also displays activity resembling that of chlorpromazine. Janssen revealed that it is possible to eliminate the morphine type of activity and simultaneously to accentuate the chlorpromazine type of activity in the butyrophenone series, provided that certain structural changes are made.

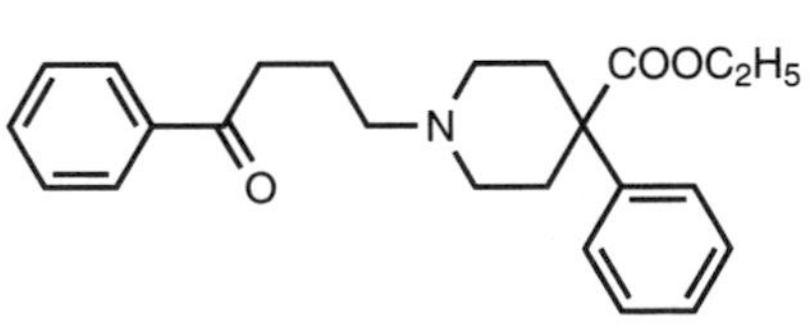

Meperidine

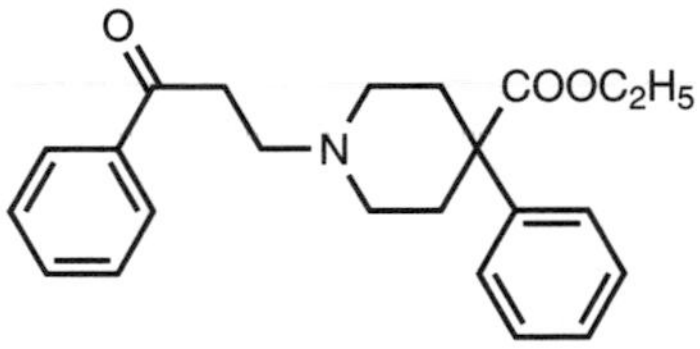

Propiophenone analog

Butyrophenone analog

Structure-Activity Relationships

All butyrophenone derivatives displaying high neuroleptic potency have the following general structure:

The attachment of a tertiary amino group to the fourth

carbon of the butyrophenone skeleton is essential for neuroleptic activity. In addition, the following structural features are noteworthy:

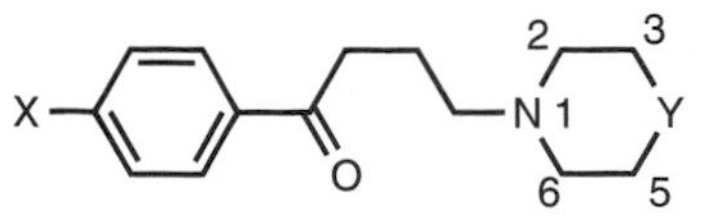

1. With one exception (anisoperidone), all potent compounds have a fluorine substituent in the para position of the benzene ring. Anisoperidone has a methoxy group in the para position.
2. Replacement of the keto group by a thioketone group (the transition of a butyrophenone to a butyrothienone derivative) decreases neuroleptic potency. Replacement of the keto group by olefinic or phenoxy groups or reduction of the carbonyl group also decrease potency.
3. Lengthening, shortening, or branching of the three-carbon (propyl) chain decreases neuroleptic potency.
4. Variations are possible in the tertiary amino group without loss of neuroleptic potency. The basic nitrogen is usually incorporated into a six-membered ring (piperidine, tetrahydropyridine, or piperazine), which usually has another substituent in position 4.

Haloperidol was introduced for the treatment of psychoses in Europe in 1958 and the United States in 1967. It has proved to be an effective alternative to more familiar antipsychotic phenothiazine drugs, being effective both in the manic phase of manic-depressive illness and in schizophrenia. Haloperidol decanoate has been introduced as depot maintenance therapy. When injected every 4 weeks, the drug appears to be as effective as orally administered haloperidol.[34]

Other currently available (mostly in Europe) butyrophenones include the most potent neuroleptic yet discovered, spiperone (spiroperidol), as well as trifluperidol and droperidol. Droperidol, a short-acting, sedating butyrophenone, is used in anesthesia for its neuroleptic and antiemetic effects and sometimes in psychiatric emergencies as a sedative-neuroleptic. Droperidol is often used in combination with the potent narcotic analgesic fentanyl (see Chapter 14). The combination (Innovar) is administered for preanesthetic sedation and anesthesia.

In most respects, the pharmacologic effects of haloperidol differ in degree but not in kind from those of the piperazine phenothiazines. As previously discussed, these drugs block the effects of dopamine.

Haloperidol is readily absorbed from the gastrointestinal tract. Peak plasma levels occur 2 to 6 hours after ingestion. The drug is concentrated in the liver and CNS, about 15% of a given dose is excreted in the bile, and about 40% is eliminated through the kidney. Fig. 12–13 shows the typical oxidative metabolic pathway of butyrophenones.[35]

Haloperidol produces a high incidence of extrapyramidal reactions, but its sedative effect in moderate doses is less than that observed with chlorpromazine. It blocks apomorphine-induced emesis. Haloperidol has less

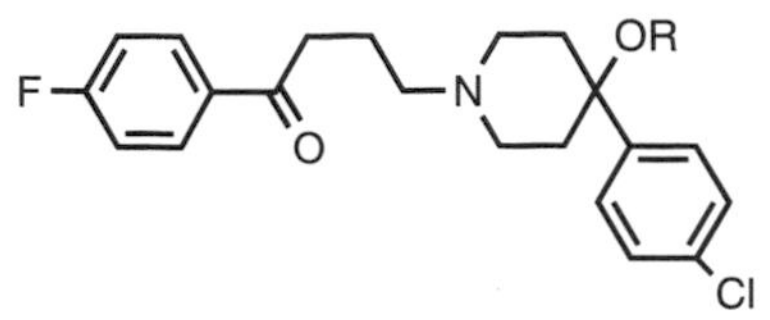

R = H Haloperidol (Haldol®)
R = $CO(CH_2)_8CH_3$, Haloperidol decanoate (Haldol® decanoate)

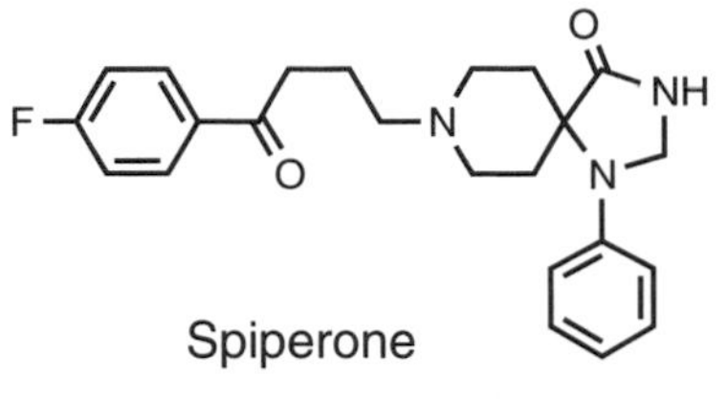

Spiperone

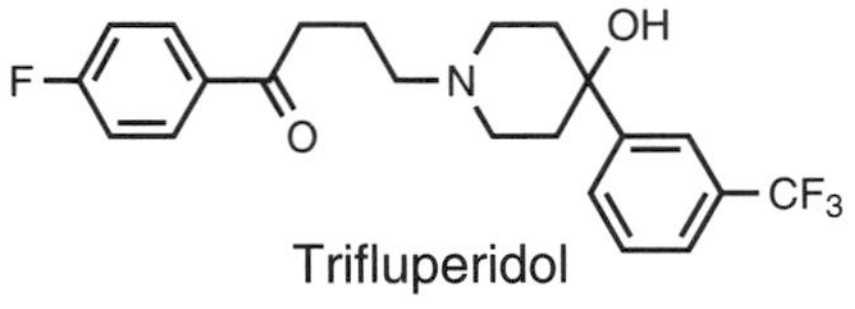

Trifluperidol
Triperidol

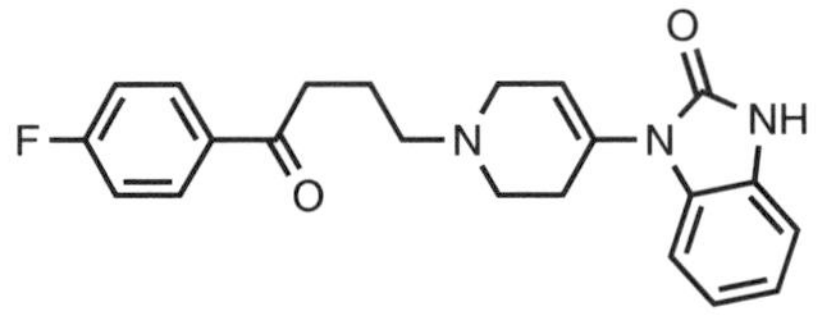

Droperidol (Inapsine®)

prominent autonomic effects than do the other antipsychotic drugs. Only mild hypotension occurs with the use of haloperidol even in high parenteral doses. A long-acting form of haloperidol, haloperidol decanoate, which is effective for 3 to 4 weeks, is also available.

Diphenylbutylpiperidines

Modification of the butyrophenone side chain (i.e., haloperidol) by replacement of the keto function with a 4-flurophenylmethane moiety resulted in the production of diphenylbutylpiperidine derivatives such as pimozide, penfluridol, and fluspirilene. The diphenylbutyl piperidines differ from butyrophenone drugs in their longer duration of action. All are effective in the control of schizophrenia, and, in particular, pimozide has been shown to be useful in treating acute exacerbations in schizophrenia and in reducing the rate of relapse in chronic schizophrenic patients when administered orally in dosages of 2 to 6 mg daily.[29,36] Pimozide has been approved as an "orphan drug" in the United States for the treatment of Tourette's syndrome, which is estimated to afflict approximately 100,000 Americans. The bizarre symptoms that characterize the disease, including facial tics, grimaces,

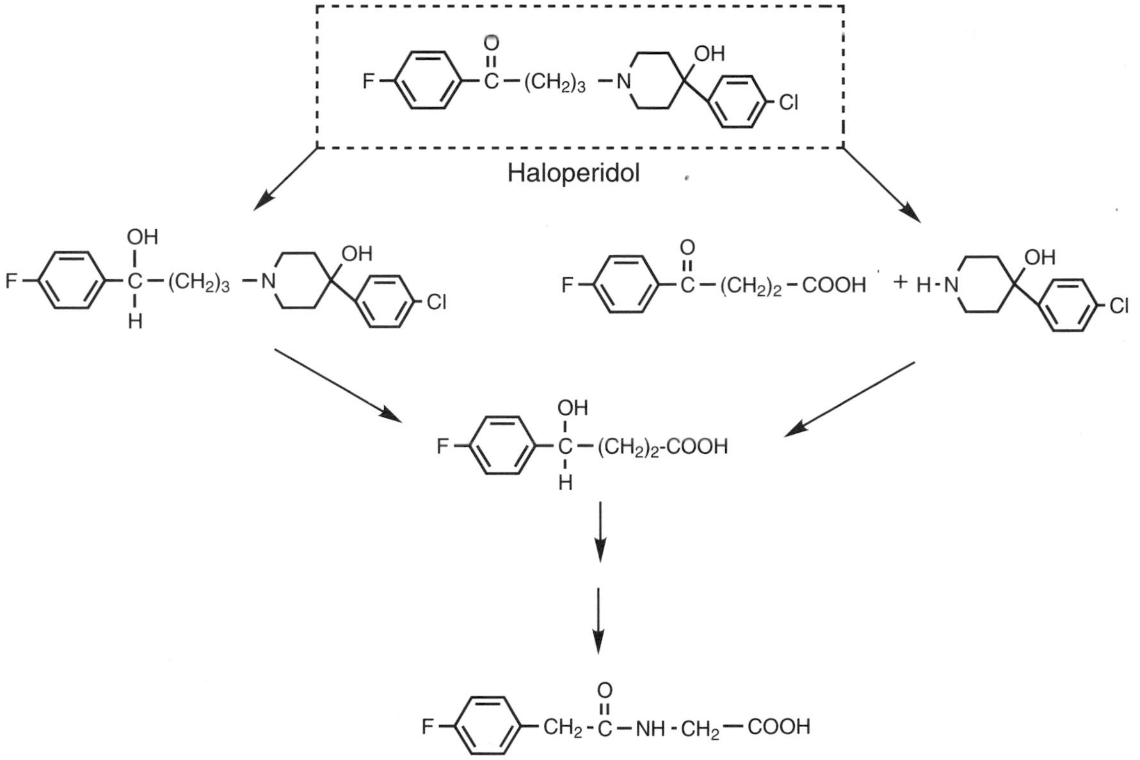

Haloperidol

Fig. 12–13. Metabolism of haloperidol.

strange uncontrollable sounds, and sometimes the involuntary shouting of obscenities, are often misdiagnosed by clinicians as schizophrenia. Typically the onset of Tourette's syndrome occurs at age 10. Standard treatment for Tourette's syndrome has been the neuroleptics, often haloperidol. Similar to other neuroleptic drugs, pimozide can induce potentially irreversible tardive dyskinesia but can also induce impaired cardiac conduction.

The long duration of action of penfluridol (orally) and fluspiriline (intramuscularly) allows their use as a once-weekly maintenance drug treatment. Neither of these agents is available in the United States. Fluspirilene differs from the other long-acting agents in that its formulation for injection is as a micronized suspension rather than a fatty acid ester, as for example, in fluphenazine decanoate (see Table 12–4). The usual duration of action

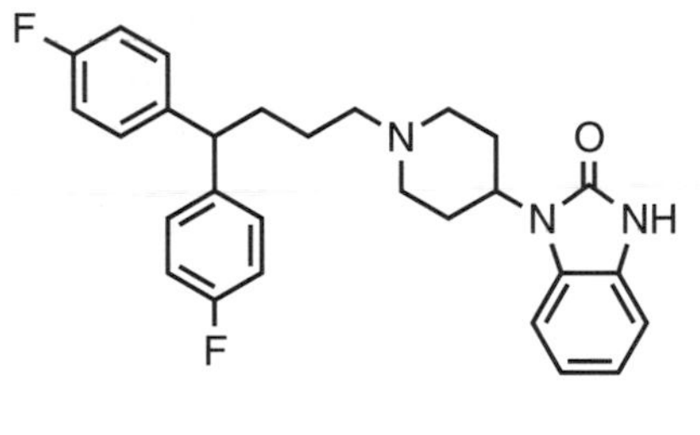

Pimozide (Orap®)

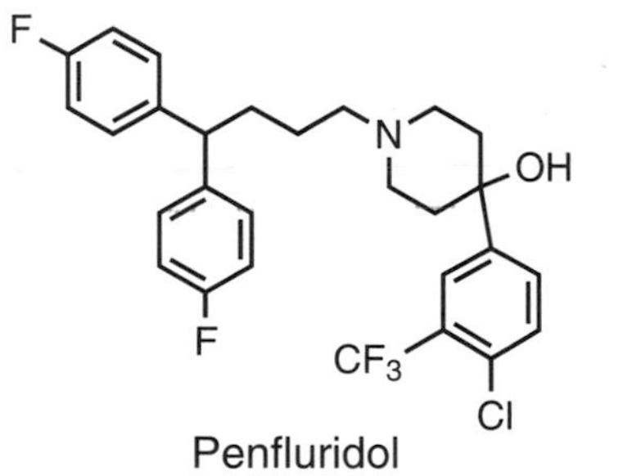

Penfluridol

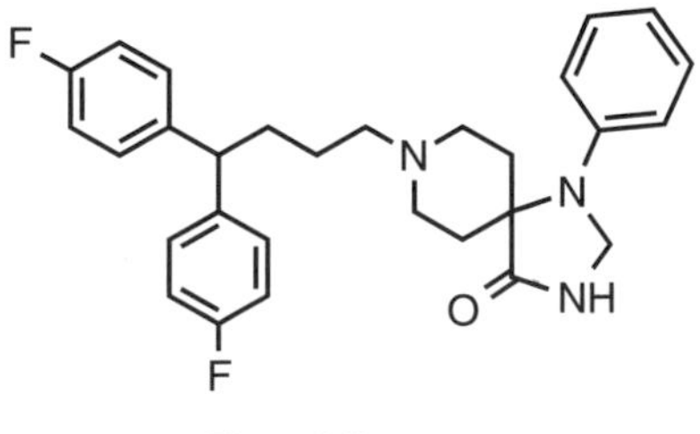

Fluspirilene

is 1 week, with maintenance doses of 1 to 10 mg weekly. Penfluridol is also a long-acting agent that has the advantage of being effective at dosages of 20 to 160 mg when given orally at weekly intervals. The drug has been effective in the treatment of acute psychosis, in severely ill schizophrenic patients, and as maintenance therapy for chronic schizophrenic patients. The side effects appear to be similar to those of other neuroleptics, i.e., primarily extrapyramidal.

Dibenzazepines

Dibenzazepine derivatives are another class of neuroleptic drugs. Loxapine succinate (Loxitane) is available in the United States for treatment of psychosis. An advantage of this compound is that it appears to have a limited number of metabolites, and methods for its blood level estimation are available.[29] Loxapine produces sedation and pronounced extrapyramidal reactions, increases the convulsive threshold, and has some antiadrenergic and weak anticholinergic effects. Metiapine and clothiapine (Table 12–5) are two members of the group of antipsychotics belonging to the dibenzothiazepine class and are active neuroleptic agents.[29] Neither of these drugs is currently available in the United States. The fourth member of this family, clozapine, is a dibenzodiazepine and an unusual neuroleptic agent that produces virtually no extrapyramidal symptoms. As previously discussed, the lack of extrapyramidal side effects may reflect its powerful anticholinergic effects relative to its weak antidopamine properties. Clozapine also has the ability to suppress symptoms of tardive dyskinesia. Findings of a high incidence of agranulocytosis have greatly limited clinical use of this agent.[29]

Pharmacologic tests have demonstrated that, in contrast to the classic neuroleptics, clozapine does not cause catalepsy, does not antagonize apomorphine, and is only a weak antagonist of apomorpine-induced stereotyped behavior in rats.[37]

Table 12–5. Antipsychotic 11-Piperazinyldibenzazepines

Generic Name (Proprietary Name)	Chemical Type	X	R_8	R_2
Clozapine (Leponex)	Dibenzodiazepine	NH	Cl	H
Loxapine (Loxitane)	Dibenzo-oxazepine	O	H	Cl
Clothiapine	Dibenzothiazepine	S	H	Cl
Metiapine	Dibenzothiazepine	S	H	CH_3

Reduced Indolones

Many indole derivatives have been synthesized and tested for neuroleptic potency. Oxypertine (Integrin), an indolylalkyl-phenylpiperazine, is a potent central nervous system depressant.[38] In the conditioned avoidance blocking test in rats and dogs, a subcutaneous dose of oxypertine was as potent as chlorpromazine; orally, oxypertine is about three times more potent than chlorpromazine. Oxypertine has not been approved for use in the United States.

Oxypertine

Molindone hydrochloride (Moban) is a neuroleptic agent that is structurally unrelated to any of the other marketed neuroleptics. This antipsychotic agent is a tetrahydroindolone derivative, 3-ethyl-6,7-dihydro-2-methyl-5-(morpholinomethyl)indol-4(5H)-one hydrochloride, but is less potent than the piperazine phenothiazines and is indicated in the management of schizophrenia.[39]

Molindone hydrochloride
Muban®

Molindone, similar to the other neuroleptics, can cause extrapyramidal reactions. In addition, molindone antagonizes the depression caused by the tranquilizing agent tetrabenazine. Metabolism studies in humans show molindone to be rapidly absorbed and metabolized when given orally. There are 36 recognized metabolites with less than 2% to 3% unmetabolized molindone being excreted in urine and feces. The usual antipsychotic dose range is 50 to 100 mg/day.

Benzamides

Substituted benzamides were first derived from the synthesis of derivatives of para-aminobenzoic acid and the analogous derivatives of para-aminosalicyclic acid (Fig. 12–14).[40] This produced ortho-methoxyprocainamide, a compound with local anesthetic properties and potent antiemetic action. Further structural alteration produced metoclopramide, which possesses antiemetic activity but limited local anesthetic properties. Metoclopramide was found to possess central dopamine antagonist properties and subsequently has been shown to be neuroleptic.[6] Further derivatives of these molecules led to the pyrrolidinyl-containing benzamides sulpiride and sultopride, which

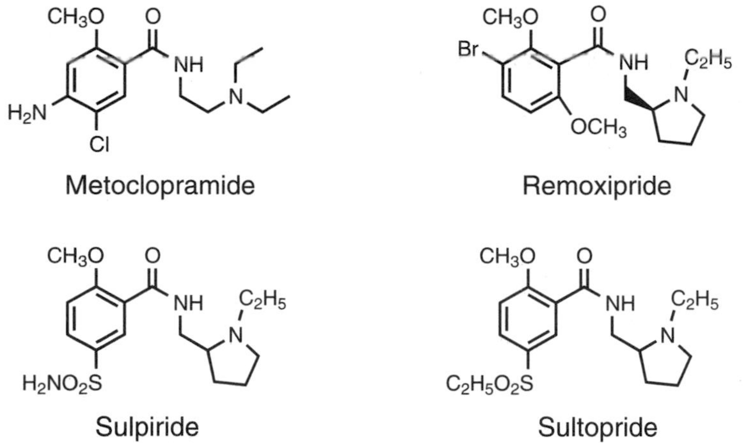

Metoclopramide

Remoxipride

Sulpiride

Sultopride

Fig. 12–14. Structural formulas of some substituted benzamide neuroleptic drugs.

also are neuroleptic agents.[6] Sultopride more closely resembles phenothiazine-type neuroleptics in that it has pronounced sedative properties and produces an appreciable incidence of extrapyramidal disturbances. In contrast, sulpiride has little sedative effect in humans and may produce disinhibitory effects in low doses.[6] It has been suggested that the lack of extrapyramidal side effects by sulpiride is due to a preferential effect on limbic in comparison with striatal tissue.[41] The hydrophilic properties of sulpiride may also account for their limited penetration into the central nervous system and their low potency. The active enantiomer of sulpiride and sultopride is the (−) enantiomer.[41] In clinical trials, emoscipride was shown to be efficacious in the treatment of schizophrenia and had less incidence of extrapyramidal side effects and elevation of serum prolactin than haloperidol.[23] Currently, no substituted benzamides are available for use as neuroleptics in the United States.

Tetrabenazine

Because the ipecacuanha and protoberberine alkaloids that are noted for pharmacologic effects contain the 1,3,4,6,7,11b-hexahydro-2H-benzo[a]quinolizine skeleton, many compounds containing it have been synthesized.

Tetrabenazine, the 9,10-dimethoxy-1,3,4,6,7,11b-hexahydro-3-isobutyl-2H-benzo[a]quinolizine-2-one, resembles reserpine in its sedative effects.

Tetrabenazine appears to be one of the most potent benzoquinolizine derivatives and has the longest duration of action, although it is still shorter than that of reserpine. The amine-depleting potency of tetrabenazine in rabbits is only $\frac{1}{20}$ that of reserpine. Tetrabenazine (Nitoman, an experimental drug in the United States), in doses up to 300 mg daily improved tardive dyskinesia in several trials.[42]

Miscellaneous Compounds

Butaclamol, a compound having a benzocycloheptapyridoisoquinoline ring system, was developed as a neuroleptic agent,[43] and was evaluated clinically but withdrawn from the market because of its toxicity. Currently this compound is of interest as a pharmacologic tool, as a stereoselective dopamine receptor antagonist. Psychopharmacologic studies have shown that activity resides solely in the (+) enantiomer, which is at least 100 times more active than the (−) enantiomer.

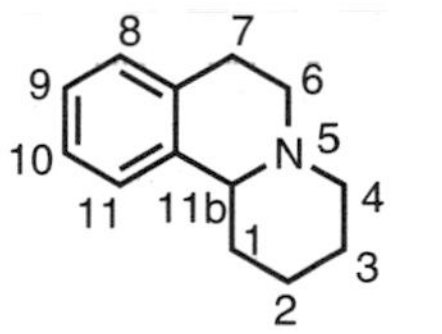

Benzoquinolizine

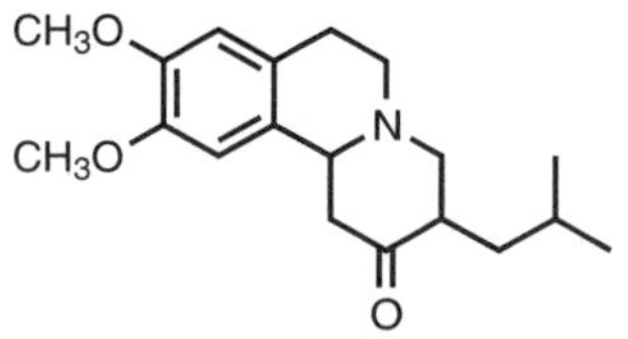

Tetrabenazine

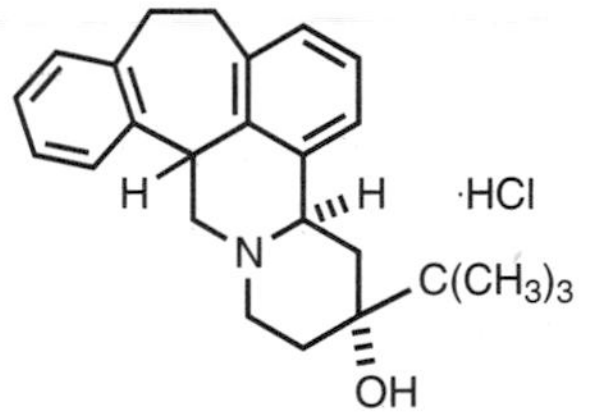

Butaclamol Hydrochloride

R(−) Apomorphine [(R) APO], used for more than 100 years as a centrally acting emetic, has also been used in the treatment of chronic schizophrenic patients in subcutaneous administration (3 mg).[44] (The dopamine receptor agonist effects of apomorphine are discussed in Chapter 13.) These clinical observations suggest that, at certain doses, (R) APO may exert a presynaptic dopamine receptor activation, inhibiting dopamine synthesis and release. These clinical observations are consistent with studies showing that at low doses, RAPO preferentially activates presynaptic D_2-type autoreceptors to inhibit dopamine synthesis and release.[44a] Thus, in contrast to agents, which primarily affect postsynaptic dopamine receptors to facilitate dopaminergic neurotransmission, selective dopamine autoreceptor agonists may be clinically useful to inhibit dopamine-mediated transmission. Accordingly, dopamine receptor agonists that preferentially affect presynaptic receptor sites may inhibit, not facilitate, dopamine-mediated transmission.

Reserpine and Related Alkaloids

Reserpine, the principal alkaloid of Rauwolfia serpentina, is now mainly of historical interest in psychiatry. It is less effective as a neuroleptic than the other drugs already discussed, and it is now used only on occasions when patients cannot tolerate other classes of neuroleptic drugs. Reserpine is more likely to produce depression, hypotension, and diarrhea as well as other serious side effects than are the other antipsychotic agents.

Rauwolfia serpentina, a climbing shrub named after the German botanist Rauwolf, is indigenous to India. Extracts of the plant were used in ancient Hindu medicine. New interest was generated in 1931 by Sen and Bose, who claimed that the root of rauwolfia may be useful in the treatment of psychoses and hypertension.[45] As a result, a systematic study to isolate the active components was undertaken by several groups.[46,47]

Reserpine was finally isolated by Mueller and co-workers in 1952.[48] Soon afterward, Bein reported that reserpine has tranquilizing as well as hypotensive potencies, the same activities displayed earlier by the extracts of rauwolfia.[49] Pure reserpine, therefore, is the active component of rauwolfia extracts. Several groups elucidated the chemical structure,[50–52] and it was confirmed by x-ray crystallography.[52] A total synthesis was completed by Woodward et al. in 1956.[53,54] The amine-depleting action of reserpine, most likely responsible for its action, is greatest for three neurohumoral transmitters: norepinephrine, dopamine, and 5-hydroxytryptamine.

Table 12–6 lists the naturally occurring reserpine-type alkaloids and semisynthetic products.

DRUGS USED IN THE TREATMENT OF ANXIETY

Antianxiety agents are used to control moderate or severe anxiety and tension in patients with anxiety disorders and mild depressive states. They may also be indicated in normal individuals in situations or during illness producing unusual stress. Such sedative-antianxiety drugs are

Table 12–6. Naturally Occurring Reserpine-type Alkaloids and Semisynthetic Products

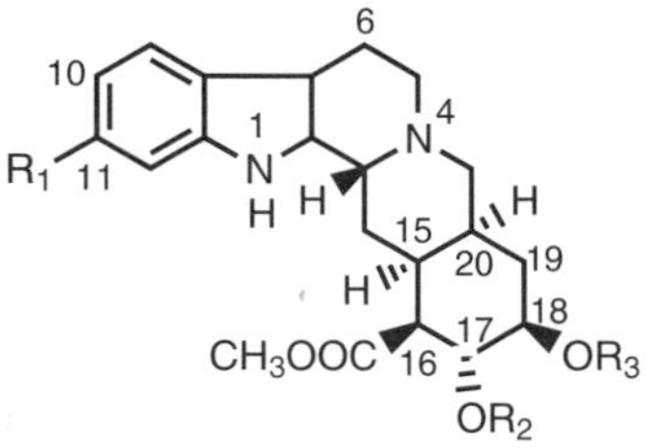

Alkaloid	Proprietary Name	R_1	R_2	R_3
Reserpine	Serpasil	OCH_3	CH_3	TMB*
Deserpidine	Harmonyl	H	CH_3	TMB*
Raunescine		H	H	TMB*
Isoraunescine		H	TMB*	H
Pseudoreserpine		OCH_3	H	TMB*
Raugustine		OCH_3	TMB*	H
Rescinnamine	Moderil	OCH_3	CH_3	TMC†
Rescidine		OCH_3	H	TMC†

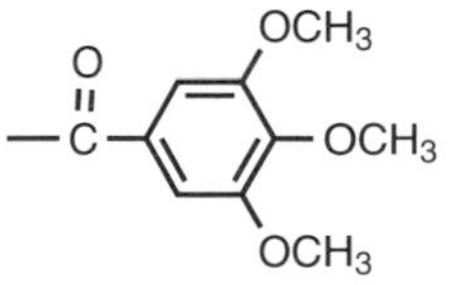

3,4,5-Trimethoxybenzoyl

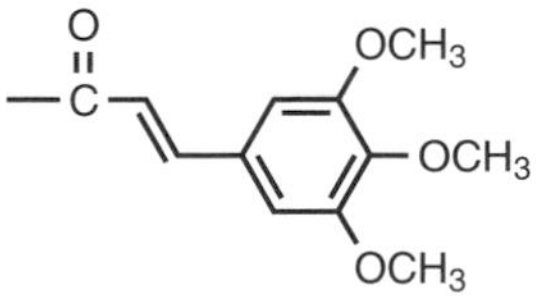

3,4,5-Trimethoxycinnamoyl

* TMB = 3,4,5-Trimethoxybenzoyl.
† TMC = 3,4,5-Trimethoxycinnamoyl (semisynthetic).
Products
1. Reserpine (Serpasil, Reserpoid).
2. Powdered Rauwolfia Serpentina (e.g., Raudixin).
3. Rauwiloid—fat-soluble fraction from whole root.

Table 12–7. Currently Available Antianxiety Agents

Generic Name	Proprietary Name	Usual Daily Dose (mg)	Half-life* (h)	Active Metabolites	Other Uses
Benzodiazepines					
Alprazolam	Xanax	0.75–4	12	Minor	
Chlordiazepoxide hydrochloride	Librium	15–60	18	Yes	IM or IV for alcohol withdrawal
Clorazepate dipotassium	Tranxene	30	100	Yes	
Diazepam	Valium	4–10	60	Yes	
Halazepam	Paxipam	60–160	50	Yes	
Lorazepam	Ativan	2–6	15	No	
Oxazepam	Serax	30–60	8	No	
Prazepam	Centrax	20–40	100	Yes	
Clonazepam	Clonopin	1.5–10	34	Yes	Primarily used as anticonvulsant
Flurazepam†	Dalmane	15–30	74†	Yes	Primarily used as hypnotic
Temazepam	Restoril	15–30	11	No	
Triazolam	Halcion	0.25–0.5	2	No	
Propanediol Carbamates					
Meprobamate	Equanil, Miltown	400–1200	10		
Tybamate	Solacen, Tybatran	500–1500	3		
Antihistaminic-Diphenylmethanes					
Hydroxyzine hydrochloride	Atarax	75–400 oral 25–100 IM			
Hydroxyzine pamoate	Vistaril	75–400			
Miscellaneous‡					
Buspirone	Buspar	10–60	2–11		

* Data from Refs. 5, 30, and 83.
† Flurazepam is essentially a prodrug for desalkylflurazepam. Half-life values are for the active metabolites.
‡ Data from Ref. 83.

prescribed more frequently than in any other group of therapeutic agents. Their misuse may be harmful and may prevent recognition of the underlying source of anxiety. By far, the most widely used antianxiety drugs (Table 12–7) are the benzodiazepines (diazepam and congeners). Propanediol carbamates (meprobamate and congeners), also discussed in Chapter 13; the diphenylmethane type antihistaminic sedatives, discussed only briefly in this chapter and in Chapter 22; and the barbiturate members of the group, described in detail in Chapter 10, are largely of historical importance. It is important to bear in mind, however, that patients who improve following treatment with antianxiety agents may also respond equally well to properly selected doses of barbiturates or other sedatives. Because of their wide margin of safety, the benzodiazepines are preferred over the barbiturates for patients with suicidal intentions. Benzodiazepines are also preferable because they produce fewer drug interactions than the barbiturates, which induce drug-metabolizing hepatic microsomal enzymes.

Chlordiazepoxide, diazepam, and oxazepam may overcome psychomotor hyperexcitability and are standard treatment for alcohol withdrawal symptoms. Diazepam and clonazepam have achieved wide acceptance as anticonvulsants (see Chapter 11). Many of the antianxiety agents also produce skeletal muscle relaxant effects. They are being used in many neurologic and musculoskeletal disorders in humans (see Chapter 13). Diazepam is effective in partially relieving spasticity in cerebral palsy. Antianxiety agents are not effective in the long-term management of chronic schizophrenia or other psychoses.

Currently the most useful drugs for the treatment of anxiety are the benzodiazepines. The chemical classification, preparations, and dosage of these major antianxiety drugs are shown in Table 12–7.

Benzodiazepines

Discovery of 1,4-Benzodiazepines as Antianxiety Drugs

The discovery of the two main classes of antianxiety agents, the propanediol carbamates and the benzodiazepines, are examples of the serendipitous discovery of new drugs based on an almost random screening of chemicals synthesized in the laboratory. Their discovery was also based on the observations made by alert pharmacologists who quickly recognized significant animal signs during general biologic screening of random chemicals. Berger in 1946 and Randall in 1957, independently and for different series of compounds, observed an unusual and characteristic paralysis and relaxation of voluntary muscles in laboratory animals. By the mid-1950s, psychopharmacology had developed to the point at which the treatment of ambulatory anxious patients with meprobamate and psychotic patients with one of the aminoalkyl-phenothiazine drugs was possible. There was a need for drugs of greater selectivity in the treatment of anxiety because of the side effects often encountered with phenothiazines. Thus, the chemist Sternbach, working in the pharmaceutical research laboratories of Hoffman-LaRoche in Nutley, New Jersey, decided to reinvestigate a relatively unexplored class of compounds, which he had studied in the 1930s while he was a postdoctoral fellow at the University of Cracow, Poland. During his synthetic studies, he obtained several compounds thought to belong to the class of 4,5-benzo-(hept-1,2,6-oxadiazines) or 3,1,4-benzoxadiazepines.[55] Reinvestigation of these compounds 20 years later revealed that they did not contain a seven-membered oxadiazine ring but were quinazoline-3-oxides.[56,57]

The biologic properties of these substances had not been investigated, so Sternbach set out to restudy them, stating, ". . . it is a fact known to medicinal chemists that basic groups frequently impart biological activity."[56] He therefore synthesized such compounds by treating chloromethylquinazoline N-oxides with various amines, but dose-response studies of about 40 of these amino derivatives were disappointing, and the project was abandoned. The last of these compounds had been synthesized using the primary amine, methylamine, instead of the secondary amines used in all other case. In 1957, during a clean-up of the laboratory, the product prepared 2 years earlier was finally submitted for pharmacologic testing as RO 5-0690. Careful study of this compound revealed that it was not a quinazoline-3-oxide similar to the other congeners with tertiary amino groups but the result of an unexpected reaction caused by the primary amine used in its synthesis. On treatment with methylamine, 6-chloro-2-chloromethyl-4-phenylquinazoline-3-oxide did not furnish the expected 2-(*N*-methyl aminomethyl) derivative, but instead rearranged to 7-chloro-2-(*N*-methylamino)-5-phenyl-1, 4-benzodiazepine 4-oxide.[56,57] Shortly thereafter, Randall reported that this compound was hypnotic, was sedative, and had antistrychnine effects similar to those of meprobamate.[58] The compound was given the name chlordiazepoxide and marketed as Librium in 1960. Structural modifications of benzodiazepine derivatives were undertaken, and a compound 5 to 10 times more potent than chlordiazepoxide was synthesized in 1959[59] and marketed as diazepam (Valium) in 1963. Diazepam was the first potent benzodiazepine derivative that did not contain a basic amino group, which was the original basis for the preparation of these compounds. The synthesis of many other experimental analogs soon followed, and by 1983 about 35 benzodiazepine drugs were available for therapy (Fig. 12–15). Benzodiazepines are the drugs of choice in the pharmacotherapy of anxiety and related emotional disorders, sleep disorders, status epilepticus, and other convulsive states; they are used as centrally acting muscle relaxants, for premedication, and as inducing agents in anesthesiology.

Mechanism of Action

Efforts to elucidate the mechanism of action of the benzodiazepines have made considerable progress. In

6-Chloro-2-(N,N-dimethylaminoethyl)-4-phenylquinazoline-3-oxide

CH_3NH_2

6-Chloro-2-chloromethyl-4-phenylquinazoline-3-oxide

7-Chloro-2-(N-methylamino)-5-phenyl-3H-1,4-benzodiazepin-4-oxide
Chlordiazepoxide

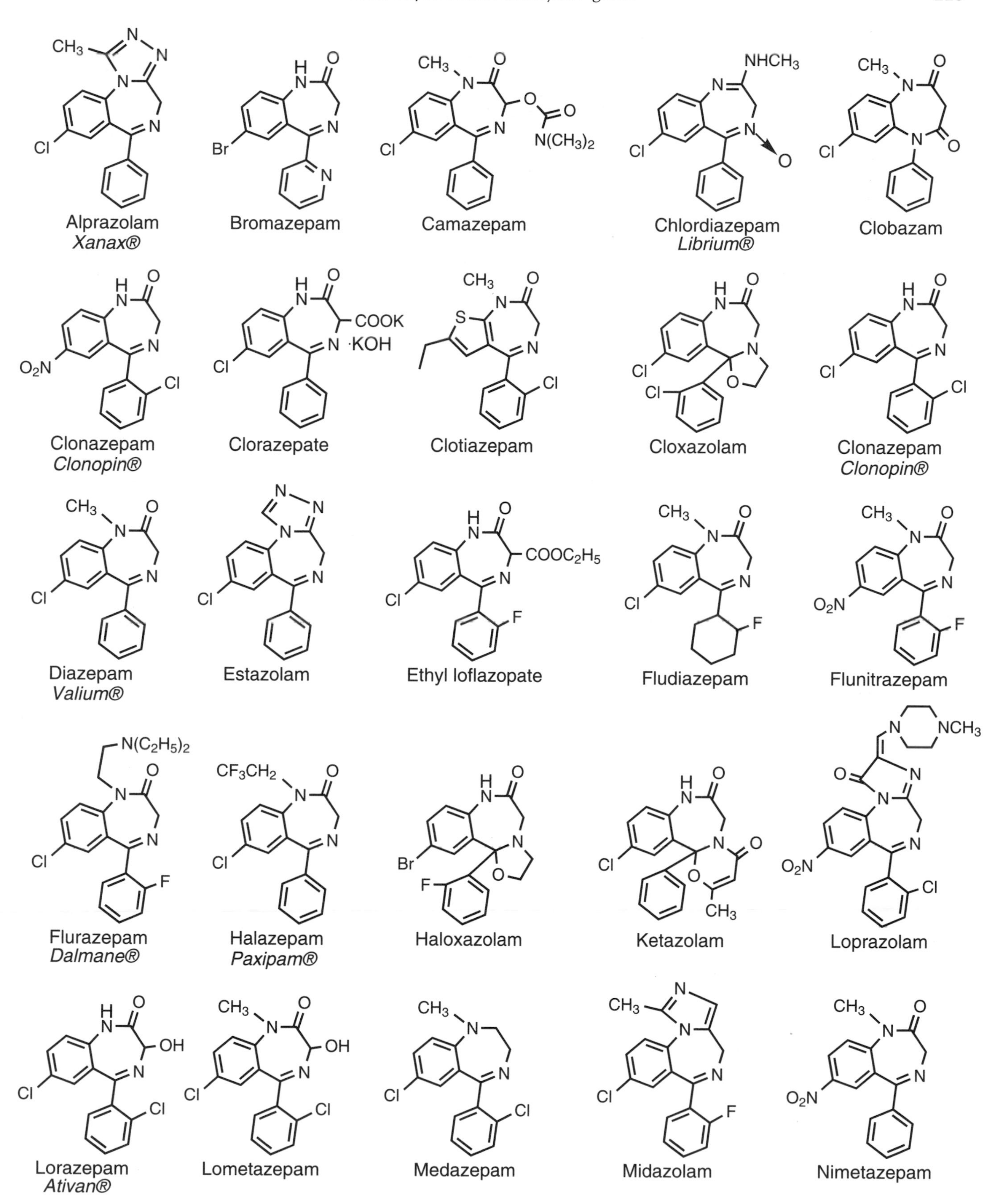

Fig. 12–15. Structural formulas and generic names of commercially available benzodiazepines. The proprietary names of agents used in the United States are also included.

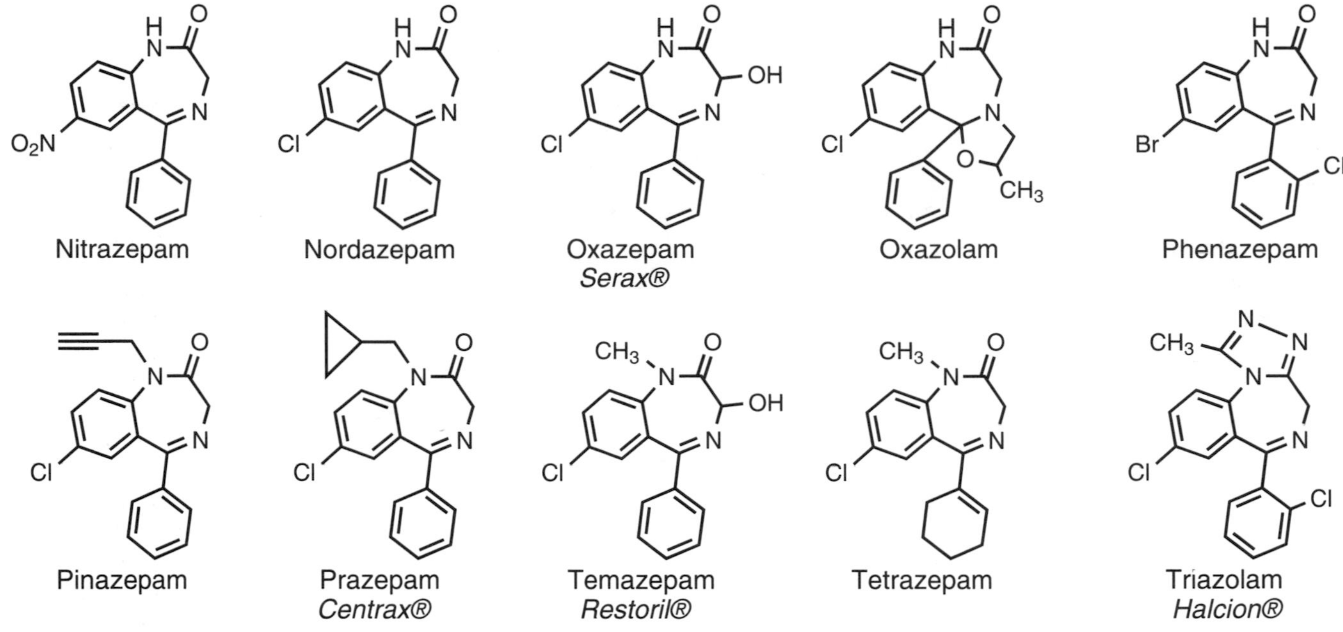

Nitrazepam

Nordazepam

Oxazepam
Serax®

Oxazolam

Phenazepam

Pinazepam

Prazepam
Centrax®

Temazepam
Restoril®

Tetrazepam

Triazolam
Halcion®

Fig. 12–15. *(Continued)*

1974, sufficient evidence from behavioral, electrophysiologic, and biochemical experiments was accumulated to indicate that benzodiazepines act quite specifically at synapses in which GABA is a neurotransmitter.[60] Benzodiazepine receptors in GABAergic synapses have been localized by autoradiography. Using tissue-binding studies with radiolabeled agents, specific, saturable, low-capacity, high-affinity binding sites for tritiated diazepam or its analogs have been identified in homogenates of mammalian brain tissue.[61–63] These studies led to increasing interest in benzodiazepine receptors. Several investigators found that GABA enhances high-affinity binding of benzodiazepines, which led to the proposal that benzodiazepines increase GABA binding to its receptor and that a close association exists between the actions of GABA and benzodiazepines. The existence of highly specific benzodiazepine binding sites led to the suggestion that these sites may be receptors for endogenous modulators. Several endogenous compounds with inhibitory activity on tritiated benzodiazepine binding have been extracted from the central nervous tissue. Braestrup, in 1980, reported the presence in normal human urine of the ethyl ester of β-carboline-3-carboxylic acid (βCCE), which had an extremely high affinity for benzodiazepine binding sites.[64] It was quickly shown that βCCE is formed by heating urine extract with ethanol at pH 1, a condition favoring the formation of the ethyl ester from β-carboline-3-carboxylic acid, which itself derives from tryptophan. Numerous derivatives of β-carboline-3-carboxylic acid have since been synthesized and investigated for their interaction with benzodiazepine binding sites. Such β-carboline derivatives with affinity for benzodiazepine binding sites are agonists, competitive antagonists, inverse agonists, or mixed forms of agonists and inverse agonists.[66]

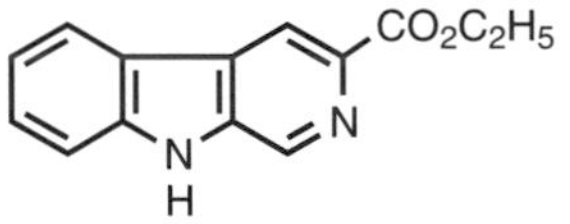

Ethyl β-carboline-3-carboxylate (β-CCE)

In 1981, the first report on the imidazolebenzodiazepine derivative RO 15-1788 was given by Hunkeler et al.[65] With this compound, it was shown that the benzodiazepine molecule could be varied in such a way that these agents would bind potently to benzodiazepine receptors and block all pharmacologic effects of the classic benzodiazepines. It was also found that some of the antagonistic

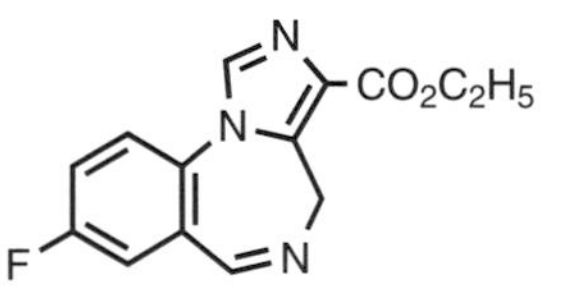

RO15-1788

β-carbolines produced effects opposite to the classic benzodiazepine effects, such as severe anxiety, convulsions, muscle rigidity, and sleep disturbance. Such benzodiazepine antagonists are under investigation for possible use in the treatment of benzodiazepine overdosage.

GABAergic Synapse as the Primary Site of Action of Benzodiazepines

The benzodiazepines have been found to affect virtually all known neurotransmitters in the central nervous system, at least at high doses. It was obvious in the early

1970s that these changes could not be due to a direct action of benzodiazepines on neurons that use catecholamines or acetylcholine as a transmitter or to an action on their receptors. The proposal made by Haefly[67] and Costa[68] that benzodiazepines may produce their effects primarily by enhancing GABAergic transmission provided an explanation for the various secondary alterations induced by these drugs in other transmitter systems. GABA acts on at least two different receptor types. The action of benzodiazepines seems to be restricted to synaptic effects of GABA, which are mediated by so-called $GABA_A$ receptors.

The $GABA_A$ receptor has been cloned,[69] and the oligomeric glycoprotein isolated from mammalian central nervous system has a molecular weight of about 250 kD. By homology with other ligand-gated ion channel receptors (e.g., the nicotinic acetylcholine receptor), the $GABA_A$ macromolecular complex probably comprises five subunits. The cDNAs encoding multiple isoforms of α-, β-, and γ-type subunits have been cloned.[69a] It is currently believed that the α subunit defines the recognition characteristics of the benzodiazepine receptor, whereas the β subunit binds GABA. In addition to binding sites for GABA and benzodiazepines, the $GABA_A$ oligomer contains sites for barbiturate-type anticonvulsants as well as convulsants such as picrotoxin.[70] Although evidence for benzodiazepine receptor heterogeneity was apparent a decade ago, and molecular cloning studies have confirmed this hypothesis, currently only two different benzodiazepine receptor subtypes, BZ_1 and BZ_2, can be distinguished pharmacologically. The nonbenzodiazepine compounds, CL 218,872 (a triazolopyridazine) and zolpidem (an imidazopyridine), both have high affinity for BZ_1 sites, and low affinity for BZ_2 sites.

The consequence of $GABA_A$ receptor stimulation by GABA seems to be an increase of the permeability of neuronal membranes for anions, mainly for Cl^- anions. It is currently accepted that benzodiazepines are able to enhance transmission at all GABAergic synapses in the mammalian central nervous system, in particular when normal GABAergic transmission is experimentally depressed.[71,72] This should not be taken to mean that in an individual treated with appropriate doses of benzodiazepines, transmission at all GABAergic synapses (estimated to constitute one-third of all synapses in the brain) is actually enhanced. Indeed, binding to the receptor may lead to different biologic effects, e.g., agonists may be anxiolytic or anticonvulsant, whereas inverse agonists produce anxiety or seizures; antagonists block effects of both agonists and inverse agonists.[66]

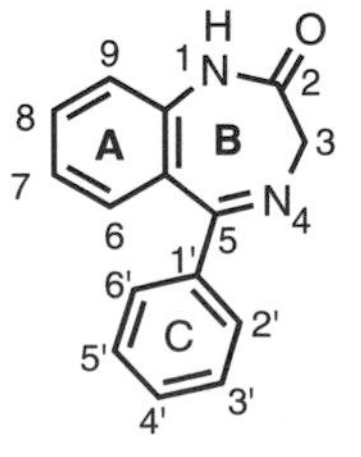

5-Phenyl-1,4-benzodiazepin-2-one: ring system of "classical" benzodiazepines

Chemistry and Structure-Activity Relationships

In their pioneering synthetic work with the classic 1,4-benzodiazepine derivatives, Sternbach and collaborators described the fundamental structure-activity relationships (SAR) of this series.[56,57,73] They recognized early that the presence of the seven-membered imino-lactam ring (ring B) was essential and that substitution was advantageous only in positions 1, 3, 7, and 2′. In particular, electronegative substituents in 7 and 2′ markedly increased activity following this simple qualitative substitution pattern. Thousands of derivatives have been synthesized all over the world, and more than a dozen have found their way into current therapy. The SAR of benzodiazepine agonists and antagonists have been thoroughly reviewed, and the reviews should be consulted for details.[66] The discovery of high-affinity binding sites for benzodiazepines and the introduction of a simple and reliable in vitro receptor binding assay greatly facilitated such SAR studies. In addition, binding studies clearly showed that structures other than 1,4-benzodiazepines may act at the same receptor complex and have a similar mechanism of action.

The three types of benzodiazepine receptor ligands can be explained according to the change in conformation they induce in the receptor from its unoccupied resting state. After primary recognition of a pharmacophore, the benzodiazepine receptor undergoes a shift in conformation to either an agonist or an inverse agonist state, based on the electronic, hydrophobic, and steric characteristics of the ligand. This conformational change then allosterically modulates the binding of GABA to the receptor. An agonist induces a conformational change that facilitates the binding of GABA (observed biologic response is anticonvulsant), whereas an inverse agonist induces a conformational change that inhibits the binding of GABA to the receptor (observed biologic response is proconvulsant). An antagonist binds to the benzodiazepine receptor but produces no conformational change.

Defining the SAR for ligand binding to the benzodiazepine receptor is a prerequisite to differentiating the structural features of molecules that may induce conformational changes in the receptor to produce functional activity (agonist or inverse agonist) or no conformational change (antagonist). Those ligands that bind to the benzodiazepine receptor with nanomolar (rather than micromolar) affinity are more likely to possess physiologically relevant biologic activity. The minimal requirements for binding at the benzodiazepine receptor have been determined to be relatively simple. Molecular modeling studies have determined that all benzodiazepine ligands share the presence of an aromatic or heteroaromatic A ring, believed to undergo π/π stacking within the receptor (probably with aromatic amino acid residues), as well as a proton-accepting group that probably exists in the same plane of the aromatic A ring and interacts with a histidine residue on the receptor. Although not required for receptor binding, the presence of a 5-phenyl aromatic group (found in diazepam-type compounds) may contribute steric or hydrophobic interactions with the receptor; however, the spatial orientation of this group with respect to the A ring is not clear. Substitution of the 4′-(para)-position of an appended 5-phenyl ring is sterically unfavorable for agonist activity. Substitution of the 2′ (ortho)-position of an appended 5-phenyl ring, however, is not detrimental to agonist activity, suggesting that the limitations at the para position are steric, rather than electronic, in nature. Substituents on the A ring have varied effects, and these are not predictable based on electronic or (within reasonable limits) steric properties. Electronegative groups in the 7-position increase functional activity; however, compounds with no substituents on the A ring may still have nanomolar affinity for the benzodiazepine receptor and have agonist, inverse agonist, or antagonist activity. Neither the amide nitrogen nor its methyl substituent are required for in vitro binding. Likewise, the 4,5-(methyleneimino) group is not required for in vitro binding of ligands. Finally, substitution of the methylene 3-position or the imine nitrogen is sterically unfavorable for agonist activity but has no effect on antagonist activity.

Stereochemical Considerations

Most of the benzodiazepines in therapeutic use have no chiral center; however, their seven-membered ring B may adopt one of two possible, energetically preferred boat conformations, i.e., *a* and *b*, which are enantiomeric (mirror images) to each other (Fig. 12–16). Nuclear magnetic resonance studies revealed that, at room temperature, the two conformations easily interconvert.[74] It is therefore impossible to predict a priori which of the two will bind best to the receptor complex. To prove the stereospecificity of the interaction with the receptor, a few examples of enantiomeric pairs possessing a chiral center at position 3 have been tested. The in vitro and in vivo activity was restricted to the enantiomers possessing the S configuration.[66]

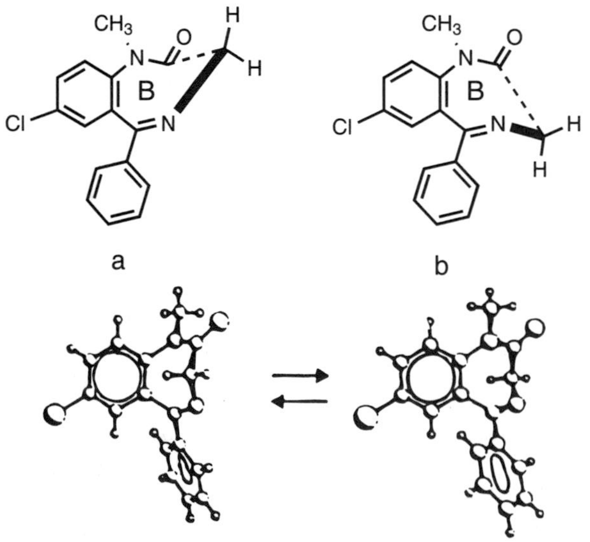

a b

Fig. 12–16. The two enantiomeric conformations of diazepam.[66]

The introduction of the methyl group in position 3 results in stabilization of conformation *a* for ring B of the S-enantiomers and to that of the opposite conformation *b* for the R-enantiomers. The methyl group is preferentially located in a pseudoequatorial position. This conformation is present in the crystalline state,[75] and also in solution as shown for the S-enantiomer of 3-methyl-diazepam.[76]

These observations indicate that it is the conformation of ring B that is determinant for the affinity for the benzodiazepine receptor and not the absolute configuration as such. The study of conformationally rigid structures starting from benzodiazepine antagonists eventually provided conclusive evidence for the importance of the conformation for receptor affinity.

1,2-Annelated 1,4-benzodiazepine. The condensation of an additional ring on the 1,2-bond of the classic benzodiazepine nucleus resulted in interesting new compounds, which stimulated new chemical and biologic work on the benzodiazepines. Among the various heterocycles that have been fused in this position, s-triazole and imidazole are of prime interest: three triazolo-benzodiazepines (estazolam, triazolam, and alprazolam) and two imidazo-benzodiazepines (midazolam, loprazolam) (see Fig. 12–15) have been marketed.

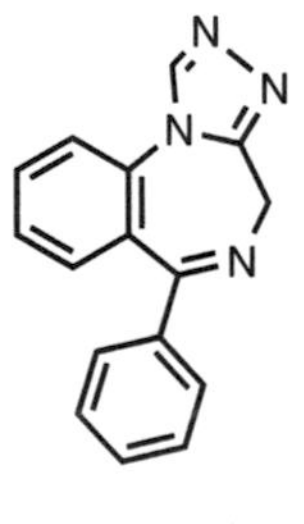

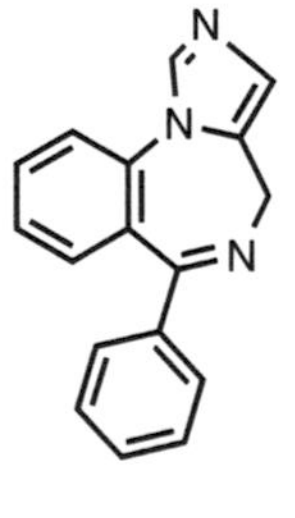

s-Triazolo[4,3a][1,4]benzodiazepine

Imidazo[1,5a][1,4]benzodiazepine

Pharmacologic Properties

Chlordiazepoxide, diazepam, and lorazepam can be considered prototypical drugs for their class. They are among the most widely used antianxiety agents. Effects of the benzodiazepines that may contribute to the relief of anxiety can readily be demonstrated in laboratory animals. The behavioral effects of antianxiety agents have been reviewed.[77]

Difficulties in evaluating the therapeutic efficacy of psychotropic drugs in humans are particularly great in the case of the antianxiety agents, largely because of the contribution of nonpharmacologic factors to the treatment of anxiety. The clinical popularity of these drugs is apparently the result of a combination of their pharmacologic actions, their relative safety, and the demand for agents of this type by both physicians and patients.

In common with barbiturates, chlordiazepoxide blocks electroencephalographic arousal from stimulation of the brain stem reticular formation. Central depressant actions of diazepam and other benzodiazepines on spinal reflexes occur and are partly mediated by the brain stem reticular system. Similar to meprobamate and the barbiturates, chlordiazepoxide depresses the duration of electrical afterdischarge in the limbic system, including the septal region, the amygdala, the hippocampus, and the hypothalamus. These and other limbic and autonomic effects are presently the focus of particular theoretical interest. Other effects include the following:

1. *Muscle relaxation.* Numerous clinical studies have shown that doses of benzodiazepines produce objective muscle relaxation, in both healthy subjects and patients with neuromuscular diseases.[78] It has been suggested that the most important action of the benzodiazepines is facilitation of brain-stem inhibitory interneurons. The benzodiazepines, however, probably have more than one site of muscle-relaxant activity.
2. *Anticonvulsant effects.* Benzodiazepines have strong anticonvulsant activity in both animals and humans. They are most effective in preventing or arresting generalized seizure activity produced by electric shock or by systematic administration of analeptic agents or local anesthetics. The use of the benzodiazepines as anticonvulsants is also discussed in Chapter 11.
3. *Circulation and respiration.* In contrast to most sedative-hypnotic drugs when given in sufficiently large doses, benzodiazepines are relatively safe and rarely produce cardiovascular and respiratory depression. Equally sedating doses of barbiturates can cause hypoventilation, hypotension, and reduced cardiac output.[78]

Absorption, Fate, and Excretion.[78,79] Chlordiazepoxide is well absorbed after oral administration, reaching peak blood levels within 4 hours. The absorption of intramuscularly administered chlordiazepoxide, however, is slow and erratic. Biotransformation rates of single doses of chlordiazepoxide vary considerably among healthy subjects, in most instances the half-life is between 6 and 30 hours. Demethylation of the 2-methylamino side chain yields an initial metabolite, desmethylchlordiazepoxide, which slowly undergoes deamination to form a 2-keto derivative, demoxepam (Fig. 12–17). Further transformation leads to two more active metabolites, desoxydemoxepam (*N*-desmethyldiazepam) and further hydroxylation at the 3-position yields oxazepam, which is rapidly glucuronidated and excreted in the urine. As shown in Figure 12–17, four distinct metabolic pathways are in operation in the biotransformation of demoxepam. Hydrolysis to the "opened lactam" and aromatic hydroxylation in the A and C rings yield inactive phenolic metabolites; formation of desoxydemoxepam can also form inactive phenolic metabolites. Less than 1% of a dose of chlordiazepoxide is excreted as demoxepam and the remainder as glucuronide conjugates of oxazepam and the other hydroxylated metabolites.[80] Repeated doses of chlordiazepoxide can produce cumulative clinical effects owing to accumulation in the body of the drug and its active metabolites.[81]

Diazepam, after oral administration, is rapidly and completely absorbed, reaching maximum blood concentrations within 2 hours. Metabolism of diazepam proceeds slowly with a half-life of 20 to 50 hours. The major metabolic product is *N*-desmethyldiazepam (Fig. 12–17), which still has considerable pharmacologic activity and is biotransformed even more slowly than diazepam. Hydroxylation at the 3-position yields oxazepam, which is rapidly glucuronidated and excreted in the urine as the major urinary metabolite of diazepam. Repeated dosage of diazepam leads to accumulation of diazepam and desmethyldiazepam in the blood. After termination of chronic administration, one or both compounds can be detected in the blood for a week or more (see Table 12–7).

The metabolism of oxazepam is facilitated by the 3-hydroxy function, which allows for direct conjugation as the glucuronide and excretion in the urine. Because oxazepam is so readily metabolized to a psychopharmacologically inactive product (half-life = 8 hours), cumulative effects during long-term therapy are much less important than with chlordiazepoxide or diazepam.[79] The short half-life of lorazepam (half-life = 2 to 6 hours) can be similarly explained.

Flurazepam is formulated as a water-soluble dihydrochloride salt for oral administration. The drug is rapidly metabolized by removal of the alkyl side chain on the 1-nitrogen, to give the 2′-fluoro analog of desoxydemoxepam whose biotransformation follows that of the other benzodiazepines (Fig. 12–17). Its half-life is approximately 7 hours. Flurazepam could produce cumulative clinical effects during repeated dosage as well as residual effects after termination of therapy.

Clorazepate is rapidly decarboxylated to give *N*-desmethyldiazepam, which is slowly biotransformed. Thus, as in the case of chlordiazepoxide and diazepam, long-lasting and cumulative effects can be anticipated during treatment with clorazepate. In contrast to barbiturates, benzodiazepines do not produce clinically important hepatic microsomal enzyme inductions. Benzodiazepines can be used more effectively when dosage schedules are guided by pharmacokinetic knowledge.

The actual kinetics for some benzodiazepines are complex and not easily analyzed by mathematical models. Thus, the usually stated half-life for the elimination phase of the drug does not adequately depict the kinetics of the early distributive phase, which can be important clinically. For example, the distributive (alpha) half-life of diazepam is about 1 hour, whereas the elimination (beta) half-time is about 1.5 days initially and even longer after

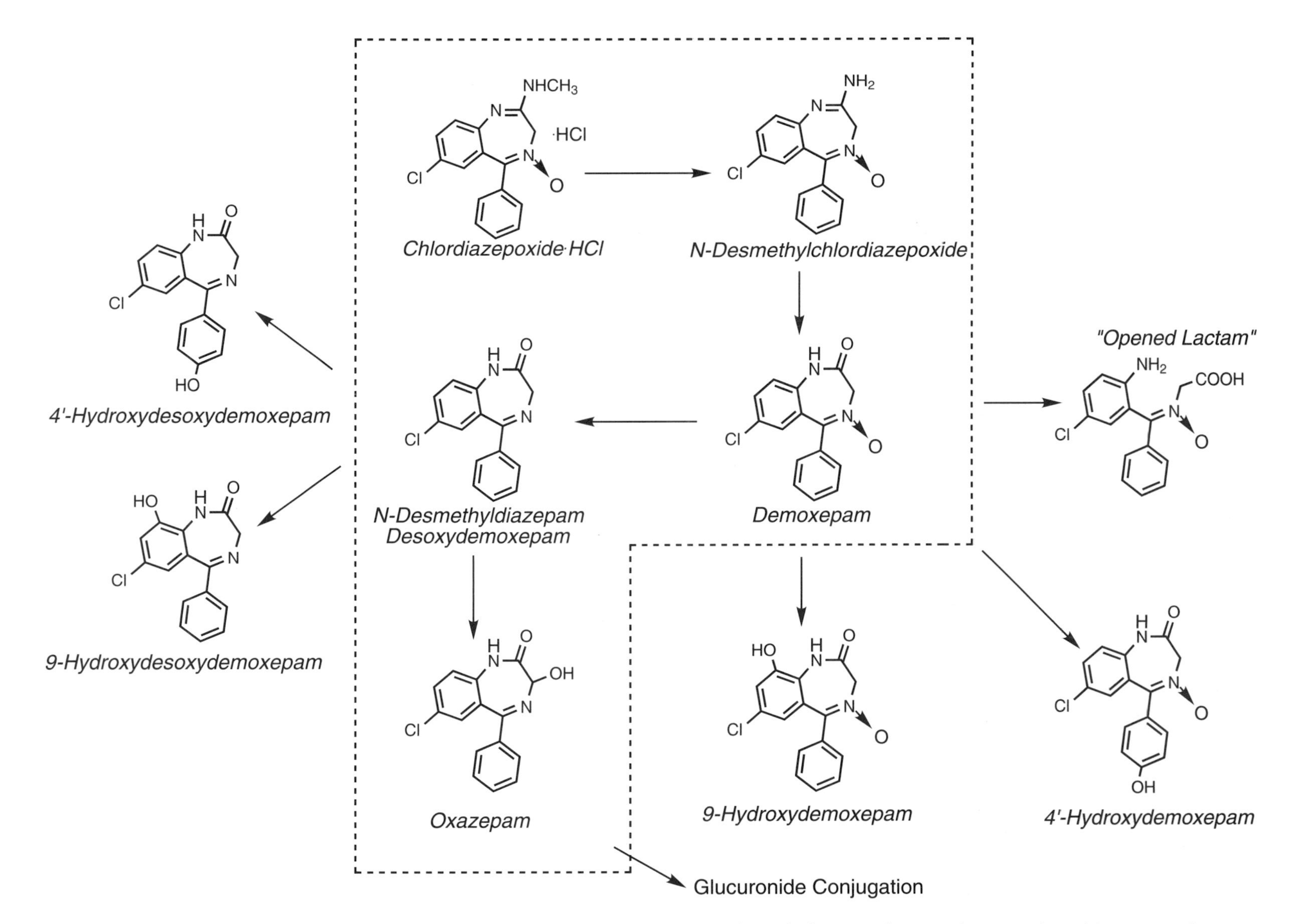

Fig. 12–17. Metabolic pathway of chlordiazepoxide hydrochloride. Active drug and metabolites are shown in boxes. (Adapted from L. Farah, *Int. Rec. Med. Gen. Prac. Clin., 169,* 370[1956].)

prolonged treatment. Moreover, although correlations between plasma concentrations of benzodiazepines and clinical effects are imperfect, it is apparent that concentrations in plasma that border on twice the values usually considered effective are associated with undesirable degrees of sedation. For this reason, the benzodiazepines are not effectively or safely given once a day, despite their relatively long elimination half-lives; doses should be divided into two to four portions for the treatment of daytime anxiety.[5] The benzodiazepines or their active metabolites can accumulate in the body during several days of therapy. Thus, therapeutic or toxic effects not apparent after 1 or 2 days of treatment may appear after 5 to 10 days of repeated administration. Conversely the clinical effects can persist for several days after the drug is discontinued. Oxazepam is more rapidly biotransformed to inactive metabolites, which makes it of value for indications where a short-acting, nonaccumulating drug is desired (see Table 12–7), such as hepatic failures and in elderly patients.[80]

Propanediol Carbamates

The two propanediol carbamates, meprobamate and tybamate (see Table 12–7), little used as antianxiety agents in current clinical practice, are chemically related and have similar activities. The chemistry and pharmacologic action of these drugs are discussed in greater detail in Chapter 13.

Meprobamate is now rarely used in the treatment of anxiety and tension and appears somewhat less effective than the benzodiazepines. The half-life is about 10 hours, with peak plasma concentrations about 3 hours after oral ingestion. The substitution of a butyl group in place of a hydrogen on one of the carbamoyl nitrogen atoms produces tybamate, a shorter-acting antianxiety agent (half-life = 3 hours). Isopropyl substitution at the same position results in carisoprodol (see Table 13–2), a muscle relaxant.

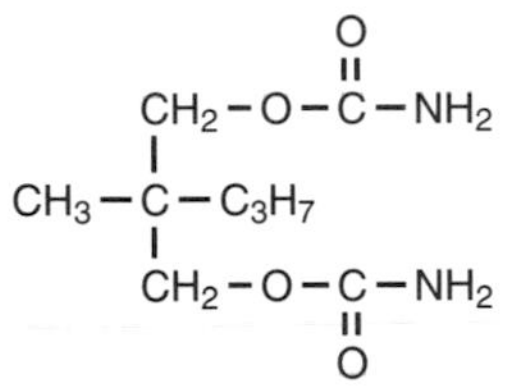

$$CH_3-C(C_3H_7)(CH_2-O-C(=O)-NH_2)_2$$

Meprobamate

$$CH_3-C(C_3H_7)(CH_2-O-C(=O)-NH_2)(CH_2-O-C(=O)-NHC_4H_7)$$

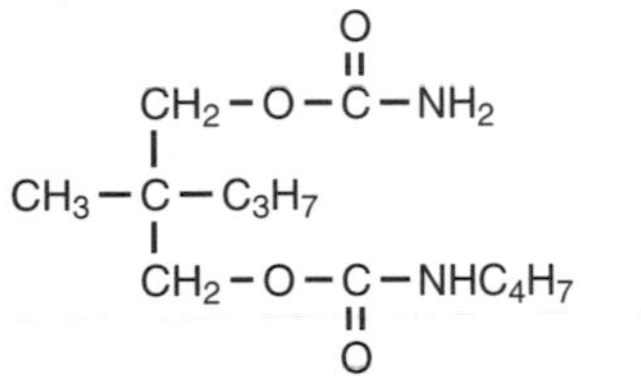

Tybamate

The demise of these agents for the treatment of anxiety is due to their tendency to produce undesirable degrees of sedation or intoxication at the dosage required to alleviate anxiety. Meprobamate has a tendency to produce tolerance, physical dependence and severe withdrawal reactions after use in doses that overlap the clinical range.

Antihistaminic Diphenylmethanes

Hydroxyzine (Atarax) and its pamoate ester (Vistaril pamoate) have found use in the treatment of anxiety, tension, and agitation. A diphenylmethane derivative, hydroxyzine, may also be regarded as an *N*-benzhydryl-*N′*-substituted piperazine having the chemical name (1-p-chlorbenzhydryl-4-[2-(2-hydroxyethoxy)ethyl]-piperazine).

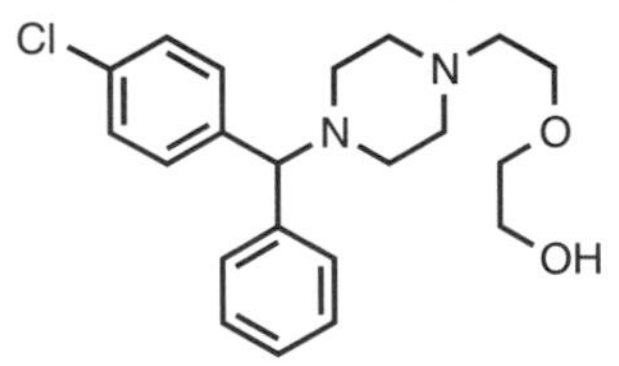

Cl N N O OH

Hydroxyzine

Pharmacology. Hydroxyzine is useful in anxiety and neurotic conditions when given in doses of 400 mg per day.[8] At such doses, it produces marked sedation. Hydroxyzine has been used for psychosomatic disturbances, nervous tension, and emotional stress. Because of its antianxiety effect, hydroxyzine has been used before and after surgery and during delivery.

In contrast to reserpine, hydroxyzine does not exert its action through its effect on the metabolism in the brain. There is no evidence that it depletes the biogenic amine content of the brain. In contrast to meprobamate and chlordiazepoxide, it lacks muscle-relaxant activity. Its mode of tranquilizing action is similar to that of meprobamate and chlordiazepoxide, as shown by the similarity of the three compounds in reducing spontaneous or induced activity in experimental animals.

Hydroxyzine is a potent antihistamine with a long duration of action. It has been used in the treatment of a variety of allergic conditions, such as asthma and pruritus. It has a weak atropine type of antispasmodic effect against cholinergic agents. It was proposed that, similar to atropine, hydroxyzine exerts its central nervous system depressant effect by an anticholinergic mechanism in the reticular activating system.

Hydroxyzine has been used in the treatment of alcoholism and is the only antianxiety agent among the many diphenylmethane compounds that has a piperazine skeleton. Slight variations in the structure of hydroxyzine produce great changes in pharmacologic properties. Substitution of a benzyl side chain for the hydroxyethoxyethyl side chain produces potent antihistamines, such as meclizine and buclizine.

Azaspirodecanediones

An entirely new type of drug with antianxiety effects but limited sedative properties is buspirone.[82]

Buspirone is a novel anxiolytic agent unrelated to the benzodiazepines in structure or pharmacologic properties. Extensive clinical studies have shown buspirone to be effective in the treatment of anxiety, with efficacy comparable to that of diazepam or clorazepate. Buspirone exhibits a unique pharmacologic profile in that it alleviates anxiety without causing sedation or functional impairment and does not promote abuse or physical dependence.[83] Furthermore, preclinical studies have shown that buspirone does not possess anticonvulsant or muscle

Meclizine

Buclizine

relaxant properties and does not interact significantly with central nervous system depressants. Biochemical and electrophysiologic studies indicate that buspirone alters monoaminergic and GABAergic systems in a manner different from that of the benzodiazepines. The uniform depressant action of the benzodiazepines upon serotoninergic, noradrenergic, and dopaminergic cell firing may result from their facilitatory effect on GABA and its known inhibitory influence in these monoaminergic areas. In contrast to the benzodiazepines, buspirone exerts a differential influence on monoaminergic neuronal activity, suppressing serotoninergic activity while enhancing dopaminergic and noradrenergic cell firing. The mechanism of action of buspirone challenges the notion that only one neurotransmitter mediates anxiety. The interaction with multiple neurotransmitters at multiple brain sites suggests that buspirone may alter diverse activities within a "neural matrix of anxiety."

Buspirone (Buspar®)

REFERENCES

1. S. H. Snyder, *Sci. Am., 253,* 132(1985); M. J. Berridge, *Sci. Am., 253,* 142(1985).
2. G. A. Robinson, et al., *Cyclic Amp,* New York, Academic Press, 1971.
3. P. Seeman, *Pharmacol. Revs., 32,* 229(1981).
4. A. Carlsson and M. Lindqvist, *Acta Pharmacol. Toxicol., 20,* 140(1963).
5. R. J. Baldessarini, *Chemotherapy in Psychiatry: Principles and Practice,* Cambridge, Harvard University Press, 1985, pp. 36–46.
6. P. Jenner and C. D. Marsden, in *Drugs in Central Nervous System Disorders.* D. C. Horwell, Ed., Marcel Dekker, New York, 1985, pp. 149–262.
7. J. W. Kebabian and D. B. Caine, *Nature, 285,* 93(1979).
8. O. Civelli, et al., *Eur. J. Pharmacol., 207,* 277(1991).
9. P. Sokoloff, et al., *Biochem. Pharmacol., 43,* 659(1992).
10. C. Kaiser, P. A. Dandrige, E. Garvey, R. A. Hahn, H. M. Sarav, P. E. Setter, L. S. Bass, and J. Clardy, *J. Med. Chem., 25,* 697(1982).
11. T. W. Lovenberg, et al., *Eur. J. Pharmacol., 166,* 111(1989).
12. M. J. Kuhar, *Trends Neurosci., 14,* 299(1991).
13. M. J. Kuhar, *Neurosci. Facts, 3,* 49(1992).
14. B. K. Madras, et al., *Mol. Pharmacol., 36,* 518(1989).
15. J. L. Neumeyer, et al., *J. Med. Chem., 34,* 3144(1991).
16. R. B. Innis, J. P. Seibyl, et al., *Proc. Natl. Acad. Sci. U.S.A., 90,* 11965(1993).
17. W. H. Oldendorf and W. G. Dewhurst, *Principles of Psychopharmacology,* 2nd ed., Wm. G. Clark and J. Del Giudice, Eds., New York, Academic Press, 1978, pp. 183–193.
18. W. M. Pardrige, *Annual Reports in Medicinal Chemistry, Vol. 20,* D. M. Bailey, Ed., 1985, pp. 305–313.
19. N. Bodor and H. H. Farag, *J. Med. Chem., 26,* 528(1983).
20. S. Snyder and H. Yamamura, *Arch. Gen. Psychiatry, 34,* 236(1977).
21. S. H. Snyder, et al., *Science, 184,* 1243(1974).
22. J. M. Walker, et al., *Pharmacol. Rev., 42,* 355(1990).
23. L. D. Wise and T. G. Heffner, in *Annual Reports in Medicinal Chemistry,* Vol. 27, 49–57 (1992).
24. D. Bovet and A. M. Staup, *C. R. Soc. Biol. (Paris), 124,* 547(1937).
25. H. Laborit, et al., *Presse Med., 60,* 206(1952).
26. P. Charpentier, U.S. Pat. 2,519,886; 2,530,451(1950); P. Charpentier, et al., *C. R. Acad. Sci. (Paris), 325,* 59(1952).
27. J. Delay, et al., *Ann. Med. Psychol. (Paris), 110,* 112(1952).
28. A. S. Horn and S. H. Snyder, *Proc. Natl. Acad. Sci. U.S.A., 68,* 2325(1971).
29. G. M. Simpson and J. H. Lee, in *Psychopharmacology: A Generation of Progress.* M. A. Lipton, et al., Eds. New York, Raven Press, 1978, pp. 1131–1137.
30. R. J. Baldessarini, in *The Pharmacological Basis of Therapeutics,* 8th ed., A. G. Gilman, et al., Eds., New York, Pergamon Press, 1990, pp. 383–435.
31. S. Fielding and H. Lal, in *Handbook of Psychopharmacology,* Vol. 10, L. L. Iverson, et al., Eds., Plenum Press, New York, 1978, pp. 91–128.
32. E. Usdin, in *Psychopharmacology: A Generation of Progress,* M. A. Lipton, et al., Eds., New York, Raven Press, 1978, pp. 895–903.
33. P. A. J. Janssen, in *Psychopharmacologica Agents,* Vol. 1, M. Gordon, Ed., New York, Academic Press, 1964, pp. 199–248.
34. R. Debert, et al., *Acta Psychiatr. Scand., 62,* 356(1980).
35. P. A. J. Janssen and F. M. Van Bever, in *Psychotherapeutic Drugs, Part II—Applications,* E. Usdin and I. S. Forrest, Eds., New York, Marcel Decker, 1977, pp. 839–921.
36. I. Faloon, et al., *Psychol. Med., 8,* 59(1978).
37. A. C. Sayers and A. A. Amsler, in *Pharmacological and Biochemical Properties of Drug Substances,* M. E. Goldberg, Ed., Washington, American Pharmaceutical Association, 1977, pp. 1–31.
38. D. W. Wyhe and S. Archer, *J. Med. Pharm. Chem., 5,* 932(1962).
39. F. J. Ayd, *Dis. Nerv. Sys., 35,* 447(1974).
40. B. M. Augrist, in *The Benzamides: Pharmacology, Neurobiology*

and Clinical Effects, J. Rotrosen and M. Stanley, Eds., Raven Press, New York, 1982, p. 1.
41. J. H. Hyttel, et al., in *CRC Handbook of Stereoisomers: Drugs in Psychopharmacology,* D. F. Smith, Ed., Boca Raton, CRC Press, 1984, pp. 143–214.
42. R. J. Baldessarini and D. Tarsy, *Eur. J. Pharmacol., 37,* 993(1976).
43. L. G. Humber, et al., *Mol. Pharmacol., 11,* 833(1975).
44. C. A. Tamminga, et al., *Science, 200,* 567(1978).
44a. R. G. Booth, et al., *Mol. Pharmacol., 38,* 92(1990).
45. G. Sen and K. C. Bose, *Indian Med. World, 2,* 194(1931).
46. S. Siddiqui and R. H. Siddiqui, *J. Indian Chem. Soc., 8,* 667(1931).
47. L. van Itallic and A. J. Steenhauer, *Arch. Pharm. (Weinheim), 270,* 313(1932).
48. J. M. Mueller, E. Schlittler, and H. J. Bein, *Experientia (Basel), 8,* 338(1952).
49. H. J. Bein, *Experientia (Basel), 9,* 107(1953).
50. P. A. Diassi, et al., *J. Am. Chem. Soc., 77,* 4687(1955).
51. E. E. van Tamelen and P. D. Hance, *J. Am. Chem. Soc., 77,* 4692(1955).
52. R. Pepinski, et al., *Acta Crystallogr., 10,* 811(1957).
53. R. B. Woodward, et al., *J. Am. Chem. Soc., 78,* 2023(1956).
54. R. B. Woodward, et al., *Tetrahedron, 2,* 1(1958).
55. K. Dziewonski and L. H. Sternbach, *Bull Int. Acad. Pol. Sci., Classe Aci. Math. Natl. Ser. A.,* 333(1935).
56. L. H. Sternbach, in *The Benzodiazepines,* S. Garattini, et al., Eds., New York, Raven Press, 1973, pp. 1–25.
57. L. H. Sternbach, *J. Med. Chem., 22,* 1(1979).
58. L. O. Randall, et al., *J. Pharmacol. Exp. Ther., 129,* 163(1960).
59. L. H. Sternbach, *J. Med. Chem., 22,* 1(1979).
60. E. Costa, et al., *Adv. Biochem. Pharmacol., 14,* 113(1975).
61. H. B. Bossmann, et al., *FEBS Lett., 82,* 368(1975).
62. H. Mohler and T. Okada, *Science, 198,* 849(1977).
63. R. F. Squires and C. Braestrup, *Nature, 266,* 732(1977).
64. C. Braestrup, et al., *Proc. Natl. Acad. Sci. USA, 77,* 2288(1980).
65. W. Hunkeler, et al., *Nature, 290,* 514(1981).
66. W. Haefely, et al., in *Advances in Drug Research, Vol. 14,* London, Academic Press, 1985, pp. 166–322.
67. W. Haefely, et al., *Adv. Biochem. Psychopharmacol., 14,* 131(1975).
68. E. Costa, et al., *Adv. Biochem. Pharmacol., 14,* 113(1975).
69. P. R. Schofield, et al., *Nature, 328,* 221(1987).
69a. A. Dobie and I. L. Martin, *Trends Pharmacol. Sci., 13,* 76(1992).
70. R. W. Olsen, *J. Neurochem., 37,* 1(1981).
71. P. Skolnick and S. M. Paul, *Int. Rev. Neurobiol., 23,* 103(1982).
72. R. W. Olsen, *Ann. Rev. Pharmacol. Toxicol., 22,* 245(1982).
73. R. I. Fryer, in *Comprehensive Medicinal Chemistry,* Vol. 3, C. Hansch, Ed., New York, Pergamon Press, 1990, pp. 539–566.
74. P. Linscheid and J. M. Lehn, *Bull. Soc. Chim. France,* 992(1967).
75. J. Blount, et al., *Mol. Pharmacol., 24,* 425–428(1983).
76. V. Sunjic, et al., *J. Heterocycl. Chem., 16,* 757(1979).
77. J. Sepinwall and L. Cook, *Handbook of Psychopharmacology, Vol. 13,* L. L. Iversen, et al., Eds., New York, Plenum Press, 1978, pp. 345–393.
78. W. Schallek, et al., *Adv. Pharmacol. Chemother., 10,* 119(1972).
79. D. J. Greenblatt and R. I. Shader, *N. Engl. J. Med., 291,* 1011(1974).
80. M. A. Schwartz, in *The Benzodiazepines,* S. Garattini, et al., Eds., New York, Raven Press, 1973, pp. 52–74.
81. L. Farah, *Int. Rec. Med. Gen. Prac. Clin., 169,* 370(1956).
82. A. S. Eison and D. L. Temple, *Am. J. Med., 80 (Suppl 3B),* 1(1986).
83. R. A. Gammans, et al., *Am. J. Med., 80 (Suppl 3B),* 41(1986).

SUGGESTED READINGS

G. Biggio and E. Costa, Eds. *GABAergic transmission and anxiety,* in *Adv. Biochem. Pharmacol.,* Vol. 41, New York, Raven Press, 1986.

W. Haefely, et al., *Recent advances in the molecular pharmacology of benzodiazepine receptors and in the structure-activity relationships of their agonists and antagonists,* in *Adv. Drug Research,* Vol. 14, London, Academic Press, 1985, pp. 165–322.

Chapter 13

DRUGS USED TO TREAT NEUROMUSCULAR DISORDERS: ANTIPARKINSONISM AGENTS AND SKELETAL MUSCLE RELAXANTS

John L. Neumeyer and Raymond G. Booth

Drugs described in this chapter have in common the ability to reduce muscle tone by virtue of their action on the central nervous system. These drugs, however, have distinctly different mechanisms of action, therapeutic uses, and chemical properties. One group includes drugs useful in treating Parkinson's disease and related syndromes. The drugs in the other group, the skeletal muscle relaxants, depress neuronal synapses controlling muscle tone and are used for treating acute muscle spasm, tetanus, and spasticity.

DRUGS USED FOR PARKINSON'S DISEASE

For more than a century, drugs with central anticholinergic properties, used empirically and with limited success, constituted the major approach to the pharmacotherapy of Parkinson's disease (paralysis agitans), first described by James Parkinson in 1817. The clinical usefulness of the belladonna alkaloids was believed to be due to their central anticholinergic properties, and based on this reasoning, the supposition of increased cerebral cholinergic activity in parkinsonism was made. The application of this concept led to the introduction of several synthetic anticholinergic drugs for the treatment of Parkinson's disease (Table 13–1). Such therapy was often supplemented with antihistamines. It was not until sophisticated biochemical assays and histochemical techniques were developed, however, that the pathophysiology of Parkinson's disease was understood. The discovery that large doses of levodopa produce dramatic improvement in the symptoms of this disease led to the use of this drug and the combination of levodopa and decarboxylase inhibitors in therapy.

Although the anticholinergic agents are still useful for parkinsonism, they are less effective than levodopa. The chemistry and mechanism of action of the anticholinergics are discussed in greater detail in Chapter 17 and those of the antihistamines are discussed in Chapter 22. This chapter is primarily a discussion of the pathophysiology of Parkinson's disease and the events that led to the use of dopamine agonists for the treatment of this disease.

Pathophysiology of Parkinson's Disease

In addition to its many other functions, the extrapyramidal system maintains posture and muscle tone and modulates voluntary movement. In patients with parkinsonism, it has been observed that dopamine levels in the striatum are depleted or severely reduced. Dopamine is believed to act as a neurotransmitter at certain striatal synapses that are concerned with mediating inhibition in the nigrostriatal pathway (Fig. 13–1). (See also Chapter 12.) The discovery of Ehringer and Hornykiewicz[1] of a gross depletion of striatal dopamine in the brains of patients with idiopathic and postencephalic parkinsonism examined post mortem led to the present hypotheses regarding the role of dopamine in the pathogenesis of parkinsonism. It is currently believed that acetylcholine is an excitatory transmitter for neurons running between the substantia nigra, the pallidum, and the striatum, and that in parkinsonism there is an imbalance between these two transmitter systems in the direction of cholinergic dominance (Fig. 13–2).

In parkinsonism, the dopaminergic input to the corpus striatum is deficient, whereas the cholinergic input remains unchanged. The resulting imbalance explains the apparent increased effect of acetylcholine in parkinsonism and justifies the former empiric use of anticholinergic drugs, although their usefulness is now limited.

It is now known that interconnections between the striatum and the substantia nigra and the distributions of D_1- and D_2-type (including perhaps D_3, D_4, D_5, and others) receptors in these brain regions are much more complex than previously thought.[2] In the striatum, D_2-type receptors are located presynaptically on dopamine neurons (autoreceptors) and postsynaptically on cholinergic and γ-aminobutyric acid (GABA)ergic neurons. It is thought that stimulation of these striatal D_2 receptors by dopamine agonists or by dopamine formed from levodopa or indirect dopamine agonists negatively modulates the excess cholinergic transmission apparent in Parkinson's disease to provide relief from extrapyramidal symptoms. In the substantia nigra, D_2-type receptors are located on the dendrites of dopamine cell bodies and, again, postsynaptically on GABAergic neurons. D_1-type receptors have been shown to be present on cell bodies of GABA and substance-P–containing neurons in the striatum and on the nerve terminals of these neurons in the substantia nigra. Evidence has accumulated to suggest that in addition to its well-characterized release in the striatum, dopamine also is released by dendrites of substantial nigra

Table 13–1. Drugs Used for Parkinsonism

Class and Nonproprietary Name	Chemical Structure	Trade Name	Single Oral Dose (initial)
Synthetic anticholinergic agents			
Benztropine mesylate	N·CH_3; CHO—; ·CH_3SO_3H	Cogentin	0.5–1.0 mg
Trihexyphenidyl hydrochloride	+; H; NCH_2CH_2·C-OH Cl^-	Artane Pipanol Tremin	1.0–2.0 mg
Procyclidine hydrochloride	+; H; NCH_2CH_2·C-OH Cl^-	Kemadrin	2.0–5.0 mg
Biperiden hydrochloride	+; H; NCH_2CH_2·C—OH Cl^-	Akineton	1.0–2.0 mg
Antihistamines			
Diphenhydramine hydrochloride	$CHOCH_2CH_2$·N(CH_3)(CH_3) Cl^-	Disipal	50.0 mg
Orphenadrine hydrochloride	$CHOCH_2CH_2$—N(CH_3)(CH_3) Cl^-; CH_3	Benadryl, others	25.0 mg
Phenothiazines (Antihistamines)			
Ethopropazine hydrochloride	S; N·CH_2 CH(CH_3)—N(CH_3)(CH_3)	Parsidol	50.0 mg

(continued)

Table 13–1. *(Continued)*

Class and Nonproprietary Name	Chemical Structure	Trade Name	Single Oral Dose (initial)
Dopamine agonists			
Levodopa	$CH_2-CH(NH_2)-COOH$; HO, OH	Bendopa Dopar Larodopa	0.1–1.0 g
Decarboxylase inhibitors			
Carbidopa (MK 486)	$CH_2-C(CH_3)(COOH)-NH-NH_2$; HO, OH	Sinemet (contains levodopa)	
Benserazide hydrochloride	$CH_2-NH-NH-C(=O)-CH(NH_2)-CH_2-OH \cdot HCl$; HO, OH, OH		
Dopamine-releasing agents			
Amantadine hydrochloride	$NH_2 \cdot HCl$	Symmetrel	100 mg

dopamine neurons, and a striatonigral pathway may play an important role in the actions of levodopa. It has been proposed that activation of D_1-type receptors in the substantia nigra may explain some of the synergistic effects of D_1 and D_2 agonists observed in animal models of Parkinson's disease. Thus, the control over motor movement exerted by the basal ganglia dopaminergic system appears to be achieved by dopamine release at both axon and dendritic terminals in the striatum and substantia nigra, to act on the D_1 and D_2 receptors located in these regions.

Symptoms of Parkinson's disease may be induced by drugs or toxicants that cause a relative central nervous system dopamine–acetylcholine imbalance by one of the following mechanisms: (1) by the depletion of dopamine from intraneuronal stores; (2) by rendering the dopamine receptor less accessible to dopamine; (3) by increas-

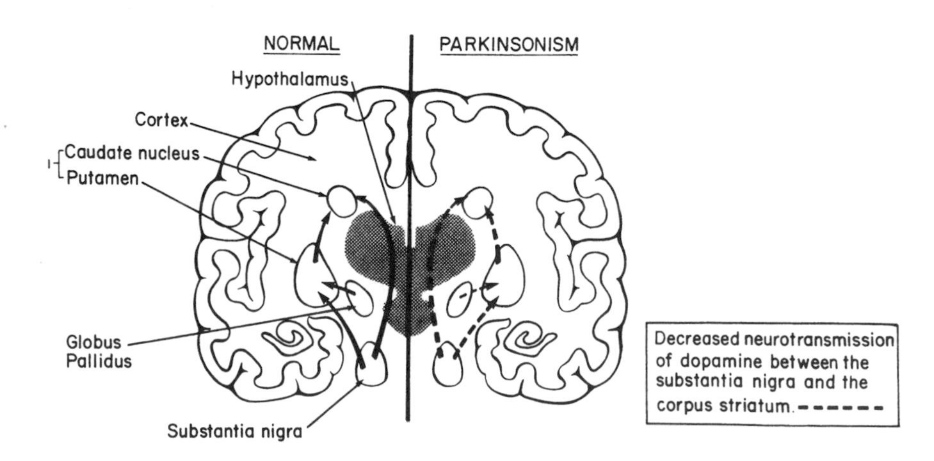

Fig. 13–1. Schematic representation of the extrapyramidal system and the nigrostriatal dopaminergic pathways.

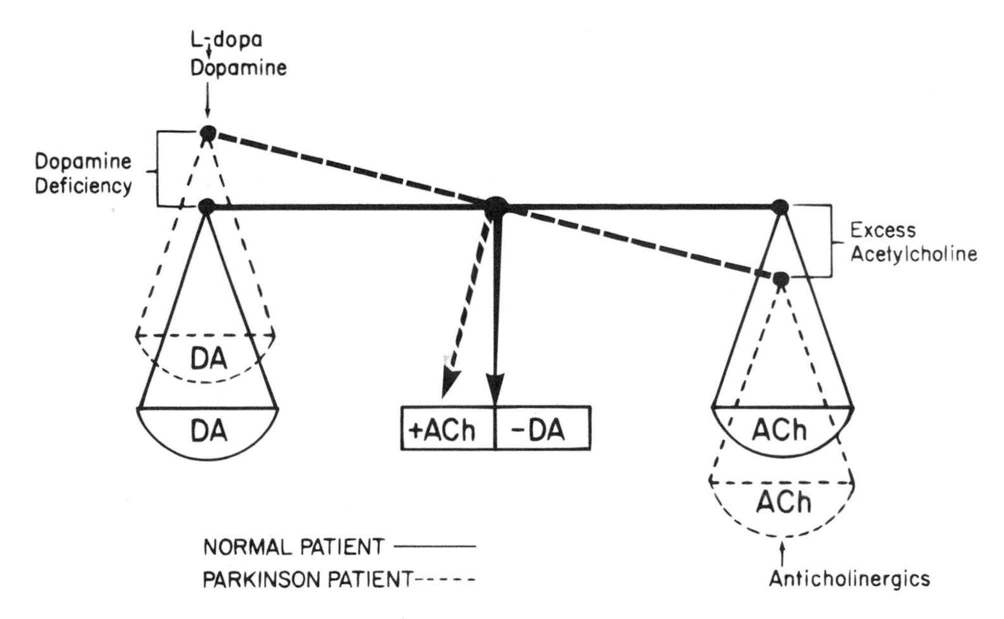

Fig. 13–2. Schematic representation of the imbalance between the excitatory neurotransmitter acetylcholine and the inhibitory neurotransmitter dopamine in the basal ganglia of parkinsonian patients.

ing acetylcholine levels; or (4) by dopamine neurotoxins such as MPTP (N-methyl-4-phenyl-1,2,3,6-tetrahydropyridine) and 6-hydroxydopamine causing degeneration of the substantia nigra. Reserpine is an example of a drug in the first category. Parkinsonian symptoms induced by such drugs can be reversed with exogenous levodopa. Among the drugs that render the dopaminergic receptor less accessible to dopamine are the neuroleptic agents (see Chapter 12), such as the butyrophenones, thioxanthines, dibenzodiazepines, and many of the phenothiazines. It is believed that these drugs selectively block the dopamine receptors, preventing the interaction between dopamine and its receptor. Drugs that increase acetylcholine levels, such as the cholinesterase inhibitor physostigmine and the cholinergic agent carbachol, aggravate parkinsonism in humans.

Etiology of Parkinson's Disease

The cause of the neuronal degeneration in Parkinson's disease remains unclear. The pattern of catecholamine abnormalities is important in understanding the cause of this disease. Dopamine neurons degenerate more with advancing age than other neuronal systems in the brain. In normal adults, levels of dopamine in the corpus striatum decline by about 13% per decade. Parkinsonian symptoms become apparent when striatal dopamine levels decline by about 70%.[3] Conceivably the symptoms of parkinsonism are produced by two processes, a specific disease-related insult combined with changes of normal aging.[4] This two-pronged pathophysiology may explain why Parkinson's disease is a progressive disorder of late onset. Perhaps the continuous deterioration of parkinsonian patients is not simply due to active disease but involves the effects of aging superimposed on the original lesion. The postulated lesion may take place in early life, even at birth, producing a loss of 20% to 40% of dopamine neurons. No symptoms are expected until late middle age, when dopamine neurons begin to die from the normal aging process. Because many neurologic conditions are genetically determined, researchers have investigated a possible genetic influence in Parkinson's disease. It is currently believed that most cases do not appear to be genetically determined and may, instead, involve environmental factors. Because all members of a family are usually exposed to similar environments, a primary role of the environment cannot be ruled out, even in the early-onset cases. Twin studies provide a more rigorous genetic analysis and have failed to reveal a genetic component of Parkinson's disease. There is direct evidence that environmental toxins may cause some types of Parkinson's disease.[5] For instance, manganese miners in South America are at risk of developing a form of Parkinson's disease characterized by increased muscle tone, tremor, shuffling gait, and masklike facies.[6]

One of the best characterized epidemiologic findings in Parkinson's disease is its lower incidence in cigarette smokers than in nonsmokers.[7–10] Although some studies did not find an altered incidence in smokers,[11] the negative studies generally had fewer subjects. Perhaps the inverse correlation of smoking and Parkinson's disease only reflects decreased smoking after patients develop parkinsonian symptoms. The lower incidence in smokers, however, is apparent as early as 20 years before the onset of parkinsonian symptoms.[10] Something in cigarette smoke may protect against a toxin, endogenous or exogenous, that is relevant to parkinsonian neuropathology. The carbon monoxide in cigarette smoke may detoxify free radicals from environmental toxins or oxidation products of dopamine.[4] It also has been suggested that compounds present in cigarette smoke or metabolites of these compounds may inhibit monoamine oxidase (MAO)–B activity,[12] the main enzyme responsible for metabolism of biogenic amines. Studies with the nigrostriatal toxin MPTP (see later), suggest that MAO-B may be involved in the pathophysiology of Parkinson's disease and cigarette smoke attenuates MPTP-induced neurotoxicity.[13]

Parkinsonism Caused by MPTP

The identification of the pyridine derivative MPTP (N-methyl-4-phenyl-1,2,3,6-tetrahydropyridine) (Fig. 13–3) as a cause of parkinsonism, similar in neuropathology and motor abnormalities to idiopathic Parkinson's disease, has provoked a serious attempt to seek specific environmental toxins that might be involved in the cause of idiopathic Parkinson's disease. The role of MPTP in parkinsonian disorders was revealed by a serendipitous series of events. In 1977, a 23-year-old college student suddenly developed parkinsonian symptoms with severe bradykinesia, rigidity, and mutism. The abrupt and early onset of symptoms was so atypical that the patient was initially labeled a catatonic schizophrenic. The subsequent diagnosis of parkinsonism was substantiated by a therapeutic response to L-dopa, whereupon the patient was referred to the National Institutes of Mental Health in Bethesda, Maryland. He admitted having synthesized and used several illicit drugs. The psychiatrist who had elicited the patient's history visited his home and collected glassware that had been used for chemical syntheses. Chemical analysis revealed several pyridines, including MPTP, formed as by-products in synthesizing the reversed ester of meperidine known as MPPP, "designer heroin," or "synthetic heroin" (N-methyl-4-propionoxy-4-phenylpiperidine). This substance is also the desmethyl analog of alphaprodine (Fig. 13–4). It was unclear, however, whether MPTP or other constituents of the injected mixture accounted for the neurotoxicity.

After the patient returned home, he continued to abuse drugs and died of an overdose; autopsy revealed degeneration of the substantia nigra.[14] Subsequently, Langston et al.[15] identified other patients with virtually identical clinical symptoms who had also been receiving intravenous injections of preparations of MPPP containing large quantities of MPTP. Because in several patients MPTP was the principal or sole constituent injected, Langston et al. provided the first definitive evidence that MPTP is a parkinsonian neurotoxin. More than 400 people are now known to have self-administered MPTP, but only a few have developed parkinsonian symptoms. Many of these individuals who are presently asymptomatic, however, may be at risk for developing parkinsonism as they age.

Exposure to the high levels of MPTP by intravenous injection is not needed to develop parkinsonism. A 37-year-old industrial chemist developed parkinsonism after working with MPTP as a synthetic intermediate without

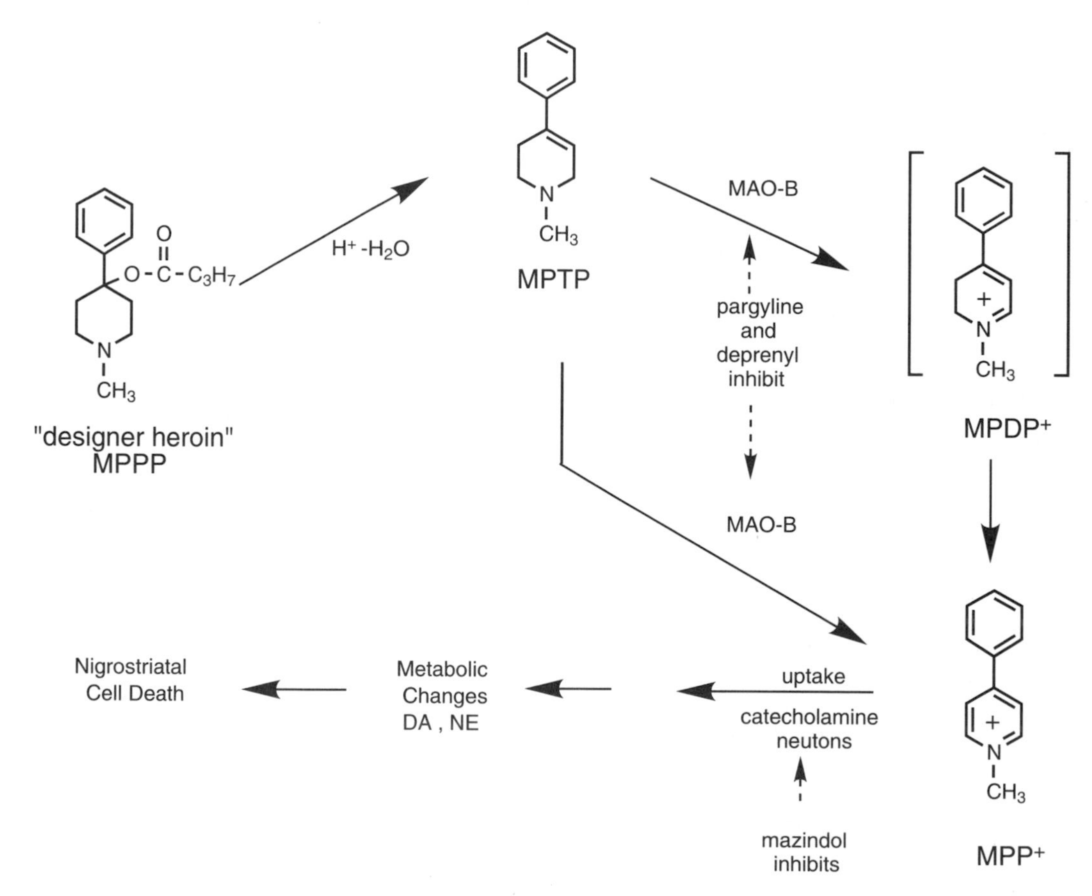

Fig. 13–3. Chemical conversion of MPPP and probable mechanism of MPTP neurotoxicity. (Adapted from J. A. Javitch, R. J. D'Amato, S. M. Strittmatter, and S. H. Snyder, *Proc. Natl. Acad. Sci. USA, 82,* 2173[1985].)

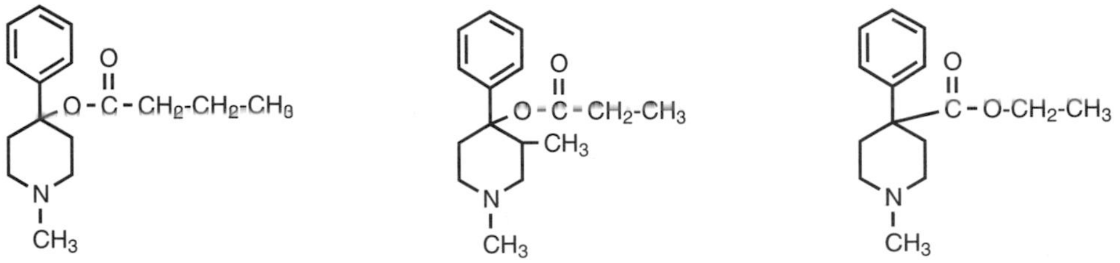

"designer heroin" (MPPP)

alphaprodine (Nisentil)

meperidine (Demerol)

Fig. 13–4. Phenylpiperidine synthetic analgesics.

ever ingesting it. Inhalation or skin contact with small amounts may be enough to cause the condition.

Both the clinical and neuropathologic features of MPTP-induced parkinsonism resemble idiopathic Parkinson's disease more closely than any previous animal or human disorder elicited by toxins, metals, viruses, or other means. Accordingly, understanding the molecular pathophysiology of MPTP neurotoxicity may shed light on the mechanisms in idiopathic parkinsonism.

Role of MAO in MPTP Neurotoxicity. An important question that needs to be answered in explaining MPTP neurotoxicity is why this substance selectively causes degeneration of dopamine neurons. Are there specific receptor sites for MPTP, analogous to high-affinity drug receptors? Castagnoli et al.[17] showed that MAO in brain mitochondria converts MPTP to 1-methyl-4-phenylpyridine (MPP^+) and that only type B MAO inhibitors prevent MPTP neurotoxicity.[18] This indicated that MPTP itself is not inherently toxic and must be bioactivated in vivo to a neurotoxic metabolite(s) by MAO-B. The localization of this enzyme accounts in part for the sensitivity of catecholamine-containing structures to MPTP. Thus, how does MAO chemically alter MPTP? As mentioned previously, in the presence of MAO-B, the partially saturated tetrahydropyridine ring in MPTP is oxidized at the allylic α-carbon to give the two-electron oxidation product, 1-methyl-4-phenyl-2,3-dihydropyridinium species ($MPDP^+$), which subsequently undergoes a further two-electron oxidation to the 1-methyl-4-phenylpyridinium species (MPP^+) via autooxidation, disproportionation, and enzyme-catalyzed mechanisms (see Fig. 13–3).[19] MPP^+ is currently believed to be the active metabolite responsible for the destruction of dopamine neurons, although a role for the unstable dihydropyridinium species $MPDP^+$ has not been ruled out.

The relationship of MAO and MPTP has neurobiologic relevance beyond MPTP neurotoxicity. MAO has been well-known for more than 50 years and has been thoroughly characterized in terms of substrate specificity. MAO oxidatively removes the amine grouping from biogenic amines, specifically those with only a single amine, such as the catecholamines (Fig. 13–5) and serotonin. Oxidation of a heterocyclic tertiary amine (e.g., MPTP) by this enzyme is unprecedented and suggests a novel physiologic role for MAO. MAO could be important in regulating the oxidation state of pyridine systems, such as those involving NADH and nucleic acids.[20]

The mechanism of MPP^+ formation in the brain may proceed via the unchanged free base form of $MPDP^+$ (see Fig. 13–3), which can diffuse across glial membranes and disproportionate extracellularly to MPP.$^+$[21] Presumably, extracellular MPP^+ must then enter the dopamine neurons to destroy the cells.

Several factors seem to be essential in the selective damage of the substantial nigra by MPTP. First, MPTP binds selectively to MAO-B, which is highly concentrated in human substantia nigra and corpus striatum. Within glia and serotonin nerve terminals, MAO transforms MPTP to $MPDP^+$, which can diffuse out and disproportionate extracellularly to MPP^+. Nigral cells, dopamine terminals, and norepinephrine terminals then accumulate MPP^+ through the catecholamine uptake system. The dense catecholamine terminal innervation of the locus coeruleus protects the neurons in this nucleus from MPP^+ accumulation. Within the cell bodies of nigral neurons, MPP^+ is bound to neuromelanin and gradually released in a depot-like fashion, maintaining a toxic intracellular concentration of MPP^+.[20]

It was suggested that the mechanism by which MPP^+ exerts its toxicity is by the inhibition of NADH dehydrogenase in the mitochondrial respiratory chain[22] This observation was later confirmed by the finding[23] that a variety of pyridine derivatives were capable of inhibiting NADH dehydrogenase. Further, MPP^+ has been shown to be rapidly accumulated by the mitochondria via an energy-dependent carrier that concentrates the compound well in excess of the concentration required for complete inhibition of NADH dehydrogenase.[24] Apparently then, inside dopamine neurons, MPP^+ produces inhibition of mitochondrial respiration leading to ATP depletion and cell death. In this regard, there is increasing evidence for a defect of mitochondrial respiratory chain function in Parkinson's disease and specific NADH CoQ_1 reductase (complex I) deficiency, which has been identified in the substantia nigra of parkinsonian brains.[25]

Thus, the study of MPTP, the product inadvertently produced from synthetic heroin, and the serendipitous finding that this substance causes Parkinson's disease have increased the understanding of catecholamine sys-

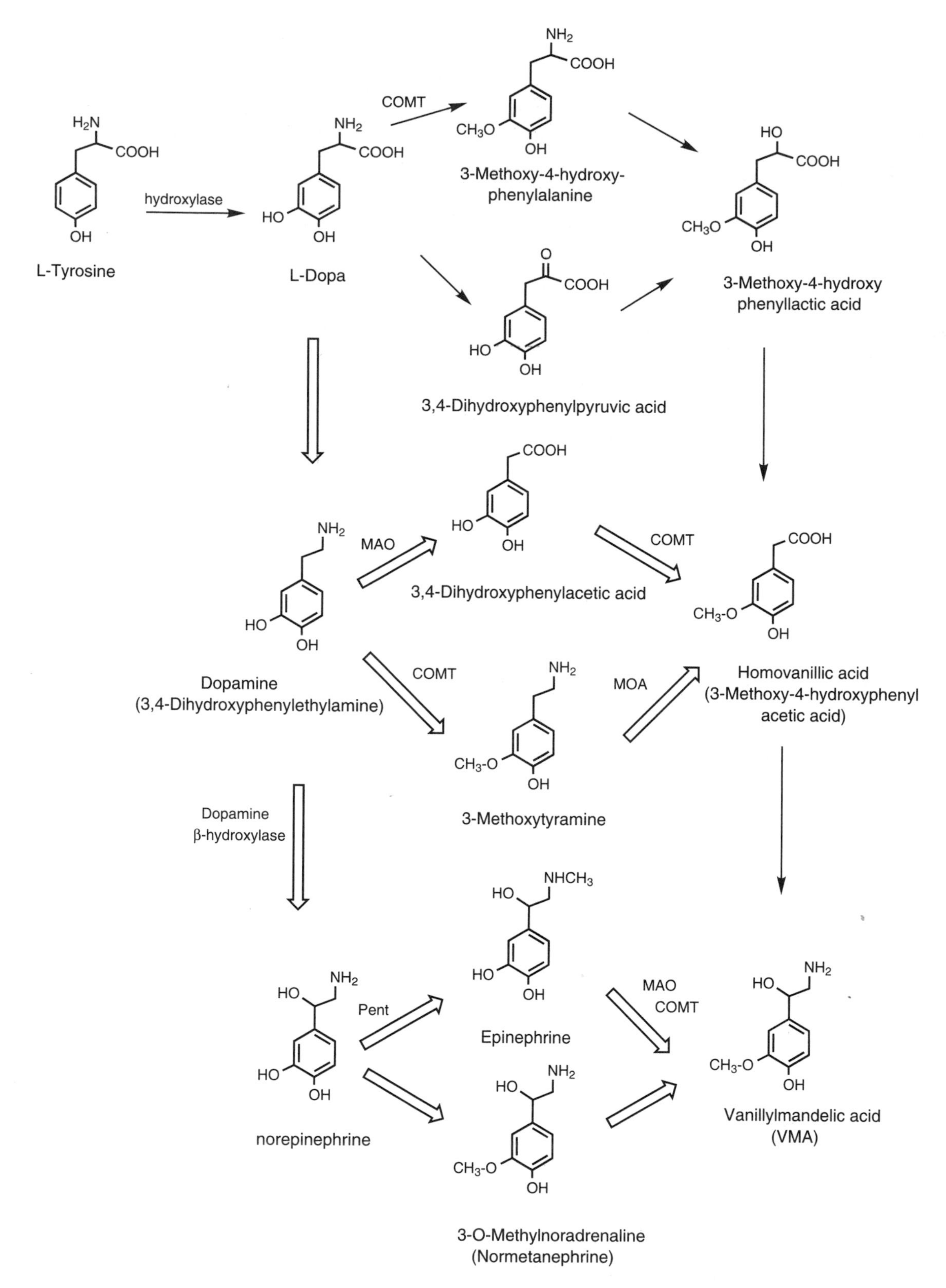

Fig. 13–5. Important pathways in the biosynthesis and metabolism of levodopa. Major pathways are shown by heavy arrows. COMT, catechol-O-methyltransferase; MAO, monoamine oxidase; PENT, phenylethanolamine N-methyl transferase.

tems in the brain and focused attention on the possible role of MPTP and MPP^+-like endogenous or environmental toxins in the cause of Parkinson's disease.

Anticholinergic Therapy

For many years, drug therapy for parkinsonism depended primarily on the limited efficacy of the belladonna alkaloids. With the newer synthetic alkaloids (see Table 13–1), attempts were made to increase the central anticholinergic effects as well as to reduce their undesirable peripheral effects. Unfortunately the side effects of these agents are troublesome, the most severe being their action on the central nervous system, including delusions, hallucinations, somnolence, ataxia, and dysarthria. The peripheral effects, such as dry mouth, blurred vision, constipation, urinary retention, and tachycardia, are also disturbing deterrents to their use.

In general, anticholinergic drugs rarely produce more than 20% improvement, and, despite continued use, the symptoms of the disease continue to progress. The most important present usage of the anticholinergic agents is with L-dopa, which, in addition to diminishing the cholinergic striatal effects, may inhibit the reuptake and storage of dopamine at the striatum.[26] The antihistamines, particularly those with central anticholinergic effects, are generally better tolerated in the elderly and may produce slightly greater relief from tremor,[27] but this therapy is rarely used today.

Dopaminergic Therapy

Several approaches may be used to decrease the dopaminergic deficiency at the striatum: (1) augmentation of the synthesis of brain dopamine, (2) stimulation of dopamine release from presynaptic sites, (3) direct stimulation of dopamine receptors, (4) decreasing reuptake of dopamine at presynaptic sites, and (5) decreasing dopamine catabolism (see Fig. 13–5).

Augmentation of Dopamine Synthesis—Levodopa Therapy

The initial report by Cotzias and co-workers in 1967,[27] describing dramatic symptomatic improvement of parkinsonian patients given high oral doses of racemic dopa, was followed by more clinical trials that confirmed the efficacy and safety of the *levo* isomer. The effectiveness of levodopa required penetration of the drug into the central nervous system and its subsequent enzymatic decarboxylation to dopamine. Dopamine does not cross the blood–brain barrier, probably because of its basicity [pK_a 8.9 (OH), 10.6 (NH_2)],[28] because it exists primarily in its protonated form under physiologic conditions. It requires the precursor amino acid, which is less basic [pK_a 2.32 (COOH), 8.72 (NH_2), 9.96 (OH), 11.79 (OH)][28] and can thus penetrate the central nervous system.

Biosynthesis and Metabolism of Levodopa. Levodopa is formed from L-tyrosine as an intermediary metabolite in the biosynthesis of catecholamines (see Fig. 13–5). Dopamine is formed from levodopa by the cytoplasmic enzyme aromatic L-amino acid decarboxylase. The effects observed following systemic administration of levodopa have been attributed to its catabolites, dopamine, norepinephrine, and epinephrine, acting at various sites in the periphery and in the brain. The principal metabolic pathways for levodopa are shown in Figure 13–5. A small amount is methylated to 3-O-methyldopa, which accumulates in the central nervous system owing to its long half-life. Most is converted to dopamine, small amounts of which are metabolized to norepinephrine and epinephrine. Metabolism of dopamine proceeds rapidly to the principal excretion products 3,4-dihydroxyphenylacetic acid (DOPAC) and 3-methoxy-4-hydroxyphenylacetic acid (homovanillic acid, HVA). The first step, which appears to be the main metabolic pathway, is the decarboxylation of levodopa by the pyridoxine-dependent enzyme aromatic-L-amino acid decarboxylase (dopa decarboxylase) to form dopamine. In humans, dopa decarboxylase activity is greater in the liver, heart, lungs, and kidneys than in the brain.[29] Therefore, ingested levodopa is converted to dopamine in the periphery in preference to the brain. It is thought that, in humans, levodopa thus enters the brain only when administered in dosages high enough to overcome losses caused by peripheral metabolism (3 to 6 g daily). Inhibition of peripheral decarboxylase markedly increases the proportion of levodopa that crosses the blood–brain barrier. This has been accomplished by the co-administration of such decarboxylase inhibitors as carbidopa and benserazide (Fig. 13–6).[30]

Side Effects of Levodopa. The actions of dopamine are complex because the amine acts on several different receptors located both centrally and peripherally. In the periphery, β-adrenoceptor stimulation enhances heart rate, α-adrenoceptor stimulation causes vasoconstriction, and dopaminergic receptor stimulation causes renal and mesenteric vasodilation (Fig. 13–6).[31] One of the most common side effects of levodopa therapy is gastric upset with nausea and vomiting. This appears to be the result of direct gastrointestinal irritation as well as stimulation of the chemoreceptor trigger zone (CTZ) in the medulla. The CTZ, which lies outside the blood–brain barrier, also has dopamine receptors. One of the results of combining levodopa with a peripheral decarboxylase inhibitor such as carbidopa (Sinemet, see Table 13–1) is a significant decrease in the incidence of and severity of nausea and vomiting as well as diminished effects on the cardiovascular system.

Because combination therapy permits a 75 to 80% reduction of the dosage of levodopa, the risk of side effects is reduced. Pyridoxine (a coenzyme for dopa decarboxylase) reverses the therapeutic effects, if administered with levodopa, by increasing the decarboxylase activity, which results in more levodopa being converted to dopamine in the periphery and consequently less being available for penetration into the central nervous system. When peripheral dopa decarboxylation is blocked with an agent such as carbidopa, the pyridoxine effect on peripheral levodopa metabolism is negligible.

Parkinsonian patients not previously treated with levodopa are usually started on a combination therapy with Sinemet, which is available in a fixed ratio of 1 part carbi-

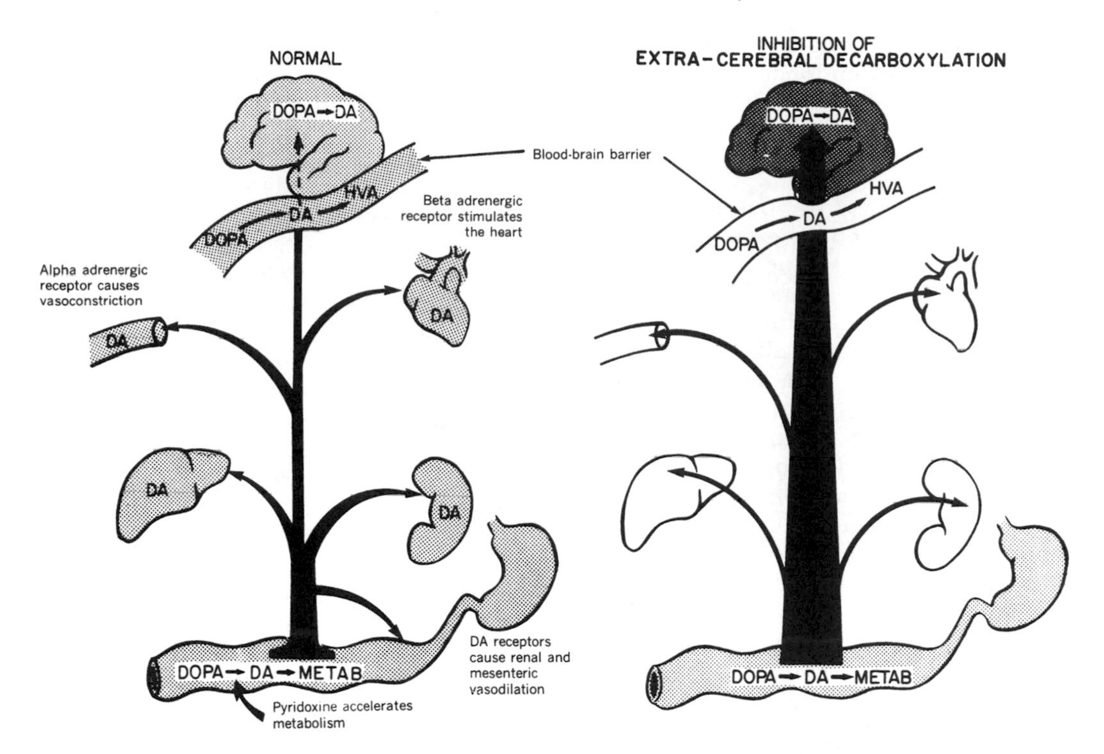

Fig. 13–6. Diagrammatic representation of the peripheral decarboxylation of levodopa to form dopamine (DA) and the mode of action of extracerebral decarboxylase on levodopa metabolism and distribution in vivo. The concurrent administration of levodopa and a decarboxylase inhibitor decreases the amount of levodopa required to elicit a therapeutic response in parkinsonism. HVA, homovanillic acid.

dopa and 10 parts levodopa, either 10/100 mg or 25/250 mg.

Another means of increasing brain dopamine levels has been suggested. The approach is based on the administration of such MAO inhibitors as nialamide or iproniazid, which promote the brain uptake of amines that otherwise do not enter the brain or enter it poorly.[32]

Selegiline (Eldepryl) the levo isomer of N, α-dimethyl-N-2-propynylbenzenethanamine, is a highly selective inhibitor of MAO-B. In contrast to

$C_6H_5-CH_2-CH(CH_3)-N(CH_3)-CH_2-C\equiv CH$

Selegiline [Eldepryl®]

the known nonspecific MAO inhibitors (e.g., pargyline, phenelzine, and isocarboxazid) selegiline does not cause profound and potentially lethal potentiation of the effects of catecholamines when administered concurrently with a centrally active amine. When selegiline is added to the therapeutic regimen, the dose of levodopa can be reduced without loss of therapeutic benefit.[33] In addition to its inhibition of MAO-B, selegiline also inhibits reuptake of dopamine and norepinephrine into presynaptic nerve terminals and increases the turnover of dopamine, thus adding to its potentiation of the pharmacologic effects of levodopa.[34] The current recommended regimen for selegiline in the treatment of Parkinson's disease is 10 mg per day.

Fusaric acid, a potent inhibitor of the enzyme dopamine β-hydroxylase, which mediates the final step in norepinephrine synthesis, has also been tried in Parkinson's disease in an attempt to increase central dopamine levels. Even though this seemed like a rational approach to therapy, fusaric acid failed to alter the severity of parkinsonism in otherwise untreated patients.[33]

$CH_3(CH_2)_3$ — pyridine ring (N) — COOH

Fusaric acid

Stimulation of Dopamine Release—Amantadine

Agents that can release dopamine from neuronal storage sites should be beneficial in the treatment of parkinsonism. Amantadine (also used as an antiviral agent) is currently the only such drug that has clinically significant antiparkinsonian effects, which are enhanced in the presence of levodopa. Amantadine has been shown to delay the reuptake of dopamine by nerve terminals and may have anticholinergic effects as well.[35] Thus, patients who respond to amantadine (indicating that their dopamine receptors are still capable of responding to dopamine) generally respond to treatment with levodopa. The rapid onset of optimal benefits from amantadine, within 2 weeks, affords an advantage over levodopa. Amantadine also produces fewer side effects than levodopa or the anticholinergic drugs. Amantadine for parkinsonism is usually given orally, 100 to 200 mg daily, and is readily absorbed from the gastrointestinal tract.

Amantadine is a primary amine with a pK_a of 10.8. The amino group is more basic than the amino group of dopamine (pK_a 10.6). Because most of the drug is in the protonated form at physiologic pH, how does it pass through the blood–brain barrier to exert its effect on dopaminergic neurons? A feasible explanation is that the cage-like structure of amantadine not only increases its lipophilicity, but also precludes its catabolism by oxidative enzymes, making more of the drug available for penetration into the central nervous system, even though it may be in minute amounts. Metabolism studies have shown that amantadine is excreted in the urine unchanged.

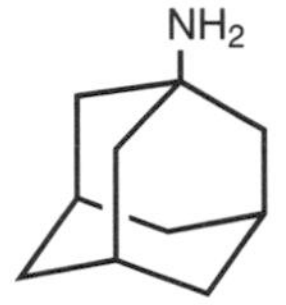

Amantadine [Symadine®, Symmetrel®]

Direct Stimulation of Dopamine Receptors

Drugs that act by directly stimulating the dopamine receptors have the advantage over such drugs as levodopa that their effect is independent of striatonigral degeneration, whereas levodopa depends on the ability of the remaining neurons to decarboxylate levodopa to dopamine.

Among the drugs that act predominantly at the dopamine receptor are bromocriptine,[36] a derivative of an ergot alkaloid, apomorphine, and its homolog N-propylnorapomorphine (NPA).[37] Bromocriptine is an ergopeptide derivative and dopamine agonist that predominantly stimulates the striatal D_2 dopamine receptors (see Chapter 12 for further discussion of dopamine receptors). Bromocriptine, in contrast to many other dopamine agonists, has mixed agonist–antagonist properties at these receptors. Bromocriptine mesylate (Parlodel), in low doses (5 to 30 mg/day), is effective in patients with mild-to-moderate Parkinson's disease, whereas bromocriptine in higher doses (31 to 100 mg/day) is needed in patients with advanced disease. In both low and high doses, bromocriptine combined with levodopa is usually more effective than bromocriptine alone.[36] Bromocriptine mesylate was originally introduced in the United States in 1978 for the short-term treatment of amenorrhea/galactorrhea associated with hyperprolactinemia owing to varied causes, excluding demonstrable pituitary tumors.

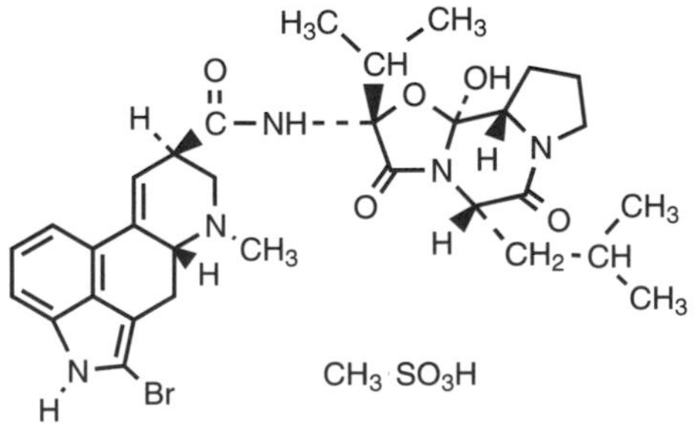

Bromocriptine mesylate [Parlodel®]
2-Bromo-α-ergocryptine

Structure-Activity Relationships of Dopamine Receptor Agonists

The discovery that dopamine-rich areas of the central nervous system contain an adenylate cyclase uniquely responsive to dopamine provides a single biochemical model system in which to test a wide range of chemicals as potential agonists or antagonists at central nervous system dopamine receptors in vitro (see Fig. 12–4). Many studies have been carried out on the actions of dopamine analogs in stimulating adenylate cyclase activity in homogenates of striatal brain tissue.[38] Such in vitro test systems, coupled with a number of in vivo behavioral tests, indicate a high degree of specificity for agonists at central nervous system dopamine receptors and reveal a spectrum of activity different from that at either α- or β-adrenoceptors. Among simple β-phenethylamine analogs of dopamine, only the N-methyl derivative epinine is equipotent with the parent compound. Although norepinephrine is effective in stimulating adenylate cyclase, it is about 20 times less potent than dopamine; the α-methyl analog of dopamine is also considerably less potent than dopamine. Because the side chain of dopamine possesses unlimited flexibility and unrestricted rotation about the β-carbonphenyl bond, little information can be obtained concerning the conformational requirements of the dopamine receptor.

Various compounds in which the catechol ring and the amino group of dopamine are held in rigid conformation can be prepared to assess more accurately the structural and conformational requirements of the dopamine receptor. The aporphine alkaloid, apomorphine, obtained by the acid-catalyzed rearrangement of morphine and recognized for years as a powerful centrally acting emetic

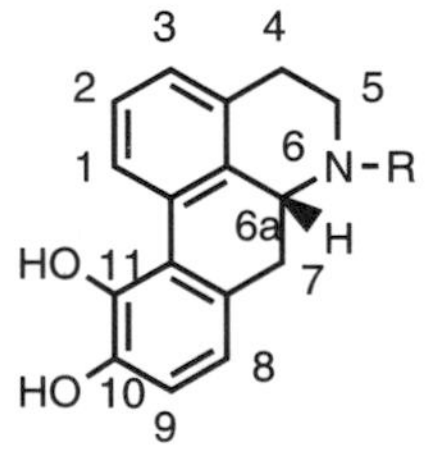

Apomorphine (R = CH_3)
N-n-Propylnorapomorphine NPA (R = C_3H_7)

by its action on the medullary CTZ, was found to produce effects similar to those of dopamine by direct stimulation of central dopamine receptors.[39] Evidently, apomor-

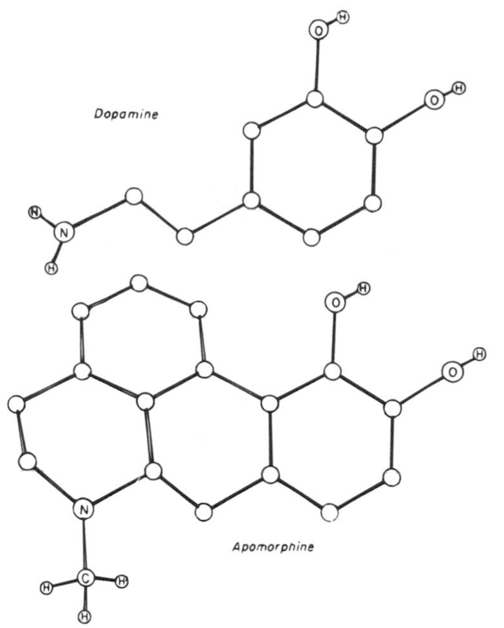

Fig. 13–7. Model of apomorphine molecule as determined by the published x-ray data of Giesecke[42] showing the structural relationship to dopamine in the *trans* α-rotameric conformation.

phine can pass through the blood–brain barrier, whereas dopamine cannot. Apomorphine is a potent D_1 and D_2 agonist that can be administered parenterally or by application to mucosal membranes. Apomorphine produces an antiparkinsonian effect equivalent to that of levodopa, and its administration of subcutaneous minipump to reverse "off" periods occurring during levodopa therapy is currently under investigation.[40] Of other apomorphine analogs examined, N-n-propylnorapomorphine was found to be 2 to 90 times more active than apomorphine as a dopamine agonist.[41] Such aporphines have enabled medicinal chemists to examine the structural requirements of the dopamine receptor(s) in a rigid system. Examination of the molecular model of apomorphine (Fig. 13–7) shows that it contains within its molecule the elements of the structure of dopamine or, more accurately, its N-methyl derivative, epinine.

Isoapomorphine contains the structure of N-methyldopamine in a β-rotameric conformation (Fig. 13–8); this compound was inactive as a dopamine agonist. Similarly, when the dopamine portion of the molecule was incorporated into 1,2-dihydroxyaporphine,[43] the catechol moiety and the amino group are forced into a *cis* or *gauche* conformation. Again, this compound resulted in an inactive dopamine agonist. It has also been shown that the preferred configuration of apomorphine for dopamine agonist activity is the *levo* (6a,R) isomer.[44] Racemic apomorphine produced by total synthesis[45] rather than by the acid-catalyzed rearrangement of morphine that yields the *levo* (6a,R) isomer is only half as potent. It has been shown the S(+) isomer of *N*-n-propylnorapomorphine (NPA) can distinguish between postsynaptic D_2-type receptors, at which it is not active as an agonist, and presynaptic D_2-type autoreceptors, at which it acts as a potent agonist to negatively modulate dopamine synthesis.[46] Studies with the semirigid aminotetralins represent the fully extended *trans* conformation of dopamine.[47] In 2-amino-5,6-dihydroxy-1,2,3,4-tetrahydronaphthalene (A-5,6-DTN), a *trans* conformation (α-rotamer) between the benzene ring and the amino side chain exists. In 2-amino-6,7-dihydroxy-1,2,3,4-tetrahydronaphthalene (A-6,7-DTN), a *trans* conformation (β-rotamer) between the benzene ring and the amino side chain can be visualized. As shown in Figure 13–8, A-5,6-DTN is structurally related to apomorphine, whereas A-6,7-DTN is more closely related to the conformation of isoapomorphine. Apomorphine, besides being active on central dopamine receptors, is also active on vascular dopamine receptors, although it is weaker than dopamine. Isoapomorphine, however, is inactive as a vascular dopamine agonist.[48]

SKELETAL MUSCLE RELAXANTS

The antispastic drugs are diverse in both their chemical structure and their site and mechanism of action. The first drug recognized to exhibit antispastic activity was Antodyne or 3-phenoxy-1,2-propanediol. In guinea pigs and rabbits, Antodyne produced prolonged paralysis without impairing consciousness. Antodyne was introduced into clinical medicine in 1910 as an analgesic and antipyretic. The duration of its skeletal muscle relaxant effect, however, was too short-lived to be clinically useful. In 1943, SAR studies of a series of simple glyceryl ethers related to Antodyne, by Berger and Bradley, identified the most active compound, which following extensive research and development led to the introduction of mephenesin (3-(o-toloxy)-1,2-propanediol) in 1946.[49] Pharmacologic studies revealed that mephenesin selectivity depressed polysynaptic while sparing monosynaptic spinal cord reflexes. Because of its safety and selective action on the spinal cord, mephenesin became the first widely prescribed centrally acting skeletal muscle relaxant. Accordingly, mephenesin is the prototype of the interneuronal blocking type of muscle relaxant. Several other compounds with mephenesin-like pharmacologic profiles, some longer-lasting, were developed and marketed as muscle relaxants. These include carisoprodol, chlorphenesin carbamate, chlorzoxazone, metaxalone, methocarbamol, and orphenadrine (Table 13–2).

The general pharmacology and neuropharmacology of propanediol carbamates have been reviewed.[50] The efficacy of these compounds is difficult to assess because of the lack of well-controlled clinical studies. These agents have been largely replaced by diazepam, sedatives, or analgesics. The pharmacology of the mephenesin-like muscle relaxants is remarkably similar to that of the sedative-hypnotics. Indeed, the only apparent difference is the greater selectivity for the spinal cord and, consequently, lesser sedation for antispastics. Both classes produce a reversible nonspecific depression of the central nervous system.[51]

The skeletal muscle relaxants are used to relieve muscular spasticity, a condition characterized by exaggerated resting tone of a muscle. Muscle hypertonus is usually

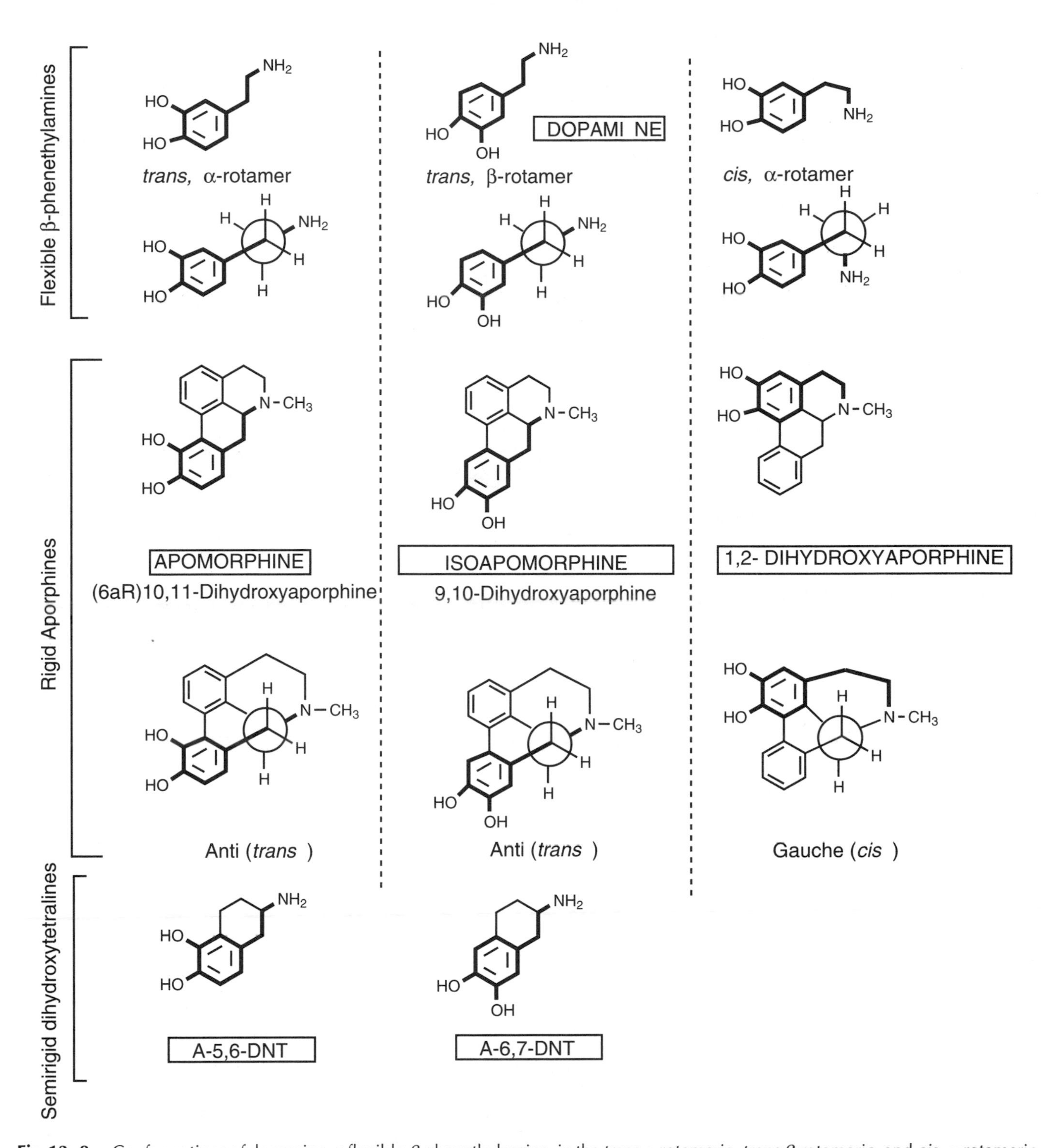

Fig. 13–8. Conformations of dopamine, a flexible *β*-phenethylamine, in the *trans α*-rotameric, *trans β*-rotameric, and *cis, α*-rotameric forms and their structural relationships to the rigid analogs of dopamine, the aporphines, apomorphine, isoapomorphine, and 1,2-dihydroxyaporphine. Also shown are the corresponding semirigid analogs of dopamine, the dihydroxytetralins, A-5, 6-DTN and A-6, 7-DTN.

Table 13–2. Skeletal Muscle Relaxants

Class and Nonproprietary Name	Chemical Structure	Trade Name	Single Dose
Glycerol monoethers and derivatives			
Mephenesin	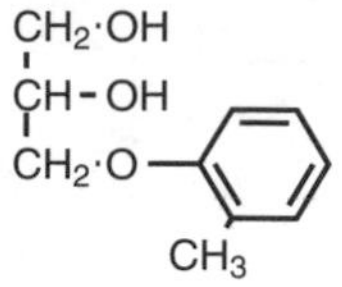 $CH_2 \cdot OH$; $CH-OH$; $CH_2 \cdot O-$(ring); CH_3	—	1–2 g (oral)
Chlorophensin carbamate	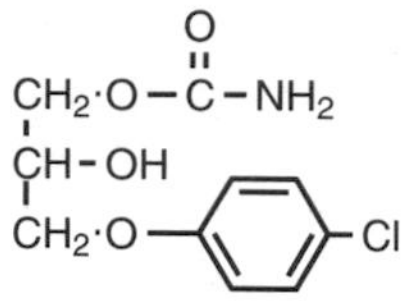 O; $CH_2 \cdot O-C-NH_2$; $CH-OH$; $CH_2 \cdot O-$(ring)$-Cl$	Maolate	800 mg (oral)
Methocarbamol	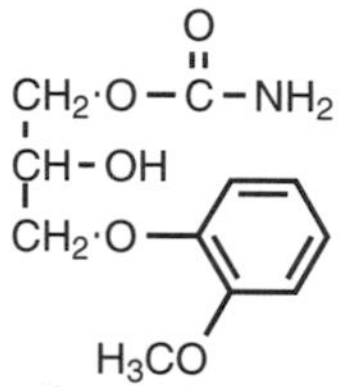 O; $CH_2 \cdot O-C-NH_2$; $CH-OH$; $CH_2 \cdot O-$(ring); H_3CO	Robaxin	1–2 g (oral) 500 mg IM 1–3 g daily (IV)
Substituted alkanediols and derivatives			
Meprobamate	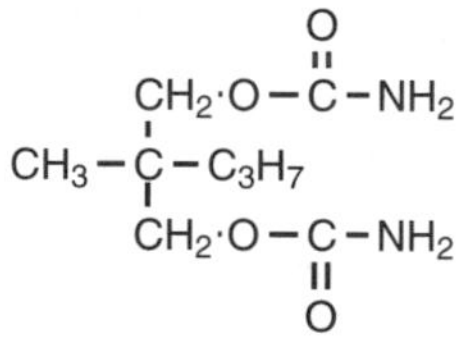 O; $CH_2 \cdot O-C-NH_2$; $CH_3-C-C_3H_7$; $CH_2 \cdot O-C-NH_2$; O	Equanil Miltown	400 mg
Carisoprodol	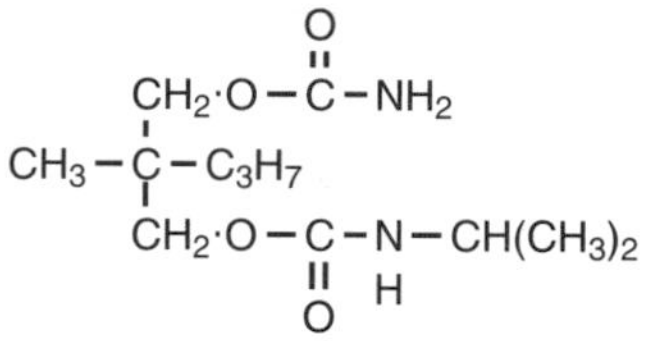 O; $CH_2 \cdot O-C-NH_2$; $CH_3-C-C_3H_7$; $CH_2 \cdot O-C-N-CH(CH_3)_2$; O; H	Rela Soma	250–350 mg
Metaxalone	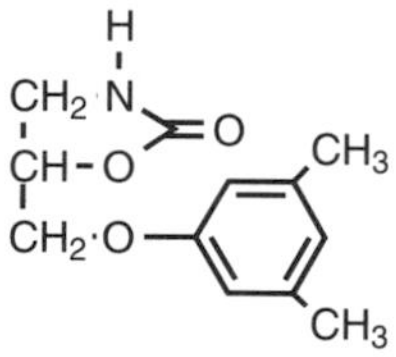 H; CH_2 N; =O; $CH-O$; $CH_2 \cdot O-$(ring); CH_3; CH_3	Skelaxin	800 mg
Benzazoles			
Chlorzoxazone	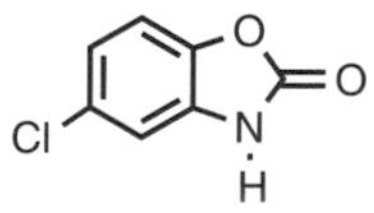 O; =O; Cl; N; H	Paraflex	250–750 mg
Miscellaneous			
Orphenadrine citrate	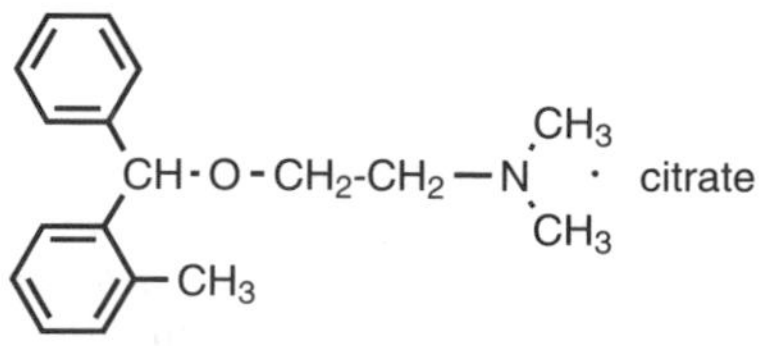 $CH-O-CH_2-CH_2-N$; CH_3; CH_3; · citrate; CH_3	Norflex	100 mg (oral) 60 mg IM or IV

accompanied by increased resistance to passive stretch. Spasticity may be caused by musculoskeletal or spinal cord trauma, brain lesions, or brain diseases. Regardless of the cause, the mechanisms essential for the expression of spasticity are contained in the spinal cord.

Clinical Evaluation of Spasticity

There is a serious lack of quantitative methodology to demonstrate neurophysiologic improvement in the drug treatment of spastic states in humans.[52] Simple tests of muscle tone or reflex latency have been impossible or useless. So far, global clinical assessments, such as the number of painful spasms per day, have been more useful. Even the combined subjective impressions of improvement by patient, family, and physician, however, cannot establish that the drug is working. A valuable strategy to establish whether a drug produces any benefit is gradual withdrawal of the drug; however, abrupt withdrawal should be avoided.[53] Because of the enormous diversity of neurologic disorders that culminate in spasticity and the subjectivity of many of the measurements, information is presently inadequate to establish the superiority of any one of the antispastic drugs. Furthermore, spasm frequently coexists with pain. Drug efficacy may be related to both skeletal muscle relaxation and analgesia.[54]

The second group of antispastic drugs to be developed includes the benzodiazepines, diazepam in particular. Diazepam exerts its skeletal muscle relaxant effect by binding to specific receptors that enhance the efficacy of GABA receptor binding and consequently synaptic inhibition by opening chloride channels in nerves. Diazepam probably has both supraspinal and spinal sites of antispastic action. Spinal actions are sufficient to relieve spasticity in patients with complete spinal cord transections. The ability of diazepam to enhance GABA-mediated inhibition in the spinal cord of the cat is well documented.[52,55] Sedation may accompany the antispastic effect of benzodiazepines. The molecular mechanism of action of benzodiazepines is discussed in Chapter 12. Similar to the interneuronal blockers, diazepam selectively depresses polysynaptic over monosynaptic spinal reflexes. Diazepam is effective in the treatment of spasticity as well as nonspastic disorders, such as tetanus and stiff man syndrome.

In an attempt to elevate GABA activity in the brain more directly, the GABA structural analog baclofen [(β-(4-chlorophenyl)GABA] was developed. In contrast to GABA, baclofen is lipophilic and penetrates the blood–brain barrier. Baclofen, however, fails to meet important criteria to be GABA-mimetic. Although it depresses neural activity, its effects cannot be blocked by the GABA receptor antagonist bicuculline. Furthermore, it has no effect on inhibitory postsynaptic potentials. The most important site of action of baclofen is the primary afferent fiber terminal, where baclofen blocks the release of excitatory neurotransmitter. In contrast to the interneuronal blockers, baclofen depresses both monosynaptic and polysynaptic activity in the spinal cord. Baclofen hyperpolarizes primary afferent terminals, in contrast to GABA, which depolarizes them.[56] Similar to diazepam, baclofen probably acts primarily in the spinal cord because it is an effective antispastic drug in complete spinal transections. Baclofen also shows analgesic activity in the whole animal, which may also enhance relief. Baclofen is effective in the treatment of spasticity in paraplegic or quadriplegic patients with multiple sclerosis or traumatic lesions to the cord.[52,55]

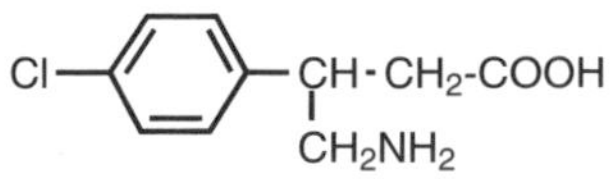

Baclofen [Lioresal®]

Since the development of dantrolene in 1974, no longer are all muscle relaxants centrally acting. Dantrolene, a peripherally acting agent, relaxes skeletal muscle by suppressing the release of calcium by the sarcoplasmic reticulum. With appropriate dose adjustment, dantrolene reduces spasticity by a graded reduction in the force of muscle contraction. Dantrolene has no other pharmacologically significant site of action in the stretch reflex arc. It does, however, have other central nervous system depressant effects, such as euphoria, lightheadedness, dizziness, and drowsiness early in treatment, but tolerance develops for these effects, and they subside with time. No tolerance develops to the antispastic action of dantrolene. Interestingly, dantrolene is also valuable in alleviating the signs of malignant hyperthermia. This is a rare genetically determined condition that can be precipitated by administration of neuromuscular blocking drugs and inhalation anesthetics. Muscular contraction and hyperthermia are apparently precipitated by release of excess calcium ions from the sarcoplasmic reticulum. Dantrolene is an effective antispastic drug, but because it weakens skeletal muscles, it is more likely to diminish voluntary motor activity than the centrally acting muscle relaxants.[52,55]

Dantrolene sodium (Dantrium), 1-(5-(p-nitrophenyl) furfurylideneamino)-hydantoin sodium salt, is slowly absorbed from the gastrointestinal tract. The mean half-life of the drug in adults is about 9 hours after a 100-mg dose. It is slowly metabolized by the liver to give the 5-hydroxy and acetamido metabolites as well as unchanged drug excreted in the urine.

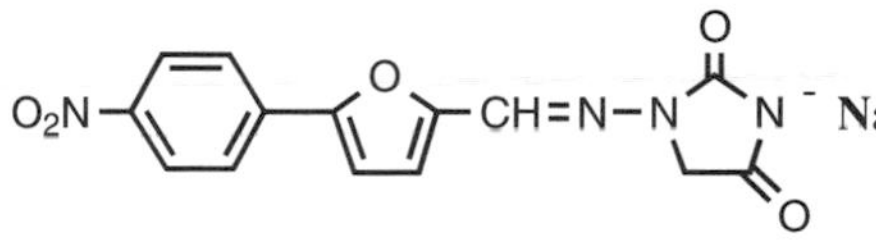

Dantrolene sodium [Dantrium®]

REFERENCES

1. H. Ehringer and O. Hornykiewicz, *Klin. Wochenschr.*, *38*, 1236(1960).
2. H. A. Robertson, *Trends Neurosci.*, *6*, 201(1992).
3. P. Riederer and S. Wuketich, *J. Neural Transm.*, *38*, 277(1976).

4. D. Calne and J. W. Langston, *Lancet, 2,* 1457(1983).
5. C. D. Ward, et al., *Neurology (Cleveland), 33,* 815(1983).
6. P. S. Spencer and H. H. Schaumburg, (Eds.), *Experimental and Clinical Neurotoxicology,* Baltimore, Williams & Wilkins, 1980, pp 618–619.
7. I. Kessler and E. L. Diamond, *Am. J. Epidemiol., 94,* 16(1971).
8. R. J. Bauman, et al., *Neurology (NY), 30,* 839(1980).
9. R. J. Marttila and U. K. Rinne, *Acta Neurol. Scand., 62,* (1980).
10. R. B. Godwin-Austen, et al., *J. Neurol. Neurosurg. Psychiatry, 45,* 577(1982).
11. A. H. Rajput, *Can. J. Neurol. Sci., 11,* 156(1984).
12. P. H. Yu and A. A. Boulton, *Life Sci., 41,* 675(1987).
13. L. A. Carr and P. P. Rowell, *Neuropharmacology, 29,* 311(1990).
14. G. C. Davis, et al., *Psychiatry Res., 11,* 156(1979).
15. J. W. Langston, et al., *Science, 219,* 979(1983).
16. J. A. Javitch, et al., *Proc. Natl. Acad. Sci. USA, 82,* 2173(1985).
17. K. Chiba, et al., *Biochem. Biophys. Res. Commun., 120,* 574(1984).
18. R. E. Heikkila, et al., *Nature, 311,* 467(1984).
19. S. Markey, et al., *Nature, 311,* 464(1984).
20. S. H. Snyder and R. J. D'Amato, *Neurology, 36,* 250(1986).
21. N. Castagnoli Jr., et al., *Life Sci., 36,* 225(1985).
22. W. J. Nicklas, et al., *Life Sci., 36,* 2503(1985).
23. R. R. Ramsay, et al., *Biochem. Biophys. Res. Commun., 146,* 53(1987).
24. R. R. Ramsay and T. P. Singer, *J. Biol. Chem., 261,* 7585(1986).
25. A. H. V. Schapira, et al., *Ann. Neurol., 32,* S116(1992).
26. M. D. Yahr, *Med. Clin. North Am., 57,* 1377(1972).
27. G. C. Cotzias, e al., *N. Engl. J. Med., 276,* 374(1967).
28. Toshiharu Nagatssu, in *Biochemistry of Catecholamines,* Baltimore, University Park Press, 1973, p. 289.
29. W. H. Vogel, et al., *Biochem. Pharmacol., 19,* 618(1970).
30. C. D. Marsden, et al., *Lancet, 2,* 1459(1973).
31. L. I. Goldberg, and T. L. Whitsett, *JAMA 218,* 1921(1971).
32. G. C. Cotzias, et al., *Proc. Natl. Acad. Sci. U.S.A., 71,* 2715(1974).
33. T. Eisler, et al., *Neurology, (NY), 31,* 19, New York (1981).
34. E. H. Heinonen and R. Lammintausta, *Acta Neurol. Scand., 84,* Suppl. 136, 44(1991).
35. A. E. Lang, *Can. J. Neurol. Sci., 11,* Suppl 1, 210(1984).
36. A. N. Lieberman and M. Goldstein, *Pharmacol. Rev., 37,* 217(1985).
37. G. C. Cotzias, et al., *N. Engl. J. Med., 294,* 567(1976).
38. L. L. Iversen, *Science, 188,* 1084(1975).
39. A. M. Ernst, *Psychopharmacologia, 7,* 391(1965).
40. J. G. Nutt, *Neurosciecne Facts, 3(22),* 1(1992).
41. M. K. Menon, et al., *Eur. J. Pharmacol., 52,* 1(1978).
42. J. Giesecke, *Acta Cryst., B29,* 785(1973); *B33,* 302(1977).
43. J. L. Neumeyer, et al., *J. Med. Chem., 16,* 1228(1973).
44. W. S. Saari, et al., *J. Med. Chem., 16,* 171(1973).
45. J. L. Neumeyer, et al., *J. Med. Chem., 16,* 1223(1973).
46. R. G. Booth, et al., *N.Y. Acad. Sci., 604,* 592(1990).
47. J. G. Cannon, in *Adv. Neurol.,* Vol. 9, D. Calne, et al., Eds. New York, Raven Press, 1975, pp. 177–183.
48. L. I. Goldberg, et al., *Annu. Rev. Pharmacol. Toxicol., 18,* 57(1978).
49. F. M. Berger and W. Bradley, *Br. J. Pharmacol. Chemother., 1,* 265(1946).
50. W. Haefely, et al., in *Handbook of Experimental Pharmacology, Vol. 55/II, Psychotropic Agents, Part II. Anxiolytics, Gerontopsychopharmacological Agents, and Psychomotor Stimulants,* E. Hoffmeister and G. Stille, Eds., Berlin, Springer, 1981, p. 263.
51. C. M. Smith, *The Nervous System—Part B: Central Nervous System Drugs,* Vol. 2, W. S. Root and F. G. Hofman, Eds., New York, Academic Press, Inc., 1965, p. 2–96.
52. R. A. Davidoff, *Neurology (Minneapolis), 28,* 46(1978).
53. R. R. Young and P. J. Delwaide, Drug therapy: spasticity (first of two parts), *N. Engl. J. Med., 304,* 28(1981).
54. J. K. Elenbaas, *Am. J. Hosp. Pharm., 37,* 1313(1980).
55. R. A. Young and P. J. Delwaide, *N. Engl. J. Med., 304,* 96(1981).
56. R. A. Davidoff, in *Spasticity: Disordered Motor Control,* R. G. Feldman et al., Eds., Chicago, Yearbook Medical Publishers, 1980.

SUGGESTED READINGS

W. Haefely, et al., in *Handbook of Experimental Pharmacology, Vol. 55/II, Psychotropic Agents, Part II. Anxiolytics, Gerontopychopharmacologic Agents, and Psychomotor Stimulants.* E. Hoffmeister and G. Stille, Eds., Berlin, Springer, 1981.

O. Hornykiewicz in *The Neurosciences: Paths of Discovery, II.* F. Samson and G. Adelman, Eds., Cambridge, Mass, Birkhauser, 1992.

Chapter 14

ANALGESICS

David S. Fries

Agents that decrease pain are referred to as analgesics or as analgetics. Although analgetic is grammatically correct, common use has made the term analgesic preferable to analgetic for the description of pain-killing drugs. Pain-relieving agents are also called antinociceptives. A number of classes of drugs are used to relieve pain. The nonsteroidal anti-inflammatory agents have primarily a peripheral site of action, are useful for mild-to-moderate pain, and often have an anti-inflammatory effect associated with their pain-killing action. Anesthetics (general or local) inhibit pain transmission by nonreceptor-mediated actions on nerve cell membranes. Dissociative anesthetics (ketamine) cause an analgesic effect by inhibiting N-methyl-D-aspartate (NMDA) receptors in the brain. All central nervous system (CNS) depressants (e.g., ethanol, barbiturates, and antipsychotics) cause some decrease in pain perception. Severe acute or chronic pain generally is treated most effectively with opioid agents.

Opioid analgesics historically have been called narcotic analgesics. Narcotic analgesic literally means that the agents cause sleep or loss of consciousness in conjunction with their analgesic effect. The term narcotic has become associated with the addictive properties of opioids and other CNS depressants. Because the great therapeutic value of the opioids is their ability to cause analgesia without causing sleep (narcosis), the term narcotic analgesic is not used further in this chapter.

The use of the juice (opium in Greek) or gum from the unripe seed pods of the poppy Papaver somniferum is among the oldest recorded medications. The writings of Theophrastus around 200 B.C. describe the use of opium in medicine; however, there is evidence that opium was used in the Sumerian culture as early as 3500 B.C. The initial use of opium was as a tonic, or it was smoked. The pharmacist Surtürner first isolated an alkaloid from opium in 1803. He named the alkaloid morphine, after the Greek god of dreams. Codeine, thebaine, and papaverine are other medically important alkaloids that were later isolated from opium gum.

Morphine was among the first compounds to undergo structure modification. Ethylmorphine (the 3-ethyl ether of morphine) was introduced as a medicine in 1898. Diacetylmorphine, which may be considered to be the first synthetic prodrug, was synthesized in 1874 and introduced as a nonaddicting analgesic, antidiarrheal, and antitussive agent in 1898.

The use of the terms opiate and opioid requires some clarification. The term opiate was used extensively until the 1980s to describe any natural or synthetic agent that was derived from morphine. One could say an opiate was any compound that was structurally related to morphine. The discovery, in the mid-1970s, of peptides in the brain that had pharmacologic actions similar to those of morphine prompted a change in nomenclature. The peptides were not easily related to morphine structurally; yet, their actions were similar to the actions of morphine. At this time, the term opioid, meaning opium-like or morphine-like in terms of pharmacologic action, was introduced. The broad group of opium alkaloids, synthetic derivatives related to the opium alkaloids, and the many naturally occurring and synthetic peptides with morphine-like pharmacologic effects is called opioids. In addition to having pharmacologic effects similar to those of morphine, a compound must be antagonized by an opioid antagonist such as naloxone to be classed as an opioid. The neuronally located proteins to which opioid agents bind to initiate a biologic response are called opioid receptors.

ENDOGENOUS OPIOID PEPTIDES AND THEIR PHYSIOLOGIC FUNCTIONS

Scientists had postulated for some time, based on structure-activity relationships, that opioids bind to specific receptor sites to cause their actions. It was also reasoned that morphine and the synthetic opioid derivatives were not the natural ligands for the opioid receptors and that some analgesic substance must exist in the brain. Techniques to prove these two points were not developed until the mid-1970s. Hughes and co-workers[1] used the electrically stimulated contractions of guinea pig ileum (GPI) and the mouse vas deferens (MVD), which are sensitive to inhibition by opioids, as bioassays to follow the purification of compounds with morphine-like activity from mammalian brain tissue. These researchers were able to isolate and identify the structures of two pentapeptides, Tyr-Gly-Gly-Phe-Met (met-enkephalin) and Tyr-Gly-Gly-Phe-Leu (leu-enkephalin), that caused the opioid activity (Figure 14–1). The compounds were named enkephalins after the Greek word kaphale that translates "from the head."

At about the same time as Hughes and co-workers were making their discoveries, three other laboratories, using a different assay technique, were able to identify endogenous opioids and opioid receptors in the brain.[2–4] These laboratories used radiolabeled opioid compounds (radioligands) with high specific activity to bind to opioid

PROOPIOMELANOCORTIN (POMC)

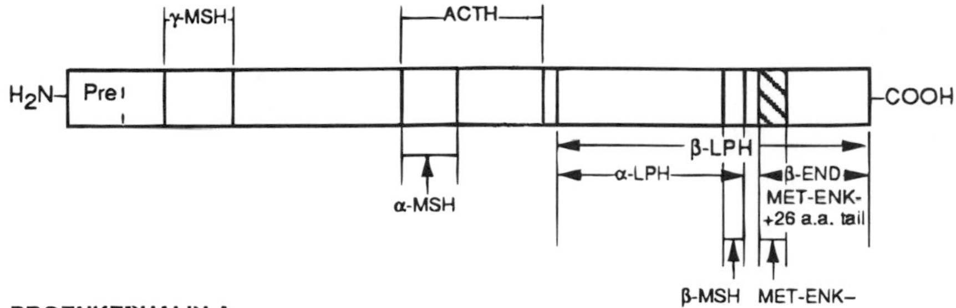

PROENKEPHALIN A

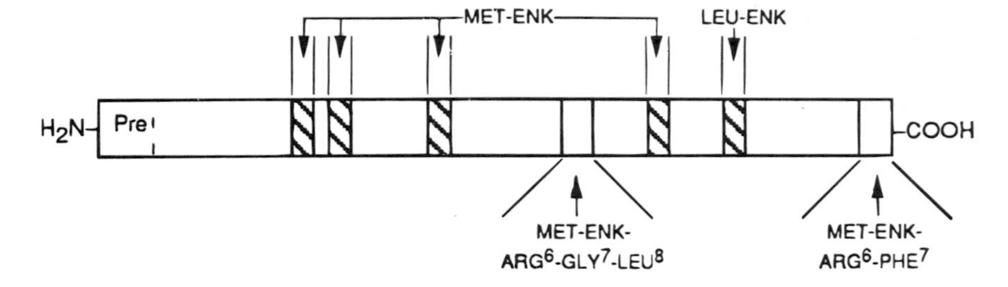

PROENKEPHALIN B (PRODYNORPHIN)

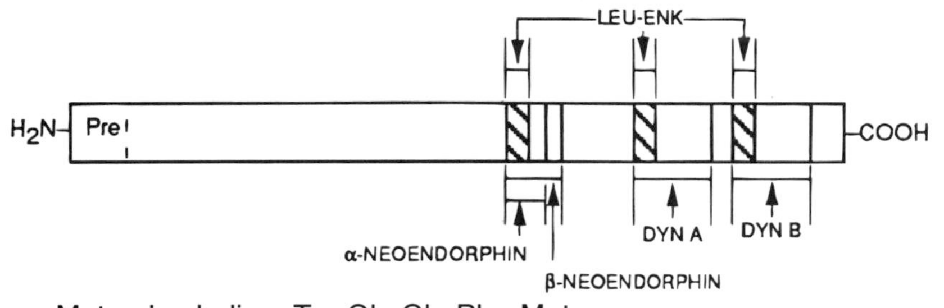

Met-enkephalin = Tyr-Gly-Gly-Phe-Met

Leu-Enkephalin = Tyr-Gly-Gly-Phe-Leu

β-Endorphin = Tyr-Gly-Gly-Phe-Met-Thr-Ser-Glu-Lys-Ser^{10}-Gln-
Thr-Pro-Val-Thr-Leu-Phe-Lys- Asn^{20}-Ala-Ile-Ile-Lys-
Asn-Ala-Tyr-Lys-Lys--Gly-Gly^{31}

Dynorphin A($dyn^{1\text{-}17}$) = Tyr-Gly-Gly-Phe-Leu-Arg-Arg-Ile-Arg-Pro-
Lys-Lys-Trp-Asp-Asn-Gln

Dynorphin B($dyn^{1\text{-}8}$) = Tyr-Gly-Gly-Phe-Leu-Arg-Arg-Ile

Dynorphin($dyn^{1\text{-}13}$) = Tyr-Gly-Gly-Phe-Leu-Arg-Arg-Ile-Arg-Pro-
Lys-Lys-Trp

α-Neoenodorphin = Tyr-Gly-Gly-Phe-Leu-Arg-Lys-Tyr-Pro-Lys

β-Neoendorphin = Tyr-Gly-Gly-Phe-Leu-Pro-Lys

Fig. 14–1. Precursor proteins to the opioid peptides.

receptors in brain homogenates.[5] One can demonstrate saturable binding (i.e., all binding sites are occupied) of the radioligands, and the bound radioligands can be displaced stereoselectively by nonradiolabeled opioids. Discovery of the enkephalins was soon followed by the identification of other endogenous opioid peptides, including β-endorphin[6] and dynorphin[7] (Fig. 14–1). Since its application to opioids, the radioligand-receptor binding technique has been used extensively for the identification of active compounds from endogenous and exogenous sources and for the characterization of neurotransmitter receptors.

The opioid peptides isolated from mammalian tissue are known collectively as endorphins, a word derived from a combination of *endo*genous and m*orphine.* The opioid alkaloids and all of the synthetic opioid derivatives are called exogenous opioids. Interestingly the isolation of morphine and codeine in small amounts has been reported from mammalian brain.[8] There is no evidence for a function for the endogenous morphine.

The endogenous opioid peptides are synthesized as part of the structures of large precursor proteins.[9] There is a different precursor protein for each of the three major types of opioid peptides (see Fig. 14–1). Proopiomelanocortin (POMC) is the precursor for β-endorphin, proenkephalin A is the precursor for met-enkephalin and leu-enkephalin, and proenkephalin B (prodynorphin) is the precursor for dynorphin and α-neoendorphin. All of the pro-opioid proteins are synthesized in the nucleus and transported to the terminals of the nerve cells from which they are released. The active peptides are hydrolyzed from the large proteins by processing proteases that recognize double basic amino acid sequences positioned just before and after the opioid peptide sequences.

Peptides with opioid activity have been isolated from sources other than mammalian brain. The hexapeptide β-casomorphin (Tyr-Pro-Phe-Pro-Gly-Pro-Ile), found in cow's milk, is a mu opioid agonist.[10] Dermorphin (Tyr-D-Ala-Phe-Gly-Tyr-Pro-Ser-NH_2), a mu selective peptide isolated from the skin of South American frogs, is about 100 times more potent than morphine in in vitro tests.[11]

The endogenous opioids exert their analgesic action at spinal and supraspinal sites (Fig. 14–2). The opioids exert an inhibitory action on afferent pain neurons in the dorsal horn of the spinal cord and on interconnecting neuronal pathways for pain signals within the brain. In the brain, the periaqueductal gray and the thalamic areas are especially rich in opioid receptors and are sites where opioids exert an analgesic action. All of the endogenous opioid peptides and the three major classes of opioid receptors appear to be at least partially involved in the modulation of pain. The actions of opioids at the synaptic level are shown in Fig. 14–3.

Analgesia that results from acupuncture or is self-induced by a placebo or biofeedback mechanism is caused by release of endogenous endorphins. Both of these types of analgesia can be prevented by the prior dosage of a patient with an opioid antagonist. Electrical stimulation from electrodes properly placed in the brain causes endorphin release and analgesia. This procedure has been used for the "self-stimulated" release of endorphins in a few chronic pain patients who did not respond to any

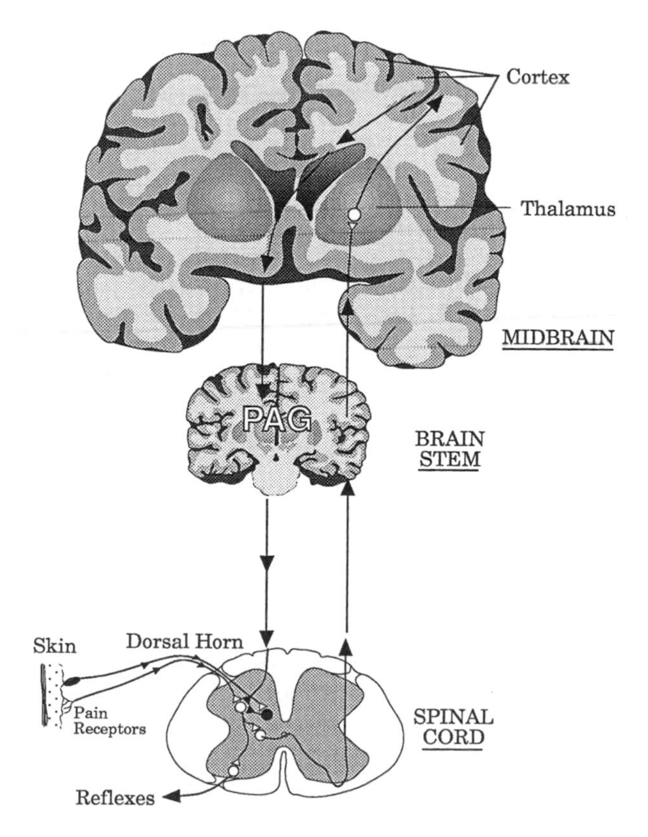

Fig. 14–2. Location of endogenous opioid nerve tracts in the central nervous system. Endorphins and opioid receptors in the dorsal horn of the spinal cord, thalamus, and periaqueductal gray (PAG) areas are associated with the transmission of pain signals.

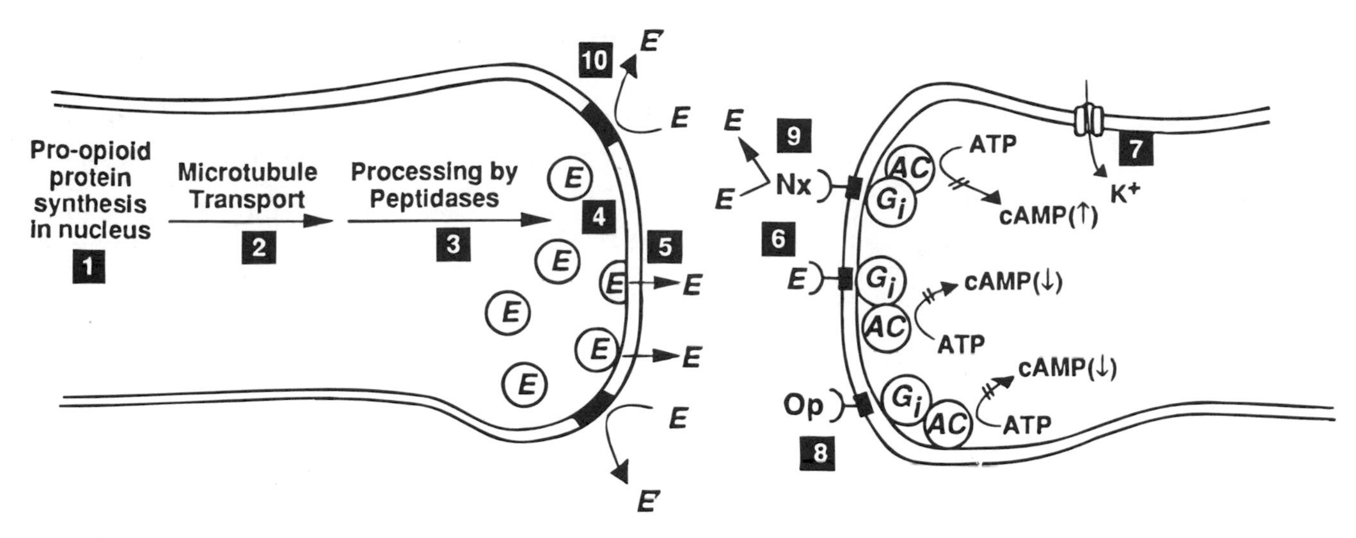

Fig. 14–3. Schematic representation of a mu (μ) enkephalinergic nerve terminal. (1) Pro-opioid proteins are synthesized in the cell nucleus. (2) Pro-opioid proteins undergo microtubular transport to the nerve terminal. (3) Active endogenous opioids (E) are cleaved from the pro-opioid proteins by the action of "processing" proteases. (4) The active peptides (E) are taken up and stored in presynaptic vesicles. (5) The peptides are released when the presynaptic neuron fires. (6) The endogenous opioid peptides bind to postsynaptic receptors and activate second message systems. (7) For mu opioid receptors, the second messenger effect is mediated by a G-inhibitory (G_i) protein complex, which promotes the inactivation of adenyl cyclase, a decrease in intracellular cyclic-adenosine-3′,5′-monophosphate (cAMP), and finally an influx of potassium ions (K^+) into the cell. The net effect is the hyperpolarization of the postsynaptic neuron and inhibition of cell firing. (8) Exogenous opioids (Op) such as morphine combine with opioid receptors and mimic the actions of E. (9) Opioid antagonists such as naloxone (Nx) combine with opioid receptors and competitively inhibit the actions of E or Op. (10) The action of E is terminated by a membrane-bound protease, which hydrolyses the Gly^3-Tyr^4 peptide bond of enkephalin.

other medical treatment. As with exogenously administered opioid drugs, tolerance develops to these procedures that release endogenous opioids.

OPIOID RECEPTORS

Opioid receptors have been divided into the three major types: mu (μ), kappa (κ)[12] and delta (δ).[13] There is good evidence for subtypes of each of these receptors. Table 14–1 lists the opioid receptor types and subtypes, their known physiologic functions, and selective agonists and antagonists for each of the receptors. All three of the opioid receptor types are located in human brain or spinal cord tissues, and each has a role in the mediation of pain. At this time, only mu and kappa agonists are available for clinical use. Delta receptor selective compounds have until recently all been peptides that do not distribute well to the CNS. A clinical use for delta agonists is yet to be discovered.

Identification of multiple opioid receptors has depended on the discovery of selective agonists and antagonists and the use of sensitive assay techniques.[14] The techniques that have been especially useful are the radioligand binding assays on brain tissues and the electrically stimulated peripheral muscle preparations. Rodent brain tissue contains all three opioid receptor types, and special evaluation procedures (computer-assisted line fitting) or selective blocking (with reversible or irreversible binding agents) of some of the receptor types must be used to determine the receptor selectivity of test compounds. The myenteric plexus containing longitudinal strips of GPI contains mu and kappa opioid receptors. The contraction of these muscle strips is initiated by electrical stimulation and is inhibited by opioids. The vas deferens from mouse contains mu, delta, and kappa receptors and reacts similarly to the GPI to electrical stimulation and to opioids. Homogeneous populations of opioid receptors are found in rat (μ), hamster (δ), and rabbit (κ) vas deferentia.

The signal transduction mechanism for mu and delta receptors is through G_i proteins and is linked to an inhibition of adenylyl cyclase activity. The resultant decrease in cyclic adenosine monophosphate (cAMP) production causes an increase in intracellular potassium ions and hyperpolarization of the cells.[15,16] Kappa receptors are also coupled to G_i proteins, but the cellular effect may be linked to Ca^{++} channels.[17]

A number of compounds have been found that are selective for mu opioid receptors (Fig. 14–4). All of the opioid alkaloids and most of their synthetic derivatives are mu selective agonists. Morphine, normorphine, and dihydromorphone have 10-fold to 20-fold mu receptor selectivity and were particularly important in early studies in differentiating the opioid receptors. Sufentanil and the peptides DAMGO and dermorphin, all with greater than 100-fold selectivity for mu over other opioid receptors, are now commonly used in the laboratory for the mu selective induction of tolerance, receptor binding assays, and activation of isolated muscle preparations.

Naloxone and naltrexone are antagonists that have marginal (5-fold to 10-fold) selectivity for mu receptors. Cyprodime is the most selective mu antagonist (about 30-

Table 14–1. Opioid Receptor Types and Subtypes

Receptor Type (Natural Ligand)	Selective Agonists	Agonist Properties	Selective Antagonists
μ (enkephalins) (β-endorphin)*	Morphine Sufentanil DAGO(Tyr-d-Ala-Gly-MePhe-NH-$(CH_2)_2$-OH)(also DAMGO) PLO17 (Tyr-Pro-MePhe-d-Pro-NH_2) DALDA (Tyr-d-Arg-Phe-Lys-NH_2) BIT (affinity label)	Analgesia (morphine-like) Euphoria Increased gastrointestinal transit time Tolerance and physical dependence Immune suppression Respiratory depression (volume) Emetic effects	Naloxone Naltrexone CPT (d-Phe-Cys-Trp-Lys-Thr-NH_2) Cyprodime β-FNA (affinity label)
μ_1 (high affinity)	N-(2-pyrazinyl)-N-(1-phenethyl-4-piperidinyl)-2-furamide		Naloxonazine
μ_2 (low affinity)	?		N-(2-pyrazinyl)-N-(1-phenethyl-4-piperidinyl)-2-furamide
κ (dynorphins) (β-endorphin)*	EKC Bremazocine Mr 2034 dyn (1-17) Trifluadom	Analgesia Sedation Miosis Diuresis Dysphoria	TENA nor-BNI
κ_1 (high affinity)	U-50,488 Spiradoline (U-62,066) U-69,593 PD 117302 UPHIT (affinity label)		
κ_2	dyn (1-17)		
κ_3	?		
δ (enkephalins) (β-endorphin)*	DADLE (d-Ala^2-d-Leu^5-enkephalin) DSLET (Tyr-d-Ser-Gly-Phe-Leu-Thr) DPDPE (d-Pen^2-d-Pen^5-enkephalin) FIT (affinity label) SUPERFIT (affinity label)	Analgesia Immune stimulation Respiratory depression (rate)	ICI 174864 Naltrindole

* Binds nonselectively to the three major opioid types.

fold selective for mu over kappa and 100-fold selective for mu over delta) available for laboratory use.[18] There is evidence that mu$_1$ receptors are high-affinity binding sites that modulate pain neurotransmission, whereas mu$_2$ receptors control respiratory depression. Naloxonazine is a selective inhibitor of mu$_1$ opioid receptors.[19]

Ethylketazocine was the primary kappa selective agonist (Fig. 14–5) used in early studies. Bremazocine is another 6,7-benzomorphan derivative with high kappa receptor selectivity. More recently, a number of arylacetamide derivatives, with high kappa activity but devoid of mu or delta activity, have been discovered. The first of these compounds (±) U50488 (κ:μ selectivity ≃ 50) has been extremely important in the characterization of kappa opioid activity.[20] Other important agents in this class are (±) PD117302[21] and (−)C1977.[22] Each of these agents has greater than 1000-fold selectivity for kappa over mu or delta receptors. There is evidence that the arylacetamides bind to a subtype of kappa receptors. Kappa agonists, in general, produce analgesia in animals including humans. Other prominent effects are diuresis, sedation, and dysphoria. Compared with mu agonists, kappa agonists lack respiratory depressant, constipating, and strong addictive (euphoria and physical dependence) properties. It was hoped that kappa agonists would become useful strong analgesics that lacked addictive properties, but it now appears that the sedative and dysphoric side effects are too strong to allow successful commercialization of these agents. The scientific evidence suggesting κ_1, κ_2, and κ_3 subtypes of kappa receptors leaves some researchers still hopeful of finding nonaddicting analgesics related to this class of opioid agents.[23] The different physiologic effects of the three kappa receptor subtypes are not currently known.

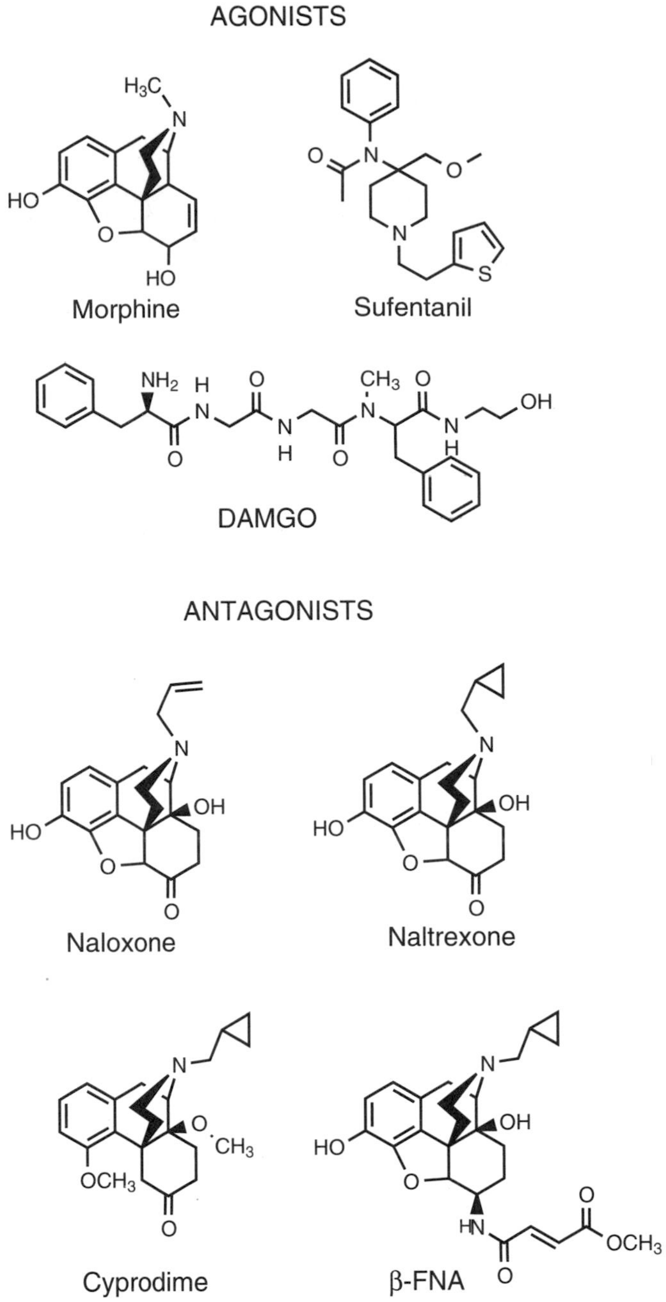

AGONISTS

Morphine Sufentanil

DAMGO

ANTAGONISTS

Naloxone Naltrexone

Cyprodime β-FNA

Fig. 14–4. Structures of opioid (*μ*) receptor selective compounds.

The peptides related to dynorphin are the natural agonists for kappa receptors. Their selectivity for kappa over mu receptors is not high. Surprisingly little work has been done on synthetic analogs of dynorphin.

The only antagonist with good selectivity for kappa receptors is nor-binaltorphimine.[24] This compound has about a 30-fold selectivity for blocking kappa over mu receptors and an even greater selectivity for kappa versus delta receptors. No medical uses for a kappa antagonist have been found; however, there is some evidence that these agents may be useful for the treatment of stroke victims and for the prevention of some forms of drug abuse.

Enkephalins, the natural ligands at delta receptors, are only slightly selective for delta over mu receptors. Changes in the amino acid composition of the enkephalins can give compounds with high potency and selectivity for delta receptors. The peptides most often used as selective delta receptor ligands (Fig. 14–6) are [d-Ala^2, d-Leu^5]enkephalin (DADLE),[25] [d-Ser^2, Leu^5]enkephalin-Thr (DSLET),[26] and the cyclic peptide [d-Pen^2, d-Pen^5]-enkephalin (DPDPE).[27] These and other delta receptor selective peptides have been useful for in vitro studies, but their metabolic instability and poor distribution properties (i.e., penetration of the blood–brain barrier is limited by their hydrophilicity) have limited their usefulness

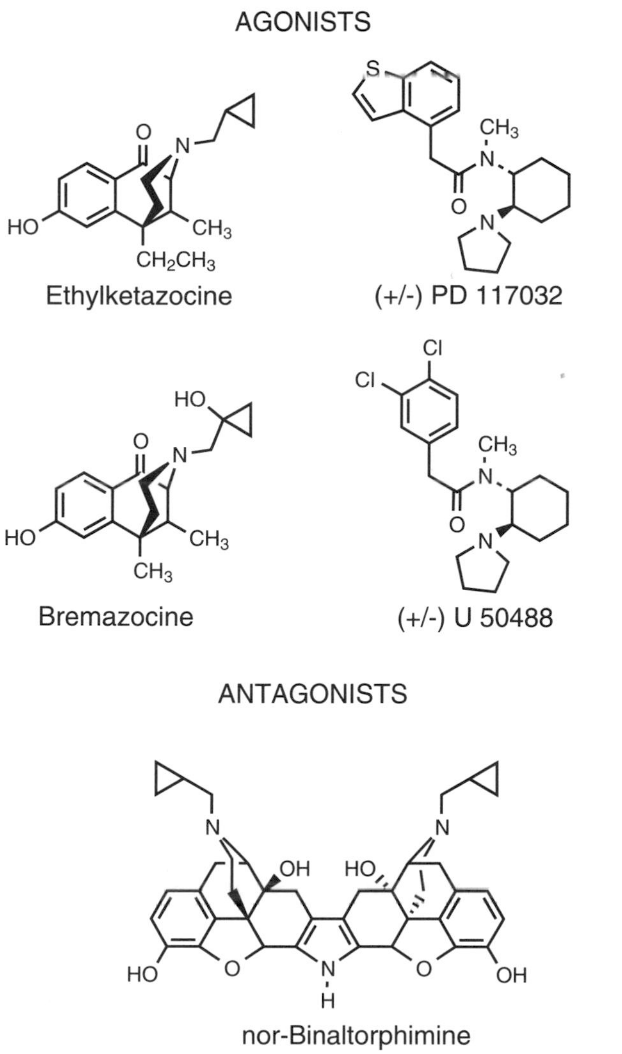

AGONISTS

Ethylketazocine

(+/-) PD 117032

Bremazocine

(+/-) U 50488

ANTAGONISTS

nor-Binaltorphimine

Fig. 14–5. Kappa (κ) receptor selective agents.

for in vivo studies. Nonpeptide agonists that are selective for delta receptors have been difficult to find. Derivatives of morphindoles (nor-OMI) show delta selectivity in in vitro assays, but in vivo the degree of receptor selectivity is not clear.[28] Radioligand binding studies in rodent brain tissue and in electrically stimulated vas deferentia have provided evidence of δ_1 and δ_2 receptors.[29] The functional significance of this differentiation has not been determined.

Naltrindole is the most selective antagonist for delta receptors.[30] Naltrindole penetrates the CNS and displays delta receptor antagonist selectivity in vitro and in vivo. Peptidyl antagonists (e.g., ICI154129) selective for delta receptors are known, but their usefulness for in vivo studies is limited by their instability and pharmacodynamic properties.

A number of opioid receptor selective affinity labeling agents (i.e., compounds that form an irreversible covalent bond with the receptor protein) have been developed (Fig. 14–7). These compounds have been important in the characterization and isolation of the opioid receptor types. Each of the affinity labeling agents contains a pharmacophore that allows initial reversible binding to the receptor. Once reversibly bound to the receptor, an affinity labeling agent must have an electrophilic group positioned so that it can react with a nucleophilic group on the receptor protein. The receptor selectivity of these agents depends on (1) the receptor type selectivity of the pharmacophore, (2) the proper location of the electrophile within the pharmacophore structure that positions it near a nucleophile on the receptor binding site and (3) the relative reactivities of the electrophilic and nucleophilic groups.

Examples of important affinity labeling agents are β-CNA which, owing to its highly reactive 2-chloroethylamine electrophilic group, irreversibly binds to all three opioid receptor types.[31] The structurally related compound β-FNA has a less reactive furamamide electrophilic group and reacts irreversibly with only mu receptors.[32] Derivatives of the fentanyl series, FIT and SUPERFIT, bind mu and delta receptors, but only the delta receptor is bound irreversibly.[33,34] Apparently, when these agents are bound to mu receptors, the electrophilic isothiocyanate group is not oriented in proper juxtaposition to a receptor nucleophile for covalent bond formation to occur. Incorporation of the electrophilic isothiocyanate into the structure of the highly kappa receptor selective benzacetamides has provided affinity labeling agents (UPHIT) for this receptor type.[35]

STRUCTURE-ACTIVITY RELATIONSHIPS OF OPIOID PEPTIDES

Structure-Activity Relationships of Mu Receptor Agonists

Morphine is the prototype opioid selective for the mu opioid receptors. The structure of morphine is composed of five fused rings, and the molecule has five chiral centers with absolute stereochemistry (5R, 6S, 9R, 13S, and 14R). The naturally occurring isomer of morphine is levo-[(ℓ) or (−)] rotatory. (+)-Morphine has been synthesized, and it is devoid of analgesic and other opioid activities.[36]

Table 14–2 shows the structure, numbering, and selected structure-activity relationships (SAR) for (−)-morphine.

It is important to recognize that changes in the structure of morphine (or any other opioid) are likely to cause a change in the affinity and intrinsic activity of the new compound for each of the opioid receptor types. The opioid receptor selectivity profile of the new compound may be different from the type of structure from which it was made or modeled (i.e., a selective mu agonist may shift to become a selective kappa agonist). In addition, the new compound has different physical properties than its parent. The different physical properties (e.g., solubility, partition coefficient, pK_a) result in different pharmacokinetic characteristics for the new drug and can affect its in vivo activity profile. For example, a new drug that is more lipophilic than its parent may distribute better to the brain and thus appear to be more active, whereas it may have lower affinity or intrinsic activity for the recep-

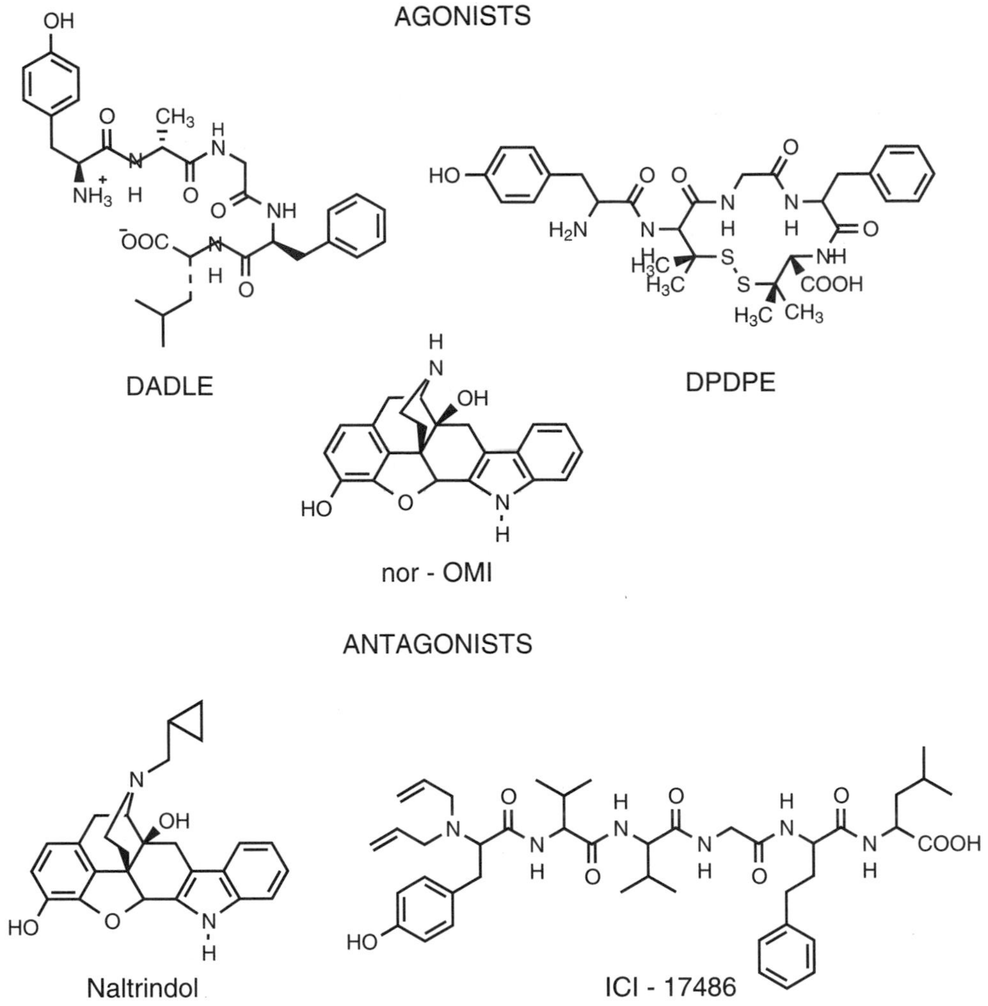

AGONISTS

DADLE

DPDPE

nor - OMI

ANTAGONISTS

Naltrindol

ICI - 17486

Fig. 14–6. Delta *(δ)* selective agents.

tor. The SARs discussed in the following paragraphs describe the relative therapeutic potencies of the compounds and are a combination of pharmacokinetic and receptor activation properties of the drugs.

The A-ring and the basic nitrogen (mostly in the protonated, ionized form at physiologic pH) are the two most common structural features found in compounds displaying opioid analgesic activity. The aromatic A ring and the nitrogen may be connected either by an ethyl linkage (9,10-positions of the B ring) or a propyl linkage (either edge of the piperidine ring that forms the D ring). The A ring and the basic nitrogen are necessary components in every potent mu agonist known; however, these two structural features alone are not sufficient for mu opioid activity. In compounds having rigid structures (i.e., fused A, B, and D rings), the 3-hydroxy group and a tertiary nitrogen either greatly enhance or are essential for activity. A summary of other important SAR features for morphine are given in Figure 14–8.

The substituent on the nitrogen of morphine and morphine-like structures is critical to the degree and type of activity displayed by an agent. A tertiary amine is usually necessary for good opioid activity. The size of the N-substituent can dictate the compound's potency and its agonist versus antagonist properties. N-Methyl substitution generally results in a compound with good agonist properties. Increasing the size of the N-substituent to three to five carbons (especially where unsaturation or small carbocyclic rings are included) results in compounds that are antagonists at some or all opioid receptor types. Larger substituents on nitrogen returns agonist properties to the opioid. An N-phenylethyl substituted opioid is usually on the order of 10 times more potent as an agonist than the corresponding N-methyl analog.

Thousands of derivatives of morphine and other mu agonists have been prepared and tested.[37,38] The objective of most of the synthetic efforts has been to find an analgesic with improved pharmacologic properties over known mu agonists. Specifically, one would like to have an orally active drug that retains the strong analgesic properties of morphine yet lacks its ability to cause tolerance, physical dependence, and respiratory depression. The success of this search has been limited. Many compounds that are more potent (on a molar comparison

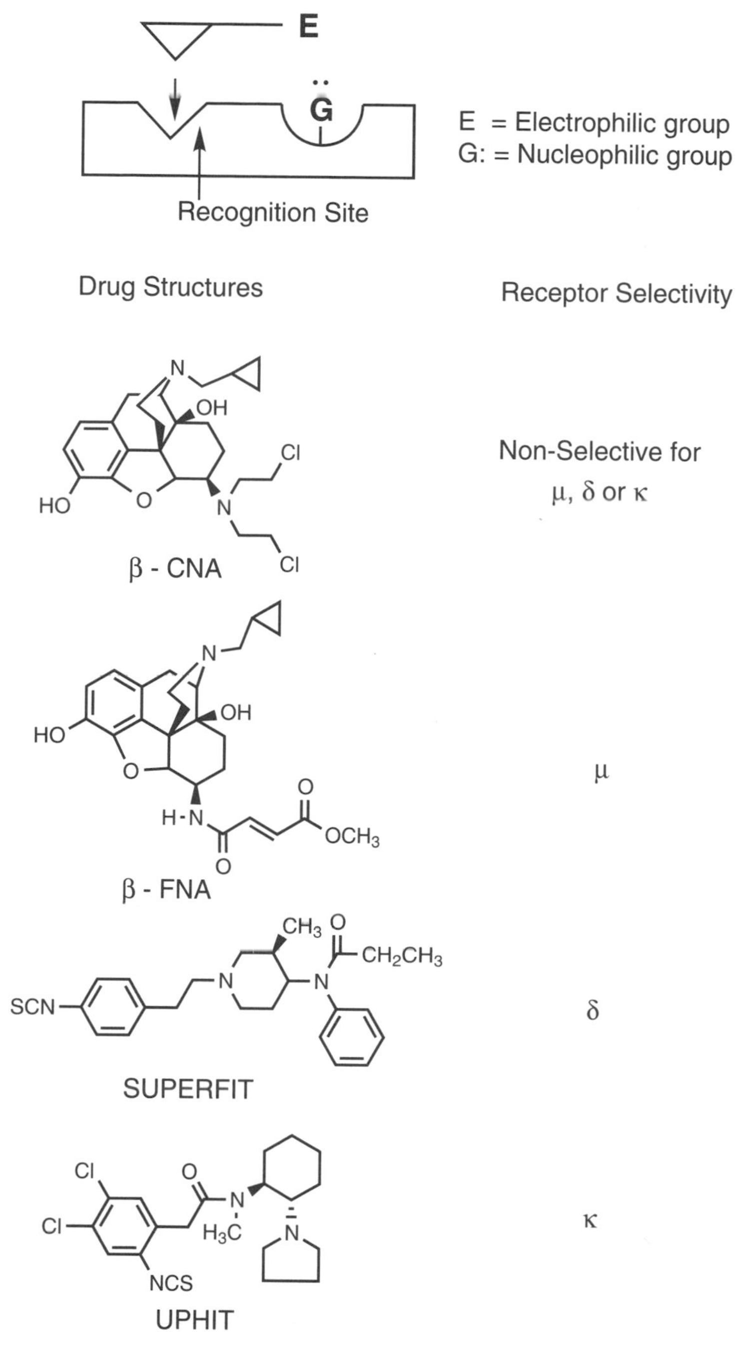

E

G

E = Electrophilic group
G: = Nucleophilic group

Recognition Site

Drug Structures — Receptor Selectivity

β - CNA — Non-Selective for μ, δ or κ

β - FNA — μ

SUPERFIT — δ

UPHIT — κ

Fig. 14–7. A representation of the concept of affinity labeling of receptors and affinity labeling agents for opioid receptors.

basis) than morphine have been discovered. Also, compounds with pharmacodynamic properties different from morphine have been discovered, and some of these compounds are preferred to morphine for selected medical uses. The ideal analgesic drug, however, is yet to be discovered. Research to find new centrally acting analgesics has turned away from classic mu agonists and now is focused on agents that act through other types or subtypes of opioid receptors or through nonopioid neurotransmitter systems.

The SARs of compounds structurally related to morphine are outlined in Table 14–2. A number of the structural variations on morphine gives compounds that are available as drugs in the United States. The most important of these agents, in terms of prescription volume, is the alkaloid codeine. Codeine, the 3-methoxy derivative of morphine, is a relatively weak mu agonist, but it undergoes slow metabolic O-demethylation to morphine, which accounts for much of its action. Codeine is also a potent antitussive agent, and it is used extensively for this purpose.

The 3,6-diacetyl derivative of morphine is known commonly as heroin. Heroin was synthesized from morphine in 1874 and was introduced to the market in 1898 by the Friedrich Bayer Co. in Germany. The 1906 Squibb's Materia Medica lists 10-mg tablets of heroin at $1.20/1000.

Table 14–2. Structure, Numbering, and Selected SAR for (−)-Morphine

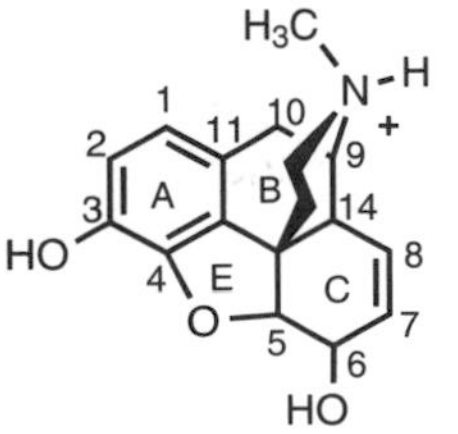

	Substituent Change	Analgesic Activity
(-) - Morphine	3-H for -OH	10× Decrease
	6-OH to 6-keto	Decrease activity or increase w/6,7-dihydro
	6-OH to 6-H	Increase activity
	7,8-dihydro	Increase activity
	14 β-OH	Increase activity
	3-OCH_3 for OH	Decrease activity
	CH_3CO- ester at 3	Decrease activity
	CH_3CO- ester at 6	Increase activity
	NCH_2CH_2Ph for NCH_3	Increase activity (10×)
	$NCH_2CH{=}CH_2$ for NCH_3	Becomes a μ antagonist

Heroin is described as "preferable to morphine because it does not disturb digestion or produce habit readily." Heroin itself has relatively low affinity for mu opioid receptors; however, its high lipophilicity compared with morphine results in enhanced penetration of the blood–brain barrier. Once in the body (including the brain), serum and tissue esterases hydrolyze the 3-acetyl group to produce 6-acetylmorphine. This latter compound has mu agonist activity in excess of morphine. The combination of rapid penetration by heroin into the brain after intravenous dose and rapid conversion to a potent mu agonist provide an 'euphoric rush" that makes this compound a popular drug of abuse. Repeated use of heroin results in the development of tolerance, physical dependence, and the acquisition of a drug habit that is often destructive to the user and to society. In addition, the use of unclean or shared hypodermic needles for self-administering heroin often results in the transmission of the acquired immunodeficiency syndrome (AIDS), hepatitis, and other infections.

Changes in the C-ring chemistry of morphine or codeine can lead to compounds with increased activity. Hydromorphone is the 7,8-dihydro-6-keto derivative of morphine, and it is 8 to 10 times more potent than morphine on a weight (mg/mg) basis. Hydrocodone, the 3-methoxy derivative of hydromorphone, is considerably more active than codeine.

The opium alkaloid thebaine can be synthetically converted to 14β-hydroxy-6-keto derivatives of morphine. The 14β-hydroxy group generally enhances mu agonist properties and decreases antitussive activity, but activity varies with the overall substitution on the compound. Oxycodone, the 3-methoxy-N-methyl derivative, is about as potent as morphine when given parenterally, but its oral-to-parenteral dose ratio is better than for morphine. Oxymorphone is the 3-hydroxy-N-methyl derivative, and it is 10 times as potent as morphine on a weight (mg/mg) basis. Substitution of a N-cyclobutylmethyl for N-methyl and reduction of the 6-keto group to 6α-OH of oxymorphone gives nalbuphine. Nalbuphine acts through kappa receptors to produce about one-half the analgesic potency as morphine. Nalbuphine is an antagonist at mu receptors. Interestingly, N-allyl-(naloxone) and N-cyclopropylmethyl-(naltrexone) noroxymorphone are "pure" opioid antagonists. Naloxone and naltrexone are slightly mu receptor selective but act as antagonists at all opioid receptor types.

Figure 14–8 contains some of the diverse chemical structures that produce mu agonist activity. The structures shown in the figure illustrate that the morphine structure may be built up or broken down to yield compounds that produce potent agonist activity. Reaction of thebaine with dienophiles (i.e., Diels-Alder reactions) results in 6,14-endo-ethenotetrahydrothebaine or oripavine derivatives.[39] Some of the oripavine derivatives are extremely potent mu agonists. The best-known of these derivatives are etorphine and buprenorphine. Etorphine is about 1000 times more potent than morphine as a mu agonist. Etorphine has a low therapeutic index in humans, and its respiratory depressant action is difficult to reverse with an opioid antagonist; thus, the compound is not useful in medical practice. Etorphine is available for use in veterinary medicine for the immobilization of large animals. Buprenorphine is a partial agonist at mu receptors with a potency of 20 to 30 times that of morphine. The compound's uses and properties are described in the section on clinically available agents.

Removal of 3,4-epoxide bridge in the morphine structure results in compounds that are referred to as morphinans. Synthetically, one cannot just remove the epoxide ring from the morphine structure; rather the morphinans are prepared synthetically from a procedure described by Grewe and Mandon.[40] The synthetic procedure yields compounds as racemic mixtures, but only the levo (−) isomers possess opioid activity. The dextro isomers have useful antitussive activity. The two morphinan derivatives that are marketed in the United States are levorphanol and butorphanol. Levorphanol is about eight times more potent as an analgesic in humans than morphine. Levorphanol's increased activity is due to an increase in affinity for mu opioid receptors and its greater lipophilicity, which allows higher peak concentrations to reach the brain. Butorphanol is a mu antagonist and a kappa agonist. These mixed agonist/antagonists are described in more detail later.

Synthetic compounds that lack both the epoxide ring and the C ring of morphine retain opioid activity. Compounds having only the A, B, and D rings are named chemically as derivatives of 6,7-benzomorphan or as 2,6-methano-3-benzazocine. In common use, they are simply

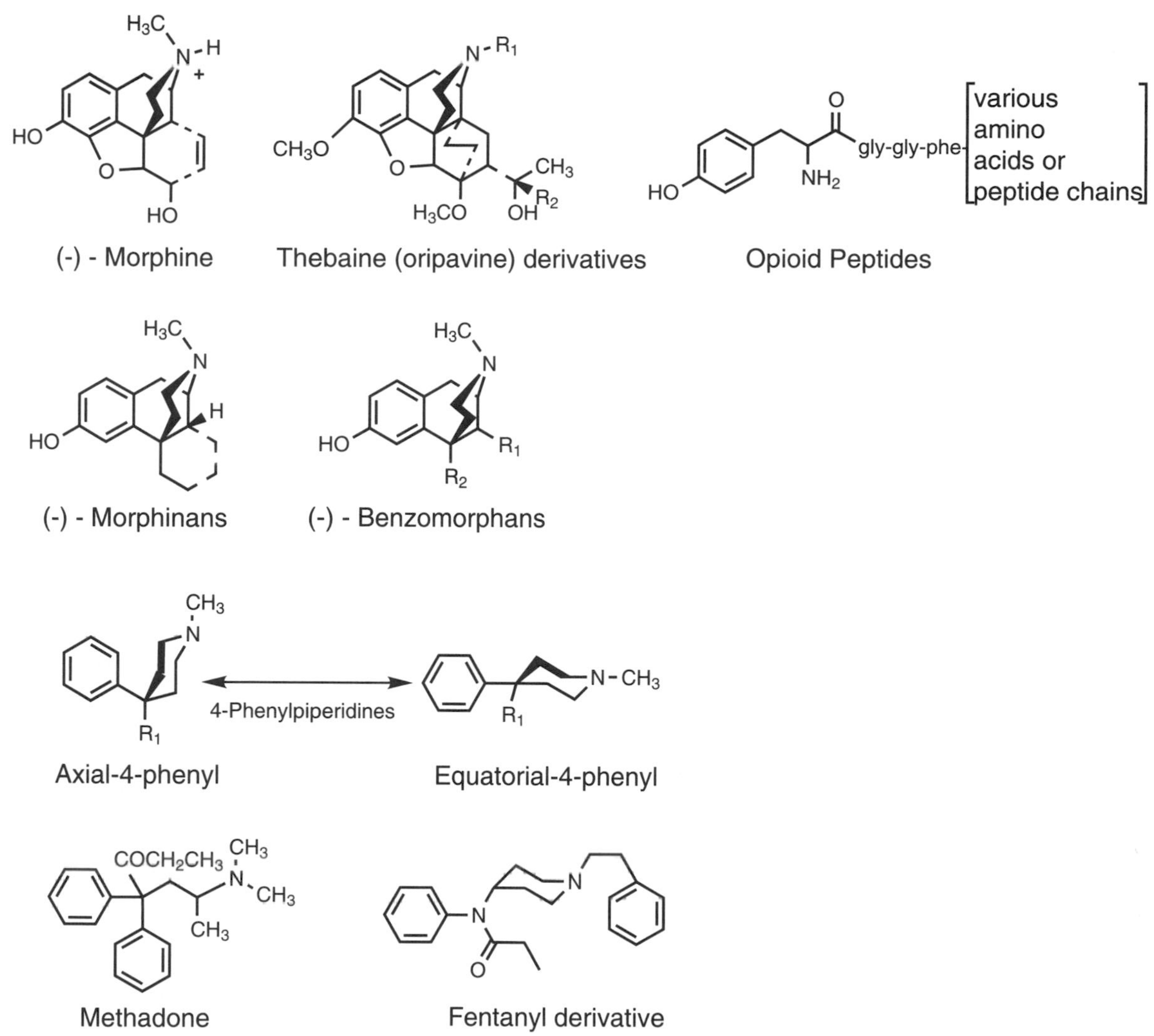

(-) - Morphine

Thebaine (oripavine) derivatives

Opioid Peptides

(-) - Morphinans

(-) - Benzomorphans

Axial-4-phenyl

4-Phenylpiperidines

Equatorial-4-phenyl

Methadone

Fentanyl derivative

Fig. 14–8. Diverse structural families that yield potent opioid agonists.

called benzomorphans. The only agent from this structural class that is marketed in the United States is pentazocine. Pentazocine has an agonist action on kappa opioid receptors—an effect that produces analgesia. Pentazocine is a weak antagonist at mu receptors. The dysphoric side effects produced by higher doses of pentazocine are due at least in part to actions at phencyclidine sites on the NMDA receptors. The benzomorphan derivative phenazocine (N-phenylethyl) is about 10 times as potent as morphine as a mu agonist and is marketed in Europe.

Aminotetralins represent A, B ring analogs of morphine. A number of active compounds in this class have been described, but only dezocine, a mixed agonist/antagonist, has been marketed.

Analgesic compounds in the 4-phenylpiperidine class may be viewed as A, D ring analogs of morphine. The opioid activity of these agents was discovered serendipitously. The first of these agents, meperidine, was synthesized in 1937 by Eisleb and Schaumann,[40] who were attempting to prepare antispasmodic agents. The compound produced an S-shaped tail (Straub tail) in cats, an effect that was known to be typical of morphine and its derivatives. Meperidine proved to be a typical mu agonist with about one-fourth the potency of morphine on a weight (mg/mg) basis. It is particularly useful in certain medical procedures because of its short duration of action. Reversed esters of meperidine have greater potency, and several of these derivatives have been marketed. The 3-methyl reversed ester derivatives of meperidine, α- and β-prodine, were available in the United States but have been removed from the market because of their low prescription volume and their potential to undergo elimination reactions to compounds that resemble the neurotoxic agent MPTP (see Chapter #13). Trimeperidine or γ-prodinol, the 1,2,5-trimethyl reserved ester of meperidine, is still heavily used in Russia as an analgesic.

Structural modification of the 4-phenylpiperidines has led to discovery of the 4-anilidopiperidine or the fentanyl group of analgesics. Fentanyl and its derivatives are mu agonists, and they produce typical morphine-like analgesia and side effects. Structural variations of fentanyl that have yielded active compounds are substitution of an isosteric ring of the phenyl group, addition of a small oxygen-containing group at the 4-position of the piperidine ring, and introduction of a methyl group onto the 3-position of the piperidine ring. Newer drugs that illustrate some of these structural changes are alfentanil and sufentanil. Both of these drugs have higher safety margins than other mu agonists. For unknown reasons, the compounds produce analgesia at much lower doses than is necessary to cause respiratory depression.

In the period just before or during World War II, German scientists synthesized another series of open chain compounds as potential antispasmodics. In an analogous manner to meperidine, animal testing showed some of the compounds to possess analgesic activity. Methadone was the major drug to come from this series of compounds. Methadone is especially useful for its oral activity and its long duration of action. These properties make methadone useful for maintenance therapy of opioid addicts and for pain suppression in the terminally ill (i.e., hospice programs). Methadone is marketed as a racemic mixture but the (−)-isomer possesses almost all of the analgesic activity. Many structural variations on the methadone structure have been made, but little success in finding more useful drugs in this class has been achieved. Reduction of the keto and acetylation of the resulting hydroxyl group gives the acetylmethadols (see later). Variations of the methadone structure have led to the discovery of the useful antidiarrheal opioids diphenoxylate and loperamide.

Propoxyphene is an open chain compound that was discovered by structural variation of methadone. Propoxyphene is a weak mu opioid agonist, having only one-fifteenth the activity of morphine. The (+)-isomer is the active enantiomer.

The SAR for mu antagonists is relatively simple if one focuses just on marketed compounds. All of the marketed rigid-structured opioid analogs that have the 3-phenolic group and a N-allyl, N-cyclopropylmethyl (CPM), or N-cyclobutylmethyl (CBM) substituent replacing the N-methyl are mu antagonists. Compounds behaving as mu antagonists may retain agonist activity at other opioid receptor types. The only exception to this rule is buprenorphine (N-CPM), which is a potent partial agonist (or partial antagonist) at mu receptors. Only two compounds are pure antagonists (i.e., act as antagonists at all opioid receptors). These compounds are the N-allyl (naloxone) and N-CPM (naltrexone) derivatives of noroxymorphone. The 14-β-hydroxyl group is believed to be important for the pure antagonistic properties of these compounds. It is not understood how the simple change of an N-methyl to an N-allyl group can change an opioid from a potent agonist into a potent antagonist. The answer may lie in the ability of opioid receptor protein to couple effectively with signal transduction proteins (G-proteins) when bound by an agonist but not to couple with the G-proteins when bound by an antagonist. This explanation infers that an opioid having an N-substituent of three to four carbon size induces a conformational change in the receptor or blocks essential receptor areas that prevents the interaction of the receptor and the signal transduction proteins.

Those interested in an in-depth understanding of the SAR for mu receptor antagonists should be aware that properly substituted N-methyl-4-phenylpiperidines, N-methyl-6,7-benzomorphans and even nonphenolic opioid derivatives that have good antagonist activity are known.[37,38]

Structure-Activity Relationships of Kappa Receptor Agonists

The SAR for kappa agonists is somewhat related to that of mu antagonists. All of the marketed kappa agonists have structures related to the rigid opioids and N-allyl, N-CPM, or N-CBM substitutions. The compounds are all mu receptor antagonists and kappa receptor agonists. The kappa agonist activity is enhanced if there is an oxygen group placed at the 8-position (e.g., ethylketazocine) or into the N-substituent (e.g., bremazocine). The oxygen group in an N-furanylmethyl substituent also enhances kappa activity.

Potent and selective kappa agonists that lack antagonistic properties at any of the opioid receptors are found in a number of *trans*-1-arylacetamido-2-aminocyclohexane derivatives. There are not enough compounds reported in this class to develop strong trends in SAR. The relative mode of receptor binding for the morphine-related versus the arylacetamide kappa agonists is not known. Evidence exists for the selective binding of the arylacetamides at a different kappa receptor subtype than the morphine-related compounds.

Structure-Activity Relationship of Delta Receptor Agonists

Structure-activity relationships for delta receptor agonists are the least developed among the opioid compounds. Nonpeptide delta selective agonists are just beginning to be discovered. Peptides with high selectivity for delta receptors are known. The SARs for some of these peptides are discussed in the following paragraphs. No selective delta agonist appears to be headed for quick entry into medical practice.

Structure-Activity Relationships of Endogenous Opioid Derivatives

Thousands of derivatives related to the endogenous opioid peptides have been prepared since the discovery of the enkephalins in 1975.[42] A thorough discussion of the SARs of these peptides would be a major task; however, there are some major trends that have emerged and easily can be discussed. Some selected general SAR points for peptide opioids are:

1. All of the endogenous peptides have Leu- or Met-enkephalin as their first five amino acid residues (see Fig. 14–2).
2. The tyrosine at the first amino acid residue position of all the endogenous opioid peptides is essential for activity. Removal of the phenolic hydroxyl group or the basic nitrogen (amino terminus group) abolishes activity. The Tyr^1 free amino group may be alkylated (methyl or allyl groups to give agonists and

antagonists), but it must retain its basic character. The structure resemblance between morphine and the Tyr[1] group of opioid peptides is especially obvious.

3. In addition to the phenol and amine groups of Tyr[1], the next most important moiety in the enkephalin structure is the phenyl group of Phe[4]. Removal of this group or changing its distance from Tyr[1] can result in full or substantial loss in activity.
4. The enkephalins have several low-energy conformations, and it is likely that different conformations are bound at different opioid receptor types and subtypes.
5. The replacement of the natural L-amino acids with unnatural d-amino acids can make the peptides resistant to the actions of several peptidases that generally rapidly degrade the natural endorphins. A d-Ala in place of Gly[2] has been especially useful for protecting the peptides from the action of nonselective aminopeptidases. The placement of bulky groups into the structure (e.g., the addition of N-Me to Phe[4]) also slows the action of peptidases. When evaluating new peptides for opioid activity, it is often difficult to tell if changes are due to metabolic stability or receptor affinity.
6. Conversion of the terminal carboxyl group into an alcohol or an amide protects the compound from carboxy peptidases.
7. Any introduction of unnatural d- or l-amino acids or bulky groups into the enkephalin structure affects its conformational stability. The resultant peptides have an increase or decrease in affinity for each of the opioid receptor types. The right combination of receptor affinity increases or decreases results in selectivity for a receptor type.
8. Structural changes that highly restrict the conformational mobility of the peptides (e.g., substitution of proline for Gly[2] or cyclization of the peptide) have been especially useful for the discovery of receptor selective opioid peptides.

For examples of the above-mentioned SARs, see the structures of the peptides given in Figs. 14–1 and 14–6.

The effect of lengthening the amino acid chain of the enkephalin peptides deserves special consideration. As previously noted, all of the endogenous opioids found in mammals have Leu- or Met-enkephalin at their amino terminus end. Lengthening the carboxyl terminus can give the peptide greater affinity or selectivity for an opioid receptor type. This effect can be illustrated by the dynorphins (see Fig. 14–1), in which incorporation of the basic amino acids (especially Arg[7]) into the C-terminus chain results in a marked increase in affinity for kappa receptors. The message-address analogy has been used to describe this effect. The first four amino acids [Tyr-Gly-Gly-Phe] are essential for peptide ligands to bind to and to activate all opioid receptor types. The N-terminus amino acids can then be referred to as carrying the "message" to the receptors. Adding additional amino acids to the C-terminus can "address" the message to a specific receptor type. The additional peptide chain may be affecting the address (selectivity) by providing new and favorable binding interactions to one of the receptor types. Alternatively the additional peptide could be inducing a conformational change in the message portion of the peptide that favors interaction with one of the receptor types.

METABOLISM

A knowledge of the metabolism of the opioid drugs is essential to the understanding of the uses of these agents.

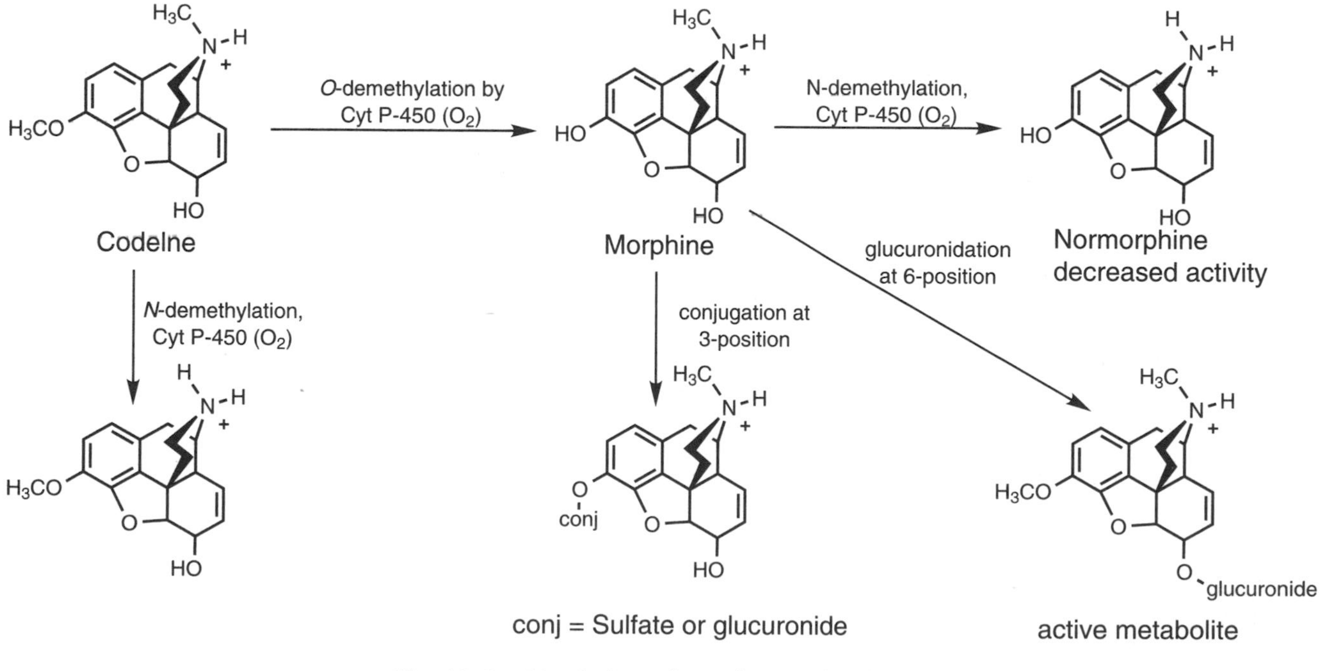

Codeine —*O*-demethylation by Cyt P-450 (O_2)→ Morphine —N-demethylation, Cyt P-450 (O_2)→ Normorphine (decreased activity)

Codeine —*N*-demethylation, Cyt P-450 (O_2)→ norcodeine

Morphine —conjugation at 3-position→ 3-O-conj (conj = Sulfate or glucuronide)

Morphine —glucuronidation at 6-position→ 6-O-glucuronide (active metabolite)

Fig. 14–9. Metabolism of morphine and codeine.

The poor oral versus parenteral dose ratio (about 6:1) of morphine is caused by extensive first-pass metabolic conjugation of morphine at the phenolic (3-OH) position (Fig. 14–9). The metabolism occurs predominantly in the liver but also in the intestinal mucosa, requiring the action of a sulfotransferase or glucuronosyltransferase enzyme. The 3-conjugates have low activity and poor distribution properties. The 3-glucuronide does undergo enterohepatic cycling, which explains the need for high initial oral doses of morphine followed by lower maintenance doses. Glucuronidation of morphine at the 6-OH position results in the formation of an active metabolite. Morphine is also N-demethylated to give normorphine, a compound that has decreased opioid activity and decreased bioavailability to the CNS. Normorphine undergoes N- and O-conjugations and excretion. Geriatric patients metabolize morphine at a slower rate than normal adult patients; thus, they are likely to show greater sensitivity to the drug and require lower doses.

In humans, about 10% of an oral dose of codeine is

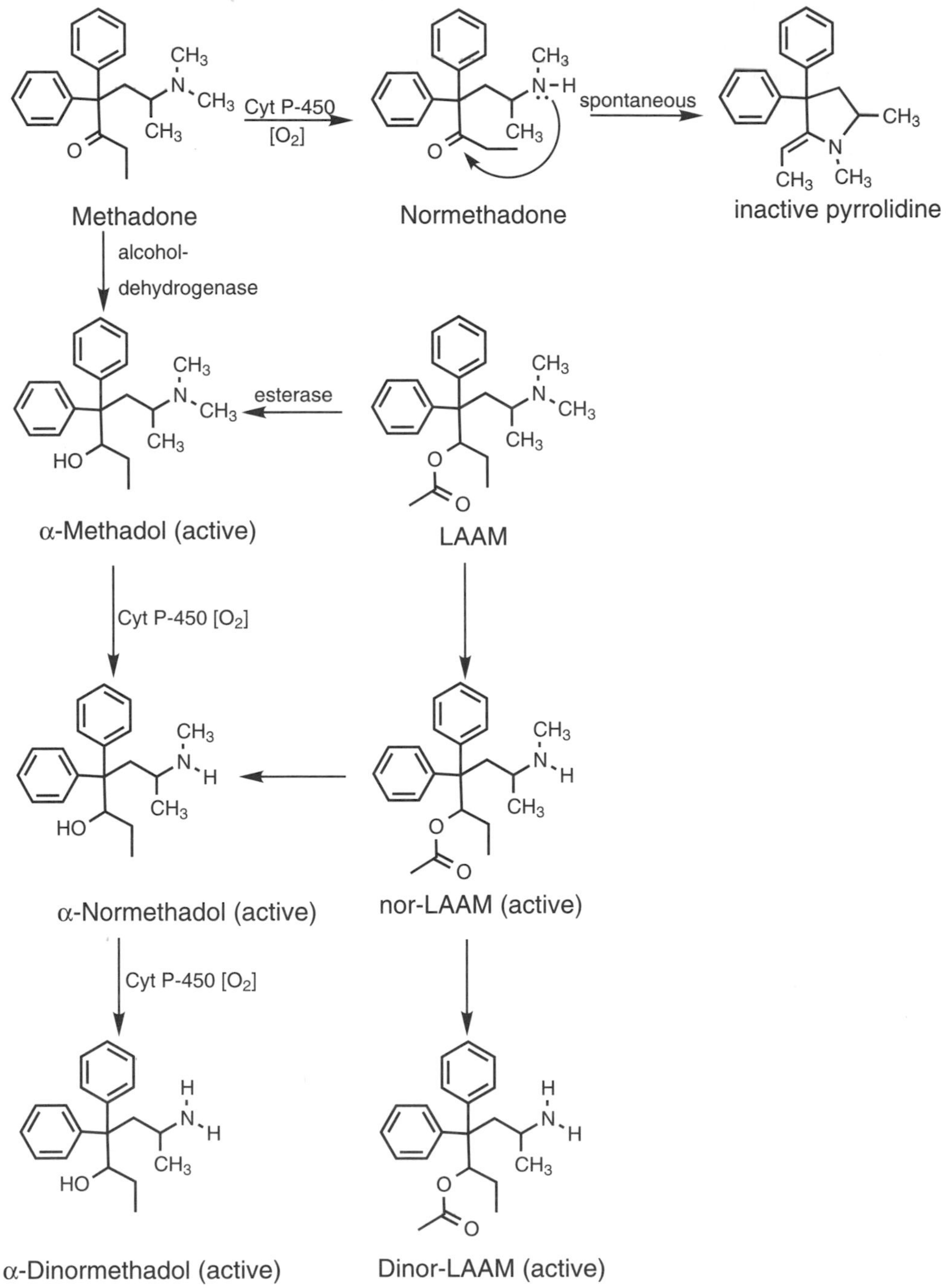

Fig. 14–10. Metabolism of methadone and α-acetylmethadol (LAAM).

O-demethylated to produce morphine. The morphine produced as a metabolite of codeine plays an important role in the analgesic effect. The bioactivation of codeine versus the bioinactivation of morphine results in an oral to parenteral dose ratio for codeine of 1.5:1.

Other rigid-structured opioid analogs undergo similar routes of metabolism as morphine. The amount of first-pass 3-O-conjugation varies from compound to compound; thus, the relative oral/parenteral usefulness of the agents varies. In general, compounds more potent and lipophilic than morphine (e.g., levorphanol) tend to have better oral activity. Compounds with N-alkyl groups larger than methyl get N-dealkylated as a major route of inactivation.

The short duration of action of meperidine is the result of rapid metabolism. Esterases cleave the ester bond to leave the inactive 4-carboxylate derivative. Meperidine also undergoes N-demethylation to give normeperidine. Normeperidine has little analgesic activity, but it contributes significantly to the toxicity of meperidine.

The metabolism of methadone, as outlined in Figure 14–10, is important to its action. The major route of inactivation results from N-demethylation and cyclization of the secondary amine to an inactive pyrrolidine derivative. If the keto group is reduced by alcohol dehydrogenase to give methadol, the demethylation product can no longer cyclize to the pyrrole derivatives. Methadol is less active than methadone as an analgesic, but the N-demethylation products of methadol, normethadol and dinormethadol, are active analgesics with increased half-lives compared with methadone. The buildup of these metabolites is responsible for the long duration of action and mild prolonged withdrawal symptoms produced by methadone.

Levo-alpha-acetylmethadol (LAAM) is longer acting than methadone. Its slow onset of action after oral dose, and the isolation of at least three active metabolites, suggests that LAAM itself is a prodrug. The relative contributions of LAAM and its active metabolites to the analgesic and addiction maintenance properties in humans have not been determined. It is clear that a 75- to 100-mg oral dose of this agent suppresses withdrawal symptoms in opioid addicts for 3 to 4 days.

MU OPIOID RECEPTOR MODELS

A number of models have been proposed to represent the bonding interactions of agonists at mu opioid receptors. These models are reflections of complementary bonding interactions of mu agonists to the receptor as revealed from SAR studies. Beckett and Casy made the first such receptor drawing in 1954.[43] They studied the configurations and conformations of the mu agonists known at that time, and proposed that all opioids could bind to the template (receptor model) shown in Figure 14–11. The model presumed that nonrigid opioids (e.g., meperidine and methadone) took a shape similar to that of morphine when binding to the receptor. It soon became apparent that the most stable conformations of meperidine and methadone were not superimposable on the structure of morphine. New compounds that could not assume the shape of morphine were also being discovered, and it became apparent that the Beckett and Casy model could not explain the activity of all mu agonists.

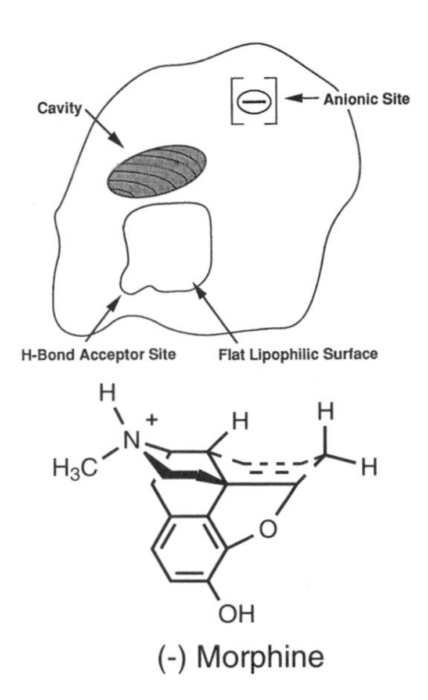

Fig. 14–11. A representation of the original model for the opioid receptor as proposed by Beckett and Casy.[43] The morphine structure would have to rotate 180° about a vertical axis before it could fit to the receptor site. The model is only good for mu selective agents.

In the mid-1960s, Portoghese attempted to correlate the structures and analgesic activities of rigid and nonrigid opioids that contained the same series of N-substituents.[44] He reasoned that if all opioids bound the receptor in the same conformation, a substituent at a similar position on any of the compounds should fall on the same surface area of the receptor. One would expect the same structure modification on any opioid structure to give the same type and degree of bonding interaction and thus the same contribution to analgesic activity. Portoghese found that parallel changes of the N-substituent on rigid (morphine, morphinan, or benzomorphan) analgesic parent structures gave parallel changes in activity. This finding supported the notion that rigid-structural opioid compounds bound to the receptor for analgesia in the same manner. When the same test was applied to nonrigid (meperidine-like) opioid structures, however, varying the N-substituent did not produce an activity change paralleling that seen for the rigid-structured series. Apparently the N-substituents in the rigid and nonrigid opioid series were falling on different surfaces of a receptor and thus making different contributions to analgesic activity. Portoghese concluded that the rigid and nonrigid series of compounds were either binding to different receptors or were interacting with the same receptor by different binding modes. He introduced the bimodal receptor binding model (Fig. 14–12) as one possible explanation of the results. Later it was discovered that the activity of the rigid opioid compounds (series 1) was enhanced by a 3-OH substituent on the aromatic ring, whereas a similar substituent in some nonrigid opioids (series 2) caused a loss of activity. Again, similar substituents pro-

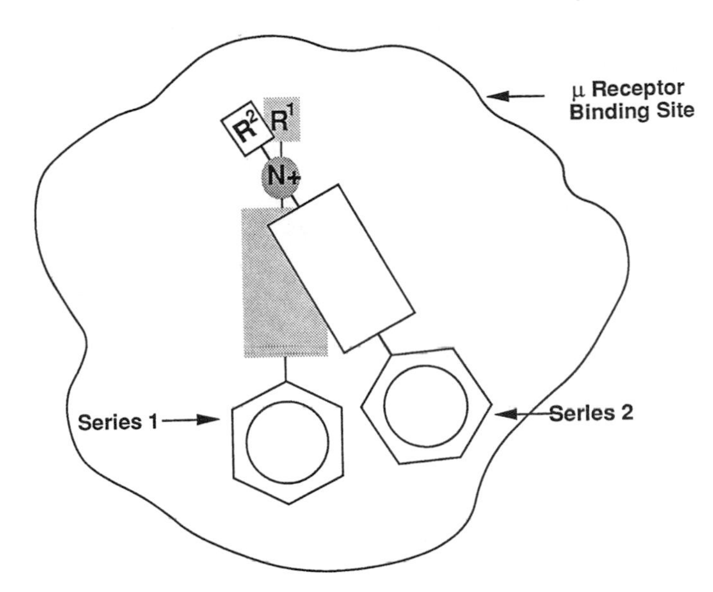

Fig. 14–12. A representation of the bimodal binding model of the mu opioid receptor as proposed by Portoghese.[44] Different opioid series bind to different surface areas of the same receptor protein.

duced nonparallel changes in activity, indicating that the aromatic rings in the two series were not binding to the same receptor site. To provide an explanation for these results, the bimodal binding model was modified to incorporate the structure of the enkephalins (Fig. 14–13).[45] The rigid-structured opioids benefiting from the inclusion of a phenolic hydroxyl group were proposed to bind the mu receptor in a manner equivalent to the tyrosine (Tyr^1 or T-subsite) of enkephalin. The nonrigid-structure opioids, which lose activity on introduction of a phenolic hydroxyl group into their structure, were proposed to interact with the receptor in a manner equivalent to the phenylalanine (Phe^4 or P-subsite) of enkephalin. The free amino group of Tyr^1 occupies the anionic binding site of the receptor that is the common binding point of both opioid series. This model closely resembles the original bimodal binding proposal. Models that attempt to explain the ability of Na^+ to decrease the binding affinity of agonists but not antagonists for the opioid receptor have been made.[46] Sodium ions also protect the receptor from alkylation by nonselective alkylating agents.

The Beckett and Casy model was extended to explain the increased potency of the oripavine analogs such as etorphine (Fig. 14–14).[39] The affinity of the oripavines for the mu opioid receptor can be much greater than that seen for morphine. It is likely that the increased receptor affinity results from auxiliary drug-receptor bonding interactions similar to those depicted in the receptor model.

Martin has proposed a receptor model for kappa opioid receptors.[47] Martin's model considers just the binding of rigid morphine-related opioid structures. The relationship of how rigid morphine-related agents interact with the kappa receptor compared with the arylacetamide kappa agonist derivatives has not been well studied. Models for the delta receptors have not been proposed.

All three opioid receptor types have been solubilized from their membrane-bound natural state, and extensive efforts have been made to clone the proteins.[48–50] There has been some success in cloning each of the receptor types. In time, the exact chemical and physical characteristics of the receptor proteins will be known, and the di-

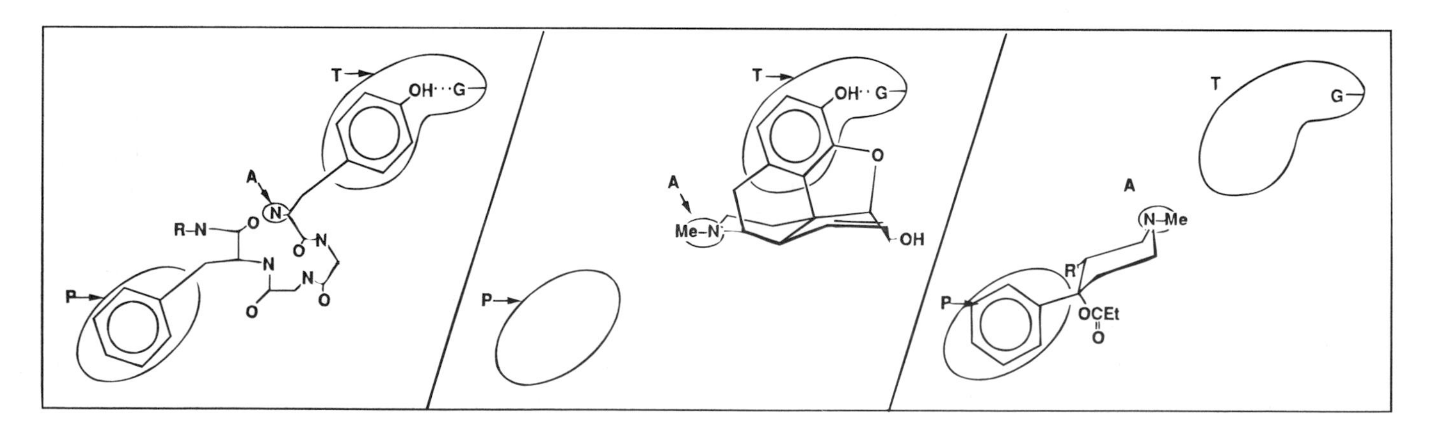

Fig. 14–13. A representation of the enkephalin binding site of mu opioid receptors.[45] *A,* An enkephalin bound to the receptor. *B,* Morphine binding the receptor by using the T-subsite (i.e., the tyrosine binding site). *C,* A Meperidine-type opioid binding the receptor by using the P-subsite (i.e., the phenylalanine binding site).

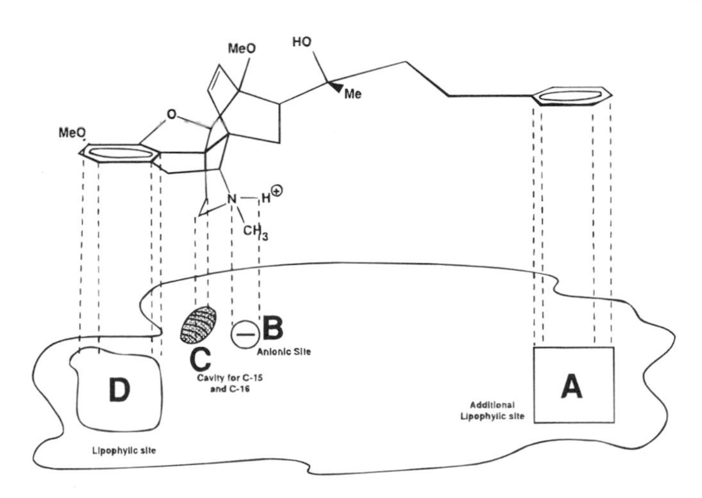

Fig. 14–14. A representation of the binding of an oripavine-type analgesic to the mu opioid receptor.[39] The hydroxyl and phenyl groups in the side chain are believed to form additional bonding interactions with the receptor compared with the Beckett and Casy receptor model.

rect bonding interactions of selective drugs with their receptors will be studied.

CLINICALLY AVAILABLE AGENTS

Mu Agonists

(−)-Morphine Sulfate: Generic Preparations

Morphine sulfate is the most often used analgesic for severe, acute, and chronic pain. Morphine is a mu agonist and is a Schedule II drug. It is available in intramuscular, subcutaneous, oral, rectal, epidural and intrathecal dosage forms. The epidural and intrathecal preparations are formulated without a preservative. Morphine is three to six times more potent when given intramuscularly than when it is given orally. The difference in activity is due to extensive first-pass 3-O-glucuronidation of morphine—an inactive metabolite. The half-life of intramuscularly dosed (10 mg) morphine is about 3 hours. The dose of morphine, by any dosage route, must be reduced in patients with renal failure and in geriatric and pediatric patients. The enhanced effects of morphine in renal failure is believed to be due to a buildup of the active 6-glucuronide metabolite, which depends on renal function for elimination.

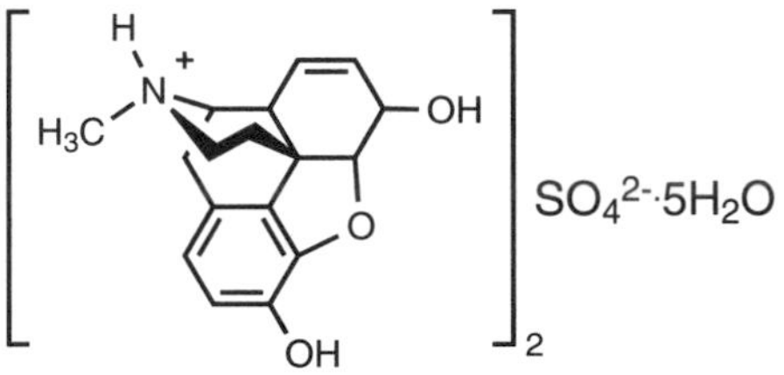

The analgesic effect of orally dosed morphine can equal that obtained by parenteral administration, if proper doses are given. When given orally, the initial dose of morphine is usually 60 mg, followed by maintenance doses of 20 to 30 mg every 4 hours. Addiction to clinically used morphine by the oral route is generally not a problem. Overdoses of morphine as well as all mu agonists in this section can be effectively reversed with naloxone.

(−)-Codeine Phosphate: Generic Preparations and Mixtures

Codeine is used extensively to treat moderate-to-mild pain. Codeine is a weak mu agonist, but about 10% of an oral dose (30 to 60 mg) is metabolized to morphine (see under Metabolism), which contributes significantly to its analgesic effect. The plasma half-life of codeine after oral dose is 3.5 hours. The parenteral dosage of codeine needed to produce analgesia causes release of histamine sufficient to produce hypotension, pruritus, and other allergic responses. Thus, administration of codeine by the parenteral route is not recommended.

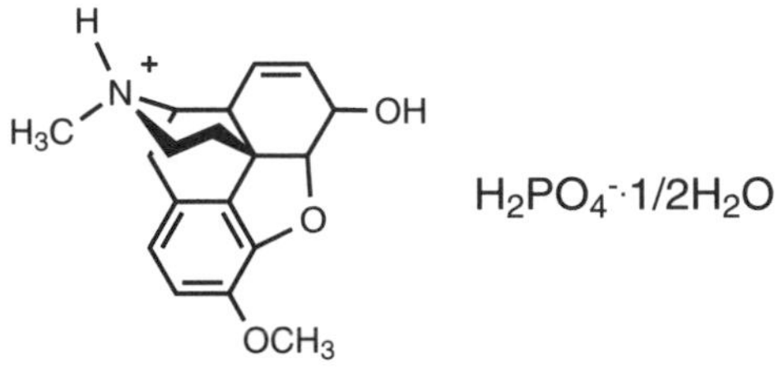

(−)-Hydromorphone Hydrochloride: Dilaudid and Generic Preparations

Hydromorphone is a potent mu agonist (eight times morphine) used to treat severe pain. It is available in intramuscular, intravenous, subcutaneous, oral, and rectal dosage forms. Similar to all strong mu agonists, hydromorphone is addicting and is a Schedule II drug. Hydromorphone has an oral-to-parenteral potency ratio of 5:1. The plasma half-lives after parenteral and oral dosage are 2.5 and 4 hours.

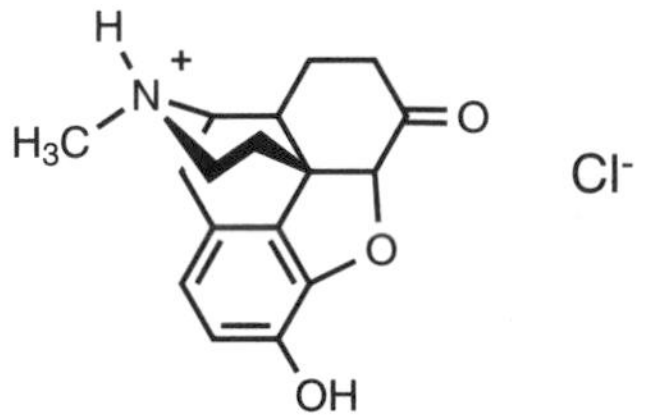

(−)-Oxymorphone Hydrochloride: Numorphan

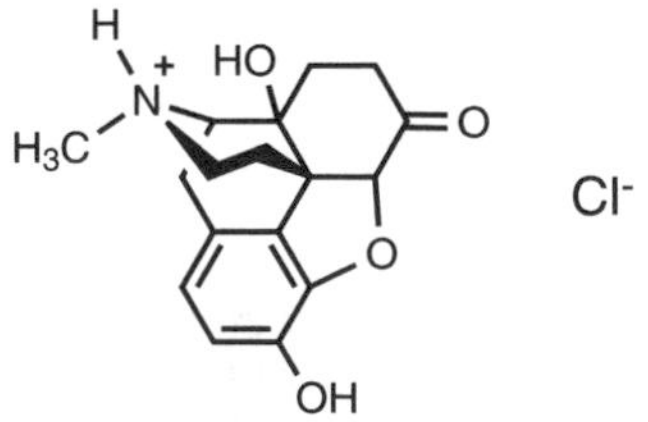

Oxymorphone is a potent mu agonist (10 times morphine) used to treat severe pain. Administration is by intramuscular, subcutaneous, intravenous, and rectal routes. The intramuscular dose of oxymorphone (1 mg) has a half-life of 3 to 4 hours. It is a Schedule II drug. Oxymorphone, because of its 14β-hydroxy group, has low antitussive activity.

(−)-Levorphanol Bitartrate: Levo-Dromoran and Generic Preparations

Levorphanol is a potent mu agonist (about six times morphine), and its uses, side effects, and physical dependence liability are similar to oxymorphone or hydromorphone. Levorphanol is available in oral, subcutaneous, and intravenous dosage forms. The oral dose of levorphanol is about twice the parenteral dose. This drug is unique among the mu agonists in that its analgesic duration of action is 4 to 6 hours, whereas its clearance half-life is 11.4 hours. Thus, effective analgesic doses of this agent can lead to an accumulation of the drug in the body and result in excessive sedation.

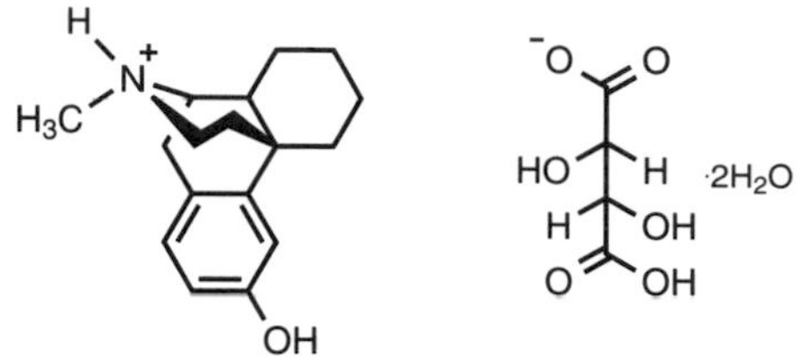

(−)-Hydrocodone Bitartrate: Lortab, Vicodin, and Generic Preparations (all in mixtures with acetaminophen)

Hydrocodone is a schedule III drug that is used to treat moderate pain. It is used mostly by the oral route (5 mg tablets and solutions) in combination with acetaminophen. The compound has good oral bioavailability and is metabolized in a manner similar to codeine.

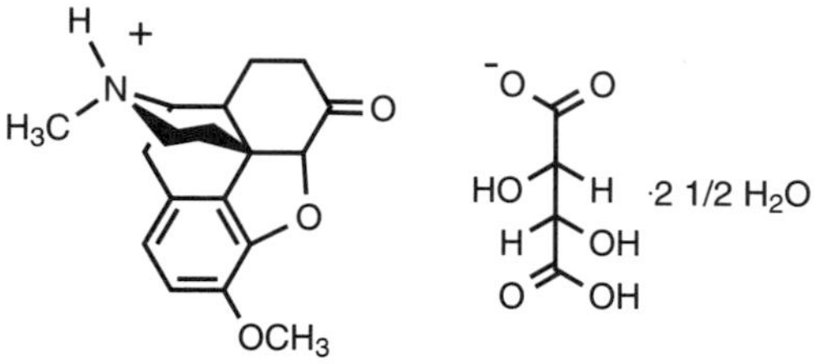

(−)-Oxycodone Hydrochloride: Roxicodone (single agent), Percocet, Percodan, Tylox, and Generic Preparations (all mixtures)

Oxycodone is about equipotent with morphine, but because of the 3-OCH_3 group, it has a much lower oral-to-parenteral dose ratio. Thus, oxycodon is used orally to treat severe-to-moderate pain. It is a Schedule II drug as a single agent and when it is combined in strong analgesic mixtures. Oxycodone has a plasma half-life of about 4 hours and requires dosing every 4 to 6 hours. Metabolism of this agent is comparable to that of codeine.

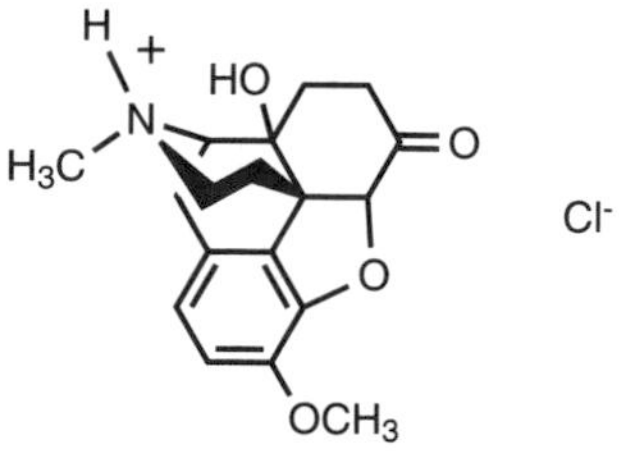

Meperidine Hydrochloride: Demerol and Generic Preparations

Meperidine is a mu agonist with about one-tenth the potency of morphine after intramuscular dose. Meperidine produces the analgesia, respiratory depression, and euphoria caused by other mu opioid agonists, but it causes less constipation, and it does not inhibit cough. When given orally, meperidine has 40 to 60% bioavailability as a result of significant first-pass metabolism. Because of the limited bioavailability, it is one-third as potent after an oral dose compared with a parenteral dose.

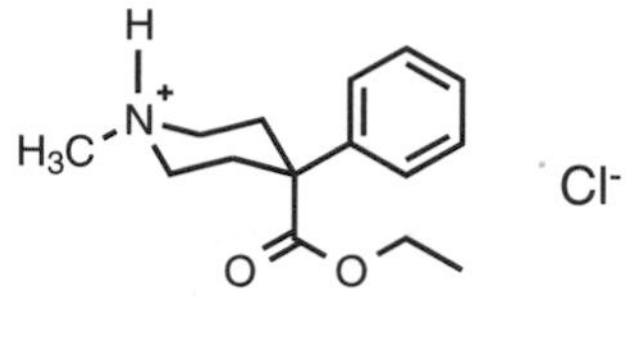

Meperidine has received extensive use in obstetrics because of its rapid onset, short duration of action and lack of antioxytocic activity. In contrast to morphine, meperidine does not prolong labor. When it is given intravenously in small (25-mg) doses during delivery, the respiratory depression in the newborn child is minimized. Meperidine is used as an analgesic in a variety of nonobstetric anesthetic procedures. Meperidine is extensively metabolized in the liver, with only 5% of the drug being excreted unchanged. Prolonged dosage of meperidine may cause an accumulation of the metabolite normeperidine (see under Metabolism). Normeperidine has only weak analgesic activity, but it causes CNS excitation and can initiate grand mal seizures. It is recommended that meperidine be discontinued in any patient who exhibits signs of CNS excitation.

The elimination half-life of meperidine is 3 to 4 hours, and it can double in patients with liver disease. Acidification of the urine causes an enhancement of the clearance of meperidine, but there is a lesser effect on the clearance of the toxic metabolite normeperidine.

Meperidine has a strong adverse reaction when given to patients receiving a monoamine oxidase (MAO) inhibitor. This drug interaction has been seen in Parkinson's disease patients taking the MAO-β selective inhibitor selegiline (Eldepryl).

(±) Methadone Hydrochloride: Dolophine Hydrochloride

Methadone is a synthetic agent with about the same mu opioid potency as morphine. The drug is used as a racemic mixture in the United States, but nearly all of the activity is due to the (−)-isomer. Methadone's usefulness is a result of its greater oral potency and longer duration of action compared with most other mu agonists. When given orally, a 20-mg dose every 8 to 12 hours can give effective analgesia. Methadone is an excellent analgesic for use in cancer patients, and it is often used in hospice programs. Oral doses of 40 mg are commonly used for 24-hour suppression of withdrawal symptoms (addiction maintenance) in opioid addicts. When given parenterally in doses of 2.5 to 10 mg, methadone (Schedule II drug) has all of the effects of morphine and other mu agonists.

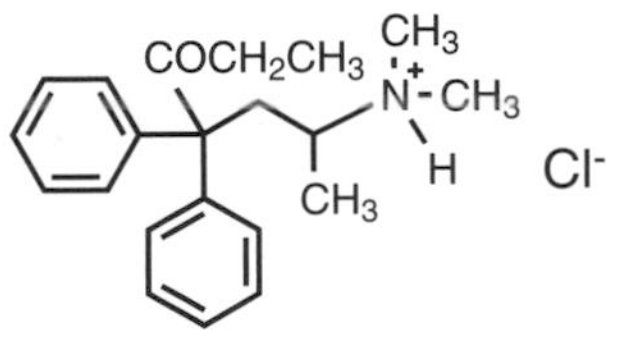

The metabolism of methadone is extremely important in determining its long duration of action. The elimination of methadone depends on liver function and urinary pH. The typical half-life is 19 hours, but if urinary pH is raised from normal values of 5.2 to 7.8, the half-life becomes 42 hours. At the higher pH, a lower percentage of methadone exists in the ionized form, and there is more renal reabsorption of the drug. The metabolism of methadone by liver enzymes is extensive and complex (see under Metabolism). There are at least two active metabolites. Enzyme inducers (e.g., phenytoin, rifampin) can lead to withdrawal in patients using methadone for maintenance of addiction. Toxic doses can build up in patients with liver disease or in geriatric patients who have a decreased oxidative metabolism capacity.

Although methadone is a good drug for maintenance of addiction, it is not ideal. Methadone requires once-a-day dosing, usually at a clinic, to suppress withdrawal symptoms. Once-a-day dosing is expensive and sometimes logistically difficult to achieve. LAAM is available and is used in some treatment programs to overcome the problems of methadone. LAAM is more potent than methadone, and it has a longer duration of action. A single oral dose of this agent can suppress abstinence withdrawal for up to 3 days.

Propoxyphene Hydrochloride or Napsylate: Darvon, Dolene, Darvon-N, and Generic Preparations

Propoxyphene is a weak mu agonist that is used as a single agent and in mixtures with nonsteroidal anti-inflammatory agents to treat mild or moderate pain. The active (+)-isomer has (2S, 3R) absolute configuration. Propoxyphene is available only in oral dosage forms. Propoxyphene has about one-twelfth the potency of morphine, and most studies show it to be equally or less effective than aspirin as an analgesic. Doses of propoxyphene that approach the analgesic efficacy of morphine are toxic. Propoxyphene's popularity is due to the fact that physicians prescribe it for its lower abuse potential (Schedule IV) compared with codeine.

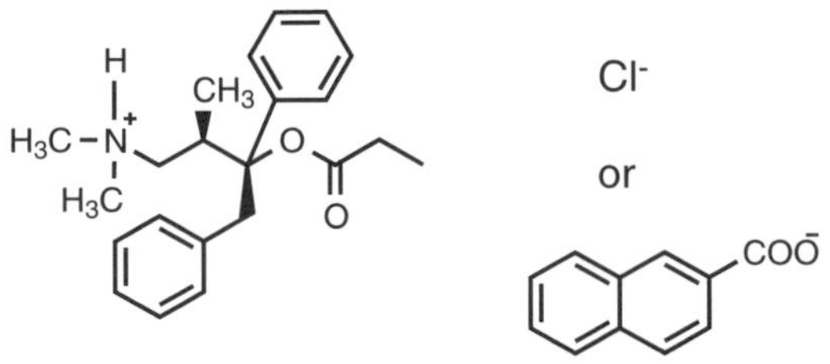

Fentanyl Citrate: Sublimaze plus Generic Preparations, also in Combination with Droperidol

Fentanyl is a mu agonist with about 80 times the potency of morphine. Fentanyl has been used in combination with nitrous oxide for balanced anesthesia and in combination with droperidol for neuroleptanalgesia. The advantages of fentanyl over morphine for anesthetic procedures are its shorter duration of action (1 to 2 hours) and the fact that it does not cause histamine release on intravenous injection.

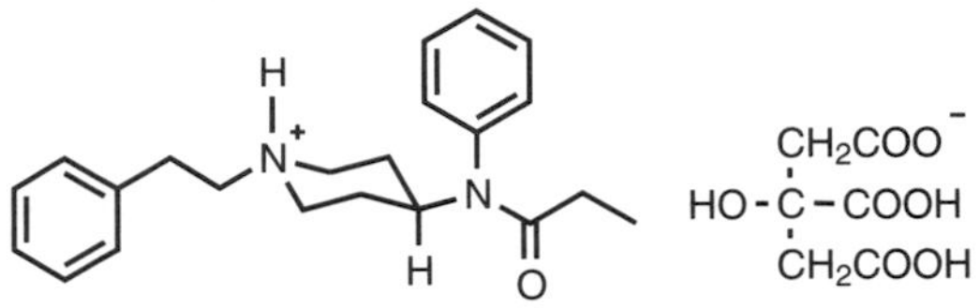

A fentanyl patch has been released for the treatment of severe chronic pain. This dosage form delivers fentanyl transdermally and provides effective analgesia for periods up to 10 hours.

Fentanyl's short duration of action after parenteral dosing is due to redistribution, rather than to metabolism or excretion. Repeated doses of fentanyl can result in accumulation and toxicities. Elderly patients are usually more sensitive to fentanyl and require lower doses.

Sufentanil Citrate: Sufenta

Addition of the 4-methoxymethyl group and bioisosteric replacement of the phenyl with a 2-thiophenyl on the fentanyl structure results in a 10-fold increase in mu opioid activity. The resultant compound, sufentanil, is 600 to 800 times more potent than morphine. Despite its greater sedative and analgesic potency, sufentanil produces less respiratory depression at effective anesthetic doses. Sufentanil is available in an intravenous dosage form, and it is used for anesthetic procedures. It has a faster onset and shorter duration of action than fentanyl. The short duration is due to redistribution from brain tissues.

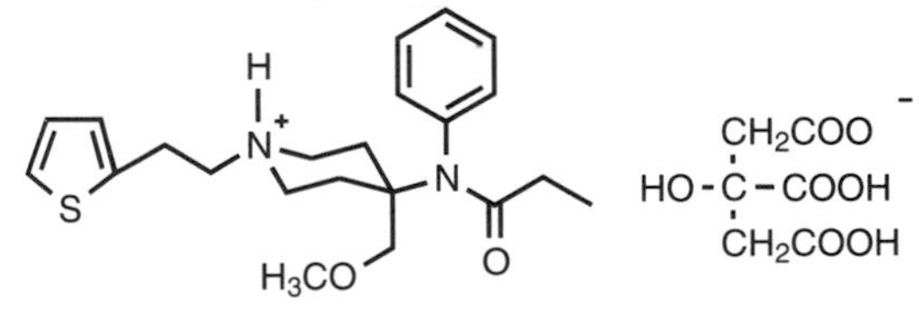

Alfentanil Hydrochloride: Alfenta

Substitution of tetrazol-5-one for the thiophene ring in sufentanil results in a decrease in potency (approximately 25 times morphine) and a decrease in the pK_a of the resultant compound, alfentanil. The lower pKa (6.50) of alfentanil results in a lower percentage of the drug existing in the ionized form at physiologic pH. Being more unionized, alfentanil penetrates the blood–brain barrier even faster than other fentanyl derivatives and has a faster onset and shorter duration of action. In addition, alfentanil is metabolized 99% in the liver and has a half-life of only 1.3 hours. Alfentanil is available as an intravenous dosage form for use in ultrashort anesthetic procedures.

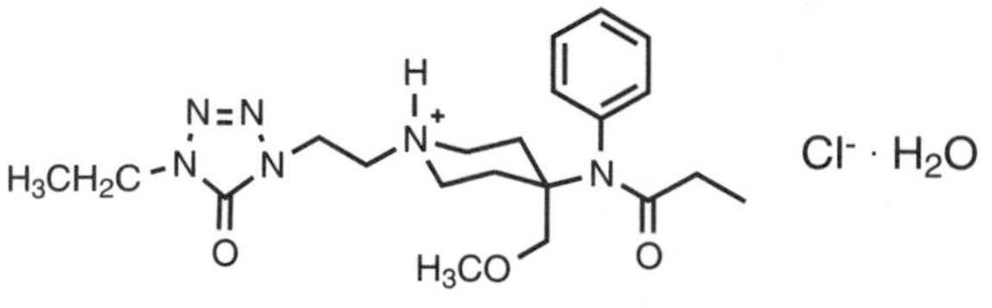

Mixed Agonist-Antagonists

(−)-Buprenorphine Hydrochloride: Buprenex

Buprenorphine is 20 to 50 times more potent than morphine in producing an ED_{50} analgesic effect in animal studies; however, it cannot produce an ED_{100} in these tests. Thus, buprenorphine is a potent partial agonist at mu opioid receptors. It is also a partial agonist at kappa receptors but more of an antagonist at delta receptors. Buprenorphine, at 0.4 mg intramuscular dose, produces the same degree of analgesia as 10 mg of morphine. Because of its partial agonist properties, it has a lower ceiling on its analgesic action but also produces less severe respiratory depression. It does not produce tolerance and addiction like full mu agonists. In fact, buprenorphine's partial agonist action, high affinity for opioid receptors, and high lipophilicity combine to give buprenorphine a tolerance, addiction, and withdrawal profile that is unique among the opioids. When given by itself to opioid-naive patients, little tolerance or addictive potential (Schedule V) is observed. A mild withdrawal can occur some 2 weeks after the last dose of buprenorphine. Buprenorphine precipitates withdrawal symptoms in highly addicted individuals, but it suppresses symptoms in individuals who are undergoing withdrawal from opioids. It effectively blocks the effect of high doses of heroin. Because of these properties, buprenorphine has been used in opioid addiction treatment programs. It has also been reported to suppress cocaine use and addiction.

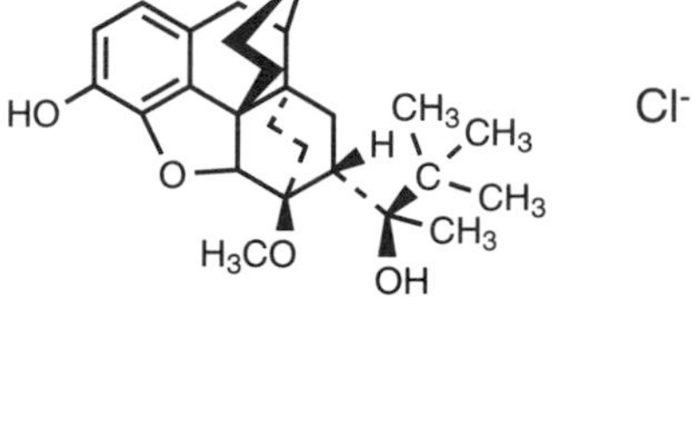

Buprenorphine undergoes extensive first-pass 3-O-glucuronidation, which negates its usefulness after oral dose. It is available in parenteral dosage forms in the United States and as a sublingual preparation in Europe. Its typical dose is 0.3 to 0.6 mg three times per day by intramuscular injection. The duration of analgesic effect is 4 to 6 hours. After parenteral dose, about 70% of the drug is excreted in the feces, and the remainder appears as N-dealkylated and conjugated metabolites in the urine.

Naloxone is not an effective antagonist to buprenorphine because of the latter's high binding affinity for opioid receptors.

(−)-Butorphanol Tartrate: Stadol

Butorphanol is an antagonist at mu opioid receptors with about one-sixth the potency of naloxone. If given to a person addicted to a mu agonist, butorphanol induces an immediate onset of abstinence syndrome. Butorphanol is a strong agonist at kappa opioid receptors, and through this interaction it is five times more potent than morphine as an analgesic. Kappa agonists have a lower ceiling analgesic effect than full mu agonists; thus, they are not as effective in treating severe pain.

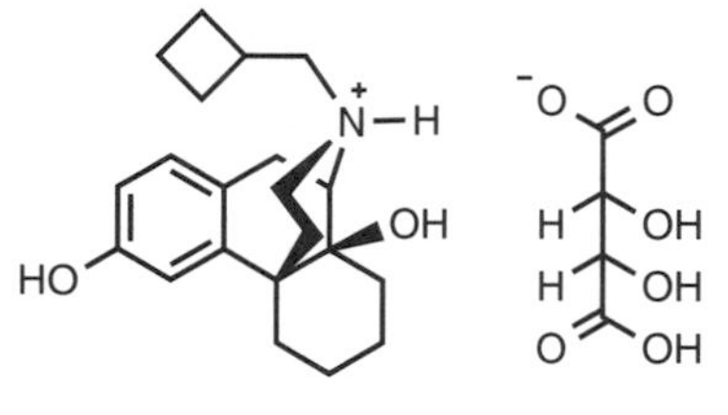

Butorphanol has a different spectrum of side effects than mu opioid analgesics. Respiratory depression occurs, but there is a lower ceiling on this effect, and it is not generally lethal, as is the case with high doses of mu agonists. Major side effects after normal analgesic doses are sedation, nausea, sweating, and dysphoric (hallucinogenic) effects at higher doses. Butorphanol causes an increase in pulmonary arterial pressure and pulmonary vascular resistance. There is an overall increased workload on the heart, and it should not be used in patients with congestive heart failure or to treat pain from acute myocardial infarction. Butorphanol has low abuse potential and is not a scheduled drug.

Because of first-pass metabolism, butorphanol is not used in an oral dose form. Given parenterally, it has a plasma half-life and a duration of analgesic effectiveness of 3 to 4 hours. The major metabolite of butorphanol is the inactive hydroxylated product, which is excreted primarily in the urine.

Nalbuphine Hydrochloride: Nubain

Nalbuphine is an antagonist at mu receptors and an agonist at kappa receptors. As an antagonist, it has about one-fourth the potency of naloxone, and it does produce withdrawal when given to addicts. On a weight basis (mg/

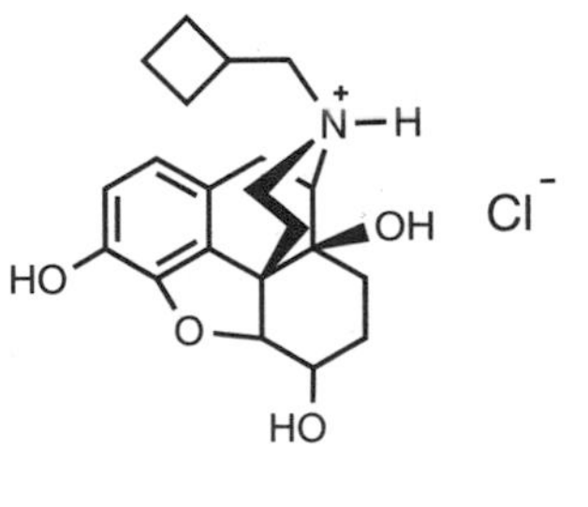

mg), the analgesic potency of nalbuphine approaches that of morphine. An intramuscular injection of 10 mg gives about the same degree of duration of analgesia as an equivalent dose of morphine.

Side effects of nalbuphine are similar to those of other kappa agonists. Dysphoria is not as common as with pentazocine. Sedation is the most common side effect. Nalbuphine does not have the adverse cardiovascular properties found with pentazocine and butorphanol. Nalbuphine has low abuse potential and is not listed under the Controlled Substances Act.

Nalbuphine is available only for parenteral dosage. Its elimination half-life is 2 to 3 hours. The primary route of metabolism is by conjugations, and greater than 90% of drug is excreted as conjugates in the feces.

(−)-Pentazocine Hydrochloride and Lactate: Talwin Nx and Talwin

Pentazocine is a weak antagonist (one-thirtieth naloxone) at mu receptors and an agonist at kappa receptors.

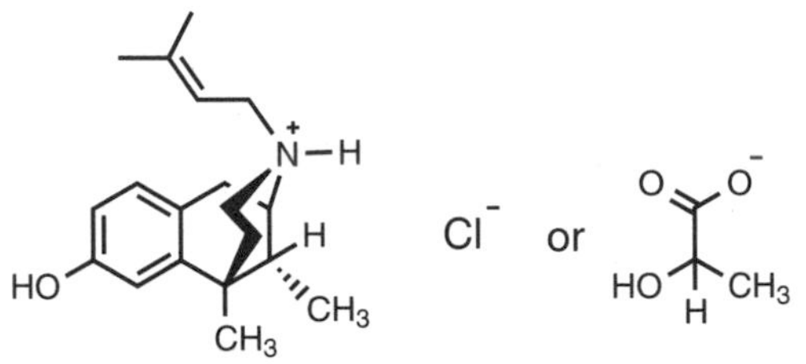

Pentazocine is one-sixth as potent as an analgesic compared with morphine after parenteral doses. Pentazocine also is dosed orally and has an oral-to-parenteral dose ratio of about 2:1. It is used to treat moderate pain. The mu antagonist properties of pentazocine are sufficient to produce abstinence signs in opioid addicts.

The side effects of pentazocine are similar to those of other kappa agonists. It has a greater tendency to produce dysphoric episodes, and it causes an increase in blood pressure and heart rate similar to that caused by butorphanol. Pentazocine is a Schedule IV drug. The major abuse of pentazocine has been its injection along with the antihistaminic drug tripelennamine (the T's and blues). Inclusion of the antihistaminic drug supposedly causes an increase in the euphoric while decreasing the dysphoric effects of the pentazocine. The manufacturers of pentazocine have attempted to thwart this use by including naloxone in the oral dose formulation of pentazocine. When taken orally, as prescribed, the naloxone has no bioavailability, and the pentazocine is able to act as intended. When the tablet is dissolved and injected, the naloxone effectively blocks all of the pharmacologic actions of the pentazocine.

The elimination half-life of pentazocine is about 4 hours after parenteral dosage and 3 hours after oral dosage. Bioavailability after oral dosage is only 20 to 50% owing to first-pass metabolism. Pentazocine is metabolized extensively in the liver and excreted via the urinary tract. The major metabolites are 3-O-conjugates and hydroxylation of the terminal methyl groups of the N-substituent. All metabolites are inactive.

Dezocine: Dalgan

Dezocine is classified as a mixed agonist/antagonist. The SAR of dezocine is unique among the opioids. It is a primary amine, whereas all other nonpeptide opioids are tertiary amines. Its exact receptor selectively profile has not been reported; however, its pharmacology is most similar to that of buprenorphine. It seems to be a partial agonist at mu receptors, to have little effect at kappa receptors, and to exert some agonist effect at delta receptors. On a weight (mg/mg) basis, it is about equipotent with morphine, and similar to morphine, it is useful for the treatment of moderate-to-severe pain. It is available for intramuscular and intravenous dose. The drug is indicated for postoperative and cancer-induced pain.

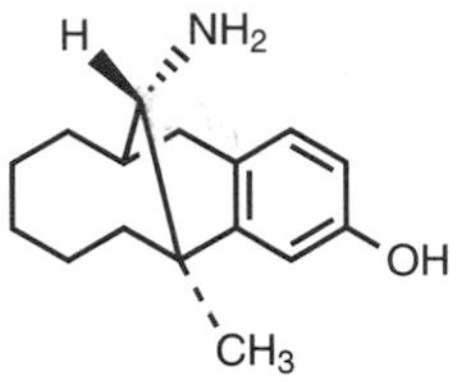

Dezocine has a half-life of 2.6 to 2.8 hours in normal patients and 4.2 hours in patients with liver cirrhosis. The onset of action of dezocine is faster (30 minutes) than equivalent analgesic doses of morphine, and its duration of action is longer (4 to 6 hours). Dezocine is extensively metabolized by glucuronidation of the phenolic hydroxyl group and by *N*-oxidation. Metabolites are inactive and excreted mostly via the renal tract.

Dezocine causes respiratory depression, but similar to buprenorphine, there is a ceiling to this effect. Presumably, there is also a ceiling to the analgesic effect of dezocine, but this point is not well documented. Dezocine does not have the high affinity for mu receptors that buprenorphine has, and its respiratory depressant effect can be reversed readily by naloxone.

The major side effects of dezocine are dizziness, vomiting, euphoria, dysphoria, nervousness, headache, pruritus, and sweating. Normal volunteers and recovered addicts report the subjective effects of single doses of dezocine to be similar to morphine. Because of the partial agonist mechanism of dezocine, one would not expect it to have a high abuse potential.

Opioids Used as Antidiarrheal Agents

Structure modification of 4-phenylpiperidines has led to the discovery of opioid analogs that are used extensively as antidiarrheal agents. Opioid agonists that act on mu and delta receptors have a strong action to inhibit the peristaltic reflex on the intestine. This action occurs because endogenous opioid tracts innervate the intestinal wall, where they synapse onto cholinergic neurons. When opioids are released onto cholinergic neurons, they inhibit the release of acetylcholine and thus inhibit peristalsis. Any mu agonist used in medicine causes constipation as a side effect. Most mu agonists are not used as antidiarrheal agents because of their potential for abuse and addiction.

Opium tincture and camphorated opium tincture (paregoric) have long been used as effective antidiarrheal agents. The bad taste of these liquid preparations and their abuse potential (Schedule II and III) serve to limit their use and to favor newer agents. Codeine sulfate or phosphate salt, as a single agent, is sometimes used for the short-term treatment of mild diarrhea. Synthetic agents that are structural combinations of meperidine and methadone are used extensively as antidiarrheal agents. Structures and uses of these agents are given next.

Diphenoxylate Hydrochloride with Atropine Sulfate: Lomotil and Generic Preparations

Diphenoxylate hydrochloride (2.5 mg) and atropine (0.025 mg) are combined in tablets or 5 ml liquid and used effectively as symptomatic treatment for diarrhea. The typical dose is two tablets, or 10 ml every 3 to 4 hours. The combination with atropine enhances the block of acetylcholine-stimulated peristalsis, and the adverse effects of atropine helps to limit the abuser of the opioid. The combination is Schedule V under the Controlled Substance Act. Diphenoxylate itself has low mu opioid agonist activity. It is metabolized rapidly by ester hydrolysis to the free carboxylate (difenoxin), which is five fold more potent after oral dosing. The high polarity of difenoxin probably limits its penetration into the CNS and explains the low abuse potential of this agent. High doses of diphenoxylate (40 to 60 mg) cause euphoria and addiction.

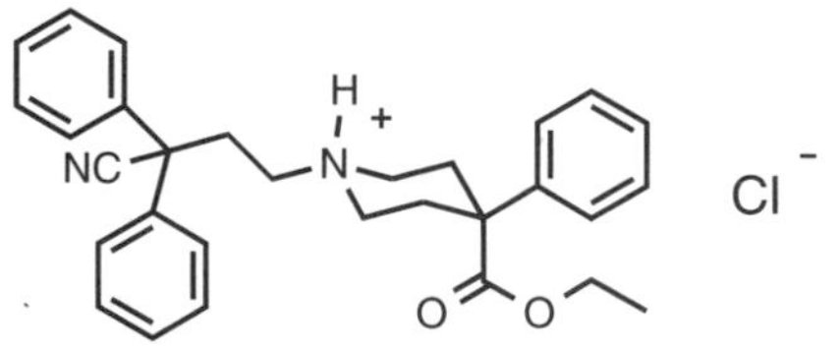

Difenoxin HCl with Atropine Sulfate: Motofen

Difenoxin, the active metabolite of diphenoxylate (as described previously), is also used as an antidiarrheal agent. Tablets contain 1 mg of difenoxin and 0.025 mg of atropine sulfate. Dosage, uses, and effectiveness are similar to that of diphenoxylate.

Loperamide Hydrochloride: Immodium

Loperamide is a safe, effective opioid-derived antidiarrheal agent, and it is not listed under the Controlled Substance Act. This medication is now available as a nonprescription item in the United States. It is used extensively for traveler's diarrhea. Loperamide is marketed as capsules (2 mg) and liquid preparations (1 mg/5ml). The recommended dosage is 4 mg initially and an additional 2 mg following each diarrheal stool. The dosage should not exceed 16 mg/day. The reason for the low abuse potential of loperamide has not been determined. The compound is highly lipophilic and undergoes slow dissolution, thus limiting the bioavailability of the agent to about 40% of the dose. The combination of a slow absorption rate, poor overall bioavailability, and first-pass metabolism may explain the low abuse potential after an oral dose. The compound is too lipophilic to dissolve for an intravenous dosage form.

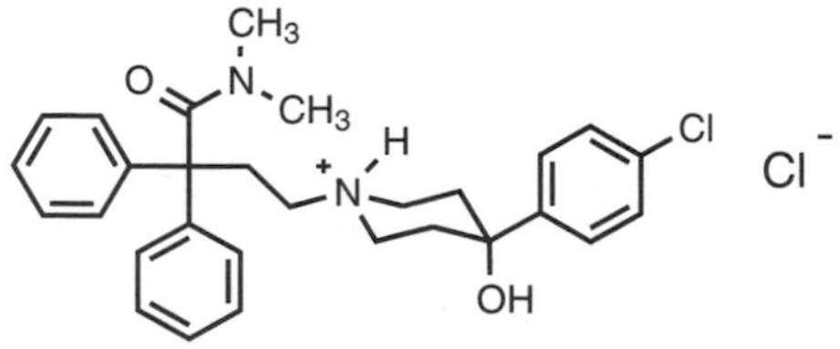

Opioid Agents Used as Cough Suppressants (Antitussives)

Many of the rigid-structured opioids have cough-suppressant activity. This action is not a true opioid effect in that it is not antagonized by opioid antagonists, and the (+)-isomers are equally effective with the analgesic (−)-isomers as cough suppressants. The 3-methoxy derivatives of morphine (codeine and hydrocodone) are nearly as effective antitussive agents as free phenolic agents. The better oral activity and decreased abuse potential of the methoxy derivatives make them preferred as antitussive agents. Incorporation of the 14β-hydroxyl into the structure (oxycodone) greatly decreases antitussive activity. If no cough suppression is desired in a patient being treated for pain, meperidine is the preferred agent.

Codeine is used extensively as a cough suppressant. It is available as a single agent or as mixtures in a variety of tablet and liquid cough-suppressant formulations. As a single agent, codeine is Schedule II, and in mixtures it is Schedule V under the Controlled Substance Act. When used properly as a cough suppressant, codeine has little abuse potential; however, cough formulas of codeine are often abused.

Hydrocodeine bitartrate is about three times more effective on a weight (mg/mg) basis compared with codeine as an oral antitussive medication. Hydrocodone also has greater analgesic activity and abuse potential then codeine. Hydrocodone is available only as a Schedule III prescription agent in combination formulations for cough suppression.

Dextromethorphan HBr is the (+)-isomer of the 3-methoxy form of the synthetic opioid levorphanol. It lacks the analgesic, respiratory depressant, and abuse potential of mu opioid agonists but retains the centrally acting antitussive action. Dextromethorphan is not an opioid and is not listed in the Controlled Substance Act. Its effectiveness as an antitussive is less than that of codeine. Dextromethorphan is available in a number of nonprescription cough formulations.

REFERENCES

1. J. Hughes, et al., *Nature (Lond.)*, *258*, 577(1975).
2. L. Terenius, *Acta. Pharmacol. Toxicol.*, *32*, 317(1973).
3. C. B. Pert and S. H. Snyder, *Proc. Natl. Acad. Sci. U.S.A.*, *70*, 2243(1973).
4. E. J. Simon, J. M. Hiller, and I. Edelman, *Proc. Natl. Acad. Sci. U.S.A.*, *70*, 1947(1973).
5. A. Goldstein, L. I. Lowney, and B. K. Pal, *Proc. Natl. Acad. Sci. U.S.A.*, *68*, 1742(1971).
6. H. H. Loh, et al., *Proc. Natl. Acad. Sci. U.S.A.*, *73*, 2895(1976).
7. A. Goldstein, et al., *Proc. Natl. Acad. Sci. U.S.A.*, *76*, 6666(1979).
8. A. Goldstein, et al., *Proc. Natl. Acad. Sci. U.S.A.*, *82*, 5203(1985).
9. H. Akil, et al., *Annu. Rev. Neurosci.*, *7*, 233(1984).
10. V. Brantl and H. Teshemacher, *Arch. Pharmacol.*, *306*, 301(1979).
11. P. C. Montecucchi, et al., *Int. J. Pept. Protein Res.*, *17*, 316(1981).
12. P. E. Gilbert and W. R. Martin, *J. Pharmacol. Exp. Ther.*, *198*, 66(1976).
13. J. A. H. Lord, et al., *Nature (Lond.)*, *267*, 495(1977).
14. F. M. Leslie, *Pharmacol. Rev.*, *39*, 197(1987).
15. W. F. Simonds et al., *Proc. Natl. Acad. Sci. U.S.A.*, *82*, 4974(1985).
16. M. Ui, *Trends Pharmacol. Sci.*, *5*, 277(1984).
17. G. DiChiara and A. North, *Trends Pharmacol. Sci.*, *13*, 185(1992).
18. H. Schmidhammer, et al., *J. Med. Chem.*, *32*, 418(1989).
19. D. Paul and G. W. Pasternak, *Eur. J. Pharmacol.*, *149*, 403(1988).
20. J. Szmuszkovick and P. F. Von Voitlander, *J. Med. Chem.*, *25*, 1125(1982).
21. C. R. Clark, et al., *J. Med. Chem.*, *31*, 831(1988).
22. J. C. Hunter, et al., *Br. J. Pharmacol.*, *101*, 183(1990).
23. R. B. Rothman, et al., *Peptides*, *11*, 311(1990).
24. P. S. Portoghese, A. W. Lipowski, and A. E. Takemori, *Life Sci.*, *40*, 1287(1987).
25. I. F. James and A. Goldstein, *J. Mol. Pharmacol.*, *25*, 337(1984).
26. G. Gacel, et al., *J. Med. Chem.*, *24*, 1119(1981).
27. H. I. Mosberg, et al., *Proc. Natl. Acad. Sci. U.S.A.*, *80*, 5871(1983).
28. P. S. Portoghese, et al., *J. Med. Chem.*, *35*, 4325(1992).
29. Q. Jiang, et al., *J. Pharmacol. Exp. Ther.*, *255*, 636(1990).
30. P. S. Portoghese, M. Sultana, and A. E. Takemori, *J. Med. Chem.*, *33*, 1714(1990).
31. P. S. Portoghese, et al., *J. Med. Chem.*, *22*, 168(1979).
32. P. S. Portoghese, et al., *J. Med. Chem.*, *23*, 233(1980).
33. K. C. Rice, et al., *Science*, *220*, 314(1983).
34. T. R. Burke, et al., *J. Med. Chem.*, *29*, 1087(1986).
35. B. R. deCosta, et al., *FEBS Lett.*, *249*, 178(1989).
36. I. Iigima, et al., *J. Org. Chem.*, *43*, 1462(1978).
37. A. F. Casy and R. T. Parfitt, *Opioid Analgesics: Chemistry and Receptors*, New York, Plenum Press, 1986.
38. D. C. Rees and J. C. Hunter, in *Comprehensive Medicinal Chemistry: The Rational Design, Mechanistic Study and Therapeutic Application of Chemical Compounds*, Vol. 3, *Membranes and Receptors*. J. C. Emment, Ed., Oxford, Pergamon Press, 1990, pp. 805–846.
39. J. W. Lewis, K. W. Bently, and A. Cowan, *Annu. Rev. Pharmacol.*, *11*, 241(1971).
40. R. Grewe and A. Mandon, *Chem. Ber.*, *81*, 279(1948).
41. O. Eisleb and O. Schaumann, *Dtsch. Med. Wschr.*, *65*, 967(1939).
42. P. W. Schiller, in *Progress in Medicinal Chemistry*, Vol. 28, G. P. Ellis and G. B. West, Eds., Amsterdam, Elsevier, 1988, pp. 301–340.
43. A. H. Beckett and A. F. Casy, *J. Pharm. Pharmacol.*, *6*, 986(1954).
44. P. S. Portoghese, *J. Med. Chem.*, *8*, 609(1965).
45. P. S. Portoghese, B. D. Alreja, and D. L. Larson, *J. Med. Chem.*, *24*, 782(1981).
46. A. P. Feinberg, I. Creese, and S. H. Snyder, *Proc. Natl. Acad. Sci. U.S.A.*, *73*, 4215(1976).
47. W. R. Martin, *Pharmacol. Rev.*, *35*, 283(1984).
48. C. J. Evens, et al., *Science*, *258*, 1952(1992).
49. G-X. Xie, A. Miyajima, and A. Goldstein, *Proc. Natl. Acad. Sci. U.S.A.*, *89*, 4124(1992).
50. J. B. Wang, et al., *Proc. Natl. Acad. Sci. U.S.A.*, *90*, 10230(1993).

SUGGESTED READINGS

F. E. Bloom, *Annu. Rev. Pharmacol. Toxicol.*, *23*, 151(1983).

H. O. J. Collier, J. Hughes, M. J. Rance, M. B. Tyers, Eds., *Opioids: Past, Present and Future*, London, Tayler & Frances Ltd., 1984.

M. C. Fournie-Zaluski, G. Gacel, B. Maigret, S. Premilat, and B. P. Roques, *Mol. Pharmacol.*, *20*, 484(1981).

D. S. Fries, in *CNS Drug-Receptor Interactions*, Vol. 1, J. G. Cannon, Ed., Greenwich, CT, JAI Press, 1991, pp. 1–21.

L. Haynes, *Trends Pharmacol. Sci.*, *9*, 309(1988).

V. Höllt, *Annu. Rev. Pharmacol. Toxicol.*, *26*, 59(1986).

V. J. Hruby and C. A. Gehrig, *Med. Res. Rev.*, *9*, 343(1989).

G. F. Koob and F. E. Bloom, *Science*, *242*, 715(1988).

G. R. Lenz, S. M. Evans, D. E. Walters and A. J. Hopfinger, *Opiates*. Orlando, Academic Press, 1986.

E. L. May, *J. Med. Chem.*, *35*, 3587(1992).

P. S. Portoghese, *J. Med. Chem.*, *35*, 1929(1992).

N. E. S. Sibinga and A. Goldstein, Opioid peptides and opioid receptors in cells of the immune system. *Annu. Rev. Immun.*, *6*, 219(1988).

E. J. Simon, *Med. Res. Rev.*, *11*, 357(1991).

W. F. Simonds, *Endocr. Rev.*, *9*, 200(1988).

D. M. Zimmerman and J. D. Leander, *J. Med. Chem.*, *33*, 895(1990).

Chapter 15

CENTRAL NERVOUS SYSTEM STIMULANTS

Karl A. Nieforth and Gerald Gianutsos

Societal living dictates rigid norms to which all members must conform to retain approval. Deviants from this set of standards may be sociologically classified as mentally abnormal or ill. Physiologically, such deviation and the loss of individuals' central control and integration system have not yet been correlated with any specific cellular or biochemical aberration.

Although society may define mental illness in terms of nonconformity, it is extremely difficult to diagnose the disease. Patients' complaints may resemble those of organic diseases involving many physiologic or biochemical systems. Once the diagnosis is made, few options are available—exorcism from the group, restricted confinement, or conformity assisted by medical therapy and drugs. Based on data from state mental hospitals, psychopharmacology has reduced hospitalization more successfully than any other form of therapy. Despite this, drug therapy often does not reflect the underlying cause of the illness as either a central biochemical alteration or an environmental incompatibility. Physicians have been accused of using psychoactive drugs to retain a sense of mastery in the undefined situation of mental illness. Only when mental alteration is chemically induced, as in the depressant effect of alcohol intoxication, barbiturate overdose, or glue inhalation, do we see a common thread of therapy among clinicians.

Mental malfunctioning encompasses a broad scope of anomalies ranging from overt psychoses to changes in mood or perception. In many cases, these phenomena can be categorized as either stimulation or depression. Compounds that increase an initial low level of physiologic activity are generally classified as stimulants. Central stimulation appears to be closely related on a molecular level to central depression because central stimulation can revert to depression by variations in species, drug dose, and minor chemical modifications of a drug. Stimulants discussed herein are grouped according to use as respiratory stimulants for overdosage of depressants, as mood elevators in functional or psychogenic depression, and finally as psychotomimetics or hallucinogens. Other terms encountered in psychiatry include psychostimulants and nootropics, which share central escalatory and disinhibitory effects with analeptics. Psychomotor stimulants produce a generalized activation often followed by sedation. Nootropics, in the presence of impaired central function, produce an activation of reduced adaptation functions. In the case of analeptics, the stimulation is generally focused on the respiratory and circulatory systems, and high doses are associated with convulsions. Within the group, only psychostimulants have shown a potential for abuse.[1]

ANALEPTICS

More than 100 years have passed since reflex lung expansion was demonstrated to inhibit inspiration and promote expiration as long as vagal integrity is maintained. Both inspiratory and expiratory centers are located in the reticular formation of the medulla and together are responsible for spontaneous respiration.

Respiratory activity is subject to a variety of chemical and nonchemical controls. Chemoreceptors in the carotid and aortic bodies are stimulated by increases in pCO_2, decreases in pH, and decreases in pO_2. Medullary chemoreceptors are also sensitive to increases in pCO_2 and stimulate hyperventilation after denervation of other chemoreceptors. Similar receptors in pulmonary and myocardial vessels initiate short-term respiratory changes on stimulation by drugs such as nicotine. Nonchemical controls, such as joint movement, temperature, blood flow, irritation of respiratory passages, gagging, vomiting, swallowing, and Hering-Breuer reflexes, complete the factors that alter the rate and depth of respiration.

In addition to this physiologic control, certain drugs stimulate or depress normal and abnormal rates of respiration. Such drugs may be used to counteract respiratory depression resulting from overdosage of general anesthetics, narcotic analgesics, or sedative hypnotics.

The relative lack of respiratory center specificity and unacceptable toxicity of most respiratory stimulants have discouraged therapeutic use. Complete discussions of these agents are found in earlier editions of this text.

Strychnine

Strychnine was probably introduced by Arabian physicians as a treatment for sores, boils, ulcers, and abscesses (*nux vomica* = ulcer nut).[2] It was used in Europe from 1500 to 1800 to poison crows, dogs, and vermin. The strychnine panic of 1852 was caused by a false announcement that it was being used as a bitter principle in certain kinds of English beers. Some of strychnine's actions have been duplicated by drugs such as morphine, thebaine, codeine, and brucine (a less potent alkaloid from Strychnos nux vomica) as well as the following synthetic compounds. The striking similarity to meperidine of the last compound in the group reemphasizes the interrelationship between strychnine and morphine.

Analeptic effects of agents such as fentanyl[3] and codeine[4] are blocked by atropine but not methyl atropine, suggesting a central cholinergic mechanism. In a related fashion, the analeptic effects of SKF38393,[5] a selective D_1 agonist, and cocaine[6] are also blocked by centrally acting anticholinergics. Inhibition of the analeptic effect by selective D_1 antagonism indicates a dopaminergic/cholinergic modulation.

Strychnine, similar to most convulsants, inhibits the activity of inhibitory neurotransmitters. In contrast to many other drugs that antagonize γ-aminobutyric acid (GABA), however, strychnine inhibits the binding of glycine. The strychnine-sensitive glycine binding site, however, is distinct from the glycine site coupled to excitatory N-methyl-D-aspartate (NMDA) receptors.[7]

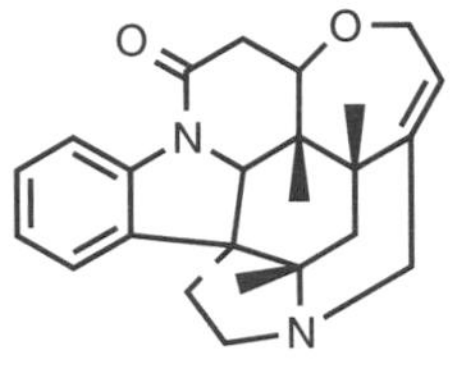

Strychnine

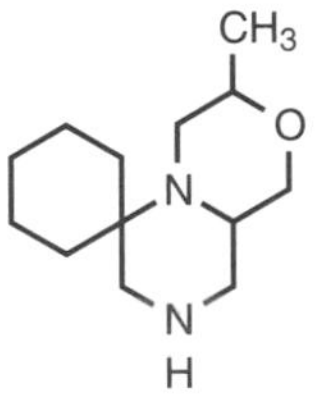

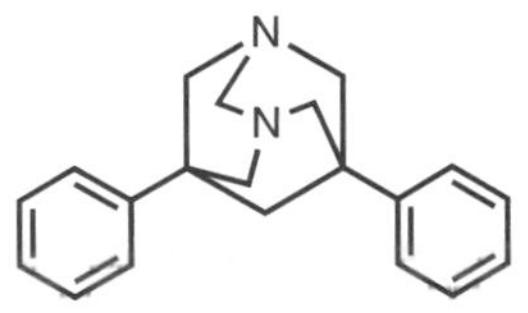

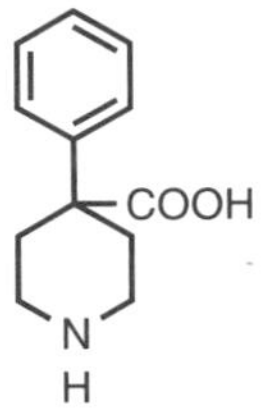

Strychnine differs from other analeptics in that it produces a tonic type of extensor convulsion. It apparently augments sensory input so that various types of nonconvulsive stimuli evoke convulsions. Strychnine is the least potent analeptic, and small doses do not stimulate respiration. Adequate absorption occurs after oral ingestion, and activity is initiated within 45 minutes. Presumably, excretion of strychnine and its metabolites is through the kidneys. Because of its weak analeptic activity, it is rarely used for any purpose.

The fatal dose for humans is approximately 60 to 90 mg, but fatalities have been reported with doses as small as 15 mg. Symptoms, which may be delayed as long as 2 hours, first appear as stiffening of face and neck muscles, hyperexcitability of reflexes, and sensory-induced tonic seizures. Later stages include full tetanic convulsions, with marked arching of the back. All muscles, including those of respiration, are in full painful contraction. This produces apnea, cyanosis, and eventually death.

Picrotoxin

Picrotoxin, a bitter, nonnitrogenous principle obtained from the dried berries of Anamirta cocculus, consists of two materials, picrotoxinin and picrotin. Natives of India and Malaya used to throw the crushed berries into the water to paralyze fish.

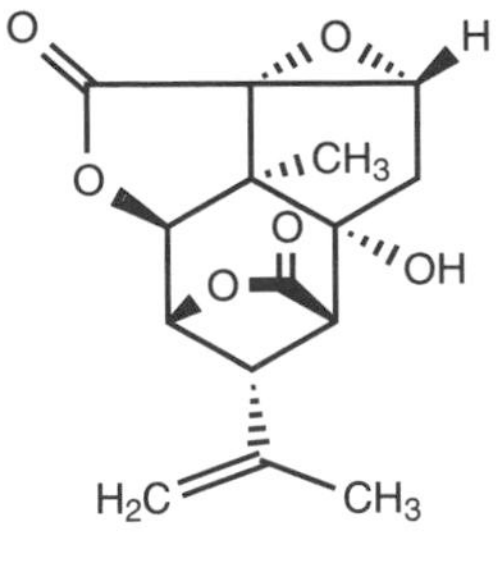

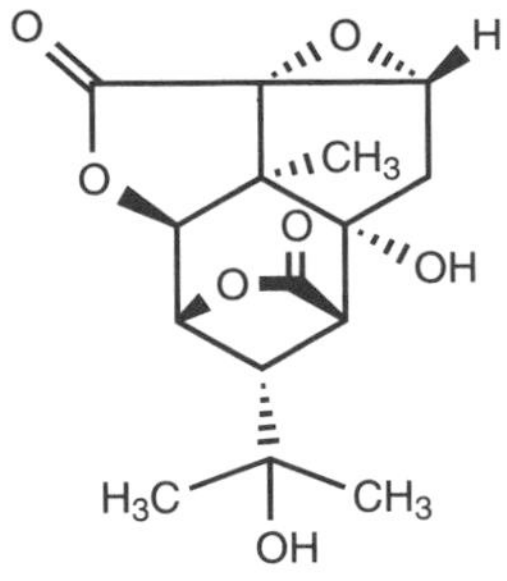

Picrotoxinin

Picrotin

Although its action is duplicated by pentylenetetrazole, few synthetic analogs of picrotoxinin have been prepared. Its activity seems to depend on the presence of an unsubstituted bridgehead hydroxyl group and a lactone ring connecting carbons 3 and 5 of picrotoxane.[8]

Picrotoxinin appears to be the most active analeptic agent, although its action as a respiratory stimulant occurs most strikingly with depressed respiration and with doses that cause convulsions. Similar doses also produce sympathetic and parasympathetic autonomic effects, such as sweating, mydriasis, salivation, increased or decreased urination, vomiting, and increased or decreased gastrointestinal motility. It is well absorbed when administered orally and is excreted with unknown metabolites in the urine.

GABA is an inhibitory amino acid neurotransmitter within the brain and spinal cord. The effects of GABA are mediated by a receptor complex that contains binding sites for GABA and the benzodiazepines and that is associated with a chloride channel that can modulate the binding at these other binding sites. Bicuculline can antagonize the effects of GABA by competitively inhibiting its binding to sites in the receptor complex. In contrast, picrotoxin binds to the receptor complex at or near the chloride channel and can block the effects of GABA and the benzodiazepines in a fashion that is not competitive. A more complete and extensive understanding of the interactions within this receptor complex may permit a greater understanding of the mechanism by which both picrotoxin and bicuculline are central stimulants.

The remaining analeptics may be chemically grouped into derivatives of bemegride, pentylenetetrazole, nikethamide, and caffeine. The questionable therapeutic value of these compounds perhaps explains why so few have reached clinical status. These drugs act by agonistic stimulation rather than by antagonism of depression. In all groups, however, there is a fine line between stimulation and depression, which may be crossed by slight molecular modification.

Other Analeptics

Nikethamide, which may be considered a model for the substituted amide type of analeptics, is weaker than compounds used as barbiturate antagonists. Diethylamide substitution of an aromatic or heterocyclic acid appears to be the only prerequisite for activity (Table 15–1).[9] The *N*-unsubstituted derivatives are weak or may have anesthetic activity. A few heterocyclic derivatives have a depressant component and are synergistic with barbiturates, which limits their usefulness as antidotes. Because of this component, nikethamide has a greater therapeutic index than do other analeptics. Doxapram has been used to treat idiopathic apnea spells in infants who show resistance to methylxanthines.[10]

Bemegride (4-ethyl-4-methyl-2,6-piperidinedione) was synthesized in 1901 and reported as a barbiturate antagonist in 1954.[11] It is not a specific antagonist, although structural similarity to glutethimide suggests such specificity. Changes in substitution at the third carbon result in transformation from central nervous system (CNS) stimulation to depression. Computer graphic–based pattern recognition analysis of similar series of compounds has revealed conformational regions likely to be responsible for convulsant and anticonvulsant activity.[12]

Tetrazole has no marked pharmacologic value, but when it is suitably substituted, powerful stimulants or moderate depressants are produced. Pentylenetetrazole serves as a model of tetrazole analeptics.

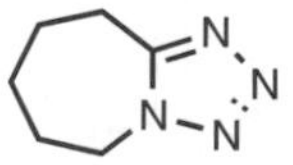

Pentylenetetrazole

It has been structurally modified to improve its analeptic effectiveness by increasing separation of respiratory stimulatory and convulsive doses.[13] Most substitutions increase potency with a concomitant decrease in effectiveness. Several derivatives that do not possess a pentylene ring are depressants but of no clinical significance.[14]

This agent was used in the past for certain depressive states to induce an electroshock-like seizure, but it has been largely replaced by other antidepressant agents. It has also been used as an electroencephalographic activator to induce mild convulsions in diagnosis of possible epileptic patients.

Table 15–2 lists proposed mechanisms for CNS stimulation.

Table 15–1. Substituted Amides

Structure	Name	Comments
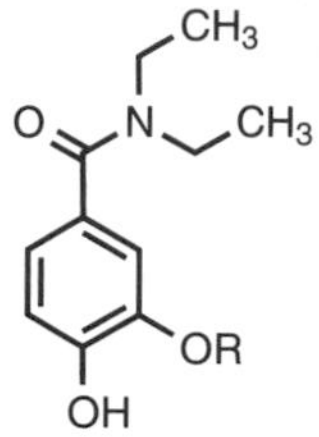	Ethamivan (Vandid, Emivan): R = CH_3 Anacardiol: R = C_2H_5	
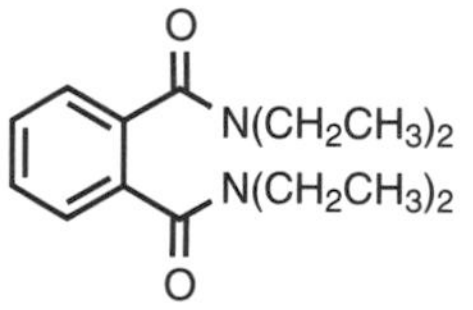	N,N,N′,N′-Tetraethylphthalamide (Neospiran)	
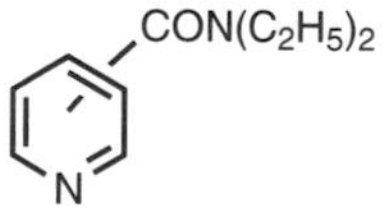	3-Isomer: Nikethamide (Coramine)	Most potent of three isomers
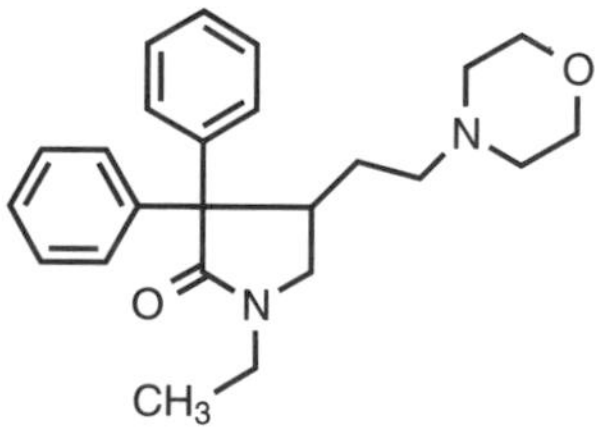	Doxapram (Dopram)	
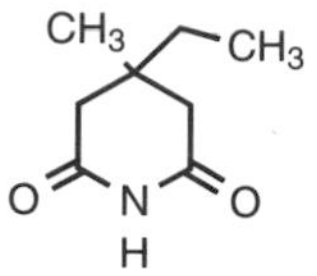	Bemegride	

Table 15–2. Proposed Mechanisms for Central Nervous System Stimulation

Possible Mechanism	Agent(s) Implicated
Blockade of postsynaptic inhibition	Strychnine
Blockade of presynaptic inhibition	Picrotoxin
Antagonism of GABA	Picrotoxin, bemegride
Cholinergic augmentation	Strychnine, pentylenetetrazole, nikethamide
Decreased energy levels	Pentylenetetrazole
Prostaglandin release	Strychnine, picrotoxin, pentylenetetrazole
Low surface activity	Picrotoxin, bemegride, pentylenetetrazole

XANTHINES

As a group, the xanthine alkaloids, which differ primarily in potency (Table 15–3), are included in many common beverages and patent medicines (Table 15–4). The mild stimulatory effects produced by these drugs account for their widespread popularity. Perhaps this group of drugs more than any other provides an example of society's ability to cultivate and control drug use with some degree of moderation, making legal restraints unnecessary. The xanthines share many of the pharmacologic properties summarized in Table 15–5.

Caffeine is the most potent xanthine, producing cortical and medullary stimulation and even spinal stimulation in large doses. Theobromine is the least potent CNS stimulant and may even be inactive in humans.[16]

Controversy exists on the extent of cortical stimulation, although caffeine can stimulate mental alertness and overcome fatigue. It also prolongs the length of time an individual is able to do physically exhausting work.[17] One proposed mechanism is the production of euphoria, which delays the development of a negative attitude toward exhausting work. Along with changing attitude, caffeine may actually improve performance beyond the improvement caused by release from fatigue and lack of motivation by direct stimulation of skeletal muscle. Improvement in intellectual capabilities has been suggested, although this is most likely best explained by caffeine's action on fatigue or boredom because no evidence indicates that caffeine enhances intelligence or learning.

Medullary stimulation owing to large doses of these agents increases the rate and depth of respiration but is less significant than that produced by other analeptics. Xanthines also stimulate the medullary vasomotor center, increasing peripheral resistance and blood pressure. Medullary vagal stimulation is responsible for the decreased pulse rate sometimes observed. After large amounts of caffeine have been ingested, spinal cord stimulation may occur, along with reflex and motor excitability.

Cardiac effects of xanthines include direct stimulation of myocardial tissue as well as indirect activation of the

Table 15–3. Relative Pharmacologic Properties of Xanthines

Structure	Central Nervous System, Respiratory, and Skeletal Muscle Stimulation	Cardiac Stimulation, Coronary Dilatation, Smooth Muscle Relaxation, Diuresis, and Goitrogenic Potential
Caffeine	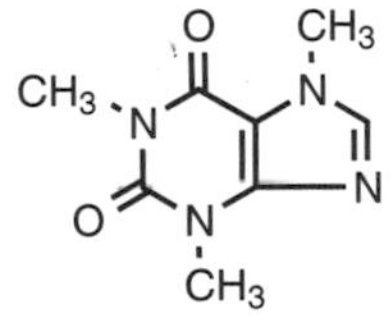 1	3
Theophylline	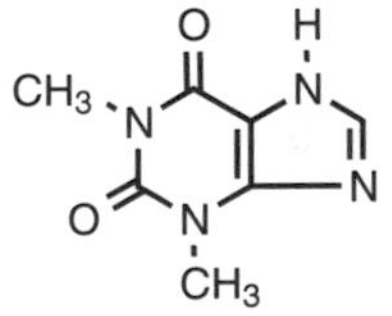 2	1
Theobromine	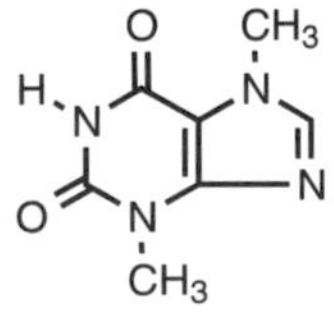 3	2

Data from A. Goodman and L. Gilman, *The Pharmacological Basis of Therapeutics,* 4th ed., New York, Macmillan, 1971, p. 359.

Table 15–4. Approximate Caffeine Content

Beverage	Caffeine Content
Brewed coffee	100–150 mg/cup
Instant coffee	50–90 mg/cup
Decaffeinated coffee	2–10 mg/cup
Tea	50–150 mg/cup
Cola drinks	35–55 mg/12 oz
Cocoa	230–280 mg theobromine/cup
OTC* analgesics	30–60 mg/dose
OTC stimulants	100 mg/dose

* OTC, Over the counter.

sympathoadrenal system, initiating increases in rate and force of contraction; large doses cause arrhythmias. Theophylline appears to cause the greatest amount of cardiac stimulation and caffeine the least. Cardiac effects are not completely predictable and may be complicated by bradycardia as a result of medullary vagal stimulation. Although therapeutic doses decrease peripheral resistance by direct vaosdilation, net effects on blood pressure are variable owing to multiple effects on the circulatory system. Coronary arteries are also dilated by xanthines, which explains their use in anginal pains. Xanthines produce vasoconstriction of cerebral arteries and have been used in combination with ergot to relieve migraine headaches. Their use in bronchial asthma is mediated by relaxation of bronchial smooth muscle and respiratory stimulation. Relaxation of biliary tone and gastrointestinal movement has also been reported.

All xanthines, especially theophylline, produce diuresis by increasing glomerular filtration and blocking tubular reabsorption of sodium ions. These drugs also stimulate acid and pepsin secretion in the stomach, which should limit their use in peptic ulcer patients.

Xanthines are well absorbed after oral administration and are distributed throughout the body in proportion to tissue water content. Protein binding occurs with caffeine and theophylline and, to a lesser extent, with theobromine. The in vitro half-life of caffeine is 3½ hours, whereas that of theophylline is about 5 hours.[20] Urinary products of the xanthines in humans are summarized in Table 15–6. Because caffeine is not appreciably metabolized by xanthine oxidase, uric acid levels are not increased, and thus the potential for gout is unlikely. Caffeine can alter the liver microsomal enzyme system, suggesting the possibility for interactions with other drugs metabolized in the liver.

Table 15–5. Pharmacologic Effects of Xanthines

Tissues	Action
Central nervous system	
Cortex	Euphoria, antifatigue, mild stimulation
Medulla	Increased respiration, vasomotor stimulation, vagal stimulation
Cardiac	Increased heart rate, increased stroke volume
Smooth muscles	
Blood vessels	Vasodilation, peripheral and coronary Vasoconstriction, cerebral
Bronchioles	Dilation
Skeletal muscle	Stimulation
Kidneys	Diuretic
Gastrointestinal tract	Increased gastric secretion and metabolic rate

Caffeine is relatively nontoxic, and theobromine is the least toxic of all three xanthines. The estimated fatal dose of caffeine in humans is approximately 10 g. Although fatal overdoses are rare, unpleasant symptoms may occur with large doses (250 mg or greater according to Drill's *Pharmacology in Medicine*).[21] Central effects resemble anxiety states[22] and include symptoms of insomnia, irritability, tremulousness, nervousness, hyperexcitability, hyperthermia, and headache. Toxic sensory disturbances of hyperesthesia, ear ringing, and visual flashes have occurred.[22] Cardiac irregularities and arrhythmias may occur as well as marked hypotension as a result of direct vasodilation. A relationship between excessive coffee drinking and increased incidence of heart attacks has been suggested but is still controversial.[27] Caffeine elevates low-density lipoprotein cholesterol[24] and increases the frequency of ventricular ectopic beats.[25]

This potential link of coffee to heart attacks, however, may result from substances in coffee other than caffeine because coffee may contain cholinomimetic vasoconstrictor substances in addition to caffeine.[26] High caffeine intake can cause increased respiration, diuresis, and gastric upset. Toxic effects may occur with theophylline and its congener, aminophylline (theophylline ethylenediamine). Predisposing cardiovascular and renal problems enhance toxicity. These agents should be injected slowly when given intravenously.

With regular use, complete tolerance develops to the effects of caffeine within a few days.[27] Low-grade tolerance develops to the diuretic and salivary-stimulating actions of xanthines, and cross-tolerance to the diuretic effect has been demonstrated for caffeine, theophylline, and theobromine. Nonusers of caffeine show greater caffeine-induced sleep disturbances than users, suggesting tolerance to certain of the central effects. Dysphoric withdrawal symptoms of irritability, nervousness, restlessness, and headache have been described.[22] Caffeine-withdrawal headaches (about 12 hours after last intake) are a frequent complaint, suggesting that certain headaches (especially those occurring on weekends or during abstinence from coffee) may best be treated with aspirin products containing low doses of caffeine.

Because of the structural similarity of the xanthines to purine nucleotides, much research has been directed at determining the mutagenic effects of caffeine. Caffeine may be mutagenic in microorganisms, fungi, plants, fruit flies, mice, and in in vitro human cells[28] and may enhance the effects of mutagenic chemicals or radiation. Caffeine ingestion, however, does not represent a significant muta-

Table 15–6. Xanthine Metabolites*

Metabolite	Theobromine	Theophylline	Caffeine
1-Methyluric acid	—	x	x
7-Methyluric acid	x	—	x
1,3-Dimethyluric acid	—	x	x
1-Methylxanthine	—	—	x
3-Methylxanthine	x	x	—
7-Methylxanthine	x	—	x
1,7-Dimethylxanthine	—	—	x
Theophylline	—	x	—
Theobromine	x	—	—
Caffeine	—	—	x

* From Cornish et al.[18] and Sved et al.[19]

genic hazard to humans based on current information.[29] Although animal studies have demonstrated some teratogenic potential of caffeine, there is no report linking caffeine to teratogenic effects in humans, suggesting that consumption of caffeine by humans probably does not result in any teratogenic danger.

It is naive to attempt to explain the pharmacologic actions of the xanthines based on one biochemical mechanism. A critical biochemical role has been established for the cyclic nucleotide adenosine 3′,5′-monophosphate (cAMP) in many metabolic and hormonal actions. Cyclic AMP is formed from ATP by the membrane-bound enzyme adenylyl cyclase and is inactivated by phosphodiesterase. Increases in cAMP levels in various tissues are responsible for many metabolic and cellular actions of the cell, most notably glycogenolysis and lipolysis. Xanthines, primarily theophylline, competitively inhibit phosphodiesterase and thereby increase cAMP levels in the cells. Because of the multitude of actions ascribed to cAMP and possibly cyclic GMP, this is an appealing method of explaining the diverse actions of xanthines. Relatively large concentrations of theophylline (0.2 mM), however, are required to elevate cAMP in vitro.[30]

Other cellular actions of the xanthines have been demonstrated, however, and should not be disregarded. Adenosine, present in most cells, exerts pronounced cardiac, vascular, metabolic, and gastrointestinal effect. Methylxanthines are competitive inhibitors at certain receptors stimulated by adenosine at therapeutically active concentrations.[31] This effect of the methylxanthines may contribute to the adverse effects seen after methylxanthine administration, particularly in systems in which adenosine (A1 receptors) inhibits adenylyl cyclase. An increased functional sensitivity to adenosine may underlie caffeine tolerance.[32]

Central release of norepinephrine by caffeine and theophylline has also been demonstrated[33] and must be considered a potential mechanism for stimulatory actions of these agents. Again, no direct evidence linking the release of norepinephrine by these agents to central stimulation is available.

Calcium involvement has also been implicated as a possible mechanism for stimulation of skeletal muscle. A time-dependent stimulation of calcium release from skeletal muscle sarcoplasmic reticulum has been demonstrated for caffeine. Inhibition of calcium uptake into the sarcoplasmic reticulum fragments has also been observed. It has been suggested that these two actions of caffeine on calcium ions may be responsible for the increased force of skeletal muscle contraction because calcium is involved in contractile processes.

The most common therapeutic indication for xanthine derivatives is in the treatment of asthma because of their bronchodilating properties. Theophylline and aminophylline (see Chapter 40) are used predominantly as bronchodilators. The bronchodilatory actions of the methylxanthines are beneficial in premature infants suffering from episodes of prolonged apnea. The xanthines appear to be useful in preventing apneic episodes that might otherwise be lethal in these infants.[34] The diuretic (see Chapter 21) and cardiac stimulatory actions have not achieved wide therapeutic usefulness. Caffeine has been examined for its central stimulatory efficacy in relieving the symptoms of minimal brain dysfunction in children. Results have been disappointing compared with those obtained with methylphenidate and amphetamine treatment.

Pentoxifylline, a methyl xanthine derivative, enhances deformability of red blood cells, thereby permitting more effective blood flow through occluded pathways in cases of peripheral arterial insufficiency. Pentoxifylline also modifies the action of interleuken-1 and tumor necrosis factor on neutrophil chemotaxis.[35]

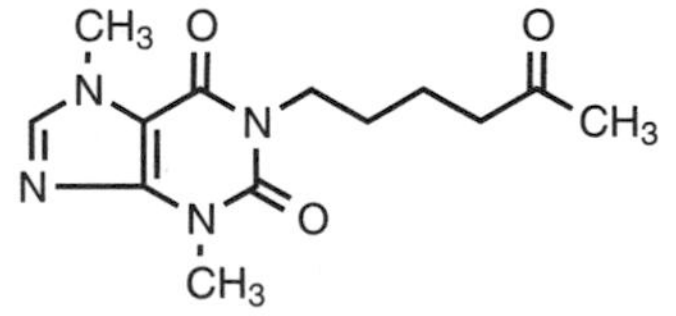

Pentoxifylline (Trental)

PHENETHYLAMINE ANALEPTICS

Barger and Dale's classic studies of structure-activity relationships with adrenergic compounds revealed the most potent adrenergic activity in derivatives of phenethyl-

Table 15–7. Phenethylamine Derivatives

Compound	Structure	Chemical Name
Methylphenidate (Ritalin)	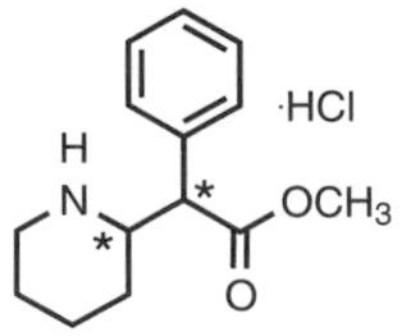	Methyl α-phenyl-α-2-piperidylacetate hydrochloride
Pipradrol (Meratran)	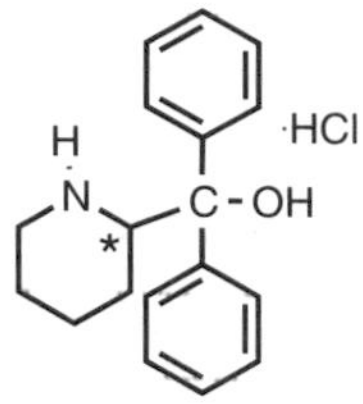	α,α-Diphenyl-2-piperidinemethanol hydrochloride
Phendimetrazine (Plegine)	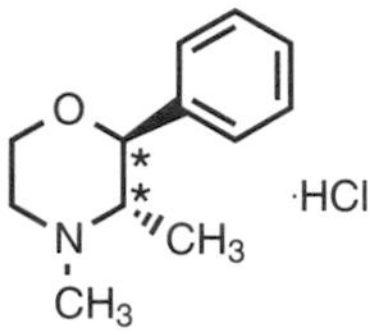	2S,3S-3,4-Dimethyl-2-phenylmorpholine hydrochloride
Phenmetrazine (Preludin)	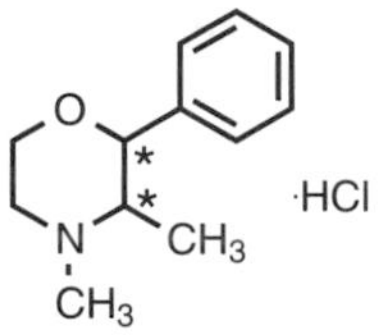	dl-3-Methyl-2-phenylmorpholine hydrochloride
Amphetamine	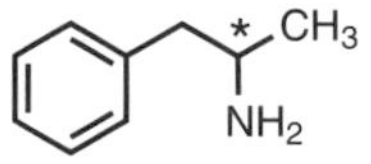	dl-α-Methylphenethylamine

amine.[36] Several of these compounds are clinically used as sympathomimetics (see Chapter 18).

Derivatives of 2-benzylpiperidine and 2-phenyl-morpholine have been developed as substitutes for the more potent amphetamines because of the toxicity and abuse potential of amphetamine (Table 15–7). Phenethylamine derivatives are used in the management of obesity, narcolepsy, and attention deficit/hyperactivity disorder and with doubtful efficacy in cases of mild depression.

Stereochemical investigations suggest different areas or mechanisms of activity.[37] Within a series of biologically active compounds having centers of asymmetry, the more active enantiomers are usually superimposable. The active isomers of pipradrol (R) and methylphenidate (2R:2R′) are not superimposable on the active amphetamine isomer (S). Amphetamine antipodal potency ratios (ratio of activities of enantiomers), however, vary considerably based on the test receptor system.

Any decrease in distance between the aromatic ring and the heterocyclic nitrogen decreases activity in this series as anticipated because of the relationship with other sympathomimetics. Activity in the phenidate series is maximized at the methyl ester and decreases with any modification not related to the hydrolysis rates. In the morphine series, substitutions on the aromatic ring as well as aromatic ring replacement by heterocyclic groups substantially decrease potency. The piperidine ring may be replaced by other heterocyclic groups, such as 3-tetrahydroisoquinolyl, 3-thiomorpholino, and 2-pyrrolidinyl functions, with only a slight loss of activity.

Phenethylamine derivatives are mild stimulants that are thought to act similar to amphetamines, although peripheral sympathetic side effects may be less obvious. For example, blood pressure, heart rate, and respiration are not appreciably increased after methylphenidate. Increased alertness, euphoria, mood elevation, and enhanced performance occur, however. Because of these actions, agents such as methylphenidate have been used as antidepressants, although the efficacy of this treatment is doubtful. In fact, symptoms of anxiety, irritability,[38–40] anorexia, and insomnia in some depressed patients can be aggravated by amphetamine and methylphenidate.

The metabolic fate of phenethylamine derivatives has been incompletely studied. Methylphenidate has a rapid onset and reaches maximum activity in 4 to 6 hours with a calculated half-life in animals of 7 to 8 hours. Amphetamine plasma half-life in humans is about 30 hours. More is known about the metabolic degradation products of amphetamine in humans (Fig. 15–1). Certain of these metabolites may have activity that contributes to the effects seen after amphetamine. Para-hydroxynorephendrine may serve to displace norepinephrine from

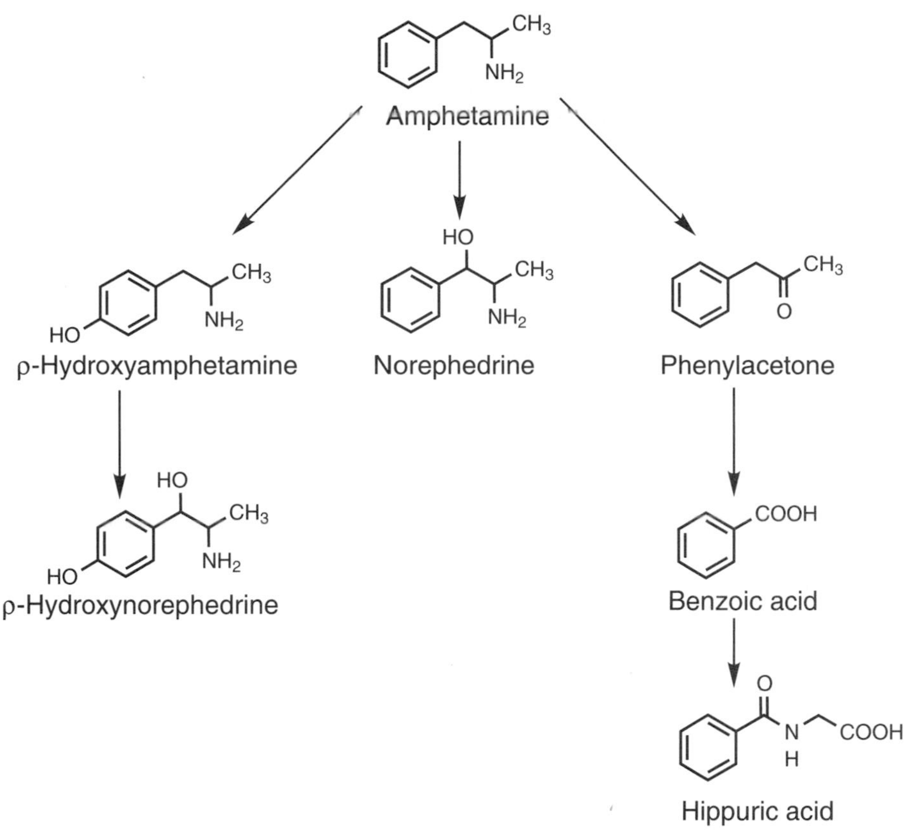

Fig. 15–1. Urinary metabolites of amphetamine in humans. (Data from J. Caldwell and P. Server, *Clin. Pharmacol. Ther., 16,* 625 [1974].)

neuronal storage sites and to function as a false neurotransmitter. Although only a small amount of amphetamine is converted to para-hydroxynorephedrine in humans, this mechanism may partially account for the tolerance observed after repeated amphetamine administration.[41] Beta-hydroxy amphetamines have also been implicated in the toxic psychosis seen after high doses of amphetamine. Amphetamine excretion is increased by acidifying urine because amphetamine as a base would be totally ionized under acid conditions, thus preventing tubular reabsorption. This property suggests that ammonium chloride acidification of urine may be useful in treating amphetamine toxicity. Both amphetamine and methylphenidate can interact with hepatic microsomal enzymes, suggesting numerous possible drug interactions.

Amphetamine, and to a lesser extent its congeners, exerts four major effects on nerves: (1) inhibition of reuptake mechanisms for several biogenic amines, (2) enhancement of neuronal release of catecholamines, (3) direct α-adrenergic receptor stimulation, and (4) inhibition of monoamine oxidase (MAO) in higher concentrations.[42] Because amphetamine is metabolized by dopamine β-hydroxylase, the same enzyme that converts dopamine to norepinephrine, it is not surprising that alterations in both these endogenous amines occur in the brain. There is controversy regarding the importance of each of these mechanisms in the biologic actions of amphetamine, although many of the actions of amphetamine are now thought to result from release or inhibition of reuptake of dopamine except for anorexia.

Side effects of the stimulants include nervousness, insomnia, irritability, anorexia, and the possibility of cardiac irregularities. Toxicity from overdosage or intravenous administration (in cases of abuse) is well documented. Attention should be drawn to the possibility of lethal cerebral hemorrhage and convulsions. Additionally, chronic amphetamine use has been associated with acute psychosis of a paranoid and delusional nature. Tolerance to the loss of appetite and euphoric effects is greater than that to insomnia. Physicial dependence on amphetamine is documented by the signs of depression, fatigue, and increased appetite observed primarily after withdrawal of chronic, high-dose amphetamine use.

Therapeutically, amphetamine has been used in narcolepsy, a problem characterized by uncontrolled episodes of sleep accompanied by cataplexy (sudden loss of muscle tone frequently caused by emotion) and hallucinations. Excessive rapid eye movement (REM) sleep is a prominent feature of narcolepsy. Stimulants have been used in narcolepsy, sometimes in combination with tricyclic antidepressant drugs.[43]

ANORECTICS

Phenethylamine derivatives and a few miscellaneous compounds are widely prescribed as anorectics to be used in combination with reduced food intake. Realization of such control has traditionally been approached through attempted modulation of noradrenergic, serotoninergic, or dopaminergic pathways. Several peptidergic pathways, using somatostatin, endorphin, insulin, glucagon, cholecystokinin, and neurotensin, have been implicated in the regulation of food intake. Other related efforts have involved inhibition of nutrient intestinal absorption, inhibition of lipid biosynthesis, enhancement of lipolysis, and increased gastric emptying. Because of complexities of central mechanisms controlling appetite and adverse effects related to central drug actions, major efforts are directed to the design of peripherally acting agents.

Pharmacotherapeutic intervention in obesity encourages sustained weight loss during the active drug and appetite control regimen. Termination of therapy is generally associated with gradual, partial, or complete regain of the lost weight. Behavioral therapy alone or in combination with anorectic agents does not significantly improve initial weight loss but, at least in one study, reduced the rate of weight regain.[44] Appetite control is an essential component of any diet regimen, for in its absence, anorectic agents effect few permanent benefits.

Appetite control, or the fine balance between caloric intake and energy expenditure, is a poorly understood, complex physiologic process. Obesity, now defined as a weight 20% above the weight considered ideal, presently affects a large and increasing percentage of the U.S. population. Vigorous therapeutic intervention has been discouraged by highly negative publicity of some historic weight loss regimens (e.g., rainbow diet, lactose diet) and the recognized side effects and abuse of the classic anorectic, amphetamine.

Each of the major neurotransmitter systems has been implicated in the reduction of food intake. Classic anorectics (phentermine, phenylpropanolamine, phenmetrazine, benzphetamine, and diethylpropion) are structural modifications of amphetamine and are, perhaps simplistically, believed to induce transient weight reduction through interactions with central biogenic amines, particularly norepinephrine (Table 15–8). Newer agents

Table 15–8. Anorectics

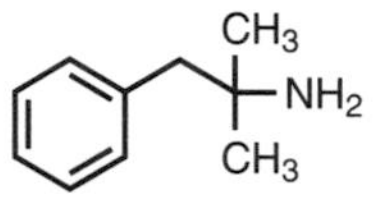

Phentermine	Phenylpropanolamine
Phendimetrazine	Benzphetamine
Diethylpropion	d-Fenfluramine
Mazindol	Sertraline (Zoloft)
(-)-*Threo* Chlorocitric acid	(+/-)-*Threo* Epoxyaconitic acid

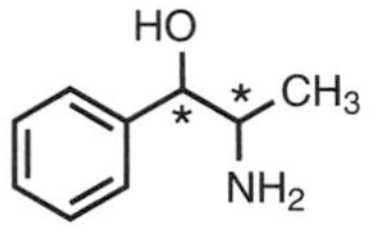

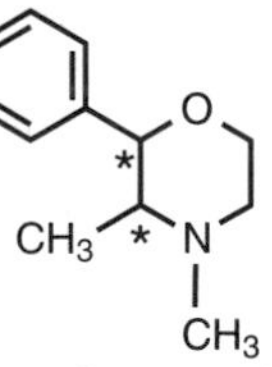

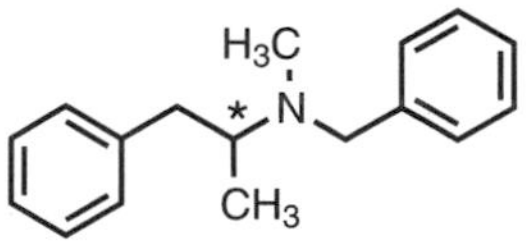

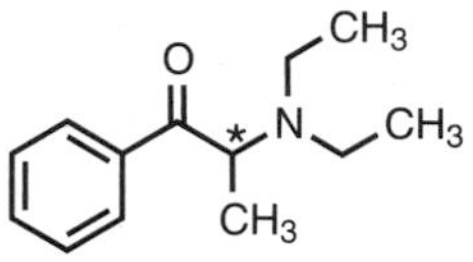

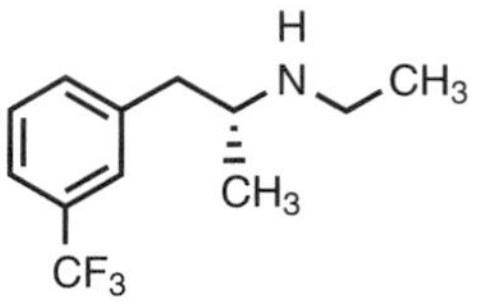

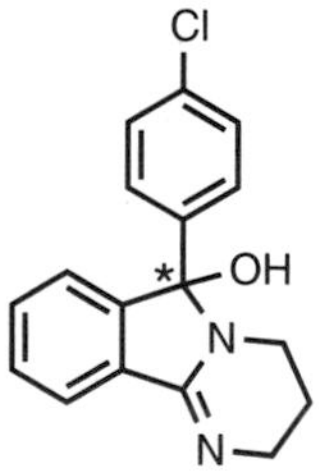

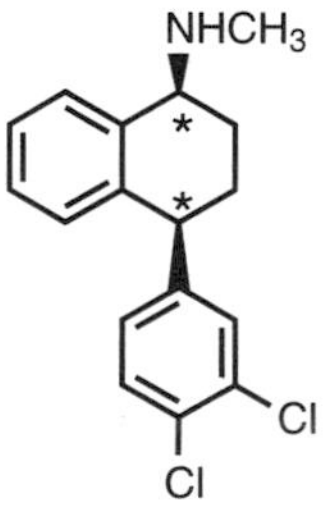

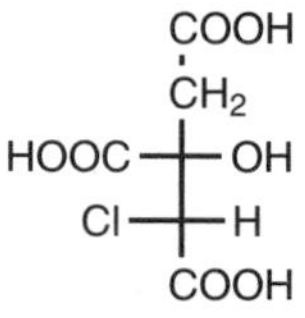

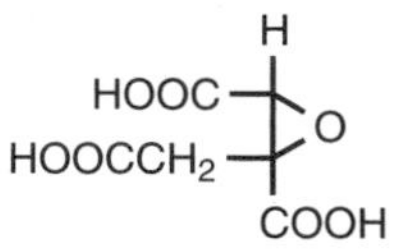

such as mazindol have been implicated with dopaminergic systems as well.[45] Fenfluramine and the more potent metabolite norfenfluramine require an intact serotoninergic system for effective anorexia.[46] Relative potencies for this compound were found to be:

$$\text{d-norfenfluramine} > \text{d-fenfluramine} > \text{dl-fenfluramine} > \ell\text{-norfenfluramine} \simeq \ell\text{-fenfluramine}.$$ [47]

Direct injection of serotonin reduces food intake. This effect is duplicated by nonspecific serotonin agonists, $5HT_{1B}$ serotonin agonists, and indirect serotonin agonists d-fenfluramine, fluoxetine, and sertraline.[48] Serotoninergically mediated anorectic drugs have been shown to spare protein at the expense of fat and carbohydrate.[49]

Peripheral agents that do not enter the brain, such as (−)-*threo*-chlorocitric acid and (±)-*threo*-epoxy aconitic acid[50] delay gastric emptying, induce satiety, and thereby reduce food intake. Similar effects are noted through ingestion of a variety of hydrophilic granules, which swell in the stomach and improve diet control.[51] Such agents offer the distinct advantage of minimizing the adverse central effects, abuse potential, and tolerance that can occur with centrally active anorectics.

Phenylpropanolamine (PPA, dl-norephedrine) is the only nonprescription anorectic presently approved by the Food and Drug Administration. Although ℓ-norephedrine is slightly more toxic than the d-isomer, neither induces noticeable stimulation unless doses approaching lethality are given.[52] The low toxicity of PPA alone (approximately one-twentieth that of amphetamine) is paralleled by the lesser incidence of CNS or cardiovascular side effects. PPA, however, is currently available in several over-the-counter drugs in combination with caffeine and pseudoephedrine or ephedrine. Such over-the-counter medications are used as diet aids, nasal decongestants, or for stimulant properties and have been associated with symptoms such as restlessness, irritability, hypertension, and in some cases psychic disturbance, headache, seizure, and stroke. The increased use of PPA in combination with other stimulants may be implicated in more severe side effects than previously recognized. PPA has a half-life of 3 to 4 hours and is cleared in the urine within 24 hours.

Tolerance through continued use of anorectics has often been observed and explained by both behavioral and pharmacologic mechanisms. A simple reduction in customary food intake increases the drive to eat and decreases anorectic efficacy. Neuronal changes have been suggested as inducing tolerance through cross-tolerance studies of drugs with primarily dopaminergic or serotoninergic mechanisms. One study demonstrated that tolerance to dopaminergic agents results in tolerance to serotoninergic agents, but that the reverse is not true. Clinically, this suggests initial treatment with an agent that enhances serotoninergic function, followed by an agent that enhances dopaminergic function if tolerance occurs.[53] Another theory suggests that anorectic agents

Table 15–9. Characteristics of Attention Deficit/Hyperactivity Disorder

Attention Deficit/Hyperactivity Disorder	Test
Intellectual functioning	Wechsler Intelligence Scale for Children
Average or superior intelligence	Revised Stanford-Binet Test
	Draw-A-Man Test
Academic achievement	Standard achievement tests
Failure to learn	
Reading problems	
Problem with abstraction	
Gross behavior	Standardized teacher report
Short attention span	Porteus Maze
Lack of ability to concentrate	Rosenzweig Picture Frustration Test
Easy distractibility	
Restless and impulsive behavior	
Explosive or aggressive outbursts	
Irritability	
Visual and auditory perception	Bender Visual-Motor Gestalt Designs
Inability to copy designs	Auditory discrimination test
Inability to reproduce designs from memory	Test of auditory synthesis
Difficulties in auditory processing of language	Sentence memory test
Motor function	Actometer
Hyperactivity	Rhomberg
Posture	Hand-nose
Fine motor control	Pronation-supination
Balance	Unilateral winking
Skilled sequential movement	

reduce the body weight set point so that weight loss occurs only long enough to obtain a new weight set point. Discontinuation of the anorectic results in a restoration of the old weight set point, and weight gain recurs.[54]

Attention deficit/hyperactivity disorder is defined as a syndrome affecting "children of near average, average or above average intelligence with certain learning or behavioral disabilities (ranging from mild to severe) associated with deviations of function of the central nervous system." This syndrome affects fewer than 4% of the population of school-age children, predominantly boys. Such children may suffer from any or all of the characteristics listed in Table 15–9 to varying degrees, hyperactivity (or hyperkinesia) being only one manifestation of this complex syndrome. Because of the complexity of this syndrome, proper identification requires a battery of physiologic, psychologic, and neurologic tests by trained personnel. The symptoms generally disappear between the ages of 12 and 18 years, although some residual signs may persist into adulthood.[55] Causes of this syndrome are not known, although hypotheses include theories that children with attention deficit disorder suffer from underactivity of the CNS that may occur from diminished effectiveness of brain norepinephrine, serotonin, or dopamine.

Stimulants such as amphetamine and methylphenidate are effective in reducing the behavioral signs of attention deficit/hyperactivity disorder.[56] Pemoline has been approved for use. Stimulants do not cure the syndrome, but serve to alleviate certain of the disturbing symptoms so that function approaches normal with overall improvement in achievement, learning, and behavior. Because treatment may cause gastric upset, irritability, and insomnia, stimulants are generally not given at night. Weight loss and moderate reduction in growth have been documented, with a rebound in growth and weight gain after termination of treatment.[57] In some, but not all, children, heart rate and blood pressure are elevated. Reported hypersensitivity to methylphenidate includes urticaria, conjunctivitis, and tic development.[58] Because stimulants are not effective in all children and because tolerance may occur after prolonged treatment, a wide variety of drugs (antidepressants, antihistamines, phenothiazines, and caffeine) have been used with variable success. Barbiturates may exacerbate the symptoms of attention deficit/hyperactivity disorder.

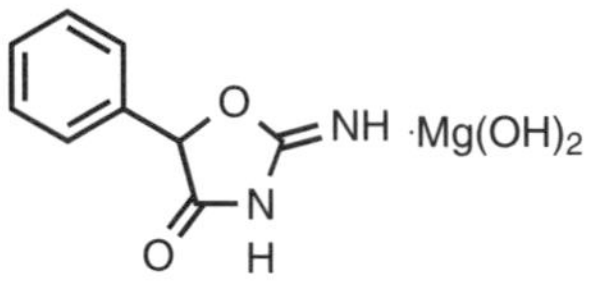

Pemoline magnesium (Cylert)

As the symptoms of the syndrome diminish with age, stimulant therapy can be reduced and eventually withdrawn with no difficulty, suggesting the lack of physical dependence in such children. The potential for abuse of stimulants is not a problem with children, but the household availability of stimulants may promote parental abuse.

Although methylphenidate possesses central anticholinergic activity, release of biogenic amines and inhibition of neuronal uptake constitute the predominant mechanisms of action. Stimulation of postsynaptic neurons is less significant than with amphetamine and is therefore not a major contributor to the biologic actions of methylphenidate.

Since effects and contraindications for this drug parallel those of other stimulants and include nervousness, insomnia, irritability, anorexia, and cardiac irregularities. Hypersensitivity is rare, although bulbous dermatologic allergic rashes have occurred. Illegitimate injection of methylphenidate tablets that have been ground with water has resulted in lung tumors because of the insoluble tablet fillers.

MONOAMINE OXIDASE INHIBITORS

Inhibitors of MAO [monoamine:oxygen oxidoreductase (deaminating) ECI.4.3.4.] were among the first drugs used to treat depression. They differ from the tricyclic antidepressant drugs in chemical structure and pharmacologic spectrum. Most of these agents inhibit not only MAO, but also other enzymes and interfere with hepatic metabolism of many drugs. Thus, these agents are not widely used.

MAO is a family of flavin-containing enzymes located primarily in the outer membranes of mitochondria. These enzymes inactivate biogenic amines, such as norepinephrine, dopamine, serotonin, tryptamine, and tyramine, by conversion of these amines to aldehydes and by their subsequent oxidation or reduction to an acid or alcohol (Table 15–10). Catechol-O-methyltransferase (COMT) is another enzyme involved in catecholamine degradation that methylates phenolic groups on aromatic rings. Neuronal norepinephrine is first metabolized by MAO, whereas circulating norepinephrine is methylated first by COMT.

The family of MAO enzymes may be subdivided into at least two isozymes. MAO-A is characterized by its substrate preference for serotonin and norepinephrine and exhibiting irreversible inhibition by chlorgyline and reversible inhibition by meclobemide. In contrast, MAO-B is characterized by its preferential location in human platelet, substrate preference for more hydrophobic amines such as phenethylamine and benzylamine, and irreversible inhibition by pargyline and selegiline.[59]

In the human brain, dopamine is predominantly a substrate of MAO-B,[60] so selective MAO-B inhibitors have found usefulness in dopamine deficiency states such as Parkinson's disease.[61]

Both enzyme subtypes are found in neuronal tissue, but their relative ratio differs with species and brain regions. MAO-B is also located in glial cells.[62] Based on the differing substrate specificity and cellular localization of these enzymes, selective inhibitors have been developed that may have distinct clinical effects.

Oxidation of amines by oxidized MAO (Table 15–11) has been proposed to proceed initially through enzymatic

Table 15–10. Classification of antidepressants

Tricyclics

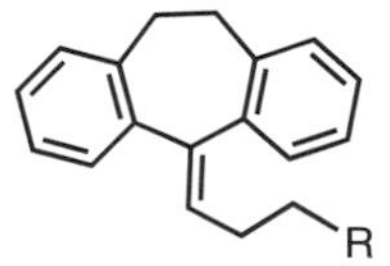

Amitriptyline (Elavil, Endep): R = $N(CH_3)_2$
Nortriptyline (Aventyl, Pamelor); R = $NHCH_3$

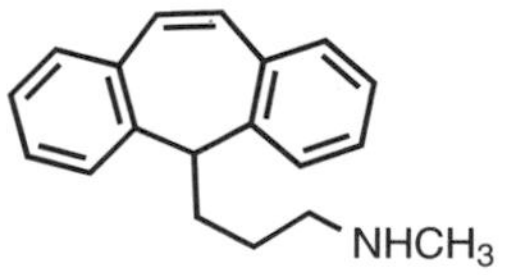

Protriptyline (Vivactil)

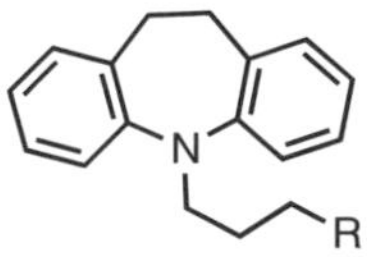

Imipramine (Tofranil): R = $N(CH_3)_2$
Desipramine (Pertrofrane): R = $NHCH_3$

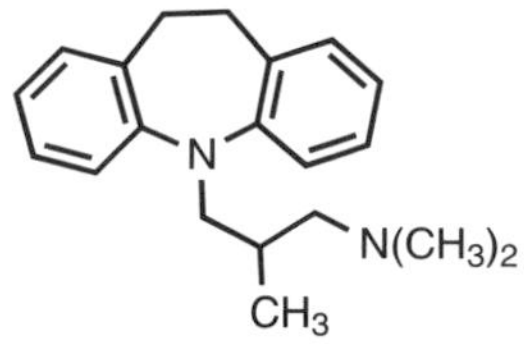

Trimipramine (Surmontil)

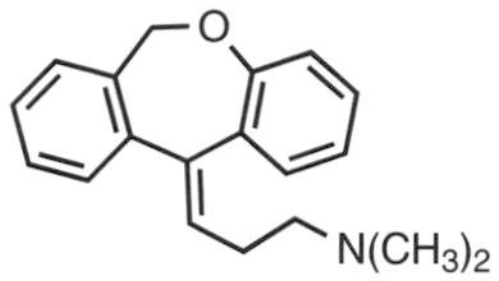

Doxepin (Adapin, Sinequan)

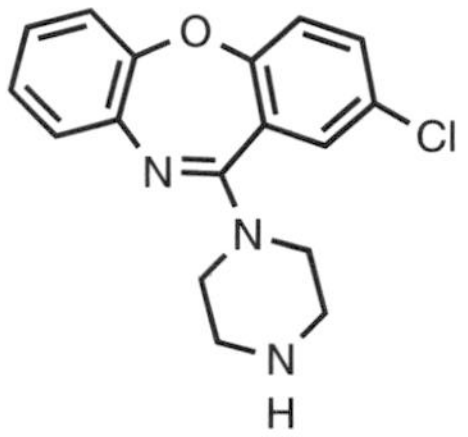

Amoxapine (Asendin)

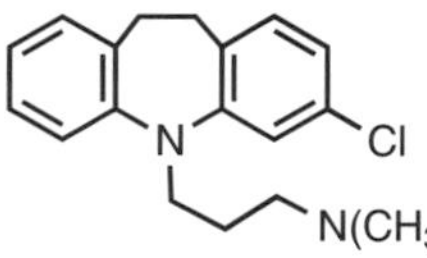

Clomipramine (Anafranil)

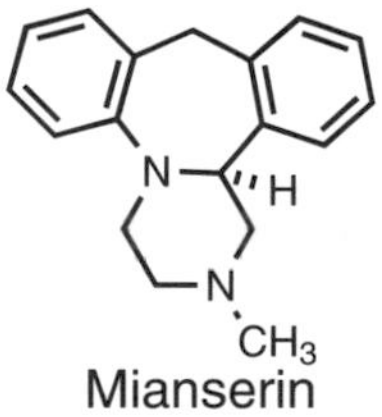

Mianserin

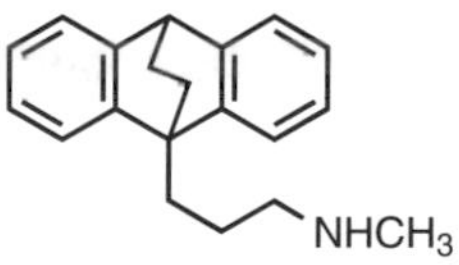

Maprotiline (Ludiomil)

Aryl- and Aryloxyalkylamines

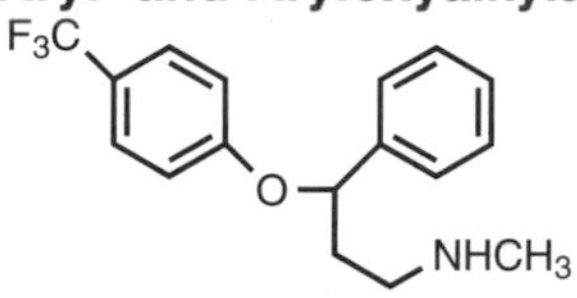

Fluoxetine (Prosac)

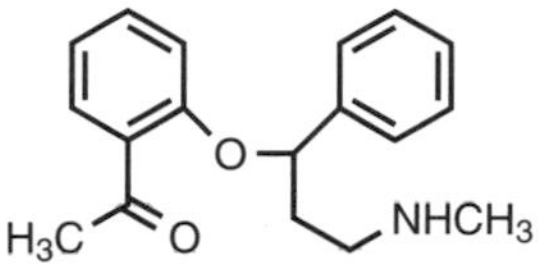

Nisoxetine

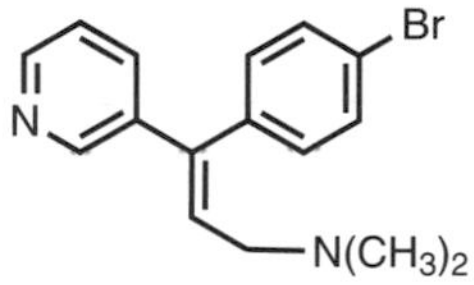

Zimeldine (Zelmid)

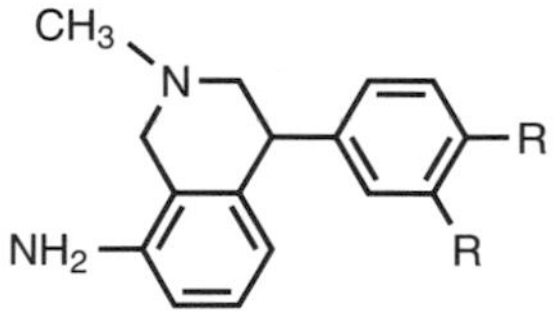

Nomifensine (Mertial): R = H
Diclofensine: R = Cl

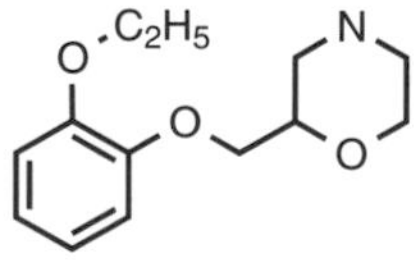

Viloxazine (Vivalan)

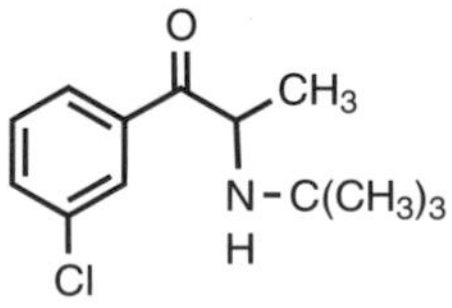

Bupropion (Wellbutrin)

(continued)

Table 15–10. *(Continued)*

Miscellaneous

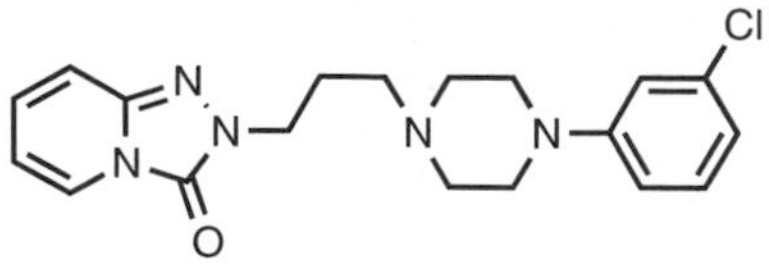

Trazodone (Desyrel)

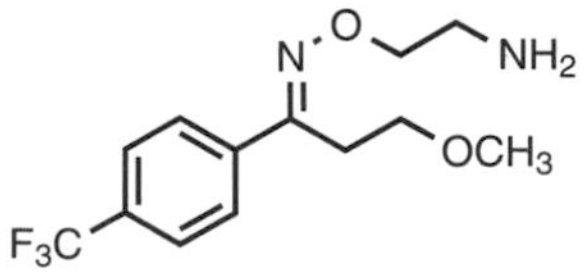

Fluvoxamine (Fluoxyfral)

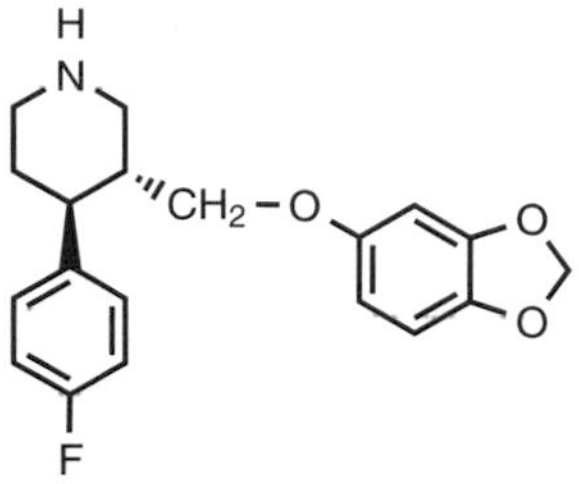

Paroxerine (Paxil)

Table 15–11. Biogenic Amine Metabolism

Central nervous system norepinephrine $\xrightarrow{MAO}$

Aldehyde intermediate
- Oxidation → Dihydroxymandelic acid $\xrightarrow{COMT}$ 3-Methoxy-4-hydroxymandelic acid
- Reduction → Dihydroxyphenylglycol $\xrightarrow{COMT}$ 3-Methoxy-4-hydroxyphenylglycol

Circulating norepinephrine $\xrightarrow{COMT}$ Normetanephrine $\xrightarrow{MAO}$ 3-Methoxy-4-hydroxymandelic acid

Dopa → Dopamine $\xrightarrow{MAO}$ Dihydroxyphenylacetaldehyde $\xrightarrow{Oxidation}$ Dihydroxyphenylacetic acid
Dopamine → Norepinephrine

Serotonin $\xrightarrow{MAO}$ 5-Hydroxyindoleacetaldehyde
- Oxidation → 5-Hydroxyindole acetic acid
- Reduction → 5-Hydroxytryptophol

Tryptamine $\xrightarrow{MAO}$ Indoleacetaldehyde $\xrightarrow{Oxidation}$ Indoleacetic acid

Tyrosine → Tyramine $\xrightarrow{MAO}$ p-Hydroxyphenylacetaldehyde $\xrightarrow{Oxidation}$ p-Hydroxyphenylacetic acid
Tyramine → Octopamine

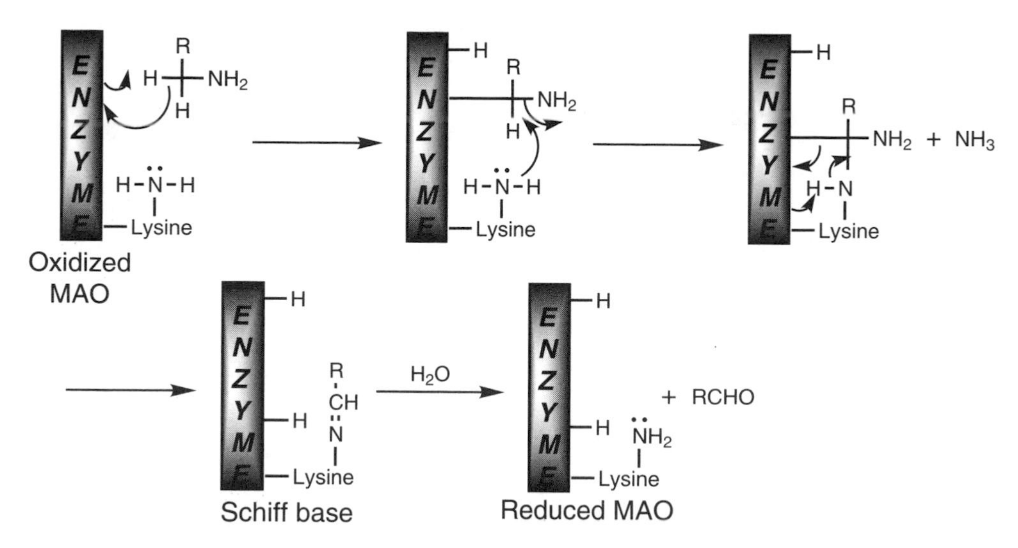

Fig. 15–2. Proposed mechanism of metabolic oxidation of amines (original diagram by C. Beecher). (Modified from C. H. Williams, *Biochem. Pharmacol., 23,* 627 [1974].)

abstraction of a proton from the α carbon of the amine. A lysine amine group on the enzyme forms a Schiff base at the α carbon of the substrate displacing the amino group from the substrate molecule. This also provides the second hydrogen to complete reduction of the enzyme. Hydrolysis of the Schiff base produces the aldehyde metabolite and the free lysylamine function. The enzyme must then be oxidized before subsequent amine oxidation (Fig. 15–2).[63] Certain compounds such as pargyline and selegiline irreversibly inhibit flavin-linked MAO through covalent attachment to flavin.[64]

Despite the diversity of compounds exhibiting MAO inhibition, including cocaine, mescaline, chlorpromazine, harmaline, and p-chloromercuribenzoate, only hydrazines and aralkylamine compounds have achieved therapeutic significance (Table 15–12).

Primary interaction of MAO and inhibitory drugs occurs through the side-chain amine with secondary interactions through the aromatic ring. Potency is increased by ring substitution with electron-withdrawing groups or replacement by certain heterocyles (2-thienyl, 2-pyridyl, 2-furyl, 2-pyrazinyl).[65] Minimal structural requirements include an electron-rich functional group (amino nitrogen or acetylenic carbon) in the plane of and separated from the center of an aromatic ring by approximately 5.25Å.[65]

Stereoselectivity of the active enzyme site is indicated by the threefold potency of *trans*-2-phenylcyclopropylamine as compared to the *cis* isomer and the antipodal potency ratio of four of the *cis* enantiomers (1S:2R/1R:2S). Similar stereochemistry (S) is possessed by the more active MAO inhibitory isomers of amphetamine and pheniprazine, suggesting a common site and binding mode for the three compounds.[66] A different site or binding mode is suggested for tranylcypromine and amphetamine at the CNS catecholamine uptake site because the stereochemistries of the more active isomers are opposite.[67]

In the treatment of depression, electroconvulsive shock therapy has been shown to be more effective than drug treatment, although drugs are widely used. In clinical studies, tricyclic antidepressants may result in significant improvement over placebo therapy, whereas many studies have been unable to demonstrate any difference between the effect of MAO inhibitors and placebo. MAO inhibitors may be recommended to a limited extent for reactive depressive illnesses with prominent phobic and hysterical symptoms. Because of the nonspecific effects of first-generation MAO inhibitors, however, potential toxicity, numerous food and drug interactions, and dubious efficacy in treating depression, these compounds are not widely prescribed and are usually used therapeutically only if no improvement occurred with tricyclic antidepressants.[68] When MAO inhibitors are tried, previous antidepressant medication should be terminated for at least 1 week before beginning therapy with a MAO inhibitor. This prevents cumulative drug effects and decreases the potential for unpleasant and toxic drug interactions. Clinical support does exist for the use of MAO inhibitors in atypical depression, a subclass that responds better to this class of drugs than to the tricyclics or electroconvulsant shock therapy.[69]

Most attempts to explain the proposed antidepressant action of MAO inhibitors have centered on their action on the enzyme. Inhibition results in increased levels of norepinephrine, dopamine, normetanephrine, tyramine, and serotonin in brain neurons and various other tissues. Urinary excretion products also vary, showing a decrease in deaminated metabolites (5-hydroxyindoleacetic acid and 3-methoxy-4-hydroxymandelic acid) and an increase in normetanephrine. Increased excretion of tyramine may also occur.

It is hypothesized that increased intracellular levels of the biogenic amines as a result of MAO inhibitors produce an overflow of these potential transmitters from the neuron and subsequent CNS stimulation.[70] Elevations in dopamine levels, the precursor to norepinephrine in some neurons, may result in increased norepinephrine synthesis that further augments neuronal overflow of nor-

Table 15–12. Monoamine Oxidase Inhibitors

Compound (Trade Name)	Structure	Chemical Name	Isozyme Selectivity
First generation: Irreversible, nonselective			
Iproniazid	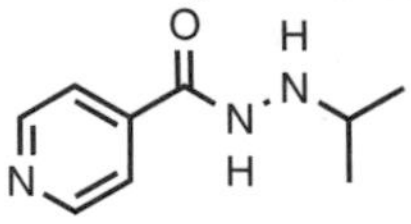	1-Isonicotinoyl-2-isopropylhydrazine	
Nialamide (Niamid)	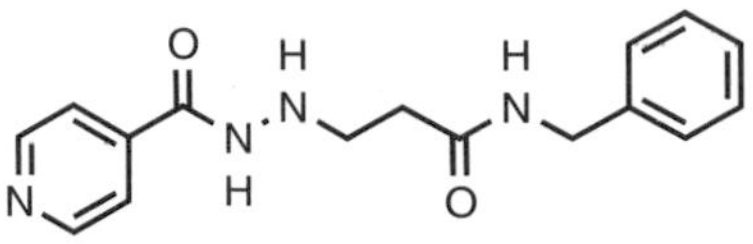	N-Benzyl-β-(isonicotinoyl-hydrazino) propionamide	
Isocarboxazid (Marplan)	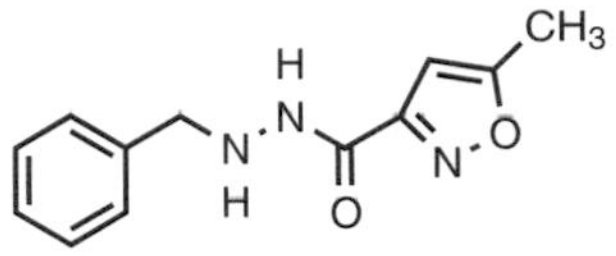	1-Benzyl-2-(5-methyl-3-isoxazolylcarbonyl) hydrazine	
Pargyline (Eutonyl)	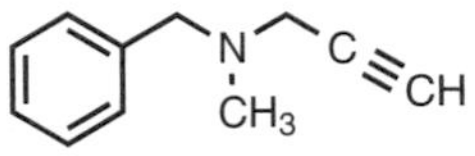	N-Methyl-N-2-propynylbenzylamine	
Phenelzine (Nardil)	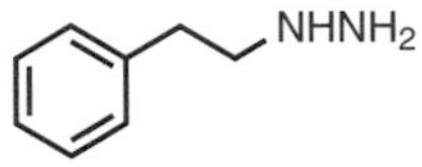	2-Phenylethylhydrazine	
Tranylcypromine (Parnate)	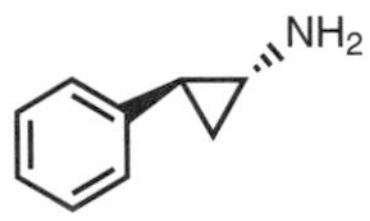	*trans*(±)-2-Phenylcyclopropylamine	
Second generation: Irreversible, preferential			
Chlorgyline	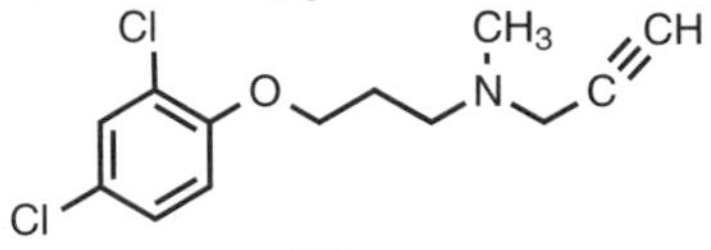	N-methyl-N-2-propynyl-3-(2,4-dichlorophenoxy) propylamine	MAO-A
Deprenyl, selegiline (Eldepryl)	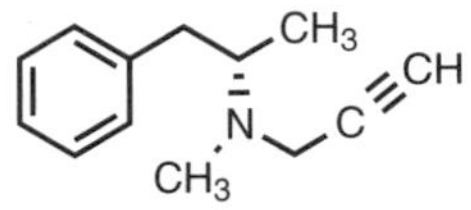	R-(−)-N,2-dimethyl-N-2-propynylphenethylamine	MAO-B
Third generation: Reversible, preferential (RIMA)			
Bromfaromine	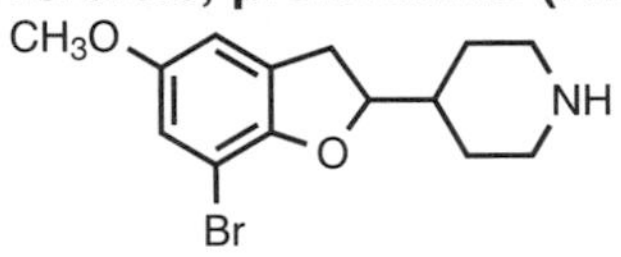	7-Bromo-5-methoxy-2-(4-piperidyl)-2,3-dihydrobenzo[b]furan	MAO-A
Meclobemide	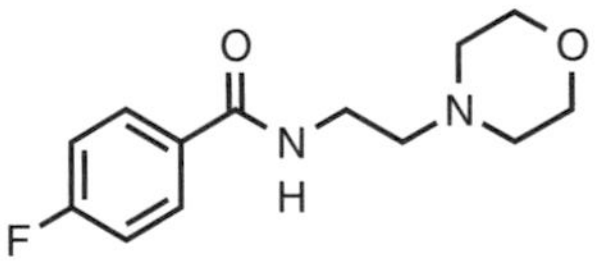	4-Fluoro-N-[2-(N-morpholino)ethyl] benzamide	MAO-A
Toloxatone	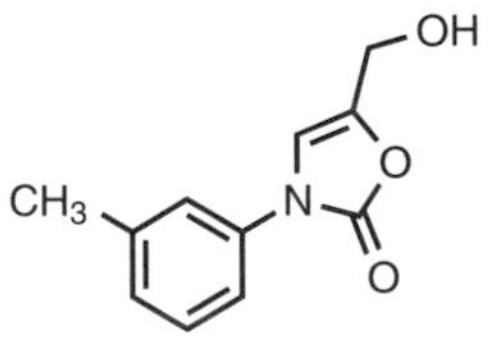	3-m-Tolyl-5-hydroxymethyl-2-oxazolidinone	MAO-A
RO-19-6327	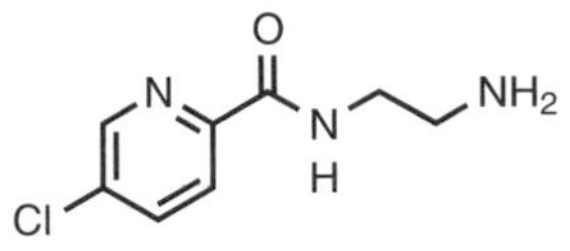	N-(2-aminoethyl)-5-chloropicolinamide	MAO-B

epinephrine after MAO inhibitors.[71] Inhibition of MAO does not potentiate the effects of nerve stimulation, however, nor are the peripheral effects of norepinephrine prolonged or intensified.

Although most explanations of antidepressant activity have been based on neuronal inhibition of MAO, investigation of other possible mechanisms indicates that some MAO inhibitors block many amine oxidases, showing a lack of specificity. Because MAO inhibitors may also block drug microsomal enzymes in the liver, the action of many drugs, such as barbiturates, narcotic analgesics, and amphetamine, may be potentiated after a MAO inhibitor. Interference with metabolism of sympathomimetic drugs, such as tyramine and amphetamine, coupled with increased levels of intraneuronal amines as a result of MAO inhibition, can initiate hypertensive crises in combination with indirectly acting sympathomimetics. The interaction of MAO inhibitors with tyramine is important because tyramine is present in many foods, such as cheese, wine, beer, snails, chicken liver, and pickled herring. The ingestion of such foods during treatment with a MAO-A inhibitor can precipitate a hypertensive crisis (see Chapter 19). Additionally, certain MAO inhibitors can stimulate postsynaptic adrenergic neurons and, similar to the tricyclic antidepressants, may block reuptake of catecholamines.

More recently, MAO inhibitors, with the exception of selegiline, have been shown, after long-term administration to induce down-regulation of β-adrenergic receptors and hence a subsensitivity to some of the actions of norepinephrine, an action that must be considered when discussing the mechanism of MAO.

Isozyme relative specificity has elevated interest in and potential for therapeutic utility of second-generation and third-generation MAO inhibitors. Selegiline, an irreversible suicide inhibitor of MAO-B, is used as an adjunct in parkinsonian patients in concert with levodopa or with the combination, levodopa/carbidopa. It is indicated for those patients not completely controlled by single-drug therapy. Selegiline does not exhibit any tyramine effect at therapeutic doses because of its weak effect on MAO-A.[73]

Reversible inhibitors of MAO (RIMAs) also demonstrate a reduced enhancement of tyramine's pressor effect.[74]

A mechanism for the paradoxic action of MAO inhibitors to produce adrenergic malfunction and postural hypotension has not been resolved. One proposed mechanism involves aberrant tyramine metabolism in adrenergic neurons. Tyramine, resulting from decarboxylation of tyrosine, is normally metabolized by MAO. Inhibition of oxidative deamination permits an alternate route of β-hydroxylation that results in the formation of octopamine, which is then stored by norepinephrine granules. Octopamine is postulated to be a false neurotransmitter with weaker pressor activity than that of norepinephrine, which it replaces. A more recent proposal for the hypotensive effects of MAO inhibitors is that increases in central norepinephrine levels may act on central α-adrenergic receptors in the nucleus tractus solitarii to lower blood pressure.[75] Pargyline, a benzylamine derivative, is a MAO inhibitor that has been used clinically for its antihypertensive properties (see Chapter 19).

When an antidepressant effect is observed, 2 to 4 weeks are commonly necessary for the onset. The effects of MAO inhibitors are cumulative and outlast the presence of the drugs. Once irreversible blockade of MAO has been established, action may not be terminated until 1 to 2 weeks after cessation of therapy. Caution must therefore be employed when changing antidepressant therapy, permitting at least 1 week without drugs to decrease the probability of drug interaction. The MAO inhibitors are noted for their mood-elevating component, but they also stimulate wakefulness and produce euphoria, respiratory stimulation, and excitement. Appetite is increased with MAO inhibitors, in contrast to anorexia caused by amphetamine.

Side effects may be related to sympathetic malfunction (postural hypotension, flushing, hyperthermia, headache, edema, and impotence), to sympathetic excess (hypertension, constipation, urinary retention, dry mouth, tachycardia, sweating, glaucoma, and blurred vision), or to general CNS effects (euphoria, insomnia, hyperreflexia, restlessness, anxiety, increased appetite, and tremors). Major consideration must be given to possible toxicity of these agents to the liver and blood-forming systems. Tranylcypromine has induced hypertensive crises even in the absence of other drugs that produce this effect in combination with MAO inhibitors. Foods containing tyramine or dopa are particular problems.

Hepatotoxicity of hydrazine MAO inhibitors may be a result of drug metabolism by acetylation. Enzymatic acetylations are genetically controlled, and people may be classified as fast, intermediate, or slow acetylators.[76] The rapid biosynthesis of acetyl hydrazines has been postulated as the mechanism for severe hepatotoxic reactions observed in fast acetylators.[77]

NOOTROPICS

The treatment of Alzheimer's disease and related dementias has historically focused on restoring lost brain cholinergic function.[78] More recently, piracetam and related pyrrolidinoacetamides have been tested as alternative modes of therapy to enhance cognition and represent the first of a group of nootropic (affect the mind favorably) agents.[79] The pyrrolidinoacetamides were originally studied because of their ability to increase the metabolism of neuronal cells by increasing the synthesis of ATP from ADP. The stimulatory effect of piracetam on adenylyl kinase may maintain tissue ATP stores under ischemic conditions.[80] These agents, however, have widespread biochemical effects, including enhancement of presynaptic and postsynaptic mechanisms of cholinergic and dopaminergic neurotransmission. Therefore, it is unclear which action is responsible for their beneficial effects.[81,82]

In animal studies, piracetam and related drugs produce a diverse array of positive effects. These include enhanced maze learning[83] and the ability to provide some protection against the decrement in performance produced by hypoxia and electroconvulsive shock in young and old rats.[84] In clinical studies, these agents are reported to improve information acquisition and facilitate information retrieval.[85]

Piracetam

ANTIDEPRESSANTS

Discovery of the conventional tricyclic antidepressant agents (formerly referred to as thymoleptics) resulted from several fortuitous observations beginning with the synthesis of aminodibenzyl derivatives in 1951. These phenothiazine bioisosteres were initially developed for their sedative and analgesic activity in animals to be used as antihistamines, sedatives, analgesics, or antiparkinsonian agents. Because these agents shared certain properties with the phenothiazine, chlorpromazine, which was being developed in the 1950s as an antipsychotic agent, and because clinical evaluation of the aminodibenzyl derivatives in 1957 showed them to be effective in some depressive states but not in severely agitated patients, interest in pursuing this former activity evolved. Many tricyclic phenothiazine isosteres have since been synthesized and, until the early 1980s, only six of these were marketed in the United States: amitriptyline, nortriptyline, imipramine, desipramine, doxepin, and protriptyline (see Table 15–10).

Because tricyclic antidepressant agents have been available for a relatively long time, research has been directed toward using the common features of these agents in an effort to understand the mechanism underlying their effectiveness in the treatment of depression. Although several theories for the biochemical basis of depression have evolved over the years, none is completely compatible with all the current information available on antidepressant agents. With the introduction and demonstration of clinical efficacy of some of the newer antidepressant agents that do not share biochemical mechanisms with the tricyclic antidepressants, newer theories have emerged that must be subjected to the test of time.

A common feature of many of the tricyclic antidepressants is the inhibition of the neuronal reuptake of neurotransmitters (see Table 15–15). The traditional tricyclics are relatively nonselective in their action, being effective inhibitors themselves or being metabolized to effective inhibitors of the reuptake of both norepinephrine and serotonin. The newer tricyclics and related analogs offer an an increasing degree of selectivity, having a primary effect on the reuptake of norepinephrine, serotonin, or dopamine (see Table 15–16). Several of the newer serotonin selective inhibitors, notably clomipramine, have also found efficacy in obsessive-compulsive disorders and panic attacks.

Depression has a variety of symptoms. Generally the common characteristics of a depressed state include feelings of sadness and guilt and, in some cases, suicidal preoccupation. In fact, depressive states may be subdivided into various syndromes, each possibly with its own biochemical basis. Depressed states can show varying symptoms depending on the frequency of occurrence, amount of anxiety, apathy, agitation, and psychosis. Depression is clearly episodic (i.e., it has exacerbations and remissions involving a cyclical occurrence). The frequency of depressive episodes can vary from individual to individual. Thus, the expression of the depressed state varies greatly among patients. Obviously the subjective nature of the disease reflects conflicting clinical reports of the agents used to treat depression and may be indicative of multiple biochemical mechanisms involved. Nevertheless, the use of antidepressant agents has clearly been established in several depressive states, although these agents are not as effective as electroconvulsive shock therapy.

Structure-Activity Relationships

A sufficient number of similar compounds have been synthesized to draw some interesting structure-activity relationships. One author[86] has arranged the tricyclic antidepressants available in the United States and Europe according to chemical and clinical characteristics, as shown in Table 15–10.

Related but separate studies have outlined general properties shared by tricyclic antidepressants that distinguish them from antipsychotics.[87,88] Each author points to the lack of coplanarity in the tricyclic structure of antidepressants as compared with the relative planarity of the two aromatic rings in antipsychotics (Table 15–13). The qualitative activity is determined by the tricyclic structure. Various side chains are noted in the entire class of antidepressants that represent modifications of substituted ethyl and propyl amine. Relative potencies and neuronal specificities of the chains relate to the nature of the aromatic structure. In general, the monomethyl derivatives are more potent and have a more rapid onset of activity than the corresponding disubstituted compounds. In contrast to the structurally similar antipsychotics, ring substitution has little positive impact on antidepressant activity.

These features have been related to the ability of the tricyclic drugs to block reuptake of D(−) norepinephrine into nerve endings.[87] Trimethylquaternary nitrogen derivatives of imipramine have in vitro activity comparable to that of the dimethyl and primary amine derivatives, which suggests that the active form is protonated. Marked activity of the secondary methylamine derivatives, which are proposed to be active metabolites of the dimethylamine derivatives, has been attributed to an important steric effect of the methyl group. Secondary ethyl and propylamine derivatives are less potent. Activity as inhibitors of neuronal reuptake mechanism may be related to spatial similarities to norepinephrine. Accordingly, one aromatic ring and the aliphatic nitrogen of the drug and norepinephrine should be spatially similar.

Both tricyclic groups share a lack of coplanarity of aromatic rings and differ in the degree of twisting (Table 15–14). Antidepressant tricyclic drugs (dibenzocyclo-

Table 15–13. Psychoactive Ring Structures

Tranquilizers

S S O O
N N

Neuroleptic-thymoleptics

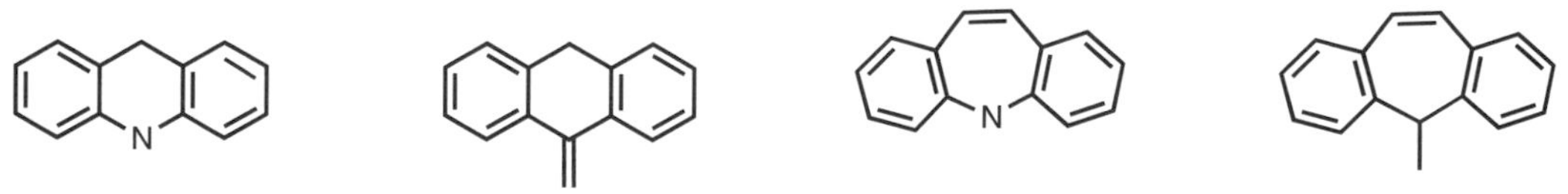

N N

Antidepressants

X X
N N

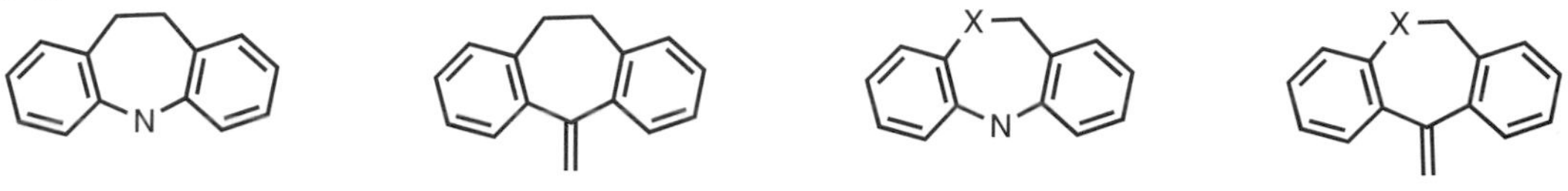

heptadiene and dihydrodibenzazepine) are characterized by considerable skewing of the two aromatic rings, whereas less thymoleptic potency is shown by the more symmetric tricyclic compounds (phenothiazine and dibenzocycloheptatriene).

Structural similarities among the nontricyclic antidepressants are less obvious despite certain spatial relationships. Zimeldine is a reasonably selective serotonin uptake inhibitor in the Z configuration but not in the E. The rigid mianserin structure is suggestive of appropriate spatial placement. The S(+) configuration is 300 times as potent as a norepinephrine uptake inhibitor as the R(−) isomer. The R(−) isomer is a more potent (α) adrenolytic.[89]

The (R) and (S) isomers of fluoxetine are equally potent as serotonin uptake inhibitors, as is the (S) isomer of norfluoxetine. R-Norfluoxetine is much less effective.[90]

Viloxazine, a specific inhibitor of norepinephrine uptake, is devoid of traditional anticholinergic and sedative side effects. Inhibition is restricted to the (S) isomer, in contrast to its serotonin releasing activity, which is not stereoselective.[90]

Structure-activity relationships determined for fluoxetine point out the exquisite sensitivity of the serotonin receptor to minor modifications. Relocation of the trifluoromethyl group to the meta or ortho position brought about decreases in receptor affinity of 10-fold and 90-fold. No other group could effectively replace the p-trifluoromethyl substituent without decreasing potency or selectivity. The selectivity of the trifluoromethyl compound for the serotonin site was not modified by the degree of N-methylation.[91] The low affinity of fluoxetine for norepinephrine uptake sites was completely reversed in the O-methoxy derivative (nisoxetine), which is a selective inhibitor or norepinephrine neuronal uptake.[92]

The absolute configuration of 3S,4R(−)-*trans*-paroxetine has been reported. Preliminary structure-activity relationships suggest that even small substitutions on the nitrogen, changing the conformation to *cis* or modifying the phenoxy ring substituents, decrease activity. Fluoro substitution enhanced potency in (−)-*trans* series and diminished potency in the (+)-*trans* series.[93]

Marketed tricyclic antidepressants share the ability to inhibit neuronal monoamine uptake with several other diverse classes. Compounds that are not considered antidepressants but do inhibit monoamine uptake include chlorpromazine, benztropine, chlorpheniramine, and methadone. Tertiary amine antidepressants, useful in patients with psychomotor agitation, are more effective inhibitors of serotonin uptake and have greater affinities for α-noradrenergic binding sites in brain tissue than secondary amines. Secondary amines, useful in retarded depression owing to a lower incidence of sedation and hypotension, are more effective inhibitors of norepinephrine uptake.[94] Although a direct relationship between inhibition of amine uptake and antidepressant activity has not been established, such pharmacologic action is being used to identify compounds for further clinical study.

Minimal structural features for monoamine uptake inhibition are an aromatic ring and an aliphatic nitrogen separated by an aliphatic chain of approximately four atoms. These minimal features characterize but do not distinguish inhibitors of dopamine, norepinephrine, and serotonin uptake.[95] Several compounds now under clinical or preclinical study do show preference for one of the three amine uptake receptors, although not to the exclusion of interaction with the other uptake sites in vitro (Table 15–15).

Biochemical Basis of Depression and the Mechanism of Action of Antidepressant Drugs

The primary hypothesis for the mechanism of action of antidepressants, known as the biogenic amine hypothesis,[98] was formulated in the mid-1960s. Although it has been largely supplanted by more recent theories, this hypothesis provided valuable insights into the mechanism of action of antidepressant drugs. The hypothesis suggests that pathologic depression is due to a deficiency of biogenic amines at postsynaptic sites in the brain and

Table 15–14. Tricyclic Ring Structures*

Ring system	Structure	Relative Potency	
Potent uptake inhibitors			
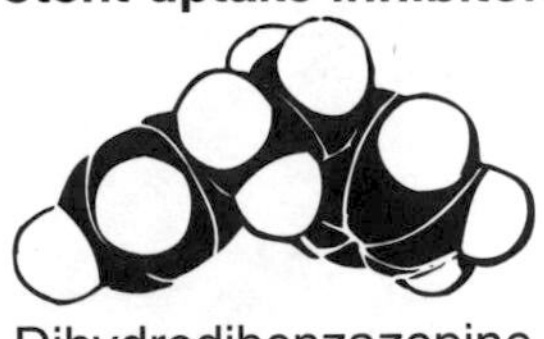 Dihydrodibenzazepine	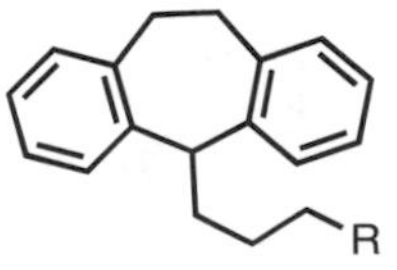	R = $N(CH_3)_2$ 38 (Imipramine)	R = $NHCH_3$ 3000 (Desmethyl imipramine, DMI)
 Phenothiazine	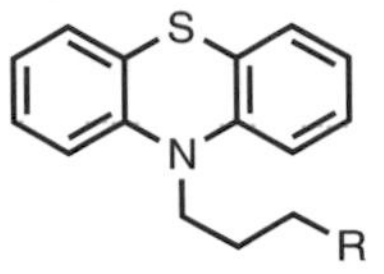	40 (Promazine) 53 2-Cl (Chlorpromazine)	1200 2-Cl
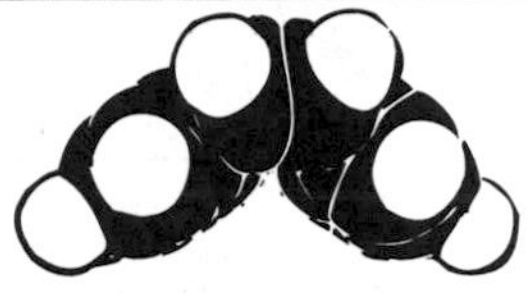 Dibenzocycloheptatriene	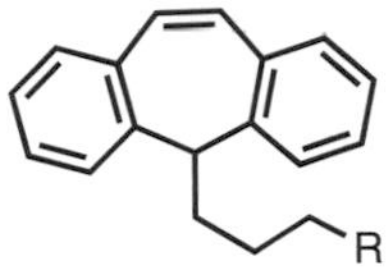		280 (Protriptyline)
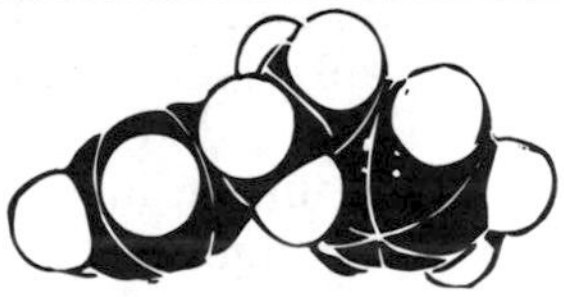 Dibenzocycloheptadiene	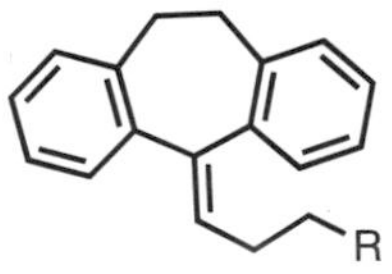	16 (Amitriptyline)	120 (Nortriptyline)
Weak uptake inhibitors			
 Phenoxazine	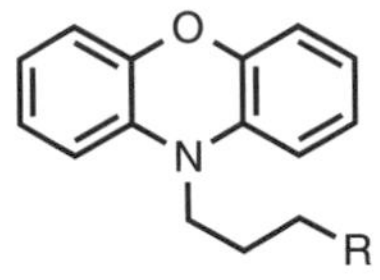	R = $N(CH_3)_2$ 2.0	
 Carbazole	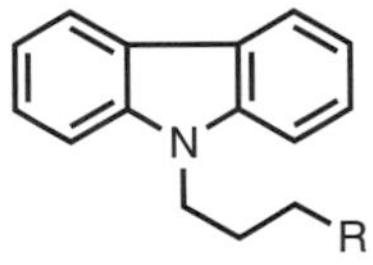	0.5	
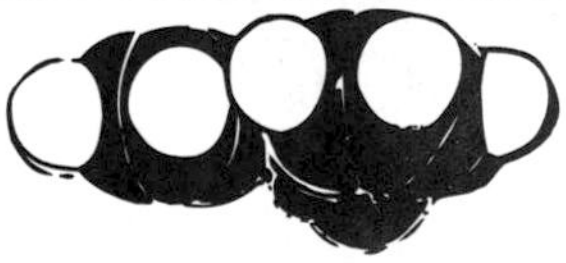 Phenanthridone	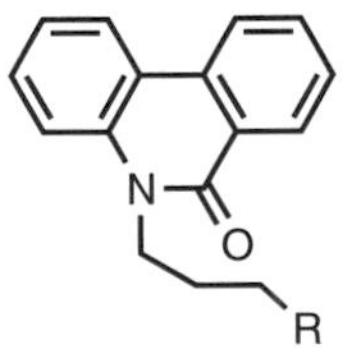	0.5	
 Dihydrophenanthridine	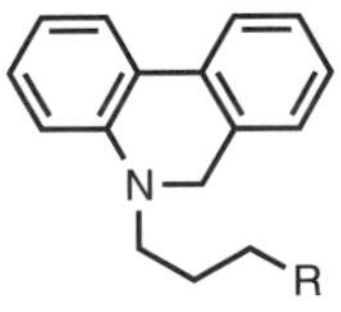	0.01	

* Molecular conformations of the ring systems of tricyclic inhibitors of the uptake of norepinephrine by rabbit aortic strips. The first column represents the Corey, Pauling, Koltan molecular model of the ring system shown in the second column. The models represent the structural formulas as viewed in the plane of the page, looking from the top of the page down. Note the well-defined angles between the two phenyl rings. Relative potencies of derivatives of each as inhibitors of the uptake of norepinephrine are given in the third column.

Table 15–15. Selective Amine Uptake Inhibitors

Compound	IC_{50} (UM) 5-HT	NE	DA	Selectivity IC_{50} Ratio NE/5-HT	DA/5-HT
Sertraline	0.058	1.2	1.1	21.0	19.0
Fluroxamine	0.54	1.9	45.0	3.5	83.0
Zimelidine	4.5	12.0	43.0	2.7	9.6
Norzimelidine	0.45	0.36	21.0	0.8	47.0
Fluoxetine	0.27	0.74	12.0	2.7	44.0
Clomipramine	0.099	0.11	8.1	1.1	82.0
Imipramine	0.81	0.066	20.0	0.081	25.0
Desipramine	3.4	0.0056	21.0	0.0016	6.2
Amitriptyline	1.2	0.13	13.0	0.11	11.0
Nortriptyline	1.7	0.025	11.0	0.0014	6.5
Paroxetine*	0.0011	0.35	—	35.0	—

* From B. Thomas et al., Psychopharmacology, *93,* 193 (1987).

is based on two pharmacologic observations. First, both groups of traditional antidepressant drugs, the MAO inhibitors and the tricyclic antidepressants, share an ability to elevate synaptic concentrations of neurotransmitter amines, especially norepinephrine and serotonin. The MAO inhibitors act by inhibiting the catabolism of these amines, whereas the tricyclic antidepressants block neuronal reuptake by inhibiting the activity of neuronal transporters,[99] the primary mechanism responsible for reducing the amount of transmitter available in the synapse. Second, early clinical observations that reserpine induced depressive-like effects were consistent with a synaptic deficiency underlying depression because reserpine acts to deplete these same amines from presynaptic storage sites. Furthermore, electroconvulsive shock therapy is also capable of increasing the turnover of certain neurotransmitters in the brain, providing a mechanism for increasing postsynaptic stimulation, which is consistent with this formulation.

In addition, based on this theory, mania is thought to be due to an excess of biogenic amine available for postsynaptic interaction. Lithium treatment has proven effective in the reversal of the manic phase of manic depression and increases the net reuptake of certain biogenic amines into nerves (an effect opposite to that of the tricyclic antidepressants) and is capable of reducing nerve-stimulated release of biogenic amines, both effects consistent with a net decrease in the amount of biogenic amines available for postsynaptic interaction in the brain. Thus, considerable data exist to support the catecholamine theory of depression.

This theory, along with the idea that more than one biochemical mechanism may be responsible for depression and that agents selective for interacting with one or more of the biogenic amines may provide more specific effects in certain depressed patients, has led to the development of antidepressant agents that can block the neuronal uptake mechanism of only selected biogenic amines. In this regard, desipramine, a conventional tricyclic antidepressant, and some of the newer antidepressants, such as amoxapine, a tricyclic dibenzoxazepine, maprotiline, and nomifensine, have been developed as selective inhibitors of the neuronal uptake for norepinephrine, exerting minimal or no effect on the neuronal uptake mechanism for serotonin. In contrast, agents such as amitriptyline, clomipramine, fluoxetine, and fluvoxamine are selective inhibitors of the neuronal uptake for serotonin. Bupropion is an agent that was under development as a selective inhibitor of the neuronal uptake for dopamine. It is hoped that, with the development of these selective agents for the different neuronal pathways, it may be possible to identify subtypes of depressed patients to be preferentially treated with these selective agents.

Although some of these newer, more selective agents provide continued evidence in support of the biogenic amine theory of depression, some agents, such as mianserin and trazodone, and not potent inhibitors of biogenic amine neuronal uptake mechanisms. Mianserin, however, by virtue of its α_2-receptor antagonist activity, may enhance the neuronal release of biogenic amines, and trazodone, by virtue of its metabolism to metachlorophenylpiperazine, a known agonist at central serotonin receptors, may serve to activate postsynaptic central biogenic amine receptors. Thus, certain actions of these antidepressants are consistent with the biogenic amine theory of depression.

Nevertheless, the fact that pharmacologic reversal of depression with antidepressant drugs normally takes from 7 to 14 days, after initiating therapy, to become apparent, whereas the biochemical effects on transmitter uptake and metabolism are apparent almost immediately, has caused a reassessment of the original hypothesis. The emphasis has been shifted from immediate presynaptic effects to more slowly occurring adaptive effects involving postsynaptic receptors.[100] The primary focus has been on β-adrenergic receptors coupled to adenylyl cyclase/cyclic AMP generating systems,[101] but changes in α-adrenergic,[102] serotoninergic,[103] dopaminergic,[102] and GABAergic[101] receptors have also stimulated interest, along with evidence of interactive effects involving coincident changes in more than one receptor system.[106] By way of illustration, administration of a variety of tricyclic antidepressants has been shown initially to stimulate cyclic AMP production (presumably mediated by stimulation of β-receptors by norepinephrine). With repeated administration, however, cyclic AMP production is reduced, and there is a gradual reduction in the number of available postsynaptic β-adrenergic receptor ligand binding sites.

This shifting emphasis on the cellular site of action of antidepressants has altered the understanding of the possible action of these drugs and led to the consideration of therapeutic agents with novel mechanisms.

Side Effects and Contraindications of Antidepressants

Autonomic side effects of the tricyclic antidepressants are primarily attributed to the peripheral anticholinergic activities of these agents. Dry mouth, decreased visual accommodation, constipation, and urinary retention are the most frequent problems associated with this atropine-type action. Sweating is also common. Cardiovascular effects include postural hypotension, tachycardia and, with larger doses, arrhythmias. The frequency of cardiac irregularities limits the usefulness of these antidepressants in elderly patients prone to cardiovascular problems. Sudden unexplained deaths have been correlated with amitriptyline use in patients predisposed to cardiovascular disease.

Major claims for the newer antidepressant agents are that they induce fewer anticholinergic side effects and less sedation and that their potential for cardiotoxicity, either in overdose or at therapeutic doses, is less than that found for tricyclic antidepressants. The activities and claims for some of the second-generation antidepressants are listed in Table 15-16. Selective serotonin reuptake inhibitors do not cause the sedative, anticholinergic, and cardiovascular side effects characteristic of tricyclic antidepressants, but common complaints include nausea, nervousness, and dizziness.

Table 15–16. Second-generation Antidepressants

Compound	Comments
Amoxapine	Weak postsynaptic dopamine receptor antagonist Active metabolites identified More rapid onset is questionable
Maprotiline	Selective inhibitor of norepinephrine neuronal uptake Active metabolite identified Less anticholinergic, sedative, and cardiotoxic activity reported Increased incidence of skin rashes and seizures
Mianserin	Little effect on serotoninergic or noradrenergic uptake systems Combines α_1, α_2, and histamine and serotonin receptor blocking activity Less anticholinergic activity
Trazodone	Little effect on serotoninergic or noradrenergic uptake systems Combines α_1 and serotonin receptor blocking activity Major metabolite, m-chlorophenylpiperazine, is a serotonin agonist Less anticholinergic activity Sedation is a major side effect
Buproprion	Selective inhibitor of dopamine neuronal uptake Dry mouth is major side effect Withdrawn due to high incidence of seizures
Nomifensine	Selective inhibitor of norepinephrine and dopamine neuronal uptake Less anticholinergic activity Withdrawn due to hemolytic anemia
Alprazolam	A benzodiazepine Little effect on biogenic amine receptors or uptake systems
Zimeldine	Selective inhibitors of serotonin neuronal uptake mechanisms
Fluoxetine	Active metabolites identified Less anticholinergic, sedative action and cardiotoxicity
Sertraline	Shorter action than fluoxetine Nonsedating
Paroxetine	No apparent anticholinergic activity Potent serotonin uptake inhibitor

Central actions of the tricyclic antidepressants include a predisposition to convulsions and seizures, disturbance in motor function, transient insomnia, fine tremors, and activation of schizophrenic symptoms. With depressive patients, these agents can produce a transition into the manic state. Other less frequent and less predictable side effects include allergic reactions, photosensitivity, jaundice, and blood dyscrasias, such as agranulocytosis. Drug contraindications are numerous (Table 15-17). Although a teratogenic potential has been suggested, no firm evidence for such an effect is available in humans. Withdrawal symptoms have been reported after stopping therapy with imipramine given for 2 months at 150 mg/daily. Symptoms include nausea, headache, malaise, vomiting, dizziness, chills, cold sweat, cramps, diarrhea, insomnia, anxiety, restlessness, and irritability.

Elderly patients are also more sensitive to sedative side effects, and balance problems may occur during initial dosage. This may be exacerbated by orthostatic hypotension, seen frequently at therapeutic levels in hypertensive patients. Age-related decreases in antidepressant metabolic capacity have been noted with amitriptyline, imipramine, maprotiline, and mianserin.

Pharmacokinetics

Conventional tricyclic antidepressants are well absorbed orally and are normally given in initial doses of 75 to 100 mg/day; the patient is then titrated until doses of 200 to 300 mg/day are achieved for maintenance. After

Table 15–17. Contraindications and Precautions with Tricyclic Antidepressant Use

Contraindication or Precaution	Action of Antidepressant
Diseases	
Glaucoma	Anticholinergic action
Constipation	
Urine retention	
Cardiovascular disease	Cardiovascular disturbances
Geriatrics	
Hyperthyroidism	
Epilepsy	CNS actions
Parkinsonism*	
Insomnia or restlessness	
Schizophrenia	
Pregnancy	Teratogenic potential
Children	Hyperpyrexia
Liver disease	Jaundice and metabolic impotence
Pharmacologic Agents	
Anticholinergic agents	Anticholinergic synergism
Indirect acting sympathomimetics	Potentiation of response
Antihypertensive agents Guanethidine and reserpine	Antagonism of action
MAO inhibitors	Potentiation of psychomotor activatlon
Alcohol	Potentiation of depression (sedation)
Phenothiazines	Schizophrenic activation
Thyroid medications	Synergism of cardiovascular arrhythmias
Barbiturates	Potentiation of sedation

* Small doses have been efficacious in parkinsonism.

a few weeks, the dosage may be reduced by about 50% if tolerated. Once therapy has begun, 1 to 4 weeks may be necessary to observe clinical improvement, which limits effectiveness in the treatment of suicidal depression. In the elderly, in whom metabolism and hepatic clearance may be reduced, dosage should be reduced.

High concentrations of the tricyclic antidepressants are found in the liver, lungs, and brain of humans, similar to the distribution of the phenothiazines. Extensive metabolism occurs with marked species and sex differences. For example, 13 unconjugated metabolites have been isolated from human urine after imipramine administration, representing side chain demethylation, ring hydroxylation, N-oxide formation, and side chain removal (Table 15–18). Glucuronide conjugates of these metabolites are the major excretory products. Of these metabolites, the desmethyl (norimipramine) is an active metabolite and may contribute to the activity seen after imipramine. Because metabolism of this group of antidepressant agents is so extensive, it is not surprising to consider the possibility that several of these agents may be metabolized to active antidepressant agents. In addition, one of the newer antidepressant agents, trazodone, is known to be metabolized to meta-chlorophenylpiperazine, an agent that activates central serotonin receptors. This mechanism may contribute to the antidepressant efficacy seen after trazodone administration. Because metabolism of antidepressant agents is extensive and contributes in some cases to their activity, factors affecting hepatic function may have a pronounced effect on the activity and duration of these compounds.

Approximately 5% of the general population are slow hydroxylators of tricyclics and may develop toxic levels. This variability in metabolism between individuals is called genetic polymorphism and may be identified by challenge with debrisoquin and determination of the rate of 4-hydroxylation (see Chapter 8).[107]

Debrisoquin

The pharmacokinetic data for selected second-generation antidepressants are listed in Table 15–19.

HALLUCINOGENS

According to Goode[109] the labeling of hallucinogens involves either prodrug or antidrug sentiments. The word *psychedelic* has prodrug connotations, implying that the mind works best under the influence of a drug. The words *hallucinogenic* and *psychotomimetic* have negative overtones because hallucinations are unreal and therefore not acceptable to society, and psychoses indicate an equally undesirable state of madness. The terms are used interchangeably in this chapter.

A psychotomimetic drug may be defined as any agent that can produce in humans a temporarily altered state

Table 15–18. Unconjugated Imipramine Urinary Metabolites

R = $(CH_2)_3N(CH_3)_2$	R = $(CH_2)_3NHCH_3$	R = $(CH_2)_3NH_2$	R = H
Imipramine	Nor_1imipramine	Nor_2imipramine	Iminodibenzyl
Imipramine-N-oxide	2-OH-Nor_1imipramine	2-OH-Nor_2imipramine	2-OH-Iminodibenzyl
2-OH-Imipramine	10-OH-Nor_1imipramine	10-OH-Nor_2imipramine	10-OH-Iminodibenzyl
10-OH-Imipramine			

Table 15–19. Pharmacokinetics of Selected Second Generation Antidepressants

Drug	Elimination Half-Life (hr)	Active Metabolites	Metabolite Elimination Half-Life (hr)	Daily Dose (mg)
Zimelidine	5	Yes	15	100–300
Fluroxamine	15	No	—	100–300
Fluoxetine	70	Yes	330	20–80
Sertraline	25	No	—	50–200
Paroxetine	12–20	No	—	10–50

Data from K. Rickels and E. J. Schweizer, *J. Clin. Psychiatry, 51,* Suppl. B, 9 (1990).

of mood, thinking, perception, and behavior. Obviously an altered state of mental behavior is a subjective phenomenon and therefore unpredictable with each agent and each individual.

College student surveys report that alcohol remains the campus "drug of choice." Over the past decade, there has been a decrease in the frequency of use of the second-choice drug, marijuana. Cocaine use remains somewhat higher than it was at the beginning of the decade. Hallucinogen use has been low throughout the decade.

This section is concerned with three chemical classes of compounds currently classified as hallucinogens: phenethylamines, indoles, and cannabinols. Other pharmacologic classes include drugs to which hallucinogenic properties have been attributed. These include veterinary anesthetics (phencyclidine), anticholinergics (scopolamine), analgesic antagonists (nalorphine), insecticides (parathion), and cholinesterase inhibitors (diisopropyl fluorophosphate). Anticholinergics deserve particular mention because they are frequently adulterants or substitutes for other street drugs. In addition to atropine and scopolamine, several esters of benzilic acid or bioisosteric acids (JB336) and phencyclidine (PCP, THC) are reported to induce a hallucinatory state or, perhaps more accurately, delirium.[111]

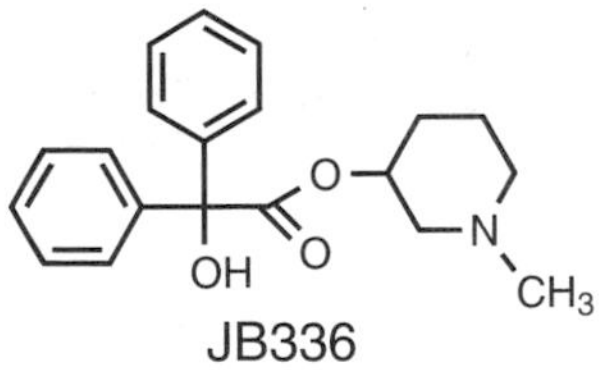

OH O CH3 N O

JB336

N

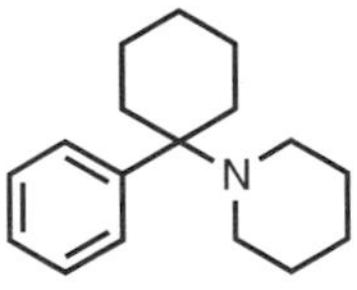

Phencyclidine

Derivatives of allylbenzene and propenylbenzene are widely distributed throughout the plant kingdom. Many are found in the aromatic fraction of oils of anise, japonicum, citronella, sassafras, calamus, clove, elemi, crowei, nutmeg, parsley seed, and dill. Reported incidences of unusual mental effects after ingestion of nutmeg[112] and sweet calomel[113] prompted further study of methoxy-substituted and methylenedioxy-substituted propenylbenzenes. The ring substitutions of myristicin, elemicin, asarone, safrole and several other constituents are reminiscent of highly potent hallucinogenic phenylisopropylamines exemplified by 2,5-dimethoxy-3,4-methylenedioxyphenylisopropylamine, a compound reported to be 12 times as potent as mescaline.

In vivo potency of these volatile oil constituents may be due to metabolic conversion of isopropylamine derivatives through either direct addition of ammonia or oxidation to a ketone and transamination to the amine. Although no direct evidence for this biotransformation exists, urine analysis after administration of myristicin

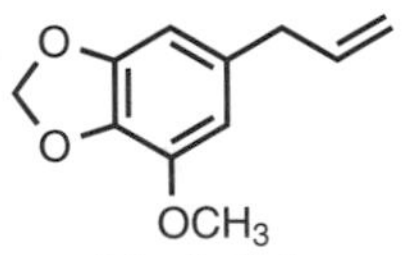

O O OCH_3

Myristicin

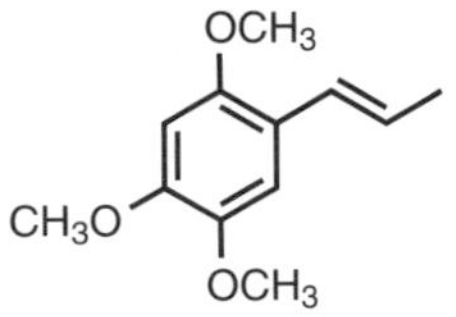

OCH_3 CH_3O OCH_3

Asarone

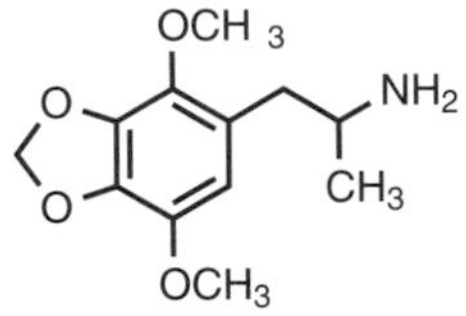

OCH_3 O O NH_2 CH_3 OCH_3

2,5-Dimethoxy-3,4-methylenedioxyphenylisopropylamine

and related agents has revealed ninhydrin-positive compounds, identified as tertiary aminopropiophenones which are competitive inhibitors of MAO.[114] The hallucinogenic activity of nutmeg has not yet been related to one constituent, and all workers do not agree that it is hallucinogenic.

Despite efforts by many states to ban the sale of clove cigarettes (30 to 40% clove, 60 to 70% tobacco), imports from Indonesia increased 10-fold between 1980 and 1985. Eugenol, a component of clove used as a dental anesthetic, has been suggested as a facilitator of deep and prolonged inhalation by anesthetizing the respiratory tract. During 1985, 12 cases of hospitalization following clove cigarette smoking were due to severe pulmonary edema, bronchospasm, and hemolysis.[115]

After the initial surge of popularity of lysergic acid diethylamide (LSD), individuals who enjoyed a chemically induced escape from reality became concerned about possible chromosomal and mutagenic effects. Attention was directed briefly toward methedrine (speed), but then came the warning that "speed kills."

On learning that mescaline was a legal sacrament for members of the Native American Church, many drug-oriented persons shifted their allegiance to this simple compound.

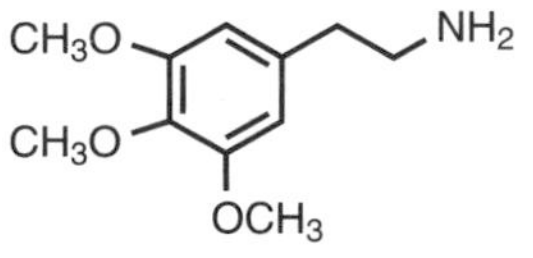

Mescaline

Mescaline occurs in several cacti and has been used sacramentally by Indian tribes from the southwestern United States to southern Canada. It has been isolated from the cacti Lophoplora williamsii, Trichocereus pachanoi Br. and R., Trichocereus bridgessi (SD) Br. and R., Trichocereus macrogenus, and Pelecyphora aselliformis Ehrenberg.

Monomethoxylated phenethylamine derivatives are generally inactive in doses up to 25 mg/kg. All disubstituted compounds are also inactive, although some reports have indicated psychotomimetic activity of 3,4-dimethoxyphenethylamine (the "pink spot" of schizophrenia).[116] Only mescaline among the trisubstituted and 2,3,4,5-tetramethoxyphenethylamine among the tetrasubstituted compounds are active. Mescaline units (MU) are used to compare the potency of these derivatives,[117] and some have been estimated to be as much as 150 times as potent as mescaline (1 MU). The pentasubstituted compound is the most potent (7 MU) polymethoxylated phenethylamine. The close relationship between potency and group position is demonstrated by the 3,4-methylenedioxy (0.2 MU), 3-methoxy-4,5-methylenedioxy (IMU), and 2-methoxy-3,4-methylenedioxy (5MU) compounds.

Ring-methoxylated analogs of amphetamine are considerably more potent than the corresponding phenethylamines (Table 15–20). Optimal activity is found in compounds bearing a p-methoxy or p-alkyl group. The addition of an o-methoxy group usually increases activity, whereas as m-methoxy group generally decreases it unless converted with an adjacent methoxy group to a methylenedioxy group. Other ortho or para groups (amino, iodo, and bromo functions) may substantially increase potency.

Stereoselectivity of binding to the 5-HT_2 receptors has been determined to reside in the R(−) isomers, although enantiomeric potency ratios are relatively small.[119] Optimum affinity for the 5-HT_2 receptor is exhibited by molecules such as DOB and DOI. Structural modifications of these two molecules decrease both affinity and selectivity.[118]

Analogs of mescaline have been said to mimic the A/B or A/C rings of lysergic acid.[120,121] Theoretical calculations[122] and spectroscopic studies have suggested that the *trans* conformation is preferred above other alternatives.[123]

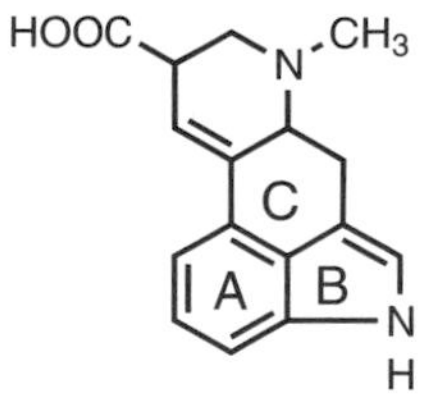

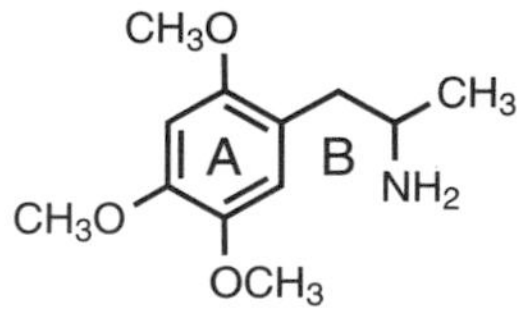

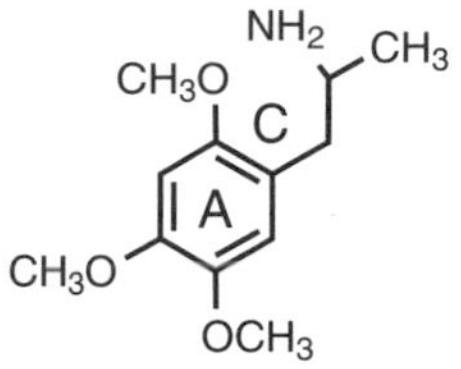

Other physical properties, including molecular orbital studies,[124] ionization potentials,[125] partition coefficients,[126] and molecular connectivity,[127] have been used in an effort to gain more insight into the structural requirements of phenethylamine and other hallucinogenic compounds.

The relatively low activity of compounds with three flanking methoxyl substituents has been related to hindrance of receptor binding by out of plane methoxyl conformations.[128] Phenylisopropyl amine is presently serving as the backbone of a diverse group of abused drugs collectively identified as designer drugs. These are illicitly synthesized compounds with substitution patterns selected to provide hallucinogenic activity and compounds not yet classified as illegal by the U.S. Drug Enforcement Agency. One such compound, 3,4-methylenedioxymethamphetamine (MDMA, Ecstasy), was designated a controlled drug by emergency scheduling because more restrictive legislation was not framed. Psychiatrists who have used MDMA as an adjunct to chemotherapy claim that it is not hallucinogenic but rather removes barriers to effective communication.

The reversed ester derivatives of meperidine, 1-methyl-4-phenyl-propionoxypiperidine (MPPP) and 1-(2-phenylethyl)-4-phenyl-4-acetoxypiperidine (PEPAOP), have also been identified as designer drugs. Improper control of temperature and pH during synthesis results in the formation of the tetrahydropyridine derivatives MPTP and PEPTP (Fig. 15–3).

MPTP has been identified as a neurotoxin with high affinity for the substantia nigra and is capable of precipitating signs of Parkinson's disease. It is rapidly metabolized by MAO B to the *N*-methyl-4-phenylpyridinium ion (MPP^+), and during this, oxidation generates the neurotoxin.[129] PEPTP is under study as a potential neurotoxin associated with patients exhibiting symptoms of Huntington's chorea.

Indoles

Indole is the base for a diverse group of hallucinogenic agents that may be subdivided into simple indole, harmine, and polycyclic derivatives, including yohimbine

Table 15–20. Phenylisopropylamine Derivatives

Substitution Pattern 2	3	4	5	6	Potency* (MU)	Ki Values† $5HT_2$ nM
H	H	OCH_3	H	H	5	33600
OCH_3	H	OCH_3	H	H	5	
OCH_3	H	H	OCH_3	H	8	5200
H	OCH_3	OCH_3	H	H	<1	43300
H	OCH_3	OCH_3	OCH_3	H (TMA)	2.2	
OCH_3	H	OCH_3	OCH_3	H (TMA-2)	17	1650
OCH_3	OCH_3	OCH_3	H	H	<2	
OCH_3	OCH_3	H	OCH_3	H	4	
OCH_3	OCH_3	H	H	OCH_3	13	
OCH_3	H	OCH_3	H	OCH_3	10	
H	—O—CH_2—O—		H	H(MDA)	3	R(−) 3420
H	OCH_3	—O—CH_2—O—		H(MMDA)	3	
OCH_3	H	—O—CH_2—O—				
OCH_3	—O—CH_2—O—		H	H	10	
—O—CH_2—O—		OCH_3	H	H	3	
OCH_3	—O—CH_2—O—		OCH_3	H	12	
OCH_3	OCH_3	—O—CH_2—O—		H	5	
H	OCH_3	—O—$(CH_2)_2$—O—		H	<1	
OCH_3	OCH_3	OCH_3	OCH_3	H	6	
OC_2H_5	H	OCH_3	OCH_3	H	<7	
OCH_3	H	OC_2H_5	OCH_3	H	15	
OCH_3	H	OCH_3	OC_2H_5	H	<7	
OCH_3	H	CH_3	OCH_3	H (DOM)	80	R(−) 60
OCH_3	H	C_2H_5	OCH_3	H	150	
OCH_3	H	Br	OCH_3	H (DOB)		R(−) 24
OCH_3	H	I	OCH_3	H (DOI)		R(−) 10

TMA = 3,4,5-trimethoxyamphetamine.
TMA-2 = 2,4,5-trimethoxyamphetamine.
MDA = 3,4-methylenedioxyamphetamine.
MMDA = 4,5-methylenedioxy-3-methoxyamphetamine.
DOM = 1-(2,5-dimethoxy-4-methylphenyl)-2-aminopropane.
DOB = 1-(2,5-dimethoxy-4-bromophenyl)-2-aminopropane.
DOI = 1-(2,5-dimethoxy-4-iodophenyl)-2-aminopropane.
* From Shulgin.[117]
† From Glennon.[118]

and lysergic acid derivatives. Many simple indole derivatives may be viewed as possible metabolites of normal physiologic compounds, such as tryptophan, serotonin, and norepinephrine, and therefore enhanced formation of such derivatives has been implicated as a cause of mental disease.

With the exception of adrenolutin, all hallucinogenic indoles are substituted in the 3 position with a 2-aminoethyl function, as shown in Table 15–21. (Aminomethyl compounds are active only when suitably substituted on the ring.) Dialkylaminoethyl derivatives are superior in activity to other indoles. Many are used ritually and recreationally by Central and South American Indians in the form of snuff. These contain 5-methoxydimethyltryptamine, dimethyltryptamine, 5-methoxy-N-methyltryptamine, N-methyltryptamine and β-carbolines. Diethyl and dipropylamine analogs are less frequently encountered than dimethyl compounds but still are considered street drugs.

Adrenolutin

Plants in the Middle East (Peganum harmala), South America (Banisteriopsis), and Africa (Leptactinia densiflora) containing harmala alkaloids are used as spices and intoxicants. Hallucinogenic effects of these alkaloids

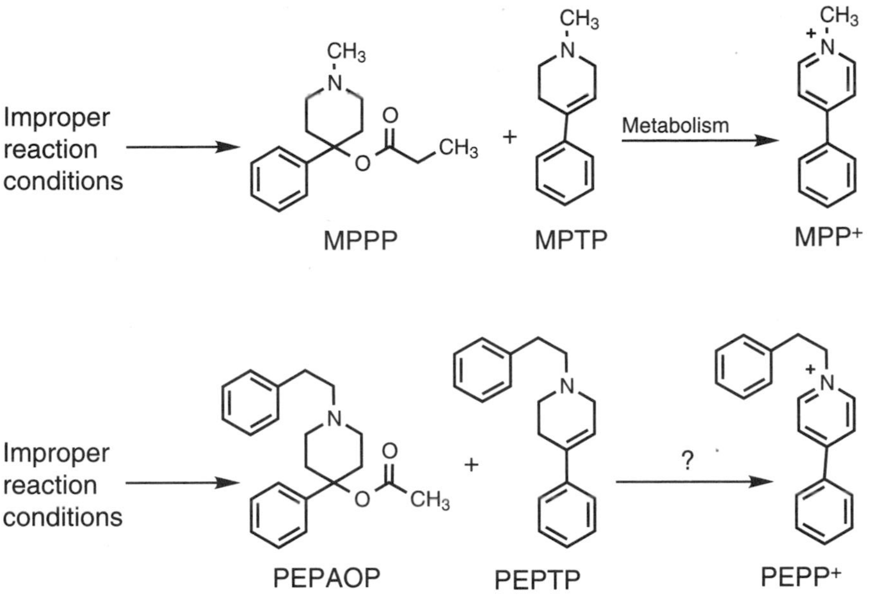

Improper reaction conditions → MPPP + MPTP —Metabolism→ MPP^+

Improper reaction conditions → PEPAOP + PEPTP —?→ $PEPP^+$

Fig. 15–3. Proposed mechanism for neurotoxin generation. (Data from K. Chiba, et al., *Neurosci. Lett., 48,* 1984.)

appear to be more closely related to mescaline than to LSD-25. Even mescaline, which is claimed to be the drug that most affects vision in its group, does not precipitate effects as numerous and realistic at those experienced with harmala alkaloids (Table 15–22). The drug 5-methoxytetrahydroharman has been isolated from animals as an in vitro product of melatonin and is currently under investigation for possible formation in schizophrenia.[131]

Tetracyclic and pentacyclic indole derivatives include the most potent hallucinogenic compounds, derivatives of lysergic acid. Lysergic acid diethylaminde (LSD) was first synthesized by Sandoz Research Laboratories in Basel, Switzerland, in 1938 but was not identified as a hallucinogen until 1943. Following this report, it was studied as a therapeutic aid in the treatment of chronically depressed persons, alcoholics, and terminal cancer patients.

Although lysergic acid has been synthesized, it is more conveniently available from naturally occurring ergot alkaloids. Claviceps purpurea, a fungus that parasitizes certain grains, such as rye and wheat, is the major source, although alkaloids may be obtained from varieties of morning glory (Rivea corymbosa, Ipomoea sidaefolia, and Convolvulus). Ingestion of bread containing high concentrations of these alkaloids results in gangrenous or convulsive ergotism. Two major epidemics in the tenth and twelfth centuries killed more than 40,000 people and were responsible for legislation limiting concentrations of ergot-contaminated grains. Ergotism has been reported as late as 1953 in France and may be encountered in the United States as a result of the fad of eating morning glory seeds. Fungus-free sources have yielded isolysergic acid amide, lysergic acid amide, and more complex alkaloids. These materials are generally less potent than

Table 15–21. Hallucinogenic Indoles

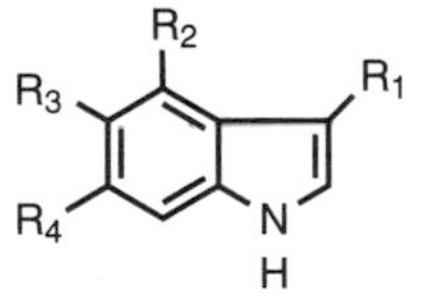

Compound	Structure				Potency (Mescaline units)
	R_1	R_2	R_3	R_4	
Bufotenine*	$(CH_2)_2N(CH_3)_2$	H	OH	H	±
Psilocin	$(CH_2)_2N(CH_3)_2$	OH	H	H	31
6-Hydroxydiethyltryptamine	$(CH_2)_2N(C_2H_5)_2$	H	H	OH	25
Dimethyltryptamine	$(CH_2)_2N(CH_3)_2$	H	H	H	4
Diethyltryptamine	$(CH_2)_2N(C_2H_5)_2$	H	H	H	>4
5-Methoxydimethyltryptamine	$(CH_2)_2N(CH_3)_2$	H	OCH_3	H	>31

* Activity controversial.
Data from P. Brawley and J. C. Duffield, *Pharm. Rev., 24,* 31 (1972).

Table 15–22. Carboline Derivatives

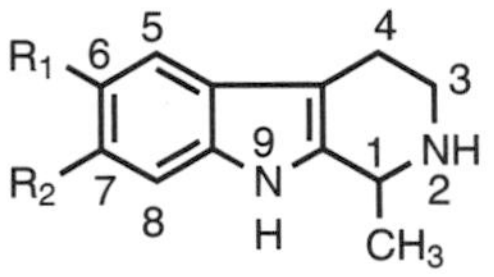

Compound	Structure R₁	Structure R₂	Position of Double Bond	Hallucinogenic Dose in Humans
Harmine	H	CH_3O	1,3	150–200 mg (IV)
Harmaline	H	CH_3O	1	1 mg/kg (IV); 4 mg/kg (orally)
6-Methoxyharmalan	CH_3O	H	1	1.5 mg/kg (orally)
6-Methoxytetrahydroharman	CH_3O	H		1.5 mg/kg (orally)
1,2,3,4-Tetrahydroharmine	H	CH_3O		50–100 mg (IV)

diethylamide derivatives, yet have been used illegitimately to produce intoxication.

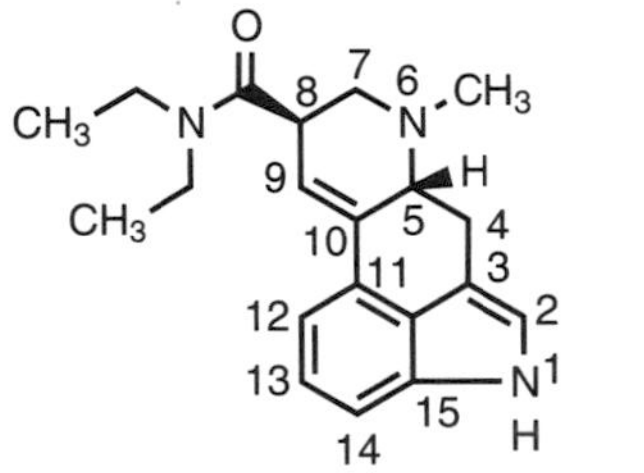

d-Lysergic acid diethylamide

Asymmetry of carbons 5 and 8 results in four diastercoisomers, d-lysergic acid, ℓ-lysergic acid, d-isolysergic acid, and ℓ-isolysergic acid, but only d-lysergic acid analogs have hallucinogenic potency. Psychotomimetic activity is closely related to structural modifications, which are limited to minor changes (Table 15–23). Exceptional activity of N-acetyl derivatives may be explained based on the ease of hydrolysis to the parent compounds. Hydrogenation of LSD to the corresponding 2,3-dihydro derivative decreases activity about eightfold, whereas hydrogenation of the 9,10-double bond destroys activity. Other compounds such as 2-bromolysergic acid diethylamide have been synthesized and lack hallucinogenic potency.

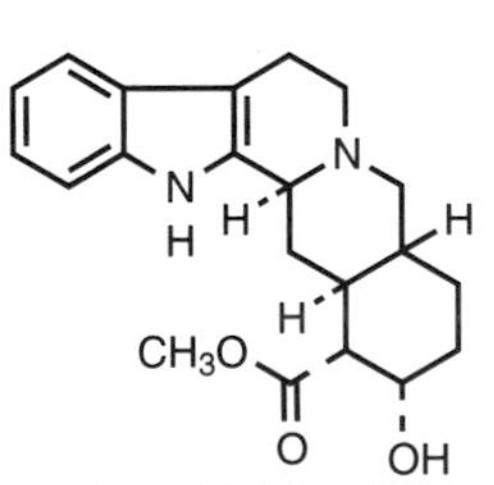

(+/-) Yohimbine

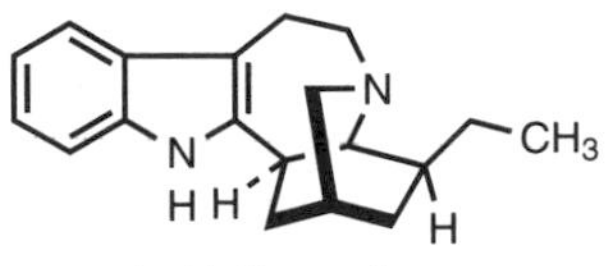

(+/-) Ibogaine

Yohimbine and ibogaine are claimed to produce hallucinogenic as well as other interesting effects, including aphrodisia and a cocaine type of euphoria.

Both compounds are CNS stimulants and cause hallucinations (yohimbine, 0.5 mg/kg; ibogaine, 3 mg/kg, both intravenously). Yohimbine may activate latent psychoses in schizophrenics[133] and produce anxiety in normal subjects. A similar phenomenon has been suggested for ibogaine, although the compounds exhibit different pharmacologic properties.

Marijuana

Marijuana plants (Cannabis sativa) contain a variety of cannabinoid derivatives, which together with alcohol constitute the world's most widely abused drugs. Concentrations of cannabinoids vary from plant to plant,[134] although cannabidiol, Δ^9-tetrahydrocannabinol (THC), cannabidiolic acid, and Δ^9-tetrahydrocannabinolic acid are considered to be primarily responsible for biologic activity.

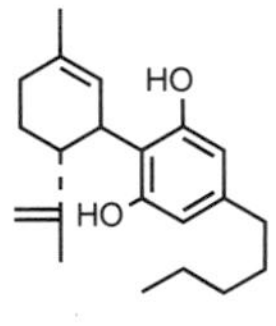

Cannabidiol

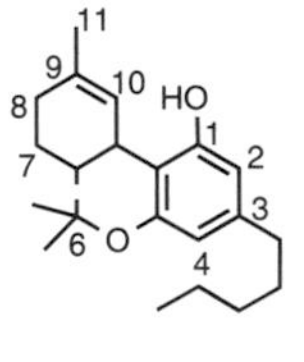

Δ^9-Tetrahydrocannabinol

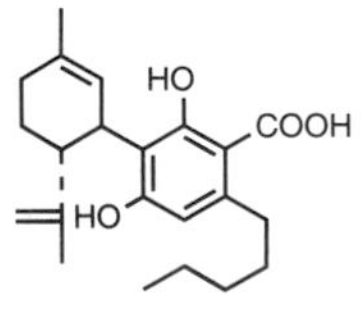

Cannabidiolic acid

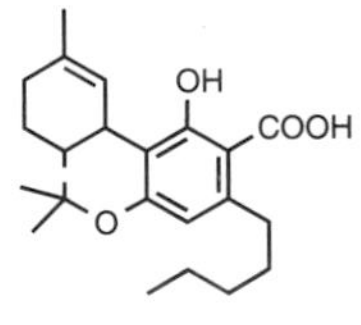

Δ^9-Tetrahydrocannabinolic acid A

Cannabinoid rings may be numbered two ways; one is based on the pyran system, which has been generally accepted, and one is based on the monoterpenoid system,

Table 15–23. Hallucinogenic Lysergic Acid Derivatives

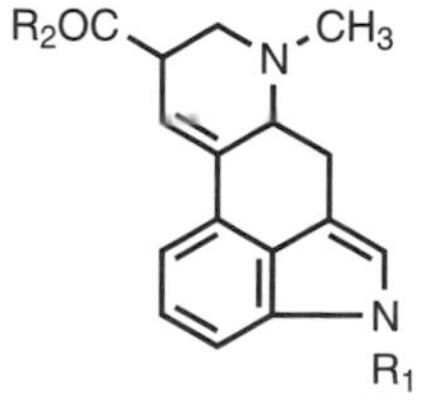

Compound	R_1	R_2	Relative Potency (LSD = 100)
d-Lysergic acid diethylamide	H	$N(C_2H_5)_2$	100
dl-Methyllysergic acid diethylamide	CH_3	$N(C_2H_5)_2$	40
dl-Acetyllysergic acid diethylamide	$COCH_3$	$N(C_2H_5)_2$	100
d-Lysergic acid amide	H	NH_2	10
d-Lysergic acid ethylamide	H	NHC_2H_5	5
d-Lysergic acid dimethylamide	H	$N(CH_3)_2$	10
d-Lysergic acid pyrrolidide	H	C_4H_8N	10
d-Lysergic acid morpholide	H	C_4H_8NO	20
dl-Methyllysergic acid monoethylamide	CH_3	NHC_2H_5	5
dl-Acetyllysergic acid monoethylamide	$COCH_3$	NHC_2H_5	5
dl-Methyllysergic acid pyrrolidide	CH_3	C_4H_8N	7

From A. Hoffer and H. Osmond, *The Hallucinogens,* New York, Academic Press, 1967, p. 94.

which is more convenient when viewing biosynthesis of the derivatives. According to the numbering in this chapter, Δ^8-tetrahydrocannabinol, would be the same as $\Delta^{1(6)}$-tetrahydrocannabinol, using the monoterpenoid system.

Pyran Nomenclature — Δ^8-Tetrahydrocannabinol

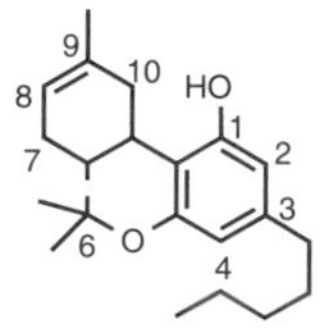

Monoterpenoid Nomenclature — $\Delta^{1(6)}$-Tetrahydrocannabinol

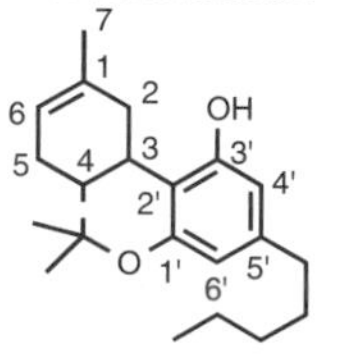

Biologic effects of marijuana smoking are duplicated by THC, which suggests that this is the primary active compound. The first cannabinol tested on human volunteers that retained the effects of the plant was Δ^8-tetrahydrocannabinol, and it was theorized to be produced from THC during smoking. Subsequent studies proved this assumption erroneous. Cannabidiol, however, is converted in small amounts to Δ^9 THC. This may explain the physiologic activity observed after smoking marijuana containing only small amounts of Δ^9 THC.[135]

Early attempts to synthesize Δ^9 THC resulted in unnatural isomers, including 7,8,9,10-tetrahydrocannabinol, which had the same amount of activity as Δ^9 THC in animals. Changes in the alkyl chain attached to the aromatic rings brought about considerable changes in activity (Table 15-24).

Despite the emotional furor surrounding marijuana, certain aspects of its pharmacology indicate potential use for its constituents or derivatives. For example, THC has been shown to reduce intraocular pressure, suggesting potential usefulness in the treatment of glaucoma. Bronchodilator, anticonvulsant, and hypotensive activity may result in novel antiasthmatic, antiepileptic, and antihypertensive therapy. The ability of tetrahydrocannabinol to reduce nausea and vomiting may provide an effective antiemetic compound particularly useful in combination with cancer chemotherapy (treatment that often results in nausea and vomiting). Also, analgesia has been identified as a component of cannabinoid activity.[136]

Δ^9-Tetrahydrocannabinol has been approved by the Food and Drug Administration under the generic name dronabinol for use in nausea and vomiting associated with cancer chemotherapy.

Direct therapy with Δ^9 THC (Marinal) suffers from the disadvantages of central dysphoric effects, the potential for tolerance, the poor water solubility, and the restriction to inhalation as the most effective route of administration (half-life greater than 2 hours). Thus, synthetic analogs are being designed in efforts to minimize the undesirable characteristics of marijuana but to retain the low toxicity and lack of physical dependence. Nabilone, a synthetic cannabinoid, and certain nitrogen and sulfur-containing benzopyrans are currently being clinically evaluated for a number of therapeutic indications. Nabilone, thought to have less propensity to produce hallucinogenic-like effects than marijuana, is currently available as Cesamet (half-life 4 hours) for use in nausea and vomiting associated with cancer chemotherapy.

Few synthetic analogs have been adequately tested in humans, although the ones that have been tested have the same potency as in animals. Tests with animals indicate the following relative potencies: dimethylheptyl-Δ^9-tetrahydrocannabinol greater than Δ^9 THC greater than Δ^8-tetrahydrocannabinol greater than 3(1,2-dimeth-

Table 15–24. Cannabinol Derivatives

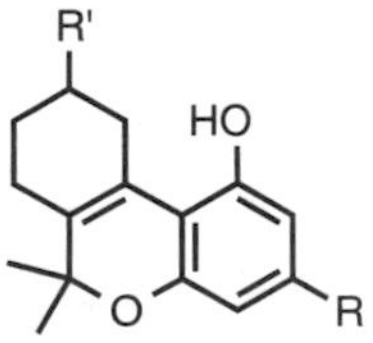

Compound	R	R′
7,8,9,10-Tetrahydrocannabinol	C_5H_{11}	CH_3
Synhexyl	C_6H_{13}	CH_3
3-(1,2-Dimethylheptyl)-7,8,9,10-tetrahydrocannabinol	$CH(CH_3)CH(CH_3)C_5H_{11}$	CH_3
3-(1-Methyloctyl)-7,8,9,10-tetrahydrocannabinol	$CH(CH_3)C_7H_{15}$	CH_3
Nabilone (Cesamet)	$C(CH_3)_2C_6H_{13}$	O

ylheptyl)-7,8,9,10-tetrahydrocannabinol greater than 3(1-methyloctyl)-7,8,9,10-tetrahydrocannabinol greater than synhexyl greater than 7,8,9,10-tetrahydrocannabinol. These tests are complicated by poor water solubility. Attempts to solubilize a compound by molecular modification of the phenolic group reduced activity except with water-soluble diethylaminobutyric acid esters. Smoking delivers between 25% and 75% of active ingredient, indicating cold nebulization as a suitable dosage form.

Structure-activity relationships for the cannabinols are listed as follows:[137]

1. The 1-hydroxyl substituted benzopyran nucleus substituted with a 3-alkyl or alkoxyl is necessary for activity.
2. Esterification of the phenol or replacement by NH_2 retains activity, whereas etherification or replacement with SH eliminates it.
3. Activity is retained by aromatic ring substitution with alkyl and OH but is destroyed by electronegative groups such as COOH.
4. Activity is related to a minimum length of the 3′-alkyl side chain and is increased with branching.
5. The 6-*gem*-dimethyl is optimum for activity in the pyran ring, but it may be expanded with a one methylene or converted to the nitrogen isostere without activity loss. Opening of the pyran ring decreases activity.
6. The alicyclic ring may be unsaturated at the 8, 9 or 10a positions, and for optimum activity the 10a–10b ring juncture should be trans.
7. The alicyclic ring may be substituted by N or S insertion in place of C-9 or C-7. A methyl substitution on C-9 enhances activity.
8. Planarity of the alicyclic ring is not essential for activity. Coplanarity of all three rings has been suggested as a factor in the low potency of the $\Delta^{9(11)}$ derivative[138]

Cocaine

Cocaine is extracted from the leaves of the coca plant, *Erythroxylon coca,* grown in the mountains of South America where it has been used by the natives for centuries to relieve fatigue and promote a feeling of well-being. Originally part of Inca culture, cocaine has become a major drug abuse problem in the United States because of its powerful stimulant and reinforcing properties.

In earlier times, the natives obtained cocaine by chewing the coca leaves, but modern use is normally by intravenous injection, intranasal application ("snorting"), or smoking the freebase form (crack). The half-life of cocaine is 1 hour or less, because of its rapid metabolism. Cocaine is metabolized by de-esterification to ecgonine and benzoylecgonine and by *N*-demethylation to norcocaine. Ester hydrolysis by plasma and tissue esterases accounts for its major metabolic route in humans.[139] A condensation product with ethanol, cocaethylene, has also been reported,[140] which is a potentially lethal metabolite when abuse of these two drugs is combined.[141]

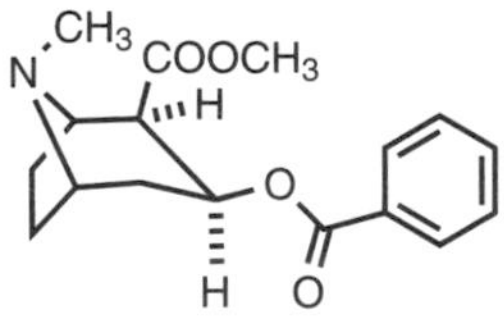

Cocaine produces an elevation of mood and self-esteem and a sense of increased mental and physical capacity. Experienced users are generally unable to distinguish between the subjective effects of cocaine and amphetamine.[142] The toxicity of cocaine affects several organ systems. Hyperexcitability of the CNS results in hyperactivity, elevated body temperature, and seizures. The cardiovascular system is also susceptible, inducing hypertension, cardiac arrhythmia, cerebrovascular ischemia, cardiomyopathy, and cardiac arrest.

The mechanism of action of cocaine is complex. Some of the toxicity is likely related to its local anesthetic properties. The rewarding effects are likely due to blockade of neuronal reuptake of biogenic amines in the CNS. Extensive structure activity studies indicate that an action on the transporter for dopamine is correlated with self-administration and abuse potential.[143] Repeated administration of cocaine in laboratory animals has been associated with a reverse tolerance to certain behaviors, probably as a result of a dopaminergic supersensitivity.[144] Clinically, it has been suggested that the long-term use of cocaine produces a dopamine depletion in reward areas, which may be responsible for psychologic dependence.[145]

Pharmacology

Variations in response to psychotomimetics and conflicting reports on effects are numerous. Many psychotomimetic studies employ statistics derived from the illicit use of drugs of uncertain chemical composition and dosage. Because street-drug contaminants are difficult to identify, these studies must be cautiously compared with responses from controlled laboratory administration of known agents. Responses in some studies are obtained after initial drug administration and in others after prolonged administration. Because these agents are usually taken more than once, studies based on prolonged administration should be more enlightening.

Drug responses have been related to subjects' previous experiences with drugs and differ in naive and experienced subjects. Set and setting of drug use present yet another difficulty. Set refers to the psychologic attitude and mood of the person at administration. If he is upset or disturbed, psychotomimetic drugs produce an effect different from that produced when he is happy, content, and exhibiting a positive attitude toward the outcome. External circumstances of setting also have an impact, depending on whether the drugs are taken with peers or in a laboratory. These effects are indicative of a significant placebo response.

Table 15-25 summarizes the physiologic effects of the psychotomimetic drugs with specific reference to prototype drugs LSD, mescaline, and marijuana. The responses indicated are those most prominent in the literature, although some studies may report an opposing result. Mescaline and LSD have strikingly similar psychic, somatomotor, autonomic, and direct actions, although mescaline appears to cause a greater alteration of visual perception.

Marijuana, classified as a psychotomimetic because of psychic actions similar to those of LSD and mescaline, differs considerably in physiologic actions. Marijuana use may be characterized by increase in pulse rate, reddening of eyes, dry mouth, and increase in appetite. CNS effects are also considerably different. Marijuana depresses the electroencephalogram when the subject is awake, and although it can initially produce excitation, the predominant effect is sedation and sleep. Despite marijuana's effect on heart rate, many investigators believe that it strongly resembles pharmacologically the sedative-hypnotics because of the sedative and anticonvulsant actions. Marijuana has been compared with morphine[146,147] and phenytoin with the suggestion that they have a common site of action.

In general, all psychotomimetics exhibit the full range of psychic alterations with only quantitative differences. By far the most potent is LSD, which humans take in an oral dose of 0.5 to 1.5 μg/kg and a street dose of 125 μg. In order of decreasing potency, these agents may be listed as follows: LSD, psilocybin, 2,5-dimethoxy-4-methylamphetamine, dimethyl-tryptamine, THC, and mescaline. With the exception of marijuana, oral administration of the psychotomimetics results in predictable effects and is preferred over other dosage forms. Effects of marijuana when taken orally are unpredictable. Inhalation of marijuana is approximately three times as effective as oral administration, and even though at least half of the drug may be destroyed by smoking, the total effect is stronger. Inhaled doses of 2 to 20 mg of THC produce almost immediately psychotomimetic symptoms that may persist 1 to 4 hours depending on the dose. Synhexyl, a synthetic marijuana derivative, has a slower onset and longer duration of activity.

The psychotomimetics are rapidly removed from the blood and distributed to various body tissues (Table 15–26). Mescaline and LSD are rapidly distributed to the liver, kidneys, and brain. Marijuana achieves a high concentration in lung tissue after both intravenous injection and inhalation. All these agents cross the blood–brain and the placental barrier. Distribution in select areas of the brain varies. Mescaline and LSD are found in high concentrations in the pituitary gland and in visual and auditory areas. High concentration of LSD is also found in the pineal gland (a site of high serotonin concentration).

The major metabolism of hallucinogens occurs in the liver, although some detoxification of LSD occurs in muscle and brain. Hydroxylation with frequent glucuronidation is the major route for LSD and THC, resulting in active metabolites of THC. Mescaline is poorly absorbed by the CNS and readily undergoes oxidative deamination to 3,4,5-trimethoxyphenylacetic acid and several O-demethylated metabolites.

Metabolic hydroxylation of marijuana may depend on frequency of use.[148] A different plasma half-life is observed in nonsmokers (56 hours) and smokers (28 hours), suggesting a more rapid metabolism in subjects accustomed to its use. One metabolite, 11-hydroxytetrahydrocannabinol, is as active as THC and accounts for nearly 25% of the metabolic products. Little unchanged THC is excreted by either experienced or naive subjects. Other possible metabolites have been identified in animals and include 7-hydroxy-THC, 7-hydroxy-Δ^8-tetrahydrocannabinol, and 8α,11-dihydroxy-THC, all of which are biologically active. More sensitive isolation procedures have revealed many additional polar metabolites that have yet to be identified. Marijuana is rapidly removed from the plasma and then slowly metabolized, resulting in a biphasic half-life, initially about 30 minutes and then 56 hours. A shorter half-life of 27 hours is observed with chronic marijuana use.

Marijuana-induced "highs" are associated with plasma levels of 150 to 200 mg/ml, although direct correlations are not possible because of considerable lag time between peak levels and "highs."[149] Extension of this observation to metabolite 11-hydroxy-THC is plausible and might contribute to delayed effects of oral administration.[150] Bioavailability of THC varies with mode of administration between 8% to 24% (smoking) and 2% to 19% (oral) owing to a considerable first-pass effect.[151]

Table 15–25. Physiologic Effects of Psychotomimetics

	LSD	Mescaline	Marijuana
Psychic			
Excitation	+	+	+ (initial)
Euphoria/depression	+	+	+
Hallucinations	+ +	+ +	+ (HD)
Depersonalization	+ +	+ +	+ (HD)
Altered perception			
Visual and tactile	+	+ +	+
Time and space	+	+	+
Electroencephalographic wakefulness	+	+	0/−
Psychotic states	+	+	+ (HD)
Sedation	0/−	0/−	+
Somatomotor			
Spontaneous activity	+	+	−
Ataxia	+	+	+ (HD)
Convulsions	+/−	+	−
Paralysis	+	+	Not reported
Autonomic			
Heart rate	+/−	+/−	+ +
Blood pressure	+/−	+	0/−
Nausea	+	+ +	−
Micturition	0/−	0/−	+
Secretions	+/−	Not reported	−
Mydriasis	+ +	+ +	0
Conjunctival reddening	0	0	+
Respiration	− (HD)	− (HD)	− (HD)
Temperature	+ +	+ +	+
Appetite	0	0	+
Analgesia	+	+ (HD)	+
Metabolic			
Blood glucose	+	0	0
17-Ketosteroid excretion	+	+	0
Free fatty acid blood levels	+	+	0
Direct			
Vasoconstriction	+ (HD)	+	−
Uterine contraction	+	+	−
Bronchial constriction	+ (HD)	Not reported	−

HD = High dose.

Tissue distribution rates are similar for both naive and experienced subjects and are rapid for the first 30 minutes. The subsequent decrease in rate of disappearance from the plasma probably represents the true rate of metabolism.

Many agents acting on the CNS cause tolerance, necessitating the use of increasingly larger doses to achieve pharmacologic responses. The mechanism of tolerance is generally unknown and is probably a combination of behavioral and biochemical factors. Tolerance to LSD occurs rapidly and is lost within a few days. In humans, tolerance to mescaline develops more slowly than to LSD. Tolerance to certain effects of marijuana has been reported in rats, dogs, pigeons, and humans.[152]

Pharmacokinetic differences discerned in light and heavy users of marijuana are not of sufficient significance to explain behavioral and pharmacologic differences.[153]

Reverse tolerance has been suggested to occur in humans, partly because of a learned behavioral response after repeated doses in which the subject needs less drug to initiate a recognizable response and partly because of conversion of Δ^9-tetrahydrocannabinol to an active metabolite. Enzyme activation and prolonged storage have also been implicated in the reverse tolerance.

Cross-tolerance occurs with LSD, mescaline, psilocybin, and bufotenin, although there is no cross-tolerance between marijuana and these agents. The drug THC has exhibited cross-tolerance with Δ^8-tetrahydrocannabinol, synhexyl, and a dimethylheptyl derivative. There is little evidence to indicate physical dependence on any of these

Table 15–26. Psychotomimetic Pharmacokinetics

	LSD	Mescaline	Marijuana
Route of administration	Oral	Oral	Inhalation
Oral dose in humans	100 μg	250–500 mg	20–50 mg tetrahydrocannabinol
Onset			
Oral	15–120 min	15–120 min	30–120 min
Intravenous	Immediate	Immediate	Immediate
Inhalation	Not reported	Not reported	Immediate
Duration	8–12 hr	24 hr	3–4 hr
Metabolism	Oxidation	Deamination	Hydroxylation
Mechanism	Hydroxylation Conjugation	O-Demethylation N-Acetylation	
Active metabolites	Not reported	Not reported	11-Hydroxytetrahydrocannabinol 7-Hydroxytetrahydrocannabinol
Inactive metabolites	2-Keto-LSD	Trimethoxyphenylacetic acid	8α,11-Dihydroxy-Δ^9-tetrahydrocannabinol
Half-life in humans	2 hr and 55 min	1–6 hr	25–36 hr (including metabolites)
Plasma-protein binding	Yes	Weak	Yes
Excretion route	Bile, urine	Urine	Bile, urine
Cross placental barrier	Slightly	Yes	Yes

agents. Dramatic opiate-like withdrawal signs observed by some investigators are not commonly observed in most studies.[154]

Relatively few instances of fatal overdoses of hallucinogens have been found. Death, when it does occur, is a result of respiratory depression. There have been frequent reports of psychic toxicity, paranoia, and confusion following initial administration of LSD, mescaline, and marijuana. An amotivational syndrome in users of the hallucinogens has also been described, although a causal relationship to the drug is not conclusive. Recurrent effects are most common for LSD and mescaline and include visual and auditory flashbacks as well as a momentary return to the depersonalized state. Reports of prolonged toxicity, although rare, include the production of chronic anxiety, mental deterioration, and psychoses.

Mutagenicity, teratogenicity, and carcinogenicity are of concern because of widespread illicit use of these drugs. Despite reported LSD-induced chromosomal damage, there is no evidence that pure normal doses cause these aberrations. Data on Drosophila and fungi suggest that in extremely high doses it is weakly mutagenic, although this dose level is unlikely to be reached in humans. Reports of marijuana-induced or mescaline-induced chromosomal changes are conflicting and inconclusive.

Evidence of teratogenic potential of LSD, mescaline, and marijuana has been obtained from observed congenital malformations in rodents. There is an indication of increased risk of spontaneous abortion in users of illicit LSD. There is no substantial evidence, however, that pure LSD causes birth defects in humans and no evidence that marijuana or mescaline alters fetal development. The use of any agent of unknown teratogenic or mutagenic capabilities during pregnancy is unwise. Preliminary evidence suggests that marijuana may suppress the immune system, an action that has potential therapeutic implications in the development of immunosuppressive and antineoplastic drugs. No evidence has yet been validated concerning possible carcinogenicity of any hallucinogen; of course, inhalation of marijuana smoke may have effects similar to those of tobacco smoke inhalation. Chronic smoking of marijuana cigarettes has been associated with increased incidence of bronchitis and asthma, most likely due to smoke inhalation because THC is a direct bronchodilator.

To date, many theories have been proposed to explain psychotomimetic effects. Changes in cerebral blood flow and permeability of cerebral capillaries have been suggested. Alterations in levels of adrenal cortical and thyroid hormones have been implicated without substantial evidence. Most research has centered on neurohormonal hypotheses that suggest changes in synthesis or metabolism of serotonin, norepinephrine, acetylcholine, or other potential transmitters.

LSD interacts with several transmitter receptors, but the most convincing evidence involves interaction with receptors for the indole neurotransmitter serotonin. Notably, LSD binds with high affinity to the $5HT_2$ receptor and demonstrates a drug discrimination pattern consistent with $5HT_2$-like activity,[155] although controversy exists whether it is acting as a full or partial agonist at this receptor. Activity at the $5HT_{1a}$ and the $5HT_{1c}$ receptors has also been proposed in addition to a mechanism involving receptors for dopamine.[156,157] Dopamine and other catecholamine neurotransmitters appear to play a more predominant role in the action of phenethylamine hallucinogens. Amphetamine and its analogs produce complex effects on dopaminergic neuronal mechanisms, including interference with release, uptake, and degradation. Several of the amphetamine analogs, in high doses, cause damage to dopaminergic and serotoninergic neurons, perhaps by promoting auto-oxidation.[158]

PCP, at high concentrations, produces an enhancement of the release of a variety of neurotransmitters, but considerable emphasis has been placed on its actions on presynaptic dopaminergic mechanisms.[159] A selective binding site,[160] probably related to the sigma receptor,[161] may be responsible for its psychotomimetic and indirect dopaminergic effects.[161] THC has a distinct mechanism from the other hallucinogens. Traditionally, THC's effects have been related to a nonspecific transmitter or membrane perturbing effects.[162] A selective binding site for THC, however, has been discovered in the brain that is negatively coupled to adenylyl cyclase activity.[163]

Various abnormal metabolic pathways for neurohormones that result in production of endogenous psychotogens have been proposed. Hallucinogenic compounds, adrenochrome or adrenolutin, 3,4-dimethoxyphenethylamine, 6-methoxyharmalan, and taraxein are postulated to result from epinephrine, dopamine, melantoin, and protein synthesis.

Cerebral function is extremely dependent on the use of energy in the form of ATP. Energy is necessary for the membrane pump to maintain neuronal ionic gradients and hence electrical excitability in cerebral neurons as well as for the termination of adrenergic transmitter activity and may be involved in release and indirect regulation of neurotransmitter levels. Psychotomimetics may disrupt cerebral energy production or use in such a fashion that it alters behavior.

REFERENCES

1. H. Caper and W. M. Herrmann, *Pharmacopsychiatry, 21,* 211(1988).
2. T. D. Turner, *Pharm. J., 189,* 151(1962).
3. A. Horita, et al., *Clin. Neuropharmacol., 28,* 481(1989).
4. A. Horita, et al., *Pharmacol. Biochem. Behav., 115*(1988).
5. A. Horita, and M. A. Carino, *Pharmacol. Biochem. Behav., 39,* 449(1991).
6. M. Yabase, et al., *Pharmacol. Biochem. Behav., 36,* 375(1990).
7. J. A. Kemp and P. D. Leeson, *Trends Pharmacol. Sci., 14,* 20(1993).
8. C. H. Jarboe, *J. Med. Chem., 11,* 729(1968).
9. P. K. Knoefel, *J. Pharmacol. Exp. Ther., 84,* 26(1945).
10. M. G. Romeo, et al., *Pediatr. Med. Chir., 13,* 77(1991).
11. F. H. Shaw, et al., *J. Pharm. Pharmacol., 9,* 666(1957).
12. P. R. Andrews, et al., *J. Med. Chem., 26,* 1223(1983).
13. E. G. Gross, *J. Pharmacol. Exp. Ther., 87,* 291(1946).
14. E. G. Gross, *J. Pharmacol. Exp. Ther., 89,* 330(1948).
15. A. Goodman and L. Gilman, *The Pharmacological Basis of Therapeutics,* 4th ed., New York, Macmillan, 1971, p. 359.
16. L. J. Dorfman, et al., *Clin. Pharmacol. Ther., 11,* 869(1970).
17. B. Weiss, et al., *Pharm. Rev., 14,* 1(1962).
18. H. H. Cornish, et al., *J. Biol. Chem., 228,* 315(1957).
19. S. Sved, et al., *Res. Commun. Chem. Path. Pharmacol., 13,* 185(1976).
20. J. W. Jenne, et al., *Clin. Pharmacol. Ther., 13,* 399(1972).
21. E. B. Truitt, Jr., in *Drill's Pharmacology in Medicine,* 4th ed. J. R. DiPalma, Ed. New York, McGraw-Hill, 1971.
22. J. F. Greden, *Am. J. Psychiatr., 131,* 1089(1974).
23. A. Z. LaCroix, et al., *N. Engl. J. Med., 315,* 977(1986).
24. B. R. Davis, et al., *Am. J. Epidemiol., 128,* 124(1988).
25. D. J. Sutherland, et al., *Chest, 87,* 319(1985).
26. S. Kalsner, *Life Sci., 20,* 1689(1977).
27. D. Robertson, et al., *J. Clin. Invest., 67,* 1111(1981).
28. W. Kuhlmann, et al., *Cancer Res., 28,* 2375(1968).
29. P. S. Thayer and P. E. Palm, *CRC Crit. Rev. Toxicol., 3,* 345(1975).
30. O. H. Choi, et al., *Life Sci., 43,* 387(1988).
31. J. W. Daly, et al., *Life Sci, 28,* 2083(1981).
32. R. M. Green and G. L. Stiles, *J. Clin. Invest., 77,* 222(1986); see S. G. Holtzman, et al., *J. Pharmacol. Exp. Ther., 256,* 62(1991).
33. B. A. Berkowitz et al., *Eur. J. Pharmacol., 10,* 64(1970).
34. *Anti-Depressant Drugs of Non-MAO Inhibitor Type.* Washington, D. C., Workshop on Anti-Depressant Drugs, 1966, p. 140.
35. Mandell, G. L., et al., *Infect Immunol., 56,* 1722(1988).
36. G. Barger and H. H. Dale, *J. Physiol., 41,* 19(1910).
37. A. Shafiee and G. Hite, *J. Med. Chem., 12,* 266(1969).
38. K. Rickels, et al., *Clin. Pharmacol. Ther., 11,* 698(1970).
39. *Med. Lett. Drug Ther., 11,* 47(1969).
40. W. L. Smith, Ed., *Drugs and Cerebral Function, Cerebral Function Symposium.* First Aspen Institute for Humanistic Studies, Springfield, IL, Charles C. Thomas, 1969, p. 136.
41. J. Caldwell and P. Server, *Clin. Pharmacol. Ther., 16,* 625(1974).
42. L. Lemberger and A. Rubin, *Physiologic Disposition of Drugs of Abuse,* New York, Spectrum Publications, Inc., 1976.
43. Editorial: Narcolepsy and cataplexy, *Lancet, 1,* 845(1975).
44. A. J. Stunkard, in *Anorectic Agents: Mechanisms of Action and Tolerance,* S. Garattini and R. Samanin, Eds., New York, Raven Press, 1981, p. 191.
45. F. Zambotti, et al., *Eur. J. Pharmacol., 36,* 405(1976).
46. V. B. Clineschmidt, et al., *Eur. J. Pharmacol, 27,* 313(1974).
47. J. A. Hirsch and R. J. Wurtman, in *Anorectic Agents: Mechanisms of Action and Tolerance,* S. Garattini and R. Ramanin, Eds., New York, Raven Press, 1981, p. 159.
48. S. Garattini, et al., *Am. J. Clin. Nutr., 55,* 160S(1922).
49. T. Silverstone, and E. Goodall, *Am. J. Clin. Nutr., 55,* 212S(1992).
50. A. C. Sullivan, et al., *Recent Advances in Obesity Research, II,* G. Bray, Ed., London, Newman, 1978, p. 442.
51. J. C. Valle-Jones, *Br. J. Clin. Pract., 34,* 72(1980).
52. M. D. Fairchild and G. A. Alles, *J. Pharmacol. Exp. Ther., 158,* 135(1967).
53. M. O. Carruba, et al., in *Anorectic Agents: Mechanisms of Action and Tolerance,* S. Garattini and R. Samanin, Eds., New York, Raven Press, 1981, p. 101.
54. A. J. Stunkard, *Life Sci., 30,* 2043(1982).
55. P. H. Wender, et al., *Arch. Gen. Psychiatry, 38,* 449(1981).
56. L. A. Stroufe and M. A. Stewart, *N. Engl. J. Med., 289,* 407(1973).
57. D. J. Safer, et al., *J. Pediatr., 86,* 113(1975).
58. M. B. Denckla, *JAMA, 235,* 1349(1976).
59. S-W. Kwan, et al., *Psychopharmacology, 106,* S1(1992).
60. V. Glover, et al., *Nature, 265,* 80(1977).
61. J. W. Tetrud and J. W. Langston, *Science, 245,* 519(1989).
62. P. Levitt, et al., *Proc. Natl. Acad. Sci. USA, 79,* 6835(1982).
63. C. H. Williams, *Biochem. Pharmacol., 23,* 627(1974).
64. A. L. Maycock, et al., *Biochemistry, 15,* 115(1976).
65. C. L. Johnson, *J. Med. Chem., 19,* 600(1976).
66. T. R. Riley and C. G. Brier, *J. Med. Chem., 15,* 1187(1972).
67. A. S. Horn and S. H. Snyder, *J. Pharmacol. Exp. Ther., 180,* 523(1972).
68. D. F. Klein and J. M. Doors, *Diagnosis and Drug Treatment of Psychiatric Disorders.* Baltimore, Williams & Wilkins Co., 1969.
69. E. K. Silberman and J. L. Sullivan, *Psychiatr. Clin. North Am., 7,* 535(1984).
70. J. Marks and C. M. B. Pore, Eds., *The Scientific Basis of Drug Therapy in Psychiatry. A Symposium at St. Bartholomew's Hospital, London,* New York, Pergamon Press, 1965, p. 115.

71. J. Cooper, F. Bloom, and R. Roth, *The Biochemical Basis of Neuropharmacology,* New York, Oxford University Press, 1970.
72. W. Haefly, et al., *Psychopharmacology, 106,* S6(1992).
73. L. I. Globe, et al., *Clin. Neuropharmacol., 11,* 45(1988).
74. W. Burkhard, et al., *Psychopharmacology, 106,* S35(1992).
75. P. A. Van Zwieten, in *Prog. Pharmacol.,* Vol. 1, H. Grobecker et al., Eds. Stuttgart, Gustav Fischer Verlag, 1975, pp. 33–35.
76. D. J. Chapron, et al., *J. Pharm. Sci., 67,* 1018(1978).
77. S. D. Nelson, et al., *Science, 193,* 901(1976).
78. J. T. Coyle, et al., *Science, 219,* 1184(1983).
79. C. U. Giurgea, *Drug. Dev. Res., 2,* 441(1982).
80. M. W. Vernon and E. M. Sorkin, *Drugs Aging, 1,* 17(1991).
81. R. T. Bartus, et al., *Neurobiol. Aging, 2,* 105(1981).
82. M. Sarter, *Trends Pharmacol. Sci., 12,* 456(1991).
83. M. Koupilova, et al., *Activas Nervosa, 22,* Suppl 193(1980).
84. A. Lenegre, et al., *Pharmacol. Biochem. Behav., 29,* 625(1988).
85. C. D. Nicholson, *Psychopharmacology, 101,* 147(1990).
86. W. Poldinger, *Wien. Z. Nervenhellk., 24,* 128(1966).
87. T. DePaulis, et al., *Mol. Pharmacol., 14,* 596(1978).
88. D. Bente, *Arzneim, Forsch, 14,* 486(1964).
89. B. E. Leonard, *Typical and Atypical Antidepressants: Molecular Mechanisms,* E. Costa and G. Racagni, Eds. New York, Raven Press, 1980, p. 30.
90. R. Howe, et al., *J. Med. Chem., 19,* 1074(1976).
91. D. T. Wong and F. P. Bymaster, *Biochem. Pharmacol., 25,* 1979(1976).
92. D. T. Wong, et al., *J. Pharmacol. Exp. Ther., 193,* 804(1975).
93. P. Plege, et al., *J. Pharm. Pharmacol., 39,* 877(1987).
94. D. C. U'Prichard, et al., *Science, 199,* 197(1978).
95. G. L. Grunewald, et al., *Biochem. Pharmacol., 28.,* 417(1979).
96. B. K. Koe, et al., *J. Pharmacol. Exp. Ther., 226,* 686(1983).
97. Thomas, B. et al., *Psychopharmacology, 93,* 193(1987).
98. J. J. Schildkraut, *Am. J. Psychiatry, 122,* 509(1965).
99. J. Glowinski and J. Axelrod, *J. Pharmacol. Exp. Ther., 149,* 43(1965).
100. M. F. Sugrue, *Pharmacol. Ther., 21,* 1(1983).
101. F. Sulser, et al., *Ann. N. Y. Acad. Sci., 430,* 91(1984).
102. M. Rehavi, et al., *Brain Res., 194,* 443(1980).
103. A. Maggi, et al., *Eur. J. Pharmacol., 61,* 91(1980).
104. L. A. Nielsen, *Neurochem. Intl., 9,* 61(1986).
105. A. Pilc and K. G. Lloyd, *Life Sci., 35,* 2149(1984); P. D. Suzdak and G. Gianutsos, *Neuropharmacology, 24,* 217(1985).
106. A. S. Eison, et al., *J. Pharmacol. Exp. Ther., 246,* 571(1988); D. H. Manier, et al., *Biochem. Pharmacol., 36,* 3308(1987).
107. F. Sjoqvist and L. Bertilsson, in *Frontiers in Biochemical and Pharmacological Research in Depression,* E. Usdin et al., eds., New York, Raven Press, 1984, p. 359.
108. K. Rickels and E. J. Schweizer, *Clin. Psychiatry, 51,* Suppl. B, 9(1990).
109. E. Goode, *The Marihuana Smokers,* New York, Basic Books, 1970.
110. P. W. Meilman, et al., *Int. J. Addict., 25,* 1025(1990).
111. L. G. Abood, in *Neuropsychopharmacology,* P. B. Bradley, P. Deniker, and C. Radauco-Thomas, Eds. New York, Elsevier, 1959, p. 433.
112. A. T. Shulgin, et al., in *Ethnopharmacological Search for Psychoactive Drugs,* D. H. Efron, Ed., Public Health Service Publication No. 1645. Washington, D. C., U. S. Government Printing Office, 1967, p. 202.
113. A. Hoffer and H. Osmond, *The Hallucinogens.* New York, Academic Press, 1967, p. 55.
114. J. E. Forrest and R. A. Heacock, *Lloydia, 35,* 440(1972).
115. F. G. Schechter, et al., *Morb. Mortal. Wkly Rep., 34,* No. 21(May 31, 1985).
116. C. R. Creveling, et al., *Nature, 216,* 190(1967).
117. A. T. Shulgin, *Nature, 221,* 537(1969).
118. R. A. Glennon, et al., *J. Med. Chem., 30,* 1(1987).
119. R. A. Glennon, et al., *J. Med. Chem., 29,* 194(1986).
120. S. H. Snyder, in *Psychopharmacology, A Review of Progress,* D. H. Efron, Ed. Washington, D.C., U.S. Government Printing Office, 1968, p. 1199.
121. R. W. Baker, et al., *Mol. Pharmacol., 9,* 23(1973).
122. H. J. R. Weintraub and D. E. Nichols, *Int. J. Quant. Chem. Quant. Biol. Symp., 5,* 321(1978).
123. K. Bailey, *J. Pharm. Sci., 60,* 1232(1971).
124. S. H. Snyder, in *Molecular Orbital Studies in Chemical Pharmacology,* L. B. Kier, Ed., Berlin, Springer, 1970, p. 238.
125. L. N. Domelsmith, *J. Med. Chem., 20,* 1346(1977).
126. C. F. Barfnecht and D. E. Nichols, *J. Med. Chem., 18,* 208(1975).
127. L. B. Kier and L. H. Hall, *J. Med. Chem., 20,* 1631(1977).
128. J. J. Knittel and A. Makriyannis, *J. Med. Chem., 24,* 906(1981).
129. K. Chiba, et al., *Neurosci. Lett., 48,* 87(1984).
130. P. Brawley and J. C. Duffield, *Pharm. Rev., 24,* 31(1972).
131. G. Farrel, *Arch. Biochem. Biophys., 94,* 443(1961).
132. A. Hoffer and H. Osmond, *The Hallucinogens,* New York, Academic Press, 1967, p. 94.
133. S. Gershon, *Arch. Int. Pharmacodyn., 135,* 31(1962).
134. P. S. Fetterman, *J. Pharm. Sci., 60,* 1246(1971).
135. F. Mikes, *Science, 172,* 1158(1971).
136. *Marihuana and Health,* Annual Reports to the Congress by the Secretary of Health, Education and Welfare, 1982.
137. R. K. Razdan, in *The Cannabinoids: Chemical, Pharmacologic and Therapeutic Aspects,* S. Agurell, W. L. Dewey and R. E. Willette, Eds., New York, Academic Press, 1984, p. 75.
138. A. Makriyannis and R. N. Kriwacki, *N. Am. Med. Chem. Sym. (Toronto),* 1982.
139. T. Inaba, et al., *Clin. Pharmacol. Ther., 23,* 547(1978).
140. R. M. Smith, *J. Anal. Tox., 8,* 38(1984).
141. R. W. Foltin and M. W. Fischman, *FASEB J., 5,* 2735(1991).
142. M. D. Fischman and C. R. Schuster, *Fed Proc., 41,* 241(1982).
143. M. C. Ritz, et al., *Science, 237,* 1219(1987).
144. M. M. Kilbey and E. H. Ellinwood, *Life Sci., 20,* 1063(1977).
145. C. A. Dackis and M. S. Gold, *Neurosci. Behav. Rev., 9,* 469(1985).
146. P. Lomax, *Proc. West. Pharmacol. Soc., 14,* 10(1971).
147. I. Tamir, et al., *J. Med. Chem., 23,* 220(1980).
148. L. Lemberger, et al., *Science, 173,* 72(1971).
149. M. Perez-Reyes, et al., *Clin. Pharmacol. Ther., 31,* 617(1982).
150. L. E. Hollister, et al., *J. Clin. Pharmacol., 21,* 171S(1981).
151. A. Ohlsson, et al., *Clin. Pharmacol. Ther., 28,* 409(1980).
152. L. E. Hollister, *Science, 172,* 21(1971).
153. J. E. Lindgren, et al., *Psychopharmacology, 74,* 208(1981).
154. R. T. Jones and N. Benowitz, in *Pharmacology of Marihuana,* M. C. Braude and S. Szara, Eds., New York, Raven Press, 1976, p. 627.
155. R. A. Glennon, *Neuropsychopharmacology, 3,* 509(1990).
156. B. L. Jacobs, Ed., *Hallucinogens,* New York, Raven, 1984.
157. E. Sanders-Bush and M. Breeding, *Psychopharmacology, 105,* 340(1991).
158. M. E. Morgan and J. W. Gibb, *Neuropharmacology, 19,* 989(1980); G. W. Ricuarte, et al., *Brain Res, 193,* 153(1980); L. S. Seiden and G. Vosmer, *Pharmacol. Biochem. Behav., 21,* 29(1984).
159. K. M. Johnson, *Fed. Proc., 42,* 2579(1983).
160. S. R. Zukin and R. S. Zukin, *Proc. Natl. Acad. Sci. USA, 76,* 5372(1979).

161. J. M. Walker, et al., *Pharmacol. Rev., 42,* 355(1990).
162. B. R. Martin, *Pharmacol. Rev., 38,* 45(1986).
163. W. A. Devane, et al., *Mol. Pharmacol., 34,* 605(1988).

SELECTED READINGS

S. Agurell, et al., Eds., *The Cannabinoids: Chemical, Pharmacological and Therapeutic Aspects,* New York, Academic Press, 1984.

T. C. Daniels and E. G. Jorgenson, *Central Nervous Stimulants in Wilson and Gisvold's Textbook of Organic Medicinal and Pharmaceutical Chemistry,* 8th ed., R. F. Doerge, Ed., Philadelphia, J. B. Lippincott, 1982.

R. W. Fuller, *Adv. Drug Res., 17,* 350(1988).

S. Garattini and R. Samanin, Eds., *Anorectic Agents: Mechanisms of Action and Tolerance,* New York, Raven Press, 1981.

A. G. Gilman, et al., Eds., *The Pharmacological Basis of Therapeutics,* 8th ed., New York, Macmillan, 1990.

L. E. Hollister, Ed., *Second Generation Antidepressants, Rational Drug Therap., 16(M12),* 1(1982).

Pharmacological Treatment of Obesity: Satellite Symposium to the 6th International Congress of Obesity. *Am. J. Clin. Nutr., 55,* Suppl. 1(1992).

R. M. Pinder and J. H. Wieringa, *Med. Res. Rev., 13,* 259(1993).

A. C. Sullivan and R. K. Gruen, *Fed. Proc., 44,* 139(1985).

F. Sulser, in *Frontiers in Biochemical and Pharmacological Research in Depression,* E. Usdin, et al., Eds., New York, Raven Press, 1984.

Chapter 16

LOCAL ANESTHETICS

Matthias C. Lu

A local anesthetic agent is a drug that, when given either topically or parenterally to a localized area, produces a state of local anesthesia by reversibly blocking the nerve conductances that transmit the feeling of pain from this locus to the brain. For this reason, local anesthetics play an important role clinically in dentistry and in other minor surgery for temporary relief of pain.

What Is Anesthesia? The term anesthesia is defined as a loss of sensation with or without loss of consciousness. According to this definition, a wide range of drugs with diverse chemical structures are anesthetics. The list includes not only the classic anesthetic agents, such as the general and local anesthetics, but also many central nervous system (CNS) depressants, such as analgesics, barbiturates, and muscle relaxants.

What Is the Difference Between a Local Anesthetic and Other Anesthetic Agents? How Does It Work at the Molecular Level? Both general anesthetic and local anesthetic drugs produce anesthesia by blocking nerve conductance in both sensory and motor neurons. This blockade of nerve conductances leads to a loss of pain sensation as well as impairment of motor functions. The anesthesia produced by local anesthetics, however, is generally without loss of consciousness or impairment of vital central functions. It is generally accepted that a local anesthetic blocks nerve conductance by binding to selective site(s) on the sodium channels in the excitable membranes, thereby reducing sodium passage through the pores and thus interfering with the action potentials. Thus, a local anesthetic decreases the excitability of nerve membranes without affecting the resting potential. By contrast, a general anesthetic agent alters physical properties of nerve membranes through rather nonspecific interactions with the lipid bilayer or the receptor/ionic channel proteins. These, in turn, reduce membrane excitability through a number of possible mechanisms, including changes in membrane fluidity, permeability, and receptor/channel functions. Furthermore, local anesthetics, in contrast to analgesic compounds, do not interact with the pain receptors or inhibit the release or the biosynthesis of pain mediators.

HOW MODERN LOCAL ANESTHETICS ARE DISCOVERED

Similar to many modern drugs, the initial leads for the design of clinically useful local anesthetics were derived from natural sources. As early as 1532, the anesthetic properties of coca leaves (Erythroxylon coca Lam) became known to Europeans from the natives of Peru, who chewed the leaves for a general feeling of well-being and to prevent hunger. Saliva from chewing the leaves was often used by the natives to relieve painful wounds. The active principle of the coca leaf, however, was not discovered until 1860 by Niemann, who obtained a crystalline alkaloid from the leaves to which he gave the name cocaine and noted the anesthetic effect on the tongue. Although Moréno y Maiz in 1868 first asked the question whether cocaine could be used as a local anesthetic, it was Von Anrep, in 1880, who recommended that cocaine be used clinically as a local anesthetic after many animal experiments. The first report of successful surgical use of cocaine appeared in 1884 by Koller, an Austrian ophthalmologist. This discovery led to an explosive development of new anesthetic techniques and local anesthetic agents.[1]

Although the structure of cocaine was not known until 1924, many attempts were made to prepare new analogs of cocaine without its addicting liability and other therapeutic shortcomings, such as allergic reactions, tissue irritations, and poor stability in aqueous solution. Also, cocaine is easily decomposed when the solution is sterilized (Fig. 16–1). Initially, analogs of ecgonine and benzoic acid, the hydrolysis products of cocaine, were prepared. When the chemical structure of ecgonine became known, the preparation of active compounds accelerated. It was soon realized that a variety of benzoyl esters of amino alcohols, including benzoyltropine, exhibited strong local anesthetic properties. Removal of the carbomethoxy

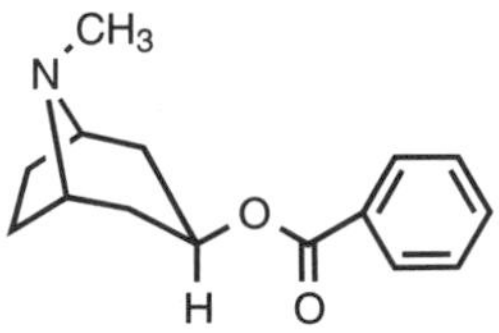

Benzoyltropine

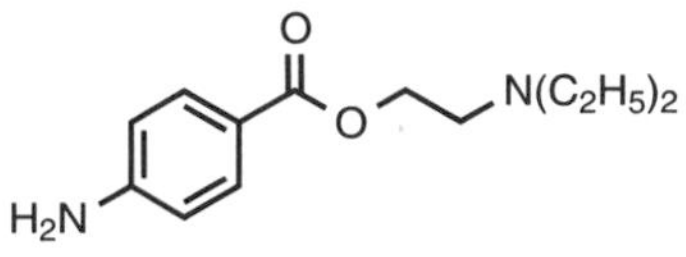

Procaine

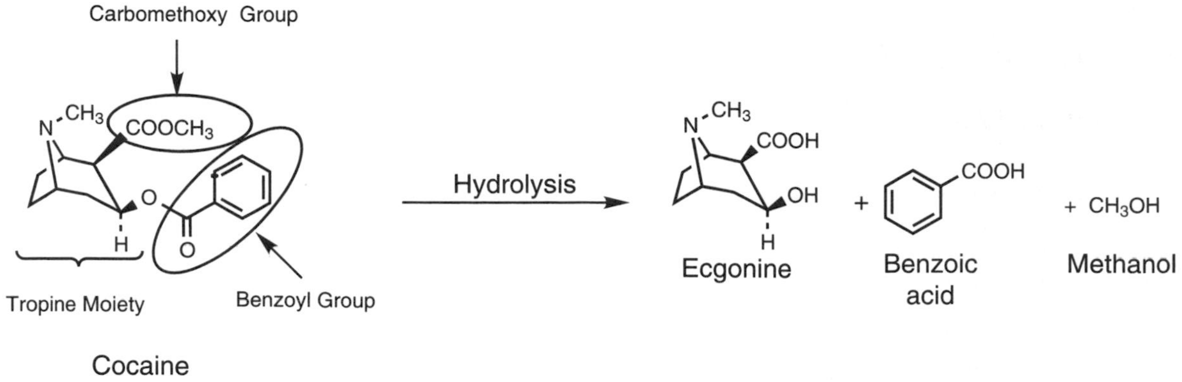

Carbomethoxy Group

Tropine Moiety

Benzoyl Group

Cocaine

Hydrolysis

Ecgonine

Benzoic acid

Methanol

Fig. 16–1. Structures of cocaine and its hydrolysis products.

group of cocaine also abolished the addicting liability. This discovery eventually led to the synthesis of procaine (N,N-diethylaminoethyl ester of p-aminobenzoic acid) in 1905, which became the prototype for local anesthetics for nearly half a century largely because it did not have the severe local and systemic toxicities of cocaine.

Although the intrinsic potency of procaine was low and the duration of action was relatively short as compared with cocaine, it has been found that these deficiencies could be remedied when it was combined with a vasoconstrictor compound such as epinephrine. Apparently a vasoconstrictor agent reduces the local blood supply and thereby prolongs the residence time of this agent.

Following the introduction of procaine, hundreds of structurally related analogs were prepared and their local anesthetic properties examined. Most of these compounds were prepared for the purposes of enhancing the intrinsic potency and the duration of action of procaine. Among these compounds, tetracaine still is the most widely used of the ester-type local anesthetic agents.

The next major turning point in the development of clinically useful local anesthetic agents was the serendipitous discovery of the local anesthetic activity of another natural alkaloidal product, isogramine, in 1935 by von Euler and Erdtman. This observation led to the synthesis of lidocaine (Xylocaine) by Löfgren in 1946, the first nonirritating, amide-type local anesthetic agent with good local anesthetic properties and yet less prone to allergic reactions than procaine analogs. A further practical advantage of lidocaine is its stability in aqueous solution because of the more stable amide functionality. Structurally, lidocaine can be viewed as an open-chain analog of isogramine and thus is a bioisosteric analog of isogramine.

In the years since 1948, extensive progress has been achieved primarily in the fields of neurophysiology and neuropharmacology rather than that of synthetic medicinal chemistry. Most of this research has significantly increased our understanding of how nerve conduction occurs and how compounds interact with the neuronal membranes to produce local anesthesia. It should be pointed out, however, that although a number of current clinically useful local anesthetic agents have been introduced into the market, unfortunately an ideal local anesthetic drug has not yet been realized.

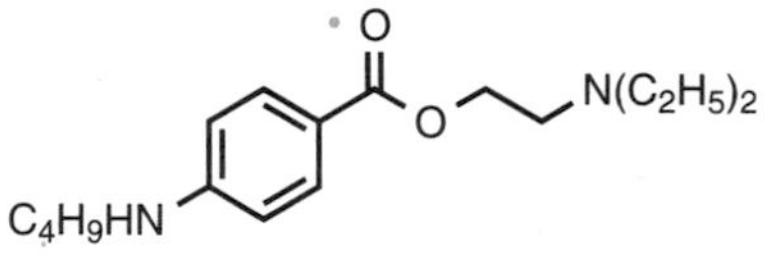

Tetracaine

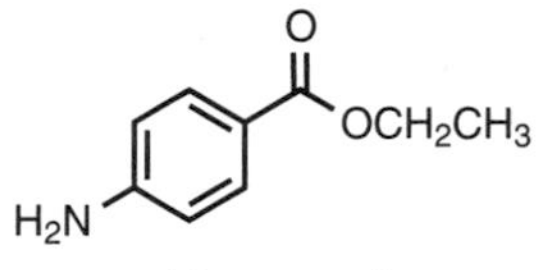

Benzocaine

It should also be mentioned that benzocaine, an effective topical anesthetic agent, was synthesized by Ritsert in 1890 and found to have good anesthetizing properties and low toxicity. Benzocaine, however, has limited water solubility except at low pH values owing to the lack of a basic aliphatic amino group, thereby disallowing the preparation of pharmaceutically acceptable parenteral solutions.

Characteristics for an Ideal Local Anesthetic

An ideal drug should produce reversible blockade of sensory nerve fibers with a minimal effect on the motor

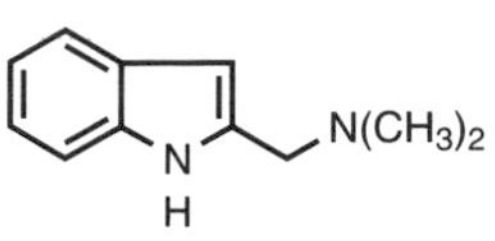

Isogramine

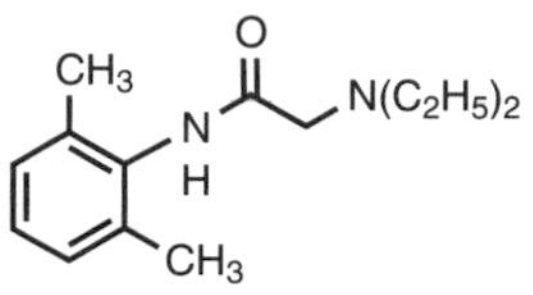

Lidocaine

fibers. It should also possess a rapid onset and have a sufficient duration of action for the completion of major surgical procedures without any systemic toxicities.

Finding an Ideal Local Anesthetic Agent

The realization of this goal, however, can only be attained through further structure-activity relationship studies, particularly with regard to their selective actions on the voltage-sensitive sodium channels. Additional leads for the design of ideal agents could also come from a more systematic metabolic and toxicity study of current available agents. To understand the chemical aspects of local anesthetics and thus provide a proper background for practical uses of these compounds, it is necessary to have a working knowledge of basic neuroanatomy and electrophysiology of the nervous system.

NEUROANATOMY AND ELECTROPHYSIOLOGY OF THE NERVOUS SYSTEM

Neuroanatomy

As can be seen in Figure 16–2, the sensory fibers (afferent neuron) course together in bundles with the motor fibers (efferent neuron) from the periphery to the spinal cord.[2] The cell bodies of the sensory fibers are found at the point at which the nerve enters the vertebra and are seen as enlargements on the nerve bundles. The cell bodies of the motor fibers are found within the spinal cord. The bundles of sensory and motor fibers outside the spinal cord are wrapped in a connective tissue sheath, the epineurium. Groups of fibers are found within this "nerve" in small bundles, each of which is surrounded by connective tissue known as perineurium and in even smaller tubes of connective tissue called endoneurium.

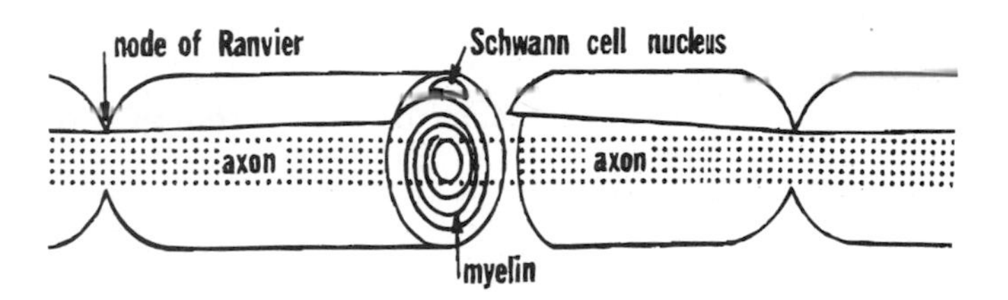

Fig. 16–3. Single myelinated nerve fiber.

Figures 16–2 and 16–3 also show that each nerve axon has its own membranous covering, often called the nerve membrane, tightly surrounded by a myelin sheath called a Schwann cell covering. The myelin is not continuous along the fiber. The interruptions are the nodes of Ranvier, which are of great importance for nerve functioning.

Electrophysiology of Nerve Membrane

Resting Potential

Most nerves have resting membrane potentials of about −70 to −90 mV as a result of a slight imbalance of electrolytes across the nerve membranes (i.e., between the cytoplasm and the extracellular fluid).[3] The origin of this membrane potential has been of great interest to neurophysiologists. The main electrolytes of nerve axons and cell bodies are sodium, potassium, calcium, magnesium, and chloride. At resting potential, the nerve membrane was believed to be impermeable to sodium because of the low sodium ion concentration in the excitable cell.

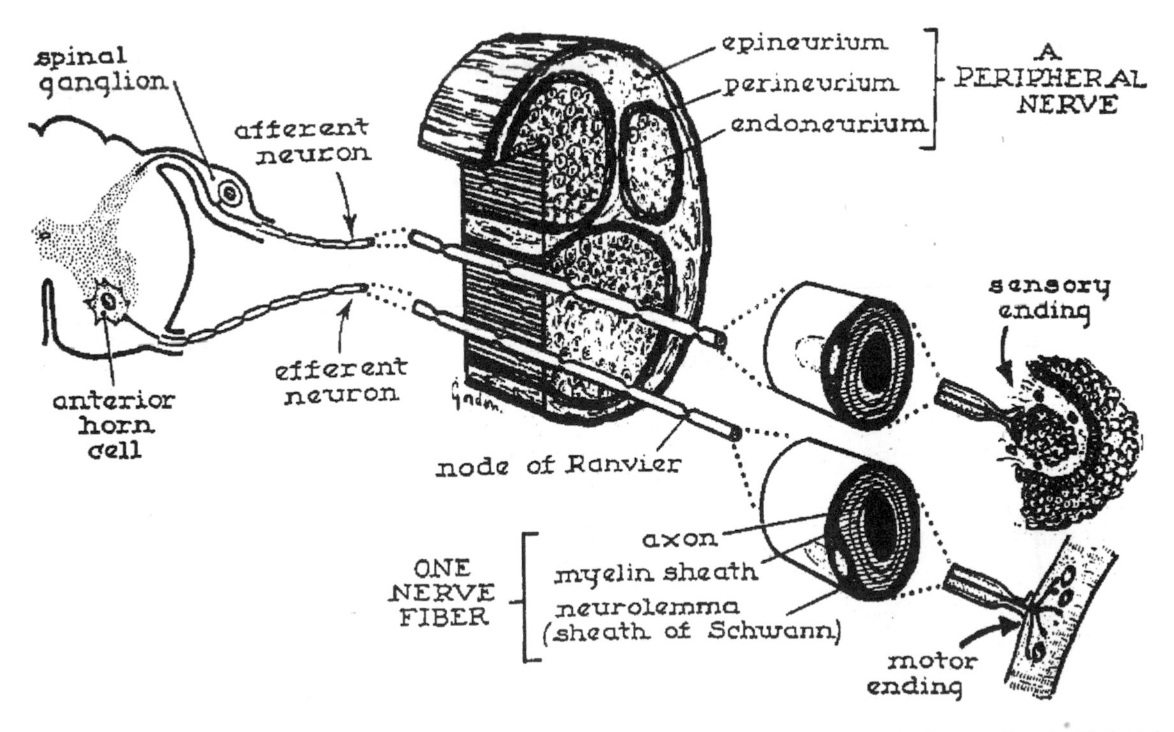

Fig. 16–2. Diagram showing the various parts of a peripheral nerve. (From A. W. Ham, *Histology,* 6th ed., Philadelphia, J. B. Lippincott.)

Potassium ions may flow in and out of the cell with ease, indicating that the membrane is highly permeable to potassium ions. A high potassium ion concentration is retained intracellularly by the attractive forces provided by the negative charges on the protein molecules. Thus, the predominant intracellular cation is potassium (~110 to 170 mmol/L), and the predominant extracellular ions are sodium (~140 mmol/L) and chloride (~110 mmol/L). It would appear that changes in the intracellular or extracellular concentration of potassium ions markedly alter the resting membrane potential. For this reason, an excitable cell was treated by neurophysiologists as if it were an electrochemical, or Nernst, cell. The resting potential for one permeant species could therefore be explained by the familiar Nernst equation:

$$E = -RT/zF \ln [K^+]_i/[K^+]_o$$

in which E = membrane potential, inside minus outside; R = gas constant; T = temperature; z = valence of ion; F = Faraday's constant; $[K^+]_i$ = activity of potassium intracellularly, and $[K^+]_o$ = activity of potassium extracellularly.

Action Potential

Action potentials are transient membrane depolarizations that result from influx of sodium ions through a brief opening of the voltage-gated sodium channels on excitation of the cells.[3] The transmembrane potential during an action potential goes from −70 to about +40 mV (a total net change of 110 mV) and promptly returns to the resting potential; the event lasts about 1 msec (Fig. 16–4).

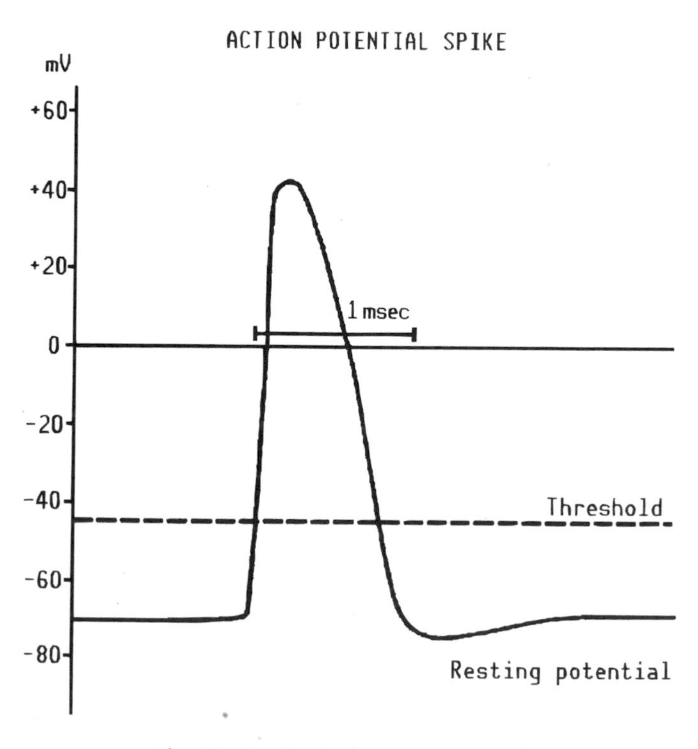

Fig. 16–4. Typical action potential.

The transmembrane potential at the peak of the action potential can be predicted from the Nernst equation by substituting appropriate sodium ion concentrations for those of potassium ions. It therefore appears that the excitable membrane can be transformed from a potassium electrode to a sodium electrode during the active process.[4] As the cell approaches its peak action potential, the permeability to sodium again decreases (sodium inactivation or repolarization). If no other event occurred, this cell would slowly return to its resting potential, but the cell again becomes highly permeable to potassium ions, allowing potassium ions to flow out and quickly restore the membrane potential. After an action potential, the cell would therefore be left with a small increase in sodium ions and a decrease in potassium ions. To explain how the nerve is restored to its original electrolyte composition at the resting potential, it was necessary to postulate a mechanism by which sodium ions could be extruded and potassium ions could probably be accumulated. It has been suggested that by using the energy derived from splitting adenosine triphosphate (ATP), an ATPase system could serve this function and act as a sodium pump.[5] Other investigators believe that, during excitation, the membrane goes from one stable state to another, functioning more like an ion exchanger.[6] Koketsu suggests that during excitation the membrane goes from a calcium-associated to calcium-dissociated state, triggered by the depolarizing pulse, and that during recovery the membrane returns to the resting state by the reassociation as a result of outward movement of potassium ions.[7]

Threshold

An electric stimulus of less than a certain voltage can result in only local electronegativity and cannot elicit a propagated action potential. The voltage necessary to change localized electronegativity into a propagated action potential is called the threshold voltage, which is closely related to the stimulus duration—the longer the stimulus, the lower the threshold voltage.

Refractoriness

Immediately after an impulse has been propagated, the axon is *absolutely refractory,* or completely inexcitable, and no stimulus, no matter how strong or long, can excite it. Shortly thereafter, the axon becomes *relatively refractory;* it responds with a propagated impulse only to stimulation that is greater than the normal threshold. The length of the refractory period is affected by the frequency of stimulation and by many drugs (Fig. 16–5).

Conductance Velocity

This is the velocity at which an impulse is conducted along the nerve and is proportional to the diameter of the fiber.

Nodal Conduction

Because longitudinal resistance is inversely proportional to cross-sectional area, impulses are conducted faster in large-diameter fibers. The squid axon is unmy-

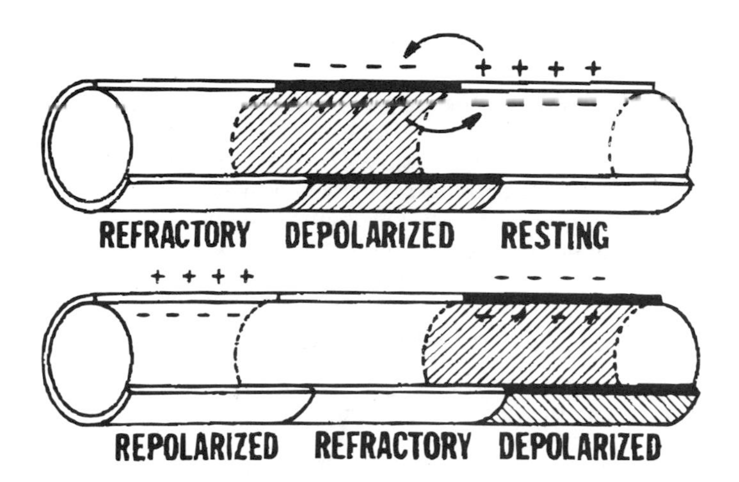

Fig. 16–5. Impulse propagation. *A,* The wave of depolarization passes down the nerve, followed by a wave of refractoriness. *B,* The wave of refractoriness is followed by a wave of repolarization. (After R. H. De Jong and F. G. Freund, *Int. Anesthesiol. Clin., 8,* 35[1970].)

elinated and exceptionally large (approximately 800 μ); impulses are therefore conducted rapidly along it. Contraction of the mantle of a squid, however, is an uncomplicated procedure that does not require a complex sensorimotor system. Perhaps, during evolution, vertebrates developed a complicated input-output system of many fibers collected in bundles, as shown in Figure 16–2. Conduction in these fibers would be slow if they were not insulated with a myelin coat, interrupted at intervals by the nodes of Ranvier where current enters and exits. Ionic fluxes occur at these nodes. The impulse jumps along the fiber from node to node faster than in unmyelinated fibers.[8]

Sodium Channel

The voltage-sensitive sodium channels are discrete membrane-bound glycoproteins that mediate sodium permeability.[9,10] It consists of an aqueous pore spanning the axon membrane, which is narrow at one point (known as selective filter) to discriminate sodium ions from other ions (i.e., sodium ions pass through this pore about 12 times faster than the potassium ions). Sodium channels open and close as they switch between several conformational states, i.e., the resting/closed form (nonconducting state), the open channel (conducting), and the inactivated form (nonconducting state). It is generally agreed that at resting potential, the sodium channels are in a rest/closed state and are impermeable to the passage of sodium ions. On activation, the channels undergo conformational changes to an open state, thus allowing rapid influx of sodium ions across the axonal membrane. Thus, when threshold potential is exceeded, most of the sodium channels are in an open or conducting state. At the peak of the action potential, the open channels spontaneously convert to an inactivated state (i.e., nonconducting, nonactivatable), leading to a decrease in sodium permeability. When a sodium channel is in the inactivated state, it cannot be opened without first transforming to the normal resting/closed form. The channel gating kinetics have been extensively studied with the use of the selective blockers of sodium channels such as tetrodotoxin (TTX, a neurotoxin isolated from puffer fish) and saxitoxin (STX, a toxic compound produced by certain marine dinoflagellates, which are known to cause environmental conditions commonly referred to as red tides." These compounds bind stoichiometrically to the outer opening of the channels and are detected with patch-clamp electrophysiologic techniques on the cut-open squid giant axon.[11] As a matter of fact, both Neher and Sakmann, two German scientists, were awarded the 1991 Nobel Prize in physiology and medicine for their work on ion channels.

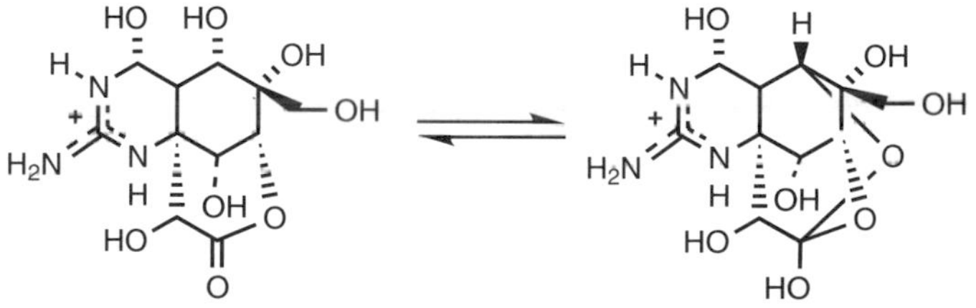

(Lactone form) (Hemiacetal form)

Tetrodotoxin (TTX)

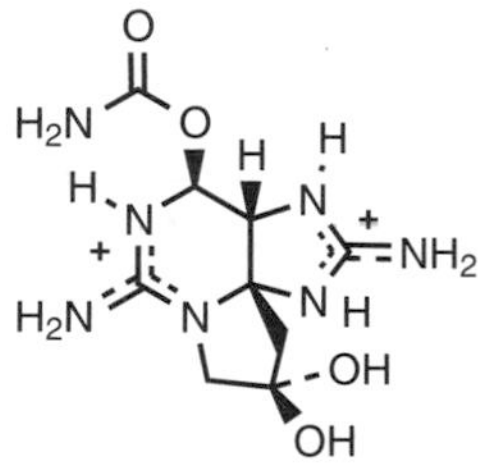

Saxitoxin (STX)

MOLECULAR MECHANISM OF ACTION OF LOCAL ANESTHETICS

Local anesthetics decrease the excitability of cells without affecting the resting potential. Because the action potential, or the ability of nerve cells to be excited, seems to be associated with the movement of sodium ions across the nerve membranes, anything that interferes with sodium movement interferes with excitability. For this reason, many hypotheses have been suggested to explain how local anesthetics regulate the sodium permeability changes that underlie the nerve impulse. These hypotheses include a direct action on ionic channels that interferes with ionic fluxes, interaction with phospholipids and calcium that reduces membrane flexibility and responsiveness to changes in electrical fields, and competition with acetylcholine for the membrane-bound cholinergic receptor as well as rather nonspecific actions on the membrane structure (i.e., membrane expansion theory of anesthesia). The nonspecific membrane actions of local anesthetics can be easily ruled out because most clinically useful agents, in contrast to general anesthetics, possess a defined set of structure-activity relationships. As mentioned earlier, local anesthetic agents block nerve conductances and produce anesthesia as a result of their selective actions on membrane-bound sodium channels. Before discussing the where and how local anesthetics bind to the sodium channel to exert its action, the other hypotheses and the reason(s) why they may not be the molecular mechanism of action for the clinically useful local anesthetics are briefly reviewed.

Interaction with Phospholipids and Calcium

Calcium exists in the membrane in a bound state. Many investigators believe that the release of the bound calcium is the first step in membrane depolarization and that this release leads to the ionic permeability changes previously described. It has been suggested that local anesthetics displace the bound calcium from these sites and form more stable bonds, thereby inhibiting ionic fluxes. The following evidence has been offered in support of this theory. Both calcium and local anesthetics bind to phospholipids in vitro, reducing their flexibility and responsiveness to changes in electric fields.[12,13] Also, membrane excitability and instability increase in calcium-deficient solutions. Local anesthetics counteract this abnormal increase in excitability, and more local anesthetic is necessary to block excitation in calcium-poor solutions.[14] Direct proof of this hypothesis, however, is lacking because of the difficulty in measuring calcium movements in vivo. It is also possible that the aforementioned cause-and-effect relationship between intracellular free calcium and membrane excitability is the result of a sodium-calcium exchange reaction; i.e., the influx of sodium ions displaces the membrane-bound calcium, which leads to an increase of intracellular free calcium and thereby increases cellular excitability.

Interaction with Acetylcholine System

A quick comparison of the chemical structures of the clinically useful local anesthetics with the known cholinergic agents (Fig. 16–6) led various authors to suggest that perhaps a local anesthetic competes with acetylcholine for the cholinergic receptor as a mechanism of action of local anesthetics.[15] Support for this hypothesis is derived from the observations that local anesthetics antagonize the depolarizing action of acetylcholine on the nerve membrane. Lidocaine also modifies the kinetics of acetylcholine interactions at the myoneural junction. Again, there is no direct support for this hypothesis. The fact that most potent local anesthetic agents have only a weak anticholinergic activity weakens the argument for such a hypothesis. Perhaps the strongest evidence against such a mechanism can be obtained from the fact that attempts to increase anticholinergic activity by insertion of a methylene group between the aromatic ring and the carbonyl function of the procaine molecule produces a compound with only weak local anesthetic activity.

Action on Voltage-Sensitive Sodium Channels

As mentioned before, the voltage-sensitive sodium channels are membrane-bound glycoproteins that mediate sodium permeability. On excitation, these channels undergo conformational changes from a closed state to an open state, thus allowing a rapid influx of sodium. The movement of sodium ions is blocked by neurotoxins TTX and STX and by local anesthetics.[16] Most electrophysiologists and neuropharmacologists have now agreed that the mechanism of action of local anesthetics is due primarily to their binding to a site(s) within the sodium channels, thus blocking the sodium conductance.[17] The exact location of the binding site(s), however, and whether all local anesthetics interact with a common site remain a matter of dispute.

How Local Anesthetics Block Sodium Conductance

Local anesthetics block sodium conductance by two possible modes of action; the tonic and the phasic inhibition.[18,19] Tonic inhibition results from the binding of local anesthetics to nonactivated closed channels and thus is independent of channel activation. Phasic inhibition may be accomplished when local anesthetics bind to either activated, open states (conducting) or inactivated (nonconducting) states of the channels. Thus, it is not surprising that a greater phasic inhibition is usually obtained with repetitive depolarization. Two reasons have been suggested to explain this observation: Channel inactivation during depolarization increases the number of binding sites normally inaccessible to local anesthetics at resting potential, or both the open and the inactivated channels possess binding sites with higher affinity and thereby bind local anesthetics more tightly and result in stronger nerve block. Furthermore, it is generally agreed that most of the clinically useful local anesthetics exert their actions by binding to the inactivated forms of the channels and thus prevent their transition to the original rest state.[19] Because most of these drugs exhibit both tonic and phasic inhibitions, however, the question as to whether tonic and phasic block results from drug interaction at the same or different sites remains unclear.

Where Local Anesthetics Bind to Sodium Channels

Most of the clinically useful local anesthetics are tertiary amines with a pK_a value of 7.5 to 9.0. Thus, under physiologic conditions, both protonated forms (onium ions) and the unionized, molecular forms are available for binding to the channel proteins. In fact, the ratio between the onium ions $[BH^+]$ and the unionized mole-

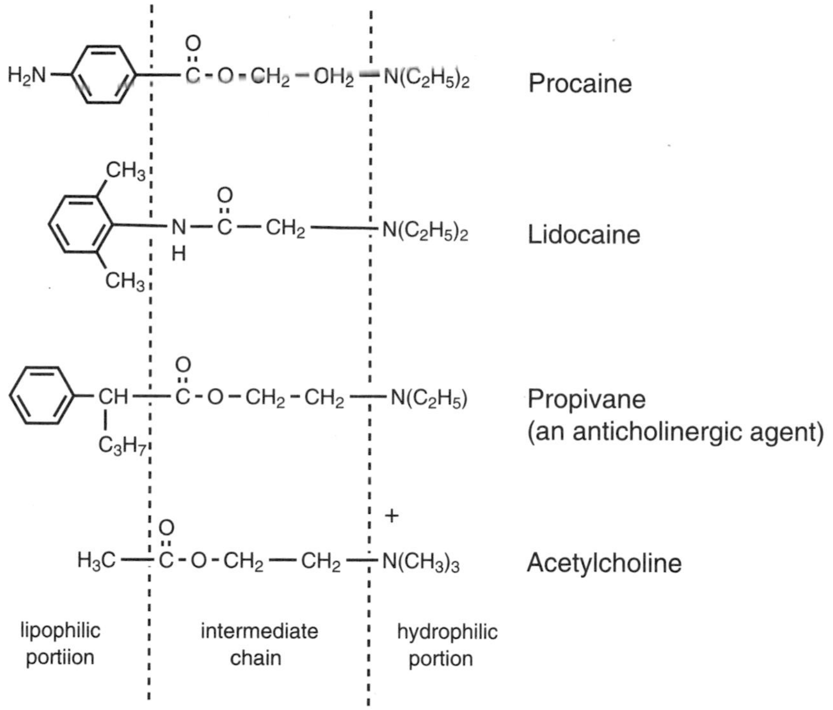

Procaine

Lidocaine

Propivane (an anticholinergic agent)

Acetylcholine

lipophilic portiion | intermediate chain | hydrophilic portion

Fig. 16–6. Comparison of local anesthetics with cholinergic agents.

cules [B] can be easily calculated based on the pH of the medium and the pK_a of the drug molecule by the Henderson-Hasselbalch equation:

$$pH = pK_a - \log [B]/[BH^+]$$

The effect of pH changes on the potency of local anesthetics has been extensively investigated.[20] Based on these studies, it was concluded that local anesthetics block action potential by first penetrating the nerve membrane in their unionized forms and then binding to a site within the channels in their onium forms. Perhaps the most direct support for this hypothesis comes from the experimental results of Narahashi and co-workers,[21,22] who studied the effects of internal and external perfusion of local anesthetics (both tertiary amines and quaternary ammonium compounds), at different pH values, on the sodium conductance of the squid axon. The observation that both tertiary amines and quaternary ammonium compounds produce greater nerve blockage when applied internally indicates an axoplasmic site for these compounds. Furthermore, only the tertiary amines exhibit a reduction in their local anesthetic activities when the internal pH was raised from 7.0 to 8.0. Because the increase of internal pH to 8.0 favors the existence of the unionized forms; this result again suggests that the onium ions are required for binding to the channel receptors. Narahashi and Frazier[23] further estimated that approximately 90% of the blocking actions of lidocaine may be attributed to onium forms of the drug molecule, whereas only about 10% may be due its unionized molecule and perhaps at a site other than the primary binding site. Benzocaine, owing to lack of a basic amine group, and other neutral anesthetics such as benzyl alcohol have been suggested to bind to this site.

In 1984, Hille proposed a unified theory involving a single receptor in the sodium channels for both onium ions (protonated tertiary amines and quaternary ammonium compounds) and unionized forms of local anesthetics.[24] As depicted in Figure 16–7, a number of pathways are available, depending on the size, the pK_a and the lipid solubility of the drug molecules, and voltage and frequency-dependent modulation of the channel states, for a drug to reach its receptor binding site. Protonated anesthetic molecules $[BH^+]$ and quaternary ammonium compounds reach their target sites via the hydrophilic pathway externally, which is available only during chan-

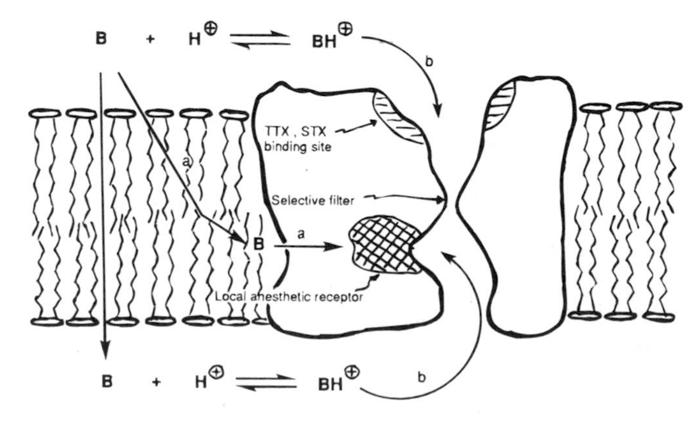

Fig. 16–7. Model of a sodium channel, as suggested by Hille,[24] depicting a hydrophilic pathway (denoted by "b") and a hydrophobic pathway (denoted by "a") by which local anesthetics may reach their receptor site(s).

nel activation. The lipid-soluble anesthetic molecules can interact with the same receptors from either the hydrophilic pathway from the internal aqueous medium in their onium ions [BH^+] or via the hydrophobic pathway in their unionized forms. Benzocaine and other nonbasic local anesthetic molecules use this hydrophobic pathway and thus bind to the same receptor binding site to produce their actions. Again, this hypothesis is purely speculative, and its acceptance is still open for further debate.

Toxicity and Side Effects

The side effects and toxicity of local anesthetics seem to be related to their actions on other excitable membranes, such as in the heart, the neuromuscular junctions, and the CNS. In general, neuromuscular junctions and the CNS are more susceptible to the toxic effects of local anesthetics than the cardiovascular system. The actions on skeletal muscles are transient and reversible, whereas the CNS side effects can be deleterious. The primary effect of the toxicity seems to be convulsions, followed by severe CNS depression, particularly of the respiratory and cardiovascular centers. This may be related to an initial depression of inhibitory neurons such as GABAergic systems, causing convulsions, followed by depression of other neurons, leading to general depression of the CNS.

Amino amide–type local anesthetics (i.e., lidocaine derivatives) are, in general, more likely to produce CNS side effects than the amino ester–type compounds (procaine analogs). It should be pointed out, however, that the toxic effects observed depend heavily on the route and site of administration as well as the lipid solubility and metabolic stability of a given local anesthetic molecule.

In contrast, allergic reactions to local anesthetics, even though rare, are known to occur exclusively with *p*-aminobenzoic ester-type local anesthetics. Whether the formation of *p*-aminobenzoic acid on ester hydrolysis is responsible for this hypersensitivity remains to be investigated.

It is commonly known that amide-type local anesthetics (e.g., procainamide and lidocaine) also possess antiarrhythmic activity when given parenterally and at a subanesthetic dosage. Although this action can also be attributed to their actions on sodium channels in cardiac tissues, current evidence suggests a distinctly different mechanism of action with respect to the modulation of channel receptors and the location of binding sites for these compounds.[25,26]

CHEMICAL ASPECTS OF LOCAL ANESTHETIC AGENTS

Structure-Activity Relationships

Since the discovery of cocaine in 1880 as a surgical local anesthetic, several thousand new compounds have been tested and found to produce anesthesia by blocking nerve conductance. Among these agents, only about 30 are clinically available in the United States as local anesthetic preparations (Table 16–1).

Table 16–2 contains chemical structures of the different types of agents in current or recent use. A quick perusal of Table 16–2 reveals that many diverse chemical structures possess local anesthetic properties, i.e., amino esters (procaine analogs), amino amides (lidocaine analogs), amino ethers (dimethisoquin and pramoxine), amino carbamates (diperodon), amino ketones (dyclonine), amidines (phenacaine), alcohols (benzyl alcohol and chlorobutanol), and phenols (eugenol and phenol). It would seem that there is no obvious structure-activity relationship among these agents. As mentioned earlier, however, most of the clinically useful local anesthetics are tertiary amines with pK_a values of 7.5 to 9.0. These compounds exhibit their local anesthetic properties by virtue of the binding of the onium ions to a selective site within the sodium channels. For this reason, any structural modifications that alter the lipid solubility, pK_a, and metabolic inactivation definitely have a pronounced effect on the ability of a drug molecule to reach or bind to the hypothetical receptor site(s), thus modifying its local anesthetic properties.

A brief discussion of known structure-activity relationships is presented according to the following structural characteristics according to Löfgren's classification[27] (see Fig. 16–6):

lipophilic portion—intermediate chain—
hydrophilic portion

Lipophilic Portion

The lipophilic portion of the molecule is essential for local anesthetic activity. For most of the clinically useful local anesthetics, this portion of the molecule consists of either an aromatic group directly attached to a carbonyl function (the amino ester series) or a 2,6-dimethylphenyl group attached to a carbonyl function through an —NH— group (the amino amide series). Both of these groups are highly lipophilic and appear to play an important role in the binding of local anesthetics to the channel receptor.

Structural modification of this portion of the molecule is known to have a profound effect on its physical and chemical properties, which, in turn, alters its local anesthetic properties. In the amino ester series, an electron-donating substituent in the *ortho* or *para* or both positions increases local anesthetic potency. Such groups as an amino (procaine, butacaine, chloroprocaine, and propoxycaine), an alkylamino (tetracaine), or an alkoxy (proparacaine, cyclomethycaine and propoxycaine) group can contribute electron density to the aromatic ring by both resonance and inductive effects, thereby enhancing local anesthetic potency over nonsubstituted analogs (hexylcaine and meprylcaine). As illustrated in Figure 16–8, using tetracaine as an example, through resonance we could expect that the structure of tetracaine is really between the two hybrid forms; i.e., the electrons from the amino group can be resonance delocalized onto the carbonyl oxygen to form the zwitterionic form as shown. Although neither structure may accurately represent the structure of tetracaine when it binds to the channel receptors, the greater the resemblance to the zwitterionic form, the greater the affinity for the receptor. Thus, it is reasonable to assume that any aro-

Table 16–1. Local Anesthetics in Current or Recent Use

Generic Name	Proprietary Name	Recommended Application
Benoxinate*	Dorsacaine	Mainly in ophthalmology
Benzocaine†	Americaine	Topical
Bupivacaine†	Marcaine, Sensorcaine	Parenteral
Butacaine†	Butyn	Topical
Butamben†	Butesin	Topical
Chloroprocaine†	Nesacaine	Parenteral
Cocaine†		Topical
Cyclomethycaine‡	Surfacaine	Topical
Dibucaine†	Nupercainal	Parenteral, topical
Dimethisoquin	Quotane	Topical
Diperodon*	Diothane	Topical
Dyclonine†	Dyclone	Topical
Etidocaine†	Duranest	Parenteral, topical
Hexylcaine*	Cyclaine	Parenteral, topical
Lidocaine†	Xylocaine, Dalcaine, Dilocaine, Nervocaine, Nulicaine	Parenteral, topical
Mepivacaine†	Carbocaine, Polocaine, Isocaine	Parenteral
Meprylcaine*	Oracaine	Parenteral, especially in dentistry
Phenacaine§	Holocaine	Mainly in ophthalmology
Pramoxine†	Tronothane	Topical
Prilocaine†	Citanest	Parenteral
Procaine†	Novocain	Parenteral
Proparacaine†	Ophthaine, Ak-Taine, Alcaine	Mainly in ophthalmology
Propoxycaine†	Blockaine	Parenteral
Pyrrocaine‖		Parenteral, especially in dentistry
Tetracaine†	Pontocaine, Prax	Parenteral, topical
Benzyl alcohol¶		Topical, mainly in combination
Chlorobutanol¶		Topical, especially in dentistry
Eugenol*		Topical, especially in dentistry
Phenol*		Mainly topical (and for irreversible blocks)
Ethyl choride*		Extracutaneous, temperature decreasing

* United States Pharmacopoeia XXII (1990).
† USP DI, 13th ed. (1993).
‡ USP DI, 6th ed. (1986).
§ United States Pharmacopoeia XXI (1985).
‖ National Formulary XIV (1975).
¶ National Formulary XVII (1990).

matic substitution that can enhance the formation of this zwitterionic form through resonance or inductive effects will produce more potent local anesthetic agents. Electron-withdrawing groups such as nitro (—NO_2), carbonyl (—CO—), and nitrile (—CN) reduce the local anesthetic activity. When an amino or an alkoxy group is attached to the *meta* position of the aromatic ring, however, no resonance delocalization of their electrons is possible. The addition of this function only increases (alkoxy group) or decreases (amino group) the lipophilicity of the molecule.

Insertion of a methylene group between the aromatic moiety and the carbonyl function in the procaine molecule, which prohibits the formation of the zwitterion, has led to a procaine analog with greatly reduced anesthetic potency. This observation lends further support for possible involvement of the zwitterionic form when a local anesthetic binds to the receptor. It should be pointed out that a similar zwitterionic form can also be envisioned in the amino amide series without the need for a direct attachment of the aromatic ring to the carbonyl function (see Fig. 16–8).

Furthermore, tetracaine, the most widely used analog of procaine, is approximately 50-fold more potent than procaine. This increase in potency cannot be correlated experimentally solely with the increase of lipid solubility of the n-butyl group. Perhaps part of this potentiation of local anesthetic activity can be attributed to the electron-releasing property of the n-butyl group via the inductive effect, which indirectly enhances the electron density of the *p*-amino group, which, in turn, increases the formation of zwitterion.

Another important aspect of aromatic substitution has been observed from structure-activity relationship studies. In the amino amides (lidocaine analogs), the *o,o*-dimethyl groups are required to provide suitable protection

Table 16–2. Structures of Local Anesthetics

1. Amino Esters

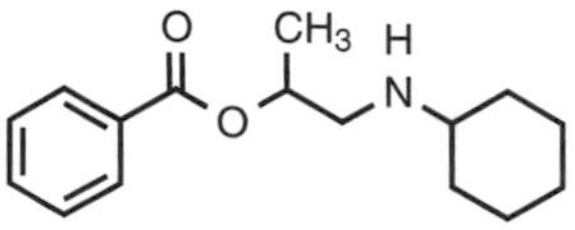

O CH_3 H N O

Hexylcaine ($pK_a = 9.3$)

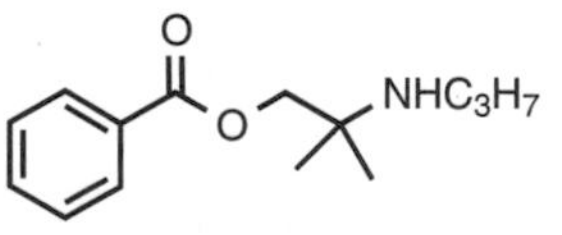

O NHC_3H_7 O

Meprylcaine ($pK_a = 7.8$)

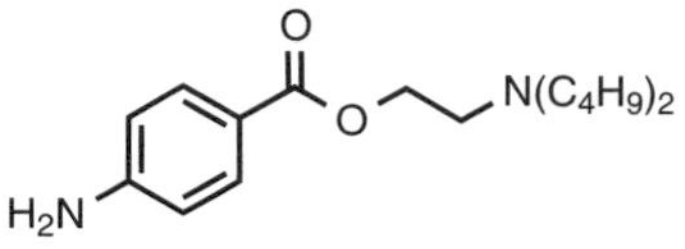

O $N(C_4H_9)_2$ O H_2N

Butacaine ($pK_a = 8.3$)

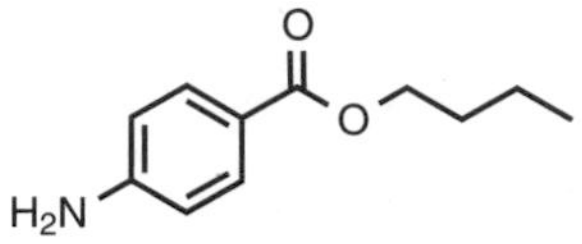

O O H_2N

Butamben ($pK_a = 2.5$)

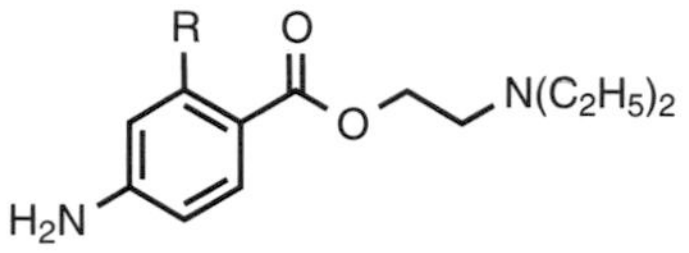

R O $N(C_2H_5)_2$ O H_2N

R =Cl, Chloroprocaine ($pK_a = 9.0$)
R = OC_3H_7, Propoxycaine ($pK_a = 9.1$)

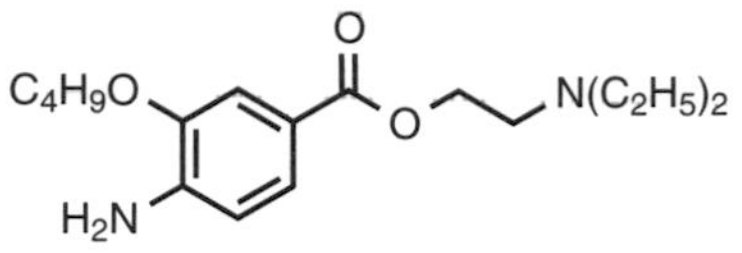

C_4H_9O O $N(C_2H_5)_2$ O H_2N

Benoxinate ($pK_a = 9.0$)

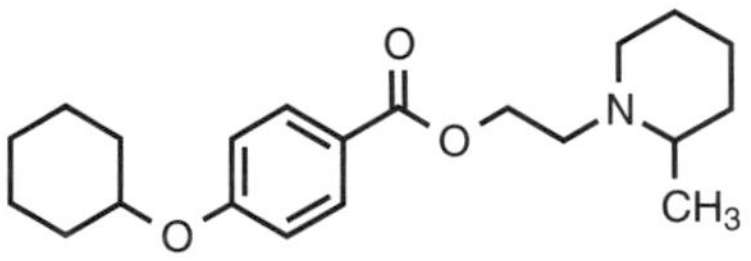

O N O O CH_3

Cyclomethycaine ($pK_a = 8.6$)

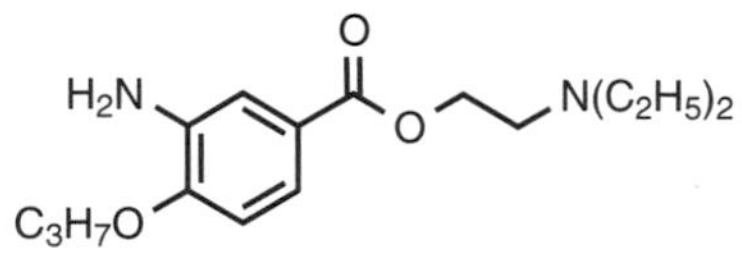

H_2N O $N(C_2H_5)_2$ O C_3H_7O

Proparacaine ($pK_a = 9.1$)

2. Amino Amides

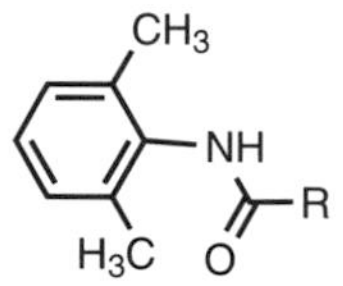

CH_3 NH R H_3C O

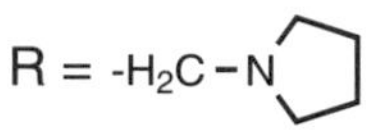

R = $-H_2C-N$ Pyrrocaine

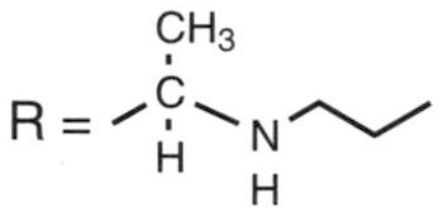

R = CH_3 C H N H Prilocaine ($pK_a = 7.9$)

R = R' N

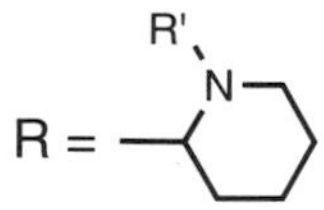

R' = C_4H_9 Bupivacaine ($pK_a = 8.1$)
R' = CH_3 Mepivacaine ($pK_a = 7.6$)

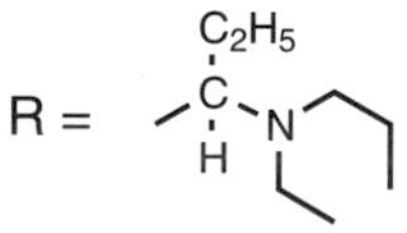

R = C_2H_5 C H N Etidocaine ($pK_a = 7.7$)

N OC_4H_9 O N H $N(C_2H_5)_2$

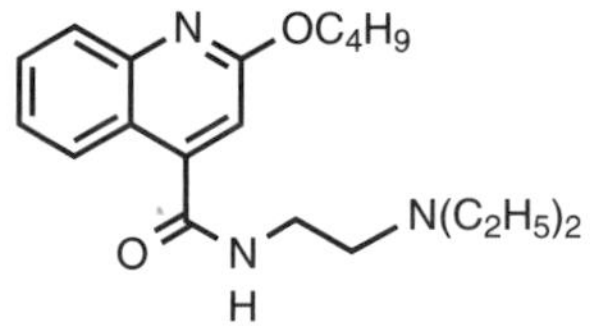

Dibucaine ($pK_a = 8.8$)

(Continued)

Table 16–2. *(Continued)*

3. Amino Ethers

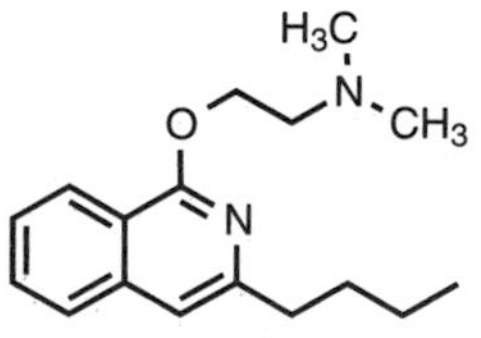

H_3C, N, CH_3, O, N

Dimethisoquin (pK_a = 7.7)

O, N, O

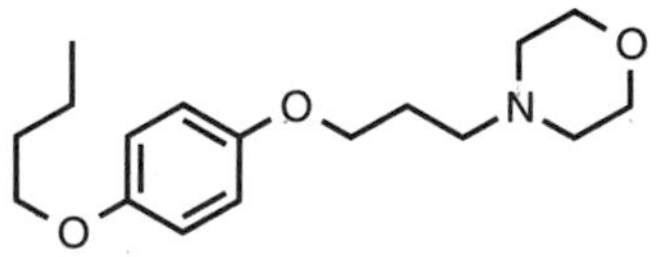

Pramoxine (pK_a = 7.1)

4. Amino Ketones

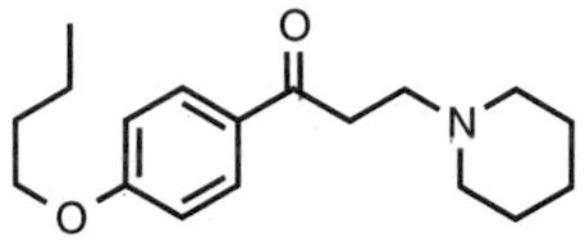

O, N, O

Dyclonine (pK_a = 8.2)

5. Amino Carbamates

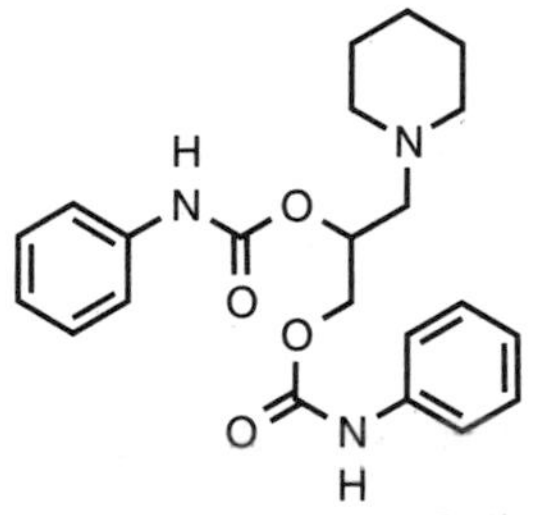

N, H, N, O, O, O, O, N, H

Diperodon (pK_a = 8.4)

6. Amidines

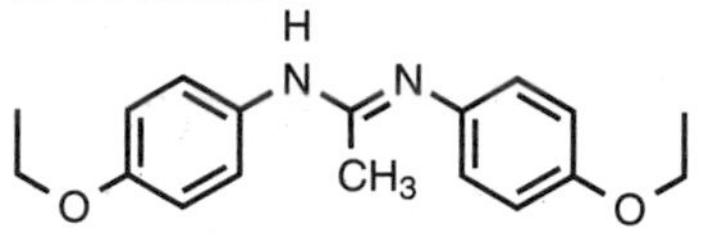

H, N, N, CH_3, O, O

Phenacaine

7. Alcohols

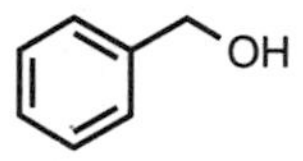

OH

Benzyl alcohol

CH_3
$Cl_3C-C-OH$
CH_3

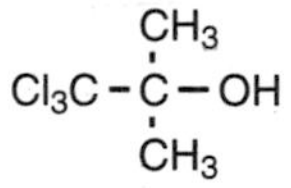

Chlorobutanol

8. Phenols

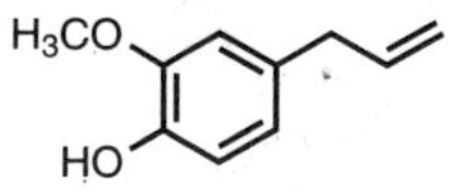

H_3CO, HO

Eugenol

HO

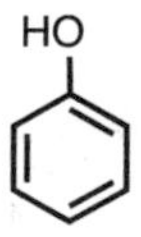

Phenol

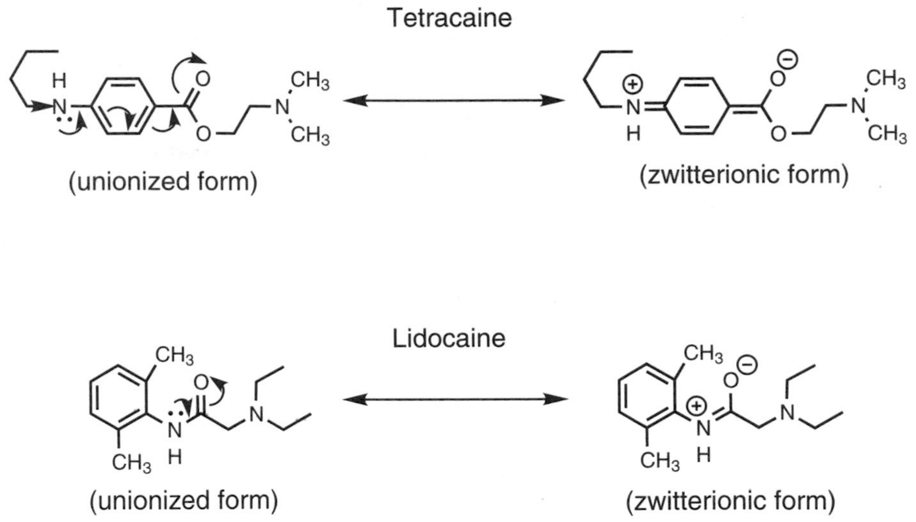

Tetracaine

(unionized form) (zwitterionic form)

Lidocaine

(unionized form) (zwitterionic form)

Fig. 16–8. Possible zwitterionic forms for procaine and lidocaine analogs.

from amide hydrolysis to ensure a desirable duration of action. Similar conclusions can be made to rationalize the increase in the duration of action of propoxycaine by the *o*-propoxy group. The shorter duration of action, however, observed with chloroprocaine when it was compared with that procaine can only be explained by the inductive effect of the *o*-chloro group, which pulls the electron density away from the carbonyl function, thus making it more susceptible for nucleophilic attack by the esterases.

Intermediate Chain

The intermediate chain almost always contains a short alkylene chain of one to three carbons in length linked to the aromatic ring via several possible organic functional groups. The nature of this intermediate chain determines the chemical stability of the drug. It also influences the duration of action and relative toxicity. In general, amino amides and amino carbamates are more resistant to metabolic inactivation than the amino esters and thus are longer-acting local anesthetics. Placement of small alkyl groups (i.e., branching), especially around the ester function (such as hexylcaine and meprylcaine) or the amide function (such as etidocaine and prilocaine), also hinders ester or amide hydrolysis and prolongs the duration of action. It should be mentioned, however, that prolonging the duration of action of a compound usually also increases its systemic toxicities, and this is discussed in a later section under biochemical distribution.

In the lidocaine series, lengthening of the alkylene chain from one to two to three increases the pK_a of the terminal tertiary amino group from 7.7 to 9.0 to 9.5. Thus, lengthening of the intermediate chain effectively reduces local anesthetic potency as a result of a reduction of onium ions under physiologic conditions. As mentioned earlier, the onium ions are required for effective binding to the channel receptors.

Hydrophilic Portion

Most clinically useful local anesthetics have a tertiary alkylamine, which readily forms water-soluble salts with mineral acids, and this portion is commonly considered as the hydrophilic portion of the molecule. The necessity of this portion of the molecule for local anesthetic activity is still a matter of debate. The strongest opposition for requiring a basic amino group for local anesthetic action comes from the observation that benzocaine, which lacks the basic aliphatic amine function, has potent local anesthetic activity. For this reason, it is often suggested that the tertiary amine function is needed only for the formation of water-soluble salts suitable for pharmaceutical preparations. With the understanding of the voltage-activated sodium channel and the possible mechanism of action of local anesthetics discussed previously, however, it is quite conceivable that the onium ions produced by protonation of the tertiary amine group are also required for binding to the receptors.

From Table 16–2, the hydrophilic group present in most of the clinically useful drugs can be in the form of a secondary or tertiary alkyl amine or as part of a nitrogen heterocycle (such as pyrrolidine, piperidine, morpholine). As mentioned earlier, most of the clinically useful local anesthetics have pK_a values of 7.5 to 9.0. As we learned in organic chemistry, the effects of an alkyl substituent on the pK_a depends on the size, length, and hydrophobicity of the group; it is difficult to see a clear structure-activity relationship among these structures. It is generally accepted, however, that local anesthetics having higher lipid solubility and lower pK_a values appear to exhibit more rapid onset and lower toxicity.

Stereochemistry

Are there any stereochemical requirements of local anesthetic compounds when they bind to the sodium channel receptors? Although a number of clinically used local anesthetics do contain a chiral center (i.e., hexylcaine, cyclomethycaine, bupivacaine, mepivacaine, etidocaine, prilocaine, and diperodon), in contrast to cholinergic drugs, the effect of optical isomerism on isolated nerve preparations revealed a lack of stereospecificity. In a few cases (e.g., prilocaine, bupivacaine, and etidocaine), when they have been administered in vivo, however, small differences in total pharmacologic profile of optical isomers have been noted. Whether these differences are due to differences in uptake, distribution, and metabolism or due to direct binding to the receptor has not been determined. When structural rigidity has been imposed on the molecule, however, as in the case of some aminoalkyl spirotetraline succinimides,[28] differences in local anesthetic potency of the enantiomers have been observed, ranging from 1:2 to 1:10. Although these differences in enantiomers are clearly not as pronounced as those in other pharmacologic agents, such as adrenergic blocking agents or anticholinergic drugs, steric requirements are necessary for effective interaction between a local anesthetic agent and its proposed channel receptors.

local anesthetic activity and the associated toxicities. Extraneuronal blood vessels near the site of drug application greatly affect the amount of drug that reaches the nerve trunk to establish anesthesia; thus, a larger dosage is required to elicit local anesthesia, which also affects the observed toxicity. After the establishment of anesthesia, however, a major contributing factor to the loss of the drug from the nerve may be the intraneuronal blood vessels. Thus, limitation of blood flow in the area of injection may substantially increase the amount of drug available to the nerve and thus improve the duration of successful anesthesia. A reduction in local blood flow also slows systemic uptake of the drug from the injection site and thus minimizes any potential toxicity.

Protein binding affinity of local anesthetic drugs seems to be correlated to the duration of local anesthetic activity. If the binding affinity is too great, however, the ability of the drug to reach its target site could be impeded, thereby leading to less active agents.

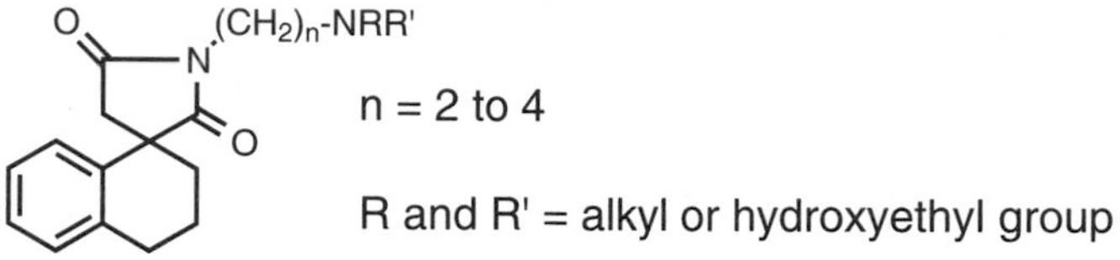

$(CH_2)_n$-NRR'

n = 2 to 4

R and R' = alkyl or hydroxyethyl group

N-Aminoalkyl spirotetralin succinimides

Biochemical Distribution

A solution of a local anesthetic is commonly administered by injection to a site near the nerve trunk to allow maximum diffusion to the nerve ending to effect desirable local anesthesia. Thus, it is reasonable to assume that any diffusion of the drug from this site into the blood circulation only contributes to the toxicity and side effects of the administered drug. In the design of local anesthetic agents, many attempts have been made to maximize penetration of nerve membranes while minimizing the loss of drug to the systemic circulation by modifying the structure of a known drug. This has not been an easy task, however, even though the ability of a drug molecule to penetrate a membrane appears to correlate well with its lipid solubility, and this distribution coefficient can be measured experimentally. The difficulty arises from the fact that the in vivo system is normally more complex than the laboratory experiments used to assess the local anesthetic potency (i.e., isolated frog sciatic nerves or rabbit cornea). For example, increasing lipid solubility of a compound may well result in facilitated penetration of a nerve membrane, but it can also enhance the ability of the drug to pass through a blood vessel wall. This increased lipid solubility, therefore, reduces the anesthetic potency owing to the more rapid removal of the agent from the site, which thereby increases its systemic toxicity. Another complicating factor is the presence of adipose tissues in the area of drug deposition. Because of the lipoidal nature of the adipose tissue, any distribution of the drug into this lipid depot or site of loss also reduces local anesthetic activity.

Furthermore, the degree of vascularization or rate of blood flow at the site of application and the total dosage administered also play an important role in governing

Metabolism

In discussing the metabolic fates of local anesthetics, most of the clinically available local anesthetics may be divided into two broad categories: the esters, of which procaine is an example, and the nonesters, exemplified by lidocaine.

The ester-type local anesthetics are hydrolyzed by esterases, which are widely distributed in body tissues. These compounds can therefore be metabolized in the blood, kidneys, and liver and, to a lesser extent, at the site of administration. For example, both procaine and benzocaine are easily hydrolyzed by esterases into *p*-aminobenzoic acid (PABA) and the corresponding alcohols. It is not surprising that potential drug interactions exist between the ester-type local anesthetics and other clinically important drugs, such as cholinesterase inhibitors or atropine-like anticholinergic drugs. These compounds either inhibit or compete with local anesthetics for esterases, therefore prolonging local anesthetic activity or toxicity. Another potential drug interaction with clinical significance may also be envisioned between benzocaine and sulfonamides; i.e., the hydrolysis of benzocaine to PABA may antagonize the antibacterial activity of sulfonamides.

The nonester-type local anesthetics, however, are primarily metabolized in the liver involving microsomal enzymes. A general metabolic scheme for lidocaine is shown in Figure 16–9. Marked species variations occur in the quantitative urinary excretion of these metabolites. For example, rats produce large quantities of the 3-hydroxy derivatives of both lidocaine and monoethylglycinexylidide,

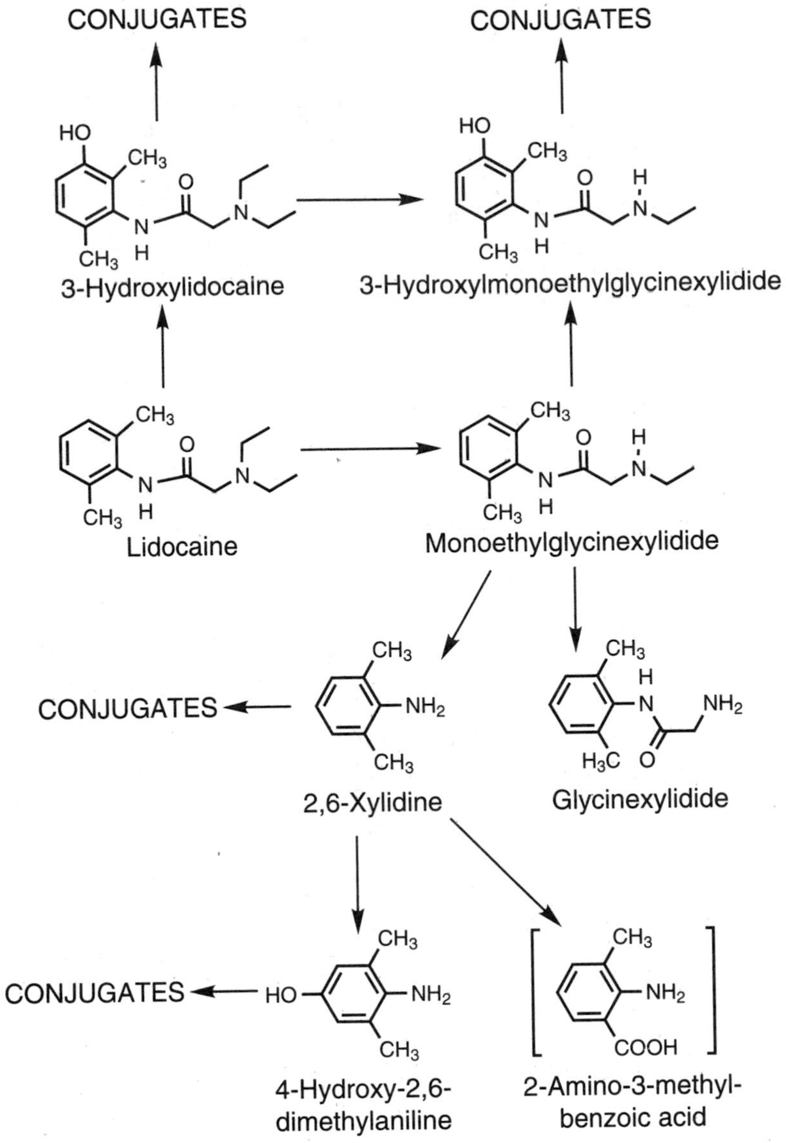

CONJUGATES
CONJUGATES
3-Hydroxylidocaine
3-Hydroxylmonoethylglycinexylidide
Lidocaine
Monoethylglycinexylidide
CONJUGATES
2,6-Xylidine
Glycinexylidide
CONJUGATES
4-Hydroxy-2,6-dimethylaniline
2-Amino-3-methylbenzoic acid

Fig. 16–9. Metabolic scheme for lidocaine. (From J. B. Keenaghan and R. N. Boyes: *J. Pharmacol. Exp. Ther.,* 180, 454(1972).)

which are subsequently conjugated and recycled in the bile. Significant quantities of these two metabolites, however, are not produced by guinea pigs, dogs, or humans. It is therefore unlikely that biliary excretion is a major pathway of excretion of these species. Species variability is important primarily when the acute and chronic toxicity of nonester-type local anesthetic agents is being evaluated.

It has been suggested that both monoethylglycinexylidide and glycinexylidide may contribute to some of the CNS side effects of lidocaine. Both of these compounds are derived from the removal of the ethyl groups of lidocaine after crossing the blood–brain barrier. To minimize these unwanted side effects, tolcainide and tolycaine have been prepared and found to possess good local anesthetic activity without any appreciable CNS side effects. Tolcainide, which lacks the vulnerable N-ethyl group but has an α-methyl group to prevent degradation of the primary amine group from amine oxidase, has desirable local anesthetic properties. Tolycaine has a *o*-carbomethoxy substituted for one of the *o*-methyl group of lidocaine. The carbomethoxy group is fairly stable in tissues but is rapidly hydrolyzed in the blood to the polar carboxylic function and is thus unable to cross the blood-brain barrier. For this reason, tolycaine lacks any CNS side effects even though it still contains the N-ethyl groups. It should be pointed out, however, that both tolcainide and tolycaine are primarily used clinically as antiarrhythmic agents.

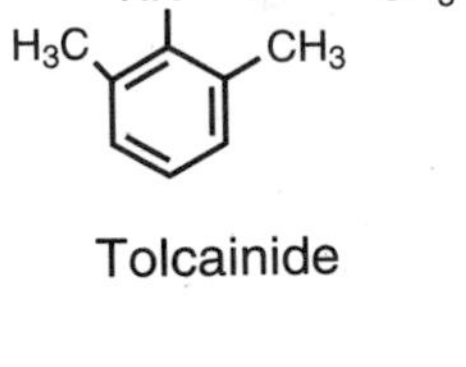

Tolcainide

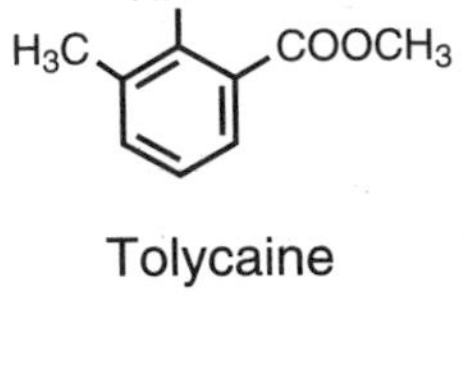

Tolycaine

Furthermore, nonester-type drugs, especially lidocaine derivatives, are also known to be more prone to enzyme induction or inhibition of other medications (e.g., cimetidine, barbiturates).

Pharmaceutical Considerations

Local anesthetic agents are generally prepared in various dosage forms: aqueous solutions for parenteral injection and creams and ointments for topical applications. Thus, chemical stability and aqueous solubility become primary factors in the preparations of suitable pharmaceutical dosage forms. As a rule, compounds containing an amide linkage have greater chemical stability than do the ester types. In this regard, an aqueous solution of an amino ester-type local anesthetic is more likely to decompose under normal conditions and cannot withstand heat sterilization because of base-catalyzed hydrolysis of the ester.

As stated before, local anesthetic activity usually increases with increasing lipid solubility. Unfortunately, this increase in lipid solubility is often inversely related to water solubility. For this reason, a suitable parenteral dosage form may not be available for these agents owing to poor water solubility under acceptable conditions. For example, benzocaine, which lacks an aliphatic amino group needed for salt formation, is practically insoluble in water at a neutral pH. Protonation of the aromatic amino group in benzocaine results in a salt with a pK_a = 2.78, which is too acidic and therefore unsuitable for the preparation of a parenteral dosage form for injection. For this reason, benzocaine and its closely related analog, butamben, are used mostly in creams or ointments to provide topical anesthesia of accessible mucous membranes or skin for burns, cuts, or inflamed mucous surfaces.

Many attempts have been made to substitute oils, fats, or fluid polymers for the aqueous vehicle commonly used in injectable local anesthetics. Unfortunately the pharmacologic results of these experiments have been quite disappointing, often as a result of the undesirable toxicity of the nonaqueous vehicle. Efforts have also been made to find additives that will potentiate the action of the classic local anesthetics. Substances such as quinine derivatives, caffeine, theobromine, antipyrine, aspirin, certain vitamins, and some quaternary ammonium compounds potentiate the action of local anesthetic drugs to various degrees. Unfortunately, these combinations are often associated with unacceptable levels of tissue irritation. The only commonly accepted organic additives to local anesthetics are vasoconstrictors, such as epinephrine. These compounds often increase the frequency of successful anesthesia and, to a limited degree, increase the duration of activity by reducing the rate of drug loss from the injection site. These agents are believed to function by constricting arterioles that supply blood to the area of the injection. The effect of these vasoconstrictors is less pronounced if the agents are added to a local anesthetic solution that is to be injected in an area that has profuse venous drainage but is remote from an arterial supply.

Because it is believed that local concentration of potassium in the vicinity of a nerve can affect the conduction of the nerve, mixtures of local anesthetic drugs enriched with potassium ions have been evaluated. The results of these experiments have also been disappointing. Ammonium sulfate, which is often used to denature proteins reversibly, has also been studied in combination with local anesthetics. Again, no substantial advantage has been established, and indications of possible irritation have been noted.

It has been reported that administration of a local anesthetic in a carbonic acid–carbon dioxide solution rather than the usual solution of a hydrochloride salt appreciably improves the time of onset and duration of action. This change in solution form is apparently not associated with local or systemic toxicity. Current theories suggest that carbon dioxide potentiates the action of local anesthetics by initial indirect depression of the axon, followed by diffusion trapping of the active form of the local anesthetic within the nerve. Use of the carbonate salt appears to be one pharmaceutic modification of the classic local anesthetic agents that may result in significant clinical advantages.

REFERENCES

1. G. Liljestrand, The historical development of local anesthesia in local anesthetics, in *International Encyclopedia of Pharmacology and Therapeutics,* Vol. I, Sect. 8, P. Lechat, Ed., Oxford, Pergamon Press, 1971.
2. L. B. Arey, *Developmental Anatomy,* 7th ed., Philadelphia, W. B. Saunders, 1965.
3. R. H. De Jong and F. G. Freund, *Int. Anesthesiol. Clin., 8,* 35(1970).
4. A. L. Hodgkin and A. F. Huxley, *J. Physiol. (Lond), 116,* 497(1952).
5. J. C. Skou, *Biochim. Biophys. Acta, 42,* 6(1960).
6. I. Tasaki, et al., *J. Gen. Physiol., 48,* 1095(1965).
7. K. Koketsu, *Neurosci. Res., 2,* 2(1969).
8. I. Tasaki, *Nerve Transmission.* Springfield, IL, Charles C Thomas, 1953.
9. W. A. Catterall, *Science, 242,* 50(1988).
10. J. S. Timmer and W. S. Agnew, *Annu. Rev. Physiol., 51,* 401(1989).
11. O. P. Hamill, et al., *Pflügers Arch., 391,* 85(1981).
12. M. B. Feinstein, *J. Gen. Physiol., 48,* 357(1964).
13. M. P. Blaustein and D. E. Goldman, *J. Gen. Physiol., 49,* 1043(1966).
14. J. M. Ritchie and P. Greengard, *Annu. Rev. Pharmacol., 6,* 405(1966).
15. W. D. Dettbarn, *Ann. N. Y. Acad. Sci.., 144,* 483(1967).
16. M. M. Tamkun, et al., *J. Biol. Chem., 259,* 1676(1984).
17. J. F. Butterworth IV and G. R. Strichartz, *Anesthesiology, 72,* 711(1990).
18. G. R. Strichartz, *J. Gen. Physiol., 62,* 37(1973).
19. K. R. Courtney, *J. Pharmacol. Exp. Ther., 195,* 225(1975).
20. T. Narahashi and D. T. Frazier, Site of action and active form of local anesthetics, in *Neurosciences Research,* Vol. 4, S. Ehrenpreis and O. C. Solnitzky, Eds., New York, Academic Press, 1971, pp. 65–99.
21. T. Narahashi, et al., *Nature, 223,* 748(1969).
22. T. Narahashi, et al., *J. Pharmacol. Exp. Ther., 171,* 32(1970).
23. T. Narahashi and D. T. Frazier, *J. Pharmacol. Exp. Ther., 194,* 506(1975).
24. B. Hille, *Ionic Channels of Excitable Membranes,* Sunderland, MA, Sinauer Associates, Inc., 1984, pp. 272–302.

25. B. P. Bean, et al., *J. Gen. Physiol., 81,* 613(1983).
26. J. C. Makielski, et al., *Biophys. J., 53,* 540(1988).
27. N. Löfgren, *Studies on Local Anesthetics: Xylocaine, a New Synthetic Drug,* Stockholm, Haegstroms, 1948.
28. S. B. A. Åkerman, et al., *Eur. J. Pharmacol., 8,* 337(1969).

SUGGESTED READINGS

M. Akeson and D. W. Deamer, Anesthetics and membranes: A critical review, in *Drug and Anesthetic Effects on Membrane Structure and Function,* R. C. Aloia, et al., Eds., Wiley-Liss, New York, 1991, pp. 71–89.

G. R. Arthur, Pharmacokinetics of local anesthetics, in *Local Anesthetics. Handbook of Experimental Pharmacology,* Vol. 81, G. R. Strichartz, Ed., Berlin, Springer, 1987, pp. 165–186.

J. F. Butterworth IV and G. R. Strichartz, *Anesthesiology, 72,* 711(1990).

K. R. Courtney and G. R. Strichartz, Structural elements which determine local anesthetic activity, in *Local Anesthetics. Handbook of Experimental Pharmacology,* Vol. 81, G. R. Strichartz, Ed., Berlin, Springer, 1987, pp. 53–94.

B. G. Covino, Local anesthetics, in *Drugs in Anaesthesia: Mechanisms of Action,* S. A. Feldman, et al., Eds., London, Edward Arnold, 1987, pp. 261–291.

B. G. Covino, Toxicity and systemic effects of local anesthetic agents, in *Local Anesthetics. Handbook of Experimental Pharmacology,* Vol. 81, G. R. Strichartz, Ed., Berlin, Springer, 1987, pp. 187–212.

A. O. Grant, *Clin. Invest. Med., 14,* 447(1991).

B. Hille, *Ionic Channels of Excitable Membranes,* Sunderland, MA, Sinauer Associates, 1984.

L. M. Hondeghem and B. G. Katzung, *Annu. Rev. Pharmacol. Toxicol., 24,* 387(1984).

P. Lechat, Ed., Local anesthetics, in *International Encyclopedia of Pharmacology and Therapeutics,* Vol. I, Sect. 8, Oxford, Pergamon Press, 1971.

K. W. Miller, General anesthetics, in *Drugs in Anesthesia: Mechanisms of Action,* S. A. Feldman, et al., Eds., London, Edward Arnold, 1987, pp. 133–159.

T. Narahashi and D. T. Frazier, Site of action and active form of local anesthetics, in *Neurosciences Research,* Vol. 4, S. Ehrenpreis and O. C. Solnitzky, Eds., New York, Academic Press, 1971, pp. 65–99.

P. Seeman, *Pharmacol. Rev., 24,* 583–655 (1972).

G. R. Strichartz and J. M. Ritchie, The action of local anesthetics on ion channels of excitable tissues, in *Local Anesthetics. Handbook of Experimental Pharmacology,* Vol. 81, G. R. Strichartz, Ed., Berlin, Springer, 1987, pp. 21–52.

L. D. Vandam, Some aspects of the history of local anesthesia, in *Local Anesthetics. Handbook of Experimental Pharmacology,* Vol. 81, G. R. Strichartz, Ed., Berlin, Springer, 1987, pp. 1–19.

Chapter 17

CHOLINERGIC AGONISTS, ACETYLCHOLINESTERASE INHIBITORS, AND CHOLINERGIC ANTAGONISTS

Danny L. Lattin

The autonomic nervous system consists of the parasympathetic and sympathetic divisions, which are referred to as the parasympathetic nervous system and the sympathetic nervous system. Medicinal chemists and other health scientists frequently subdivide the autonomic nervous system according to the chemical neurotransmitters that mediate autonomic nerve impulses. Acetylcholine is the chemical neurotransmitter in the cholinergic division; norepinephrine is the principal chemical neurotransmitter in the adrenergic division. The diagram of the autonomic nervous system in Figure 17–1[1] shows the sites of action of these neurotransmitters. Acetylcholine is the chemical neurotransmitter released by preganglionic neurons in the autonomic nervous system, all parasympathetic postganglionic neurons, some sympathetic postganglionic neurons (e.g., sweat and salivary glands), all somatic neuromuscular junctions, and some neurons in the central nervous system.

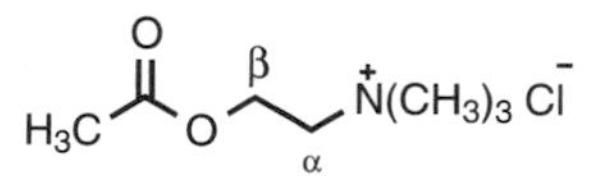

Acetylcholine chloride

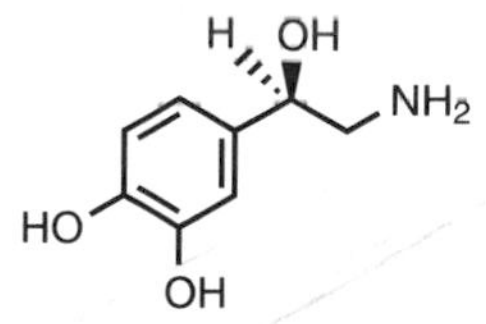

R(-) - Norepinephrine

In serving as a chemical neurotransmitter, acetylcholine is released from the presynaptic nerve ending into the synapse. The acetylcholine traverses the synaptic space and interacts with a specific receptor at a postsynaptic site. This interaction leads to a receptor-mediated response. Synaptic acetylcholinesterase catalyzes the rapid hydrolysis of acetylcholine to afford acetate and choline. This hydrolytic inactivation of acetylcholine serves as the physiologic mechanism for terminating its effects.

The parasympathetic nervous system innervates both smooth and cardiac muscles as well as certain exocrine glands (e.g., salivary and sweat glands). Parasympathetic nerve impulses are responsible for stimulating contractions of smooth muscles in the gastrointestinal tract, the urinary tract, and the eye, but decreases the rate of heart contractions. The parasympathetic nervous system causes relaxation of smooth muscle in the vasculature producing vasodilation.

Organic medicinal compounds that stimulate the parasympathetic nervous system are called cholinomimetic or parasympathomimetic agents. Cholinergic agonists are cholinomimetic agents that act directly at receptors for acetylcholine. Compounds inhibiting acetylcholinesterase, the enzyme responsible for the hydrolysis of acetylcholine, are also cholinomimetic but are not receptor agonists. Those compounds that block or inhibit parasympathetic nervous system responses by possessing affinity but without intrinsic activity for cholinergic receptors are termed cholinergic antagonists or parasympatholytic agents. This chapter is devoted to the discussion of both types of compounds as well as to the biochemistry of the parasympathetic nervous system.

Perhaps no other mammalian system, and no other chemical neurotransmitter, has been studied as exhaustively as the parasympathetic nervous system and acetylcholine. Scientists of all disciplines have been involved in this research, and their discoveries have found application to studies of all other systems in the human body. Studies of the parasympathetic nervous system and cholinergic agents led to the concept of neurochemical transmission, were instrumental in describing stereochemical influences on drug action, and pioneered the development of early drug receptor hypotheses and models. The history of research developments from the latter part of the nineteenth century to about 1940 makes interesting reading. An excellent summary of this history is presented in the first chapter of Waser.[2]

In 1914, Dale defined the two fundamental subdivisions of the parasympathetic nervous system when he observed that ethers and esters (including acetylcholine) of choline produced effects similar to either muscarine (muscarinic effects) or nicotine (nicotinic effects) in different pharmacologic preparations.[2] The initial experiments were performed using an ergot extract contaminated with acetylcholine, although Dale was unaware of this contamination. Ewins, a chemist who collaborated with Dale, isolated acetylcholine from the ergot extract and subsequently synthesized acetylcholine. This permitted Dale to prove that the unexpected muscarine-like effects he observed with the ergot preparation were due to

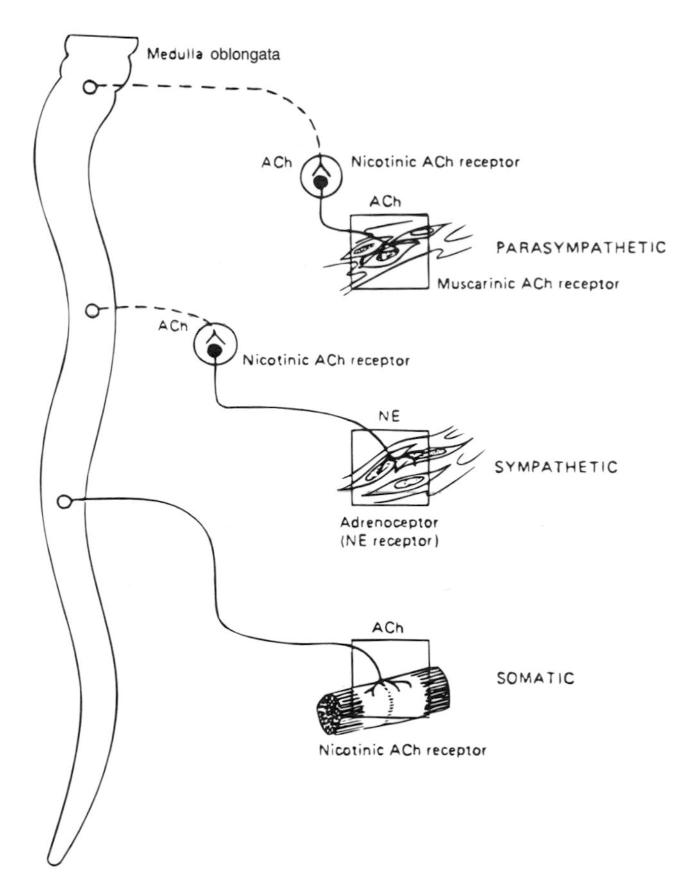

Fig. 17–1. The sites of action of acetylcholine (ACh) and norepinephrine (NE) in autonomic and somatic nervous systems. (From B. G. Katzung, Introduction to autonomic pharmacology, in *Basic and Clinical Pharmacology,* B. G. Katzung, Ed., Norwalk, CT, Appleton & Lange, 1992.)

acetylcholine. He proposed the term parasympathomimetic to describe the ability of acetylcholine to produce the same effects as electrical stimulation of parasympathetic nerves and suggested that acetylcholine was a chemical neurotransmitter in the parasympathetic nervous system. Dale also observed that the action of acetylcholine in his preparations was short-lived and proposed that the tissues contained an esterase that hydrolyzed acetylcholine to inactive products. Loewi's elegant experiments in 1921 were the first unequivocal demonstration that a chemical compound mediated impulses between nerves; he referred to the chemical substance in his preparation as *vagusstoff.*[3] In 1926, Loewi and Navratil provided experimental evidence suggesting that vagusstoff was acetylcholine.[4]

(-) - Muscarine chloride

S(+) - Nicotine

These classic studies are the foundation for current understanding of the role of acetylcholine in cholinergic nerve transmission and the recognition of muscarinic and nicotinic cholinergic receptors. These discoveries provided the stimulus for the subsequent studies of acetylcholine biochemistry, the synthesis of new organic compounds such as cholinergic and anticholinergic drugs, and the purification of cholinergic receptors.

The concept of muscarinic and nicotinic receptors to explain the different physiologic responses produced by acetylcholine was derived from the early research of Dale and Loewi. It is currently recognized that there may be many classes of muscarinic receptors, and perhaps of nicotinic receptors, but the general classification of two types of cholinergic receptors, muscarinic and nicotinic, continues to explain effectively the different physiologic responses produced by acetylcholine. Muscarinic receptors are those at which muscarine is the classic cholinergic agonist; nicotine is the classic cholinergic agonist at nicotinic receptors. Muscarinic receptors mediate cholinergic responses at all postganglionic nerve terminals and on autonomic presynaptic membranes. On the other hand, nicotinic receptors mediate responses on autonomic presynaptic nerve membranes, on autonomic ganglia, and at somatic neuromuscular junctions.

Because of the important role of acetylcholine as a chemical neurotransmitter in the autonomic nervous system, an imbalance in parasympathetic tone can lead to serious consequences and physiologic difficulties. Deficiencies of acetylcholine could conceptually be treated by administering the neurotransmitter itself, but acetylcholine is a poor therapeutic agent. It is nonselective in its actions, producing effects at all cholinergic receptor sites and leading to undesirable side effects, which could result in serious consequences for the patient. Acetylcholine is poorly absorbed across biologic membranes because it is a quaternary ammonium salt; this provides for poor bioavailability regardless of the route of administration. Furthermore, it is chemically labile owing to rapid hydrolysis of its ester functional group in aqueous solutions, in the gastrointestinal tract, and in blood, where hydrolysis is catalyzed by esterases. For these reasons, medicinal chemists have vigorously sought alternatives to acetylcholine as therapeutic agents from the time it was demonstrated to be a chemical neurotransmitter in the autonomic nervous system.

Most therapeutic cholinomimetic agents are those possessing muscarinic effects. Muscarinic cholinergic agents are used postsurgically to reestablish smooth muscle tone of the gastrointestinal tract and the urinary tract to relieve abdominal distention and urinary retention. They are also used to treat some forms of glaucoma by enhancing the outflow of aqueous humor and reducing intraocular pressure. Cholinomimetic compounds possessing activity in the central nervous system are being evaluated for the treatment of cognitive disorders such as Alzheimer's disease. Cholinomimetic drugs having nicotinic effects are commonly used to treat myasthenia gravis.

The largest number of medicinal agents used to modify the effects of acetylcholine are the cholinergic antagonists or anticholinergic drugs. Cholinergic muscarinic antagonists are sometimes referred to as antispasmodics because of their ability to reduce smooth muscle spasms resulting from overstimulation of the gastrointestinal smooth muscles.

One of the goals of medicinal chemistry research and drug discovery is to provide a rational basis for the design of new medicinal agents. As a result, many synthetic cholinergic agonists were designed using structure-activity relationships (SAR) based on the structure of acetylcholine. To design cholinergic agents selective for specific cholinergic receptors, it is necessary to have a more complete understanding of acetylcholine neurochemistry as well as of the chemical nature and role of cholinergic receptors.

ACETYLCHOLINE NEUROCHEMISTRY

The neurochemistry of acetylcholine includes its biosynthesis, storage, release, and metabolism, which are summarized in Figure 17–2. Acetylcholine is biosynthesized in cholinergic neurons by the enzyme-catalyzed transfer of the acetyl group from acetyl coenzyme A (ace-

HO–CH₂–CH(NH₂)–COOH → (Serine Decarboxylase) → HO–CH₂–CH₂–NH₂

↓ Choline N-methyltransferase / S-Adenosylmethionine

H_3C–C(=O)–O–CH₂–CH₂–$\overset{+}{N}(CH_3)_3$ ← (Acetyl-S-CoA; Choline Acetyltransferase) ← HO–CH₂–CH₂–$\overset{+}{N}(CH_3)_3$

Biosynthesis of acetylcholine.

tyl CoA) to choline, a quaternary ammonium alcohol.[5] The enzyme catalyzing this reaction, choline acetyltransferase, is also biosynthesized by the cholinergic neuron. Some choline is biosynthesized from the amino acid serine, but most of the choline used to form acetylcholine is recycled following the enzymatic hydrolysis of acetylcholine in the synaptic space. Extracellular choline is ac-

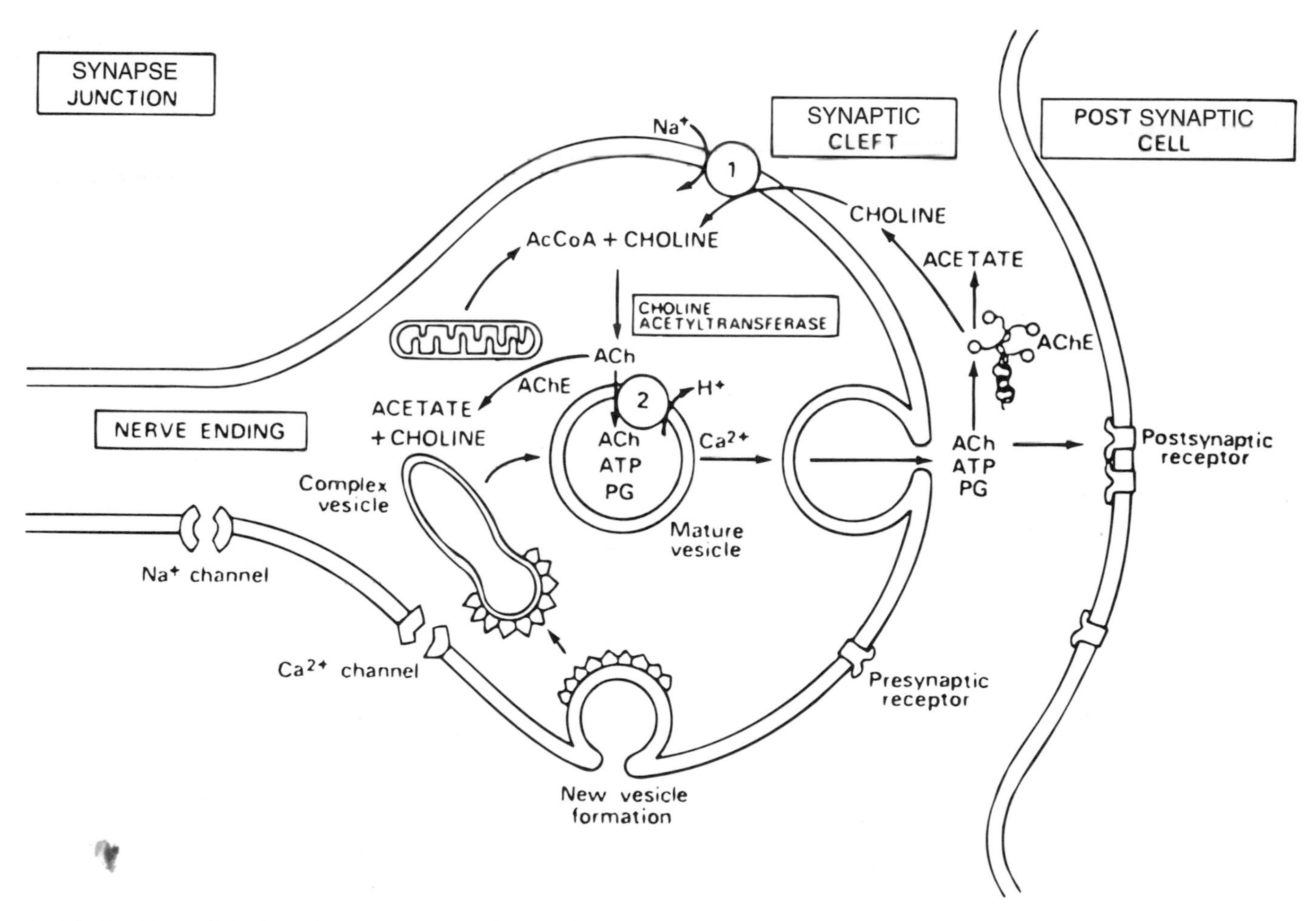

Fig. 17–2. General cholinergic nerve junction showing location of receptor sites and biosynthesis, storage, release, and hydrolysis of acetylcholine. (From B. G. Katzung, Introduction to autonomic pharmacology, in *Basic and Clinical Pharmacology,* B. G. Katzung, Ed., Norwalk, CT, Appleton & Lange, 1992.)

tively transported into the presynaptic nerve terminal by both high-affinity and low-affinity uptake sites. The high-affinity site, inhibited by hemicholinium, is probably re-

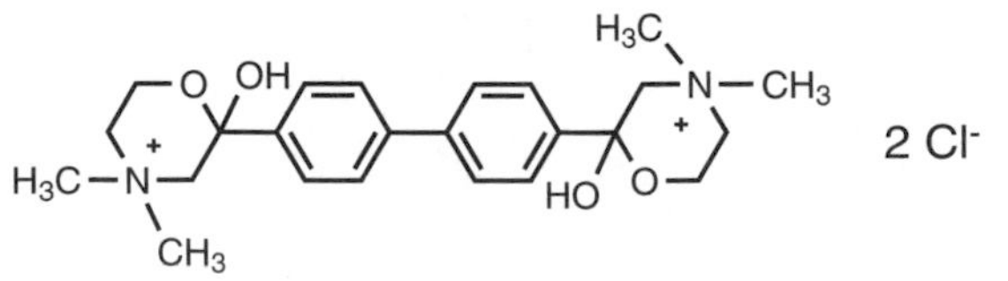

Hemicholinium chloride

sponsible for most of the choline recycled from the synapse and used to biosynthesize acetylcholine. The active uptake of choline is considered to be the rate-determining step in the biosynthesis of acetylcholine.

Efforts to develop therapeutic agents based on regulation of acetylcholine biosynthesis have not been successful. Dexpanthenol, the dextrorotatory enantiomer of the alcohol derived from pantothenic acid (a vitamin), was once used as a cholinomimetic agent to help reestablish normal smooth muscle tone in the gastrointestinal tract following surgery. Pantothenic acid is essential for the biosynthesis of coenzyme A. The apparent rationale for

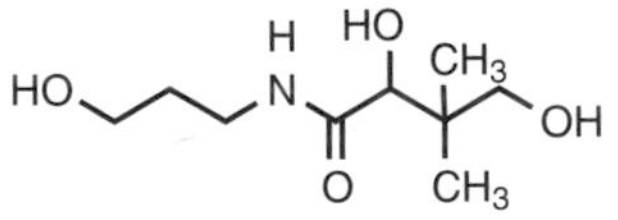

Dexpanthenol

the therapeutic use of dexpanthenol was that it would be biotransformed to pantothenic acid, which would be incorporated into coenzyme A. This would lead to increased intracellular levels of acetyl CoA, which would facilitate increased biosynthesis of acetylcholine. Limited therapeutic success of dexpanthenol, difficulty with administration, and the effectiveness of synthetic cholinergic agonists led to its discontinuation. The quaternary pyridinium salt, *trans*-N-methyl-4-(1-naphthylvinyl)pyridinium (MNPV) iodide, is an effective inhibitor of cho-

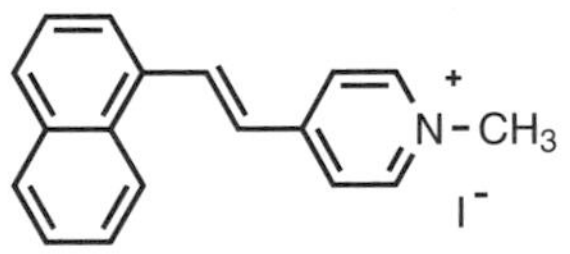

trans-N-Methyl-4-(1-naphthylvinyl)pyridinium iodide (MNPV)

line acetyltransferase in vitro. It has proven to be a poor inhibitor in whole animal experiments. Despite the efforts to design cholinergic agents based on the mechanism of biosynthesis of acetylcholine, such agents would be expected to have nonselective effects because it is currently thought that acetylcholine is biosynthesized by this same mechanism in all cholinergic neurons.

Most newly biosynthesized acetylcholine is actively transported into cytosolic storage vesicles located in the presynaptic nerve endings, where it is maintained until it is released.[6] Some acetylcholine remains in the cytosol and is eventually hydrolyzed to acetate and choline. Only the stored form of acetylcholine serves as the functional neurotransmitter.

Acetylcholine release from the storage vesicles is initiated by an action potential that has been carried down the axon to the presynaptic nerve membrane. This action potential leads to the opening of voltage-dependent calcium channels affording an influx of Ca^{++} and an exocytotic release of acetylcholine into the synapse. The increase in intracellular Ca^{++} may induce fusion of the acetylcholine storage vesicles with the presynaptic membrane before release of the neurotransmitter. Each synaptic vesicle contains a quantum of acetylcholine; one quantum represents between 12,000 and 60,000 molecules of acetylcholine. A single action potential causes the release of several hundred quanta of acetylcholine into the synapse.

Acetylcholine in the synapse can bind with receptors on the postsynaptic or presynaptic membranes to produce a response. Free acetylcholine, that which is not bound to a receptor, is hydrolyzed by acetylcholinesterase. This hydrolysis is the physiologic mechanism for terminating the action of acetylcholine. Acetylcholinesterase can hydrolyze approximately 3×10^8 molecules of acetylcholine in 1 msec; thus, there is adequate enzyme activity to hydrolyze all of the acetylcholine (approximately 3×10^6 molecules) released by one action potential. A number of useful therapeutic cholinomimetic agents have been developed based on the ability of the compounds to inhibit acetylcholinesterase. These agents are addressed later in this chapter.

ACETYLCHOLINE RECEPTORS

Medicinal chemists and other scientists have devoted a great deal of effort to understanding how cholinergic receptors carry out the two primary functions of all receptors—molecular recognition and signal transduction. A complete understanding of these phenomena is essential to achieve the desired goals of rational, efficient, and selective drug therapy.

Knowledge of the structure and function of acetylcholine receptors has increased substantially in the 80 years since the concept of distinct muscarinic and nicotinic receptors was first postulated. The earliest efforts to describe these receptors were hindered by the fact that the receptors were only a concept. The location and fundamental chemical nature of the receptor was unknown because no one had isolated a receptor. Indeed, the existence of receptors was not established until 1973, when Snyder and Pert[7] provided demonstrable evidence for the existence of opiate receptors.

The earliest attempts by medicinal chemists to characterize cholinergic receptors were based on SAR and stereochemical studies of cholinergic agonists and antagonists. This research led to the synthesis of agonists and antagonists with exceptionally high affinity and selectivity for cholinergic receptors as well as to the synthesis of radiolabeled cholinergic ligands possessing high specific radioactivity. These chemical advances were paralleled by advances in biochemistry, molecular pharmacology, and molecular biology, making possible the purification and sequencing of small quantities of protein, the measurement of ligand binding to cell membranes and subcellular components of cells, and the cloning and base sequencing of genes. These scientific and technologic advances culminated in the isolation, purification, and deduced amino acid sequencing of one of the nicotinic acetylcholine receptors—the first acetylcholine receptor, and the first neurotransmitter receptor, to be this completely characterized.[8,9] Subsequently, muscarinic receptor have been isolated and purified, as well as sequenced, using these techniques.

Current evidence from pharmacologic and molecular biology research indicates that there are multiple muscarinic and nicotinic acetylcholine receptor subtypes.[10,11] The traditional classification of muscarinic receptors and nicotinic receptors, however, is adequate to describe the actions of most cholinergic medicinal agents and is used throughout this chapter. Furthermore, most of the current therapeutic agents acting at muscarinic receptors exhibit little selectivity for the receptor subtypes.

MUSCARINIC RECEPTORS

Prior to the isolation and structural characterization of muscarinic receptors, SAR and stereostructure-activity relationship studies of cholinergic agonists and antagonists provided clues to the structure of the muscarinic receptor. This pioneering research led to the development of effective cholinergic medicinal agents and has provided an invaluable foundation for more recent discoveries.

SAR were the basis for early models of the receptor structure that would account for the affinity and efficacy of cholinergic agonists. An early model of the muscarinic receptor was that depicted in Figure 17–3, which accounts for the importance of muscarinic agonists having an ester functional group and a quaternary ammonium group separated by two carbons. This model depicts ionic binding of acetylcholine to the receptor by an electrostatic interaction between the positive charge of the quaternary nitrogen and a negative charge at the anionic site of the receptor. The negative charge was suggested to be due to a carboxylate ion from the free carboxyl group of a dicarboxylic amino acid (e.g., aspartate or glutamate) located at the binding site of the receptor protein. This model also involves a hydrogen bond between the ester oxygen of acetylcholine and a hydroxyl group contributed by the esteratic site of the receptor.

Although this early muscarinic receptor model accounted for two important SAR requirements for muscarinic agonists, it failed to explain the following: (1) at least two of the alkyl groups bonded to the quaternary nitrogen must be methyl groups; (2) the known stereochemical requirements for agonist binding to the receptor; and (3) the fact that all potent cholinergic agonists have only five atoms between the quaternary nitrogen and the terminal hydrogen atom (Ing's rule of five[12]).

Subsequent models of the cholinergic muscarinic receptor depicted the receptor as a binding site on a protein molecule and explained more completely the structural and stereochemical requirements for cholinergic agonist activity. Some scientists proposed that the muscarinic receptor and acetylcholinesterase were the same entity. This proposal was dispelled by experiments that demonstrated that interaction of cholinergic ligands with the muscarinic receptor did not lead to a chemical change (hydrolysis) of the ligand. None of these models, however, could explain completely the diverse pharmacologic effects produced by all the muscarinic agonists and antagonists.

At this point in muscarinic receptor research, a number of important developments began to play a major role. One of these was the strong evidence that muscarinic receptor effects, like those of adrenergic receptors, are mediated by second messengers. Muscarinic receptors mediate at least two important biochemical events leading to second messengers: (1) inhibition of adenylyl cyclase and (2) increased turnover of phosphatidyl inositol. Both of these biochemical events involve a guanosine triphosphate (GTP)–dependent mechanism. Two other important developments were the synthesis of highly selective muscarinic ligands and the utilization of molecu-

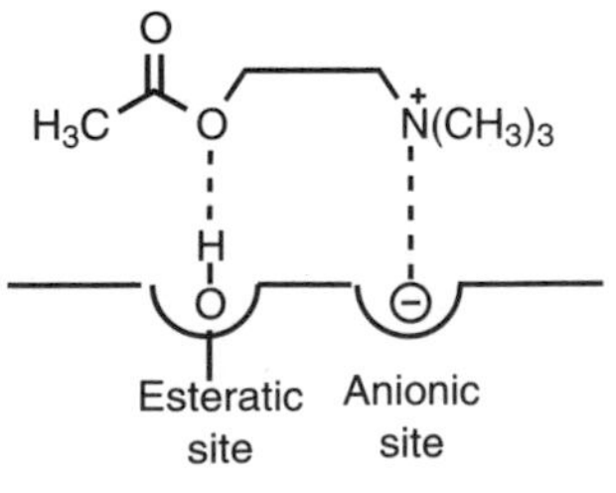

Esteratic site

Anionic site

Fig. 17–3. Original representation of the muscarinic receptor.

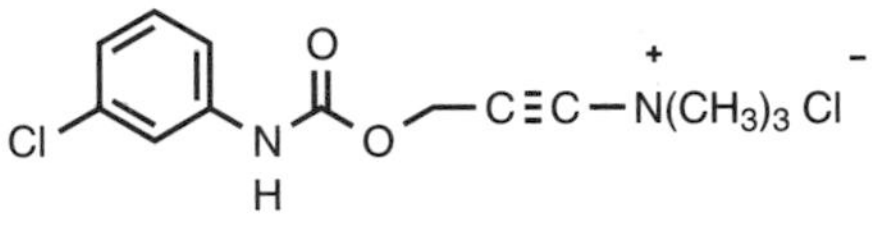

McN-A-343

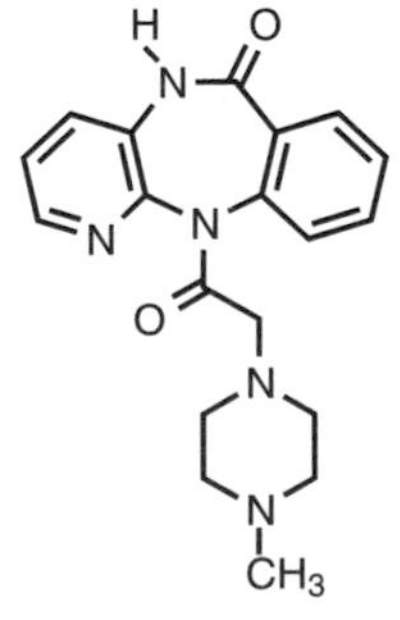

Pirenzepine

lar biology techniques in the study of muscarinic receptors.

It has become well accepted that the muscarinic receptors are not a homogeneous group of receptors located at different anatomic sites but actually consist of at least two pharmacologically distinct subtypes referred to as M1 and M2, each responsible for different muscarinic responses.[13,14] It is important to note that muscarinic receptors identified by pharmacologic and radioligand binding experiments are designated with an upper case "M." Those receptors identified in molecular biology studies are designated with a lower case "m." This pharmacologic distinction of receptor subtypes was made possible by the availability of synthetic muscarinic ligands possessing a high degree of affinity and differential selectivity for these receptors as well as by in vitro techniques for measuring radioligand binding to membrane-bound receptors. Muscarinic M1 receptors, located on presynaptic membranes in autonomic ganglia and in the brain, are characterized by their high affinity for the muscarinic agonist McN-A-343 and the muscarinic antagonist pirenzepine. The M2 muscarinic receptors, located primarily on postganglionic membranes in the autonomic nervous system (smooth muscle, heart, exocrine glands), have much lower affinity for both of these ligands. Further investigations with these ligands as well as with other muscarinic ligands possessing receptor selectivity and differential affinity indicate that there may be a third muscarinic receptor subtype, M3.

This apparent multiplicity of muscarinic receptor subtypes has been supported by research using molecular biology techniques for cloning and base sequencing of the genes for muscarinic receptors. These studies indicate that there may be at least five muscarinic receptor subtypes, m1–m5, most of which exist in the brain. Cloning and base sequencing of the genes encoding for muscarinic receptors have been major advances in the understanding of the chemical nature and function of these receptors.[12,15–17] These experiments demonstrated that the muscarinic receptors belong to a group of receptors known as G-protein coupled receptors (GPCR); the β-adrenergic receptors are also members of the GPCR family. The guanine nucleotide regulatory protein to which these receptors are coupled serves to link the receptor to second messengers. Binding of acetylcholine and other muscarinic agonists to the GPCRs leads to a variety of effector responses, which includes inhibition of adenylyl cyclase, stimulation of guanylyl cyclase, regulation of phos-

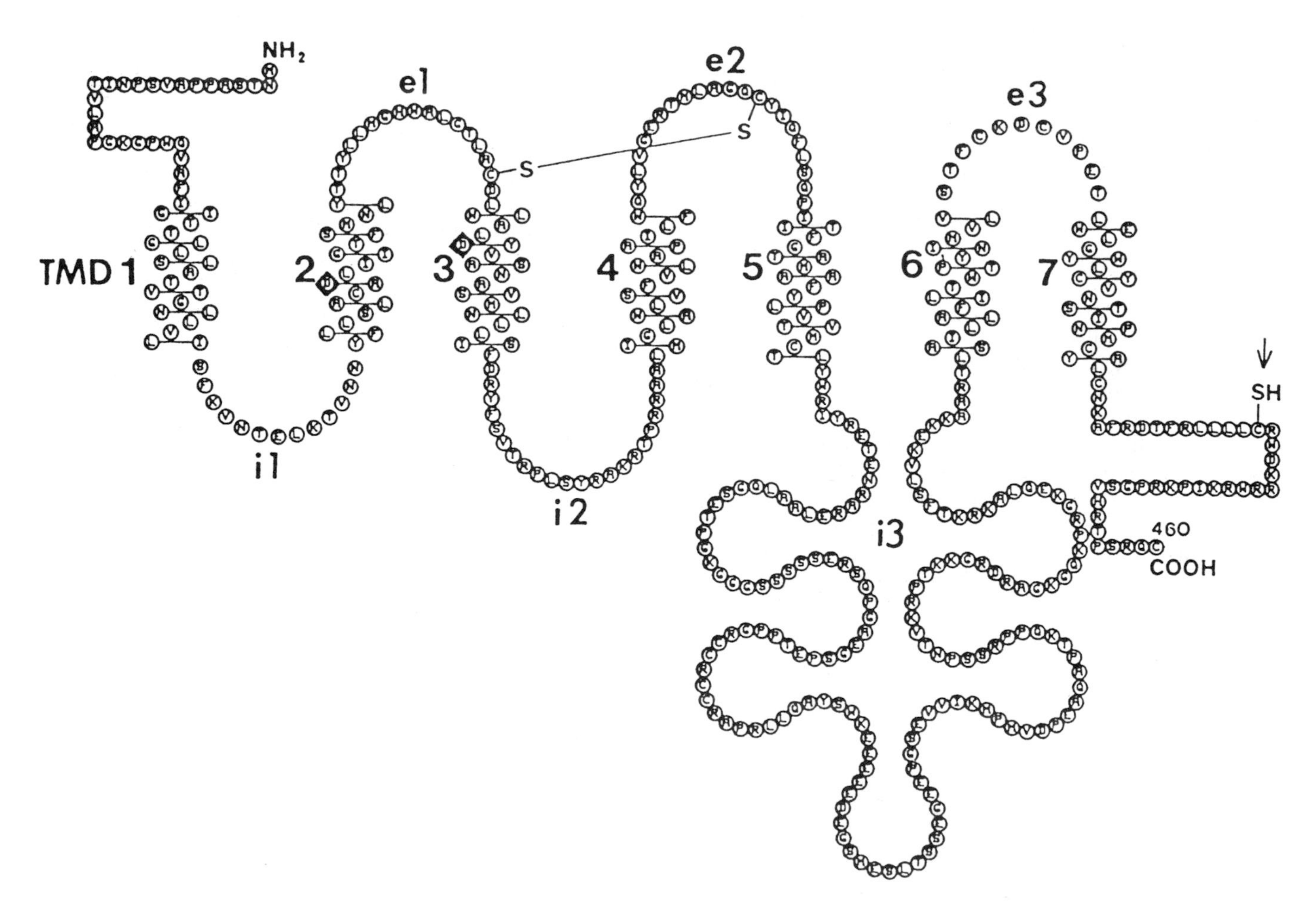

Fig. 17–4. Deduced amino acid sequence of human muscarinic acetylcholine receptor M1 and putative arrangement of the seven transmembrane domains, three intracellular domains (i1–i3), and three extracellular domains (e1–e3). (From J. Lameh, et al., *Pharm. Res., 7,* 1213[1990].)

phatidyl inositol turnover, and regulation of potassium and calcium ion channel activity. The ultimate observable response is a function of the tissue where the receptor is located.

The amino acid sequences (primary structures) of the muscarinic receptor proteins expressed by the cloned genes for the GPCRs have been deduced from the base sequence of the respective genes. The application of molecular modeling programs to the deduced structures of the muscarinic GPCR, as well as other GPCRs, suggests that GPCRs are components of the cell membrane and consist of seven transmembrane helical domains that are hydrophobic as well as three extracellular and three intracellular loops that are hydrophilic.[18] The N-terminus of the GPCR protein is extracellular, and the C-terminus is intracellular. This proposed arrangement for the human type-1 muscarinic receptor (Hm1), including its deduced amino acid sequence, is illustrated in Figure 17–4.[15] Computer-assisted molecular modeling has also made it possible to obtain three-dimensional models of the muscarinic receptor; a proposed top-view model of the m1 muscarinic receptor is shown in Figure 17–5.[18] It is interesting to observe that this model suggests that the quaternary nitrogen of acetylcholine participates in an ionic bond with the free carboxylate group of an aspartate residue (D105)—one of the receptor functional groups that was hypothesized to be involved in receptor binding of acetylcholine almost 50 years ago using only SAR data and the powers of deduction.

The current model for muscarinic receptors is much more descriptive than earlier models of the relationship between ligand binding to the receptor molecule (molecular recognition) and the resulting effect (signal transduction). Figure 17–6 illustrates this relationship.[17] In this model, acetylcholine (H) binds to the muscarinic receptor located in the cell membrane, and this ligand-receptor interaction is translated, presumably by a conformational perturbation, through the receptor protein to the receptor-coupled guanine nucleotide regulatory protein (G-protein). A proposal for the relationship between the guanine nucleotide regulatory protein and the effector is shown in Figure 17–7.[19] In this proposal, a GTPase cycle controls the guanine nucleotide regulatory protein (G-protein) coupled with the receptor protein. The G-protein is maintained in an inactive state by tightly bound GDP. The GDP, possessing high affinity for G-protein, dissociates slowly from the G-protein; this results in minimal spontaneous activation of the G-protein. When an agonist ligand, such as acetylcholine, binds to its receptor, the receptor causes the G-protein to become activated by releasing GDP and binding GTP. Although GTP has lower binding affinity than GDP, its intracellular concentration is much higher. Binding of GTP to the G-protein changes the protein to its active form, which, in turn, interacts with its effector and either inhibits or activates the effector; this leads to regulation of the second messenger. The active form of the G-protein reverts to its inactive, GDP-bound form following hydrolysis of GTP to GDP by GTPase.

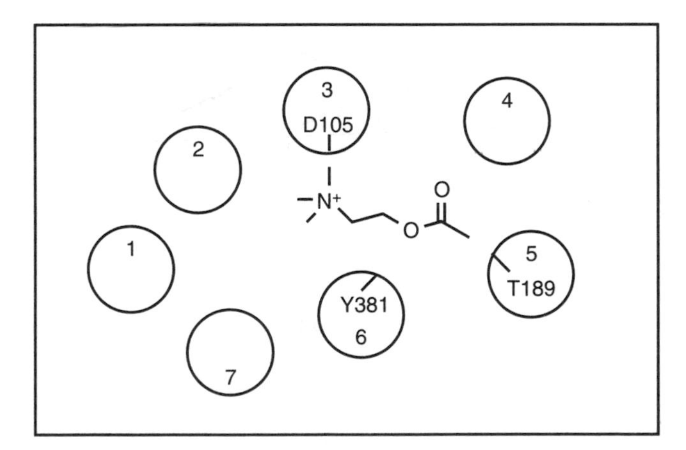

Fig. 17–5. Model of acetylcholine interaction with muscarinic M1 receptor. Circles represent seven transmembrane domains; D105, T189, and Y381 are aspartate, threonine, and tyrosine residues. (Adapted from reference 18.)

NICOTINIC RECEPTORS

Nicotinic receptors have been the focus of intensive research interest even though the majority of clinically effective cholinergic medicinal agents have been designed as either agonist or antagonist ligands for muscarinic receptors. This interest is due both to the ready availability of nicotinic receptor protein from the electric organs of the electric eel (Electrophorus electricus) and the marine ray (Torpedo californica) and to the important role that nicotinic receptors play in myasthenia gravis, an autoimmune disease.

Pharmacologic and medicinal chemical evidence has long supported the concept of multiple nicotinic receptors based on the different anatomic sites (autonomic ganglia and skeletal neuromuscular junction) of these receptors and on the different structural requirements for nicotinic agonists and antagonists acting at these two receptor populations. Multiplicity of nicotinic receptors is also supported by more recent molecular biology research.[6]

Acetylcholine nicotinic receptors belong to a group of receptors classified as ligand-gated ion channel receptors. The receptor creates a transmembrane ion channel (the gate), and acetylcholine (the ligand) serves as a gatekeeper by interacting with the receptor to modulate passage of ions, principally K^+ and Na^+, through the channel.

A nicotinic receptor was the first neurotransmitter receptor to be isolated and purified in an active form using the same molecular biologic techniques described for the isolation and characterization of the muscarinic receptors. The primary sequence of nicotinic receptors has been deduced from the cloning and sequencing of the genes that encode the nicotinic receptor proteins.[20,21]

The nicotinic receptor of electric organs and muscle tissue is a glycoprotein consisting of four distinct subunits—the α, β, γ, and δ-subunits. In a single receptor molecule, there is a pentameric arrangement of two α-subunits in combination with one each of the β, γ, and δ-subunits; this is usually abbreviated $\alpha_2\beta\gamma\delta$. The five sub-

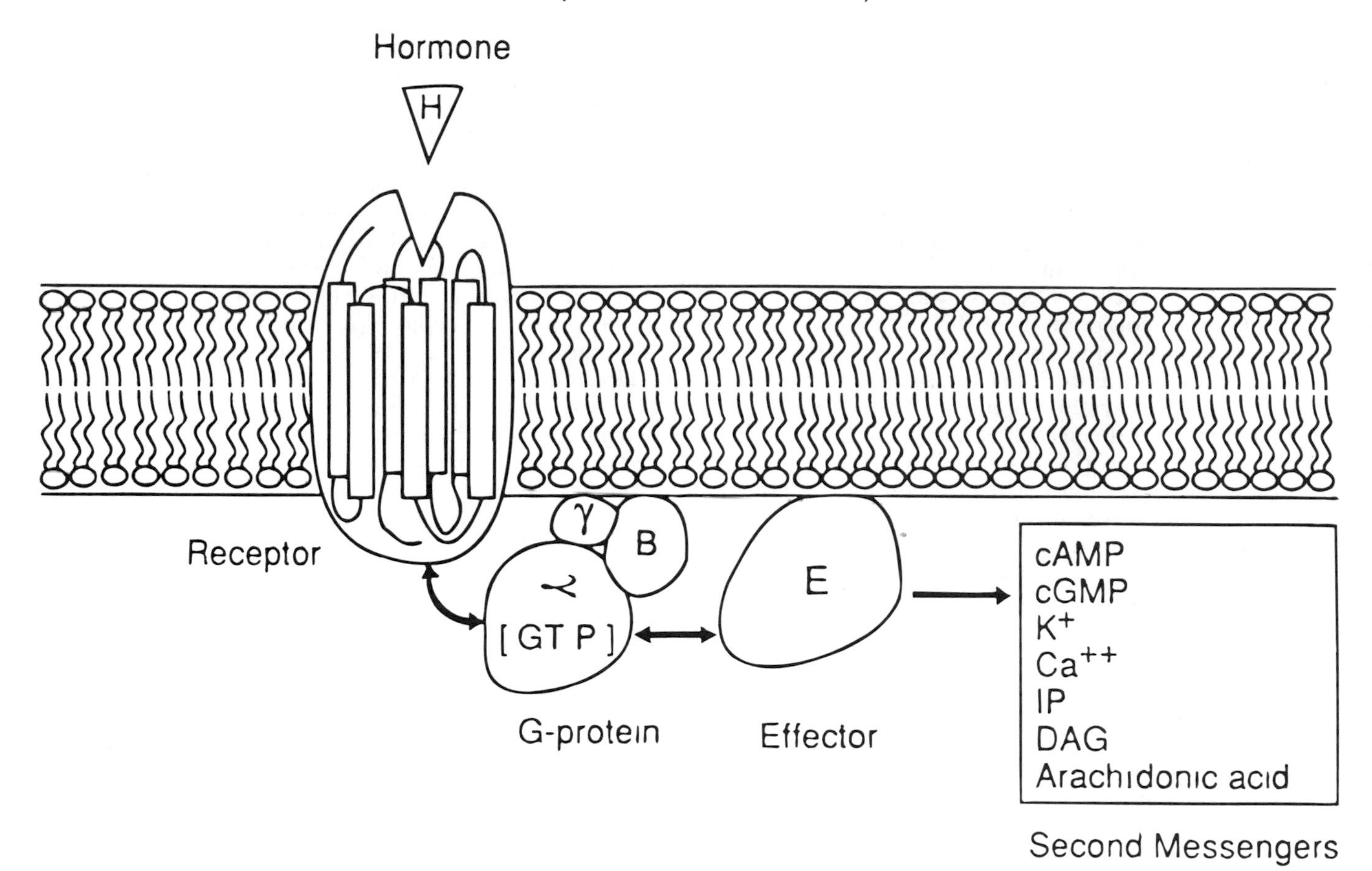

Fig. 17–6. Model of signal transduction by a G-protein coupled receptor. This illustrates a proposed relationship between receptor, G-protein, the effector, and various second messengers. (From R. A. F. Dixon, et al., *Ann. Rep. Med. Chem., 23,* 221[1988].)

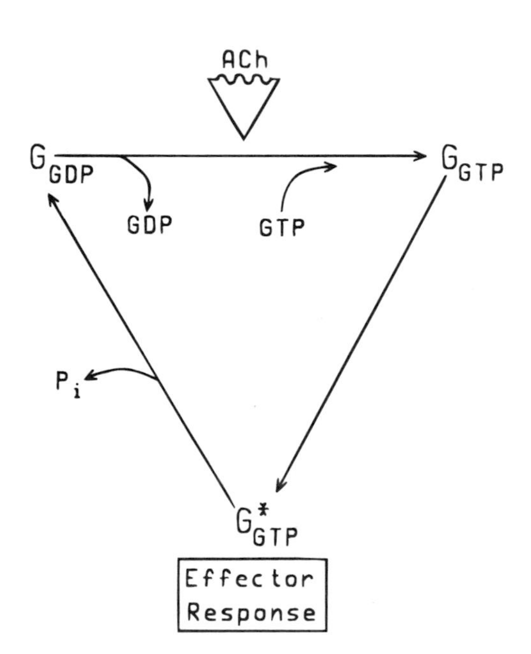

Fig. 17–7. Diagram of a GTPase cycle proposed to control signal transduction between muscarinic G-protein coupled receptors and the effector. The active form of G-protein is designated G^*_{GTP}. (Adapted from A. M. Spiegel, *Ann. Rep. Med. Chem., 23,* 235[1988].)

units of each receptor protein are arranged around a central pore that serves as the ion channel. Based on molecular modeling of the deduced primary structure of the individual subunits, it is proposed that each subunit possesses an extracellular N-terminus that is hydrophilic, an intracellular C-terminus that is also hydrophilic, four alpha helical hydrophobic domains that are in the cell membrane, and an amphipathic alpha helical domain that may be located either in the membrane or extend into the cytoplasm (Fig. 17–8).[21] It has been suggested that these five amphipathic domains make up the walls of the ion channel in the pentameric arrangement of the receptor subunits. There are two acetylcholine binding sites on the extracellular domain of each receptor molecule. One binding site is located on each α-subunit; the binding sites possess a positive cooperativity even though the two α-subunits are not adjacent to each other in the pentameric receptor.

Current evidence indicates that the receptor found in the electric organ is the same as that found at the mammalian neuromuscular junction, but this is based entirely on deduced protein primary structures from the cloned genes and awaits confirmation by actual amino acid sequencing of the receptor protein. Cloning experiments, however, indicate that receptor heterogeneity does exist. Nicotinic-like receptors in the central nervous system appear to be composed of only α- and β-subunits. Receptors

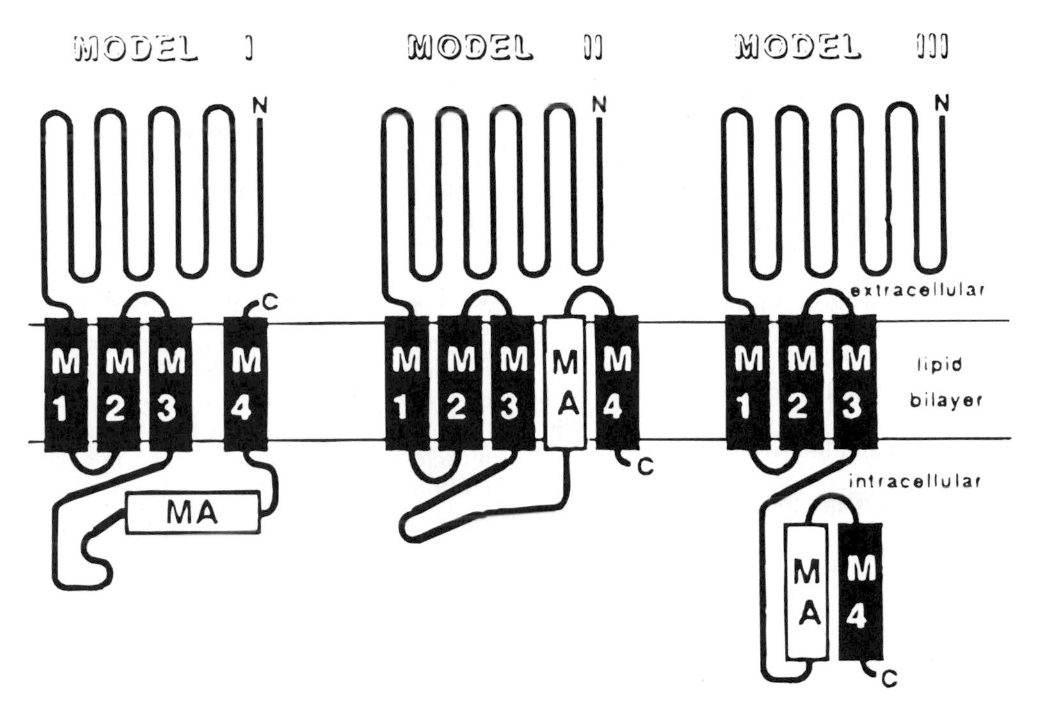

Fig. 17–8. Three models of the nicotinic (n) receptor showing amino terminus (N), carboxy terminus (C), four hydrophobic transmembrane domains (M1–M4), and the amphipathic domain (MA). (From W. H. M. L. Luyten and S. G. Heinemann, *Ann. Rep. Med. Chem., 22,* 281 [1987].)

in muscle tissue differ between embryo-derived tissues and that derived from adult tissue; adult muscle nicotinic receptors exhibit a ε subunit found in place of the γ-subunit in embryonic muscle tissue receptors.

Our knowledge and understanding of the muscarinic and nicotinic receptors has advanced tremendously from the time these receptors were only ethereal concepts, thanks to the dedicated efforts of many scientists. This understanding of cholinergic receptors provides the basis for the rational design of new selective therapeutic agents to treat diseases associated with cholinergic neurons.

STEREOCHEMISTRY OF ACETYLCHOLINE

One shortcoming of all the early models for cholinergic receptors was that they did not account for the observed stereoselectivity of the receptors for agonist and antagonist ligands. Even though acetylcholine does not exhibit optical isomerism, many of the synthetic and naturally occurring agonists and antagonists are optical isomers; usually one of the enantiomers is many times more active than the other. It was apparent to early receptor investigators that the stereochemistry of cholinergic ligands is important for receptor binding. In this regard, the stereochemical-activity relationships of cholinergic ligands has been studied extensively to provide a rational basis for the design of cholinergic drugs as well as to describe the properties and functions of cholinergic receptors.

The stereochemistry of acetylcholine resides in the different arrangements in space of its atoms by virtue of rotation about sp^3 covalent bonds, i.e., conformational isomerism. Because of the relatively unrestricted rotation about these single covalent bonds, acetylcholine can exist in a number of conformations. Most of the studies on the conformational isomerism of acetylcholine have focused on the torsion angles between the ester oxygen atom and the quaternary nitrogen resulting from rotation about the C_α–C_β bond. Four of these conformations are illustrated by Newman projections in Figure 17–9.

Techniques used to determine the thermodynamically preferred conformation of acetylcholine have included nuclear magnetic resonance spectrometry (NMR), x-ray crystallography (x-ray), and molecular orbital calculations. The NMR studies provided the preferred conformation of acetylcholine in aqueous solution and indi-

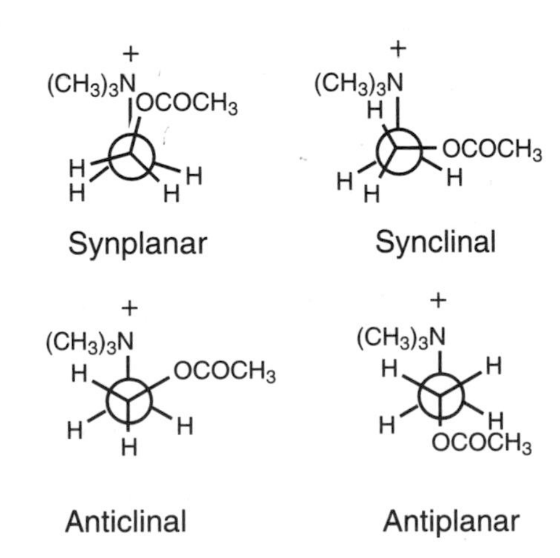

Fig. 17–9. Conformational isomers of acetylcholine.

cated that the molecule assumes an anticlinal (gauche) conformational relationship between the ester oxygen and the quaternary nitrogen. This conformation is supported by the x-ray crystallographic data, which established that the conformation of acetylcholine in the solid, crystalline state is also anticlinal. Molecular orbital calculations of the preferred conformation of acetylcholine also resulted in the conclusion that the preferred conformation is anticlinal, in keeping with the conformation derived from NMR and x-ray determinations. These experimental and theoretical determinations of the acetylcholine conformation differ from the antiplanar conformation that might be expected from molecular models by minimizing bond overlap. There is an intramolecular interaction, most probably an electrostatic attraction, between the quaternary nitrogen and the ester carbonyl oxygen, which stabilizes the anticlinal conformation.

It must be emphasized that the experimentally determined conformations of acetylcholine are only those measured in aqueous solution (NMR) or the crystalline state (x-ray). These may not be the conformation preferred by the receptors; indeed, the conformation of receptor-bound acetylcholine could be much different and might not be a thermodynamically preferred conformation.

In recognition of this possibility, medicinal chemists synthesized and tested conformationally restricted acetylcholine analogs. These compounds represent some of the possible conformations of acetylcholine. The most significant study in this regard is that of Armstrong and colleagues.[22] They synthesized and evaluated the muscarinic and nicotinic activity of the *cis*- and *trans*-isomers of a conformationally rigid model of acetylcholine, *cis*- and

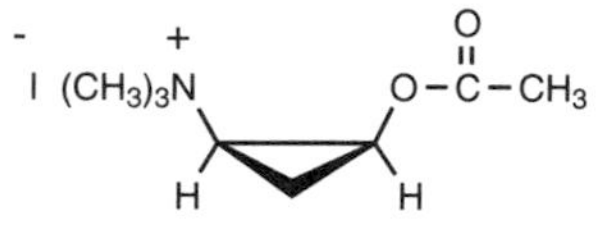

cis-ACTM

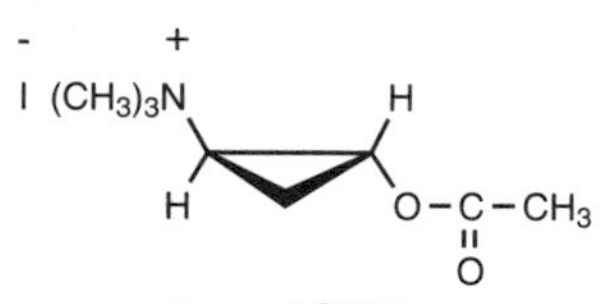

trans-ACTM

trans-ACTM. Because this model is based on the cyclopropane ring, the ester and quaternary ammonium functional groups cannot change their relative positions by bond rotation. The *cis*- and *trans*-isomers are rigidly constrained to the conformations shown. The *cis*-isomer is similar to the synplanar conformation of acetylcholine, and the *trans*-isomer approximates the synclinal conformation. The (+)-*trans*-isomer was observed to be equally as potent or more potent depending on the pharmacologic test used than acetylcholine at muscarinic receptors; it was much more potent than the (−)-*trans*-enantiomer. The racemic *cis*-compound had almost no activity in the same muscarinic receptor test system, and all compounds were weak nicotinic agonists.

The important conclusion drawn from this study was that acetylcholine would most probably interact with muscarinic receptors in its synclinal conformation. The most active isomer, the (+)-*trans* enantiomer, of these cyclopropane analogs was found to have a torsion angle of 137° between the ester oxygen and the quaternary nitrogen. This is significantly different from the 60° torsion angle in the anticlinal conformation found by NMR and x-ray determinations to be the preferred conformation.

The stereochemistry of cholinergic ligands and stereoselectivity of the receptors has played an important role in the design of cholinergic ligands as therapeutic agents. This role becomes apparent in subsequent sections.

ACETYLCHOLINE MIMETICS

Muscarinic Agonists

The interaction of cholinergic agonists with muscarinic receptors leads to well-defined pharmacologic responses depending on the tissue or organ in which the receptor is located. These responses include contractions of smooth muscle, vasodilation of the vascular system, increased secretion from exocrine glands, miosis, and a decrease in both the cardiac rate and the force of contractions of the heart.

Acetylcholine is the prototypical muscarinic (and nicotinic) agonist because it is the physiologic chemical neurotransmitter. It is a poor therapeutic agent, however, because of the chemical and physicochemical properties associated with its ester and quaternary ammonium salt functional groups. Acetylcholine is quite stable in the solid crystalline form, but it undergoes rapid hydrolysis in aqueous solution. This hydrolysis is accelerated in the presence of catalytic amounts of either acid or base. For this reason, acetylcholine cannot be administered orally owing to rapid hydrolysis in the gastrointestinal tract. Even when administered by parenteral routes, its pharmacologic action is fleeting as a result of enzyme-catalyzed hydrolysis by serum and tissue esterases. The quaternary ammonium functional group imparts excellent water solubility to acetylcholine, but quaternary ammonium salts are poorly absorbed across lipid membranes because of the highly hydrophilic character imparted by the ionic quaternary ammonium functional group. Therefore, even if acetylcholine were chemically stable enough to be administered orally, it would be poorly absorbed. Finally, the pharmacologic effects of acetylcholine are not selective, inasmuch as it is the neurotransmitter at both muscarinic and nicotinic receptors.

The necessity to design compounds that would serve as alternatives to acetylcholine as therapeutic agents and as probes to study the role of acetylcholine in neurotransmission led to an exhaustive study of the structural features that are required for the action of acetylcholine. The SAR that developed from these studies have provided the basis for the design of all the muscarinic agonists currently used as therapeutic agents.

To review these SAR, it is logical to divide the structure of acetylcholine into the three components shown below to examine the effects of chemical modification of each group.

Modification of the Quaternary Ammonium Group

Analogs in which the nitrogen atom was replaced by arsenic, sulfur, or selenium were among the first synthetic compounds studied as substitutes for acetylcholine.[23] Although they had some of the activity of acetylcholine, these compounds were less active and not used clinically.

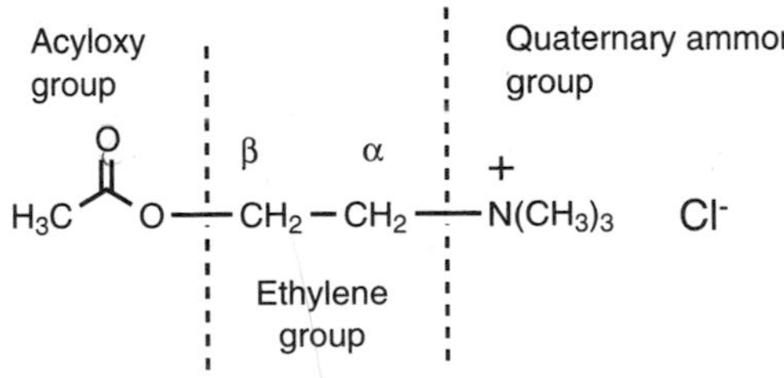

Acyloxy group | Ethylene group | Quaternary ammonium group

It was concluded that only compounds possessing a positive charge on the atom in the position of the nitrogen had appreciable muscarinic activity.

Compounds in which all three methyl groups on the nitrogen were replaced by larger alkyl groups were inactive as agonists. When the methyl groups are replaced by three ethyl groups, the resulting compound is a cholinergic antagonist. Replacement of only one methyl group by an ethyl or propyl group affords a compound that is active but much less so than acetylcholine.[24,25] Furthermore, successive replacement of the methyl groups with hydrogen atoms to afford a tertiary, secondary, or primary amine leads to successively diminishing muscarinic activity.

Modification of the Ethylene Bridge

The synthesis of acetic acid esters of quaternary ammonium alcohols of greater length than choline led to a series of compounds with activity that was rapidly reduced as the chain length increased. This observation led Ing to postulate his rule of five. This rule suggests that there should be no more than five atoms between the nitrogen and the terminal hydrogen atom for maximal muscarinic potency.[12] Present concepts suggest that the muscarinic receptor cannot successfully accommodate molecules larger than acetylcholine and still produce its physiologic effect. Although larger molecules may bind to the receptor, they lack efficacy and demonstrate antagonist properties.

Replacement of the hydrogen atoms of the ethylene bridge by alkyl groups larger than methyl affords compounds that are much less active than acetylcholine. The introduction of a methyl group on the carbon beta to the quaternary nitrogen affords acetyl-*β*-methylcholine (methacholine), which has muscarinic potency almost equivalent to that of acetylcholine; it has selectivity for muscarinic receptors in that it possesses much greater muscarinic potency than nicotinic potency. Methacholine has been used therapeutically to treat some forms of glaucoma. A methyl group on the carbon alpha to the quaternary nitrogen affords acetyl-*α*-methylcholine, a compound having greater nicotinic than muscarinic potency; this compound is not used as a therapeutic agent.

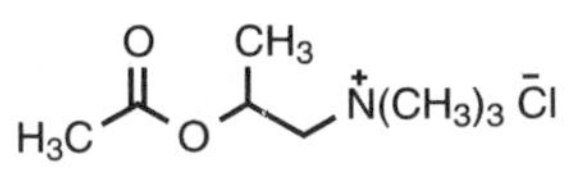

Acetyl-β-methylcholine chloride

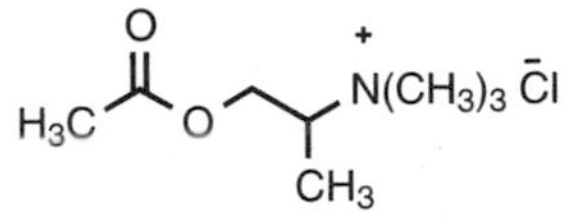

Acetyl-α-methylcholine chloride

The addition of methyl groups to either one or both of the ethylene carbons results in asymmetric molecules exhibiting optical isomerism. The muscarinic receptors and acetylcholinesterase (AChE) display stereoselectivity for the enantiomers of acetyl-*β*-methylcholine. The S(+) enantiomer is eqipotent with acetylcholine, and the R(−) enantiomer is about 20-fold less potent. Acetylcholinesterase hydrolyzes the S(+) isomer much slower (about half the rate) than acetylcholine. The R(−) isomer is not hydrolyzed by AChE but is a weak competitive inhibitor of the enzyme. This stability to AChE hydrolysis as well as the AChE inhibitory effect of the R(−) enantiomer may explain why methacholine (racemic acetyl-*β*-methylcholine) produces a longer duration of action than acetylcholine. The nicotinic receptor and AChE exhibit little stereoselectivity for the optical isomers of acetyl-*α*-methylcholine.

Modification of the Acyloxy Group

As would be predicted by Ing's rule of five,[12] when the acetyl group is replaced by higher homologs (i.e., the propionyl or butyryl groups), the resulting esters are less potent than acetylcholine. Choline esters of aromatic or higher molecular-weight acids possess cholinergic antagonist activity.

The fleeting pharmacologic action and chemical instability of acetylcholine are due to its rapid hydrolysis. A logical approach to the development of better muscarinic therapeutic agents was to replace the acetyloxy functional group with a functional group were resistant to hydrolysis. This observation led to the synthesis of the carbamic acid ester of choline (carbachol), which is a potent cholinergic agonist possessing both muscarinic and nicotinic activity. Esters derived from carbamic acid are referred to as carbamates and are more stable than carboxylate esters to hydrolysis because the carbonyl carbon is less electrophilic. Carbachol is less readily hydrolyzed in the gastrointestinal tract or by acetylcholinesterase than acetylcholine and can be administered orally. Owing to its erratic absorption and pronounced nicotinic effects, however, its use has been limited to the treatment of glaucoma.

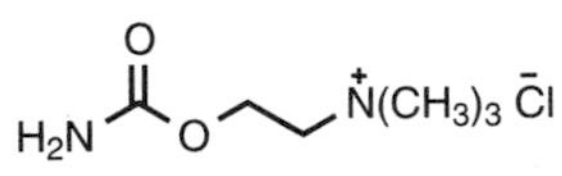

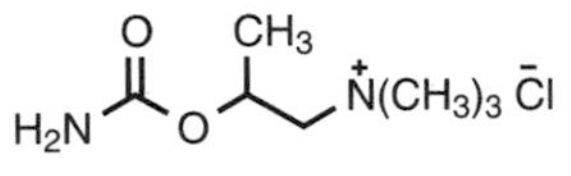

Carbachol

Bethanechol

This same chemical logic was extended to acetyl-*β*-methylcholine, which led to the synthesis of its carbamate ester, bethanechol. Bethanechol is an orally effective potent muscarinic agonist. Therapeutically, it possesses almost no nicotinic activity and is used to treat postsurgical urinary retention and abdominal distention. As would be expected, muscarinic receptors exhibit a stereoselectivity for the two optical isomers of bethanechol. The S(+) enantiomer exhibits much greater binding to muscarinic receptors than the R(−) enantiomer in isolated receptor preparations.

The profound muscarinic activity for the alkaloid muscarine provided substantial rationale for synthesizing ethers of choline. Muscarine, obtained from the red variety of mushroom, Amanita muscaria and other mushrooms, is one of the oldest known cholinergic agonists and is the compound for which muscarinic receptors were named. It was used in many pharmacologic experiments in the latter nineteenth century and early part of the twentieth century, and its use preceded the discovery and chemical characterization of acetylcholine.[2] The chemical structure of muscarine, however, was not completely characterized until 1957. Muscarine possesses three chiral centers (C2, C3, and C5), or eight optical isomers (four enantiomeric pairs). Of these, only the naturally occurring alkaloid, (2S, 3R, 5S)(+)-muscarine, is correctly referred to as muscarine. The C5 carbon of (+)-muscarine has the same absolute configuration as the analogous chiral beta carbon in S(+)-methacholine.

Other choline ethers as well as alkylaminoketones have been synthesized and evaluated for muscarinic activity.

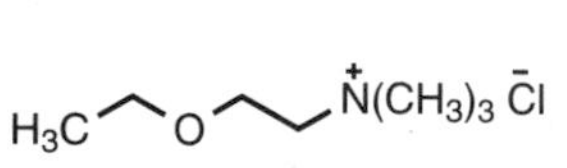

Choline ethyl ether 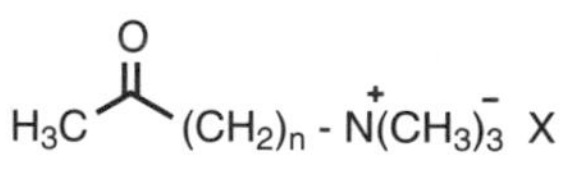   Alkylaminoketones

Choline ethyl ether exhibits significant muscarinic activity and is chemically quite stable, but it has not been used clinically. The most potent ketone derivatives possess the carbonyl on the carbon delta to the quaternary nitrogen; this is the same relative position as the carbonyl in acetylcholine. This suggests that these carbonyl groups bind by either a hydrogen bond or other dipole-dipole interaction with an appropriate group on the muscarinic receptor. Furthermore, the activity of these ethers and ketones demonstrates that neither the ester functional group nor a carbonyl is required for muscarinic agonist activity.

The classic SAR for muscarinic agonist activity can be summarized as follows:

1. The molecule must possess a nitrogen atom capable of bearing a positive charge, preferably a quaternary ammonium salt.
2. For maximum potency, the size of the alkyl groups substituted on the nitrogen should not exceed the size of a methyl group.
3. There should be an oxygen atom, preferably an ester-like oxygen, capable of participating in a hydrogen bond.
4. There should be a two-carbon unit between the oxygen atom and the nitrogen atom.

It is important to note that this SAR was based on pharmacologic evaluations using isolated tissues and sometimes whole animals during the first 60 years of research on the cholinergic nervous system. Scientists conducting this research did not have the luxury of modern, highly refined biologic testing systems (i.e., protein binding assays, cell membrane binding assays, and single-cell models) that are considered state-of-the-art today for pharmacologic evaluation of new medicinal agents. This is why some classic muscarinic agonists and many of the more modern agents do not adhere to this SAR. Indeed, SAR rules are not meant to be static; they are expected to change as new experimental data refine the structural and stereostructural requirements for muscarinic agonist activity.

Pilocarpine hydrochloride, an alkaloid obtained from Pilocarpus jaborandi, is an example of a muscarinic agonist that does not adhere to the traditional SAR. Its structure was described in 1901, although Langley reported in 1876 that extracts containing the alkaloid stimulated the end organs of parasympathetic neurons. Pilocarpine is subject to chemical degradation by hydrolysis of the lactone (a cyclic ester) to afford the pharmacologically inactive pilocarpic acid. In addition, pilocarpine undergoes base-catalyzed epimerization at C3 in the lactone to isopilocarpine, an inactive stereoisomer of pilocarpine. Pilocarpine is still used in ophthalmology for the treatment of glaucoma.

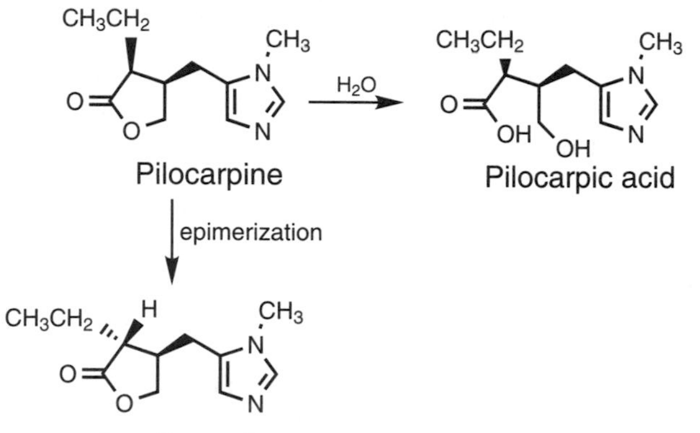

Pilocarpine → (H_2O) Pilocarpic acid

Pilocarpine → (epimerization) Isopilocarpine

Future Muscarinic Agonists

Current research interest in the design and synthesis of new muscarinic agonists is focused on discovering agents that might be effective in the treatment of Alzheimer's disease and other cognitive disorders.[28,29] In this regard, investigators are actively searching for muscarinic agonists that exhibit selectivity for muscarinic receptors in the brain. Among these compounds are analogs of arecoline, oxotremorine, and McN-A-343 as well as many other novel chemical structures possessing muscarinic agonist

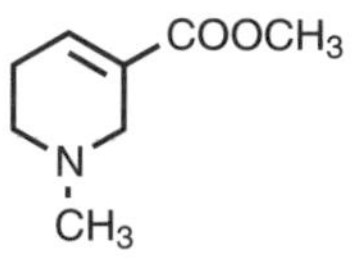

Arecoline 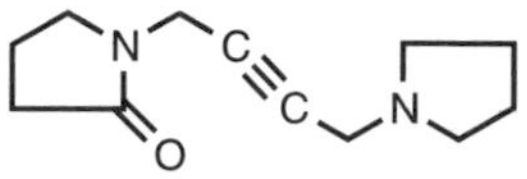   Oxotremorine

activity. Arecoline is of historical interest because its structure, similar to those of many other early medicinal agents, was determined and confirmed by a nineteenth-century German pharmacist, E. Jahns.[2] Some of these new analogs of older muscarinic agents exhibit a high degree of muscarinic receptor selectivity and demonstrate effectiveness in laboratory models of cognitive disorders.

ACETYLCHOLINESTERASE INHIBITORS

Another means of producing an autonomic response is to interfere with the mechanism by which the action of the neurotransmitter is terminated. In the parasympathetic nervous system, the action of acetylcholine is terminated by its rapid, AChE-catalyzed hydrolysis to acetic acid and choline.

$$H_3C{-}C(=O){-}O{-}CH_2CH_2{-}\overset{+}{N}(CH_3)_3 \xrightarrow{H_2O,\ AChE} H_3C{-}C(=O){-}OH + HO{-}CH_2CH_2{-}\overset{+}{N}(CH_3)_3$$

The inhibition of AChE increases the concentration of acetylcholine in the synapse and results in the production of both muscarinic and nicotinic responses.

Acetylcholinesterase inhibitors (AChEI), sometimes referred to as anticholinesterases, are classified as indirect cholinomimetics because their principle mechanism of action does not involve binding to cholinergic receptors. These agents are used therapeutically to treat myasthenia gravis and glaucoma. AChEIs are being investigated for the treatment of Alzheimer's disease and similar cognitive disorders.[28,30] Acetylcholinesterase inhibitors are used extensively as insecticides and have been used as chemical warfare agents.

Extensive studies of AChE have resulted in the purification and amino acid sequencing of the enzyme from several sources as well as the description of its quaternary structure from x-ray crystallographic and molecular modeling studies.[31] To understand the mode of action of AChEIs, it is necessary to examine the mechanism by which AChE catalyzes the hydrolysis of acetylcholine. This enzymatically controlled hydrolysis parallels the two chemical mechanisms for the hydrolysis of esters. The first mechanism is acid-catalyzed hydrolysis, in which the initial step involves protonation of the carbonyl oxygen.

$$R{-}C(=O){-}O{-}R' + H^+ \rightleftharpoons \left[R{-}\overset{+}{C}(OH){-}O{-}R'\right] \rightleftharpoons \left[R{-}C(OH)(OR')(\overset{+}{O}H_2)\right] \longrightarrow R{-}C(=O){-}O{-}H + H^+ + R'{-}OH$$

The transition state is formed by the attack of a molecule of water at the electrophilic carbonyl carbon atom. Collapse of the transition state affords the carboxylic acid and the alcohol.

$$R{-}C(=O){-}O{-}R' + OH^- \rightleftharpoons \left[R{-}C(O^-)(OR')(O{-}H)\right] \rightleftharpoons R{-}C(=O){-}O^- + R'OH$$

The second mechanism, base-catalyzed hydrolysis, involves nucleophilic attack of hydroxide anion on the electrophilic carbonyl carbon.

Both mechanisms for ester hydrolysis are proposed to be involved in the mechanism for AChE-catalyzed hydrolysis of acetylcholine. Figure 17–10 is a schematic illustration of the binding of acetylcholine to the catalytic (active) site of AChE, which consists of an esteratic binding site and a anionic binding site. This figure reflects ionic binding of the quaternary nitrogen of acetylcholine to the anionic site on the enzyme. The negative charge at the anionic site is proposed to be contributed by the free carboxylate group of a glutamate residue. In this illustration, there is a concerted protonation of the ester carbonyl oxygen by a imidazole proton from a histidine residue and a nucleophilic attack on the partially positive carbon of the carbonyl group by the hydroxyl group of a serine residue. The remainder of the hydrolysis mechanism is described in the following reaction. Transition

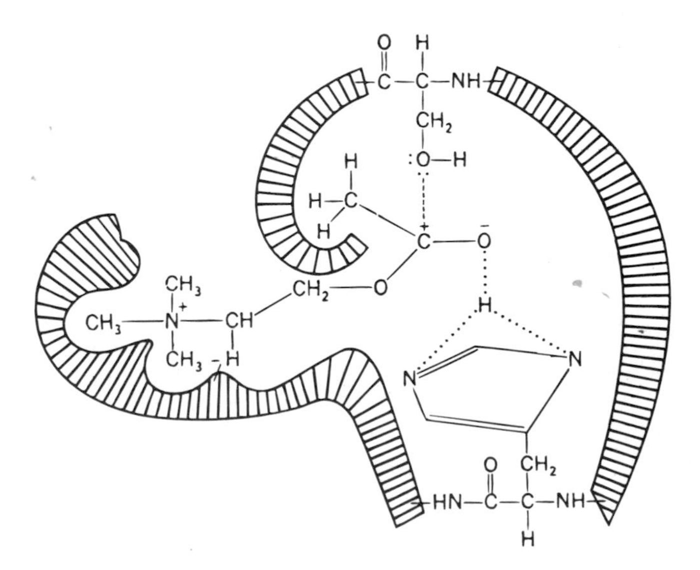

Fig. 17–10. Binding of acetylcholine to catalytic site of acetylcholinesterase; role of serine and histidine residues is illustrated.

AChE-Ser-OH + H_3C–C(=O)–O–CH_2CH_2–$\overset{+}{N}(CH_3)_3$ ⇌ [H_3C–C(OH)(O-Ser-AChE)–O–CH_2CH_2–$\overset{+}{N}(CH_3)_3$]

A ⇌ B

B → H_3C–C(=O)–O-Ser-AChE (C) + HO–CH_2CH_2–$\overset{+}{N}(CH_3)_3$

H_3C–C(=O)–O-Ser-AChE (C) + H_2O → AChE-Ser-OH + CH_3–C(=O)–OH (D)

state B is unstable and collapses to form choline and acetylated AChE (C); this form of the enzyme is referred to as the acetylated form. As long as the enzyme is in the acetylated form, it cannot bind another molecule of acetylcholine; the enzyme is inactive. The acetylated enzyme undergoes rapid hydrolysis to regenerate the original, active form of AChE (D) and a molecule of acetic acid.

The latter step in the mechanism, the regeneration of the active enzyme, is important in the development of AChEIs. If the enzyme becomes acylated by a functional group (i.e., carbamyl or phosphate) that is more stable to hydrolysis than a carboxylate ester, the enzyme remains inactive for a longer period of time. This chemical principle regarding the rates of hydrolysis led to the discovery and design of two classes of AChEIs, the reversible inhibitors and the irreversible inhibitors.

Reversible Inhibitors of Acetylcholinesterase

Reversible AChEIs are those compounds that either (1) are substrates for and react with AChE to form an acylated enzyme more stable than the acetylated enzyme but still capable of undergoing hydrolytic regeneration or (2) bind to AChE with greater affinity than acetylcholine but do not react with the enzyme as substrates. The clinically useful inhibitors are those of the first type and include the aryl carbamates, e.g., esters of carbamic acid and phenols, such as physostigmine, the classic AChEI. Alkyl carbamates (esters of carbamic acid and alcohols), such as carbachol and bethanechol, both structurally related to acetylcholine, are also substrates for and competitively inhibit AChE but are hydrolyzed very slowly by AChE. For reasons previously discussed, carbachol and bethanechol are more resistant to AChE-catalyzed hydrolysis than acetylcholine.

When aryl carbamate AChEIs, such as physostigmine and its analogs, bind to the catalytic site of AChE, hydrolysis of the carbamate occurs, which transesterifies the serine residue with carbamic acid; this is referred to as the carbamylated enzyme.

R_2N–C(=O)–O-Ser-AChE

Carbamylated AChE

Regeneration of active AChE by hydrolysis of the carbamylated enzyme is much slower than hydrolysis of the acetylated enzyme. The rate for regeneration of the carbamylated AChE is measured in minutes (e.g., half-life for methyl carbamates is about 15 minutes); the rate of regeneration of acetylated AChE is measured in milliseconds (half-life for acetylcholine is about 0.2 msec). Despite the longer time required to regenerate the carbamylated enzyme, the fact remains that the active form of AChE is eventually regenerated. Therefore, these inhibitors are considered to be reversible.

R_2N–C(=O)–O-Ser-AChE + H_2O → R_2N–C(=O)–OH + HO-Ser-AChE

Aryl carbamates are superior to alkyl carbamates as AChEIs because they have better affinity for, and therefore carbamylate, AChE more efficiently. Physostigmine and other aryl carbamates exhibit inhibition constants (K_i) on the order of 10^{-9} to 10^{-8} M and are three to four orders of magnitude more effective than alkyl carbamates such as carbachol (approximate K_i of 10^{-5} M). This is to be expected owing to the fact that phenoxide anions are more stable than alkoxide anions, inasmuch as phenoxide anions are stabilized through resonance with the aromatic ring. Thus, the therapeutically effective carbamate inhibitors of AChE are derived from phenols.

The classic AChEI, physostigmine, is an alkaloid obtained from seeds of Physostigma venenosum.[30] Its parasympathomimetic effects were recognized long before its structure was elucidated in 1923. In 1929, Stedman found that the mechanism of the parasympathomimetic effects of physostigmine was due to its inhibition of AChE. It inhibits AChE by acting as its substrate and carbamylating the enzyme. AChE is carbamylated at a slow rate but physostigmine has exceptionally high affinity (approximate K_i of 10^{-9} M) for the catalytic site of the enzyme; for comparison, the K_s for acetylcholine is on the order of 10^{-4} M. Thus, physostigmine is classified as a reversible AChEI that carbamylates the enzyme, but at a slow rate; the carbamylated AChE is also regenerated quite slowly.

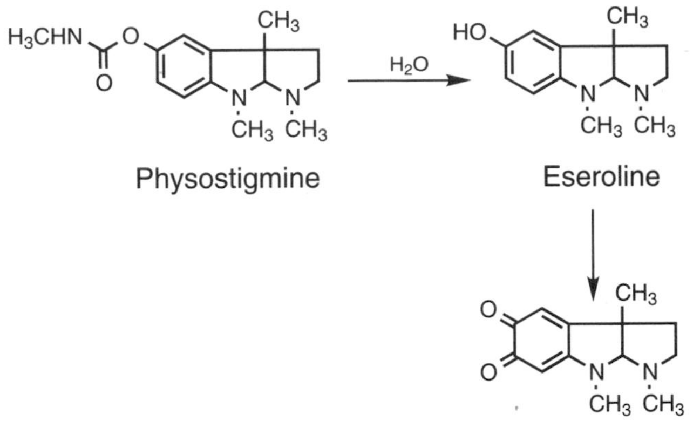

Physostigmine → (H_2O) → Eseroline → Rubreserine

Physostigmine undergoes hydrolytic decomposition in aqueous solutions to form eseroline, which is subject to light-catalyzed oxidation to form rubreserine, a red colored compound. Both degradation products are inactive as AChEIs. Physostigmine has been used for many years in ophthalmology for the treatment of glaucoma. More recently, the salicylate salt has been used in hospital emergency rooms to treat overdoses of compounds possessing significant anticholinergic effects (depression) in the central nervous system, such as atropine and tricyclic antidepressants. Physostigmine, in contrast to many other AChEIs, is not a quaternary ammonium salt and is therefore more lipophilic and can readily diffuse across the blood-brain barrier into the central nervous system to inhibit AChE in the brain and reverse the effects of anticholinergic compounds. The ability of physostigmine to cross the blood–brain barrier has led to renewed interest in this molecule. It is one of a number of centrally active AChEIs being investigated as indirect cholinomimetics for use in the treatment of Alzheimer's disease and other cognitive disorders.[28]

The discovery that physostigmine and other aryl carbamates inhibit AChE reversibly led to research to find other AChEIs possessing this activity. Most of this research involved the incorporation of the required structural features of both physostigmine and acetylcholine into the new molecules. This research led to the synthesis of neostigmine, a compound resembling physostigmine but having a much simpler structure. Neostigmine retains the substituted carbamate group, the benzene ring and the nitrogen atom of the first heterocyclic ring of physostigmine. The distance between the ester and the quaternary ammonium group is approximately the same as that found in acetylcholine and physostigmine. Neostigmine is a reversible AChEI by the same mechanism as physostigmine but is chemically more stable with a longer duration of action. It is an effective agent following oral administration and is commonly used to treat myasthenia gravis because it is not centrally active. Pyridostigmine is another therapeutically useful reversible AChEI that is structurally related to physostigmine and neostigmine.[32]

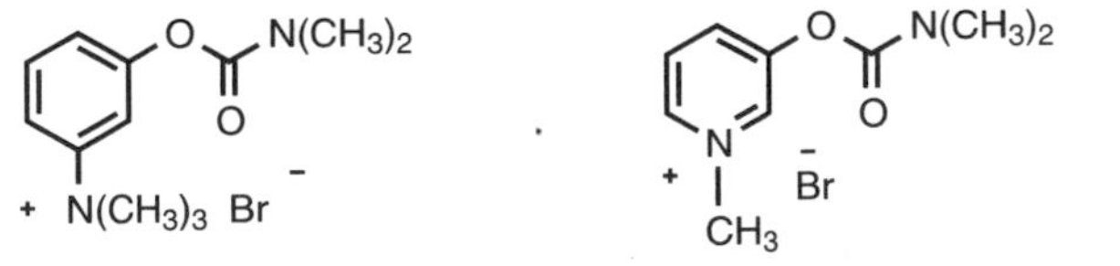

Neostigmine bromide   Pyridostigmine bromide

The aminoacridine tacrine, and a series of related aminoacridines including velacrine, represent a series of nonclassic AChEIs currently under investigation for the treatment of cognitive disorders.[28] Tacrine, synthesized in the 1930s, has shown some promise in clinical trials, and was approved for the treatment of Alzheimers disease as Cognex.

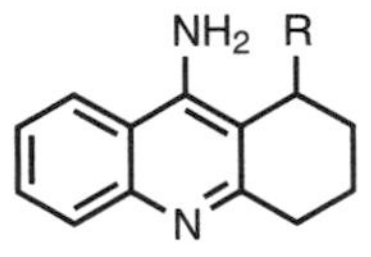

Tacrine, R = H
Velnacrine, R = OH

Carbaryl is a reversible carbamate-derived AChEI that

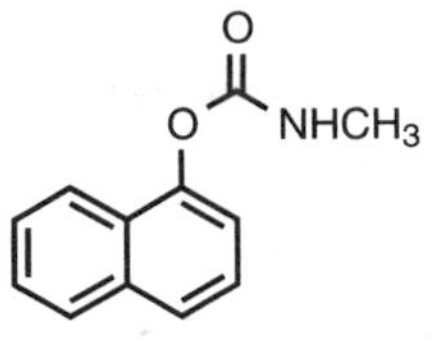

Carbaryl

has tremendous economic impact as an insecticide for use on house plants and vegetables as well as for control of fleas and ticks on pets. Its structural relationship to physostigmine and neostigmine is readily apparent. A number of other carbamate AChEIs are also commercially available for this use.

Irreversible Inhibitors of Acetylcholinesterase

The chemical logic involved in the development of effective AChEIs was to synthesize compounds that would be substrates for AChE and result in an acylated enzyme more stable to hydrolysis than a carboxylate ester. Phosphate esters are very stable to hydrolysis, being even more stable than many amides. Application of this chemical property to the design of AChEI compounds led to derivatives of phosphoric, pyrophosphoric, and phosphonic acids that are effective inhibitors of AChE. These act as inhibitors by the same mechanism as the carbamate inhibitors except that they leave the enzyme esterified as phosphate esters. The rate of hydrolyic regeneration of the phosphorylated enzyme is much slower than that of the carbamylated enzyme, and its rate is measured in hours (e.g., half-life for diethyl phosphates is about 8 hours). Because the duration of action of these compounds is much longer than that of the carbamate esters, they are referred to as irreversible inhibitors of AChE.

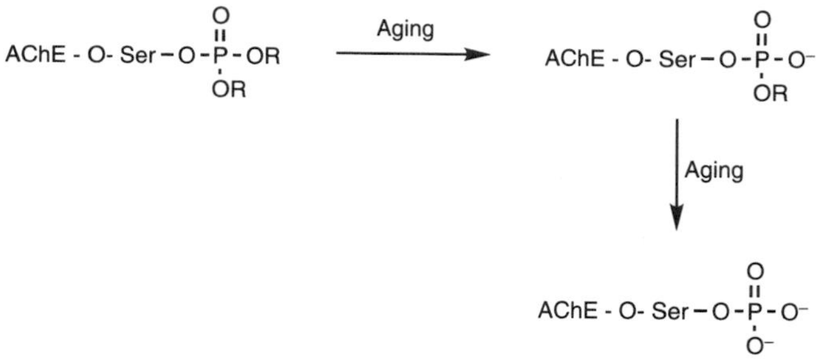

AChE - O- Ser – O – P(=O)(OR) – OR —Aging→ AChE - O- Ser – O – P(=O)(OR) – O⁻ —Aging→ AChE - O- Ser – O – P(=O)(O⁻) – O⁻

An important difference between the irreversible phosphoester-derived AChEIs and the reversible AChEIs is that the phosphorylated AChE can undergo a process known as aging. This aging process, illustrated above, plays an important role in the toxicity of these irreversible AChEIs. Aging is the hydrolytic cleavage of the phosphoester bond while the AChE is phosphorylated. This reaction affords an anionic phosphate that possesses a phosphorus atom, which is much less electrophilic and therefore much less likely to undergo further hydrolysis than the original phosphoester. Thus, the aged phosphorylated enzyme does not undergo nucleophilic attack and regeneration by antidotes (see next section) for phosphate ester AChEIs. This aging process occurs over a period of time, which depends on the rate of the hydrolysis reaction; during this time, the antidotes to phosphate ester poisoning may be effective.

Only those phosphorus-derived AChEIs that possess at least one phosphoester group undergo this aging process. Knowledge of the chemical mechanisms associated with irreversible inhibition of AChE and the aging process led to the development of deadly phosphorus-derived chemical warfare agents, one of which is sarin (GB). When this compound phosphorylates AChE, only one aging reaction takes place, and then the enzyme becomes completely refractory to regeneration by the currently available antidotal agents.

Two of these phosphate ester AChEIs have found therapeutic application: diisopropylfluorophosphate and echothiophate iodide. The selectivity of echothiophate for the AChE catalytic site was enhanced by incorporating in the molecule a quaternary ammonium salt functional group two carbons removed from the phosphoryl group. Because of their toxicity, these compounds are not used for their systemic action; they are applied topically to the eye to reduce intraocular pressure in the treatment of glaucoma. One novel phosphate ester AChEI, anatoxin-a(s), has been evaluated for its effects in animal models of cognitive disorders.

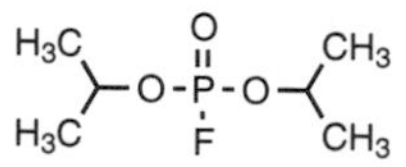

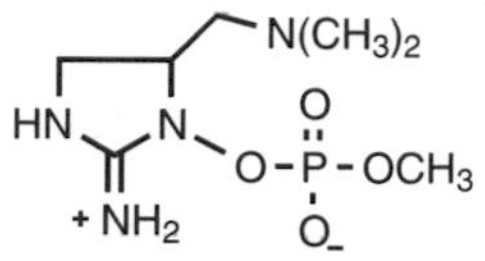

Diisopropylfluorophosphate (DFP)

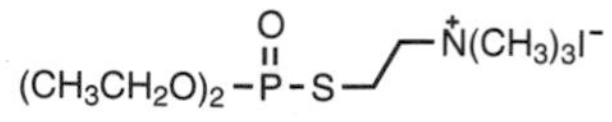

Echothiophate iodide

Anatoxin - a(s)

A number of lipophilic derivatives of phosphoester AChEIs have been designed as insecticides, the structures of some are shown in Figure 17–11. This group of irreversible AChEI insecticides is beneficial to agricultural production throughout the world. In addition to being extremely lipophilic, another physicochemical property common to these compounds is a high vapor pressure. This combination of physicochemical properties makes it imperative that these compounds be used with extreme caution to prevent inhalation of the vapors and absorption of the compounds through the skin. Both routes of exposure cause a number of poisoning accidents, some of which are fatal, every year.

Some of these irreversible AChEI insecticides possess a sulfur atom bonded to the phosphorus atom with a coordinate-covalent bond. These compounds exhibit little AChEI activity, but they are rapidly bioactivated (desulfuration) by microsomal oxidation to afford the corresponding oxo derivatives, which are quite potent. A good example of this bioactivation phenomenon is illustrated by parathion and paraoxon. Parathion is the commercially available insecticide and it is bioactivated to paraoxon.

The marked toxicity of the phosphate ester irreversible AChEIs, their widespread use as insecticides, and their proliferation as chemical warfare agents posed serious problems that stimulated research to develop antidotes for these agents. The solution of this problem required the rational use of reaction kinetics, organic reaction mechanisms, and synthetic organic chemistry. Water is a strong enough nucleophile to hydrolyze acetylated AChE rapidly and regenerate the active enzyme. Phosphorylated AChE (irreversibly inhibited), however, was known to involve a phosphate ester of serine. It was well established from reaction kinetic studies that the rate of hydrolysis was much slower for organic phosphate esters than for carboxylate esters and that a significantly stronger nucleophile than water was required for the efficient hydrolysis of phosphate esters. Therefore, the problem resolved itself to one of designing reagents capable of efficiently catalyzing phosphate ester hydrolysis and regenerating active AChE, while being safe

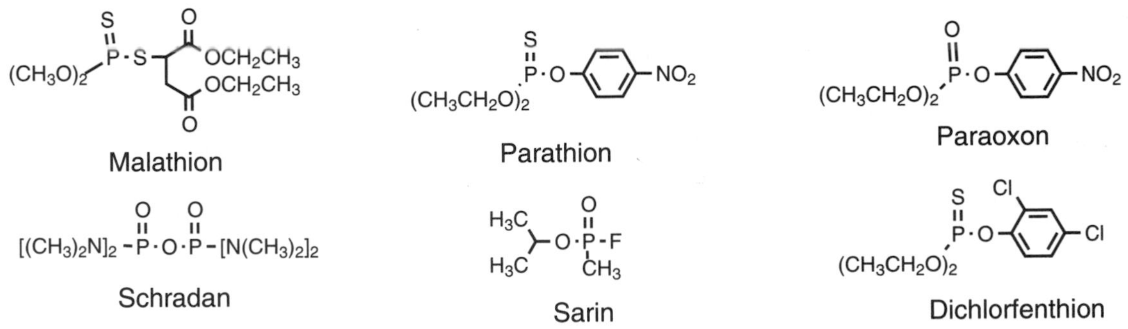

Malathion

Parathion

Paraoxon

Schradan

Sarin

Dichlorfenthion

Fig. 17–11. Irreversible acetylcholinesterase inhibitors.

enough for use as therapeutic agents. The resolution of this problem is an elegant example of the application of chemical principles to the solution of a therapeutic problem.[34–36]

Organic chemists had found that hydroxylamine (NH_2-OH) is a strong nucleophilic compound that efficiently cleaves phosphate esters. Hydroxylamine was demonstrated to increase significantly the rate of hydrolysis of phosphorylated AChE, but only at concentrations that were toxic.[33] This prompted the development of a number of structurally related compounds in the hope of eliminating toxicity. The toxicity inherent in hydroxylamine would most probably be present in any structurally related compound, although it might be minimized if the hydrolytic action could be achieved with smaller doses. It would be logical to design a compound that would (1) have a high degree of selectivity and strong binding affinity for AChE and (2) carry a hydroxylamine-like nucleophile into close proximity to the phosphorylated serine residence. This was achieved by the synthesis of hydroxylamine derivatives of organic compounds, which possessed a functional group bearing a positive charge.

The well-known reaction of hydroxylamine with aldehydes or ketones affords oximes, which possess a nucleophilic oxygen atom. A pyridine ring is an attractive carrier for the oxime function because it is present in a number of biochemicals (e.g., NAD, NADP), which would indicate a low order of toxicity. Furthermore, there are three readily available positional isomers of pyridine aldehyde that can be converted easily to oximes. Finally, the nitrogen atom of the pyridine ring can be converted to a quaternary ammonium salt by treatment with methyl iodide. This cationic charge would be expected to increase affinity of the compound for the anionic binding site of the phosphorylated AChE.

The three isomeric pyridine aldoxime methiodides were synthesized and biologically evaluated. Of these three, the most effective is the isomer derived from 2-pyridinyladehyde. This compound, known as 2-PAM (pralidoxime) for 2-pyridine aldoxime methyl chloride, is the only currently available agent proven to be clinically effective as an antidote for poisoning by phosphate ester AChEIs. The proposed mechanism for regeneration of AChE by 2-PAM is illustrated in the following reaction: The initial step involves binding of the quaternary ammo-

H_2N-OH

Hydroxylamine

Oxime

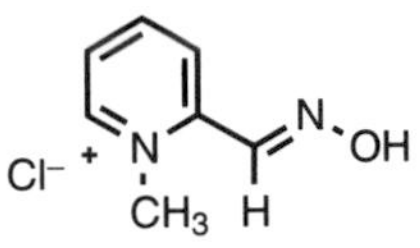

Pralidoxime chloride (2-PAM)

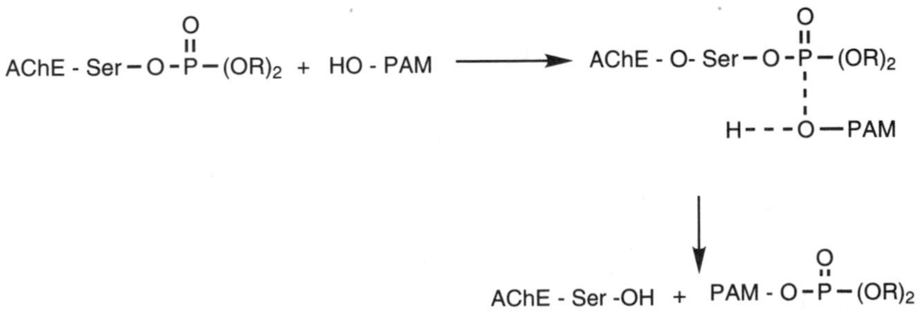

$$\text{AChE - Ser}-O-P(=O)-(OR)_2 + \text{HO - PAM} \longrightarrow \text{AChE - O- Ser}-O-P(=O)-(OR)_2 \cdots H\text{- - -}O-\text{PAM}$$

$$\downarrow$$

$$\text{AChE - Ser -OH} + \text{PAM - O}-P(=O)-(OR)_2$$

nium nitrogen of 2-PAM to the anionic binding site of phosphorylated AChE. This places the nucleophilic oxygen of 2-PAM in close proximity to the electrophilic phosphorus atom. Nucleophilic attack of the oxime oxygen results in breaking of the ester bond between the serine oxygen atom and the phosphorus atom. The final products of the reaction are the regenerated active form of AChE and phosphorylated 2-PAM.

Pralidoxime must be given within a short period of time, generally a few hours, for it to be effective because of the aging process of the phosphorylated enzyme. After the aging process has occurred, 2-PAM will not regenerate the enzyme. For this reason, as well as because new phosphate ester AChEIs capable of aging rapidly are being developed as insecticides and chemical warfare agents, there is a continuing research effort to discover new and better substitutes for 2-PAM. This research is focused on finding substitutes that are better nucleophiles than 2-PAM, and therefore more effective generators of active AChE, as well as compounds that cross the blood-brain barrier to regenerate phosphorylated AChE in the brain.

ACETYLCHOLINE ANTAGONISTS

Muscarinic Antagonists

Muscarinic antagonists are compounds that have high binding affinity for muscarinic receptors but have no intrinsic activity. When the antagonist binds to the receptor, it is proposed that the receptor protein undergoes a conformational perturbation that is different from that produced by an agonist. Therefore, antagonist binding to the receptor produces no response. Muscarinic antagonists are commonly referred to as anticholinergics, antimuscarinics, cholinergic blockers, antispasmodics, or parasympatholytics. The term anticholinergic refers in a pure sense to medicinal agents that are antagonists at both muscarinic and nicotinic receptors. Common usage of the term, however, has become synonymous with muscarinic antagonist, and it is used as such in this section.

Muscarinic antagonists are frequently employed as both prescription drugs and over-the-counter medications. Because they act as competitive (reversible) antagonists of acetylcholine, these compounds have pharmacologic effects that are the opposite of the muscarinic agonists. The responses of muscarinic antagonists include decreased contractions of the smooth muscle of the gastrointestinal and urinary tracts, dilation of the pupils, reduced gastric secretion, and decreased secretion of saliva. It follows that these compounds have therapeutic value in treating smooth muscle spasms, in ophthalmologic examinations, and in treatment of gastric ulcers. Compounds possessing muscarinic antagonist activity are common components of cold and flu remedies acting to reduce nasal and upper respiratory tract secretions.

The earliest known anticholinergic agents are alkaloids found in the family Solanaceae, a large family of plants that includes potatoes. Atropa belladonna (deadly nightshade), Hyoscymus niger (black henbane), and Datura stramonium (jimsonweed, thorn apple) are plants that have significant historical importance to our understanding of the parasympathetic nervous system. Pharmacologic effects of extracts from these plants have been recognized since the Middle Ages, although these effects were not associated with the autonomic nervous system until the latter part of the nineteenth century.

Atropine, an alkaloid isolated from A. belladonna and observed to block the effects of electrical stimulation and muscarine on the parasympathetic nervous system, became the first anticholinergic compound to be recognized as such. Scopolamine, another of the belladonna alkaloids, is chemically and pharmacologically similar to

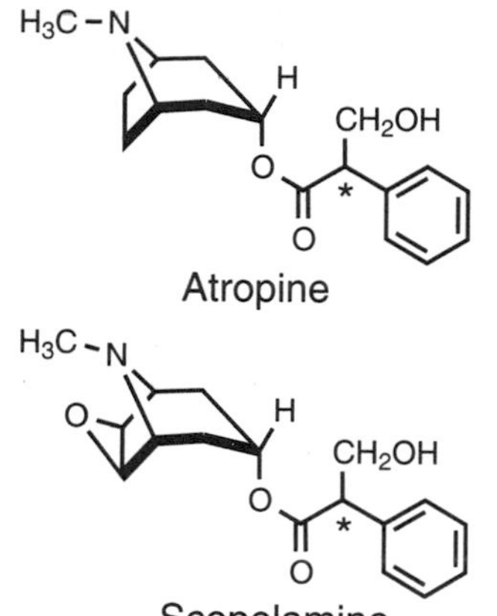

Atropine

Scopolamine

atropine. Atropine is (±)-hyoscyamine, the tropic acid ester of tropine and the naturally occurring alkaloid is (−)-hyoscyamine. Atropine results from the base-catalyzed racemization of the chiral carbon of tropic acid, which occurs during the isolation process. Scopolamine is the common name given to (−)-hyoscine, the naturally occurring alkaloid. The racemic compound isolated during extraction of the alkaloid from plants is atroscine.

The structural characterization of atropine and scopolamine and the discovery that these belladonna alkaloids are antagonists of acetylcholine were followed by the introduction of a great many novel synthetic anticholinergic drugs. This research was stimulated by the need for additional chemical tools to probe the phenomenon of neurochemical transmission as well as the desire to develop the therapeutic potential of these agents. Efforts to discover new muscarinic antagonists remain unabated today, but the thrust is to design antagonists that are specific for the different muscarinic receptor subtypes.

This effort has achieved some success. Muscarinic antagonists are the agents of choice to study muscarinic receptors because of their high degree of binding affinity. In this regard, antagonists selective for muscarinic receptor subtypes were the radioligands used to elucidate two pharmacologically distinct (as opposed to molecular biologically distinct) muscarinic receptor subtypes.[37]

Atropine, the prototype anticholinergic agent, provided the structural model that guided the design of synthetic muscarinic antagonists for almost 70 years. The circled portion of the atropine molecule shown below

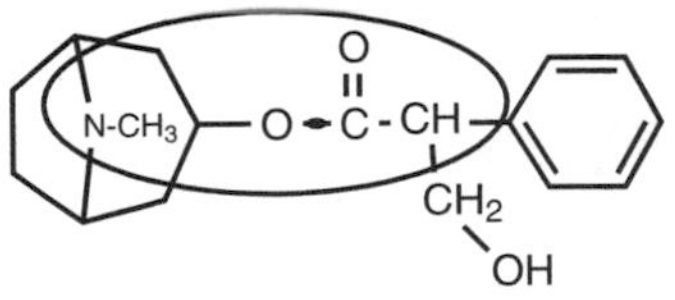

depicts the segment resembling acetylcholine. Although the amine functional group is separated from the ester oxygen by more than two carbons, the conformation assumed by the tropanol ring orients these two atoms such that the intervening distance is similar to that in acetylcholine. One of the important structural differences between atropine and acetylcholine, both esters of amino alcohols, is the size of the acyl portion of the molecules. Based on the assumption that size was a major factor in blocking action, many substituted acetic acid esters of amino alcohols were prepared and evaluated for biologic activity.

It became apparent that the most potent compounds were those that possessed two lipophilic ring substituents on the acyl carbon atom. This is the first of the classic SAR for muscarinic antagonist activity, and this SAR became more precisely defined as research on these antagonists continued. The SAR for muscarinic antagonists can be summarized as follows:

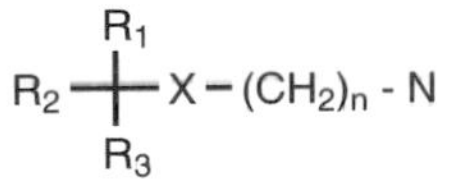

$$R_2-C(R_1)(R_3)-X-(CH_2)_n-N$$

1. Substituents R_1 and R_2 should be carbocyclic or heterocyclic rings for maximal antagonist potency. The rings may be identical, but the more potent compounds have different rings. Generally, one ring is aromatic and the other saturated or possessing only one olefinic bond. R_1 and R_2, however, may be combined into a fused aromatic tricyclic ring system such as that found in propantheline (Table 17–1). The size of these substituents is limited. For example, substitution of naphthalene rings for R_1 and R_2 affords compounds that are inactive apparently owing to steric hindrance of binding of these compounds to the muscarinic receptor.
2. The R_3 substituent may be a hydrogen atom, a hydroxyl group, a hydroxymethyl group, or a carboxamide, or it may be a component of one of the R_1 and R_2 ring systems. When this substituent is either a hydroxyl group or a hydroxymethyl group, the antagonist is usually more potent than the same compound without this group. The hydroxyl group presumably increases binding strength by participating in a hydrogen bond interaction at the receptor.
3. The X substituent in the most potent anticholinergic agents is an ester, but an ester functional group is not an absolute necessity for muscarinic antagonist activity. This substituent may be an ether oxygen, or it may be absent completely.
4. The N substituent is a quaternary ammonium salt in the most potent anticholinergic agents. This is not a requirement, however, because tertiary amines also possess antagonist activity presumably by binding to the receptor in the cationic (conjugate acid) form. The alkyl substituents are usually methyl, ethyl, propyl, or isopropyl.
5. The distance between the ring-substituted carbon and the amine nitrogen is apparently not critical, inasmuch as the length of the alkyl chain connecting these may be from two to four carbons. The most potent anticholinergic agents have two methylene units in this chain.

Muscarinic antagonists must compete with agonists for a common receptor. Their ability to do this effectively is because the large groups R_1 and R_2 enhance binding to the receptor. Because antagonists are larger than agonists, this suggests that groups R_1 and R_2 bind outside the binding site of acetylcholine. It has been suggested that the area surrounding the binding site of acetylcholine is hydrophobic in nature.[38] This accounts for the fact that in potent cholinergic antagonists, the groups R_1 and R_2 must be hydrophobic (usually phenyl, cyclohexyl, or cyclopentyl). This concept is also supported by the current models for muscarinic receptors.

Tables 17–1 and 17–2 include some of the anticholinergic agents that have found clinical application. These compounds reflect the SAR features that have been described. All of these compounds are effective when administered orally or by parenteral routes. Anticholinergic agents possessing a quaternary ammonium functional group are generally not well absorbed from the gastrointestinal tract because of their ionic character. These drugs are useful primarily in treatment of ulcers or other conditions in which a reduction in gastric secretions and reduced motility of the gastrointestinal tract are desired. Those antagonists having a tertiary nitrogen are much better absorbed and distributed following all routes of administration and are especially useful when systemic distribution is desired. The amino ether-derived and amino alcohol-derived anticholinergic agents readily cross the blood-brain barrier. These have proven particu-

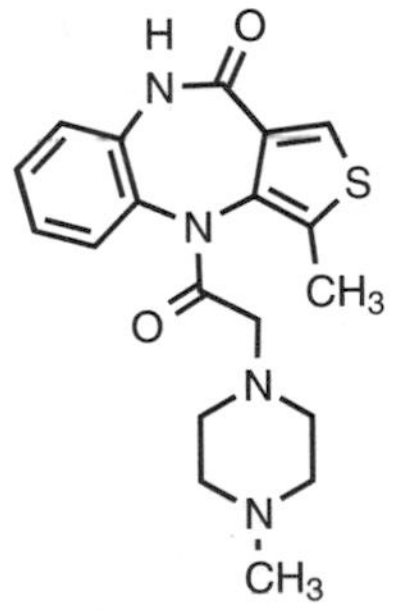

Telenzepine

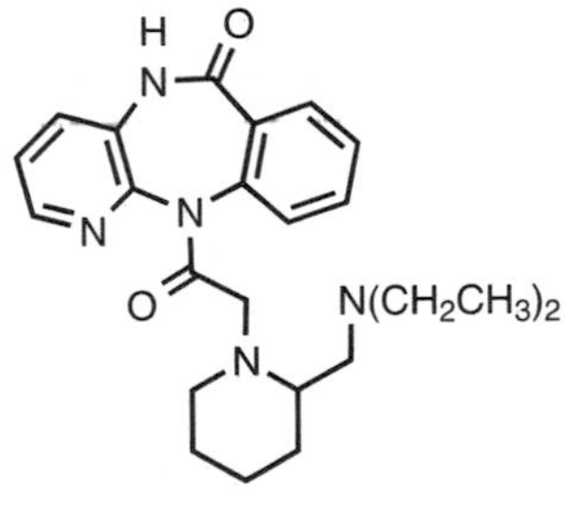

AFDX - 116

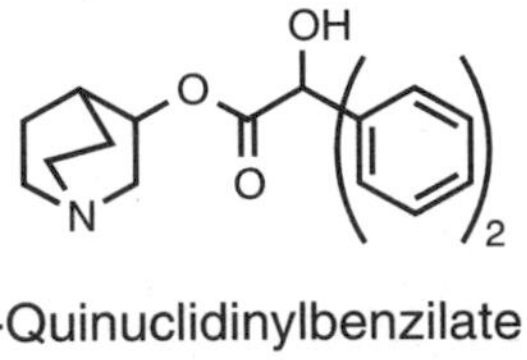

3-Quinuclidinylbenzilate (QNB)

Table 17–1. Amino Alcohol Esters of General Structure, RCOOR′

R	R′	Generic Name
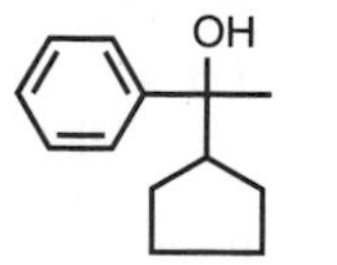	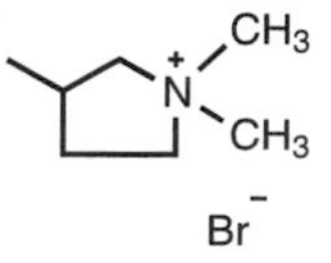	Glycopyrrolate
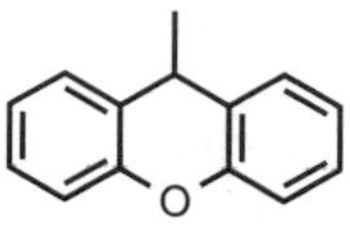	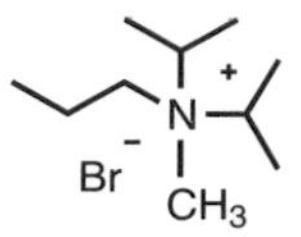	Propantheline
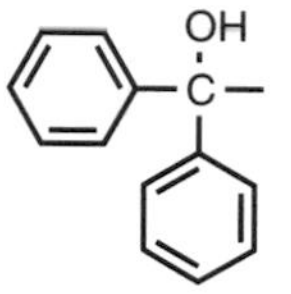	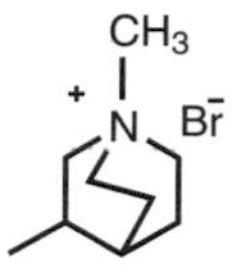	Clidinium
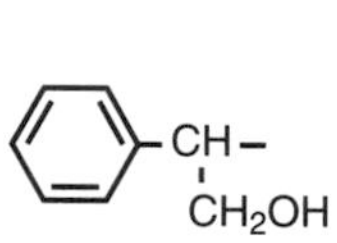	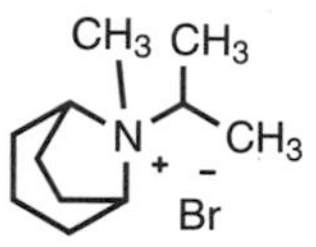	Ipratropium
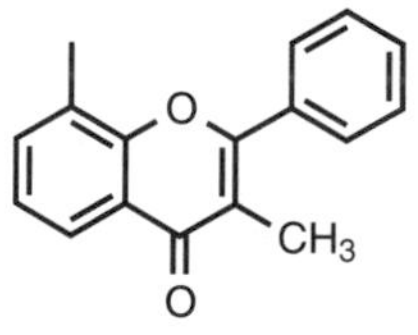	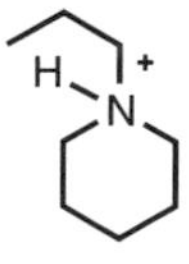	Flavoxate
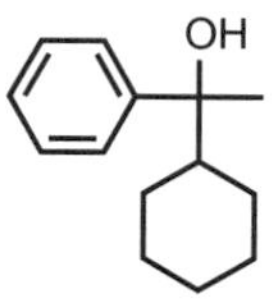	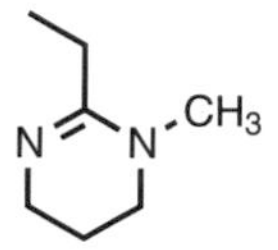	Oxyphencyclimine

larly beneficial in the treatment of Parkinson's disease and other diseases requiring a central anticholinergic effect.

All these drugs display pronounced selectivity for muscarinic receptors; however, some of those possessing the quaternary ammonium functional group exhibit nicotinic antagonist activity at high doses. For the most part, these display no marked selectivity for either the M1 or M2 subtypes of muscarinic receptors.

More recently discovered muscarinic antagonists display higher affinity for the receptors than the older agents as exemplified by quinuclidinylbenzilate (QNB), which possesses structural features common to the classic anticholinergic agents. Radiolabeled QNB was instrumental in the development of muscarinic receptor labeling techniques as well as the discovery of subtypes of muscarinic receptors. This latter research depended as well on the highly M1-selective antagonist pirenzepine, a compound having a novel structure for muscarinic antagonist activity. A number of compounds structurally related to pirenzepine have been demonstrated to be selective M1 muscarinic antagonists; among these is telenzepine.[39] Owing to their high degree of selectivity for muscarinic M1 receptors, pirenzepine and telenzepine have been evaluated in clinical trials for the treatment of duodenal ulcers. It is of interest to note that AFDX-116, structurally similar to pirenzepine, is a muscarinic antagonist exhibiting selectivity for cardiac M2 receptors.

Nicotinic Antagonists

Nicotinic antagonists are chemical compounds that bind to cholinergic nicotinic receptors but have no efficacy. All therapeutically useful nicotinic antagonists are competitive antagonists; i.e., the effects are reversible with acetylcholine. There are two subclasses of nicotinic

Table 17–2. Amino Alcohols and Amino Ethers

Amino Alcohols of the General Structure, $R\text{-}CH_2\text{-}CH_2\text{-}R'$

R	R′	Generic Name
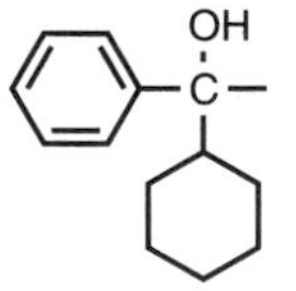 OH, C (phenyl, cyclohexyl)	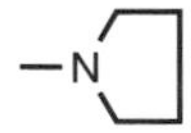 N (pyrrolidinyl)	Procyclidine
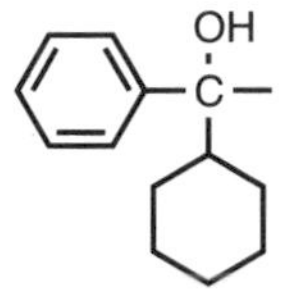 OH, C (phenyl, cyclohexyl)	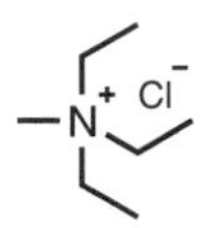 N^+ Cl^- (triethyl)	Tridihexethyl chloride

Amino Ethers of the General Structure, R-O-R′

R	R′	Generic Name
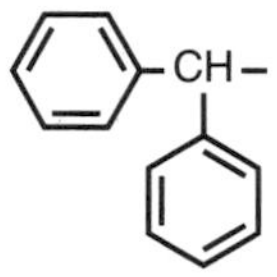 CH (diphenyl)	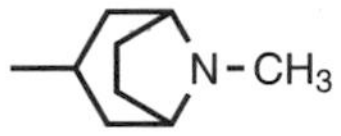 N-CH_3 (tropane)	Benztropine
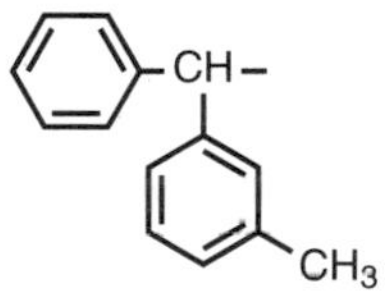 CH (phenyl, methylphenyl), CH_3	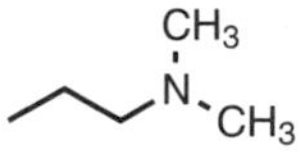 CH_3, N, CH_3	Orphenadrine

antagonists, skeletal neuromuscular blocking agents and ganglionic blocking agents, classified according to the two populations of nicotinic receptors. This section emphasizes nicotinic antagonists used clinically as neuromuscular blocking agents. These medicinal agents should not be confused with those skeletal muscle relaxant compounds that produce their effects through the central nervous system.

In terms of the historical perspective, tubocurarine, the first known neuromuscular blocking drug, was as important to the understanding of nicotinic antagonists as atropine was to muscarinic antagonists. The neuromuscular blocking effects of extracts of curare were first reported as early as 1510, when explorers of the Amazon River region of South America found natives using these plant extracts as arrow poisons. Early research with these crude plant extracts indicated that the active components caused muscle paralysis by effects on either the nerve or the muscle (the reader must remember that the concept of neurochemical transmission was not introduced until the late nineteenth century). In 1856, however, Bernard described the results of his experiments, which demonstrated unequivocally that curare extracts prevented skeletal muscle contractions by an effect at the neuromuscular junction and not on either the nerve innervating the muscle or the muscle itself.[40]

Much of the early literature concerning the effects of curare is confusing and difficult to interpret. This is not at all surprising when it is realized that this research was performed using crude extracts, many of which came from different plants. It was not until the late 1800s that scientists recognized curare extracts contained quaternary ammonium salts. This knowledge prompted the use of other quaternary ammonium compounds to explore the neuromuscular junction. In the meantime, curare extracts continued to be used to block the effects of nicotine and acetylcholine at skeletal neuromuscular junctions and explore the nicotinic receptors.

In 1935, King[41] isolated a pure alkaloid which he named *d*-tubocurarine from a tube curare of unknown botanical origin. The word tube refers to the container in which the South American natives transported their plant extract. It was almost 10 years later that the botanical source for *d*-tubocurarine was clearly identified as Chondodendron tomentosum. The structure that King assigned to tubocurarine possessed two nitrogen atoms, both of which were quaternary ammonium salts (a *bis*-quaternary ammonium compound). This structure was incorrect, but the correct structure was not reported until 1970 by Everett and colleagues.[42] The correct structure shown here has only one quaternary ammonium nitrogen; the other nitrogen is a tertiary amine salt. Nevertheless, the incorrect structure of tubocurarine served as the model for the synthesis of all the neuromuscular blocking agents in use today. These compounds have been of immense therapeutic value for surgical and orthopedic procedures and have been essential to the research that led to the isolation and purification of nicotinic receptors. It is interesting to contemplate the consequences if no one had ever questioned the structure of tubocurarine as originally determined.

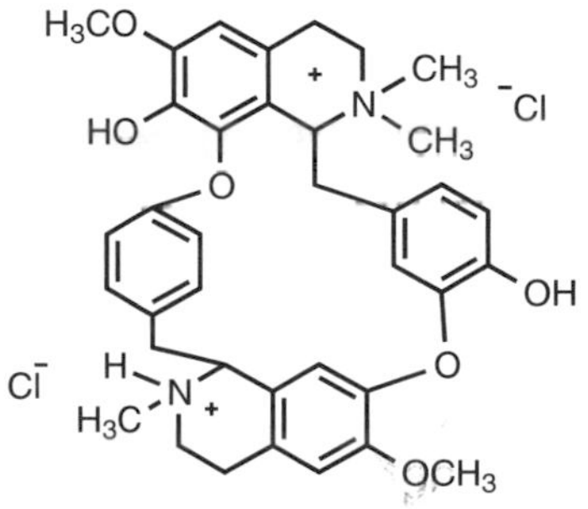

d-Tubocurarine

The potential therapeutic benefits of the neuromuscular blocking effects of tubocurarine as well as the difficulty in obtaining pure samples of the alkaloid encouraged medicinal chemists to design structurally related compounds possessing nicotinic antagonist activity. Using the *bis*-quaternary ammonium structure of tubocurarine (as reported by King) as a guide, a large number of compounds were synthesized and evaluated. It became apparent that a *bis*-quaternary ammonium compound, having two quaternary ammonium salts separated by 10 to 12

Table 17–3. Clinically Useful Neuromuscular Blocking Agents

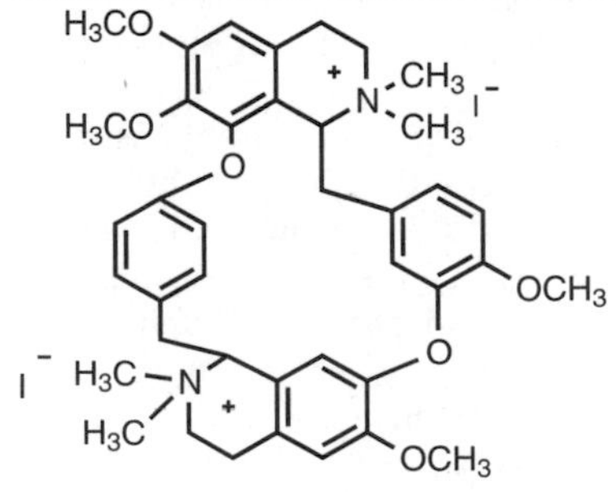

Metocurine iodide

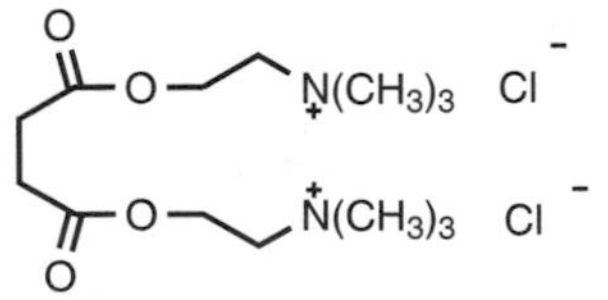

Succinylcholine chloride

$(CH_3)_3\text{-}\overset{+}{N}\text{-}(CH_2)_{10}\text{-}\overset{+}{N}\text{-}(CH_3)_3$ Cl^- Cl^-

Decamethonium chloride

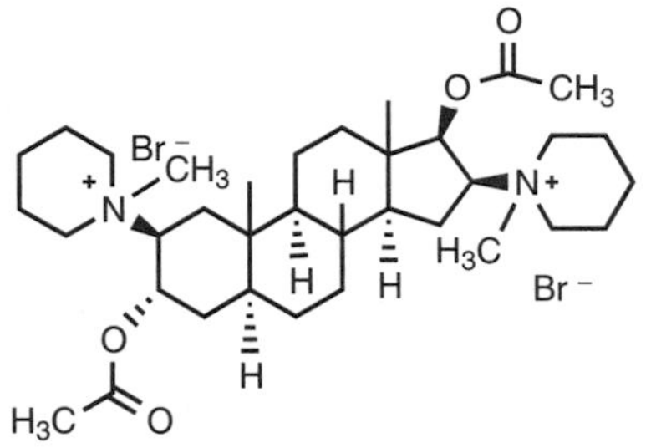

Pancuronium bromide

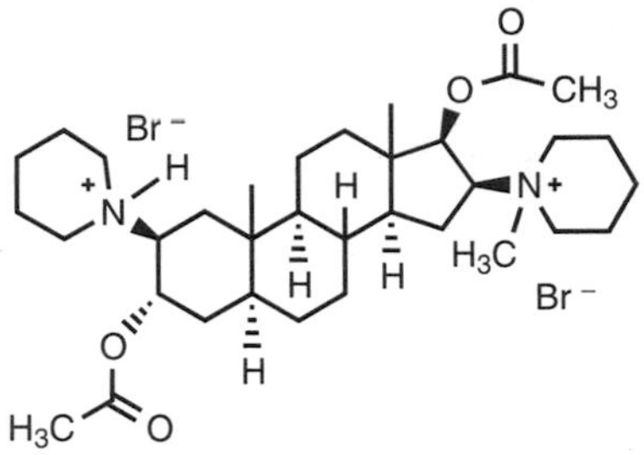

Vecuronium bromide

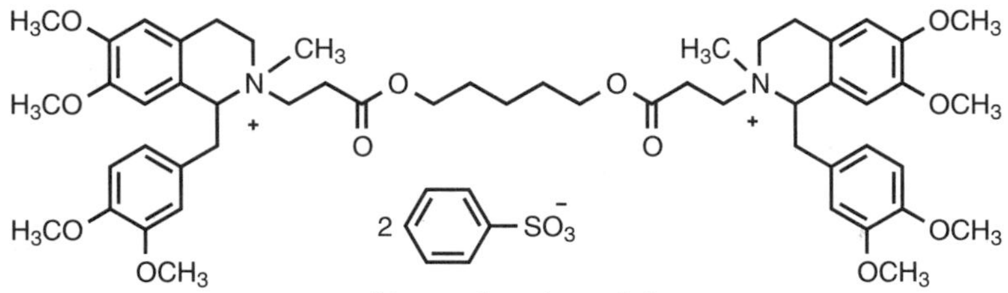

Atracurium besylate

carbon atoms (similar to the distance between the nitrogen atoms in tubocurarine), was a requirement for neuromuscular blocking activity. The rationale for this structural requirement was that, in contrast to muscarinic receptors, nicotinic receptors possessed two anionic binding sites, both of which had to be occupied for a neuromuscular blocking effect. It is important to observe that the current transmembrane model for the nicotinic receptor protein has two anionic sites in the extracellular domain.

Some of the new *bis*-quaternary ammonium agents produced a depolarization of the postjunctional membrane at the neuromuscular junction before causing a blockade; other compounds, such as tubocurarine, did not produce this depolarization. Thus, the structural features of the remainder of the molecule determined whether the nicotinic antagonist was a depolarizing or a nondepolarizing neuromuscular blocker. Table 17–3 shows the structures and generic names of some of the clinically useful neuromuscular blocking agents.

Depolarizing Neuromuscular Blocking Agents

One of the first neuromuscular blocking agents to be synthesized was decamethonium. A structure-activity study on a series of *bis*-quaternary ammonium compounds with varying numbers of methylene groups separating the nitrogen atoms demonstrated that maximal neuromuscular blockade occurred with 10 to 12 unsubstituted methylene groups. Activity diminished as the number of carbons was either decreased or increased. The compound with six methylene groups, hexamethonium, is a nicotinic antagonist in autonomic ganglia (ganglionic blocking agent). All the compounds in this series that possessed neuromuscular blocking activity also caused depolarization of the postjunctional membrane.

Another depolarizing neuromuscular blocking agent is succinylcholine, which represents two molecules of acetylcholine connected at the carbons of the acyl group. The manner in which this compound is drawn in Table 17–3 gives the impression that the quaternary ammonium functional groups are not separated by the distance represented by decamethonium or tubocurarine. The molecule can exist, however, in an extended conformation as shown in the Newman projection. This might ac-

$CO_2CH_2CH_2\text{-}\overset{+}{N}(CH_3)_3\ Cl^-$

H H H H

$CO_2CH_2CH_2\text{-}\underset{+}{N}(CH_3)_3\ Cl^-$

count for the appropriate separation of the quaternary nitrogens. Succinylcholine is rapidly hydrolyzed and rendered inactive, both in aqueous solution and by plasma esterases. This chemical instability must be considered when preparing solutions for parenteral administration. This same chemical property, however, gives the compound a brief duration of action. As a result, succinylcholine is used frequently for the rapid induction of neuromuscular blockade and when blockade of short duration is desired. The depolarizing property is undesirable in neuromuscular blockers, so most research efforts have been directed toward the design of agents that are nondepolarizing.

Nondepolarizing Neuromuscular Blocking Agents

The reaction of *d*-tubocurarine with methyl iodide affords metocurine iodide, in which the two phenolic hydroxyl groups of tubocurarine are changed to the methyl ethers, and the tertiary amine becomes quaternary. This agent is about four times more potent than tubocurarine in neuromuscular blocking activity. Both agents have a long duration of action and are eliminated (predominantly unchanged) via the kidney.

An ideal neuromuscular blocking agent would be a nondepolarizing compound that is inactivated by metabolism and eliminated rapidly, although possessing a duration of action longer than succinylcholine. The efforts to design ideal neuromuscular blocking agents have resulted in the development of pancuronium, vecuronium, and atracurium. Pancuronium and vecuronium are derivatives of malouetine, an alkaloid purportedly used as an arrow poison by African natives. Pancuronium has a duration of action only slightly shorter than that of tubocurarine and is more active. Both pancuronium and vecuronium, a slightly less potent agent, depend on renal elimination for terminating their effects. Atracurium has a duration of action slightly longer than that of succinylcholine. It is metabolized, and inactivated, by hydrolysis of the ester functional groups that connect the two quaternary nitrogens. Thus, termination of the effects of atracurium are independent of renal elimination.

REFERENCES

1. B. G. Katzung, Introduction to autonomic pharmacology, in *Basic and Clinical Pharmacology,* B. G. Katzung, Ed., Norwalk, CT, Appleton & Lange, 1992.
2. G. Holmstedt, Pages from the history of research on cholinergic mechanisms, in *Cholinergic Mechanisms,* P. G. Waser, Ed., New York, Raven Press, 1975, pp. 1–21.
3. O. Loewi, *Ges. Physiol., 189,* 239(1921).
4. O. Loewi and E. Navratil, *Arch. Ges. Physiol., 214,* 689 (1926).
5. S. Tucek, Choline acetyltransferase and synthesis of acetylcholine, in *The Cholinergic Synapse, Handbook of Experimental Pharmacology,* V. P. Whittaker, Ed., Berlin, Springer-Verlag, 1988, pp. 125–166.
6. R. J. Lefkowitz, et al., Neurohumoral transmission: The autonomic and somatic motor nervous systems, in *Goodman and Gilman's The Pharmacological Basis of Therapeutics,* 8th Ed., A. G. Gilman, T. W. Rall, A. S. Nies, and P. Taylor, Eds., New York, Pergamon Press, 1990, pp. 84–120.
7. S. H. Snyder and C. B. Pert, *Science, 179,* 1011(1973).
8. J. P. Changeaux, et al., Acetylcholine receptor: An allosteric protein. *Science, 225,* 1335(1984).
9. M. Mishina, et al., *Nature, 307,* 604(1984).
10. S. R. Kerlavage, et al., *Trends Pharmacol. Sci., 87,* 426 (1987).
11. R. Baker and J. Saunders, *Ann. Rep. Med. Chem., 24,* 31(1989).
12. H. R. Ing, *Science, 109,* 264(1949).
13. R. Hammer, et al., *Nature, 283,* 90(1980).
14. L. J. Ignarro and P. J. Kadowitz, *Ann. Rev. Pharmacol. Toxicol., 25,* 171(1985).
15. J. Lameh, et al., *Pharm. Res., 7,* 1213(1990).
16. V. Drubbisch, et al., *Pharm. Res., 9,* 1644(1992).
17. R. A. F. Dixon, et al., *Ann. Rep. Med. Chem., 23,* 221 (1988).
18. C. Humblet T. Mirzadegan, *Ann. Rep. Med. Chem., 27,* 291(1992).
19. A. M. Spiegel, *Ann. Rep. Med. Chem., 23,* 235(1988).
20. J.-P. Changeux, *Trends Pharmacol Sci., 11,* 485(1990).
21. W. H. M. L. Luyten and S. G. Heinemann, *Ann. Rep. Med. Chem., 22,* 281(1987).
22. P. D. Armstrong, et al., *Nature, 220,* 65(1968).
23. A. Welch and M. H. Roepke, *J. Pharmacol. Exp. Ther., 55,* 118(1935).
24. R. L. Stehle, et al., *J. Pharmacol. Exp. Ther., 56,* 473(1936).
25. P. Hotlon and H. R. Ing, *Br. J. Pharmacol., 4,* 190(1949).
26. A. Simonart, *J. Pharmacol. Exp. Ther., 46,* 157(1932).
27. J. H. Welsh and R. Taub, *J. Pharmacol. Exp. Ther., 103,* 62(1951).
28. M. R. Pavia, et al., *Ann. Rep. Med. Chem., 25,* 21(1989).
29. R. Baker and J. Saunders, *Ann. Rep. Med. Chem., 24,* 31(1989).
30. A. G. Karczmar, Anticholinesterase agents, in *International Encyclopedia of Pharmacology and Therapeutics,* Section 13, Vol. 1, C. Raduoco-Thomas, Ed., New York, Pergamon Press, 1970.
31. J. L. Sussman, et al., *Science, 253,* 872(1991).
32. J. E. Tether, *JAMA, 160,* 156(1956).
33. S. Hestrin, *J. Biol. Chem., 180,* 879(1949).
34. I. B. Wilson and S. Ginsburg, *Biochim. Biophys. Acta., 18,* 168(1955).
35. I. B. Wilson, *Fed. Proc., 18,* 752(1959).
36. I. B. Wilson, *J. Biol. Chem., 190,* 111(1951).
37. N. M. Nathanson, *Ann. Rev. Neurosci., 10,* 195(1987).
38. E. J. Ariens, *Adv. Drug Res., 3,* 235(1966).
39. G. Mihm and B. Wetzel, *Ann. Rep. Med. Chem., 23,* 81(1988).
40. A. R. McIntyre, History of curare, in *Neuromuscular Blocking and Stimulating Agents,* Vol. 1, *International Encyclopedia of*

Pharmacology and Therapeutics, Sect. 14, J. Cheymol, Ed., Oxford, Pergamon Press, 1972, pp. 187–203.
41. H. King, *J. Chem. Soc.,* 1381(1935).
42. A. J. Everett, et al., *Chem. Commun.,* 1020(1970).

SUGGESTED READINGS

R. Baker and J. Saunders, *Ann. Rep. Med. Chem., 24,* 31 (1989).

B. Belleau, in *Advances in Drug Research,* Vol. 2, N. J. Harper and A. R. Simmons, Eds., New York, Academic Press, 1965, p. 143.

J. G. Cannon, Cholinergics, in *Burger's Medicinal Chemistry,* Part III, 4th Ed., M. E. Wolff, New York, John Wiley & Sons, 1981.

M. Willians and S. J. Enna, *Ann. Rep. Med. Chem., 21,* 211 (1986).

N. M. Nathanson, *Ann. Rev. Neurosci., 10,* 195(1987).

Chapter 18

ADRENERGIC DRUGS

Robert K. Griffith and Elizabeth A. Johnson

Adrenergic drugs are agents that exert their principal pharmacologic effects on that portion of the autonomic nervous system, the sympathetic, which employs norepinephrine as the chemical mediator between a neuron and its effector cell. The autonomic nervous system controls the unconscious or involuntary functions of the body and includes nerves and ganglia that provide innervation to the heart, blood vessels, glands, respiratory system, visceral organs, and smooth muscles. The efferent autonomic nervous system consists of two major divisions, the sympathetic and the parasympathetic, which have structural and physiologic differences. Each of the two divisions of the autonomic system makes use of multineuronal connections in which the neurons exit the spinal cord but, rather than making direct connections with their target organ, synapse first in a ganglion. Ganglia are small but complex structures that contain synapses between preganglionic and postganglionic neurons in both the sympathetic and the parasympathetic nervous system. Both systems employ acetylcholine as the neurotransmitter between the preganglionic and postganglionic neurons, but the sympathetic nervous system largely employs norepinephrine to transmit a signal from the postganglionic fiber to its effector cell. Closely related to norepinephrine and having similar pharmacologic effects is the hormone epinephrine, synthesized and released into circulation by the chromaffin cells of the adrenal medulla. Both epinephrine and norepinephrine are members of a chemical class known as catecholamines because they contain within their structures *ortho*-dihydroxybenzene, known by the common chemical name of catechol.

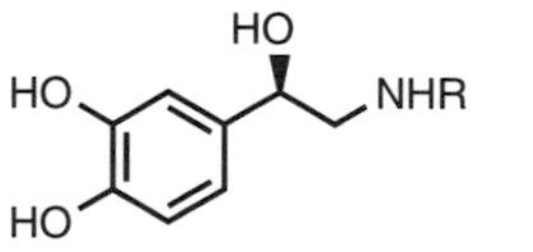

HO, HO, HO, NHR

R (–) Norepinephrine: R = H
R (–) Epinephrine: R = CH_3

The term sympathetic nervous system is an anatomic description for a set of nerve fibers that dates from the time of Galen.[1] The use of the term adrenergic nervous system stems from the discovery in the early part of the twentieth century that the administration of adrenaline, (epinephrine), produced in the adrenal medulla, caused effects similar to stimulation of the sympathetic nervous system.[2] For a number of years, adrenaline was thought to be the neurotransmitter in the adrenergic (sympathetic) nervous system, but it was also recognized that although they were similar, the effects of administered epinephrine were not quite identical to sympathetic stimulation. Finally in the 1940s, norepinephrine (noradrenaline) was identified as the true neurotransmitter at the terminus of the sympathetic nervous system.[3] Today the terms adrenergic nervous system and sympathetic nervous system are generally used interchangeably.

General stimulation of the sympathetic nervous system causes physiologic changes in keeping with a fight-or-flight response. These effects include increased rate and force of heart contractions, a rise in blood pressure, shift of blood flow to skeletal muscles, dilation of bronchioles and pupils, and an increase in blood glucose levels through gluconeogenesis and glycogenolysis. Thus, the large class of adrenergic drugs includes numbers of widely used agents, ranging from drugs for treating hypertension, cardiac arrhythmias, and asthma to nonprescription nasal decongestants acting through localized vasoconstriction. There are a variety of mechanisms of action by which these drugs exert their influence, and an understanding of these mechanisms requires knowledge of the details of norepinephrine biosynthesis, storage, release, and fate after release.

BIOSYNTHESIS, STORAGE, AND RELEASE OF NOREPINEPHRINE

Biosynthesis of norepinephrine takes place within adrenergic neurons near the terminus of the axon and the neuroeffector junction. The biosynthetic pathway (Fig. 18–1) begins with the amino acid L-tyrosine, which is actively transported into the neuron cell.[2] In the first step, the cytoplasmic enzyme tyrosine hydroxylase (tyrosine-3-monooxygenase) oxidizes the 3′ position of tyrosine to form the catechol-amino-acid L-dopa. This is the rate-limiting step in norepinephrine biosynthesis, and the activity of tyrosine hydroxylase is carefully controlled.[4] The enzyme is under feedback inhibition control by product catecholamines and is controlled through a complex pattern of phosphorylation/dephosphorylation, in which phosphorylation by protein kinases activates the enzyme, and dephosphorylation by phosphatases decreases activity.

In the second step, L-dopa is decarboxylated to dopamine by aromatic-L-amino acid decarboxylase, another cytoplasmic enzyme. Aromatic-L-amino acid decarboxylase was discovered in the 1930s and originally named

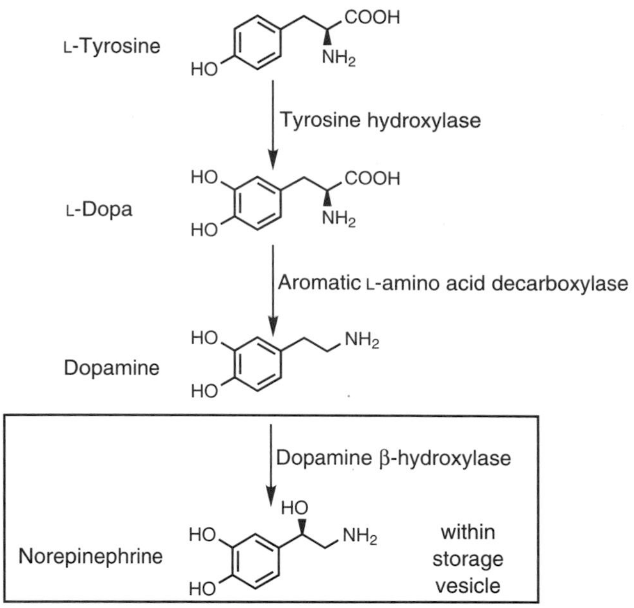

L-Tyrosine

Tyrosine hydroxylase

L-Dopa

Aromatic L-amino acid decarboxylase

Dopamine

Dopamine β-hydroxylase

Norepinephrine

within storage vesicle

Fig. 18–1. Biosynthesis of norepinephrine. Within the cytoplasm of an adrenergic neuron, the amino acid L-tyrosine is hydroxylated by the enzyme tyrosine hydroxylase to form L-dopa, which is, in turn, converted to dopamine by aromatic L-amino acid decarboxylase. Dopamine is taken up into a storage vesicle, where it is converted by the enzyme dopamine β-hydroxylase into the neurotransmitter norepinephrine.

dopa decarboxylase. Researchers subsequently discovered that dopa decarboxylase is not specific for dopa and decarboxylates other aromatic amino acids having the L(S) absolute configuration, such as 5-hydroxytryptophan, tryptophan, and tyrosine. Nevertheless the enzyme is still commonly referred to by the older name.

The dopamine formed in the cytoplasm by decarboxylation of L-dopa is then taken up by active transport into storage vesicles or granules located near the terminus of the adrenergic neuron. Within these vesicles, the enzyme dopamine β-hydroxylase stereospecifically introduces a hydroxyl group in the *R* absolute configuration on the carbon atom beta to the amino group to generate the neurotransmitter norepinephrine. Norepinephrine is stored in the vesicles in a 4:1 complex with adenosine triphosphate (ATP) in such quantities that each vesicle in a peripheral adrenergic neuron contains between 6,000 and 15,000 molecules of norepinephrine.[5] (The pathway for epinephrine biosynthesis in the adrenal medulla is the same with the additional step of conversion of norepinephrine to epinephrine by the enzyme phenylethanolamine-N-methyltransferase.) The norepinephrine remains in the vesicles until released into the synapse during signal transduction. When a wave of depolarization reaches the terminus of an adrenergic neuron, it triggers the transient opening of voltage-dependent calcium channels causing an influx of calcium ions. This influx of calcium ions triggers fusion of the storage vesicles with the neuronal cell membrane, spilling the norepinephrine and other contents of the vesicles into the synapse through exocytosis. The exact mechanism by which the calcium influx triggers fusion of the storage vesicle and cell wall is not clearly understood.[6] A summary view of the events involved in norepinephrine biosynthesis, release, and fate is given in Figure 18–2.

FATE OF NOREPINEPHRINE FOLLOWING RELEASE

Following release, norepinephrine diffuses through the intercellular space to bind reversibly to adrenergic receptors (alpha or beta), on the effector cell, triggering a biochemical cascade that results in a physiologic response by the effector cell. In addition to the receptors on effector cells, there are also adrenoreceptors that respond to norepinephrine (α_2-receptors) on the presynaptic neuron, which, when stimulated by norepinephrine, act to inhibit the release of additional neurotransmitter into the synapse. Classification of various types of adrenoceptors and the nature of their interaction with norepinephrine and other drugs are discussed more fully later. Once the neurotransmitter has been released and is stimulating various receptors, there must be mechanisms for removing the norepinephrine from the synapse and terminating the adrenergic impulse. By far the most important of these mechanisms for removing the norepinephrine is transmitter recycling through active transport uptake into the presynaptic neuron. This process, called uptake-1, is efficient, and in some tissues, up to 95% of released norepinephrine is likely removed from the synapse by this mechanism.[7] Part of the norepinephrine taken into the presynaptic neuron by uptake-1 is metabolized to 3,4-dihydroxyphenylglycolaldehyde (DOPGAL) by mitochondrial monoamine oxidase (MAO) (see Fig. 18–2), and part of it is sequestered in the storage vesicles to be used again as neurotransmitter. A less efficient uptake process, uptake-2, operates in a variety of other cell types but only in the presence of high concentrations of norepinephrine. That portion of released norepinephrine which escapes uptake-1 diffuses out of the synapse and is metabolized in extraneuronal sites by catechol-O-methyltransferase, COMT, which methylates the meta hydroxyl group. Norepinephrine is also metabolized to DOPGAL by MAO present at extraneuronal sites, principally the liver and blood platelets. Both DOPGAL and normetanephrine are subject to further metabolism, as outlined in Figure 18–3.[8]

CHARACTERIZATION OF ADRENERGIC RECEPTOR SUBTYPES

The discovery of subclasses of adrenergic receptors and the ability of relatively small molecule drugs to stimulate differentially or block these receptors represented a

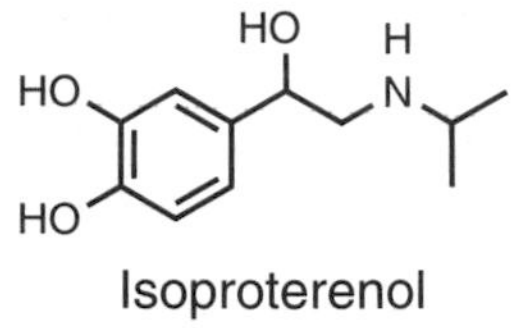

Isoproterenol

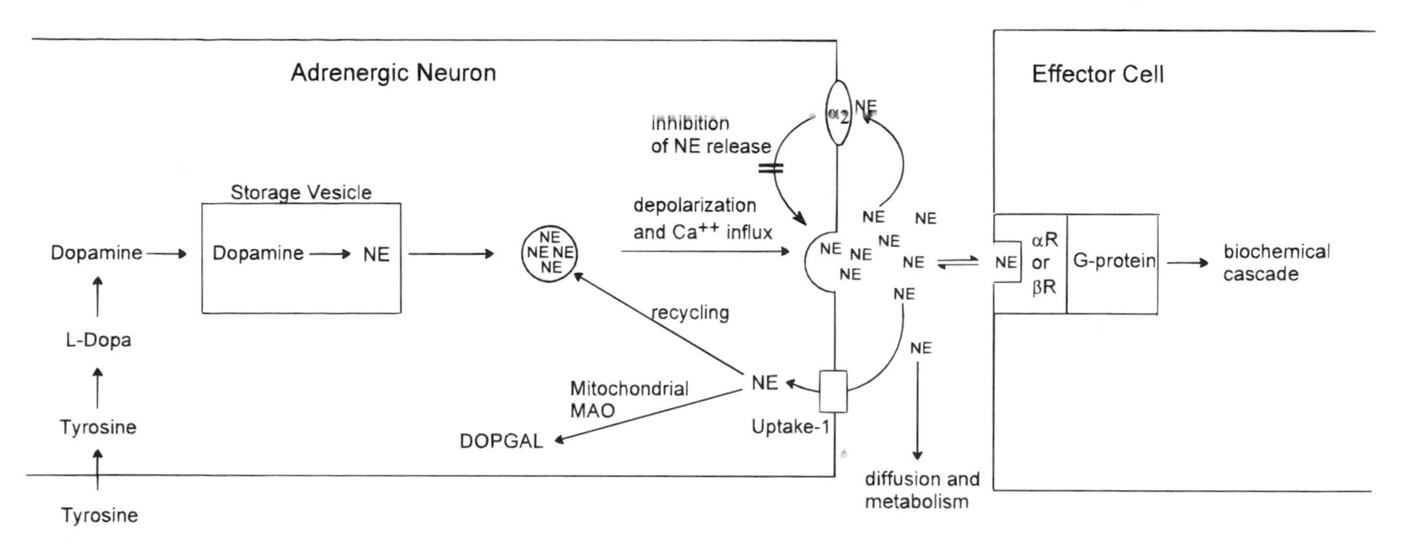

Fig. 18–2. Schematic of neurotransmission events in an adrenergic neuron and effector cell. Following biosynthesis and storage in a vesicle, norepinephrine release into the synapse is triggered by depolarization-induced calcium influx. The norepinephrine in the synapse interacts with postsynaptic G-protein–linked α or β receptors on the effector cell, triggering effector cell response, or with presynaptic α_2 receptors on the neuron, which inhibit release of more norepinephrine. Most of the synaptic neurotransmitter is taken back into the presynaptic neuron by uptake-1 active transport. Some of the norepinephrine is metabolized by monoamine oxidase and the remainder stored in a vesicle to be used again. That portion not captured by uptake-1 diffuses out of the synapse and is metabolized. NE, norepinephrine; αR, alpha adrenoceptor; βR, beta adrenoceptor; MAO, monoamine oxidase; DOPGAL, 3,4-dihydroxyphenylglycolaldehyde.

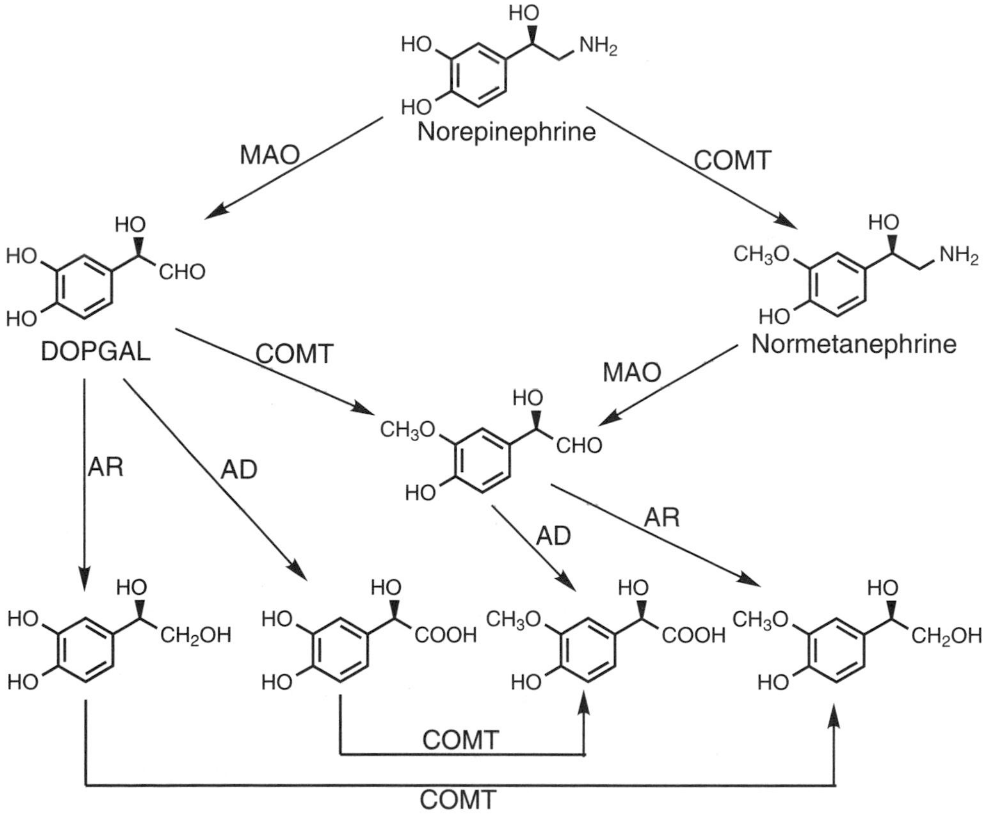

Norepinephrine
MAO
COMT
DOPGAL
Normetanephrine
COMT
MAO
AR
AD
AD
AR
COMT
COMT

Fig. 18–3. Metabolism of norepinephrine. Norepinephrine is initially metabolized to DOPGAL by monoamine oxidase to normetanephrine by catechol-O-methyltransferase. Each of the initial products may, in turn, be further metabolized, and these metabolites are subject to further metabolism as shown. All of the structures shown are subject to phase II conjugations. MAO, Monoamine oxidase; COMT, catechol-O-methyltransferase; AR, aldehyde reductase; AD, aldehyde dehydrogenase.

major advance in several areas of pharmacotherapeutics. Adrenergic receptors were subclassified by Ahlquist in 1948 into alpha and beta adrenoreceptor classes according to their responses to different adrenergic receptor agonists, principally norepinephrine, epinephrine, and isoproterenol.[9] these catecholamines are able to stimulate α-adrenoceptors in the descending order of sensitivity epinephrine > norepinephrine > isoproterenol. In contrast, β-adrenoceptors are stimulated most potently by isoproterenol > epinephrine > norepinephrine. In the years since Ahlquist's original classification, additional small molecule agonists and antagonists have been used to allow further subclassification of alpha and beta receptors into α_1 and α_2 subtypes of alpha receptors and β_1 and β_2 subtypes of beta adrenoceptors. Most recently, the powerful tools of molecular biology have been used to clone and identify an additional beta receptor, β_3, and even more subtypes of alpha receptors for a total of six, three subtypes of α_1 and three subtypes of α_2.[10] This provides a wealth of information on the structures and biochemical properties of both alpha and beta receptors. Intensive research continues in this area, and the coming years may provide evidence of additional subtypes of both alpha and beta receptors. At this time, however, only the α_1, α_2, β_1, and β_2 receptor subtypes are sufficiently well differentiated by their small molecule binding characteristics to be considered clinically significant in pharmacotherapeutics. Therefore, this chapter concentrates on these four receptor subtypes.

The adrenoceptors, both alpha and beta, are members of a receptor superfamily of membrane-spanning proteins, including muscarinic, serotonin, and dopamine receptors, which are coupled to intracellular GTP-binding proteins (G-proteins), which determine the cellular response to receptor activation.[11] All of these receptors exhibit a common motif of a single polypeptide chain that is looped back and forth through the cell membrane seven times with an extracellular N-terminus and intracellular C-terminus. One of the most thoroughly studied of these receptors is the human β_2-adrenoreceptor (Fig. 18–4).[12] The seven transmembrane segments, I–VII, are composed primarily of lipophilic amino acids arranged in α-helices connected by regions of hydrophilic amino acids. The hydrophilic regions form loops on the intracellular

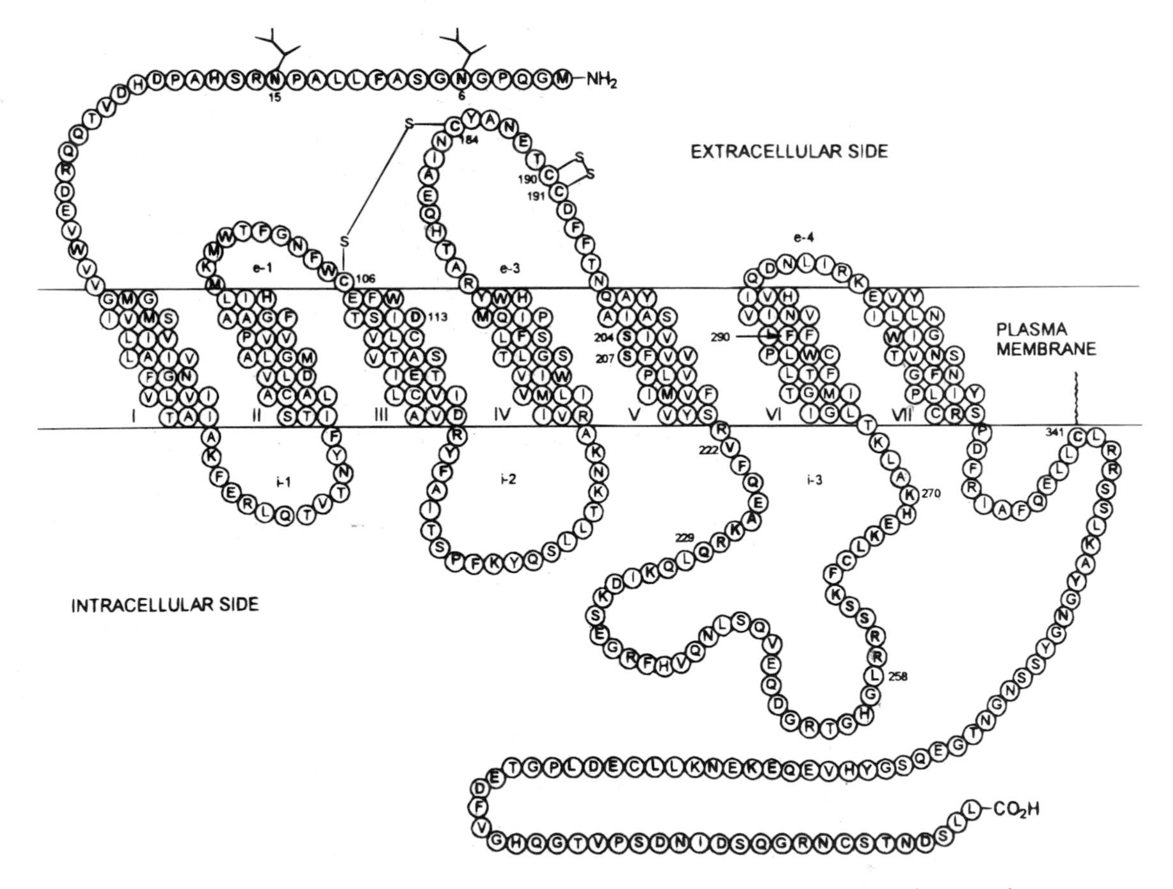

Fig. 18–4. Human β_2 adrenergic receptor. Amino acid sequence of the human β_2 receptor showing the seven transmembrane domains, I–VII, the connecting intracellular and extracellular loops, extracellular glycosylation sites at asparagines 6 and 15, and intrachain disulfide bonds between cysteines 106–184 and 190–191. Also indicated are the amino acids identified as participating in neurotransmitter binding—aspartate 113 in transmembrane domain III, which binds the positively charged amine of the neurotransmitter, and serines 204 and 207 of transmembrane domain V, which form H-bonds with the catechol hydroxyls. Phenylalanine 290 may also participate in agonist binding. Amino acids 222–229 and 258–270 of the third intracellular loop are critical for G-protein coupling, and palmitoylated cysteine 341 is critical for proper adenylyl cyclase activation. (Adapted from J. Ostrowski, et al., *Annu. Rev. Pharmacol. Toxicol., 32,* 167–183, [1992]).

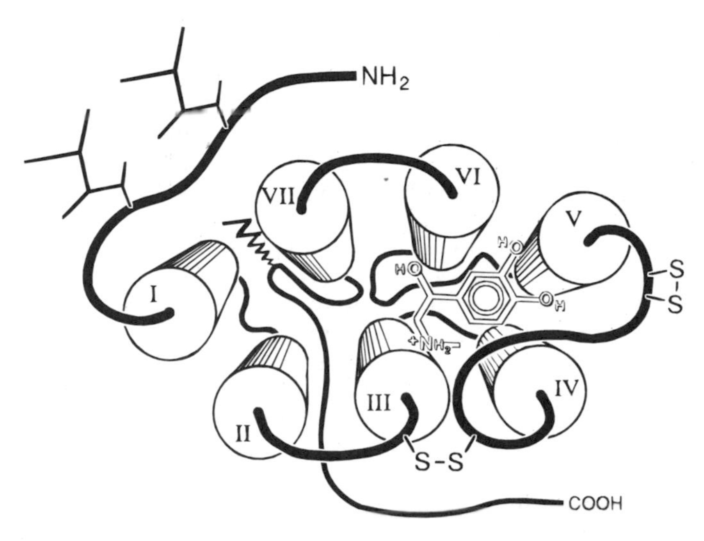

Fig. 18–5. Proposed arrangement for the transmembrane helices of the β_2-adrenergic receptor depicting the binding site for epinephrine as viewed from the extracellular side. (From J. Ostrowski, et al., *Annu. Rev. Pharmacol. Toxicol., 32,* 167–183 [1992]).

and extracellular faces of the membrane. In all of the adrenoceptors, the agonist/antagonist recognition site is located within the membrane-bound portion of the receptor. This binding site is within a pocket formed by the membrane spanning regions of the peptide, as illustrated in Figure 18–5 for epinephrine bound to the human β_2-receptor. All of the adrenoceptors are coupled to their effector systems through a G-protein, which is linked through reversible binding interactions with the third intracellular loop of the receptor protein.

Salient features of the extensively studied β_2-adrenoreceptor are indicated in Figure 18–4. Binding studies with selectively mutated β_2-receptors have provided strong evidence for binding interactions between agonist functional groups and specific residues in the transmembrane domains of adrenoceptors. Such studies indicate that Asp_{113} in transmembrane domain III of the β_2-receptor is the acidic residue that forms a bond, presumably ionic or a salt bridge, with the positively charged amino group of catecholamine agonists. An aspartic acid residue is also found in a comparable position in all of the other adrenoceptors as well as other known G-protein coupled receptors that bind substrates having positively charged nitrogens in their structures. Elegant studies with mutated receptors and analogs of isoproterenol demonstrated that Ser_{204} and Ser_{207} of transmembrane domain V are

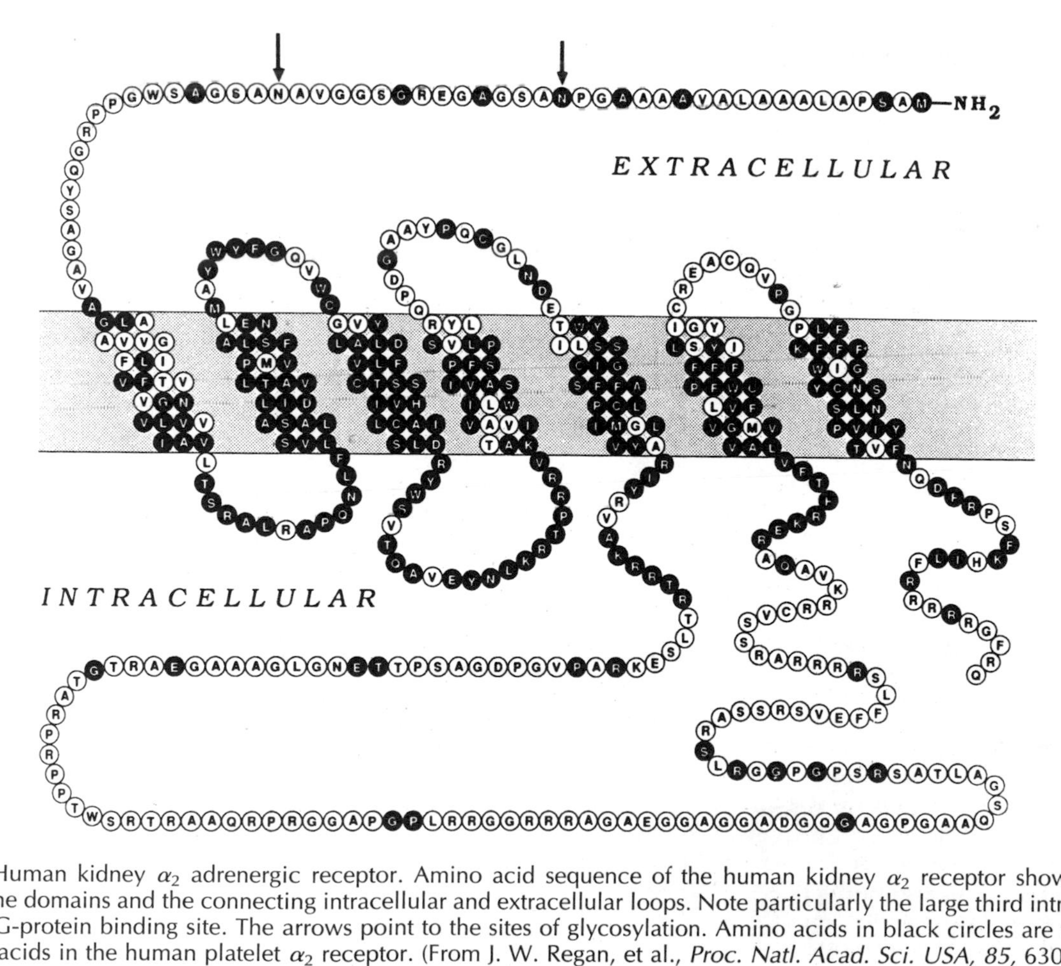

Fig. 18–6. Human kidney α_2 adrenergic receptor. Amino acid sequence of the human kidney α_2 receptor showing the seven transmembrane domains and the connecting intracellular and extracellular loops. Note particularly the large third intracellular loop, which is the G-protein binding site. The arrows point to the sites of glycosylation. Amino acids in black circles are those identical to the amino acids in the human platelet α_2 receptor. (From J. W. Regan, et al., *Proc. Natl. Acad. Sci. USA, 85,* 6301 [1988]).

the residues that form hydrogen bonds with the catechol hydroxyls of β_2-agonists.[13] Furthermore the evidence indicates that Ser_{204} interacts with the *meta* hydroxyl group of the ligand, whereas Ser_{207} interacts specifically with the *para* hydroxyl group. Serine residues are found in corresponding positions in the fifth transmembrane domain of the other known adrenoceptors. Evidence indicates that the phenylalanine residue of transmembrane domain VI is also involved in ligand-receptor bonding with the catechol ring. Studies such as these and others that indicated the presence of specific disulfide bridges between cysteine residues of the β_2-receptor led to the binding scheme shown in Figure 18–5.

Structural differences exist among the various adrenoceptors with regard to their primary structure, including the actual peptide sequence and length. Each of the adrenoceptors is encoded on a distinct gene, and this information was considered crucial to the proof that each adrenoreceptor is indeed distinct although related. The amino acids that make up the seven transmembrane regions are highly conserved among the various adrenoreceptors, but the hydrophilic portions are quite variable. The largest differences occur in the third intracellular loop connecting transmembrane domains V and VI, which is the site of linkage between the receptor and its associated G-protein. Compare the diagram of the β_2-receptor in Figure 18–4 with that of the α_2-receptor in Figure 18–6.[14]

EFFECTOR MECHANISMS OF ADRENERGIC RECEPTORS

The receptors each are coupled through a G-protein to an effector mechanisms. Effector mechanisms are proteins that are able to translate the conformational change caused by activation of the receptor into a biochemical event within the cell. Both of the beta adrenoceptors are coupled via specific G-proteins (G_s) to the activation of adenylyl cyclase, as outlined in Figure 18–7.[15] Thus when the receptor is stimulated by an agonist, adenylyl cyclase is activated to catalyze the formation of cyclic-adenosine monophosphate (cAMP) from ATP. cAMP, called a second messenger for the beta adrenoceptors, is known to function as a second messenger for a number of other receptor types. cAMP is considered a messenger because it can diffuse through the cell for at least short distances to modulate biochemical events remote from the synaptic cleft. Modulation of biochemical events by cAMP includes a phosphorylation cascade of other proteins. cAMP is rapidly deactivated by hydrolysis of the phosphodiester bond by the enzyme phosphodiesterase. The α_2-receptor may use more than one effector system depending on the location of the receptor; however, to date the best understood effector system of the α_2-receptor appears to be similar to that of the β-receptors except that linkage via a G-protein (G_i) leads to inhibition of adenylyl cyclase instead of activation.

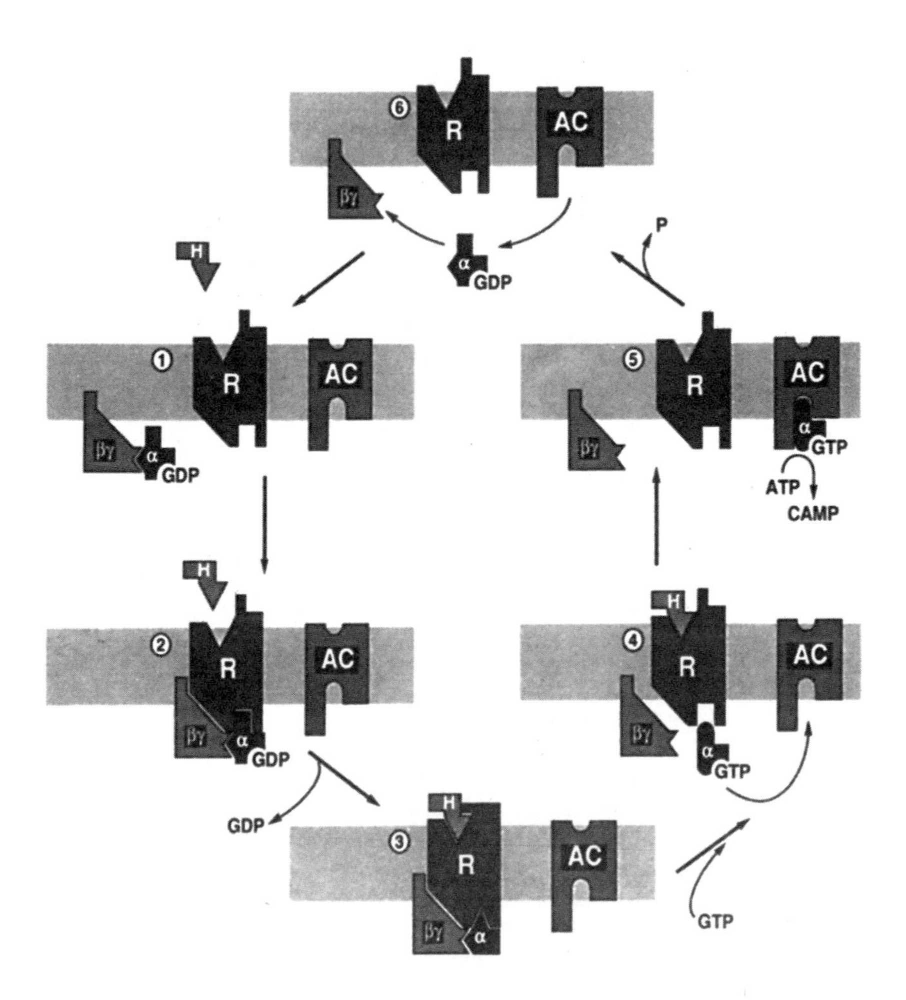

Fig. 18–7. β-adrenoceptor interactions with G-proteins. *1,* The adrenoceptor has low affinity for neurohormone agonist in the absence of GDP–G-protein complex. *2,* GDP–G-protein complex binds receptor increasing affinity for agonist. *3,* Agonist binds receptor causing conformational change in α-subunit of G-protein, which no longer binds GDP but now has affinity for GTP. *4,* GTP–α-subunit complex loses affinity for receptor, separates from receptor, and binds adenylyl cyclase. *5,* GTP–α-subunit activates adenylyl cyclase leading to conversion of ATP to cyclic-AMP, which initiates phosphorylation cascade. *6,* Following hydrolysis of GTP to GDP, α-subunit separates from adenylyl cyclase, terminating impulse transduction. $\beta\gamma$, Beta and gamma subunits of G-protein; α, alpha subunit of G-protein; R, beta adrenoceptor; AC, adenylyl cyclase; H, neurohormone epinephrine. (Adapted from B. Kobilka, *Annu. Rev. Neurosci., 15,* 87–114 [1992]).

ATP → Adenylyl Cyclase → cAMP → Phosphodiesterase → AMP

The α_1-adrenoreceptor is linked through yet another G-protein to a complex series of events involving hydrolysis of polyphosphatidylinositol.[16] The first event set in motion by activation of the α_1-receptor is activation of the enzyme phospholipase C. Phospholipase C catalyzes the hydrolysis of phosphatidylinositol-4,5-biphosphate (PIP_2). This hydrolysis yields two products, each of which has biologic activity as second messengers of the α_1-receptor. These are 1,2-diacylglycerol (DAG) and inositol-1,4,5-triphosphate (IP_3). IP_3 causes the release of calcium ions from intracellular storage sites in the endoplasmic reticulum resulting in an increase in free intracellular calcium levels. Increased free intracellular calcium is correlated with smooth muscle contraction. DAG is thought to activate cytosolic protein kinase C, which may induce slowly developing contractions of vascular smooth muscle. The end result of a complex series of protein interactions triggered by agonist binding to the α_1-receptor includes increased intracellular free calcium, which leads to smooth muscle contraction. When the smooth muscle innervated by α_1-receptors is in vascular walls, stimulation leads to vascular constriction.

RECEPTOR LOCALIZATION

The generalization made in the past about synaptic locations of adrenoreceptor subtypes were that all α_1, β_1, and β_2-receptors are postsynaptic receptors that are linked to stimulation of biochemical processes in the postsynaptic cell. The α_2-receptor has been traditionally viewed as a presynaptic receptor that resides on the outer membrane of the nerve terminus or presynaptic cell and reacts with released neurotransmitter. The α_2-receptor serves as a sensor and modulator of the quantity of neurotransmitter present in the synapse at any given moment. Thus, during periods of rapid nerve firing and neurotransmitter release, the α_2-receptor is stimulated and causes an inhibition of further release of neurotransmitter. This is a well-characterized mechanism of modulation of neurotransmission, but it is now known that not all α_2-receptors are presynaptic.[17] At this time, the physiologic significance of postsynaptic α_2-receptors is less well understood.

THERAPEUTIC RELEVANCE OF ADRENERGIC RECEPTOR SUBTYPES

When one considers the adrenoreceptor subtypes and effector responses of only a few organs and tissues innervated by the sympathetic nervous system, the clinical utility of receptor selective drugs becomes obvious. A small but therapeutically relevant sample is listed in Table 18–1. For example, the predominant response to adrenergic stimulation of smooth muscle of the peripheral vasculature is constriction causing a rise in blood pressure. Because this response is mediated through α_1-receptors, an α-antagonist would be expected to cause relaxation of the blood vessels and a drop in blood pressure with clear implications for treating hypertension. A smaller number of β-receptors on vascular smooth muscle mediate arterial dilation, particularly to skeletal muscle, and a few antihypertensives act through stimulation of these β-receptors. Adrenergic stimulation of the lungs causes smooth muscle relaxation and bronchodilation mediated through β_2-receptors. Drugs acting as β-agonists are useful for alleviating respiratory distress in persons with asthma or other obstructive pulmonary diseases. Activation of β_2-receptors in the uterus also causes muscle relaxation, and so some β_2-agonists are used to inhibit uterine contractions in premature labor. Adrenergic stimulation of the heart causes an increase in rate and force of contraction, which is mediated primarily by β_1-receptors. Drugs with β-blocking activity slow the heart rate and decrease the force of contraction; these drugs have utility in treating hypertension, angina, and certain cardiac arrhythmias. More details on therapeutic relevance of each class of drugs are provided with the discussions of the structure-activity relationships (SAR) of the various classes.

STRUCTURE-ACTIVITY RELATIONSHIPS OF ADRENERGIC DRUGS

From the preceding discussions of the biosynthesis, storage, release, and fate of norepinephrine, one can readily conceive of a number of possible sites of drug action, and there are indeed many sites of action for adrenergic drugs. There are drugs that act directly on the receptors as agonists and antagonists, drugs that affect storage and release from vesicles, drugs that affect neurotransmitter biosynthesis, and drugs that affect uptake and catabolism of norepinephrine and epinephrine. These categories are discussed in turn. Many adrenergic drugs fit into well-defined classes with readily defined SAR. Unfortunately, some adrenergic drugs do not permit such straightforward structural definition of their activity. We begin with a discussion of phenylethanolamine (or phenethanolamine) agonists, which do have reasonably clear SAR. Although many of these drugs directly stimulate ad-

Table 18–1. Selected Tissue Responses to Stimulation of Adrenoceptor Subtypes

Organ or Tissue	Receptor Type	Response
Arterioles	α_1	Constriction
	β_2	Dilation
Eye (radial muscle)	α_1	Contraction (pupillary dilation)
Fat cells	α,β	Lipolysis
Heart	β_1	Increased rate and force
		Increased conduction velocity
Intestine	α,β	Decreased motility
Liver	α,β_2	Increased gluconeogenesis and glycogenolysis
Lungs	β_2	Relaxation (bronchial dilation)
Uterus	α_1	Contraction (pregnant uterus)
	β_2	Relaxation

renergic receptors, others exhibit what is termed indirect activity. Indirect agonists do not directly bind to and activate adrenergic receptors; rather, they are taken up into the presynaptic neuron, where they cause the release of norepinephrine, which can diffuse to the receptor and cause the observed response. Mixed acting drugs have both a direct and an indirect component to their action, and the relative amount of direct versus indirect activity for a given drug varies considerably with the tissue preparation examined and experimental animal species.

PHENYLETHANOLAMINE AGONISTS

The SAR of many clinically useful phenylethanolamine-type adrenergic agonists are summarized in Table 18–2. Agents of this type have been extensively studied over the years since the discovery of the naturally occurring prototypes, epinephrine and norepinephrine, and the structural requirements and tolerances for substitutions at each of the indicated positions have been established.[18] In general, a primary or secondary aliphatic amine separated by two carbons from a substituted benzene ring is minimally required for high agonist activity in this class. Because of the basic amino group, these agents are highly charged at physiologic pH. Most agents in this class have a hydroxyl group on C1 of the side chain, β to the amine, as in epinephrine and norepinephrine. This hydroxyl substituted carbon must be in the *R* absolute configuration for maximal direct activity as in the

Table 18–2. Phenylethanolamine Adrenergic Agonists

Drug	R_1	R_2	R_3	Receptor Activity
Norepinephrine	H	H	3′,4′-diOH	$\alpha + \beta$
Epinephrine	CH_3	H	3′,4′-diOH	$\beta \geq \alpha$
α-Methylnorepinephrine	H	CH_3	3′,4′-diOH	$\alpha + \beta$
Ethylnorepinephrine	H	CH_2CH_3	3′,4′-diOH	$\beta > \alpha$
Isoeproterenol	$CH(CH_3)_2$	H	3′,4′-diOH	β
Colterol	$C(CH_3)_3$	H	3′,4′-diOH	β_2
Isoetharine	$CH(CH_3)_2$	CH_2CH_3	3′,4′-diOH	β_2
Metaproterenol	$CH(CH_3)_2$	H	3′,5′-diOH	β_2
Terbutaline	$C(CH_3)_3$	H	3′,5′-diOH	β_2
Albuterol	$C(CH_3)_3$	H	3′-CH_2-OH,4′-OH	β_2
Mabuterol	$C(CH_3)_3$	H	3′-CF_3,4′-NH_2,5′-Cl	β_2
Ritodrine	$-CH_2CH_2-C_6H_4-OH$	CH_3	4′-OH	β_2
Phenylephrine	CH_3	H	3′-OH	α
Metaraminol	H	CH_3	3′-OH	α
Methoxamine	H	CH_3	3′,5′-diOCH_3	α
Ephedrine	CH_3	CH_3	H	Mixed
Phenylpropanolamine	H	CH_3	H	Mixed

natural neurotransmitter, although most drugs are currently sold as mixtures of both (R) and (S) stereoisomers at this position. Given these features in common, the nature of the other substituents determines receptor selectivity and duration of action. In the following discussions, one must keep in mind that a drug selective for a given receptor does not mean it has zero activity at other receptors and that the clinically observed degree of selectivity is frequently dose dependent.

R^1, N-Substitution

We have already seen that as R^1 is increased in size from hydrogen in norepinephrine to methyl in epinephrine to isopropyl in isoproterenol, activity at α-receptors decreases, and activity at β-receptors increases. These compounds were used to define alpha and beta activity long before receptor proteins could be isolated and characterized. The activity at both α and β-receptors is maximal when R^1 is methyl as in epinephrine, but α-activity is dramatically decreased when R^1 is larger than methyl and is negligible when R^1 is isopropyl as in isoproterenol, leaving only β-activity. In fact, the β-activity of isoproterenol is actually enhanced over the natural neurotransmitters. Presumably the β-receptor has a large lipophilic binding pocket adjacent to the amine-binding aspartic acid residue, which is absent in the α-receptor. As R^1 becomes larger than butyl, affinity for α_1-receptors returns, but not intrinsic activity, which means large lipophilic groups can afford compounds with α_1-blocking activity, (e.g., see labetalol under β-antagonists). In addition, the N-substituent can also provide selectivity for different β-receptors, with a t-butyl group affording selectivity for β_2-receptors. For example, with all other features of the molecules being constant, colterol is a selective β_2-agonist, whereas isoproterenol is a general β-agonist. When considering use as a bronchodilator, a general β-agonist such as isoproterenol has undesirable cardiac stimulatory properties owing to its β_1-activity that are diminished or absent in a selective β_2-agonist. A variety of other bulky N-substituents can be used to provide β-selectivity, as seen, for example, in ritodrine, a selective β_2-agonist used to inhibit uterine contractions.

R^2, Substitution α to N

Small alkyl groups, methyl or ethyl, may be present on the carbon alpha to the amino nitrogen, carbon 2 in Table 18–2. Such substitution slows metabolism by monoamine oxidase (MAO) but has little overall effect on duration of action in catecholamines because they remain substrates for catechol-O-methyltransferase (COMT). Resistance to MAO activity is more important in noncatechol indirect acting phenylethylamines. An ethyl group in this position diminishes α-activity far more than β-activity, affording compounds with β-selectivity such as ethylnorepinephrine. Substitution on this carbon also introduces another asymmetric center into these molecules producing pairs of diastereomers, which can have significantly different biologic and chemical properties. For example, maximal direct activity in the stereoisomers of α-methylnorepinephrine resides in the *erythro* stereoisomer with the (1R,2S) absolute configuration.[19] The configuration of C2 has a great influence on receptor binding because the (1R,2R) diastereomer of α-methylnorepinephrine has primarily indirect activity even though the absolute configuration of the hydroxyl-bearing C1 is the same as in norepinephrine. In addition, with respect to α-activity, this additional methyl group also makes the direct-acting (1R,2S) isomer of α-methylnorepinephrine selective for α_2-adrenoceptors over α_1-adrenoceptors. This has important consequences in the antihypertensive activity of α-methyldopa, which is discussed later. The same stereochemical relationships hold for metaraminol and other phenylethanolamines, in which stereochemical properties have been investigated.

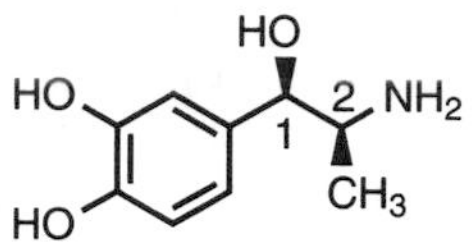

1R,2S
α-Methylnorepinephrine
direct acting stereoisomer

C1 Substitution

As previously mentioned, maximal direct agonist activity is obtained when C1 bears a hydroxyl group in the (R) absolute configuration as in epinephrine and norepinephrine. There is a clear stereospecific binding interaction with adrenoceptors because substitution with a hydroxyl in the (S) absolute configuration affords compounds with lowered activity, approximately equal to those with no hydroxyl group.

R^3, Ring Substitution

The natural 3′,4′-dihydroxy substituted benzene ring present in norepinephrine provides excellent receptor activity for both α and β-sites, but such catechol-containing compounds have poor oral activity because they are rapidly metabolized by COMT. Alternative substitutions have been found that retain good activity but are more resistant to COMT metabolism. In particular, 3′,5′-dihydroxy compounds are not good substrates for COMT and, in addition, provide selectivity for β_2-receptors. Thus, because of its ring substitution pattern, metaproterenol is an orally active bronchodilator having little of the cardiac stimulatory properties possessed by isoproterenol.

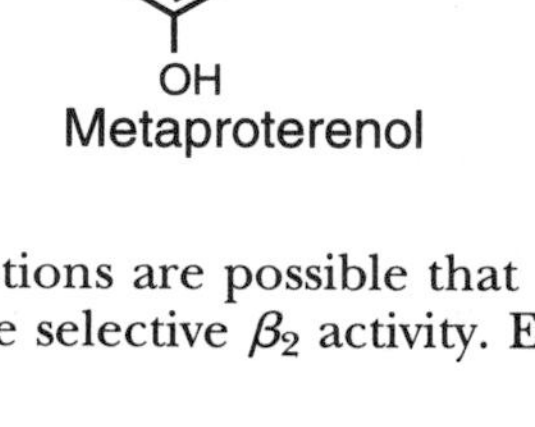

Metaproterenol

Other substitutions are possible that enhance oral activity and provide selective β_2 activity. Examples include the 3′-hydroxymethyl, 4′-hydroxy substitution pattern of albuterol and the 3′-trifluoromethyl, 4′-amino, 5′-chloro pattern of mabuterol. At least one of the groups must be capable of forming hydrogen bonds, and if there is only one, it should be at the 4′ position to retain β-activity. For example, ritodrine has only a 4′-OH for R^3, yet retains good β activity.

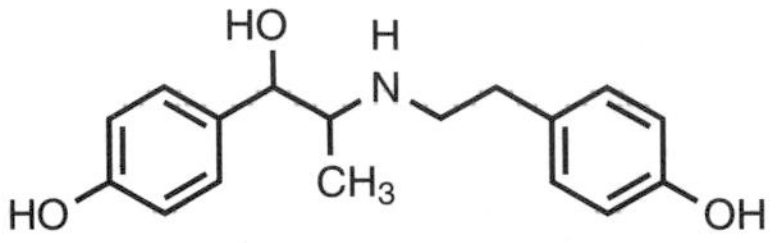

Ritodrine

If R^3 is only a 3′-OH, however, activity is reduced at α sites but almost eliminated at β sites, affording selective α agonists such as phenylephrine and metaraminol. Further indication that α sites have a wider range of substituent tolerance for agonist activity is shown by the 2′,5′-dimethoxy substitution of methoxamine, which is a selective α-agonist that also has β-blocking activity at high concentrations.

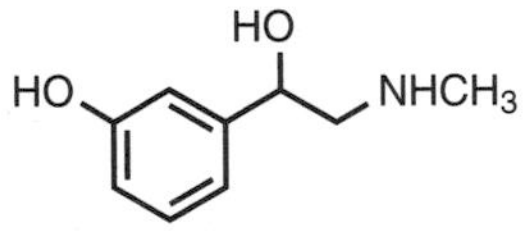

Phenylephrine

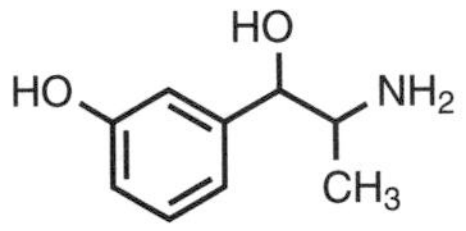

Metaraminol

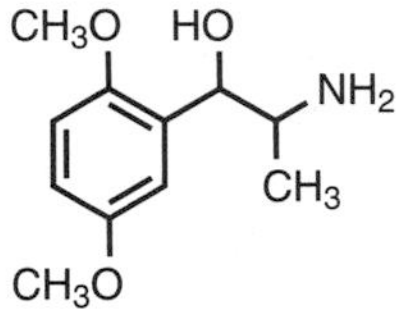

Methoxamine

Selected Applications of Phenylethanolamine Agonists

The mixed α- and β-agonist norepinephrine has limited clinical application because of the nonselective nature of its action, which causes both vasoconstriction and cardiac stimulation. It has no oral activity and because of its rapid metabolism by COMT and MAO must be given intravenously. Rapid metabolism limits its duration of action to only 1 or 2 minutes even when given by infusion. The drug is used to counteract various hypotensive crises and as an adjunct treatment in cardiac arrest. Metaraminol and methoxamine are selective for α receptors and so have little cardiac stimulatory properties. Because they are not substrates for COMT, their duration of action is significantly longer than norepinephrine, but their primary use is limited to treating hypotension during surgery or shock. The β-blocking activity of methoxamine, which is seen at high concentrations, affords some use in treating tachycardia. Phenylephrine, also a selective α-agonist, is used similar to metaraminol and methoxamine for hypotension but also has widespread use as a nonprescription nasal decongestant in both oral and topical preparations. Constriction of dilated blood vessels in mucous membranes shrinks the membranes and reduces nasal congestion. Phenylephrine preparations applied topically to the eye constrict the dilated blood vessels of bloodshot eyes and in higher concentrations are used to dilate the pupil during eye surgery.

Epinephrine is far more widely used clinically than norepinephrine, although it also lacks oral activity. Epinephrine, similar to norepinephrine, is used to treat hypotensive crises and, because of its greater β-activity, is used to stimulate the heart in cardiac arrest. Epinephrine's β_2 activity, lacking in norepinephrine, leads to its administration intravenously and in inhalers to relieve bronchoconstriction in asthma and to application in inhibiting uterine contractions. Because it has significant α activity, epinephrine also is used in nasal decongestants.

Most of the β-selective agents listed in Table 18–2 are used primarily as bronchodilators in asthma and other constrictive pulmonary conditions. Isoproterenol is a general β-agonist, and the cardiac stimulation caused by its β_1-activity and its lack of oral activity have led to diminished use in favor of selective β_2-agonists. Noncatechol selective β_2-agonists, such as albuterol, metaproterenol, and terbutaline, are available in oral dosage forms as well as in inhalers. All have similar activities and durations of action. Pirbuterol is an interesting analog of albuterol in which the benzene ring has been replaced by a pyridine ring. Similar to albuterol, pirbuterol is a selective β_2-agonist but currently is available only for administration by inhalation.

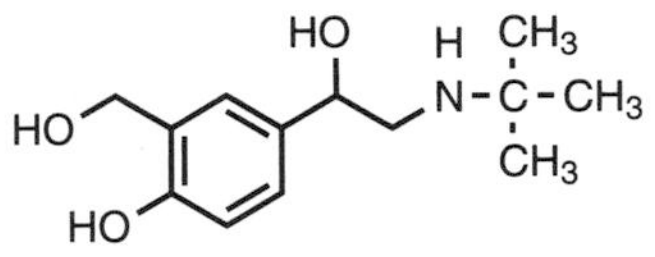

Albuterol

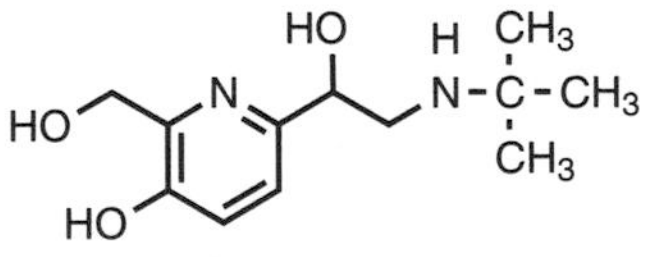

Pirbuterol

Bitolterol is a prodrug form of colterol in which the catechol hydroxyl groups have been converted to 4-methylbenzoic acid esters, which provide increased lipid solubility and prolonged duration of action. Bitolterol is administered by inhalation, and the ester groups are hydrolyzed by esterases to liberate the active drug, colterol. Colterol is then subject to metabolism by COMT, but the duration of action of a single dose of the produg bitolterol, 8 hours, is twice that of a single dose of colterol,

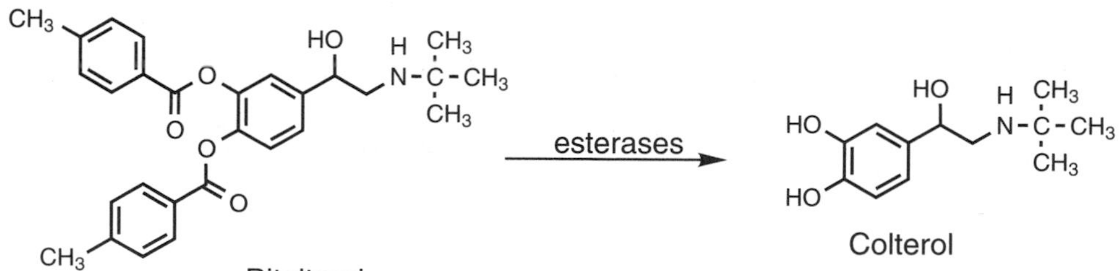

Bitolterol → esterases → Colterol

permitting less frequent administration and greater convenience to the patient.

As previously mentioned, ritodrine is a selective β_2-agonist that is used exclusively for relaxing uterine muscle and inhibiting the contractions of premature labor. Terbutaline has also been widely used for this purpose.

Ephedrine is a natural product isolated from several species of ephedra plants, which were used for centuries in folk medicines in a variety of cultures worldwide.[20] Pure ephedrine was first isolated and crystalized from a Chinese herbal medicine called Ma Huang in 1887. Its sympathomimetic activity was not recognized until 1917, and the pure drug was used clinically even before epinephrine and norepinephrine were isolated and characterized. Ephedrine does not have any substituents on the phenyl ring, giving it good oral activity because it is not a substrate for COMT. Lacking hydrogen-bonding phenyl substituents, ephedrine is less polar than the other compounds discussed thus far and crosses the blood–brain barrier far better than catechols. Because of its ability to penetrate the central nervous system, ephedrine has been used as a stimulant and exhibits side effects related to its action in the brain. Ephedrine has two asymmetric centers and therefore has four stereoisomers in a pair of diastereomers. The drug ephedrine is a mixture of the erythro enantiomers (1R,2S) and (1S,2R). The *threo* pair of enantiomers is pseudoephedrine (ψ-ephedrine). As discussed for α-methylnorepinephrine, the ephedrine stereoisomer with the (1R,2S) absolute configuration has direct activity on the receptors, both α and β, as well as an indirect component. The (1S,2R) enantiomer has primarily indirect activity. Ephedrine is widely used for many of the same indications as epinephrine, including use as a bronchodilator, vasopressor, cardiac stimulant, and nasal decongestant.

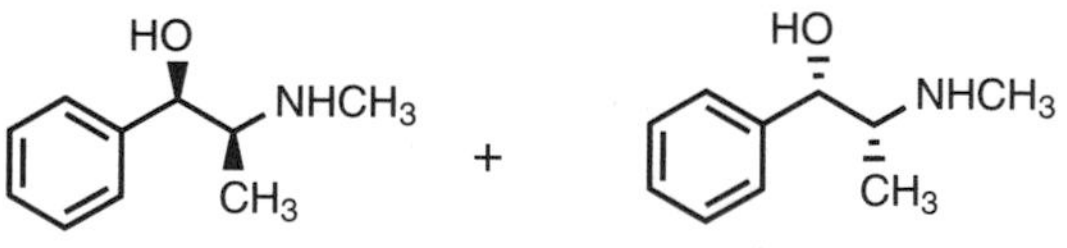

(1R:2S) + (1S:2R)

Ephedrine

(1R:2R) + (1S:2S)

Pseudoephedrine

Pseudoephedrine, the *threo* diastereomer of ephedrine, has virtually no direct activity and fewer central nervous system side effects than ephedrine. Pseudoephedrine is widely used as a nasal decongestant in a variety of dosage forms.

Phenylpropanolamine (see Table 18–2) is a N-desmethyl analog of ephedrine that has many of the same properties. Lacking the N-methyl group, however, phenylpropanolamine has none of the β_2-agonist activity of ephedrine, is slightly less lipophilic, and therefore does not enter the central nervous system as well as ephedrine. Phenylpropanolamine, similar to ephedrine, is a mixture of *erythro* enantiomers with mixed direct and indirect activity. The drug is the active ingredient of a number of nasal decongestants and is also used as a nonprescription appetite suppressant (anorexiant).

Not all the adrenergic agonists with direct activity have an aliphatic hydroxyl group such as the agents discussed so far. One of these is the catechol dobutamine. Dobutamine is a dopamine analog with a bulky arylalkyl group on the nitrogen and one asymmetric center. Dobutamine has direct activity on both α- and β-receptors, but because of some unusual properties of its two enantiomers, the overall pharmacologic response looks similar to that of a selective β_1-agonist.[21] The (−)-isomer of dobutamine is a powerful α_1-agonist and vasopressor. The (+)-isomer is a powerful α_1-antagonist; thus, when the racemate is used clinically, the α-effects of the enantiomers cancel out each other. The stereochemistry of the methyl substituent does not affect the ability of the drug to bind to the α_1-receptor but does affect the ability of the molecule to activate the receptor. That is, the stereochemistry of the methyl group affects intrinsic activity but not affinity. Because both stereoisomers are β-agonists with the (+)-isomer about 10 times as potent as the (−)-isomer, the net effect is β stimulation. Dobutamine is used as a cardiac stimulant after surgery or congestive heart failure. As a catechol, dobutamine is readily metabolized by COMT and has a short duration of action with no oral activity.

Dopamine, although not strictly an adrenergic drug, is a catechol with properties related to the cardiovascular activities of the other agents in this chapter. Dopamine acts on specific dopamine receptors to dilate renal vessels, increasing renal blood flow. Dopamine also stimulates cardiac β_1-receptors through both direct and indirect mechanisms. It is used to correct hemodynamic imbal-

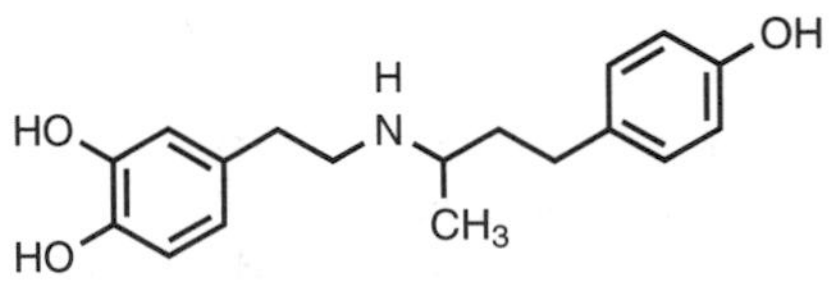

Dobutamine

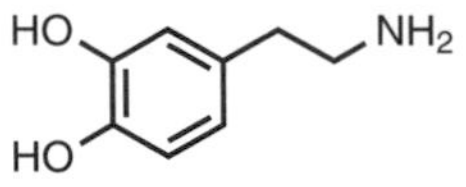

Dopamine

ances induced by conditions such as shock, myocardial infarction, trauma, or congestive heart failure. As a catechol and primary amine, dopamine is rapidly metabolized by COMT and MAO and, similar to dobutamine, has a short duration of action with no oral activity.

Other phenylethylamines, such as amphetamine, which lack both ring substituents and a side chain hydroxyl, are sufficiently lipophilic to cross the blood–brain barrier readily and have dramatic central nervous system effects, some of which give them serious abuse potential. The clinical utility of amphetamine and its derivatives is entirely based on central nervous system stimulant and central appetite suppressant effects. These agents are discussed in the chapter on central nervous system stimulants (see Chapter 15).

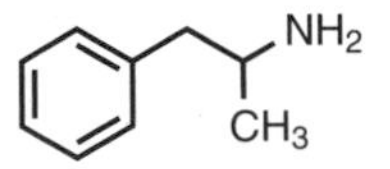

Amphetamine

IMIDAZOLINE α_1-AGONISTS

Although nearly all β-agonists are phenylethanolamine derivatives, α-receptors accommodate a more diverse assortment of structures. The imidazole derivatives in Figure 18–8 are selective α_1-agonists and therefore vasoconstrictors. They all contain a one-carbon bridge between C2 of the imidazoline ring and a phenyl substituent, and therefore the general skeleton of a phenylethylamine is contained within the structures. Lipophilic substitution on the phenyl ring ortho to the methylene bridge appears to be required for agonist activity at both types of α-receptor.[22] Presumably the bulky lipophilic groups attached to the phenyl ring at the meta or para positions provide selectivity for the α_1-receptor by diminishing affinity for α_2-receptors. These compounds are used in topical preparations as nasal decongestants and in eye drops.

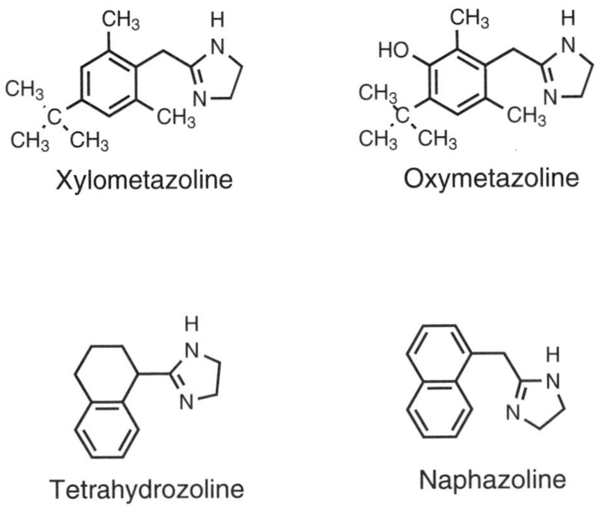

Xylometazoline

Oxymetazoline

Tetrahydrozoline

Naphazoline

Fig. 18–8. Imidazoline α_1 agonists.

IMIDAZOLINE AND OTHER α_2-AGONISTS

Closely related structurally to the imidazoline nasal decongestants is clonidine, which was originally synthesized as a vasoconstricting nasal decongestant but in early clinical trials was found to have dramatic hypotensive effects in contrast to all expectations for a vasoconstrictor.[23] Subsequent pharmacologic investigations showed not only that clonidine does have some α_1-agonist (vasoconstrictive) properties in the periphery, but also that clonidine is a powerful agonist at α_2-receptors in the central nervous system. Stimulation of central postsynaptic α_2-receptors leads to a reduction in sympathetic neuronal output and a hypotensive effect. Because of its peripheral activity on extraneuronal vascular postsynaptic α_2-receptors, initial doses of clonidine may produce a transient vasoconstriction and an increase in blood pressure that is soon overcome by vasodilation as clonidine penetrates the blood–brain barrier and interacts with central nervous system α_2-receptors.

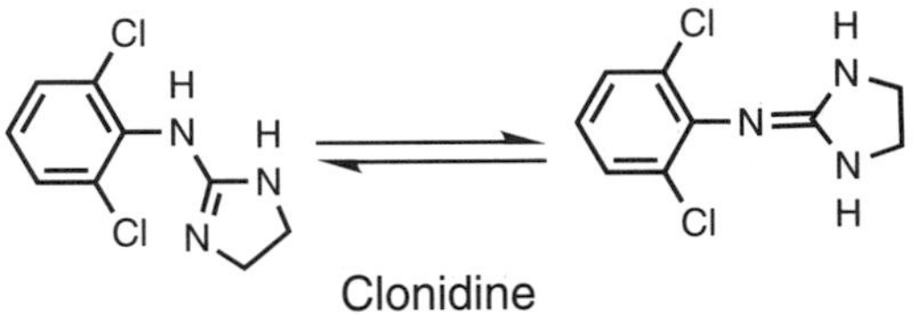

Clonidine

Similar to the imidazoline α_1-agonists, clonidine has lipophilic ortho substituents on the phenyl ring. Chlorines afford better activity than methyls at α_2 sites. The most readily apparent difference between clonidine and the α_1-agonists in Figure 18–8, is the replacement of the CH_2 on C1 of the imidazoline by an amine NH. This makes the imidazoline ring part of a guanidino group, and the uncharged form of clonidine exists as a pair of tautomers as shown. Clonidine has a pKa of 8.05 and at physiologic pH is about 82% ionized. The positive charge is shared over all three nitrogens, and the two rings are forced out of coplanarity by the bulk of the two ortho chlorines, as illustrated in Figure 18–9.

Following the discovery of clonidine, extensive research into the SAR of central α_2 agonists showed that the imidazoline ring was not necessary for activity in this class, but the phenyl ring required at least one ortho chlorine or methyl group. It has been found in this research that other amidine containing functions can replace the guanidino group, but none of them are yet in clinical use.

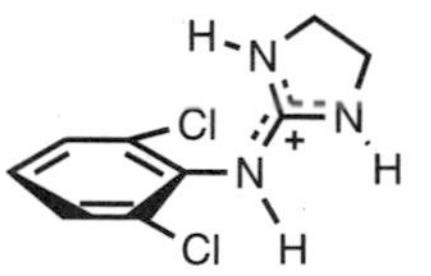

Fig. 18–9. Protonated clonidine. The positive charge is shared through resonance by all three nitrogens of the guanidino group. Steric crowding by the ortho chlorines does not permit a coplanar conformation of the two rings.

Two clinically useful agents resulting from this effort are guanfacine and guanabenz. These are ring-opened analogs of clonidine, and their mechanism of action is the same as that of clonidine.

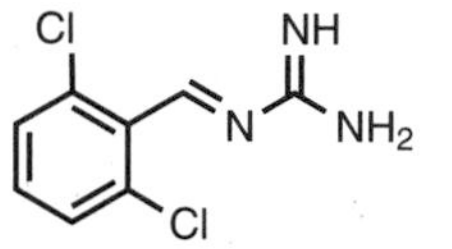

Guanabenz

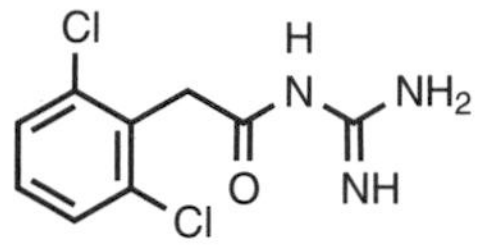

Guanfacine

Although structurally unrelated to clonidine, guanabenz and guanfacine, another antihypertensive agent acting as an α_2 agonist in the central nervous system via its metabolite, α-methyl-norepinephrine is L-α-methyldopa (methyldopa) (Fig. 18–10), originally synthesized as a

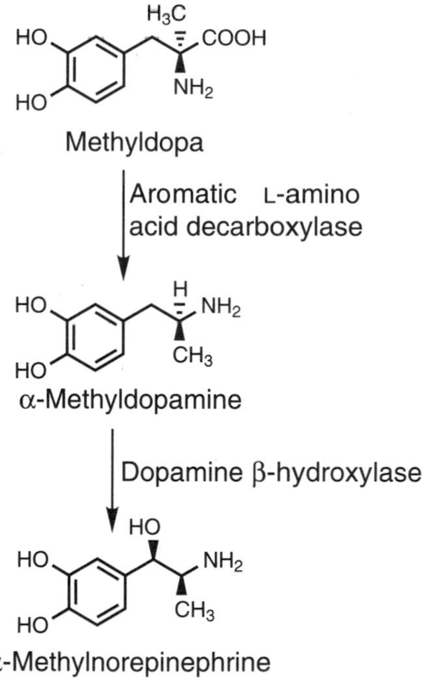

Methyldopa

Aromatic L-amino acid decarboxylase

α-Methyldopamine

Dopamine β-hydroxylase

α-Methylnorepinephrine

Fig. 18–10. Methyldopa bioactivation. Methyldopa (L-α-methyldopa) crosses the blood–brain barrier, where it is decarboxylated by aromatic L-amino acid decarboxylase to α-methyldopamine, which is converted stereospecifically to (1R,2S)-α-methylnorepinephrine. This acts as an α_2 agonist to inhibit sympathetic neural activity and lower blood pressure. Of the three chemicals shown, only methyldopa can cross the blood–brain barrier.

norepinephrine biosynthesis inhibitor.[24] When originally introduced, methyldopa was thought to act through a combination of inhibition of norepinephrine biosynthesis through dopa decarboxylase inhibition and metabolic decarboxylation to generate α-methylnorepinephrine. The latter would replace norepinephrine in the nerve terminal and, when released, have less intrinsic activity than the natural neurotransmitter. This latter mechanism is an example of the concept of a false neurotransmitter. The antihypertensive mechanism of methyldopa, however, has since been shown to be due to activity in the central nervous system. Methyldopa (the drug is the L(S)-isomer) is decarboxylated to α-methyldopamine followed by stereospecific β-hydroxylation to the (1R,2S) stereoisomer of α-methylnorepinephrine. As mentioned previously, this stereoisomer is an α_2-agonist and acts much like clonidine to cause a decrease in sympathetic output from the central nervous system.

ADRENERGIC ANTAGONISTS

α Antagonists

Because α agonists cause vasoconstriction and raise blood pressure, one would expect α antagonists to have antihypertensive activity. An old but powerful drug in this class is phenoxybenzamine, a β-haloalkylamine that alkylates α-receptors. Beta-haloalkylamines are present in nitrogen mustard anticancer agents and are highly reactive alkylating agents. The acid salt of phenoxybenzamine is stable, but at physiologic pH, at which there is equilibrium between protonated drug and free base. The unprotonated amino group is nucleophilic and displaces the chlorine atom in an intramolecular reaction to form a highly reactive aziridinium ion (Fig. 18–11). The aziridinium ion reacts with nucleophiles, X, to form a covalent bond between the nucleophile and the drug. The substituents attached to the haloalkylamine provide selectivity for binding to α-adrenoceptors so that the nucleophile is generally part of the target receptor. The nucleophile X is presumably part of an amino acid side chain, such as a cysteine thiol, serine hydroxyl, or lysine amino group, but the specific site of covalent attachment to the α receptor has not been determined. Because the reaction in which phenoxybenzamine forms covalent bonds with the receptors is irreversible, new receptors must be synthesized before the effects can be overcome. The α-blockade is therefore long lasting. Unfortunately, other biomolecules besides the target α-receptor are also alkylated. Because of its nonselectivity and toxicity, the use of phenoxybenzamine is limited to alleviating the effects of pheochromocytoma. This tumor of chromaffin cells of the adrenal medulla produces large amounts of epinephrine and norepinephrine, which are released into the blood stream producing hypertension and generalized sympathetic stimulation.

Imidazoline α Antagonists

Tolazoline and phentolamine are two imidazoline α-antagonists that also have antihypertensive activity, although they have been replaced in general clinical use

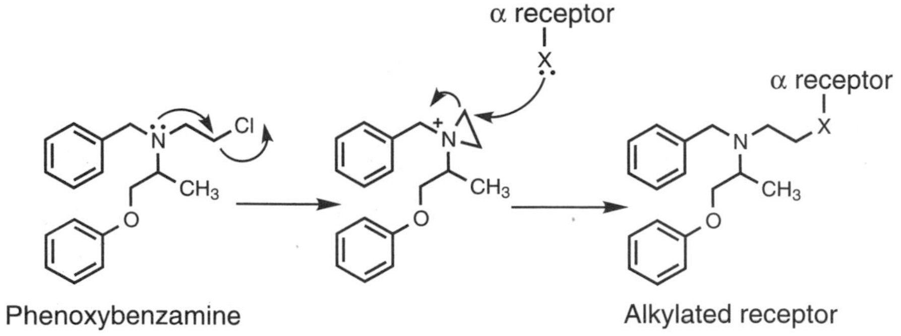

α receptor
X
Cl
N
CH3
O
Phenoxybenzamine
α receptor
N
X
CH3
O
Alkylated receptor

Fig. 18–11. Phenoxybenzamine alkylation of α adrenoceptors. The unshared pair of electrons on the nitrogen of phenoxybenzamine free base displaces the β-chlorine to form a reactive aziridinium ion. If this occurs in the vicinity of an α receptor, a nucleophile group X on the receptor can open the aziridinium ion in a nucleophilic reaction to form a covalent bond between the receptor and the drug. X is a nucleophile, such as S, N, or O.

by far better agents. Tolazoline has clear structural similarities to the imidazoline α-agonists, such as naphazoline and xylometazoline (see Fig. 18–8), but does not have the substituents required for agonist activity. The resemblance of phentolamine is not as readily apparent, but extensive molecular modeling studies have provided a topologic scheme for α_1-antagonist SAR.[25] This pattern, however, cannot be readily visualized without computer graphics and is beyond the scope of this chapter. Both phentolamine and tolazoline are potent but rather nonspecific α-antagonists. Both drugs stimulate gastrointestinal smooth muscle, an action blocked by atropine, which would indicate cholinergic activity, and they both stimulate gastric secretion, possibly through release of histamine. Because of these and other side effects, the clinical applications of tolazoline and phentolamine are limited to treating the symptoms of pheochromocytoma.

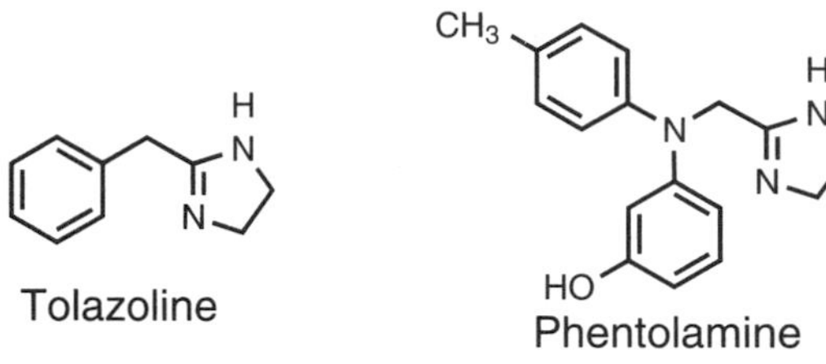

Tolazoline

Phentolamine

Selective α_1 Antagonists

Prazosin, the first known selective α_1 blocker, was discovered in the late 1960s [26] and is now one of a family of quinoxaline antihypertensives with selective α_1-blocking activity, which also includes terazosin, doxazosin, and trimazosin (Fig. 18–12). These agents lower peripheral resistance both at rest and during exercise by reversing elevated vasoconstrictor tone through α_1 blockade.

All of these agents contain a 4-amino-6,7-dimethoxyquinazoline ring system attached to a piperazine nitrogen. The only structural differences are in the groups attached to the other nitrogen of the piperazine, and the differences in these groups afford dramatic differences in some of the pharmacokinetic properties of these agents. For example, when the furan ring of prazosin is reduced to form the tetrahydrofuran ring of terazosin, the compound becomes significantly more water-soluble,[27] as would be expected because tetrahydrofuran is much more water-soluble than furan. Some of the important clinical parameters of the quinazolines are shown in Table 18–3. Perhaps most significant are the long half-

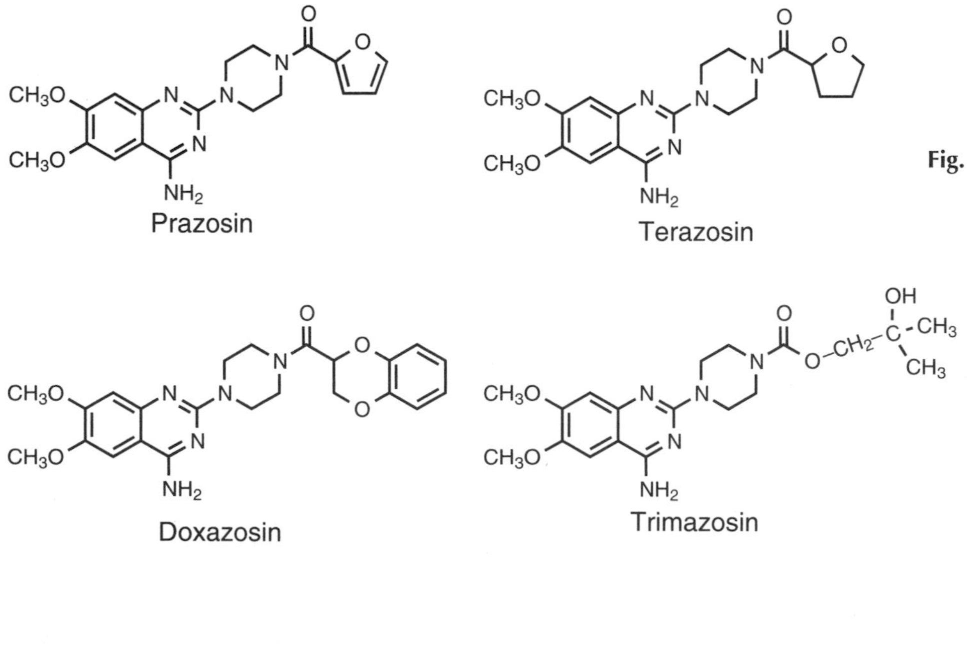

Prazosin

Terazosin

Doxazosin

Trimazosin

Fig. 18–12. Quinoxaline α_1 antagonists.

Table 18–3. Selected Clinical Parameters of α_1-antagonist Antihypertensives

Drug	Therapeutic Dose (mg)	Half-life (h)	Duration of Action (h)	Frequency of Administration	Bioavailability (%)
Prazosin	2–20	2–3	4–6	BID–TID	45–65
Terazosin	1–40	12	>18	QD–BID	90
Doxazosin	1–16	19	18–36	QD–BID	65
Trimazosin	100–900	3	3–6	BID–TID	60–80
Indoramin	50–150	5	>6	BID–TID	30

QD, Once daily; BID, twice daily; TID, three times daily.

lives and durations of action of terazosin and doxazosin, which permit once-a-day dosing and generally lead to increased patient compliance.

Another α_1-antagonist is indoramin, which is an indole derivative having a piperidine in the side chain instead of a piperazine. Although indoramin is highly selective for α_1 over α_2-adrenoceptors, it also blocks histamine H_1 and serotonin receptors.[27] It has other pharmacologic properties in the cardiovascular system that are not well understood. Indoramin's half-life and duration of action are also listed in Table 18–3.

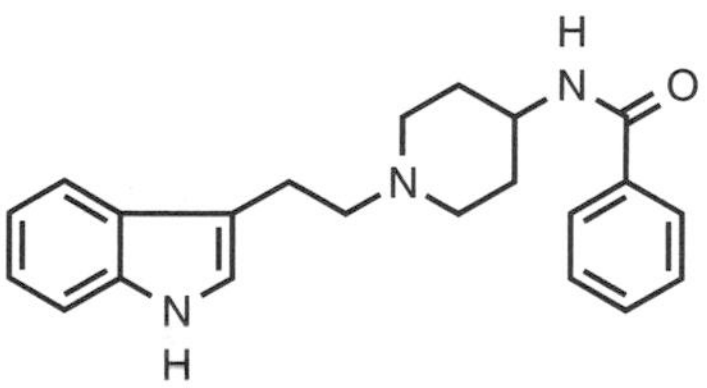

Indoramin

α_2-Antagonists, Yohimbine

Yohimbine, an indole alkaloid isolated from Pausinystlia yohimbe bark and Rauwolfia roots is an α_2-antagonist selective for α_2 over α_1 adrenoceptors, but it is also a serotonic antagonist. It has actions both in the central nervous system and in the pheriphery inducing hypertension and increases in heart rate. Yohimbine has no indications sanctioned by the Food and Drug Administration in the United States but has been used to treat male impotence and postural hypotension. Yohimbine has also been used in research to induce anxiety.

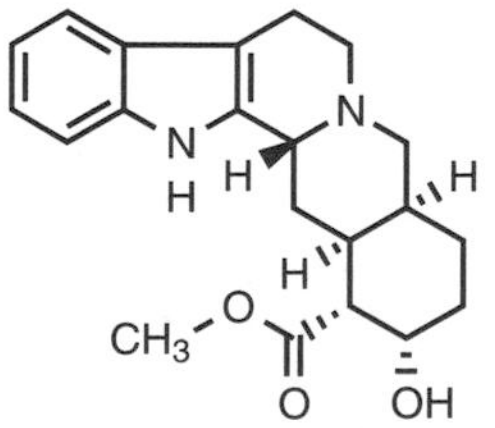

Yohimbine

β-Antagonists

In the 1950s, a derivative of isoproterenol in which the catechol hydroxyls had been replaced by chlorines, dichloroisoproterenol (DCI), was discovered to be a β-antagonist that blocked the effects of sympathomimetic amines on bronchodilation, uterine relaxation, and heart stimulation.[28] Although DCI had no clinical utility, replacement of the 3,4-dichloro substituents with a carbon bridge to form a naphthylethanolamine derivative did afford a clinical candidate, pronethalol, introduced in 1962 only to be withdrawn in 1963 because of tumor induction in animal tests.

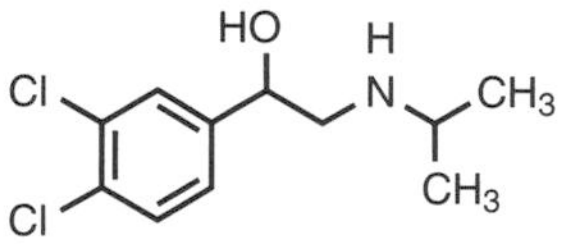

Dichloroisoproterenol

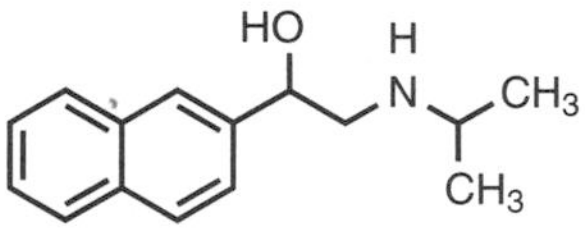

Pronethalol

Shortly thereafter, a major innovation was introduced when it was discovered that an oxymethylene bridge, OCH_2, could be introduced into the arylethanolamine structure of pronethalol to afford propranolol, an aryloxypropanolamine and the first clinically successful β-blocker. Note that, along with the introduction of the oxymethylene bridge, the side chain has been moved from C2 of the naphthyl group to the C1 position. In general, the aryloxypropanolamines are more potent β-blockers than the corresponding arylethanolamines.

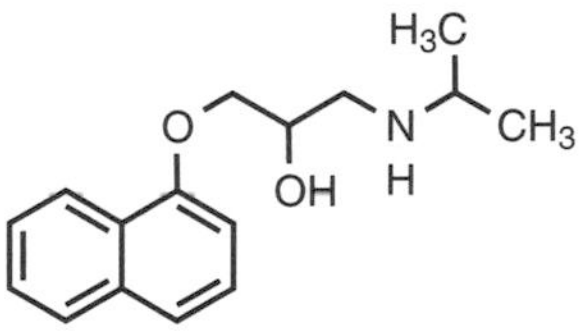

Propranolol

Initially, it might appear that lengthening the side chain would prevent appropriate binding of the required functional groups to the same receptor site. But simple molecular models show that the side chains of aryloxypropanolamines can adopt a conformation that places the hydroxyl and amine groups into approximately the same position in space (Fig. 18–13). The simple two-dimensional drawing in Figure 18–13 exaggerates the true degree of overlap; however, elaborate molecular modeling studies confirm that the aryloxypropanolamine side

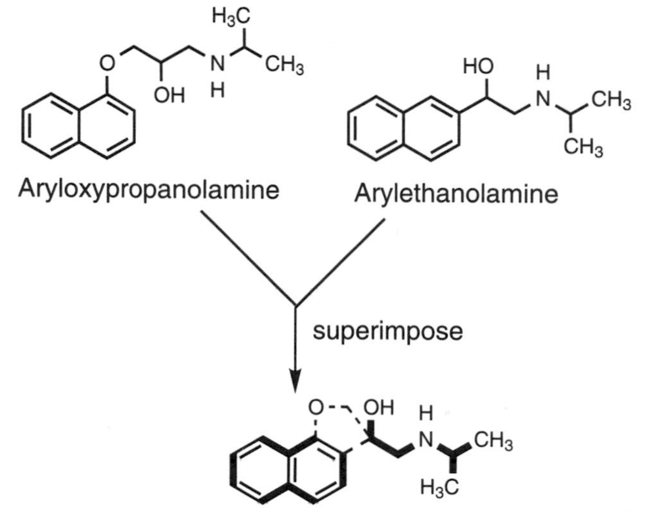

Aryloxypropanolamine Arylethanolamine

superimpose

Fig. 18–13. Overlap of aryloxypropanolamines and arylethanolamines. The structures of prototype β antagonists propranolol and pronethalol may be superimposed so the critical functional groups occupy the same approximate regions in space as indicated by the bold lines in the superimposed drawings. The dotted lines are those parts that do not overlap but are not necessary to receptor binding.

chain can adopt a low-energy conformation that permits close overlap with the arylethanolamine side chain.[29]

Propranolol was initially introduced for the treatment of angina pectoris, followed by trials as an antiarrhythmic. During clinical trials as an antianginal, propranolol was discovered to have antihypertensive properties. Propranolol rapidly became widely used for a variety of cardiovascular disorders.[30]

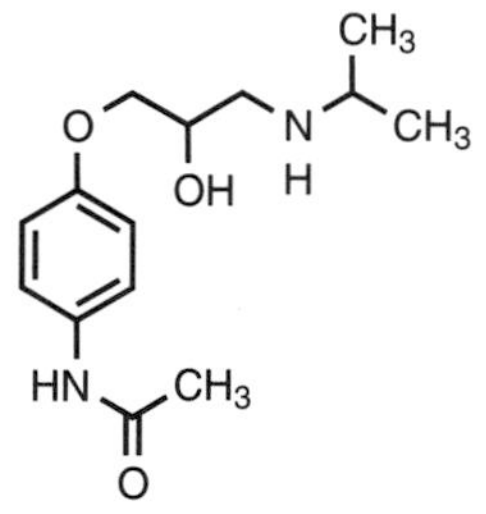

Practolol

At approximately this same time, a new series of 4-substituted phenyloxypropanololamines emerged, such as practolol, which selectively inhibited sympathetic cardiac stimulation. These observations led to the recognition that not all β-receptors were the same and to the introduction of β_1 and β_2 nomenclature to differentiate cardiac β-receptors from others.[31] Development of β-blockers proceeded rapidly, and there are now a large number of additional drugs available on the world market, both general β-antagonists (Fig. 18–14), and selective β-antagonists (Fig. 18–15). All of the drugs shown in Figure 18–14 and 18–15 are aryloxypropanolamines. Metipranolol (Fig. 18–14) is an exception to the general rule that 4-substituted aryloxypropanolamines are selective β_1-blockers.

Sotalol is an arylethanolamine general β-blocker, whereas labetalol is a unique arylethanolamine antihypertensive with α_1, β_1, and β_2-blocking activity. In terms of its SAR, you will recall from the earlier discussion of phenylethanolamine agonists that although groups such as isopropyl and t-butyl eliminated α-receptor activity, large groups could bring back α_1-affinity but not intrinsic activity.

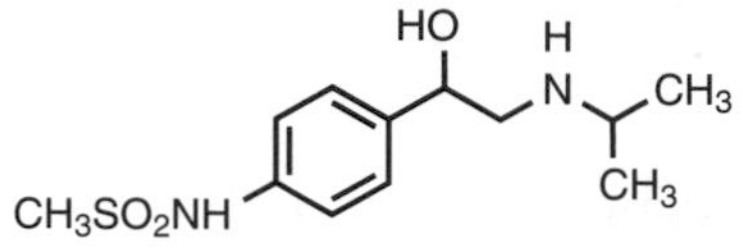

Sotalol

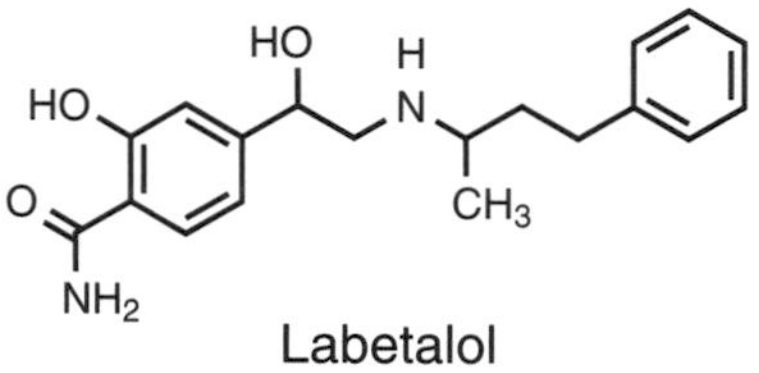

Labetalol

A factor that sometimes causes confusion when comparing the structures of arylethanolamines with aryloxypropanolamines is the stereochemical nomenclature of the side chain carbon bearing the hydroxyl group. For maximum effectiveness in receptor binding, the hydroxy group must occupy the same region in space as it does for the phenylethanolamine agonists in the *R* absolute configuration. Because of the insertion of an oxygen atom in the side chain of the aryloxypropanolamines, the priority of the substituents around the asymmetric carbon changes, and the isomer with the required special arrangement now has the *S* absolute configuration. This is an effect of the nomenclature rules, and the groups still have the same spatial arrangements (Fig. 18–16).

β-blockers are widely used in the treatment of hypertension, angina, cardiac arrhythmias, myocardial infarction, and glaucoma. The ability of β-blockers to slow the heart rate and decrease force of contraction lowers the workload on the heart and relieves angina. Although the exact mechanism is less straightforward, these same effects may be the principal cause of the hypotensive effects of β-blockers. Lowered output from the heart is thought eventually to lead to relaxation of the peripheral vasculature, although an effect on renin output may be involved. The ability to decrease automaticity leads to their use in treating arrhythmias. The action of β-blockers in treating glaucoma is more difficult to explain.[32] The β-blockers lower intraocular pressure by decreasing the amount of aqueous humor fluid produced in the eye by the ciliary body, and β_2-receptors have been found in that tissue. Observations, however, that the ciliary body has no adrenergic innervation; the effect is not stereoselective, and a correlation of activity exists with decreased ciliary blood flow and decreased dopamine levels indicate that the

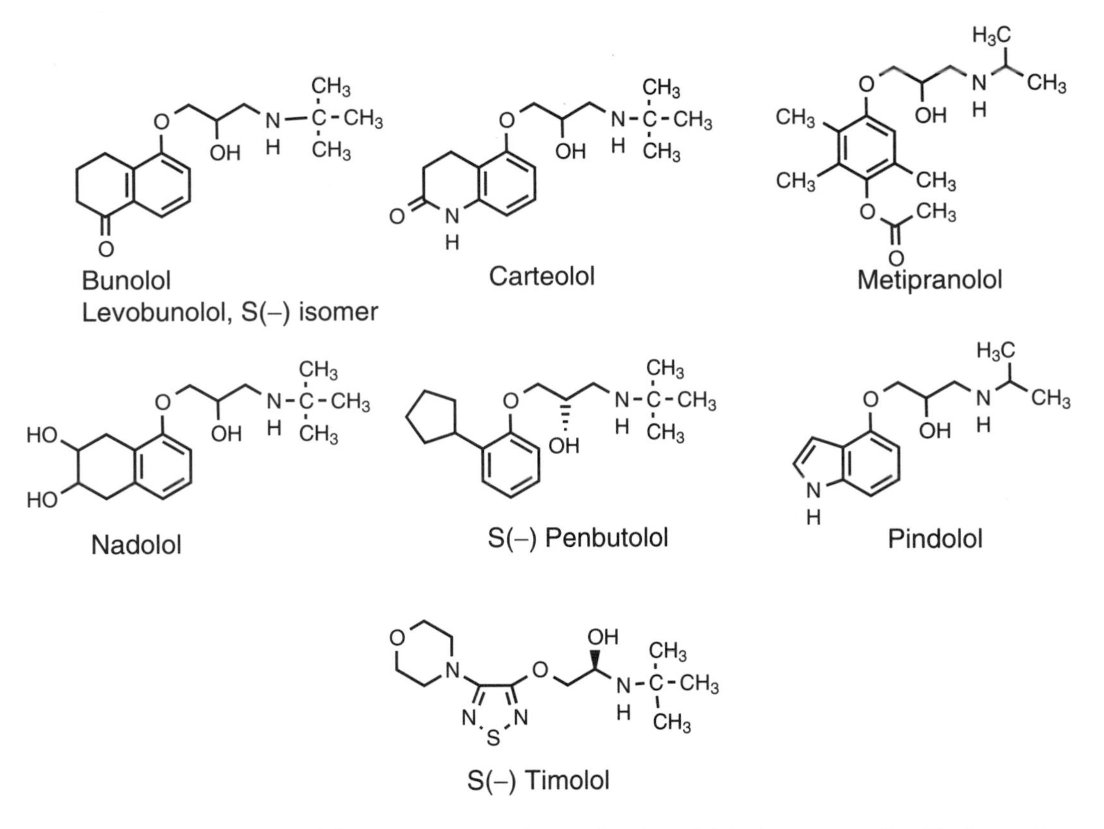

Fig. 18–14. General β antagonists. Bunolol is a racemate, whereas levobunolol is the more active S(−) enantiomer of the same structure. Penbutolol and timolol are marketed as the single (S)-enantiomer as shown. The hydroxyl groups on the ring of nadolol are cis to each other.

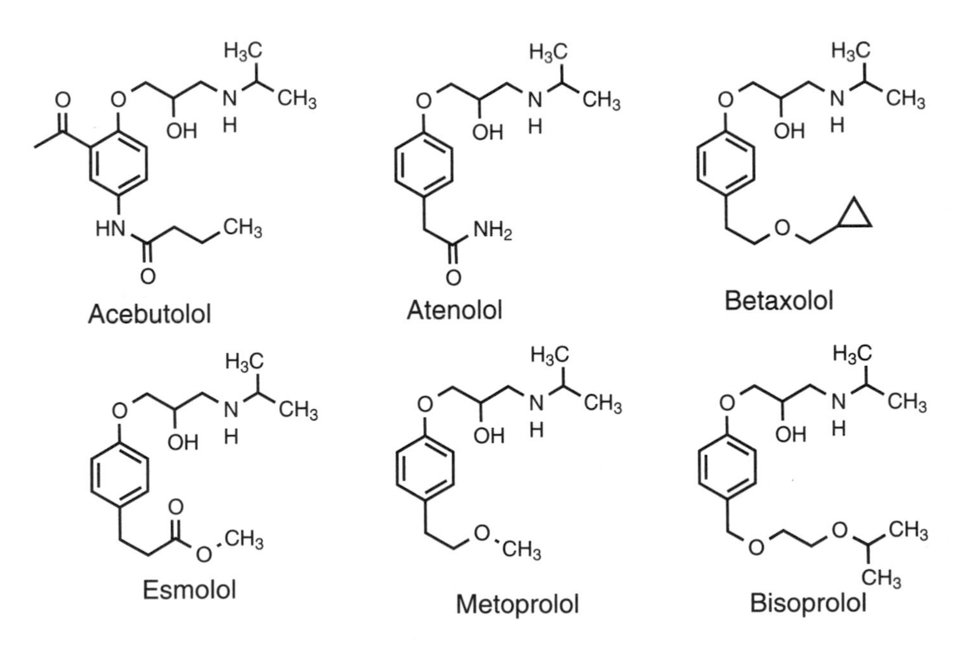

Fig. 18–15. Selective β_1 antagonists.

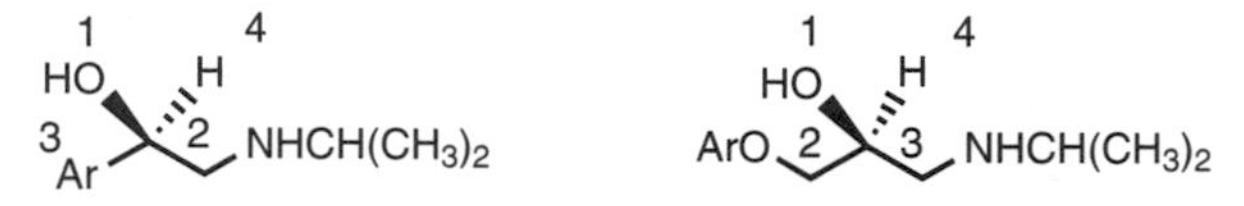

(R) absolute configuration (S) absolute configuration

Fig. 18–16. Stereochemical nomenclature for arylethanolamines versus aryloxypropanolamines. The relative positions in space of the four functional groups are the same in the two structures; however, one is designated (*R*) and the other (*S*). This is because the introduction of an oxygen atom into the side chain of the aryloxypropanolamine changes the priority of two of the groups used in the nomenclature assignment.

mechanism of action of β-blockers in treating glaucoma is unusual and may not be β-blockade in the classic sense.[32]

Even though β-blockers are widely used for a variety of ailments, certain precautions must be taken in the use of these agents in certain patient populations. Because the selectivity of β_1-antagonists is not absolute, at the higher range of recommended doses the compounds also block β_2-receptors. Therefore, extreme caution is advised in using these agents in persons with asthma or other pulmonary disease. Caution is also required with diabetics because the β-blockers inhibit catecholamine-induced glycogenolysis in response to insulin-induced hypoglycemia and may thus aggravate the condition.

Other than β_1-selectivity of 4-substituted aryloxypropanolamines, there is little obvious structural pattern to relate β-blockers to specific clinical applications with the exception of esmolol. Esmolol has a methyl ester at the end of the 4-substituent, which makes it susceptible to hydrolysis by serum esterases, and the acid generated by hydrolysis has little β-blocking activity. For this reason, esmolol has a half-life of only about 8 minutes and is used to control supraventricular tachycardia during surgery when a short-acting agent is desirable.

Another physicochemical parameter that has some clinical correlation is relative lipophilicity of different agents. Propranolol is by far the most lipophilic of the available β-blockers and enters the central nervous system far better than less lipophilic agents, such as atenolol or nadolol. Lipophilicity as measured by octanol/water partitioning also correlates with primary site of clearance as seen in Table 18–4.[33] The more lipophilic drugs are primarily cleared by the liver, whereas, the more hydrophilic agents are cleared by the kidney. This could have an influence on choice of agents in cases of renal failure or liver disease.

Table 18–4. Lipophilicity and Clearance of β Antagonists

Drug	Octanol/Water Partition Coefficient	Primary Clearance Site
Propranolol	20.40	Hepatic
Metoprolol	0.89	Hepatic
Pindolol	0.82	Hepatic plus renal
Atenolol	0.008	Renal
Nadolol	0.006	Renal

DRUGS AFFECTING NOREPINEPHRINE/EPINEPHRINE BIOSYNTHESIS

Hypothetically, inhibitors of any of the three enzymes involved in the conversion of L-tyrosine to norepinephrine could be used as drugs to moderate adrenergic transmission. Inhibitors of the rate-limiting enzyme tyrosine hydroxylase would be the most logical choice. One inhibitor of tyrosine hydroxylase, metyrosine or α-methyl-L-tyrosine, is in limited clinical use to help control hypertensive episodes and other symptoms of catecholamine overproduction in patients with the rare adrenal tumor pheochromocytoma.[34] Metyrosine, a competitive inhibitor of tyrosine hydroxylase, inhibits the production of catecholamines by the tumor. Although metyrosine is useful in treating hypertension caused by excess catecholamine biosynthesis in pheochromocytoma tumors, it is not useful for treating essential hypertension. The drug metyrosine is the L(S) stereoisomer of α-methyltyrosine. The enantiomer, D(R)-α-methyltyrosine, does not bind to the active site of tyrosine hydroxylase and thus has no useful activity.

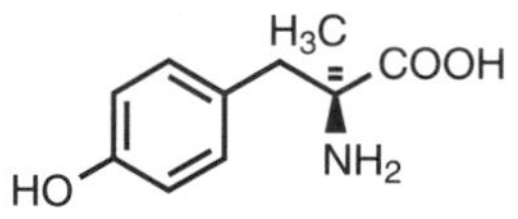

Metyrosine

Of far greater clinical significance in the treatment of hypertension is methyldopa, which is an inhibitor of aromatic L-amino acid decarboxylase through its ability to serve as an alternative substrate for the enzyme. As previously discussed, however, the mechanism of antihypertensive activity of methyldopa is not due to norepinephrine biosynthesis inhibition but rather to metabolism to an α_2-agonist in the central nervous system. Other more powerful inhibitors of aromatic L-amino acid decarboxylase (e.g., carbidopa) have proven to be clinically useful, but not as modulators of peripheral adrenergic transmission. Rather these agents are used to inhibit the metabolism of exogenous L-dopa administered in the treatment of Parkinson's disease and are discussed in Chapter 13.

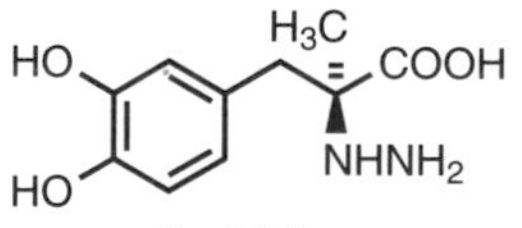

Carbidopa

The next enzyme in the biosynthetic pathway to norepinephrine and epinephrine, dopamine β-hydroxylase, has been the subject of extensive research into its chemical mechanism and the subject of many enzyme inhibition studies. The inhibitors known to date, however, are primarily of basic biochemical research interest and have no therapeutic relevance. The same is true of phenylethanolamine-N-methyltransferase, the last enzyme in the biosynthesis of epinephrine in the adrenal medulla.

DRUGS AFFECTING STORAGE VESICLES

An old and historically important drug that affects the storage and release of norepinephrine is reserpine.[35] Reserpine is one of several indole alkaloids isolated from the roots of Rauwolfia serpentina, a plant whose roots were used in India for centuries as a remedy for snake bites and as a sedative. The antihypertensive effects of the root extracts were first reported in India in 1918 and in the West in 1949. Shortly thereafter, reserpine was isolated and identified as the principal active agent. Reserpine was the first effective antihypertensive drug introduced into Western medicine, although it has largely been replaced in clinical use by agents with fewer side effects.

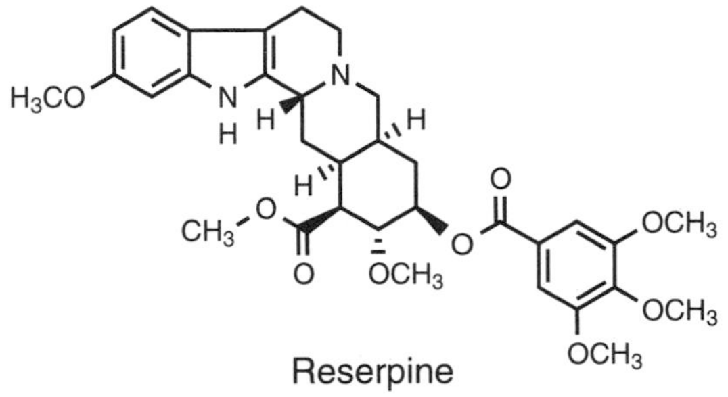

Reserpine

Reserpine acts to deplete the adrenergic neurons of their stores of norepinephrine by inhibiting the active transport Mg-ATPase responsible for sequestering norepinephrine and dopamine within the storage vesicles. The norepinephrine and dopamine that are not sequestered in vesicles are destroyed by MAO. Because the storage vesicles contain so little neurotransmitter, adrenergic transmission is dramatically inhibited, resulting in decreased sympathetic tone and vasodilation. Reserpine has the same effect on epinephrine storage in the adrenals. Reserpine readily enters the central nervous system, where it also depletes the stores of norepinephrine and serotonin. The central nervous system neurotransmitter depletion led to the use of reserpine in treating certain mental illnesses.

Bretylium, guanethidine, and guanadrel are three other drugs whose similar mechanisms involve norepinephrine storage granules. They are taken up into adrenergic neurons by uptake-1, where they bind to the storage vesicles and prevent release of neurotransmitter in response to a neuronal impulse. Guanethidine and guanadrel are orally active antihypertensives that cause a slow decrease in the amount of norepinephrine in the vesicles, but their principal mechanism is a poorly understood inhibition of neurotransmitter release. Bretylium is a quaternary ammonium salt and must be given intravenously because it has poor oral absorption. Initially, it can cause a release of norepinephrine and a transient rise in blood pressure. Its clinical utility is limited to cardiac arrhythmias.

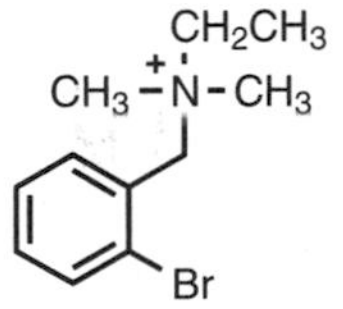

Bretylium

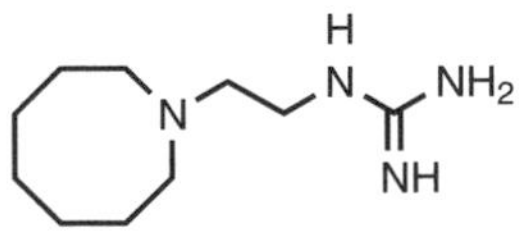

Guanethidine

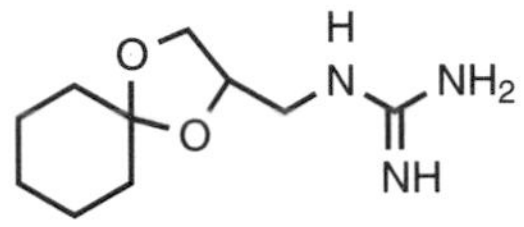

Guanadrel

ERGOT ALKALOIDS

The ergot alkaloids are a large group of indole alkaloids isolated from the ergot fungus, Claviceps purpurea, which is a plant parasite principally infecting rye. Eating grain contaminated with ergot caused a severely debilitating and painful disease during the Middle Ages called St. Anthony's Fire, but in small doses, ergot was known to midwives for centuries for its ability to stimulate uterine contraction. The pharmacology of the various ergot alkaloids is complex, involving actions on the adrenergic nervous system as well as a number of others. The structures of several ergot alkaloids are shown in Figure 18–17.

Ergotamine is a mixed agonist/antagonist of various peripheral and central adrenergic receptors. It is a strong inducer of contractions in the pregnant uterus, which appears to be partially an α-effect because it is blocked by phentolamine, but other receptors may be involved. The drug is used to contract the postpartum uterus to prevent excessive bleeding. Ergotamine is also used to treat migraine headache. Ergonovine and methyergonovine are also strong inducers of uterine contractions and because of better oral absorption have largely replaced ergotamine for this purpose. Methylsergide, structurally identical to methyergonovine except for the addition of a methyl group to the indole nitrogen, has far less of the uterine stimulatory properties of the other agents and is instead used exclusively for treatment of migraine headache. This is believed to be a serotonin antagonist effect.

All of these ergot alkaloids are amide derivatives of lysergic acid, but only the diethylamide LSD produces the profound hallucinatory effects for which it is so well-known.

XANTHINE BRONCHODILATORS

Although they are not truly adrenergic drugs, methylxanthines are extensively used as bronchodilators in asthma and so are included here (Fig. 18–18). The most widely used of the xanthines is theophylline and its salts.

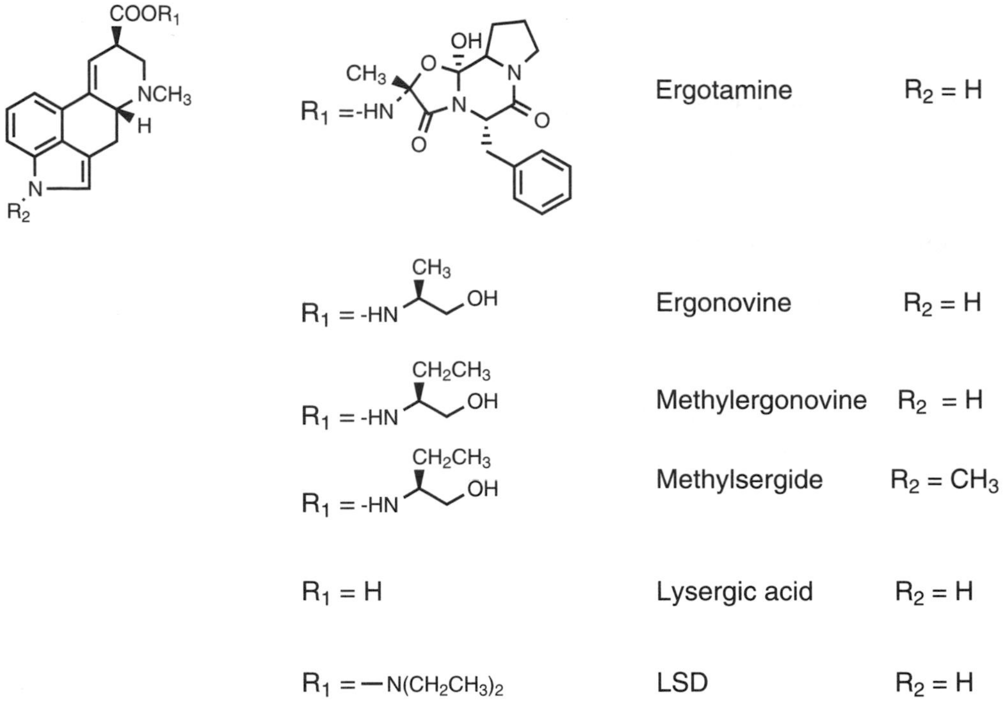

R_1		R_2
	Ergotamine	$R_2 = H$
R_1 = -HN–CH(CH_3)–CH_2OH	Ergonovine	$R_2 = H$
R_1 = -HN–CH(CH_2CH_3)–CH_2OH	Methylergonovine	$R_2 = H$
R_1 = -HN–CH(CH_2CH_3)–CH_2OH	Methylsergide	$R_2 = CH_3$
$R_1 = H$	Lysergic acid	$R_2 = H$
R_1 = —N(CH_2CH_3)$_2$	LSD	$R_2 = H$

Fig. 18–17. Ergot alkaloids.

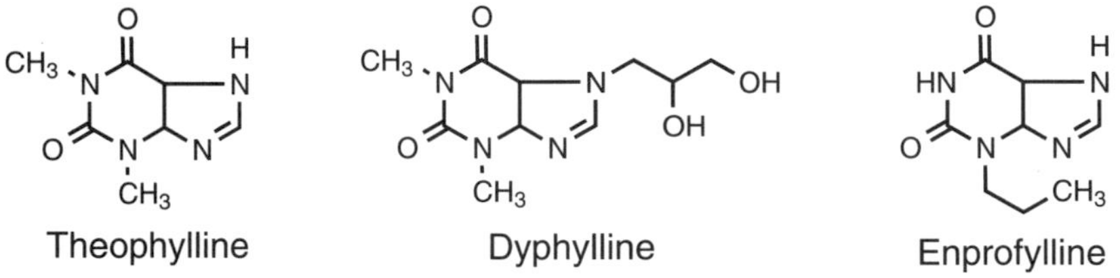

Theophylline Dyphylline Enprofylline

Fig. 18–18. Xanthine bronchodilators.

Aminophylline is the ethylemediamine salt of theophylline, and oxtriphylline is the choline salt. Enprofylline is widely used outside the United States.

These xanthine derivatives are effective bronchodilators; however, their pharmacologic mechanism of action remains controversial despite many years of intensive study.[37] Among the mechanisms of bronchodilation induction that have been suggested are inhibition of phosphodiesterase hydrolysis of cAMP, adenosine receptor antagonism, stimulation of increased secretion of endogenous catecholamines, inhibition of prostaglandins, and reduction of intracellular calcium ion concentrations. Adenosine receptor antagonism was highly favored but is currently not considered a likely mechanism.[38] It is possible that the xanthines act by the sum of several actions or by an as yet undiscovered mechanism.

REFERENCES

1. O. Appezeller, The vegetative nervous system, in *Handbook of Clinical Neurology,* Vol. I, P. J. Vinken and G. W. Bruyn, Eds., Amsterdam, North-Holland, 1969, pp. 427–428.
2. U. S. von Euler, Synthesis, uptake and storage of catecholamines in adrenergic nerves, the effect of drugs, in *Catecholamines, Handbook of Experimental Pharmacology,* Vol. 33, H. Blaschko and E. Marshall, Eds., New York, Springer, 1972, pp. 186–230.
3. D. J. Triggle, Adrenergics: Catecholamines and related agents, in *Burger's Medicinal Chemistry,* 4th Ed., Part III, M. E. Wolff, Ed., New York: John Wiley & Sons, 1981, pp. 225–283.
4. S. Kaufman and T. J. Nelson, Studies on the regulation of tyrosine hydroxylase activity by phosphorylation and dephosphorylation, in *Progress in Catecholamine Research, Part A: Basic Aspects and Peripheral Mechanisms,* A. Dahlstrom, R. H. Belmaker, and M. Sandler, Eds. New York, Alan R. Liss, 1988, pp. 57–60.
5. A. Philippu and H. Matthaei, Transport and storage of catecholamines in vesicles, in *Catecholamines I, Handbook of Experimental Pharmacology,* Vol. 90, U. Trendelenburg and N. Weiner, Eds., New York, Springer, 1988, pp. 1–42.
6. R. B. Kelley, et al., *Ann. Revu. Neurosci., 2,* 399(1979).
7. U. Trendelenburg, Factors influencing the concentration of catecholamines at the receptors, in *Catecholamines, Handbook of Experimental Pharmacology,* Vol. 33, H. Blaschko and E. Marshall, Eds., New York, Springer, 1972, pp. 726–761.
8. I. J. Kopin, Metabolic degradation of catecholamines. The

relative importance of different pathways under physiological conditions and after administration of drugs, in *Catecholamines, Handbook of Experimental Pharmacology,* Vol. 33, H. Blaschko and E. Marshall, Eds., New York, Springer, 1972, pp. 270–282.
9. R. P. Ahlquist, *Amer. J. Physiol., 153,* 586(1948).
10. J. K. Harrison, et al., *Trends Pharmacol. Sci., 12,* 62(1991).
11. S. Trumpp-Kallmeyer, et al., *J. Med. Chem., 35,* 3448(1992).
12. J. Ostrowski, et al., *Annu. Rev. Pharmacol. Toxicol., 32,* 167(1992).
13. C. D. Strader, et al., *J. Biol. Chem., 264,* 13572(1989).
14. J. W. Regan, et al., *Proc. Natl. Acad. Sci. USA, 85,* 6301(1988).
15. B. Kobilka, *Annu. Rev. Neurosci., 15,* 87(1992).
16. A. J. Nichols, α-Adrenoceptor signal transduction mechanisms, in α-Adrenoceptors: Molecular Biology, Biochemistry and Pharmacology, R. R. Ruffolo, Jr., Ed., New York, Karger, 1991, pp. 44–74.
17. P. B. M. W. M. Timmermans and P. A. van Zwieten, *J. Med. Chem., 25,* 1389(1982).
18. D. J. Triggle, Adrenergics: Catecholamines and related agents, in *Burger's Medicinal Chemistry,* 4th Ed., Part III, M. E. Wolf, Ed., New York, John Wiley & Sons, 1981, pp. 225–283.
19. P. N. Patil and D. Jacobowitz, *J. Pharmacol. Exp. Ther., 161,* 279(1968).
20. K. K. Chen and C. F. Schmidt, *Medicine, 9,* 1(1930).
21. R. R. Ruffolo, Jr., et al., *J. Pharmacol. Exp. Ther., 219,* 447(1981).
22. A. J. Nichols and R. R. Ruffolo, Jr., Structure-activity relationships for α-adrenoceptor agonists and antagonists, in *α-Adrenoceptors: Molecular Biology, Biochemistry and Pharmacology,* R. R. Ruffolo, Jr., Ed., New York, Karger, 1991, pp. 75–114.
23. W. Kobinger, Central α-adrenergic systems as targets for hypotensive drugs., *Rev. Physiol. Biochem. Pharmacol., 81,* 39(1978).
24. M. Henning and P. A. Van Zwieten, *J. Pharm. Pharmacol., 20,* 409(1968).
25. R. M. DeMarinis, et al., Structure-activity relationships for α_1-adrenoceptor agonists and antagonists, in *The Alpha-1 Adrenergic Receptors,* R. R. Ruffolo, Jr., Ed., Clifton, Humana Press, 1987, pp. 211–265.
26. A. Scriabine, et al., *Experientia, 24,* 1150(1968).
27. L. X. Cubedda, *Am. Heart J., 116,* 133(1988).
28. N. C. Moran, *Ann. N.Y. Acad. Sci., 139,* 649(1967).
29. T. Jen and C. Kaiser, *J. Med. Chem., 20,* 693(1977).
30. D. B. Evans, et al. *Ann. Rep. Med. Chem., 14,* 81(1979).
31. A. M. Lands, et al., *Nature, 214,* 597(1967).
32. T. S. Lesar, *Clin. Pharm., 6,* 451(1987).
33. F. M. Gengo and J. A. Green, Beta-blockers, in *Applied Pharmacokinetics: Principles of Therapeutic Drug Monitoring,* W. E. Evans, et al., Eds., Spokane, Applied Therapeutics, 1986, pp. 735–781.
34. R. N. Brogden, et al., *Drugs, 21,* 81(1981).
35. W. T. Comer, et al., Antihypertensive agents, in *Burger's Medicinal Chemistry,* 4th Ed., Part III, M. E. Wolf, Ed., New York, John Wiley & Sons, 1981, pp. 285–337.
36. T. W. Rall, Oxytocin, prostaglandins, ergot alkaloids, and other drugs: Tocolytic agents, in *Goodman and Gilman's The Pharmacologic Basis of Therapeutics,* 8th Ed., A. G. Gilman, et al., Eds. New York, Pergamon Press, 1990, pp. 933–953.
37. P. J. Barnes, Bronchodilator mechanisms, in *Asthma, Clinical Pharmacology and Therapeutic Progress,* A. B. Kay, Ed., Oxford, Blackwell Scientific Publications, 1986, pp. 146–160.
38. C. G. A. Persson and R. Pauwels, Pharmacology of antiasthma xanthines, in *Pharmacology of Asthma, Handbook of Experimental Pharmacology,* Vol. 98, C. P. Page and P. J. Barnes, Eds., New York, Springer, 1991, pp. 207–225.

SUGGESTED READINGS

B. B. Hoffman and R. J. Lefkowitz, Catecholamines and sympathomimetic drugs, in *Goodman and Gilman's The Pharmacologic Basis of Therapeutics,* 8th Ed., A. G. Gilman, et al., Eds., New York, Pergamon Press, 1990, pp. 187–220.

B. B. Hoffman and R. J. Lefkowitz, Adrenergic receptor antagonists, in *Goodman and Gilman's The Pharmacologic Basis of Therapeutics,* 8th Ed., A. G. Gilman, et al., Eds., New York, Pergamon Press, 1990, pp. 221–243.

R. J. Lefkowitz, et al., Neurohumoral transmission: The autonomic and somatic motor nervous systems, in *Goodman and Gilman's The Pharmacologic Basis of Therapeutics,* 8th Ed., A. G. Gilman, et al., Eds., New York, Pergamon Press, 1990, pp. 84–121.

J. Ostrowski, et al., *Annu. Rev. Pharmacol. Toxicol., 32,* 167(1992).

D. J. Triggle, Adrenergics: Catecholamines and Related Agents, in *Burger's Medicinal Chemistry,* 4th Ed., Part III. Edited by M. E. Wolff. New York, John Wiley & Sons, 1981, pp. 225–283.

Chapter 19

CARDIAC AGENTS: CARDIAC GLYCOSIDES, ANTIANGINALS, AND ANTIARRHYTHMIC DRUGS

Ahmed S. Mehanna

Heart diseases are grouped into three major disorders: cardiac failure, ischemia (with angina as its primary symptom), and cardiac arrhythmia.

CARDIAC GLYCOSIDES IN THE TREATMENT OF HEART FAILURE

Congestive Heart Failure

Cardiac failure can be described as the inability of the heart to pump blood effectively at a rate that meets the needs of metabolizing tissues. This is found to be a direct result of a reduced contractility of the cardiac muscles, especially those of the ventricles. The cardiac output decreases, and the blood volume of the heart increases (hence the name congested). As a result, the systemic blood pressure and the renal blood flow are both reduced, which often lead to the development of edema in the lower extremities and the lung (pulmonary edema) as well as renal failure. A group of drugs known as the cardiac glycosides were found to reverse most of these symptoms and complications.

Cardiac Glycosides

The cardiac glycosides are an important class of naturally occurring drugs whose actions include both beneficial and toxic effects on the heart. The desirable cardiotonic action is of particular benefit in the treatment of congestive heart failure and associated edema. Cardiac glycoside preparations have been used as medicinal agents as well as poisons since 1500 B.C., and this dual application serves to highlight the toxic potential of this class of life-saving drugs. Despite the extended use and obvious therapeutic benefits of the cardiac glycosides, it was not until the famous monograph by Withering in 1785, entitled "An Account of the Foxglove and Some of its Medical Uses," that cardiac glycoside therapy started to become more standardized and rational.[1,2] The therapeutic use of purified cardiac glycoside preparations has occurred only over the last century. Today the cardiac glycosides represent one of the most important drug classes available to treat heart failure.

Chemistry of the Cardiac Glycosides

Cardiac glycosides, similar to other glycosides, are composed of two portions: the sugar and the nonsugar (the aglycone) moiety.

Aglycones. The aglycone portion of the cardiac glycosides is a steroid nucleus with a unique set of fused rings, which makes these agents easily distinguished from other steroids. Rings A-B and C-D are *cis* fused, and ring B-C has a *trans* configuration. Such ring fusion gives the aglycone nucleus of cardiac glycosides the characteristic U shape as shown in Figure 19–1. The steroid nucleus also carries, in most cases, two angular methyl groups at C-10 and C-13. Hydroxyl groups are located at C-3, the site of the sugar attachment, and C-14. The latter hydroxyl is normally unsubstituted. Additional hydroxyl groups may be found at C-12 and C-16, the presence or absence of which distinguishes the important genins: digitoxigenin, digoxigenin, and gitoxigenin (Fig. 19–2). These additional hydroxyl groups have significant impact on the partitioning and kinetics of each glycoside as discussed later. The lactone ring at C-17 is a major structural feature of the cardiac aglycones. The size and degree of unsaturation of the lactone ring varies with the source of the glycoside. In most cases, cardiac glycosides of plant origin, the cardinolides, possess a five-membered, α,β-unsaturated lactone ring, whereas those derived from animal origin, the bufadienolides, possess a six-membered lactone ring with two conjugated double bonds (generally referred to as α-pyrone) (see Fig. 19–1).

Sugars. The hydroxyl group at C-3 of the aglycone portion is usually conjugated to a monosaccharide or a polysaccharide with β-1,4-glucosidic linkages. The number and identity of sugars vary from one glycoside to another as detailed subsequently. The most commonly found sugars in the cardiac glycosides are D-glucose, D-digitoxose, L-rhamnose, and D-cymarose (Fig. 19–3). These sugars predominately exist in the cardiac glycosides in the β-conformation. In some cases, the sugars exist in the acetylated form. The presence of an acetyl group on a sugar greatly affects the lipophilic character and kinetics of the entire glycoside as discussed subsequently.

Sources and Common Names of Cardiac Glycosides

The cardiac glycosides occur mainly in plants and in rare cases in animals, such as poisonous toads. Digitalis purpurea or the foxglove plant, Digitalis lanata, Strophanthus gratus, and Strophanthus kombé are the major plant sources of the cardiac glycosides. Based on the nature and number of sugar molecules and the number of hydroxyl groups on the aglycone moiety, each combina-

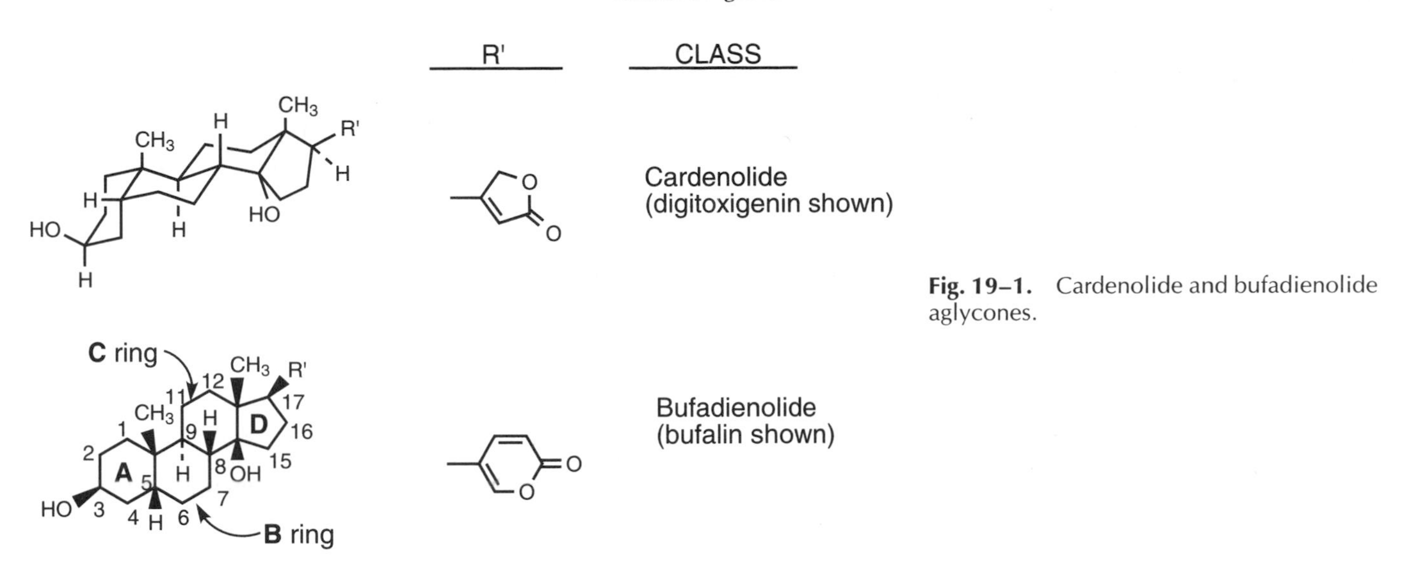

Fig. 19–1. Cardenolide and bufadienolide aglycones.

tion of sugars and aglycones assumes different generic names. The site of the glycosides concentration in the plant, types of glycosides, and the names of the structural components of these glycosides are summarized in Table 19–1.

Digitalis lanata. Lanatoside A is composed of the aglycone digitoxigenin (genin indicates no sugar) connected to three digitoxose molecules, the third of which carries an acetyl group, and a terminal glucose molecule; i.e., the structure sequence is glucose-digitoxose (3-acetyl)-digitoxose-digitoxose-digitoxigenin.

Lanatoside B has an identical sugar portion to lanatoside A, but the aglycone unit has an extra hydroxyl group at C-16 and is given the name gitoxigenin. The structure sequence is glucose-digitoxose(3-acetyl)-digitoxose-digitoxose-gitoxigenin.

Lanatoside C also has the same sugars found in both lanatosides A and B; however, the aglycone has the nucleus of lanatoside A plus an additional hydroxyl group at C-12 and is named digoxigenin. The structure sequence is glucose-digitoxose(3-acetyl)-digitoxose-digitoxose-digoxigenin.

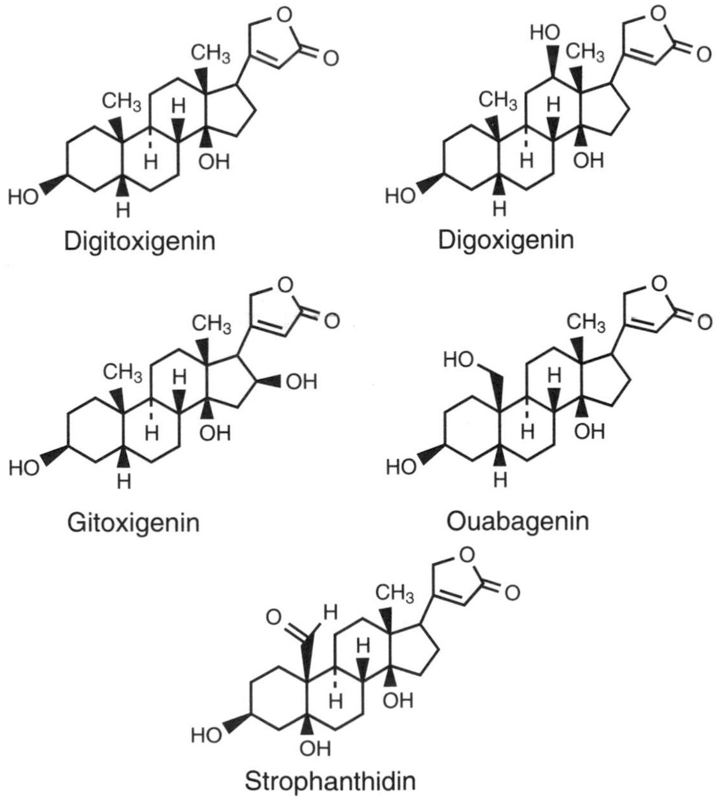

Digitoxigenin

Digoxigenin

Gitoxigenin

Ouabagenin

Strophanthidin

Fig. 19–2. Major cardenolide aglycones.

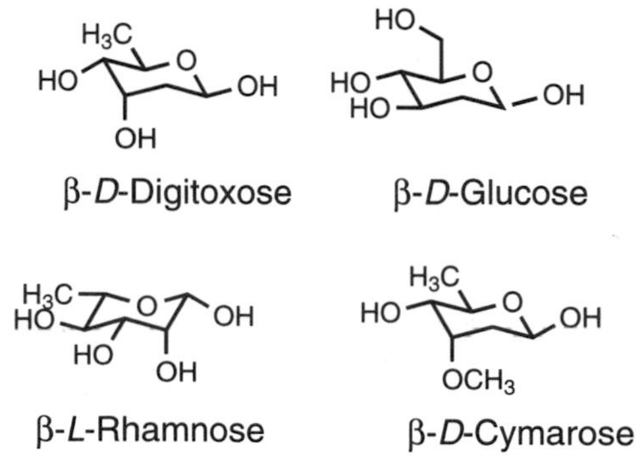

β-*D*-Digitoxose β-*D*-Glucose

β-*L*-Rhamnose β-*D*-Cymarose

Fig. 19–3. Selected sugars found in naturally occurring cardiac glycosides.

Partial hydrolysis of the glucose molecule and the acetyl group off the digilanides A and C produces two new, important cardiac glycosides known as digitoxin and digoxin with the following sequences: digitoxin—(digitoxose)$_3$-digitoxigenin—and digoxin—(digitoxose)$_3$-digoxigenin.

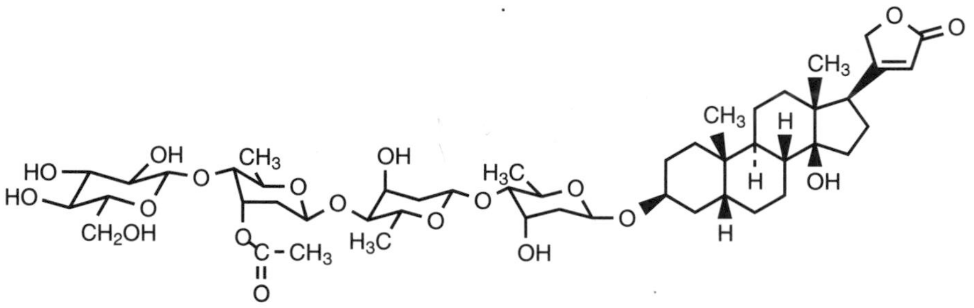

digilanides A and B. Their sequences are as follows: purpurea glycoside A—glucose-(digitoxose)$_3$-digitoxigenin—and purpurea glycoside B—glucose-(digitoxose)$_3$-gitoxigenin. There is no purpurea glycoside C.

Strophanthus gratus and Strophanthus kombé. The glycosides extracted from the plants Strophanthus gratus and Strophanthus kombé are called g-strophanthin (or ouabain) and k-strophanthoside. The corresponding aglycone for ouabain is ouabagenin and for k-strophanthoside is strophanthidin. The former has a polyhydroxylated steroidal nucleus, and the latter has an additional hydroxyl group at C-5 and an angular aldehyde group at C-10, replacing the traditional methyl group at that position (see Fig. 19–2). Ouabagenin is conjugated only to a single molecule of L-rhamnose, whereas strophanthidin is conjugated to a molecule of cymarose, which is further linked to two molecules of glucose.

The medicinally used preparations are mainly obtained from Digitalis purpurea and Digitalis lanata plants. These glycosides are generally referred to as digitalis glycosides or cardiac glycosides or simply as cardinolides. Strophanthus glycosides are not as widely used as digitalis preparations. Cardiac glycosides from animal sources (generally referred to as bufadienolides) are rare and of far less medicinal importance because of high toxicity. Pharmaceutical preparations of whole plants and partially hydrolyzed glycosides of Digitalis lanata and Digitalis purpurea

Lanatoside A

Digitalis purpurea. Purpurea glycosides A and B have identical structures to the lanatosides A and B but with no acetyl group on the third digitoxose. Therefore, purpurea glycosides A and B are sometimes called desacetyl

Table 19–1. Selected Natural Cardiac Glycosides and Their Sources

Source	Glycoside	Aglycone	Sugar*
Digitalis lanata (leaf)	Lanatoside A (digilanide A)	Digitoxigenin	Glucose
	Lanatoside B (digilanide B)	Gitoxigenin	Digitoxose (3-acetyl)
	Lanatoside C (digilanide C)	Digoxigenin	Digitoxose
			Digitoxose
Digitalis purpurea (leaf)	Purpurea glycoside A (desacetyl digilanide A)	Digitoxigenin	Glucose
	Purpurea glycoside B (desacetyl digilanide B)	Gitoxigenin	Digitoxose
			Digitoxose
			Digitoxose
Strophanthus gratus (seed)	G-Strophanthin	Ouabagenin	Rhamnose
Strophanthus kombé (seed)	K-Strophanthoside	Strophanthidin	Glucose
			Glucose
			Cymarose

* Conjugated with the C-3 hydroxyl of the aglycone via the bottom sugar. All sugars are conjugated via β-1,4-glucosidic bond.

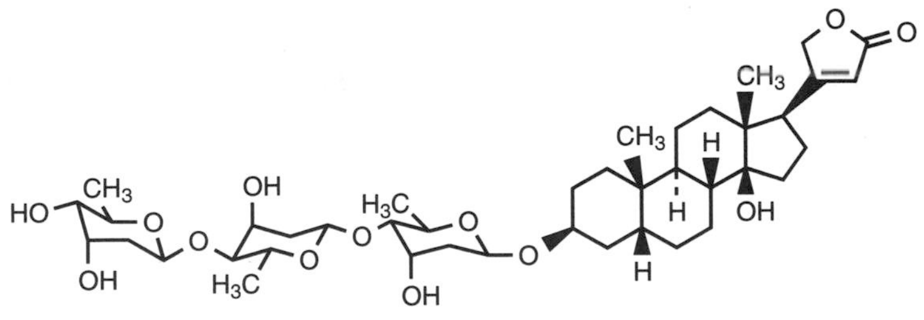

Digitoxin

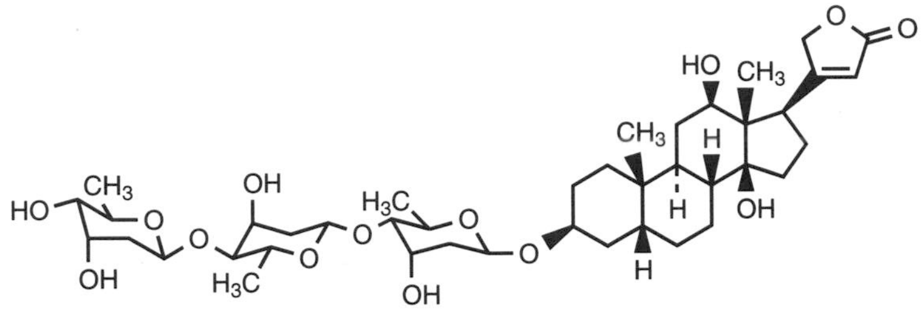

Digoxin

have been widely used clinically. The advancement in isolation and purification techniques, however, has made it possible to obtain highly purified digitoxin and digoxin preparations.

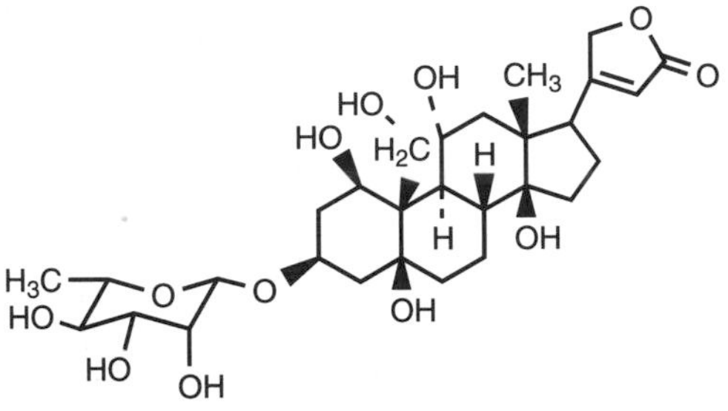

Ouabain

Pharmacology

Cardiac glycosides affect the heart in a dual fashion, both directly (on the cardiac muscle and the specialized conduction system of sinoatrial (S-A) node, atrioventricular (A-V) node, and His-Purkinje system), and indirectly (on the cardiovascular system mediated by the autonomic nerve reflexes). The combined direct and indirect effects of the cardiac glycosides lead to changes in the electrophysiologic properties of the heart, including alteration of the contractility; heart rate; excitability; conductivity; refractory period; and automaticity of the atrium, ventricle, Purkinje fibers, A-V node, and S-A node. The heart response to the cardiac glycosides is a dose-dependent process and varies considerably between normal and the congestive heart failure (CHF) diseased heart. The effects observed after the administration of low doses (therapeutic doses) differ considerably from those observed at high doses (cardiotoxic doses). The pharmacologic effects discussed subsequently relate mainly to therapeutic doses administered to CHF patients. The effects of cardiac glycosides on the properties of the heart muscle and different sites of the conduction system are summarized in Table 19–2. The increased force and rate of myocardial contraction (positive inotropic effect) and the prolongation of the refractory period of the A-V node are the most relevant effects on the CHF problem. Both of these effects are due to the direct action of the cardiac glycosides on the heart. The indirect effects are manifested as increased vagal nerve activity, which probably results from the glycoside-induced sensitization of the baroreceptors of the carotid sinus to changes in the arterial pressure; i.e., any given increase in the arterial blood pressure results in an increase in the vagal activity (parasympathetic) coupled with a greater decrease in the sympathetic activity. The vagal effect with uncompensated sympathetic response results in decreased heart rate and decreased peripheral vascular resistance (afterload). Therefore, cardiac glycosides reverse most of the symptoms associated with CHF as a result of increased sympathetic system activity, including increased heart rate, vascular resistance, and afterload. The administration of cardiac glycosides to a patient with CHF increases cardiac muscle contraction, reduces heart rate, and decreases both edema and the heart size.

Biochemical Mechanism of Action

The mechanism whereby cardiac glycosides cause a positive inotropic effect and electrophysiologic changes

Table 19–2. Effects of Cardiac Glycosides on the Heart*

	Atrium	Ventricle	Purkinje Fiber	A-V Node	A-S Node
Contractility	↑	↑	–	–	–
Excitability	0	Variable	↑	–	–
Conductivity	↑	↑	↓	↓	–
Refractory period	↓	↓	↑	↑	–
Automaticity	–	–	↑	–	↓

* Modified from *Drug Facts and Comparisons,* in *Facts and Comparisons,* 47th ed., St. Louis, C.V. Mosby, 1993.
↑ = Increased action; ↓ = decreased action; 0 = no action; – = no data available.

is still not completely known despite years of active investigation. Several mechanisms have been proposed, but the most widely accepted mechanism involves the ability of cardiac glycosides to inhibit the membrane-bound Na^+, K^+–adenosine triphosphatase (Na^+, K^+-ATPase) pump responsible for sodium/potassium exchange. To understand better the correlation between the pump and the mechanism of action of cardiac glycosides on the heart muscle contraction, one has to consider the sequence of events associated with cardiac action potential that ultimately lead to muscular contraction. The process of membrane depolarization/repolarization is controlled mainly by the movement of the three ions, Na^+, K^+, and Ca^{++}, in and out of the cell.

At the resting state (no contraction), the concentration of sodium is high outside the cell. On membrane depolarization, sodium fluxes-in, leading to an immediate elevation of the action potential. Elevated intracellular sodium triggers the influx of Ca^{++} which occurs slowly and is represented by the plateau region of the cardiac action potential. The influx of calcium results in efflux of potassium out of the myocardium. The Na^+/K^+ exchange occurs at a later stage of the action potential to restore the membrane potential to its normal level (for further detail, see antiarrhythmic agents and their classification at the end of this chapter). The Na^+/K^+ exchange requires energy and is catalyzed by the enzyme K^+, Na^+-ATPase. Cardiac glycosides are proposed to inhibit this enzyme with a net result of reduced sodium exchange with potassium, i.e., increased intracellular sodium, which, in turn, results in increased intracellular calcium. Elevated intracellular calcium concentration triggers a series of intracellular biochemical events that ultimately result in an increase in the force of the myocardial contraction, or a positive inotropic effect. The events that lead to muscle contraction are discussed in further detail elsewhere in this chapter under the mechanism of action of the calcium channel blockers.

The above-mentioned mechanism of the cardiac glycosides via inhibiting the Na^+, K^+-ATPase pump is in agreement with the fact that the action of the cardiac glycosides is enhanced by low extracellular potassium and inhibited by high extracellular potassium. The cardiac glycosides–induced changes in the electrophysiology of the heart can also be explained based on inhibition of Na^+, K^+-ATPase. It has been suggested that the intracellular loss of potassium owing to inhibition of the pump causes a decrease in the cellular transmembrane potential approaching zero. This decrease in the membrane potential is sufficient to explain the increased excitability and other electrophysiologic effects observed following cardiac glycosides administration.

Structural Requirements for Intrinsic Activity

Many hypotheses have been put forth to explain the cardiac glycoside structure-activity relationships (SAR). Some of the difficulty in arriving at a universally acceptable SAR model has been attributed to the early method of testing cardiac glycoside preparations and the lack of well-characterized cardiac glycosides "receptors." Until the early 1970s, nearly all cardiac glycosides were evaluated based on their cardiac toxicity rather than on more therapeutically relevant criteria. This was partly because of the belief that the cardiac toxicity was, in fact, an extension of the desired cardiotonic action. Thus, comparisons of cardiac glycoside preparations were based on the amount of drug required to cause cardiac arrest in test animals, usually anesthetized cats. More recently, most SAR studies rely, at least initially, on results obtained with isolated cardiac tissue or whole-heart preparations. In these models, inotropic activity, contractility, and so forth can be directly assessed. In addition, the recognition of cardiac Na^+, K^+-ATPase as the probable receptor for the cardiac glycosides has made the inhibition of this enzyme system an important criterion for the cardiac glycosides activity.

Much of the interest in the effects of structural modification on activity is due to the desire to develop agents with less toxic potential. Early studies based primarily on cardiac toxicity testing data suggested the importance of the steroid "backbone" shape, the 14-*β*-hydroxyl and the 17-unsaturated lactone for activity. More recent studies have been directed toward characterizing the interaction of the cardiac glycosides with Na^+, K^+-ATPase, the putative cardiac glycosides receptor. Using this enzyme model with enzyme inhibition as the biologic end point, a number of hypotheses for cardiac glycosides receptor binding interactions have been put forth. Many of these suggested that the 17-lactone plays an important role in receptor binding. Using synthetic analogs, it was found that unsaturation in the lactone ring was important, with the saturated lactone analog showing diminished activity.[3,4] Further investigations of synthetic compounds in which the lactone was replaced with open chain structures of varying electronic and steric resemblance to the lactone showed that, in fact, the α,β-unsaturated lactone ring at C-17 was not an absolute requirement and that several α,β-unsaturated open chain groups could be replaced with little or no loss in activity.[3,4] For example, analogs possessing an α,β-unsaturated nitrile at the 17-β position had high activity. In light of this, most current theories point to a key interaction of the carbonyl oxygen (or nitrile nitrogen) with the cardiac glycoside binding site on Na^+, K^+-ATPase.[5,6] Some controversy, however, exists over this point.

The importance of the "rest" of the cardiac glycoside molecule must not be ignored. Despite the apparently dominant role of the 17-substituent, it is the steroid (A-B-C-D) ring system that provides the lead structure for cardiac glycosides activity. Lactones alone, when not attached to the steroid ring system, show no Na^+, K^+-ATPase inhibitory activity. Some important steroid structural features have become apparent. The C-D *cis* ring juncture appears critical for activity in compounds possessing the unsaturated butyrolactone in the normal 17-β position. This apparent requirement may be a reflection of changes in the spatial orientation of the 17-substituent.[6] Moreover, the 14-β-OH is now believed to be dispensable. The contribution to activity previously attributed to this group is now thought to be related to the need to retain the sp^3 and *cis* character of the C-D ring juncture. The earlier interpretation arose from the fact that 14-deoxy analogs often had unsaturation in the D ring in place of the 14-OH. This double bond markedly influenced the position of the C-17 substituent, thereby complicating interpretation of 14-OH group importance. Finally, the A-B *cis* ring juncture appears not to be mandatory for cardiac glycosides activity. This feature, however, is characteristic of all clinically useful cardiac glycosides, and conversion to an A-B *trans* ring system generally leads to a marked drop in activity, unless compensating modifications are made elsewhere in the molecule.

Pharmaceutical Preparations

The clinically used digitalis preparations range from the powdered whole plants to purified individual glycosides, including lanatoside C, its partially hydrolyzed product deslanatoside C, and digitoxin and digoxin, which are the most commonly used digitalis preparations. Dosage forms and adult dosages of the cardiac glycosides are summarized in Table 19–3.

To arrive at an effective plasma concentration, often a large initial dose (i.e., digitalizing or loading dose) is given (Table 19–3). The purpose of this large initial dose is to achieve a therapeutic blood and tissue level in the shortest possible time. Depending on the condition of the patient and the desired therapeutic goal, the loading dose may be much less than the dose likely to cause toxicity, or almost equal to it. Once the desired effect is obtained, the amount of drug lost from the body per day is replaced with a maintenance dose. For digoxin, this is approximately 35% of the total body store, and for digitoxin it is about 10% (Table 19–3).

Absorption, Metabolism, and Excretion of Digitoxin and Digoxin

The therapeutic effects of all cardiac glycosides on the heart are qualitatively similar; however, the glycosides

Table 19–3. Cardiac Glycosides and Their Dosage Forms

Name	Dosage Forms	Typical Adult Dosages*
Digitoxin	Tablets: 0.05, 0.1, 0.15, 0.2 mg; injection: 1 ml ampules, 0.2 mg/ml	Loading dose: 0.8–1.2 mg total oral or IV (starting with 0.6 mg, then 0.4 mg at 6 h and 0.2 mg every 6 h thereafter as needed); maintenance: 0.05–0.2 mg orally per day
Digitalis (powdered digitalis leaf)	Tablets: 32.5, 48, 75, 65, 100 mg; capsules: 100 mg	Loading: 1.2 g total in divided doses at 6-h intervals; maintenance: 100–200 mg/day
Gitalin	Tablets: 0.5 mg	Loading: initially 2.5 mg, then 0.75 mg every 6 h for 24 h if needed; maintenance: 0.25–1.25 mg/day
Lanatoside C	Tablets: 0.5 mg	Loading: 3.5 mg on first day, 2.5 mg on second day, 2 mg on third day, then 1.5 mg/day until digitalized; maintenance: 0.5–1.5 mg/day
Deslanoside (desacetyl lanatoside C)	Injection: 2 ml ampules, 0.2 mg/ml	Loading: 1.6 mg IV or 1.6 mg IM at two different sites; not appropriate for maintenance therapy
Digoxin	Tablets: 0.125, 0.25, 0.5 mg; elixir, pediatric: 0.05 mg/ml; injection: 2-ml ampules, 0.25 mg/ml; injection, pediatric: 1-ml ampules, 0.1 mg/ml	Loading: IV, 0.5–1.0 mg total in increments of 0.25 mg at 6-h intervals; or loading: oral, 0.75–1.5 mg total in increments of 0.25–0.5 mg cautiously every 8 h; maintenance, IV, 0.125–0.5 mg/day; or maintenance, oral, 0.125–0.5 mg/day
Ouabain (G-strophanthin)	Injection: 2-ml ampules, 0.25 mg/ml	Loading: IV, 0.25 mg followed by 0.1 mg hourly until desired effect. Ouabain is not suitable for maintenance therapy

* All dosages are subject to modification according to patient condition and response.

largely differ in their pharmacokinetic properties. The latter are greatly influenced by the lipophilic character of each glycoside. In general, cardiac glycosides with more lipophilic character are absorbed faster and exhibit longer duration of action as a result of a slower urinary excretion rate. The lipophilicity of an organic compound is measured by its partitioning between two immiscible solvents (in most cases, chloroform and water or water mixed with an alcohol). The higher the concentration of the compound in the chloroform phase, the higher its partition coefficient, and the more lipophilic it is. The partition coefficients of five cardiac glycosides are listed in Table 19–4. It is evident that the lipophilicity is markedly influenced by the number of sugar molecules and the number of hydroxyl groups on the aglycone part of a given glycoside. Lanatoside C has a partition coefficient of 16.2, which is far less than that of acetyldigoxin (98), which differs only in lacking the terminal glucose molecule. Likewise, comparison of digitoxin and digoxin structures reveals that they differ only by an extra hydroxyl in digoxin at C-12. This seemingly minor difference results in a significant decrease in the partition coefficient from 96.5 to 81.5 for digitoxin and digoxin. This results in notable differences in the pharmacokinetic behavior of these agents. Table 19–4 also illustrates that the presence of an acetyl group on the sugar greatly enhances the lipophilic character of the glycoside (compare digoxin with acetyldigoxin with partition coefficients of 81.5 and 98). The glycoside G-strophanthin (ouabain) possesses a low lipophilic character owing to the presence of five free hydroxyl groups on the steroid nucleus of the aglycone ouabagenin.

Digitoxin and digoxin are the most frequently used therapeutic agents of the cardiac glycosides family. Absorption of digoxin and digitoxin from the gastrointestinal tract is a passive process that depends on the lipid solubility of the drug. Digitoxin has a greater lipid solubility than digoxin and is therefore more readily absorbed (see Table 19–4 to compare the lipophilicity of the two drugs). Absorption of digoxin following oral administration is variable and can range from 70% to 85% of an administered dose, depending on the size of the digoxin crystals. Therefore, different digoxin preparations may have different absorption characteristics. For this reason, it is important to establish carefully the effective dose of digoxin for each patient.

Table 19–4. Effect of Glycoside Structure on Partition Coefficient

Glycoside	Partition Coefficient ($CHCl_3$/ 16% aq.MeOH)
Lanatoside C (glucose-3-acetyldigitoxose-digitoxose$_2$-digoxigenin)	16.2
Digoxin (digitoxose$_3$-digoxigenin)	81.5
Digitoxin (digitoxose$_3$-digitoxigenin)	96.5
Acetyldigoxin (3-acetyldigitoxose-digitoxose$_2$-digoxigenin)	98.0
G-Strophanthin (rhamnose-ouabagenin)	Very low

Table 19–5. Pharmacokinetic Comparison of Digoxin and Digitoxin

	Digoxin	Digitoxin
Gastrointestinal absorption	70–85%	95–100%
Average half-life	1–2 days	5–7 days
Protein binding	25–30%	90–95%
Enterohepatic circulation	5%	25%
Excretion	Kidneys; largely unchanged	Liver metabolism
Therapeutic plasma level	0.5–2.5 ng/ml	20–35 ng/ml
Digitalizing dose (mg)	Oral: 0.75–1.5 IV: 0.5–1.0	Oral: 0.8–1.2 IV: 0.8–1.2
Maintenance dose (oral; mg)	0.125–0.5	0.05–0.2

Once the cardiac glycosides are absorbed, they bind to plasma protein. Digitoxin has 90% protein binding, whereas digoxin has only 30% binding. The half-life of digoxin in patients with normal renal function is 1.5 to 2 days, whereas that of digoxin ranges between 5 and 7 days. Approximately 25% of an absorbed dose of digitoxin is excreted in the bile unchanged and is reabsorbed. This enterohepatic circulation of digitoxin contributes to its long half-life. Biliary excretion of digoxin is minimal.

Digoxin is eliminated primarily unchanged by both glomerular filtration and tubular secretion by the kidney. Digitoxin, however, is extensively metabolized by the liver to a variety of metabolites, including small amounts of digoxin. The pharmacokinetic data of digoxin and digitoxin are summarized in Table 19–5.

Drug Interactions

The absorption of digoxin and digitoxin after oral administration can be significantly altered by other drugs concurrently present in the gastrointestinal tract. For example, laxatives may interfere with the absorption of digoxin because of increased intestinal motility. The presence of the drug cholestyramine, an agent used to treat hyperlipoproteinemia, decreases the absorption of both digoxin and digitoxin by binding to these drugs and retaining them in the gastrointestinal tract. Cholestyramine also alters the half-life duration of digitoxin by interrupting the entero/hepatic circulation. Antacids, especially magnesium trisilicate, and antidiarrheal adsorbent suspensions may also inhibit the absorption of the glycosides. Concurrent use of the cardiac glycosides with antiarrhythmics, calcium salts, sympathomimetics, β-adrenergic blockers, and calcium channel blockers may result in different degrees of arrhythmias. Potassium-depleting diuretics, such as thiazides, may increase the possibility of digitalis toxicity owing to the additive hypokalemia. Several other drugs that are known to bind to plasma proteins, such as thyroid hormones, have the potential to displace the glycosides, especially digitoxin, from their plasma-binding sites and thereby increase the free drug concentration to a toxic level. The metabolism of digitoxin may be accelerated by drugs that induce microsomal enzymes, such as phenobarbital, phenytoin, or phenylbutazone. Induction of the rate of metabolism results in a decreased digitoxin half-life.

Therapeutic Uses

Although the primary clinical use of the digitalis glycosides is for the treatment of CHF, these agents are used also in cases of atrial flutter or fibrillation and paroxysmal atrial tachycardia.

Toxicity

All cardiac glycosides preparations have the potential to cause toxicity. Because the minimal toxic dose of the glycosides is only two to three times the therapeutic dose, intoxication is quite common. In mild-to-moderate toxicity, the common symptoms are anorexia, nausea and vomiting, muscular weakness, bradycardia, and ventricular premature contractions. The nausea is a result of excitation of the chemoreceptor trigger zone (CTZ) in the medulla. In severe toxicity, the common symptoms are blurred vision, disorientation, diarrhea, ventricular tachycardia, and A-V block, which may progress into ventricular fibrillation. It is generally accepted that the toxicity of the cardiac glycosides is due to inhibition of the Na^+, K^+-ATPase pump, which results in increased intracellular levels of Ca^{++}. Hypokalemia (decreased potassium), which can be induced by coadministration of thiazide diuretics, glucocorticoids, or by other means can be an important factor in initiating a toxic response. It has been shown that low levels of extracellular K^+ partially inhibit the Na^+, K^+-ATPase pump. In a patient stabilized on a cardiac glycoside, the Na^+, K^+-ATPase pump is already partially inhibited, and the hypokalemia only further inhibits the pump, causing an intracellular buildup of sodium, which leads to an increase in intracellular calcium levels. The high levels of calcium are responsible for the observed cardiac arrhythmias characteristic of cardiac glycosides toxicity.

A common procedure used in treating cardiac glycosides toxicity is to administer potassium salts to increase extracellular potassium level, which stimulates the Na^+, K^+-ATPase pump, resulting in decreased intracellular sodium levels and thus decreased intracellular calcium. In treating any cardiac glycoside–induced toxicity, it is important to discontinue administration of the drug, in addition to administering a potassium salt. Other drugs that may be useful in treating the tachyarrhythmias present during toxicity are lidocaine, phenytoin, and propranolol. Specific antibodies directed toward digoxin (Dig-Bind) have been used experimentally and proven effective.

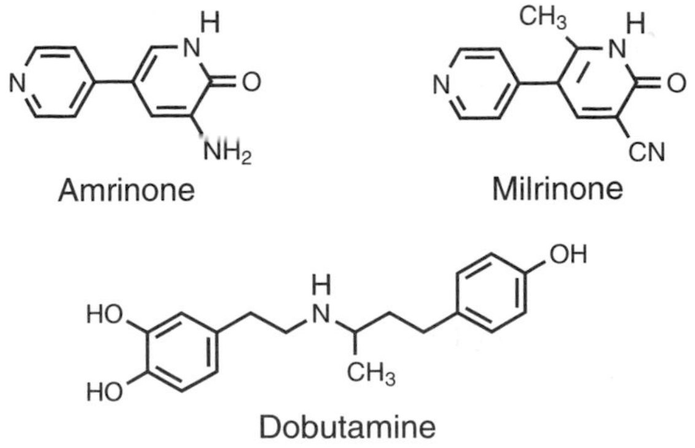

Fig. 19–4. Miscellaneous inotropic agents.

Additional Inotropic Agents

Although the digitalis glycosides are the principal therapeutic agents for the treatment of CHF, they are not the only positive inotropic agents available. Among several types of "nonglycoside" inotropic agents, a few have emerged as potentially useful drugs.[7] The first of these new agents was amrinone (Inocor) (Fig. 19–4), introduced in 1978. Amrinone produced both positive inotropic and concentration-dependent vasodilatory effects. Despite similar positive inotropic action of the cardiac glycosides, amrinone appears to act through a distinctly different mechanism. This agent and related compounds are thought to elicit their effects by the inhibition of a specific phosphodiesterase (phosphodiesterase fraction III) in the myocardium. This inhibition leads to elevated levels of cyclic adenosine monophosphate (cAMP), which through a complex chain of biochemical events leads to an increase in the intracellular Ca^{++} and ultimately an increase in muscle contractility. Amrinone was approved in 1984 by the Food and Drug Administration (FDA) for short-term intravenous administration in patients with severe heart failure refractory to other measures. The compound undergoes some conjugative metabolism in the liver and is excreted in the urine. Although amrinone is orally active, several adverse side effects have dampened enthusiasm for long-term oral amrinone therapy. These effects include gastrointestinal disturbances, thrombocytopenia, and impairment of the liver function.

The promising, although limited, success of amrinone led to the development of structurally related newer agents such as milrinone (Primacor) (see Fig. 19–4). The latter produces similar pharmacologic effects and probably acts through the same mechanism as amrinone. Milrinone, however, is of an order of magnitude more potent than amrinone. Furthermore, preliminary reports show it to be better tolerated, with no apparent thrombocytopenia or gastrointestinal disturbances. Milrinone is excreted largely unchanged in the urine, and accordingly patients with impaired renal function require reduced dosages.

Another promising area for the development of new positive inotropic agents is that of β-adrenergic receptor agonists. Adrenergic nervous system activity is mediated by two distinct receptor types, designated α and β. In addition, these receptors are further differentiated into α_1 and α_2 and β_1 and β_2 subtypes.[8] The distribution of these receptor subtypes depends greatly on the tissue in question. The myocardium has mostly β-adrenergic receptors of the β_1 subtype. It has been found that stimulation of these receptors by a variety of β-adrenergic agonists produces a potent positive inotropic response. The mechanism underlying this effect is believed again to involve elevation of cAMP levels, this time through the indirect stimulation of the enzyme adenylate cyclase. Elevated cAMP levels lead to a cascade of events ultimately producing an increase in intracellular Ca^{++} and thereby increased myocardial contractility. Although many agents possess β-adrenergic agonist activity, most have side effects that make them inappropriate for the treatment of CHF. For example, the well-known sympathetic amines, norepinephrine and epinephrine, are potent β-receptor agonists. Because the actions of these agents are not limited to the myocardial β-receptors, however, they produce undesirable positive chronotropic effects, exacerbate arrhythmias, and produce vasoconstriction. These effects limit their utility in the treatment of CHF.

Among the most promising β-adrenergic agonists are those derived from dopamine, the endogenous precursor to norepinephrine. Dopamine itself is a potent stimulator of the β-receptors, but it results in many of the undesirable side effects described previously. The new analogs of dopamine that have been developed retain the potent inotropic effect but possess fewer effects on heart rate, vascular tone, and arrhythmias. Dobutamine (Dobutrex) (see Fig. 19–4) is a prime representative of this group of agents. Dobutamine is a potent β_1-adrenergic agonist. Its beneficial effects, although largely attributed to its β_1 agonistic activity on the myocardium, are likely the composite of a variety of actions on the heart and the peripheral vasculature. Dobutamine is active only by the intravenous route because of the rapid first-pass metabolism. Therefore, its use is limited to critical care situations. Nonetheless, its success has led to the development of several new orally active drugs presently in clinical trials. One of the major limitations associated with β_1 agonists is the phenomenon of myocardial β-receptor desensitization. This lowered responsiveness of the receptors appears to be due to a decrease in the number of β_1 receptors and partial uncoupling of the receptors from adenylate cyclase. Table 19–6 lists dosage forms and typical adult doses of nonglycosidic inotropic agents.

DRUGS FOR THE TREATMENT OF ANGINA

Disease State

Angina pectoris is the disease of the coronary artery. The latter is the supply route of blood carrying oxygen from the left ventricle to all heart tissues, including the ventricles themselves. When the coronary artery becomes less efficient in supplying blood and oxygen to the heart, the heart is said to be ischemic (short in oxygen). Angina is the primary symptom of ischemic heart disease, characterized by a sudden, severe pain originating in the chest, often radiating to the left shoulder and down the left arm. Angina is further subclassified into typical or variant angina based on the precipitating factors and the electrophysiologic changes observed during the attack. Typical angina usually is the result of an advanced state of atherosclerosis and is provoked by food, exercise, and emotional factors. It is characterized by low S-T segment of the electrocardiogram. Variant or acute angina results from sudden spasm in the coronary artery, unrelated to atherosclerotic narrowing of the coronary circulation, and can occur at rest. It is characterized by an increase in the S-T segment of the electrocardiogram.

Antianginal Drugs

Therapy of angina is directed mainly toward alleviating and preventing anginal attacks by dilating the coronary artery. Three classes of drugs are found to be efficient in this regard, although via different mechanisms. These include organic nitrates, calcium channel blockers, and β-adrenergic blockers.

Organic Nitrates

Organic nitrates have dominated the treatment of acute angina over the last 100 years. Although the introduction of the calcium channel blockers and the β-blockers as antianginal agents has expanded the physician's therapeutic arsenal, organic nitrates are still the class of choice to treat acute anginal episodes.

Chemistry. Organic nitrates are esters of simple organic alcohols or polyols with nitric acid. This class was developed after the antianginal effect of amyl nitrite (ester of isoamyl alcohol with nitrous acid) was first observed in 1857. Five members of this class are in clinical use today: amyl nitrite (amyl nitrite inhalant USP), nitroglycerin (e.g., Nitro-Bid, Deponit, Nitro-Dur, Nitrogard),

Table 19–6. Nonglycosidic Positive Inotropic Agents

Name	Dosage Forms	Typical Adult Dosage*
Amrinone lactate	Injection: 20 ml ampules, 5 mg/ml	Initially 0.5 mg/kg IV over 2–3 min, then 5–10 μg/kg/min IV infusion (not to exceed 10 mg/kg per day)
Dobutamine hydrochloride	Injection: 20 ml vials, 12.5 mg/ml (250 mg total)	IV infusion: 2.5–10 μg/kg/min
Milrinone lactate	Injection: 5 ml, 1 mg/ml	Initially 0.05 mg/kg every 10 min Maintenance 0.59–1.13 mg/kg over 24 hr

* All dosages are subject to modification according to patient condition and response.

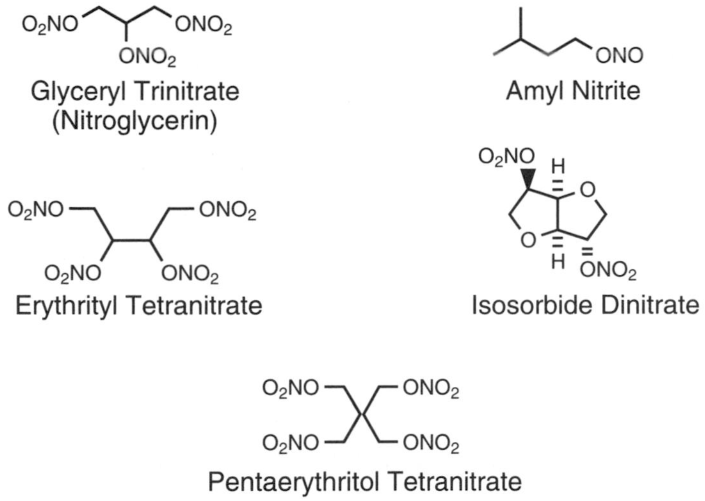

O_2NO ONO_2 ONO_2
Glyceryl Trinitrate (Nitroglycerin)

ONO
Amyl Nitrite

O_2NO ONO_2 O_2NO ONO_2
Erythrityl Tetranitrate

O_2NO H O O H ONO_2
Isosorbide Dinitrate

O_2NO ONO_2 O_2NO ONO_2
Pentaerythritol Tetranitrate

Fig. 19–5. Organic nitrates and nitrites.

isosorbide dinitrate (e.g., Iso-Bid, Sorbitrate, Novosorbide, Isordil, Dilatrate), erythrityl tetranitrate (Cardilate), and pentaerythritol tetranitrate (e.g., Duotrate, Pentylan, Peritrate). The chemical structures of these agents are shown in Figure 19–5. This class is usually referred to as organic nitrates because all of these agents (except amyl nitrites are nitrate esters. It should be noted that the generic names do not always precisely describe the chemical nature of the drug but are used for simplicity. For example, the drug nitroglycerin is not really a nitro compound because nitro means a nitro group attached to a carbon atom, i.e., NO_2-C; the correct chemical name of nitroglycerin is glyceryltrinitrate. Another example is amyl nitrite, whose structure indicates that it is an ester of isoamyl alcohol with nitrous acid; the correct chemical name of this drug is isoamyl nitrite.

The chemical nature of these molecules as esters constitutes some problems in formulating these agents for clinical use. The small nonpolar ester character makes them volatile. Volatility is an important concern in drug formulation owing to the potential loss of the active principle from the dosage form. In addition, moisture should be avoided during storage to minimize the hydrolysis of the ester bond, which can lead to a decrease in the therapeutic effectiveness. Lastly, because these agents are nitrate esters, they possess explosive properties, especially in the pure concentrated form. Dilution in a variety of vehicles and excipients eliminates this potential hazard. The nonpolar nature of these esters, however, makes these agents efficient in emergency treatment of anginal episodes as a result of rapid absorption through biomembranes.

Pharmacologic Actions. The oxygen requirements of the myocardial tissues are related to the workload of the heart, which is, in part, a function of the heart rate, the systolic pressure, and the peripheral resistance of the blood flow. Myocardial ischemia occurs when the oxygen is insufficient to meet the myocardial workload. This can occur, as explained previously, because of atherosclerotic narrowing of the coronary circulation (typical) or vasospasm of the coronary artery (variant). The nitrates have been shown to be effective in treating angina resulting from either cause. The vasodilating effect of organic nitrates on the veins leads to pooling of the blood in the veins and decreased venous return to the heart (decreased preload), whereas vasodilation of the arteries decreases the resistance of the peripheral tissues (decreased afterload). The decrease in both preload and afterload results in a generalized decrease in the myocardial workload, which translates into a reduced oxygen demand by the myocardium. Organic nitrates restore the balance between oxygen supply (by vasodilating the coronary artery) and oxygen demand (by decreasing the myocardial workload).

Biochemical Mechanism of Action. Although the physiologic effects of the organic nitrates are clearly due to its vasodilating effect on the vascular smooth muscles, the underlying molecular mechanism remains elusive. The presence of nitrate receptors on vascular smooth muscle has been postulated by Needleman and Johnson.[9] These investigators suggested that the nitrate-receptor interaction is accompanied by the oxidation of a critical receptor sulfhydryl group, which initiates vascular relaxation. The involvement of free tissue sulfhydryl groups was supported by experimental evidence showing that prior administration of *N*-acetylcysteine, which should increase the availability of free sulfhydryl groups, resulted in an increase in the vasodilating effect of organic nitrates. Similarly, pretreatment with reagents that react with free sulfhydryl groups, such as ethacrynic acid, blocked glyceryl trinitrate vasodilation in vitro.[10] A more complex mechanism for nitrate vasodilation, however, was proposed by Ignarro and co-workers.[11] It was suggested that the nitrates act indirectly by stimulating the enzyme guanylate cyclase, thereby producing elevated levels of cyclic guanosine monophosphate (cGMP), which, in turn, leads to vasodilation. The initial stimulation of guanylate cyclase is believed to be mediated by a nitrate-derived nitrosothiol metabolite produced intracellularly. In support of this mechanism is the observation that a variety of synthetic nitrosothiols were found to increase markedly guanylate cyclase activity and produce vasodilation in vitro.[12–16]

Table 19–7. Antianginal Nitrates and Nitrites

Name	Dosage Forms	Typical Adult Dosage*
Amyl nitrite (isoamylnitrite)	Inhalant: 0.18, 0.3 ml	0.18 or 0.3 ml inhaled as needed
Nitroglycerin (glyceryl trinitrate)	Sublingual tablets: 0.15, 0.3, 0.4, 0.6 mg; extended-release capsules/tablets; 2.5, 2.6, 6.5, 9.0, 13 mg; ointment: 2%; transdermal systems: many (e.g., 26, 51, 77, 104, 154-mg disks that release 2.5, 5.0, 7.5, 10, 15 mg/24 h)	0.15–0.6 mg sublingually; 2.5–9 mg orally, extended-release every 8–12 hours; 1.25–5 cm topical ointment every 4–8 hours; one transdermal disk per day (2.5–15 mg/day)
Isosorbide dinitrate	Sublingual tablets: 2.5, 5, 10 mg; sustained-release tablets/capsules: 40 mg; chewable tablets: 5, 10 mg; oral tablets: 5, 10, 20, 30 mg	2.5–10 mg sublingually every 4–6 hours; 40 mg sustained-release every 6–12 hours; 5–10 mg chewable every 2–3 hours; 5–30 mg orally every 6 hours
Erythrityl tetranitrate	Oral/sublingual tablets: 5, 10, 15 mg; chewable tablets: 10 mg	5–10 mg oral/sublingual three times a day; 10 mg chewable three times a day
Pentaerythritol tetranitrate	Oral tablets: 10, 20, 40 mg; sustained-release tablets/capsules: 30, 45, 60, 80 mg	10–40 mg orally four times a day; 30–80 mg sustained-release every 12 hours

* All dosages are subject to modification according to patient condition and response.

Such a mechanism is consistent with the requirement for free sulfhydryl groups described previously. A unifying mechanism suggests that the organic nitrates through the release of nitrous oxide activate guanylate cyclase, causing an increase in cGMP, which, in turn, reduces the Ca^{++} catalyzed vascular contraction.[17–20] It is clear that much research is required to clarify this complex process.

Pharmaceutical Preparations and Dosage Forms. Organic nitrates are administered by inhalation; by injection; as sublingual, chewable, and sustained-release tablets; as capsules; as transdermal disks; and as ointments. Table 19–7 summarizes different dosage forms and typical adult dosages of organic nitrates as antianginal drugs.

Absorption, Metabolism, and Therapeutic Effects. Organic nitrates are used for both treatment and prevention of painful anginal attacks. The therapeutic approaches to achieve these two goals, however, are distinctly different. For the treatment of acute anginal attacks, that is, attacks that have already begun, a rapid-acting preparation is required. In contrast, preventive therapy requires a long-acting preparation with more emphasis on duration and less emphasis on onset. The onset of organic nitrate action is influenced not only by the specific agent chosen, but also by the route of administration. Sublingual administration is used predominantly for a rapid onset of action. The duration of nitrate action is strongly influenced by metabolism. All of the organic nitrates are subject to fairly rapid first-pass metabolism, not only in the liver by the action of glutathione-nitrate reductase, but also in extrahepatic tissues, such as the blood vessels themselves.[21,22] In addition, avid uptake into the vessel walls plays a significant role in the rapid disappearance of organic nitrates from the blood stream. Sublingual, transdermal, and bucal administration routes have been used in an attempt to avoid at least some of the hepatic metabolism.

Acute angina is most frequently treated with sublingual glyceryl trinitrate. This sublingual preparation is rapidly absorbed from the sublingual, lingual, and buccal mucosa and provides relief usually within 2 minutes. The duration of action is also short, lasting about 30 minutes. Other treatments include amyl nitrite by inhalation and sublingual isosorbide dinitrate. Amyl nitrite is by far the fastest-acting preparation, with an onset of about 15 to 30 seconds, but lasts only about 1 minute. Isosorbide dinitrate, although usually used as a long-acting agent, may be used to treat acute angina. Sublingually administered isosorbide dinitrate has a somewhat slower onset than glyceryl trinitrate (about 3 minutes), but its action may last for 4 to 6 hours. Although the onset appears to be almost as rapid as that of glyceryl trinitrate, to wait an additional minute for relief may be deemed unacceptable by some patients.

To prevent recurring angina, long-acting organic nitrate preparations are used. Several agents fall into this category, such as orally administered isosorbide dinitrate, pentaerythritol tetranitrate, and erythrityl tetranitrate. In addition, a number of long-acting glyceryl trinitrate preparations are available. These include oral sustained-release forms, glyceryl trinitrate ointment, transdermal patches, and buccal tablets. Of these therapeutic options, isosorbide dinitrate and glyceryl trinitrate preparations are by far the most frequently used. The whole concept of prophylactic nitrate use was at first met with skepticism by many physicians because early studies indicated that oral nitrates were nearly completely broken down by first-pass metabolism,[21,22] and blood levels of the parent drug appeared to be virtually nil. These findings, in conjunction with several clinical studies showing equivocal efficacy, led Needleman and co-workers[22] to conclude that "[T]here is no rational basis for the use of 'long-acting' nitrates (administered orally) in the prophylactic therapy of angina pectoris." More recent studies, however, suggest that oral prophylactic nitrates may be effective if appropriate doses are used.[23] Moreover, some metabolites of long-acting nitrates are active as vasodilators, albeit less potent than the parent drug. Isosorbide dinitrate is an example of this. Isosorbide dinitrate is metabolized primarily in the liver by glutathione-nitrate reductase. This enzyme, which also participates in the metabolism of other organic nitrates, catalyzes the denitration of the parent drug to yield two metabolites, 2- and 5-isosorbide mononitrate.[24] Of these, the 5-isomer is still a potent vasodilator, and its plasma half-life of about 4.5 hours is much longer than that of isosorbide dinitrate itself. The extended half-life, owing to the metabolite's resistance to further metabolism, has suggested that it may be contributing to the prolonged duration of action associated with isosorbide dinitrate use.[24]

Adverse Effects. Most patients tolerate the nitrates fairly well. Headache and postural hypotension are the most common side effects of organic nitrates. Dizziness, nausea, vomiting, rapid pulse, and restlessness are among the additional side effects reported. These symptoms may be controlled by administering low doses initially and gradually increasing the dose. Fortunately, tolerance to nitrate-induced headache develops after a few days of therapy. Because postural hypotension may occur in some individuals, it is wise to advise the patient to sit down when taking a rapid-acting nitrate preparation for the first time. An effective dose of nitrate usually produces a fall in upright systolic pressure of 10 mm Hg and a reflex rise in heart rate of 10 beats per minute.

Another concern associated with prophylactic nitrate use is the development of tolerance.[23,25] Tolerance, usually in the form of shortened duration of action, is commonly observed with long-term nitrate use. The clinical importance of this tolerance is, however, a matter of controversy. Because tolerance to nitrates has not been reported to lead to a total loss of activity, some physicians believe it is not clinically relevant. In addition, an adjustment in dosage can compensate for the reduced response.[23] It has also been reported that intermittent use of long-acting and sustained release preparations may limit the extent of tolerance development.

Drug Interactions. The most significant interactions of organic nitrates are with those agents that cause hypotension, such as other vasodilators, alcohol, and tricyclic antidepressants, in which a potential of orthostatic hypotension may arise. Concurrent administration with sympathomimetic amines, such as ephedrine and norepinephrine, may lead to a decrease in the antianginal efficacy of the organic nitrates.

Calcium Channel Blockers

The second major therapeutic approach to the treatment of angina is the use of calcium channel blockers.[26,27] The recognition that inhibition of calcium ion (Ca^{++}) influx into myocardial cells may be advantageous in preventing angina occurred in the 1960s. Three classes of calcium channel blockers are currently approved for use in the prophylactic treatment of angina, the dihydropyridines nifedipine (Adalat, Procardia), nicardipine (Cardene), and amlodipine (Norvasc); the benzothiazepine derivative diltiazem (Cardizem); and the aralkyl amine derivatives verapamil (Calan, Isoptin) and bepridil (Vascor) (Fig. 19–6). The last-mentioned are reserved for treatment failures in that serious arrhythmias may occur.

Chemistry. The structural dissimilarity of these agents is apparent and serves to emphasize the fact that each is distinctly different from the others in its profile of effects. Although nifedipine and similar drugs belong to the so-called dihydropyridine family, diltiazem belongs to the benzo[*b*-1,5]thiazepine family. Verapamil is structurally characterized by a central basic nitrogen to which alkyl and aralkyl groups are attached. It is noteworthy that these latter two compounds are chiral, possessing asymmetric centers. In each case, the dextrorotatory, i.e., the (+), isomer is about an order of magnitude more potent as a calcium channel blocker than the levorotatory, i.e., the (−), isomer.

Pharmacologic Effects. Calcium ions are known to play a critical role in many physiologic functions. Physiologic calcium is found in a variety of locations, both intracellular and extracellular. Because calcium plays such a ubiquitous role in normal physiology, the overall therapeutic effect of the calcium channel blockers is often the composite of numerous pharmacologic actions in a variety of tissues. The most important of these tissues associated with angina are the myocardium and the arterial vascular bed. Because of the dependency of the myocardium contraction on calcium, these drugs have a negative inotropic effect on the heart. Vascular smooth muscle also depends on calcium influx for contraction. Although the underlying mechanism is somewhat different, inhibition of calcium channel influx into the vascular muscles by the calcium channel blockers leads to vasodilation, particularly in the arterial smooth muscles. The venous beds appear to be less affected by the calcium channel blockers. The negative inotropic effect and arterial vasodilation result in decreased heart workload and afterload. The preload is not affected because of lesser sensitivity of the venous bed to the calcium channel blockers.

Mechanism of Action. The depolarization and contraction of the myocardial cells are mediated in part by calcium influx. As explained before, the overall process consists of two distinct inward ion currents. The first of these is the rapid flow of sodium ions into the cell through the "fast channel." Then calcium enters more slowly through the "slow channel." The calcium ions trigger contraction indirectly by binding and inhibiting troponin, a natural suppressor of the contractile process. Once the inhibitory effect of troponin is removed, actin and myosin can interact to produce the contractile response. The calcium channel blockers produce negative inotropic effect by interrupting the latter process. In vascular smooth muscles, calcium causes constriction by binding to calmodulin, a specific intracellular protein, to form a complex that initiates the process of vascular constriction. The calcium channel blockers inhibit vascular smooth muscle contraction by depriving the cell of the calcium ions.

The effects of the three classes of calcium channel blockers on the myocardium and the arteries varies from one class to the other. Although verapamil and diltiazem affect both the heart and the arterial bed, the dihydropyridines have much less effect on the cardiac tissues and

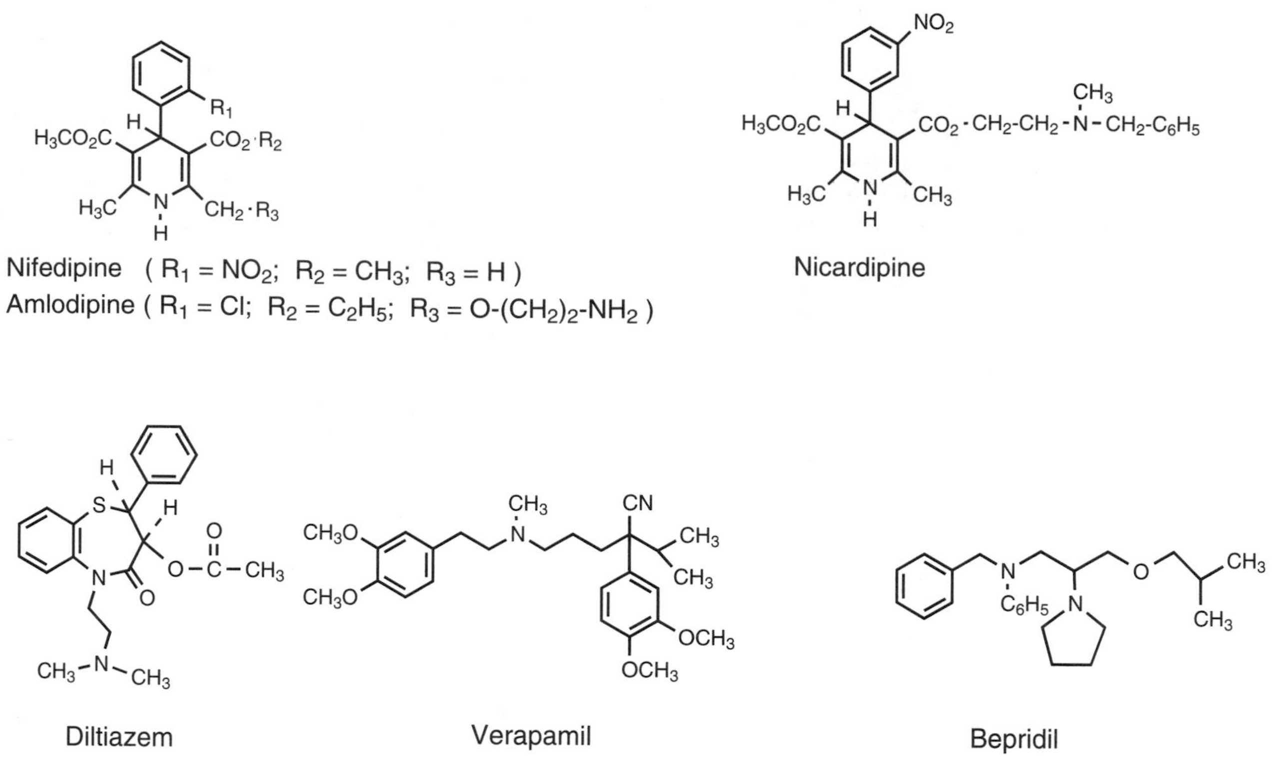

Nifedipine ($R_1 = NO_2$; $R_2 = CH_3$; $R_3 = H$)
Amlodipine ($R_1 = Cl$; $R_2 = C_2H_5$; $R_3 = O\text{-}(CH_2)_2\text{-}NH_2$)

Nicardipine

Diltiazem

Verapamil

Bepridil

Fig. 19–6. Calcium channel blockers.

higher specificity for the vascular bed. Therefore, both verapamil and diltiazem are clinically used in the management of angina, hypertension, and cardiac arrhythmia, whereas the dihydropyridines are more frequently used as antianginal and antihypertensive agents. Because nicardipine has a less negative inotropic effect than nifedipine, it may be preferred over nifedipine for patients with angina pectoris or hypertension who also have CHF dysfunction.

The recognition of the pivotal role of calcium flux on biologic functions led to the reexamination of several therapeutic agents already in clinical use to determine if their effects were also mediated through calcium-dependent mechanisms. Interestingly, many drugs were found to influence calcium movement and availability. In many cases, however, this effect was not found to contribute significantly to the desirable pharmacologic activity, with other mechanisms playing more dominant roles.

Pharmaceutical Preparations. Calcium channel blockers are administered as oral tablets and capsules. Preparations are available as regular and sustained-release forms. Verapamil is also administered by injection. The pharmaceutical dosage forms and the corresponding adult doses are summarized in Table 19–8.

Absorption, Metabolism, and Excretion. The calcium channel blockers are rapidly and completely absorbed after oral administration. First-pass metabolism occurs extensively, especially for verapamil, leading to low bioavailability (verapamil, 10 to 30%, diltiazem 30 to 60%, nicardipine 15 to 40%, nifedipine 50 to 60%, amlodipine 60 to 65%). Verapamil is metabolized by *N*-demethylation to norverapamil, which retains approximately 20% of ver-

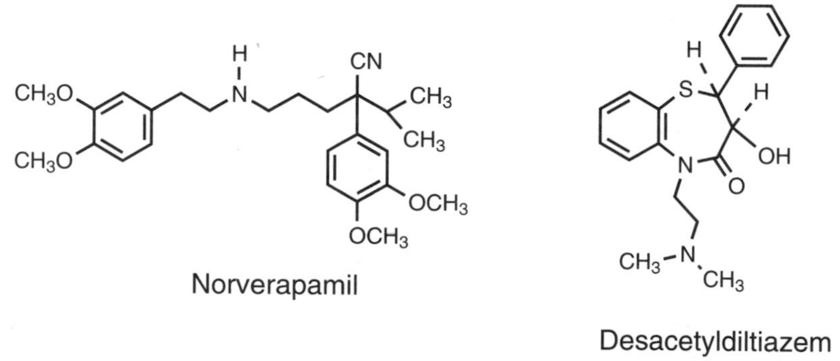

Norverapamil

Desacetyldiltiazem

Table 19–8. Calcium Channel Blockers and Their Dosage Forms

Name	Dosage Forms	Typical Adult Dosage*
Verapamil	Tablets: 80, 120 mg; injection: 5 mg/2 ml	IV: 5–10 mg over 2 minutes; orally: 80 mg three or four times a day (usually 320–480 mg/day)
Diltiazem	Tablets: 30, 60 mg	30 mg orally four times a day, then if necessary, increased to 60 mg, three or four times a day
Nifedipine	Capsules: 10 mg	10 mg orally three times a day, may be increased to 20 mg three times daily, not to exceed 180 mg/day
Nicardipine	Capsules: 20 mg	20 mg three times a day; allow three days before increasing the dose to 40 mg three times a day
Amlodipine	Tablets: 2.5, 5, 10 mg	5–10 mg/day
Bepridil	Tablets: 200, 300, 400 mg	Initially 200 or 300 mg/four times a day with maximum 400 mg/day

* All dosages are subject to modification according to patient condition and response.

apamil activity, and by O-demethylation into inactive metabolites. Diltiazem is metabolized to the desacetyl derivative, which has approximately 25 to 50% of the diltiazem activity. The dihydropyridines are metabolized largely to a variety of inactive metabolites. Binding to plasma protein is high, ranging from 70 to 98% depending on the individual agent: verapamil 90%, diltiazem 70 to 80%, nifedipine 92 to 98%, nicardipine greater than 90%, amlodipine greater than 90%. Little of these agents is excreted unchanged in urine (0 to 4%). The duration of action of the calcium channel blockers ranges from 4 to 8 hours (verapamil 4 hours, diltiazem 6 to 8 hours, nifedipine 4 to 8 hours, nicardipine 6 to 8 hours). Amlodipine has a 24-hour duration of action. Thus, it is the only calcium channel blocker that can be given once daily as non–sustained-release product.

Adverse Effects. The most common side effects of the calcium channel blockers include dizziness, hypotension, headache, and peripheral and pulmonary edema. These symptoms are mainly related to the excessive vasodilation, especially with the dihydropyridines. Verapamil was reported to cause constipation in some patients.

Drug Interactions. With other vasodilators, antihypertensive drugs, and alcohol, excessive hypotension may arise owing to an additive effect. The high protein-binding nature of these drugs precipitates a potential of mutual plasma displacement with other drugs known to possess the same property, such as oral anticoagulants, digitalis glycosides, oral hypoglycemic agents, sulfa drugs, and salicylates. Dose adjustment may be necessary in some cases.

Therapeutic Uses. Calcium channel blockers are used clinically as antianginal, antiarrhythmic, and antihypertensive agents.

β-Adrenergic Blocking Agents

The use of β-adrenergic blockers as antianginal agents is limited to the treatment of exertion-induced angina. Propranolol is the prototype drug in this class, but several newer agents have been approved for clinical use in the United States (see Chapter 18). Although these agents may be used alone, they are often used in combination therapy with nitrates, calcium channel blockers, or both. In several instances, combination therapy was found to provide more improvement than did either agent alone. This, however, is not always the case.

Miscellaneous Coronary Vasodilators

Another approach to the treatment of myocardial insufficiency is the use of the coronary vasodilators dipyridamole, papaverine, and cyclandelate. Dipyridamole (Fig. 19–7) causes a long-acting and selective coronary vasodilation, presumably through inhibition of adenosine uptake by the red blood cells and vasculature. Adenosine is a natural vasodilatory substance released by the myocardium during hypoxic episodes. Inhibition of its uptake is therefore believed to prolong its vasodilatory effects. Some structural similarity of adenosine to dipyridamole is apparent and substantiates this mechanism. Dipyridamole is generally used prophylactically, but its efficacy in reducing the incidence and severity of anginal attacks is not universally accepted. Cyclandelate and papaverine (see Fig. 19–7), although structurally unrelated to one another, appear to act similarly by the relaxation of vascular smooth muscle. The exact mechanism of action of these agents is unclear. Despite few studies supporting their efficacy, they still see clinical use.

DRUGS FOR THE TREATMENT OF CARDIAC ARRHYTHMIA

Disease State

Arrhythmia is an alteration in the normal sequence of electrical impulse activation that leads to the contraction of the myocardium. It is manifested as an abnormality in the rate, in the site from which the impulses originate, or in the conduction through the myocardium. This process is controlled by the so-called pacemaker cells in the A-V and S-A nodes; however, both the atria and the ventricles are also involved.

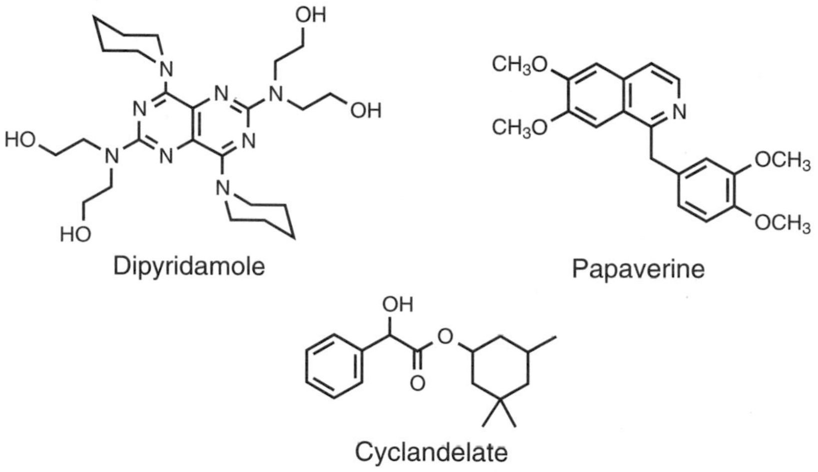

Dipyridamole

Papaverine

Cyclandelate

Fig. 19–7. Miscellaneous coronary vasodilators.

Causes of Arrhythmias

Many factors influence the normal rhythm of electrical activity in the heart. Arrhythmias may occur because pacemaker cells fail to function properly or because of a blockage in transmission through the A-V node. Underlying diseases such as atherosclerosis, hyperthyroidism, or lung disease may also be initiating factors. Some of the more common arrhythmias are those termed ectopic, which occur when electrical signals spontaneously arise in regions other than the pacemaker and then compete with the normal impulses. Myocardial ischemia, excessive myocardial catecholamine release, stretching of the myocardium, and cardiac glycoside toxicity have all been shown to stimulate ectopic foci. A second mechanism for the generation of arrhythmias is from a phenomenon called reentry. This occurs when the electrical impulse does not die out after firing but continues to circulate and reexcite resting heart cells into depolarizing. The result of this reexcitation may be a single, premature beat or runs of ventricular tachycardia. Reentrant rhythms are common in the presence of coronary atherosclerosis.

Antiarrhythmic Agents and Their Classification

It is widely accepted that most currently available antiarrhythmic drugs may be classified into four categories, grouped based on their effects on the cardiac action potential (Table 19–9) and consequently on the electrophysiologic properties of the heart. To understand the basis of classification and the pharmacology of these agents, an understanding of normal cardiac electrophysiology is necessary.

Normal cardiac contractions are largely a function of the action of a single atrial pacemaker, a fast and generally uniform conduction in predictable pathways, and a normal duration of the action potential and refractory period. Figure 19–8 depicts a normal cardiac action potential from a Purkinje fiber. The resting cell has a membrane potential of approximately −90 mV, with the inside of the cell being electronegative relative to the outside of the cell. This is called the transmembrane resting potential. On excitation, the transmembrane potential reverses, and the inside of the membrane rapidly becomes positive with respect to the outside. On recovery

Table 19–9. Classification of Antiarrhythmic Agents

Class	Agent	Primary Pharmacologic Effect
IA	Quinidine Procainamide Disopyramide	Decrease maximal rate of depolarization; increase duration of action potential
IB	Lidocaine Phenytoin Tocainide Mexiletine	Decrease maximal rate of depolarization; decrease duration of action potential
IC	Flecainide	Decrease maximal rate of depolarization; no change in duration of action potential
II	Propranolol	Inhibition of sympathetic activity
III	Bretylium Amiodarone	Prolongation of duration of action potential
IV	Verapamil	Inhibition of inward slow calcium current

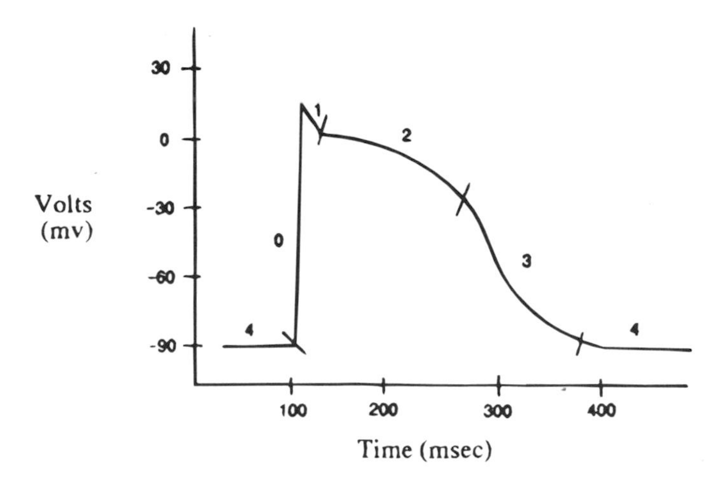

Fig. 19–8. Cardiac action potential recorded from a Purkinje fiber.

from excitation, the resting potential is restored. These changes have been divided into five phases: Phase 0 represents depolarization and reversal of the transmembrane potential, phases 1 to 3 represent different stages of repolarization, and phase 4 represents the resting potential. During phase 0, also referred to as rapid depolarization, the permeability of the membrane for sodium ions increases, and sodium rapidly enters the cell, causing it to become depolarized. Phase 1 results from the ionic shift, which creates an electrochemical and concentration gradient that reduces the rate of sodium influx but favors the influx of chloride and efflux of potassium. Phase 2, the plateau phase, results from the slow inward movement of calcium, which is triggered by the rapid inward movement of sodium in phase 0. During this time, there is also an efflux of potassium that balances the influx of calcium, thus resulting in little or no change in membrane potential. Phase 3 is initiated by a slowing of the calcium influx coupled with a continued efflux of potassium. This continued efflux of potassium from the cell restores the membrane potential to normal resting potential levels. During phase 4, the Na^+, K^+-ATPase pump restores the ions to their proper local concentrations. The action potential is a coordinated sequence of ion movements in which sodium initially enters the cell, followed by a calcium influx, and finally a potassium efflux that returns the cell to its resting state. Several antiarrhythmic agents exert their effects by altering these ion fluxes.

Class I drugs are generally local anesthetics acting on nerve and myocardial membranes to slow conduction. Myocardial membranes show the greatest sensitivity. Class I drugs decrease the maximal rate of depolarization without changing the resting potential. They also increase the threshold of excitability, increase the effective refractory period, decrease conduction velocity, and decrease spontaneous diastolic depolarization in pacemaker cells. The decrease in diastolic depolarization tends to suppress ec-

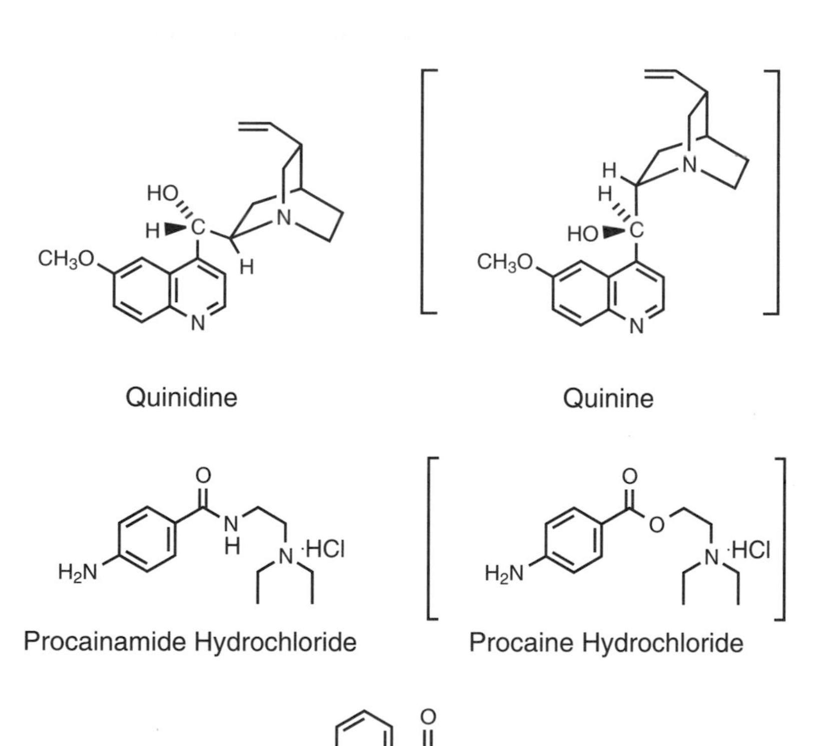

Fig. 19–9. Class IA antiarrhythmics (excluding agents shown in parentheses).

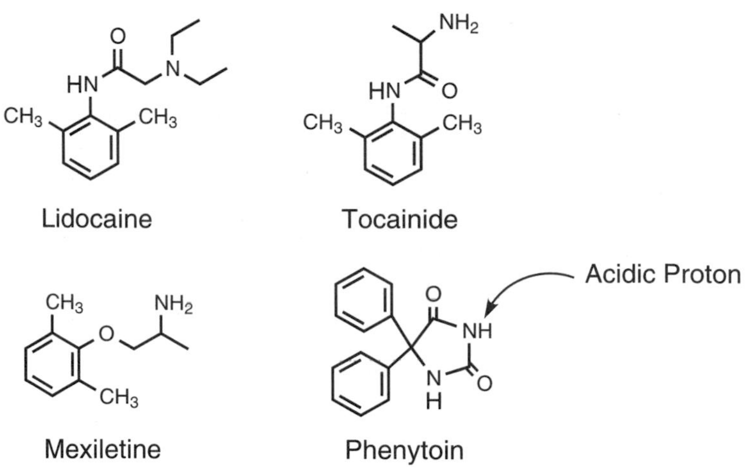

Lidocaine Tocainide

Mexiletine Phenytoin

Acidic Proton

Fig. 19–10. Class IB antiarrhythmics.

topic foci activity. Prolongation of the refractory period tends to abolish reentry arrhythmias. This class is further subclassified into classes IA, IB, and IC based on the primary pharmacologic effect. Table 19–9 summarizes those effects, and Figures 19–9 to 19–11 illustrate the chemical structures of members of each subclass. Quinidine is considered the prototype drug for class I.

Class II antiarrhythmic drugs are β-adrenergic receptor blocking agents, which block the role of the sympathetic nervous system in the genesis of certain cardiac arrhythmias. Their dominant electrophysiologic effect is to depress adrenergically enhanced phase 4 depolarization through β-receptor blockade. Drugs in this class decrease neurologically induced automaticity at normal therapeutic doses. At higher doses, these drugs may also exhibit anesthetic properties, which cause decreased excitability, decreased conduction velocity, and a prolonged effective refractory period. In normal therapeutic situations, the β-blocking effects are more important than any local anesthetic effects these drugs may have. Propranolol is the prototype drug for class II (see Chapter 17).

Class III drugs cause a homogeneous prolongation of the duration of the action potential. This results in a prolongation of the effective refractory period. Figure 19–12 illustrates the chemical structures of members of class III. Bretylium is the prototype drug for this class.

Class IV antiarrhythmic drugs comprise a group of agents that selectively block the slow inward current carried by calcium, i.e., calcium channel blockers. The slow inward current in cardiac cells has been shown to be important for the normal action potential in pacemaker cells. It has also been suggested that this inward current is involved in the genesis of certain types of cardiac arrhythmias. Administration of a class IV drug causes a prolongation of the refractory period in the A-V node and the atria, a decrease in A-V conduction, and a decrease in spontaneous diastolic depolarization. These effects block conduction of premature impulse at the A-V node and thus are effective in treating supraventricular arrhyth-

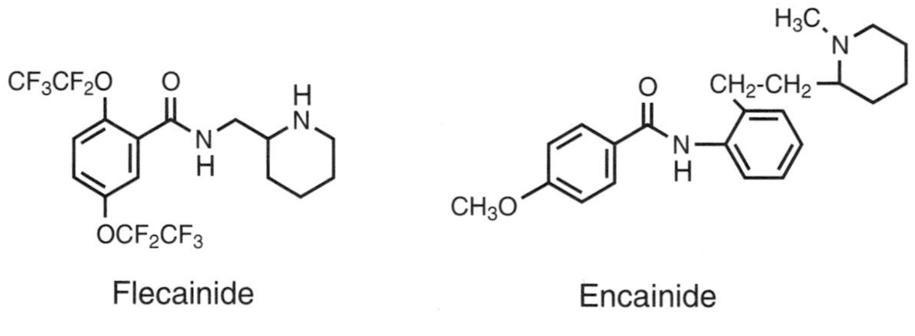

Flecainide Encainide

Fig. 19–11. Class IC antiarrhythmics.

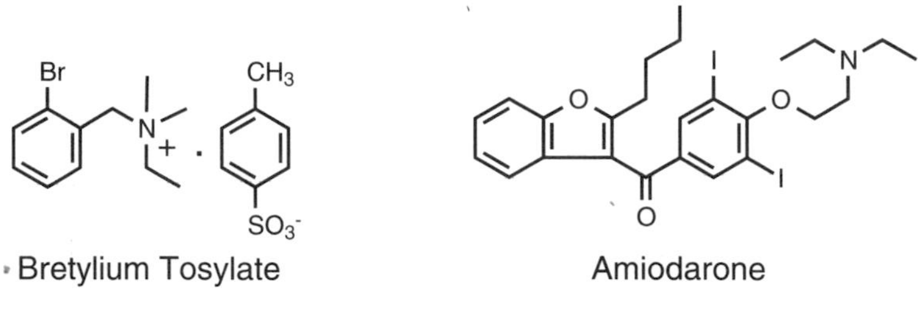

Bretylium Tosylate Amiodarone

Fig. 19–12. Class III antiarrhythmics.

Table 19–10. Class I Antiarrhythmic Drugs and Their Dosage Forms

Name	Dosage Forms	Typical Adult Dosage*
Quinidine sulfate, USP	Tablets: 100, 200, 300 mg; tablets, extended-release 300 mg; capsules: 200, 30 mg; solution for injection, 200 mg/ml	200–300 mg orally three or four times daily; a loading dose of 600–1000 mg may be used
Quinidine gluconate, USP	10-ml ampules, 80 mg/ml	IV: slow infusion, 0.3–0.4 mg/kg/min
Procainamide hydrochloride, USP	Capsules: 250, 375, 500 mg; tablets: 250, 375, 500 mg; 10-ml ampules, 100, 500 mg/ml	Initially, 1.25 g orally then 0.5–1.0 g every 4 h; 0.5–1.0 g IM; 0.2–0.5 g IV, not exceeding 50 mg/min, then maintenance of 2–6 mg/min
Disopyramide phosphate, USP	Capsules: 100, 500 mg	100–200 mg four times a day
Lidocaine hydrochloride, USP	Ampules: 5 and 50 ml, 10, 20, 40, 100, 200 mg/ml	50–100 mg IV, then 1–4 mg/min infusion
Phenytoin (diphenylhydantoin sodium, USP)	Capsules: 30, 100 mg; vials (to be reconstituted): 100, 200 mg; oral suspension: 125 mg/5 ml	50–100 mg IV over 5 min, not to exceed 50 mg/min; 5–6 mg/kg/day orally in one or two divided doses
Tocainide hydrochloride	Tablets: 400, 600 mg	400–600 mg orally three times a day
Mexiletine hydrochloride	Capsules: 100, 200, 250 mg	200 mg orally every 8 h with incremental increase of 50 mg each if needed
Flecainide	Tablets: 100 mg	100 mg orally every 12 h with incremental increases of 50 mg each as needed to a maximum of 400 mg/day

* All dosages require individual adjustment according to patient condition and response.

mias. Verapamil is the prototype drug for this class (see Fig. 19–6). Dosage forms and typical adult dosages of class I drugs are summarized in Table 19–10, and those of classes II through IV are summarized in Table 19–11.

Quinidine (e.g., Cardioquin, Cin-Quin, Duraquin, Quinora)

Quinidine is widely used for acute and long-term treatment of ventricular and supraventricular arrhythmias, especially supraventricular tachycardia. It is a member of a family of alkaloids found in Cinchona bark (Cinchona officinalis L.) and is a close relative of quinine. In fact, quinidine and quinine are simply diastereomers of one another, as shown in Figure 19–9. Despite their structural similarity, quinidine and quinine differ markedly in their effects on the cardiac muscles, with quinidine's effects being much more pronounced. Structurally, quinidine is composed of a quinoline ring and the bicyclic quinuclidine ring system with a hydroxymethylene bridge connecting these two components. Examination of quinidine reveals two basic nitrogens, with the quinuclidine nitrogen (pK_a 10) being the stronger of the two. Because of quinidine's alkaline character, it is always used in water-soluble salt forms. These salts include quinidine sulfate, gluconate, and polygalacturonate. Good absorption (approximately 95%) is observed with each of these forms after oral administration. In special situations, quinidine may be administered intravenously as the gluconate salt. The use of intravenous quinidine, however, is rare. The gluconate salt is particularly suited for parenteral use because of its high water solubility and lower irritant potential.

Quinidine's bioavailability appears to depend on first-pass metabolism in the liver. The bioavailabilities of quini-

Table 19–11. Class II–IV Antiarrhythmic Drugs and Their Dosage Forms

Name	Dosage Forms	Typical Adult Dosage*
Propranolol, USP	Tablets: 10, 40, 80 mg; ampules: 2 ml, 2.5 mg/ml	20–80 mg orally per day; 0.5–3 mg IV
Bretylium tosylate	Ampules: 10 ml, 50 mg/ml	5–10 mg/kg IV infusion over 10–30 min, repeated every 6 h
Amiodarone	Tablets: 200 mg	800–1600 mg orally per day, gradually decreased over 1–3 weeks to 600–800 mg/day, then further decreased to 400 mg/day
Verapamil	Tablets: 80, 120 mg; ampules: 2 ml, 2.5 mg/ml	5–10 mg IV over 2 min, then 80–120 mg orally four times a day

* All dosages require individual adjustment according to patient condition and response.

dine sulfate and gluconate are 80 to 85% and 70 to 75%. Once absorbed, quinidine is largely (approximately 85%) plasma protein bound, with an elimination half-life of about 6 hours. Quinidine is metabolized mainly in the liver, and renal excretion of unchanged drug is also significant (approximately 10 to 50%). The metabolites are hydroxylated derivatives at either the quinoline ring through the first pass, O-demethylation, or at the quinuclidine ring through oxidation of the vinyl group, oxydihydroquinidine. These metabolites possess only about one-third the activity of quinidine. Their contribution to quinidine's overall therapeutic effect is unclear.

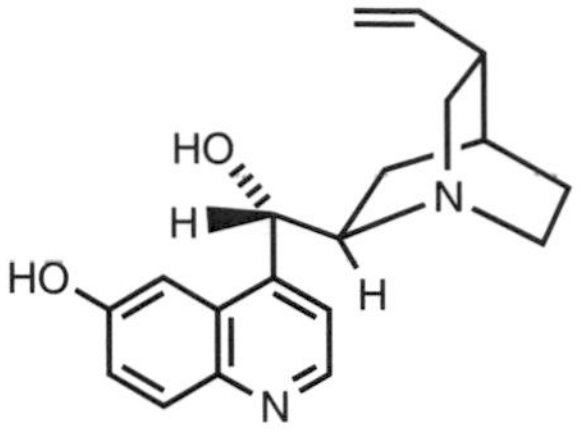

O-Demethylquinidine

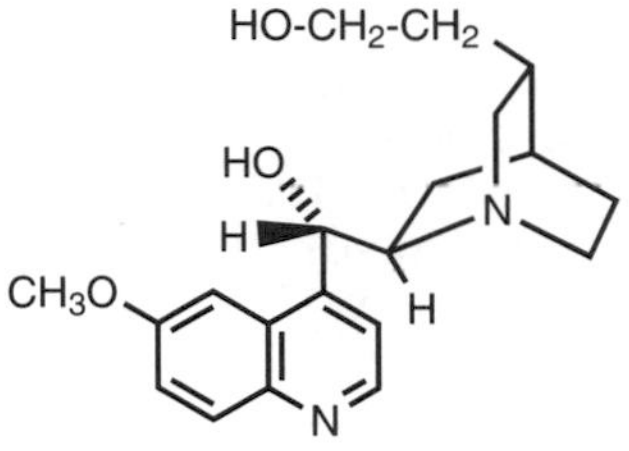

Oxydihydroquinidine

In addition, a common contaminant in quinidine preparations, dihydroquinidine (derived from reduction of the quinuclidine vinyl group at C-3 to an ethyl group), may also contribute to activity.[28] Although similar to quinidine in pharmacodynamic and pharmacokinetic behavior, this contaminant is both more potent as an antiarrhythmic and more toxic. Thus, levels of this contaminant may contribute to variability between commercial preparations. The most frequent adverse effects associated with quinidine therapy are gastrointestinal disturbances, such as nausea, diarrhea, and vomiting.

Procainamide (Procan, Promine, Pronestyl, Rhythmin)

Procainamide is effective in the treatment of several types of cardiac arrhythmias. Its actions are similar to those of quinidine, and yet procainamide may be effective in patients who are unresponsive to quinidine. The initial development of procainamide was stimulated by the observation that the local anesthetic procaine (see Fig. 19–9), when administered intravenously, produced significant although short-lived antiarrhythmic effects. Unfortunately, considerable central nervous system toxicity, in addition to the short duration, limited the usefulness of this agent. Moreover, procaine was not active orally. Because the short duration of action was attributable to hydrolysis, both chemical and enzymatic (catalyzed by plasma esterases), a logical modification of this molecule was the isosteric replacement of the ester with an amide group. This produced procainamide, which was more resistant to plasma esterase enzymes and chemical hydrolysis. In addition, procainamide was orally active. Peak plasma levels of procainamide are observed within 45 to 90 minutes after oral administration, and about 70 to 80% of the dose is bioavailable. About half of this dose is excreted unchanged, and the remaining half undergoes metabolism in the liver. Metabolites of procainamide include *p*-aminobenzoic acid and *N*-acetylprocainamide.

Interestingly the acetylated metabolite is active as an antiarrhythmic. Its formation accounts for up to one-third of the administered dose and is catalyzed by the liver enzyme, *N*-acetyl transferase. Because acetylation is strongly influenced by an individual's genetic background, marked variability in the amounts of this active metabolite may be observed from patient to patient. Renal excretion dominates, with about 90% of a dose excreted as unchanged drug and metabolites. The elimination half-life is about 3.5 hours. A substantial percentage (60 to 70%) of patients on procainamide show elevated levels of antinuclear antibodies after a few months. Of these patients, between 20 and 30% develop a drug-induced lupus syndrome if therapy is continued. These adverse effects are observed more frequently and more rapidly in "slow acetylators." Usually the symptoms associated with procainamide-induced lupus syndrome subside fairly rapidly after the drug is discontinued. These problems, however, have discouraged long-term procainamide therapy.

Disopyramide (Norpace)

Disopyramide phosphate is used orally for the treatment of certain ventricular and atrial arrhythmias. Despite its structural dissimilarity to quinidine and procainamide (see Fig. 19–9), its cardiac effects are similar. Disopyramide is rapidly and completely absorbed from the gastrointestinal tract. Peak plasma level is usually reached within 1 to 3 hours, and a plasma half-life of 5 to 7 hours is common. About half of an oral dose is excreted unchanged in the urine. The remaining drug undergoes hepatic metabolism, principally to the corresponding *N*-dealkylated form. This metabolite retains about half of disopyramide's antiarrhythmic activity and is also subject to renal excretion. Adverse effects of disopyramide are frequently observed. These effects are primarily anticholinergic in nature and include dry mouth, blurred vision, constipation, and urinary retention.

Lidocaine (Baylocaine, LidoPen, Xylocaine)

Lidocaine, similar to procaine, is an effective, clinically used local anesthetic (see Fig. 19–10). Its cardiac effects, however, are distinctly different from those of procainamide or quinidine. Lidocaine is normally reserved for the treatment of ventricular arrhythmias and is, in fact, usually the drug of choice for emergency treatment of ventricular arrhythmias. Its utility in these situations is due to the rapid onset of antiarrhythmic effects on intravenous infusion. In addition, these effects cease soon after the infusion is terminated. Thus, lidocaine therapy may be rapidly modified in response to changes in the patient's status. Lidocaine is effective as an antiarrhythmic only when given parenterally, and the intravenous route is the most common. Antiarrhythmic activity is not observed after oral administration because of the rapid and efficient first-pass metabolism by the liver. Parenterally administered lidocaine is about 60 to 70% plasma protein bound. Hepatic metabolism is rapid (plasma half-life is about 15 to 30 minutes) and primarily involves deethylation to yield monoethylglycinexylide, followed by amidase-catalyzed hydrolysis into *N*-ethylglycine and 2,6-dimethylaniline (2,6-xylidine) (see Fig. 19–13).

Monoethylglycinexylide has good antiarrhythmic activity. It is not clinically useful, however, because it undergoes rapid enzymatic hydrolysis and has emetic and convulsant properties. Lidocaine's adverse effects predominantly involve the central nervous system and heart. The central nervous system effects may begin with dizziness and paresthesia and, in severe cases, ultimately lead to epileptic seizures.

Tocainide (Tonocard)

Tocainide (see Figure 19–10) is an *α*-methyl analog structurally related to monoethylglycinexylide, the active metabolite of lidocaine. It possesses similar electrophysiologic effects to lidocaine. In contrast to lidocaine, tocainide is orally active, and oral absorption is excellent. Similar to lidocaine, it is usually reserved for the treatment of ventricular arrhythmias. The *α*-methyl group is believed to slow metabolism and thereby contributes to oral activity. The plasma half-life of tocainide is about 12 hours, and nearly 50% may be excreted unchanged in the urine. Adverse effects associated with tocainide are similar to those observed with lidocaine, specifically, gastrointestinal disturbances and central nervous system effects.

Mexiletine (Mexitil)

Mexiletine (see Fig. 19–10) is similar to both lidocaine and tocainide in its effects and therapeutic application. It is used principally to treat and prevent ventricular arrhythmias. Similar to tocainide, mexiletine has good oral activity and absorption properties. Clearance depends on metabolism and renal excretion. A relatively long plasma half-life of about 12 to 16 hours is common. Adverse effects are similar to those experienced with tocainide and lidocaine.

Phenytoin

For 50 years, phenytoin (Dilantin, diphenylhydantoin sodium; see Fig. 19–10) has been used clinically in the treatment of epileptic seizures. During this time, it was noticed that phenytoin also produced supposedly adverse cardiac effects. On closer examination, these adverse effects were actually found to be beneficial in the treatment of certain arrhythmias. Currently, phenytoin is used in the treatment of atrial and ventricular arrhythmias resulting from digitalis toxicity. It is, however, not officially approved for this use.

Phenytoin may be administered either orally or intravenously. It is absorbed slowly after oral administration, with peak plasma levels achieved after 3 to 12 hours. It is extensively plasma protein bound (approximately 90%), and the elimination half-life is between 15 and 30 hours. These large ranges reflect the considerable variability observed from patient to patient. Parenteral administration of phenytoin is usually limited to the intravenous route. Phenytoin for injection is dissolved in a highly alkaline vehicle (pH 12). This alkaline vehicle is required because phenytoin is weakly acidic and has poor solubility in its protonated form. Intramuscular phenytoin is generally avoided because it results in tissue necrosis at the site of injection and erratic absorption. In addition, intermittent intravenous infusion is required to reduce the incidence of severe phlebitis.

Phenytoin metabolism is relatively slow and predominantly involves conversion to hydroxylated, inactive metabolites. Metabolism is also subject to large interindividual variability. The major metabolite, 5-*p*-hydroxyphenyl-5-phenylhydantoin, accounts for about 75% of a dose. This metabolite is excreted through the kidney as the *β*-glucuronide conjugate. Phenytoin clearance is strongly influenced by metabolism, and therefore agents that affect phenytoin metabolism may cause intoxication. In addition, because phenytoin is highly plasma protein bound, agents that liberate phenytoin may also cause toxicity.

Flecainide (Tambocor)

Flecainide is a relatively new drug that exhibits properties distinctly different from those of class IA (e.g., quinidine) or class IB (e.g., lidocaine) antiarrhythmic agents. Flecainide is a fluorinated benzamide derivative (see Fig. 19–11), available as the acetate salt. Flecainide has been approved by the FDA for the treatment of ventricular arrhythmias. Clinical studies suggest that this new agent may be more effective than either quinidine or disopyramide in suppressing premature ventricular contractions. Oral flecainide is well absorbed, and the plasma half-life is about 14 hours. About half of an oral dose is metabolized in the liver, and a third is excreted unchanged in the urine. As with other antiarrhythmics, flecainide may produce adverse effects. The most severe is flecainide's tendency occasionally to aggravate existing arrhythmias or induce new ones. Although fewer than 10% of patients experience this effect, it may be life-threatening. Accordingly, it may be desirable to start therapy in the hospital. Other less serious side effects include blurred vision, headache, nausea, and abdominal pain.

Encainide (Enkaid)

Encainide (see Fig. 19–11) represents another benzamide derivative, with similar pharmacologic properties to encainide, however, with less negative inotropic effect.

Propranolol (Inderal, Ipran, Betachron)

Propranolol, an adrenergic *β*-receptor blocker, is the prototype for the class II antiarrhythmics. Its pharmacology is discussed in detail in Chapter 18. It is generally

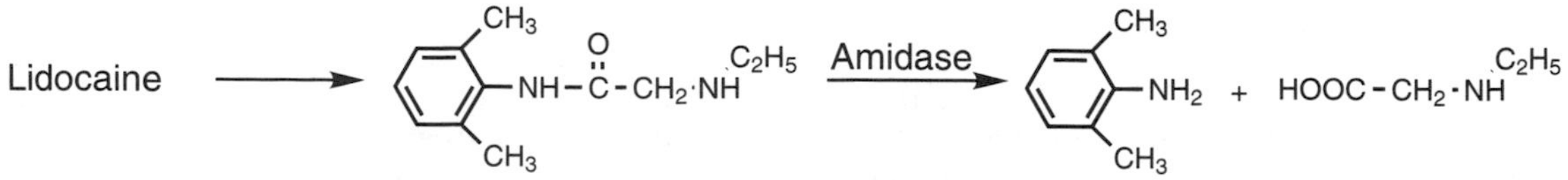

Fig. 19–13. Metabolism of lidocaine.

used in the treatment of supraventricular arrhythmias, including atrial flutter, paroxysmal supraventricular tachycardia, and atrial fibrillation. Propranolol is also reported to be effective in the treatment of digitalis-induced ventricular arrhythmias. Moreover, beneficial results may be obtained when propranolol is used in combination with other agents. For example, in certain cases, quinidine and propranolol together have proved more successful in alleviating atrial fibrillation than either agent alone. Few serious adverse effects are associated with propranolol therapy. Another β-blocker, sotalol (Betapace, Sotacor), has been approved by the FDA for use as an antiarrhythmic agent.

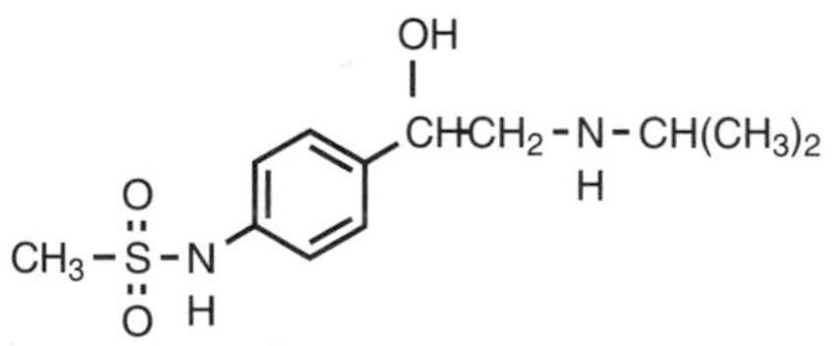

Sotalol

Bretylium Tosylate (Bretylol)

Bretylium tosylate is a quaternary ammonium salt derivative (Fig. 19–12), originally developed for use as an antihypertensive. Its antiarrhythmic use is limited to emergency life-threatening situations in which other agents, such as lidocaine and procainamide, have failed. Generally, bretylium is used only in intensive care units. It may be administered either intravenously or intramuscularly and is eliminated largely unchanged in the urine. The plasma elimination half-life is usually about 10 hours. The major adverse effect associated with bretylium tosylate is hypotension, including orthostatic hypotension. This effect may be severe.

Amiodarone (Cordarone)

Amiodarone (Fig. 19–12), initially developed as an antianginal agent (coronary vasodilator), has antiarrhythmic effects that are somewhat similar to those of bretylium. It is approved by the FDA for the treatment of life-threatening ventricular arrhythmias refractory to other drugs. Its cardiac effects are not well characterized, but clinical studies indicate that it is a promising new class III agent. Severe toxicity, however, makes it the drug of last choice. As with bretylium tosylate, use of this agent should be initiated in a hospital setting.

Verapamil (Calan, Isoptin)

Verapamil (see Fig. 19–6), similar to amiodarone, was originally conceived as a coronary vasodilator for angina. It is, however, also effective and widely used for the treatment of supraventricular arrhythmias. Therapeutic effects after oral administration are observed within 2 hours. Considerable variability may be observed in the elimination half-life of 1.5 to 7 hours. In addition, the plasma half-life may not always accurately predict the duration of action owing to the presence of active metabolites. Because of verapamil's extensive first-pass metabolism, nearly all of the drug is excreted in the urine as metabolites. Few serious adverse effects are encountered with oral verapamil. Intravenous administration, however, may lead to hypotension, bradycardia, or asystole in individuals with atrioventricular block.

REFERENCES

1. T. W. Smith, et al., *Prog. Cardiovasc. Dis., 26,* 413; *27,* 21(1984).
2. N. Rietbrock and B. G. Woodcock, *Trends Pharmacol. Sci., 6,* 267(1985).
3. R. Thomas, et al., *J. Pharm. Sci., 63,* 1649(1974).
4. R. Thomas, in *Burger's Medicinal Chemistry,* 4th ed., Part III, M. E. Wolff, Ed., *Cardiac Drugs,* New York, John Wiley & Sons, 1981.
5. K. Repke, *Trends Pharmacol. Sci., 6,* 275(1985).
6. D. S. Fullerton, et al., *Trends Pharmacol. Sci., 6,* 279(1985).
7. W. S. Colucci, et al., *N. Engl. J. Med., 314,* 290(1986).
8. B. F. Hoffman and R. J. Lefkowitz, in *Goodman and Gilman's The Pharmacological Basis of Therapeutics,* 8th ed., A. G. Gilman, et al., Eds., New York, *Catecholamines and Symnpathomimetic Drugs,* Pergamon Press, 1990.
9. P. Needleman and E. J. Johnson, Jr., *J. Pharmacol. Exp. Ther., 184,* 709(1973).
10. P. Needleman, et al., *J. Pharmacol. Exp. Ther., 187,* 324(1973).
11. L. J. Ignarro, et al., *J. Pharmacol. Exp. Ther., 218,* 739(1981).
12. L. J. Ignarro, et al., *Biochem. Biophys. Res. Commun., 94,* 93(1980).
13. L. J. Ignarro, et al., *Biochim. Biophys. Acta, 673,* 394(1981).
14. L. J. Ignarro, et al., *FEBS Lett., 110,* 275(1980).
15. L. J. Ignarro and C. A. Gruetter, *Biochim. Biophys. Acta, 631,* 221(1980).
16. L. J. Ignarro, et al., *Arch. Biochem. Biophys., 208,* 75(1981).
17. S. Moncada, et al., *Pharmacol. Rev., 43,* 109(1991).
18. S. H. Snyder and D. S. Bredt, *Trends Pharmacol. Sci., 12,* 125(1991).
19. T. McCall and P. Vallance, *Trends Pharmacol. Sci., 13,* 1(1992).
20. P. L. Feldman, et al., *Chem. Eng. News,* 26(1993).
21. H. L. Fung, et al., *J. Pharmacol. Exp. Ther., 228,* 334(1984).

22. P. Needleman, et al., *J. Pharmacol. Exp. Ther., 181,* 489(1972).
23. J. Abrams, *Am. J. Cardiol., 56,* 12A(1985).
24. H. L. Fung, *Am. J. Med., 72*(Suppl.), 13(1983).
25. S. Corwin and J. A. Reiffel, *Arch. Intern. Med., 145,* 538(1985).
26. R. G. Rahwan, et al., *Ann. Rep. Med. Chem., 16,* 257(1981).
27. K. C. Yedinak, *Am. Pharm, 33,* 49(1993).
28. H. L. Conn, Jr., and R. J. Luchi, *Am. J. Med., 37,* 685(1964).

SUGGESTED READINGS

J. T. Bigger, Jr., and B. F. Hoffman, Antiarrhythmic drugs, in *Goodman and Gilman's The Pharmacological Basis of Therapeutics,* 8th ed., A. G. Gilman, et al., Eds., New York, Pergamon Press, 1990.

H. A. Fozzard and M. F. Sheets, *J. Am. Coll. Cardiol., 5,* 10A(1985).

P. D. Hansten, Ed., *Drug Interactions: Clinical Significance of Drug-drug Interactions,* 5th ed., Philadelphia, Lea & Febiger, 1985.

B. F. Hoffman and J. T. Bigger, Jr., Digitalis and allied glycosides, in *Goodman and Gilman's The Pharmacological Basis of Therapeutics,* 8th ed., A. G. Gilman, et al., Eds., New York, Pergamon Press, 1990.

H. R. Kaplan, *Fed. Proc., 45,* 2184(1986).

A. M. Katz, *J. Am. Coll. Cardiol. 5,* 16A(1985).

R. J. Mangini, Ed., *Drug Interaction Facts,* Philadelphia, J. B. Lippincott, 1983.

F. Murad, Drugs used for the treatment of angina, in *Goodman and Gilman's the Pharmacological Basis of Therapeutics,* 8th ed., A. G. Gilman, et al., Eds., New York, Pergamon Press, 1990.

R. E. Thomas, Cardiac drugs, in *Burger's Medicinal Chemistry,* 4th ed., Part III, M. E. Wolff, Ed., New York, John Wiley & Sons, 1981.

Chapter 20

ANTICOAGULANTS, COAGULANTS, AND PLASMA EXTENDERS

Richard H. Hammer

Cardiovascular diseases, especially various forms of thrombosis, such as coronary, embolic, venous, and traumatic thromboses, account for a large number of deaths per year. In fact, it is estimated by the American Heart Association that 54% of all deaths in the United States can be attributed to cardiovascular disease. It is therefore important that the pharmacist be familiar with physical, chemical, and clinical aspects of the drugs that are frequently used to treat these forms of thrombosis.

Before studying the coagulant and anticoagulant drugs used in health care, it is important to appreciate and understand the biochemical steps and factors involved in clot formation. Clinical and experimental findings indicate that intravascular clotting or thrombosis can be caused by vascular injury, blood hypercoagulability, or blood stasis.

The first coagulation scheme was proposed by Morawitz in 1903. Since that time, 12 coagulation factors as well as the Fletcher and Fitzgerald factors have been identified (Table 20–1).

When an injury to the subendothelial cells of a blood vessel or to tissue occurs, an immediate vasoconstrictive reflex reduces the volume of blood flow, and platelets begin to adhere to the injured cells through a process called "platelet adhesion." At this time, the platelets are physically transformed from a disklike shape into irregularly shaped forms with filaments projecting from the edges that help the platelets to adhere to the cells or tissue. The platelets subsequently release adenosine diphosphate (ADP) and prostaglandin endoperoxides and adhere to one another, forming a platelet plug. These processes are referred to as "platelet release" and "platelet aggregation." The release of biochemicals from the platelets can lead to thromboxane A_2 synthesized by the platelet, which induces platelet aggregation, or to prostacyclin (PGI_2), synthesized by the blood vessel cells, which inhibits platelet aggregation. It may therefore be a delicate balance between the amounts of thromboxane A_2 synthesized by the platelets and prostacyclin formed by the vessels as to whether or not a thrombus is formed. The coagulation factors (Table 20–1) interact through a series of reactions to form fibrin, which intertwines, strengthens, and reinforces the platelet plug (Figs. 20–1 and 20–2). The coagulation factors are proteins except for calcium and platelet phospholipids. They are inactive proenzymes until converted to active protease enzymes by cleavage of one or two peptide bonds. The active enzymes then activate another proenzyme in a cascade effect to form fibrin (Fig. 20–2).

The coagulation process is divided into an intrinsic system (all factors present in blood) and an extrinsic system, which depends on the release of thromboplastin from injured cells and tissues (Fig. 20–2). Factor XIIa serves as an activator of factor XI and activates prekallikrein along with high-molecular-weight kininogen. The initiating enzyme that activates the coagulation process by converting prekallikrein to kallikrein, which, in turn, activates factor XII, is not known. Both intrinsic and extrinsic coagulation pathways proceed through a common pathway by forming activated factor X (e.g., Xa). Figure 20–3 illustrates the mechanism of protein activation by showing prothrombin, a single-chain glycoprotein, being activated by enzymatic degradation to thrombin, a two-chain glycoprotein. Thrombin, in turn, cleaves fibrinogen to form fibrin by removing two pairs of peptides from fibrinogen. Prothrombin is cleaved at two points by factor Xa to yield thrombin. Ten γ-carboxyglutamic acid residues are located on the N-terminal end of the prothrombin molecule. Vitamin K is required in the liver biosynthesis of the prothrombin γ-carboxyglutamic groups by participating in the carboxylation of the γ-carbon of glutamic acid. These carboxy groups are required for binding calcium to prothrombin, which induces a conformation change in prothrombin enabling it to bind to cofactors on the phospholipid surfaces during its conversion to thrombin by factor Xa, factor V, and platelet phospholipids in the presence of calcium.[3–6]

The three groups of drugs used to prevent or treat thromboses and emboli—anticoagulants, antiplatelet drugs, and fibrinolytic agents—are discussed here, along with coagulants and plasma extenders.

ANTICOAGULANTS

Since the first clinical use of heparin as an anticoagulant in 1937, and the subsequent discovery in 1941 that bishydroxycoumarin had anticoagulant activity when administered orally, many compounds have been synthesized and tested, although only a few have reached the market.

Anticoagulants are usually administered to patients with myocardial infarction, venous thrombosis and its extension, peripheral arterial emboli, and pulmonary em-

Table 20–1. Coagulation Factors

Factor	Name	Molecular Weight	Source	Plasma Half-Life (hours)
I	Fibrinogen	Protein (340,000) (six chains)	Liver	72–96
II	Prothrombin	α-2-globulin protein (70,000) (one chain)	Liver (requires vitamin K)	60 (contains γ-carboxyglutamic acid)
III	Tissue thromboplastin, thrombokinase, tissue factor	Glycoprotein (45,000) (one chain)	Liver (may require vitamin K)	—
IV	Calcium (Ca^{++})	—	—	—
V	Proaccelerin, labile factor	Globulin protein (330,000) (one chain)	Liver	15
VI	Deleted factor	—	—	—
VII	Proconvertin, stable factor	β-globulin protein (55,000) (one chain)	Liver (requires Vitamin K)	5 (factor VII disappears rapidly in the presence of coumarin drugs)
VIII	Antihemophilic A factor (AHF), antihemophilic globulin (AHG)	α-2 or β-globulin protein (1–2,000,000), (6–10 subunits)	Liver	10 (contains γ-carboxyglutamic acid)
IX	Antihemophilic B factor, plasma thromboplastin component (PTC), Christmas factor	β-globulin protein (57,000) (one chain)	Liver (requires vitamin K)	25 (contains γ-carboxyglutamic acid)
X	Stuart or Stuart-Prower factor	α-globulin protein (59,000) (two chains)	Liver (requires vitamin K)	40 (contains γ-carboxyglutamic acid)
XI	Plasma thromboplastin antecedent (PTA)	β_2-globulin protein (124,000) (two chains)	Unknown	45–65
XII	Hageman factor, contact factor	Protein (80,000) (one chain)	Unknown	60
XIII	Fibrin stabilizing factor, fibrinase	Protein (300,000) (four chains)	Unknown	150
	Fletcher factor, prekallikrein factor	γ-globulin protein (80,000) (one chain)	Liver	—
	Fitzgerald factor, high-molecular-weight kininogen	α-globulin protein (120,000) (one chain)	Liver	156

boli; anticoagulants have also been used successfully in patients with mechanical prosthetic heart valves and in atrial fibrillation. They are used to prevent transient ischemic attacks and to reduce the risk of recurrent myocardial infarction, although data for the latter two indications are conflicting. Several types of infarction and the rationale for use of anticoagulants are discussed in the following paragraphs.

1. *Acute myocardial infarction.* Anticoagulant therapy for this condition may decrease the mortality rate by 3 to 4%, especially in poor risk cases (e.g., those with previous infarction, shock, and complicating diseases, such as diabetes or obesity). Generally, heparin is the drug of choice.
2. *Long-term therapy after acute myocardial infarction.* The effect of long-term therapy on reducing reinfarction is still disputed. Approximately 50% of patients benefit slightly from this therapy. The balance have shown no significant benefit. Oral anticoagulants are the preferred agents for long-term therapy. It is important, however, that routine laboratory control of anticoagulation prothrombin times be available.
3. *Prophylaxis and treatment of pulmonary and venous thrombosis.* When thrombosis has occurred, anticoagulant therapy may prevent extension of the formed clot and secondary thromboembolic complications.

Heparin

Heparin, a mucopolysaccharide with a molecular weight ranging from 6,000 to 30,000, was first isolated

1. Platelet adhesion

   a. Platelets adhere to injured cells and tissue by being transformed from a disklike shape to a spiny shape.

   b. Platelets adhere to exposed collagen basement membrane and/or to microfibrils from damaged tissue. Ca^{++} and von Willebrand factor believed to be involved.

2. Platelet aggregation

   a. Platelets adhere to one another.

   b. Platelets $\overset{ADP}{\rightleftharpoons}$ Primary aggregation $\xrightarrow{ADP}$ Secondary aggregation

   and/or

   Platelet prostaglandins (e.g., thromboxane A_2)

3. Platelet release reaction

   a. Release contents of platelet granules to the outside (exocytosis) induced in part by thromboxane A_2.

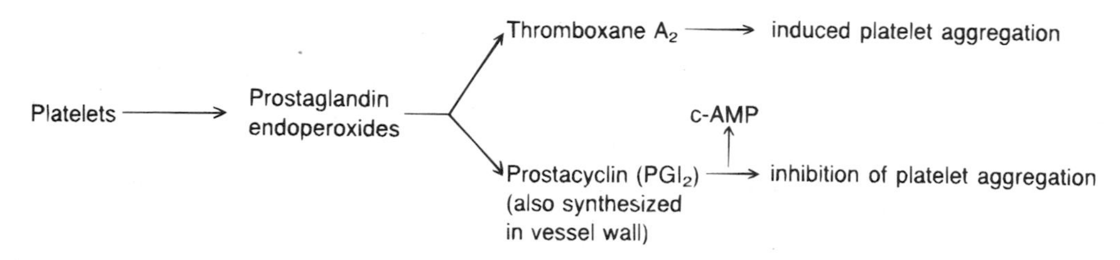

4. Blood coagulation—Platelet plug strengthened by fibrin formation

Stage 1: Intrinsic and extrinsic generation of activated Factor X

Stage 2: Formation of thrombin from prothrombin

Stage 3: Formation of fibrin from fibrinogen

Fig. 20–1. Steps in the hemostatic process.[1–6]

from liver tissue by McLean in 1916. He was searching for coagulating substances and discovered that heparin had anticoagulating properties. Subsequent work by Best, Charles, and Scott demonstrated that heparin occurs in many other tissues in the body, with the lungs having an especially abundant supply. By 1936, heparin had been sufficiently purified to be used in humans, and its first use as an anticoagulant was in 1937. Today, heparin is produced commercially by selective processing of pig intestines. Commercial heparin contains both low-molecular-weight and high-molecular-weight heparins and is infractionated.

Chemistry and Mode of Action

Because of its acidic properties, heparin is also referred to as heparinic acid. Chemically it is similar to hyaluronic acid, chondroitin, and chondroitin sulfate A and B. It is a large, dextrorotatory, highly acidic polysaccharide molecule composed of sulfated D-glucosamine and D-glucuronic acid residues. The polymer chain contains a repeating disaccharide unit of D-glucosamine that has an O-sulfate at carbon 6 and an N-sulfate at carbon 2 combined with D-glucuronic acid that has an O-sulfate at carbon 2. These monosaccharide units are linked by α,1→4 bonds.

COOH, OH, O, H, OSO_3^-, α-1-4 linkage, ^-O_3SO, OH, O, $NHSO_3^-$, n

D-Glucuronic acid + O-sulfate at carbon 2

D-glucosamine N-sulfate at carbon 2 + O-sulfate at carbon 6

Because of its highly acidic sulfate groups, heparin exists as the anion at physiologic pH and is usually administered as the sodium salt.

Heparin acts as an anticoagulant by combining with antithrombin III, an α-2-glycoprotein, which is synthe-

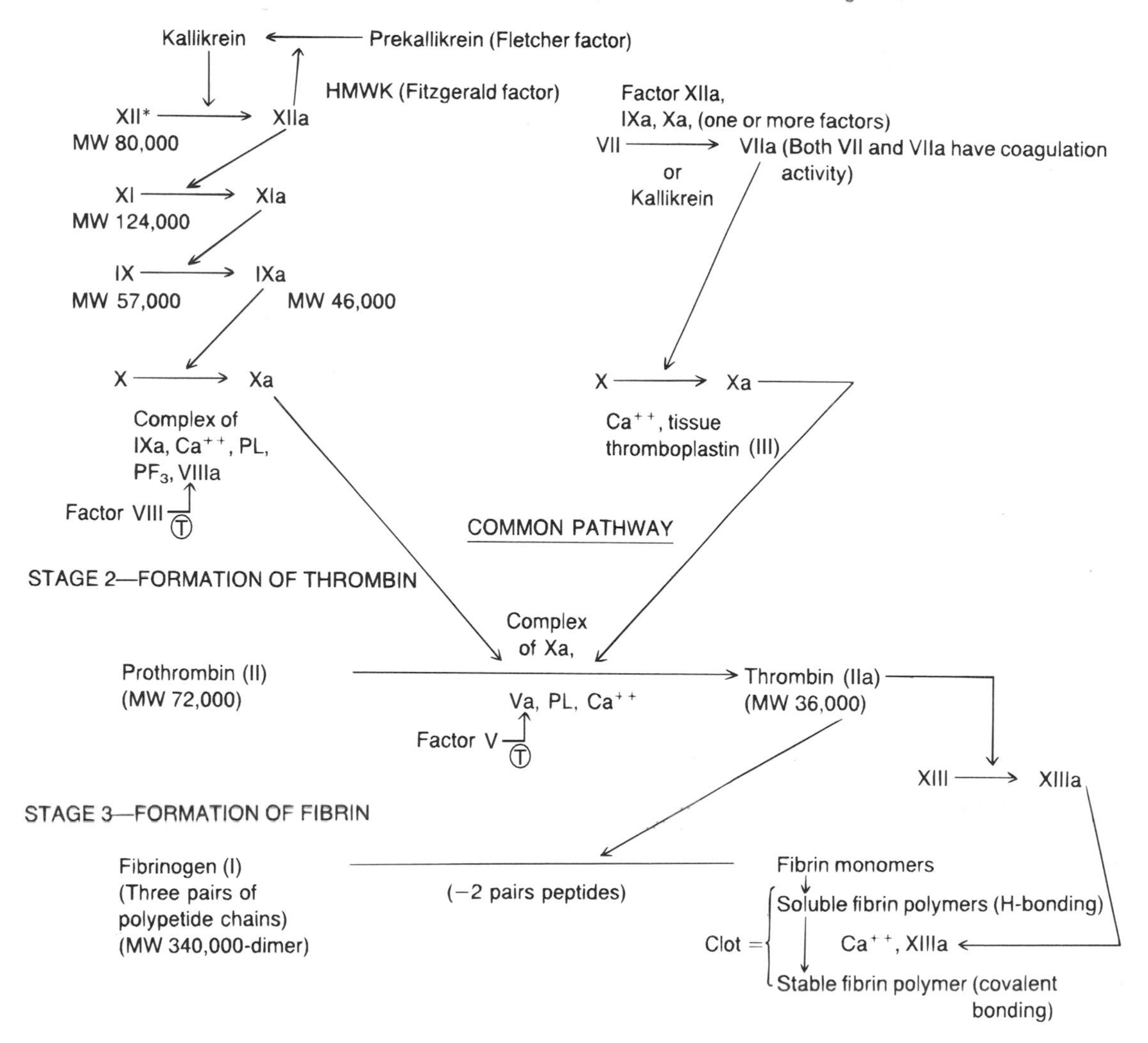

Fig. 20–2. Three stages of blood coagulation. Ⓣ = Thrombin; PF_3 = platelet factor 3; PL = phospholipid; MMWK = high-molecular-weight kininogen.[1–3]

sized in the liver. The heparin–antithrombin III complex inactivates factor Xa and therefore inhibits the generation of thrombin (equation 1). The heparin–antithrombin III complex can accelerate the inactivation of thrombin (IIa), which is required to convert fibrinogen to fibrin. Antithrombin III normally inactivates thrombin but at a slower rate than the heparin-antithrombin III complex, which is 100 to 1000 times faster (equation 2). To a lesser extent, heparin–antithrombin III complex also inactivates factors IXa, XIIa, and XIIIa.[1] When factor XIIIa is inhibited, the formation of a stable, covalently bonded fibrin clot is inhibited (Fig. 20–2).

Two mechanisms:

Heparin-Antithrombin III Complex + Factor Xa $\rightleftharpoons$ Inactivated Factor Xa Complex

Equation 1

Thrombin + Antithrombin III $\xrightleftharpoons{\text{Slow}}$ Inactivated Thrombin

Thrombin + Heparin-Antithrombin III $\xrightleftharpoons{\text{Fast}}$ Inactivated Thrombin

Equation 2

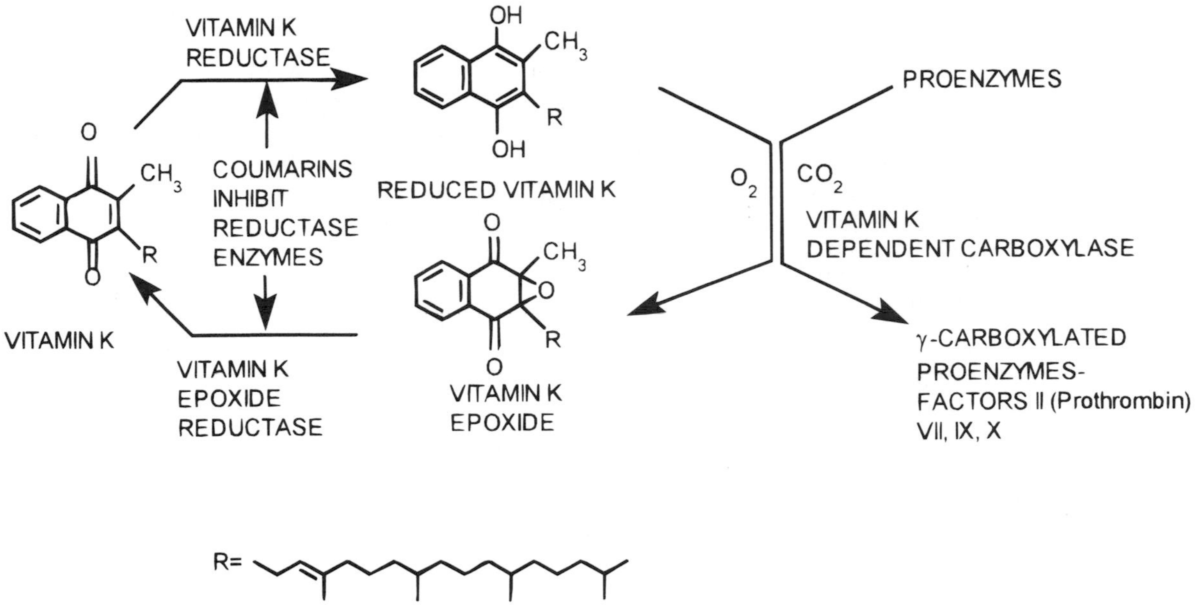

VITAMIN K REDUCTASE

REDUCED VITAMIN K

PROENZYMES

COUMARINS INHIBIT REDUCTASE ENZYMES

O_2 CO_2

VITAMIN K DEPENDENT CARBOXYLASE

VITAMIN K

VITAMIN K EPOXIDE REDUCTASE

VITAMIN K EPOXIDE

γ-CARBOXYLATED PROENZYMES- FACTORS II (Prothrombin) VII, IX, X

R=

Fig. 20–3. Mechanism of vitamin K–dependent γ-carboxylation of proenzymes and inhibition by coumarins. (Modified from B. Furie and B. C. Furie, *Blood 75,* 1753[1990].)

Heparin is partially metabolized in the liver by heparinase to uroheparin, which has only slight antithrombin activity. Twenty percent to 50% of heparin is excreted unchanged. The heparin polysaccharide chain is degraded in gastric acid and must therefore be administered intravenously or subcutaneously. Heparin should not be given intramuscularly because of the danger of hematoma formation.

A commercial product called Antithrombin III (Human) (ATnativ.) obtained from pooled human plasma has been marketed. It inactivates thrombin and factors IX, X, XI, and XII. Antithrombin III (Human) is used to treat patients with hereditary antithrombin III deficiency when they are undergoing surgical or obstetrical procedures or to treat thromboembolism.

Clinical Pharmacology

Heparin doses are measured in USP units because of variation in potency from one preparation to another. The USP unit of heparin is defined as the quantity that prevents 1.0 ml of citrated sheep plasma from clotting for 1 hour on addition of 0.2 ml of a 1/100 calcium chloride solution. Standardized preparations of sodium heparin,

Table 20–2. Commercial Preparations of Heparin (Unfractionated) and Low-Molecular-Weight Heparin

Generic Name	Trade Name	Dosage Forms (units/ml)	Route of Administration
Heparin sodium injection	Liquaemin, Lipo-Hepin, Hepathrom	1,000, 5,000, 10,000, 20,000, and 40,000	Parenteral
Heparin sodium and sodium chloride (0.9%)	—	1000 and 2000	Parenteral
Heparin sodium and sodium chloride (0.45%)	—	12,500 and 25,000	Parenteral
Heparin calcium injection	Calciparine	5000	Parenteral
Heparin sodium lock flush solution	Hep-Lock	10 and 100 units	IV flush for indwelling IV catheter
Enoxaparin sodium (low-molecular-weight heparin)	Lovenox	30 mg/0.3 ml (~3000 IU of low-molecular-weight heparin reference standard)	Subcutaneously (prefilled syringes)

USP, must contain at least 120 USP units of anticoagulant activity per milligram. See Table 20–2 for the commercial preparations that are available.

Many people believe heparin to be the anticoagulant of choice, but parenteral administration precludes its long-term use. It is generally given to postoperative patients and to those with acute infarctions requiring immediate anticoagulant activity. Heparin is commonly used as an anticoagulant in the prophylaxis and treatment of thromboembolic conditions, such as pulmonary embolism, venous thrombosis, and its extension thrombophlebitis, and to prevent thromboembolic complications emanating from cardiovascular surgery, frostbite, and dialysis techniques. Because the drug does not cross the placental membrane, it is the best anticoagulant to use during pregnancy.

Heparin is relatively nontoxic for short-term therapy except for hemorrhage. Hypersensitivity reactions are uncommon. Temporary alopecia may occur 3 or 4 months after therapy.

The most satisfactory dosage regimen is 5,000 to 10,000 USP units every 4 hours. Four hours after injection, the dose should produce an activated partial thromboplastin time (APTT) 1.5 to 2 times the control value.

If hemorrhage occurs, the anticoagulant effects of heparin can be reversed in minutes by administration of protamine sulfate. Protamines are low-molecular-weight proteins found in the sperm of various fish. They are rich in arginine, which confers a basic property to the protein. Other common amino acids found in protamines are alanine and serin.

Because of the basic properties of the protamines and the acidic properties of heparin, these agents have a strong ionic attraction for each other, and the interaction results in inactivation of the anticoagulant activity of heparin. Approximately 1.0 mg of protamine sulfate neutralizes 100 USP units of heparin within 5 minutes after intravenous injection. No more than 50 mg of protamine should be given because amounts greater than 100 mg can produce an anticoagulant effect by inhibiting the thrombin-fibrinogen reaction.

Low-Molecular-Weight Heparin

The use of low-molecular-weight heparins (molecular weight 2000 to 8000) may well prove to be the heparin products of the future. They are made by enzymatic or chemical controlled hydrolysis of unfractionated heparin.[4,5] In a randomized, double-blind study in patients with thrombosis in the proximal veins of lower limbs, those receiving low-molecular-weight heparin subcutaneously once a day had 60% less recurrent vein thrombosis or pulmonary embolism and 91% fewer major bleeding complications compared with patients receiving regular, unfractionated heparin by continuous intravenous infusion. Even though the low-molecular-weight heparins have been tested in 20,000 patients, differences of opinion exist, and only time will determine whether they are safer and more efficacious than unfractionated heparin. The first low-molecular-weight heparin, enoxaparin (Lovenox), has been approved for preventing blood clots following hip replacement surgery. Eventually, it should be approved for treating other potential thrombotic conditions, such as knee replacement surgery. Enoxaparin binds more strongly to factor Xa than unfractionated heparin and acts in a similar way by inactivating factors IIa (thrombin) and Xa.

Coumarin Derivatives

Coumarin derivatives and 1,3-indandiones are the principal orally active anticoagulants. The first one of these, bishydroxycoumarin, was isolated in 1934 by Link and Campbell. It had been known since 1921 that spoiled sweet clover hay caused a bleeding disease, referred to as sweet clover disease, in cattle. These and other findings eventually led to its use in humans in 1941 as the first orally effective anticoagulant drug.

Chemistry and Mode and Action

Coumarin, 4-hydroxycoumarin, and their derivatives are water-insoluble lactones.

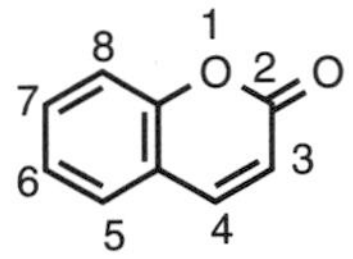

Coumarin

Substitution of a hydroxyl group in the 4 position of coumarin confers weakly acidic properties (pK_a = 4.1 for 4-hydroxycoumarin) to an otherwise neutral compound. The hydroxyl group of coumarin and its derivatives reacts with appropriate bases to form water-soluble salts, as shown in equation 3. An exception is cyclocumarol, which is a neutral compound because the acidic hydroxyl is part of a cyclic ketal ring system.

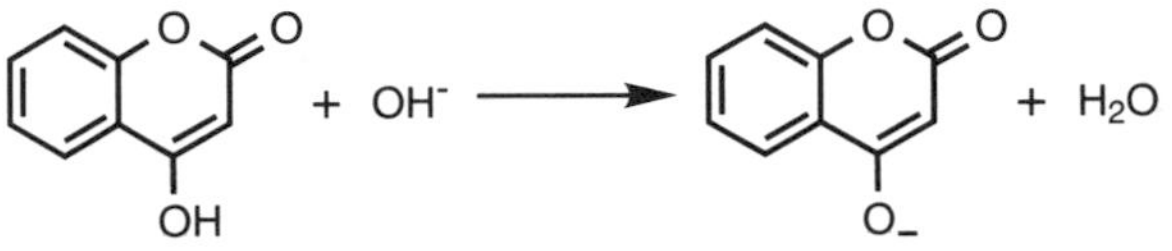

Equation 3

Warfarin is marketed as the sodium salt. It has one asymmetric carbon atom, as indicated in Table 20–3. Both the (R)(+) and the (S)(−) enantiomers of warfarin have been tested for anticoagulant activity in rats, and the (S)(−) isomer was shown to be five to eight times more potent.[8] Commercially available warfarin for use in humans, however, is the racemic mixture.

Structural features of coumarin derivatives essential to oral anticoagulant activity are based on the groups substituted in the 3 and 4 positions of the lactone ring. Link, who pioneered the isolation and characterization of bishydroxycoumarin from sweet clover, concluded after examination of typical coumarins that the minimal requirements for anticoagulant activity are a 4-hydroxy group, a 3-substituent, and a bis molecule.[9] Bishydroxycoumarin

Table 20–3. Coumarin Derivatives

Generic Name (Trade Name)	Chemical Name	Structure	Onset (hours)	Duration (days)	Dosage Forms
—	4-Hydroxycoumarin		—	—	—
Bishydroxycoumarin (Dicumarol)	3,3′-Methylenebis(4-hydroxycoumarin)		17–120	2–10	Tablets: 25, 50, and 100 mg
Warfarin sodium (Coumadin, Panwarfin)	3-(α-Acetonylbenzyl)-4-hydroxycoumarin		36–72	2–5	Tablets: 2, 2.5, 5, 7.5, and 10; 50-mg injectable vial with 2-ml ampules sterile water for injection

* Asymmetric center.

was a bis type of coumarin but is no longer used in medicine.

Subsequent investigators compared the activity of the coumarins with that of vitamin K because the oral anticoagulants are competitive inhibitors of vitamin K in the biosynthesis of prothrombin (factor II) in the liver. They also are believed to depress the hepatic synthesis of other vitamin K–dependent clotting factors, VII, IX, and X. Warfarin inhibits vitamin K reductase and vitamin K epoxide reductase enzymes to decrease the availability of reduced vitamin K, which serves as a substrate for the carboxylase enzyme in the γ-carboxylation of the proenzymes (Figs. 20–3 and 20–4).

Coumarins exert their effects in vivo only after a latent period of 12 to 48 hours. Their effects last for 1.5 to 5 days. The observed slow onset may be due to the time required to decrease predrug prothrombin blood levels, (half-life = 60 hours), whereas the long duration of action observed with warfarin (4 to 5 days) may be due to the lag time required for the liver to resynthesize pro-

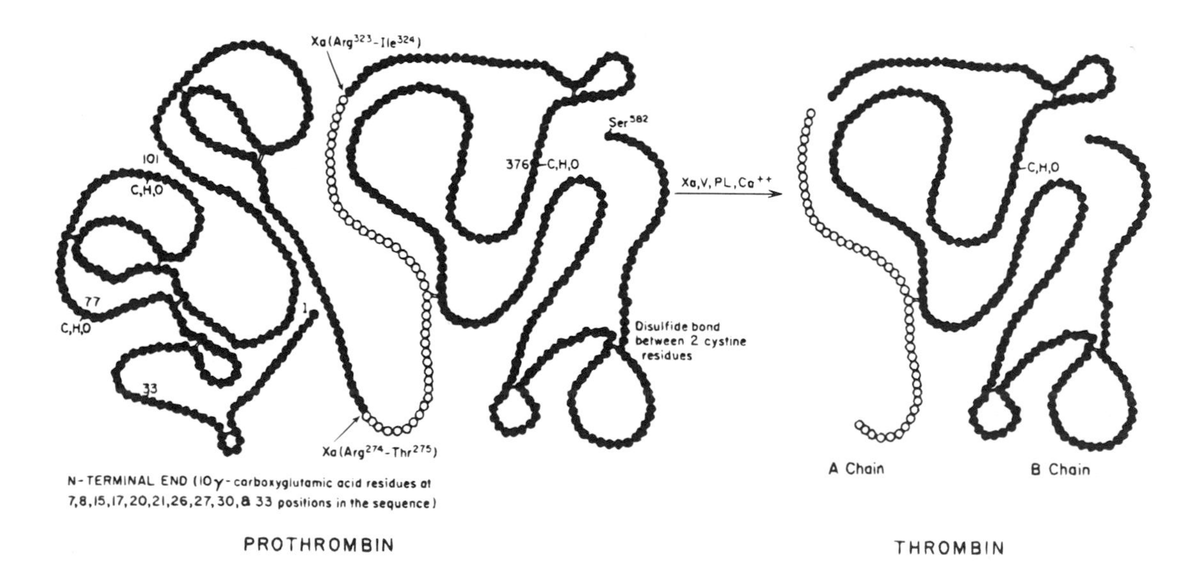

Fig. 20–4. Biosynthesis of thrombin from prothrombin. Cleavage by factor Xa occurs at arginine^{274}–threonine^{275} and at arginine^{323}–isoleucine^{324} to produce the two-chain thrombin molecule. (Modified from M. P. Esnouf, Prothrombin and related proteins, in *Human Blood Coagulation, Haemostasis and Thrombosis,* R. Biggs and C. R. Rizza, Eds., Boston, Blackwell Scientific Publications, 1984, p. 34.)

thrombin to predrug blood levels. One postulation is that vitamin K is converted to an active hemiketal form in the body, as shown by equation 4. Oral anticoagulants that compete with vitamin K or function as an antimetabolite could be converted to similar cyclic hemiketals, as shown in equation 5.

In the presence of vitamin K antagonists, an abnormal prothrombin glycoprotein with little, if any, biologic activity is biosynthesized. Abnormal prothrombin is identical to normal prothrombin except that the 10 γ-carboxyglutamic acid residues are replaced partly by glutamic acid when oral anticoagulants are present. The normal vitamin K–dependent coagulation proteins contain 10 to 13 glutamic acid residues that are γ-carboxylated during the vitamin K–dependent carboxylase conversion reaction (see Fig. 20–3). Prothrombin molecules with fewer than six γ-carboxylated glutamic acid residues have only 2% of the normal prothrombin activity and are fraudulent (abnormal) prothrombin molecules. Prothrombin molecules with nine γ-carboxylated glutamic acid residues have 70% of the normal activity.[3] The abnormal prothrombin does not bind calcium as readily, which reduces its ability to bind with phospholipid and other biochemicals involved in thrombin formation[3] (Fig. 20–4).

Other studies have shown that methoxy groups increase the activity, especially when substituted at position 8.[10] Nuclear magnetic resonance studies show that warfarin exists in three forms in solution: two cyclic diastereomeric hemiketals and one open-chain form.[11]

The commercially available coumarin derivative is summarized in Table 20–3.

Clinical Pharmacology

The coumarin derivatives commonly used for thrombophlebosis, pulmonary embolism, and coronary thrombosis vary considerably in their activity, which is reflected in different onsets and durations of action. They are normally administered once daily in the late afternoon. Absorption and response to the drug vary considerably from one patient to another, and adjustments in dosage may be required from time to time. Close supervision of ambulatory patients through biweekly checks of prothrombin times may also be necessary. Thromboplastin, however, employed in the "Quick One-Stage" prothrombin method, varies in purity, and this has led to a reassessment of prothrombin times. Studies now suggest that therapeutic prothrombin times should be maintained at 1.5 to 2.0 times the normal value of 11 to 13 seconds.

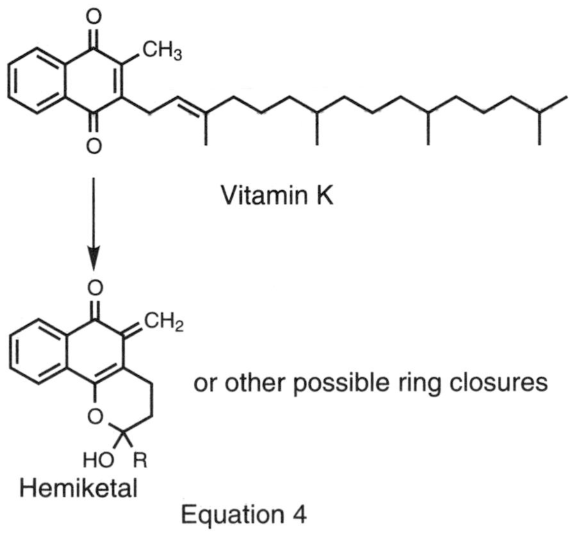

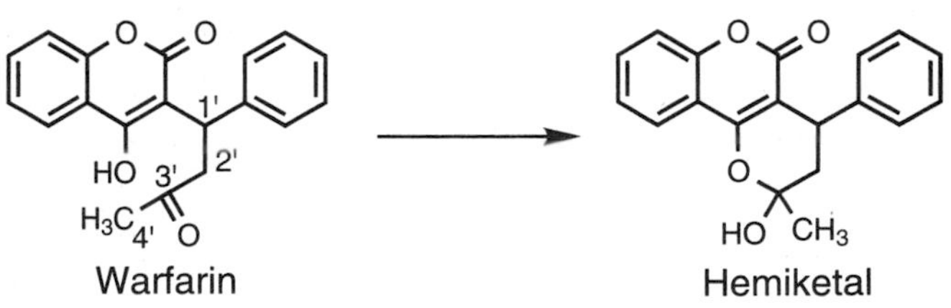

Vitamin K

or other possible ring closures

Hemiketal

Equation 4

Warfarin

Hemiketal

Equation 5

Even with careful monitoring of prothrombin times with warfarin therapy, however, bleeding complications still occur in 20% of the patients. Studies show that the measurement in blood of the native prothrombin antigen (normal fully carboxylated prothrombin) by an immunoassay method may be more accurate in assessing warfarin therapy. For example, it was found that maintaining a therapeutic window of 12 to 24 μg/ml of native prothrombin antigen (normal prothrombin levels are 110 μg/ml) leads to an 85% reduction in bleeding or thrombotic complications in patients treated with warfarin. Bleeding complications occurred at 12 μg/ml or lower, and thrombotic complications occurred at 24 μg/ml or higher. Therefore the native prothrombin antigen test may more accurately define the patient's proper warfarin dosage, which would lead to fewer side effects. The native prothrombin antigen test could well be the assay of the future.[3]

Because of the slow onset of action of oral anticoagulants (12 to 48 hours), therapy should be initiated at the same time as heparin, if the latter drug is necessary. High heparin blood levels prolong the prothrombin time, which makes it difficult to adjust oral anticoagulant dosage. This can be minimized by drawing blood for prothrombin times immediately before a dose of heparin.

Various factors can produce changes in the prothrombin time, which may be difficult to control. For example, variations in intake and absorption of dietary vitamin K may change the prothrombin time. In addition, the dietary intake of fats can modify absorption of vitamin K and influence the therapeutic regimen. Because coumarins are metabolized primarily to inactive hydroxylated products in the liver, patients with decreased hepatic function are more susceptible to anticoagulant drugs.

Coumarins and 1,3-indandiones interact with certain concurrently administered drugs, and pharmacologic activity is thereby altered. For example, the action of oral anticoagulants can be enhanced by drugs such as phenylbutazone and salicylates and antagonized by barbiturates and vitamin K (Table 20–4). Salicylate enhancement of anticoagulant activity is believed to occur through salicylate molecules displacing coumarin molecules from their protein-binding sites and therefore increasing the amount of available drug, as evidenced by higher blood

Table 20–4. Changes in Drug Activity with Anticoagulants

Factors That Enhance Oral Anticoagulant Activity	Factors That Antagonize Oral Anticoagulant Activity	Drugs Enhanced by Oral Anticoagulants
ACTH	Barbiturates (e.g., phenobarbital)	Sulfonylureas
Broad-spectrum antibiotics (suppress intestinal flora required for vitamin K synthesis)	Carbamazepine	Diphenylhydantoin
Cimetidine	Chloral hydrate	
Clofibrate	Glutethimide	
Corticosteroids (e.g., hydrocortisone)	Griseofulvin	
Guanethidine	Meprobamate	
Methylphenidate	Penicillins	
Methylthiouracil	Rifampin	
Metronidazole	Vitamin K	
Omeprazole	Diarrhea	
Oxyphenylbutazone		
Phenylbutazone		
Phenylramidol		
Quinidine		
Quinine		
Salicylates (e.g., aspirin)		
Sulfinpyrazone		
Sulfonamides		
D-Thyroxine		
Fever		
Stress		
Radioactive compounds		
Ticlopidine		
X-rays		

levels of unbound anticoagulant. In addition, other factors, such as fever and stress, can enhance the anticoagulant effect, whereas diarrhea can antagonize it. The dosage of anticoagulant drug may therefore have to be reduced or increased, depending on whether enhancement or antagonism is present.

In view of these adverse effects on oral anticoagulant activity, the physician must exercise caution when prescribing drugs to be taken concurrently. Patients taking coumarin drugs, however, can still be given acetaminophen as an antipyretic and acetaminophen, d-propoxyphene, codeine, and occasionally salicylates for relief of mild pain.

The safest and most commonly prescribed coumarin anticoagulant is warfarin. Usually 40 to 60 mg of warfarin is given the first day, none on the second, and the dosage administered every third day thereafter as based on the prothrombin time. Maintenance doses usually are 5 to 10 mg. Peak action for each dose of warfarin occurs at 36 to 72 hours, with duration of action of 4 to 5 days.

Toxic side effects that may develop are rash, nausea, vomiting, diarrhea, fever, jaundice, leukopenia, thrombocytopenia, and vasculitis. Coumarin drugs can cross the placental membrane and possibly cause lethal neonatal hemorrhage. Significant bleeding occurs in skin, mucous membranes, and gastrointestinal and genitourinary tracts. Antidotes for bleeding and excessively long prothrombin times are vitamin K and whole blood. Depending on the severity of the disorder, vitamin K may be given orally, intramuscularly, or by slow intravenous infusion.

Coumarins such as coumachlor are used principally as rodenticides.

1,3-Indandione Derivatives

Chemistry and Mode of Action

The 1,3-indandiones have been recognized for their anticoagulant activity since the 1940s.

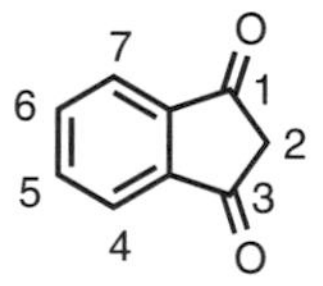

1,3-Indandione

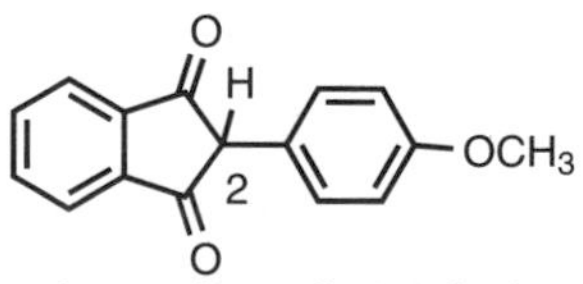

2-(ρ-methoxyphenyl)-1,3-indandione (Anisidione, Miradon)

A commercially available indandione is anisindione, which is substituted with a lipophilic 2-(p-methoxyphenyl) group. Because of the weakly ionizable proton on the second carbon atom, it is soluble in alkaline solutions as shown in Equation 6. This proton becomes activated because of the electron withdrawing effects of the 1,3-dicarbonyl groups and the p-substituted phenyl ring in anisindione. Anisindione has an onset of 48 to 72 hours with a duration of 1 to 3 days.

The mechanism of action of 1,3-indandiones is similar to that of coumarin derivatives in that the synthesis of plasma prothrombin and factors VII, IX, and X are inhibited, thereby lengthening the prothrombin times.

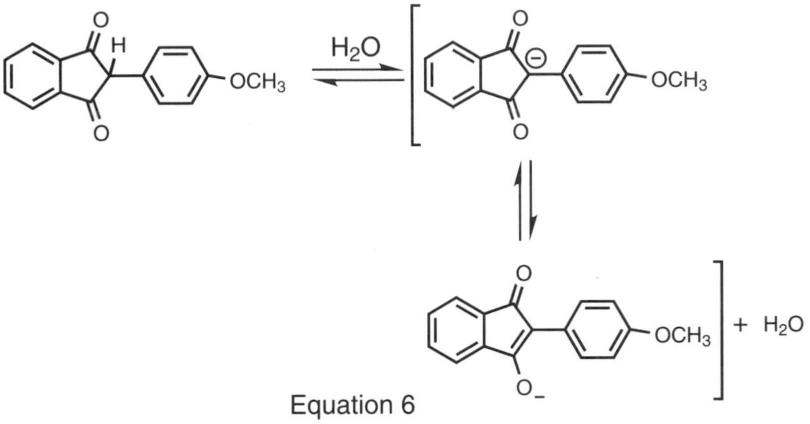

Equation 6

Little is known about the metabolism of this class of oral anticoagulants except for phenindione and its analogs, which produce metabolites that color the urine red-orange. This may be alarming to the patient who is unable to distinguish this phenomenon from hematuria. Acidification of the urine removes the color, thus differentiating it from hematuria. The patient receiving this drug should be forewarned by either the physician or the pharmacist concerning this red-orange color.

Clinical Pharmacology

The onset and duration of action of anisindione are shorter than those of bishydroxycoumarin and about the same as those of warfarin. The chief disadvantage of indandiones is their side effects. Approximately 1.5 to 3.0% of patients are hypersensitive to it and develop a rash, pyrexia, and leukopenia, accompanied by malaise and headache. Anisindione has been reported to have fewer toxic side effects than other drugs of this class; only dermatitis has been observed.

Considering the toxicity and side effects of indandiones, it would appear that warfarin would be the drug of choice.

Miscellaneous Anticoagulants

Citric Acid

Citric acid is a water-soluble, tricarboxylic acid widely distributed in plants and in animal tissues and fluids. It is produced through fermentation of molasses and strains of Aspergillus niger and occurs in citrus fruit and pineapple waste. Lemon juice, for example, contains 5 to 8%. Citric acid has many industrial and pharmaceutical uses, mostly as an acidulant in beverages, confectionery, pharmaceutical syrups, elixirs, tablets, and effervescent powders.

Medicinally, citric acid is usually used as sodium citrate (trisodium citrate). When this drug is taken orally as the sodium salt, the citrate anion is metabolized in vivo to the bicarbonate anion, ultimately producing systemic alkalinization, which is of value for treatment of acidosis. The oral alkalinizing dose is normally 1.0 g. The principal use for sodium citrate is as an in vitro anticoagulant. The citrate anion (Ci^{-3}) forms a soluble complex with calcium by sequestering the calcium ion (factor IV), one of the agents required to convert prothrombin to thrombin. Clot formation is therefore inhibited.

Sodium citrate is commonly found in official preparations, such as anticoagulant sodium citrate solution, USP (4%; 50 ml added to 500 ml of blood or plasma) and anticoagulant acid citrate dextrose solution, USP (0.73 or 0.44%). Both solutions are used as anticoagulants for whole blood. It is recommended that either 15 ml of the 0.73% solution or 25 ml of the 0.44% solution be added to 100 ml of whole blood to prevent clotting.

The use of sodium citrate in vivo as an anticoagulant

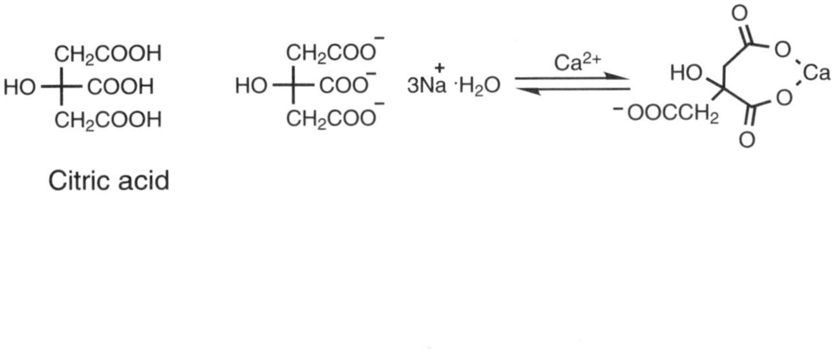

Citric acid

is not possible because of the toxic manifestations of sequestering systemic calcium.

Fibrinolysis

It has been estimated that 200,000 deaths annually are caused by pulmonary embolism, and 676,000 hospital admissions for acute myocardial infarction were recorded in 1983. Research offers new approaches for combating thromboembolic circulatory disorders through the use of four peptide enzymes, urokinase, streptokinase, alteplase, and antistreplase. Although anticoagulant drugs such as heparin have been used as adjuncts in the treatment of thromboembolic disorders, they have not been successful in dissolving formed coronary and pulmonary clots. The discovery of fibrinolytic agents gives the clinician major therapeutic alternatives to heparin or surgical treatment.

The proteolytic enzymes (Table 20–5) act as amidases by splitting a peptide bond of plasminogen, converting it to plasmin, which breaks down clots formed in situ (thrombi) and those carried in the blood to other sites (emboli) (equations 7, 8, and 9). Urokinase is believed to cleave the arginine^{560}–valine bond of single-chain plasminogen to form plasmin, a two-chain protein. Streptokinase requires a two-step reaction. First, a stoichiometric activated complex is formed between plasminogen and streptokinase that, in turn, cleaves the arginine^{560}–valine bond, as urokinase does. Streptokinase generates about 10 times more plasmin than urokinase.[13] Both are used to treat acute massive pulmonary embolism and coronary artery thrombosis. The Food and Drug Administration (FDA) warns that these agents must be used cautiously because of the danger of life-threatening cerebral and other hemorrhages during therapy and because they degrade coagulation proteins indiscriminately. Overdoses of fibrinolytic agents can be reversed by injections of aminocaproic acid.[14] Streptokinase can be used by intravenous infusion or by intracoronary infusion. Because streptokinase contains no preservatives, it should be reconstituted immediately before use.[15]

On intravenous administration, alteplase binds to fibrin in the thrombus and reacts with the entrapped plasminogen, converting it to plasmin. The plasmin degrades the fibrin, which limits alteplase's effects somewhat to systemic proteolysis. Alteplase is produced using the complementary DNA for natural human tissue-type plasminogen activator obtained from a human melanoma cell line.

Anistreplase is an inactive prodrug that is synthesized in vitro from lys-plasminogen and streptokinase. When streptokinase combines with plasminogen systemically to form activated plasminogen-streptokinase, the plasminogen is predominantly glu-plasminogen compared with lys-plasminogen in the anistreplase complex. By forming the p-anisoyl derivative of lys-plasminogen streptokinase activator complex, the drug is inactive until deacylation of the p-anisoyl group occurs in vivo. The fibrinolytic half-life of anistreplase is 94 minutes.

Plasminogen (profibrinolysin) (inactive glycoprotein found in clots and blood; mol. wt. 90,000, 790 amino acids) + Urokinase (UK) (mol. wt. 55,000) —Splits ARG^{560} - Val bond and possibly at a second site→ Plasmin (fibrinolysin) lyses fibrin clot

Equation 7

Plasminogen + Streptokinase(SK) (mol. wt. 48,000 ⇄ Plasminogen-SK activated complex (1:1)

Plasminogen + Plasminogen-SK activated complex (1:1) —Splits ARG^{560} - Val bond and possibly at a second site→ Plasmin

Equation 8

Clot (fibrin) + Alteplase or Anistreplase ⇄ Fibrin-Alteplase complex

Fibrin-Alteplase complex ⇅ Plasminogen → Fibrin-Alteplase-plasminogen complex

Fibrin-Alteplase-plasminogen complex → Plasmin (acts on fibrin in the clot)

Equation 9

Table 20–5. Proteolytic Drugs

Generic Name (Trade Name, Common Name)	Source and Properties	Indications
Alteplase, recombinant (Activase, tissue plasminogen activator, tPA)	Produced by recombinant DNA; molecular weight ~60,000 (527 amino acids)	Acute myocardial infarction; pulmonary embolism; possible use for unstable angina pectoris
Anistreplase (Eminase, anisoylated plasminogen streptokinase activator complex, APSAC)	Inactive p-anisoylated derivative of the lys-plasminogen streptokinase activator complex; produced in vitro from lys-plasminogen and streptokinase	Acute myocardial infarction
Urokinase (Abbokinase)	Human urine; peptide; molecular weight 55,000	Pulmonary emboli; coronary artery thrombosis; IV catheter clearance
Streptokinase (Kabikinase, Streptase)	β-Hemolytic streptococci; peptide; molecular weight 48,000	Pulmonary emboli; coronary artery thrombosis (intracoronary and IV infusion); clearing A-V cannulae

ANTIPLATELET DRUGS

Considerable interest and research have focused on the use of platelet inhibitor drugs in preventing or modifying coronary artery disease. Because platelet aggregation is implicated in thrombus formation, particularly in the arterial system, and in general in the pathogenesis of atherosclerosis, drugs that inhibit platelet aggregation should be able to modify or prevent atherosclerotic disease. The four drugs that show the most promise are aspirin, sulfinpyrazone (Anturan), dipyridamole (Persantine), and ticlopidine (Ticlid).

A review by the FDA of seven large studies (11,000 people) involving aspirin concluded that one aspirin tablet a day (0.324 g = 5 g) reduces by 20% the risk of a second heart attack in those who have had one attack already. One aspirin tablet (5 g) per day reduced by 51% the risk of a heart attack in men with unstable angina in a study involving 1266 men.[16] A 5-year study involving 22,000 healthy physicians taking an aspirin tablet (325 mg) every other day establishes its preventive everyday use.[17] There were 139 myocardial infarctions among those assigned to aspirin and 239 among those assigned to aspirin placebo, which represents a 44% reduction in risk. A slightly increased risk of stroke among those taking aspirin was not statistically significant. The reduction in the risk of myocardial infarction was shown only among those who were 50 years of age or older. The benefit was present at all levels of cholesterol but appeared the highest at low levels. Ulcer occurrence in the aspirin group was 169 participants compared with 138 in the placebo group. The American Heart Association guidelines now suggest that 160 mg of aspirin every other day or 80 mg every day is effective for preventing myocardial infarctions and strokes, while decreasing the frequency of hemorrhagic strokes.

The FDA has estimated that 30,000 to 50,000 lives could be saved each year, considering that 5 million people in the United States have a history of heart disease, angina, or both. Another clinical study shows that aspirin may be effective in high doses in treating the complications of Kawasaki disease, a disorder in children (6 months to 4 years) that produces coronary aneurysms leading to heart attack and death.[18]

Aspirin acts as an antiplatelet drug by irreversibly inhibiting platelet aggregation by acetylating cyclooxygenase, a platelet enzyme. This, in turn, inhibits the synthesis of thromboxane A_2, a powerful vasoconstrictor and inducer of the platelet release reaction and platelet aggregation. The irreversible effect lasts for the life of the acetylated platelet (4 to 7 days).

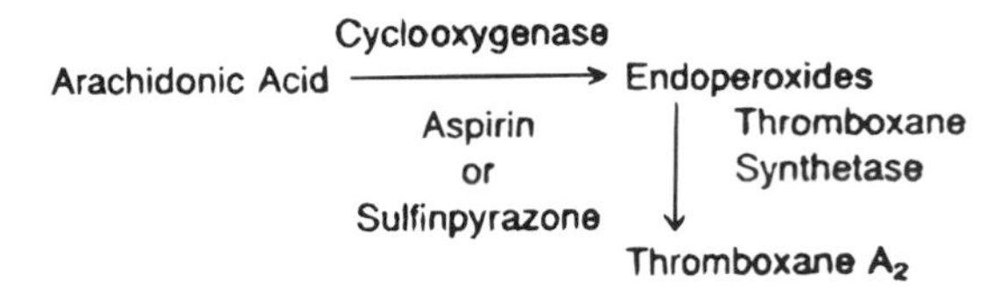

Sulfinprazone (Anturane), a uricosuric drug chemically related to phenylbutazone, has antiplatelet activity because of its thioether metabolite. It inhibits cyclooxygenase but is less active than aspirin. It reversibly inhibits the enzyme, in contrast to aspirin, which has irreversible inhibitory activity (see Table 20–6). Clinical studies in 1629 post–myocardial infarction patients indicate that 200 mg four times daily reduces the risk of sudden heart attack death.

A third drug that has been studied for its antiplatelet activity alone or in combination with aspirin is dipyridamole (Persantin), a coronary vasodilator. Preliminary results indicate that dipyridamole therapy may be beneficial in preventing myocardial infarction and sudden death when given soon after infarction has occurred and may be beneficial in preventing occlusion of coronary artery bypass grafts. Dipyridamole is indicated as an adjunct to coumarin anticoagulants to prevent postoperative thromboembolic complications of cardiac valve replacement. Its antiplatelet mechanism is different from that of aspirin or sulfinpyrazone. It blocks the platelet phosphodiester-

Table 20–6. Antiplatelet Drugs

Name	Structure	Dose	Indications
Acetylsalicylic acid (aspirin)	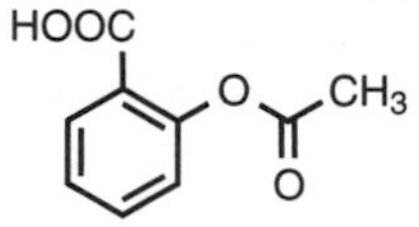 HOOC, O, CH3, O	0.324 g (doses as low as 80 mg every day or 160 mg every other day are effective with fewer side effects)	Prophylactic protective effects in 0.324 g (5 g) daily dose against acute myocardial infarction in men with unstable angina and in preventing a second heart attack.[16–18] Prophylactic effect against myocardial infarction in healthy men (0.325 g every other day)[17]
Sulfinpyrazone (Anturan)	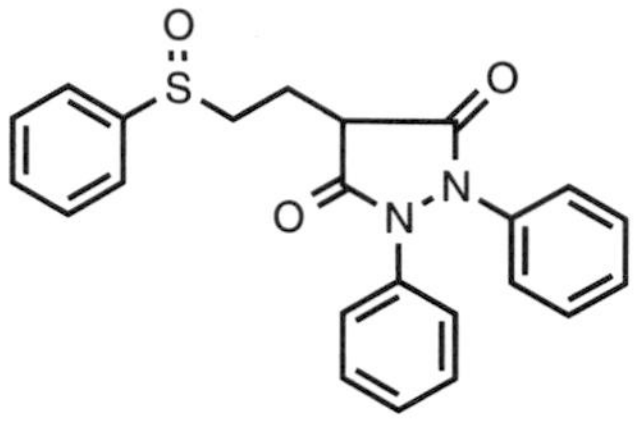 O, S, O, O, N, N	600–800 mg daily (200 mg four times a day)	May reduce mortality (sudden death) during period after a myocardial infarction
Dipyridamole (Persantin)	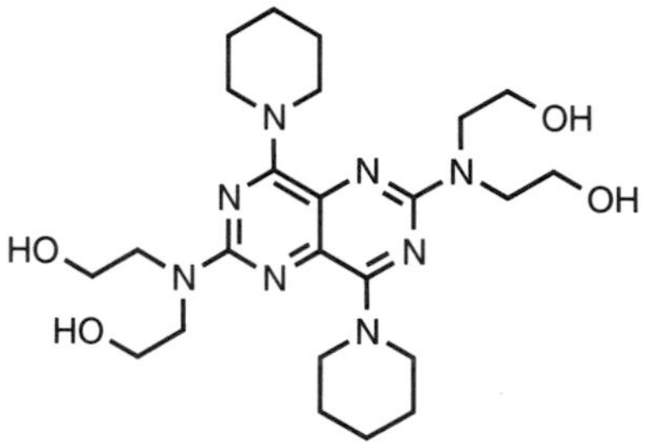 N, OH, N, N, N, OH, HO, N, N, N, HO, N	Aspirin (32.5 mg daily) plus dipyridamole (75 mg three times daily)	Alone and with aspirin to prevent myocardial reinfarction and reduction of mortality post–myocardial infarction; in combination with aspirin to prevent coronary bypass graft occlusion[19,20]
Ticlopidine HCl (Ticlid)	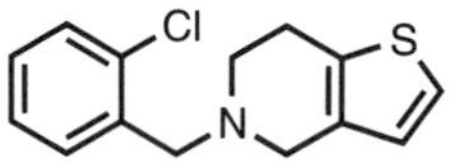 Cl, S, N	250-mg tablets two times daily with food	To reduce the risk of fatal or nonfatal thrombotic strokes in patients who have had stroke precursors or a thrombotic episode. Owing to potential neutropenia agranulocytosis toxicity, ticlopidine should be reserved for aspirin-intolerant patients

ase enzyme, therefore leading to higher cyclic adenosine monophosphate (cAMP) levels. The increase in cAMP levels by dipyridamole inhibits platelet release and aggregation and augments the rise in cAMP and platelet inhibition induced by vessel wall prostacyclin.

Ticlopidine was approved by the FDA in 1991. It inhibits both platelet aggregation and the release of platelet granule substances. Ticlopidine inhibits ADP-induced fibrinogen-platelet binding and the resultant platelet-platelet interactions. Ticlopidine reduced the risk of fatal and nonfatal stroke by 24% compared with aspirin in a study in patients who were experiencing stroke warning signs. Aspirin is considered the drug of choice, unless the patient is intolerant to aspirin. Ticlopidine has been associated with a risk of neutropenia and agranulocytosis, which can be life-threatening.

All four drugs—aspirin, sulfinpyrazone, dipyridamole, and ticlopidine—may cause one or more side effects, such as gastrointestinal disturbances, bleeding complications, headaches, and skin rash.

COAGULANTS

The coagulant drugs are classified according to their chemical structures and discussed according to whether the effect is systemic or local. Generally, coagulants include vitamin K and its analogs, proteins and amino acids, plasma coagulants, and miscellaneous coagulants such as cellulose and carbazochrome salicylate. Vitamin K and its analogs are discussed in Chapter 29.

Proteins and Amino Acids

Thrombin, fibrinogen, collagen, and gelatin are the proteins and aminocaproic acid is the one amino acid used systemically or locally for coagulating effects (Table 20–7).

Table 20–7. Coagulant Drugs

Generic Name (Trade Name)	Molecular Weight	Activity	Solubility in Water	Dosage Forms	Stability and Storage
Proteins: Thrombin, topical (Throminar, Thrombostat)	33,580	Topical	Freely soluble	Vials: 1,000, 5,000 and 10,000, 20,000, 50,000 units (1 unit = amount necessary to coagulate 1 ml of standard fibrinogen in 15 sec)	2–8°C; 3 years in solid state. (Use within a few hours after reconstitution and only topically)
Fibrinogen, human	400,000	Systemic	Sparingly soluble	1 and 2 g; dose = 2–6 g IV	2–8°C; 5 years
Antihemophilic factor, factor VIII (AHF) (Hemofil M, Koate, Monoclate)	—	Systemic	Sparingly soluble	80, 125, 175, 225, 250, 275, 1000 units (1 unit = antihemophilic factor activity present in 1 ml of normal fresh pooled plasma less than 1 h old)	2–8°C; 1–2 years
Antihemophilic factor, recombinant, rFVIII (Recombinate)	—	Systemic	—	—	—
Factor IX complex, human (Konyne, Proplex, Profilnine, Alphanine)	—	Systemic	Soluble	500 units (1 unit = average factor activity present in 1 ml of normal fresh pooled plasma less than 1 h old)	2–8°C; 2 years
Coagulation Factor IX (Mononine, Monoclonal Factor IX)	—	Systemic	—	—	—
Anti-inhibitor coagulant complex (autoplexT, Feiba VH-Immuno)	—	Systemic	—	Each vial labeled with units of Factor VIII Correctional Activity. One unit of Factor VIII Correctional Activity is the quantity of activated prothrombin complex that, on addition to an equal volume of Factor VIII deficient or inhibitor plasma, will correct the clotting time (ellagic acid-activated partial thromboplastin time) to 35 sec (normal)	2–8°C; (unreconstituted)
Absorbable gelatin sponge (Gelfoam)	—	Topical and by implantation	—	Sponges, packs, dental packs, prostatectomy cones	—
Absorbable gelatin film, sterile (Gelfilm, Gelfilm Ophthalmic)	—	Topical and by implantation	—	Film (100 mm × 125 mm); ophthalmic (25 mm × 50 mm)	—
Absorbable gelatin powder, sterile (Gelfoam)	—	Topical and by implantation	—	Powder (1-g jars)	—
Microfibrillar collagen hemostat (Avitene)	—	Topical	Used dry	Fibrous form (1, 5 g jars); nonwoven web form (70 mm × 70 mm × 1 mm and 70 mm × 35 mm × 1 mm)	—
Amino acids: Aminocaproic acid (Amicar)	6-Amino-hexanoic acid	Systemic	Freely soluble	Tablets: 500 mg; syrup: 25%; injection vial: 250 mg/ml, 20 ml (dosage = 10 g/day orally; or 6 g IV initially and then 6 g over 24 h up to 0.2 g/kg)	—
Tranexamic acid	4-Aminomethyl-cyclohexane carboxylic acid	Systemic	Freely soluble	Tablets: 500 mg; injection: 100 mg/ml, 10 ml	—
Miscellaneous coagulants: Oxidized cellulose (Surgicel, Oxycel)	—	Topical	Insoluble	Gauze pads, strips, Foley cones, cotton pledgets	Protect from sunlight; store in refrigerator

Thrombin, fibrin, microfibrillar collagen, and gelatin are used topically to clot oozing blood. Fibrin and thrombin play a role in the clotting process and possibly speed up clot formation because of their participation in the reaction shown in Figure 20–1. Gelatin is used alone or in combination with thrombin and is absorbed in 3 to 5 weeks after topical application. In addition to promoting clotting, these proteins may mechanically aid in stopping the flow of blood, depending on the form in which they are used (e.g., fibrin foam or absorbable gelatin sponge).

Aminocaproic acid is an ε-substituted amino acid that is believed to inhibit plasminogen, a precursor to plasmin. It is used orally or intravenously to prevent bleeding in hemophilia and after heart and prostate operations in which plasminogen may be activated. Tranexamic acid (4-aminomethylcyclohexane carboxylic acid) is a cyclic carboxylic acid that is a competitive inhibitor of plasminogen's activation to plasmin. It is about 10 times more potent in vitro than aminocaproic acid.

Human antihemophilic factor VIII and human factor IX complex are lyophilized concentrates of fresh plasma used in treating hemophilia A (deficiency in factor VIII) and hemophilia B (deficiency in factor IX). They are administered by slow intravenous injection after reconstitution with sterile water for injection to prevent bleeding episodes or before surgical or dental procedures are performed. These agents are standardized to have greater factor activity than normal human plasma (e.g., antihemophilic factor [human] must have a factor VIII activity 15 times that of normal human plasma). Other products are recombinant antihemophilic factor VIII and monoclonal coagulation factor IX (Table 20–7).

Anti-inhibitor coagulant complex is obtained from pooled human plasma and is used in patients with factor VIII inhibitors who are bleeding or who must undergo surgery. About 10% of patients with hemophilia A have inhibitors to factor VIII that can be measured.

Miscellaneous Coagulants

Miscellaneous coagulants include oxidized cellulose (Table 20–7). Oxidized cellulose is the result of oxidation of gauze or cotton (90% cellulose) so that 16 to 24% of the molecular weight is attributable to carboxyl groups. Cellulose is a glucose polymer involving β,1→4 linkages. The molecular weight has been estimated to be between 50,000 and 400,000. Application of oxidized cellulose to oozing wounds stops the bleeding by producing an artificial clot. It is absorbed in 2 days to 6 weeks.

PLASMA EXTENDERS

Development of shock from the loss of blood during trauma, wounds, or surgery may lead to a decrease in total blood volume of 15 to 20% in mild shock and as much as 40% in severe shock. The loss of blood may be due to hemorrhage, pooling of blood, and loss of plasma into extracellular compartments or to the outside. It is essential to the patient's well-being that therapy with whole blood, one of the blood derivatives, or one of the plasma extenders be initiated as quickly as possible.

The use of whole blood to restore blood volume, espe-

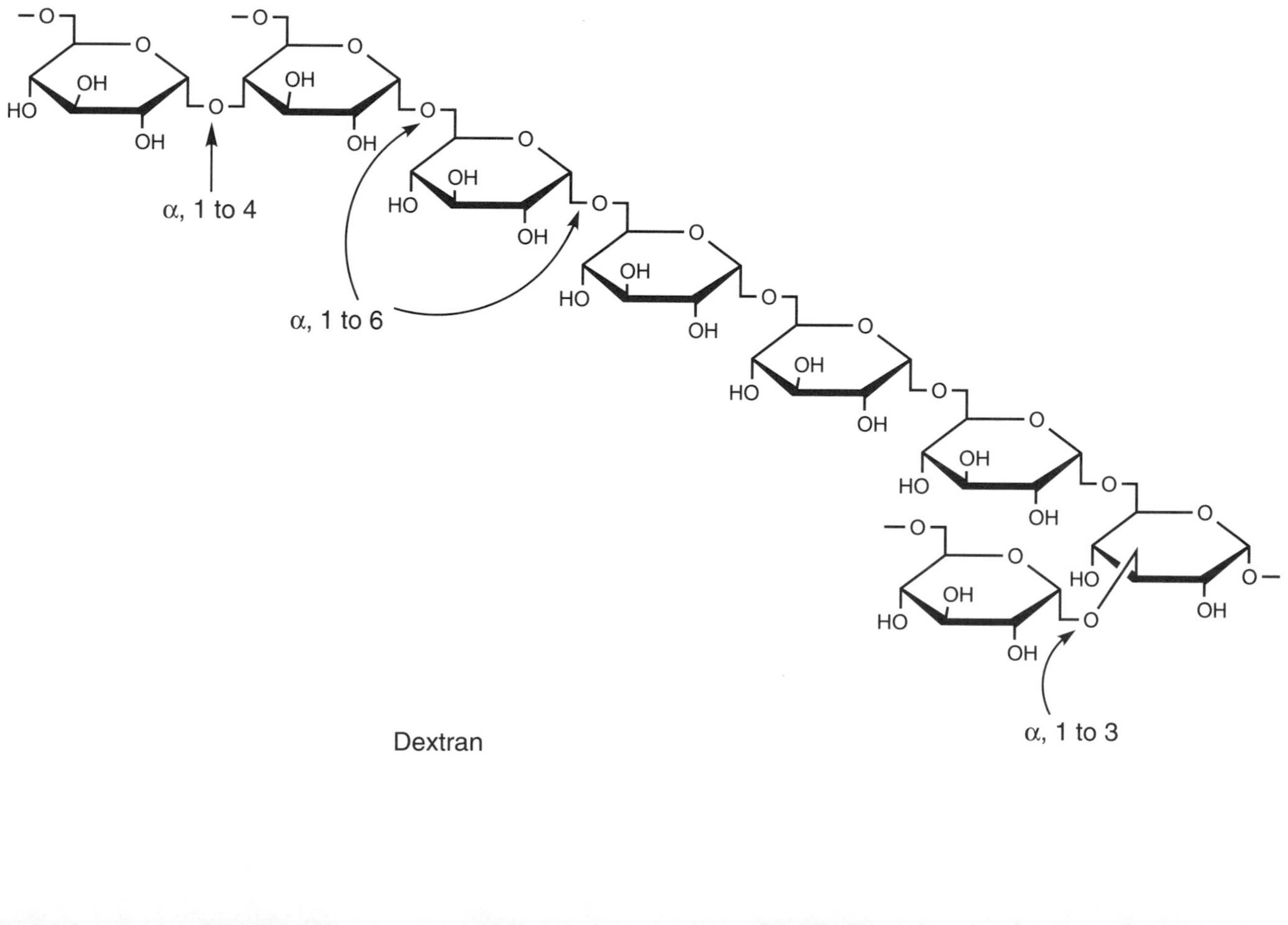

Dextran

cially in shock owing to bleeding, is most desirable. Disadvantages of the use of this and other blood products, such as plasma and albumin, are their expense and lack of availability. In addition, whole blood may cause pyrogenic, hemolytic, or sensitization reactions, serum hepatitis, and difficulties in cross-matching. Human plasma is easy to store, but it produces a higher incidence of hepatitis because of pooling of plasma. Albumin does not present the danger of viral hepatitis but is expensive. When available and properly used, these substances can be lifesaving to the patient in shock.

Blood products used in the regulation of plasma volume are human albumin and plasma protein fraction, human (PPF). Human albumin (albumisol) is administered by infusion in 5 or 25% solutions. It is prepared from pooled normal human blood, plasma, or serum, and at least 90% of the total protein must be albumin. Human PPF contains 5 g of protein per 100 ml, of which 83 to 90% is albumin. Similar to human albumin, PPF is administered by intravenous infusion as a plasma volume expander in the treatment of shock, e.g., from burns, surgery, or hemorrhage.

The plasma extenders that are currently used include polymers, such as the dextrans, and salt solutions that include one or more of the naturally occurring salts in aqueous solution.

Polymers

The dextrans with molecular weights of 40,000, 70,000, and 75,000 and hetastarch (hydroxyethyl starch) are used as plasma extenders.

Chemistry and Mode of Action

Dextran is a large, water-soluble polymer of glucose formed by the action of bacteria (Leuconostoc species) on sucrose substrate (molecular weight ≈ 4 million) that is partially hydrolyzed in vitro to yield dextrans with average molecular weights of 40,000, 70,000, and 75,000.

Structurally, they are long, slender molecules with α,1→6 linkages and α,1→2, α,1→3, or α,1→4 branching, depending on the strain and species of bacteria employed. The branching is believed to occur every five glucose units.

The dextran polymers function in vivo by an osmotic effect similar to that of albumin. The intravascular blood volume incresaes through the migration of water into the vascular system to alleviate the hypertonicity caused by the intravenously administered dextran.

Hetastarch is a complex mixture of ethoxylated amylopectin molecules (polysaccharide) with molecular weights ranging from 10,000 to greater than 1 million and an average molecule weight of 450,000. The activity of 6% hetastarch approximates that of human albumin.

Clinical Pharmacology

Dextran 70 (Dextran, Gentran), with a molecular weight of 70,000, is administered intravenously in doses of 500 ml of a 6% solution in saline solution. Dextran 40 (Gentran, Dextran), with a molecular weight of 40,000, is used as a plasma extender at 10% concentration. Because of its low molecular weight, Dextran 40 minimizes the sludging of blood and rouleau formation in shock. Caution must be exercised with the dextrans because of their antigenicity. Both polymers are used as substitutes for blood or blood products in the treatment of shock. Side effects are few, but if dextran molecules with molecular weights greater than 100,000 are present, the probability of sensitivity reactions increases.

To reduce the antigenicity of Dextran 40 and Dextran 70, Dextran 1 (Promit) (molecular weight 1000) can be given before intravenous administration of Dextran 40 or 70 to minimize the formation of immune complexes with these dextrans and to reduce the occurrence of anaphylaxis by a factor of 15 to 20 times.

Salt Solutions

The salts may be used temporarily until whole blood, plasma, or one of the polymers can be employed. The most commonly employed salt solutions are (1) normal saline, or physiologic salt solution, containing 0.9% sodium chloride in water; (2) Ringer's injection, containing 0.86% sodium chloride, 0.03% potassium chloride, and 0.033% calcium chloride; (3) lactated Ringer's injection, containing 0.3% sodium lactate (racemic mixture), 0.6% sodium chloride, 0.03% potassium chloride, and 0.02% calcium chloride.

The use of salt solutions in the treatment of shock is only a temporary or initial method of increasing blood volume through an osmotic effect because of renal excretion of the salts within several hours. The usual intravenous dose of any one of the three salt solutions is 500 ml.

REFERENCES

1. B. A. Brown, *Hematology: Principles and Procedures,* 4th ed., Philadelphia, Lea & Febiger, 1984, p. 179.
2. E. J. Bowie and C. A. Owen, The hemostatic mechanism, in *Thrombosis,* H. C. Kwaan and E. J. W. Bowie, Eds., Philadelphia, W. B. Saunders, 1982, p. 7.
3. B. Furie and B. C. Furie, *N. Engl. J. Med. 326,* 800(1992).
4. J. Hirsh and M. N. Levine, *Blood 9,* 1(1992).
5. R. D. Hull, et al., *N. Engl. J. Med. 326,* 975(1992).
6. B. Furie and B. C. Furie, *Blood 75,* 1753(1990).
7. M. P. Esnouf, Prothrombin and related proteins, in *Human Blood Coagulation, Haemostasis and Thrombosis,* R. Biggs and C. R. Rizza, Eds., Boston, Blackwell Scientific Publications, 1984, p. 34.
8. J. N. Eble, et al., *Biochem. Pharmacol., 15,* 1003(1966).
9. K. P. Link, *Harvey Lect. 39,* 162(1943).
10. R. B. Arora and C. N. Mathur, *Br. J. Pharmacol., 20,* 29(1963).
11. E. J. Valente, et al., *J. Med. Chem., 20,* 1489(1977).
12. D. Pennica, et al., *Nature, 301,* 214(1983).
13. K. C. Robbins, Fibrinolysis, in *Thrombosis,* H. C. Kwaan and E. J W. Bowie, Eds., Philadelphia, W. B. Saunders, 1982, p. 23.
14. *FDA Drug Bulletin, 8,* 4(1978).
15. The TIMI Study Group, *N. Engl. J. Med., 312,* 932(1993).
16. H. D. Lewis, Jr., et al., *N. Engl. J. Med., 309,* 396(1983).
17. C. H. Hennekens, et al., *N. Engl. J. Med., 321,* 129(1989).
18. G. Koren, et al., *JAMA 254,* 767(1985).

19. J. H. Chesebro, et al., *N. Engl. J. Med., 307,* 73(1982).
20. The American-Canadian Co-operative Study Group, *Stroke, 14,* 99(1983).

SUGGESTED READINGS

R. Biggs and C. R. Rizza, Eds., *Human Blood Coagulation, Haemostasis and Thrombosis,* Boston, Blackwell Scientific Publications, 1984.

I. Chanarin, et al., Eds., *Blood and Its Diseases,* 3rd ed., New York, Churchill Livingston, 1984.

A. Lazlo, Ed. *Blood Platelet Function and Medicinal Chemistry,* New York, Elsevier Biomedical, 1984.

R. W. Colman, et al., *Hemostasis and Thrombosis, Basic Principles and Clinical Practice,* 2nd ed., Philadelphia, J. B. Lippincott, 1987.

O. D. Ratnoff and C. D. Forbes, Eds., *Disorders of Hemostasis,* 2nd ed., Philadelphia, W. B. Saunders, 1991.

Chapter 21

DIURETICS

Gary O. Rankin

Diuretics are chemicals that increase the rate of urine formation. By increasing the urine flow rate, diuretic usage leads to the increased excretion of electrolytes (especially sodium and chloride ions) and water from the body. These pharmacologic properties have led to the use of diuretics in the treatment of edematous conditions resulting from a variety of causes (e.g., congestive heart failure, nephrotic syndrome, chronic liver disease) and in the management of hypertension. Diuretics also are useful as the sole agent or as adjunctive therapy in the treatment of a wide range of clinical conditions, including hypercalcemia, diabetes insipidus, acute mountain sickness, primary hyperaldosteronism, and glaucoma.

The primary target organ for diuretics is the kidney, where these compounds interfere with the reabsorption of sodium and other ions from the lumina of the nephrons, the functional units of the kidney. The amount of ions and accompanying water that are excreted as urine following administration of a diuretic, however, is determined by many factors, including the chemical structure of the diuretic, the site or sites of action of the agent, and the amount of extracellular fluid present. In addition to the direct effect of diuretics to impair solute and water reabsorption from the nephron, diuretics can also trigger physiologic events that have an impact on either the magnitude or the duration of the diuretic response. Thus, it is important to be aware of the normal mechanisms of urine formation and renal control mechanisms to understand clearly the ability of chemicals to induce diuresis.

URINE FORMATION

Two important functions of the kidney are to maintain a homeostatic balance of electrolytes and water and to excrete water-soluble end products of metabolism. The kidney accomplishes these functions through the formation of urine by the nephrons (Fig. 21–1). Each kidney contains approximately 1 million nephrons and is capable of forming urine independently. The nephrons are composed of a specialized capillary bed called the glomerulus and a long tubule divided anatomically and functionally into the proximal tubule, loop of Henle, and distal tubule. Each component of the nephron contributes to the normal functions of the kidney in a unique manner, and thus all are targets for different classes of diuretic agents.

Urine formation begins with the filtration of blood at the glomerulus. Approximately 1200 ml of blood per minute flows through both kidneys and reaches the nephron by way of afferent arterioles. About 20% of the blood entering the glomerulus is filtered into Bowman's capsule to form the glomerular filtrate. The glomerular filtrate is composed of blood components with a molecular weight less than albumin (approximately 69,000) and not bound to plasma proteins. The glomerular filtration rate (GFR) averages 125 ml/minute in humans but can vary widely even in normal functional states.

The glomerular filtrate leaves Bowman's capsule and enters the proximal convoluted tubule, where the majority (50 to 60%) of filtered sodium is reabsorbed isosmotically. Sodium reabsorption is coupled electrogenically with the reabsorption of glucose, phosphate, and amino acids and nonelectrogenically with bicarbonate reabsorption. Glucose and amino acids are completely reabsorbed in this portion of the nephron, whereas phosphate reabsorption is between 80 and 90% complete. The early proximal convoluted tubule is also the primary site of bicarbonate reabsorption (80 to 90%), a process that is mainly sodium dependent and coupled to hydrogen ion secretion. The reabsorption of sodium and bicarbonate is facilitated by the enzyme carbonic anhydrase which is present in proximal tubular cells and catalyzes the formation of carbonic acid from water and carbon dioxide. The carbonic acid provides the hydrogen ion, which drives the reabsorption of sodium bicarbonate. Chloride ions are reabsorbed passively in the proximal tubule, where they follow actively transported sodium ions into tubular cells.

Reabsorption of electrolytes and water also occurs isosmotically in the straight proximal tubule or pars recta. By the end of the straight segment, between 65 and 70% of water and sodium, chloride, and calcium ions; 80 to 90% of bicarbonate and phosphate; and essentially 100% of glucose, amino acids, vitamins, and protein have been reabsorbed from the glomerular filtrate. The proximal tubule is also the site of active secretion of weakly acidic and weakly basic organic compounds. Thus, many of the diuretics can enter luminal fluid not only by filtration at the glomerulus, but also by active secretion.

The descending limb of the loop of Henle is impermeable to ions, but water can freely move from the luminal fluid into the surrounding medullary interstitium, where the higher osmolality draws water into the interstitial space and concentrates luminal fluid. Luminal fluid continues to concentrate as it descends to the deepest portion of the loop of Henle, where the fluid becomes the most concentrated. The hypertonic luminal fluid next

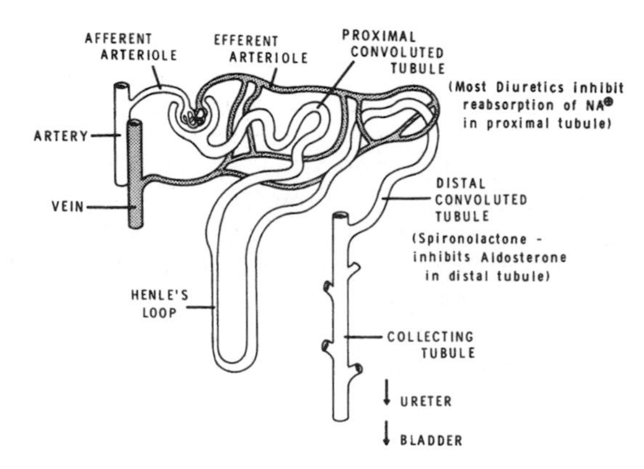

Fig. 21–1. The nephron. (From C. H. Best and N. D. Taylor, *The Living Body,* 3rd ed., New York, Henry Holt & Co., 1952, p. 295.)

enters the water-impermeable thick ascending limb of the loop of Henle. In this segment of the nephron, approximately 20 to 25% of the filtered sodium and chloride ions are reabsorbed via a cotransport system ($Na^+/K^+/2\ Cl^-$) on the luminal membrane. Reabsorption of sodium and chloride in the medullary portion of the thick ascending limb is important for maintaining the medullary interstitial concentration gradient. Reabsorption of sodium chloride in the cortical component of the thick ascending limb and the early distal tubule contributes to urinary dilution, and as a result these two nephron sections are sometimes called the cortical diluting segment of the nephron.

Luminal fluid leaving the early distal tubule next passes through the late distal tubule and cortical collecting tubule, where sodium is reabsorbed in exchange for hydrogen and potassium ions. This process is partially controlled by mineralocorticoids (e.g., aldosterone) and accounts for the reabsorption of between 2 and 3% of filtered sodium ions. Although the reabsorption of sodium ions from these segments of the nephron is not large, this sodium-potassium/hydrogen ion exchange system determines the final acidity and potassium content of urine. Several factors, however, can influence the activity of this exchange system, including the amount of sodium ions delivered to these segments, the status of the acid-base balance in the body, and the levels of circulating aldosterone.

The urine formed during this process represents only about 1 to 2% of the original glomerular filtrate, with more than 98% of electrolytes and water filtered at the glomerulus being reabsorbed during passage through the nephron. Thus, a change in urine output of only 1 or 2% could double urine volume. Urine leaves the kidney through the ureters and travels to the bladder, where the urine is stored until urination removes the urine from the body.

NORMAL REGULATION OF URINE FORMATION

The body contains several control mechanisms that regulate the volume and contents of urine. These systems are activated by changes in solute or water content of the body, by changes in systemic or renal blood pressure, and by a variety of other stimuli. Activation of one or more of these systems by diuretic drugs can modify the effectiveness of these drugs to produce their therapeutic response and may require additional therapeutic measures to ensure a maximal response.

The kidney has the ability to respond to changes in the GFR through the action of specialized distal tubular epithelial cells called the macula densa. These cells are in close contact with the glomerular apparatus of the same nephron and detect changes in the rate of urine flow and luminal sodium chloride concentration. An increase in the urine flow rate at this site (as can occur with the use of some diuretics) activates the macula densa cells to communicate with the granular cells and vascular segments of the juxtaglomerular (JG) apparatus. Stimulation of the JG apparatus causes renin to be released, which leads to the formation of angiotensin II and subsequent renal vasoconstriction. Renal vasoconstriction leads to a decrease in GFR and possibly a decrease in the effectiveness of the diuretic. Renin release also can be stimulated by factors other than diuretics, including decreased renal perfusion pressure, increased sympathetic tone, and decreased blood volume.

Another important regulatory mechanism for urine formation is antidiuretic hormone (ADH), also known as vasopressin. ADH is released from the posterior pituitary in response to reduced blood pressure and elevated plasma osmolality. In the kidney, ADH acts on the collecting tubule to increase water permeability and reabsorption. As a result, the urine becomes more concentrated, and water is conserved.

DIURETIC DRUG CLASSES

Compounds that increase the urine flow rate have been known for centuries. One of the earliest substances known to induce diuresis is water, an inhibitor of ADH release. Calomel (mercurous chloride) was used as early as the sixteenth century as a diuretic, but because of poor absorption from the gastrointestinal tract and toxicity, calomel was replaced clinically by the organomercurials (e.g., chlormerodrin). The organomercurials represented the first group of highly efficacious diuretics available for clinical use. The need to administer these drugs parenterally, the possibility of tolerance, and their potential toxicity, however, soon led to the search for newer, less toxic diuretics. Today the organomercurials are no longer used as diuretics, but their discovery began the search for many of the diuretics used today. Other compounds previously used as diuretics include the acid-forming salts (ammonium chloride) and methylxanthines (theophylline).

The diuretics currently in use today (Table 21–1) are classified by their chemical class (thiazides), mechanism of action (carbonic anhydrase inhibitors, osmotics), site of action (loop diuretics), or effects on urine contents (potassium-sparing diuretics). These drugs vary widely in their efficacy (ability to increase the rate of urine formation) and their site of action within the nephron. Efficacy

Table 21–1. Diuretics: Sites and Mechanisms of Action

Class of Diuretic	Site of Action	Mechanism of Action
Osmotics	Proximal tubule	Osmotic effects decrease sodium and water reabsorption
	Loop of Henle	Increases medullary blood flow to decrease medullary hypertonicity and reduce sodium and water reabsorption
	Collecting tubule	Sodium and water reabsorption decreases because of reduced medullary hypertonicity and elevated urinary flow rate
Carbonic anhydrase inhibitors	Proximal convoluted tubule	Inhibition of renal carbonic anhydrase decreases sodium bicarbonate reabsorption
Thiazides and thiazide-like	Cortical portion of the thick ascending limb of loop of Henle and distal tubule	Inhibition of sodium chloride reabsorption
Loop or high-ceiling	Thick ascending limb of the loop of Henle	Inhibition of the luminal $Na^+/K^+/2\ Cl^-$ cotransport system
Potassium-sparing	Distal tubule and collecting duct	Inhibition of sodium and water reabsorption by: Competitive inhibition of aldosterone (spironolactone); Blockade of sodium uptake at the luminal membrane (triamterene and amiloride)

is often measured as the ability of the diuretic to increase the excretion of sodium ions filtered at the glomerulus (i.e., the filtered load of sodium) and should not be confused with potency, which is the amount of the diuretic required to produce a specific diuretic response.

Efficacy is determined in part by the site of action of the diuretic. Drugs (e.g., carbonic anhydrase inhibitors) that act primarily on the proximal tubule to induce diuresis are weak diuretics because of the ability of the nephron to reabsorb a significant portion of the luminal contents in latter portions of the nephron. Likewise, drugs (potassium-sparing diuretics) that act at the more distal segments of the nephron are weak diuretics because most of the glomerular filtrate has already been reabsorbed in the proximal tubule and ascending limb of the loop of Henle before reaching the distal tubule. Thus, the most efficacious diuretics discovered so far, the high-ceiling or loop diuretics, interfere with sodium chloride reabsorption at the ascending limb of the loop of Henle, which is situated after the proximal tubule but before the distal portions of the nephron and collecting tubule.

OSMOTIC DIURETICS

Osmotic diuretics are low-molecular-weight compounds that are not extensively metabolized and are passively filtered through Bowman's capsule into the renal tubules. Once in the renal tubule, they have a limited reabsorption. They form a hypertonic solution and cause water to pass from the body into the tubule, producing a diuretic effect.

Polyols such as mannitol, sorbitol, and isosorbide provide this effect. Sugars such as glucose and sucrose can also have this diuretic effect. Although not a polyol, urea has a similar osmotic effect and has been used as a diuretic.

Osmotic diuretics are not frequently used in medicine today except in the prophylaxis of acute renal failure to inhibit water resorption and to maintain urine flow. They may be helpful in cases in which urinary output is diminished because of severe bleeding or traumatic surgical experiences. They are not considered primary diuretic agents in ordinary edemas. Isosorbide is an osmotic diuretic that is used primarily to reduce intraocular pressure in glaucoma cases.

Mannitol is the agent most commonly used as an osmotic diuretic. Sorbitol can also be used for similar reasons. These compounds can be prepared by the electrolytic reduction of glucose or sucrose.

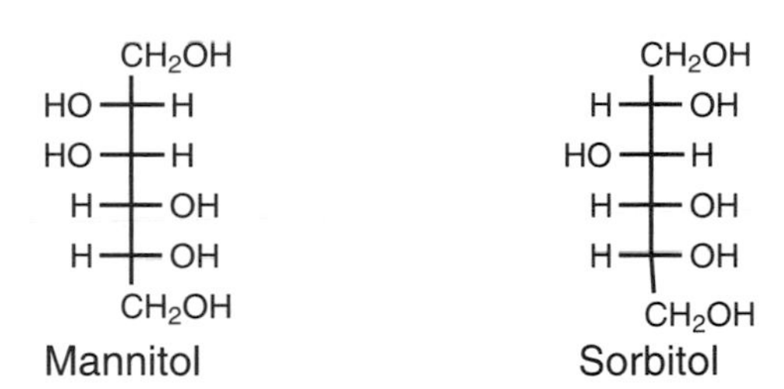

These products are administered intravenously in solutions of 5 to 50%. The rate of administration is adjusted to maintain the urinary output at 30 to 50 ml per hour. Mannitol is filtered at the glomerulus and is poorly reabsorbed by the kidney tubule. The osmotic effect of mannitol in the tubule inhibits the reabsorption of water, and the rate of urine flow can be maintained.

Isosorbide is basically a bicyclic form of sorbitol, used orally to cause a reduction in intraocular pressure. Although a diuretic effect is noted, its ophthalmologic properties are its primary value.

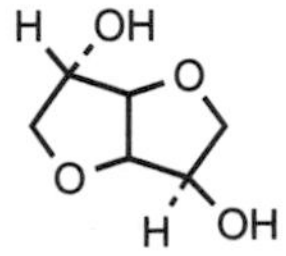

H OH
O
O
H OH
Isosorbide

CARBONIC ANHYDRASE INHIBITORS

It was proposed in 1937 that the normal acidification of urine was caused by secretion of hydrogen ions by the tubular cells of the kidney. These ions were provided by the action of the enzyme carbonic anhydrase, which catalyzes the formation of carbonic acid (H_2CO_3) from carbon dioxide and water.

$$CO_2 + H_2O \longleftrightarrow H_2CO_3 \longleftrightarrow H^+ + HCO_3^-$$

It was also observed that sulfanilamide rendered the urine of dogs alkaline because of the inhibition of carbonic anhydrase. This inhibition of carbonic anhydrase resulted in a lesser exchange of hydrogen ions for sodium ions in the kidney tubule. The sodium ions, along with bicarbonate ions, were then excreted, and a diuretic effect was noted. The large doses required and the side effects of sulfanilamide prompted a search for more effective carbonic anhydrase inhibitors as diuretic drugs.

It was soon learned that the sulfonamide portion of an active diuretic molecule could not be monosubstituted or disubstituted. It was reasoned that a more acidic sulfonamide would bind more tightly to the carbonic anhydrase enzyme. Synthesis of more acidic sulfonamides produced compounds more than 2500 times more active than sulfanilamide. Acetazolamide was introduced in 1953 as an orally effective diuretic drug. Previous to that time, the organic mercurials, which commonly required intramuscular injection, were the principal diuretics available.

Carbonic anhydrase inhibitors induce diuresis by inhibiting the formation of carbonic acid within proximal and distal tubular cells to limit the number of hydrogen ions available to promote sodium reabsorption. For a diuretic response to be observed, more than 99% of the carbonic anhydrase must be inhibited. Although carbonic anhydrase activity in the proximal tubule regulates the reabsorption of about 20 to 25% of the filtered load of sodium, the carbonic anhydrase inhibitors are not highly efficacious diuretics. An increased excretion of only 2 to 5% of the filtered load of sodium is seen with carbonic anhydrase inhibitors owing to increased reabsorption of sodium ions by the ascending limb of the loop of Henle and more distal nephron segments.

With prolonged use of the carbonic anhydrase inhibitor diuretics, the urine becomes more alkaline, and the system becomes more acidic. When acidosis occurs, the carbonic anhydrase inhibitors lose their effectiveness as diuretics. They remain ineffective until normal acid-base balance in the body has been regained. For this reason, this class of compounds is limited in its diuretic use. Today they are most commonly used in the treatment of glaucoma. Acetazolamide reduces the rate of aqueous humor formation, and the intraocular pressure is subsequently reduced. These compounds have also found some limited use in the treatment of absence seizures, to alkalinize the urine and prophylactically to reduce acute mountain sickness.

The following compounds are of interest as carbonic anhydrase inhibitors.

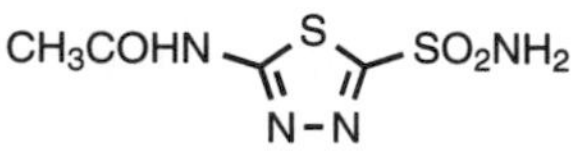

CH_3COHN S SO_2NH_2
N–N
Acetazolamide

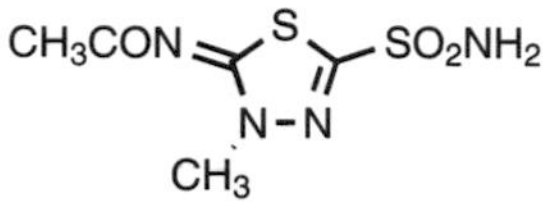

CH_3CON S SO_2NH_2
N–N
CH_3
Methazolamide

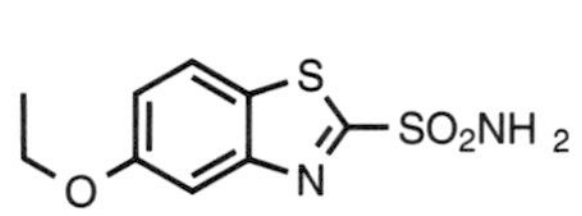

S SO_2NH_2
O N
Ethoxzolamide

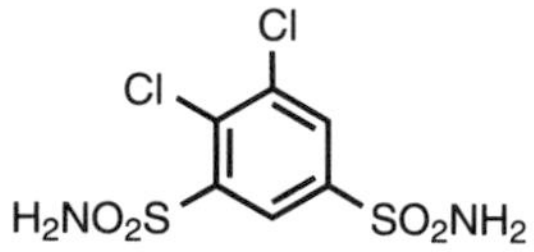

Cl
Cl
H_2NO_2S SO_2NH_2
Dichlorphenamide

Acetazolamide (Diamox) was the first of the carbonic anhydrase inhibitors to be introduced as an orally effective diuretic. It produces a diuretic effect that lasts about 8 to 12 hours. As mentioned earlier, its diuretic action is limited because of the systemic acidosis it produces. It is used primarily in the treatment of glaucoma and absence seizures. The dose is 250 mg to 1 g per day.

Methazolamide is a derivative of acetazolamide in which one of the active hydrogens has been replaced by a methyl group. This decreases the polarity and permits a greater penetration into the ocular fluid, where it acts as a carbonic anhydrase inhibitor, reducing intraocular pressure. Its dose for glaucoma is 50 to 100 mg two to three times a day.

Ethoxzolamide is another carbonic anhydrase inhibitor whose properties and uses resemble those of acetazolamide. Dichlorphenamide is a disulfonamide derivative that shares the same pharmacologic properties and clinical uses as the previously discussed compounds. The dose of dichlorphenamide is 25 to 100 mg one to three times a day.

Further study of these benzene disulfonamide derivatives provided some compounds with a high degree of diuretic activity. Chloro and amino substitution gave compounds with increased activity, but these compounds were weak carbonic anhydrase inhibitors. When the amino group was acylated, an unexpected ring closure took place. These compounds possessed a diuretic activity independent of the carbonic anhydrase inhibitory activity, and a new series of diuretics called the benzothiadiazines was discovered.

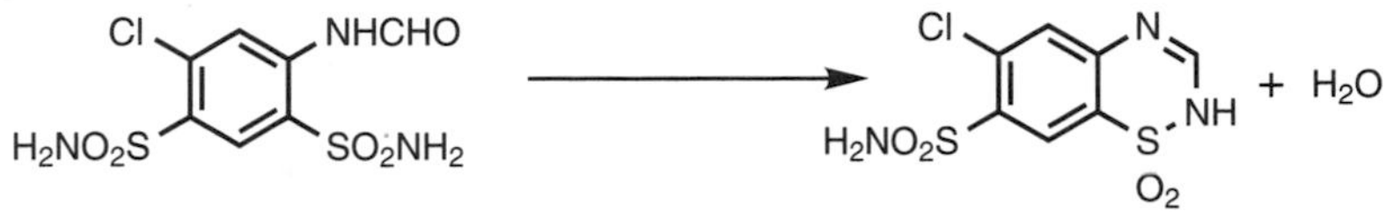

BENZOTHIADIAZINE OR THIAZIDE DIURETICS

The benzothiadiazine or thiazide diuretics have the following nucleus:

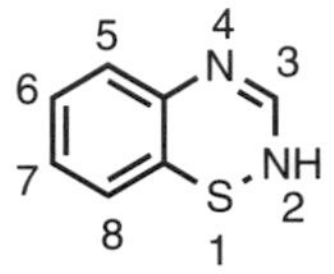

Chlorothiazide is the simplest member of this series.

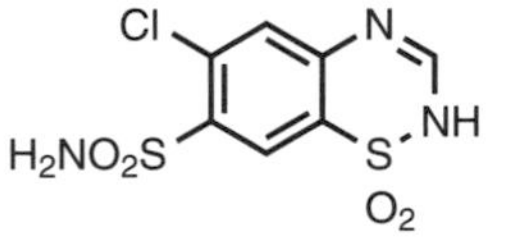

Chlorothiazide

These compounds are weakly acidic. The hydrogen atom at the 2-N is the most acidic because of the electron-withdrawing effects of the neighboring sulfone group. The sulfonamide group that is substituted at C-7 provides an additional point of acidity in the molecule but is less acidic than the 2-N proton These acidic protons make possible the formation of a water-soluble sodium salt that can be used for intravenous administration of the diuretics.

derivative produces a diuretic that is 10 times more active than the unsaturated derivative. Substitution with a lipophilic group at position 3 gives a marked increase in the diuretic potency. Haloalkyl, aralkyl, or thioether substitution increases the lipoid solubility of the molecule and yields compounds with a longer duration of action. Alkyl substitution on the 2-N position also decreases the polarity and increases the duration of diuretic action.

Although these compounds do have carbonic anhydrase activity, there is no correlation of this activity with their saluretic activity (excretion of sodium and chloride ions). Relative activities of the thiazides are shown in Table 21–2.

The mode of action of the benzothiadiazine diuretics is not completely understood. These diuretics are actively secreted in the proximal tubule and are carried to the loop of Henle and to the distal tubule. The major site of action of these compounds is in the distal tubule, with some action possible in the loop of Henle. Here the thiazide diuretics inhibit the reabsorption of sodium and chloride ions. For this reason, they are referred to as saluretics. They also inhibit the reabsorption of potassium and bicarbonate ions but to a lesser degree.

The thiazide diuretics are administered once a day or in divided daily doses. Some have a duration of action that permits administration of a dose every other day. The compounds are rapidly absorbed orally and can show their diuretic effect in an hour. These compounds are not extensively metabolized and are excreted unchanged in the urine. Thiazide diuretics are used to treat edemas caused by cardiac decompensation as well as in hepatic or renal disease. They are also commonly used in the treatment of hypertension. Their effect may be attributed to a reduction in blood volume and a direct relaxation of vascular smooth muscle.

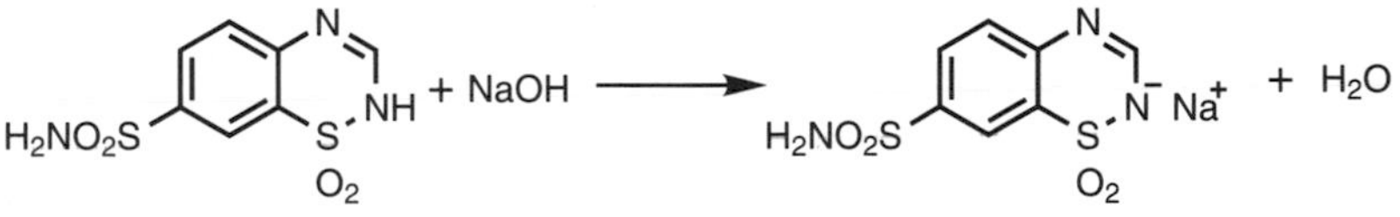

An electron-withdrawing group is necessary at position 6. Little diuretic activity is seen with a hydrogen atom at position 6, whereas compounds with a chloro or trifluoromethyl substitution are highly active. The trifluoromethyl-substituted diuretics are more lipoid-soluble and have a longer duration of action than their chloro-substituted analogs. When electron-releasing groups, such as methyl or methoxyl, are placed at position 6, the diuretic activity is markedly reduced.

Replacement or removal of the sulfonamide group at position 7 yields compounds with little or no diuretic activity. Saturation of the double bond to give a 3,4-dihydro

When the thiazide diuretics are administered for long periods, they can cause an increase in the elimination of potassium ions as well as sodium and chloride ions. Hypokalemia may result. Potassium supplements may be administered in such cases, but use of potassium supplements is controversial. These supplements are usually administered as potassium chloride, potassium gluconate, or potassium citrate. The salts are administered as solutions or timed-release tablets. Generally, about 20 mEq of potassium is given daily. In cases of hypokalemia, 40 to 100 mEq per day may be administered.

Thiazide diuretics may induce side effects, including

Table 21–2. Thiazides

Generic Name (Trade Name)	Structure	Relative Oral Natriuretic Potency in Humans*	50% Inhibition of Carbonic Anhydrase in Vitro	Duration of Action (h)	Dosage Forms (Effective Oral Dose)
1,2,4-Benzothiadiazines Chlorothiazide, U.S.P. (Diuril)	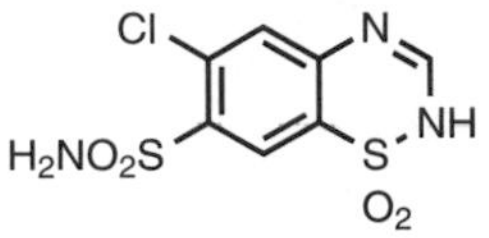	0.8	2×10^{-6}	6–10	250-mg and 500-mg tablets (500–2000 mg)
Benzthiazide, U.S.P. (Exna)	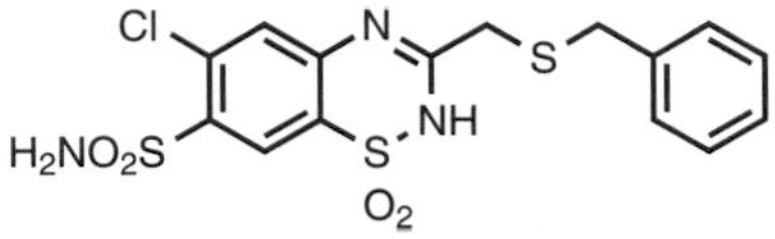	1.3	$\approx 10^{-7}$	12–18	50-mg tablets (25–50 mg)
Flumethiazide (Ademol)	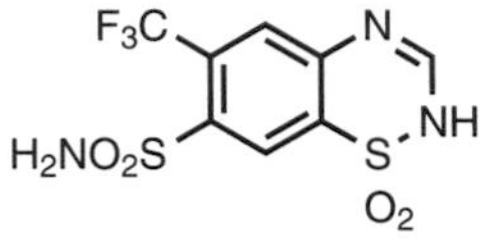	0.7	4×10^{-5}	—	500-mg tablets (500–2000 mg)
3,4-Dihydro-1,2,4-benzothiadiazines Hydrochlorothiazide, U.S.P. (Hydro-Diuril, Esidrix, Oretic)	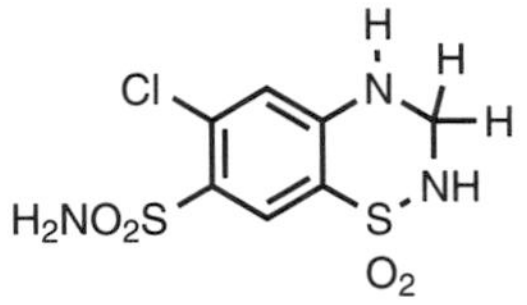	1.4	2×10^{-5}	8–12	25-mg and 50-mg tablets (25–100 mg)
Hydrobentizide	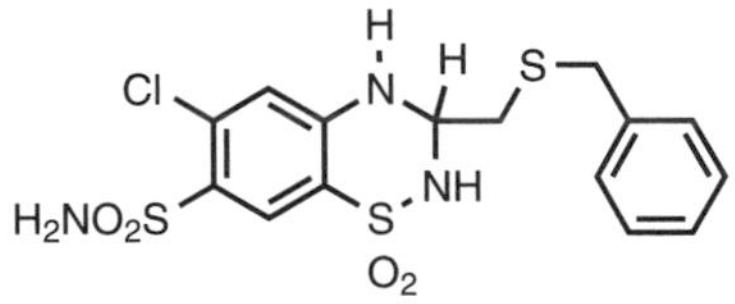	1.3	7×10^{-7}	—	(10–100 mg)
Trichlormethiazide, U.S.P. (Naqua, Metahydrin)	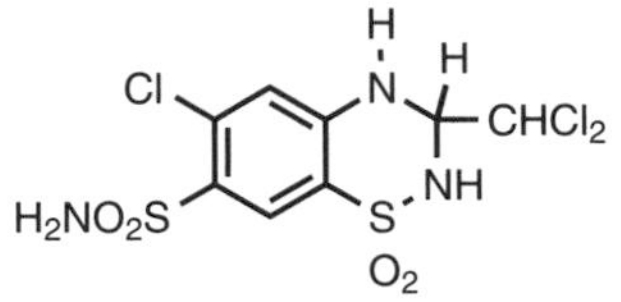	1.7	6×10^{-5}	24–36	2 mg and 4 mg tablets (4–8 mg)
Methyclothiazide, U.S.P. (Enduron)	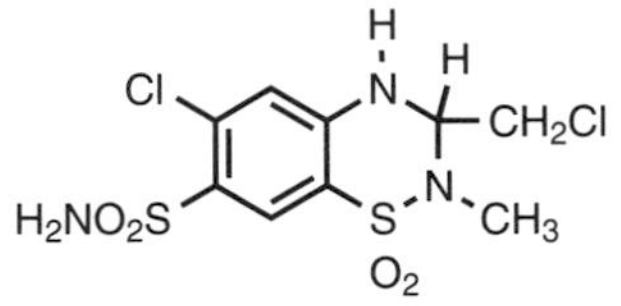	1.8	—	24+	2,5-mg and 5-mg tablets (5–10 mg)
Polythiazide, U.S.P. (Renese)	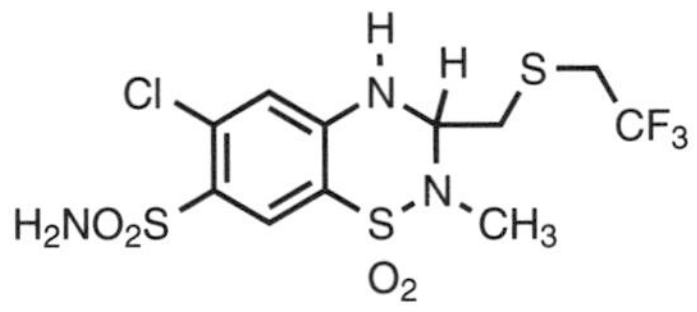	2.0	5×10^{-7}	24–36	1-mg, 2-mg, and 4-mg tablets (4–8 mg)
Cyclothiazide, U.S.P. (Anhydron)	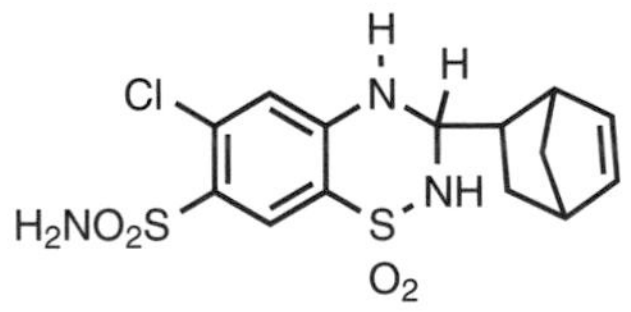	—	—	18–24	2-mg tablets (1–6 mg)
Hydroflumethiazide, U.S.P. (Saluron)	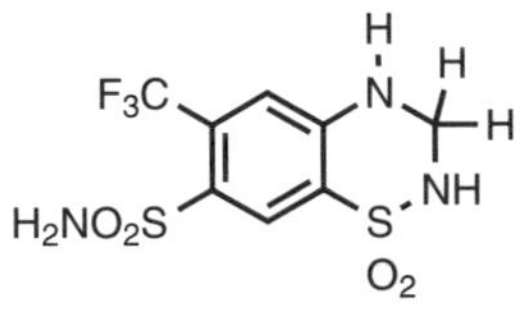	1.3	2×10^{-4}	18–24	50-mg tablets (25–50 mg)

Table 21–2. *(Continued)*

Generic Name (Trade Name)	Structure	Relative Oral Natriuretic Potency in Humans*	50% Inhibition of Carbonic Anhydrase in Vitro	Duration of Action (h)	Dosage Forms (Effective Oral Dose)
Bendroflumethiazide, U.S.P. (Naturetin, Benuron)	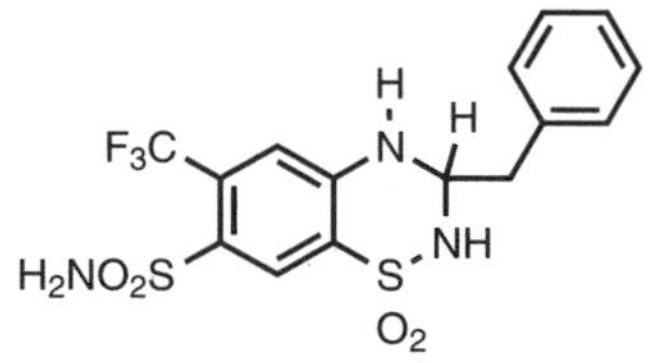	1.8	3×10^{-4}	18–24	2.5-mg and 5-mg tablets (2–5 mg)

* The numerical values refer to potency ratios with the natriuretic response to that of a standard dose of meralluride, given a value of 1.
Data from G. N. Mudge and I. M. Weiner, Drugs affecting renal function and electrolyte metabolism, in *Goodman and Gilman's The Pharmacological Basis of Therapeutics,* 7th ed., A. G. Gilman, et al., Eds., New York, Macmillan, 1985.

hypersensitivity reactions, gastric irritation, nausea, and electrolyte imbalances such as hypokalemia and hypochloremic alkalosis. Individuals who exhibit hypersensitivity reactions to one thiazide are likely to have a hypersensitivity reaction to other thiazides and sulfamoyl-containing diuretics (e.g., thiazide-like and some high-ceiling diuretics).

Quinazolinone Derivatives

The quinazolin-4-one molecule has been structurally modified in a manner similar to the modification of the thiazide diuretics.

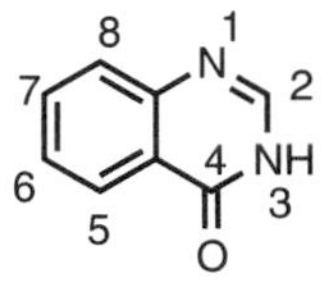

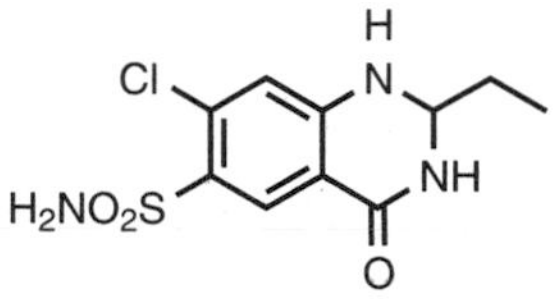

Quinethazone

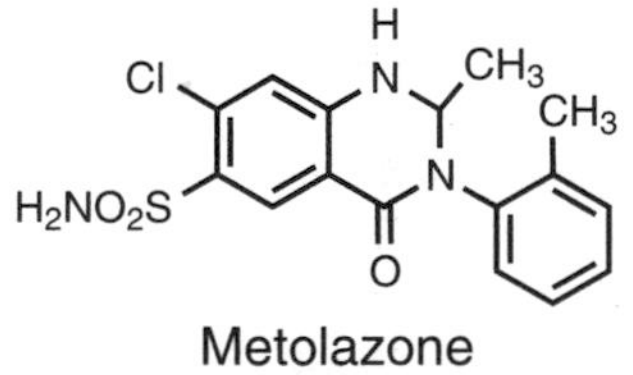

Metolazone

Quinethazone and metolazone are examples of this class. The real difference is the replacement of the sulfone group (—SO_2—) with a carbonyl group (—CO—). Because of their similar structures, it is not surprising that the quinazolinones have a diuretic effect similar to that of the thiazides. The side effects are also the same.

The quinazolinone diuretics have a long duration of action. Although chlorothiazide has a duration of action of 6 to 10 hours, quinethazone has a duration of 18 to 24 hours, and metolazone has a duration of 12 to 24 hours. Metolazone also has an increased potency. The mode of action of both compounds is similar to that of the thiazide derivatives. The dose of quinethazone is 50 to 100 mg daily, and that of metolazone is 2.5 to 20 mg given as a single oral dose.

Phthalimidine Derivatives

Chlorthalidone is an example of a diuretic in this class of compounds that bears a structural analogy to the quinazolinones.

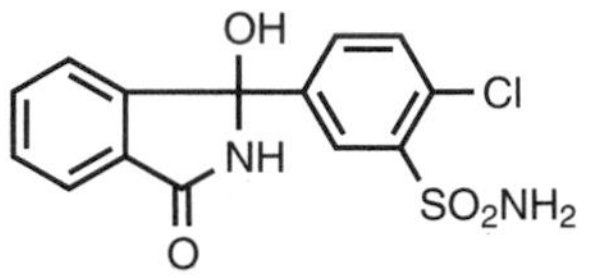

Chlorthalidone

This compound may be named as a-1-oxo-isoindoline or a phthalimidine. Although the molecule exists primarily in the phthalimidine form, the ring may be opened to form a benzophenone derivative.

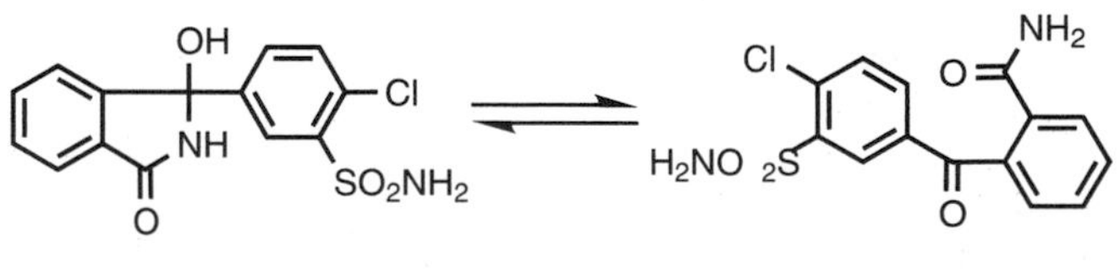

The benzophenone form illustrates the relationship to the quinazolinone series of diuretics. It may be regarded as an open ring variation.

Chlorthalidone has a long duration of action, 48 to 72 hours. Although quinethazone and metholazone are administered daily, chlorthialidone may be administered in doses of 25 to 100 mg three times a week.

INDOLINES

The prototypic indoline diuretic is indapamide, reported as a diuretic in 1984. Indapamide contains a polar chlorobenzamide moiety and a nonpolar methylindoline group. In contrast to the thiazides, indapamide does not contain a thiazide ring, and only one sulfonamide group is present within the molecular structure.

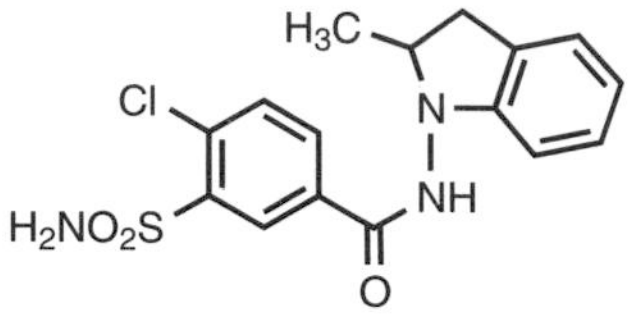

Indapamide

Uses of indapamide include the treatment of essential hypertension and edema owing to congestive heart failure. The duration of action is approximately 24 hours with the normal oral adult dosage starting at 2.5 mg given each morning. The dose may be increased to 5.0 mg per day, but doses beyond this level do not appear to provide additional results.

HIGH-CEILING OR LOOP DIURETICS

This class of drugs is characterized more by its pharmacologic similarities than its chemical similarities. These diuretics produce a peak diuresis much greater than that observed with the other commonly used diuretics, hence the name high-ceiling diuretics. Their main site of action is believed to be on the thick ascending limb of the loop of Henle, where they inhibit the luminal $Na^+/K^+/2Cl^-$ cotransporter. These diuretics are commonly referred to as loop diuretics. Additional effects on the proximal and distal tubules are also possible. High-ceiling diuretics are characterized by a quick onset and short duration of activity. Their diuretic effect appears in about 30 minutes and lasts for about 6 hours.

Furosemide is an example of a high-ceiling diuretic.

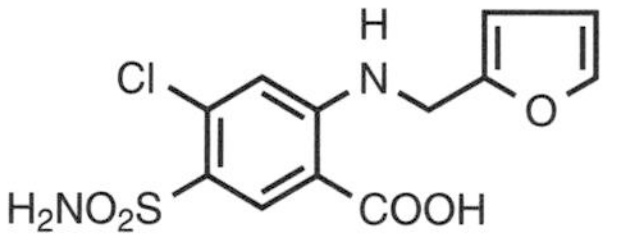

Furosemide

This molecule may be regarded as a derivative of anthranilic acid, or o-aminobenzoic acid. Research on 5-sulfa-

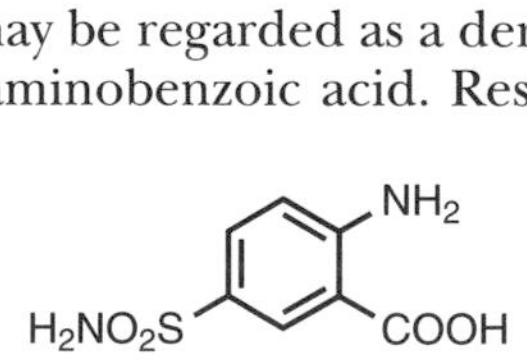

5-Sulfamoyl-anthranilic acid

moylanthranilic acids at the Hoechst Laboratories in Germany showed them to be effective diuretics. The most active of a series of variously substituted derivatives was furosemide.

The chlorine and sulfonamide substitutions are features seen also in previously discussed diuretics. Because the molecule possesses a free carboxyl group, furosemide is a stronger acid than the thiazide diuretics. This drug is excreted mostly unchanged. A small amount of metabolism, however, can take place on the furan ring, which is substituted on the aromatic amine.

Furosemide has a saluretic effect 8 to 10 times that of the thiazide diuretics. It has a short duration of action, about 6 to 8 hours. Furosemide causes an excretion of sodium, chloride, potassium, calcium, magnesium, and bicarbonate ions. It is effective for the treatment of edemas connected with cardiac, hepatic, and renal sites. Because it lowers the blood pressure similar to the thiazide derivatives, one of its uses is in the treatment of hypertension.

Furosemide is orally effective but may be used parenterally when a more prompt diuretic effect is desired. The dosage of furosemide, 20 to 80 mg per day, may be given in divided doses because of the short duration of action of the drug and carefully increased up to a maximum of 600 mg per day. Clinical toxicity of furosemide involves abnormalities of fluid and electrolyte balance. Hyperuricemia and gastrointestinal side effects are also commonly observed.

A diuretic structurally related to furosemide is bumetanide.

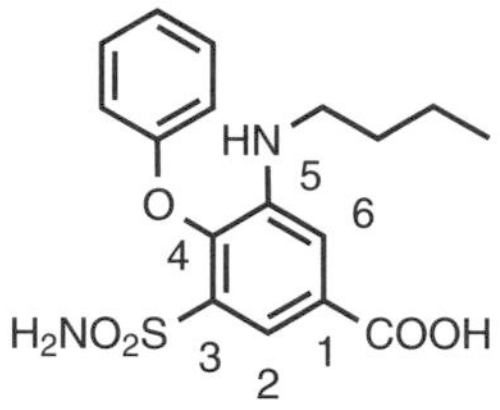

Bumetanide

This compound also functions as a high-ceiling diuretic in the ascending limb of the loop of Henle. It has a duration of action of about 4 hours. The uses of this compound are similar to those described for furosemide. The dose of bumetanide is 0.5 to 2 mg per day, given as a single dose.

In this compound, a phenoxy group has replaced the customary chloro or trifluoromethyl substitutions seen in other diuretic molecules. The phenoxy group is an electron-withdrawing group similar to the chloro or trifluoromethyl substitutions. The amine group that had been customarily seen at position 6 has been moved to position 5. These minor variations from furosemide produced a compound with a mode of action similar to that of furosemide, but with a marked increase in diuretic potency. The short duration of activity is similar, but the compound is about 50 times more potent. Replacement of the phenoxy group at position 4 with a C_6H_5NH— or C_6H_5S— group also gives compounds with a favorable activity. When the butyl group on the C_5 amine is replaced with a furanylmethyl group, such as in furosemide, however, the results are not favorable.

Further modification of furosemide-like structures has led to the development of torsemide.

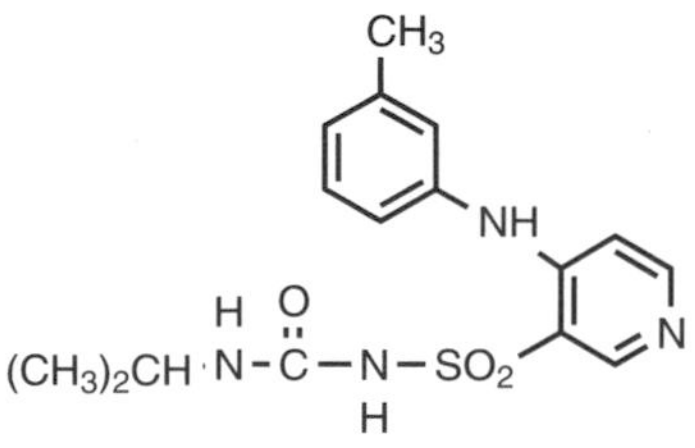

Torsemide

Instead of the sulfonamide group found in furosemide and bumetanide, torsemide contains a sulfonylurea moiety. Similar to other high-ceiling diuretics, torsemide inhibits the luminal $Na^+/K^+/2\ Cl^-$ cotransporter in the ascending limb of the loop of Henle to promote the excretion of sodium, potassium, chloride, calcium, and magnesium ions and water. An additional effect on the peritubular side at chloride channels may enhance the luminal effects of torsemide. Torsemide, however, does not act at the proximal tubule, in contrast to furosemide and bumetanide, and therefore does not increase phosphate or bicarbonate excretion. Peak diuresis is observed in 1 to 2 hours following oral or intravenous administration with a duration of action of about 6 hours. Torsemide was recommended for approval by the Food and Drug Administration's Cardiovascular and Renal Drug's Advisory Committee for the treatment of hypertension and edema associated with congestive heart failure and cirrhosis.

Another major class of high-ceiling diuretics is the phenoxyacetic acid derivatives. These compounds were developed at about the same time as furosemide but were designed to act mechanistically similar to the organomercurials (i.e., via inhibition of sulfhydryl-containing enzymes involved in solute reabsorption). Optimal diuretic activity was obtained when an oxyacetic acid group was positioned *para* to an α,β-unsaturated carbonyl (or other sulfhydryl-reactive group) and chloro or methyl groups were placed at the 2- or 3-position of the phenyl ring. In addition, hydrogen atoms on the terminal alkene carbon also provided maximum reactivity. Thus, a molecule with a weakly acidic group to direct the drug to the kidney and an alkylating moiety to react with sulfhydryl groups and lipophilic groups seemed to provide the best combination for a diuretic in this class. These features led to the development of ethacrynic acid as the prototypic agent in this class.

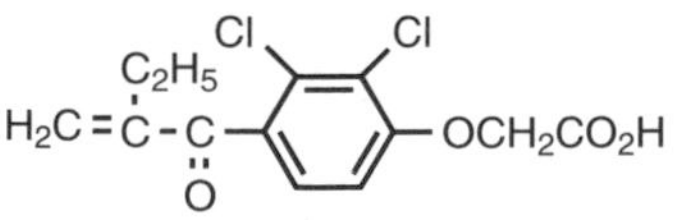

Ethacrynic acid

The mode of action of ethacrynic acid appears to be more complex than the simple addition of sulfhydryl groups of the enzyme to the drug molecule. When the double bond of ethacrynic acid is reduced, the resultant compound is still active, although the diuretic activity is diminished. The sulfhydryl groups of the enzyme would not be expected to add to the drug molecule in the absence of the α, β-unsaturated ketone.

In 1984, a new series of diuretics was reported.[3] The following formula is representative of this series.

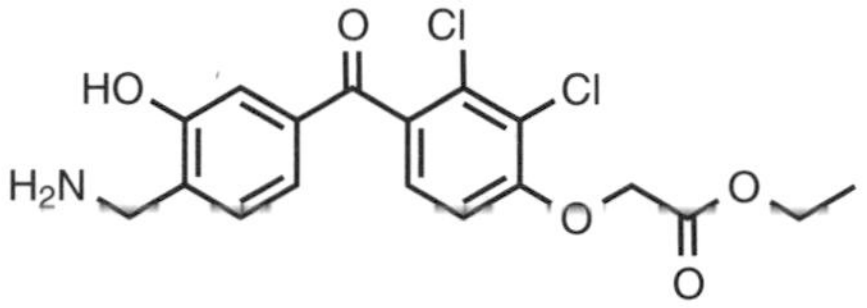

These compounds are potent high-ceiling diuretics that resemble ethacrynic acid in their mechanism of action. The ethyl ester group represents a prodrug that can be easily hydrolyzed to the free carboxyl group. As in ethacrynic acid, a 2,3-dichloro substitution is necessary. In addition, a para-hydroxyl group and an unsubstituted aminomethyl group on the benzene ring are highly beneficial. The carbonyl group can be replaced with an ether or sulfide group. These compounds have no ability to add the sulfydryl groups of the kidney enzymes. The complete mechanism of action of these compounds remains in doubt.

Similar to the other high-ceiling diuretics, ethacrynic acid inhibits the $Na^+/K^+/2Cl^-$ cotransporter in the ascending limb of the loop of Henle to promote a marked diuresis. Sodium, chloride, potassium, and calcium excretion are increased following oral or intravenous administration of ethacrynic acid. Oral administration of ethacrynic acid results in diuresis within 1 hour and a duration of action of 6 to 8 hours. Toxicity induced by ethacrynic acid is similar to that induced by furosemide and bumetanide. Ethacrynic acid, however, is not widely used because it induces a greater incidence of ototoxicity and more serious gastrointestinal effects than furosemide or bumetanide.

Two other high-ceiling diuretics reported are muzolimine and etozolin.

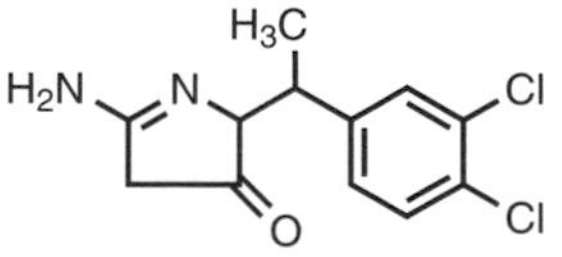

Muzolimine

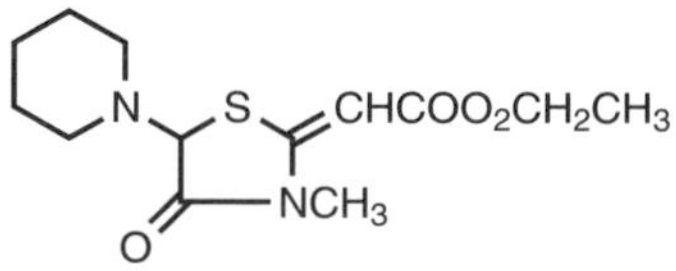

Etozolin

As can be seen by these varied structures, the high-ceiling diuretics are characterized more by their pharmacologic similarities than by their chemical similarities.

POTASSIUM-SPARING DIURETICS

Antihormone Diuretics

The adrenal cortex secretes a potent ADH called aldosterone. Other corticosteroids

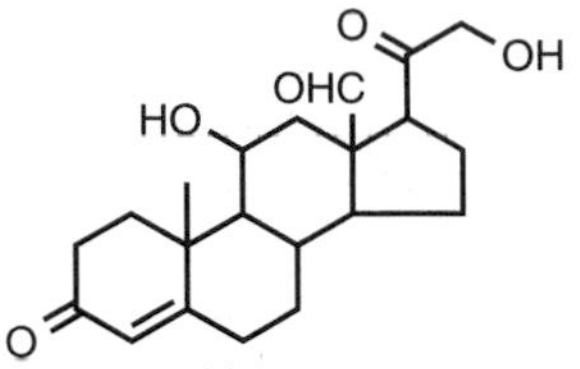

Aldosterone
(aldol from)

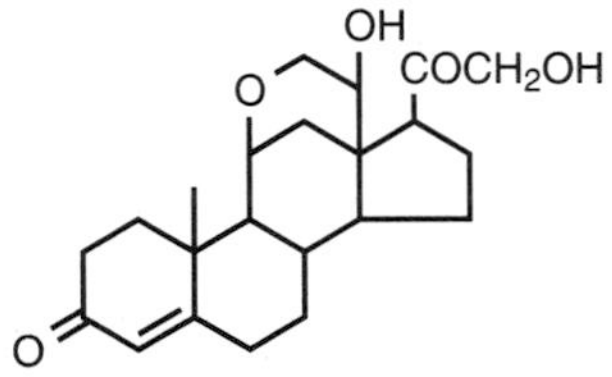

Aldosterone
(hemiacetal from)

have an effect on the electrolytic balance of the body, but aldosterone is the most potent. Its ability to cause increased reabsorption of sodium and chloride ion and increased potassium ion excretion is about 3000 times that of hydrocortisone. A substance that antagonizes the effects of aldosterone could conceivably be a good diuretic drug. Spironolactone is such an antagonist.

Spironolactone is a competitive antagonist to the mineralocorticoids such as aldosterone. The aldosterone receptor is a protein that appears to exist in two different configurations. Only one of these configurations is active

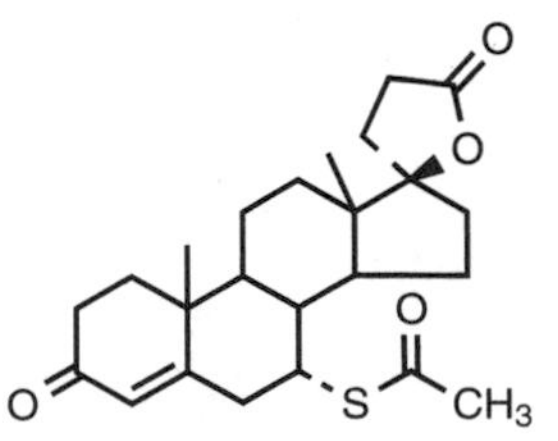

Spironolactone

in binding aldosterone. Spironolactone apparently binds to the inactive configuration of the receptor and prevents its conversion to the active configuration. Thus, aldosterone is prevented from binding to the receptor and cannot cause the reabsorption of sodium and chloride ions with the accompanying water. The most important site of these receptors is in the late distal tubule and collecting system.

On oral administration, about 70% of the dose of spironolactone is absorbed. The drug is significantly metabolized during its first passage through the liver. The major metabolite is canrenone, which can easily be converted to the canrenoate anion. Canrenone is an antagonist to aldosterone. The canrenoate anion is not active per se but acts as aldosterone antagonist because of its conversion to canrenone, which exists in the lactone form. Canrenone has been suggested as the active form of spironolactone as an aldosterone antagonist. The formation of canrenone, however, cannot account fully for the total activity of spironolactone. Both canrenone and potassium canrenoate are used as diuretics in other countries, but they are not yet available in the United States.

The most serious side effect of spironolactone is hyperkalemia because it has a potassium-sparing effect. Potassium levels should be monitored during the use of this drug. The dose of spironolactone is 100 mg per day given in single or divided doses. Spironolactone can also be

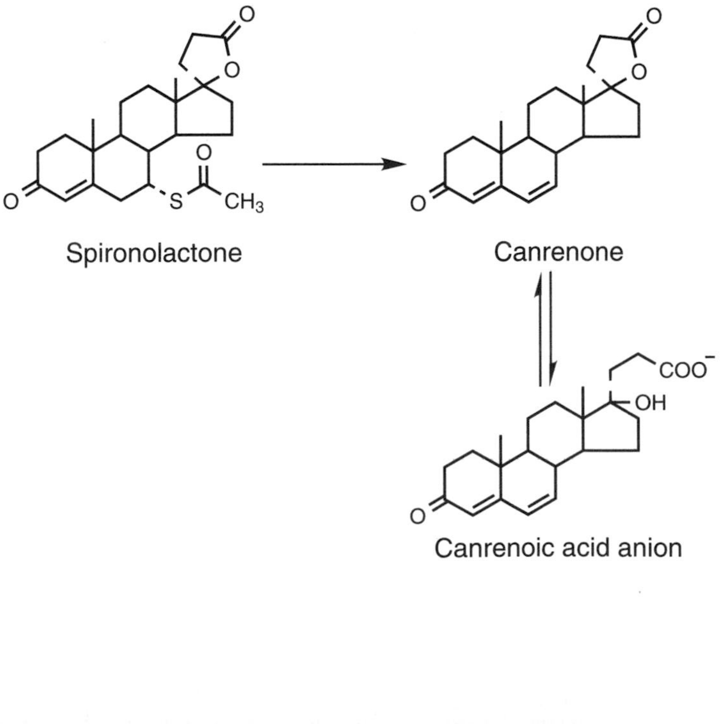

Spironolactone

Canrenone

Canrenoic acid anion

administered in a fixed-dose combination with hydrochlorothiazide.

Spironolactone has been implicated in tumor production in chronic toxicity studies in rats.

Pteridines

Pteridines have a marked potential for influencing biologic processes.

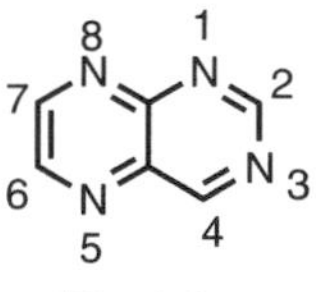

Pteridine

Early screening of pteridine derivatives revealed that 2,4-diamino-6,7-dimethylpteridine was a fairly potent diuretic. Further structural modification led to the development of triamterene.

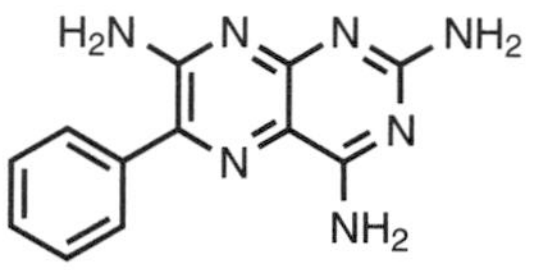

Triamterene

Triamterene interferes with the process of cationic exchange in the distal tubule. It blocks the reabsorption of sodium ion and blocks the secretion of potassium ion. This is done by a mechanism other than aldosterone antagonism. The net result is increased sodium and chloride ion excretion in the urine and almost no potassium excretion. In fact, hyperkalemia may result from the use of triamterene. Triamterene is about 50% absorbed on oral administration. The diuretic effect occurs rapidly, in about 30 minutes, and reaches a peak in about 6 hours. Triamterene is extensively metabolized, and some of the metabolites are active as diuretics. Both the drug and its metabolites are excreted in the urine.

Triamterene is administered initially in doses of 100 mg twice a day. A maintenance dose for each patient should be individually determined. This dose may vary from 100 mg a day to as low as 100 mg every other day.

The most serious side effect associated with the use of triamterene is hyperkalemia. For this reason, potassium supplements are contraindicated, and serum potassium levels should be checked regularly. Triamterene is also used in combination with hydrochlorothiazide. Here the hypokalemic effect of the hydrochlorothiazide counters the hyperkalemic effect of the triamterene. Other side effects seen with the use of triamterene are nausea, vomiting, and headache.

Modifications of the triamterene structure are not usually beneficial in terms of diuretic activity. Activity is retained if an amine group is replaced with a lower alkylamine group. Introduction of a para-methyl group on the phenyl ring decreases the activity about one-half. Introduction of a para-hydroxyl group on the phenyl ring yields a compound that is essentially inactive as a diuretic.

Aminopyrazines

Amiloride, another potassium-sparing diuretic, is an aminopyrazine structurally related to triamterene as an

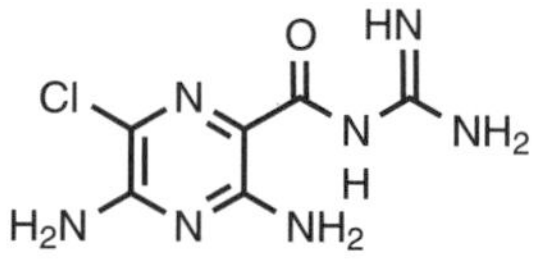

Amiloride

open-chain analog. Similar to triamterene, it interferes with the process of cationic exchange in the distal tubule. It blocks the resorption of sodium ion and the secretion of potassium ion. It has no effect on the action of aldosterone. Oral amiloride is about 50% absorbed. The duration of action is about 10 to 12 hours, slightly longer than that of triamterene. Although triamterene is extensively metabolized, amiloride is not. As with triamterene, the most serious side effect associated with amiloride is hyperkalemia, and it also has the other side effects associated with triamterene. The dose of amiloride is 5 to 10 mg per day.

REFERENCES

1. C. H. Best and N. D. Taylor, *The Living Body,* 3rd ed., New York, Henry Holt & Co., 1952, p. 295.
2. G. N. Mudge and I. M. Weiner, Drugs affecting renal function and electrolyte metabolism, in *Goodman and Gilman's The Pharmacological Basis of Therapeutics,* 7th ed., A. G. Gilman et al., Eds., New York, Macmillan, 1985.
3. Cheuk-Man Lee, et al., *J. Med. Chem., 27,* 1579 (1984).

SUGGESTED READINGS

M. A. Acara, Renal pharmacology—diuretics, in *Textbook of Pharmacology,* C. M. Smith and A. M. Reynard, Eds., Philadelphia, W. B. Saunders, 1992.

W. O. Berndt and R. E. Stitzel, Water, electrolyte metabolism, and diuretic drugs, in *Modern Pharmacology,* 3rd ed., C. R. Craig and R. E. Stitzel, Eds., Boston, Little, Brown & Co., 1990.

B. M. Brenner and F. C. Rector, Jr., Eds., *The Kidney,* 4th ed., Philadelphia, W. B. Saunders, 1991.

J. Breyer and H. R. Jacobson, *Annu. Rev. Med., 41,* 265(1990).

Diuretics, in *Drug Evaluations, Annual 1993,* D. R. Bennett, Ed.-in-chief, Chicago, American Medical Association, 1993.

I. M. Weiner, Diuretics and other agents employed in the mobilization of edema fluid, in *The Pharmacological Basis of Therapeutics,* 8th ed., A. G. Gilman, et al., Eds., New York, Pergamon Press, 1990.

Chapter 22

ANTIALLERGIC AND ANTIULCER DRUGS

Maria Di Bella, Daniela Braghiroli, and Donald T. Witiak

Histamine [1H-imidazole-4-ethanamine or 2-(imidazol-4-yl)ethylamine], which is biosynthesized by decarboxylation of the basic amino acid histidine, is widespread in nature and fundamental to many of the conditions described in this chapter. This compound is found in ergot and other plants, many venoms, bacteria, and all organs and tissues of the human body.

Histamine numbering system

Histamine is an important chemical messenger communicating information from one cell to another, and is involved in a variety of complex biologic actions. Histamine is distributed within mast cells and almost all mammalian tissues. This messenger is mainly stored in an inactive bound form from which it is released as a result of an antigen-antibody reaction initiated by different stimuli such as venoms, toxins, proteolytic enzymes, detergents, foods, and numerous chemicals. After mast cell membrane alteration by the antigen-antibody reaction, histamine is released from its ionic binding sites on granular heparin-protein complexes. These complexes are found within the mast cell; a passive exchange of histamine for cations, possibly Na^+, into the extracellular fluid takes place. The release of histamine depends on the presence of two effectors, Ca^{2+} and guanosine triphosphate (GTP).[2]

Histamine also makes an important physiologic contribution as a stimulant of gastric secretion from parietal cells, and is widely distributed within mammalian brain in both neuronal and non-neuronal compartments. Although concrete evidence for specific functions of this central histaminergic system is lacking, a likely role for histamine is as a neurotransmitter or neuromodulator. Consistent with this wide distribution, neuronal histamine seems to be involved in a variety of central nervous system (CNS) functions, including alertness, hormone release, cerebral glycogenolysis, feeding, drinking, sexual behavior, autonomic function, and analgesia.[3]

Histamine exerts its biologic functions by interacting with at least three distinctly specific receptors. These receptors have not been identified and characterized using physico-chemical methods, but their presence is inferred pharmacologically by employment of synthetic agonists and antagonists. The three receptor types are designated H_1,[4] H_2,[5] and H_3.[6]

Activation of H_1 receptors stimulates the contraction of smooth muscles in many organs such as gut, uterus, and bronchi. Activation of these receptors also causes relaxation of capillaries, resulting in increased permeability leading to edema. Such effects are readily blocked by pyrilamine and related structures known as H_1 antagonists.[4]

Gastric secretion results from activation of H_2 receptors and, accordingly, is inhibited by H_2 antagonists.[5] Hypotension, resulting from vascular dilatation, is mediated by both H_1 and H_2 receptors.[7] H_3 receptors, detected in some peripheral organs, seem to be most important in the CNS.[3]

Histamine has two basic centers; fully protonated histamine is a dication, and the first stoichiometric ionization constant (pK_{a1} = 5.80 at 37°C) corresponds to dissociation of the ring NH to give the monocation. The second ionization constant (pK_{a2} = 9.40 at 37°C) corresponds to dissociation of the side chain NH_3^+ group to give the uncharged molecule. At high pH, the ring NH ionizes (pK_{a3} = 14) to give an anion,[8] but at physiologic pH, the major species is the monocation.

Histamine has no therapeutic application. The compound is mainly used as a diagnostic agent (histamine phosphate U.S.P.) to test for secretory action of the stomach, in the diagnosis of pheochromocytoma[9,10] and as a positive control in allergy skin testing.[9] Simple molecular modification of the histamine structure generally results in a less active agonist rather than an antagonist. Only β-aminoethyl-substituted heterocyclic compounds closely related to histamine exhibit appreciable agonist activity. The related pyrazole analog betazole (1H-pyrazole-3-ethanamine) hydrochloride was previously but is no longer used to diagnose impairment of the acid-producing cells of the stomach. Betazole is less potent than histamine in this regard, and has even less ability to act as an agonist at other sites of histamine interaction.

Another histamine analog, betahistine (N-methyl-2-pyridineethanamine) dihydrochloride, is claimed to improve the microcirculation and thus is used to reduce the symptoms of Meniere's disease.[11]

Hundreds of compounds have been synthesized and evaluated for histamine agonist activity. Only N^α-methyl- and N^α,N^α-dimethylhistamine are more active than histamine as stimulants of gastric acid secretion in the dog.[12]

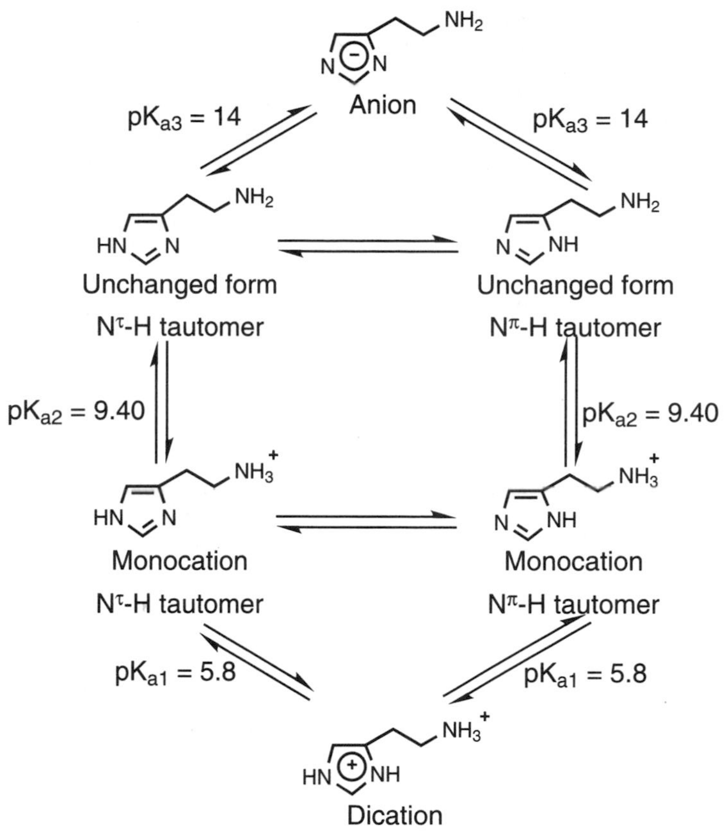

Anion

$pK_{a3} = 14$ $pK_{a3} = 14$

Unchanged form
N^{τ}-H tautomer

Unchanged form
N^{π}-H tautomer

$pK_{a2} = 9.40$ $pK_{a2} = 9.40$

Monocation
N^{τ}-H tautomer

Monocation
N^{π}-H tautomer

$pK_{a1} = 5.8$ $pK_{a1} = 5.8$

Dication

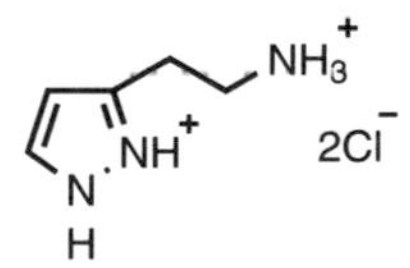

Betazole hydrochloride

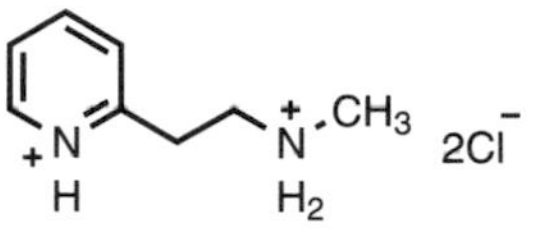

Betahistine

ANTIALLERGENIC AGENTS

When a previously sensitized individual produces an adverse response to a foreign chemical or to a physical condition, that individual is said to have an allergy. Such hypersensitivity reactions cause many chronic and acute illnesses, including hay fever, pruritus, contact dermatitis, drug rashes, urticaria, atopic dermatitis, and anaphylactic shock.

An immediate (anaphylactic type) allergenic reaction may develop in seconds when a presensitized individual comes in contact with an allergen. Anaphylaxis is an intense, systemic allergic reaction that, in some cases, is fatal. In delayed allergies, the reaction is mediated by cells (i.e., lymphocytes), and symptoms may take up to 24 hours to appear. In this case, a reaction does not occur between an antigen and a circulating antibody. Whether or not an allergic reaction is of the immediate or delayed type, the precipitating factor involves interaction of two macromolecules, an antigen (immunogen) and an antibody.

As part of the allergic response to an antigen, reaginic antibodies (IgE) are generated, and these bind to the surface of mast cells and basophils. The IgE molecules function as receptors and interact with signal transduction systems in the membranes of sensitized cells.[7] On subsequent exposure, the antigen bridges the IgE molecules and causes activation of phospholipase C. This action leads to the generation of inositol-1,4,5-*tris*(dihydrogen phosphate) and diacylglycerols and to an elevation of intracellular Ca^{2+},[13] which triggers the release of secretory granule contents by exocytosis.

Dale and Laidlaw first suggested that histamine was the mediator associated with allergic manifestations.[14,15] This substance is still accepted as the cause of the wheal-and-flare response in the skin and is thought to play a major role in producing the symptoms of allergic rhinitis. Subsequent studies by numerous investigators, however, have revealed that histamine alone cannot account for all of the symptoms of allergy.

These pharmacologic effects of histamine and structurally related compounds are referred to as actions of the agonist. Antiallergic agents that block some of the actions of histamine are known as antihistamines or, more specifically, as histamine H_1 antagonists. Structural requirements for histamine (agonist) activity are considerably different from those required for antagonist activity.

H^1 Antagonists

In 1933, Fourneau and Bovet,[16] working at the Pasteur Institute in France, observed that a compound designated 933F, namely piperoxan [1-[2,3-dihydro-1,4-benzodioxin-2-yl)methyl]piperidine], protected animals from bronchial spasm caused by aerosolized histamine.

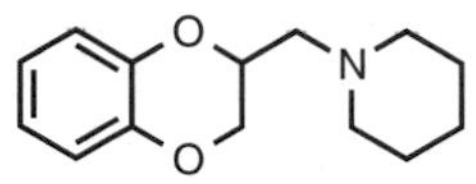

Piperoxan

This observation, the result of a general investigation of such compounds for pharmacologic activity, initiated H_1 antihistaminic research.

Ethylenediamine Derivatives

Four years later, Staub (of the same institute) reported that Fourneau's compound 1571F (N-phenyl-N,N′,N′-triethyl-1,2-ethanediamine) is superior to piperoxan.[17]

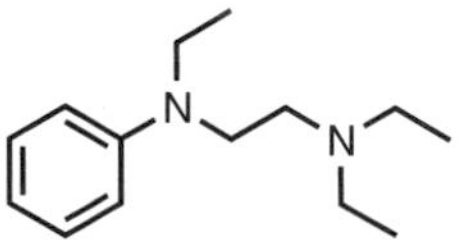

N-Phenyl-N,N'N'-triethyl-1,2-ethanediamine

The structural relationship of the two compounds is apparent. The piperidino group of piperoxan is replaced by the diethylamino function and the benzodioxane ring system is replaced by the phenylethylamino group.

Following these pioneering studies, Halpern in 1942, reported 24 derivatives of N-phenyl-N,N′,N′-triethyl-1,2-ethanediamine.[18] Of these, phenbenzamine (N,N-dimethyl-N′-phenyl-N′-phenylmethyl-1,2-ethanediamine), was one of the most potent compounds in the series and the first to become a clinically useful H_2 antagonist.[18,19]

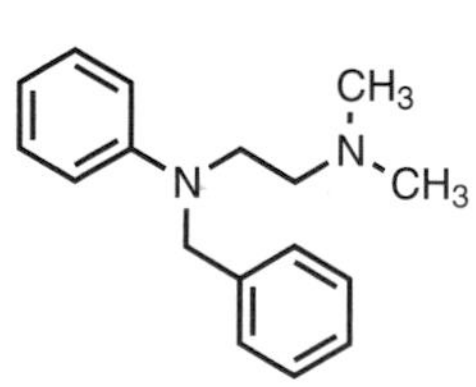

Phenbenzamine

Phenbenzemine served as the model for compounds having the general structure ArAr′N-CH_2CH_2-NRR′. Synthesis and biologic evaluation of analogs possessing this structure produced a multitude of clinically useful H_1 antagonists.

A common method of obtaining biologically active drugs is to replace functional groups of the model compound with other groups of similar size. This type of substitution is referred to as isosteric replacement. Staub had suggested that one of the amino groups should contain small alkyl groups, such as $(CH_3)_2N$-, for maximal H_1 antagonist activity. Phenbenzamine therefore was mainly modified by isosteric replacement of phenyl or phenylmethyl groups with other ring systems; introduction of small *p*-substituents into the benzyl ring also resulted in compounds with potent H_1 antagonist activity.

Clinically useful ethylenediamines were introduced between 1944 and 1950. Pyrilamine [N-[(4-methoxyphenyl)methyl]-N′,N′-dimethyl-N-2-pyridinyl-1,2-ethanediamine] maleate (U.S.P.), also known as mepyramine, is obtained when the phenylmethyl group of phenbenzamine is replaced by a 4-methoxybenzyl group and the phenyl ring is replaced by a 2-pyridyl group. Mono salts of dibasic acids such as maleic acid are acidic in solution; a 10% solution of pyrilamine maleate has a pH of 5.1.[20]

The H_1 antagonist methaphenilene, no longer marketed, which was less toxic but also less potent, has the benzyl group of phenbenzamine replaced by a 2-thenyl function. The term 2-thenyl refers to the 2-thienylmethyl functionality and is isosteric with the benzyl group; i.e., -S- is sterically equivalent to C═C.

Tripelennamine ([N,N-dimethyl-N′-(phenylemethyl)-N′-2-pyridinyl-1,2-ethanediamine] was the first ethylenediamine developed in the United States. This popular H_1 antagonist was introduced in 1946 and is related to phenbenzamine by the simple isosteric replacement of the phenyl by a 2-pyridyl group.[21] The citrate (U.S.P.) and hydrochloride (U.S.P.) salts are available; a 1% solution of tripelennamine citrate has a pH of 4.3. Solutions of the hydrocholoride salt have a pH closer to neutrality.[22]

Heterocyclic ring systems previously used in place of the phenyl group are 2-thiazolyl, 2-pyrimidyl, and 2-pyridyl (zolamine, thonzylamine, methapyrilene, thenyldiamine). All H_1 antagonists are dispensed as water-soluble hydrochloride salts.

Aminoalkyl Ether Analogs

Structurally, the aminoalkyl ethers are closely related to the ethanediamines: Ar($ArCH_2$)-N- is replaced with $(Ar)_2$CHO-. Research with these compounds from the early 1940s to the middle 1950s was simultaneous with the development of the ethanediamine series. Rieveschl and Huber carried out the earliest syntheses and reported the first important antihistamine in the series.[23,24] The compound diphenhydramine, available as a hydrochloride salt (U.S.P.) and identified as 2-diphenylmethoxy-N,N-dimethylethanamine, represents the first efforts of investigators in the United States to block the effects of histamine on what are now termed H_1 receptors.

Diphenhydramine is two to four times more effective than piperoxan and compound 1571F, and is readily synthesized by condensing benzhydryl bromide with dimethylaminoethanol in the presence of sodium carbonate. The nitrogen analogs may be prepared similarly.

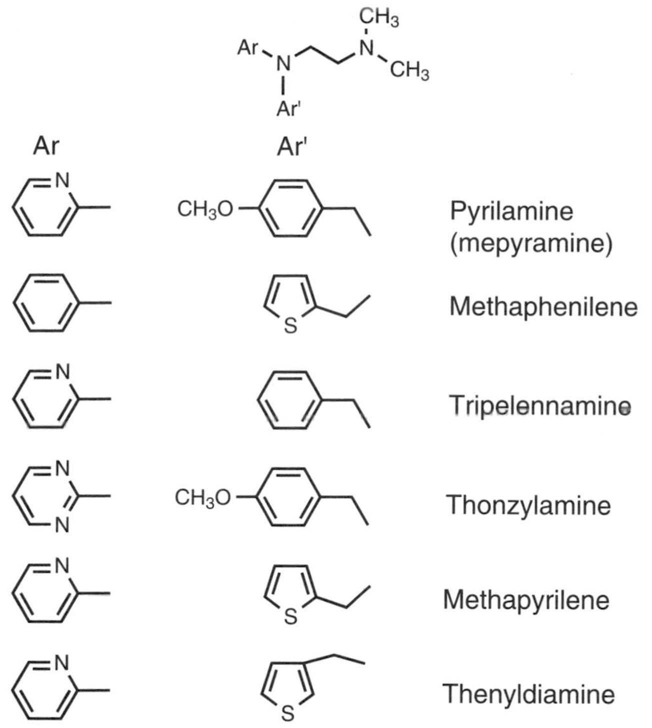

Ar Ar'

Pyrilamine (mepyramine)

Methaphenilene

Tripelennamine

Thonzylamine

Methapyrilene

Thenyldiamine

All antihistamines produce side effects such as drowsiness or nervousness. Because purines such as caffeine stimulate the CNS, investigators studied salts of diphenhydramine with centrally stimulating purine derivatives. The compound 8-chlorotheophylline is the only purine sufficiently acidic to form a stable salt with diphenhydramine. The resulting salt, dimenhydrinate (U.S.P.), still exhibits CNS depressant activity, but it is used widely in the treatment of motion sickness. This pharmacologic property is not related to antihistaminic activity.

Other antihistamines, including diphenhydramine hydrochloride, that are not salts of 8-chlorotheophylline are also effective for motion sickness. Buclizine, cyclizine, meclizine, and the phenothiazines are examples of antihistamines with this activity.[25] Sedation is the most common side effect, but anticholinergic activity and consequently the symptoms of dry mouth, dizziness, blurred vision, and fatigue are often experienced.

Clinically useful analogs of diphenhydramine are obtained through the use of functional groups similar to those investigated in the ethylenediamine series. For example, medrylamine [2-[(4-methoxyphenyl)-phenylmethoxy]-N,N-dimethylethanamine], chlorodiphenhydramine [2-[(4-chlorophenyl)-phenylmethoxy]-N,N-dimethylethanamine], and bromodiphenhydramine [2-[(4-bromophenyl)-phenylmethoxy]-N,N-dimethylethanamine] (bromodiphenhydramine hydrochloride U.S.P.) are obtained by *p*-substitution of one of the phenyl groups of diphenhydramine with CH_3O, Cl, and Br, respectively, and may have advantages in some patients because of fewer side effects.[26–30] In the diphenhydramine series, steric factors seem to have an effect on H_1 antihistamine potency. Use of a *p*-methyl group affords the active antihistamine *p*-methyldiphenhydramine, whereas an *o*-methyl group causes loss of H_1 antihistaminic properties and enhances atropine-like activity.[31,32]

Doxylamine (N,N-dimethyl-2-[1-phenyl-1-(2-pyridinyl)ethoxy]ethanamine), dispensed as a succinate salt (U.S.P.), results from replacement of the benzhydryl hydrogen of diphenhydramine with a methyl group and isosteric replacement of one phenyl group with a 2-pyridyl group. A 1% aqueous solution of doxylamine succinate has a pH of about 5[25] and exhibits an efficacy comparable to that of diphenhydramine and tripelennamine preparations.[33,34]

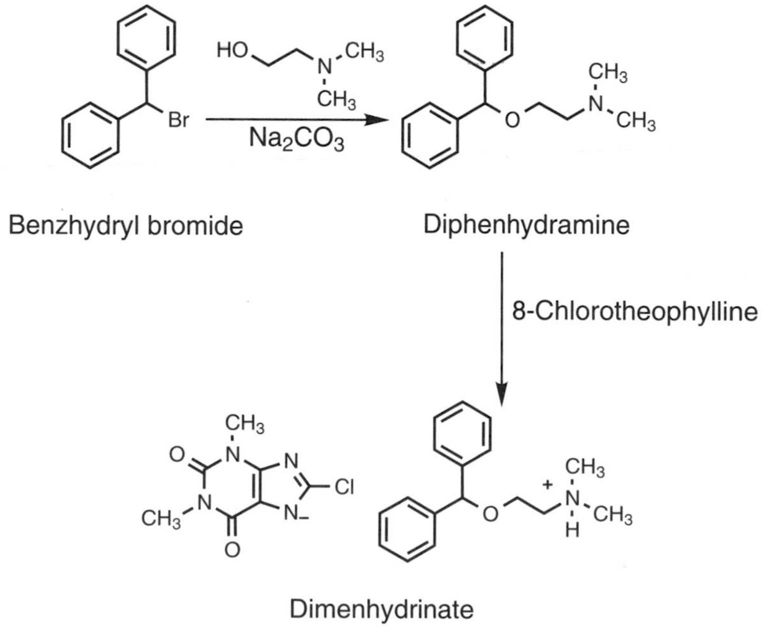

Na_2CO_3

Benzhydryl bromide

Diphenhydramine

8-Chlorotheophylline

Dimenhydrinate

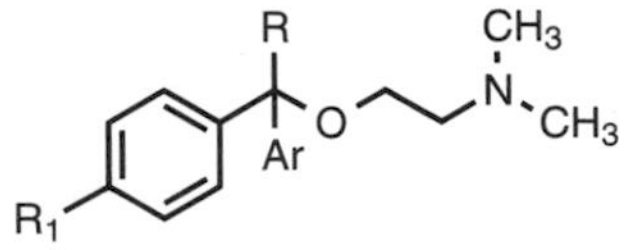

Ar	R_1	R	
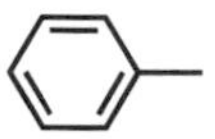	H	H	Diphenhydramine
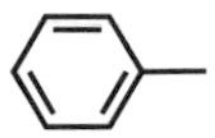	OCH_3	H	Medrylamine
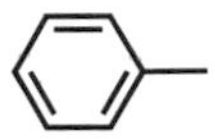	Cl	H	Chlorodiphenhydramine
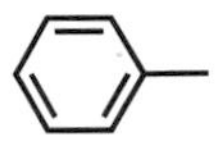	Br	H	Bromodiphenhydramine
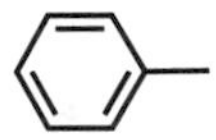	CH_3	H	Methyldiphenhydramine
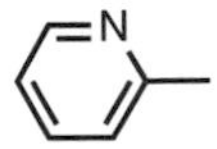	H	CH_3	Doxylamine
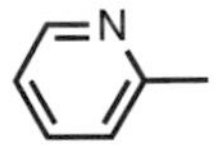	Cl	H	Carbinoxamine

Carbinoxamine maleate (U.S.P.), which is obtained by *p*-substitution of a Cl on one phenyl ring of diphenhydramine and replacement of the second ring by a 2-pyridyl group, is of particular interest. Resolution of carbinoxamine (2-[(4-chlorophenyl)-2-pyridinylmethoxy]-N,N-dimethylethanamine) produced the more active levorotatory isomer (rotoxamine), which was also dispensed as a tartrate salt (Twiston).[35] The significance of this observation is discussed under structure-activity relationships.

Another compound that is comparable in efficacy to the classic aminoalkyl ethers and ethylenediamines is phenyltoloxamine, which is used as the less hygroscopic dihydrogen citrate. This compound, an N,N-dimethyl-2-[2-(phenylmethyl)phenoxy]ethanamine, is a position isomer of diphenhydramine.[36,37]

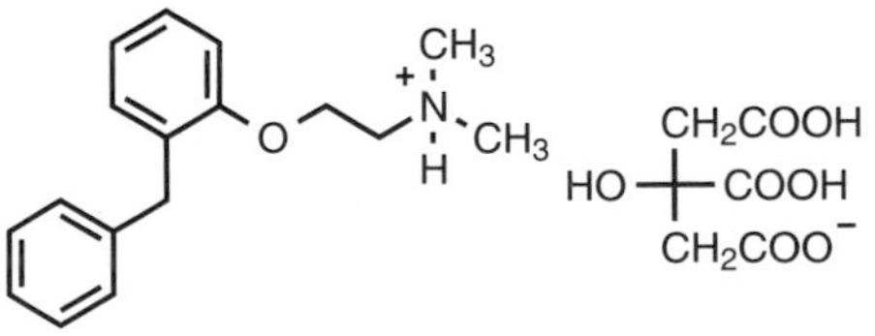

Phenyltoloxamine citate

Cyclic Basic Chain Analogs

Hundreds of ethylenediamine derivatives have been prepared. The search for active analogs has not been limited to compounds containing the dimethylamino function. Clemizole [1-[(4-chlorophenyl)methyl]-2-(1-pyrrolidinylmethyl)-1H-benzimidazole] antazoline [4,5-dihydro-N-phenyl-N-(phenylmethyl)-1H-imidazole-2-methanamine], and thenaldine [1-methyl-N-phenyl-N-(2-thienylmethyl)-4-piperidinamine] represent molecular modifications of the general ethylenediamine structure in which the basic dimethylamino group is replaced by a small basic heterocyclic ring. Careful examination of these structures reveals that they contain the same struc-

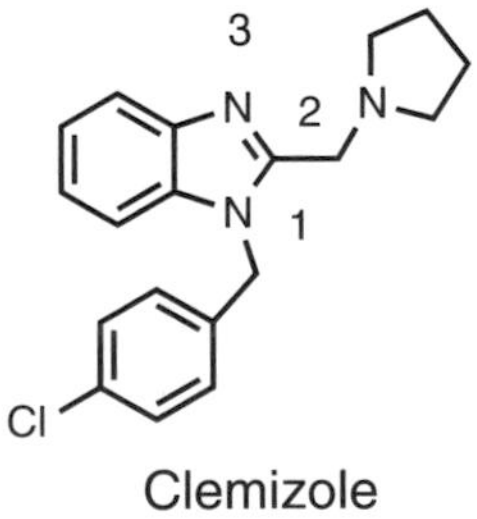

Clemizole

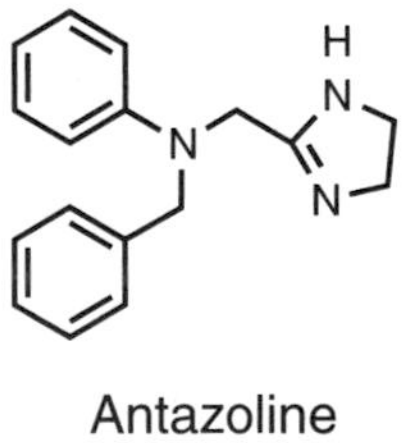

Antazoline

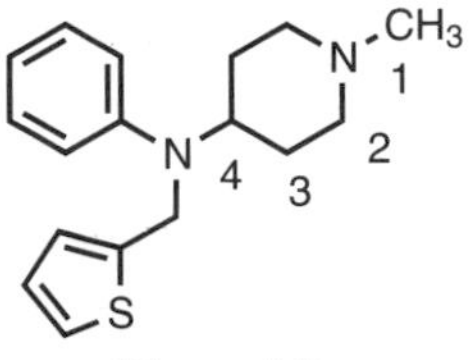

Thenaldine

tural elements as those of simpler ethylenediamines. Only in thenaldine is potent H_1 antagonist activity observed when the two nitrogen atoms are separated by three rather than two carbons.

These substituted ethylenediamine antihistamines contain aliphatic amino groups that are sufficiently basic to form stable salts with mineral acids. The phosphate salt of antazoline (U.S.P) is less irritating than the hydrochloride salt when applied to the cornea of the eye. The nitrogen atoms to which the aromatic rings are bonded are considerably less basic, in part because of delocalization of the free electrons on the aryl nitrogen into the aromatic ring. The resonance structure illustrating this electron delocalization is as follows:

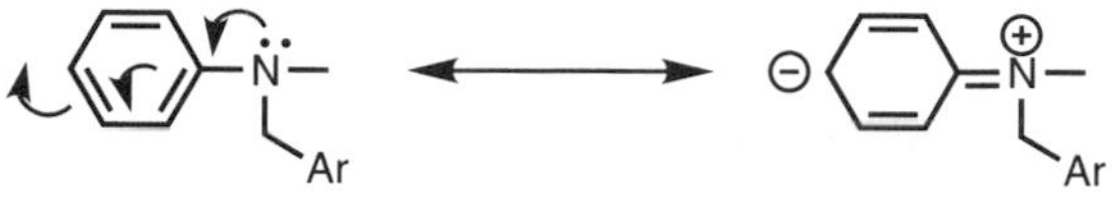

Because electron density on nitrogen is decreased, protonation at this position takes place less readily; i.e., dissociation is favored.

Piperazine derivatives, exemplified by cyclizine, chlorcyclizine, hydroxyzine, meclizine, buclizine, cinnarizine, and oxatomide, are also ethylenediamine analogs. These potent antihistamines, which have CNS depressant effects, have a slow onset of action and a prolonged duration of activity, and are synthesized from ethylenediamine. Condensation with 1,2-dichloroethane produces piperazine. Alkylation using 1 mol of the appropriate alkyl bromide yields the respective N-alkyl piperazine; reaction with substituted benzhydryl bromides then results in cyclizine analogs.[38–41]

Because both nitrogen atoms in the piperazine series are replaced with aliphatic groups, both are readily protonated. Cyclizine (1-diphenylmethyl-4-methylpiperazine) (cyclizine, cyclizine hydrochloride, and cyclizine lactate U.S.P.) and chlorcyclizine [1-[(4-chlorophenyl)phenylmethyl]-4-methylipiperazine] are dispensed as monohydrochloride salts, but meclizine [1-[(4-chlorophenyl) phenylmethyl]-4-[(3-methylphenyl)methyl]piperazine] hydrochloride (U.S.P.) is a dihydrochloride in which both N_1 and N_4 nitrogens are protonated. Although meclizine is a potent H_1 antagonist with a long duration of action,[42] it is used primarily as an antinauseant for the prevention of motion sickness.[25] The compound is effective in the treatment of nausea and vomiting associated with vertigo and radiation sickness.

Hydroxyzine, 2-[2-[4-[(4-chlorophenyl)phenylmethyl]-1-piperazinyl]ethoxy]ethanol, is available as the pamoate (U.S.P.) and hydrochloride (U.S.P.) salts, and because of its sedative action is used mainly as a minor tranquillizer. It is also used as an antiemetic and in allergic conditions, particularly pruritus.[25]

Replacement of the N-methyl group in cyclizine with a β-styrylmethyl moiety affords cinnarizine, 1-(diphenylmethyl)-4-(3-phenyl-2-propenyl)-piperazine.[43] In both a competitive and noncompetitive manner, cinnarizine antagonizes histamine on isolated tissue.[44] The mechanism of action seems to involve inhibition of Ca^{2+} transfer

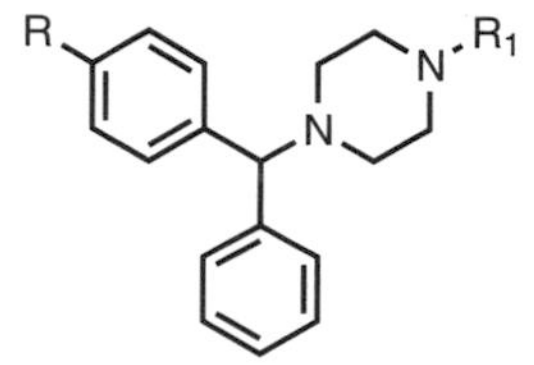

R	R_1	
H	$-CH_3$	Cyclizine
Cl	$-CH_3$	Chlorcyclizine
Cl	$-CH_2CH_2-O-CH_2CH_2OH$	Hydroxyzine
Cl	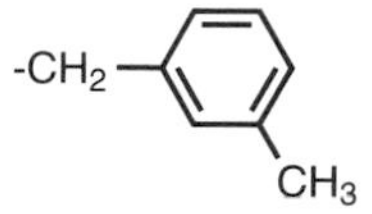 $-CH_2-$(3-methylphenyl)	Meclizine
Cl	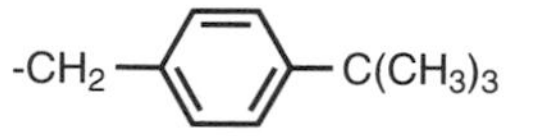 $-CH_2-$(4-$C(CH_3)_3$-phenyl)	Buclizine
H	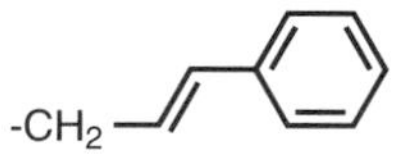 $-CH_2-CH=CH-$phenyl	Cinnarizine
H	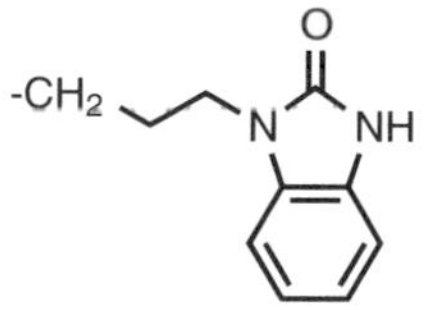 $-CH_2CH_2CH_2-$(2-oxobenzimidazolyl)	Oxatomide

from the outside to the inside of the cell; such Ca^{2+} migration takes place during contraction of smooth muscle following stimulation with spasmogenic agents.[45,46] Although its H_1 antagonist potency is relatively modest, it has proved valuable for the symptomatic management of nausea and vertigo in labyrinthine disturbance, and for the prevention of motion sickness. Cinnarizine also is reported to possess smooth muscle relaxant effects and is used in the treatment of vascular disorders.[47]

Oxatomide, (1-[3-[4-(diphenylmethyl)-1-piperazinyl]-propyl]-1,3-dihydro-2H-benzimidazole-2-one), in addition to its antihistamine activity, displays some antiserotonin and antimuscarinic properties, and possibly antagonizes slow-reacting substance of anaphylaxis. In common with some other antihistamines, oxatomide has also shown some mast cell stabilizing activity, although the contribution this action makes to the clinical activity of the drug is unknown.[47]

Clemastine, (2-[2-[1-(4-chlorophenyl)-1-phenylethoxy] ethyl]-1-methyl pyrrolidine), is one of four isomers obtained by condensing 4-chloro-α-methylbenzhydryl alcohol with 2)(2-chloroethyl)-1-methylpyrrolidine. Clemastine has the R configuration at both its benzylic and pyrrolidino chiral centers.[48] It is marketed as the fumarate salt (U.S.P.), is an excellent H_1 antagonist, and has minimal anticholinergic and CNS depressant effects.[49] Importantly, the CNS effects of clemastine do not appear until 6 to 7 hours after administration.[50] The R(benzylic) S(pyrrolidino) diastereoisomer is also an effective H_1 antagonist, but the SS- and the SR-isomers have only low potency. When chirality is proximal to the aromatic system, stereoselectivity is observed. When chirality is proximal to the amino function, however, little or no stereoselectivity is observed.

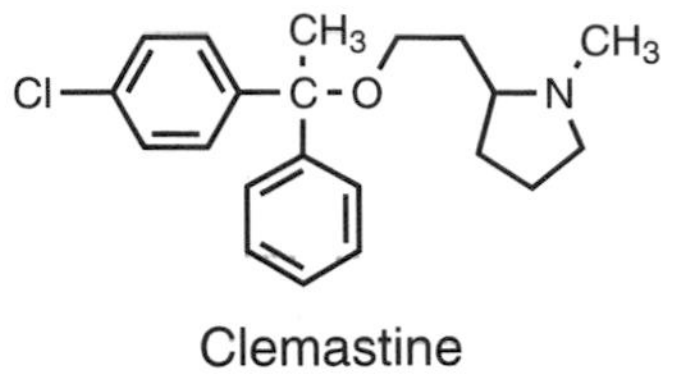

Clemastine

Diphenylpyraline (4-(diphenylmethoxy)-1-methylpiperidine) represents a molecular modification in which the dimethylaminoethyl group of diphenhydramine is replaced by a piperidine ring.

Many different heteroxyclic rings have been investigated; the 1-methyl-4-piperidyl group seems particularly effective even though the N and the O are separated by three carbon atoms.

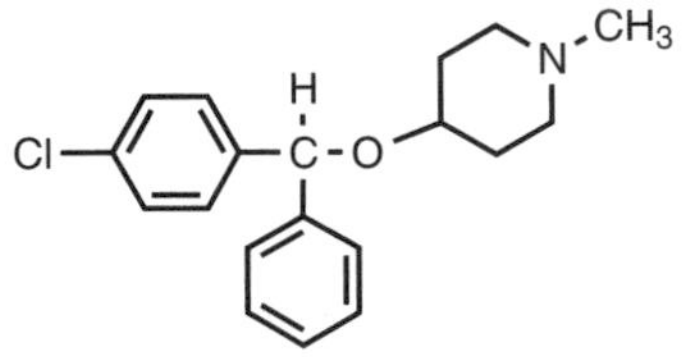

Diphenylpyraline

Monoaminopropyl Analogs

Replacement of $Ar(ArCH_2)$-N- of the ethylenediamine series with $(Ar)_2CH$- affords the monoaminopropyl antihistamines. Pheniramine (N,N-dimethyl-γ-phenyl-2-pyridinepropanamine), chlorpheniramine [γ-(4-chlorophenyl)-N,N-dimethyl-2-pyridinepropanamine] (chlorpheniramine maleate, U.S.P.), and brompheniramine [γ-(4-bromophenyl)-N,N-dimethyl-2-pyridinepropanamine] (brompheniramine maleate, U.S.P.) are compounds of the monoaminopropyl series synthesized between 1948 and 1952.[51–54] Chlorpheniramine and brompheniramine have a long duration of action; the compounds are effective in doses 50 times smaller than those of tripelennamine.

Insertion of a halogen into the para position of the phenyl ring of pheniramine increases potency twentyfold but does not appreciably increase toxicity. Like carbinoxamine of the aminoalkyl ether series, the pheniramines have asymmetric carbon atoms. The dextrorotatory isomers of chlorpheniramine and brompheniramine exhibit greater potency than do the levorotatory isomers, i.e., stereoselective H_1 antagonist activity is observed with

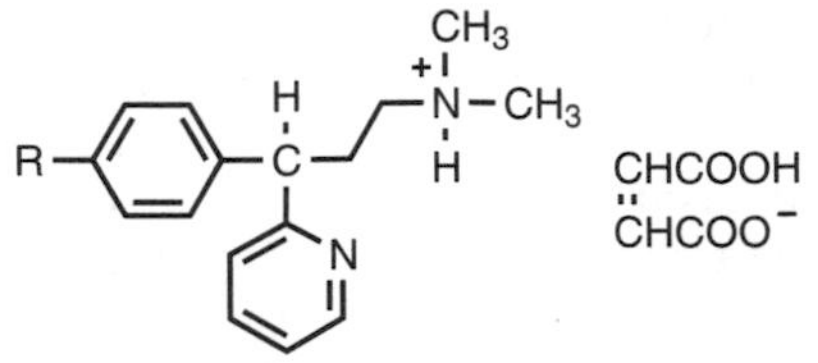

R	
H	Pheniramine maleate
Cl	Chlorpheniramine maleate
Br	Brompheniramine maleate

these compounds. Acute toxicity with the (+)-isomers of chlorpheniramine and brompheniramine (dexchlorpheniramine maleate and dexbrompheniramine maleate, U.S.P.) is no greater than with the (±)-mixture.

Other important compounds in this series may be defined as unsaturated (olefinic) analogs of monoaminopropyl antihistamines. The readily absorbed pyrrobutamine phosphate is a 1-[4-(4-chlorophenyl)-3-phenyl-2-butenyl]pyrrolidine.

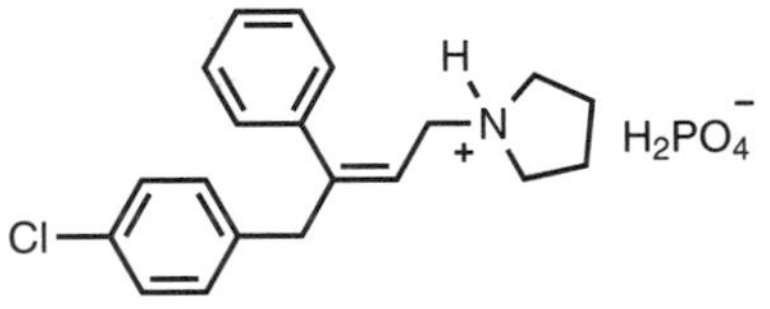

Pyrrobutamine phosphate

Triprolidine hydrochloride (U.S.P.) is most active as the *(E)*-2-[1-(4-methylphenyl)-3-(1-pyrrolidinyl)-1-propenyl]-pyridine isomer.

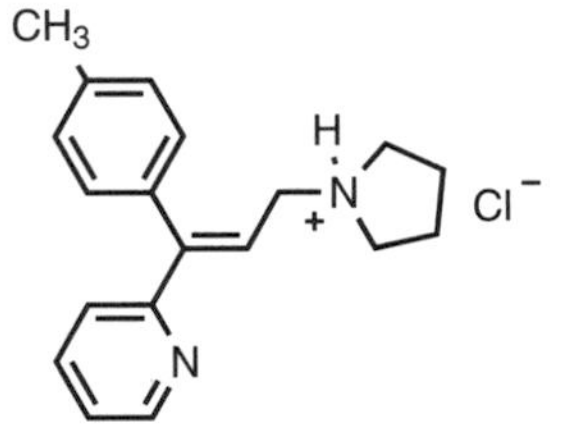

Triprolidine hydrochloride

Dimethindene maleate is most potent as the levorotatory isomer of N,N-dimethyl-3-[1-(2-pyridinyl)ethyl]-1H-indene-2-ethanamine.[55–57]

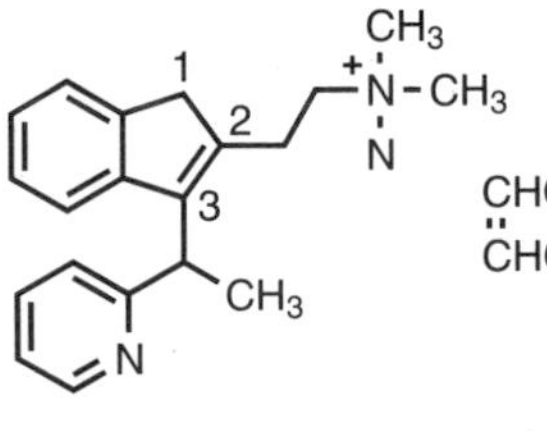

Dimethindene maleate

These differences in activity of isomers may prove helpful in determining the nature of the antihistamine receptor site.

The aminoalkyl side chain becomes part of a tetrahydropyridindene ring in phenindamine tartrate.

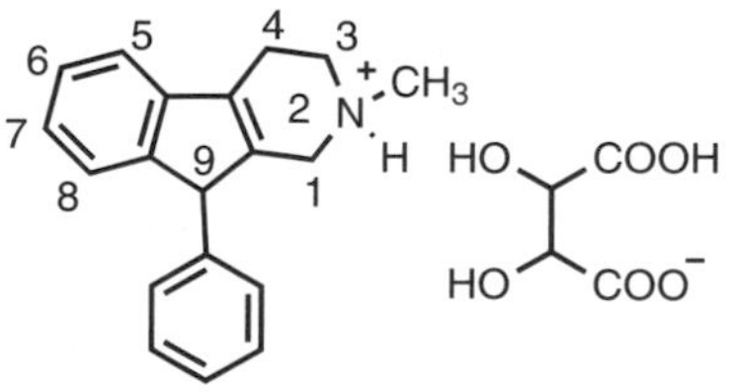

Phenindamine tartrate

This compound is identified as a 2,3,4,9-tetrahydro-2-methyl-9-phenyl-1H-indeno[2,1-c]pyridine and is dispensed as hydrogen tartrate. It is related to mebhydrolin, a 1,3,4,5–tetrahydro-2-methyl-5-(phenylmethyl)-2H-pyrido[4,3-b]indole, available as the 1,5-naphthalenedisulfonate, which also contains the aminopropyl side chain within the carboline heterocyclic system.[58]

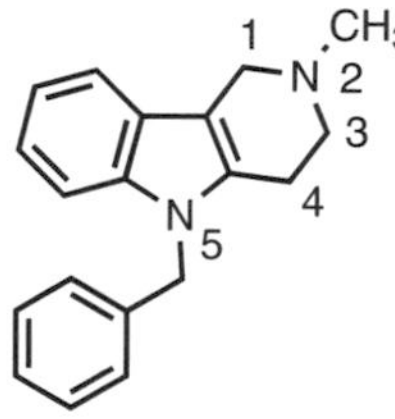

Mebhydrolin

These compounds are of particular interest because their structures are rigid. This rigidity enables measurement of distances between important pharmacophores and helps to define structural requirements for H_1 antagonistic activity.

Tricyclic Ring System

Phenindamine and mebhydrolin contain a tricyclic ring system, but another way of obtaining tricyclic analogs designed to possess H_1 antagonistic activity is to connect the ortho positions of the aromatic rings of the general antihistamine structure. Such a connection, Y, in which Y = carbon, a heteroatom, —CH═CH—, or carbon-heteroatom, results in tricyclic H_1 antagonists. There are many potent compounds in this class.

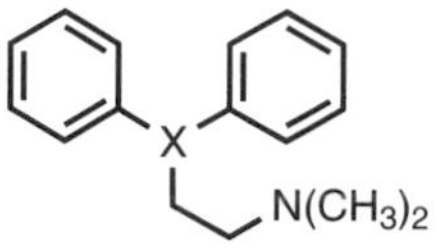

an H_1 antagonist

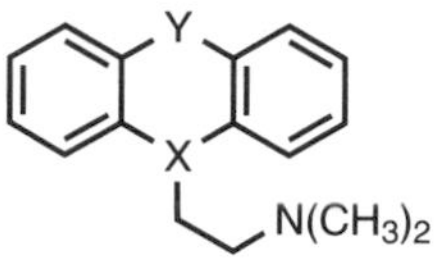

a tricyclic H_1 antagonist

When X = N and Y = S in the tricyclic H_1 antagonist structure, we have the phenothiazine series. These compounds, the first members of the tricyclic ring series, were introduced in 1945 by Halpern.[59,60] The first of the 10-aminoalkylphenothiazine group, fenethazine hydrochloride (N,N-dimethyl-10H-phenothiazine-10-ethanamine hydrochloride) is capable of counteracting 700 lethal doses of histamine in the guinea pig. Branching in the side chain as in promethazine (N,N,α-trimethyl-10H-phenothiazine-10-ethanamine) (promethazine hydrochloride, U.S.P.) improves H_1 antagonistic potency.[61] Parathiazine (pyrathiazine) hydrochloride, a 2-pyrrolidinoethyl analog [10-[2-(1-pyrrolidinyl)ethyl]-10H-phenothiazine], trimeprazine, an N,N,β-trimethyl-10H-phenothiazine-10-propanamine (tartrate, U.S.P.), and methdila-

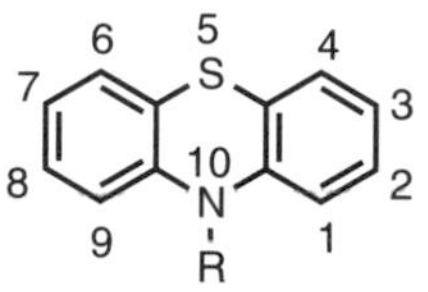

R	
$—CH_2CH_2\overset{+}{N}H(CH_3)_2\ Cl^-$	Fenethazine hydrochloride
$—CH_2CH(CH_3)\overset{+}{N}H(CH_3)_2\ Cl^-$	Promethazine hydrochloride
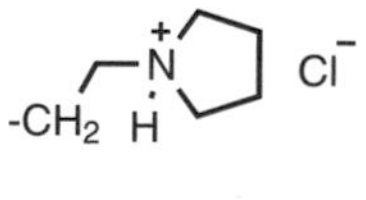	Parathiazine hydrochloride
$—CH_2CH(CH_3)CH_2NH(CH_3)_2$ 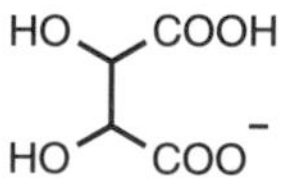	Trimeprazine tartrate
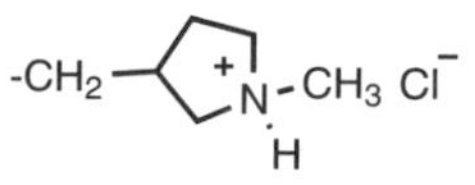	Methdilazine hydrochloride

zine, a 1-methyl-3-pyrrolidinylmethyl analog, [10-[(1-methyl-3-pyrrolidinyl)methyl]-10H-phenothiazine] (methdilazine and methdilazine hydrochloride, U.S.P.), are members of this class.[61–63]

In addition to H_1 antagonistic activity, these compounds have tranquilizing and antiemetic activities; they potentiate analgesics and sedatives. Generally, insertion of halogen or trifluoromethyl (CF_3) at the 2 position and/or the use of a 3-dimethylaminopropyl rather than of an ethyl side chain make the compounds more effective tranquilizers and less potent H_1 antagonists.[60,64]

The nitrogen-containing isostere of phenothiazine, 1-azaphenothiazine, was used successfully in the preparation of isothipendyl (N,N,α-trimethyl-10H-pyrido[3,2-b][1,4]benzothiazine-10-ethanamine) hydrochloride.[65]

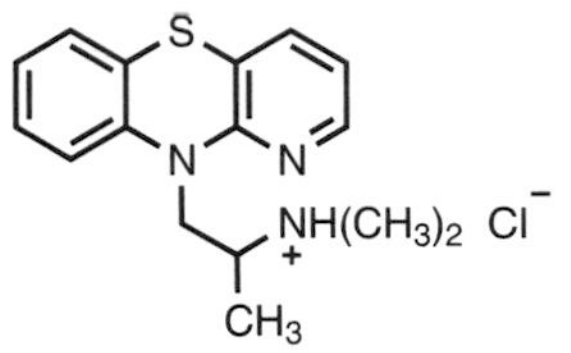

Isothipendyl hydrochloride

Other azaphenothiazines have been investigated, but isothipendyl is the only one of commercial significance.[63,66]

The nitrogen atom in the phenothiazine ring system may be replaced by an sp^2 or an sp^3 carbon atom resulting in the tricyclic thioxanthene series without loss of biologic activity.[67] Pimethixene [1-methyl-4-(9H-thioxanthen-9-ylidene)-piperidine], a potent H_1 antagonist, is an example of a thioxanthene compound with an sp^2 hybridized carbon atom bonded to a cyclic 1-methyl-4-piperidylidene ring system.

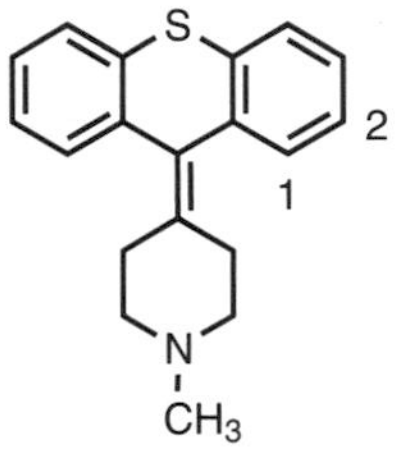

Pimethixene

Methixene, a 1-methyl-3-(9H-thioxanthen-9-ylmethyl)-piperidine, exhibits moderate H_1 antagonist and marked antimuscarinic activity and is used for the symptomatic treatment of parkinsonism.[68,69] As with phenothiazines, introduction of chloride at the 2 position of thioxanthenes weakens H_1 antagonist activity relative to other actions.

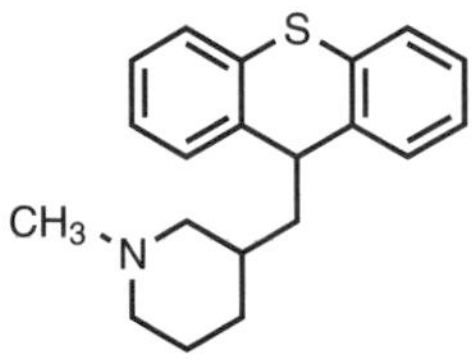

Methixene

Cyproheptadine hydrochloride (U.S.P.), a 4-(5H-dibenzo[a,d]cyclohepten-5-ylidene)-1-methylpiperidine, is also related to the phenothiazines. With this agent, the S in the tricyclic ring is replaced by —CH═CH— and the N is replaced by an sp^2 carbon atom.

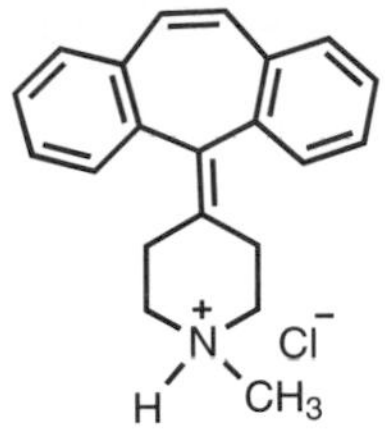

Cyproheptadine hydrochloride

The compound exhibits both H_1 antagonist and antiserotonin activity.[70] Replacement of S by —CH═CH— is an example of bioisosterism, i.e., S and —CH═CH— are similar in size, and similar biologic activities are observed. Saturating the 10,11 double bond of cyproheptadine does not affect H_1 antagonist activity, but lowers antiserotonin potency slightly.

Azatadine (6,11-dihydro-11-(1-methyl-4-piperidinylidene)-5H-benzo[5,6] cyclohepta[1,2-*b*]pyridine), as the maleate salt (U.S.P.), is the most potent analog in this series. This tricyclic compound protects mice from fatal anaphylaxis and inhibits the effect of histamine on the guinea-pig ileum; the compound is three to four times more potent than chlorpheniramine maleate. Other studies have established its advantage over pheniramine in treating allegic rhinitis[71] and its low level of CNS side effects.[72,73]

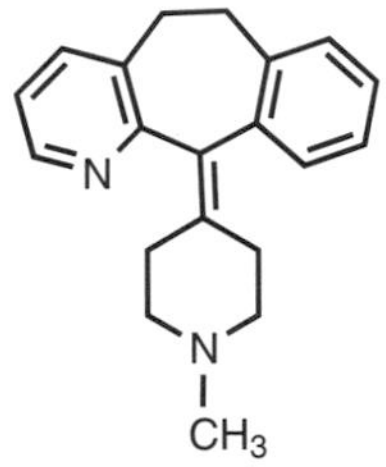

Azatadine

Variously substituted tricyclic ring systems have been studied for both antiallergic and psychotherapeutic activity. Generally, tricyclic compounds with the $(CH_3)_2NCH_2CH_2$- side chain or its equivalent exhibit marked H_1 antagonist activity. Some analogs also have antiserotonin properties; methixene is also a noncompetitive antagonist of bradykinin and is equally as active against acetylcholine and histamine in small animals.[74]

Among the ring systems investigated, ketotifen (4,9-dihydro-4-(1-methyl-4-piperidinylidene)-10H-benzo-[4,5]cyclohepta[1,2*b*]-thiophen-10-one) is particularly interesting. This compound is a potent inhibitor of the H_1-histamine receptor. In addition, it has an action on mast cells resembling that of sodium cromoglycate, but only achieves its maximal effect after it has been taken for several weeks.[75] Ketotifen is dispensed as the hydrogen fumarate salt and is completely absorbed through the gut. It is used for the commonly accepted indications of an H_1 antagonist and in the prophylactic treatment of asthma,[75] although its efficacy in the management of asthma is still debated.[76]

In contrast to other tricyclic H_1 antagonists, the introduction of a 7-chloro moiety affords a more potent, long-lasting drug.[77]

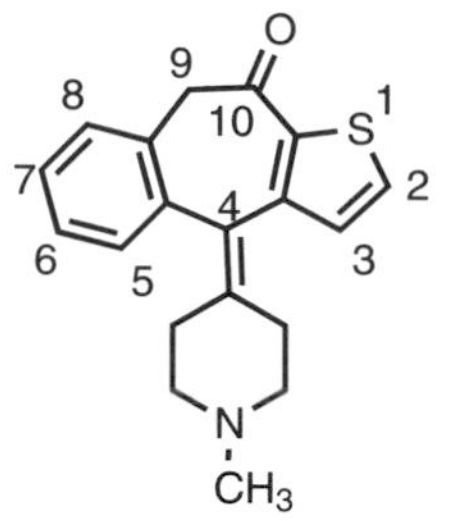

Ketotifen

Pizotyline (pizotifen), 4-(9,10-dihydro-4H-benzo[4,5]-cyclohepta[1,2-b]thien-4-ylidene)-1-methylpiperidine, is also a potent H_1 and serotonin antagonist. The compound, available as hydrogen maleate salt, is used for the prophylaxis of recurrent vascular headache, including migraine.[47,75]

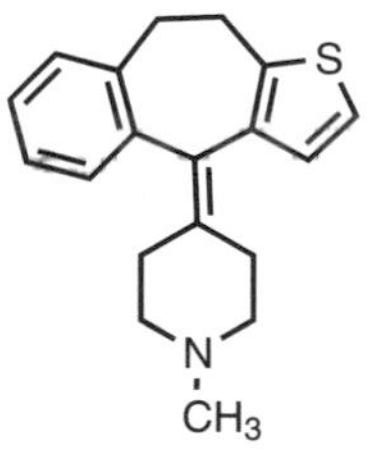

Pizotyline

Compounds of the 4,9-dihydrothieno[2,3-c][2]benzothiepin series exhibit H_1 antagonist effects that exceed those of promethazine in guinea pigs in the histamine aerosol test.[77,78] *Cis* isomers are less potent than *trans* isomers, which are an order of magnitude more potent than promethazine and cyproheptadine.[79]

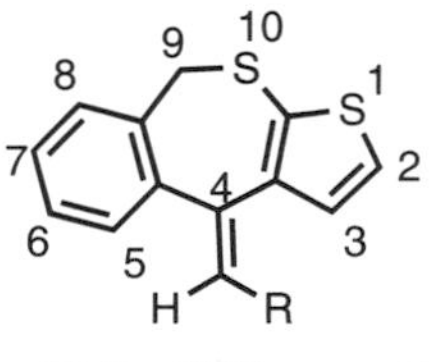

4,9-Dihydrothieno[2,3-c][2]benzothiepin derivatives

Mianserin (1,2,3,4,10,14b-hexahydro-2-methyldibenzo[c,f]pyrazino[1,2-a]azepine) hydrochloride may be regarded as a double-bridged semirigid analog of phenbenzamine. This compound possesses a range of actions, has high affinities for both histamine and 5-hydroxytryptamine sites,[80] and is used clinically as an antidepressant.

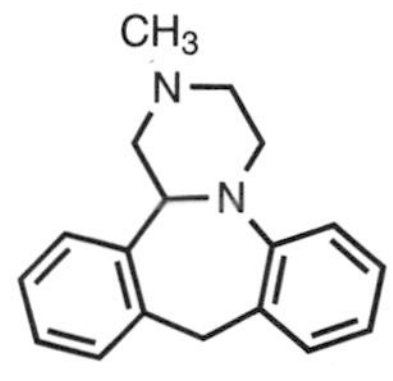

Mianserin

H[1] Antagonists with Decreased Sedative Properties

The major side effect of the classic H_1 antagonists is sedation, possibly attributable to occupation of cerebral H_1 receptors.[81,82] Central nervous system effects are to be anticipated because most classic antihistamines really cross the blood-brain barrier.

During the 1980s, H_1 receptor antagonists with better clinical properties were discovered. These substances have a relative low affinity for central H_1 receptors, a limited ability to penetrate the CNS, or both. Consequently, they are largely free of side effects, especially sedation.

A particularly interesting modification of cyclic basic chains of a cyclizine analog led to the development of terfenadine (U.S.P.), α-[4-(1,1-dimethylethyl)phenyl]-4-hydroxydiphenylmethyl)-1-piperidinebutanol, an H_1 antagonist with virtually no CNS side effects[83,84] such as dizziness, drowsiness, and fatigue, which are common with classic antihistamines.[7,85,86] In contrast, the sedative effect of one early antihistamine, promethazine, is so intense that the suggestion was made to use it as an antipsychotic agent.[7] The relationship between promethazine, the N-methyl derivative of the antihallucinatory agent azacyclonol, and the general structure of H_1 antagonists (note structure-activity relationships) led to studies of the properties of azacyclonol derivatives. Modification of this structure produced fluorobutyrophenones related to haloperidol, which has weak antipsychotic activity with potent H_1 antagonist properties.[83,84] Replacement of the fluorine atom by *t*-butyl and reduction of the ketone to an alcohol resulted in the H_1 antagonist terfenadine. This nonsedating compound has enjoyed wide popularity in the treatment of allergic manifestations.

In a series of 4-(diarylhydroxymethyl)-1-[3-(aryloxy)-propyl]piperidines, the 4-fluorophenyl derivative AHR-5333B is more effective than terfenadine[87] and has undergone clinical investigation.[80]

Ebastine (1-[4-(1,1-dimethylethyl)phenyl]-4-[4-(diphenylmethoxy)-1-piperidinyl]-1-butanone), commercialized in Spain (1990), may be viewed as a structural hybrid of diphenylpyraline, a potent classic H_1 antagonist, and terfenadine.[88] As such, it combines potent, long-lasting, and selective blockade of histamine H_1 receptors with an absence of sedative side effects, and like terfenadine, undergoes biotransformation to active carboxylic acid metabolites by oxidation of one methyl group of the *t*-butyl moiety. Whereas the acidic metabolite of terfenadine is less potent than its precursor as an H_1 antagonist,[89] the equivalent metabolite derived from ebastine possesses considerably more antihistaminic activity in vivo.

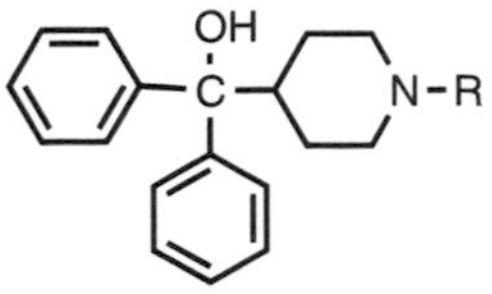

R	
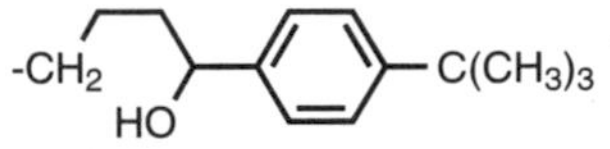 $-CH_2CH_2CH_2CH(OH)C_6H_4C(CH_3)_3$	Terfenadine
-H	Azacyclonol
$-CH_3$	N-Methylazacyclonol
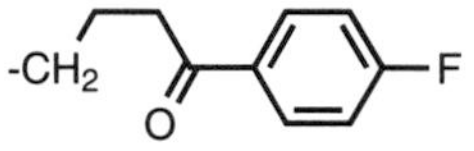 $-CH_2CH_2CH_2C(=O)C_6H_4F$	Weakly active experimental antipsychotic

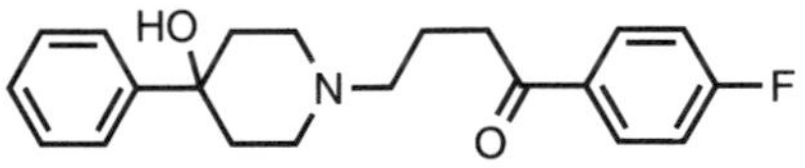

Haloperidol

Cetirizine ([2-[4-[(4-chlorophenyl)phenylmethyl]-1-piperazinyl]ethoxy]-acetic acid) is the main metabolite of hydroxyzine. Because of its higher polarity, this zwitterionic compound less readily penetrates the blood-brain barrier than does the parent compound, and, accordingly, this drug also has reduced sedative side effects.[12,80]

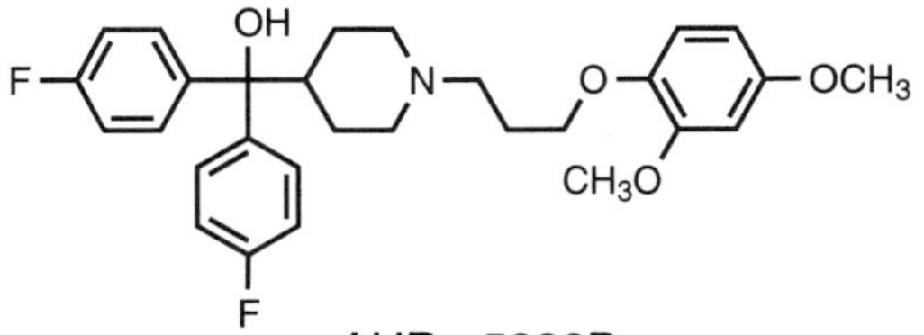

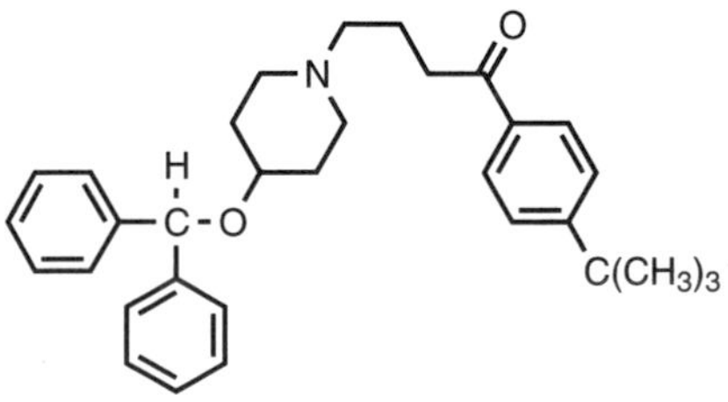

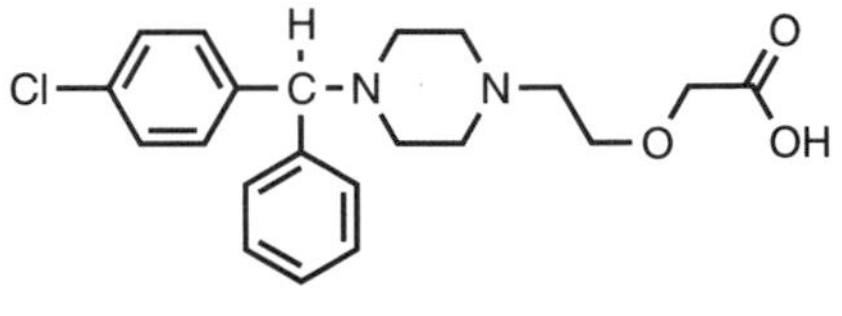

AHR - 5333B

Ebastine

Cetirizine

Acrivastine ((*E,e*)-3-[6-[1-(4-methylphenyl)-3-(1-pyrrolidinyl)-1-propenyl]-2-pyridinyl]-2-propenoic acid) is derived from the classic antihistamine triprolidine by insertion of a carboxyethenyl function at the 2 position of the pyridine ring. This peripheral H_1 receptor antagonist

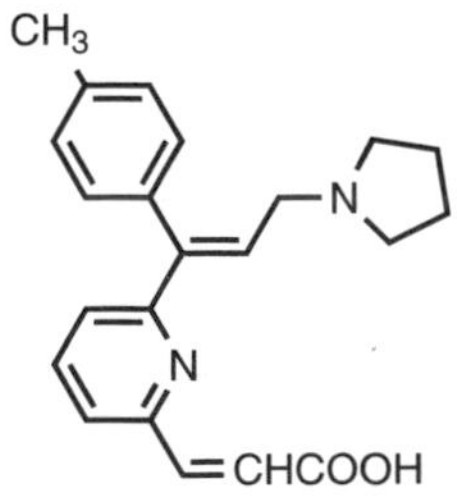

Acrivastine

has activity similar to that of triprolidine. Animal tests show this zwitterionic compound only slowly passes the blood-brain barrier and is less sedative than the parent drug.[76,90]

Astemizole (1-[(4-fluorophenyl)methyl]-N-[1-[2-(4-methoxyphenyl)ethyl]-4-piperidinyl]-1H-benzimidazole-2-amine), a nonsedative antihistamine structurally resembling clemizole and developed during the investigation of derivatives of 2-aminobenzimidazole,[91] is a potent H_1 receptor antagonist with a long duration of action. This drug has no anticholinergic or local anesthetic properties. Astemizole is metabolized slowly and extensively, mainly by aromatic hydroxylation and oxidative dealkylation. The desmethyl metabolites are pharmacologically active and likely contribute to the extended duration of antihistaminic action.[76] From studies in vitro and in vivo, it appears that the reduced level of CNS activity of astemizole is related to its poor penetration of the blood-brain barrier.[92,93]

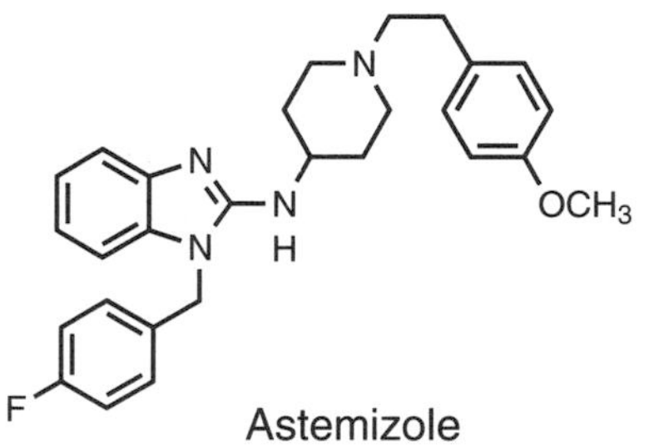

Astemizole

Two additional H_1 antagonists, made available during the middle to late 1980s, are mequitazine and loratadine. Mequitazine (10-(1-azabicyclo[2.2.2]oct-3-ylmethyl)-10H-phenothiazine) is a tricyclic analog in which the basic side chain is positioned within a quinuclidinylmethyl substituent. This drug has less sedative activity than promethazine.[47,94]

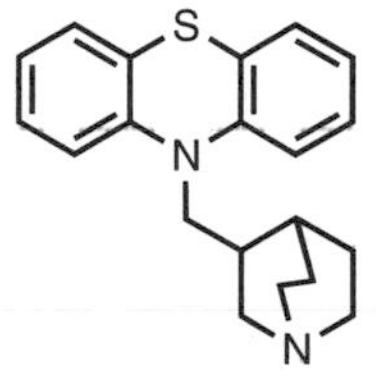

Mequitazine

Loratadine (4-(8-chloro-5,6-dihydro-11H-benzo[5,6]-cyclohepta[1,2-b]pyridin-11-ylidene)-1-piperidinecarboxylic acid ethyl ester) is a hydrophobic analog of azatadine obtained by addition of an aryl chloro group and derivatization of the otherwise basic piperidino nitrogen as the neutral carbamate. H_1 antagonist potency is comparable to that of terfenadine. The drug has a long duration of action and little or no CNS effects.[76] Studies in vitro show a higher affinity for peripheral H_1 receptors than for CNS receptors.[92]

Other nonsedating H_1 antihistamines, such as temelastine and epinastine, are under clinical evaluation. Temelastine (2-[[4-(5-bromo-3-methyl-2-pyridinyl)butyl]amino]-5-[(6-methyl-3-pyridinyl)methyl]-4(1H)-pyrimidinone) is a potent and selective H_1 receptor antagonist with insignificant antimuscarinic activity, and with negligible ability to penetrate the CNS.[95] Interestingly, this drug differs chemically from most H_1 receptor antagonists. The compound is not a tertiary amine and is mainly noncationic at physiologic pH.[96]

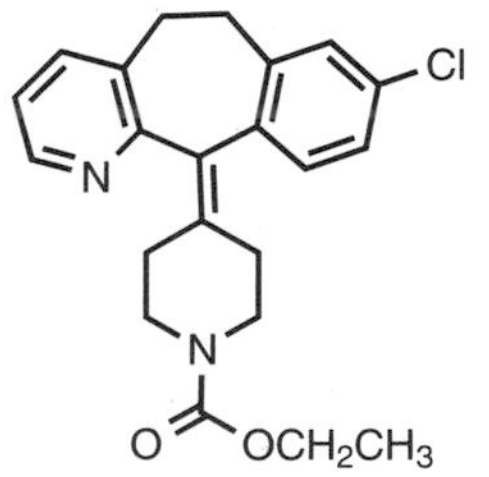

Loratadine

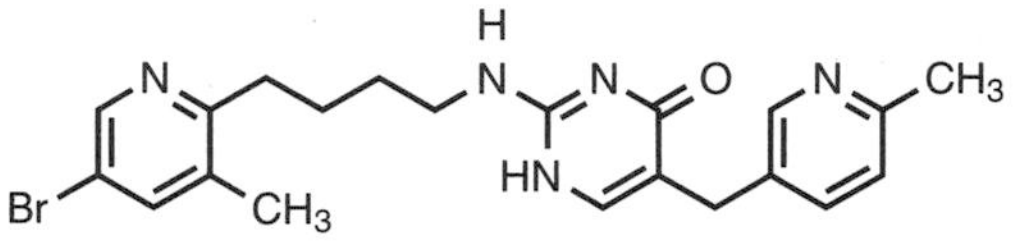

Temelastine

Epinastine (3-amino-9,13b-dihydro-1H-dibenz[c,f]imidazo[1,5-a]azepine)hydrochloride is a tetracyclic compound the structure of which differs from that of other known H_1 antagonists.[97] Epinastine is a peripherally active antagonist that, in the dose range required to give a clear antihistaminic effect, is practically devoid of side effects. Epinastine has also been found to be effective in reducing bronchospasm induced by histamine.[98]

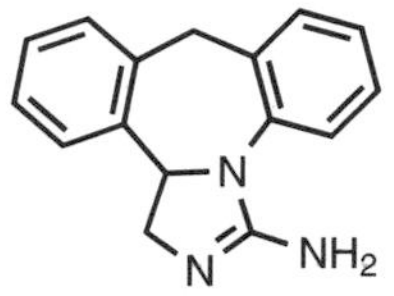

Epinastine

Distribution, Metabolism, and Excretion

H_1 antagonists generally are orally active amine salts that are not degraded appreciably in the gastrointestinal tract.[99] These compounds are not readily absorbed in the stomach, where they exist in the protonated ionic form. When they reach the small intestine, which is alkaline, the equilibrium shifts to favor the free base, which is lipid soluble and readily absorbed through the intestinal wall.

$$\underset{\text{Protonated ionic form}}{Ar{\sim}\overset{+}{N}H(R)_2} \rightleftharpoons \underset{\text{Free base}}{Ar{\sim}N(R)_2} + H^+$$

In the blood, where H_1 antagonists exist in equilibrium free and bound to serum proteins, such as albumin, they are transported to various tissues. These drugs are found in highest concentration in the lungs, kidneys, and spleen of small animals. The liver and other tissues contain enzymes that metabolize these drugs. When hepatic function is impaired, H_1 antagonists remain in the body for long periods.[100] Normally, they are excreted in the urine as degradation products or secondary metabolites. Only small quantities (approximately 10 to 20%) of unchanged drug are usually found.[101]

Metabolite formation depends on the chemical properties of the antagonist and the animal in which the drug is studied. Small animals have fairly simple excretion patterns in contrast to humans.[102] Typical metabolites for antihistamines result from N-dealkylation, deamination, side chain degradation, ring hydroxylation, and oxidation.[103] The metabolism of specific drugs follows, but a more thorough treatment of antihistamine metabolism is available.[58,104,105]

Diphenhydramine and its o-methyl analog are metabolized by rat liver microsomes, yielding formaldehyde through N-demethylation in vitro.[106] In the rhesus monkey, only a small amount of the unchanged drug is excreted in the urine, whereas the major portion of the administered dose is excreted as the glutamine conjugate of α,α-(diphenylmethoxy)acetic acid.[107] This acid is formed either by direct enzymatic oxidation or from the secondary and primary amines after successive N-demethylations. With some antihistamines, N-demethylation may occur through a tertiary amine N-oxide intermediate that is also excreted in the urine. Small quantities of the N-oxide are found in the urine of the rhesus monkey.

In contrast, the small quantities of chlorcyclizine N-oxide formed in vivo are not excreted, but are reduced back to chlorcyclizine, which is subsequently N-demethylated in vivo to norchlorcyclizine.[108,109] Other possible N-dealkylation mechanisms have been proposed wherein N-oxide formation is not a requisite intermediate.[109] Tricyclic antihistamines generally undergo ring hydroxylation as well as side-chain N-dealkylation.[58,104] For example phenothiazines are hydroxylated at positions 3 and 7 and also undergo N-demethylation. No ring-hydroxylated metabolites are reported for the tricyclic antihistamine cyproheptadine, but the principal metabolite in humans is identified as a quaternary ammonium glucuronide-like conjugate.[110] Aryl thiol ethers of the phenothiazine type are also oxidized to their corresponding, more polar sulfoxide metabolites.

H_1 antagonists are noted for their tendency to cause drowsiness. Children are most susceptible to toxic effects resulting from ingestion of large quantities of these drugs.[111] The result is sedation, followed by restlessness, irritability, and muscular twitching. Convulsive seizures may develop, followed by coma and respiratory failure. Adults are more resistant to these severe toxic effects. The difference in susceptibility of young and fully grown animals is a result of increased absorption in the young rather than inability to metabolize the drug.[112]

Prolonged administration of some H_1 antagonists (e.g., diphenhydramine) enhances the activity of hepatic microsomal enzymes. This results in increased metabolism of the drug and of many unrelated medications administered concurrently.[105,113]

Mode of Action

On a molecular level, H_1 antagonists block the agonist histamine from interaction with a receptor site. Although high concentrations of antagonists do inhibit histamine release in vitro, these drugs probably do not work by this mechanism in vivo.[114] Antagonist concentrations necessary to inhibit histamine release in vitro are higher than those likely to be produced with protective doses of antagonist in vivo. Instead, these drugs interact with histamine receptors, thereby preventing histamine from eliciting responses.

The competitive antagonism of histamine and antihistamines on H_1 receptors is studied by use of dose ratios ($[A_2]/[A_1]$), in which $[A_2]$ is the concentration of agonist needed in the presence of a given concentration of antagonist [B] that is required to produce the same response as the initial concentration of agonist $[A_1]$. This dose ratio is related to [B] and K_B by the following equation:

$$\frac{[A_2]}{[A_1]} = 1 + [B]\, K_B$$

The K_B is the association (affinity) constant of the receptor-antagonist complex for the equation:

$$\text{Receptor} + \text{B} \rightleftharpoons \text{Receptor—B}$$

For agonists that have intrinsic activity, such as histamine, the negative log of the dose that causes 50% of the maximal effect ($-\log[A]_{50}$) is defined as the pD_2 value. Because H_1 antagonists have no intrinsic activity, they cannot be defined in terms of pD_2 values. For antagonists, Schild introduced the term pA_2; here, $pA_2 = -\log[B]$ in which [B] is the dose of antagonist that necessitates doubling the dose of agonist to compensate for the action of the antagonist.[115,116] In other words, when the dose ratio ($[A_2]/[A_1]$) is 2, $BK_B = 1$ and $K_B = 1/[B]$. Therefore, the $\log K_B = \log 1/[B] = -\log[B] = pA_2$. If a tenfold instead of a twofold increase in agonist is necessary, the designation becomes pA_{10}. Generally, then, for $[B_X]$, which produces a dose ratio of X, $\log(1/[B_X])$ is called pA_X.

If we know [B] and the dose ratio, we can calculate K_B. If pA_2 values (or dose ratios) for two different agonists of a given antagonist are the same, the inference is that the agonists are operating on the same receptor site. This kind of analysis, using pA_2 values, can be made only if the antagonism is competitive; therefore, the dose-response curves must be parallel in the presence and absence of antagonist. This is the case for the histamine-antihistamine relationship, which is competitive on H_1 receptors. On these receptors of the guinea pig ileum in vitro, most common antagonists have pA_2 values between 8.5 and 10.0. Cholinergic antagonists such as atropine have much lower pA_2 values than antihistamines. The pA_2 value obtained in vitro is not linearly related to the potency of the drug in vivo. A single pA_2 value is usually insufficient to characterize an antagonist fully. For accuracy, it is necessary to state both the time-action and the concentration-action relationships of the test system. Longer contact times give higher pA_2 values and usually indicate approximate equilibrium conditions.[115]

When drugs competitively inhibit the action of histamine at H_1 receptors, it is usually assumed that the drug binds to the same site as does histamine; when the antagonist binds, it has no intrinsic activity. This molecular mechanism may be a grossly simplified explanation. Indeed, to date the histamine pharmacologic receptor has not been isolated. Furthermore, to determine what happens (on a molecular level) after histamine reacts with the receptor requires considerably more research. It is likely, however, that histamine interacts with cell membranes, leading to an increase in membrane permeability to inorganic ions, which causes the variety of pharmacologic actions. Facilitation of Ca^{2+} entry explains the stimulant effect of histamine on smooth muscle contraction.[7]

Pharmacology, Side Effects, and Clinical Applications

The Magnus procedure is one method of evaluating these drugs in vitro; H_1 antagonists are evaluated in terms of their ability to relax histamine-induced spasms in an isolated strip of guinea pig ileum. Antagonists may be evaluated in vivo in terms of their ability to protect animals from the effects of intravenously injected or aerosolized histamine. They may be evaluated for protection against anaphylaxis-induced death in small animals resulting from administration of foreign protein; they are generally less effective against anaphylaxis. H_1 antagonists may also be evaluated in terms of their ability to inhibit local skin reactions to intradermally injected histamine.[117]

Hives, drug rashes, contact dermatitis, and other allergic manifestations that result from histamine-induced dilation of arterioles and capillaries are effectively treated with H_1 antagonists. These drugs are of limited value in the treatment of asthma and neither prevent nor cure the common cold.[7] Their ineffectiveness in the treatment of asthma is attributed to their inability to block the effects of other compounds released during the attack (e.g., leukotrienes and platelet activating factor).

The H_1 antagonists provide relief from allergic seasonal rhinitis (i.e., hay fever), but are less effective in the treatment of chronic perennial rhinitis. These drugs are of some value in controlling allergic dermatoses, serum sickness, blood transfusions, reactions of the nonhemolytic nonpyrogenic type, and drug reactions attributable to an allergic phenomenon.

Side effects with H_1 antagonists vary in incidence and severity with each patient as much as with each drug, although some of the drugs give rise to more side effects than others. The CNS stimulation observed in some patients results in restlessness, nervousness, and insomnia; overdosage may result in convulsions, which can be controlled with diazepam intravenously, but sedatives should be avoided.[47]

Dizziness, drowsiness, and fatigue are other frequently observed side effects of H_1 antagonists in man, but as stated previously, newer analogs are devoid of these side effects. The depressant activity also renders some of these drugs useful in the treatment of insomnia. Clearly, taking these antagonists before driving is contraindicated.

H_1 antagonists also potentiate the CNS depression caused by barbiturates and ethyl alcohol.[118] Patients should be warned of this effect.

The central depressant activity of H_1 antagonists accounts for their use in motion sickness and nausea of pregnancy.[119] Some drugs are more effective than others; aminoalkyl ethers, such as diphenhydramine, may be particularly useful. The phenothiazines and other antihistamines are used for treatment of postoperative vomiting and nausea of pregnancy.

Extrapyramidal symptoms, such as akathisia, dystonia, tremors, and rigidity, may develop with phenothiazine derivatives. These effects have also been reported with other H_1 antagonists, including oxatomide.[47]

Any antiallergic action obtained from H_1 antagonists used in cold remedies is far outweighed by their anticholinergic properties. The anticholinergic properties of these drugs have a drying effect on the respiratory tree, frequently resulting in a dry, painful sensation of the paranasal sinus mucosa.[7] Although these drugs are more potent antagonists of histamine than acetylcholine, their structural similarity to cholinergic blocking agents enables one to predict that they will also be effective antagonists of the pharmacologic action of acetylcholine.

H_1 antagonists, particularly diphenhydramine, have been used in paralysis agitans (parkinsonism); they lessen rigidity and tremor and improve movement and speech, probably because of their ability to antagonize acetylcholine in the CNS. Other effects common to these drugs include urinary frequency and dysuria, headache, tingling sensation, loss of appetite, nausea, vomiting, constipation, and diarrhea.

The H_1 antagonists, promethazine and pyrilamine, have local anesthetic activity. Again, these agents have a structural resemblance to local anesthetics. Tripelennamine and diphenhydramine have been used as local anesthetics during tooth extraction.[120] This activity does not parallel their ability to antagonize acetylcholine or histamine; some of these drugs are more potent local anesthetics than procaine.[7] Because of this activity, they are incorporated in creams and lotions to relieve itching and eczema. Caution is advised in their use; H_1 antagonists, like many drugs, may cause hypersensitivity reactions, especially when applied topically. Photosensitivity has also been reported, particularly with the phenothiazines.

Various H_1 antagonists have been associated with fetal abnormalities when taken during pregnancy, but authors of major studies have failed to demonstrate any strong correlations. Overdose may be fatal, especially in infants and children. The elderly are more susceptible to the CNS depressant and hypotensive effects, even at therapeutic doses.[47]

Structure-Activity Relationships

In general, whenever a compound contains the function

$$\begin{matrix} Ar_1 \\ \quad\diagdown \\ \quad X-(CH_2)_nN(CH_3)_2 \\ \quad\diagup \\ Ar_2 \end{matrix}$$

in which Ar^1 = a phenyl group, Ar^2 = a phenyl, benzyl, pyridyl, or thenyl group, X = O, C, or N, and n is a number (usually 2) so that the distance between the centers of the aromatic rings and the aliphatic N may be 5 to 6 Å, the compound will have competitive histamine H_1 antagonist activity in vitro.[58]

Substitutions on the Ar groups, replacement of the aliphatic dimethylamino group with small basic heterocyclic rings, increased branching on $(—CH_2—)_n$, and substitution between X and N all serve to modify the potency, metabolism, ability to reach the site of action, toxicity, and side reactions in vivo. This distance of 5 to 6 Å is easily achieved when antihistamines exist in the *trans* conformation; more rigid antihistamines have a similar distance between aromatic rings and the basic nitrogen. It seems, however, that there is no strict requirement for the fully extended *trans*-N-C-C-N conformation for H_1 receptor blockade. Rather, a range of values for this torsional angle is suitable for antagonist-receptor interaction.[121]

For maximal H_1 antagonist activity, both rings of diphenhydramine cannot be in the same plane (coplanar). The fluorene analog, in which both rings are coplanar, is 100 times less active. In potent tricyclic systems, rings A and C are not in the same plane; for example, the B ring of a phenothiazine H_1 antagonist is in the shape of a boat.

If an assumption is made that a basic amino group is

trans Conformation of diphenhydramine

Fluorene analog of diphenhydramine

Boat-shaped phenothiazine ring

necessary for attachment to an anionic site on the H_1 histamine receptor (just as the amino group of histamine seems necessary for such an interaction), the aromatic rings of the antagonist may bind to sites outside the histamine receptor. It seems that sites outside the histamine receptor are asymmetric, because stereoselectivity is observed with carbinoxamine, pheniramine, chlorpheniramine, and brompheniramine. The more active levo isomer of carbinoxamine has been shown to have the (S) absolute configuration[122] and to be superimposable on the more active dextro isomer (S configuration[123]) of chlorpheniramine. The more active enantiomorphs of pheniramine and brompheniramine have been shown to have the S absolute configuration.[123]

S absolute configuration of substituted pheniramines

For a compound to exhibit stereoselective H_1 antagonist activity, the asymmetric center must be located on the carbon to which the aromatic functions are bonded. If the asymmetric center is located on the carbon to which the dimethylamino group is bonded, stereoselective antagonism is not observed.[124] It therefore seems reasonable to propose that sites to which the aromatic rings bind are asymmetric. Studies of asymmetric drugs with known absolute configurations help to clarify structural requirements for drug-receptor interaction and provide insight into the nature of the H_1 receptor.[125,126]

Geometric isomers, namely *cis* and *trans* N,N-dimethyl-1,5-diphenyl-3-pyrrolidinamine, were synthesized to evaluate stereochemical requirements for H_1 antagonist activity.[121,127,128] Both isomers are potent H_1 antagonists, but the *trans* compound is the more potent and longer acting isomer.[121] The active conformation of several H_1 antagonists are being investigated to define further the biologically active conformation required for histamine H_1 receptor interaction.[129]

Cis *Trans*

N,N-Dimethyl-1,5-diphenyl-3-pyrrolidinamine

Quantitative structure-activity relationships (QSAR) and molecular modeling studies are available for histamine H_1 antagonists.[130] These studies provide considerable insight into the physicochemical and conformational features that define receptor binding of H_1 antagonists. The QSAR analyses indicate that at least seven classes of H_1 antagonists bind to the same site on the receptor. For these compounds, a basic amino group, an aromatic ring (the so-called *cis* ring) and a hydrophobic group (at the position of the so-called *trans* ring) are essential for receptor binding. Similarities in the QSAR equations of diphenhydramines, mono phenyl analogs of diphenhydramines, and benzimidazoles show that hydrophobic and steric factors are important for binding at the *trans* ring location. Modeling studies reveal the optimal distance between the basic nitrogen atom and one of the aromatic rings is approximately 6 Å. Additionally, cyproheptadine, which has a piperidylidene ring in the boat conformation, is a useful template for modeling studies.

Chemical Mediators in Hypersensitivity Reactions

The classic histamine hypothesis provides only a partial explanation for the spectrum of effects that results from immediate hypersensitivity reactions. In addition to activation of phospholipase C and the hydrolysis of inositol phospholipids, stimulation of IgE receptors also activates phospholipase A_2. This action leads to the production of a host of mediators, including platelet activating factor (PAF) and metabolites of arachidonic acid (AA)[7] (Figs. 22–1 and 22–2).

In the cyclooxygenase pathway (Fig. 22–1) of arachidonic acid (AA) metabolism, AA is oxidized to a 9,11-endoperoxide and 15α-hydroperoxide known as PGG_2. PGG_2 is reduced to PGH_2, which serves as precursor to prostacyclin (PGI_2), thromboxanes (TXs), and the prostaglandins, such as PGE_2 and $PGF_{2\alpha}$.

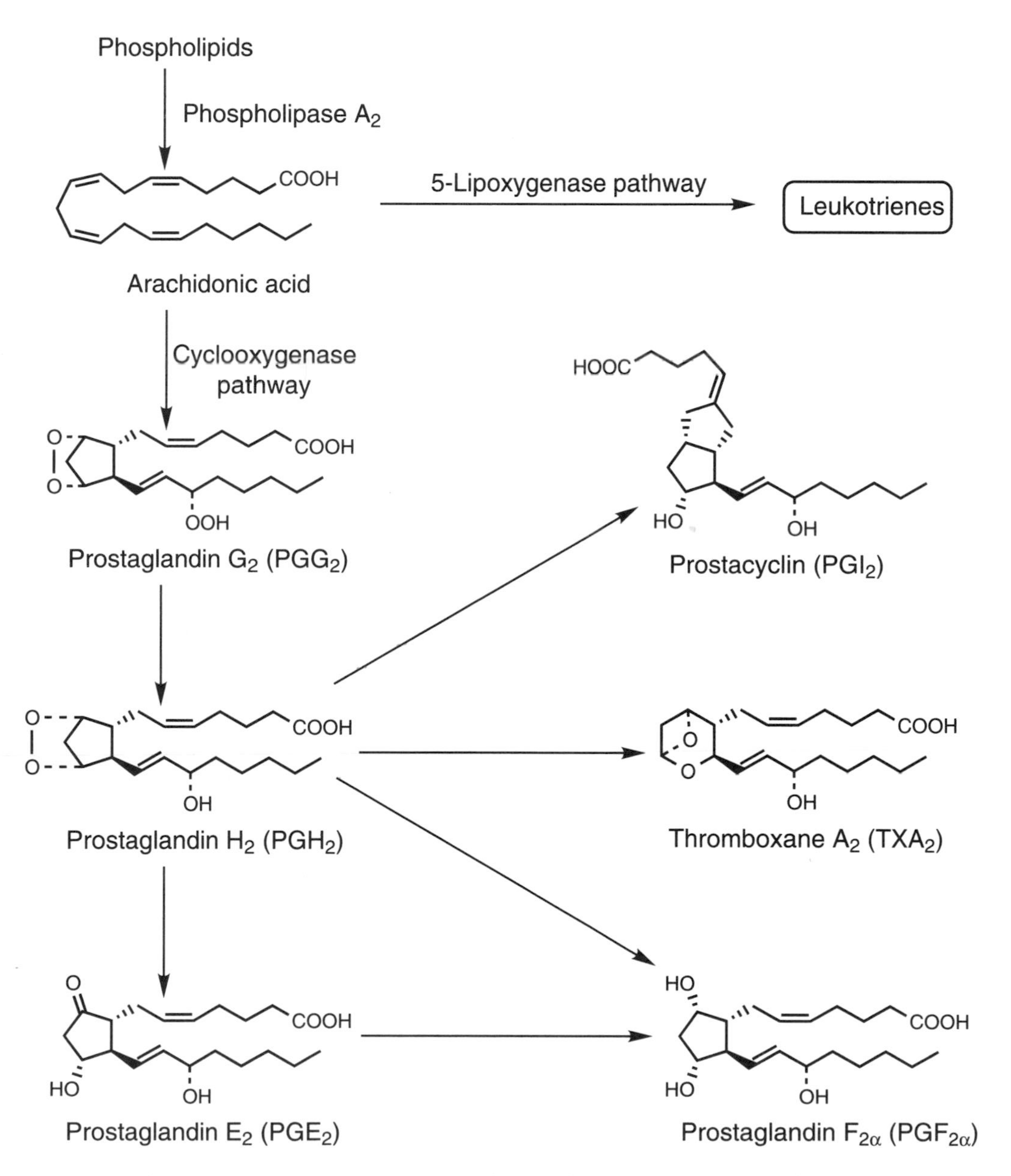

Fig. 22–1. The cyclooxygenase pathway of arachidonic acid metabolism.

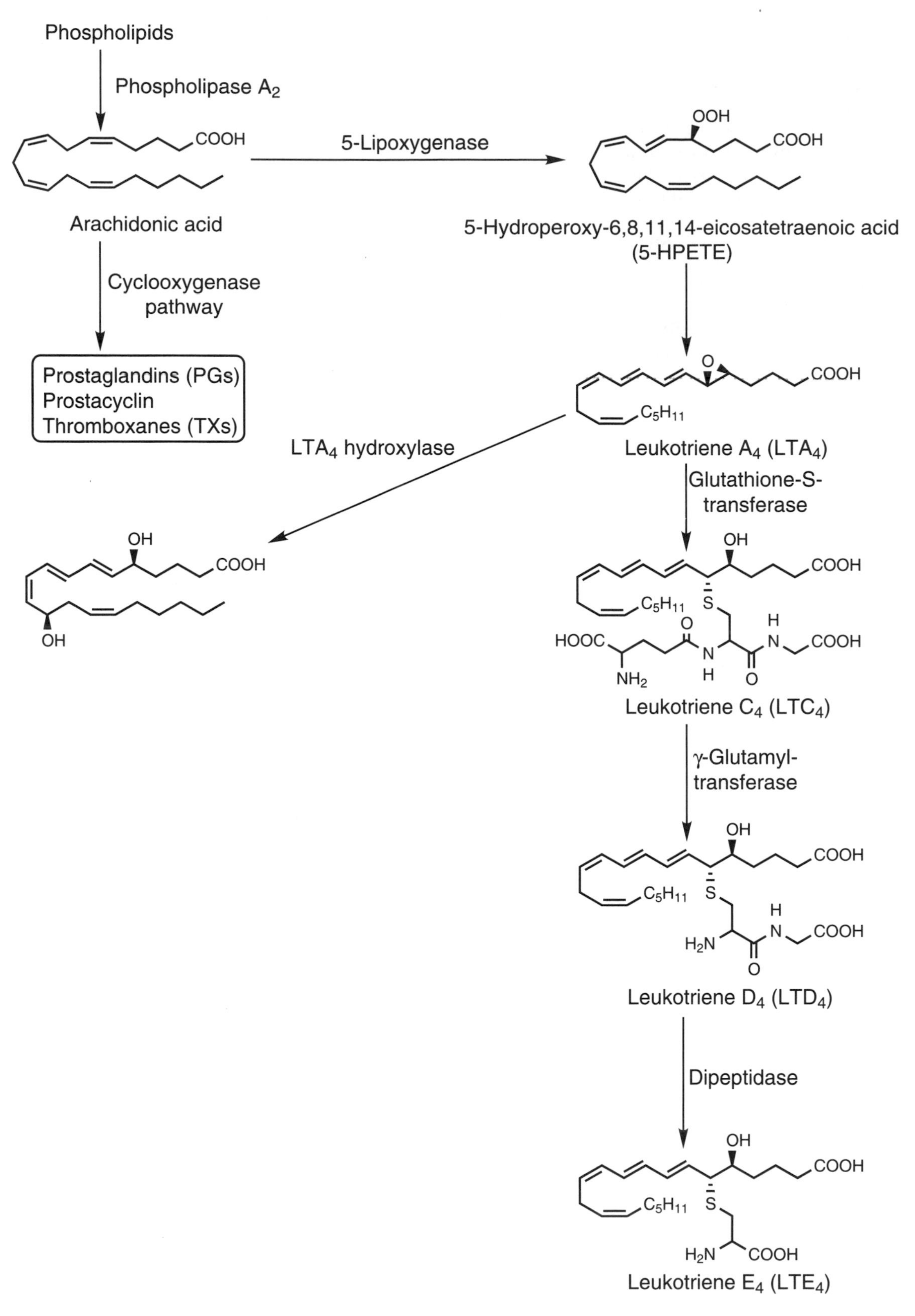

Fig. 22–2. The 5-lipoxygenase pathway of arachidonic acid metabolism.

The 5-lipoxygenase pathway (Fig. 22–2). produces 5-hydroperoxy-6,8,11,14-eicosatetraenoic acid (5-HPETE) from AA, and this metabolite is intermediate to a series of leukotrienes (LTA_4, LTB_4, LTC_4, LTD_4 and LTE_4).

Leukotriene D_4, generated as in Figure 22–2, stimulates smooth muscle contraction of the bronchial tree and may be a dominant factor in allergic conditions, such as asthma.[131,132] Kinins are also generated during some allergic responses. Thus, mast cells secrete a variety of inflammatory compounds in addition to histamine, and each contributes to the major symptoms of the allergic response.

The wide variety of mediators released during the allergic manifestation explains the ineffectiveness of drug therapy that involves use of a single antagonist. Research directed at the regulation of mediator release from mast cells and basophils is important. These cells have receptors linked to signaling systems that may enhance or block IgE-induced mediator release.

Inhibition of the Release of Mediators

bis(Chromones) and Related Drugs

In the late 1960s, a new class of compounds, referred to as bis(chromones), was introduced by Fisons Corporation in England. These antiallergenic drugs were synthesized as part of an attempt to improve on the bronchodilator activity of khellin, a chromone (benzopyrone) derived from the plant Amni visnaga, previously used by the ancient Egyptians for its spasmolytic properties.[133]

The most representative compound of this class is cromolyn sodium, which, unlike the natural product prototype khellin, has no muscle relaxant properties, but is effective in preventing, but not reversing, antigen-induced bronchospasms.[134–136]

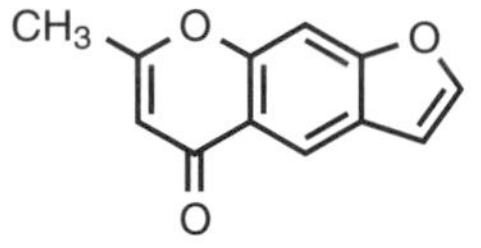

Khellin

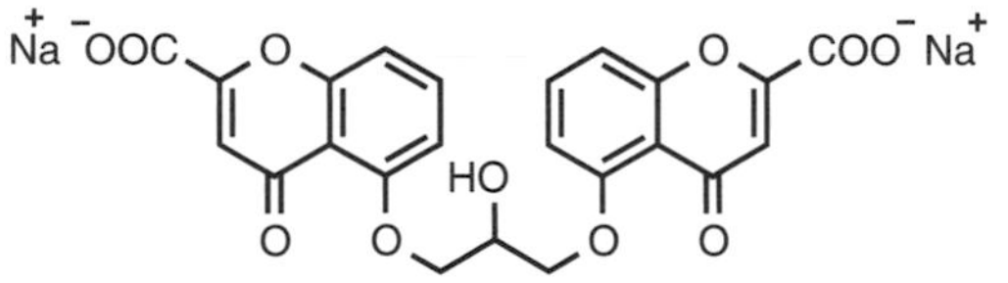

Cromolyn sodium

Cromolyn sodium, disodium 5,5′-[(2-hydroxy-1,3-propanediyl)bis-(oxy)]bis[4-oxo-4H-1-benzopyran-2-carboxylate] (U.S.P.), is poorly absorbed orally and thus is administered intranasally, topically onto the eye, and by oral inhalation.[137] The drug is prophylactic and is not a histamine antagonist. Cromolyn appears to act mainly through a local effect on the lung mucosa, nasal mucosa, and eyes. Cromolyn prevents release of mediators of allergic reactions, including histamine and slow-reacting substance of anaphylaxis (SRS-A) from sensitized mast cells after the antigen-antibody union has taken place. The drug does not inhibit the binding of IgE to mast cells nor the interaction between cell-bound IgE and the specific antigen; instead, cromolyn suppresses the release of substances (e.g., histamine, SRS-A) in response to this reaction. The drug may block calcium channels in mast cell membranes and may inhibit histamine release from mast cells by regulating phosphorylation of a specific mast cell protein involved in secretory mechanisms. The specific mechanism(s) of action of the drug on mast cells, however, remains to be established.[137,138]

Proper administration of cromolyn sodium is necessary to obtain optimal results. The maximal effect is observed when the interval between the drug (predose) and antigen (challenge dose) is between 5 and 30 minutes. The effectiveness of this therapy depends on its administration at regular intervals. Usually, one 20-mg capsule is inhaled four times daily. Excessive use of this drug leads to a rapid, induced tolerance. Although cromolyn sodium is poorly absorbed orally, such administration is indicated for the symptomatic treatment of systemic mastocytosis.[137]

Unlike the relationship of the aryl rings in H_1 antagonists, coplanarity of the two chromone rings is a most important requirement for biologic activity in the bis(chromone)series.[139] The length and position of the linking chain in bis(chromone) structures does not seem to be critical for antiallergic activity unless the chain is bonded to the 8 and 8′ positions of the chromone functions. Thus, 8,8′-substituted bis(chromones) and 8-substituted monochromone-2-carboxylic acids are inactive. No antiallergic activity is observed with bis(chromone) structures bonded through a single methylene group or when the linking chain [O-$(CH_2)_n$-O] is longer than $n = 6$. Subsequent to the introduction of cromolyn sodium, several oral antiallergic agents have become available that also inhibit the release of chemical mediators, but none are equal to the prototype in the clinic.

Nedocromil sodium, disodium 9-ethyl-6,9-dihydro-4,6-dioxo-10-propyl-4H-pyrano[3,2-g]quinoline-2,8-dicarboxylate, a compound possessing a structure as well as an action on mast cells that resembles that of cromolyn sodium, has been investigated in asthmatic patients.[140] This drug is administered by inhalation for the prophylactic treatment of asthma.

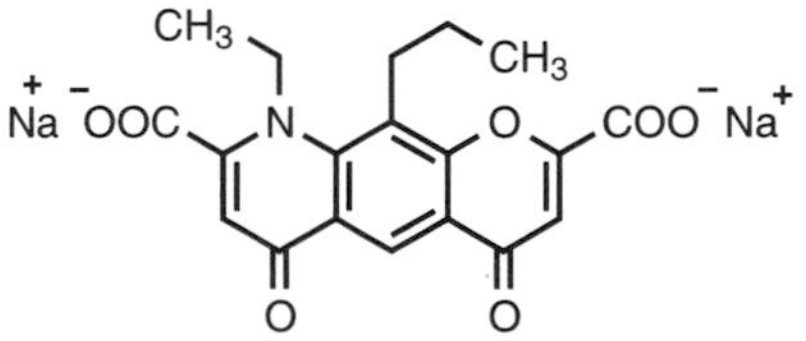

Nedocromil sodium

Peptidoleukotriene Antagonists

Peptidoleukotrienes (pLT) are metabolites of AA, which are enzymatically liberated from membrane phospholipids. Arachidonic acid is converted to prostaglandins, prostacyclin, and thromboxanes via PGG_2 (see Fig. 22–1). Such biosynthesis is blocked by the nonsteroidal anti-inflammatory drugs (NSAIDs), which inhibit cyclooxygenase. Additionally, AA is transformed by 5-lipoxygenase to produce 5-HPETE, which serves as precursor to the epoxide leukotriene A_4 (LTA_4), from which all other leukotrienes are derived (see Fig. 22–2).

Local production of pLT is likely a major factor in the pathogenesis of asthma and related conditions. Constriction of bronchial smooth muscle and excessive production of bronchial mucus are characteristic symptoms evoked by pLT. This response results in a decreased lung volume associated with an asthma attack.[141] Consequently, potent and selective pLT antagonists should be useful therapeutic agents for the treatment of this disease.

Twenty years of research in this field have generated a number of biologically active compounds from different structural classes. The first generation of pLT antagonists was disappointing in asthmatic patients, however, because of insufficient potency.[142]

The structures of the pLT, collectively known as SRS-A, were defined in the late 1970s and early 1980s.[143,144] Peptidoleukotriene antagonist research evolved using the natural mediators as a starting point for chemical modification. This effort was highly sophisticated because of chemical lability and the complicated lead structures. Even so, four antagonists (SK&F 104353, SK&F 106203, LY 170680, and CGP 45715A) have been identified for testing in humans.[142]

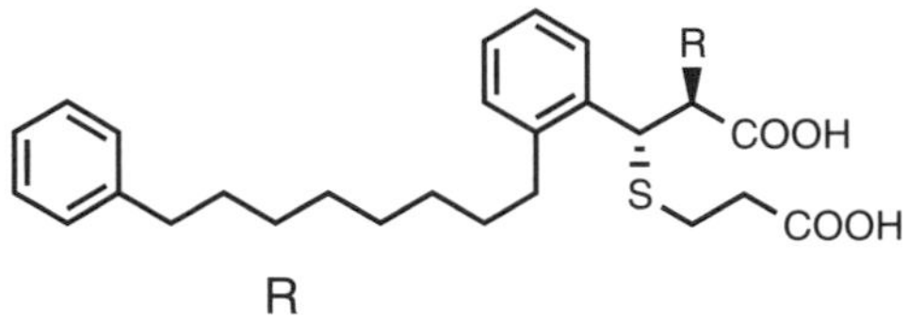

R	
OH	SK&F 104353
H	SK&F 106203

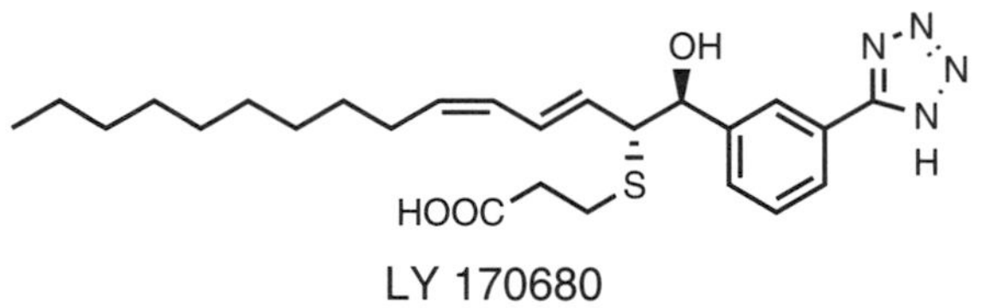

LY 170680

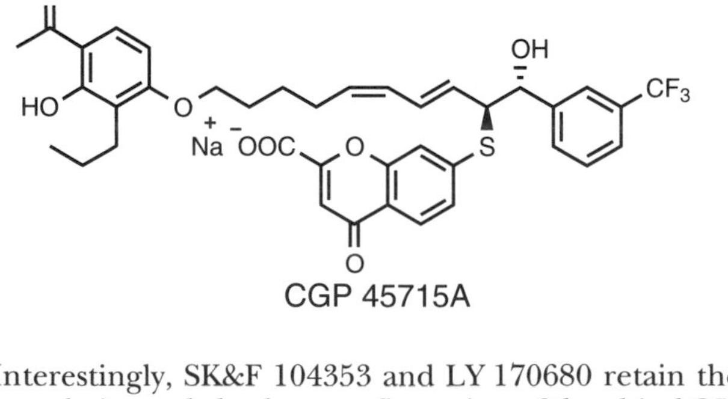

CGP 45715A

Interestingly, SK&F 104353 and LY 170680 retain the same relative and absolute configuration of the chiral OH and thioether centers of LTD_4, whereas in CGP 45715A, the opposite diastereomer is required for potent pLT antagonism. The SK&F compounds as well as LY 170680 also retain the acidic function equivalent to the carboxylic group in the eicosanoid backbone of LTD_4. Clinically, SK&F 104353 is the most advanced pLT analog and may be one of the first pLT antagonists to reach the market. Problems with an irritant aerosol formulation need to be resolved. SK&F 106203 (the dehydroxy derivative of SK&F 104353), although ten times less potent in vitro, has improved oral activity when compared to the parent compound. Accordingly, an oral formulation of SK&F 106203 is being developed. The most potent orally active antagonist of the pLT analog type is CGP 45715A. Nonetheless, in the guinea pig, CGP 45715A is three or four orders of magnitude more potent when given as an aerosol, and thus aerosol formulations are of interest clinically.

The substituted indole derivative ICI 204219 is among the most potent LTD_4 antagonists known to date and is the one to have reached the most advanced stage of clinical investigation. It is effective against allergen-[145,146] or exercise-[147] induced bronchoconstriction in asthmatics and also improves lung function.[148]

The tetrazolylderivative ONO-1078, 8-[4-(4-phenylbutoxy)benzamido]-2-(tetrazol-5-yl)-4H-1-benzopyran-4-one, is also well advanced in clinical development and is reported to show beneficial effects in asthmatic patients.[149]

Among the many compounds examined as potential antagonists of SRS-A, some unusual properties are observed in a new isopropylpyrazolopyridine known as ibudilast (KC-404). Ibudilast (2-methyl-1-[2-(1-methylethyl)-pyrazolo[1,5-a]-pyridin-3-yl]-1-propanone) is a "pLT antagonist" on the market in Japan. The compound does not bind to the LTD_4 receptor and has weak functional pLT properties.[150] The antiasthmatic activity observed for this drug in man may be attributed to a number of other biologic activities.[142]

Pemirolast, a new antiallergic agent with a unique chemical structure has been marketed in Japan.[151] This potassium salt of 9-methyl-3-(1H-tetrazol-5-yl)-4H-pyrido[1,2-a]pyrimidin-4-one inhibits the release of histamine, pLTs, PGD_2, TXB_2, and other mediators following the antigen-antibody reaction.[152–154]

Bradykinin Antagonists

Bradykinin, an endogenous linear nonapeptide hormone, and kallidin are collectively known as "kinins." These substances are generated from kininogens by proteolysis catalyzed by kallikrein enzymes. Such enzymes are present in most tissues and fluids of the body. After release, kinins contribute to many pathophysiologic responses, including inflammation, allergic reactions, arthritis, asthma, pancreatitis, and viral rhinitis.[155–158]

$$\text{R-Arg}^1\text{-Pro}^2\text{-Pro}^3\text{-Gly}^4\text{-PHe}^5\text{-Ser}^6\text{-Pro}^7\text{-Phe}^8\text{-Arg}^9$$

R = H Bradykinin

R = Lys Kallidin

Kinin receptors are categorized into several subtypes.

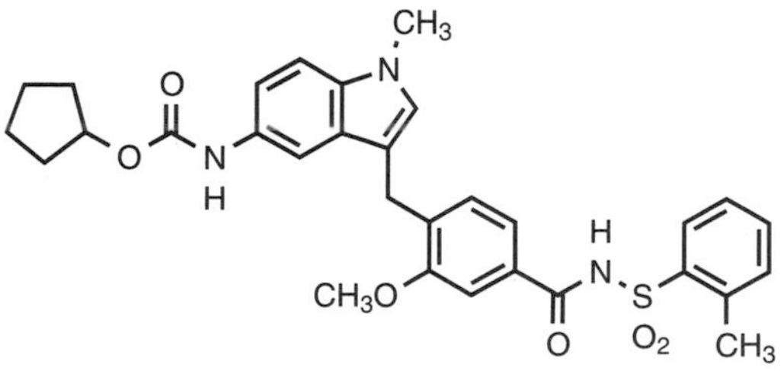

ICI 204219

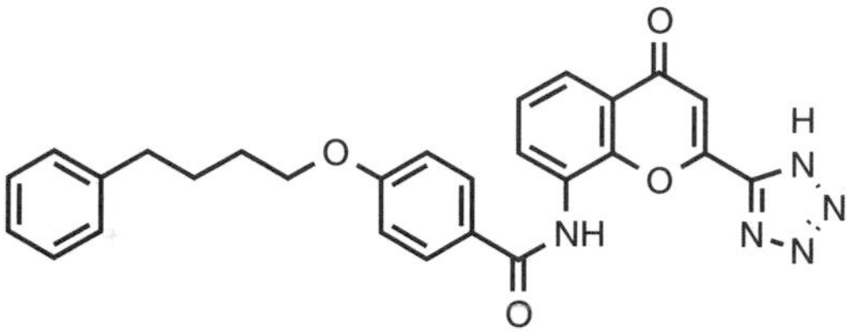

ONO 1078

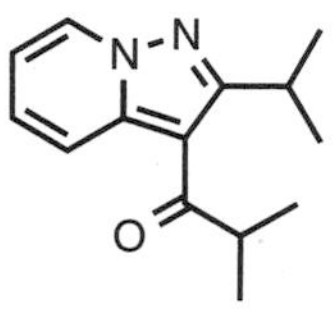

Ibudilast (KC 404)

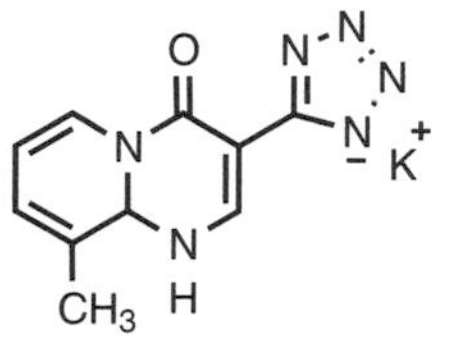

Pemirolast

The major receptor subtypes are terms B_1 and B_2, but additional numeric classifications may include B_3, B_4, and B_5.[159] Development of bradykinin analogs predates receptor subtype classification, and it continued for a period of more than two decades[160–163] before the discovery of the first antagonist, [D-Phe^7]-bradykinin.[164] Subsequently, antagonists that block B_1 and B_2 receptors or are selective for the B_1 receptor became available. The most selective and potent B_1 antagonists are desArg^9[Leu^8]-bradykinin and desArg^{10}[Leu^9]kallidin.[165] One example of an indiscriminate B_1 and B_2 receptor antagonist is D-Arg[4-hydroxy-Pro^3-D-PHe^7]bradykinin. Analogs of this type block most of the B_2 receptor-mediated biologic effects of kinins.[165] Such peptides have not been clinically useful, however, because of their short biologic half-lives and poor oral bioavailability. In initial trials, the nonselective bradykinin antagonist D-Arg[4-hydroxy-Pro^3-D-Phe^7]bradykinin produced encouraging results in the relief of cold symptoms caused by rhinovirus, in the suppression of pain from burns, and in the treatment of allergic asthma.[165,166]

Because peptides are expensive to prepare, difficult to deliver, usually short-lived in vivo, and orally inactive, much interest has been generated in the development of nonpeptide kinin receptor antagonists. A comprehensive review of such compounds is available.[167] Despite substantial research efforts, however, there are no potent and selective nonpeptide bradykinin antagonists. Even so, interest in bradykinin receptor antagonists continues. Most exciting are reports describing second-generation peptides that are significantly more potent and of longer duration of action. For many research peptides, conformational constraints introduced via unnatural amino acids provide information regarding the more likely receptor-bound conformation. With certain hydroxyproline ether antagonists, receptor binding topographic insites are available despite the lack of x-ray crystallographic or nuclear magnetic resonance-derived structural data.[159]

Several second-generation peptides are 500 to 1000 orders of magnitude more potent than first-generation compounds, and some of these have longer half-lives. For these reasons, clinical evaluation continues. HOE140 is one such agent of interest in allergic rhinitis[168] and NPC17761 is a second agent of interest in rhinovirus infections, asthma, and sepsis.[169]

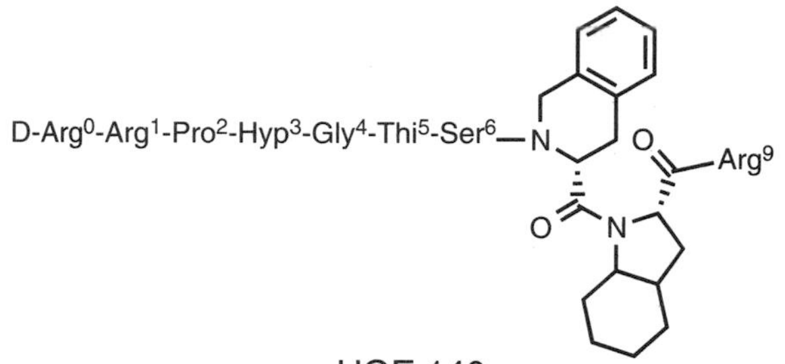

D-Arg^0-Arg^1-Pro^2-Hyp^3-Gly^4-Thi^5-Ser^6— ... Arg^9

HOE 140

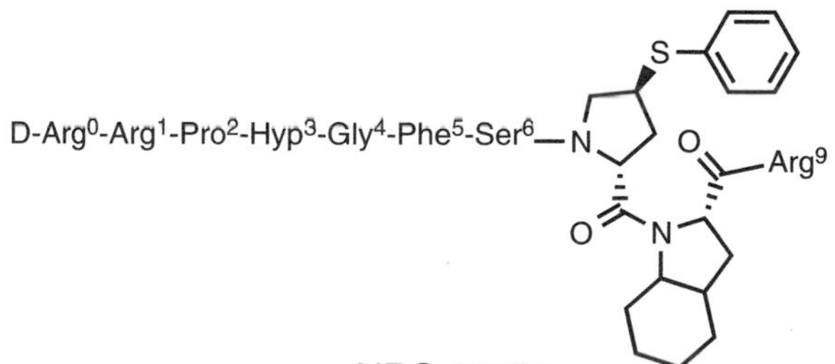

D-Arg^0-Arg^1-Pro^2-Hyp^3-Gly^4-Phe^5-Ser^6— ... Arg^9

NPC 17761

ANTIULCER AND H_2 RECEPTOR INTERACTIVE DRUGS

H_2 Agonists

In 1966, Ash and Schild defined histamine sites blocked by mepyramine as H_1 receptors.[4] Because mepyramine does not antagonize histamine-stimulated gastric acid secretion or atrial and rat uterus contraction, applicable receptors most likely differ from those of the H_1 type and were named H_2. In the late 1960s, Black and Ganellin began their search for an H_2 antagonist that would inhibit the release of gastric acid caused by the action of histamine. Such a drug should be useful for the treatment of gastric ulcers.[170]

After the H_2 receptor was defined,[5] most research was directed toward the discovery of H_2 antagonists.[171] Less attention was given to H_2 agonists, compounds potentially useful in the treatment of severe congestive heart failure.[172,173] Thus far, potent and highly selective H_2 agonists are not available. Generally, modification of the histamine molecule leads to loss of activity. Only the 5-

methyl analog is a selective H_2 agonist with about the same potency as histamine.[171] Two other H_2 agonists are dimaprit and impromidine.

Dimaprit (S - [3 - (N,N - dimethylamino) propyl] isothiourea) is selective, but not very potent (about 0.5 times histamine). Impromidine (N-[3-(1H-imidazol-4-yl)-propyl] - N′ - [2 - [[(5-methyl - 1H - imidazol - 4 - yl)methyl]-thio]ethyl] guanidine) is a potent H_2 agonist (30 to 50 times histamine), but is nonselective. This compound also is a potent H_3 antagonist.[174] Impromidine induces hemodynamic improvement in patients with idiopathic congestive cardiomyopathy in end-stage heart failure,[175] although this compound is not used routinely in the clinic because of side effects.

Arpromidine (N-[3-(4-fluorophenyl)-3-(2-pyridinyl)-propyl]-N′-[3-(1H-imidazol-4-yl)propyl]guanidine) is a possible chemical lead for the development of cardiohistaminergics.[176] Such a compound, while related to impromidine, does not contain the imidazolylmethylthio group found in cimetidine, an H_2 antagonist. Compounds possessing two phenyl-ring halogen substituents, such as 3,4-difluoro and 3,4-dichloro, are up to 160 times more potent than histamine and may have therapeutic value in the treatment of congestive heart failure.[177] Analysis of impromidine structure-activity relationships,[171] coupled with the use of computational techniques,[178,179] resulted in the generation of several new potent and selective H_2 agonists. For example, thiapromidine, the 2-amino-5-methylthiazole analog of impromidine, is a potent and selective H_2 agonist with about the same potency as impromidine. VUF 9012 is the thiapromidine analog in which the 5-methylimidazole group is replaced by a bis-(aryl)methyl grouping of the type found in H_1 interactive drugs. VUF 9012 is a potent H_2 agonist with a profile comparable to arpromidine.[180]

Structure-activity relationships for a series of 65 H_2 agonists of the impromidine (phenyl analog) and arpromidine type were investigated by Free-Wilson analysis.[181] These compounds have in common an imidazolylpropylguanidine moiety that is connected to 1 or 2 aromatic rings through a flexible aliphatic chain that may or may not contain heteroatoms. In one of the rings, substituents are varied in all positions. Free-Wilson analysis revealed that the properties of the side chain and the arrangement of rings are of primary importance for receptor affinity, whereas substituents show an irregular pattern of marginal contributions. Substituent effects are not constant but vary with changes in the chain.

Quantum chemical calculations comparing histamine and dimaprit reveal similarities in electron distributions and molecular electrostatic potentials (MEP). These analyses led to the synthesis of rigid (cyclic) analogs of dimaprit possessing the 4- and 5-(2-aminoethyl)thiazole functionality with small alkyl substitutions. One of these, amthamine (2-amino-5-(2-aminoethyl)-4-methylthiazole) is a potent histamine H_2 receptor agonist that may be useful as a pharmacologic tool to investigate subtypes of histamine receptors.[179]

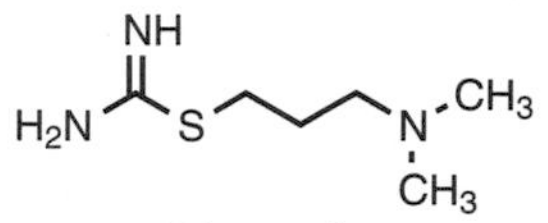

Dimaprit

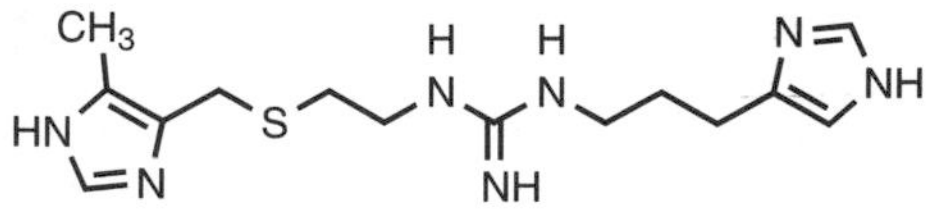

Impromidine

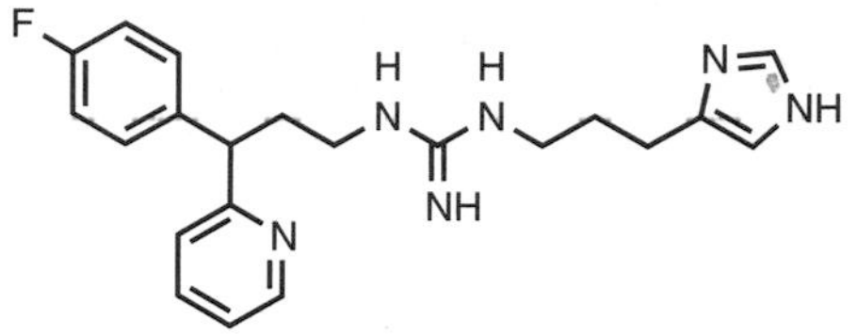

Arpromidine

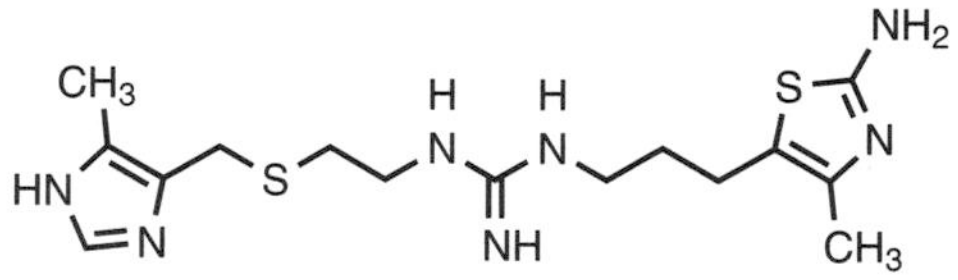

Thiapromidine

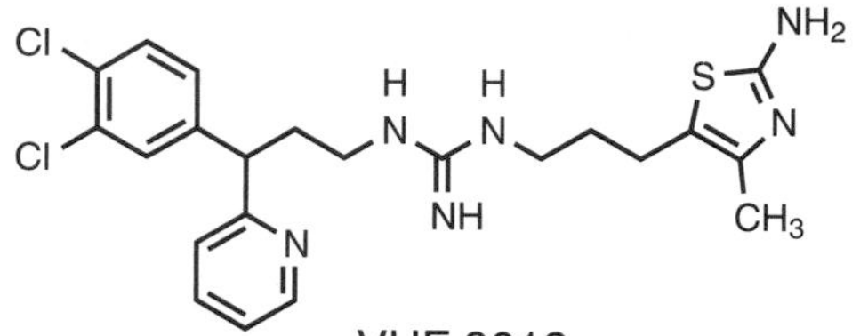

VUF 9012

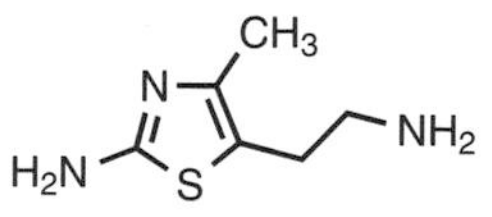

Amthamine

H_2 Antagonists

Unlike H_1 receptor antagonists, which were discovered by chance, the discovery of H_2 antagonists resulted from an extensive multidisciplinary-multinational research initiative. This activity represents an approach to rational drug design.

Different physico-chemical properties of histamine analogs determine selectivity for either the H_1 or H_2 receptor subtype. Burimamide (N-[4-(1H-imidazol-5-yl)butyl]-N′-methylthiourea), an early discovery, possesses a chemical structure different from those of the classic H_1 antagonists. This antagonist does not have affinity for H_1 receptors and served, in part, as impetus to define the H_2

receptor subtype.[5] For histamine, the mono cationic N^{τ}-H tautomer seems to be the active form at both H_1 and H_2 receptors. Potent antagonists are imidazoles of the N^{τ}-H type, a form proposed to be required for maximal H_2 antagonist activity.

Burimamide is not useful clinically. Marginal potency is attributable to the electron-donating effect of the alkyl side chain, which favors the N^{π}-H tautomeric form, a species less favorable for H_2 receptor binding. Modification of the burimamide structure with electron-withdrawing groups shifts the equilibrium toward the N^{τ}-H form, the tautomer presumed necessary for H_2 receptor interaction.

Substitution of one methylene group in the side chain of burimamide with an isosteric sulfur atom and addition of one methyl group to position 5 on the imidazole ring gives the H_2 receptor active compound methiamide (N-methyl-N′-[2-[[5-methyl-1H-imidazol-4-yl)methyl]thio]ethyl]thiourea), the dominating neutral tautomer of which is the N^{τ}-H form. Metiamide is not marketed, however, because of side effects, which include agranulocytosis, uncovered during preclinical trials. Such side effects are attributed to the thiourea function.

Interestingly, the cyanoguanidine functionality is isoelectronic with the thiourea group. This substitution furnishes cimetidine (N-cyano-N′-methyl-N″-[2-[[(5-methyl-1H-imidazol-4-yl)methyl]thio]ethyl]guanidine) which is devoid of the undesirable side effects possessed by metiamide and related thioureas.

The N^{τ}-H tautomer of cimetidine also predominates in solution, and this drug competitively inhibits the action of histamine at H_2 receptors. Thus, betazole- or histamine-induced gastric acid secretion is inhibited by cimetidine. Duodenal ulcers represent a chronic recurrent condition, and cimetidine inhibits most of the night-time basal and food-stimulated secretion of gastric acid in patients, thereby permitting ulcer healing. Cimetidine, first marketed in 1976, is used worldwide for the treatment of conditions associated with gastric hyperacidity.

Drugs such as cimetidine are synthesized from 4,5-disubstituted imidazoles. Reaction of precursor alcohols with β-mercaptoethylamine hydrochloride generates the β-imidazolylmethylthio intermediate, which on condensation with N-cyano-N′,S-dimethylisothiourea, generates the desired compound.

Biological activity that depends on tautomeric forms

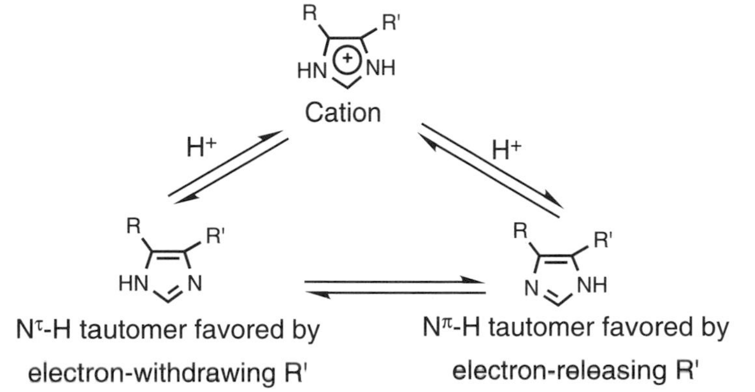

Cation

H^+ H^+

N^{τ}-H tautomer favored by electron-withdrawing R′

N^{π}-H tautomer favored by electron-releasing R′

	R	R′
Histamine	H	$-CH_2CH_2NH_2$
Burimamide	H	$-CH_2CH_2CH_2CH_2NHC(S)NHCH_3$
Metiamide	CH_3	$-CH_2SCH_2CH_2NHC(S)NHCH_3$

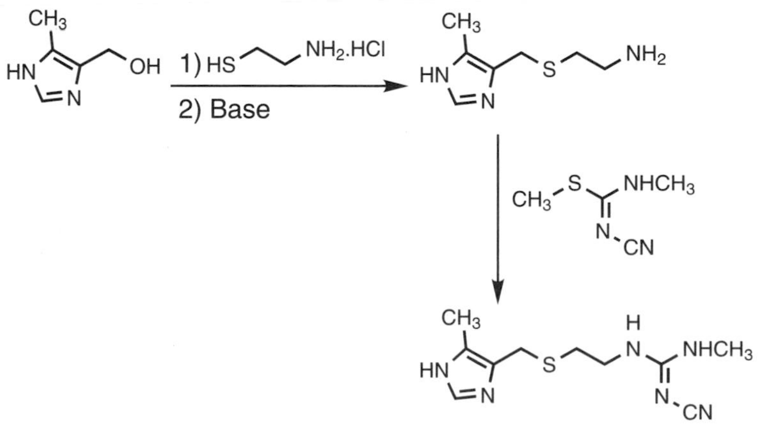

1) HS–CH_2CH_2–NH_2.HCl
2) Base

Cimetidine

is only significant for imidazole-containing antagonists. Several second-generation H_2 antagonists, introduced in the 1980s, use aromatic and other heteroaromatic groups in place of the imidazole functionality. These agents include ranitidine, nizatidine, famotidine, roxatidine.

In ranitidine (N-[2-[[[5-[(dimethylamino)methyl]-2-furanyl]methyl]thio]-ethyl]-N′-methyl-2-nitro-1,1-ethenediamine) (U.S.P.), the methylimidazolyl ring of cimetidine is replaced by a dimethylaminomethylfuryl ring, and the cyanoguanidine functionality is replaced by an isosteric methylnitroethenediamine moiety. Ranitidine is some 4 to 10 times more potent than cimetidine.[182]

Syntheses for such a compound parallel the syntheses for cimetidine, but in this case, N-methyl-1-methylthio-2-nitroetheneamine is used in the last step, and an appropriately substituted 2-furanylmethyl alcohol serves as starting material.

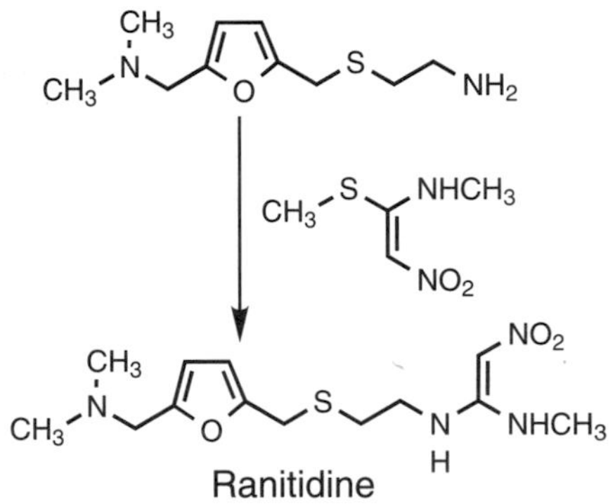

Ranitidine

Replacement of the furan ring in ranitidine by a thiazole group generates nizatidine (N-[2-[[[2-[(dimethylamino)methyl]-4-thiazolyl]methyl]thio]ethyl]-N′-methyl-2-nitro-1,1-ethenediamine), which is 5 to 18 times as potent as cimetidine.[183]

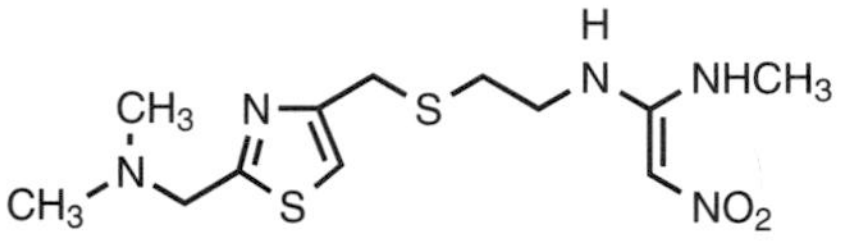

Nizatidine

Famotidine (3-[[[2-[(diaminomethylene)amino]-4-thiazolyl]methyl]thio]-N-(aminosulfonyl)propanimidamide) is a histamine H_2 receptor antagonist in which the N-methyl-2-nitro-1,1-ethenediamine group of nizatidine is replaced by the aminosulfonylimidamide function, and the basic dimethylaminomethyl group is replaced by the basic guanidino functionality.[184]

Roxatidine (2-(acetyloxy)-N-[3-[3-(1piperidinylmethyl)phenoxy]propyl]acetamide) differs most dramatically from other known H_2 histamine antagonists in that it possesses an aryloxypropyl connecting chain bonded to an acetyloxyacetamide moiety.

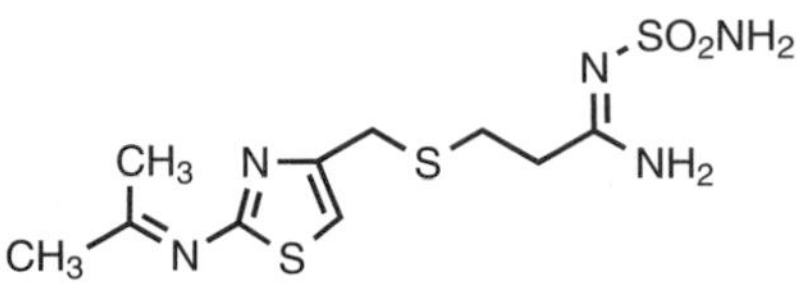

Famotidine

On the basis of stereostructure-inhibitory activity relationships of a series of cimetidine[185,186] and famotidine[187] analogs, it has been shown that the molecular conformation of neutral antagonists revealed by x-ray crystal analysis reflects, to some extent, the folded form suitable for binding to the receptor, although the crystal structure of the cationic molecule such as the hydrochloride salt always shows the nonactive extended conformation. The large conformational change caused by the acid (i.e., by the protonation of the polar atom) appears to be a common feature of the antagonists cimetidine, ranitidine, and famotidine, all of which have three fundamental substructures: (1) the substituted heteroaromatic ring, (2) the connection via a methylthioethyl chain to an end group, and (3) an essentially neutral urea-like end group. Roxatidine does not exhibit such behavior; the effect of the hydrochloride salt on the molecular conformation is not significant, and both the free and cationic molecules have similar extended forms.[188] Possibly, roxatidine is a new type of H_2 receptor antagonist, differing from cimetidine, ranitidine, and famotidine. In contrast to cimetidine, ranitidine, and famotidine, roxatidine is not a competitive inhibitor of ^{3}H-cimetidine binding to the H_2 receptor.[189] Likely, the binding site for roxatidine on the H_2 receptor is different from that of the other antagonists.[188]

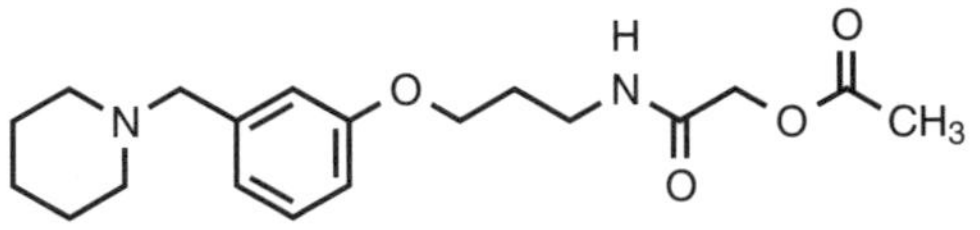

Roxatidine

In summary, compounds possessing clinically useful H_2 antagonist properties have a variety of structural possibilities. Most, but not all, can be represented by the following general formula:[12,171,184]

basic-heterocycle group	flexible chain or aromatic ring system	polar group

Antiulcer Agents Working by Mechanisms Other Than H_2 Receptor Blockade

Antiulcer agents with mechanisms of action other than H_2 receptor antagonism are known or are under investigation. Pirenzepine (5,11-dihydro-11-[(4-methyl-1-piperazinyl)acetyl]-6H-pyrido[2,3-b][1,4]benzodiazepin-6-one),dispensed as the dihydrochloride, is a selective antimuscarinic agent. This drug displays preferential binding to the receptors of the gastric mucosa and causes a reduction in the secretion of both gastric acid and pepsin. The pirenzepine analog telenzepine [4,9-dihydro-3-methyl-4-[(4-methyl-1-piperazinyl)acetyl]-10H-thieno[3,4-b][1,5]

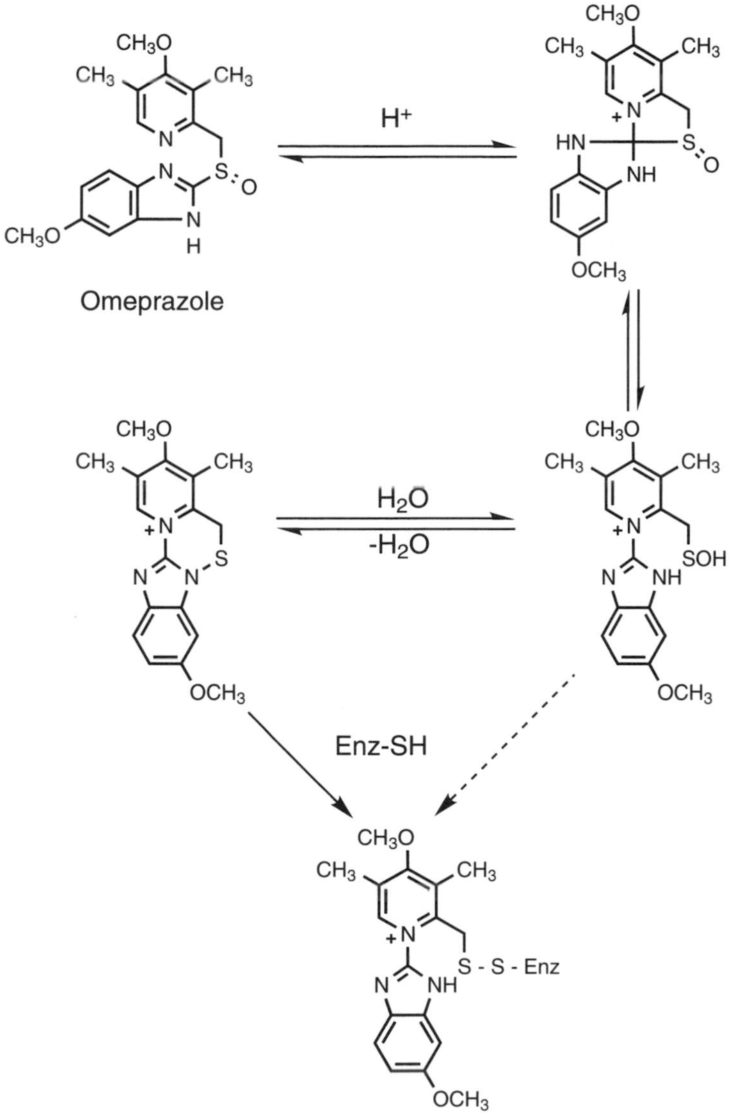

Omeprazole

Fig. 22–3. Mechanism of activation of omeprazole in acid. (From P. Lindberg, et al., *J. Med. Chem., 29,* 1327(1986).

benzodiazepin-10-one] is currently undergoing clinical investigation in the United States.

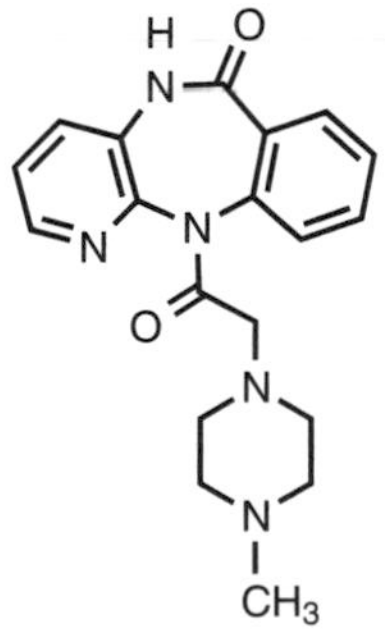

Pirenzepine

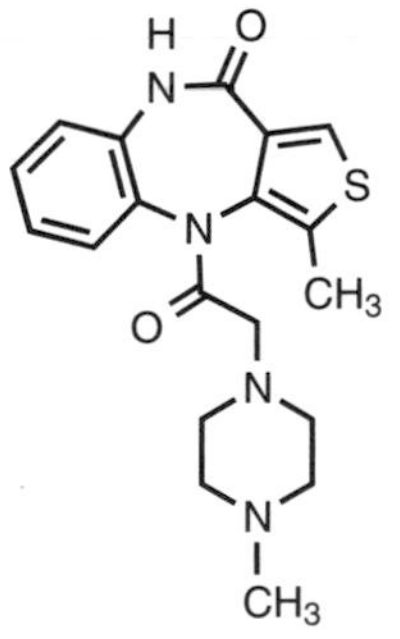

Telenzepine

Proton-pump inhibitors represent another class of drugs likely useful for control of gastric acidity and the treatment of peptic ulcers. Omeprazole, released for clinical use, is the most widely studied member of this group. Chemically known as 5-methoxy-2-[[(4-methoxy-3,5-dimethyl-2-pyridinyl)methyl]sulphinyl]-1H-benzimidazole, omeprazole blocks gastric acid secretion by irreversibly inhibiting hydrogen/potassium adenosine triphosphatase.[190] Omeprazole does not undergo reaction with enzymes at neutral pH and is activated within the acidic compartment of the parietal cells. A cationic spiro intermediate is generated that in turn rearranges to form a reactive sulfenic acid or a sulfenamide (Fig. 22–3). Either the sulfenic acid or the sulfenamide undergoes reaction with the enzyme (Enz-SH) to form an active benzimidazolpyridinium species.

Sustained suppression of gastric acid secretion, either by long-acting histamine H_2 receptor antagonists or by irreversible proton pump inhibitors, is associated with the formation of gastric carcinoids in long-term carcinogenicity studies.[191–193] For these reasons, reversible (H^+/K^+)-ATPase antagonists and thus shorter acting inhibitors of gastric acid secretion are attracting attention for the treatment of acid-related gastrointestinal disorders.[194] Acting on the final stage of secretion, such compounds have the potential to combine inhibition of multistimuli-initiated acid secretion and the dosing flexibility available with short-acting H_2 receptor antagonists. Such compounds may have a reduced potential for the formation of gastric carcinoids in long-term therapy.

These studies led to the identification of the potent, orally active, reversible (H^+/K^+)-ATPase inhibitor 3-butyryl-8-methoxy-4-(2-methylphenylamino) quinoline (SK&F 96067), which was selected for development after the evaluation of a series of analogs.[195] The nature of the 3-substituent is important for significant potency, and is associated with effects on both molecular conformation and quinoline pK_a. A 3-butyryl substituent provides for high oral activity and an absence of nephrotoxicity associated with the 3-carbethoxy group of the lead compound. SK&F 96067 is currently in clinical trials.[196]

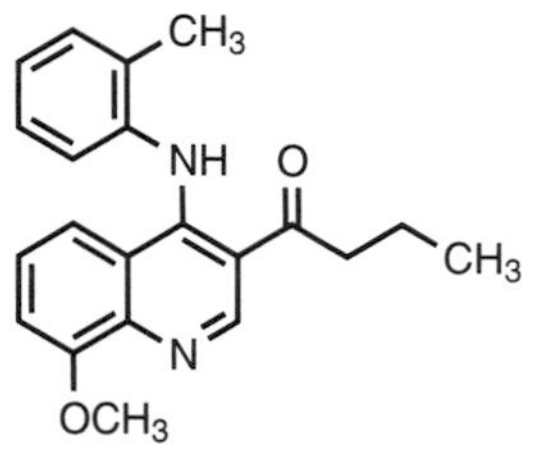

SK&F 96067

Prostaglandins are known to be involved in the preservation of the integrity of gastric epithelium. Furthermore, their prior administration protects the gastric mucosa of animals against various ulcerogenic agents.[197] To produce more useful drugs, stable, orally active prostaglandin analogs have been developed for use in the treatment of peptic ulcer in humans. One compound, misoprostol ((11α,13E)-($\pm$)-11,16-dihydroxy-16-methyl-9-oxoprost-13-en-1-oic acid methyl ester) is used in the United States as a gastric antisecretory agent with protective effects on the gastroduodenal mucosa.[198]

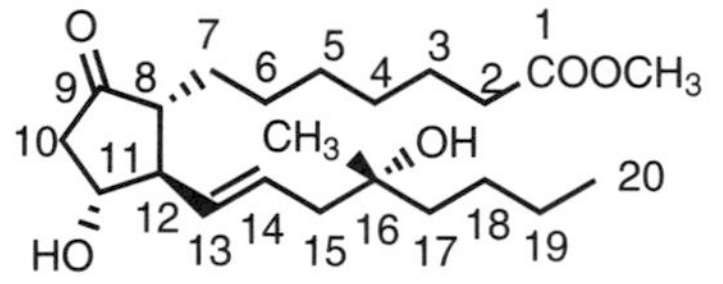

(+/-) -S-form
Misoprostol

Misoprostol is a synthetic analog of prostaglandin E_1 (PGE_1; alprostadil) and differs structurally from PGE_1 by the presence of a methyl ester at C-1, a methyl group at C-16, and a hydroxy group at C-16 rather than at C-15. The methyl ester increases both the antisecretory potency and duration of action, whereas positioning the hydroxyl group at C-16 with addition of the methyl group improves oral activity, increases duration of action, and provides a safer drug when compared to PGE_1.

During the manufacture of this drug, approximately equal amounts of four isomers are formed, but it appears that the 11R,16S-isomer is principally responsible for the gastric acid inhibitory activity. The remaining three possibilities have little gastric antisecretory activity.

Other drugs, the mechanism of action of which is unclear, are also used in antiulcer therapy; among these are sucralfate and carbenoxolone. Sucralfate is a complex formed from sucrose octasulfate and polyaluminum hydroxide. A primary structure for sucralfate consists of $Al_2(OH)_5$ groups combined to each of seven or eight sulfate groups on a sucrose moiety with a secondary polymerization producing intermolecular aluminum hydroxide bridges.[199]

Sucralfate has more stable glycosidic linkages than most disaccharides or polysaccharides because of the sulfate groups. When the pH is less than 4, extensive polymerization and cross-linking occurs. Continued reaction in acid gradually frees aluminum hydroxide until some sucrose octasulfate moieties are completely free of the metal. The reaction is slow and incomplete in the stomach. Sucralfate has no practical acid-neutralizing capacity. Even though the pH $\gg$ 4 in the duodenum, the gel retains its viscid, demulcent properties. Binding to ulcer craters probably represents the main therapeutic action of the polymer. The gel adheres strongly to epithelial cells and to the base of ulcer craters. Affinity for the crater base is higher than that for the epithelial surface. It is difficult to wash the gel from the crater. In man, the gel remains bound to ulcerated epithelium for over 6 hours. Duodenal binding is favored. The incidence and severity of side effects from this drug are low.[200]

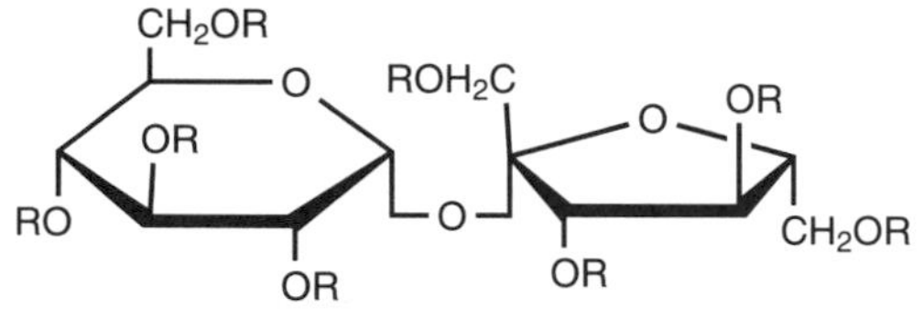

R = $SO_3[Al(OH)_5]$

Sucralfate

Carbenoxolone (3-(3-carboxy-1-oxopropoxy)-11-oxo-olean-12-en-29-oic acid), obtained from licorice extract, has been in general use as an antiulcer drug in Europe since 1962. The drug has a steroid-like structure and possesses significant mineralocorticoid activity. Other mineralocorticoids do not have antiulcer properties, but con-

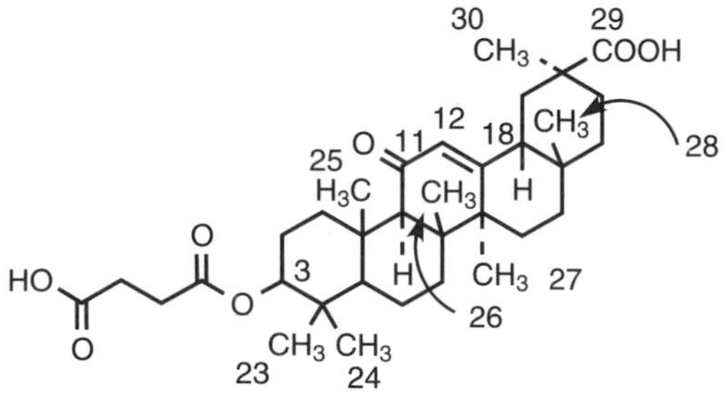

Carbenoxolone

current administration of spironolactone interferes with the therapeutic effect of carbenoxolone.[201] Carbenoxolone alters the composition of mucus and enhances the mucosal barrier to diffusion of acid. Suggested therapeutic effects include augmentation of glycoprotein synthesis, inhibition of enzymes that inactivate prostaglandins, and suppression of the activation of pepsinogen. Although carbenoxolone appears to be as effective as cimetidine in some trials, frequent adverse effects resulting from its mineralocorticoid actions are major deterrents to the use of this drug.

H_3 RECEPTOR INTERACTIVE COMPOUNDS

H_3 receptor antagonists were discovered within a group of H_2 antagonists.[174] Burimamide, the first H_2 receptor antagonist,[5] is 100-fold more potent at H_3 than at H_2 receptors. Impromidine, a potent and specific H_2 receptor agonist, also exhibits high H_3 antagonistic potency.[12] A related compound with H_3 antagonist properties is thioperamide, a burimamide analog wherein the side chain is a piperidine ring and a hydrophobic cyclohexyl group is substituted for methyl on the terminal thioureido nitrogen atom. Thioperamide is a high-affinity selective H_3 receptor antagonist.[6]

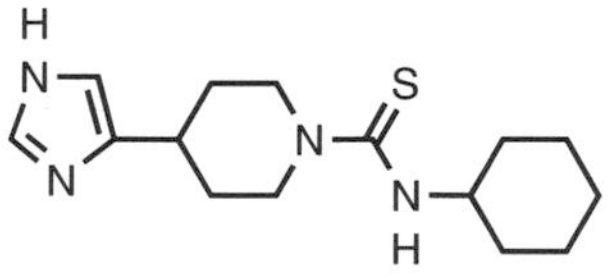

Thioperamide

Several other histamine derivatives are reported to be H_3 receptor agonists or antagonists.[202,203] It seems that H_3 receptors are part of a general regulatory system that may serve as target for the design of new therapeutics. Imetit(S-[(2-imidazol-4-yl)ethyl]isothiourea) is one example of a novel, potent, and selective H_3 agonist.[204] Several isothiourea analogs of histamine serve to define the structural specificity at H_3 receptors as well as structural requirements for the transition from agonists to antagonists.[205–207] A discussion of such structure-activity relationships is available.[205,206,208]

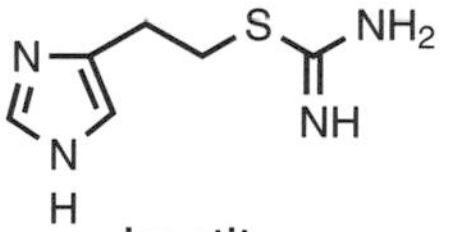

Imetit

REFERENCES

1. J. W. Black and C. R. Ganellin, *Experientia, 30,* 111(1974).
2. B. D. Gomperts, in *Handbook of Experimental Pharmacology,* Vol. 97, B. Uvnäs, Ed., Berlin, Springer, 1991.
3. A. Yamatodani, et al., in *Handbook of Experimental Pharmacology,* Vol. 97, B. Uvnäs, Ed., Berlin, Springer, 1991.
4. A. S. F. Ash and H. O. Schild, *Br. J. Pharmacol., 27,* 427(1966).
5. J. W. Black, et al. *Nature, 236,* 385(1972).
6. J. M. Arrang, et al., *Nature, 327,* 117(1987).
7. J. C. Garrison in *The Pharmacological Basis of Therapeutics,* 8th ed., A. G. Gilman, et al., Eds., New York, Pergamon Press, 1990.
8. T. B. Paiva, et al., *J. Med. Chem., 13,* 689(1970).
9. J. E. F. Reynolds, Ed., *Martindale,* 29th ed., London, The Pharmaceutical Press, 1989, p. 941.
10. G. K. McEvoy, Ed., *A. H. F. S. Drug Information,* Bethesda, Am. Soc. of Hospital Pharmacists, Inc, 1992, p. 1416.
11. J. E. F. Reynolds, Ed., *Martindale,* 29th ed., London, The Pharmaceutical Press, 1989, p. 1494.
12. D. G. Cooper, et al., in *Comprehensive Medicinal Chemistry,* Vol. 3, C. Hansch, Ed., New York, Pergamon Press, 1990.
13. J. R. Cunha-Melo, et al., *J. Biol. Chem., 262,* 11455(1987).
14. H. H. Dale and P. P. Laidlaw, *J. Physiol. (Lond.), 41,* 318(1910).
15. H. H. Dale and P. P. Laidlaw, *J. Physiol. (Lond.), 43,* 182(1911).
16. E. Fourneau and D. Bovet, *Arch. Int. Pharmacodyn. Ther., 46,* 178(1933).
17. A. M. Staub, *Ann. Inst. Pasteur, 63,* 400(1939).
18. B. N. Halpern, *Arch. Inst. Pharmacodyn., 68,* 339(1942).
19. M. M. Mosnier, French patent 913, 161(1943).
20. S. Budavari, Ed., *The Merck Index,* 11th ed., Rahway, N.J., Merck and Co., Inc., 1989, p. 7996.
21. C. P. Huttrer, et al., *J. Am. Chem. Soc., 68,* 1999(1946).
22. S. Budavari, Ed., *The Merck Index,* 11th ed., Rahway, N.J., Merck and Co., Inc., 1989, p. 9651.
23. G. R. Rieveschl, U.S. Patent 2,421,714 (1947).
24. G. R. Rieveschl and W. F. Huber, *Abstracts,* 109th Meeting, American Chemical Society, 1946, p. 50K.
25. G. K. McEvoy, Ed., *A. H. F. S. Drug Information,* Bethesda, American Society of Hospital Pharmacists, Inc., 1992, p. 1.
26. J. C. Drach and J. P. Howell, *Biochem. Pharmacol., 17,* 2125(1968).
27. R. Kuhn, *Psychopharmacologia, 8,* 201(1965).
28. J. B. Wyngaarden and M. H. Seeves, *JAMA, 145,* 277(1951).
29. C. C. Lee, *Toxicol. Appl. Pharmacol., 8,* 210(1966).
30. J. J. Burns, et al., *Ann. N. Y. Acad. Sci., 104,* 881(1963).
31. I. Mota and W. deSilva, *Br. J. Pharmacol., 15,* 396(1960).
32. H. O. Schild, *Br. J. Pharmacol., 2,* 189(1947).
33. C. H. Tilford, et al., *J. Am. Chem. Soc., 70,* 4001(1948).
34. J. Cany and H. Huidoboro, *Thérapie, 15,* 159(1960).
35. J. E. F. Reynolds, Ed., *Martindale,* 28th ed., London, The Pharmaceutical Press, 1982, p. 1319.
36. J. Fakstorp and E. Ifversen, *Acta Chem. Scand., 4,* 1610(1950).
37. J. B. Hoekstra, et al., *J. Am. Pharm. Assoc., 42,* 587(1953).
38. B. H. Chase and A. M. Downes, *J. Chem. Soc.,* 3874(1953).
39. R. Baltzly, et al., *J. Org. Chem., 14,* 775(1949).
40. K. E. Hamlin, et al., *J. Am. Chem. Soc., 71,* 2731(1949).
41. H. Morren, et al., *Bull. Soc. Chim. Belge, 60,* 282(1951).
42. E. A. Brown, et al., *Ann. Allergy, 8,* 32(1950).
43. B. N. Halpern, et al., *Arch. Int. Pharmacodyn. Ther., 142,* 170(1963).
44. J. M. Van Nueten and P. A. J. Janssen, *Arch. Int. Pharmacodyn. Ther., 204,* 37(1973).
45. T. Godfraind and A. Kaba, *Br. J. Pharmacol., 36,* 549(1969).
46. T. Godfraind and A. Kaba, *Arch. Int. Pharmacodyn. Ther., 196,* 35(1972).
47. J. E. F. Reynolds, Ed., *Martindale,* 29th ed., London, The Pharmaceutical Press, 1989, p. 443.
48. A. Ebnöther and H. P. Weber, *Helv. Chim. Acta, 59,* 2462(1976).
49. D. Roemer and H. Weidmann, *Med. Welt, 27,* 2794(1966).
50. A. W. Peck, et al., *Eur. J. Clin. Pharmacol., 8,* 455(1975).

51. A. LaBelle and R. Tislow, *Fed. Proc., 7,* 236(1948).
52. R. Tislow, et al., *Fed. Proc., 8,* 338(1949).
53. N. Sperber, D. Papa, and E. Schwenk, U.S. Patent 2,676,964 (1954).
54. S. Saijo, *J. Pharm. Soc. Jpn., 72,* 1529(1952).
55. D. W. Adamson, et al., *Nature, 168,* 204(1951).
56. A. F. Green, *Br. J. Pharmacol. Chemother., 8,* 171(1953).
57. C. F. Huebner, et al., *J. Am. Chem. Soc., 82,* 2077(1960).
58. D. T. Witiak and R. C. Cavestri, in *Burger's Medicinal Chemistry,* 4th ed., Vol. 3, M. E. Wolff, Ed., New York, John Wiley & Sons, 1981.
59. B. N. Halpern, *JAMA, 129,* 1219(1945).
60. B. N. Halpern and R. Ducrot, *C. R. Soc. Biol (Paris), 140,* 361(1946).
61. W. B. Reid, et al., *J. Am. Chem. Soc., 70,* 3100(1948).
62. M. J. VanderBrook, et al., *J. Pharmacol. Exp. Ther., 94,* 197(1948).
63. H. W. Wahner and C. A. Peters, *Mayo Clin. Proc., 35,* 161(1960).
64. C. P. Huttrer, *Enzymologia, 12,* 277(1948).
65. K. Stach and W. Poldinger, *Arzneimittelforschung, 9,* 148(1966).
66. A. Von Schilchtegroll, *Arzneimittelforschung, 7,* 237(1957); *8,* 489(1958).
67. P. V. Petersen, et al., *Arzneimittelforschung, 8,* 395(1958).
68. R. Caviezel, et al., *Arch. Int. Pharmacodyn. Ther., 141,* 331(1963).
69. J. E. F. Reynolds, Ed., *Martindale,* 29th ed., London, The Pharmaceutical Press, 1989, p. 539.
70. C. A. Stone, et al., *J. Pharmacol. Exp. Ther., 131,* 73(1961).
71. S. R. Rege and S. D. Sharma, *Curr. Ther. Res., 39,* 378(1986).
72. B. Biehl, *Curr. Med. Res. Opin., 6,* 62(1979).
73. D. K. Luscombe, P. J. Nicholls, and P. S. Spencer, *Br. J. Clin Pract., 34,* 75(1980).
74. H. vanRiezen, *J. Pharm. Pharmacol., 18,* 688(1966).
75. J. E. F. Reynolds, Ed., *Martindale,* 29th ed., London, The Pharmaceutical Press, 1989, p. 1419.
76. C. Tester-Dalderup, *Drugs of Today, 24,* 117(1988).
77. J. Metys and J. Metysova, *Acta Biol. Med. Ger., 15,* 871(1965).
78. M. Rajsner, et al., *Collect. Czech. Chem. Commun., 32,* 2854(1967).
79. M. Rajsner, et al., *Collect. Czech. Chem. Commun., 39,* 1366(1974).
80. A. F. Casy, in *Handbook of Experimental Pharmacology,* Vol. 97, B. Uvnäs, Ed., Berlin, Springer, 1991.
81. T. T. Quach, et al., *Eur. J. Pharmacol., 60,* 391(1979).
82. T. T. Quach, et al., *Neurosci. Lett., 17,* 49(1980).
83. A. A. Carr and D. R. Meyer, *Arzneimittelforschung, 32,* (II)9a, 1157(1982).
84. A. A. Carr, et al., *Pharmacologist, 15,* 221(1973).
85. J. P. Green, et al., *Psychopharmacology: A Generation of Progress,* New York, Raven Press. 1978, p. 319.
86. H. C. Cheng and J. K. Woodward, *Drug Dev. Res., 2,* 181(1982).
87. D. A. Walsh, et al., *J. Med. Chem., 32,* 105(1989).
88. *Drugs of the Future, 15,* 674(1990).
89. D. A. Gartiez, et al., *Arzneimittelforschung, 32,* 1185(1982).
90. A. F. Cohen, et al., *Eur. J. Clin. Pharmacol., 28,* 197(1985).
91. F. Janssens, et al., *J. Med. Chem., 28,* 1925, 1934, 1943(1985).
92. H. S. Ahn, and A. Barnett, *Eur. J. Pharmacol., 127,* 153(1986).
93. C. J. E. Niemegeers, et al., *Agents Actions, 18,* 141(1986).
94. G. Le Fur, et al., *Life Sci., 29,* 547(1981).
95. *Drugs of the Future, 10,* 209(1985).
96. G. J. Durant, et al., *Br. J. Pharmacol., 82(Suppl),* 232(1984).
97. *Drugs of the Future, 12,* 1106(1987).
98. W. S. Adamus, et al., *Arzneimittelforschung, 37,* 569(1987).
99. P. Naranjo and E. Naranjo, *Arch. Int. Pharmacodyn. Ther., 94,* 383(1953); *J. Allergy, 24,* 442 (1953).
100. T. Iwasaki, *Nippon Yakurigaku Zasshi, 53,* 375(1957).
101. M. Kikkawa, et al., *Osaka City Med. J., 3,* 69(1956).
102. H. Goldenberg and V. Fishman, *Biochem. Biophys. Res. Commun., 14,* 404(1964).
103. A. J. Glazko and W. A. Dill, *J. Biol. Chem., 179,* 409(1949); A. J. Glazko, et al., *J. Biol. Chem., 179,* 417(1949).
104. D. T. Witiak and N. J. Lewis, in *Handbook of Experimental Pharmacology,* Vol. 18/2, M. Rocha and E. Silva, Eds., Berlin, Springer, 1978, p. 513–560.
105. D. M. Paton and D. R. Webster, *Clin. Pharmacokinet, 10,* 477(1985).
106. R. C. Roozemond, *Biochem. Pharmacol., 14,* 699(1965).
107. D. Robinson, *Biochem. J., 68,* 584(1958).
108. R. Kuntzman, et al., *J. Pharmacol. Exp. Ther., 149,* 29(1965).
109. R. Kuntzman, et al., *J. Pharmacol. Exp. Ther., 155,* 337(1967).
110. M. Strolin-Benedetti and J. Caldwell, in *Comprehensive Medicinal Chemistry,* Vol. 5, C. Hansch, Ed., New York, Pergamon Press, 1990, p. 482.
111. J. B. Wyngaarden and M. H. Seeves, *JAMA, 145,* 277(1951).
112. C. C. Lee, *Toxicol. Appl Pharmacol., 8,* 210(1966).
113. F. E. R. Simons and K. J. Simons, *J. Allergy Clin. Immunol., 81,* 975(1988).
114. I. Mota and W. deSilva, *Br. J. Pharmacol., 15,* 396(1960).
115. J. W. Black, et al., *Nature, 236,* 385(1972).
116. H. O. Schild, *Br. J. Pharmacol, 2,* 189(1947).
117. F. Leonard and C. P. Huttrer, *Histamine Antagonists.* Washington, D.C., National Research Council, 1950; E. R. Loew, *Boston Med. Q., 3,* 1(1952).
118. C. A. Winter, *J. Pharmacol. Exp. Ther., 94,* 7(1948).
119. T. M. Glaser and G. R. Hervey, *Lancet, 262,* 490(1952).
120. R. A. Meyer and W. Jakubowski, *J. Am. Dent. Assoc., 69,* 32(1964).
121. P. E. Hanna and A. E. Ahmed, *J. Med. Chem., 16,* 963(1973).
122. V. Barouh, et al., *J. Med. Chem., 14,* 834(1971).
123. A. Shafi'ee and G. Hite, *J. Med. Chem., 12,* 266(1969).
124. F. E. Roth, *Chemotherapia, 3,* 120(1961).
125. D. T. Witiak, et al., *J. Med. Chem., 14,* 24(1971).
126. A. F. Casy and R. R. Ison, *J. Pharm. Pharmacol., 22,* 270(1970).
127. P. E. Hanna, et al., *J. Pharm. Sci, 62,* 512(1973).
128. P. E. Hanna and R. T. Borchardt, *J. Med. Chem., 17,* 471(1974).
129. M. J. Van Drooge, et al., *J. Comput. Aided Mol. Des., 5,* 357(1991).
130. A. M. ter Laak, et al., *Quant. Struct.-Act. Relat., 11,* 348(1992).
131. K. P. Mathews, *JAMA, 248,* 2587(1982).
132. S. E. Dahlen, et al., *Proc. Natl. Acad. Sci. U.S.A., 80,* 1712(1983).
133. G. G. Shapiro, and P. Köning, *Pharmacotherapy, 5,* 156(1985).
134. J. S. G. Cox, *Nature, 216,* 1328(1967).
135. J. S. G. Cox, et al., *Adv. Drug Res., 5,* 115(1970).
136. R. N. Brogden, et al., *Drugs, 7,* 164(1974).
137. G. K. McEvoy, Ed., *A. H. F. S. Drug Information,* Bethesda, Am. Soc. of Hospital Pharmacists, Inc., 1992, p. 2263.
138. T. W. Rall, in *The Pharmacological Basis of Therapeutics,* 8th ed., A. G. Gilman, et al., Eds., New York, Pergamon Press, 1990, p. 631.
139. H. Cairns, et al., *J. Med. Chem., 15,* 583(1972).
140. J. P. Gonzales and R. N. Brodgen, *Drugs, 34,* 560(1987).

141. J. Rokach, Ed., *Bioactive Molecules,* Vol. 11, Amsterdam, Elsevier, 1989.
142. A. Von Sprecher, et al., *Chimia, 46,* 304(1992).
143. S. Hammarström, et al., *Biochem. Biophys. Res. Commun., 91,* 1266(1979).
144. E. J. Corey, et al., *J. Am. Chem. Soc., 102,* 1436(1980).
145. R. A. Nathan, et al., *J. Allergy Clin. Immunol., 87,* Pt 2,256(1991).
146. I. K. Taylor, et al., *Lancet, 337,* 690(1991).
147. J. P. Finnerty, et al., *Am. Rev. Respir. Dis., 145,* 746(1992).
148. K. P. Hui and N. C. Barnes, *Lancet, 337,* 1062(1991).
149. T. Nakagawa, et al., *Adv. Prostaglandin Thromboxane Leukotriene Res., 21A,* 465(1991).
150. S. Etoh, et al., *Br. J. Pharmacol., 100,* 564(1990).
151. *Drugs of Today, 28,* 29(1992).
152. T. Kawashima, et al., *Jpn. J. Allergy, 37,* 438(1988).
153. J. Yanagihara, et al., *Jpn. J. Pharmacol., 48,* 103(1988).
154. J. Yanagihara et al., *Jpn. J. Pharmacol., 51,* 83(1989).
155. D. Proud and A. P. Kaplan, *Annu. Rev. Immunol., 6,* 49(1988).
156. R. W. Colman and P. Y. Wong, in *Handbook of Experimental Pharmacology,* Vol. 25 (Suppl.), E. G. Erdos, Ed., Berlin, Springer, 1979, p. 567.
157. M. W. Greaves, *Br. J. Dermatol., 119,* 419(1988).
158. Kinins and their antagonists (editorial), *Lancet, 338,* 287(1991).
159. D. J. Kyle and R. M. Burch, *Drugs of the Future, 17,* 305(1992).
160. M. Rocha e Silva, *Cien. Cult. (Sao Paulo), 1,* 32(1949).
161. E. Schröder, in *Handbook of Experimental Pharmacology,* Vol. 25, E. G. Erdos, Ed., Berlin, Springer, 1979.
162. J. M. Steward, in *Handbook of Experimental Pharmacology,* Vol. 25(Suppl.), E. G. Erdos, Ed., Berlin, Springer, 1979, p. 225.
163. J. M. Stewart and R. J. Vavrek, in *Bradykinin Antagonists: Basic and Clinical Research,* R. M. Burch, Ed., New York, Marcel Dekker, 1991, p. 51.
164. R. J. Vavrek and J. M. Stewart, *Peptides, 6,* 161(1985).
165. R. M. Burch, et al., *Med. Res. Rev., 10,* 143(1990).
166. L. R. Steranka, et al., *FASEB J., 3,* 2019(1989).
167. J. B. Calixto, et al., in *Bradykinin Antagonists: Basic and Clinical Research,* R. M. Burch, Ed., New York, Marcel Dekker, 1991, p. 97.
168. P. B. M. W. M. Timmermans, et al., *Trends Pharmacol. Sci., 12,* 55(1991).
169. L. R. Steranka, *Int. Kinin Conf. (Munich),* 108(1991).
170. C. R. Ganellin and G. J. Durant, in *Burger's Medicinal Chemistry,* 4th ed., Vol. 3, M. E. Wolff, Ed., New York, John Wiley & Sons, 1981.
171. H. Van der Goot, et al., in *Handbook of Experimental Pharmacology,* Vol. 97, B. Uvnäs, Ed., Berlin, Springer, 1991.
172. G. Baumann, et al., *Pharmacol. Ther., 24,* 165(1984).
173. P. Mörsdorf, et al., *Drugs of the Future, 15,* 919(1990).
174. J. M. Arrang, et al., *Nature, 302,* 832(1983).
175. G. Baumann, et al., *Pharmacol. Ther., 24,* 165(1984).
176. A. Buschauer, *J. Med. Chem., 32,* 1963(1989).
177. A. Buschauer and G. Baumann, *XIIth International Symposium on Medicinal Chemistry,* Basel, 1992, p. 135.
178. H. Timmerman, *Quant. Struct.-Act. Relat., 11,* 219(1992).
179. J. C. Eriks, et al., *J. Med. Chem., 35,* 3239(1992).
180. A. Buschauer and G. Baumann, *Arch. Pharm. (Weinheim), 324,* 736(1991).
181. R. Franke and A. Buschauer, *Eur. J. Med. Chem., 27,* 443(1992).
182. J. Bradshaw, et al., *Br. J. Pharmacol., 66,* 464P(1979).
183. T. M. Lin, et al., *Gastroenterology, 84,* 1231(1983).
184. T. H. Brown and R. C. Young, *Drugs of the Future, 10,* 51(1985).
185. M. Shibata, et al., *J. Pharm. Sci, 72,* 1436(1983).
186. M. Shibata, et al., *Acta Crystallogr., C39,* 1255(1983).
187. T. Ishida, et al., *Mol. Pharmacol., 31,* 410(1987).
188. Y. In, et al., *Chem. Pharm. Bull, 36,* 2295(1988).
189. A. Tanaka, et al., *Igaku No Auymi, 130,* 433 (1985).
190. P. Lindberg, et al., *J. Med. Chem., 29,* 1327(1986).
191. D. Poynter, et al., *Gut, 26,* 1284(1985).
192. E. Carlsson, et al., *Scand. J. Gastroenterol, 21(Suppl. 118),* 31(1986).
193. R. Hakanson and F. Sunder, *Trends Pharmacol Sci., 7,* 386(1986).
194. R. J. Ife, et al., in *Annual Reports in Medicinal Chemistry,* Vol. 25, J. A. Bristol, New York, Academic Press, 1990, p. 159.
195. R. J. Ife, et al., *J. Med. Chem., 35,* 3413(1992).
196. *Drugs of the Future, 17,* 796(1992).
197. B. J. R. Whittle and J. R. Vane, in *Physiology of the Gastrointestinal Tract,* L. R. Johnson, Ed., New York, Raven Press, 1987, p. 143.
198. G. K. McEvoy, Ed., *A. H. F. S. Drug Information,* Bethesda, American Society of Hospital Pharmacists, Inc., 1992, p. 1774.
199. R. Nagashima and N. Yoshida, *Arzneimittelforschung, 29,* 1668(1979).
200. G. K. McEvoy, Ed., *A. H. F. S. Drug Information,* Bethesda, American Society of Hospital Pharmacists, Inc., 1992, p. 1784.
201. L. L. Brunton, in *The Pharmacological Basis of Therapeutics,* 8th ed., A. G. Gilman, et al., New York, Pergamon Press, 1990, p. 911.
202. R. Lipp, et al., Poster presentation, *10th International Symposium on Medicinal Chemistry,* Budapest, 1988, Abst. p. 119.
203. J. M. Arrang, et al., Eur. Pat 0197840(1986).
204. J. C. Schwartz, et al., in *Trends in Receptor Research,* Amsterdam, Elsevier, 1992, p. 141.
205. C. R. Ganellin, et al., *11th International Symposium on Medicinal Chemistry,* Basel, 1992, p. 64.
206. H. Van der Goot, et al., *11th International Symposium on Medicinal Chemistry,* Basel, 1992, p. 185.
207. H. Van der Goot, et al., *Eur. J. Med. Chem., 27,* 511(1992).
208. J. C. Schwartz, et al., in *Handbook of Experimental Pharmacology,* Vol. 97, B. Uvnäs, Ed., Berlin, Springer, 1991.

SUGGESTED READINGS

D. T. Witiak and R. C. Cavestri, in *Burger's Medicinal Chemistry,* 4th ed., Vol. 3, M. E. Wolff, Ed., New York, John Wiley & Sons, 1981.
C. R. Ganellin and G. J. Durant, in *Burger's Medicinal Chemistry,* 4th ed., Vol., 3, M. E. Wolff, Ed., New York, John Wiley & Sons, 1981.
D. G. Cooper, et al., in *Comprehensive Medicinal Chemistry,* Vol. 3, C. Hansch, Ed., New York, Pergamon Press, 1990.
A. G. Gilman, et al., *The Pharmacological Basis of Therapeutics,* 8th ed., Chaps. 23–25, 37, New York, Pergamon Press, 1990.
B. Uvnäs, Ed., *Handbook of Experimental Pharmacology,* Vol. 97, Berlin, Springer, 1991.
A. M. ter Laak, et al., *Quant. Struct.-Act. Relat., 11,* 348(1992).

Chapter 23

CHOLESTEROL, ADRENOCORTICOIDS, AND SEX HORMONES

Robert W. Brueggemeier, Duane D. Miller, and Donald T. Witiak

Cholesterol, adrenocorticoids, and sex hormones have much in common. All are steroids, and consequently the rules that define their structures, chemistry, and nomenclature are the same. The rings of these biochemically dynamic and physiologically active compounds have a similar stereochemical relationship. Changes in the geometry of the ring junctures generally result in inactive compounds regardless of the biologic category of the steroid. Similar chemical groups are used to render some of these agents water soluble or active when taken orally or to modify their absorption.

In addition, the adrenocorticoids and the sex hormones, which include the estrogens, progestins, and androgens, are mainly biosynthesized from cholesterol, which, in turn, is synthesized from acetyl-CoA. Cholesterol and steroid hormone catabolism takes place primarily in the liver. Although the products found in the urine and feces depend upon the hormone undergoing catabolism, many of the metabolic reactions are similar for these compounds. For example, reduction of double bonds at positions 4 and 5 or 5 and 6, epimerization of 3β-hydroxyl groups, reduction of 3-keto groups to the 3α-hydroxyl function, and oxidative removal of side chains are transformations common to these agents.

Despite the similarities in chemical structures and stereochemistry, each class of steroids demonstrates unique and distinctively different biologic activities. Adrenocorticoids are composed of two classes of steroids, the glucocorticoids, which regulate carbohydrate, lipid, and protein metabolism and the mineralocorticoids, which influence salt balance and water retention. The sex hormones include the female sex hormones, progestins and estrogens, and the male sex hormones, androgens. Minor structural modifications to the steroid nucleus, such as changes in or insertion of functional groups at different positions, cause marked changes in physiologic activity. The first part of this chapter focuses on the similarities among the steroids and describes steroid nomenclature, stereochemistry, cholesterol biosynthesis and metabolism, and the general mechanism of action. The second portion of the chapter focuses on each class of steroids and discuss the unique chemistry, biochemistry, medicinal chemistry, and pharmacology of endogenous steroid hormones, synthetic agonists, and synthetic antagonists.

NOMENCLATURE, STRUCTURE, AND CONFORMATIONAL ASPECTS OF STEROIDS

Steroids consist of four fused rings (A, B, C, and D). Chemically, these hydrocarbons are cyclopentanoperhydrophenanthrenes; they contain a five-membered cyclopentane (D) ring plus the three rings of phenanthrene.

A perhydrophenanthrene (rings A, B, and C) is the completely saturated derivative of phenanthrene.

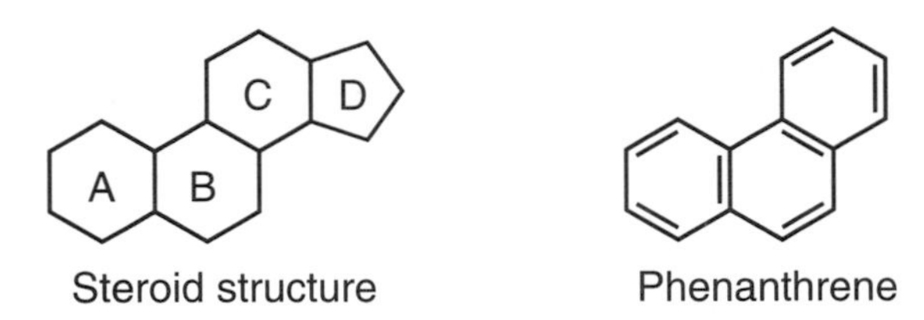

Steroid structure

Phenanthrene

The polycyclic hydrocarbon known as 5α-cholestane will be used to illustrate the numbering system for a steroid. The 5α notation is used because the hydrogen atom at position 5 is on the opposite side of the rings from the angular methyl groups at positions 18 and 19, which are assigned the β side of the molecule. The term *cholestane* refers to a steroid with 27 carbons that includes a side chain of eight carbons at position 17 on the β side. Functional groups on the β side of the molecule are denoted by solid lines; those on the α side are designated by dotted lines. Side chains at position 17 are always β unless indicated by dotted lines or in the nomenclature of the steroid (e.g., 17α).

5α-Cholestane

Just as cyclohexane can be drawn in a chair conformation, the three-dimensional representation for 5α-cholestane is shown by the following conformational formula. Although cyclohexane may undergo a flip in conformation, steroids are rigid structures because they generally have at least one *trans* fused ring system, and these rings must be diequatorial to each other.

If one is aware that the angular methyl groups at positions 18 and 19 are β and have an axial orientation (i.e., perpendicular to the plane of the rings) rather than an equatorial orientation (i.e., peripheral to the plane of the

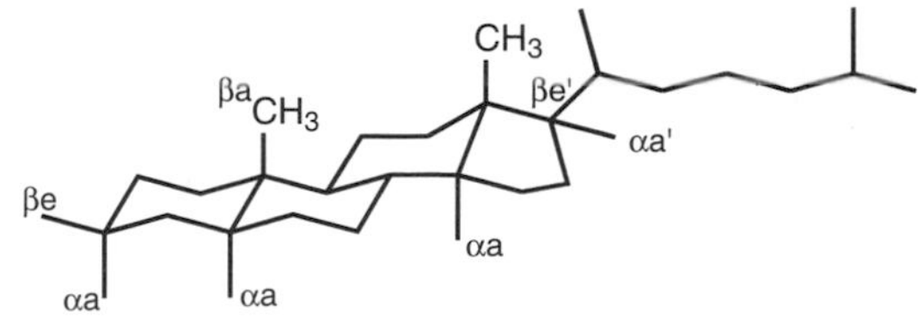

Conformational representation of 5α-cholestane
a = axial a' = quasi-axial
e = equatorial e' = quasi-equatorial

rings), the conformational orientation of the remaining bonds of a steroid can be assigned easily. For example, in 5*α*-cholestane the C-19 methyl group at position 10 is always *β* and axial; the two bonds at position 1 must be either *β*-equatorial or *α*-axial, as indicated.

The orientation of the remaining bonds on a steroid may be determined if one recalls that groups on a cyclohexane ring that are positioned on adjacent carbon atoms (vicinal, $-\overset{1}{\text{C}}\text{H}-\overset{2}{\text{C}}\text{H}-$) of the ring (i.e., 1, 2 to each other) are *trans* if their relationship is 1,2-diaxial or 1,2-diequatorial; they are *cis* to each other if their relationship is 1,2-equatorial-axial.

Steroid chemists often refer to the series of carbon-carbon bonds shown with heavy lines as the backbone of the steroid.

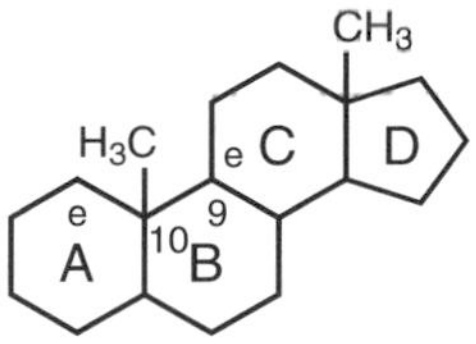

The *cis* or *trans* relationship of the four rings may be expressed in terms of the backbone. The compound 5*α*-cholestane is said to have a *trans-anti-trans-anti-trans* backbone. In this structure, all the fused rings have *trans* (diequatorial) stereochemistry, i.e., the A/B fused ring, the B/C fused ring, and the C/D fused ring are *trans*. The term *anti* is used in backbone notation to define the orientation of rings that are connected to each other and have a *trans*-type relationship. For example, the bond equatorial to ring B, at position 9, which forms part of ring C, is *anti* to the bond equatorial to ring B, at position 10, which forms part of ring A. 5*β*-Cholestane has a *cis-anti-trans-anti-trans* backbone, in which the A/B rings are fused *cis*. The term *syn* is used in a similar fashion as *anti* to define a *cis*-type relationship. No natural steroids exist with a *syn*-type geometry, although such compounds can be chemically synthesized. Thus, the conventional drawing of the steroid nucleus is the natural configuration and does not show the hydrogens at 8*β*, 9*α*, or 14*α* positions. If the carbon at position 5 is saturated, the hydrogen is always drawn, either as 5*α* or 5*β*. Also, the conventional drawing of a steroid molecule has the C-18 and C-19 methyl groups shown only as solid lines.

The stereochemistry of the rings markedly affects the biologic activity of a given class of drugs. Nearly all biologically active steroids have the cholestane-type backbone. In most of the important steroids discussed in this chap-

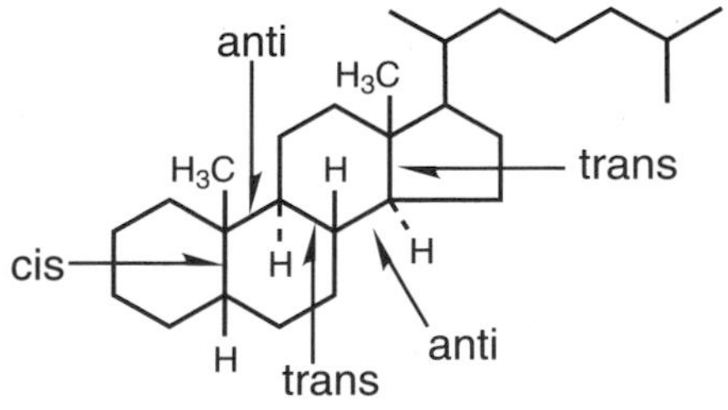

5β-Cholestane

ter, a double bond is present between positions 4 and 5 or 5 and 6, and consequently there is no *cis* or *trans* relationship between rings A and B. The symbol Δ is often sued to designate a C=C bond in a steroid. If the C=C is between the 4 and 5 position, the compound is referred to as a Δ^4-steroid; if the C=C is between positions 5 and 10, the compound is designated a $\Delta^{5(10)}$-steroid.

Cholesterol (cholest-5-en-3*β*-ol) is a Δ^5 steroid or, more specifically, a Δ^5-sterol because it is an unsaturated alcohol.

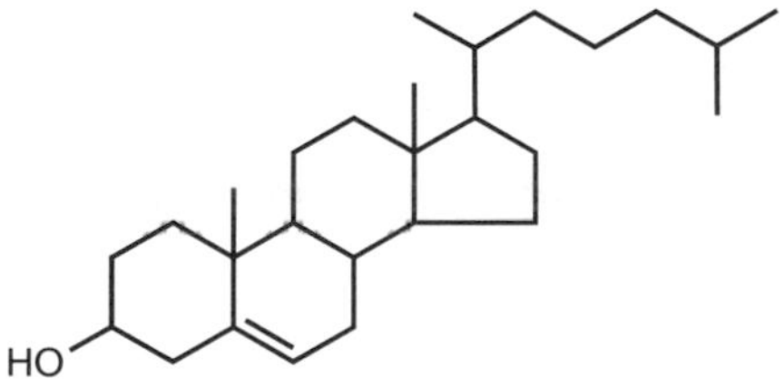

Cholesterol

The bile acids, on the other hand, contain no C=C and belong to the 5*β* series. For example, cholic acid, which is an important cholesterol metabolite in man and animals, has the 5*β*-cholestane backbone and is named 3*α*,7*α*,12*α*-trihydroxy 5*β*-cholan-24-oic acid.

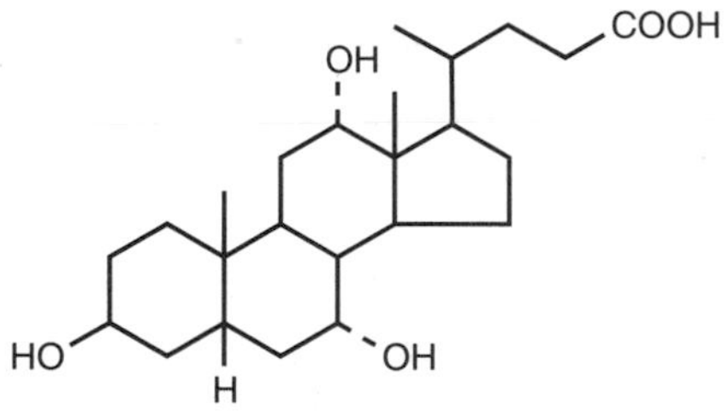

Cholic acid

Most of the cardiac glycosides also belong to the 5*β* series.

Three other steroid hydrocarbons having the 5*α*-cholestane configuration are mentioned throughout this chapter. Biologically active compounds discussed are members of the 5*α*-pregnane, 5*α*-androstane, and 5*α*-estrane steroid classes. Pregnanes are steroids with 21 carbon atoms, androstanes have 19 carbon atoms, and estranes have 18 carbon atoms, with the C-19 angular methyl group at C-10 replaced by hydrogen.

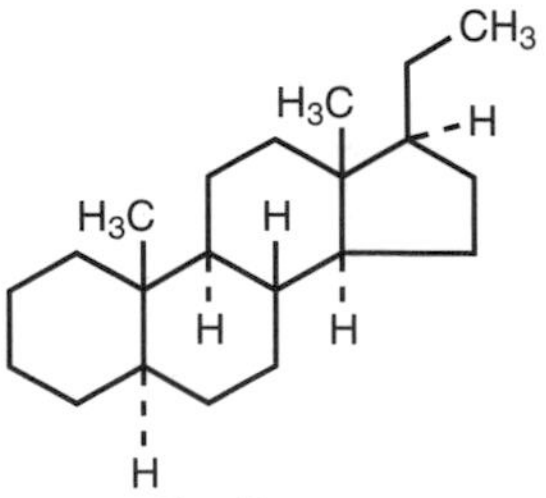

5α-Pregnane

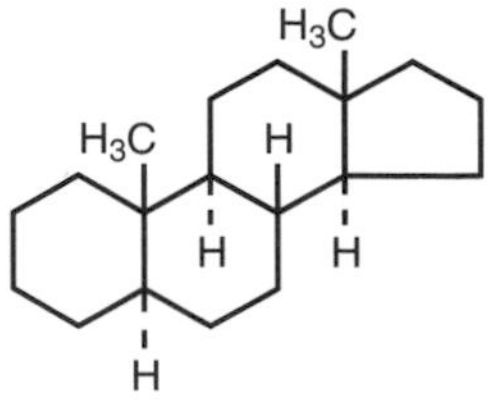

5α-Androstane

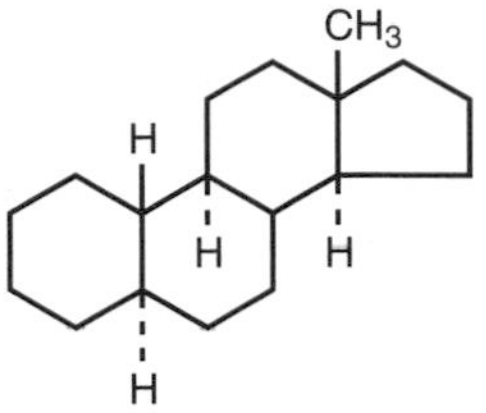

5α-Estrane

Numbering is the same as in 5α-cholestane.

The adrenocorticoids (adrenal cortex hormones) are pregnanes and are exemplified by cortisone, which is a 17,21-dihydroxypregn-4-ene-3,11,20-trione. The acetate ester is named 12,71-dihydro-oxypregn-4-ene-3,11,20-trione 21-acetate.

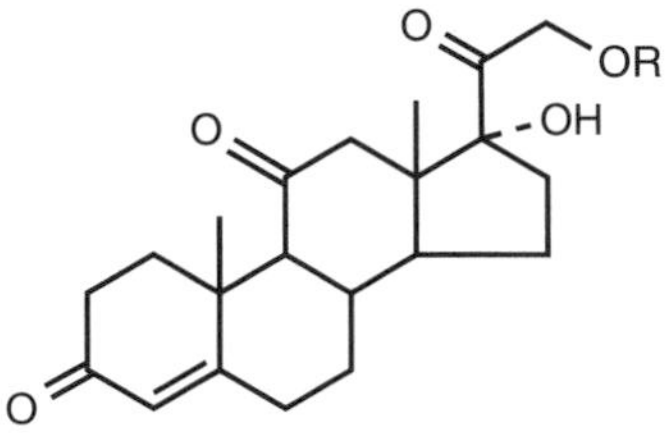

Cortisone: R = H
Cortisone acetate: R = $COCH_3$

Progesterone (pregn-4-ene-3,20-dione), a female sex hormone synthesized by the corpus luteum, is also a pregnane analog.

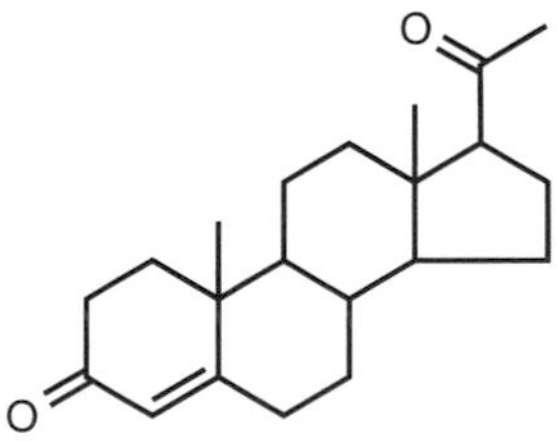

Progesterone

The male sex hormones (androgens) are based on the structure of 5α-androstane. Testosterone, an important naturally occurring androgen, is named 17β-hydroxy-4-androsten-3-one.

The estrogens, which are female sex hormones synthesized by the graafian follicle, are estrane analogs containing an aromatic A ring. Although the A ring does not

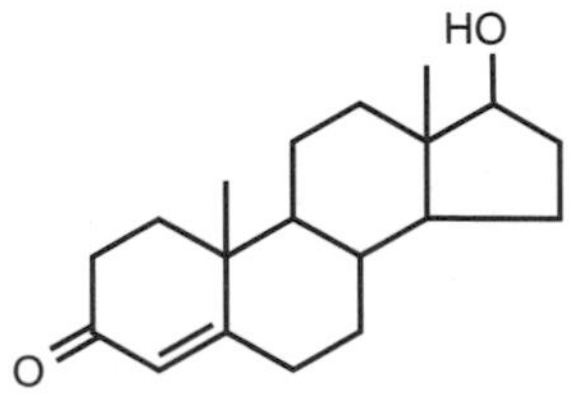

Testosterone

contain isolated C═C groups, these analogs are named as if the bonds were in the positions shown in 17β-estradiol. Hence, 17β-estradiol, a typical member of this class of drugs, is named estra-1,3,5,(10)-triene-3,17β-diol.

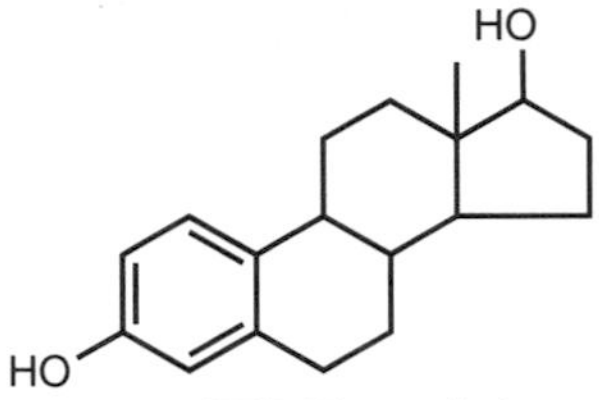

17β-Estradiol

Other examples of steroid nomenclature are found throughout this chapter.

Aliphatic side chains at position 17 are always assumed to be β when cholestane or pregnane nomenclature is employed; hence, the notation 17β need not be used when naming these compounds. If a pregnane has a 17α chain, however, this should be indicated in the nomenclature. Finally, the final e in the name for the parent steroid hydrocarbon is always dropped when it precedes a vowel, regardless of whether a number appears between the two parts of the word. Note the nomenclature for cholesterol and testosterone versus that for cortisone, for example. For a more extensive discussion of steroid nomenclature, consult the literature.[1]

CHOLESTEROL BIOSYNTHESIS AND METABOLISM

Biosynthesis

The biosynthesis of cholesterol in man is particularly important because this sterol is implicated in many diseases and because it is the precursor for other endogenous steroids, i.e., the adrenocorticoids and sex hormones. Cholesterol is formed in many tissues, including the liver, skin, intestines, arteries, and glands that produce the steroid hormones.

This sterol and its metabolites are required for growth of almost all life forms. The currently accepted scheme for the first stage of biosynthesis of cholesterol involves the anaerobic conversion of acetate to squalene, as shown in equation 1 in Figure 23–1.[2] Under physiologic conditions, phosphate and carboxyl groups in the intermediates are ionized.

Cholesterol is derived from two-carbon units by a complex enzymatic process and uses 18 molecules of acetyl-CoA in the biosynthesis. The first enzymatic step is the condensation of two acetyl-coenzyme A (acetyl-CoA) mol-

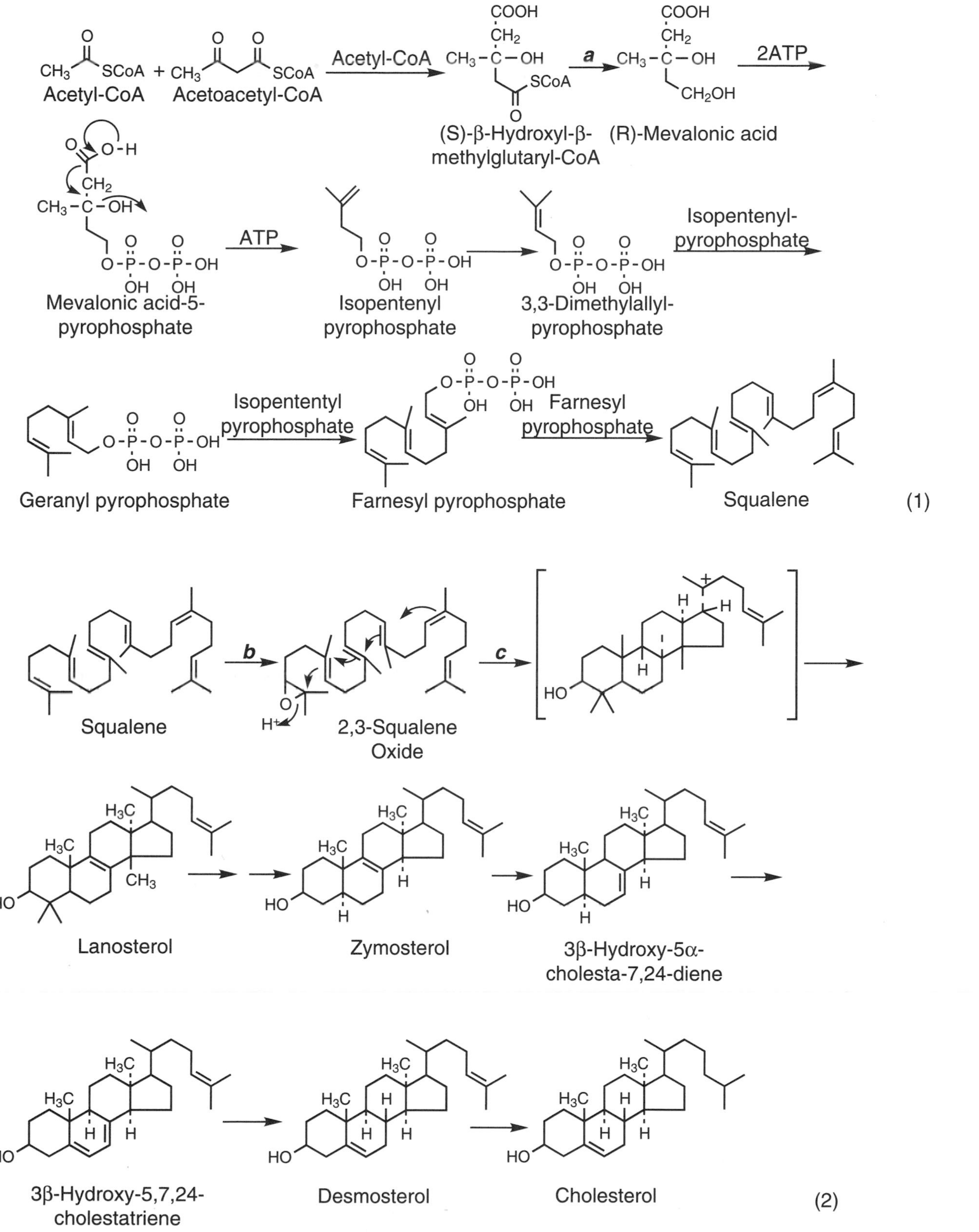

Fig. 23–1. Cholesterol biosynthesis. The biosynthesis is illustrated in two parts: (1) Formation of squalene from acetyl-CoA units and (2) conversion of squalene to cholesterol. Key enzymatic steps in the biosynthesis of cholesterol are catalyzed by HMG-CoA reductase (a), squalene epoxide (b), and 2,3-oxidosqualene-sterol cyclase (c).

ecules, catalyzed by a thiolase, to give acetoacetyl-CoA. A third unit of acetyl-CoA is then added to acetoacetyl-CoA to produce (S) 3-hydroxy-3-methylglutaryl-CoA (abbreviated HMG-CoA) by the enzyme HMG-CoA synthase. These initial steps of cholesterol biosynthesis are reversible.

The first irreversible enzymatic step is the reduction of HMG-CoA by the enzyme HMG-CoA reductase to yield (R) mevalonic acid. This enzymatic step is also the rate-limiting step in cholesterol biosynthesis[3] and has become an important target for therapeutic control of cholesterol production.[4] Mevalonic acid is subsequently phosphorylated by mevalonate kinase utilizing two molecules of adenosine triphosphate (ATP) to from mevalonic acid-5-pyrophosphate.

A third molecule of ATP is required for the dehydrodecarboxylation of mevalonic acid-5-pyrophosphate; this results in isopentenyl pyrophosphate. During this reaction the terminal phosphate of ATP accepts the leaving tertiary hydroxyl group of mevalonic acid-5-pyrophosphate.[5] In addition, CO_2, inorganic phosphate, and adenosine diphosphate (ADP) are formed. Isopentenyl pyrophosphate undergoes reversible isomerization to 3,3-dimethylallylpyrophosphate. Condensation of this compound with isopentenyl pyrophosphate units in successive steps leads to geranyl pyrophosphate and farnesyl pyrophosphate. Two farnesyl pyrophosphate units condense in a tail-to-tail fashion and produce squalene. This reductive process is catalyzed by a membrane-bound enzyme squalene synthetase and requires reduced nicotinamide adenine dinucleotide phosphate (NADPH).[6]

The second stage in cholesterol biosynthesis is initiated by oxidative cyclization of squalene by the enzyme squalene epoxidase, as shown in Figure 23–1. Squalene first undergoes epoxidation with molecular O_2, yielding the corresponding 2,3-squalene epoxide. This epoxide cyclizes[7–10] by the action of 2,3-oxidosqualene-sterol cyclase, a microsomal enzyme, and rearranges to the tetracyclic triterpenoid (lanosterol), which is the first compound in the sequence to have a steroidal structure. Other possible pathways for the formation of lanosterol from squalene are discussed in the literature.[3,10]

Lanosterol is sometimes classified as a $\Delta^{8,24}$-C-30 sterol. After its formation, the scheme of cholesterol biosynthesis is less well defined. Conversion of lanosterol to cholesterol involves (a) oxidative removal of the two *gem*-dimethyl groups at position 4 and the single methyl group at position 14, (b) removal of the 8,9-double bond and insertion of a 5,6-double bond, and (c) reduction of the side chain.

Reduction of the 24,25-double bond in the side chain may occur at many different points in the conversion of lanosterol to cholesterol.[11,12] It has been suggested that the pathway from lanosterol to cholesterol in various tissues involves either early or late saturation of the 24,25-double bonds as the preferred routes.[13–19] For convenience, the route in equation 2 (see Fig. 23–1) is illustrated for the unsaturated side chain; a similar route may be shown for the reduced side chain. Bach[20] has presented a more advanced and detailed discussion of the relationship between pathways involving a reduced versus unsaturated side chain.

The two methyl groups at C-4 and the methyl group at C-14 are removed by cytochrome P450-mediated monooxygenase reactions. The 14α-methyl group is removed in three consecutive oxidation steps catalyzed by 14α-demethylase.[21] Regarding the two methyl groups at C-4, evidence suggests that the 4α-methyl group is removed first.[22,23] This proceeds by stepwise oxidation to the CO_2H group followed by decarboxylation and loss of CO_2. The resulting zymosterol serves as a precursor of 3β-hydroxy-5α-cholesta-7,24-diene, which results from an enzyme-catalyzed migration of the 8,9-double bond of zymosterol to the 7,8 position by an addition-elimination mechanism.[24] Dehydrogenation results in the formation of 3β-hydroxy-5,7,24-cholestatriene; reduction of the 7,8-double bond affords desmosterol. This sterol, cholest-5,24-dien-3β-ol, is one of the immediate precursors of cholesterol. The coenzyme NADPH is necessary for reduction of the 7,8-double bond in the triene and the 24,25 double bond in desmosterol.

Cholesterol has two primary biochemical roles in the body—as an integral component of the plasma membrane in all cells and as a biosynthetic precursor in steroidogenesis in endocrine cells of the adrenal gland, ovary, testes, and placenta. All cells have the capability to biosynthesize cholesterol, and yet the major source of cholesterol is derived from circulating low density lipoproteins (LDL), which contain primarily cholesterol esters. In the elegant studies of Brown and Goldstein,[25,26] cells incorporate LDL by the binding of LDL to specific, high affinity receptors located on the external portion of the cell membrane. The LDL-receptor complex is internalized via receptor-mediated endocytosis, and the complex is degraded by lysosomal enzymes, releasing the cholesterol in the cell. The lipoprotein-derived cholesterol down-regulates endogenous cholesterol biosynthesis and is converted to cholesteryl oleate for storage in lipid droplets in the cell. This stored cholesterol is the biosynthetic precursor for steroid hormones; more details are provided in later sections of this chapter. The relationships between hormones and lipoproteins have been reviewed.[27] Chapter 24 discusses cholesterol, lipoproteins, and antilipidemic agents.

Metabolism: Conversion to Pregnenolone and Steroid Hormones

Cholesterol is converted by enzymatic cleavage of its side chain to pregnenolone (3β-hydroxypregn-5-en-20-one), which serves as the common biosynthetic precursor of the adrenocorticoids and sex hormones. The enzyme that catalyzes this biotransformation is a mitochondrial cytochrome P450 enzyme complex referred to as side-chain cleavage. This enzyme complex found in the mitochondrial membrane consists of three proteins—the chcytochrome $P450_{SCC}$, adrenodoxin, and adrenodoxin reductase.[28] Three oxidation steps are involved in the conversion by this single enzyme complex, and three moles of NADPH and molecular oxygen are consumed for each mole of cholesterol converted to pregnenolone. The first oxidation results in the formation of cholest-5-ene-3β,22R-diol, followed by the second oxidation yielding cholest-5-ene-3β,20R,22R-triol. The third oxidation

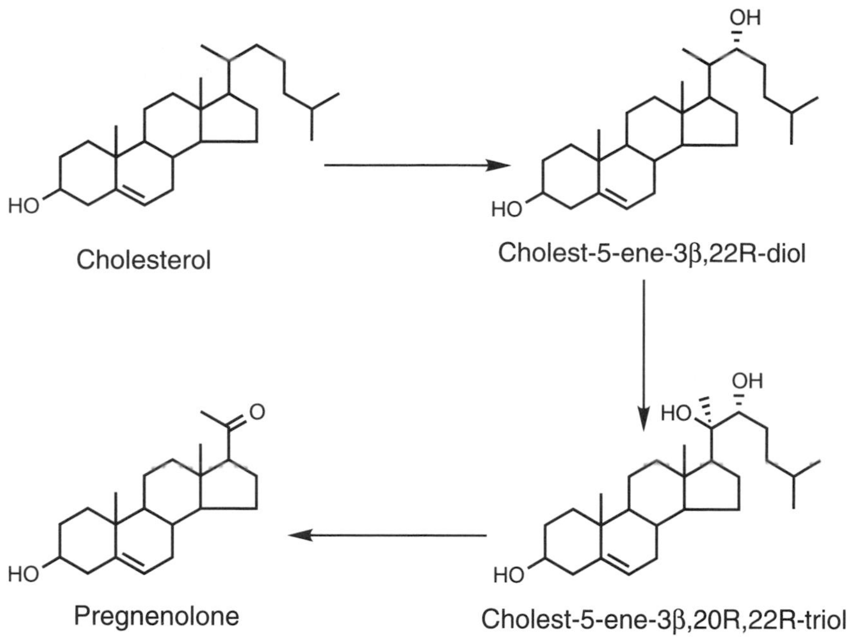

Cholesterol

Cholest-5-ene-3β,22R-diol

Cholest-5-ene-3β,20R,22R-triol

Pregnenolone

Fig. 23–2. Biosynthesis of pregnenolone from cholesterol.

step catalyzes the cleavage of the $C_{20}-C_{22}$ bond to release pregnenolone and isocaproic aldehyde.

Pregnenolone serves as the common precursor in the formation of the steroid hormones. This C_{21} steroid is converted via enzymatic oxidations and isomerization of the double bond to a number of physiologically active C_{21} steroids, including the female sex hormone progesterone and the adrenocorticoids hydrocortisone (cortisol), corticosterone, and aldosterone. Oxidative cleavage of the two carbon side chain of pregnenolone and subsequent enzymatic oxidations and isomerization lead to C_{19} steroids, including the androgens testosterone and dihydrotestosterone. The final group of steroids, the C_{18} female sex hormones, are derived from oxidative aromatization of the A ring of androgens to produce estrogens. More detailed information on these biosynthetic pathways are described later in the chapter under the particular class of steroid hormones.

Conversion to Bile Acids

Cholesterol is secreted as neutral sterol esters or as bile acid metabolites. The direct transformation of cholesterol to one of the most important bile acids, cholic acid, was first reported by Block and co-workers in 1943.[29] In the liver, the bile acids produced from the metabolism of cholesterol are converted to their conjugates with either glycine ($H_2NCH_2CO_2H$) or taurine ($H_2NCH_2CH_2SO_3H$). The sodium salts of these bile acid conjugates (bile salts) are secreted by the liver and are found in bile.[30] When they enter the large intestine, they are hydrolyzed back to bile acids by the intestinal flora. A portion of these steroid acids is returned to the liver by way of the portal blood; in the liver, bile acids regulate their own production.[31] They do so by interfering with enzymes needed for their synthesis from cholesterol; feedback mechanisms are involved. Cholic acid, the main cholesterol metabolite in man, has an average biologic half-life of 2.8 days.[32]

A major route for the catabolism of cholesterol, the obligatory precursor of bile acids, is seen in Figure 23–3. As much as 90% of administered cholesterol is converted to bile acids by way of this partially elucidated scheme.[33–35] Hydroxylation at the 7α position of cholesterol by the microsomal 7α-hydroxylase cytochrome P450 enzyme complex produces 5-cholestene-$3\beta,7\alpha$-diol; this represents the initial and rate-limiting step in cholesterol catabolism to the bile acids.[31] Steroid ring hydroxylation must precede side-chain oxidation. Subsequent 12α hydroxylation, epimerization of the 3β-OH group to the 3α configuration, and reduction of the 5,6-double bond result in $3\alpha,7\alpha,12\alpha$-trihydroxy-5β-cholestane. Carbon atoms 25, 26, and 27 are lost by stepwise oxidation of one of the methyl groups (26 or 27) followed by β-oxidation of the resulting acid.[36–38] This ultimately results in cholic acid.

The taurine and glycine conjugates formed in the liver are secreted into the large intestine, where they are subsequently hydrolyzed. In humans, cholesterol is mainly converted to glycocholic acid and some taurocholic acid. Other bile acids, such as chenodesoxycholic acid, desoxycholic acid, and lithocholic acid, are found as conjugates in mammals.

Desoxycholic acid, in humans, is produced from cholic acid through the action of intestinal microorganisms. It is found in the bile because it is reabsorbed from the intestine; it is not a precursor of cholic acid.[38] Lithocholic acid may be a precursor of cholic acid, but the exact order of hydroxylation still needs to be determined; species differences do exist.[30]

Bile salts are surface-active agents that act as anionic detergents. They emulsify fats and fat-soluble vitamins, thereby promoting absorption through the intestinal mucosa. Emulsified fats have an increased surface area and

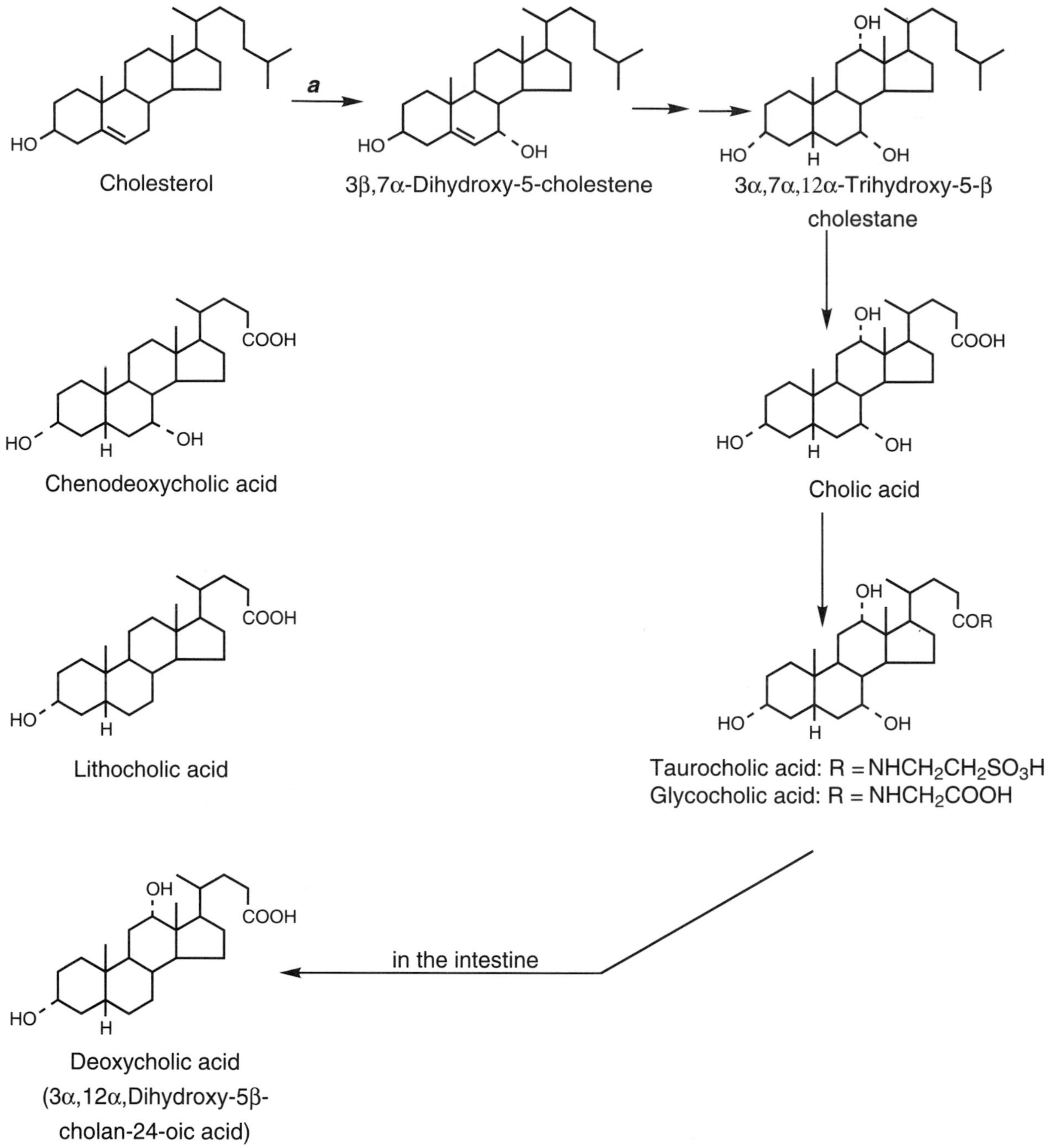

Cholesterol → *a* → 3β,7α-Dihydroxy-5-cholestene → → 3α,7α,12α-Trihydroxy-5-β cholestane

Chenodeoxycholic acid

Cholic acid

Lithocholic acid

Taurocholic acid: R = $NHCH_2CH_2SO_3H$
Glycocholic acid: R = $NHCH_2COOH$

in the intestine

Deoxycholic acid (3α,12α,Dihydroxy-5β-cholan-24-oic acid)

Fig. 23–3. Biosynthesis of bile acids. The rate-limiting enzyme in the biosynthesis pathway is cholesterol 7a-hydroxylase (a).

are consequently more readily hydrolyzed by the lipolytic enzyme lipase. Ox-bile extracts (containing at least 45% cholic acid) are used as choleretics; such agents increase excretion of bile by the liver. Other preparations, such as Fairchild's bile salts (consisting of sodium glycolate and sodium taurocholate) and concentrated ox bile that is free of bile pigment (called glycotauro bile salts), are used for similar purposes.

Dehydrocholic acid National Formulary (N.F.) (3,7,12-triketo-5*β*-cholan-24-oic acid), which is administered orally, and sodium dehydrocholate N.F., which is water soluble and administered by injection, are also used as choleretics. These compounds, prepared by oxidation of cholic acid, alleviate hepatic functional insufficiency and product diuresis. They are less effective diuretics than the organomercurials.

BIOCHEMICAL MECHANISM OF STEROID HORMONE ACTION

Although the steroid hormones—adrenocorticoids, estrogens, progestins, and androgens—are present in the body only in extremely low concentrations (e.g., 0.1 to

Dehydrocholic acid: R = H
Sodium dehydrocholate: R = Na

1.0 nM), they exert potent physiologic effects on sensitive tissues. Extensive research activities directed at elucidation of the general mechanism of steroid hormone action have been performed for over three decades, and several reviews have appeared.[39–44]

The steroid hormones act on *target* cells to regulate gene expression and protein biosynthesis via the formation of steroid-receptor complexes, as outlined in Figure 23–4. The lipophilic steroid hormones are carried in the bloodstream, with the majority of the hormones reversibly bound to serum carrier proteins and a small amount of free steroids. The free steroids can diffuse through the cell membrane and enter cells. Those cells sensitive to the particular steroid hormone (referred to as *target* cells) contain high-affinity steroid receptor proteins capable of interacting with the steroid. Early studies suggested that the steroid receptor proteins were located in the cytosol of target cells[45] and, following formation of the steroid-receptor complex, the steroid-receptor complex translocated into the nucleus of the cell. More recent investigations on estrogen, progestin, and androgen action indicate that active, unoccupied receptor proteins are present only in the nucleus of the cell, whereas the majority of the glucocorticoid receptor is located in the cytoplasm.[39,44,46] In the current model, the steroid enters the nucleus of the cell and binds to the nuclear steroid receptor protein. This binding initiates a conformational change or activation of the steroid-nuclear receptor complex to form a dimer, and the dimer interacts with particular regions of the cellular deoxyribonucleic acid (DNA), referred to as hormone-responsive elements (HRE), and with various nuclear transcriptional factors. Binding of the nuclear steroid-receptor complex to DNA initiates transcription of the DNA sequence to produce messenger ribonucleic acid (mRNA). Finally, the elevated levels of mRNA lead to an increase in protein synthesis in the endoplasmic reticulum; these proteins include enzymes, receptors, and secreted factors that subsequently result in the steroid hormonal response regulating cell function, growth, and differentiation.

The primary amino acid sequences of the various steroid hormone receptors have been deduced from cloned

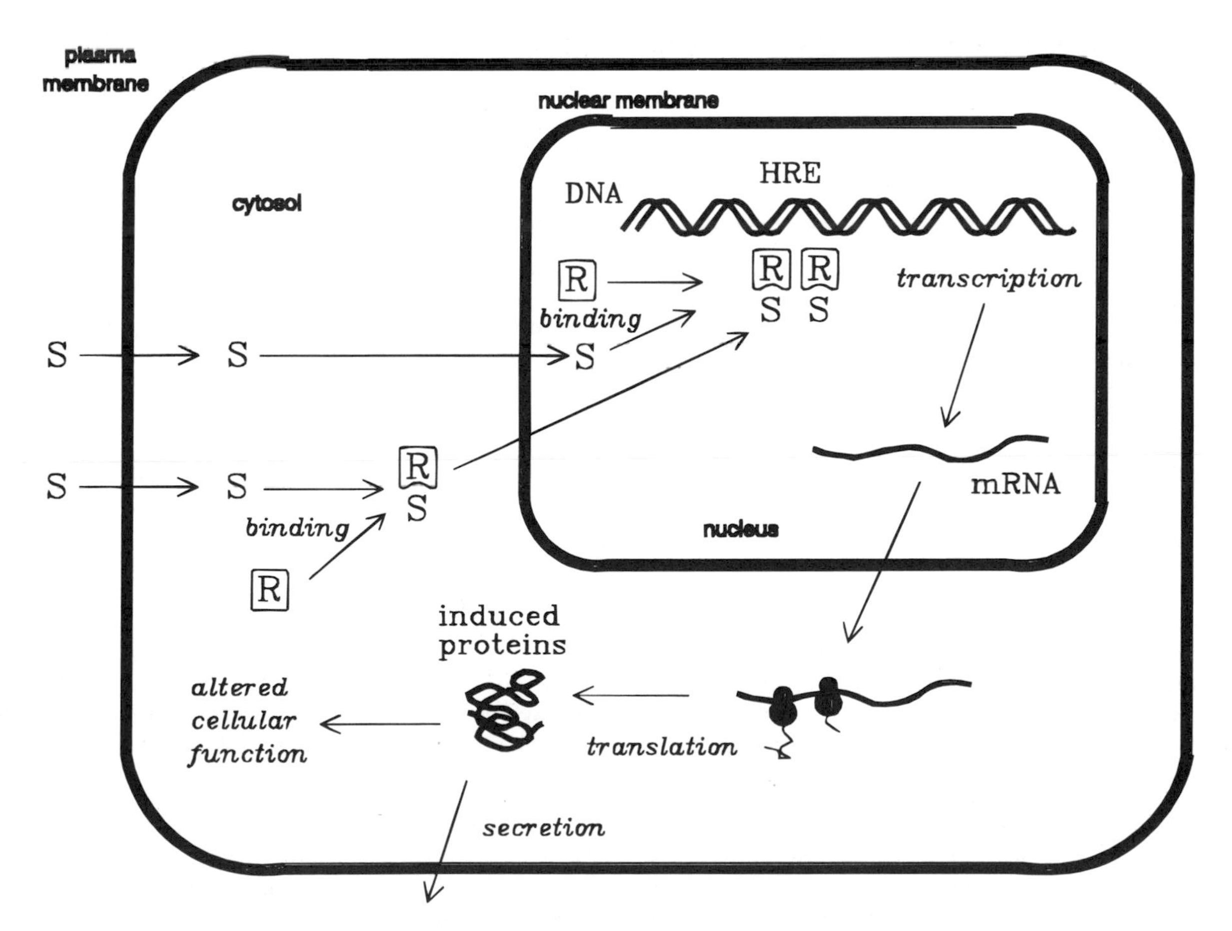

Fig. 23–4. Mechanism of steroid hormone action.

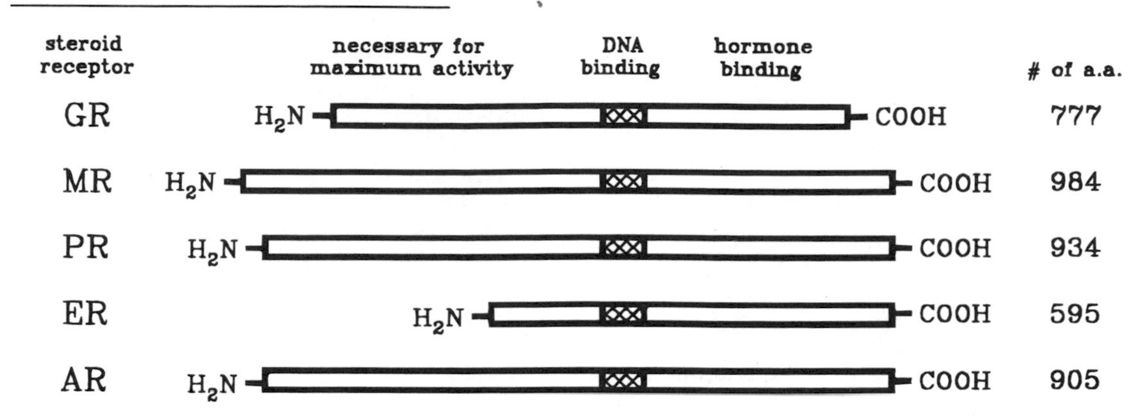

HORMONE RESPONSIVE ELEMENT (HRE)

	nucleotide sequence
GRE	G G T A C A n n n T G T T C T
MRE	"
PRE	"
ARE	"
ERE	A G G T C A n n n T G A C C T

Fig. 23–5. Structural features of steroid hormone receptors and hormone responsive elements. Schematic comparison of the amino acid sequences for steroid receptors (GR, glucocorticoid; MR, mineralocorticoid; PR, progesterone; ER, estrogen; AR, androgen) with high homology in the DNA binding region. HRE sequences are also compared (GRE, glucocorticoid; MRE, mineralocorticoid; PRE, progesterone; ARE, androgen, ERE, estrogen)

complementary DNA (cDNA).[40,42] The steroid receptor proteins are part of a larger family of nuclear receptor proteins that also include receptors for vitamin D, thyroid hormones, and retinoids. The overall structures of the receptors have strong similarities (Fig. 23–5), with the receptor protein–containing regions that bind to the DNA and bind to the steroid hormone ligand. A high degree of homology (sequence similarities) in the steroid receptors are found in the DNA binding region that interacts with the HRE. The DNA binding region is rich in cysteine amino acids and are thought to chelate zinc ions, forming finger-like projections called zinc fingers that bind to the DNA. Structure-function studies of cloned receptor proteins also identify regions of the molecules that are important for interactions with nuclear transcriptional factors and activation of gene transcription. The interactions necessary for formation of the steroid-receptor complexes and subsequent activation of gene transcription are complicated, involve multi-stage processes, and leave many unanswered questions.

The basic mechanism of steroid hormone action on target cells is similar for the various classes of agents. Differences in the actions of adrenocorticoids, estrogens, progestins, and androgens arise from the specificity of the particular receptor proteins, the particular genetic processes initiated, and the specific cellular proteins produced. These topics are discussed in greater detail in the sections dealing with each class of steroid hormones.

Adrenocorticoids

The adrenal glands are flattened, caplike structures located above the kidneys. The inner core (medulla) of the gland secretes catecholamines, while the shell (cortex) of the gland synthesizes certain steroids known as the adrenocorticoids.

The importance of the adrenal glands has been recognized for some time; adrenalectomy in small animals causes death in a few days. Addison's disease, Cushing's disease, and Conn's syndrome are pathologic conditions related to the adrenal cortex and the hormones produced by the gland.

Addison's disease was named after Thomas Addison, who in 1855 described a syndrome in which the physiologic significance of the adrenal cortex was emphasized.[47] This disease is characterized by extreme weakness, anorexia, anemia, nausea and vomiting, low blood pressure, hyperpigmentation of the skin, and mental depression caused by decreased secretion of steroid hormones by the adrenal cortex.

Conditions of this type, generally referred to as hypoadrenalism, may result from several causes, including destruction of the cortex by tuberculosis or atrophy or decreased secretion of adrenocorticotropin (ACTH) because of diseases of the anterior pituitary (adenohypophysis). Cushing's disease, or hyperadrenalism, on the other hand, many result from adrenal cortex tumors or increased production of ACTH; increased corticotropin levels may be caused by pituitary carcinoma.

Conn's syndrome is apparently due to an inability of the adrenal cortex to carry out 17α-hydroxylation in the biosynthesis of the hormones from cholesterol. Consequently, the disease is characterized by a high secretory level of aldosterone, which has no 17α-hydroxyl function. In addition, hypernatremia, polyuria, alkalosis, and hypertension are observed.[48]

The importance of the corticoids is most dramatically observed in adrenalectomized animals. There is an increase of urea in the blood, muscle weakness (asthenia), decreased liver glycogen and decreased resistance to insulin, reduction in resistance to trauma, such as cold and mechanical or chemical shock, and electrolyte disturbances. Potassium ions are retained, and excretion of Na^+, Cl^-, and water is increased.

Biosynthesis

The glucocorticoids and mineralocorticoids are secreted under the influence of peptide hormones secreted by the hypothalamus and anterior pituitary (adenohypophysis). Removal of the pituitary results in atrophy of the adrenal cortex and a marked decrease in the rate of glucocorticoid secretion. In hypophysectomized animals, the rate of secretion of aldosterone is only slightly decreased or remains unchanged. Consequently, the electrolyte balance remains nearly normal. Aldosterone secretion rates are influenced to a greater extent by the octapeptide angiotensin II.

The peptide hormone in the anterior pituitary that influences glucocorticoid biosynthesis is adrenocorticotropic hormone (ACTH; corticotropin), whereas the peptide hormone in the hypothalamus is corticotropin-releasing factor (CRF). The production of both ACTH and CRF is regulated by the central nervous system (CNS) and by a negative corticoid feedback mechanism. CRF is released by the hypothalamus into the portal vein and is transported to the anterior pituitary, where it stimulates the release of ACTH in the bloodstream. ACTH is transported to the adrenal glands, where it stimulates the biosynthesis and secretion of the glucocorticoids. The circulating levels of glucocorticoids act on the hypothalamus and anterior pituitary to regulate the release of both CRF and ACTH. As the levels of glucocorticoids rise, smaller amounts of CRF and ACTH are secreted and a negative feedback is observed. A variety of stimuli, including pain, noise, and emotional reactions, increases the secretion of CRF, ACTH, and consequently, the glucocorticoids. Once the stimulus is alleviated or removed, the negative feedback mechanism inhibits further production and helps return the body to a normal hormonal balance.[49,50]

ACTH acts at the adrenal gland by binding to a receptor protein on the surface of the adrenal cortex cell to stimulate the biosynthesis and secretion of glucocorticoids. The only steroid stored in the adrenal gland is cholesterol, found in the form of cholesterol esters stored in lipid droplets. ACTH stimulates the conversion of cholesterol esters to glucorticoids by initiating a series of biochemical events through its surface receptor. The ACTH receptor protein is coupled to a G-protein and to adenylyl cyclase, located on the inner side of the plasma membrane. Binding of ACTH to its receptor leads to activation of adenylyl cyclase via the G-protein. The result is an increase in intracellular cyclic adenosine monophosphate (cAMP) levels. One of the processes influenced by elevated cAMP levels is the activation of cholesterol esterase, which cleaves cholesterol esters and liberates free cholesterol.

This free cholesterol is then converted during mitochondria to pregnenolone via the side-chain cleavage reaction described earlier. Pregnenolone is converted to adrenocorticoids by a series of enzymatic oxidations and isomerization of the double bond (Fig. 23–6). The next several enzymatic steps in the biosynthesis of glucocorticoids occur in the endoplasmic reticulum of the adrenal cortex cell. Hydroxylation of pregnenolone at position 17 by the enzyme 17α-hydroxylase produces 17α-hydroxypregnenolone; this hydroxyl group in the final adrenocorticoid potentiates hormonal action. In one step, 17α-hydroxypregnenolone is oxidized to a 3-keto intermediate by the action of the enzyme 5-ene-3β-hydroxysteroid dehydrogenase and isomerized to 17α-hydroxyprogesterone by the enzyme 3-oxosteroid-4,5-isomerase. Another hydroxylation occurs by the action of 21-hydroxylase to give rise to 11-deoxycortisol, which contains the physiologically important ketol (—$COCH_2OH$) side chain. The final step in the biosynthesis is catalyzed by the enzyme 11β-hydroxylase, a mitochondrial cytochrome P450 enzyme complex, and results in the formation of cortisol (hydrocortisone), the most potent endogenous glucocorticoid secreted by the adrenal cortex. Several reviews[49,51–54] provide more detailed discussions about the enzymology and regulation of adrenal steroidogenesis.

The pathway for the formation of the potent mineralocorticoid molecule, aldosterone, is similar to that for cortisol and uses several of the same enzymes. The preferred pathway involves the conversion of pregnenolone to progesterone by 5-ene-3β-hydroxysteroid dehydrogenase and 3-oxosteroid-4,5-isomerase. Hydroxylation at position 21 of progesterone by 21-hydroxylase results in 21-hydroxyprogesterone (deoxycorticosterone). Again, these first conversions occur in the endoplasmic reticulum of the cell, whereas the next enzymatic steps occur in the mitochondria. 11β-Hydroxylase catalyzes the conversion of deoxycorticosterone to corticosterone, which exhibits mineralocorticoid activity. The final two oxidations involve hydroxylations at the C-18 methyl group and are catalyzed by 18-hydroxylase. These reactions produce first 18-hydroxycorticosterone and then aldosterone, the most powerful endogenous mineralocorticoid secretion of the adrenal cortex. The aldehyde at C-18 of aldosterone exists in equilibrium with its hemiacetal form.

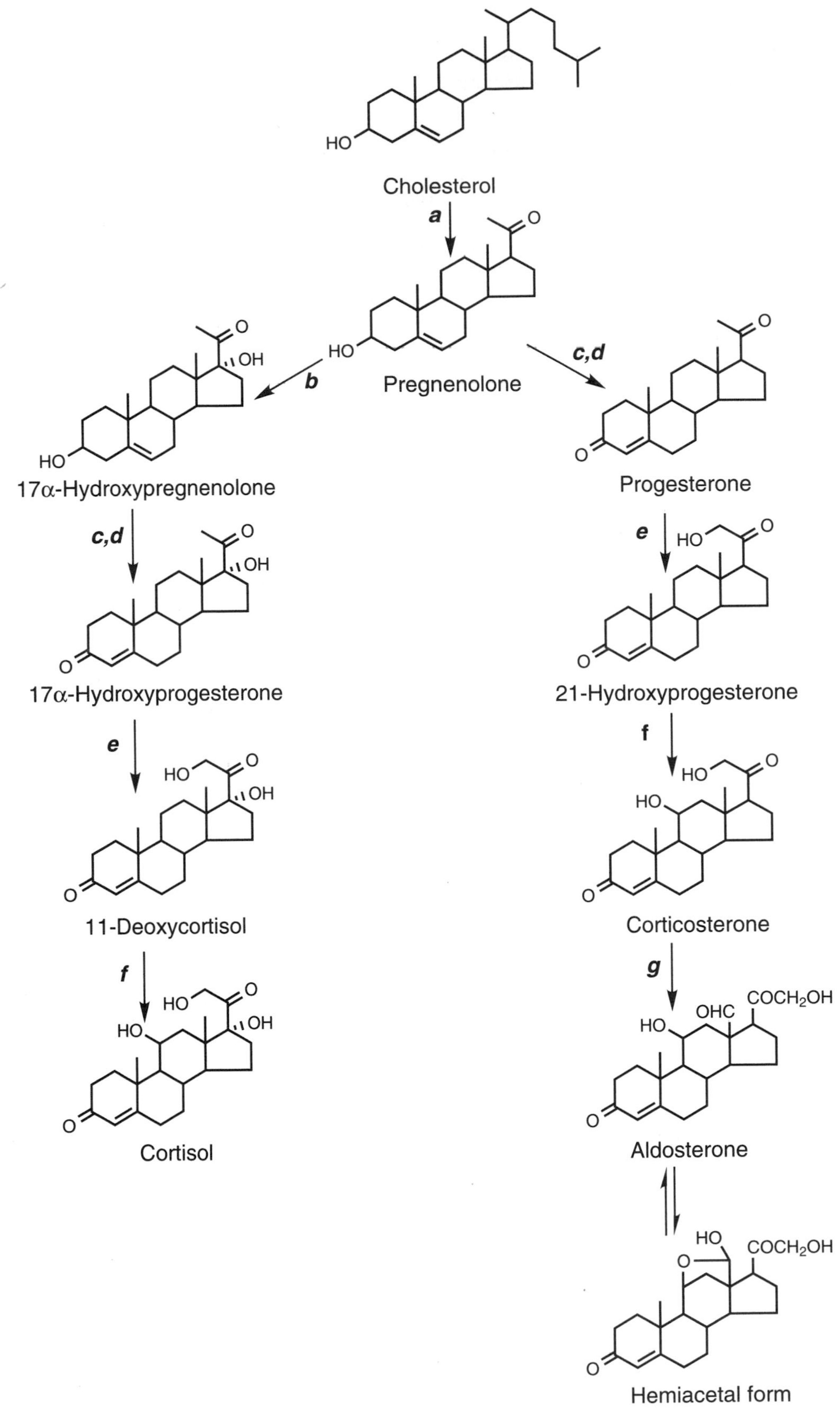

HO
Cholesterol
a
O
HO
Pregnenolone
b
c,d
O
OH
HO
17α-Hydroxypregnenolone
Progesterone
c,d
17α-Hydroxyprogesterone
e
HO
21-Hydroxyprogesterone
f
11-Deoxycortisol
Corticosterone
g
OHC
$COCH_2OH$
Cortisol
Aldosterone
Hemiacetal form

Fig. 23–6. Biosynthesis of Adrenocorticoids. The enzymes involved in glucocorticoid bisynthesis are (a) side chain cleavage, (b) 17α-hydroxylase, (c) 5-ene-3β-hydroxysteroid dehydrogenase, (d) 3-oxosteroid-4,5-isomerase, (e) 21-hydroxylase, (f) 11β-hydroxylase, and (g) 18-hydroxylase.

Metabolism

The adrenocorticoids are metabolized by several routes in many tissues, including the liver, muscles, and red blood cells.[52,53,55] The liver metabolizes them most rapidly. Catabolic products can be isolated from the urine and bile and can be formed in tissue preparations *in vitro*.[56,57] In urine they are found primarily in conjunction with glucuronic acid; conjugation with this or sulfuric acid renders the steroid metabolite more water soluble. Metabolites of the endogenous adrenocorticoids are similar; one reason is that cortisone and hydrocortisone are biochemically interconvertible.[58]

Metabolites may be produced by reduction of the 4,5-double bond by 5β- or 5α-reductase and reduction of the 3-ketone by 3α-hydroxysteroid dehydrogenase or 3β-hydroxysteroid dehydrogenase. The tetrahydro metabolite, the major metabolite formed, has the 5β-pregnane geometry and 3α-hydroxyl function. This is similar to the geometry of the bile acids. Several compounds of this type have been isolated.[59,60] These are exemplified in Figure 23–7 by the two major ones, urocortisol and urocortisone, which are named after cortisol (hydrocortisone) and cortisone.

Minor metabolites result from reduction of the C-20 ketone of urocortisol or urocortisone, producing 20-hydroxyl analogs. Alternatively, loss of the 17-ketol side chain results in C-19 steroids with the geometry of either 5β-androstane or 5α-androstane.[51–65] This system is illustrated by two metabolites, 3α,11β-dihydroxy-5α-androstan-17-one and 3α,11β-dihydroxyl-5β-androstan-17-one. In addition, some ring A aromatic adrenocorticoid metabolites that resemble the estrogens have been isolated.[66]

Development of Adrenocorticoid Drugs

After Addison's observations in 1855, physiologists, pharmacologists, and chemists from many countries contributed to our understanding of adrenocorticoids. It was not until 1927, however, that Rogoff and Stewart found that extracts of adrenal glands, administered by intravenous injection, kept adrenalectomized dogs alive.

Since this discovery, similar experiments have been repeated many times; it was thought that the biologic activity of the extract was due to a single compound. Later, 47 compounds were isolated from such extracts, and some were highly active. Among the biologically active corticoids isolated, hydrocortisone, corticosterone, aldosterone, cortisone (see equations 3 and 7 for structures), 11-desoxycorticosterone, 11-dehydrocorticosterone, and 17α-hydroxy-11-desoxycorticosterone, were found to be most potent.[67]

The 11-desoxycorticosterone was the first naturally occurring corticoid to be synthesized; it was prepared, before its isolation from the adrenal cortex, by Steiger and Reichstein.[68] This synthesis, shown in Figure 23–8, involved conversion of 3β-hydroxy-Δ^5-etienic acid 3-acetate to the diazoketone, hydrolysis, and reaction with acetic acid to give a ketol acetate. The dibromide addition product was then oxidized and the bromine atoms removed with zinc, producing the 21-acetate ester of 11-desoxycorticosterone.

Because of this and other early work with corticoids,

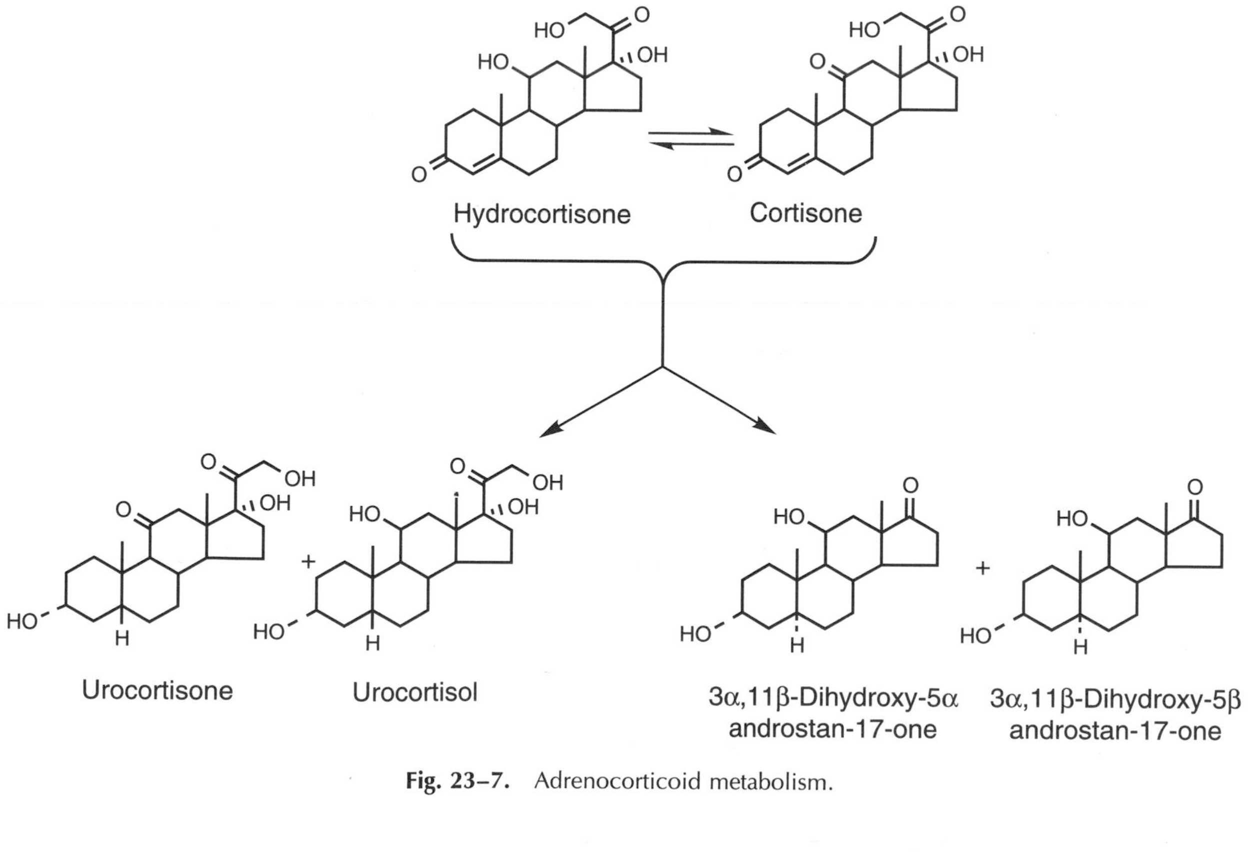

Hydrocortisone ⇌ Cortisone

Urocortisone + Urocortisol

3α,11β-Dihydroxy-5α androstan-17-one + 3α,11β-Dihydroxy-5β androstan-17-one

Fig. 23–7. Adrenocorticoid metabolism.

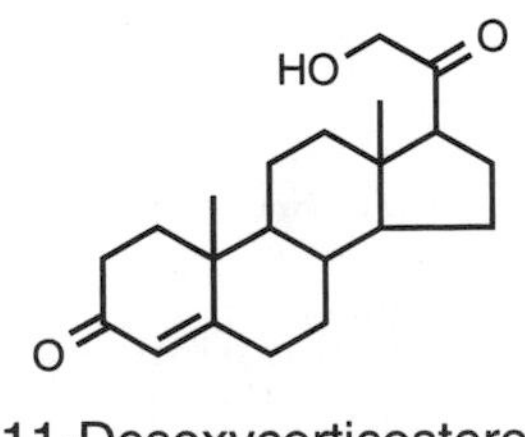

11-Desoxycorticosterone (cortexone)

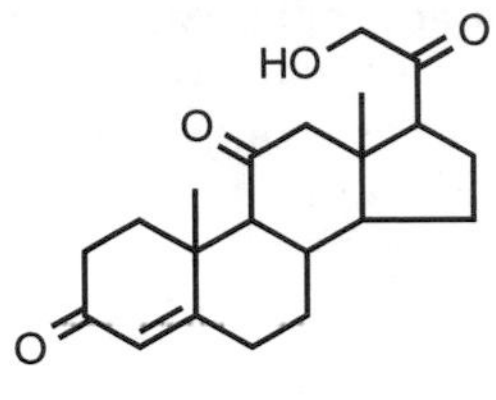

11-Dehydrocorticosterone

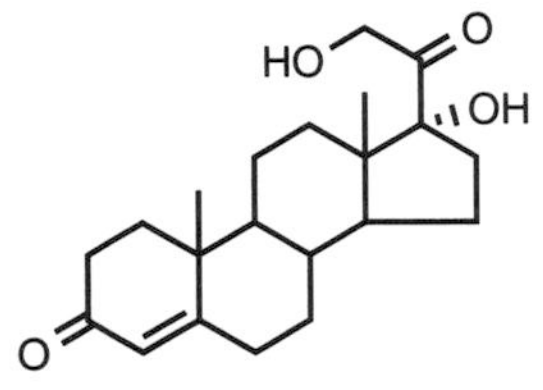

17α-Hydroxy-11-desoxycorticosterone

Reichstein later shared the Nobel Prize with Kendall, another chemist who was instrumental in carrying out early steroid syntheses, and with Hench, a rheumatologist who in 1929 discovered that cortisone is effective in the treatment of rheumatoid arthritis. Kendall's basic research ultimately led to the synthesis of cortisone from naturally occurring bile acids.[69]

After the synthesis of 11-desoxycorticosterone in 1937, all the corticoids were synthesized and their structures confirmed. The first synthesis of cortisone from methyl *βα*-hydroxy-11-ketobisnorcholanate was reported by Sarett in 1946.[70] Earlier work of Kendall and co-workers involving its preparation from the methyl ester of desoxycholic acid was used in his research.[71] Later, several chemists, including Sarett,[72] Kendall, and Tishler, found ways to improve the yields and to decrease the labor involved in the multistep conversion of bile acids to cortisone acetate. In 1949 Merck sold limited quantities of this glucocorticoid to physicians at $200 per gram for treating rheumatoid arthritis. Subsequent improvements in the methods of synthesis reduced the price to $10 per gram by 1951; in 1955 Upjohn used an efficient process involving the synthesis of cortisone acetate from progesterone, the latter steroid being prepared from diosgenin, and further reduced the price to $3.50 per gram.[69] Other pharmaceutical companies also began to sell cortisone synthesized from bile acids by a well-developed but lengthy procedure, and by 1958 sales of this and other corticoids reached $100 million per year in the United States.[69]

Cortisone (17,21-dihydroxypregn-4-ene-3,11-20-trione) is used primarily as its 21-acetate (cortisone acetate U.S.P.) because of its increased stability; the acetate derivative also has a longer duration of action when administered by injection, and smaller doses can be used. Similarly, hydrocortisone (U.S.P.), also known as cortisol (11*β*,17,21-trihydroxypregn-4-ene-3,20-dione), may be dispensed as its 21-acetate (hydrocortisone acetate U.S.P.), which is superior to cortisone acetate when injected intra-articularly.

Cortisone acetate is available in tablets (U.S.P.) or in ophthalmic suspension (U.S.P.). The usual oral or intra-

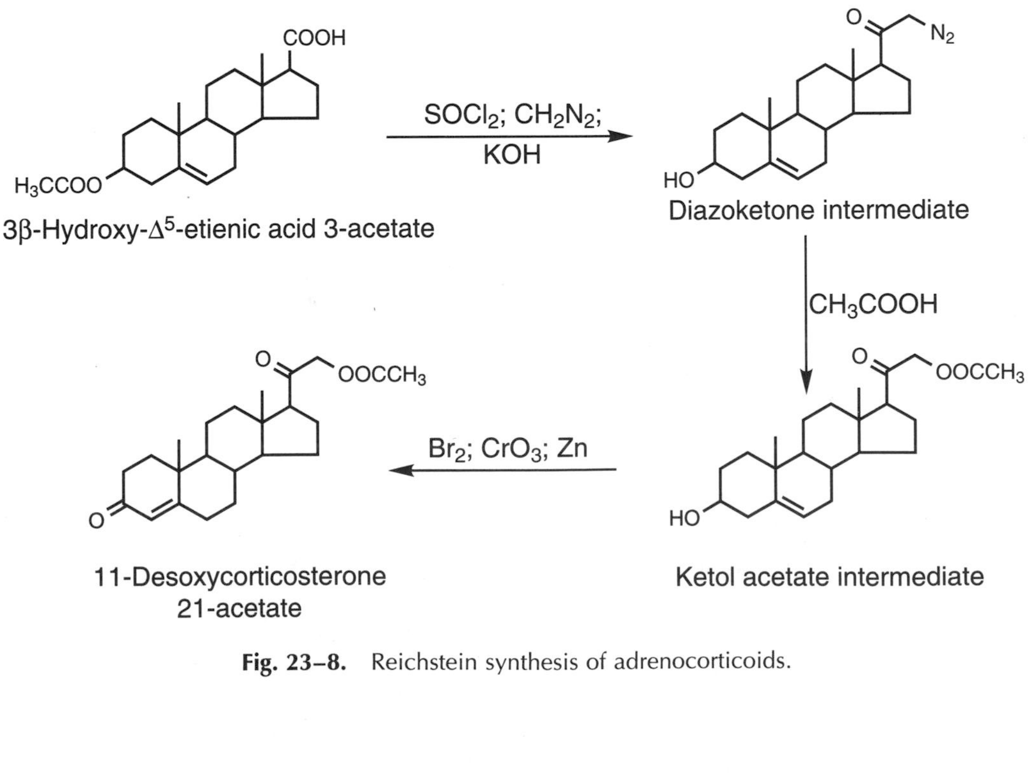

3β-Hydroxy-Δ5-etienic acid 3-acetate → ($SOCl_2$; CH_2N_2; KOH) → Diazoketone intermediate → (CH_3COOH) → Ketol acetate intermediate → (Br_2; CrO_3; Zn) → 11-Desoxycorticosterone 21-acetate

Fig. 23–8. Reichstein synthesis of adrenocorticoids.

muscular dose is between 10 and 400 mg daily; usually 25 mg four times daily, administered orally. For topical anti-inflammatory action, 1 to 2.5% lotions and ointments are available. The usual intra-articular dose of hydrocortisone acetate is 25 mg; as a topical anti-inflammatory agent, the drug is used in 0.5 and 2.5% lotions and ointments.

Other 21-ester derivatives that are available are hydrocortisone cypionate, which is a 21-(3-cyclopentylpropionate) ester, hydrocortamate hydrochloride, which is the hydrochloride salt of a 21-dimethylcarbamate ester, hydrocortisone sodium succinate U.S.P., which is a 21-sodium succinate ester, and hydrocortisone sodium phosphate U.S.P.

The water-insoluble hydrocortisone cypionate is used in doses expressed in terms of hydrocortisone for slower absorption from the gastrointestinal tract. The water-soluble hydrocortamate hydrochloride is used in 0.5% ointment and is claimed to be twice as potent as hydrocortisone acetate owing to better absorption and distribution. The extremely water-soluble 21-sodium succinate ester is used in solution for intravenous or intramuscular injection in the management of emergency conditions that can be treated with anti-inflammatory steroids. The usual intramuscular dosage ranges from 100 to 500 mg daily.

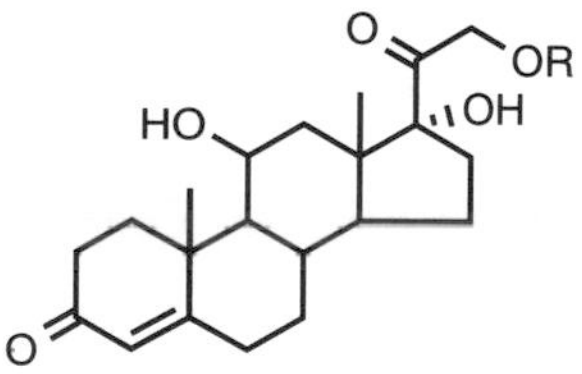

Cortisol (hydrocortisone): R = H

Hydrocortisone acetate: R = CH_3CO-

Hydrocortisone cypionate: R = (cyclopentyl)CH_2CH_2CO-

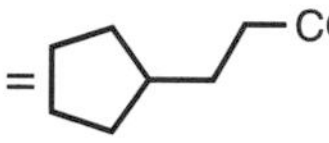

Hydrocortamate hydrochloride: R = Cl^- $(CH_3)_2\overset{+}{N}HCO$-

Hydrocortamate sodium succinate: R = Na^+ $^-OOCCH_2CH_2CO$-

Hydrocortisone sodium phosphate: R = Na_2O_3P-

After the introduction of cortisone (1948) and later hydrocortisone (1951) for the treatment of rheumatoid arthritis, many investigators began to search for superior agents having fewer side effects. When these drugs are used in doses necessary to suppress symptoms of rheumatoid arthritis, they also affect other metabolic processes. Side effects such as excessive sodium retention and potassium excretion, negative nitrogen balance, increased gastric acidity, edema, and psychosis are exaggerated manifestations of the normal metabolic functions of the hormones.

It was hoped that a compound with high glucocorticoid and low mineralocorticoid activity could be synthesized. Because it was recognized early that a carbonyl group at C-3, a double bond between carbons 4 and 5, an oxygen (C═O or β-OH) at carbon 11, and a β-ketol side chain at position 17 are necessary for superior glucocorticoid activity, investigators began to synthesize analogs containing these functions. Insertion of additional groups into other parts of the steroid might modify the glucocorticoid and mineralocorticoid activities of the parent drugs.

The first potent analogs discovered, however, did not result from a concentrated effort to find a better drug, but rather from basic chemical research concerned with the preparation of hydrocortisone from 11-epicortisol.

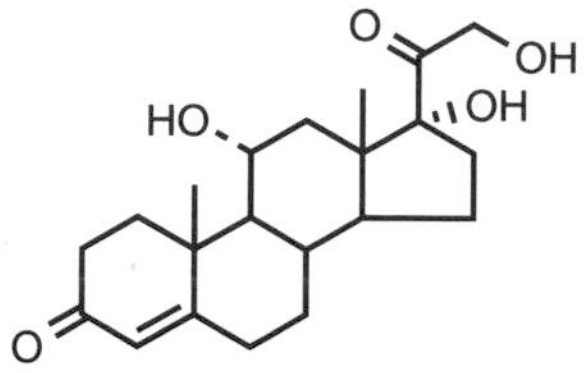

11-Epicortisol

A 9α-bromo analog that had one-third the glucocorticoid activity of cortisone acetate was prepared in these investigations.[73] Other halogens were introduced into the 9α position, and it was soon observed that glucocorticoid activity is inversely proportional to the size of the halogen at carbon 9; the 9α-fluoro analog (fludrocortisone) is approximately 11 times as potent as cortisone acetate.

Several modifications have been devised for the preparation of fludrocortisone; a key intermediate is the 9β,11β-oxide, which can be prepared from 17,21-dihydroxypregna-4,9(11)-diene-3,20-dione by addition of HOBr followed by reaction with CH_3CO_2Na. Reaction of the 9β,11β-oxide intermediate with HF in 5% C_2H_5OH/$HCCl_3$ results in fludrocortisone as illustrated in Figure 23–9.[74]

This 9α-fluoro steroid, administered as its 21-acetate derivative, was tested clinically in patients with rheumatoid arthritis and was found to be effective in about one-tenth the dose of cortisone acetate. Although glucocorti-

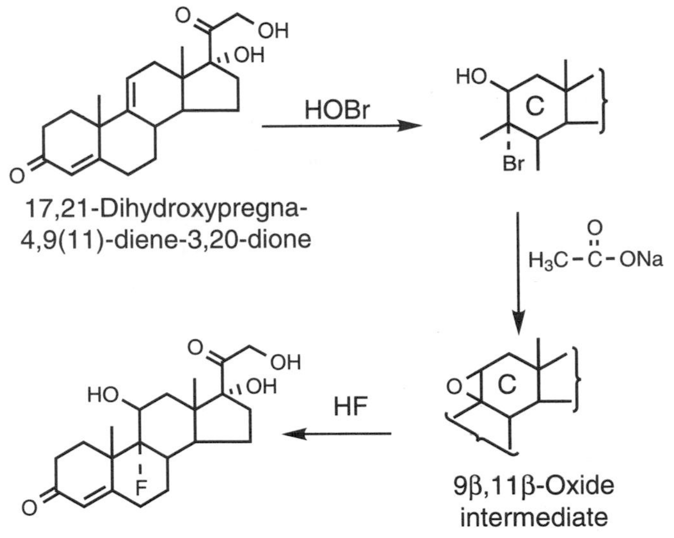

17,21-Dihydroxypregna-4,9(11)-diene-3,20-dione

HOBr

$H_3C-C(=O)-ONa$

9β,11β-Oxide intermediate

HF

Fig. 23–9. Synthesis of fluorinated adrenocorticoids.

coid activity is increased elevenfold by insertion of the 9α-fluoro substituent, mineralocorticoid activity is increased 300 to 800 times.[75,76] This property is objectionable be-

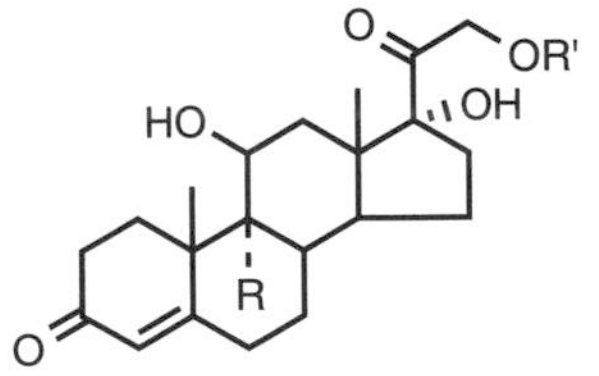

Fludrocortisone: R = F, R' = H
Fludrocortisone acetate: R = F, R' = CH_3CO-
9α-Bromo analog of fludrocortisone: R = Br, R' = H

cause it leads to edema. Consequently, fludrocortisone acetate, which has 10 times the anti-inflammatory activity of cortisone acetate, is used topically in ointments and lotions for this action. It is also available in tablet form to maintain patients with Addison's disease. This drug, introduced in 1954, helped to provide the impetus for the synthesis and biologic evaluation of newer halogen-containing analogs.

One year after the introduction of fludrocortisone, the Δ-corticoids were brought forth into clinical medicine. Investigators at Schering observed that the 1-dehydro derivatives of cortisone and hydrocortisone, namely prednisone and prednisolone, are more potent antirheumatic and anti-allergenic agents than the parent compounds and produced fewer undesirable side effects. These compounds are known as Δ-corticoids because they contain an additional double bond between positions 1 and 2 (i.e., Δ^1).

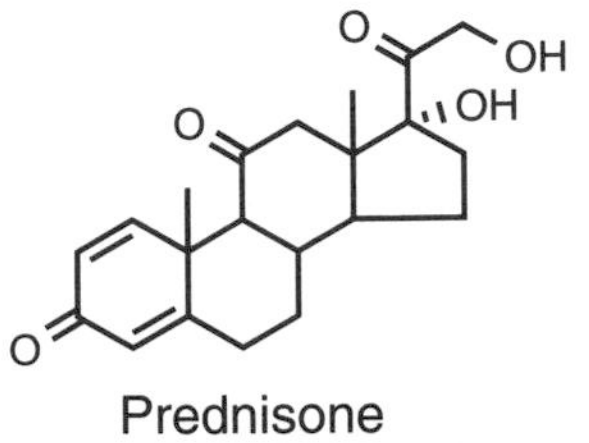

Prednisone

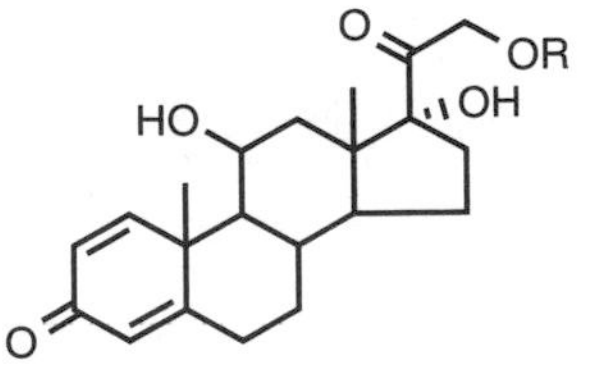

Prednisolone: R = H
Prednisolone acetate: R = CH_3CO-
Prednisolone tert-butylacetate: R = $(CH_3)_3CCH_2CO$-
Prednisolone sodium phosphate: R = Na_2O_3P-

The increased potency may reflect the change in geometry of ring A caused by the introduction of >C=C< functions. Although the remaining portions of the steroid are essentially unchanged (except for less easily visualized molecular perturbations), the conformation of ring A changes from a chair, as in 5α-pregnan-3-one, to a half-chair (pregn-4-en-3-one) and to a flattened boat (pregna-1,4-dien-3-one) upon introduction of such unsaturation.

The Δ-corticoids, which can be prepared by microbial dehydrogenation of cortisone or hydrocortisone with Corynebacterium simplex[77] and by several synthetic methods,[69] represent the first chemical innovation leading to the creation of a modified compound that could be prescribed for rheumatoid arthritis. One high-yield route involves oxidation of 5α-pregnane or 5β-pregnane precursors that have appropriate oxygen substitutions with selenium dioxide.[78,79]

Both prednisone and prednisolone were found to have adrenocortical activity (measured by eosinopenic response, liver glycogen decomposition, and thymus involution in adrenalectomized mice). In these tests, prednisone and prednisolone were found to be three or four times more potent than cortisone and hydrocortisone. Antiphlogistic strengths in human subjects were similarly augmented, but their electrolyte activities were not proportionately increased.

When prednisone or prednisolone is used in the treatment of rheumatoid arthritis, smaller doses are required than of hydrocortisone; the usual dose is 5 mg two to four times a day. For practical purposes, prednisone and prednisolone are equally potent and may be used interchangeably.

Prednisolone acetate is available in suspension (U.S.P.) and ointment forms for use externally. Just as with hydrocortisone, several other 21-esters of prednisolone are available. Prednisolone *tert*-butylacetate (3,3-dimethylbutyrate) is used in suspension form and by injection for the same reasons the 21-ester derivatives of hydrocortisone are employed. The butylacetate ester, which is suitable only for use by injection, has a long duration of action owing to low water solubility and a slow rate of hydrolysis. The drug is administered in doses of 4 to 20 mg.

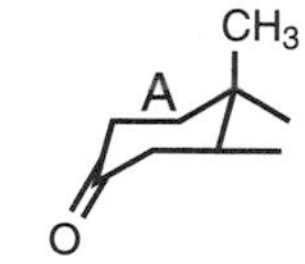

Ring A of 5α-pregnan-3-one

Ring A of pregn-4-en-3-one

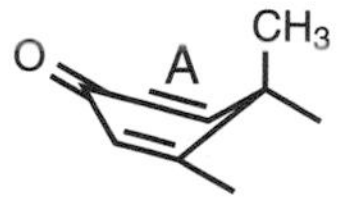

Ring A of pregna-1,4-dien-3-one

Prednisolone sodium phosphate U.S.P., in which the 21-hydroxyl group is esterified with disodium phosphate, is a water-soluble derivative having a rapid onset and a short duration of action when administered by injection (usual dose of 20 mg intravenously or intramuscularly). Topically, 1 or 2 drops of a 0.5% solution may be used four to six times daily for its anti-inflammatory action in the eye.

When doses of equivalent antirheumatic potency are given to patients not treated with steroids, the Δ-corticoids promote the same pattern of initial improvement as hydrocortisone. Statistical results of the improvement status during the first few months of therapy have been similar with prednisolone, prednisone, and hydrocortisone; the results of longer-term therapy have been significantly better with the modified compounds. Studies indicate that the Δ-corticoids may be used continuously in patients with rheumatoid arthritis without undue gastrointestinal hazard.

Methylprednisolone N.F. dosage ranges from 2 to 60 mg. Generally, 4 mg are administered four times daily in the treatment of rheumatoid arthritis. Dispensed as the 21-acetate analog, it is found in suspension (N.F.) for external use; as the sodium succinate analog, it is available as a water-soluble form for injection (N.F.)

A natural extension of corticoid research involved examination of compounds containing both a 9α-fluoro group and a double bond between positions 1 and 2. Triamcinolone U.S.P. (9-fluoro-11β,16α,17,21-tetrahydroxypregna-1,4-diene-3,20-dione) introduced in 1958, is an example of such a system. This drug combines the structural features of a Δ-corticoid and 9α-fluoro corticoid.

As mentioned previously, the 9α-fluoro group increases the anti-inflammatory potency, but it also markedly increases the mineralocorticoid potency. This is undesirable if the drug is to be used internally for the treatment of rheumatoid arthritis. By inserting a 16α-hydroxy group into the molecule, one can decrease the mineralocorticoid activity.

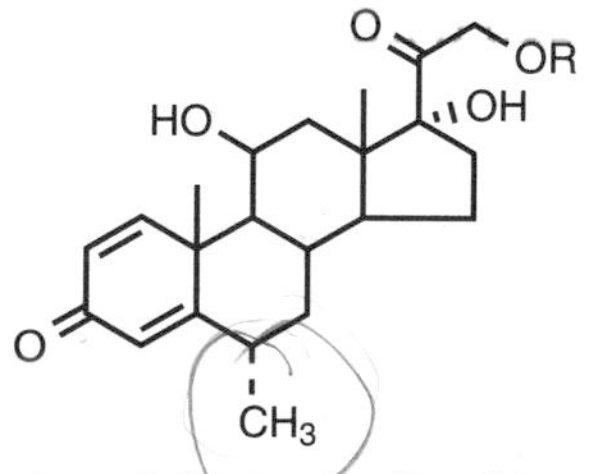

Methylprednisolone: R = H
Methylprednisolone 21- acetate: R = CH_3CO-
Prednisolone sodium succinate: R = Na^{+} $^{-}OOCCH_2CH_2CO$-

Between 1953 and 1962, many derivatives of the Δ-corticoids and the halogen-containing analogs (especially fluorinated compounds) were synthesized, and some became useful clinical agents. Studies with methylcorticoids revealed 2α-methyl derivatives to be inactive, whereas the 2α-methyl-9α-fluoro analogs had potent mineralocorticoid activity. It was thought that 6α-methyl substitution would slow metabolism, i.e., slow reduction of the A ring. These compounds potentiated glucocorticoid activity with negligible salt retention.[80]

Methylprednisolone (6α-methyl-11β,17,21-trihydroxypregna-1,4-diene-3,20-dione) was synthesized in 1956 and introduced into clinical medicine. In human subjects, the metabolic effects did not differ appreciably from those of prednisolone. Its activities with respect to nitrogen excretion, ACTH suppression, and reduction of circulating eosinophils were similar to those of prednisolone. The sodium retention and potassium loss were slightly less than with the parent compound.[81]

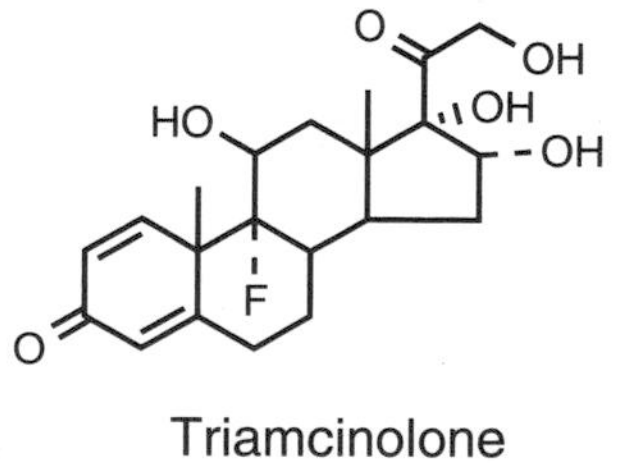

Triamcinolone

The original interest in 16α-hydroxy corticoids stemmed from their isolation from the urine of a boy with an adrenal tumor, the desire of chemists to synthesize them, and the hope that such analogs might have potent biologic activity.[82–85] In fact, 16α-hydroxy analogs of natural corticoids retain glucocorticoid activity and have a considerably reduced mineralocorticoid activity.

They may cause sodium excretion rather than sodium retention.

On a weight-for-weight basis, the antirheumatic potency of triamcinolone is greater than that of prednisolone (about 20%) and about the same as that of methylprednisolone. Initial improvement following administration of triamcinolone is similar to that noted with other compounds; reports in the literature, however, indicate that the percentage of patients maintained satisfactorily for long periods has been distinctly smaller than with prednisolone.

Even though triamcinolone has an apparent decreased tendency to cause salt and water retention and edema and may induce sodium and water diuresis, it causes other unwanted side effects, including anorexia, weight loss, muscle weakness, leg cramps, nausea, dizziness, and a general toxic feeling.[86] Intramuscular triamcinolone is reportedly effective and safe in the treatment of dermatoses, and in combination with folic acid antagonists, it is effective in the treatment of psoriasis.[87,88]

Triamcinolone is generally dispensed as the more potent acetonide, a 16α,17α-cyclic ketal or isopropylidene derivative, and is effective in the treatment of psoriasis and other dermatologic conditions.[89] The peculiar side effects of the drug, however, have occurred with sufficient frequency to discourage its routine use for rheumatoid patients requiring steroid therapy. The drug may be employed advantageously as a special-purpose steroid in instances in which salt and water retention (due to other corticoids, hypertension, or cardiac compensation) or excessive appetite and weight gain are problems in management. A newer acetonide analog, desonide, is a nonfluorinated derivative of triamcinolone acetonide. Desonide is a potent topical glucocorticoid and used for various skin disorders.[90]

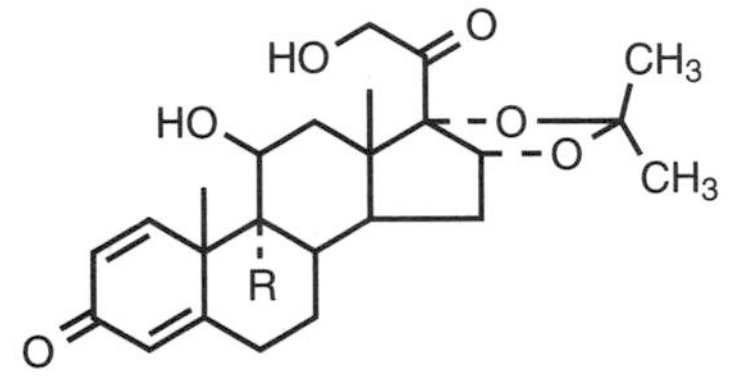

Triamcinolone acetonide: R = F
Desonide: R = H

Research with 16-methyl substituted corticoids was initiated in part because investigators hoped to stabilize the 17β-ketol side chain to metabolism in vivo. A 16α-methyl group does decrease the reactivity of the 20-keto group to carbonyl reagents and increases the stability of the drug in human plasma in vitro.[91,92] Unlike 16α-hydroxylation, a methyl group increases the anti-inflammatory activity; like the 16α-hydroxyl group, the methyl group appears to reduce markedly the salt-retaining properties of the compound.

These studies led to the development of dexamethasone (9-fluoro-16α-methyl-11β,17,21-trihydroxypregna-1,4-diene-3,20-dione), which was introduced for clinical trial.

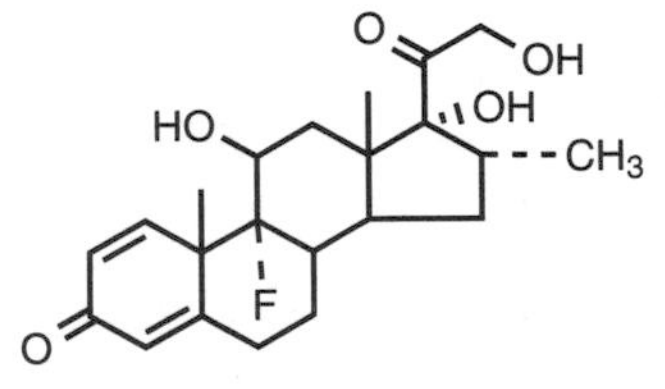

Dexamethasone

The data from these investigations were published in 1958.[93–97] The activity of dexamethasone, as measured by glycogen deposition, is 20 times greater than that of hydrocortisone. It has five times the anti-inflammatory activity of prednisolone. Clinical data indicate that this compound has seven times the antirheumatic potency of prednisolone. It is roughly 30 times more potent as hydrocortisone.

In practical management, 0.75 mg of dexamethasone promotes a therapeutic response equivalent to that from 4 mg of triamcinolone or methylprednisolone, 5 mg of prednisolone, and 20 mg of hydrocortisone. Clinical investigations with small groups of patients indicate that this compound could control patients who did not respond well to prednisolone. Over long periods, the improved status of some patients deteriorated.

In summarizing the biologic properties of this drug, it seems clear that, with doses of corresponding antirheumatic strength, this steroid has approximately the same tendency as prednisolone to produce facial mooning, acne, and nervous excitation. Peripheral edema is uncommon (7%) and mild. The more common and most objectionable side effects are excessive appetite and weight gain, abdominal bloating, and distention. The frequency and severity of these symptoms vary with the dose (1 mg maximum for females and 1.5 mg maximum for males).

The striking increase in potency does not confer a general therapeutic index on dexamethasone that is higher than that of prednisolone. Again, this drug is probably best employed as a special-purpose corticoid. It may be useful when other steroids are no longer effective or when increased appetite and weight gain are desirable.[98–104] Its efficacy may be increased when it is used in combination with cyproheptadine as an antiallergenic, antipyretic, and anti-inflammatory agent.[93,94]

Shortly after the introduction of dexamethasone, betamethasone U.S.P., which differs from dexamethasone only in configuration of the 16-methyl group, was made available for the treatment of rheumatic diseases and der-

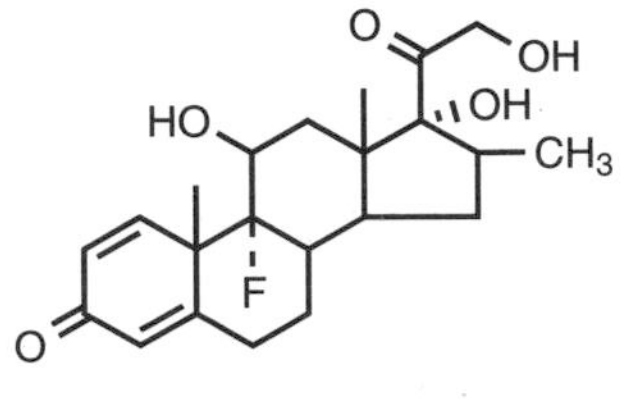

Betamethasone

matologic disorders. This analog, which contains a 16β-methyl group, has received sufficient clinical trial to indicate that it is as effective as dexamethasone or perhaps slightly more active. Although this drug has been reported to be less toxic than other steroids, some clinical investigators suggest it is best used for short-term therapy; toxic side effects, such as increased appetite, weight gain and facial mooning, occur with prolonged use. Generally, a 0.5-mg tablet of betamethasone is equivalent to a 5.0-mg tablet of prednisolone, and except for isolated instances, this drug is apparently on a par with dexamethasone.[105,106]

Paramethasone acetate N.F. (6α-fluoro-16α-methyl-11β-17,21-trihydroxypregna-1,4-diene-3,20-dione 21-acetate), synthesized in 1960, is also a modification of dexamethasone.

Dispensed as a 21-acetate ester, this drug retains the 16α-methyl group, but the 9α-fluoro substituent has been moved to the 6α position. It was thought that this manipu-

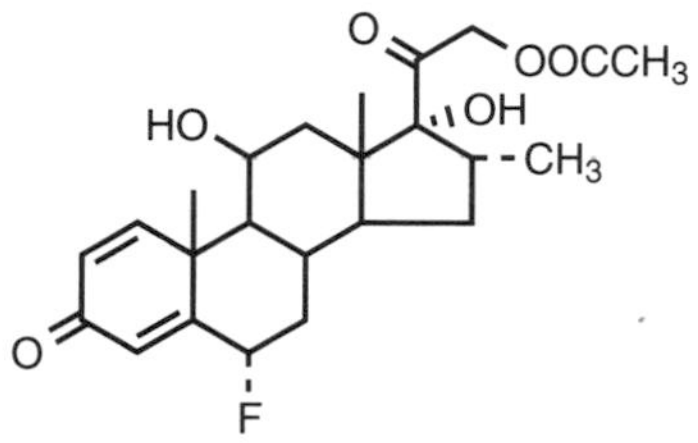

Paramethasone

lation would reduce the electrolyte loss sometimes found with dexamethasone. Although reports in the literature indicate that, in humans, paramethasone causes a slight loss of sodium and chloride with little or no loss of potassium (with doses as large as 15 mg), an analysis of continuous therapy over 9 months indicates that the therapeutic efficacy of paramethasone does not differ greatly from that of fluprednisolone.[107]

Fluprednisolone (6α-fluoro-11β,17,21-trihydroxy-pregna-1,4-diene-3,20-dione) is actually a 16-desmethyl para-

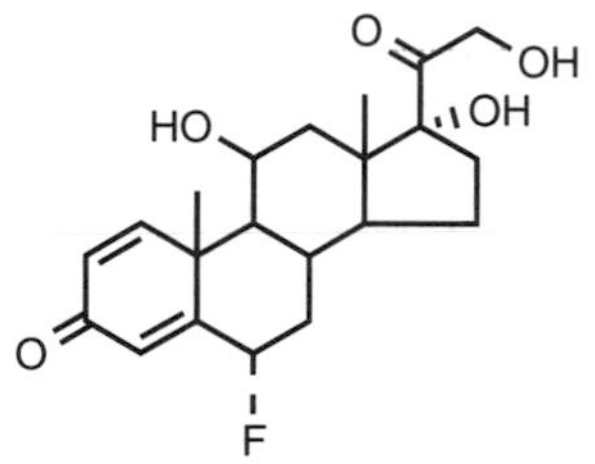

Fludprednisolone

methasone. Its activity is similar to that of paramethasone, and it has about 2½ times the antirheumatic potency of prednisolone. With doses of 2 to 7 mg, evidence of salt and water retention has not been noted. The therapeutic index, however, is probably the same or only a little greater than that for prednisolone. Because no new or strange adverse reactions have been noted during the administration of this drug, at least on a short-term basis, it seems that a 6α-fluoro group does not deleteriously affect the activity of prednisolone.[81,108]

Additional fluorinated analogs, flurandrenolone (6α-fluoro-11β,16α,17,21-tetrahydroxypregna-1,4-diene-3,20-dione), fluorometholone (6α-methyl-9-fluoro-11β-17,21-trihydroxypregna-1,4-diene-3,20-dione), and fluocinolone (6α,9-difluoro-11β,16α,17,21-tetrahydroxypregna-1,4-diene-3,20-dione), are dispensed only for topical administration.

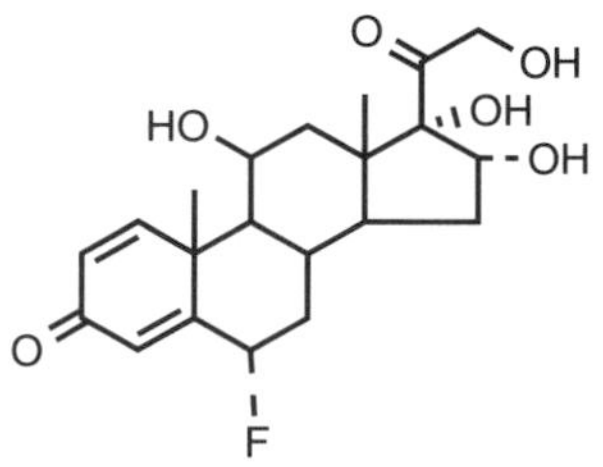

Flurandrenolone

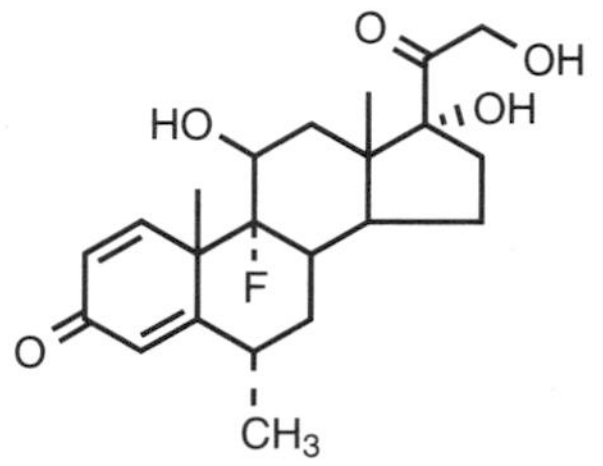

Fluorometholone

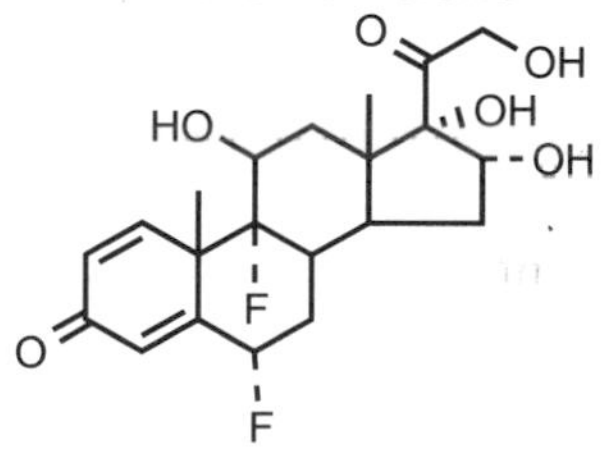

Fluocinolone

These compounds are potent anti-inflammatory agents. Flurandrenolide N.F. is the 16,17-acetonide derivative of flurandrenolone and is used in creams (N.F.) in a concentration of 0.025 to 0.05%.

Psoriasis is one of the few inflammatory dermatoses that has not responded to routine topical steroid therapy, but these more potent steroids appear to work if a special occlusive dressing is used. In this technique, a thin layer of cream or ointment containing flurandrenolone, usually as the acetonide derivative, is applied to the individual patch of psoriasis. The area is then covered with plastic food wrap or a similar pliable plastic film.[109–115]

Clinical investigations generally show flurandrenolone (0.05%) to be more effective than 1% hydrocortisone and about the same in activity as 0.1% triamcinolone. Fluorometholone is claimed to be as effective as 1% hydrocortisone in a concentration of 0.025%. Some investigators believe the greater activity to be due to an increased biologic half-life. In other words, these analogs are not metabolized as readily. It seems that fluocinolone, which may be used as the acetonide derivative (N.F.), has about the same anti-inflammatory activity as fluorometholone.[109–115]

Newer synthetic glucocorticoids have incorporated chlorine atoms onto the steroid molecule. Beclomethasone, a 9α-chloro analog of betamethasone, is a potent glucocorticoid, and the dipropionate derivative is utilized in inhalation aerosol therapy for asthma and rhinitis.[116] Two potent topical glucocorticoids that contain chlorine atoms are alclometasone, with a 7α-chloro group,[117] and halometasone, which contains a 2-chloro functionality.[118]

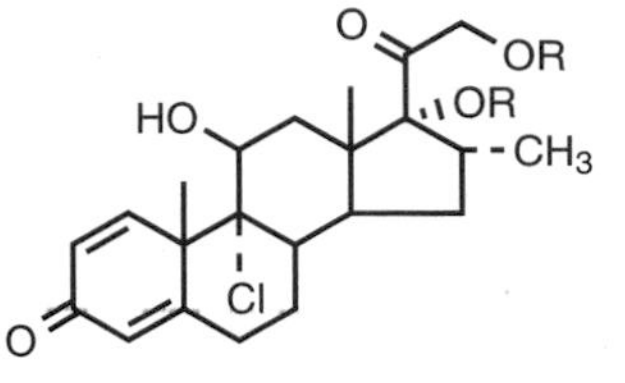

Beclomethasone: R = H
Beclomethasone dipropionate: R = C_2H_5CO-

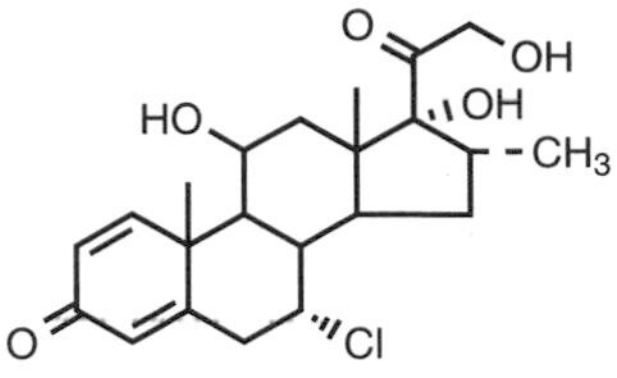

Alclomestanone

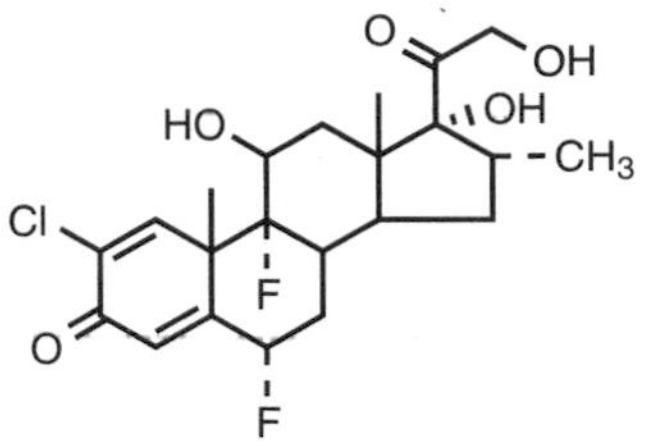

Halometasone

Substitution of a chlorine atom for the 21-hydroxyl group on the glucocorticoids provides topical anti-inflammatory agents with similar pharmacologic activities.[119] Clobetasol and halcinonide are two examples of the 21-chlorocorticoids.[120–122]

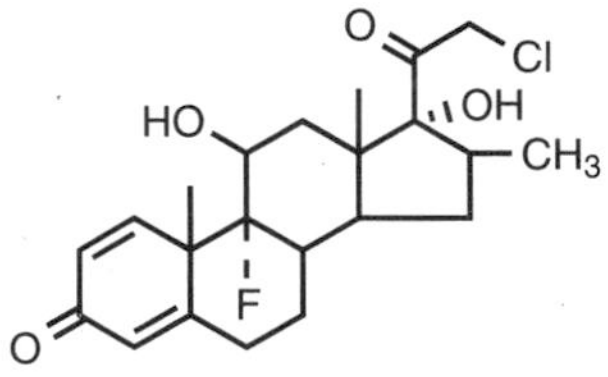

Clobetasol

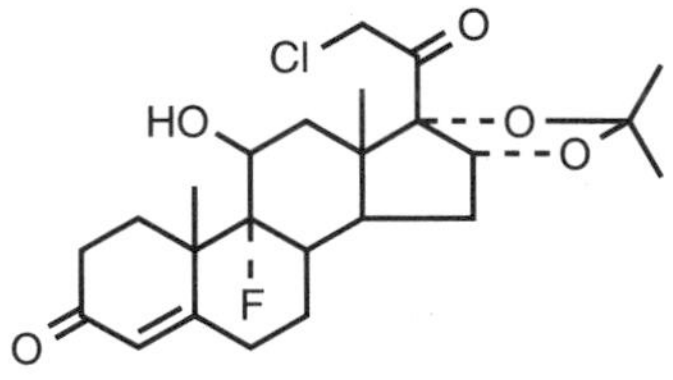

Halcinonide

Modes of Action

Corticosteroids influence all tissues of the body and produce numerous and varying effects in cells. These steroids regulate carbohydrate, lipid, and protein biosynthesis and metabolism (glucocorticoid effects), and they influence water and electrolyte balance (mineralocorticoid effects). Cortisol is the most potent glucocorticoid secreted by the adrenal gland, and aldosterone is the most potent endogenous mineralocorticoid. Both naturally occurring glucocorticoids and related semisynthetic analogs can be evaluated in terms of their ability to sustain life,[123,124] to stimulate an increase in blood glucose concentrations and a deposition of liver glycogen,[125] to decrease circulating eosinophils,[126] and to cause thymus involution in adrenalectomized animals.[127,128] In addition, corticosteroids can affect immune system functions, inflammatory responses, and cell growth.

The biochemical mechanism of corticosteroid action is the regulation of gene expression and subsequent induction of protein biosynthesis via specific, high affinity corticosteroid receptors, as described in general terms earlier in the chapter. Glucocorticoid action is mediated through the glucocortocoid receptor, which is found primarily in the cytosol of the cell when not bound to glucocorticoids. The glucocorticoid receptor is stabilized in the cytosol by complexation with phosphorylated proteins, including a 90 Kd protein referred to as a heat-shock protein (hsp90).[129] The steroid molecule binds to the glucocorticoid receptor, resulting in a conformational change of the receptor to dissociate the other proteins and initiate translocation of the steroid-receptor complex into the nucleus. The steroid-nuclear glucocorticoid receptor complex interacts with particular HRE regions of the cellular DNA, referred to as glucocorticoid responsive elements (GREs), and initiates transcription of the DNA sequence to produce mRNA. Finally, the elevated levels of mRNA lead to an increase in protein synthesis in the endoplasmic reticulum; these proteins then mediate glucocorticoid effects on carbohydrate, lipid and protein metabolism. Some of the specific proteins induced by glucocorticoids have been identified and are discussed later in this section. Mineralocorticoid effects are observed in several tissues and specific mineralocorticoid receptors have been characterized that mediate mineralocorticoid functions.[130]

The primary physiologic function of glucocorticoids is to maintain blood glucose levels and thus ensure glucose-dependent processes critical to life, particularly brain functions. Cortisol and related steroids accomplish this by stimulating the formation of glucose, by diminishing glucose use by peripheral tissues, and by promoting glycogen synthesis in the liver in order to increase carbohydrate stores for later release of glucose. For glucose formation, glucocorticoids mobilize amino acids and promote amino acid metabolism and gluconeogenesis. These steroids, acting via the glucocorticoid receptor mechanism, induce the production of a variety of enzymes important for glucose formation. The synthesis of tyrosine aminotransferase increases within 30 minutes of glucocorticoid exposure.[131–134] This enzyme promotes the transfer of amino groups from tyrosine to α-ketoglutarate to form glutamate and hydroxyphenylpyruvate. Another amino-acid metabolizing enzyme induced rapidly by glucocorticoids is tryptophan oxidase.[135] This enzyme oxidizes tryptophan to formylkynurenine, which is subsequently converted to alanine. Alanine transaminase is also induced by glucocorticoids.[136,137] Alanine, and to a lesser extent glutamate, is important for gluconeogenesis in the liver.

Several other enzymes important in gluconeogenesis and glycogen formation are elevated for several hours following glucocorticoid administration, including glycogen synthetase, pyruvate kinase, phosphoenol pyruvate carboxykinase, and glucose-6-phosphate kinase.[138–141] The delayed increases in these enzymes suggest their biosyntheses are not regulated directly by glucocorticoids.[140] In peripheral tissues, glucocorticoid-induced inhibition of phosphofructokinase is observed.[138] This enzyme catalyzes the formation of D-fructose-1,6-diphosphate from D-fructose-6-phosphate during glycolysis. Inhibition of this enzyme decreases glucose utilization by peripheral tissues and results in maintaining blood glucose levels. Reviews of the multiple effects of glucocorticoids on carbohydrate metabolism have been published.[140,141]

Additional effects of glucocorticoids in the body are preventing or minimizing inflammatory reactions and suppressing immune responses. These steroids interfere with both early events in inflammation (such as release of mediators, edema, and cellular infiltration) and later stages (capillary infiltration and collagen formation). Only a few of the mechanisms involved in glucocorticoid suppression of inflammation are known. Cortisol will induce the production of lipocortin and related proteins by increasing gene expression through the glucocorticoid receptor mechanism.[142] Lipocortin inhibits the activity of phospholipase A_2, which liberates arachidonic acid and leads to the biosynthesis of eicosanoids (such as prostaglandins and leukotrienes). Lipocortin also mediates the decreased production and release of platelet activating factor.[143] Glucocorticoids can also suppress the expression of interleukin-1 (IL-1) and tumor necrosis factor (TNF).[144–146] These eicosanoids and peptide factors are important as mediators in the inflammatory response. Some of these factors also play important roles in cellular infiltration and capillary permeability in the inflamed region. Suppression of the immune responses are also mediated by inhibition of the synthesis and release of important mediators. In macrophages, glucocorticoids inhibit IL-1 synthesis and thus interfere with proliferation of B-lymphocytes, important for antibody production.[145] IL-1 is also important for activation of resting T-lymphocytes, important for cell-mediated immunity. The activated T-cells produce interleukin-2 (IL-2); its biosynthesis is also reduced by glucocorticoids.[147]

The primary physiologic function of mineralocorticoids is to maintain electrolyte balances in the body by enhancing Na^+ reabsorption and increasing K^+ and H^+ secretion in the kidney. Similar effects on cation transport are observed in a variety of secretory tissues, including the salivary glands, sweat glands, and mucosal tissues of the gastrointestinal tract and the bladder. Aldosterone is the most potent endogenous mineralocorticoid. Deoxycorticosterone is approximately 20 times less potent than aldosterone; cortisol exhibits weak mineralocorticoid activity in vivo due to rapid metabolism of cortisol by 11β-hydroxysteroid dehydrogenase. The mechanism of action of aldosterone involves binding of the steroid to the aldosterone receptor and initiation of gene transcription, mRNA biosynthesis, and protein production. A protein referred to as *aldosterone-induced protein* (AIP) is produced through this mechanism and is thought to aid in Na^+ retention. One possible mode of action of AIP is to act as a *permease* to increase the permeability of the cell membrane to Na^+.[148] This results in an accelerated rate of Na^+ influx and elevated activity of Na^+,K^+-ATPase to pump Na^+ into extracellular space.[149]

Pharmacology, Side Effects, and Clinical Applications

In addition to their natural hormonal actions, the adrenocorticoids have many clinical uses. Glucocorticoids and mineralocorticoids may be used for the treatment of adrenal insufficiency (hypoadrenalism), which results from failure of the adrenal glands to synthesize adequate amounts of the hormones. Adrenocorticoids are also used to maintain patients who have had partial or complete removal of their adrenal glands or adenohypophysis (adrenalectomy and hypophysectomy, respectively).

Two major uses of glucocorticoids are in the treatment of rheumatoid diseases and allergic manifestations. Their use in the treatment of severe asthma is well documented.[150] They are effective in the treatment of rheumatoid arthritis, acute rheumatic fever, bursitis, spontaneous hypoglycemia in children, gout, rheumatoid carditis, sprue, allergy, including contact dermatitis, and other conditions. The treatment of chronic rheumatic diseases and allergic conditions with glucocorticoids is symptomatic and continuous; symptoms return after withdrawal of the drug.

In addition, these drugs are moderately effective in the treatment of ulcerative colitis, dermatomyositis, periarteritis nodosa, idiopathic pulmonary fibrosis, idiopathic thrombocytopenic purpura, regional ileitis, acquired hemolytic anemia, nephrosis, cirrhotic ascites, neurodermatitis, and temporal arteritis. Some of the newer potent analogs, such as flurandrenolone, fluometholone, and fluocinolone, are effective topically in the treatment of psoriasis. Glucocorticoids may be combined with antibiotics to treat pneumonia, peritonitis, typhoid fever, and meningococcemia.[151]

When dosages with equivalent antirheumatic potency are given to patients not treated with steroids, the Δ-corticoids (prednisone and prednisolone) promote the same pattern of initial improvement as hydrocortisone. Statistical results of improvement during the first few months of therapy have been similar with prednisone, prednisolone, and hydrocortisone; the results of longer term therapy have been significantly better with the modified compounds.

Satisfactory rheumatic control, lost after prolonged cortisone or hydrocortisone therapy, may be regained in an appreciable number of patients by changing to prednisone, prednisolone, or other modified drugs. Of patients whose conditions deteriorate below adequate levels during hydrocortisone administration, nearly half reach their previous level of improvement after Δ-corticoids (in doses slightly larger in terms of anti-rheumatic strength) are used. With further prolongation of steroid therapy, improvement again wanes in some patients; in others such management remains successful for longer than 2 years. In some instances, the improvement is attributed to increased effectiveness of the drug due to correction of salt and water retention; in other instances, there is no adequate explanation.

When these drugs are administered in doses that have similar antirheumatic strengths, the general incidence of adverse reactions with prednisone and prednisolone is about the same as with hydrocortisone. The compounds differ, however, in their tendencies to induce individual side effects. The incidence and degree of salt and water retention and blood pressure elevation are less with the Δ-corticoids. Conversely, these analogs are more likely to promote digestive complaints, peptic ulcer, vasomotor symptoms, and cutaneous ecchymosis.

Although these analogs have unwanted side effects, most clinical investigators prefer the Δ-corticoids to cortisone and hydrocortisone for rheumatoid patients who require steroid therapy. The reasons are that these drugs have less tendency to cause salt and water retention and potassium loss, and that they restore improvement in a significant percentage of patients whose therapeutic control has been lost during cortisone and hydrocortisone therapy.

It seems desirable to administer prednisone and prednisolone in conjunction with nonabsorbable antacids; this affords improvement of long-term therapy. It appears that the therapeutic indices of these two analogs, especially when used in conjunction with nonabsorbable antacids, are higher than for the naturally occurring glucocorticoids.

Although side effects and toxicities vary with the drug and sometimes with the patient, facial mooning, flushing, sweating, acne, thinning of the scalp hair, abdominal distention, and weight gain are observed with most glucocorticoids. Protein depletion (with osteoporosis and spontaneous fractures), myopathy (with weakness of muscles of the thighs, pelvis, and lower back), and aseptic necrosis of the hip and humerus are other side effects. These drugs may cause psychic disturbances, headache, vertigo, and peptic ulcer, and may suppress growth in children.

Patients with well-controlled diabetes must be closely watched and their insulin dosage increased if glycosuria or hyperglycemia ensues during or following glucocorticoid administration. Patients should also be watched for signs of adrenocorticoid insufficiency after discontinuation of glucocorticoid therapy, and individuals with a history of tuberculosis should receive prophylactic doses of antituberculosis drugs.

Most important, glucocorticoids should not be withdrawn abruptly in cases of acute infections or severe stress, such as surgery or trauma. Myasthenia gravis, peptic ulcer, diabetes mellitus, hyperthyroidism, hypertension, psychic disturbances, pregnancy (first trimester), and infections may be aggravated by glucocorticoid administration; hormone therapy is contraindicated in these conditions and should be used only with the utmost precaution.

Semisynthetic analogs exhibiting high mineralocorticoid activity are not employed in the treatment of rheumatic disorders because of toxic side effects resulting from a disturbance of electrolyte and water balance. Some newer synthetic steroids (Table 23–1) are relatively free of sodium-retaining activity; they may show other toxic manifestations, however, and eventually have to be withdrawn.

Some modified compounds have been recommended for use when other analogs are no longer effective or when it is desirable to promote increased appetite and weight gain. Triamcinolone may be used advantageously when salt and water retention (from other glucocorticoids, hypertension, or cardiac compensation) or excessive appetite and weight gain are problems in management.

Adrenal insufficiency may result from adrenal atrophy owing to prolonged use of large doses of these drugs. Glucocorticoids inhibit ACTH production by the adenohypophysis and this, in turn, reduces endogenous glucocorticoid production. With time, atrophy of the adrenal glands takes place.

Glucocorticoids are sometimes used in the treatment of scleroderma, discoid lupus, acute nephritis, osteoarthritis, acute hepatitis, hepatic coma, Hodgkin's disease, multiple myeloma, lymphoid tumors, acute leukemia, metastatic carcinoma of the breast, and chronic lymphatic leukemia.[151] Glucocorticoids may be more or less effective in these diseases depending on the clinical condition.

One factor must not be overlooked when applying potent anti-inflammatory agents with high mineralocorticoid activity to the skin. Consideration must be given to percutaneous absorption. Sodium retention and edema occur in patients with dermatitis who apply as much as 75 mg of fludrocortisone acetate (i.e., 30 ml of a 0.25% lotion) to the skin in 24 hours. The relative rate of percutaneous absorption, administered as a cream in rats, was triamcinolone acetate $\geqslant$ hydrocortisone $>$ dexamethasone, but dexamethasone was deposited in skin longer than the other two drugs; hydrocortisone disappeared most rapidly.[152]

An important factor in sodium retention following topical application is the concentration of drug. It is estimated that in three applications daily a patient applies approximately 0.5 ml of lotion to each 1% of skin surface. In a patient with 40% of the skin involved in a dermatitis, 20 ml of the lotion would be used in 24 hours. With 0.25% fludrocortisone acetate lotion, this would amount to application of 50 mg of drug.

Clinical edema has resulted from ingestion of as little as 2 mg of fludrocortisone acetate; if as little as 1% of the applied drug were absorbed percutaneously, sodium retention and clinical edema could result. Furthermore, on skin involved in acute or subacute dermatitis, it is likely that much more than 1% would be absorbed percutaneously. Application to the vulva, scrotum, and perianal areas also results in greater absorption because of the moisture in these areas.

Table 23–1. Characteristics of Adrenocorticoids*

	Activity Relative to Hydrocortisone		Equivalent Dose in Milligrams† (Glucocorticoid Action)	Plasma Biologic Half-life (min)	
Adrenocorticoid	**Mineralocorticoid Activity**	**Glucocorticoid Activity**		**Human**	**Dog**
Hydrocortisone	1.0	1.0	20	101.6	57
Cortisone	0.8	0.8	25		
Prednisone	0.6	3.5	5		
Prednisolone	0.6	4.0	5	200	69
Methylprednisolone	0	5.0	4		91
Triamcinolone	0	5.0	4	300	116.7
Paramethasone	0	10.0	2		
Dexamethasone	0	30.0	0.75	200	
Betamethasone	0	35.0	0.6		
Fludrocortisone	800	10.0	Not employed		
Fluprednisolone	0	10.0	1.6		
Aldosterone	800	0.2	Not employed		
11-Desoxycorticosterone	40	0	Not employed		
Corticosterone	5	0.5	Not employed		

* Abstracted from Sciuchetti, L. A.: *Pharmindex, 5,* 7 (1963).
† Based on the oral dose of an anti-inflammatory agent in rheumatoid arthritis.

It has been suggested that patients using a lotion or ointment containing these drugs be instructed to apply them sparingly and to spread them lightly over the affected areas. The extent and frequency of applications should be carefully considered. Apparently, a lotion vehicle is more effective when treating a dermatitis, but a greater degree of percutaneous absorption occurs than when ointments are used.[153,154]

Structure-Activity Relationships

The following structure depicts the absolute configuration of hydrocortisone. The all *trans* backbone is necessary for activity.

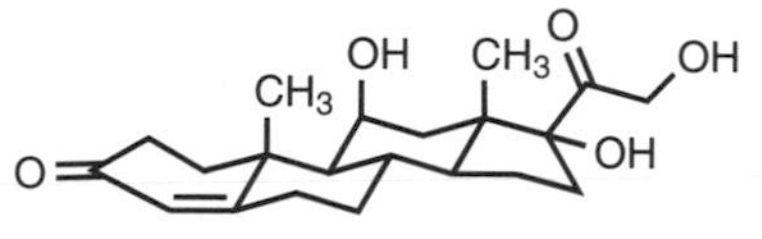

Hydrocortisone

As previously pointed out, the adrenocorticoids are generally classified as either glucocorticoids, which affect intermediary metabolism and are associated with inhibition of the inflammatory process, or mineralocorticoids. In fact, most naturally occurring and semisynthetic analogs exhibit both of these actions. The 17β-ketol (—$COCH_2OH$) side chain and the Δ^4-3-ketone functions are found in clinically used adrenocorticoids, and these groups do contribute to the potency of the agents. Modifications of these groups may result in derivatives that retain biologic activity. For example, replacement of the 21-OH group with fluorine increases glucocorticoid and sodium-retaining activities, whereas, substitution with chlorine or bromine abolishes activity. Some compounds that do not contain the Δ^4-3-ketone system have appreciable activity. It has been suggested that this group makes only a minor contribution to the specificity of action of these drugs or to the steroid-receptor association constant.[155]

If the concept of a hormone-receptor interaction is accepted, it seems that, on the basis of structure-activity studies, the C and D rings, involving positions 11, 12, 13, 16, 17, 18, 20, and 21, are more important for receptor binding than are the A and B rings.[155] Generally, insertion of bulky substituents on the β-side of the molecule abolishes glycogenic activity, while insertion on the α-side does not. It has been suggested that association of these steroids with receptors involves β-surfaces of rings C and D and the 17β-ketol side chain.[155] It is possible, however, that association with the α-surface of rings A, C, and D, as well as with the ketol side chain, is essential for sodium-retaining activity. Many functional groups, such as 17α-OH, 17α-CH_3, 16β-CH_3 and 16α-CH_3, 16α-CH_3O and 16α-OH substituents, abolish or reverse this activity in 11-desoxycorticosterone and 11-oxygenated steroids. Discussions of exceptions of these generalities are found in the literature.[155]

Although some steroids retain sodium, many have glucocorticoid and either sodium-retaining or sodium-excreting action. Difficulties in correlating the structures of adrenocorticoids with biologic action are compounded because of differences in assay methods, species variation, and the mode of drug administration. For example, whereas liver glycogen and anti-inflammatory assays in the rat correlate well, some drugs show high anti-inflammatory action in the rat but little or no antirheumatic activity in humans.[156] The 9α-F analog, fludrocortisone

Table 23–2. Biologic Potencies of Modified Adrenocorticoids in the Rat and Humans*

	Potency Relative to Cortisol				
	Rat		Humans		
Adrenocorticoid	Thymus Involution	Liver Glycogen Deposition	Eosinopenic Potency	Hyperglycemic Potency	Antirheumatic Potency
Corticosterone	—	0.8	0.06	0.06	<0.1
Prednisone	—	3	—	4	4
Prednisolone	2	3.9	4	4	4
Methylprednisolone	10	11	5	5	5
Triamcinolone	4	47	5	5	5
Paramethasone	—	150	12	12	11
Dexamethasone	56	265	28	28	29
Fludrocortisone acetate	6	9	8	8	10
Fluprednisolone	6	81	9	9	10
Triamcinolone acetonide	33	242	3	3	3
Flurandrenolone	4	—	1	—	2
Fluorometholone	25	115	10	10	—
Fluocinolone	19	112	5	6	9

* Data from I. Ringler, in *Methods of Hormone Research, Vol. 3, Part A,* R. I. Dorfman, New York, Academic Press, 1964.

acetate, is more active than the 9α-Cl analog in terms of sodium retention in the dog; the reverse is true in the rat. While 16α-methylation and 16β-methylation enhance glucocorticoid activity, anti-inflammatory action is increased disproportionately to glycogenic action in both series.

In humans, eosinopenic and hyperglycemic potencies are essentially the same; there is a close correlation in efficacy ratios derived from these tests and antirheumatic potency (Table 23–2). Because the eosinopenic-hyperglycemic activity and antirheumatic potency show excellent agreement, it has been suggested that these assays afford advantages in the preliminary estimation of anti-inflammatory potency; the methods are simple and require minimal amounts of steroid.[156]

Structure-activity studies of glucocorticoids have mainly been carried out in animals and are not necessarily applicable to clinical efficacy in man. Relative activity and dose correlations for the clinically useful drugs are found in Table 23–1.

Several other compounds have been studied in animals; from these several structure-activity relationships have been derived. For example, insertion of a double bond between positions 1 and 2 in hydrocortisone increases glucocorticoid activity. Such analogs have a much longer half-life in the blood than hydrocortisone; ring A is much more slowly metabolized.[156] If, however, a double bond is inserted between positions 9 and 11 (no oxygen function at 11), a decrease in glucocorticoid activity is observed. Except for cortisone, which results in an analog with decreased glucocorticoid activity when a double bond is inserted between position 6 and 7, such modification of other glucocorticoids generally produces no change in activity.[156]

Insertion of α-CH_3 groups at positions 2 (in 11β-OH analogs), 6, and 16 increases glucocorticoid activity in animals. Again, insertion of a 2α-CH_3 group into the glucocorticoid almost completely prevents reduction of the Δ^4-3-ketone system in vivo and in vitro. Substitution at positions 4α, 7α, 9α, 11α, and 21 decreases activity.[156]

Although some analogs, such as 16α, 17α-isopropylidinedioxy-6α-methylpregna-1,4-diene-3,20-dione and the 1,2-dihydro derivative, are 11-desoxysteroids and are biologically active, the 11β-OH group of hydrocortisone does seem to be involved in the drug-receptor interaction.[155] It has been proposed that cortisone, which contains an 11-keto function, must be reduced in vivo to hydrocortisone. The drug 2α-methylhydrocortisone exhibits high glucocorticoid activity; this is probably because of steric hindrance to reduction (i.e., C=O → C-β-OH) by the methyl group, thus rendering the analog inactive.[157–159] Insertion of α-OH groups into most other positions (1, 6, 7, 9, 14, and 16) or reduction of the 20-ketone, however, decreases glucocorticoid activity.

The 9α-F group increases glucocorticoid activity and nearly prevents metabolic oxidation of the 11β-OH group to a ketone.[155] Metabolism of such steroids is mainly restricted to the Δ^4-3-ketone and 17β-ketol side chains. The 9α-F group may increase activity by an inductive effect, which increases the acidic dissociation constant of the 11β-OH group and thereby increases the ability of the drug to hydrogen bond to biologic receptors.

A 6α-F group also increases glucocorticoid activity, but it has less effect than the 9α-F function on sodium retention. Insertion of 2α, 11α (no OH group at 11), or 21-F groups decreases glucocorticoid activity. Of particular interest is a 12α-F group. When this function is inserted into corticosterone, which has no 17α-OH group, it potentiates activity to the same extent as a 9α-F group. Insertion of a 12α-F group into a 16α,17α-dihydroxy steroid, however, renders the compound inactive; a 9α-F group potentiates activity in such analogs.

It has been proposed that hydrogen bonding between the 12α-F and 17α-OH groups renders the analog inac-

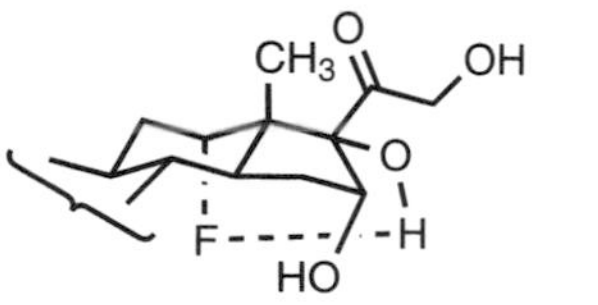

16α,17α-Dihydroxy steroid

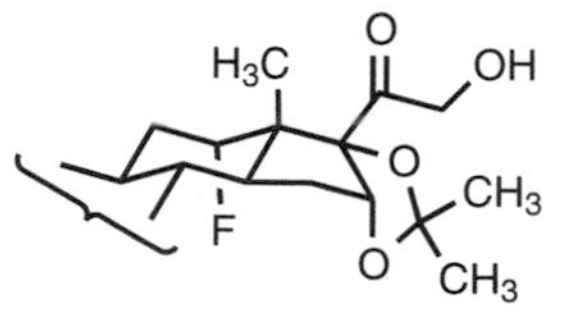

16α,17α-Isopropylidenedioxy steroid

tive; conversion to the 16α,17α-isopropylidinedioxy derivative, which cannot hydrogen bond, restores biologic activity.[160]

The mineralocorticoid activity of adrenocorticoids is another action of major significance. Many toxic side effects, making it necessary to withdraw steroid therapy in rheumatoid patients, are a result of this action. Highly active naturally occurring mineralocorticoids have no OH function in positions 11 and 17. In fact, OH groups in any position reduce the sodium-retaining activity of the adrenocorticoid.

Generally, 9α-F, 9α-Cl, and 9α-Br substitution causes increased retention of urinary sodium with an order of activity in which F > Cl > Br, but species differences do exist. For these reasons, such compounds are not used internally in the treatment of diseases such as rheumatoid arthritis. Insertion of a 16α-OH group into the molecule affects the sodium retention activity so markedly that it not only negates the effect of the 9α-F atom, but also causes sodium excretion.

A double bond between positions 1 and 2 (Δ-corticoids) also reduces the sodium retention activity of the parent drug. This functional group, however, contributes to the parent drug only about one-fifth the sodium-excreting tendency of a 16α-OH group.[161]

12α-F and 2α-CH_3 substitution contributes as much to sodium retention as does a 9α-Cl. A 21-OH group, found in all these drugs, contributes to this action to the same degree; because 21-OH groups also contribute to glucocorticoid activity, it is easy to understand why it is difficult to develop compounds with only one major action.

A 2α-CH_3 group is about three times and a 21-F substituent two times as effective as unsaturation between positions 1 and 2 in reducing sodium retention. Other substituents reported to inhibit sodium retention include 16α-CH_3 and 16β-CH_3, 16α-CH_3O, and 6α-Cl functions. A 17α-OH group, present in naturally occurring and semisynthetic analogs, reduces sodium retention to about the same extent as does unsaturation between positions 1 and 2.

Adrenocorticoid Antagonists

Antagonists of adrenocorticoids include agents that compete for binding to steroid receptors (antiglucocorticoids or antimineralocorticoids) and inhibitors of adrenosteroid biosynthesis. The action of adrenal steroids can be blocked by antagonists that compete with the endogenous steroids for binding sites on their respective cytosolic receptor proteins. The antagonist-receptor complexes are unable to stimulate the production of new mRNA and protein in the target tissues and are thus unable to elicit the biologic responses of the hormone agonist. Spironolactone (SC 9420) and related analogs bind to the aldosterone receptor in the kidney and result in the diuretic response of increased Na^+ excretion and K^+ retention. The 3-keto-4-ene A-ring is essential for this antagonistic activity and the opening of the lactone ring dramatically reduces activity. The 7α-substituent increases both intrinsic activity and oral activity.[162,163] Progesterone also has shown antimineralocorticoid activity at 10^{-4} molar concentrations.

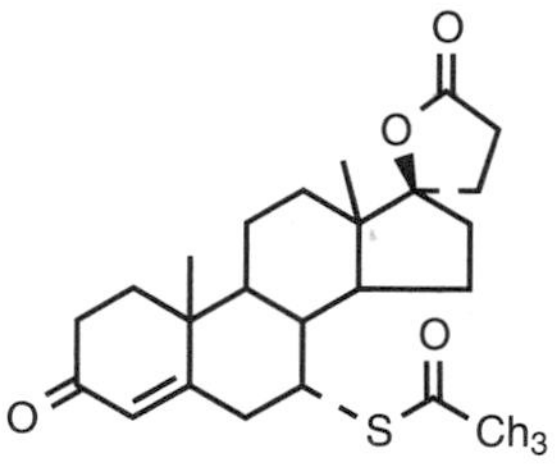

Spironolactone

Receptor antagonists of glucocorticoids have been described; they are derivatives of 19-nortestosterone.[164,165] Mifepristone (RU 486) was originally developed as an antiprogestin and is described in more detail later in the chapter. Mifepristone also exhibits very effective antagonism of glucocorticoids.

Several inhibitors of adrenocorticoid biosynthesis have been described, with the majority nonsteroidal agents inhibiting one or more of the cytochrome P450 enzyme complexes involved in adrenosteroid biosynthesis. Metyrapone reduces cortisol biosynthesis by primarily inhibiting mitochondrial 11β-hydroxylase; it also inhibits to a lesser degree 18-hydroxylase and side chain cleavage. This agent is used to test pituitary-adrenal function and the ability of the pituitary to secrete ACTH.[166] Aminoglutethimide inhibits side chain cleavage[167] and has been utilized as a *medical adrenolectomy*. Several azole antifungal drugs inhibit adrenocorticoid biosynthesis; ketoconazole is one example. It inhibits fungal sterol biosynthesis at low concentrations; however, at higher doses, ketoconazole inhibits several cytochrome P450 enzymes in adrenosteroid biosynthesis.[168] Trilostane (4α,5α-epoxy-17β-hydroxy-3-oxoandrostan-2α-carbonitrile) is a steroidal inhibitor of 3β-hydroxysteroid dehydrogenase[169] and has been used to treat Cushing's syndrome.

FEMALE SEX HORMONES

Reproductive Cycle

Among the key events in the female reproductive process is ovulation, which is regulated by the endocrine and

Metyrapone

Ketoconazole

Aminoglutethimide

Trilostane

central nervous systems.[170,171] The female reproductive cycle is controlled through an integrated system involving the hypothalamus, pituitary gland, ovary, and reproductive tract (Fig. 23–10). The hypothalamus exerts its action on the pituitary gland through a decapeptide called luteinizing hormone-releasing hormone (LHRH), also referred to as gonadotropin-releasing hormone (GnRH), which is released by the hypothalamus and initiates the release of gonadotropins, follicle stimulating hormone (FSH) and luteinizing hormone (LH).

The two main gonadotropins, FSH and LH, regulate the ovary and its production of sex hormones. As the name implies, FSH promotes the initial development of the immature graafian follicle in the ovary. This hormone cannot induce ovulation but must work in conjunction with LH. The combined effect is to promote follicle growth and increase secretion of estrogens. Through a negative feedback system the estrogens inhibit production of FSH and stimulate output of LH. The level of LH rises to a sharp peak at midpoint in the menstrual cycle and acts on the mature follicle to bring about ovulation; LH levels are low during the menses. In contrast, FSH reaches its high level during menses, falls to a low level during and after ovulation, and then increases again toward the onset of menses.

Once ovulation has taken place, LH induces luteinization of the ruptured follicle, which leads to corpus luteum formation. After luteinization has been initiated, there is an increase in progesterone from the developing corpus luteum, which in turn suppresses production of LH. Once the corpus luteum is complete, it begins to degenerate toward menses, and the levels of progesterone and estrogen decline. The major events are summarized in Figure 23–11.

The endometrium, a component of the genital tract, passes through different phases, which depend on the steroid hormones secreted by the ovary. During the development of the follicle, which takes approximately 10 days and is referred to as the follicular phase, the endometrium undergoes proliferation owing to estrogenic stimulation. The luteal phase follows ovulation, lasts about 14 days, and ends at menses. During the luteal phase, the endometrium shows secretory activity, and cell proliferation declines.

In the absence of pregnancy, the levels of estrogen and progesterone decline; this leads to sloughing of the endo-

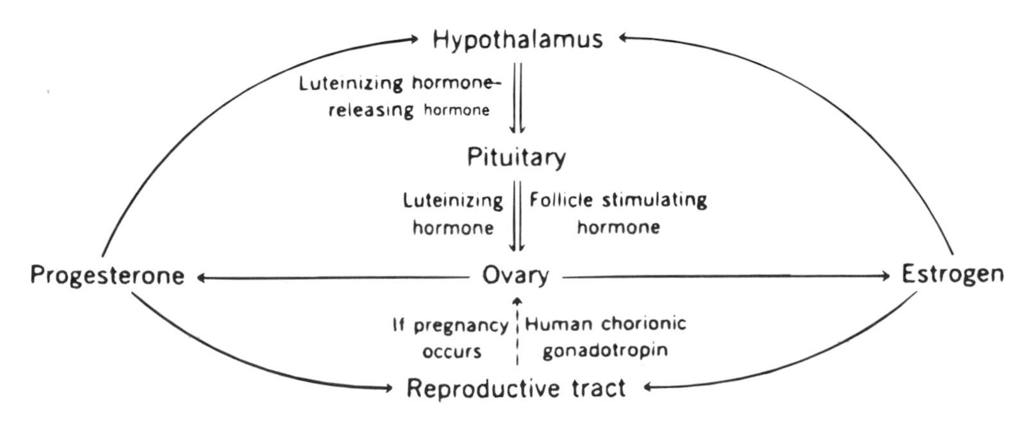

Fig. 23–10. Hypothalamic, ovarian, and reproductive tract interrelationships.

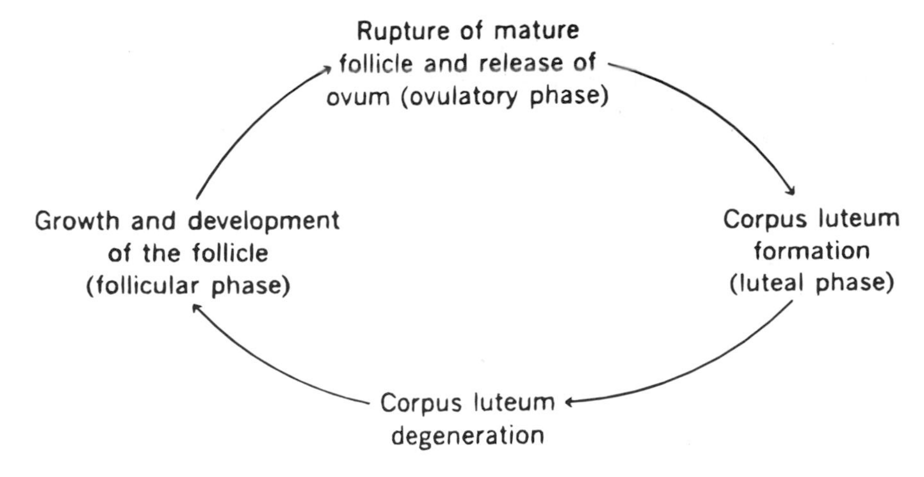

Fig. 23–11. Ovarian cycle.

metrium. This, together with the flow of interstitial blood through the vagina, is called menses and lasts for 4 to 6 days. Because the estrogen and progesterone levels are now low, the hypothalamus releases more LHRH, and the cycle begins again. The reproductive cycle in the female extends from the onset of menses to the next period of menses, with a regular interval varying from 20 to 35 days; the average length is 28 days.

If pregnancy occurs, the menstrual cycle is interrupted because of the release of a fourth gonadotropin. In human pregnancy, the gonadotropin produced by the placenta is referred to as human chorionic gonadotropin (HCG). HCG maintains and prolongs the life of the corpus luteum. The HCG level in the urine rises to the point where it can be detected after 14 days and reaches a maximum around the seventh week of pregnancy. After this peak, the HCG concentration falls to a constant level, which is maintained throughout pregnancy.

The corpus luteum, because of HCG stimulation, provides an adequate level of the steroidal hormones to maintain pregnancy during the first nine weeks. After this period, the placenta can secrete the required level of estrogen and progestational hormones to maintain pregnancy. The levels of estrogen and progesterone increase during pregnancy and finally reach their maximal concentrations a few days before parturition. Because the level of HCG in the urine rises rapidly after conception, it has served as the basis for many pregnancy tests.

Sexual maturation, or the period in which cyclic menstrual bleeding begins to occur, is reached between the ages of 10 and 17; the average age is 13. The period of irregular menstrual cycles before the cessation of menses (usually between the ages of 40 and 50) is commonly known as menopause.

Estrogens

The two principal classes of female sex hormones that are steroids are estrogens and progestins. Estrogens are substances that induce estrus in various mammalian species. They are important because they bring about the secondary sex characteristics in females.

Chemically, the naturally occurring estrogens have in common an unsaturated A ring (which is planar) with a resulting phenolic function in the 3 position that aids in separation and purification from nonphenolic substances.

Allen and Doisy[172] showed in 1923 that an extract of ovaries can produce estrus. Soon thereafter, it was found that a good source of estrogens is the urine of pregnant

Estrone: R = H
Estrone sodium sulfate: R = $SO_3^-\ Na^+$
Piperazine estrone sulfate: R = $SO_3^-\ H_2N^+$ NH

17β-Estradiol

Estratriol

women. Estrone N.F. [estra-1,3,5(10)-trien-3-ol-17-one] was the first crystalline estrogen to be isolated from such a source. Two other C-18 estrogen steroids, 17β-estradiol and estriol, were later isolated and characterized to round out what are considered to be the three classic estrogens.[173,174]

Biosynthesis. Estrogens are biosynthesized from cholesterol, primarily in the ovary in mature, premenopausal women. During pregnancy, the placenta is the main source of estrogen biosynthesis and pathways for production change.[173–176] Small amounts of these hormones are also synthesized by the testes in the male and by the adrenal cortex, the hypothalamus, and the anterior pituitary in both sexes. The major source of estrogens in both postmenopausal women and men occurs in extraglandular sites, particularly in adipose tissue.[177]

In endocrine tissues, cholesterol is the steroid that is stored and is converted to estrogen, progesterone, or androgen when the tissue is stimulated by a gonadotropic hormone. The major pathways for the biosynthesis of sex steroid hormones are summarized in Figure 23–12. In the ovary, FSH acts on the preovulatory follicle to stimulate the biosynthesis of estrogens. The thecal cells of the preovulatory follicle convert cholesterol into androgens, whereas the granulosa cells convert androgens to estrogens. After ovulation, LH acts on the corpus luteum to

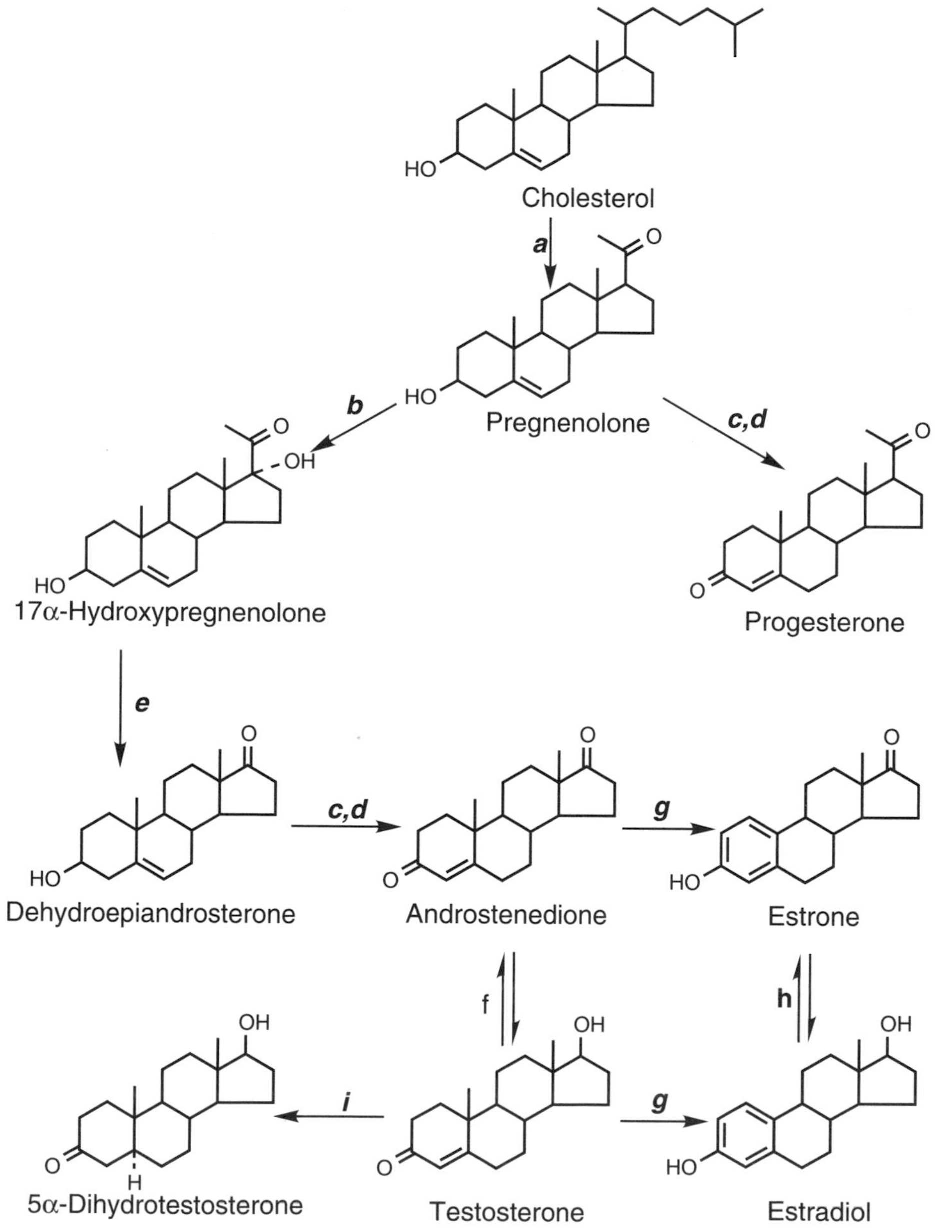

Fig. 23–12. Biosynthesis of sex steroid hormones. The enzymes involved in this biosynthesis are **(a)** side chain cleavage, **(b)** 17α-hydroxylase, **(c)** 5-ene-3β-hydroxysteroid dehydrogenase, **(d)** 3-oxosteroid-4,5-isomerase, **(e)** 17,20-lyase, **(f)** 17β-hydroxysteroid dehydrogenase, **(g)** aromatase, **(h)** estradiol dehydrogenase, and **(i)** 5α-reductase.

stimulate both estrogen and progesterone biosynthesis and secretion.

Cholesterol is converted by side chain cleavage into pregnenolone, which can be converted to progesterone or by several steps to the ring A aromatic system found in estrogens.[178] Pregnenolone is converted by 17α-hydroxylase to 17α-hydroxypregnenolone. The next biosynthetic step involves the cleavage of the C_{17}–C_{20} carbon-carbon bond and is referred to as the 17-20-lyase reaction. This step is catalyzed by the same cytochrome P450 enzyme, 17α-hydroxylase.[179] Loss of the 17β-acetyl side chain of 17α-hydroxypregnenolone affords the C_{19} intermediate dehydroepiandrosterone (3β-hydroxyandrost-5-en-17-one, DHA). DHA is converted by 5-ene-3β-hydroxysteroid dehydrogenase and 3-oxosteroid-4,5-isomerase to the 17-ketosteroid, androstenedione (androst-4-ene-3,17-dione). This compound is interconvertible with the reduced 17β-hydroxy intermediate (testosterone) by 17β-hydroxysteroid dehydrogenase. Loss of the C-19 angular methyl group and aromatization of the A ring of testosterone or androstenedione results in 17β-estradiol or estrone, respectively. 17β-estradiol and estrone are metabolically interconvertible, catalyzed by estradiol dehydrogenase.

The final step in the biosynthesis is the conversions of the C_{19} androgens to the C_{18} estrogens. This is catalyzed by the microsomal cytochrome P450 enzyme complex termed aromatase. Research interests in the aromatization reaction continue to expand from basic endocrinology and reproductive biology studies to aromatase inhibition for the treatment of estrogen-dependent cancers, as illustrated in several conferences and reviews.[180–183] Androstenedione is the preferred substrate for aromatization and three molecules of NADPH and three molecules of oxygen are necessary for conversion of one molecule of androgen to estrogen.[184,185] Aromatization proceeds via three successive steps catalyzed by a single enzyme complex, consisting of the cytochrome $P450_{arom}$ protein and NADPH-cytochrome P450 reductase, present in the endoplasmic reticulum of the cell. The mechanism of aromatization is illustrated in Figure 23–13. The first step involves oxidation of the angular C-19 methyl group to provide 19-hydroxyandrostenedione (19-hydroxyandrost-4-ene-3,17-dione). 19,19-Dihydroxyandrostenedione, isolated as 19-oxoandrostenedione, is formed by the second oxidation step. The exact mechanism of the last oxidation remains to be fully determined.[186–188] Following the last oxidation, the C_{10}–C_{19} bond is cleaved and aromatization of the A ring occurs. In this last step, the C-19 carbon atom is lost as formic acid and oxygens from the first and last step are incorporated into the formic acid. The 1β-hydrogen and 2β-hydrogen atoms are also lost during aromatization. A summary of the proposed mechanisms and the biochemistry of aromatase has recently been reviewed.[189]

Metabolism. The endogenous estrogens, estrone and 17β-estradiol, are biochemically interconvertible by the enzyme estradiol dehydrogenase and yield the same metabolic products.[190–192] These hormones are metabolized mainly in the liver and largely excreted as water-soluble glucuronide and sulfate conjugates. Other tissues, such as those in the kidneys and intestines, may also be involved, but unlike the liver, they apparently contain no steroid sulfokinases.[193] Renal and intestinal preparations affect sulfate conjugation with low-molecular-weight phenols, but not with estrogens.[58]

Most of the accountable metabolites appear in the urine, even though 50% of a dose of 17β-estradiol first goes to the kidney through the liver, and the remaining 50% from the liver is found in bile fluid.[190] The reason is that much of the material found in bile subsequently enters the intestine, is reabsorbed there, and is returned to the liver; only 10% of the administered dose is found in the feces.[173]

Many metabolites have been isolated from urine, but to date no one has been able to account for all the radioactivity of an administered dose of ^{14}C-labeled 17β-estradiol. The major metabolites are shown in Figure 23–14. Both 17β-estradiol and estrone are converted by the cytochrome P450 enzyme complex 16α-hydroxylase to yield estriol (estra-1,3,5[10]-triene-3,16α,17β-triol). Estriol is found in the urine as the glucuronide conjugate. Estro-

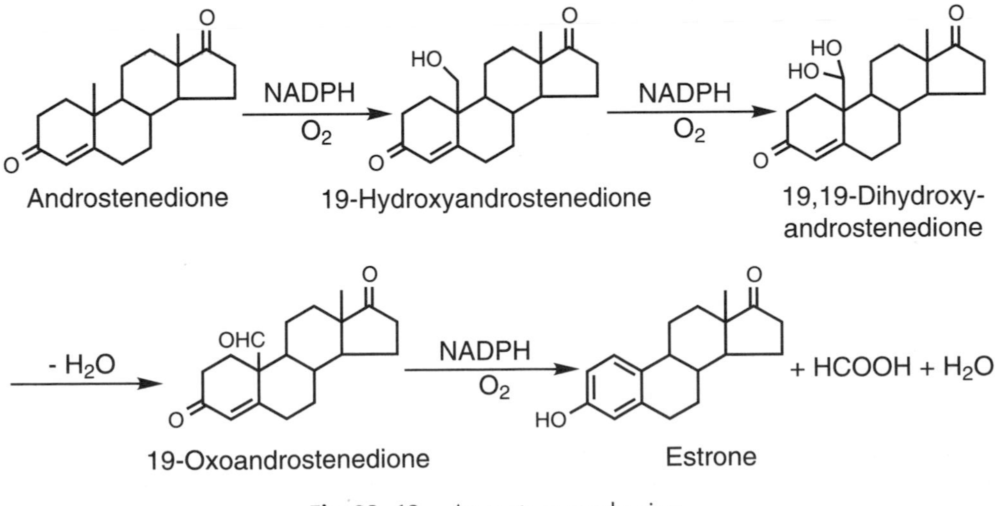

Fig. 23–13. Aromatase mechanism.

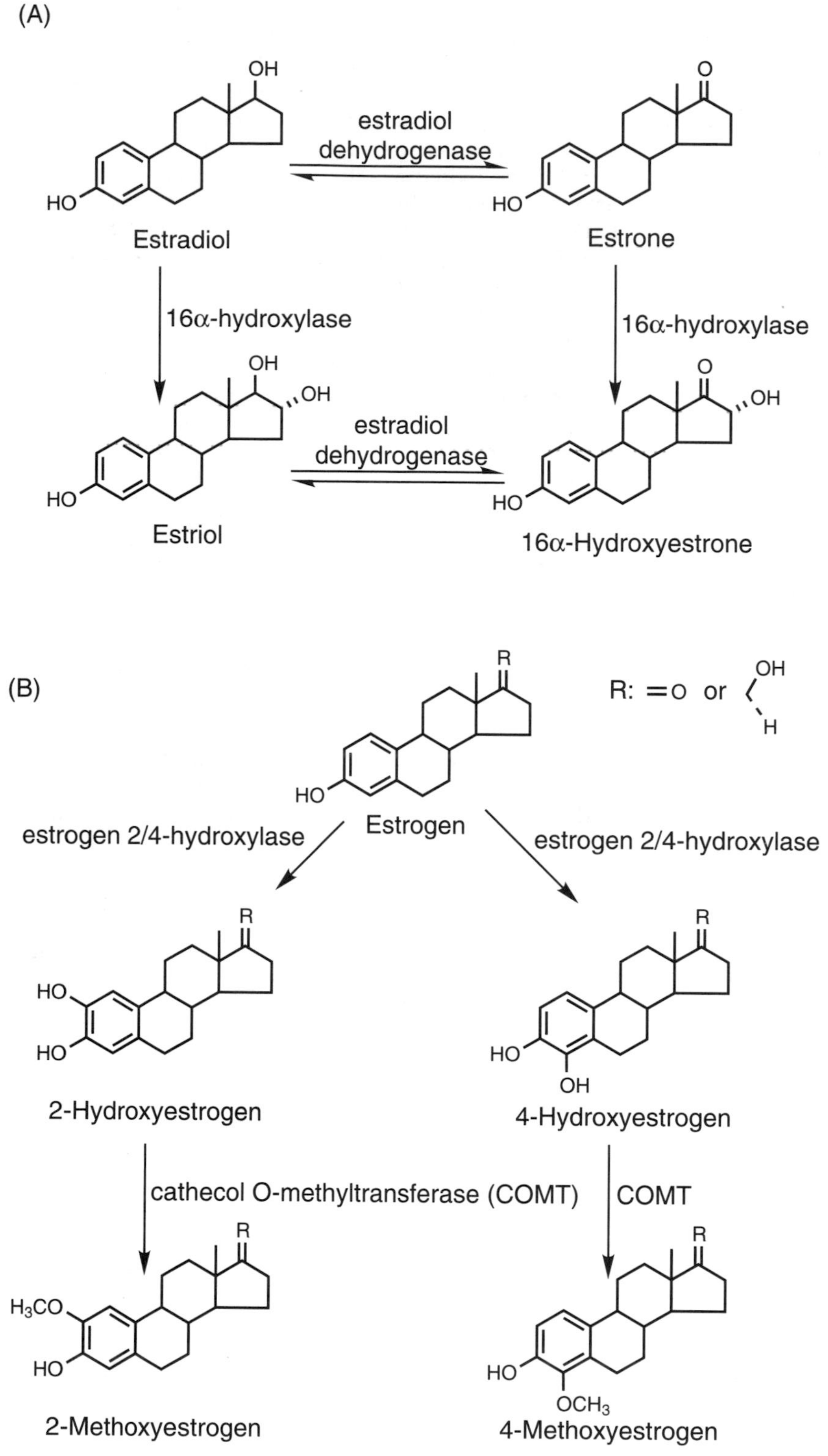

Fig. 23–14. Estrogen metabolism.

gens are also metabolized via oxidation at positions ortho to the 3-phenolic group. Metabolism by estrogen 2/4-hydroxylase provides the catechol estrogens, 2-hydroxyestrogens and the 4-hydroxyestrogens. These metabolites are unstable in vivo and are rapidly converted to 2-methoxyestrogens and 4-methoxyestrogen metabolites and also to glucuronide, sulfate, and glutathione conjugates. These compounds are also found in comparatively large amounts in the urine.[190–192,194–196] The catechol estrogens will bind to estrogen receptors and produce weak to moderate estrogenic effects. These steroids have also been shown to be produced in certain CNS tissues such as those in the pituitary and hypothalamus, suggesting possible neuroendocrine functions. The formation, metabolism, and biologic effects of catechol estrogens have been reviewed.[197]

Steroidal Estrogens. Estradiol is the most potent endogenous estrogen, exhibiting high affinity for the estrogen receptor and potency when administered parenterally. The naturally occurring estrogens are only weakly active when administered orally. They are thought to be rapidly absorbed from the intestine, but can be degraded by microorganisms in the gastrointestinal (GI) tract. The endogenous estrogens are promptly metabolized by the liver to a great extent, resulting in the observed low therapeutic effectiveness when administered orally. Comparative dosage ranges for endogenous estrogens (estradiol, estrone, and estriol) and synthetic preparations are shown in Table 23–3.

One successful method of overcoming this rapid inactivation of 17β-estradiol by the liver has been to stabilize the alcoholic function at C-17 with an appropriate substitute. Treatment of estrone with potassium acetylide in liquid ammonia results in ethinyl estradiol U.S.P. [17α-ethynyl-1,3,5(10)-estratriene-3,17β-diol], an estrogenic substance that is potent when taken orally.[198]

Another semisynthetic estrogen, synthesized by a similar route, is mestranol U.S.P. [17α-ethynyl-3-methoxy-1,3,5(10)-estratrien-17-ol]. This methyl ether analog is used to a great extent in oral contraceptives.[199] One of the most potent of all orally active estrogens used clinically is quinestrol, [3-cyclopentyloxy-17α-ethinylestra-1,3,5(10)-trien-17β-ol], shown in Figure 23–15. This estrogen ether is stored in body fat and is slowly released over several days.[200,201]

Esters of 17β-estradiol and other estrogens have been used to prolong estrogenic action. The ester derivatives of estrogens are usually given intramuscularly in contrast to the ethynyl derivatives, which are given orally. Among the esters of estradiol that have been used to the greatest extent in estrogen therapy are the 3-benzoate (N.F.), the 3,17-dipropionate (N.F.), the 17-valerate (U.S.P.), and the 17-cyclopentylpropionate (cypionate) (N.F.).

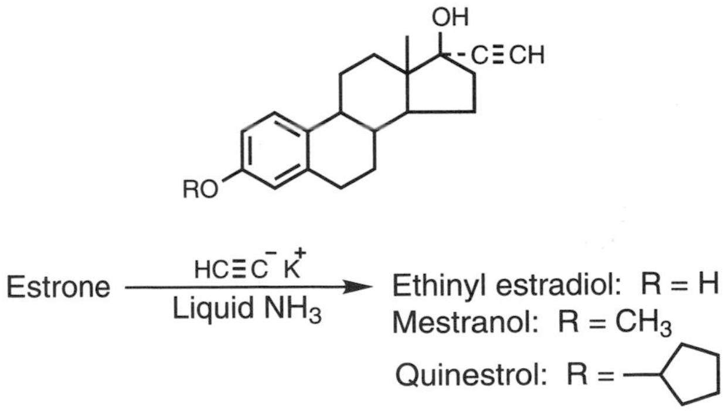

OH
-C≡CH
RO

Estrone —($HC\equiv C^- K^+$ / Liquid NH_3)→ Ethinyl estradiol: R = H
Mestranol: R = CH_3
Quinestrol: R = cyclopentyl

Fig. 23–15. Synthesis of 17-ethinyl derivatives.

Slow hydrolysis of the ester releases the free estrogen over a prolonged period. These derivatives may be

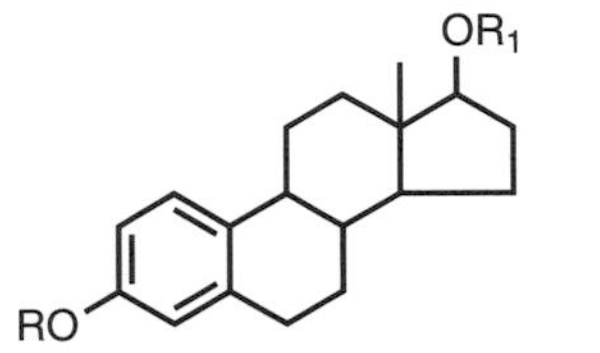

OR_1
RO

Estradiol-3-benzoate: R = C_6H_5CO-; R_1 = H
Estradiol-3,17-dipropionate: R = R_1 = C_2H_5CO-
Estardiol-17-valerate: R = H; R_1 = $CH_3(CH_2)_3CO$-
Estradiol-17-cyclopentylpropionate: R = H; R_1 = (cyclopentyl)-CH_2CH_2CO-

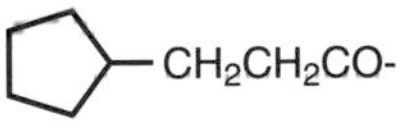

thought of as prodrugs; when they are administered intramuscularly, their actions may persist as long as 4 weeks.

Conjugated, water-soluble forms of naturally occurring estrogens obtained from the urine of pregnant mares are

Table 23–3. Dosages for Estrogens*

Estrogen	Route	Frequency	Usual Dosage Range (mg)
Estradiol	Intramuscular	Daily	0.02–0.2
Estrone	Intramuscular	Daily	0.1–3
Estriol	Oral	Daily	0.2–2
Ethinyl estradiol	Oral	Daily	0.02–0.2
Diethylstilbestrol	Oral	Daily	0.1–3
	Intramuscular	Daily	0.1–3
	Vaginal	Daily	0.1–0.5
Diethylstilbestrol dipropionate	Intramuscular	Semiweekly	0.1–3
Estradiol benzoate	Intramuscular	Weekly	0.5–5
Estradiol cypionate	Intramuscular	Biweekly	0.5–5
Conjugated estrogenic substances	Oral	Daily	0.3–1.25
	Intramuscular	Daily	0.1–3
	Vaginal	Daily	0.2–5

* Abstracted from *American Hospital Formulary Service.* Drug Information, American Society of Hospital Pharmacists, Inc., Bethesda, MD, 1987.

also utilized as estrogen preparations, referred to as conjugated estrogenic substances. Horses produce two unique estrogenic compounds, equilin and equilenin, and secrete them in the urine as sodium sulfate conju-

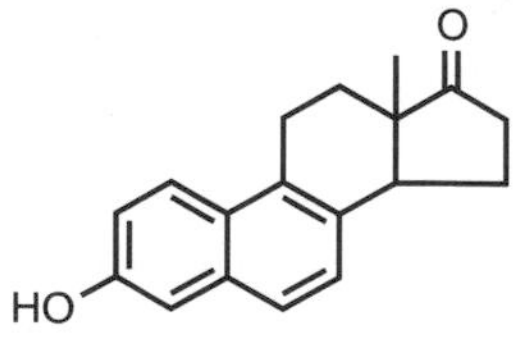

Equilenin

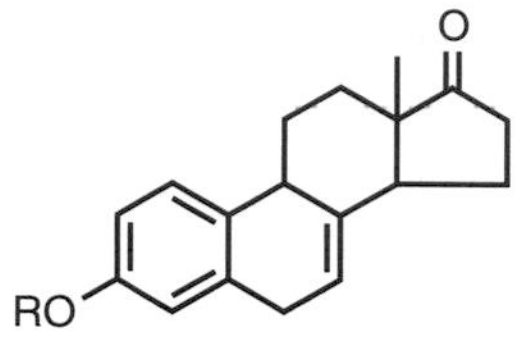

Equilin: R = H
Equilin sodium sulfate: R = SO_3^- Na^+

gates. Because the compounds in estrogenic substance are the sodium sulfate conjugates of equilin [3-hydroxyestra-1,3,5(10),7-tetraen-17-one] and estrone, the preparation is incompatible with acidic solutions. Equilin has not been detected in urine of humans. Another sulfate conjugate that is soluble in warm water and orally effective is piperazine estrone sulfate. This derivative has the same actions and uses as the conjugated naturally occurring estrogens.

A recent formulation of estradiol utilized for the treatment of menopause symptoms is Estraderm, a patch which is placed on the abdomen for local release of estradiol. This estrogen patch provides sufficient estradiol concentrations for physiologic effects on the uterus, whereas the systemic levels of estradiol are low because of rapid liver metabolism of the hormone once it enters the circulatory system.[202,203]

Nonsteroidal Estrogens. The steroid nucleus is not required for estrogenic activity. Many derivatives of stilbene, which is considerably more stable as the *trans* isomer, are potent estrogenic substances and are used therapeutically. One of the most important synthetic estrogens is diethylstilbestrol U.S.P. (α,α'-diethyl-4,4'-stilbenediol). This drug is significantly cheaper than naturally occurring estrogens and yet can produce all the same effects. It has been reported that neutral or alkaline aqueous solutions of the compound decompose over a 2 week period and that the rate of decomposition can be increased by oxidizing agents.[204]

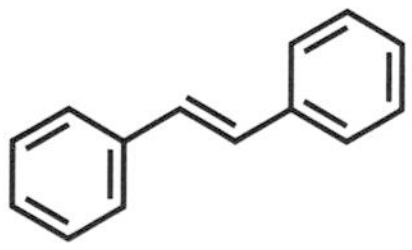

Stilbene

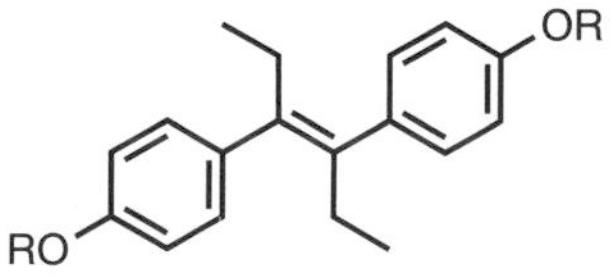

Diethylstilbestrol: R = H
Diethylstilbestrol dipropionate: R = C_2H_5CO-
Diethylstilbestrol dipalmitate: R = $C_{15}H_{31}CO$-

The official diethylstilbestrol is the *trans* isomer, which has 10 times the estrogenic potency of the *cis* isomer because it resembles more closely the natural estrogen, estradiol.[205]

Diethylstilbestrol, taken during pregnancy, is believed to be the cause of neoplasias in the offspring. Also, a higher incidence of uterine cancers has been linked to clinical use of estrogens as replacement therapy in menopausal women.

Hexestrol is the meso form of 3,4-bis(p-hydroxyphenyl)-n-hexane.

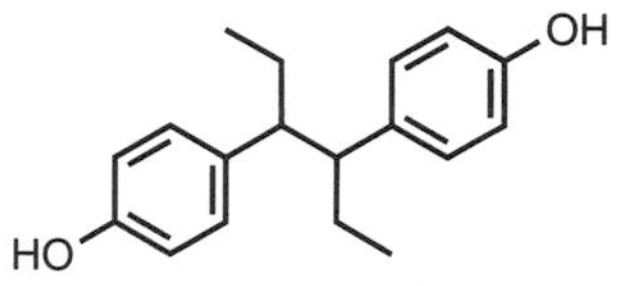

Hexestrol

It has the greatest estrogenic potency of the three stereoisomers of the dihydro analog of diethylstilbestrol; however, it is less potent than diethylstilbestrol. Other orally active substances related to diethylstilbestrol are dienestrol N.F. [3,4-bis(p-hydroxyphenyl)-2,4-hexadiene], benz-

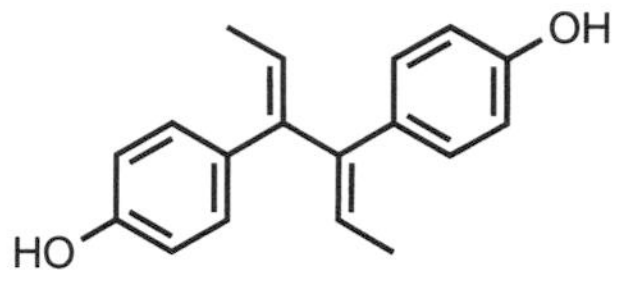

Dienestrol

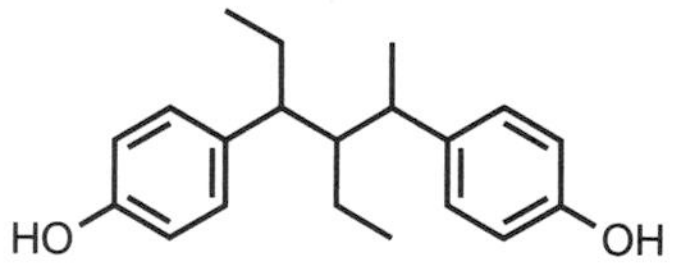

Benzestrol

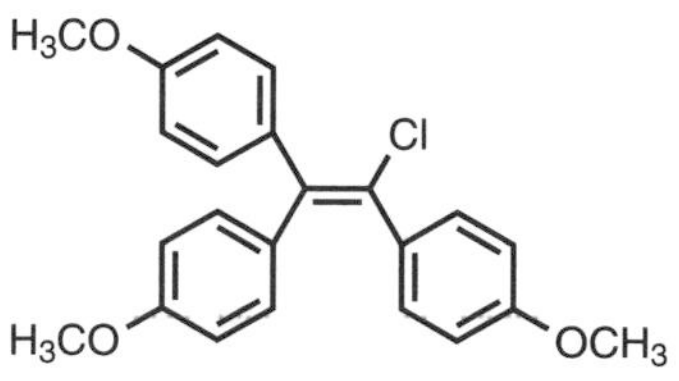

Cholotrianisene

estrol N.F. [3-ethyl-2,4-bis(p-hydroxyphenyl)hexane], and chlorotrianisene N.F. [chlorotris-(p-methoxyphenyl)-ethylene].

Chlorotrianisene can be considered a proestrogen; it is thought to be stored in adipose tissue, liberated slowly over a period of time, and then metabolized to a more potent estrogen than the parent substance.

Although there are several synthetic estrogens that are related to diethylstilbestrol, none has gained wide therapeutic acceptance. A synthetic estrogen that differs drastically from the others is methallenestril [3-(6-methoxy-2-naphthyl)-2,2-dimethylpentanoic acid]. The potency of diethylstilbestrol administered orally appears to be 10 times that of methallenestril.

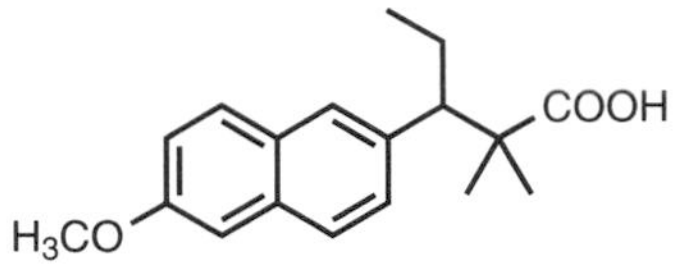

Methallenestril

Mode of Action. Estrogens act on many tissues, such as those of the reproductive tract, breast, and CNS. The growth and development of tissues in the reproductive tract of animals, in terms of actual weight gained, are not seen for as long as 16 hours after administration of the estrogen, although some biochemical processes in the cell are affected immediately. The growth response produced in the uterus by estrogens is temporary, and the maintenance of such growth requires the hormone to be available almost continuously. The initial growth induced by the estrogen is therefore of limited duration, and atrophy of the uterus occurs if the hormone is withdrawn.

Because estrogens localize and produce dramatic and selective responses in female reproductive tissues, the search for a biochemical mechanism of action of estradiol focused on these tissues. Using labeled estradiol, Jensen and Jacobson[45] showed that uptake of estradiol is rapid; the estradiol is retained to a high degree in the uterus and vagina.

The biochemical mechanism of estrogen action is the regulation of gene expression and subsequent induction of protein biosynthesis via specific, high affinity estrogen receptors, as described in general terms earlier in the chapter. Receptor binding sites for estradiol are located in the nucleus of target cells[46,206] and exhibit both high affinity ($K_D = 10^{-11}$ to 10^{-10} M) and low capacity. Binding to the estrogen receptor is specific. The uptake of [2,4,6,7-^{3}H]-17β-estradiol by the receptors can be inhibited by pretreatment with unlabeled 17β-estradiol or diethylstilbestrol. Pretreatment with testosterone, cortisol, or progesterone does not inhibit binding of radiolabeled estradiol. The binding is stereospecific because 17α-estradiol, which differs from 17β-estradiol in the configuration of the hydroxyl group at carbon 17, does not prevent the binding of 17β-estradiol to the estrogen receptor.[207]

When estradiol binds to the estrogen receptor, a conformational change of the estradiol-receptor complex occurs and results in interactions of the estradiol-receptor complex with particular HRE regions of the cellular DNA, referred to as estrogen responsive elements (EREs). Binding of the complex to ERE elements results in initiation of transcription of the DNA sequence to produce mRNA. Finally, the elevated levels of mRNA lead to an increase in protein synthesis in the endoplasmic reticulum.

Estrogens produce their effects upon the mammalian uterus by increasing synthesis of RNA in the target cells. Within 24 hours after administration of estradiol to mice, there is an increase in uterine RNA concentrations without a similar increase in DNA. The stimulatory effect on nuclear RNA synthesis occurs within 2 minutes after administration of estrogens.[206] It has been observed that RNA synthesized by estrogen-stimulated nuclei differs from that produced by nonstimulated nuclei.[206,208,209] Estradiol, for example, increases the frequency of the consecutive pairs guanine and uracil (GpU) and adenine and uracil (ApU) and decreases frequency of the consecutive pairs uracil and uracil (UpU) and cytosine and uracil (CpU) in nearest-neighbor nucleotide analysis of RNA synthesized by rat uterine nuclei.

Metabolic inhibitors have been used to delineate the role of estrogens in the activation of RNA synthesis. Actinomycin D, an inhibitor of DNA-dependent RNA synthesis, prevents the acceleration of protein and phospholipid synthesis induced by estradiol.[210] Puromycin, an inhibitor of protein synthesis, blocks the response of administered estrogen.[211] The RNA extracted from a uterus that has been stimulated by estrogens appears to produce several uterine effects seen with normal estrogen stimulation. When the RNA extract obtained from estrogenized uterine tissue is treated with RNAase or diphosphoesterase, it loses most estrogenic activity. In addition, RNA taken from other tissues, such as from the liver, treated with estrogens has no estrogenic effect on the uterus. The increased RNA synthesis eventually leads to an increase in the synthesis of proteins. Several proteins produced in target cells by the action of estrogens have been identified. In the chick oviduct the transport protein ovalbumin is induced by estrogen.[206] One protein initially termed IP (Induced Protein) and later identified as the brain-type creatine kinase[212,213] is produced in rat uterine tissue on E_2 stimulation. Also, levels of progesterone receptors increase in uterine tissue, thus preparing the tissue for the actions of progesterone on the uterus during the later half of the reproductive cycle.[214]

Estrogens affect the activity of various enzyme systems. The activity of enzyme systems increases with the proper concentration of estrogens[215] including lactic dehydrogenase, β-glucuronidase, alkaline phosphatase (from the uterus), peroxidase,[216,217] glucose-6-phosphate dehydrogenase, and plasminogen activator.[214]

Another physiologic effect of estrogens, observed 1 hour after their administration, is edema in the uterus. During this period, vasodilation of the uterine pre- and postcapillary arterioles occurs, and there is an increase in permeability to plasma proteins. These effects appear to occur predominantly in the endometrium and not to any great extent in the myometrium.[218]

Another target of estrogens is breast tissue. Estrogens can stimulate the proliferation of breast cells and promote the growth of hormone-dependent mammary carcinoma. Because the breast is the primary site for cancer in women, considerable research has been focused on understanding breast cancer and the factors that influence its growth. Estradiol will stimulate gene expression and the production of several proteins in breast cancer cells via the estrogen receptor mechanism. These proteins include both intracellular proteins important for breast cell function and growth, and secreted proteins that can influence tumor growth and metastasis. Intracellular proteins include enzymes needed for DNA synthesis, such as DNA polymerase, thymidine kinase, thymidylate synthetase, and dihydrofolate reductase.[219,220] Progesterone receptors are induced in breast cells by estrogens,[221] and the content of both estrogen receptors and progesterone receptors is utilized clinically as markers for hormone responsiveness of the breast cancer in determining hormonal therapy.[222] Another prominent protein made in breast cells is a 7-kd protein, whose function is yet unknown, and is derived from estrogen-induced mRNA referred to as pS2.[223] Several secreted proteins induced via the estrogen receptor are potential cellular growth factors.[224] Stimulatory growth factors induced by estrogen treatment include transforming growth factor-α (TGFα)[225–227] and insulin-like growth factor I (IGF-I).[228] A secreted 52-kd glycoprotein, a cathepsin D protein that exhibits protease activity, is also induced by estrogens and is thought to be important in breast cancer cell metastasis.[229] The levels of another growth factor protein, transforming growth factor-β (TGFβ), are decreased in breast cancer cells by estrogen treatment and are induced by antiestrogens,[230] suggesting that it may exert a negative effect on cell growth and serve as an endogenous regulator.

Pharmacology, Side Effects, and Clinical Applications. Administration of estrogens to immature animals can increase the rate of sexual maturity. One of the principal actions of estrogens is to promote the development of secondary sex characteristics in the female. The feminizing characteristics include growth of hair, softening of skin, growth of breasts, and accretion of fat in the thighs, hips, and buttocks. Estrogens also stimulate the growth and development of the female reproductive tract, including the uterine oviduct, cervix, and vagina. The proliferative changes that occur in the endometrium and myometrium on estrogen administration resemble those that take place naturally. Bleeding often follows withdrawal of the estrogens.

These hormones appear to prevent coronary atherosclerosis in women before menopause because of an alteration in the composition of circulating lipids.[27] Because of feminizing effects, estrogen therapy in males is limited. One of the primary therapeutic uses is in the treatment of menopausal symptoms such as hot flashes, chilly sensations, dizziness, fatigue, irritability, and sweating. For many women, menopause does not cause much discomfort; in some, however, both physical and mental discomfort may occur and can usually be prevented through estrogen therapy.

Estrogens are used in a wide variety of menstrual disturbances, such as amenorrhea, dysmenorrhea, and oligomenorrhea. They are also effective in failure of ovarian development, acne, and senile vaginitis. After childbirth, estrogens have been used to suppress lactation. In nonresectable prostate carcinoma, the treatment of choice is combined estrogen therapy and castration. One of the most widespread uses of estrogens is in birth control. These hormones are used in osteoporosis because it is thought that an estrogen deficiency in postmenopausal women can lead to this serious disorder of the bone.

Nausea appears to be the main side effect; other adverse effects include vomiting, anorexia, and diarrhea. If small doses are used to initiate therapy, and the dose is gradually increased, most of the side effects can be avoided. Excessive doses of estrogens inhibit the development of bones in young patients by accelerating epiphyseal closure.

When estrogens are given in large doses over long periods of time, they can inhibit ovulation because of their feedback action; inhibition of the release of FSH from the adenohypophysis results in inhibition of ovulation. Administration of these drugs may promote sodium chloride retention; the result is retention of water and subsequent edema. This effect, however, is less pronounced than with glucocorticoids. More detailed information on the pharmacology and toxicology of estrogens can be found in published reviews.[173,174,231]

Structure-Activity Relationships. The aromatic A-ring and the hydroxyl substituent at C–3 of the steroid nucleus are the structural features essential for estrogenic activity. The 17β-hydroxyl functional group, the distance between the two hydroxyl groups, and a planar hydrophobic molecule are also important for imparting full estrogenic activity to the molecule. In vitro investigations with subcellular fractions containing estrogen receptors have provided a more specific analysis of the structure-activity relationships for interacting with the estrogen receptor[231,232] and demonstrated the high affinity and specificity of the most potent endogenous estrogen, estradiol.

The observed in vivo biologic activity varies with the mode of administration of the estrogens. The order of activity of the three naturally occurring steroids when administered subcutaneously is estradiol > estrone > estriol. The order changes to estriol > estradiol > estrone, however, when the drugs are administered orally.[233] Although the naturally occurring steroids have activity when administered orally, chemical modifications have led to better orally effective estrogens.

Therapeutically, ethinyl estradiol is a more effective oral estrogen than many others because of its resistance to microbiologic degradation in the gastrointestinal tract and to metabolism in the liver. Also, metabolic conjugation may produce estrogens that retain some of their activity; it has been reported that estrone sulfate is actually more active than the original hormone when administered orally to rats. The modified drug also retains some activity in humans.[234,235]

Ester derivatives (acetates and benzoates) of the naturally occurring and synthetic estrogens have a prolonged action. The free hormone is released slowly because of hydrolysis of the ester in vivo. Another attempt to prepare highly active estrogens has led to the development of labile ethers of various estrogens.[236] Two potent derivatives of estradiol are the 3-(2-tetrahydropyranyl) and 17-(2-tet-

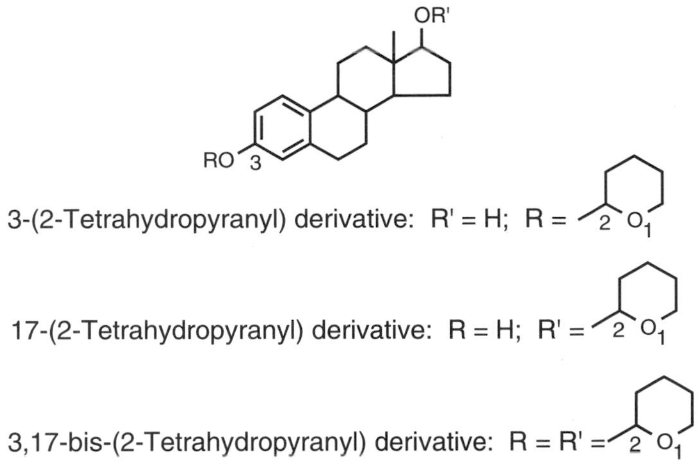

3-(2-Tetrahydropyranyl) derivative: R' = H; R = 2-tetrahydropyranyl

17-(2-Tetrahydropyranyl) derivative: R = H; R' = 2-tetrahydropyranyl

3,17-bis-(2-Tetrahydropyranyl) derivative: R = R' = 2-tetrahydropyranyl

rahydropyranyl) derivatives. These drugs proved to be 12 and 15 times as active, respectively, as estradiol. The estrogenic activity was greatest when these drugs were administered orally. The remarkably high ratio of oral to subcutaneous potency of 403 observed with 17β-estradiol-3-(2-tetrahydropyranyl) ether may be contrasted with the 3,17-bis-(2-tetrahydropyranyl) derivative, which is less active than estradiol.

Substituents on the estrone nucleus significantly modify estrogenic activity. Insertion of hydroxyl groups at positions 6, 7, and 11 reduces activity: Removal of the oxygen function from position 3 or 17 or epimerization of the 17-hydroxyl group of estradiol to the α-configuration results in a less active estrogenic substance.[237] Introduction of unsaturation into the B ring similarly reduces potency. Enlargement of the D ring of both estradiol and estrone (i.e., D-homoestradiol and D-homoestrone) greatly reduces estrogenic activity.

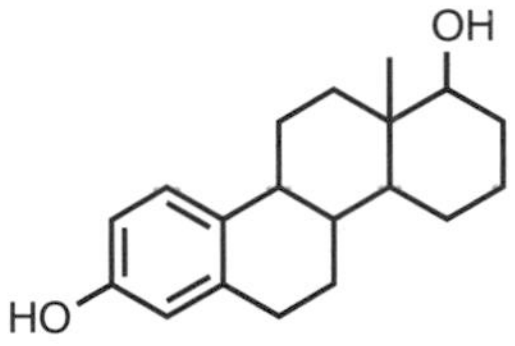

D-Homoestradiol

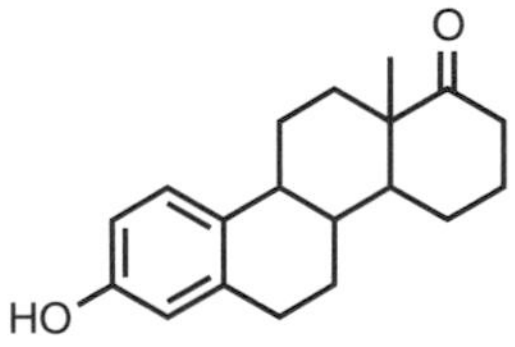

D-Homoestrone

Many compounds have estrogenic activity; this includes substances that do not have the normal steroid nucleus.[238] Estrogenic activity is unique in that it does not require a strict structural configuration as do the other sex hormones and the adrenocorticoids. The constituents of many plants have estrogenic activity, and one of these substances, genistein, found in subterranean clover, causes infertility in Australian sheep.

Another naturally occurring estrogen found in a number of legumes is coumestrol, which is even more potent than genistein.[239]

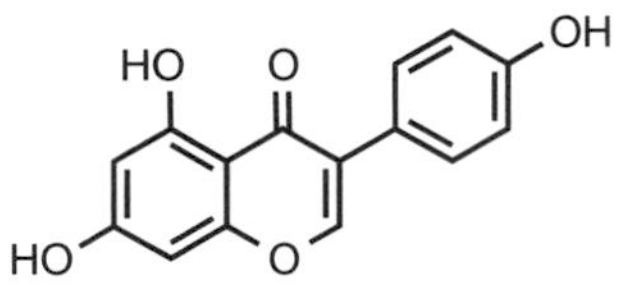

Genistein

The Schueler hypothesis[240] is that molecules with a distance of 8.55 Å between groups that can form hydrogen bonds (e.g., ketones, and phenolic and alcoholic hydroxyl groups) on a large, inert skeleton have optimal estrogenic activity. Diethylstilbestrol and estrone were the basis for this hypothesis; both compounds have such a distance between the hydrogen bonding groups, as shown by x-ray crystallography. Many other potent steroidal and nonsteroidal estrogens conform to this hypothesis.

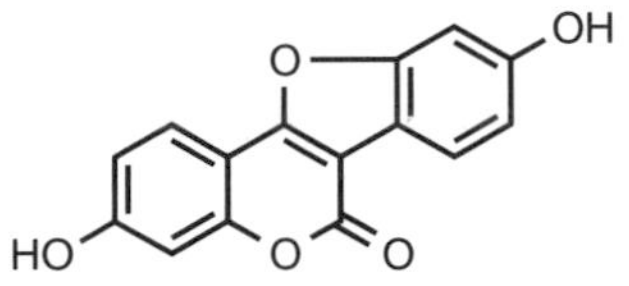

Coumestrol

An intact D ring is not required for estrogenic activity in rats; doisynolic acid, obtained from cleavage of estrone by a strong base, retains a high degree of hormonal activity. (Note the structural similarity of doisynolic acid and methallenestril.)

When estrogens were first obtained, the scarcity of the natural hormones led chemists to prepare synthetic materials that could be used as substitutes. Diethylstilbestrol, prepared in 1939, is the most successful drug in the nonsteroidal class. Although many stilbene derivatives have been prepared, the *trans* isomer of diethylstilbestrol

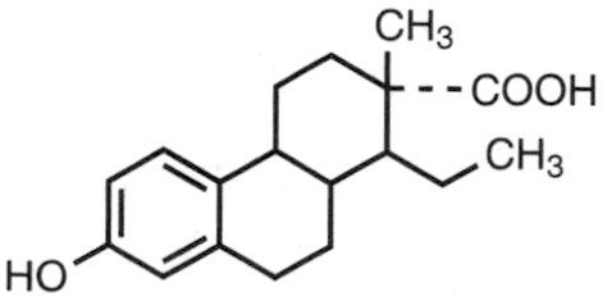

Doisynolic acid

remains one of the most active nonsteroidal estrogens.[241–243]

Although the estrogenic activity of triphenylethylene derivatives, such as chlorotrianisene, is low compared to that of diethylstilbestrol, these derivatives do exert their action for prolonged periods. The synthetic estrogens have the same spectrum of effects as the natural steroidal estrogens.

Estrogen Antagonists

Agents that antagonize the actions of estrogens are of particular interest for their ability to modify reproductive processes and for the treatment of estrogen-dependent breast cancer. Three groups of agents are identified as estrogen antagonists—impeded estrogens, triphenylethylene antiestrogens, and aromatase inhibitors.

The impeded estrogens interact with the estrogen receptor in target tissues but dissociate from the receptor too rapidly to produce any strong estrogenic effect. If, however, the impeded estrogens are present in a high *local* concentration, these agents compete with or *impede* the access of estradiol to the receptor site, thus decreasing the estradiol's effect on the cell.[162,173] The classical impeded estrogen is estriol.

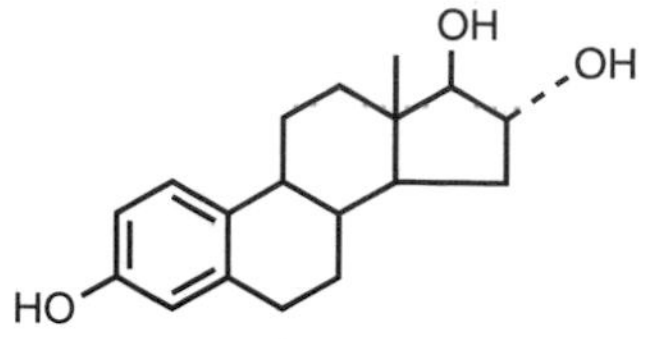

Estriol

The triphenylethylene antiestrogens are structurally related to the nonsteroidal estrogens and the antilipidemic agent, triparanol. They exhibit a strong and persistent binding to the estrogen receptor, producing antiestrogen-receptor complexes that are either unable to translocate into the nucleus of target cells or, if translocated, do not bind properly to the acceptor site of the chromatin to produce an estrogenic response.[244,245] Clomiphene is used for the treatment of infertility and produces an increase in the secretion of gonadotropins.[173] This effect is the result of the binding of clomiphene to estrogen receptors in the hypothalamus, leading to a blockade of the feedback inhibition exhibited by estrogens. Tamoxi-

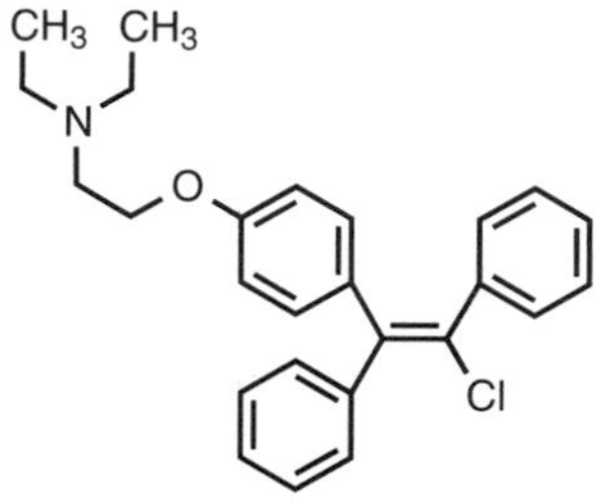

Chlomiphene

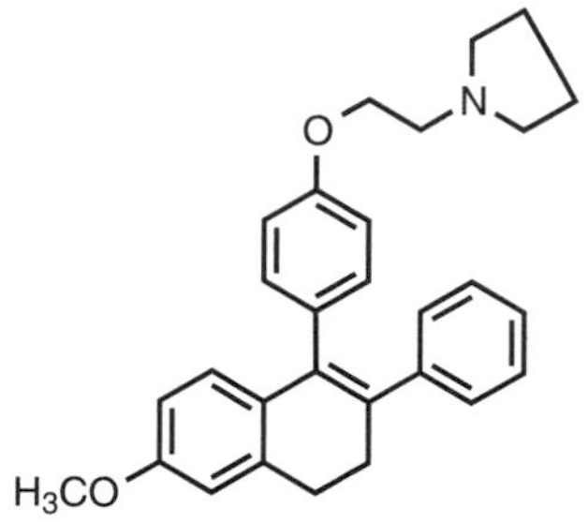

Nafoxidine

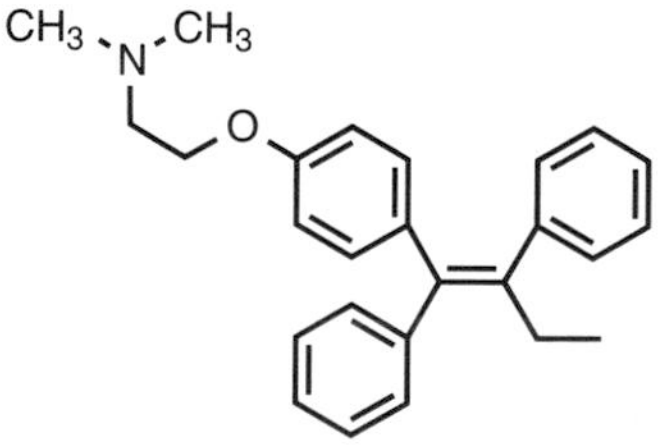

Tamoxifen

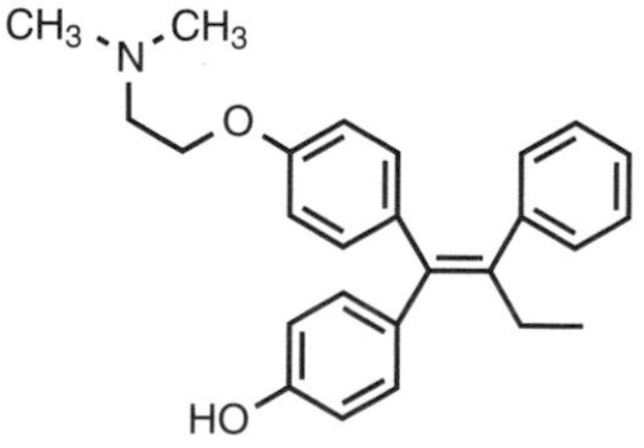

Monohydroxytamoxifen

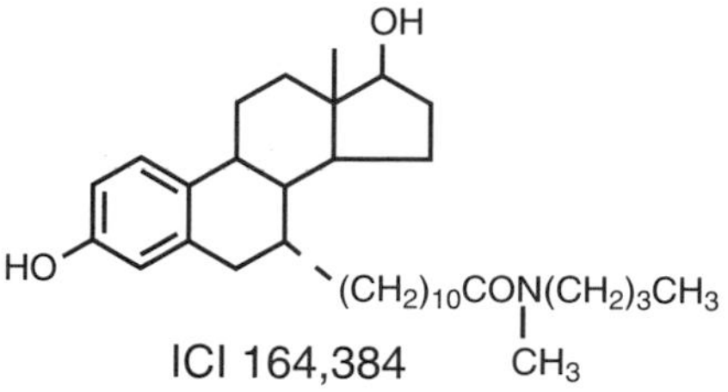

ICI 164,384

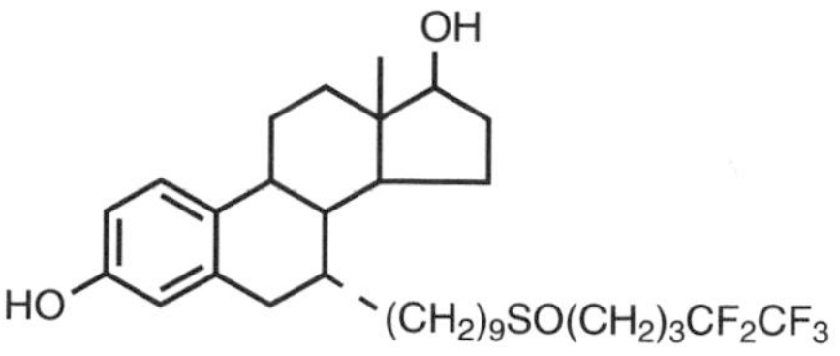

ICI 182,780

fen and nafoxidine are antiestrogens used in the treatment of estrogen-dependent breast cancer. Evidence suggests that these compounds are most active after metabolism to the para-hydroxyl derivatives.[246,247] The aminoethyl ether side chain of the triphenylethylene antiestrogens is critical for the antiestrogenic activity of these agents. Recently, steroidal derivatives (ICI 164,384 and 182,780) have been described and are referred to as *pure antiestrogens* because they bind tightly to estrogen receptors, effectively block estradiol binding, and are completely devoid of estrogenic activity.[248,249]

Inhibitors of aromatase, the cytochrome P450 enzyme complex responsible for the conversion of androgens to estrogens, have therapeutic potential in the control of reproductive functions and in the treatment of estrogen-dependent cancers such as breast cancer. Steroidal agents under investigation include 4-hydroxyandrostenedione,[250] various 7α-substituted androst-4-ene-3,17-diones[251] and androsta-1,4-diene-3,17-diones,[252] 10β-propynylestr-4-ene-3,17-dione,[253] 1-methylandrost-4-ene-3,17-dione,[254] and 6-methyleneandrost-4-ene-3,17-dione.[255] These steroidal agents compete with androstenedione for binding to the active site of the aromatase enzyme. In addition, 4-hydroxyandrostenedione, several androsta-1,4-diene-3,17-diones, and 10β-propynylestr-4-ene—3,17-dione act as enzyme-activated irreversible inhibitors (suicide substrates) in vitro. The structure-activity relationships of steroidal aromatase inhibitors indicate that the best inhibitors are analogs of the substrate, with only small structural changes made on the A-ring and at C-19. Incorporation of aryl functionalities at the 7α-position results in inhibitors with enhanced affinity for the enzyme. Nonsteroidal inhibitors of aromatase include aminoglutethimide[256] (which is also an inhibitor of adrenal steroidogenesis), fadrazole (CGS 16949A),[257] and virazole (R 76713).[258] These agents all contain heterocyclic rings and are thought to produce inhibition by binding of the nitrogen atom of the heterocycle with the heme

Steroidal Aromatase Inhibitors

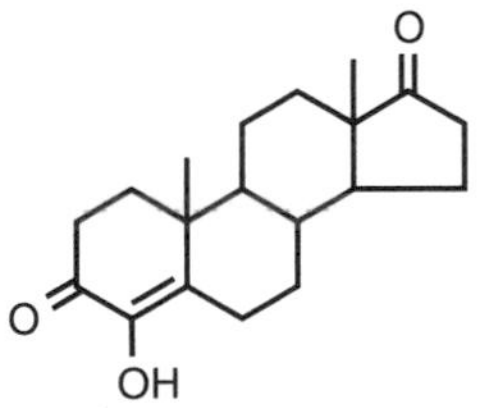

4-Hydroxyandrostenedione

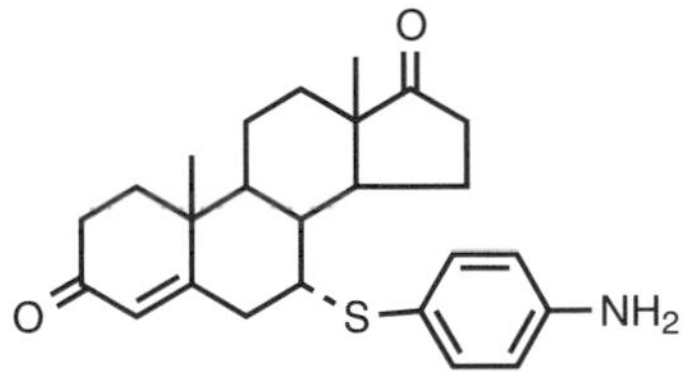

7α-Aminophenylthioandrost-4-ene-3,17-dione

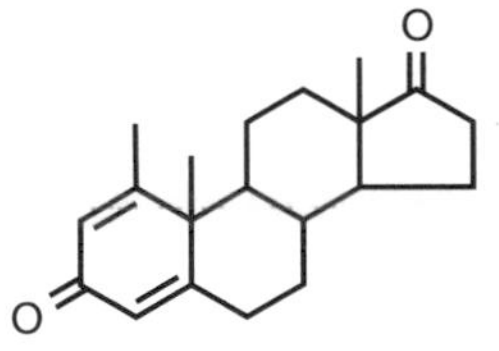

1-Methylandrosta-1,4-diene-3,17-dione

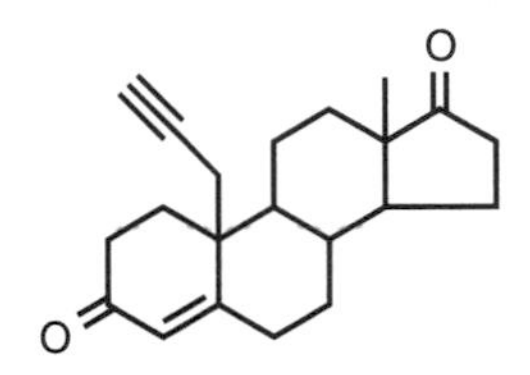

10β-Propynylest-4-ene-3,17-dione

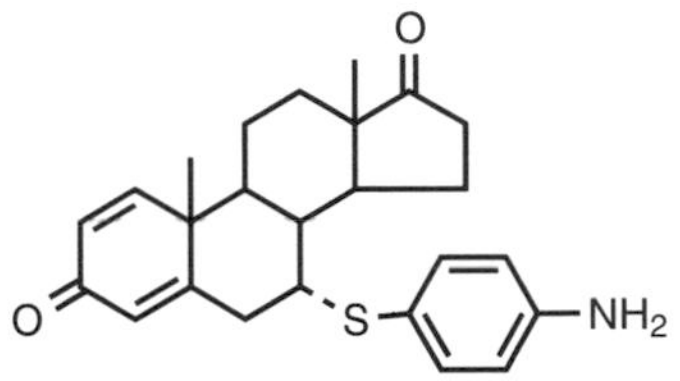

7α-Aminophenylthioandrost-1,4-diene-3,17-dione

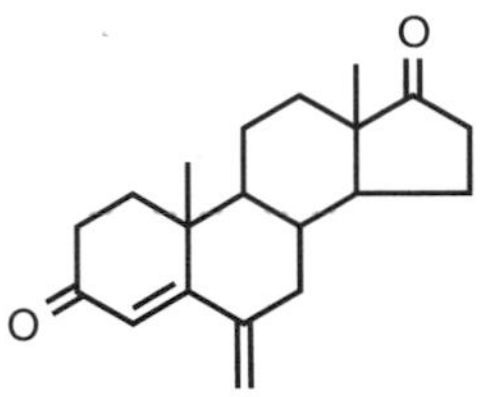

6-Methyleneandrost-4-ene-3,17-dione

Nonsteroidal Aromatase Inhibitors

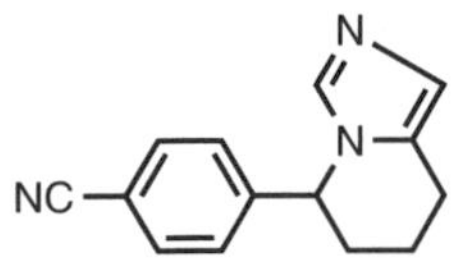

Fadrazole (CGS 16949A)

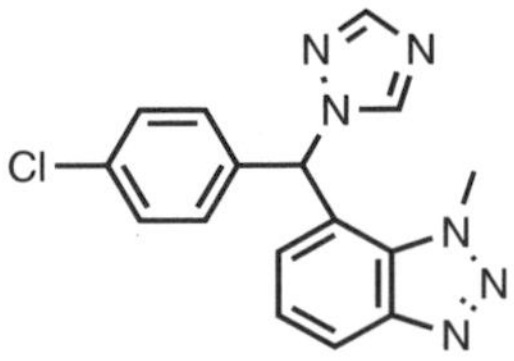

Virazole (R 76713)

iron of the cytochrome P450 protein. The structure-activity relationships of these inhibitors has recently been reviewed.[189,259–262]

Progestins

Once ovulation has taken place, the tissue remaining from the ruptured follicle forms the corpus luteum, which has the important function of preparing for and maintaining pregnancy if it occurs. Fraenkel[263] first observed in 1903 that removal of the corpus luteum shortly after conception results in termination of the pregnancy. In 1914, Pearl and Surface[264] showed that the corpus luteum can prevent ovulation in animals. In 1929, Corner and Allen[265] developed a method of assay for progestational activity.

By 1934, the progestational hormone progesterone had been isolated by several research groups.[69] It was shown in 1937 that pure progesterone alone can maintain pregnancy in animals.[266] Besides performing important functions in the reproductive system, this hormone serves as a precursor to androgens, estrogens, and adrenocorticoids.

The most abundant pregnane steroid found in the urine during pregnancy is 5β-pregnanediol (5β-pregnane-3α,20α-diol) glucuronide, which is also the main excretory product of exogenous progesterone. This conjugated substance can serve as an index of the corpus luteum placenta activity and a premature drop in its level in the urine may be a warning of possible miscarriage. Pregnanediol appears to have no progestational activity, but it apparently can antagonize the action of progesterone in certain animals.[267]

Biosynthesis. Progesterone is biosynthesized and secreted by the corpus luteum of the ovary during the luteal phase of the reproductive cycle. Luteinizing hormone (LH), the anterior pituitary glycoprotein hormone, binds to the LH receptor on the surface of the cells to initiate progesterone biosynthesis. As in other endocrine cells such as adrenal cortical cells, the binding of LH results in an increase in intracellular cAMP levels via activation of a G-protein and adenylyl cyclase. One of the processes influenced by elevated cAMP levels is the activation of cholesterol esterase, which cleaves cholesterol esters and liberates free cholesterol. The free cholesterol is then converted in mitochondria to pregnenolone via the side-chain cleavage reaction described earlier, and progesterone is formed from pregnenolone by the action of 5-ene-3β-hydroxysteroid dehydrogenase and 3-oxosteroid-4,5-isomerase (see Fig. 23–12).

Metabolism. Progesterone, the female sex hormone synthesized by the corpus luteum, is metabolized in part by routes similar to those for the adrenocorticoids. Progesterone is mainly excreted as conjugates of 5β-pregnanediol. Reduction of the 20-ketone to an alcohol, reduction of the 4,5-double bond resulting in 5β geometry, and reduction of the 3-ketone to the 3α configuration characterize the production of 5β-pregnanediol from progesterone. Minor pathways of metabolism in the liver can occur, with the side-chain at position 17 removed and pathways similar to those for the metabolism of androgens observed.

Synthetic Progestins. Progesterone, being an important hormone for maintaining pregnancy and normal menstrual bleeding, was used to correct disorders in these areas. The naturally occurring hormone was recognized for its ability to prevent ovulation during pregnancy; it can be considered a natural contraceptive.[268] A good source of progesterone was therefore desired.

Sequences were devised for the synthesis of progesterone from naturally occurring steroids, including diosgenin, ergosterol, and bile acids. The natural hormone, however, has many drawbacks, including a relatively low potency when orally administered; the hormone is administered parenterally for therapeutic effects.

Because progesterone had to be injected repeatedly over relatively short periods for best results and because in some instances this method of administration produced local irritation and pain, orally active derivatives were desired.

The first synthetic progestin to be used to any extent was synthesized from male sex hormones (androstanes). Ethisterone N.F. (17-ethynyl-17β-hydroxyandrost-4-en-3-one) was prepared in 1937 in an attempt to find an orally active androgen.[269]

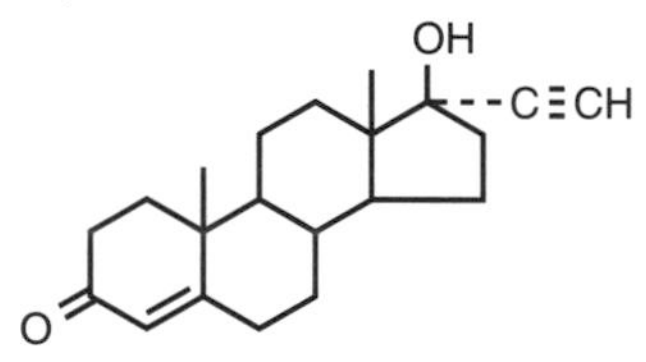

Ethisterone

The substance later proved to be an effective oral progestin and became useful in the treatment of menstrual dysfunctions.[270]

Several molecular modifications of ethisterone have enhanced progestational activity; introduction of methyl groups in the C-6α and C-21 positions, as in dimethisterone N.F., provided active analogs.[174]

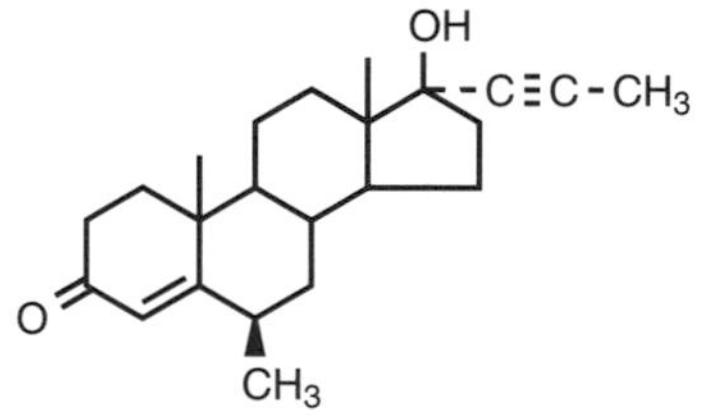

Dimethisterone

Ethisterone, therefore, paved the way for the synthesis of other progestins that did not have a typical progesterone-type C-17 side chain.

A second breakthrough was made in 1944 when Ehrenstein[271] discovered that the C-19 methyl group on steroids is not necessary for progestational activity. In fact, his work showed that loss of the C-19 methyl (from a compound he thought was progesterone, but later turned out to be a stereoisomer of progesterone, namely, 19-nor-14β,17α-pregn-4-ene-3,20-dione) results in activity equal to or greater than that of parenterally administered progesterone.

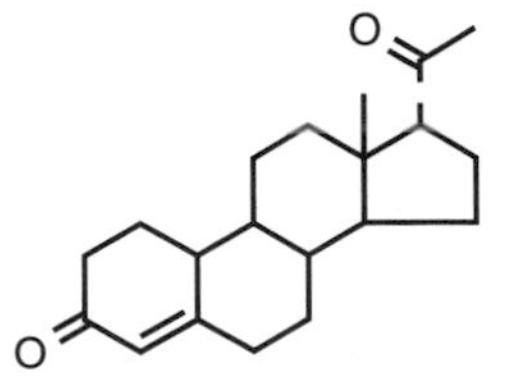

19-Nor-14β,17α-pregn-4-ene-3,20-dione

Ehrenstein's procedure for converting strophanthidin (a cardiac aglycone) to a stereoisomer of 19-norprogesterone using a 12-step process was too complicated and unsuited for further preparations of C-19 nor steroids. This work, however, did lead to intensive attempts to prepare orally active progestins that were devoid of estrogenic and androgenic activities.

Eventually, in 1953, Djerassi and co-workers[272] did synthesize 19-norprogesterone.

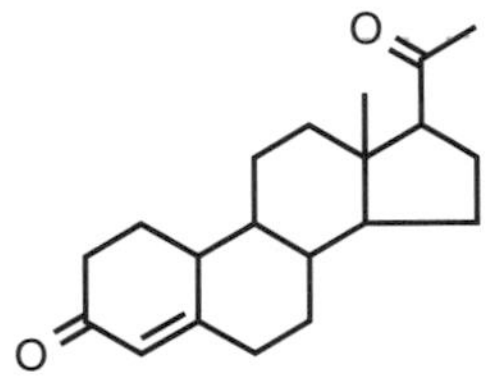

19-Norprogesterone

This drug differed from the natural hormone only in replacement of the C-19 angular methyl group by hydrogen. This analog was 8 times as active as progesterone when administered parenterally (Clauberg's test) and was the most potent progestin known. This work culminated in the synthesis of two potent, orally active progestins, namely, norethindrone U.S.P. (17-ethynyl-17β-hydroxyestr-4-en-3-one) and norethynodrel U.S.P. [17-ethynyl-17β-hydroxyestr-5(10)-en-3-one].[171]

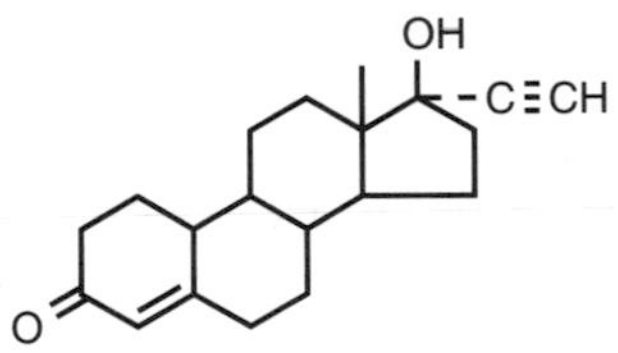

Norethindrone

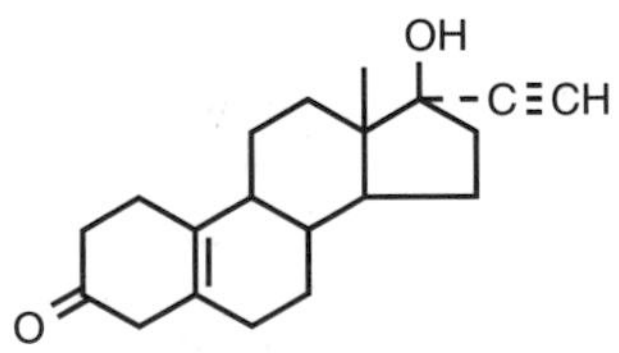

Norethynodrel

The progestational activity of norethynodrel is about one-tenth that of norethindrone; both compounds appear to have weak estrogenic activity. These two substances were among the first 19-nor steroids to be used clinically for progesterone hormonal disorders. They also afforded a method, when used with estrogens such as mestranol, for control of conception. The first oral contraceptive, Enovid, is a combination of mestranol and norethynodrel introduced in 1960.

The usefulness of norethynodrel and norethindrone for therapy of irregular menses and as contraceptive agents provided the impetus to continue research in the area of 19-nor steroids. Another 19-nor steroid reported to be effective and exhibit few side effects is ethynodiol diacetate U.S.P. (17-ethynylestr-4-ene-3,17β-diol 3,17-diacetate). This drug has been used as an oral progestin.[273]

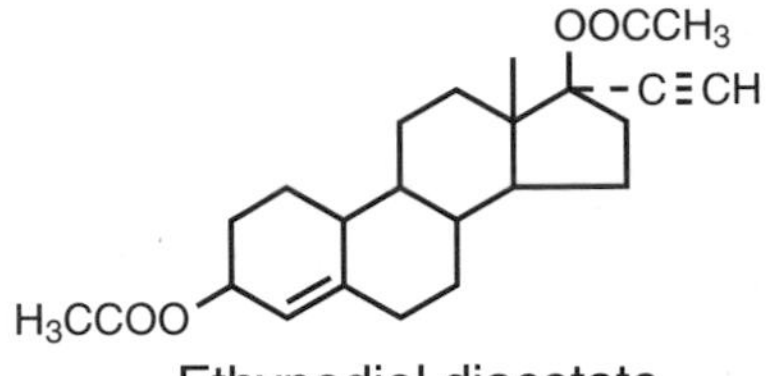

Ethynodiol diacetate

Although many of the early structural modifications of progesterone led to weakly active or inactive progestational agents, it was eventually shown that 17α-acetoxyprogesterone had some activity when administered orally even though the parent compound, 17α-hydroxyprogesterone, was inactive.[274] A second derivative of 17α-hydroxyprogesterone, the 17-caproate ester, is used extensively in therapeutics; because of its long duration of action, the drug is administered parenterally for locating problems related to deficiency of natural progesterone.[275]

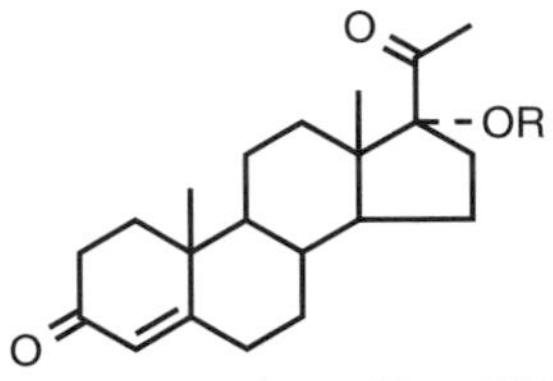

17α-Acetoxyprogesterone: R = CH_3CO-

17α-Hydroxyprogesterone 17-caproate: R = $C_5H_{11}CO$-

Further structural modifications of 17α-acetoxy-progesterone have enhanced its oral contraceptive action. In most instances, these modifications are carried out at the sixth carbon position. Substituents in this position hinder catabolism of the compounds and increase their lipid solubility; the result is an enhanced biologic effect.[276]

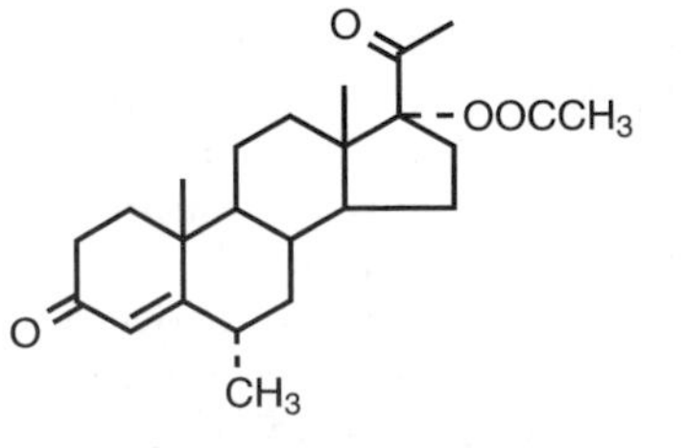

Medroxyprogesterone acetate

Among the first of these interesting analogs of 17α-acetoxyprogesterone to be used in progestational therapy was medroxyprogesterone acetate U.S.P. (6α-methyl-17-hydroxy-4-pregnene-3,20-dione 17-acetate).[277] This drug is 25 times as active as ethisterone; the compound has low estrogenic and no androgenic activity.

Progestational activity is further enhanced with 6-substituted 17α-acetoxyprogesterones when a double bond is introduced between positions 6 and 7. Megestrol acetate (6-methyl-17-hydroxy-4,6-pregnadiene-3,20-dione 17 acetate) and chlormadinone acetate N.F. (6-chloro-17-hydroxy-4,6-pregnadiene-3,20-dione 17-acetate) are typical examples of clinically useful progestins.

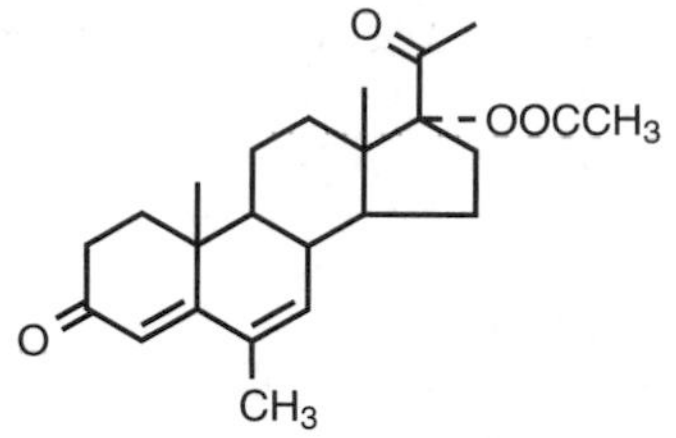

Megestrol acetate

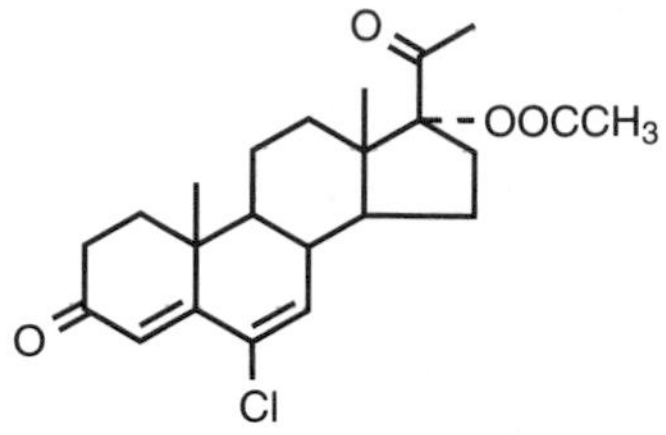

Chlormadinone acetate

Progestational activity of a steroid does not have to be related to its antiovulatory or antigonadotropic activity. Dydrogesterone (N.F.), which has a *cis* juncture between rings B and C and is named 9β,10α-4,6-pregnadiene-3,20-dione, is a progestational agent that can maintain pregnancy in spayed rats when administered subcutaneously. This isomer does not inhibit ovulation (even in large doses), however, and has no androgenic, estrogenic, or masculinizing effects on the female fetus. It also has no thermogenetic activity.

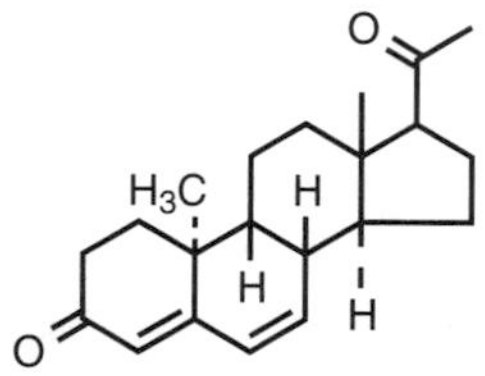

Dydrogesterone

Mode of Action. The uterus is the primary site of progesterone action in the female. Once the endometrium proliferates and becomes dense under the influence of estrogens, the levels of progesterone rise. This hormone inhibits the proliferation of, and initiates a secretory phase of, the reproductive cycle. During this stage, the endometrium becomes edematous and glycogen increases in the epithelium of the endometrium.

In an attempt to understand the cellular transformations induced by progesterone that involve gene expression, O'Malley, Shrader, and colleagues studied the effects of progesterone on the chick oviduct, a particularly useful biologic system for the examination of the mechanism of action of progesterone.[278] These studies on progesterone receptors extend into mammalian systems as well.[279] The progesterone receptor consists of two hormone binding proteins, receptors A and B.[280,281] Biologically active progesterone receptors are present in the nucleus of target cells, whereas inactive receptors have been found in the cytosol as a complex with hsp90,[282] similar to inactive glucocorticoid complexes. The nuclear progesterone receptor heterodimer binds progesterone with high affinity, resulting in a conformational change in the complex. The steroid-receptor complex interacts with particular HRE regions of the cellular DNA, referred to as progesterone responsive elements (PRE), and initiates transcription of the DNA sequence to produce mRNA. Finally, the elevated levels of mRNA lead to an increase in protein synthesis in the endoplasmic reticulum.

Administration of progesterone to estrogen-stimulated chicks resulted in the synthesis of the specific oviduct protein, avidin. In mammals, uteroglobin (a small secretory protein of the uterus) and the enzyme estradiol dehydrogenase have been identified as proteins induced by progesterone.

Extragenital effects of progesterone, except when secreted in large amounts, are slight. Progesterone is natriuretic, probably because of antagonism of aldosterone. Subsequently, increased sodium excretion stimulates secretion of aldosterone, which affects sodium retention. Progesterone is also catabolic, because it increases the total nitrogen excretion brought about by catabolism of amino acids.[283]

The main feedback effects of progesterone in the central nervous system are thought to occur in the hypothalamus; this causes inhibition of pituitary secretion.[284] Progesterone receptors have been identified in the hypothalamus and are involved in this feedback inhibition. Prior administration of estrogens or progestins does not appear to inhibit ovulation induced by exogenous gonadotropins. Progesterone appears to have a biphasic feedback effect on ovulation. During the first few hours after administration of this hormone, ovulation is produced and the effects are inhibited. It appears that the effects of progesterone are reversed as time passes.

Recent investigations have identified additional actions of progesterone and progesterone metabolites in the central nervous system. The identification of various C_{21} and C_{19} steroids and enzymatic processes for their production in brain tissues led investigators to suggest that these steroids have a possible function in the CNS.[285] Two 5α-reduced metabolites of progesterone, pregnanolone (3α-hydroxy-5α-pregnan-20-one) and its hydroxy derivative (3α,21-dihydroxy-5α-pregnan-20-one), have been shown to bind to the gamma amino butyric acid A ($GABA_A$) receptor complex at 10^{-8} M concentrations and potentiate GABA responses.[286,287] Another C_{21} metabolite found in CNS tissues is pregnenolone sulfate, which demonstrates an inhibitory activity on the $GABA_A$ receptor complex.[288] The physiologic relevance of these progestin metabolites in CNS function remains to be determined.

It has been suggested that the temperature-raising effect of progesterone may be due to increased body heat resulting from reduced sweating. This effect is not unique to progesterone; other steroids in the pregnane and androstane series can also produce it.[289]

Pharmacology, Side Effects, and Clinical Applications. The mechanism controlling ovarian secretion of progesterone involves the release of LH from the anterior pituitary during ovulation. The LH induces progesterone secretion from the corpus luteum during the second half of the menstrual cycle.

Progesterone has many biologic functions. The hormone acts on both the endometrium (inner mucous lining) and the myometrium (muscle mass) of the uterus. It acts on the endometrium, which has been primed by the estrogens, to induce the secretory phase, during which the endometrial glands grow and secrete large amounts of carbohydrates that will possibly be used by the fertilized ovum as a source of energy.

The primary function of progesterone with respect to the myometrium is to stop the spontaneous rhythmic contractions of the uterus. The effects of progesterone on the uterus are to prepare the endometrium for reception, implantation, and maintenance of the fertilized ovum and to suppress the myometrial contractions so that the embryo is not dislodged from the uterus.

The corpus luteum is the primary source of progesterone for the first third of pregnancy; subsequently, the developing placenta is the major source of progesterone and estrogens. Both hormones are secreted continually in large amounts until parturition. The high levels of progesterone produced by the corpus luteum and placenta during pregnancy act upon the hypothalamus through the negative feedback system to prevent the formation of new ova.

This information led to studies involving progesterone and its analogs as contraceptives.[173,290] If conception does not occur, the corpus luteum regresses and progesterone production decreases. This finally leads to sloughing of part of the endometrium during menstruation.

Development of the alveolar sacs in the mammary gland during pregnancy is stimulated by progesterone and estrogens, but lactation does not occur until after the levels of these hormones fall at parturition. Progesterone also increases the basal temperature and decreases the motility of the fallopian tubes. In large doses, progesterone can produce weak analgesia and general anesthesia.

Progesterone, and more recently its synthetic analogs, have been used to treat dysmenorrhea, endometriosis, functional uterine bleeding, and amenorrhea. Progesterone and its derivatives have been used to treat habitual abortions, although not always successfully.[291] This seems to be a reasonable use because progesterone is considered a pregnancy-supporting hormone. Because abortion is not always due to a hormonal deficiency, however, progestin treatment has not been as successful as predicted.

An early pregnancy can be diagnosed by giving the combination of an estrogen and a progestin for several days and then withdrawing it. If bleeding occurs in a few days, the patient is not pregnant. Progesterone has also been used in the treatment of carcinoma of the endometrium. The major use of progestins is in combination with estrogens as a contraceptive.

Among the side effects seen with progestins are nausea and vomiting, drowsiness, spotting and irregular bleeding; these may occur when these drugs are taken for a short time. With prolonged therapy, a greater incidence of side effects may be seen, including edema and weight gain, breast discomfort, breakthrough bleeding, decreased libido, and masculinization of the female fetus.

Structure-Activity Relationships. Progestational activity appears to be restricted to molecules with a steroid nucleus. Klimstra has pointed out that it is difficult to compare progestins on the basis of studies reported in the literature because there are many ways to do so.[292] Two of the most common methods of measuring uterine glandular development are Clauberg's and McGinty's tests. Other biologic evaluations of the progestins include their effect on uterine carbonic anhydrase, inhibition of gonadotropin hormones, and delay of parturition, and their ability to maintain pregnancy in a spayed female animal. Substances should be evaluated in the same laboratory; the resulting data are more informative.

The synthetic progestins can generally be divided into two classes of compounds, namely, the androgens (19-norandrostane or estrane derivatives) and the 17α-hydroxyprogesterones.[171,215,276,292] In the androgen series, a 17α-substituent, such as ethynyl, methyl, ethyl, and variations of these, renders the molecule active when administered orally.

Ethisterone, the first androgenic compound found to be effective, has about one-third the activity of progesterone in women when taken subcutaneously but is 15 times as active when taken orally. Because this analog is closely related to testosterone, it has androgenic activity. Removal of the CH_3 group at position 19 leads to norethindrone (norethisterone), which has five to ten times more progestational activity. The activity of norethindrone may be increased further by substituting a chlorine atom at position 21 or by adding a methyl group at carbon 18 (norgestrel).

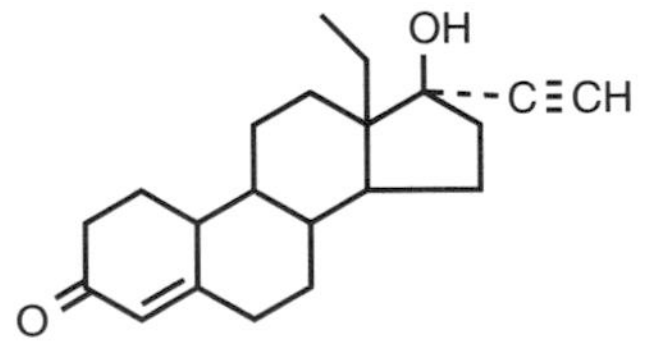

Norgestrel

Further unsaturation of the B or C ring of 19-androstane derivatives usually enhances progestational activity. Introduction of halogen or methyl substituents in the 6α-position or 7α-position generally increases hormonal activity. Acetylation of the 17β-OH of norethindrone results in a longer duration of action. Removal of the keto function of norethindrone at carbon 3 gives lynestrenol (17-ethynylestr-4-en-17β-ol), which retains potent progestational activity and is free of androgenic effects. This hormone is used in combination with an estrogen as a contraceptive agent.

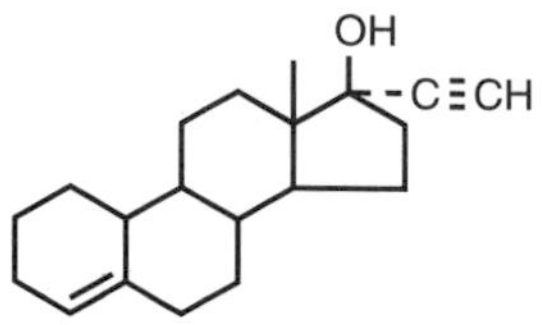

Lynestrenol

Activity of 17α-hydroxyprogesterones is enhanced by unsaturation at positions 6 and 7 and substitution of a methyl group or a halogen at position 6. This activity may be further increased by introducing a CH_3 group at position 11. These substitutions on the progesterone molecule probably prevent metabolic reduction of the two carbonyl groups and metabolic oxidation at position 6. Substitution of a fluoro group at position 21 apparently prevents hydroxylation at this point and enhances the oral effectiveness. Some of the potent orally administered progestins belong to this series of compounds.

Ethynodiol acetate is an extremely potent oral progestin; it is more active orally than parenterally and is effective as an oral contraceptive when combined with an estrogen.

A progestin with a prolonged duration of action is 16α,17-dihydroxyprogesterone acetophenide. When given parenterally, this agent appears to be devoid of both androgenic and estrogenic activities.

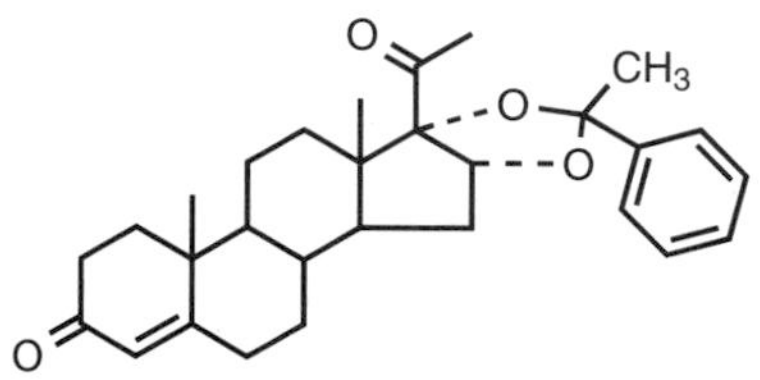

16α,17-Didydroxyprogesterone acetophenide

Inversion of the configuration at positions 9 and 10 in progesterone leads to retroprogesterone, which is more active parenterally and orally than progesterone. Further unsaturation at positions 6 and 7 of retroprogesterone gives dydrogesterone, which has previously been discussed. The adrenocortical hormone 21-hydroxyprogesterone and the precursor of progesterone, pregnenolone, have minimal or no progestational activity.

Progestin Antagonists

An antiprogestin, a compound that antagonizes the actions of progesterone by competing for its receptor, would be an important agent for interfering with the early phases of pregnancy. This area of research has received considerable attention but few results. In 1982, the first antiprogestin, mifepristone (RU 38,486 or abbreviated RU 486), was reported.[293] Mifepristone, 17β-hydroxy-11β-(4-dimethylaminophenyl-1-)-17α-(prop-1-ynl) estra-4,9-dien-3-one, blocked implantation of fertilized eggs in rats and was shown to interrupt early stages of implantation and pregnancy in humans.[294] Mifepristone also demonstrates antiglucocorticoid activity.[165] Additional antiprogestin analogs, such as onapristone (ZK 98,299), have been developed that exhibit lowered antiglucocorticoid activity.[295] These antiprogestins have also demonstrated therapeutic potential for the treatment of hormone-dependent breast cancer.[296]

Female Contraceptives and Abortifacients

Pincus and his colleagues initiated the use of steroidal hormones in oral contraception in the early 1950s.[297] Early findings in animals were extended to human subjects in Haiti and Puerto Rico; such investigations showed that a combination of an estrogen and a progestin prevents conception.[298] At about the same time, Greenblatt and Goldzieher developed a sequential method of contraception.[299] At present, the combination estrogen-progestin pill is used in various forms, but the sequential method is no longer being used. A summary of oral contraceptive preparations available is given in Table 23–4.

In the combination method, both an estrogen and a progestin are administered for 20 or 21 days and then stopped for 7 or 8 days (including the 5-day menstrual period); administration is then repeated. In some instances, a 28-day regimen is given that includes 6 or 7 inert tablets or 7 tablets of 75 mg of ferrous fumarate. In the obsolete sequential method, the estrogen was first administered alone, and then a progestin was added toward the end of each cycle. In some cases a placebo was administered (in the sequential method) between the estrogen-progestin sequence and the starting estrogen. The discontinuation of sequential agents, e.g., Norquen (Syntex Laboratories, Inc., Palo Alto, Calif), Oracon (Mead Johnson, Evansville, Ind), and Ortho-Novum SQ (Ortho Pharmaceutical Corporation, Raritan, NJ) was due to an increased complication rate, linkage to endometrial cancer, and decreased efficiency in preventing

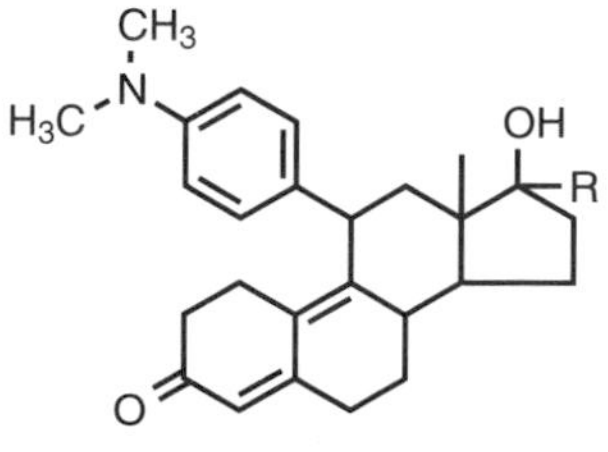

Mifepristone (RU 38,486): R = —C≡C-CH_3
Onapristone (ZK 98,299): R = -$CH_2CH_2CH_2OH$

Table 23–4. Oral Contraceptives

Trade Name	Progestin	(mg)	Estrogen	(mg)
Combination Oral Contraceptives:				
Brevicon (Syntex)	Norethindrone	0.5	Ethinyl estradiol	0.035
Demulen 1/35 (Searle)	Ethynodiol diacetate	1.0	Ethinyl estradiol	0.035
Demulen 1/50 (Searle)	Ethynodiol diacetate	1.0	Ethinyl estradiol	0.050
Enovid 5 mg (Searle)*	Norethynodrel	5.0	Mestranol	0.075
Enovid-E (Searle)*	Norethynodrel	2.5	Mestranol	0.100
Genora 1/35 (Rugby)	Norethindrone	1.0	Ethinyl estradiol	0.035
Genora 1/50 (Rugby)	Norethindrone	1.0	Mestranol	0.050
Levlen (Berlex)	Levonorgestrel	0.15	Ethinyl estradiol	0.030
Loestrin 1/20 (Parke-Davis)	Norethindrone acetate	1.0	Ethinyl estradiol	0.020
Loestrin 1.5/30 (Parke-Davis)	Norethindrone acetate	1.5	Ethinyl estradiol	0.030
Lo/Ovral (Wyeth-Ayerst)	Norgestrel	0.3	Ethinyl estradiol	0.030
Modicon (Ortho)	Norethindrone	0.5	Ethinyl estradiol	0.035
Nelova 0.5/35E (Warner-Chilcott)	Norethindrone	0.5	Ethinyl estradiol	0.035
Nelova 1.0/35E (Warner-Chilcott)	Norethindrone	1.0	Ethinyl estradiol	0.035
Nelova 1/50M (Warner-Chilcott)	Norethindrone	1.0	Mestranol	0.050
Nordette (Wyeth-Ayerst)	Levonorgestrel	0.15	Ethinyl estradiol	0.030
Norethin 1/35E (Searle)	Norethindrone	1.0	Ethinyl estradiol	0.035
Norethin 1/50M (Searle)	Norethindrone	1.0	Mestranol	0.050
Norinyl 1 + 35 (Syntex)	Norethindrone	1.0	Ethinyl estradiol	0.035
Norinyl 1 + 50 (Syntex)	Norethindrone	1.0	Mestranol	0.050
Norinyl 1 + 80 (Syntex)*	Norethindrone	1.0	Mestranol	0.080
Norinyl 2 mg (Syntex)	Norethindrone	2.0	Mestranol	0.100
Norlestrin 1/50 (Parke-Davis)	Norethindrone acetate	1.0	Ethinyl estradiol	0.050
Norlestrin 2.5/50 (Parke-Davis)	Norethindrone acetate	2.5	Ethinyl estradiol	0.050
Ortho-Novum 1/35 (Ortho)	Norethindrone	1.0	Ethinyl estradiol	0.035
Ortho-Novum 1/50 (Ortho)	Norethindrone	1.0	Mestranol	0.050
Ortho-Novum 10 mg (Ortho)*	Norethindrone	10.0	Mestranol	0.060
Ortho-Novum 1/80 (Ortho)*	Norethindrone	1.0	Mestranol	0.080
Ortho-Novum 2 mg (Ortho)*	Norethindrone	2.0	Mestranol	0.100
Ovcon-35 (Mead-Johnson Nutrition)	Norethindrone	0.4	Ethinyl estradiol	0.035
Ovcon-50 (Mead-Johnson Nutrition)	Norethindrone	1.0	Ethinyl estradiol	0.050
Ovral (Wyeth-Ayerst)	Norgestrel	0.5	Ethinyl estradiol	0.050
Ovulen (Searle)*	Ethynodiol diacetate	1.0	Mestranol	0.100
Biphasic Oral Contraceptives:				
Nelova 10/11 (Warner-Chilcott)	Norethindrone	0.500	Ethinyl estradiol	0.035
	Norethindrone	1.000	Ethinyl estradiol	0.035
Ortho-Novum 10/11 (Ortho)	Norethindrone	0.500	Ethinyl estradiol	0.035
	Norethindrone	1.000	Ethinyl estradiol	0.035
Triphasic Oral Contraceptives:				
Ortho-Novum 7/7/7/ (Ortho)	Norethindrone	0.500	Ethinyl estradiol	0.035
	Norethindrone	0.750	Ethinyl estradiol	0.035
	Norethindrone	1.000	Ethinyl estradiol	0.035
Tri-Levlen (Berlex)	Levonorgestrel	0.050	Ethinyl estradiol	0.030
	Levonorgestrel	0.075	Ethinyl estradiol	0.040
	Levonorgestrel	0.125	Ethinyl estradiol	0.030
Tri-Norinyl (Syntex)	Norethindrone	0.500	Ethinyl estradiol	0.035
	Norethindrone	1.000	Ethinyl estradiol	0.035
	Norethindrone	0.500	Ethinyl estradiol	0.035
Triphasil (Wyeth-Ayerst)	Levonorgestrel	0.050	Ethinyl estradiol	0.030
	Levonorgestrel	0.075	Ethinyl estradiol	0.040
	Levonorgestrel	0.125	Ethinyl estradiol	0.030
Progestin Oral Contraceptives:				
Micronor (Ortho)	Norethindrone	0.35		
Nor-Q.D. (Syntex)	Norethindrone	0.35		
Ovrette (Wyeth-Ayerst)	Norgestrel	0.075		

* Combination oral contraceptives with estrogen content greater than 0.050 mg are no longer commercially available in the U.S.

pregnancies.[300] The new biphasic and triphasic formulations are designed to simulate more closely the normal menstrual cycle and minimize breakthrough bleeding.[301]

A third method of contraception is to administer continuously a small dose of a progestin (Table 23–4). In what is referred to in some instances as the mini pill, no estrogen is given at any time. This is thought to reduce some of the risks associated with the use of higher doses of estrogens. A major disadvantage is that irregular bleeding is usually observed during the first 18 months of therapy. After this period, it has been reported, abnormal bleeding does not occur.[302] Progestin contraceptive therapy is thought to cause less interference with the endocrine system. There is no pituitary inhibition, and this method does not produce the thromboembolic episodes reported with the regular combination and sequential methods.

Various long-term steroidal contraceptives such as once-a-week and once-a-month preparations are being investigated. One of these longer-term products uses an injectable form of medroxyprogesterone acetate every 3 months. Prolonged infertility or permanent infertility, however, may occur after the use of such a depot contraceptive. Norplant is another product that has been developed as a long-term steroidal contraceptive. It consists of slender Silastic tubes containing a synthetic progestin, norethistrone, in a powdered form. These tubes are implanted under the skin and release the progestin for 3 or 5 years.[303] Another long-term approach uses a progestin-containing intrauterine device (IUD), sometimes called a uterine therapeutic system (UTS). The Progestasert IUD contains 38 mg of progesterone dispersed in silicone oil within a flexible T-shaped polymer that releases 6 μg/day of progesterone into the uterine area for 1 year. The use of the nonsteroidal estrogenic substance diethylstilbestrol as a postcoital contraceptive or "morning after" contraceptive is not recommended because of its risk for causing reproductive carcinomas in offspring.

Approximately 55 million women use oral contraceptives, the most popular form of contraception used in the world today.[304] The oral contraceptive steroids appear to be superior to any other method of preventing pregnancies. The failure rate of oral contraceptives is 0.2 to 0.6 per 100 woman-years; these data include absorption abnormalities, e.g., those produced by diarrhea. The more recent contraceptive steroid combination or progestin-only products produce fewer side effects than the original agents. The safety of the oral contraceptives has been a subject of keen interest. One of the more studied uses of estrogens has been the treatment of postmenopausal women, and reports have concentrated on the increased risk of cancer of the endometrium in such therapy.[305,306] Among the side effects more commonly encountered are nausea, vomiting, menstrual disturbances, breakthrough bleeding, decreased hepatic function, breast tenderness, weight gain, changes in libido, and headaches.

Evidence indicates an increased risk of a more serious nature—thromboembolic and vascular problems in women using oral contraceptives. Other serious conditions are increased risk of myocardial infarction, hepatic adenomas, hypertension, gallbladder disease, breast cancer, and altered carbohydrate metabolism.[302,304,307–309] Most of these studies were done with oral contraceptives containing relatively high doses of estrogen and that the trend has been toward lowering the estrogen component. Apparently, the estrogenic substance is the prime component in oral contraceptives that is responsible for producing thromboembolisms, and lowering the dose of the estrogen has provided a lower incidence of this problem.[308]

Other toxic effects that may be induced by the estrogen include altered carbohydrate metabolism and cancer.[302,309] Oral contraceptives are contraindicated in pregnancy because of the possibility of masculinization of the fetus.[310] The effects of prolonged use of these agents are still under investigation.

The primary mode of action of the combination and sequential oral contraceptives seems to involve blocking ovulation by inhibiting secretion of hypothalamus-releasing factors through a sex steroid feedback mechanism. Other mechanisms may also play a major role; for example, the contraceptive action of various steroids may be related to an increase in viscosity of the cervical mucus.

In recent years, two prostaglandins, PGE_2 and $PGF_{2\alpha}$, have been used as abortifacients. $PGF_{2\alpha}$ is injected into the amniotic sac, whereas PGE_2 is given by vaginal suppository to induce abortion. Saline-induced abortions have also been used previously.

ANDROGENS

The mechanisms controlling growth and development of the male gonads are similar to those in the female. The hypothalamus controls the adenohypophysis through the same releasing factor as for females, namely, LHRH. This substance then brings about the release of follicle-stimulating hormone (FSH) and luteinizing hormone (LH), or, as the latter is more commonly called in the male, interstitial cell-stimulating hormone (ICSH), from the adenohypophysis.

The two gonadotropins appear to have separate functions. FSH promotes sperm development, or spermatogenesis, by stimulating the seminiferous tubules. The pri-

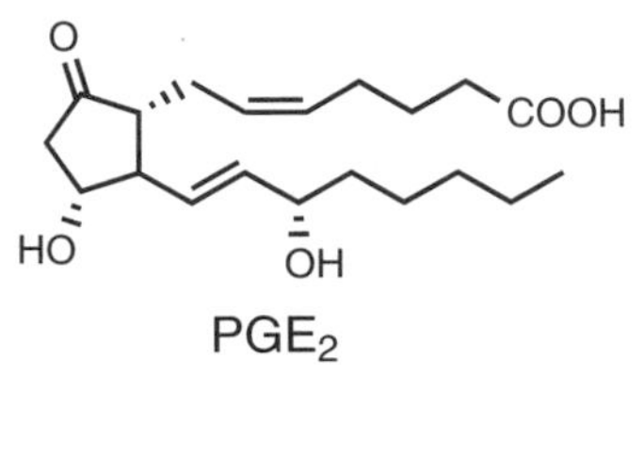

PGE_2

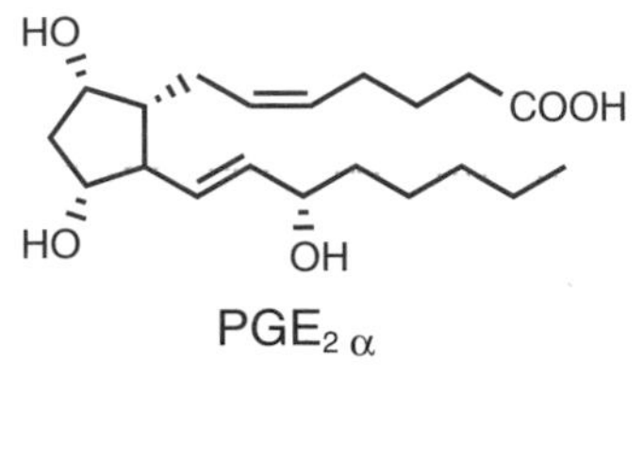

$PGE_{2\alpha}$

mary action of ICSH, on the other hand, is to stimulate the interstitial Leydig's cells to secrete androgens.

The feedback system between the testes and hypothalamus is not well understood. If the androgens, primarily testosterone, reach a certain level, they cause a decrease in LHRH released from the hypothalamus with a resultant decrease in FSH secretion from the adenohypophysis. Castration causes a deficiency of endogenous testosterone; the result is an increase in secretion of the gonadotropins from the adenohypophysis along with an increase in urinary excretion.

The primary testicular androgen, testosterone, has important sexual and metabolic activities. It controls the development and maintenance of the sex organs, including the vas deferens, prostate, seminal vesicles, and penis. Spermatogenesis also depends on testosterone. Functional sperm, therefore, depends on both the gonadotropins and testosterone.

This androgen is also needed for the development of secondary sex characteristics. The male voice deepens because of thickening of the laryngeal mucosa and lengthening of the vocal cords. It plays a role first in stimulating the growth of hair on the face, arms, legs, and pubic areas and later in the recession of the male hairline. The fructose content of human semen and the size and secretory capacity of the sebaceous glands also depend on the levels of testosterone.

Testosterone causes nitrogen retention by increasing the rate of protein synthesis while decreasing the rate of protein catabolism. The positive nitrogen balance therefore results from both decreased catabolism and increased anabolism of proteins that are used in male sex accessory apparatus and muscle. The thickness and linear growth of bones are stimulated and later limited by testosterone because of closure of the epiphyses.

Biosynthesis

Androgens (male sex hormones) are synthesized from cholesterol in the testes and adrenal cortex. The major pathway for the biosynthesis of testosterone, the most important androgen, is shown in Figure 23–12.[311–313] In the liver, androgens are formed from C-21 steroids. Small amounts are also secreted by the ovary. This is not surprising because androgens are intermediates in the biosynthesis of estrogens. Testosterone is the major circulating androgen in males; the primary source of testosterone are the Leydig's cells of the testes. LH binds to its receptor on the surface of the Leydig's cells to initiate testosterone biosynthesis. As in other endocrine cells, the binding of the gonadotropin results in an increase in intracellular cAMP levels via activation of a G-protein and adenylyl cyclase.

One of the processes influenced by elevated cAMP levels is the activation of cholesterol esterase, which cleaves cholesterol esters and liberates free cholesterol. The free cholesterol is then converted in mitochondria to pregnenolone via the side-chain cleavage reaction, as described earlier. Pregnenolone is converted by 17α-hydroxylase to 17α-hydroxypregnenolone and then to dehydroepiandrosterone (DHA) via 17-20 lyase. DHA is converted by 5-ene-3β-hydroxysteroid dehydrogenase and 3-oxosteroid-4,5-isomerase to the 17-ketosteroidal androgen, androstenedione (androst-4-ene-3,17-dione). Testosterone is formed by reduction of the 17-ketone of androstenedione by 17β-hydroxysteroid dehydrogenase.[314,315] Testosterone and androstenedione are metabolically interconvertible. Testosterone is secreted by the Leydig's cells and can act in a negative feedback fashion in the hypothalamus and pituitary to decrease the release of gonadotropins.

The most potent endogenous androgen is the 5α-reduced steroid, 5α-dihydrotestosterone (5α-DHT,17β-hydroxy-5α-androstan-3-one). This molecule is formed in androgen target tissues by the enzyme 5α-reductase, which has been located in both the microsomal fraction and the nuclear membrane of homogenized target tissues. The 5α-reductase enzyme catalyzes an irreversible reaction and requires NADPH as a cofactor, which provides the hydrogen at C-5.[316,317]

Metabolism

The metabolism of testosterone can lead either to physiologically active steroids or to inactive molecules, as shown in Figure 23–16.[311–313] Both reductive and oxidative pathways of metabolism of testosterone can produce important, physiologically active, steroidal compounds. In androgen target tissues such as the prostate gland, testosterone is converted by 5α-reductase to 5α-DHT, as described previously. In ovarian and adipose tissues, the angular C-19 methyl group of testosterone is oxidized by aromatase to lead to the cleavage of the methyl group, aromatization of the A ring, and the production of the estrogen estradiol. In CNS tissues, both pathways are active, and research suggests that these biotransformations influence brain differentiation and function.[318,319]

As with cholesterol and the other physiologically active steroids, the liver is a major site of metabolic inactivation of androgens.[311–313] In the urine, metabolites of the major androgenic agent, testosterone, are found as their more water-soluble glucuronide conjugates and to a lesser extent as sulfate conjugates or in the free form.[320] Androsterone (3α-hydroxy-5α-androstan-17-one) and its 5β-diastereoisomer etiocholanolone are the predominant metabolites of testosterone. Just as 17β-estradiol and estone are interconvertible, testosterone is interconvertible with androstenedione. These compounds are precursors of 17β-estradiol and estrone. This conversion is important in the ovary and testes, but of minor significance in the adrenal glands. A number of minor metabolites of testosterone have also been isolated from urine and identified as 5α-androstanes and 5β-androstanes with a 3α-hydroxyl function. Most 17-ketosteroids isolated from the urine result from catabolism of the adrenocorticoids rather than from metabolism of androgens.

Naturally Occurring Androgens

One of the early and unusual experiments with testicular extracts was carried out in 1889 by the French physiologist Brown-Séquard. He administered such an extract to himself and reported that he felt an increased vigor and

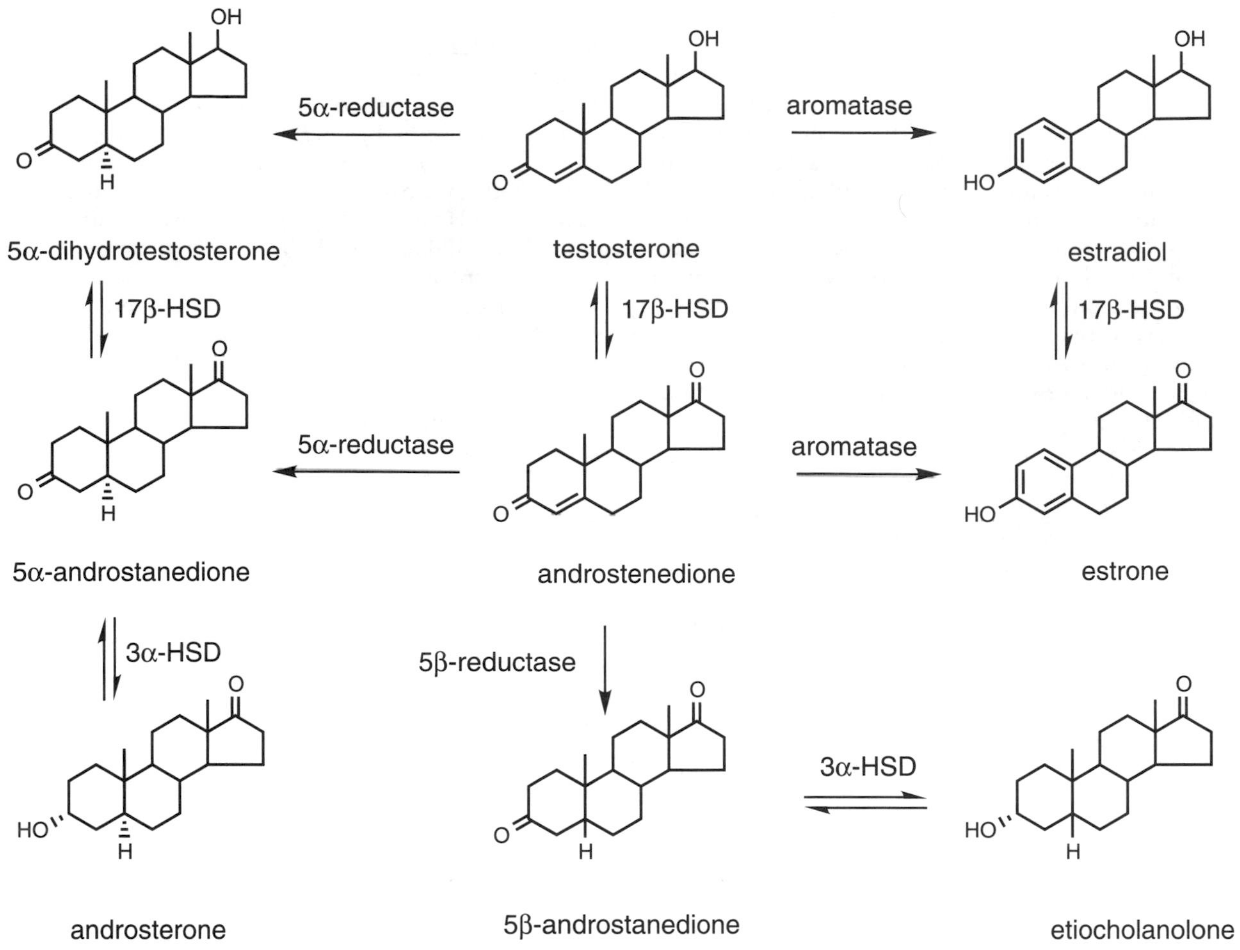

Fig. 23–16. Androgen metabolism.

capacity for work.[321] In 1911, Pézard[322] showed that extracts of testicular tissue increase comb growth in capons. Early attempts to isolate pure male hormones from the testes failed because only small amounts are present in this tissue.

The earliest report of an isolated androgen was presented by Butenandt in 1931.[323] He isolated 15 mg of crystalline androsterone from 15,000 liters of human male urine. A second crystalline compound, dehydroepiandrosterone, which has weak androgenic activity, was isolated by Butenandt and Dannenberg in 1934.[324] In the following year, testosterone N.F. (17β-hydroxy-4-androsten-3-one) was isolated from bull testes by David and associates.[325] This hormone was shown to be 6 to 10 times as active as androsterone.

Shortly after testosterone was isolated, Butenandt and Hanisch[326] reported its synthesis. In the same year, extracts of urine from males were shown to cause nitrogen retention as well as the expected androgenic effects.[327] Many steroids with androgenic activity have subsequently been synthesized. Steroid hormones may have many potent effects on various tissues, and slight chemical alterations of androgenic steroids may increase some of these effects without altering others.[328]

Testosterone was the first androgen to be used clinically for its anabolic activity, but new sources of the hormone were needed because only 270 mg could be isolated from a ton of bull testes.[329] Commercially, testosterone is prepared from various steroids, including sarsasapogenin, diosgenin, and certain androgens found in stallion urine. Owing to its androgenic action, testosterone is limited in its use in humans as an anabolic steroid. Many steroids were synthesized in an attempt to separate the androgenic and the anabolic actions. Because testosterone had to be given parenterally, it was also desirable to find orally active agents.

Synthetic Androgen and Anabolic Agents

Some of the early studies with androgens included structural modifications of the naturally occurring hormones. As with other hormones, esterification of testosterone with acids, affording the propionate (U.S.P.), heptanoate (U.S.P.), and phenylacetate, at the 17β-position resulted in agents with an increased duration of action when given parenterally.

The synthesis of 17α-methyltestosterone (N.F.) made available a compound that was orally active[330] in daily doses between 10 and 50 mg.

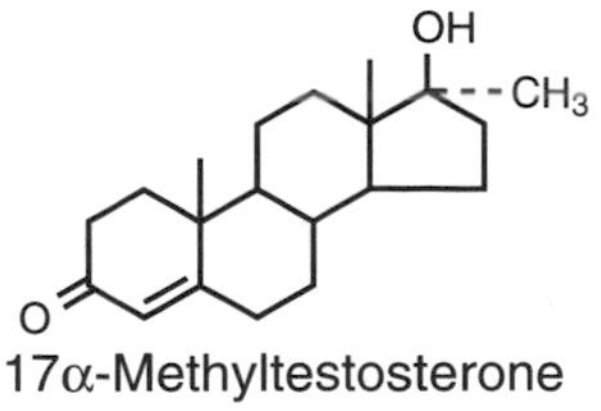

17α-Methyltestosterone

This drug has the androgenic and anabolic activities of testosterone. Although orally active, it is more effective when administered sublingually.

Increasing the length of the alkyl side chain at the 17α-position, however, resulted in decreased activity, and the incorporation of other substituents, such as the 17α-ethynyl group, produced compounds with useful progestational activity. Several modifications of 17α-methyltestosterone lead to potent, orally active anabolic agents. Two

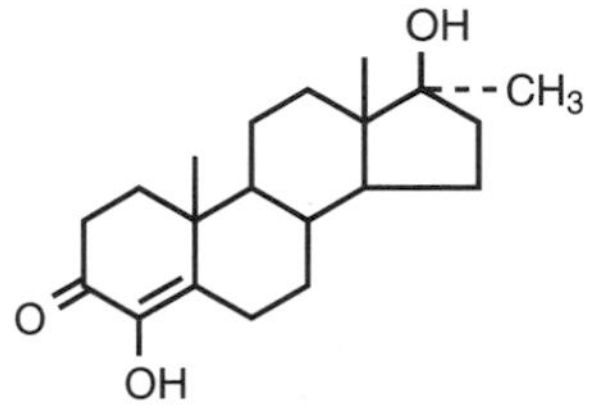

Oxymesterone

hydroxylated derivatives include oxymesterone (4,17β-dihydroxy-17-methyl-4-androsten-3-one) and oxymetholone (2-hydroxymethylene-17β-hydroxy-17-methyl-5α-androstan-3-one).

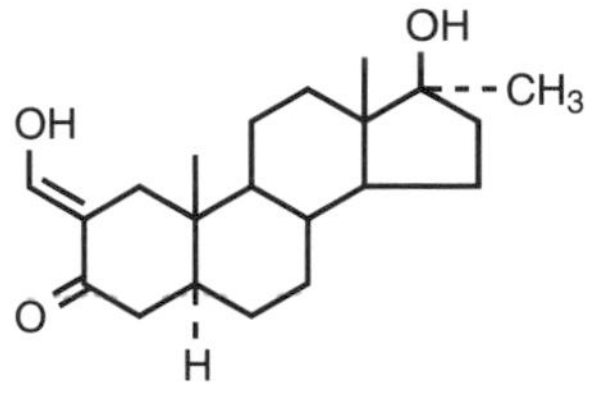

Oxymetholone

These drugs have at least three times the anabolic and half the androgenic activity of the parent compound.[311,312]

Selenium dioxide dehydrogenation of 17α-methyltestosterone yields methandrostenolone (N.F.), known chemically as 17β-hydroxy-17-methylandrost-1,4-dien-3-one.

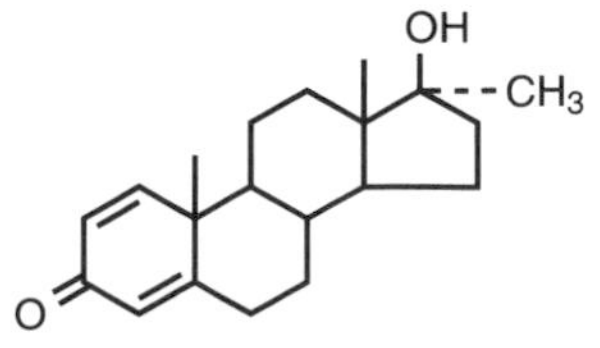

Methandrostenolone

This drug has several times the anabolic activity of the starting material. It has low androgenic activity but can apparently produce mammogenic effects in men. These effects are thought to result from metabolites.

Fluoxymesterone (U.S.P.), also called 9-fluoro-11β,17β-dihydroxy-17-methyl-4-androsten-3-one, has 20 times the anabolic and 10 times the androgenic activity of 17α-methyltestosterone.[330]

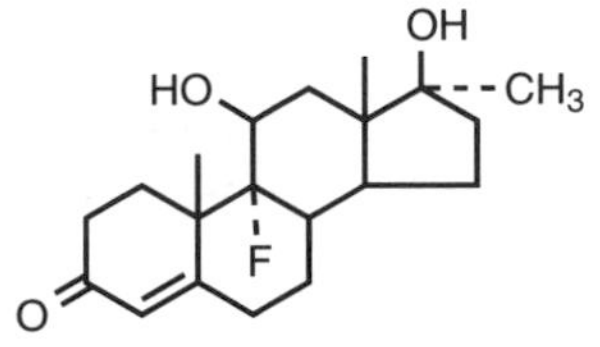

Fluoxymesterone

It is used clinically as an androgen in doses of 4 to 10 mg per day.

A heterocyclic analog of 17α-methyltestosterone is oxandrolone (N.F.), named 17β-hydroxy-17-methyl-2-oxa-5α-androstan-3-one.

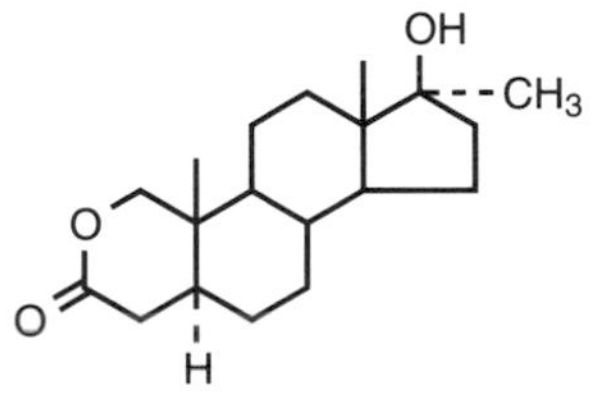

Oxandrolone

This drug, which contains a lactone in the A ring and is therefore susceptible to hydrolysis, has three times the anabolic activity of 17α-methyltestosterone; it exhibits slight androgenic activity. Clinically, oxandrolone is used in dosages of 7.5 to 20 mg per day for its anabolic action.

Another heterocyclic compound used for its anabolic effects is stanozolol (N.F.), or 17β-hydroxy-17-methylandrostano-[3,2-c]-pyrazole. Usually, 2 mg of this material are administered three times a day for its anabolic action.

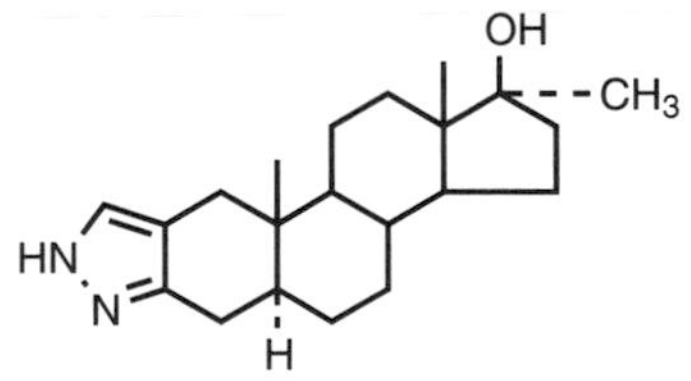

Stanozolol

Danazol, 17α-pregna-2,4-diene-20-yn[2,3-d]isoxazol-17-ol, has been marketed for use in endometriosis. Previous treatment of endometriosis had been surgical or with progestins or a combination of estrogen and progestin. Danazol is a weak androgenic substance that inhibits release of gonadotropins from the pituitary gland.[331]

Klimstra[328] reported that alkylation in the 1, 2, 7, and 18 positions of the androstane molecule generally increases anabolic activity. One of these derivatives, methe-

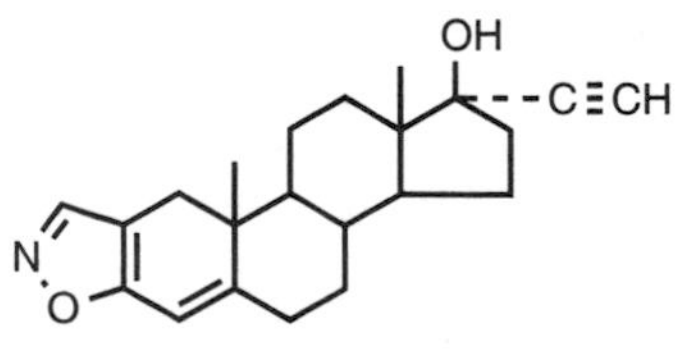

Danazol

nolone acetate (1-methyl-17β-hydroxy-5α-androst-1-en-3-one 17-acetate) is an example of a potent anabolic agent that does not have an alkyl substituent at the 17α-position.

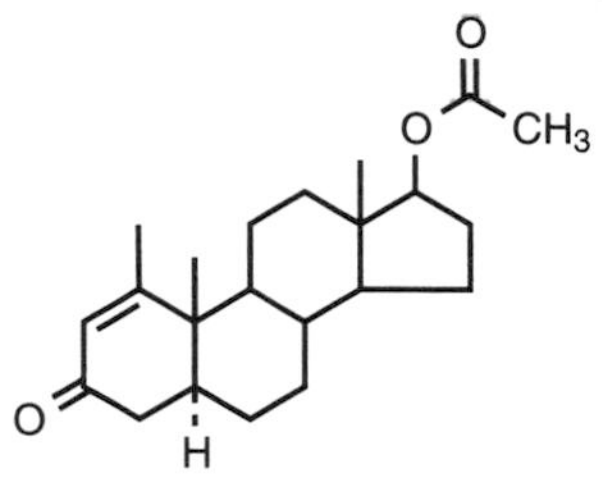

Methenolone acetate

This compound is administered parenterally in a dose of 20 mg per week for its anabolic action.

A halogenated anabolic agent used in about the same dosage is chlortestosterone (4-chloro-17β-hydroxy-4-androsten-3-one 17-acetate).

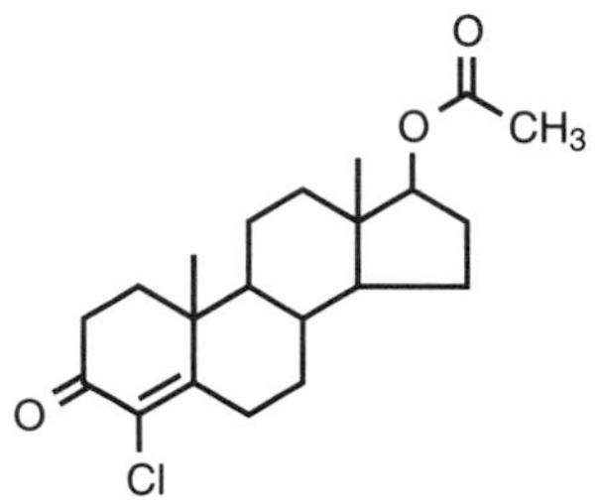

Chlortestosterone acetate

Androgens, having no methyl group in the 10 position of the steroid nucleus, are an important class of anabolic agents often referred to as the 19-norandrogens. These compounds can be synthesized by the Birch reduction illustrated in Figure 23–17. The aromatic A ring of a 3-methoxy estrogen is reduced to a diene (3-methoxy-17β-hydroxy-2,5(10)-estradiene). Cleavage of the enol ether with HCl results in 19-nortestosterone (17β-hydroxy-4-estren-3-one).

In animal assays, 19-nortestosterone has about the same anabolic activity as the propionate ester of testosterone, but its androgenic activity is much lower. Because 19-nortestosterone showed some separation of anabolic and androgenic activities, related analogues were synthesized and biologically investigated. Two of the more potent members of the series are norethandrolone (N.F.), a 17α-ethyl-19-nortestosterone (17β-hydroxy-17-ethyl-4-estren-3-one), and ethylestrenol (17β-hydroxy-17-ethyl-4-estrene).

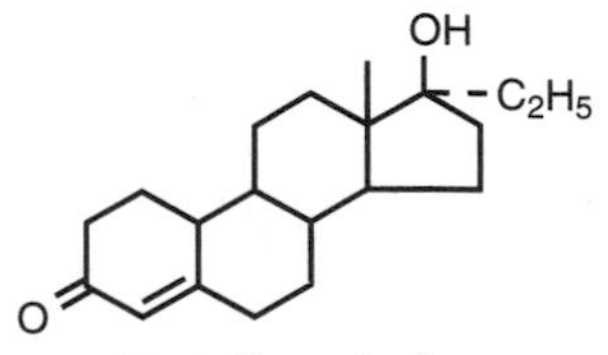

Norethandrolone

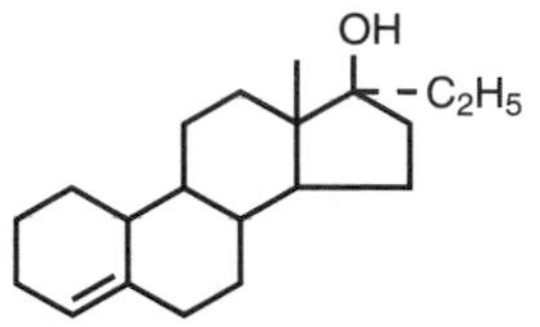

Ethylestrenol

Norethandrolone has a better ratio of anabolic to androgenic activity than does either 19-nortestosterone or 17α-methyl-19-nortestosterone. The usual dose of norethandrolone is 30 to 50 mg per day administered orally or parenterally. Both androgenic and progestational side effects have been observed with this agent. Ethylestrenol is more potent than norethandrolone as an anabolic agent and is used in a dosage of 4 mg per day orally.

Testolactone (N.F.), D-homo-17α-oxoandrosta-1,4-diene-3,17-dione, possesses some anabolic but no androgenic effects. It is used as an antineoplastic agent and is thought to depress ovarian function, thus reducing the level of estrogens that would stimulate the growth of breast tissue. Testolactone is available in both parenteral and oral forms.

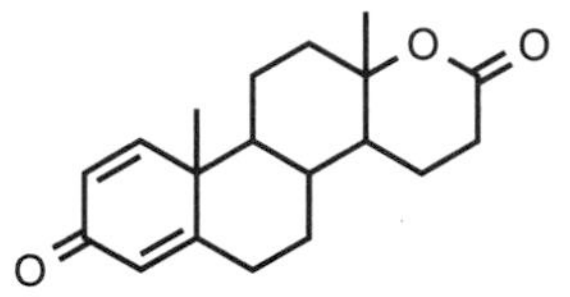

Testolactone

Mode of Action

Androgens produce various physiologic effects. Nevertheless, research has shown the mechanism of action of androgens to be through regulation of protein synthesis in target cells by the formation of a steroid-receptor complex. Extensive reviews on the mechanism of androgen action have appeared.[206,331–335] The current concept of this mechanism of action has evolved from studies using androgen-sensitive rat ventral prostate. Testosterone was found to be rapidly converted to 5α-dihydrotestosterone (DHT) in the prostate, and this reduced metabolite was selectively retained.[316,336] Thus, the mechanism of action of testosterone in the prostate begins with the conversion of testosterone to DHT by 5α-reductase. DHT binds to proteins present in the nucleus of prostate cells with extremely high affinity and specificity for DHT.[337] This DHT-receptor complex then undergoes a conformational change and the steroid-receptor complex then in-

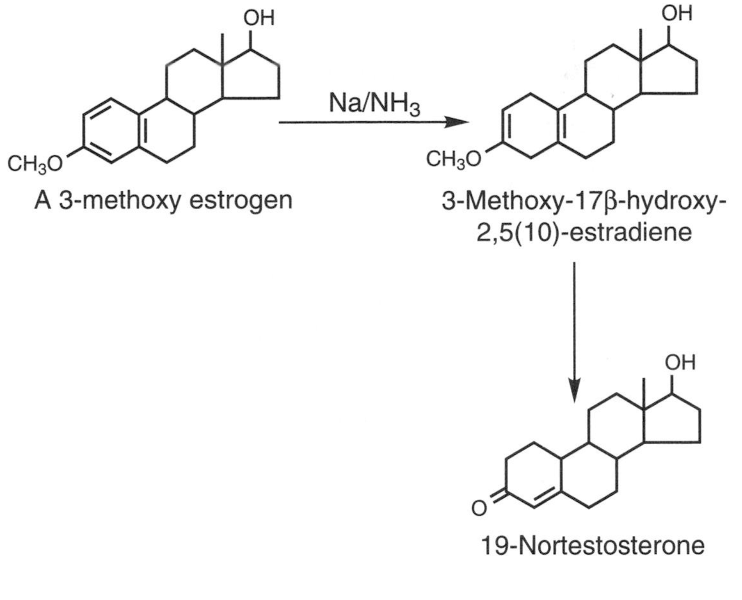

Na/NH$_3$

A 3-methoxy estrogen

3-Methoxy-17β-hydroxy-2,5(10)-estradiene

19-Nortestosterone

Fig. 23–17. Synthesis of 19-nortestosterone by the Birch reduction.

teracts with the chromatin present in the target cell and results in the increased production of messenger RNA (mRNA).[337–340] These elevated levels of mRNA result in increased protein synthesis and subsequent stimulation of cell growth and differentiation.

Androgen receptors have been identified in other tissues, such as seminal vesicles,[341,342] testis,[343,344] epididymis,[343,345] kidney,[346] brain,[347–349] liver,[350] and androgen-sensitive tumors.[351,352] DHT, however, is not the only functioning form of androgen in other androgen-sensitive tissues. DHT is rapidly biosynthesized in tissues such as those of the prostate, but is not readily formed in tissues such as those of the kidney. The mechanism of anabolic action of the androgens also appears to involve the formation of a steroid-receptor complex. Testosterone has greater activity than DHT in the androgen-mediated growth of muscle both in vivo and in tissue culture.[353–355] Receptor proteins specific for testosterone have been identified in muscle tissues.[356–362] Thus, the involvement of specific receptor proteins and the resulting increase in protein synthesis are the common mechanism of action of the androgens in various target tissues.

Pharmacology, Side Effects, and Clinical Applications

The male sex hormones are responsible for normal growth and development of the male sex organs and for the retention of nitrogen and certain inorganic substances, such as potassium, calcium, chloride, phosphorus, and sodium. The anabolic effect of androgens is responsible for the fast growth of males at puberty.

Among the important therapeutic applications of the androgenic hormones is replacement therapy in patients with deficient endogenous androgen production. Various preparations of testosterone and its congeners include aqueous suspensions and oil solutions for parenteral administration, tablets for oral administration, and tablets for buccal and sublingual absorption. Solid pellets are also available for implantation under the skin.

The various androgen preparations can be used in the treatment of eunuchism and eunuchoidism to restore or maintain secondary sexual characteristics. Gonadotropins may be administered when hypogonadism exists to determine if the testes are capable of producing the needed hormones. Androgens have been used alone and in combination with gonadotropins for cryptorchidism (failure of the testes to descend), and are also used for the treatment of faulty spermatogenesis, benign prostatic hypertrophy, and impotency.

In older men who are undergoing the male climacteric, which is analogous to menopause in women, androgens may be beneficial. In women the androgens are used for breast engorgement, inoperable mammary carcinoma, hypolibido, and chronic cystic mastitis. At one time, the androgens were used in menstrual disorders, but this treatment has been supplanted by the progestins and estrogens. In small amounts, androgens, along with estrogens, are used in the treatment of menopause.

The androgens have been used in both sexes as anabolic agents, to increase weight in both adults and children, and to reverse the loss of protein resulting from trauma, prolonged immobilization, and wasting diseases. Because androgens retain calcium, they are used in the treatment of osteoporosis, which often occurs in the elderly.

The use of androgens is limited because of the side effects they can produce in humans. In women, masculinization may occur, particularly during long-term therapy. This leads to growth of hair on the face, voice changes, and the development of a more muscular body. In both sexes, there may be increased retention of electrolytes and water during extended periods of androgen therapy, leading to edema.

Hepatic dysfunction may occur in patients taking the 17α-alkylated androgens, but this appears to be a transitory effect because hepatic function returns to normal on cessation of therapy. Premature epiphyseal closure and virilization limit the use of anabolic agents in children. Caution must also be exercised when using androgens in conjunction with other drugs. The anticoagulant response of coumarin-type anticoagulants can be increased by norethandrolone.[363]

Structure-Activity Relationships

For a substance to have androgenic activity, it must contain a steroid skeleton.[364] Oxygen functional groups normally occurring at positions 3 and 17 are not essential because the basic nucleus, 5α-androstane, has androgenic activity. This appears to be the minimal structural requirement for hormonal activity. For derivatives of etiocholane, in which the hydrogen is in the 5β-position, thereby affording a *cis*-A/B ring juncture, no active androgens and anabolic agents are known.[365] Generally, both ring expansion (to form homo derivatives by inserting a methylene group into one of the rings in the steroid nucleus) or ring contraction (by removing a methylene group) significantly reduces or destroys the androgenic and anabolic activities.

Introduction of a 3-ketone function or a 3α-OH group enhances androgenic activity. A hydroxyl group in the 17α-position of androstane contributes no androgenic or anabolic activity; no known substituent can approach the effectiveness of a 17β-OH group. Evidence indicates that the longer-acting esters of the 17β-OH compounds are hydrolyzed in vivo to the free alcohol, which is the active species.[366] It is thought that the 17β-oxygen atom is important for attachment to the receptor site[365] and that 17α-alkyl groups are important for preventing metabolic changes at this position. Such 17α-substituents render the compounds orally active.

Ordinarily, halogen substitution produces compounds with decreased activity except when inserted into positions 4 or 9. Replacement of a carbon atom in position 2 by oxygen has produced the only clinically successful heterocyclic steroid among a number of azasteroids and oxasteroids. Some of the 2-oxasteroids are potent anabolic agents.

Introduction of an sp^2 hybridized carbon atom into the A ring renders the ring more planar, and this in turn may be responsible for greater anabolic activity. The 19-norsteroids are of interest because these agents seem to produce a more favorable ratio of anabolic to androgenic activity. Vida[365] has extensively reviewed the replacement of various hydrogens on the androgen steroid skeleton by other functional groups. It appears that certain substitutions at positions 1, 2, 7, 17, and 18 may result in compounds with favorable activities that will be of clinical importance.[330]

Bioassays used in determining the androgenic activity of these hormones include measurements of capon comb growth (in terms of size and weight), increased weight of seminal vesicles, and increased weight of the ventral part of the prostate of castrated rats. Measurements of anabolic activity are based on an increase in nitrogen retention or an increase in mass of specific muscles in animals.

Although much effort has concentrated on the development of female contraception, some work has been carried out in males.[367] A surgical method of male contraception, vasectomy, is reported to be the most effective method and one of the most widely used, especially in India and China. It is a rather simple surgical procedure that disrupts spermatozoal exit and produces few side effects. The irreversible nature of vasectomy and the subsequent psychologic barrier raised in some men has prompted a continuing search for reversible approaches to vasectomy and drugs.[368]

It has been shown that androgens, progestins, estrogens, and many chemically nonrelated substances have the ability to interfere with the formation of mature ejaculated spermatozoa.[367] Unfortunately, none of the steroid preparations is without side effects. It has been reported that, in a village in China, not a single childbirth occurred over a 10-year period from 1930 to 1940, whereas before and after this period, fertility was normal.[369] During those years, it was noted that the villagers had switched from soybean oil to cottonseed oil for cooking. Tests in animals showed that cottonseed oil had an antispermatogenic effect. The active principle responsible for the infertility effect is gossypol, the main side effects of which are fatigue, gastrointestinal upsets, weakness, hypokalemia, low libido, and carcinogenic effects in the skin of mice. The finding of the activity of gossypol is an important milestone in the search for male antifertility agents.[370] Gossypol has been tested in men in China as a male contraceptive.

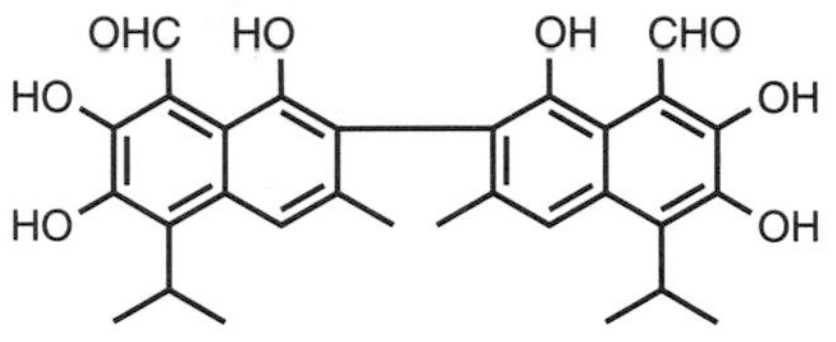

Gossypol

Androgen Antagonists

Antagonists of androgens include agents that block androgen receptors (antiandrogens) and inhibitors of androgen biosynthesis. An antiandrogen is a substance that antagonizes the actions of dihydrotestosterone at the androgen receptor and, when administered with an androgen, blocks or diminishes the effectiveness of androgens in androgen-sensitive tissues. Such compounds have shown potential therapeutic use in the treatment of acne, virilization in women, and hyperplasia and neoplasia of the prostate.[371] Several steroidal and nonsteroidal agents have demonstrated antiandrogenic activity. Cyproterone acetate suppresses gonadotropin release[372] and binds with high affinity to the androgen receptor.[373] Oxendolone also acts by competing for the receptor binding sites.[374] Recently, a novel androgen receptor antagonist, WIN 49,596, was described;[375] this agent contains a fused pyrazole ring at carbons 2 and 3 of the steroid nucleus. A potent nonsteroidal antiandrogen, flutamide, has been shown to compete with DHT for the androgen receptor.[376] Its hydroxylated metabolite is a more powerful antiandrogen in vivo, and has a higher affinity for the receptor than the parent compound.[377]

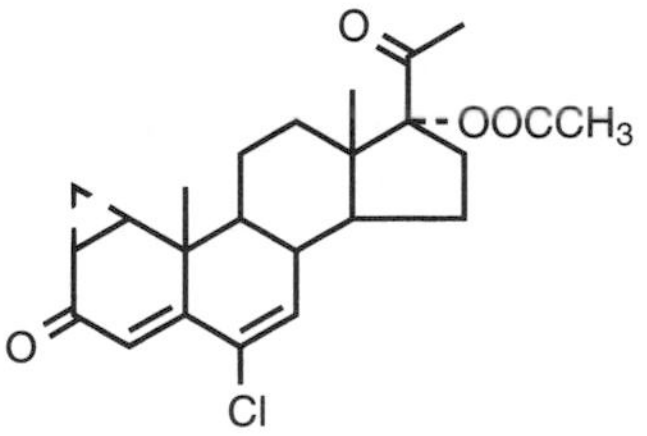

Cyproterone acetate

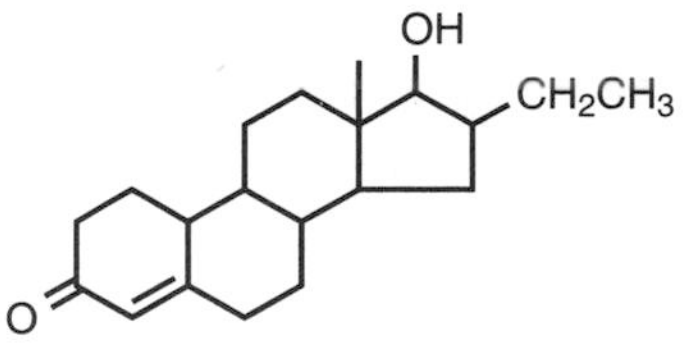

Oxendolone

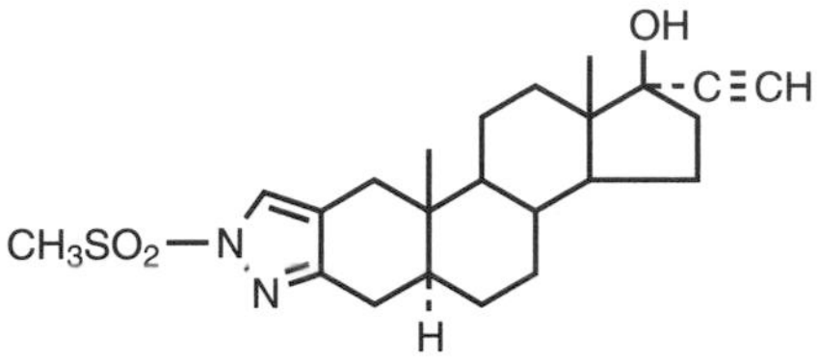

WIN 49,596

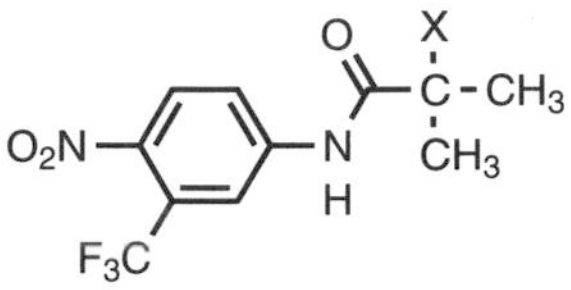

Flutamide: X = H
Metabolite: X = OH

Inhibitors of androgen biosynthesis result in a decrease in active androgen concentrations in target tissues and thus antagonize androgen action. The critical enzyme targeted for inhibition is 5α-reductase, which converts testosterone to the most potent endogenous androgen, 5α-dihydrotestosterone. The first agent to demonstrate 5α-reductase inhibition was a progestin analog, medrogesterone.[378] The azasteroid finasteride (Proscar) is a potent inhibitor of 5α-reductase[379,380] and effectively decreases 5α-DHT concentrations in both plasma and in prostate tissues.[380–382] This drug is approved for treatment of benign prostatic hyperplasia. A second enzyme system targeted for inhibition is 17α-hydroxylase/17,20-lyase, which converts pregnenolone to DHA and subsequently testosterone. Since testosterone has significant androgenic activity by itself, inhibition of its biosynthesis would be useful in treating androgen-dependent diseases such as prostate cancer. The antifungal agent, ketoconazole, inhibits 17α-hydroxylase at high concentrations and demonstrated clinical activity in metastatic prostate cancer patients.[383] The steroidal compound, MDL 27,302,[384] and the nonsteroidal agent, R 75,251,[383] are under development as potentially more selective inhibitors of 17α-hydroxylase/17,20-lyase.

REFERENCES

1. IIUPAC-IUB Commission on Nomenclature, *Eur. J. Biochem., 186,* 429(1989).
2. J. S. Baran, *Intra-Sci. Chem. Rep., 3,* 35(1969).
3. V. W. Rodwell, et al., *Adv. Lipid Res., 14,* 1(1976).
4. T.-K. Lee, *Trends Pharmacol. Sci., 8,* 442(1987).
5. M. Lindberg, et al., *Biochemistry, 1,* 182(1962).

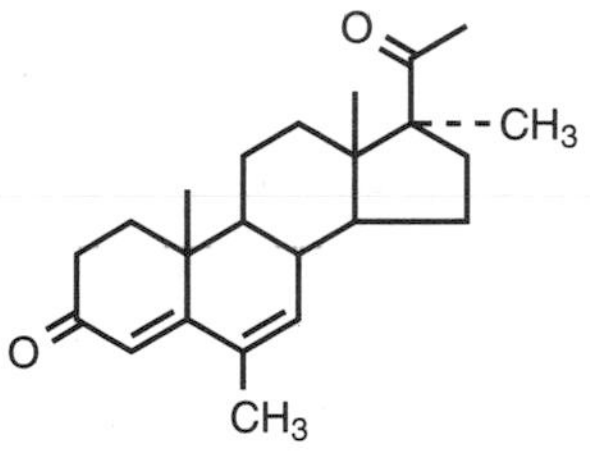

Medrogesterone

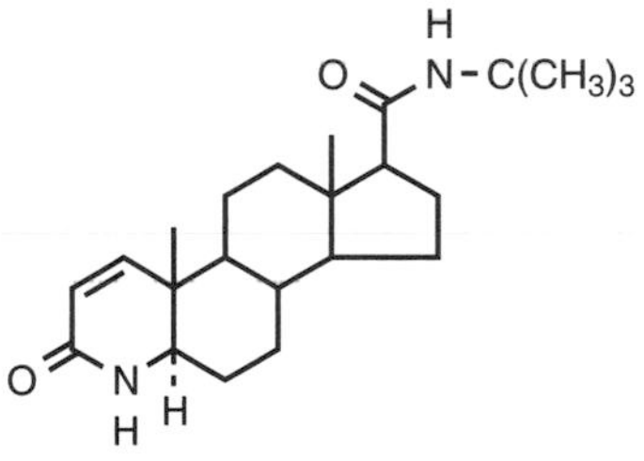

Finasteride

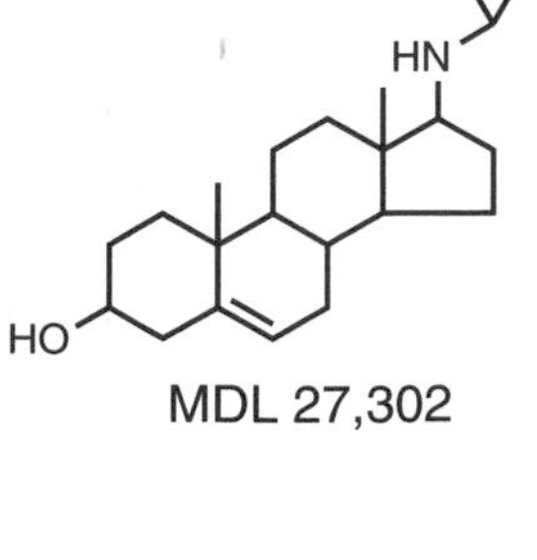

MDL 27,302

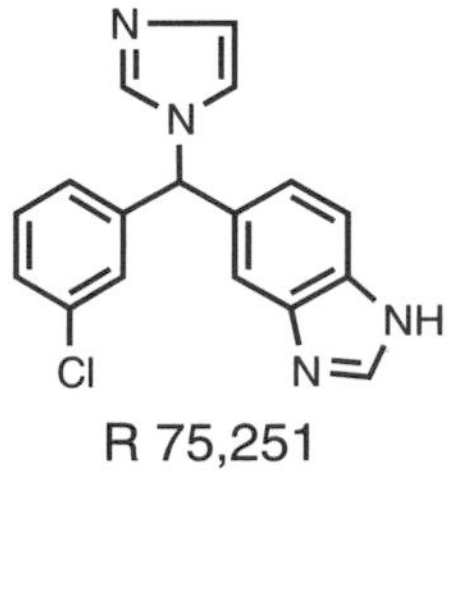

R 75,251

6. C. D. Poulter and H. C. Rilling, in *Biosynthesis of Isoprenoid Compounds, Vol. 1,* J. W. Porter and S. L. Spurgeon, Eds., New York, Wiley, 1981, p. 161.
7. J. D. Willett, et al., *J. Biol. Chem., 242,* 4182(1967).
8. F. J. Corey, W. E. Russey, and P. R. Ortiz de Montellano, *J. Am. Chem. Soc., 88,* 4750(1966).
9. E. E. van Tamelen, et al., *J. Am. Chem. Soc., 88,* 4752(1966).
10. E. E. van Tamelen, *Accts. Chem. Res., 1,* 111(1968).
11. J. Avigan, et al., *J. Biol. Chem., 238,* 1283(1963).
12. J. L. Gaylor, *Arch. Biochem. Biophys., 101,* 108(1963).
13. I. D. Frantz and G. J. Schroepfer, *Annu. Rev. Biochem., 36,* 691(1967).
14. R. B. Clayton, et al., *J. Lipid Res., 4,* 166(1963).
15. L. Horlick and J. Avigan, *J. Lipid Res., 4,* 160(1963).
16. L. Horlick, *J. Lipid Res., 7,* 116(1966).
17. A. A. Kandutsch and A. E. Russell, *J. Biol. Chem., 235,* 2256(1960).
18. W. A. Fish, et al., *J. Biol. Chem., 237,* 334(1962).
19. R. Fumagalli, et al., *J. Am. Oil Chem. Soc., 42,* 1018(1965).
20. F. L. Bach, Chapter 42, in *Medicinal Chemistry,* A. Burger, Ed., New York, Wiley, 1970.
21. J. Tzaskos, et al., *J. Biol. Chem., 259,* 1655(1984).
22. K. B. Sharpless, et al., *J. Am. Chem. Soc., 91,* 3394(1962).
23. K. B. Sharpless, et al., *J. Am. Chem. Soc., 90,* 6874(1968).
24. M. Akhtar and A. D. Rahimtula, *Chem. Commun.,* 259(1968).
25. M. L. Brown and J. L. Goldstein, *Science, 191,* 150(1976).
26. J. L. Goldstein and M. L. Brown, *Annu. Rev. Biochem., 46,* 897(1977).
27. R. W. Brueggemeier and P.-K. Li, Hormonal regulation of lipoproteins and lipoprotein metabolism, in *Antilipidemic Drugs, Medicinal, Chemical and Biochemical Aspects,* D. L. Witiak, H. A. I. Newman, and D. R. Feller, Eds., Amsterdam, Elsevier Science Publishers, 1992, pp. 493–526.
28. E. R. Simpson, *Mol. Cell. Endocrinol., 13,* 213(1979).
29. K. Block, et al., *J. Biol. Chem., 149,* 511(1943).
30. D. Kritchevsky, *Cholesterol.* New York, Wiley, 1958, pp. 118–123.
31. E. H. Mosback, in *Adv. Exp. Med. Biol. (1968),* Vol. 4, W. L. Holmes, et al., Eds., New York, Plenum Press, 1969, p. 421.
32. S. Lindstedt, *Acta Physiol. Scand., 40,* 1(1957).
33. M. D. Siperstein, et al., *Proc. Soc. Exp. Biol. Med., 81,* 720(1952).
34. M. D. Siperstein and A. W. Murray, *J. Clin. Invest., 34,* 1449(1955).
35. S. Dayton, et al., *Fed. Proc., 14,* 460(1955).
36. S. Bergstrom, *Rec. Chem. Prog., 16,* 63(1955).
37. S. Bergstrom, *Proc. 4th Int. Cong. Biochem., 4,* 161(1959).
38. S. Linstedt, *Arkh. Kemi., 11,* 145(1957).
39. J.-A. Gustaffson, et al., *Endocr. Rev., 8,* 185(1987).
40. R. M. Evans, *Science, 240,* 889(1988).
41. G. Ringold, Ed., *Steroid Hormone Action,* New York, Liss, 1988.
42. M. Beato, *Cell, 56,* 335(1989).
43. B. O'Malley, *Mol. Endocrinol., 4,* 363(1990).
44. M. A. Carson-Jurica, et al., *Endocr. Rev., 11,* 201(1990).
45. E. V. Jensen and H. I. Jacobsen, *Recent Prog. Horm. Res., 18,* 387(1962).
46. J. Gorski, et al., *Rec. Prog. Horm. Res., 42,* 297(1986).
47. T. Addison, *On the Constitutional and Local Effects of Disease of the Suprarenal Capsules.* London, Samuel Higley, 1855.
48. P. J. Murison, *Med. Clin. North Am., 51,* 883(1967).
49. L. T. Samuels and D. H. Nelson, *Handbook of Physiology, 6,* 55(1975).
50. R. C. Haynes, Chapter 60 in *Goodman and Gilman's The Pharmacological Basis of Therapeutics,* 8th ed., A. G. Gilman, et al., Eds., New York, Pergamon Press, 1990, pp. 1431–1462.
51. P. Kremers, *J. Steroid Biochem. Mol. Biol., 7,* 571(1976).
52. H. L. J. Makin, Ed., *Biochemistry of Steroid Hormones, Second Edition,* Oxford, Blackwell Scientific, 1984.
53. E. R. Simpson and M. R. Waterman, *Annu. Rev. Physiol., 50,* 427(1988).
54. W. L. Miller, *Endocrine Rev., 9,* 295(1988).
55. E. D. Robbins, et al., *J. Clin. Endocrinol., 17,* 111(1957).
56. W. Stevens, et al., *Endocrinology, 68,* 875(1961).
57. J. H. Glick, *Endocrinology, 60,* 368(1957).
58. C. H. Gray and J. B. Lunnon, *J. Endocrinol., 3,* 19(1956).
59. L. P. Romanoff, et al., *J. Clin. Endocrinol., 21,* 1413(1961).
60. D. K. Fukushima, et al., *J. Biol. Chem., 212,* 449(1955).
61. T. F. Gallagher, *Cancer Res., 17,* 520(1957).
62. A. D. Kemp, et al., *J. Biol. Chem., 210,* 123(1954).
63. S. Burstein, et al., *Endocrinology, 52,* 448(1953).
64. S. Burstein, et al., *Endocrinology, 53,* 88(1953).
65. G. Birke and L. O. Plantin, *Acta Med. Scand., 146,* 184(1953).
66. E. Chang and T. I. Dao, *Biochim. Biophys. Acta, 57,* 609(1962).
67. C. W. Shoppee, *Chemistry of Steroids.* 2nd ed. London, Butterworth, 1964, pp. 245–246.
68. M. Steiger and T. Reichstein, *Helv. Chim. Acta, 20,* 1164(1937).
69. L. F. Fieser and M. Fieser, Chapter 19, in *Steroids.* New York, Reinhold, 1959.
70. L. H. Sarett, *J. Biol. Chem., 162,* 601(1946).
71. B. F. McKenzie, et al., *J. Biol. Chem., 173,* 271(1948).
72. L. H. Sarett, *J. Am. Chem. Soc., 70,* 1454(1948); L. H. Sarett, *J. Am. Chem. Soc., 71,* 2443(1949).
73. J. Fried and E. F. Sabo, *J. Am. Chem. Soc., 75,* 2273(1953).
74. J. Fried and E. F. Sabo, *J. Am. Chem. Soc., 79,* 1130(1957).
75. H. M. Robinson, *Bull. Sch. Med. Univ. Maryland, 40,* 72(1955).
76. D. Stuart, *Pharmindex, 1,* 6(1959).
77. A. Nobile, et al., *J. Am. Chem. Soc., 77,* 4184(1955).
78. C. Myestre, et al., *Helv. Chim. Acta, 39,* 734(1956).
79. S. A. Szpilfogel, et al., *Rec. Trav. Chim., 75,* 402(1956).
80. G. B. Spero, et al., *J. Am. Chem. Soc., 79,* 1515(1957).
81. E. W. Boland, *Ann. Rheum. Dis., 21,* 176(1962).
82. H. Hirshmann, et al., *Am. Chem. Soc., 75,* 4862(1953).
83. B. Ellis, et al., *J. Chem. Soc., 77,* 4383(1955).
84. W. S. Allen and S. Bernstein, *J. Am. Chem. Soc., 77,* 1028(1955).
85. W. S. Allen and S. Bernstein, *J. Am. Chem. Soc., 78,* 1909(1956).
86. E. W. Boland, *Ann. N.Y. Acad. Sci., 82,* 887(1959).
87. A. L. Weiner, *Antibiot. Chemother., 12,* 360(1962).
88. W. L. Dobes, *South. Med. J., 56,* 187(1963).
89. M. M. Nierman, *Clin. Med., 70,* 771(1963).
90. L. Mantica, et al., *Arnzeimittelforschung, 30,* 1543(1980).
91. G. E. Arth, et al., *J. Am. Chem. Soc., 80,* 3160(1958).
92. E. P. Oliveto, et al., *J. Am. Chem. Soc., 80,* 4428(1958).
93. P. A. Sperber, *Curr. Ther. Res., 4,* 70(1962).
94. A. L. Welsh and M. Ede, *J. New Drugs, 2,* 223(1962).
95. R. H. Silber, *Ann. N.Y. Acad. Sci., 82,* 821(1959).
96. S. Tolksdorf, *Ann. N.Y. Acad. Sci., 82,* 829(1959).
97. K. M. West, et al., *Arth. Rheumat., 3,* 129(1960).
98. M. M. Nierman, *Clin. Med., 69,* 1311(1962).
99. A. Cohen and J. Coldman, *Penn. Med. J., 65,* 347(1962).
100. A. I. Cohen, *Antibiot. Chemother., 12,* 91(1962).
101. D. L. Unger and R. Bartolomei, *Ann. Allerg., 19,* 1312(1961).
102. J. H. Glyn and D. B. Fox, *Br. Med. J., 1,* 876(1961).
103. J. H. Glyn and D. B. Fox, *Br. Med. J., 2,* 650(1961).

104. D. S. Wilkinson, *Br. Med. J., 1,* 1319(1961).
105. G. W. Irwin, et al., *Metabolism, 10,* 852(1961).
106. S. W. Simon, *Ann. Allerg., 20,* 460(1962).
107. C. Stritzler, et al., *Arch. Derm., 85,* 505(1962).
108. S. M. Feinberg, et al., *J. New Drugs, 1,* 268(1961).
109. M. M. Cahn and E. J. Levy, *Clin. Med., 70,* 571(1963).
110. R. O. Noojin, et al., *Clin. Med., 70,* 747(1963).
111. A. Rostenberg, *J. New Drugs, 1,* 118(1961).
112. M. M. Cahn and E. J. Levy, *J. New Drugs, 1,* 262(1961).
113. M. P. Lazar, *Illinois Med. J., 121,* 552(1962).
114. J. M. Fox, *J. Indiana Med. Assoc., 55,* 1162(1962).
115. H. S. Appell and L. W. Koster, *Conn. Med., 26,* 579(1962).
116. R. N. Brogden, et al., *Drugs, 28,* 99(1984).
117. M. J. Green, et al., *J. Steroid Biochem., 11,* 61(1979).
118. H. Asche, et al., *Pharm. Acta Helv., 60,* 232(1985).
119. N. Bodor, et al., *J. Med. Chem., 26,* 318(1983).
120. T. L. Popper, et al., *J. Steroid Biochem. Mol. Biol., 27,* 837(1987).
121. W. L. Shapiro, et al., *J. Med. Chem., 30,* 1581(1987).
122. F. K. Bagatell and M. A. Augustine, *Curr. Ther. Res., 16,* 748(1974).
123. R. I. Dorfman, in *Hormone Assay,* C. W. Emmens, Ed., New York, Academic Press, 1950, p. 326.
124. K. Junkman, *Arch. Exp. Pathol., 227,* 212(1955).
125. R. M. Reinecke and E. C. Kendall, *Endocrinology, 31,* 573(1942).
126. R. S. Speirs and R. K. Meyer, *Endocrinology, 48,* 316(1951).
127. I. Ringler and R. Brownfield, *Endocrinology, 66,* 900(1960).
128. R. I. Dorfman and A. S. Dorfman, *Endocrinology, 69,* 283(1961).
129. W. B. Pratt, *J. Cell Biochem., 35,* 51(1987).
130. D. Marver, *Vitam. Horm., 38,* 57(1980).
131. F. Sereni, et al., *J. Biol. Chem., 234,* 609(1959).
132. D. Kupfer, *Arch. Biochem. Biophys., 127,* 200(1968).
133. J. D. Baxter and G. M. Tomkins, *Proc. Natl. Acad. Sci. U.S.A., 65,* 709(1970).
134. K. L. Lee, et al., *J. Biol. Chem., 245,* 5806(1970).
135. P. Feigelson, et al., *Recent Prog. Horm. Res., 31,* 213(1975).
136. P. Felig, et al., *Science, 167,* 1003(1970).
137. K. L. Lee and F. T. Kenney, *Biochem. Biophys. Res. Commun., 40,* 469(1970).
138. B. R. Landau, *Vitam. Horm., 23,* 1(1965).
139. P. L. Ballard, in *Glucocorticoid Hormone Action,* J. D. Baxter and G. G. Rousseau, Eds., New York, Springer, 1979, pp. 493–515.
140. F. D. Sistare and R. C. Haynes, Jr., *J. Biol. Chem., 260,* 12754(1985).
141. M. McMahon, et al., *Diabetes Metab. Rev., 4,* 17(1988).
142. M. DiRosa, et al., *Agents Actions, 17,* 284(1985).
143. L. Parente and R. J. Flower, *Life Sci., 36,* 1225(1985).
144. C. A. Dinarello and J. W. Mier, *N. Engl. J. Med., 317,* 940(1987).
145. W. Lew, et al., *J. Immunol., 140,* 1895(1988).
146. B. Beutler and A. Cerami, *N. Engl. J. Med., 316,* 379(1987).
147. J. S. Goodwin, et al., *J. Clin. Invest., 77,* 1244(1986).
148. G. W. G. Sharp and A. Leaf, *Recent Prog. Horm. Res., 22,* 431(1966).
149. B. M. Koeppen, et al., *Am. J. Physiol., 244,* F35(1983).
150. A. F. Wilson in *Fortschritte de Arzneimittelforschung,* H. von Redige and E. Jucker, Eds., Vol. 28, Basel-Boston-Stuttgart, Birkhauser-Verlag, 1968, p. 122.
151. L. A. Sciuchetti, *Pharmindex, 5,* 7(1963).
152. M. Suzuki, *Nippon Hifuka Gakkai Zasshi, 92,* 757(1982).
153. T. B. Fitzpatrick, et al., *J. Am. Chem. Soc., 158,* 1149(1955).
154. C. S. Livingood, et al., *Arch Dermatol., 72,* 313(1955).
155. I. E. Bush, *Pharmacol. Rev., 14,* 317(1962).
156. I. Ringler, in *Methods of Hormone Research,* Vol. 3, R. I. Dorfman, Ed., Part A. New York, Academic Press, 1964, pp. 227–349.
157. I. E. Bush and V. B. Mahesh, *Biochem. J., 71,* 718(1959).
158. E. M. Glenn, et al., *Endocrinology, 61,* 128(1957).
159. W. E. Dulin, et al., *Proc. Soc. Exp. Biol. Med., 94,* 303(1957).
160. J. Fried, in *Inflammation and Diseases of Connective Tissue,* L. C. Mills and J. H. Moyer, Eds., Philadelphia, W. B. Saunders, 1961, pp. 353–355.
161. J. Fried and A. Borman, *Vitam. Horm., 16,* 303(1958).
162. I. W. Funder, et al., *Biochem. Pharmacol., 23,* 1493(1974).
163. M. Peterfalvi, et al., *Biochem. Pharmacol., 29,* 353(1980).
164. D. Duval, et al., *J. Steroid Biochem., 20,* 283(1984).
165. M. K. Agarwal, et al., *FEBS Lett., 217,* 221(1987).
166. J Napoli and R. E. Counsell, *J. Med. Chem., 20,* 762(1977).
167. M. A. Shaw, et al., *J. Steroid Biochem. Mol. Biol., 31,* 137(1988).
168. N. Sonino, *N. Engl. J. Med., 317,* 812(1987).
169. G. O. Potts, et al., *Steroids, 32,* 257(1978).
170. G. W. Harris and F. Naftolin, *Br. Med. Bull., 26,* 1(1970).
171. R. B. Jaffe, et al., Chapter 1, in *Contraception: The Chemical Control of Fertility,* D. Lednicer, Ed. New York, Marcel Dekker, 1969.
172. E. Allen and E. A. Doisy, *J.A.M.A., 81,* 819(1923).
173. F. Murad and J. A. Kuret, Chapter 58, in *Goodman and Gilman's The Pharmacological Basis of Therapeutics,* 8th ed., A. G. Gilman, et al., Eds., New York, Pergamon Press, 1990, pp. 1384–1412.
174. R. Deghenghi and M. L. Givner, Chapter 29, in *Burger's Medicinal Chemistry, Part II, 4th Ed.,* M. E. Wolff, Ed., NY, Wiley, 1979.
175. J. Fishman, et al., *J. Biol. Chem., 237,* 1487(1962).
176. E. Gurpide, et al., *J. Clin. Endocrinol., 22,* 935(1962).
177. E. R. Simpson, et al., *Endocr. Rev., 10,* 136(1989).
178. K. J. Ryan and O. W. Smith, *Recent Prog. Horm. Res., 21,* 367(1965).
179. S. Najakin and P. F. Hall, *J. Biol. Chem., 256,* 3871(1981).
180. H. A. Harvey, et al., Eds., *Cancer Res., 42,* 3267s–3468 (1982).
181. R. J. Santen, Ed., *Steroids, 50,* 1–655(1985).
182. R. J. Santen and A. M. H. Brodie, Eds., Third International Aromatase Conference, *J. Steroid Biochem. Mol. Biol., 44* (4–6) (1993).
183. A. M. H. Brodie, Ed., *J. Enzyme Inhib., 4,* 75–200(1990).
184. K. J. Ryan, *J. Biol. Chem., 234,* 268(1959).
185. E. A. Thompson and P. K. Siiteri, *J. Biol. Chem., 249,* 5364(1974).
186. J. Goto and J. Fishman, *Science, 195,* 80(1977).
187. M. Akhtar, et al., *Biochem. J., 201,* 569(1982).
188. P. A. Cole and C. H. Robinson, *J. Am. Chem. Soc., 110,* 1284(1988).
189. R. W. Brueggemeier, *J. Enzyme Inhibition, 4,* 101(1990).
190. J. B. Brown, *Adv. Clin. Chem., 3,* 157(1960).
191. H. Breuer, *Vitam. Horm., 20,* 285(1962).
192. E. Diczfalusy and C. Lauritzen, *Oestrogen bein Menschen.,* Berlin, Springer, 1961.
193. Y. Nose and F. Lipmann, *J. Biol. Chem., 233,* 1348(1958).
194. C. J. Migeon, et al., *J. Clin Invest., 38,* 619(1959).
195. B. T. Brown, et al., *Nature, 182,* 50(1958).
196. J. Fishman, et al., *J. Biol. Chem., 235,* 3104(1960).
197. G. R. Merriam and M. B. Lipsett, *Catechol Estrogens,* New York, Raven Press, 1983.
198. H. H. Inhoffen and W. Hohlweg, *Naturwissenschaften, 29,* 96(1938).
199. L. B. Colton, et al., *J. Am. Chem. Soc., 79,* 1123(1957).
200. A. Ercoli and R. Gardi, *Chem. Ind. (Lond.),* 1037(1961).
201. F. R. Zuleski, et al., *J. Pharm. Sci., 67,* 1138(1978).
202. R. J. Chetkowski, et al., *N. Engl. J. Med., 314,* 1615(1986).

203. H. Judd, *Am. J. Obstet. Gynecol., 156,* 1326(1987).
204. R. E. Vanderlinde, et al., *J. Am. Chem. Soc., 77,* 4176(1955).
205. U. V. Solmssen, *Chem. Rev., 37,* 481(1945).
206. L. Chan and B. W. O'Malley, *N. Engl. J. Med., 294,* 1322–1328, 1372–1381, 1430–1437(1976).
207. W. P. Noteboom and J. Gorski, *Arch. Biochem. Biophys., 111,* 559(1965).
208. S. Liao and A. H. Lin, *Proc. Natl. Acad. Sci. U.S.A., 57,* 379(1967).
209. B. W. O'Malley, et al., *Recent Prog. Horm. Res., 25,* 121(1969).
210. H. Wi and G. C. Mueller, *Proc. Natl. Acad. Sci. U.S.A., 50,* 256(1963).
211. G. C. Mueller, et al., *Proc. Natl. Acad. Sci. U.S.A., 47,* 164(1961).
212. B. S. Katzenellenbogen and J. Gorski, *J. Biol. Chem., 247,* 1299(1972).
213. A. M. Kaye, *J. Steroid Biochem., 19,* 33(1983).
214. S. J. Segal and S. S. Koide, in *Pharmacology of Estrogens, International Encyclopedia of Pharmacology and Therapeutics, Section 106,* Pergamon Press, Oxford, 1981, pp. 113–150.
215. E. Heftmann, *Steroid Biochemistry.* New York, Academic Press, 1970, p. 140.
216. P. H. Jellinck and A. M. Newcombe, *J. Endocrinol., 74,* 147(1977).
217. C. R. Lyttle and E. R. DeSombre, *Nature, 268,* 337(1977).
218. O. Hechter and I. D. K. Halkerston, *Hormones, 5,* 799(1964).
219. S. C. Aitken and M. E. Lippman, *Cancer Res., 43,* 4681(1983).
220. S. C. Aitken and M. E. Lippman, *Cancer Res., 45,* 1611(1985).
221. K. B. Horwitz and W. L. McGuire, *J. Biol. Chem., 253,* 2223(1978).
222. M. E. Lippman, in *Williams' Textbook of Endocrinology,* 7th ed., J. D. Wilson and D. W. Foster, Eds., Philadelphia, W. B. Saunders, 1985, pp. 1309–1326.
223. S. B. Jakolew, et al., *Nucleic Acids Res., 12,* 2861(1984).
224. R. B. Dickson and M. E. Lippman, *Endocr. Rev., 8,* 29(1987).
225. S. E. Bates, et al., *Cancer Res., 46,* 1707(1986).
226. R. B. Dickson, et al., *Science, 232,* 1540(1986).
227. R. B. Dickson, et al., *Endocrinology, 118,* 138(1986).
228. K. K. Huff, et al., *Cancer Res., 46,* 4613(1986).
229. F. Vignon, et al., *C. R. Acad. Sci. III, 296,* 151(1983).
230. D. Bronzert, et al., *Cancer Res., 47,* 1234(1987).
231. P. J. Bentley, *Endocrine Pharmacology,* Cambridge, Cambridge University Press, 1980.
232. J.-P. Raynaud, et al., *Mol. Pharmacol., 9,* 520(1973).
233. L. F. Fieser and M. Fieser, in *Steroids.* New York, Reinhold, 1959, p. 477.
234. G. A. Grant and D. Beall, *Recent Prog. Horm. Res., 5,* 307(1950).
235. H. S. Kupperman, et al., *J. Clin. Endocrinol., 13,* 688(1953).
236. A. D. Cross, et al., *Steroids, 4,* 423(1964).
237. J. S. Baran, *J. Med. Chem., 10,* 1188(1967).
238. A. A. Albanese, et al., *N. Y. State J. Med., 65,* 2116(1965).
239. E. Heftmann, *Steroid Biosynthesis.* New York, Academic Press, 1970, p. 141.
240. F. W. Schueler, *Science, 103,* 221(1946).
241. M. Rubin and H. Wiskinsky, *J. Am. Chem. Soc., 66,* 1948(1944).
242. E. C. Dodds, *Nature, 142,* 34(1938).
243. G. W. Harris and F. Naftolin, *Br. Med. Bull., 26,* 197(1970).
244. B. S. Katzenellenbogen and E. R. Ferguson, *Endocrinology, 97,* 1(1975).
245. J. H. Clark, et al., *Nature, 251,* 446(1974).
246. V. C. Jordan, et al., *J. Endocrinol., 75,* 305(1977).
247. V. C. Jordan and C. S. Murphy, *Endocrine Rev., 11,* 578(1990).
248. A. E. Wakeling and J. Bowler, *J. Steroid Biochem., 30,* 141(1988).
249. A. E. Wakeling, et al., *Cancer Res., 51,* 3867(1991).
250. A. M. H. Brodie, et al., *Cancer Res., 42,* 3360s(1982).
251. R. W. Brueggemeier, et al., *Cancer Res., 42,* 3334s(1982).
252. R. W. Brueggemeier, et al., *J. Steroid Biochem. Mol. Biol., 37,* 379(1990).
253. J. O. Johnston, et al., *Endocrinology, 115,* 776(1984).
254. D. Henderson, et al, *Endocrinology, 115,* 776(1986).
255. D. Giudici, et al., *J. Steroid Biochem. Mol. Biol., 30,* 391(1988).
256. H. A. Salhanick, *Cancer Res., 42,* 3315s(1982).
257. A. S. Bhatnagar, et al., *J. Steroid Biochem. Mol. Biol., 37,* 363(1990).
258. R. De Coster, et al., *J. Steroid Biochem. Mol. Biol., 37,* 335(1990).
259. J. A. Johnston and B. W. Metcalf, in *Novel Approaches to Cancer Chemotherapy,* P. Sunkara, Ed., New York, Academic Press, 1984, p. 307.
260. L. Banting, et al., *J. Enzyme Inhib., 2,* 215(1988).
261. D. F. Covey, in *Sterol Biosynthesis Inhibitors,* D. Berg and M. Plempel, Eds., Chichester, UK, Ellis Horwood, 1988, p. 534.
262. L. Banting, et al., *Prog. Med. Chem., 26,* 253(1989).
263. S. Fraenkel, *Arch. Gynaek., 68,* 438(1903).
264. R. Pearl and F. M. Surface, *J. Biol. Chem., 19,* 263(1914).
265. G. W. Corner and W. M. Allen, *Am. J. Physiol., 88,* 326(1929).
266. A. W. Makepace, et al., *Am. J. Physiol., 119,* 512(1937).
267. E. Heftmann and E. Mosetlig, in *Biochemistry of Steroids.* New York, Reinhold, 1960, p. 106.
268. C. Djerassi, *Science, 151,* 1055(1966).
269. L. Ruzicka and K. Hofmann, *Helv. Chim. Acta, 20,* 1280(1937).
270. P. D. Klimstra, *Am. J. Pharm. 34,* 630(1970).
271. M. Ehrenstein, *J. Org. Chem., 9,* 435(1944).
272. C. Djerassi, et al., *J. Am. Chem. Soc., 75,* 4440(1953).
273. G. Pincus, *Science, 138,* 439(1962).
274. M. E. Davis and G. L. Wied, *J. Clin. Endocrinol., 15,* 923(1955).
275. I. Siegel, *Obstet. Gynecol., 21,* 666(1963).
276. A. Klopper, *Br. Med. Bull., 26,* 39(1970).
277. J. C. Babcock, et al., *J. Am. Chem. Soc., 80,* 2904(1958).
278. B. W. O'Malley, et al., *Recent Prog. Horm. Res., 25,* 105(1969).
279. M. A. Carson-Jurica, et al., *Endocr. Rev., 11,* 201(1990).
280. H. Gronemeyer, et al., *EMBO J., 6,* 3985(1987).
281. M. Misrahi, et al., *Biochem. Biophys. Res. Commun., 143,* 740(1987).
282. M. A. Carson-Jurica, et al., *J. Steroid Biochem. Mol. Biol., 34,* 1(1989).
283. K. Fotherby, *Vitam. Horm., 22,* 153(1964).
284. G. W. Harris and O. F. Naftolin, *Br. Med. Bull., 26,* 3(1970).
285. E. E. Baulieu, in *Steroid Hormone Regulation of the Brain,* K. Fuxe, et al., Eds., Oxford, Pergamon Press, 1981, pp. 3–14.
286. M. D. Majewska, et al., *Science, 222,* 1004(1986).
287. M. D. Morrow, et al., *Eur. J. Pharmacol., 142,* 483(1987).
288. E. E. Baulieu and P. Robel, *J. Steroid Biochem. Mol. Biol., 37,* 395(1990).
289. I. Rothchild, in *Metabolic Effects of Gonadal Hormones and Contraceptive Steroids,* H. A. Salhanick, et al., Eds., New York, Plenum Press, 1969, p. 668.
290. V. A. Drill, *Oral Contraceptives.* New York, McGraw-Hill, 1966, p. 3.

291. J. Crossland, Chapter 38, in *Lewis's Pharmacology,* London, E. & S. Livingstone, 1970.
292. P. D. Klimstra, *Am. J. Pharm. Ed., 34,* 630(1970).
293. D. Philibert, R. Deraedt, et al., *64th Annual Meeting of The Endocrine Society,* San Francisco, June 1982, Abstract No. 668.
294. E. E. Baulieu, *Science, 245,* 1351(1989).
295. W. Elger, et al., *J. Steroid Biochem. Mol. Biol., 25,* 835(1986).
296. K. B. Horwitz, *Endocr. Rev., 13,* 146(1992).
297. G. Pincus, *Science, 153,* 493(1966).
298. G. Pincus, et al., *Science, 130,* 81(1959).
299. V. Petrow, *Chem. Rev., 70,* 713(1970).
300. H. W. Berendes, in *Pharmacology of Steroid Contraceptive Drugs,* S. Garattini and H. W. Berendes, Eds., New York, Raven Press, 1977, p. 223.
301. G. V. Upton, *Int. J. Fertil., 28,* 121(1983).
302. C. Dodds, *Clin. Pharmacol. Ther., 10,* 147(1969).
303. D. Shoupe and D. R. Mishell, *Am. J. Obstet. Gynecol., 60,* 1286(1989).
304. R. Wiechert, *Angew, Chem. Int. Ed. Engl., 16,* 506(1977).
305. F. D. A. Consumer, *Women and Estrogens,* April 1976, p. 4.
306. F. D. A. Drug Bulletin, *Estrogens and Endometrial Cancer, 6,* 18(1976).
307. Facts and Comparison, *Sex Hormones,* St. Louis, Mo., Facts and Comparison Inc., 1975, p. 107d.
308. Med. Lett., *18,* 21(1976).
309. A. Klopper, *Br. Med. J., 26,* 39(1970).
310. M. M. Grumbach and J. R. Ducharme, *Fertil. Steril., 11,* 157(1960).
311. R. E. Counsell and R. W. Brueggemeier, Chapter 28, in *Burger's Medicinal Chemistry, Part II, 4th Ed.,* M. E. Wolff, Ed., New York, Wiley, 1979.
312. R. W. Brueggemeier, in *Handbook of Hormones, Vitamins, and Radiopaques,* M. Verderame, Ed., Boca Raton, FL, CRC Press, 1986, pp. 1–49.
313. R. I. Dorfman and F. Ungar, *Metabolism of Steroid Hormones.* New York, Academic Press, 1965, p. 123.
314. J. M. Rosner, et al., *Endocrinology, 75,* 299(1964).
315. E. E. Baulieu, et al., *Steroids, 2,* 429(1963).
316. N. Bruckovsky and J. D. Wilson, *J. Biol. Chem., 243,* 2012(1968).
317. P. Ofner, *Vit. Horm., 26,* 237(1968).
318. L. Martini, in *Subcellular Mechanisms in Reproductive Endocrinology,* F. Naftolin, et al., Eds., Amsterdam, Elsevier, 1976, p. 327.
319. F. Naftolin, et al., in *Subcellular Mechanisms in Reproductive Endocrinology,* F. Naftolin, et al., Eds., Amsterdam, Elsevier, 1976, p. 347.
320. A. E. Kellie and E. R. Smith, *Biochem. J., 66,* 490(1957).
321. C. E. Brown-Séquard, *C. R. Seanc. Soc. Biol., 1,* 420(1889).
322. A. Pézard, *C. R. H. Acad. Sci., 153,* 1027(1911).
323. A. Butenandt, *Angew. Chem., 44,* 905(1931).
324. A. Butenandt and H. Dannenberg, *Z. Physiol. Chem., 229,* 192(1934).
325. K. David, et al., *Z. Physiol. Chem., 233,* 281(1935).
326. A. Butenandt and G. Hanisch, *Chem. Ber., 68B,* 1859(1935).
327. C. D. Kochakian and J. R. Murlin, *J. Nutr., 10,* 437(1935).
328. J. A. Vida, Chapter 1, in *Androgens and Anabolic Agents.* New York, Academic Press, 1969.
329. F. C. Kock, *Bull. N. Y. Acad. Med., 14,* 655(1938).
330. P. D. Klimstra, Chapter 8, in *The Chemistry and Biochemistry of Steroids,* Vol. 3. Los Altos, Calif., Geron-X, 1969.
331. B. W. O'Malley and A. R. Means, *Receptors for Reproductive Hormones,* New York, Plenum Press, 1974.
332. R. J. B. King and W. P. Mainwaring, *Steroid-Cell Interactions,* Baltimore, University Park Press, 1974.
333. S. Liao, *Int. Rev. Cytol., 41,* 87(1975).
334. H. G. Williams-Ashman and A. H. Reddi, in *Biochemical Actions of Hormones, Vol. 2,* G. Litwack, Ed., New York, Academic Press, 1972, 257.
335. W. I. P. Mainwaring, *The Mechanism of Action of Androgens,* New York, Springer-Verlag, 1977.
336. W. I. P. Mainwaring, *J. Endocrinol., 44,* 323(1969).
337. S. Fang and S. Liao, *J. Biol. Chem., 246,* 16(1971).
338. W. I. P. Mainwaring and B. M. Peterken, *Biochem. J., 125,* 285(1971).
339. J. L. Tymoczko and S. Liao, *Biochem. Biophys. Acta, 252,* 607(1971).
340. W. I. P. Mainwaring, et al., *Biochem. J., 137,* 513(1974).
341. K. J. Tvter and O. Unhjem, *Endocrinology, 84,* 963(1969).
342. J. M. Stern and A. J. Eisenfeld, *Science, 166,* 233(1969).
343. W. I. P. Mainwaring and F. R. Mangan, *J. Endocrinol., 59,* 121(1973).
344. V. Hansson, et al., *Steroids, 23,* 823(1974).
345. D. J. Tindall, et al., *Biochem. Biophys. Res. Commun., 49,* 1391(1972).
346. E. M. Ritzin, et al., *Endocrinology, 19,* 116(1972).
347. P. Jovan, et al., *J. Steroid Biochem., 4,* 65(1973).
348. M. Sar and W. E. Stumpf, *Endocrinology, 92,* 251(1973).
349. D. P. Cardinali, et al., *Endocrinology, 95,* 179(1974).
350. A. K. Roy, et al., *Biochem. Biophys. Acta, 354,* 213(1974).
351. N. Bruchovsky and J. W. Meakin. *Cancer Res., 33,* 1689(1973).
352. N. Bruchovsky, et al., *Biochem. Biophys. Acta, 381,* 61(1975).
353. R. L. Gloyna and J. D. Wilson, *J. Clin. Endocrinol., 29,* 970(1969).
354. V. Hansson, et al., *J. Steroid Biochem., 3,* 427(1972).
355. G. Giannopoulos, *J. Biol. Chem., 248,* 1004(1973).
356. M. L. Powers and J. R. Florini, *Endocrinology, 90,* 1043(1975).
357. I. Jung and E. E. Baulieu, *Nature [New Biol.], 237,* 24(1972).
358. G. Michel and E. E. Baulieu, *C. R. Acad. Sci. III, 279,* 421(1974).
359. M. Krieg, *Steroids, 28,* 261(1976).
360. M. Krieg and K. D. Voigl, *Acta Endocrinol. (Copenh), Suppl., 214,* 43(1977).
361. M. Snochowski, et al., *Eur. J. Biochem., 111,* 603(1980).
362. M. Snochowski, et al., *J. Steroid Biochem., 14,* 765(1981).
363. J. J. Schrozie and H. M. Solomon, *Clin. Pharmacol. Ther., 8,* 797(1967).
364. A. Segaloff and R. B. Grabbard, *Endocrinology, 67,* 887(1960).
365. J. A. Vida, Chapter 3, in *Androgens and Anabolic Agents.* New York, Academic Press, 1969.
366. J. Van der View, *Acta Endocrinol, 49,* 271(1965).
367. R. I. Dorfman, Chapter 30, in *Burger's Medicinal Chemistry, Part II,* 4th ed., M. E. Wolff, Ed., New York, Wiley, 1979.
368. S. Grombe, *East Afr. Med. J., 60,* 203(1983).
369. S. Qian and Z. Wang, *Annu. Rev. Pharmacol. Toxicol., 24,* 329(1984).
370. M. R. N. Prasad and E. Siczfalusy, *Int. J. Androl., Suppl. 5,* 53(1982).
371. L. Martini and M. Motta, Eds., *Androgens and Antiandrogens.* New York, Raven Press, 1977.
372. R. O. Neri, *Adv. Sex. Horm. Res., 2,* 233(1976).
373. S. Fang and S. Liao, *Mol. Pharmacol., 5,* 240(1969).
374. G. Goto, et al., *Chem. Pharm. Bull., 13,* 1294(1965).
375. B. W. Snyder, et al., *J. Steroid Biochem. Mol. Biol., 33,* 1127(1989).
376. S. Laio, et al., *Endocrinology, 94,* 1205(1974).
377. A. E. Wakeling, et al., *J. Steroid Biochem., 15,* 355(1981).
378. S. Y. Tan, et al., *J. Clin. Endocrinol. Metab., 39,* 936(1974).
379. G. H. Rasmussen, et al., *J. Med. Chem., 29,* 2298(1986).

380. T. Liang, et al., *Endocrinology, 117,* 571(1985).
381. A. Vermeulen, et al., *Prostate, 14,* 45(1989).
382. J. D. McConnell, et al., *J. Urol., 141,* 239A(1989).
383. J. P. Van Wauwe and P. A. J. Janssen, *J. Med. Chem., 32,* 2231(1989).
384. M. R. Angelastro, et al., *Biochem. Biophys. Res. Commun., 162,* 1571(1989).

SELECTED READINGS

M. Beato, *Cell, 56,* 335(1989).

P. J. Bentley, *Endocrine Pharmacology,* Cambridge, Cambridge University Press, 1980.

R. W. Brueggemeier, in *Handbook of Hormones, Vitamins, and Radiopaques,* M. Verderame, Ed., Boca Raton, FL, CRC Press, 1986, pp. 1–49.

R. E. Counsell and R. W. Brueggemeier, Chapter 28 in *Burger's Medicinal Chemistry, 4th ed., Part II,* M. E. Wolff, Ed., New York, John Wiley & Sons, 1979, pp. 873–916.

R. Deghenghi and M. L. Givner, Chapter 29 in *Burger's Medicinal Chemistry, 4th ed., Part II,* M. E. Wolff, Ed., New York, Wiley, 1979, pp. 917–939.

R. M. Evans, *Science, 240,* 889(1988).

D. W. Fullerton, Chapter 18 in *Wilson and Gisvold's Textbook of Organic Medicinal and Pharmaceutical Chemistry, Ninth Edition,* J. N. Delgado and W. A. Remers, Eds., Philadephia, J. B. Lippincott Co., 1991, pp. 675–765.

R. C. Haynes, Jr., Chapter 60 in *Goodman and Gilman's The Pharmacological Basis of Therapeutics, 8th ed.,* A. G. Gilman, et al., Eds., New York, Pergamon Press, 1990, pp. 1431–1462.

R. A. Magarian, et al., in *Handbook of Hormones, Vitamins, and Radiopaques,* M. Verderame, Ed., Boca Raton, FL, CRC Press, 1986, pp. 51–92.

H. J. L. Makin, Ed., *Biochemistry of Steroid Hormones, Second Edition,* Oxford, Blackwell Scientific Publications, 1984.

F. Murad and J. A. Kuret, Chapter 58 in *Goodman and Gilman's The Pharmacological Basis of Therapeutics, 8th ed.,* A. G. Gilman, et al., Eds., New York, Pergamon Press, 1990, pp. 1384–1412.

J. D. Wilson, Chapter 59 in *Goodman and Gilman's The Pharmacological Basis of Therapeutics, 8th ed.,* A. G. Gilman, et al., Eds., New York, Pergamon Press, 1990, pp. 1413–1430.

R. F. Witzman, *Steroids: Keys to Life,* New York, Van Nostrand-Reinhold, 1981.

M. E. Wolff, Chapter 63 in *Burger's Medicinal Chemistry, Fourth Edition, Part III,* M. E. Wolff, Ed., New York, Wiley, 1979, pp. 1273–1316.

F. J. Zeelen, *Medicinal Chemistry of Steroids,* Amsterdam, Elsevier, 1990.

Chapter 24

ANTILIPIDEMIC DRUGS

Dennis R. Feller, Larry M. Hagerman, Howard A. I. Newman, and Donald T. Witiak

Coronary heart disease (CHD) is the leading cause of death in the United States. More than 650,000 people die each year from CHD in this country and about 20% of the population will develop the first symptoms before age 60.[1,2] In 1988, the American Heart Association estimated that CHD costs the United States more than $88 billion, a figure that greatly underestimates the costs of lost economic productivity.[3]

Although nearly everyone has some degree of coronary atherosclerosis, the underlying cause of CHD, this condition is most common in affluent nations. Diets high in fat, elevated serum cholesterol levels, lack of exercise, emotional stress, hypertension, and smoking all contribute to the production of this disease.[4–10]

Recognition and treatment of different forms of atherosclerosis and the development of antilipidemic drugs have taken many centuries. Our understanding of atherosclerosis and its treatment is far from complete. Many major advances have been made only recently. Indeed, in his opening remarks to the Third International Symposium on Drugs Affecting Lipid Metabolism, in Milan, in September, 1968, Steinberg recalled that in 1960, discussions of cholesterol, triglycerides, and free fatty acids (FFA) could not be found in well-known pharmacology textbooks.[11]

In fact, Ruffer's treatise on "Arterial Lesions Found in Egyptian Mummies" describes arteriosclerosis as a common disease among the Egyptians in 1500 B.C.[12] Although a set of pathologic conditions termed "coronary sclerosis" had been known before 1833, it was not until then that the term arteriosclerosis was introduced.[13] Drelincort is credited with the first observations on the pathologic changes in coronary sclerosis in 1700.

Most of the clinical and experimental studies of the disease during the 19th century were conducted in Germany. Rokitansky, a leading pathologist, described arteriosclerosis as an excessive formation and deposition of endogenous products from the blood (fatty globules, cholesterin crystals, and calcium salts) and fibrin in the membrane lining the artery.[14] Another 19th century investigator, Virchow, disagreed with Rokitansky's proposal that fibrin deposition is basic to arteriosclerosis, and until Winternitz, and later Duquid, reinvestigated the problem in the early 1900s, Rokitansky's work was in disfavor.[15,16]

Atherosclerosis occurs in stages in human arteries. First, juvenile fatty streaks form along the intima. These initial lesions contain foam cells presumably derived from lipid-engorged macrophages. Later, the deposition of lipoproteins, cholesterol, and phospholipids causes the formation of softer, larger plaques. Associated with this lipid deposition is the proliferation of arterial smooth muscle cells into the intima and the laying down of collagen, elastin, and glycosaminoglycans leading to fibrous plaques. Ultimately the surface of the plaque deteriorates and an atheromatous ulcer is formed with a fibrous matrix, accumulation of necrotic tissue, and appearance of cholesterol and cholesterol ester crystals. A complicated lesion also shows calcification and hemorrhage with the formation of organized mural thrombi. Patients with these complicated lesions may have stroke, gangrene, aneurysm, and myocardial infarction. Thrombosis results from changes in the arterial walls and in the blood-clotting mechanism.

By the 1960s, an association between elevated serum cholesterol and increased risk to CHD was widely recognized. It was not proven, however, that diet or drug interventions that lowered serum cholesterol also lowered risk to disease. The results of the first major drug trial, *The Coronary Drug Project*, which used clofibrate, nicotinic acid, D-thyroxine or estrogen in men with symptomatic disease were disappointing.[17,18] Estrogen and D-thyroxine were dropped before the planned 5-year treatment period because of excess cardiovascular events or other adverse effects. Subsequently, the World Health Organization conducted a primary prevention trial using clofibrate in hypercholesterolemic men without symptomatic disease. This trial also was disappointing and, in fact, overall mortality in the clofibrate treated group was about 25% higher than that taking a placebo.

Finally, in 1984, the National Institutes of Health (NIH) concluded its $150 million, 7-year, primary prevention trial in 3806 hypercholesterolemic males given cholestyramine or placebo.[19,20] This landmark study was the first to provide convincing evidence that lowering serum cholesterol carried with it a clear benefit with respect to reduction of risk to CHD and associated symptoms. The results suggested that for each 1% reduction in serum cholesterol level there was a 2% reduction in CHD risk. The conclusions of the NIH primary prevention trial were reinforced by the Helsinki Heart Study, which also was a primary prevention trial comparing effects of gemfibrozil versus placebo in hypercholesterolemic males.[21] The results of this study clearly suggested that both elevating high-density lipoprotein (HDL) and

Table 24–1. Clinical Trials Showing Atheroma Regression With Cholesterol Lowering Drugs

Study	Therapy	Date Published
US NHLBI Type II Coronary Intervention Study[23]	Cholestyramine	1984
Cholesterol Lowering Atherosclerosis Study (CLAS I & II)[24,25]	Colestipol plus niacin	1987, 1990
Familial Atherosclerosis Treatment Study (FATS)[26]	Colestipol plus niacin or colestipol plus lovastatin	1990
University of California, San Francisco, Arteriosclerosis Specialized Center of Research (CSF-SCOR)[27]	Colestipol plus niacin or colestipol plus lovastatin	1990
St. Thomas Atherosclerosis Regression Study (STARS)[28]	Cholestyramine	1992
Mevinolin (lovastatin) Atherosclerosis Regression Study (MARS)[29]	Lovastatin	1992

lowering low-density lipoprotein (LDL) cholesterol levels are effective in the primary prevention of CHD.[22]

In addition to the primary prevention trials, in several secondary trials, the effects of a variety of cholesterol-lowering drugs on progression of established coronary disease were determined using serial angiography. Six secondary prevention trials (Table 24–1) showed that serum lipid-lowering drugs not only halted the progression of coronary atherosclerosis, but in some patients also induced regression of pre-existing disease.

Thus, now there is widespread confidence that coronary atherosclerosis is not only a preventable disease, but also that clinical benefit can be achieved using cholesterol lowering agents in individuals with established, symptomatic disease.

BIOCHEMISTRY OF LIPOPROTEINS

Lipids are one of a group of organic compounds with a greasy feel, freely soluble in fat solvents (e.g., ethanol, isopropanol, chloroform, diethyl ether) but insoluble in water. The major lipids found in the bloodstream are: triglycerides (triacylglycerols, simple fatty acid [FA], esters of glycerol), phospholipids (glycerophosphatides, FA-substituted phosphoric acid esters of glycerol), cholesterol, and cholesterol esters.

All lipids found in the bloodstream are solubilized by their association with proteins. These macromolecular aggregates are called lipoproteins. Proteins associated with lipoproteins are known as apoproteins. Apoproteins are divided into four major groups: A, B, C, and E. Figure 24–1 summarizes the interrelationships among the various lipoproteins and their biosynthesis and degradation. The largest of these particles, through its endowment of triglycerides, is the chylomicron. Chylomicra contain exogenous lipids primarily as triglycerides (86 to 94%) with small amounts of cholesterol (0.5 to 1%), cholesterol esters (1 to 3%), phospholipids (3 to 8%), and apolipoproteins A, C, E, and B-48 (1 to 2%).[30] These large lipid particles are formed through extrusion of resynthesized triglycerides from the mucosal cells into the intestinal lacteals. The chylomicra flow through the thoracic ducts into the subclavian veins. During circulation, they are degraded into remnants by the action of plasma membrane lipoprotein lipase (LpL) located on capillary endothelial cell surfaces. The remnants interact with and are taken up predominantly by liver parenchymal cells due to apoE-III and apoE-IV isoform recognition sites.[31]

Next in size are the very low density lipoproteins (VLDL), made up of 55 to 65% triglyceride, 6 to 8% cholesterol, 12 to 14% cholesterol esters, 12 to 18% phospholipids, and 5 to 10% apoproteins. B-100 and C apolipoproteins predominate.[30] These macromolecular aggregates primarily carry endogenously synthesized or modified lipids from the liver into the bloodstream. Nascent VLDL containing triglycerides and only apoE and apoB, secreted from the liver, interact with HDL to generate mature VLDL with added cholesterol esters, apoC-II, and apoC-III.[32] Again, LpL catalyzes triglyceride degradation to generate VLDL remnants, which are further degraded by hepatic triglyceride lipase (HTGL) to generate LDL.[33]

LDL contains 5 to 10% cholesterol, 35 to 40% cholesterol esters, 20 to 25% phospholipid, 8 to 12% triglyceride and 20 to 24% apoB-100 apoprotein.[30] LDL is taken up by and is the primary source of cholesterol for both liver and peripheral cells. The apoB-100 is degraded to amino acids by lysosomal proteases in these cells. The cholesterol from LDL inhibits both the rate-limiting factor in cholesterol biosynthesis, 3-hydroxy-3-methylglutaryl CoA (HMGCoA) reductase, and through 25-hydroxycholesterol, the production of LDL receptor.[34] 25-Hydroxycholesterol combines with an inhibitory protein which, in turn, displaces an activator of transcription for the LDL receptor.[35] In other metabolic pathways, LDL, modified through reaction with malondialdehyde and glucose, or oxidized through free radical reactions catalyzed by transitional metals or by enzymes, can be taken up by specific acetyl-LDL or LDL_{ox} receptors on macrophage plasma membranes.[36]

HDL contains 14 to 18% cholesterol esters, 3 to 5% cholesterol, 20 to 30% phospholipids, 3 to 6% triglycerides and 45 to 50% apoA, C, and E. E apoprotein is composed of several subspecies (apoE-II, apoE-III, apoE-IV). Discoidal HDL, a rapidly evolving form of HDL, contains cholesterol, phospholipid, apoE, and is disk shaped.[30] HDL may be separated into subsets differing in composition, density, and function. HDL_3 is composed of choles-

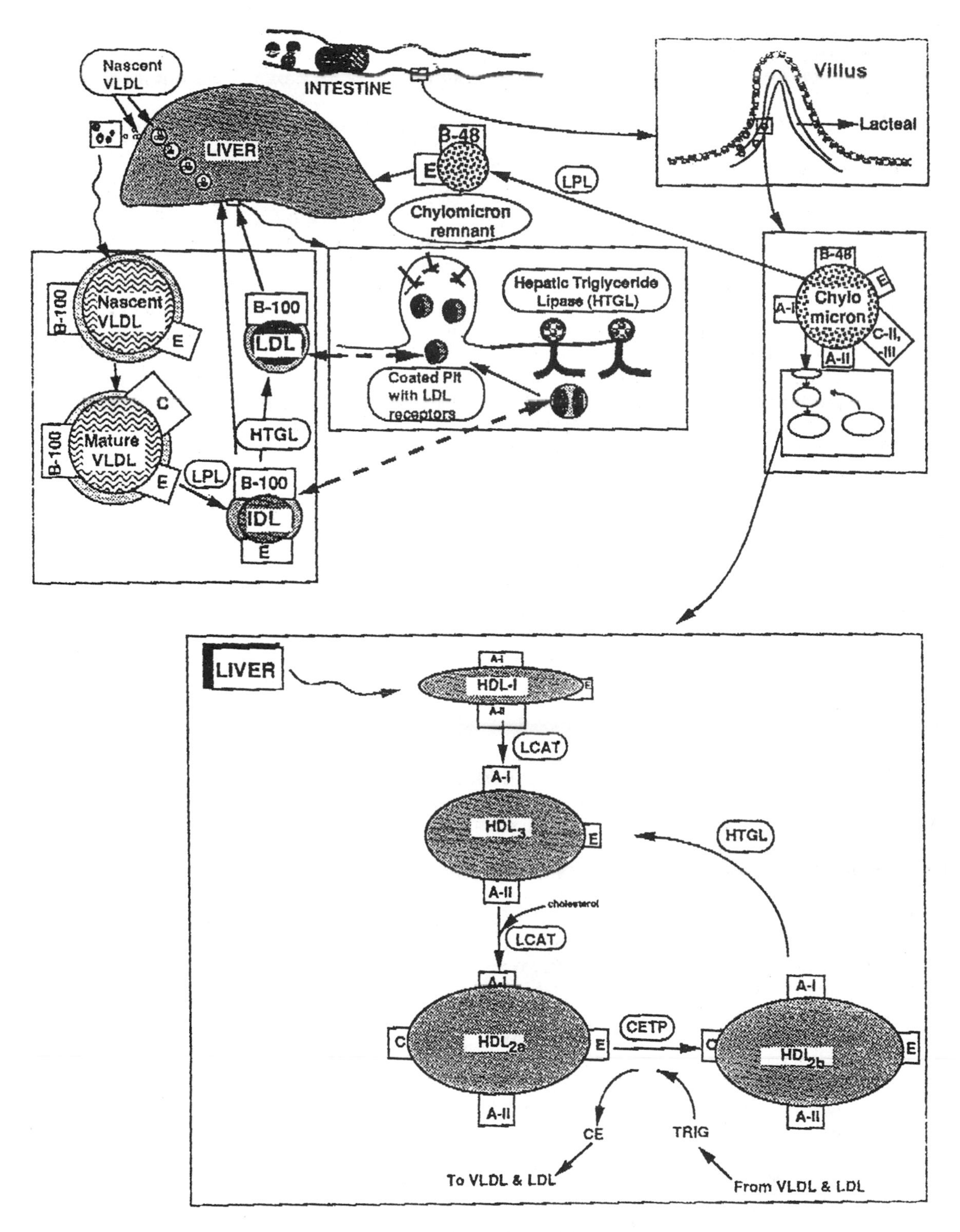

Fig. 24–1. Interrelationships of lipoproteins.

terol, cholesterol ester, phospholipid, and apoA and apoE. HDL_2 has a composition similar to that of HDL_3 but also contains apoC. HDL_2 has been further subdivided into HDL_{2a}, which is cholesterol ester-enriched, and HDL_{2b}, which is triglyceride-enriched.[37] Discoidal HDL, containing apoA-I, apoA-II and possibly apoE, is synthesized in the liver and intestine and then converted to HDL_3 by interaction with chylomicra remnants and lecithin cholesterol acyl transferase (LCAT). HDL_3 interacts with cell plasma membranes to remove free cholesterol. Reaction with LCAT converts HDL_3 into HDL_{2a}, an HDL with a high apoE and cholesterol ester content.

HDL_1 can competitively displace LDL at the apoB and apoE receptors of peripheral cells. Thus, it is possible that HDL_1 can effectively block uptake of LDL into peripheral cells.[38] Cholesterol ester-rich HDL_{2a} is converted to triglyceride-rich HDL_{2b} by concomitant transfer of HDL cholesterol esters to VLDL and VLDL triglycerides to HDL. Ultimately the VLDL cholesterol esters enter the liver through either VLDL or LDL uptake, and the HDL triglycerides are hydrolyzed by HTGL.

Lipoprotein nomenclature is based on mode of separation. When preparative ultracentrifugation is used, lipoproteins are identified as VLDL, intermediate density lipoprotein (IDL), LDL, and HDL. When electrophoresis is employed in the separation, lipoproteins are designated as pre-β, β, and α. IDL is mainly found in the pre-β fraction as a second electrophoretic band and is currently believed to be an intermediate lipoprotein in the catabolism of VLDL to LDL. IDL and chylomicron remnants may show similar electrophoretic and ultracentrifugation separation characteristics. In general, VLDL, LDL and HDL corresponds to pre-β-, β-, and α-lipoproteins, respectively. LDL is characterized as an atherogenic lipoprotein, whereas HDL is protective against atheromatosis.[39] LP(a)s are another even more atherogenic family of lipoproteins.[40] These combine LDL and (a), the protein designated "a" or "(a)," which contains multiple copies of kringle IV, a single kringle V and the protease of plasminogen.[41] Kringle proteins are named after the shape of a pastry made in Denmark.[42]

CLASSIFICATION OF HYPERLIPIDEMIAS

The presence of excess lipids in the blood is known as hyperlipemia or hyperlipidemia. Because atherosclerotic plaques contain lipid material, such as phospholipids, lipoproteins, and cholesterol, plaque formation is thought to be related to the content of these materials in the bloodstream. Hyperlipemia is divided into primary and secondary types. The primary disease may respond to antilipidemic drugs, but in the secondary type, which is caused by renal diseases (lipid nephrosis), diabetes, or hypothyroidism, treatment of the original disease rather than the hyperlipemia is necessary.[43]

Primary hyperlipemic conditions have been classified into six types by Fredrickson and co-workers.[44] Table 24–2 summarizes characteristics related to each of the hyperlipoproteinemias. Recommended diets in all six types can be obtained from the National Heart and Lung Institute (NHLI), Bethesda, Maryland. Although the phenotyping of lipoprotein disorders has the advantage of providing a practical basis for dietary and drug therapy for the physician, it does not provide information to establish genetic forms of hyperlipemia because multiple phenotypes may occur in one genotype. Furthermore, the phenotype may vary with drug treatment, dietary changes, and physiologic variability. Also, phenotyping by Fredrickson's protocol does not take into consideration the concentrations of HDL and other epidemiologically derived risk factors such as total cholesterol, cigarette smoking, systolic blood pressure, glucose tolerance, and left ventricular hypertrophy.[45] Studies have also supported the substitution of serum LDL cholesterol for serum total cholesterol and of apoB and apoA for LDL and HDL cholesterol, respectively. In clinical studies, positive correlation is often better between CHD risk and ratio of LDL to HDL cholesterol or of apoB to A-I.[46] Additionally, the ratio of HDL_2 to HDL_3 and the absolute concentration of apoA-I particles are inversely correlated with the severity of coronary occlusion.[47]

Antilipemic drugs are not effective in all six types of primary hyperlipidemia, as classified by Fredrickson. For example, no drugs are beneficial in the treatment of type I hyperlipemia, a rare condition characterized by the detection of high concentrations of chylomicrons in blood of patients who fasted for 14 hours. This disease is characterized by an absence of activity of LpL, an enzyme that catalyzes the hydrolysis of triglycerides in chylomicrons and VLDL. Free FA generated are resynthesized into triglycerides, which are stored in adipose (fat) tissue.

Type IIa hyperlipemia is characterized by high LDL and cholesterol levels and a slight increase in triglyceride content of the blood. Type IIb hyperlipemia is characterized by elevated serum cholesterol and triglyceride concentrations and by high LDL and VLDL. Antilipemic drugs are of benefit in types IIa and IIb hyperlipemia, both common in the United States.

Dietary treatment of type III hyperlipemia has been unusually gratifying; types IV and V also may be partly controlled by diet. Type III hyperlipemia is characterized by high cholesterol and triglyceride levels and by increased amounts of chylomicron remnants in the bloodstream. Types IV and V are related in that they both may result (in part) from faulty carbohydrate metabolism. Type IV can also be characterized by rapid triglyceride synthesis (faster than catabolism), reduced catabolism, and a high VLDL content. In addition to dietary controls, some drugs are effective against Type IV hyperlipemia.

Type V hyperlipemia may also result from faulty carbohydrate metabolism, but is further complicated by a diet high in fat; it is often seen secondary to acute metabolic disorders, such as diabetic acidosis, pancreatitis, alcoholism, and nephrosis. This disease, which may be controlled in part through a proper diet, is characterized by increased amounts of chylomicrons, VLDL and low concentrations of HDL and LDL.

Even with a normal lipoprotein pattern, atherosclerosis may develop. This may be a function of hypoalphalipoproteinemia,[48] small, dense, triglyceride-rich LDL[49] and Lp(a) > 40 mg/dl.[50] Gemfibrozil may be useful in managing the first two conditions[51,52] whereas nicotinic acid has efficacy in treatment of all three.[53–56]

In 1986, the National Cholesterol Education Program (NCEP) was formed. It subsequently published guidelines for diagnosis and management of hypercholesterolemia.[57] A stratification of CHD risk according to serum levels of total, LDL, and HDL cholesterol and total triglycerides is shown in Table 24–3.

The NCEP recommends dietary modification as a first step in the management of patients whose lipid and lipoprotein profile places them in the moderate to high risk groups. Those who remain at high risk after diet modification should be treated with appropriate cholesterol lower-

Table 24–2. Characteristics Related to Hyperlipoproteinemias

Hyperlipoproteinemia	Type I	Type IIA	Type IIB	Type III*	Type IV	Type V
Occurrence	Very rare	Common familial	Less common than IIA	Infrequent	Common	Rare
Etiology	Familial occurrence; lipoprotein lipase deficiency	Hypercholesterolemia, monogenic, polygenic; familial combined hyperlipoproteinemia	Familial hypercholesterolemia; familial hypertriglyceridemia; familial combined hyperlipidemia	Familial "Broad B" disease, same as in IIB	Familial hypertriglyceridemia; combined hyperlipidemia	Possibly familial lipoprotein lipase deficiency, "Broad B" disease
Cholesterol	N or ↑	Markedly ↑	Markedly ↑	Variable ↑	Essentially No	↑ (250–500 mg/dl)
Triglyceride	Markedly ↑ (2000 mg/dl)	No	Markedly ↑	Variable ↑	Markedly ↑	↑
Serum or plasma findings	Creamy layer upon standing in refrigerator; chylomicrons present in large amount above clear serum	Clear appearance; no chylomicrons; β-lipoproteins	Clear appearance; no chylomicrons, β-lipoproteins, pre-β-lipoprotein	Clear or cloudy; chylomicrons present sporadically; β-lipoprotein, broad, β-VLDL, pre-β-lipoprotein N or ↑	Clear or cloudy; no chylomicrons; β-lipoprotein N; pre-β-lipoprotein ↑	Creamy layer over turbid serum in refrigerator; pre-β-lipoprotein chylomicrons present
Secondary causes	Dysproteinemias, diabetes mellitus, acute alcoholism, pancreatitis	Dysproteinemia, excess cholesterol intake, hypothyroidism	Same as Type IIA	Dysproteinemia, hypothyroidism (myxedema)	Diabetes mellitus, pregnancy, oral contraceptives, alcoholism, dysproteinemia, nephrotic syndrome	Dysproteinemia, nephrotic syndrome, decompensated diabetes mellitus, alcoholism
Carbohydrate sensitivity	No	No	No	Yes	Yes	Possibly yes
Dietary treatment	Low fat diet with glycerides containing short chain fatty acids	Reduction in dietary saturated fats; low cholesterol	Low caloric intake; reduction in dietary saturated fats Weight reduction; maintenance of ideal body weight, low cholesterol, control carbohydrate intake	Weight reduction and maintenance of ideal body weight, control of carbohydrate and fat intake	Weight reduction and maintenance of ideal body weight, control of carbohydrate intake	Weight reduction and maintenance of ideal body weight
Drug treatment	—	Resins; probucol; nicotinic acid; lovastatin	Nicotinic acid; resins; probucol; lovastatin	Clofibrate; nicotinic acid	Nicotinic acid; gemfibrozil	Nicotinic acid; gemfibrozil

* Classification of Type III requires confirmation by ultracentrifugation before treatment.
↑ = increased.

ing drugs. Generally diets are planned for reduction of obesity and replacement of meat and saturated fat products with foods containing nonsaturated fats and high-protein, non-meat substances.[58] Concepts concerning appropriate diets were developed largely during the 20th century, and extensive clinical investigations continue.

The general consensus is that three dietary changes are necessary to lower cholesterol concentrations. First, decreasing food consumption in the obese patient; second, reducing consumption of cholesterol-containing foods; and third, decreasing saturated fat intake. More information is provided by Conner and Conner.[59]

Table 24–3. Stratification of CHD Risk According to Serum Lipid and Lipoprotein Levels

Total Cholesterol (mg/dl)	LDL Cholesterol (mg/dl)	HDL Cholesterol (mg/dl)	Triglycerides (mg/dl)	Risk for CHD
≤200	≤130	—	—	Low
200–239	130–159	≥35	—	Moderate
≥240	≥160	<35	≥250	High

HORMONAL REGULATION OF LIPOPROTEIN METABOLISM

A more detailed discussion of hormones and biochemical events influenced by these agents was provided by Brueggemeier and Li.[60]

Estrogens

Interest in the use of estrogens as antilipidemic drugs was stimulated by the observation that men and women exhibit different serum lipid levels; and that there is also a lower incidence of CHD in females than in males.

During the menstrual cycle, the relative amounts of LDL and VLDL in serum vary. For example, estrogen secretion is greatest before ovulation and during the luteal phase of the menstrual cycle. At this time the ratio of HDL to LDL cholesterol is increased and the ratio of cholesterol to phospholipids decreases.

On the other hand, during pregnancy there is a continual increase in the serum cholesterol, and the ratio of cholesterol to phospholipids increases until, in the final months, the serum lipid content resembles that in patients with CHD. During pregnancy, no correlation of serum estrogen and lipid levels is observed.[61–63]

Estrogens are also effective in males and cause regression of experimentally induced atherosclerosis in small animals. A major research objective has been to prepare compounds with high antilipidemic activity but few feminizing side effects. At present such analogs are not used clinically.

Equine estrogens were investigated in the national Coronary Drug Project[17,18] at daily doses of 2.5 and 5.0 mg. The 5-mg-per-day estrogen regimen was discontinued in 1970, chiefly because of an excessive number of nonfatal cardiovascular events in this group compared to those in the placebo group.[17] In 1973, 2.5-mg-per-day estrogen therapy was discontinued for lack of therapeutic efficacy.[18]

Estradiol increases the concentration of apoA-I, the major protein in HDL, and this is thought to lead to an elevation in HDL subfraction known as HDL_2.[60] Estrogens also suppress the activity of HTGL, which catalyzes the hydrolysis of triglycerides in HDL and VLDL.[64] The antiestrogen, tamoxifen, lowers LDL concentrations in breast cancer patients with a tendency toward elevated HDL levels.[65,66]

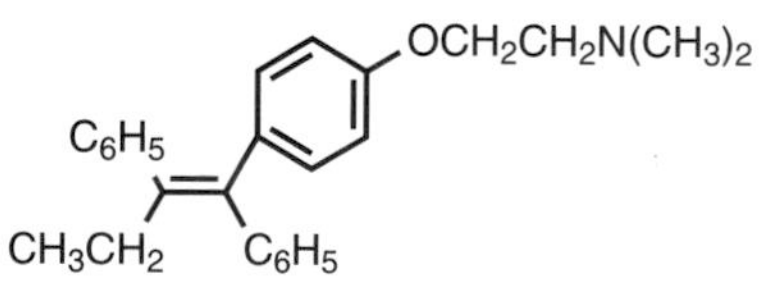

Tamoxifen

Progestins

The biochemical effects of progesterone are more or less opposite to those of the estrogens.[60] Thus, progesterone lowers HDL_2 levels and stimulates the activity of HTGL. Although progesterone may be protective in myocardial infarctions,[67] in oral contraceptives the increased risk of atherosclerosis is a function of this steroid,[60] but risk in thrombosis is associated with the estrogenic compound.[60] Effects of oral contraceptives on lipoprotein levels depend on the formulation, dose, and potency of the preparation.[60]

Androgens and Anabolic Steroids

HDL cholesterol levels fall in males at puberty and this correlates with the rise in plasma testosterone concentrations,[60,68] but in adults such correlations are conflicting.[60] Prostate carcinoma patients treated with luteinizing hormone-releasing hormone (LHRH) with or without the antiandrogen, flutamide, show increases in HDL cholesterol and apoA-I.[69] Drug abuse by weight lifters is of significance;[60] those taking 6 mg/day of stanozolol have their HDL cholesterol reduced by 33% and the HDL_2 subfraction reduced by 71% at the end of week six,[70] but apparently less drastic effects are noted with testosterone.[60] The stimulation of HTGL activity correlates with decreased plasma HDL_2 levels.[60] Additionally, these compounds inhibit apoA-I synthesis and LCAT activity. LCAT transfers unsaturated fatty acids from position 2 of phosphatidylcholine to cholesterol in HDL to increase the number of dense HDL_2 particles.[60]

Glucocorticoids

After 14 days of prednisolone at 30 mg/day, VLDL triglyceride and cholesterol, HDL cholesterol, apoA-I and apoE increase.[71] Long-term administration of glucocorticoids also results in VLDL-triglyceride and cholesterol increases and increases in LDL levels, although the mechanisms are not clearly understood.[60] The high incidence of CHD observed in people with Type A personalities or in Cushing's syndrome also is related to higher serum corticosteroid levels.[60]

Thyroid Hormones

Hypothyroid animals, including humans, synthesize and metabolic (catabolize) cholesterol at a slower rate than do normal or hyperthyroid animals. L-Thyroxine stimulates catabolism to a greater extent than anabolism of cholesterol.

Hypothyroid individuals exhibit an increase in LDL cholesterol.[72–75] Administration of L-thyroxine to patients who do not have a hypothyroid condition (myxedema) is dangerous and undesirable because it increases basal and myocardial metabolism; for these reasons, administration to patients with CHD is contraindicated.

The drug D-thyroxine was investigated in males using 6.0-mg daily doses. This enantiomorph of the naturally occurring hormone has marked hypocholesterolemic activity (attributed to increased cholesterol catabolism), but apparently has less influence on basal and myocardial metabolism in vivo.[76–78] The difference in action of the isomers is probably related to differences in distribution in vivo. Because high mortality was associated with this drug, D-thyroxine was withdrawn from the Coronary Drug Project.[17] The compound is still used clinically as an anti-

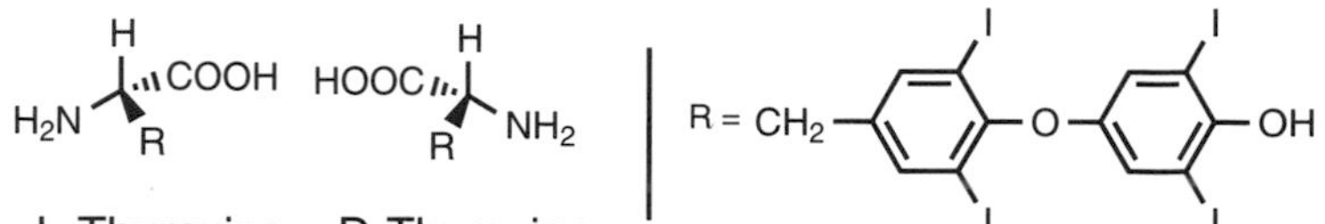

L-Thyroxine D-Thyroxine

lipidemic agent, but should be used with caution in patients having evidence of cardiac arrhythmia and other heart disorders.

Mechanisms are complex and not well understood, but in hyperthyroidism the low HDL-, LDL- and total cholesterol levels are likely a function of increased elimination. In contrast, elevated VLDL-triglycerides and LDL cholesterol fractions in hypothyroid patients result from decreased catabolism and elimination.[60] These hormones also stimulate LDL receptor function and enhance catabolism of these lipoproteins.[60] The decrease in HTGL activity in hyperthyroidism correlates with an increase in HDL cholesterol levels.[60,79]

Insulin

The increase in VLDL cholesterol and triglyceride induced by corticosteroids is likely secondary to the induced hyperinsulinemia,[80] but in diabetic patients the role of insulin in lipoprotein metabolism is more complex.[60] In Type 1 patients (those with an insulin deficiency) there is an increase in free FA with concomitant increases in the synthesis and secretion of VLDL-triglyceride, a decrease in VLDL-triglyceride catabolism because of decreased LpL activity, and an increase in LDL cholesterol levels because of a decrease in LDL receptors in liver and peripheral tissues.[60,81]

Insulin increases the number of LDL receptors in Type 1 diabetic patients as well as the clearance of such lipoproteins in non-diabetic individuals.[82] Additionally, LDL apoB glycation (nonenzymatic glycosylation) at lysine residues, results from hyperglycemia in Type 1 individuals and this ultimately leads to decreases in LDL-receptor mediated uptake because it is the apoB of LDL that binds to the receptor.[83,84]

Most Type 2 patients (those who are insulin resistant) have high plasma VLDL-triglyceride and free FA levels and enhanced secretion and synthesis of such lipids.[60] Hyperglycemia may also result in increased glycation of apoB and decreased receptor mediated uptake of LDL. Decreases in HDL levels may be a function of both increases in HTGL activity,[60,85] and glycation of apoA-I in HDL resulting in enhanced catabolism.[60,86]

Growth Hormone

Serum triglyceride and cholesterol levels are generally normal or slightly increased in growth hormone-deficient patients,[60,87] but again the situation is complex. In acromegaly, serum cholesterol levels are lower and triglyceride levels higher.[60] Familial combined hyperlipoproteinemia (FCH) and growth hormone deficiency lead to increased levels of serum triglyceride and cholesterol. Because growth hormone stimulates cholesterol 7α-hydroxylase activity, the rate-limiting step in bile acid biosynthesis, it is by this mechanism that the hormone is thought to increase cholesterol catabolism and lower serum total and LDL cholesterol levels.[60,88]

DRUGS OF HISTORICAL INTEREST

Plant Sterols

During the mid-1900s, considerable interest arose in plant sterols as serum cholesterol-lowering agents. Mixtures of plant sterols, including soybean sterols and β-sitosterol (22,23-dihydrostigmasterol), are among the preparations that are absorbed to a limited extent from the intestinal tract and compete for the absorption of both endogenous and exogenous cholesterol.[89,90]

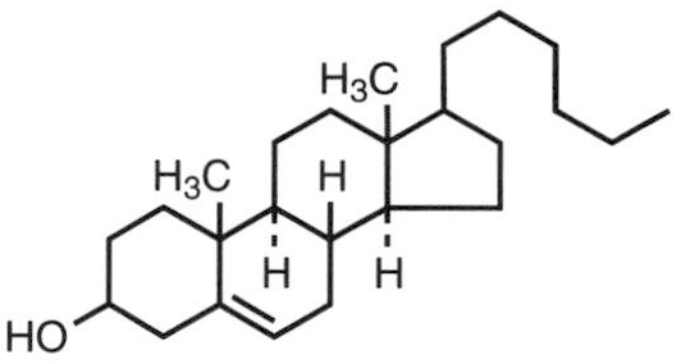

β-Sitosterol

In animals, β-sitosterol also increases the rate of cholesterol and lipid metabolism.[91] Long-term studies with β-sitosterol show no detectable toxicity or tissue accumulation, but the hypocholesterolemic effect is variable and unpredictable.[92,93] For these reasons, this sterol, which is expensive and only of research interest, has gained limited status in the treatment of hypercholesterolemia and atherosclerosis.

Triparanol

Blohm and co-workers reported that triparanol (Mer-29) was a potent inhibitor of cholesterol biosynthesis.[94,95] This drug and related compounds resulted from a search

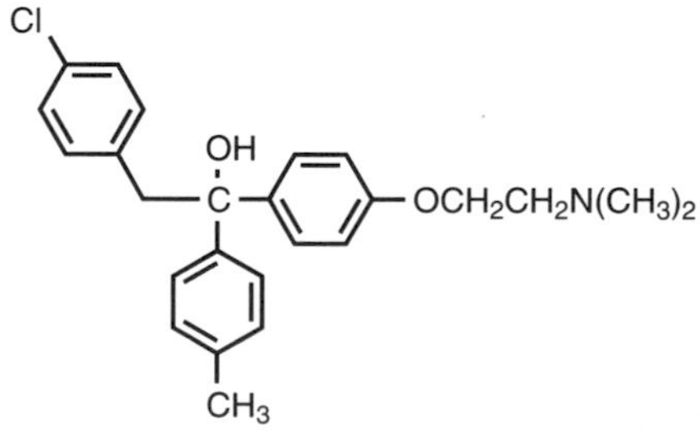

Triparanol

for nonsteroidal agents devoid of the undesirable side effects of steroids; for example, feminization may be produced by certain antilipemic estrogens. Although initial clinical trials with triparanol showed promise, toxic manifestations, including vaginal bleeding, ichthyosis, loss of hair, and a cataractogenic effect were subsequently reported.[96,97] Clinical investigations were consequently terminated.

Large doses of this inhibitor of cholesterol biosynthesis also reduced adrenocorticoid production; doses of 2 g of triparanol daily have been used to induce medical adrenalectomy in patients with bilateral hyperplasia and Cushing's syndrome.[98]

Although it was disappointing that triparanol did not become a drug of clinical significance, research with this compound alerted investigators to hazards associated with agents that block at late steps in cholesterol biosynthesis. Moreover, triparanol played an important part in understanding of the enzymes involvement in the process. This drug blocks reduction of 7,8- and 24,25-double bonds of the intermediate products in cholesterol biosynthesis (see Chapter 23); the enzymes are referred to as Δ^7- and Δ^{24}-reductases, respectively.[99]

In rats, desmosterol, the immediate cholesterol precursor, accumulates in serum, liver, and other tissues.[100,101] Tissue studies in a 16-year-old boy who received triparanol for 2 months before dying also showed desmosterol accumulation.[102,103]

Diazacholesterols

It was originally suggested that diazacholesterols and related analogs, which look sufficiently like cholesterol, would bind to negative feedback enzymes because of electron-rich nitrogens. Such binding would mimic cholesterol binding and inhibit cholesterol biosynthesis. Compounds such as 20,25-diazacholesterol, however, generally inhibit reduction of 7,8- and 24,25- double bonds,[104–111] and in vivo, HMG-CoA reductase activity is stimulated even though serum cholesterol levels are reduced.[112] Unlike the structurally unrelated triparanol molecule, which is a noncompetitive antagonist, 20,25-diazacholesterol is probably a competitive inhibitor[99] that binds to Δ^7-reductase and Δ^{24}-reductase.

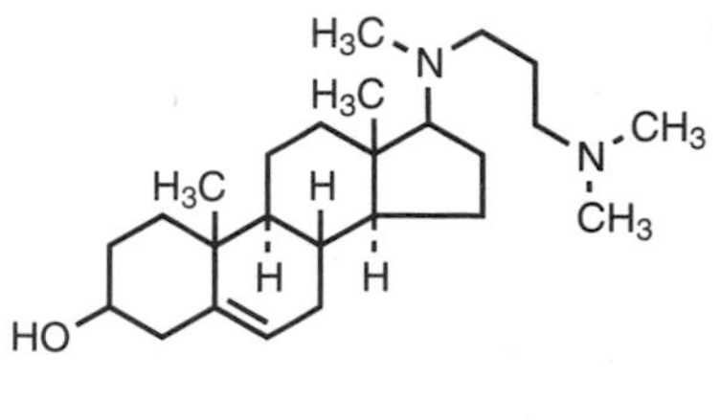

20,25-Diazacholesterol

Subsequently, it was observed that 20,25-diazacholesterol blocked egg laying capabilities in hens. The compound (Ornitrol, Searle, Skokie, IL) was used to control pigeons by supervised baiting on roof-tops and in bell towers and grain elevators.[113] The conversion of desmosterol to cholesterol is also blocked in the larvae of the tobacco hornworm[114] and in animals and humans, the compound causes myotonia.[115] For these reasons, this compound and related analogs are of experimental interest.

DRUGS OF CLINICAL SIGNIFICANCE

Statins (Lovastatin, Pravastatin, and Simvastatin)

Statins (or vastins) represent a new class of lipid-lowering compounds that inhibit HMG-CoA reductase, the first and rate-limiting step in cholesterol biosynthesis in cells.[116–118] The hypocholesterolemic actions of statins are related to both the inhibition of liver cholesterol biosynthesis and depletion of the cholesterol pool. This cholesterol-depleting effect results in the induction of messenger ribonucleic acid (mRNA) for the LDL-receptor and increased synthesis of hepatic LDL receptors. Increased numbers of cellular LDL receptors enhance the clearance of circulating serum IDL and LDL. This produces a fall in serum cholesterol, LDL particles, and LDL cholesterol.[119–122] Thus, the effectiveness of these agents depends on LDL receptor regulation, and this class of compounds is unable to lower LDL in patients who lack gene expression (homozygous familial hypercholesterolemia) of LDL receptors, or those who are LDL receptor negative.

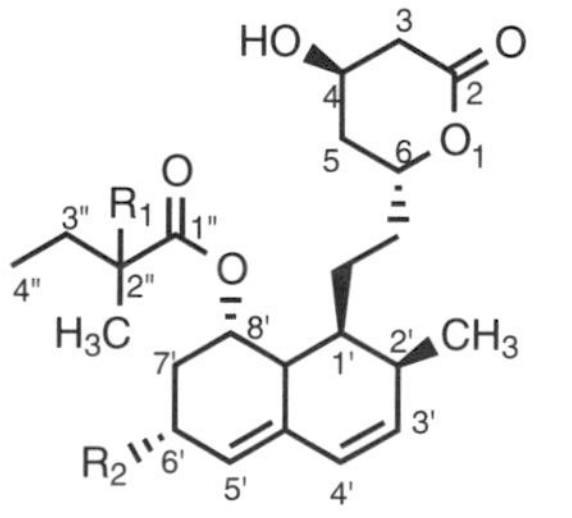

	R_1	R_2
Mevastatin	H	H
Lovastatin	H	CH_3
Simvastatin	CH_3	CH_3

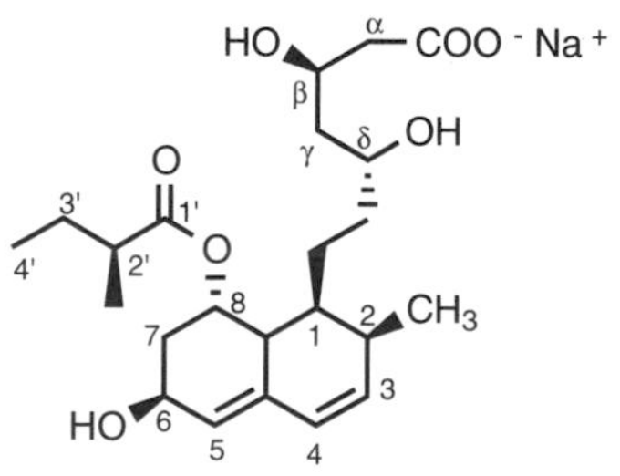

Pravastatin

Mevastatin (compactin: 6-demethylmevinolin) and lovastatin (mevinolin) were isolated as fungal metabolites from cultures of Penicillium brevicompactum (or P. citrinum) and Aspergillus terreus, respectively.[123–125] Other closely related analogs include pravastatin (pravastatin sodium) and simvastatin (synvinolin).

Lovastatin (Mevacor) was the first drug of this chemical class to be approved (1987) in the United States, and pravastatin sodium (Pravachol) and simvastatin (Zocor) were approved in 1991. Lovastatin, pravastatin, and simvastatin are approved for use in the reduction of total and LDL cholesterol blood levels in patients with primary hypercholesterolemia when the response to dietary and other nonpharmacologic measures have been inadequate.

Statins are generally more effective than probucol,[126–129] bile acid sequestrants,[130–135] and fibrates[136–150] at lowering total and LDL cholesterol. These drugs now represent the major class of compounds for monotherapy treatment of patients with Types IIa and IIb hypercholesterolemia. Each of these compounds is a potent competitive inhibitor of HMG-CoA reductase.[116–118] Unlike triparanol (MER-29), which inhibits a late step in cholesterol biosynthesis leading to desmosterol accumulation and toxicity,[96,97] these drugs selectively block HMG-CoA reductase activity. Thus, the buildup of sterol intermediates is not seen with mevastatin or lovastatin administration.[117,123,151] Whereas pravastatin (pravastatin sodium) is given as the active hydroxy acid, the remaining statins are given as prodrug lactone forms, which are enzymatically converted by serum or tissue esterases to the corresponding hydroxy acid.[117] These hydrolysis products inhibit HMG-CoA reductase activity in cultured cell types[119,152,153] and in animals.[117,120,154,155] Lovastatin, pra-

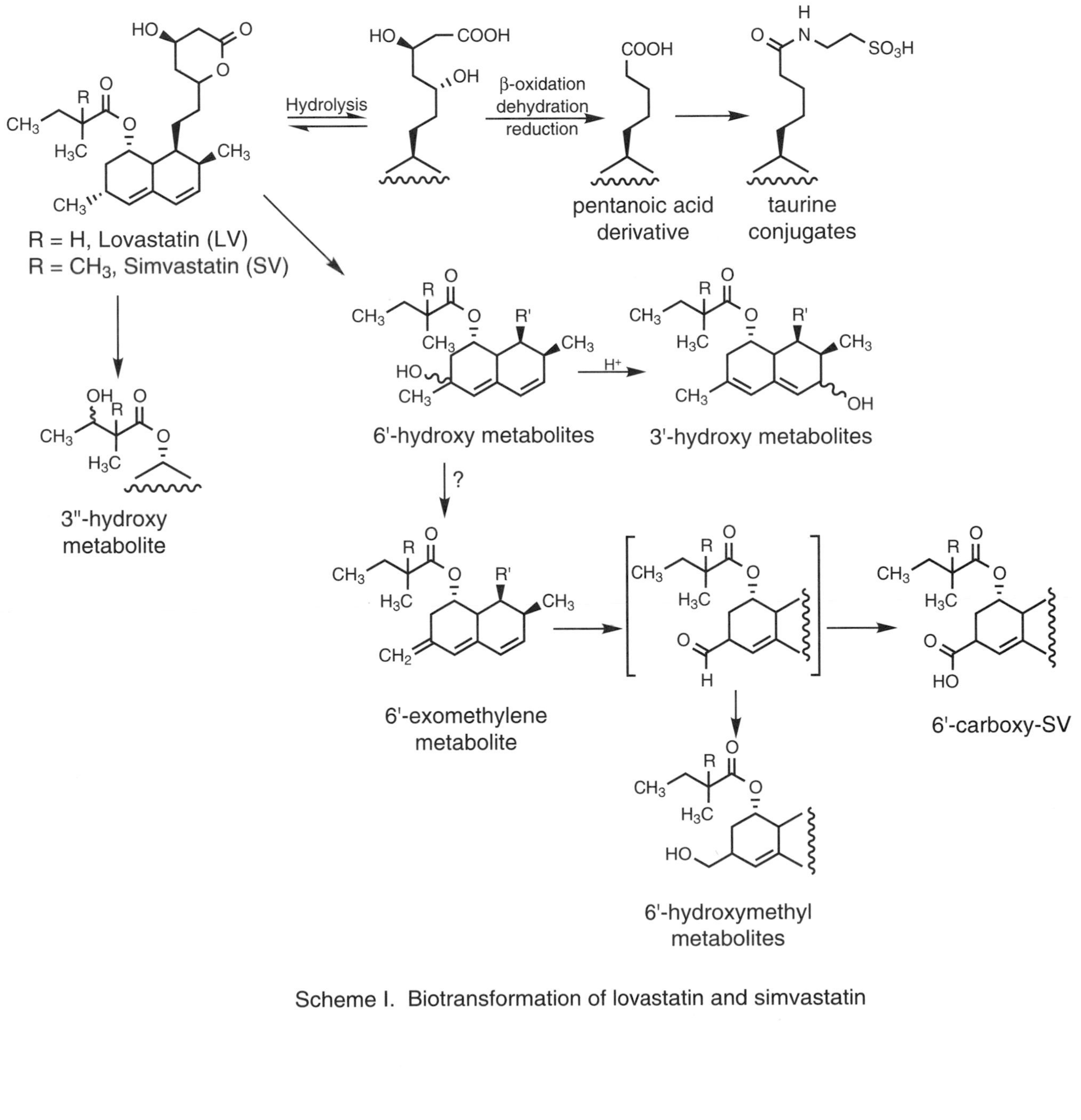

Scheme I. Biotransformation of lovastatin and simvastatin

vastatin, and simvastatin are more potent inhibitors than mevastatin of HMG-CoA reductase and cholesterogenesis in vitro and in vivo.[117,154] Pravastatin is a metabolite of mevastatin and simvastatin differs from lovastatin by the presence of an additional methyl group on the 2-methylbutylyl side chain (2,2-dimethylbutylyl group). Other analogs have been prepared; however, they generally are less potent than mevastatin or lovastatin or are less selective inhibitors of lipid synthesizing enzymes.[156,157] In other studies, isosimvastatin-6-one,[158] and the 6β-methyl-5-oxa and 5α-hydroxy of 3,4,4a,5-tetrahydro analogs of simvastatin,[159] possess potent HMG-CoA reductase inhibitory activity.

Lovastatin[160,161,162] and simvastatin[163] are extensively biotransformed in vivo and in vitro (scheme I). The prodrug statins also undergo reversible equilibrium between the lactone and open chain hydroxy acid forms in vivo, and statin metabolites are mainly eliminated into bile.[164,165] Oxidized phase I metabolites of lovastatin and simvastatin include 6′β-hydroxy, 3″-hydroxy and 6′-exomethylene derivatives. Liver microsomal metabolism of stereoselective 6′β-hydroxy lovastatin is blocked by cimetidine and famotidine.[162] Further conversion of lovastatin and simvastatin acidic forms by the fatty acid β-oxidizing pathway of intermediary metabolism leads to the formation of pentanoic acid and taurine conjugate derivatives. These metabolites are formed in rodents but not in humans.[160] Like lovastatin, simvastatin is converted to the 6′-hydroxy and 3′-hydroxy metabolites as major metabolites in vivo and in vitro. 6′-Hydroxy statins undergo rapid nonenzymatic conversion to corresponding 3′-hydroxy derivatives under acidic conditions. Therefore, a non-enzymatic acid catalyzed formation of statins to various metabolites may occur prior to absorption. Prodrug statins are converted to 3′-exomethylene derivatives, in vitro. The 6′-hydroxymethyl and 6′-carboxyl statin derivatives are formed in vivo from 3′-exomethylene statin through a postulated 6-aldehyde intermediate.[164] Both lactone and hydroxy acid forms of prodrug statin metabolites accumulate in tissues, in vivo.[160,164,166,167]

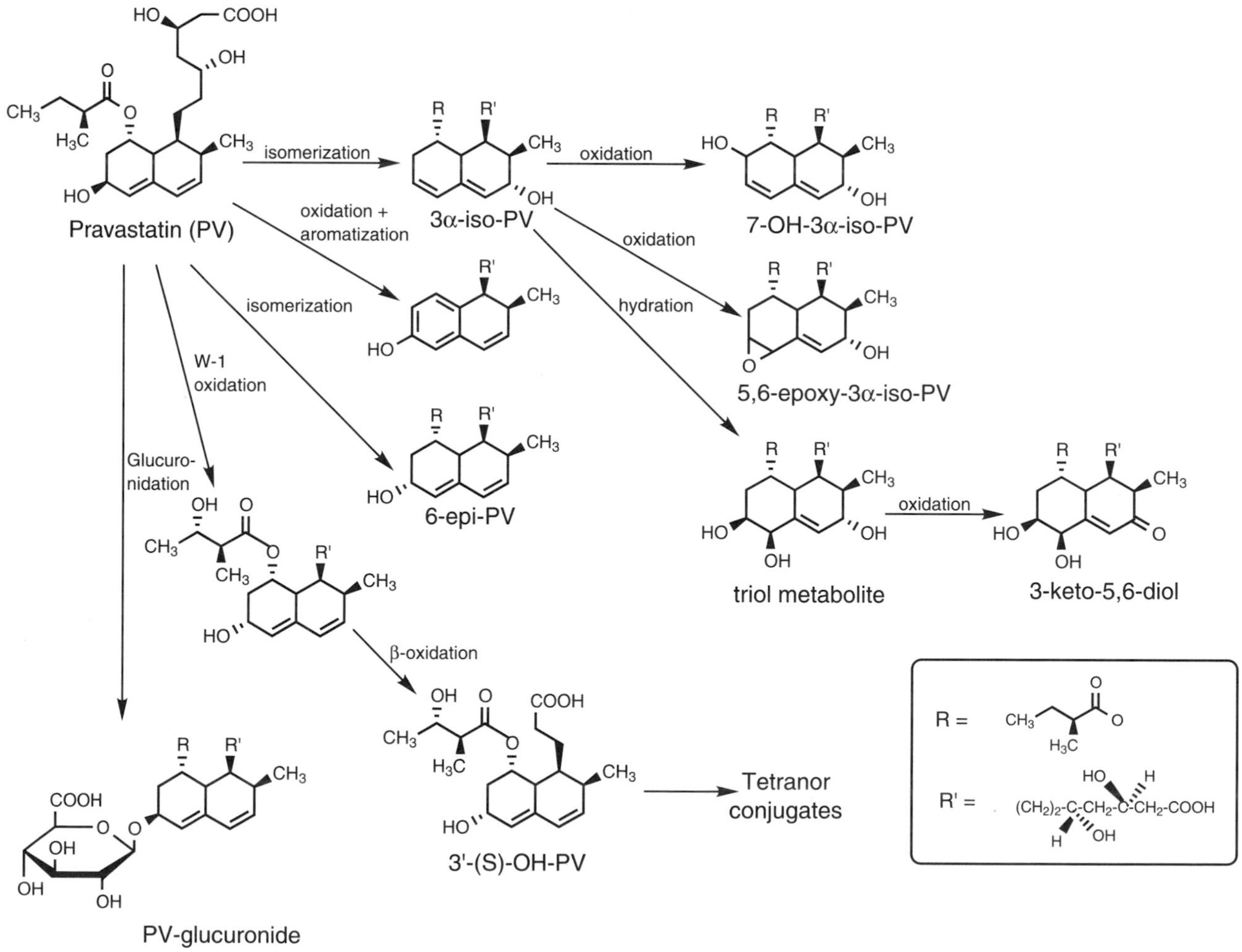

Scheme II. Biotransformation of pravastatin.

A large number of pravastatin metabolites arising from oxidation (including ring, ω-1 and β-oxidation pathways), hydration, conjugation, and isomerization reactions (scheme II) have been identified.[168] Pravastatin, like lovastatin[160] and simvastatin,[164] undergoes ω-1 hydroxylation of the methylbutyryl side chain [3′(S)-hydroxypravastatin] and isomerization to form 3α-isopravastatin. In human plasma, urine, and feces, intact pravastatin was the major component, and 3α-iso- and 3α-5,6-diol- (triol) pravastatin are the main metabolites.[166,168] Other metabolites identified were pravastatin glucuronide, desacyl pravastatin, 6-epipravastatin, and tetranor derivatives.[168] 6-Epi and 3α-isopravastatin metabolites result from a nonenzymatic acid-catalyzed isomerization reaction. Oxidation of the 1.2.6.7.8.8a-hexahydronapthalene ring of 3α-isopravastatin produces the corresponding 5,6-epoxy and 7-hydroxy metabolites. Desacylpravastatin is produced by oxidation of 6-hydroxypravastatin to a keto intermediate followed by spontaneous aromatization. The triol derivative is derived via a 4α,5β,-epoxide intermediate and is further oxidized to the 3-keto-5,6-diol analog of pravastatin. The tetranor metabolite undergoes conjugation, and its formation from either 3′-hydroxypravastatin or pravastatin may arise by β-oxidation and ω-1 pathways of fatty acid intermediate metabolism. Similar to lovastatin and simvastatin, a pentanoic acid derivative of pravastatin has been identified in dog, rodent, and monkey plasma, but not in that of humans.[168]

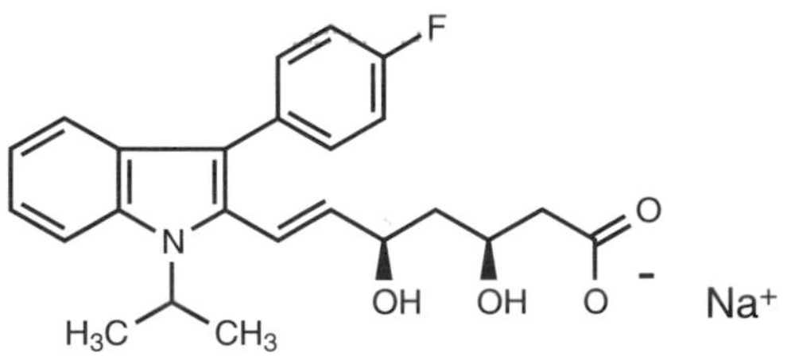

Fluvastatin sodium

Fluvastatin sodium is the most recently approved statin in the United States. It is a white to pale yellow, hygroscopic powder that is freely soluble in water, ethanol, and methanol. It is supplied as 20-mg and 40-mg capsules for oral administration.

Fluvastatin sodium is structurally distinct from the other statins, having fluorophenyl and indole ring systems. This drug, a close structural derivative of HMG-CoA, is a water-soluble agent that directly inhibits liver HMG-CoA reductase. Although bioavailability is reduced by food intake as with other statins, fluvastatin is almost completely absorbed (98%) in fasting patients. Fluvastatin sodium is highly plasma protein bound (>98%) and is rapidly extracted (68%) by first pass effect into the liver.

Biotransformation pathways for fluvastatin involve hydroxylations at the 5- and 6-positions of the indole ring, N-dealkylation, and β-oxidation. The hydroxylated metabolites possess HMG-CoA reductase inhibitory activity, but are not widely distributed to peripheral tissues. Both enantiomers of fluvastatin are metabolized in a similar manner, and unchanged fluvastatin and its N-deschloropropionic acid inactive metabolite are the major circulating compounds in plasma. Following administration of [^{3}H]-fluvastatin, 90% was cleared into feces and 5% into urine. Unchanged fluvastatin accounted for less than 2% of the eliminated radioactivity.

Twenty to 40 mg of fluvastatin sodium is recommended for the treatment of primary hypercholesterolemia (Type IIA and Type IIB hyperlipoproteinemias). Like the other statins, this drug lowers blood LDL-cholesterol, total cholesterol, and Apo B levels in patients with Type IIA, IIB hyperlipoproteinemias. Efficacy for blood cholesterol lowering is comparable to other statins; a 20 to 25% decrease in blood LDL-cholesterol levels is found after a 20-mg dose (QPM) over 9 weeks. Similar to the other statins, administration to familial hypercholesterolemia (FH) patients may produce small increases in blood HDL-cholesterol and decreases in blood triglycerides. No effect on LP(a) levels were observed in patients receiving fluvastatin sodium therapy.

The contraindications to fluvastatin sodium use are the same as for other statins. Side effects with statins, including fluvastatin, are rare. Patients may exhibit symptoms of myotonia with concomitantly administered niacin, fibrates, or other statins; hepatotoxicity, which should be monitored on initiation of fluvastatin therapy or in patients with compromised liver function; and drug interactions with cyclosporin and erythromycin. Neurotoxicity (or insomnia) is not a major consideration because this water-soluble statin does not easily penetrate the blood-brain barrier. In comparison to the other fungus-derived statins, fluvastatin sodium (a synthetic drug) may offer patients a savings in cost to achieve the same extent of LDL-cholesterol lowering.

Hydrophobic prodrug statins undergo a greater first pass extraction by the liver than pravastatin. Conversion of lovastatin and simvastatin to the active hydroxy acid forms (lovastatin and simvastatin acids) occurs rapidly.[160,164] Pravastatin, as the hydroxy acid form, is more hydrophilic than the prodrug statins, hepatic levels of pravastatin are 50% lower, and plasma and nonhepatic tissue concentrations are considerably higher than either lovastatin or simvastatin.[169] Pharmacokinetic studies indicate that hydroxy acid metabolites of these prodrug statins contribute significantly to the overall inhibitory activity against HMG-CoA reductase and cholesterol lowering response, in vivo.[166,167] Relative to lovastatin acid (100%), the hydroxy acid forms of 3″-hydroxy lovastatin (15%); 6′-exomethylene lovastatin (50%), and 6′β-hydroxy lovastatin (10%) were less potent.[160] In contrast, the hydroxy acid form of 3′-hydroxy lovastatin, and the pentanoic acid and taurine conjugate derivatives of lovastatin were inactive. Only 6-epipravastatin, a minor in vivo metabolite, produced significant inhibitory activity (60% that of pravastatin) against HMG-CoA reductase activity; the remaining metabolites of pravastatin were inactive or had negligible (2–6%) activity. Although pravastatin, as the active hydroxy acid form, is the major inhibitor of HMG-CoA reductase activity after dosing, in vivo; hydroxy acid forms of statin metabolites other than lovastatin acid and simvastatin acid appear to contribute significantly to the overall LDL-cholesterol lowering response of prodrug statins.[166]

After oral administration, statins are moderately absorbed (about 30%) and both prodrug and active acid forms are extensively bound (>95%) to plasma proteins.[166,167] Between 70 and 80% is eliminated in feces, and 10 to 20% of an oral dose of statins is eliminated in urine. In human volunteers given an oral dose (9.2 mg) of radiolabeled pravastatin, 71 and 20% recoveries of total dose were found in feces and urine, respectively. The systemic bioavailability of total inhibitors (active form and potentially active metabolites) for lovastatin was higher than that for pravastatin. Pravastatin, however, was more rapidly absorbed (T_{max} of 1 hour for pravastatin versus 4 hours for lovastatin) and achieves much higher concentrations of the active inhibitor (C_{max} of 64 ng for pravastatin versus 9 ng for lovastatin) following the administration of 40 mg to healthy volunteers.[166] Lovastatin and simvastatin both undergo a high hepatic extraction rate leading to the accumulation of the prodrug lactone forms and active open ring metabolite(s) in this target tissue. The lactone forms of prodrug statins are in reversible equilibrium with the active acid forms (relactonization) in vivo.[165] Only small amounts of the lactone are observed in human plasma of following pravastatin treatment; thus, intact pravastatin is the major component in blood and various tissues.[166] Whereas lovastatin, simvastatin, and mevastatin are all virtually insoluble in water, the greater hydrophilicity of pravastatin may explain its increased tissue selectivity as an inhibitor of hepatic HMG-CoA reductase.[170,171] All statins are extracted preferentially into the liver, but pravastatin accumulates at 3 to 6 times greater concentrations than simvastatin and lovastatin in peripheral tissues.[169]

Initial studies indicated that lovastatin and mevastatin significantly reduce plasma total cholesterol and LDL levels with a minimal effect on plasma triglycerides in dogs,[117,120,154] whereas little or no cholesterol lowering effect is known in normolipemic rats or mice given mevastatin.[117,172] By contrast, treatment of normolipemic volunteers or heterozygous familial hypercholesterolemic (FH) patients with lovastatin[121,173–177] or heterozygous FH or combined hyperlipidemic patients with mevastatin[178–180] gave marked decreases in serum total cholesterol (19 to 32%) and LDL cholesterol (27 to 40%). Treatment of heterozygous FH patients with mevastatin also produces significant reductions in serum IDL triglyceride, cholesterol, LDL cholesterol and triglyceride without a change in VLDL cholesterol, triglyceride or in HDL cholesterol levels.[180] In addition, serum LDL and IDL phospholipids and total serum phospholipid concentrations decrease with mevastatin therapy in these patients.[178] Likewise, lovastatin therapy appears to produce little effect on HDL, HDL cholesterol, and serum triglyceride concentrations in humans.[174,175] Reports of serum triglyceride lowering in heterozygous FH patients given mevastatin correlate to a reduction in serum LDL or IDL levels,[178,180] which may be explained by increased removal of these apoB-containing lipoproteins by liver.

Few studies compare the effects of statin drugs within the same regimen. The effects of statins on total and LDL cholesterol lowering in hypercholesterolemic patients are comparable. Lovastatin and simvastatin appear to produce greater increases in HDL cholesterol and lowering of serum triglycerides, VLDL triglycerides, and VLDL cholesterol than pravastatin.[181–183] Clinically, the adverse effects of the statins are also similar, and thus a clear recommendation for one over the other is difficult.[184]

Evidence also indicates that lovastatin and simvastatin accumulate in plasma lipoproteins,[185] and that LDL of patients on lovastatin therapy (20 mg/day) are less susceptible to oxidation.[186] An antioxidant role is proposed for lovastatin, although this potential antiatherogenic effect is considerable less than that observed for probucol or vitamin E.

Statins[187–193] produce consistent dose-related reductions in LDL cholesterol along with modest decreases in triglycerides and increases in HDL cholesterol levels. Patients with initially low HDL cholesterol and high triglyceride levels have greater therapeutic responses. Statins do not, however, consistently elevate HDL cholesterol or HDL particles in normolipemic patients with hypoalphalipoproteinemia.[194] Statins offer an attractive alternative therapy for primary hypercholesterolemia (Types IIa, IIb) and especially heterozygous FH. Statins are not effective in LDL-receptor negative patients (homozygous FH) but LDL-receptor defective patients and homozygous FH patients undergoing LDL plasmaphoresis may be responsive.[195–198] The observed reduction in IDL and VLDL cholesterol levels after mevastatin therapy suggests that these drugs may also be beneficial in the treatment of familial dysbetaliproteinemia (Type III hyperlipoproteinemia) alone or in combination with fibrates.[180,199–201] Patients with familial defect apoB-100 or cholesterol ester storage disorders may also respond to lovastatin therapy.[202–204] Lovastatin and pravastatin represent safe and effective treatments for hypercholesterolemic patients undergoing renal and cardiac transplantation,[205–207] provided low doses are given and that changes in renal and skeletal function and blood levels of immunosuppressants are monitored.[208,209]

A greater serum cholesterol-lowering effect of combined therapy with lovastatin and colestipol,[121,173,210–212] mevastatin and colestipol,[213] pravastatin and colestipol,[131] or statins and cholestyramine.[131,178,214–218] Similarly, simvastatin and combined therapy with a bile acid sequestrant or nicotinic acid produce a greater LDL cholesterol-lowering activity than the statin alone in heterozygous FH patients.[192,219,220] Lovastatin and gemfibrozil therapy are also beneficial for treatment of hypertriglyceridemias in noninsulin-dependent diabetes mellitus[221,222] and in heterozygous FH and mixed hyperlipidemic patients.[223–225] Polytherapy in hypercholesterolemic FH patients indicates that colestipol plus lovastatin give a maximal LDL cholesterol-lowering effect (52%).[226] The addition of probucol produced little additional benefit.[226] In another triple drug regimen study, the combined administration of a bile sequestrant, statin, and nicotinic acid reduced LDL cholesterol to a maximum of 59 to 67%.[227] In patients with initially high apoB levels, lesion progression (using quantitative arteriography) was reduced, frequency of regression was increased, and clinical events (myocardial infarctions, death, revascularization for worsening symptoms) were reduced by the combination of lovastatin and colestipol therapy.[228] Therefore, the combination of bile acid sequestrants and statins may be most useful for treatment of more severe or refractory types of primary hypercholesterolemias. Considering the use of these two cholesterol-lowering agents, the statins are more cost effective than cholestyramine and compare favorably with other medication strategies (nicotinic acid), especially if treatment is initiated at an early age.[229–231] The statins either increased or did not lower circulating plasma levels of the atherogenic lipoprotein, Lp(a).[232–236]

Adverse effects of statin therapy in humans are few and may include abdominal pain and gastrointestinal (GI) symptoms (cramps, bloating, diarrhea, constipation, and nausea), insomnia, skin rashes, and muscular pain (myositis, myalgia) with moderate increases in serum CPK, SGOT or SGPT.[174,175,178–180] The frequency of side effects of simvastatin is similar to that of lovastatin.[237] Sleep disturbances are observed with lovastatin but not pravastatin therapies (40 mg/day), and are reversible.[238] Lovastatin is more hydrophobic than pravastatin and penetrates more readily into the CNS.[239] Potentially life-threatening, but uncommon, side effects of lovastatin therapy are acute renal failure and rhabdomyolysis, and this may occur with statins alone or during combined therapy with immunosuppressant drugs (cyclosporine, azathioprine),[208,240] erythromycin,[241] fibrates,[242] and nicotinic acid.[243] There are reports of liver and skeletal muscle toxicities with statins alone, and these adverse effects may increase in patients given higher doses[244,245] or in combination therapy with immunosuppressants and fibrates.[246] The mechanism of myopathy is related to a reduced clearance of the statins with significant accumulations in skeletal muscle.[247] A possible risk of lens opacities or cataracts in patients given lovastatin (20 to 80 mg/day) was proposed; however, recent studies do not confirm this adverse effect for lovastatin, pravastatin or simvastatin.[218,248–250]

Use of statins is also contraindicated during pregnancy and lactation. These drugs accumulate in breast milk and produce embryotoxic effects, such as skeletal muscle abnormalities, in laboratory animals. Statins also are reported to produce carcinogenicity in animals at doses that greatly exceed the therapeutic dose range.[251]

One concern is that these HMG-CoA reductase inhibitors may interfere with the formation of nonsterol and other sterol compounds derived from mevalonic acid. The effect of statins on formation of ubiquinones (coenzyme Q),[252,253] dolichol,[254] sterols,[255] and farnesyl intermediates, however, are necessary for protein isoprenylation. These prenylated proteins are of concern because they are associated with signal transduction pathways related to cell growth or function.[256–262] Generally, at doses of statins that interfere with cholesterol biosynthesis, little effect is observed on isoprenylated proteins such as $p21^{ras}$ and prelamin A.[262] However, urinary mevalonic acid levels[263–265] and coenzyme Q levels are reduced[266] on patients given statins. No significant changes in blood steroids such as vitamin D_3, ACTH, cortisol, LH, FSH, aldosterone, progesterone, testosterone, prolactin, and plasma renin activity,[267–274] whole body cholesterol metabolism,[275] or growth of enterocytes[276] are observed. Modest effects in leukocytes[277,278] are observed in patients on statin therapy. Reductions in dolichol and ubiquinone levels in FH patients on pravastatin paralleled the lowering of LDL cholesterol. No changes in hepatic cholesterol, dolichol, or ubiquinone concentrations are observed.[279] Thus, the low therapeutic doses of statins preclude adverse effects that may be associated with the inhibition of nonsterol and other sterol products in vivo.

Due to the high plasma protein binding of statins, patients receiving warfarin and statin therapies should be monitored to prevent hypoprothrombinemia and bleeding episodes.[280] Propranolol coadministered with statins reduces the mean plasma concentrations and area under the curve (AUC) of active and total inhibitors by 16 to 23%; however, the clinical significance of these changes is minimal.[281]

In contrast to fibrates, statins do not increase the biliary cholesterol saturation index nor the frequency of gallstone formation in hypercholesterolemic patients.[282] Lovastatin lowers the biliary cholesterol saturation index by reducing bile acid synthesis and reducing cholesterol secretion into bile.[282,283] Statin therapy normalizes platelet hyperreactivity in hypercholesterolemic patients[284–287] and produces little effect on whole body rheology or on factors of the coagulation pathway.[288,289] The elevations of fibrinogen and whole blood viscosity in hypocholesterolic patients are normalized by pravastatin.[290] Increases in fibrinogen level and platelet aggregability and count may also be observed with lovastatin therapy.[291] Lovastatin shows promise in the prevention of restenosis in arteries after successful coronary angioplasty.[292]

Fibrates (Gemfibrozil, Clofibrate and Newer Analogs)

Gemfibrozil

Gemfibrozil [2,2-dimethyl-5-(2,5-xylyloxy)valeric acid, Lopid, Warner-Lambert] was introduced in 1981 after the statins, and is the second most frequently prescribed antilipemic drug. This drug is a white solid that is nearly insoluble in water and acidic solutions, and soluble at greater than 1% in dilute base.[293] It is stable at room temperature and the recommended daily dose in adults is 1.2 g (range of 0.9 to 1.5 g daily). Gemfibrozil, like clofibrate (Atromid-S), is a phenoxy acid derivative.

Clinical trials have shown that gemfibrozil decreases total serum cholesterol, triglyceride, VLDL triglyceride, VLDL cholesterol, and HDL triglyceride.[294–302] It increases serum HDL cholesterol concentrations in hyperlipidemic patients.[296–302] Thus, the main effect of gemfibrozil, like that of clofibrate, is to lower serum triglycerides; however, this drug appears more effective than clofibrate at increasing HDL cholesterol in certain patients.[300–302]

Clinical comparison trials with clofibrate and gemfibrozil show that the overall lipid-lowering efficacies in Type II hyperlipemias[300,303–305] and hypertriglyceridemic patients[304,306] are similar. In one study,[299] clofibrate and

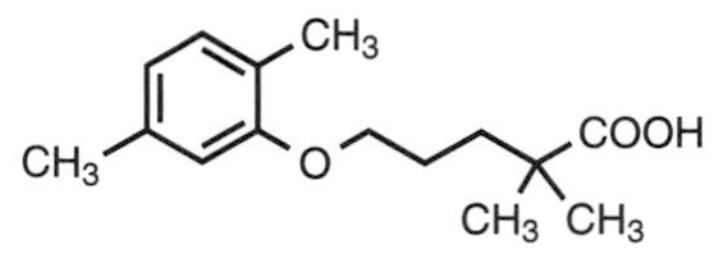

Gemfibrozil

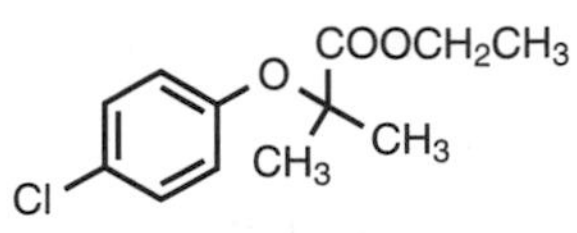

Clofibrate

gemfibrozil treatments caused 32 and 51% reductions, respectively, in total serum triglycerides. VLDL triglycerides were lowered 38 and 57% by clofibrate and gemfibrozil, respectively, and gemfibrozil increased HDL cholesterol levels by 31%.[299] Overall, gemfibrozil effectively lowers plasma triglycerides, VLDL cholesterol, and VLDL-apoB concentrations, increases HDL cholesterol and HDL-apoA-I concentrations in Type IIa and IIb hyperlipoproteinemic patients, and produces a greater reduction in LDL cholesterol levels in Type IIa patients.[307,308] In comparison to fenofibrate, gemfibrozil produces greater reductions in serum triglycerides and LDL cholesterol, and elevations in HDL cholesterol in patients with IIa, IIb, and IV hyperlipoproteinemias.[309,310] Gemfibrozil and bezafibrate produces greater responses than fenofibrate, etofibrate, and etofylline clofibrate in patients with Type IIb and Type IV hyperlipoproteinemias.[310] Bezafibrate produces a more favorable lipid-lipoprotein lowering response than gemfibrozil.[311,312] Although the newer fibrates (beclobrate, bezafibrate, ciprofibrate, and fenofibrate) are more effective than gemfibrozil in lowering serum LDL cholesterol levels in hypercholesterolemic patients, fibrates are not the drugs of first choice for the treatment of primary hypercholesterolemias.[313–315] A combination of fibrate with either a bile acid sequestrant or statin represents an excellent regimen for the treatment of Type IIb hyperlipoproteinemias.[137,141]

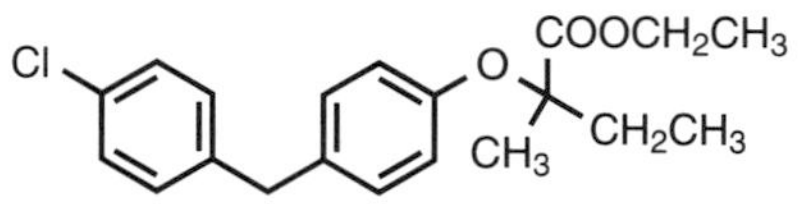

Beclobrate

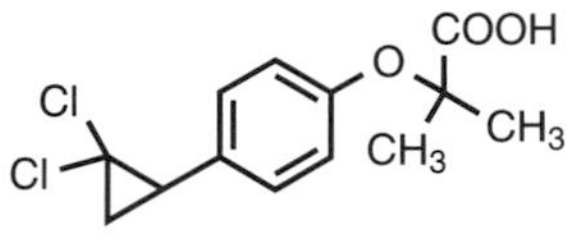

Bezafibrate

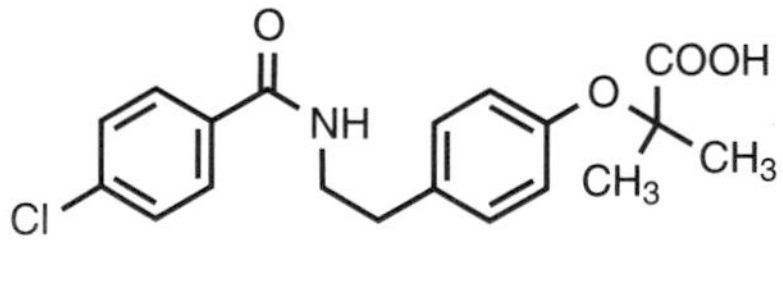

Ciprofibrate

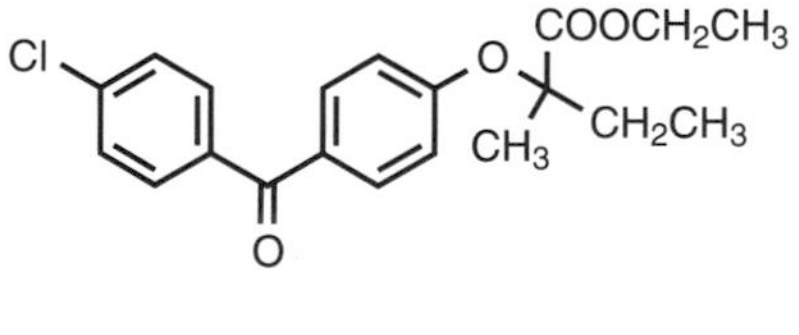

Fenofibrate

Gemfibrozil is rapidly and well absorbed in humans with a peak plasma level occurring 1 to 2 hours after dosing.[293,316,317] This drug has a circulating plasma $t_{1/2}$ of 1 to 2 hours and a biologic $t_{1/2}$ of 8.5 to 35 hours.[293] Plasma levels of gemfibrozil are related to dose, and maximal steady-state plasma concentrations are seen 14 days after 0.8 or 1.6 g daily dosing. Using radiolabeled gemfibrozil, 66 and 6% of the total dose are eliminated in urine and feces, respectively, after 5 days.[318] Gemfibrozil is extensively metabolized in humans and experimental animals.[318] Unchanged drug in urine accounts for only 5% of the dose and this percentage is irrespective of dose (200, 400, or 800 mg/day) or duration of treatment (1 or 7 days).[317] Conjugation of gemfibrozil to form an ester glucuronide and phase I oxidation of gemfibrozil to form phenolic, benzylic alcohol, and benzoic acid metabolites represent the major biotransformation products in humans and experimental animals (scheme III).[318] Like clofibrate, gemfibrozil is highly protein bound to serum albumin and undergoes extensive enterohepatic circulation.[319,320]

Treatment of animals[320–322] and humans[299,323] with gemfibrozil produces some effects that differ from those of clofibrate. Only gemfibrozil treatment increases serum HDL cholesterol levels in rats[320] and decreases hepatic VLDL production in humans.[299,323] The net total fecal excretion of acidic and neutral steroids by gemfibrozil is unchanged,[299] whereas clofibrate produces an increase in the net excretion of steroids.[324–326] The effects of these two drugs on hepatic cholesterogenesis in rats are also qualitatively and quantitatively different[322] and only clofibrate stimulates activity of hepatic mitochondrial α-glycerophosphate dehydrogenase.[321] Despite the metabolic differences between these two drugs, their overall lipid-count lowering effects in humans are similar.

Gemfibrozil is indicated for the treatment of adult Type IV hyperlipemic patients with high serum triglyceride levels (>750 mg/dl) who are at risk of abdominal pain and pancreatitis and do not respond to diet alone.[327]

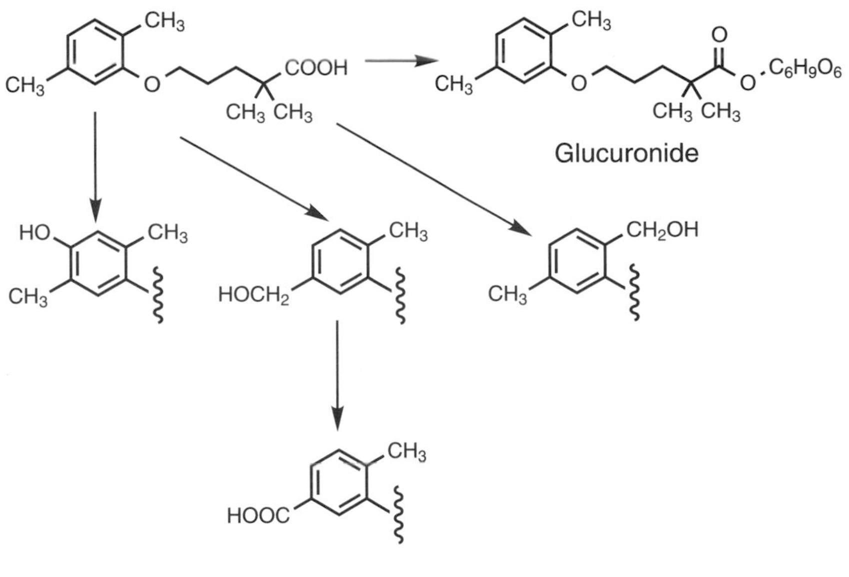

Scheme III. Biotransformation of gemfibrozil

The Helsinki Heart Study provided evidence of a 34% reduction in the incidence of CHD with gemfibrozil therapy, and further indicated that Types IIb and IV, and patients with low initial HDL cholesterol levels may benefit from this therapy.[328] Gemfibrozil also lowers serum triglyceride and cholesterol concentrations in diabetic patients,[293,329] and lowers VLDL and increases HDL levels in Types III[297] and V[299,330,331] hyperlipemias. Thus, gemfibrozil, like clofibrate, may be useful for treatment of patients with Types III and V hyperlipemias. Gemfibrozil also produces marked elevations in apoA-I, apoA-II and HDL cholesterol concentrations of familial hypertriglyceridemic patients with HDL deficiency,[332,333] and patients with Types IIa, IIb and IV hyperlipoproteinemia also respond favorably.[334–336] Although several clinical studies indicate that gemfibrozil lowers LDL cholesterol and raises HDL cholesterol levels in Type IIa and IIb patients,[298,300,337] it is not as effective as bile acid sequestrants[338] or statins at lowering total and LDL cholesterol concentrations in patients with primary hypercholesterolemia.[132,136–138,147,148,224,308,339] In patients with hypertriglyceridemia with borderline high cholesterol and elevated apoB, gemfibrozil only reduces VLDL and IDL cholesterol levels whereas lovastatin also lowers apoB, LDL- and total cholesterol concentrations.[149] Gemfibrozil more greatly lowers LDL cholesterol and LDL cholesterol/HDL cholesterol in normotriglyceridemic Type IIa patients than in Type IIb hyperlipoproteinemic patients. Presumably this is related to the increased catabolism of VLDL to LDL in the latter group of hypertriglyceridemic patients.[333] Gemfibrozil may prove beneficial alone or in drug combination to hypercholesterolemic patients who are nonresponsive to diet and bile acid sequestrants[334,338,340] or to statins alone.[306] A combination of gemfibrozil plus colestipol therapy in severely hyperlipemic[341] and nephrotic[342] patients shows a dramatic lipid-lowering response. Patients with non-insulin dependent diabetes mellitus respond with reductions of VLDL cholesterol and triglycerides to gemfibrozil, and a lowering of total and LDL cholesterol with combination with a statin.[221] Hypertriglyceridemic elderly patients given gemfibrozil also demonstrate an improvement in cerebral blood flow and cognition.[343]

Several substitutions in the aromatic ring and side chain of gemfibrozil have been examined for lipid lowering actions in animals.[344] Lipid-lowering effects are seen in analogs with three or more carbons separating the phenoxy group from the isobutyric acid function and the presence of α,α-dimethyl substitution on the propyl side chain. 2,5- or 3,5-Disubstitution of methyl, methoxy, or chloro groups on the aromatic ring lower triglyceride levels well. Dimethyl substitution, as in gemfibrozil, produces the most active compound and trimethyl, mixed methylalkyl, or chloromethyl analogs are less active. The presence of a small alkyl carboxylic ester group or conversion of the carboxylic group to an oxygenated function does not markedly reduce the triglyceride lowering activity in comparison to the parent compound. The gemfibrozil analog, 1,4-phenylenebis(oxy)-bis[2,2-dimethylpentanoic acid] (CI-924), is more potent than gemfibrozil at lowering plasma total cholesterol and increasing plasma HDL cholesterol levels in cholesterol-cholic acid-fed rats,[345] and in patients with Type II and IV hyperlipoproteinemias.[346] This compound was the most potent in a series of 100 gemfibrozil analogs tested in animals.[345]

Patients on gemfibrozil, like those on clofibrate, may exhibit GI distress (abdominal or epigastric pain), dermatitis, myositis, blurred vision, diarrhea, flatulence, and headache. Myopathy, including a severe form (rhabdomyolysis) accompanied by renal failure rarely occurs with gemfibrozil alone but may be observed during combina-

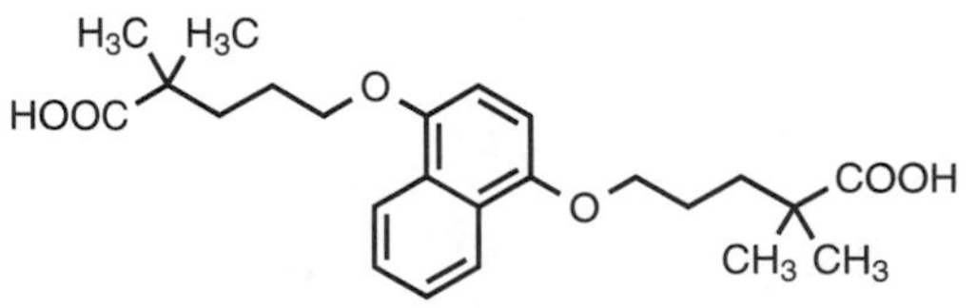

CI - 924

tion therapy with statins, nicotinic acids, and other fibrates.[347–349] Other symptoms include anemia and leukopenia.[293,306,337,350,351] Periodic blood counts are recommended during the first year of drug treatment. Abnormal liver function, including elevations in levels of serum glutamic acid-pyruvic acid transaminase (SGPT), lactic acid dehydrogenase (LDH), and alkaline phosphatase, are noted occasionally and are reversible on removal of gemfibrozil.

Gemfibrozil causes a modest increase in biliary cholesterol saturation in healthy volunteers[352] and shows increased incidence of new or enlarged gallstones in patients after treatment for one year.[306] Gemfibrozil decreases the biliary cholesterol saturation index, however, as well as the hepatic bile acid secretion rate, fecal bile acid excretion, and bile acid pool size in hyperlipidemic patients.[353] This potentially adverse effect is less for gemfibrozil than clofibrate[353] and severe symptoms (cholecystitis) may require discontinuation of treatment.

Teratologic studies of gemfibrozil in animals do not reveal any drug-induced malformations.[354] As with clofibrate, gemfibrozil treatment of rodents produces hepatomegaly, peroxisomal proliferation, and hepatocarcinomas.[323,355–357] The degree of hepatic peroxisome proliferation is correlated to elevations in HDL cholesterol levels in rodents.[358] No evidence of gemfibrozil-induced malignancy or hepatic peroxisomal proliferation in biopsies of treated patients are found, however.[357,359,360] Because of high risk for the fetus or newborn, gemfibrozil is contraindicated in females during pregnancy and lactation.

Like clofibrate, gemfibrozil potentiates the action of oral anticoagulants[306,307,361] and prothrombin clotting times should be monitored in all patients. Maintenance dose of anticoagulants may be reduced by more than one half during gemfibrozil therapy. Although gemfibrozil does not alter plasminogen activator inhibitory activity in hypertriglyceridemic patients with a history of thrombosis,[362] treatment of hyperlipidemic patients produces an increase in plasma fibrinogen levels, which is a potentially adverse effect.[363] In contrast, gemfibrozil therapy in Type IV hyperlipoproteinemic patients reduces plasma fibrinogen and α-2-antiplasmin levels and normalizes times for antithrombin III and euglobulin lysis.[364] Combination therapy with resins and gemfibrozil should be given at least 2 hours apart to avoid interference with gemfibrozil absorption.[365]

Clofibrate and Newer Analogs (Benzafibrate, Ciprofibrate, Fenofibrate)

Clofibrate (Atromid-S) was the first fibrate introduced in the United States. Clofibrate [ethyl 2-(4-chlorophenoxy)-2-methylpropionic acid] is a nearly colorless, oily liquid, insoluble in water, and miscible with organic solvents.[366] This drug is sensitive to oxidation and light, and it hydrolyzes to form clofibric acid.[367] Clofibrate is sold commercially under at least 80 trade names worldwide.[367] 4-Chlorophenol, acetone, and chloroform are starting materials for the synthesis of clofibric acid, which is esterified with ethanol to yield clofibrate.[367]

Clofibrate is rapidly and completely absorbed after oral administration in humans and undergoes hydrolysis to the active form, clofibric acid.[368–371] The mean absorptive half-life of clofibrate (1 to 2 g dose) in humans is 1.7 hours and maximal serum concentrations of clofibric acid (100 to 180 μg/ml) are usually seen within 3 to 6 hours. Clofibric acid is highly plasma protein-bound (93 to 98%) and possesses a small volume of distribution (5 to 9 L). The elimination half-life of clofibric acid from plasma at therapeutic doses is usually between 13 and 17 hours with reported variations of 6 to 24 hours.[370,372] A 1-g dosing regimen, twice daily, in humans is appropriate for maintaining maximal steady state concentrations of clofibric acid of 100 to 160 μg/ml.

Clofibric acid is conjugated to form the 1-O-acyl glucuronide as the major metabolite (scheme IV).[368–372] Clofibric acid and principally the glucuronide metabolite of clofibric acid are eliminated in urine (>97% of the dose) within 72 hours. The 1-O-acyl glucuronide of clofibric acid undergoes enterohepatic circulation in the GI tract after hydrolysis by β-glucuronidase and reabsorption of clofibric acid in to the bloodstream. Serum of patients with renal insufficiency or Gilbert's syndrome shows higher levels of the glucuronide metabolite.[370,372,373] Isomeric 2-, 3-, and 4-O-acyl glucuronides of clofibric acid are found in the urine of patients and are related to a nonenzymatic pH dependent intramolecular migration of glucuronic acid from the 1-O-acyl glucuronide conjugate of clofibric acid.[374,375] These isomeric ester glucuronides, unlike the 1-O-acyl glucuronide, do not undergo enzymatic hydrolysis by β-glucuronidase or enterohepatic circulation. The taurine conjugate of clofibric acid has been isolated in several species but not in humans.[376]

Clofibrate is used in the treatment of primary (Type III or broad β-VLDL dyslipoproteinemia) that does not respond to diet alone. Lovastatin alone or in combination with clofibrate lowers total LDL cholesterol, VLDL cholesterol concentrations in Type III hyperlipoproteinemia.[199] Beneficial effects of clofibrate are found for the treatment of endogenous hypertriglyceridemias (Types IV and V), of FCH (Types IIa, IIb and IV), either alone or in drug combination, and of diabetic retinopathy.[137,377] Clofibrate may also be indicated as an adjunctive therapy in Types IIb, IV, and V hyperlipemias when diet and drug therapy with nicotinic acid (niacin) or gemfibrozil are not effective,[307] and may be useful to decrease elevated fibrinogen levels and to normalize thrombogenic potential in high risk patient populations.[378]

Antitriglyceridemic effects of clofibrate in humans are related to an increased catabolism of serum triglyceride-rich lipoproteins (VLDL and VLDL remnants), but not to any effect on hepatic triglyceride or VLDL synthesis and release from the liver.[379–383] The action of clofibrate

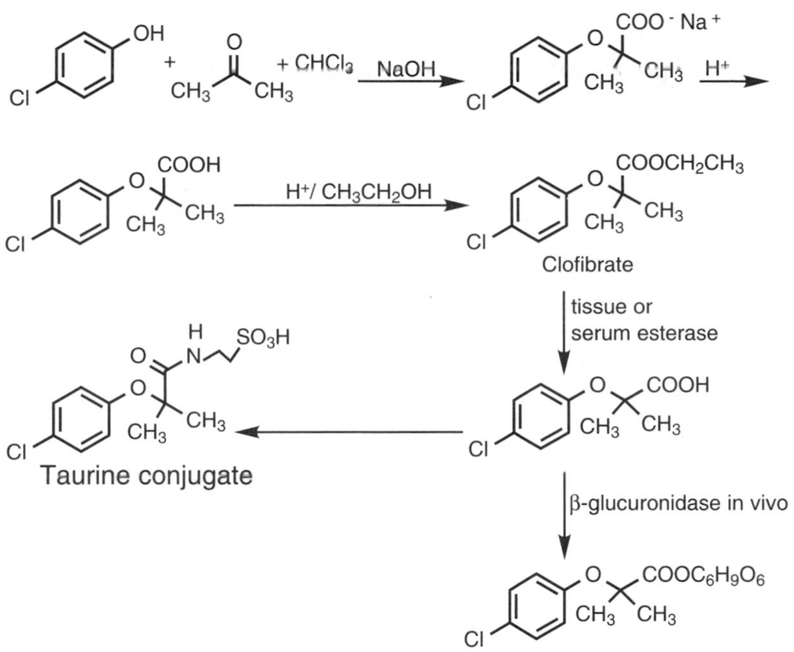

Scheme IV. Clofibrate metabolic pathways

is related to an increase in adipose tissue or muscle LpL activity which accelerates the rate of intravascular catabolism of VLDL to IDL and LDL. Accordingly, clofibrate-mediated increases in LDL cholesterol levels of hyperlipoproteinemic patients (Types IIb, IV, and V hyperlipemias) may be caused by the increased breakdown of VLDL and conversion to LDL.[379,384–386]

Serum cholesterol lowering by clofibrate in animals is related to changes in liver cholesterol through blockade of HMGCoA reductase.[387–389] Serum cholesterol catabolism remains unchanged in humans after clofibrate treatment,[324] however, acidic and neutral sterol excretion into bile is increased by clofibrate treatment.[324–326] The net loss of cholesterol into bile and subsequently into feces is presumably achieved by mobilization of cholesterol from extrahepatic tissues by clofibrate. This action explains the increased lithogenicity of bile, cholecystitis, and gallstone formation in patients seen during long-term administration.[367,390,391] The mechanism underlying cholesterol mobilization from extrahepatic tissues may involve the elevations in HDL concentrations by clofibrate[384,385,392–394] and activation of cholesterol removal from peripheral tissues to the liver for subsequent elimination in bile.

Administration of clofibrate to animals produces a marked hepatomegaly associated with a proliferation of mitochondria, endoplasmic reticulum, and peroxisomes (microbodies).[367–397] These hepatic effects are usually reversible.[396] In humans, liver enlargement is related to increased mitochondria, hypertrophy of the endoplasmic reticulum, and either increased or unchanged peroxisome levels.[367,397,398] Elevation in serum HDL cholesterol is associated with proliferation of the endoplasmic reticulum.[399–401] This effect may explain the modest rise in serum HDL concentrations by clofibrate in hyperlipidemic patients.[384,385,394]

Biochemical changes of enzymes of FA (fatty acid) metabolism and triglyceride synthesis in liver[367,395,402–405] may contribute to the hypotriglyceridemic action of clofibrate. FA oxidation in isolated hepatocytes[405] or perfused liver[402,403] of animals is increased by clofibrate. Peroxisomes contain catalase, hydrogen peroxide-generating enzymes, and lipid-metabolizing enzymes of the FA spiral, including fatty acyl CoA oxidase (FACO), which functions in the catabolism of long-chain FAs.[355] A variety of hypolipidemic agents structurally similar to clofibrate cause peroxisomal proliferation and markedly stimulate the activity of FACO.[355,356,406,407] Increased FA β-oxidation may be related to the hypolipidemic action of these compounds.[408] Increased hepatic FA oxidation can reduce triglyceride and VLDL biosynthesis or release, and this may explain the inhibitory action of clofibrate on hepatic triglyceride synthesis in experimental animals.[395,403,409] Although clofibrate apparently does not modify triglyceride synthesis or VLDL release in humans,[379–383,395] the increased intravascular catabolism of VLDL may lead to an elevated FA uptake in liver with an enhanced capacity of mitochondria and peroxisomes to oxidize FA. Thus, the elevation in hepatic FA oxidation may account indirectly for part of the antitriglyceridemic action of clofibrate.

Clofibrate treatment decreases the sensitivity of platelets of Type IIb patients to aggregatory agents,[410] prolongs platelet survival time,[411] and produces a benefit in patients without showing a serum lipid-lowering effect.[412] Even though the antiaggregatory effects of clofibrate are not consistently observed in humans,[395] the inhibitory action of this drug has been repeatedly demonstrated in isolated human platelet preparations.[395,413–415] Clofibric acid blocks the release of arachidonic acid from platelet membrane phospholipids and acts as an inhibitor of prostaglandin biosynthesis in vitro.[415] Analogs of clofibric acid also manifest greater antiaggregatory activity than that of the parent drug[415,416] and thus have potential therapeutic utility.

Many acyclic and cyclic analogs of clofibric acid have been tested in experimental animals.[395] Only a few have reached the stage of clinical evaluation in humans.[156,417] Three groups of clofibric acid analogs can be identified: prodrugs or salts of clofibric acid; newer congeners of clofibric acid, which possess varying lipophilic substituents on the aryl ring; and gemfibrozil types.

Salts (aluminum, calcium and magnesium), ester or diester prodrugs (clofenpyride, clofibride, clofibrate, etofibrate, etofylline clofibrate, pyridoxine clofibrate, pirifibrate, simfibrate, and tiafibrate) or amide (timofibrate) analogs, which share the same antilipidemic activity profile of clofibric acid are represented in Table 24–4. Many of these analogs have been tested with positive findings and although some differences in bioavailability and other pharmacokinetic parameters may exist,[156,369,372,417] their efficacy can be primarily related to the liberation in vivo of clofibric acid and/or nicotinic acid (etofibrate, binifibrate, and ronafibrate). Etofibrate decreases serum cholesterol and triglyceride levels and the LDL:HDL cholesterol ratio in hypercholesterolemic patients,[418] and improves pain-free walking distances in hyperlipidemic patients with arteriosclerosis obliterans.[419] Etofylline clofibrate liberates clofibric acid in vivo and produces a serum triglyceride and VLDL-lowering profile similar to clofibrate.[420] Bezafibrate treatment, but not etofylline clofibrate, lowers plasmas apoB levels, and total and LDL cholesterol concentrations in hyperlipoproteinemic patients.[421]

Selected newer commercially available hypolipidemic analogs of clofibrate, which are being evaluated in humans or are currently used in Europe, generally exhibit the same profile of hypotriglyceridemic activity (increased VLDL catabolism by stimulating LpL)[422] as clofibrate, but may have advantages over clofibrate for certain phenotypes (IIa and IIb). These fibrates produce modest reductions in total and LDL cholesterol levels in Type IIa hyperlipoproteinemia, produce less consistent reductions in LDL cholesterol and significant decreases in triglyceride levels in Type IIb hyperlipoproteinemia, and significant reductions in VLDL- and serum-triglycerides in Type IV hyperlipoproteinemia.[423] Ciprofibrate and fenofibrate reduce total cholesterol, LDL cholesterol, VLDL cholesterol, and apoB concentrations. They increase HDL cholesterol and apoA levels in Type II hypercholesterolemic patients.[424] Ciprofibrate produces greater increases in HDL cholesterol and apoA levels than fenofibrate, and possesses a long half-life so that a single daily dose may be administered.

Combining clofibrate with other commercially available hypolipidemic agents has generally produced additive or synergistic effects in hyperlipidemic patients.[156,395,417] Positive results are reported with clofibrate and colestipol,[425–430] cholestyramine,[395,431,432] nicotinic acid and its prodrugs,[433–435] and Secholex (DEAE-Sephadex)[432,436,437] in hyperlipidemic patients. Combinations of nicotinic acid and fenofibrate[438,439] and of colestipol and fenofibrate[440,441] also are beneficial in lowering serum lipids and lipoproteins. A severe hypoalphalipoproteinemia has been observed in hyperlipidemic patients receiving a combination of clofibrate and probucol;[442] such a deficiency in HDL may increase risk for CHD in this patient population.

Hepatomegaly and peroxisomal proliferation persist during long-term administration of clofibrate and related analogs in animals.[355,443] An increased number of hepatic tumors occurs with a variety of agents (structurally related and unrelated to clofibrate) that cause peroxisomal proliferation,[355,356,444] and several reports on the tumorigenicity of clofibrate, gemfibrozil, fenofibrate, bezafibrate, and ciprofibrate in rats have appeared.[355,445–447] Hepatic peroxisomal proliferators may represent a unique class of chemical carcinogens.[448] More recent evidence indicates these compounds interact with a ligand-specific peroxisomal receptor, called a peroxisome proliferation activated receptor (PPAR) that belongs to the steroid hormone receptor superfamily.[449]

An increased incidence of non-CHD related deaths in hyperlipidemic patients[391] is associated with a possible increase in malignancies of the liver, pancreas, and intestine of clofibrate-treated patients,[391] and liver tumorigenicity has been confirmed in animals.[355,445,446] Clinicians have strongly advised the evaluation of benefit versus risk in patients undergoing clofibrate treatment.[391,450] The FDA has issued a warning about the potential tumorigenicity associated with clofibrate therapy in humans.[451] Similar concerns exist for other fibrates, especially for newer and more potent analogs, such as fenofibrate, beclobrate, and ciprofibrate.

Gastrointestinal symptoms (abdominal discomfort, diarrhea and constipation) are the most common complaints associated with fibrate therapy.[452] Skin rashes, fatigue, headache, loss of libido, impotence, or insomnia occur with less frequence. Hepatotoxicity and severe liver dysfunctions (hepatitis) are infrequently reported with fibrate therapy.[453] Diseases of the kidney (nephrotic syndrome or renal insufficiency) and liver (including biliary cirrhosis) may produce elevations in the retention of clofibric acid so that plasma levels increase.[370,372] Such patients may exhibit myositis and severe renal failure requiring a reduction or termination of fibrate (clofibrate, gemfibrozil or ciprofibrate) therapy.[347,454,455] Gallstone enlargement and cholelithiasis and cholecystitis are also potential adverse effects observed with fibrates.[456]

Increased displacement of drugs from plasma protein-binding sites and concomitant enhancement of the pharmacologic response or toxicity for such compounds as phenytoin, tolbutamide, and coumarin anticoagulants are reported.[370,379] The increased anticoagulant effect of coumarin-like drugs by clofibric acid necessitates a frequent monitoring of prothrombin clotting times and may require a reduction in the dose of the anticoagulant by one-half or more.[379] Coadministration of rifampin and

Table 24–4. Salts and Prodrugs of Clofibric Acid

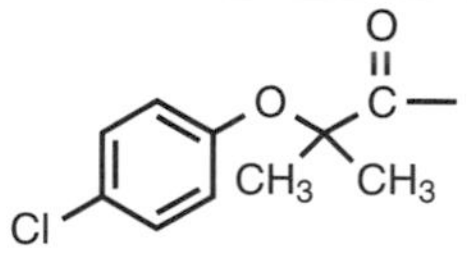

Compound	R
I. Salts	
Alufibrate	- O)$_3$Al
Calcium clofibrate	- O)$_2$Ca
Magnesium clofibrate	- O)$_2$Mg
II. Prodrugs	
Binifibrate	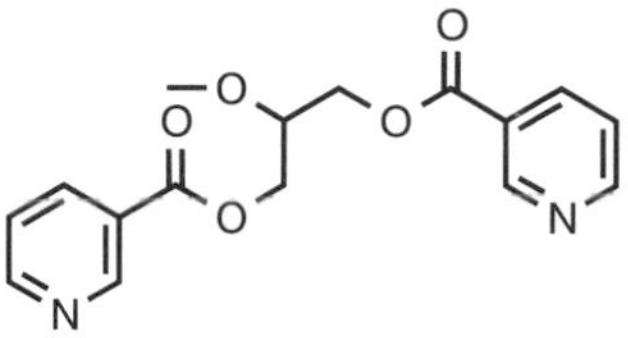
Clofenpyride	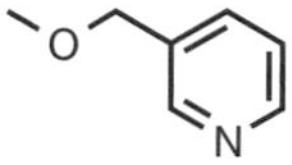
Clofibride	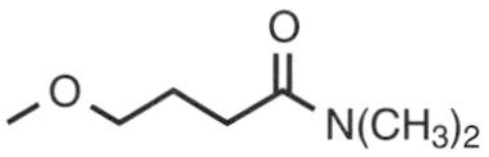 N(CH$_3$)$_2$
Etofibrate	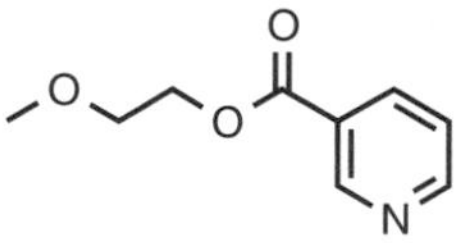
Etofylline clofibrate	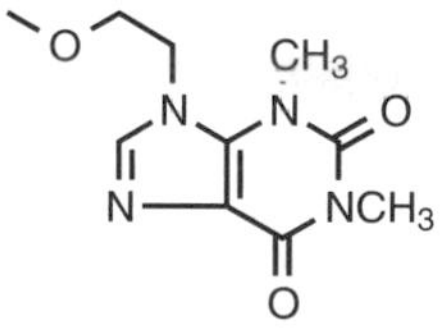 CH$_3$; NCH$_3$
Pirifibrate	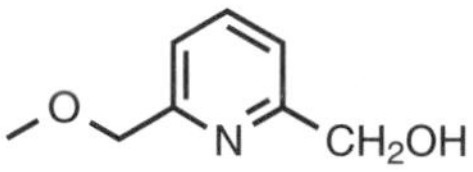 CH$_2$OH
Pyridoxine clofibrate	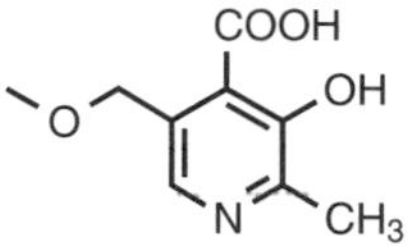 COOH; OH; CH$_3$
Ronafibrate	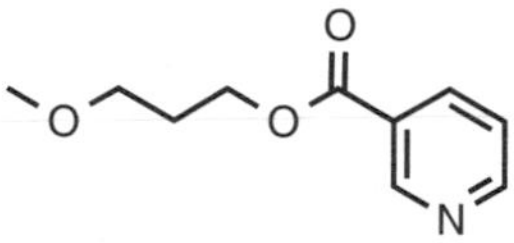
Simifibrate	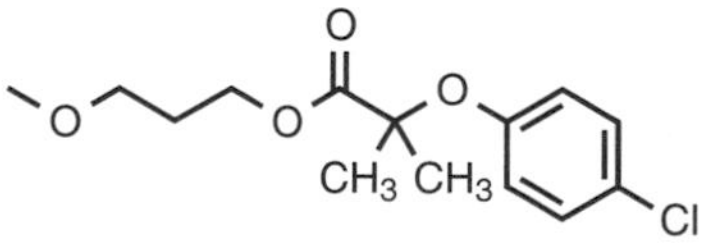 CH$_3$ CH$_3$; Cl
Tiafibrate	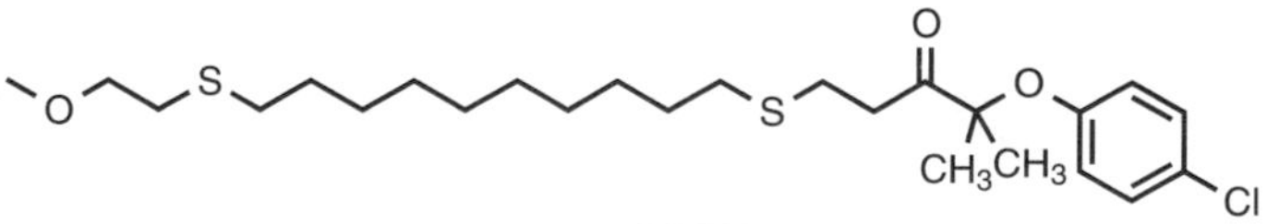 CH$_3$CH$_3$; Cl
Timofibrate	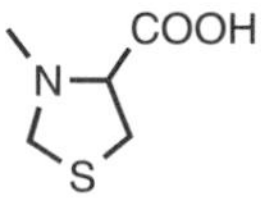 COOH

clofibrate to patients reduces serum clofibric acid concentrations and increases the dose of clofibrate required.[370] To prevent risk to fetuses and infants, clofibrate is contraindicated in women during pregnancy, lactation, or in those of childbearing age who are not practicing birth control.[367,307,379] Embryotoxicities are observed with clofibrate, etofylline clofibrate, and fenofibrate in animal studies at doses much higher than those used therapeutically.[457]

Bile Acid Sequestrants (Cholestyramine and Colestipol)

Cholestyramine and colestipol hydrochloride are anion-exchange resins that were approved by the FDA in 1973 and 1977, respectively, for reduction of elevated serum cholesterol levels in patients with primary hypercholesterolemia who do not adequately respond to diet. Both resins are water insoluble, inert to digestive enzymes in the intestinal tract, and are not absorbed.[458,459]

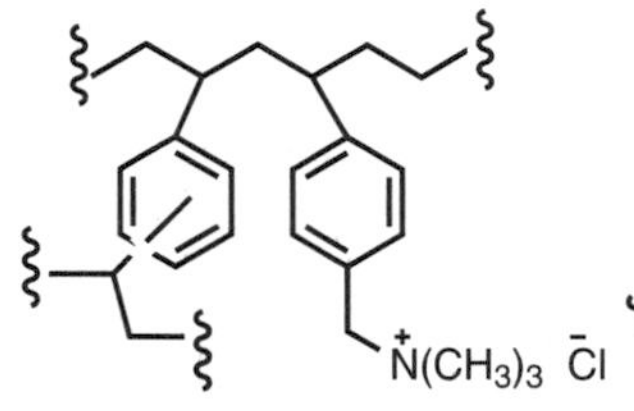

Cholestyramine

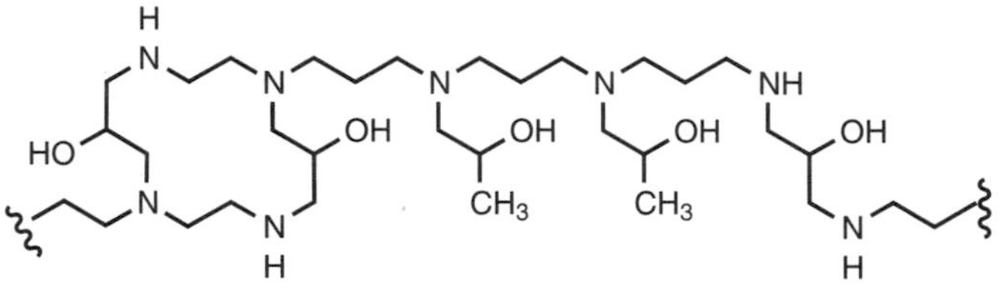

Colestipol

Cholestyramine is a high molecular weight (>1,000,000) copolymer of 98% polystyrene and 2% divinylbenzene (DVB) containing ~4 meq of fixed quaternary ammonium groups per gram of dry resin. The resin is administered as the chloride salt, which exchanges for other anions in the intestinal tract with a greater affinity for the positively charged functional groups on the resin.[460] One gram of cholestyramine binds ~2 g (4 meq) of glycocholate when equilibrated for 2 hours with the bile salt. Thus, virtually all of the binding sites on the resin are accessible by bile salt. Increasing the cross linkage of cholestyramine from 2% to 4% to 8% DVB reduces the porosity of the resin and the amount of bile salt bound is reduced by a *sieve exclusion* effect, which prevents binding to interior binding sites.[460] The rate of binding of bile salt to resin is increased by reducing resin particle size[461] or by increasing equilibration temperature.[462,463]

Colestipol is the hydrochloride salt of a copolymer of diethylenetriamine and 1-chloro-2,3-epoxypropane. The functional groups on colestipol are secondary and tertiary amines. Although the total nitrogen content of colestipol is greater than that of cholestyramine, the functional anion exchange capacity of the resin depends on the pH in the intestinal tract and may be less than cholestyramine.[464] Recent studies in vitro indicate that in buffered solutions with phosphate anions competing with bile salt for binding sites, cholestyramine has a higher adsorption capacity than colestipol for bile salts.[465] Quaternization of colestipol with methyl iodide increases the capacity in vitro for glycocholate.[466]

The serum cholesterol lowering efficacy in humans of cholestyramine and colestipol is well established, and the importance of these resins for treating primary hypercholesterolemia is widely recognized.[467,468] All of the cholesterol lowering effect is attributable to LDL lowering. Serum levels of triglycerides and HDL cholesterol are either unchanged or increased.[469] The usual cholesterol lowering dose is 8 to 24 g/day for cholestyramine and 10 to 30 g/day for colestipol. In some studies the effect obtained with 8 g of cholestyramine[470,471] or 10 g of colestipol[472] is similar to that with higher doses of either resin. Administration of resin two times each day is as effective as three or four doses per day.[350] Typical responses in hypercholesterolemic patients to a therapeutic dose are a 15 to 25% serum cholesterol lowering and a 20 to 35% LDL cholesterol reduction.[467,468] Most of the decrease occurs during the first week of treatment, and a maximal effect is usually obtained within 2 to 3 weeks.

Serial coronary angiography is used to monitor resin treatment effects on human coronary atherosclerosis.[23–28,46,473,474] Two studies have shown that treatment with cholestyramine not only slows or halts the progression of coronary atherosclerosis, but also produces regression associated with a net increase in coronary lumen diameter. Cardiovascular signs and angina and exercise stress test performance results improve. The clinical benefit seen with cholestyramine is correlated with reduction of IDL and a large molecular weight subset of LDL close in size and density to IDL.[475] Regression of established coronary lesions is also achieved with colestipol in combination with nicotinic acid or lovastatin.[24–27]

Cholestyramine and colestipol lower plasma LDL levels by increasing the rate at which LDL is cleared from the bloodstream.[476] Because the sequestrants are not absorbed from the intestinal tract, their effects are indirect and mediated by their ability to reduce the production or absorption of gut-derived substances, which may, in turn, regulate the removal of LDL from blood. Resins bind bile salts in the intestinal tract, thereby reducing reabsorption of bile salts by an active transport process in the ileum and increasing excretion in feces.[477] The liver compensates for the increased fecal bile salt loss by increasing de novo synthesis of bile salts from cholesterol.[478] To compensate for this cholesterol-depleting effect, the hepatic synthesis of cholesterol is increased.[479,480]

Increased fractional clearance of raidoiodinated (^{125}I)-LDL in cholestyramine-treated patients occurs by the high-affinity receptor pathway.[481] Removal of LDL by the scavenger (non-receptor-mediated) pathway is not affected by resin treatment. These findings may explain why resins fail to reduce LDL levels in homozygous Type II patients who have a complete absence of high-affinity receptors.[351,482] Subsequent studies in rabbits[483] and dogs[484] show that cholestyramine increases hepatic synthesis of high-affinity LDL receptors on liver membranes. Thus, the serum LDL-lowering effect of cholestyramine may be mediated by binding bile salts, which then lower intracellular hepatic cholesterol concentration and induce the synthesis of the high-affinity LDL receptors on the hepatocyte plasma membrane.

It has not been conclusively proven that LDL-lowering effects of resins are mediated by intestinal bile salt binding activity. For example, oral administration of bile salts (cholic acid or deoxycholic acid) to humans does not increase serum LDL or LDL cholesterol concentrations.[485–487]

Because the resins increase the rate of clearance and catabolism of LDL, they work well in combination with drugs that suppress the synthesis of VLDL and LDL, such as nicotinic acid,[488–491] or the cholesterol synthesis inhibitors, lovastatin[120,173] and mevastatin.[178] With combination therapy, a 50% reduction in serum LDL levels is often achieved. Other studies[492–495] indicate that probucol and resin combinations are also more effective than either agent alone. A mechanistic explanation of why probucol works well in combination with cholestyramine may be provided by a study showing that probucol alters the composition and properties of LDL in a way that increases its clearance by both receptor and nonreceptor pathways.[496]

Studies of the effects of clofibrate in combination with resin have generally not produced the degree of serum cholesterol lowering reported for other drug combinations,[120,173,178,488–495] but serum triglyceride levels in patients given clofibrate and resin were 30 to 50% lower than those given resin alone.[426,432,497] Other agents that have been used in combination with resin are neomycin,[498] gemfibrozil,[341] fenofibrate[440,441] and pectin.[499]

Probucol

Probucol {4, 4′-(isopropylidenedithio) bis[(2,6-di-tertiarybutyl)phenol]} gained approval by the FDA in 1977 as therapy adjunctive to diet for the reduction of elevated serum cholesterol levels in patients with primary hypercholesterolemia. Probucol is a hydrophobic white powder that poorly dissolves in water but is readily solubilized in most organic solvents. Originally synthesized as an antioxidant for plastics and rubber, probucol has biologically significant antioxidant activity and scavenges reactive oxygen species (ROS).[500] The ultraviolet (UV) absorbance spectrum of a methanolic solution of probucol has a maximum at 242 ηm.[501]

Only about 1 to 10% of the usual 1-g dose of probucol (500 mg twice daily) is absorbed from the gastrointestinal tract.[502,503] Probucol associates with digestive lipid micelles and is transported by chylomicrons and VLDL via lymphatics to the systemic circulation where transfer to different lipoprotein classes takes place. The plasma $t_{1/2}$ for probucol in humans is about 2 days.[502] Plasma probucol levels range from 5 to 50 μg/ml (10-100 μM), and no significant relationship appears to exist between total plasma cholesterol and probucol levels. When ^{14}C-radiolabeled probucol is incubated in whole human blood, about 90% partitions into plasma.[504] After ultracentrifugal fractionation, 38%, 44%, and 13% of the drug appears in VLDL, LDL and HDL, respectively. Molar uptake of probucol positively correlates with lipid content of the lipoprotein particle, suggesting that probucol essentially dissolves in the particle lipid core.

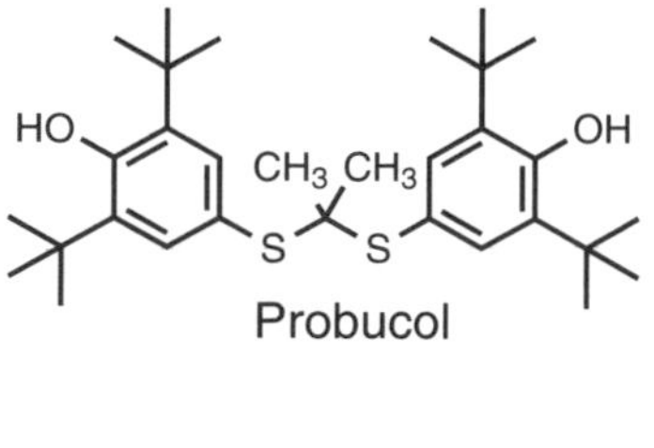

Probucol

Probucol accumulates in plasma lipoproteins that serve as vehicles for delivery of the drug to body tissues. With continuous oral administration for 3 months at 500 mg twice daily, probucol accumulates in adipose tissue where it may remain for 6 months.[505] The major excretion pathway is via bile to feces with a small renal clearance. The drug has also been detected in animal milk, thus precluding its use by nursing human mothers. Monkey adipose tissue, adrenal glands, and liver retain this drug at concentrations 100, 25 and 4 times, respectively, greater than those in plasma. Although highly hydrophobic, probucol is not detected in the brain.[502]

Clinical studies provide probucol concentrations in plasma[506–508] and are typified by the data of Tedeschi et al.[506] Probucol concentrates in VLDL, LDL, and HDL to the extent of 3 to 24, 7 to 15, and 3 to 6 μg/ml, respectively.

Probucol lowers plasma cholesterol levels in Types IIa, IIb, III, IV, and V[502,506,509–514] and is one of the few drugs to reduce both cholesterol levels[515–517] and xanthomata[516] in homozygous FH patients. This drug should be seriously considered for heterozygous FH patients[518] or for those with primary moderate hypercholesterolemia.[512,519] Elderly patients[520] and children with FH[521] also benefit. Probucol is also effective in normolipemic subjects.[522] The responsiveness to probucol-induced cholesterol concentration lowering in heterozygous FH is greater than those with the an apoE polymorphism with an ε4 allele.[523] Long-term use of probucol reduces stroke incidence without decreasing CHD.[512] Probucol's effects on serum triglyceride levels are inconsistent[513] and it is not a first-line drug employed with hypertriglyceridemia as the abnormality of most concern.[524–526]

Probucol in combination with clofibrate produces a severe hypoalphalipoproteinemia.[527] Combined colestipol and probucol therapy reduces both LDL and HDL and also blocks coronary artery lesion progression.[528] Combining lovastatin and colestipol provides no advantage over lovastatin and colestipol alone.[529] Niacin[530] and Nicterol[531] in combination with probucol also reduce LDL and HDL, but no clear advantage is established with this regimen over monotherapy.

Most of the cholesterol-lowering effect is attributable to LDL and apoB-100 concentration reduction,[520] but in one study cholesterol lowering occurs without apoB-100 reduction.[532] In mice, probucol lowers total hepatic HMG-CoA reductase; thus, serum-cholesterol lowering may in part be a function of reduced cholesterol synthesis.[522]

Numerous inconsistencies exist between animals and humans in probucol responses. For example, probucol inhibits cholesterol absorption in rats, but not in humans. Fecal excretion of both neutral steroids and bile acids are increased in rats, whereas in humans only bile acid excretion increases.[533] In properly designed experiments, fractional catabolic rates of LDL increase.[519,524,534,535] Although no simple explanation for the LDL cholesterol lowering can be provided,[536] when there is an effect, it appears to be caused by a change in lipoprotein structure or composition rather than acceleration of a tissue-based receptor-dependent or independent pathways.[507,535]

Probucol does reduce the temperature of phase transitions of phosphatidylcholine liposomes at concentrations somewhat higher than those which prevent liposome arachidonic acid oxidation.[537] Further, the drug interferes in the association of apoC-III to dimyristylphosphatidylcholine.[538] At low concentrations, the phase transition assayed by differential scanning calorimetry indicates probucol association with the glycerol backbone, whereas at high concentrations the association is with acyl chains.[537]

Probucol treatment decreases ApoA-I,[511,539–541] whereas ApoA-II levels remain unchanged[505] or decrease.[520] Plasma CETP and apoE concentrations increase with a concomitant reduction in HDL cholesterol. These changes observed during probucol treatment correspond with an increased reverse cholesterol transport via the remnant pathway,[542] and the decreased HDL-cholesterol and apolipoprotein A-I relate predominantly to a reduced HDL_{2b} subfraction.[522,543,544] This increase in reverse cholesterol transport may function to reduce the size or number of xanthelasmas in some patients.[545,546] This regression of lesions is highly correlated to HDL reduction.[547] The reduction may be related to probucol treatment causing a reduction in the size of HDL particles (HDL cholesterol is reduced without concomitant decreases in ApoA-I).[522] Because probucol suppresses LpL activities in postheparin plasma, it is likely that the decreases in HDL_2 may be caused by disturbance in VLDL conversion to IDL and HDL_3 conversion to HDL_2.[548] In cholesterol-fed rabbits a combination of probucol and pantethine reduces total cholesterol concentrations without lowering HDL cholesterol; further, the drug decreases atheromatosis.[549]

Part of this process of enhanced removal of cholesterol may be because probucol modulates the physical state of cellular cholesterol esters through increased enthalpy to convert anisotropic cholesterol ester crystals in rat hepatoma cells to an isotropic structure. This change is coupled to the increased rate of efflux of cholesterol from these cells,[550] a phenomenon also observed in adult rat hepatocytes[551,552] and human skin fibroblasts.[553]

Although it may not yet be clear how the HDL concentration reduction effect of probucol relates to its potential use in the prevention or treatment of atherosclerosis and xanthoma regression, the antioxidant properties of probucol may play a significant role in reducing atherosclerosis progression based on the concept that oxidized lipoproteins participate in the generation of foam cells in atheroma.[554] Probucol inhibits LDL oxidation by Cu^{2+} catalysis in endothelial and smooth muscle cell cultures.[555] Further, LDL from probucol-treated hypercholesterolemic[556] and diabetic patients[557] resists oxidation. Probucol within LDL modifies copper-catalyzed LDL_{ox} and concurrently, during oxidation, the drug is converted to a spiroquinone. A further reaction converts the spiroquinone to diphenoquinone and then to bisphenol (scheme V).[558]

A water-soluble probucol derivative, diglutaryl probucol, prevents cell-induced LDL oxidation.[559] Barnhart, Wagner, and Jackson have reviewed probucol hypolip-

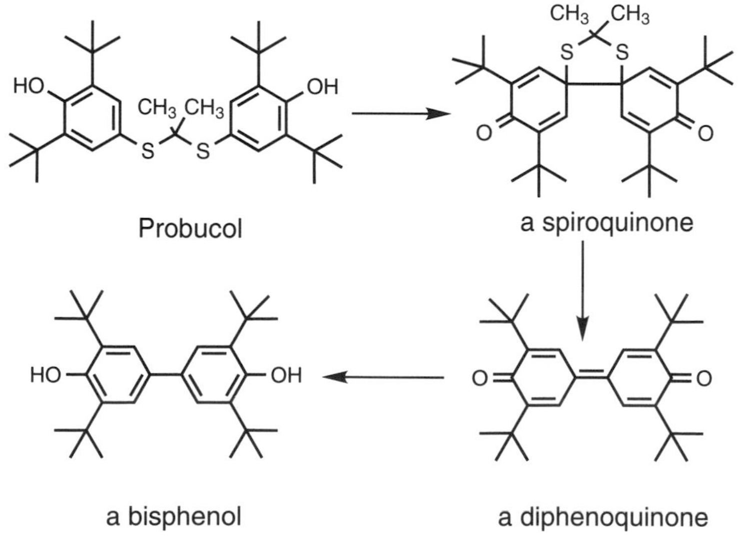

Probucol

a spiroquinone

a bisphenol

a diphenoquinone

Scheme V. Oxidation of Probucol

idemic and antioxidant properties.[560] In Watanabe Heritable Hyperlipidemic (WHHL) rabbits, substitutions at the disulfide-linked carbon and at the phenolic ring *tert*-butyl group of probucol eliminates cholesterol concentration reduction, but all analogs inhibit LDL oxidation in proportion to their concentrations in LDL.[561] In most human studies, the maximum effect of probucol occurs after 1 to 3 months of treatment.[502,506] After treatment, plasma cholesterol levels return to pre-existing levels only after several months. Probucol does not preserve endogenous α-tocopherol, γ-tocopherol and β-carotene in LDL during oxidative modification as ascorbate does,[562] but ascorbate does act synergistically with probucol to extend the lag phase for LDL oxidation.[563] The incorporation of probucol into the LDL does reduce the los of β-structure in LDL undergoing oxidation, and prevents the loss of reactive amino groups and highly reactive heparin binding.[564] In other systems probucol inhibits the nicotinamide-adenine dinucleotide phosphate (NADPH)-dependent microsomal lipid peroxidation and the iron-doxorubicin complex-induced phospholipid peroxidation at about half the concentration of the powerful antioxidant, 2-hydroxyestrone.[565]

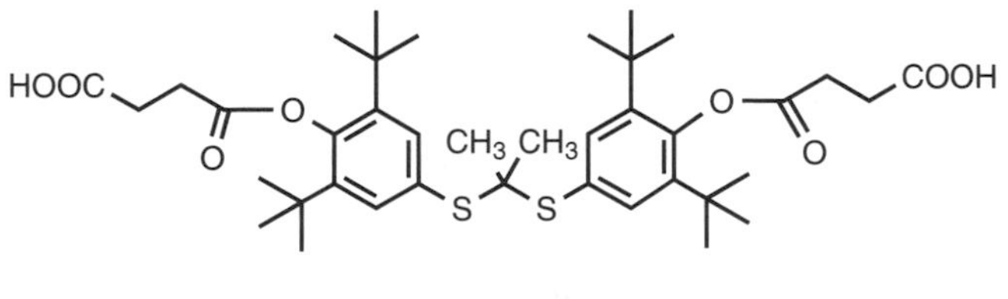

HOOC O O CH3 CH3 S S O O COOH

Diglutaryl Probucol

The effects of probucol on macrophage utilization of modified LDL are controversial. Incubation of macrophage cell lines with acetyl LDL and probucol suppress conversion to foam cells and enhance the secretion of apoE and cholesterol.[566] The uptake of LDL_{ox} and acetyl LDL, however, is not inhibited by probucol treatment in vivo or in vitro of peritoneal WHHL rabbit[567] or murine peritoneal macrophages.[568] Treatment of WHHL rabbit peritoneal macrophages with probucol enhance chemotaxis of murine peritoneal macrophages to native and oxidized LDL and to HDL thus possibly increasing egress of macrophages from the arterial wall.[569] Probucol also inhibits the production of IL-1, a potent smooth muscle cell proliferator, in lipopolysaccharide-primed human monocytic leukemic cells,[570] THP-1 monocytic cells,[571] as well as in zymosan-primed mice.

Although probucol reduces severity and extent of atherosclerosis in one set of cholesterol-fed rabbits,[572] the drug does not do so in another experiment in which probucol-treated and control rabbits have comparable serum cholesterol concentrations.[573] WHHL rabbits, however, uniformly have reductions in lesions.[574–577] Clotting factors may also participate in thromboatherogenesis in WHHL rabbits because probucol-treated homozygous animals have reductions in fibrinogen and Factor VIII after 8 months of treatment.[578]

The increased LDL_{ox} with resulting cytotoxicity occurring in strepzotocin-induced diabetes in rats attenuates with probucol treatment.[579] This treatment also delays the development of diabetes in BB-Wistar rats.[580]

Patients generally tolerate probucol; only about 3% discontinue treatment because of side effects. Diarrhea, the most frequently reported side effect, may occur in about one-third of treated patients.[502,581] Less frequently reported side effects include flatulence, abdominal pain, nausea, and vomiting.[582]

Toxicity studies were conducted in rats, mice, dogs, and monkeys.[502,583–587] The oral LD_{50} of probucol in rats and mice is greater than 5000 mg/kg, and oral administration of 800 mg/kg to rats for 2 years shows no adverse effects. Probucol is toxic in dogs given high doses for 90 days.[89] In this species, the drug appears to sensitize the canine myocardium to epinephrine slowly. Ventricular fibrillation and sudden death occur in about one-third of treated dogs. In monkeys on an atherogenic diet, electrocardiographic (ECG) QT intervals increase following probucol treatment.[587] Probucol treatment also significantly increases the QT interval in humans, but has no effect on premature ventricular complexes.[588]

Nicotinic Acid and Analogs

Nicotinic acid (niacin), first produced by oxidation of nicotine,[589] is a stable, nonhygroscopic, white, crystalline powder, freely soluble in alkaline solution ($pK_a = 4.76$) but insoluble in ether. The pH of a saturated solution is 2.7 and its UV absorbance spectrum exhibits a maximum at 283 ηm. The amide metabolite, nicotinamide, has no antilipidemic activity.[590] The amount of nicotinic acid and another metabolite, nicotinuric acid, found in the urine depends on the dose, rate, and route of administration.[590]

Nicotinic acid and nicotinamide are important cofactors required to prevent pellagra. In 1955, Altschul and co-workers[591] observed that high doses of nicotinic acid lower cholesterol levels in humans, an activity unrelated to its property as a vitamin. Nicotinic acid also lowers serum triglyceride levels and is effective against hyperlipidemias Types II, III, IV, and V.[379,590,592,593] To develop tolerance to the so-called flushing and gastrointestinal

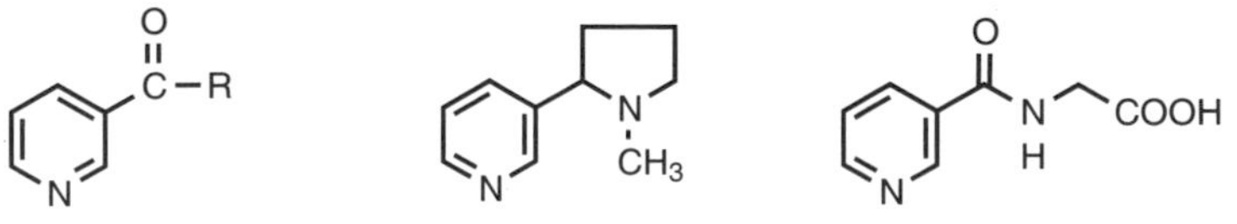

O C–R N | N CH3 N | O N H COOH N

R = OH; Nicotinic Acid
R = NH_2; Nicotinamide

Nicotine

Nicotinuric Acid

side effects associated with high doses, nicotinic acid therapy is initially given in small doses of three 100-mg tablets per day. The dosage is gradually increased over 2 or 3 weeks until a therapeutic dose (2 to 9 g/day) is reached.

A typical lipid lowering response to nicotinic acid would be an approximate 15% reduction in serum cholesterol and 30% reduction in triglyceride levels.[594] In the different hyperlipoproteinemic phenotypes, however, a variable response of the plasma lipoprotein classes is often seen, and the overall effects of nicotinic acid may be a composite of multiple interactions on a variety of enzymatic pathways of lipid metabolism.[595–598] In normolipidemic men and women, nicotinic acid (3 g/day for 21 days) decreases VLDL cholesterol by 30%, LDL cholesterol by 36%, but raises HDL cholesterol by 23%.[594] The effect on HDL is associated with an increase in plasma apoA-I levels and a striking 345% increase in the HDL_2:HDL_3 ratio.[594] Nicotinic acid is the most commonly used drug for lowering Lp(a). This is important because this lipoprotein is very atherogenic because of its high affinity for the arterial wall and its interference in fibrinolysis.[40] In patients who have received angioplasty, Lp(a) may be involved in restenosis.[599] High serum concentrations (>40 mg/dl) of this atherogenic lipoprotein are a genetic trait in some individuals.[600]

Nicotinic acid lowers VLDL levels, inhibits lipolysis in adipose tissue, decreases liver triglyceride esterification, and increases LpL activity.[590,601–603] Nicotinic acid also stimulates the oxidative metabolism of cholesterol in rat liver preparations,[604] but in humans, the cholesterol lost from the blood is not accounted for as fecal bile acids or neutral steroids.[605–607] The increase in fecal neutral steroids observed with Type II hypercholesterolemic patients[608] is probably a function of cholesterol mobilization from tissue.[609–611]

Nicotinic acid decreases mobilization of FA from adipose tissue and reduces plasma FFA levels.[601] Hepatic triglyceride synthesis is reduced resulting in a smaller VLDL particle size[609] and a decreased production of chylomicrons.[612] Decreases in plasma LDL levels also are not related to clearance or catabolism, but rather to decreased synthesis of the lipoprotein.[476] Thus, combinations of nicotinic acid and resin have a profound effect on plasma LDL levels.[488–491]

Clinical benefits have been reported in patients treated with nicotinic acid in whom eruptive, tuboeruptive, and tuberous xanthomata have regressed.[379] In the U.S Coronary Drug Project, nicotinic acid reduced serum cholesterol levels by 10% and triglyceride levels by 26% in 1,110 men treated for 5 years.[18] The incidence of non-fatal myocardial infarctions decreased relative to the placebo-control group, but there were no significant effects on total or CHD mortality during 5 years of nicotinic acid treatment. In a 10-year follow up to the Coronary Drug Project, however, cardiovascular mortality in the group that had been given nicotinic acid was 11% lower ($p = 0.0004$) than expected on the basis of mortality in the placebo group.[613]

The most common, and often dose-limiting, side effects of nicotinic acid treatment are cutaneous vasodilation (flushing) and gastrointestinal intolerance, which may occur in 20 to 50% of treated patients. Flushing and pruritus are apparently mediated via prostaglandin E_1[614] and prostacyclin[615] and may be prevented by aspirin or indomethacin given before nicotinic acid.[488,616]

Glucose tolerance may worsen in diabetes, and serum uric acid levels may increase, precipitating attacks of gout in hyperuricemic patients.[379,617] Plasma aspartate transaminase (AST), alanine transaminase (ALT), lactate dehydrogenase (LDH), and alkaline phosphatase levels are often elevated by nicotinic acid, but return to normal when therapy is discontinued for 1 to 2 weeks.[617]

Cutaneous flushing and plasma FFA rebound associated with nicotinic acid therapy prompted the development of various prodrugs and dosage forms designed to sustain blood levels of the parent drug and produce a longer duration of action. Less flushing is evident with xanthinol nicotinate[618] and tetranicotinoyl fructose.[619] Less FA rebound occurs with D-glucitol hexanicotinate,[620] β-pyridylcarbinol,[621,622] and nicotinic acid esterified to a polysaccharide polymer.[623]

The properties of many other nicotinic acid analogs have been reviewed.[590] In a series of ring substituted nicotinic acid derivatives, 5-fluoronicotinic acid was the most potent with respect to antilipolytic activity in vitro and serum FFA lowering in rats.[624] In clinical studies, the FA lowering effects of 5-fluoronicotinic acid is 5 to 10 times greater than those of the parent drug. The fluoro analog, however, is no more effective than nicotinic acid in lowering serum VLDL and LDL.[625]

Acipimox (5-methylpyrazine carboxylic acid 4-oxide), like 5-fluoronicotinic acid, shows greater antilipolytic activity than nicotinic acid in rats[626] and man[627] and produces a beneficial elevation in the HDL_2 subfraction.[628] However, the cholesterol and triglyceride-lowering activities of Acipimox do not appear to be greater than those of nicotinic acid.[629]

Many examples of prodrug nicotinic acid esters have been researched.[590] One, Niceritrol, a pentaerythritol derivative, does not lower serum triglyceride concentrations in Type IIa hyperlipidemic patients but lowers serum cholesterol in a dose-related manner.[630] This and many other esters of nicotinic acid undergo hydrolysis by plasma esterases. The fructose tetranicotinate (Nicofuranose) slowly releases nicotinic acid, and the flushing is better tolerated.[631] The methyl groups on Cyclonicate sterically hinder enzymatic hydrolysis and the release of nicotinic acid. Thus, decreases in FFA and triglycerides are observed without the usual side effects (flushing, arrhythmias). Nicomol, a tetraester, also increases the HDL_2:HDL_3 ratio and increases the total HDL concentration.[632] L-44 is a sterically hindered antilipidemic nicotinic acid ester, wherein 47% of the parent drug is found unaltered in the feces.[633] Additional molecular modifications include the ethanolamine analog 2-(N-isopropyl-N-nicotinoyl)ethylnicotinate, which undergoes metabolism at the more easily hydrolyzed ester linkage generating a more favorable toxicity/activity profile than for nicotinic acid.[634] Etofibrate is a bis(ester) of nicotinic and clofibric acids with ethylene glycol (a questionable choice because of toxicity associated with this alcohol). Binifibrate is a glycerol tri(ester) with two equivalents of nicotinic acid and one equivalent of clofibric acid, and ronafibrate is a related bis(ester) of propylene glycol.[635–638] Although these substances have other useful properties, nicotinic acid itself remains the drug of choice in this series.[590]

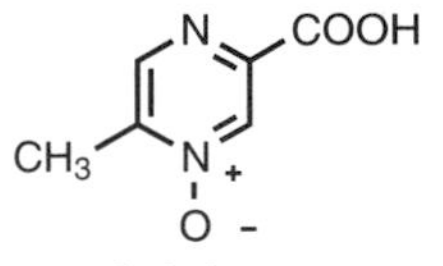

Acipimox

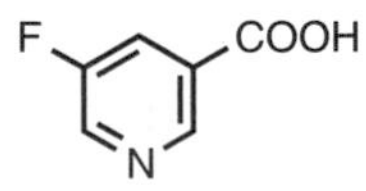

5-Fluoronicotinic acid

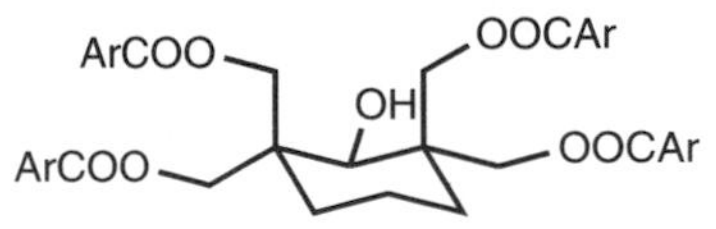

Nicomol

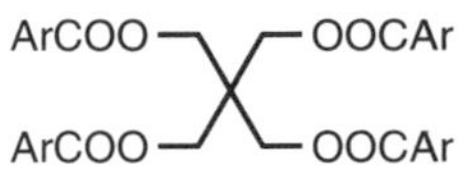

Niceritrol

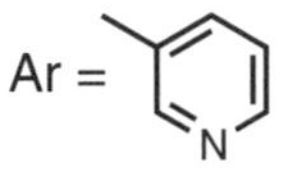

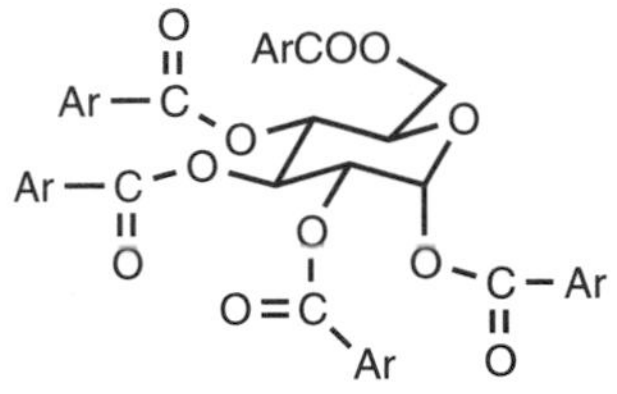

Ar =

L-44

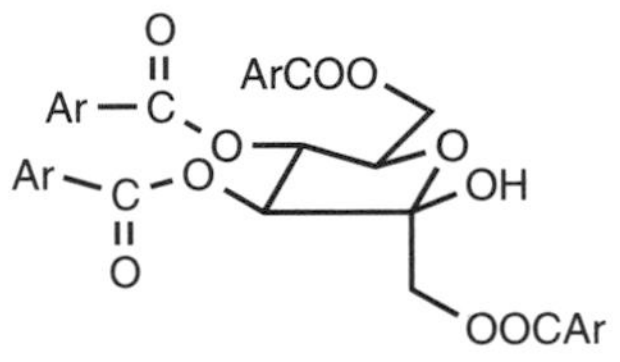

Nicofuranose

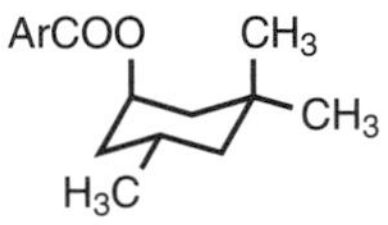

Cyclonicate

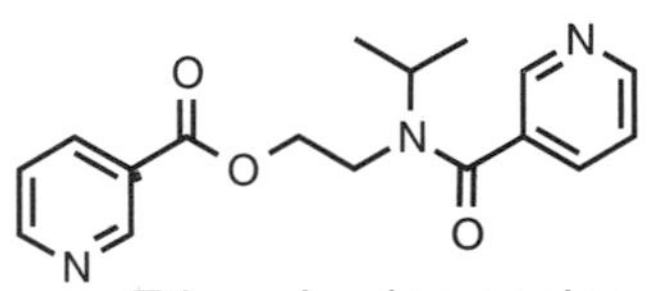

Ethanolamine analog

ANTIATHEROSCLEROSIS DRUGS IN DEVELOPMENT

Antioxidants

Atherosclerotic fatty streaks occur through cholesterol ester engorgement by cells. This takes place after uptake of LDL_{ox} or modified LDL. This lesion proceeds to a fibrous plaque and then to a complicated thrombus-involved atheroma, which may block arterial blood flow and produce clinically manifested CHD.[639]

LDL_{ox}, identified in both human[640] and atherosclerotic animal model[640] arteries and in the bloodstream[641] results from a free radical reaction sequence with LDL lipids.[642] This produces endothelial cytotoxicity, modified platelet functionality, increased monocyte adherence to and chemotaxis through the endothelium. This results in accumulation of arterial foam cells and enhanced smooth muscle proliferation.[639] Such an oxidative process, in vitro, may involve either transition metal ions, endothelial cells, monocytes, macrophages, or smooth muscle cells.[643] Additional systems that can oxidize or otherwise modify LDL are platelets,[644] a peroxyl radical generating system [2,2′-azobis(2-amidinopropane) hydrochloride or 2,2′-azobis(2,4-dimethylvaleronitrile)][645] or UV irradiation.[646] Many comprehensive reviews of this field and of the LDL antioxidants are available.[643,647–651]

Superoxide anion radicals ($O_2^{\cdot -}$) are continuously generated in normal cells by reduction of dioxygen (O_2). Defense mechanisms against $O_2^{\cdot -}$ include radical scavengers such as ascorbate and vitamin E and enzymes, superoxide dismutase (SOD), and catalase, among others. $O_2^{\cdot -}$ is the anionic form of $HOO\cdot$ having a $pK_a = 4.8$. Both $O_2^{\cdot -}$ and $HOO\cdot$ are reactive species but are not as reactive as their metabolites. $O_2^{\cdot -}$ upon dismutation catalyzed by SOD yields H_2O and hydrogen peroxide (H_2O_2). Preferably, H_2O_2 converts to H_2O and O_2 through the action of catalase.

When normally protective mechanisms cannot handle accumulation of H_2O_2, a Fenton's reaction with ferrous ion (Fe^{2+}) may cause further reduction to produce both hydroxyl radical ($HO^{\cdot}$) and hydroxy anion (HO^-). H_2O_2 and especially $HO^{\cdot}$ are highly toxic substances which carry out covalent bond breaking reactions with macromolecules such as DNA, enzymes, proteins, and lipoproteins (i.e., LDL). Such free radical reactions may result in pathologic conditions such as CHD, cancer, liver damage, rheumatoid arthritis, immunologic incompetence, among others.

Lipid peroxidation involves highly reactive radicals. Thus, radicals ($R^{\cdot}$) generated under conditions previously described undergo reaction with lipid side chains (LH) to generate new radicals ($L^{\cdot}$). When $L^{\cdot}$ is an unsaturated chain, reaction with O_2 produces allylic peroxides ($RCH = CH\text{-}CH_2OOH$) and epoxides (oxiranes, $R\text{-}CH\overset{O}{—}CH$) which are also reactive species. For example, peroxides may react with LH to generate $L^{\cdot}$, which produces more $LOO^{\cdot}$ and LOOH. Additionally, $O_2^{\cdot -}$ may react with LH by an insertion mechanism to generate LOOH or may, by reductive mechanisms, lead to H_2O_2 and $HO^{\cdot}$. LOOH may undergo reduction by Fe^{2+} or oxidation by ferric ion (Fe^{3+}) in reinitiation reactions, in which $LO^{\cdot}$ and $LOO^{\cdot}$, respectively, are produced. Glutathione and other reducing agents, such as ascorbate, may remove LOOH through generation of LOH and H_2O. Unfortunately, ascorbate can also reduce Fe^{3+} generating Fe^{2+}, which serves to increase concentrations of HO· in Fenton's reaction. Ascorbate also can reduce other transition metal ions. Termination of these chain reactions may involve radical scavengers, oxidation or reduction to other oxygen containing species such as aldehydes, ketones, and alcohols; or the less likely coupling of two radicals to generate less reactive species.[642]

Lipid and lipoprotein oxidation and inhibition may be followed by:

- the generation of malondialdehye, nonaldehyde, and many other aldehydes as detected by thiobarbituric acid or HPLC with a luminescence detection system
- carotene decolorization
- increased diene conjugation
- increased electrophoretic mobility of LDL_{ox}
- increased uptake of $^{125}I\text{-}LDL_{ox}$ into macrophages
- HPLC of phospholipids with chemilluminescence detection
- fluorescence at 430 ηm with excitation at 355 ηm, a putative measure of peroxy aldehydes reacting with apoB
- free amino groups
- spin label traps
- iodimetric peroxide measurement
- oxygen consumption[649]

Many drugs discussed in this chapter and throughout the book may serve by scavenging these species or chelating with transition metals interrupting free radical pathology. Such antioxidants can function at each of these steps to prevent the generation of LDL_{ox}.[649]

Antioxidants are present in foodstuffs,[649] dietary supplements,[652–654] and endogenously in LDL.[649] Further, many drugs have antioxidant properties.[649,655] Ascorbate protects the endogenous antioxidants in LDL, α-tocopherol, γ-tocopherol, ubiquinols, α- and β-carotene, lycopene, cryptoxanthin, cantoxanthin, lutein, zeoxanthin and phytofluene[649] and, in turn, urate protects ascorbate.[656] The ready oxidation of LDL with a high content of polyunsaturated FA[657] and low concentration of tocopherols,[649] however, particularly in the presence of low concentrations of HDL, HDL apoA-I, and HDL paroxaonase, known protectors of HDL oxidation,[658–660] makes it important to have powerful antioxidants available to preserve these endogenous antioxidants and the nativity of LDL. Ascorbic acid preserves the endogenous antioxidants, carotenes and tocopherols,[661] and the hypolipidemic drug, probucol.[662] Probucol, however, fails to provide protection for the endogenous antioxidants present in LDL.[661] 4-(4-Chlorophenyl)-2-hydroxytetronic acid (CHTA) and related hydrophobic *aci*-reductones are very effective hypolipidemic drugs in vivo, and antioxidants in vitro.[663] In the future, such experimental antiatherosclerotic drugs may more effectively fulfill the antioxidant ascorbate function of protecting endogenous LDL antioxidants.

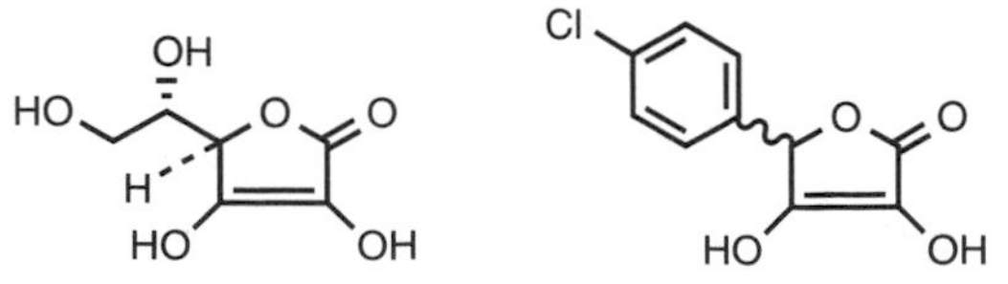

Ascorbic Acid

CHTA

AcylCoA: Cholesterol Acyltransferase (ACAT) Inhibitors

ACAT, a microsomal enzyme found in several tissues from a variety of species,[664–667] catalyzes the intracellular esterification of cholesterol with CoA-activated FA. The products are cholesterol esters and CoA. An extensive review of ACAT is available.[668]

To maximize ACAT activity, both the sterol nucleus β-hydroxyl group and the aliphatic side chain must be present. Neither the presence nor the position of the double bond found in cholesterol is crucial. Steroids with either shorter or more branched side chains than those found in cholesterol are poor substrates for ACAT.

The role of ACAT in the atherosclerotic process is significant. In hypercholesterolemic rabbits with endothelial-denuded iliac-femoral arteries, ACAT inhibition by the amide CI-976 reduces atheroma foam cell area by 28%, lowers esterified cholesterol by 46%, and blunts the fatty streak lesion development with no reduction in lesion free cholesterol.[669] ACAT activity in arterial microsomes is elevated 61-fold with cholesterol feeding and remains elevated during lesion development or regression.[670] Thus, inhibition of ACAT is important for lowering serum cholesterol concentrations and slowing atheroma formation. ACAT serves as an aid in cholesterol absorption and reduces intracellular free cholesterol.

Compounds of many diverse structures perform as ACAT inhibitors. On the basis of either functionality or structure similarity, they are classified as

- Antihypertensive agents at pharmacologic dose levels
- CNS agents involved in microsomal membrane perturbation
- Fibrates
- Cholesterol synthesis inhibitors
- Agents developed as specific ACAT inhibitors
    - Fatty acyl amides
    - Disubstituted ureas
    - Trisubstituted ureas
    - Cetaben sodium
    - Fatty acid anilides

Inhibitory potencies in vitro and pharmacologic effects in vivo are listed in Table 24–5.

In a large series of N,N-dialkyl-N′-arylureas, N′-(2,4-dimethylphenyl)-N-benzyl-N-butylurea is particularly effective as an ACAT inhibitor and is undergoing extensive biologic evaluation.[671] Within a series of fatty acid anilides with non-branched acyl compounds, inhibitory potency is optimal with bulky 2,6-dialkyl substitution. Generally, with α-aryl substituted acyl analogs there is little dependence on potency in vitro of anilide substitution, but 2,4,6-trimethoxy anilides are uniquely preferred.[672] Such

Table 24–5. ACAT Inhibitor Characteristics and Testing Protocols

	In Vitro Studies		In Vivo Studies	
Inhibitor	**Tissue**	**IC_{50} or Concentration Used***	**Animal Model**	**Effect Observed**
2-Isopropyl,6-methyl oleic acid anilide*	Cholesterol-fed rabbit intestinal microsomes[671]	0.023 μM	Cholesterol-cholic acid-peanut oil fed rat[671]	45% decrease in total plasma cholesterol
2,4,6-Trimethoxy-α-methyl heptanoic anilide†	Cholesterol-fed rabbit intestinal microsomes[671]	0.059 μM	Cholesterol-cholic acid-peanut oil fed rat[671]	53% decrease in total plasma cholesterol and 86% increase in plasma HDL-C
Trimethylcyclohexanyl mandelate	Rat hepatic microsomes[673]	80 μM		
	Transformed mouse macrophage J774 microsomes[673]	30 μM		
[a] CI-976	Rabbit intestinal microsomes[674]	0.075 μM		
	Mouse peritoneal microsomes[674]	0.62 μM		
	CaCO-2 cell microsomes[675]	0.51 μM		
[b] 58-035	CaCO-2 cell microsomes[676]	4 μM	Hypercholesterolemic rat[680]	30 mg/kg oral dose reduced intestinal microsomal acyl-CoA: cholesterol acyltransferase (ACAT) by 75% and completely prevented increase in serum cholesterol induced by cholesterol/cholic acid feeding over night.
	Mouse peritoneal macrophages[677]	11 μM*		
	Rat intestinal microsomes[678]	0.5 μM		
	Fu5AH rat hepatoma cells[679]	<1 μM		
N-[1-(4-benzyloxyphenyl)-2-phenylethyl]-N-benzyl-N-(3-trifluorotolyl)urea	Rat intestinal microsomes[678]	80 μM		
1-Benzylidene-4,4-diphenylthiosemicarbazone	Rat intestinal microsomes[678]	2.1 μM		
Verapamil	CT2 cell microsomes[681]	No effect on ACAT.		
[c] CI-277,082	Intestinal microsomes[682]	0.14 μM		
	Liver microsomes[682]	0.74 μM		
	Adrenal microsomes[682]	1.18 μM		
Chlorpromazine	Intact Fu5AH cells[679]	17 μM		
Propranolol	Intact Fu5AH cells[679]	>57 μM		

(continued)

Table 24–5. *(Continued)*

	In Vitro Studies		In Vivo Studies	
Inhibitor	**Tissue**	**IC_{50} or Concentration Used***	**Animal Model**	**Effect Observed**
Dibucaine	Intact Fu5AH cells[679]	39 μM		
[d] U-73482	Intact Fu5AH cells[679]	1.1 μM		
Lovastatin	Rabbit intestinal mucosal microsomes[683]	36 μM		
	CaCo-2 microsomes[684]	10 μg/mL		
Simvastatin	Rabbit intestinal mucosal microsomes[683]	20 μM	Hypercholesterolemic rabbit[685]	70% reduction in intestinal mucosa ACAT with 0.13% fed with 1% cholesterol diet.
[e] Pentacyclic triterpene ester (PTE)	Rabbit and rat liver microsomes[686]	50 μM		
[f] Purpactin A	Intact J774 macrophages[687]	1.2 μM		
	J774 microsomes[687]	121 μM		
Clofibrate			Rat	0.3% of diet gave 50–70% reduction of ACAT in liver microsomes[688]
Bezafibrate			Rat	0.1% of diet gave 50–70% reduction of ACAT in liver microsomes[688]
Ciprofibrate			Rat	0.016% of diet gave 50–70% reduction of ACAT in liver microsomes[688]
5α-Cholest-8(14)-en-3β-ol-15-one			Rat	0.1% of diet causes 77% reduction in ACAT[689]

[a] 2,2-Dimethyl-N-(2,4,6-trimethoxyphenyl)dodecanamide; [b] 3-(3-[decylmethylsilyl]-N-[2-(4-methylphenyl)-1-phenylethyl]propanamide; [c] (2,4-difluorophenyl)-N-{[4-(2,2-dimethylpropyl)phenyl]methyl}-n-hepturea; [d] 7-7′-[1,2-ethanediylbis(4,1-piperidinediyl methylene)]-bis-4,9-dimethoxy-5H-furo-[3,2-g][1]benzopyran-5-one; [c] Member of the olean-12-ene triterpene family; [f] 3-1′-acetoxy-11-hydroxy-4-methoxy-9-methyl-3′-methylbutyl-5H-dibenzo[b,g][1,5]dioxicin-5-one

* One of the most potent in vitro and in vivo in a series of alkyl substituted oleic acid analides.

† One of the most potent in vitro and in vivo in a series of α-substituted saturated 2,4,6 trimethoxyanilides.

compounds have IC_{50}s in the nanomolar range and produce significant reductions in total serum cholesterol concentrations in cholesterol fed rats. A number of compounds have been identified that inhibit ACAT in vitro, reduce plasma total cholesterol, and elevated HDL cholesterol in animals.

Based on the results in Table 24–5 and the activities found in vivo, it is apparent that no simple relationship exists between ACAT activity and lowered cholesterol levels. Additional evidence showing ACAT inhibitors are useful for prevention of atherosclerosis is needed before this novel mechanism will be widely accepted clinically.

α- and β-Adrenergic Agents

Adrenergic receptor selective agonists and antagonists regulate metabolic effects of cellular lipolysis and affect circulating plasma lipoproteins.[690] In adipocytes, β-adrenoceptor agonists stimulate triglyceride breakdown through an adenylate cyclase coupled receptor, whereas α-adrenoceptor activation inhibits both β-adrenoceptor-induced and basal lipolysis. Brown adipose tissue, which plays a role in diet-induced thermogenesis, responds in a similar way to α- and β-adrenergic agonists, and may be a target for weight-reducing agents.[691] Clinical trials with adipocyte selective β-adrenoceptor agonists such as BRL

26830[692] enhance weight loss under caloric restriction, and oral administration of the selective β_2-agonist, terbutaline,[693] elevates HDL cholesterol without any influence on plasma total cholesterol concentrations in humans. In contrast, β-adrenoceptor antagonists increase plasma triglycerides, VLDL/LDL ratio and plasma cholesterol, often accompanied by a fall in the HDL cholesterol levels.[694–696] β_1-Selective antagonists such as atenolol and metoprolol may produce less undesirable lipoprotein modifications than nonselective antagonists such as propranolol.[697] Promising results on the regression of atherosclerotic plaques are observed in animal models with β-antagonists.[698]

Norepinephrine produces an α_2-adrenoceptor-mediated antilipolytic action and blockade of these receptors represents an approach to the treatment of obesity. To date, treatment of humans with yohimbine (an α_2-adrenoceptor antagonist) has been variable, perhaps due to the short half-life of this agent. Guanabenz (a centrally acting, nonselective, α_2-adrenoceptor agonist) reduces plasma cholesterol levels in humans,[699] suggesting that more studies with selective α_1- and α_2-adrenergic agonists is desirable.

α_1-Selective adrenoceptor antagonists such as prazosin, doxazosin, terazosin, trimazosin, and indoramin generally reduce plasma total cholesterol and triglyceride levels, with marginal increases or no changes in lipoproteins, such as HDL cholesterol.[695,700,701] Because doxazosin reduces the density of foam cells in the aortic arch in hamsters,[702] and prazosin inhibits smooth muscle proliferation in denuded endothelium of rabbits,[703] these agents may have a beneficial effect on reduction of atheroma formation in humans.

Gene Therapy

The two main approaches to gene therapy for atherosclerosis are modifying liver cell genes to induce expression of LDL receptors as a means for reducing plasma cholesterol concentrations, and inhibiting cell proliferation and pathology through direct molecular manipulation of the coronary artery lesion. Both techniques use the transfection protocol that introduces foreign genetic material via retroviral vectors. These protocols can either transfect ex vivo or in vivo. Balloon catheter injection of the genetically altered material provides a direct access to the coronary artery wall, but the current technique limits transfection with the LDL receptor gene even with a high titer of transfecting agent.[704] Hepatocyte-directed transfer of LDL receptor gene, in vivo, in WHHL rabbits leads to transient reduction in LDL cholesterol,[705] but long-term cholesterol lowering takes place with ex vivo introduction of retroviral vectors containing the LDL receptor gene into WHHL hepatocytes[706] or fibroblasts.[707,708] This technique, ex vivo, can also transfect human hepatocytes with the LDL receptor gene.[709] Although these approaches remain tentative, human trials of gene therapy for homozygous familial hypercholesterolemic patients will soon take place.

Gugulipid

Gugulipid, at a dose of 500 mg, produces modest reductions in serum cholesterol (17%) and triglycerides (11%). HDL cholesterol concentrations are elevated 60% in responders.[710] In a crossover study, clofibrate produces similar reductions in serum lipids, but does not change HDL cholesterol levels in these patients.

Squalestin 1

Squalestin 1, a fermentation product derived from Phoma species (Coelomycetes), a potent inhibitor of squalene synthase, produces a marked serum cholesterol (75%) and apoB lowering effect without changing apoA-I concentrations in marmosets.[711] The serum cholesterol lowering is not attenuated after 8 weeks of administration. Squalene synthase, which catalyzes a reaction between two molecules of farnesyl diphosphate in the pathway of sterol synthesis (see Chapter 23), does not interfere with formation of nonsterol products (dolichol, ubiquinone, and protein isoprenylation) derived from mevalonate. The molecular mechanism of action differs from the statins, and squalestin 1 or its analogs may represent an alternative clinical therapy.

REFERENCES

1. E. H. Wittels, et al., *Am. J. Cardiol., 65,* 432(1990).
2. L. Goldman, *Am. Heart J., 119,* 733(1990).
3. W. B. Statson, *Am. Heart J., 119,* 718(1990).
4. C. W. Frank, *Bull. N.Y. Acad. Med., 44,* 900(1968).
5. F. H. Epstein, *Bull. N.Y. Acad. Med., 44,* 916(1968)
6. S. H. Rinzler, *Bull. N.Y. Acad. Med., 44,* 936(1968).
7. S. M. Fox and W. L. Haskell, *Bull. N.Y. Acad. Med., 44,* 950(1968).
8. Q. B. Deming, *Bull. N.Y. Acad. Med., 44,* 968(1968).
9. J. Stamler, *Bull. N.Y. Acad. Med., 44,* 1476(1968).
10. L. Garfinkel, *Bull. N.Y. Acad. Med., 44,* 1495(1968).
11. D. Steinberg, in: *Adv. Exp. Med. Biol.* (1968), Vol. 4, W. L. Holmes, L. A. Carlson and R. Paoletti, Eds., New York, Plenum Press, 1969, p. 1.
12. M. A. Ruffer, *J. Pathol., 15,* 453(1911).
13. A. D. Morgan, *The Pathogenesis of Coronary Occlusion,* Springfield, IL, Charles C Thomas, 1956, pp. 5–9.
14. C. Moses, *Atherosclerosis, Mechanisms as a Guide to Prevention,* Philadelphia, Lea & Febiger, 1963, pp. 29–31.
15. M. C. Winternitz, et al., *The Biology of Arteriosclerosis,* Springfield, IL, Charles C Thomas, 1938, p. 22.
16. J. B. Duquid, *J. Pathol., 58,* 207(1946).
17. The Coronary Drug Project Research Group, *JAMA, 220,* 996(1972).
18. The Coronary Drug Project Research Group, *JAMA, 226,* 652(1973).
19. Lipid Research Clinics Program, *JAMA, 251,* 365(1984).
20. Lipid Research Clinics Program, *JAMA, 251,* 361(1984).
21. M. H. Frick, et al., *N. Engl. J. Med., 317,* 1237(1987).
22. V. Manninen, et al., *JAMA, 260,* 641(1988).
23. J. F. Brensike, et al., *Circulation, 69,* 313(1984).
24. D. H. Blankenhorn, et al., *JAMA, 257,* 3233(1987).
25. L. Cashin-Hemphill, et al., *JAMA, 264,* 3013(1990).
26. G. Brown, et al., *N. Engl. J. Med., 323,* 1289(1990).
27. J. P. Kane, et al., *JAMA, 264,* 3007(1990).
28. G. F. Watts, et al., *Lancet, 339,* 563(1992).
29. Presented at 59th European Atherosclerosis Society Congress, Nice, France, May 17–21, 1992.
30. G. Assman, in: *Lipid Metabolism and Atherosclerosis,* Stuttgart, F. K. Schattauer, 1982, p. 16.
31. R. E. Pitas, et al., *Biochemistry, 19,* 4359(1980).

32. N. B. Myant in *Cholesterol Metabolism, LDL and the LDL Receptor,* San Diego, Academic Press, 1990, pp. 190–191.
33. N. B. Myant in *Cholesterol Metabolism, LDL and the LDL Receptor,* San Diego, Academic Press, 1990, pp. 194–195.
34. L. Orci, et al., *Cell, 36,* 835(1984).
35. P. A. Dawson, et al., *J. Biol. Chem., 263,* 3372(1988).
36. D. Steinberg, et al., *N. Engl. J. Med., 320,* 915(1989).
37. R. Zechner, et al., *Biochem. Biophys. Acta, 918,* 27(1987).
38. G. Assmann, in *Lipid Metabolism and Atherosclerosis,* Stuttgart, F. K. Schattauer, 1982, p. 51.
39. N. E. Miller, et al., *Lancet, 1,* 1741(1981).
40. A. Rosengren, L., et al., *Br. Med. J., 301,* 1248(1990).
41. J. W. McLean, et al., *Nature, 330,* 132(1987).
42. P. G. Lerch, et al., *Eur. J. Biochem., 107,* 7(1980).
43. K. B. Sharpless, et al., *J. Am. Chem. Soc., 91,* 3394(1969).
44. D. S. Fredrickson, et al., *N. Engl. J. Med., 276,* 32(1967).
45. W. B. Kannel and A. Schatzkin, *Prog. Cardiovasc. Dis., 26,* 309(1983).
46. R. I. Levy, et al., *Circulation, 69,* 325(1984).
47. N. E. Miller, et al., *Lancet, 1,* 1741(1981).
48. W. P. Castelli, et al., *JAMA, 256,* 2835(1986).
49. M. A. Austin, et al., *JAMA, 260,* 1917(1988).
50. D. J. Rader and H. B. Brewer, Jr., *JAMA, 267,* 1109(1992).
51. B. D. Roth and R. S. Newton in *Antilipidemic Drugs, Medicinal, Chemical, and Biochemical Aspects,* D. T. Witiak, et al., Eds., Amsterdam, Elsevier/North Holland, 1991, p. 241.
52. B. D. Roth and R. S. Newton in *Antilipidemic Drugs, Medicinal, Chemical, and Biochemical Aspects,* D. T. Witiak, et al., Eds., Amsterdam, Elsevier/North Holland, 1991, p. 240.
53. G. Assmann, in *Lipid Metabolism and Atherosclerosis,* Stuttgart, F. K. Schattauer, 1982, p. 53.
54. H. R. Superko and R. M. Krauss, *Atherosclerosis, 95,* 69(1992).
55. A. Gurakar, et al., *Atherosclerosis, 57,* 293(1985).
56. L. A. Carlson, et al., *J. Intern. Med., 226,* 271(1989).
57. Anon., *Arch. Intern. Med., 148,* 36(1988).
58. G. Christakis and S. K. Rinzler, in *Atherosclerosis,* F. G. Schettler and G. S. Boyd, Eds., New York, Elsevier, 1969.
59. S. L. Conner and W. E. Conner, in *Antilipidemic Drugs, Medicinal, Chemical, and Biochemical Aspects,* D. T. Witiak, et al., Eds., Amsterdam, Elsevier/North Holland, 1991, pp. 455–491.
60. R. W. Brueggemeier and P.-K. Li, in *Antilipidemic Drugs, Medicinal, Chemical, and Biochemical Aspects,* D. T. Witiak, et al., Eds., Amsterdam, Elsevier/North Holland, 1991, pp. 493–526.
61. N. Applezweig, *Steroid Drugs,* New York, McGraw-Hill, 1962, p. 226.
62. L. N. Katz, et al., *Nutrition and Atherosclerosis, Current Status of the Problem,* Philadelphia, Lea & Febiger, 1958.
63. J. Stamler, et al., *Ann. N.Y. Acad. Sci., 64,* 596(1956).
64. D. Applebaum-Bowden, et al., *J. Clin. Invest., 59,* 601(1977).
65. J. D. Bagdade, et al., *J. Clin. Endocrinol. Metab., 70,* 1132(1990).
66. J. A. Dewar, et al., *Br. Med. J., 305,* 225(1992).
67. M. F. Kalin and B. Zumoff, *Steroids, 55,* 330(1990).
68. R. R. French, et al., *Am. J. Epidemiol., 108,* 486(1978).
69. S. Moorjani, et al., *J. Clin. Endocrinol. Metab., 66,* 314(1988).
70. P. D. Thompson, et al., *JAMA, 261,* 1165(1989).
71. W. H. Ettinger and W. R. Hazzard, *Metabolism, 37,* 1055(1988).
72. S. B. Weiss and W. Marx, *J. Biol. Chem., 213,* 349(1955).
73. S. O. Byers and M. Friedman, *Am. J. Physiol., 168,* 297(1952).
74. J. C. Thompson and H. M. Vars, *Proc. Soc. Exp. Biol. Med., 83,* 246(1953).
75. N. B. Myant, in *Lipid Pharmacology,* Vol. 2, R. Paoletti, Ed., New York, Academic Press, 1964, p. 229.
76. M. M. Levy and E. Levy, *Presse Med., 40,* 240(1932).
77. S. P. Asper, et al., *Bull. Johns Hopkins Hosp., 93,* 164(1953).
78. H. A. Lardy and G. F. Maley, *Recent Prog. Horm. Res., 10,* 129(1954).
79. S. Valdemarsson, et al., *Scand. J. Clin. Lab. Invest., 44,* 183(1984).
80. Z. Chap, et al., *J. Clin. Invest., 78,* 1355(1986).
81. D. J. Betteridge, *Br. Med. Bull., 45,* 285(1989).
82. T. Mazzone, et al., *Diabetes, 33,* 333(1984).
83. J. L. Witztum, et al., *Diabetes, 31,* 283(1982).
84. B. Gonen, et al., *Diabetes, 30,* 875(1981).
85. J. J. Abrams, et al., *Diabetes, 31,* 903(1982).
86. C. C. T. Smith, et al., *Diabetes Res., 2,* 277(1985).
87. M. Muggeo, et al., *J. Clin. Endocrinol. Metab., 48,* 17(1979).
88. G. Gacs and L. Romics, *Exp. Clin. Endocrinol., 90,* 227(1987).
89. O. J. Pollak, *Circulation, 7,* 702(1953).
90. H. H. Hernandez, et al., *Proc. Soc. Exp. Biol. Med., 83,* 498(1953).
91. T. Gerson and F. B. Shorland, *Nature, 200,* 579(1963).
92. R. E. Shipley, et al., *Circ. Res., 6,* 373(1958).
93. A. H. Levere, et al., *Metabolism, 7,* 338(1958).
94. T. R. Blohm and R. D. MacKenzie, *Arch. Biochem. Biophys., 85,* 245(1959).
95. T. R. Blohm, et al., *Arch. Biochem. Biophys., 85,* 250(1959).
96. R. W. P. Achor, et al., *Mayo Clin. Proc., 36,* 217(1961).
97. R. C. Laughlin and T. F. Carey, *JAMA, 181,* 339(1962).
98. H. S. Seltzer and J. C. Melby, *J. Lab. Clin. Med., 58,* 957(1961).
99. R. Niemiro and R. Furnagalli, *Biochem. Biophys. Acta, 98,* 624(1965).
100. J. Avigan, et al., *J. Biochem., 235,* 3123(1960).
101. D. Steinberg, et al., *J. Clin. Invest., 40,* 884(1961).
102. D. H. Blankenhorn, et al., *Circulation, 24,* 889(1961).
103. D. H. Blankenhorn and O. Kuzma, *Metabolism, 10,* 763(1961).
104. R. E. Counsell, et al., *J. Med. Pharm. Chem., 5,* 720(1962).
105. R. E. Counsell, et al., *J. Med. Pharm. Chem., 5,* 1224(1962).
106. P. D. Klimstra, et al., *J. Med. Chem., 9,* 323(1966).
107. R. E. Counsell, et al., *J. Med. Chem., 8,* 45(1965).
108. K. Irmscher, et al., *Steroids, 7,* 557(1966).
109. L. G. Humber, et al., *J. Med. Chem., 9,* 329(1966).
110. N. J. Doorenbos and M. T. Wu, *J. Pharm. Sci., 54,* 1290(1965).
111. M. Martin Smith, *Rep. Prog. Appl. Chem., 52,* 146(1967).
112. R. Langdon, et al., *J. Lipid Res., 18,* 24(1977).
113. R. E. Counsell, Professor of Pharmacology, University of Michigan. Personal communication.
114. J. A. Svododa and W. E. Robbins, *Science, 156,* 1637(1967).
115. J. Ashraf, et al., *Biosci. Rep., 4,* 1115(1984).
116. A. Endo, et al., *FEBS Lett., 72,* 323(1976).
117. A. W. Alberts, et al., *Proc. Natl. Acad. Sci., 77,* 3957(1980).
118. V. W. Rodbell, et al., *Adv. Lipid Res., 14,* 1(1976).
119. M. S. Brown, et al., *J. Biol. Chem., 253,* 1121(1978).
120. P. T. Kovanen, et al., *Proc. Natl. Acad. Sci., 78,* 1194(1981).
121. D. W. Bilheimer, et al., *Clin. Res., 31,* 544A(1983).
122. G. Haba, et al., *J. Clin. Invest., 67,* 1532(1981).
123. Anon, *Drugs of the Future, 9,* 197(1984).
124. A. Endo, et al., *J. Antibiot., 29,* 1346(1976).
125. A. G. Brown, et al., *J. Chem. Soc. Perkin Trans., 1,* 1165(1976).
126. The Lovastatin Study Group IV, *Am. J. Cardiol., 66,* 22B(1990).
127. E. Helve and M. J. Tikkanen, *Atherosclerosis, 72,* 189(1988).

128. J. C. Sienra Perez, et al., *Arch. Inst. Cardiol. (Mexico), 61,* 365(1991).
129. D. A. Pietro, et al., *Am. J. Cardiol., 63,* 682(1989).
130. The Lovastatin Study Group III, *JAMA, 260,* 359(1988).
131. D. McTavish and E. M. Sorkin, *Drugs, 42,* 65(1991).
132. H. H. Ditchuneit, et al., *Med. Klin., 86,* 142(1991).
133. J. M. Bard, et al., *Metabolism, 39,* 269(1990).
134. E. Stein, et al., *Arch. Intern. Med., 150,* 341(1990).
135. J. Molgaard, et al., *Eur. J. Clin. Pharmacol., 36,* 455(1989).
136. R. B. D'Agostino, et al., *Am. J. Cardiol., 69,* 28(1992).
137. M. S. Gavelli, et al., *Clin. Ter., 136,* 31(1991).
138. J. P. Ojala, et al., *Cardiology, 77,* Suppl. 4, 39(1990).
139. F. U. Beil, et al., *Cardiology, 77,* Suppl. 4, 22(1990).
140. J. Sanchez-Dominguez, et al., *Rev. Esp. Cardiol., 44,* 251(1991).
141. M. J. Tikkanen, et al., *Eur. J. Clin. Pharmacol., 40,* Suppl. 1, S23(1991).
142. R. Goldberg, et al., *Am. J. Cardiol., 66,* 16B(1990).
143. G. Crepaldi, et al., *Arch. Intern. Med., 151,* 146(1991).
144. J. M. Bard, et al., *Atherosclerosis, 91,* S29(1991).
145. E. Bruckert, et al., *Ann. Med. Interne (Paris), 142,* 505(1991).
146. D. H. Smith, et al., *S. Afr. Med. J., 77,* 500(1990).
147. F. Valles, et al., *Atherosclerosis, 91,* S3(1991).
148. S. Berioli, et al., *Cardiologia, 35,* 335(1990).
149. G. L. Vega and T. P. Gross, *Arch. Intern. Med., 150,* 2169(1990).
150. G. Francechini, et al., *J. Lab. Clin. Med., 114,* 250(1989).
151. A. Endo, et al., *Eur. J. Biochem., 77,* 31(1970).
152. T. Habo, et al., *J. Clin. Invest., 67,* 1532(1981).
153. I. Kaneko, et al., *Eur. J. Biochem., 87,* 313(1978).
154. Y. Tsujita, et al., *Atherosclerosis, 32,* 307(1979).
155. M. Kuroda, et al., *Lipids, 14,* 585(1979).
156. R. Fears, *Drugs of Today, 20,* 257(1984).
157. T. J. Lee, et al., *J. Med. Chem., 34,* 2474(1991).
158. H. Hoshua, et al., *J. Antibiot. (Tokyo), 44,* 366(1991).
159. M. E. Duggan, et al., *J. Med. Chem., 34,* 2489(1991).
160. K. P. Vyas, et al., *Drug Metab. Dispos., 18,* 203(1990).
161. K. P. Vyas, et al., *Drug Metab. Dispos., 18,* 218(1990).
162. K. P. Vyas, et al., *Biochem. Pharmacol., 39,* 67(1990).
163. K. P. Vyas, et al., *Biochem. Biophys. Res. Commun., 166,* 1155(1990).
164. S. Vickers, et al., *Drug Metab. Dispos. Biol. Fate Chem. 18,* 476(1990).
165. S. Vickers, et al., *Drug Metab. Dispos., 18,* 138(1990).
166. H. Y. Pan, et al., *J. Clin. Pharmacol., 30,* 1128(1990).
167. H. Y. Pan, *Eur. J. Clin. Pharmacol., 40,* Suppl. 1, S15(1991).
168. D. W. Everett, et al., *Drug Metab. Dispos. Biol. Fate Chem. 19,* 740(1991).
169. J. I. Germershausen, et al., *Biochem. Biophys. Res. Commun., 158,* 667(1989).
170. A. T. Serajuddin, et al., *J. Pharm. Sci., 80,* 830(1991).
171. T. Koga, et al., *Biochim. Biophys. Acta, 1045,* 115(1990).
172. A. Endo, et al., *Biochim. Biophys. Acta, 575,* 266(1979).
173. D. W. Bilheimer, et al., *Proc. Natl. Acad. Sci. USA, 80,* 4124(1983).
174. J. A. Tobert, *Atherosclerosis, 41,* 61(1982).
175. J. A. Tobert, *J. Clin. Invest., 69,* 913(1982).
176. S. M. Grundy and G. L. Vega, *Circulation, 70,* II-168(1984).
177. B. Lewis, et al., *Br. Med. J., 287,* 21(1983).
178. H. Mabuchi, et al., *N. Engl. J. Med., 308,* 609(1983).
179. A. Yamamoto, et al., *Atherosclerosis, 35,* 259(1980).
180. H. Mabuchi, et al., *N. Engl. J. Med., 305,* 478(1981).
181. W. O. Richter, et al., *Int. J. Tissue React., 13,* 107(1991).
182. P. L. Malini, et al., *Clin. Ther., 13,* 500(1991).
183. H. H. Ditschuneit, et al., *Eur. J. Clin. Pharmacol., 40,* Suppl. 1, S27(1991).
184. V. F. Mauro and J. L. MacDonald, *Drug Intelligence and Clinical Pharmacy, Ann. Pharmacother., 25,* 257(1991).
185. M. Aviram, et al., *Eur. J. Chem. Clin. Biochem., 29,* 657(1991).
186. M. Aviram, et al., *Metabolism, 41,* 229(1992).
187. H. Y. Pan, et al., *Clin. Ther., 13,* 368(1991).
188. M. Rubenfire, et al., *Arch. Intern. Med., 151,* 2234(1991).
189. C. L. Shear, et al., *Circulation, 85,* 1293(1992).
190. V. M. Maher and G. R. Thompson, *Q. J. Med., 74,* 165(1990).
191. J. Johansson, et al., *Atherosclerosis, 91,* 175(1991).
192. J. J. Brocard, et al., *Schweiz. Med. Wochenschr., 121,* 977(1991).
193. H. Y. Pan, et al., *Clin. Pharmacol. Ther., 48,* 201(1990).
194. G. L. Vega and S. M. Grundy, *JAMA, 262,* 3148(1989).
195. A. Ausina-Gomez, et al., *Ann. Esp. Pediatr., 35,* 327(1991).
196. J. Thiery, et al., *Eur. J. Pediatr., 149,* 716(1990).
197. R. Uauy, et al., *J. Pediatr., 113,* 387(1988).
198. E. A. Stein, *Arteriosclerosis, 9,* 1145(1989).
199. D. R. Illingworth and J. P. O'Malley, *Metabolism, 39,* 403(1990).
200. P. M. Stuyt, et al., *J. Intern. Med., 230,* 151(1991).
201. P. M. Stuyt, et al., *Am. J. Med., 88,* 42N(1990).
202. D. R. Illingworth, et al., *Lancet, 339,* 8793(1992).
203. M. D. Tarantino, et al., *J. Pediatr., 118,* 131(1991).
204. A. M. DiBisceglie, et al., *Hepatology, 11,* 764(1990).
205. B. L. Kasiske, et al., *Transplantation, 49,* 95(1990).
206. P. C. Kuo, et al., *Am. J. Cardiol., 64,* 631(1989).
207. N. Yoshimura, et al., *Transplantation, 53,* 94(1992).
208. C. L. Corpier, et al., *JAMA, 260,* 239(1988).
209. J. A. Kobashigawa, et al., *Circulation, 82,* Suppl. 5, IV281(1990).
210. J. L. Goldstein and M. S. Brown, *Clin. Res., 30,* 417(1982).
211. M. Uusitupa, et al., *J. Cardiovasc. Pharmacol., 18,* 496(1991).
212. G. Brown, et al., *N. Engl. J. Med., 33,* 1289(1990).
213. W. V. Brown, *Lowering Blood Cholesterol to Prevent Heart Disease,* in: Washington, DC, NIH Consensus Development Conference, 1984, pp. 41–45.
214. G. L. Vega, et al., *Arteriosclerosis, 91,* 1135(1989).
215. T. P. Leren, et al., *Atherosclerosis, 73,* 135(1988).
216. W. Schwartzkopff, et al., *Arzneimittelforschung, 40,* 1322(1990).
217. N. Hoogerbrugge, et al., *J. Intern. Med., 228,* 261(1990).
218. J. Molgaard, et al., *Atherosclerosis, 91,* S21(1991).
219. R. C. O'Brien, et al., *Med. J. Aust., 152,* 480(1990).
220. J. Emmerich, et al., *Eur. Heart J., 11,* 149(1990).
221. A. Garg and S. M. Grundy, *Diabetes, 38,* 364(1989).
222. G. Paolisso, et al., *Eur. J. Clin. Pharmacol., 40,* 27(1991).
223. R. B. D'Agostino, et al., *Am. J. Cardiol., 69,* 28(1992).
224. G. L. Vega and S. M. Grundy, *Arch. Intern. Med., 150,* 1313(1990).
225. D. R. Illingworth and S. Bacon, *Circulation, 79,* 590(1989).
226. J. L. Witxtum, et al., *Circulation, 79,* 16(1989).
227. E. W. Erkelens, *Cardiology, 77,* Suppl. 4, 33(1991).
228. G. Brown, et al., *N. Engl. J. Med., 323,* 1289(1990).
229. L. L. Martens, et al., *Am. J. Cardiol., 65,* 27F(1990).
230. F. A. Lederle and E. M. Rogers, *J. Gen. Intern. Med., 5,* 459(1990).
231. J. W. Hay, et al., *Am. J. Cardiol., 67,* 789(1991).
232. M. J. Mol., et al., *Neth. J. Med., 36,* 182(1990).
233. G. M. Kostner, et al., *Circulation, 80,* 1313(1989).
234. J. Thiery, et al., *Klin. Wochenschr., 66,* 462(1988).
235. K. Berg and T. P. Leren, *Lancet, II,* 812(1989).
236. D. Crook, et al., *Lancet, 339,* 313(1992).
237. D. W. Bilheimer, *Cardiology,* 77, Suppl. 4, 58(1990).
238. A. N. Vgontzas, et al., *Clin. Pharmacol. Ther., 50,* 730(1991).
239. R. E. Botti, et al., *Clin. Neuropharmacol., 14,* 256(1991).

240. J. A. Tolbert, *Am. J. Cardiol., 62,* 28J(1988).
241. D. H. Spach, et al., *West. J. Med., 154,* 213(1991).
242. L. R. Pierce, et al., *JAMA, 264,* 71(1990).
243. C. S. Wallace and B. A. Mueller, *Ann. Pharmacother., 26,* 190(1992).
244. C. A. Dujovne, et al., *Am. J. Med., 91,* 25S(1991).
245. G. Mantell, et al., *Am. J. Cardiol., 66,* 11B(1990).
246. K. Jensen, *Ugeskr, Laeger, 153,* 862(1991).
247. P. F. Smith, et al., *J. Pharmacol. Exp. Ther., 257,* 1225(1991).
248. M. Ulbig and T. Schneider, *Fortschr. Ophthalmol., 88,* 431(1991).
249. J. Schmitt, et al., *Fortschr. Ophthalmol., 88,* 843(1991).
250. B. L. Lundh and S. E. Nilsson, *Act. Ophthalmol. (Copenhagen), 68,* 658(1990).
251. Mevacor Monograph in *Physicians' Desk Reference,* Montvale, NJ, Medical Economics Data, 1992.
252. R. A. Willis, et al., *Proc. Natl. Acad. Sci. USA, 87,* 8928(1990).
253. Y. Nagata, et al., *Jpn. J. Pharmacol., 54,* 315(1990).
254. E. A. Porta, et al., *Adv. Exp. Med. Biol., 169,* (1989).
255. M. Sinensky, et al., *J. Biol. Chem., 265,* 19937(1990).
256. F. G. Fenton, et al., *J. Cell Biol., 117,* 347(1992).
257. R. E. Law, et al., *Mol. Cell. Biol., 12,* 103(1992).
258. T. S. Vincent, et al., *Biochem. Biophys. Res. Commun., 180,* 1284(1991).
259. J. Linna, et al., *Adv. Exp. Med. Biol., 288,* 269(1991).
260. M. Jakobisiak, et al., *Proc. Natl. Acad. Sci. USA, 88,* 3628(1991).
261. M. Jakobisiak, et al., *Proc. Natl. Acad. Sci. USA, 88,* 3628(1991).
262. M. Sinensky, et al., *J. Biol. Chem., 32,* 19937(1990).
263. F. U. Beil, et al., *Cardiology, 77,* Suppl. 4, 44(1990).
264. A. S. Pappu and D. R. Illingworth, *J. Lab. Clin. Med., 114,* 554(1989).
265. A. S. Pappu, et al., *Metabolism, 38,* 542(1989).
266. K. Folkers, et al., *Proc. Natl. Acad. Sci. USA, 87,* 8931(1990).
267. W. Sturmer, et al., *Klin. Wochenschr., 69,* 307(1991).
268. K. Purvis, et al., *Eur. J. Clin. Pharmacol., 42,* 61(1992).
269. C. Azzarito, et al., *Metabolism, 41,* 148(1992).
270. J. Prihoda, et al., *J. Clin. Endocrinol. Metab., 72,* 567(1991).
271. M. J. Mol, et al., *Clin. Endocrinol. (Oxford), 31,* 679(1989).
272. R. H. Jay, et al., *Br. J. Clin. Pharmacol., 32,* 417(1991).
273. A. S. Dobs, et al., *Metabolism, 40,* 524(1991).
274. H. Ide, et al., *Clin. Ther., 12,* 410(1990).
275. I. J. Goldberg, et al., *J. Clin. Invest., 86,* 801(1990).
276. R. Geghard, et al., *Lipids, 26,* 492(1991).
277. B. G. Stone, et al., *J. Lipid Res., 30,* 1943(1989).
278. D. Owens, et al., *Biochim. Biophys. Acta, 1082,* 303(1991).
279. J. Elmberger, et al., *J. Lipid Res., 32,* 935–1991).
280. S. Ahmad, *Arch. Intern. Med., 150,* 2407(1990).
281. H. Y. Pan, et al., *Br. J. Clin. Pharmacol., 31,* 665(1991).
282. R. Nitsche, et al., *Z. Gastroenterol., 29,* 242(1991).
283. J. Mitchell, et al., *J. Lipid Res., 32,* 71(1991).
284. S. E. Barrwo, et al., *Br. J. Clin. Pharmacol., 32,* 127(1991).
285. K. Schor, *Eicosanoids, 3,* 67(1990).
286. K. Schror, et al., *Eicosanoids, 2,* 39(1989).
287. G. Davi, et al., *Atherosclerosis, 79,* 79(1989).
288. M. Bo, et al., *Angiology, 42,* 106(1991).
289. A. V. Wever, et al., *Hypertension, 17,* 203(1991).
290. R. H. Jay, et al., *Atherosclerosis, 85,* 249(1990).
291. Y. Beigel, et al., *J. Clin. Pharmacol., 31,* 512(1991).
292. R. Sahni, et al., *Am. Heart J., 121,* 1600(1991).
293. J. T. Pento, *Drugs of Today, XVIII,* 585(1982).
294. A. Eisalo and V. Mannien, *Proc. Roy. Soc. Med., (Suppl. 2), 69,* 49(1976).
295. A. G. Olsson, et al., *Proc. R. Soc. Med., (Suppl. 2), 69,* 38(1976).
296. P. Schwandt, et al., *Artery, 5,* 117(1979).
297. B. J. Hoogwerf, et al., *Atherosclerosis, 51,* 251(1984).
298. S. Kaukola, et al., *Acta Med. Scand., 209,* 69(1981).
299. Y. A. Kesaniemi and S. M. Grundy, *J. Amer. Med. Assoc., 251,* 2241(1984).
300. D. T. Nash, *J. Med., 11,* 107(1980).
301. E. R. Nye, et al., *N. Z. J. Med., 92,* 345(1980).
302. B. Vessby, et al., *Proc. R. Soc. Med. (Suppl. 2.), 69,* 32(1976).
303. A. N. Howard and P. Ghosh, *Proc. R. Soc. Med. (Suppl. 2), 69,* 88(1976).
304. J. Tuomilehto, et al., *Proc. R. Soc. Med. (Suppl. 2), 69,* 32(1976).
305. S. W. Rabkin, et al., *Atherosclerosis, 73,* 233(1988).
306. Anon., *The Medical Letter, 24,* 59(1982).
307. R. I. Levy, in: *Proc. 2nd World Conference Clinical Pharmacology,* L. Lemberger and M. M. Riedenberg, Eds., Bethesda, MD, ASPET, 1984, p. 916.
308. P. J. Lupien, et al., *Can. J. Cardiol., 7,* 27(1991).
309. D. Manojlovic, et al., *Srp. Arh. Celok. Lek., 119,* 22(1991).
310. L. Klosiewicz-Latoszek and W. B. Szostak, *Eur. J. Clin. Pharmacol., 40,* 33(1991).
311. E. Nakandakare, et al., *Atherosclerosis, 85,* 211(1990).
312. P. Weisweiler, *Arzneimittelforschung, 38,* 925(1988).
313. P. Zimetbvaum, et al., *J. Clin. Pharmacol., 31,* 25(1991).
314. D. R. Illingworth and S. Bacon, *Arteriosclerosis, 9,* 1121 (1989).
315. P. Kremer, et al., *Curr. Med. Res. Opin., 11,* 293(1989).
316. M. N. Cayen, *Drug Metab. Rev., 11,* 291(1980).
317. T. C. Smith, *Proc. R. Soc. Med. (Suppl. 2), 69,* 24(1976).
318. R. A. Okerholm, et al., *Proc. R. Soc. Med. (Suppl. 2), 69,* 11(1976).
319. R. A. Okerholm, et al., *Fed. Proc., 35,* 327(1976).
320. R. E. Maxwell, et al., *Artery, 4,* 303(1978).
321. M. T. Kahonen and R. H. Ylikahri, *Atherosclerosis, 32,* 47(1979).
322. R. E. Maxwell, et al., *Atherosclerosis, 98,* 195(1983).
323. A. H. Kissebah, et al., *Atherosclerosis, 24,* 199(1976).
324. S. M. Grundy, et al., *J. Lipid Res., 13,* 531(1972).
325. L. Horlick, et al., *Circulation, 43,* 299(1971).
326. P. J. Nestel, et al., *J. Clin. Invest., 44,* 891(1965).
327. Anon., *F.D.C. Reports,* January 4, 1982.
328. V. Mannienen, et al., *Am. J. Cardiology, 63,* 42H(1989).
329. I. de Salcedo, et al., *Proc. R. Soc. Med. (Suppl. 2), 69,* 64(1976).
330. P. Varthakavi, et al., *J. Assoc. Physicians (India), 38,* 860 (1990).
331. D. A. Leaf, et al., *JAMA, 262,* 3154(1989).
332. M. L. Kashyap and K. Saku, *Adv. Exp. Med. Biol., 285,* 233(1991).
333. P. J. Lupien, et al., *Can. J. Cardiol., 7,* 27(1991).
334. E. Ros, et al., *Arch. Intern. Med., 151,* 301(1991).
335. P. Pauciullo, et al., *J. Intern. Med., 228,* 425(1990).
336. S. C. Kundu, et al., *J. Assoc. Physicians (India), 38,* 156(1990).
337. H. D. Peabody, Jr., in: *Hyperlipoproteinemia and Coronary Artery Disease—The Atherogenic Connection,* New York, Parke-Davis, 1982, p. 52.
338. P. Zimetbaum, et al., *J. Clin. Pharmacol., 31,* 25(1991).
339. M. J. Tikkanen, et al., *Am. J. Cardiol., 62,* 35J(1988).
340. B. Odman, et al., *Eur. J. Clin. Invest., 21,* 344(1991).
341. D. T. Nash, *Postgrad. Med., 73,* 75(1983).
342. G. C. Groggel, et al., *Kidney Int., 36,* 266(1989).
343. R. L. Rogers, et al., *Angiology, 40,* 260(1989).
344. P. L. Creger, et al., *Proc. R. Soc. Med. (Suppl. 2), 69,* 3(1976).
345. I. Sicar, et al., *J. Med. Chem., 26,* 1020(1983).
346. C. N. Corder, et al., *Eur. J. Clin. Pharmacol., 37,* 477(1989).
347. G. J. Magarian, et al., *Arch. Intern. Med., 151,* 1873(1991).
348. L. Pierce, et al., *JAMA, 264,* 71(1990).

349. G. E. Marais and K. K. Larson, *Ann. Intern. Med., 112,* 228(1990).
350. C. B. Blum, et al., *Ann. Intern. Med., 85,* 287(1976).
351. A. K. Khachadurian, *J. Athero. Res., 8,* 177(1968).
352. M. J. Hall, et al., *Atherosclerosis, 39,* 511(1981).
353. G. Mazzella, et al., *Scand. J. Gastroenterol., 25,* 1227(1990).
354. S. M. Kurtz, et al., *Proc. R. Soc. Med. (Suppl. 2), 69,* 15(1976).
355. J. K. Reddy and N. D. Lalwani, *CRC Crit. Rev. Toxicol., 12,* 1(1984).
356. N. D. Lalwani, et al., *Hum. Toxicol., 2,* 27(1983).
357. R. H. Gray and F. A. de la Iglesia, *Hepatology, 4,* 520(1984).
358. E. J. McGuire, et al., *Am. J. Pathol., 139,* 217(1991).
359. F. A. de la Iglesia, et al., *Atherosclerosis, 43,* 19(1982).
360. F. A. de la Iglesia, et al., *Micron, 12,* 97(1981).
361. S. Ahmad, *Chest, 98,* 1041(1990).
362. W. D. Haire, *Thromb. Res., 64,* 493(1991).
363. M. D. Stringer, et al., *Curr. Med. Res. Opin., 12,* 207(1990).
364. G. Avellone, et al., *Int. Angiol., 7,* 270(1988).
365. S. C. Forland, et al., *J. Clin. Pharmacol., 30,* 29(1990).
366. *Merck Index,* 10th ed, Rahway, NJ, Merck, 1983.
367. World Health Organization, *IARC Monograph on the Evaluation of Carcinogenic Risk of Chemicals (Some Pharmaceutical Drugs), Vol. 24,* New York, UN/WHO, 1980, pp. 39–58.
368. J. M. Thorp, *Lancet, 1,* 1323(1962).
369. M. N. Cayen, et al., *J. Pharmacol. Exp. Ther., 200,* 33(1977).
370. R. Gugler, et al., in *Drugs Affecting Lipid Metabolism,* R. Fumagalli, D. Kritchevsky and R. Paoletti, Eds., Amsterdam, Elsevier/North Holland Press, 1980, p. 183.
371. J. R. Baldwin, et al., *Biochem. Pharmacol., 29,* 3143(1980).
372. M. N. Cayen, *Drug Metab. Rev., 11,* 291(1980).
373. K. Kutz, et al., *Gastroenterology, 73,* 1229(1977).
374. C. E. Hignite, et al., *Life Sci., 28,* 2077(1981).
375. K. A. Sinclair and J. Caldwell, *Biochem. Soc. Trans., 9,* 215(1981).
376. J. Caldwell, et al., *Br. J. Pharmacol., 66,* 421(1979).
377. H. Keen, et al., *Lancet, 2,* 1241(1980).
378. K. G. Green, et al., *Int. J. Epidemiol., 18,* 355(1989).
379. R. I. Levy, in: *The Pharmacological Basis of Therapeutics,* 6th Ed., A. G. Gilman, et al., Eds., New York, MacMillan, 1980, pp. 834–847.
380. E. A. Nikkila, et al., *Metabolism, 26,* 179(1977).
381. K. G. Taylor, et al., *Lancet, 2,* 1106(1977).
382. H. Lithell, et al., *Eur. J. Clin. Pharmacol., 8,* 64(1978).
383. A. H. Kissebah, et al., *Eur. J. Clin. Invest., 4,* 163(1974).
384. P. J. Nestel, in: *Drugs Affecting Lipid Metabolism,* R. Fumagalli, et al., Eds., Amsterdam, Elsevier/North Holland, 1980, p. 159.
385. L. A. Carlson, et al., *Atherosclerosis, 26,* 603(1977).
386. R. Pichardo, et al., *Atherosclerosis, 26,* 573(1977).
387. R. G. Gould, et al., *J. Athero. Res., 6,* 555(1966).
388. C. W. White, *J. Pharmacol. Exp. Ther., 178,* 361(1971).
389. B. I. Cohen, et al., *Biochem. Biophys. Acta, 369,* 79(1974).
390. M. C. Bateson, et al., *Am. J. Dig. Dis., 23,* 623(1978).
391. Committee of Principal Investigators, WHO Cooperative Trial, *Lancet, 2,* 379(1980).
392. J. R. Patch, et al., *Am. J. Med., 63,* 1001(1977).
393. D. B Hunningshake, in *Atherosclerosis,* A. M. Gotto, L. C. Smith and B. Allen, Eds., New York, Springer-Verlag, 1980, p. 74.
394. J. M. Falko, et al., *Am. J. Med., 66,* 303(1979).
395. D. T. Witiak, et al., in *Medicinal Research Series* Vol. 7, "Clofibrate and Related Analogs, A Comprehensive Review," New York, Marcel Dekker, 1977.
396. R. Hess, et al., *Nature, 208,* 856(1965).
397. D. J. Svoboda and D. L. Azarnoff, *Fed. Proc., 30,* 841(1971).
398. M. Hanefeld, et al., *Atherosclerosis, 36,* 159(1980).
399. C. R. Sirtori, et al., *Atherosclerosis, 30,* 45(1978).
400. M. Kaste, et al., *Acta Neurol. Scand., 66,* 18(1982).
401. S. Kaukola, et al., *J. Cardiovasc. Pharmacol., 3,* 207(1981).
402. C. R. Mackerer, *Biochem. Pharmacol., 26,* 2225(1977).
403. M. E. Laker and P. A. Mayes, *Biochem. Pharmacol., 28,* 2813(1979).
404. T. P. Krishnakantha and C. K. R. Kurup, *Biochem. J., 130,* 167(1972).
405. J. Kim, et al., *Exp. Mol. Pathol., 25,* 263(1976).
406. J. K. Reddy and T. P. Krishnakantha, *Science, 190,* 787(1975).
407. P. B. Lazarow, *Science, 197,* 580(1977).
408. P. B. Lazarow and C. de Duve, *Proc. Natl. Acad. Sci. USA., 273,* 2043(1976).
409. J. M. Odonker and M. P. Rodgers, *Biochem. Pharmacol., 33,* 1337(1984).
410. R. W. Colman, et al., *Stroke, 5,* 299(1974).
411. J. B. Gilbert and J. F. Mustard, *J. Atheroscler., 3,* 623(1963).
412. J. F. Mustard, et al., in *Atherosclerosis,* Proc. 3rd Intl. Sym., G. Schlettler and A. Weizel, Eds., New York, Springer-Verlag, 1973, p. 253.
413. M. A. Packham and J. F. Mustard, *Circulation, 62,* V-26(1980).
414. C. Y. Lin and S. Smith, *Life Sci., 18,* 563(1976).
415. Huzoor-Akbar, et al., *Biochem. Pharmacol., 30,* 2013(1981).
416. D. T. Witiak, et al., *J. Med. Chem., 25,* 90(1982).
417. L. M. Hagerman, et al., in *Handbook of Cardiovascular and Anti-inflammatory Agents,* M. Verderame, Ed. CRC Press, Boca Raton, 1986, pp. 225–260.
418. J. J. Series, et al., *Atherosclerosis, 69,* 233(1988).
419. A. Dembinsk Kiec, et al., *Fortshr. Med., 107,* 450(1989).
420. R. Ceska, et al., *Vnitr. Lek., 36,* 363(1990).
421. L. Arsenio and A. Rossi, *Clin. Ter., 131,* 403(1989).
422. H. N. Ginsberg, *Am. J. Med., 83,* 66(1987).
423. D. B. Hunninghake and J. R. Peters, *Am. J. Med., 83,* 44(1987).
424. J. Rouffy, et al., *Atherosclerosis, 54,* 273(1985).
425. D. S. Goodman, et al., *J. Clin. Invest., 52,* 2646(1973).
426. R. Fellin, et al., *Atherosclerosis, 29,* 24(1978).
427. D. B. Hunningshake, et al., *Metabolism, 30,* 605(1981).
428. J. P. Kane, et al., *N. Engl. J. Med., 304,* 251(1981).
429. C. A. Dujove, et al., *Clin. Pharmacol. Ther., 16,* 291(1974).
430. P. J. Nestel, et al., *Aust. N. Z. J. Med., 3,* 630(1973).
431. A. N. Howard and D. E. Hyams, *Br. Med. J., 263,* 25(1971).
432. B. J. Hoogwerf, et al., *Metabolism, 34,* 978(1985).
433. A. G. Olsson, et al., *Atherosclerosis, 22,* 91(1975).
434. G. Rosenhamer and L. A. Carlson, *Atherosclerosis, 37,* 129(1980).
435. L. A. Carlson and G. Rosenhamer, *Acta Med. Scand., 223,* 405(1988).
436. R. J. C. Evans, et al., *Angiology, 24,* 22(1973).
437. S. Ritand, et al., *Scand. J. Gastroenterol., 10,* 791(1975).
438. A. G. Olsson, et al., *Lancet, 1,* 1311(1982).
439. S. Rossner and A. G. Olsson, *Atherosclerosis, 35,* 413(1980).
440. F. R. Heller, et al., *Metabolism, 30,* 67(1981).
441. A. Lehtonen and J. Viikari, *Artery, 10,* 353(1982).
442. J. Davignon, et al., *Adv. Exp. Med. Biol., 201,* 111(1986).
443. D. E. Moody and J. K. Reddy, *J. Cell Biol., 71,* 768(1976).
444. J. K. Reddy, in *Drugs Affecting Lipid Metabolism,* R. Fumagalli, et al., Eds., Amsterdam, Elsevier/North Holland, 1980, p. 301.
445. J. K. Reddy and S. A. Qureshi, *Br. J. Cancer, 38,* 537(1979).
446. D. J. Svoboda and D. L. Azarnoff, *Cancer Res., 39,* 3419(1979).
447. P. I. Eacho and D. R. Feller, In *Antilipidemic Drugs,* D. T. Witiak, et al., Eds., Elsevier, Amsterdam, Holland, p. 375(1991).

448. J. K. Reddy, et al., *Nature, 283,* 397(1980).
449. I. Issemann and S. Green, *Nature, 347,* 645(1990).
450. Editorial, *N. Z. Med. J., 92,* 315(1980).
451. Anon. *Drug Therapy,* Sept. 16(1979).
452. G. F. Blane, *Am. J. Med., 83,* 26(1987).
453. G. Migneco, et al., *Minerva. Med., 77,* 799(1986).
454. N. Hattori, et al., *Jpn. J. Med., 29,* 545(1990).
455. A. Baglin, et al., *Therapie, 42,* 247(1987).
456. W. V. Brown, *Am. J. Med., 83,* 85(1987).
457. E. Ujhazy, et al., *Pharmacol. Toxicol., 64,* 286(1989).
458. D. G. Gallo and A. L. Sheffner, *Proc. Soc. Exp. Biol. Med., 120,* 91(1965).
459. R. C. Thomas, et al., *Atherosclerosis, 29,* 9(1978).
460. J. Blanchard and J. G. Nairn, *J. Phys. Chem., 72,* 1204(1968).
461. L. M. Hagerman, et al., *Proc. Soc. Exp. Biol. Med., 139,* 248(1972).
462. W. H. Johns and T. R. Bates, *J. Pharm. Sci., 58,* 179(1969).
463. W. H. Johns and T. R. Bates, *J. Pharm. Sci., 59,* 788(1970).
464. N. E. Miller, et al., *Med. J. Aust., 1,* 1223(1973).
465. X. X. Zhu, et al., *J. Pharm. Sci., 81,* 65(1992).
466. S. D. Clas, *J. Pharm. Sci., 80,* 128(1990).
467. R. C. Heel, et al., *Drugs, 19,* 161(1980).
468. C. E. Day, *Low Density Lipoproteins,* New York, Plenum Press, 1976, p. 421.
469. R. I. Levy, *Ann. Intern. Med., 79,* 51(1973).
470. B. Angelin and K. Einarsson, *Atherosclerosis, 38,* 33(1981).
471. J. R. Farah, *Lancet, 1,* 59(1977).
472. G. Schlierf, et al., *Atherosclerosis, 41,* 133(1982).
473. P. T. Kuo, et al., *Circulation, 59,* 199(1979).
474. D. Nash, et al., *Int. J. Cardiol., 2,* 43(1982).
475. R. M. Krauss, et al., *Circulation, 70, Suppl. II,* II-269(1984).
476. R. I. Levy and T. Langer, *Drugs Affecting Lipid Metabolism: Pharmacological Control of Lipid Metabolism,* New York, Plenum Press, 1972, p. 155.
477. K. Einarsson, et al., *Eur. J. Clin. Invest., 4,* 405(1974).
478. I. Bjorkhem, et al., *Biochem. Biophys. Res. Commun., 61,* 934(1974).
479. D. G. Gallo, et al., *Proc. Soc. Exp. Biol. Med., 122,* 238(1966).
480. E. Bosisio, et al., *Life Sci., 34,* 2075(1984).
481. J. Shepherd, et al., *N. Engl. J. Med., 302,* 1219(1980).
482. C. D. Montafis, et al., *Atherosclerosis, 26,* 329(1977).
483. H. R. Slater et al., *J. Biol. Chem., 255,* 10210(1980).
484. P. T. Kovanen, et al., *Proc. Natl. Acad. Sci., 78,* 1194(1981).
485. J. T. Garbutt and T. J. Kenny, *J. Clin. Invest., 51,* 2781(1972).
486. J. L. Thistle and L. J. Schoenfield, *Gastroenterology, 61,* 488(1971).
487. H. E. Gallo-Torres, et al., *Am. J. Clin. Nutr., 32,* 1363(1979).
488. J. P. Kane, et al., *N. Engl. J. Med., 304,* 251(1981).
489. C. J. Packard, et al., *Artery, 7,* 281(1980).
490. P. T. Kuo, et al., *Chest, 79,* 286(1981).
491. D. R. Illingworth, et al., *Lancet, 1,* 296(1981).
492. J. M. Jackson and H. A. Lee, *Atherosclerosis, 51,* 189(1984).
493. C. A. Dujovne, et al., *Am. J. Cardiol., 53,* 1514(1984).
494. C. A. Dujovne, et al., *Ann. Int. Med., 100,* 477(1984).
495. R. Pasquali, et al., *Lancet, 1,* 1368(1981).
496. M. Naruszewicz, et al., *J. Lipid Res., 25,* 1206 (1984).
497. A. H. Seplowitz, et al., *Atherosclerosis, 39,* 35(1981).
498. T. A. Miettinen, *J. Clin. Invest., 64,* 1485(1979).
499. P. Schwandt, et al., *Atherosclerosis, 44,* 379(1982).
500. A. B. Bridges, et al., *Atherosclerosis, 89,* 263(1991).
501. P. Pignard, *Presse Med., 9,* 2973(1980).
502. R. C. Heel, et al., *Drugs, 15,* 409(1978).
503. J. A. Arnold, et al., *Fed. Proc., 29,* Abstract #622(1970).
504. S. Urien, et al., *Mol. Pharmacol., 26,* 322(1984).
505. T. E. Strandberg, et al., *Gen. Pharmac., 19,* 317(1988).
506. R. E. Tedeschi, et al., *Artery, 110,* 22(1982).
507. R. Fellin, et al., *Atherosclerosis, 59,* 47(1986).
508. C. Dachet, et al., *Atherosclerosis, 58,* 261(1985).
509. A. A. Polachek, et al., *Curr. Med. Res. Opin., 1,* 323(1973).
510. W. B. Parsons, *Am. Heart J., 96,* 213(1978).
511. M. J. Mellies, et al., *Metabolism, 29,* 956(1980).
512. T. A. Miettinen, et al., *Am. J. Cardiol., 57,* 49H(1986).
513. R. Naumova, et al., *Vutr. Boles. 25,* 73(1986).
514. J. P. Paez Moreno and G. Gonzalez, *Curr. Med. Res. Opin., 11,* 523(1984).
515. S. G. Baker, et al., *S. Afr. Med. J., 62,* 7(1982).
516. A. J. Yamamoto, et al., *Atherosclerosis, 48,* 157(1983).
517. M. J. Tikkanen, et al., *Eur. Heart J., 8* (Suppl. E), 97(1987).
518. P. N. Durrington and J. P. Miller, *Atherosclerosis, 55,* 187(1985).
519. P. Nestel and T. Billington, *Atherosclerosis, 38,* 203(1981).
520. N. Morisaki, et al., *J. Am. Geriatr. Soc., 38,* 15(1990).
521. P. Sanjurjo, et al., *Acta. Paediatr. Scand., 77,* 132(1988).
522. A. Berg, et al., *Atherosclerosis, 72,* 49(1988).
523. A. C. Nestruck, et al., *Metabolism, 36,* 743(1987).
524. T. A Miettinen, *Atherosclerosis, 15,* 163(1972).
525. T. S. Danowski, et al., *Clin. Pharmacol. Ther., 12,* 929(1971).
526. Z. Kalams, et al., *Curr. Ther. Res., 13,* 692(1971).
527. J. Davignon, et al., *Adv. Exp. Med. Biol., 201,* 111(1986).
528. P. T. Kuo, et al., *Am. J. Cardiol., 57,* 43H(1986).
529. J. L. Witztum, et al., *Circulation, 79,* 16(1989).
530. L. Cohen and J. Morgan, *J. Fam. Pract., 26,* 145(1988).
531. K. Yamamoto, et al., *Artery, 18,* 133(1991).
532. A. Berg, et al., *Atherosclerosis, 72,* 49(1988).
533. A. Beynen, *Atherosclerosis, 61,* 249(1986).
534. Y. A. Kesaniemi and S. M. Grundy, *J. Lipid Res., 25,* 780(1984).
535. D. Steinberg, *Am. J. Cardiol., 57,* 16H(1966).
536. M. F. Baudet, et al., *Atherosclerosis, 62,* 65(1986).
537. L. R. McLean and K. A. Hagaman, *Biochim. Biophys. Acta., 1029,* 161(1990).
538. L. R. McLean and K. A. Hagaman, *Biochim. Biophys. Acta, 959,* 201(1988).
539. R. F. Atmeh, et al., *Biochim. Biophys. Acta, 751,* 175(1983).
540. L. A. Simons, et al., *Atherosclerosis, 40,* 299(1981).
541. C. Glueck, *Am. J. Cardiol., 52,* 28B(1983).
542. R. McPherson, et al., *Arterioscler. Thromb., 11,* 476(1991).
543. E. Helve and M. J. Tikkanen, *Atherosclerosis, 72,* 189(1988).
544. C. R. Sirtori, et al., *Am. J. Cardiol., 62,* 73B(1988).
545. F. Canosa, et al., *Clin. Pharmacol. Ther., 17,* 230(1975).
546. A. Yamamoto, et al., *Am. J. Cardiol., 57,* 29H(1986).
547. Y. Matsuzawa, et al., *Am. J. Cardiol., 62,* 66B(1988).
548. Y. Homma, et al., *Atherosclerosis, 88,* 175(1991).
549. K. Tawara, et al., *Jpn. J. Pharmacol., 41,* 211(1986).
550. L. R. McLean, et al., *J. Biol. Chem., 267,* 12291(1992).
551. J. W. Barnhart, et al., *Am. J. Cardiol., 62,* 52B(1988).
552. F. M. La Vega and T. Mendoza-Figueroa, *Biochim. Biophys. Acta, 1081,* 293(1991).
553. R. B. Goldberg and A. Mendez, *Am. J. Cardiol., 62,* 57B(1988).
554. D. Steinberg, et al., *N. Engl. J. Med., 320,* 915(1989).
555. S. Parthasarathy, et al., *J. Clin. Invest., 77,* 641(1986).
556. J. Regnstrom et al., *Atherosclerosis, 82,* 43(1990).
557. A. V. Babiy, et al., *Biochem. Pharmacol., 43,* 995(1992).
558. R. L. Barnhart, et al., *J. Lipid Res., 30,* 1703(1989).
559. S. Parthasarathy, *J. Clin. Invest., 89,* 1618(1992).
560. J. W. Barnhart, et al., in *Antilipidemic Drugs, Medicinal, Chemical and Biochemical Aspects,* D. T. Witiak, et al., Eds., Amsterdam, Elsevier/North Holland, 1991, pp. 277–299.
561. S. J. T. Mao, et al., *J. Med. Chem., 34,* 298(1991).
562. I. Jialal and S. M. Grundy, *J. Clin. Invest., 87,* 597(1994).
563. B. Kalyanarman, et al., *J. Biol. Chem., 267,* 6789(1992).

564. L. R. McLean and K. A. Hagaman, *Biochemistry, 28,* 321(1984).
565. H. Minakami, et al., *Arzneimittelforschung, 39,* 1090(1984).
566. A. Yamamoto, et al., *Am. J. Cardiol., 62,* 31B(1988).
567. Y. Nagano, et al., *Arterosclerosis, 9,* 453(1989).
568. G. Ku, et al., *Atherosclerosis, 80,* 191(1990).
569. S. Hara, et al., *Arterioscler. Thromb., 12,* 593(1992).
570. G. Ku, *FASEB J., 4,* 1645(1990).
571. A. L. Akeson, et al., *Atherosclerosis, 86,* 261(1991).
572. J. Henahan, *JAMA, 246,* 2309(1981).
573. Y. Stein, et al., *Atherosclerosis, 75,* 145(1989).
574. T. Kita, et al., *Proc. Natl. Acad. Sci. USA, 84,* 5928(1984).
575. A. Daugherty, et al., *Br. J. Pharmacol., 103,* 1013(1991).
576. B. Finckh, et al., *Eur. J. Clin. Pharmacol., 40,* Suppl 4 S77(1991).
577. K. O'Brien, et al., *Arterioscler. Thromb., 11,* 751(1991).
578. Y. Mori, et al., *Thromb. Haemostasis, 61,* 140(1989).
579. G. M. Chisholm III and D. W. Morel, *Am. J. Cardiol., 62,* 20B(1988).
580. A. L. Drash, et al., *Am. J. Cardiol., 62,* 27B(1988).
581. H. L. Taylor, et al., *Clin. Pharm. Ther., 23,* 131(1978).
582. D. McCaughan, *Artery, 10,* 56(1982).
583. J. A. Molello, et al., *Toxicol. Appl. Pharmacol., 24,* 594(1973).
584. F. N. Marshall and J. E. Lewis, *Toxicol. Appl. Pharmacol., 24,* 594(1979).
585. J. E. LeBeau, *Presse Med., 9,* 3001(1980).
586. F. N. Marshall, *Artery, 10,* 7(1982).
587. M. Barber, et al., *Circulation, 64,* Suppl. IV, 123(1981).
588. K. F. Browne, et al., *Am. Heart J., 107,* 680(1984).
589. C. Huber, *Ann. der Chemie Pharmacie, 65,* 271(1867).
590. R. A. Parker and M. Weiner, in *Antilipidemic Drugs: Medicinal, Chemical and Biochemical Aspects,* D. T. Witiak, et al., Eds., Pharmacochemistry Library, vol. 17, Amsterdam, Elsevier, 1991, Chapt. 9.
591. R. Altschul, et al., *Arch. Biochem., 54,* 558(1955).
592. D. Kritchevsky, in *Metabolic Effects of Nicotinic Acid and Its Derivatives,* K. F. Gey and L. A. Carlson, Eds., Hans Huber, Bern, 1971, p. 541.
593. S. M. Grundy, et al., *J. Lipid Res., 22,* 24(1981).
594. J. Shepherd, et al., *J. Clin. Invest., 63,* 858(1979).
595. L. A. Carlson and G. Wahlberg, *Atherosclerosis, 31,* 77(1978).
596. L. A. Carlson, et al., *Atherosclerosis, 26,* 603(1977).
597. C. J. Packard, et al., *Biochim. Biophys. Acta, 618,* 53(1980).
598. J. G. Yovos, et al., *J. Clin. Endocrinol. Metab., 54,* 1210(1982).
599. G. L. Cushing, et al., *Arteriosclerosis, 9,* 593(1989).
600. G. Uttermann, et al., *J. Clin. Invest., 80,* 458(1987).
601. L. A. Carlson and L. Oro, *Acta Med. Scand., 172,* 641(1962).
602. L. A. Carlson, in *Metabolic Effects of Nicotinic Acid and Its Derivatives,* L. A. Carlson and K. F. Gey, Eds., Bern, Hans Huber, 1971, p. 163.
603. W. Holtz, *Adv. Lipid Res., 20,* 195(1983).
604. D. Kritchevsky, et al., in *Metabolic Effects of Nicotinic Acid and Its Derivatives,* K. F. Gey and L. A. Carlson, Eds., Bern, Hans Huber, 1961, p. 585.
605. G. A. Goldsmith, et al., *Arch. Intern. Med., 105,* 512(1960).
606. O. N. Miller, et al., *Am. J. Clin. Nutr., 8,* 480(1960).
607. O. N. Miller, et al., *Am. J. Clin. Nutr., 10,* 285(1962).
608. T. A. Miettinen, *Metabolic Effects of Nicotinic Acid and Its Derivatives,* K. F. Gey and L. A. Carlson, Eds., Bern, Hans Huber, 1971, p. 677.
609. S. M. Grundy, et al., *J. Lipid Res., 22,* 24(1981).
610. B. J. Kudehodkar, et al., *Clin. Pharmacol. Ther., 24,* 354(1978).
611. H. S. Sodhi, et al., *Atherosclerosis, 17,* 1(1973).
612. I. B. Marsh, in *Plasma Lipoproteins,* R. M. S. Smellie, Ed., New York, Academic Press, 1971, p. 89.
613. K. G. Berge and P. L. Canner, *Eur. J. Clin. Pharmacol., 40(Suppl 1),* S49(1991).
614. B. Eklund, *Prostaglandins, 17,* 821(1979).
615. M. Weiner and J. van Eys, in *Nicotinic Acid—Nutrient, Cofactor, Drug,* New York, Marcel Dekker, 1983.
616. N. Svedmyr, et al., *Acta Pharmacol. Toxicol., 41,* 397(1977).
617. R. I. Levy and B. M. Rifkind, in *Cardiovascular Drugs,* Sydney, ADIS Press, 1977, p. 1.
618. P. L. Sharma, et al., *J. Clin. Pharmacol., 7,* 299(1973).
619. H. A. Salmi and H. Frey, *Curr. Ther. Res., 16,* 669(1974).
620. P. Avogaro, et al., *Pharm. Res. Comm., 10,* 127(1978).
621. L. A. Carlson and L. Oro, *Acta Med. Scand., 186,* 337(1969).
622. P. Schraepler, et al., *Med. Welt., 28,* 1900(1977).
623. L. Puglisi, et al., *Pharmacol. Res. Commun., 8,* 379(1976).
624. C. Dalton, et al., in *Metabolic Effects of Nicotinic Acid and Its Derivatives,* K. F. Gey and L. A. Carlson, Eds., Bern, Hans Huber, 1971, p. 331.
625. L. A. Carlson, *Adv. Exp. Med. Biol., 109,* 225(1978).
626. P. O. Lovisolo, et al., *Pharmacol. Res. Comm., 13,* 163(1981).
627. L. M. Fuccella, et al., *Clin. Pharmacol. Ther., 28,* 790(1980).
628. D. Sommariva, et al., *Curr. Therap. Res., 37,* 363(1985).
629. C. R. Sirtori, et al., *Atherosclerosis, 38,* 267(1981).
630. A. G. Olsson, et al., *Atherosclerosis, 19,* 61(1974).
631. M. E. Benaim and H. A. Dewar, *J. Int. Med. Res., 3,* 423(1975).
632. S. Ikeda, et al., *Acta Med. Okayama, 35,* 143(1981).
633. G. Quack, et al., in *Drugs Affecting Lipid Metabolism,* R. Paoletti, et al., Eds., Berlin, Heidelberg Springer-Verlag, 1987, p. 376.
634. K. Seki, et al., *Chem. Pharm. Bull., 31,* 4116(1983).
635. D. P. Mertz, et. al., *Med. Welt., 33,* 405(1982).
636. E. Herrara, et al., *Metabolism, Biochim. Biophys. Acta, 963,* 42(1988).
637. C. Figols, *Drugs of the Future, 4,* 681(1979).
638. G. Buzzelli, et al., *Clin. Ther., 89,* 251(1979).
639. C. J. Schwartz, et al., *Clin. Cardiol., 14,* 11(1991).
640. S. Ylä-Hertuala, et al., *J. Clin. Invest., 84,* 1086(1989).
641. P. Avagaro, et al., *Arteriosclerosis, 8,* 79(1988).
642. N. I. Kinsky, *Proc. Soc. Exp. Biol. Med., 200,* 248(1992).
643. H. Esterbauer, et al., *Chem. Res. Tox., 3,* 77(1990).
644. M. Avirim, et al., *J. Clin. Chem. Clin. Biochem., 27,* 3(1989).
645. R. Stocker, et al., *Proc. Natl. Acad. Sci. USA, 88,* 1646(1991).
646. N. Dousset, et al., *Biochim. Biophys. Acta., 1045,* 219(1990).
647. H. A. I. Newman, et al., in *Antilipidemic Drugs, Medicinal, Chemical and Biochemical Aspects,* D. T. Witiak, et al., Eds., Amsterdam, Elsevier Science Publishers, 1991, pp. 345–374.
648. U. P. Steinbrecher, et al., *Free Radical Biol. Med., 9,* 155(1990).
649. H. Esterbauer, et al., *Free Radic. Biol. Med., 13,* 341(1992).
650. D. Steinberg, et al., *N. Engl. J. Med., 320,* 915(1989).
651. J. L. Witztum, *Adv. Exp. Med. Biol., 285,* 353(1991).
652. H. M. Sharma, et al., *Pharmacol. Biochem. Behav., 43,* 1175(1992).
653. K. Ondrias, et al., *Free Radic. Res. Commun., 16,* 227(1992).
654. C. V. de Whalley, et al., *Biochem. Pharmacol., 39,* 1743(1990).
655. A. N. Hanna, et al., *Biochem. Pharmacol.,* in press (1992).
656. A. Sevanian, et al., *Am. J. Clin. Nutr., 54,* 1129S(1991).
657. E. M. Berry, et al., *Am. J. Clin. Nutr., 53,* 899(1991).
658. S. Parthasarathy, et al., *Biochim. Biophys. Acta, 1044,* 275(1990).
659. T. Ohta, et al., *FEBS Lett., 257,* 435(1989).
660. M. I. Mackness, et al., *FEBS Lett., 286,* 152(1991).
661. I. Jialal and S. M. Grundy, *J. Clin. Invest., 82,* 597(1991).
662. B. Kalyanaraman, et al., *J. Biol. Chem., 267,* 6789(1992).
663. D. T. Witiak, et al., *Trends in Medicinal Chemistry '90,* S.

Sarel, R. Mechoulam and I. Agranat, Oxford, Blackwell Scientific Publications, 1992, pp. 243–256.
664. A. A. Spector, et al., Eds., *Prog. Lipid Res., 18,* 31(1979).
665. T. Y. Chang and G. M. Doolittle, *Enzymes, 16,* 523(1983).
666. J. G. Heidler, in *Pharmacological Control of Hyperlipidemia,* Barcelona, Spain, J. R. Prous Science Publishers, 1986, pp. 423–428.
667. J. T. Billheimer, *Methods in Enzymology, 3,* 286(1985).
668. F. G. Kathawala and J. G. Heider, in *Antilipidemic Drugs, Medicinal, Chemical and Biochemical Aspects,* D. T. Witiak, et al., Eds., Amsterdam, Elsevier/North Holland, 1991, pp. 241.
669. T. A. Bocan, et al., *Arterioscler. Thromb., 11,* 1830(1991).
670. P. J. Gillies, et al., *Atherosclerosis, 83,* 177(1990).
671. V. G. Devries, et al., *J. Med. Chem., 32,* 2318(1989).
672. B. D. Roth, et al., *J. Med. Chem., 35,* 1609(1992).
673. F. Heffron, et al., *Biochem. Pharmacol., 39,* 575(1990).
674. B. R. Krause, et al., *Drugs Affecting Lipid Metabolism, 10,* 3(1989).
675. F. J. Field, et al., *Lipids, 26,* 1(1991).
676. N. T. P. Kam, et al., *J. Lipid Res., 30,* 371(1989).
677. L. Dory, *J. Lipid Res., 30,* 809(1989).
678. L. Gallo, et al., *J. Lipid Res., 28,* 381(1987).
679. F. P. Bell, et al., *Atherosclerosis, 92,* 115(1992).
680. R. J. Williams, et al., *Biochim. Biophys. Acta, 1003,* 213(1989).
681. O. Stein and Y. Stein, *Arterio. Thromb., 7,* 578(1987).
682. E. E. Largis, et al., *J. Lip. Res., 30,* 681(1989).
683. F. Ishida, et al., *Chem. Pharm. Bull., 37,* 1635(1989).
684. N. T. P. Kam, et al., *Biochem. J., 272,* 427(1990).
685. F. Ishida, et al., *Biochim. Biophys. Acta, 148,* 920(1987).
686. I. Tabas, et al., *J. Biol. Chem., 265,* 8042(1990).
687. H. Tomoda, et al., *J. Antibiotics, 44,* 136(1991).
688. D. Stahlberg, et al., *J. Lipid Res., 30,* 953(1989).
689. D. H. Needleman, et al., *Biochim. Biophys. Acta, 148,* 920(1987).
690. J. P. Hieble and R. R. Ruffolo, Jr., in *Antilipidemic Drugs, Medicinal, Chemical and Biochemical Aspects,* Eds., D. T. Witiak, H. Newman and D. Feller, Amsterdam, Elsevier/North Holland, 1991, pp. 301–344.
691. R. T. Jung, et al., *Nature, 279,* 322(1979).
692. J. R. Arch, et al., *Nature, 309,* 163(1984).
693. P. L. Hooper, et al., *N. Engl. J. Med., 35,* 1455(1981).
694. M. H. Weinberger, *Am. J. Med., 80 (Suppl. 2A),* 64(1986).
695. J. J. Rohlfing and J. D. Brunzell, *West. J. Med., 145,* 210(1986).
696. L. G. Howes and H. Krum, *J. Auton. Pharmacol., 9,* 293(1989).
697. R. Fogari, et al., *J. Cardiovasc. Pharmacol., 14 (Suppl. 7),* S28(1989).
698. J. R. Kaplan, et al., *Eur. Heart J., 8,* 928(1987).
699. N. M. Kaplan, *J. Cardiovasc. Pharmacol., 6 (Suppl. 6),* S841(1984).
700. L. X. Cubeddu, *Am. Heart. J., 116,* 133(1988).
701. H. R. Superko and P. D. Wood, *Am. J. Med., 86 (Suppl. 1B),* 57S(1989).
702. M. C. Kowala and R. J. Nicolosi, *J. Cardiovasc. Pharmacol., 13 (Suppl. 3),* S45(1989).
703. M. K. O'Malley, et al., *Br. J. Surg., 76,* 629(1989).
704. M. Y. Flugelman, et al., *Circulation, 85,* 1110(1992).
705. J. M. Wilson, et al., *J. Biol. Chem., 267,* 963(1992).
706. J. R. Chowdhury, et al., *Science, 254,* 1802(1991).
707. D. A. Dichek, et al., *Trans. Assoc. Am. Physicians, 103,* 73(1990).
708. A. Miyanohara, et al., *Proc. Natl. Acad. Sci., 85,* 6538(1988).
709. S. E. Raper, et al., *Surgery, 119,* 457(1992).
710. S. Nityanand, et al., *J. Assoc. Physicians (India), 37,* 323(1989).
711. A. Baxter, et al., *J. Biol. Chem., 267,* 11705(1992).

SUGGESTED READING

D. T. Witiak, et al., in *Clofibrate and Related Analogs: A Comprehensive Review,* New York, Marcel Dekker, Inc., 1977.
N. B. Myant, in *Cholesterol Metabolism, LDL and the LDL Receptor,* San Diego, Academic Press, 1990, pp. 190–191.
G. Assmann, in *Lipid Metabolism and Atherosclerosis,* Stuttgart, Schattauer, 1982.
N. M. Kaplan and J. Stamler, in *Prevention of Coronary Heart Disease. Practical Management of the Risk Factors,* Philadelphia, W. B. Saunders, 1983.
L. M. Hagerman, et al., in *Handbook of Cardiovascular and Anti-inflammatory Agents,* M. Verderame, Ed., Boca Raton, CRC Press, 1986, pp. 225–260.
M. S. Brown and J. L. Goldstein, *Sci. Am., 58,* Nov.(1984).
D. T. Witiak, et al., *Antilipidemic Drugs: Medicinal, Chemical, and Biochemical Aspects,* Amsterdam, Elsevier/North Holland, 1991.

Chapter 25

NONSTEROIDAL ANTI-INFLAMMATORY DRUGS

Ronald F. Borne

The classification of drugs covered in this chapter as nonsteroidal anti-inflammatory agents (NSAIAs) is somewhat misleading because many of these entities possess antipyretic and analgetic properties in addition to anti-inflammatory properties useful in the treatment of certain rheumatic disorders. There are, however, other agents which possess analgetic-antipyretic properties, but are essentially devoid of anti-inflammatory activity. Additionally, agents that possess uricosuric properties useful in the treatment of gout will also be discussed in this chapter. The prototype agent of this class is acetylsalicylic acid, or aspirin, which has therapeutically useful analgetic, antipyretic, and anti-inflammatory actions; other agents to be covered may possess only one or two of these properties. Steroidal anti-inflammatory agents are covered in Chapter 23.

The medicinal agents considered in this chapter are commonly used in both prescription and nonprescription forms. Other than caffeine or ethyl alcohol, aspirin may be the most widely used drug in the world.[1] Prescription sales of NSAIAs for 1993 have been estimated at $2.3 billion with an annual growth rate of 11%.[2] Over 75 million prescriptions are written annually for NSAIAs, including aspirin.[3] Rheumatic diseases, which have been classified by the American Rheumatism Association (Table 25–1),[4] are inflammatory disorders affecting more individuals than any other chronic illness. Approximately 15% of the U.S. population, or 37 million individuals, suffer from a rheumatoid disease, with females being affected about twice as much as males. Almost 7 million U.S. citizens suffer from arthritis in its most debilitating forms.[5] Rheumatoid arthritis is thought to affect 2.5 million U.S. citizens, 1.8 million of whom are females, whereas juvenile arthritis affects 71,000 children under 16 years of age, 61,000 of whom are females. In addition, nonrheumatoid osteoporosis affects 24 million females (50% of all women over age 45 and 90% of all women over 75)[5] and 16 million males. Because more than 80% of the U.S. population over the age of 55 have joint abnormalities detectable radiographically,[6] the use of NSAIAs will increase as Americans live longer. It has been estimated that $35 billion is spent on medical costs and economic losses associated with arthritis.[5] It is not surprising, therefore, that the development of new NSAIAs continues at a rapid pace as evidenced by the introduction of seven new agents on the U.S. market from 1988 to 1992. For general properties of NSAIAs (dose, peak blood level, pK_a, and plasma binding) see Table 25–3.

The diseases mentioned are considered to be host-defense mechanisms. Inflammation is a normal and essential response to any noxious stimulus which threatens the host and may vary from a localized response to a more generalized one.[7] The inflammation sequence can be summarized as follows: (1) initial injury causing release of inflammatory mediators (e.g., histamine, serotonin, leukokinins, SRS-A, lysosomal enzymes, lymphokinins, and prostaglandins); (2) vasodilation; (3) increased vascular permeability and exudation; (4) leukocyte migration, chemotaxis and phagocytosis; and (5) proliferation of connective tissue cells. The most common sources of chemical mediators include neutrophils, basophils, mast cells, platelets, macrophages, and lymphocytes.[7] The etiology of inflammatory and arthritic diseases has received a great deal of recent attention but remains, for the most part, unresolved, hindering the development of new cures. Agents to relieve the symptoms of the disease are now available, but are not curative.

The currently accepted pathogenesis of these disorders can be summarized as follows: an unknown antigen gains access to the patient's tissues and combines with an antibody in the joint activating the complement sequence. An antigen-complement-antibody immune complex then precipitates in the synovium and joint fluid, generating the release of chemical mediators which subsequently cause the migration of numerous polymorphonuclear leukocytes phagocytizing the immune complexes. Lysosomal membranes become unstable and discharge hydrolytic enzymes (proteases, collagenases, etc.) from the leukocytes and synovial cells. Tissue damage ensues with continuing inflammation, tissue destruction, collagen depolymerization, and loss of physical properties of the connective tissue and joints. Anti-inflammatory agents may thus act by interfering with any one of several mechanisms including immunologic mechanisms such as antibody production or antigen-antibody complexation, activation of complement, cellular activities such as phagocytosis, interference with the formation and release of the chemical mediators of inflammation, or stabilization of lysosomal membranes.

The role of complement in inflammation is of considerable current interest.[8–10] The complement system is one component of the host-defense system, which aids in the elimination of various microorganisms and antigens from blood and tissues. Although complement normally plays a functional role in the development of disease states, by promoting inflammation locally, excessive com-

Table 25–1. Classification of Rheumatic Diseases

I. Polyarthritis of unknown etiology
  A. Rheumatoid arthritis
  B. Juvenile rheumatoid arthritis
  C. Ankylosing spondylitis
  D. Reiter's syndrome
  E. Others
II. Connective tissue disorders
  A. Systemic lupus erythematosus
  B. Progressive scleroderma
  C. Polymyositis and dermatomyositis
  D. Mixed connective disease
  E. Necrotizing arthritis and other forms of vasculitis
  F. Others
III. Acute rheumatic fever
IV. Degenerative joint disease (osteoarthritis)
V. Nonarticular rheumatism
VI. Diseases with which arthritis is frequently associated
  A. Sjögren's syndrome
  B. Others
VII. Associated with known infectious agents
  A. Bacterial
  B. Rickettsial
  C. Viral
  D. Fungal
  E. Parasitic
VIII. Traumatic neurogenic disorders
IX. Associated with known biochemical or endocrine abnormalities
  A. Gout
  B. Others
X. Neoplasms
XI. Allergy and drug reactions
XII. Inherited and congenital disorders
XIII. Miscellaneous disorders

Modified from *Primer on the Rheumatic Diseases,* 9th edition, copyright 1988. Used by permission of the Arthritis Foundation.

plement activation is detrimental. Individuals with a deficiency of individual complement proteins, though, either acquired or hereditary, are more susceptible to infection caused by pyrogenic bacteria and diseases resulting from the generation of autoantibodies and immune complexes. Complement proteins are numbered C1–C9 and their cleavage products indicated by suffixes a, b, and so forth. The complement system consists of two activating pathways (an antibody-mediated classical pathway and a nonimmunologically activated alternate pathway), a single termination pathway, regulatory proteins, and complement receptors, and it involves approximately 30 membrane and plasma proteins.[9] A major function of complement is to mark antigens and microorganisms with C3 fragments that direct them to cells containing C3 receptors, such as phagocytic cells.[9] Complement has been implicated in many diseases including allergic, hematologic, dermatologic, infectious, renal, hepatic inflammatory (rheumatoid arthritis [RA] and systemic lupus erythematosus [SLE]), pulmonary, and others, such as multiple sclerosis and myasthenia gravis. Complement activation can induce the synthesis or release of inflammatory mediators such as interleukin-1 (a potent pro-inflammatory cytokine) and certain prostaglandins (PGE_2). Leukotrienes (LTB_4) and thromboxanes are also released. Complement also aids the immigration of phagocytes associated with inflammation. Thus, inhibition of the complement system by controlling its activation or inhibiting those active fragments that are produced should be beneficial in reducing or eliminating tissue damage associated with inflammatory diseases.

Connective tissue diseases include the following states: RA, ankylosing spondylitis, SLE, polyarteritis nodosa, gout, rheumatic fever, and osteoarthritis, most common forms of which are RA, osteoporosis, and gout. Rheumatoid arthritis is a subacute or chronic, nonsuppurative inflammatory disease of unknown cause affecting primarily peripheral synovial joints. The onset is usually insidious with immunologic reactions playing a major role. The pathogenesis of RA has been summarized as follows:[11] (1) an unknown initiation factor causes the production of antigenic IgG, which stimulates the synthesis of the rheumatoid factors IgM and IgG forming immune complexes; (2) IgG aggregates activating the complement system leading to the generation of chemotactic factors, which attract polymorphonuclear leukocytes into the articular cavity; and (3) the polymorphonuclear leukocytes ingest immune complexes to become RA cells which discharge hydrolases from lysosomal granules which, in their turn, degrade extracellular tissue components, polysaccharides, and collagens in cartilage, thus provoking an inflammatory response in rheumatoid joints. Clinical symptoms characteristically include symmetric swelling of joints accompanied by tenderness, erythema, stiffness, and pain. The joints primarily involved are those of the extremities and the patient often suffers a low grade fever accompanied by malaise, anorexia, and fatigue. Serum protein irregularities are common. It is often difficult to distinguish RA from other connective tissue diseases. Treatment usually involves a program of rest, exercise, and a balanced diet with the initial use of salicylates. If conservative management fails, drug therapy, including NSAIAs, corticosteroids, or the antirheumatic gold salts are employed.

Osteoarthritis, also known as degenerative joint disease, is the most common form of arthritis and is characterized by degeneration of cartilage and hypertrophy of bone at the articular margin. Secondary inflammation of synovial tissue is common. The most common symptoms involve joint pain associated with movement and, sometimes, bone enlargement. Weight-bearing joints and those of the hands and fingers are usually involved. Abnormalities in laboratory tests are generally not observed. Treatment usually involves exercise and salicylates. Aspirin is considered first-choice therapy, with NSAIAs being employed in patients who do not tolerate salicylates.

Gout is a metabolic disease characterized by recurrent episodes of acute arthritis, usually of the monoarticular type, and is associated with abnormal levels of uric acid in the body, particularly the presence of monosodium urate crystals in synovial fluid. Primary gout is a hereditary disease in which hyperuricemia is caused by an error in uric acid metabolism—either overproduction or an inability to excrete uric acid. The term secondary gout refers to those cases in which hyperuricemia is due to an acquired disease or disorder such as chronic renal disease, lead poisoning, or myeloproliferative disorders. Gout generally occurs in mid-life and affects males significantly more than females (9:1). Treatment usually involves the use of uricosuric agents, colchicine, NSAIAs, or corticosteroids.

The search for new and effective treatment requires the availability of adequate screening tests. Although no model adequately reflects the events which occur in human arthritic conditions, several in vivo and in vitro assays are used. The most common in vivo animal assays measure the ability of anti-inflammatory agents to inhibit edema induced in the rat paw by carrageenan (a mucopolysaccharide derived from a sea moss of the Chondrus species), to inhibit adjuvant arthritis in rats induced by Mycobacterium butyricum or M. tuberculosis, to inhibit granuloma formation usually induced by the implantation of a cotton pellet beneath the abdominal skin of rats, or to inhibit erythema of guinea pig skin as a result of exposure to ultraviolet (UV) radiation. In vitro techniques include the ability of NSAIAs to stabilize erythrocyte membranes or, more commonly, to inhibit the biosynthesis of prostaglandins, particularly in cultured human synoviocytes and chondrocytes, and monocyte culture fluid stimulated bovine synoviocytes and chondrocytes.

ROLE OF CHEMICAL MEDIATORS IN INFLAMMATION

As indicated previously, several chemical mediators have been postulated to play important roles in the inflammatory process. Prior to 1971, the proposal by Shen[12,13] that the NSAIAs exert their effects by interacting with a hypothetical ant-inflammatory receptor was widely accepted. The topography of the proposed receptor was based upon known structure-activity relationships primarily within the series of indole acetic acid derivatives of which indomethacin was the prototype. Most NSAIAs, whether salicylates, pyrazolidinediones, arylalkanoic acids, oxicams, or anthranilic acid derivatives, possess the common structural features of an acidic center, an aromatic or heteroaromatic ring, and an additional center of lipophilicity in the form of either an alkyl chain or an additional aromatic ring. The proposed receptor to which indomethacin was postulated to bind consisted of a cationic site to which the carboxylate anion would bind, a flat area to which the indole ring would bind through

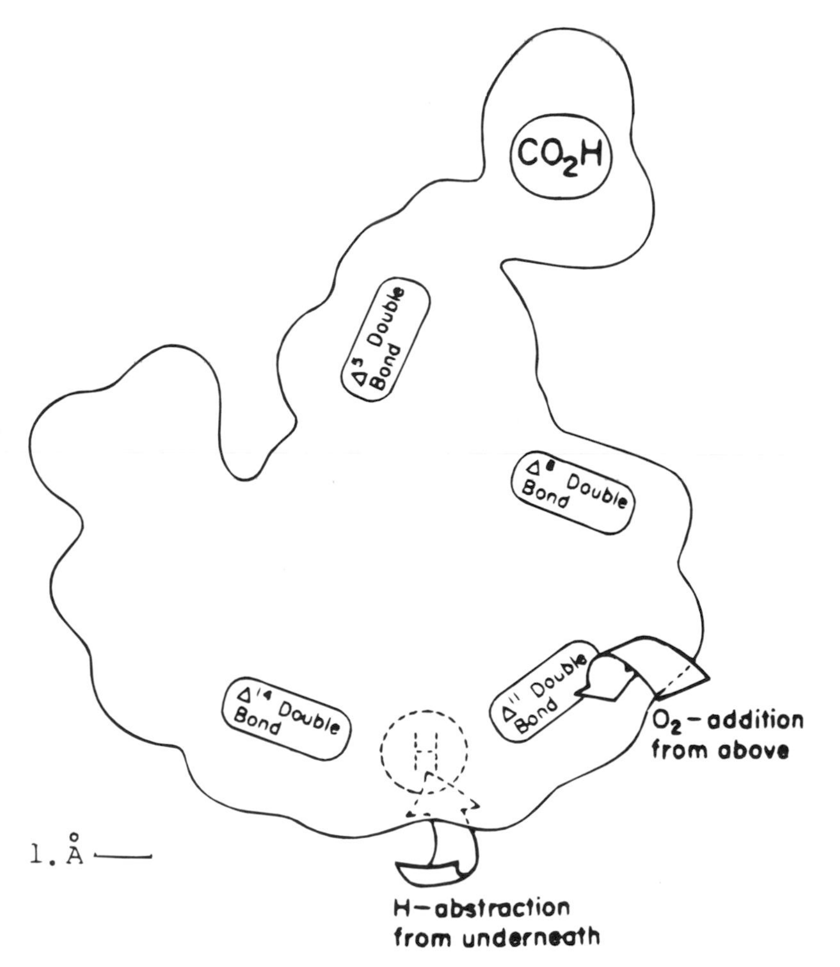

Fig. 25–1. Shen's proposed model of the fatty acid substrate binding site of prostaglandin synthetase. (Modified, with permission, from R. F. Borne, Handbook of Cardiovascular and Anti-Inflammatory Agents, M. Verame, Ed., Boca Raton, FL, CRC Press, 1986).

Van der Waals' forces, and an out-of-the-plane trough to which the benzene ring of the ρ-chlorobenzoyl group binds through hydrophobic or charge-transfer interactions. Additional binding sites for the methoxy and carbonyl groups were also suggested. In 1971, Vane[14] published a classic paper in which he reported that indomethacin, aspirin, and salicylate, in this descending order of potency, inhibited the biosynthesis of prostaglandins from arachidonic acid using cell-free preparations of guinea pig lung and further suggested that the clinical actions of these agents were caused by this inhibition. This theory has become the most widely accepted explanation of the mechanism of action of NSAIAs. Shen subsequently[15] modified his hypothesis and proposed that his earlier anti-inflammatory receptor model actually described the active site of the key enzyme in prostaglandin biosynthesis, that is, prostaglandin cyclooxygenase (Fig. 25–1).

PROSTAGLANDINS, THROMBOXANES, PROSTACYCLIN, AND LEUKOTRIENES

Prostaglandins are naturally occurring 20-carbon cyclopentano-fatty acid derivatives produced in mammalian tissue from polyunsaturated fatty acids. They belong to the class of eicosanoids, a member of the group of autocoids derived from membrane phospholipids. The eicosanoids are derived from unsaturated fatty acids and include the following groups of compounds: prostaglandins, thromboxanes, prostacyclin, and leukotrienes. They have been found in essentially every area of the body. In 1931, Kurzrok and Lieb[16] reported that human semen possessed potent contractile and relaxant effects on uterine smooth muscle. Shortly thereafter, Goldblatt[17] in England and von Euler[18] in Sweden independently reported vasodepressor and smooth muscle contracting properties in seminal fluid; von Euler identified the active constituent as a lipophilic acidic substance he termed prostaglandin. These observations attracted little attention during the years of World War II, but shortly thereafter, primarily through the efforts of Samuelsson and Bergstrom,[19] it was realized that von Euler's prostaglandin was actually a mixture of a number of structurally-related fatty acids. The structure of the first prostaglandins was reported in 1962, which initiated several studies relating to the chemical and biologic properties of these potent substances.

The general structure of the prostaglandins (PGs) is shown in Figure 25–2. All naturally occurring PGs possess a 15α-hydroxy group and a *trans* double bond at C-13. Unless a double bond occurs at the C-8, C-12 positions, the two side chains (the carboxyl-bearing chain termed

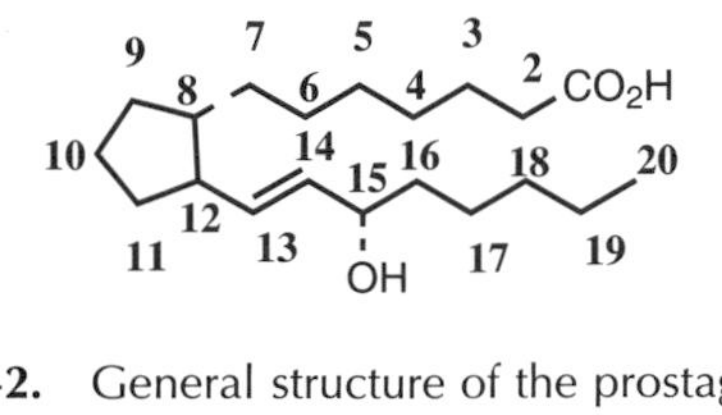

Fig. 25–2. General structure of the prostaglandins.

the α-chain and the hydroxyl-bearing chain termed the β-chain) are of the *trans* stereochemistry depicted. The PGs are classified by the capital letters A, B. C, D, E, F, G, H, and I depending on the nature and stereochemistry of oxygen substituents at the 9- and 11- positions. For example, members of the PGE series possess a keto function at C-9 and an α-hydroxyl group at C-11, whereas members of the PGF series have α-hydroxyl groups at both of these positions. Members of the PGG and PGH series are cycloendoperoxide intermediates in the biosynthesis of prostaglandins as depicted in Figure 25–3. The number of double bonds in the side chains connected to the cyclopentane ring is designated by subscripts 1, 2, or 3, indicative of the nature of the fatty acid precursor. The subscript 2 indicates an additional *cis* double bond at the C-5, C-6 positions whereas the subscript 3 indicates a third double bond of *cis* stereochemistry at the C-17, C-18 positions.

Prostaglandins are derived biosynthetically from unsaturated fatty acid precursors. The number of double bonds contained in the naturally occurring PGs reflects the nature of the biosynthetic precursors. Those containing one double bond are derived from 8, 11, 14-eicosatrienoic acid, those with two double bonds from arachidonic acid (5,8,11,14-eicosatetraenoic acid), and those with three double bonds from 5,8,11,14,17-eicosapentenoic acid. The most common of these fatty acids in humans is arachidonic acid and hence PGs of the 2 series play an important biologic role. Arachidonic acid is derived from dietary linoleic acid or is ingested from the diet[20] and esterified to phospholipids (primarily phosphatidylethanolamine or phosphatidylcholine) in cell membranes. Various initiating factors interact with membrane receptors coupled to G proteins (guanine nucleotide-binding regulatory proteins) activating phospholipase A_2 which, in turn, hydrolyzes membrane phospholipids resulting in the release of arachidonic acid. Other phospholipases (e.g., phospholipase C) are also involved. Phospholipase C differs from phospholipase A_2 by inducing the formation of 1,2-diglycerides from phospholipids with the subsequent release of arachidonic acid by the actions of mono- and diglyceride lipases on the diglyceride.[18] Interleukin-1 (IL 1), a polypeptide, produced by leukocytes, that mediates inflammation, increases phospholipase activity and thus activates PG biosynthesis. The steroidal anti-inflammatory agents (corticosteroids) appear to act, in part, by inhibiting these phospholipases. The liberated arachidonic acid may then be acted on by two major enzyme systems: arachidonic acid cyclooxygenase (prostaglandin endoperoxide synthetase) to produce prostaglandins, thromboxanes, and prostacyclin, or by lipoxygenases to produce leukotrienes.

Interaction of arachidonic acid with cyclooxygenase in the presence of oxygen and heme produces first the cyclic endoperoxide, PGG_2 and thence, through its peroxidase activity, to PGH_2, both of which are chemically unstable and decompose rapidly ($t_{1/2}$ ~ five minutes). PGE_2 is formed by the action of PGE isomerase and PGD_2 by the actions of isomerases or glutathione-S-transferase on PGH_2, whereas $PGF_{2\alpha}$ is formed from PGH_2 via an endoperoxide reductase system (see Fig. 25–3). It is at the

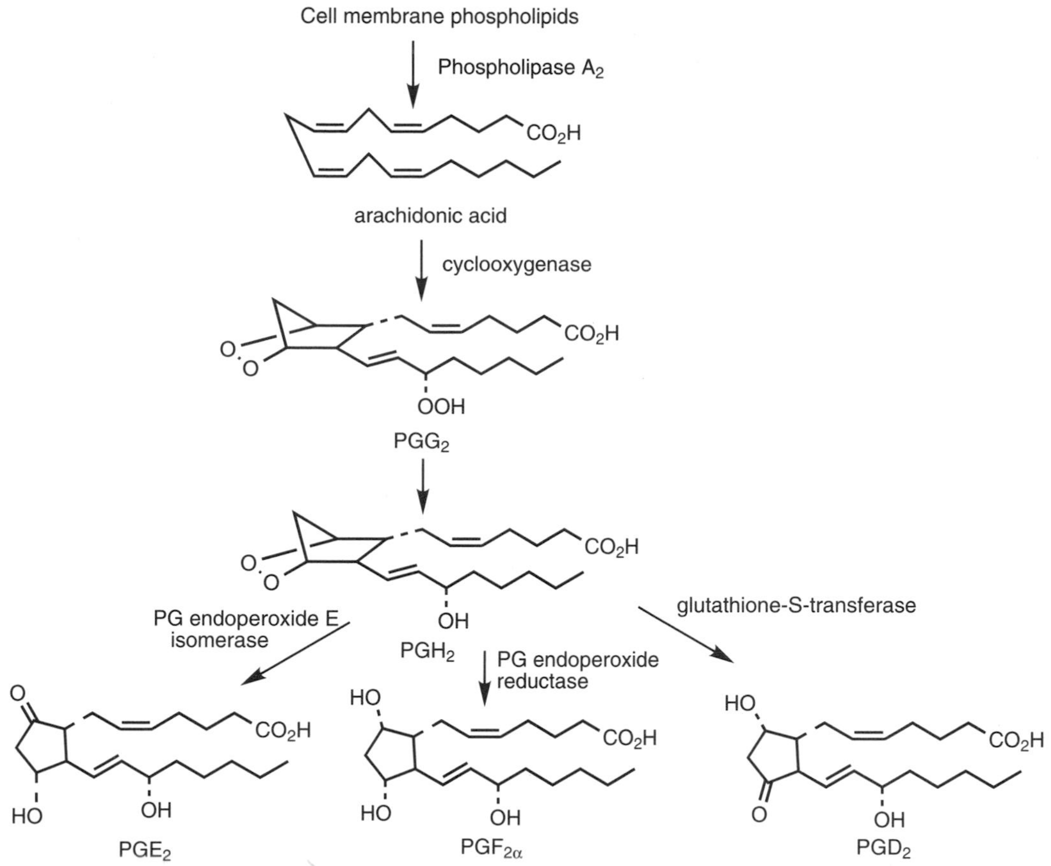

Cell membrane phospholipids
Phospholipase A_2
CO_2H
arachidonic acid
cyclooxygenase
CO_2H
OOH
PGG_2
CO_2H
PG endoperoxide E isomerase
OH
PGH_2
PG endoperoxide reductase
glutathione-S-transferase
PGE_2
$PGF_{2\alpha}$
PGD_2

Fig. 25–3. Biosynthesis of prostaglandins from arachidonic acid.

cyclooxygenase step at which the NSAIAs inhibit PG biosynthesis preventing inflammation. Because PGG_2 and PGH_2 themselves may possess the ability to mediate the pain responses and to produce vasoconstriction, and because PGG_2 may mediate the inflammatory response, cyclooxygenase inhibition would profoundly reduce inflammation.

Prostaglandins are rapidly metabolized and inactivated by various oxidative and reductive pathways. The initial step involves rapid oxidation of the 15α-OH to the corresponding ketone by the prostaglandin specific enzyme 15-hydroxy-prostaglandin dehydrogenase. This is followed by reduction of the C13–14 double bond by prostaglandin Δ^{13}-reductase to the corresponding dihydro ketone, which for PGE_2 represents the major metabolite in plasma. Subsequently, enzymes normally involved in β- and ω-oxidation of fatty acids more slowly cleave the α-chain and oxidize the C-20 terminal methyl group to the carboxylic acid derivative, respectively. Hence, dicarboxylic acid derivatives containing only 16 carbon atoms are the major metabolites of PGE_1 and PGE_2, which are excreted.

The actions of the various PGs are diverse. When administered intravaginally, PGE_2 will stimulate the endometrium of the gravid uterus to contract in a manner similar to uterine contractions observed during labor. Thus, PGE_2 is therapeutically available as dinoprostone (Prostin E2, Upjohn) for use as an abortifacient used at 12 to 20 weeks gestation and for evacuation of uterine content in missed abortion or intrauterine fetal death useful up to 28 weeks gestation. PGE_2 is also a potent stimulator of smooth muscle of the gastrointestinal (GI) tract and can elevate body temperature in addition to possessing potent vasodilating properties in most vascular tissue and also possessing constrictor effects at certain sites. PGEs usually cause pain when administered intradermally. Many of these properties are shared by $PGF_{2\alpha}$ which is also therapeutically available as an abortifacient at 16 to 20 weeks gestation and is marketed as dinoprost tromethamine (Prostin F2 alpha, Upjohn). The synthetic 15-methyl derivative of $PGF_{2\alpha}$, carboprost, is also available as the tromethamine salt (Prostin 15/M, Upjohn) as an abortifacient used at 13 to 20 weeks gestation. $PGF_{2\alpha}$ differs from PGE_2, however, in that it does not significantly alter blood pressure in humans. PGD_2 causes both vasodilation and vasoconstriction. Whereas the PGEs produce a relaxation of bronchial and tracheal smooth muscle, PGFs and PGD_2 cause contraction. PGE_1 is available as alprostadil (Prostin VR Pediatric, Upjohn) to maintain patency of the ductus arteriosus in neonates until surgery can be performed to correct congenital heart defects.

The effects of prostaglandins on the GI tract deserve special mention. PGEs and PGI_2 inhibit gastric secretion which may be induced by gastrin or histamine. PGs ap-

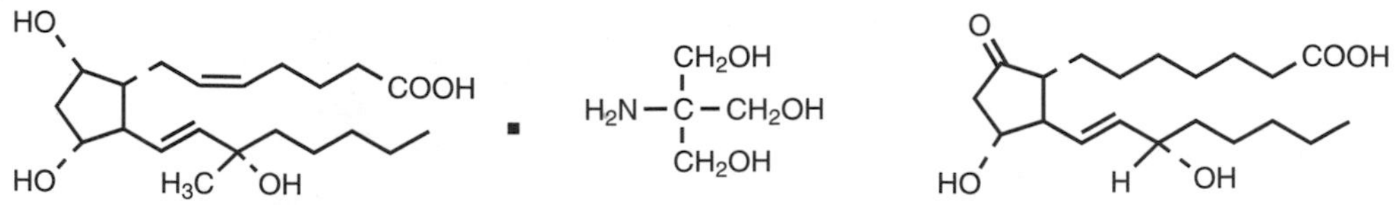

Carboprost tromethamine

Alprostadil

pear to play a major cytoprotecive role in maintaining the integrity of gastric mucosa. PGE_1 exerts a protective effect on gastroduodenal mucosa by stimulating secretion of an alkaline mucus and bicarbonate ion and also by maintaining or increasing mucosal blood flow. Thus, inhibition of PG formation in joints produces favorable results as indicated by a reduced fever, pain, and swelling, but inhibition of PG biosynthesis in the GI tract is unfavorable because it may cause disruption of mucosal integrity resulting in peptic ulcer disease which, as will be discussed later in this chapter, is commonly associated with the use of NSAIAs and aspirin.

Alternatively, nonprostanoids can also be formed from PGH_2 as illustrated in Figure 25–4. Thromboxane synthetase acts on PGH_2 to produce thromboxane A_2 (TxA_2) whereas prostacyclin synthetase converts PGH_2 to prostacyclin (PGI_2), both of which possess short biologic half-lives. TxA_2, a potent vasoconstrictor and inducer of platelet aggregation, has a biologic half-life of about 30 seconds, being rapidly nonenzymatically converted to the more stable, but inactive, TxB_2. Prostacyclin, a potent hypotensive and inhibitor of platelet aggregation, has a half-life of about 3 minutes and is nonenzymatically converted to 6-keto-$PGF_{1\alpha}$. Platelets contain primarily thromboxane synthetase whereas endothelial cells contain primarily prostacyclin synthetase. Considerable research efforts are being expended in the development of stable prostacyclin analogs and thromboxane antagonists as car-

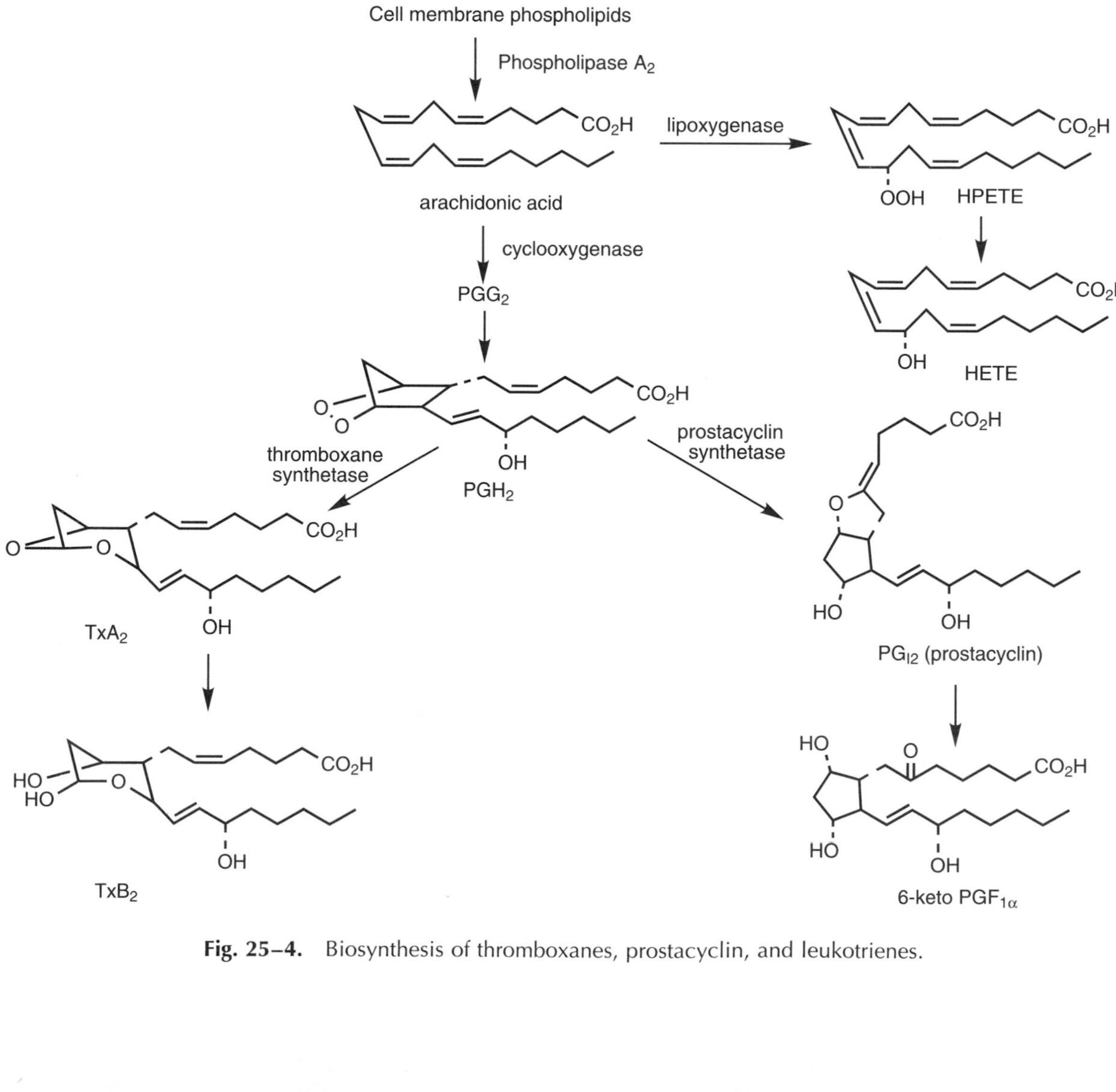

Fig. 25–4. Biosynthesis of thromboxanes, prostacyclin, and leukotrienes.

Table 25–2. Pharmacologic Properties of Prostaglandins, Thromboxanes and Prostacyclin

	PGE_2	$PGF_{2\alpha}$	PGI_2	TxA_2
Uterus	Oxytocic	Oxytocic		
Bronchi	Dilation	Constriction		
Platelets			Inhibits aggregation	Aggregation
Blood vessels	Dilation	Constriction	Dilation	Constriction

Modified from T. Nogrady, *Medicinal Chemistry—A Biochemical Approach,* 2nd ed., New York, Oxford University Press, 1988, p. 330, with permission.

diovascular agents. The pharmacologic effects of some prostaglandins, thromboxane A_2 and prostacyclin are summarized in Table 25–2.

The existence of distinct prostaglandin receptors may explain the broad spectrum of action displayed by the prostaglandins. The nomenclature of these receptors is based on the affinity displayed by natural prostaglandins, prostacyclin or thromboxanes at each receptor type. Thus, EP receptors are those receptors for which the PGEs have high affinity, FP receptors for PGFs, DP receptors for PGDs, IP receptors for PGI_2, and TP receptors for TxA_2. These receptors are coupled through G-proteins to effector mechanisms, which include stimulation of adenyl cyclase, and hence increased cyclic adenosine monophosphate levels (cAMP), and phospholipase C, which results in increased levels of IP_3 (inositol 1,4,5-triphosphate). Three distinct receptors for leukotrienes have also been identified.

Lipoxygenases are a group of enzymes which oxidize polyunsaturated fatty acids possessing two *cis* double bonds separated by a methylene group to produce lipid hydroperoxides.[20] Arachidonic acid is thus metabolized to a number of hydroperoxyeicosatetraenoic acid derivatives (HPETEs). These enzymes differ in the position at which they peroxidize arachidonic acid and in their tissue specificity. For example, platelets possess only a 12-lipoxygenase but leukocytes possess both a 12-lipoxygenase and a 5-lipoxygenase.[21] The HPETE derivatives are not stable, being rapidly converted to different metabolites. Leukotrienes are products of the 5-lipoxygenase pathway and are divided into two major classes, hydroxylated eicosatetraenoic acids (LTs) represented by LTB_4 and peptidoleukotrienes (pLTs) such as LTC_4, LTD_4 and LTE_4. The enzyme 5-lipoxygenase will produce leukotrienes from 5-HPETE as shown in Figure 25–5. LTA synthetase converts 5-HPETE to an unstable epoxide called LTA_4 which may

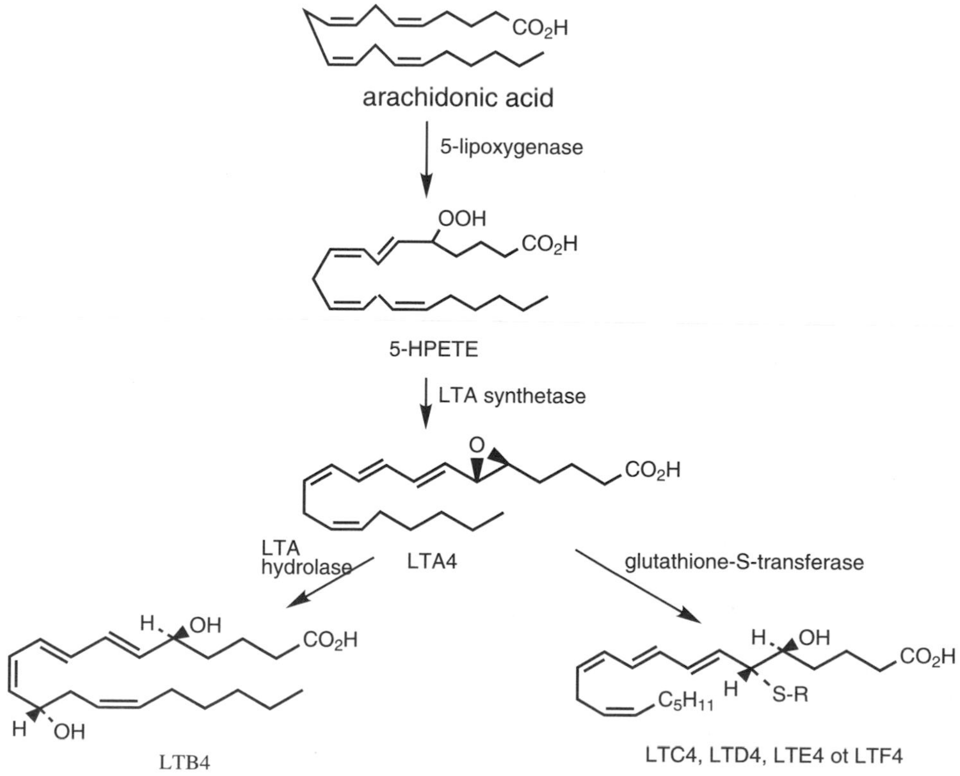

arachidonic acid

5-lipoxygenase

5-HPETE

LTA synthetase

LTA hydrolase

LTA4

glutathione-S-transferase

LTB4

LTC4, LTD4, LTE4 ot LTF4

Fig. 25–5. Biosynthesis of leukotrienes.

be converted by the enzyme LTA hydrolase to the leukotriene LTB_4 or by glutathione-S-transferase to LTC_4. Other leukotrienes (e.g., LTD_4, LTE_4, and LTF_4) can then be formed from LTC_4 by the removal of glutamic acid and glycine and then re-conjugation with glutamic acid, respectively. One mediator of inflammation known as SRS-A (an acronym for slow-reacting substance of anaphylaxis) is primarily a mixture of two leukotrienes, LTC_4 and LTD_4. The physiologic roles of the various leukotrienes is becoming better understood. LTB_4 is a potent chemotactic agent for polymorphonuclear leukocytes and causes the accumulation of leukocytes at inflammation sites leading to the development of symptoms characteristic of inflammatory disorders. LTC_4 and LTD_4 are potent hypotensives and bronchoconstrictors. Because of the role played by LTs and pLTs in inflammatory conditions and asthma, it is not surprising that intensive research is being conducted in the area of inhibitors of leukotriene biosynthesis.

Based on the generally accepted hypothesis that NSAIAs produce their anti-inflammatory effects primarily through inhibition of PG biosynthesis at the cyclooxygenase step, Shen[22] represented the structural relationships of arylalkanoic acids (the largest group of NSAIAs) and the ability of these agents to structurally superimpose with the structure of arachidonic acid to demonstrate the ability of both agents to accommodate to a computer-modeled hypothetical binding site of arachidonic acid cyclooxygenase. These relationships are illustrated in Figure 25–6.

Therapeutic Approach to Arthritic Disorders

The management of arthritic disorders involves a stepwise approach to the use of therapeutic agents currently available. Relief of pain and reduction of inflammation are immediate goals because of the severity of symptoms most frequently encountered in arthritics. Longer-term goals would be to halt the progression of the disease and preserve the functions of muscles and joints in order to permit the patient to lead a more productive life. Fortunately, a large number of NSAIAs are therapeutically

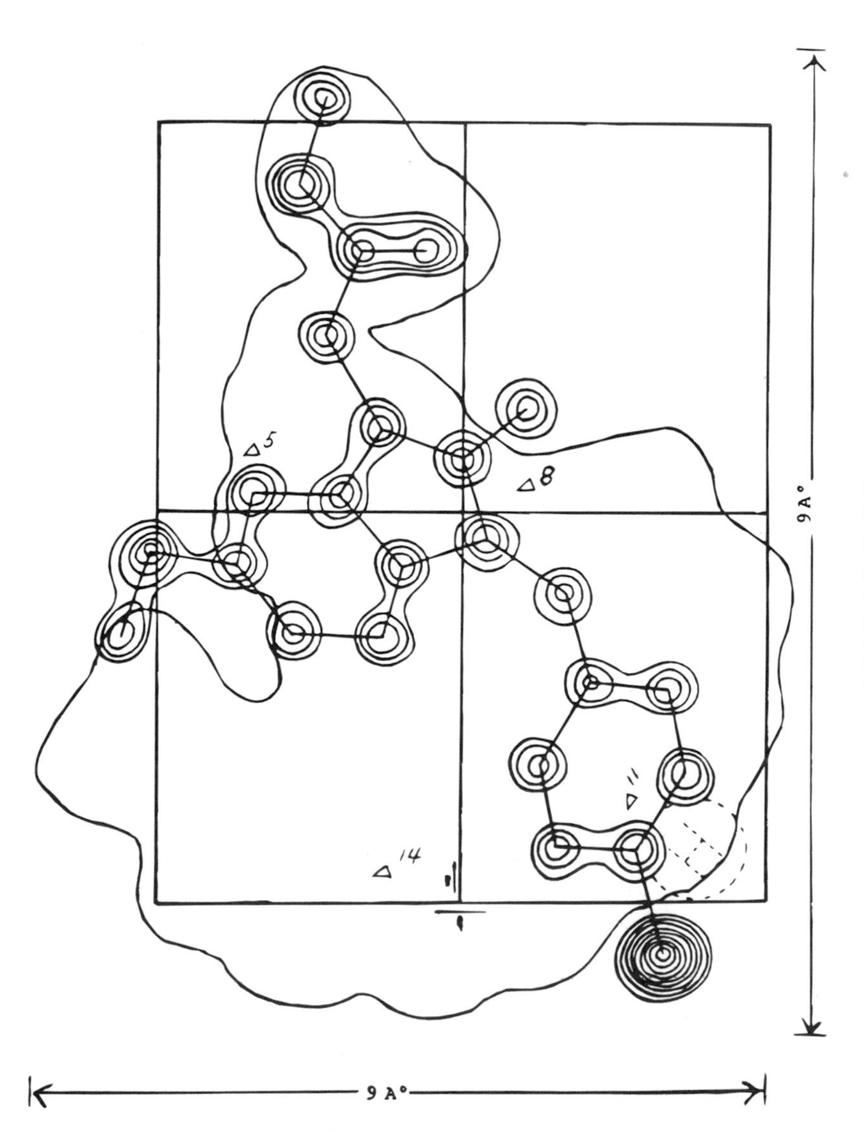

Fig. 25–6. The superposition of the x-ray projection of a noncoplanar anti-inflammatory molecule over a computer modeled hypothetical binding site of arachidonic acid in cyclooxygenase. The relative positions of four double bonds at the 5-, 8-, 11-, and 14- positions in arachidonic acid are indicated. (Reprinted with permission from R. W. Bennett, US Pharmacist, June, 1990.)

available, differing in efficacy, but perhaps more importantly, differing also in overall toxicity. As a group, NSAIAs can cause GI toxicity such as dyspepsia, abdominal pain, heartburn, gastric erosion, peptic ulcer formation, bleeding, diarrhea, renal disorders such as acute renal failure, tubular necrosis, and analgesic nephropathy, and other effects such as tinnitus and headache. A recent report of the Arthritis, Rheumatism, and Aging Medical Information System Post-Marketing Surveillance Program ranked the overall toxicity of NSAIAs studies in the following decreasing order: indomethacin > tolmetin > meclofenamate > ketoprofen > fenoprofen > piroxicam > sulindac > naproxen > ibuprofen > salsalate > aspirin. If these agents prove ineffective, alternate treatments include the use of disease-modifying antirheumatic agents, corticosteroids and immunosuppressive agents.

A generally accepted stepwise approach to treatment is:

- Step 1 Physical Therapy, Rest, Patient Education, and Counseling
- Step 2 Salicylates (aspirin)
- Step 3 Nonsteroidal Anti-inflammatory Agents (NSAIAs)
- Step 4 Disease-Modifying Antirheumatic Drugs (DMARDs)
- Step 5 Corticosteroids
- Step 6 Immunosuppressive Agents

ANTIPYRETIC ANALGESICS: ANILINES AND ρ AMINOPHENOLS

Agents are included in this class which possess analgesic and antipyretic actions but lack anti-inflammatory effects. Antipyretics interfere with those processes by which pyrogenic factors produce fever, but do not appear to lower body temperature in afebrile subjects. It had been historically accepted that the antipyretics exert their actions within the central nervous system (CNS), primarily at the hypothalamic thermoregulatory center but more recent evidence suggests that peripheral actions may also contribute. Endogenous leukocytic pyrogens may be released from cells which have been activated by various stimuli and antipyretics may act by inhibiting the activation of these cells by an exogenous pyrogen or by inhibiting the release of endogenous leukocytic pyrogens from the cells once they have been activated by the exogenous pyrogen. Substantial evidence exists suggesting a central antipyretic mechanism, an antagonism which may result from either a direct competition of a pyrogen and the antipyretic agent at CNS receptors or an inhibition of prostaglandins in the CNS.[23] Despite extensive use of acetaminophen, its mechanism of action has not been fully elucidated. Acetaminophen may inhibit pain impulses by exerting a depressant effect on peripheral receptors; an antagonistic effect on the actions of bradykinin may play a role. The antipyretic effects may not result from inhibition of release of endogenous pyrogen from leukocytes but by inhibition of the action of released endogenous pyrogen on hypothalamic thermoregulatory centers. The fact that acetaminophen is an effective antipyretic-analgesic but an ineffective anti-inflammatory agent may be due to its greater inhibition of prostaglandin biosynthesis in the CNS than in the periphery.

Acetanilide was introduced into therapy in 1886 as an antipyretic-analgetic agent under the name antifebrin but was subsequently found to be too toxic, having been associated with methomeoglobinemia and jaundice, particularly at high doses, to be useful. Phenacetin was introduced the following year and was widely used but was withdrawn recently because of persistent reports of nephrotoxicity. Phenacetin is longer acting than acetaminophen despite the fact that it is metabolized to acetaminophen but is a weaker antipyretic. Acetaminophen (paracetamol) was subsequently introduced in 1893 but remained unpopular for over 50 years until it was observed that it is a metabolite of both acetanilide and phenacetin. It remains the only useful agent of this group and is widely used as a nonprescription antipyretic-analgesic under a variety of trade names (Tylenol [McNeil], Datril [Bristol-Myers], Tempra [Mead Johnson]). Whereas the analgesic activity of acetaminophen is comparable to aspirin's, it lacks useful anti-inflammatory activity. Its major advantage over aspirin as an analgetic, however, is that individuals who are hypersensitive to salicylates generally respond well to acetaminophen.

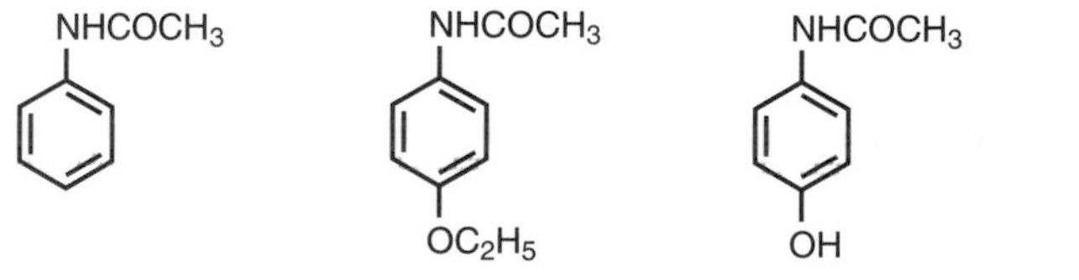

$NHCOCH_3$ $NHCOCH_3$ $NHCOCH_3$

OC_2H_5 OH

acetanilide phenacetin acetaminophen

Structure-Activity Relationships

The structure-activity relationships of ρ-aminophenol derivatives has been widely studied. Based on the comparative toxicity of acetanilide and acetaminophen, aminophenols are less toxic than the corresponding aniline derivatives, although ρ-aminophenol itself is too toxic for therapeutic purposes. Etherification of the phenolic function with methyl or propyl groups produces derivatives with greater side effects than the ethyl derivative. Substituents on the nitrogen atom which reduce basicity reduce activity unless that substituent is metabolically labile, e.g., acetyl. Amides derived from aromatic acids, e.g., N-phenylbenzamide, are less active or inactive.

Preparations Available

Acetaminophen, U.S.P. Acetaminophen is a weakly acidic ($pK_a = 9.51$), colorless powder which can be synthesized by the acetylation of ρ-aminophenol. It is weakly bound to plasma proteins (18 to 25%). Acetaminophen is indicated for use as an antipyretic-analgesic, particularly in individuals displaying an allergy or sensitivity to aspirin. Although it does not possess anti-inflammatory activity, it will produce analgesia in a wide variety of arthritic and musculoskeletal disorders. It is available in various formulations including suppositories, tablets, capsules, granules, and solutions. The usual adult dose is 325 to 650 mg every 4 to 6 hours. Doses above 2.6 g per day are not recommended for long-term therapy. Acetaminophen, unlike aspirin, is stable in aqueous solution making liquid formulations readily available, a particular advantage in pediatric cases.

Metabolism and Toxicity

The metabolism of acetanilide, acetaminophen, and phenacetin is illustrated in Figure 25–7.[24] As indicated earlier, both acetanilide and phenacetin are metabolized to acetaminophen. Additionally, both undergo hydrolysis to yield aniline derivatives which produce directly, or through their conversion to hydroxylamine derivatives, significant methemoglobinemia and hemolytic anemia, primary factors which resulted in the drugs' removal from the U.S. market. On the other hand, acetaminophen is metabolized primarily by conjugation reactions, the O-sulfate conjugate being the primary metabolite in children and the O-glucuronide in adults. A minor, but significant, metabolic product of both acetaminophen and phenacetin is the N-hydroxyamide produced by a cytochrome P_{450} mixed function oxidase system. The hydroxyamide is then converted to a reactive toxic metabolite, an acetimidoquinone, which has been suggested[25] to produce the nephrotoxicity and hepatotoxicity associated with acetaminophen and phenacetin. Normally this quinone is detoxified by conjugation with hepatic glutathione. In cases of ingestion of larger doses or overdoses of acetaminophen, however, hepatic stores of glutathione may be depleted by more than 70%, thus allowing the reactive quinone to interact with nucleophilic functions,

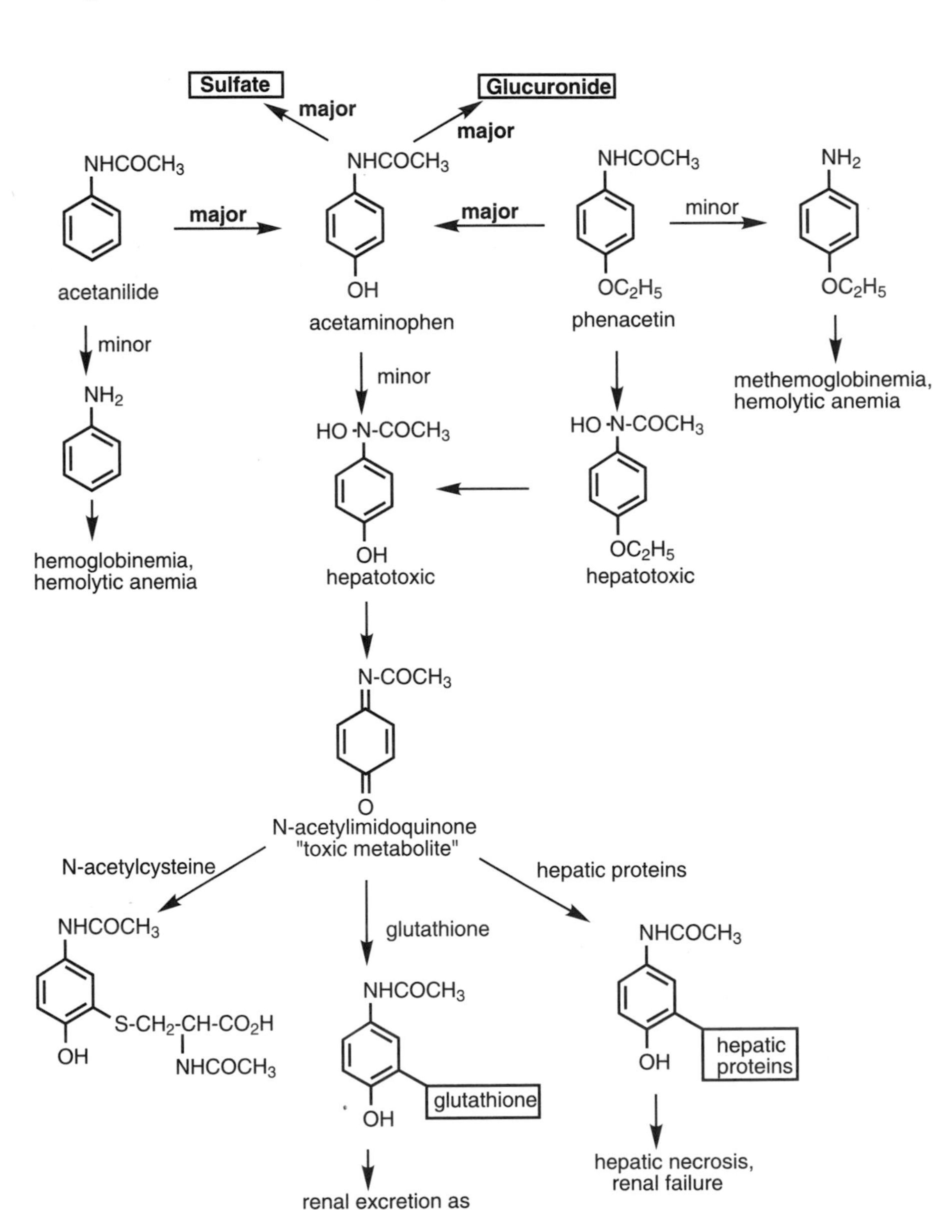

Fig. 25–7. Metabolism of acetaminophen. Modified from, and used with permission of, W. G. Clark, *Gen. Pharmacol., 10,* 71(1979).

primarily -SH groups, on hepatic proteins resulting in the formation of covalent adducts which produce hepatic necrosis. Overdoses of acetaminophen can produce potentially fatal hepatic necrosis, renal tubular necrosis, and hypoglycemic coma. Various sulfhydryl-containing compounds were found to be useful as antidotes to acetaminophen overdoses. The most useful of these, N-acetylcysteine (Mucomyst [Mead Johnson]), substitutes for the depleted glutathione by enhancing hepatic glutathione stores and by enhancing disposition by nontoxic sulfate conjugation.[26] N-Acetylcysteine may also inhibit the formation of the toxic imidoquinone metabolite.[27] In cases of overdoses N-acetylcysteine is administered as a 5% solution in water, soda, or juice. The recommended dose is 140 mg/kg followed by 17 maintenance doses of 70 mg/kg every 5 hours.

Drug Interactions

Hepatic necrosis develops at much lower doses of acetaminophen in some heavy drinkers than would be expected,[24] due, perhaps, to the induction of the cytochrome P_{450} system, depletion of glutathione stores, or by aberrations in the primary sulfate and glucuronide conjugation pathways. Acetaminophen has been reported to potentiate the response to oral anticoagulants although the effect on prothrombin time is not clear. Interactions with warfarin, dicumarol, anisindione, and diphenadione have been suggested. The mechanism of these interactions has not been fully elucidated but may be associated with competition for plasma protein binding sites because acetaminophen is a weak acid and is weakly bound, but may also be due to the induction of hepatic microsomal enzymes. The effects of acetaminophen are reduced in the presence of microsomal enzyme inducers such as barbiturates and are enhanced by metoclopramide and salicylamide. The absorption of acetaminophen is enhanced by polysorbate and sorbitol and is reduced by anticholinergics and narcotic analgesics. Chemical incompatibilities have also been reported based on hydrolysis by strong acids or bases or phenolic oxidation in the presence of oxidizing agents. Acetaminophen forms ''sticky'' mixtures with diphenhydramine HCl and discolors under humid conditions in the presence of caffeine or codeine phosphate.

ANTI-INFLAMMATORY AGENTS

Salicylates

The use of salicylates dates back to the 19th century when salicylic acid was first obtained in 1838 from salicin, a glycoside which is present in most willow and poplar bark. Interestingly, Hippocrates prescribed chewing willow bark for pain relief in the fifth century B.C.E. In 1860, Kolbe synthesized salicylic acid from sodium phenoxide and carbon dioxide, using a method which produced large quantities of salicyclic acid inexpensively. Derivatives of salicylic acid began to receive medical attention shortly thereafter. Sodium salicylate was employed as an antipyretic-antirheumatic agent in 1875 and the phenyl ester was used in 1886. Acetylsalicylic acid was prepared in 1853 but was not used medicinally until 1899. The name Aspirin was given to acetylsalicyclic acid by Dreser, director of pharmacology at Frederich Bayer and Company in Germany as a contraction of the letter *a* from acetyl and *spirin,* an older name given to salicylic acid (spiric acid) that was derived from a natural source in spirea plants. Since then, numerous derivatives of salicylic acid have been synthesized and evaluated pharmacologically, yet only a relatively few derivatives have achieved therapeutic utility.

In addition to possessing antipyretic, analgesic, and anti-inflammatory properties, salicylates possess other actions which have been proven to be therapeutically beneficial. Because salicylates promote the excretion of uric acid they are useful in the treatment of gouty arthritis. More recently, attention has been given to the ability of salicylates to inhibit platelet aggregation which may contribute to heart attacks and stroke. Aspirin appears to inhibit prostaglandin cyclooxygenase in platelet membranes thus blocking formation of the potent platelet aggregating factor thromboxane A_2 in an irreversible manner. The Physicians' Health Study recently concluded that in a group of 22,071 participants, there was a 44% reduction in the risk of myocardial infarction in the group taking a single 325 mg aspirin tablet taken every other day vs. the placebo group.[28] The role of aspirin in reducing cardiac mortality has been recently reviewed.[29] Also, a recent study suggested that aspirin and other NSAIAs may be protective against colon cancer.[30] Thus the therapeutic utility of aspirin continues to increase. Unfortunately, a number of side effects are associated with the use of salicylates, most notably GI disturbances such as dyspepsia, gastroduodenal bleeding, gastric ulcerations, and gastritis.

A number of possible mechanisms of action have been proposed for salicylates over the years. Among those suggested have been inhibition of the biosynthesis of histamine, antagonism of the actions of various kinins, inhibition of mucopolysaccharide biosynthesis, inhibition of lysosomal enzyme release, and inhibition of leukocyte accumulation. Currently, the most widely accepted mechanism of action is the ability of these agents to inhibit the biosynthesis of prostaglandins at the cyclooxygenase stage discussed earlier.

Structure-Activity Relationships

Despite the vast effort expended in the search to find a ''better'' aspirin (i.e., one possessing fewer gastrointestinal side effects, with greater potency, longer duration of action, but yet inexpensive, being antipyretic, analgesic, and anti-inflammatory), none superior to aspirin has yet to be discovered. But as a result of this search, the following structure-activity relationships have been established.

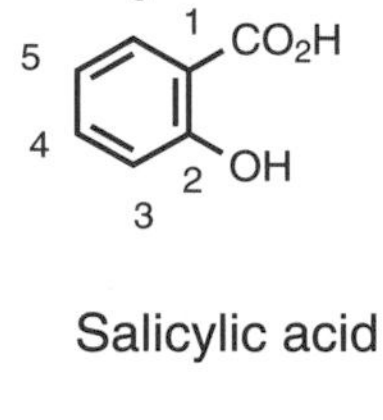

Salicylic acid

The active moiety appears to be the salicylate anion. Side effects of aspirin, particularly the GI effects, appear to be associated with the carboxylic acid function. Reducing the acidity of this function, e.g., the corresponding amide (salicylamide), maintains the analgesic actions of salicylic acid derivatives but eliminates the anti-inflammatory properties. Substitution on either the carboxyl or phenolic hydroxyl groups may affect potency and toxicity. Benzoic acid itself has only weak activity. Placing the phenolic hydroxyl group *meta*- or *para*- to the carboxyl group abolishes activity. Substitution of halogen atoms on the aromatic ring enhances potency and toxicity. Substitution of aromatic rings at the 5-position of salicylic acid increases anti-inflammatory activity (e.g., diflunisal).

Absorption, Metabolism, and Toxicity

Most salicylates are rapidly and effectively absorbed on oral administration with the rate of absorption and bioavailability depending on several factors including the dosage formulation, gastric pH, food contents in the stomach, gastric emptying time, the presence of buffering agents or antacids, and particle size. Because salicylates are weak acids (acetylsalicylic acid $pK_a = 3.49$), absorption generally takes place primarily from the small intestine and to a lesser extent from the stomach by the process of passive diffusion of unionized molecules across the epithelial membranes of the GI tract. Thus, gastric pH is an important factor in the rate of absorption of salicylates. Any factor which increases gastric pH, e.g., buffering agents, will slow the rate of absorption because more of the salicylate will be in the ionized form, but this may be counteracted by an increased solubility of salicylates which enhances absorption. The differences in the rates of absorption of aspirin, salicylate salts, and the buffered preparations of salicylates are actually quite small with absorption half-times in humans ranging from approximately 20 minutes for buffered preparations to 30 minutes for aspirin itself. The presence of food in the stomach slows the rate of absorption. Formulation factors may contribute to the differences in absorption rates of the various brands of plain and buffered salicylate preparations. Tablet formulations consisting of small particles are absorbed faster than those of larger particle size. The bioavailability of salicylate from enteric-coated preparations may be inconsistent. Absorption of salicylate from rectal suppositories is slower, incomplete, and not recommended when high salicylate levels are required. Topical preparations of salicylic acid are effective in that the rate of salicylate absorption from the skin is rapid.

Salicylates are highly bound to plasma protein albumin with binding being concentration dependent. At low therapeutic concentrations of 100 μg/ml, approximately 90% is plasma protein bound whereas at higher concentrations of about 400 μg/ml, only 76% binding is observed. Plasma protein binding is a major factor in the drug interaction observed for salicylates.

The major metabolic routes of esters and salts of salicylic acid are illustrated in Figure 25–8. The initial route of metabolism of these derivatives is their conversion to salicyclic acid which may be excreted in the urine as the free acid (10%) or undergo conjugation with either glycine to produce the major metabolite, salicyluric acid (75%), or with glucuronic acid to form the glucuronide ether and ester (15%). In addition, small amounts of metabolites resulting from microsomal aromatic hydroxylation are found. The major hydroxylation metabolite, gentisic acid, was once thought to be responsible for the anti-inflammatory actions of the salicylates but its presence in trace quantities would rule out a major role for gentisic acid, or the other hydroxylation metabolites, in the pharmacologic action of salicylates. The metabolism and pharmacokinetic properties of salicylates has been extensively reviewed.[31]

The most commonly observed side effects associated with the use of salicylates relate to disturbances of the GI tract. Nausea, vomiting, epigastric discomfort, intensification of symptoms of peptic ulcer disease, such as dyspepsia and heartburn, gastric ulcerations, erosive gastritis, and gastrointestinal hemorrhage occur in individuals on high doses of aspirin. The incidence of these side effects is rarer at lower doses but a single dose of aspirin can cause GI distress in 5% of individuals. Gastric bleeding induced by salicylates is generally painless but can lead to fecal blood loss and may cause a persistent iron deficiency anemia. At dosages that generally are useful in anti-inflammatory therapy, aspirin may lead to a loss of 3 to 8 ml of blood per day. The mechanism by which salicylates cause gastric mucosal cell damage may be due to a number of factors including gastric acidity, the ability of salicylates to damage the normal mucosal barrier which protects against the back diffusion of hydrogen ions, the ability of salicylates to inhibit the formation of prostaglandins, particularly those of the PGE series which normally inhibit gastric acid secretion, and inhibition of platelet aggregation leading to an increased tendency toward bleeding. Thus, salicylate use prior to surgery or tooth extraction is contraindicated.

Reye's syndrome is an acute condition that may follow influenza and chicken pox infections in children from infancy to their late teens with the majority of cases seen between the ages of 4 and 12 years. It is characterized by symptoms including sudden vomiting, violent headaches, and unusual behavior in children who appear to be recovering from an often mild viral illness. Although a rare condition (60 to 120 cases per year or an incidence of 0.15 per 100,000 population of 18 years of age or younger), it can be fatal with a death rate of between 20 to 30%. Fortunately, the number of cases is declining, due mainly to the observations that over 90% of children with Reye's syndrome were on salicylate therapy during a recent viral illness. Based on these observations, the FDA has proposed that aspirin and other salicylates be labeled with a warning against their use in children under 16 years of age with influenza, chicken pox, or other flu-like illness. Acetaminophen would appear to be the drug of choice in children with these conditions.

Salicylates account for approximately 25% of all accidental poisonings in children in the United States.

Drug Interactions

Because of the widespread use of salicylates, it is not surprising that interactions with many other drugs used

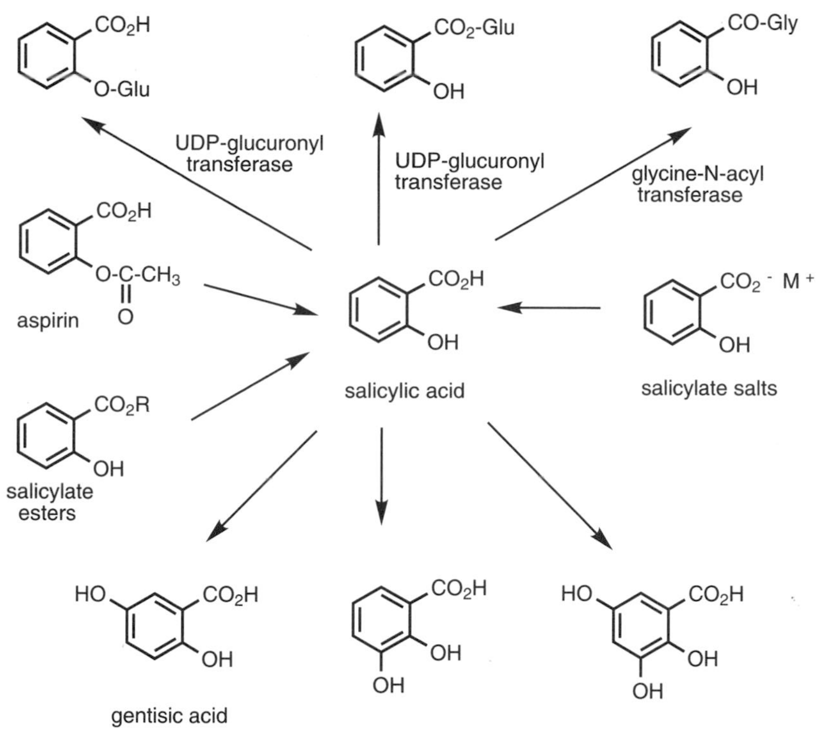

Fig. 25–8. Metabolism of salicylic acid derivatives. (glu, glucoronide conjugate; gly, glycine conjugate.)

in therapeutic combinations have been observed, several of which are clinically significant. More data is available for aspirin than any other specific salicylate product. As mentioned previously, acetylsalicylic acid is a weak acid that is highly bound to plasma proteins (50–80%) and will compete for these plasma protein binding sites with other drugs which are also highly bound to these sites. The interaction that results from the combination of salicylates with oral anticoagulants represents one of the most widely documented clinically significant drug interactions reported to date. The plasma concentration of free anticoagulant increases in the presence of salicylates, necessitating a possible decrease in the dosage of anticoagulant required to produce a beneficial therapeutic effect. The ability of salicylates to produce GI ulcerations and bleeding coupled with the inhibition of the clotting mechanism results in a clinically significant drug interaction. In addition, salicylates may inhibit the synthesis of prothrombin by antagonizing the actions of vitamin K. NSAIAs can also produce these interactions. The competition for plasma protein binding sites can also lead to an increase in free methotrexate levels (thus enhancing the toxicity of methotrexate), enhanced toxicity of long-acting sulfonamides, and a hypoglycemic effect resulting from displacement of oral hypoglycemic agents. In large doses, salicylates given concomitant with uricosuric agents such as probenecid and sulfinpyrazone may lead to a retention of uric acid and thus antagonize the uricosuric effect, despite the fact that salicylates alone increase urinary excretion of uric acid. The diuretic activity of aldosterone antagonists, such as spironolactone, may be antagonized by salicylates. Corticosteroids may decrease blood levels of salicylates because of their ability to increase the glomerular filtration rate. The incidence and severity of GI ulcerations may be increased if corticosteroids, salicylates, and NSAIAs are administered together. The GI bleeding induced by salicylates may be enhanced by the ingestion of ethanol. Other interactions have been reported but their clinical significance has not been fully established.

Salicylate hypersensitivity, particularly to aspirin, is relatively uncommon but must be recognized because severe and potentially fatal reactions may occur. Signs of aspirin hypersensitivity appear soon after administration and include skin rashes, watery secretions, urticaria, vasomotor rhinitis, edema, bronchoconstriction, and anaphylaxis. Less than 1% of the U.S. population may experience aspirin hypersensitivity; this group consists primarily of middle-aged individuals. Females are more likely to experience aspirin intolerance or hypersensitivity. Aspirin-sensitive asthmatics are especially at high risk. Mild salicylism may occur after repeated administration of large doses. Symptoms include dizziness, tinnitus, nausea, vomiting, diarrhea, and mental confusion. Doses of 10 to 30 g have been known to cause death in adults but some individuals have ingested up to 130 g without fatality. Over 10,000 cases of serious salicylate toxicity occur in the United States each year.

Preparations Available

The structures of the marketed preparations of salicylic acid are presented in Figures 25–9.

Aspirin, U.S.P. Acetylsalicylic acid, or aspirin, is a white powder which is stable in a dry environment but which is hydrolyzed to salicylic acid and acetic acid under humid or moist conditions. Hydrolysis can also occur

acetylsalicylic acid salicylamide salsalate (Disalcid®)

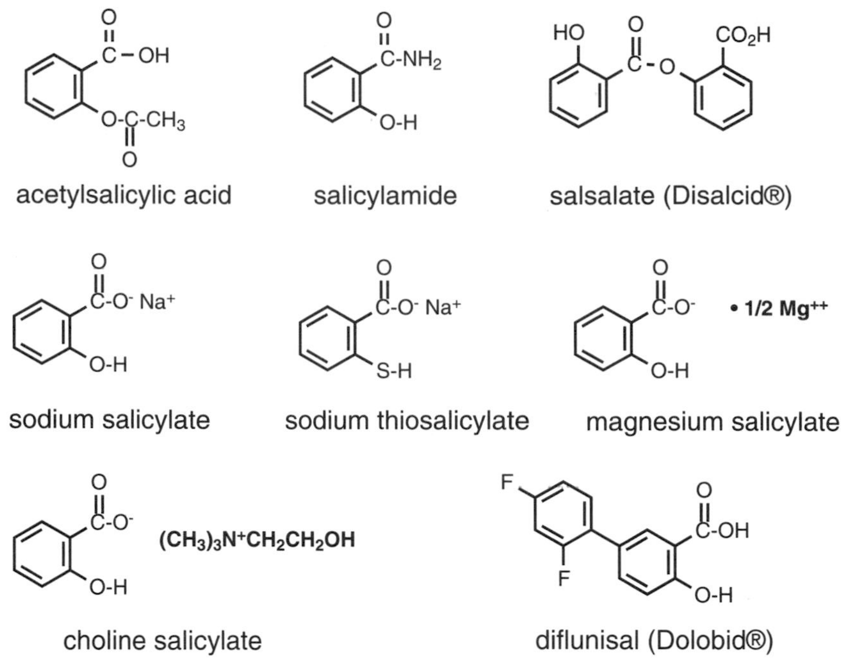

sodium salicylate sodium thiosalicylate magnesium salicylate • 1/2 Mg^{++}

choline salicylate $(CH_3)_3N^+CH_2CH_2OH$ diflunisal (Dolobid®)

Fig. 25–9. Structures of marketed derivatives of salicylic acid.

when aspirin is combined with alkaline salts or with salts containing water of hydration. Stable aqueous solutions of aspirin are thus unobtainable despite the addition of modifying agents which tend to decrease hydrolysis. Aspirin is rapidly absorbed, largely intact from the stomach and upper small intestine upon oral administration but is rapidly hydrolyzed by plasma esterases. Peak plasma levels are usually achieved within 2 hours after administration. Increasing the pH of the stomach by the addition of buffering agents may affect absorption because the degree of ionization would be increased.

Aspirin is indicated for the relief of minor aches and mild to moderate pain (325 to 650 mg every 4 hours), for arthritis and related arthritic conditions (3.2 to 6 g per day), to reduce the risk of transient ischemic attacks in men (1.3 g per day), and for myocardial infarction prophylaxis (as little as 40 mg/day to as much as 325 mg every other day are being used). It is available in a large number of dosage forms and strengths as tablets, suppositories, capsules, enteric-coated tablets, and buffered tablets.

Salicylamide. Salicylamide is a white crystalline powder which is much less acidic than other salicylic acid derivatives. Although poorly soluble in water, stable solutions can be formed at pH 9. It is absorbed from the GI tract on oral administration and rapidly metabolized to inactive metabolites by intestinal mucosa, but not by hydrolysis. Activity appears to reside in the intact molecule. It is approximately 40 to 55% plasma protein bound. Salicylamide competes with other salicylates and acetaminophen for glucuronide conjugation decreasing the extent of conjugation of these other agents. Excretion occurs rapidly, primarily in the urine. The major advantages of salicylamide are its general lack of gastric irritation, compared with aspirin's, and its use in individuals who are hypersensitive to aspirin. Salicylamide enters the CNS more rapidly than other salicylates and will case sedation and drowsiness when administered in large doses. Whereas salicylamide is reported to be as effective as aspirin as an analgesic-antipyretic and is effective in relieving pain associated with arthritic conditions, it does not appear to possess useful anti-inflammatory activity.[32] Thus indications for its use in the treatment of arthritic disease states are unwarranted and its use is restricted to the relief of minor aches and pain at a dosage of 325 to 650 mg, three or four times per day. Its effects in humans are not reliable however, and its use is not widely recommended.

Salicylate Salts. Several salts of salicylic acid, sodium salicylate, U.S.P., choline salicylate, U.SP., and magnesium salicylate, U.S.P., and one salt of thiosalicylic acid, sodium thiosalicylate, U.S.P., are available. These salts are used primarily because their use decreases the incidence of GI disturbances or because they form stable aqueous solutions. Sodium salicylate is half as potent, on a weight basis, as aspirin as an analgesic and antipyretic but produces less GI irritation and equivalent blood levels, and is useful in patients exhibiting hypersensitivity to aspirin. It generates salicylic acid in the GI tract accounting for some gastrointestinal irritation and sodium bicarbonate is sometimes given concomitantly to reduce acidity. Sodium salicylate, unlike aspirin, does not affect platelet function, although prothrombin times are increased. It is available as tablets, enteric-coated tablets, and as a solution for injection.

Choline salicylate has lower GI side effects than aspirin and has been shown to be particularly useful in treating juvenile rheumatoid arthritis where aspirin was ineffective. It is absorbed more rapidly than aspirin and produces higher salicylate plasma levels. It is available as a mint-flavored liquid.

Magnesium salicylate has a low incidence of GI side effects. Both sodium salicylate and magnesium salicylate should be used cautiously in individuals in whom excessive amounts of these electrolytes might be detrimental. The possibility of magnesium toxicity in individuals with renal insufficiency exists. It is available as tablets but its safety in children under 12 years of age has not been fully determined.

Sodium thiosalicylate is indicated for rheumatic fever, muscular pain and acute gout and is available as a solution for intramuscular injection.

Salsalate (Disalcid [Riker]). Salsalate, salicylsalicylic acid, is a dimer of salicylic acid. It is insoluble in gastric juice but is soluble in the small intestine where it is partially hydrolyzed to two molecules of salicylic acid and absorbed. On a molar basis it produces 15% less salicylic acid than aspirin. It does not cause GI blood loss and can be given to aspirin-sensitive patients. Salsalate is available as capsules and tablets.

Diflunisal (Dolobid [Merck]). Diflunisal [5-(2,4-difluorophenyl)salicylic acid] was introduced in the United States) in 1982 and has gained considerable acceptance as an analgesic and in the treatment of rheumatoid arthritis and osteoarthritis. It currently is one of the 200 most-used prescription drugs. It is a white, odorless, crystalline powder, which is practically insoluble in water at neutral or acidic pH but is soluble in most organic solvents and aqueous alkaline solutions, and is stable to both heat and light. Diflunisal is metabolized primarily to ether and ester glucuronide conjugates. No metabolism involving changes in ring substituents has been reported. It is more potent than aspirin but produces fewer side effects, and has a biologic half-life 3 to 4 times greater than that of aspirin. It is rapidly and completely absorbed on oral administration with peak plasma levels being achieved within 2 to 3 hours of administration. It is highly bound (99%) to plasma proteins after absorption. Side effects most frequently reported include GI disturbances (nausea, dyspepsia and diarrhea), dermatologic reactions, and CNS effects such as dizziness and headache.

Diflunisal is a moderately potent inhibitor of prostaglandin biosynthesis but differs from the manner in which aspirin inhibits the cyclooxygenase system in that the inhibition is competitive and reversible in nature. Diflunisal does not have an appreciable effect on platelet aggregation, however, and does not significantly produce gastric or intestinal bleeding. It is available as 250-mg and 500-mg tablets.

3,5-Pyrazolidinediones

In 1884, during a search for antipyretic agents, the German chemist Ludwig Knorr attempted to synthesize quinoline derivatives as structural analogs of quinine, whose structure was unresolved at that time. Instead, he obtained a pyrazole derivative, more specifically a 5-pyrazolone derivative, antipyrine, which was found to have marked antipyretic and analgesic activities. The 4-dimethylamino derivative, aminopyrine, was subsequently found to be more potent but slower acting than antipyrine (Fig. 25–10). Both antipyrine and aminopyrine have analgesic, antipyretic, and antirheumatic potencies similar to those

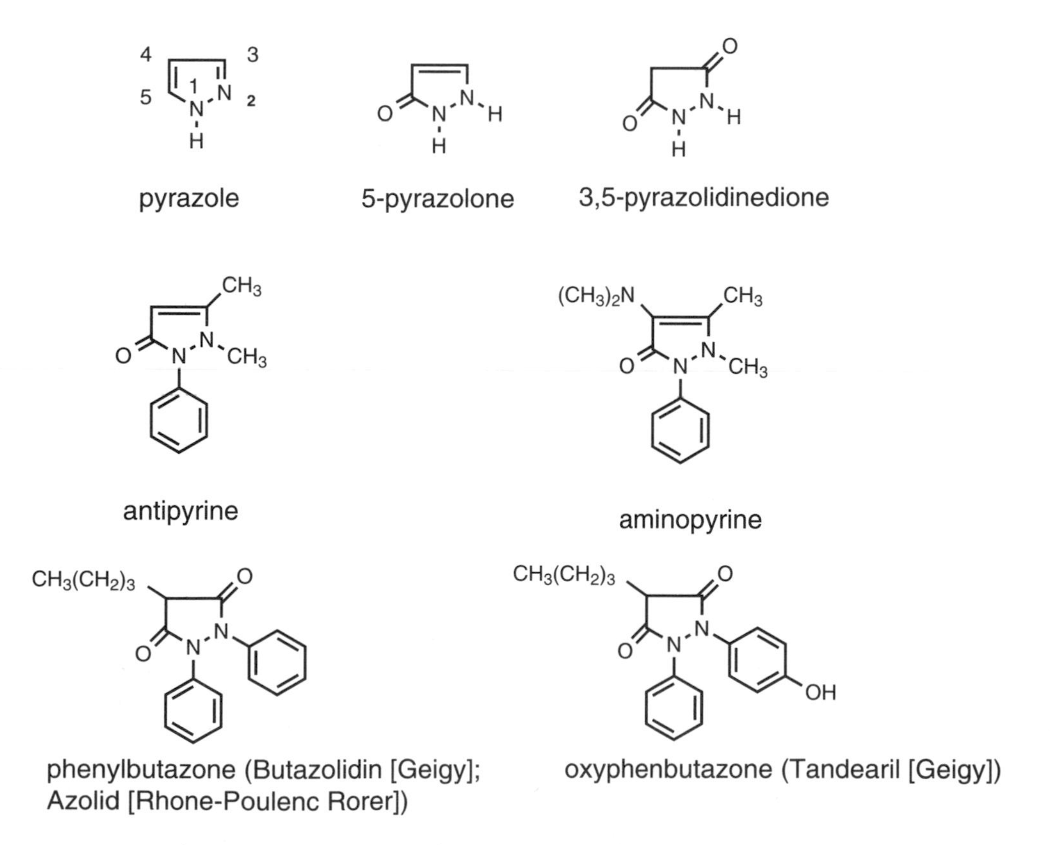

Fig. 25–10. Structures of 5-pyrazones and 3,5 pyrazolidinediones.

of aspirin and sodium salicylate and so were widely used in the United States and Europe for those purposes. Reports of fatal agranulocytosis diminished interest in these agents, however. The antipyretic properties of antipyrine were so widely accepted that it was used in in various nonprescription ophthalmic and analgesic preparations until recently when the Food and Drug Administration OTC (over-the-counter) panel on Topical Analgesic, Antirheumatic, Otic, Burn, and Sunburn Treatment Products determined that it was neither safe nor effective for OTC use. Interest in the pyrazolones had begun to decline before that ruling until the mid-1940s when, in an effort to increase the analgesic properties of 5-pyrazolones, a series of 3,5-pyrazolidinediones was synthesized. One member of this group, phenylbutazone, was shown to possess antipyretic and analgesic properties equipotent with antipyrine and, being acidic, could enhance the solubilization of other drugs. The sodium salt of phenylbutazone was initially introduced for use as a solubilizing agent for aminopyrine. A few years later a combination of aminopyrine and phenylbutazone was found to be beneficial in patients with various rheumatic disorders and eventually phenylbutazone itself was evaluated for anti-inflammatory activity. The confirmation of the anti-inflammatory activity of phenylbutazone represented a major breakthrough in arthritis therapy in 1952 when phenylbutazone was introduced as an antirheumatic agent.

Like most NSAIAs, the 3,5-pyrazolidinediones possess a number of pharmacologic and biochemical properties. Among the biochemical properties are the ability to uncouple oxidative phosphorylation, stabilize lysosomal membranes, inhibit the biosynthesis of various mucopolysaccharides, and, perhaps most important, also inhibit prostaglandin biosynthesis at the cyclooxygenase stage. Phenylbutazone possesses three major pharmacologic properties: anti-inflammatory activity (as measurable in the carrageenan-induced rat paw, cotton pellet granuloma, and adjuvant-induced arthritis assays), analgesic activity, and uricosuric activity. The anti-inflammatory activity is greater than that of the salicylates in the rat cotton pellet granuloma assay. Aspirin at a dose of 50 mg/kg produces a 15% inhibition whereas phenylbutazone at 30 mg/kg produces a 35% inhibition.[33] Analgesic activity, however, is less than that of the salicylates and the routine use of phenylbutazone for this purpose is not recommended. Phenylbutazone has been widely used as an analgesic in animals, particularly in race horses. At relatively high doses (600 mg) phenylbutazone displays measurable uricosuric effects resulting in decreased tubular reabsorption of uric acid while low concentrations appear to promote uric acid retention by inhibiting its secretion. In addition to these actions, phenylbutazone will cause the retention of sodium and the accompanying decrease in volume of urine can lead to edema. Oxyphenbutazone possesses equipotent anti-inflammatory and sodium-retaining effects as phenylbutazone and is somewhat better tolerated because of decreased GI irritation.

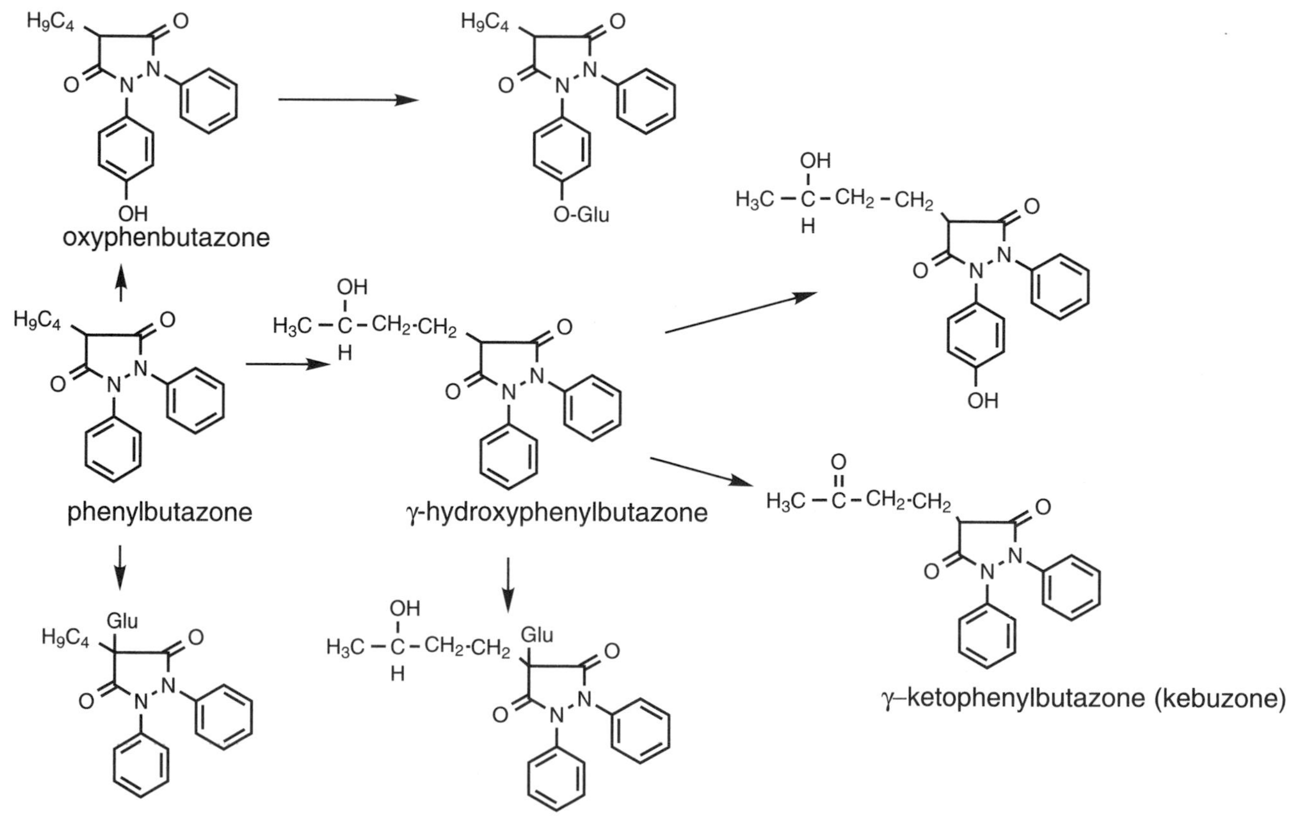

Fig. 25–11. Metabolism of phenylbutazone in humans. (Glu, glucuronide.)

Structure-Activity Relationships

Perhaps the most striking chemical feature of the 3,5-pyrazolidinediones is the fact that their acidity approaches that of carboxylic acids and the pharmacologic activity of 3,5-pyrazolidinediones is closely related to their acidity. The dicarbonyl functions at the 3 and 5 positions enhance the acidity of hydrogen atoms at the 4-position resulting in the following equilibrium:

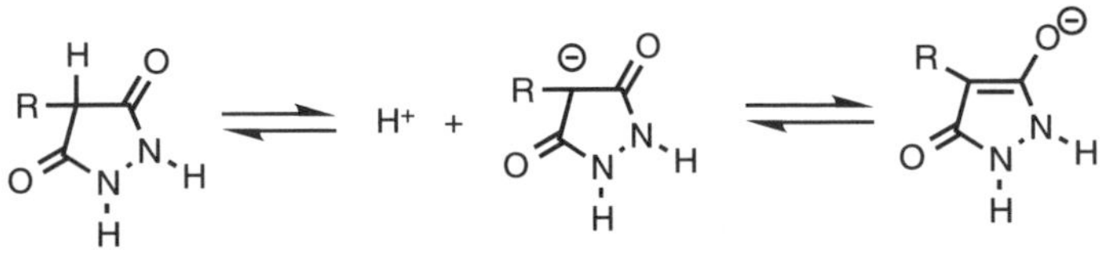

Decreasing or eliminating acidity by removing the acidic proton at the 4-position (e.g., 4,4-dialkyl derivatives) abolishes anti-inflammatory activity. Thus, if the hydrogen atom at the 4-position of phenylbutazone is replaced by a substituent such as a methyl group, anti-inflammatory activity is abolished. If acidity is enhanced too much anti-inflammatory and sodium-retaining activities decrease (presumably through an enhanced excretion rate as evidenced by a decreased serum half-life), while other properties such as the uricosuric effect increase. A single alkyl group at the 4-position enhances anti-inflammatory activity. Although the n-butyl group enhances activity most, propyl and allyl analogs also possess anti-inflammatory activity. Introduction of polar functions on these alkyl groups gives mixed results. The γ-hydroxy-n-butyl derivative (a metabolite of phenylbutazone) possesses pronounced uricosuric activity but gives few anti-inflammatory effects. The corresponding γ-keto-n-butyl derivative (kebuzone) possesses significant anti-inflammatory activity—it has been used in Europe for that purpose. Substitution of a 2-phenylthioethyl group at the 4-position produces interesting effects. The corresponding sulfoxide derivative is marketed as a uricosuric agent called sulfinpyrazone. The most active derivatives are those in which the substituents on both ring nitrogen atoms is a phenyl group. The presence of both phenyl groups is essential for neither anti-inflammatory nor analgesic activity, however. The monophenyl analog, mofebutazone, possesses anti-inflammatory activity. Various substituents in the *para-* position of one or both aromatic rings do not drastically affect activity. A ρ-hydroxy group, present in oxyphenbutazone, the major metabolite of phenylbutazone, contributes to therapeutically useful anti-inflammatory activity. Other derivatives such as methyl, chloro, or nitro groups also possess activity. The importance of a pyrazole ring system is less readily apparent. Corresponding pyrrole and isoxazole analogs retain the anti-inflammatory properties of phenylbutazone, but carbocyclic analogs such as the cyclopentane or cyclopentene derivatives and the corresponding acyclic derivatives do not.

The structures of the 5-pyrazolones and 3,5-pyrazolidinediones are presented in Figure 25–10.

Absorption, Metabolism, and Toxicity

Both phenylbutazone and oxyphenbutazone are rapidly and completely absorbed from the GI tract with peak plasma levels being achieved within 2 hours. At therapeutic doses, approximately 98% of phenylbutazone and 95% of oxyphenbutazone is bound to serum albumin. Both drugs possess relatively long biologic half lives, 72 to 84 hours for phenylbutazone and 48 to 72 hours for oxyphenbutazone. Significant concentrations of phenylbutazone may persist in joints up to three weeks after therapy has been terminated.

As suggested earlier, metabolism plays an important role in the actions of 3,5-pyrazolidinediones. The metabolism of phenylbutazone is illustrated in Figure 25–11. Phenylbutazone is slowly metabolized by aromatic hydroxylation via liver microsomal enzymes to oxyphenbutazone, a pharmacologically active metabolite displaying a similar pharmacologic profile to the parent drugs. Oxyphenbutazone is excreted primarily as the O-glucuronide conjugate. Liver microsomal oxidation of the n-butyl group via ω-1 oxidation produces a second major metabolite, γ-hydroxyphenylbutazone, which possesses significant uricosuric activity. Subsequent oxidation of γ-hydroxyphenylbutazone yields ρ, γ-dihydroxyphenylbutazone and γ-keto-phenylbutazone (kebuzone) with the latter being widely used in Europe as an anti-inflammatory agent. Phenylbutazone and γ-hydroxyphenylbutazone form C-glucuronides in humans as a result of conjugation of glucuronic acid to the carbon atom at the 4-position, early examples of this novel route of metabolism.[34] Phenylbutazone is also an inducer of liver microsomal enzymes and will induce its own metabolism.

Phenylbutazone is a relatively potent drug and is poorly tolerated in many individuals. Among the most frequently reported side effects are abdominal discomfort, nausea, peptic ulcerations and skin rashes. Less common side effects include vomiting and diarrhea, CNS disturbances (headache and drowsiness), and blurred vision. Susceptibility to these toxic effects increases with age and thus prolonged use in patients over 60 years of age is not advised. Although incidences are rare, phenylbutazone can produce fatal blood dyscrasias and such symptoms of dyscrasias (lesions in the mouth, skin rash, fever) should be immediately reported. The use of phenylbutazone in children under 14 is also contraindicated. Whereas doses as low as 200 to 499 mg may cause death in children, other children have survived doses as high as 5000 mg. There is no specific antidote for overdosage and emergency treatment generally consists of inducing emesis or performing gastric lavage and supportive treatment. Oxyphenbutazone shares many of the toxic effects of phenylbutazone with the major exception that it causes less gastric irritation and may thus be better tolerated than phenylbutazone.

Drug Interactions

Both phenylbutazone and oxyphenbutazone, like the salicylates, are highly bound to plasma proteins. Thus, like the salicylates, they will displace other plasma protein-bound drugs from these binding sites leading to increased serum levels of the displaced drug enhancing the pharmacologic activity, duration of effect, and toxicity of the displaced drug. Enhanced effects of oral anticoagulants (such as warfarin), oral hypoglycemics (such as tolbutamide), sulfonamides, other NSAIAs, and other acidic drugs have been reported. Inducers of hepatic microsomal enzymes, such as barbiturates, may decrease the half-life of phenylbutazone. Cholestryamine may bind phenylbutazone in the gut and may thus retard its absorption on oral administration. For this reason, it is recommended that phenylbutazone be administered at least 1 hour before the administration of the sequestering agent. Alcohol administered concomitantly with phenylbutazone or oxyphenbutazone may cause an impairment of psychomotor skills.

Preparations Available

Phenylbutazone (Butazolidin [Geigy], Azolid [USV], [Rhone-Poulenc Rorer]). Phenylbutazone (4-n-butyl-1,2-diphenyl-3,5-pyrazolidinedione) is a white, odorless powder which forms water-soluble salts when treated with base. It can be prepared in a relatively straightforward manner by treating either diethyl n-butylmalonate or the corresponding malonyl dichloride with hydrazobenzene in the presence of base. It is indicated for use only after other therapeutic measures, including other NSAIAs, have proven unsatisfactory for the relief of symptoms of active rheumatoid arthritis and acute ankylosing spondylitis, acute gouty arthritis, and acute attacks of degenerative bone disease. It is available as 100 mg tablets and capsules. Because phenylbutazone may cause GI upset, it is recommended that it be taken with food or milk.

Oxyphenbutazone, U.S.P. Oxyphenbutazone [4-n-1-(*p*-hydroxyphenyl)-2-phenyl-3,5-pyrazolidinedione] is white and odorless, a powder which forms water soluble salts when treated with base. It is prepared by condensing *p*-benzyloxyhydrazobenzene with diethyl n-butylamine in the presence of base and subsequent hydrogenolysis. It is a major metabolite of phenylbutazone which has the same indications and recommended dosages as the parent drug. It was formerly available as Tandearil (Geigy), but was removed from the U.S. market in 1985 for safety considerations.

Arylalkanoic Acids

The largest group of NSAIAs is represented by the class of arylalkanoic acids as typified by the following general chemical structure:

```
      R  O
      |  ||
AR—C—C—OH
      |
      H
```

R = H, CH_3 or Alkyl

AR = aryl or heteroaryl

Several factors have caused this to be one of the most active areas of drug development in recent years. The impact that the introduction of phenylbutazone in the 1950s had on arthritis therapy was more than matched by the interest generated by the introduction of indomethacin in the mid-1960s. As a result of a study designed to investigate the anti-inflammatory activity of 350 indole acetic acid derivatives related structurally to serotonin and metabolites of serotonin, the Merck group, led by T. Y. Shen,[35] reported the synthesis and antipyretic and anti-inflammatory activity of the most potent compound in the series, indomethacin. The observation that indomethacin possessed 1085 times the anti-inflammatory activity and 20 times the antipyretic activity of phenylbutazone (and 10 times the antipyretic activity of aminopyrine) generated considerable interest in the development of other aryl and heteroaryl acetic acid and propionic acid derivatives. The marketplace was ripe for new anti-inflammatory agents and most pharmaceutical companies joined in the search for new arylalkanoic acids. The introduction of ibuprofen in the 1970s by Upjohn was quickly followed by the appearance of fenoprofen calcium, naproxen, and tolmetin. Sulindac, a prodrug analog of indomethacin was introduced in the late 1970s. The 1980s produced zomepirac, benoxaprofen, ketoprofen, flurbiprofen, suprofen, and diclofenac sodium. Thus far, the 1990s have already produced ketorolac, etodolac, and nabumetone. This rapid development has been accompanied by some setbacks, however. Zomepirac, introduced in 1980 as an analgesic, was withdrawn in 1983 because of severe anaphylactoid reactions, particularly in patients sensitive to aspirin. Benoxaprofen was withdrawn within 6 months of its introduction in 1982 because of several deaths caused by cholestatic jaundice in Europe and the United States. In addition, benoxaprofen produced photosensitivity reactions in patients when they were exposed to sunlight and onycholysis (loosening of the fingernails) in some patients. Suprofen, introduced as an analgesic in 1985, was removed from the market 2 years later because of flank pain and transient renal failure. It was re-introduced in 1989 for ophthalmic use. Numerous other arylalkanoic acids are currently being evaluated in various stages of clinical trials.

As discussed earlier, most NSAIAs manifest several biochemical and pharmacologic actions. As was the case for the salicylates and 3,5-pyrazolidinediones, the arylalkanoic acids, to various extents, share the property of inhibition of prostaglandin biosynthesis by blocking the arachidonic acid cyclooxygenase system.

Structure-Activity Relationships

Agents of this class share certain common structural features. These general structure-activity relationships will be discussed here as they pertain to the proposed mechanism of action. Specific structure-activity relationships for each drug or drug class are presented separately, as appropriate.

All agents possess a center of acidity which can be represented by a carboxylic acid function, an enolic function, a hydroxamic acid function, a sulfonamide, or a tetrazole ring. The relationship of this acid center to the carboxylic acid function of arachidonic acid is obvious. The activity of ester and amide derivatives of carboxylic acids is generally attributed to the metabolic hydrolysis products. One nonacidic drug, nabumetone, has recently been introduced in the U.S. but, as will be discussed later, its activity is attributed to its bioactivation to an active acid metabolite. The center of acidity is generally located one carbon atom adjacent to a flat surface represented by an aromatic or heteroaromatic ring. The distance between these centers is critical because increasing this distance to two or three carbons generally diminishes activity. Derivatives of aryl or heteroaryl acetic or propionic acids are most common. The aromatic ring system appears to correlate with the double bonds at the 5- and 8-positions of arachidonic acid. Substitution of a methyl group on the carbon atom separating the acid center from the aromatic ring tends to increase anti-inflammatory activity. The resulting α-methyl acetic acid, or 2-substituted propionic acid, analogs have been given the class name "profens" by the U.S. Adopted Name Council. Groups larger than methyl decrease activity, but incorporation of this methyl group as part of an alicyclic ring system does not drastically affect activity. Introduction of a methyl group creates a center of chirality. Anti-inflammatory activity in those cases wherein the enantiomers have been separated and evaluated, whether determined in vivo or in vitro by cyclooxygenase assays, is associated with the (S)-(+)-enantiomer. Interestingly, in those cases where the propionic acid is administered as a racemic mixture, in vivo conversion of the R-enantiomer to the biologically active S-enantiomer is observed to varying degrees. A second area of lipophilicity which is generally noncoplanar with the aromatic or heteroaromatic ring enhances activity. This second lipophilic area may correspond to the area of the double bond in the 11-position of arachidonic acid. This lipophilic function may consist of an additional aromatic ring or alkyl groups either attached to, or fused to, the aromatic center.

Metabolism

Essentially all of the arylalkanoic acid derivatives which are therapeutically available are extensively metabolized. Metabolism occurs primarily through hepatic microsomal enzyme systems and may lead to deactivation or bioactivation of the parent molecules. Metabolism of each drug will be treated separately.

Drug Interactions

All of the arylalkanoic acids are highly bound to plasma proteins and may thus displace other drugs from protein binding sites resulting in an enhanced activity and toxicity of the displaced drugs. Interestingly, despite the high degree of plasma protein binding, indomethacin does not display this characteristic drug interaction. The most commonly observed interaction is that between the arylalkanoic acid and oral anticoagulants, particularly warfarin. Coadministration may prolong prothrombin time. Potential interactions with other acidic drugs, such as hydantoins, sulfonamides, and sulfonylureas should be monitored. Concomitant administration of aspirin decreases plasma levels of arylalkanoic acids by as much as 20%. Probenecid, on the other hand, tends to increase these plasma levels. Interactions with drugs which may induce hepatic microsomal enzyme systems (such as phenobarbital) may enhance or diminish anti-inflammatory activity depending on whether the arylalkanoic acid is metabolically bioactivated or inactivated by this enzyme system. Certain diuretics, such as furosemide, inhibit the metabolism of prostaglandins by 15-hydroxy-prostaglandin dehydrogenase and the resulting increase in PGE_2 levels induce plasma renin activity. Because the arylalkanoic acids block the biosynthesis of prostaglandins, the effects of furosemide can be antagonized, in part, offering a potentially significant drug interaction.

Preparations Available

The structures of the aryl- and heteroarylacetic acid derivatives and the aryl- and heteroarylpropionic acids ("profens") available are presented in Figures 25–12 and 25–13, respectively.

Aryl- and Heteroarylacetic Acids.

Indomethacin (Indocin [Merck, Sharp & Dohme]). Indomethacin [1-(*p*-chlorobenzoyl)-5-methoxy-2-methylindole 3-acetic acid] is a yellow-tan crystalline powder that is odorless, has a bitter taste, and is light-sensitive. It is water soluble and although soluble in base, alkaline solutions are not stable due to the ease of hydrolysis of the *p*-chlorobenzoyl group. The original synthesis of indomethacin by Shen[35] involved the formation of 2-methyl-5-methoxyindole acetic acid and subsequent acylation after protection of the carboxyl group as the t-butyl ester. It was introduced in the United States in 1965 and has remained a popular drug over the years. It is still one of the most potent NSAIAs in use, possessing approximately 25 times the activity of phenylbutazone. It is also a more potent antipyretic than either aspirin or acetaminophen and possesses about 10 times the analgesic potency of aspirin, although the value of the analgesic effect is widely overshadowed by concern over the frequency of side effects.

Structure-Activity Relationships. Replacement of the carboxyl group with other acidic functionalities decreases activity. Anti-inflammatory activity generally increases as the acidity of the carboxyl group increases and decreases as the acidity is decreased. Amide analogues are inactive. Acylation of the indole nitrogen with aliphatic carboxylic acids or aralkylcarboxylic acids results in amide derivatives which are less active than those derived from benzoic acid. N-Benzoyl derivatives substituted in the *para*-position with fluoro, chloro, trifluoromethyl, or thiomethyl groups are the most active. The 5-position of the indole

Indomethacin (Indocin [Merck, Sharp & Dohme])

Sulindac (Clinoril [Merck & Co.])

Tolmetin sodium (Tolectin [McNeil Pharm.])

Diclofenac sodium (Voltaren [Geigy])

Etodolac (Lodrine [Wyeth-Ayerst])

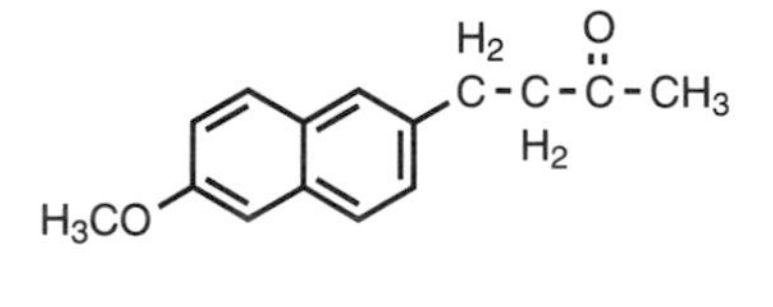

Nabumetone (Relafen [SK Beecham])

Fig. 25–12. Structures of aryl- and heteroacylacetic acid derivatives.

ring is most flexible with regard to the nature of substituents which enhance activity. Substituents such as methoxy, fluoro, dimethylamino, methyl, allyloxy, and acetyl are more active than the unsubstituted indole ring. The presence of an indole ring nitrogen is not essential for activity because the corresponding 1-benzylidenylindene analogs (e.g., sulindac) are active. Alkyl groups, especially methyl, at the 2-position are more active than aryl substituents. Substitution of a methyl group at the α-position of the acetic acid side chain (to give the corresponding propionic acid derivative) leads to equiactive analogs. The resulting chirality introduced in the molecules is important. Anti-inflammatory activity is displayed only by the (S)(+)-enantiomer. The conformation of indomethacin appears to play a crucial role in its anti-inflammatory actions. The acetic acid side chain is flexible and can assume a large number of different conformations. The preferred conformation of the N-*p*-chlorobenzoyl group is one in which the chlorophenyl ring is oriented away from the 2-methyl group (or *cis* to the methoxyphenyl ring of the indole nucleus) and is non-coplanar with the indole ring because of steric hindrance produced by the 2-methyl group and the hydrogen atom at the 7-position. This conformation may be represented as follows:

Absorption and Metabolism. Absorption of indomethacin occurs rapidly on oral administration and peak plasma levels are obtained within 2 to 3 hours. Being an acidic substance (pKa = 4.5), it is highly bound to plasma proteins (97%). Indomethacin is converted to inactive metabolites, approximately 50% of a single dose being converted to the O-demethylated metabolite and 10% conjugated with glucuronic acid. Nonhepatic enzyme systems hydrolyze indomethacin to N-deacylated metabolites. The metabolism of indomethacin is illustrated in Figure 25–14.

The ability of indomethacin to potently inhibit prostaglandin biosynthesis may account for its anti-inflammatory, antipyretic, and analgesic actions. Pronounced side effects are frequently observed at anti-rheumatic doses.

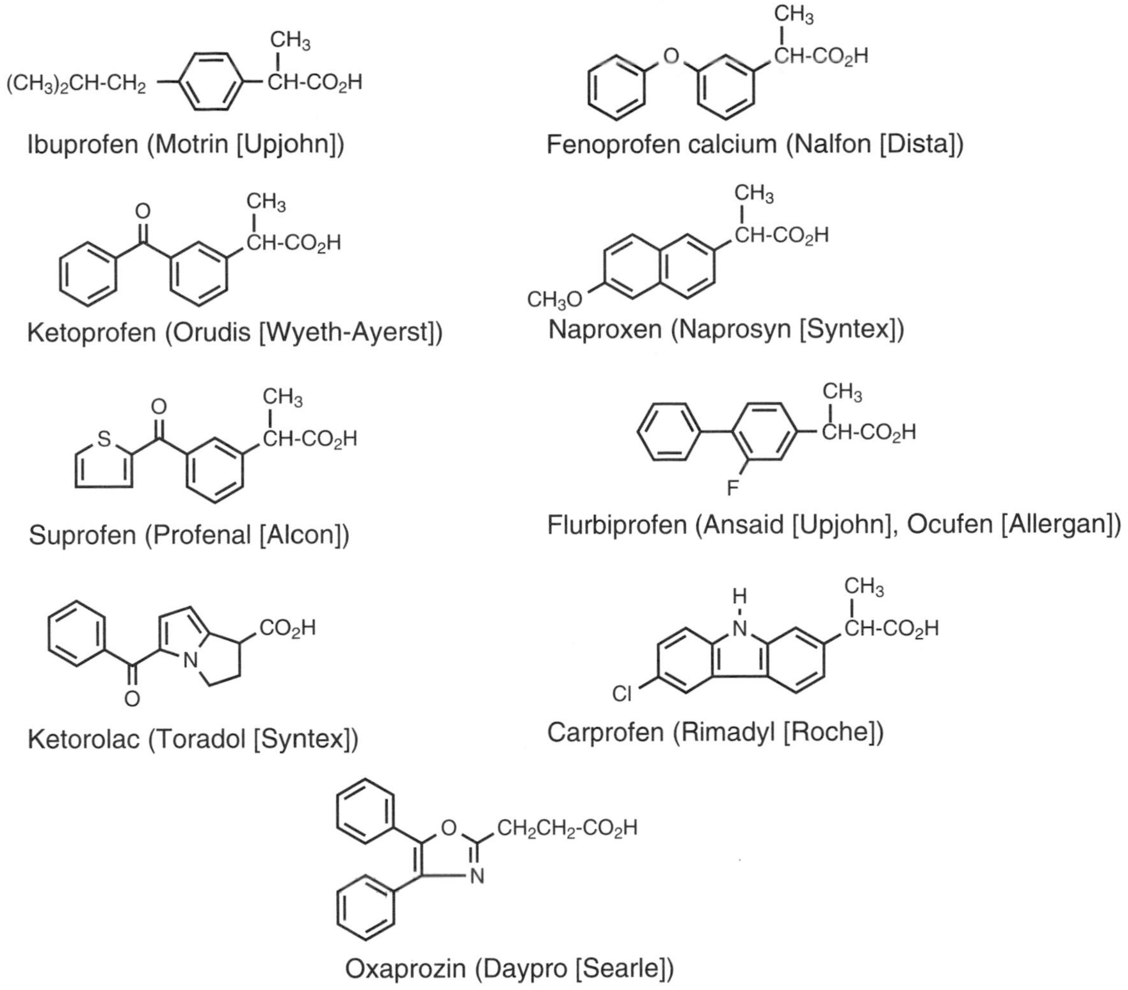

Ibuprofen (Motrin [Upjohn])

Fenoprofen calcium (Nalfon [Dista])

Ketoprofen (Orudis [Wyeth-Ayerst])

Naproxen (Naprosyn [Syntex])

Suprofen (Profenal [Alcon])

Flurbiprofen (Ansaid [Upjohn], Ocufen [Allergan])

Ketorolac (Toradol [Syntex])

Carprofen (Rimadyl [Roche])

Oxaprozin (Daypro [Searle])

Fig. 25–13. Structures of aryl- and heteroacylacetic acid derivatives.

A large number of individuals taking indomethacin experience undesirable effects of the GI tract (nausea, dyspepsia, and diarrhea), the CNS (headache, dizziness, and vertigo) and the ears (tinnitus): many must discontinue its use. As with other arylalkanoic acids, administration of indomethacin with food or milk decreases GI side effects.

Indomethacin is available for the treatment of rheumatoid and acute gouty arthritis, ankylosing spondylitis, and moderate to severe osteoarthritis in a number of dosage forms including 25-mg and 50-mg capsules, 75-mg sustained release capsules, a 25 mg per 5 ml oral suspension, and 50-mg suppositories. An injectable form is also available as the sodium trihydrate salt for intravenous use in premature infants with patent ductus arteriosus. Because of its ability to suppress uterine activity by inhibiting prostaglandin biosynthesis, indomethacin also has an unlabeled use to prevent premature labor.

Sulindac (Clinoril [Merck & Co.]). Sulindac {(Z)-5-fluoro-2-methyl-1-[(4-methylsulfinyl)phenylmethylene]-1H-indene-3-acetic acid} is a yellow crystalline powder which is soluble in water only at alkaline pH. It is stable in alkaline aqueous solutions and in air at 100° C. It was introduced in the United States in 1978 by the same company as indomethacin as a result of chemical studies designed to produce an analog free of the side effects commonly associated with the use of indomethacin, particularly GI irritation. It achieved wide popularity and remains one of the more commonly used NSAIAs. Its synthesis was also reported by Shen's group.[36] Sulindac is a "prodrug" and is converted to a metabolite which appears to inhibit the cyclooxygenase system about eight times as effectively as aspirin. In anti-inflammatory and antipyretic assays it is only about half as potent as indomethacin but is equipotent in analgesic assays.

Structure-Activity Relationships. The use of classical bioisosteric changes in medicinal chemistry drug design were invoked in the design of sulindac. The isosteric replacement of the indole ring with the indene ring system resulted in a derivative with therapeutically useful anti-inflammatory activity and fewer CNS and GI side effects but which possessed other undesirable effects, particularly poor water solubility and resulting crystalluria. The replacement of the N-p-chlorobenzoyl substituent with a benzylidiene function resulted in active derivatives. When the 5-methoxy group of the indene isostere was replaced with a fluorine atom, enhanced analgesic effects were ob-

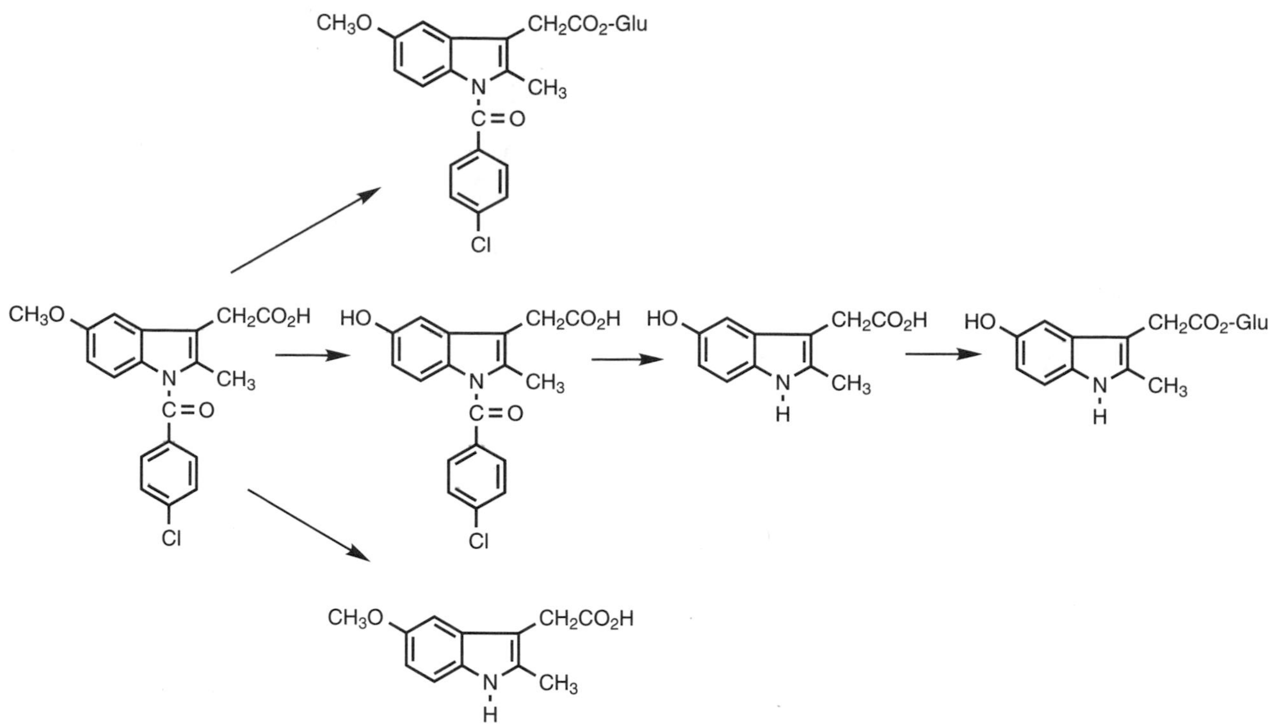

Fig. 25–14. Metabolism of indomethacin. (Glu, glucuronide.)

served. The decreased water solubility of the indene isostere was alleviated by replacing the chlorine atom of the phenyl substituent with a sulfinyl group. The importance of stereochemical features in the action of sulindac, introduced by the benzylidene double bond, is evidenced by the observation that the (Z)-isomer is a much more potent anti-inflammatory agent than the corresponding (E)-isomer. This *cis*-relationship of the phenyl substituent to the aromatic ring bearing the fluoro substituent (see Fig. 25–12) is similar to the proposed conformation of indomethacin suggesting that both indomethacin and sulindac assume similar conformations at the active site of arachidonic acid cyclooxygenase.

Absorption and Metabolism. Sulindac is well absorbed on oral administration (90%), reaches peak plasma levels within 2 to 4 hours and, being acidic (pK_a = 4.5), is highly bound to serum proteins (93%). The metabolism of sulindac plays a major role in its actions because all the pharmacologic activity is associated with its major metabolite. Sulindac is, in fact, a prodrug, the sulfoxide function being reduced to the active sulfide metabolite. Sulindac is absorbed as the sulfoxide that is not an inhibitor of prostaglandin biosynthesis in the GI tract. As discussed earlier, prostaglandins exert a protective effect in the GI tract and inhibition of their synthesis here leads to many of the GI side effects noted for most NSAIAs. Once sulindac enters the circulatory system, it is reduced to the sulfide which is an effective inhibitor of prostaglandin biosynthesis in the joints. Thus, sulindac produces fewer GI side effects, such as bleeding, and ulcerations, than indomethacin and many other NSAIAs. In addition, the active metabolite has a plasma half-life approximately twice that of the parent compound (~16 hours vs. 8 hours), which favorably affects the dosing schedule. In addition to the sulfide metabolite, sulindac is oxidized to the corresponding sulfone, which is inactive. A minor product results from hydroxylation of the benzylidene function and the methyl group at the 2-position. Glucuronides of several metabolites are also found. Sulindac, the sulfide, and the sulfone metabolites are all highly protein-bound. Despite the fact that the sulfide metabolite is a major activation product and is found in high concentration in human plasma, it is not found in human urine, perhaps because of its high degree of protein binding. The major excretion product is the sulfone metabolite and its glucuronide conjugate. The complete metabolism of sulindac is illustrated in Figure 25–15.

Whereas the toxicity of sulindac is lower than that observed for indomethacin and other NSAIAs, the spectrum of adverse reactions is very similar. The side effects most frequently reported are associated with irritation of the GI tract (nausea, dyspepsia, and diarrhea), although these effects are generally mild. Effects on the CNS (dizziness and headache) are less common. Dermatologic effects are less frequently encountered.

Sulindac is indicated for long-term use in the treatment of rheumatoid arthritis, osteoarthritis, ankylosing spondylitis, and acute gouty arthritis. It is available as 150-mg and 200-mg tablets. The usual maximum dosage is 400 mg per day with starting doses recommended at 150 mg twice a day. It is recommended that sulindac be administered with food.

Tolmetin sodium (Tolectin [McNeil Pharmaceuticals]). Tolmetin sodium [1-methyl-5-(4-methylbenzoyl)-1H-pyrrole-

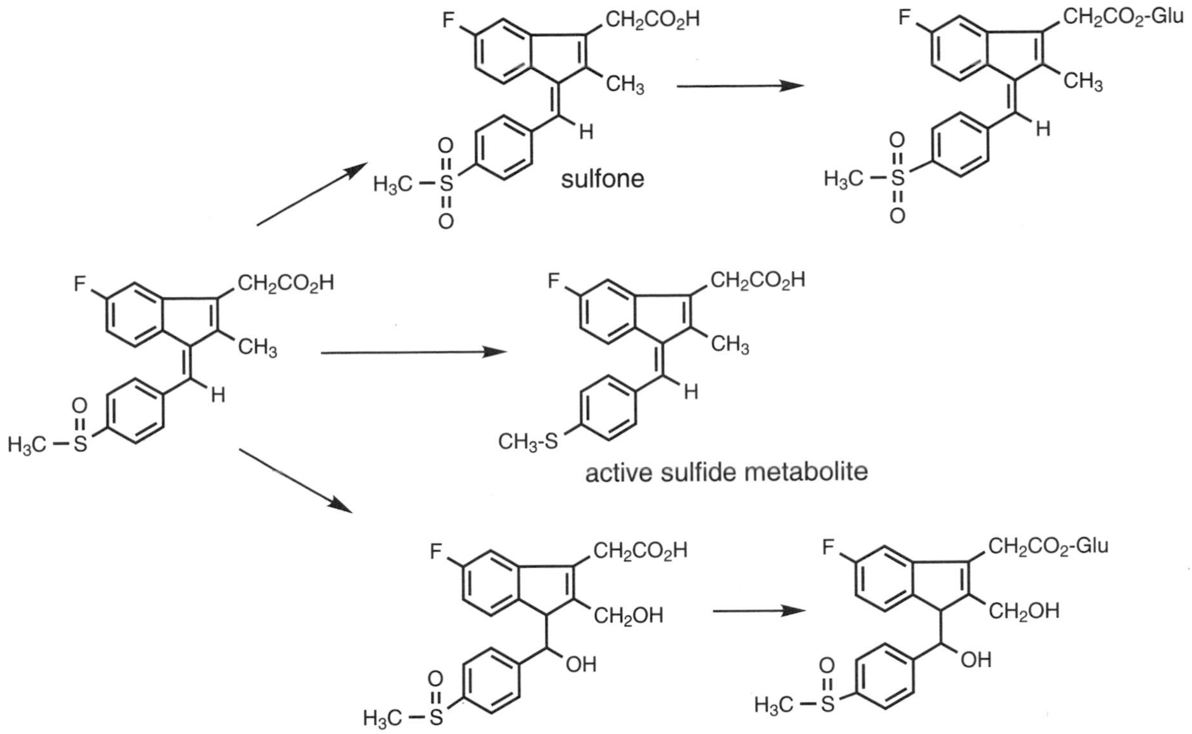

Fig. 25–15. Metabolism of sulindac. (Glu, glucuronide.)

2-acetic acid sodium salt] is a light yellow crystalline solid that is very water soluble. The free acid form, however, is virtually water insoluble. Tolmetin is synthesized straightforwardly from 1-methylpyrrole.[37] It was introduced in the United States in 1976, and like other NSAIAs, tolmetin inhibits prostaglandin biosynthesis. Tolmetin also inhibits polymorph migration and decreases capillary permeability, however. Its anti-inflammatory activity, as measured in carrageenan-induced rat paw edema and cotton pellet granuloma assays, is intermediate between that of phenylbutazone and indomethacin.

Structure-Activity Relationships. The relationship of tolmetin to indomethacin is clear, each containing a *p*-substituted benzoyl group and an acetic acid function. Tolmetin possesses a pyrrole ring instead of the indole ring in indomethacin. Replacement of the 5-p-toluoyl group with a p-chlorobenzoyl moiety produced little effect on activity whereas introduction of a methyl group in the 4-position of the pyrrole ring produced interesting results. The 4-methyl-5-p-chlorobenzoyl analog is approximately four times as potent as tolmetin. This compound was marketed by McNeil in 1980 as zomepirac, an analgesic that was removed from the market in 1983 because of severe anaphylactic reactions, particularly in patients sensitive to aspirin. Unlike the previous structure-activity relationships discussed for arylalkanoic acids, the propionic acid analog is slightly less potent than tolmetin.

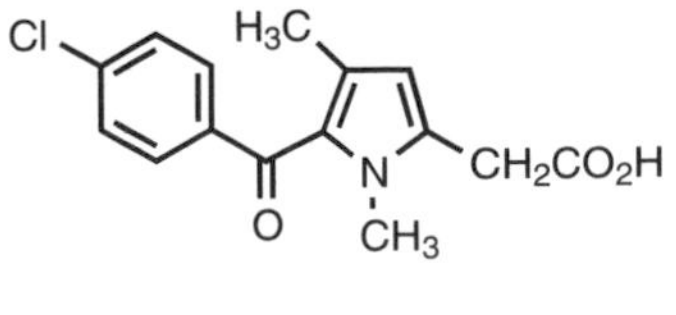

Zomepirac

Absorption and Metabolism. Tolmetin sodium is rapidly and almost completely absorbed on oral administration with peak plasma levels being attained within the first hour of administration. It has a relatively short plasma half-life of about 1 hour. The free acid (pK_a = 3.5) is highly bound to plasma proteins (99%) and excretion of tolmetin and its metabolites occurs primarily in the urine. Tolmetin is extensively metabolized with approximately 70% of the drug being metabolized to the dicarboxylic acid shown below. This metabolite is inactive in standard in vivo anti-inflammatory assays. Approximately 15 to 20% of an administered dose is excreted unchanged and 10% as the glucuronide conjugate of the parent drug. Conjugates of the dicarboxylic acid metabolite account for the majority of the remaining administered drug.

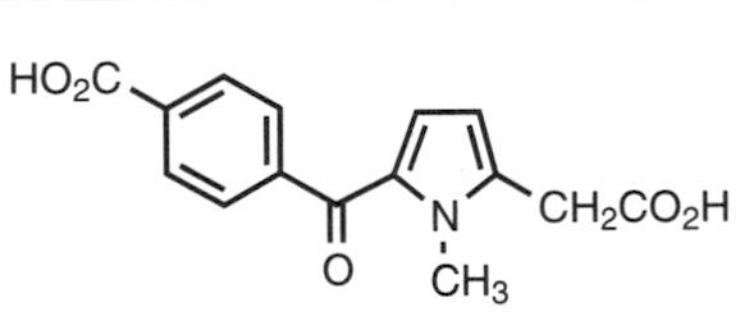

The most frequently adverse reactions are those involving the GI tract (abdominal pain, discomfort, and nausea) but appear to be less than those observed with aspirin. Some CNS effects (dizziness and drowsiness) are also observed. Few cases of overdose have been reported but in such cases recommended treatment involves elimination of the drug from the GI tract by emesis or gastric lavage and elimination of the acidic drug from the circulatory system by enhancing alkalinization of the urine with sodium bicarbonate.

Tolmetin sodium is indicated for the treatment of rheumatoid arthritis, juvenile rheumatoid arthritis, and osteoarthritis. The recommended dosage is 400 mg, 3 times a day. It is available as 200-mg and 600-mg tablets and as 400-mg capsules.

Diclofenac sodium (Voltaren [Geigy]). Diclofenac sodium {2-[(2,6-dichlorophenyl)amino]benzene acetic acid sodium salt} is a faintly yellow-white to light beige, odorless, slightly hygroscopic crystalline powder, which is sparingly soluble in water. It is synthesized from N-phenyl-2,6-dichloroaniline.[38] Diclofenac is available in 120 different countries and is perhaps the most widely used NSAIA in the world, being the eighth largest selling drug overall.[39] It was introduced in the U.S. in 1989, but was first marketed in Japan in 1974. It ranks among the top 30 prescription drugs in the United States. Diclofenac possesses structural characteristics of both the arylalkanoic acid and the anthranilic acid classes of anti-inflammatory agents and displays anti-inflammatory, analgesic, and antipyretic properties. In the carrageenan-induced rat paw edema assay, it is twice as potent as indomethacin and 450 times as potent as aspirin. As an analgesic, it is 6 times more potent than indomethacin and 40 times more potent than aspirin in the phenyl-benzoquinone-induced writhing assay in mice. As an antipyretic it is twice as potent as indomethacin and over 350 times as potent as aspirin in the yeast-induced fever assay in rats. Diclofenac is unique among the NSAIAs in that it possesses three possible mechanisms of action: (1) inhibition of the arachidonic acid cyclooxygenase system (3–1000 times more potent than other NSAIAs on a molar basis) resulting in a decreased production of prostaglandins and thromboxanes; (2) inhibition of the lipoxygenase pathway resulting in decreased production of leukotrienes, particularly the pro-inflammatory leukotriene B_4; and (3) inhibition of arachidonic acid release and stimulation of its re-uptake resulting in a reduction of arachidonic acid availability. Its properties and place in therapy have been extensively reviewed.[39]

Structure-Activity Relationships. Structure-activity relationships in this series have not been extensively studied. It does appear that the function of the two o-chloro groups is to force the anilino-phenyl ring out of the plane of the phenylacetic acid portion, this twisting effect being important in the binding of NSAIAs to the active site of the cyclooxygenase enzyme, as previously discussed.

Absorption and Metabolism. It is rapidly and completely (~100%) absorbed on oral administration with peak plasma levels being reached within 1.5 to 2.5 hours. The free acid ($pK_a = 4.0$) is highly bound to serum proteins(99.5%), primarily albumin. Only 50 to 60% of an oral dose is bioavailable because of extensive hepatic metabolism. Four major metabolites resulting from aromatic hydroxylation have been identified. The major metabolite, the 4′-hydroxy derivative, accounts for 20 to 30% of the dose excreted whereas the three others, the 5-hydroxy, the 3′-hydroxy, and the 4′,5-dihydroxy metabolites each account for 10 to 20% of the excreted dose. The remaining drug is excreted in the form of sulfate conjugates. Although the major metabolite is much less active than the parent compound, it may contribute to the overall biologic activity because it accounts for 30 to 40% of all of the metabolic products. The metabolism of diclofenac is illustrated in Figure 25–16.

Diclofenac sodium is indicated for the treatment of rheumatoid arthritis, osteoarthritis, and ankylosing spondylitis. Recommended doses range from 100–200 mg per day in divided doses depending on the indication. It is available as 25-mg, 50-mg, and 75-mg enteric-coated tablets.

Etodolac (Lodine [Wyeth-Ayerst]). Etodolac (1,8-diethyl-1,3,4,9-tetrahydropyrano-[3,4-b]indole-1-acetic acid) is a white, crystalline compound, which is insoluble in water but soluble in most organic solvents. It is promoted as the first of a new chemical class of anti-inflammatory agents, the pyranocarboxylic acids. Although not strictly an arylacetic acid derivative (because there is a two-carbon atom separation between the carboxylic acid function and the hetero-aromatic ring), it still possesses structural characteristics similar to those of the hetero-arylacetic acids and is classified here as such. It was introduced in the United States in 1991 for acute and long-term use in the management of osteoarthritis, and as an analgesic. It is registered in 28 other countries around the world, and it also possesses antipyretic activity. Etodolac is marketed as a racemic mixture although only the (S)(+)-enantiomer possesses anti-inflammatory activity in animal models. Etodolac also displays a high degree of enantioselectivity in its inhibitory effects on the arachidonic acid cyclooxygenase system. With regard to its anti-inflammatory actions, etodolac was about 50 times more active than aspirin, 3 times more potent than sulindac, and one-third as active as indomethacin. The ratio of the anti-inflammatory activity to the ED_{50} for gastric ulceration or erosion, however, was more favorable for etodolac ($ID_{50}/ED_{50} = 10$) than for aspirin, naproxen, suldinac, or indomethacin ($ID_{50}/ED_{50} = 4$). At 2.5 to 3.5 times the effective anti-inflammatory dose, etodolac was reported to produce less GI bleeding than indomethacin, ibuprofen, or naproxen. The primary mechanism of action appears to be inhibition of the biosynthesis of prostaglandins at the cyclooxygenase step, with no inhibition of the lipoxygenase system. Etodolac, however, possesses a more favorable ratio of inhibition of prostaglandin biosynthesis in human rheumatoid synoviocytes and chondrocytes than by cultured human gastric mucosal cells compared to ibuprofen, indomethacin, naproxen, diclofenac, or piroxicam. Thus, although etodolac is no more potent an NSAIA than many others, the lower incidence of GI side effects represents a potential therapeutic advantage.

Structure-Activity Relationships. During a search for newer, more effective antiarthritic agents in the 1970s, the Ayerst group led by Humber investigated a series of pyranocarboxylic acids of the general structure shown below.[40] Structure-activity relationships studies indicated

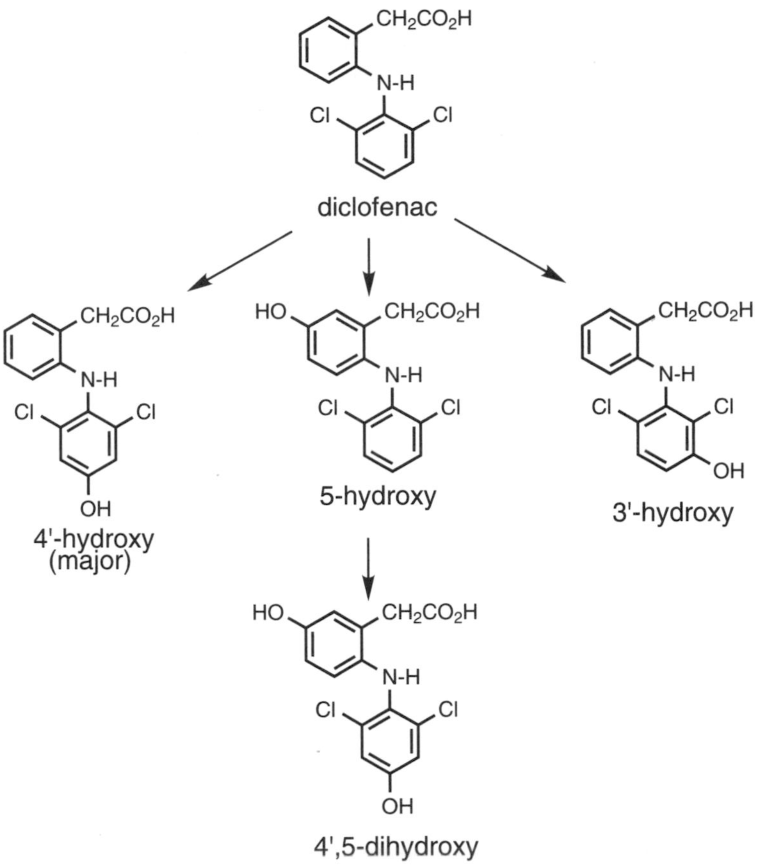

diclofenac

4'-hydroxy (major)

5-hydroxy

3'-hydroxy

4',5-dihydroxy

Fig. 25–16. Metabolism of diclofenac.

that alkyl groups at R_1 and an acetic acid function at R_2 enhanced anti-inflammatory activity. Lengthening the acid chain, or ester or amide derivatives, gave inactive compounds. The corresponding α-methylacetic acid derivatives were also inactive. Increasing the chain length of the R_1 substituent to ethyl or n-propyl gave derivatives which were 20 times more potent than methyl. A number of substituents on the aromatic ring were evaluated and substituents at the 8-position proved most beneficial. Among the most active were the 8-ethyl, 8-n-propyl, and 7-fluoro-8-methyl derivatives. Etodolac was found to possess the most favorable anti-inflammatory to gastric distress properties among these analogs.

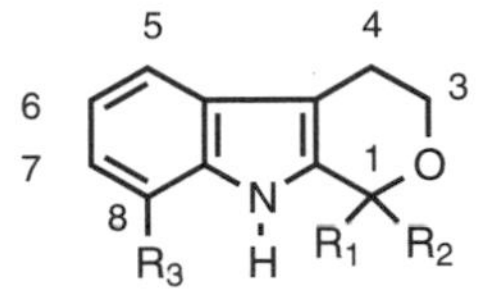

Absorption and Metabolism. Etodolac is rapidly absorbed following oral administration with maximum serum levels being achieved within 1 to 2 hours and is highly bound to plasma proteins (99%). The penetration of etodolac into synovial fluid is greater than or equal to tolmetin, piroxicam, or ibuprofen. Only diclofenac appears to provide greater penetration. Etodolac is metabolized to three hydroxylated metabolites and to glucuronide conjugates, none of which possesses important pharmacologic activity. Metabolism appears to be the same in the elderly as in the general population, so no adjustment of dosage appears necessary.

Etodolac is indicated for the management of the signs and symptoms of osteoarthritis and for the management of pain. The usual recommended dosage 800 to 1200 mg per day for osteoarthritis and 200 to 400 mg every 6 to 8 hours as needed for pain and is available as 200 mg and 300 mg capsules.

Nabumetone (Relafen [SK Beecham]). Nabumetone [4-(6-methoxy-2-naphthalenyl)-2-butanone] is a nonacidic, white to off-white, crystalline compound which is insoluble in water but soluble in most organic solvents. It is unique among the NSAIAs in that it represents a new class of nonacidic prodrugs, being rapidly metabolized after absorption to form a major active metabolite. It is synthesized from 2-acetyl-6-methoxynaphthalene[41] and was introduced in the United States in 1992. Gastric damage produced by NSAIAs generally involves a dual insult mechanism (Fig. 25–17). Most NSAIAs are acidic substances which produce a primary insult by direct acid damage, an indirect contact effect, and a back diffusion of hydrogen ions. The secondary insult results from inhibition of prostaglandin biosynthesis in the GI tract where prostaglandins exert a cytoprotective effect. The dual insult leads to gastric damage. Nabumetone, being nonacidic, does not produce a significant primary insult and is an ineffectual inhibitor of prostaglandin cyclooxygenase in gastric mucosa, thus producing minimum second-

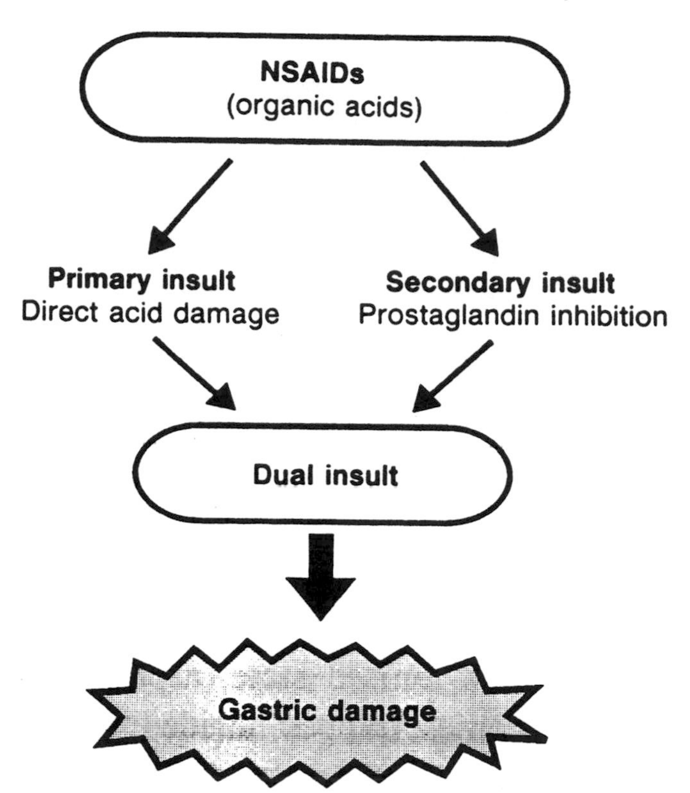

Fig. 25–17. NSAIA-induced production of gastric damage by a dual insult mechanism. (Reprinted, with permission from A. Goudie et al., *J. Med. Chem., 21,* 1260[1978].)

ary insult. The result is that gastric side effects of nabumetone appear to be minimized. Once the parent drug enters the circulatory system, however, it is metabolized to an active metabolite, 6-methoxynaphthalene-2-acetic acid (6MNA), which is an effective inhibitor of prostaglandin synthesis in joints. Nabumetone thus represents a classic example of the prodrug approach in drug design.

In the carrageenan-induced rat paw assay, nabumetone is approximately 13 times more potent than aspirin, a third as active as indomethacin and half as active as diclofenac. It is only half as active as aspirin as an analgesic as measured by the phenylquinone-induced writhing assay in mice. Despite its lower potency, advantages to nabumetone may reside in the favorable gastric irritancy profile. The ratio of gastric irritancy in rats (GTD_{50}) to anti-inflammatory activity in rats (ED_{50}) for nabumetone is 21.25, whereas this ratio is 0.41 for aspirin, 0.55 for indomethacin, 0.72 for diclofenac, 3.00 for tolmetin, and 7.85 for zomepirac.

Structure-Activity Relationships. Introduction of methyl or ethyl groups on the butanone side chain greatly reduced anti-inflammatory activity. The ketone function can be converted to a dioxolane with retention of activity while converting the ketone to an oxime reduced activity. Removal of the methoxy group at the 6-position reduced activity, but replacement of the methoxy with a methyl or chloro group gave active compounds. Replacement of the methoxy with hydroxyl, acetoxy, or N-methylcarbamoyl groups, or positional isomers of the methoxy group at the 2- or 4-positions, greatly reduced activity. The active metabolite, 6MNA, is closely related structurally to naproxen, differing only by the lack of an α-methyl group. The ketone precursor [4-(6-methoxy-2-naphthyl)-pentan-2-one], however, which would be expected to produce naproxen as a metabolite, was inactive in chronic models of inflammation.

Absorption and Metabolism. Nabumetone is absorbed primarily from the duodenum. Milk and food increase the rate of absorption and the bioavailability of the active metabolite. Plasma concentrations of unchanged drug are too low to be detected in most subjects after oral administration, so most pharmacokinetic studies have involved the disposition of the active metabolite. Pharmacokinetic properties are altered in elderly patients with higher plasma levels of the active metabolite having been noted. Nabumetone undergoes rapid and extensive metabolism in the liver with a mean absolute bioavailability of the active metabolite of 38%. The metabolism of nabumetone is illustrated in Figure 25–18. The major and most active metabolite is 6MNA but the initial alcohol metabolite, a minor product, and its esters, also possess anti-inflammatory properties.

Nabumetone is indicated for the acute and chronic treatment of the signs and symptoms of osteoarthritis and rheumatoid arthritis. The recommended starting dosage is 1000 mg as a single dose with or without food. More symptomatic relief of severe or persistent symptoms may be obtained at doses of 1500 mg or 2000 mg per day. It is available as 500-mg and 750-mg tablets.

Aryl- and Heteroarylpropionic Acids.

Ibuprofen (Motrin [Upjohn], Rufen [Boots]). Ibuprofen [2-(p-isobutylphenyl)propionic acid] is a colorless, crystalline solid which is only very slightly soluble in water, but is soluble in most organic solvents. The synthesis of ibuprofen was originally reported in 1964 from *p*-isobutylacetophenone[42] but the drug was not marketed in the United States until 1974 despite the fact that it had been available for several years in Europe. It was the first NSAIA introduced since indomethacin and was immediately accepted in therapy. Its success was a factor in the introduction of many new agents in the 1970s. Ibuprofen was the first arylpropionic acid derivative to be marketed in the United States. This chemical class currently comprises the largest group of NSAIAs under investigation with as many as 25 derivatives in various stages of development. It recently became the first prescription NSAIA to become available as an over-the-counter analgesic in almost 30 years and is available under a number of trade names, Advil (Whitehall) and Nuprin (Bristol-Myers) perhaps being among the more widely used. The continuing popularity of ibuprofen is evidenced by the appearance of proprietary and nonproprietary forms in the list of top 200 prescription drugs in the United States. It is marketed as the racemic mixture although biologic activity resides almost exclusively in the (S)(+)-isomer. Ibuprofen is more potent than aspirin but less potent than indomethacin in anti-inflammatory and prostaglandin biosynthesis inhibition assays. It produces moderate degrees of gastric irritation.

Structure-Activity Relationship. The substitution of an α-methyl group on the alkanoic acid portion of many

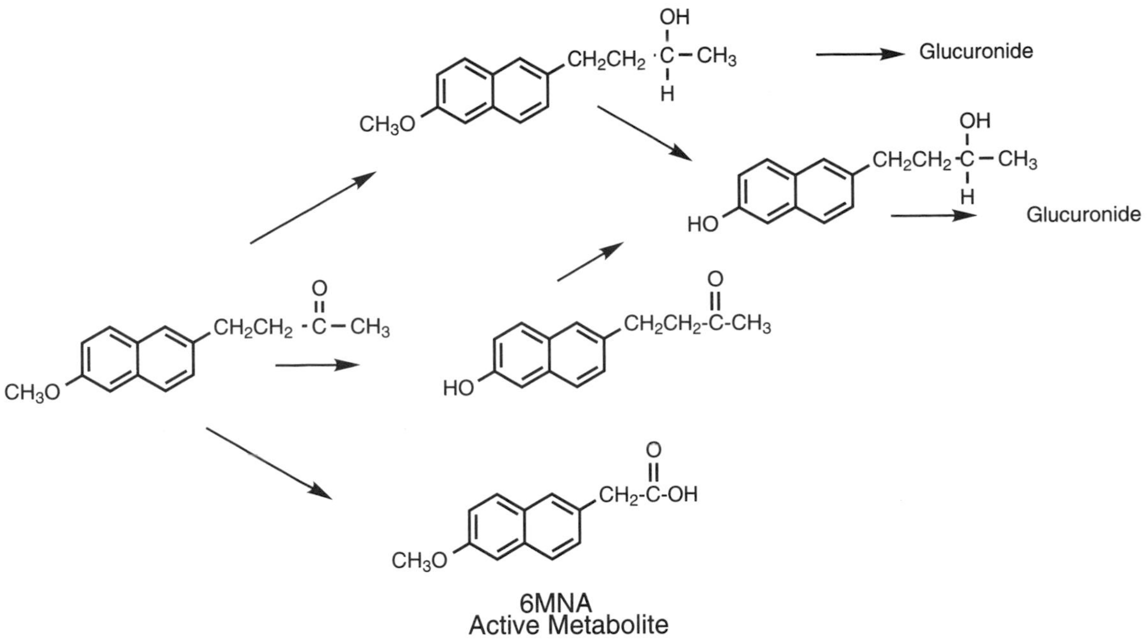

6MNA
Active Metabolite

Fig. 25–18. Metabolism of nambumetone.

acetic acid derivatives enhances anti-inflammatory actions and reduces many side effects. For example, the acetic acid analog of ibuprofen, ibufenac (ρ-isobutylphenylacetic acid), is less potent and more hepatotoxic than ibuprofen. The stereochemistry associated with the chiral center in the arylpropionic acids, but lacking in the acetic acid derivatives, plays an important role in both the in vivo and in vitro activities of these agents. As indicated earlier, although marketed as a racemic mixture, the (+)-enantiomer of ibuprofen possesses greater activity in vitro than the (−)-isomer. The eudismic (S/R) ratio for the inhibition of bovine prostaglandin synthesis is approximately 160, but in vivo, the two enantiomers are equiactive (see section on metabolism). The (+)-enantiomer of ibuprofen, and of most of the arylpropionic acids under investigation, has been shown to possess the (S)-absolute configuration.

Absorption and Metabolism. Ibuprofen is rapidly absorbed on oral administration with peak plasma levels being generally attained within 2 hours. As with most of these acidic NSAIAs, ibuprofen (pK_a = 4.43) is extensively bound to plasma proteins (99%) and will interact with other acidic drugs which are protein bound. Metabolism occurs rapidly and the drug is nearly completely excreted in the urine as unchanged drug and oxidative metabolites within 24 hours following administration. Metabolism involves primarily ω-1 and ω-2 oxidation of the *ρ*-isobutyl side chain, followed by alcohol oxidation of the primary alcohol resulting from ω-oxidation to the corresponding carboxylic acid. All metabolites are essentially inactive. When ibuprofen is administered as the individual enantiomers, the major metabolites isolated are the (+)-isomers regardless of the configuration of the administered enantiomer. Interestingly, the (R)(−)-enantiomer is inverted to the (S)-(+)-enantiomer in vivo, accounting for the observation that the two enantiomers are bioequivalent in vivo. This is a metabolic phenomenon that has also been observed for other arylpropionic acids such as ketoprofen, benoxaprofen, fenoprofen and naproxen.[43] The metabolism of ibuprofen is shown in Figure 25–19.

Ibuprofen is indicated for the relief of the signs and symptoms of rheumatoid arthritis and osteoarthritis, the relief of mild to moderate pain, the reduction of fever, and the treatment of dysmenorrhea. The recommended anti-arthritic dose is 1.2 to 3.2 g per day and the analgesic dose is 400 mg every 4 to 6 hours as necessary. It is available as 200, 300, 400, 600, and 800-mg tablets and as a fruit-flavored suspension of 100 mg per 5 ml. It is also available in a number of over-the-counter products.

Fenoprofen calcium (Nalfon [Dista]). Fenoprofen calcium [2-(3-phenoxyphenyl)propionic acid calcium salt] is a white, crystalline powder that is slightly soluble in water. The calcium and sodium salts of fenoprofen possess similar bioavailability, distribution, and elimination characteristics, however, it is the calcium salt that is marketed because it has the advantage of being less hygroscopic. Its original synthesis was reported in 1970[44] and it was marketed in the United States in 1976. Fenoprofen is less potent in anti-inflammatory assays than ibuprofen, indomethacin, ketoprofen, or naproxen. As an inhibitor of prostaglandin biosynthesis it is much less potent than indomethacin, more potent than aspirin, and about equipotent as ibuprofen. It also possesses analgesic and antipyretic activity. It possesses other pharmacologic properties such as inhibition of phagocytic and complement functions and stabilization of lysosomal membranes. Fenoprofen is marketed as a racemic mixture because no differences have been observed in the in vivo anti-inflammatory or analgesic properties of the individual enanti-

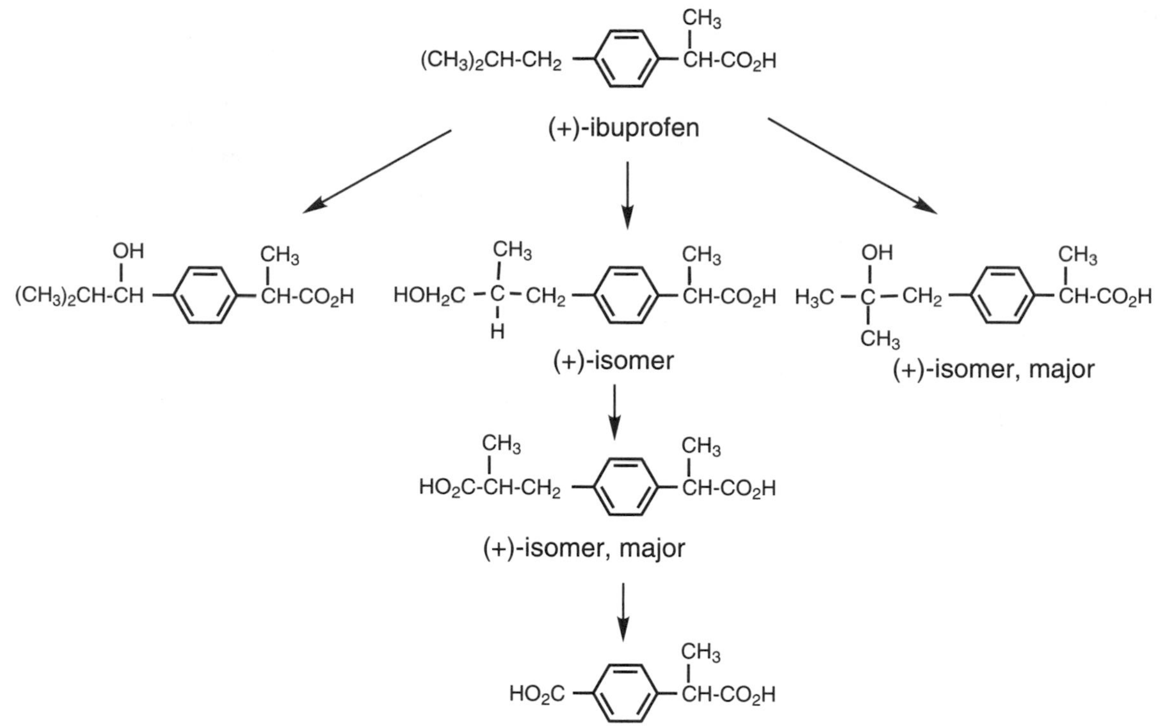

(+)-ibuprofen

(+)-isomer

(+)-isomer, major

(+)-isomer, major

Fig. 25–19. Metabolism of ibuprofen.

omers. The ability of (R)(−)-arylpropionic acids to undergo inversion to the (S)(+)-enantiomers may be involved, however. Like other NSAIAs, in vitro prostaglandin synthesis assays indicate that the (S)(+)-enantiomer is more potent than the (R)(−)-isomer.

Structure-Activity Relationship. Placing the phenoxy group in the ortho- or para-position of the arylpropionic acid ring markedly decreases activity. Replacement of the oxygen bridge between the two aromatic rings with a carbonyl group yields an analog (ketoprofen) which is also marketed.

Absorption and Metabolism. Fenoprofen is readily absorbed (85%) upon oral administration and is highly bound (99%) to plasma proteins. Peak plasma levels are attained within 2 hours of administration. The free acid has a pK_a of 4.5, which is within the range of the other arylalkanoic acids. Fenoprofen is rather extensively metabolized, primarily through glucuronide conjugation with the parent drug and the 4′-hydroxy metabolite.

Fenoprofen calcium is indicated for treatment of osteo- and rheumatoid arthritis and the relief of mild to moderate pain. The usual recommended antirheumatic dose is 300 to 600 mg, 3 or 4 times a day and the analgesic dose is 200 mg every 4 to 6 hours as needed. Total doses should not exceed 3.2 g/day. Fenoprofen calcium is available as 200-mg and 300-mg capsules and 600-mg tablets.

Ketoprofen (Orudis [Wyeth-Ayerst]). Ketoprofen [2-(3-benzoylphenyl)propionic acid] is a white or off-white, odorless, fine to granular powder that is practically insoluble in water but is soluble in most organic solvents. It was introduced in 1986 and has gained wide acceptance. It is synthesized from 2-(p-aminophenyl)propionic acid via a thiaxanthone intermediate.[45] Ketoprofen, unlike many other NSAIAs, inhibits the synthesis of leukotrienes and leukocyte migration into inflamed joints in addition to inhibiting the biosynthesis of prostaglandins. It stabilizes the lysosomal membrane during inflammation resulting in decreased tissue destruction. Antibradykinin activity has also been observed. Bradykinin is released during inflammation and can activate peripheral pain receptors. In addition to anti-inflammatory activity, ketoprofen also possesses antipyretic and analgesic properties. Although it is less potent than indomethacin as an anti-inflammatory agent and as an analgesic, its ability to produce gastric lesions is about the same.[46]

Absorption and Metabolism. Ketoprofen is rapidly and nearly completely absorbed on oral administration, reaching peak plasma levels within 0.5 to 2 hours. It is highly plasma protein-bound (99%) despite a lower acidity (pK_a = 5.94) than some other NSAIAs. Wide variation in plasma half-lives have been reported. It is metabolized by glucuronide conjugation of the carboxylic acid, hydroxylation of the benzoyl ring, and reduction of the keto function.

Ketoprofen is indicated for the long-term management of rheumatoid arthritis and osteoarthritis, for mild to moderate pain, and for primary dysmenorrhea. The usual recommended dosage for the arthritic indications is 150 to 300 mg per day divided into 3 or 4 doses, not to exceed 300 mg per day. The dose range for primary dysmenorrhea is 25 to 50 mg every 6 to 8 hours as needed. It is available as 25-mg, 50-mg, and 75-mg capsules.

Naproxen (Naprosyn [Syntex]). Naproxen [(+)-2-(6-methoxy-2-naphthyl)propionic acid] is a white to off-white, crystalline powder possessing a slightly bitter taste. Solubility in water is attained only at alkaline pH. The racemic mixture is synthesized from 2-methoxynaphthalene and the (+)-isomer obtained by resolution with cinchonidine.[47] It was introduced in the United States in 1976 and has consistently been among the more popular NSAIAs, ranking in the top 15 prescription drugs almost since its introduction. It is marketed as the (S)(+)-enantiomer but, interestingly, the sodium salt of the (−)-isomer is also on the market as Anaprox (Syntex), which itself ranks among the top 200 prescription drugs in the United States. As an inhibitor of prostaglandin biosynthesis, it is 12 times more potent than aspirin, 10 times more potent than phenylbutazone, 3 to 4 times more potent than ibuprofen and 4 times more potent than fenoprofen, but is about 300 times less potent than indomethacin. In vivo anti-inflammatory assays are consistent with this relative order of potency. In the carrageenan rat paw edema assay, it is 11 times more potent than phenylbutazone and 55 times as potent as aspirin but only 0.7 times as potent as indomethacin. In the phenylquinone writhing assay for analgesia, it is 9 times as potent as phenylbutazone and 7 times as potent as aspirin but only 10% as potent as indomethacin. In the yeast-induced pyrexia assay for antipyretic activity, it is 7 times as potent as phenylbutazone, 22 times as potent as aspirin, and 1.2 times as potent as indomethacin. The order of gastric ulcerogenic activity is sulindac < naproxen < aspirin, indomethacin, ketoprofen, and tolmetin.

Structure-Activity Relationships. In a series of substituted 2-naphthylacetic acids, substitution in the 6-position led to maximum anti-inflammatory activity. Small lipophilic groups such as Cl, CH_3S, and CHF_2O were active analogs with CH_3O being the most potent. Larger groups were found to be less active. Derivatives of 2-naphthylpropionic acids are more potent than the corresponding acetic acid analogs. Replacing the carboxyl group with functional groups capable of being metabolized to the carboxyl function (e.g., $-CO_2CH_3$, $-CHO$ or $-CH_2OH$) led to a retention of activity. As with other propionic acid derivatives, the (S)(+)-isomer is the more potent enantiomer. Naproxen is the only arylalkanoic acid NSAIA marketed as optically active isomers.

Absorption and Metabolism. Naproxen is almost completely absorbed following oral administration. Peak plasma levels are achieved within 2 to 4 hours following administration. Like most of the acidic NSAIAs (pK_a = 4.15), it is highly bound (99.6%) to plasma proteins. Approximately 70% of an administered dose is eliminated as either unchanged drug (60%) or as conjugates of unchanged drug (10%). The remainder is converted to the 6-desmethyl metabolite (5%) and the glucuronide conjugate of the demethylated metabolite (22%). The 6-desmethyl metabolite lacks anti-inflammatory activity. Like most of the arylalkanoic acids the most common side effect associated with the use of naproxen is irritation to the GI tract. The most frequent other adverse reactions are associated with CNS disturbances, such as nausea and dizziness.

Naproxen is indicated for the treatment of rheumatoid arthritis, osteoarthritis, juvenile arthritis, ankylosing spondylitis, tendinitis, bursitis, acute gout, primary dysmenorrhea, and for the relief of mild to moderate pain. The recommended dose for arthritic indications (except juvenile arthritis where the total daily dose is 10 mg/kg in 2 divided doses) is 250 to 500 mg twice a day with a maximum dose of 1.5 g/day for limited periods. As an analgesic and for primary dysmenorrhea, the recommended dose is 500 mg followed by 250 mg every 6 to 8 hours not to exceed 1.25 g total daily dose. It is available as 250-mg, 375-mg, and 500-mg tablets and as an oral suspension (125 mg per 5 ml).

As stated, the sodium salt of the (−)-enantiomer is available as Anaprox, which generally has the same indications and dosage regimens. Anaprox is not indicated in juvenile arthritis and daily doses should not exceed 1.375 g. It is available as 275-mg and 550-mg tablets.

Suprofen (Profenal [Alkon]). Suprofen [α-methyl-4-(2-thienylcarbonyl)benzeneacetic acid] is a white, microcrystalline powder which is slightly soluble in water. It was originally synthesized from thiophene in 1974[48] and was introduced in the United States in 1985 for the treatment of dysmenorrhea and as an analgesic for mild to moderate pain. Reports of severe flank pain and transient renal failure appeared, however, the syndrome being noted abruptly within several hours after one or two doses of the drug, and suprofen was removed from the U.S. market in May, 1987. Obviously, clinical trials are not sufficient to determine a drug's safety and postmarketing surveillance becomes most important. Suprofen was reintroduced in the United States in 1990 as a 1% ophthalmic solution for the prevention of surgically induced miosis during cataract extraction. Miosis complicates the removal of lens material and implantation of a posterior chamber intraocular lens which thus increases the risk of ocular trauma. The mechanism of this action also involves inhibition of prostaglandin synthesis because prostaglandins constrict the iris sphincter independent of a cholinergic mechanism. Additionally, prostaglandins also break down the blood-aqueous barrier allowing the influx of plasma proteins into aqueous humor resulting in an increase in intraocular pressure. There is a possibility of suprofen being re-introduced as an anti-inflammatory agent.

Flurbiprofen (Ansaid [Upjohn], Ocufen [Allergan]). Flurbiprofen [2-(2-fluoro-4-biphenylyl)propionic acid] is a white or slightly yellow, crystalline powder that is soluble in water at pH 7.0 and readily soluble in most polar organic solvents. Its synthesis was originally reported in 1973.[49] During a study of the pharmacologic properties of a large number of substituted phenylalkanoic acids, including ibuprofen and ibufenac, the most potent were found to be substituted 2-(4-biphenyl)propionic acids. Further toxicologic and pharmacologic studies indicated that flurbiprofen possessed the most favorable therapeutic profile, so it was selected for further clinical development. It was not marketed until 1987 when its sodium salt was introduced as Ocufen as the first topical NSAIA indicated for ophthalmic use in the United States. The indication of Ocufen is the same as that of Profenal, that is, to inhibit intraoperative miosis induced by prostaglandins in cataract surgery. Thus, flurbiprofen is an inhibitor of prostaglandin synthesis. The free acid oral form was introduced in 1988 as Ansaid (Upjohn) (*a*nother *n*on-*s*teroidal *a*nti-*i*nflammatory *d*rug) and gained immediate acceptance. An estimated 3.2 million prescriptions were dispensed in 1989 (3.2% of the total NSAIA prescriptions) thrusting it immediately into the top 200 prescription drugs. It currently ranks among the top 80 most often prescribed drugs in the United States. In acute inflammation assays in adrenalectomized rats, flurbiprofen was found to be 536 times more potent than aspirin and 100 times more potent than phenylbutazone. Orally, it was half as potent as methylprednisolone. As an antipyretic, it was 403 times as potent as aspirin in the yeast-induced fever assay in rats and was 26 times more potent than ibuprofen as an antinociceptive.

Absorption and Metabolism. Flurbiprofen is well absorbed after oral administration with peak plasma levels being attained within 1.5 hours. Food alters the rate of absorption but not the extent of its bioavailability. It is extensively bound to plasma proteins (99%), and has a plasma half-life of 2 to 4 hours. Metabolism is extensive with 60 to 70% of flurbiprofen and its metabolites being excreted as sulfate and glucuronide conjugates. Flurbiprofen shows some interesting metabolic patterns with 40 to 47% produced as the 4′-hydroxy metabolite, 5% as the 3′,4′-dihydroxy metabolite, 20 to 30% as the 3′-hydroxy-4′-methoxy metabolite and the remaining 20 to 25% of the drug being excreted unchanged. None of these metabolites demonstrates significant anti-inflammatory activity. The metabolism of flurbiprofen is presented in Figure 25–20.

Flurbiprofen is indicated as an oral formulation for the short- or long-term treatment of rheumatoid arthritis and osteoarthritis and as an ophthalmic solution for the inhibition of intraoperative miosis. The recommended oral dose is 200 to 300 mg in divided doses 2, 3, or 4 times a day. The ophthalmic dose is one drop approximately every half hour beginning 2 hours before surgery. It is available as 50-mg and 100-mg tablets and as a 0.03% solution.

Ketorolac tromethamine (Toradol [Syntex]). Ketorolac (5-benzoyl-2,3-dihydro-1H-pyrrolizine-1-carboxylic acid tromethamine salt) represents a cyclized heteroarylpropionic acid derivative with the α-methyl group being fused to the pyrrole ring. It was introduced in 1990 and is indicated for use only as a peripheral analgesic for short-term use although it exhibits anti-inflammatory and antipyretic activity as well. It was initially introduced only in an injectable form, but recently an oral formulation has been made available. Its analgesic activity resembles that of the centrally-acting analgesics with 15 to 30 mg of ketorolac producing analgesia equivalent to a 12-mg dose of morphine and it has become a widely accepted alternative to narcotic analgesia. Ketorolac inhibits prostaglandin synthesis. Although the analgesic effect is achieved within 10 minutes of injection, peak analgesia lags behind peak plasma levels by 45 to 90 minutes. The free acid has a pK_a of 3.54 and is highly plasma protein bound > 99%). Ketorolac is metabolized to the ρ-hydroxy derivative and to conjugates that are excreted primarily in the urine.

Ketorolac tromethamine is available as 15-mg/ml and 30-mg/ml injectable solutions and as 10-mg tablets. The recommended intramuscular dose is 30 or 60 mg as a loading dose followed by half the loading dose every 6 hours as needed. The oral dose regimen is 10 mg as needed every 4 to 6 hours.

Carprofen (Rimadyl [Roche]). Carprofen was approved by the FDA as a NSAIA in 1987 but Roche has not yet introduced the drug on the U.S. market. It appears to be an effective anti-inflammatory agent with analgesic and antipyretic properties and a low incidence of serious GI side effects but does not appear to offer an advantage over currently marketed agents.

Oxaprozin (Daypro [Searle]). Oxaprozin (4,5-diphenyl-2-oxazole propionic acid) was approved by the FDA in late 1992 for the short- and long-term treatment of osteoarthritis and rheumatoid arthritis, but not marketed until 1993. Although not formally a propionic acid derivative of the α-methylacetic acid type, it appears to be similar to the propionic acid derivatives considered here. The recommended dose is two 600-mg caplets once a day.

N-Arylanthranilic Acids (Fenamic Acids)

The anthranilic acid class of NSAIAs result from the application of classic medicinal chemistry bioisosteric

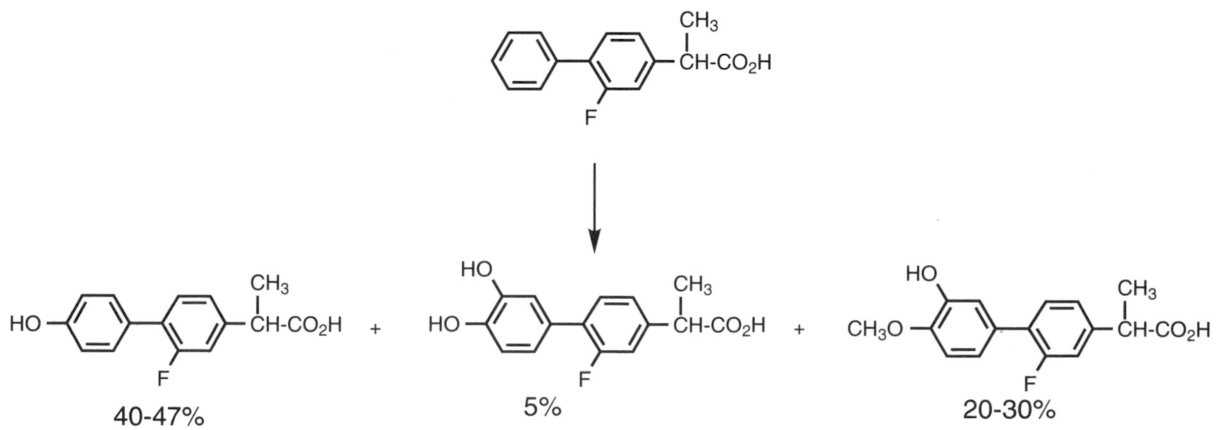

Fig. 25–20. Metabolism of flurbiprofen.

drug design concepts as these derivatives are nitrogen isosteres of salicylic acid. In the early 1960s, the Parke-Davis research group reported the development of a series of N-substituted anthranilic acids that have since been given the chemical class name of fenamic acids. The fact that this class of compounds possess little advantage over the salicylates with respect to their anti-inflammatory and analgesic properties has diminished interest in their development on a scale relative to that of the arylalkanoic acids. Mefenamic acid was introduced in the United States in 1967 as an analgesic—this remains its primary indication despite its having modest anti-inflammatory activity. Flufenamic acid has been available in Europe as an antirheumatic agent, but there are no apparent plans to introduce this drug in the United States. With regard to anti-inflammatory activity, mefenamic acid is approximately 1.5 times as potent as phenylbutazone and half as potent as flufenamic acid. Meclofenamic acid was introduced in the United States as its sodium salt in 1980 primarily as an antirheumatic agent and analgesic. The structures of these fenamic acids are shown in Figure 25–21.

The fenamic acids share a number of pharmacologic properties with the other NSAIAs. Because these agents are potent inhibitors of prostaglandin biosynthesis, it is tempting to speculate that this represents their primary mechanism of action. Scherrer, like Shen, had proposed a hypothetical receptor for NSAIAs and later modified[50a] the receptor to represent the active site of arachidonic acid cyclooxygenase. Structurally, the fenamic acids fit the proposed active site of arachidonic acid cyclooxygenase proposed by Shen[15] (See Fig. 25–1) because they possess an acidic function connected to an aromatic ring along with an additional lipophilic binding site, in this case the N-aryl substituent. The greater anti-inflammatory activity of meclofenamic acid compared to that of mefenamic acid correlates well with their ability to inhibit prostaglandin synthesis. Scherrer[50b] compared the in vivo anti-inflammatory activities, clinical anti-inflammatory doses in humans, and the in vitro inhibition of prostaglandin synthesis activities of mefenamic acid, meclofenamic acid, phenylbutazone, indomethacin, and aspirin and suggested an important role of prostaglandin synthesis inhibition in the production of therapeutic effects of the fenamic acids.

Side effects are those primarily associated with GI disturbances (dyspepsia, discomfort, and especially diarrhea), some CNS effects (dizziness, headache, and drowsiness), skin rashes, and transient hepatic and renal abnormalities. Isolated cases of hemolytic anemia have been reported.

Structure-Activity Relationships

Substitution on the anthranilic acid ring generally reduces activity whereas substitution on the N-aryl ring can lead to conflicting results. In the UV erythema assay for anti-inflammatory activity, the order of activity is generally $3' > 2' >> 4'$ for monosubstitution with the $3'$-CF_3 derivative (flufenamic acid) being particularly potent. The opposite order of activity was observed, however, in the rat paw edema assay, the $2'$-Cl derivative being more potent than the $3'$-Cl analog. In disubstituted derivatives, where the nature of the two substituents is the same, $2',3'$-disubstitution appears to be the most effective. A plausible explanation may be found in an examination of the proposed topography of the active sites of arachidonic acid cyclooxygenase using either the Shen or Sherrer models. Proposed binding sites include a hydrophobic trough to which a lipophilic group, non-coplanar with the ring bearing the carboxylic acid function binds. Substituents on the N-aryl ring which force this ring to be non-coplanar with the anthranilic acid ring should enhance binding at this site and thus enhance activity. This may account for the enhanced anti-inflammatory activity of meclofenamic acid (which has two ortho-substituents

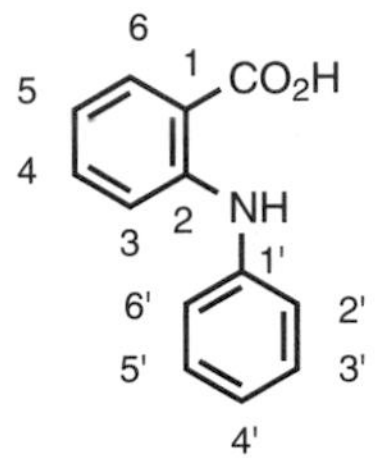

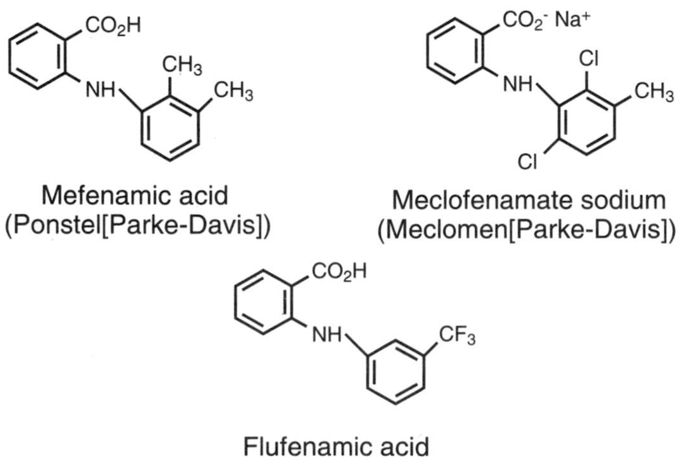

Mefenamic acid (Ponstel[Parke-Davis])

Meclofenamate sodium (Meclomen[Parke-Davis])

Flufenamic acid

Fig. 25–21. Structures of N-arylanthranilic acids (fenamic acids).

forcing this ring out of the plane of the anthranilic acid ring) over flufenamic acid (no ortho-substituents) and mefenamic acid (one ortho-substituent). Meclofenamic acid possesses 25 times greater anti-inflammatory activity than mefenamic acid. The NH-moiety of anthranilic acid appears to be essential for activity since replacement of the NH function with O, CH_2, S, SO_2, N-CH_3, or N-$COCH_3$ functionalities significantly reduces activity. Finally, the position, rather than the nature, of the acidic function is critical for activity as is the case for the salicylates. Anthranilic acid derivatives are active whereas the m- and ρ-aminobenzoic acid analogs are not. Replacement of the carboxylic acid function with the isosteric tetrazole moiety has little effect on activity.

Drug Interactions

The pK_a's of the N-arylanthranilic acids (4.0–4.2) resemble those of the arylalkanoic acids, thus it is not surprising that they are strongly bound to plasma proteins with interactions with other drugs which are highly protein-bound being very probable. The most common interactions reported are those of mefenamic acid and meclofenamic acid with oral anticoagulants. Concurrent administration of aspirin results in a reduction of plasma levels of meclofenamic acid.

Preparations Available

Mefenamic acid (Ponstel [Parke-Davis]). Mefenamic acid [N-(2,3-xylyl)anthranilic acid] is a white to off-white crystalline solid with a bitter aftertaste. It will darken if exposed to light for long periods but is otherwise stable at room temperature. It is virtually water insoluble except at an alkaline pH. It is synthesized from o-chlorobenzoic acid and 2,3-dimethylaniline under catalytic conditions.[51] Mefenamic acid is the only fenamic acid derivative which produces analgesia centrally and peripherally.

Absorption and Metabolism. Mefenamic acid is rapidly absorbed following oral administration with peak plasma levels being attained within 2 to 4 hours. It is highly bound to plasma proteins (78.5%) and has a plasma half-life of 2 to 4 hours. Metabolism occurs through regioselective oxidation of the 3′-methyl group and glucuronidation of mefenamic acid and its metabolites. Urinary excretion accounts for approximately 50 to 55% of an administered dose with unchanged drug accounting for 6%, the 3′-hydroxymethyl metabolite (primarily as the glucuronide) accounting for 25%, and the remaining 20% as the dicarboxylic acid (of which 30% is the glucuronide conjugate). These metabolites are essentially inactive. The metabolism of mefenamic acid is illustrated in Figure 25–22.

Mefenamic acid is indicated for the short term relief of moderate pain and for primary dysmenorrhea. The recommended dose is 500 mg initially followed by 250 mg every 6 hours as needed and should be given with food. It is available as 250-mg capsules.

Meclofenamate sodium (Meclomen [Parke-Davis]). Meclofenamate sodium [N-2-(2,6-dichloro-m-tolyl)anthranilic acid sodium salt] is a white powder that is highly water soluble. The pH of an aqueous solution is 8.7. It is synthesized from o-methyl-ρ-hydroxyacetophenone.[52]

Absorption and Metabolism. Meclofenamate sodium is rapidly and almost completely absorbed following oral administration reaching peak plasma levels within 2 hours. It is highly bound to plasma proteins (99.8%) and has a plasma half-life of 2 to 4 hours. Metabolism involves oxidation of the methyl group, aromatic hydroxylation, monodehalogenation and conjugation. Urinary excretion accounts for approximately 75% of the administered

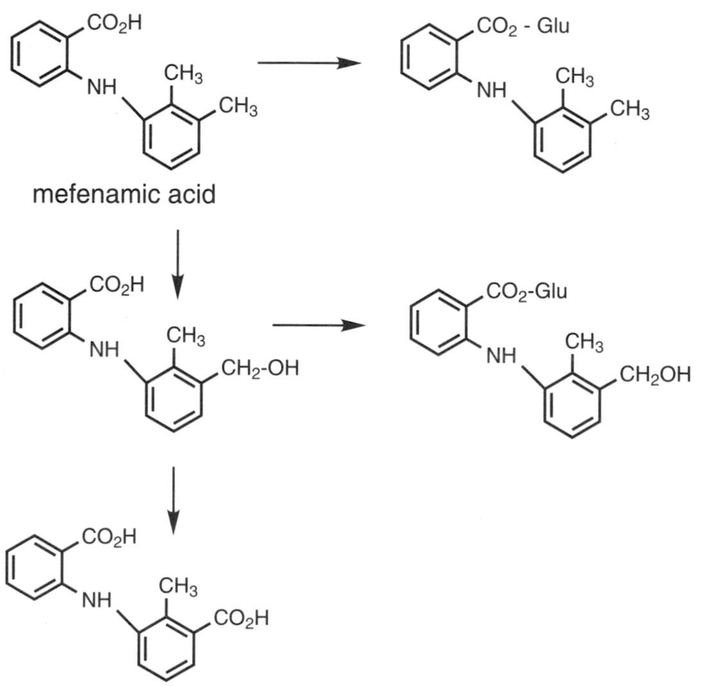

Fig. 25–22. Metabolism of mefenamic acid (glu, glucuronide).

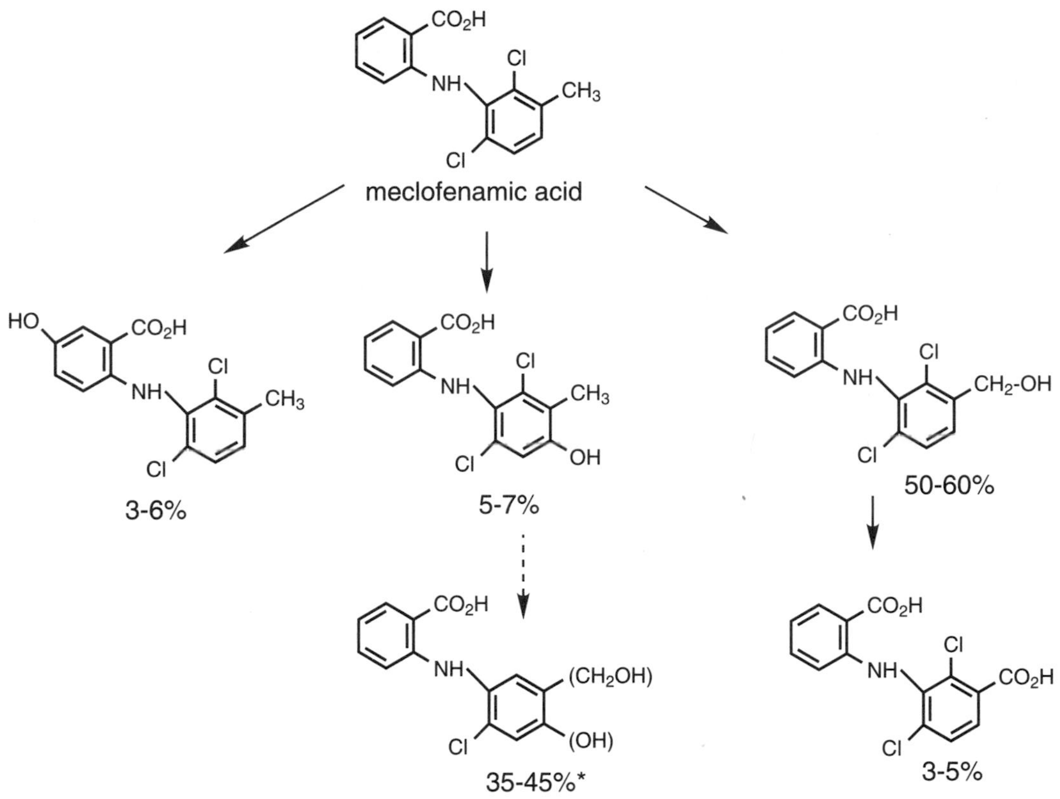

Fig. 25–23. Metabolism of meclofenamic acid of the dehalogenated metabolite (the asterisk indicates that hydroxylation occurs on either the methyl group or the aromatic ring).

dose. The major metabolite is the product of methyl oxidation and has been shown to possess anti-inflammatory activity. The metabolism of meclofenamic acid is illustrated in Figure 25–23.

Meclofenamate sodium is indicated for relief of mild to moderate pain, short- and long-term treatment of rheumatoid arthritis and osteoarthritis, treatment of primary dysmenorrhea, and treatment of idiopathic heavy menstrual blood loss. The analgesic dose is 50 mg every 4 to 6 hours whereas the arthritic dose is 200 to 400 mg per day in 3 or 4 equal doses. It is available as 50 mg and 100 mg capsules.

Oxicams

The term oxicams has been adopted by the USAN Council to describe the relatively new enolic acid class of 4-hydroxy-1,2-benzothiazine carboxamides with anti-inflammatory and analgesic properties. These structurally distinct agents resulted from extensive studies by the Pfizer group in an effort to produce non-carboxylic acid, potent, and well-tolerated anti-inflammatory agents. As part of this study, several series were prepared and evaluated including 2-aryl-1,3-indanediones, 2-arylbenzothiophen-3-(2H)-one 1,1-dioxides, dioxoquinoline-4-carboxamides, and 3-oxa-2H-1,2-benzothiazine-4-carboxamide 1,1-dioxides. These results, combined with the previously known activity of 1,3-dicarbonyl derivatives such as phenylbutazone, led to the development of the oxicams. The first member of this class, piroxicam, was introduced in the United States in 1982 as Feldene (Pfizer) and gained immediate acceptance in the United States where it was among the top 50 prescription drugs for several years. Although piroxicam is the only agent of this class in therapeutic use today, several others are being clinically evaluated, and thus the oxicams represent a potentially growing class of NSAIAs.

Piroxicam is quite active in standard in vivo assays, being 200 times more potent than aspirin and at least 10 times as potent as any other standard agent in the UV erythema assay, equipotent with indomethacin, more potent than phenylbutazone or naproxen in the carrageenan rat paw edema assay, and equipotent with indomethacin and 15 times more potent than phenylbutazone in the rat adjuvant arthritis assay. It is less potent than indomethacin, equipotent with aspirin, and more potent than fenoprofen, ibuprofen, naproxen, and phenylbutazone as an analgetic in the phenylquinone writhing assay. Piroxicam inhibits the migration of polymorphonuclear cells into inflammatory sites and also inhibits the release of lysosomal enzymes from these cells. It also inhibits collagen-induced platelet aggregation. It is an effective inhibitor of arachidonic acid cyclooxygenase, being almost equipotent with indomethacin and more potent than ibuprofen, tolmetin, naproxen, fenoprofen, phenylbutazone, and aspirin in the inhibition of prostaglandin biosynthesis by methylcholanthrene-transformed mouse fibroblasts (MC-5) assay. A template for designing anti-inflammatory agents based upon CPK space filling models of the peroxy radical precursor of PGG and inhibitors of cyclooxygenase was proposed,[53] the ability of oxicams, particularly piroxicam, to inhibit this enzyme was subsequently rationalized on the ability of oxicams to assume a conformation resembling that of the peroxy radical precursor.[54]

Approximately 20% of individuals on piroxicam report adverse reactions. Not unexpectedly, the greatest incidence of side effects are those resulting from GI disturbances. The incidence of peptic ulcers reported is less than 1%, however.

Structure-Activity Relationships

Within the series of 4-hydroxy-1,2-benzothiazine carboxamides represented by the following structure, optimum activity was observed when R_1 was a methyl substituent. The carboxamide substituent, R, is generally an aryl or heteroaryl substituent because alkyl substituents are less active. Oxicams are acidic compounds with pK_as that range between 4 and 6. N-heterocyclic carboxamides are generally more acidic than the corresponding N-aryl carboxamides and this enhanced acidity was attributed[55] to stabilization of the enolate anion by the pyridine nitrogen atom as illustrated in tautomer A and additional stabilization by tautomer B:

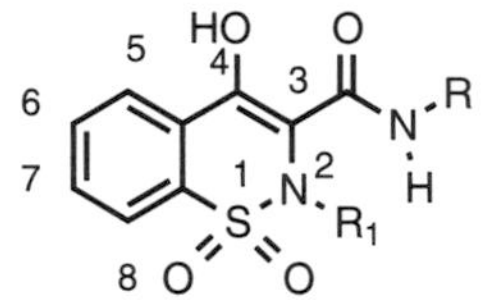

Piroxicam (Feldene{Pfizer})

B

A

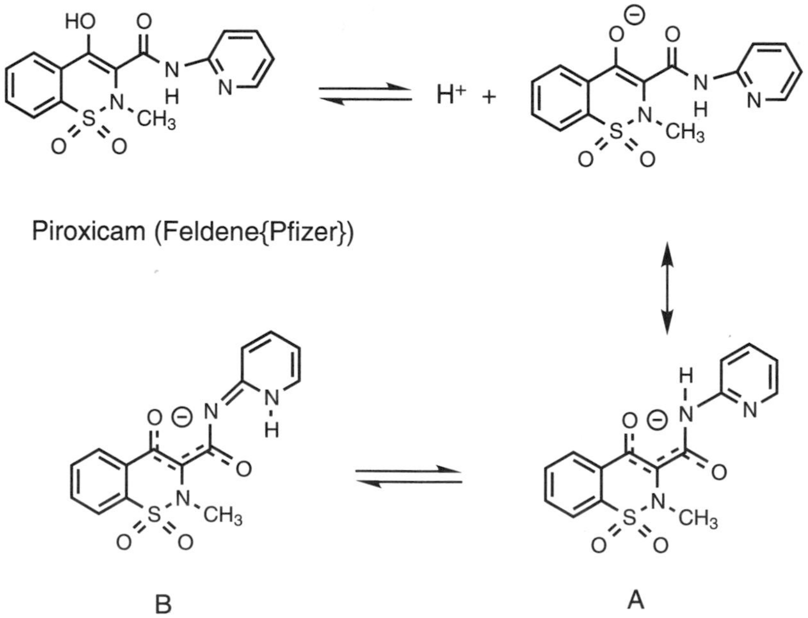

This explains the observation that primary carboxamides are more potent than the corresponding secondary derivatives because no N-H bond would be available to enhance the stabilization of the enolate anion. When the aryl group is o-substituted variable results were obtained, whereas m-substituted derivatives are generally more potent than the corresponding ρ-isomers. In the aryl series, maximum activity is observed with a m-chloro substituent. No direct correlations were observed between acidity and activity, nor with partition coefficient, electronic or spatial properties in this series. Two major differences, however, are observed when R = heteroaryl rather than aryl: pK_as are generally between 2 and 4 units lower and anti-inflammatory activity increases as much as sevenfold. The greatest activity is associated with the 2-pyridyl (as in piroxicam), 2-thiazolyl, or 3-(5-methyl)isoxazolyl ring systems, the latter derivative (isoxicam) having been withdrawn from the European market in 1985 following several reports of severe skin reactions. In addition to possessing activity equal to or greater than indomethacin in the carrageenan rat paw edema assay, the heteroaryl carboxamides also possess longer plasma half-lives providing an improvement in dosing scheduling regimens.

Metabolism

Although the metabolism of piroxicam varies quantitatively from species to species, qualitative similarities are found in the metabolic pathways found in humans, rats, dogs, and rhesus monkeys. It is extensively metabolized in humans with less than 5% of an administered dose being excreted unchanged. The major metabolites in humans result from hydroxylation of the pyridine ring and subsequent glucuronidation, other metabolites being of lesser importance. Aromatic hydroxylation at several positions of the aromatic benzothiazine ring also occurs; two hydroxylated metabolites have been extracted from rat urine. On the basis of NMR deuterium-exchange studies, hydroxylation at the 8-position was ruled out indicating that hydroxylation occurs at two of the remaining positions. Other novel metabolic reactions occur. Cyclodehydration gave a tetracyclic metabolite (the major metabo-

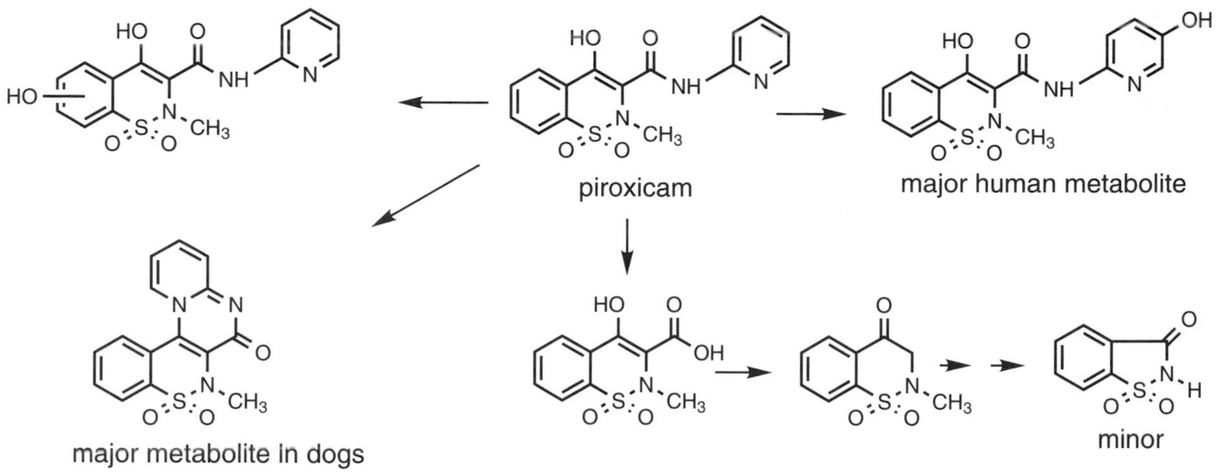

Fig. 25–24. Metabolism of piroxicam.

lite in dogs), whereas ring contraction following amide hydrolysis and decarboxylation eventually yields saccharin. All known metabolites of piroxicam lack anti-inflammatory activity. For example, the major human metabolite is 1000 less effective as an inhibitor of prostaglandin biosynthesis than piroxicam itself. Related oxicams undergo different routes of metabolism. For example, sudoxicam (the N-2-thiazolyl analogue) undergoes primarily hydroxylation of the thiazole ring followed by ring-opening whereas isoxicam undergoes primarily cleavage reactions of the benzothiazine ring. The metabolism of piroxicam is illustrated in Figure 25–24.

Drug Interactions

Few reports of therapeutically significant interactions of oxicams with other drugs have appeared. Concurrent administration of aspirin has been shown to reduce piroxicam plasma levels by about 20%, although the anticoagulant effect of acenocoumarin is potentiated, presumably as a result of plasma protein displacement.

Preparations Available

Piroxicam (Feldene [Pfizer]). Piroxicam [4-hydroxy-2-methyl-N-2-pyridinyl-2H,1,2-benzothiazine-3-carboxamide 1,1-dioxide] is a white crystalline solid that is sparingly soluble in water. The parent ring system is synthesized by ring expansion reactions of saccharin derivatives.[56] Piroxicam is readily absorbed on oral administration reaching peak plasma levels in about 2 hours. Peak plasma levels appear to be lower when given with food at low doses (30 mg) with no differences appearing at a 60 mg dose but, in general, food does not markedly affect bioavailability. Being acidic ($pK_a = 6.3$), the drug is highly bound to plasma proteins (99.3%). Piroxicam possesses an extended plasma half-life (38 hours) making single daily dosing possible.

Piroxicam is indicated for long-term use in rheumatoid arthritis and osteoarthritis. The initial recommended maintenance dose is a single 20 mg dose that may be divided. It is available in 10-mg and 20-mg capsules.

NSAIA-INDUCED GASTROENTEROPATHY

The effectiveness and popularity of the NSAIAs in the United States and Europe make this class one of the most commonly used classes of therapeutic entities. Over 75 million prescriptions are written annually for the NSAIAs, including aspirin. Unfortunately, almost all of the current agents share the undesirable property of producing damaging effects to gastric and intestinal mucosa resulting in erosion, ulcers, and gastrointestinal bleeding. These major adverse reactions are drawbacks to the use of NSAIAs. As the use of these agents increases, so does the incidence of these side effects. These acute and chronic injuries to gastric mucosa result in lesions referred to as NSAIA gastropathy, which differs from peptic ulcer disease by the localization of these lesions more frequently in the stomach rather than in the duodenum. NSAIA-induced lesions also occur more frequently in the elderly than typical peptic ulcers. Normally the stomach protects itself from the harmful effects of hydrochloric acid and pepsin through protective mechanisms collectively referred to as the gastric mucosal barrier, which consists of epithelial cells, the mucus and bicarbonate layer, and mucosal blood flow. Gastric mucosa is actually a gel consisting of polymers of glycoprotein which limit the diffusion of hydrogen ions. These polymers reduce the rate at which hydrogen ions (produced in the lumen) and bicarbonate ion (secreted by the mucosa) mix and thus a pH gradient is created across the mucus layer. Normally, gastric mucosal cells are rapidly repaired when they are damaged by factors such as food, ethanol, or acute ingestion of NSAIAs. Among the cytoprotective mechanisms is the ability of prostaglandins of the PGE series, particularly PGE_1, to increase the secretion of bicarbonate ion and mucus and to maintain mucosal blood flow. PGs also decrease acid secretion permitting the gastric mucosal barrier to remain intact.

As mentioned earlier, Figure 25–17 illustrates the ability of aspirin and NSAIAs to induce gastric damage by a dual insult mechanism. Aspirin and the NSAIAs are acidic substances that can damage the GI tract, even in the absence of hydrochloric acid, by changing the permeability of cell membranes allowing a back diffusion of hydrogen ions. These weak acids remain unionized in the stomach but the resulting lipophilic nature of these agents allows an accumulation or concentration in gastric mucosal cells. Once inside these cells, however, the higher pH of the intracellular environment causes the acids to dissociate and become trapped within the cells. The permeability of the mucosal cell membrane is thus altered and the accumulation of hydrogen ions causes mucosal cell damage. This gastric damage results, therefore, from the primary insult of acidic substances. As detailed earlier in this chapter, the primary mechanism of action of the NSAIAs is to inhibit the biosynthesis of PGs at the cyclooxygenase step. The resulting inhibition of PG biosynthesis in the GI tract prevents the PGs from exerting their protective mechanism on gastric mucosa and thus the NSAIAs induce gastric damage through this secondary insult mechanism.

The use of PGE_1 itself to reduce NSAIA-induced gastric damage is limited by the fact that it is ineffective orally and degrades rapidly on parenteral administration, primarily by oxidation of the 15-hydroxy group. To overcome these limitations, misoprostol was synthesized. Misoprostol is a PG prodrug analogue in which oral activity was achieved by administering the drug as the methyl ester allowing the bioactive acid to be liberated after absorption. Oxidation of the 15-hydroxy group of the PGs was overcome by moving the hydroxy group to the 16-position thus "fooling" the enzyme prostaglandin 15-OH dehydrogenase. Oxidation was further limited by the introduction of a methyl group at the 16-position producing a tertiary alcohol which is more difficult to oxidize than the secondary alcohol group of the PGs. Misoprostol exists as a mixture of stereoisomers at the 16-position. It

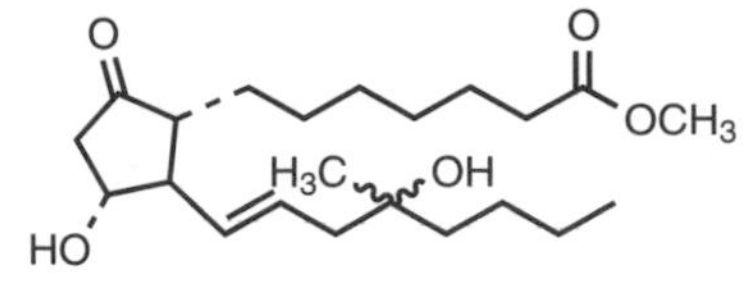

misoprostol

was introduced in 1989 in Cytotec (Searle) for the prevention of NSAIA-induced gastric ulcers (but not duodenal ulcers) in patients at high risk of complications from a gastric ulcer, particularly the elderly and patients with concomitant debilitating disease and in individuals with a history of gastric ulcers (see Table 25–3).

DISEASE-MODIFYING ANTIRHEUMATIC DRUGS (DMARDs)

The agents previously discussed as NSAIAs have proven to be beneficial in the symptomatic treatment of arthritic disorders and are a popular therapeutic regimen. Despite their effectiveness and popularity, however, it should be remembered that none of these agents is effective in preventing or inhibiting the underlying pathogenic, chronic, inflammatory processes. Recent interest has been generated by agents that are effective in the treatment of arthritic disorders yet fail to demonstrate significant activity in the standard screening assays for antiarthritic agents.

Table 25–3. Properties of NSAIAs

Drugs	Date Introduced	Antiinflammatory Dose (mg/day)	Peak Blood Level (hrs)	pK_a	Plasma Binding (%)
Aspirin	1899	3200–6000	2	3.5	90
Diflunisal	1982	500–1000	2–3	3.3	99
Phenylbutazone	1952	300–600	2–2.5	4.5	98
Oxyphenylbutazone				4.7	92
Indomethacin	1965	75–150	2–3	4.5	97
Sulindac*	1978	400	2–4	4.5	93
Tolmetin	1976	1200	<1	3.5	99
Diclofenac	1989	100–200	1.5–2.5	4.0	99.5
Etodolac	1991	800–1200	1–2		
Nabumetone* 6MNA	1992	1500–2000	2.5–4	Neutral	99
Ibuprofen	1974	1200–3200	2	4.43	99
Fenoprofen Ca	1976	1200–2400	2	4.5	99
Ketoprofen	1986	150–300	0.5–2	5.94	99
Naproxen	1976	500–1000	2–4	4.15	99.6
Flurbiprofen	1988	200–300	1.5	4.22	99
Mefenamic acid	1967	1000	2–4	4.2	78.5
Meclofenamate Na	1980	200–400	2	4.0	99.8
Piroxicam	1982	20	2	6.3	99.3

* Prodrugs

Disease-modifying antirheumatic drugs (DMARDs) differ from the previously discussed drugs in that they retard or halt the underlying progression of arthritis while lacking anti-inflammatory and analgesic effects observed for the NSAIAs. They are much slower acting, taking as long as 3 months for measurable clinical benefits to be observed. These agents also possess potentially dangerous adverse side effects which in many cases, limit their long-term use. Yet DMARDs are effective in reducing joint destruction and the progression of arthritic disorders in patients. The group of drugs that comprise the DMARDs include the gold compounds, some antimalarial agents, sulfhydryl compounds such as penicillamine, and immunosuppressive agents.

Gold compounds

At the end of the 19th century, the chemotherapeutic applications of heavy metal derivatives were gaining considerable interest. Among those metals receiving great attention were gold compounds, or gold salts. The first of these, gold cyanide, was found to be effective in vitro against Mycobacterium tuberculosis. This discovery prompted an extension of the use of gold compounds to other disease states, thought to be tubercular in origin. Early clinical observations had suggested some similarities in the symptoms of tuberculosis and rheumatoid arthritis and some thought rheumatoid arthritis to be an atypical form of tuberculosis. In 1927, aurothioglucose was found to relieve joint pain when used to treat bacterial endocarditis. The era of chyrsotherapy had begun. Subsequent investigations led to an extensive study of gold compounds in Great Britain by the Empire Rheumatism Council, which reported in 1961 that sodium aurothiomalate was effective in slowing the development of progressive joint diseases. Both aurothioglucose and sodium aurothiomalate are orally ineffective and have to be administered by injection. In 1985 the first orally effective gold compound for arthritis, auranofin, was introduced in the United States. Several other gold compounds have been evaluated clinically but do not appear to offer efficacy or toxicity advantages.

The biochemical and pharmacologic properties shared by gold compounds are quite diverse. The mechanisms by which they produce their antirheumatic actions has not been totally determined. The earlier observations that gold compounds were effective in preventing arthritis induced by hemolytic streptococci and by pleuropneumonia-like organisms led to the postulation that they acted through an antimicrobial mechanism. The inability of gold compounds to consistently inhibit mycoplasma growth in vitro while inhibiting the arthritic process independent of microbial origins, however, suggested that they did not directly produce their effects by this mechanism. The involvement of immunologic processes in the pathogenesis of arthritis suggested that a direct suppression of the immunologic response by gold compounds was involved. Available evidence, however, suggests that although enzymatic mediators released as a result of the immune response may be inhibited, no direct effect on either immediate or delayed cellular responses is evident to suggest any immunosuppressive mechanism. Suggestions have been made that protein denaturation and macroglobulin formation cause the proteins to become antigenic, thus initiating the immune response producing biochemical changes in connective tissue, which ultimately lead to rheumatoid arthritis. The possibility that gold compounds inhibit the aggregation of macroglobulins and in turn inhibit the formation of immune complexes may account for their ability to slow connective tissue degradation. Interaction with collagen fibrils and thus reduction of collagen reactivity that alters the course of the arthritic process has also been postulated. Perhaps the most widely accepted mechanism of action is related to the ability of gold compounds to inhibit lysosomal enzymes, the release of which promotes the inflammatory response. The lysosomal enzymes glucuronidase, acid phosphatase, collagenase and acid hydrolases are inhibited, presumably through a reversible interaction of gold with sulfhydryl groups on the enzymes. Gold thiomalate inhibits glucosamine-6-phosphate synthetase, a rate-limiting step in mucopolysaccharide biosynthesis, a property shared to a lesser extent by several NSAIAs. Gold sodium thiosulfate is a potent uncoupler of oxidative phosphorylation. Gold sodium thiomalate is also a fairly effective inhibitor of prostaglandin biosynthesis in vitro, but the relationship of this effect to the antiarthritic actions of gold compounds has not been clarified.

Toxic side effects have been associated with the use of gold compounds with the incidence of reported adverse reactions in patients on chrysotherapy being as high as 55%. Serious toxicity occurs in 5 to 10% of reported cases. The most common adverse reactions include dermatitis (erythema, papular, vesicular, and exfoliative dermatitis), mouth lesions (stomatitis preceded by a metallic taste and gingivitis), pulmonary disorders (interstitial pneumonitis), nephritis (albuminuria and glomerulitis) and hematologic disorders (thrombocytopenic purpura, hypoplastic and aplastic anemia, and eosinophilia; blood dyscrasias are less common but can be severe). Less commonly reported reactions are GI disturbances (nausea, anorexia, and diarrhea), ocular toxicity (keratitis with inflammation and ulceration of the cornea and subepithelial deposition of gold in the cornea), and hepatitis. In those cases in which severe toxicity occurs, excretion of gold can be markedly enhanced by the administration of chelating agents, the two most common of which are dimercaprol (British Anti-Lewisite, BAL) and penicillamine. Corticoids also suppress the symptoms of gold toxicity and the concomitant administration of dimercaprol and corticosteroids has been recommended in cases of severe gold intoxication.

Structure-Activity Relationships

Structure-activity relationships of gold compounds have not received a great amount of attention. Two important relationships have been established, however: (1) monovalent gold (aurous ion, Au^{+}) is more effective than trivalent gold (auric ion, Au^{+++}) or colloidal gold and (2) only those compounds in which aurous ion is attached to a sulfur-containing ligand are active. The nature of the

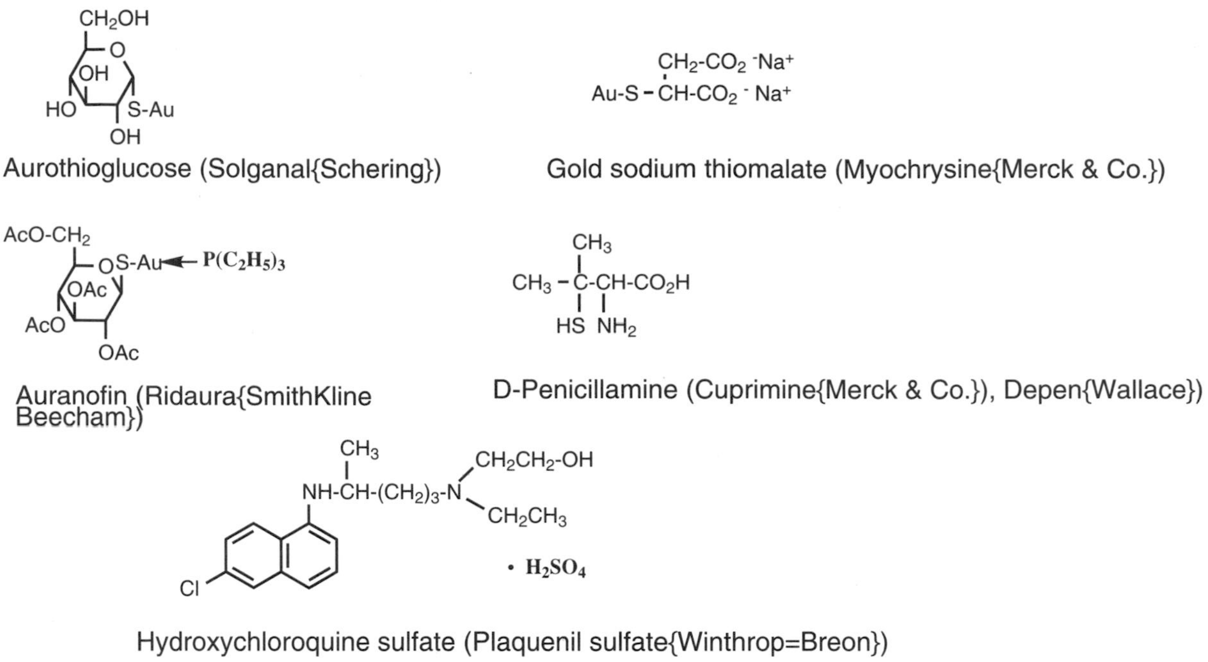

Aurothioglucose (Solganal{Schering})

Gold sodium thiomalate (Myochrysine{Merck & Co.})

Auranofin (Ridaura{SmithKline Beecham})

D-Penicillamine (Cuprimine{Merck & Co.}), Depen{Wallace})

Hydroxychloroquine sulfate (Plaquenil sulfate{Winthrop=Breon})

Fig. 25–25. Structures of disease-modifying antirheumatic drugs (DMARDs).

ligands affects tissue distribution and excretion properties and are usually highly polar, water-soluble functions. Aurous ion has only a brief existence in solution and is rapidly converted to metallic gold or to auric ion. Aqueous solutions decompose on standing at room temperature posing a stability problem for the two injectable gold compounds therapeutically available (aurothioglucose and gold sodium thiomalate). Complexation of Au^+ with phosphine ligands stabilizes the reduced valence state and results in nonionic complexes that are soluble in organic solvents and an enhancement of oral bioavailability. Other changes also occur. In the phosphine-Au-S compounds, gold has a coordination number of 2 and the molecules are nonconducting monomers in solution. The injectable gold compounds are monocoordinated. Whereas non-gold phosphine compounds are ineffective in arthritic assays, the nature of the phosphine ligand in the gold coordination complexes appears to play a greater role in antiarthritic activity than the other groups bound to gold. Within a homologous series, the triethylphosphine gold derivatives provide greatest activity.

The structures of the three therapeutically available gold compounds in the United States are shown in Figure 25–25.

Absorption and Metabolism

Gold compounds are generally rapidly absorbed following intramuscular injection and the gold is widely distributed in body tissues with the highest concentrations being found in the reticuloendothelial system and in the adrenal and renal cortices. Binding of gold from orally administered gold to red blood cells is higher than in injectable gold. Gold accumulates in inflamed joints where high levels persist for at least 20 days following injection. Although gold is excreted primarily in the urine, the bulk of injected gold is retained.

Drug Interactions

The only significant drug interactions reported are the concurrent administration of drugs that also produce blood dyscrasias (most notably phenylbutazone and the antimalarial and immunosuppressive drugs) and one report suggesting that phenytoin blood levels may be increased when auranofin and phenytoin are coadministered.

Preparations Available

Gold sodium thiomalate (Myochrysine [Merck & Co.]). Gold sodium thiomalate (mercaptobutanedioic acid, monogold, disodium salt)—actually a mixture of mono- and disodium salts of gold thiomalic acid—is a white to yellow-white, odorless powder that is very water soluble. It is available as an light-sensitive, aqueous solution of pH 5.8–6.5. The gold content is approximately 50%. It is administered intramuscularly because it is not absorbed on oral administration and is highly bound (95%) to plasma proteins.

Gold sodium thiomalate is indicated in the treatment of active adult and juvenile rheumatoid arthritis as one part of a complete therapy program. It is recommended that injections be given to patients only when the patients are supine. They must remain so for 10 minutes following injection. The recommended dosage schedule is 10 mg initially followed by gradual increases in weekly injections to 25 to 50 mg until either toxic symptoms or signs of clinical improvement appear. If neither appear, weekly injections should continue until the cumulative administered dose reaches 1 g. It is available as sterile aqueous solutions of 25 mg/ml in 1-ml ampules and 50 mg/ml in 1-ml ampules, and 10-ml vials.

Aurothioglucose (Solganal [Schering-Plough]). Aurothioglucose [(1-thio-D-glucopyranosato)gold] is a nearly odorless, yellow powder. Although highly water soluble, aqueous solutions decompose on long standing. It is therefore available as a suspension in sesame oil. Gold content is approximately 50%. Following intramuscular injection it is highly protein bound (95%) and peak plasma levels are generally achieved within 2 to 6 hours. Following a single 50-mg dose, the biologic half-life ranges from 3 to 27 days, but following successive weekly doses, the half-life increases to 14 to 40 days after the third dose. The therapeutic effect does not correlate with serum plasma gold levels but appears to depend on total accumulated gold.

Aurothioglucose is indicated for the adjunctive treatment of adult and juvenile rheumatoid arthritis. The usual dosage schedule consists of an initial 10-mg dose with subsequent doses increased to weekly 50-mg injections until a total of 800 mg to 1000 mg has been administered. If no toxicity is observed, the 50-mg dose may be continued at 3 to 4-week intervals. It is available as a suspension of 50 mg/ml in 10-ml vials, each milliliter containing 50 mg of aurothioglucose in sterile sesame oil with 2% aluminum monostearate and 1 mg of propylparaben as a preservative.

Auranofin (Ridaura [Smith, Kline Beecham]). Auranofin [(2,3,4,6-tetra-O-acetyl-1-thio-β-D-glucopyranosato-S)(triethylphosphine)gold] is a crystalline substance that is essentially insoluble in water but is soluble in organic solvents. Gold content is approximately 29%. The carbohydrate portion of auranofin assumes a chair conformation and as such all substituents occupy equatorial positions. It is the first orally effective gold compound used to treat rheumatoid arthritis. On a mg gold/kg basis, it is reported to be as effective in the rat adjuvant arthritis assay as the parenterally effective agents. Daily oral doses produce a rapid increase in kidney and blood gold levels for the first 3 days of treatment with a more gradual increase on subsequent administration. Plasma gold levels are lower than those attained with parenteral gold compounds. The major route of excretion is via the urine. Auranofin may produce fewer adverse reactions than parenteral gold compounds but its therapeutic efficacy may also be less.

Auranofin is indicated in adults with active rheumatoid arthritis who have not responded sufficiently to one or more NSAIAs. The usual dosage is 6 mg per day either as a 3 mg twice daily dose or as a 6-mg dose once daily. The dose may be increased to 9 mg per day if a satisfactory response is not achieved within 6 months. It is available as 3-mg capsules.

Sulfhydryl Compounds

In the mid-1950s, certain sulfhydryl compounds were reported to dissociate macroglobulins that may be involved in the rheumatic process. In 1960, D-penicillamine was investigated as an antirheumatic agent and was subsequently found to lower circulating levels of rheumatoid factors when administered over a period of several weeks. This effect persisted for several months after discontinuation of penicillamine thus discounting macroglobulin dissociation as a primary mechanism of action. A double-blind, multicenter British clinical trial in 1973 concluded that the drug was of value in the treatment of acute, severe rheumatoid disease. Further studies demonstrated the equivalent efficacy of penicillamine and gold compounds, although D-penicillamine was found to produce more frequent side effects. These results prompted the investigation of other sulfhydryl compounds as antirheumatic agents including pyridoxine analogs such as 5-thiopyridoxine, the corresponding disulfide, and the disulfide of 3-hydroxy-4-mercaptomethyl-5-vinylpyridine.

The ability of D-penicillamine to chelate metals (particularly copper, lead, mercury, and zinc) is the basis of its use as an antidote for heavy metal poisoning and in Wilson's disease (a disorder characterized by elevated copper levels caused by hereditary deficiency of copper metabolism). It is also of use in treating cystinuria, an inherited defect of the renal tubules in which resorption of cystine in impaired, urinary excretion is increased and cystine calculi are often formed in the urinary tract. D-penicillamine appears to be involved in sulfhydryl-disulfide exchange forming a mixed disulfide with cysteine, which is more soluble than cystine, thus reducing overall levels of cystine. A number of other in vitro effects of D-penicillamine have been demonstrated including inhibition of protein and DNA synthesis, inhibition of collagen cross-linking, and inhibition of pyridoxal metabolism. D-penicillamine influences those models of inflammation and rheumatoid diseases that contain a cell-mediated immune component. Its ability to dissociate IgM rheumatoid factors in synovial fluid, to depress T-cell activity, to influence phagocytosis of particulate matter by polymorphonuclear leukocytes, to inhibit the migration of inflammatory cells into the synovium, to improve lymphocyte function, to affect the plasma complement system, and to provide copper to copper-dependent enzymes (such as superoxide dismutases) involved in the arthritic process have all received considerable attention. Despite the knowledge of the biochemical and pharmacologic actions of D-penicillamine, its mechanism of action in rheumatoid arthritis has not been totally elucidated.

Preparations Available

D-Penicillamine (Cuprimine [Merck & Co.], Depen [Wallace]). D-Penicillamine (D-3-mercaptovaline) is a white crystalline powder possessing a slight odor and a bitter taste. It is freely soluble in water and aqueous solutions are stable at pH 2–4, but it undergoes oxidation to the disulfide at higher pH. Evidence exists to demonstrate the presence of polymorphs. The D-form (levorotatory) of penicillamine is the natural form initially isolated as a product of the acid hydrolysis of penicillin. It is now primarily obtained by synthesis[57] and resolution. D-penicillamine is readily absorbed on oral administration with peak plasma levels being attained within 1 to 2 hours.

Excretion is primarily urinary. Its use as an antiarthritic agent is limited because of the high incidence of side effects, some of which are potentially fatal. The major adverse reactions include GI effects (anorexia, nausea, diarrhea, and stomatitis or oral disorders), allergic reactions (rashes, urticaria, and exfoliative dermatitis), hematologic effects (leukopenia, bone marrow depression, thrombocytopenia, and aplastic anemia) and renal disturbances (proteinuria and hematuria). Tinnitus and diminution or total loss of taste perception can also occur. D-penicillamine is less toxic than the L-isomer or the racemic mixture.

D-Penicillamine is indicated for the treatment of rheumatoid arthritis in patients who have not responded to conventional therapy, for Wilson's disease and for cystinuria. Because improvement of the clinical symptoms of rheumatoid arthritis may not be noted for the first 2 to 3 months of drug therapy, administration should be carefully monitored. The initial recommended dose is 125 mg or 250 mg daily as a single dose (on an empty stomach at least 1 hour before meals), increasing the dose at 1 to 3 month intervals by 125 mg or 250 mg as patient response and tolerance indicate. Drug administration should cease if no response has been achieved after 3 to 4 months of therapy at daily doses of 1 to 1.5 g daily. It is available as 125-mg and 250-mg capsules and 125 mg tablets.

Antimalarial agents

The 4-aminoquinoline class of antimalarial agents has been known to possess pharmacologic actions which are beneficial in the treatment of rheumatoid arthritis. Two of these agents, chloroquine and hydroxychloroquine, have been used as antirheumatics since the early 1950s. The corneal and renal toxicity of chloroquine has resulted in its discontinuance for this purpose, however, although it is still indicated as an antimalarial agent and as an amebicide. Whereas hydroxychloroquine is less toxic, it is also less effective than chloroquine as an antirheumatic. The mechanism of action of these agents as an antirheumatic remains unresolved. Interestingly, most of the data available relates to chloroquine rather than hydroxychloroquine, but is assumed to be applicable to the latter. The spectrum of action of the 4-aminoquinolines differs from the NSAIAs in that chloroquine appears to be an antagonist of certain preformed prostaglandins. This effect, however, would indicate an acute, rather than a chronic, antirheumatic effect whereas chloroquine has been shown to be similar to gold compounds in that it possesses a slow onset of action—beneficial effects are noted only after 1 to 2 months of administration. Chloroquine inhibits chemotaxis of polymorphonuclear leukocytes in vitro but not in vivo. Its effects on collagen metabolism in connective tissue is also unclear. The most widely accepted mechanism of action of chloroquine, and presumably hydroxychloroquine, is related to its ability to accumulate in lysosomes. Although evidence indicating stabilization of lysosomal membranes is not convincing, it may inhibit the activity of certain lysosomal enzymes such as cartilage chondromucoprotease and cartilage cathepsin B. There does not appear to be a correlation of the antirheumatic effects of 4-aminoquinolines with their antimalarial activity.

Preparations Available

Hydroxychloroquine sulfate (Plaquenil Sulfate [Sanofi Winthrop]). Hydroxychloroquine sulfate {7-chloro-4-[5-(N-thyl-N-2-hydroxyethylamino)-2-pentyl]aminoquinoline sulfate salt} is a white to off-white, odorless, bitter-tasting crystalline powder. The sulfate salt is highly water soluble and exists in two different forms of different melting points. It is readily absorbed on oral administration reaching peak plasma levels within 1 to 3 hours. It concentrates in organs such as the liver, spleen, kidneys, heart, lung, and brain, thereby prolonging elimination. Hydroxychloroquine is metabolized by N-dealkylation of the tertiary amines followed by oxidative deamination of the resulting primary amine to the carboxylic acid derivative. In addition to possessing corneal and renal toxicity, hydroxychloroquine may also cause CNS, neuromuscular, GI and hematologic side-effects.

Hydroxychloroquine sulfate is indicated for the treatment of rheumatoid arthritis, lupus erythematosus, and malaria. The initial anti-arthritic dose is 400 mg to 600 mg daily taken with milk. It is available as 200-mg tablets.

IMMUNOSUPPRESSIVE AGENTS

The area of drugs that modify the immune response, whether as immunoregulatory, immunostimulatory or immunosuppressive agents, has been the focus of much recent research activity. Several agents that suppress the immune system have been explored as antirheumatic agents because the cause of rheumatoid arthritis may involve a destructive immune response. Thus, unlike agents previously discussed, immunosuppressive agents may act at the steps involved in the pathogenesis of rheumatic disorders, an attractive approach to antirheumatic therapy. These agents are quite cytotoxic, however, as a group, as evidenced by the initial development of these agents as anticancer drugs. Among the more widely employed immunosuppressives are azathioprine (Imuran [Burroughs-Wellcome]), methotrexate (Rheumatrex [Lederle]) and cyclophosphamide (Cytoxan [Bristol-Meyers/Mead Johnson Oncology]). All of these agents are quite toxic and generally are indicated for rheumatoid arthritis only in those patients with severe, active disease who have not responded to full dose NSAIAs therapy and at least one DMARD. Interestingly, whereas aspirin and NSAIAs are effective in only one-third of children with juvenile arthritis, methotrexate, given only once a week at low doses to minimize side effects, appears to offer promise. More recently, cyclosporine (Sandimmune [Sandoz]) has been investigated in rheumatoid arthritis and appears to offer short-term benefits, although its long-term use is limited by its toxic effects as well. Cyclosporine appears to inhibit activation of T-helper/inducer lymphocytes involved in the etiology of rheumatoid arthritis. Several other immunosuppressive agents are being explored and future developments in this area appear promising.

AGENTS USED TO TREAT GOUT

Gout is an inflammatory disease characterized by elevated levels of uric acid (as urate ion) in the plasma and urine and may take both acute and chronic forms. Acute gouty arthritis results from the accumulation of needle-like crystals of monosodium urate monohydrate within the joints, synovial fluid, and periarticular tissue, and usually appears without warning signs. Initiating factors may be minor trauma, fatigue, emotional stress, infection, overindulgence in alcohol or food, or by drugs such as penicillin or insulin. Chronic gout symptoms develop as permanent erosive joint deformity appears. The increase in extracellular urate may be caused by increased uric acid biosynthesis, decreased urinary excretion of uric acid, or perhaps a combination of both. The formation of uric acid from adenine and guanine is illustrated in Figure 25–26. Uric acid is formed by the oxidation of xanthine by the enzyme xanthine oxidase. Xanthine is a metabolic product of adenine (via hypoxanthine) and guanine, formed by the enzymes adenine deaminase and guanine deaminase, respectively. Thus, uric acid is an excretory product of purine metabolism in humans. In mammals, uric acid is hydrolyzed to allantoin by the enzyme uricase that is then subsequently hydrolyzed by allantoinase to allantoic acid. Hydrolysis of allantoic acid by allantoicase yields as final products urea and glyoxylic acid.

Normal total body pool of uric acid is approximately 1000 to 1200 mg in males, which is twice that of females. In patients suffering from gout these levels may be as high as two or three times above the normal levels. Uric acid is a weak acid (pK_a = 5.7 and 10.3) with very low water solubility. At physiologic pH it exists primarily as the monosodium salt, which is somewhat more soluble in aqueous media than the free acid. In humans, uric acid is reabsorbed primarily via renal tubular absorption. When levels of uric acid in the body increase either as a result of decreased excretion or increased formation, the solubility limits of sodium urate are exceeded and precipitation of the salt from the resulting supersaturated solution causes deposits of urate crystals to form. It is the formation of these urate crystals in joints and connective tissue which initiates attacks of gouty arthritis. The control of gout has been approached from the following therapeutic strategies: (1) control of acute attacks by drugs to reduce inflammation caused by the deposition of urate crystals (these drugs may possess only an anti-inflammatory component such as colchicine or both anti-inflammatory and analgesic actions such as in indomethacin, phenylbutazone, and naproxen); (2) increasing the rate of uric acid excretion (by definition these drugs are termed uricosuric agents and include probenecid and sulfinpyrazone); and (3) inhibiting the biosynthesis of uric acid by inhibiting the enzyme xanthine oxidase with drugs such as allopurinol.

Treatment of Acute Attacks of Gout

Preparations Available (Fig. 25–27)

Colchicine. Colchicine{(S)-N-(5,5,7,9-tetrahydro-1,2,3,10-tetramethoxy-9-oxobenzo[a]heptalen-7-yl)acetamide} is a pale yellow, odorless powder that is obtained from various species of Colchicum, primarily Colchicum autumnale Liliaceae. Its total chemical synthesis has been achieved but the primary source of colchicine currently remains alcohol extraction of the alkaloid from the corm and seed of C. autumnale L. It darkens on exposure to light and possesses moderate water solubility. Colchicine has a pK_a of 12.35. Its use in the treatment of gout dates back to the 6th century C.E. Unlike those agents that are discussed next, colchicine does not alter serum levels of uric acid. It does appear to retard the inflammation process initiated by the deposition of urate crystals. Acting on polymorphonuclear leukocytes and diminishing phagocytosis, it inhibits the production of lactic acid thus increasing the pH of synovial tissue and decreasing urate deposition because uric acid is more soluble at the higher pH. Additionally, colchicine inhibits the release of lysosomal enzymes during phagocytosis, which also contributes to the reduction of inflammation. Because colchicine does not lower serum urate levels, it has been found

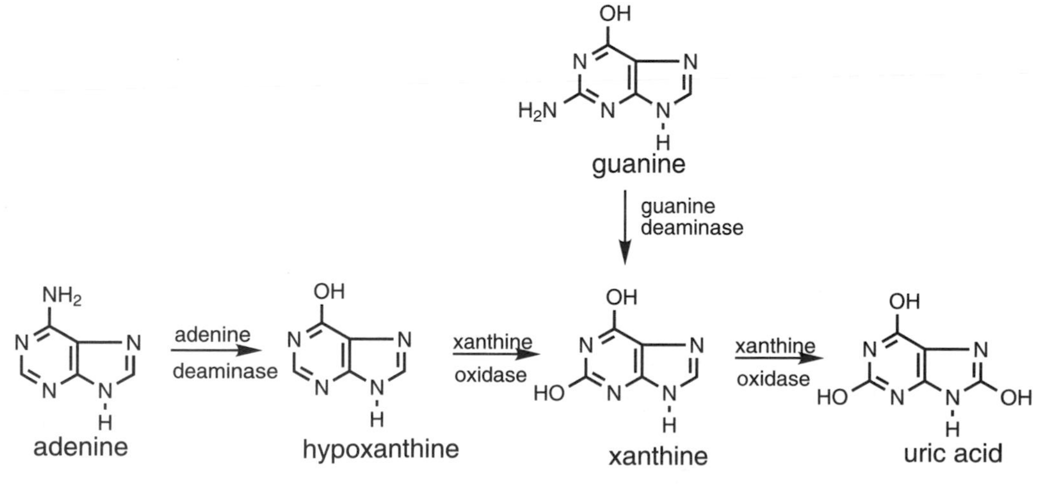

Fig. 25–26. Formation of uric acid.

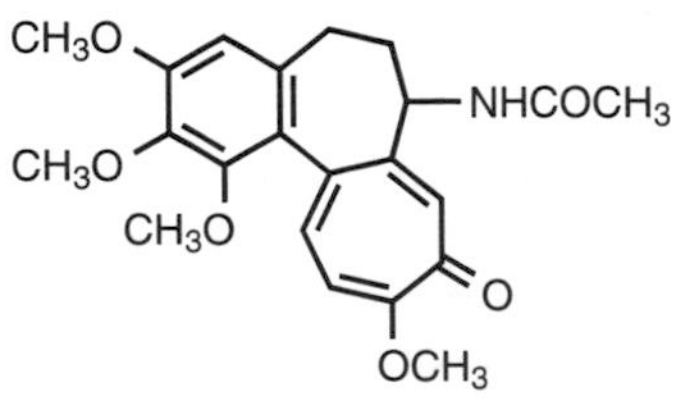

Colchicine

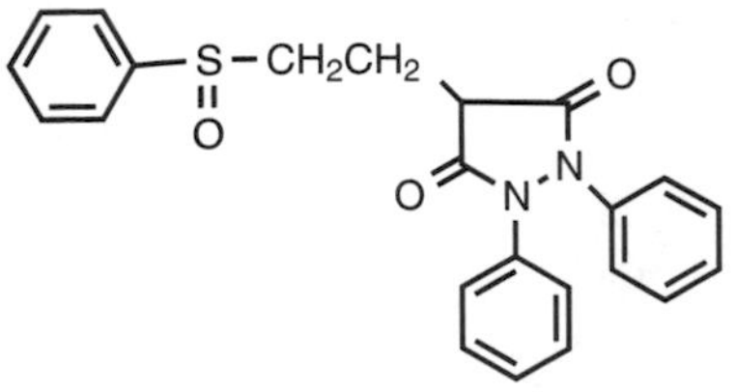

Sulfinpyrazone (Anturane{Ciba})

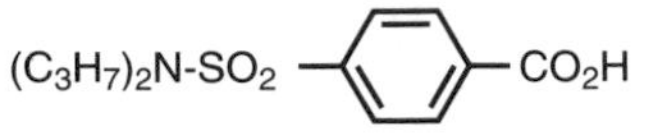

$(C_3H_7)_2N\text{-}SO_2\text{-}C_6H_4\text{-}CO_2H$

Probenecid (Benemid{Merck & Co.})

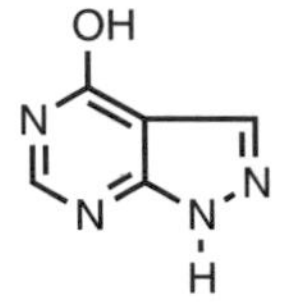

Allopurinol (Zyloprim{Burrough=Wellcome})

Fig. 25–27. Structures of agents used to control gout.

beneficial to combine colchicine with a uricosuric agent, particularly probenecid. It is a potent drug, being effective at doses of about 1 mg with doses of as small as 7 mg causing fatalities.

Colchicine is absorbed upon oral administration with peak plasma levels being attained within 0.5 to 2 hours after dosing. Plasma protein binding is only 31%. It concentrates primarily in the intestinal tract, liver, kidney, and spleen and is excreted primarily in the feces, with only 20% of an oral dose being excreted in the urine. It is retained in the body for considerable periods, being detected in the urine and leukocytes for 9 to 10 days following a single dose. Metabolism occurs primarily in the liver with the major metabolite being the amine resulting from amide hydrolysis. Colchicine may produce bone marrow depression with long-term therapy resulting in thrombocytopenia or aplastic anemia. Gastrointestinal disturbances (nausea, diarrhea, and abdominal pain) may occur at maximum dose levels. Acute toxicity is characterized by GI distress, including severe diarrhea, resulting in excessive fluid loss, respiratory depression, and kidney damage. Treatment normally involves measures to prevent shock, and morphine and atropine to diminish abdominal pain. Several drug interactions have been reported. In general, the actions of colchicine are potentiated by alkalinizing agents and inhibited by acidifying drugs, factors consistent with its mechanism of action of increasing the pH of synovial fluid. Responses to CNS depressants and to sympathomimetic agents appear to be enhanced. Clinical tests may be affected, most notably elevated alkaline phosphatase and serum glutamic-oxaloacetic transaminase (SGOT) values, and decreased thrombocyte values may be obtained.

Colchicine is indicated for the effective treatment of acute attacks of gout. The usual dose is 1.0 to 1.2 mg followed by 0.5 mg to 1.2 mg every 1 to 2 hours until either pain relief or symptoms of GI distress is observed. When a rapid response is required, or if GI reactions warrant discontinuance of oral administration, intravenous administration (usually 2 mg initially) may be indicated. It is available as 0.5 mg and 0.6 mg tablets and as an injectable solution of 1 mg in 2 ml ampules. It is often given in combination with probenecid. Combination products of the two are available in tablets containing 500 mg probenecid and 0.5 mg of colchicine.

Uricosuric Agents

Preparations Available

Probenecid (Benemid [Merck & Co.]). Probenecid {4-[di-(n-propylamino) sulfonyl] benzoic acid} is a white, practically odorless, crystalline powder that possesses a slightly bitter taste, but pleasant aftertaste. It is insoluble in water and acidic solutions but soluble in alkaline solutions buffered to pH 7.4. Probenecid was initially synthesized as a result of studies in the 1940s on sulfonamides, which indicated that the sulfonamides decreased the renal clearance of penicillin, extending the half-life of penicillin as war-time supplies were endangered. It was thus initially used, and is still indicated, for that purpose. Probenecid promotes the excretion of uric acid by decreasing the reabsorption of uric acid in the proximal tubules. The overall effect is to decrease plasma uric acid concentrations and decreasing the rate and extent of urate crystal deposition in joints and synovial fluids. Within the series of N-dialkylsulfamylbenzoates from which probenecid is derived, renal clearance of these compounds is decreased as the length of the N-alkyl substituents increases. Uricosuric activity increases with increasing size of the alkyl group in the series methyl, ethyl, and propyl.

Probenecid is essentially completely absorbed from the GI tract on oral administration with peak plasma levels observed within 2 to 4 hours. Like most acidic compounds, probenecid (pK_a = 3.4) is extensively plasma protein bound (93 to 99%). The primary route of elimination of probenecid and its metabolites is in urine. It is extensively metabolized in humans with only 5 to 10% being excreted as unchanged drug. The major metabolites detected result from glucuronide conjugation of the carboxylic acid, ω-oxidation of the n-propyl side chain and subsequent oxidation of the resulting alcohol to the carboxylic acid derivative, ω-1 oxidation of the n-propyl group, and N-dealkylation. Those metabolites possessing a free carboxylic acid function generally possess some uricosuric activity. Probenecid appears to be generally well-tolerated with few adverse reactions. The major side effects are related to GI distress (nausea, vomiting, and anorexia) but occur in only 2% of patients at low doses. Other adverse effects include headache, dizziness, frequency of urination, hypersensitivity reactions, sore gums, and anemia. Overdosages do not appear to present major difficulties; a case of a 49-year-old man who recovered from the ingestion of 47 g in a suicide attempt has been reported. Should overdosage occur, treatment consists of emesis or gastric lavage, short-acting barbiturates (if CNS excitation occurs), and epinephrine (for anaphylactic reactions). Certain drug interactions have been reported. Despite the high degree of plasma protein binding, displacement interactions with other drugs bound to plasma proteins does not appear to occur to any significant extent. Salicylates counteract the uricosuric effects of probenecid. Increased plasma levels of the following drugs may be observed because probenecid inhibits their renal excretion: aminosalicylic acid, methotrexate, sulfonamides, dapsone, sulfonylureas, naproxen, indomethacin, rifampin, and sulfinpyrazone. The effects on penicillin plasma levels were discussed previously.

Probenecid is indicated for the treatment of hyperuricemia associated with gout and gouty arthritis and for the elevation and prolongation of plasma levels of penicillins and cephalosporins. In gout, treatment should not begin until after an acute attack has subsided. It is not recommended in individuals with known uric-acid kidney stones or blood dyscrasias nor for children under 2. The recommended dose in gout is 250 mg twice a day for 1 week followed by 500 mg twice a day. As an adjunct for antibiotic therapy, the recommended dose is 2 g per day in divided doses. It is available as 500-mg tablets.

Sulfinpyrazone (Anturane [Ciba]). Sulfinpyrazone {1,2-diphenyl-4-[2-(phenylsulfinyl)ethyl]-3,5-pyrazolidinedione} is a white crystalline powder which is only slightly soluble in water but soluble in most organic solvents and alkaline solutions. Its synthesis is similar to that of phenylbutazone.[58] It produces its uricosuric effect in a manner similar to that of probenecid. A dose of 35 mg produces a uricosuric effect equivalent to that produced by 100 mg of probenecid while 400 mg/day of sulfinpyrazone produces an effect comparable to that obtained with doses of 1.5 to 2 g of probenecid. It also possesses, not surprisingly, some of the properties of phenylbutazone. It is an inhibitor of human platelet prostaglandin synthesis at the cyclooxygenase step, resulting in a decrease in platelet release and a reduction in platelet aggregation. This antiplatelet effect suggests a role for sulfinpyrazone in reducing the incidence of sudden death that can occur in the first year following a myocardial infarction. It lacks the analgesic and anti-inflammatory effects of phenylbutazone, however.

Sulfinypyrazone is a strong acid ($pK_a = 2.8$), an important factor in the production of the uricosuric effect because, within a series of pyrazolidinedione derivatives, the stronger the acid, the more potent the uricosuric effect. Polar substitution on the side chain also influences uricosuric activity as discussed previously with regard to the pyrazolidinediones.

Oral administration results in rapid and essentially complete absorption, peak plasma levels being attained within 1 to 2 hours of administration. It is highly bound (98 to 99%) to plasma proteins and it is excreted in the urine primarily (50%) as unchanged drug. Metabolites produced result from sulfoxide reduction, sulfur and aromatic oxidation, and C-glucuronidation of the heterocyclic ring in a manner similar to that for phenylbutazone. The metabolite resulting from ρ-hydroxylation of the aromatic ring possesses uricosuric effects in humans. The sulfide metabolite, a major metabolic product, may contribute to the antiplatelet effects of sulfinpyrazone but not the uricosuric effects. The most frequent adverse reactions are GI disturbances, however, their incidence is relatively low. Sulfinpyrazone is a much weaker inhibitor of prostaglandin synthesis in bovine stomach microsomes than either aspirin or indomethacin, a factor which may account for its gastric tolerance. There are much rarer reports of blood dyscrasias and rash. Overdosage produces symptoms that are primarily gastrointestinal in nature (nausea, diarrhea, and vomiting) but may also involve impaired respiration and convulsions. Treatment consists of emesis or gastric lavage and supportive treatment. Like probenecid, its uricosuric effects are antagonized by salicylates, and probenecid markedly inhibits the renal tubular secretion of sulfinpyrazone. It potentiates the actions of other drugs that are highly plasma-protein bound such as coumarin-type oral anticoagulants, antibacterial sulfonamides, and hypoglycemic sulfonylureas.

Sulfinpyrazone is indicated for the treatment of chronic and intermittent gouty arthritis. The initial recommended dose is 200 to 400 mg daily in two divided doses taken with meals and maintenance doses between 200 and 800 mg daily. Sulfinpyrazone is available as 100-mg tablets and 200-mg capsules.

Inhibitors of Uric Acid Synthesis

Figure 25–26 illustrates the biosynthesis of uric acid from the immediate purine precursor xanthine, which results from either adenine, via the intermediate hypoxanthine, or from guanine. The enzyme xanthine oxidase is involved in two steps, the first being the conversion of hypoxanthine to xanthine, and the final step, the conversion of xanthine to uric acid. Allopurinol was originally designed as an antineoplastic antimetabolite to antagonize the actions of key purines, inasmuch as it differs from normal purines only by the structural inversion of the nitrogen and carbon atoms at the 7- and 8-positions of the purine ring system, but was found to have little or no effect on experimental tissues. It was subsequently found that allopurinol serves as a substrate for xanthine oxidase (15 to 20 times greater the affinity of xanthine) and reversibly inhibits that enzyme. Normally, uric acid is a major metabolic end product in humans but, when allopurinol is administered, xanthine and hypoxanthine are elevated in the urine and uric acid levels decrease. When the synthesis of uric acid is inhibited, plasma urate levels decrease, supersaturated solutions of urate are no longer present, and the urate crystal deposits dissolve, eliminating the primary cause of gout. The increased plasma levels of hypoxanthine and xanthine pose no real problem because they are more soluble than uric acid and are readily excreted.

Preparations Available

Allopurinol (Zyloprim [Burroughs-Wellcome]). Allopurinol (1H-pyrazolol[3,4-d]pyrimidin-4-ol) is a white, fluffy tasteless powder possessing a slight odor. It is only very slightly soluble in water but is soluble in polar organic solvents and alkaline aqueous solutions. It was synthesized in 1956 as part of a study of purine antagonists.[59] It is well absorbed on oral administration with peak plasma concentrations appearing within 1 hour. Decreases of uric acid can be observed within 24 to 48 hours. Excretion of allopurinol and its major metabolite occurs primarily in the urine with about 20% of a dose being excreted in the feces. Allopurinol is rapidly metabolized via oxidation and the formation of numerous ribonucleo-

allopurinol —Xanthine oxidase→ alloxanthine (oxypurinol)

side derivatives. The major oxidation metabolite, alloxanthine or oxypurinol, has a much longer half-life (18 to 30 hours vs. 2 to 3 hours) than the parent drug and is an effective, although less potent, inhibitor of xanthine oxidase. The longer plasma half-life of alloxanthine results in its accumulation in the body during chronic administration, thus contributing significantly to the overall therapeutic effects of allopurinol. The major adverse effects are primarily dermatologic: skin rash and exfoliative lesions. Other effects such as GI distress (nausea, vomiting, and diarrhea), hematopoietic effects (aplastic anemia, bone marrow depression, and transient leukopenia), neurologic disorders (headache, neuritis and dizziness) and ophthalmologic effects (cataracts) are less common. Allopurinol may also initiate attacks of acute gouty arthritis in the early stages of therapy and may require the concomitant administration of colchicine. Drug interactions include those drugs which are normally also metabolized by xanthine oxidase. For example, the oxidation of 6-mercaptopurine, a useful antineoplastic agent, is inhibited, permitting a reduction in therapeutic dose. Allopurinol also has an inhibitory effect on liver microsomal enzymes, thus prolonging the half-lives of drugs such as oral anticoagulants, which are normally metabolized and inactivated by these enzymes, although this effect is quite variable. The incidence of ampicillin-related skin rashes increases with the concurrent administration of allopurinol.

Allopurinol is indicated for the treatment of primary and secondary gout, for malignancies such as leukemia and lymphoma, and the management of patients with recurrent calcium oxalate calculi. The average dose in gout is 200 to 300 mg per day for mild gout and 400 to 600 mg per day for more severe cases. It is available as 100-mg and 300-mg tablets.

FUTURE DEVELOPMENTS

Future development of new approaches and agents to treat arthritic processes is promising. Among the most intensive research areas is the investigation of inhibitors of leukotrienes as anti-inflammatory agents. Both leukotrienes and prostaglandins are involved in the inflammatory processes. Currently available NSAIAs all appear to act through the same mechanism of action, that is, inhibition of prostaglandin biosynthesis at the cyclooxygenase stage. This results in a buildup of arachidonic acid that permits the other major arachidonic acid pathway, the biosynthesis of leukotrienes, to function more significantly and to produce leukotrienes such as LTB_4 and slow-reacting substance of anaphylaxis (primarily a mixture of LTC_4 and LTD_4). LTB_4 is a potent chemotactic agent for polymorphonuclear leukocytes and plays a major role in inflammation as a pro-inflammatory agent while LTC_4 and LTD_4 are potent hypotensives and bronchoconstrictors. The enzyme 5-lipoxygenase is the key enzyme in the process, specific for the biosynthesis of leukotrienes; its inhibition may block their biosynthesis. Thus, continued development of anti-inflammatory agents as novel dual inhibitors of the cyclooxygenase and leukotriene pathways may lead to the development of "safer" NSAIAs, agents which possess a lower side-effect profile than currently available agents. Several experimental compounds are being investigated which specifically affect the leukotriene pathway by inhibiting 5-lipoxygenase and are receiving considerable clinical attention. Agents that inhibit the activation of 5-lipoxygenase, rather than inhibit the enzyme, are also being investigated as are mixed 5-lipoxygenase/leukotriene antagonists.

The role of complement and complement receptors in the immune response is also an intense area of research. Complement is one of the major mediators of the acute inflammatory response, therefore the control of the activation of complement or the inhibition of active complement fragments are potential ways to reduce tissue damage in inflammatory diseases. Several approaches are being explored, including the use of endogenous inhibitors of the complement system prepared by recombinant DNA technology and complement receptor inhibitors. The recent cloning of the C5a receptor[60] should facilitate developments in the latter area.

The use of glucocorticosteroids in the treatment of inflammatory disorders is generally accompanied by a number of adverse reactions making their use secondary to the employment of NSAIAs. These agents prevent the biosynthesis of prostaglandins and leukotrienes by inhibiting phospholipase A_2 and inhibiting the release of arachidonic acid and preventing the cascade from operating. This helps to explain the use of corticosteroids in allergic and asthmatic conditions in addition to inflammatory conditions, because NSAIAs are ineffective in allergic and asthmatic disorders. Several non-steroidal compounds have been discovered recently[61] that are effective in anti-inflammatory assays and do not affect the activities of either cyclooxygenase or 5-lipoxygenase but exhibit potent, glucocorticoid-like anti-inflammatory profiles. These agents appear to produce their anti-inflammatory actions in part by increasing serum levels of endogenous glucocorticoids thus representing a potential new class of anti-inflammatory agents.

There have been recent reports of anti-inflammatory agents that produce their effects independent of the cyclooxygenase or lipoxygenase pathways. Some compounds possess a broad spectrum of anti-inflammatory activity, while possessing no cyclooxygenase activity and very weak lipoxygenase inhibitory activity.[62] Another interesting group of compounds are related to the pungent natural ingredient in peppers of the Capsicum family, capsaicin, which is known to parenterally inhibit neurogenic inflammation by blocking neurokinin release. Capsaicin itself is orally inactive but analogs, which are orally effective have been reported.[63] The potential use of capsaicin analogs include roles as analgesic and anti-inflammatory agents.

Interleukins play an important role in the immune response and are also receiving a great deal of attention. IL-1 is one of a class of pro-inflammatory proteins (cytokines) produced by stimulated mononuclear phagocytes that plays a major role in inflammatory and immunologic processes of infection and tissue damage. It has been implicated in various chronic inflammatory disorders including rheumatoid arthritis. Thus, inhibitors of IL-1 synthesis, or release, or inhibition of the interaction of IL-1 with its "receptors" represent novel approaches to the control of inflammatory diseases. A natural peptide IL-1 receptor antagonist has been isolated and its primary structure determined.[64] This antagonist, known as IL-1ra, does not appear to be immunosuppressive or immunomodifying in nature but may act by blocking granulocyte extravasation and activating tissue cells to produce proteases. The development of specific inhibitors of IL-1 are being intensively sought. This represents a most promising area for the development of new agents effective in the battle against arthritic diseases.

Selectins comprise a group of glycoproteins involved in leukocyte adhesion to platelets or vascular endothelium, which is an early step in the extravasation of leukocytes associated with several disease states including inflammation, some cancers, and thrombosis. The involvement of leukocyte adhesion in inflammation and the immune system has been reviewed.[65] Three protein receptors, termed E-selectin, L-selectin and P-selectin, have been identified and have been shown to contain several complement binding protein-like domains. Each selectin receptor possesses a native carbohydrate ligand. Sialyl Lewis^{X}, a tetrasaccharide that forms the terminus of some cell-surface glycolipids and which apparently plays a major role in the inflammatory process, is a ligand for E-selectin and P-selectin. Once tissue is injured, cytokines are released to signal endothelial cells to change conformation and express selectins on their surface. E-selectin recognizes sialyl Lewis^{X} and the sialyl Lewis^{X} groups on leukocytes then bind to selectins on endothelial cells. The sialyl Lewis^{X-} containing white blood cells adhere to leukocytes, thus allowing them to migrate to the site of tissue injury. Several biotechnology research groups are exploring the development of complex oligosaccharides such as sialyl Lewis^{X} and related compounds through enzymatic methods as potential anti-inflammatory agents on the basis that these compounds may occupy binding sites on blood vessel wall receptors and thus prevent the binding of white blood cells to these sites.

REFERENCES

1. R. F. Borne, in *Handbook of Cardiovascular and Anti-Inflammatory Agents,* M. Verderame, Ed., Boca Raton, FL, CRC Press, 1986, p. 27.
2. S. C. Stinson, *Chem. Eng. News,* Oct. 16, 1989, p. 38.
3. S. H. Fuller and C. McKenzie, *U. S. Pharmacist,* May, 1992, p. 35.
4. G. P. Rodnam and H. R. Schumacher, Eds., *Primer on the Rheumatic Diseases,* 8th ed., Atlanta, GA, Arthritis Foundation, 1983.
5. A. Schwartz, *U.S. Pharmacist,* June, 1990, p. 8.
6. R. W. Bennett, *U.S. Pharmacist,* June, 1990, p. 30.
7. G. H. Hamor, in *Principles of Medicinal Chemistry,* 3rd ed., W. O. Foye, Ed., Philadelphia, Lea & Febiger, 1989, p. 503.
8. M. Walport, in *Immunology,* 2nd ed., I. M. Roitt, et al., Eds., St. Louis, C. V. Mosby Company, 1989.
9. T. Kinoshita, *Immunol. Today, 12,* 291(1991).
10. M. M. Frank and L. F. Fries, *Immunol. Today, 12,* 322(1991).
11. T. Y. Shen, in *Burger's Medicinal Chemistry, Part III,* 4th ed., M. Wolff, Ed., New York, John Wiley & Sons, 1981, p. 1212.
12. T. Y. Shen, in *Proc. Int. Symp.,* Milan, 1964, S. Grattini and M. N. G. Dukes, Eds., Excerpta Medica, Amsterdam, 1965, p. 18.
13. T. Y. Shen, *Top. Med. Chem., 1,* 29(1967).
14. J. R. Vane, *Nature, 231,* 232(1971).
15. P. Gund and T. Y. Shen, *J. Med. Chem., 20,* 1146(1977).
16. R. Kurzrok and C. Lieb, *Proc. Soc. Exp. Biol. N.Y., 28,* 268(1931).
17. M. W. Goldblatt, *J. Physiol. (London), 84,* 208(1935).
18. U. S. von Euler, *J. Physiol. (London), 88,* 213(1936).
19. S. Bergstrom and B. Samuelsson, *Endeavour, 27,* 109(1968).
20. W. B. Campbell, in *Goodman and Gilman's The Pharmacological Basis of Therapeutics,* 8th ed., A. Gilman, et al., New York, Pergamon Press, 1990, p. 600–601.
21. T. Okazaki, et al., *J. Biol. Chem., 256,* 7316(1981).
22. T. Y. Shen, in *Burger's Medicinal Chemistry,* Part III, 4th ed., M. E. Wolff, Ed., New York, Wiley, 1981, p. 1222.
23. W. G. Clark, *Gen. Pharmacol., 10,* 71(1979).
24. D. B. Kunkel, *Emergency Medicine,* Geigy Pharmaceuticals, July 15, 1985.
25. I. C. Calder, et al., *J. Med. Chem., 16,* 499(1973).
26. M. J. Smilkstein, et al., *N. Engl. J. Med., 319,* 1557(1988).
27. A. R. Buckpitt, et al., *Biochem. Pharmacol., 28,* 2941(1979).
28. C. H. Hennekens, et al., *N. Engl. J. Med., 321,* 129(1989).
29. T. A. Gossel, *U.S. Pharmacist,* February, 1988, p. 34.
30. M. J. Thun, et al., *N. Engl. J. Med., 325,* 1393(1991).
31. C. Davison, *Ann. N.Y. Acad. Sci., 179,* 249(1971).
32. H. E. Paulus and M. W. Whitehouse, in *Search for New Drugs,* Vol. 6, A. Rubin, Ed., New York, Marcel Dekker, 1972, p. 28.
33. C. A. Winter, in *Non-Steroidal Anti-Inflammatory Drugs,* Proc.

Int. Symp., Milan, 1964, S. Garattini and M. N. G. Dukes, Eds., Amsterdam, Exercepta Medica, 1965, p. 191.
34. W. Dieterle, et al., *Arzneimittel forschung, 26,* 571(1976).
35. T. Y. Shen, et al., *J. Am. Chem. Soc., 85,* 488(1963).
36. T. Y. Shen, et al. *U.S. Patent 3,654,349,* 1971; *Chem. Abstr., 74,* 141379v(1971).
37. J. R. Carson, et al., *J. Med. Chem., 14,* 646(1971).
38. A. Sallman and R. Pfister, *Ger. Patent,* 1,815,802(1969).
39. V. A. Skoutakis, et al., *Drug. Intell. Clin. Pharm., 22,* 850(1988).
40. C. A. Demerson, et al., *J. Med. Chem., 18,* 189(1975); C. A. Demerson, et al., *J. Med. Chem., 19,* 391(1976).
41. A. C. Goudie, et al., *J. Med. Chem., 21,* 1260(1978).
42. J. S. Nicholson and S. S. Adams, *Br. Patent 971,700*(1964); *Chem. Abstr. 61,* 14591d(1964).
43. A. J. Hutt and J. Caldwell, *J. Pharm. Pharmacol., 35,* 693(1983).
44. W. S. Marshall, *French Patent 2,015,728* (1970); *Chem. Abstr., 75,* 48707m(1971).
45. D. Farge, et al., *U.S. Patent 3,641,127* (1972); *Chem. Abstr., 68,* 524(1968).
46. K. Ueno, et al., *J. Med. Chem., 19,* 941(1976).
47. I. T. Harrison, et al., *J. Med. Chem., 13,* 203(1970).
48. P. A. J. Janssen, et al., *Ger. Patent 2,353,375*(1974).
49. S. S. Adams, et al., *U.S. Patent 3,755,427*(1975).
50. R. A. Scherrer, in *Anti-Inflammatory Agents,* Vol. 1, R. A. Scherrer and M. W. Whitehouse, Eds., New York, Academic Press, 1974, p. 35; R. A. Scherrer, in *Anti-Inflammatory Agents,* Vol. 1, R. A. Scherrer and M. W. Whitehouse, Eds., New York, Academic Press, 1974, p. 56.
51. C. V. Winder, et al., *J. Pharmacol. Exp. Ther., 138,* 405(1962).
52. P. F. Juby, T. W. Hudyma, and M. Brown, *J. Med. Chem., 11,* 111(1968).
53. R. A. Appleton and K. Brown, *Prostaglandins, 18,* 29(1979).
54. T. J. Carty, et al., *Prostaglandins, 19,* 671(1980).
55. J. G. Lombardino and E. H. Wiseman, *Med. Res. Rev., 2,* 127(1982).
56. J. G. Lombardino, et al., *J. Med. Chem., 16,* 493(1973).
57. K. Savard, et al., *Can. J. Res., 24B,* 28(1946).
58. R. Pfister and F. Haflinger, *Helv. Chim. Acta, 44,* 232(1961).
59. R. K. Robins, *J. Am. Chem. Soc., 78,* 784(1956).
60. C. Gerard and N. P. Gerard, *Nature, 349,* 614(1991).
61. F. Suzuki, et al., *J. Med. Chem., 35,* 2863(1992).
62. R. M. Burch, et al. *Proc. Natl. Acad. Sci., U.S.A., 88,* 355(1991).
63. L. M. Brand, et al. *Agents Actions, 31,* 329(1990).
64. C. H. Hannum, et al., *Nature, 343,* 336(1990); S. P. Eisenberg, et al., *Nature, 343,* 341(1990).
65. T. A. Springer, *Nature, 346,* 425(1990).

Chapter 26

DRUGS AFFECTING SUGAR METABOLISM

Edmund J. Hengesh

Although the symptoms of diabetes mellitus were described in the Ebers Papyrus nearly 3500 years ago and were associated with the pancreas by von Mering and Minkowski in 1889,[1] little progress was made in managing the disease until 1921. In that year, two young investigators, Fred Banting and Charles Best, demonstrated that an extract from beef pancreata could successfully lower blood glucose levels in pancreatectomized dogs.[2] Furthermore, daily injections of the pancreatic extract were able to ameliorate the diabetic symptoms "indefinitely," i.e., 70 days for "Marjorie," one of Banting and Best's test animals. Shortly thereafter, in January, 1922, the administration of a pancreatic extract to the first human diabetic subject, Leonard Thompson, marked the advent of the use of the pancreatic antidiabetic principle, insulin, in the treatment of diabetes mellitus. Mr. Thompson survived for 15 years on insulin, dying at age 29 of bronchopneumonia.

The ability of insulin to bring about such a dramatic reversal in the symptoms of diabetes, with a return to a "near normal" life expectancy, led the medical community to conclude that the problems of etiology and treatment had been resolved. Viewed retrospectively these conclusions were premature, whereas insulin does return control of the blood glucose level and does offset the development of ketoacidosis, it does not appear to rectify all of the metabolic defects identifiable in the diabetic. These metabolic defects, as well as a wide variety of diabetic complications, also cannot be adequately controlled by the use of any of the currently available oral hypoglycemic agents. At present, aside from pancreas transplantation,[3] we are able to treat only the symptoms of diabetes mellitus—not the disease itself.

The search for the ideal antidiabetic agent has been hampered by a lack of fundamental knowledge of the basic biochemical abnormalities of diabetes. Some of the questions to be answered include: What is the primary lesion of diabetes? Do both insulin-dependent and noninsulin-dependent diabetics have the same lesion? How is the lesion transmitted from generation to generation? Can it be detected in the prediabetic state? What is the relationship between diabetes and its vascular and neurologic complications? In other words, what is diabetes? Until the answers to these questions are found, diabetes mellitus will continue to be characterized on the basis of its more obvious symptoms, and the medicinal chemist and the pharmacologist will be forced to continue a random search for more effective antidiabetic drugs.

DIABETES MELLITUS—THE DISEASE

Etiology

Diabetes mellitus is divided into two classes depending on whether or not the patient is dependent upon exogenous insulin. Insulin-dependent diabetes mellitus (IDDM) is called type I, whereas noninsulin-dependent diabetes mellitus (NIDDM) is called type II. Both types have been recognized as genetic diseases for centuries, yet the contribution of genetic factors to the development of the diseases remains obscure. The complexity of the problem can be appreciated when one inspects the rates at which twins, born of families with a history of diabetes, both become diabetic. Normally, if one twin of an identical pair develops a genetic disease, the remaining twin can also be expected to develop the disease, because they have the same genotype. This corresponds to a concordance rate of 100%. Even though, however, diabetes mellitus is considered a genetic disease, the concordance rate observed with identical twins is only about 50% (36% for type I diabetes and up to 100% for type II[4]). Furthermore, the concordance rate for fraternal twins is 3 to 37%. This is approximately the rate expected for the population at large (10% treated, undiagnosed, and total potential) and suggests that fraternal twins may have little more potential to develop diabetes mellitus than any other pair of individuals. Although the information derived from these studies does support the concept that there is a genetic component to diabetes, it also implies that other factors, perhaps environmental and/or immunologic, may participate in the development of the disease. This is consistent with the now widely accepted hypothesis that type I diabetes is an autoimmune disease.[5]

Efforts to explain the lack of concordance have led to the postulation of many different modes of inheritance, ranging from simple autosomal recessive to much more elaborate inheritance models. None of the proposed modes of inheritance adequately explains all the patterns of diabetic inheritance that have been observed.[6] It is generally agreed that diabetes is not a single genetic defect; it probably involves multiple genetic loci and a number of contributing factors. Diabetes is most likely a heterogeneous group of diseases sharing nothing more than a real or potential abnormality of glucose metabolism.

Attempts to identify individuals who are predisposed to diabetes have been most fruitful in the investigation of the development of type I diabetes. Family studies indicate that genetic susceptibility factors play a role.[7] Attention has focused on a group of genes associated with the major histocompatibility complex located on chromosome 6. These genes, known as the human leukocyte antigens (HLA), must be compatible for tissue transplants or skin grafts to be successful. It has been found that 93 to 98% of type I diabetic patients of northern European ancestry have HLA-DR3 or HLA-DR4. The presence of both DR3 and DR4 antigens yields an even higher risk than the additive risk with either antigen alone. The predictive value of this relationship, however, is lessened by the observation that the frequency of these antigens in the control population is also high, averaging greater than 50%. Another HLA antigen, DR2, is found less frequently in diabetic patients than in the general population, thus suggesting some type of protective effect. More recent studies present evidence that alleles at the DQ rather than the DR locus may be more intimately related to increased susceptibility with the DQw8 allele associated with a twofold to sixfold greater risk.[8] Models to explain the mechanisms of the increased susceptibility and the protective effects have been proposed by Nepom[9] and Sheehy.[10] The lack of correlation between the HLA antigens and noninsulin-dependent diabetes provides further support for the concept of diabetic heterogeneity.

Because type I diabetes is considered to be an autoimmune disorder resulting in the destruction of the insulin-secreting pancreatic β-cells, early identification of patients in whom autoimmunity has been activated may be of value in preventing or minimizing the course of the disease. To date, three markers with sufficient specificity to be predictive have been identified[11]: insulin autoantibodies (IAA), cytoplasmic islet cell antibodies (ICA), and antibodies to a 64K protein—which could be glutamic acid decarboxylase.[12] IAA and antibodies to a 64K protein have an advantage over ICA in that they are islet β-cell specific. Due to concerns regarding the sensitivity and specificity of the analysis and side effects associated with the therapy, the American Diabetes Association has proposed that intervention be restricted to high risk individuals (mainly first degree relatives of patients with IDDM) in whom immune markers are persistently present, in high titer and in combination, and in whom β-cell function is extremely low.[13]

Symptoms and Complications

Because the biochemical basis of diabetes is unclear, the disease is frequently referred to by its symptoms. All tissues have an energy requirement that is usually met by metabolizing glucose. The entry of glucose from the blood into the cells of liver, skeletal muscle, and adipose tissue is promoted by the presence of insulin. In the diabetic, the insulin-dependent tissues cannot assimilate glucose normally, and glucose accumulates within the blood (hyperglycemia). As the blood glucose concentration increases, osmotic forces come into play that tend to increase the blood volume and urine output (polyuria). As the blood glucose level exceeds its renal threshold, glucose appears in the urine (glucosuria). The increased loss of water from the body triggers compensatory adjustments that lead to an increase in thirst (polydipsia). The inability of glucose to enter some tissues increases the need for alternate sources of energy, such as ketone bodies, which are synthesized by the liver. Increased concentrations of ketone bodies appear in the blood (ketonemia) and ultimately in the urine (ketonuria). Muscular weakness and weight loss also occur.

Apart from these rapidly developing symptoms, which are usually reversed by appropriate therapy, diabetes can also be characterized by its long-term complications. These involve the cardiovascular, renal, neural, and visual systems. The development of these complications appears to be somewhat related to the duration of the disease and might involve prolonged exposure to abnormally high concentrations of glucose or its metabolites. The chronic complications did not become apparent until after diabetic control was realized, but they now represent the major causes of death in the diabetic. For example, from 1914 to 1922, only 24.6% of diabetics died of cardiovascular and renal disease, whereas in the 2-year span from 1966 to 1968, 74.2% of diabetics died of these disorders. Similarly, diabetic retinopathy, which may take 10 or more years to develop, now ranks as a leading cause of blindness in the United States. It is hoped, but remains to be demonstrated, that the development of late complications can be prevented through more judicious control of the blood glucose level.

Diabetes mellitus can be divided symptomatically into two basic syndromes: type I (formerly juvenile onset), and type II (formerly adult or maturity onset). In type I diabetes, the hyperglycemia, glucosuria, polyuria, and polydipsia develop rapidly. These diabetics lack the ability to synthesize and secrete insulin and are prone to ketoacidosis. The type I diabetic then depends on the administration of insulin for survival and tends to develop chronic problems at an earlier age. Approximately 95% of patients who develop clinical diabetes before age 30 years have type I diabetes.

Type II typically afflicts people over 40, and the symptoms develop more slowly. Obesity is often a contributing factor. For the most part, the type II diabetic can synthesize and secrete insulin, although not in sufficient quantity to sustain normal metabolism. These patients do not normally exhibit severe ketosis. Because type II diabetes is a milder form of the disease, the individuals in this group are usually more easily controlled—many by dietary regulation, with or without the use of an oral hypoglycemic agent.

Detection and Diagnosis

The detection and diagnosis of diabetes mellitus vary significantly with the severity of the disease. In the more overt cases, the disease can be readily detected by measuring urinary glucose and ketone bodies using any of the readily available, self-administered tests. In less developed cases, the urinary tests are often negative, and the diabetologist must look for fasting hyperglycemia.

To confirm the diagnosis, patients are frequently subjected to an oral glucose tolerance test (OGTT). The OGTT measures the ability of an individual to assimilate a glucose load. The normal response to the test is a rapid, substantial increase in the blood glucose level within the first hour, followed by a decline to a near-fasting level within 3 hours. In the diabetic, the blood glucose level rises much higher initially and takes much longer to return to normal. Although clinicians differ in opinion as to how the results of the OGTT should be interpreted and as to the conditions under which the test should be run, the OGTT still plays a prominent role in the classification of diabetics.

Prediabetics are individuals who are genetically predisposed to diabetes, but have normal fasting glycemia and a normal OGTT response. Subclinical diabetics have normal fasting blood glycemia and exhibit an abnormal OGTT only under stress conditions. Latent or chemical diabetes is the more common undiagnosed form of the disease and is characterized by normal fasting blood glucose but an abnormal OGTT. Overt diabetes represents the symptomatic form and is subdivided into the two types discussed previously.

Hormonal Interrelationships

Classically, diabetes mellitus has been considered a *unihormonal* disease. All of the metabolic defects were thought to be primarily or secondarily due to deficiency of effective insulin. More recently, some investigators have advanced the hypothesis that diabetes is a *bihormonal* disease, characterized by an excess of glucagon in conjunction with a lack of insulin.[14] Such a hormonal relationship was not anticipated. Glucagon is normally present only during times of hypoglycemia, and the blood glucose level is elevated in diabetes. In every form of spontaneous or experimentally induced diabetes examined to date, however, hyperglucagonemia has been found to coexist with hyperglycemia.

Both glucagon and insulin originate from the islets of Langerhans in the pancreas. Glucagon is produced by the α-cells, and insulin is produced by the β-cells. Each of these hormones can alter the secretion of the other. Insulin suppresses the secretion of glucagon, whereas glucagon increases the release of insulin. Proponents of the *bihormonal* hypothesis argue that the primary defect of diabetes may be a loss of glucose-sensing ability by the islets, which is expressed as an abnormal hormonal secretion pattern.[15] The relative hyposecretion of insulin accounts for the underutilization of glucose, while the hypersecretion of glucagon promotes glucose production at a rate exceeding its utilization, thus resulting in hyperglycemia.

Evidence in support of this *double trouble* hypothesis came mainly from studies using somatostatin. Somatostatin is a recently discovered polypeptide hormone produced by the "D" cells of the islets. It has the ability to suppress the release of many hormones, including insulin and glucagon.

When somatostatin is administered to insulin-dependent diabetics not receiving insulin therapy, the hyperglycemia in both the fasting and postprandial states is attenuated. The decrease in the blood glucose level is attributed to the somatostatin-induced suppression of glucagon release. Furthermore, an insulin-dependent diabetic suddenly deprived of insulin will rapidly develop hyperglycemia; however, if the removal of insulin is countered by the infusion of somatostatin, the rise of the blood glucose concentration is much more gradual and not nearly as extreme. Somatostatin has been used with insulin in the treatment of type I diabetes. The use of these agents, in combination, allowed the dose of insulin to be decreased while the same degree of blood glucose control was provided.

Controversy surrounds the *double trouble* hypothesis. Those opposed to the concept contend that diabetes is solely caused by a deficiency of effective insulin, and that, although glucagon may aggravate the consequences of insulin lack, it is not necessary for the development of the disease. These investigators are quick to note that fasting hyperglycemia, although markedly diminished when glucagon levels are suppressed with somatostatin, *does* develop on withdrawal of insulin. Even the mechanism by which somatostatin reduces hyperglycemia has been challenged. Felig et al. suggest that somatostatin reduces hyperglycemia by decreasing and/or delaying carbohydrate absorption.[16]

The controversy has, however, encouraged a search for a series of new medicinal agents designed to control plasma glucagon levels. Potential agents may accomplish this purpose by neutralizing glucagon, i.e., glucagon antibodies, or by suppressing glucagon release using somatostatin-like compounds with selected biologic activity and a reasonable life span in the plasma.[17] Somatostatin itself has a biologic half-life of less than 2 minutes, so long-term administration is not now feasible. These drugs are presently theoretic in nature and are of little practical value. The current approach to diabetes therapy still centers around the administration of insulin and the use of the oral hypoglycemic agents.

Insulin

Irrespective of any other abnormality, diabetes mellitus is characterized by a deficiency of effective insulin, a proteinaceous hormone secreted by the β-cells of the islets of Langerhans of the pancreas.

The first investigators to connect the pancreas with diabetes were von Mering and Minkowski who, in 1889, demonstrated that the symptoms of diabetes could be induced in laboratory animals by removal of the pancreas.[1] De Meyer, in 1909, coined the name "insuline" to refer to the then hypothetic substance believed responsible for preventing the occurrence of the diabetic symptoms. A means of controlling these symptoms was not available until 1921, when Banting and Best successfully treated pancreatectomized dogs with crude pancreatic extracts.[2] These extracts were later found to contain insulin. From this monumental discovery evolved the first rational treatment of diabetes, and the work of Banting and Best was acknowledged with a Nobel Prize. It has been said that, under present circumstances, Banting and Best could not have obtained research support because what they were attempting to do had been tried by numerous other investigators, who had failed.

Chemistry

Insulin was isolated in crystalline form by Abel in 1926, and its chemical structure was elucidated by Sanger and his co-workers in the early 1950s. Sanger found that the insulin molecule is composed of two polypeptide chains: an A chain, consisting of 21 amino acid residues, and a B chain, containing 30 residues. The two chains are connected by two disulfide bonds, and there is an additional disulfide linkage within the A chain.

The amino acid sequences of insulin for at least 28 species have been reported. Although most of these insulins are markedly similar in amino acid composition and molecular weight, insulin from hagfish has been found to differ from human insulin in almost 50% of the residues. The two types of insulin of greatest importance to us, due to their therapeutic role, are those isolated from the pig and cow. Porcine insulin differs from human insulin in that it contains a C-terminal alanine in the B chain, whereas human insulin has a C-terminal threonine. Because they are so similar in structure, porcine insulin can be converted into human insulin. Human insulin (Novo) is prepared in this manner. The sequential differences between bovine insulin and human insulin are more striking in that, in addition to the porcine change, there are amino acid differences at positions 8 and 10 of the A chain. At position A-8, alanine replaces threonine, and at A-10, valine is substituted for isoleucine. Although the biologic activity is retained with these amino acid changes, it should be kept in mind that the more dissimilar the sequences, the greater is the potential for antigenicity. Predictably, bovine insulin is more antigenic than porcine insulin in humans.

The three-dimensional arrangement of atoms in insulin has been determined by x-ray crystallography. From this work, we can appreciate that insulin looks like a typical globular protein, despite its small size. Insulin crystallized at neutral pH is composed of six molecules of insulin organized into three identical dimers. The three dimers are arranged around two zinc ions, each of which is coordinated to the imidazole nitrogens of three B-10 histidines. Presumably, insulin is stored in granules in the β-cell as the hexamer, but the physiologically active form of insulin is the monomer.

The chemical synthesis of insulin was achieved independently by three groups during the mid-1960s.[17–19] All three used the method of separate synthesis of the A and B chains followed by random combination. The process involved more than 200 steps and took several years to accomplish. Using the modern-day methods of solid-state synthesis developed by Merrifeld and Marglin, the same results have been obtained in just a few days.

The major difficulty encountered in the synthesis of insulin was the correct positioning of the three disulfide bonds. This was because of the many ways in which the sulfhydryl groups could combine. The problem of poor yields was partially resolved by treating the sulfhydryl form of one chain with the S-sulfonated derivative of the other. Proper positioning of the disulfide bonds has been further facilitated through conformational-directed disulfide bond formation. By reversible cross-linking of amino acids known to be in juxtaposition in the three-dimensional structure, the formation of the correct disulfide bonds can be enhanced.

Another novel approach to the synthesis of insulin has been to circumvent the problem of disulfide bond formation. The appropriate disulfide bond was first formed between fragments of the A and B chains. Then the fragments, with the disulfide bonds correctly positioned, were condensed in an orderly fashion to obtain the final product. Human insulin synthesized in this manner has been shown to be biologically equivalent to the natural hormone. Although the fragment condensation approach is a significant improvement over other methods for synthesizing proteins containing disulfide bonds, it is not likely to become a commercial source of insulin due to the rapid advances made in recombinant DNA technology.

From a rather humble beginning in the late 1970s, when Ullrich and co-workers successfully incorporated the gene for rat insulin into bacterial plasmids and Gilbert's team from Harvard reported the successful synthesis of rat proinsulin from another bacterial clone, recombinant DNA technology, as applied to insulin, has flourished (for a review see ref. 18). Collaborative arrangements between Genentech and Eli Lilly and Company resulted in the commercial production of human insulin by recombinant means. Although the product was initially made by joining the A and B chains of human insulin produced by separate strains of E. coli K12, currently human insulin is obtained from an E. coli clone capable of producing proinsulin directly. This approach offsets the difficulties encountered in obtaining the proper orientation of the disulfide bonds. This achievement ensures an almost limitless supply of insulin for the world's 60 million diabetics.

The availability of insulin analogs and chemically and enzymatically modified insulin derivatives has provided a means of studying the relationship between the chemical structure and the biologic activity of the insulin molecule. Neither the A chain nor the B chain is active in the fully reduced form.

There is no apparent loss of activity in the insulin molecule when the C-terminal alanine is removed from the B chain of porcine or bovine insulin, but the biologic activity is diminished appreciably when both the C-terminal alanine and the C-terminal asparagine are removed. The removal of the second and third amino acids from the C-terminus of the B chain decreases the potency slightly, and desoctapeptide insulin (insulin minus eight amino acids from the C-terminal end of the B chain) has almost no activity and retains little ability to dimerize. Removing or chemically modifying the N-terminal glycine of the A chain substantially decreases biologic activity, but activity is retained if the B-chain N-terminal phenylalanine is deleted. Chemical modification of the side chain carboxyl groups or the tyrosine residues also leads to inactivity.

The information obtained from studies such as these has confirmed that the conformation of the insulin molecule is critical to its activity, and has provided a basis for Pullen and collaborators to propose a receptor-binding region within the insulin molecule.[19] The receptor-binding region, as proposed, consists largely of invariant amino acids located both in the A chain (residues A-1 gly, A-5 gln, A-19 tyr, and A-21 asn) and in the B chain (residues B-24 phe, B-25 phe, B-26 tyr, B-12 val and B-16 tyr). Modification of insulin or the receptor may decrease biologic activity by reducing the hormone: receptor affinity or by decreasing the ability of the complex, once formed, to elicit a response.

Biosynthesis

Information regarding the biosynthesis of insulin within the β-cells of the islets of Langerhans has been accumulating rapidly.[20,21] Thanks to the studies of Steiner and associates, insulin is known to originate as part of a high-molecular weight, metabolically inactive protein known as proinsulin. Proinsulin is enzymatically converted into insulin within the pancreatic β-cell.

Chance and co-workers determined the amino acid sequence of porcine proinsulin (Fig. 26–1) and found that the insulin A chain constitutes the C-terminal end of the proinsulin sequence, while B chain is at the N-terminal end. Separating the two insulin chains is a polypeptide (the connecting or C-peptide) that has a pair of basic amino acids at each end of its sequence. The basic amino acids identify the cleavage sites for the conversion reaction.

The number of amino acid residues and their appearance within the C-peptide are highly variable and species-dependent. Human C-peptide contains 31 residues, porcine 29, and bovine 26. A comparison of one avian and nine mammalian C-peptide sequences shows that only nine amino acid positions are invariant. Such variability in sequence results from a high mutational frequency and suggests that the C-peptide has no active center. The presence of the C-peptide within the proinsulin polypeptide chain allows the molecule to assume a conformational state that favors the formation of the correct disulfide bonds. Steiner and Clark demonstrated that the disulfide bonds of proinsulin, reduced to the sulfhydryl form, reform correctly under physiologic conditions with yields greater than 70%. Thus, nature avoids the problem of random disulfide bond formation encountered by chemists who attempted to combine the separate A and B chains.

The transformation of the proinsulin molecule into insulin requires protease activity similar to that of trypsin and carboxypeptidase B. The products formed during the conversion reaction include the insulin molecule, four basic amino acids (arg-31, arg-32, lys-62, and arg-63 in the case of porcine proinsulin), and the C-peptide. Each molecule of proinsulin yields one molecule of insulin, containing an A chain and a B chain, and one molecule of C-peptide.

The actual synthesis of proinsulin occurs on the ribosomes of the endoplasmic reticulum. Evidence, substantiated by the mapping of the human insulin gene to the distal short arm of chromosome II, indicates that proinsulin originates from an even larger "preproinsulin" protein. Preproinsulin turns over rapidly and may be responsible for sequestering proinsulin within the channels of

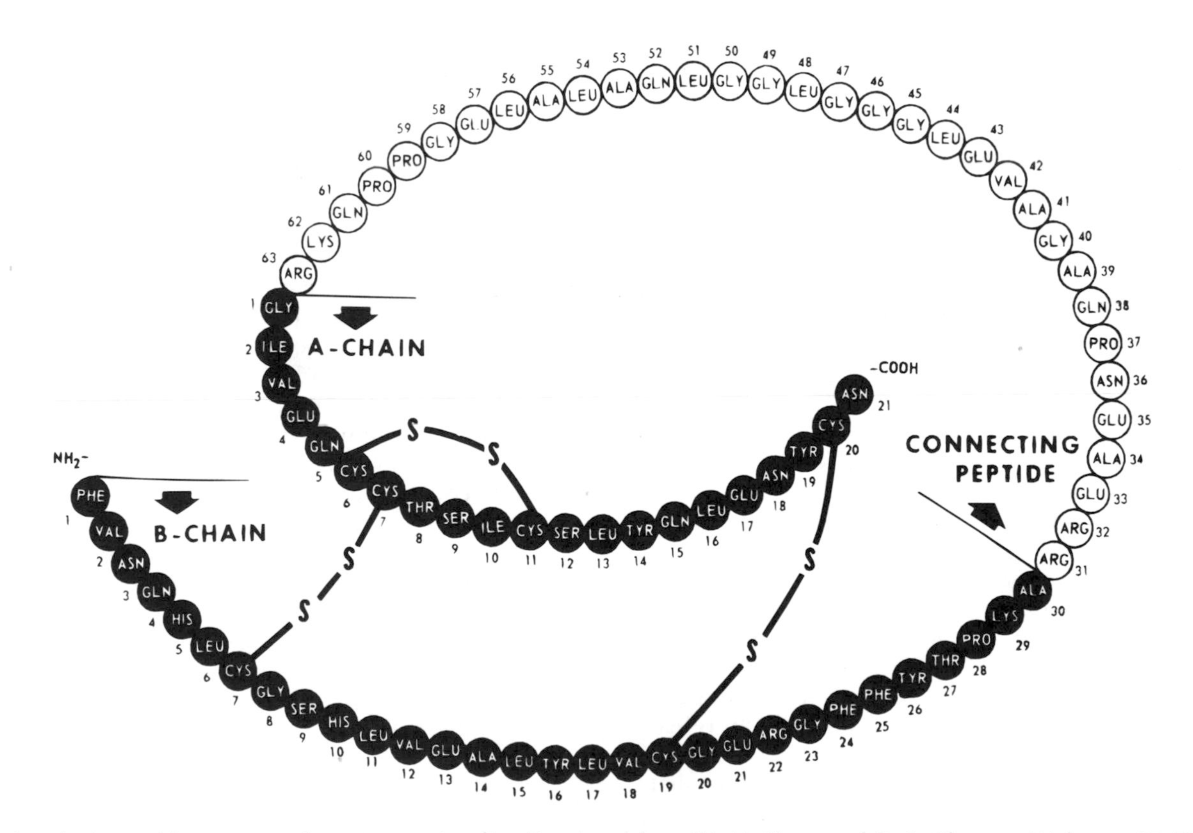

Fig. 26–1. Amino acid sequence of porcine proinsulin. (Reprinted from W. N. Shaw and R. E. Chance, *Diabetes, 17,* 737(1968) with permission of author and publisher.)

the endoplasmic reticulum. From there, proinsulin is translocated to the Golgi apparatus, where it is packaged into new secretory granules. The conversion of proinsulin to insulin occurs as these newly formed granules *mature* in the cytosol. The half-time for conversion is about 1 hour, and the conversion continues for many hours. By the fifth hour, only a low percentage of proinsulin remains, and, perhaps at this time, the contents of the granules crystallize with zinc. The contents of each mature granule represent one "quantum" of secretory product and consist, to the extent of 94%, of equimolar quantities of insulin and C-peptide. The remaining 6% is largely proinsulin with trace amounts of other conversion products that were entrapped during the crystallization process.

Secretion

The release of pancreatic β-cell secretory products is stimulated by a large number of endogenous and exogenous substances. Secretagogues include glucose, mannose, ribose, arginine, glucagon, and gastrointestinal hormones (secretin, pancreozymin, and gastric inhibitory protein), and the sulfonylureas. Some of these have a primary effect in that they stimulate insulin secretion directly, whereas others are potentiators that require the presence of some other compound. Substances known to inhibit the release of insulin include the catecholamines and somatostatin. The major determinant of β-cell function appears to be glucose, which stimulates both insulin synthesis and release.[21]

The glucose-provoked release of insulin is biphasic. An *early,* acute phase occurs almost immediately and peaks within 5 to 10 minutes. The second or *late* phase of insulin release occurs 15 to 20 minutes later, and it persists for as long as the glucose concentration remains elevated. Inhibitors of protein synthesis, e.g., puromycin, have no effect upon the early secretory phase but do reduce the *late* phase response by about 50%. No newly synthesized insulin is released from the pancreas within the first 2 hours. This observation has been taken to indicate that there are at least two targets for glucose action within the β-cell. Glucose given by mouth is much more effective than infused glucose in eliciting insulin release. This may be explained by the fact that oral glucose promotes the release of gastrointestinal hormones that enhance the glucose response. This is not the case with glucose infused intravenously.

Although the precise mechanism by which glucose causes insulin secretion is not known, it is generally accepted that glucose-induced insulin secretion involves closure of K^+ channels, depolarization of the β-cells, opening of the Ca^{++} channels, influx of Ca^{++}, and activation of exocytosis. Guanine nucleotide-binding proteins (G-proteins), which are critically important mediators in many signal transduction systems, have been implicated in both the inhibition and stimulation of insulin release from the β-cells through their ability to decrease or increase the intracellular cyclic adenosine monophosphate (cAMP) concentration, respectively.[22] Once exocytosis is activated, the secretory granules migrate toward the periphery of the cell, where they fuse with the plasma membrane before releasing their contents into the extracellular space.

Many diabetics with impaired response to glucose challenges retain their ability to respond to other secretagogues. This suggests that there could be many functional disorders, occurring at any of the steps in the sequence leading from β-cell insulin biosynthesis to secretion, all of which are collectively known as diabetes mellitus.

The development of a sensitive method for immunoassay of insulin has made it possible to study pancreatic secretions in normal and diabetic subjects. Under normal fasting conditions, the pancreas secretes about 20 μg of insulin per hour into the portal circulation. This results in a portal insulin concentration of 2 to 4 ng/ml. About 50% of this insulin is destroyed at the liver before entering the general circulation. Glutathione : insulin transhydrogenase catalyzes the reductive cleavage of insulin into the A and B chains, and each of these polypeptides is then hydrolyzed by proteases to the constituent amino acids. Proteases using insulin as a substrate are often called insulinases. Insulin escaping catabolism by the liver enters the general circulation and is present in concentrations ranging around 0.5 ng/ml. Systemically circulating insulin is further degraded by the kidney. Insulin is small enough to be filtered by the glomeruli, and the fraction that leaves the plasma is catabolized by the tubules upon resorption. As one might predict, insulin has a relatively short plasma half-life of 3 to 4 minutes.

After the patient ingests and absorbs a carbohydrate meal, plasma insulin levels normally increase 10- to 20-fold. No such increase in plasma insulin levels is seen in type I diabetics, who secrete little, if any, insulin. In type II diabetics, the increase in plasma insulin is delayed, but the increase may eventually exceed that of normal subjects under identical conditions. The delayed response may contribute to abnormal glucose tolerance and may be a factor in the pathogenesis of the disease.

The coexistence of hyperinsulinemia and abnormal glucose tolerance suggests insulin antagonism or resistance; the molecular basis for this lack of *effective* insulin is not known. Some conditions, however, are normally characterized by elevated basal plasma insulin levels. For example, in obesity there is a linear correlation between plasma insulin levels and the degree of the obesity. Furthermore, the plasma insulin levels in the obese are increased proportionately upon absorption of a carbohydrate meal. Apparently, obese individuals require more insulin to control lipolysis and also additional insulin for proper glucose uptake. The relationship between hyperinsulinemia and obesity may be significant in light of the evidence that obesity is often associated with the development of type II diabetes. Hyperinsulinemia could also result from a defect involving insulin catabolism in the liver and kidney; but there is no convincing evidence to suggest that the enzyme levels of glutathione: insulin transhydrogenase and insulinase are elevated or that their activities are enhanced.

The plasma insulin levels measured in insulin-treated diabetics are somewhat difficult to interpret due to the presence of the circulating antibodies found in virtually all patients within a few weeks of starting injections. Some conclusions, however, can be drawn as to how the insulin levels resulting from therapy differ from those measured under normal physiologic conditions. First, the amount of insulin that must be injected into diabetics to achieve near-normal glycemia results in plasma insulin levels that are practically always above normal. Second, the plasma insulin levels of diabetics are not nearly as variable as those of normal individuals, who may experience a 10- to 20-fold variation in going from the fasting to the fed state. Finally, because insulin normally enters the portal circulation en route to the liver, whereas the insulin injected by the diabetic enters the general circulation, it follows that the liver of the insulin-dependent diabetic never experiences normal insulin concentrations. The full ramifications of these problems are not known, but they will become of increasing concern as clinicians attempt to approximate more closely normal physiologic conditions.

Insulin immunoassays cannot readily distinguish between endogenously secreted insulin and exogenously administered bovine or porcine insulin. Therefore, insulin immunoassays are of practically no value for determining residual β-cell activity in diabetics undergoing insulin therapy. But residual β-cell activity can be assessed through use of the C-peptide immunoassay. The interpretation of the results obtained from this assay is based upon the fact that C-peptide and insulin are synthesized in equimolar quantities. Hence, any C-peptide measured in the plasma of an insulin-dependent diabetic should reflect the synthesis of an equal molar quantity of insulin. The assay has been useful in measuring the recovery of β-cell function after the correction of ketoacidosis with insulin therapy. It may also be of prognostic value. When C-peptide levels in stable and unstable insulin-requiring diabetics were compared, the stable group had higher C-peptide levels in both the basal state and after stimulation with glucose.

Mechanism of Action

Insulin exerts a variety of effects on responsive cells; all are mediated through interaction of insulin with specific receptors. The surface concentration of receptors ranges from as few as 40 on circulating erythrocytes to more than 200,000 on adipocytes and hepatocytes.

The gene encoding the human insulin receptor is composed of 22 exons and 21 introns and spans a region in excess of 120 kilobases on chromosome 19.[23] The product is a single polypeptide that undergoes post-transcriptional modifications including proteolytic cleavage, glycosylation, and fatty acylation. The basic structure of the mature insulin receptor is a heterodimer formed by α- and β subunits, which undergoes further dimerization to produce a heterotetrameric $\alpha_2\beta_2$ product. The four subunits are linked together by disulfide bonds to give a symmetric complex having the configuration (α-s-s-β)-s-s-(α-s-s-β). Mutations in the insulin receptor gene have been identified and found to be associated with insulin resistance.[24] Because insulin resistance contributes greatly to the pathogenesis of type II diabetes, the relationship between it and receptor mutations is of great interest.

The α-subunit consists of either 719 or 731 amino acids depending on splicing alternatives involving exon 11, which codes for a 12 amino acid segment at the COOH-terminal end. It is believed that individual cells express only one form of the α-subunit. When the receptor is positioned in the membrane, the α-subunits are extracellular and provide cysteine-rich regions that participate in the binding of insulin. α-Subunits with the extra 12 amino acid domain have a two- to threefold lower binding affinity for insulin than those without.

The β-subunits consist of 620 amino acid residues that provide an extracellular region of 194 residues to which the α-subunits are attached, a transmembrane anchoring domain of 23 amino acids, and a 403 amino acid cytoplasmic extension. The cytoplasmic extension has tyrosine kinase activity, which, through a series of autophosphorylations, is stimulated by the interaction of insulin with the α-subunits. Activation of tyrosine kinase activity thus provides a means for the internalization of the signal produced through the insulin-receptor interaction.[25]

In addition to the signal, the insulin-receptor complex is also internalized through endocytotic uptake. After internalization, insulin dissociates from its receptor to be degraded or released from the cell through a regulated process termed retroendocytosis.[23] The receptor can also be recycled back to the membrane, sequestered within the cell, or it can undergo degradation. Internalization can be looked at as providing a mechanism by which insulin can regulate the number of its own receptors on the cell's surface.[26]

A major, unanswered question is the identity of the immediate cellular targets of the insulin-stimulated tyrosine kinase activity. Although many potential protein substrates have been identified, none has been firmly established as being physiologically relevant. There is, however, mounting evidence that phosphorylation/dephosphorylation cascades do play a role in insulin's mechanism of action.[25] Other candidates for an insulin signaling system include the hydrolysis of a specific phospholipid derivative to yield diacylglycerol and the release of inositol phosphate from a phosphatidylinositol compound.[27] It is conceivable that these systems are not mutually exclusive and that a variety of divergent pathways are ultimately responsible for the attainment of insulin's multiple effects.

Metabolic Effects

The effects produced by insulin on carbohydrate, lipid, protein, and nucleic acid metabolism depend on the tissue involved and its metabolic state. Which of these are primary insulin effects, and which are secondary to one or more intracellular messengers, is not known. The metabolic effects of insulin are generally anabolic and promote the synthesis of energy reserves. The effects of insulin are antagonized by catabolic hormones such as glucagon and epinephrine. In the type I diabetic, insulin-mediated effects are severely attenuated or missing altogether.

Effects on Carbohydrate Metabolism

The most familiar of all of insulin's actions is its ability to promote glucose uptake by the cells of responsive tissues. Glucose uptake in virtually all cells occurs through the process of facilitated diffusion mediated by proteins known as glucose transporters. Five different, but structurally related, transporters having distinctive tissue distributions and responsiveness to insulin have been characterized and are identified as GLUT1 through 5.[28,29] GLUT1 is found in the membrane of most cells, has a high affinity (low K_m) for glucose and is responsible for basal glucose uptake by both insulin-dependent and insulin-independent tissues. The insulin-dependent tissues—adipose, cardiac, and skeletal muscle, in addition to GLUT1—have an intracellular pool of GLUT4, which is mobilized to the cell's membrane upon insulin stimulation. The additional glucose transported by this mobilized pool of transporters accounts for the 10- to 40-fold increase in glucose uptake attributable to the action of insulin. The liver, although being classified as being insulin sensitive, contains GLUT2, which does not require insulin for glucose uptake. GLUT2 has a high capacity and low affinity (K_m = 17–30 mM) for glucose. Consequently, glucose uptake by the liver is proportionate to the level of glucose in the blood with a net efflux occurring at blood glucose levels below 5 mM and a net influx occurring above 5 mM. GLUT2 is also found in the pancreatic β-cells where its responsiveness to the blood glucose level allows the β-cell to proportionately adjust its output of insulin.[30]

Glucose in the cell can be metabolized in many ways. It can be oxidized by the Embden-Meyerhof pathway and the Krebs's cycle to produce energy. It can be diverted through the hexose monophosphate shunt to obtain the reduced cofactors necessary for biosynthetic reactions, or, as in the skeletal muscles and the liver, it can be added to glycogen deposits (glycogenesis) to be stored for later retrieval (glycogenolysis). In adipose cells, glucose can be metabolized to acetyl CoA, which can then be used to synthesize fatty acids. Esterification of fatty acids with glycerol yields triglycerides, which are an energy storage form.

Insulin affects the preceding processes in a number of ways. In skeletal muscles and the liver, insulin favors glycogen formation by promoting glycogenesis while decreasing glycogenolysis. Glucose oxidation is accelerated by insulin, and glucogenesis is decreased. The glycogen-sparing effect of insulin in the liver, along with the decrease in glucogenesis, results in a reduction of hepatic glucose output. In the adipose cells, glucose oxidation is increased, as is the formation of reduced cofactors. The net effect of these insulin actions is to increase fatty acid synthesis and triglyceride formation.

Effects on Lipid Metabolism

Insulin promotes lipogenesis by producing conditions conducive to the synthesis of fatty acids and triglycerides from carbohydrate precursors. In addition to promoting their formation, insulin also inhibits their breakdown. Through these combined actions, insulin favors energy storage and gradually reduces the plasma levels of free fatty acids. As long as enough insulin is available to maintain the blood glucose level within the normal range, the synthesis of ketone bodies by the liver occurs at a minimal rate. Correspondingly, acetoacetic acid, acetone, and β-hydroxybutyric acid are present in the plasma in low concentrations. During severe hypoglycemia or in the untreated type I diabetic, however, ketogenesis is stimulated markedly, and a thousandfold increase in the concentrations of the ketone bodies may occur. High plasma concentrations of the ketone bodies are a major factor in the development of diabetic acidosis.

Effects on Protein and Nucleic Acid Metabolism

Insulin stimulates the transport of certain amino acids into the tissues and generally favors a positive nitrogen balance. In the liver, insulin decreases the levels of the glucogenic enzymes (pyruvate carboxylase, PEP carboxykinase, fructose-1,6-diphosphatase, and glucose-6-phosphatase) and increases the synthesis of the glycolytic enzymes (glucokinase, phosphofructokinase, and pyruvate kinase). These hormonal effects, no doubt, require both RNA and protein synthesis and are initiated through the interaction of insulin with its membrane receptor.[31] The increased rate of protein synthesis does appear to be independent of the stimulatory effects of insulin on amino acid and glucose transport. Insulin also has been shown to suppress the release of amino acids from the skeletal muscles, thereby accentuating its ability to lower plasma amino acid levels.

Insulin Therapy

The primary goal in the treatment of diabetes mellitus is to prevent the development of the metabolic abnormalities experienced by diabetics even in the presence of adequate food consumption. For many diabetics, the goal can be realized only by supplemental or replacement insulin therapy. Included in this group are all type I diabetics and all diabetics who are ketosis-prone, with or without the stress due to trauma, surgery, or infection. Insulin is also required by type II diabetics who cannot be maintained adequately on oral hypoglycemic agents or dietary regulation.

No single insulin preparation or combination of preparations can successfully meet the demands of such a diverse group. Consequently, a large number of insulin preparations have been developed, each of which has certain advantages and disadvantages. Although all of these preparations consist primarily of insulin and exhibit the biologic effects of insulin, they do differ in their onset and duration of action. Accordingly, they are classified on the basis of their duration of action into short (usual onset 0.5–2 hours; usual duration 3–6 hours), intermediate (usual onset 3–6 hours; usual duration 12–20 hours), and long-acting categories (usual onset 6–12 hours; usual duration 18–36 hours). The time course of action of any insulin may vary considerably in different individuals, or at different times of day in the same individual. Consequently, these time courses should be considered only as general guidelines.

The insulin contained in insulin preparations is of animal origin (either pork or beef or mixtures of the two) or is human insulin that has been prepared semisynthetically or has been isolated from bacteria that have been modified using recombinant DNA technology. It is anticipated that in the future, human insulin preparations will replace all other types. To minimize antigenicity, all insulin preparations available in the United States are highly purified in that they contain less than 10 parts per million proinsulin with many containing less than 1 part per million. Due to differences in the amino acid sequences of the animal insulins as compared to human insulin, however, preparations containing beef insulin (three amino acid differences) are still more antigenic than those containing pork insulin (one amino acid difference), which are more antigenic than those containing human insulin. In general, human insulin may act more quickly, peak earlier, and last a shorter time than animal insulins.

All insulin preparations are marketed in a strength containing 100 units/ml (U-100). For patients demonstrating extreme insulin resistance, e.g., a daily insulin requirement of 200 or more units, a U-500 preparation is available. Insulin units used to be based on biologic activity, but are now based on an absolute weight of insulin prepared from a recrystallized composite sample.

Insulin Preparations

The number and types of insulin preparations available in the United States is constantly changing and the student is referred to a current edition of the *Physicians' Desk Reference* (PDR)[32] for a current list. In general, they can be grouped into three basic types: regular, the protamine products, and the lentes, each of which differs from the others in physical and chemical properties.

Regular insulin is a solution of crystalline zinc insulin. It is the only insulin that is a solution, all others are suspensions. Regular insulin can be mixed with modified insulin preparations if one is directed to do so. Regular insulin has a rapid onset and a short duration of action, which generally preclude its routine use in the daily treatment of diabetes because of the number of injections required. It is used instead for treating medical emergencies, such as ketoacidosis, diabetic coma, and surgical complications. Due to the increasing interest in achieving fine control over the blood glucose level, regular insulin is often incorporated into multiple daily injection protocols in combination with a modified insulin product. Buffered insulin preparations are also available for use in the insulin pumps required for continuous subcutaneous infusion of insulin (CSII)—another method to achieve fine blood glucose control.

The insulin products containing protamine were originally developed in 1936 during a search for a longer acting insulin preparation. In the first product, insulin was combined with zinc and an excess of protamine, a basic protein obtained from the testes of fish, to form a protamine zinc insulin (PZI) complex. The complex has an isoelectric point close to physiologic pH; hence, it is less soluble in extracellular fluids than is insulin, and it leaves the site of injection more slowly. This accounts for the longer onset and the prolonged duration of action (longer than 36 hours). Because of concerns about too little effect in the daytime and too much at night, PZI lacked popularity, and was discontinued in the United States in 1991. A related product, Neutral Protamine Hagedorn insulin (NPH, also called isophane), however, developed at Nordisk Insulinlaboratorium in Denmark in 1946, achieved wide acceptance and products continue to be readily available.

Like PZI, NPH insulin is a complex of insulin, zinc, and protamine, but it contains less protamine than does PZI. The protamine and zinc insulin crystals are in stoichiometric quantities. PZI, NPH insulin has a quicker onset of action than PZI, and it has a shorter duration of action (24 to 28 hours). Because of the more rapid onset of action, it is usually not necessary to supplement NPH insulin with regular insulin, as was often the case with PZI. The 24-hour duration of NPH insulin is more compatible with day-to-day living than the 36-hour duration of PZI. Recently, premixed insulins having a fixed ratio of NPH insulin to regular insulin have been brought to market. Preparations available in the United States consist of 70% NPH and 30% regular insulin (70/30) and 50% NPH and 50% regular insulin (50/50). Other premixed preparations of 90/10 and 80/20 are available in Europe. NPH insulin is more stable than insulin mixtures due to concern about the potential antigenicity of fish protamine.

The lente insulins were developed in 1951. These preparations are unique in that they do not contain a protein modifier to prolong their action. Lente insulins are prepared by precipitating insulin from an acetate buffer by the addition of zinc. At relatively low zinc concentrations, an amorphous powder (semilente insulin) precipitates, and at higher zinc concentrations, a crystalline material (ultralente insulin) precipitates.

Semilente insulin has an onset and duration of action similar to those of regular insulin, while ultralente insulin is comparable to protamine zinc insulin. A mixture of 70% ultralente crystals and 30% semilente insulin is available as lente insulin, which is similar to NPH insulin in both onset and duration of action. Because of its rapid onset and intermediate duration, lente insulin is well suited for once-a-day administration and has been well accepted for use in this manner.

The lente insulins should not be mixed with insulin preparations containing phosphate buffers, because phosphates alter the solubility characteristics of the lente crystals and influence the onset and duration of action. Protamine zinc insulin, if available, and NPH insulin are both marketed as suspensions in pH 7.2 phosphate buffer and should not be mixed with the lente preparations.

The pharmacokinetics and therapeutic applications of the human insulins have been reviewed and compared to those of the animal insulins.[33]

Problems with Insulin Therapy

Insulin has been used for over 70 years to treat the symptoms of diabetes mellitus. During this time, the life expectancy of a newly diagnosed type I diabetic has increased from mere months to a life expectancy approaching that of a nondiabetic of similar age. The problems of the diabetic, however, are far from being solved. Insulin-treated diabetics are at risk to develop cardiovascular and kidney problems, and are prone to cataracts and visual impairment, including blindness. Some clinicians feel that these problems occur, not because of insulin therapy itself, but because of the poor blood glucose control obtained by the way in which insulin is administered. They argue that the tissues of an insulin-treated diabetic rarely, if ever, experience physiologic levels of insulin and that a method of delivering insulin physiologically must be developed. Ideally, the insulin should be introduced into the portal circulation to mimic normal pancreatic secretion. Evidence is accumulating that good control of the blood glucose level may lessen the likelihood of diabetic complications.

Improved methods for the delivery of insulin are being developed to improve glycemic control.[34] Perhaps the best method involves a pump that functions as an "artificial" pancreas. The device monitors the blood glucose concentration through a glucose sensor, and through several feedback mechanisms can inject controlled amounts of either insulin or glucagon according to the ambient glucose level. Although such a system is available (the Biostator from Miles Laboratories), its size and technical limitations limit its use to short periods in patients who are confined to the hospital. Until the technical problems, particularly in regard to the glucose sensor, can be overcome, there will be no widespread use of this device. A much simpler device, the portable insulin infusion pump, however, is rapidly gaining in popularity. Although the pump does not respond to changes in the concentration of blood glucose, it can be programmed to provide a continuous supply of insulin, as well as premeal bursts, into the subcutaneous tissue. Patients using this method of insulin delivery (CSII) must be carefully chosen and are required to self-monitor their blood glucose. To minimize the risk of hypoglycemia and ketoacidosis, an implantable form of the insulin infusion pump is available with the added advantage that it can deliver insulin intravenously or intraperitoneally, thereby mimicking more closely the physiologic delivery route.

Another alternative has been the transplantation of biologically active tissue that has retained its insulin secretory ability. Transplanted tissues vary from intact pancreata to pancreatic islets and β-cells. Transplantation of the intact gland is, in fact, a crude procedure in that 97% of the gland is not necessary. Nevertheless, through mid-1990, more than 2700 pancreatic transplantations have been carried out. The number is progressively increasing with about 300 operations a year being performed in the United States and Europe.[3] Whereas the transplant can restore normoglycemia, the operation is technically difficult and the patients must receive long-term immunosuppressive therapy. Consequently, the operation is usually reserved for those patients who have had a kidney transplant in the past and are receiving immunosuppressive therapy or are currently candidates for a dual transplant. Four-year patient survival from the operation has improved to approximately 90% with the survival of the functioning graft approaching 60%.[35]

Many of the problems associated with pancreatic transplants can be eliminated with the transplantation of islets. But of the approximately 160 attempts around the world since 1970, only one of the recipients was considered insulin-independent.[36] Islets are also susceptible to rejection mechanisms and are difficult to isolate, particularly from the human pancreas. It is hoped, however, that purified islets obtained by new techniques in combination with the latest methods to suppress rejection will be successful in man.

A novel approach to avoiding tissue rejection has been to culture β-cells on the outside of hollow, permeable, synthetic capillaries inserted into an iliac artery-to-vein shunt. The pore size of the capillaries is adequate to allow insulin to pass through into the blood, but too small to permit antibodies and lymphocytes to pass out. Dramatic short-term results have been demonstrated using this approach in alloxan diabetic rats, but the procedure is still highly experimental and of little practical value. The current status and future direction of this approach to insulin delivery is summarized in a recent review.[37]

Perhaps the most attractive and direct approach to administering insulin into the portal circulation is through the oral route. Insulin is a protein and is destroyed by digestion, but clinical trials have suggested that it can be absorbed in the duodenum. Doses 20 times higher than those effective subcutaneously are required, however. Efforts are being directed toward the derivation or compartmentalization of insulin in hope of decreasing the dose required.[38] Perhaps the increasing availability and possible lower cost of insulin produced by recombinant DNA technology may allow for such lavish expenditures of insulin. Oral administration of insulin would also eliminate the problem of lipodystrophy that may develop at the injection site. These tissue changes are generally benign, but may slow down insulin absorption or make it unpredictable. Alternatives to the oral administration of insulin under investigation include the nasal and rectal routes. The rectal route is considered socially unacceptable, but the nasal route may have clinical application.[34]

Because of its proteinaceous nature, insulin can cause allergic reactions. Hypersensitivity to insulin is common, but, fortunately, most of the dermal reactions are fairly mild and disappear within a week or two after initiation of therapy. Generalized allergic reactions occur much less frequently. Anaphylactic shock has also been reported in conjunction with insulin therapy, but it is extremely rare.

It is not always possible to attain good control of blood sugar levels by insulin administration. The brittle diabetic is particularly difficult to regulate, and in this case less than good control must be accepted. Also, because insulin is an extremely potent hormone, it can cause dangerous hypoglycemia when the dosage is not regulated to conform to conditions of diet and exercise.

Insulin resistance is another problem. Although the insulin requirement after pancreatectomy is approximately 50 units daily, some diabetics require much more than this to control blood glucose levels. Individuals are classified as insulin-resistant if their daily insulin requirement exceeds 200 units. Extreme cases have been reported in which patients require more than 2000 units of insulin daily to control hyperglycemia. Resistance does not usually occur until several months after the initiation of insulin therapy. This may be related to the production of circulating antibodies, which bind insulin and decrease its effectiveness.

Insulin antibodies are normally found in virtually all patients within a few weeks of starting injections. If insulin resistance, mediated by elevated levels of antibodies, occurs with conventional therapy, an alternative would be to convert the patient to more purified preparations. Newly diagnosed diabetics treated from the start with human insulin generally produce less antibodies than patients treated with the animal insulins.

ORAL ANTIDIABETIC AGENTS

It is evident that insulin therapy, as currently practiced, is not a panacea for diabetes mellitus. This realization has promoted a great deal of research toward the development of more effective ways of treating the disease and has led to the discovery of several orally active agents, including the sulfonylureas and the biguanides. These oral agents have received wide acceptance and are primarily used for the treatment of the noninsulin-dependent type of diabetes. The therapeutic value of these antidiabetic agents has undergone extensive reappraisal in the wake of the University Group Diabetes Program (UGDP).

The UGDP was commissioned by the National Institutes of Health (NIH) in 1959. It was to be a multicenter, long-term study designed to evaluate the natural history of type II diabetes, which includes about 80% of all cases of diabetes, and to compare the relative effectiveness of the sulfonylureas, specifically tolbutamide, and insulin in reducing vascular complications. The patients in the study had all been diagnosed as diabetic no more than 1 year before, and all were controllable without insulin. They were randomly assigned to one of four treatment groups: (1) fixed dose of tolbutamide (1.5 g/day); (2) fixed dose of insulin (10 to 16 units, based on body surface area); (3) variable dose of insulin adjusted to control the blood glucose; and (4) placebo. Eighteen months after the study began, a fifth group was added in which patients were given a fixed dose of a biguanide, phenformin hydrochloride (100 mg per day). There were approximately 200 patients in each group.

By 1969, an unexpected observation was being reported with increasing regularity. Cardiovascular disease was the designated cause of death in 12.7% of the patients from the tolbutamide group as compared to 4.9% from the placebo group, 6.2% from the fixed-insulin group, and 5.9% from the variable-insulin group. Based on these findings and the conclusion that tolbutamide was no more effective than diet in treating the symptoms of diabetes, investigators discontinued tolbutamide treatment.[39] Shortly thereafter, in May of 1971, the UGDP investigators moved to discontinue treatment of the phenformin group because of an observed cardiovascular mortality rate of 13.2%. A month later, the FDA outlined its proposed changes for the labeling of all sulfonylurea drugs.

Since then and continuing for more than a decade, the design, conduct, analysis and interpretation of the UGDP were the subjects of controversy. Despite certain shortcomings of the study brought to light by two independent audits of the data, one by the Biometric Society (an international organization of biostatisticians) and one by the FDA itself, the FDA is now requiring (since April 11, 1984) that all oral hypoglycemic agents of the sulfonylurea class be labeled with a specific warning on the increased risk of cardiovascular mortality. The FDA thought it prudent from a safety standpoint to include all sulfonylureas, although only tolbutamide was investigated.

Salicylates

The hypoglycemic effect of salicylates was first observed over 100 years ago by German physicians who noted a decrease in glycosuria in patients ingesting large quantities of sodium salicylate. Consequently, many of these compounds have since been studied as potential antidiabetic agents. The exact mechanism by which they lower blood sugar levels is not known, although they do enhance glucose-stimulated insulin secretion in normal and diabetic humans, probably by a mechanism medicated by prostaglandin synthesis inhibition. They also exert other effects, which may contribute to the hypoglycemic activity. Woods and associates demonstrated that sodium salicylate inhibits glucogenesis from lactate and alanine and suggested that this effect may be due to the uncoupling of oxidative phosphorylation and to the inhibition of alanine aminotransferase.[40] Certain salicylates decrease plasma levels of free fatty acids, triglycerides, and cholesterol. Because elevated levels of plasma-free fatty acids inhibit glucose utilization, a decrease in their concentration could contribute to the hypoglycemic action.

The clinical usefulness of salicylates has been limited because of the large doses (5 g daily) necessary to maintain adequate control of blood sugar levels. At these dosage levels, numerous side effects, such as gastric irritation, nausea, vomiting, and tinnitus, occur frequently. Attempts to find new salicylates that have hypoglycemic activity when administered in smaller doses and do not produce these side effects, have been unsuccessful. Salicylates have not been precisely evaluated as hypolipidemic agents for the same reasons.

Diguanidines

Guanidine was found to lower blood sugar levels in animals in 1918, but its relatively high toxicity precluded its use in clinical medicine. The guanidine derivatives, Synthalin A and Synthalin B, however, were introduced into diabetes therapy in the 1920s.

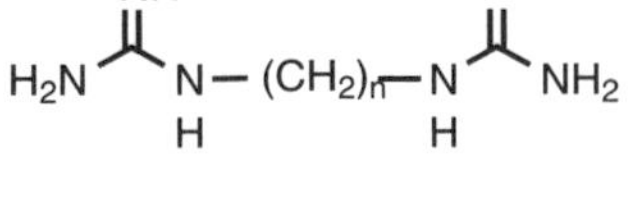

$$H_2N-C(=NH)-N(H)-(CH_2)_n-N(H)-C(=NH)-NH_2$$

Synthalin A: n = 10
Synthalin B: n = 12

These agents are substantially less toxic than guanidine, but they have no particular advantages over insulin. Furthermore, prolonged use of these compounds can cause renal and hepatic damage. Accordingly, their use was discontinued in the early 1930s.

Biguanides

Guanidine derivatives were largely ignored until the 1950s, when it was discovered that phenformin (DBI, Meltrol) had antidiabetic properties.

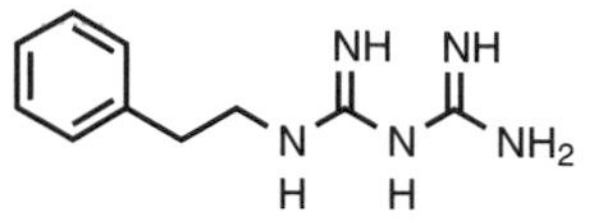

Phenformin

Because of the earlier experiences with Synthalins, acceptance of phenformin was slow, and two other biguanides, metformin (N-1,1-dimethyl biguanide) and buformin (N-butylbiguanide) were never released on the market in the United States, although they are used in other countries.

Soon after its introduction, phenformin was found to increase blood lactic acid levels, but the clinical syndrome, lactic acidosis, was considered to be a rare complication. Based on the data accumulated on lactic acidosis in 1977, however, the FDA estimated the true incidence at 0.25 to 4 cases per 1000 users per year with death occurring in approximately 50% of the patients.[41] In the United States, this mortality rate among phenformin users would amount to 50 to 700 deaths per year. Consequently, on July 25, 1977, the Secretary of Health, Education, and Welfare suspended the New Drug Application for phenformin and ordered an end to general marketing within 90 days.

In addition to lactic acidosis, cardiovascular complications are a risk in patients receiving phenformin therapy. As identified by the UGDP study, the cardiovascular mortality rate of the phenformin treatment group exceeded that of all other treatment groups and was almost three times the rate obtained with the placebo group. Although the validity of the results of the UGDP has been challenged, phenformin has been stigmatized, and when these results were considered along with the increased risks of lactic acidosis, the effect has been a decline in phenformin use worldwide, with metformin emerging as the biguanide of choice.

Metformin usually causes only a small increase in peripheral lactate with the reported incidence of lactic acidosis being 0–0.084 cases per 1000 patient-year; death occurs in 30% of patients.[42] This incidence, when compared to that stated above for phenformin, is approximately one order of magnitude less. Almost all reported cases of metformin-associated lactic acidosis have occurred in patients in whom the drug was contraindicated, or in cases of attempted suicide.[43] By avoiding use of the drug in patients with renal insufficiency, liver disease, and cardiopulmonary disorders, the incidence of lactic acidosis can be minimized. Metformin has been used for more than 20 years in Canada, accounting for about 25% of the orders for hypoglycemic agents, and not a single case of drug-associated lactic acidosis has been reported.[44] In a comparison of the risk of lactic acidosis caused by metformin to the risk of severe hypoglycemia caused by the sulfonylurea agents, it was determined that metformin was safer.[45] Metformin is currently undergoing clinical testing in the United States and could reach the United States' market in 1995.[46] In general, the biguanides should be used only in stable type II diabetics who are free of liver, kidney, and cardiovascular disease and cannot be adequately controlled with diet. Whereas it has been shown that metformin can reduce the insulin requirements of type I diabetic patients, the side effects of metformin coupled with the need for good renal function discourage this practice.[42] Biguanides can be used either alone or in combination with sulfonylureas. Particular attention has been shown to a fixed combination tablet containing phenformin and glyburide,[42] but most studies have involved the use of glyburide or chlorpropamide in combination with metformin.[47] Although the pharmacokinetic characteristics of the biguanides are a function of their structures, there are no studies addressing the potential of pharmacokinetic interactions between the sulfonylureas and the biguanides.

Mechanistically, the biguanides might better be called antidiabetic agents than hypoglycemic agents because they have relatively little effect in normal individuals. They have no direct stimulatory effect on insulin release, and in fact actually reduce the serum insulin level in both the basal and stimulated states. This decrease is believed to be secondary to the biguanide-induced decrease in blood glucose concentration and is usually attributed to one or more influences, working individually or in combination, that include: decreased lumen-to-plasma glucose transport, suppression of hepatic glucogenesis coupled with a net decrease in hepatic glucose output, and increased anaerobic glucose utilization in tissues—especially the intestine.[42,47] At the subcellular level, metformin has been shown to increase insulin binding to its receptor both in vitro and in vivo. Poor correlations, however, between improvements in glucose homeostasis in diabetic states without change in insulin binding and increased insulin binding in nondiabetic states without measurable change in glucose metabolism suggest that the effects of metformin may not only be directed at the receptor but also may be directed against intracellular postreceptor events.[42,44] For example, in rat adipocytes, metformin increases insulin-stimulated glucose transport by potentiating insulin-induced translocation of glucose transport proteins from an intracellular pool to the plasma membrane.[48] A similar effect has been observed with skeletal muscle.[49] Other effects of metformin include a decrease in circulating triglycerides, a decrease in plasma cholesterol, and an increase in high density lipoproteins. Patients receiving metformin often achieve a modest weight loss, which is particularly beneficial when the patient is obese.

The normal therapeutic dose of phenformin ranges from 25 to 125 mg per day. With oral administration, about 50% of the drug is absorbed from the gastrointestinal tract, and the peak serum concentration is reached within 2 to 4 hours. About one-third of the drug reaching the circulation is rendered biologically inactive via para-hydroxylation by the liver. Less than 20% of the drug is protein-bound and the plasma half-life is 7 to 15 hours. The rate of hydroxylation appears to be genetically determined and poor hydroxylators may be at increased risk to develop lactic acidosis. Excretion is partially biliary, with the majority being eliminated through the kidney into the urine. The usual duration of action is 4 to 6 hours.

The therapeutic dose of metformin is considerably higher than that of phenformin. Generally, 1000 to 2500 mg per day is given, divided into 2 or 3 doses taken with meals. The estimated absorption half-time is 0.9 to 2.6 hours, and the bioavailability is 50 to 60%.[42] Plasma half-life is 1.7 to 4.5 hours and 100% of the unchanged drug is eliminated by the kidneys within 24 hours (90% within 12 hours). The clearance rate is greater than the glomerular filtration rate (GFR), suggesting that the drug is secreted by the proximal convoluted tubules.[42] Metformin's most common side effects are gastrointestinal: anorexia, nausea, abdominal discomfort, and diarrhea. They occur initially in 5 to 20% of patients but are usually transient.[47]

Sulfonamides

The compound 2-(p-aminobenzenesulfonamido)-5-isopropylthiadiazole (IPTD) was used in the treatment of typhoid fever in the early 1940s.

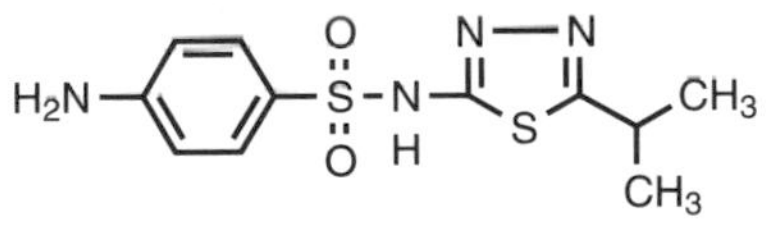

2-(p-Aminobenzenesulfonamido)-5-isopropylthiadiazole

Many patients died from obscure causes while being treated with large doses of the drug. These deaths were eventually attributed to acute and prolonged hypoglycemia, caused by stimulation of pancreatic insulin release. Relatively little attention was paid to the potential significance of IPTD in the treatment of diabetes until several years later when clinical studies were reported. The drug never gained wide acceptance as an antidiabetic agent, however, because at the same time as the clinical findings were reported, other workers published the results of studies conducted on carbutamide (BZ55), a sulfonylurea.

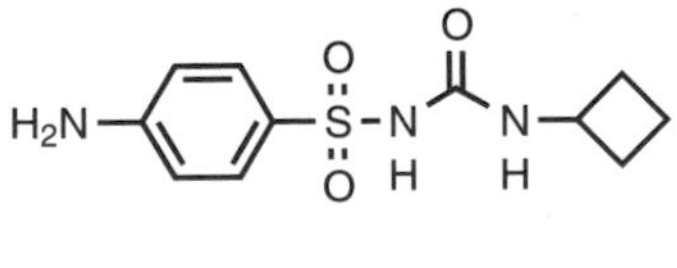

Carbutamide

Sulfonylureas

Carbutamide was found substantially more active than IPTD and was the first sulfonylurea hypoglycemic agent to be marketed. Approximately 12,000 sulfonylureas have since been synthesized as potential antidiabetic agents, and many have been found active. The development of sulfonylureas had a tremendous impact on diabetes treatment and research. Six sulfonylureas are currently marketed in the United States: tolbutamide, chlorpropamide, tolazamide, and acetohexamide comprise the first generation whereas glyburide and glipizide make up the more recently introduced second generation. As can be seen in Table 26–1, the sulfonylurea hypoglycemic agents are structurally similar, differing primarily with respect to the substituent in the para position of the benzene ring R_1 and the group attached to the terminal urea nitrogen (R_2).

Structure-Activity Relationships

The relationship between chemical structure and hypoglycemic activity is summarized as follows. The benzene ring should contain one substituent, preferably in the para position. Some substituents that seem to enhance hypoglycemic activity are methyl, amino, acetyl, chloro, bromo, iodo, methylthio, and trifluoromethyl groups. Compounds containing the p-(β-arylcarboxamidoethyl) substituent are orders of magnitude more powerful than the original (first generation) compounds. It is believed that the high activity of these derivatives is a function of the specific distance between the nitrogen atom of the substituent and the sulfonamide nitrogen atom.

The group attached to the terminal urea nitrogen should be of a certain size and should impart lipophilic properties to the molecule. The N-methyl derivatives are usually inactive, while N-ethyl derivatives show low levels of activity. Optimal activity is usually found in compounds containing three to six carbons in the N-substituent, while activity is lost if the N-substituent contains 12 or more carbons. Some active compounds contain an alicylic ring (with five, six, or seven members) on the terminal nitrogen; others, such as tolazamide, contain a heterocyclic ring.

Mechanism of Action

The hypoglycemic action of the sulfonylureas is usually attributed to their ability to stimulate the release of insulin from the pancreatic islets. It is proposed that the sulfonylureas bring about their increase in insulin release through binding to receptors on the islet β-cell membrane that are linked to closure of the channels that facilitate the passive efflux of K^+ from the cell.[50,51] These K^+ channels normally are responsive to the cytoplasmic ATP/ADP ratio and close when the ratio increases because of an increase in glucose metabolism.[52] Whether the closure of the channels is in response to the increase in the ATP/ADP ratio or to the interaction of sulfonylurea with a receptor, the plasma membrane undergoes depolarization. The reduced K^+ conductance opens voltage-dependent Ca^{++} channels that allow for an influx of Ca^{++} into the cytoplasm. The increase in cytosolic Ca^{++} activates the effector system that leads to the translocation of secretory granules to the exocytotic sites at the plasma membrane at which insulin is released. The binding ca-

Table 26–1. Currently Marketed Sulfonylureas

$R_1-C_6H_4-SO_2-NH-C(=O)-NH-R_2$

Compound	R_1	R_2
Tolbutamide	CH_3—	—$CH_2CH_2CH_2CH_3$
Chlorpropamide	Cl—	—$CH_2CH_2CH_3$
Tolazamide	CH_3—	—N (azepane ring)
Acetohaxamide	$CH_3C(=O)$—	cyclohexyl
Glyburide	5-Cl-2-OCH_3-C_6H_3-C(=O)-NH-CH_2CH_2—	cyclohexyl
Glipizide	5-H_3C-pyrazinyl-C(=O)-NH-CH_2CH_2—	cyclohexyl

pacity of various sulfonylureas to the putative sulfonylurea receptor closely correlates with their ability to stimulate insulin release.[53]

The released insulin caused by the interaction of sulfonylureas with their receptors is largely preformed insulin because these agents have little immediate effect on insulin synthesis. The impact on first and second phase insulin release appears to depend somewhat on the status of the patient and the agent used. Using glipizide, Groop was unable to demonstrate any appreciable increase in first phase insulin secretion in patients with manifest type II diabetes. Second phase insulin secretion, however, increased linearly with the ambient plasma glucose.[53] In general, the secretory pattern obtained with the sulfonylureas is similar to but distinct from that obtained with glucose stimulation. Because the sulfonylureas apparently stimulate insulin secretion by a mechanism that differs from that of glucose, they can be used in the treatment of diabetics who have lost their ability to respond to a glycemic stimulus, but who have retained residual pancreatic β-cell function, e.g., type II diabetics. These agents are of no use in the treatment of type I diabetics or in alloxan-diabetic animals, provided that β-cell destruction is complete. Studies have been done to evaluate the therapeutic potential of combination insulin-sulfonylurea treatment in type I diabetic patients. With two exceptions, the studies do not support the contention that such an approach is clinically useful.[54]

The hyperinsulinemia realized with these agents is transient. When the sulfonylureas are administered chronically, the plasma insulin levels usually, but not always, decline to pretreatment levels or below. The decline has been shown to be associated with decreases in proinsulin synthesis, pancreatic insulin content, and insulin secretion.

Although the serum insulin levels of patients receiving sulfonylureas gradually return to near normal (glipizide appears to be an exception; increased insulin secretory responses of 100 to 1500% have been maintained in some cases for longer than 2 years), the blood glucose concentration continues to be suppressed. To explain this phenomenon, investigators have postulated that the prolonged administration of the sulfonylureas improve the responsiveness of the β-cells to the glycemic stimulus. Lesser quantities of glucose[53] are required to provoke insulin release, and this more appropriate release of insulin improves glucose tolerance in the absence of exaggerated plasma insulin levels.

In addition to the direct effects of the sulfonylureas on the pancreas, other, extrapancreatic actions have been attributed to their use. Most can be explained via a sulfonylurea-induced potentiation of, or a decrease in resistance to, the actions of insulin on liver, skeletal muscle, and adipose tissue and, potentially, a reduction in hepatic insulin clearance.[53] Whereas the sulfonylureas have been shown to affect insulin receptor number or affinity, the extrapancreatic effects appear to be more directly related to post-receptor activities.

At the liver, sulfonylurea therapy is associated with a decrease in the hepatic glucose output (HGO), known to be a major contributor to the hyperglycemia characteristic of untreated type II diabetes.[55] The HGO decrease is predominantly caused by reduced hepatic glucogenesis rather than an increased glycogen mobilization rate.[56] Because both hyperglucagonemia and increased oxidation of free fatty acids are thought to stimulate glucogenesis, and because both conditions occur in type II diabetes,[57] it is possible that the sulfonylureas may decrease HGO through their normalization. It is also possible that the sulfonylureas may lower blood glucose by stimulating glycolysis by the liver. Tolbutamide favors production of fructose 2,6 bisphosphate, a potent activator of phosphofructokinase—the rate-limiting step.[58]

The sulfonylureas have also been shown to increase glucose disposal by skeletal muscle. In the type II diabetic patient, glucose conversion into glycogen is impaired by the extent of impairment correlated with a decrease in glycogen synthase activity.[56] Activation of glycogen synthase, such as has been demonstrated with the sulfonylurea, gliclazide,[59] promotes glucose uptake and favors a decrease in the blood glucose level.

In summary, although the mechanistic details have yet to be fully elucidated, the net effect of the sulfonylureas on blood glucose levels appears to be the result of a combination of factors and depends on the time course of therapy. Early in the course of treatment, insulin secretion is stimulated, resulting in an increase in circulating insulin levels which promote a move toward hypoglycemia. Later on in the course of therapy, the hypoglycemic effect is maintained in the face of declining insulin levels by an increase in insulin effectiveness mediated through an enhancement of insulin's membrane effects.

Pharmacology

The sulfonylureas are most likely to be effective in the treatment of obese type II diabetics who are over 40 years of age with diabetes of less than 10 years duration and whose fasting blood glucose level is less than 200 mg/dl and can be controlled on less than 20 units of insulin per day. They should not be used to treat patients who are prone to ketoacidosis. If adequate control over a patient's hyperglycemia cannot be achieved with one sulfonylurea (primary failure), it might be possible to gain control by using a different agent. Likewise, clinical success can sometimes be accomplished by switching a patient whose hyperglycemia was adequately controlled but now has become refractory to a sulfonylurea (secondary failure) over to another, more potent agent. Previously, secondary failures in sulfonylurea therapy were often treated successfully by combining a biguanide with the sulfonylurea. Now that metformin is undergoing clinical testing for the United States market, such a combination may once again become possible. An alternative strategy that is emerging due to the evidence that the sulfonylureas potentiate the action of insulin is to combine a sulfonylurea with insulin.[60] This approach, however, needs to be more fully evaluated before it becomes common practice.

The sulfonylureas are rapidly absorbed from the gastrointestinal tract and transported in the blood as highly protein-bound complexes. Binding to proteins by first generation agents is primarily ionic whereas binding to proteins by second generation agents is primarily nonionic. They vary mainly in potency and duration of action (Table 26–2). The variations are largely attributable to differences in mode and rate of metabolism. The pharmacokinetic parameters of the sulfonylureas have been extensively reviewed.[43,53,61–63]

Carbutamide is the parent compound in the series and was the first to be used clinically. It did, however, exhibit marked hepatic toxicity when administered in large doses, and, consequently, it was never marketed in the United States. Carbutamide is metabolized by N^4-acetylation, as are the sulfa drugs, and the metabolite has no hypoglycemic activity.

Tolbutamide is metabolized by the liver to p-hydroxytolbutamide, which has about 35% of the hypoglycemic activity of the parent. The metabolite is short-lived, however, because oxidation to the inactive carboxyl derivative ensues rapidly. Because of this, tolbutamide is the least potent and shortest acting of the currently available sulfonylurea hypoglycemic agents. The drug is usually prescribed in doses given two to three times a day.

Tolazamide is also metabolized to a carboxylic acid through the same two-stage oxidation, but it can also be converted into 4-hydroxy and hydroxymethyl derivatives. These hydroxylated forms have less hypoglycemic activity than tolazamide but are more potent than tolbutamide. Consequently, tolazamide has increased potency and a

Table 26–2. Pharmacokinetic Characteristics of Sulfonylureas

Agent	Equivalent Therapeutic Dose (mg)	% Serum Protein Binding	Mode of Metabolism	Biologic Half-life (hr)	Duration (hr)	Renal Excretion (%)
First generation						
Tolbutamide	1000	95–97	Carboxylation	4.5–6.5	6–12	100
Chlorpropamide	250	88–96	Cleavage, hydroxylation	36	Up to 60	80–90
Tolazamide	250	94	Carboxylation, hydroxylation	7	12–14	85
Acetohexamide	500	65–88	Reduction, hydroxylation	6–8	12–18	60
Second generation						
Glyburide	5	99	Hydroxylation	1.5–3.0	Up to 24	≈50
Glipizide	5	92–97	Hydroxylation	4	Up to 24	68

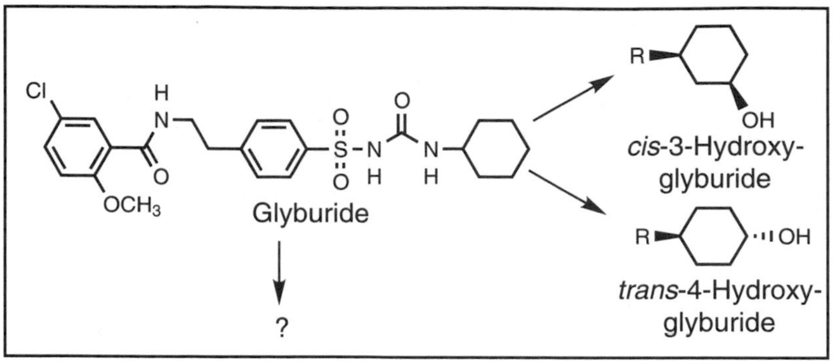

Fig. 26–2. Metabolism of glyburide.

longer duration of action. It is usually given in 1 or 2 doses per day.

Acetohexamide is rapidly metabolized by reduction of the ketone to a secondary alcohol. The metabolite, 1-hydroxyhexamide, has 2.5 times the hypoglycemic activity of acetohexamide and accounts for acetohexamide's longer duration of action. Acetohexamide is also subject to hydroxylation to yield the 4-hydroxy derivative. 4-Hydroxyacetohexamide is inactive and accounts for only 15% of the excreted dose. Once-a-day administration of acetohexamide is adequate to control many diabetics.

Chlorpropamide, despite reports to the contrary, does undergo significant metabolism, but slowly, which accounts for its long half-life. Primary metabolites are 2-hydroxy and 3-hydroxychlorpropamide and p-chlorobenzenesulfonylurea. Because of its long half-life, blood levels gradually increase on a once-a-day dose schedule, reaching a plateau in 7 to 10 days. The dosage of the drug should not be increased during this transitional period, because hypoglycemia is apt to occur.

The second-generation agents that have been introduced to the U.S., although they have been on foreign markets much longer, include glyburide (glibenclamide) and glipizide. Glyburide is totally metabolized by the liver (Fig. 26–2) with the main metabolite being the *trans*–4–hydroxy derivative. *Trans*–4–hydroxyglyburide has approximately 15% of the hypoglycemic activity of the parent compound. Besides urinary excretion, about one-half of an administered dose of glyburide undergoes biliary excretion. Glipizide is metabolized to at least three derivatives (Fig. 26–3) that are essentially devoid of hypoglycemic activity. *Trans*–4–hydroxyglipizide is the one that is found in the greatest concentration in the urine after 24 hours. Both glyburide and glipizide are orders of magnitude more powerful than the original series and can be given in single daily doses. While these agents are generally considered to function according to the same mechanisms as the other sulfonylureas, it is not clear whether their administration in such low doses subjects the patient to the same hazards and side effects.

Other second generation sulfonylureas, not available in the United States, are glibornuride and gliclazide (see formula). Glibornuride is approximately 40 times more potent than tolbutamide in lowering blood glucose concentrations. Glibornuride has an average plasma half-life of 8 hours and a relatively short duration of action. It is almost completely metabolized to six inactive metabolites that are excreted in the urine.

Gliclazide was introduced to the foreign market in 1972 and was the subject of a symposium that was recently published.[64] In addition to the usual actions attributed to the sulfonylureas, gliclazide is of particular interest due to its

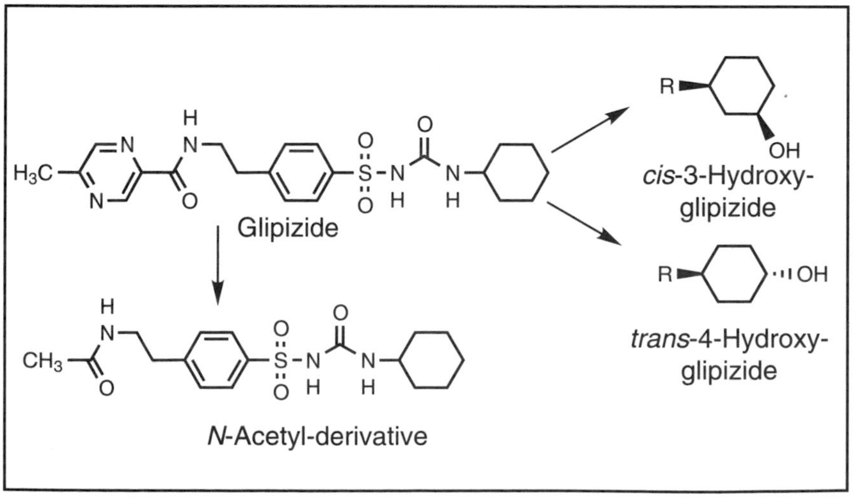

Fig. 26–3. Metabolism of glipizide.

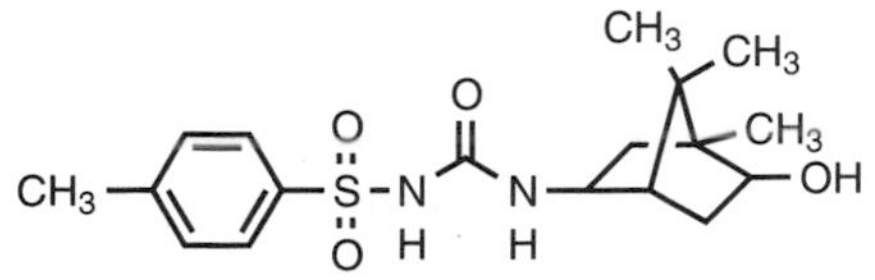

Glibornuride

having an antiplatelet-aggregating activity and a potential action in preventing or retarding the progression of diabetic retinopathy.[65] Given in divided doses ranging from 80 to 320 mg per day, gliclazide reaches peak plasma concentrations within 2 to 4 hours, has a plasma half-life of 8 to 11 hours, and a duration of action up to 24 hours. It is primarily metabolized by the liver, resulting in the formation of eight inactive metabolites that are eliminated in the urine.[61,65]

Hazards and Side Effects

A great deal of attention has been directed toward the cardiovascular effects of the sulfonylureas since the findings of the UGDP were released. The study was designed to determine whether or not control over the blood glucose levels helps to prevent or delay vascular disease, but instead, it concluded that patients receiving oral hypoglycemic agents appear to face an increased risk of cardiovascular mortality, that tolbutamide plus diet was no more effective than diet alone in prolonging life and that insulin treatment was no better than diet alone in altering the course of vascular complications in type II diabetes.[66] To establish the validity of the UGDP results, many prospective studies have been undertaken. The results of some of these studies tend to contradict the UGDP results, while the results of others are supportive of the UGDP conclusions. Although none of these studies was as extensive as the UGDP and all have been subjected to varying degrees of criticism, they do point out the need for careful evaluation of therapy and the continuing need for studies designed to answer the questions posed by the UGDP.

The UGDP has since published a supplementary report on the nonfatal events experienced by patients taking part in the program.[67] Some of the findings discussed in the report agree with the original conclusions reached by the UGDP; other findings raise questions concerning the appropriateness of any of the current modes of diabetic therapy. For instance, none of the four different treatment schedules had much effect on body weight, incidence of hypertension, or occurrence of myocardial infarction, although favorable changes in blood glucose levels were observed, particularly in the variable insulin treatment group. These results are interpreted by the UGDP to suggest that normalization of the blood glucose levels of diabetic patients may not alter the incidence of renal impairment, retinal changes, or any of the other common complications of diabetes. Furthermore, the UGDP has determined that the mortality rate among patients suffering a myocardial infarction is somewhat higher in the tolbutamide treatment group (50%) than in the placebo treatment group (18%), the variable insulin group (40%), or the standard insulin group (35%). The UGDP considered this observation to be consistent with the findings of others[68,69] who have demonstrated that the sulfonylureas have a positive inotropic effect on the heart and also increase cardiac irritability. Hence, patients who are susceptible to cardiovascular problems and are receiving sulfonylurea drugs may have their condition aggravated by these agents.

The incidence of reversible side effects from sulfonylurea therapy is low, involving less than 5% of the patients with fewer than 2% of patients discontinuing therapy,[43] which may produce facial flushing in susceptible individuals after alcohol intake. The more common complaints involve the gastrointestinal tract (1.4%) and include anorexia, nausea, vomiting, diarrhea, and abdominal pain. Skin reactions occur in 2 to 3% of the patients receiving sulfonylureas, and, in the case of chlorpropamide, may persist for several days after the drug has been withdrawn.

Chlorpropamide is known to cause water retention and hyponatremia. Chlorpropamide exerts this effect by increasing both the release of antidiuretic hormone (ADH) from the posterior pituitary and the sensitivity of the renal tubules to the hormone. In contrast, acetohexamide, tolazamide, and glyburide have diuretic properties and should be considered if water intoxication is a potential problem.

The more serious side effects associated with sulfonylurea therapy, apart from the potential cardiovascular problems, are fortunately rare. These agents, particularly chlorpropamide, have some hepatotoxicity and may cause a transient form of cholestatic jaundice (0.5%). Blood dyscrasias have been reported (0.24%) and include transient leukopenia, granulocytosis, and fatal aplastic anemia. There are no definitive changes in the thyroid function with the short-term use of these agents, but, apparently, sulfonylurea-induced hypothyroidism can occur after long-term administration.

All six of the commonly used sulfonylureas have caused severe hypoglycemic reactions, but with varying frequencies. The true incidence of severe hypoglycemia is difficult to determine because of different interpretations of the definition and underreporting. Data from Sweden report only 0.22 episodes per 1000 patient years,[53] whereas 20.2% of the patients in a U.K. study experienced hypoglycemic symptoms during the previous 6 months.[70] Nevertheless, most cases of severe and fatal hypoglycemia have been associated with the use of the long-acting sulfonylureas, glyburide and chlorpropamide. Risk factors for hypoglycemia to be considered before sulfonylurea therapy included: age greater than 70 years, alcohol intake, poor nutrition, intercurrent gastrointestinal disease, impaired renal function, and drug interactions.[53]

Interactions with Other Drugs

Several interactions between sulfonylurea hypoglycemic agents and other drugs have been observed. Some of the more commonly occurring interactions will be briefly discussed, but for a more detailed treatment of the subject, the reader is referred to compendia such as Drug Interactions & Updates[71] or Drug Interaction Facts.[72]

All of the currently used thiazide diuretics, as well as many of the thiazide-like diuretics, antagonize the hypoglycemic effects of the sulfonylureas. The antagonism leads to an increase in blood glucose levels and is believed to occur through drug interference of the sulfonylurea-induced release of insulin, possibly due to depletion of tissue potassium. The effect is usually reversed by discontinuing use of the diuretic, supplementing potassium levels, or increasing the dose of the hypoglycemic agent. The loop diuretics, e.g., furosemide, may also promote hyperglycemia by decreasing glucose tolerance.

An elevation in the blood glucose level as a response to antagonism of the hypoglycemic effect of the sulfonylureas is also a possibility with epinephrine, isoproterenol, and other adrenergic agents. The ability of the β-adrenergic stimulants to promote glycogenolysis is probably responsible for this effect. On the other hand, the β-adrenergic blocking agent propranolol may potentiate the hypoglycemic action of the sulfonylureas, probably by inhibiting glycogenolysis. The β-blockers may also attenuate the hypoglycemic action of the sulfonylureas by decreasing peripheral glucose uptake, decreasing patient's sensitivity to insulin, and inhibiting glucose-stimulated insulin release.[71]

Antagonism of the hypoglycemic activity of the sulfonylureas is also a problem associated with hormone therapy, including the use of oral contraceptives and the glucocorticoids. Both of these types of hormones favor hyperglycemia by promoting peripheral insulin resistance. In addition, the glucocorticoids generally oppose the metabolic effects of insulin by increasing the hepatic synthesis of glucose and decreasing peripheral glucose utilization. Maintenance of control over the diabetic state when these drugs are administered concurrently may require that the dose of the sulfonylurea be adjusted upward.

Monoamine oxidase inhibitors, both of the hydrazide and nonhydrazide types, potentiate the hypoglycemic effect of insulin and the sulfonylureas. Although the interaction is well documented, no exact mechanism has been determined. It has been postulated that a false neurotransmitter may be generated that does not have adrenergic effects and does not induce compensatory hyperglycemia. Because of the potentially dangerous hypoglycemia that might result from this drug interaction, care must be taken when a monoamine oxidase inhibitor is administered in combination with a sulfonylurea.

Some of the other drug interactions involve displacement of the sulfonylurea from plasma protein binding sites or prolongation of their biologic half-life. Displacement is associated with clofibrate and halofenate coadministration. The sulfonamides may impair hepatic metabolism of the sulfonylureas and may alter plasma protein binding.

Sulfaphenazole, a sulfonamide not currently available in the United States, has been shown to displace first-generation sulfonylureas from protein-binding sites on human serum albumin. This increases the concentration of the free (active) drug and produces a more intense reaction that may cause hypoglycemia. Furthermore, it is believed that sulfaphenazole may either diminish the hepatic metabolism or delay the renal clearance of the sulfonylureas; five- to sixfold increases in the half-life of

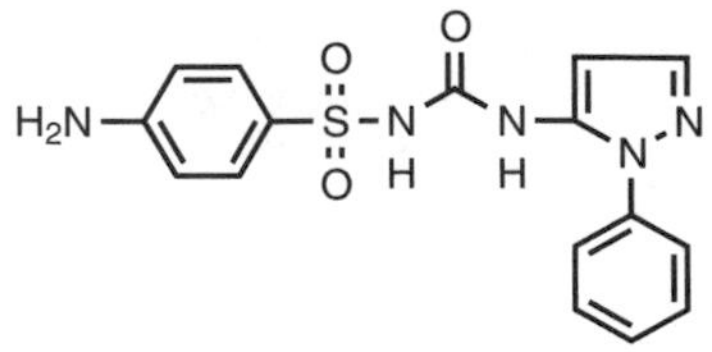

Sulfaphenazole

the sulfonylureas have been reported. Although this drug interaction between the sulfonylureas and sulfaphenazole may have some clinical significance, it appears not to pertain to all of the other sulfonamides. The second-generation agents are less susceptible to protein displacement because their binding is more dependent on hydrophobic than charge interactions.

Phenylbutazone is another drug that may enhance the hypoglycemic activity of the sulfonylureas by protein displacement as well as by increasing their half-life. Phenylbutazone appears to exert its effect by interfering with the renal excretion of both the sulfonylureas and their metabolites. Because this drug interaction is considered a "definite possibility," the therapy should be supervised closely and the dose adjusted if necessary.

The administration of some drugs to patients receiving sulfonylureas must be approached cautiously if the drugs are known to have hypoglycemic activity of their own. Examples include aspirin and other salicylates and alcohol. Although the potential for this drug interaction is low with the dosage of aspirin used to treat headaches and minor pains, it may become significant with the large doses required required to obtain an anti-inflammatory effect.

Besides its inherent hypoglycemic effect, alcohol interacts with the sulfonylureas in many other ways that make the ultimate effect upon the blood glucose level difficult to predict. For instance, the infrequent ingestion of large amounts of alcohol by patients receiving sulfonylureas may result in potentiation of their hypoglycemic responses, due to decreased metabolism of the sulfonylureas. At the other extreme, chronic alcohol consumption by patients receiving sulfonylureas may actually diminish the hypoglycemic response because of the induction of the hepatic enzymes responsible for the metabolism of these agents.

Although not related to a change in blood glucose levels, another drug interaction occurs between alcohol and sulfonylureas. Within 10 minutes of taking alcohol, many patients on sulfonylurea therapy may experience facial flushing, headache, rapid breathing, and a rapid heart rate. This disulfiram-like effect is most common with chlorpropamide.

It is evident that a great deal of information about the interactions of sulfonylureas with other drugs has been accumulated. It is equally apparent, however, that the significance and the fundamental basis of many of these observations are poorly understood. Much more research is necessary before this subject can be viewed in its proper perspective.

Other Hypoglycemic Agents

Hypoglycemic activity has been observed with compounds of widely diverse chemical structures, but the clinical usefulness of these compounds has yet to be established. In the mid-1960s, there was a great deal of optimism with regard to a number of pyrazoles and isoxazoles as potential oral antidiabetic agents, but animals developed resistance to these compounds in a relatively short time.

Hypoglycin, p-tert-butylbenzoate, methyl 2-tetradecylglycidate, and other compounds that bring about a decrease in hyperglycemia due to their ability to inhibit fatty acid oxidation have been studied, but their inherent toxicities preclude their clinical usefulness.[73] Hypoglycemic activity has also been observed in indole derivatives, indans, monoamine oxidase inhibitors, benzoquinones, naturally occurring amino acids, and other classes of compounds. At present, however, evidence is insufficient that any of these compounds is equal to the currently available antidiabetic agents. The same arguments apply equally well to glicodiazine, a sulfapyrimidine, and other "second-generation" sulfonylureas that have found their way into the foreign market. Although these compounds may differ quantitatively in their action as compared to currently available sulfonylureas, their mechanism appears to be essentially the same and, accordingly, they appear to offer no particular advantages.

The observation that diabetes, with rare exceptions, is associated with elevated plasma glucagon levels in combination with decreased or absent insulin levels and the knowledge that glucagon promotes hyperglycemia have triggered a search for agents that can selectively suppress glucagon release. Toward this goal, the amino acid sequence of somatostatin is being systematically altered. D-cysteine[14] somatostatin, D-tryptophan[8]-D-cysteine[14] somatostatin, and alanine[2]-D-tryptophan[8]-D-cysteine[14] somatostatin have all been shown to suppress arginine-induced glucagon release with a concomitant decrease in blood glucose. These peptides, however, also retain varying levels of suppressor effect on other hormones and have short plasma half-lives. More recently, a selective action on glucagon suppression with no effect on insulin secretion in isolated rat islets was demonstrated using a substituted sulfonamido-benzamide (M&B 39890A).[74] Whereas this combination of hormonal effects is the one desired, research with the drug is preliminary. Yet another approach has been the search for analogs of glucagon that could antagonize the effects of glucagon on target tissues. Two such antagonists have been developed,[75] but further research is needed. It is apparent that there will be no clinical application of this concept in the foreseeable future.

A novel approach to the treatment of both type I and type II diabetes may be the use of intestinal α-glucosidase inhibitors as adjuncts. When α-glucosidase inhibitors, e.g., acarbose, are ingested with sucrose or starch-containing meals, they lower the postprandial rise in blood glucose.[76] These agents are being evaluated to determine which patients are the best candidates and the degree of long-term benefit.

Other agents that have demonstrated hypoglycemic effects but are still in the research stage are the thiazolidinediones (ciglitazone, pioglitazone, and englitazone), the substituted guanidines (linogliride and pirogliride), dichloroacetate (DCA), and a thiopyranopyrimidine derivative (MTP-3631). The thiazolidinediones have similar pharmacologic effects and appear to act by sensitizing target tissues to the effects of insulin without stimulating insulin secretion from the pancreas.[75,77] These agents are currently in phase I or phase II clinical testing. Linogliride has been shown to stimulate insulin release in the absence of glucose either in the presence or absence of a physiologic mixture of amino acids. Its mechanism of action appears to be similar to that of tolbutamide.[78] DCA reduces blood glucose by inhibiting hepatic glucose synthesis and stimulating glucose clearance and use by peripheral tissues.[79] It is not known whether DCA and its analogs are safe and effective agents for long-term use in humans. MTP-3631 is the latest in a series of thiopyranopyrimidine derivatives that have demonstrated hypoglycemic activity. Although the mechanism for the blood glucose lowering effect is not established, it appears to require the presence of insulin and is distinct from that of the sulfonylureas and the biguanides.[80]

REFERENCES

1. R. Luft, *Diabetologia, 32,* 399(1989).
2. R. B. Tattersall and D. A. Pyke, *Lancet, 2,* 1120(1972).
3. D. Pyke, *Diabetes Metab. Rev., 7,* 3(1991).
4. S. S. S. Lo, et al., *Diabetes Metab. Rev., 7,* 223(1991).
5. D. W. Drell and A. L. Notkins, *Diabetologia, 30,* 132(1987).
6. S. S. Rich, *Diabetes, 39,* 1315(1990).
7. J. R. Baker, *JAMA, 268,* 2899(1992).
8. J. M. Baisch, et al., *N. Engl. J. Med., 322,* 1836(1990).
9. G. T. Nepom, *Diabetes, 39,* 1153(1990).
10. M. J. Sheehy, *Diabetes, 41,* 123(1992).
11. J. P. Palmer and D. K. McCulloch, *Diabetes, 40,* 943(1991).
12. S. Baekkeskov, et al., *Nature, 347,* 151(1990).
13. Anon., American Diabetes Association, *Diabetes, 39,* 1151(1990).
14. R. H. Unger and L. Orci, *Lancet, 1,* 14(1975).
15. R. H. Unger, *Diabetes, 25,* 136(1976).
16. P. Felig, et al., *Diabetes, 25,* 1091(1976).
17. R. R. Davies, et al., *Diabet. Med., 6,* 103(1989).
18. Various authors, *Diabetes Care, 5(Suppl 2),* 1(1982).
19. R. A. Pullen, et al., *Nature, 259,* 369(1976).
20. J. C. Hutton, *Diabetologia, 32,* 271(1989).
21. P. A. Halban, *Diabetologia, 34,* 767(1991).
22. R. P. Robertson, et al., *Diabetes, 40,* 1(1991).
23. J. M. Olefsky, *Diabetes, 39,* 1009(1990).
24. S. I. Taylor, et al., *Diabetes Care, 13,* 257(1990).
25. J. H. Exton, *Diabetes, 40,* 521(1991).
26. J. L. Carpentier, *Diabetologia, 32,* 627(1989).
27. A. R. Saltiel, *Diabetes Care, 13,* 244(1990).
28. L. J. Elsas and N. Longo, *Annu. Rev. Med., 43,* 377(1992).
29. W. T. Garvey, *Diabetes Care, 15,* 396(1992).
30. M. Mueckler, *Diabetes, 39,* 6(1990).

31. R. M. O'Brien and D. K. Granner, *Biochem. J., 278,* 609(1991).
32. A. L. Dowd, Ed., *Physicians' Desk Reference,* 47th ed. Montvale, NJ, Medical Economics Data, 1993.
33. R. N. Brogden and R. C. Heel, *Drugs, 34,* 350(1987).
34. F. P. Kennedy, *Drugs, 42,* 213(1991).
35. R. P. Robertson, *Diabetes, 40,* 1085(1991).
36. C. Hellerstrom, et al., *Diabetes Care, 11(Suppl 1),* 45(1988).
37. G. Reach, *Int. J. Artif. Organs, 13,* 329(1990).
38. R. S. Spangler, *Diabetes Care, 13,* 911(1990).
39. Anonymous, A study of the effects of hypoglycemic agents on vascular complications in patients with adult-onset diabetes. *Diabetes, 19(Suppl 2),* 747(1970).
40. H. F. Woods, et al., *Clin. Exp. Pharmacol. Physiol., 1,* 534(1974).
41. Anonymous, Phenformin: New labeling and possible removal from market, *FDA Drug Bulletin, 7,* 6(1977).
42. C. J. Bailey, *Diabetes Care, 15,* 755(1992).
43. J. E. Gerich, *N. Engl. J. Med., 321,* 1231(1989).
44. R. Vigneri and I. D. Goldfine, *Diabetes Care, 10,* 118(1987).
45. I. W. Campbell, *Metformin and sulphonylurea derivatives: the comparative risks.* In *Diabetes and Metformin: A Research and Clinical Update.* H. M. J. Krans, Ed. London, Royal Society of Medicine, 1985, p. 45.
46. J. A. Colwell, *Diabetes Care, 16,* 653(1993).
47. L. S. Hermann, *Diabetes Care, 13,* 37(1990).
48. S. Matthaei, et al., *Diabetes, 40,* 850(1991).
49. A. Klip and L. A. Leiter, *Diabetes Care, 13,* 696(1990).
50. L. Siconolfi-Baez, et al., *Diabetes Care, 13(Suppl 3),* 2(1990).
51. W. J. Malaisse and P. Lebrun, *Diabetes Care, 13(Suppl 3),* 9(1990).
52. C. Schwanstecher, et al., *Mol. Pharmacol., 41,* 480(1991).
53. L. C. Groop, *Diabetes Care, 15,* 737(1992).
54. H. E. Lebovitz and R. M. Pasmantier, *Diabetes Care, 13,* 667(1990).
55. S. D. Prato, et al., *Am. J. Med., 90(Suppl. 6A),* 29S(1991).
56. R. A. DeFronzo, et al., *Diabetes Care, 15,* 318(1992).
57. A. Consoli, *Diabetes Care, 15,* 430(1992).
58. M. Aoki, et al., *Diabetes, 41,* 334(1992).
59. A. B. Johnson, et al., *Diabet. Med., 8,* 243(1991).
60. T. S. Bailey and N. H. E. Mezitis, *Diabetes Care, 13,* 687(1990).
61. R. E. Ferner and S. Chaplin, *Clin. Pharmacokinet., 12,* 379(1987).
62. A. Melander, et al., *Drugs, 37,* 58(1989).
63. P. Marchetti, et al., *Clin. Pharmacokinet., 21,* 308(1991).
64. Various authors, Oral treatment of diabetes mellitus: The contribution of gliclazide. *Am. J. Med., 90(Suppl 6A),* 1S(1991).
65. H. Rifkin, *Am. J. Med., 90(Suppl. 6A),* 3S(1991).
66. Anonymous, Effects of hypoglycemic agents on vascular complications in patients with adult-onset diabetes VIII. Evaluation of insulin therapy: final report. *Diabetes, 31(Suppl. 5),* 1(1982).
67. Anonymous, A study of the effects of hypoglycemic agents on vascular complications in patients with adult-onset diabetes VI. Supplementary report on nonfatal events in patients treated with tolbutamide. *Diabetes, 25,* 1129(1976).
68. K. C. Lasseter, et al., *J. Clin. Invest., 51,* 2429(1972).
69. B. H. Tan, et al., *Diabetes, 33,* 1138(1984).
70. A. M. Jennings, et al., *Diabetes Care, 12,* 203(1989).
71. P. D. Hansten and J. R. Horn, *Antidiabetic Drug Interactions.* in *Drug Interactions and Updates,* Philadelphia, Lea & Febiger, 1990.
72. D. S. Tatro, Ed., *Drug Interaction Facts.* St. Louis, Facts and Comparisons, 1993.
73. J. E. Foley, *Diabetes Care, 15,* 773(1992).
74. M. Tadayyon, et al., *Diabetologia, 30,* 41(1987).
75. R. Bressler and D. Johnson, *Diabetes Care, 15,* 792(1992).
76. G. M. Reaven, et al., *Diabetes Care, 13,* 32(1990).
77. D. K. Kreutter, et al., *Diabetes, 39,* 1414(1990).
78. P. Ronner, et al., *Diabetes, 40,* 878(1991).
79. P. W. Stacpoole and Y. J. Greene, *Diabetes Care, 15,* 785(1992).
80. H. Ogawa, et al., *Life Sci., 50,* 375(1992).

SUGGESTED READINGS

1. M. B. Davidson, *Diabetes Mellitus: Diagnosis & Treatment,* 3rd ed. New York, Churchill-Livingstone, 1991.
2. M. A. Sperling, Ed., *Physicians' Guide to Insulin–Dependent (Type I) Diabetes: Diagnosis & Treatment,* Alexandria, VA, American Diabetes Association, 1988.
3. H. Rifkin, Ed., *Physicians' Guide to Non-Insulin–Dependent (Type II) Diabetes: Diagnosis & Treatment,* 2nd ed., Alexandria, VA, American Diabetes Association, 1988.
4. J. A. Galloway, et al., Eds., *Diabetes Mellitus,* 9th ed., Indianapolis, Eli Lilly & Company, 1988.
5. Symposium, H. E. Lebovitz and A. M. Melander, Eds., *Diabetes Care, 13(Suppl 3),* 1(1990).
6. P. Marchetti and R. Navalesi, *Clin. Pharmacokinet., 16,* 100(1989).
7. L. C. Groop, *Diabetes Care, 15,* 737(1992).
8. C. J. Bailey, *Diabetes Care, 15,* 755(1992).
9. J. E. Gerich, *N. Engl. J. Med., 321,* 1231(1989).
10. R. A. DeFronzo, *Diabetes Reviews, 1,* 1(1993).

Chapter 27

AMINO ACIDS, PEPTIDES, AND PROTEINS

Roy L. Kisliuk

The important amino acids common to proteins include alanine, arginine, aspartic acid, cysteine, glutamic acid, glycine, histidine, isoleucine, leucine, lysine, methionine, phenylalanine, proline, serine, threonine, tryptophan, tyrosine, and valine. They are analogous to letters of the alphabet because when ordered in appropriate sequence in peptide or protein polymers or when appropriate derivatives are formed, they communicate information of biologic significance.

The structures, abbreviations, and important derivatives of these amino acids are given in Table 27–1.[1] With the exception of proline, which has a cyclic structure, all these amino acids have the general formula depicting an α-aminocarboxylic acid with various side chains.

$$RCHNH_2COOH$$

Two stereoisomers are possible for all these amino acids except glycine, threonine, and isoleucine. Glycine does not have an asymmetric carbon, whereas threonine and isoleucine have two each. Only one form is therefore possible for glycine, whereas four stereoisomeric forms are possible for threonine and for isoleucine. Amino acids are designated D or L depending on whether the configuration about the a carbon corresponds to D-glyceraldehyde or L-glyceraldehyde, respectively. The amino acids so far found in proteins and peptide hormones discussed in this chapter all have the L-configuration. The D-amino acids are found in bacterial cell walls, antibiotics, and frog skin.[2]

Racemization is the process by which D or L amino acids are converted to DL mixtures. This process occurs to some extent spontaneously and is accelerated by alkali treatment. D-Aspartic acid accumulates with age in human-brain white matter and may play a role in the dysfunction of aging brain.[2] Amino acids may be classified according to the properties of their side chains, as outlined in Table 27–2. The 20 amino acids listed are specified in protein biosynthesis by the genetic code. In addition there are hundreds of amino acids formed by post-translational modifications including methylation, formylation, sulfation, adenylation, acetylation, iodination, phosphorylation, hydroxylation, carboxylation, amidation, and isoprenylation.[4] Such modifications are often vital for protein function. Phosphorylation of serine, threonine, and tyrosine residues in enzymes is an important mechanism, regulating the synthesis and breakdown of glycogen and signal transduction. The vitamin K-dependent carboxylation of glutamate residues in prothrombin results in the formation of γ-carboxyglutamic acid residues, which are involved in the binding of calcium ions that, in turn, is necessary for proper blood clotting. Collagen, the most abundant protein in the body, matures by forming cross-links. Specific lysine and hydroxylysine residues are oxidized to aldehydes that condense with residues on adjacent chains. Collagen cross-linking is related to age-associated alterations such as wrinkled skin and loss of elasticity in the cardiovascular system.[5]

PROPERTIES OF PEPTIDES

Peptide Bond

Peptides are formed from amino acids by linking the α-carboxyl group of one to the α-amino group of another with the elimination of water. The C–N bond thus formed is called a peptide or amide bond. Peptides are written with the residue having the free amino group on the left and named as the carboxy terminal amino acid, e.g., alanylglycine, $CH_3CHNH_2CONHCH_2CO_2H$.

The C–N peptide bond is a partial double bond substituted in a manner described as the *trans* configuration because the carbon substituents are on opposite side of the bond.

Amide bond

The six atoms (a, b, c, d, e, and f) are in the same plane. The structure of this bond is important in determining the structure of peptides and proteins.[5]

The Charge on the α-Amino and α-Carboxyl Groups

In aqueous solution near a pH of 7.0, amino acids exist as dipolar ions in which both the carboxyl group and the amino group are charged (zwitterionic). For example, the structure of glycine at pH 6 is $^{+}H_3N\text{-}CH_2CO_2^{-}$. As amino acids are incorporated into peptides of increasing length, the charges on the terminal amino and carboxyl moieties are separated. In many naturally occurring

Table 27–1. Amino Acids Common to Proteins

Nomenclature			Structure	
Common Name	**Chemical Name**	**Abbreviation**	**R CH-COOH NH_2 R–**	**Initially Isolated From (in)**
Alanine	α-aminopropionic acid	Ala, A	CH_3-	Silk hydrolysates (1888)
Arginine	α-amino-δ-guanidino valeric acid	Arg, R	$H_2N-C(=NH)-NH-CH_2-CH_2-CH_2-$	Found in histones
Asparagine	α-aminosuccinamic acid	Asn, N	$H_2N-C(=O)-CH_2-$	Asparagus (1806)
Aspartic acid	α-aminosuccinic acid	Asp, D	$HO-C(=O)-CH_2-$	Hydrolysis of asparagine
Cysteine	α-amino-β-mercaptopropionic acid	Cys, C	$HSCH_2$-	Reduction of cystine (1884) Component of glutathione
Cystine	β,β′-dithiodialanine	Cys-S-S-Cys		Urinary calculi (1810)
Glutamic acid	α-aminoglutaric acid	Glu, E	$HO-C(=O)-CH_2-CH_2-$	Wheat gluten hydrolysate (1886)
Glutamine	α-aminoglutaramic acid	Gln, Q	$H_2N-C(=O)-CH_2-CH_2-$	Beets (1883)
Pyroglutamic acid	5-oxo-2pyrrolidine	carboxylic acid	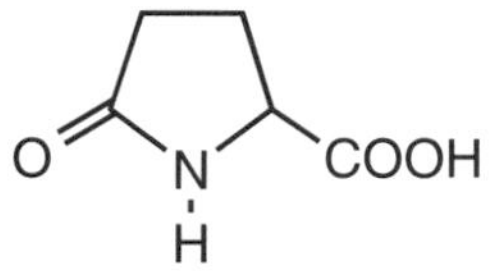 O, N, H, COOH	
Glycine	aminoacetic acid	Gly, G	H-	Gelatin hydrolysate (1820)
Histidine	α-amino-β-imidazole proprionic acid	His, H	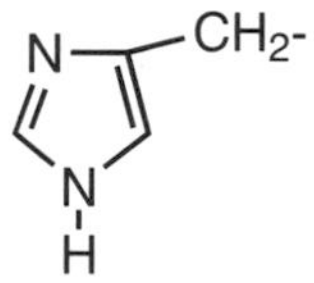 N, CH_2-, N, H	Sturgeon sperm (1896)
Isoleucine	α-amino-β-methylvaleric acid	Ile, I	$C_2H_5-CH(CH_3)-$	Beet sugar molasses (1903)
Leucine	α-aminoisocaproic acid	Leu, L	$C_2H_5-CH(CH_3)-CH_2-$	Muscle and wool hydrolysate
Lysine	α,ε-diaminocaproic acid	Lys, K	$H_2N(CH_2)_4$-	Casein hydrolysate
Methionine	α-amino-γ-methylthiobutyric acid	Met, M	$CH_3-S-CH_2CH_2-$	Casein hydrolysate (1928)
Phenylalanine	α-amino-β-phenyl propionic acid	Phe, F	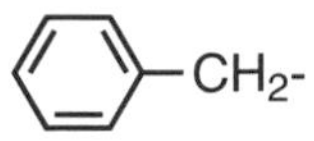 CH_2-	Lupine sprouts (1879)

Table 27–1. *(Continued)*

Nomenclature			Structure $R\,CH\text{-}COOH$ (NH_2)	
Common Name	**Chemical Name**	**Abbreviation**	**R–**	**Initially Isolated From (in)**
Proline	pyrrolidine-2-carboxylic acid	Pro, P	(pyrrolidine ring) N–H, COOH	Casein hydrolysate (1901)
Serine	α-amino-β-hydroxypropionic acid	Ser, S	$HO{-}CH_2{-}$	Silk (1865)
Threonine	α-amino-β-hydroxybutyric acid	Thr, T	$CH_3{-}CH(OH){-}$	Fibrin hydrolysate (1935)
Tryptophan	α-amino-3-indolepropionic acid	Trp, W	(indole ring, N–H) $CH_2{-}$	Pancreatic digest of casein (1901)
Tyrosine	α-amino-β-(ρ-hydroxy phenyl)propionic acid	Tyr, Y	$HO{-}C_6H_4{-}CH_2{-}$	Casein hydrolysate (1846)
Valine	α-aminoisovaleric acid	Val, V	$CH_3{-}CH(CH_3){-}$	Pancreatic extract (1856)

amino acids and peptides, the terminal amino or carboxyl group is substituted in a manner which abolishes the formal charge. In most instances these substitutions are essential for biologic activity. For example, thyrotropin-releasing hormone, pyroglutamylhistidylprolinamide, is a peptide in which both the terminal amino and the carboxyl groups are substituted. Another example is gastrin, which has a pyroglutamyl moiety at the amino terminal and phenylalanineamide at the carboxy terminal.

Table 27–2. Classification of Amino Acids Based on Characteristics of the Side Chain

- I. Aliphatic
  - A. Nonpolar
    - Glycine, Leucine, Alanine, Isoleucine, Valine, Proline (imino acid)
  - B. Hydroxy
    - Serine, Threonine
  - C. Sulfur-containing
    - Cysteine, Methionine
  - D. Acidic
    - Aspartic acid, Glutamic acid
  - E. Basic
    - Arginine, Lysine
  - F. Amides
    - Asparagine, Glutamine
- II. Aromatic
  - Phenylalanine, Tyrosine, Tryptophan, Histidine

In the biosynthesis of many neurohormones and related compounds derived from single amino acids such as γ-aminobutyric acid, histamine, norepinephrine, and serotonin, decarboxylation removes the negative charge, leaving positively charged amino groups.

The Charge on the Side Chains

Ionic bonds of the type: $-COO^{-}{-}{-}{}^{+}H_3N-$ formed between the charged side chains of aspartate, glutamate, arginine, and lysine as well as C-terminal and N-terminal amino acid residues are important in maintaining protein structure. In hemoglobin, for example, x-ray crystallographic studies show an ionic bond between the NH_3^+ of lysine 10 in the α subunit and the COO^- of the C-terminal arginine of the β-subunit. These bonds are not present in oxyhemoglobin, but form as oxygen dissociates and serve to stabilize deoxyhemoglobin.[5]

In many proteins the charged side chains interact with the surrounding solvent rather than with each other. For example, x-ray crystallographic studies on myoglobin show that the charged groups are on the outer surface of the molecule with hydrophobic amino acids found most commonly in the interior. A similar situation occurs with lysozyme and ribonuclease.

Ionic bonds are important in the interaction of histones (basic proteins rich in arginine and lysine found in chromosomes), with the phosphoryl groups of DNA.

Table 27–3. Hydrophobicity of Amino Acids

Amino Acid	Free Energy of Transfer (Calories/mole)
Tryptophan	3400
Phenylalanine	2500
Tyrosine	2300
Leucine	1800
Valine	1500
Methionine	1300
Histidine	500
Alanine	500
Threonine	400
Serine	−300

Hydrophobic Bonding

The side chains of amino acids bearing nonpolar hydrocarbon groups have a tendency to coalesce in the interior of proteins. Hydrophobic bonds are of great significance in the stabilization of the structure and in their function. They are also vital in binding the α-peptide chains of hemoglobin to the β-chains. In this case, hydrophobic residues on the surface of each chain interact with each other. Hydrophobic interactions play an important role in the binding of peptide hormones to their cellular receptors.

Exclusion of water from the interior of lysozyme and ribonuclease is advantageous to their catalytic function. Water acts as an insulator, preventing the electronic interactions necessary for the reactions to occur. The respective substrates for these enzymes fit into a cleft surrounded by a hydrophobic region.

A quantitative estimate of the hydrophobicity of amino-acid side chains may be obtained by comparing their solubility in organic solvents to that in water. By this measure, tryptophan has the most hydrophobic side chain (Table 27–3).[6]

Hydrogen Bonds

Hydrogen bonding occurs between a hydrogen atom covalently bonded to an electronegative atom (nitrogen or oxygen) and a second electronegative atom. Both elec-

$$>N{-}H \; + \; O{=}C< \;\rightleftharpoons\; >N{-}H{-}{-}{-}O{=}C<$$

Hydrogen bond

trostatic and covalent interactions contribute to the hydrogen bond. These bonds are relatively weak, being about $\frac{1}{20}$ as strong as the O-H covalent bond in water, however, if a molecule contains many such bonds, they contribute to the formation of a stable structure. Because of their bonding orbitals, hydrogen bonds are restricted as to the direction in which they can form. Peptide chains exist as helices, sheets, or irregular forms, depending on the arrangement of hydrogen bonds. Boiling or treatment with urea breaks the hydrogen bonds (denaturation), and the chain assumes a random form.

The α-Helix

The α-helix commonly found in globular proteins is a right-handed helix in which the H atom on the amino group is bonded with the oxygen of the carbonyl on the fourth amino acid behind it.[5] The hydrogen bonds are oriented parallel to the long axis of the helix. Each peptide bond participates in hydrogen bonding. The amino-acid side chains extend outward from the helix axis. There are 3.6 amino acids in each turn of the helix. The α-helix is so named because this type of structure was

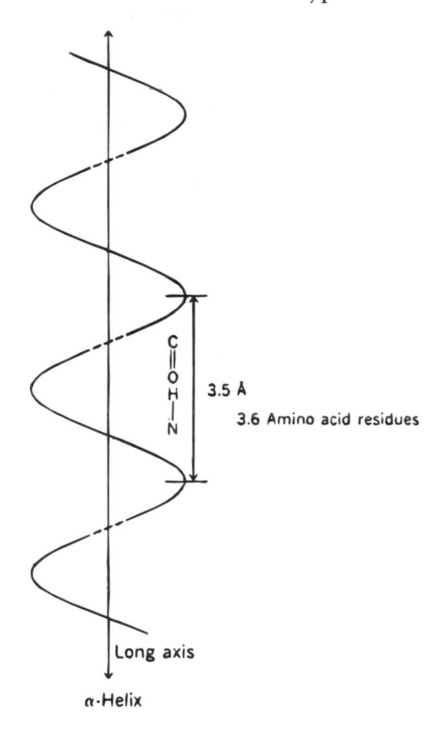

α-Helix

first demonstrated by x-ray crystallography in α-keratin, a protein derived from wool.

Individual amino acids differ in their tendency to form helices. Alanine, glutamic acid, leucine, lysine, and methionine have a strong tendency to promote helix formation, whereas proline and glycine are considered helix breakers.

β-Sheets

Another arrangement of polypeptide chains is in β-sheets, in which hydrogen bonds are formed with groups in a neighboring, rather than in the same, chain, as in the α-helix. The hydrogen bonds are almost perpendicular to the long axis of the peptide chain. They are called β-sheets because this type of structure is found in β-keratin, a substance produced from α-keratin by heating and stretching wool fibers in a moist atmosphere.

Examples of proteins containing chains arranged in the β-sheet form are ribonuclease, chymotrypsin, and papain in which the sheets are antiparallel (strands run in opposite directions) and lysozyme and carboxypeptidase, which contain both parallel (strands run in the same direction) and antiparallel segments. Silk fibroin consists almost entirely of chains in the antiparallel β-sheet form.

Hormones Derived from Single Amino Acids

The simplest biologically active compounds derived from amino acids are formed by decarboxylation as in the formation of histamine from histidine and γ-aminobutyric acid from glutamic acid. In many instances, the decarboxylated amino acids are further modified by hydroxylation, methylation, and acetylation. Examples include the conversion of tyrosine to dopamine, norepinephrine, and epinephrine, the conversion of serine to acetylcholine, and the conversion of tryptophan to 5-hydroxytryptamine (serotonin) (Fig. 27–1). Unmodified

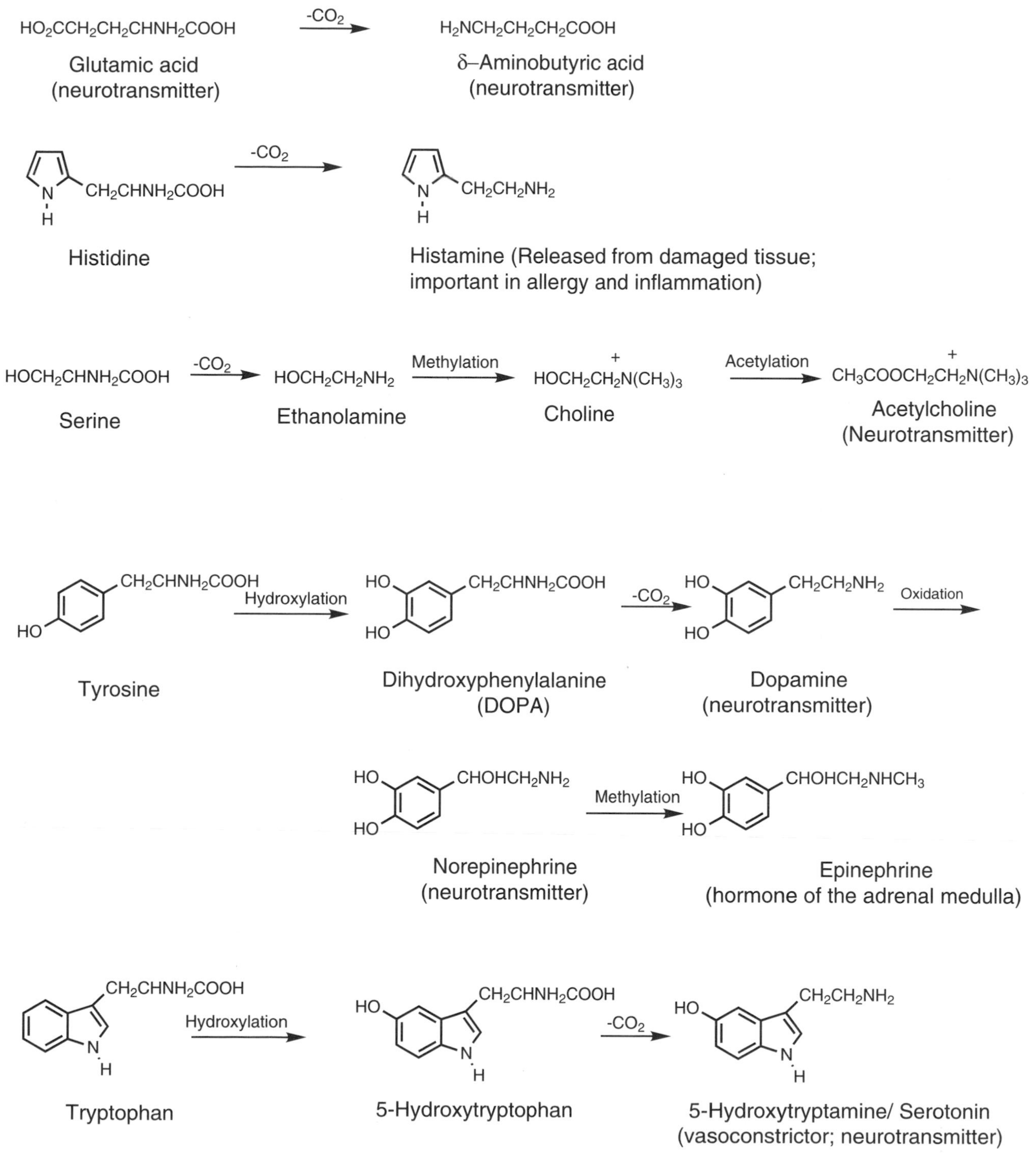

$HO_2CCH_2CH_2CHNH_2COOH$ $\xrightarrow{-CO_2}$ $H_2NCH_2CH_2CH_2COOH$

Glutamic acid (neurotransmitter) → δ–Aminobutyric acid (neurotransmitter)

CH_2CHNH_2COOH $\xrightarrow{-CO_2}$ $CH_2CH_2NH_2$

Histidine → Histamine (Released from damaged tissue; important in allergy and inflammation)

$HOCH_2CHNH_2COOH$ $\xrightarrow{-CO_2}$ $HOCH_2CH_2NH_2$ $\xrightarrow{\text{Methylation}}$ $HOCH_2CH_2\overset{+}{N}(CH_3)_3$ $\xrightarrow{\text{Acetylation}}$ $CH_3COOCH_2CH_2\overset{+}{N}(CH_3)_3$

Serine → Ethanolamine → Choline → Acetylcholine (Neurotransmitter)

CH_2CHNH_2COOH $\xrightarrow{\text{Hydroxylation}}$ CH_2CHNH_2COOH $\xrightarrow{-CO_2}$ $CH_2CH_2NH_2$ $\xrightarrow{\text{Oxidation}}$

Tyrosine → Dihydroxyphenylalanine (DOPA) → Dopamine (neurotransmitter)

$CHOHCH_2NH_2$ $\xrightarrow{\text{Methylation}}$ $CHOHCH_2NHCH_3$

Norepinephrine (neurotransmitter) → Epinephrine (hormone of the adrenal medulla)

CH_2CHNH_2COOH $\xrightarrow{\text{Hydroxylation}}$ CH_2CHNH_2COOH $\xrightarrow{-CO_2}$ $CH_2CH_2NH_2$

Tryptophan → 5-Hydroxytryptophan → 5-Hydroxytryptamine/ Serotonin (vasoconstrictor; neurotransmitter)

Fig. 27–1. Hormones derived from single amino acids.

glutamic acid, aspartic acid, and glycine serve as neurotransmitters in the central nervous system (CNS).[7]

Triiodothyronine and thyroxine are formed in the thyroid gland by iodination of tyrosine in thyroglobulin, followed by transfer of an iodinated phenol moiety to a second tyrosine residue. There are 120 tyrosine residues in each molecule of thyroglobulin, some of which are iodinated much more readily than others. The hormones are released from thyroglobulin by proteolysis and then are secreted into the blood.[8]

Peptide Hormones

An ever increasing number of peptide hormones are being identified and synthesized. They have a fascinating role in the control of life processes. Some of these hormones, arranged in order of increasing size, are listed in Table 27–4. Peptides with a molecular weight exceeding 5000 are usually called proteins, but this distinction is not rigidly maintained.

Table 27–4. Peptide Hormones

Hormone	Amino Acid Residues	Molecular Weight
Thyrotropin-releasing hormone	3	363
Tuftsin	4	501
Met-enkephalin	5	645
Angiotensin II	8	1,031
Oxytocin	8	986
Vasopressin	8	1,029
Sleep peptide	9	992
Bradykinin	9	1,069
Gonadotrophin-releasing hormone	10	1,182
Substance P	11	1,527
Somatostatin	14	1,876
Gastrin	17	2,110
Secretin	27	2,876
Ser-Leu-Arg-Arg-Atriopeptin III	28	3,547
Vasoactive intestinal peptide	28	3,809
Glucagon	29	3,374
β-Endorphin	31	3,476
Calcitonin	32	3,415
Cholecystokinin	33	4,492
Corticotropin	39	4,600
Corticotropin-releasing factor	41	5,400
Growth hormone-releasing factor	44	5,800
Relaxin	48	5,500
Insulin	51	5,700
Parathyroid hormone	83	8,500
β-Lipotropin	91	10,000
Nerve growth factor	115	14,000
Growth hormone	191	21,000
Prolactin	198	22,800

Greek prefixes are often used to designate the number of amino acid residues in a peptide, i.e., tuftsin is a tetrapeptide and somatostatin a tetradecapeptide. These prefixes become unwieldy as the number of residues increases. Arabic numerals are used throughout this chapter. For example, tuftsin is designated as a 4-peptide and somatostatin as a 14-peptide.

Thyrotropin-Releasing Hormone

The synthesis and release of triiodothyronine and thyroxine are regulated by thyrotropin (thyroid-stimulating hormone, TSH), a glycoprotein elaborated in the anterior pituitary gland. Thyrotropin stimulates the formation of thyroid cells, increases the uptake of iodide from the blood, and increases the proteolysis of thyroglobulin.[8]

At least two mechanisms control the level of thyrotropin. One is feedback by which increased levels of triiodothyronine and thyroxine reduce thyrotropin excretion. The second, mediated through the hypothalamus, is a response to external stimuli such as food intake or anxiety, which raise thyroxine levels. The hypothalamus contains thyrotropin-releasing hormone, a 3-peptide secreted into blood vessels leading directly to the anterior pituitary, where it causes the release of thyrotropin. The hog, sheep, cow, rat, and human hormones all have the same structure, pyroglutamylhistidylprolinamide. A precursor protein of molecular weight 29,000 has been identified in rat brain.[8] It contains five Gln-His-Pro-Gly sequences potentially capable of forming five molecules of thyrotropin-releasing hormone. Gln is a good precursor of the amino terminal pyroglutamate and the amide of the carboxy terminal prolineamide arises from the degradation of glycine.[9]

One nanogram of thyrotropin-releasing hormone injected into a mouse increases the uptake of iodide from the blood into the thyroid gland. The compound is used to test the response of the thyroid gland. When 3-methylhistidine replaces histidine, hormone activity is enhanced tenfold.

Tuftsin

Tuftsin (Thy-Lys-Pro-Arg) arises from leukokinin, a protein found in the γ-globulin fraction of blood. Leukokinin is a substrate for a spleen endocarboxypeptidase that catalyzes cleavage of an Arg-Glu bond between residues 292 and 293. The nicked leukokinin binds to the plasma membrane of phagocytic cells that release another peptidase which catalyzes the formation of tuftsin by cleaving the protein between Lys^{288} and Thr^{289}.[10]

Tuftsin stimulates phagocytosis and promotes antibody-dependent cellular cytotoxicity. A potent peptidase-resistant agonist of tuftsin is obtained when the —CONH— bond between Thr and Lys is changed to —NHCO—[retro-inverso-tuftsin].[10]

Met-enkephalin

Met-enkephalin (Tyr-Gly-Gly-Phe-Met) isolated from brain and small intestine, acts as morphine does, in that it binds to the same receptor and has analgesic activity.[11]

Therefore, along with endorphins, dynorphins, and Leu-enkephalin, it is classified as an endogenous opioid. Met-enkephalin arises from two larger peptides, proenkephalin and pro-opiomelanocortin. Enkephalin is a model for the design of analgesic agents. An interesting example is the analog in which the Gly-2 residue is replaced with D-alanine, the N of Phe-4 is methylated, and the Met residue is converted to the sulfoxide derivative of methionine alcohol.[12] This compound is 30,000 times as active as met-enkephalin and 1000 times as active as morphine as an analgesic when administered intracerebroventricularly. It also shows some activity after oral administration, which is unusual for a peptide. Dermorphin, a potent opiod peptide (Tyr-D-Ala-Phe-Gly-Tyr-Pro-Ser-NH_2) isolated from frog skin also has D-alanine as the second amino acid.[3]

Angiotensin

When the flow of blood to the kidneys is restricted, these organs respond by releasing the peptidase renin (molecular weight 40,000) into the blood. Renin catalyses the cleavage of the 10-peptide angiotensin I from angiotensinogen, a blood globulin.[13] Angiotensin I is inactive but is cleaved to an 8-peptide, angiotensin II by the action of another peptidase, called angiotensin converting enzyme (ACE). Angiotensin II is an extremely powerful pressor agent. Infusion of 2 ng/kg/min increases blood pressure by constricting blood vessels as well as by stimulating the adrenal gland to release aldosterone, which causes retention of sodium leading to fluid retention. Angiotensin II has also been implicated in the development of thirst. Angiotensin is a substrate for aminopeptidase which converts it to angiotensin III a peptide that retains much of the activity of angiotensin II. Further metabolism by aminopeptidase, carboxypeptidases, and endopeptidases leads to inactive fragments.

The renin-angiotensin system is of great interest in studies of hypertension. Captopril, D-3-mercaptomethylpropanoyl-L-proline, specifically designed as an inhibitor of ACE,[14] is widely used in the treatment of essential hypertension. Non-peptidic angiotensin II antagonists cause the relaxation of vascular smooth muscle.[15]

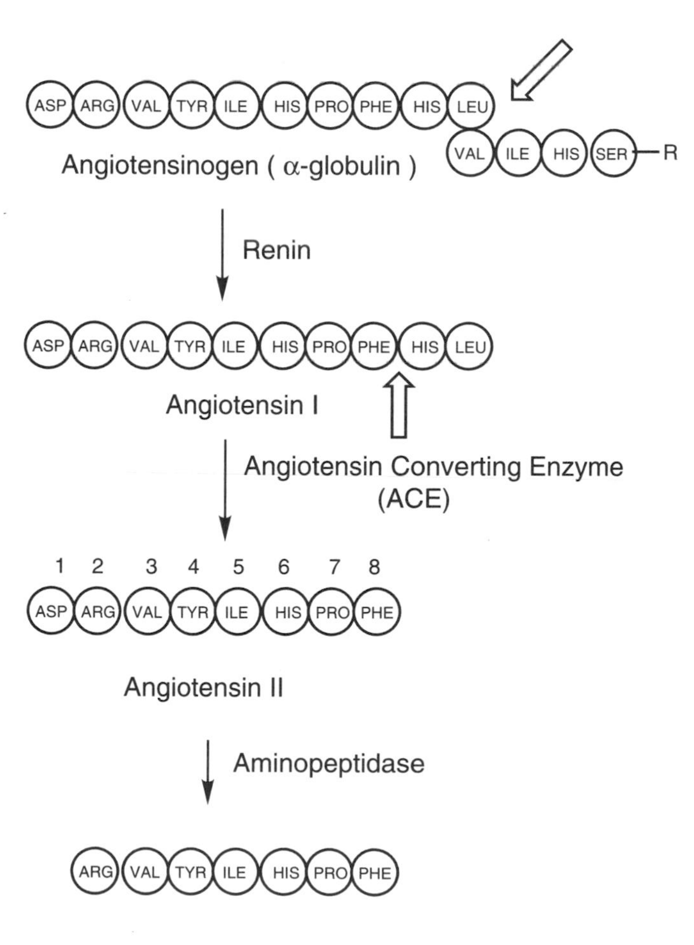

Oxytocin and Antidiuretic Hormone (ADH)

Biologic Activity

The posterior pituitary gland stores these two important peptide hormones. ADH acts on the membranes of the distal convoluted tubules and collecting ducts of the kidney, causing them to become permeable to water. This permits uptake of water by osmosis because the interstitial side of the membrane has a high concentration of electrolytes. ADH also acts as a vasoconstrictor of vascular smooth muscle. Recent work has opened new horizons for ADH as a neurohormone in the CNS controlling cardiovascular, renal, and thermoregulatory systems.[16]

Oxytocin intensifies uterine contractions during parturition and causes the ejection of milk from the mammary gland. It may also play an important role in the CNS.[17]

Each hormone has slight activity in systems affected by the other. Oxytocin has detectable antidiuretic activity, and ADH can cause uterine contractions at relatively high concentrations. Release of the hormones from the posterior pituitary is triggered by reflex action of the CNS. Dilation of the birth canal at term or suckling causes oxytocin release, whereas an increase in the osmotic pressure of blood causes the secretion of ADH. ADH is used to treat diabetes insipidus, a disease in which there is an excessive flow of dilute urine. Oxytocin is used to induce labor.

Both hormones occur in the posterior pituitary bound to a protein, neurophysin. The linkage to this protein is not covalent, because the hormone-protein complex can be disrupted by boiling it or by filtering it through a dextran column. Both hormones are synthesized in the hypothalamus in neurons distinct for each hormone.

Chemical Activity

Both oxytocin and ADH contain a cyclic 6-peptide with a 3-peptide side chain. Reduction of the cystine disulfide bond yields, in each case, a linear 9-peptide containing two cysteine residues. ADH differs from oxytocin in having a phenylalanine residue at position 3 and an arginine residue at position 8 in place of the corresponding isoleucine and leucine residues of oxytocin.

Congeners of these hormones have been discovered in

nature. Swine ADH contains lysine instead of arginine at position 8 and is referred to as lysine ADH. It is more stable than the arginine-containing hormone and is therefore preferred for clinical use. An interesting compound called arginine vasotocin has been isolated from chickens, frogs, and fish. It has the ring structure of oxytocin and the side chain of arginine ADH. When tested in mammals, it shows activity of both oxytocin and ADH. Both oxytocin and ADH are cleaved by the action of peptidases located in brain synaptic membranes.

The synthesis of oxytocin in 1954 in the laboratory of du Vigneaud was a milestone in medicinal chemistry—the first synthesis of a peptide hormone.[18] An excellent tribute to the work of du Vigneaud has been prepared by Hofmann.[19] The use of sodium in liquid ammonia, a technique du Vigneaud learned from Audrieth, was instrumental to his success. "The observation that the S-benzyl group was removed from S-benzylcysteine by sodium in liquid ammonia represents a significant contribution to peptide chemistry since it made possible the transient protection of the thiol group of cysteine during peptide synthesis."[19] Hundreds of analogs of posterior pituitary hormones have been synthesized. Arginine vasotocin was prepared in the laboratory prior to its discovery in nature.

Structure-Activity Relationship

The structure-activity relationship of both oxytocin and vasopressin have been extensively studied. Most alterations decrease activity, some enhance activity, whereas others result in activity qualitatively different from that of either parent hormone. Two of the many biologic tests used to evaluate these compounds are based on oxytocic activity (the ability to cause contraction of isolated uterine muscle obtained from a rat in estrus) and antidiuretic activity (the ability to diminish urine excretion in the hydrated rat).

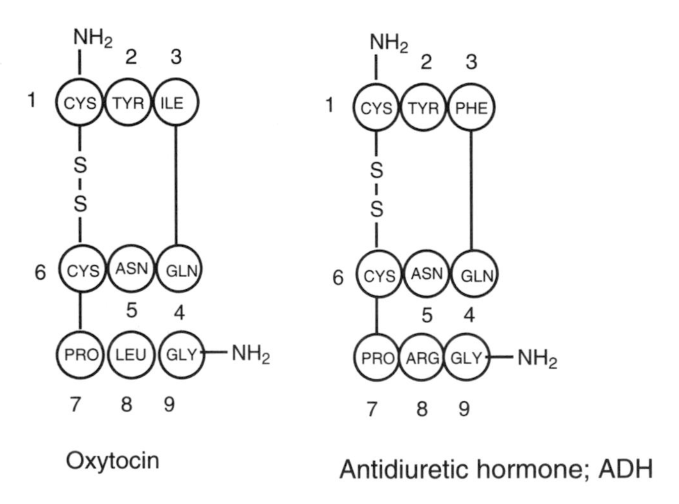

Oxytocin

Antidiuretic hormone; ADH

Alterations That Decrease Activity

The ring structure involving amino acid residues 1 through 6 contains 20 atoms. If this is decreased by one atom, as in I-δ–mercaptoacetic acid oxytocin, or increased by one atom, as in 4-β-alanine oxytocin, the oxytocic activity is greatly diminished.[20] The I-δ-mercaptovaleric acid oxytocin, which has 22 atoms in the ring, has only slight oxytocic activity and inhibits the effects of oxytocin. If the two sulfur atoms of the ring are replaced by methylene groups (dicarbaoxytocin), activity is reduced to one-fifth that of the parent compound.[21] The significant oxytocic activity retained after this alteration proves that the sulfur atoms are not essential for activity.

If glycine is attached to the free amino group of the cysteine residue on the amino acid at position 1 (1-glycyl oxytocin), the oxytocic potency is decreased, but the duration of action is increased due to the gradual release of oxytocin by peptidase action.

Alterations That Increase Activity

If the free terminal amino group is removed, i.e., the cysteine residue at position 1 is replaced by β-mercaptoproionic acid (1-desamino oxytocin), the activity of both oxytocin and ADH is enhanced approximately twofold.[22]

Desmopressin, 1-deamino-8-D-arginine ADH, has greater antidiuretic activity than ADH because it is resistant to peptidase degradation; it also has lower pressor activity and is therefore preferred for the treatment of diabetes insipidus.

An important goal in the synthesis of analogs of peptide hormones is to produce specific antagonists to be used to assess the importance of a given agonist. A specific antagonist for the antidiuretic action of ADH results when Cys-1 is replaced with β-mercapto-β,β-cyclopentamethylene propionic acid and Tyr^2 by D-Phe.[23] D-Cys^6 oxytocin is a potent oxytocin antagonist.[24]

Bradykinin

Like angiotensin II, this substance results from the enzymatic cleavage of a protein in blood.

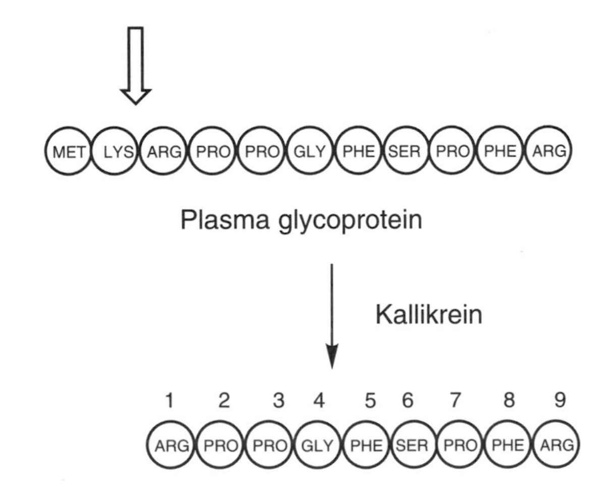

Bradykinin

The complex events leading to bradykinin formation begin by activation of the Hageman's factor. Hageman's factor is a glycoprotein that is also involved in the clotting of blood. When activated, it catalyzes the conversion of plasma dilution factor to its active form, which in turn catalyzes the conversion of prekallikrein to kallikrein, a proteolytic enzyme that catalyzes the release of bradykinin from a plasma globulin.

The biologic activity of bradykinin is similar to that of histamine. It is also involved in inflammation, but its action is not blocked by antihistamines. It causes dilation and increased permeability of blood vessels, increased migration of leukocytes, and production of pain. Peptides having an amino terminal 9-peptide identical to that in bradykinin occur in wasp venom. Bradykinin is inactivated by cleavage of the carboxy terminal dipeptide catalyzed by angiotensin converting enzyme, the same enzyme that converts angiotensin I to angiotensin II. This enzyme therefore catalyzes the formation of a pressor substance and the removal of a vasodilator.

Many analogs of bradykinin have been synthesized.[25] If alanine replaces arginine at position 1 or 9, phenylalanine at position 5 or 8, proline at position 2 or 7, or glycine at position 4, activity is lost. A specific antagonist of bradykinin is (D-Phe^7) bradykinin.

Physical studies employing optical rotatory dispersion and circular dichroism indicate that no helical segments are present in bradykinin. The present of three proline residues prevents helix formation.

Gonadotrophin-Releasing Hormone (Luteinizing Hormone-Releasing Hormone [LHRH])

A peptide hormone produced in the hypothalamus elicits the release of both luteinizing hormone and follicle-stimulating hormone from the pituitary. The structure of the human hormone, a 10-peptide, has been determined and its synthesis achieved. The amino terminal pyroglutamyl histidine is the same as in thyrotropin-releasing factor, which also has a carboxy terminal amide.

Gonadotropin-releasing hormone

Gonadotrophin-releasing hormone elicits the release of luteinizing hormone and follicle-stimulating hormone in vitro from pituitary glands derived from male as well as female rats. It is probable that the sex steroids play an important role in controlling the amount of hormone released in vivo. Studies on gonadotrophin-releasing hormone and its analogs may yield new possibilities for the control of fertility.

Substance P

Substance P is an 11-peptide found in sensory neurons and is associated with pain transmission and inflammation.

Substance P

Substance P is representative of a group of peptides called tachykinins or neurokinins all of which have Phe-X-Gly-Leu-Met-NH_2 as their carboxy terminal sequence.[26]

Somatostatin

This is a 14-peptide, isolated from the hypothalamus, which inhibits the release of growth hormone from the pituitary gland.

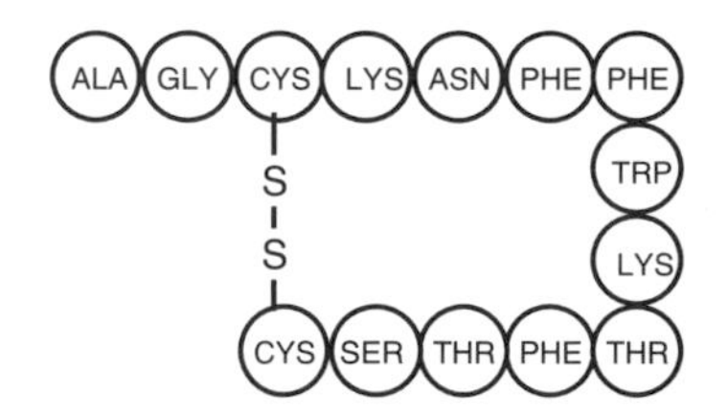

Somatostatin

Somatostatin is also found in the stomach, intestine, and pancreas and inhibits the release of insulin, glucagon, gastrin, secretin, and other hormones. It has a biologic half-life of less than 4 minutes. Synthetic analogs have been prepared with prolonged activity and greater selectivity than the natural hormone. D-Cysteine-14-somatostatin, for example, inhibits the release of growth hormone and glucagon to a greater extent than it inhibits insulin release.[27]

Gastrin

This substance is produced in the gastric mucosa near the juncture of the stomach and the small intestine. It is released into the blood in response to stimulation by cholinergic nerves, which are activated by the presence of food in the stomach, and it promotes acid and pepsin secretion therein. The flow of pancreatic fluid is also stimulated by this hormone.

Early workers confused the action of gastrin with that of histamine, which has a similar effect on the stomach. Gastrin is more selective in its action than histamine. Overproduction of gastrin by tumor cells occurs in the Zollinger-Ellison syndrome. This leads to the production of acid in the stomach even during fasting, which in turn leads to severe ulcers. Effective antagonists of gastrin would therefore be of clinical interest.

Gastrin was isolated in 1962 from hog stomach and synthesized in 1964; it is a 17-peptide. It occurs in tissues in two forms—one with and the other without a sulfate group on tyrosine. Both forms are equally potent.[28]

Structure-Activity Relationship

Hormonal activity resides with the carboxyl terminal 4-peptide amide Trp-Met-Asp-Phe-NH_2. This substance has $\frac{1}{12}$ the activity of the intact hormone. The carboxy terminal 3-peptide and 2-peptide amides are not active. As appropriate amino acids are added to the 4-peptide fragment proceeding toward the amino terminal position, full activity is attained when there are 12 amino

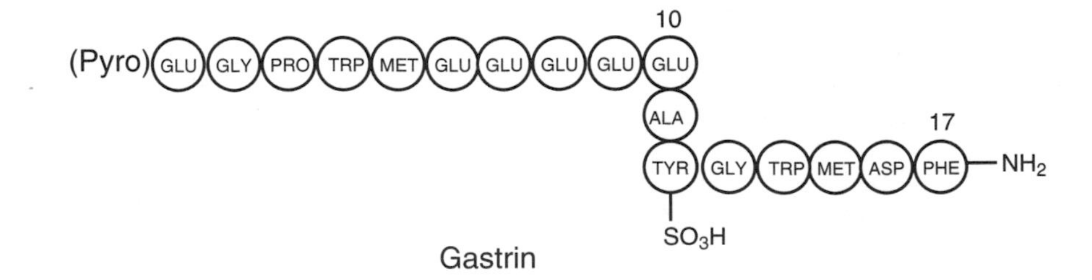

Gastrin

acids, 5 less than the number found in the natural hormone. The synthetic derivative pentagastrin (tert-butyloxycarbonyl-β-Ala-Trp-Met-Asp-Phe-NH_2) has several times the activity of the 4-peptide.

The penultimate aspartyl residue is the most sensitive of the four amino acids to alteration. Replacement of the aspartic acid by glutamic acid results in an inactive derivative. The amide group is essential, but it can be methylated without loss of activity. Methionine can be replaced with nor-leucine, tryptophan with 5-HO-tryptophan, and phenylalanine with o-methylphenylalanine without diminishing activity.

Gastrin isolated from man, hog, cat, dog, and cow differs in only one or two amino acids. The 33-peptide cholecystokinin has the same carboxy terminal pentapeptide as gastrin. This explains the fact that each of these hormones has observable effects on targets characteristic of the other. Another naturally occurring compound having this carboxy terminal sequence is cerulein, a 10-peptide isolated from toad skin.

Endothelin

Endothelin-1 is a 21-peptide isolated from cultured aortic endothelial cells. It is an extremely potent vasoconstrictor that also acts as a mitogen (i.e., a substance that induces mitosis), and as a modulator of neurotransmitter release.[31]

Secretin

This substance was discovered in 1902 when it was observed that an extract of duodenal mucosa stimulates pancreatic secretion.[32] The term hormone was coined soon thereafter and applied first to secretin and then to other substances that acted as chemical messengers among tissues. Secretin release is triggered by the presence of acid in the duodenum. Secretin elicits a pancreatic alkaline secretion with relatively few digestive enzymes. It neutralizes the acid introduced from the stomach and thus provides the appropriate pH for the pancreatic digestive enzymes to function. The pancreatic secretion elicited by cholecystokinin has a higher enzyme content.

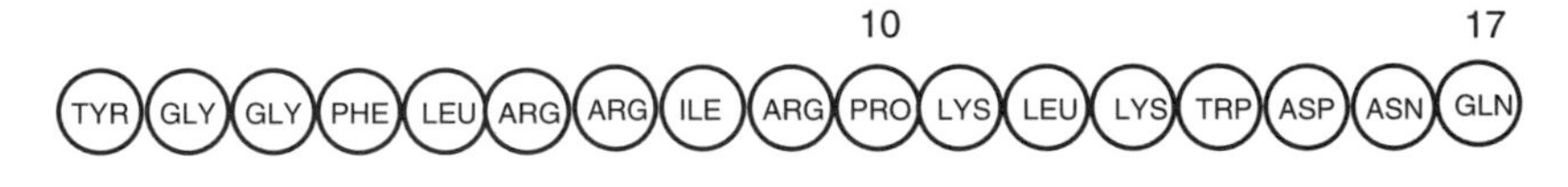

Dynorphin

Gastrin-releasing peptide is a 27-peptide that plays a role in secretion, smooth muscle contraction, and growth regulation. Antagonists of this hormone are being investigated for their ability to block the growth of small cell lung carcinoma.[29]

Dynorphin

Dynorphin is a 17-peptide opiod found in neurons. It can modulate the excitatory effects of glutamate in the guinea pig hippocampus.[30]

Hog secretin is a 27-peptide. It was synthesized by Bodanszky and co-workers in 1967.[33] There is a high degree of homology between the amino acid sequence of secretin and that of glucagon, a hormone involved in the regulation of carbohydrate metabolism. Fourteen of the amino acid residues of secretin occupy the same positions as in glucagon. Secretin also shows homology with growth hormone and placental lactogen. These homologies sug-

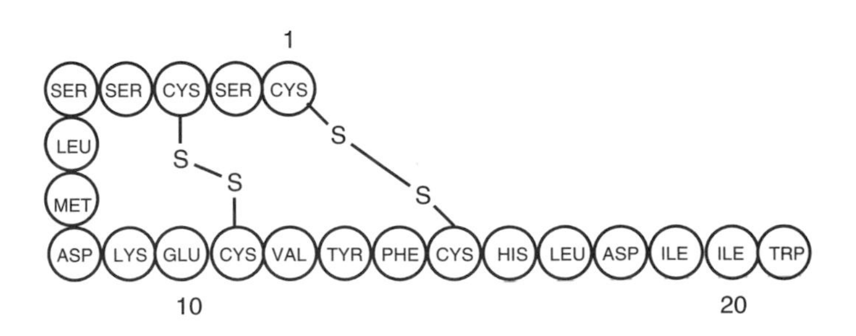

Endothelin-1

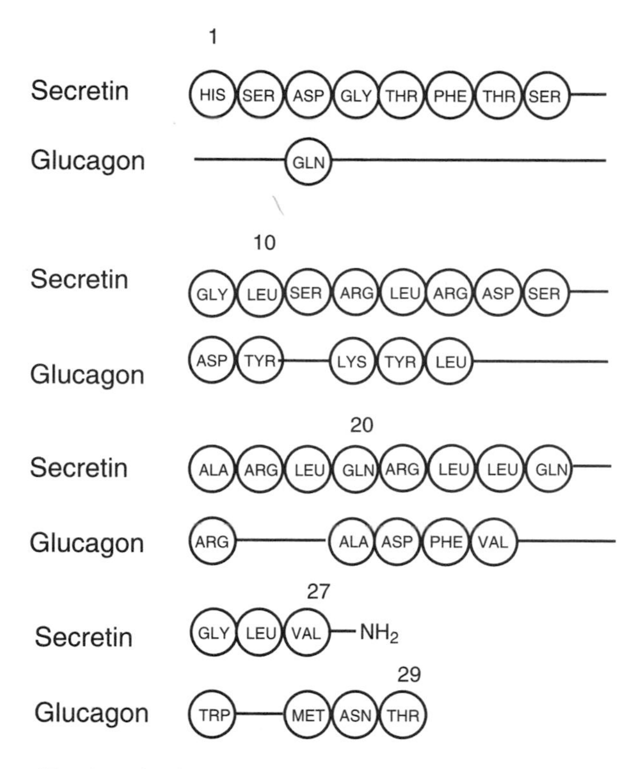

(Dashes indicate same residues in both hormones.)

gest that peptide hormones evolved from preexisting hormones by gene duplication and mutation.

No fragment of secretin has been found that has activity approaching that of the whole molecule. Removal of the amino terminal histidine drastically reduces activity. The proteolytic enzyme thrombin cleaves secretin between Arg^{14} and Asp^{15}. The resulting fragments are inactive, even when tested together in the same solution.

Glucagon

Glucagon is produced in the pancreas and released into the blood in response to low blood glucose. It stimulates the production of glucose from glycogen and from amino acids.

Glucagon was discovered as a contaminant in preparations of insulin, a pancreatic hormone that lowers blood glucose. It arises from a larger peptide, preproglucagon, by proteolytic processing and is assayed by its ability to stimulate the enzyme adenyl cyclase found in liver cell membrane particles. Adenyl cyclase catalyzes the conversion of ATP to cyclic 3′,5′-adenylic acid, which in turn stimulates the action of phosphorylase, the enzyme that catalyzes the conversion of glycogen to glucose-1-phosphate.

Structure-Activity Relationship

The structure of glucagon was determined in 1956, and it was synthesized in 1967. It has 29 amino acids arranged in a sequence remarkably similar to that in secretin. The target tissues and physiologic responses brought about by the two hormones are different. With both hormones the entire molecule is necessary for activity. Removal of N terminal histidine of glucagon causes complete loss of activity. des-His^1-[Glu^9] glucagon amide is an antagonist of glucagon in hepatic plasma membranes.[34]

Nuclear magnetic resonance and optical rotatory dispersion studies of glucagon indicate that the structure in solution is largely random but contains a helical region involving residues 22 through 27. When the compound is placed in a nonpolar solvent or when it is treated with detergents, its helicity increases. A similar increase in helicity would be expected when glucagon interacts with its receptor sites in lipid-rich membranes. Based on these considerations, an analog with enhanced helical potential, (Lys-17,18,Glu-21), glucagon was synthesized and shown to be five times more active than the parent hormone.

Atrial Natriuretic Factors

The atria of mammalian hearts contain a group of peptides that causes natriuresis, diuresis, and lower arterial blood pressure.[35] They are of great interest as potential agents for the control of blood pressure. A circulating form that appears to predominate is the 28-peptide ser-leu-arg-arg-atriopeptin III.[36]

It arises from a large peptide and requires a disulfide bridge between the two cysteine residues for activity.

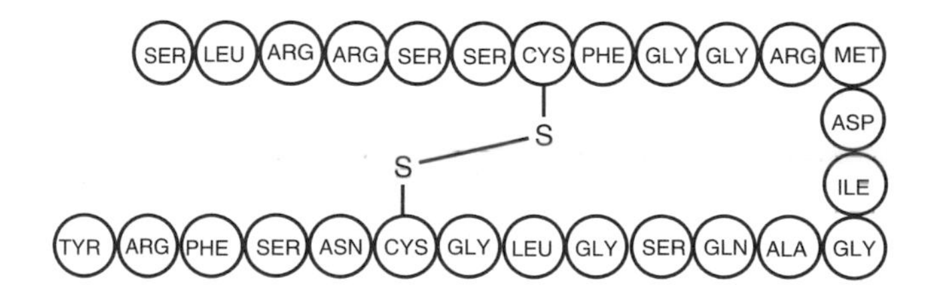

Ser-Leu-Arg-Arg-atriopeptin III

Vasoactive Intestinal Peptides (VIP)

This 28-peptide was originally isolated from hog duodenum but is also found in the nervous system and in endocrine cells where it acts as a neurotransmitter and a hormone.[37] Nine of its residues are positioned identically to those in secretin, and six are identical to those of glucagon.

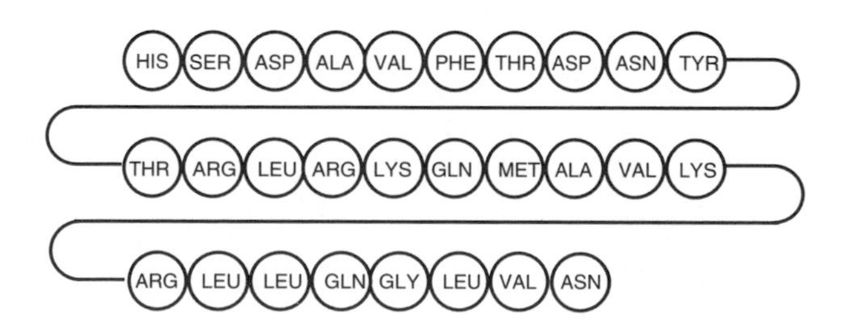

Vasoactive Intestinal Peptide

Its physiologic effects include vasodilation, increased cardiac contractility, and increased glycogen breakdown. It also acts as a growth factor for cultured mouse embryos.[38]

β-Endorphin

β-Endorphin is a 31-peptide isolated from hypothalamus. The five N-terminal amino acids are the same as those found in met-enkephalin. *β*-Endorphin is more active than met-enkephalin in its analgesic effects and in its ability to promote the release of growth hormone and prolactin.

A synthetic derivative of *β*-endorphin in which Gly^2 has been replaced with D-Ala is 60% more active in the mouse vas deferens assay. Contraction of this tissue is inhibited by morphine and other opiate analgesics. Both *β*-endorphin and enkephalin are probably neurotransmitters. A physiologic role for endorphin is suggested by work showing that mice exposed to repeated attacks by other mice show decreased sensitivity to pain, an effect which is blocked by opiate antagonists.[39]

A high-molecular-weight precursor of met-enkephalin and *β*-endorphin called pro-opiomelanocortin is present in rat pituitary extracts. This protein also gives rise to adrenocorticotropic hormone.

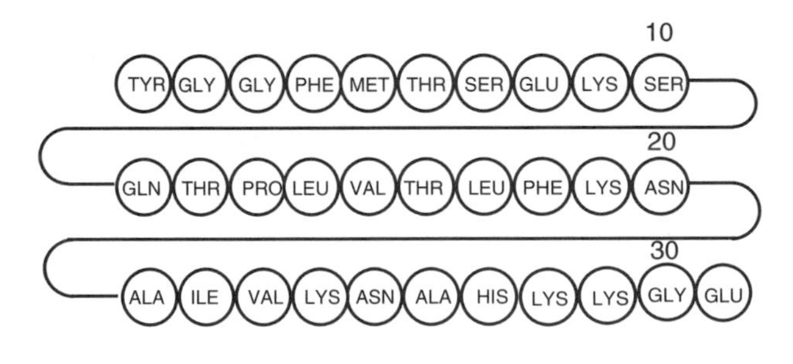

β–Endorphin

Calcitonin

Calcium is essential for muscle contraction and membrane function; its level in the blood is closely regulated. Calcitonin, discovered in 1962, is produced by specialized cells in the thyroid gland and is released in response to high levels of blood calcium. It decreases the amount of calcium entering the blood from bone. Parathyroid hormone has an opposing action in that it is released in response to low blood calcium and causes dissolution of bone.

The structure of calcitonin from humans, hog, and salmon has been elucidated. In each case there are 32 amino acids.[40] At the amino terminal there is a ring composed of cystine and five other amino acids. The carboxyl terminal amino acid is prolinamide. Only 9 of the 32 residues are the same in these three species. No immunologic cross reaction occurs between the hormone from man and that from other species because of the variability in amino acid composition at the center of the molecule. Despite these differences, all three compounds show hormonal activity in humans. Salmon calcitonin has been synthesized by recombinant DNA-directed synthesis of a precursor in Escherichia coli followed by in vitro enzymatic amidation.[40]

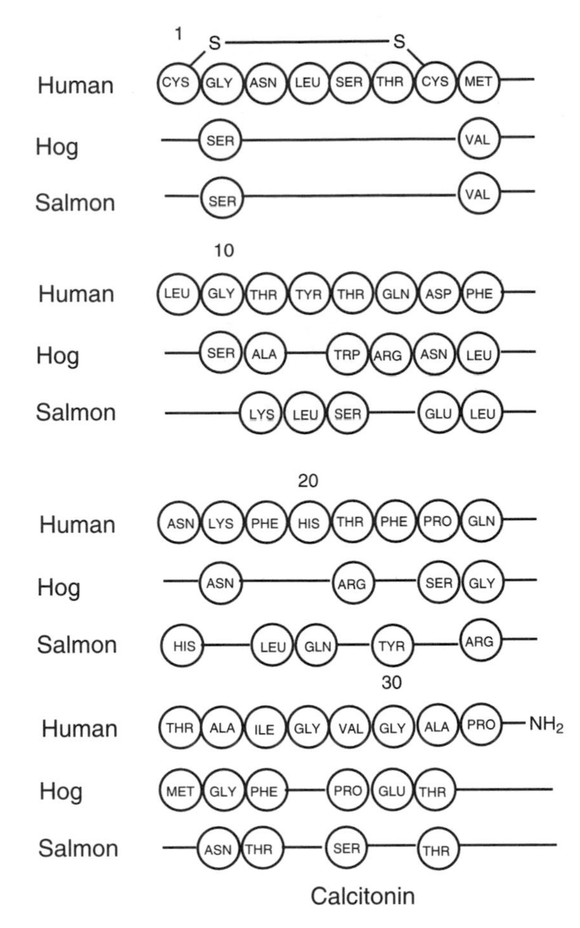

Calcitonin

(Dashes indicate homology with the hormone in humans.)

Cholecystokinin

In 1928, the name cholecystokinin was applied to material present in certain extracts of duodenal mucosa, which caused gallbladder contractions.[37] In 1943, it was found that similar extracts also stimulate the release of pancreatic enzymes. The active principle was termed pancreozymin. Subsequently, both activities were found to reside in the same chemical entity. It is released into the blood in response to stimulation of cholinergic nerves, which in turn are stimulated by the presence of food in the duodenum. It is also found in the CNS where it plays a role in the suppression of food intake. Cholecystokinin can abolish morphine-induced analgesia in the spinal cord and may play a role as a safety signal under conditions where the sensation of pain could aid survival.[41] The gene encoding cholecystokinin in rats has been sequenced,[42] as has a gene for a human brain cholecystokinin receptor.[43]

Cholecystokinin is a 33-peptide. The carboxyl terminal 5-peptide amide is the same as that in gastrin.

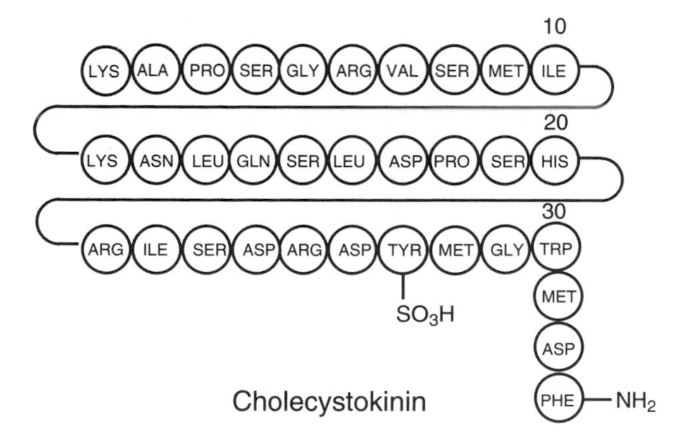

Cholecystokinin

Structure-Activity Relationship

The carboxyl terminal 7-peptide amide is the smallest fragment that shows the activity of the intact hormone, but is less potent. Addition of one more amino acid, to form the 8-peptide amide, yields a compound as potent as the intact hormone. A remarkable enhancement of activity is seen with the 10-peptide amide, which is 10 times as active as the 8-peptide amide. As with gastrin, the penultimate Asp is important for activity.

Removal of the sulfate from the tyrosine of the 8-peptide amide causes loss of most of the activity. This differs from gastrin, in which the absence of sulfate causes little change in activity. The position of the tyrosine sulfate in the chain is also important for activity. If this moiety is interchanged with the amino acid on either side, activity is lost. The 4-peptide amide Asp-Tyr(SO_3H)-Met-Gly-NH_2 is devoid of activity, showing that the tyrosine sulfate and its immediate surroundings are not sufficient to activate the receptor. A sulfated peptide with remarkable sequence homology with gastrin and cholecystokinin has been isolated from head extracts of the cockroach.[44]

Corticotropin (Adrenocorticotropic Hormone [ACTH])

Corticotropin is released from the anterior pituitary in response to signals from the CNS elicited by stress. It is also released when the blood level of corticosteroids is diminished. Both the nervous and chemical stimuli act on the hypothalamus causing release of corticotropin-releasing factor, a 41-peptide that causes corticotropin release from the anterior pituitary.

Corticotropin stimulates the adrenal cortex to convert cholesterol to steroid hormones, such as aldosterone and cortisol, which promote sodium retention and liver glycogen deposition, respectively. Synthesis of these hormones involves the introduction of oxygen into the steroid nucleus, a biochemical process facilitated by ascorbic acid, which is present in high concentration in the adrenal cortex. Corticotropin decreases the adrenal ascorbic acid concentration, a phenomenon that is a basis for the bioassay of the hormone.

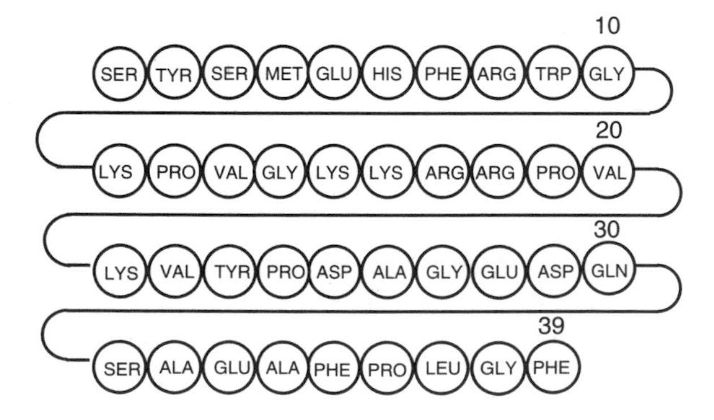

Corticotropin

Structure-Activity Relationship

Corticotropin is a 39-peptide discovered in 1933.[45] It arises from pro-opiomelanocortin (which also gives rise to met-enkephalin) by proteolytic cleavage. Removal of the amino terminal serine causes complete loss of activity. If this residue is acetylated or replaced by D-serine, however, enhanced and prolonged activity results because of resistance of the analogs to enzymatic hydrolysis. Sequential removal of amino acids from the carboxyl terminal does not result in loss of activity until residue 20 is deleted. The resulting 19-peptide has 30% of the activity of the intact hormone. No activity is detectable after residue 16 is removed. If Arg^{17} and Arg^{18} are replaced by norvaline residues, activity is sharply reduced. The basic amino acids Lys^{16}, Arg^{17}, and Arg^{18} are therefore important in determining potency.

Corticotropin is able to correct behavioral impairment in rats caused by removal of the anterior pituitary. Fragments containing only residues 1–10 also show the ability to modify behavior.

Relaxin

Relaxin is a 53-peptide produced in the corpora lutea of pregnant mammals that aids the passage of the fetus by causing dilation of the birth canal and relaxation of the pubic joints. It consists of two polypeptide chains linked by two disulfide bridges placed in the same positions as are found in insulin. Significant homology with insulin is lacking in the other amino acid residues of this hormone. Solid phase chemical synthesis of relaxin and congeners has been achieved.[46]

Insulin

The story of the isolation of insulin from the pancreas is an especially dramatic chapter in the history of medicine. Insulin alleviates the symptoms (hyperglycemia, lipemia, ketonuria, and azoturia) of diabetes mellitus, a disease in which insulin secretion is diminished or absent. Isolation proved difficult because proteolytic enzymes in pancreas extracts destroyed hormonal activity. After the enzyme producing cells were caused to degenerate by closing off the pancreatic duct, insulin could be isolated.

Insulin is released in response to an increased concentration of glucose, fatty acids, or amino acids in the blood. Insulin increases the uptake and storage of these metabolites by cells. Stimulation of the vagus nerve to the pancreas also causes insulin release. In contrast, norepinephrine blocks insulin release, causing glucose, fatty acids, and amino acids to be utilized for energy rather than storage.

Insulin is synthesized in the pancreas as the 104-peptide preproinsulin.[47] The first 23 residues contain a high proportion of hydrophobic amino acids that aid in secretion by directing the protein across the microsomal membrane into the intracisternal space. Preproinsulin is then cleaved to proinsulin. Both preproinsulin and proinsulin have negligible hormone activity. Proteolytic activity in the pancreas cleaves proinsulin between residues 30 and 31 and 60 and 61, liberating insulin and the connecting peptide.[48]

The structure of proinsulin resembles that of nerve growth factor, a protein isolated from submaxillary glands and other tissues, that causes rapid production of neurotubules when applied to nerves in tissue culture.[49]

Insulin has two peptide chains, the A chain, consisting of 21 amino acids, and the B chain, having 30 (Fig. 27–2). The chains are joined by two disulfide bridges, one between residues A_7 and B_7 and the other between A_{20} and B_{19}. The A chain has a third disulfide bond connecting residues A_6 and A_{11}. The amino acid sequence was determined by the pioneering effort of Sanger.[50] His development of fluorodinitrobenzene as a reagent for determining amino terminal amino acids aided greatly. The structure has been confirmed by synthesis, a monumental effort carried out independently in the United States, People's Republic of China, and Germany.[51,52]

The disulfide bonds of insulin can be reduced by treatment with mercaptoethanol and urea, after which the chains can be separated. Neither chain taken alone or in a mixture of chains shows hormonal activity. Reoxidation of a mixture of A chains and B chains led to the formation of insulin in 2% yield. The same experiment performed with proinsulin yielded 70% recovery of native proinsulin.[48] A function of the connecting peptide is to augment the correct alignment of sulfhydryl groups, preventing formation of isomers on oxidation. In addition to the two disulfide bonds, the A and B chains are held together by hydrophobic bonds, hydrogen bonds, and salt linkages.[53]

Comparison of the amino acid composition of human, cow, and hog proinsulin reveals that the variation in the connecting segments (17 of 35 residues) is greater than that in the corresponding insulins (3 of 51 residues). In most instances, substitution with similar amino acids occurs, e.g., glycine for alanine, and leucine for valine.

The tyrosine residues of insulin can be iodinated with I^{125} without loss of activity making tracer studies of the insulin receptor possible. The receptor is a membrane

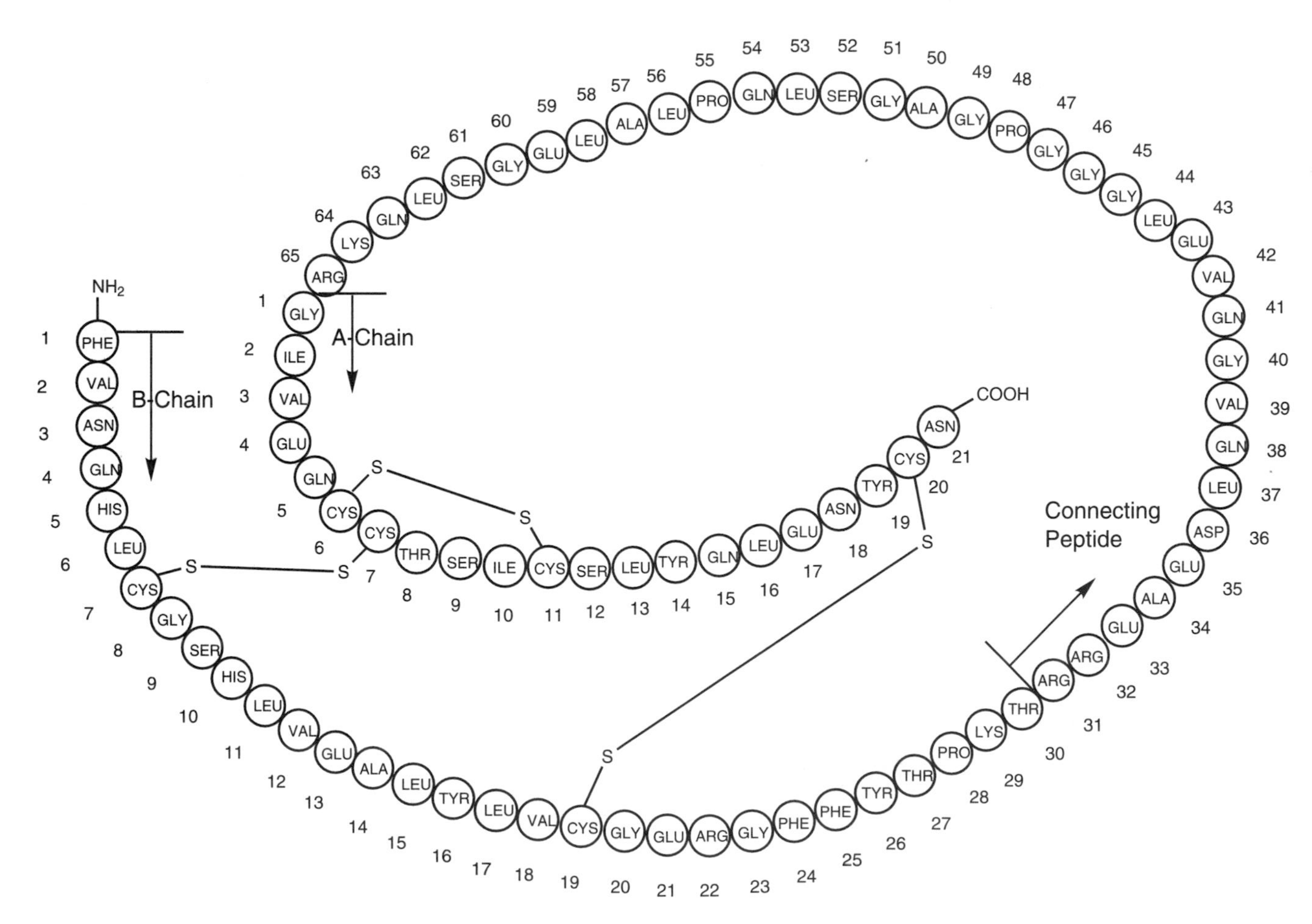

Fig. 27–2. Human proinsulin and insulin (Chain A plus Chain B, Humulin [Lilly]).

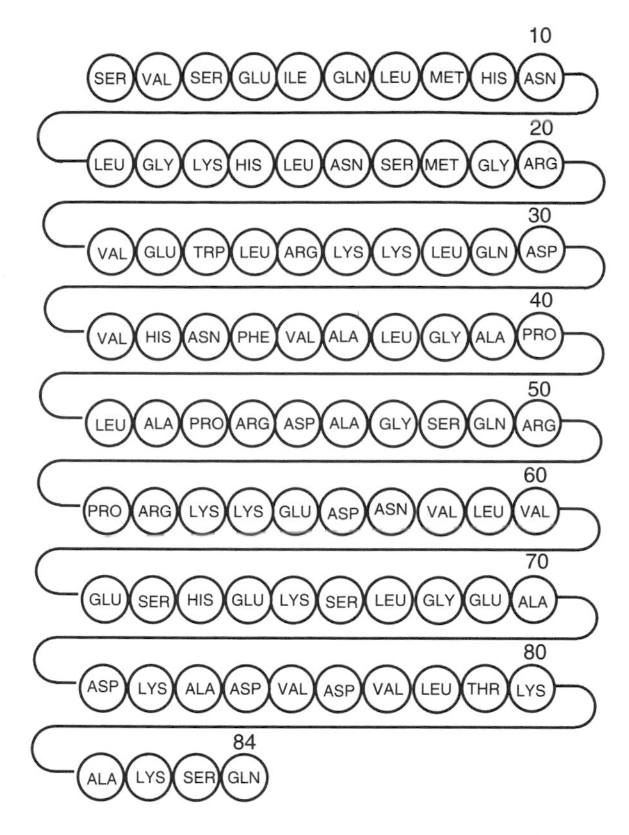

Fig. 27–3. Human parathyroid hormone.[54]

glycoprotein composed of two α- and two β-subunits. The β-subunit has tyrosine kinase activity, which is stimulated by the binding of insulin to the α receptor. This phosphorylation plays a role in signal transduction.

Parathyroid Hormone

Parathyroid hormone is an 84-peptide first noted in acid extracts of parathyroid glands in 1925 (Fig. 27–3). It serves to maintain constant calcium ion levels in body fluids by mobilization from bone, increased absorption from the intestine, and increased recovery from the kidney. Vitamin D derivatives also play a role in these processes.

A fragment composed of residues 1 through 34 has been synthesized and shown to have full activity. The function of the 50 additional residues of the native hormone is not clear.

Growth Hormone (Somatotrophin, Humatrope [Lilly])

Growth hormone is a 191-peptide elaborated in the adenohypophysis and released in response to fasting, hypoglycemia, exercise, and sleep. It is not understood why these stimuli control release of the hormone. Production is regulated by the hypothalamus, which secretes a 44-peptide, growth hormone releasing factor.

Overproduction of growth hormone brought about by a tumor of the adenohypophysis causes giantism in young individuals whose growth has not yet ceased and acromegaly in mature individuals. In acromegaly, the bones of the hands, feet, and face grow, but the other bones of the body do not. Underproduction of growth hormone leads to adults 3 to 4 feet tall. Administration of the hormone to young people with this condition increases their growth.

The metabolic actions of growth hormone are complex. It affects most tissues of the body and acts directly on these tissues as well as by potentiating the effects of other hormones. It stimulates protein synthesis and mobilizes carbohydrate and fat stores. Although human growth hormone stimulates the growth of rats, growth hormone shows greater species specificity than most other peptide hormones. Bovine growth hormone for example, is inactive in humans and monkeys.

Human growth hormone resembles human chorionic somatomammotrophin, a hormone with similar activity elaborated in the placenta. Of the 191 residues, 161 are in the same position in these two hormones.

Prolactin

Prolactin is a 198-peptide produced in the pituitary gland that is essential for induction of lactation at parturition. It acts synergistically with estrogen to promote mammary gland proliferation. It has a 16% sequence homology with human growth hormone. Homology is most pronounced at the amino and carboxyl terminal regions.

Hormone Receptors

Peptide hormones function by interacting with specific cellular membrane receptors.[55] The amino terminal domain of these receptors extends to the outside of the cell, whereas the carboxy terminal domains are located in the cytoplasm. Between the amino and carboxy domains, hydrophobic and hydrophilic segments of the protein loop in and out of the cell membrane. When a peptide hormone binds to its receptor, the signal is transmitted to the cytoplasm by G-proteins, on/off switches controlled by the reversible binding of GTP and GDP. In the "on" mode, when GTP is bound, the G-proteins activate protein kinases, which alter cellular processes by the phosphorylation of specific enzymes.

Proteolysis

The cleavage of the amide bond in peptides and proteins catalyzed by proteolytic enzymes controls the formation and duration of action of peptide hormones. Some of the enzymes involved in the formation of peptide hormones have been cloned and sequenced.[56] They are serine proteases containing the catalytic triad Ser, His, Asp at their active sites. The design of effective hormone agonists and antagonists involves synthesizing compounds that are resistant to proteolysis. Proteolysis also plays a vital role in fertilization, cell growth, blood clotting, virus assembly, and many other processes including clot lysis where the serine protease tissue plasminogen activator (t-PA, Activase Genentech) stimulates the breakdown of thrombi by catalyzing the conversion of plasminogen to plasmin, a serine protease that cleaves peptide bonds in thrombi.

Purification of Peptides

The most favorable conditions for extraction of peptide hormones from tissues must be determined empirically. They will vary with the hormone. For example, dilute acid is used to extract secretin from the duodenal mucosa, whereas dilute alkali is used to obtain growth hormone from the adenohypophysis.

Purification procedures have improved greatly through the introduction of ion exchange chromatography, gel filtration, and high performance liquid chromatography.

Commonly used ion exchange materials include sulfonated polystyrene, diethylaminoethylcellulose, and carboxymethyl cellulose. Peptides that attach themselves to these materials may be eluted by using a gradient of increasing or decreasing pH or by increasing the ionic strength.

Gel filtration employs dextran beads that separate materials based on molecular size. The beads are porous and exclude materials of high molecular weight, which are eluted first, followed by materials of lower molecular weight, which are retarded because they penetrate the beads.

Determination of Structure

Amino acid composition is determined by hydrolyzing the peptide in 6N HCl or by incubation with proteolytic enzymes. The individual amino acids may be separated by paper chromatography, ion exchange chromatography, high performance liquid chromatography, or gas chromatography. Molecular weights of amino acids, peptides, and proteins can be determined accurately by mass spectrometry.[57]

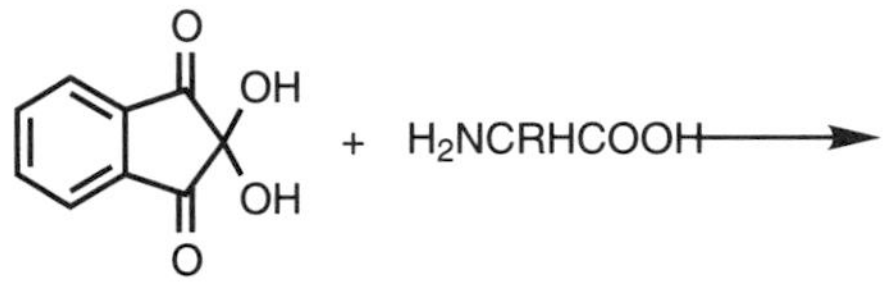

Ninhydrin
(Triketohydrindene hydroate)

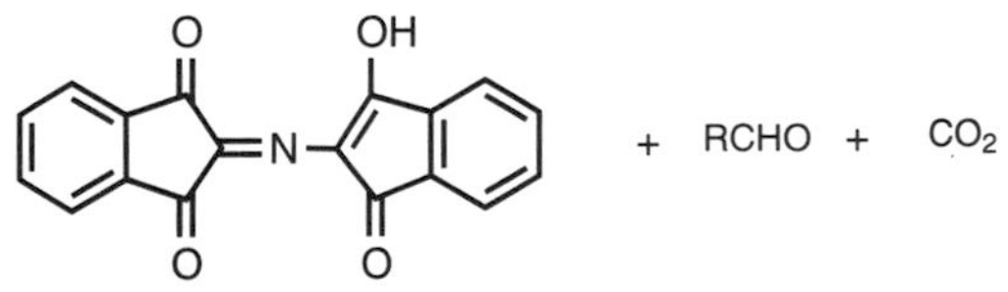

Diketohydrindylidenediketohydrindamine

The ninhydrin reaction may be used to determine the quantity of each amino acid. Reaction with fluorescamine is also used to detect amino acids. It forms fluorescent derivatives that can be detected at the picomole level.

Tryptophan is destroyed by acid hydrolysis and may be isolated after alkaline hydrolysis, a procedure that destroys cysteine, cystine, serine, threonine, and arginine, and also causes racemization. The NH_2-terminal residue

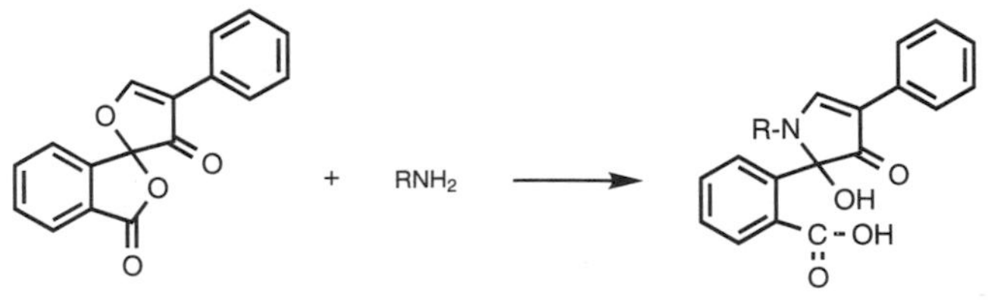

Fluorescamine

can be identified by reaction with fluorodinitrobenzene, with 1-dimethylaminonaphthalene-5-sulfonyl chloride

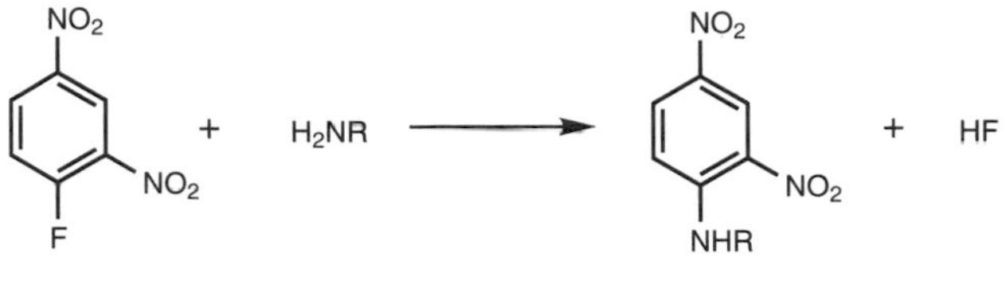

Fluorodinitrobenzene

(dansyl chloride) (equation 4), with phenylisothiocyanate (Edman's reagent) (equation 5) or by enzymatic degradation catalyzed by aminopeptidase.

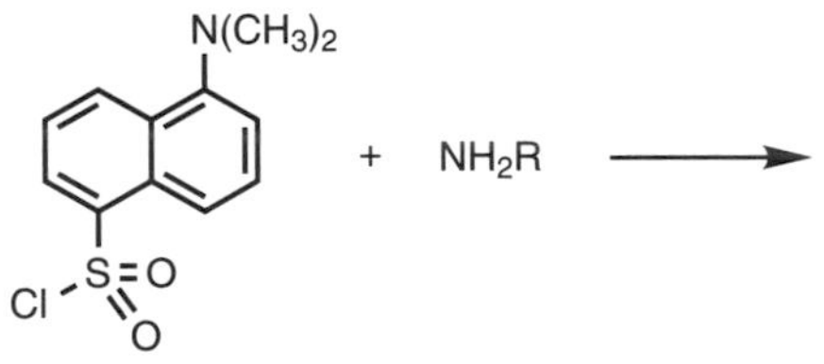

1-Dimethylaminonaphthalene-
5-sulfonyl chloride
(Dansyl chloride)

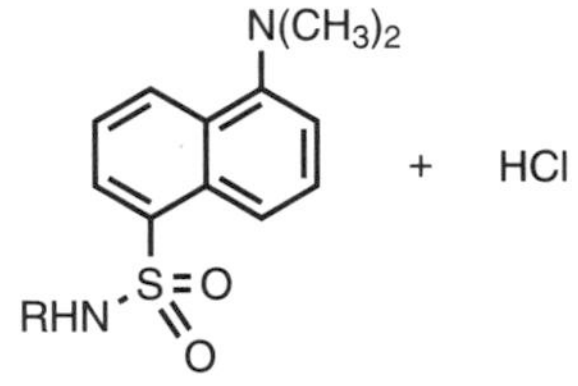

After reaction with fluorodinitrobenzene, the peptide is hydrolyzed in acid and the yellow 2,4-dinitrophenyl derivative of the amino terminal residue is identified chromatographically. The ϵ-NH_2-group of lysine also reacts with fluorodinitrobenzene, but the derivative formed is readily distinguished from the α-amino derivative. The dansyl chloride procedure is similar but has an advantage—the dansyl amino acids are fluorescent and small amounts can be detected. The Edman's phenylisothiocyanate procedure has the advantage that it may be used to degrade a peptide in a stepwise manner, isolating the phenylthiohydantoin of the NH_2-terminal amino acid as well as the residual peptide after each step.

Aminopeptidase specifically cleaves the NH_2-terminal amino acid from the chain. Because the NH_2-terminal residue in the peptide remaining after initial cleavage will

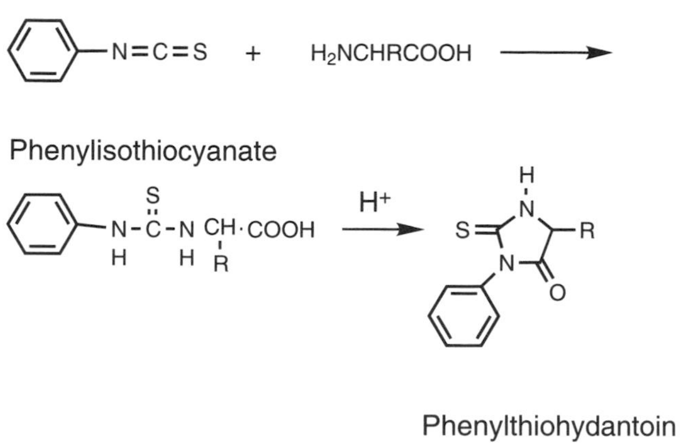

$$C_6H_5-N=C=S + H_2NCHRCOOH \longrightarrow$$

Phenylisothiocyanate

$$C_6H_5-NH-C(=S)-NH-CH(R)-COOH \xrightarrow{H^+}$$ Phenylthiohydantoin

Edman's degradation

also be cleaved as it becomes exposed, it is necessary to determine the order in which the amino acids are released.

The COOH-terminal amino acid residue may be determined after reduction of its carboxyl group to an alcohol with sodium borohydride. Acid hydrolysis liberates an amino alcohol, which is isolated by paper chromatography, high performance liquid chromatography, or electrophoresis. Carboxypeptidase, an enzyme that specifically cleaves the carboxy-terminal residue, is also used for this purpose in a manner analogous to that described for aminopeptidase.

Large peptides are cleaved at specific residues and the resulting small peptides placed in the appropriate order by obtaining overlapping sequences. Cyanogen bromide, which cleaves peptide chains specifically at methionine residues, is used in this type of study as is trypsin, which catalyses the cleavage of the peptide chain on the carboxyl side of arginine or lysine.

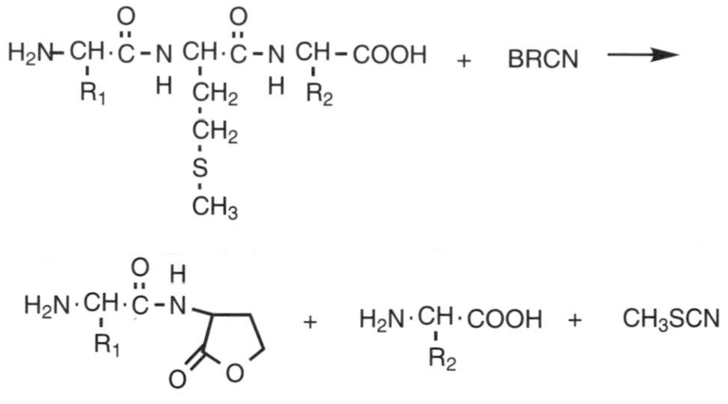

$$H_2N-CH(R_1)-CO-NH-CH(CH_2CH_2SCH_3)-CO-NH-CH(R_2)-COOH + BRCN \longrightarrow$$

$$H_2N\cdot CH(R_1)\cdot CO-NH-\text{(homoserine lactone)} + H_2N\cdot CH(R_2)\cdot COOH + CH_3SCN$$

Peptide of homoserinelactone

SYNTHESIS OF PEPTIDES

Chemical Synthesis

The simplest dipeptide, glycylglycine, was synthesized in 1901 by Fischer, who appreciated the significance of being able to synthesize peptides of defined sequence.[58] Protecting groups had to be introduced to prevent amino acids from reacting with themselves or with side chains. Some common reagents used in peptide synthesis are given in Table 27–5.

As an example of the synthesis of a protein, the solution synthesis of His-Ser-Asp-Gly-Thr-Phe, the first six amino acids of secretin, is discussed.[59] The synthesis consists of a series of condensations of protected amino acids followed by removal of protecting groups. The benzyloxycarbonyl group is used as a protecting group for primary amines by treating the amino acid with benzyloxycarbonyl chloride.

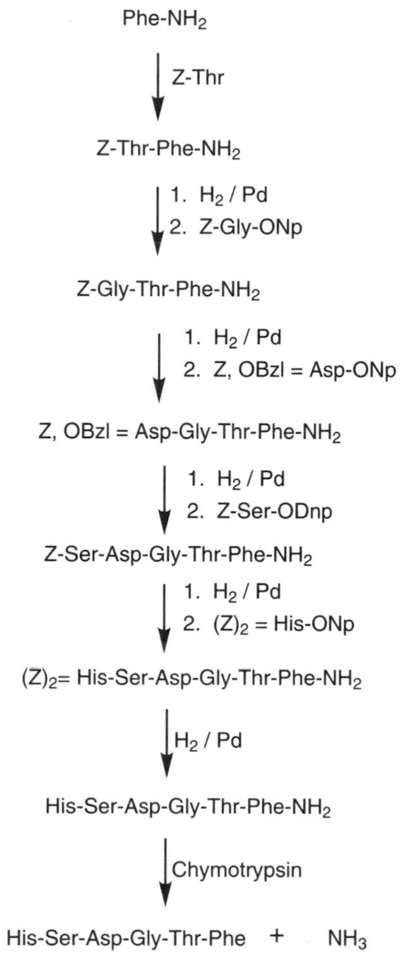

Phe-NH2
↓ Z-Thr
Z-Thr-Phe-NH2
↓ 1. H2 / Pd 2. Z-Gly-ONp
Z-Gly-Thr-Phe-NH2
↓ 1. H2 / Pd 2. Z, OBzl = Asp-ONp
Z, OBzl = Asp-Gly-Thr-Phe-NH2
↓ 1. H2 / Pd 2. Z-Ser-ODnp
Z-Ser-Asp-Gly-Thr-Phe-NH2
↓ 1. H2 / Pd 2. (Z)2 = His-ONp
(Z)2= His-Ser-Asp-Gly-Thr-Phe-NH2
↓ H2 / Pd
His-Ser-Asp-Gly-Thr-Phe-NH2
↓ Chymotrypsin
His-Ser-Asp-Gly-Thr-Phe + NH3

$$C_6H_5-CH_2OCOCl + H_2NR \longrightarrow C_6H_5-CH_2OCONHR + HCl$$

Benzyloxycarbonyl chloride

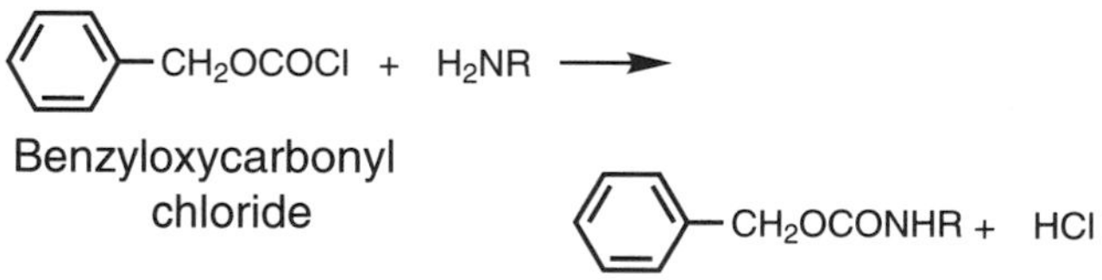

Nitrophenol is used to activate the carboxyl group by treating the amino acid with *p*-nitrophenol in the presence of dicyclohexylcarbodiimide. The latter aids in the removal of water during condensation to form the ester. This reaction is shown in the following equation:

Table 27–5. Reagents Used in Peptide Synthesis

Name	Structure	Use	Abbreviation
Benzyloxycarbonyl	C_6H_5–CH_2–O–C(=O)–	N protection	Z
t-Butyloxycarbonyl	$(CH_3)_3C$–O–C(=O)–	N protection	Boc
9-Fluorenylmethoxycarbonyl	Fluorenyl–CH_2·O–C(=O)–	N protection	Fmoc
Benzyl ester	C_6H_5–CH_2·O–	COOH protection	OBzl
Benzyl	C_6H_5–CH_2–	S protection Imidazole protection	Bzl
p–Nitrophenoxy ester	O_2N–C_6H_4–O–	COOH activation	ONp
2,4-Dinitrophenoxy ester	O_2N–$C_6H_3(NO_3)$–O–	COOH activation	ODnp
N-Hydroxysuccimido ester	Succinimide N–O–	COOH activation	HOSu
Dicyclohexylcarbodimide	C_6H_{11}–N=C=N–C_6H_{11}	Condensing agent	DCC

C_6H_{11}–N=C=N–C_6H_{11} + HNZ·CH(R)·COOH + HO–C_6H_4–NO_2

Dicyclohexylcarbodiimide — p-Nitrophenol

↓

C_6H_{11}–NH–C(=O)–NH–C_6H_{11} + HNZ·CH(R)·C(=O)–O–C_6H_4–NO_2

Dicyclohexylurea — Nitrophenyl ester

Initial condensation of phenylalanine amide with benzyloxycarbonyl threonine was brought about with the aid of N-ethyl-5-phenylisoxazolium-3′-sulfonate, a specific carboxyl activating group.[60] This compound forms an ester after reaction with the carboxyl groups of benzyloxycarbonyl threonine to activate the carboxyl. All subsequent condensations were carried out as nitrophenoxy esters. The benzyloxycarbonyl and benzyl ester protecting groups were removed by hydrogenolysis with palladium catalyst. Cleavage of the carboxy terminal amide was catalyzed by chymotrypsin, which does not cause hydrolysis of any other linkage in this peptide.

Secretin has been synthesized by a step-by-step addition of amino acids to the carboxy terminal residue and by fragment condensation in which residues 1 to 4, 5 to 8, 9 to 13, and 14 to 27 are prepared individually and then condensed together.[33] The synthesis of glucagon was much more difficult, although it is structurally similar to secretin. The introduction of N-hydroxysuccimido esters minimized unwanted side reactions.

The reagents most commonly used to block α-amino groups are Boc[61] and Fmoc,[62] both introduced by L. A. Carpino. The Boc group is removed by treatment with trifluoroacetic acid and the Fmoc group by treatment with piperidine.

Introduction of solid phase peptide synthesis in 1963 by Merrifield[63] proved a giant step forward. In this technique, the carboxy terminal amino acid is attached to polystyrene resin beads. After each amino acid is added, excess reagents are washed away and the necessity of isolating each intermediate is eliminated. When the chain is complete, the peptide is cleaved from the resin. The entire process has been automated. The 9-peptide bradykinin can be synthesized in a few hours.

A functional group is introduced into the resin by converting benzene residues to benzyl chloride residues via

$ClCH_2OCH_3$ + C_6H_5—resin ⟶ $ClCH_2$—C_6H_4—resin + CH_3OH

Chlormethylmethyl ether

reaction with chloromethyl methyl ether. The carboxy terminal amino acid is anchored to the resin through formation of a benzyl ester by treating the triethylamine salt of a t-butyloxycarbonyl amino acid with the benzyl chloride residues.

$(CH_3)_3C-O-CO-NH-CHR-COO^-$ $(C_2H_5)_3\overset{+}{N}H$ + $ClCH_2$—C_6H_4—resin $\xrightarrow[80^\circ C]{C_2H_5OH}$

Triethylamine salt of t-butyloxycarbonyl amino acid

$(CH_3)_3C-O-CO-NH-CHR-CO-O-CH_2$—$C_6H_4$—resin + $(C_2H_5)_3\overset{+}{N}H$ Cl^-

The t-butyloxycarbonyl group is removed by treatment with acid (deprotection) and the next protected amino acid condensed to the amino acid bound to the resin. The completed peptide chain is removed from the resin by treatment with HBr in trifluoroacetic acid.

The t-butyloxycarbonyl amino acids are prepared by reaction with t-butyl-*p*-nitrophenylcarbonate, as shown in the following equation.[61]

O_2N—C_6H_4—$OCOOC(CH_3)_3$ + $H_2NCHRCOONa$ $\xrightarrow{Na^+}$

t-Butyl-p-nitrophenylcarbonate

$(CH_3)_3COCOHNCHRCOONa$ + O_2N—C_6H_4—ONa

Synthesis of the 53-peptide murine epidermal growth factor by the solid phase method[64] is a fine illustration of the state-of-the-art production of a fully active synthetic.

> . . . the assembly of the 53-amino acid residue protected peptide proceeded smoothly with no indication of synthetic difficulties. The final peptide resin was obtained in 99% yield based on weight gain of the resin, the ninhydrin test, and amino acid analysis. The incorporation of (^{3}H) Leu at position 52 greatly facilitated monitoring and quantitation of the synthetic peptides. [Success is attributed to attention to detail] . . . i.e., use of clean, well-characterized reagents, proper synthetic protocols, and monitoring of amino acid incorporation . . .

Another outstanding example is the total chemical synthesis of both the D- and L-enantiomers of the 99-peptide HIV protease.[65,66] The enzyme enantiomers showed reciprocal chiral specificity on peptide substrates.[66]

Synthesis Directed by Recombinant DNA

In this method genes are cloned in Escherichia coli and their expression maximized. Among proteins synthesized in large quantities by this method are human insulin, growth hormone, α- and γ-interferon, t = PA, erythropoietin, colony-stimulating factor, and interleukin-2. Site-directed mutagenesis affords the possibility of producing materials in which specific amino acid substitutions are made.

Design of Drugs Based on the Structure of Peptide Hormones

The ability of peptide hormones to interact with membrane bound receptors depends on their ability to assume shapes complementary to that of the receptor. Once bound, hormones transmit a signal that leads to the biologic response. This is followed by dissociation of the hormone from the receptor. Peptide hormones have a short duration of action and are rapidly degraded. They may interact with more than one type of receptor. They usually show low activity when administered orally. Taking these factors into account, drug design is directed to the preparation of compounds with high receptor specificity, and prolonged activity, as well as oral activity. This may be achieved by designing peptidomimetics whose conformation is restricted.

Current strategies of drug design are illustrated by the following examples:

N-CH_3 HO O OH

Morphine

1. Morphine is an orally active peptidomimetic alkaloid of restricted conformation that interacts with receptors for endorphin and enkephalin. Whereas the phenolic entity of morphine mimics the amino

terminal phenolic side chain of Tyr in endorphin and enkephalin, the complete definition, in chemical terms, of the features of morphine that enable it to mimic the natural hormones has not been achieved.[67] The design of conformationally restricted peptidomimetics is an active field of investigation, however.[68]

2. Intense current interest focuses on the design of inhibitors of the protease involved in the replication of the human immunodeficiency virus (HIV).[69] These studies employ "structure-based drug design" in which the structure of a ligand bound to its macromolecular target (ligate), as determined by x-ray crystallography and nuclear magnetic resonance, is used to direct the search for lead compounds.
3. ^{15}N and ^{1}H nuclear magnetic resonance studies on the interaction of peptide boronic acid inhibitors of the serine protease trypsin show that the conformation of the ligand as well as the ligate are involved in determining the type of complex formed.[70] The ligate is not a rigid mold but adapts itself according to the ligand. These studies have important implications for the design of specific inhibitors of proteolytic enzymes as well as for the development of specific agonists and antagonists of peptide hormone receptors.
4. Orally active peptide-based renin inhibitors have been developed by incorporating solubilizing as well as lipophilic groups.[71,72]
5. Methods are being developed for the selection of ligands for any given ligate from large peptide libraries. One method involves the display of peptides on the surface of a bacteriophage.[73] The peptides are generated from a randomly mutated region of the phage genome. A given ligate is used to select phage bearing active peptide ligands that are identified by sequencing the corresponding bacteriophage DNA. Chemical methods involving tagging peptide libraries with synthetic deoxynucleotide[74] or peptide sequences[75] have also been described. Selection of tight binding ligands should reveal leads for drug development.

REFERENCES

1. A. Meister, *Biochemistry of the Amino Acids,* 2nd ed., New York, Academic Press, 1965.
2. A. Mor, et al., *Trends Biochem. Sci., 17,* 481(1992).
3. E. Culotta and D. E. Koshland, *Science, 258,* 1862(1992).
4. S. Clarke, *Annu. Rev. Biochem., 61,* 335(1992).
5. O. K. Mathews and K. E. van Holde, *Biochemistry,* Redwood City, CA, Benjamin/Cummings, 1990.
6. Y. Nazaki and C. Tanford, *J. Biol. Chem., 246,* 2211(1971).
7. J. W. Olney, *Annu. Rev. Pharmacol. Toxicol., 30,* 47(1990).
8. R. C. Haynes, Jr., in *Pharmacological Basis of Therapeutics,* A. G. Gilman, et al., Eds., New York, Pergamon Press, 1990, p. 1361.
9. R. C. Bateman, Jr., et al., *J. Biol. Chem., 260,* 9088(1985).
10. A. S. Verdini, et al., *J. Med. Chem., 34,* 3372(1991).
11. J. R. Jaffe and W. R. Martin, in *Pharmacological Basis of Therapeutics,* A. G. Gilman, et al., Eds., New York, Pergamon Press, 1990, p. 481.
12. D. Roemer, et al., *Nature, 268,* 547(1977).
13. J. C. Garrison and M. J. Peach, in *Pharmacological Basis of Therapeutics,* A. G. Gilman, et al., Eds., New York, Pergamon Press, 1990, p. 749.
14. E. W. Petrillo and M. A. Ondetti, *Med. Res. Rev., 2,* 1(1982).
15. W. J. Greenlee and P. K. S. Siegl, *Annu. Rep. Med. Chem., 27,* 59(1992).
16. R. M. Hays, in *Pharmacological Basis of Therapeutics,* A. G. Gilman, et al., Eds., New York, Pergamon Press, 1990, p. 732.
17. T. W. Rall, in *Pharmacological Basis of Therapeutics,* A. G. Gilman, et al., Eds., New York, Pergamon Press, 1990, p. 933.
18. V. du Vigneaud, et al., *J. Am. Chem. Soc., 76,* 3115(1954).
19. K. Hofmann, in *Peptides,* E. Gross and J. Meienhofer, Eds., Rockford, Illinois, Pierce Chemical Company, 1979, p. 5.
20. W. H. Sawyer, *Pharmacol. Rev., 13,* 255(1961).
21. T. Yamanaka, et al., *Mol. Pharmacol., 6,* 474(1970).
22. R. Walter, et al., *Am. J. Med., 42,* 653(1967).
23. W. H. Sawyer and M. Manning, *FASEB J., 43,* 87(1984).
24. G. Flouret, et al., *J. Med. Chem., 36,* 747(1993).
25. J. M. Bathon and D. Proud, *Annu. Rev. Pharmacol. Toxicol., 31,* 129(1991).
26. M. E. Logan, et al., *Annu. Rep. Med. Chem., 26,* 43(1991).
27. A. V. Schally, et al., *Annu. Rev. Biochem., 47,* 89(1978).
28. P. C. Emson and B. E. B. Sandberg, *Annu. Rep. Med. Chem., 18,* 31(1983).
29. J. J. Leban, et al., *Proc. Natl. Acad. Sci. USA, 90,* 1922(1993).
30. M. G. Weisskopf, et al., *Nature, 362,* 423(1993).
31. A. M. Doherty, *J. Med. Chem., 35,* 1493(1992).
32. E. Jorpes, *Gastroenterology, 55,* 157(1968).
33. A. Marglin and R. B. Merrifield, *Annu. Rev. Biochem., 39,* 841(1970).
34. S. R. Post, et al., *Proc. Natl. Acad. Sci. USA, 90,* 1662 (1993).
35. R. W. Lappe and R. L. Wendt, *Annu. Rep. Med. Chem., 21,* 273(1986).
36. P. Needleman, *FASEB J., 45,* 2096(1986).
37. V. Mutt, *Ann. N.Y. Acad. Sci., 527,* 1(1988).
38. P. Gressens, et al., *Nature, 362,* 155(1993).
39. K. A. Miezek, et al., *Science, 215,* 1520(1982).
40. M. V. L. Ray, et al., *Biotechnology, 11,* 64(1993).
41. E. P. Wiertelak, et al., *Science, 256,* 830(1992).
42. R. J. Deschenes, et al., *J. Biol. Chem., 260,* 1280(1985).
43. Y.-M. Lee, et al., *J. Biol. Chem., 268,* 8164(1993).
44. R. J. Nachman, et al., *Science, 234,* 71(1986).
45. R. Schwyzer, in *Protein and Polypeptide Hormones,* M. Margoulies, Ed., Amsterdam, Exerpta Medica, 1969, p. 201.
46. E. E. Bullesbach and C. Schwabe, *J. Biol. Chem., 266,* 10754(1991).
47. P. T. Lomedico, et al., *J. Biol. Chem., 252,* 7971(1977).
48. C. Nolan, et al., *J. Biol. Chem., 246,* 2780(1971).
49. K. A. Thomas, *Annu. Rep. Med. Chem., 17,* 219(1982).
50. F. Sanger, *Br. Med. Bull., 16,* 183(1960).
51. P. G. Katsoyannis and J. Z. Ginos, *Annu. Rev. Biochem., 38,* 881(1969).
52. K. Lubka and H. Klostermeyer, *Adv. Enzymol., 33,* 445(1970).
53. T. L. Blundell, et al., *Contemp. Physics, 12,* 209(1971).
54. H. T. Keutmann, et al., *Biochemistry, 17,* 5723(1978).
55. C. J. Evans, et al., *Science, 258,* 1952(1992).
56. D. F. Steiner, et al., *J. Biol. Chem., 267,* 23,435(1992).
57. K. Biemann, *Annu. Rev. Biochem., 61,* 977(1992).
58. J. S. Fruton, *Adv. Protein Chem., 5,* 41(1949).
59. M. A. Ondetti, et al., *Biochemistry, 7,* 4069(1968).
60. R. B. Woodward, et al., *J. Am. Chem. Soc., 83,* 1010(1961).
61. L. A. Carpino, *J. Am. Chem. Soc., 79,* 4427(1957).

62. L. A. Carpino and G. Y. Han, *J. Org. Chem., 37,* 3404(1972).
63. B. Merrifield, *Science, 232,* 341(1986).
64. W. F. Heath and R. B. Merrifield, *Proc. Natl. Acad. Sci. USA, 83,* 6367(1986).
65. R. F. Nutt, et al., *Proc. Natl. Acad. Sci. USA, 85,* 7129(1988).
66. R. C. deL. Milton, et al., *Science, 256,* 1445(1992).
67. M. Barinaga, *Science, 258,* 1882(1992).
68. J. Rizo and L. M. Gierasch, *Annu. Rev. Biochem., 61,* 387(1992).
69. J. W. Erickson and S. W. Fesik, *Annu. Rep. Med. Chem., 27,* 271(1992).
70. E. Tsilikounas, et al., *Biochemistry, 31,* 12839(1992).
71. S. H. Rosenberg, et al., *J. Med. Chem., 36,* 460(1993).
72. H. D. Kleinert, et al., *Science, 257,* 1940(1992).
73. J. K. Scott, *Trends Biochem. Sci., 17,* 241(1992).
74. S. Brenner and A. Lerner, *Proc. Natl. Acad. Sci. USA, 89,* 531(1992).
75. J. M. Kerr, et al., *J. Am. Chem. Soc., 115,* 2529(1993).

Chapter 28

PHARMACEUTICAL BIOTECHNOLOGY

Robert D. Sindelar

The early 1980s saw the products of modern pharmaceutical biotechnology come to the marketplace as the Food and Drug Administration (FDA) approved recombinant DNA–produced insulin in 1982, and second-generation home pregnancy test kits containing monoclonal antibodies were developed. In an article entitled "Biotechnology: Are you ready for it?", appearing as the cover story in the May 1990 issue of *Drug Topics,* Conlan[1] suggested that "pharmacists will skillfully ride the coming biotechnology drug wave into the twenty-first century, where they'll reign as the unchallenged drug therapy experts, designing, dispensing, counseling, and monitoring medicines in the brave new world of genetic engineering." Although yet to be fulfilled, this is certainly an attractive scenario and represents an exciting opportunity for pharmacists.[2] Biotechnology generates useful products and the basic scientific knowledge. There is little doubt that the techniques of biotechnology have led to the development and marketing of new and novel therapies residing on pharmacists' shelves today; improved methods of manufacture of pharmaceuticals; and significant contributions to our better understanding of disease etiology, pathophysiology, and biochemistry. Obviously, advances in biotechnology are going to have an increasing impact on pharmacy practice in the 1990s and into the future. Although many of the first biotechnology-derived therapeutics were initially used in acutely ill hospital patients, products of modern pharmaceutical biotechnology will increasingly have an impact on the chronic disease patient populations constituting much of ambulatory care practice.

Biotechnology, innovation harnessing nature's own biochemical tools, uses living organisms and their cellular, subcellular, and molecular components. In its broadest definition, Sumerian and Babylonian beer brewing around 6000 B.C., the baking of leavened bread by the Egyptians (by 4000 B.C.), cheese making, and even Alexander Fleming's discovery of penicillin in 1928 are products of biotechnology. Therefore, biotechnology is an evolutionary technology, not a revolutionary one. What we may think of as modern biotechnology is enabled by a collection of techniques arising out of developments in molecular and cell biology, microbiology, genetics, biochemistry, protein chemistry, organic chemistry, and immunology in the 1970s and 1980s (Fig. 28–1). The FDA defines biotechnology as a technique that uses living organisms or a part of a living organism to produce or modify a product, to improve a plant or animal, or to develop a microorganism to be used for a specific purpose. Applications of biotechnology are found in medicine, pharmaceutical sciences, agriculture, environmental science, food science, forensics, and materials science. As applied to pharmaceuticals, the techniques of biotechnology discover, develop, and produce useful therapeutic agents and diagnostics. Pioneering strategies for new drug design are products of pharmaceutical biotechnology.

Several textbooks on the pharmaceutical aspects of biotechnology are available.[3–5] Additionally, dictionaries of biotechnology[6,7] and other biotechnology texts[8–11] have appeared in print. Most biochemistry and molecular biology textbooks are fine sources of biotechnology-related material.[12–14] Excellent chapters on biotechnology in medicinal chemistry[15,16] and numerous reviews of biotechnology for pharmacists[1,2,17–41] are also available. In addition, listings of biotechnology-produced pharmaceuticals in development[42,43] and a biotechnology reference resource catalog[44] have been published.

IMPACT OF BIOTECHNOLOGY ON PHARMACEUTICAL CARE

The biotechnology industry has grown explosively over the past two decades, with estimated sales for fiscal 1993 of $7.0 billion.[45] Therapeutic agents (41%) and diagnostic products (27%) compose the greatest share of the biotechnology industry market. Nearly half of all the 4446 biotechnology patents issued by the U.S. Patent and Trademark Office in 1992 were for health care. Approximately 75% of the nearly $5.9 billion in worldwide biotechnology sales in 1992 came from biotechnology-produced pharmaceuticals.[46] As reported in the 1993 survey of *Biotechnology Medicines in Development* from the Pharmaceutical Manufacturers Association (PMA),[42] a consulting group found that 33% of research projects in major pharmaceutical companies were biotechnology based in 1993, as compared with only 2% in 1980. The survey showed that 143 biotechnology-produced medicines and vaccines are in clinical trials or at the FDA awaiting final approval. This number represents a nearly 77% increase over the past 5 years (Fig. 28–2). Of the 143 medicines, 59 products are for cancer or cancer-related conditions, 23 are in the development for human immunodeficiency virus (HIV) infection and acquired immunodeficiency syndrome (AIDS), and 11 are for the treatment of sepsis.

Improved manufacturing of pharmaceuticals was the first major contribution of biotechnology to pharmaceutical care in the 1980s. Biotechnology-produced human

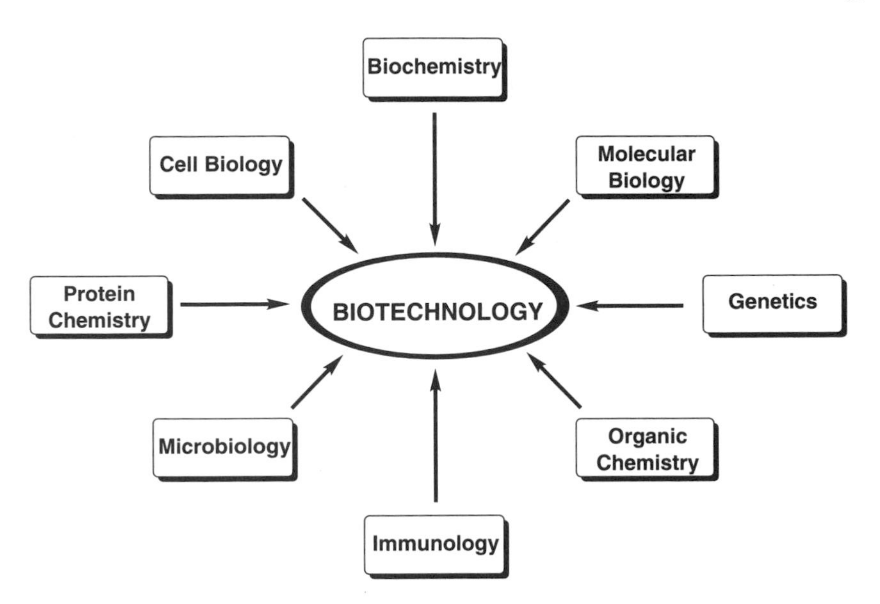

Fig. 28–1. Enabling technologies of biotechnology.

insulin, growth hormone, and erythropoietin, all replacements of highly specific, endogenous molecules, are major advances in therapy. Since the late 1980s, however, pharmaceutical biotechnology has also helped to identify new compounds with new mechanisms of action. Significant contributions to the understanding of the mechanism of disease at the molecular level continue to be made by biotechnology researchers and will translate into newer, better pharmaceuticals. The impact of biotechnology on pharmaceutical care will exponentially increase as advances in technology continue to yield novel medicinal agents such as the colony-stimulating factors (CSFs), tissue-type plasminogen activator (tPA), new vaccines, and DNase. As Abrams[47] stated, "the dream for biotechnology, however, was not merely to provide replacements for deficiency states, but to treat disease with molecules that, because they are designed by nature, had highly specific actions and a paucity of side effects." Biotechnology will help to target drug therapy toward the underlying cause of diseases, not just the treatment of disease symptoms. Products of biotechnology are playing a critical role in the discovery and design as well as the production of treatments for life-threatening diseases, such as cancer, AIDS, and cardiovascular disease.

Pharmaceutical care using biotechnology-derived products requires:[38]

1. An understanding of how the handling and stability of biopharmaceuticals differs from other drugs pharmacists dispense.
2. A preparation of the product for patient use, including reconstitution or compounding, if required.
3. Patient education on their disease, benefits of the prescribed biopharmaceutical, potential side effects or drug interactions to be aware of, and the techniques to self-administer the biotechnology drug.
4. Patient counseling on the reimbursement issues involving an expensive product.
5. Monitoring of the patient for compliance.

TECHNIQUES OF BIOTECHNOLOGY

The techniques made available by advances in biotechnology that have provided medicinal agents fall into two broad areas. First, recombinant DNA (rDNA) technology,

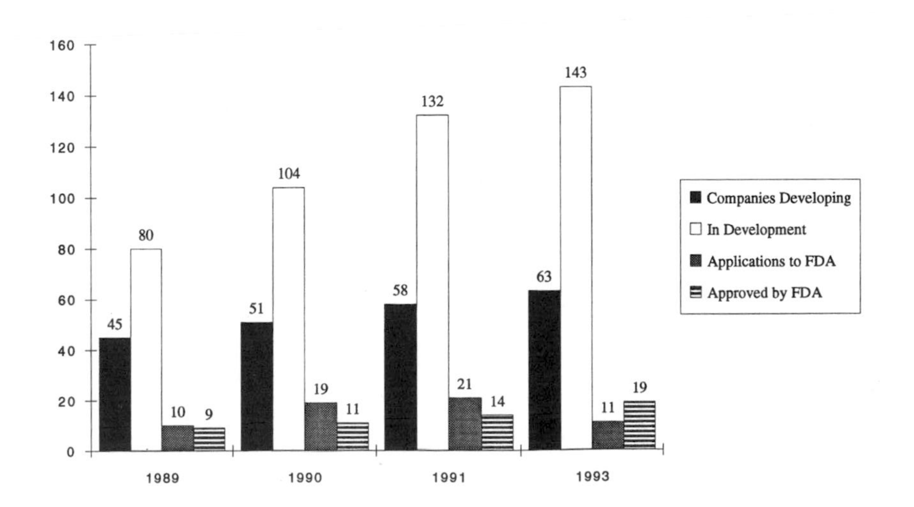

Fig. 28–2. Number of technology-produced pharmaceuticals. (Data from *Biotechnology Medicines in Development, 1993 Survey,* Pharmaceutical Manufacturers Association, Washington, D.C., 1993.)

the ability to manipulate the genetic information inherent within the nucleus of living cells, provides the ability to take identified gene sequences from one organism and place them functionally into another to permit the production of protein medicines. Second, hybridoma techniques permit the production of monoclonal antibodies (MoAb). MoAbs are ultrasensitive hybrid immune system–derived proteins designed to recognize specific antigens and are used as diagnostic agents for laboratory and home kits and site-directed therapeutics.

Additionally the development of technologies to study DNA-DNA and DNA-RNA interactions has led to the formation of DNA probes (antisense technology) for a variety of research purposes with potential uses as diagnostics and therapeutics.

RECOMBINANT DNA TECHNOLOGY

The revolution in biology and genetics that has occurred over the past 20 years, affecting both the basic research and its practical aspects, has been fueled by rDNA technology. Sometimes also referred to as genetic engineering, gene cloning, or in vitro genetic manipulation, rDNA technology provides the ability to introduce genetic material from any source into cells (bacterial, fungal, plant, or animal) or whole plants and animals.[48] A general understanding of the technologies involved should readily help the pharmacist better gain insight into and comprehension of a biotechnology drug's use, stability, handling, side effects, and potential toxicity: in other words, how these agents differ from traditional drugs. Also, such an understanding should readily demonstrate the impact that rDNA technology is having on both current and future pharmacy practice.

Normal Process of Protein Synthesis by the Cell

To understand the impact of rDNA technology, a general understanding of the process of protein synthesis by a cell is essential. Sections on protein synthesis in any biochemistry or molecular biology[12–14] textbook furnish a detailed review. A general review of gene expression and the process of protein synthesis follows and is schematically represented in Figure 28–3.

The active working components of the cellular machinery are proteins, which constitute more than half the total dry weight of a cell. Proteins carry out most biologic activities of cells. A protein's particular function is determined by the exact linear order of amino acids composing that protein, its primary structure. Therefore, protein synthesis is critical to cell development, maintenance, and growth. The information or coding for protein synthesis and the chemical synthesis of the protein are the central dogma of biology: DNA to RNA to protein.[49,50]

The genetic information necessary for a cell to live and function, i.e., to encode for the synthesis of specific proteins, is stored in discrete genes within the linear molecules of 2′-deoxyribonucleic acid (DNA) making up the chromosomes in the cell nucleus. The genes consist of double helical strands of DNA. The exact sequence of building blocks of the cell's DNA, the nucleic acid bases, contain the genetic code to synthesize a specific protein. Each base is linked through a phosphate bond at the 5′-position of a 2′-deoxyribose sugar to the 3′-end of the deoxyribose portion on the next nucleotide. A nucleoside is a base + sugar, whereas a nucleotide is a base + sugar + phosphate. The purines adenine (A) and guanine (G) and the pyrimidines thymine (T) and cytosine (C) are the only nucleic acid bases found in DNA. The hydrogen bonding interactions of complementary bases on each strand hold the two strands of DNA together to form the regularly appearing structure of double helical DNA with the sugar-phosphate backbones directed toward the outside. Owing to the specificity of the hydrogen bonding, A can only bond with T, whereas G can only pair with C. All four nucleotide bases can exist in all possible permutations of sequence along any given DNA chain. The particular combination of A, G, T, and C found in one of the single strands of DNA and complementary to those in

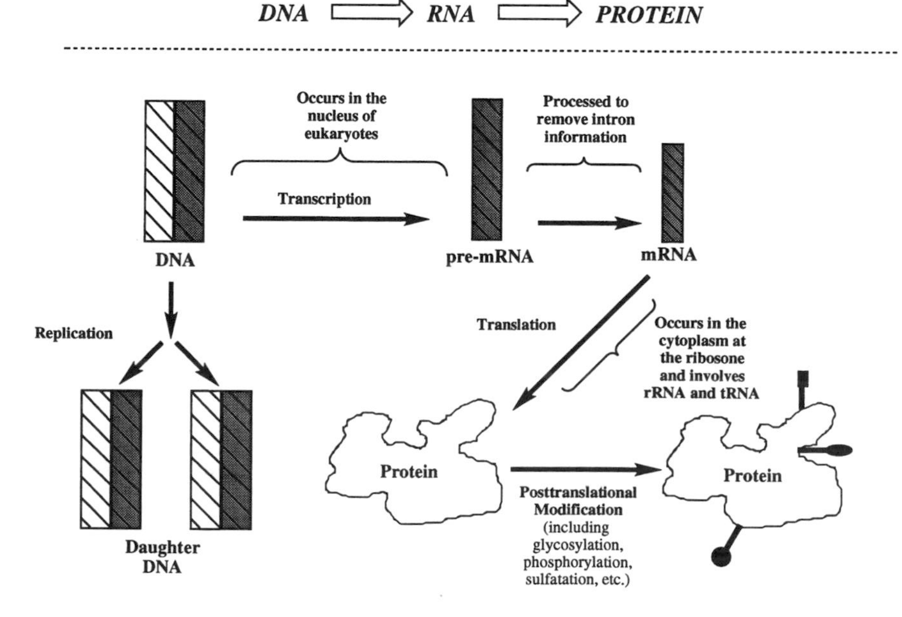

Fig. 28–3. Gene expression: the synthesis of proteins.

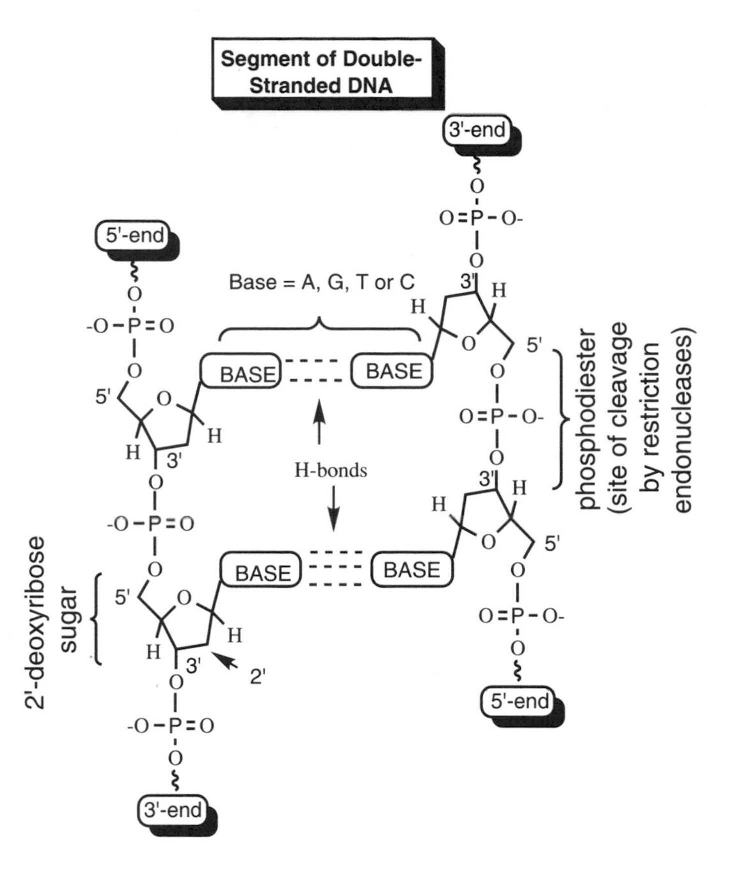

the other strand of the double-stranded DNA helix dictates the genetic message.

Three types of ribonucleic acid (RNA) work together to convert the genetic code stored in a cell's nucleus as DNA into a protein. Messenger RNA (mRNA), by the process of transcription, encodes the genetic information in the base pair sequences. Although each hydrogen bonding interaction holding the two DNA strands together is relatively weak, each DNA molecule contains so many base pairs that the two strands never spontaneously separate under physiologic conditions.[51] During transcription, the enzyme RNA polymerase[52] initiates the unraveling of the two strands of helical DNA (the sense strand and the antisense strand) to create temporarily single-stranded DNA chains. The sense strand, containing the specific genetic code sequence of base pairs, is the template for the forming mRNA. Transcription proceeds in the 5′-to-3′ direction. mRNA is produced as a complementary sequence of base pairs to the single-stranded DNA template. Nucleotides G and C are complementary. In mRNA, however, the pyrimidine base uracil (U) is substituted for T. Therefore, a T in the DNA complements with an A in mRNA, whereas an A in the DNA template complements with a U in the newly formed mRNA. In eukaryotes, further processing of this mRNA, more properly termed pre-mRNA, occurs.[53] Eukaryotic DNA contains regions of several hundred base pairs that are not expressed in the protein molecule known as introns (or nonsense DNA, or silent DNA) that alternate with the coding sequences or exons. The pre-mRNA corresponding to the whole gene is processed enzymatically by special intracellular enzymes and spliced to remove the introns producing a mRNA that codes correctly for the amino acid sequence of the protein.

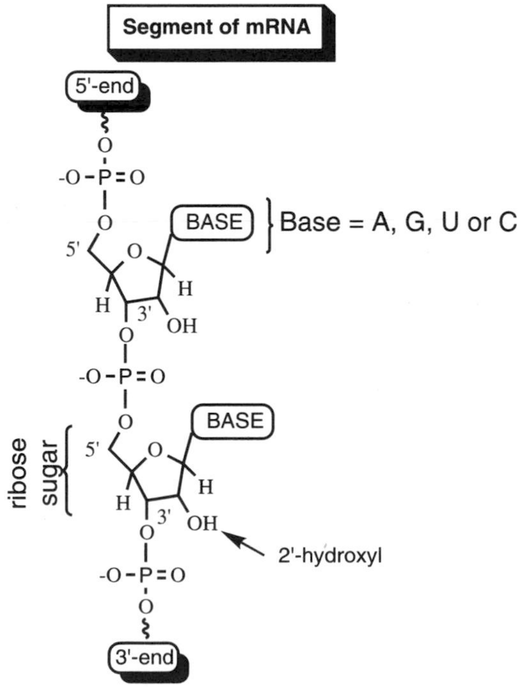

Segment of mRNA

5'-end
BASE
Base = A, G, U or C
BASE
ribose sugar
2'-hydroxyl
3'-end

The mRNA relays the genetic information obtained from the corresponding DNA into a protein by the process of translation occurring at the ribosome. Only the processed mRNA contains the necessary leader sequence of nucleotides that initiates the translation process. Ribosomal RNA (rRNA) interacts with a set of proteins to form the ribosome, providing sites of interaction for all the molecules necessary for the translation process. The genetic information contained in the DNA is carried by mRNA in a three-letter code. Amino acids are added one at a time by a specialized type of RNA, transfer RNA (tRNA), to make the specific protein coded for by the cell's original DNA. Each sequence of three nucleotide bases along the length of mRNA is a distinct codon or message that specifies which amino acid is to be added next to the end of the growing protein molecule (Table 28–1). The four nucleotides, A, G, U, and C in all possible three-nucleotide combinations create 64 possible codes representing the 20 amino acids, starts and stops. For example, the codon (5′) AUG (3′) on the mRNA codes specifically for methionine and is the most common initiator codon for protein synthesis. Stop or termination codons include UAA, UGA, and UAG. Each tRNA acts as a translator molecule that converts the information in each mRNA codon so that only the appropriate amino acid is added to the growing polypeptide chain. Twenty enzymes called aminoacyl tRNA synthetases, each corresponding to an amino acid, add the correct amino acid to tRNA, which recognizes the codon and adds the amino acid to the chain.

As the amino acid addition occurs, the newly synthesized protein begins to assume its characteristic three-dimensional shape (secondary and tertiary structure). In addition, post-translational modifications of the protein can commonly occur in the endoplasmic reticulum or Golgi apparatus after completion of the protein chain. In most cases, these enzymatic chemical reactions alter the side chains of some of the amino acids in the protein. Some examples of post-translational modifications include O-glycosylation of threonine or tyrosine; N-glycosylation of asparagine; sulfate conjugation of thyrosine; hydroxylation of proline or lysine; and phosphorylation of tyrosine, serine, or threonine.[15] Although the codon system is consistent among all organisms (thus allowing rDNA technology), it is important to note that many post-translational modifications occur only in higher organisms (not bacteria).

Table 28–1. Genetic Code: mRNA to Amino Acid

Position 1 (5′ end)	Position 2				Position 3 (3′ end)
	U	C	A	G	
U	Phe	Ser	Tyr	Cys	U
	Phe	Ser	Tyr	Cys	C
	Leu	Ser	Stop	Stop	A
	Leu	Ser	Stop	Trp	G
C	Leu	Pro	His	Arg	U
	Leu	Pro	His	Arg	C
	Leu	Pro	Gln	Arg	A
	Leu	Pro	Gln	Arg	G
A	Ile	Thr	Asn	Ser	U
	Ile	Thr	Asn	Ser	C
	Ile	Thr	Lys	Arg	A
	Met/Start	Thr	Lys	Arg	G
G	Val	Ala	Asp	Gly	U
	Val	Ala	Asp	Gly	C
	Val	Ala	Glu	Gly	A
	Val/Met	Ala	Glu	Gly	G

Process of Recombinant DNA

Useful reviews detailing the process of recombinant DNA are available.[3–5,8,10,15,16,50] Applicable sections in any biochemistry or molecular biology[12–14] textbook provide more detailed reviews. Several reviews of rDNA technology have been written for practicing pharmacists.[18,26,27–31,37,39] In addition, a biotechnology resource catalog listing references of various aspects of rDNA has been published.[44] A general summary of the typical rDNA production of a protein follows and is schematically presented in Figure 28–4.

In theory, one can produce any protein desired as long as a copy of the corresponding gene is made available. Reproducing the DNA involved is called cloning. There are two major methods to obtain the necessary gene. The first is genomic cloning, in which identification and isolation of the DNA coding for the protein is achieved by breaking up the entire genome into fragments and using DNA probes to screen the resulting genomic library for the desired gene. DNA probes are the standard method for identifying a DNA sequence within a mixture.[54] The probes are labeled (radioactive or fluorescent) specific nucleotide sequences that bind only to their complementary or copy DNA (cDNA) in the mixture of DNA fragments to form a double helix. The resulting double helix is detected using the label. Probes are the basis for DNA fingerprinting, the method of making a pattern of the DNA of an individual unique from any other individual.

The second method, cDNA cloning, involves the isolation of the mRNA that codes for the amino acid sequence of the protein and is achieved using the viral enzyme reverse transcriptase to "retro"-synthesize the cDNA. Reverse transcriptase, also called RNA-dependent DNA polymerase, is an enzyme capable of converting mRNA into complementary single-stranded DNA. The enzyme DNA polymerase synthesizes a complementary second strand of DNA on the created single strand of DNA.

An additional method to obtain DNA for use in rDNA technology is the automated synthesis of the gene through biochemical means, if the amino acid sequence of the protein is known so the codon sequence can be deduced. At present, however, this method is practical only for relatively small proteins.

Once the gene coding for the desired protein has been identified and isolated, the genetic material is introduced into cells using restriction endonucleases and a vector to

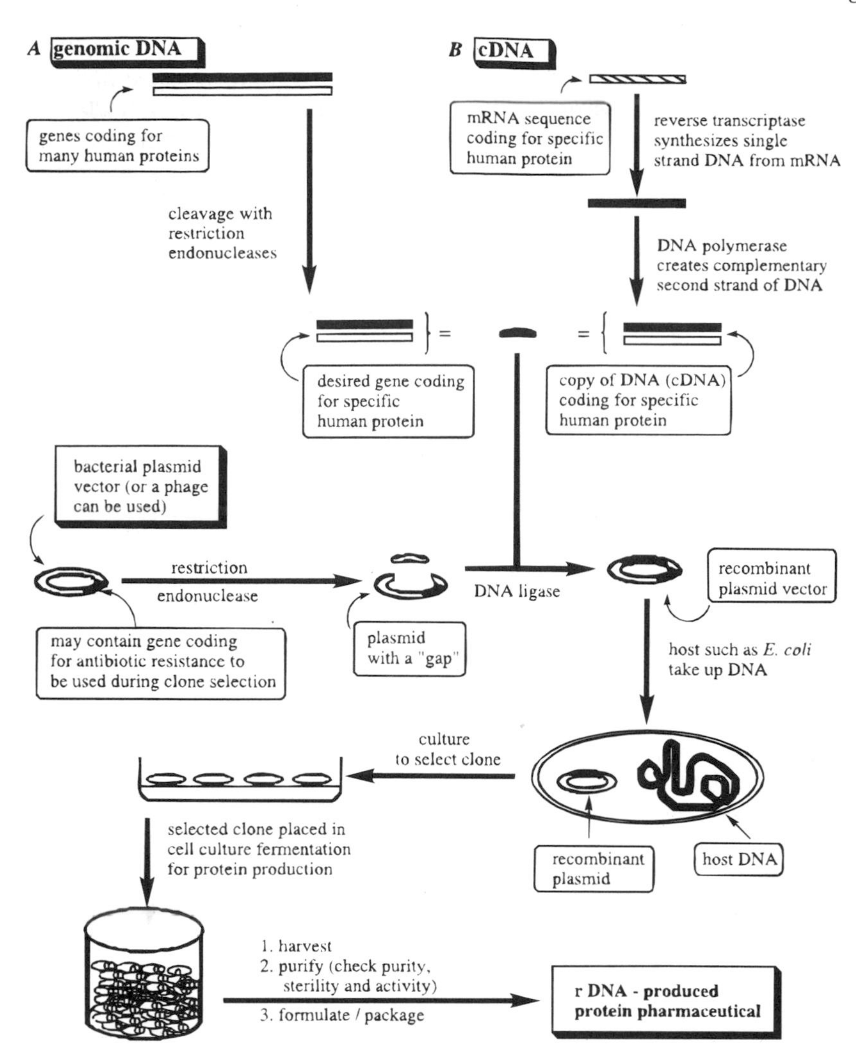

Fig. 28–4. Summary of typical rDNA production of a protein from either *(A)* genomic DNA or *(B)* cDNA.

Table 28–2. Some Representative Examples of Restriction Endonucleases

Restriction Endonucleases	Source	Recognition Sequence*	Cleavage Products	
Alu I	Arthrobacter luteus	5′-TT/CGAA-3′	-TT-	-CGAA-
		3′-AAGC/TT-5′	-AAGC-	-TT-
Asu II	Ananaena subcylindria	5′-AG/CT-3′	-AG-	-CT-
		3′-TC/GA-5′	-TC-	-GA-
Bal I	Brevibacterium albidum	5′-TGG/CCA-3′	-TGG-	-CCA-
		3′-ACC/GGT-5′	-ACC-	-GGT-
Eco RI	Escherichia coli	5′-G/AATTC-3′	-G-	-AATTCT-
		3′-CTTAA/G-5′	-CTTAA-	-G-
Eco RV	Escherichia coli	5′-GAT/ATC-3′	-GAT-	-ATC-
		3′-CTA/TAG-5′	-CTA-	-TAG-
Hha I	Haemophilus haemolyticus	5′-GCG/C-3′	-GCG-	-C-
		3′-C/GCG-5′	-C-	-GCG-

Data from S. W. Zito, Ed., *Pharmaceutical Biotechnology—A Programmed Text,* Lancaster, PA, Technomic Publishing, 1992; and J. D. Watson, et al., *Recombinant DNA,* 2nd ed., New York, Scientific American Books, 1992.
* Illustrates the position where the DNA strands are cleaved.

enable that DNA to replicate and produce protein. The discovery of restriction endonuclease enzymes that recognize explicit sequences of bases in double-stranded DNA and precisely hydrolyze the phosphodiester bonds of the nucleic acids at specific sites along the DNA strands offered a way of isolating predictable fragments of any DNA molecule. This family of "scissors-like" enzymes is used to provide the DNA fragment coding for the protein of interest by "cutting" or "clipping" the DNA. The bacterial enzymes, numbering greater than 100 variations isolated (see Table 28–2 for representative examples), have led to the powerful techniques of DNA sequencing and rDNA technology. Most restriction endonucleases create two single-strand breaks, one in each strand generating a 5′-phospho*mono*ester and 3′-hydroxyl group from each cleavage. The breaks are not necessarily opposite one another as can be seen in Table 28–2 with examples ASU II, Bal I, and Eco RV For instance, Eco RI (in Table 28–2) causes breaks that are not opposite one another creating sticky (or cohesive or complementary) ends to the DNA strands. The DNA fragments can be isolated, purified, and identified.

Other vectors include the constructed plasmids pBR322 and pUC18 and bacteriophages. Bacteriophages are bacterial viruses modified to accept large pieces (7 to 20 kilobases in length) of exogenous DNA without altering their ability to infect and replicate inside the bacteria.[56] Large genomic libraries have been created by fragmenting all of the embryonic or sperm cell DNA of an organism, inserting them into bacteriophage lambda, and screening with DNA probes to identify the gene sequences.

The DNA fragments coding for the desired protein can be cloned from genomic DNA or cDNA as described previously. The piece of DNA coding for the protein of interest is then inserted into the vector that carries the code to synthesize the protein into the host. The plasmid or bacteriophage DNA is opened with a restriction enzyme, and the exogenous gene is pasted into it by an enzyme called DNA ligase with the assistance of special sections of DNA called linkers. DNA ligase seals the single-strand breaks in double-stranded DNA. Also, promoter or enhancer DNA sequences are added to increase plasmid replication and increase protein synthesis by the gene. A promoter is a short DNA sequence that amplifies the expression of protein by the adjacent target gene. An enhancer is a viral DNA sequence that dramatically increases the level of transcription of adjacent DNA. A gene providing antibiotic resistance (as a selection tool) may also be placed in a plasmid that is inserted into a bacterial host. This gene confers a particular antibiotic resistance on the clone and may be used in the clone selection later in the rDNA process. This vector is now rDNA molecule consisting of the gene, linker, promoter/enhancer, and vector DNA.

Cleavage by Restriction Endonucleases

phosphodiester (site of cleavage by restriction endonucleases)

restriction endonuclease

phosphomonoester

hydroxyl

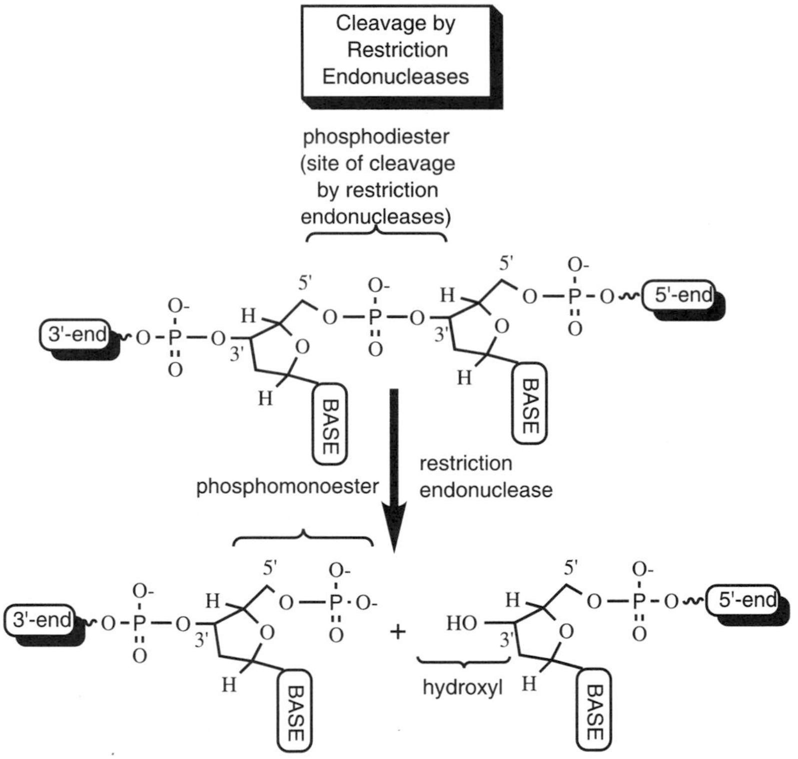

A cloning vector is a carrier molecule, the vehicle that is used to insert foreign DNA into a host cell. Typically, vectors are genetic elements that can be replicated in a host cell separately from that cell's chromosomes. Bacterial plasmids, circular DNA of only a few thousand base pairs outside of the nucleus that replicate freely within the cell, are ideal to carry the gene into the host organism. In the first such rDNA experiments in early 1973, Cohen and associates[55] used the small Escherichia coli plasmid pSC101 containing only a single Eco RI recognition site as the vector to insert foreign plasmid DNA into E. coli.

The vector containing the code for the target protein

is then inserted into the host. Host cells are typically bacteria (e.g., E. coli), eukaryotic yeast such as Saccharomyces cerevisiae (baker's yeast), or mammalian cell lines. Examples of mammalian cell lines include CHO (Chinese hamster ovary), VERO (African green monkey kidney), and BHK (baby hamster kidney). The choice of host system is influenced primarily by the type of protein to be expressed and the key differences among the various host cells.[15,57] Bacterial and yeast cells are more easily cultured in large fermentors. Overall protein yields are generally much lower in mammalian cells, but, in some cases, this may be the only system that produces some mammalian proteins. Another difference is that yeast and mammalian cells do not form toxins, whereas gram-negative bacteria produce endotoxins. Finally, an important distinction is that posttranslational modification reactions such as glycosylation do not occur in bacteria.

The host cells containing the vector are grown in small-scale culture to select only for the correct clone containing the desired gene and able to express the best yields of the protein. The selected cloned cells (or cell bank cells) are used as inoculum first for a small-scale cell culture/fermentation, which is then followed by larger fermentations in bioreactors. The medium is carefully controlled to enhance cell reproduction and protein synthesis. The host cells divide, and the vectors within the hosts multiply. The host produces its natural proteins along with the desired protein, which may be secreted into the growth medium. The protein of interest can then be isolated from the fermentation, purified and formulated to give a potential rDNA-produced pharmaceutical.

Protein Isolation and Purification

The isolation and purification of the final protein product from the complex mixture of cells, cellular debris, medium nutrients, and other host metabolites is a challenging task. The structure, purity, potency, and stability of the recombinant protein must be assayed and taken into consideration in this process. Often, sophisticated filtrations, phase separations, precipitation, and complex multiple-column chromatographic procedures are required to obtain the desired protein. Although isolation of the recombinant protein, produced in culture in relatively large amounts, is generally easier than isolating the native protein, ensuring the stability and retention of the bioactive three-dimensional structure (correct protein folding) of any biopharmaceutical is a more arduous task. In addition, recombinant proteins from bacterial hosts require removal of endotoxins, whereas viral particles may need to be removed from mammalian cell culture products. A discussion of these techniques is beyond the scope of this chapter, however. Useful reviews on the extraction and purification,[58–60] analysis,[61] and chromatography[62–64] of biotechnology products are available as a resource for further information.

SOME GENERAL PROPERTIES OF BIOTECHNOLOGY-PRODUCED MEDICINAL AGENTS

Although a majority of traditional medicinal agents are relatively small organic molecules, rDNA and hybridoma technologies have made it possible to produce large quantities of highly pure, therapeutically useful proteins. rDNA-derived proteins and MoAb are not dissimilar to the other protein pharmaceuticals or biopharmaceuticals that pharmacists have dispensed in the past. As proteins, polymers of amino acids joined by peptide bonds, their properties generally differ from small organic molecule pharmaceuticals. An overview of the general properties of biotechnology-produced medicinal agents is actually a review of the general physicochemical properties of pro-

The unique chemical structure of a protein results from the particular combination of 20 natural amino acids (19 chiral molecules of L-configuration) forming the peptide amide bond (***primary structure***).

1st amino acid of L-configuration + 2nd amino acid of L-configuration ⟹ 1st — 2nd (rotatable bonds; rigid, labile peptide bond; rotatable bond)

PROTEIN

The protein assumes a unique physical structure (***secondary and tertiary structure***) based on: 1) its ability to form intramolecular and intermolecular interactions (via hydrogen bonding amide functional groups and also some R side chains); and 2) conformations resulting from rotation about rotatable bonds.

teins. Therefore, to study the stability, handling, storage, route of administration, and metabolism of biotechnology-produced pharmaceuticals, it is valuable to understand the chemical nature of proteins. Chapter 27, reviews the physical biochemistry of protein drugs,[65–68] and related chapters in any biochemistry textbook all review this topic.

Stability of Protein Pharmaceuticals

Several detailed reviews on the stability of proteins and protein pharmaceuticals written for pharmaceutical scientists are available.[66,69–73] A brief overdose of these resources follows and provides additional information. The instability of proteins, including protein pharmaceuticals, can be separated into two distinct classes. Chemical instability results from bond formation or cleavage yielding a modification of the protein and a new chemical entity. Physical instability involves a change to the secondary or higher-order structure of the protein rather than a covalent bond-breaking modification.

Chemical Instability

A variety of reactions give rise to the chemical instability of proteins, including hydrolysis, oxidation, racemization, β-elimination, and disulfide exchange. Each of these changes may cause a loss of biologic activity. Proteolytic hydrolysis of peptide bonds results in fragmentation of the protein chain. It is well established that in dilute acids, aspartic acid (Asp) residues in proteins are hydrolyzed at a rate of at least 100 times faster than other peptide bonds owing to the mechanism of the reaction. An additional hydrolysis reaction is the deamidation of the neutral residue of asparagine (Asn) and glutamine (Gln) side-chain linkages forming the ionizable carboxylic acid residues aspartic acid (Asp) and glutamic acid (Glu). This conversion may be considered primary sequence isomerization.

Oxidative degradative reactions can occur to the side chains of sulfur-containing methionine (Met) and cysteine (Cys) residues and the aromatic amino acid residues histidine (His), tryptophan (Trp), and tyrosine (Tyr) in

hydrolysis

n=1, Asn residue
n=2, Gln residue

n=1, Asp residue
n=2, Glu residue

proteins during their isolation and storage. The weakly nucleophilic thioether group of Met ($R\text{-}S\text{-}CH_3$) can be oxidized at low pH by hydrogen peroxide as well as by oxygen in the air to the sulfoxide ($R\text{-}SO\text{-}CH_3$) and the sulfone ($R\text{-}SO_2\text{-}CH_3$). The thiol (sulfhydryl, R-SH) group of Cys can be successively oxidized to the corresponding sulfenic acid (R-SOH), disulfide (R-SS-R), sulfinic acid ($R\text{-}SO_2H$), and, finally, sulfonic acid ($R\text{-}SO_3H$). A number of factors, including pH, influence the rate of this oxidation. Oxidation of His, Trp, and Tyr residues is believed to occur with a variety of oxidizing agents resulting in the cleavage of the aromatic rings.

Base-catalyzed racemization reactions may occur in any of the amino acids except achiral glycine (Gly) to yield residues in proteins with mixtures of L- and D-configurations. The α-methine hydrogen is removed to form a carbanion intermediate. The degree of stabilization of this intermediate controls the rate of this reaction. Racemization generally alters the proteins' physicochemical properties and biologic activity. Also, racemization generates nonmetabolizable D-configuration forms of the amino acids. Generally, most amino acid residues are relatively stable to racemization, with a notable exception. Aspartate residues in proteins racemize at 10^5-fold faster rate than when free, in contrast to the twofold to fourfold increase for the other residues. The facilitated rate of racemization for Asp residues is believed to result from the formation of a stabilized cyclic imide.

Proteins containing cysteine (Cys), serine (Ser), threonine (Thr), phenylalanine (Phe), and lysine (Lys) are prone to β-elimination reactions under alkaline condi-

L-amino acid residue | carbanion intermediate | D-amino acid residue

-H+ / +H+

α-hydrogen

OH-

Asp if R = CH_2CO_2H

side-chain carbonyl

peptide carbonyl

L-Asp as cyclic imide | cyclic imide carbanion intermediate | D-Asp as cyclic imide

tions. The reaction proceeds through the same carbanion intermediate as racemization. It is influenced by a number of additional factors, including temperature and the presence of metal ions.

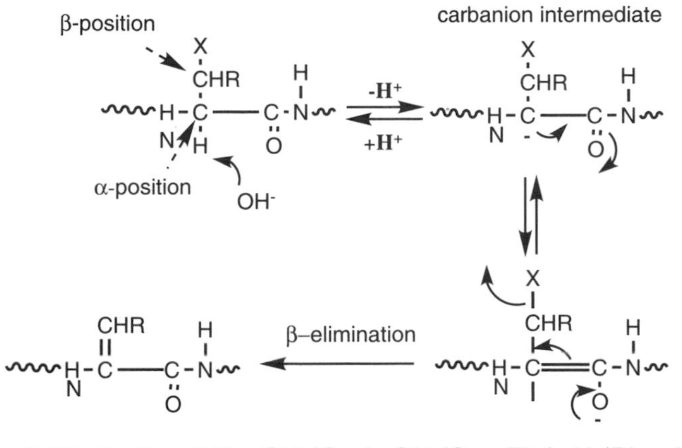

β-Elimination if X = SH (Cys), OH (Ser, Thr), H (Phe, Lys)

The interrelationships of disulfide bonds and free sulfhydryl groups in proteins are important factors influencing the chemical and biologic properties of protein pharmaceuticals. Disulfide exchange can result in incorrect pairings and major changes in the higher-order structure (secondary and above) of proteins. The exchange may occur in neutral, acidic, and alkaline media.

Physical Instability

Generally not encountered in most small organic molecules, physical instability is a consequence of the polymeric nature of proteins. Proteins adopt secondary, tertiary, and quaternary structures, which influence their three-dimensional shape and, therefore, their biologic activity. Any change to the higher-order structure of a protein may alter both. Physical instability includes denaturation, adsorption to surfaces, and noncovalent self-aggregation (soluble and precipitation). The most widely studied aspect of protein instability is denaturation. Noncovalent aggregation, however, is one of the primary mechanisms of protein degradation.[66]

A protein, in principle, can be folded into a virtually infinite number of conformations. The combination of spatial arrangements and noncovalent intramolecular interactions of nearby amino acid residues providing the lowest energy conformation is the most stable secondary structure. Longer-distance interactions cause the globular nature of proteins (tertiary structure), including their ability to fold so that hydrophilic amino acid side chains are directed toward the exterior surface of the protein exposed to an aqueous environment. In general, all molecules of any protein species adopt the same conformation, or native state. Denaturation occurs by disrupting the weaker noncovalent interactions that hold a protein together in its secondary and tertiary structures. Temperature, pH, and the addition of organic solvents and solutes may cause denaturation. The process can be reversible or irreversible. In general, denaturation affects the protein by decreasing aqueous solubility, altering three-dimensional molecular shape, increasing susceptibility to enzymatic hydrolysis, and causing the loss of the native protein's biologic activity.

Handling and Storage of Biotechnology-Produced Products

The preparation and administration of drugs of recombinant or hybridoma origin are in contrast to the small organic molecule pharmaceutical preparations and, yet, are not dissimilar to the other protein pharmaceuticals that pharmacists have been dispensing in the past. As proteins, they generally have a more limited shelf stability. The average shelf life for a biotechnology product is 12 to 18 months versus more than 36 months for a small molecule drug. Although each individual biotechnology drug may be different, several generalizations can be made.

Proper storage of the lyophilized and the reconstituted drug is essential. As most are expensive drugs, special care must be taken not to inactivate the therapeutic protein during storage and handling. The human proteins have limited chemical and physical stability, which is shortened on reconstitution. Expiration dating ranges from 2 hours to 30 days. The self-association of either native state or misfolded protein subunits may readily occur under certain conditions. This can lead to aggregation and precipitation and results in a loss of biologic activity. Self-association mechanisms depend on the conditions of formulation and may occur as a result of hydrophobic interactions.

Many of the biotechnology-produced drugs are stored refrigerated, *but not frozen,* between 2° and 8°C. Freezing or exposure to excessive heat decreases the physical stability of the protein. Anything that causes denaturation or self-aggregation, even though labile peptide bonds are not broken, may inactivate the protein. If the patient must travel any distance home after receiving the medication, the pharmacist should help package the biotechnology product according to the manufacturer's directions. This may mean supplying a reusable cooler for the patient's use. Because the protein drug should not be frozen, the cooler should contain an ice pack rather than dry ice.

Some rDNA-derived pharmaceuticals, in particular, the cytokines (such as the interferons, interleukin-2, and colony-stimulating factors), require human serum albumin in their formulation to prevent adhesion of the protein drug to the glass surface of the vial, which results in loss of protein. The vials should not be shaken in order to prevent foaming of the albumin, which causes protein loss or inactivation of the biotechnology-derived proteins. Care must be exercised in reconstituting protein pharmaceuticals. The diluent used for reconstitution of biotechnology drugs varies with the product and is specified by the manufacturer. Diluents can include normal saline, bacteriostatic water, and 5% dextrose. Several reviews of biotechnology drugs written for pharmacists contain additional information on the subjects of handling and storage.[33,36–41,74]

Biotechnology Drug Delivery

Protein-based pharmaceuticals, whether produced by biotechnology or isolated from traditional sources, present challenges to drug delivery owing to the unique demands imposed by their physicochemical and biologic properties. Although a detailed discussion of this topic is beyond the scope of this chapter, a brief overview follows. Useful reviews are available for further information.[70,73,75–77]

Delivery of large-molecular-weight, biotechnology-produced drugs into the body is difficult because of the poor absorption of these compounds, the acid lability of peptide bonds, and their rapid enzymatic degradation in the body. In addition, protein pharmaceuticals are susceptible to physical instability, complex feedback control mechanisms, and peculiar dose-response relationships.[78]

Given the limitations of today's technology, the strongly acidic environment of the stomach, peptidases in the gastrointestinal tract, and the barrier to absorption presented by gastrointestinal mucosal cells preclude successful oral administration of most protein drugs. Therefore, administration of all of the biotechnology-produced protein drugs is currently parenteral (by intravenous, subcutaneous, or intramuscular injections) to provide a better therapeutic profile. Manufacturers supply most of these drugs as sterile solutions without a preservative. In such cases, it is recommended that only one dose be prepared from each vial to prevent bacterial contamination. Novel solutions to overcome delivery problems associated with biotechnology protein products are being explored. Oral drug delivery approaches in development for various biotechnology-derived drugs include conjugated systems (such as with polyethylene glycol [PEG]), liposomes, microspheres, erythrocytes as carriers, and viruses as drug carriers.[73] Specialized delivery methods being examined are transdermal systems, pulmonary delivery, intranasal sprays, buccal administration, ocular delivery systems, rectal administration, iontophoresis, phonophoresis, metered pumps, protein prodrugs, lymphatic uptake, coadministration of peptidase inhibitors, and penetration enhancers.

Metabolism

The plasma half-life of most administered proteins is relatively short because they are susceptible to a wide variety of metabolic reactions. Rapid hydrolytic degradation of peptide bonds by both nonspecific enzymes and highly structurally selective aminopeptidases, carboxypeptidases, deamidases, and proteinases occurs at the site of administration, while crossing the vascular endothelia, at the site of action, in the liver, in the blood, in the kidney, and, in fact, in most tissues and fluids of the body. Metabolic oxidation reactions may occur to the side chains of sulfur-containing residues similar to that observed for in vitro chemical instability. Methionine can be oxidized to the sulfoxide, whereas metabolic oxidation of cysteine residues forms a disulfide. Metabolic reductive cleavage of disulfide bridges in proteins may occur yielding free sulfhydryl groups.

Adverse Effects

An important consideration in the pharmaceutical care of a patient being administered a biotechnology-produced medicinal agent is the potential for adverse reactions. Many of the protein agents are biotechnology-derived versions of endogenous human proteins normally present, on stimulus, in minute quantities near their specific site of action. Therefore, the same protein administered in much larger quantities may cause adverse effects not commonly observed at normal physiologic concentrations.[28,79] Careful monitoring of patients administered biotechnology-produced drugs is critical for the health care team.

Immunogenicity

The immune system may respond to an antigen such as a protein pharmaceutical triggering the production of antibodies. Biotechnology-derived proteins may possess a different set of antigenic determinants (regions of a protein recognized by an antibody) owing to structural differences between the recombinant protein and the natural human protein.[66,70] Factors that can contribute to this immunogenicity include incorrect or lack of glycosylation, amino acid modifications, and amino acid additions and deletions. A number of recombinant proteins produced with bacterial vectors contain an N-terminal methionine in addition to the natural human amino acid sequence. Bacterial vector–derived recombinant protein preparations may also contain small amounts of immunoreactive bacterial polypeptides as possible contaminants. Also, immunogenicity may result from proteins that are misfolded, denatured, or aggregated.

RECOMBINANT DNA–PRODUCED MEDICINAL AGENTS

rDNA technology provides a powerful tool for new pharmaceutical product development and production. Table 28–3 lists the rDNA-produced drugs and vaccines approved by the FDA through 1993. The 20 products include hormones, enzymes, cytokines, hematopoietic growth factors, blood clotting factors, and vaccines.

Hormones

Insulin

Biotechnology has provided hormone replacement therapy with the introduction of human insulin,[80–82] the

Table 28–3. Recombinant-DNA–Produced Drugs and Vaccines Approved by the Food and Drug Administration

Year	Generic name	Trade Name (Company)	Indication
1982	Human insulin	Humulin (Eli Lilly)	Diabetes*
1985	Somatrem (rhGH)	Protropin (Genentech)	hGH deficiency in children*
1986	Hepatitis B vaccine	Recombivax HB (Merck)	Hepatitis B prevention*
	Interferon alfa-2a (rINF-α-2a)	Roferon-A (Hoffmann-LaRoche	Hairy cell leukemia* AIDS-related Kaposi's sarcoma (1988)[+]
	Interferon alfa-2b (rINF-α-2b)	Intron-A (Schering-Plough)	Hairy cell leukemia* Genital warts (1988)[+] AIDS-related Kaposi's sarcoma (1988)[+] Hepatitis C (1991)[+] Hepatitis B (1992)[+] Use in combination with podophyllin in treating genital warts (1993)[+]
1987	Alteplase (rt-PA, tissue-type plasminogen activator)	Activase (Genentech)	Acute myocardial infarction* Acute pulmonary embolism (1990)[+]
	Somatropin (rhGH)	Humatrope (Eli Lilly)	hGH deficiency in children*
1989	Epoetin alfa (rEPO, erythropoietin)	Epogen (Amgen)	Treatment of anemia associated with chronic renal failure including patients on dialysis and not on dialysis, anemia in Retrovir-treated HIV-infected patients* Treatment of anemia caused by chemotherapy in patients with nonmyeloid malignancies (1993)[+]
	Hepatitis B vaccine	Engerix-B (SmithKline-Beecham)	Hepatitis B prevention*
1990	Epoetin alfa (rEPO, erythropoietin)	Procrit (Ortho Biotech) (marketed under Amgen's epoetin alfa PLA)	Treatment of anemia associated with chronic renal failure including patients on dialysis and not on dialysis, anemia in Retrovir-treated HIV-infected patients* Treatment of anemia caused by chemotherapy in patients with nonmyeloid malignancies (1993)[+]
	Interferon gamma-1b	Actimmune (Genentech)	Management of chronic granulomatous disease*
1991	Filgrastim (rG-CSF, granulocyte colony-stimulating factor, r-metHuGCSF,)	Neupogen (Amgen)	Chemotherapy-induced neutropenia*
	Human insulin	Novolin (Novo Nordlsk)	Diabetes*
	Sargramostim, rGM-CSF granulocyte-macrophage colony-stimulating factor	Leukine (Immunex)	Myeloid reconstitution after autologous bone marrow transplantation*
1992	Aldesleukin (IL-2, interleukin-2)	Proleukin (Chiron)	Metastatic renal cell carcinoma in adults*
	Antihemophiliac factor (rAHF, Factor VIII)	Recombinate (Baxter)	Treatment of hemophilia A*
1993	Antihemophiliac factor (rAHF, factor VIII)	KoGENate (Miles)	Treatment of hemophilia A
	Dornase alfa (rhDNase, deoxyribonuclease I)	Pulmozyme (Genentech)	Treatment of symptoms of cystic fibrosis*
	Interferon beta-1b	Betaseron (Berlex)	Relapsing, remitting multiple sclerosis*
	Somatropin (rhGH)	Nutropin (Genentech)	Treatment of growth failure associated with chronic renal insufficiency up to the time of renal transplantation*

* Denotes first approved indication(s).
[+] Denotes additional indications (year of FDA approval).
hGH = human growth hormone; AIDS = acquired immunodeficiency syndrome; HIV = human immunodeficiency virus.
Some data from *Biotechnology Medicines in Development, 1993 Annual Survey,* Washington, D.C., Pharmaceutical Manufacturers Association, 1993.

first FDA-approved rDNA drug in 1982, for the treatment of insulin-dependent diabetes. The human insulin molecule has the structural characteristics of a large protein yet is only the size of a polypeptide totaling 51 amino acid residues. Two disulfide bonds (cysteine [Cys] A7 to Cys B7 and Cys A20 to Cys B19) link two polypeptide chains, the A-chain consisting of 21 amino acids and the 30-residue B-chain. An additional disulfide loop is found in the A-chain between Cys A6 and Cys A11. Before the availability of rDNA-produced human insulin, porcine and bovine insulin were the most commonly used pharmaceutical preparations. Both porcine and bovine insulin differ in primary structure from human with alanine (Ala) replacing Thr at the C-terminal of the B-chain (B30). Bovine insulin also differs from human insulin by Ala replacing Thr at A8 and valine (Val) substituting for isoleucine (Ile) at A10. These subtle differences can result in immunologic responses to the nonhuman insulins requiring a modification of therapeutic regimen. The biotechnology solution has several advantages over insulin derived from animal sources: (a) It should have potentially fewer serious immune reactions; (2) it is pyrogen free; (3) it is not contaminated with other peptide hormones, such as glucagon, somatostatin, and proinsulin found in isolated products; and (4) it can be produced in larger amounts. The first successful attempts to tailor a protein hormone for therapy by rDNA techniques has yielded interesting insulin analogs.[83,84]

Human insulin rDNA origin is available as Humulin (Eli Lilly) and Novolin (Novo Nordisk). The two products are produced using genetically modified strains of two different microorganisms. Humulin is prepared using recombinant E. coli bacteria. The pharmaceutical preparation is reported to contain less than 4 ppm of immunoreactive bacterial polypeptides that act as possible contaminants. S. cerevisiae, a yeast, serves as the recombinant organism for the production of Novolin. Before 1986, Humulin was produced by chemically joining together the separately rDNA-derived A-chain and B-chain. Today the product is prepared by enzymatically cleaving the connecting peptide in recombinant proinsulin. Humulin is available in R (regular) and BR (buffered regular) injection, N isophane insulin suspension (NPH), 70/30 and 50/50 isophane insulin and regular insulin suspension, L insulin zinc suspension (lente), and U ultralente insulin zinc suspension, extended formulations. Novolin is available in R and R PenFill injection, N and N PenFill isophane insulin suspension (NPH), 70/30 and 70/30 PenFill isophane insulin suspension and injection, and L insulin zinc suspension forms.

Studies in anmals, healthy adults, and patients with type I diabetes mellitus have shown human insulin to have identical pharmacologic effects as purified porcine insulin. A comparable pharmacokinetic profile has also been shown. Human insulin, however, administered intramuscularly or intravenously may have a slightly faster onset and slightly shorter duration of action when compared with purified porcine insulin in patients with diabetes. The usual precautions concerning toxic potentials observed with insulin of animal origin should be followed with rDNA human insulin. As would be expected the recombinant product has been shown to be less immunogenic than animal insulins.

Growth Hormones

The introduction of rDNA human growth hormone (hGH),[85–87] previously isolated from cadaver pituitaries, greatly improved the long-term treatment of children who have growth failure caused by a lack of adequate endogenous hGH. Currently, three hGH products have been approved by the FDA (see Table 28–3).

Pituitary hGH is a globular protein of 191 amino acids in a single polypeptide chain with a molecular weight of 22,000 daltons. Pituitary hGH is a roughly spherical protein with a hydrophobic interior. Degradation pathways of hGH include typical proteolysis reactions; deamidation of Asn and Gln residues; oxidation of Met, Trp, His, and Tyr residues; disulfide exchange; and aggregation. All three recombinant pharmaceuticals are produced in special laboratory strains of E. coli.

Somatrem (Protropin, Genentech), introduced in 1985, contains the identical 191–amino acid sequence found in the pituitary-derived hGH plus the addition of a methionine amino acid at the N-terminus of the peptide chain (resulting in a 192–amino acid protein). Both rDNA somatotropin products, Humatrope (Eli Lilly, approved in 1987) and Nutropin (Genentech, approved in 1993), contain the 191–amino acid sequence identical to that of hGH of pituitary origin. Protropin and Humatrope are each available in vials of 5 mg of lyophilized powder for subcutaneous or intramuscular injection after reconstitution with Bateriostatic Water for Injection, USP (benzyl alcohol preserved). Individualized doses are administered three times per week. Nutropin differs from the other two rhGHs only in that it is also supplied as a 10-mg vial and the approved route of administration is subcutaneous. All three products before and after reconstitution must be stored at 2° to 8°C with freezing avoided. It is recommended that reconstituted vials of somatrem be used in 7 days, whereas somatropin is stable up to 14 days.

Both Protropin and Humatrope are indicated for the long-term treatment of children who have growth failure as a result of a lack of adequate endogenous growth hormone secretion. Protropin is currently indicated only for the treatment of children who have growth failure associated with chronic renal insufficiency up to the time of renal transplantation. Actions of rDNA-derived hGHs include an increase in the linear growth of the patient, increased skeletal growth, increased protein synthesis, reduction in body fat stores, and increased organ growth. Clearance and bioavailability do not appear to be clinically or statistically significantly different for somatrem and somatropin. Although the direct clinical comparison of the efficacy of the two forms of hGH have yet to be performed, separate controlled studies of growth hormone–deficient patients were similar. Recombinant hGHs are safe and effective therapies with relatively few side effects.

Enzymes

Tissue Plasminogen Activator

The fibrinolytic system is activated in response to the presence of an intracellular thrombus or clot. The process of clot dissolution is initiated by the conversion of plasminogen to plasmin. Plasminogen activation is catalyzed by two endogenous highly specific serine proteases, urokinase-type plasminogen activator (u-PA) and tissue-type plasminogen activator (t-PA).[88–99]

The mature human t-PA is a glycoprotein consisting of a single chain of 527 amino acids. Its molecular weight is about 70,000 daltons. Human t-PA contains 35 cysteines assigned to 17 disulfide bonds. A serine protease domain of about 260 residues is located at the carboxy-terminal end of this protein. A fibronectin "finger" domain, two kringle domains, and an epidermal growth factor domain are also present. The t-PA protease domain is approximately 35 to 40% homologous with typical serine protease, such as bovine trypsin and chymotrypsin.

Mammalian cells produce two t-PA variants of N-linked glycosylation, type 1 (at asparagines 117, 184, and 448) and type 2 (only as asparagines 117 and 448). The rate of fibrin-dependent plasminogen activation is two to three times faster for type 2 compared with type 1. The cDNA obtained from a human melanoma cell line was expressed in CHO cells to achieve glycosylation and a protein identical to the natural protein. Protein engineering studies have produced variant t-PA molecules with modified pharmacokinetics, affinity for fibrin, catalytic activity, and side effects.

The sole rDNA thrombolytic agent approved in the United States is alteplase (Activase, Genentech). It is supplied as 20-mg and 50-mg vials of a white to off-white lyophilized powder for injection. The powder should be stored at a room temperature of 15° to 30°C or refrigerated at 2° to 8°C. The expiration date is 2 years after manufacture. Reconstituted solutions (diluent supplied) contain no preservatives and should be stored at 2° to 30°C for no more than 8 hours.

Alteplase is indicated for the treatment of acute myocardial infarction (administered as a bolus) and acute massive pulmonary embolism (administered by intravenous infusion). The mechanism of action of t-PA is unlike that of streptokinase and urokinase. It is the first fibrin-selective thrombolytic agent preferentially activating fibrinogen bound to fibrin. Thus, the thrombolytic effect is localized to a blood clot and avoids systemic activation of fibrinogen, preventing bleeding elsewhere in the body. Plasma t-PA concentrations are proportional to the rate of infusion. Alteplase is rapidly cleared from circulating plasma, with 50% cleared within 5 minutes after termination of infusion. The mechanisms for clearance of t-PA from the blood are poorly understood. Detectable levels of antibody against alteplase have been found in patients receiving the drug, although 12 days to 10 months later antibody determinations were negative.

DNase

According to the Cystic Fibrosis Foundation, cystic fibrosis (CF) is the most common fatal genetic disorder, afflicting approximately 30,000 patients, most of whom die before the age of 30. They develop thick mucous secretions and suffer from severe, frequent lung infections. Studies during the 1950s and 1960s determined that CF-related secretions in the lungs contained large amounts of DNA. Mucus-thickening DNA release resulted from an inflammatory response and ensuing white blood cell death. The enzyme deoxyribonuclease I (DNase I) specifically cleaves extracellular DNA such as that found in the mucous secretion of CF patients and has no effect on the DNA of intact cells. The FDA has approved a recombinant human DNase.[87,91,92]

DNase I is a glycoprotein containing 260 amino acids with an approximate molecular weight of 37,000 daltons. The recombinant protein is expressed by genetically engineered CHO cells encoding for the native enzyme, although DNase I was not purified or sequenced from human sources at the time. A degenerate sequence, based on the sequence of bovine DNase (263 amino acids), was used to synthesize probes and screen a human pancreatic DNA library. The primary amino acid sequence of rhDNase is identical to native human DNase I.

The only FDA-approved DNase product, dornase alfa (Pulmozyme inhalation solution, Genentech), has been developed as a therapeutic agent for the management of CF. The product is supplied in single-use ampules delivering 2.5 ml of a sterile, clear, colorless solution containing 1.0 mg/ml dornase alfa with no preservative. Administration is by nebulizer aerosol delivery systems.

Dornase alfa is indicated for daily administration in conjunction with standard CF therapies to reduce the frequency of respiratory infections requiring parental antibiotics to improve pulmonary function. The breakdown of DNA in infected sputum results in improved air flow in the lung and reduced risk of bacterial infection. The medicinal agent has been shown in clinical trials to have a positive effect on pulmonary function, which returned to baseline on stopping therapy. Although effective for the management of the respiratory symptoms of CF, dornase alfa is not a replacement for antibiotics, bronchodilators, and daily physical therapy. Two short-term studies have reported no adverse reactions. The agent is also in early clinical trials for the treatment of chronic bronchitis, a disease afflicting 400,000 patients in the United States alone.

Cytokines

Cytokine is a generic term for the soluble protein molecules released by participating and interacting cells in the immune system to communicate in a dynamic cellular network during an immune/inflammatory response to an antigen. Lymphokine and monokine are the terms used for a cytokine derived from lymphocytes and macrophages. Cytokines, usually released and targeted to produce a localized effect, regulate the growth, differentiation, and activation of the hematopoietic cells responsible for the maintenance of the immune response. A wide array of glycoproteins, including interferons, interleukins, hematopoietic growth factors, and tumor necrosis factor, are cytokines. Among the many reviews discussing the cytokines, several written for pharmacists or pharmaceutical scientists are excellent sources for a readable overview of the area.[93–98]

Interferons

The interferons[98] are a family of cytokines discovered in the late 1950s, with broad-spectrum antiviral and potential anticancer activity making them biologic response modifiers (BRM). Biotherapy (the therapeutic use of any substance of biologic origin) of cancer is different than standard chemotherapy. That is, biotherapy agents belong to a group of compounds that enhanced normal immune interactions (therefore, they are also immunomodulators) with cells in a specific or nonspecific fashion. Chemotherapeutics interact directly with the cancer cells themselves. Three types of naturally occurring interferons have been found present in small quantities: leukocyte interferon (interferon alfa, INF-α), produced by lymphocytes and macrophages; fibroblast interferon (interferon beta, INF-β) produced by fibroblasts, epithelial cells, and macrophages; and immune interferon (interferon gamma, INF-γ) synthesized by T4, T8, and natural killer lymphocytes. Molecules in INF-α and INF-β, also known as type I interferons, exhibit approximately 30% primary sequence homology but no structural similarity to INF-γ, a type II interferon. All three are glycoproteins. Previously only available in low yields by chemical synthesis or isolation, four rDNA interferon pharmaceuticals now have been marketed in the United States: two INF-αs, an INF-β, and an INF-γ (see Table 28–3).

Interferon Alfa. At least 23 different subtypes of INF-α,[99,102] with slight structural variations, are known. Human INF-α proteins generally are composed of either 165 or 166 amino acids. INF-α-2a and INF-α-2b, the two primary subtypes, both contain 165 amino acids differing only at position 23, with α-2a containing a lysine group and α-2b an arginine group at this position. They have molecular weights of approximately 19,000 daltons. Although cultures of genetically modified E. coli produce two recombinant FDA-approved interferon alfas, INF-α-2a and INF-α-2b, their method of purification differs. INFα-2a's purification includes affinity chromatography using a murine MoAb, whereas INFα-2b's does not.

INFα-2a (Roferon-A, Hoffmann-La Roche) is commercially available as a sterile solution or a sterile white-to-beige lyophilized powder to reconstitute for subcutaneous or intramuscular injection. INFα-2b (Intron A, Schering-Plough) is available only as a sterile white to-cream-colored lyophilized powder to reconstitute for subcutaneous or intramuscular injection. Storage of the lyophilized powders or the reconstituted solutions should be at 2° to 8°C. INFα-2a and INFα-2b, as lyophilized powders, have expiration dates of 36 and 24 months after manufacture. Reconstituted solutions, if stored properly, are stable for up to 30 days. During the manufacture of the rDNA interferon-alfas, human albumin is added to minimize adsorption to glass and plastic by these cytokines. Solutions, therefore, should not be shaken.

Interferon-alfa possesses complex antiviral, antineoplastic, and immunomodulating activities. Both products are approved for the treatment of hairy cell leukemia and AIDS-related Kaposi's sarcoma. INFα-2b is also indicated for the treatment of Condylomata acuminata (genital warts) alone and in combination with podophyllin, the treatment of hepatitis B, and the treatment of hepatitis C. Although the precise mechanism of action of interferon alfas is not known, they are believed to interact with cell surface receptors to produce their biologic effects. The actions appear to result from a complex cascade of biologic modulation and pharmacologic effects that include the modulation of host immune responses; cellular antiproliferative effects; cell differentiation, transcription, and translation processes; and reduction of oncogene expression.

Interferon alfa is filtered through the glomeruli in the kidney and undergoes rapid proteolytic degradation during tubular reabsorption. Toxicity is generally dose and time dependent with fever, fatigue, myalgia, chills, and anorexia, all flu-like symptoms, generally occurring within 2 to 8 hours after administration of high doses.

Interferon Beta. Human interferon beta[103,104] was first cloned and expressed in 1980; however, its instability made it unsuitable for clinical use. The more stable recombinant interleukin beta-1b, a 165–amino acid analog of human interferon beta, differs from the native protein with a serine residue substituted for cysteine at position 17. The highly purified product of biotechnology has a molecular weight of 18,500 daltons.

Approved in December 1993, INFβ-1b (Betaseron, Berlex) is indicated for the treatment of patients with exacerbating-remitting multiple sclerosis (MS). The National Multiple Sclerosis Society says that 250,000 to 300,000 Americans have this disease, with greater than 60% of the patients falling into the exacerbating-remitting category. A vial of recombinant INFβ-1b contains 0.3 mg of protein with dexrose and human albumin as stabilizers. The sterile white lyophilized powder is reconstituted without preservative for single-use subcutaneous injection every other day. Before and after reconstitution, the preparation should be refrigerated at 2° to 8°C. The refrigerator solution should be administered within 3 hours of reconstitution.

Results from a large double-blind, placebo-controlled study have found that 8 million IU subcutaneously (the low-dose treatment group) of interferon beta every other day brought the most promising results, with a one-third reduction in exacerbations and a decrease by half in the frequency of severe attacks. Cranial magnetic resonance imaging scans showed that treated patients did not develop new lesions in the central nervous system and had less active lesions than placebo patients. The exact mechanism of action of IFNβ-1b is not known. Its immunomodulating effects, however, may benefit MS patients by decreasing the levels of endogenous interferon gamma. Levels of INF-gamma are believed to rise before and during acute attacks in MS patients.

As observed with other interferons, flu-like symptoms are common on administration. During clinical trials, a suicide and four attempted suicides were reported. Depression and suicide have previously also been reported with patients receiving interferon alfa.

Interferon Gamma. Human interferon gamma[98,101,105,106] is a single-chain glycoprotein with a molecular weight of approximately 15,500 daltons. The cytokine mainly exists as a noncovalent dimer of differentially glycosylated chains in solution in vivo. Glycosylation does not appear to be necessary for biologic activity. The 140–amino acid INFγ-1b is produced by fermentation of a recombinant E. coli. Approved in 1990, INFγ-1b (Actimmune, Genentech) is supplied as a sterile solution containing 100 μg of the drug for subcutaneous injection. Vials must be placed in a refrigerator at 2° to 8°C and neither frozen nor shaken.

INFγ-1b possesses biologic activity identical to the natural human gamma interferon derived from lymphoid cells. Although all the interferons share certain biologic effects, gamma interferon differs distinctly from alfa and beta interferons by its potent capacity to activate phagocytes involved in host defense. These activating effects include the ability to enhance the production of toxic oxygen metabolites within phagocytes, resulting in a more efficient killing of various microorganisms. This activity is the basis for the use of INFγ-1b in the management of chronic granulomatous disease. Chronic granulomatous disease is a group of rare X-linked or autosomal genetic disorders of the phagocytic oxygen metabolite generating system leaving patients susceptible to severe infections. The drug extends the time patients spend without being hospitalized for infectious episodes. Investigational applications of gamma interferon include the treatment of renal cell carcinoma, small-cell lung cancer, infectious disease, trauma, atopic dermatitis, asthma, allergies, rheumatoid arthritis, and venereal warts. Adverse reactions are similar to those reported for interferon alfa.

Interleukin-2

Interleukins are cytokines involved in immune cell communication. Synthesized by monocytes, macrophages, and lymphocytes, interleukins serve as soluble messengers between leukocytes. Currently, at least 12 interleukins have been observed. One of the most studied cytokines is interleukin-2 (IL-2),[107–109] originally called T cell growth factor owing to its ability to stimulate growth of T lymphocytes.

Human IL-2 is a 133–amino acid, 15,400-dalton protein O-glycosylated at a threonine in position 3. An intramolecular disulfide bond between cysteine 58 and cysteine 105 is essential for biologic activity. A recombinant version of IL-2 is marketed as aldesleukin (Prokeukin, Chiron). Aldesleukin differs from the native protein by the absence of glycosylation, a lack of the N-terminal alanine residue at position 1 (132 amino acids), and the replacement of cysteine with serine at position 125 of the primary sequence. Sequence changes were accomplished by site-directed mutagenesis to the IL-2 gene before cloning and expression. Aldesleukin exists as noncovalent microaggregates with an average size of 27 recombinant IL-2 molecules. The recombinant drug does possess the biologic activity of the native protein.

Aldesleukin is supplied as a lyophilized formulation of protein admixed with mannitol to provide bulk in the vial. Sodium dodecyl sulfate is present to ensure sufficient water solubility on reconstitution. The drug is administered as an intravenous infusion. Handling and storage considerations for aldesleukin are consistent with the other cytokines. Recombinant IL-2 should be administered within 48 hours after reconstitution.

Aldesleukin is used in cancer biotherapy (therapeutic use of any substance of biologic origin) as a biologic response modifier for the treatment of metastatic renal cell carcinoma. Although the exact mechanism of action is not established, IL-2 is known to bind to an IL-2 receptor that has been well studied. In vitro, IL-2 induces killer cell activity, enhances lymphocyte mitogenesis and cytotoxicity, and induces interferon gamma production. The extent of its antitumor effect is directly proportional to the amount of IL-2 administered. Side effects are the major dose-limiting factor because aldesleukin is an extremely toxic drug. The manufacturer's labeling should be consulted for full details. Careful patient selection and thorough patient monitoring are essential. The incidence of nonneutralizing anti-IL-2 antibodies in patients treated on an every-8-hour regimen is quite high (76% in one clinical study).

Hematopoietic Growth Factors

Hematopoiesis is the complex series of events involved in the formation, proliferation, differentiation, and activation of red blood cells, white blood cells, and platelets. Hematopoietic growth factors are cytokines that regulate these events.[110–113] Investigators have identified and cloned at least 20 factors, including IL-3 (or multi-CSF), IL-4, IL-5, IL-6, IL-7, erythropoietin (EPO), granulocyte-macrophage colony-stimulating factor (GM-CSF), granulocyte colony-stimulating factor (G-CSF), macrophage colony-stimulating factor (M-CSF), and stem cell factor (SCF). Figure 28–5 summarizes the elaborate hematopoietic cascade. All blood cells originate within the bone marrow from a single class of pluripotent stem cells. In response to various external and internal stimuli, regulated by hematopoietic growth factors, stem cells give rise to additional new stem cells (self-renewal) and differentiate into mature specialized blood cells.

Erythropoietin EPO,[101,102,114–116] a 30,000 to 34,000-molecular weight glycoprotein produced by the kidney, stimulates the division and differentiation of erythroid progenitors in the bone marrow, increasing the production of red blood cells. Epoetin alfa (rHuEPO-α), a recombinant EPO prepared from cultures of genetically engineered mammalian CHO cells, consists of the identical 165–amino acid sequence of endogenous EPO. The molecular weight is approximately 30,400 daltons. The protein contains two disulfide bonds (linking cysteine 7 with 161 and 29 with 33) and four sites of glycosylation (one O-site and three N-sites); the disulfide bonds and glycosylation are necessary for the hormone's biologic activity. Deglycosylated natural EPO or bacterial-derived EPO (without glycosylation) have greatly decreased in vivo activity, although in vitro activity is largely conserved.

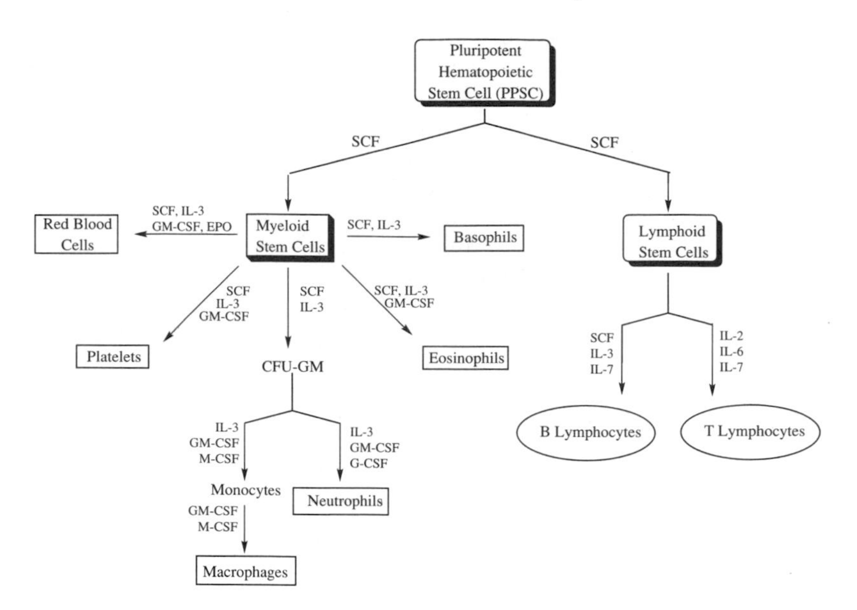

Fig. 28–5. Schematic overview of hematopoiesis. SCF = stem cell factor; IL-3, interleukin-3; GM-CSF = granulocyte-macrophage colony-stimulating factor; EPO = erythropoietin; CFU-GM = colony-forming unit—granulocyte-macrophage; G-CSF = granulocyte colony-stimulating factor; M-CSF = macrophage colony-stimulating factor. (Adapted from S. P. Smith and G. C. Yee, *Pharmacotherapy, 12,* 114[1992].)

The sugars may play a role in thermal stability or the prevention of aggregation in vivo.

Epoetin alfa is marketed as both Epogen (Amgen) and Procrit (Ortho Biotech). These products are formulated as a sterile, colorless, preservative-free liquid for intravenous or subcutaneous administration. Each 1-ml vial contains 2000, 3000, 4000, or 10,000 units of epoetin alfa formulated in an isotonic NaCl/Na citrate buffer at neutral pH. The vial should not be shaken, or the glycoprotein may become denatured rendering it inactive.

Epoetin alfa is indicated for the treatment of anemia associated with chronic renal failure, including patients on or not on dialysis; the treatment of anemia related to therapy with azidothymidine (AZT) in human immunodeficiency virus (HIV)–infected patients; and treatment of anemia caused by chemotherapy in patients with nonmyeloid malignancies. Epoetin alfa represents a major scientific advance in the treatment of patients with chronic renal failure, serving as a replacement therapy for inadequate production of endogenous EPO by failing kidneys. Epoetin alfa may decrease the need for infusions in dialysis patients. By several mechanisms related to elevating the erythroid progenitor cell pool, epoetin alfa increases the production of red blood cells.

The manufacturer's full prescribing information should be consulted for dosing regimens because the dose is titrated individually to maintain the patient's target hematocrit. The circulating half-life is 4 to 13 hours in patients with chronic renal failure. Peak serum levels are achieved within 5 to 24 hours following subcutaneous administration.

Granulocyte Colony-Stimulating Factor. Colony-stimulating factors are glycoprotein cytokines that promote progenitor proliferation, differentiation, and some functional activation. Recombinant DNA-derived G-CSF[110,111,117,118] or filgrastim (Neupogen, Amgen), was approved in 1991 to decrease the incidence of infection in patients with nonmyeloid malignancies receiving myelosuppressive anticancer drugs. Filgrastim is a 175–amino acid, single-chain protein with a molecular weight of 18,800 daltons. Filgrastim, produced by a recombinant bacteria, differs from the endogenous human protein by the addition of a methionine at the N-terminus (recombinant methionyl G-CSF is sometimes called r-metHuG-CSF) and the lack of glycosylation. Glycosylation, however, does not appear to be necessary for the biologic activity.

Filgrastim injectable solution is supplied in 1-ml-vials containing 300 μg/ml or 1.6-ml-vials containing 480 μg (at 300 μg/ml). The product should be refrigerated at 2° to 8°C and is packaged with a patented indicator that turns red at temperatures below -4°C. The vials should never be shaken before the dose is withdrawn (contains human albumin). Administration is by intravenous infusion or by subcutaneous injection or infusion. The drug is rapidly absorbed with perk serum concentrations in 4 to 5 hours. Elimination half-life is approximately 3.5 hours.

Filgrastim is lineage selective for the neutrophil lineage type of white blood cells, whereas GM-CSF is multilineage stimulating progenitors of neutrophils, monocytes, basophils, and eosinophils. The drug reduces the period of neutropenia, the number of infections, and the number of days the patient is on antibiotics. Filgrastim is generally well tolerated, with medullary bone pain being the most frequently encountered side effect.

Granulocyte-Macrophage Colony-Stimulating Factor. GM-CSF[119,120] has been produced by rDNA technology in the yeast S. cerevisiae. Sargramostim (Leukine, Immunex) is a glycoprotein of 127 amino acids differing from endogenous human GM-CSF by substituting leucine at position 23. Also, glycosylation may be different from the native protein.

Sargramostim is supplied in 250μg and 500-μg vials of lyophilized powder (with mannitol) for intravenous infusion. Reconstituted with 1 ml of Sterile Water for Injection, USP (without preservative), the drug should be administered within 6 hours. Vials are intended for single use only. The powder, reconstituted solution, and diluted solution requires refrigeration at 2° to 8°C without being frozen or shaken.

Sargramostim is indicated for the acceleration of myeloid recovery in patients after autologous bone marrow transplantation. Cellular division, maturation, and activation are induced through GM-CSF's binding to specific receptors expressed on the surface of target cells. On 2-hour intravenous infusion, the alpha half-life is 12 to 17 minutes followed by a slower decrease (beta half-life) of 2 hours. The manufacturer's label should be consulted for precautions. Additional indications for GM-CSF under study are as an adjuvant to chemotherapy and an adjuvant to AIDS therapy.

Clotting Factors

Antihemophiliac factor (AHF), or factor VIII, is required for the transformation of prothrombin (factor II) to thrombin by the intrinsic clotting pathway. Hemophilia A, a life-long bleeding disorder, results from a deficiency of factor VIII. Conventional biotherapy for the treatment of hemophilia A includes protein concentrates from human plasma collected by transfusion services or commercial organizations.[121] Therefore, the concentrates may possibly contain other native human proteins and microorganisms, such as viruses (e.g., HIV, hepatitis), derived from infected blood. Two versions of recombinant factor VIII, highly purified, microorganism-free proteins, are now available. Antihemophiliac factor (Recombinate, Baxter) was approved in 1992, and antihemophiliac factor (KoGENate, Miles) was approved in 1993. Both products are produced by the insertion of cDNA encoding for the entire factor VIII protein into mammalian cells. The mature, heavily glycosylated protein is composed of 2332 amino acids (1 to 2 million daltons) and contains sulfate groups. Stability of the large protein is a concern. The products have proved to be safe and effective in reducing bleeding time in patients. There is the possibility of induction of inhibitors in previously untreated patients.

Vaccines

Vaccines[102,122–124] enable the body to resist infection by diseases. In response to an injection of vaccine, the immune system makes antibodies, which recognize surface antigens found in the vaccine. If the subject is later exposed to a virulent form of the virus, the immune system is primed and ready to eliminate it. Many viral vaccines are produced from the antigens isolated from pooled human plasma of virus carriers. Although generally safe, the minimal risk of vaccine-produced infections can be eliminated by administration of highly purified vaccine antigens of recombinant origin. Two hepatitis B vaccines, Recombivax HB (Merck Sharp & Dohme) and Engerix-B (SmithKline-Beecham) were marketed in 1986 and 1989. Both are derived from a hepatitis B surface antigen and are produced in yeast cells. The primary difference between the two appears to be in exact dosing regimens. The immunization regimen consists of three injections: initial, at 1 month, and at 6 months. The immune response and clinical reactions for both intramuscular and subcutaneous administration are comparable. Vials containing the vaccine in solution should be stored at 2° to 8°C; freezing destroys potency.

RECOMBINANT DNA–PRODUCED PHARMACEUTICALS IN DEVELOPMENT

According to the PMA 1993 survey of *Biotechnology Medicines in Development,* a total of 63 companies are developing 143 biotechnology pharmaceuticals. Table 28–4 lists some interesting examples of rDNA drugs in development.

Superoxide dismutase (SOD) is an enzyme that destroys oxygen free radicals. Recombinant SOD has the potential to be useful in the treatment of oxygen toxicity in premature infants, myocardial infarction, organ transplantation, and stroke.

Growth factors are cytokines responsible for regulating cell proliferation, differentiation, and function. Each cell type's response is specific for each particular growth factor and differs from growth factor to growth factor. Several rDNA-produced growth factors are now undergoing clinical trials. Epidermal growth factor (EGF) stimulates epidermal cell proliferation. Indications being explored include tissue repair in corneal and cataract surgeries, angiogenesis (new capillary formation), and improvement in wound healing including burns and tendon repair. Because EGF assists in nerve regeneration, the potential may exist for the treatment of damaged spinal cords sometime in the future. Fibroblast growth factor (FGF), a member of a family of heparin-binding growth factors, is similar to EGF in its potential therapeutic uses. Platelet-derived growth factor (PDGF) is in early clinical trials for wound healing and ulcer repair. Other exciting growth factors being studied by the pharmaceutical industry include nerve factor factor, insulin-like growth factor, and stem cell growth factor.

The CD4 antigen is a plasma membrane glycoprotein found on the surface of CD4 helper T cells and serves as the receptor for the HIV during the process of infecting helper T cells. Recombinant soluble CD4 molecules, the extracellular portions of the CD4 glycoprotein, may serve as decoys or sponges to prevent binding of the virus to the surface of helper T cells, thus preventing HIV infection.

Recombinant vaccines in human clinical trials are directed against HIV, melanoma, herpes simplex 2 virus, and malaria. Each of the vaccines uses a recombinant surface protein or protein fragment. The HIV proteins used in vaccine development include the gp120 envelope glycoprotein (gp), the entire gp 160 glycoprotein, and the p24 core protein.

IL-1, released in response to microbial challenge, interacts with receptors on lymphocytes, macrophages, and other cell types and is one of the most powerful cytokines driving inflammatory responses. Thus, blocking IL-1 synthesis or receptor interaction could be helpful in treating autoimmune diseases. Anakinra or IL-1 receptor is a recombinant derivative of a naturally occurring IL-1 recep-

Table 28–4. Some rDNA-Derived Medicines in Development

Group	Product	Indication	Status
Colony-stimulating factors	Macrophage colony-stimulating factor (M-CSF)	Fungal disease	Phase II
		Cancer	Phase II
	PIXY321 (combined GM-CSF and IL-3)	Chemotherapy-induced neutropenia and thrombocytopenia	Phase III
Dismutases	Superoxide dismutase	Oxygen toxicity in premature infants	Phase I/II
Growth factors	Epidermal growth factor	Corneal and cataract surgeries	Phase III
	Fibroblast growth factor	Neuropathic ulcers and pressure sores	Phase II
	Nerve growth	Peripheral neuropathies	Phase I
	Platelet-derived growth factor	Wound healing	Phase III
Interferons	Consensus interferon	Hepatitis C	Phase III
Interleukins	Interleukin-1 receptor	Allergy and asthma, rheumatoid arthritis	Phase II
		HIV, sepsis	Phase I
	Interleukin-3	Autologous bone marrow transplants, adjuvant to chemotherapy	Phase III
Recombinant soluble CD4s (rCD4)	sCD4-PE40	AIDS	Phase II
	rsCD4	ARC, AIDS	Phase II
Tumor necrosis factors	Tumor necrosis factor	Cancer	Phase II
Vaccines	Herpes vaccines	Herpes simplex 2, genital herpes	Phase II/III
	Malaria vaccines	Malaria	3 in Phase II
	Melanoma therapeutic vaccine	Stage III–IV melanoma	Phase III
	gp160	AIDS	Phase III as a therapeutic Phase II as a vaccine
Others	Anakinra (interleukin-1 receptor antagonist)	Severe sepsis	Phase III
		Rheumatoid arthritis, ocular inflammation	Phase II
	Atrial natriuretic peptide	Acute kidney failure	Phase III
	Ciliary neurotrophic factor (CNTF)	Amyotrophic lateral sclerosis (Lou Gehrig's disease)	Phase II/III

HIV = human immunodeficiency virus; AIDS = acquired immunodeficiency syndrome; ARC = AIDS-related complex.
Adapted from *Biotechnology Medicines in Development, 1993 Annual Survey,* Washington, D.C., Pharmaceutical Manufacturers Association, 1993.

tor antagonist, currently undergoing clinical trials for the treatment of sepsis arthritis, ocular allergies, asthma, and inflammatory bowel disease.

MONOCLONAL ANTIBODIES

Antibodies

The human immune system is composed of two major branches or arms: the cell-mediated immune system (which includes macrophages, lymphocytes, and granulocytes) and the humoral immune system (which includes the antibody-secreting B cells or plasma cells). In contrast to the cell-mediated nature of the immune actions carried out by most lymphocytes, antibodies (Abs) or immunoglobulins (Igs) are soluble proteins that are produced in response to an antigenic stimulus. Antibodies are proteins. As part of the normal immune system, each B cell produces as many as a hundred million antibody proteins directed against bacteria, viruses, and other foreign invaders.[37] Antibodies act by binding to a particular antigen (Ag), thereby "tagging" it for removal or destruction by other immune system components. The selectivity of any antibody for a particular antigen is determined by its structure and specifically by the variable or antigen-binding regions (Fig. 28–6). Enzymatic digestion of the antibody with papain yields the Fab fragment, which contains the antigen binding sites, and the Fc fragment, which specifies the other biologic activities of the molecule.

Hybridoma Technology

MoAbs are ultrasensitive hybrid immune system–derived proteins designed to recognize specific antigens. Nobel Laureates Kohler and Milstein first reported

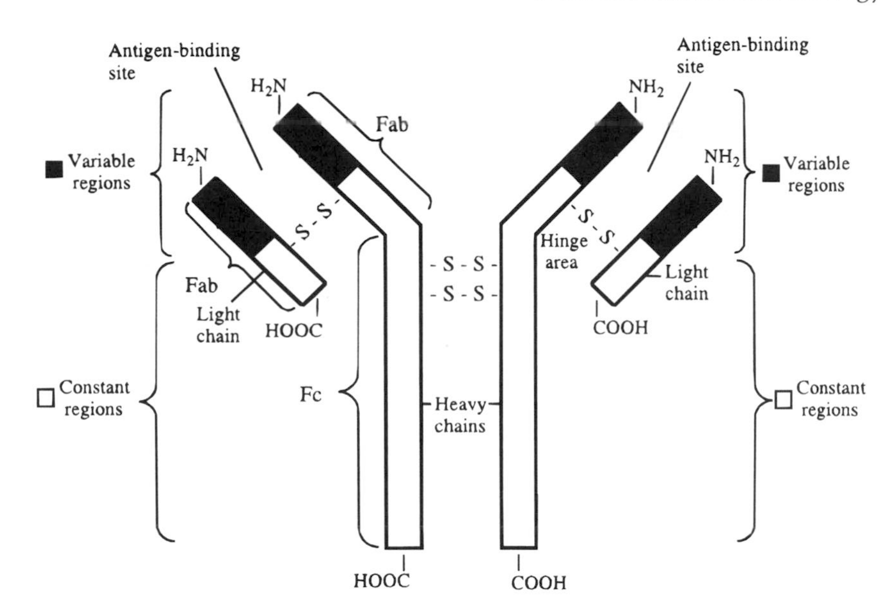

Fig. 28–6. Schematic model of an antibody molecule. (Adapted from R. A. Hudson and C. D. Black, *Biotechnology—The New Dimension in Pharmacy Practice: A Working Pharmacist's Guide,* Toledo, Council of Ohio Colleges of Pharmacy, 1992; T. Comstock, *Pharmaceutical Biotechnology Monitor, 1:* 2[1991]; M. M. E. Klegerman, in *Biotechnology and Pharmacy,* J. M. Pezzuto, et al., Eds., New York, Chapman & Hall, 1993; R. D. Mayforth, *Designing Antibodies,* San Diego, Academic Press, 1993, p. 54.)

MoAbs in 1975.[125] MoAbs have been used in laboratory diagnostics, site-directed drugs, and home test kits.[126,127] The B lymphocyte produces a wide range of structurally diverse antibody proteins with varying degrees of specificity in response to a single antigen stimulus. Because of their structural diversity, these antibodies would be called polyclonal antibodies. MoAbs are homogeneous hybrid proteins produced by a selected single clone of an engineered B lymphocyte. They are designed to recognize specific sites or epitopes on antigens.[128]

Hybridoma technology, the technology used to produce MoAbs, consists of combining or fusing two different cell lines: a myeloma cell (generally from a mouse) and a plasma spleen cell (B lymphocyte) capable of producing an antibody that recognizes a specific antigen (Fig. 28–7).[128,129] The resulting fused cell or hybridoma possesses some of the characteristics of both original cells: the myeloma cell's ability to survive and reproduce in culture (immortality) and the plasma spleen cell's ability to produce antibodies to a specific antigen.

Two myeloma variant cell lines with defects in nucleotide synthesis pathways are commonly used as fusion partners.[128–130] One lacks the HGPRT gene coding for an essential enzyme in purine biosynthesis. The other lacks the Tk gene coding for a pyrimidine biosynthetic enzyme. Following fusion, the cells are cultured in a medium containing hypoxanthine, aminopterin, and thymidine (HAT medium). Correct hybridomas (1 myeloma plus 1 spleen cell), although missing the gene from the myeloma partner, possess the gene from its spleen cell partner. Fused myeloma hybrids do not survive because they lack the essential gene, and fused spleen cell hybrids do not grow in culture. Hybridomas can be grown in large quantities, and clones producing antibodies with the appropriate specificity for the original antigen can be isolated from culture. Various techniques have been developed to select the single hybridoma clone producing the desired antibody (thus MOAb).[129,130] Hybridomas are grown in in vitro cell culture or in vivo in mouse (murine) ascites to yield large amounts of MoAbs (1 to 100 μg/ml in culture and 1 mg/ml in ascites).

MoAbs are more attractive than polyclonal antibodies for diagnostic and therapeutic applications because of their increased specificity of antigen recognition. Thus, they can serve as target-directed "homing devices" to find and attach to the targeted antigen. Developments in hybridoma technology have led to highly specific diagnostic agents for home use in pregnancy testing and ovulation prediction kits, laboratory use in colorectal and ovarian cancer detection, and in the design of site-directed therapeutic agents such as muromonab-CD3 for kidney transplant rejection.[128,129]

Monoclonal Antibody–Based In-Home Diagnostic Kits

The strong trend toward self-care coupled with a heightened awareness by the public of available technology and an emphasis on preventive medicine has increased the use of in-home diagnostics. Sales of in-home diagnostics are predicted to exceed $1 billion by the mid-1990s. MoAb specifically minimizes the possibilities of interference from other substances that might yield false-positive test results. The antigen being selectively detected by MoAb-based pregnancy test kits is human chorionic gonadotropin (hCG), the hormone produced if fertilization occurs and that continues to increase in concentration during the pregnancy.[131–135] Table 28–5 lists some examples of MoAb-containing in-home pregnancy test kits. Luteinizing hormone present in the urine is the antigen detected by MoAb ovulation prediction home test kits.[131–135] These test kits can help determine when a woman is most fertile because ovulation occurs 20 to 48 hours after the luteinizing hormone surge. Table 28–5 provides some examples of MoAb-based in-home ovulation prediction kits.

Other Monoclonal Antibody Diagnostics and Therapeutics

Hybridoma technology has also led to two highly specific MoAb diagnostic agents for laboratory use in colo-

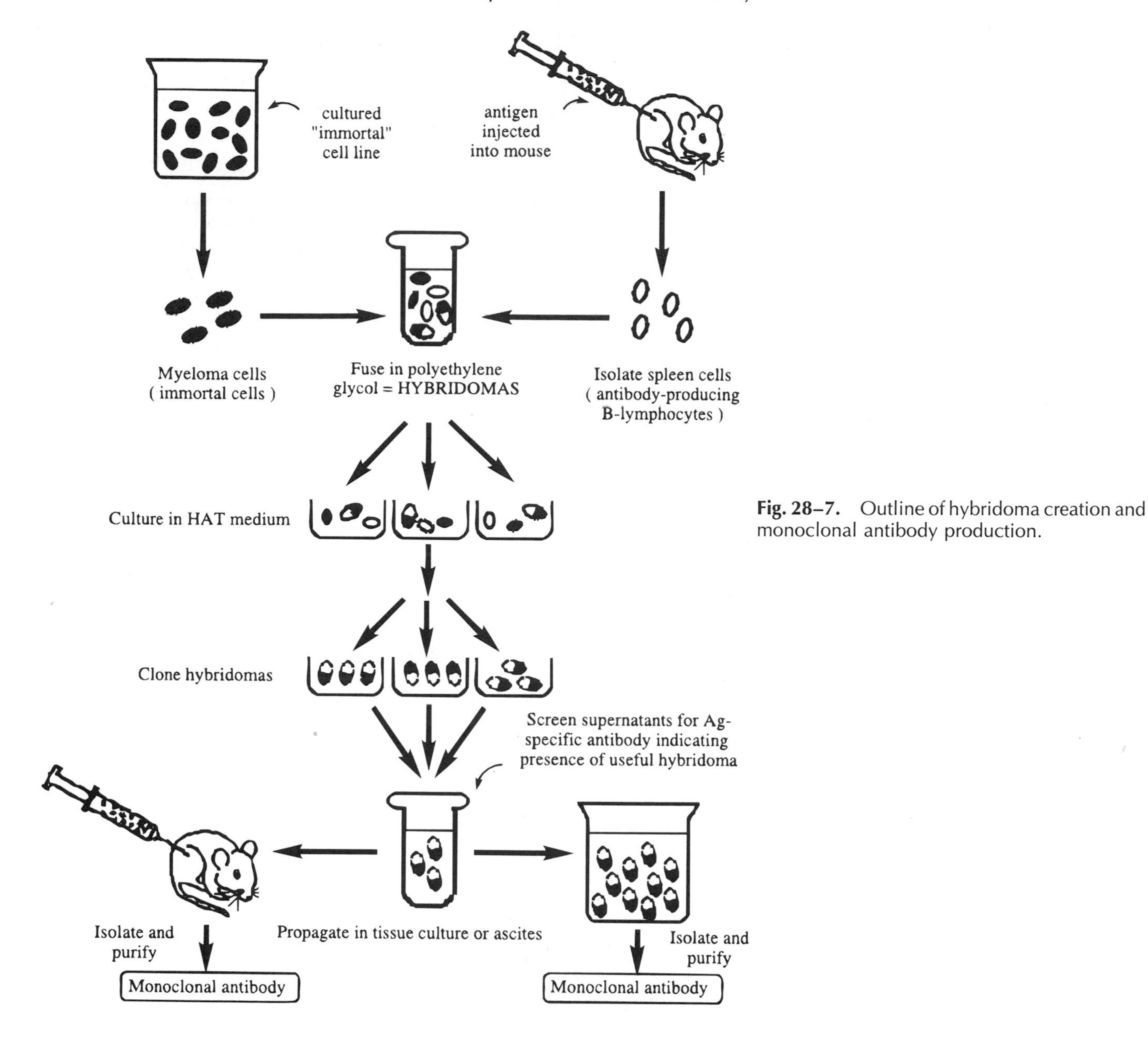

Fig. 28–7. Outline of hybridoma creation and monoclonal antibody production.

rectal (OncoScint CR Cytogen) and ovarian cancer (OncoScint OV, Cytogen) detection and in the design of site-directed therapeutic agents such as muromonab-CD3 (ORTHOCLONE OKT-3, Ortho Biotech) for the treatment of kidney, liver, and heart transplant rejection.[136]

Although difficulties in showing efficacy in the treatment of septic shock have slowed the progress of some therapeutic hybridoma-produced agents such as Xomen-E5 (Xoma/Ortho) and HA-1A (Centoxin, Centocor), several diagnostic MoAb-based products have enjoyed greater success and are currently awaiting FDA approval (Table 28–6). These diagnostic products include a variety of imaging agents for the detection of blood clots and cancer cells.[136] In addition to the progress being made in MoAb-based imaging agents, immunotoxins (hybrid molecules consisting of a MoAb attached to a cytotoxic entity), "humanized" MoAb (MoAb engineered to prevent the formation of human antimouse antibodies) and Fab fragments (still contains the antigen-binding site but less immunogenic mouse protein) are also progressing through clinical trials for the treatment of various immunologic, infectious, and neoplastic diseases.[31,42,136,137]

INFLUENCE OF BIOTECHNOLOGY ON DRUG DISCOVERY

The search for novel, efficacious, and safer medicinal agents is an increasingly costly and complex process. In a review of the impact of biotechnology on drug discovery, Venuti[138] has suggested that medicinal chemists have refined, "without major obstacles," the process that optimizes the pharmacologic properties of an identified novel chemical lead with a useful biological property. While taking advantage of any opportunity or technique

Table 28–5. Some Monoclonal Antibody–Based In-Home Test Kits

Product	Manufacturer or Distributor	Indicator Method (#) *or* Days Tested (*)
Pregnancy		
Advance	Advanced Care Products	Color change (#)
Answer 2	Carter Products	Hemagglutination inhibition (#)
Answer Plus	Carter Products	Color change (#)
Answer Plus 2	Carter Products	Color change (#)
Answer Quick & Simple	Carter Products	Color change (#)
Answer	Carter Products	Hemagglutination inhibition (#)
C&T	Healthcheck Corp.	Color appearance (#)
Clearblue	Whitehall Labs	Color change (#)
Clearblue Easy	Whitehall Labs	Color change (#)
Daisy 2	Advanced Care Products	Color appearance (#)
e.p.t. Stick	Parke-Davis Consumer Products	Color appearance (#)
Fact Plus	Advanced Care Products	Color appearance (#)
First Response 5-minute test for pregnancy	Hygeia Sciences	Color appearance (#)
Q Test for pregnancy	Becton Dickinson	Color appearance (#)
Ovulation		
Answer Ovulation	Carter Products	6-day kit (*)
Clearplan Easy	Whitehall Labs	5-day kit (*)
First Response Ovulation Predictor	Hygeia Sciences	5-day kit, 3-day refill (*)
Fortel	Biomerica	9-day kit (*)
OvuKIT Self-Test	Monoclonal Antibodies, Inc.	6-day kit and 9-day kit (*)
OvuQUICK Self-Test	Monoclonal Antibodies, Inc.	6-day kit and 9-day kit (*)
Q Test for ovulation	Becton Dickinson	5-day kit (*)

Data from references 131–135.

available to aid in the drug discovery process, the challenge for medicinal chemists remains lead identification.[139] Traditionally, medicinal chemistry research programs aimed at discovering new therapeutic agents relied heavily on random screening followed by analog synthesis and lead optimization. In discovering a lead, medicinal chemists encounter a problem. What biologic property does one screen for that directly correlates with the desired therapeutic outcome?

Biotechnology's contribution to pharmaceutical care in the 1980s was an improvement in the methods used to manufacture peptide-based and protein-based biophar-

Table 28–6. Some Monoclonal Antibodies in Development

Product	Indication	Status
Capiscint	Atherosclerotic plaque imaging agent	Phase II
Centara (chimeric anti-CD4 antibody)	Rheumatoid arthritis, multiple sclerosis	Phase II
CentroRx	Antiplatelet and prevention of blood clots	Phase III
Centoxin (HA-1A)	Sepsis and septic shock	Submitted
Xomen-E5	Gram-negative sepsis	Phase III
Fibriscint (anti-fibrin antibody)	Blood clot imaging agent	Submitted
ImmuRAID-CEA	Imaging colorectal cancer	Submitted
LYM-1	Lymphoma	Phase III
Myoscint	Imaging agent to diagnose heart attacks	Submitted
Oncolysin B (anti-B4-blocked ricin)	Intravenous treatment of B-cell leukemias and lymphomas	Phase III
OncoScint PR356 (CYT-356-In-111)	Detection, staging, and follow-up of prostate adenocarcinoma	Phase III
OncoTrac	Imaging metastatic and small cell lung cancer	Submitted
Orthozyme (CD5 Plus)	Immunoconjugate for treatment of graft-versus-host disease in bone marrow transplants	Submitted
Panorex (17-1A)	Therapeutic directed against colorectal cancer	Phase III

Adapted from *Biotechnology Medicines in Development, 1993 Annual Survey,* Washington, D.C., Pharmaceutical Manufacturers Association, 1993.

maceuticals, including insulin and hGH. Manufacturing shifted to highly controlled, well-characterized microbial fermentation or culture techniques from extraction from blood or other biologic materials. This improved the availability, safety, and, to some extent, the efficacy of widely used hormones and vaccines. Since the late 1980s, advances in biotechnology have not only affected manufacturing, but also have contributed to a greater understanding of the cause and progression of disease and have identified new therapeutic targets forming the basis of novel drug screens. These advances have facilitated the discovery of new agents with novel mechanisms of action for diseases that were previously difficult or impossible to treat. Having entered all areas of the drug discovery process, there is little doubt that biotechnology is playing a major role in shaping the medicinal agents that pharmacists will dispense in the future.[16]

With the evolution of biotechnology and, in particular, recombinant DNA techniques, detailed structural information about therapeutic targets is becoming increasingly available. The application of the techniques of biotechnology "to the identification of proteins and other macromolecules as drug targets, and their production in meaningful quantities as discovery tools thus can provide an answer to at least one of the persistent problems of lead detection."[138]

In Vitro Screening

Enzymes, membrane-bound proteins, and other binding proteins can serve as receptors for molecules acting as stimulators or inhibitors, agonists or antagonists. rDNA technology has provided the ability to clone, express, isolate, and purify these target proteins in larger quantities than ever before. Instead of using receptor proteins isolated from animal tissue for screening, in vitro bioassays can use the exact human protein target. This area has been extensively reviewed.[140,141] Examples of the application of biotechnology to in vitro screening include (1) immobilized preparations of receptors, antibodies, and other ligand-binding proteins; (2) soluble enzymes and extracellular cell-surface expressed protein receptors; (3) cloned membrane-bound receptors expressed in cell-lines carrying few endogenous receptors; and (4) whole-cell binding assays in which the production of the target protein is associated with a phenotypic change in the organism.

Receptor Structure Determination

Recombinant receptors can be crystallized and their three-dimensional structure determined by the well-established technique of x-ray crystallography. Macromolecules such as recombinant proteins require specialized procedures to obtain suitable crystals for x-ray analysis. At present, it is the only available technique that elucidates the complete three-dimensional structure of a molecule in high-resolution detail, including bond distances, angles, stereochemistry, and absolute configuration.[142] X-ray crystallography can provide atomic resolution structures of protein receptor targets and protein-ligand interactions with other molecules, such as substrates, cofactors, or inhibitors.[143] The structure of these complexes can be obtained by co-crystallizing the protein and the ligand or by soaking the ligand into an existing protein crystal.

Rational drug design (sometimes called mechanistic drug design or computer-assisted drug design) is an iterative process in which three-dimensional structure determination, analysis, design, synthesis, and bioassay form a dynamic feedback cycle. Rational drug design requires a structure for the target receptor, a demand satisfied by the availability of macromolecular crystal structures, including recombinant proteins. Drug design approaches can use crystal structures in a priori design (the terms de novo design or ab initio design have sometimes been used in the same context) and a posteriori analysis.[143] A priori design, a rational drug design approach, employs the crystal structure to create the initial, structurally novel lead compound with the aid of molecular graphics. A retrospective method, a posteriori analysis rationalizes existing structure-activity relationship data with x-ray structure and proposes design improvements.

HIV encodes for a protease (HIV-1 PR) required for viral replication. An example of the use of x-ray crystallography in drug design is the intensive study of inhibitors of HIV-1 PR as potential anti-AIDS drugs.[144,145] Medicinal chemists have applied a priori design and a posteriori analysis to the problem. Crystallographic studies of both recombinant expressed and synthetic HIV-1 PR[146–148] suggested a substrate binding site.[149] In addition, the crystal structure of a complex of an inhibitor bound to HIV-1 PR was determined, resulting in a closest contact map detailing hydrogen bonding interactions that serves as an aid to future HIV-1 PR inhibitor design (Fig. 28–8).[150] Also, high-throughput enzyme inhibition assays based on recombinant HIV-1 PR have facilitated the drug design studies.

Although x-ray crystallography may not reflect the three-dimensional molecular structure of a recombinant receptor protein under biologic conditions in solution, multidimensional nuclear magnetic resonance (NMR) spectroscopy may be the most powerful technique for such examinations. NMR can often provide structural information faster than x-ray crystallography and more easily study the dynamic interaction of ligand and receptor because the technique does not require crystals.[143,151]

rDNA technology has provided a reproducible, less expensive, and, in principle, inexhaustible source of pure human receptor subtypes. With the number of cloned receptors steadily growing, an illustrative example would be the families of cell surface receptors. Table 28–7 provides selected examples of cloned families of cell surface receptors, receptor fragments, and/or ligands.[152] The number of different receptor classes as well as the number of genes encoding receptor subtypes within each class is continually growing and is far greater than previously expected. Detailed structural analysis of cloned receptors and receptor subtypes has provided insight into receptor function through comparisons of sequence and molecular modeling.[152] Cloned and expressed receptors have proved helpful in highlighting the pharmacologic differences between humans and animals, in developing ligand binding assays for drug discovery, and in reevaluating the mechanism of action of known medicinal agents.

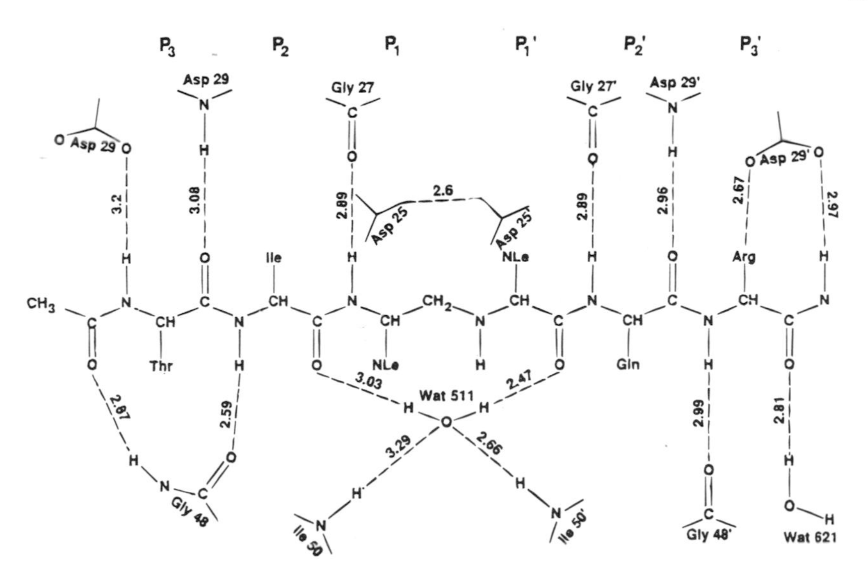

Fig. 28–8. Schematic representation of the hydrogen bond interactions of an inhibitor with HIV-1 PR as determined by x-ray crystallography. (From M. Miller, et al., *Science, 246,* 1149[1989], with permission of the American Association for the Advancement of Science, 1989.)

Protein Engineering

Protein engineering is the rational design and construction of unique proteins by rDNA techniques with enhanced or novel properties.[153–156] As a technique, protein engineering has tremendous theoretical and practical implications in medicinal chemistry for studying protein structure and function; examining properties of enzymes and protein and polypeptide folding and stability; introducing spectroscopically detectable reporter groups into proteins as an analytical tool; and producing second-generation, clinically useful proteins.[57] Site-directed mutagenesis is a protein engineering technique that allows one to specifically (site-directed) change (mutate) an amino acid found in a protein by modifying the base sequence of the DNA coding for that protein.

Table 28–7. Some Examples of Cloned Families of Cell-Surface Receptors, Receptor Fragments, and/or Ligands

Cytokine receptor superfamily
Erythropoietin
Granulocyte colony-stimulating factor (G-CSF)
Granulocyte macrophage colony-stimulating factor (GM-CSF)
Growth hormone
Interleukin-2 (IL-2)
Interleukin-3 (IL-3)
Interleukin-4 (IL-4)
Interleukin-5 (IL-5)
Interleukin-6 (IL-6)
Interleukin-7 (IL-7)
G-protein–coupled receptors
Adenosine
Angiotensin
Bradykinin
GABA-B
Histamine
Interleukin-8 (IL-8)
Muscarinic acetylcholine
Platelet-activating factor (PAF)
Vasopressin
Ion channel receptors
GABA-A
Nicotinic acetylcholine
Serotonin 5-HT_3

Data from W. H. M. L. Luyten and J. E. Leysen, *TIB TECH, 11,* 247(1993).

The exact amino acid sequence (primary structure) affects the conformation of a protein. This, in turn, affects the protein's complex three-dimensional structure.[156] The conformational preference of the specific amino acid sequence determines the formation of α-helices, β-sheets, or turns (secondary structure). The local secondary structures are folded into three-dimensional tertiary structures made up of domains. The domains are not only structural units, but also are functional units often containing intact ligand binding or catalytic sites. Therefore, protein engineering techniques, such as site-directed mutagenesis, can aid in the examination of the structure and function of proteins of interest to the medicinal chemist at the molecular level in three dimensions.

OTHER TECHNOLOGIES

Polymerase Chain Reaction

Polymerase chain reaction (PCR) is an in vitro gene amplification technique invented by 1993 Nobel Prize winner Mullis.[157,158] The technique involves taking a single copy of a specific double-stranded nucleic acid sequence and using it to create millions or billions of copies of itself. An extremely specific, accurate reaction, PCR is a fast, highly sensitive detection system that can find a single gene "needle" sequence in a proverbial "haystack" of other DNA sequences. Although straightforward in methodology, only the discovery of restriction enzymes has had as much of an impact on molecular biology as PCR has had.[159] The method uses a repeating three-cycle sequence: denature, anneal, and extend.[159–161] Specific oligonucleotide primer strands complementary to the DNA flanking the template sequence of interest are synthesized and on heat denaturation (separating the segment of DNA to be amplified into the sense and the antisense strand) annealed at the 3′ ends of each separated strand pointing toward each other. The DNA polymerase catalyzed synthesis of a complementary copy to each strand of the sense and antisense sequence, in effect, extends each chain. This completes the first cycle. The now extended product of each primer can anneal to the opposite strand primer and serve as the template for another round of extensions. One cycle produces twice the relative number of copies of a specific PCR product, 2 cycles 4 times, 3 cycles 8 times, 10 cycles 1024 times, and 30 cycles 1,073,741,824 times, and so forth. PCR is a valuable tool in direct DNA sequencing, genomic cloning, and site-directed mutagenesis. The first PCR-based diagnostic test to be marketed in the United States gained approval on June 15, 1993, for the detection of chlamydia, a sexually transmitted disease affecting 4 million people annually in the United States.[162]

Gene Therapy

Since the development of rDNA technology, methods to deliver genes to mammalian cells have stimulated great interest in treating human disease by gene therapy or gene transplantation. Although initially defining gene therapy as the insertion of a gene, or genes, into the cells of a patient with a genetic disease to correct the function of defective cells, scientists now recognize it as a more powerful technology.[163] In broader terms, gene therapy is simply the transfer of new genetic material to an individual to change that individual's genetic make-up.[164] By 1993, gene transfer clinical trials had been approved for use in treating more than 50 patients in the United States. Among the protocols used to date, gene therapy has found use in treating rare genetic diseases, such as adenosine deaminase (ADA) deficiency, CF, and familial hypercholesterolemia due to a defective low-density lipoprotein receptor protein. The technology has also found application in cancer biotherapy.[165–171]

Catalytic Antibodies

In 1948, Pauling[172] suggested that an enzymatic reaction accelerates if the catalytic site is more structurally complementary to the high-energy transition state of the reaction than to its substrate or product ground states. Normal antibodies, although generally ligand-specific, lack catalytic activity owing to binding their ligands in low-energy ground states.[172] Jencks[173] first directly hypothesized the concept of catalytic antibodies in 1969 when he suggested that antibodies that selectively bind the transition states of reactions would be catalytic. The development of hybridoma technology permits the production of homogeneous, MoAbs in the sufficient quantities necessary to purify and characterize reproducibly the properties of catalytic antibodies. First reported independently by the Tramontano[174] and Pollack[175] groups in 1986, catalytic antibodies or "abzymes" have been considered as a new class of designer enzymes catalyzing reactions for which no natural enzyme exists. Numerous reviews are available in this field.[176–189]

Chemists have developed general strategies to induce catalytic activity into antibody binding sites by introducing reactive groups (nucleophilic, electrophilic, acidic, or basic amino acid side chains) precisely in locations optimal for reaction with the the substrate. A hapten specifically designed to resemble the transition state is used to form the catalytic antibody. Accelerations in reaction rates of 10^3- to 10^6-fold were observed for antibodies raised against stable analogs of the transition states in the hydrolysis of esters and carbonates (Fig. 28–9). Significantly higher binding affinities were observed between the catalytic antibody and the transition state analogs. One limitation of the antibody catalysts is that they function only in predominantly aqueous environments as observed with enzymes. Organic solvents may denature antibodies, which are proteins. The literature reports a wide range of additional reactions, including enantioselective reactions, proton transfers, Claisen rearrangements, bimolecular amide bond formation, lactone ring formation, redox reactions, and Diels-Alder reactions.[182–189] Abzymes may have potential as therapeutic agents that

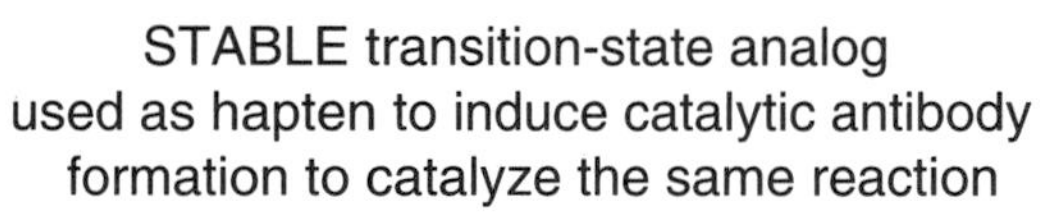

STABLE transition-state analog
used as hapten to induce catalytic antibody
formation to catalyze the same reaction

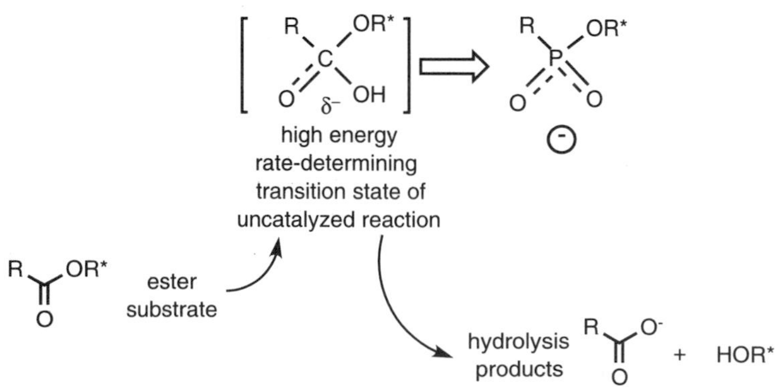

Fig. 28–9. Hapten specifically designed to mimic transition states in hydrolysis of esters.

could enzymatically cleave specific surface proteins or sugars of microbes or tumor cells, thereby destroying the invading cells.

Antisense Technology

During the process of transcription, the double-stranded DNA separates into the sense DNA strand, which then serves as the template for the mRNA and the antisense DNA strand. Antisense sequences were first described as a naturally occurring event in which an endogenous antisense RNA can be formed complementary to a cellular mRNA resulting in a repressor of gene expression.[185–189] Antisense binding occurs when the bases adenine, guanine, thymine, and cytosine of the antisense oligonucleotide align in a precise, sequence-specific manner with a complementary series of bases in the target RNA. Antisense RNAs inhibit gene expression and thus protein synthesis, through the action of an enzyme that recognizes the RNA-RNA duplex, disrupts the base pairing, and changes many of the adenosine residues to inosine.[190,191] Inhibition of gene expression occurs because the modified mRNA is no longer competent for translation and resultant protein synthesis.

Although most traditional pharmaceuticals act by targeting enzymes or other receptor proteins, antisense technology involves the blocking of genetic messages to turn off the production of disease-producing proteins at the source.[192,193] These potential genetic-code blocking drugs might control disease by inhibiting deleterious or malfunctioning genes, differing from gene therapy, which inserts needed genetic information. Among the potential therapeutic targets are viruses (HIV, herpes simplex, human papillomavirus) and oncogenes *(c-myb, c-ras)*.[194] A related approach is triple helix (triplex) technology in which short oligodeoxynucleotides (ODNs) of 15 to 27 nucleotides in length can bind sequence specifically to complementary segments of duplex DNA. The resulting triple helices can inhibit DNA replication.

Hurdles to antisense drug development include difficulties in cell permeation of the negatively charged, highly lipophobic molecules and problems regarding sensitivity to degradation of the phosphodiester linkages in ODNs by endogenous cellular nucleases.[194,195] Chemical modification of the oligonucleotides by substituting for the negatively charged oxygen on the phosphodiester linkages with a sulfur, methyl, or nitrogen results in phosphorothiolate, methyl phosphonate, and phosphoamidate analogs, each possessing an additional chiral center, which increase resistance to nucleases and increase lipophilicity. Derivatives containing 3′–5′ formacetal linkers have an added advantage in not creating additional chiral centers that make purification difficult (Fig. 28–10).[192,194,196]

adenosine residue → inosine residue

Glycobiology

Glycobiology is the study of sugars, and their role in biology. Sugars or oligosaccharides are an integral component of many biologically important macromolecules, such as structural proteins, enzymes, hormones, immunoglobulins, cell adhesion molecules, and lipids.[197,198] Scientists have only recently recognized the importance of sugars in biology. Most biologic proteins are glycosylated. Attached oligosaccharides modify the structure, physical properties, and dynamics of the proteins. They are attached to proteins via covalent O-glycosyl linkages to serine, threonine, hydroxyproline, or hydroxylysine and via N-glycosyl linkages to asparagine residues. Glycoforms are the variants or subsets resulting form different oligosaccharides linked to a protein at the same site or several sites on the same molecule. A glycobiologist would describe this variation as "microheterogeneity."

One cell can synthesize a series of glycoforms of the same protein that may differ in physiochemical and biochemical properties. For example, EPO has three N-linked and one O-linked glycosylation sites. The removal of the terminal sialic acids destroys in vivo activity, and desialylation results in a more rapid clearance of the molecule and a shorter circulatory half-life.[199–201] Yet, deglycosylation of GM-CSF has the opposite effect, increasing the specific activity sixfold.[202] Glycobiology is also important to the medicinal chemist because the sugars of glycoproteins are thought to be responsible for the recognition and binding of these biomolecules to other molecules in disease states, such as asthma, rheumatoid arthritis, HIV infection, cancer, and several diseases associated with pregnancy.[202] Identification of which oligosaccharides are involved may lead to new molecular targets for possible therapeutic intervention.[203] For example, strong evidence exists that metastasis involves specific cell-surface carbohydrates present on the surface of colonic, pancreatic, and other tumors.[204]

Transgenic Animals

A transgenic animal is one that has been altered to contain a gene (transgene) from another organism, usu-

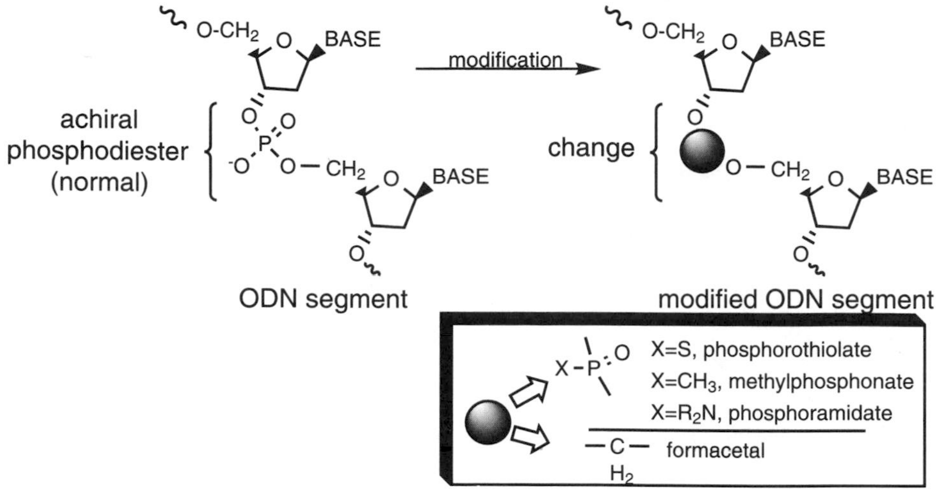

Fig. 28–10. Chemical modifications of oligonucleotides in antisense technology.

ally another species, within its genome.[140] The most common method for generating transgenic animals is the microinjection of the foreign DNA into the male pronucleus of a fertilized ovum, which is then transferred back into the female and allowed to develop. Species of transgenic animals produced include mice, rats, rabbits, sheep, pigs, and cows. Medicinal chemistry research may employ these genetically engineered species in the production of protein biopharmaceuticals, the overexpression of drug receptors for bioassay and drug side-effects models, and the creation of animal models of human disease (such as genetic deficiencies, cancer, type I diabetes, and familial Alzheimer's disease). Disease models provide reproducible and renewable pharmacologic tests for the screening of potential therapeutic agents directed against the disease.[205]

Biosensors

Biosensors are a synergistic combination of biochemistry, membrane technology, and microelectronics.[206–208] A biosensor is an extremely sensitive, selective analytical device that converts a biologic action into an electrical signal that can be quantified. The biologic component, immobilized to a transducer, can be an enzyme, multienzyme system, DNA probe, antigen, antibody, organelle, or whole organism capable of creating a product of or forming an intermolecular interaction with the analyte. The advantage of biosensors as analytical tools to detect particular molecules present in a highly complex mixture is the high selectivity resulting from the integral biologic element used as a part of the sensor. The major application area, accounting for a 53% market share of this biotechnology, is in clinical diagnostics, with a commercialized example being the use of a glucose oxidase–based biosensor as a test for blood glucose levels.

CONCLUSION

Many advances have occurred in biotechnology since Watson and Crick determined the structure of DNA (Fig. 28–11). Pharmaceutical biotechnology and, in particular, recombinant DNA technology and hybridoma methodology have provided improved pharmaceuticals, novel therapeutic agents, unique diagnostic products, and new drug design tools. Pharmacists and medicinal chemists are poised to be among the great beneficiaries of the advances made in the future developments of the techniques and products of biotechnology. Wade and Levy[2] state that "these pivotal developments may catapult the pharmacy profession into new roles and may provide many opportunities for pharmacists to enhance the

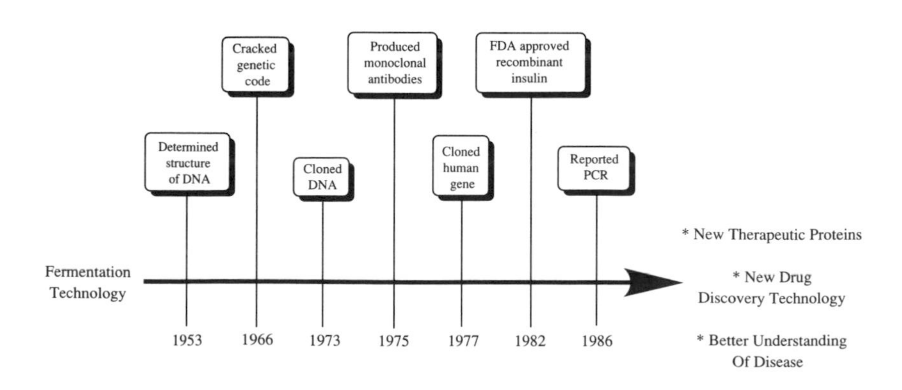

Fig. 28–11. Some major advances in biotechnology since Watson and Crick discovered the double helical nature of the DNA molecule.

professional dimension of their practice—but only if they are prepared."

REFERENCES

1. M. F. Conlan, *Drug. Top.*, 34(1990).
2. D. A. Wade and R. A. Levy, *Am. Pharm., NS32,* 33(1992).
3. S. W. Zito, Ed., *Pharmaceutical Biotechnology—A Programmed Text,* Lancaster, PA, Technomic Publishing, 1992.
4. J. M. Pezzuto, et al., Eds., *Biotechnology and Pharmacy,* New York, Chapman & Hall, 1993.
5. M. E. Klegerman and M. J. Groves, Eds., *Pharmaceutical Biotechnology—Fundamentals and Essentials,* Buffalo Grove, IL, Interpharm Press, 1992.
6. W. Bains, *Biotechnology from A to Z,* Oxford, Oxford University Press, 1993.
7. J. Coombs, *Dictionary of Biotechnology,* 2nd Ed., New York, MacMillan Press, 1992.
8. F. Franks, Ed., *Protein Biotechnology,* Totowa, NJ, Humana Press, 1993.
9. V. Moses and R. E. Cape, Eds., *Biotechnology—The Science and the Business,* Chur, Switzerland, Harwood Academic Publishers GmbH, 1991.
10. J. M. Walker and E. B. Gingold, Eds., *Molecular Biology and Biotechnology,* 3rd Ed., Cambridge, Royal Society of Chemistry, 1993.
11. J. Davies and W. S. Reznikoff, Eds., *Milestones in Biotechnology: Classic Papers on Genetic Engineering,* Stoneham, MA, Butterworth-Heinemann, 1992.
12. B. Alberts, et al., *Molecular Biology of the Cell,* 2nd Ed., New York, Garland Publishing, 1989.
13. J. Darnell, et al., *Molecular Cell Biology,* 2nd Ed., New York, Scientific American Books, 1990.
14. D. Freifelder and G. M. Malacinski, Eds., *Essentials of Molecular Biology,* 2nd Ed., Boston, Jones & Bartlett Publishers, 1993.
15. S. Carlsen and H. Dalboge, *Biotechnology in pharmaceutical research and production,* in *A Textbook of Drug Design and Development,* P. Krogsgaard-Larsen and H. Bundgaard, Eds., Chur, Switzerland, Harwood Academic Publishers GmbH, 1991, p. 244.
16. T. J. R. Harris, *The current and future impact of molecular biology in drug discovery,* in *Medicinal Chemistry, The Role of Organic Chemistry in Drug Research,* 2nd Ed., C. R. Ganellin and S. M. Roberts, Eds., San Diego, Academic Press, 1993, p. 103.
17. D. Brixner, *Pharm. Times, Nov.,* 39(1991).
18. J. Demuth, *An Introduction to Pharmaceutical Biotechnology,* University of Wisconsin–Madison, Extension Services in Pharmacy, School of Pharmacy, 1990.
19. L. P. Gage, *Am. J. Pharm. Educ., 50,* 360(1986).
20. E. T. Herfindal, *Am. J. Hosp. Pharm., 46,* 2520(1989).
21. L. Hood, *JAMA, 259,* 1844(1988).
22. T. O'Connor, *Pharm. Times, Apr.,* 39(1990).
23. M. M. Piascik, *Am. J. Hosp. Pharm., 48* (Suppl. 1), S4(1991).
24. M. L. Sethi, *Pharm. Times, Jul.,* 122(1990).
25. M. K. Speedie, *Am. J. Pharm. Educ., 54,* 55(1990).
26. C. F. Stewart, *Biotechnol. Inform. Issues Pharm., 1,* 1(1991).
27. C. F. Stewart and R. A. Fleming, *Am. J. Hosp. Pharm., 46* (Suppl. 2), S3(1989).
28. C. F. Stewart and R. A. Fleming, *Am. J. Hosp. Pharm., 46* (Suppl. 2), S4(1989).
29. J. A. Tami, *Am. Pharm., NS31,* 131(1991).
30. R. D. Sindelar, *Drug. Top., special insert, Apr.,* 1(1992).
31. R. A. Hudson and C. D. Black, *Biotechnology—The New Dimension in Pharmacy Practice: A Working Pharmacist's Guide,* Toledo, Council of Ohio Colleges of Pharmacy, 1992.
32. R. K. Miyahara and D. Nykamp, *U.S. Pharm., Dec.,* 50(1990).
33. B. J. Kane and J. G. Kuhn, *Drot. Top., special insert, Oct.,* 1(1992).
34. Anonymous, *Am. Pharm., NS31,* 28(1991).
35. M. Montague, *Am. J. Pharm. Educ., 53,* 21s(1989).
36. S. Fields, *Am. Pharm., NS33,* 28(1993).
37. P. Hall, *Pharmaceutical Biotechnology,* New York, Global Medical Communications, 1992.
38. R. D. Sindelar, *Drug Top., May 3,* 66(1993).
39. R. D. Sindelar, *Drug Store News, 3,* 31(1993).
40. R. D. Sindelar, *Am. Pharm., NS33,* 27(1993).
41. *Biotech Rx: Opportunities in Therapy Management,* Washington, D.C., American Pharmaceutical Association, 1993.
42. *Biotechnology Medicines in Development, 1993 Annual Survey,* Washington, D.C., Pharmaceutical Manufacturers Association, 1993.
43. *Med Ad News, 12,* (1993).
44. *Biotechnology Resource Catalog,* Philadelphia, Philadelphia College of Pharmacy and Science, 1993.
45. J. Impoco, *U.S. News & World Report, Oct. 18,* 58(1993).
46. A. Klausner, *BIO/TECH., 11,* S35(1993).
47. P. Abrams, *BIO/TECH., 11,* 450(1993).
48. E. B. Gingold, *An introduction to recombinant DNA technology,* in *Molecular Biology and Biotechnology,* 3rd Ed., J. M. Walker and E. B. Giingold, Eds., Cambridge, Royal Society of Chemistry, 1993, p. 23.
49. F. H. C. Crick, *Symp. Soc. Exp. Biol., 12,* 138(1958).
50. J. D. Watson, et al., *Recombinant DNA,* 2nd Ed., New York, Scientific American Books, 1992.
51. *Ibid,* p. 22.
52. P. H. von Hippel, et al., *Annu. Rev. Biochem., 53,* 389(1984).
53. C. K. Mathews and K. E. van Holde, *Biochemistry,* Redwood City, CA, Benjamin/Cummings Publishing Company, 1990.
54. W. Bains, *Biotechnology from A to Z,* Oxford, Oxford University Press, 1993, p. 108.
55. S. N. Cohen, et al., *Proc. Natl. Acad. Sci. U.S.A., 70,* 3240(1973).
56. L. G. Davis, *Background to recombinant DNA technology,* in *Biotechnology and Pharmacy,* J. M. Pezzuto, et al., Eds., New York, Chapman & Hall, 1993, p. 3.
57. L. A. Fothergill-Gilmore, *Recombinant protein technology,* in *Protein Biotechnology,* F. Franks, Ed., Totowa, NJ, Humana Press, 1993, p. 467.
58. M. D. Scawen, et al., in *Molecular Biology and Biotechnology,* 3rd Ed., J. M. Walker and E. B. Gingold, Eds., Cambridge, Royal Society of Chemistry, 1993, p. 327.
59. R. Seetharam and S. K. Sharma, Eds., *Purification and Analysis of Recombinant Proteins,* New York, Marcel Dekker, 1991.
60. R. R. Burgess, *Protein purification,* in *Protein Engineering,* D. L. Oxender and C. F. Fox, Eds., New York, Alan R. Liss, 1987, p. 71.
61. R. J. Prankerd and S. G. Schulman, *Analytical methods in biotechnology,* in *Biotechnology and Pharmacy,* J. M. Pezzuto, et al., Eds., New York, Chapman & Hall, 1993, p. 71.
62. J. L. Dwyer, in *Protein Biotechnology,* F. Franks, Ed., Totowa, NJ, Humana Press, 1993, p. 49.
63. V. B. Lawlis and H. Heinsohn, *LC-GC, 11,* 720(1993).
64. C. Horvath and L. S. Ettre, Eds., *Chromatography in Biotechnology,* Washington, D.C., American Chemical Society, 1993.
65. J. M. Samanen, in *Peptide and Protein Drug Delivery,* V. H. L. Lee, Ed., New York, Marcel Dekker, 1991, p. 137.
66. J. Q. Oeswein and S. J. Shire in *Peptide and Protein Drug Delivery,* V. H. L. Lee, Ed., New York, Marcel Dekker, 1991, p. 167.

67. F. Franks, in *Protein Biotechnology,* F. Franks, Ed., Totowa, NJ, Humana Press, 1993, p. 133.
68. S. W. Zito, *Medicinal chemistry and pharamceutical biotechnology,* in *Pharmaceutical Biotechnology—A Programmed Text,* S. W. Zito, Ed., Lancaster, PA, Technomic Publishing, 1992, pp. 17–49.
69. M. C. Manning, et al., *Pharmaceut. Res., 6,* 903(1989).
70. Indra K, Reddy, in *Pharmaceutical Biotechnology—A Programmed Text,* S. W. Zito, Ed., Lancaster, PA, Technomic Publishing, 1992, p. 77.
71. Y. J. Wang and R. Pearlman, Eds., *Stability and Characterization of Protein and Peptide Drugs, Case Histories,* New York, Plenum Press, 1993.
72. T. L. Lemke, *Review of Organic Functional Groups,* 3rd Ed., Malvern, PA, Lea & Febiger, 1992.
73. D. J. Burgess, In *Biotechnology and Pharmacy,* J. M. Pezzuto, et al., New York, Chapman & Hall, 1993, p. 116.
74. J. Koeller and S. Fields, *Contemp. Pharm. Issues, April,* 4(1991).
75. V. H. L. Lee, Ed., *Peptide and Protein Drug Delivery,* New York, Marcel Dekker, 1991.
76. H. C. Ansel and N. G. Popovich, *Pharmaceutical Dosage Forms and Drug Delivery Systems,* 5th Ed., Malvern, PA, Lea & Febiger, 1990.
77. R. A. Hutchinson and C. D. Black, *Am. Pharm., NS 33,* 64(1993).
78. V. H. L. Lee, *Pharm. Int., 7,* 208(1986).
79. D. R. Parkinson, *Semin. Oncol., 15,* 10(1988).
80. *Facts and Comparisons,* St. Louis, MO, 1993, Facts and Comparisons, p. 129f.
81. *AHFS Drug Information,* Bethesda, MD, American Society of Hospital Pharmacists, 1993, p. 1958.
82. J. Brange and L. Langkjaer, in *Stability and Characterization of Protein and Peptide Drugs, Case Histories,* J. Wang and R. Pearlman, Eds., New York, Plenum Press, 1993, p. 315.
83. J. Brange, et al., *Nature, 333,* 679(1988).
84. J. Markussen, et al., *Protein Eng., 2,* 157(1988).
85. *Facts and Comparisons,* St. Louis, MO, 1993, Facts and Comparisons, p. 116b.
86. R. Pearlman and T. A. Bewley, in *Stability and Characterization of Protein and Peptide Drugs, Case Histories,* J. Wang and R. Pearlman, Eds., New York, Plenum Press, 1993, p. 1.
87. Personal communication, January 14, 1994, Genentech, Inc.
88. *Facts and Comparisons,* St. Louis, MO, 1993, Facts and Comparisons, p. 88o.
89. *AHFS Drug Information,* Bethesda, MD, American Society of Hospital Pharmacists, 1993, p. 869.
90. D. L. Higgins and W. F. Bennett, *Annu. Rev. Pharmacol. Toxicol., 30,* 91(1990).
91. *Facts and Comparisons,* St. Louis, MO, 1993, Facts and Comparisons, p. 884.
92. S. M. Edgington, *BIO/TECH., 11,* 580(1993).
93. D. L. Urdal, *Ann. Rep. Med. Chem., 26,* 221(1991).
94. K. Cooper and H. Masamune, *Ann. Rep. Med. Chem., 27,* 209(1992).
95. J. P. Dutcher, *Hosp. Formul., 27,* 694(1992).
96. W. J. McIntyre and J. A. Tami, *Drug. Top., special insert, Aug.,* 1(1992).
97. L. A. Hansen, in *Concepts in Immunology and Immunotherapeutics,* J. Koeller and J. Tami, Eds., Bethesda, MD, American Society of Hospital Pharmacists, 1990.
98. M. E. Klegerman and N. P. Plotnikoff, in *Biotechnology and Pharmacy,* J. M. Pezzuto, et al., Eds., New York, Chapman & Hall, 1993, p. 53.
99. *Facts and Comparisons,* St. Louis, MO, 1993, Facts and Comparisons, p. 682c.
100. *AHFS Drug Information,* Bethesda, MD, American Society of Hospital Pharmacists, 1993, p. 579.
101. M. H. Schwenk, in *Pharmaceutical Biotechnology—A Programmed Text,* S. W. Zito, Ed., Lancaster, PA, Technomic Publishing Co., 1992, p. 123.
102. D. Brixner, in *Biotechnology and Pharmacy,* J. M. Pezzuto, et al., New York, Chapman & Hall, 1993, p. 381.
103. H. S. Panitch, *Drugs, 44,* 946(1993).
104. *Facts and Comparisons,* St. Louis, MO, 1993, Facts and Comparisons, p. 476c.
105. *Ibid,* p. 476.
106. P. A. Todd and K. L. Goa, *Drugs, 43,* 111(1992).
107. *Facts and Comparisons,* St. Louis, MO, 1993, Facts and Comparisons, p. 685e.
108. *AHFS Drug Information,* Bethesda, MD, American Society of Hospital Pharmacists, 1993, p. 523.
109. J. Geigert, et al., in *Stability and Characterization of Protein and Peptide Drugs, Case Histories,* J. Wang and R. Pearlman, Eds., New York, Plenum Press, 1993, p. 249.
110. S. L Huber, Ed., *New Product Bulletin—Filgastrim,* Washington, D.C., American Pharmaceutical Association, 1993.
111. *Colony-Stimulating Factors,* New York, Triclinica Communications, 1990.
112. S. P. Smith and G. C. Yee, *Pharmacotherapy, 12,* 11S(1992).
113. A. Wlodawer, A. Pavlovsky, and A. Gustchina, *Protein Sci., 2,* 1373(1993).
114. *Facts and Comparisons,* St. Louis, MO, 1993, Facts and Comparisons, p. 84b.
115. *AHFS Drug Information,* Bethesda, MD, American Society of Hospital Pharmacists, 1993, p. 847.
116. S. E. Graber, *Hematol. Oncol. Clin. North Am., 3,* 369(1989).
117. *Facts and Comparisons,* St. Louis, MO, 1993, Facts and Comparisons, p. 84i.
118. *AHFS Drug Information,* Bethesda, MD, American Society of Hospital Pharmacists, 1993, p. 859.
119. *Ibid,* p. 864.
120. *Facts and Comparisons,* St. Louis, MO, 1993, Facts and Comparisons, p. 84m.
121. C. V. Prowse, In *Plasma and Recombinant Blood Products in Medical Therapy,* C. V. Prowse, Ed., Chichester, England, John Wiley & Sons, 1992, p. 89.
122. *Facts and Comparisons,* St. Louis, MO, 1993, Facts and Comparisons, p. 467.
123. J. E. Ciardi, et al., Eds., *Genetically Engineered Vaccines,* New York, Plenum Press, 1992.
124. R. W. Ellis, Ed., *Vaccines,* Stoneham, MA, Butterworth-Heinemann, 1992.
125. G. Kohler and C. Milstein, *Nature, 256,* 495(1975).
126. T. Comstock, *Pharmaceutical Biotechnology Monitor, 1,* 2(1991).
127. A. Gringauz, *U. S. Pharm., Oct.,* 38(1985).
128. M. Webb, in *Molecular Biology and Biotechnology,* 3rd Ed., J. M. Walker and E. B. Gingold, Eds., Cambridge, Royal Society of Chemistry, 1993, p. 357.
129. M. M. E. Klegerman, in *Biotechnology and Pharmacy,* J. M. Pezzuto, et al., Eds., New York, Chapman & Hall, 1993, p. 39.
130. R. D. Mayforth, *Designing Antibodies,* San Diego, Academic Press, 1993, p. 54.
131. T. A. Gossel and M. P. Mahalik, *U. S. Pharm., July,* 36(1993).
132. G. D. Newton, *Am. Pharm., NS33,* 22(1993).
133. S. M. Meyer, in *Handbook of Nonprescription Drugs,* T. R. Covington, et al., Eds., Washington, D.C., American Pharmaceutical Association, 1993, p. 39.
134. S. J. Coons, *Am. Pharm., NS29,* 46(1989).
135. T. A. Gossel, *U. S. Pharm., Apr.,* 26, (1990).

136. J. M. Brown, in *Biotechnology and Pharmacy,* J. M. Pezzuto, et al., Eds., New York, Chapman & Hall, 1993, p. 227.
137. L. M. Hinman and G. Yarranton, *Ann. Rep. Med. Chem., 28,* 237(1993).
138. M. C. Venuti, *Ann. Rep. Med. Chem., 25,* 289(1990).
139. K. R. Freter, *Pharm. Res., 5,* 397(1988).
140. C. W. Dykes, in *Molecular Biology and Biotechnology,* 3rd Ed., J. M. Walker and E. B. Gingold, Eds., Cambridge, Royal Society of Chemistry, 1993, p. 164.
141. C. K. Angerhofer and J. M. Pezzuto, in *Biotechnology and Pharmacy,* J. M. Pezzuto, et al., Eds., New York, Chapman & Hall, 1993, p. 312.
142. D. J. Abraham, in *Computer-Aided Drug Design, Methods and Applications,* T. J. Perun and C. L. Propst, Eds., New York, Marcel Dekker, 1989, p. 93.
143. J. W. Erickson and S. W. Fesik, *Ann. Rep. Med. Chem., 27,* 271(1992).
144. J. R. Huff, *J. Med. Chem., 34,* 2305(1991).
145. D. W. Norbeck and D. J. Kempf, *Ann. Rep. Med. Chem., 26,* 141(1991).
146. M. A. Navia, et al., *Nature, 337,* 615(1989).
147. A. Wlodawer, et al., *Science, 245,* 616(1985).
148. P. Lapatto, et al., *Nature, 342,* 299(1989).
149. I. T. Weber, et al., *Science, 243,* 928(1989).
150. M. Miller, et al., *Science, 246,* 1149(1989).
151. S. W. Fesik, in *Computer-Aided Drug Design, Methods and Applications,* T. J. Perun and C. L. Propst, Eds., New York, Marcel Dekker, 1989, p. 133.
152. W. H. M. L. Luyten and J. E. Leysen, *TIBTECH, 11,* 247(1993).
153. D. L. Oxender and C. F. Fox, Eds., *Protein Engineering,* New York, Alan R. Liss, 1987.
154. D. L. Oxender, Ed., *Protein Structure, Folding and Design,* Vols. I and II, New York, Alan R. Liss, 1985–86.
155. S. A. Narang, Ed., *Protein Engineering, Approaches to the Manipulation of Protein Folding,* Stoneham, MA, Butterworth Publishers, 1990.
156. D. L. Oxender and T. J. Graddis, in *Biotechnology—The Science and the Business,* V. Moses and R. E. Cape, Eds., Chur, Switzerland, Harwood Academic Publishers GmbH, 1991, p. 153.
157. R. K. Saiki, et al., *Science, 230,* 1350(1985).
158. K. B. Mullis, *Sci. Am., 262,* 36(1990).
159. C. Buffery, in *Molecular Biology and Biotechnology,* 3rd Ed., J. M. Walker and E. B. Gingold, Eds., Cambridge, Royal Society of Chemistry, 1993, p. 51.
160. K. B. Mullis and F. A. Faloona, *Methods in Enzymology, 155,* 263(1986).
161. E. Daniell, in *Biotechnology—The Science and the Business,* V. Moses and R. E. Cape, Eds., Chur, Switzerland, Harwood Academic Publishers GmbH, 1991, p. 145.
162. *Med Ad News, Aug.* 1993.
163. R. M. Blaese, *Pharmaceut. Biotech. Monitor, 2,* 2(1992).
164. W. Bains, *Biotechnology from A to Z,* Oxford, Oxford University Press, 1993, p. 147.
165. J. W. Larrick and K. L. Burck, *Gene Therapy, Application of Molecular Biology,* Amsterdam, Elsevier, 1991.
166. W. F. Anderson, *Science, 256,* 808(1992).
167. A. D. Miller, *Nature, 357,* 455(1992).
168. R. C. Mulligan, *Science, 260,* 926(1993).
169. R. A. Morgan, *Biopharm, Jan.-Feb.,* 1993, p. 32.
170. T. Friedmann, *TIBTECH, 11,* 159(1993).
171. G. J. McGarrity and Y. Chiang, *Ann. Rep. Med. Chem., 28,* 267(1993).
172. L. Pauling, *Am. Sci., 36,* 51(1948).
173. W. Jencks, *Catalysis in Chemistry and Enzymology,* New York, McGraw-Hill, 1969.
174. A. Tramontano, et al., *Science, 234,* 1566(1986).
175. S. J. Pollack, et al., *Science, 234,* 1570(1986).
176. P. G. Schultz, *Science, 240,* 426(1988).
177. R. A. Lerner and S. J. Benkovic, *Bioessays, 9,* 107(1988).
178. P. G. Schultz, *Acc. Chem. Res., 22,* 287(1989).
179. P. G. Schultz, *Angew. Chem. Int. Ed. Eng., 28,* 1283(1989).
180. B. S. Green, *Biotechnol. Processes, 11,* 359(1989).
181. G. A. Krafft and G. T. Wang, *Ann. Rep. Med. Chem., 25,* 299(1990).
182. P. G. Schultz, et al., *Chem. Eng. News, May 28,* 1990, p. 26.
183. R. D. Mayforth, *Designing Antibodies,* San Diego, Academic Press, 1993, p. 167.
184. P. G. Schultz and R. A. Lerner, *Acc. Chem. Res., 26,* 391(1993).
185. W. Bains, *Biotechnology from A to Z,* Oxford, Oxford University Press, 1993, p. 19.
186. P. C. Zamecnik and M. L. Stephenson, *Proc. Natl. Acad. Sci., 75,* 280(1978).
187. H. M. Weintraub, *Sci. Am., 262,* 40(1990).
188. C. Helene and J. Toulme, *Biochim. Biophys. Acta, 1049,* 99(1990).
189. P. J. Green, et al., *Annu. Rev. Biochem., 55,* 569(1986).
190. B. L. Bass and H. M. Weintraub, *Cell, 55,* 1089(1988).
191. R. W. Wagner, et al., *Proc. Natl. Acad. Sci. U.S.A., 86,* 5586(1989).
192. A. M. Thayer, *Chem. Eng. News, Dec. 3,* 1990, p. 17.
193. E. Uhlmann and A. Peymann, *Chem. Rev., 90,* 544(1990).
194. J. F. Milligan, et al., *J. Med. Chem., 36,* 1923(1993).
195. J. Toulme, in *Antisense RNA and DNA,* J. A. H. Murray, Ed., New York, Wiley-Liss, NY, 1992, p. 175.
196. P. Mohan in *Biotechnology and Pharmacy,* J. M. Pezzuto, et al., Eds., New York, Chapman & Hall, 1993, p. 256.
197. H. J. Allen and E. C. Kisalus, *Glycoconjugated—Composition, Structure and Function,* New York, Marcel Dekker, 1992.
198. A Varki, *Glycobiology, 3,* 97(1993).
199. M. S. Dordal, et al., *Endocrinology, 116,* 2293(1985).
200. K. Yamaguchi, et al., *J. Biol. Chem., 266,* 20434(1991).
201. S. Dube, et al., *J. Biol. Chem., 268,* 399(1993).
202. D. McCormick, in *Biotechnology—The Science and the Business,* V. Moses and R. E. Cape, Eds., Chur, Switzerland, Harwood Academic Publishers GmbH, 1991, p. 408.
203. J. H. Musser, *Ann. Rep. Med. Chem., 27,* 301(1992).
204. T. Imura, et al., *Semin. Cancer Biol., 2,* 129(1991).
205. J. D. Coombes and M. Evans, *Ann. Rep. Med. Chem., 24,* 207(1989).
206. M. Gronow, in *Biotechnology—The Science and the Business,* V. Moses and R. E. Cape, Eds., Chur, Switzerland, Harwood Academic Publishers GmbH, 1991, p. 355.
207. E. A. H. Hall, *Biosensors,* Buckingham, Open University Press, 1990.
208. A. E. G. Cass, *Biosensors, A Practical Approach,* Oxford, Oxford University Press, 1990.

Chapter 29

VITAMINS AND COENZYMES

Raymond W. Doskotch and Robert W. Curley, Jr.

Vitamins are organic substances that, along with carbohydrates, proteins, fats, and inorganic salts, are required in the diet for normal health and growth of an organism. This need results from the inability of cells to produce these compounds, although not all organisms require vitamins, and some need more than others. At one end of the scale are the blue-green algae, which survive and function solely on inorganic salts. Other living systems require varying degrees of preformed nutrients, some of which are classified as vitamins.

The minute quantities involved indicate a catalytic role in the cell; many function as coenzymes or part of coenzymes. The function of others at the biochemical level is undetermined. The diseases associated with vitamins result, therefore, from inadequate supply or uptake, and the associated pathologic conditions have their origin in the biochemical lesions.

The early history of vitamins can be traced to observations that certain diets produced diseases that could be cured by a supplement of another foodstuff. For example, the ancient Arab physicians and their successors valued the use of liver to cure night blindness (nyctalopia), and the North American Indian knew that scurvy could be treated by a tea made from cedar leaves (*Thuja occidentalis* L.).

In the seventeenth century, it was standard procedure for the British Navy to provide lemon juice for its sailors to prevent scurvy. The Japanese Navy settled on a more balanced diet when it was shown that this disease was due to the inordinate use of rice. The taking of fish liver oil to ward off or cure rickets, a deficiency characterized by faulty bone formation, was well-known by the end of the nineteenth century, but despite this evidence, the vitamin concept was still not formulated. The work on polyneuritis in fowl and the general study of animal nutrition by Hopkins helped to formulate it clearly.

Eijkman, a Dutch physician working in Java at the end of the nineteenth century, observed that chickens fed polished rice developed a disorder involving the peripheral nervous system. This condition, he noticed, was similar to that of patients with beriberi. The addition of rice bran to the food not only prevented the disease, but also cured the afflicted. Rice bran gave an active extract, which was initially, but erroneously, concluded to contain an antidote that neutralized a toxin postulated to be present in the polished kernel. This was not unreasonable because only starch diets caused the disease; meat, for example, did not.

In 1901, Grijns, Eijkman's successor and collaborator, recognized the true nature of the discovery—that they were dealing with an organic material necessary for health, specifically for the nervous system, that was not a protein, a carbohydrate, or a fat. This material was named vitamine by Funk in 1912 because it had basic properties and was essential for life (*vita,* Latin). The term was also used for three other food factors that cured pellagra, rickets, and scurvy, which were likewise believed to be caused by faulty nutrition. Subsequently, the *e* was dropped from vitamine when it was evident that not all such substances were basic.

Another phase of the vitamin story resulted from the evaluation of foods in experiments with animals. Toward the end of the nineteenth century, the Swiss school of Lunin and Socin showed that mice on diets containing artificial mixtures of the separate bulk constituents of milk failed to survive. The proposed explanation concentrated on the lack of inorganic constituents held in organic combination—a form necessary for metabolism—that were separated. Pekelharing of Holland performed similar experiments in 1905 and recognized the importance of the unknown minor substance in milk and other foods. He even attempted to isolate it.

Hopkins, unaware of this work, published in 1912 the classic paper on accessory growth factors. By careful quantitative experiments, in which food consumption was measured and the growth curves determined, he showed that the growth rate of rats was indeed related to the presence of unknown factors when food intake was quantitatively adequate. This work was of tremendous importance in directing a concentrated effort in this field. As a result, Hopkins shared with Eijkman the 1929 Nobel Prize in Medicine.

Extending Hopkins' study, McCollum and Davis in 1915 found evidence for at least two growth factors, one a fat-soluble component designated A and another a water-soluble component B. Component A cured an eye disease caused by a nutritional deficiency, and component B was effective in treating beriberi and was designated the antiberiberi factor, or vitamin B. The two concepts were therefore formally combined. As additional factors were recognized, other letters of the alphabet were assigned; when the original factor was shown to be a mixture of several, numerical subscripts were added. The antiberiberi factor became vitamin B_1. These terms are today used in reference to the physiologic activity of the vitamin, and a trivial name is reserved for a specific compound having that activity; for example, thiamine has vitamin B_1 activity.

In 1935, the cofactor, or coenzyme, required by Warburg's yellow enzyme was shown to be a derivative of vitamin B_2, or riboflavin, and was the first evidence for the long-suspected catalytic role for vitamins. As other coenzymes were discovered, they too were shown to contain vitamins. These findings were not always in this order because nicotinamide was shown to be a part of a coenzyme 2 years before it was discovered to be a vitamin. Complete knowledge of the vitamins, therefore, must include the understanding of their related coenzymes (the biologically active forms) as well as their biochemical role.

Extensive studies have been made to determine the daily requirements for each vitamin necessary to maintain proper nutrition in healthy people, and these results are made known periodically in the United States through the Food and Nutrition Board. These values reflect a primary dietary need and may be inadequate for other conditions. Additional amounts of vitamins are generally required under stress conditions, such as during pregnancy (folic acid), during the metabolism of large quantities of ethanol by an alcoholic, or when inadequate absorption of the fat-soluble vitamins occurs from lack of bile. In addition, some suggest that healthy individuals may benefit from so-called megavitamin therapy with doses substantially in excess of the recommended daily requirements. Vitamins C and E and niacin have received particular attention in this regard.

THIAMINE

The antiberiberi vitamin, or thiamine, was the first to be isolated after Eijkman's early observations on polyneuritis in fowl. Although many attempts were made and numerous concentrates prepared, it was not until 1926, more than 30 years later, that the crystalline material was obtained.[1] Success was due to the use of a large amount of rice bran, an adsorption step on fuller's earth to concentrate the chemically sensitive material, and a quick 10-day bioassay using small rice birds to follow the purification.

Structure and Synthesis

The structure of thiamine was determined to be 3-(4-amino-2-methyl-5-pyrimidinylmethyl)-5-(β-hydroxyethyl)-4-methylthiazolium chloride hydrochloride by Williams and co-workers in 1936.[2] Contributors from European laboratories also played an important role.

The key to the problem was the facile cleavage of thiamine at room temperature with sodium sulfite to produce 4-amino-2-methyl-5-pyrimidinyl-methyl sulfonic acid and 5-(β-hydroxyethyl)-4-methylthiazole, which were then characterized individually. Within 1 year after the structure was established, syntheses of thiamine were announced from three laboratories. The first report involved preparing the pyrimidine and thiazole moieties separately and then a final joining.[3] This procedure served as the basis, with subsequent modifications, for the commercial production of the vitamin (Fig. 29–1).

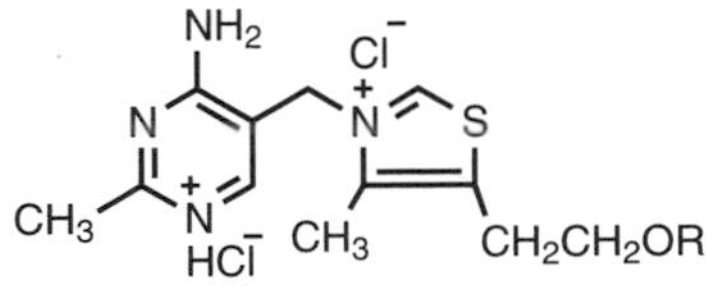

Thiamine: R = H

Thiamine pyrophosphate: R = $P_2O_6^{-3}$

(or Thiamine diphosphate)

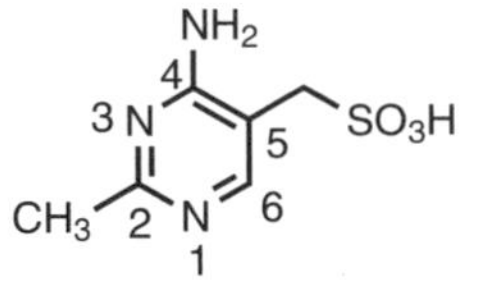

4-Amino-2-methyl--5-pyrimidinyl methyl sulfonic acid

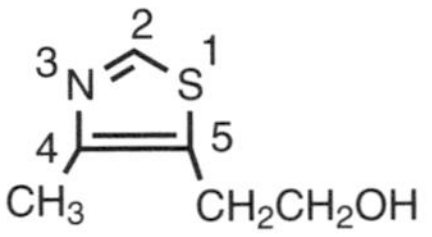

5-(β-Hydroxyethyl)-4-methylthiazole

After thiamine is absorbed by the organism, it must be converted to the biologically active thiamine pyrophosphate (or diphosphate), first isolated as cocarboxylase, the coenzyme for a decarboxylating enzyme. The monophosphate and triphosphate are not biologically active. Thiamine pyrophosphate functions in carbohydrate metabolism in the decarboxylation of α-keto acids, such as pyruvic and α-ketoglutaric acids, and in the transfer of acyl groups in the transketolase and phosphotransketolase reactions (Fig. 29–2).

These conversions have been extensively studied and can be grouped together as involving a transfer of aldehyde.[4] For example, pyruvic acid, through a nucleophilic attack by the ylid of thiamine pyrophosphate, forms the active acetaldehyde intermediate after the loss of carbon dioxide.

The active acetaldehyde can be transferred to an acceptor or broken down to yield acetaldehyde. Anaerobic glycolysis leading to the production of ethanol results from reduction of the released acetaldehyde by alcohol dehydrogenase. Acetoin, a product of microbial fermentation, is produced when acetaldehyde is the acceptor; in the pyruvic dehydrogenase reaction, lipoic acid is the temporary acceptor, which passes on the group to coenzyme A for utilization in the citric acid cycle.

In a similar manner, thiamine pyrophosphate serves to transfer glycolaldehyde ($HOCH_2CHO$) in the transketolase reaction, the intermediate being the 2-(1,2-dihydroxyethyl)thiazole analog.[5] The function of the pyrimidine moiety of thiamine in these conversions is to enhance the activity of the thiazole unit by inductive electron withdrawal. There is evidence, although unclear and

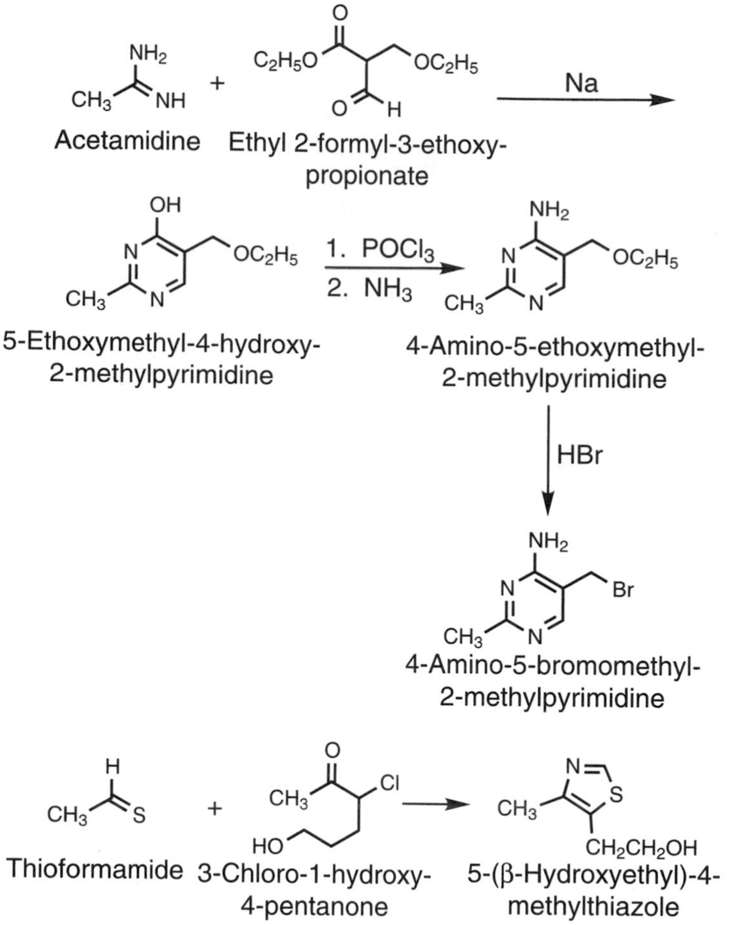

Fig. 29–1. Synthesis of pyrimidine and thiazole moieties of thiamine.

equivocal, that thiamine diphosphate or triphosphate plays a biophysical role in neural conduction.

Many analogs of thiamine were prepared, and in this way the structural features necessary for the vitamin activity were determined. For example, the methylene bridge between the thiazole and pyrimidine groups is essential, as are the hydroxyethyl group of the thiazole and the amino group in the pyrimidine ring. Replacement of the hydrogen at carbon 2 of the thiazole ring by a methyl group causes total loss of activity and is consonant with

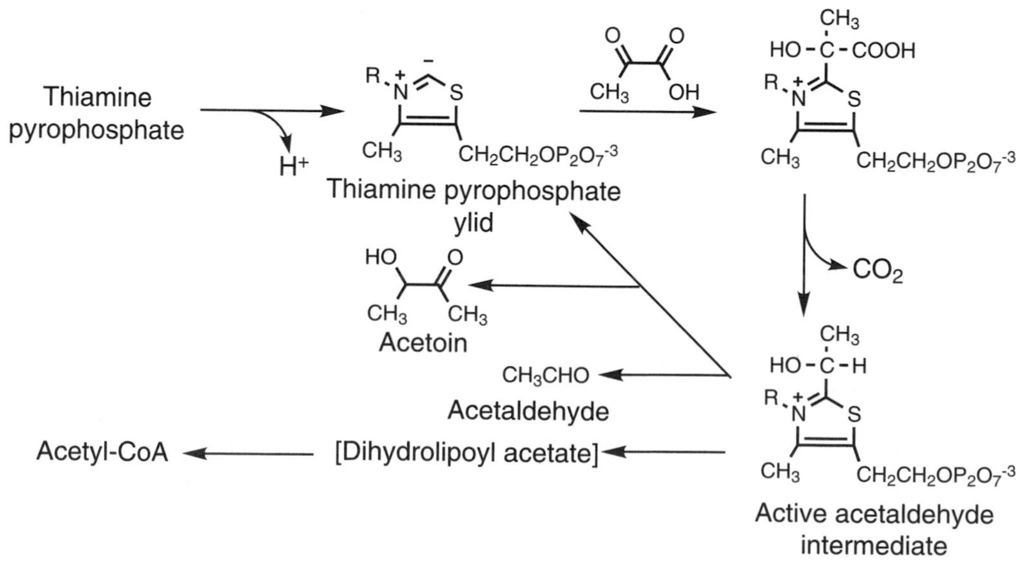

Fig. 29–2. Metabolic rate of thiamine pyrophosphate.

the mechanism of thiamine action, but the methyl group at carbon 2 of the pyrimidine is not required because the ethyl analog is active. The propyl derivative is inactive, and the butyl derivative is an antagonist.

Two other potent antivitamins that have been useful in studying thiamine action are oxythiamine and neopyrithiamine. When added to diets containing sufficient amounts of thiamine, these compounds produce symptoms of thiamine deficiency in animals. It is believed that oxythiamine competitively inhibits thiamine pyrophosphate and becomes active after phosphorylation, whereas neopyrithiamine prevents the conversion of thiamine to thiamine pyrophosphate.

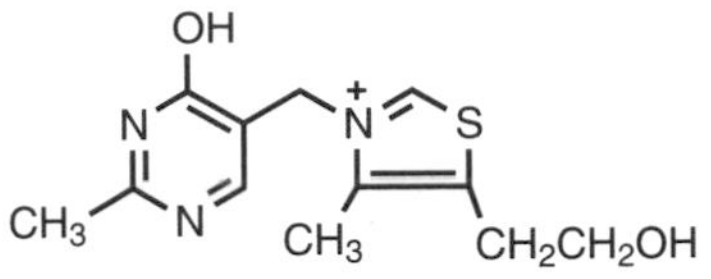

Oxythiamine

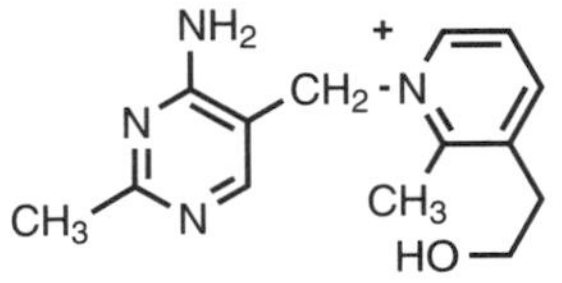

Neopyrithiamine

In the organisms that make thiamine, the pyrimidine and thiazole portions are synthesized separately, but the exact steps are yet to be determined. Salmonella typhimurium uses 5-aminoimidazole 3-ribonucleotide, a purine intermediate, as a precursor of the pyrimidine moiety.[6] In baker's yeast, these intermediates are suitably phosphorylated by the appropriate kinases—the pyrimidine to the pyrophosphate and the thiazole to the phosphate. The product of the two units is thiamine monophosphate, phosphorylation of which yields the coenzyme thiamine pyrophosphate. Phosphatases dephosphorylate the monophosphate to the vitamin, which can pass freely across cell membranes, but once the pyrophosphate group is added, movement out of the cell ceases.

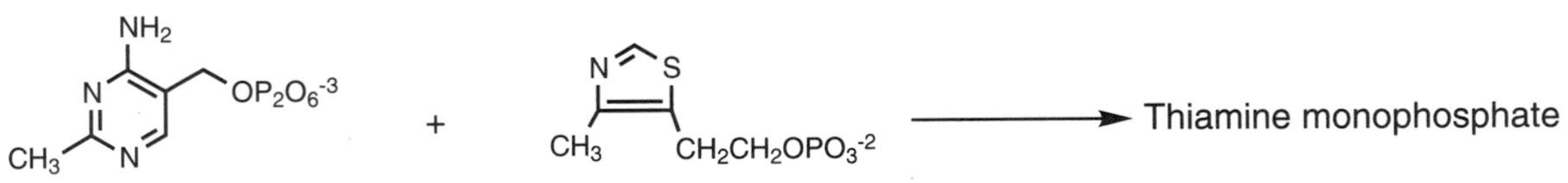

4-Amino-5-hydroxymethyl-2-methyl-pyrimidine pyrophosphate + 5-(β-Hydroxyethyl)-4-methyl-thiazole phosphate → Thiamine monophosphate

Chemical and Biologic Assays

Thiamine can be assayed by a variety of chemical and biologic methods. The thiochrome method, which can measure a few nanograms of the vitamin, is based on the case of conversion of thiamine by alkaline ferricyanide to thiochrome, a blue fluorescent compound.

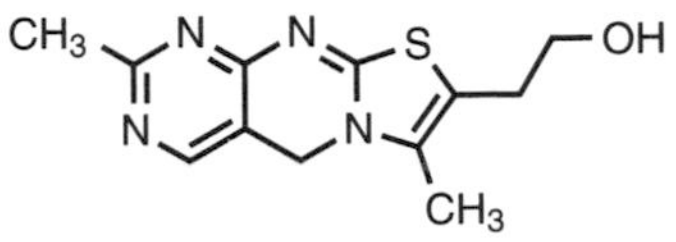

Thiochrome

The fluorescence is measured after an extraction with isobutanol to remove the thiamine pigment from interfering substances. The phosphate esters of thiamine are not extracted, but a test on the enzymatically dephosphorylated sample indicates the total thiamine content; the procedure can therefore give values for the free and esterified forms.

The biologic tests with animals, which are generally time-consuming and costly, can involve either a curative or a protective (prevention of symptoms of deficiency) end point, or the growth rate can be compared to standards. Many biologic methods that overcome some of the drawbacks with animals take advantage of microorganisms that require thiamine for growth. A direct relationship between the amount of thiamine and the quantity of cell growth can be easily established. A manometric method (CO_2 release) employing yeast carboxylase is used if only the quantity of thiamine pyrophosphate must be determined.

Deficiency

Beriberi, the thiamine deficiency syndrome, is rare in the western world but is found in parts of Asia, where polished rice is a dietary staple. The disease occurs in the wet form or in the dry form, which is more severe. Pyruvate and lactate accumulate in the tissues, especially the blood and brain. Recovery is rapid after daily treatment with thiamine (20 to 100 mg orally), provided that irreversible damage has not occurred.

A form of polyneuritis common to alcoholics results from decreased thiamine intake and a simultaneous high intake of calories, which requires an increased amount of thiamine.

Daily Requirements

The normal daily requirement for the vitamin is 0.5 mg/1000 calories up to 3000 calories and then 0.3 mg/additional 1000 calories. Foods rich in thiamine are liver, pork, uncooked green vegetables, and whole cereals.

RIBOFLAVIN

With the isolation of pure thiamine and its addition to a basal animal diet, it was found that inadequate growth and poor health still resulted; supplements of yeast extracts, however, prevented this condition. A second nutritional factor was therefore proposed and designated vitamin B_2 in Britain and vitamin G in the United States. In contrast to thiamine, this factor was heat-stable and appeared at that time to be a cure for a rat dermatitis that was believed to be similar to pellagra in humans. Purification of the thermostable factor employing the rat dermatitis cure as the bioassay resulted in recognition of still other food factors because the purer fractions of vitamin B_2 were still inadequate for health and growth. These were grouped together as the vitamin B_2 complex.

Progress leading to the isolation of riboflavin was made when its growth-promoting activity was studied rather than its pellagra-curing properties. Yeast, liver, heart, and kidney extracts all provided purified fractions, but egg-white extracts, having fewer contaminants, yielded the crystalline vitamin named ovaflavin (the flavin from egg).[7] Active fractions were noted to be greenish yellow and to fluoresce a yellowish green; this observation considerably simplified the isolation problem. Subsequently, other sources yielded the vitamin, notably whey, which provided lactoflavin in quantities required for chemical study.

The name riboflavin was adopted when it was shown that a ribityl moiety is part of the structure. The vitamin is an optically active, light-sensitive amphoteric substance that decomposes at about 280°C, shows only slight solubility in neutral polar solvents, and decomposes in alkaline solutions, in which it dissolves more readily. It is relatively stable under acid conditions.

Structure and Synthesis

Riboflavin, $C_{17}H_{20}N_4O_6$, has the structure 7,8-dimethyl-10-(1′-D-ribityl)isoalloxazine.

Photochemical cleavage of riboflavin under alkaline conditions produces lumiflavin, $C_{13}H_{12}N_4O_2$, which furnishes urea and a bicyclic carboxylic acid on alkaline hydrolysis.[8] On further hydrolysis, the carboxylic acid produces 4-amino-1,2-dimethyl-5-methylaminobenzene.[9] This compound, with alloxan, regenerates lumiflavin.

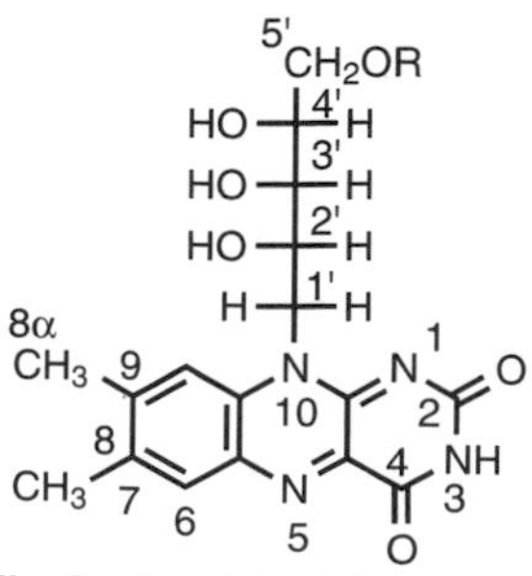

Riboflavin: R = H

Riboflavin-5′-phosphate: R = PO_3^{-2}

In the formation of lumiflavin, a $C_4H_8O_4$ unit is lost, and because riboflavin forms a tetraacetate on acetylation, and formaldehyde on lead tetraacetate cleavage, a linear four-carbon polyol was proposed. Its attachment must be to the methyl carbon of lumiflavin because riboflavin does not contain an N-methyl group. The synthesis of riboflavin from 4-amino-5-carbomethoxyamino-1,2-dimethylbenzene, D-ribose, and alloxan settled the structural problem, including the stereochemistry of the polyhydroxy unit.[10]

Several syntheses of riboflavin have been developed (Fig. 29–3), and because the ribose unit is the most difficult to obtain, only the processes successfully solving this problem are of commercial importance. The Merck procedure begins with D-gluconic acid, which is converted to D-ribonic acid and then to D-ribose tetraacetate.[11] The ribose tetraacetate is condensed with m-xylidine, and the product is reduced to 1,2-dimethyl-4-tetraacetyl-D-ribitylaminobenzene and coupled with a ρ-nitrobenzenediazonium salt. The azobenzene product is treated with barbituric acid and, after methanolysis, yields riboflavin.

Microbiologic synthesis by bacteria (Clostridium acetobutylicum), yeasts (Candida flaveri), or fungi (Eremothecium ashbyii and Ashbya gossypii) are also of commercial importance.

The biologic function of riboflavin was established within a few years after its discovery, when Warburg and Christian in 1932 isolated the "old yellow enzyme" (a

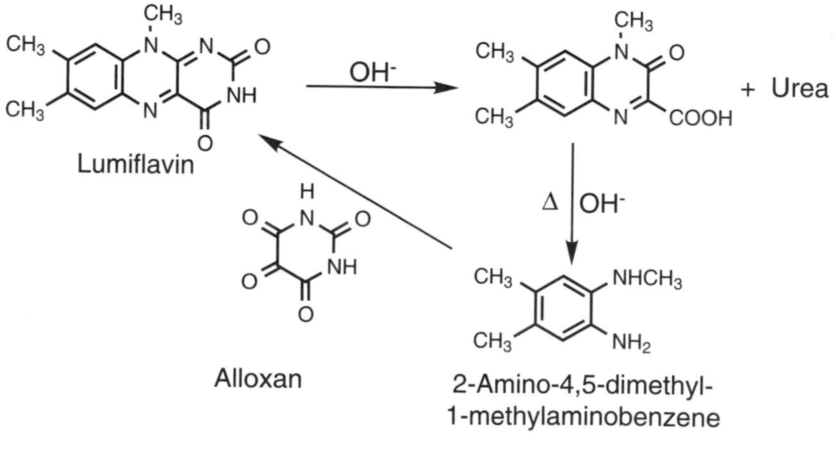

Lumiflavin

+ Urea

Alloxan

2-Amino-4,5-dimethyl-1-methylaminobenzene

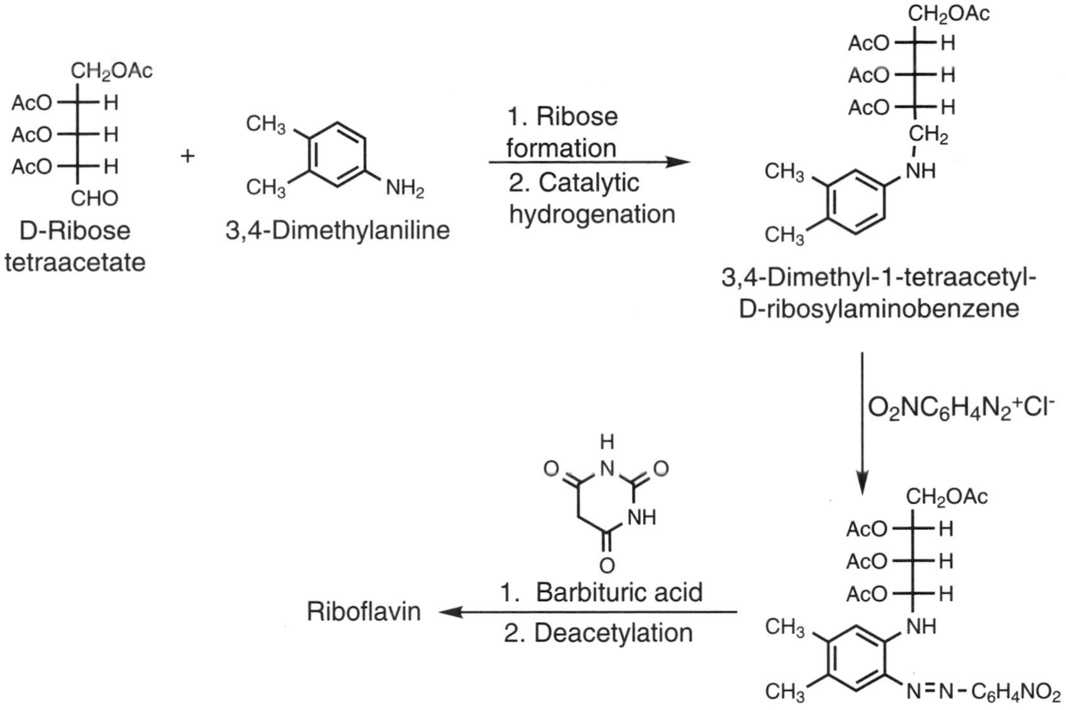

D-Ribose tetraacetate + 3,4-Dimethylaniline → (1. Ribose formation; 2. Catalytic hydrogenation) → 3,4-Dimethyl-1-tetraacetyl-D-ribosylaminobenzene → ($O_2NC_6H_4N_2^+Cl^-$) → azo compound → (1. Barbituric acid; 2. Deacetylation) → Riboflavin

Fig. 29–3. Synthesis of riboflavin.

flavoprotein) from erythrocytes involved in the oxidation of glucose-6-phosphate. The enzyme was separated into a colorless protein and a yellow prosthetic group, which was spectroscopically similar to riboflavin but contained a phosphate group. By 1936, riboflavin-5′-phosphate, or flavin mononucleotide (FMN), as it is now known, was synthesized and shown to be the colored component of the yellow enzyme. This was the first synthesis of a prosthetic group.[12] Strictly speaking, it is misnamed as a nucleotide because it is not derived from ribose but from ribitol. Riboflavin is converted to flavin mononucleotide by an enzyme that requires adenosine triphosphate (ATP).

Other flavoproteins contain yet another riboflavin coenzyme, flavin adenine dinucleotide (FAD), which is composed of flavin mononucleotide and adenosine-5′-phosphate. This coenzyme is biosynthesized from riboflavin-5′-phosphate and ATP with release of inorganic pyrophosphate.[13,14] The riboflavin coenzymes function as hydrogen-transferring cofactors, accepting hydrogen from a substrate (as in the succinic dehydrogenase reaction) or serving as a carrier between the nicotinamide coenzymes and the cytochromes in the electron transport system. In succinic dehydrogenase, the FAD coenzyme is covalently linked through the 8α-carbon to the 3-nitrogen of a protein histidyl residue.[15] A similar covalent form is present in liver monoamine oxidase but with the bond to the sulfur of a cysteine group.[16] The hydrogen is accepted by the isoalloxazine moiety of the molecule in two separate one-electron transfers to form first a free radical semiquinone intermediate, which can be detected by electron paramagnetic resonance spectroscopy and then

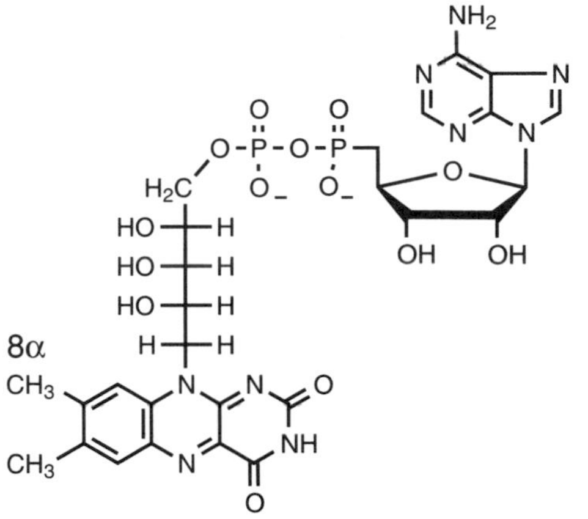

Flavin adenine dinucleotide

is converted to the colorless dihydroriboflavin derivative (Fig. 29–4).

Singularly, riboflavin and its coenzymes have uniform ultraviolet absorption spectra, fluorescence spectra, and oxidation-reduction potentials, but when they are combined with the enzyme protein, these properties are changed, suggesting a binding at the isoalloxazine ring. Another point of attachment is the phosphate group. The 3-substituted riboflavins are nonfluorescent, as are the protein-bound coenzymes; this suggests that the NH group is required for protein binding. A supporting observation is that the 3-substituted coenzymes do not form enzyme complexes.

Analogs of riboflavin delineate the structural require-

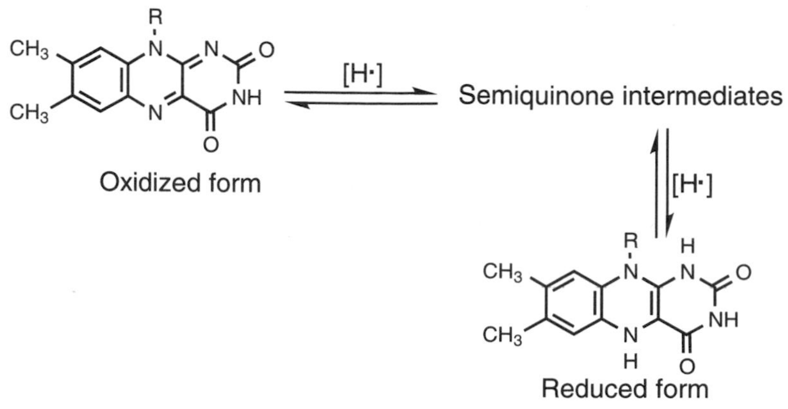

Fig. 29–4. Riboflavin in one-electron transfer reaction.

ments. To promote growth in animals, the flavin must have a D-ribityl group and a methyl group at position 7 or 8, or at both, although an ethyl group at position 7 is possible. Isoriboflavin (the 6,7-dimethyl analog) is an antagonist, and the absence of methyl groups resulted in a toxic substance. Of the glycityl derivatives tested, the L-ribityl, L-arabityl, D-xylityl, and D-lyxityl derivatives were inactive, whereas the D-arabityl and D-dulcityl (or galactoflavin) compounds were antagonists.

Riboflavin is biosynthesized by green plants and most microorganisms, with the initial stages consisting of the pathway for the formation of purines. Subsequently the purine nucleoside guanosine is converted to 6,7-dimethyl-8-ribityllumazine through 5-amino-2,6-dihydroxy-4-ribityl-aminopyrimidine, which is also a by-product of the final step to riboflavin (Fig. 29–5). Two acetates or a four-carbon carbohydrate unit are required in this conversion. In a unique reaction catalyzed by riboflavin synthetase, two lumazine units serve as substrate; one is split at the diazine ring to release the pyrimidine by-product with the remaining four-carbon unit becoming attached to the second lumazine at the methyl groups to form riboflavin.[17] The cross-addition of this unit, as illustrated in the diagram, should be noted. The four different marks (○, ■, Δ, *) show the origin of the eight-carbon unit of riboflavin from two lumazine precursors.

Chemical and Biologic Assays

The assaying of riboflavin content in materials is performed almost exclusively by two methods: a microbiologic procedure (using Lactobacillus casei) of Snell and Strong[18] or a modification thereof or a chemical assay based on the fluorescence of riboflavin. Although animal growth tests can be used, the success of other methods has all but eliminated their routine use.

Deficiency

Riboflavin deficiency disease in humans has been found only in association with deficiencies of other B complex vitamins, especially niacin and thiamine. Volunteers on riboflavin-free diets develop a lesion of the legs (chelosis) and a seborrheic dermatitis in the facial folds,

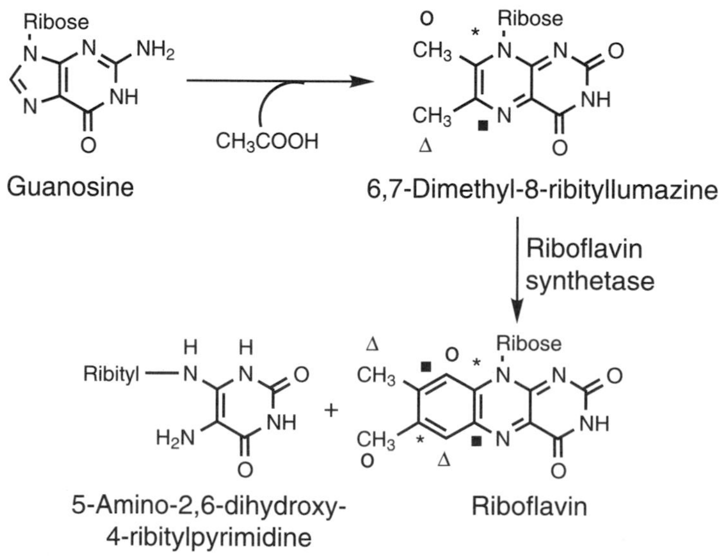

Fig. 29–5. Biosynthesis of riboflavin.

especially about the nose, ears, and eyelids. Ocular disturbances, such as corneal inflammation and photophobia, are also characteristic symptoms.

Daily Requirements

Daily requirements of riboflavin are from 0.6 mg for children to about 1.8 mg for adults, with increased amounts needed during pregnancy and lactation. Good dietary sources are heart, liver, kidney, milk, eggs, and green vegetables.

NICOTINIC ACID AND NICOTINAMIDE

Nicotinic acid, also known as niacin, or its amide is the antipellagra vitamin.Pellagra was recognized in 1706 as a disease prevalent in areas of the world in which corn was the principal component of the diet. Studies in the United States by Goldberger[19] established that it was dietary in origin because supplements of milk and eggs prevented its outbreak and cured the stricken. Certain amino acids, especially tryptophan, were found to be helpful.

Isolation of the vitamin was hampered for many years by the lack of a proper animal assay that reflected the condition in humans. It was only when dogs fed pellagra-producing diets developed the canine black tongue disease that an assay was available. As a result, nicotinamide was isolated from liver extracts.[20,21] Along with nicotinic acid, it was shown to cure black tongue in dogs and pellagra in humans.

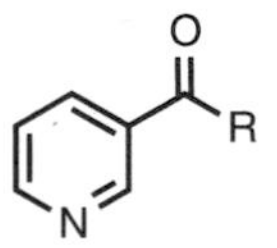

Nicotinamide: R = NH_2
Nicotinic acid: R = OH

Nicotinic acid has been known since 1867, when it was obtained by nitric acid oxidation of nicotine, the major alkaloid of tobacco. Its structure was determined a number of years later, and it was obtained coincidentally as an isolate in the search for the antiberiberi factor, but its significance was not recognized at that time.

Structure and Synthesis

Nicotinic acid and nicotinamide are synthesized commercially by several methods. One procedure involves the formation of 3-cyanopyridine from pyridine followed by hydrolysis to yield the acid; another involves oxidation with alkaline hydrogen peroxide to give the amide.

Nicotinamide is a component of two important coenzymes, nicotinamide adenine dinucleotide (NAD), formerly known as diphosphopyridine nucleotide (DPN) or coenzyme I, and nicotinamide adenine dinucleotide phosphate (NADP), also called triphosphopyridine nucleotide (TPN) or coenzyme II. In 1909, NAD was first encountered as the dialyzable, heat-stable cofactor named cozymase necessary for the formation of ethanol from glucose by a cell-free enzyme preparation.[22] In 1936, NADP was recognized as the cofactor from erythrocytes that was necessary for the oxidation of glucose-6-phosphate. Shortly thereafter, NAD was also shown to contain nicotinamide. Chemical and enzyme degradation studies established the structure for these coenzymes, and final confirmation came with the total synthesis of NAD many years later.[23,24]

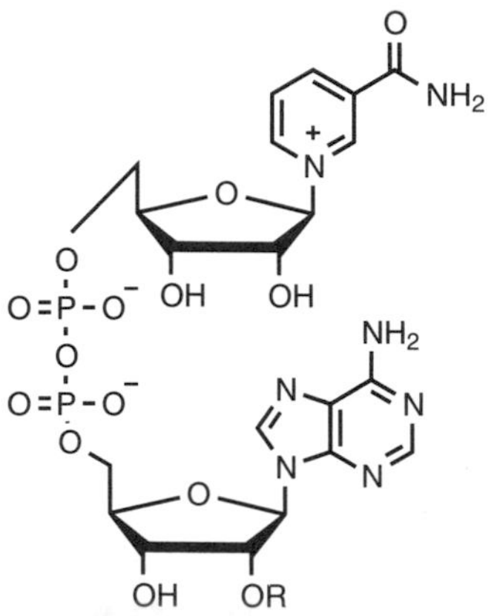

Nicotinamide adenine dinucleotide: R = H
Nicotinamide adenine dinucleotide phosphate: R = PO_3^{-2}

The nicotinamide coenzymes participate in hydrogen-transfer reactions, functioning as the hydride ion carriers of biologic systems. In the dehydrogenation of the substrate, a hydride transfer to position 4 of the nicotinamide ring occurs,[25] and a proton, the other element of the hydrogen molecule, is released into the medium. The reduced form shows an absorption maximum at 340 nm, which is used extensively in enzyme studies to follow its presence.

Because the dihydronicotinamide group has a prochiral carbon at position 4, a stereochemical addition of hydride is suggested. Indeed, enzymes catalyze this transfer stereospecifically to either side (A or B, or *pro-R* or *pro-S*) but not to both. The reduced coenzymes can be used for synthesis if required, or they may transfer the reducing power via a flavoprotein through the electron

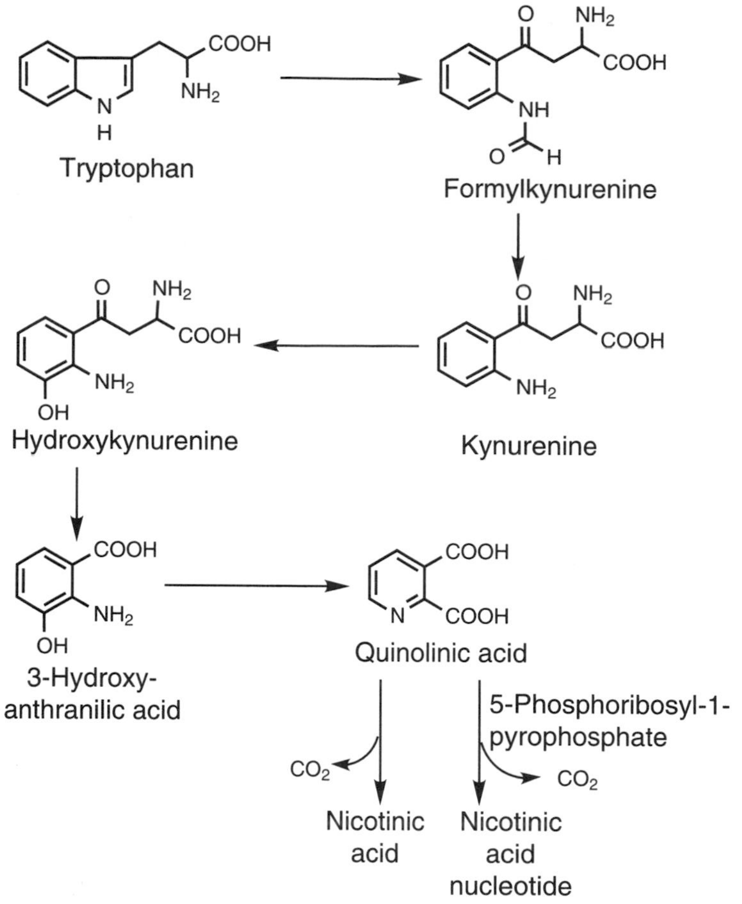

Fig. 29–6. Biosynthesis of nicotinic acid from tryptophan.

transport system to generate ATP. Reversible hydrogen transfer between NADH and NADP is catalyzed by transhydrogenase enzymes.

Many mammalian species do not require nicotinic acid as such but can form it from tryptophan. This explains the beneficial effect of tryptophan when added to the pellagra-producing corn diet because zein, the corn protein, has only a small amount of this amino acid. The formation of nicotinic acid is believed to proceed as shown in Figure 29–6. Quinolinic acid could yield nicotinic acid directly or the nucleotide with 5-phosphoribosyl-1-pyrophosphate. In humans, about 1 mg of nicotinic acid is produced from about 60 mg of tryptophan.

Nicotinic acid is not converted directly to nicotinamide; instead, desamido-NAD is formed first from nicotinic acid nucleotide and amidated in a step requiring ATP and glutamine. Nicotinamide forms nicotinamide nucleotide and then, with ATP, yields NAD (Fig. 29–7).

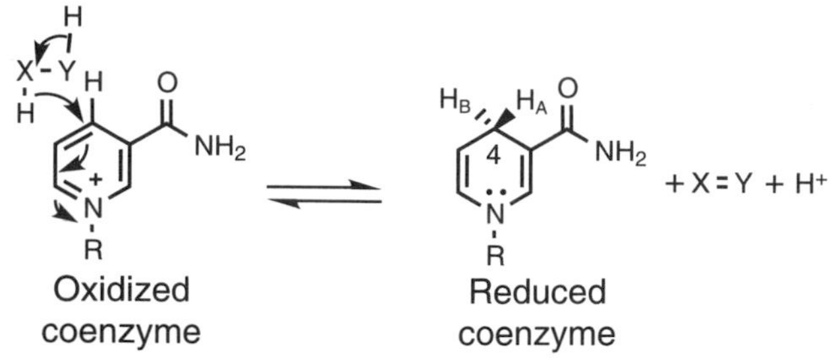

Chemical and Biologic Assays

Nicotinic acid and its amide can be assayed chemically, microbiologically, or biologically. The amount of growth of Lactobacillus arabinosus is directly related to the amount of vitamin present. The growth can be measured turbidimetrically, and by comparison with standards, the amount of vitamin is determined. Chicks, dogs, and rats on purified diets can be used to evaluate the pellagra-preventing properties of the test material, but these tests are less convenient to conduct routinely. The chemical methods, although simple, lack specificity because other pyridine derivatives interfere, but if suitable chromatographic separations are first conducted, a reliable value can be obtained.

Deficiency

Pellagra is characterized by bilaterally symmetric lesions on the sides of the body and hands. There is hyperpigmentation and thickening of the skin, inflammation of the tongue and mouth, and an alimentary disorder as indicated by indigestion, anorexia, and diarrhea. In later stages, irritability, insomnia, amnesia, and delirium appear.

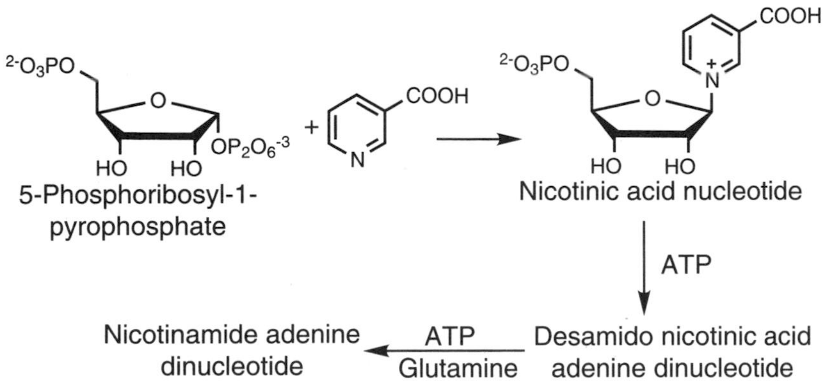

5-Phosphoribosyl-1-pyrophosphate + nicotinic acid → Nicotinic acid nucleotide → (ATP) Desamido nicotinic acid adenine dinucleotide → (ATP, Glutamine) Nicotinamide adenine dinucleotide

Fig. 29–7. Biosynthesis of nicotinamide adenine dinucleotide from nicotinic acid.

Daily Requirements

The minimum daily requirement for nicotinic acid is about 20 mg, but in the treatment of pellagra, doses of 50 to 500 mg of nicotinamide are used because it is better tolerated as a therapeutic agent than nicotinic acid.

PYRIDOXINE

Animals fed purified diets supplemented with concentrates of thiamine and riboflavin did not show good growth and developed a dermatitis characterized by redness and swelling of the extremities. This condition, called acrodynia, was prevented by the addition of yeast and was traced to the vitamin B complex of the water-soluble fraction. It was specifically designated vitamin B_6. In 1938, no less than five reports from different laboratories documented the isolation of this vitamin from yeast and rice bran.[26–31] This substance was given the trivial name pyridoxine.

Structure and Synthesis

Pyridoxine is a 4,5-di(hydroxymethyl)-3-hydroxy-2-methylpyridine, and it has been renamed pyridoxol. Its

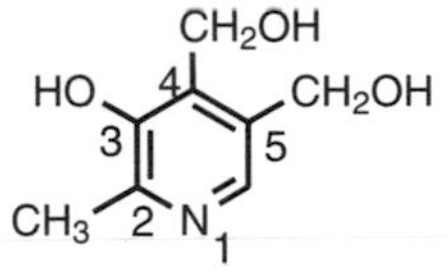

Pyridoxol

structure was established from chemical and spectroscopic evidence,[32] in particular, the ultraviolet spectrum, which was similar to that of several 3-hydroxypyridines. Methylation of the phenolic group and oxidation of the

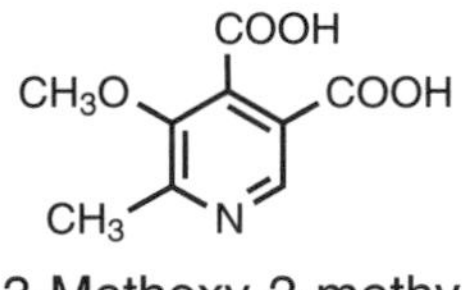

3-Methoxy-2-methyl-pyridine-4,5-dicarboxylic acid

product to the dicarboxylic acid located the hydroxymethyl groups at positions 4 and 5, based on a characteristic color test ($FeCl_2$) not given by 2-carboxypyridine or 6-carboxypyridine. The synthesis of the diacid confirmed the test and located the C-methyl group at position 2.

Numerous syntheses of the vitamin have been recorded, and commercial production is based on a sequence beginning with ethyl N-formyl-d,l-alaninate to form an oxazole, which accepts the dienophile 1,4-diacetoxybut-2-ene (Fig. 29–8). Hydrolysis of the adduct gives pyridoxol.[33]

For certain lactic acid bacteria, other compounds were shown by Snell and co-workers[34] to have greater vitamin B_6 activity than pyridoxol. These were identified as pyridoxal and pyridoxamine, and their constitution was proved by synthesis. In humans, the three pyridoxins (the general name for these compounds) are readily interchangeable and have the same activity. The 4-carboxy analog, however, is inactive and is found as a product of excretion.

In the cell, pyridoxal and pyridoxamine are present in the coenzyme form as the corresponding 5-phosphates. An ATP-requiring kinase phosphorylates the three pyridoxins, and pyridoxal-5-phosphate is formed from both pyridoxal-5-phosphate and pyridoxamine-5-phosphate by an oxidase.

Vitamin B_6 coenzymes function mainly in amino acid metabolism. The reactions proceed with pyridoxal-5-phosphate and the amino acid to form, reversibly, a Schiff base generally stabilized by coordination with a metal ion.

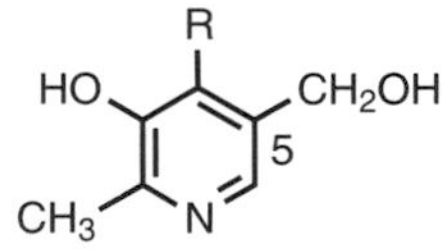

Pyridoxal: R = CHO
Pyridoxamine: R = CH_2NH_2
4-Carboxypyridoxine: R = COOH

The α-carbon bonds of the complexed amino acid become labilized, and model studies involving various metal ions have provided an understanding of the enzyme-directed cleavages.[35,36]

Decarboxylations result when bond *a* is broken (equation 1); racemizations (equation 2) and formation of α-

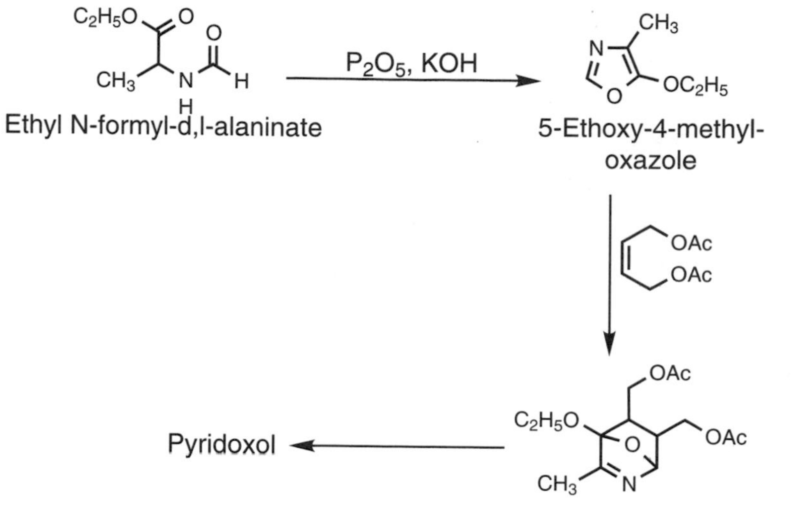

Fig. 29–8. Commercial synthesis of pyridoxol.

keto acids, as in the aminotransferase reaction (equation 3) are a consequence of bond *b* cleavage; and aldol-type cleavage (equation 4) occurs when bond *c* is involved. Reactions in the side chain such as β-elimination and γ-elimination also require the pyridoxal coenzyme. In addition, phosphorylases, the enzymes that cleave $\alpha[1 \rightarrow 4]$polyglucans by phosphorolysis to glucose-1-phosphate, contain pyridoxal phosphate, but the mechanism of action is unknown. It does not, however, involve a Schiff base.

Pyridoxal or the 5-phosphate, although stable when pure, can be readily destroyed when present in a mixture because the aldehyde function can react with amino acids, amines, and sulfhydryl groups, especially if heavy metal ions are present. Cooking of foods can, therefore, result in considerable loss of vitamin B_6 activity.

The biosynthesis of pyridoxol has been studied by use of radiolabeled precursors in a mutant of Escherichia coli that is blocked between pyridoxine and pyridoxal. The carbon skeleton is derived from glycerol and a two-carbon unit at the oxidation level of acetaldehyde.[37]

Chemical and Biologic Assays

Vitamin B_6 activity can be assayed enzymatically by the glutamate-aspartate aminotransferase reaction; microbiologically by use of a number of microorganisms; and chemically by the use of colorimetric, spectrophotometric, and fluorometric techniques. It can also be assayed biologically by using test animals on purified diets, but this is time-consuming and expensive and consequently used infrequently. The microbiologic methods are preferred. The major difficulty with the other methods concerns accurate assessment of B_6 activity because quantita-

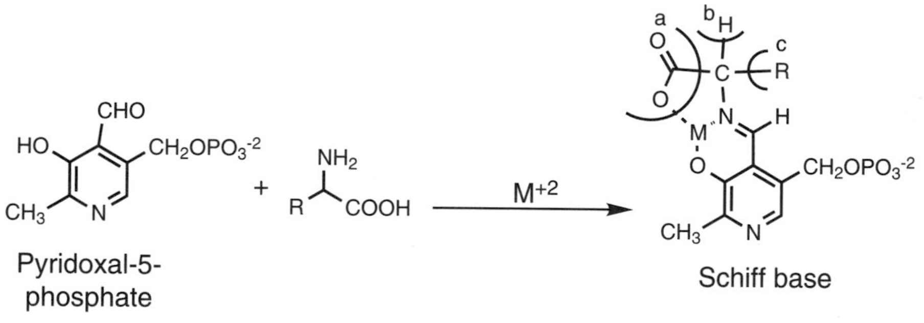

		Equations
Histidine ⟶	CO_2 + Histamine	1
L-Alanine ⟶	D-Alanine	2
α-Ketoglutarate + Alanine ⟶	Glutamate + Pyruvate	3
Serine ⟶	Glycine + Formaldehyde	4

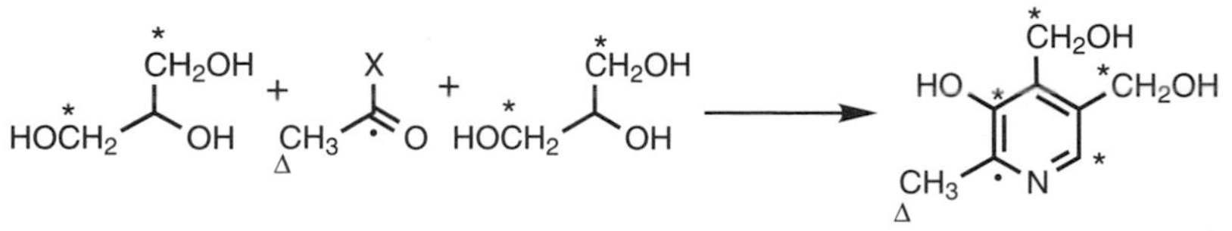

tive extraction of the pyridoxins from mixtures, especially blood, is not always possible.

Deficiency

The need for pyridoxins in human nutrition is clear. Subjects with deficiencies induced by antagonists, such as 4-desoxypyridoxol, show lesions about the eyes, nose, and mouth similar to those observed in riboflavin and niacin deficiencies, together with an altered blood picture. Infants on restricted pyridoxol intake develop nervous disorders, which are corrected by pyridoxol therapy. A sensitive test for detecting early vitamin B_6 deficiency is based on the level of xanthurenic acid in the urine, as the result of a decrease in the kynureninase activity. (Pyridoxal phosphate is a coenzyme.) Kynurenine, a breakdown product of tryptophan, is normally converted to kynurenic acid, but in vitamin B_6 deficiency it is shunted to form xanthurenic acid.

Daily Requirements

The daily requirement of pyridoxol is estimated at 1 to 2 mg, and because it is widely distributed in meat, liver, vegetables, eggs, and unmilled cereals, ordinary diets should supply the need. An increase is required, however, during pregnancy and when the antitubercular agent isonicotinic acid hydrazide is taken. The latter depletes the vitamin by hydrazone formation.

PANTOTHENIC ACID

The discovery of pantothenic (Greek, meaning from everywhere) acid resulted from investigations of growth requirements for certain microorganisms and of the curing and preventive factor of a chick dermatitis. In the first instance, it was a component of bios, the name given to the material present in yeast, beef, and malt extracts that stimulated the growth of yeast when added to a simple medium of sugar and salts. In the second, it was the filtrate factor, also known as vitamin B_3, that was part of the heat stable components of the vitamin B complex but, in contrast to pyridoxol, was not adsorbed on fuller's earth. A short history of pantothenic acid is given by Williams.[38]

Structure and Synthesis

The structure of pantothenic acid was established indirectly by examining the partially purified yeast-stimulating factor for characteristic structural features and then by synthesizing the postulated substance. It was found that the molecular weight was near 200, that hydroxyl groups were present, and that basic properties were lacking, as were sulfhydryl, ketonic, olefinic, aldehydic, aromatic, and carbohydrate groups.

Hydrolysis yielded β-alanine (a yeast growth stimulator) and an impure, partially characterized lactone that, when treated with a β-alanine ester, gave pantothenic acid after hydrolysis of the ester group. The lactone was

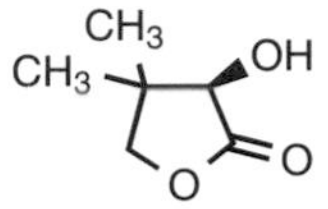

α-Hydroxy-β,β-dimethyl-γ-butyrolactone

identified as α-hydroxy-β,β-dimethyl-γ-butyrolactone and synthesized. Resolution into its optical antipodes showed that only the (R)-isomer yielded fully active pan-

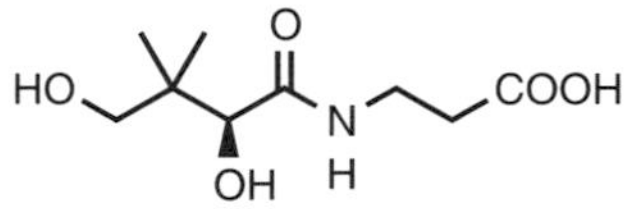

Pantothenic acid

tothenic acid; the other was completely inactive. The vitamin is a viscous oil that is relatively unstable to heat, acid, and base and is best handled as the calcium salt.

The commercial synthesis of calcium pantothenate

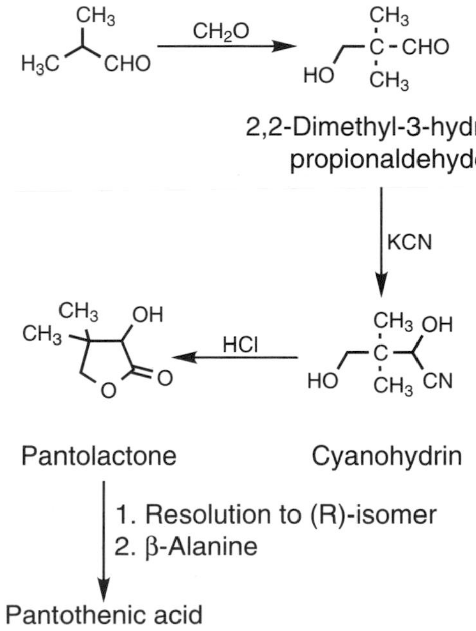

2,2-Dimethyl-3-hydroxy-propionaldehyde

KCN

HCl

Pantolactone

Cyanohydrin

1. Resolution to (R)-isomer
2. β-Alanine

Pantothenic acid

Fig. 29–9. Commercial synthesis of pantothenic acid.

starts with isobutyraldehyde and formaldehyde to form 2,2-dimethyl-3-hydroxypropionaldehyde, which with potassium cyanide yields the cyanohydrin. Racemic pantolactone, formed on acid hydrolysis of the cyanohydrin, is treated with *β*-alanine, after resolution to the (*R*)-isomer, to give pantothenic acid (Fig. 29–9).

The coenzyme form of pantothenic acid is coenzyme A (abbreviated CoA), which was originally discovered as the cofactor necessary for acetylation reactions. Purified fractions of this coenzyme were cleaved by degradative enzymes, and from the reaction products a structure was proposed that was proved correct by synthesis.[39–41] Coenzyme A functions as an acyl carrier, the acyl unit being attached to the sulfhydryl group in a thioester linkage.

Acetyl-CoA formed from pyruvate or some other donor initiates the Krebs cycle by the addition of the acetyl

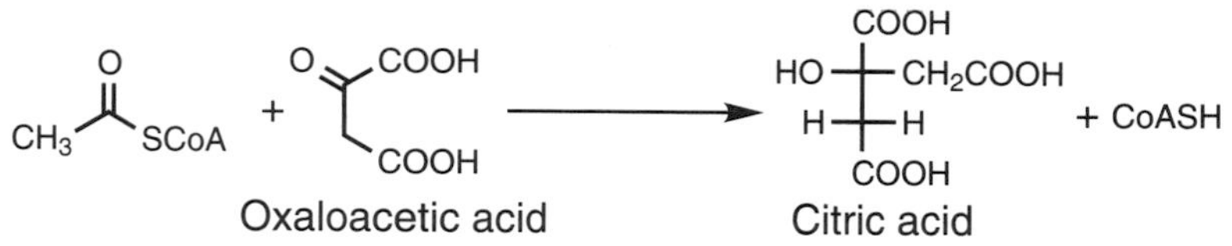

$CH_3COSCoA$ + Oxaloacetic acid → Citric acid + CoASH

group to oxoloacetic acid or is directed to the formation of fatty acids or terpenes, in which case the chain-extending unit is malonyl-CoA formed from acetyl-CoA and car-

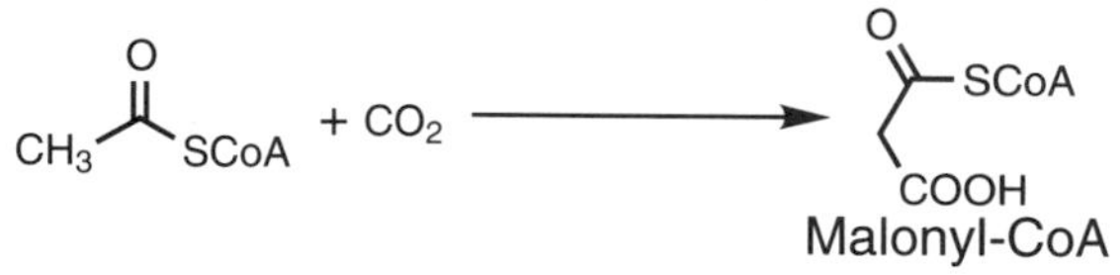

$CH_3COSCoA + CO_2$ → Malonyl-CoA

bon dioxide. Derivatives of coenzyme A are also involved in fatty acid breakdown, amino acid metabolism, and heme biosynthesis, to name a few.

The cytoplasmic fatty acid synthesizing system uses a protein analog of CoA (instead of CoA), which is called acyl carrier protein. In CoA, pantotheine-4-phosphate is attached to adenosine-5′-phosphate; in acyl carrier protein from E. coli, it is joined to a serine hydroxyl of the heat-stable protein that has 77 amino acids (but no cystine), the sequence of which has been determined.[42] The importance of pantotheine derivatives, briefly illustrated here, explains the reason for the presence of pantothenic acid in all cells.

Pantoyltaurine and *ω*-methylpantothenic acid are antagonists that have been used to block pantothenic acid utilization.

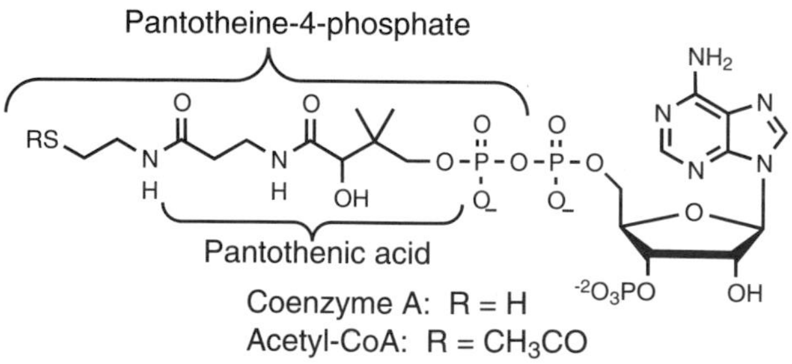

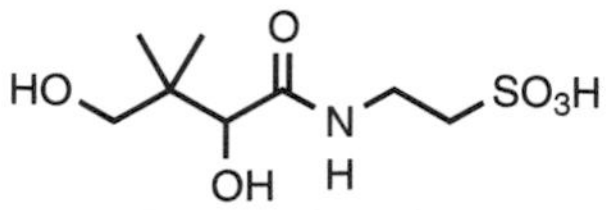

Pantotheine-4-phosphate

Pantothenic acid

Coenzyme A: R = H

Acetyl-CoA: R = CH_3CO

Pantoyltaurine

The *ω*-methylpantothenic acid was tried in humans, but the resulting symptoms could not be alleviated by administration of pantothenic acid alone. Addition of cortisone did produce remission.

Pantothenic acid is synthesized from valine by microorganisms by the pathway shown in Figure 29–10 to yield pantoic acid, which is condensed with *β*-alanine derived from aspartic acid.

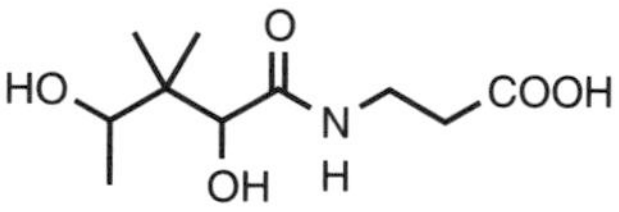

ω-Methylpantothenic acid

The formation of CoA proceeds by phosphorylation of pantothenic acid with ATP to the 4-phosphate, which with cysteine gives 4-phosphopantothenylcysteine; on decarboxylation, this product yields 4-phosphopantotheine. With ATP, dephospho-CoA is formed which is then phosphorylated at the 3′ position of the ribose to generate coenzyme A.

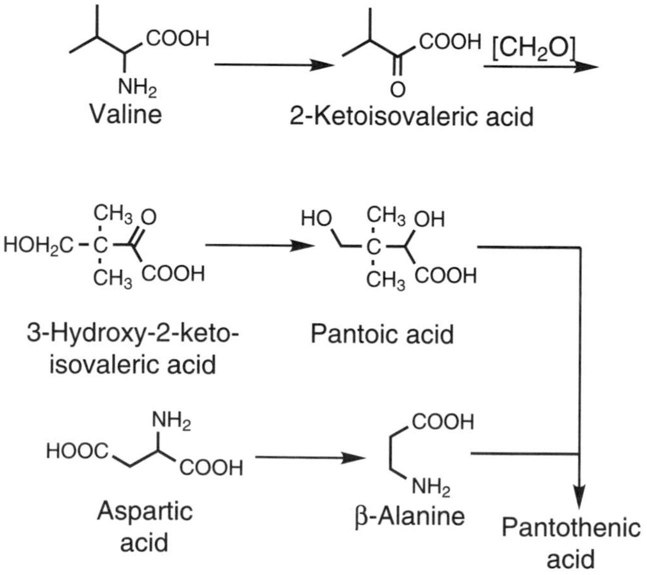

Valine → 2-Ketoisovaleric acid → $[CH_2O]$ → 3-Hydroxy-2-keto-isovaleric acid → Pantoic acid

Aspartic acid → β-Alanine

Pantoic acid + β-Alanine → Pantothenic acid

Fig. 29–10. Synthesis of pantothenic acid.

Chemical and Biologic Assays

Pantothenic acid can be easily and accurately assayed microbiologically by a number of organisms, one of which is Lactobacillus plantarum. The chemical methods are most successful with relatively pure mixtures, whereas the tests with animals, although indicating the total vitamin activity, are less often used because of the inconvenience and cost.

Deficiency

There is no record of pantothenic acid deficiency disease in humans, but the need for the vitamin has been shown by using synthetic diets lacking only pantothenic acid and including an antagonist to deplete the system further. Various symptoms resulted, involving the cardiovascular, gastrointestinal, respiratory, and nervous systems.

Daily Requirements

Pantothenic acid is abundant in most foods, especially liver, eggs, beef, cabbage, peas, and skim milk. Natural sources with the greatest amounts are royal jelly and codfish ovaries. Cooking does not appreciably decrease its content. A minimal dietary requirement has not been established, but the amount in an average American diet, which appears adequate, is about 10 mg.

BIOTIN

The isolation of biotin as the methyl ester from egg yolk was reported by Kögl and Tönnis[43] in 1936 as the growth factor for yeast, one of the constituents of bios. Four years later, vitamin H, or the "egg white injury factor," was isolated from liver concentrates and found to be identical to biotin. This unusual injury, a deficiency disease, is induced by a diet that consists of a substantial amount of uncooked egg white, which contains avidin, a protein that tenaciously binds biotin, thereby depriving the animal of this vitamin. By providing a biotin supplement in excess of the binding capacity of avidin, the disease is alleviated.

Structure and Synthesis

Biotin has the empiric formula $C_{10}H_{16}O_3N_2S$. The structure has been elucidated.[44,45] On hydrolysis, it yields a diaminocarboxylic acid that with phosgene regenerates biotin, suggesting a cyclic urea-type structure (Fig. 29–11). The amino groups are primary, as established by a Van Slyke determination. Oxidation of the diamine produces the six-carbon dicarboxylic acid, adipic acid. The sulfur atom is disubstituted because it forms a sulfone with hydrogen peroxide. Double bonds are not present; the molecular formula therefore requires a bicyclic structure.

Hofmann degradation of the diamine produces δ-(2-thienyl)valeric acid; formation of a quinoxaline with the diamine and phenanthraquinone establishes the 1,2-arrangement of the amino groups. Preparation of the quinoxaline from desthiobiotin confirmed the biotin structure except for stereochemistry. With three asymmetric centers, eight isomers are possible. These have been made synthetically.

The Merck synthesis[46] gave access to three of the four racemates, and only one isomer, the positive rotating, all-*cis* biotin, had biologic activity. The fourth racemate, obtained by another route, was likewise inactive. The commercial synthesis (Hoffman-LaRoche) (Fig. 29–12) differs from the others, which start with the thiophene ring, by forming first the imidazole portion, avoiding in this way the formation of the *trans* ring junction isomers. The compound *meso*-bisbenzylamine succinic acid is converted to the bicyclic *cis*-thiolactone by a series of steps. The side chain was added through the Grignard reagent and the product dehydrated and then hydrogenated stereospecifically to give the all-*cis*-ethyl ether. Cleavage of

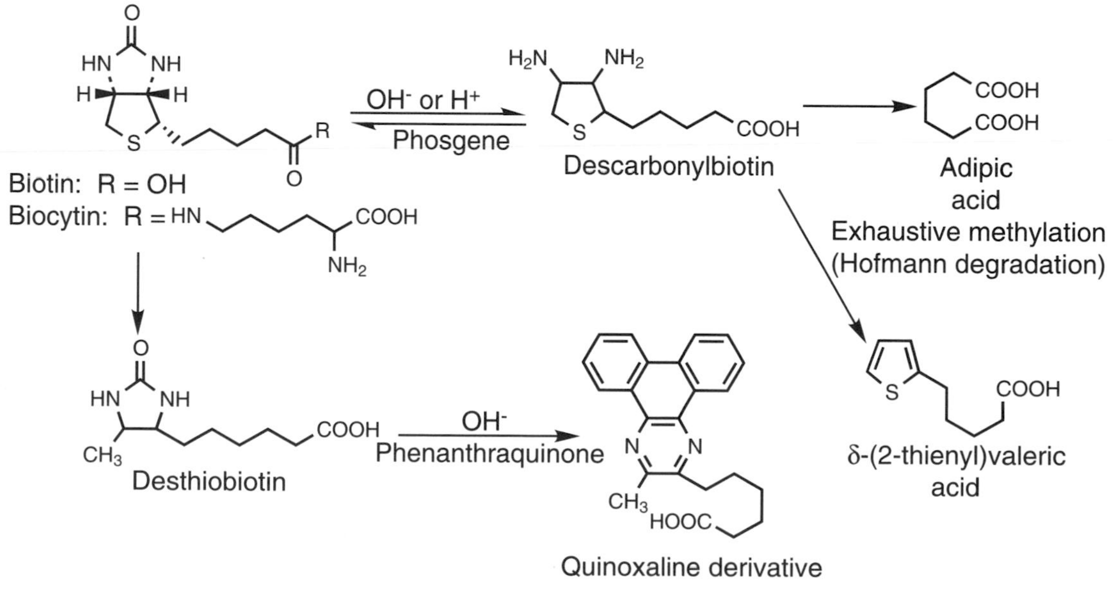

Biotin: R = OH
Biocytin: R = HN–(CH₂)₄–CH(NH₂)–COOH

OH⁻ or H⁺ / Phosgene — Descarbonylbiotin — Adipic acid

Exhaustive methylation (Hofmann degradation) — δ-(2-thienyl)valeric acid

Desthiobiotin — OH⁻, Phenanthraquinone — Quinoxaline derivative

Fig. 29–11. Chemical identification of biotin.

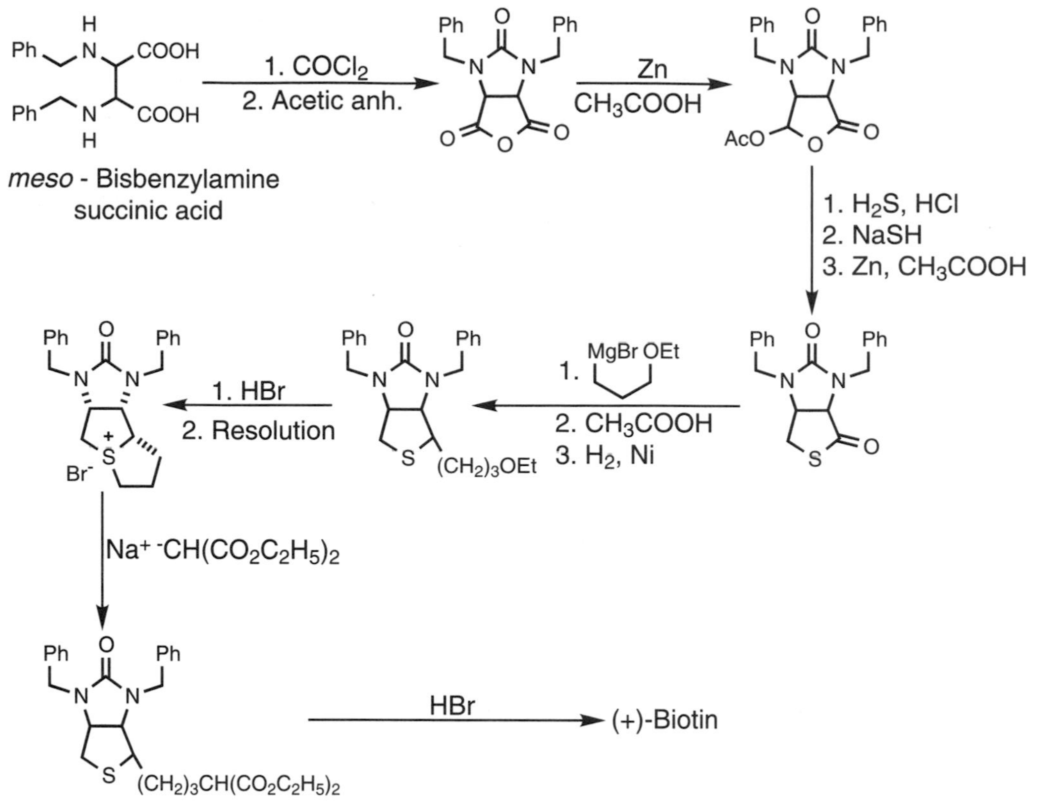

Fig. 29–12. A commercial synthesis of biotin.

the ether formed the transient alkyl bromide that generated the racemic sulfonium salt, which was resolved with d-camphorsulfonate anion. Sodium diethyl malonate opened the ring, hydrolysis of the malonic ester with hydrobromic acid removed the protecting groups, and concomitant decarboxylation formed (+)-biotin. There has been a renewed interest in synthetic methods[47] for biotin generated by new applications in the area of nutrition and growth promotion.

The biologic function of biotin is concerned with the nonphotosynthetic fixation of carbon dioxide, by enzymes called carboxylases, with the transfer of carboxy groups by transcarboxylases and with decarboxylations. Tissues from biotin-deficient animals show, for example, a reduction in the conversion of propionic acid to succinic acid, which is reversed by biotin supplementation. Although the presence of biotin or a dissociable coenzyme form has not been demonstrated, avidin can, however, inhibit the reaction stoichiometrically. In these enzymes, biotin is bound to the protein by an amide linkage through its carboxyl group to that of the ϵ-amino group of lysine. Activation of carbon dioxide proceeds by the following reaction with formation of a biotin-CO_2 intermediate, which then transfers the activated group to the substrate in another step.

$$\text{Biotin-enzyme} + \text{ATP} + CO_2 \xrightarrow{Mg^{2+}} CO_2\text{-biotin enzyme} + \text{ADP} + \text{Pi}$$

Characterization of the biotin-CO_2 complex was made with the beef liver enzyme propionyl carboxylase, which converts propionyl-CoA to methylmalonyl-CoA. The carboxylated enzyme was degraded and yielded ϵ-N-(1′-N-carboxyl-(+)-biotinyl)-lysine.[48] Similar enzymes, such as β-methylcrotonyl-CoA carboxylase (in leucine degradation), acetyl-CoA carboxylase (in fatty acid synthesis), and pyruvate carboxylase (in formation of oxoloacetic acid), likewise involve N-carboxybiotin. The transcarboxylase and decarboxylase reactions, however, do not require ATP to form the N-carboxybiotin intermediate.

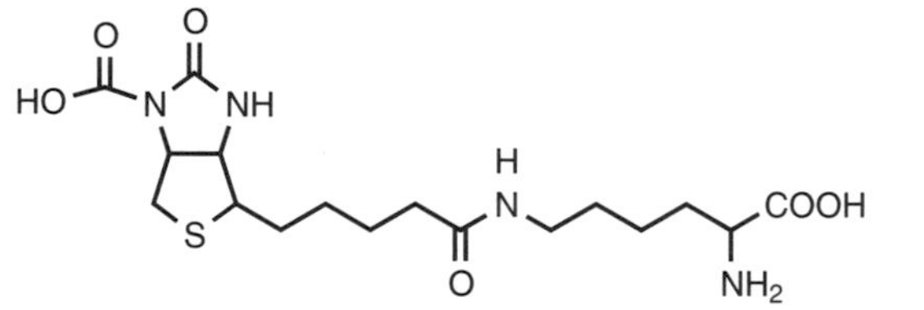

ε-N-(1'-N-Carboxyl-(+)-biotinyl)-lysine

Of the biotin analogs that have been prepared, desthiobiotin is of special interest because it can have vitamin activity or no activity, or be an antagonist, depending on the microorganism under study.

Oxybiotin, in which the sulfur is replaced by oxygen, can replace biotin in microorganisms and animals but has less activity. Biotin sulfoxide, which was first obtained from milk, and also prepared by oxidation of biotin with 1 mol of hydrogen peroxide, has two isomers (because

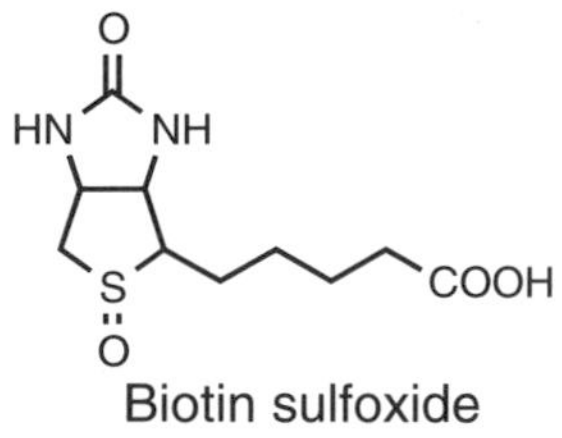

Biotin sulfoxide

of asymmetry of the sulfoxide group); the dextrorotatory form has high biotin activity, but the levorotatory form is inactive.

Biotin sulfone is either a potent antagonist or at best only weakly active.

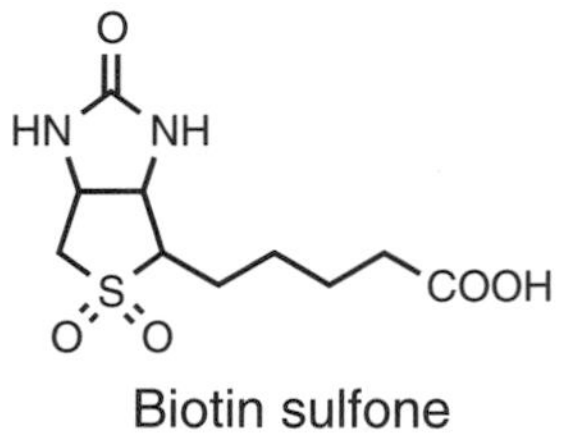

Biotin sulfone

Variation in the length of the side chain affects activity. Norbiotin (with three methylene groups in the side chain), homobiotin (with five methylene groups), and other with longer chains are antagonists.

Avidin, the classic biotin antagonist, is a protein (molecular weight 70,000) from egg white, which has four subunits and four specific binding sites for the vitamin or its derivatives.[49] A natural function for this material is not known. The tight avidin-biotin complex can be broken by denaturation (steaming) to release biotin, but it is surprisingly stable to mild heat, acid, and proteolytic enzymes; this accounts for its passage through the alimentary tract without destruction.

Biosynthesis of biotin has been studied with mutant microorganisms and purified enzymes; it proceeds from pimeloyl CoA and L-alanine to give 7-keto-8-aminopelargonic acid. The second amino group is obtained from S-adenosylmethionine and the carbonyl of the ureido group from carbon dioxide. Introduction of the sulfur to desthiobiotin in the final step is not completely understood, but methionine and inorganic sulfur compounds are more effective donors than cysteine or its derivatives.[50,51]

Chemical and Biologic Assays

The estimation of biotin in foods is most conveniently done by microbiologic methods, of which the Wright and Skeggs method[52] with lactic acid bacteria is the most widely employed. Because none of the bound forms of biotin, of which biocytin (ϵ-N-biotinyl-L-lysine)[53] is a well-known example, is active in this test, prior hydrolysis with a strong acid is required for estimation of full potency. Chemical and physical methods are not available, and animal tests are not convenient.

Deficiency

Biotin deficiency is unknown in humans except as induced in four volunteers consuming a diet low in biotin and fortified with egg white and in one individual whose unusual eating habits included the intake of two to six dozen raw eggs per week for several years. The symptoms in these cases involve a mild dermatitis, muscle aches, lethargy, anorexia, and nausea. A daily dose of 150 μg of biotin for about a week brought about complete recovery.

Daily Requirements

The richest dietary sources are yeast, liver, and egg yolk. Although no minimal daily requirements have been determined for biotin, between 150 and 300 μg appear to suffice. Intestinal flora can be a source of this vitamin but to what extent is an open question.

FOLIC ACID

The history of the folic acid group of vitamins is complicated and was made clear only after isolation and charac-

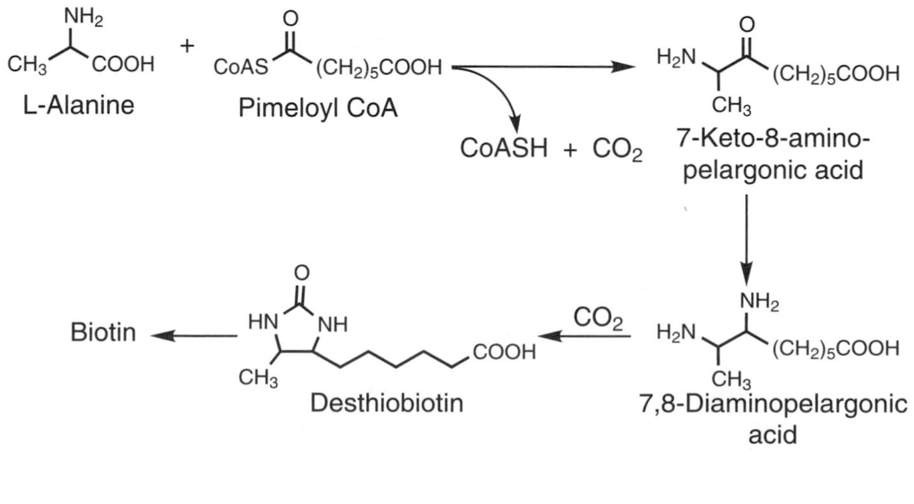

L-Alanine + Pimeloyl CoA → 7-Keto-8-aminopelargonic acid ($CoASH + CO_2$)

7-Keto-8-aminopelargonic acid → 7,8-Diaminopelargonic acid

7,8-Diaminopelargonic acid + CO_2 → Desthiobiotin → Biotin

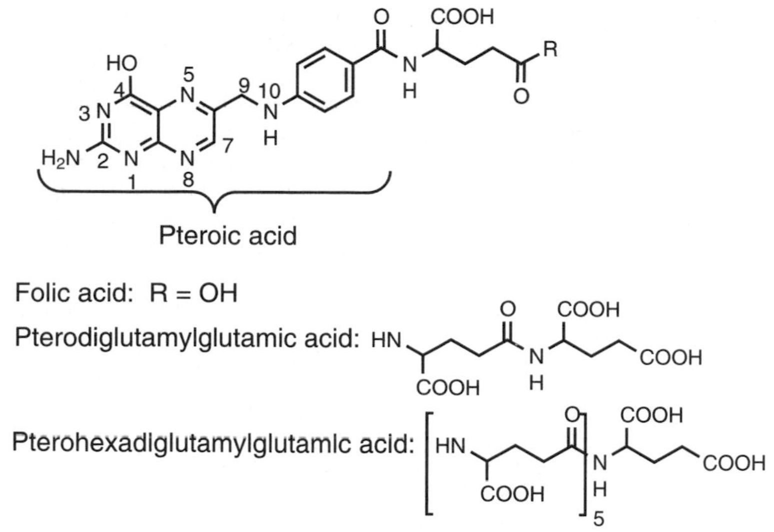

Pteroic acid

Folic acid: R = OH

Pterodiglutamylglutamic acid:

Pterohexadiglutamylglutamic acid:

terization of the active compounds. The first member, folic acid (Latin: *folium,* or leaf), was obtained from spinach leaves as the growth factor for Streptococcus faecalis R., although it was also effective for Lactobacillus casei.[54] Extracts from liver, yeast, and other sources also stimulated growth in lactic acid bacteria but showed quantitative differences from one test organism to another, indicating the existence of several forms of the vitamin. Concurrent studies with animals showed the microbiologically active factors to prevent a macrocytic anemia in monkeys and deficiency diseases in chicks, guinea pigs, and rats. Folic acid has been called a hemopoietic vitamin because it is involved in the development of blood cells.

Structure and Synthesis

Folic acid, or pteroylglutamic acid, is an optically active, slightly water-soluble, yellow-orange crystalline solid with the systematic name of N-[4-{[(2-amino-4-hydroxy-6-pteridinyl)methyl]amino}benzoyl]glutamic acid. Its structure was established in 1948.[55]

Alkaline hydrolysis of folic acid in air (oxygen) gives two products: a fluorescent dibasic acid, $C_7H_5N_5O_3$, and a diazotizable amine. The dibasic acid can be decarboxylated and on treatment with hypochlorous and gives guanidine. From this and other information, the pteridine structure was proposed for the dibasic acid and confirmed by synthesis from 4-hydroxy-2,5,6-triaminopyrimidine and diethyl mesoxalate.

Extended studies, including behavior of model compounds, established the carboxyl carbon as the point of attachment to the other product of the oxidative cleavage reaction and that it was originally present as a methylene group because anaerobic alkaline hydrolysis does not cause cleavage. The diazotizable amine was shown to be ρ-aminobenzoylglutamic acid, forming ρ-aminobenzoic acid and L-glutamic acid on acid hydrolysis.

Two other substances with activity similar to that of

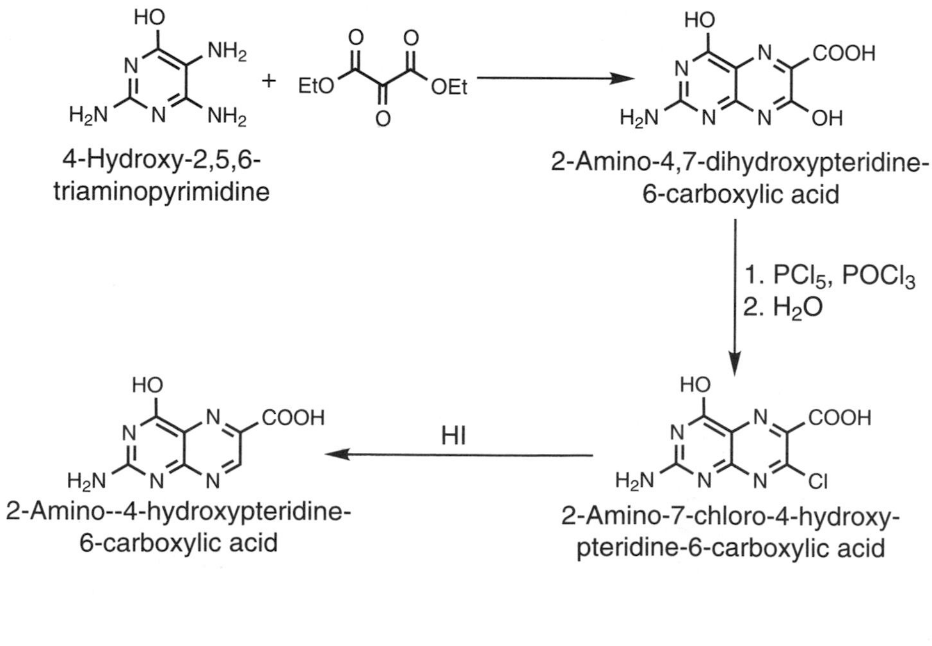

4-Hydroxy-2,5,6-triaminopyrimidine

2-Amino-4,7-dihydroxypteridine-6-carboxylic acid

1. PCl_5, $POCl_3$
2. H_2O

HI

2-Amino--4-hydroxypteridine-6-carboxylic acid

2-Amino-7-chloro-4-hydroxy-pteridine-6-carboxylic acid

folic acid were characterized from their acid hydrolysis products. The yeast-derived material that was only half as active as the vitamin in S. faecalis furnished, in addition to racemic folic acid, 2 mol of L-glutamic acid. Aerobic alkaline hydrolysis gave ρ-aminobenzoylglutamic acid, the pteridine, and 2 mol of glutamic acid. This factor, pteroyldiglutamylglutamic acid, has the glutamyl units in amide linkage through the γ-carboxyl group. The second factor, also from yeast and named vitamin B_C conjugate, is active in chicks but ineffective in L. casei and S. faecalis. It was determined to be pteroylhexaglutamylglutamic acid. Species unable to use the folic acid conjugates lack the enzyme conjugase (usually found in the intestinal mucosa) that liberates free folic acid.

Folic acid has been synthesized in a one-step procedure that requires three reagents—4-hydroxy-2,5,6-triaminopyrimidine, α,β-dibromopropionaldehyde, and ρ-aminobenzoyl-L-glutamic acid—and a buffer at pH 4. No improvement in the overall yield was observed when dibromopropionaldehyde reacted first with the pyrimidine or ρ-aminobenzoylglutamic acid and then with the third component.

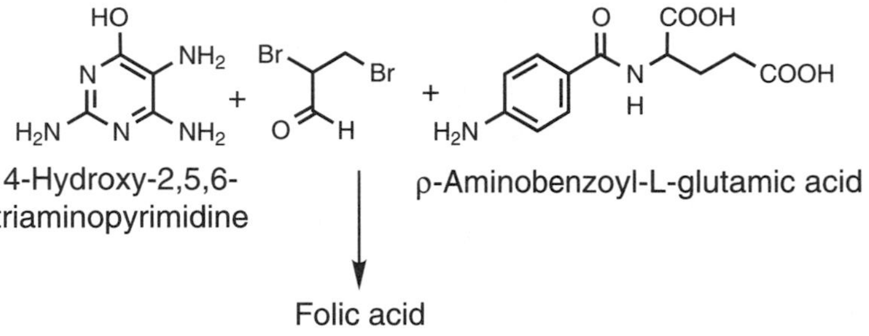

4-Hydroxy-2,5,6-triaminopyrimidine + ρ-Aminobenzoyl-L-glutamic acid → Folic acid

The various coenzyme forms of folic acid are all involved in the metabolism of one-carbon units. Before folic acid can function, however, it must be reduced by the enzyme dihydrofolate reductase to 5,6(S),7,8-tetrahydrofolic acid. The one-carbon units carried by this coenzyme can be at the level of oxidation corresponding to formic acid, formaldehyde, or methanol and are interrelated through a series of reactions depicted in Figure 29–13, in which only the relevant part of the structures is given. The one-carbon unit, depending on the coenzyme form, can be at position N^5 or N^{10} or bridged between N^5 and N^{10}. These coenzymes participate in the synthesis of purine and pyrimidine nucleotides and in the metabolism of the amino acids serine, histidine, and methionine.[56] The last is a methyl donor, via the coenzyme S-adenosylmethionine, employed in the formation of N-methyl and O-methyl compounds, of which choline is an important example.

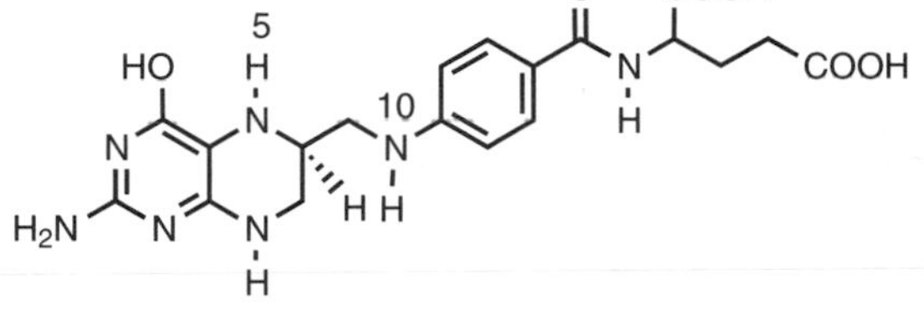

5,6(S),7,8-Tetrahydrofolic acid

Because folic acid coenzymes play a key role in metabolism, several structural analogs have been found to be important in medicine. Aminopterin and amethopterin (Methotrexate) (Fig. 29–14) both strongly inhibit the reduction of folic acid by the NADPH-linked enzyme dihydrofolate reductase. They thereby prevent formation of the one-carbon–carrying coenzymes and inhibit normal synthesis of deoxyribonucleic acid (DNA). These drugs are used in the treatment of leukemias.

The pathway of synthesis of folic acid begins with guanosine-5-triphosphate (Fig. 29–15). The carbon at position 8 is lost, and the ribose unit contributes carbons 1′, 2′, and 3′ to form the pteridine ring carbons, 6, 7, and 9 of the only characterized intermediate of this pathway, 2-amino-4-hydroxy-6-hydroxymethyl-7,8-dihydropteridine. The dihydropteridine is pyrophosphorylated with ATP and reacts with ρ-aminobenzoic acid to form dihydropteroic acid. The glutamate group is introduced next, and the product, 7,8-dihydrofolic acid, is reduced to the coenzyme form, tetrahydrofolic acid. Although the biosynthetic pathway yields the reduced folic acids, they readily oxidize in air to folic acid, the form one isolates from natural sources.

The step in which ρ-aminobenzoic acid is incorporated to yield folic acid is susceptible to competitive inhibition by sulfonamide drugs, the simplest of which is ρ-aminobenzenesulfonamide (sulfanilamide). Organisms synthesizing their own folic acid are therefore subject to growth inhibition by these agents, and infections by such pathogens can be effectively treated.

Chemical and Biologic Assays

The estimation of folic acid in foods and other mixtures is best performed microbiologically with either L. casei or S. faecalis R., although chemical methods are available.

Deficiency

In humans, folic acid deficiency causes macrocytic anemias in which the level of blood platelets and polynuclear erythrocytes is decreased. The condition is also accompanied by gastrointestinal disturbances.

Daily Requirements

The daily requirement for this vitamin is estimated at 0.5 mg, and foods such as liver, kidney, spinach, asparagus, bananas, and strawberries are good dietary sources.

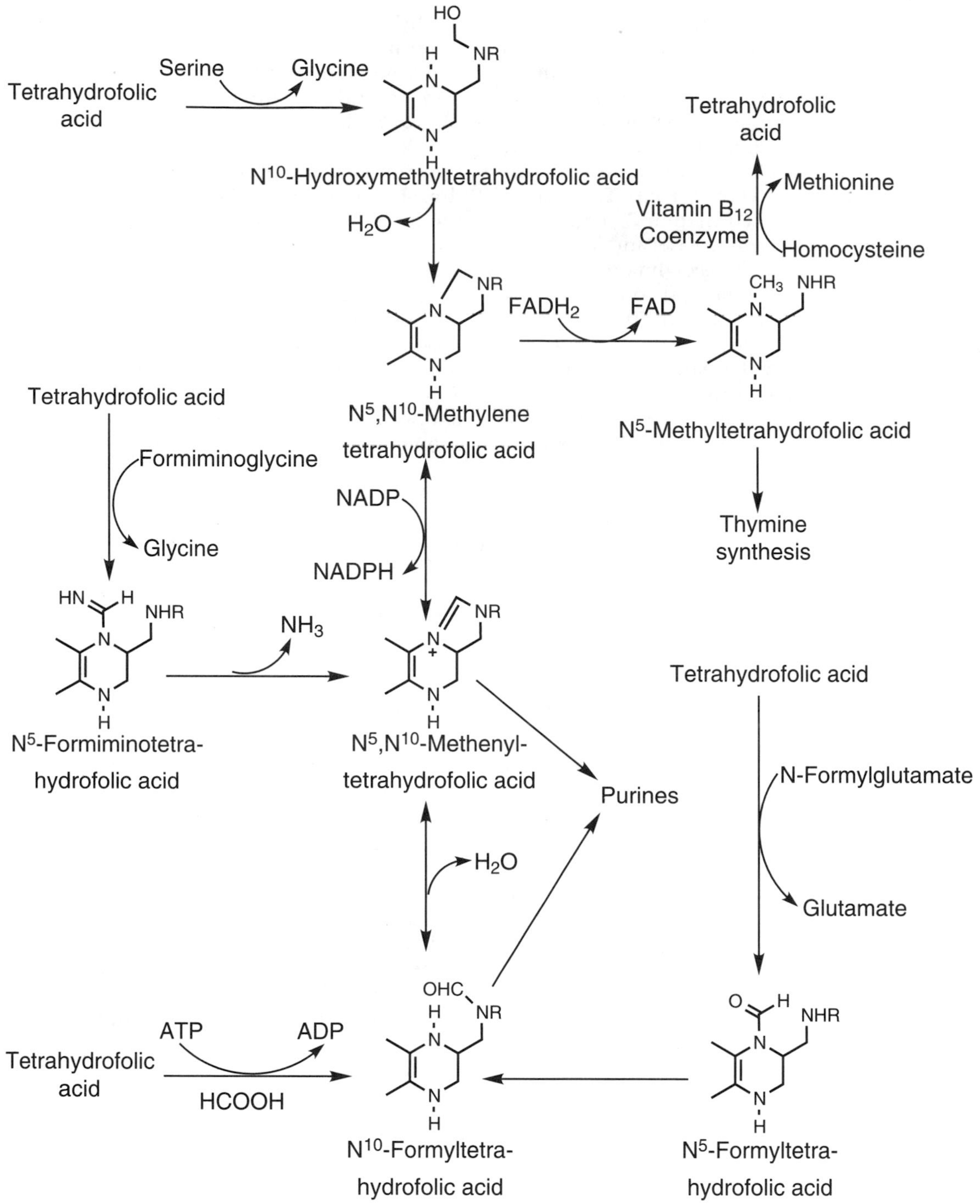

Fig. 29–13. Various one-carbon transfer reactions catalyzed by folic acid.

Folic acid is not included in some over-the-counter vitamin preparations because of its ability to mask the diagnostic signs of pernicious anemia and other diseases and to aggravate these conditions.

VITAMIN B_{12}

The discovery of vitamin B_{12} (cyanocobalamin) did not follow the usual pattern of tracing down a nutritional factor for a microorganism or an animal. Instead, it resulted from isolation of the curative agent for pernicious anemia in humans. Minot and Murphy[57] observed in 1926 that this condition, a macrocytic anemia with accompanying neurologic lesions, could be treated by a diet containing a large amount of raw liver.

For 20 years, efforts were made to isolate the active constituent that was present in minute amounts, but

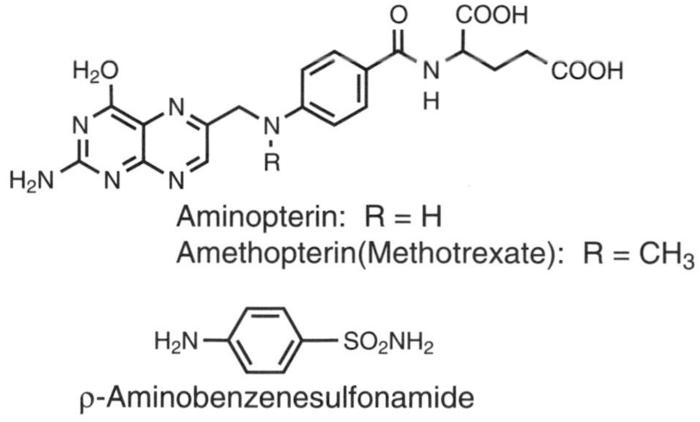

Aminopterin: R = H
Amethopterin(Methotrexate): R = CH_3

p-Aminobenzenesulfonamide

Fig. 29–14. Antimetabolites of folic acid biosynthesis.

progress was slow because the bioassay involved anemic patients. A comparable anemia could not be produced in animals. In addition, the number of anemics in any clinical area is small. Search for a microorganism that could serve in the assay eventually led to the use of L. lactis Dorner.

In 1948, a report in the United States[58] and one in Great Britain[59] announced the isolation of the therapeutic agent, which was effective in a single dose of only 3 to 6 μg. Its concentration in liver is about 1 ppm, and the isolation method used involved adsorption and partition chromatography, extraction steps, and crystallization from aqueous acetone.

Structure and Synthesis

Vitamin B_{12} is a red, cobalt-containing substance that has a characteristic ultraviolet spectrum unaffected by acid or base, although it is destroyed by extremes of pH as well as light and reducing agents. Its structure was established by x-ray analyses in a landmark study by Hodgkin and collaborators,[60] although many useful fragments were obtained from chemical degradation studies.

The molecule has a modified porphyrin unit (a class that includes heme of hemoglobin and chlorophyll of plants) called a corrin ring, coordinating a trivalent cobalt ion, and it is nearly planar. A cyanide ion is also coordinated to the cobalt. It is a product of the isolation method and can be replaced by an hydroxide, chloride, nitrite, or other anion.

The sixth coordination position is directed to a nitrogen of a 5,6-dimethylbenzimidazole ring, which is positioned perpendicular to the corrin ring. The other nitrogen of the imidazole is attached to α-D-ribofuranose-3′-phosphate. This unit is referred to as α-ribazole phosphate and bears a close resemblance to the nucleotides,

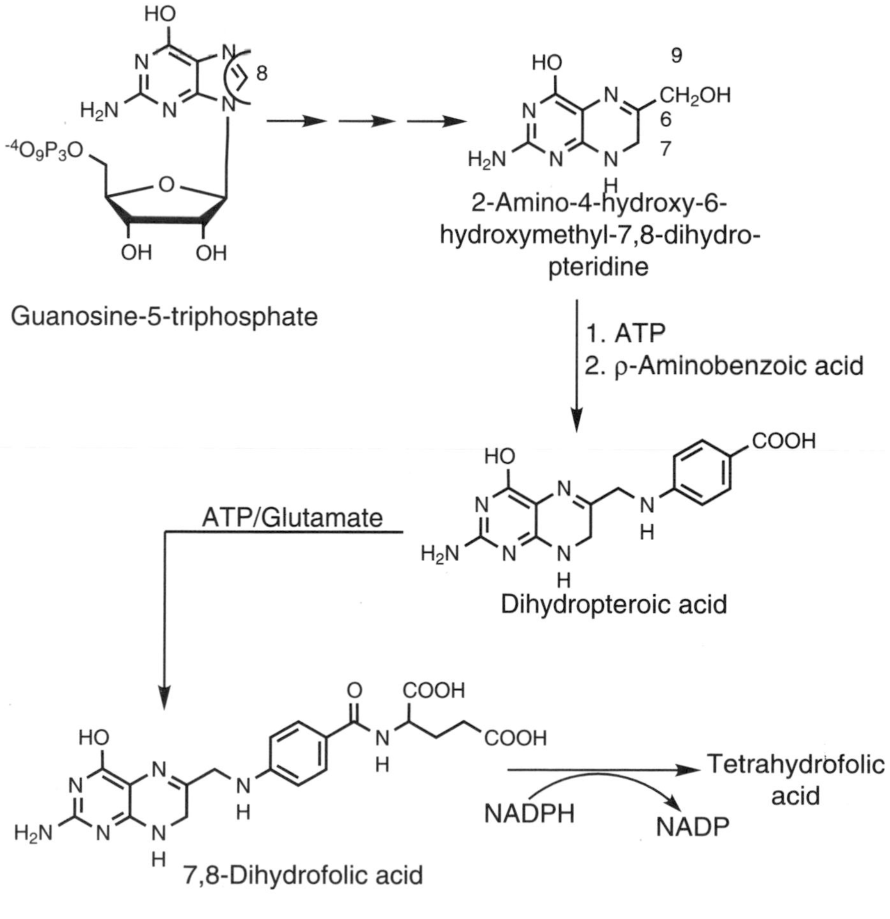

Guanosine-5-triphosphate

2-Amino-4-hydroxy-6-hydroxymethyl-7,8-dihydropteridine

1. ATP
2. p-Aminobenzoic acid

Dihydropteroic acid

ATP/Glutamate

7,8-Dihydrofolic acid

NADPH → NADP

Tetrahydrofolic acid

Fig. 29–15. Folic acid biosynthetic pathway.

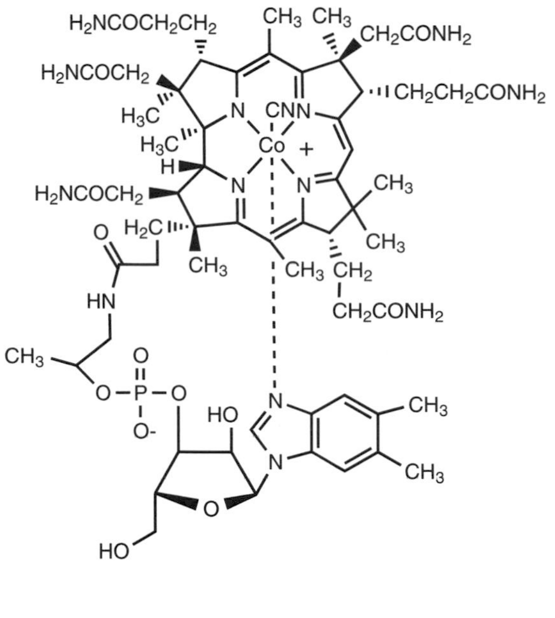

Vitamin B_{12}

with the exception that the glycosidic linkage is α rather than β.

An optically active amino alcohol, (*R*)-1-amino-2-propanol, is ester-linked to the phosphate and amide-linked to a propionate residue of the corrin. The negative charge on the phosphate neutralizes the positive charge of the cobalt, resulting in an uncharged molecule. The corrin is not a regular porphyrin, although it is derivable from four porphobilinogen units, the biogenetic precur-

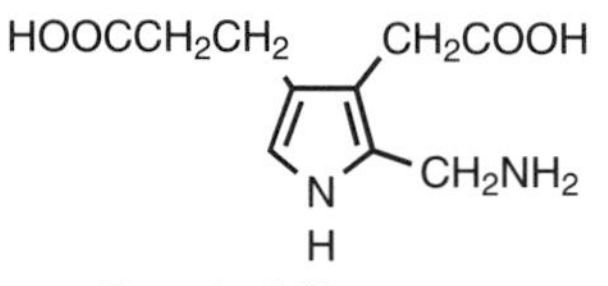

Porphobilinogen

sor of the prophyrin system. Only three units are joined through a methylene bridge; the fourth connection is directly between the rings. The seven extra methyl groups of the corrin ring, including the one at the point of attachment of the two hydropyrrole rings, originate from the methyl of methionine.

Another differing feature concerns the more reduced nature of the corrin; consequently, its ultraviolet and visible spectrum does not reveal its close relationship to the porphyrins. A porphyrin intermediate, uroporphyrinogen III, which contains the unaltered acetic and propionic acid side chains of porphobilinogen, is a precursor of both protoporphyrin IX, leading to the heme of hemoglobin, and of vitamin B_{12}.[61] The total synthesis of this vitamin was announced in 1973. It was the culmination of 11 years of international cooperation between two large research groups totalling 99 workers.[62]

The coenzyme function of vitamin B_{12} was discovered by Barker and co-workers[63,64] in their studies of glutamic acid metabolism by the anaerobe Clostridium tetanomorphum. The first step in that pathway is the isomerization to β-methyl-aspartic acid. The cofactor in this reaction is an extremely light-sensitive vitamin B_{12} coenzyme, the structure of which differs from that of the vitamin by the replacement of the cyanide group with 5′-deoxyadenosine, forming an alkyl cobalt derivative, the first-discovered covalent carbon-to-cobalt bond.[65]

Coenzyme reactions that depend on vitamin B_{12} involve the interchange of a substituent group on a carbon

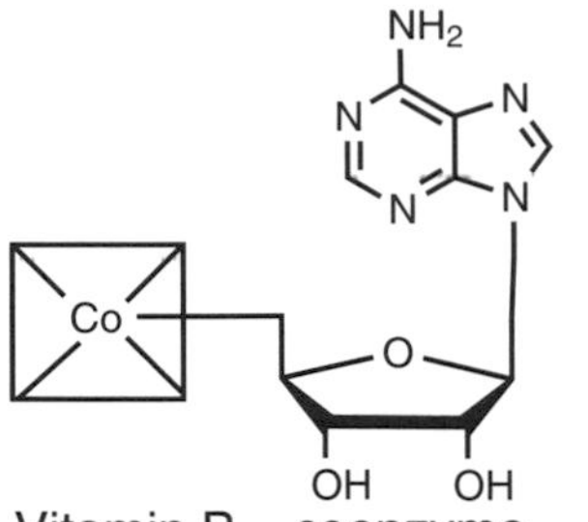

Vitamin B_{12} coenzyme
(5'-Deoxyadenosylcobalamin)

with a hydrogen from an adjacent carbon.[66] The coenzyme is the hydrogen carrier. The three illustrated reactions catalyzed by methylmalonyl CoA mutase, glycerol dehydrase, and ornithine mutase are typical. The blocked groups migrate and are replaced by hydrogen.

Formation of the coenzyme requires reduction of the vitamin by NADH through a flavoprotein and a sulfhydryl protein to the Co^{+1} form, which with ATP generates the coenzyme and tripolyphosphate.

A protein-bound vitamin B_{12} prosthetic group participates in a number of reactions. These lack 5′-deoxyadenosine, and the nature of the alkyl group, if any, is not known. The most studied enzyme of this type is methionine synthetase, which transfers a methyl group from N^5-methyltetrahydrofolic acid to homocysteine to form methionine and in which the methyl group is covalently linked to the cobalt of vitamin B_{12}. The vitamin therefore is a methyl carrier.

Various analogs of vitamin B_{12} have been isolated from sewage in which the 5,6-dimethylbenzimidazole moiety has been replaced by 5-hydroxybenzimidazole, adenine, 2-methyladenine, hypoxanthine, and other groups. Biosynthetic analogs have also been produced by the introduction of benzimidazole derivatives during fermentation to vitamin B_{12}–producing microorganisms. These are not as active in animals as a cyanocobalamin, although some microorganisms use them effectively. No modifications of the corrin ring have yet been made.

Chemical and Biologic Assays

Vitamin B_{12} content can be easily estimated microbiologically by use of a number of organisms after proper preparation of the extract to liberate the bound form and to prevent the vitamin from readily decomposing. Physical (spectrophotometry, polarography, and isotope dilution) and chemical methods (based on release of cya-

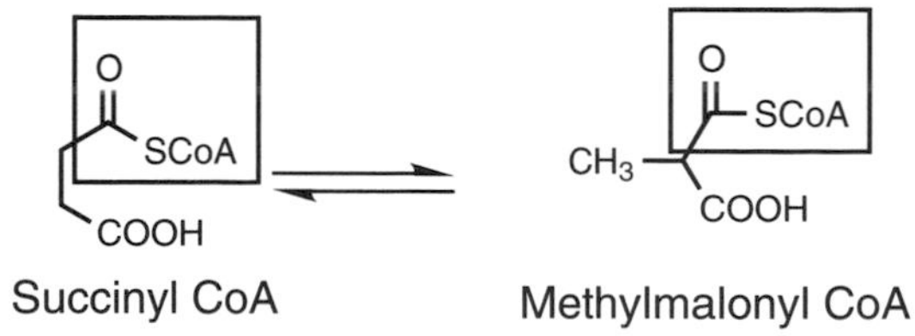

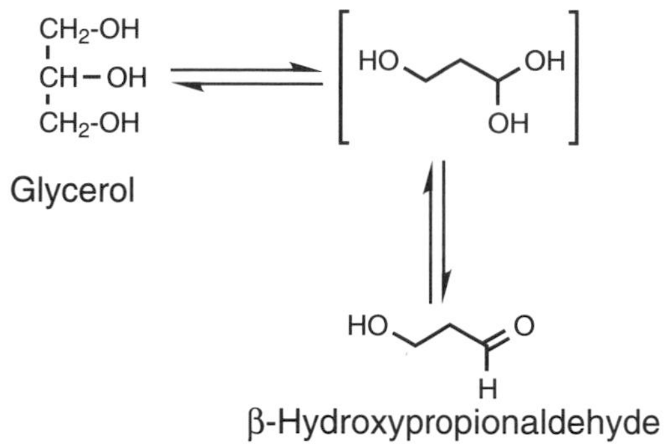

Succinyl CoA ⇌ Methylmalonyl CoA

Glycerol ⇌ β-Hydroxypropionaldehyde

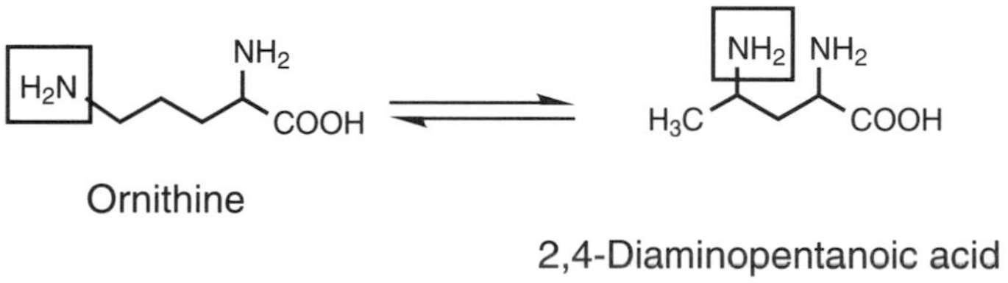

Ornithine ⇌ 2,4-Diaminopentanoic acid

nide or 5,6-dimethylbenzimidazole) can be used in certain cases, provided that interfering substances are not involved.

Deficiency

People with pernicious anemia, even if they consume an adequate supply of vitamin B_{12}, nevertheless have a deficiency because of a lack of a gastric glycoprotein, called the intrinsic factor, which is necessary for the absorption of vitamin B_{12}, the extrinsic factor. Remission in these patients occurs after injection of extremely small amounts of the vitamin or after oral administration of normal stomach protein fractions containing the intrinsic factor. Intake of a large quantity of vitamin B_{12}, such as when eating liver, results in forced uptake of enough material to satisfy the need.

It is difficult to distinguish clinically between a folic acid and a vitamin B_{12} deficiency, if one considers only the hematopoietic system. One must examine the metabolic products of other tissues. Vitamin B_{12} therapy is required for patients after total gastrectomy, for those infested with fish tapeworms (because the food source is depleted by the parasite), and for elderly people because of decreased production of intrinsic factor.

Daily Requirements

Because the primary source of vitamin B_{12} is microbiologic, high concentrations are found in sewage and other products of fermentation. Plants are poor sources, and animals derive their supply through the food chain (ruminants from rumen bacteria). Good sources for humans are liver, kidney, clams, and oysters. An estimate of 3 to 4 μg of vitamin B_{12} per day is considered adequate. This level is easily obtained from the diet or the intestinal flora.

ASCORBIC ACID

Scurvy, that dreaded disease that afflicted explorers in the Middle Ages, was traced to the lack of fresh fruits or vegetables on extended trips. The practice of including lemons, limes, and vegetables on voyages represents the first recognition of a human disease associated with a food deficiency. Isolation of the factor designated as vitamin C was made possible by the accidental finding that scurvy could be induced in guinea pigs. Various purified preparations of the antiscorbutic factor were obtained, but the first crystalline isolate was reported in 1932 from lemon juice.[67] This substance had been isolated 4 years earlier from orange and cabbage juices and from the adrenal cortex as a result of a study of biologic oxidation-reduction systems, but its vitamin C activity was not suspected. The name hexuronic acid was chosen, but after it was found to be identical to the antiscorbutic principle, ascorbic acid became the accepted term.

Structure and Synthesis

Ascorbic acid has the formula $C_6H_8O_6$ and the systematic name L-*threo*-2,3,4,5,6-pentahydroxy-2-hexenoic acid-

4-lactone. Its strong reducing property stems from the enediol system, which forms the 1,2-dione moiety of dehydroascorbic acid on oxidation.

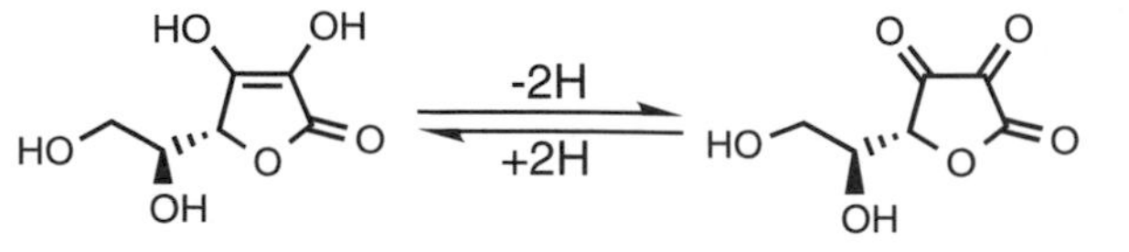

Ascorbic acid ⇌ (−2H / +2H) Dehydroascorbic acid

The dehydro substance is equally as active a vitamin as ascorbic acid and, incidentally, is not acidic. The 2,3-diketo-L-gulonic acid, the product of hydrolysis of dehydroascorbic acid, is inactive.

This reversible oxidation-reduction system accounts for the biologic function of the vitamin. The enol groups impart the strong acid properties to the molecule. The substance is stable in crystalline form and dissolves readily in water and polar solvents, but it is insoluble in organic

L-Idonic acid

solvents. In solution, it is sensitive to light, air, and heavy metals, especially under alkaline conditions.

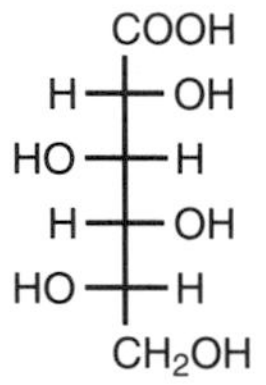

Catalytic reduction of ascorbic acid formed L-idonic acid to establish the straight chain of the carbon skeleton, whereas acetylation to a tetraacetate indicated four hydroxyl groups. Two of these are enolic and form the

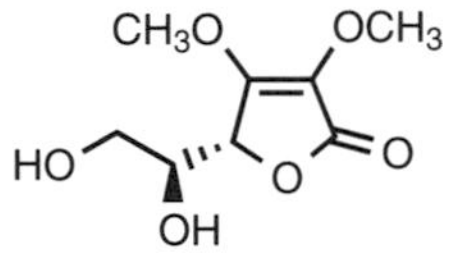

Ascorbic acid dimethyl ether

methyl ether, which no longer has reducing properties. Oxidation of ascorbic acid with sodium hypoiodite gave a quantitative yield of L-threonic acid and oxalic acid. The L-threonic acid fixed the two stereochemical centers.

Commercial synthesis of ascorbic acid begins with reduction of glucose to D-sorbitol, which is oxidized microbiologically by Acetobacter suboxydans to L-sorbose (Fig. 29–16). The diisopropylidene derivative of L-sorbose is

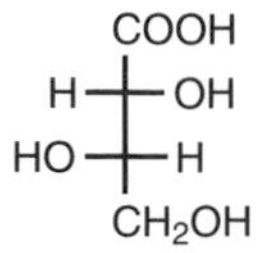

L-Threonic acid

oxidized to the acid derivative, which on treatment with acid forms L-*xylo*-2-ketohexanoic acid; this substance spontaneously enolizes and lactonizes to ascorbic acid.

The coenzyme form of ascorbic acid, if there is one, is not known, and the entire biochemical function for this vitamin remains to be uncovered. Besides a role based on its reducing and metal chelating properties, ascorbic acid is involved in hydroxylation reactions by monooxygenases. Two of these enzymes are required for hydroxylation of lysine and proline in the conversion of procollagen to collagen.[68] Tyrosine catabolism is hampered by ascorbic acid deficiency, and abnormal excretion of ρ-hydroxyphenylpyruvate results. Administration of the vitamin restores the oxygenases that normally break down tyrosine.

The adrenal glands contain a high concentration of ascorbic acid necessary for hydroxylation reactions in the biosynthesis of adrenocortical hormones, and the formation of norepinephrine from dopamine by β-hydroxylase.[69] The latter reaction is also important in the brain. An electron transport function between NADH and cytochrome b_5 in mammalian microsomes has also been implicated.

The biosynthesis of ascorbic acid, in animals that can perform it, starts with reduction of D-glucuronic acid (from D-glucose) to L-gulonic acid, which is cyclized to L-gulono-γ-lactone.[70] Oxidation of the 2-hydroxyl group to the ketone and enolization result in ascorbic acid. Carbon 6 of glucuronic acid becomes carbon 1 of ascorbic acid in this conversion.

Animals that are unable to form the vitamin lack the enzyme for oxidation of position 2 of L-gulono-γ-lactone. Plants make ascorbic acid from glucose, but by a different pathway, in which carbon 1 of glucose becomes carbon 1 of ascorbic acid.[71] Details of this pathway are not known.

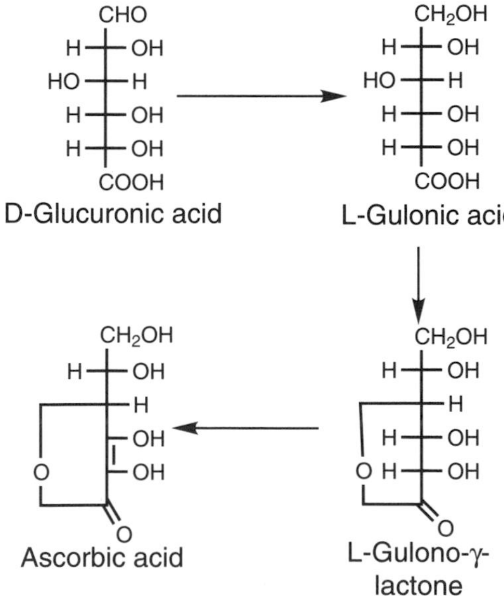

D-Glucuronic acid → L-Gulonic acid → L-Gulono-γ-lactone → Ascorbic acid

Chemical and Biologic Assays

Ascorbic acid activity can be assayed chemically, based on its reducing properties, or biologically with guinea pigs.

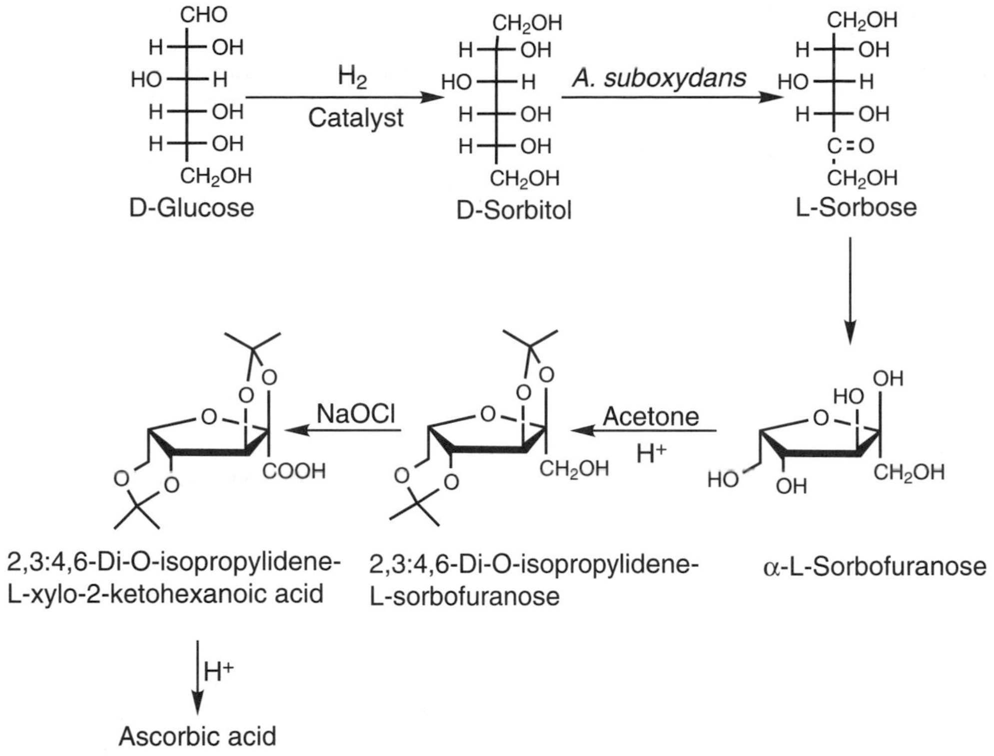

D-Glucose → (H_2, Catalyst) → D-Sorbitol → (*A. suboxydans*) → L-Sorbose → α-L-Sorbofuranose → (Acetone, H^+) → 2,3:4,6-Di-O-isopropylidene-L-sorbofuranose → (NaOCl) → 2,3:4,6-Di-O-isopropylidene-L-xylo-2-ketohexanoic acid → (H^+) → Ascorbic acid

Fig. 29–16. Commercial synthesis of ascorbic acid.

Deficiency

Scurvy is characterized by anemia, a general weakness accompanied by painful joints, and hemorrhages in the mouth and gastrointestinal tract. The gums become weakened and spongy and the teeth loosened. Weakened areas become susceptible to infection and gangrene. Wounds do not heal, and even old scars may break. Ascorbic acid is used in connection with surgery to speed up healing and in conditions of stress (infection and burns). It is believed by some that large daily doses have a prophylactic effect against the common cold and other viral attacks and also aid the body under physical and emotional stress.[72]

Daily Requirements

The daily requirement of ascorbic acid is estimated at about 60 mg.

Good food sources are liver; fresh vegetables such as green peppers, spinach, cabbage, and lettuce; and fresh citrus fruits and strawberries.

VITAMIN A

Shortly after Hopkins reported his results on the accessory food principle required in the purified diet of lard, carbohydrates, proteins, and salts, McCollum and Davis[73] showed that an ether extract of animal fats or fish oils markedly improved the growth of rats. This fat-soluble factor, later named vitamin A, was found also to prevent and cure eye conditions characterized by xerophthalmia, or a drying of the eye tissues. A keratinization of this area occurs, infection may appear, and in the end, permanent eye damage occurs. Night blindness (nyctalopia) was also traced to this vitamin deficiency.

Structure and Synthesis

Retinol (vitamin A_1) was isolated in 1931 by Karrer and co-workers[74] from fish liver oil, the richest common natural source, after saponification to release the vitamin from the ester-bound forms. The vitamin, which is sensitive to heat and oxygen, shows a characteristic absorption at 328 nm and turns blue with antimony trichloride (Carr-Price reaction), which was useful in its purification. The nonsaponifiable material was fractionated by preferential solubility steps and adsorption chromatography on alumina. Material obtained in this manner was pure but not

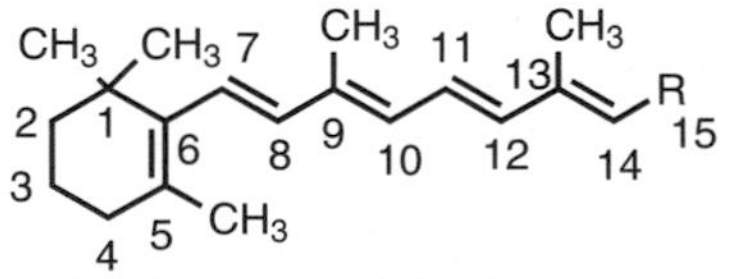

Retinol: R = CH_2OH
Retinal: R = CHO
Retinoic acid: R = COOH

crystalline and served to establish the structure. It was not until 1937 that crystalline retinol was obtained as a solvate with methanol, and in 1942 it was obtained as the pure substance.

The liver of freshwater fishes has yielded a second vitamin A, 3-dehydroretinol, or vitamin A_2.

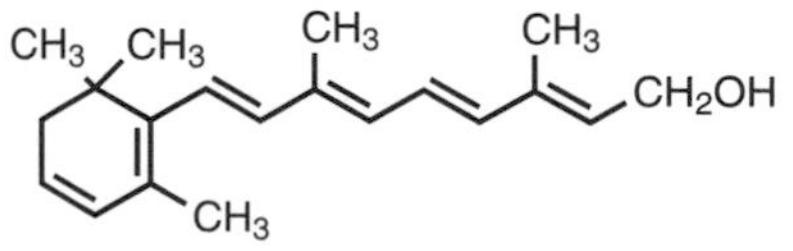

3-Dehydroretinol

The extra double bond is conjugated with the other five and causes a shift of the ultraviolet absorption maximum by 23 nm toward the visible region. This compound is one-half as potent as retinol in the rat assay.

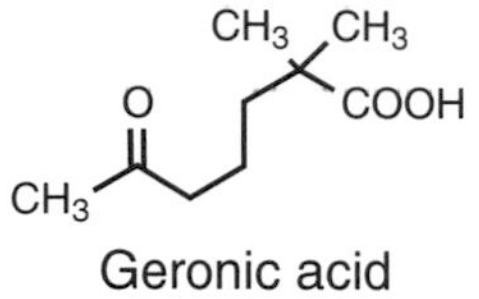

Geronic acid

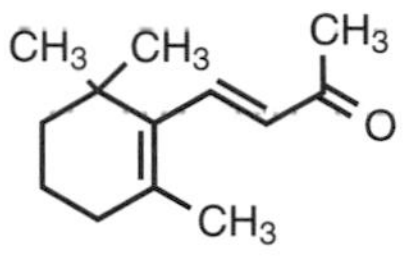

β-Ionone

The structure of retinol, $C_{20}H_{30}O$, was established by hydrogenation and oxidation studies and has the all *trans* structure. The ring system became known from the product of ozonolysis, geronic acid, a compound previously obtained from the terpene *β*-ionone. The fully reduced product, decahydroretinol, was synthesized from *β*-ionone, leaving only the location and stereochemistry of the bonds to be settled. This was done on the basis of the relationship between retinol and *β*-carotene.

Animal feeding experiments demonstrated that plant sources also contain the vitamin A growth-promoting properties but are devoid of retinol. Certain yellow pigments in these plants were found to be responsible for this provitamin activity, and administration of a tiny amount of a mixture of carotenes was equivalent to giving the animal a vitamin A supplement.

Only a few of the many carotenes have provitamin activity. The *β*-carotene is cleaved in the intestinal mucosa to give two molecules of retinal, the aldehyde derivative of retinol. The reaction requires oxygen and is catalyzed by an iron-containing dioxygenase. Reduction of the aldehyde by a NADH-requiring enzyme forms retinol, which is absorbed, esterified, and transported to the liver for storage. Retinol is transported in the plasma as a complex with a specific transport protein (molecular weight about 21,000), which is associated with prealbumin, another serum protein.[75,76]

Only carotenes that can be cleaved to yield retinol (or 3-dehydroretinol) have vitamin A activity. For example, *γ*-carotene is only one-half as active as *β*-carotene. Other compounds, such as lycopene (the red pigment of tomatoes), with no ring, cannot give retinal and are therefore inactive.

Since 1950, vitamin A has been commercially supplied almost exclusively as retinol acetate or palmitate (Fig. 29–17). Before that time, highly purified fish liver oil fractions were used, but the increased demands could no longer be met by this supply. Many synthetic pathways were developed, and at least six manufacturing processes are in operation today. As far as is known, all begin with *β*-ionone, a bulk chemical of the perfume industry, which can be made synthetically from acetone or from the monoterpenes, citral (from lemongrass oil), or *β*-pinene (from pine oil).

The synthetic route of Hoffmann-LaRoche of Basle is used as an illustration.[77,78] The side chain construction begins with methyl vinyl ketone, which when treated with sodium acetylide gives 3-hydroxy-3-methyl-1-pentene-4-yne. Allylic rearrangement of the product results in *cis* and *trans* isomers, which are separated by fractional distillation. The *trans* isomer is converted to the Grignard reagent and is condensed with a C_{14}-aldehyde, which is derived from *β*-ionone by a Darzens condensation followed by treatment of the glycidic ester with a base. The alcohol, after partial hydrogenation and selective acetylation of the primary hydroxyl group, undergoes an allylic rearrangement to the hydroxy acetate. Dehydration of this compound yields retinyl acetate, from which retinol can be obtained by alkaline hydrolysis.

A biologic function of retinol in the photoreceptor process became clear when Wald[79] in 1935 isolated retinal (vitamin A_1 aldehyde) from the retina after bleaching. The protein rhodopsin, or visual purple, present in the rods of the human eye, releases retinal after exposure to light in an important step of the visual process. After a quantum of light is absorbed by rhodopsin, the pigment undergoes a series of transformations observable as changes in the ultraviolet and visible spectrum and culminating in the dissociation of the polyene from the protein. This last step is the hydrolysis of the Schiff base between the aldehyde group of retinal and the *ε*-amino group of a lysine residue in the protein. At some stage in the transformation, a nerve impulse is initiated that registers the event.

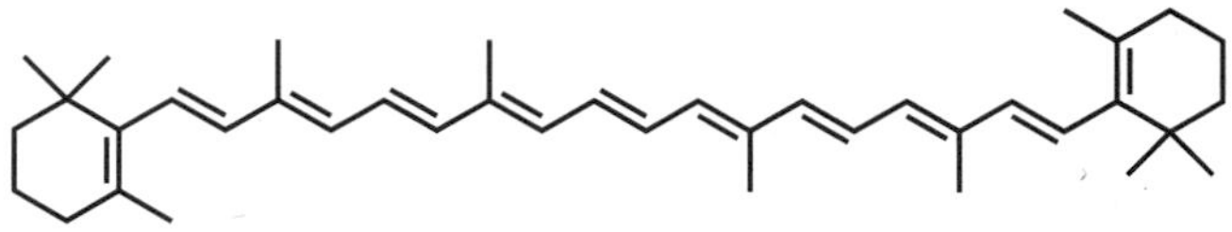

β-Carotene

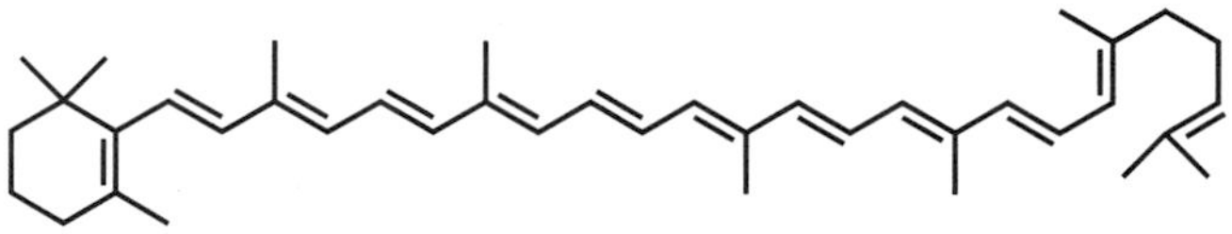

γ-Carotene

A pigment similar to rhodopsin, called iodopsin, in which only the protein part differs, is present in the cones, the other visual cell structure, which is considered to be responsible for color vision. Opsin recombines with retinal only after the carbon-11 double bond in retinal has been properly isomerized.

Retinol and retinal have four double bonds that can exist in a *cis* or *trans* form, theoretically producing 16 stereoisomers. The number known is much less because only a few isomers can form without steric hindrance.[80] Isomerization of the double bond at carbon 9 causes only a small interaction between the protons at positions 8 and

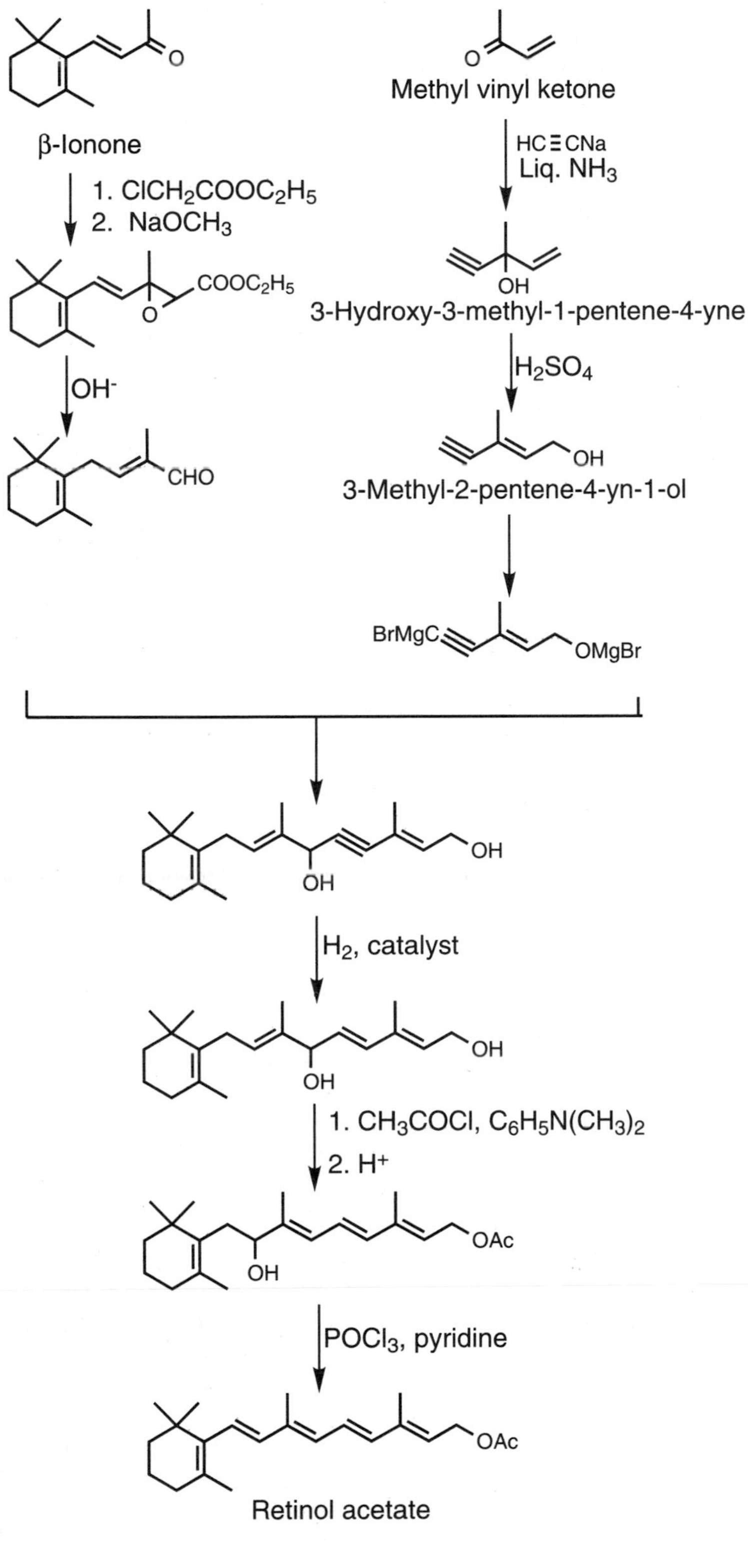

β-Ionone

1. $ClCH_2COOC_2H_5$
2. $NaOCH_3$

$COOC_2H_5$

OH^-

CHO

Methyl vinyl ketone

$HC{\equiv}CNa$
Liq. NH_3

3-Hydroxy-3-methyl-1-pentene-4-yne

H_2SO_4

3-Methyl-2-pentene-4-yn-1-ol

BrMgC OMgBr

H_2, catalyst

1. CH_3COCl, $C_6H_5N(CH_3)_2$
2. H^+

OAc

$POCl_3$, pyridine

Retinol acetate

Fig. 29–17. Synthetic route to retinol acetate.

11. A similar weak interaction occurs on isomerization of the carbon 13 double bond. The unhindered isomers 9-*cis*-retinol, 13-*cis*-retinol, and 9,13-di*cis*-retinol are known; all have less biologic activity than *trans*-retinol.

Isomerization at carbon 11 leads to the less stable or hindered retinols, 11-*cis*-retinol and 11,13-di*cis*-retinol, in which a methyl group and a hydrogen are crowded. These are likewise less active in the standard assay of vitamin A in rats. The 7-*cis*-retinol is not known, because apparently the two methyl interactions make it too labile to exist under normal conditions.

The hindered aldehyde isomer, 11-*cis*-retinal, is required to form rhodopsin. To regenerate the visual pigment, the released *trans*-retinal does not simply become isomerized but must undergo the following steps: (1) reduction to *trans*-retinol with NADH by the enzyme of the retina, retinal reductase; (2) transportation to the liver and isomerization to 11-*cis*-retinol, which is returned to the eye via the blood; and (3) reoxidation by the reductase to 11-*cis*-retinal, which recombines with opsin. A mechanism is available in the retina for converting *trans*-retinal to 11-*cis*-retinal, but it is too slow to be of primary use.

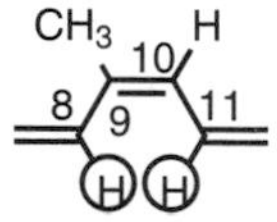

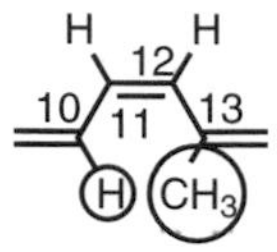

Retinol has other biologic functions besides photoreception, but these are not understood as well. For example, vitamin A–deficient animals have an impaired adrenal cortex and consequently show the effects of an inadequate adrenal hormone supply. Glycogen formation, for one, is depressed, and mucous membrane tissues are unable to make the mucopolysaccharides, but when their cell homogenates are supplemented with retinol, this ability is restored. There is evidence that vitamin A has a role in maintaining cell membranes, protein synthesis, and skeletal formation. Retinoic acid (vitamin A_1 acid), a substance that can function as the vitamin except in the visual and reproductive processes, is the active form in these cases[81] and functions by receptor-mediated regulation of gene expression.[82]

Understanding the biosynthesis of vitamin A–active compounds requires knowledge about the formation of their precursors, the carotenoids, a large class of C_{40} polyene compounds that are built up from repeating isopentene or isoprene units. Similar to other isoprene-derived hydrocarbons (squalene), β-carotene has its origin entirely in acetate units (acetyl-CoA), with mevalonic acid and isopentenyl pyrophosphate as key intermediates. The early stages of the pathway are exactly as established for steroid formation.[83] At the farnesyl pyrophosphate stage, another isopentenyl pyrophosphate unit is added in a regular manner, and two molecules of the product, geranylgeranyl pyrophosphate, are linked tail to tail to yield the first carotene, phytoene. A series of dehydrogenations produces lycopene, and subsequent cyclization steps form β-carotene (Fig. 29–18).[84,85]

Chemical and Biologic Assays

Vitamin A, which is restricted almost exclusively to animal tissues, is easily estimated chemically by the Carr-Price reaction (the blue color is measured colorimetrically) or by the intensity at the absorption maximum (328 nm) in the ultraviolet region. Assay for the carotene provitamins requires separation of the mixture by chromatography and identification of the active components. Simply measuring the yellow color of a mixture does not indicate the quantity of provitamin A carotenes. Animal assays, of course, provide a value for the total vitamin A activity.

Deficiency

The deficiency disease in humans involves an ocular disorder characterized by night blindness and conjunctivitis, a drying of the skin with accompanying hyperkeratosis of the follicles, and a general decline in health and growth.

Daily Requirements

The recommended daily allowance is 1000 μg for men and 800 μg for women. During pregnancy and lactation, the level should be raised to 1000 and 12,000 μg. Children should receive 400 to 1000 μg, depending on age and sex. Excellent dietary sources of vitamin A are liver, kidney, fish liver oils, and fruits and vegetables high in carotene content, such as carrots, spinach, peaches, prunes, and apricots.

Toxicity

Excess intake of vitamin A causes severe poisoning, which is sometimes fatal. Arctic explorers, after eating polar bear liver, have experienced acute symptoms, which include drowsiness, severe headache, and vomiting. Chronic symptoms include anorexia, hyperirritability, thickening of skin and bones, blurring of vision and diplopia, and skeletal abnormalities. Vitamin A compounds are recognized as potent teratogens. Recovery is complete after intake of the vitamin is stopped.

VITAMIN D

The use of cod liver oil for the treatment of rickets, a childhood disease connected with improper bone formation, was started more than a century ago, even though the nutritional cause was still unknown. The cause was recognized 50 years later.[86] The curative substance was found in the steroid fraction of the unsaponifiable part of the oil and was distinguishable from vitamin A by its heat stability during aeration.

The discovery that irradiation of certain foods as well as rachitic animals produced the antirachitic activity helped explain the observation that sunlight was useful in the treatment of the disease.[87,88] The curative factor was designated vitamin D, and substances that form the vitamin on ultraviolet irradiation were called provitamins D. Because it was recognized that several substances had activity, numerical subscripts after the letter D were used to distinguish them. Products derived from them maintain the subscript, and the carbons are numbered according to the steroid numbering system.

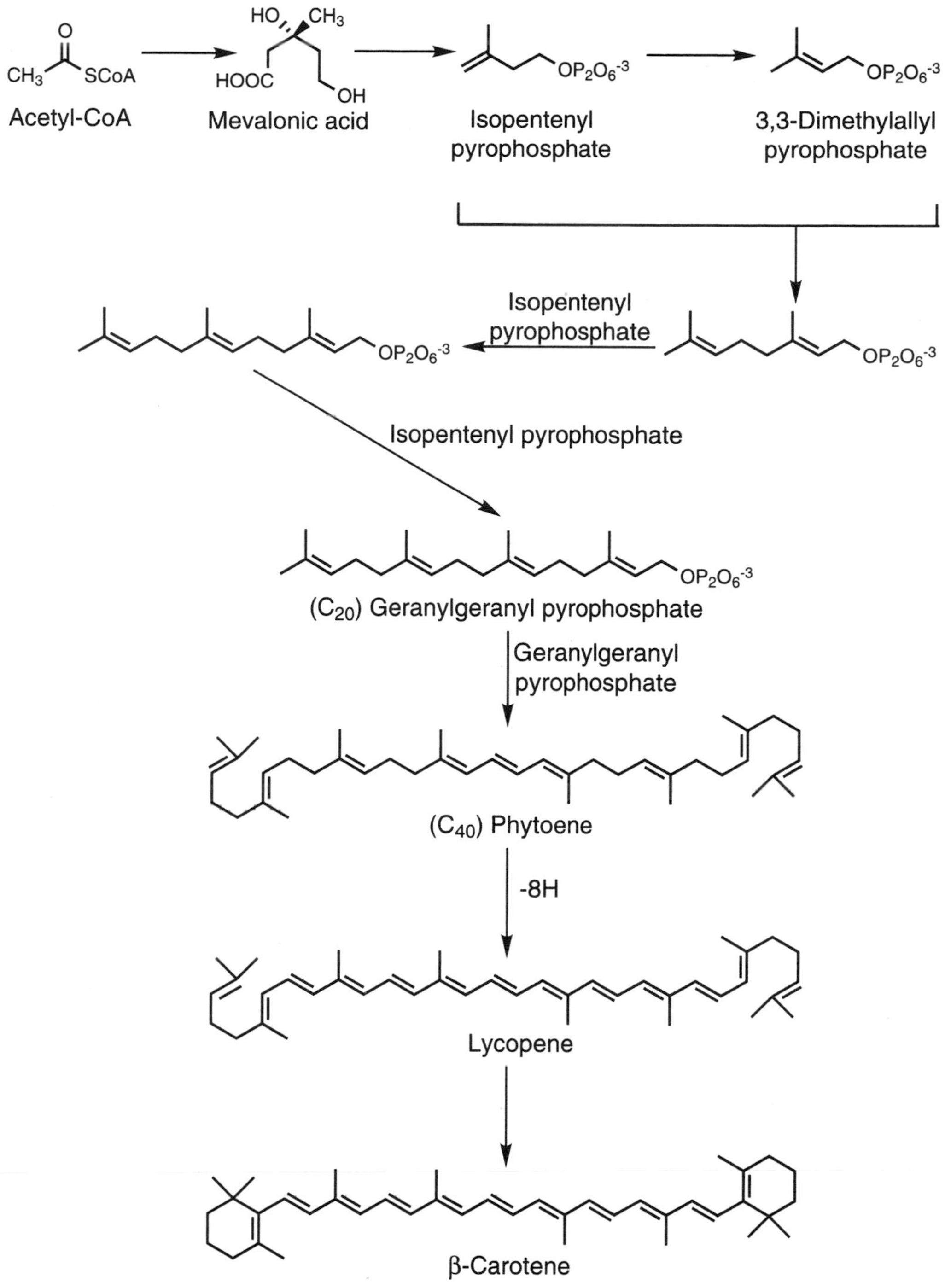

Fig. 29–18. Biosynthetic pathway leading to β-carotene.

Structure and Synthesis

Ergocalciferol, or vitamin D_2, the first product to be isolated pure, was obtained from the photolysis reaction products of ergosterol.[89,90] The isolate designated vitamin D_1 was found to be a 1 : 1 molecular complex of ergocalciferol and lumisterol$_2$ and is no longer used. In differential bioassay studies with chicks and rats, ergocalciferol was found to be less active in chicks than were crude preparations of cod liver oil or irradiated cholesterol. Following this observation, cholecalciferol, or vitamin D_3, a substance 50 to 100 times as active as ergocalciferol in chicks, was obtained from fish liver oil.[91] The compound was identical to the irradiated product of 7-dehydrocholesterol.

The provitamins are 3β-hydroxy-$\Delta^{5,7}$-steroids and differ in the type of C-17 side chain. A common feature of each is the 280-nm peak in the ultraviolet spectrum of the

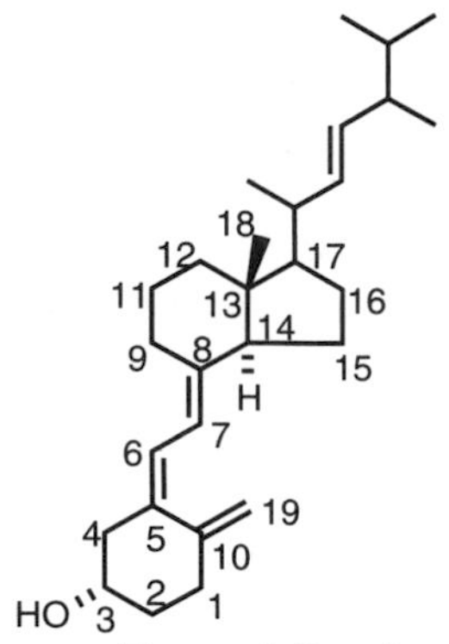

Ergocalciferol

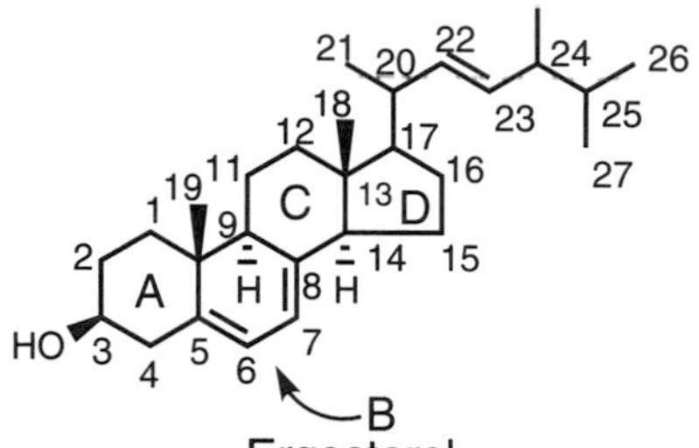

Ergosterol

homoannular diene system. Vitamin D congeners formed from plant steroids bearing varied side chains are biologically inactive or less active than cholecalciferol. The side chain, therefore, is vital.

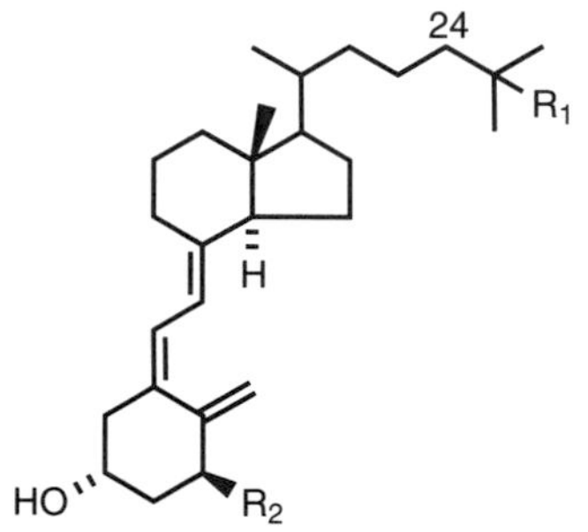

Cholecalciferol: $R_1 = R_2 = H$
25-Hydroxycholecalciferol: $R_1 = OH$, $R_2 = H$
1α,25-Dihydroxycholecalciferol: $R_1 = R_2 = OH$

The first product of irradiation of ergosterol in the formation of ergocalciferol is pre-ergocalciferol, which is isomeric with the starting material and formed in a reversible reaction in which orbital symmetry is conserved

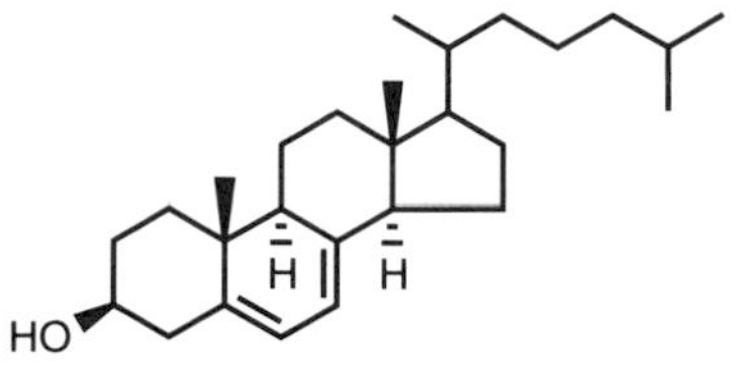

7-Dihydrocholesterol

(Fig. 29–19).[92] This conversion results in a high yield. The next step is also reversible but requires only heat. The reaction is not unidirectional, and a number of by-products are formed; one, tachysterol$_2$, is the light-induced reversibly formed 6-*trans*-isomer of pre-ergocalciferol. In another photochemical step, tachysterol$_2$ is converted to lumisterol$_2$, a 9β,10α-isomer of ergosterol, again with conservation of orbital symmetry. Lumisterol$_2$ is converted to pre-ergocalciferol by an irreversible photochemical reaction. Not included in the scheme are products of overirradiation and overheating, such as the suprasterols and the pyroergocalciferols. The pyroergocalciferols are two symmetry-allowed thermal isomers of ergosterol originating from pre-ergocalciferol in which both the 9-hydrogen and the 10-methyl groups are α or β. The yield of ergocalciferol can be maximized to avoid formation of the biologically inactive by-products by carefully controlling the conditions of irradiation. A similar sequence of steps is operational in the conversion of 7-dehydrocholesterol to cholecalciferol.

A formal total synthesis of cholecalciferol was achieved once cholesterol was synthesized because it can be readily converted to 7-dehydrocholesterol.

A second pathway (not through the steroid nucleus and directed specifically toward the vitamin structure) requires first formation of the bicyclic system (rings C and D of the steroid series) with the side chain and then attachment of the remaining nine-carbon unit.[93]

The biologic function of vitamin D is connected with bone formation—the deposition of calcium and phosphate into the cartilage region of bone as well as intestinal phosphate transport, mobilization of calcium from bone, and reabsorption of calcium by the kidney. Specifically the absorption of calcium (and phosphate with it) from the digestive tract is affected. Rachitic animals have a low calcium level in the plasma, although the diet contains an abundant supply. This condition has been traced to the lack of a transport protein for calcium. The protein has been isolated from chick duodenum.[94,95] It is not present in rachitic chicks but is formed on administration of the vitamin. Its rate of appearance parallels the rate of calcium absorption.

The location at which vitamin D acts in this process appears to be, in part, at gene expression of calcium binding proteins.[96] Involvement at the level of genetic expression appears likely, because absorption of calcium by rachitic animals on administration of vitamin D is prevented by actinomycin D or puromycin. Actinomycin D inhibits transcription of the DNA information to a messenger ribonucleic acid (mRNA), whereas puromycin interferes with the normal building of a protein chain. Addition of actinomycin D 4 to 8 hours after administration of the vitamin does not affect calcium absorption.

Calcium metabolism is not solely under the control of vitamin D. Thyrocalcitonin (a thyroid hormone), for example, lowers the levels of calcium and phosphate in the blood; the parathyroid hormone (parathormone) raises the level of calcium and phosphate in the blood; the parathyroid hormone (parathormone) raises the level of calcium.

Vitamin D$_3$ is converted in animals to 1α,25-dihydroxycholecalciferol,[97–99] now considered to be the calcium

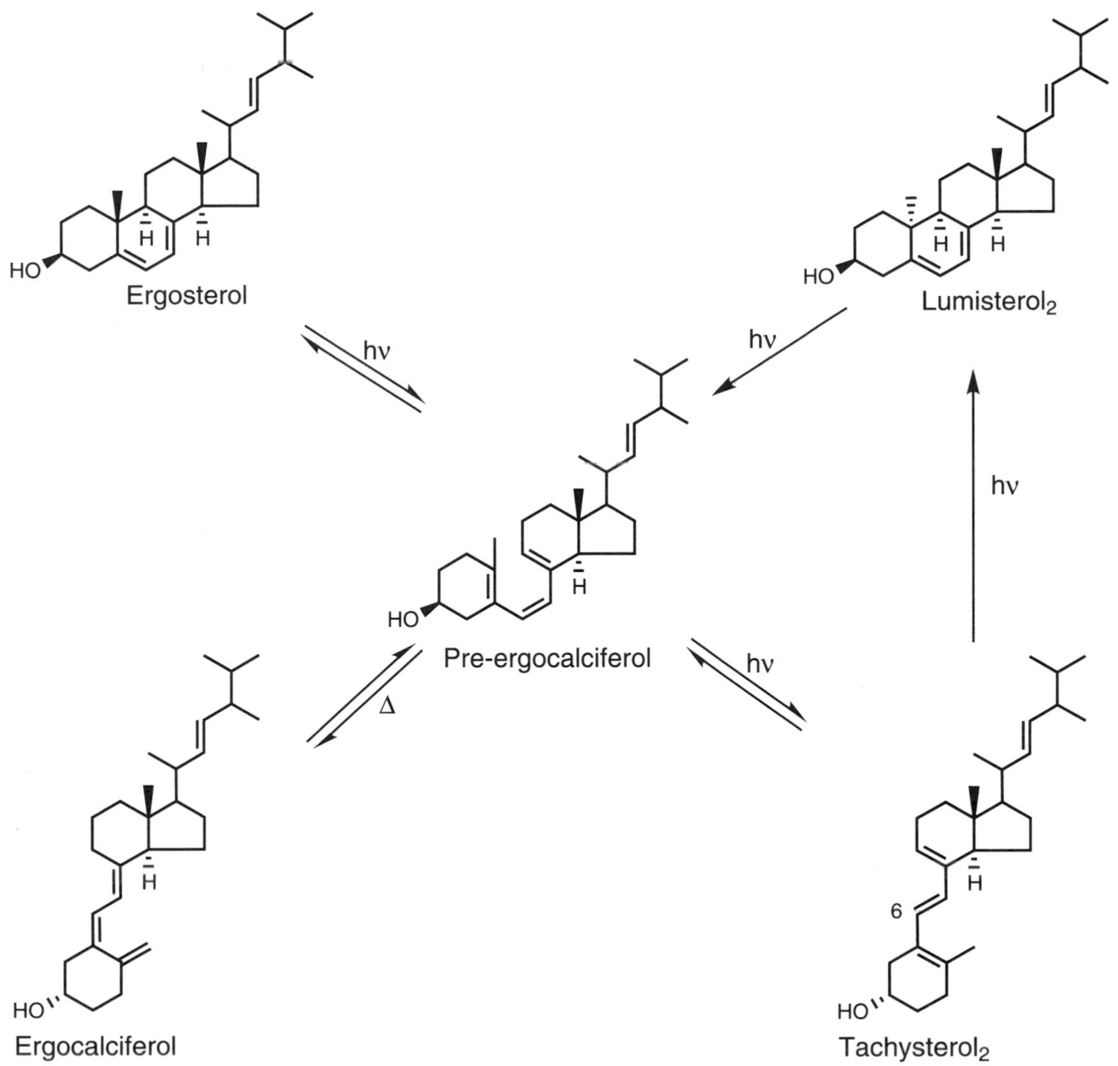

Fig. 29–19. Interconversion of vitamin D and precursor.

and phosphate-mobilizing hormone, because it performs all of the known functions of the vitamin and is a product of one organ that elicits activity in others. The hydroxylation is sequential: The first occurs in the liver at the 25-position,[100] and the second is restricted to the kidney. The kidney hydroxylase is stimulated by the parathyroid hormone. An additional kidney hydroxylase introduces a hydroxyl into position 24 and may be part of the hormone-inactivation process. Disease states in which vitamin D metabolism has been impaired, such as renal osteodystrophy, hypoparathyroidism, pseudohypoparathyroidism, vitamin D–dependency disease (a genetic disease with a defective 1-hydroxylase), and drug-induced osteomalacia,[101] have been successfully treated with 1α-hydroxycholecalciferol or 1α,25-dihydroxycholecalciferol. The former compound is hydroxylated in vivo to give the active hormone.

In examining the biosynthesis of vitamin D, one considers the formation of the steroid nucleus and specifically cholesterol.[83] The pathway is known in detail, and up to the intermediate farnesyl pyrophosphate (already mentioned in connection with formation of β-carotene), the steps are identical. At this point, branching occurs; two farnesyl pyrophosphate units are joined tail to tail to form squalene, which after epoxidation at the 2,3-double bond is cyclized by cyclase to the tetracyclic triterpene, lanosterol. In the subsequent steps, three methyl groups are lost by oxidation to carbon dioxide, the double bond at carbon 24 is reduced, and the other one is moved from carbon 8 to carbon 5 to form cholesterol.

The intermediate before cholesterol is 7-dehydrocholesterol, a provitamin. Another source of the provitamin is cholesterol. The dehydrogenation is performed by a NADP-requiring enzyme found in the intestinal wall and other tissues. The provitamin is found in the skin, which is properly situated for absorption of radiant energy to effect the isomerization to cholecalciferol. In lower plants, ergosterol is made by a similar pathway. The methyl group at carbon 24 has its origin in S-adenosylmethionine.

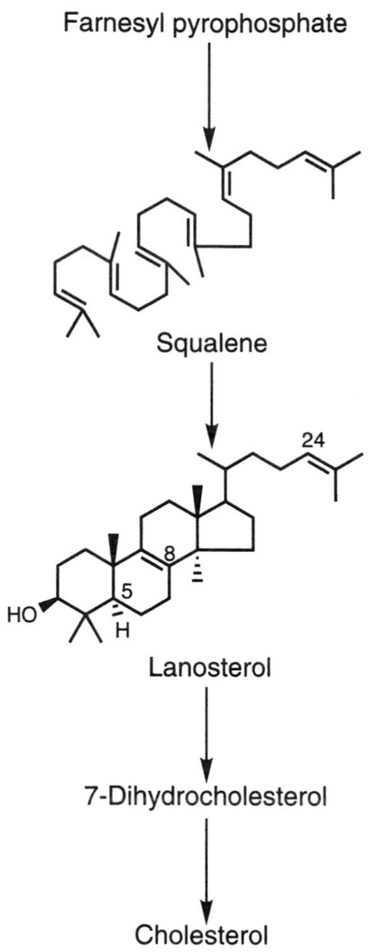

Farnesyl pyrophosphate

Squalene

Lanosterol

7-Dihydrocholesterol

Cholesterol

Chemical and Biologic Assays

A good indicator of vitamin D activity is the curing of rickets in rats; however, cholecalciferol and ergocalciferol can be assayed nonbiologically by adsorption or gas chromatographic methods.[102,103]

Daily Requirements

The average daily adult requirement of vitamin D is 400 IU* or 0.01 mg of cholecalciferol or ergocalciferol. In climates in which the skin is easily exposed, the vitamin need is met by irradiation together with the dietary source. In cold climates with little winter sun, a dietary supplement is necessary. Pregnancy increases the demand for calcium and vitamin D, which, if not met, results in osteomalacia, a mobilization of calcium from the bones of the mother.

Toxicity

Excess vitamin D causes nausea, loss of appetite, thirst, and elevation of the calcium and phosphate blood levels, resulting in their deposition in the heart, lungs, and kidneys and demineralization of the bones. If the condition is recognized in time and the vitamin is withheld, the process can be reversed.

* The International Unit (IU) is measured by the rat line test, in which calcium deposition in the radii and ulnae bones is determined.

VITAMIN E

A diet containing adequate bulk nutrients and the growth factors known in 1922 caused sterility in rats, although normal growth was observed.[104,105] Both sexes were affected; pregnant females resorbed the fetus, and males showed atrophy of the reproductive organs. A supplement of fresh lettuce, meat, milk, or yeast resulted in return to a healthy reproductive state. The factor, an antisterility substance named vitamin E (its recognition followed the discovery of vitamin D), was found in the unsaponifiable lipid fraction. Cereal seed oils were found to be especially good sources. Wheat germ oil yielded two active compounds, α-tocopherol and β-tocopherol, as crystalline allophanate esters ($NH_2CONHCO_2R$), which on hydrolysis regenerated the vitamins as heavy oils.[106] The generic name tocopherol (Greek: *tokos,* for childbirth, and *phero,* to bear) was adopted when it was evident that more than one compound possessed the activity. Eight tocopherols have been characterized.

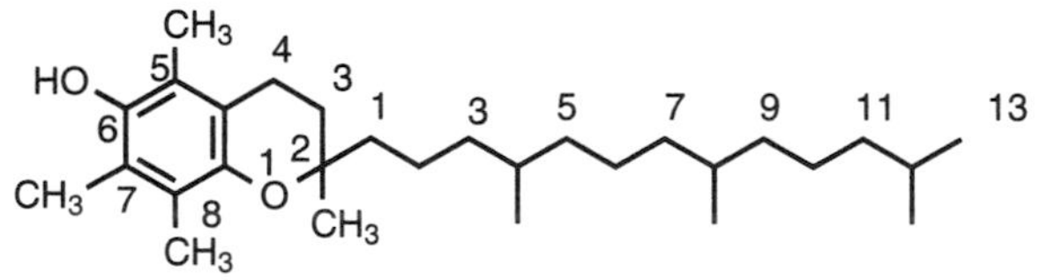

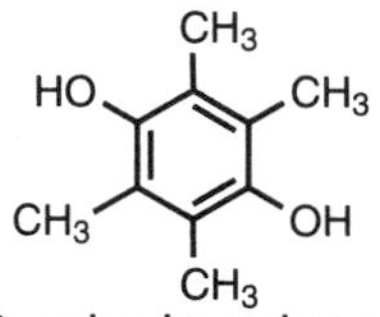

α-Tocopherol

Durohydroquinone

Structure and Synthesis

The structure of α-tocopherol, $C_{29}H_{50}O_2$, was established as 2,5,7,8-tetramethyl-2-(4,8,12-tetramethyltridecyl)-6-chromanol. Pyrolysis at a high temperature formed durohydroquinone.

Because ether-cleaving reagents did not yield the same product, the possibility that the vitamin was a simple ether of durohydroquinone was ruled out. The hydrocarbon chain was characterized from chromic acid oxidation products, one of which was a carboxylic acid with the formula $C_{16}H_{32}O_2$ and containing three C-methyl groups as determined by the Kuhn-Roth method. Application of the isoprene rule formulated the compound as 4,8,12-trimethyltridecanoic acid. Hydrogenation studies and the finding that a benzoquinone formed on silver nitrate oxidation helped deduce a cyclic ether structure, in which the aliphatic side chain is joined to the aromatic ring.

Synthesis of optically inactive α-tocopherol with po-

Trimethylhydroquinone

Phytyl bromide

Coumaran structure

tency equal to that of the natural product was accomplished by Karrer's group in Switzerland using trimethylhydroquinone and phytyl bromide, a procedure adopted for manufacture of the vitamin. The method of formation, however, could not distinguish between the coumaran and chroman structures.[107] Conclusive evidence for the chroman structure was the inability to oxidize to a ketone the hydroxyl group in the side chain of a ring-opened derivative of α-tocopherol. The hydroxyl group in this derivative must, therefore, be tertiary and the ring system in α-tocopherol six-membered.

Six of the eight known tocopherols have the 4,8,12-trimethyltridecyl side chain but differ in the number of methyl groups on the aromatic ring. Tocol is the basic unit of the tocopherols; the aromatic methyl groups are omitted. In this system, α-tocopherol is named 5,7,8-trimethyltocol. Three possible dimethyl tocols are known, but only two (substituted at positions 7 and 8) of the three monomethyl tocols have been isolated. The remaining two known tocopherols have unsaturated side chains with three unconjugated double bonds—a 4,8,12-trimethyltrideca-3,7,11-trienyl unit. For one of these, the ring has the three methyls as in α-tocopherol, whereas the other has only two ring methyls at positions 5 and 8.

Biologic potency varies with the tocopherols; for example, α-tocopherol is the most active, β-tocopherol (5,8-dimethyltocol) is less active, and γ-tocopherol (7,8-dimethyltocol) is inactive. An enzyme-related role for vitamin E has not been revealed, and animals can be raised normally on a vitamin-free diet, provided that an antioxidant is added. This antioxidant property of vitamin E is considered to be responsible for its activity, although there is evidence that some effects of tocopherol cannot be explained solely on the basis of a nonspecific role.

Chemical and Biologic Assays

Vitamin E activity can be assayed biologically by using rat fertility as an indicator or chemically by measuring its reducing property. The second method requires removal of interfering reducing agents or a procedure for subtracting their contribution.

Deficiency

The deficiency symptoms observed in animals, such as fetus resorption in rats, myocardial degeneration in cattle, and dystrophy in voluntary muscles of dogs and chicks, with chicks also showing neurologic and vascular degeneration, are believed to be secondary manifestations of a primary lesion, i.e., muscular degeneration of dystrophy. In each animal species, some muscles are more sensitive than others; these will be affected first, resulting in a myriad of physiologic abnormalities.

A vitamin E deficiency disease is not recognized in humans, but in muscular dystrophy, habitual abortion, and cardiac disease, α-tocopherol therapy has been used with varied success.

Daily Requirements

For good health, 5 to 30 mg of α-tocopherol daily is considered adequate. The larger amount is needed by individuals on a diet containing unsaturated fats (vegetable oils). Good dietary sources are vegetable oils, cereals, eggs, and butter.

VITAMIN K

In 1929, Dam,[108] a Danish scientist, found that chicks on a purified diet developed a hemorrhagic disease that was prevented by a food supplement. The active factor was named koagulations vitamin, which has since become vitamin K. The deficiency disease is characterized by an increase in the blood-clotting time and results from impaired synthesis of prothrombin, proconvertin, and other proteins required for blood clotting. The vitamin is fat-soluble, and the first pure sample (now called vitamin K_1) was isolated from alfalfa.[109] The vitamin K activity in putrefied fish meal, which has a microbiologic origin, was due to another compound, named vitamin K_2.[110]

Structure and Synthesis

Vitamin K_1, a yellow oil, has the structure 2-methyl-3-phytyl-1,4-naphthoquinone. The empiric formula $C_{31}H_{46}O_2$ was obtained from elemental analyses and the

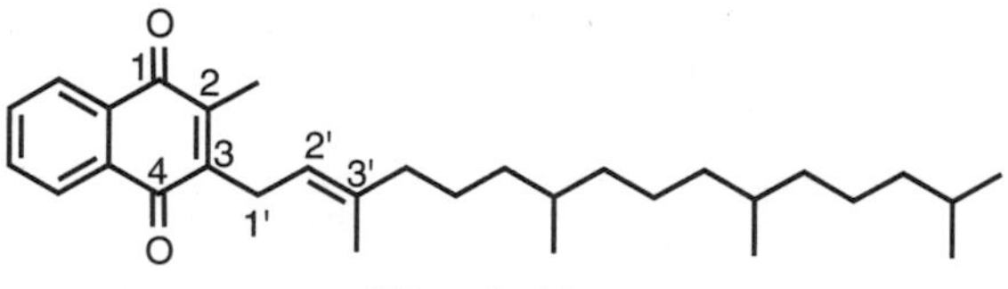

Vitamin K_1

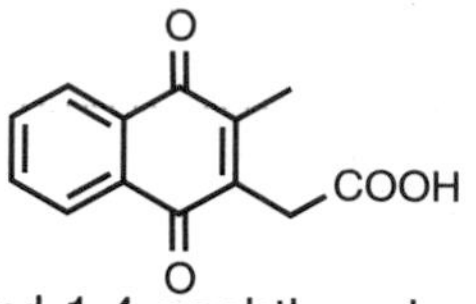

2-Methyl-1,4-naphthoquinone-3-acetic acid

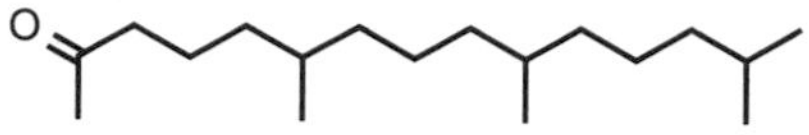

6,10,14-Trimethylpentadeca-2-one

molecular weight determined by potentiometric titration. Chronic acid oxidation yielded phthalic acid and 2-methyl-1,4-naphthoquinone-3-acetic acid; ozonolysis yielded 6,10,14-trimethylpentadeca-2-one, which is identical to the ketone (compared as the semicarbazone) obtained by oxidation of phytol. These products suggested a possible structure for vitamin K_1 that was confirmed by synthesis.

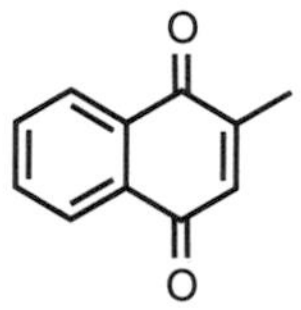

Menadione

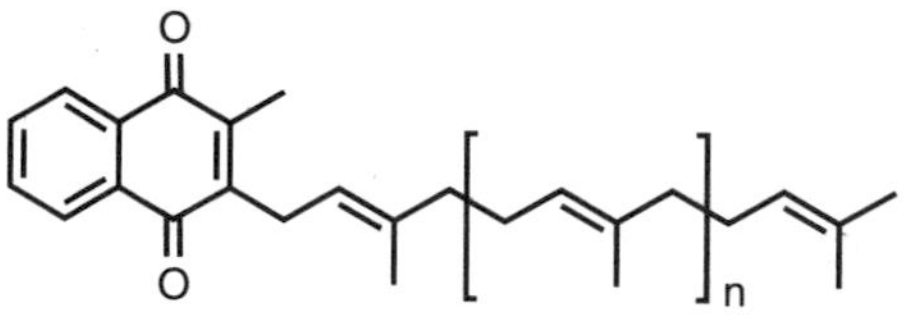

Vitamin $K_{2(35)}$ n = 5
Viatmin $K_{2(20)}$ n = 2

Vitamin $K_{2(35)}$,* with a melting point of 52° to 53°C, differs from vitamin K_1 only in the nature of the side chain. In addition to being longer (seven isoprene units instead of four), it is unsaturated to the extent of one double bond for each isoprene unit. Incidentally, vitamin $K_{2(35)}$ was originally thought to have 30 carbon atoms. Synthesis of vitamin $K_{2(30)}$, nearly 20 years after its structure was proposed, showed that an additional isoprene unit was required. Vitamin $K_{2(30)}$ is found in purified fish meal, but it is only a minor component relative to vitamin $K_{2(35)}$.

Menadione, the vitamin K naphthoquinone moiety without the side chain, has activity equivalent to that of the vitamin on a molar basis. It is a synthetic compound readily made by oxidation of 2-methylnaphthalene. When fed to animals, menadione is converted to vitamin $K_{2(20)}$, the metabolic form of the vitamin. Apparently the 20-carbon substituent is obtained from a geranylgeranyl pyrophosphate group. The other forms of vitamin K are likewise converted to vitamin $K_{2(20)}$ by formation of menadione and then its alkylation by the tetraisoprene unit.

All vitamin K–active substances must contain a 1,4-naphthoquinone moiety or a precursor readily converted to it. The 2-methyl substituent is also necessary, but increasing its size to ethyl decreases the response. The 3-substituents showing the greatest activity are isoprenoid; potency increases with chain length, with maximal activity appearing at 20 to 30 carbon atoms. The unsaturation at the 2′-position enhances the function, but other points of unsaturation are without effect.

The biologic role of vitamin K is now known to be connected with posttranslational carboxylation of proteins,[111] in particular, those involved in blood coagulation, such as prothrombin and factors VII, IX, and X. The carboxylation step occurs on glutamic acid side chains to produce γ-carboxyglutamic acid residues. This amino acid was not previously recognized as a component of proteins, and it imparts to them such characteristics as Ca^{2+} binding and association with phospholipids of membranes. Absence of these properties prevents normal blood coagulation. In the case of prothrombin, a protein of 72,000 daltons, the 10 γ-carboxyglutamyl groups are localized at the amino end up to residue 32. The other proteins likewise contain the carboxylated residues at the

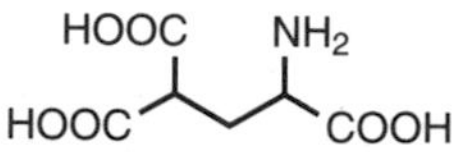

γ-Carboxyglutamic acid

* The subscript in parentheses is added to the K_2 designation to signify the carbon length of the unsaturated side chain, since other members have been isolated and differ in this respect.

amino end. There is evidence that proteins not connected with blood coagulation contain γ-carboxyglutamic acid so that the carboxylation process may be widely distributed.

Introduction of the carboxy group to the precursor protein is by a microsomal enzyme that requires vitamin K, bicarbonate, a reducing environment (dithiothreitol), and oxygen. Vitamin K reacts with oxygen while in its hydroquinone form to give a basic epoxy-monoquinone hydrate, which abstracts the γ-hydrogen of glutamate, which then becomes carboxylated via CO_2 from bicarbonate.[112] The other products are vitamin K (quinone form) epoxide and water. Regeneration of the epoxide quinone is a two-step reductive process; the first requires a dithiol to give vitamin K quinone and water, and the second uses a dithiol or NAD(P)H to form vitamin K. Both reductases appear to be inhibited by the clinically useful 4-hydroxycoumarins (e.g., warfarin).[113]

Chemical and Biologic Assay

Vitamin K activity can be estimated by measuring the clotting time of blood from chicks on controlled diets and comparing it with a standard, such as menadione. Chemical methods, which take advantage of the naphthoquinone group to produce a color, have also been used, as have the ultraviolet absorption properties for development of a physical method.

Deficiency

In normal adults, lack of vitamin K is seldom a problem because the need is met by the diet, in which leafy green vegetables are an excellent source, or by the intestinal flora. Normal adults require between 70 and 140 μg per day. A deficiency may appear, however, if absorption of the vitamin is impaired, such as in diseases that affect the level of bile salts. A water-soluble analog of vitamin K may be used in such cases.

Vitamin K is also used in conjunction with bishydroxycoumarin or related anticoagulants to maintain the prothrombin level at a safe value in patients with cardiovascular disorders. Newborn infants may show a low prothrombin level because of a lack of the vitamin because their intestinal tract is sterile and vitamin K is not available from the intestinal flora. Coupled with this is the fact that transfer of the vitamin from the mother is poor. Consequently, administration of vitamin K to the mother before delivery can ensure an adequate supply and safeguard against hemorrhage in infants, which, if it occurs in the brain, can cause irreparable damage.

REFERENCES

1. B. C. P. Jansen and W. F. Donath, *Proc. Kon. Nererl. Akad. Wet., 29,* 1390(1926).
2. R. R. Williams, *J. Am. Chem. Soc., 58,* 1063(1936).
3. R. R. Williams and J. K. Cline, *J. Am. Chem. Soc., 58,* 1504(1936).
4. R. Breslow, *J. Am. Chem. Soc., 80,* 3719(1958).
5. A. G. Datta and E. Racker, *J. Biol. Chem., 236,* 624(1961).
6. P. C. Newell and R. G. Tucker, *Biochem. J., 106,* 279(1968).
7. R. Kuhn, et al., *Chem. Ber., 66,* 317(1933).
8. R. Kuhn and T. Wagner-Jauregg, *Chem. Ber., 66,* 1577(1933).
9. R. Kuhn and H. Rudy, *Chem. Ber., 67,* 892, 1125, 1298(1934).
10. P. Karrer, et al., *Helv. Chim. Acta, 18,* 426(1935).
11. M. Tishler, et al., *J. Am. Chem. Soc., 69,* 1487(1947).
12. R. Kuhn, et al., *Chem. Ber., 69,* 2034(1936).
13. C. Deluca and N. O. Kaplan, *Biochim. Biophys. Acta, 30,* 6(1958).
14. G. W. E. Plaut, in *Comprehensive Biochemistry,* Vol. 21, M. Florkin and E. H. Stotz, Eds. New York, Elsevier, 1971, p. 11.
15. W. H. Walker and T. P. Singer, *J. Biol. Chem., 245,* 4224(1970).
16. W. C. Kenney, et al., in *Metabolic Pathways,* Vol. 7, 3rd ed., D. M. Greenberg, Ed. New York, Academic Press, 1975, p. 189.
17. R. A. Harvey and G. W. E. Plaut, *J. Biol. Chem., 241,* 2120(1966).
18. E. E. Snell and F. M. Strong, *Ind. Eng. Chem. Anal. Ed., 11,* 346(1939).
19. J. Goldberger, *JAMA, 66,* 471(1916).
20. C. A. Elvehjem, et al., *J. Am. Chem. Soc., 59,* 1767(1937).
21. C. A. Elvehjem, et al., *J. Biol. Chem., 123,* 137(1938).
22. A. Harden and W. Young, *Proc. R. Soc. Lond.* (Biol.), *81,* 528(1909).
23. L. J. Haynes, et al., *J. Chem. Soc.,* 3727(1957).
24. N. A. Hughes, et al., *J. Chem. Soc.,* 3733(1957).
25. F. A. Loewus, et al., *J. Am. Chem. Soc., 77,* 3391(1955).
26. J. C. Keresztesy and J. R. Stevens, *Proc. Soc. Exp. Biol. Med., 38,* 64(1938).
27. S. Lepkovsky, *Science, 87,* 169(1938).
28. R. Kuhn and G. Wendt, *Chem. Ber., 71,* 780(1938).
29. P. Gyorgy, *J. Am. Chem. Soc., 60,* 938(1938).
30. A. Ichiba and K. Michi, *Sci. Papers Inst. Phys. Chem. Res. (Tokyo), 34,* 623(1938).
31. A. Ichiba and K. Michi, *Chem. Abstr., 32,* 7534[8](1938).
32. E. T. Stiller, et al., *J. Am. Chem. Soc., 61,* 1237(1939).
33. E. E. Harris, et al., *J. Org. Chem., 27,* 2705(1962).
34. E. E. Snell, et al., *J. Biol. Chem., 143,* 519(1942).
35. D. E. Metzler, et al., *J. Am. Chem. Soc., 76,* 648(1954).
36. H. C. Dunathan, in *Adv. Enzymol.,* Vol. 35, A. Meister, Ed. New York, Interscience, 1971, p. 79.
37. R. E. Hill, et al., *J. Am. Chem. Soc., 93,* 518(1971).
38. R. J. Williams, in *Comprehensive Biochemistry,* Vol. 11, M. Florkin and E. H. Stotz, Eds. New York, Elsevier, 1963, Chapter 5.
39. W. H. DeVries, et al., *J. Am. Chem. Soc., 72,* 4838(1950).
40. J. Baddiley, et al, *Nature, 171,* 76(1953).
41. J. G. Moffatt and H. P. Khorana, *J. Am. Chem. Soc., 83,* 663(1961).
42. T. C. Vanaman, et al., *J. Biol. Chem., 243,* 6420(1968).
43. F. Kögl and B. Tönnos, *Z. Physiol. Chem., 242,* 43(1936).
44. L. H. Sternbach, in *Comprehensive Biochemistry,* Vol. 11, M. Florkin and E. H. Stotz, Eds. New York, Elsevier, 1963, Chapter 6.
45. J. Trotter and J. A. Hamilton, *Biochemistry, 5,* 713(1966).
46. S. A. Harris, et al., *J. Am. Chem. Soc., 66,* 1956(1944).
47. P. N. Canfalone, et al., *J. Org. Chem., 42,* 135(1977).
48. M. D. Lane and F. Lynen, *Proc. Natl. Acad. Sci. U.S.A., 49,* 379(1963).
49. N. M. Grren, *Nature, 217,* 254(1968).

50. M. A. Eisenberg, in *Metabolic Pathways,* Vol. 7, 3rd ed., D. M. Greenberg, Ed. New York, Academic Press, 1975, p. 27.
51. R. J. Parry and M. G. Kunitani, *J. Am. Chem. Soc., 98,* 4024(1976).
52. L. D. Wright and H. R. Skeggs, *Proc. Soc. Exp. Biol. Med., 56,* 95(1944).
53. L. D. Wright, et al., *Science, 114,* 635(1951).
54. H. K. Mitchell, et al., *J. Am. Chem. Soc., 63,* 2284(1941).
55. J. H. Mowat, et al., *J. Am. Chem. Soc., 70,* 14(1948).
56. E. L. R. Stokstad and J. Koch, *Physiol. Rev., 47,* 83(1967).
57. G. R. Minot and W. P. Murphy, *JAMA, 87,* 470(1926).
58. E. L. Rickes, et al., *Science, 107,* 396(1948).
59. E. L. Smith and L. F. J. Parker, *Biochem. J., 43,* viii(1948).
60. D. C. Hodgkin, et al., *Proc. R. Soc. Lond. (A), 242,* 228(1957).
61. A. I. Scott, *Tetrahedron, 31,* 2639(1975).
62. T. H. Maugh II, *Science, 179,* 266(1973).
63. H. A. Barker, et al., *Proc. Natl. Acad. Sci. U.S.A., 44,* 1093(1958).
64. H. Weissbach, J. Toohey, and H. A. Barker, *Proc. Natl. Acad. Sci. U.S.A., 45,* 521(1959).
65. P. G. Lenhert and D. C. Hodgkin, *Nature, 192,* 937(1961).
66. T. C. Stadtman, *Science, 171,* 859(1971).
67. W. A. Waugh and C. G. King, *J. Biol. Chem., 97,* 325(1932).
68. M. J. Barnes and E. Kodicek, *Vitam. Horm., 30,* 1(1972).
69. M. Goldstein, et al., *Biochemistry, 7,* 2724(1968).
70. O. Touster, *Annu. Rev. Biochem., 31,* 407(1962).
71. F. A. Loewus and R. Jang, *Biochim. Biophys. Acta, 23,* 205(1957).
72. T. W. Anderson, *Nutrition Today, (Jan–Feb),* 6(1977).
73. E. V. McCollum and M. Davis, *J. Biol. Chem., 15,* 167(1913).
74. P. Karrer, R. Morf, and K. Schopp, *Helv. Chim. Acta, 14,* 1036, 1431(1931).
75. A. Vahlquist and P. A. Peterson, *Biochemistry, 11,* 4526(1972).
76. Y. Muto, et al., *J. Biol. Chem., 247,* 2542(1972).
77. O. Isler, et al., *Helv. Chim. Acta, 30,* 1911(1947).
78. O. Isler and U. Schwieter, *Dtsch. Med. J., 16,* 576(1965).
79. G. Wald, *J. Gen. Physiol., 19,* 351, 781(1935–1936).
80. L. Pauling, *Fortschr. Chem. Org. Naturstoffe, 3,* 203(1939).
81. J. E. Dowling and G. Wald, *Vitam. Horm., 18,* 387(1960).
82. M. Petkovich, *Annu. Rev. Nutr., 12,* 443(1992).
83. R. B. Clayton, *Quart. Rev., 19,* 168(1965).
84. R. B. Clayton, *Quart. Rev., 19,* 201(1965).
85. J. W. Porter and D. G. Anderson, *Arch. Biochem. Biophys., 97,* 520(1962).
86. E. Mellanby, *Lancet, 196,* 407(1919).
87. H. Steenbock, *Science, 60,* 224(1924).
88. H. Steenbock and A. Black, *J. Biol. Chem., 61,* 405(1924).
89. A. Windaus, et al., *Justus Liebigs Ann. Chem., 492,* 226(1932).
90. F. A. Askew, et al., *Proc. R. Soc. Lond.* (Biol.), *B109,* 488(1932).
91. H. Brockmann, *Z. Physiol. Chem., 241,* 104(1936).
92. R. B. Woodward and R. Hoffmann, *The Conservation of Orbital Symmetry.* New York, Academic Press, 1970.
93. H. H. Inhoffen, *Angew. Chem., 72,* 857(1960).
94. R. H. Wasserman, et al., *J. Biol. Chem., 243,* 3978(1968).
95. R. H. Wasserman and A. N. Taylor, *J. Biol. Chem., 243,* 3987(1968).
96. H. F. DeLuca, *FASEB J., 2,* 224(1988).
97. D. E. M. Lawson, et al., *Natural, 230,* 228(1971).
98. M. F. Holick, et al., *Proc. Natl. Acad. Sci. U.S.A., 68,* 803(1971).
99. A. W. Norman, et al., *Science, 173,* 51(1971).
100. J. W. Blunt, et al., *Biochemistry, 7,* 3317(1968).
101. H. F. DeLuca and H. K. Schnoes, *Annu. Rev. Biochem., 52,* 411(1983) and references therein.
102. A. W. Norman and H. F. DeLuca, *Anal. Chem., 35,* 1247(1963).
103. P. P. Nair, et al., *Anal. Chem., 37,* 631(1965).
104. H. M. Evans and K. S. Bishop, *Science, 56,* 650(1922).
105. H. A. Matrill, *J. Biol. Chem., 50,* xliv(1922).
106. H. M. Evans, et al., *J. Biol. Chen., 113,* 319(1936).
107. P. Karrer, et al., *Helv. Chim. Acta, 21,* 520(1938).
108. H. Dam, *Biochem. Z., 215,* 475(1929).
109. S. B. Binkley, et al., *J. Biol. Chem., 130,* 219(1939).
110. R. W. McKee, et al., *J. Biol. Chem., 131,* 327(1939).
111 J. Stenflo and J. W. Suttie, *Annu. Rev. Biochem., 46,* 157(1977).
112. P. Dowd, et al., *J. Am. Chem. Soc., 114,* 7613(1992).
113. J. W. Suttie, *Annu. Rev. Biochem., 54,* 459(1985).

SUGGESTED READINGS

G. G. Birch and K. J. Parker, Eds. *Vitamin C.* New York, John Wiley and Sons, 1974.

R. L. Blakley, *The Biochemistry of Folic Acid and Related Pteridines.* Amsterdam, North-Holland Publishing Co., 1969.

G. W. Clark, *A Vitamin Digest.* Springfield, Ill., Charles C. Thomas, 1953.

M. B. Davies, et al., *Vitamin C.* Cambridge, Royal Society of Chemistry, 1991.

A. P. DeLeenheer, et al., *Modern Chromatographic Analysis of Vitamins.* 2nd ed. New York, Marcel Dekker, 1992.

D. Dolphin, B_{12}; Vol. 1, *Chemistry;* Vol. 2. *Biochemistry and Medicine.* New York, Wiley-Interscience, 1982.

H. J. Deuel, Jr., Ed., *The Lipids,* Vols. 1–3. New York, Interscience Publishers, 1951–1957.

S. F. Dyke, *The Chemistry of the Vitamins.* New York, Interscience Publishers, 1965.

M. Florkin and E. H. Stotz, Eds., *Comprehensive Biochemistry,* Vols. 9 and 11 (1963), Vol. 21 (1971), New York, Elsevier Publishing Co.

T. W. Goodwin, *The Biosynthesis of Vitamins and Related Compounds.* New York, Academic Press, 1963.

C. J. Gubler, Ed., *Thiamine.* New York, John Wiley & Sons, 1976.

P. Gyorgy, *Vitamin Methods,* Vol. 1 (1950), Vol. 2 (1951), New York, Academic Press.

P. Gyorgy and W. N. Pearson, Eds., *The Vitamins,* Vols. 6 and 7. New York, Academic Press, 1967.

R. S. Harris and K. V. Thimann, Eds., *Vitamins and Hormones,* Vols. 1–46. New York, Academic Press, 1943–1991.

L. O. Krampitz, *Thiamine Diphosphate and Its Catalytic Function.* New York, Marcel Dekker, 1970.

R. J. Kutsky, *Handbook of Vitamins and Hormones.* 2nd ed. New York, Van Nostrand Reinhold, 1981.

S. Lewin, *Vitamin C, Its Molecular Biology and Medical Potential.* New York, Academic Press, 1976.

J. M. Luck, et al., Eds., *Annual Review of Biochemistry,* Vols. 1–61. Palo Alto, Calif., Annual Reviews, 1932–1992.

L. J. Machlin, *Handbook of Vitamins,* 2nd Ed. New York, Marcel Dekker, 1990.

T. Moore, *Vitamin A.* Princeton, Van Nostrand, 1957.

R. S. Rivlin, Jr., Ed., *Riboflavin.* New York, Plenum Press, 1975.

F. A. Robinson, *The Vitamin B Complex,* New York, John Wiley & Sons, 1951.

F. A. Robinson, *The Vitamin Co-factors of Enzyme Systems.* New York, Pergamon Press, 1966.

W. H. Sebrell, Jr., and R. S. Harris, Eds., *The Vitamins,* 2nd ed., Vol. 1 (1967), Vol. 2 (1968), Vol. 3 (1971). New York, Academic Press.

E. L. Smith, *Vitamin* B_{12}. New York, John Wiley & Sons, 1965.

Subcommittee on the Tenth Edition of the RDAs, Food and Nutrition Board, National Research Council, *Recommended Dietary Allowances,* 10 ed., Washington, 1989.

A. F. Wagner and K. Folkers, *Vitamins and Coenzymes.* New York, Interscience Publishers, 1964.

R. J. Williams, et al., *The Biochemistry of B Vitamins.* New York, Reinhold, 1950.

Chapter 30

THYROID FUNCTION AND THYROID DRUGS

Alireza Banijamali

The thyroid gland is a highly vascular, flat structure located at the upper portion of the trachea, just below the larynx. It is composed of two lateral lobes joined by an isthmus across the ventral surface of the trachea. The gland is the source of two fundamentally different types of hormones, thyroxine (T_4) and triiodothyronine (T_3). Both are vital for normal growth and development and control essential functions, such as energy metabolism and protein synthesis.

The word thyroid, meaning shield-shaped, was introduced by Wharton in his description of the gland.[1] Like many before him, he attributed a solely cosmetic function to it because of the more frequent presence of enlarged glands in women, giving the throat region a more beautiful roundness. Later it was observed, however, that some characteristic symptoms for diseases always were accompanied by an obvious change in the size of the thyroid. This change was correctly interpreted as evidence that this structure plays a major role in normal body function.

An important step in the understanding of thyroid function was taken by Baumann.[2] He discovered that the thyroid gland was the only organ in mammals that had the capability to incorporate iodine into organic substances. That discovery was important in research on the phylogeny of the thyroid.

Major clues to the physiologic roles of thyroid hormones were provided when normal and abnormal thyroid function were related to oxygen uptake[3] and when thyroid hormones were found to induce metamorphosis in tadpoles.[4] The first discovery led to investigations into the role of thyroid hormone in metabolism and calorigenesis, and the second inspired research into specific receptors as points of initiation of thyroid hormone expression. A patient lacking thyroid hormones may be treated with synthetic hormones or natural preparations. Better agents to treat hyperthyroidism are still being sought. Presently available drugs, other compounds affecting thyroid function, and present approaches in the search of new drugs are presented in this chapter within the context of thyroid biochemistry and physiology.

BIOCHEMISTRY AND PHYSIOLOGY

Thyroid Follicular Cells

All vertebrates have a thyroid gland consisting of functional units, the follicles. The morphologic and functional characteristics of the follicles are essentially similar in all vertebrate groups.

The follicle is a spherical, cystlike structure about 300 μ in diameter and consists of a luminal cavity surrounded by a one-cell-deep layer of cells called follicular or acinar cells. The center of the follicles is filled with a gelatinous colloid, the main component of which is a glycoprotein called thyroglobulin. The follicular cells contain an extensive network of rough endoplasmic reticulum, a well-developed Golgi apparatus, and lysosomes of various sizes.[5] Thyroglobulin is synthesized in the rough endoplasmic reticulum of the follicle cells and transported by way of the Golgi complex to the exocytic vesicles, which empty the thyroglobulin into the follicle lumen.

The follicle cell contains two major assembly lines operating in opposite directions.[6] One line moves in an apical direction and produces thyroglobulin that is delivered to the follicle lumen; the other line begins at the apical cell surface with endocytosis of thyroglobulin and ends by delivering hormones at the basolateral cell surface. The follicle cell seems, therefore, to fulfill the functions of typically secretory and typically absorptive cells simultaneously. In addition to the functions associated with these two lines, the follicle has the specific ability to metabolize iodine, comprising the accumulation of iodide, iodination of tryosyl residues in thyroglobulin, and coupling of iodinated tyrosyls to form thyroid hormones.

Parafollicular cells, also called light cells or C cells, are located individually or in clusters between follicular cells but do not border on the colloid. These cells produce thyrocalcitonin, a peptide hormone involved in calcium homeostasis. The extrafollicular space of the gland is occupied by blood vessels, capillaries, lymphatic vessels, and connective tissue.

Hormones of the Thyroid Gland

Thyroid hormones are iodinated amino acids, synthesized in the thyroid gland and stored as amino acid residues of thyroglobulin. The thyroid hormones, tetraiodothyronine and triiodothyronine play numerous, profound roles in regulating metabolism, growth, and development and in the maintenance of homeostasis. It is generally believed that these actions result from effects of thyroid hormones on protein synthesis.

The first biologically active iodine-containing compound of the thyroid gland was isolated from thyroid extracts by Kendall[7] and named thyroxine. Later its structure was established by Harington[8] as the 3,5,3′,5′-

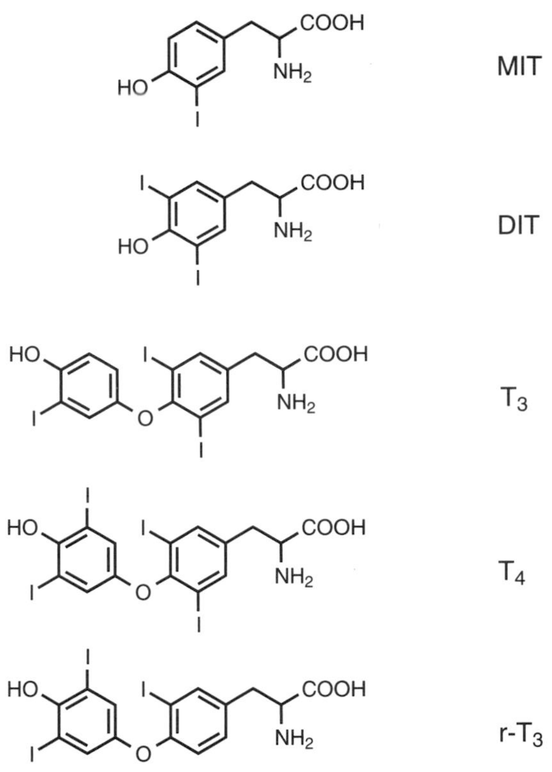

Fig. 30–1. Structure of the iodinated compounds of the thyroid glands.

tetraiodo-L-thyronine (T_4) (Fig. 30–1), and its synthesis was accomplished by Harington and Bargar.[9] Twenty-five years later, with the availability of chromatographic techniques and radioactive iodine, another thyroid hormone was characterized and identified as 3,5,3′-triiodo-L-thyronine (T_3) (Fig. 28–1) simultaneously by Gross and Pitt-Rivers[10] and Roch and co-workers.[11] The thyroid gland also contains two quantitatively important iodinated amino acids; diiodotyrosine (DIT) was isolated from thyroid tissues by Harington and Randall,[12] and monoiodotyrosine (MIT) was discovered by Fink and Fink.[13] In addition, there are small amounts of other iodothyronines, such as 3,3′-diiodothyronine (T_2) and 3,3′,5′-triiodothyronine (reverse T_3, rT_3). None of the latter compounds possess any significant hormonal activity.

Chemically, MIT is 3-iodo-L-tyrosine and DIT is 3,5-diiodo-L-tyrosine. The coupling of two DIT residues or of one DIT with one MIT residue (each with the net loss of alanine) leads to the formation of the two major thyroid hormones, T_4 and T_3.

It has been estimated that thyroglobulin, which has a molecular weight of 660,000, accounts for one-third of the weight of the thyroid gland and carries an average of 6 of its tyrosine residues as monoiodotyrosine, 5 as diiodotyrosine, 0.3 as triiodothyronine, and 1 as thyroxine.[14] From these values, it can be estimated that a 20-g gland stores roughly 10 μmol (7.8 mg) of thyroxine and 3 μmol (2.0 mg) of triiodothyronine and that the normal human thyroid gland contains enough potential T_4 to maintain a euthyroid state for 2 months without new synthesis.[15] The structures of the iodinated compounds of the thyroid gland are shown in Figure 30–1.

Formation of Thyroid Hormones

The thyroid hormones T_3 and T_4 are formed in a giant prohormone molecule, thyroglobulin, the major component of the thyroid and more precisely of the colloid. Thyroglobulin is an iodinated glycoprotein made up of two identical subunits, each with a molecular weight of 330,000 daltons. It is of special importance because it is necessary for the synthesis of thyroid hormones and represents their form of storage.

The formation of the thyroid hormones depends on an exogenous supply of iodide. The thyroid gland is unique in that it is the only tissue of the body able to accumulate iodine in large quantities and incorporate it into hormones. The metabolism of iodine is so closely related to thyroid function that the two must be considered together. The formation of thyroid hormones involves the following complex sequence of events: (1) active uptake of iodide by the follicular cells, (2) oxidation of iodide and formation of iodotyrosyl residues of thyroglobulin, (3) formation of iodothyronines from iodotyrosines, (4) proteolysis of thyroglobulin and release of T_4 and T_3 into blood, and (5) conversion of T_4 to T_3. These processes are summarized in Figure 30–2.[16]

Active Uptake of Iodide by Follicular Cells. The first step in the synthesis of the thyroid hormones is the uptake of iodide from the blood by the thyroid gland. An adequate intake of iodide is essential for the synthesis of sufficient thyroid hormone. Dietary iodine is converted to iodide and almost completely absorbed from the gastrointestinal tract. Blood iodine is present in a steady state in which dietary iodide, iodide "leaked" from the thyroid gland, and reclaimed hormonal iodide provide the input, which thyroidal uptake, renal clearance, and a small biliary excretion providing the output. The thyroid gland regulates both the fraction of circulating iodide it takes up and the amount of iodide that it leaks back into the circulation. A simplified scheme of iodide metabolism is present in Figure 30–3.

The mechanism enabling the thyroid gland to concentrate blood iodide against a gradient into the follicular cell is sometimes referred to as the iodide pump. It brings about a ratio of thyroid iodide to serum iodide (T/S ratio) of 20:1 under basal conditions but of more than 100:1 in hyperactive gland.

Iodide uptake may be blocked by several inorganic ions, such as thiocyanate and perchlorate. Because iodide uptake involves concurrent uptake of potassium, it can also be blocked by cardiac glycosides that inhibit potassium accumulation.

Oxidation of Iodide and Formation of Iodotyrosines. The second step in the process is a concerted reaction in which iodide is oxidized to an active iodine species that, in turn, iodinates the tyrosyl residues of thyroglobulin. The reaction takes place at the border of the lumen using iodide concentrated within the follicle and is catalyzed by thyroid peroxidase (TPO) in the presence of iodide and hydrogen peroxide. Although DIT residues

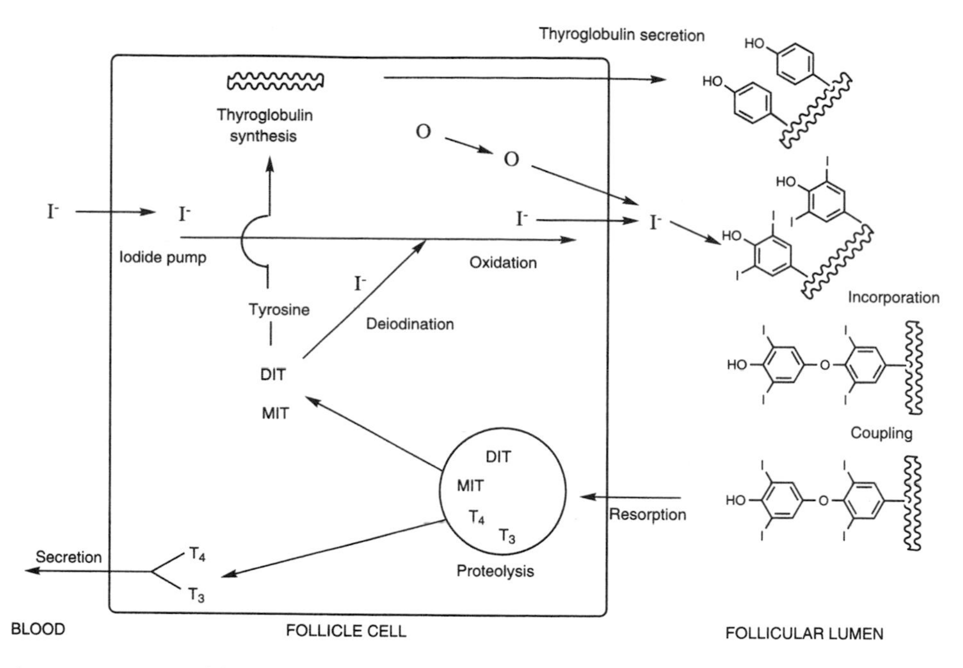

Fig. 30–2. Schematic representation of thyroid hormone biosynthesis and secretion. The protein portion of thyroglobulin is synthesized on rough endoplasmic reticulum, and carbohydrate moieties are added by the Golgi apparatus. Thyroglobulin proceeds to the apical surface in secretory vesicles, which fuse with the cell membrane and discharge their contents into the lumen. Iodide enters the cell by active transport, is oxidized by a perioxidase at the apical border, and is incorporated into tyrosine residues in peptide linkage in thyroglobulin. Two iodinated tyrosyl groups couple in ether linkage to form thyroxine, which is still trapped in thyroglobulin. For the secretory process, thyroglobulin is engulfed by pseudopods at the apical border of the follicular lumen and resolved into vesicles that fuse with lysosomes. Lysosomal protease breaks down thyroglobulin to amino acids, T_4, T_3, diiodotyrosine (DIT), and monoiodotyrosine (MIT). T_4 and T_3 are secreted by the cell into the blood. DIT and MIT are deiodinated to free tyrosine and iodide, both of which are recycled back into iodinated thyroglobulin. (From H. M. Goodman and L. Van Middlesworth, The thyroid gland, in *Medical Physiology,* Vol 2, V. B. Mooncastle, Ed., St. Louis, C. V. Mosby, 1980, p. 1500.)

constitute the major products, some MIT peptides are also produced.

In the thyroid, intracellular iodide taken up from blood is bound in organic form in a few minutes so less than 1 percent of the total iodine of the gland is found as iodide. Therefore, inhibition of the iodide transport system requires blockade of organic binding. This can be achieved by the use of antithyroid drugs, of which n-propyl-6-thiouracil and 1-methyl-2-mercaptoimidazole are the most potent.

Coupling of Iodotyrosine Residues. This reaction takes place at thyroglobulin and involves the coupling of two DIT residues or one DIT with one MIT residue (each with the net loss of alanine) to produce peptide-containing residues of the two major thyroid hormones T_4 and T_3. It is also believed that these reactions are catalyzed by the same peroxidase that effects the iodination and, therefore, can be blocked by compounds such as thiourea, thiouracils, and sulfonamides.

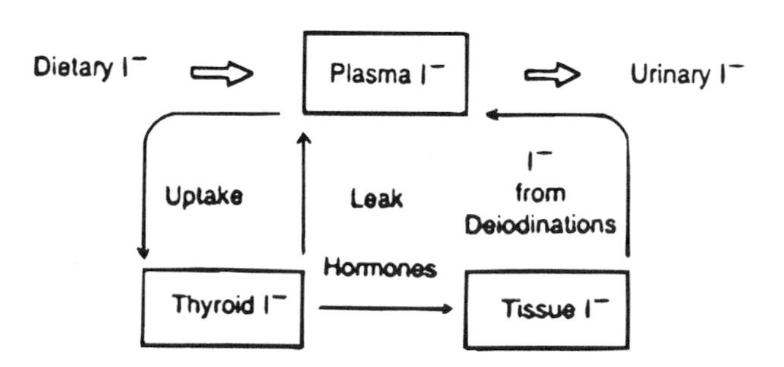

Fig. 30–3. Simplified scheme of iodide metabolism.

Proteolysis of Thyroglobulin and Release of Iodothyronines. The release of thyroid hormones from thyroglobulin is a process that involves endocytosis of colloid droplets into the follicular epithelial cells and subsequent proteolysis of the contents of these droplets by the digestive enzymes of the lysosomes of the follicular cells. The digestion yields MIT, DIT, T_3, and T_4. MIT and DIT, although formed, do not leave the thyroid. Instead, they are selectively metabolized, and the iodide liberated is reincorporated into protein. T_3 and T_4 are secreted by the cell into the circulation.

Conversion of Thyroxine to Triiodothyronine. Although T_4 is by far the major thyroid hormone secreted by the thyroid (about 8 to 10 times the rate of T_3), it is usually considered to be a prohormone. Because T_4 has a longer half-life, much higher levels of T_4 than T_3 are in the circulation. The enzymatic conversion of T_4 to T_3 is an obligate step in the physiologic action of thyroid hormones in most extrathyroidal tissues. In the peripheral tissues, about 33% of the T_4 secreted undergoes 5′-deiodination to give T_3, and another 40% undergoes deiodination of the inner ring to yield the inactive material rT_3.[17] The deiodination of T_4 is a reductive process catalyzed by a group of enzymes named iodothyronine deiodinases referred to as deiodinases and symbolized by D, found in a variety of cells.

Three types of deiodinases are currently known and are distinguished from each other primarily based on their location, substrate preference, and susceptibility to inhibitors. Type I deiodinase is found in liver and kidney and catalyzes both inner ring and outer ring deiodination (i.e., T_4 to T_3 and rT_3 to 3,3′-T_2). Type II deiodinase catalyzes mainly outer ring deiodination (i.e., T_4 to T_3 and T_3 to 3,3′-T_2) and is found in brain and the pituitary. Type III deiodinase is the principal source of rT_3 and is present in brain, skin, and placenta.[18]

Transport of Thyroid Hormones in Blood

The iodothyronines secreted by the thyroid gland into thyroid vein blood are of limited solubility. They equilibrate rapidly, however, through noncovalent association with the plasma proteins thyroxine-binding globulin (TBG), thyroxine-binding prealbumin (TBPA), and albumin. The thyroid hormones form a 1:1 complex with TBG, the major carrier protein in humans with a molecular weight of 63,000 daltons. The plasma proteins involved in thyroid hormone transport and their approximate association constants (K_a) for T_3 and T_4 are shown in Table 30–1. This table indicates that TBG has a high affinity for T_4 (K_a about 10^{10} M) and lower affinity for T_3. TBPA and albumin also transport thyroid hormones in the blood; prealbumin has K_a values of about 10^7 and 10^6 M for T_4 and T_3. The equilibrium between the free hormone and that which is protein bound determines the accessibility of the free thyroid hormone for the tissue receptors as well as to peripheral sites where biotransformations take place. The lower binding affinity for T_3 to plasma proteins may be an important factor in the more rapid onset of action and in the shorter biologic half-life for T_3.

Thyroid hormones are taken into cells by facilitated diffusion or by active transport secondary to a sodium gradient.[15] Once in the cell, thyroid hormones bind to cytosolic binding proteins and are not readily available for exchange with plasma hormones.[19] T_3 and T_4 are not evenly distributed in body cells: A great part of T_4 is stored in liver and kidney, whereas most T_3 appears in muscle and brain.[15]

Table 30–1. Plasma Proteins Involved in Thyroid Hormone Transport

		Binding of T_4		Binding of T_3	
Protein	Concentr. mg/dl	K_a	% Bound	K_a	% Bound
TBG	1.5	10^{10}	75	10^9	70
TBPA	25	10^7	15	10^6	
Albumin	4000	10^6	10	10^5	30

TBG = Thyroxine-binding globulin; TBPA = thyroxine-binding prealbumin.

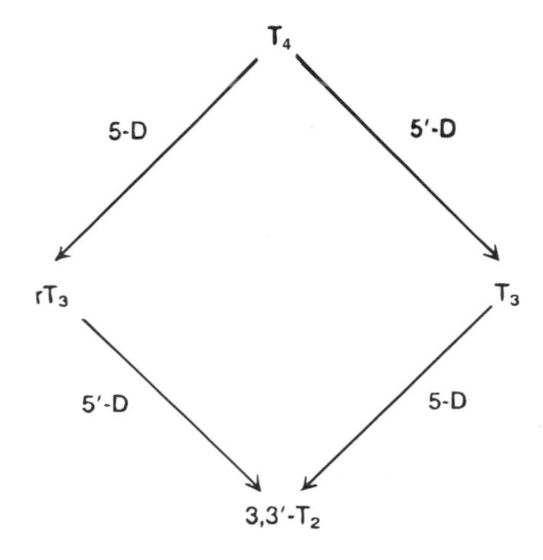

Fig. 30–4. Deiodination of T_4.

Metabolism and Excretion

As discussed earlier, T_4 is considered to be a prohormone, and its peripheral metabolism occurs in two ways: outer ring deiodination by the enzyme 5′-D, which yields T_3, and inner ring deiodination by the enzyme 5-D, which yields rT_3, for which there is no known biologic function (Fig. 30–4). In humans, deiodination is the most important metabolic pathway of the hormone, not only because of its dual role in the activation and inactivation of T_4, but also in quantitative terms.

Degradative metabolism of the thyroid hormones, apart from peripheral deiodination, occurs mainly in the liver, where both T_3 and T_4 are conjugated to form either glucuronide (mainly T_4) or sulfate (mainly T_3) through the phenolic hydroxyl group. The resulting iodothyronine conjugates are excreted via the bile into the intestine, where a portion is hydrolyzed by bacteria. It also undergoes marginal enterohepatic circulation and is excreted unconjugated in feces.

T_4 is conjugated with sulfate in kidney and liver, and the $T_4$4′-O-sulfate, an excellent substrate for 5′-D,[20] is believed to play a role in the regulation of T_4 metabolism.

Additional metabolism, involving side-chain degradation, proceeds by transamination, oxidative deamination, and decarboxylation to yield thyroacetic acid[21] and thyroethanediol[22]; also, cleavage of the diphenyl ether linkage[23] has been detected both in vitro and in vivo. The reactions through which thyroid hormone is metabolized are summarized in Figure 30–5.

Actions of Thyroid Hormones

The two most important actions of thyroid hormone are those related to oxygen consumption and those related to protein synthesis. Most effects of thyroid hormones can be related to the activation of genes following the binding of the hormone to high-affinity receptors of cell nuclei, but direct interactions of thyroid hormones with other cellular receptors cannot be excluded.[24]

Oxygen Consumption and Calorigenesis

A respiratory component of the action of thyroid hormones was first observed almost a century ago. Respira-

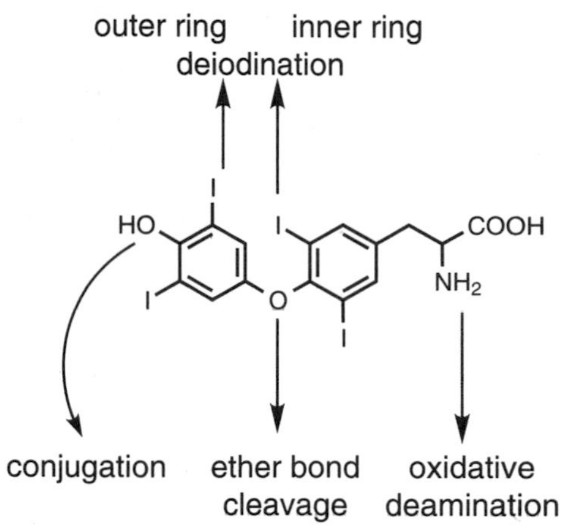

Fig. 30–5. Metabolic pathways for thyroxine.

tory exchange was depressed in patients diagnosed as hypothyroid and increased in patients diagnosed as hyperthyroid.[3] The increase in respiration that follows the administration of thyroid hormone reflects an increase in metabolic rate thyroid function has indeed long been assessed by measuring the basal or resting metabolic rate (BMR), a test in which the oxygen consumed, measured in an individual at rest, is used to calculate total body energy production. The BMR of a hyperthyroid individual is above the normal range, or positive, and that of a hypothyroid individual below the normal range, or negative.

Thyroid hormones increase the oxygen consumption of most isolated tissues but not of isolated adult brain. This suggests that the behavioral changes seen in abnormal thyroid states (anxiety and nervousness in hyperthyroidism and impaired memory in hypothyroidism) in the adult are not directly linked to overall changes in brain oxygen consumption.

Because most of the energy produced by cellular respiration eventually appears as heat, an increase in cellular respiration leads necessarily to an increase in heat production, i.e., to a thermogenic or calorigenic effect. Thus, to the degree that thyroid hormones control BMR, they also control thermogenesis.[25]

Clinically the inability to adjust to environmental temperature is symptomatic of departure from the euthyroid status. Patients with myxedema frequently have subnormal body temperature, have cold and dry skin, and tolerate cold poorly; the thyrotoxic patient, who compensates for excess heat production by sweating (warm, moist hands) does not easily tolerate a warm environment.

Thyroid hormones regulate the turnover of carbohydrates, lipids, and proteins. They promote glucose absorption, hepatic and renal gluconeogenesis, hepatic glycogenolysis, and glucose utilization in muscle and adipose tissue.[26] They increase de novo cholesterol synthesis but even more increase low-density lipoprotein (LDL) degradation and cholesterol disposal, leading to a net decrease in total and in LDL cholesterol plasma levels.[27] Thyroid hormones are anabolic when present at normal concentrations; they then stimulate the expression of many key enzymes of metabolism. Thyroid hormones at the levels present in hyperthyroidism are catabolic; they lead to the mobilization of tissue protein and especially of muscle tissue protein for gluconeogenetic processes.[28] Thus, the depletion of liver glycogen, the increased breakdown of lipids, and the negative nitrogen balance observed in hyperthyroidism represent toxic effects. The metabolic processes of hypermetabolism are wasteful. In hyperthyroidism, a smaller than normal amount of liberated energy is available for useful work, and an excessive amount is wasted as heat.

Because of the role of mitochondria in cellular respiration and energy production, efforts to elucidate the mechanism of thyroid hormone action on metabolism and calorigenesis have focused on mitochondrial studies. Thyroid hormones in vitro are known to uncouple oxidative phosphorylation in isolated mitochondria, but these effects occur at unphysiologic doses of T_4; in physiologic concentrations, T_4 increases ATP formation and the number and inner membrane surface area of mitochondria[29] but does not reduce the efficiency of oxidative phosphorylation. Further 2,4-dinitrophenol, a classic uncoupler of oxidative phosphorylation, can neither relieve hypothyroidism nor duplicate other physiologic effects of thyroid hormones.

Because in most physiologic circumstances oxygen consumption is controlled by energy metabolism, Ismail-Beigi and Edelman[30] proposed that the primary effect of thyroid hormones is to increase the amount of energy expended in translocating cations across cell membranes, probably as a response to an increased passive leak of sodium into, and potassium out of, cells. The extent to which this transport contributes to heat production and ATP utilization is uncertain. The stimulation of futile cycles by thyroid hormones[26] has been suggested to be an additional component of ATP disposal.

The effectiveness of β-blocking agents in treating the symptoms of thyrotoxicosis has led to investigations that have indicated that thyroid hormones modulate adrenergic effects by increasing the number of β-adrenergic receptors[31] and, at least in cardiac tissue, by increasing adenylyl adrenergic receptors are seen as coupled with a membrane-bound adenyl cyclase,[33] both observations appear mutually consistent. The role of thyroid hormones in adaptive thermogenesis[25] could be effected by stimulating an increase in brown adipose tissue β-adrenergic receptors, catecholamines, which are known to activate the β receptor–linked adenylyl cyclase in that tissue,[34] would then start the lipolysis cascade resulting in heat production.

Differentiation and Protein Synthesis

Stimulation of amphibian metamorphosis by thyroid hormones has been known for a long time.[4] Changes in amphibian morphogenesis, such as involution of the tail of Xenopus (toads) and metamorphosis of the axolotl (salamanders) and of rana (frogs) species, have been repeatedly used as end points for the bioassay of thyroid analogs. In young mammals, thyroid hormone is necessary not only for general growth, but also for proper differentiation of the central nervous system. The counterpart of amphibian metamorphosis in the developing mammalian brain is myelogenesis and the formation of axons and dendrites.[35] Behavioral studies in neonate hypothyroid mammals show that impairment in learning capacity[35] paralleled the impairment of proper anatomic development in the brain. A deficiency of thyroid hormone during the critical period when the developing human brain is sensitive to thyroid hormone results in an irreversible clinical entity termed cretinism, which is characterized by stunted growth and mental retardation.

A great deal of attention has been devoted to the events taking place in the roughly 48-hour interval between the administration of thyroid hormone and the manifestation of certain effects caused by that administration. After Tata[36] had observed that the anabolic effects observed in rats given T_4 could be blocked by inhibitors of protein synthesis, further investigations led Tata and Widnell[37] to the conclusion that thyroid hormones were activating protein synthesis at the ribosomal level. Subsequently, Dillmann and co-workers[38] showed that the T_3-induced formation of α-glycerophosphate dehydrogenase (E.C. 1.1.99.5) could be blocked by α-amanitin, an inhibitor of RNA polymerase II, inferring that thyroid hormones affected the transcription process. In 1972, when the partition of thyroid hormones between plasma and a number of tissues was investigated, Schadlow and associates[39] demonstrated that the uptake of T_3 by unfractionated pituitary tissue was inverse to the dose, suggesting that the pituitary was limited in its capacity to bind T_3. This finding prompted efforts to find the subcellular location of the binding sites and to search for the presence of specific binding sites in other tissues.[40] The research revealed that the binding sites were located in cell nuclei and were present also in liver and kidney.[40] Eventually, receptors were located in all mammalian tissues and eventually also in tadpole erythrocytes.[41] In cultured pituitary cells, the receptor was described as an acidic, nonhistone 50,000-dalton protein[42] with an equilibrium dissociation (Kd) constant of 2.9×10^{-11} for T_3 and of 2.6×10^{-10} for T_4, indicating that T_3 had a tenfold greater affinity than T_4 for the receptor. There were 6000 nuclear binding sites in the pituitary, 4000 in the liver, but only 16 in the testis, a tissue that is not responsive to thyroid hormone.[43] Tata's findings of the activation of one RNA polymerase[37] were extended to all RNA polymerases, suggesting that thyroid hormones were involved in all phases of gene activation.

Receptors located on the plasma membrane, in the cytoplasm, and in mitochondria have also been described.[44] Plasma membrane receptors are believed to mediate the transport of hormone into the cell and possibly to mediate nonnuclear, immediate effects of thyroid hormone.[44] The presence of a large number of low-affinity cytosol binding proteins has been known for some time,[45] but these proteins have not been assigned a specific physiologic role; it has been proposed that the cytosolic binding proteins merely serve as a large intracellular thyroid hormone reservoir delaying the metabolic disposition of thyroid hormone.

Because thyroid hormones immediately increase oxygen uptake and ATP formation and eventually lead to an increase in the number, size, and inner-membrane surface area of mitochondria as well as to the induction of many mitochondrial enzymes,[46–48] many efforts to characterize mitochondrial thyroid hormone receptors have been made. Sterling and co-workers[49] have reported the presence of very high affinity receptors in mitochondria but their work has not been duplicated, and many authors believe that most mitochondrial enzymes are synthesized on cytoplasmic ribosomes and then transferred into mitochondria.[50]

Control of Thyroid Hormone Formation

The primary role of thyroid is to produce thyroid hormones. The primary regulator of thyroid, both its function and growth, is the pituitary hormone, thyrotropin (thyroid-stimulating hormone [TSH]). Thyrotropin is a glycoprotein with a molecular weight of about 28,000, which can dissociate into two equally large polypeptide chains, an α-chain, which is also present in other pituitary peptides, and a β-chain, which is present in TSH only. TSH is produced by and released from the anterior pituitary gland causing the thyroid hormone to initiate new thyroid hormone synthesis. Increases in iodide uptake, the iodination of thyroglobulin, and endocytosis of colloid are all observed in response to stimulation of TSH. The effects of TSH on the thyroid appear to be the consequence of binding to high-affinity receptors on the capillary membrane of follicular cells and activation of adenylyl cyclase and protein kinase enzymes, found on the inner side of the membrane, with subsequent phosphorylation of cellular proteins. It is believed that most, if not all, effects of TSH are mediated through the cyclic adenosine monophosphate (cAMP) formed by the adenylyl cyclase and possibly also by cAMP-activated intracellular phosphokinases.

The amount of thyroid hormone circulating in body fluids and present in tissues remains fairly constant. Accounting for this constancy are the relatively long biologic half-life of the thyroid hormones, the regulation of gland activity by the pituitary-hypothalamic system, and the availability of iodide. Thyroid hormone research points to T_3 as the thyroid hormone, and therefore factors affecting peripheral T_3 formation by the 5′-D enzymes are highly relevant.

The relation of biological half-life to protein binding has already been alluded to. T_4 firmly bound to plasma proteins; only 0.05% of T_4 is not protein bound, and T_4 had a biologic half-life of 1 week as compared to 1 day for the less firmly bound T_3.

The biosynthesis and secretion of TSH is, in turn, regulated by thyrotropin-releasing hormone (TRH) and the quantity of thyroid hormone in circulation through feedback control. TRH, the tripeptide pyroglutamylhistidylprolylamide, is formed in the hypothalamus, reaches the thyrotrophs (i.e., the TSH-producing pituitary cells) by way of the hypophyseal portal system, and stimulates TSH release by binding to the TRH receptors of the thyrotrophs.[51] The ability of thyroid hormones to prevent the

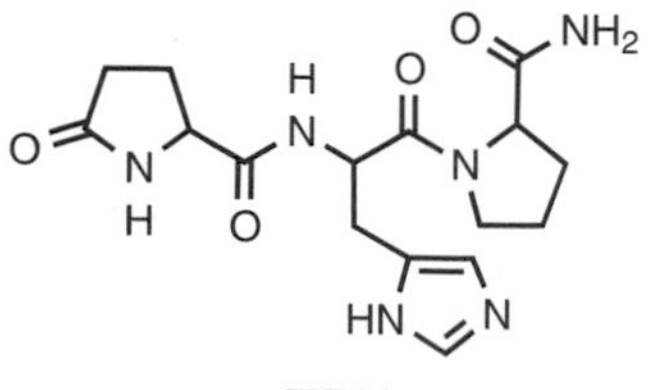

TRH

release of TSH is referred to as feedback regulation. It is believed that T_3 prevents the TRH stimulated release of TSH by stimulating the synthesis of a peptide that would compete with TRH for the TRH receptor sites on the thyrotroph.[52] Inhibition of the transcription of the α and β subunits of TSH by thyroid hormone[53] may also contribute to the feedback effect.

The amount of iodide available to the gland for hormone synthesis is also an important regulator of thyroid function. The efficiency of the thyroid pump mechanism and the rate of thyroglobulin and thyroid peroxidase (TPO) synthesis are all TSH-dependent. In iodide deficiency, the production of thyroid hormone is lowered, and TSH rises through the pituitary feedback mechanism described previously. The effect of the increased TSH is to produce more thyroid hormone by increasing efficiency of the iodide pump and increasing thyroglobulin and TPO synthesis. Thus, in iodide deficiency, there is an increased uptake of iodide and [54] larger MIT-to-DIT ratios, which leads to larger T_3-to-T_4 ratio. T_3 is the more rapidly acting hormone, mitigating the effect of iodide deficiency. When iodide deficiency is severe, a persistent rise in TSH is observed. This then results in thyroid growth.

In the presence of an excess of circulating iodide, the absolute amount of iodide taken up remains approximately constant. There is a decrease in the fraction of the total iodide taken up and an increase in the amount of iodide leaked from the thyroid gland. In addition, there may be a decrease in organification (i.e., in the formation of iodinated TG residues) and in the release of hormones from the gland. The decrease in organification that occurs at excessive physiologic doses of iodide has been called the Wolff-Chaikoff block.[55] The block may be caused by an interaction of iodide with NADPH,[56] which depletes follicular NADPH and, in turn, depletes the H_2O_2 necessary for TPO activity. The decrease in hormone release occurs when iodide is given in pharmacologic (mg) quantities; it may last for a few weeks.

The activity of deiodinases reflects thyroid status, general health, and food intake. In hypothyroidism, there is a decrease in hepatic 5′ D-I but an increase in 5′ D-II activity.[57] There is a decrease in T_3 and an increase in rT_3 after hepatic disease[58] renal damage,[59] chronic illness,[60] and starvation,[61] indicating a decrease in the activity of the quantitatively more important 5′ D-I. An increase in deiodinase activity has been observed after overfeeding subjects or after the administration of a high-carbohydrate or high-fat diet[62] to experimental animals. The mechanism of deiodinase control is presumably intricate, but the inhibiting effects of propranolol[63] on 5′ D-I and of prazosin on the 5′ D-II of BAT [64] infer the involvement of adrenergic components in the regulation. In addition, the rapid change in activity observed after asphyxia[65] points to a rapid, possibly cAMP-related control.[65]

Diseases Involving the Thyroid Gland

An enlarged, palpable thyroid gland is referred to as a goiter. When insufficient thyroid hormone is liberated from the thyroid gland, the breakdown of the thyroid-pituitary-hypothalamic feedback mechanism results in the release of excess TSH and in the formation of a thyroid hypertrophy referred to as a nontoxic goiter. The increase in size of a nontoxic goiter is a compensatory mechanism, which may lead to a normal thyroid hormone output.

Endemic goiters are those that occur in a significant segment of a given population. Goiters are most frequently caused by inadequate intake of dietary iodide in regions not reached by iodide-providing sea mists and occasionally by the prolonged intake of goitrogens derived from plant sources or aquifers.

A characteristic sign of hypothyroidism is a decrease in metabolic rate, with a reduction in calorigenic effect and defective thermoregulation. The elevated serum cholesterol level seen in hypothyroidism is the result of a decrease in cholesterol degradation that exceeds the decrease in cholesterol biosynthesis; it reflects a general slowdown in catabolic processes. The slurred speech and lethargy of adult myxedema are symptomatic of a depressed central nervous system and indicate that the adult brain, which does not respond to thyroid administration by an increase in oxygen uptake, is sensitive to thyroid hormone.[66]

Cretinism, the irreversible clinical entity characterized by defective physical and mental development, appears when thyroid hormone is not available in early childhood for normal bone formation and normal brain growth and differentiation.

Myxedema is used either as a specific term to describe the infiltration of the intercellular spaces of skin and muscle with mucopolysaccharide or as a general term, as in adult myxedema or pituitary myxedema, to denote hypothyroid status.

The increased metabolic rate of hyperthyroidism results in symptoms opposite to those seen in hypothyroidism. The increased oxygen demand in hyperthyroidism leads to an increased heart rate and increased cardiac output and thereby places strain on the heart. Exaggerated catabolic processes lead to decreased serum cholesterol and possibly also to poor glucose tolerance and glucosuria, and the excessive calorigenic effect produced by catabolic processes causes anorexia and poor thermoregulation. Excessive hormone function is expressed by the word toxic, as in toxic goiter, toxic adenoma, and thyrotoxicosis. Thyrotoxic crisis or thyroid storm is an emergency caused by a stress-triggered augmentation of the symptoms of thyrotoxicosis.[67] Graves' disease is the most frequently encountered form of hyperthyroidism. Signs of this disease may include goiter, a protrusion of the eyeballs called exophthalmos, and pretibial myxedema.

Hashimoto[68] described a goiter that, on histologic examination, revealed invasion of thyroid structures by plasma cells, lymphocytes, and fibrous tissue. Pitt-Rivers and Tata[69] suggested that the sequence of events in Hashimoto's disease was initiated by injury to some thyroid structure, following which the normally sequestered thyroglobulin would be released and exposed to immunologic mechanisms, thus setting into motion a progressive interaction between thyroglobulin and circulating autoantibodies. Hashimoto antibodies are present in adult-onset myxedema and in most persons with Graves' disease, but it is recognized that other antigenic factors are involved in these and in other autoimmune diseases of the thyroid.

In the serum of most patients suffering from Graves' disease, an immunoglobulin (IgC) is present that, after a slow onset, elicits a long-lasting stimulation of the thyroid gland. Because the release of the long-acting thyroid stimulator (LATS) is not controlled by thyroid hormone levels as is TSH, patients with LATS have a hyperplastic, hypertrophied gland, which produces and releases an excessive amount of hormone.

The indications are that the exophthalmos is caused by a substance produced by a distinct autoimmune system,[70] possibly in response to a modified TSH of pituitary origin.

MOLECULAR ASPECTS

A large number of organic and some inorganic compounds stimulate or prevent thyroid hormone formation by interfering with iodide uptake into follicular cells, inhibiting TPO, preventing thyroid hormone binding to plasma proteins, or acting as effectors of thyroid deiodinases. Some of the agents described in this section are of therapeutic or diagnostic value, some illustrate potential side effects of drugs, and some are experimental compounds designed to achieve unmet therapeutic goals or to define structural parameters necessary for thyroid hormone actions.

THERAPEUTIC AGENTS

Natural Thyroid Hormones

Hormone replacement appears to be the established therapy in the treatment of various forms of hypothyroidism, from the complete absence of thyroid function seen in myxedema to the simple goiter and cretinism. Formerly, desiccated, defatted porcine or bovine thyroid glands designated thyroid, USP, and partially purified thyroglobulin designated Proloid have been used orally. The hormones were released from the proteolytic activity of gut enzymes. Potency is based on total iodine content or bioassay and is somewhat variable with different preparations.

Thyroid Hormone Preparations

Replacement appears to be the established therapy in the hypothyroidism represented by myxedema and simple goiter. The choice in replacement therapy is between a biologic preparation, such as desiccated thyroid or partially purified thyroglobulin, and synthetic, crystalline T_4.

Desiccated thyroid preparations (thyroid, USP) are essentially acetone powders of bovine or porcine thyroid glands compressed into oral tablets. A diluent is usually present because the preparations (especially those of porcine origin) commonly exceed the 0.17 to 0.23% iodine content required by the United States Pharmacopeia. Because the iodine of desiccated thyroid is in the form of the iodinated tyrosyl and thyronyl residues of the precipitated thyroglobulin, the preparation owes its efficacy to the hormones that are eventually liberated by intestinal proteases. In desiccated preparations, T_3 and T_4 may be present in a ratio approximately that found in humans. Desiccated preparations are less expensive than synthetic hormones but have been shown to produce variable T_4/T_3 blood levels because of inconsistencies between and within animal sources of the thyroid gland. Most comments regarding desiccated thyroid apply to partially purified thyroglobulin because the two preparations differ in their total as well as in their relative amounts of T_4 and T_3.

Synthetic, crystalline thyroid hormones are more uniformly absorbed than biologic preparations and contain more precisely measured amounts of active ingredient in their dosage forms. Of present interest are T_4 (levothyroxine), T_3 (liothyronine), dT_4 (dextrothyroxine), and T_4-T_3 mixtures referred to as liotrix.[71]

Because of its firmer binding to carrier proteins, synthetic crystalline (sodium levothroxine, Synthroid) has a slower onset of action than has crystalline T_3 or than a desiccated thyroid preparation. Its administration leads to greater increase in serum T_4 but to a lesser increase in serum T_3 than that of thyroid, USP.[72]

Crystalline T_3 (liothyronine sodium, Cytomel) has a rapid onset and short duration and is the therapy of choice in circumstances in which it is desirable to have rapid onset or cessation of activity, such as in patients with heart disease.

A mixture of the sodium salts of T_4 and T_3 in a 4:1 ratio by weight is distributed as liotrix, (Euthroid, and Thyrolar).

Dextrothyroxine, the synthetic optical isomer of L-T_4, was introduced in hypocholesteremic-hypolipidemic therapy with the premise that it would be void of calorigenic effects. The possibility of trace contamination with and metabolic conversion to L-thyroxine and congeners has restricted its use, however, especially in patients with coronary heart disease.[71]

Iodide

Inhibition of the release of thyroid hormone by iodide is the basis for its use in hyperthyroidism. Iodide decreases the vascularity of the enlarged thyroid gland and also lowers the elevated BMR. It has also been suggested that excess iodide might change the conformation of thyroglobulin, making the protein less susceptible to thyroidal proteolysis.[73]

With the use of antithyroid drugs, the role of iodide in hyperthyroidism has been relegated to that of preparation for thyroid surgery. Iodide, as Lugol's solution (strong iodine solution, USP) or as saturated potassium iodide solution, is administered for about 2 weeks to ensure decreased vascularity and firming of the gland. Iodism, a side effect of iodine administration, is apparently an allergic reaction, characterized by dermatologic and common cold–like symptoms.[74]

All isotopes of iodine are rapidly taken up in thyroid follicles. So far, only the isotopes ^{131}I and ^{125}I have been used consistently. The isotope ^{131}I, which decays to ^{131}Xe mainly with the emission of 0.6 meV β-particle and of approximatively 0.3 meV γ-rays, has a half-life of 8 days. The isotope ^{125}I, with a half-life of 60 days, decays to ^{125}Te by electron capture. The major component of its decay is a 27 keV X-ray, and the minor component is a 35.5 keV γ-ray.

The γ-radiation emitted by ^{131}I can be detected by a suitably placed scintillation crystal. This is the basis for the diagnostic use of this isotope in iodine uptake and in thyroid-scanning procedures.

The absorption of ^{131}I β-radiation, which leads to the highly localized destruction of the thyroid follicles in which the isotope is taken up, has promoted radioiodine as a therapeutic alternative to surgical removal of the gland. Advantages of radioiodine therapy over surgery include the simplicity of the procedure, its applicability to patients who are poor surgical risks, and the avoidance of surgical complications such as hypoparathyroidism. The development of late hypothyroidism[75] and the fear of chromosomal damage are arguments against the use of radioiodine in patients under age 20 years and during pregnancy.

A review of the use of ^{125}I in thyrotoxicosis has indicated that the potential advantages of the ^{125}I isotope, which are based on its lower penetrability and more localized action, have not been realized in practice.[76]

Perchlorate and Pertechnetate

The ability of large anions to be taken up into the thyroid by way of the iodide pump is linearly related to molar volume.[77] The affinity of iodide for the iodide pump was equal to that of thiocyanate but much smaller than that of the larger perchlorate and pertechnetate ions.

In contrast to iodide and SCN^-, TcO_4^- and CLO_4^- do not undergo intrathyroidal metabolism after they are trapped. This property has made TcO_4^- labeled with the short-lived technetium-99^m, a widely used radioisotope for thyroid trapping and for thyroid imaging.

Perchlorate, which competitively inhibits the uptake of iodide, has been used in both diagnosis and treatment of thyroid disease. In continental Europe, perchlorate (Irenat, Anthyrium) has been used for surgical preparation and in the long-term treatment of thyrotoxicosis. In the United States, the use of perchlorate was drastically curtailed after aplastic anemia and severe renal damage were reported following its use.

Diagnostically, perchlorate is used to assess the intrathyroidal organification of iodine. When perchlorate is administered after a dose of radioactive iodine, perchlorate washes out or discharges intrathyroidal inorganic iodide but does not affect covalently bound organic iodide. When organification is inadequate, there is a sharp decrease in intrathyroidal radioactive iodine after perchlorate administration.

Methimazole, Propylthiouracil, and Related Compounds

Thionamides are the most important class of antithyroid compounds in clinical practice used in nondestructive therapy of hyperthyroidism. These agents are potent inhibitors of the thyroid peroxidase enzymes (TPO) responsible for iodination of tyrosine residues of thyroglobulin and the coupling of iodotyrosine residues to form iodothyronines. These drugs have no effect on iodide trap or on thyroid hormone release. The most clinically useful thionamides are thioureylenes, which are five- or six-membered heterocyclic derivatives of thiourea and include the thiouracil, 6-n-propyl-2-thiouracil (PTU), and the thioimidazole, 1-methyl-2-mercaptoimidazole (methimazole, Tapazole, MMI).

The study of 6-alkylthiouracil showed maximal antithyroid activity with 6-propylthiouracil. 6-Methylthiouracil has less than one-tenth the activity of PTU.

The ability of PTU to inhibit the enzyme 5′ D-I, i.e., the peripheral deiodination of T_4 to T_3 (in addition to its intrathyroidal inhibition of thyroid hormone formation) has made PTU the drug of choice in the treatment of emergency of thyroid storm.[78] Single doses of PTU in excess of 300 mg are capable of almost total blockage of peripheral T_3 production.[79]

A number of studies have defined the structure-activity relations of thiouracils and other related compounds as inhibitors of outer-ring deiodinase.[80] The C_2 thioketo/thioenol group and an unsubstituted N_1 position are essential for activity. The enolic hydroxyl group at C_4 in PTU and the presence of alkyl group at C_5 and C_6 enhance the inhibitory potency.

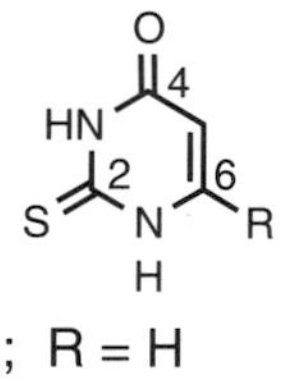

Thiouracil; R = H
Methylthiouracil; R = CH_3
Propylthiouracil (PTU); R = n-C_3H_7

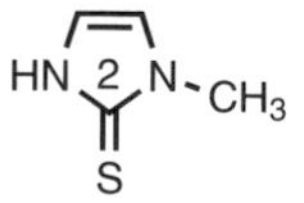

Methimazole(MMI)

Methimazole has more thyroid peroxidase inhibitory activity and is longer-acting than PTU but, in contrast to PTU, is not able to inhibit the peripheral deiodination of T_4 presumably because of the presence of the methyl group at N_1 position. The suggested maintenance dosages listed in USP DI 1986 are 50 to 800 mg daily for PTU and 5 to 30 mg daily for MMI.

Efforts to improve the taste and decrease the rate of release of MMI led to the development of 1-carbethoxy-3-methylthioimidazole (carbimazole). Carbimazole, the prodrug derivative of methimazole, gives rise to methimazole in vivo and is used in the same dosage.

The side effects of thioamides include diarrhea, vomiting, jaundice, skin rashes, and, at times, sudden onset of agranulocytosis. There does not appear to be a great difference in toxicity among the compounds currently in use.

Chemically the grouping $R{-}\overset{\overset{S}{\|}}{C}{-}N{-}$ has been referred to as thioamide, thionamide, thiocarbamide, and if R is N, as it is in PTU and MMI, as thioureylene. This structure may exist in the thioketo or thioenol tautomeric

Thioketo ⇌ Thioenol

forms. PTU and MMI are extensively taken up by the thyroid gland and act as substrates for and inhibitors of thyroid peroxidase (TPO). Taurog described the thioureylenes as potent inhibitors of thyroidal iodination and suggested that a thioureylene such as propylthiouracil (PTU.SH) would prevent the formation of a TPO-iodine complex when the thioureylene-to-iodide ratio was high and compete with thyroglobulin tyrosyl residues (TG.Tyr) when the PTU.SH-to-iodide ratio was low.[81] In the course of the reaction, the thioureylene PTU.SH would be oxidized,[82] possibly to a dimer such as PTU.SS.-PTU. Because of the rapid reaction of sulfenyl iodide with thioureylenes, a TPO sulfenyl iodide intermediate (TPO.SI) has been proposed to be a part of the reaction:

$$\text{TPO.SI} + \text{PTU.SH} \longrightarrow \text{TPO.SH} + \text{PTU.SI}$$

$$\text{PTU.SI} + \text{PTU.SH} \longrightarrow \text{PTU.SS.PTU} + \text{HI}$$

Thioureylene drugs also effectively prevent the coupling of thyroglobulin residues, which yields iodinated thyronines. This effect has been related to an alteration of the conformation of TG brought on by the binding of the thioureylene to TG (i.e., by the formation of a compound such as TPO.S.S.PTU).[83]

After the observation that PTU inhibited the peripheral deiodination of T_4,[78,84] attempts to relate deiodinase inhibitory activity to structural parameters were undertaken.[84] These studies emphasized the need for tautomerization to a thiol form and for the presence of a polar hydrogen on the nitrogen adjacent to the sulfur-bearing carbon. A study of the relation of chemical structure to 5′ D-I inhibitory activity related to similar studies of structural requirements for TPO inhibition could prove fruitful in the design of improved antithyroid drugs.

Extrathyroidal Agents

Because symptoms of thyrotoxicosis resemble those of adrenergic overstimulation, attempts to decrease such symptoms by adrenergic blockade have been undertaken. Reserpine and guanethidine, both depletors of catecholamines, and propranolol, a β-blocking agent, have been used effectively to decrease the tachycardia, tremor, and anxiety of thyrotoxicosis. Because of its less serious side effects, propranolol has become the drug of choice in this adjunctive therapy. Reports of decreased T_3 plasma levels during propranolol treatment suggest that blocking of the peripheral deiodination of T_4 may contribute to the beneficial effects of propranolol. The use of propranolol as a preventive drug in acute thyrotoxicosis has been found to be beneficial by some but not by others.

Goitrogens and Drugs Affecting Thyroid Function

The presence of environmental goitrogens was suggested by the resistance of endemic goiters to iodine prophylaxis and iodide treatment in Italy and Colombia. In the past, endemic outbreaks of hypothyroidism have pointed to calcium as a source of water-borne goitrogenicity, and it is presently believed that calcium is a weak goitrogen able to cause latent hypothyroidism to come to the surface.

Lithium salts have been used as safe adjuncts in the initial treatment of thyrotoxicosis.[85] Lithium is concentrated by the thyroid gland[86] with a thyroid-to-serum ratio of more than 2:1, suggesting active transport. Lithium ion inhibits adenylyl cyclase, which forms cAMP. cAMP is formed in response to TSH and is a stimulator of the processes involved in thyroid hormone release from the gland. Inhibition of hormone secretion by lithium has proved a useful adjunct in treatment of hyperthyroidism.[87]

In view of the role of cysteine residues in the conformation of thyroglobulin, the mode of action of TPO, and the deiodination of T_4, the effect of sulfur-containing compounds on thyroid hormone formation is hardly surprising. Most naturally occurring sulfur compounds are derived from glucosinolates[88] (formerly referred to as thioglucosides), present in foods such as cabbage, turnip, mustard seed, salad greens, and radishes (most of these are from the genus brassica of cruciferae) as well as in the milk of cows grazing in areas containing brassica weeds. Chemically, glucosinolates can give rise to many components, including thiocyanate (CNS^-), isothiocyanate (SCN^-), nitriles (RCN), and thiooxazolidones. Thiocyanate is a large anion that competes with iodide for uptake by the thyroid gland; its goitrogenic effect can be reversed by iodide intake. Goitrin, 5-R-vinyloxazolidine-2-thione, is a potent thyroid peroxidase inhibitor,[89] claimed to be more effective than PTU in humans[90] and held to be the cause of a mild goiter endemia in Finland. In rats, goitrin is actively taken up by the thyroid gland and appears to inhibit the coupling of TG diiodotyrosyl residues.[91] Many workers believe, however, that the goitrogenic effects of brassica are due to the additive effects of all goitrogenic components present.

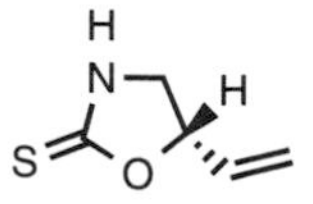

Goitrin

Other compounds affecting thyroid function include sulfonamides, anticoagulants, and oxygenated and iodinated aromatic compounds. The hypoglycemic agent carbutamide and the diuretic diamox are examples of sulfonamides. Of the anticoagulants, heparin appears to interfere with the binding of T_4 to plasma transport proteins,[92] but warfarin and dicoumarol are competitive inhibitors of the substrate T_4 or rT_3 in the 5′D reaction, with a K_i in the micromolar range.[93] Other oxygenated compounds affecting the 5′D include resorcinol, long known to be a goitrogen, and phloretin, a dihydrochalocone with an I_{50} of 4 M.

The ability of oxidation products of 3,4-dihydroxycinnamic acid to prevent the binding of TSH to human thyroid membranes[94] suggests that other oxygenated phenols may interfere with thyroid hormone function in more than one way. Examples of iodinated drugs affecting thyroid function are the antiarrhythmic agent amiodarone and the radiocontrasting agents iopanoic acid (Telepaque) and ipodoic acid (Oragrafin). All of these compounds interfere with the peripheral deiodination of T_4 and are being tested as adjuncts in the treatment of hyperthyroidism.

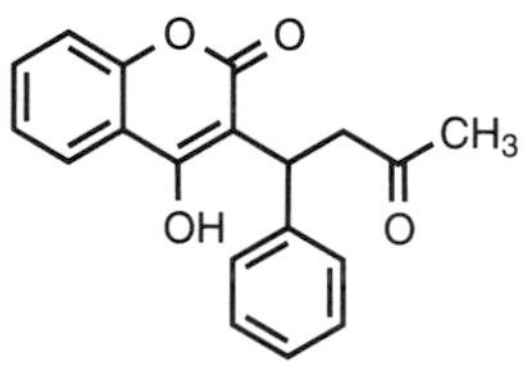

Warfarin

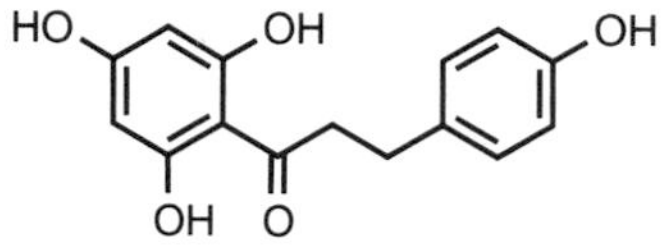

Phloretin

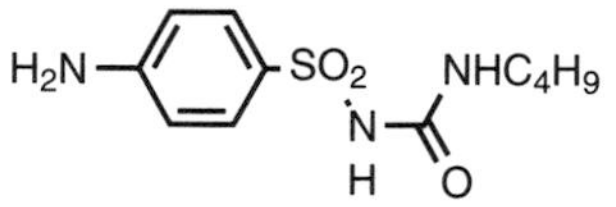

Carbutamide

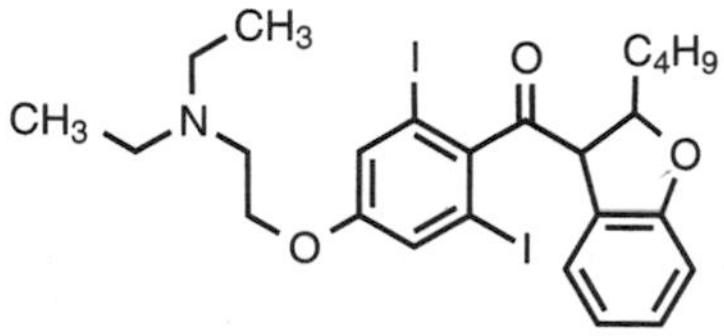

Amiodarone

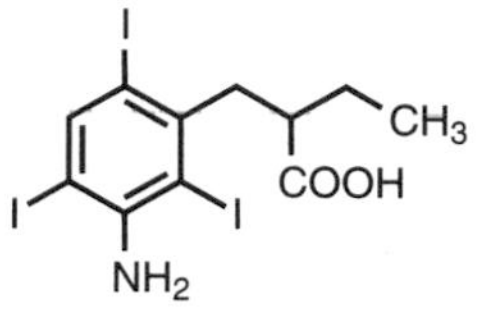

Iopanoic acid

The binding of thyroid hormones to plasma carrier proteins is affected by endogenous agents or by drugs that can change the concentration of these proteins or compete with thyroid hormones for binding sites. Examples of the first group are testosterone (and related anabolic agents) that are able to decrease and estrogens (and related contraceptive agents) that are able to increase the concentration of T_4-binding globulin. Salicylates, diphenylhydantoin, and heparin are members of the large group competing with thyroid hormones for binding sites. Alterations in the binding of T_3 and T_4 are of no large physiologic consequence because the steady-state concentrations of free hormone are rapidly restored by homeostatic mechanisms. Knowledge of the presence of agents affecting thyroid hormone binding is, however, important for the interpretation of diagnostic tests assessing the presence of free or total hormone in plasma.

Thyroid Hormone Analogs

The search for thyroid hormone analogs was prompted by the desire to establish structure-activity relationships for the hormone and by the need for a safe, specific antagonist able to block thyroid action without delay at a peripheral site. The reader will recall that TPO inhibitors such as PTU do not affect the extensive amount of hormone stored in thyroidal thyroglobulin.

Some early analogs designed as antagonists were butyl-3,5-diiodo-4-hydroxybenzoate (BHDB), which was designed as a deiodinase inhibitor and is to a degree effec-

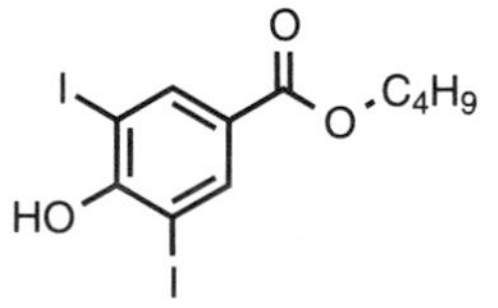

DHDB

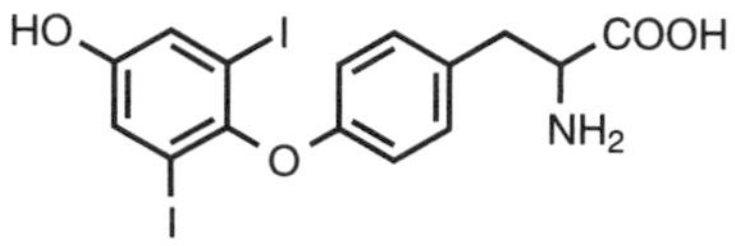

2',6'-Diiodothyronine

tive as such, and 2′6′-diiodothyronine, which has a small antithyroid effect.

When the activity of T_3 was discovered in 1955, and when none of the many analogs tabulated by Selenkow and Asper[95] demonstrated significant antithyroid activity, efforts were made to redefine the structural requirements for thyroid-like activity to provide a better rationale for the design of a peripheral thyroid antagonist. The biologic assay methods used to assay the newer compounds included measurements of in vivo oxygen uptake, of hepatic lipogenic enzyme activity, of goiter formation, of the rate of amphibian metamorphosis, and, more recently, of the binding to nuclear receptors.

The measurement of oxygen uptake provides a direct index of metabolic rate; this measurement can be done simply by placing experimental animals in a calibrated vessel maintained at constant temperature and connected to an oxygen analyzer.[96] Repeatedly used as an index of thyromimetic activity have been the in vitro assay of oxygen uptake[48] by suspended mitochondria and the spectrophotometric assays of rat liver mitochondrial α-glycerophosphate dehydrogenase[97] and of cytoplasmic malic enzyme.[98]

In the "goiter" or "antigoiter" assay, goiter formation is induced by the administration of a TPO inhibitor, such as thiouracil. The inhibitor is blended with the food or added to the drinking water and given to experimental animals for 10 to 14 days. The increase in gland weight caused by the TPO inhibitor can be reversed by T_4 or by a test compound with thyromimetic activity. The effect of a given dose of thyromimetic agent compared with that of a given dose of T_4 yields the "antigoitrogenic effect" of the compound. Further, when both a test compound and T_4 are given to an animal receiving a TPO inhibitor, the ability of the test compound to antagonize T_4, the "antithyroid effect" of the test compound, can be assessed.

Discrepancies have been seen between the results obtained with amphibian metamorphosis and those obtained with other assays. These discrepancies have been attributed to the ability of test animals such as tadpoles to concentrate test compounds, and especially lipophilic test compounds, from the medium.[99] The binding affinity of thyroid analogs to nuclear receptors correlates well with the effectiveness of these analogs in the antigoiter assay, provided that the metabolism of the analogs is taken into consideration.[99]

Structure-Activity Relationships

The synthesis and biologic evaluation of a wide variety of T_4 and T_3 analogs allowed a significant correlation of structural features with their relative importance in the production of hormonal responses. The key findings are summarized in Table 30–2. In general, only compounds with the appropriately substituted phenyl-X-phenyl nucleus (as shown on top of Table 30–2) have shown significant thyroid hormonal activities. Both single ring compounds such as DIT and a variety of its aliphatic and alicyclic ether derivatives showed no T_4-like activity in the rat antigoiter test,[100] the method most often used in determining thyromimetic activity in vivo.[101] Structure-activity relationships are discussed in terms of single structural variations of T_4 in the (1) alanine side chain, (2) 3- and 5-positions of the inner ring, (3) the bridging atom, (4) 3′- and 5′-positions of the outer ring, and (5) the 4′-phenolic hydroxyl group.

Aliphatic Side Chain. The naturally occurring hormones are biosynthesized from L-tyrosine and possess the L-alanine side chain. The L-isomers of T_4 and T_3 (Nos. 1, 3) are more active than the D-isomers (Nos. 2, 4). The carboxylate ion and the number of atoms connecting it to the ring are more important for activity than is the intact zwitterionic alanine side chain. In the carboxylate series, the activity is maximum with the two-carbon acetic acid side chain (Nos. 7, 8) but decreases with either the shorter formic acid (Nos. 5, 6) or the longer propionic and butyric acid analogs (Nos. 9 through 12). The ethylamine side chain analogs of T_4 and T_3 (Nos. 13, 14) are less active than the corresponding carboxylic acid analogs. In addition, isomers of T_3 in which the alanine side chain is transposed with the 3-iodine or occupies the 2-position were inactive in the rat antigoiter test,[102] indicating a critical location for the side chain in the 1-position of the inner ring.

Alanine Bearing Ring. The phenyl ring bearing the alanine side chain, called the inner ring or α-ring, is substituted with iodine in the 3 and 5 positions in T_4 and T_3. As shown in Table 30–2, removal of both iodine atoms from the inner ring to form 3′,5′-T_2 (No. 15) or 3′-T_1 (No. 16) produces analogs devoid of T_4-like activity primarily owing to the loss of the diphenyl ether conformation. Retention of activity observed on replacement of the 3 and 5 iodine atoms with bromine (Nos. 17, 18) implies that iodine does not play a unique role in thyroid hormone activity. Moreover, a broad range of hormone activity found with halogen free analogs (Nos. 19, 20) indicates that a halogen atom is not essential for activity. In contrast to T_3, 3′-isopropyl-3,6-dimethyl-L-thyronine (No. 20) has the capacity to cross the placental membrane and exerts thyromimetic effects in the fetus after administration to the mother. This could prove useful in treating fetal thyroid hormone deficiencies or in stimulating lung development (by stimulating lung to synthesize special phospholipids [surfactant], which ensure sufficient functioning of the infant's lungs at birth) immediately before premature birth.[103] Substitution in the 3- and 5-positions by alkyl groups significantly larger and less symmetric than methyl groups, such as isopropyl and secondary butyl moieties, produces inactive analogs (Nos. 21, 22). These results show that 3,5-disubstitution by symmetric, lipophilic groups, not exceeding the size of iodine, is required for activity.

Bridging Atom. Several analogs have been synthesized in which the ether oxygen bridge has been removed or replaced by other atoms. The biphenyl analog of thyroxine (No. 23), formed by removal of the oxygen bridge, is inactive in the rat antigoiter test. The linear biphenyl structure is a drastic change from the normal diphenyl ether conformation found in the naturally occurring hormones. Replacement of the bridging oxygen atom by sulfur (No. 24) or by a methylene group (No. 25) produces

Table 30–2. Relative Rat Antigoiter Activity of Thyroxine Derivatives

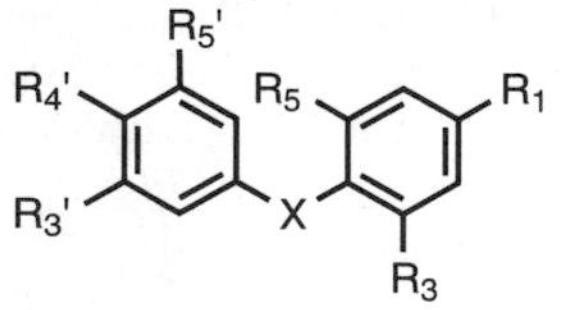

No.	R_1	R_3	R_5	X	$R_{3'}$	$R_{5'}$	$R_{4'}$	Antigoiter Activity*
1. L-T_4	L-Ala	I	I	O	I	I	OH	100
2. D-T_4	D-Ala	I	I	O	I	I	OH	17
3. L-T_3	L-Ala	I	I	O	I	H	OH	550
4. D-T_3	D-Ala	I	I	O	I	H	OH	41
5.	COOH	I	I	O	I	I	OH	0.1
6.	COOH	I	I	O	I	H	OH	0.4
7.	CH_2COOH	I	I	O	I	I	OH	50
8.	CH_2COOH	I	I	O	I	H	OH	36
9.	$(CH_2)_2COOH$	I	I	O	I	I	OH	15
10.	$(CH_2)_2COOH$	I	I	O	I	H	OH	20
11.	$(CH_2)_3COOH$	I	I	O	I	I	OH	4
12.	$(CH_2)_3COOH$	I	I	O	I	H	OH	5
13.	$(CH_2)_2NH_2$	I	I	O	I	I	OH	0.6
14.	$(CH_2)_2NH_2$	I	I	O	I	H	OH	6
15.	L-Ala	H	H	O	I	I	OH	<0.01
16.	L-Ala	H	H	O	I	H	OH	<0.01
17.	DL-Ala	Br	Br	O	I	H	OH	93
18.	L-Ala	Br	Br	O	iPr	H	OH	166
19.	L-Ala	Me	Me	O	Me	H	OH	3
20.	L-Ala	Me	Me	O	iPr	H	OH	20
21.	DL-Ala	iPr	iPr	O	I	H	OH	0
22.	DL-Ala	sBu	sBu	O	I	H	OH	0
23.	DL-Ala	I	I	—	I	I	OH	0
24.	DL-Ala	I	I	S	I	H	OH	132
25.	DL-Ala	I	I	CH_2	I	H	OH	300
26.	L-Ala	I	I	O	H	H	OH	5
27.	L-Ala	I	I	O	OH	H	OH	1.5
28.	L-Ala	I	I	O	NO_2	H	OH	<1
29.	DL-Ala	I	I	O	F	H	OH	6
30.	L-Ala	I	I	O	Cl	H	OH	27
31.	DL-Ala	I	I	O	Br	H	OH	132
32.	L-Ala	I	I	O	Me	H	OH	80
33.	L-Ala	I	I	O	Et	H	OH	517
34.	L-Ala	I	I	O	iPr	H	OH	786
35.	L-Ala	I	I	O	nPr	H	OH	200
36.	DL-Ala	I	I	O	Phe	H	OH	11
37.	DL-Ala	I	I	O	F	F	OH	2.3
38.	L-Ala	I	I	O	Cl	Cl	OH	21
39.	L-Ala	I	I	O	I	H	NH_2	<1.5
40.	DL-Ala	I	I	O	I	H	H	>150
41.	DL-Ala	I	I	O	CH_3	H	CH_3	0
42.	L-Ala	I	I	O	I	H	CH_3O	225

* See Refs. 107, 108. In vivo activity in rats relative to L-T_4 = 100% or DL-T_4 = 100% for goiter prevention.

highly active analogs. This provides evidence against the Niemann quinoid theory, which postulates that the ability of a compound to form a quinoid structure in the phenolic ring is essential for thyromimetic activity, and emphasizes the importance of the three-dimensional structure and receptor fit of the hormones. Attempts to prepare amino and carbonyl-bridged analogs of T_3 and T_4 have been unsuccessful.[104,105]

Phenolic Ring. The phenolic ring, also called the outer or *β*-ring, of the thyronine nucleus is required for hormonal activity. Variations in 3′ or 3′,5′ substituents on the phenolic ring have dramatic effects on biologic activity and the affinity for the nuclear receptor. The unsubstituted parent structure of this series L-T_2 (No. 26) possesses low activity. Substitution at 3′-position by polar hydroxyl or nitro groups (Nos. 27, 28) causes decrease in activity as a consequence of both lowered lipophilicity and intramolecular hydrogen bonding with the 4′-hydroxyl.[106] Conversely, substitution by nonpolar halogen or alkyl groups results in an increase in activity in direct relation to bulk and lipophilicity of the substituent, e.g., F < Cl < Br < I (Nos. 29 through 31–3) and CH_3 < CH_2CH_3 < $CH(CH_3)_2$ (Nos. 32 through 34). Although 3′-isopropylthyronine (No. 34) is the most potent analog known, being about 1.4 times as active as L-T_3, n-propylthyronine (No. 35) is only about one-fourth as active as isopropyl, apparently because of its less compact structure. As the series is further ascended, activity decreases with a further reduction for the more bulky 3′-phenyl substituent (No. 36). Substitution in both 3′- and 5′-positions by the same halogen produces less active hormones (Nos. 37, 38) than the corresponding 3′-monosubstituted analogs (Nos. 29, 30). The decrease in activity has been explained as due to the increase in phenolic hydroxyl ionization and the resulting increase in binding to TBG (the primary carrier of thyroid hormones in human plasma). In general, a second substituent adjacent to the phenolic hydroxyl (5′-position) reduces activity in direct proportion to its size.

Phenolic Hydroxyl Group. A weakly ionized phenolic hydroxyl group at the 4′-position is essential for optimum hormonal activity. Replacement of the 4′-hydroxyl with an amino group (No. 39) results in a substantial decrease in activity, presumably as a result of the weak hydrogen bonding ability of the latter group. The retention of activity observed with the 4′-unsubstituted compound (No. 40) provides direct evidence for metabolic 4′-hydroxylation as an activating step. Introduction of a 4′-substituent that cannot mimic the functional role of a hydroxyl group, such as a methyl group (No. 41), and that is not metabolically converted into a functional residue results in complete loss of hormonal activity. The thyromimetic activity of the 4′-methyl ether (No. 42) was ascribed to the ready metabolic cleavage to form an active 4′-hydroxyl analog. The pK_a of 4′-phenolic hydroxyl group for T_4 is 6.7 (90 percent ionized at pH 7.4) and for T_3 is 8.5 (approximately 10 percent ionized). The greater acidity for T_4 is reflective of its stronger affinity for plasma proteins and consequently its longer plasma half-life.

Conformational Properties of Thyroid Hormones and Analogs

The importance of the diphenyl ether conformation for biologic activity was first proposed by Zenker and Jorgensen.[109] Through molecular models, they showed that a perpendicular orientation of the planes of the aromatic rings of 3,5-diiodothyronines would be favored, to minimize interactions between the bulky 3,5 iodines and the 2′,6′ hydrogens. In this orientation, the 3′- and 5′-positions of the ring are not conformationally equivalent, and the 3′ iodine of T_3 could be oriented either distal (away from) or 5′ proximal (closer) to the side chain–bearing ring (Fig. 30–6). Because the activity of compounds such as 3′,5′-dimethyl-3,5-diiodothyronine had demonstrated that alkyl groups could replace the 3′- and 5′-iodine substituents, model compounds bearing alkyl groups in the 3′-position and alkyl or iodine substituents in the 5′-position (in addition to the blocking 2′-methyl group) were synthesized for biologic evaluation.[109]

Biologic evaluation of 2′,3′- and 2′,5′-substituted diiodothyronines[110] revealed that 3′-substitution was favorable for thyromimetic activity but that 5′-substitution was not. The structures of representative distal analogs, 2′,3′-dimethyl-3,5-DL-diiodothyronine (I) and O-(4′-hydroxy-1′-naphthyl)-3,5-DL-diiodotyrosine (II), and of the proximal analogs, 2′,5′-dimethyl-3,5-DL-diiodothyronine (III) and 2′-methyl-3,5,5′-DL-triiodothyronine (IV), are given in Figure 30–6. The effectiveness of these compounds in rat antigoiter assay[111] is presented in Table 30–3. These results clearly indicate that in 2′ blocked analogs, a distal 3′ substitution is favorable for thyromimetic activity, but a proximal 5′ substitution is not.

The perpendicular orientation of the rings of 3,5-diiodothyronines, which was postulated from molecular models, has been confirmed by x-ray crystallographic studies,[112] molecular orbital calculations,[113] and nuclear magnetic resonance studies.[114]

In addition to being perpendicular to the inner ring, the outer phenolic ring can adopt conformations relative to the alanine side chain, which would be *cis* or *trans*. In other words, the *cisoid* and *transoid* conformations result from the methine group in the alanine side chain being either *cis* or *trans* to the phenolic ring (Fig. 30–7). Although the bioactive conformation of the alanine side chain in thyroid hormone analogs has not yet been defined, these conformations appear to be similar in energy

Table 30–3. Effectiveness of Distal and Proximal Compounds

	Antigoiter Assay	
Compound	Dose*	% T_4 act.
I	0.025	50
II	0.013	>100
III	2.3	<1
IV	0.5	2

* mg/kg/day.

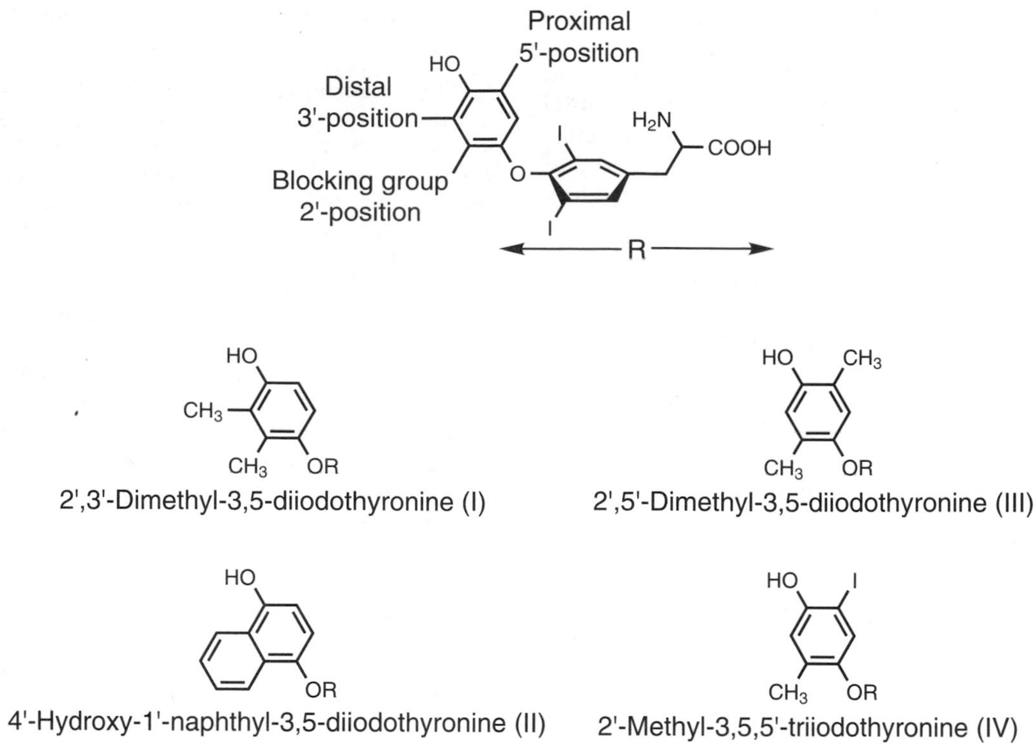

Fig. 30–6. Structures of representative distal and proximal compounds.

because both are found in thyroactive structures determined by x-ray crystallography.[115] The synthesis of conformationally fixed cyclic or unsaturated analogs may allow evaluation of the bioactivity of the two conformers.

Thyroxine-Binding Prealbumin Receptor Model. An additional tool in structural analysis and analog design has been TBPA, a plasma protein that binds as much as 27% of plasma T_3.[116] The amino acid sequence of the TBPA T_3 binding site is known, and the protein has therefore served as a model, although admittedly an approximate model, for the T_3 receptor. The TBPA model portrays the T_3 molecule as placed in an envelope near the axis of symmetry of the TBPA dimer. In this envelope, hydrophobic residues, such as those of leucine, lysine, and alanine, are near pockets accommodating the 3,5,3′- and 5′-positions of T_3, whereas the hydrophilic groups of serine and threonine, hydrogen bonded to water, are between the 3′ substituent and 4′ phenolic group. Taking this model into account, Ahmad suggested that 3′-acetyl-3,5-di-iodothyronine might be a good analog or a good inhibitor of T_3 because the carbonyl group of the 3′ acetyl substituent would form a strong hydrogen bond with the 4′ phenolic hydrogen, preventing thereby its bonding with the hydrated residue of the putative receptor.[117]

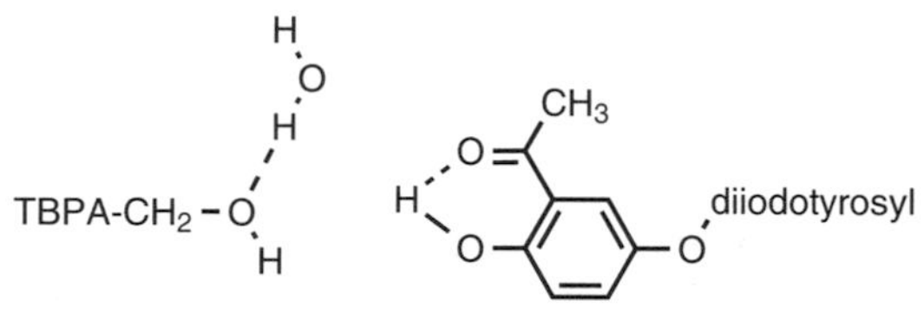

The compound, prepared by Benson and co-workers,[118] was found to be indistinguishable from T_3 in oxygen uptake and glycerophosphate activity tests and to be half as active as T_3 in displacing labeled T_3 from rat liver nuclei in specific in vivo conditions.

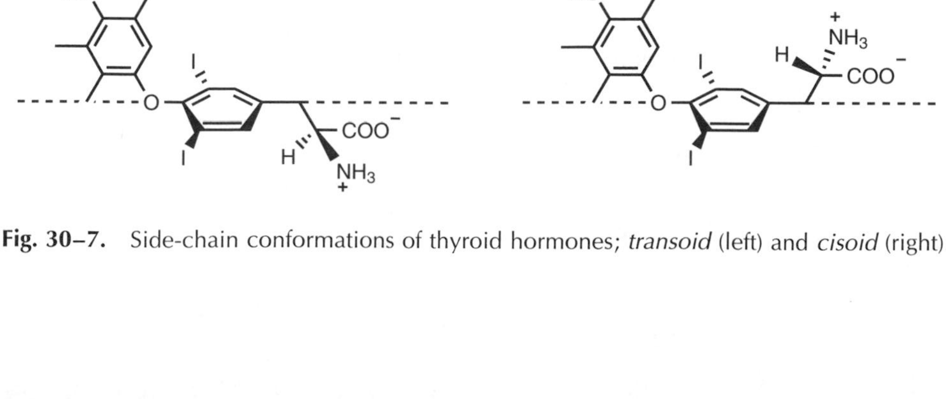

Fig. 30–7. Side-chain conformations of thyroid hormones; *transoid* (left) and *cisoid* (right).

REFERENCES

1. C. R. Harington, *Lancet, 1,* 1199(1935).
2. E. J. Baumann, *Z. Physiol. Chem., 21,* 319(1986).
3. A. Magnus-Levy, *Berl. Klin. Wocheschr., 32,* 650(1895).
4. J. F. Gudernatsch, *Arch. Entwicklungsmech. Organ, 35,* 457(1912).
5. N. J. Nadler, in *Handbook of Physiology,* M. A. Greer, D. H. Solomon, eds. Sec. 7, Vol. III, Ch. 4, Washington, D. C. Am. Physiol. Soc., 1974.
6. R. Ekholm and U. Bjorkman, in *The Thyroid Gland,* L. Martini, Ed. New York, Raven Press, 1990, pp. 38–39.
7. E. C. Kendall, *JAMA, 64,* 2042(1915).
8. C. R. Harington, *Biochem. J., 20,* 300(1926).
9. C. R. Harington and C. Barger, *Biochem. J., 21,* 169(1927).
10. J. Gross and R. Pitt-Rivers, *Biochem. J., 53,* 645(1953).
11. J. Roche, *C. R. Acad. Sci. (Paris), 234,* 1228(1952).
12. C. R. Harington and S. S. Randall, *Biochem J., 23,* 373(1929).
13. K. Fink and R. M. Fink, *Science, 108,* 358(1948).
14. C. T. Sawin, *The Hormones,* Boston, Little & Brown, p. 98, 1969.
15. L. J. de Groot, et al., *The Thyroid and its Diseases,* 5th Ed., New York, Wiley, 1984.
16. H. M. Goodman and L. Van Middlesworth, The thyroid gland, in *Medical Physiology,* V. B. Mouncastle, Ed., Vol. 2, St. Louis, Mosby, 1980, p. 1500.
17. I. J. Chopra, et al., *Recent Prog. Horm. Res., 34,* 521(1978).
18. T. J. Visser, et al., *Endocrinol. Invest., 9*(Suppl. 4), 17(1986).
19. M. J. Obregon, et al., *Clin. Endocrinol., 10,* 305(1979).
20. T. J. Visser, et al., *Endocrinology, 122,* 1547(1983).
21. R. Pitt-Rivers, *Lancet, 2,* 234(1953).
22. S. Han, et al., *Int. J. Peptide Protein Res., 30,* 656(1987).
23. J. Wynn and R. Gibbs, *J. Biol. Chem., 237,* 3499(1962).
24. M. J. Muller and H. J. Seitz, *Biochem. Pharmacol., 33,* 1579(1984).
25. J. Himms-Hagen, *Annu. Rev. Physiol., 38,* 315(1976).
26. M. J. Muller and H. J. Seitz, *Klin Wochenschr., 62,* 11(1984).
27. M. J. Muller and H. J. Seitz, *Klin. Wochenschr., 62,* 49(1984).
28. M. J. Muller and H. J. Seitz, *Klin. Wochenschr., 62,* 97(1984).
29. K. Sterling, *Bull. N. Y. Acad. Med., 53,* 260(1977).
30. F. Ismail-Beigi and I. S. Edelman, *Proc. Natl. Acad. Sci. U.S.A., 67,* 1071(1970).
31. L. T. Williams, et al., *J. Biol. Chem., 252,* 2787(1977).
32. G. S. Levey and S. E. Epstein, *J. Clin. Invest., 48,* 1663(1969).
33. L. Stryer, *Biochemistry,* 2nd Ed., San Francisco, Freeman, 1981, p. 843.
34. E. W. Sutherland and G. A. Robinson, *Pharmacol. Rev., 18,* 145(1966).
35. J. T. Ayres and W. A. Lishman, *Br. J. Anim. Behav., 3,* 17(1955).
36. J. R. Tata, *Nature, 197,* 1167(1963).
37. J. R. Tata and C. C. Widnell, *Biochem. J., 98,* 604(1966).
38. W. H. Dillmann, et al., *Endocrinology, 100,* 1621(1977).
39. A. R. Schadlow, et al., *Science, 176,* 1252(1972).
40. J. H. Oppenheimer, et al., *J. Clin. Endocrinol. Metab., 35,* 330(1972).
41. V. A. Galton, *Endocrinology, 114,* 735(1984).
42. H. H. Samuels, in *Molecular Basis of Thyroid Hormone Action,* J. H. Oppenheimer and H. H. Samuels, Eds., New York, Academic Press, 1983, pp. 35–65.
43. J. H. Oppenheimer, in *Molecular Basis of Thyroid Hormone Action.* J. H. Oppenheimer and H. H. Samuels, Eds., New York, Academic Press, 1983, pp. 1–34.
44. C. P. Barsano and L. J. DeGroot in *"Molecular Basis of Thyroid Hormone Action,* J. H. Oppenheimer and H. H. Samuels, Eds., New York, Academic Press, 1983, pp. 139–177.
45. J. Robbins and J. E. Rall, *Physiol. Rev., 40,* 415(1960).
46. Y. P. Lee, et al., *J. Biol. Chem., 234,* 3051(1959).
47. W. W. Westerfeld, et al., *Endocrinology, 77,* 802(1965).
48. N. Zenker, et al., *Science, 159,* 1102(1968).
49. K. Sterling, et al., *Science, 197,* 996(1977).
50. H. L. Schwartz and J. H. Oppenheimer, *Pharmacol. Ther. B., 3,* 349(1978).
51. N. Barden and F. Labrie, *J. Biol. Chem., 248,* 7601(1973).
52. K. Sterling and J. H. Lazarus, *Annu. Rev. Physiol., 39,* 349(1977).
53. W. W. Chin, et al., *Endocrinology, 116,* 873(1985).
54. F. S. Greenspan and P. H. Forsham, *Basic and Clinical Endocrinology,* Los Altos, Lange Medical, 1983, p. 141.
55. J. Wolff and I. L. Chaikoff, *J. Biol. Chem., 174,* 555(1948).
56. A. Virion, et al., *Eur. J. Biochem., 148,* 239(1985).
57. J. L. Leonard, et al., *Science, 214,* 571(1981).
58. J. McConnon, et al., *J. Clin. Endocrinol. Metab., 34,* 144(1972).
59. V. S. Lim, et al., *J. Clin. Invest., 60,* 522(1977).
60. J. N. Carter, et al., *Clin. Endocrinol., 5,* 587((1976).
61. S. W. Spaulding, et al., *J. Clin. Endocrinol. Metab., 42,* 197(1976).
62. M. A. Chacon, et al., *Fed. Proc., 43,* 866(1984).
63. P. Heyma, et al., *Endocrinology, 106,* 1437(1980).
64. J. E. Silva and P. R. Larsen, Abstract, Am. Thyr. Assoc. 59th Meeting, 1983.
65. N. Zenker, et al., *Life Sci., 35,* 2213(1984).
66. N. L. Eberhardt, et al., *Endocrinology, 102,* 556(1978).
67. I. N. Rosenberg, *N. Engl. J. Med., 283,* 1052(1972).
68. H. Hashimoto, *Arch. J. Klin. Chir., 97,* 219(1912).
69. R. Pitt-Rivers and J. R. Tata, *The Chemistry of Thyroid Disease,* Springfield, Ill., Charles C. Thomas, 1960, p. 38.
70. J. E. Mahaux, J. Chamla-Soumenkoff, R. Delcourt, N. Nagel, and S. Levin, *Acta Endocrinol., 61,* 400(1969).
71. H. A. Selenkow and L. I. Rose, *Pharmacol. Ther. C., 1,* 331(1976).
72. I. M. Jackson and W. E. Cobb, *Am. J. Med., 64,* 284(1978).
73. L. Lamas and S. H. Ingbar, in *Thyroid Research,* J. Robbins and L. E. Braverman, Eds., Amsterdam, Excerpta Medica, 1976, p. 213.
74. J. A. Pittman, Jr., *Diagnosis and Treatment of Thyroid Disease,* Philadelphia, F. A. Davis, 1963, p. 48.
75. J. A. Pittman, Jr., in *Diagnosis and Treatment of Common Thyroid Disease,* H. A. Selenkow and F. Hoffman, Eds., Amsterdam, Excerpta Medica, 1971, pp. 72–73.
76. W. F. Brenner, et al., *Clin. Endocrinol., 5,* 225(1976).
77. J. Wolff and J. R. Maurey, *Biochem. Biophys. Acta, 69,* 58(1963).
78. G. Morreale de Escobar and R. Escobar del Rey, *Rec. Prog. Horm. Res., 23,* 87(1967).
79. D. S. Cooper, et al., *J. Clin. Endocrinol. Metab., 54,* 101(1982).
80. T. J. Visser, et al., *FEBS Lett., 103,* 314(1979).
81. A. Taurog, *Endocrinology, 98,* 1031(1976).
82. T. Nakashima, A. Taurog and G. Riesco, *Endocrinology, 103,* 2187(1978)
83. P. D. Papapetrou, et al., *Acta Endocrinol., 79,* 248(1975).
84. I. J. Chopra, et al., *Endocrinology, 110,* 163(1982).
85. J. G. Turner, et al., *Acta Endocrinol., 83,* 86(1976).
86. S. C. Berens, J. Wolff, and D. L. Murphy, *Endocrinology, 87,* 1085(1970).
87. R. Templer, et al., *J. Clin. Invest., 51,* 2746(1972).
88. H. L. Tookey, et al., *Toxic Constituents of Foodstuffs,* I. E. Liener, Ed., New York, Academic Press, 1980, pp. 103–142.

89. P. Langer and N. Michajlovskij, *Endocrinol. Exp., 6,* 97(1972).
90. M. A. Greer, *Rec. Prog. Horm. Res., 18,* 187(1962).
91. S. Elfving, *Ann. Clin. Res., 12(Suppl. 28),* 7(1980).
92. M. Tabchnick, et al., *J. Clin. Endocrinol. Metab., 36.,* 392(1973).
93. A. Goswami, et al., *Biochem. Biophys. Res. Comm., 104,* 1231(1982).
94. M. Auf'mKolk, et al., *Endocrinology, 116,* 1677(1985).
95. H. A. Selenkow and S. P. Asper, Jr., *Physiol. Rev., 35,* 426(1955).
96. N. Zenker, et al., *Life Sci., 18,* 183(1976).
97. N. Zenker, et al., *Biochem. Pharmacol., 25,* 1757(1976).
98. A. L. Tarentino, et al., *Biochim. Biophys. Acta, 124,* 295(1966).
99. E. C. Jorgensen, *Pharmacol. Ther. B, 2,* 661(1976).
100. E. C. Jorgensen and P. A. Lehman, *J. Org. Chem., 26,* 894(1961).
101. M. V. Mussett and R. Pitt-Rivers, *Metab. Clin. Exp., 6,* 18(1957).
102. E. C. Jorgensen and J. A. W. Reid, *J. Med. Chem., 7,* 701(1964).
103. P. L. Ballard, et al., *Pediatr. Res., 12,* 1164(1978).
104. S. L. Trip, et al., *J. Med. Chem., 236,* 2891(1973).
105. R. Mukherjee and P. Block, Jr., *J. Chem. Soc.* (C) 1596(1971).
106. P. D. Leeson, et al., *J. Med. Chem., 31,* 37(1988).
107. E. C. Jorgensen in *Hormonal Proteins and Peptides,* Vol. 6, *Thyroid Hormones,* C. H. Li, Ed., New York, Academic Press, 108, 1978.
108. E. C. Jorgensen, in *Burger's Medicinal Chemistry,* Part 3, 4th ed., M. E. Wolff, Ed., New York, Wiley, 1981, p. 103.
109. N. Zenker and E. C. Jorgensen, *J. Am. Chem. Soc., 81,* 4643(1959).
110. E. C. Jorgensen, et al., *J. Biol. Chem., 235,* 1732(1960).
111. E. C. Jorgensen, et al., *J. Biol. Chem., 237,* 3832(1962).
112. V. Cody, *J. Med. Chem., 20,* 1628(1977).
113. P. A. Kollman, et al., *J. Am. Chem. Soc., 95,* 8518(1973).
114. P. A. Lehman and E. C. Jorgensen, *Tetrahedron, 21,* 363(1965).
115. V. Cody, *Rec. Prog. Horm. Res., 34,* 437(1978).
116. R. Ekins, et al., *Free Thyroid Hormones,* Amsterdam, Excerpta Medica, 1979, p. 7.
117. P. Ahmad, *Biochem. Pharmacol., 24,* 1103(1975).
118. M. G. Benson, et al., *Biochem. Pharmacol., 33,* 3143(1984).

Chapter 31

METABOLITE ANTAGONISM

Nitya Anand

HISTORICAL DEVELOPMENT

Early History

Metabolite antagonism is a term used to describe the inhibition of cellular growth/response by structural analogs of a metabolite that can bind to the enzyme/receptor but do not undergo the normal chemical reactions of the metabolite or produce the usual physiologic response. Such structural analogs are called antimetabolites.

The present concepts of metabolite antagonism and antimetabolites have grown slowly, parallel with the development of knowledge of bacterial and animal nutrition, enzyme action, and the pharmacologic antagonism of drug action. As with many other branches of science, the facts of metabolite antagonism were noted first as isolated observations, and the underlying principles and the enunciation of the basic theory were later recognized. The recognition of the structural similarity of drugs to some nutrients normally required for the growth of microorganisms and the chemical interaction of drugs with receptors were inherent in the chemoreceptor postulate of Ehrlich. In his Nobel prize address (1908), Ehrlich[1] described the receptor as special chemical groupings in the cell normally concerned with the uptake of nutrients by the cell but that could also take up specific antigens or drugs instead. The earliest examples of enzyme inhibition by a substance related in chemical structure to the substrate or the product of the reaction are those of invertase by fructose and of α-glucosidase by glucose, reported by Michaelis and associates[2,3] in 1913. The first clear examples of the phenomenon of competitive inhibition by a structural analog were provided by Quastel and associates[4] in their demonstration of the marked competitive inhibition by malonate of succinic acid dehydrogenase and of the growth of Escherichia coli on fumarate as the sole carbon source.[5] It was suggested that the active center of an enzyme is so constituted that it may combine with a variety of substances, all having similarity of chemical structure, but that only a few are actual substrates and undergo subsequent chemical change. Their picture of the mechanism of inhibition was similar to the picture held today.

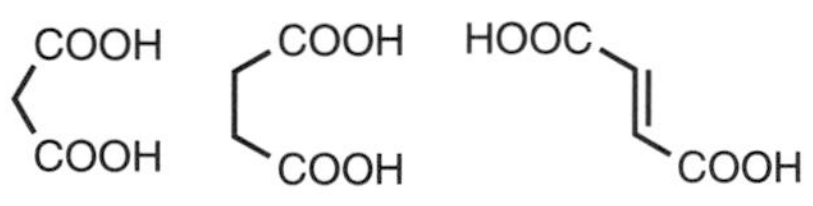

Malonic Acid | Succinic Acid | Fumaric Acid

In the period from 1920 to 1940, many instances were reported of structural analogs of a vitamin, a hormone, or an amino acid antagonizing the biologic effects of the related growth factors. One of the earliest reports was that 3-acetylpyridine and pyridine-3-sulfonic acid were toxic to dogs suffering from nicotinic acid deficiency but harmless to animals receiving a normal diet[6]; both of these substances were later shown to act as antagonists of nicotinic acid in various animal species. In 1938, Dyer[7] observed the toxic effect of ethionine in rats; ethionine was later found to be an effective antimetabolite of methionine in many biologic systems.

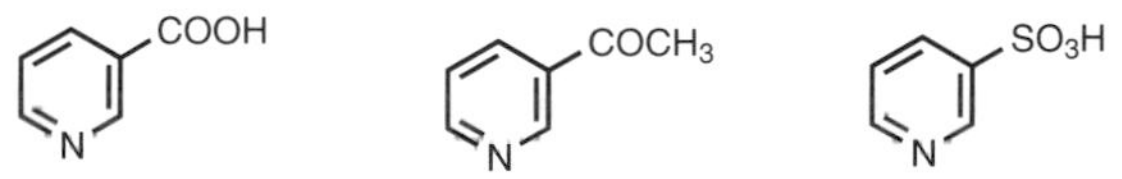

Nicotinic Acid | 3-Acetylpyridine | Pyridine-3-sulfonic Acid

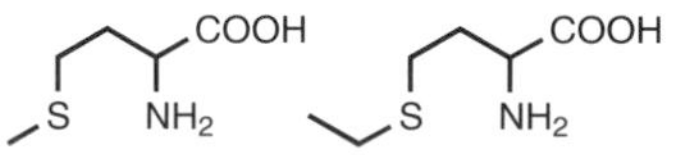

Methionine | Ethionine

Other findings of antagonism between structurally similar compounds reported during this period were those between acetylcholine and allylisopropylcholine[8] and other quaternary ammonium bases,[9,10] and between acetylcholine and atropine.[11] Clark[11] recognized the structural relationship of the drugs to the hormone and saw in this relationship the possibility of understanding the mode of action of these drugs. From a consideration of the competitive antagonism between structurally related drugs, he was able to conclude that each member of a pair of antagonistic agents probably acted on a common site in a reversible manner in the cell. His deduction of the mode of action of these drugs represented an important step in the development of the concepts of antimetabolites as applied to pharmacology; the term agonist was used for those substances that mimic the effects of the endogenous regulatory substance, antagonists for those that are themselves devoid of intrinsic activity but have affinity and compete with agonist for binding, and partial agonist for those that can produce only a submaximal response and can antagonize the agonist action.

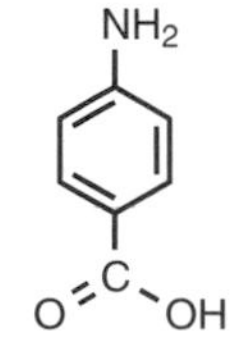

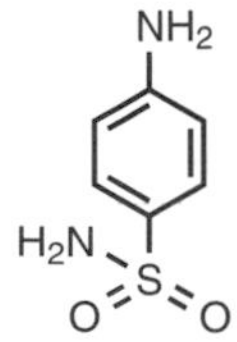

p-Aminobenzoic Acid   4-Aminobenzene sulfonamide

Sulfonamides and Fildes' Theory of Antimetabolites

The full impact and the practical value of metabolite antagonism were, however, not fully appreciated until the discovery of the sulfonamides in the early 1930s by Domagk[12] and the demonstration by Woods in 1940[13] that the bacteriostatic action of sulfanilamide was reversed completely by *p*-aminobenzoic acid (PABA). The sulfonamides were thus viewed as structural analogs of PABA, a metabolite, and were considered to owe their bacteriostatic action to this fact. This was the first definitive demonstration of metabolite antagonism as a mechanism of drug action. This led Fildes[14] to propose his classic theory of antimetabolites, which provided a logical basis for drug action. This theory had a profound influence on future thinking in this field. Although the action of sulfonamides was concerned with the chemotherapy of microbial diseases, it was soon realized that the principle of metabolite antagonism had much wider ramifications and was equally applicable in the fields of hormonal regulation, pharmacologic action, and animal nutrition.

The emergence of the Woods-Fildes antimetabolite theory verbalized the long sought after rational approach to drug design. This was followed by a period of great excitement at the prospect of rationally designing drugs; in fact, many antimetabolites were designed during this period as antimicrobials. But much of this early work proved insignificant because these agents were modeled on metabolites that were as necessary for the host as for the microbe. This resulted in attention being focused on exploiting the differences in the biochemical processes between the host and the microbe. This was also a period in which there were major advances in understanding of cellular mechanisms, biochemical pathways, and bacterial physiology, particularly of nucleic acid biosynthesis and its relationship with protein synthesis, mechanism of enzyme action, and structure of enzymes. These developments provided a fertile ground for major advances in the design of antimetabolites. In addition to looking for differences in the metabolic pathways between the host and the parasite, the demonstration of the occurrence of isoenzymes[15] (isozyme) added another useful dimension to the design of selective inhibitors. This term is used for those multiple forms of enzymes that carry out identical biocatalytic functions but have genetically determined differences in primary structure and different physicochemical characteristics, Michaelis constant, and differential response to inhibitors, which can be usefully exploited for the design of selective inhibitors. The synthesis of selective inhibitors of dihydrofolate reductases of bacteria, protozoa, and mammals has been an important practical result of this development.[16,17] Developments in molecular biology making it possible to clone and characterize the receptors and in molecular modeling (CAMM) have added another useful dimension to this field and have brought more precision to the design of receptor agonists and antagonists.

Antimetabolites were first pictured as reversible inhibitors that compete with essential metabolites but themselves remain unaltered. It was soon realized, however, that a structural analog of a metabolite may undergo similar transformation as does the metabolite, and the transformed product becomes the true inhibitor. The first instance reported of such conversion was that of fluoroacetic acid, which by itself is not toxic but is converted into fluorocitric acid by the same enzymes that use acetic acid; fluorocitric acid is a powerful inhibitor of aconitase. This phenomenon of metabolic incorporation to form the actual inhibitor has been found to be a common occurrence and has been termed lethal synthesis by Peters.[18,19] Most of the purine and pyrimidine metabolite inhibitors act after their metabolic conversion to nucleosides and nucleotides.

Active Site–Directed Irreversible Enzyme Inhibitors

As the application of the concept of metabolite antagonism progressed, greater sophistication was introduced into the design of inhibitors and antagonists. At first, only small structural changes were made in the metabolite, primarily by bioisosteric replacement of, for example, OH by NH_2, O by S, C=C by S, H by F or CH_3, COO^- and CONHR by SO_2NHR. These changes were based mainly on consideration of size or electronic effects. These classic antimetabolites generally have short and reversible action and therefore must be administered repeatedly; this is one of their limitations. To overcome this problem, Baker prepared antimetabolites with larger structural changes in the metabolite, which, although not interfering with the stereospecific requirements of the enzyme surface for complexing, would lead to stronger and longer-lasting inhibition.[20] This idea was extended to the development of irreversible inhibitors that would react with the enzyme surface (or the receptors) by covalent bond formation.[20] A major stimulus in the development of this concept was the discovery that diisopropylphosphofluoridate (DFP) was a specific antagonist for the active sites of cholinesterases and proteolytic enzymes.[20–24] Related compounds have subsequently been found to have powerful insecticidal activity owing to inhibition of cholinesterases. A study of DFP showed that it acts by forming a phosphorylserine linkage with chymotrypsin, and out of about 20 serine residues in the enzyme, only one residue at the active site reacted, thus indicating the unique character of this one serine residue.

In a comprehensive approach to the concept of active site–directed irreversible inhibitors, Baker[20,25] considered two types of such inhibitors, one that binds ionically to the enzyme much more strongly than the substrate (pseudoirreversible) and one that combines chemically with the enzyme through covalent bond formation; the

Diisopropylphosphofluoridate

4-(iodoacetamido)-salicylic Acid

latter could happen by endoalkylation, i.e., within the active site, or by exoalkylation on the enzyme surface but outside the active site (Fig. 31–1). Baker and associates described many compounds that act by one of these mechanisms, such as 4-(iodoacetamido)-salicylic acid, which inhibits both lactic dehydrogenase and glutamic dehydrogenase by exoalkylation.[25] These developments inspired many new ideas and more precise design of enzyme inhibitors.[24]

It was, however, realized that because these irreversible inhibitors carry the reactive group, they also have nonspecific interactions in the biologic system. As progress was made in the knowledge of the structures of the enzymes and their mechanism of action, more selective approaches to the design of enzyme inhibitors developed, including mechanism-based inhibitors and transition state analogs.

Mechanism-Based Enzyme Inhibitors

Mechanism-based enzyme inhibitors are structural analogs of the substrate or products of an enzyme reaction that contain a latent reactive group (chemically inert as such). In a particular microenvironment, this latent reactive group is uncovered by the specific catalytic step(s) at the active site of the enzyme to generate a reactive intermediate, which, without release, can covalently react with some residue at the catalytic site and inactivate the enzyme for subsequent reaction. The essential feature of this type of inhibitor is that it must be processed by the target enzyme to generate the actual inhibiting species. Such inhibitors have also been termed suicide substrates/inhibitors and Kcat inhibitors, although the term mechanism-based or enzyme-activated inhibitor seems more appropriate and accurate. Although the concept has been stated in clear terms only recently, many clinically useful drugs acting by enzyme inhibition have turned out, in retrospect, to be mechanism-based inhibitors.

The first report of a mechanism-based enzyme inhibitor was by Helmkamp and associates in 1969.[26] They demonstrated that β-hydroxydecanoylthiol-ester dehydrase, a key enzyme in the synthesis of unsaturated fatty acids of bacteria, was inhibited by 3-decynyl-S-pantetheine, by the specific transformation of the unreactive acetylene to a conjugated allene during the isomerization step. Since then, many cases of mechanism-based inhibition have

SUBSTRATE: β-Hydroxydecanoyl Thioester

INHIBITOR: β,γ-Decynoyl-*N*-acetylcysteamine

been reported, and several reviews have appeared.[27–29] The target enzymes investigated for the development of mechanism-based inhibitors include penicillinase, β-lactamase, and α-alanine racemase for antibacterial chemotherapy; ornithine decarboxylase as antitumor and antiparasitic agents; GABA transaminase for development of antiepileptics; and aromatase for development of antiestrogenic agents. Enzyme-catalyzed generation of electrophiles is the most important strategy in the design of mechanism-based enzyme inactivators. The functional groups most often used in these inactivators are olefins, acetylenes, fluorocarbons, diazo compounds, and cyclo-

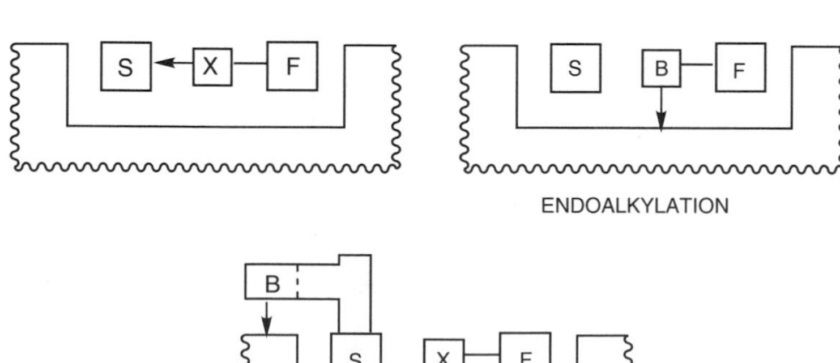

Fig. 31–1. Schematic presentation of the concepts of endoalkylation and exoalkylation according to Baker[25]; F, cofactor; S, substrate, X, group to be transferred to S; B, alkylating group, which could also be present on S; → covalent bond formation.

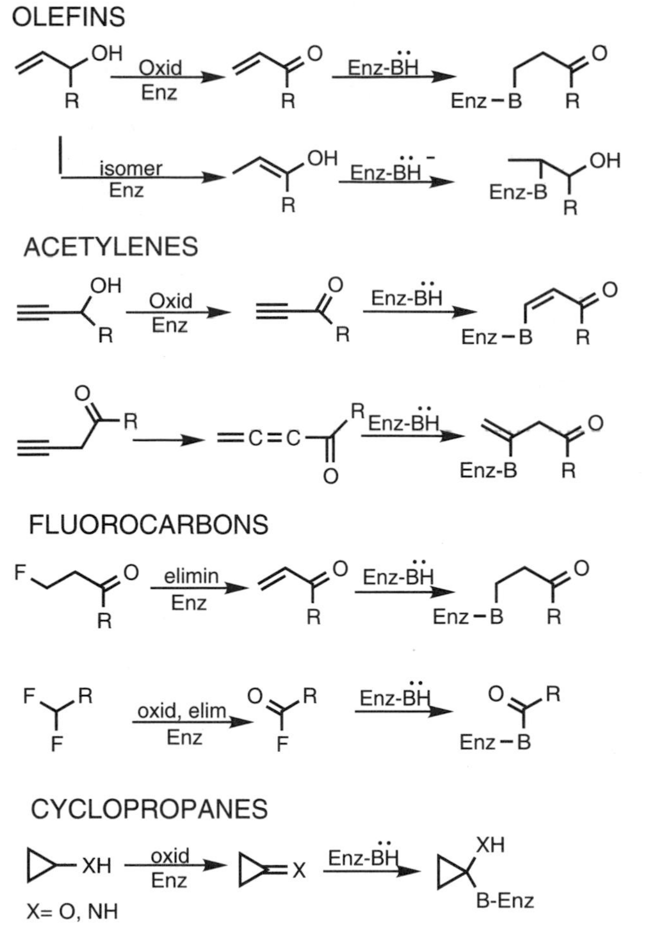

Fig. 31–2. Illustrative mechanism-based enzyme inactivation processes. (Adapted from R. R. Rando, *Pharmacol. Rev., 36,* 111[1984]).

propanoids. In each instance, a chemically unreactive group is transformed into a reactive electrophile, which can engage in a chemical reaction with a nucleophile at or in the vicinity of the active site. Organic reactions that have been implicated in mechanism-based inhibition include elimination, isomerization, decarboxylation, oxidation, and rearrangement, followed by addition to multiple bonds. Figure 31–2 illustrates the more important types of chemical reactions that occur during mechanism-based enzyme inactivation. Criteria for detecting mechanism-based inactivation have gradually evolved, including time-dependent loss of enzyme activity showing saturation kinetics, kinetic protection by normal substrate, nonregeneration of enzyme activity by dialysis or gel filtration, and the effect of nucleophilic scavengers in solution on the rate of inactivation.[32]

The potential utility of mechanism-based enzyme inhibition for the discovery of extremely specific drugs and as probes for the study of enzyme mechanisms is obvious. Both reversible and irreversible inhibitors have several disadvantages as drugs; the former require repeated doses, whereas the latter are nonspecific in action and can produce toxic side effects. In principle, both these problems can be avoided with mechanism-based inhibitors. Although in most cases inhibition is irreversible, this is not essential, and some cases have been reported in which there is no covalent linkage, but the generated inhibitor is tightly bound to give slow regeneration of the enzyme.

Transition State Analogs

The basis for the transition state analog approach was first suggested by Pauling,[30] who speculated that the catalytic specificity of enzymes required that the active site of the enzyme and the transition state of the substrate during catalysis should be structurally complementary. Molecules that resemble the transition state structure more closely would thus be expected to bind to the active site much more tightly than the natural substrate. This concept was elaborated by Lienhard[31] and Wolfenden.[32–35]

The major aim of transition state analog design is to obtain a stable molecule with the functional groups in the right position to bind to the complementary groups in the active site of the enzyme. Although this seems difficult to achieve precisely because of the long partial bonds of transition state structures, a reasonable match can often be obtained by designing compounds that mimic the hybridization state of the proposed transition state structure while incorporating one or more atoms with larger covalent radii than those in the substrate.[36]

Classically the transition state has been visualized by chemical intuition based on accepted ideas of organic reaction mechanisms and metastable intermediates. More recently, quantum mechanical methods are being used, and with the advent of computer graphic techniques, it has become possible to screen a number of three-dimensional structures as potential analogs of a calculated transition state structure. With these techniques, analogs that best resemble the proposed transition state structures are identified for synthesis and testing.

An often-quoted example of a transition state analog is that of adenosine deaminase, which catalyzes the hydrolysis of adenosine to inosine through a tetrahedral-like transition state formed by addition of water. It has been shown that the antibiotic coformycin is a strong inhibitor of adenosine deaminase and is an apparent transition state analog for this enzyme; it binds approximately a million times tighter than inosine.[37]

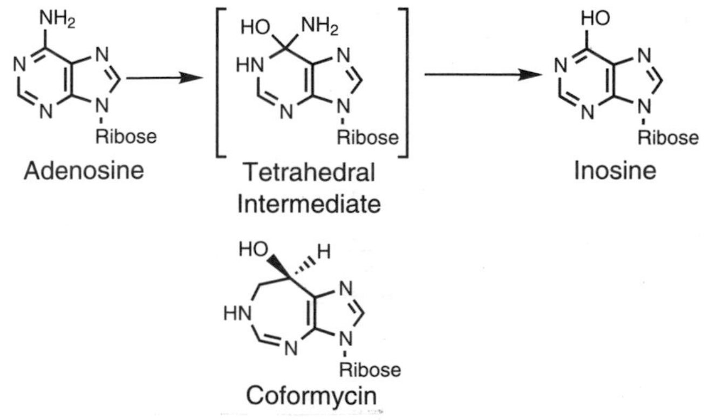

ANTIFOLATES

Folic Acid Metabolism

Folic acid is a widely distributed vitamin of the B group that was detected in the course of studies on tropical macrocytic anemia (tropical sprue), a disease cured by liver extract. It was isolated in crystalline form from liver by Pfiffner et al.[38] in 1943, and its structure was established[39] and confirmed by total synthesis as pteroylglutamic acid.[40] The pathway of folate biosynthesis was worked out using an extract of E. coli by Brown[41]; a schematic representation is given in Figure 31–3. The folate-synthesizing system is specific to microorganisms (and some plants) that cannot absorb preformed dihydrofolic acid; animals, including humans, do not make their own dihydrofolic acid but obtain it from food. The selectivity of action of sulfonamides is due to this difference in folic acid metabolism between humans and microorganisms.

Folic acid and dihydrofolic acid have no coenzyme activity. The coenzyme form of folic acid is 5,6,7,8-tetrahydrofolic acid, which is formed by a two-step reduction through dihydrofolate as an intermediate. N^5-Formimino tetrahydrofolate is formed by transfer of formyl from formimino acids catalyzed by formimino transferases. The pool of one-carbon donors consists of free formate and its combinations with folate coenzymes at all levels of oxidation, such as methyl, methylene, hydroxymethyl, formyl, or formimino groups, and the biosynthesis of purines, pyrimidines, coenzymes, and some essential amino acids occurs by the reaction of these donors with one carbon acceptors. The folate to folate–coenzyme converting system, however, is present both in microorganisms and mammals, and inhibitors of this system do not offer the selectivity that folate synthesis offers. It has, however, been found that dihydrofolate reductase enzymes from bacteria, protozoa, and mammals have different affinity characteristics for different inhibitors, and this difference has provided a basis for the design of selective inhibitors. The folate pathways thus provide a suitable system for interference, for both infectious and neoplastic conditions, and have been successfully used for the design of chemotherapeutic agents. The fact that different steps in the same pathway can be inhibited by different types of compounds offers a rational approach to synergism, which has been clinically exploited.[99]

Sulfonamides and Sulfones

The discovery of the antibacterial activity of sulfonamides in the early 1930s marked the beginning of the present era of chemotherapy. Soon after the announcement of the marked antibacterial activity of prontosil in vivo by Domagk[12] in 1935, it was established that the activity was due to its metabolic product 4-aminobenzenesulfonamide (sulfanilamide), and it was shown that, in susceptible organisms, sulfanilamide was active both in vitro

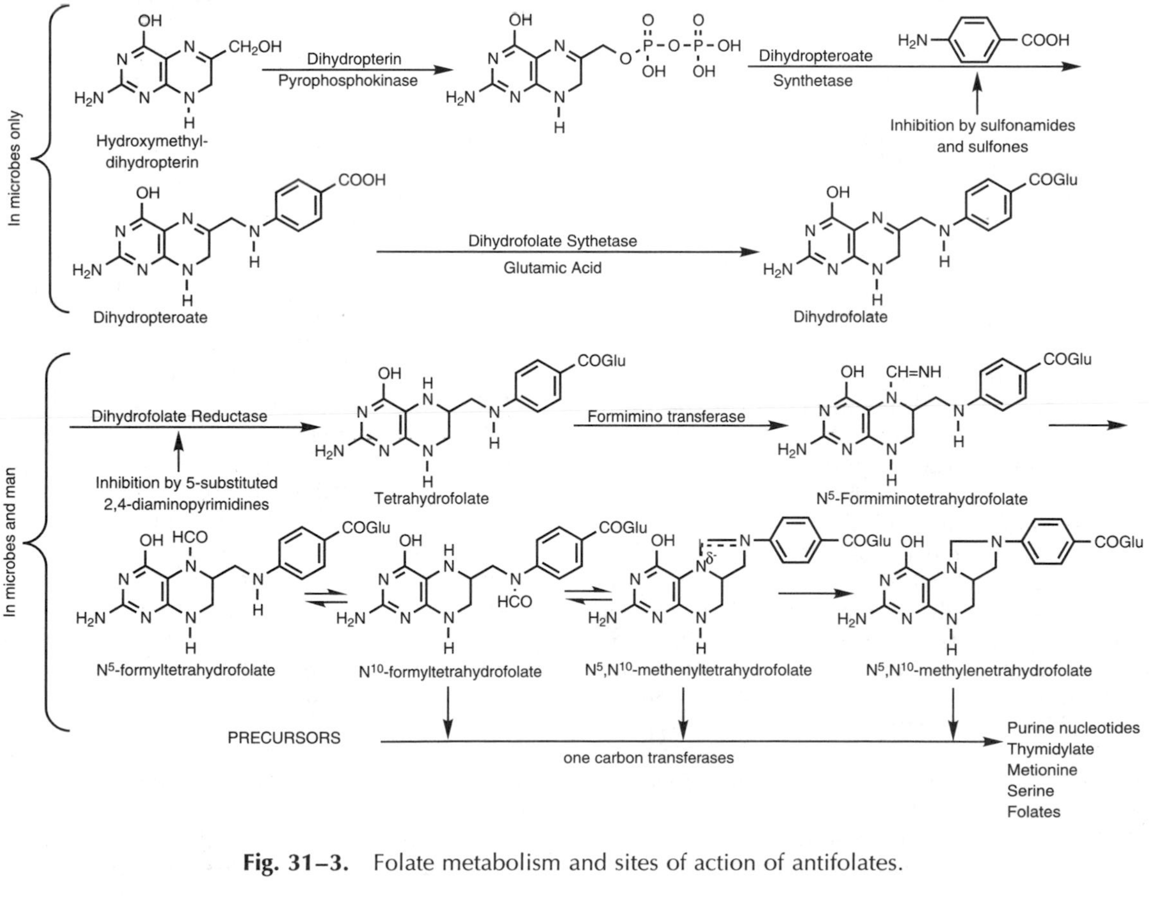

Fig. 31–3. Folate metabolism and sites of action of antifolates.

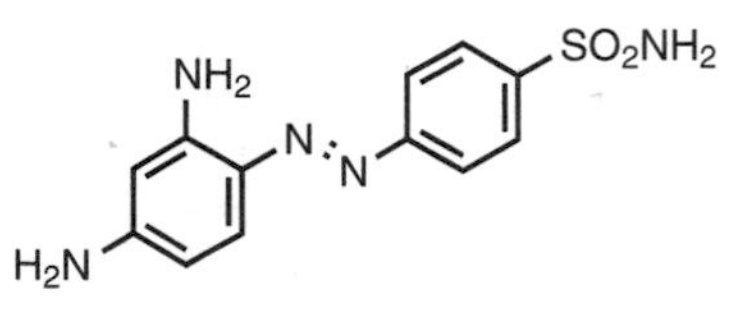

Prontosil 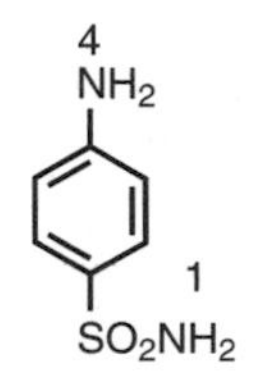 Sulfanilamide

and in vivo.[42,43] When the potential of sulfonamides was recognized, research programs were initiated worldwide to prepare analogs and derivatives of sulfanilamide, particularly with a view toward improving its antimicrobial spectrum, therapeutic ratio, and pharmacokinetic properties and delaying the emergence of resistant forms. The analogs studied included position isomers, compounds having substituents on the functional group, isosteres, and ring annelates of the benzene ring.[44,45] The following generalizations regarding structure-activity relationships, arrived at early in the development of sulfonamides, guided subsequent work on molecular modification and still hold true.

1. The amino and sulfonyl radicals on the benzene ring should be in a 1,4 disposition for the compounds to show activity; the amino group should be unsubstituted or have a substituent that is readily removed in vivo.
2. Replacement of the benzene ring by other ring systems, or the introduction of additional substituents on it, decreases or abolishes the activity.
3. Exchange of the SO_2NH_2 by $SO_2C_6H_4$—p—NH_2 retains the activity, whereas exchange by CONH, COC_6H_4—p—NH_2 markedly reduces it.
4. N^1-Monosubstitution (on the amide N) results in more active compounds with greatly modified pharmacokinetic properties; N^1-disubstitution in general leads to inactive compounds.
5. These generalizations also apply to diphenyl sulfones, so that only N-monosubstituted dapsone derivatives retain full activity.

The presence of a p-aminobenzenesulfonyl radical (H_2N—C_6H_4—SO_2—) therefore seems essential for maintaining good activity, and practically all later attention has been focused on N^1-substituents. The more important N^1-substituted derivatives of sulfonamides that were found clinically useful belonged to N^1-heterocycle and N^1-acyl classes. Alteration of the N^1-substituent made it possible to obtain sulfonamides with widely varying physicochemical and pharmacokinetic characteristics, greatly enlarging the clinical usefulness of sulfonamides as a class. The sulfonamides commonly used in clinical practice and their important characteristics are listed in Table 31–1. Their half-lives vary from 2.5 to 150 hours. These sulfonamides can be grouped based on their absorption and half-life. Sulfonamides of one group remain largely unabsorbed after oral administration and hence are considered useful for gastrointestinal infections; those of another group are characterized by high solubility, quick absorption, and rapid excretion (half-life up to 10 hours) mainly in the unaltered form and are used widely for urinary tract infections. Sulfonamides of yet another group are absorbed rapidly but excreted slowly, resulting in maintenance of adequate levels in the blood for long periods; they require less frequent administration and are useful for chronic infections and for prophylaxis. Sulfonamides with half-lives between 10 and 24 hours are termed medium-acting and those with half-lives longer than 24 hours long-acting. This wide range of pharmacokinetic properties of sulfonamides, coupled with their ease of administration, broad spectrum of antimicrobial activity, noninterference with host-defense mechanisms, and relative freedom from the problem of superinfection, is responsible for their continued, widespread use in clinical practice five decades after their discovery. The antimicrobial activity of sulfonamides (and sulfones) extends to a number of microbial species having a folic acid pathway, which includes many gram-positive and gram-negative cocci and bacilli, mycobacteria, and some large viruses, protozoa, and fungi. In all cases, their action is related to PABA antagonism. The use of sulfonamides and sulfones now includes the treatment of acute and chronic gram-positive and gram-negative bacterial infections, including leprosy, trachoma, lymphogranuloma venereum, malaria, nocardiosis, coccidiosis, and toxoplasmosis.

Physicochemical Properties

Since the discovery of sulfonamides, studies have continued to find a correlation between their physicochemical properties and bacteriostatic activity. The parameters that have attracted the most attention are degree of ionization, protein binding, and lipid/water solubility. It was established early that the primary aromatic amino group was essential for biologic activity and that the dissociation constant of this group, which is close to that of PABA, was not greatly affected by a change in the N^1-substituent. Most attention has thus been focused on the acid dissociation constant of the SO_2NHR group, which varies widely (from 3 to 11) in various clinically used sulfonamides. As early as 1942, Bell and Roblin,[46] in an extensive study of the relationship between the pK_a of a series of sulfonamides and their antibacterial activity in vitro, found that the plot of log 1/MIC against pK_a was a parabolic curve and that the highest point of this curve lay between pK_a 6.0 and 7.4; the maximal activity was thus observed in compounds whose pK_a approximated the physiologic pH. Because the pK_a values are related to the nature of the N^1-substituent, the investigators emphasized the value of this relationship for predicting the MIC of new sulfonamides. The pK_a of most of the active sulfonamides discovered since then, particularly the long-acting ones, falls within this range. Bell and Roblin correlated Woods and Fildes' hypothesis regarding the structural similarity between a metabolite and its antagonist with the observed facts of ionization. They emphasized the need for polarization of the sulfonyl group of the active sulfonamide for it to resemble as closely as possible the geometrical and electronic characteristics of the *p*-aminobenzoate ion and postulated that the more negative the SO_2 group of an N^1-substituted sulfanilamide, the more bacteriostatic it

Table 31–1. Characteristics of Some Commonly Used Sulfonamides and Sulfones*

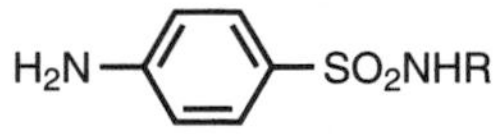

$H_2N-C_6H_4-SO_2NHR$

Generic Name	R	In Vitro Activity† Against E. coli, μmol/L	pK_a	Liposolubility‡ %	Protein Binding at 1.0 mol/ml, % Bound	Plasma Half-Life (in Humans)	% N^4-Metabolite in Urine§ (in Humans)
Phthalylsulfathiazole‖	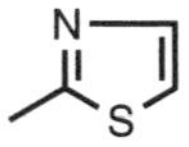		Acid				
Sulfamethizole	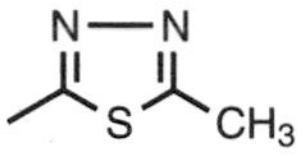		5.5		22	2.5	6
Sulfisoxazole	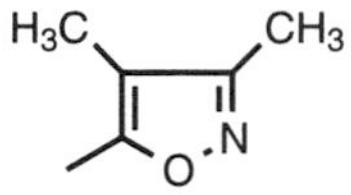	2.15	5.0	4.8	76.5	6.0	16 (30)
Sulfamethazine	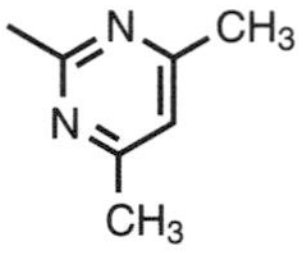	1.7	7.4	82.6	66	7	60
Sulfisomidine	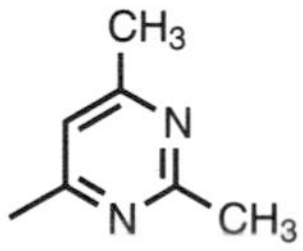	1.5	7.4	19.0	67	7.5	4
Sulfacetamide	$-COCH_3$	2.3	5.40	2.0	9.5	7	5
Sulfanilamide	H	128.0	10.50	71.0	9	9	—
Sulfaphenazole	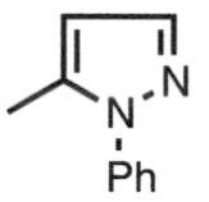	1.0	6.09	69.0	87.5	10	20 (80)
Sulfamethoxazole	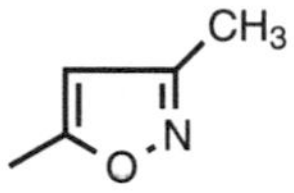	0.8	6.0	20.5	60	11	60 (14)
Sulfadiazine	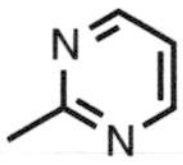	0.9	6.52	26.4	37.8	17	25
Sulfamethoxydiazine	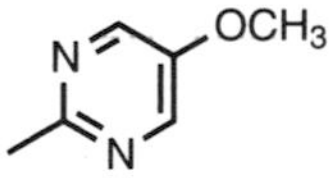	2.0	7.0	64.0	74.2	37	20‡,ii (30)
Sulfamethoxypyridazine	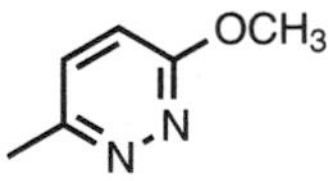	1.0	7.2	70.4	77	37	50 (15)
Sulfadimethoxine	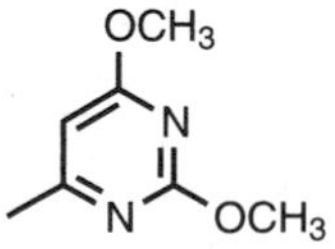	0.7	6.1	78.7	92.3	40	15 (70)
Sulfamethoxypyrazine	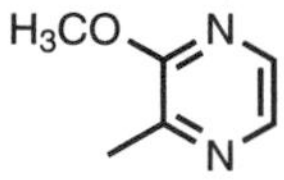	1.85	6.1		65	65	65

(continued)

Table 31–1. (Continued)

$H_2N-C_6H_4-SO_2NHR$

Generic Name	R	In Vitro Activity† Against E. coli, µmol/L	pK_a	Liposolubility‡ %	Protein Binding at 1.0 mol/ml, % Bound	Plasma Half-Life (in Humans)	% N^4-Metabolite in Urine§ (in Humans)
Sulfadoxine	5,6-dimethoxy-4-pyrimidinyl (OCH_3, H_3CO)	0.8	6.1	5	95	150	60§,ii (10)
4,4′-Diaminodiphenylsulfone		44	pk*b*13		50‡,ii	20	

* Unless otherwise stated, the data are from Rieder J., *Arzneim. Forsch., 13,* 81, 89, 95(1963).
† From Struller T., Progress in sulfonamide research, in *Progress in Drug Research,* 12, Ed. E. Jucker, Basel, Birkhäuser, 1968, p. 389.
‡ From (i) M. Windholz, Ed., *The Merck Index,* Rahway, N.J., Merck & Co., 1976; (ii) Martindale, *The Extra Pharmacopoea,* 26th ed., N. W. Blacow, Ed., London, Pharmaceutical Press, 1972.
§ From R. T. Williams and D. V. Parke, *Ann. Rev. Pharmacol., 4,* 85(1964).
|| N^4-Phthalyl.

will be. From a study of the effect of pH of the medium on the antibacterial activity of sulfonamides, Cowles[47] and Brueckner[48] suggested that sulfonamides penetrate the bacterial cell in the unionized form, but once they are inside the cell, their bacteriostatic action is due to the ionized form. Therefore, for optimum activity, the compound should have a pK_a that gives a proper balance between the activity and penetration; a half-dissociated state appears to present the best compromise between transport and activity. This also provides an explanation for the parabolic relationship between pK_a and MIC observed by Bell and Roblin.

An important factor in the chemotherapeutic activity of sulfonamides and their transport is the lipid solubility of the undissociated form. Lipid solubility of different sulfonamides varies over a wide range. These differences unquestionably influence pharmacokinetics and antibacterial activity. Long-acting sulfonamides with high tubular reabsorption are generally characterized by high lipid solubility. Although a precise relationship among these factors has not been established, it has been shown that, in general, as the lipid solubility increases so do the half-life and antibacterial activity in vitro.

In subsequent studies on correlation of physicochemical properties and biologic activity, additional parameters have been considered, such as Hammett $\sigma\rho$ values, electronic distribution calculated by molecular orbital methods, spectral characteristics, and hydrophobicity constants. The acid dissociation constant does seem to play a predominant role in determining the activity of sulfonamides because it affects their solubility, partition coefficient and transport across membranes, protein binding, tubular secretion, and reabsorption in the kidneys. Thus, intensive work in this field over four decades has fully justified the earlier view of the predominant role of ionization in the overall antimicrobial activity of sulfonamides.

Mode of Action

The mode of action of sulfonamides is characterized by competitive antagonism of certain metabolites vital to the growth of the microorganism. Evidence for this antagonism came soon after the discovery of sulfonamides. It was found that substances antagonizing their action are present in peptones, various body tissues and fluids, bacteria, and yeast extract. Woods[13] obtained evidence that PABA is the probable antagonizing agent in yeast extract and showed that PABA could completely reverse the bacteriostatic activity of sulfanilamide against various bacteria in vitro; PABA was finally isolated from these sources.[49–51] This led Woods to suggest that sulfanilamide, because of its similarity of structure to that of PABA, interferes with the utilization of the latter by the enzyme systems necessary for the growth of bacteria. In further studies, it was shown that the inhibition of growth by sulfonamides in simple media can be reversed not only competitively by PABA, but also noncompetitively by compounds not structurally related to PABA, such as L-methionine, L-serine, glycine, adenine, guanine, and thymine.[52,53] The relationship of sulfonamides to purines was uncovered by the finding that the sulfonamide-inhibited cultures accumulated 4-amino-5-imidazole-carboxamide ribotide,[54] a compound later shown by Shive[55] and by Gots[56] to be a precursor of purine biosynthesis.

With the knowledge gained of the structure, function, and biosynthesis of folic acid coenzyme, these isolated facts could be gradually fitted into a pattern. It was suggested that sulfonamides compete in the condensation of PABA with pteridine pyrophosphate to form dihydropteroate.[57] The amino acids, purines, and pyrimidines that are able to replace or spare PABA are precisely those whose formation requires one-carbon transfer catalyzed by folic acid coenzymes.

Direct evidence of the inhibition of dihydropteroate and dihydrofolate synthesis by sulfonamides was soon obtained by studies on bacterial cultures and in cell-free enzymes of bacteria,[41,58,59] and, in general, the more potent inhibitors of folate biosynthesis were also better growth inhibitors. Hotchkiss and Evans[60] suggested that the differences in response of various organisms to sulfonamides may be due to quantitative differences in the ability of individual enzymes to produce folic acid from PABA. Brown and Weisman[58,61] subsequently obtained evidence of incorporation of sulfonamides. Bock et al.,[62] using ^{35}S-labeled sulfamethoxazole, confirmed this incorporation by isolating the product formed using a partially purified enzyme system and a growing culture of E. coli and have identified the product as N_1-3-(5-methylisoxazolyl)-N^4-(7,8-dihydro-6-pteridinylmethyl)sulfanilamide (dihydropterin sulfonamide). This incorporation, however, is not of physiologic significance.[63]

The mechanism of action of diaminodiphenylsulfone (DDS) is similar to that of sulfonamides; its action against bacteria and plasmodia is antagonized competitively by PABA and noncompetitively by folic acid.[64,65] Much subsequent work has been carried out on the binding characteristics of the isolated enzymes. Based on consideration of the formal electronic charge on sulfanilamide and PABA determined by the Huckel molecular orbital approach, Moriguchi and Wada[66] proposed two binding sites on the enzyme, separated by about 6.7–7 Å, one being specific for the NH_2 group and the other nonspecific where the acidic group of PABA or sulfonamide binds. Shefter et al.,[67] on the basis of x-ray crystallographic data and study of molecular models, noted certain similarities between N^1-substituted sulfanilamides and *p*-aminobenzoylglutamate and suggested that the N^1-substituent may be competing for a site on the enzyme surface reserved for the gluatamate residue, either by directly influencing the linking of aminobenzoic acid–glutamate with the pteridine or by the coupling of glutamate to the dihydropteroic acid. Competition with PABA is thus the primary mode of action of sulfonamides. The subsequent reduction in the rate of dihydropteroate synthesis decreases the concentration of tetrahydrofolate in the cell. This prevents or slows down the formation of the raw materials of protein, RNA, and DNA biosynthesis, thereby affecting a number of synthetic processes of the organism concurrently and thus reducing the cell growth rate. Sulfonamides inhibit only growing organisms, and the bacteriostasis of the latter is preceded by a lag phase, which can now be explained as due to stored PABA or folates, or both.

Development of Resistance

Emergence of drug-resistant strains is one of the principal constraints of sulfonamide therapy. Resistance can develop as a result of one or more factors, such as overproduction of PABA,[68,69] altered permeability of the organism to sulfonamides,[70–72] and, most important of all, reduced affinity of dihydropteroate synthetase for sulfonamides while maintaining the affinity for PABA.[71,73–76] Different sulfonamides show cross-resistance, but there is no cross-resistance to other antibacterials. Resistant strains can develop by random mutation and selection[71] or by transfer of resistance factors.[72,75–78] In plasmodia, resistance can also develop by a bypass mechanism, i.e., ability of the organisms to use preformed folic acid.[79]

Dihydrofolate Reductase Inhibitors

The history of dihydrofolate reductase inhibitors (DHFRI) goes back to 1947 when 2,4-diaminopterins, with and without a *p*-aminobenzoylglutamic residue, having powerful antifolate activity, were described.[80–84] Of

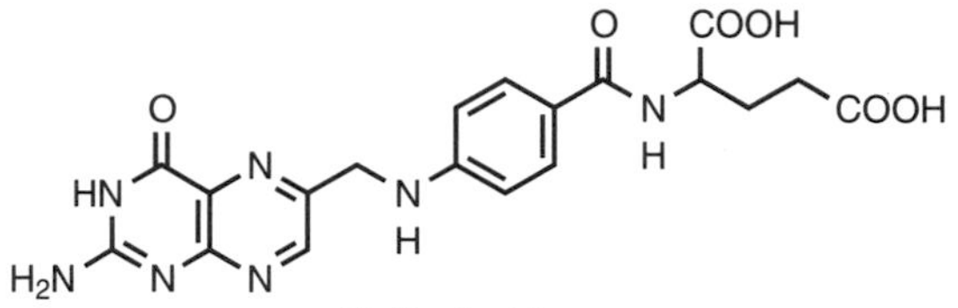

Folic Acid

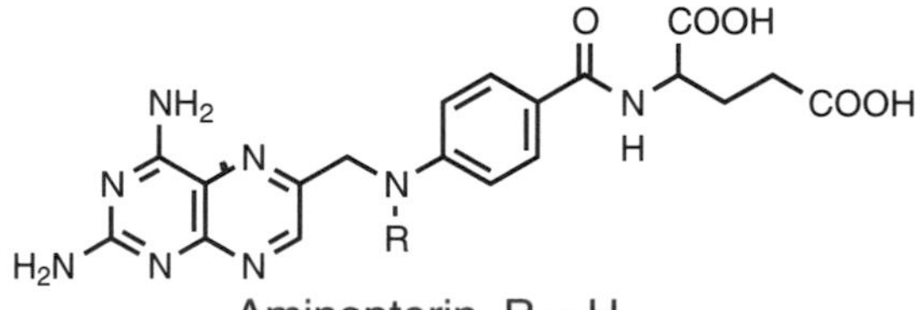

Aminopterin, R = H
Methotrexate, R = CH_3
(Amethopterin)

these the classic DHFRIs, aminopterin and amethopterin (methotrexate [MTX]),[80,85] close analogs of folic acid, proved particularly useful; MTX is still one of the most widely used antitumor drugs. Its activity has been shown as due to competitive inhibition of DHFR.[85–87] It is notable for both its high inhibition constant (1×10^{-10} M) and its lack of selectivity[88]; it is among the most tightly bound enzyme inhibitors known.

Soon after the demonstration of the strong antifolate activity of diaminopteridines, Hitchings et al.[89] reported that simple 2,4-diaminopyrimidines, such as the 5-methyl and 5,6-dimethyl derivatives, although originally synthesized as pyrimidine antagonists, inhibited the growth of L. casei through interference with the utilization of folic acid. Later, these were shown to act through inhibition of dihydrofolate reductase and were called nonclassic because they lacked the glutamate residue found in classic DHFRI such as MTX. This discovery was a trailblazer and led to the synthesis of a variety of diaminopyrimidines having potent antimicrobial activity, which was particularly strong in 5-aroyl-, 5-benzyl- and 5-aryl-2,4-diaminopyrimidines. What was even more important, however, was the selectivity of action observed against bacteria and plasmodia with a change in the substituents. For example, it was found that 5-benzylpyrimidines lacking a 6-substituent were more active against bacteria than plasmodia.[90] In contrast to MTX, these compounds showed selective

Table 31–2. Comparative Binding of Inhibitors to Dihydrofolate Reductases from Various Species

	IC_{50} ($\times 10^{-8}$ M)			
Compound	**Plasmodium berghei**	**Escherichia coli**	**Staphylococcus aureus**	**Rat Liver**
Methotrexate	—	0.01	0.02	0.02
Pyrimethamine	0.05	250	300	180
Trimethoprim	17.0	0.7	1.5	35,000
Butylphenyldihydrotriazine	—	65,000	50,000	24

From J. J. Burchall, Dihydrofolate reductase; in *Inhibition of Folate Metabolism in Chemotherapy,* G. H. Hitchings, Ed., Berlin, Springer-Verlag, 1983; and G. H. Hitchings, *Ann. N.Y. Acad. Sci., 186,* 444(1971).

antimicrobial activity without significant mammalian toxicity. Attempts to optimize these results led to the development of trimethoprim (TMP) as a powerful antibacterial agent and pyrimethamine (PYR) as an antiprotozoal.

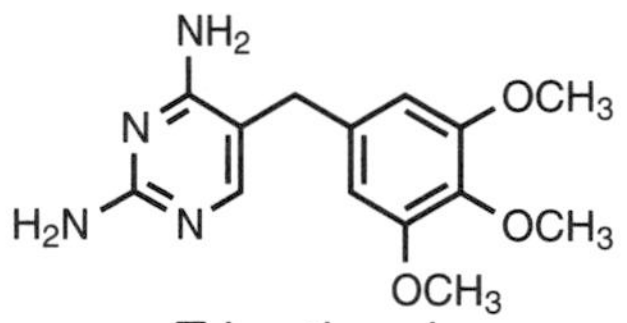

Trimethoprim

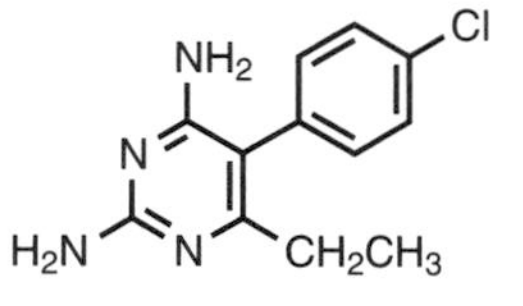

Pyrimethamine

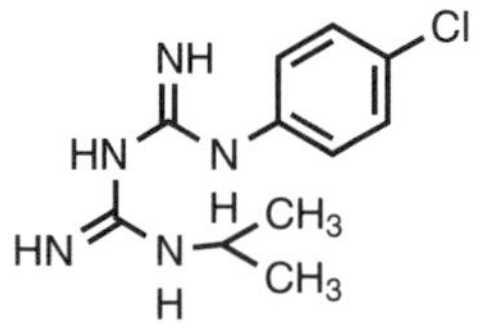

Proguanil

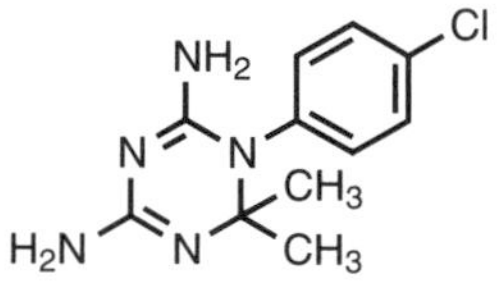

Cycloguanil

Another entry into this field was through biguanides, such as proguanil, which had been developed as an antimalarial.[91] The activity of proguanil in vivo was found to be due to its metabolization to cycloguanil, a diaminodihydrotriazine.[92] The similarity of this metabolite to diaminopyrimidines was noted by Hitchings et al.[91]; the diaminotriazines were also found to act as DHFRI. Diaminotriazines thus constitute another prototype of DHFRI.

Concurrent research carried out on dihydrofolate reductases from different sources indicated that the selectivity of these compounds may be due to their differential binding to DHFRs of different origins. The affinity data of some important compounds, when tested on reductases obtained from bacterial, protozoal, and mammalian sources (Table 31–2), show clearly that the potency and selectivity of diamino heterocyclic inhibitors can be explained satisfactorily by their differential binding to reductases of different origins.[17] For example, MTX binds tightly to all reductases tested and is lethal to any cell it can enter. TMB and PYR show a remarkable ability to distinguish between bacterial, mammalian, and plasmodial reductases. This important difference in binding is the basis of the selective chemotherapeutic spectrum of the latter two compounds.

Many DHFR from various sources have been purified, and the determination of their amino acid sequences[93] has confirmed the differences in the characteristics of DHFR from different sources that the inhibition studies had detected. The availability of x-ray diffraction pictures of crystalline enzyme complexes has added a new, important dimension to understanding the selectivity and specificity of this action at the molecular level. Baker et al.[94] provided the first information on the crystal structure of the trimethoprim–E. coli–DHFR complex. The diaminopyrimidine ring is surrounded by Asp-27, Phe-31, Ile-94, and Thr-113, whereas the trimethoxy moiety is enclosed by Phe-31, Ile-50, and Leu-28. NMR spectrometry has also been used to study the mode of binding of TMP and other ligands to DHFR.

Two second-generation nonclassic lipophilic DHFRI, trimetrexate[96] and piritrexim,[97] have found use in cancer and antiprotozoal chemotherapy and for psoriasis. Their advantage over classic inhibitors such as MTX is that they do not require the reduced folate carrier system for their transport and have a superior tumor spectrum; these are, however, prone to multidrug resistance.

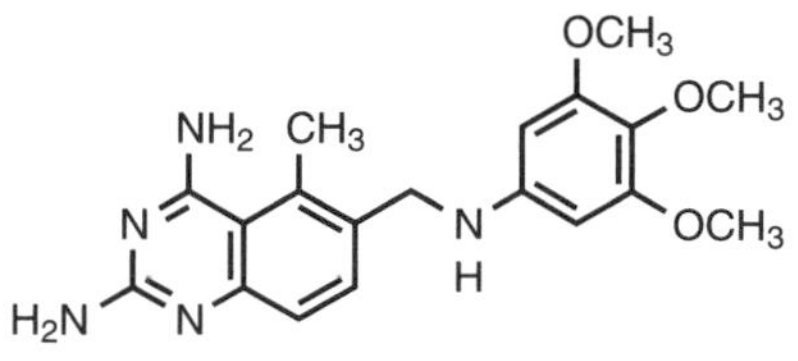

Trimetrexate

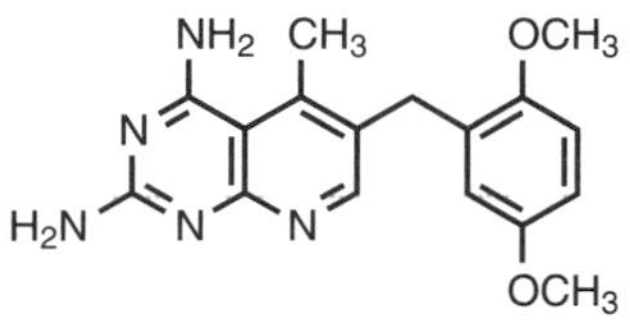

Piritrexim

DHFRI thus continue to be of interest in development of new drugs.[98] Classic and nonclassic antifolates both have their own advantages and shortcomings in regard to selectivity and spectrum of activity, resistance, and toxicity, which determine their specific clinical use.

Synergism of Sulfonamides and Dihydrofolate Reductase Inhibitors

Daniel et al.,[83] in a study of the antifolate activity of diaminopteridines, observed a synergistic effect on adding sulfonamides to the medium. They postulated that this synergism is due to two types of inhibitors competing with two different parts of the folic acid molecule and expected this to be of clinical significance because it reduced the amount of sulfonamides required. The elucidation of the folic acid pathway and of the mode of action of sulfonamides of antifolates (DHFRIs) has shown that this synergism is indeed a consequence of the sequential occurrence of inhibition at two different steps of the same pathway. TMP, for example, increases the concentration of dihydrofolate, which by the law of mass action drives the reaction to the right and produces tetrahydrofolate, which partially overcomes the induced block. This effect is minimized by including a sulfonamide, which blocks the synthesis of dihydrofolate. Such synergism is now recognized as a general occurrence and has been demonstrated for many microorganisms, both in vitro and in vivo. Combination therapy with sulfonamides and DHFRI has added a new dimension to treatment with these agents.[99] Factors that contribute to the usefulness of such a combination include a manyfold increase in therapeutic index, better tolerance, ability to delay development of resistance, and ability to produce cures when the curative effects of the individual drugs are minimal. The choice of drugs to be used in the combination is based primarily on their half-life characteristics. PYR has a long half-life (about 130 hours) and should be combined with a sulfonamide such as sulfadoxine, with a half-life of the same order for use in protozoal infections. TMP has a half-life of about 11 hours and should be combined with a sulfonamide of medium half-life, such as sulfamethoxazole or sulfadiazine.

AMINO ACID ANTAGONISTS

In view of the importance of amino acids in nutrition, metabolic processes, and translation of information, they have been an important target in the design of antimetabolites. The number of amino acid analogs, however, that have shown significant chemotherapeutic activity is rather small. This may be due, in part, to the high serum concentration use needed, which is difficult to maintain over long periods, and facile interconversion among amino acids, which helps to overcome metabolic blocks easily.

An example of an antibiotic substance produced by a streptomyces is D-cycloserine, which bears a strong structural resemblance to D-alanine, and may act as an alanine antagonist.

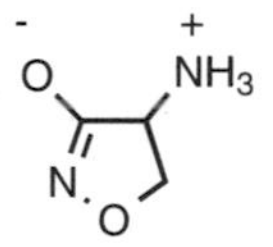

D-Cycloserine 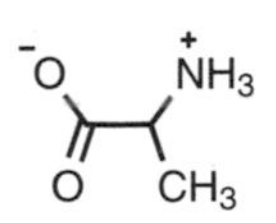 D-Alanine

Phenylalanine Antimetabolites

The modifications that have been made in the phenylalanine structure include ring substitution, replacement of the aromatic ring by an isosteric ring, and replacement of the benzyl group by a nonaromatic group having appropriate planarity and with substituents on the side chain. Among the ring-substituted analogs, 3- and 4-fluorophenylalanines have been shown to be the most effective antagonists of phenylalanine, as shown by growth inhibition studies in bacterial[100,101] and chicken heart cell cultures.[102]

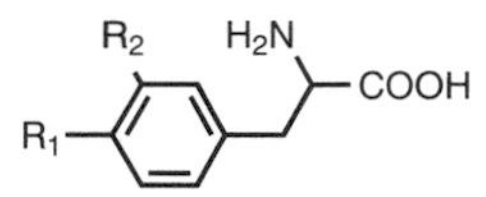

3-Fluorophenylalanine R_1=H, R_2=F
4-Fluorophenylalanine R_1=F, R_2=H

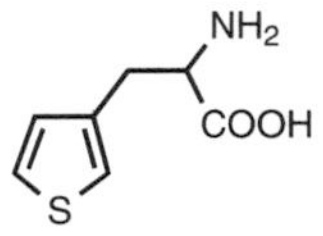

3-Thiophenealanine

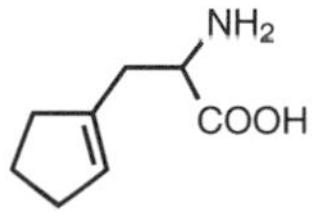

1-Cyclopentenylalanine

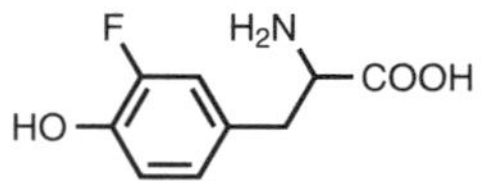

3-Fluorotyrosine

Some representatives of isosteric ring analogs found to possess significant activity were 2- and 3-thiophenealanines[103–107] and 1-cyclopentenylalanine,[108] which inhibited microbial growth and produced phenylalanine deficiency symptoms in rats[109] that could be reversed by phenylalanine.

Tyrosine Antimetabolites

Among tyrosine analogs, 3-fluorotyrosine appears to be one of the few compounds that specifically inhibit tyrosine utilization in a competitive manner. It causes inhibition of growth of N. crasa, which is competitively reversed by tyrosine.[110] It is toxic to rats[111,112] and mice.[113]

Tryptophan Antimetabolites

The most effective tryptophan analogs contain a methyl or a fluorine substituent on the indole moiety. The 4-, 5-, and 6-methyltryptophans inhibit the growth of a number of microorganisms, and the inhibition is reversed by tryptophan.[114–117] Also, 5-fluorotryptophan and 6-fluorotryptophan inhibit the use of anthranilic acid and the growth of many microorganisms.[118–120]

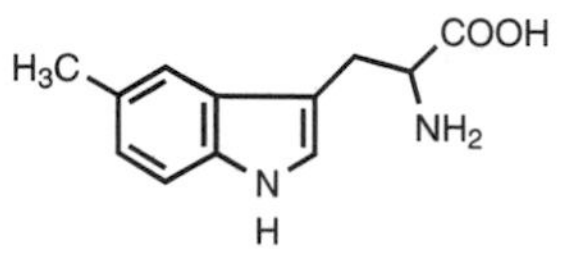

5-Methyltryptophan

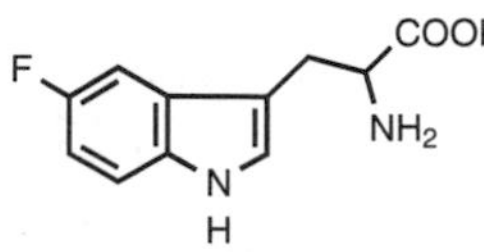

5-Fluorotryptophan

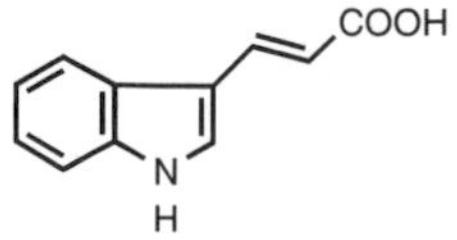

Indolylacrylic Acid

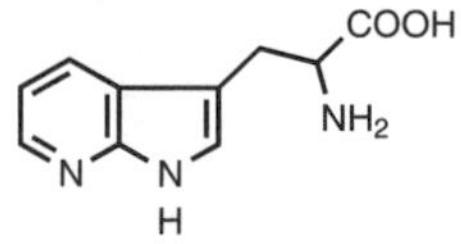

7-Azatryptophan

Replacement of the 7-carbon atom of the indole ring by nitrogen, as in 7-azatryptophan, has given a competitive antagonist of tryptophan,[121,122] which inhibits the growth of Tetrahymena pyriformins. It has been found that 3-indolylacrylic acid inhibits the growth of E. coli and B. typhosum and that tryptophan and indole reverse this inhibition; indole accumulates in the presence of the inhibitor, which appears to act by preventing the formation of tryptophan from indole.[123]

Methionine Antimetabolites

Ethionine has been found to be a potent antagonist of methionine; it produces methionine deficiency symptoms in rats,[7] causes inhibition of growth of E. coli[124] and a number of other microorganisms, and interferes in a variety of biochemical roles of methionine, such as normal protein synthesis and incorporation of the methionine sulfur into cysteine. Ethionine is converted in yeast to S-adenosylethionine. Ethionine, 50 years after its discovery, continues to be of interest as an antimetabolite of methionine; it was reviewed in 1982.[125]

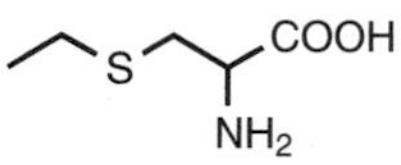

Ethionine

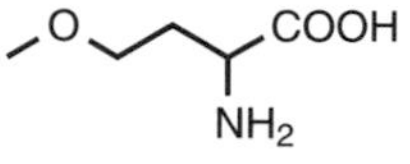

Methoxinine

Methoxinine, a methionine analog containing an oxygen in place of sulfur, is a methionine antagonist in E. coli and S. aureus[126] and is toxic to rats; the antagonism is reversed by methionine.[127] This analog also has antiviral activity.[128]

Glutamic Acid Antimetabolites

Structural modifications of glutamic acid that have resulted in effective antagonists include replacement of the γ-carboxyl group by structurally related groups and the introduction of various substituents on the α-, β-, and γ-carbon atoms. Methionine sulfoxide was inhibitory to L. arabinosus, and the inhibition was reversed by 1-glutamic acid.[129] β-Hydroxyglutamic acid and α-methylglutamic acid were also shown to inhibit the utilization of glutamic acid by L. arabinosus.[130,131]

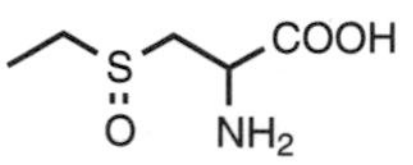

Methioninesulfoxide

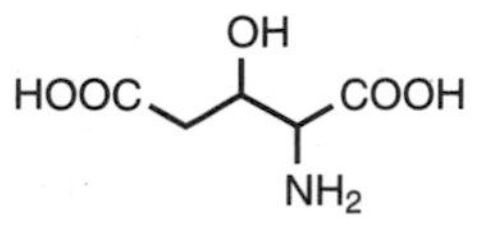

β-Hydroxyglutamic Acid

Glutamine Antimetabolites

Two types of structural modifications of glutamine that have produced glutamine antagonists include replacement of the amide group and substitution of the 3-carbon by sulfur or oxygen atoms.

γ-Glutamohydrazide inhibits S. faecalis[132,133] and some strains of mycobacteria[134] and prevents the deamination of glutamine during glycolysis; these effects are reversed by glutamine.

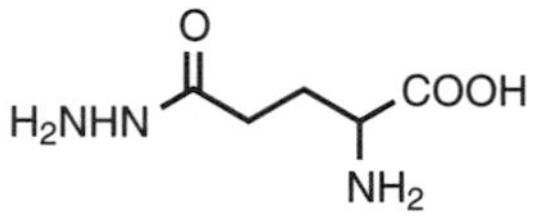

γ-Glutamohydrazide

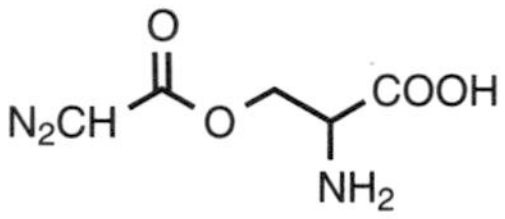

Azaserine

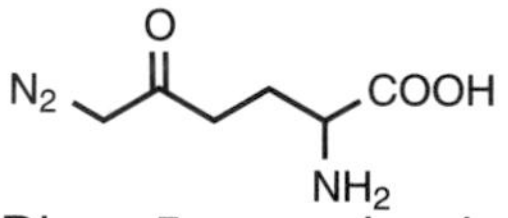

6-Diazo-5-oxonorleucine (DON)

Two naturally occurring diazo compounds closely related in structure to glutamine and function as glutamine antagonists are O-diazoacetyl-L-serine (azaserine) and 6-diazo-5-oxo-L-norleucine (DON), which find clinical use in different forms of cancer; DON is about 30 times as active as azaserine.[135,136] They block the conversion of formylglycineamide ribonucleotide to formylglycineamidine ribonucleotide, an essential step in purine biosynthesis in which glutamine acts as an amino group donor in the presence of amidotransferases. Azaserine and DON inactivate the enzyme irreversibly, and this inactivation is delayed by glutamine.[137]

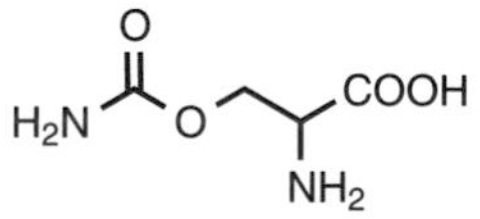

O-Carbamoylserine

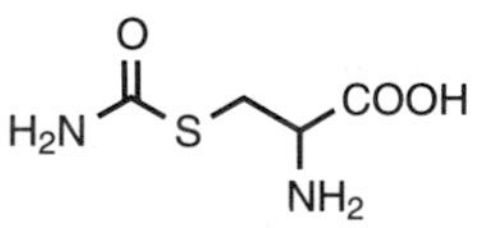

S-Carbamoylcysteine

A nonreactive analog of glutamine, L-O-carbamoylserine, has been found to be a competitive antagonist of glutamine for inhibition of microbial growth, whereas the corresponding S-carbamoylcysteine, which contains a reactive thioester group, was found to be a noncompetitive antagonist.[138] This thioester analog has antitumor activity in mice.[139]

VITAMIN ANTAGONISTS

Thiamine (Vitamin B_1) Antagonists

PYR, in which the sulfur atom of the thiazole ring of thiamine has been replaced by C═C—, was the first antimetabolite of thiamine to be studied.[140–142] It is a competitive antagonist of thiamine, able to produce the typical signs of thiamine deficiency in higher animals and to inhibit the growth of thiamine-requiring microorganisms. It causes inhibition of the cocarboxylase-synthesizing enzyme.[143,144] Some congeners of PYR have found use in the treatment of coccidiosis in chickens.[145] Some other thiamine antimetabolites studied include oxythiamine, in which the NH_2 on the pyrimidine is replaced by OH, which interferes with the action of cocarboxylase and produces in higher animals a thiamine deficiency.[146] The corresponding imidazolyl analog was inactive.

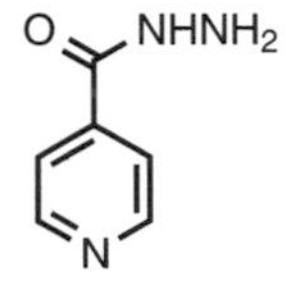

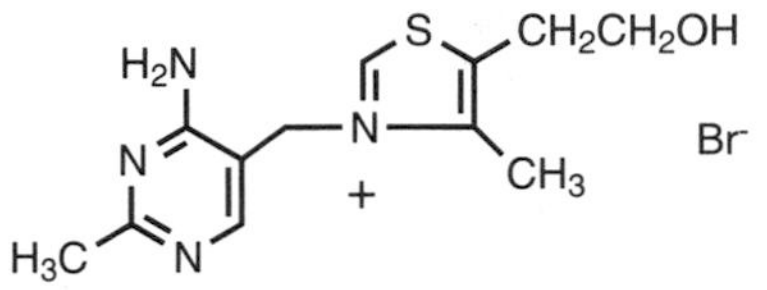

Thiamine

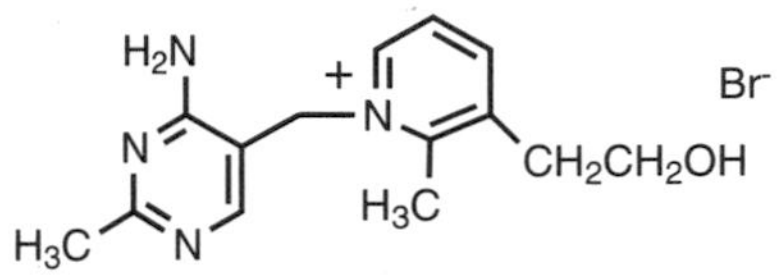

Pyrithiamine

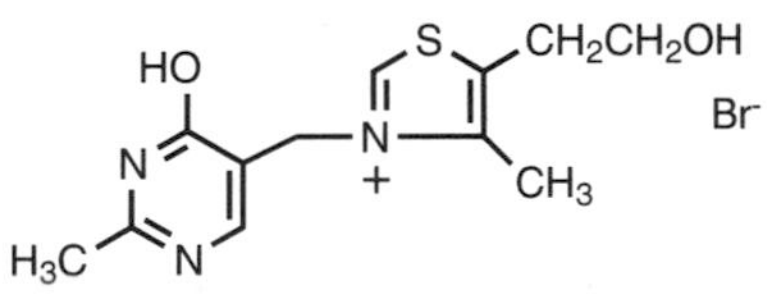

Oxythiamine

Pyridoxine (Vitamin B_6) Antagonists

There are three naturally occurring forms of vitamin B_6; pyridoxal in the form of its phosphate ester is the one whose coenzymic role is well established. The best-studied antimetabolites of B_6 are 4-deoxypyridoxine and *α*-methylpyridoxine; the former was found to inhibit the growth of many pyridoxine-requiring bacteria and fungi[147] and to produce pyridoxine deficiency in chicks.[148] In resting bacterial cells, *α*-methylpyridoxine is phosphorylated to an ester, which then competes with pyridoxal phosphate for the apoenzyme of some amino acid decarboxylases.[149] *α*-Methylpyridoxine is able to replace the vitamin in some biologic systems, whereas in others it acts as an antimetabolite[150,151]; in some organisms, it can perform both roles, inhibiting some pyridoxine-catalyzed reactions and promoting others.[151] Isonicotinic acid hydrazide (INH), a well-known antituberculosis drug, produces symptoms of pyridoxine deficiency and is considered a B_6 antimetabolite. Some of the toxic effects of INH are reversed in pyridoxine-requiring organisms.[152,153]

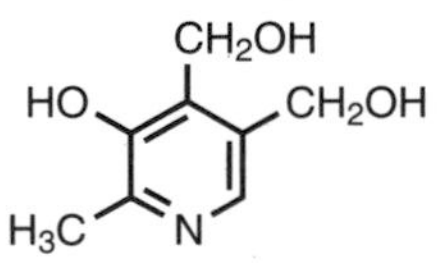

Isonicotinic acid hydrazide

Pyridoxine

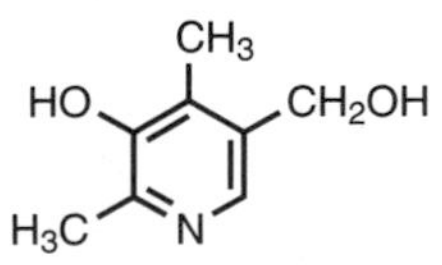

4-Deoxypyridoxine

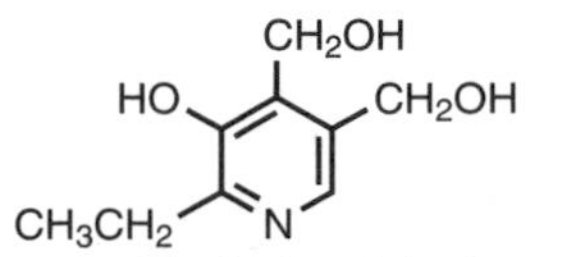

α-Methylpyridoxine

Vitamin K Antagonists

The naturally occurring K vitamins are a group of compounds with antihemorrhagic properties in birds and mammals. Chemically, all are 2-methyl-1,4-naphthoquinones, substituted in the 3-position with saturated or unsaturated polyisoprenoid side chains of varying length, and are of types K_1 and K_2. Many other compounds, most of them naphthoquinones, have been synthesized and found to exhibit vitamin K activity; the most active of them is vitamin K_3 (menadione), whose reduced dihydroxy form is menadiol.

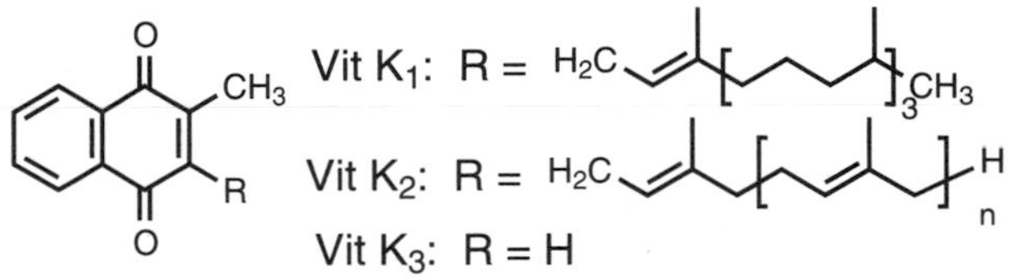

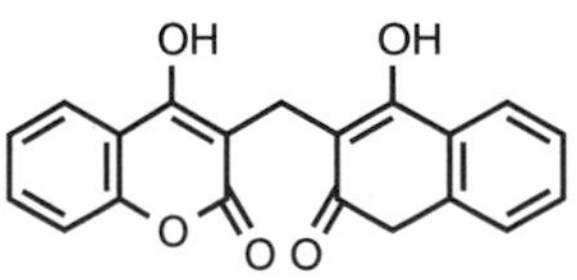

Vit K_1: R = ...

Vit K_2: R = ...

Vit K_3: R = H

Dicoumarol

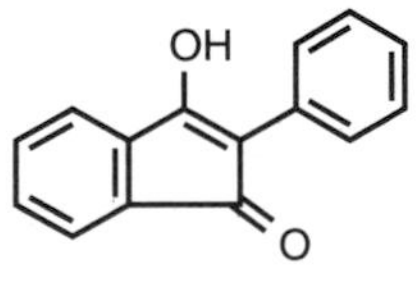

Phenylindanedione

A naturally occurring substance able to produce hemorrhagic conditions in animals was first isolated from spoiled sweet clover hay[154] and shown to be 3,3′-methylenebis-(4-hydroxycoumarin), commonly called dicoumarol. This compound rapidly produces hypothrombinemina in many species, readily reversed by vitamin K. Much work on synthetic anticoagulants has been carried out as a follow-up to the discovery of dicoumarol, and this subject has been well reviewed.[155,156] The most active anticoagulants of this chemical class are these "double" molecules, although these are not essential for anticoagulant activity. The action of all anticoagulants can be reversed by vitamin K but not to the same extent. Another potent vitamin K antagonist is 2-phenyl-indanedione,[157] which is almost as active as dicoumarol and is used clinically. Both these compounds are also important rodenticides; their pharmacologic activity is a direct result of their antivitamin effect.

Nicotinic Acid Antagonists

Several antagonists of nicotinic acid (niacin) have been found; they include 3-acetylpyridine,[158] pyridine-3-sulfonic acid,[159] 6-aminonicotinamide,[160] and 5-fluoronicotinic acid. Many of these antimetabolites have multiple actions. They probably exert their effects at several steps in the metabolic chain by which nicotinic acid is converted to the pyridine nucleotides, which then participate as coenzymes in several reactions. The analogs can thus show vitamin-like activity as well as antimetabolite effects.

attached sugars, and phosphate residues. Not withstanding early disappointments, many useful drugs have now emerged that are widely used against tumors and viral diseases; some have been found to possess useful antiprotozoal and antifungal activities, and one is widely used for the treatment of gout.

In most cases, the free bases are not active and need to be converted in vivo to nucleotides or nucleosides, which are the active forms, and compete with the normal nucleotides/nucleosides, but nucleotides cannot be administered because they fail to penetrate the cell. This dilemma creates problems in testing in cell-free systems. The principal ribotide-forming enzymes are kinases, phosphoribosyl pyrophosphate (PRPP) transferase, and ribonucleoside diphosphate reductase; each of these is a potential target for interference. For each base and riboside, there is an independent uptake mechanism that is also independent of any subsequent phosphorylation reaction.

Pyrimidine Analogs

The biosynthesis of thymine (5-methyluracil), the pyrimidine base unique to DNA, was selected by Heidelberger and co-workers[163] as a critical point for chemotherapeutic attack. Because it was known that the methyl group of thymine is inserted into 2-deoxyuridine-5′-phosphate by the enzyme thymidylate synthetase, a compound

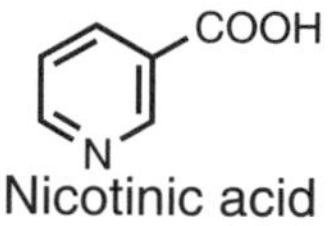

Nicotinic acid

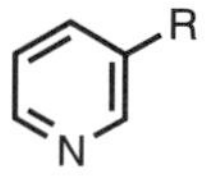

3-Acetylpyridine, R = $COCH_3$
Pyridine-3-sulfonic acid, R = SO_3H

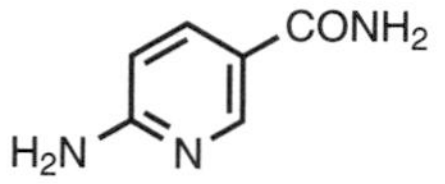

6-Aminonicotinamide

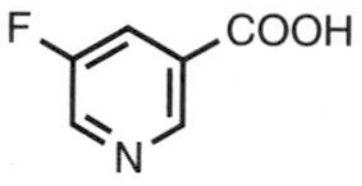

5-Fluoronicotinic acid

ANTAGONISTS DIRECTED AGAINST NUCLEIC ACIDS

Nucleic acids, in view of their central role in cellular function, have attracted much attention in the design of antimetabolites. The elucidation of the structure, function, and biosynthesis of the components of nucleic acid has resulted in targeting these components for attack. Advances in molecular biology, particularly those concerned with the processes of transcription and translation of genetic information, have uncovered many new sites amenable to interference, which have therapeutic potential and are the focus of much current interest.

Pyrimidine and Purine Antimetabolites

The design of pyrimidine and purine antimetabolites[162] has involved changes in substituents on the ring, isosteric replacement of ring atoms, changes in ring size,

with a stable fluorine substituted for the 5-hydrogen of uracil was considered as a good candidate for inhibition of this enzyme and thus inhibition of the biosynthesis of DNA. This led to the synthesis of 5-fluorouracil (5-FU) and its 2′-deoxyribonucleoside (5-FUDR), which proved to be potent antimetabolites exerting their effect mainly by blocking the synthesis of thymidylic acid. 5-FU has significant anticancer activity, and it has been found useful in treating cancers of the breast, colon, and rectum. It becomes active only after its conversion to 5-fluorodeoxyuridylic acid, whose affinity for thymidylate synthetase is several thousand times that of the natural substrate deoxyuridylic acid.[164] Thus, it is able to keep the substrate off the enzyme, causing "thymineless death." This unusually powerful inhibition of a key enzyme in the synthesis of thymidylate is sufficient to explain most of the cytotoxic effects of 5-fluorouracil. The emergence of 5-fluorouracil as an anticancer agent focused attention on other analogs

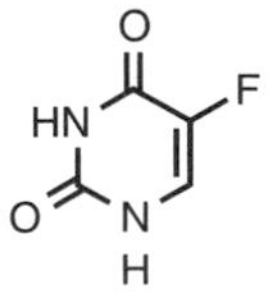

5-Fluorouracil

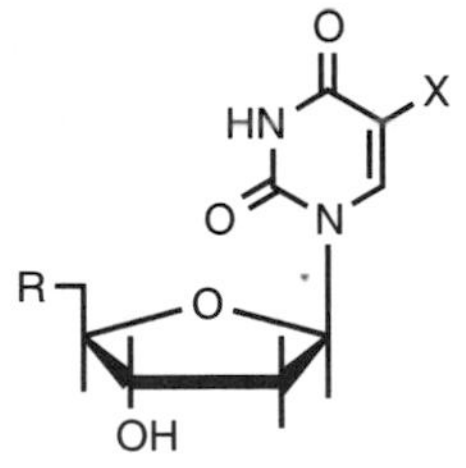

5-Iododeoxyuridine X = I, R = OH
5-Trifluoromethyldeoxyuridine X = CF_3, R = OH
5'-Amino-5-iododeoxyuridine X = I, R = NH_2

of the natural pyrimidines for nonneoplastic diseases as well. Many such analogs have since been synthesized and shown to possess marked biologic activity.

The first effective drug for a viral disease, herpetic keratitis of the eye, was 5-iodo-2-deoxyuridine (idoxuridine, IUDR).[165] It acts by being converted to 5-monophosphate, which interferes with the use of thymidine by cells. It is also incorporated into DNA in place of thymine[166]; in viruses, this leads to formation of incomplete virus particles.[167]

From the mechanistic point of view, 5-fluorouracil and idoxyuridine provide an interesting study. If one compares the van der Waals radii of the various substituents, the size of the fluorine atom is nearer that of hydrogen, whereas the size of the iodine atom is close to that of the methyl group. On this basis, their action becomes understandable. Idoxyuridine has relatively little effect on the biosynthesis of thymidylic acid. It is converted enzymatically within the cells to phosphorylated derivatives, which can be incorporated into DNA in place of thymidylic acid and can suppress the growth of both experimental and human neoplasms.[166] Thus, although fluorouracil acts as an antimetabolite of uracil and inhibits the enzyme thymidylate synthetase, idoxuridine acts as an antimetabolite of thymidine. Another effective antimetabolite of thymidine is 5-trifluoromethyldeoxyuridine (trifluridine),[167] which also inhibits thymidylate synthetase and has antitumor activity.

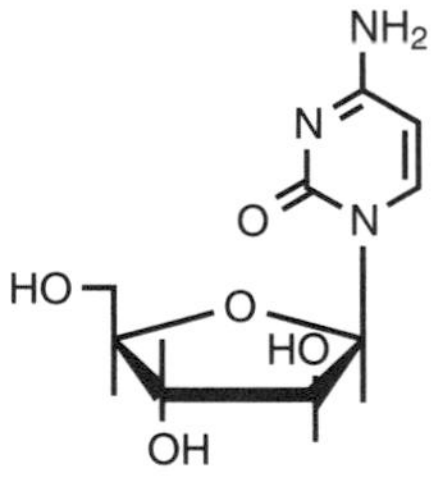

Cytosine arabinoside

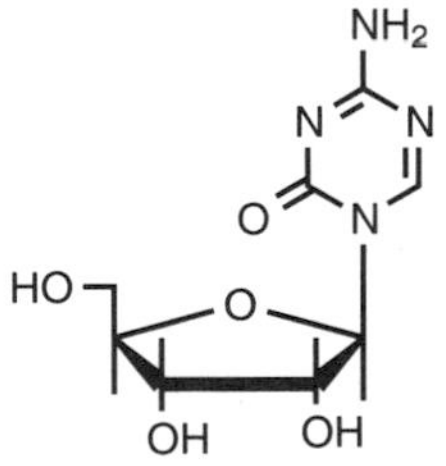

5-Azacytidine

Cytosine arabinoside (cytarabine, Ara-C), an isomer of cytidine, is only slightly incorporated into DNA. The principal action of this drug, which follows its conversion to the triphosphate by healthy cells, appears to be the inhibition of DNA polymerase; it is a specific inhibitor of the S-phase of the cell cycle.[168] Cytarabine is valued for its ability to induce remission of acute leukemia in children and adults. 5-Azacytidine (azactidine), the most active of the azapyrimidines, shows substantial activity in the treatment of human leukemias but little activity against solid tumors. It inhibits methyltransferases. It is phosphorylated in vivo to the monophosphates, diphosphates, and triphosphates, is incorporated into DNA, and affects gene expression; the synthesis of mRNA is inhibited. It is also incorporated into t-RNA, which decreases its amino acid acceptor activity, further suppressing protein synthesis.[169]

Another fruitful target for the design of antiviral nucleoside analogs is the viral DNA polymerases, which has led to the discovery of several clinically useful antiviral drugs, of which 3′-azido-3′-deoxythymidine (AZT)[170–172] and 2′,3′-dideoxycyctidine (DDC)[173,174] are of particular interest because of their strong anti-human immunodeficiency virus (HIV) activity. AZT is converted to its triphosphates, which inhibits reverse transcriptase competitively. In addition, AZT is incorporated into the growing DNA chains, which results in the termination of DNA synthesis. AZT inhibits HIV reverse transcriptase with an IC_{50} of 40 nM but is 100 to 300 times less active against mammalian DNA polymerases.[175] AZT is currently used clinically to treat acquired immunodeficiency syndrome (AIDS). Dideoxycytidine is also efficiently converted to its triphosphate form in human cells and is one of the most potent anti-HIV compounds. It has been approved for clinical use.

Some other pyrimidines that have shown promising biologic activity are 5-iodo-5′-amino-2′,5′-dideoxyuridine (aminoidoxuridine)[176] and 5-fluorocytosine (flucytosine).[177] Aminoidoxuridine displays good antiherpes activity; it is phosphorylated specifically by virus-induced thymine kinase, and thus it is potentiated only by virus-infected cells. Flucytosine is an orally active antifungal agent. It is effective in the treatment of candidiasis and cryptococcosis. In sensitive fungi, it is metabolized to 5-fluorodeoxyuridine monophosphates, which blocks thymidylate synthetase.[178]

Purine Analogs

Hitchings and associates[179] began their pioneering studies on analogs of purines as antimetabolites in the early 1940s. The first analogs found to have marked anticancer activity in experimental tumors were 8-azaguanine

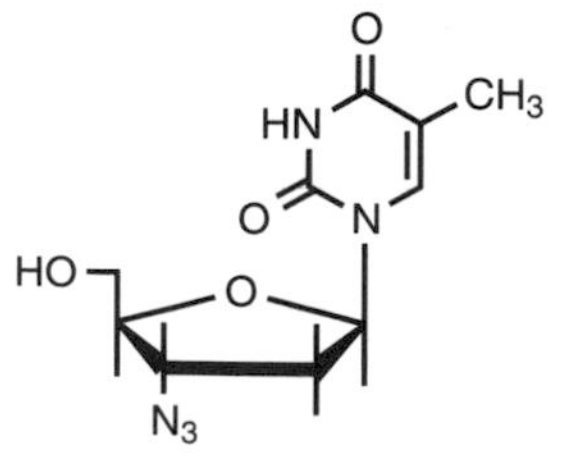

3'-Azido-3'-deoxythymidine (AZT)

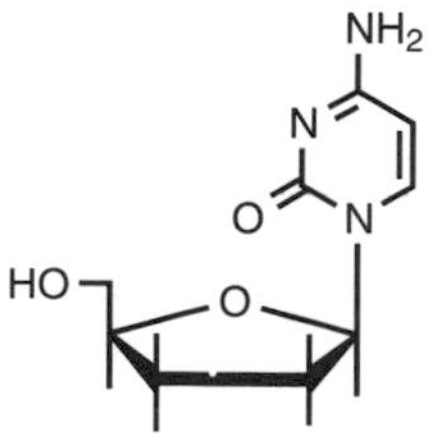

2',3'-Dideoxycytidine

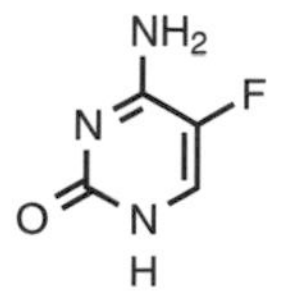

5-Fluorocytidine

and 2,6-diaminopurine, although these compounds did not prove useful clinically. The discovery of 6-mercaptopurine, which had a high order of activity against human leukemias, soon followed[179]; this discovery was an important landmark in the development of antineoplastic and immunosuppressive agents. Most of the 6-mercaptopurine is converted in the cell into 6-thioinosine 5′-phosphate,[180] which inhibits the conversion of inosine 5′-phosphate to AMP, thus bringing neogenesis of purines to a halt.[181] It also exerts feedback inhibition of the biosynthesis of phosphoribosylamine, which is involved in the early steps of purine biosynthesis.[182]

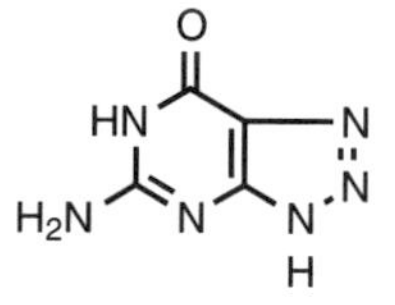

8-Azaguanine

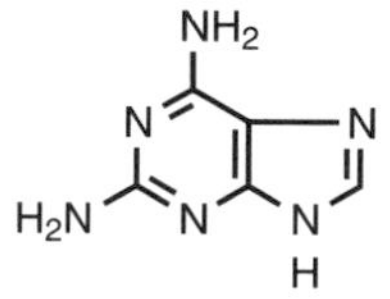

2,6-Diaminopurine

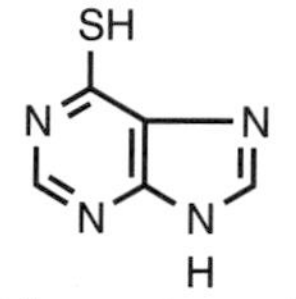

6-Mercaptopurine

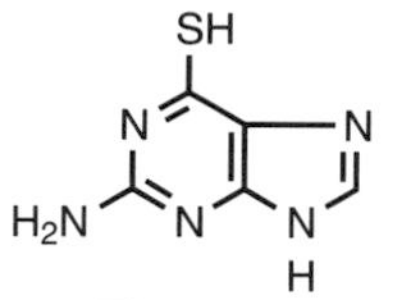

6-Thioguanine

Many analogs have since been prepared in which the purine ring, the substituents on the ring, or the sugar residue has been modified, and some of them were found to possess clinically useful activities. It has been shown that, as in the case of pyrimidines, most of the antitumor purine analogs also are not active as such but undergo transformation to ribonucleotides, which are the active species.

The 2-amino derivative of 6-mercaptopurine, thioguanine, is used to a certain extent in cancer therapy and as an immunosuppressant. Its S-(1-methyl-4-nitroimidazol-5-yl) derivative, azathioprine, is often used to prevent rejection of organ grafts, particularly those of the kidney.[183]

9-β-D-Arabinosyladenine (vidarabine, Ara-A) is a powerful antiviral agent with a high therapeutic index. Its most successful application has been in herpesvirus encephalitis and herpetic keratitis of the eye.[184,185] It is converted in vivo to the triphosphate, which inhibits DNA polymerase more strongly in the virus than in the host cells. Vidarabine is readily deaminated in the body by adenosine deaminase, and this is a major limitation in its clinical use. It has been found that simultaneous administration of erythro-9-(2-hydroxynon-3-yl) adenine (EHNA), a strong inhibitor of adenosine deaminase, which has no antiviral activity of its own, improves the therapeutic response to vidarabine. Deoxycoformycin, another strong inhibitor of adenosine deaminase, has been found useful in increasing the half-life of vidarabine and improving its therapeutic index.[186] EHNA has a selective action against red blood cell deaminase but has no activity against parasite deaminases.[187,188]

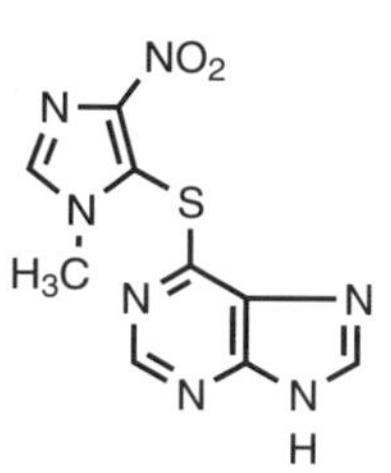

Azathioprine

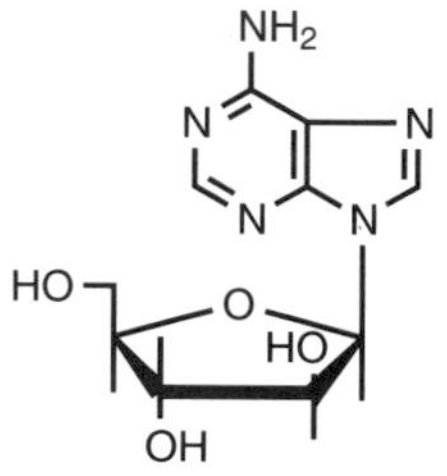

Vidarabine

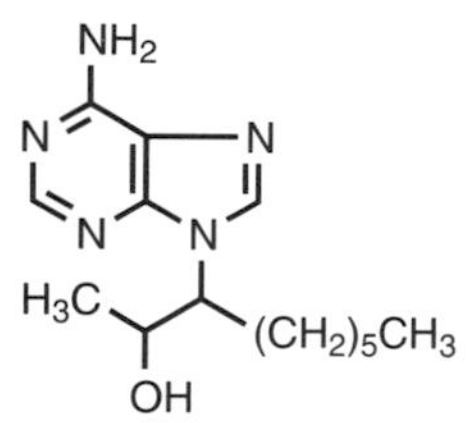

9-(2-Hydroxynon-3-yl)adenine

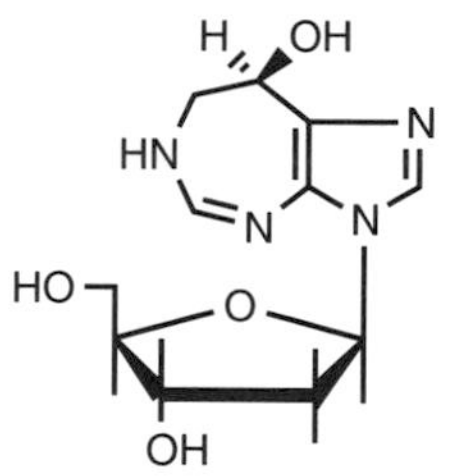

Deoxycoformycin

Another important antiviral compound discovered is 9-(2-hydroxyethoxymethyl)guanine (acyclovir), which has a high therapeutic index. It is converted to a monophosphate by a virus-specific kinase and then to a triphosphate, which inhibits DNA synthesis in the infected cells. The phosphorylation does not take place in healthy cells, which are thereby spared, and this explains its strong selectivity of action. Acyclovir is about 3000 times as toxic to herpes simplex virus as to mammalian red blood cells.[189] Clinically, it has been found effective against all types of herpes infection but not against other species of viruses.[190]

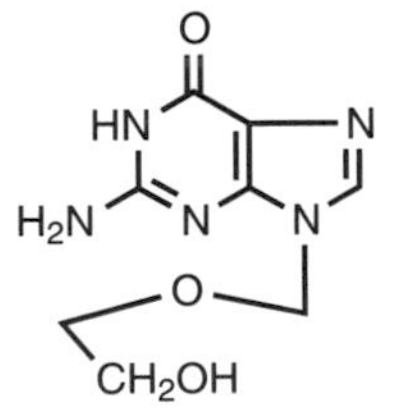

Acyclovir

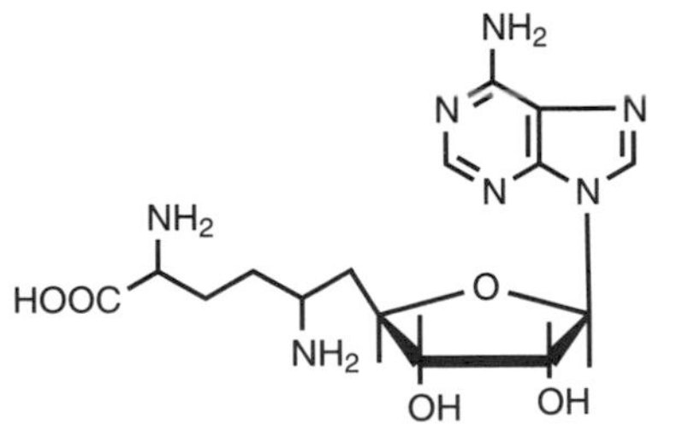

Sinefungin

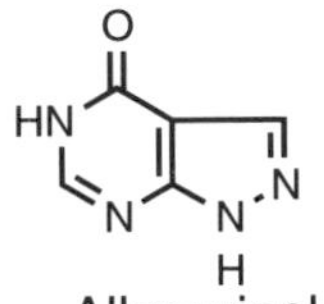

Allopurinol

A universal inhibitor of methyltransferases is S′-adenosylhomocysteine (SAH). Synthetic analogs of SAH include 5′-S-isobutylthioadenosine (SIBA) and sinefungin. SIBA has interesting antiviral and antimalarial activities.[191,192] Sinefungin has strong antifungal and antiprotozoal activities,[192–194] but it is rather toxic and cannot be used clinically. It is, however, a useful lead for further exploration.

Although originally synthesized as an antineoplastic agent, 4-oxopyrazole(3,4,-d)pyrimidine (allopurinol) was found to lack antimetabolite activity but proved an inhibitor of xanthine oxidase. Allopurinol thus delayed inactivation of 6-mercaptopurine by xanthine oxidase and also reduced the plasma concentration and renal excretion of uric acid. In clinical trials, it proved useful for the treatment of gout. It was soon found that allopurinol is both an inhibitor of the enzyme and a substrate and is converted to alloxanthine and that both allopurinol and alloxanthine are inhibitors of xanthine oxidase (Fig. 31–4). At low concentrations, allopurinol is a substrate and competitive inhibitor of the enzyme, and at high concentrations, it is a noncompetitive inhibitor; xanthine, the metabolite formed by xanthine oxidase, is a noncompetitive inhibitor of the enzyme. Inhibition of the penultimate and ultimate steps in uric acid biosynthesis reduces the plasma concentration and urinary excretion of uric acid and increases the plasma concentration and renal excretion of its more soluble oxypurine precursors, thus relieving the symptoms of gout.[195]

Antisense Oligonucleotides

Transcription of genetic information from genomic DNA by RNA polymerases to produce messenger RNA (mRNA), ribosomal RNA, or transfer RNA (tRNA) is the central process in cell metabolism, which is followed by translation of the information encoded in the RNA into protein synthesis (Fig. 31–5). The information in the nucleic acids is encoded in their base sequence and conveyed through complementary base pairing (hybridization) to the new nucleic acid strand being synthesized, whether of DNA or RNA. It has been shown that oligonucleotides (ON), which are complementary to the base sequence present on the genetic transcription element or mRNA and hybridize with these, are able to arrest the process of transcription or translation and are classified as antisense oligonucleotides (ASON). This can be achieved in the following ways: (1) Synthetic oligodeoxynucleotides can be employed, which are complementary to either the transcribing strand of DNA duplex or mRNA; (2) The genetic information for antisense RNA is cloned into the genome (antisense gene) by using a vector, which is expressed as antisense mRNA, which is capable of hybridizing with the normal mRNA.

This hybridization could be by Hoogsteen base pairing directed at the double-stranded DNA or by Watson-Crick base pairing with the single-stranded DNA or mRNA. Both of these models of intervention by oligonucleotides have the potential to regulate gene expression and their possible use in modulating a variety of human and plant diseases as well as expression of new characteristics in plants.

Because the design of oligomers is based on the uniqueness of the base sequence, the drugs have the potential for a great deal of specificity and selectivity. A num-

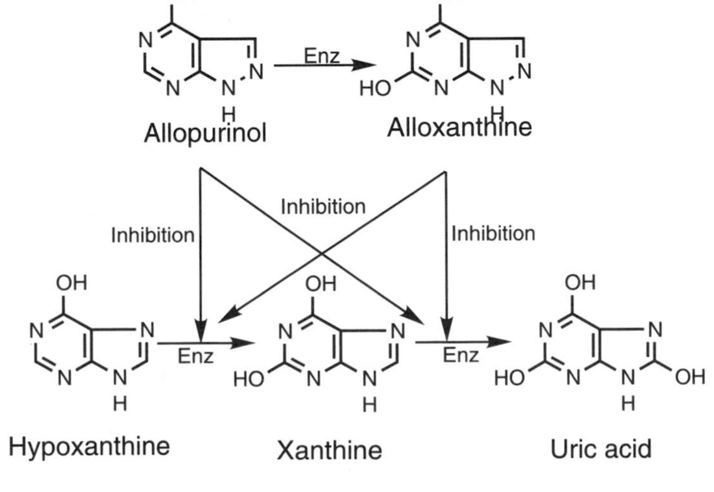

Allopurinol — Enz → Alloxanthine

Inhibition / Inhibition / Inhibition

Hypoxanthine — Enz → Xanthine — Enz → Uric acid

Fig. 31–4. Allopurinol inhibition of uric acid formation. Enz, xanthine oxidase.

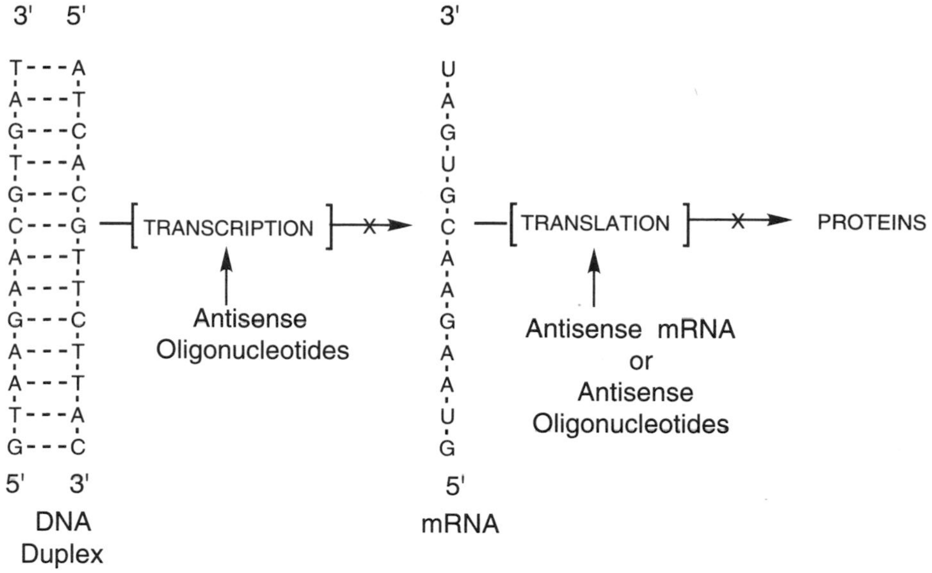

Fig. 31–5. Schematic presentation of the process of genetic transcription and translocation and sites of intervention by antisense oligonucleotides.

ber of good reviews have appeared highlighting the various aspects of the field.[196–201]

Some of the essential requirements for oligonucleotides to have effective antisense function are (1) length of the 15 to 30 mers to give them uniqueness and selectivity; (2) stability to nucleases, which is achieved by introduction of different residues at the terminal positions; (3) composition with greater than 50% CG content for good hybridization with Tm greater than 37°; (4) sequence should have 5′-initiation codon as target site; and (5) new delivery forms. Different approaches have been described to meet these requirements. Some success has already been achieved in both prokaryotes and eukaryotes against viruses[202] and in developing plants with advantageous characteristics.[203]

DOPA DECARBOXYLASE INHIBITORS

After phenylalanine is hydroxylated enzymatically to 3,4-dihydroxyphenylalanine (dopa), decarboxylation, catalyzed by the enzyme dopa decarboxylase, takes place to produce the biogenic amine dopamine, which is metabolized further to form epinephrine and norepinephrine. In view of the important physiologic roles of these amines, dopa decarboxylase inhibitors have been the subject of much interest. *α*-Methyldopa was found to be an inhibitor of this enzyme and an effective antihypertensive agent.[204,205] Detailed studies, however, showed that its antihypertensive effect could not be explained merely by the inhibition of dopa decarboxylase in the peripheral nerves. It was found that *α*-methyldopa itself was metabolized to *α*-methylnorepinephrine, which can be stored in sympathetic nerve endings; it was hypothesized that the latter displaced norepinephrine and acted as a "false transmitter."[206] Even this could not explain all the facts about the antihypertensive effect of *α*-methyldopa. It is currently accepted that the major antihypertensive action of *α*-methyldopa is on the central nervous system, and that the effect is due more to a metabolic product of *α*-methyldopa than to enzyme inhibition; *α*-methylnorepinephrine is probably the active agent.[207] The antihypertensive effect of *α*-methyldopa provides an example of the complexities of biologic systems and the difficulty in mechanism identification.

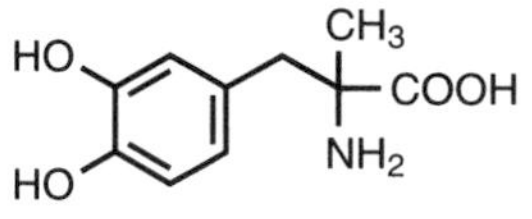

α-Methyldopa 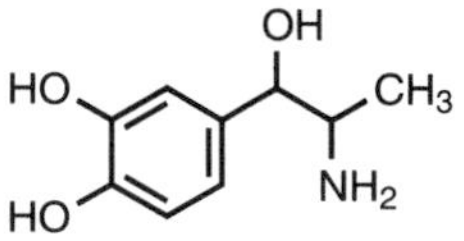 α-Methylnorepinephrine

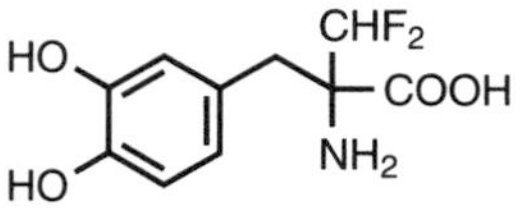

α-Difluoromethyldopa 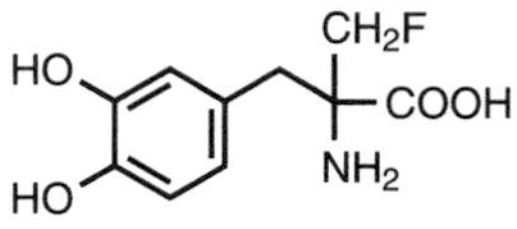 α-Fluoromethyldopa

α-Difluoromethyldopa and *α*-fluoromethyldopa have been studied as mechanism-based inhibitors of dopa decarboxylase. The former compound has a selective peripheral action and increases central levels of dopa, dopamine, and norepinephrine[208]; in contrast, the latter is a more potent inactivator, inhibits both peripheral and central activities, depletes central catecholamine levels, and has antihypertensive properties.[209,210]

DOPAMINE *β*-HYDROXYLASE INHIBITORS

Dopamine *β*-hydroxylase converts dopamine to norepinephrine and is a copper-containing enzyme that depends on oxygen and ascorbate. It is inactivated by *p*-hydroxybenzyl cyanide, which thereby undergoes conversion to *p*-hydroxymandelonitrile. It has been suggested that the inactivation may proceed through an enzymatically induced prototropic shift, leading to the ketenimine, which could alkylate an enzyme nucleophile.[211]

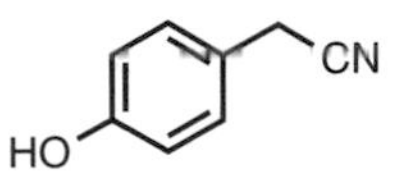

ρ-Hydroxybenzyl cyanide

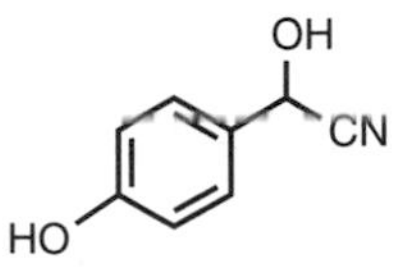

ρ-Hydroxymandelonitrile

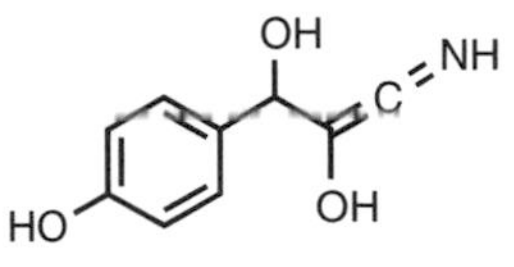

Ketenimine intermediate

GABA TRANSAMINASE INHIBITORS

GABA transaminase is responsible for the destruction of γ-aminobutyric acid (GABA), an inhibitory neurotransmitter in the central nervous system of mammals, and is thus a potential target for anticonvulsant drug action. Mechanism-based irreversible inhibitors of GABA transaminase have been developed, some of which show clinical promise. γ-Vinyl-GABA has been found to increase GABA levels in the cerebrospinal fluid in a dose-dependent manner and has promise for clinical use in several neurologic disorders.[212] Trans-2-fluoro- and 3-chloro-4-aminocrotonic acids[213,214] and 4-amino-5-fluoropentanoic acid[215] have also been reported to be mechanism-based inactivators of GABA transaminase. The proposed mechanism involves enzymatic proton abstraction, β-elimination of halogen, and generation of a reactive alkylating species (Fig. 31–6).

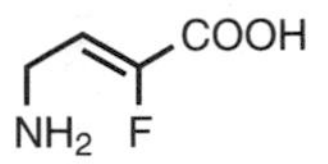

trans-2-Fluoro-4-aminocrotonic acid

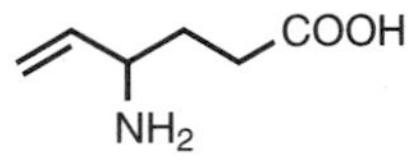

γ-Vinyl-GABA

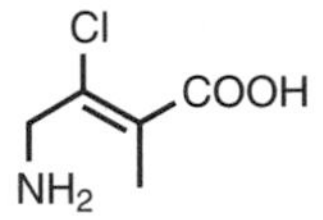

trans-3-Chloro-4-aminocrotonic acid

β-LACTAMASE INHIBITORS

β-Lactamase production is responsible for the development of resistance to β-lactam antibiotics, so its inhibitors are of great clinical importance. It is a chromosome and plasmid-encoded enzyme. Several mechanism-based inhibitors of this enzyme have been synthesized, and clavulanic acid[216] and a penicillanic sulfone, sulbactam,[217] are among the best studied of these inhibitors (see Chapter 33). It has been proposed that an initial acyl enzyme complex fragments by a β-elimination process wherein an enolate and sulfinate are the leaving groups. The remaining acyl bond is stabilized to hydrolysis by the unsaturation of the C_5—C_6 bond.[218,219] A number of carbenapenams, such as olivanic acid, also cause mechanism-based inactivation of β-lactamase.[220] The proposed mechanisms of inactivation have been discussed at length in the reviews by Rando and Walsh.[27,28]

Both clavulanic acid and sulbactam have been found to show synergism with standard penicillins such as ampicillin and amoxicillin against penicillin-resistant organisms; this observation is of clinical significance.[221]

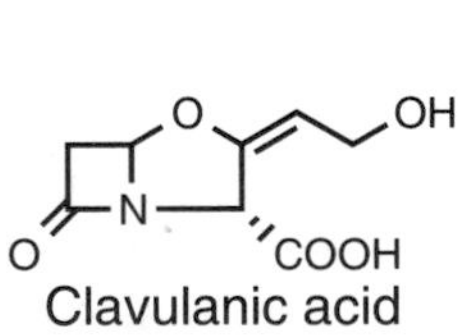

Clavulanic acid

Sulbactam

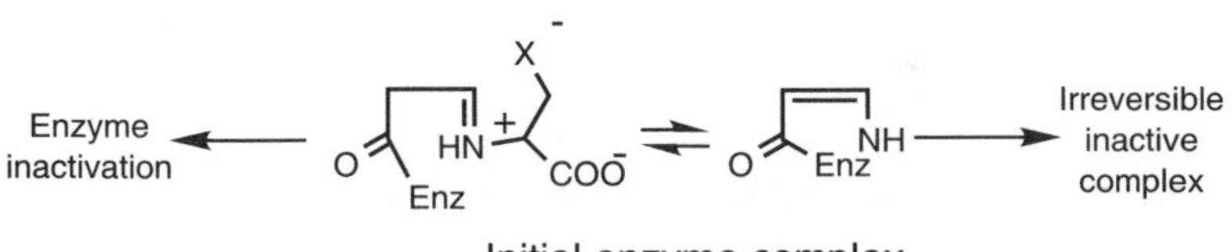

Enzyme inactivation ← Initial enzyme complex ⇌ → Irreversible inactive complex

Initial enzyme complex

β-LACTAM ANTIBIOTICS

The first step in bacterial cell wall synthesis involves the cross-linking of peptidoglycan chains. This process is regulated by various transpeptidases and carboxypeptidases, which catalyze the cleavage of the terminal D-alanyl-D-alanine bond of the glycopeptides. These enzymes are the likely sites of action of β-lactam antibiotics, which act by inhibition of bacterial cell wall synthesis[222] and are considered transition-state or mechanism-based inactivators of the enzymes.

After it was observed that penicillins cause an accumu-

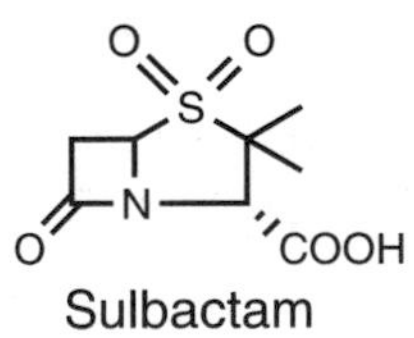

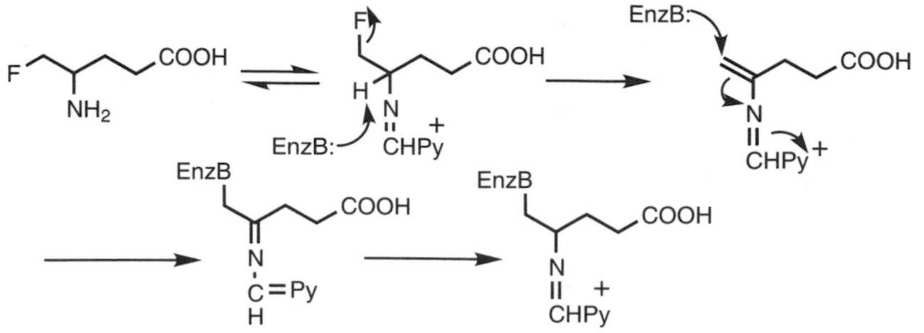

Fig. 31–6. Proposed mechanism of inactivation of GABA transaminase by 4-amino-5-fluoropentenoic acid.

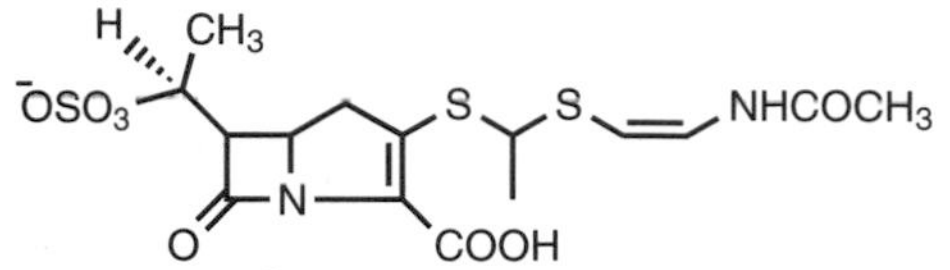

Olivanic acid

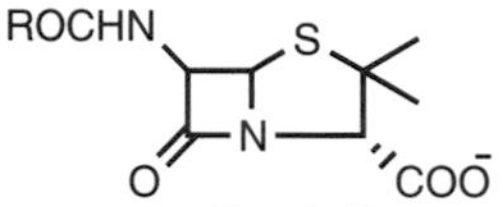

Penicillins

lation of muramyl peptides in the bacterial culture medium and that the basic lesion is produced in the cell wall, it was suggested by Tipper and Strominger[223] and by Lee[224] that the nonplanar amide bond of the penicillins could resemble a possible transition state structure involved in the cleavage of the D-alanyl-D-alanine peptide bond. Many β-lactam antibiotics have been shown to acylate functional groups in the active site of peptidases, and the extent of their activity can be correlated with the calculated reactivity of the β-lactam ring.[224] The β-lactam antibiotics thus combine to some extent the virtues of both transition-state analogs and mechanism-based inhibitors.

RENIN-ANGIOTENSIN INHIBITORS

The renin-angiotensin system is a proteolytic cascade that is found in the circulation system and plays an important role in the regulation of blood pressure and electrolyte balance. Renin is a highly specific enzyme that converts angiotensinogen to angiotensin I; the latter is then converted by angiotensin-converting enzyme (ACE) to the vasopressor octapeptide angiotensin II, which binds with the receptor for maintenance of blood pressure. All these steps have been the subjects of much current interest with the intent of developing antihypertensive agents,[225] although to date only ACE inhibitors (ACE I) have proven efficacious; in the treatment of hypertension and heart failure.

ACE also inactivates the powerful vasodepressor, bradykinin. This enzyme is, therefore, a prime target for antihypertensive drug design. Small, orally active peptides have been synthesized as inhibitors of this enzyme, and it has been proposed that these act as transition state analogs. Their development has been based on extensive studies of related peptidases. It was found that (R)-benzylsuccinic acid, whose structure was based on the corresponding dipeptide substrate by substitution of a methylene group for the imino function of the scissile peptide bond, is a powerful inhibitor of carboxypeptidase A.[226] A follow-up of this rationale and consideration of the structure-activity relationship of known ACE inhibitors led to the development of the clinically useful antihypertensive agent captopril[227] and then to enalapril[228]; the latter is claimed to have a transition-state geometry in the scissile bond region through the $CHCO_2R$ and NH groups. Many other potent ACE inhibitors have been developed, and their structure-activity relationship has been reviewed.[229]

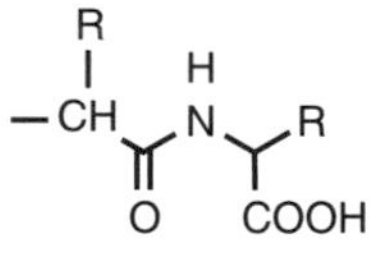

Substrate

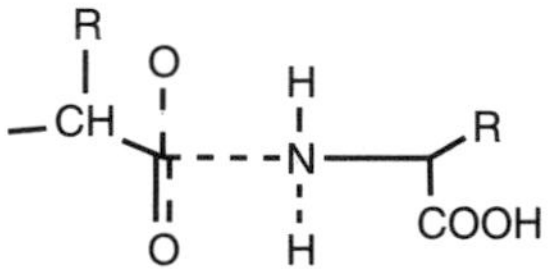

Transition State

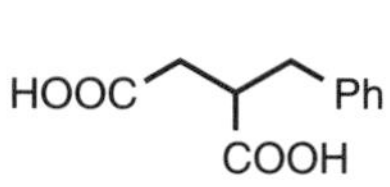

(R)-Benzylsuccinic acid

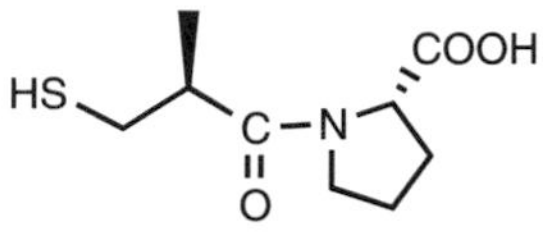

Captopril

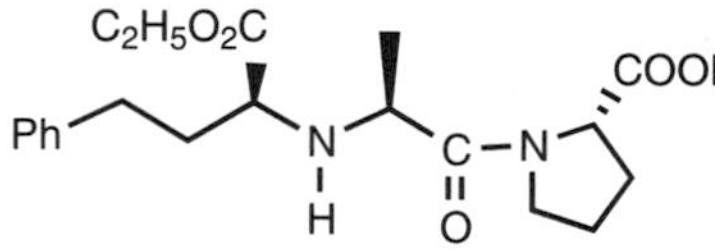

Enalapril

ORNITHINE DECARBOXYLASE INHIBITORS

Ornithine decarboxylase is the rate-limiting enzyme in polyamine biosynthesis, and its inhibition results in depletion of putrescine, spermidine, and other polyamines. In view of the importance of polyamines in cellular proliferative and differentiation processes, these inhibitors have a potential for application of antigestational, antiparasitic, antimicrobial, antitumor, and antihyperproliferative skin disease agents.[230] γ-Methylornithine,[231] α-hydrazino-δ-aminovaleric acid[232] and β, γ-dehydroornithine[233] are potent competitive inhibitors of ornithine de-

α-Methylornithine 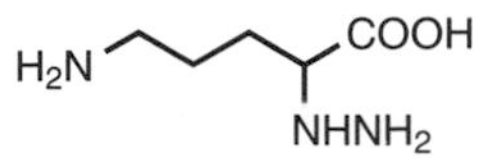 α-Hydrazino-γ-amino-valeric acid

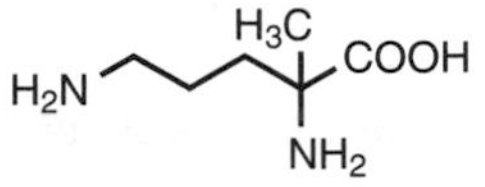

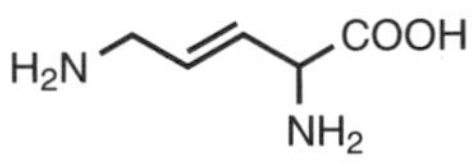

trans-β,γ-Dehydroornithine

carboxylase and inhibit the proliferation or differentiation of several cell lines in cultures. These competitive inhibitors, however, have the disadvantage of increasing the amount of ornithine decarboxylase, probably by slowing down its degradation, which results in an increased production of polyamines after the inhibitor has disappeared.[234]

α-Fluoromethylornithine and *α*-difluoromethylornithine (DFMO)[235,236] and 1,4-diaminohex-5-yne[237] have been shown to be potent irreversible inhibitors of ornithine decarboxylase. These agents are among the most well-studied mechanism-based inhibitors. The proposed mechanism of inhibition involves enzyme-induced decarboxylation of the pyridoxal phosphate–derived Schiff's base and generation of an electrophilic inactivating species that can trap a nucleophile at the active site.[237] DFMO is at present the drug of choice for sleeping sickness, although the dose needed is extremely high, which is a significant limitation.[238]

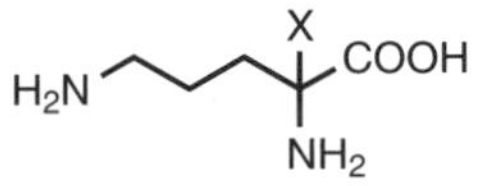

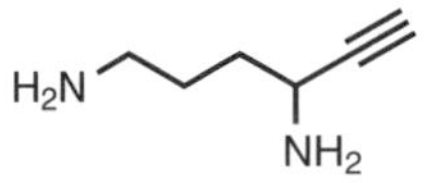

α-Fluoromethylornithine, X = CH_2F 1,4-Diaminohex-5-yne

α-Difluoromethylornithine, X = CHF_2

AROMATASE INHIBITORS

Inactivators of aromatase, a P-450–dependent enzyme that converts testosterone to estradiol by aromatizing ring A, are of considerable interest because estrogen production has been associated with gynecomastia, endometriosis, and breast and endometrial cancers. A wide range of both reversible and irreversible/mechanism-based inhibitors have been reported.[239,240] In addition to those that are substrate analogs, compounds such as 3-(4-aminophenyl)-3-ethylglutarimide have been shown to inhibit aromatase and find clinical use in treating estrogen-dependent breast cancers.[241] The more interesting compounds are the mechanism-based inhibitors. These include analogs with the C_{19} angular methyl group replaced by propynyl, propargyl, and allenyl groups.[241,242] Because androgens are successively oxygenated at the C-19 methyl group, it has been hypothesized that the acetylenic and allenic analogs are similarly processed through the catalytic cycle to generate a reactive electrophilic acetylenic ketone or vinyloxirane species, which inactivates the enzyme by enzyme alkylation (Fig. 31–7).

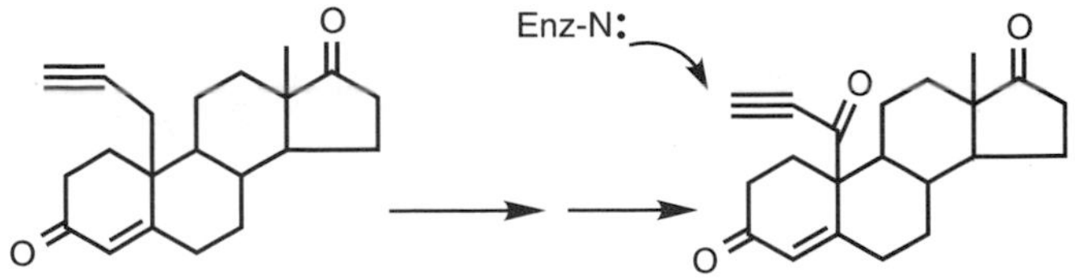

Fig. 31–7. Proposed mechanism of inactivation of aromatase by 19-ethinyl-androstenedione.

Following similar reasoning, Marcotte and Robinson[243] synthesized the 19-difluoromethyl analog and found that it inactivates human placental microsomal aromatase, presumably by hydroxylation and decomposition of the hydroxydifluorotetrahedral adduct to an electrophilic acyl fluoride group. In contrast, the corresponding monofluoroandrostenedione, which on hydroxylation and fluoride elimination changes to the aldehyde oxidation state, is not an inactivator and in fact goes through the catalytic cycle to yield estrone.

PLATELET-ACTIVATING FACTOR ANTAGONISTS

Since its discovery[245] in the early 1970s, platelet-activating factor (PAF, 1-0-alkyl-2-(R)acetyl-*syn*-glyceryl-3-phosphocholine), an endogenous phospholipid mediator produced by a variety of cells, including endothelial cells and leukocytes, has been shown to be involved in a number of pathophysiologic mechanisms, such as inflammation, immediate and delayed hypersensitivity reactions, immune-complex formation, hemostasis, and shock.[245,246] Its cellular and molecular mode of action has been shown to involve inhibition of adenylate cyclase, with a concomitant decrease in cellular cAMP, promotion of arachidonic acid release, phosphoinosinic acid turnover, increased activity of protein kinase C leading to enhanced activity of the Na^+/H^+ antiport, induction of Ca^{2+} fluxes, and an increase in the level of cytoskeletal action in various cellular systems. An important development is the recognition of PAF/cytokine interaction as one of the basic cellular mechanisms in PAF's action. These effects can be antagonized by specific PAF receptor antagonists, and some of these are candidates for clinical management of stroke, myocardial infarction, gastrointestinal ulceration, and different pathologic conditions related to the peripheral circulation. The effects of PAF antagonists on cellular Ca^{2+} metabolism deserves partic-

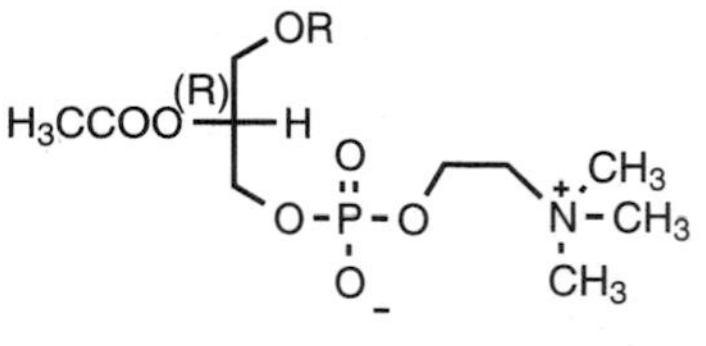

Platelet activating factor (PAF)

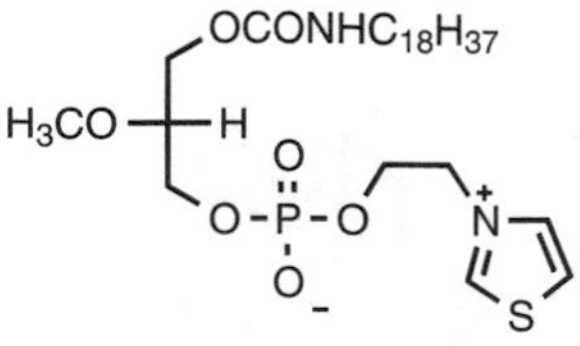

CV 3988

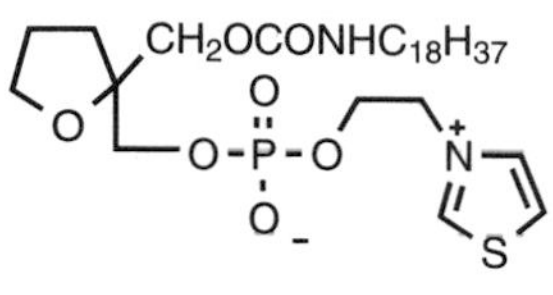

SRI 63-072

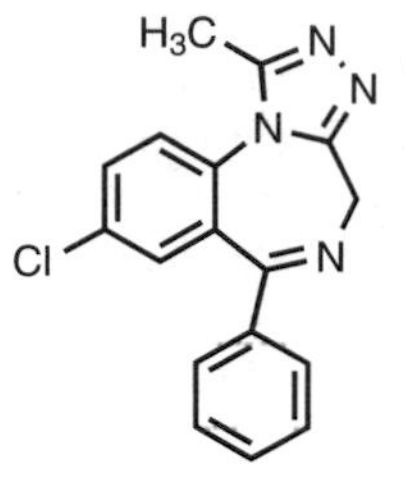

Alprazolam

ular mention and may have an important role in various ischemic states. PAF seems to have an important modulatory role in cell-to-cell interaction, and PAF/cytokine interaction may be a basic mechanism of cellular injury and a possible site for therapeutic application of PAF antagonists.

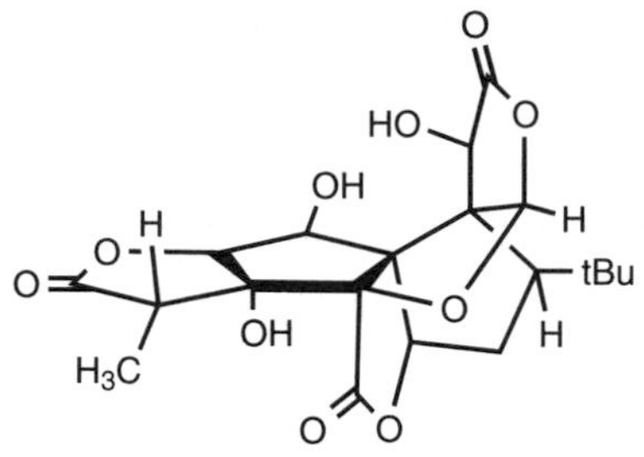

Ginkgolide

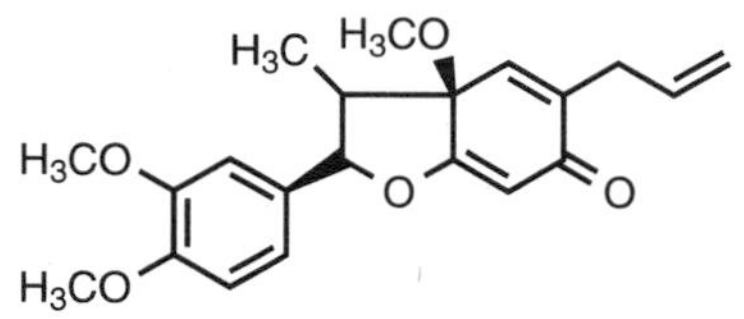

Kadsurenone

PAF antagonists can be broadly grouped as substances of synthetic or natural origin. Their chemistry has been reviewed by Hosford and Brequet.[247] The majority of the synthetic compounds are modified derivatives of the PAF framework, and these include nonconstrained analogs of the general structure related to CV 3988, conformationally restricted congeners such as SAI 63-072, and some unrelated structures that include a number of benzodiazepine-derived drugs such as alprazolam.

A number of natural products, which include ginkgolides from Ginkgo biloba, a Chinese plant, and lignans such as kadsurenone, have been shown to possess highly specific PAF antagonistic activity. The fact that different structural types show PAF antagonist activity points to the possibility of diversity of PAF receptors in various organs. The specificity of these binding sites would no doubt be revealed by the discovery of more specific antagonists, which could result in better understanding of diverse pathophysiologic events in which PAF is involved and more specific approaches to various diseases. At present, PAF antagonists appear to be appropriate for the treatment of various ischemic disorders, especially cerebral ischemia. Ginkgolide B, both in the form of a standardized extract and as a single compound is already marketed in some European countries for this purpose.

PROSPECTS

Although the concept of metabolite antagonism was originally enunciated in relation to the action of antimicrobial agents, the characterization of a large number of enzymes and their substrates, an understanding of metabolic processes of the mammalian system, and more recently of various pharmacologic receptors and their ligands have resulted in widening of the scope of metabolite antagonists. Metabolic antagonism is now considered as a general phenomenon that is valid for a variety of biologic systems or situations. The practical use of this principle, once a metabolite/regulator/ligand is identified, is the synthesis of antagonists using the well-established approaches to drug design. Another important consideration for research in this area is as a system for studying biologic recognition mechanisms.

The initial successful clinical use of metabolite antagonists to microbial infections is now extended to the treatment of parasitic and viral infections, cancers, hypertension and ischemia, and hormonal disorders. With dramatic advances in molecular biology resulting in the characterization of human genes, in the discovery of new generations of receptors and their regulators, and in computer-aided molecular modeling, the prospects of new discoveries in this field are indeed bright.

Acknowledgment

The author would like to express his deep gratitude to Drs. S. Bhattacharji and Naresh Kumar for their help in the preparation of the manuscript.

REFERENCES

1. P. Ehrlich, *Gessammelte arbeiten zur immunitatsforschung,* Les Prix Nobel, Stockholm, 1908.
2. L. Michaelis and P. Bona, *Biochem. Z., 60,* 62(1913).
3. L. Michaelis and H. Pechstein, *Biochem. Z., 60,* 79(1913).
4. J. H. Quastel and W. R. Woolridge, *Biochem. J., 22,* 689(1928).
5. J. H. Quastel and W. R. Woolridge, *Biochem. J., 23,* 115(1929).

6. D. W. Woolley, et al., *Biol. Chem., 124,* 175(1938).
7. H. M. Dyer, *J. Biol. Chem., 124,* 519(1938).
8. A. Blankart, *Festschrift,* Basel, Barell, 1936, p. 284.
9. J. Raventos, *J. Physiol., 88,* 5(1936).
10. H. R. Ing and W. M. Wright, *Proc. R. Soc.,* Series B., *109,* 337(1931).
11. A. J. Clark, Qualitative aspects of drug antagonism, in *Handbunch der Experimentellen Pharmakologie,* IV, Berlin, Springer, 1937, p. 190.
12. G. Domagk, *Dtsch. Med. Wochenschr., 61,* 250(1935).
13. D. D. Woods, *Br. J. Exp. Pathol., 21,* 74(1940).
14. P. Fildes, *Lancet, 1,* 955(1940).
15. C. L. Markert and F. Mollir, *Proc. Natl. Acad. Sci., U.S.A., 45,* 753(1959).
16. J. J. Burchall, *J. Antimicrob. Chemother., 5,*(Suppl. B.), 3(1979).
17. J. J. Burchall and G. H. Hitchings, *Mol. Pharmacol., 1,* 126(1965).
18. R. Peters, et al., *Nature, 171,* 111(1953).
19. C. Liebecq and R. A Peters, *Biochem. Biophys. Acta, 3,* 215(1949).
20. B. R. Baker, et al., *J. Med. Pharm. Chem., 2,* 633(1960).
21. M. Dixon and E. C. Webb, *Enzymes,* New York, Academic Press, 1958, pp. 376–386.
22. E. D. Adrian, et al., *Br. J. Pharmacol., 2,* 56(1947).
23. J. F. Mackworth and E. C. Webb, *Biochem. J., 42,* 91(1948).
24. A. R. Main, *Science, 144,* 992(1964).
25. B. R. Baker, *Design of Active-Site-Directed Irreversible Enzyme Inhibitors,* New York, John Wiley & Sons, 1967, pp. 1–3, 17–22.
26. G. M. Helmkamp, et al., *J. Biol. Chem., 243,* 3229(1968).
27. R. R. Rando, *Pharmacol. Rev., 36,* 111(1984).
28. C. T. Walsh, *Annu. Rev. Biochem., 53,* 493(1984).
29. R. B. Silverman and S. J. Hoffman, *Med. Res. Rev., 4,* 415(1984).
30. L. Pauling, *Chem. Eng. News, 24,* 1375(1946).
31. G. E. Lienhard, *Annu. Rev. Med. Chem., 7,* 249(1971).
32. R. Wolfenden, *Nature, 223,* 704(1969).
33. R. Wolfenden, *Acct. Chem., Res., 5,* 10(1972).
34. R. Wolfenden, *Annu. Rev. Bioplhy. Bioeng., 5,* 271(1976).
35. R. Wolfenden, in *Transition States of Biochemical Processes.* R. D. Gandour and R. L. Showen, Eds., New York, Plenum Press, 1978, pp. 555–578.
36. P. R. Andrews and D. A. Winkler, *The design and medicinal applications of transition state analogues,* in *Drug Design: Fact or Fantasy?* G. Jolles and K. R. H. Wooldridge, Eds., New York, Academic Press, 1984, pp. 145–172.
37. S. Cha, et al., *Biochem. Pharmacol., 24,* 2187(1975).
38. J. J. Pfiffner, et al., *Science, 97,* 404(1943).
39. J. J. Pfiffner, et al., *J. Am. Chem. Soc., 69,* 1476(1947).
40. R. B. Angier, et al., *Science, 103,* 667(1946).
41. G. M. Brown, *Adv. Enzymol., 35,* 35(1971).
42. J. Trefouel, et al., *C. R. Soc. Biol., 120,* 756(1935).
43. A. T. Fuller, *Lancet, 2,* 194(1937).
44. E. H. Northey, *The Sulfonamides and Allied Compounds,* Amer. Chem. Soc. Monograph Series, New York, Reinhold, 1948.
45. N. Anand, Sulfonamides and sulfones, in *Burger's Medicinal Chemistry: Part II,* 4th Ed., M. E. Wolff, Ed., New York, Wiley & Sons, 1979, p. 1.
46. P. H. Bell and R. O. Roblin, Jr., *J. Am. Chem. Soc., 64,* 2905(1942).
47. P. B. Cowles, *Yale. J. Biol. Med., 14,* 599(1942).
48. A. H. Brueckner, *Yale J. Biol. Med., 15,* 813(1943).
49. K. C. Blanchard, *J. Biol. Chem., 140,* 919(1941).
50. H. McIlwain, *Br. J. Exp. Pathol., 23,* 265(1942).
51. S. R. Rubbo and J. M. Gillepsie, *Nature, 146,* 838(1940).
52. E. A. Bliss and P. H. Long, *Bull. Johns Hopkins Hosp., 69,* 14(1941).
53. E. E. Snell and H. K. Mitchell, *Arch. Biochem., 1,* 93(1943).
54. M. R. Stetten and C. L. Fox, *J. Biol. Chem., 161,* 333(1945).
55. W. Shive, et al., *J. Am. Chem. Soc., 69,* 725(1947).
56. J. S. Gots, *Nature, 172,* 256(1953).
57. R. Tschesche, *Z. Natureforsch, (C), 2b,* 10(1947).
58. C. M. Brown, *J. Biol. Chem., 237,* 536(1962).
59. C. M. Brown, et al., *J. Biol. Chem., 236,* 2534(1961).
60. R. D. Hotchkiss and A. H. Evans, *Fed. Proc., 19,* 912(1960).
61. R. A. Weisman and G. M. Brown, *J. Biol. Chem., 239,* 326(1964).
62. L. Bock, et al., *J. Med. Chem., 17,* 23(1974).
63. S. Roland, et al., *J. Biol. Chem., 254,* 10337(1960).
64. R. Donovick, et al., *Am. Rev. Tuberc. Pulm. Dis., 66,* 219(1952).
65. J. Maier and E. Riley, *Proc. Soc. Exp. Biol. Med., 50,* 152(1942).
66. I. Moriguchi and S. Wada, *Chem. Pharm. Bull* (Tokyo), *16,* 734(1968).
67. E. Shefter, et al., *J. Pharm. Sci., 61,* 872(1972).
68. M. Landy, et al., *Science, 97,* 265(1943).
69. P. J. White and D. S. Woods, *J. Gen. Microbiol., 40,* 243(1965).
70. T. Akiba and T. Yokota, *Med. Biol., 63,* 55(1962).
71. M. L. Pato and G. M. Brown, *Arch. Biochem. Biophys., 103,* 443(1963).
72. N. Nagate, et al., *Microbiol. Immunol., 22,* 367(1978).
73. B. Wolf and R. D. Hotchkiss, *Biochemistry, 2,* 145(1963).
74. R. I. Ho, et al., *Antimicrob. Agents Chemother., 5,* 388(1974).
75. E. M. Wise, Jr., and M. M. Abou-Donia, *Proc. Natl. Acad. Sci., U.S.A., 72,* 2621(1975).
76. O. Skold, *Antimicrob. Agents Chemother., 9,* 49(1976).
77. T. Watanabe, *Bacteriol. Rev., 27,* 87(1963).
78. S. Mitsuhashi, et al., *Antimicrob. Agents Chemother., 12,* 418(1977).
79. A. Bishop, *Parasitology, 53,* 10(1963).
80. D. R. Seeger, et al., *J. Am. Chem. Soc., 69,* 2567(1947).
81. D. R. Seeger, et al., *J. Am. Chem. Soc., 71,* 1753(1949).
82. L. J. Daniel and L. C. Norris, *J. Biol. Chem., 170,* 747(1947).
83. L. J. Daniel, et al., *J. Biol. Chem., 169,* 689(1947).
84. M. F. Mallette, et al., *J. Am. Chem. Soc., 69,* 1814(1947).
85. S. F. Zakrzewski and C. A. Nichol, *Biochim. Biophys. Acta, 27,* 425(1958).
86. M. J. Osborn, et al., *Proc. Soc. Exp. Biol. Med., 97,* 429(1958).
87. W. C. Werkheiser, *J. Biol. Chem., 236,* 888(1961).
88. R. Blakeluy, *The Biochemistry of Folic Acid and Related Pteridines,* New York, John Wiley, 1969.
89. G. H. Hitchings, et al., *J. Biol. Chem., 174,* 765(1948).
90. G. H. Hitchings, and S. R. M. Busbby, 5-Denzyl-2,4-diaminopyrimidines, a new class of systemic antibacterial agents, in *Vth International Congress of Biochemistry,* N. M. Sissakian, Ed., 1961, pp. 165–171.
91. F. H. S. Curd, et al., *Ann. Trop. Med., 39,* 208(1945).
92. A. F. Crowther and A. A. Levi, *Br. J. Pharmacol. Chemother., 8,* 93(1953).
93. G. H. Hitchings and S. L. Smith, Dihydrofolate reductases as targets for inhibitors, in *Advances in Enzyme Regulation,* Vol. 18, G. Weber, Ed., Oxford, Pergamon, 1980, pp. 349–371.
94. D. I. Baker, et al., *FEBS Lett., 126,* 49(1981).
95. P. J. Cayley, et al., *Biochemistry, 18,* 3886(1979).
96. P. J. O'Dwyer, et al., *Invest. New Drugs, 3,* 71(1985).
97. N. Clendininn, et al., *Invest. New Drugs, 5,* 131(1987).
98. E. M. Berman, and L. M. Werled, *J. Med. Chem., 34,* 479(1991).
99. G. H. Hitchings, Ed., *Inhibition of Folate Metabolism in Chemo-*

therapy. The Origins and Uses of Co-trimoxazole, Berlin, Springer-Verlag, 1983.
100. H. J. Mortin and J. F. Morgan, *J. Biol. Chem., 234,* 2698(1959).
101. D. E. Atkinson, et al., *Arch. Biochem. Biophys., 31,* 205(1951).
102. H. J. Morton and J. F. Morgan, *Can. J. Biochem. Physiol., 39,* 925(1961).
103. Du. V. Vincent, et al., *J. Biol. Chem., 59,* 385(1945).
104. K. Dittmer, *J. Am. Chem. Soc., 71,* 1205(1949).
105. G. R. Garst, et al., *J. Biol. Chem., 180,* 1013(1949).
106. W. F. Drea, *J. Bacteriol., 58,* 257(1948).
107. H. Pope, *J. Bacteriol., 58,* 223(1949).
108. P. R. Pal, et al., *J. Am. Chem. Soc., 78,* 5116(1948).
109. N. Kaufman, et al., *Fed. Proc., 20,* 2(1961).
110. H. K. Mitchell and C. Niemann, *J. Am. Chem. Soc., 69,* 1232(1947).
111. P. D. Boyer, et al., *J. Pharmacol. Exp. Ther., 73,* 176(1941).
112. C. Niemann and M. M. Rapport, *J. Am. Chem. Soc., 68,* 1671(1946).
113. K. Niedner, *Z. Krebsforsch., 51,* 159(1941).
114. W. G. Gordon and R. W. Jackson, *J. Biol. Chem., 110,* 151(1935).
115. T. F. Anderson, *Science, 101,* 565(1945).
116. J. H. Marshall and D. D. Woods, *Biochem. J., 51,* ii(1952).
117. P. Files and H. N. Rydon, *Br. J. Exp. Pathol., 28,* 211(1947).
118. E. D. Bergmann, *Koninkl. Ned. Akad. Wetenschap. Proc., 57C,* 108(1954).
119. E. D. Bergmann, et al., *Bull. Res. Council, Israel, 2,* 308(1952).
120. H. S. Moyed and M. Friedman, *Science, 129,* 968(1959).
121. G. W. Kidder and V. C. Dewey, *Biochem. Biophys. Acta, 17,* 288(1955).
122. M. R. Robinson and B. L. Robinson, *J. Am. Chem. Soc., 77,* 456(1955).
123. P. Fildes, *Br. J. Exp. Pathol., 26,* 416(1945).
124. H. I. Kohn and J. S. Harris, *J. Pharmacol. Exp. Ther., 73,* 343(1941).
125. J. H. Alix, *Microbiol. Rev., 46,* 281(1982).
126. R. O. Robin, et al., *J. Am. Chem. Soc., 67,* 290(1948).
127. C. B. Shaffer and F. H. Critchfield, *J. Biol. Chem., 174,* 489(1984).
128. R. L. Thompson, *J. Immunol., 55,* 345(1947).
129. H. Waelsch, et al., *J. Biol. Chem., 166,* 273(1946).
130. E. Borek and H. Waelsch, *J. Biol. Chem., 177,* 135(1949).
131. P. Ayangar and E. Roberts, *Proc. Soc. Exp. Biol. Med., 79,* 476(1952).
132. J. A. Roper and H. McIlwain, *Biochem. J., 42,* 485(48).
133. H. McIlwain, et al., *Biochem. J., 42,* 492(1948).
134. B. Ginsburg and M. S. Dunn, *Ann. Rev. Tuberc. Pulm. Dis., 75,* 688(1957).
135. M. Rabinovitz, et al., *J. Am. Chem. Soc., 77,* 3109(1955).
136. W. M. Harding and W. Shive, *J. Biol. Chem., 206,* 401(1954).
137. B. Levenberg, et al., *J. Biol. Chem., 225,* 163(1957).
138. C. G. Skinner, et al., *J. Biol. Chem., 78,* 2412(1956).
139. C. G. Skinner, et al., *Tex. Repts. Biol. Med., 16,* 493(1958).
140. D. W. Woolley, *J. Am. Chem. Soc., 72,* 5763(1950).
141. D. W. Woolley and A. G. C. White, *J. Biol. Chem., 149,* 285(1943).
142. D. W. Woolley and A. G. C. White, *J. Exp. Med., 78,* 489(1943).
143. D. W. Woolley, *J. Biol. Chem., 191,* 43(1951).
144. S. Eich and L. R. Cerecedo, *J. Biol. Chem., 207,* 295(1954).
145. E. F. Rogers, et al., *J. Am. Chem. Soc., 82,* 2874(1960).
146. L. J. Daniel and L. C. Norris, *Proc. Soc. Exp. Biol. Med., 72,* 165(1949).
147. J. C. Rabinowitz and E. E. Snell, *Arch. Biochem. Biophys., 43,* 408(1953).
148. W. H. Ott, *Proc. Soc. Exp. Biol. Med., 61,* 125(1946).
149. W. W. Umbreit and J. G .Waddell, *Proc. Soc. Exp. Biol. Med., 70,* 293(1949).
150. M. Ikawa and E. E. Snell, *J. Am. Chem. Soc., 76,* 637(1954).
151. R. Sandnain and E. E. Snell, *Proc. Soc. Exp. Biol. Med., 90,* 63(1955).
152. F. Rosen, *Proc. Soc. Exp. Biol. Med., 88,* 243(1955).
153. H. C. Lichstein, *Proc. Soc. Exp. Biol. Med., 88,* 519(1955).
154. H. H. Campbell and K. P. Link, *J. Biol. Chem., 138,* 21(1941).
155. C. Mentzer, *Bull. Soc. Chim. Biol., 30,* 872(1948).
156. K. P. Link, et al., *J. Biol. Chem., 147,* 463(1943).
157. H. Kabat, et al., *J. Pharmacol. Exp. Therap., 80,* 160(1944).
158. D. W. Woolley, et al., *J. Biol. Chem., 124,* 715(1938).
159. J. Matti, et al., *Ann. Inst. Pasteur Immunol., 67,* 240(1941).
160. S. S. Strerberg and F. S. Philips, *Bull. N.Y. Acad. Med., 35,* 811(1959).
161. D. E. Hughes, *Biochem. J., 57,* 485(1954).
162. J. B. Hobbs, Purine and pyrimidine targets, in *Comprehensive Medicinal Chemistry,* Vol. 2, Ed. pp. 299–332.
163. C. Heidelberger, et al., *Adv. Enzymol., 34,* 58(1983).
164. P. Reyes and C. Heidelberger, *Mol. Pharmacol., 1,* 14(1965).
165. A. D. Welch and W. H. Prusoff, *Cancer Chemother. Rep., 6,* 29(1960).
166. A. S. Kaplan and T. Ben-Porat, *J. Mol. Biol., 19,* 320(1966).
167. D. Paven-Langston and R. Langston, *International Ophthalmological Clinics: Ocular Virus Disease,* Boston, Little, Brown, 1975.
168. H. E. Skipper, et al., *Cancer Chemother. Rep., 35,* 1(1964).
169. W. S. Zielinski and M. Sprinzl, *Nucleic Acid Res., 12,* 5025(1984).
170. E. DeClercq, *Trends Pharamcol. Sci., 8,* 339(1987).
171. P. A. Furman and D. W. Barry, *Am. J. Med., 85(2A),* 176(1988).
172. P. A. Furman, et al., *J. Biol. Chem., 262,* 2187(1987).
173. H. Mitsuya and S. Broder, *Nature* (London), *325,* 773(1987).
174. E. DeClercq, in *Antiviral Drug Development, A Multidisciplinary Appraoch,* Vol. 143, E. DeClercq and R. T. Walker, Eds., NATO ASI Series, Series A, New York, Plenum Press, 1988, p. 97.
175. Y. Cheng, et al., *J. Biol. Chem., 262,* 2187(1987).
176. I. Sim, et al., *Antiviral Res., 1,* 393(1982).
177. E. DeClercq, et al., *Biochem. Pharmacol., 29,* 1848(1980).
178. R. B. Diasso, et al., *Biochem. Pharmacol., 27,* 703(1978).
179. G. B. Elion, et al., *J. Am. Chem. Soc., 74,* 411(1952).
180. R. W. Brockman, *Cancer Res., 23,* 1191(1963).
181. J. S. Salser and M. E. Balis, *Cancer Res., 25,* 539(1965).
182. L. L. Bennett, et al., *Cancer Res., 23,* 1574(1963).
183. G. B. Elion and G. Hitchings, in *Antineoplastic and Immunosuppressive Agents,* Part 2, A. Sartorelli and D. Johns, Eds., Berlin, Springer, 1975, p. 404.
184. W. E. G. Muller, et al., *Ann. N.Y. Acad. Sci., 284,* 34(1977).
185. R. J. Whiteley, et al., *N. Engl. J. Med., 297,* 289(1977).
186. R. Agarwal, et al., *Biochem. Pharmacol., 24,* 693(1977).
187. C. M. Schimandle and I. W. Sherman, *Biochem. Pharmacol., 32,* 115(1983).
188. P. E. Daddona, et al., *J. Biol. Chem., 259,* 1472(1984).
189. G. B. Elion, et al., *Proc. Natl. Acad. Sci., U.S.A., 74,* 5716(1977).
190. G. B. Elion, *Adv. Enzyme Regul., 18,* 53(1980).
191. Robert-Gero, et al., Analogs of S-denosyl-homocysteine as in vitro inhibitors of transmethylases and in vitro inhibitors of vital oncogenesis nad other cellular events, in

Enzyme Inhibitors, U. Brodbeck, Ed., Weinhim, Verlag Chemie, GmbH, 1980, p. 61.
192. W. M. Trager, et al., *FEBS Lett., 85,* 264(1977).
193. F. M. Vedel, et al., *Biochem. Biophys. Res. Commun., 85,* 371(1978).
194. R. T. Borcharde, *Biochem. Biophys. Res. Commun., 89,* 919(1979).
195. G. B. Elion, et al., *Biochem. Pharmacol., 15,* 863(1966).
196. P. S. Miller and P. O. P. Ts'o, *Ann. Rep. Med. Chem., 23,* 295(1988).
197. E. Uhlmann and A. Piyman, *Chem. Rev., 90,* 554(1990).
198. Y. Uguchi, et al., *J. Ann. Rev. Biochem., 60,* 631(1991).
199. J. S. Cohen, *Pharmacol. Ther., 52,* 211(1991).
200. E. Wickstrom, *Tibtech, 10(B),* 281(1992).
201. C. Sartorius and R. M. Franklin, *Parasitology Today, 7,* 90(1991).
202. S. Agarwal, *Tibtech, 10,* 152(1992).
203. J. Gray, et al., *Plant Molecular Biol., 19,* 69(1992).
204. T. L. Sourkes, *Arch. Biochem. Biophys., 51,* 444(1954).
205. J. A. Oates, et al., *Science, 131,* 1890(1960).
206. I. J. Kopin, et al., *J. Pharmacol. Exp. Ther., 147,* 186(1965).
207. M. D. Day, et al., *Eur. J. Pharmacol., 21,* 271(1973).
208. M. G. Palfreyman, et al., *J. Neurochem., 31,* 927(1978).
209. M. J. Jung, et al., *Life Sci., 24,* 1037(1979).
210. A. L. Maycock, et al., *Biochem., 19,* 709(1980).
211. J. M. Baldoni and J. J. Villafranca, *J. Biol. Chem., 255,* 8987(1980).
212. J. Grove, et al., *Lancet, ii,* 647(1980).
213. R. D. Allan, et al., *Clin. Exp. Pharmacol. Physiol., 6,* 687(1979).
214. R. D. Allan, et al., *Aust. J. Chem., 33,* 1115(1980).
215. R. B. Silverman and M. A. Levy, *Biochem. Biophys. Res. Commun., 95,* 250(1980).
216. T. T. Howarth and A. G. Brown, *J. Chem. Soc. Chem. Comm.,* 266(1976).
217. A. R. English, et al., *Antimicrob. Agents Chemother., 14,* 414(1978).
218. J. F. Fisher and J. R. Knowles, in *Enzyme Inhibitors as Drugs,* M. Sandler, Ed., Baltimore, University Park Press, 1979, p. 209.
219. R. L. Charnas and J. R. Knowles, *Biochemistry, 20,* 3214(1981).
220. J. Hood, et al., *J. Antibiot., 32,* 295(1979).
221. M. V. Fast, et al., *Lancet, ii,* 509(1982).
222. R. R. Yocum, et al., *Proc. Natl. Acad. Sci., U.S.A., 76,* 2730(1979).
223. D. J. Tipper and J. L. Strominger, *Proc. Natl. Acad. Sci., U.S.A., 54,* 1133(1965).
224. B. Lee, *J. Mol. Biol., 61,* 463(1971).
225. W. J. Greenlee and P. K. S. Siegl, *Ann. Rep. Medicinal Chem., 27,* 59(1992).
226. L. D. Bayers and R. Wolfenden, *Biochemistry, 12,* 2070(1973).
227. D. W. Cushman, et al., *Biochemistry, 16,* 5484(1977).
228. A. A. Patchett, et al., *Nature, 288,* 280(1980).
229. E. W. Petrillo and M. A. Ondetti, *Med. Res. Rev., 2,* 1(1982).
230. J. Janne, et al., *Spec. Top. Endocrinol. Metab., 5,* 227(1983).
231. M. H. Abdel-Monem, et al., *J. Med. Chem., 17,* 447(1974).
232. S. I. Harik and S. H. Snyder, *Biochem. Biophys. Acta, 327,* 501(1973).
233. N. Relyea and R. R. Rando, *Biochem. Biophys. Res. Commun., 67,* 392(1975).
234. P. S. Mamont, et al., *Proc. Natl. Acad. Sci., U.S.A., 73,* 1626(1976).
235. J. Kollonitsch, et al., *Nature, 274,* 906(1978).
236. B. W. Metcalf, et al., *J. Am. Chem. Soc., 100,* 255(1978).
237. C. Danzin, et al., *J. Med. Chem., 24,* 16(1981).
238. A. E. Pegg and P. P. McCann, *ISI, Atlas of Science: Biochem.,* 11(1988).
239. J. Fishman and J. Goto, *J. Biol. Chem., 256,* 4466(1981).
240. M. Akhtar, et al., *Biochem. J., 201,* 569(1982).
241. D. F. Covey, et al., *J. Biol. Chem., 256,* 1076(1983).
242. B. W. Metcalf, et al., *J. Am. Chem. Soc., 103,* 3221(1981).
243. P. A. Marcotte and C. H. Robinson, *Biochemistry, 21,* 2773(1982).
244. J. Bennevist, *Nature* (London), *249,* 381(1974).
245. B. B. Vargaftig and P. Braquet, *Br. Med. Bull., 43,* 312(1987).
246. P. Braquet, et al., *Adv. Inflamm. Res., 16,* 179(1987).
247. D. Hosford and P. Brequet, *Prog. Med. Chem., 27,* 325(1990).

SUGGESTED READINGS

A. Albert, *Selective Toxicity. The Physico-Chemical Basis of Therapy,* 7th Ed., London, Chapman & Hall, 1985.

B. R. Baker, *Design of Active-Site Directed Enzyme Inhibitors,* New York, Wiley, 1967.

J. R. Bertino, *Folate Antagonists as Chemotherapeutic Agents,* Vol. 186, New York, Annals of New York Academy of Sciences, 1971.

C. Hansch, Enzymes and Other Molecular Targets. P. G. Sammes, Eds., Comprehensive Medicinal Chemistry, Vol. 2, Vol. Ed., New York, Pergamon Press, 1990.

G. H. Hitchings, Ed., *Inhibition of Folate Metabolism in Chemotherapy. The Origins and Uses of Co-trimoxazole,* Berlin, Springer-Verlag, 1983.

R. M. Hochster and J. H Quastel, Eds., *Metabolic Inhibitors. A Comprehensive Treatise. Vols. I, II.* New York, Academic Press, 1963.

R. M. Hochster, et al., Eds., *Metabolic Inhibitors, A Comprehensive Treatise,* New York, Academic Press, Vol. III, 1972, Vol. IV, 1973.

D. W. Woolley, *A Study of Antimetabolites,* New York, Wiley, 1952.

Chapter 32

PARASITE CHEMOTHERAPY

William O. Foye

Parasitic diseases are widely distributed around the world, and the morbidity they cause in human terms has a heavy impact on social and economic development in many countries. The incidence of some parasitic diseases reaches as high as 80% in some developing countries, and in some countries, more than one of the extremely serious parasitic diseases are endemic. Several international programs have been organized to combat these diseases, which are often referred to as the neglected diseases of humans. In many of the developed countries such as United States, endemic parasitic diseases are no longer among the serious problems. With modern transportation available to many segments of the world's population, however, parasitic diseases are carried by travelers across international borders.

Parasites vary in size from single-celled protozoa that cause malarial infections and amebic dysentery to larger, more complex helminths, or worms, which cause many debilitating conditions such as schistosomiasis and an invasion of flukes in the liver and lungs. This chapter discusses the protozoan parasitic diseases and then the helminthic infections. Antiprotozoan drugs are listed in Table 32–1, and anthelminthic drugs are listed in Table 32–2. Some of these drugs find application in both areas.

PROTOZOAN DISEASES

Protozoans cause more human parasitic infections than any other type of organism. Malaria, for example, is one of the most common causes of death in the world at present. It is also the most widespread of the protozoan diseases. There are approximately 10,000 protozoan parasitic species known to the literature. Protozoan species primarily infect the human gastrointestinal tract, blood, blood-forming organs, vagina, and urethra. Whereas protozoan diseases are found throughout the world, they are most common in populations living in tropical regions. Such areas have a higher incidence of malnutrition, poor health education, and inadequate sanitation. Protozoan diseases of domesticated animals and poultry also seriously diminish the food supply for humans, especially in Africa.

Discovery and development of new chemotherapeutic agents to fight protozoan diseases has been slow. The urgency to combat this problem has been heightened by the prevalence of acquired immune deficiency syndrome (AIDS) in the endemic areas, a syndrome that is complicated by opportunistic infections. Many of these are protozoan in origin. There are no successful therapeutics for many of the protozoan diseases. Parasites in many parts of the world are developing drug resistance to some of the existing drugs. Drug resistance can be defined as the ability of an organism to adapt or adjust to a drug action at a drug concentration that is equal to or exceeds the recommended effective dose. Researchers from around the world are also encountering parasites with multidrug resistance. This has made it necessary to continue the design, synthesis, and screening of potential new drugs for prophylaxis and therapy. One of the leading research organizations in parasitic diseases is the Walter Reed Army Institute of Research. This institute has studied therapeutic properties of more than 250,000 compounds.

Chemotherapeutic agents effective against the main protozoan diseases are discussed here. Numerous references and books available on parasitology and protozoology provide additional details.

Malaria

Malaria is transmitted by the infected female Anopheles mosquito. The specific protozoan organisms causing malaria are from the genus Plasmodium. Only 4 of approximately 100 species cause the disease in humans. The remaining species affect birds, monkeys, livestock, rodents, and reptiles. The four species that affect humans are: Plasmodium falciparum, P. vivax, P. malariae, and P. ovale. Concurrent infections by more than one of these species are seen in endemically affected regions of the world. Such multiple infections further complicate patient management and the choice of treatment regimens.

Malaria affects an estimated 300 million humans globally, and causes more than 2 million deaths annually. It is estimated that a third of these fatalities occurs in children under 5 years old. Although this disease is found primarily in the tropics and subtropics, it has been observed far beyond these boundaries.

Malaria has essentially been eradicated in most temperate-zone countries. However, more than 1000 cases of malaria were documented recently in United States' citizens returning from travel abroad. Today, malaria is found in most countries in Africa, Central and South America, and Southeast Asia. It is reported to be on the increase in Afghanistan, Bangladesh, Brazil, Burma, Cambodia, Colombia, China, Iran, India, Indonesia, Mexico, the Philippines, Thailand, and Vietnam. Infection from Plasmodia can cause anemia, pulmonary edema, renal failure, jaundice, shock, cerebral malaria, and, if not treated in a timely manner, can result in death.

Types of Malaria

Malarial infections are known according to the species of the parasite involved. Plasmodium falciparum infection has an incubation period (time from mosquito bite to clinical symptoms) of 1 to 3 weeks (average of 12 days). The P. falciparum life cycle in man begins with the bite of an infected female mosquito. The parasites in the sporozoite stage enter the circulatory system through which they can reach the liver in about an hour. These organisms grow and multiply 30,000 to 40,000-fold by asexual division within liver cells in 5 to 7 days. Then, as merozoites, they leave the liver to re-enter the blood stream and invade the erythrocytes, or red blood cells (RBCs), where they continue to grow and multiply further for 1 to 3 days. These infected RBCs rupture, releasing merozoites in intervals of about 48 hours. This process results in the patient exhibiting clinical symptoms such as chills, fever, sweating, headaches, fatigue, anorexia, nausea, vomiting, and diarrhea. Recurrence of these symptoms on alternate days is characteristic of tertian malaria. The P. falciparum parasite can also cause RBCs to clump and adhere to the wall of blood vessels. Such a phenomenon has been known to cause partial obstruction and sometimes restriction of the blood flow to vital organs like the brain, liver, and kidneys. Reinfection of RBCs can occur, allowing further multiplication and remanifestation of the malaria symptoms. Some merozoites develop into male and female sexual forms, called gametocytes, which can then be acquired by the female mosquito after biting the infected human. Gametocytes enter the mosquito's stomach where they are fertilized and go on to form zygotes. Zygotes migrate to the insect's salivary glands where the sporozoite form can be transmitted to another human following a mosquito bite. The life cycle of the malaria parasites is shown in Figure 32–1.

Plasmodium vivax (benign tertian) is the most prevalent form of malaria. It has an incubation period of 1 to 4 weeks (2 weeks average). This form of malaria can cause spleen rupture and anemia. Relapses (renewed manifestations of erythrocytic infection) can occur. This results from the periodic release of dormant parasites (hypnozoites) from the liver cells. Whereas the erythrocytic forms are generally considered to be susceptible to chloroquine, some chloroquine-resistant strains have been reported in parts of Indonesia and New Guinea.

Plasmodium malariae is responsible for quartan malaria. It has an incubation period of 2 to 4 weeks (average of 3 weeks). The asexual cycle occurs every 72 hours. In addition to the usual symptoms, this form also causes nephritis. This is the mildest form of malaria and does not relapse. The RBC infection associated with P. malariae can last for many years. There have been no P. malariae chloroquine-resistant cases reported.

Plasmodium ovale has an incubation period of 9 to 18 days (14 days average). Relapses have also been known to occur in people infected with this plasmodium. These relapses are indicative of ovale tertian malaria. There are no reported incidents of chloroquine resistance associated with this form.

Types of Chemotherapy

Tissue Schizonticides. These eradicate the exoerythrocytic liver-tissue stages of the parasite which prevents the parasite's entry into the blood. Drugs of this type are useful for prophylaxis. Some tissue schizonticides can act on the long-lived tissue form (hypnozoites of P. vivax and P. ovale), and thus can prevent relapses.

Blood Schizonticides. These destroy the erythrocytic stages of parasites and can cure cases of falciparum malaria or suppress relapses. This is the easiest phase to treat because drug delivery into the blood stream can be accomplished rapidly.

Gametocytocides. These kill the sexual forms of the plasmodia (gametocytes), which are transmittable to the

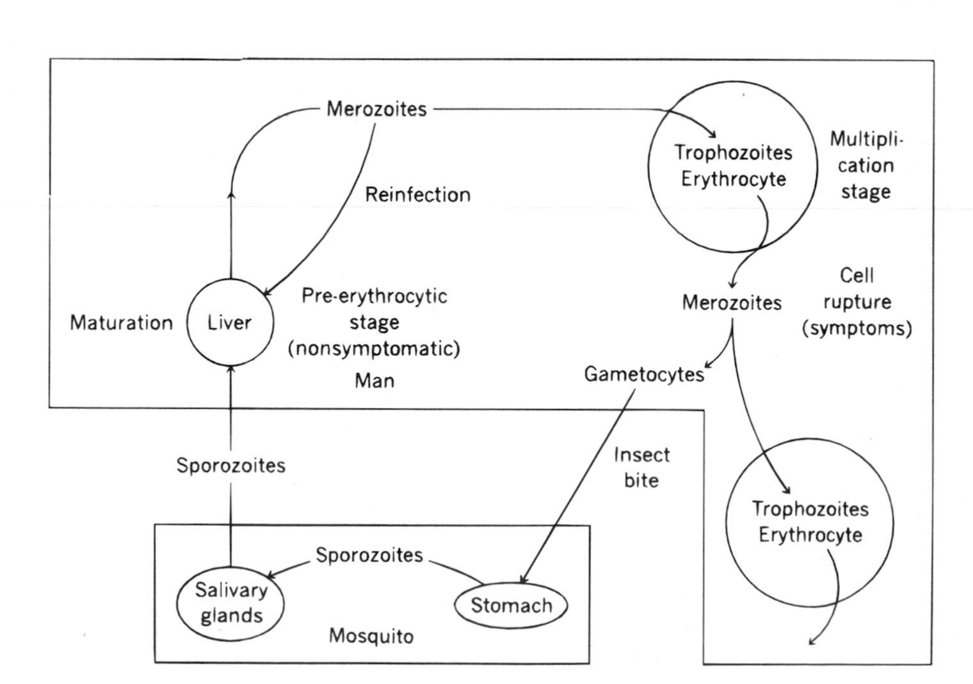

Fig. 32–1. Life cycle of malarial protozoa.

Anopheles mosquito, thereby preventing transmission of the disease.

Sporontocides (sporozoitocides). These act against sporozoites and are capable of killing these organisms as soon as they enter the bloodstream following a mosquito bite. It should be noted that antimalarials may operate against more than one form, and may be effective against one species of plasmodium but lack efficacy against others. In addition, antimalarial drugs may be classified according to their structural types.

Specific Antimalarial Agents

Quinine (1),* is the first known antimalarial. It is a 4-quinolinemethanol derivative bearing a substituted quinuclidine ring. Use of quinine in Europe began in the seventeenth century, after the Incas of Peru informed the Spanish Jesuits of the antimalarial properties of the bark of an evergreen mountain tree they called quinquina [later called cinchona after Francisca Henriquez de Ribera (1576–1639), Countess of Chinchona, and wife of the Peruvian Viceroy]. The bark, when made into an aqueous solution, was capable of curing most forms of malaria. It was listed in the *London Pharmacopoeia* of 1677. The alkaloid derived from it, quinine, was isolated in the mid-1820s. Quinine, a very bitter substance, has been used by millions of malaria sufferers. Recently it has been employed successfully to treat chloroquine-resistant strains of P. falciparum. Occasionally, however, it falls short of providing a complete cure of the infection.

One of the isomeric cinchona alkaloids is shown in the following diagram, and the four major alkaloids are listed. Quinine is the most prevalent alkaloid present in the bark extracts (about 5%). All alkaloids with the same substitution at R_1 and R_2 are diastereoisomers, differing in configuration at the third and fourth chiral centers (C–8 and C–9). Although all four alkaloids have antimalarial activity, their C–9 epimers (i.e., the *epi*-series having either 8R:9R or 8S:9S configurations) are inactive. Also, any modification of the secondary alcohol at C–9, through oxidation, esterification, or similar processes, diminishes activity. The quinuclidine portion is not necessary for activity; however, an alkyl tertiary amine at C–9 is important. These findings provided the basis for synthesis of many potential antimalarials.

Quinine is metabolized in the liver to the 2′-hydroxy derivative, followed by additional hydroxylation on the quinuclidine ring, with the 2,2′-dihydroxy derivative as the major metabolite. This metabolite has low activity and is rapidly excreted.

Quinine acts on the asexual blood forms of the plasmodia in a slower manner than that of many synthetic drugs. Overdose causes tinnitus and visual disturbances; these side effects disappear on discontinuation of the drug. Quinine can also cause premature contractions in late stages of pregnancy. Although quinine is suitable for parenteral administration, this route is considered hazardous because of its ability to cause hemolysis. **Quinidine,**

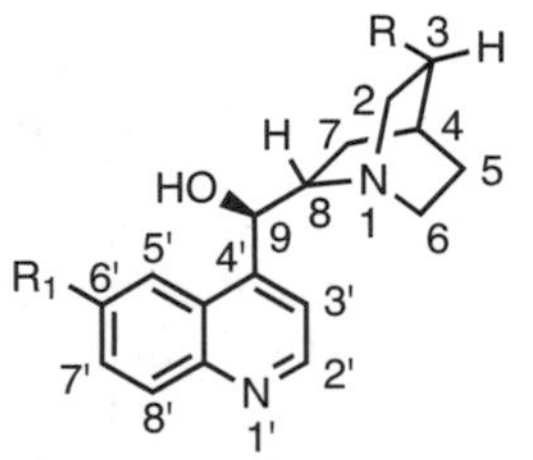

Cinchona alkaloid (configuration: 3R:4S:8S:9R)
Quinine: R_1 = OCH_3; R_2 = -CH=CH_2; (–)-8S:9R isomer
Quinidine: R_1 = OCH_3; R_2 = -CH=CH_2; (+)-8R:9S isomer
Cinchonine: R_1 = H; R_2 = -CH=CH_2; (+)-8R:9S isomer
Cinchonidine: R_1 = H; R_2 = -CH=CH_2; (–)-8S:9R isomer

which is the (+) isomer of quinine, has been shown to be more effective in combating the disease, but it has undesirable cardiac side effects. Mixtures of cinchona alkaloids, known as totaquine, have also been used in malaria treatment. Standardized totaquines contain a minimum of 15% quinine.

The usefulness of quinine motivated the search for other antimalarials. The greatest initiative for the development of synthetic drugs came during this century when the World War II interrupted the supply of cinchona bark from the East Indies. Another 4-quinolinemethanol, **mefloquine** (2, Lariam [Roche]), is now an optional drug for chloroquine-resistant P. falciparum. Mefloquine requires less than one tenth the dose of quinine to effect cures. It, too, has undesirable side effects, such as gastrointestinal upset and dizziness, but they tend to be temporary. Resistance to mefloquine has been reported, even though the drug is relatively new. With its potential central nervous system (CNS) side effects, mefloquine is not recommended for use in patients using beta-blockers, in those whose job requires fine coordination and spatial discrimination, or in those with a history of epilepsy or psychiatric disorders.

The most effective among hundreds of synthesized 4-aminoquinolines has been **chloroquine** (4). 4-Aminoquinolines are active against erythrocytic forms of P. vivax and P. falciparam. Chloroquine was the drug of choice until chloroquine resistance became widespread. It is still considered one of the most important antimalarial drugs in use for suppression and prophylaxis in most regions of the world where malaria is endemic and where sensitivity to the drug remains. Chloroquine acts as a blood schizonticidal agent and rarely produces serious side effects in persons using it prophylactically. Unfortunately, it was the prophylactic use of the drug that led to resistance to the drug. Chloroquine phosphate, when administered orally, or its chloride salt when administered parenterally, has side effects that include gastrointestinal upset, headache, dizziness, itching, and blurred vision. Discontinuing taking the drug has been found to be unnecessary for management of these relatively minor side effects. Prolonged use of chloroquine justifies occasional ophthalmic examinations to guard against retinal damage,

* Figures in parenthesis following compounds refer the reader to Tables 32–1 and 32–2.

Table 32–1. Antiprotozoan Compounds

Number	Generic Name Chemical Name	Structure
1	Quinine 6′-Methoxycinchonan-9-ol	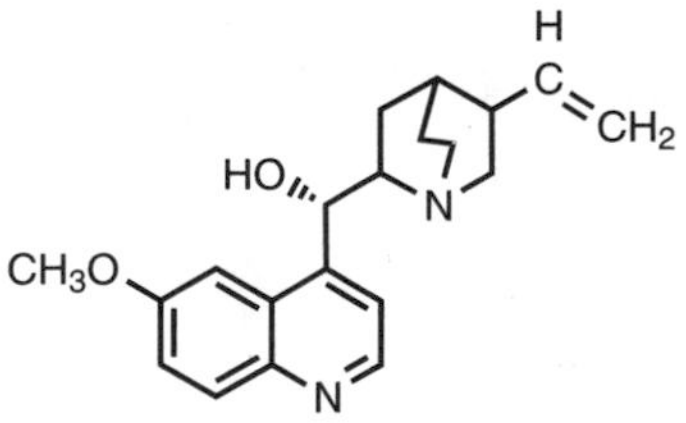
2	Mefloquine (R*,S*)-(±)-α-2-Piperidinyl-2,8-bis(trifluoromethyl)-4-quinolinemethanol	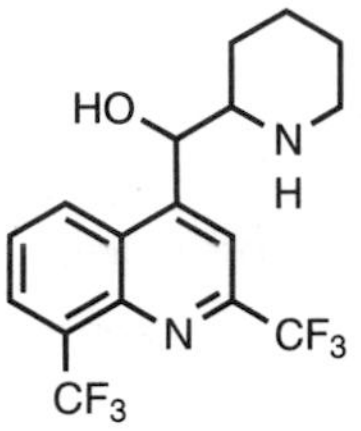
3	Halofantrine 1,3-dichloro-α-[2-dibutylamino)-ethyl]-6-(trifluoromethyl)-9-phenanthrenemethanol	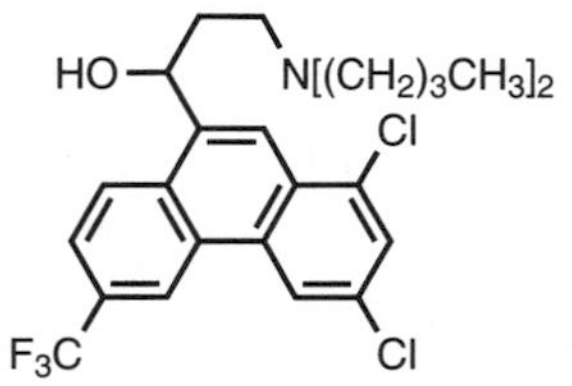
4	Chloroquine N^4-(7-Chloro-4-quinolinyl)-N^1,N^1-diethyl-1,4-pentanediamine	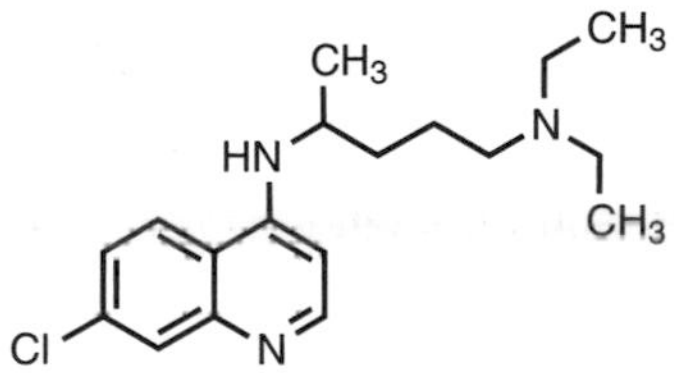
5	Amodiaquine 4-[(7-Chloro-4-quinolinyl)amino]-2-[(diethylamino)methyl]phenol	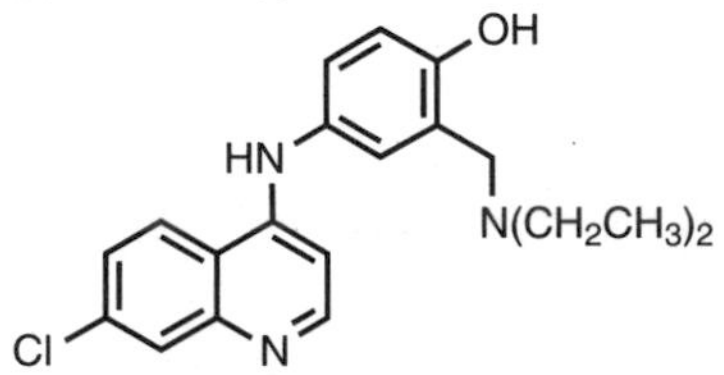
6	Hydroxychloroquine 2-[[4-(7-chloro-4-quinolinyl)-amino]pentyl]ethylamino]-ethanol	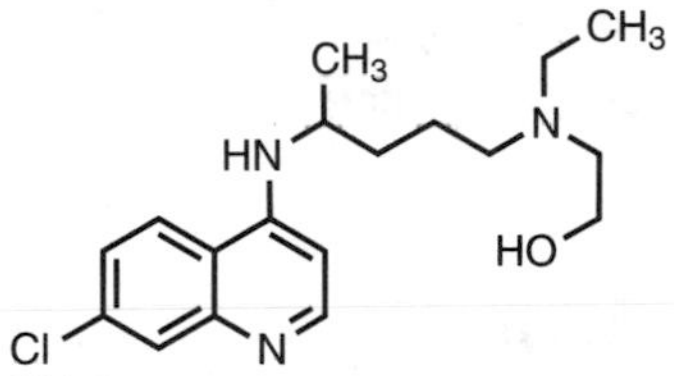
7	Pamaquine N^1,N^1-Diethyl-N^4-(6-methoxy-8-quinolinyl)-1,4-pentanediamine	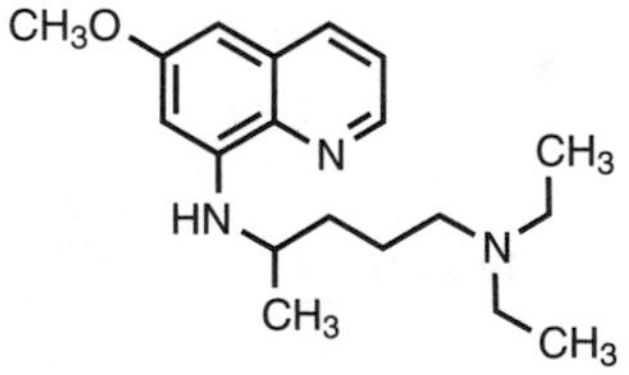
8	Primaquine N^4-(6-Methoxy-8-quinolinyl)-1,4-pentanediamine	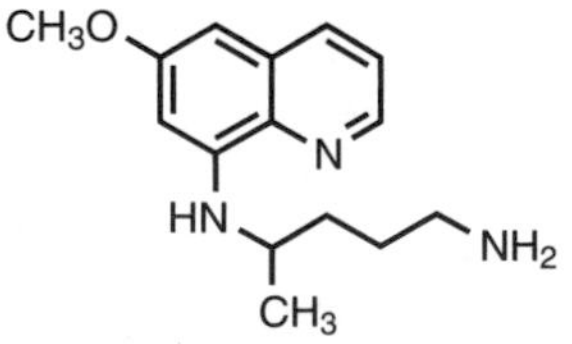

(continued)

Table 32–1. *(Continued)*

Number	Generic Name Chemical Name	Structure
9	Pyrimethamine 5-(4-Chlorophenyl)-6-ethyl-2,4-pyrimidinediamine	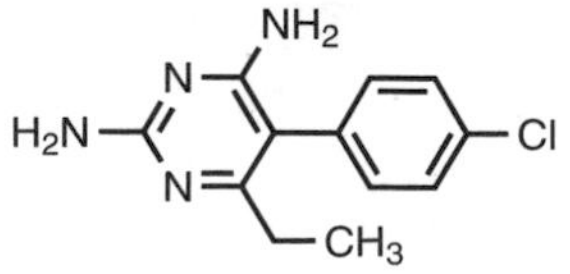
10	Dapsone 4,4′-Sulfonylbisbenzeneamine	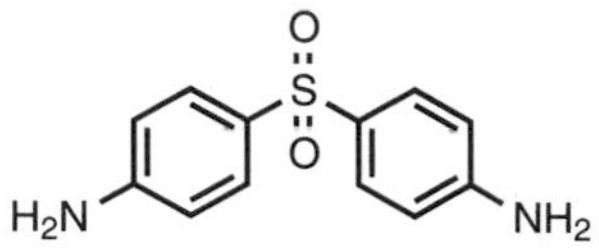
11	Sulfadoxine 4-Amino-N-(5,6-dimethoxy-4-pyrimidinyl)benzenesulfonamide	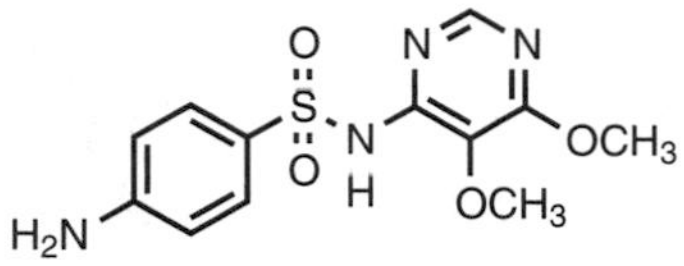
12	Sulfadiazine 4-Amino-N-2-pyrimidinylbenzenesulfonamide	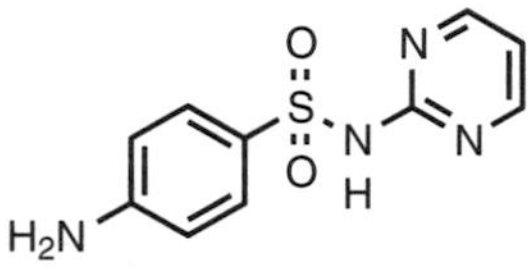
13	Sulfalene 4-Amino-N-(3-methoxypyrazinyl)-benzenesulfonamide	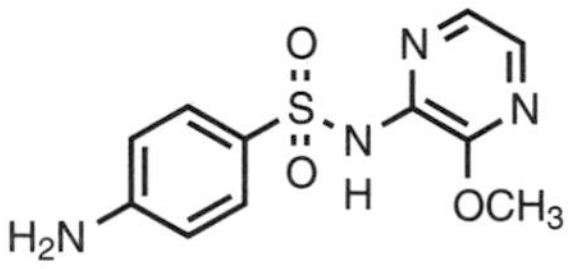
14	Trimethoprim 5-[(3,4,5-Trimethoxyphenyl)-methyl]-2,4-pyrimidinediamine	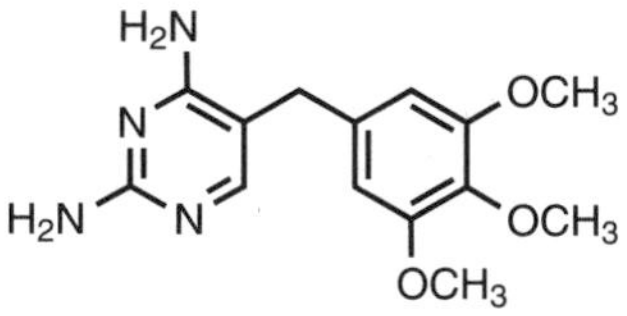
15	Chlorguanide N-(4-Chlorophenyl)-N′-(1-methyl-ethyl)imidodicarbonimidic diamide	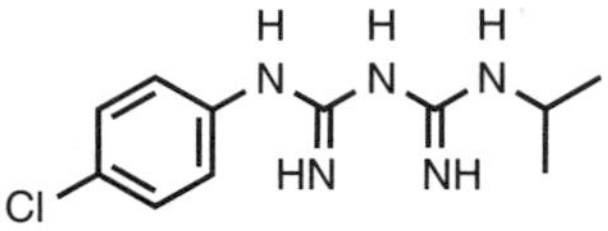
16	Cycloguanil 1-(4-Chlorophenyl)-1,6-dihydro-6,6-dimethyl-1,3,5-triazine-2,4-diamine	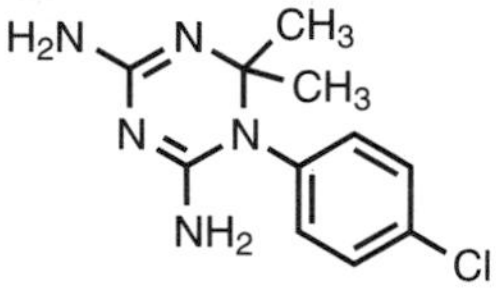
17	Menoctone 2-(8-Cyclohexyloctyl)-3-hydroxy-1,4-naphthoquinone	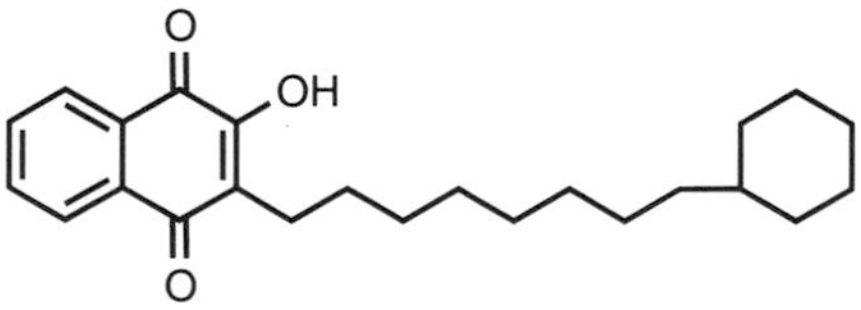
18	Metronidazole 2-Methyl-5-nitroimidazole-1-ethanol	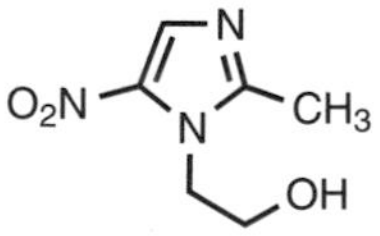
19	Iodoquinol 5,7-Diiodo-8-quinolinol	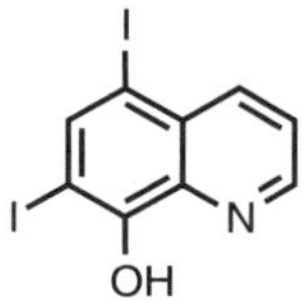

(continued)

Table 32–1. *(Continued)*

Number	Generic Name Chemical Name	Structure
20	Oxytetracycline 4-(Dimethylamino)-1,4,4a,5,5a,6,11,12a-octahydro-3,5,6,10,12,12a-hexahydroxy-6-methyl-1,11-dioxo-2-naphthacenecarboxamide	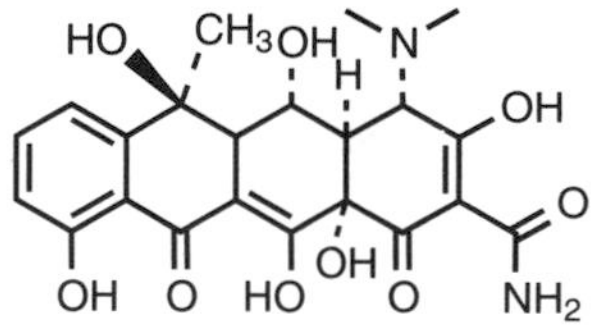
21	Emetine 6′,7′,10,11-Tetramethoxyemetan	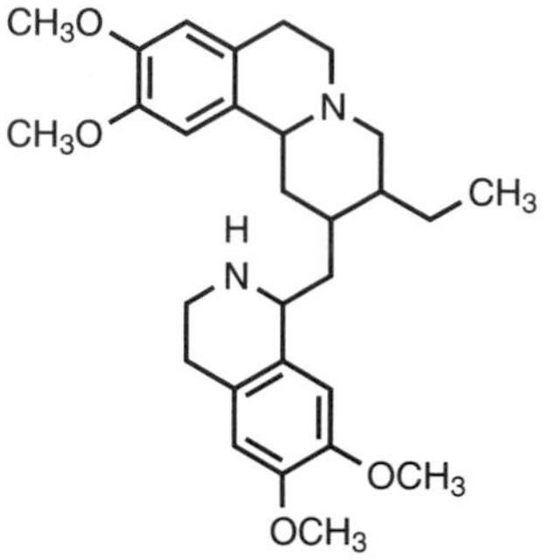
22	Paromomycin O-2-Amino-2-deoxy-α-D-glucopyranosyl-(1→4)-O-[O-2,6-diamino-2,6-dideoxy-β-L-idopyranosyl-(1→3)-β-D-ribofuranosyl-(1→5)]-2-deoxy-D-streptamine	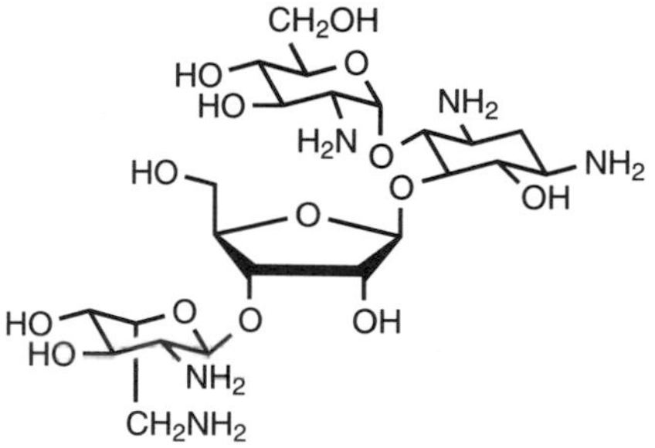
23	Diloxanide Furoate 2,2-Dichloro-N-(4-hydroxyphenyl)-N-methylacetamide-2′-furoate	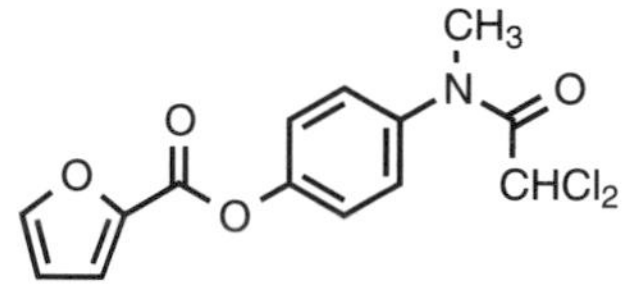
24	Tinidazole 1-[2-(Ethylsulfonyl)ethyl]-2-methyl-5-nitro-1H-imidazole	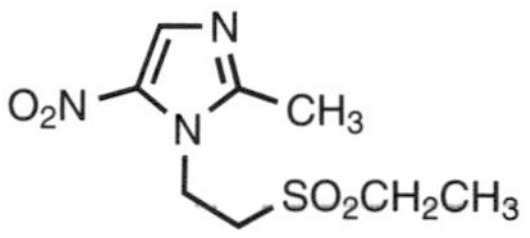
25	Bithionol 2,2′-Thiobis[4,6-dichlorophenol]	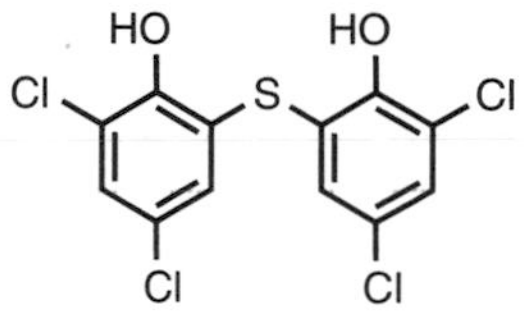
26	Amphotericin B	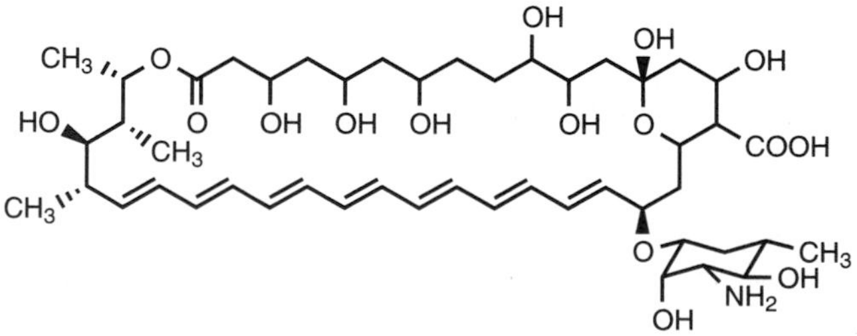

(continued)

Table 32–1. *(Continued)*

Number	Generic Name Chemical Name	Structure
27	Quinacrine N^4-(6-Chloro-2-methoxy-9-acridinyl)-N^1,N^1-diethyl-1,4-pentanediamine	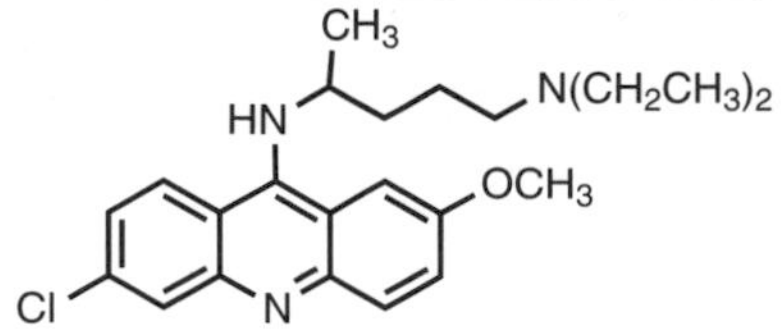
28	Furazolidone 3-[[(5-Nitro-2-furanyl) methylene]-amino]-2-oxazolidinone	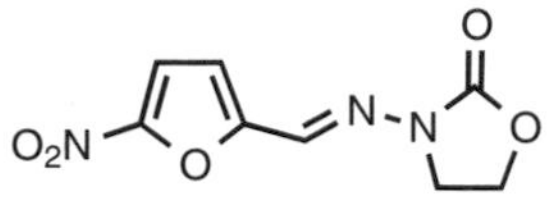
29	Acranil 1-[(6-Chloro-2-methoxy-9-acridinyl)-amino]-3-(diethylamino)-2-propanol	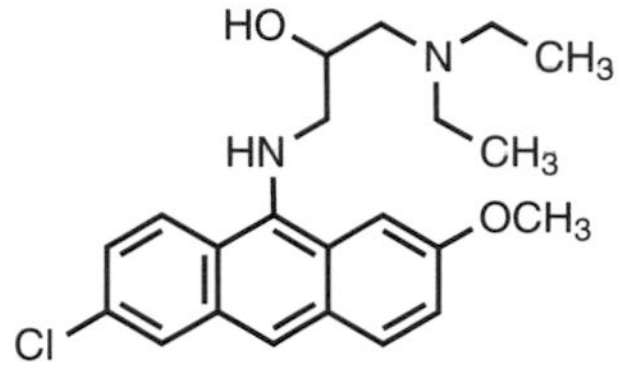
30	Sodium Stibogluconate Pentavalent sodium antimonyl gluconate	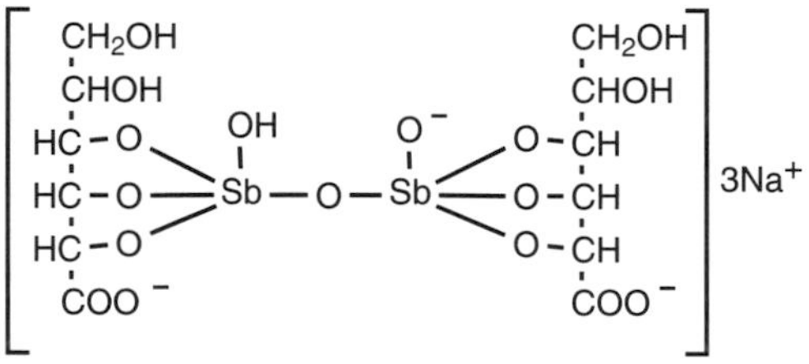
31	Stilbamidine 4,4′-(1,2-Ethenediyl) bisbenzenecarboximidamide	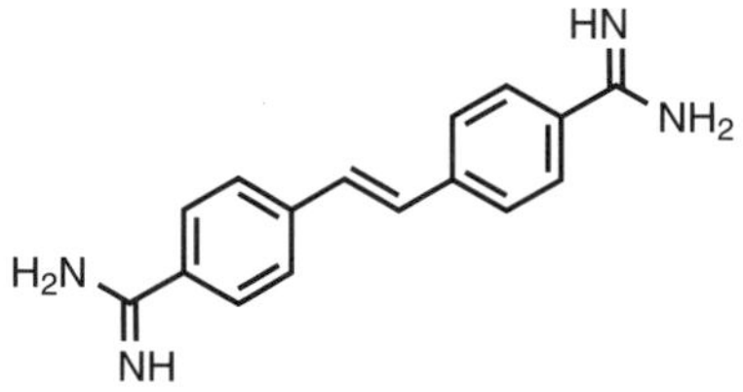
32	Allopurinol 1,5-Dihydro-4H-pyrazolo[3,4-d]pyrimidin-4-one	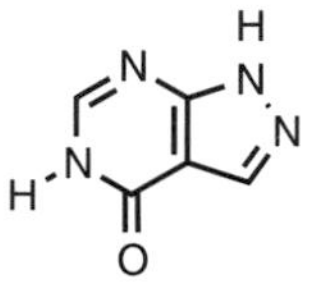
33	Ketoconazole cis-1-Acetyl-4-[r-[[2-(2,4-dichlorophenyl)-2-(1H-imidazol-1-ylmethyl)-1,3-dioxolan-4-yl]-methoxy]phenyl]piperazine	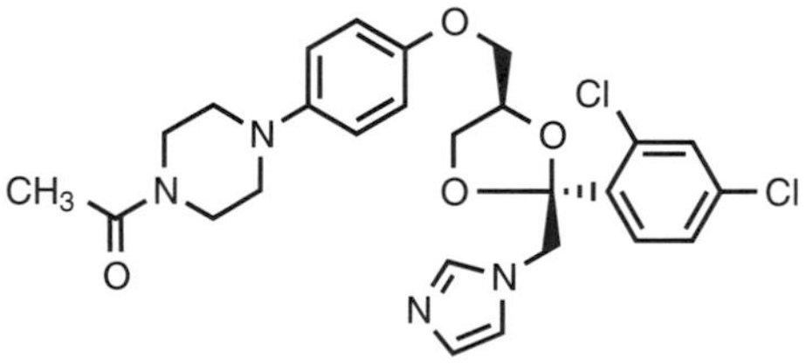
34	Sulfamethoxazole 4-Amino-N-(5-methyl-3-isoxazolyl)benzenesulfonamide	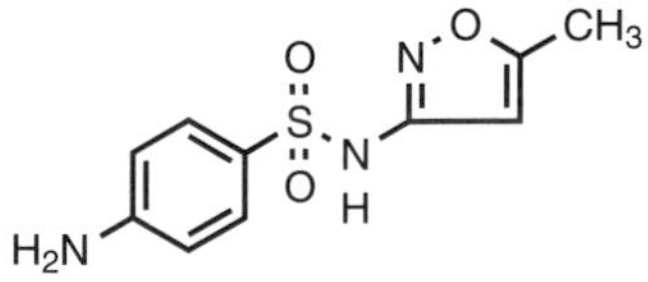
35	Eflornithine 2-(Difluoromethyl)-DL-ornithine	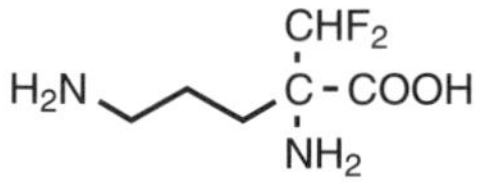

(continued)

Table 32–1. *(Continued)*

Number	Generic Name Chemical Name	Structure
36	Clindamycin (2S-trans)-Methyl-7-chloro-6,7,8-trideoxy-6-[[(1-methyl-4-propyl-2-pyrrolidinyl)carbonyl]amino]-1-thio-L-threo-α-D-galacto-octopyranoside	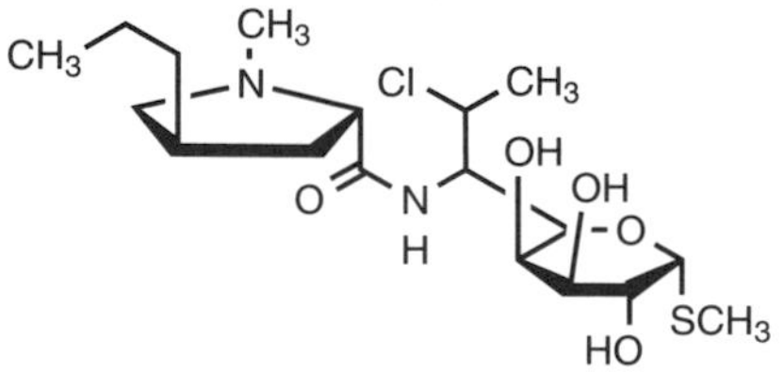
37	Suramin Sodium 8,8′-[Carbonylbis[imino-3,1-phenylenecarbonylimino(4-methyl-3,1-phenylene)-carbonylimino]]bis-1,3,5-naphthalenetrisulfonic acid hexasodium salt	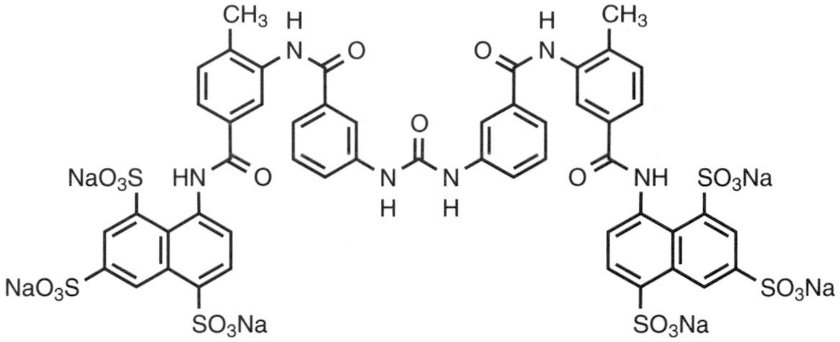
38	Tryparsamide [4-[(2-Amino-2-oxoethyl)amino]-phenyl]arsonic acid monosodium salt	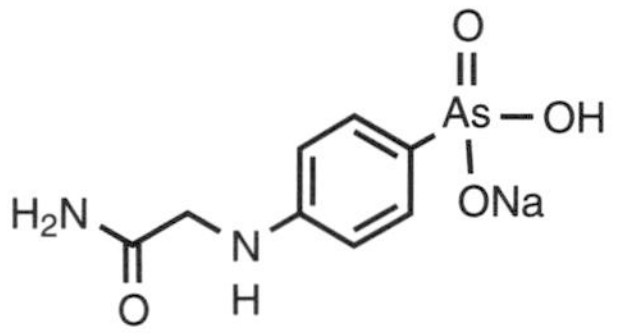
39	Pentamidine 4,4′-[1,5-Pentanediylbis(oxy)]bis-benzenecarboximidamide	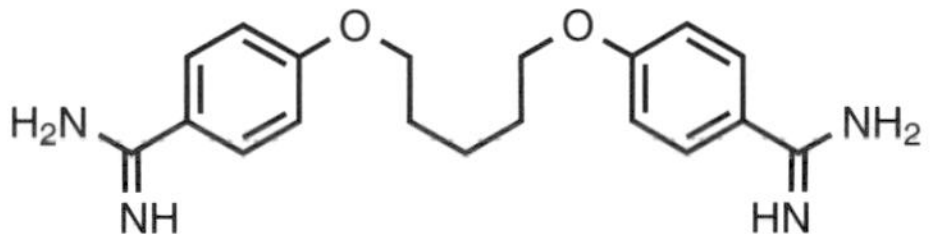
40	Melarsoprol 2-[4-[(4,6-Diamino-1,3,5-triazin-2-yl)amino]phenyl]-1,3,2-dithiarsolane-4-methanol	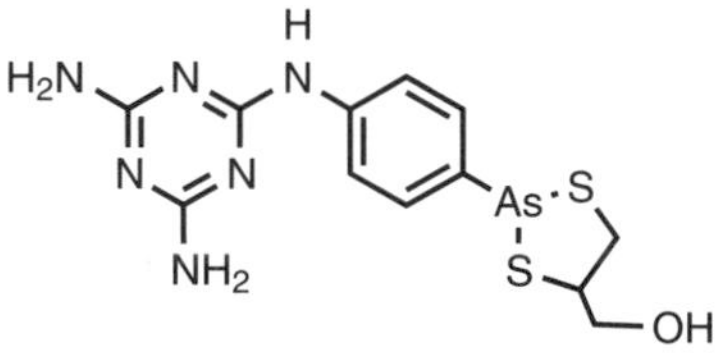
41	Nifurtimox 3-Methyl-N-[(5-nitro-2-furanyl)-methylene]-4-thiomorpholinamine 1,1-dioxide	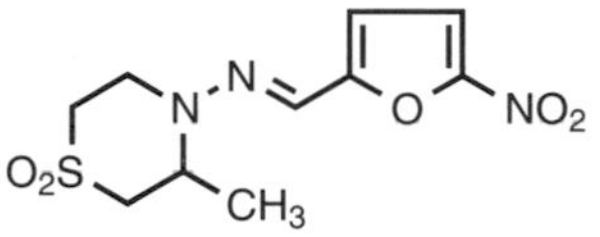
42	Benznidazole 2-Nitro-N-(phenylmethyl)-1H-imidazole-1-acetamide	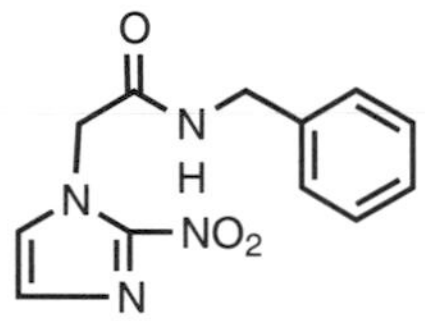

(continued)

especially when the drug is used as an anti-inflammatory for 5 or more years.

A phenanthrene-methanol derivative named **halofantrine** (3) is effective against chloroquine-resistant malaria and is now being evaluated for human use. It has a temporary side effect of gastrointestinal disturbances.

Amodiaquine (5) and **hydroxychloroquine** (6) were found to be effective as alternative drugs to chloroquine-resistant P. falciparum strains. These compounds are less effective, however, and amodiaquine is more toxic than chloroquine. The 4-aminoquinolines act as gametocytocidal agents against the various plasmodia species. It has also been demonstrated that chloroquine resistance could be reversed in vitro, and in vivo in rodents, by treating P. falciparum-parasitized erythrocytes with calcium-channel blockers and antihistamines.

Some 8-aminoquinolines are effective against both primary and secondary tissue forms and sexual blood stages of the parasite. They also have activity against asexual blood forms in humans, but only at high, toxic levels.

Table 32–2. Antihelminthic Compounds

Number	Generic Name Chemical Name	Structure
43	Praziquantel 2-(Cyclohexylcarbonyl)-1,2,3,6,7,11b-hexahydro-4H-pyrazino[2,1-a] isoquinolin-4-one	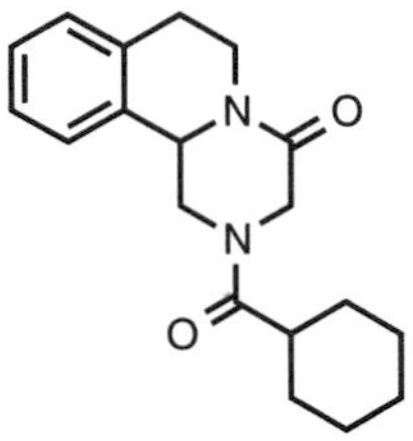
44	Oxamniquine 1,2,3,4-Tetrahydro-2-[[(1-methylethyl)amino]methyl]-7-nitro-6-quinolinemethanol	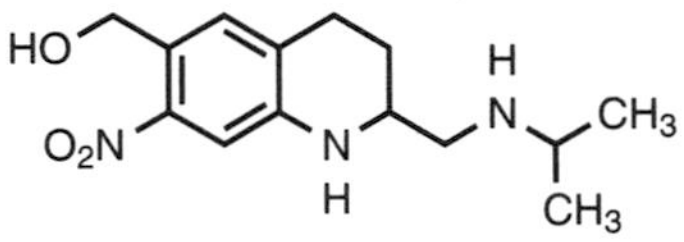
45	Metrifonate (2,2,2-Trichloro-1-hydroxyethyl)-phosphonic acid dimethyl ester	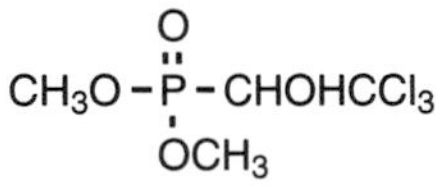
46	Tetrachloroethylene	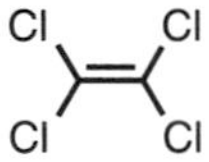
47	Niclosamide 5-Chloro-N-(2-chloro-4-nitrophenyl)-2-hydroxybenzamide	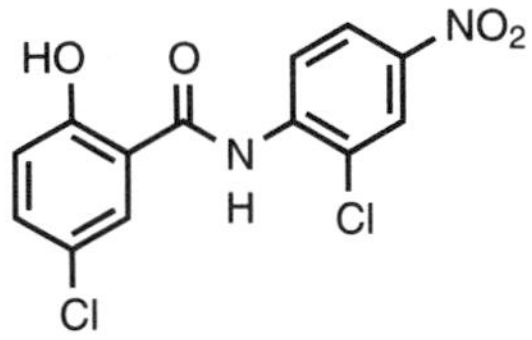
48	Diethylcarbamazine N,N-Diethyl-4-methyl-1-piperazinecarboxamide	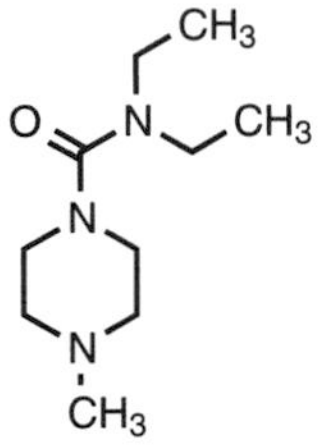
49	Pyrantel (E)-1,4,5,6-Tetrahydro-1-methyl-2-[2-(2-thienyl)ethenyl]-pyrimidine	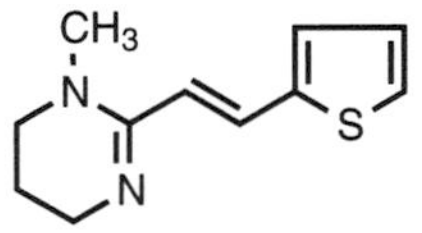
50	Mebendazole (5-Benzoyl-1H-benzimidazol-2-yl)carbamic acid methyl ester	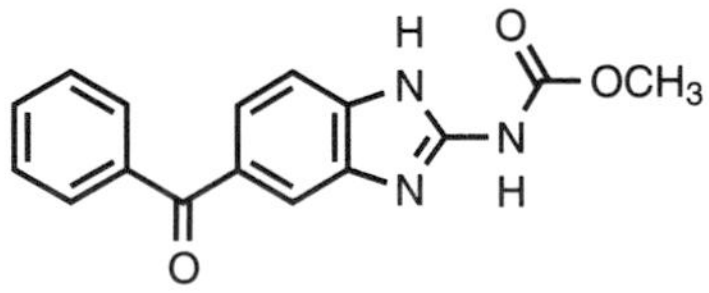
51	Thiabendazole 2-(4-Thiazolyl)-1H-benzimidazole	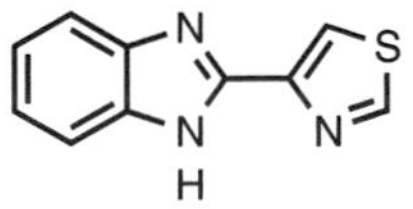

(continued)

Table 32–2. *(Continued)*

Number	Generic Name Chemical Name	Structure
52	Ivermectin	R: B_{1a} = C_2H_3; B_{1b} = CH_3

This toxicity makes their use impractical. **Pamaquine** (7, plasmochin or plasmoquine), the oldest useful member of the 8-aminoquinolines, was synthesized in Germany in the 1920s. **Primaquine** (8) is less toxic and more effective, and is the most widely used for the 8-aminoquinolines. It has gametocytocidal properties against all species of human malaria parasites and is an antirelapsing drug. It causes severe hemolysis in glucose-6-phosphate dehydrogenase (G6PD)-deficient individuals, however. It is considered inadvisable for primaquine to be taken during pregnancy because as it could be passed on to a G6PD-deficient fetus, thereby causing hemolytic anemia. Primaquine is usually coadministered with a blood schizonticidal agent such as chloroquine (4), amodiaquine (5), or **pyrimethamine** (9). It can prevent relapses caused by P. vivax and P. ovale.

Folate antagonists can also act as blood schizonticides. The parasites readily develop resistance to these drugs, however. Most antifolates have poor oral tolerance and absorption, and produce host toxicity. These drugs are divided into two types depending on their mechanisms of action.

The first type includes compounds that are competitors of p-aminobenzoic acid (PABA) and interrupt host de novo formation of the tetrahydrofolic acid required for nucleic acid synthesis. Two examples are sulfones and sulfonamides. The best-known of the sulfones is **dapsone** (10). Its toxicity has discouraged its use, however. Production of folic acid, which requires PABA, a pteridine unit, and glutamate, is disturbed by the incorporation of a sulfonamide, which is structurally similar to PABA. Those sulfonamides used as antimalarials include **sulfadoxine** (11), **sulfadiazine** (12), and **sulfalene** (13), and sulfamethoxypyrazine. Compounds of this group are rapidly absorbed, but cleared very slowly.

The second type of folate antagonists binds preferentially with, and thus selectively inhibits the enzyme dihydrofolate reductase in, the plasmodia. This inhibition interferes with the ability of the malaria parasites to convert dihydrofolate to tetrahydrofolic acid. In the host, however, dihydrofolate reductase is considerably less sensitive to these drugs, and the RBCs are capable of utilizing exogenous folate. Dihydrofolate reductase inhibitors are potent blood schizonticides that act on the asexual blood stages. Members of this class include **pyrimethamine** (9), and **trimethoprim** (14). Pyrimethamine is an effective suppressant against P. falciparum. It is, however, a slow-acting drug that is not recommended for use in acute attacks. Pyrimethamine's action is potentiated by sulfadoxine (11) (commercially, this mixture is called Fansidar [Roche]). Pyrimethamine has been combined with many other drugs, including dapsone (10), sulfalene (13), and chloroquine (4). Trimethoprim (14) has a mode of action similar to that of pyrimethamine (9). Because it is also slow-acting, it is generally used in combination with a fast-acting sulfonamide such as sulfalene (13), to give an effective treatment against P. falciparum.

The biguanides have a mechanism of action similar to that of pyrimethamine and trimethoprim. Proguanil is metabolically cyclized to **cycloguanil** (16), the active drug. Cycloguanil pamoate has a duration of action of several weeks. These compounds have low toxicity and are also effective as causal prophylactic drugs, whereby they destroy parasites upon entering the bloodstream and before they invade the RBCs.

Quinones and naphthoquinones were explored for use as antimalarial drugs during World War II, but this research was not continued. Now that chloroquine resistance is becoming a serious problem, compounds of this group, such as menoctone, (17) are being reevaluated.

Antimalarial agents can be divided into two categories depending on their mode of action. The first category includes the cinchona alkaloids, aminoquinolines, and the acridines. Their action is believed to be caused by the inhibition of nucleic acid and protein synthesis in the protozoal cell. These effects can result from the interaction of the drug and DNA. It has been proposed that the flat aromatic quinoline or acridine ring can position, or intercalate, between the base pairs in the DNA helix, and the secondary alcohol group in quinine, or the amino groups in the other compounds provide secondary binding through hydrogen-bond formation. Because these

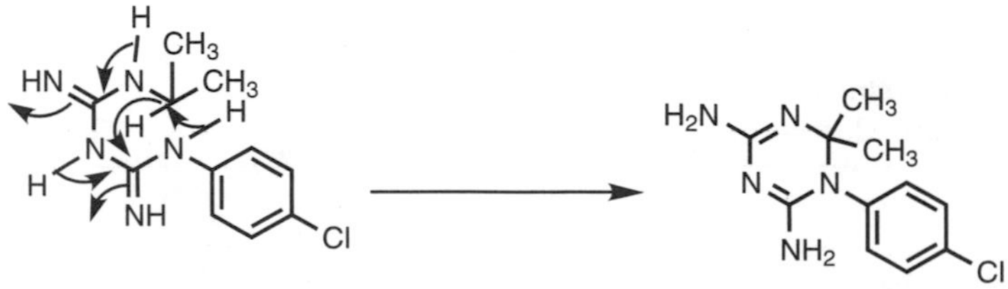

Proguanil

Cycloguanil
(Active metabolite)

events can take place in mammalian host cells as well as in parasite cells, antimalarial action depends on selective accumulation of the drugs in the parasite cell. As in chloroquine, erythrocytic schizonts can concentrate the drug to a level many times that of the host-plasma concentration. Host cells require a 100-fold greater concentration to be affected than that necessary to kill parasite cells.

The second category includes the pyrimines and biguanides, and involves interference with the synthesis of tetrahydrofolic acid. Their mode of action is described more fully in the chapter on antimetabolites. Their action occurs in host as well as parasite cells, but a greater concentration occurs in parasite cells. The sulfonamides and sulfones interfere with the synthesis of dihydrofolic acid, probably by incorporation into folic acid in place of PABA, but the mechanism is not wholly clear. This mechanism does not operate in mammalian cells.

Other Protozoan Diseases

Amebiasis

Amebiasis is caused by pathogenic strains of the protozoan Entamoeba histolytica, which are found either in the stable infective cyst form or in the more fragile trophozoite form. Amebiasis is widespread in humans with an estimated 450 million cases annually. Amebiasis occurs mainly in the tropics, but can also occur in temperate areas with poor sanitation. It is transmitted by ingestion of water and foods contaminated by fecal matter from infected persons or insects, most usually flies. The most common manifestation is dysentery, but the disease can also bring about fever, chills, and intestinal bleeding. Such symptoms may escalate to acute amebic dysentery or colitis. Many amebic infections occur without symptoms. Blood-borne organisms have been reported to cause abscesses in the liver and lungs. It is believed that amebic infections are the third leading cause of death in China. It was shown recently that E. histolytica has both pathogenic and nonpathogenic strains with characteristic isoenzyme electrophoresis patterns, which occur in symptomatic and asymptomatic individuals, respectively.

In general, therapy for amebiasis is based on several stages of the disease including an asymptomatic phase with the presence of cysts in the stool; intestinal symptoms (mild to bloody diarrhea) with the presence of cysts, trophozoites, or both; and an extraintestinal phase that most commonly causes hepatic abscesses, but can involve other tissues (e.g., lung, pericardium, skin, anal regions, or the genitourinary tract).

Acute intestinal amebic dysentery is usually treated with **metronidazole** (18,Flagyl [Searle]) but **iodoquinol** (19, diiodohydroxyquin, diiodohydroxyquinoline), has also been used successfully, particularly in combination with metronidazole or **oxytetracycline** (20, Terramycin [Pfizer]). Patients who cannot tolerate iodoquinol orally are given **emetine** (21) or dehydroemetine (2,3-dehydroemetine). Emetine is an alkaloid found in ipecac and has been in use for over 60 years. Not well tolerated orally, it is given either subcutaneously or intramuscularly. Emetine is sometimes followed by chloroquine (Aralen [Sanofi Winthrop]) phosphate or iodoquinol. Because emetine and dehydroemetine cause cardiac arrhythmias, muscle weakness, and inflammation at the injection site, these toxic compounds are used primarily to rescue patients whose lives are threatened by the disease. Antibiotics, such as tetracycline (Achromycin [Lederle]) and **paromomycin** (22, Humatin [Parke-Davis]), are also effective against moderately severe intestinal amebiasis.

Extraintestinal diseases, such as hepatic amebiasis, are treated with metronidazole. This can be followed by iodoquinol or a combination of dehydroemetine or emetine hydrochloride with chloroquine phosphate. Iodoquinol is most commonly used for asymptomatic amebiasis, and **diloxanide furoate** (23, Furamide [Clin-Comar-Byla]) has long been used to treat symptomatic and asymptomatic intestinal amebic cyst carriers.

Related ameba that cause infections of lower incidence but greater severity are the invasive organisms Naegleria fowleri and species of Acanthomoeba such as A. castellanii and A. polyphaga. These ameba live in soil or warm fresh water, and affect many animals. Entry of Naegleria into the body usually occurs through the nasal passages. Acanthamoeba can enter the host via the eye, skin, or lungs. They cause primary amebic meningoencephalitis (PAM), a usually fatal infection of the CNS. Symptoms in humans include severe headache, confusion, high fever, nausea, seizures, and coma. The olfactory tract is frequently affected. These infections are seen most often in humans, who are immunosuppressed because of AIDS, steroid use, radiation-, or chemotherapy.

Naegleria is treatable with intravenous **amphotericin B** (26, Fungizone [Squibb]), a toxic drug that must be used only with caution to avoid renal damage.

Giardiasis

Giardiasis is a water-borne enteric disease of protozoan origin that occurs throughout the world. It is the most prevalent protozoal disease found in humans in the United States; the disease is particularly highly endemic in the Rocky Mountain area. It is also common in Central America, Russia, and India. This infection is caused by the flagellated protozoan, Giardia lamblia, and infects the small intestine. The disease is transmitted by fecally contaminated food or drinking water. Hosts for this protozoan include the domesticated dog and certain wild animals, especially the beaver. The disease is more prevalent in children than in adults, and is especially common in those attending daycare centers. It is increasing in prevalence among male homosexuals.

Untreated giardiasis may produce no apparent symptoms, regardless of the duration of the infection. It may, however, continue for several weeks or months and become extremely debilitating. Symptoms include severe and chronic diarrhea, abdominal cramps, weakness, anorexia, vomiting, fatigue, and weight loss. In severe giardiasis, there is inflammation and damage to the duodenal and jejunal mucosa.

Treatment of giardiasis is usually with **quinacrine** (27), (mepacrine) or its dihydrochloride, Atabrine (Sanofi-Winthrop). Side effects from this drug include dizziness, headache, nausea, vomiting, and reversible yellowing of the skin. Metronidazole (18) is also recommended by the Centers for Disease Control, but because of its potential carcinogenicity, it should not be administered to children or pregnant women. **Furazolidone** (28) may be more effective in children than quinacrine. The aminoglycoside paromomycin (22) is recommended for pregnant women because the usual drugs are mutagenic and not recommended. **Tinidazole** (24) has been used with some success, and the disease has also been treated with **acranil** (29).

Leishmaniasis

Leishmaniasis is a protozoan infection widespread in the tropics. This disease is transmitted by a bite from an infected female sandfly of the genus Phlebotomus. The parasites typically multiply near the site of the original bite. where they cause a lesion. The sandfly itself acquires the infection along with a blood meal from an infected host. The leishmania parasite has many reservoirs including humans, dogs, cats, rodents, and horses.

Leishmaniasis affects 10 to 15 million people annually in endemic areas where 200 to 400 million humans are at risk of being infected. The two primary forms of leishmaniasis are the visceral and the cutaneous diseases, and a less prominent third form is the mucocutaneous. Leishmanial disease ranges from asymptomatic infection to one that exhibits considerable destruction of cutaneous tissue and mucous membranes. Severe visceral forms can be fatal.

Visceral leishmaniasis is a systemic form of the disease wherein the parasite invades internal organs. It is also known as kalaz-azar, meaning black fever in Hindi. It is characterized in patients by an enlarged liver and spleen, fever, weight loss, hemorrhage, leukopenia, and anemia. There is abdominal discomfort due to the spleen and liver involvement and their possible enlargement. Death caused by severe diarrhea, pneumonia, and gastrointestinal bleeding may ensue if the disease is not treated. Following treatment, patients may relapse with a leishmanial form that solely affects the skin (post-kala-azar dermal leishmaniasis).

Cutaneous leishmaniasis is characterized by one or more slow-healing superficial, but possibly painful, ulcers. These lesions are liable to secondary bacterial infection and may remain as open sores or become hard, wartlike nodules. The form of cutaneous leishmaniasis referred to as New World disease is caused by subtypes of Leishmania braziliensis (L. braziliensis, L. panamanesis, and L. peruviana) and subtypes of Leishmania mexicana (L. mexicana, L. amazonensis, and L. venezualensis) in the western hemisphere. Old World disease (oriental sore, Delhi boil) is caused by L. tropica and L. major in Asia, Africa, and southern Europe, and L. aethiopica, in eastern Africa. The incubation period for cutaneous leishmaniasis ranges from a few weeks to several months, and spontaneous cures can take place in a period ranging from one month to several years.

Certain types of leishmania (L. braziliensis) may cause mucocutaneous leishmaniasis, a form most frequently seen in South America. The affected patient initially develops skin infections that later escalate to produce highly disfiguring mucocutaneous lesions in and about the face.

Types of Chemotherapy

The primary treatments for all types of leishmaniasis are based on preparations of pentavalent antimony, following the discovery that tartar emetic had some effect on the mucocutaneous type. The two most important of these are **sodium stibogluconate** (30, Pentostam [Burroughs Wellcome]) and glucantime, (N-methyl glucamine antimonate, meglumine antimonate). The exact chemical structures of these antimonials are not known. Studies indicate that they are mixtures of compounds with widely differing molecular weights. Cardiac and hepatic toxicity have been observed when the drug is administered over the usual 3 to 4 weeks of treatment. In 1990, an estimated 5% of the cases of kala azar reported in India were unresponsive to treatment with antimonials. Some strains more resistant to antimony are found in Kenya. Although antimony-resistant strains of leishmania can be treated with diamidines, these compounds are also relatively toxic. **Pentamidine** (39), as the isethionate salt, is the most widely used diamidine, and has a high cure rate. Another diamidine, **stilbamidine** (31), is also effective.

Amphotericin B (26), an antifungal macrolide antibiotic produced by Streptomyces nodosus, has been used as an alternative drug to the antimonials. It serves as a leishmanicide against the visceral and mucocutaneous forms of the disease, but causes severe nephrotoxicity. The drug must be administered at low doses over an extended period.

Because of the problem of antimony-resistant leishmaniasis and the need to develop orally-effective therapy, many other compounds have been investigated. Those demonstrating some clinical utility are **allopurinol** (32), **ketoconazole** (33), and **paromomycin** (22).

Pneumocystis

The organism responsible for pneumocystis (pneumocystosis) in humans is Pneumocystis carinii. It has the morphologic characteristics of a protozoan and its DNA pattern resembles that of fungi. Acute pneumocystis rarely strikes healthy individuals, although the organism is harbored in a wide variety of animals and most humans without any apparent adverse effect. P. carinii becomes active only in those individuals who have a serious impairment of their immune systems. More recently, this disease has appeared in AIDS patients, 80% of whom ultimately contract P. carinii pneumonia, one of the main causes of death among AIDS patients. The disease also occurs in those receiving immunosuppressive drugs to prevent rejection following organ transplantation or for the treatment of malignant disease. Pneumocystis is also seen in malnourished infants whose immunologic systems are impaired. The disease is characterized by a severe pneumonia caused by a rapid multiplication of the organisms almost exclusively in lung tissue. Untreated, the acute form of the disease is generally fatal.

The most successful drug is co-trimoxazole (Bactrim [Roche]), a mixture of trimethoprim (14) and **sulfamethoxazole** (34) given orally or intravenously. Pyrimethamine (9) is also combined with either sulfadiazine (12) or sulfadoxine (11). Pentamidine (39) isethionate has had some use, but is generally not prescribed for prophylactic use because of the high incidence of hypotension, renal failure, pain and tissue injury at the site of injection. Prophylactic trials with the drug, however, in an aerosol form for inhalation, were successful. **Eflornithine** (35) is also effective against this disease. A combination of **clindamycin** (36) and primaquine (8) shows promise against this protozoan when administered for prophylaxis, or as a therapy in mild to moderately severe cases. Neither drug is effective separately.

Trypanosomiasis

This disease is found in two distinctly different forms: African and American trypanosomiasis. These two varieties will be discussed separately.

African trypanosomiasis affects 20 to 30 thousand people each year. It is endemic in several subSaharan countries where more than 50 million inhabitants are at risk of infection. The disease in humans is called sleeping sickness, and is transmitted by the bite of the tsetse fly. Trypanosomes travel throughout the body, eventually locating in the cerebrospinal fluid and the CNS. African trypanosomiasis is caused by Trypanosoma brucei gambiense or its sister species, Trypanosoma brucei rhodesiense, which is a more acute form of the disease with more rapid progression. Trypanosomiasis in cattle is common. The disease has a high mortality rate if untreated.

Suramin Sodium (37) is the most effective treatment for the early stages of the disease. Because this drug does not cross the blood-brain barrier, it is not effective in the later stages following CNS involvement, including the manifestation of sleeping sickness. Other early stage therapeutics include suramin in combination with either **tryparsamide** (38) or pentamidine (39). **Melarsoprol** (40), in combination with 2,3-dimercaptopropanol or British Anti Lewisite (BAL [Hynson, Westcott & Dunning]) and melarsen oxide (Arsobal [Specia]) is effective in advanced stages of the disease. This drug combination has serious side effects, including myocarditis, renal, and hepatic damage.

Eflornithine (35), a drug approved by the FDA in 1990, irreversibly inhibits ornithine decarboxylase, and inhibits the growth and multiplication of the parasite in both T. brucei gambiense and T. brucei rhodesiense. A mixture of eflornithine (35) and **nifurtimox** (41) also shows good activity against T. brucei gambiense.

American trypanosomiasis, also known as Chagas' disease, is found in Central and South America. It is caused by the protozoan parasite, Trypanosoma cruzi, which is carried and stored in dogs, cats, pigs, opossums, monkeys, bats, and other rodents. It affects 15 to 20 million people each year and is present in areas in which more than 85 million people are at risk of contracting the disease. Children are more susceptible to this infection than adults. The disease is transmitted by the blood-sucking triatoma bug, (reduviidae, cone-nosed bug). The trypanosomes migrate through the body and penetrate the cardiac muscle, and then circulate to all parts of the body through the blood stream.

This disease has two phases: acute and chronic. The acute phase has a duration of 60 days during which the parasite spreads. The infection may be asymptomatic or marked by fever, abdominal and esophageal distension, liver and spleen enlargement, acute myocarditis, and rash. In this serious stage in children, most fatalities occur. The chronic phase is marked by slow tissue loss that can persist for several years. There may be cardiac muscle and digestive tract involvement. The cardiac complications constitute the primary cause of death in either the acute or the chronic stages of Chagas' disease.

Nifurtimox (41) is the most effective drug against trypanosoma cruzi. It is active against all forms of the parasite. In combination with cortisone, it is capable of eliminating the parasite from the cardiac muscle. **Benznidazole** (42) is another effective drug that reduces inflammation and destroys parasites within the heart muscles. Primaquine (8) can destroy parasites in the bloodstream. It is ineffective, however, against cellular or tissue stages of the parasite.

HELMINTHIC DISEASES

Helminthiasis, or worm infestation, is one of the most prevalent diseases and one of the most serious public health problems in the world. Many worms are parasitic in humans and cause serious complications. Helminths that infect human hosts are divided into two categories, or phyla. These are: Platyhelminths (flatworms) and Aschelminths (roundworms). Platyhelminths include the classes Cestoda and Trematoda. Members of the class Trematoda are the flukes or schistosomes. Members of

Table 32–3. Common Worm Parasites of Man and Their Treatment

Common Name	Genus and Species	Disease	Organs Affected	Chemotherapy
Blood flukes	Schistosoma haematobium S. mansoni S. japonicum	Schistosomiasis	Intestines Bladder Liver Lungs	Niridazole Sodium antimony dimercaptosuccinate Stibophen Lucanthone Oxamniquine Praziquantel
Chinese liver fluke	Clonorchis sinensis	Clonorchiasis	Liver	Chloroquine
Tapeworms	Taenia saginata T. solium	Cysticercosis	Intestines Muscle (cysts)	Niclosamide Quinacrine Paromomycin
Roundworm	Ascaris lumbricoides	Ascariasis	Liver Lungs Intestines	Pyrantel pamoate Mebendazole Piperazine Tetramisole
Hookworms	Ancylostoma duodenale Necator americanus	Ancylostomiasis	Intestines	Mebendazole Albendazole Pyrantel pamoate Bephenium
Whipworm	Trichuris trichiura	Trichuriasis	Intestines	Mebendazole Albendazole Thiabendazole
Pinworm	Enterobius vermicularis	Enterobiasis	Intestines Perianal mucosa	Pyrantel pamoate Mebendazole Albendazole Pyrvinium pamoate Piperazine
Trichina	Trichinella spiralis	Trichinosis	Muscle	Thiabendazole
Filariasis and tropical eosinophilia	Wuchereria bancrofti	Elephantiasis Lymphangitis	Lymph	Diethyl carbamazine
	Loa loa	Loaiasis	Brain	
	Onchocerca volvulus	Erysipelas	Subcutaneous tissue Eyes	plus Suramin

the Cestoda class (tapeworms) are flat and ribbonlike. The Aschelminths include the Nematoda class, and are cylindrical with significant variations in size, proportion, and structure. The common helminths are listed in Table 32–3.

Cestoda Infections

Tapeworms

Members of this class that are of concern as parasites in humans are of the following types.

Beef Tapeworm (Taenia saginata). This worm is found world-wide and infects people who eat undercooked beef. This worm reaches a length of over 5 m and contains about 100 segments/m. Each of these segments contains its own reproductive organs.

Pork tapeworm (Taenia solium). Pork tapeworms are sometimes called bladder worms and are occasionally found in uncooked pork. The worm attaches itself to the intestinal wall of the human host. The adult worm reaches 5 m in length, if untreated, survives in the host for many years.

Dwarf tapeworm (Hymenolepis nana). This infection is transmitted directly from one human to another without an intermediate host. H. nana reaches only 3 to 4 cm in length. It is found in temperate zones, and children are most frequently infected.

Fish tapeworm (Diphyllobothrium latum). This worm reaches a length of 10 m and contains about 400 segments/m. These tapeworms attach themselves to the intestinal wall and rob the host of nutrients. They especially absorb vitamin B_{12} and folic acid. Depletion of some of these critical nutrients, especially vitamin B_{12}, can lead to pernicious anemia. Tapeworm eggs are passed in the patient's feces. Contamination of this type results in transmission of the infestation.

Chemotherapeutic Agents

Niclosamide (47). This is a halogenated salicylanilide effective in one dose against tapeworm infections. It interferes with respiration and glucose uptake of the parasite. Niclosamide also inhibits uptake of inorganic phosphate into adenosine triphosphate (ATP). It causes the tapeworm to detach itself from the intestinal wall and die. A repeat dose 1 week later is recommended to destroy newly hatched and young worms.

Quinacrine (27). This drug was used widely before the emergence of niclosamide. It is less effective and causes more side-effects than niclosamide. It is still preferred by some clinicians, however, because it is effective in causing killed parasites to be expelled intact, especially T. solium. Dead parasites are stained yellow by action of this dye-like drug. Quinacrine is administered orally and unlike nicloramide is absorbed readily from the intestinal tract. It has been found in patient tissues several weeks after administration. It can intercalate with DNA and interfere with nucleic acid synthesis.

Paromomycin (22). A broad-spectrum antibiotic that serves as an antibacterial agent in ameba-related diarrhea, it has also been used in the treatment of infections from H. nana and T. saginata.

Trematoda Infections

Blood Flukes, Schistosomiasis

There are three primary species that cause schistosomiasis in man: Schistosoma hematobium, S. mansoni, and S. japonicum. Infections result from the penetration of the normal skin by living (free-swimming) cercaria.

The cercaria develop to preadult forms in the lungs and skin. Then these parasites travel in pairs via the bloodstream and invade various tissues. The adult worm reaches approximately 2 cm. The female deposits her eggs near the capillary beds, where granulomas form. The patients might experience headache, fatigue, fever, and gastrointestinal disturbances during the early stages of the disease. Hepatic fibrosis and ascites occur in later stages. Untreated patients can harbor as many as 100 pairs of worms. Untreated worms can live 5 to 10 years within the host.

Chemotherapeutic Agents

Praziquantel (43). This is one of the leading agents used to treat schistosomiasis. It is an acylated pyrazino-isoquinoline that is preferred to over more toxic traditional trivalent antinomial drugs. Primary action of praziquantel is thought to alter the permeability of the worm's cell membrane to calcium. This causes paralysis and dislodging of the worms. Cure rate with this drug has been reported to reach as high as 80% with a single treatment regimen. Adverse reactions include headache, gastrointestinal disturbance, and dizziness.

Oxamniquine (44). A tetrahydroquinoline effective against Schistozoma mansoni infection, it is thought to exhibit anticholinergic properties, but the exact mode of action is not clearly defined. Although the drug is well tolerated, dizziness has been reported by a third of the patients receiving it. This drug also causes allergic reactions such as skin rashes, fever, and pulmonary infiltration.

Metrifonate (45). This organophosphate was one of the earliest insecticides used. As a drug, it is converted to an active metabolite (2,2-dichlorovinyl dimethylphosphate), which inhibits acetylcholinesterase of the schistosomal form and perhaps causes paralysis of the worm. This drug was found to be active against S. hematobium only. Metrifonate is given in 3 doses at 2-week intervals. Side effects of this drug include nausea, vertigo, and abdominal cramps.

Lung and Liver Flukes. These Fasciola organisms are 2 to 3 cm long. They mature in the bile ducts of the host's liver. Fibrosis of the bile duct wall may result in severe infection. Tissue reaction and damage is found throughout the peritoneal cavity due to migration of these worms. The liver flukes, like the lung and intestinal flukes, are transmitted through fecal contamination from infected individuals.

Lung flukes (Paragonimus organisms) average about 1 cm in diameter and live in cysts in the host's lung. Although they primarily infect lung tissue, they can be found in other parts of the viscera or even in brain tissues, where they cause tumor-like symptoms.

Both liver and lung flukes are treatable by **Bithionol** (25). This biphenolic compound is very effective against both the paragonimus and fasciola organisms. Bithionol interferes with the neuromuscular physiology of these worms. It inhibits egg formation, inhibits oxidative phosphorylation, and damages the protective cuticle that covers the worm. Bithionol chelates iron readily and may inactivate iron-containing enzymes. The usual treatment against both organisms is a 10-day regimen. Side effects include diarrhea, vomiting, abdominal pain, photosensitivity, and possible rash.

Chinese liver flukes (Clonorchis sinensis) can grow to 2 cm. They infect the biliary tract and cause inflammation, diarrhea, and liver enlargement. Biliary obstruction and cirrhosis can occur in severe, untreated cases. Untreated Chinese liver flukes can live in the host for up to 30 years. Praziquantel is the recommended therapy against this infection.

Intestinal flukes (Fasciolopsis buski) live by attaching themselves to the wall of the small intestine. They grow to a length of 5 to 8 cm. Severe, untreated infections result in toxemia and intestinal ulceration and are more likely to cause death in children. Praziquantel (43) and **tetrachloroethylene** (46) are effective against this organism.

Nematodoa Infections

Hookworms

The two most widespread types of hookworm in humans are the American hookworm (Necator americanus) and the common hookworm (Ancylostoma doudenale). The lifecycles of both are similar. The larva are found in the soil and are transmitted either by skin penetration or are ingested orally. The circulatory system transports these via the respiratory tree to the digestive tract where they live for 9 to 15 years if left untreated. These worms feed on intestinal tissue and blood. Infestations cause pulmonary lesions, skin reactions, intestinal ulceration, and anemia.

Pinworms (Enterobius vermicularis)

These are widespread in temperate zones, and are a common infestation of households and institutions. The pinworm lives in the lumen of the gut, attaching itself by the mouth. Mature worms reach 10 mm. The female migrates to the rectum to deposit her eggs. This event is noted by the patient, because it causes considerable itching. Eggs resist drying and can be inhaled with household dust to continue the lifecycle.

Roundworms (Ascaris lumbricoides)

Most common in developing countries, the adult roundworm, which reaches 25 to 30 cm, lodges in the small intestine. Some infections are without symptoms, but abdominal discomfort and pain are common with this infection. Roundworm eggs are ingested with contaminated food or beverages and hatch in the stomach. Some patients have reported adult worms exiting the esophagus through the oral cavity. It is not unusual for live ascaris to be expelled with a bowel movement.

Whipworms (Trichuris trichiura)

Infections by this parasite are caused by swallowing eggs with contaminated food and beverages. Eggs are passed with the feces from an infected individual. They live in the soil for many years. Adult worms, which reach about 5 cm, thread their bodies into the epithelium of the colon. They feed on tissue fluids and blood. Infections from this worm cause symptoms of irritation and inflammation of the colonic mucosa, abdominal pain, diarrhea, and distention. Infections can last 5 or more years if not treated.

Guinea worms (Dracunculus medinensis)

These string-like worms that measure about 1 cm long and 2 mm in diameter. They live under the skin surface with the head often visible in a sore or a skin ulcer, usually on the feet. These worms have a hook at the posterior end, which makes them difficult to extricate from a sore. Broken worms, as a result of an attempt to pull these out of an ulcer, can cause irritation and secondary infections. Some worms exit the wound spontaneously after 3 to 4 weeks. Metronidazole and thiabendazole have little effect on the worm, but their anti-inflammatory effect is helpful in the therapy against infections.

Chemotherapeutic Agents

Piperazine salts, especially the citrate, are effective against roundworms and pinworms. **Diethylcarbamazine** (48) is the only piperazine currently used, however. One daily dose for 2 consecutive days is sufficient to cure 90% of roundworm cases. It is possible to eradicate pinworms in a 7-day regimen. Piperazine salts' mode of action against Ascaris is thought to be the blocking of acetylcholine regeneration. This cycle disruption causes the worm to dislodge and fall into the feces where it is excreted. Side effects of piperazines include nausea, vomiting, abdominal cramps, diarrhea, headache, vertigo, occasional tremor, and lethargy.

Pyrantel (49) pamoate is effective against round worm in a single dose but, most importantly, it cures pinworm and hookworm infections. It is extremely valuable as a therapeutic agent in cases with multiple helminthic infections. A 3-day regimen is highly effective against hookworm infestations.

Mebendazole (50) is a broad-spectrum anthelminthic drug useful against ascaris, hookworms, and tapeworms. It is exceptionally effective against whipworms. This drug is administered orally for 3 to 4 days. A single dose, however, is sufficient to eradicate pinworms. The mode of action of mebendazole is believed to be inhibition of uptake of glucose and interference with glucose metabolism, resulting in depletion of ATP in the worms.

Thiabendazole (51) is an effective oral drug for the treatment of ascaris and some tissue-dwelling parasites. This was the first broad-spectrum anthelminthic drug. Although the mode of action of this drug is not clearly defined, evidence has shown that it inhibits the enzyme fumarate reductase, which is regularly found in the helminthic mitochondrium. Side effects include neurologic involvement, digestive disturbances, and various allergic responses. Thiabendazole and metronidazole are useful in treatment against guinea worm.

Filariasis

The term filiariasis denotes infections with any of the Filarioidea, although it is commonly used to refer to lymphatic-dwelling filaria, such as Wuchereria bancrofti, Brugia malayi, and Brugia timori. Other filarial infections include Loa Loa and Onchocerca volvulus. The latter two are known as the eyeworm and river blindness worm, respectively. Elephantiasis is the most common disease associated with filariasis. These parasites vary in length from 6 cm for brugia to 50 cm for onchocerca. The incubation periods also vary from 2 months for the brugia to 12 months for the bancroftian filaria.

Diethylcarbamazine (48) is effective against all lymphatic filarias. It kills the blood stages (microfilariae) of most species and has a limited effect against microfilariae of Onchocerca volvulus. **Ivermectin** (52) has gained global acclaim in recent years as a potent microfilarial agent, especially in the treatment of onchocerciasis (river blindness). Ivermectin, however, has no effect against the adult macrofilarial nematodes. Ivermectin is a semisynthetic derivative of the avermectin family of natural products. It is a mixture of two analogs: 80% dihydroavermectin B1a, in which $R=C_2H_5$, and 20% of B1b, in which $R=CH_3$. It is a broad-spectrum antiparasitic drug that acts on the parasites by binding to GABA at postsynaptic sites on the neuromuscular junction to paralyze the organisms. Ivermectin is a slow-acting drug with a long duration of action. It is administered orally for suppressive therapy every 6 to 18 months. The most common side effects are lymph node swelling with some pain, fever, and rash.

SUGGESTED READINGS

J. Becker, *Anthelminthics*, in *Kirk-Othmer Encyclopedia of Chemical Technology*, 4th ed., Vol. 3, New York, Wiley, 1992, pp. 456[n]472.

A. S. Benenson, Ed., *Control of Communicable Diseases in Man,* 15th ed., Washington, D.C., Amer. Public Health Assoc., 1990.

H. M. Bryson, *Drugs, 43*(2), 236(1992).

W. C. Campbell and R. S. Rew, *Chemotherapy of Parasitic Diseases,* New York, Plenum, 1986.

G. C. Cook, *Aust. N.Z. J. Med., 22,* 69(1992).

G. C. Cook, *Parasitic Diseases in Clinical Practice,* New York, Springer-Verlag, 1990.

J. E. Heck, *Malaria, Parasitic Diseases, Primary Care, 18*(1), 195(1991).

Informational Material for Physicians, Ivermectin, Atlanta, Centers for Disease Control, 1986.

D. L. Klayman, *Antiprotozoals,* in *Kirk-Othmer Encyclopedia of Chemical Technology,* 4th ed., Vol. 3, New York, Wiley, 1992, 99 489[n]526.

Merck Index, 11th ed., Rahway NJ, Merck, 1989.

J. A. Najera, *Parasitologia, 32,* 215(1990).

C. Naquira, et al., *Am. J. Trop. Med. Hyg., 40,* 304(1989).

E. A. Nodiff, et al., *Prog. Med. Chem., 28,* 1(1991).

E. B. Roche, et al., *Parasite Chemotherapy,* in *Principles of Medicinal Chemistry,* 3rd Ed., W.O. Foye, Ed., Philadelphia, Lea & Febiger, 1989, pp. 717–730.

J. E. Rosenblatt, *Mayo Clin. Proc., 67*(3), 276(1992).

A. F. G. Slater and A. C. Cerami, *Nature, 355*(9), 167(1992).

G. T. Strickland, Ed., *Hunter's Tropical Medicine,* 7th ed., Philadelphia, W.B. Saunders, 1991.

W. H. Wernsdorfer and D. Payne, *Pharmacol. Ther., 50,* 95(1991).

Chapter 33

ANTIMYCOBACTERIAL AGENTS

Thomas L. Lemke

MYCOBACTERIAL DISEASES

Mycobacteria are a genus of acid-fast bacilli belonging to the Mycobacteriaceae which include the organisms responsible for tuberculosis and leprosy, as well as other less common diseases.

Leprosy (Hansen's Disease)

Throughout the *Bible* one finds reference to the condition of leprosy such as that described in Leviticus: "is there any flesh in the skin of which there is a burn by fire, and the quick flesh of the burn becomes a bright spot, reddish white or white, ... and if the hair in the bright spot is turned white, and it appears deeper than the skin, it is leprosy broken out in the burn." Associated with the disease was a belief that individuals suffering from this disease were unclean. Today leprosy is recognized as a chronic granulomatous infection caused by Mycobacterium leprae. The disease may consist of lepromatous leprosy, tuberculoid leprosy, or a condition with characteristics between these two poles referred to as so-called borderline leprosy. The disease is more common in tropical countries, but is not limited to regions with a warm climate. It is thought to afflict some 10 to 20 million individuals. Whereas children appear to be the most susceptible population, the signs and symptoms usually do not occur until much later in life. The incubation period is 3 to 5 years. Whereas the disease is contagious, the infectiousness is quite low. Person-to-person contact appears to be the means by which the disease is spread with entrance into the body occurring through the skin or through the mucosa of the upper respiratory tract. Skin and peripheral nerves are the regions most susceptible to attack. The first signs of the disease consist of hypopigmented or hyperpigmented macules. Additionally, anesthetic or paresthetic patches may be experienced by the patient. Neural involvement in the extremities ultimately leads to muscle atrophy, resorption of small bones, and spontaneous amputation. When facial nerves are involved, corneal ulceration and blindness may occur. The identification of M. leprae in skin or blood samples is not always possible, but the detection of the antibody to the organism is an effective diagnostic test, especially for the lepromatous form of the disease.

Tuberculosis

Tuberculosis (TB) is a disease known to man from the earliest recorded history. It is a disease that is characterized as a chronic bacterial infection caused by M. tuberculosis, an acid-fast aerobic bacillus with an unusual cell wall. The cell wall has a high lipid content resulting in a high degree of hydrophobicity and resistance to alcohol, acids, alkali, and some disinfectants. The M. tuberculosis cell wall after staining with a dye, cannot be subsequently decolorized with acid wash and thus the characteristic of being an acid-fast bacillus (AFB). It is estimated that today a third to a half of the world's population is infected with M. tuberculosis, with tuberculosis causing 6% of all deaths worldwide.[1,2] Tuberculosis is the infectious disease with the highest worldwide death rate. Reported cases of TB had been declining in the United States from the 1950s until 1985. From 1985 until 1988, this decline leveled off, but beginning in 1989 an increase was noted. In 1991, the Center for Disease Control reported 25,701 new cases of TB. Today, the press and professional publications announce the "epidemic" spread of TB. The resurgence has been linked to urban crowding, homelessness, immigration, drug abuse, the disappearance of preventive-medicine health clinics, crowded prisons, and the AIDS epidemic. "Most alarming is the emergence of multidrug-resistant TB (MDR-TB)".[3] Prior to 1984, only 10% of the organisms isolated from patients with TB were resistant to any drug. In 1984, 52% of the organisms were resistant to at least one drug and 32% were resistant to more than one drug. MDR-TB may have a fatality rate as high as 50%. As a result of MDR-TB, isolates of M. tuberculosis should be tested for antimicrobial susceptibility.

M. tuberculosis is transmitted primarily via the respiratory route. The organism appears in water droplets expelled during coughing, sneezing, or talking. Either in the droplet form or as the desiccated airborne bacilli, the organism enters the respiratory tract. The infectiousness of an individual depends on the extent of the disease, the number of organisms in the sputum, and the amount of coughing. Usually, within 2 weeks of beginning therapy, the infected individual will no longer be infectious. Tuberculosis is a disease that mainly affects the lungs (80 to 85% of the cases), but the M. tuberculosis can spread through the bloodstream and the lymphatic system to the brain, bones, eyes, and skin (extrapulmonary TB). In pulmonary TB, the bacilli reach the alveoli and are ingested by pulmonary macrophages. Substances secreted by the macrophages stimulate surrounding fibroblasts to enclose the infection site leading to formation of granulomas or tubercles. The infection, thus contained locally, may lie dormant encapsulated in a fibrotic lesion for years only to reappear later. Extrapulmonary TB is much more common in HIV-infected patients (40 to 75%).

MAC (M. avium-intracellulare Complex)

Approximately half of all AIDS patients develop an infection caused by M. avium and M. intracellulare. The organisms are difficult to distinguish and are considered a complex, thus, the acronym MAC. The lungs are the organs most commonly involved in nonAIDS patients but the infection may involve bone marrow, lymph nodes, liver, and blood in AIDS patients.

DRUG THERAPY

Tuberculosis

First-Line Agents

Isoniazid (Nydrazid [Squibb], Laniazid [Lannett]). Isoniazid (INH) is a synthetic antibacterial agent with bactericidal action against M. tuberculosis. The drug was discovered in the 1950s as a beneficial agent effective against intracellular and extracellular bacilli and is generally considered the primary drug for treatment of M. tuberculosis. Its action is bactericidal against replicating organisms but appears to be only bacteriostatic against nonreplicating organisms. After treatment with INH, the M. tuberculosis loses its acid fastness, which may be interpreted as indicating that the drug interferes with cell wall development.

Isoniazid

Mechanism of Action (MOA). Although extensively investigated, the MOA of INH remains unknown. A variety of mechanisms have been proposed, several of which suggest a role in which INH interferes with NAD^+.[4] One mechanism involves a role in which INH affects the electron transport system and the conversion of NAD^+ to NADH. A very attractive theory is that INH is converted to isonicotinic acid, which in turn acts as an antimetabolite of nicotinic acid. Isonicotinic acid is then incorporated into NAD^+ in place of nicotinic acid. The fraudulent NAD^+ is not able to catalyze the normal oxidation/reduction reactions (Fig. 33–1). Unfortunately, the amount of data to support this mechanism is minimal. A third mechanism suggests INH as in inhibitor of the conversion of C_{24} and C_{26} saturated fatty acids to C_{24} and C_{26} unsaturated acids by blocking desaturase.[5] These acids appear to serve as precursors to mycolic acid, a key component in bacterial cell walls: The uniqueness of M. tuberculosis cell wall would account for the selectivity of INH if this mechanism were operating.

Isonicotinic acid   Nicotinic acid

Structure-Activity Relationship. An extensive series of derivatives of nicotinaldehyde, isonicotinaldehyde, and substituted isonicotinic acid hydrazide have been prepared and investigated for their tuberculostatic activity. Isoniazid hydrazones were found to possess activity but these compounds were shown to be unstable in the GI tract releasing the active isonicotinic acid hydrazide (INH). Thus, it would appear that their activity resulted from the INH and not the derivatives.[6,7] Substitution of the hydrazine portion of INH with alkyl and aralkyl substitu-

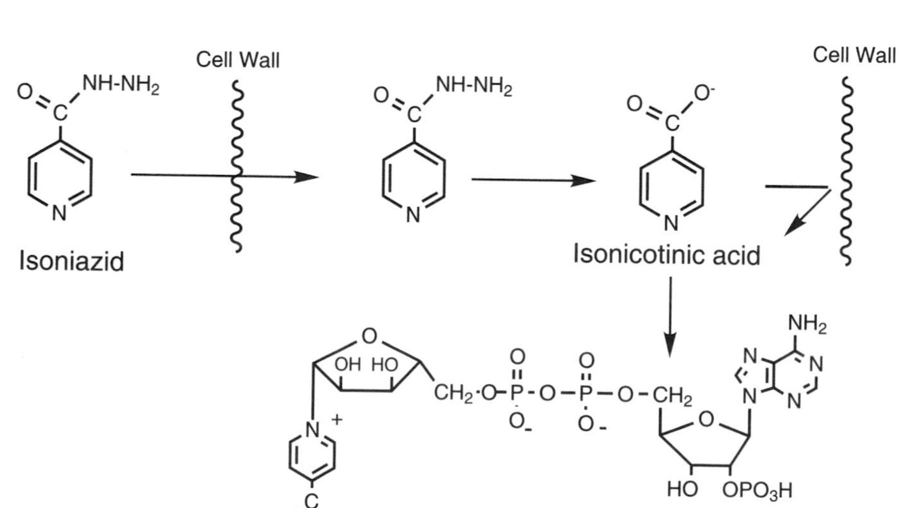

Fig. 33–1. Biosynthesis of fraudulent NAD^+ from INH.

ents, isonicotinic acid hydrazides, resulted in a series of active and inactive derivatives.[8–11] Substitution on the N_2 position resulted in active compounds (R_1 and R_2 = alkyl; R_3 = H), whereas any substitution of the N_1 hydrogen with alkyl groups destroyed the activity (R_1 and R_2 = H; R_3 = alkyl). None of these changes produced compounds with superior activity over INH.

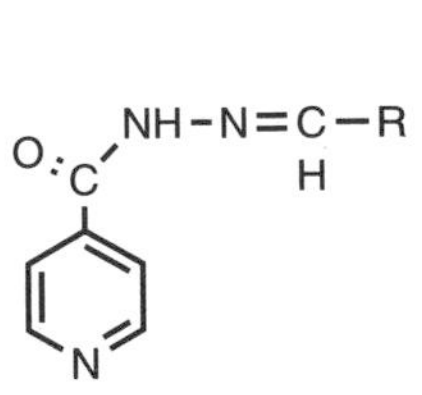

Isoniazid hydrazones 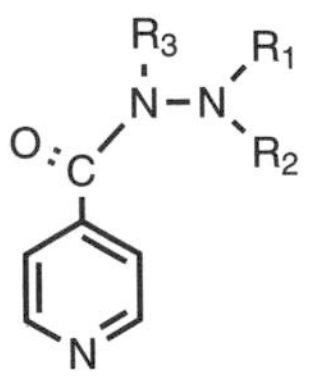   Isonicotinic acid hydrazides

Metabolism. Isoniazid is readily absorbed following oral administration. Food and various antacids, especially antacids containing aluminum, may interfere or delay absorption and therefore it is recommended that the drug be taken on an empty stomach. The drug is well distributed to body tissues including infected tissue. INH is extensively metabolized to inactive metabolites (Fig. 33–2).[12,13] The major metabolite is N-acetylisoniazid. The enzyme responsible for acetylation, cytosolic N-acetyltransferase, is produced under genetic control in an inherited autosomal fashion. Individuals who possess high concentrations of the enzyme are rapid acetylators whereas those with low concentrations are slow acetylators. This may result in a need to adjust the dosage for fast acetylators. The N-acetyltransferase is located primarily in the liver and small intestine. Other metabolites include isonicotinic acid, which is found in the urine as a glycine conjugate, and hydrazine. Isonicotinic acid may also result from hydrolysis of acetylisoniazid, but in this case, the second product of hydrolysis is acetylhydrazine. Acetylhydrazine is acetylated by N-acetyltransferase to diacetylhydrazide. This reaction occurs more rapidly in rapid acetylators. The formation of acetylhydrazine is significant in that this compound has been associated with the hepatotoxicity that occurs during isoniazid therapy. Acetylhydrazine has been postulated to serve as a substrate for microsomal P-450 resulting in the formation of a chemical capable of acetylating liver protein thus causing liver necrosis.[14] It has been suggested that a hydroxylamine intermediate, is formed, which in turn results in an active acetylating agent (Fig. 33–3). The acetyl radical/cation acylates liver protein.

Rifampin (Rifadin [Merrell Dow], Rimactane [Ciba]). Rifampin (RIF) is a semisynthetic agent prepared from rifamycin B, an antibiotic isolated from Streptomyces mediterranei. This molecule is a member of the ansamycins class of natural products. This class is characterized as molecules with an aliphatic chain forming a bridge between two nonadjacent positions of an aromatic moiety. Rifamycin B is lacking a substituent at C-3 that

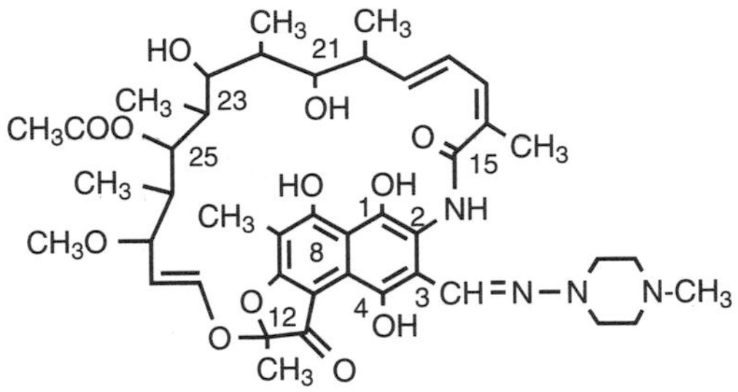

Rifampin

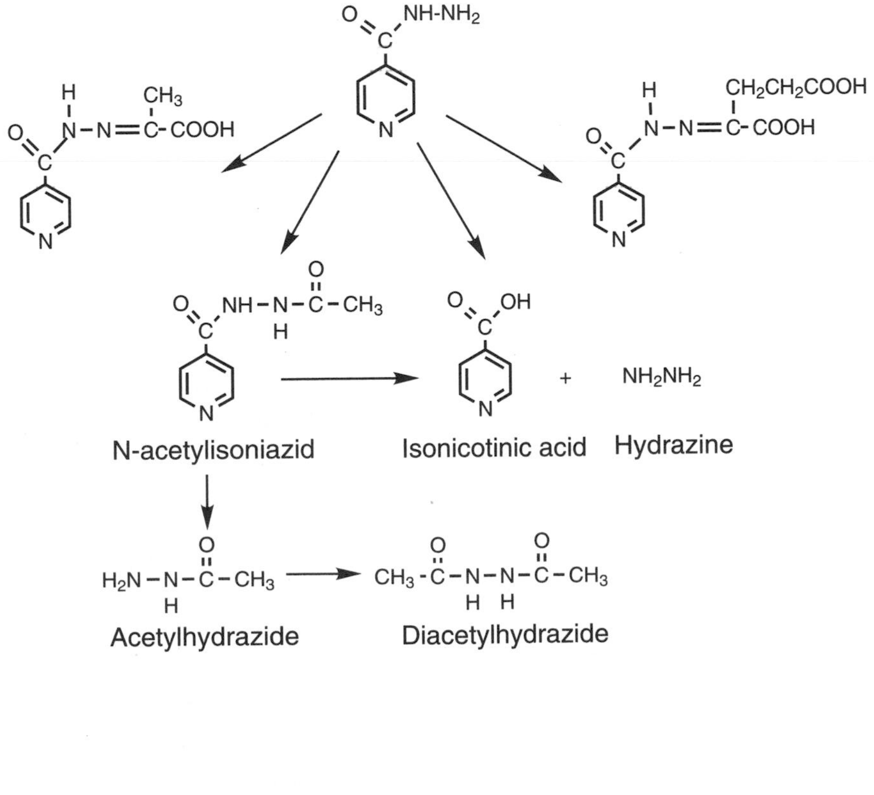

N-acetylisoniazid   Isonicotinic acid   Hydrazine

Acetylhydrazide   Diacetylhydrazide

Fig. 33–2. Metabolites of isoniazid.

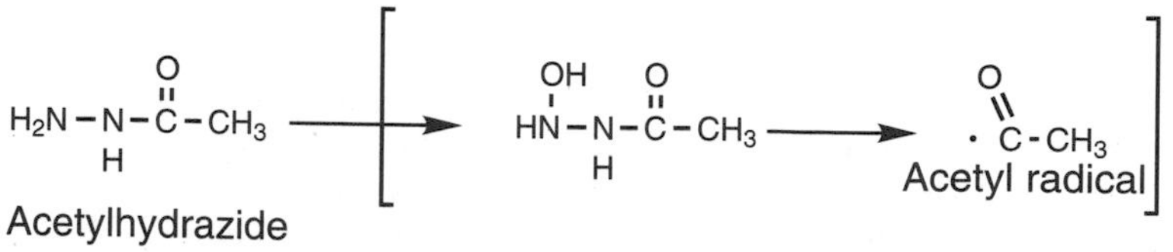

Acetylhydrazide → [ ] → Acetyl radical

Fig. 33–3. Acylating metabolite of isoniazid.

is introduced in the form of a formyl group and then derivitized as a hydrazone with a variety of hydrazines.[15] Rifampin as a derivative of rifamycin B gains the advantage of being orally active, highly effective against a variety of gram-positive and gram-negative organisms, and of high clinical efficacy in the treatment of TB.

Mechanism of Action. Rifampin inhibits bacterial DNA-dependent RNA polymerase (DDRP) and is highly active against rapidly dividing intra- and extracellular bacilli. Rifampin is active against DDRP from both gram-positive and gram-negative bacteria, but because of poor penetration of the cell wall of gram-negative organisms by rifampin, the drug has less value in infections caused by such organisms. Inhibition of DDRP leads to a blocking of the initiation of chain formation in RNA synthesis. It has been suggested that the naphthalene ring of RIF π—π bonds to an aromatic amino acid ring in the DDRP protein.[16] DDRP is a metalloenzyme that contains two zinc atoms. Oxygens at C-1 and C-8 of rifampin chelate a zinc atom that increases the binding of rifampin to DDRP and finally, the oxygens at C-21 and C-23 form strong hydrogen bonds to the DDRP. The binding of rifampin to DDRP results in the inhibition of the RNA synthesis.

Structure-Activity Relationship (SAR). Many derivatives of the naturally-occurring rifamycins have been prepared.[17] From these compounds, the following generalizations can be made concerning the SAR: (1) free-OH groups are required at C-1,8,21 and 23; (2) these groups appear to lie in a plane and appear to be important binding groups for attachment to DDRP, as previously indicated; (3) acetylation of C-21 and C-23 produces inactive compounds; (4) reduction of the double bonds in the macro ring results in a progressive decrease in activity; (5) opening of the macro ring also gives inactive compounds. These two last changes greatly affect the conformational structure of the rifamycins, which in turn decreases binding to DDRP. Substitution at C-3 or C-4 results in compounds with varying degrees of antibacterial activity. The substitution at these positions appear to affect transport across the bacterial cell wall.

Metabolism. Rifampin is readily absorbed from the intestine, although food within the tract may interfere with this absorption and therefore the drug should be taken on an empty stomach.[13] Rifampin does not interfere with the absorption of isoniazid, but there are conflicting reports on whether isoniazid affects absorption of RIF. The major metabolism of RIF is deacetylation, which occurs at the C-25 acetate (Fig. 33–4). The resulting product, desacetylrifampin, is still an active antibacterial agent. Desacetylrifampin may be found in the urine as the glucuronide. A second metabolite resulting from hydrolysis is 3-formylrifamycin SV, which possesses a broad spectrum of activity, but is not as effective as RIF.

Physical Chemical Properties. Rifampin is a red-orange crystalline compound with zwitterionic properties. The presence of the phenolic groups results in acidic properties, $pK_a = 1.7$, whereas the piperazine moiety gives basic properties, $pK_a = 7.9$. The compound is prone to acid hydrolysis giving rise to 3-formylrifamycin SV. Rifampin is also prone to air oxidation of the para phenolic groups in the naphthalene ring to give the *p*-quinone (C-1,4 quinone) (see Fig. 33–4). Rifampin and its metabolites are excreted in the urine, feces (biliary excretion), saliva, sweat, and tears. Because RIF has dye characteristics, one may note discoloration of the bodily fluids containing the drug. Notably, the tears may be discolored and permanent staining of contact lens may occur.

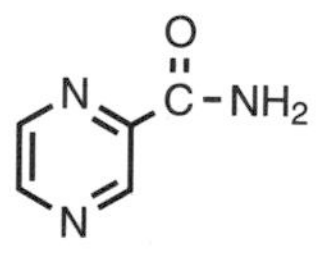

Pyrazinamide

Pyrazinamide (PZA) (Lederle). Pyrazinamide (pyrazinecarboxamide) was discovered during investigations of analogs of nicotinamide. Pyrazinamide is a bioisoster of nicotinamide and possess bactericidal action against M. tuberculosis. Pyrazinamide has become one of the more popular antituberculin agents despite the fact that resistance develops quickly to this agent. Combination therapy has proven an effective means of reducing the rate of resistant strain development. The activity of PZA appears to be pH dependent with good in vivo activity at pH 5.5, but the compound is nearly inactive at neutral pH.

Mechanism of Action. The mechanism of action of PZA is unknown, but recent findings suggest that PZA may be active totally or in part as a prodrug. Susceptible organisms produce deaminidase, which is responsible for conversion to pyrazinoic acid. Resistant strains of M. tuberculosis do not produce this amidase enzyme suggesting that the acid form of the drug is the active form. Pyrazinoic acid has been shown to possess biologic activity at a pH of 5.4 or lower, but in vitro tests show pyrazinoic acid is 8–16X less active than pyrazinamide.[18] Pyrazinoic acid may lower the pH in the immediate surroundings of the M. tuberculosis to an extent that the organism is unable to grow, but this physical-chemical property appears to account for only some of the activity.

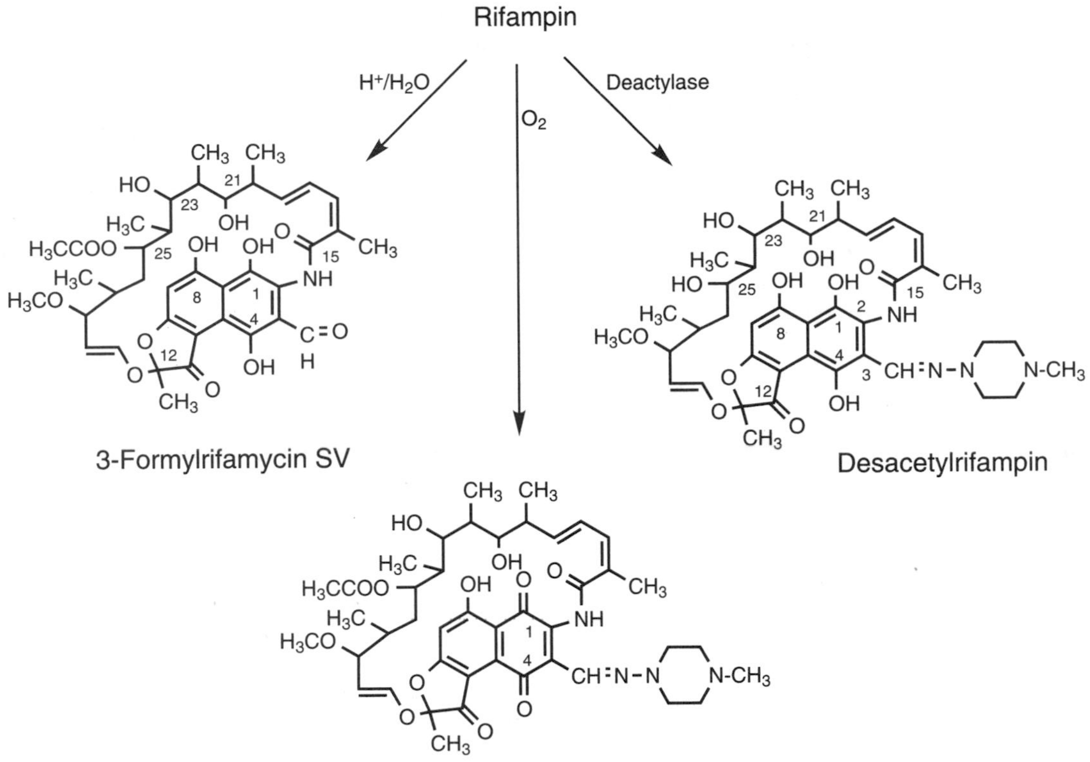

Rifampin

H^+/H_2O — 3-Formylrifamycin SV

O_2

Deactylase — Desacetylrifampin

Fig. 33–4. Metabolism and in vitro reactions of rimfampin.

Structure-Activity Relationship. Nearly all of the derivatives of PZA have proven to be either inactive or much less active than the original structure.[19] Substitution on the pyrazine ring, use of alternate heterocyclic aromatic rings, and replacement of the amide moiety with similar functional groups have proven unsuccessful, with the possible exception of the carboxylic acid analog.

Metabolism. Pyrazinamide is readily absorbed after oral administration but little of the intact molecule is excreted unchanged (Fig. 33–5). The major metabolic route consists of hydrolysis by hepatic microsomal PZA deaminidase to pyrazinoic acid, which may then be oxidized by xanthine oxidase to 5-hydroxypyrazinoic acid. The latter compound may appear in the urine as a conjugate with glycine.[13]

Ethambutol (Myambutol [Lederle]). Ethambutol, (+)-2,2′-(ethylenediimino)di-1-butanol, is utilized in its *d*-isomer form, which is 200 to 500 times more active than the *l*-isomer. The difference in activity between the two isomers suggests a specific receptor for its site of action. Ethambutol is a water-soluble, bacteriostatic agent readily absorbed (75 to 80%) following oral administration.

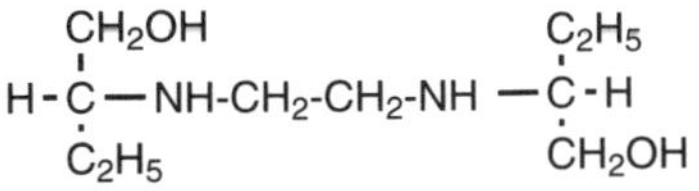

Ethambutol

Mechanism of Action. The mechanism of action of ethambutol remains unknown although there is some evidence suggesting that ethambutol may interfere with cell wall synthesis in M. tuberculosis.[20] Ethambutol has been shown to inhibit the transfer of mycolic acid into the mycobacterial cell wall. The authors have pointed out the similarity between trehalose monomycolate and ethambutol (Fig. 33–6).

Structure-Activity Relationship. An extensive number of analogs of ethambutol have been prepared, but none have been proven to be superior to ethambutol itself. Extension of the ethylene diamine chain, replacement of either nitrogen, increasing the size of the nitrogen substituents, and moving the location of the alcohol

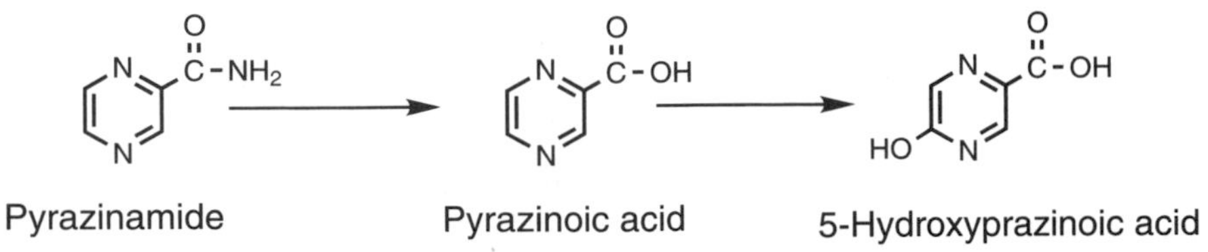

Pyrazinamide → Pyrazinoic acid → 5-Hydroxyprazinoic acid

Fig. 33–5. Metabolism of pyrazinamide.

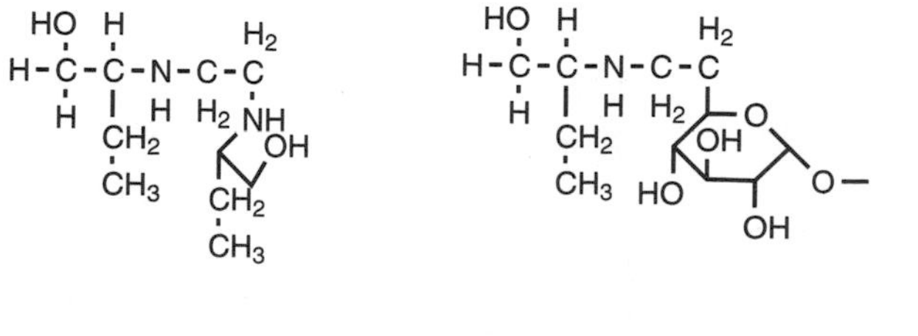

Ethambutol   Trehalose monomycolate

Fig. 33–6. Structural similarity between ethambutol and cell wall mycolates.

groups are all changes that drastically reduce or destroy biologic activity.

Metabolism. Most administered ethambutol is excreted unchanged with no more than 15% appearing in the urine as either metabolite A or metabolite B (Fig. 33–7). Both metabolites are devoid of biologic activity.

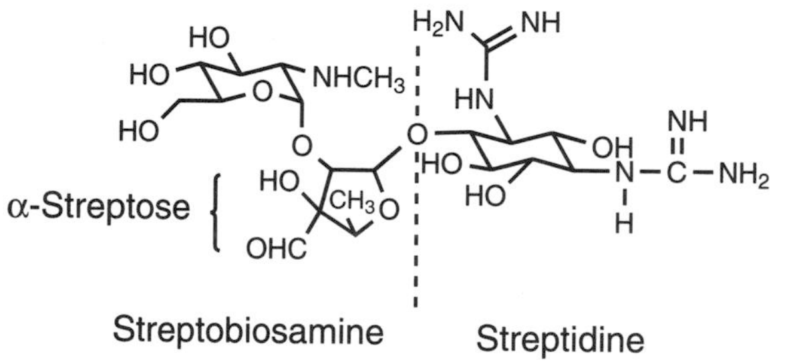

α-Streptose   Streptobiosamine   Streptidine

Streptomycin

Streptomycin (STM). Streptomycin was first isolated by Waksman and co-workers in 1944 and represented the first biologically active aminoglycoside.[21] The material was isolated from a manure-containing soil sample and was ultimately shown to be produced by Streptomyces griseus. The structure was proposed and later confirmed by Folkers and co-workers in 1948.[22] STM is water soluble with basic properties. The compound is usually available as the trihydrochloride or sesquisulfate salt, both of which are soluble in water. The hydrophilic nature of STM results in very poor absorption from the gastrointestinal tract. Orally administered STM is recovered intact from the feces, indicating that the lack of biologic activity results from poor absorption and not from chemical degradation.

Mechanism of Action. The mechanism of action of STM, and the aminoglycosides in general, has not been fully elucidated. It is known that the STM inhibits protein synthesis, but additional effects on misreading of a m-RNA template and membrane damage may contribute to the bactericidal action of STM. STM diffuses across the outer membrane of M. tuberculosis and ultimately penetrates the cytoplasmic membrane through an electron-dependent process. Binding of STM to the 30 S ribosomal subunit appears to be the primary site of action. As a result of this binding, a disruption of normal protein synthesis occurs, leading to inhibition of protein synthesis, as well as the formation of abnormal proteins.[23,24]

Structure-Activity Relationship. Although all of the aminoglycosides have very similar pharmacologic, pharmacodynamic, and toxic properties, only STM, and to a lesser extent kanamycin, are used to treat TB. This is an indication of the narrow band of structurally allowed modifications that gives rise to active analogs. Modification of the α-streptose portion of STM has been extensively studied. Reduction of the aldehyde to the alcohol results in a compound, dihydrostreptomycin, which has activity similar to STM, but with a greater potential for producing delayed severe deafness. Oxidation of the aldehyde to a carboxyl group or conversion to Schiff base derivatives (oxime, semicarbazone, or phenylhydrazone) results in inactive analogs. Oxidation of the methyl group in α-streptose to a methylene hydroxy gives an active analog but with no advantage over STM. Modification of the aminomethyl group in the glucosamine portion of the molecule by demethylation or by replacement with larger alkyl groups reduces activity, whereas removal or modification of either guanidine in the streptidine nucleus results in decreased activity.

Metabolism. No human metabolites of STM have been isolated in the urine of patients who have been administered STM with approximately 50 to 60% of the dose being recovered as unchanged drug.[13] Whereas metabolism appears insignificant on a large scale, it is implicated as a major mechanism of resistance. An early recognized problem with STM was the development of resistant strains of M. tuberculosis. Combination drug therapy was partially successful in reducing this problem, but in time, resistance has greatly reduced the value of STM as a chemotherapeutic agent for treatment of TB. Various mechanisms may lead to the resistance seen in M. tuberculosis. Permeability barriers may result in STM's not being transported through the cytoplasmic membrane, but the evidence suggests that enzymatic inactivation of STM represents the major problem. The enzymes responsible for inactivation are adenyltransferase, which catalyzes adenylation of the C-3 hydroxyl group in the N-

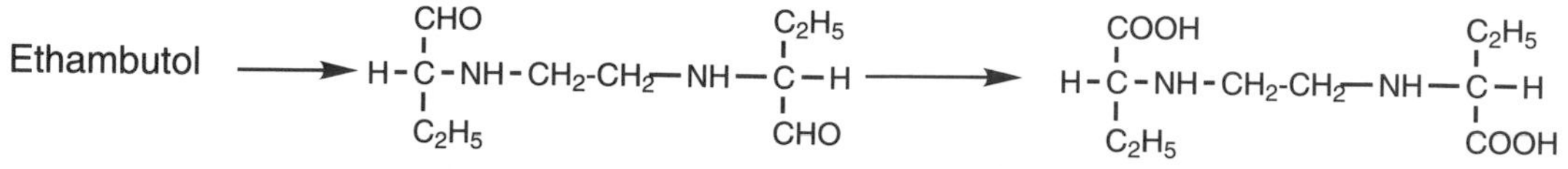

Ethambutol → Metabolite A → Metabolite B

Fig. 33–7. Metabolism of ethambutol.

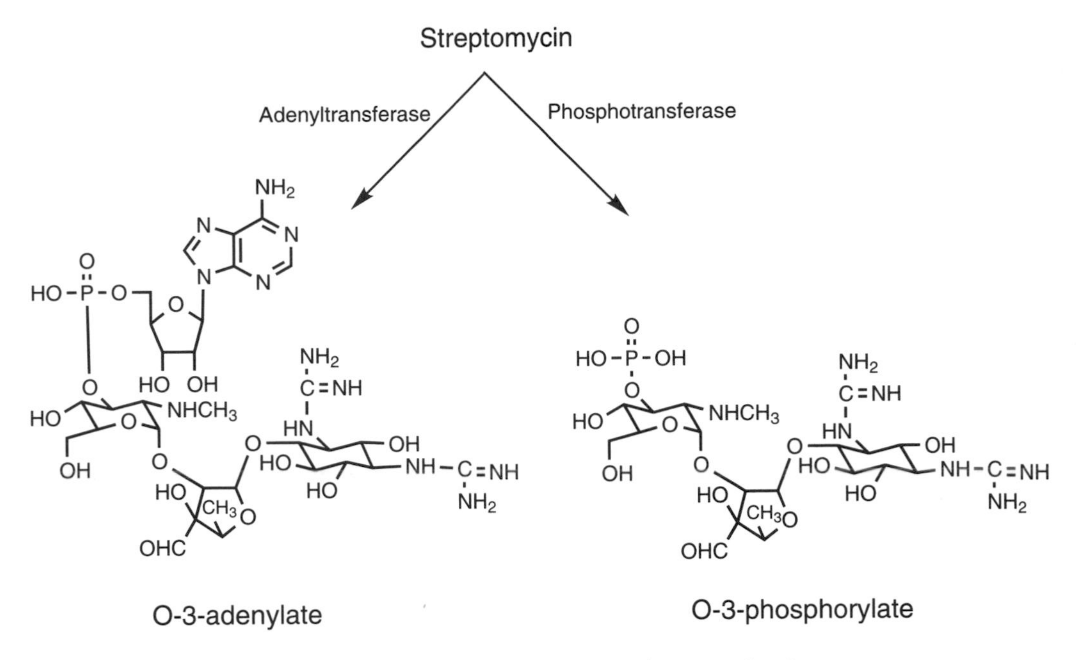

Fig. 33–8. Metabolism of STM as a mechanism of resistance.

methylglucosamine moiety to give O-3-adenylate metabolite, and phosphotransferase, which phosphorylates the same C-3 hydroxyl to give O-3-phosphorylate metabolite (Fig. 33–8). This latter reaction appears to be clinically the most significant. The result of these chemical modifications is that the resulting metabolites will not bind to ribosomes.

Second-Line Agents

Certain drugs, including ethionamide, para-aminosalicyclic acid, cycloserine, capreopmycin, and kanamycin are considered second-line agents. Although these agents are active antibacterial agents, they usually are less well tolerated or have a higher incidence of adverse effects. These agents are utilized in cases of resistance, retreatment, or intolerance to the first-line drugs.

Ethionamide

Ethionamide (Trecator-SC) [Wyeth-Ayerst]. The synthesis of analogs of nicotinamide resulted in the discovery of ethionamide and a homolog in which the ethyl group is replaced with a propyl (prothionamide). Both compounds have proven to be bactericidal against M. tuberculosis and M. leprae. Whereas it has been investigated in great depth, it has been suggested that ethionamide inhibits protein synthesis by blocking the incorporation of cysteine and methionine into proteins. Ethionamide is orally active but in a single large dose (>500 mg) is not well tolerated. The GI irritation can be reduced by administration with meals. Additional side effects may include CNS effects, hepatitis, and hypersensitivities. Less than 1% of the drug is excreted in the free form with the remainder appearing as one of six metabolites. Among these is ethionamide sulfoxide (oxide of the thiocarbonyl), which has antimicrobial activity either as such or because of reduction back to its earlier state.

Para-aminosalicylic acid

Para-aminosalicylic acid (PAS). Once a popular component in TB therapy, PAS is used only as a second-line agent today. A combination of bacterial resistance and severe side effects has greatly reduced is value. PAS, as a bacteriostatic agent, is used at a dose of up to 12 g/day, which causes considerable GI irritation, but in addition, hypersensitivity reactions occur in 5 to 10% of the patients with some of these reactions being life-threatening. The mechanism of action of PAS is thought to involve interference with para-aminobenzoic acid incorporation into folic acid. PAS, when coadministered with INH, is found to reduce the acetylation of INH, itself being the substrate for acetylation, thus increasing the plasma levels of INH. This action may be especially valuable in rapid acetylators. PAS is extensively metabolized by acetylation of the amino group, and conjugation with glucuronic acid and glycine at the carboxyl group. PAS is used primarily in cases of resistance, retreatment, and intolerance of other agents. PAS is available from the CDC.

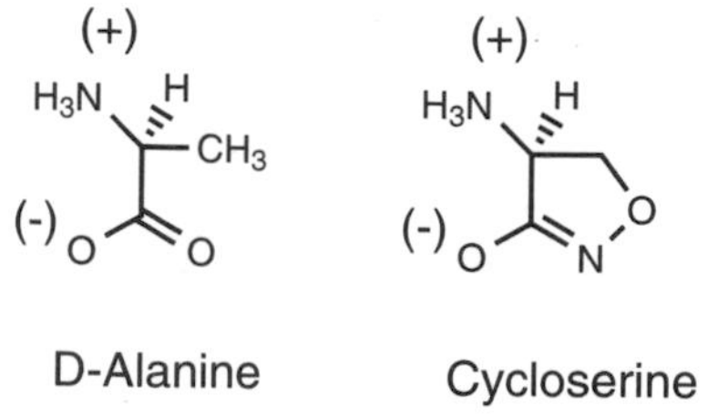

D-Alanine Cycloserine

Fig. 33–9. Structural similarity between D-alanine and cycloserine.

Cycloserine (Seromycin [Lilly] Cycloserine is a natural product isolated from Streptomyces orchidaceus as the

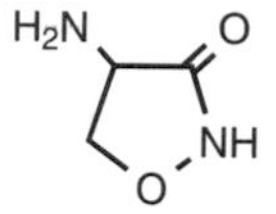

Cyclosporin

D(+) isomer. The L(−) isomer is also active and the D,L mixture is more active than the pure enantiomers. The mechanism of action is that of interfering with the role of D-alanine, probably in cell wall synthesis. Cycloserine functions as an antimetabolite of alanine (see Fig. 33–9). This is supported by the fact that organisms grown in media that contains D-alanine are not affected by cycloserine. Cycloserine is readily absorbed after oral administration and widely distributed, including in the CNS. As a second-line agent, cycloserine should only be used when retreatment is necessary or when the organism is resistant to other drugs. Cycloserine should not be used as a single drug, but must be used in combination

Capreomycin (Capreomycin 1a, R=OH; Capreomycin 1b, R=H; Capastat [Lilly]). Capreomycin is a mixture of four cyclic polypeptides of which capreomycin 1a and

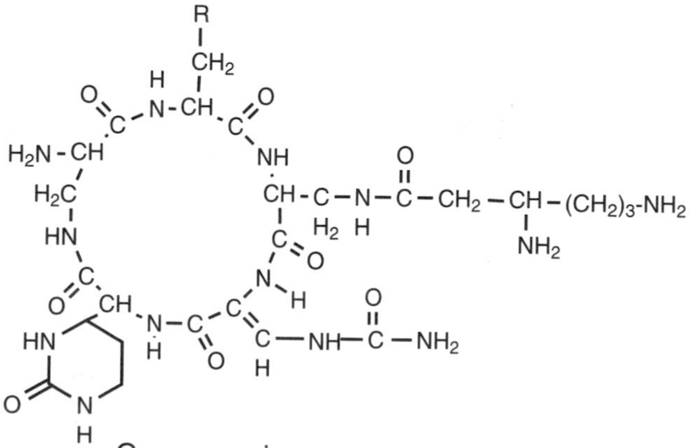

Capreomycin

1b make up 90% of the mixture. Capreomycin is produced by Streptomyces capreolus and is similar to the antibiotic viomycin. Little, if anything, is known about its mechanism of action, but if the chemical and pharmacologic similarity to viomycin carries over to mechanism of action, then one might expect a similarity. Viomycin is a potent inhibitor of protein synthesis, particularly that which depends on m-RNA at the 70S ribosome.[23] Viomycin blocks chain elongation by binding to either or both the 50S or 30S ribosomal subunits. As a polypeptide, the drug must be administered parenterally with the preferred route of administration being intramuscular. As a second-line bacteriostatic, antituberculin drug, it is reserved for so-called resistant infections and treatment-failure cases. The drug should not be given as a single agent, but rather used in combination with ethambutol or INH. Reported toxicity to capreomycin include renal and hepatic damage, hearing loss, and allergic reactions.

Kanamycin (Kanamycin A, R=OH; Kanamycin B, R=NH_2; Kantrex [Bristol]). A member of the aminoglycoside class, kanamycin is a second-line agent with only

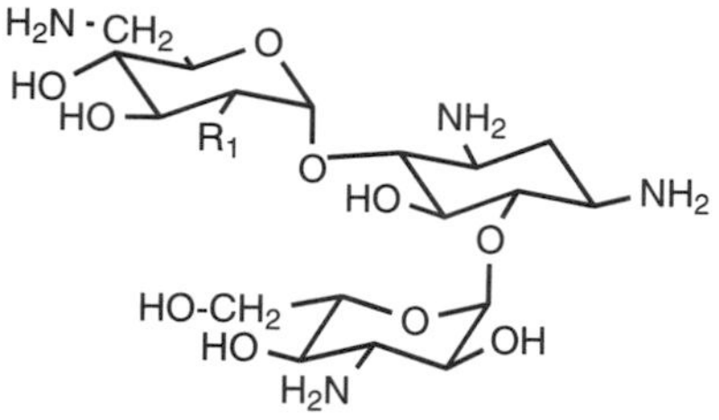

Kanamycin

limited use in the treatment of M. tuberculosis. The drug is only used to treat resistant organisms and then should be used in combination with other effective agents. The parenteral form of the drug is used, because as an aminoglycoside, the drug is poorly absorbed orally. The narrow range of effectiveness and the severe toxicity, especially if the drug is administered over a long period of time, have limited its usefulness. For additional information on kanamycin and aminoglycosides, see discussion in Chapter 34.

Unapproved Agents

The fluoroquinolones show some promise for treatment of tuberculosis. Specifically ciprofloxacin (Cipro [Mileo]) and ofloxacin (Floxin [Ortho])[25] have been shown to possess in vivo activity against M. tuberculosis.

Therapeutic Considerations

Various stages of infectious organisms that may require special consideration for chemotherapy have been identified. The organism may be in a dormant stage, which is usually impervious to drugs. In the continuously growing stage of the organism, the bacteria may be either extracellular or intracellular. A stage of the organism, which is classified as the very slowly metabolizing bacteria, exists in a relatively acidic environment. Finally, the organism may exhibit a stage in which it is dormant, followed by spurs of further growth. As noted in the discussion of specific drugs, one stage or another may be more susceptible or less to a particular drug based on the characteristics discussed. It is also recognized that organisms from some geographic regions may show a low incidence of drug resistance, whereas those from other regions have a high incidence of drug resistance. For TB patients likely to be infected with organisms suspected of showing low rates of drug resistance, the CDC and the American Thoracic Society currently recommend a 24-week regimen consisting of daily doses of isoniazid (300 mg), RIF (600 mg), and pyrazinamide (25–35 mg/kg) for 8 weeks (induction or sterilization phase). In the remaining 16 weeks, the therapy consists of isoniazid and RIF (continuation or maintenance phase). The addition of pyrazinamide to the other two drugs reduces treatment time from 9 to 6 months. Individuals on this regimen are considered noninfectious after the first 2 weeks. It is expected that modification of this regimen may be soon recommended, in that either streptomycin (15 mg/kg to a maximum of 1 g daily) or ethambutol (15 mg/kg), will be added to the initial three-drug therapy. For patients coming from locations that have a high incidence of drug resistance, the following regimen is recommended: isoniazid, RIF, pyrazinamide and ethambutol for 9 to 12 months or longer. This same group of drugs is recommended for TB patients with AIDS. The "cardinal rules" for all TB regimens is: (1) get drug susceptibility information as soon as possible; (2) always begin therapy with at least three drugs; (3) at all costs avoid a regimen employing only one effective drug; and (4) always add at least two drugs to a failing regimen.[3,26,27]

The only proven treatment for prophylaxis of TB (in patients with a positive skin test or in those with a high-risk factor) is INH used for 12 months. Adverse effects when using combination therapy over a long treatment period can be serious. Hepatotoxicity is a complication of treatment with several of the TB drugs. INH may cause severe liver damage. The drug should be removed if serum aminotransferase activity increases to three to five times normal or the patient develops symptoms of hepatitis. Peripheral neuropathy may be seen with INH therapy. This condition may be prevented by coadministration of pyridoxine. Hepatitis, thrombocytopenia, and nephrotoxicity may be seen with RIF therapy. Rifampin is thought to potentiate the hepatitis caused by INH. Gastrointestinal upset and staining effects caused by RIF are of minor importance. Hepatotoxicity is a common adverse effect of PZA, but in this case, it does not appear that PZA significantly increases the hepatotoxicity caused by combination with the other two drugs. The major adverse effects of ethambutol are optic neuritis and hyperuricemia. The former appears as changes in visual acuity and in an inability to distinguish the color green. The major toxicities reported for STM, and for aminoglycosides in general, are nephrotoxicity and ototoxicity.

Leprosy

Sulfones

The diaryl sulfones represent the major class of agents used to treat leprosy. The initial discovery of the sulfones resulted from studies directed at exploring the structure-

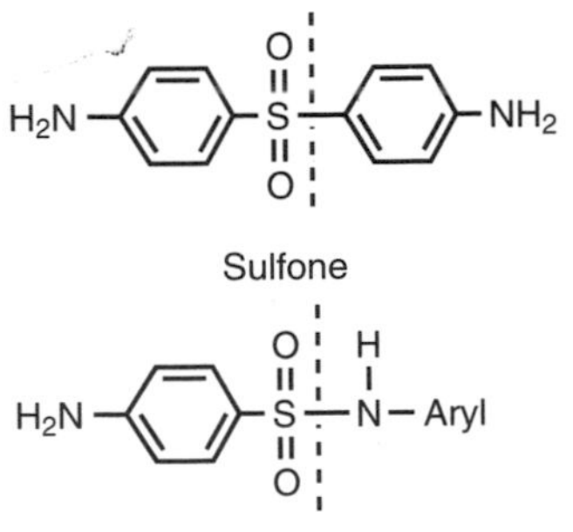

Sulfone

Sulfonamide

Fig. 33–10. Structural comparison of sulfones vs. sulfonamide.

activity relationship of sulfonamides. (Fig. 33–10). A variety of additional chemical modifications have produced several other active agents but none have proved more beneficial than the original lead, 4,4′-diaminodiphenyl sulfone. (DDS, Dapsone)

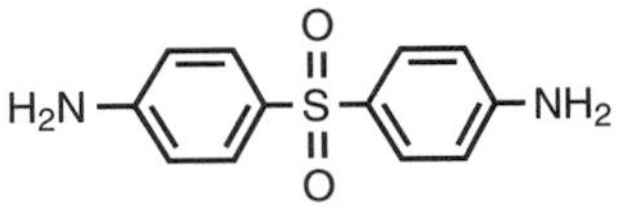

4,4'-Diaminodiphenylsulfone (Dapsone)

4,4′-Diaminodiphenyl sulfone (Dapsone, DDS). Dapsone is nearly insoluble in water and is very weakly basic ($pK_a \sim 1.0$). This lack of solubility may account in part for the occurrence of GI irritation. Despite the lack of solubility, the drug is efficiently absorbed from the GI tract. Dapsone is bound to plasma protein (at about 70%), but is distributed throughout the body.

Mechanism of Action. Dapsone, a bacteriostatic agent, is thought to act in a manner similar to that of the sulfonamides, namely, through competitive inhibition of para-aminobenzoic acid (PABA) incorporation into folic acid. Bacteria synthesize folic acid, but host cells do not. As a result, coadministration of dapsone and PABA will inactivate the dapsone. Dapsone and clofazimine have significant anti-inflammatory actions, which may or may not play a role in the antimicrobial action. The anti-inflammatory action may also be a beneficial side effect by offsetting the complication of erythema nodosum leprosum (ENL) seen in some patients. The anti-inflammatory action may come about by inhibition of myeloperoxidase catalyzed reactions.[28]

Structure-Activity Relationship. Several derivatives of DDS have been prepared in an attempt to increase the activity of DDS. Isosteric replacement of one benzene ring resulted in the formation of thiazolsulfone. Although still active, it is less effective than DDS. Substitution on the aromatic ring, to produce acetosulfone, reduces activity but increases water solubility and decreases GI irritation. A successful substitution consists of adding methanesulfinate to DDS to give sulfoxone sodium. This water-soluble form of DDS is hydrolyzed in vivo to pro-

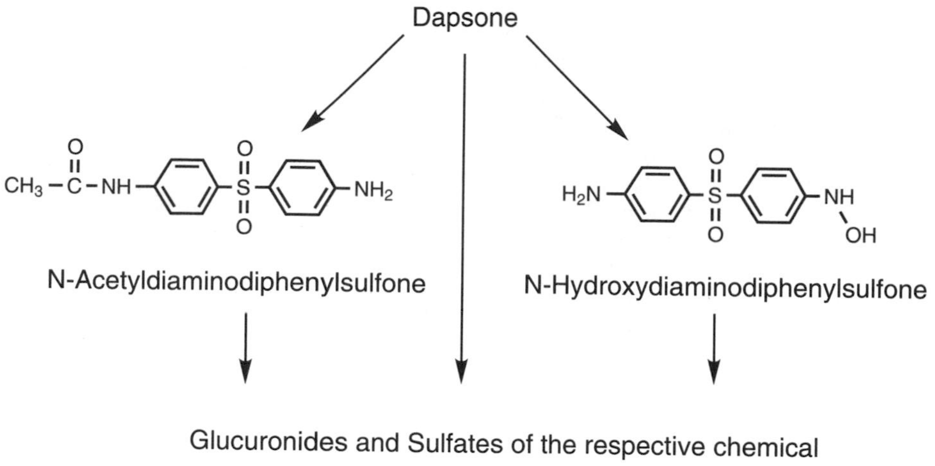

Dapsone

N-Acetyldiaminodiphenylsulfone

N-Hydroxydiaminodiphenylsulfone

Glucuronides and Sulfates of the respective chemical

Fig. 33–11. Metabolites of dapsone.

duce DDS. Sulfoxone sodium is used in individuals who are unable to tolerate DDS because of GI irritation, but it must be used in a dose three times that of DDS because of inefficient metabolism.

Metabolism. The major metabolic product of dapsone results from N-acetylation in the liver by N-acetyltransferase. Dapsone is also N-hydroxylated to a hydroxylamine

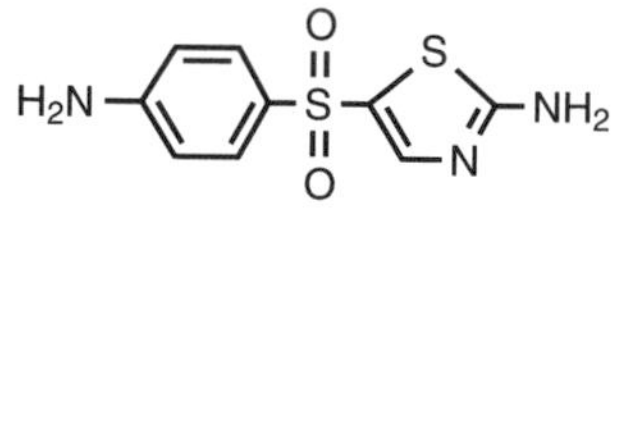

Thiazolsulfone

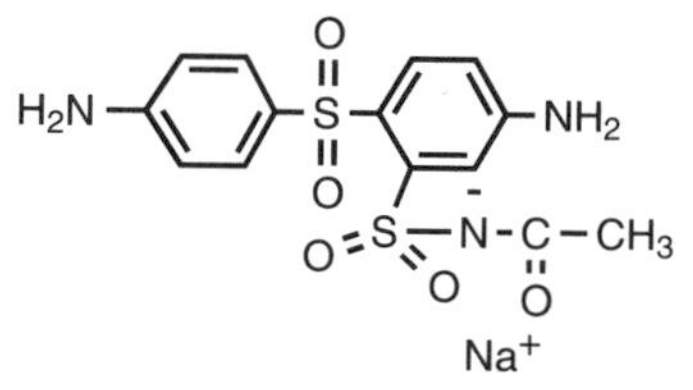

Acetosulfone

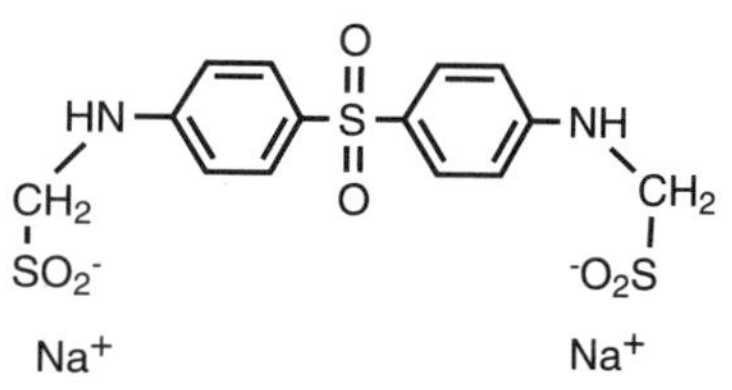

Sulfoxone sodium

derivative. Neither of these compounds possess significant biologic activity, although N-acetyldiaminodiphenylsulfone may be deacetylated back to dapsone. Products found in the urine consist of small amounts of dapsone, metabolites N-acetyldiaminodiphenylsulfone and N-hydroxydiaminodiphenylsulfone, as well as glucuronide and sulfates of each of these substances (Fig. 33–11).

Phenazines and Adjunct Agents

Clofazimine (Lamprene [Geigy]). Although classified as a secondary drug for treatment of leprosy and commonly used as a component of multiple-drug therapy, clofazimine use appears to be increasing. The chemical, a phenazine derivative, is a water-insoluble dye (dark red crystals) that causes pigmentation of the skin. In addition, discoloration of the feces, lining of the eyelids, sputum, sweat, tears, and urine is seen (pink, red, or brownish black discoloration).

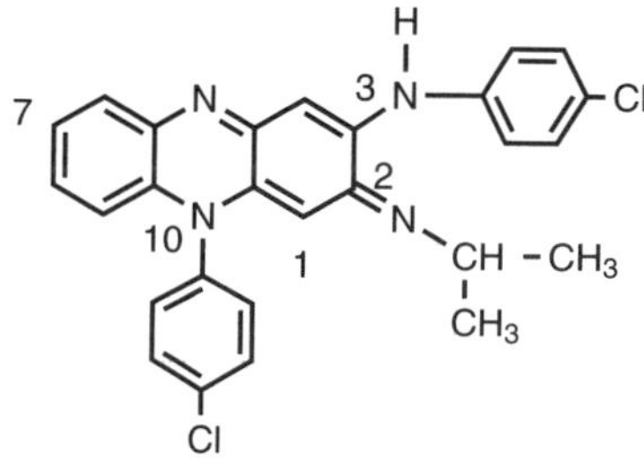

Clofazimine

Mechanism of Action. The mechanism of action remains unclear at the present time. The molecule has direct antimycobacterial and immunosuppressive properties. It has been shown that clofazimine increases prostaglandin synthesis and the generation of antimicrobial reactive oxidants from neutrophils, which may play a role in the antileprosy effects. The host cell defense may be stimulated by clofazimine, resulting in the generation of oxidants, such as the superoxide anion, which in turn could have a lethal effect on the organism.[29]

Structure-Activity Relationship. Several investigators have reported studies directed toward an understanding of the structure activity-relationship of clofazimine.[29–31] Substituents on the imino group at position 2, para-chloro substitution on the phenyls at C-3 and N-10, as well as substituents at position 7, have been investigated. The imino group at C-2 appears essential with activity increased when the imino group is substituted with alkyl and cycloalkyl groups. Chlorine substituents on the two phenyls at C-3 and N-10 enhance activity but are not essential to activity. In the analogs studied, the increased activity correlates well with pro-oxidative activities of the molecule, e.g., ability to generate superoxide anion.

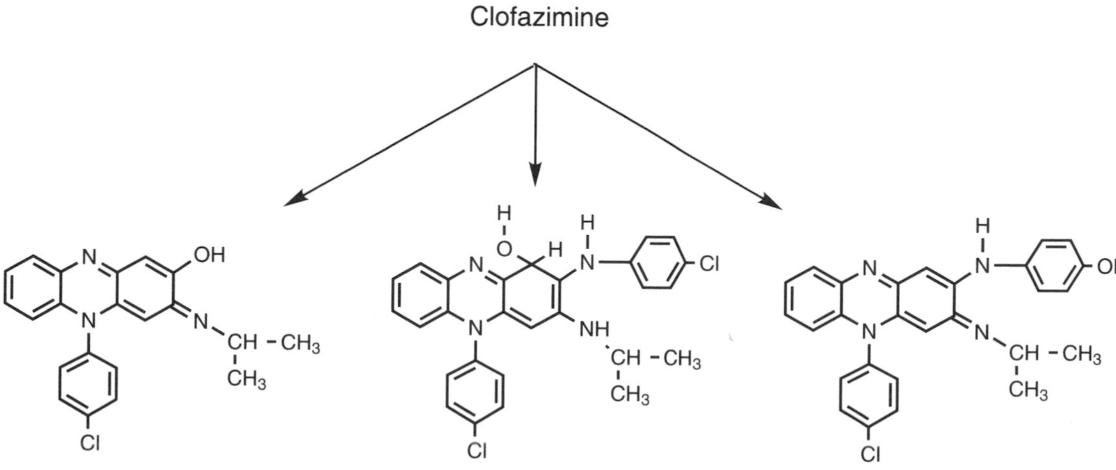

Clofazimine

Fig. 33–12. Metabolism of clofazimine.

Metabolism. Various metabolites of clofazimine have been identified, but account for only 1% of the administered dose.[32] The lack of higher concentrations of the metabolites may, in part, be caused by the very slow elimination of clofazimine from the body. Clofazimine has an estimated half-life of between 8.8 and 69 days. The lipophilic nature of clofazimine results in distribution and storage of the drug in fat tissue. The identified metabolites consist of hydrolytic deamination, hydration, and hydrolytic dehalogenation (Fig. 33–12). The former two compounds appear as glucuronide derivatives at the newly formed -OH's.

Rifampin-(Rifadin [Merrell-Dow], Rimactane [Ciba]). Rifampin, an antituberculin drug, has already been discussed. Its actions against M. leprae parallels those reported for M. tuberculosis. Today, RIF is considered an effective antileprosy agent, when used in combination with the sulfones.

Thalidomide. A possible complication of the chemotherapy of leprosy is the condition of erythema nodosum leprosum (ENL). The condition is characterized by tender, inflamed subcutaneous nodules that may last a week or two, but may reappear and last for long periods. This condition has been successfully treated with thalidomide. Thalidomide is a potent teratogenic agent, but can be used safely in men and postmenopausal women.

Therapeutic Considerations

From its introduction into the chemotherapy of leprosy in 1947, dapsone has proved to be the single most effective agent. This drug was used as monotherapeutic agent despite the recognition that resistant strains were beginning to emerge. The problems of primary and secondary drug resistance are not the only treatment complications experienced. Relapse can result from bacterial persistence. Nonmultiplying microorganisms unsusceptible to dapsone may emerge at a later date, thus causing reinfection. Since 1977, monotherapy with dapsone is no longer recognized as an acceptable method for treatment of leprosy. Today, combination chemotherapy is the method of choice. The combination consists of RIF (600 mg monthly), dapsone (100 mg daily) and clofazimine (300 mg monthly and 50 mg daily). This regimen is used for treatment of multibacillary leprosy, including lepromatous and borderline cases. Therapy is usually continued for at least 2 years or as long as skin smears are positive. The patient is kept under supervision for 5 years following completion of chemotherapy. For paucibacillary leprosy, including tuberculoid and indeterminate cases, only RIF and dapsone are used with doses as indicated above. Treatment may continue for 6 months and the patient kept under observation for an additional 2 years.[33] These new regimens allow a shortened treatment period and a reduced rate of relapse. Antileprosy agents are being investigated as a possible treatment of M. avium-intracellular complex infections. This condition is one of the most common opportunistic infection in AIDS patients.[34]

MAC

Currently, there is no recognized drug therapy for the treatment of MAC, disseminated mycobacterium avium complex. Combination drug therapy has been used without much success. The combinations normally consists

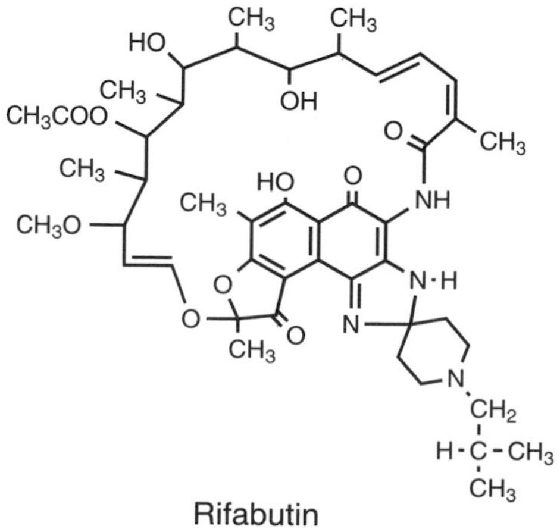

Rifabutin

of agents presently used for treatment of tuberculosis and leprosy. Recently, the Antiviral Drugs Advisory Committee has recommended approval of rifabutin (Mycobutin [Adria]) for treatment of MAC. Rifabutin, an analog, RIF, if approved, will be used as a prophylaxis against MAC.

REFERENCES

1. C. Marwick, *JAMA, 257,* 1174(1992).
2. T. M. Daniel, *Tuberculosis,* in *Harrison's Principles of Internal Medicine,* 12th ed., Jean D. Wilson, et al., Eds., New York, McGraw-Hill, 1991.
3. H. G. Taylor, *Am. Pharm., 32,* 41(1992).
4. R. P. Herman and M. M. Weber, *Antimicrob. Agents Chemother., 17,* 450(1980).
5. L. A. Davidson and K. Takayama, *Antimicrob. Agents Chemother., 16,* 104(1979).
6. E. M. Bavin, et al., *J. Pharm. Pharmacol., 7,* 1032(1955).
7. E. M. Bavin, et al., *J. Pharm. Pharmacol., 4,* 844(1952).
8. H. H. Fox and J. T. Gibas, *J. Org. Chem., 17,* 1653(1952).
9. H. H. Fox and J. T. Gibas, *J. Org. Chem., 18,* 994(1953).
10. H. H. Fox and J. T. Gibas, *J. Org. Chem., 20,* 60(1955).
11. H. H. Fox and J. T. Gibas, *J. Org. Chem., 21,* 356(1956).
12. W. W. Weber and D. W. Hein, *Clin. Pharmacokinet., 4,* 401(1979).
13. M. R. Holdiness, *Clin. Pharmacokinet., 9,* 511(1984).
14. J. A. Timbrell, et al., *J. Pharmacol. Exp. Ther., 213,* 364(1980).
15. G. Lancini, *Ansamycins,* in *Biotechnology,* Vol 4., H. Pape and H.-J. Rehm, Eds., Deerfield Beach FL, VCH, 1986.
16. S. K. Arora, *J. Med. Chem., 28,* 1099(1985).
17. G. Lancini, and W. Zanchelli, *Structure-Activity Relationship in Rifamycins,* in *Structure-Activity Relationship Among the Semisynthetic Antibiotics,* D. Perlman, Ed., New York, Academic Press, 1977.
18. L. B. Heifets, et al., *Antimicrob. Agents Chemother., 33,* 1252(1989).
19. S. Kushner, *J. Am. Chem. Soc., 74,* 3617(1952).
20. K. Takayama, et al., *Antimicrob. Agents Chemother., 16,* 240(1979).
21. A. Schatz, et al., *Proc. Soc. Exp. Biol. Med., 55,* 66(1944).
22. F. A. Kuehl, et al., *J. Am. Chem. Soc., 70,* 2325(1948).
23. E. F. Gale, et al., *The Molecular Basis of Antibiotic Action,* 2nd Ed. London, Wiley, 1981.
24. K. E. Price, et al., *Effect of Structural Modifications on the Biological Properties of Aminoglycoside Antibiotics Containing 2-Deoxystreptamine,* in *Structure-Activity Relationship Among the Semisynthetic Antibiotics,* D. Perlman, Ed., New York, Academic Press, 1977.
25. W. W. Yew, et al., *J. Antimicrob. Chemother., 26,* 227(1990).
26. *The Medical Letters.* February 7, 1992.
27. C. M. Reinke and D. H. Albrant, *U.S. Pharmacist,* October, 37, 1991.
28. J. M. van Zyl, et al., *Biochem. Pharmacol., 42,* 599(1991).
29. J. E. Savage, et al., *J. Antimicrob. Chemother., 23,* 691(1989).
30. S. G. Franzblau and J. F. O'Sullivan, *Antimicrob. Agents Chemother., 32,* 1583(1988).
31. J. F. O'Sullivan, et al., *J. Med. Chem., 31,* 567(1988).
32. V. K. Kapoor, *Clofazimine,* in *Analytical Profiles of Drug Substances,* Vol. 18., K. Florey, San Diego, Academic Press, 1989.
33. H. P. Lambert and F. W. O'Grady, *Antibiotic and Chemotherapy,* 6th ed, Edinburgh, Churchill Livingstone, 1992.
34. A. Sahu, et al., *Int. J. Immunopharmacol., 13,* 419(1991).

Chapter 34

ANTIBIOTICS AND ANTIMICROBIAL AGENTS

Lester A. Mitscher

Antibiotics are microbial metabolites or synthetic analogs inspired by them that, in small doses, inhibit the growth and survival of microorganisms without serious toxicity to the host. Selective toxicity is the key concept. Examples are the penicillins and the tetracyclines. Antibiotics are among the medications most frequently prescribed today, although their continued efficacy is threatened by microbial resistance caused by evolutionary pressures. In many cases the clinical utility of natural antibiotics has been enhanced through medicinal chemical manipulation of the original structure leading to broader antimicrobial spectrum, greater potency, lesser toxicity, and more convenient administration. Examples of such semisynthetic antibiotics are amoxicillin and doxycycline. Through customary usage, the many synthetic substances unrelated to natural products that inhibit or kill microorganisms are referred to as antimicrobial agents, examples of which are sulfisoxazole and ciprofloxacin.

Our environment, body surfaces, and cavities support rich and characteristic microbial flora which causes us neither significant illness nor inconvenience, so long as neither we nor our neighbors do not indulge in behavior that exposes us to exceptional quantities of, nor unusual strains of, microbes or introduces bacteria into parts of the body where they are not normally resident. Protection against this happening is obtained primarily through public health measures, healthful habits, intact skin and mucosal barriers, and a properly functioning immune system. All the parts of our bodies that are in contact with the environment support microbial life. All of our internal fluids, organs, and body structures are sterile under normal circumstances and the presence of bacteria, fungi, and viruses in these places is prima facie diagnostic evidence of infection. When mild microbial disease occurs, the otherwise healthy patient will often recover without treatment. Here an intact, functioning immune system is called upon to clump and phagocytize invasive microorganisms. When this is insufficient to protect us, appropriate therapeutic intervention is indicated. The chronicle of civilization before the discovery of bacteria and their role in infectious disease and, subsequently, of the discovery of antibiotics and antimicrobial agents, is punctuated by the outbreak of recurrent devastating pandemics, such as the successive waves of bubonic plague that dramatically decreased the population of Europe in the Middle Ages. Humankind was mystified both by the cause of, and what one might constructively do to find, a cure. In warfare, infections often disabled or killed more individuals than did the action of generals. Our own family histories record the premature loss of loved ones, particularly small children, to one infection or another, and in the Third World, this pattern remains all too common. This depressing picture has been altered dramatically in this century by the discovery and application of the powerful therapeutic agents described in this chapter. It is fortunately no longer common for persons to live short lives, and it is now rare for parents to bury their children. Public health measures such as purification of water supplies, routine preventive vaccination, pasteurization of milk, and avoidance of unhealthy behavior, e.g., such as spitting in public places, have also greatly diminished our exposure to infection. Considering that the first truly effective antimicrobial agents date from the middle 1930s (the sulfonamides) and the first antibiotics came into use only in the 1940s, it is amazing that we have already grown complacent, and diseases that very recently seemed on their way to extinction, such as tuberculosis and gonorrhea, are once again becoming serious public health problems because of societal changes and the emergence of resistant pathogens. Previously unknown infectious diseases, such as acquired immunodeficiency syndrome (AIDS) and Legionnaires' disease, are increasingly disturbing features of modern life. Unfortunately, we can no longer confidently depend on the discovery of increasing numbers of novel antibiotics and antimicrobial agents to keep infectious diseases under control, but must pay increased attention to neglected public health measures and to concentrate on using antibiotics safely and effectively only when these measures fail.

HISTORY

Humankind has been subject to infection by microorganisms since before the dawn of recorded history. One presumes that mankind has been searching for suitable therapy for nearly as long. This was a desperately difficult enterprise given the acute nature of most infections and the nearly total lack of understanding of their origins, which was prevalent until the last century. Although one can find indications in old medical writings of folkloric use of plant and animal preparations, soybean curd, moldy bread and cheese, counter infection with other microbes, use of extracts from fermentations, the slow development of public health measures, and an understanding of the desirability of personal cleanliness, these factors were erratically and inefficiently applied and they often failed. Until after the discovery of bacteria about

300 years ago, and subsequent understanding of their role in infection about 150 years ago, there was no hope for rational therapy.

In Germany, in the last century, Koch showed that specific microorganisms could always be isolated from the excreta and tissues of people with particular infectious diseases and that these same microorganisms were absent in healthy individuals. They could be grown on culture media and be used to reproduce in healthy individuals all of the classic symptoms of the same disease. Finally, the identical microorganism could then be isolated from this deliberately infected person. Following these rules, at long last, a chain of cause and effect was forged between certain microorganisms and specific infectious diseases.

Pasteur and Joubert reported in 1877 that if what they termed common bacteria were introduced into a pure culture of anthrax bacilli, the bacilli died, and that an injection of deadly anthrax bacillus into a laboratory animal was harmless if common bacteria were injected along with it. This did not always work but led to the Darwinian appreciation of antibiosis wherein two or more microorganisms competed with one another for survival. It was more than a half century later that the underlying mechanisms by which this phenomenon was achieved began to be appreciated and applied to routinely successful therapy.

The modern anti-infective era opened with the discovery of the sulfonamides in France and Germany in 1936 as an offshoot of Ehrlich's earlier achievements in treating infections with organometallics and his theories of vital staining. The discovery of the utility of sulfanilamide was acknowledged by the award of a Nobel prize in 1938. The well-known observation of a clear zone of inhibition (lysis) in a bacterial colony surrounding a colony of contaminating air-borne Penicillium mold by Fleming in England in 1929 and the subsequent purification of penicillin from it in the late 1930s and early 1940s by Florey, Chain, Abraham, and Heatley, provided important additional impetus. With the first successful clinical trial of crude penicillin on February 12, 1941, and the requirements of wartime, an explosion of successful activity ensued which continues over 50 years later. In rapid succession, deliberate searches of the metabolic products of a wide variety of soil microbes led to discovery of tyrothricin (1939), streptomycin (1943), chloramphenicol (1947), chlortetracycline (1948), neomycin (1949), erythromycin (1952), and more, and this ushered in the age of the miracle drugs. Microbes of soil origin remain to this day the most fruitful sources of antibiotics, although the specific means employed today for their discovery are infinitely more sophisticated than those employed 40 years ago. Initially, extracts of fermentations were screened simply for their ability to kill pathogenic microorganisms in vitro. Those that did were pushed along through ever more complex pharmacologic and toxicologic tests in attempts to reach the clinic. Today many thousands of such extracts of increasingly exotic microbes are tested each week and the tests now include sophisticated assays for agents operating through particular biochemical mechanisms or possessing particular desirable properties. As a consequence of this work, humankind now has many choices for powerful, effective and specific therapy for some of humankind's most ancient and common perils.

The 1990 worldwide commerce in antibiotics is measured in multiple tonnages and is valued in excess of $7 billion. About half of this is associated with beta-lactam antibiotics alone. More than 20% of the most frequently prescribed outpatient medications in the United States are anti-infective agents. Approximately 100 antibiotics have seen substantial clinical use, although more than 20,000 other natural antibiotics have been described in the literature, and an order of magnitude of more semi- and totally synthetic antimicrobial agents has been prepared. These agents have had a major impact on the practice of medicine and pharmacy, and on the lives of persons still living who remember clearly the perils of life before antibiotics became available.

GENERAL PRINCIPLES AND IMPORTANT DEFINITIONS

Nomenclature

The names given to antimicrobials and antibiotics are as varied as their inventor's taste, and yet some helpful unifying conventions are followed. For example, the penicillins are produced by fermentation of fungi, and their names most commonly end in the suffix -cillin, as in the term ampicillin. The cephalosporins are likewise fungal products, although their names mostly begin with the prefix cef- (or sometimes, following the English practice, spelled ceph-). The synthetic fluoroquinolones mostly end in the suffix -floxacin. Although helpful in many respects, this nomenclature does result in many substances possessing quite similar names. This can make remembering them a burden. Most remaining antibiotics are produced by fermentation of soil microorganisms belonging to various Streptomyces species. By convention these have names ending in the suffix -mycin, as in streptomycin. Some prominent antibiotics are produced by fermentation of various Micromonospora sp. These antibiotics have names ending in -micin, e.g., gentamicin. The student has to take considerable care to avoid confusing these in written communication. Whereas their pronunciation is essentially identical, they are spelled differently. In earlier times, the terms broad- and narrow-spectrum had some clinical meaning. The widespread emergence of single-agent and multiple-agent resistant microbial strains have made these terms much less meaningful. It is, nonetheless, of some value to remember that some antimicrobial families have the potential of inhibiting a wide range of bacterial genera belonging to both gram-positive and gram-negative cultures, and so are called broad-spectrum (such as the tetracyclines), whereas others inhibit only a few bacterial genera and are termed narrow-spectrum (such as the glycopeptides, typified by vancomycin, which are used almost exclusively for a few gram-positive and anaerobic microorganisms).

Gram Stain

The Gram stain was developed in the last century by Hans Christian Gram, a Danish microbiologist, in order to visualize bacteria more easily under the microscope. The basis of this differential stain lies in the chemistry of the bacterial cell walls, which causes certain bacteria to react differently with the stains employed. Gram-positive microbes stain blue. Gram-negative microbes do not retain the blue dye when washed with alcohol and are counterstained red upon treatment with a different dye. It is a convenient fact for the study of antibiotics that gram-positives reactions are generally more sensitive to comparatively nontoxic antibiotics than are gram-negative ones. Thus it is useful for the student of antibiotics to be familiar with the Gram-staining characteristics of given pathogens.

Importance of Identification of the Pathogen Before Instituting Therapy

Fundamental to appropriate antimicrobial therapy is an appreciation that individual species of bacteria are associated with particular infective diseases and that given antibiotics are more likely to be useful than others for killing them. Sometimes this can be used as the basis for successful empiric therapy. For example, first-course community-acquired urinary tract infections in otherwise healthy individuals are most commonly caused by gram-negative Escherichia coli of fecal origin. Even just knowing this much can give the physician several convenient choices for useful therapy. Likewise, skin infections, such as boils, are commonly the result of infection with gram-positive Staphylococcus aureus. In most other cases, the disease is less obvious and so is the agent that might be useful against it. It is important to determine what specific disease one is dealing with in these cases and the susceptibility patterns exhibited by the causative microorganism. Knowing these factors enables the clinician to narrow the range of therapeutic choices. The only certainty, however, is that inability of a given antibiotic to kill or inhibit a given pathogen in vitro is a virtual guarantee that the drug will fail in vivo. Unfortunately, activity in vitro all too often also results in failure to cure in vivo but here, at least, there is a significant possibility of success.

Experimentally Based Choice of Antibiotic Therapy

The modern clinical application of Koch's discoveries to the selection of an appropriate antibiotic involves sampling infectious material from a patient before instituting anti-infective chemotherapy, culturing the microorganism on suitable growth media, and identifying its genus and species. The bacterium in question is then grown in the presence of a variety of antibiotics to see which of them will inhibit its growth or survival and what concentrations will be needed to achieve this result. This is expressed in minimum inhibitory concentration units (MIC). The term MIC describes that concentration that will inhibit 99% or more of the microbe in question and represents the minimum quantity that must reach the site of the infection in order to be useful. It is usually desirable to have several multiples of the MIC at the site of infection. This brings into play an understanding of pharmacokinetic and pharmacodynamic considerations as well at the results of accumulated clinical experience. The choice of anti-infective agent is made from among those which are active. One of the most convenient experimental procedures is that of Kirby and Bauer. With this technique, filter paper disks impregnated with fixed doses of commercially available antibiotics are placed upon the seeded Petri dish. The dish is then incubated at 37°C for 12 to 24 hours. If the antibiotic is active against the particular strain of bacterium isolated from the particular patient, a clear zone of inhibition will appear around the disk. If a given antimicrobial agent is ineffective, the bacterium will grow right up to the edge of the disk. The diameter of the inhibition zone is directly proportional to the degree of sensitivity of the bacterial strain and the concentration of the antibiotic in question. This powerful methodology gives the clinician a choice of possible antibiotics to use in the particular patient. The widespread occurrence of resistance of individual strains of bacteria to given antibiotics reinforces the need to perform Kirby-Bauer susceptibility disk testing. Other, less common, laboratory methods can be employed for similar purposes.

Bactericidal versus Bacteriostatic

The student must gain an appreciation of the practical difference between the terms bacteriostatic and bactericidal. Almost all antibiotics are bactericidal, that is to say they will kill bacteria, if the concentration or dose is sufficiently high. In the laboratory, it is almost always possible to use such doses. Subsequent inoculation of fresh, antibiotic-free, media with a culture that has been so treated will not produce growth of the culture because the cells are dead. When such doses are achievable in patients, such drugs are clinically bactericidal. At somewhat lower concentrations, bacterial multiplication is prevented even though the microorganism remains viable. The spread between a bactericidal dose and a bacteriostatic dose is characteristic of given families of antibiotics. With gentamicin, for example, doubling or quadrupling the dose changes the effect on bacteria from bacteriostatic to bactericidal. Such doses are usually achievable in the clinic. The difference between bactericidal and bacteriostatic doses of tetracycline is approximately forty-fold. It is not possible to achieve such doses safely in patients, so tetracycline is referred to as bacteriostatic. If a bacteriostatic antibiotic is withdrawn prematurely from a patient, the microorganism can resume growth and the infection can reestablish itself because the culture is still alive. Subsequent inoculation of fresh laboratory media not containing the antibiotic with a culture so treated will also result in colony development. Obviously, in immunocompromised patients who are unable to contribute natural body defenses to fight their own disease, having the drug kill the bacteria is essential for recovery. When a patient is immunocompetent or the infection is not severe, however, a bacteriostatic concentration will break the fulminating stage of the infection when bacterial cell numbers are increasing at a logarithmic rate. With E. coli, for example, the number of cells doubles every 2 hours. A bacteriostatic agent will interrupt this rapid growth and will give the immune system a chance to deal with the disease. Cure usually follows. Thus, whereas it is preferred that an antibiotic be bactericidal, bacteriostatic antibiotics are widely used and are usually satisfactory. Obviously though, patients should not skip doses!

Resistance

Resistance of bacteria to the toxic effects of antimicrobial agents and to antibiotics develops fairly easily both in the laboratory and in the clinic and is an ever-increasing public health hazard. Challenging a culture in the laboratory with sublethal quantities of antibiotic kills the most intrinsically sensitive percentage of the strains in the colony. Those neither killed nor seriously inhibited continue to grow and have access to the remainder of the nutrients. A mutation to lower sensitivity also enables individual members to survive against the selecting pressure of the antimicrobial agent. If the culture is treated several times in succession with sublethal doses in this manner, the concentration of antibiotic required to prevent growth becomes ever higher. When the origin of this form of resistance is explored, it is almost always found to be caused by an alteration in the biochemistry of the colony, so that the molecular target of the antibiotic has become less sensitive, or it can lead to decreased uptake of antibiotic into the cells. This is genomically preserved and passes to the next generation by binary fission. The altered progeny may be weaker than the wild strain, so that they die out if the antibiotic is not present to give them a competitive advantage. Resistance of this type is usually expressed toward other antibiotics with the same mode of action, so is a familial affiliation—most tetracyclines, for example, show extensive cross resistance with other tetracyclines. This is very enlightening with respect to discovery of the molecular mode of action but is not very relevant to the clinical situation.

In the clinic, resistance more commonly takes place by Resistance- (R) factor mechanisms. In the more lurid examples, enzymes are elaborated that attack the antibiotic and inactivate it. The genetic material coding for this form of resistance is generally carried on extrachromosomic elements of small circular DNA molecules known as plasmids. A bacterial cell may have many plasmids or none. The plasmid may carry DNA for several different enzymes capable of destroying structurally dissimilar antibiotics. Such plasmid DNA may migrate within the cell from plasmid to plasmid or from plasmid to chromosome or from chromosome to chromosome by a process known as transposition. Such plasmids may migrate from cell to cell by conjugation (passage through a sexual pilus), transduction (carriage by a virus vector), or by transformation (excretion of DNA from cell A and its subsequent uptake by cell B). These mechanisms can convert an antibiotic-sensitive cell to an antibiotic-resistant cell. This can take place many times in a bacterium's already short generation time. The positive selecting pressure of inadequate levels of antibiotic favors explosive spread of R-factor resistance. This provides a rationale for conservative but aggressive application of appropriate antimicrobial chemotherapy.

Postantibiotic Effect

Some antibiotics exert a significant toxicity to certain microorganisms that persists after the drug is withdrawn. The microbe is termed sick.. A continuing multiple of the minimum inhibitory concentration of drug may not be essential when a postantibiotic effect (PAE) is operating. The terms used for expressing a PAE are the time required for a tenfold increase in viable bacterial colonies to occur after exposure to a single dose of the antimicrobial agent. Although the pharmacologic basis for this effect is not clear, it is speculated that adherence to the intercellular target prevents some significant quantity of the antimicrobial agent from being washed away or possibly that there are other drug-related effects that injure the bacterium and from which it only slowly recovers, or that both of these are involved, or that some yet undiscovered cause is at play. Under some conditions a postantibiotic effect can be detected for days in the chemotherapy of mycobacterial infections.

Inocculum Effect

In certain cases, microbial resistance is mediated by the production of bacterial enzymes that attack the antibiotic molecule, changing its structure to an inactive form. This often leads to a so-called innoculum effect, in which a susceptible antibiotic is apparently less potent when greater numbers of bacteria are present in the medium than when fewer cells are employed. The more bacteria taken, the more antibiotic destroying enzyme is present and the more antibiotic that is required to overcome this and achieve desired response. An antibiotic that is not enzyme-modified is comparatively free from the inocculum effect. These circumstances, when present, make quantitative comparisons of the likely effectiveness of a susceptible agent with a resistant agent tricky to interpret. Inocculum effects are fairly common in testing beta-lactam antibiotics, for example.

Use of Antibiotics in Fixed Dose Combinations

The student may suppose that use of combinations of antibiotics would be superior to the use of individual antibiotics because this would broaden the antimicrobial spectrum and make less critical the accurate identification of the pathogen. It has been found, however, by experiment, that all too often such combinations are antagonistic. A useful generalization, but one that is not always correct, is that one may often successfully combine two bactericidal antibiotics, particularly if their molecular mode of action is different. A common example is the use of a beta-lactam antibiotic, e.g., cefotaxime, and an aminoglycoside, e.g., gentamicin, for first-day empiric therapy of overwhelming sepsis of unknown etiology. Therapy must be instituted as soon after a specimen is obtained as is humanly possible, or the patient may die. This often does not allow the microbiologic laboratory sufficient time to identify the offending microorganism nor to determine its antibiotic susceptibility. An emergency resort is therefore made to what is called shotgun therapy. Both of the antibiotic families applied in this example are bactericidal in readily achievable parenteral doses. As will be detailed later in this chapter, the beta-lactams inhibit bacterial cell wall formation and the aminoglycosides interfere with protein biosynthesis and membrane function. Their modes of action are supplementary. Because of toxicity considerations and the potential for untoward side effects, this empiric therapy is replaced by suitably specific, monotherapy at the first opportunity. One may also often successfully combine two bacteriostatic antibiotics for special purposes, for example, a macrolide and a sulfonamide. These are occasionally used in combination for the treatment of an upper respiratory tract infection caused by Haemophilus influenzae, as the combination of this protein biosynthesis inhibitor and an inhibitor of RNA biosynthesis allows fewer relapses than the use of either agent alone. The use of a bacteriostatic agent, such as tetracycline, in combination with a bactericidal agent, such as a beta-lactam, is usually discouraged. The beta-lactam antibiotics are much more effective when used against growing cultures and a bacteriostatic agent interferes with bacterial growth, often giving an indifferent or antagonistic response when they are combined. Additional possible disadvantages of combination chemotherapy are higher cost, greater likelihood of side effects, and difficulties in demonstrating synergism in humans.

Serum Protein Binding

The influence of serum protein binding on antibiotic effectiveness is fairly straight-forward. It is considered in most instances that the percentage of antibiotic so bound is not available at that moment for the treatment of infections so it must be subtracted from the total blood level in order to get the effective blood level. Thus a heavily and firmly serum-protein-bound antibiotic would not generally be a good choice for the treatment of either septicemias or infections in deep tissue, even though the microorganism involved is susceptible in in vitro tests. If the antibiotic is rapidly released from bondage, this factor decreases in importance and the binding becomes a depot source. An antibiotic that is not to some extent protein-bound will normally be rapidly excreted and have a short half life. Thus, some protein binding of poorly water soluble agents is normally regarded as helpful. The student will recall that under most circumstances the urine is a protein-free filtrate so that proportion of an antibiotic that is firmly bound to serum proteins will be retained in the blood. Thus, a highly and firmly protein-bound antibiotic will be a poor choice for septicemia but will be satisfactory for mild urinary tract infection.

Preferred Dosing Methods

Under ideal circumstances it is desirable for an antibiotic to be available in both parenteral and oral forms. Whereas there is no question that the convenience of oral medication makes this ideal for outpatient and community use, very ill patients often require parenteral therapy, and it would be consistent with today's practice of discharging patients "quicker and sicker," and to send them home with an efficacious oral version of the same antibiotic that led to the possibility of discharge in the first place. In that way, the patient would not have to come back to the hospital at intervals for drug administration nor would one have to risk treatment failure by starting therapy with a new drug. For drugs with significant toxicities, the physician will prefer the injection form. The physician using this method is certain that the whole dose has been taken at the appropriate time. If the local pharmacist is adept at administration of parenteral medication, these considerations become less important.

Initiation of Therapy

As bacteria multiply rapidly—populations often double in 2 or 3 hours—it is important to institute antibiotic therapy as soon as possible. It is thus often desirable to initiate therapy with a double (loading) dose and then to follow this with a smaller (maintenance) dose. To prevent relapse, the physician must instruct the patient not to skip doses and to take all of the medication provided even though the presenting symptoms, e.g., diarrhea or fever, resolve before all of the drug is taken. Treatment failure, relapse, and the emergence of resistance are probably all too often caused by lack of compliance or premature cessation of therapy by the patient.

Prophylactic Use of Antibiotics

Antibiotics are often used prophylactically, for example in preoperative bowel sanitization and orally for treatment of viral sore throats. These are not sound practices because the patient is exposed to the possibility of both drug-associated side effects and a suprainfection by drug-resistant microorganisms, moreover the therapeutic gain from such practices is often marginal. However frustrating this may be to the infectious diseases specialist, these are common medical practices and hence difficult to stop.

Agricultural Use

Antibiotics are often used in agriculture. Their use for treatment of infections of plants and animals is not to be discouraged so long as drug residues from the treatment do not contaminate foods. In contamination, problems such as penicillin allergy or subsequent infection higher up the food chain by drug resistant microbes can occur. Animals demonstrably grow more rapidly to marketable size when antibiotics are added to their feed even though the animals have no apparent infection. This is believed to be due in large part to suppression of subclinical infections and the subsequent diversion of protein biosynthesis from muscle and tissue growth into proteins needed to combat the infection. Under appropriate conditions antibiotic feed supplementation is partly responsible for the comparative wholesomeness and cheapness of our food supplies. This practice has the potential, however, to contaminate the food we consume or to provide reservoirs of drug-resistant enteric microorganisms. Occasionally infections are traced to this cause. It might be better to use antibiotics for such purposes that are not systematically absorbed and where they are not cross-resistant with antibiotics used in clinical practice in humans. This is, in fact, the practice in England and some other countries.

Cost

Antibiotics are often expensive, but so are morbidity and mortality. For many patients, nonetheless, cost is a significant consideration. The pharmacist is in an ideal position to guide the physician and the patient on the question of possible alternative equivalent treatments that might be more affordable. The most frequent comparisons are based on the cost of the usual dose of a given agent for a single course of therapy (usually the wholesale cost to the pharmacist for 10 days worth of the drug).

In summary, when antibiotics are used intelligently they are remarkably effective, and the population explosion and the increased life span that has characterized recent generations can be traced in part to them. When used carelessly or inappropriately, they can lead to complex ecologic problems such as infection with multidrug-resistant microorganisms. This phenomenon is becoming increasingly troublesome.

SYNTHETIC ANTIMICROBIAL AGENTS

Synthetic antimicrobial agents have not been modeled after any natural product so they may not properly be called antibiotics. Some are extremely effective for treatment of infections and are widely used. Very few antibiotics are known to work in precisely the same way as these agents in killing bacteria. Also curious is that those agents whose molecular mode of action is known are at present nearly all effective against key enzymes needed for the biosynthesis of nucleic acids. Because they interrupt the biosynthesis of nucleic acids rather than attacking the finished products or substituting for them in nucleic acids they are not genotoxic but are comparatively safe to use.

Sulfonamides

Sulfonamides were discovered in the middle 1930s despite an incorrect hypothesis, but by observing the results carefully and drawing correct conclusions. Prontosil rubrum, a red dye, was one of a series of dyes examined by Domagk of Bayer of Germany in the belief that it might be taken up selectively by certain pathogenic bacteria and not by human cells, in a manner analogous to the way that the Gram-stain works, and so serve as a selective poison to kill these cells. If this were true, this would be a useful validation of theories elaborated by Ehrlich at the turn of the century. The dye, indeed, proved active in vivo against streptococcal infections in mice. Curiously, it was not active in vitro. Trefouel and Bovet in France soon showed that the urine of prontosil rubrum-treated animals was bioactive in vitro. Fractionation led to identification of the active substance as *p*-aminobenzenesulfonic acid amide (sulfanilamide), a colorless cleavage product formed by liver metabolism of the administered dye. Today we would call prontosil rubrum a prodrug. The discovery of sulfanilamide's in vivo antibacterial properties ushered in the modern anti-infective era, and these investigators shared the award of a Nobel Prize for medicine in 1938. As poor as the potency of sulfanilamide is when compared with that of the modern agents that began to succeed it shortly thereafter, its impact on medicine was enormous. For the first time in the long and weary chronicle of human struggle against infectious disease, physicians now had a comparatively safe and responsive oral drug to use. Anxious friends and family at long last had hope that loved ones, especially children, would not succumb to infectious disease at an early age. Taken along with the use of penicillin only five or so years later, the era of the so-called wonder drugs had dawned. It is one of life's great ironies to consider that although sulfanilamide had been synthesized in 1908, it had not been tested before as an anti-infective because there was no reason at that time to suppose that it would be useful to do so.

The sulfonamides are bacteriostatic when used in achievable doses. They inhibit the enzyme dihydropteroate synthase, an important biocatalyst needed for the biosynthesis of folic acid derivatives and, ultimately, DNA. They do this by competing at the active site with *p*-aminobenzoic acid (PABA), a normal structural component of folic acid derivatives. PABA is incorporated into the developing tetrahydrofolic acid molecule by condensation with a dihydropteroate diphosphate precursor under the influence of dihydropteroate synthetase. Thus sulfonamides may also be classified as antimetabolites. Indeed, the antimicrobial efficacy of sulfonamides can be reversed by adding significant quantities of PABA into the diet (in multivitamin preparations, for example) or into the culture medium. Most susceptible bacteria are unable to take up performed folic acid from their environment and convert it to a tetrahydrofolic acid but, instead, synthesize their own folates de novo. As folates are essential intermediates for the preparation of certain DNA bases, without which bacteria cannot multiply, this inhibition is strongly bacteriostatic and ultimately bactericidal. Humans are unable to synthesize folates from component parts, lacking the necessary enzymes (including dihydropteroate synthase), so, for us, folic acid is a vitamin which must be in our diet, and sulfonamides have no lethal effect upon us. The basis for the selective toxicity of sulfonamides is, thus, clear.

In a few strains of bacteria, however, the picture is more complex. Here, sulfonamides are attached to the dihydropteroate diphosphate in the place of the normal PABA. The resulting product, however, is not capable of undergoing the next necessary reaction, condensation

liver [H]

Protonsil Rubrum → **Sulfanilamide**

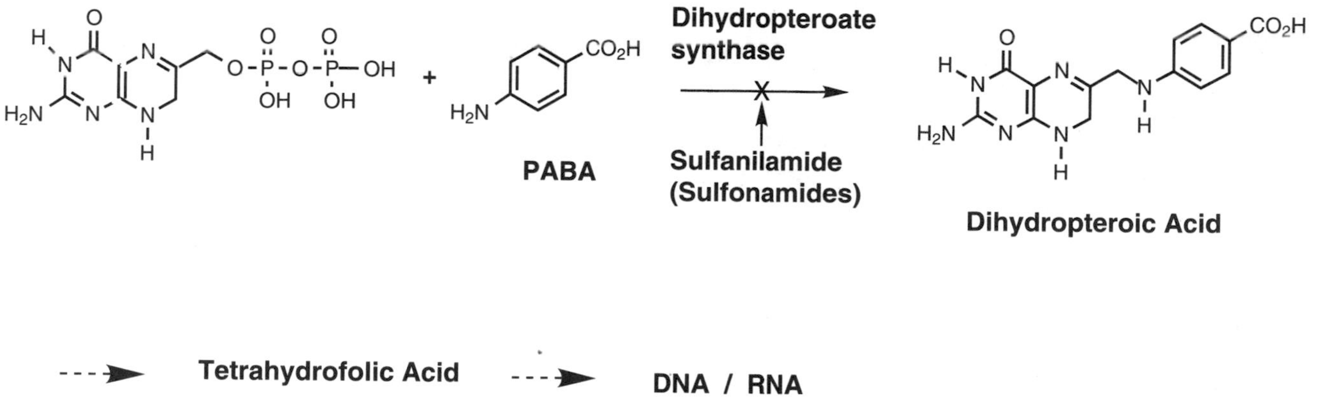

Dihydropteroate synthase

PABA

Sulfanilamide (Sulfonamides)

Dihydropteroic Acid

Tetrahydrofolic Acid → **DNA / RNA**

with glutamic acid. This false metabolite is also an enzyme inhibitor and the net result is inability of the bacteria to multiply as soon as the preformed folic acid in their cells is all used up and further nucleic acid biosynthesis becomes impossible. The net result is the same, but the molecular basis of the effect is somewhat different in these strains.

Bacteria that are able to take up preformed folic acid into their cells are intrinsically resistant to sulfonamides.

SO_2NH_2 → THFA

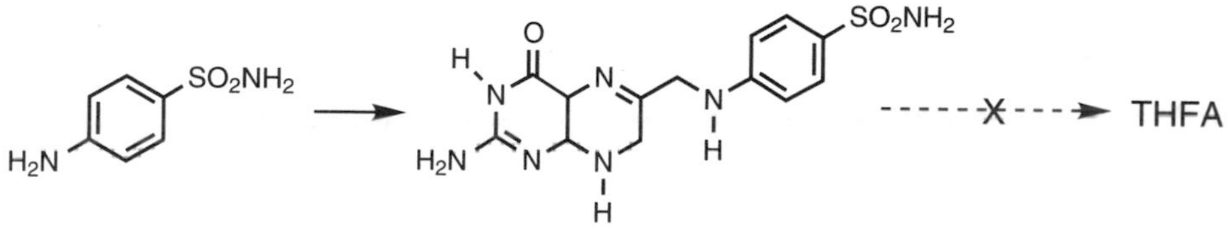

Replacement of PABA by sulfanilamide

The basis of the structural resemblance of sulfonamides to PABA that is so devastating to these bacteria is clear to chemists. The functional group that differs in the two molecules is the carboxyl of PABA and the sulfonamide moiety of sulfanilamide. The strongly electron withdrawing character of the aromatic SO_2 group makes the nitrogen atom to which it is directly attached partially electropositive. This in turn increases the acidity of the hydrogen atoms attached to the nitrogen so that this functional group is not neutral in water, as are most normal amides, but is actually feebly acidic ($pK_a \sim 10.4$). The pK_a of the carboxyl group of PABA is about 6.5 so the resemblance of the two molecules is not very close. It was soon found, following a crash synthetic program, that replacement of one of the NH_2 hydrogens by an electron-withdrawing heteroaromatic ring was not only consistent with antimicrobial activity but also greatly acidified the remaining hydrogen and dramatically enhanced potency. With suitable groups in place, the pK_a came down to the same range as that of PABA itself. Not only did this markedly increase the antibacterial potency of the product but greatly increased the water solubility under physiologic conditions. The pK_a of sulfisoxazole, one of the most popular of the sulfonamides in present use, is about 5.0. The poor solubility in water of the earliest sulfonamides led to occasional crystallization in the urine (crystallurea) and resulted in kidney damage because the molecules were un-ionized at urine pH values. It is still recommended to drink plenty of water to avoid crystallurea when taking certain sulfonamides but this form of toxicity is now comparatively uncommon with the more important agents used today.

Of the thousands of sulfonamides that have been evaluated, sulfisoxazole is currently the most popular. Along with the surviving sulfonamides, it has a comparatively broad antimicrobial spectrum in vitro but its clinical use is largely restricted to the treatment of primary uncomplicated urinary tract infections and occasionally as a back-up to other normally more preferred agents in special

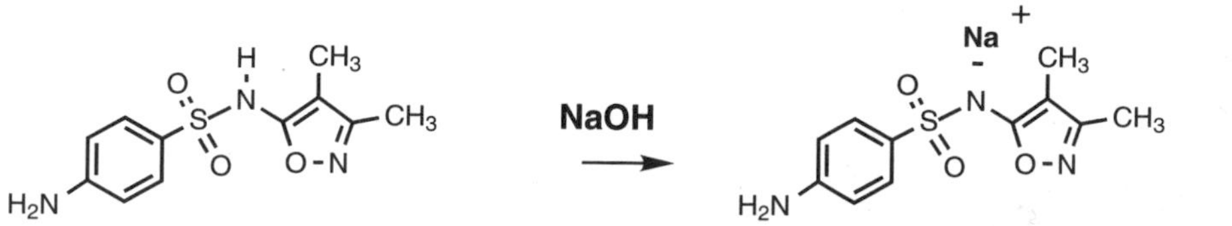

NaOH

Sulfisoxazole $pK_a = 5$ → **Sodium Sulfisoxazole**

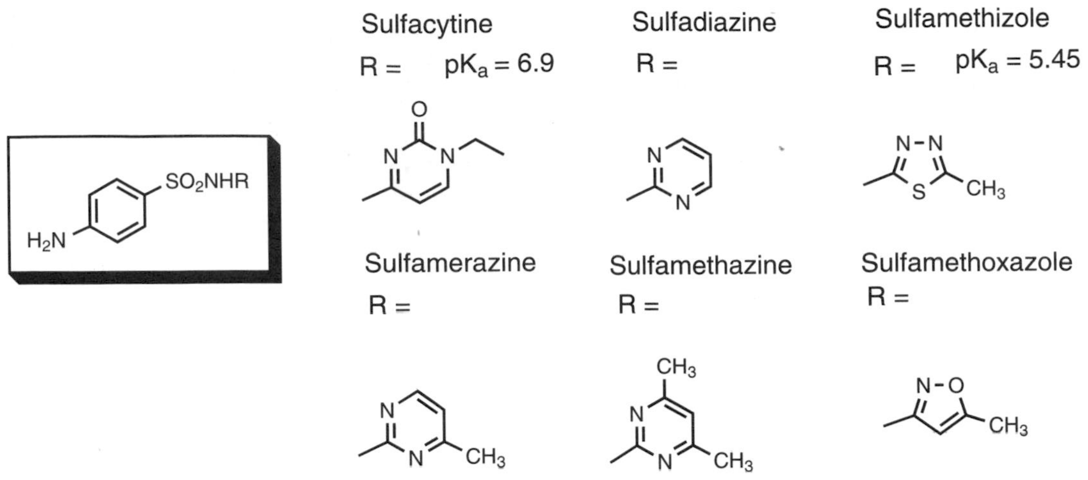

H_2N–C$_6$H$_4$–SO_2NHR

Sulfacytine R = $pK_a = 6.9$

Sulfadiazine R =

Sulfamethizole R = $pK_a = 5.45$

Sulfamerazine R =

Sulfamethazine R =

Sulfamethoxazole R =

Fig. 34–1. Structures of clinically useful sulfonamides.

situations. Sulfisoxazole is well absorbed following oral administration, distributes fairly widely, and is excreted by the kidneys. Plasmid-mediated resistance development is common, particularly among gram-negative microorganisms and usually takes the form of decreased sensitivity of dihydropteroate synthase. Sulfonamides are partly deactivated by acetylation and glucuronidation in the liver. In addition to frequent resistance, another drawback of the sulfonamides that decreases their use is the frequency of some severe side effects. Allergic reactions are the most common and take the form of rash, photosensitivity, and drug fever. Less common problems are kidney and liver damage, hemolytic anemia, and other blood problems. The most severe is onset of Stevens-Johnson syndrome characterized by sometimes fatal erythema multiforme and ulceration of mucous membranes of the eye, mouth, and urethra. Fortunately these effects are comparatively rare.

Other sulfonamides still in use include sulfacytine, sulfadiazine, sulfamethizole, sulfamethoxazole, and sulfasalazine (Fig. 34–1). Multiple (or triple) sulfas are a 1:1:1 combination of three sulfapyrimidines, namely, sulfadiazine, sulfamerazine, and sulfamethazine. The basis for use of this combination is physicochemical. A solution saturated in a given substance often can still be used as a solvent for other materials if these do not possess poorly soluble ions in common. In triple sulfas, each component contributes one-third of the potency of the combination (sum = 100%), whereas the combination is significantly more soluble in urine than would be the case if any single component were employed in a full therapeutic dose. That is, the resulting solution of the combination is not saturated in any of its components.

Of this group of less commonly employed sulfonamides, sulfasalazine stands out for other reasons. It is a prodrug in that the absence of a free *p*-aminoaromatic moiety makes it intrinsically inactive. Sulfasalazine is a red azo dye given orally. It is largely not absorbed in the duodenum so the majority of the dose is delivered to the distal bowel. Reductive metabolism by gut bacteria converts the drug to sulfapyridine and *p*-aminosalicylic acid (PAS), the active component. The liberation of PAS is the purpose for administering this drug. This agent is used to treat both ulcerative colitis and Crohn's disease because it is an anti-inflammatory. Direct administration of PAS is otherwise irritating to the gastric mucosa.

The comparative cheapness of the sulfonamides is one of their most attractive features.

Sulfones

Hansen's Disease (leprosy) is a disfiguring ancient malady of low invasiveness and slow progression. It is caused by Mycobacterium leprae and is endemic in the United States to Hawaii and the southernmost tier of contiguous states. Prolonged treatment with sulfones, such as dapsone, cures this disease. Sulfones characteristically have a fully oxygenated sulfur atom bonded directly to two

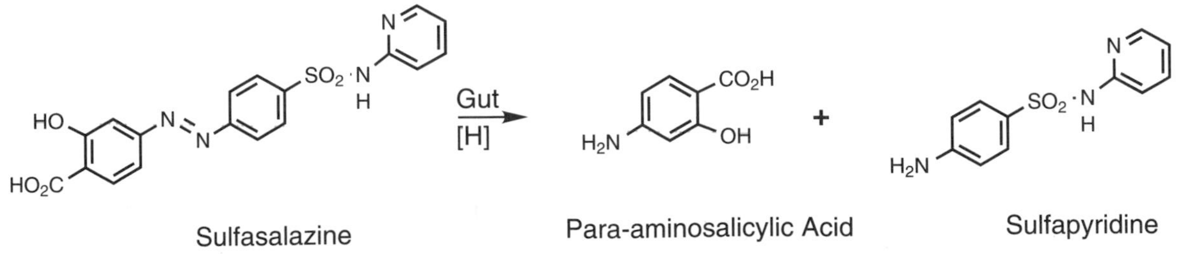

Sulfasalazine —Gut [H]→ Para-aminosalicylic Acid + Sulfapyridine

carbon atoms. The mode of action of the sulfones is apparently the same as for the sulfonamides. The reason for the particular effectiveness of dapsone and the comparative lack of utility of the other sulfonamides in treating Hansen's disease is not understood.

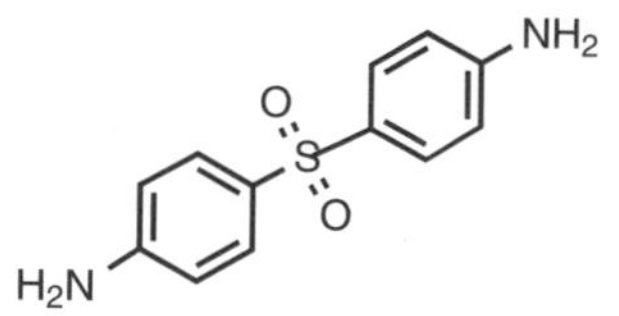

Dapsone

Trimethoprim

Trimethoprim

Somewhat further along the same pathway leading from the pteroates to folic acid derivatives and on to DNA bases one encounters the enzyme dihydrofolate reductase. Exogenous folic acid must be reduced stepwise to dihydro- and then to tetrahydrofolic acid in order to progress to nucleic acids. Endogenously produced dihydrofolate enters the pathway directly. Inhibition of this key enzyme had been widely studied in attempts to find anticancer agents by starving rapidly dividing cancer cells of needed DNA. The student will recall that methotrexate and its analogs came from such studies. Methotrexate is, however, much too toxic to be used as an antibiotic. Subsequently, however, trimethoprim was developed by Hitchings and Elion (who shared a Nobel Prize for this and other contributions to chemotherapy). This inhibitor prevents the use of folic acid analogs for DNA synthesis and the result is bacteriostasis. The student may wonder at first how this can work, because it is clear that mammals must also perform this step enzymatically. The basis for the favorable selectivity comes from subtle but significant architectural differences between the bacterial and the mammalian dihydrofolate reductases away from the active site. Whereas the bacterial enzyme and the mammalian enzyme both efficiently catalyze the conversion of dihydrofolic acid to tetrahydrofolic acid, the bacterial enzyme is sensitive to inhibition by trimethoprim by up to 40,000 times lower concentrations than is the mouse enzyme. This difference explains the useful selective toxicity of trimethoprim.

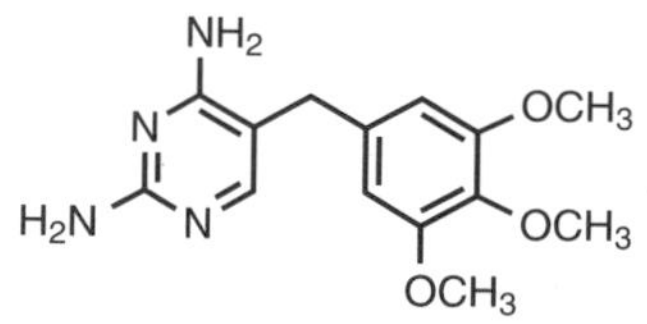

Trimethoprim is frequently used as a single agent clinically for the oral treatment of uncomplicated urinary tract infections caused by susceptible bacteria (predominantly community acquired E. coli and other gram-negative rods). It is, however, most commonly used in a 1:5 fixed ratio with the sulfonamide sulfamethoxazole. This combination is not only synergistic in vitro but is less likely to induce bacterial resistance than either agent alone. It is rationalized that microorganisms not completely inhibited by sulfamethoxazole at the pteroate condensation step will not likely be able to push substrates past a subsequent blockade of dihydrofolate reductase. Thus these agents block sequentially at two different steps in the same essential pathway, and this combination is extremely difficult for a naive microorganism to survive. It is also comparatively uncommon that a microorganism will successfully mutate to resistance at both enzymes during the course of therapy. Of course, if the organism is already resistant to either drug at the outset of therapy, which happens more and more often, much of the advantage of the combination is lost.

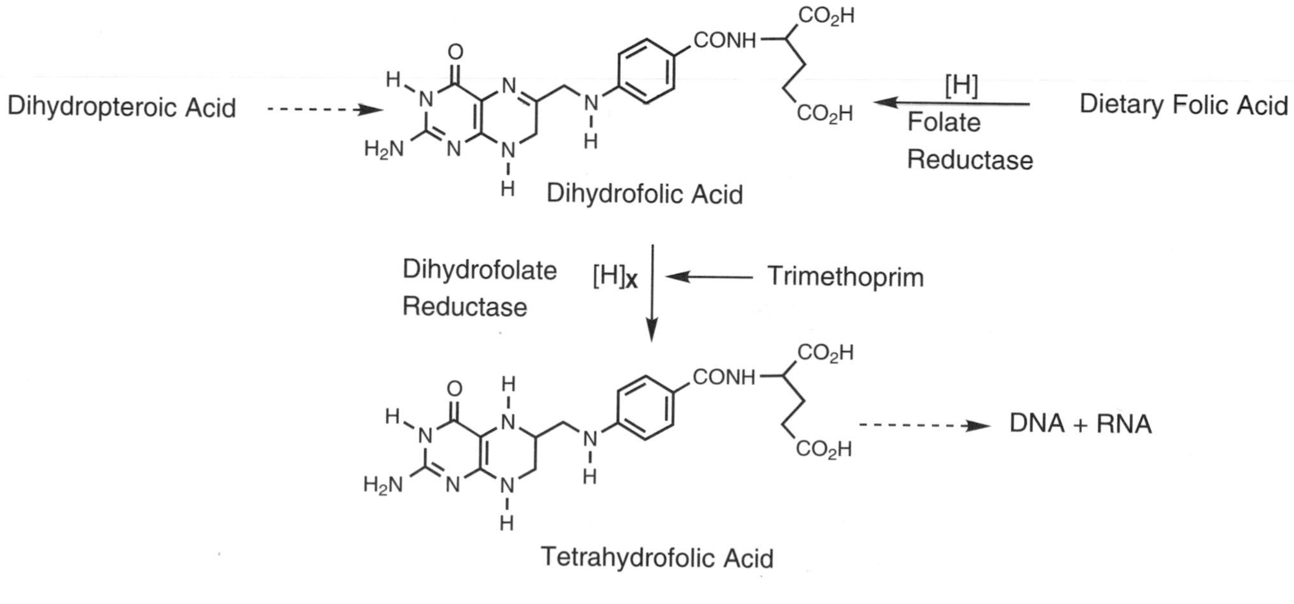

Dihydropteroic Acid → Dihydrofolic Acid ← [H] Folate Reductase ← Dietary Folic Acid

Dihydrofolic Acid → Dihydrofolate Reductase [H]x (← Trimethoprim) → Tetrahydrofolic Acid → DNA + RNA

Pairing these two particular antibacterial agents was done based on pharmacokinetic factors. For such a combination to be useful in vivo, the two agents must arrive at the infected site in the tissue compartment at the correct time(s) and in the right ratio. In this context, the optimum ratio of these two agents in vitro is 1:20! Of all of the combinations tried, sulfamethoxazole came closest to being optimal for trimethoprim.

It is easier to demonstrate synergy in vitro than in vivo, and concerns about the toxic contribution of the sulfonamide (and doubtless commercial considerations also) have led to a recent vogue for use of trimethoprim alone. This agent does often work when used alone.

Trimethoprim is broad-spectrum in vitro so it is potentially useful against many microorganisms. Combined with sulfamethoxazole, it is used for oral treatment of urinary tract infections, shigellosis, otitis media, traveler's diarrhea, and bronchitis. Among the opportunistic pathogens that afflict AIDS patients is the pneumonia-causing protozoan, Pneumocystis carinii. Immunocompetent individuals rarely become infected with P. carinii but it is a frequent pathogen in AIDS patients, and it is nearly 100% fatal in such immunocompromised individuals. The combination of sulfamethoxazole-trimethoprim has proven to be useful and comparatively non-toxic for these patients.

An injectable form is available for use in severe infections and is particularly used in AIDS patients. This treatment form leads to more frequent toxic reactions, however.

The most frequent side effects of trimethoprim-sulfamethoxazole are rash, nausea, and vomiting. Blood dyscrasias are less common as is pseudomembranous colitis (caused by nonantibiotic sensitive opportunistic gut anaerobes, often Clostridium difficile). Many broad-spectrum antimicrobials can lead to such severe drug-related diarrhea and this side effect must be monitored carefully. Severe, non-resolving diarrhea can be fatal. It is therefore, a justification for withdrawing existing therapy in favor of an antianaerobic antibiotic.

Quinolones

The quinolone antimicrobials comprise a group of synthetic substances possessing in common an N-alkylated 3-carboxypyrid-4-one ring fused to a second aromatic ring that carries other substituents. The first quinolone to be marketed, in 1965, was nalidixic acid. Nalidixic acid is primarily effective against gram-negative bacteria and its clinical spectrum is primarily reserved for oral treatment of uncomplicated urinary tract infections caused by susceptible microorganisms (usually E. coli). Over the following years, the remaining first generation quinolones were introduced. These compounds consist of more potent and somewhat broader spectrum agents such as oxolinic acid, cinnoxacin and enoxacin but they are not very popular today and are classified only as urinary tract disinfectants. The quinolones are often well absorbed following oral administration and are highly serum-protein bound. This leads to a comparatively long half-life but restricts their use mainly to protein-free compartments such as the urinary tract. The tendency, because of their comparatively low potency, is to use them in higher doses to achieve coverage, and this leads to a frequent incidence of side effects, notably gastrointestinal (GI) upset, rash, and visual disturbances. These drugs are proconvulsant and photosensitizing in susceptible individuals. They remained a small group of rarely used agents until the discovery of the fluoroquinolones, of which norfloxacin was the first to become important. Norfloxacin (1986) is broad spectrum and equivalent in potency to many of the fermentation-derived antibiotics. Over a thousand of these second generation analogs have now been made and ciprofloxacin, ofloxacin and lomefloxacin are currently marketed.

The properties of ciprofloxacin, the most widely used fluoroquinolone, are typical of those of the group. It is rapidly and often completely absorbed on oral administration and is not highly protein-bound. Lomefloxacin has a comparatively long half-life and it is conveniently administered less frequently than the other quinolones.

The quinolones are rapidly bactericidal largely as a

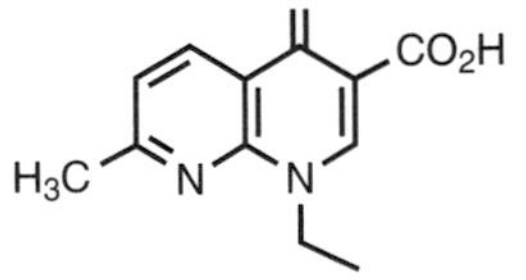

Nalidixic Acid

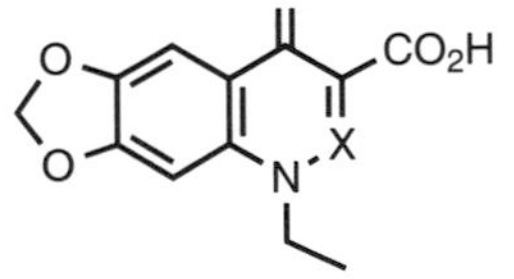

Oxolinic Acid, X = CH
Cinoxacin, X = N

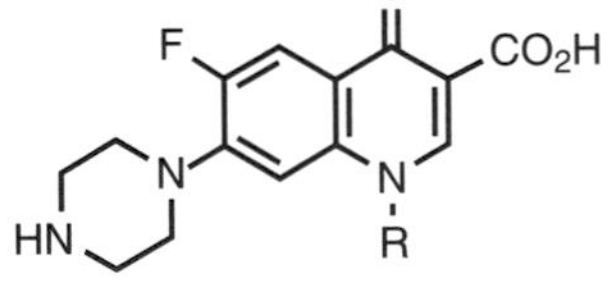

Norfloxacin, R = C2H5
Ciprofloxacin, R = cyclo-Pr

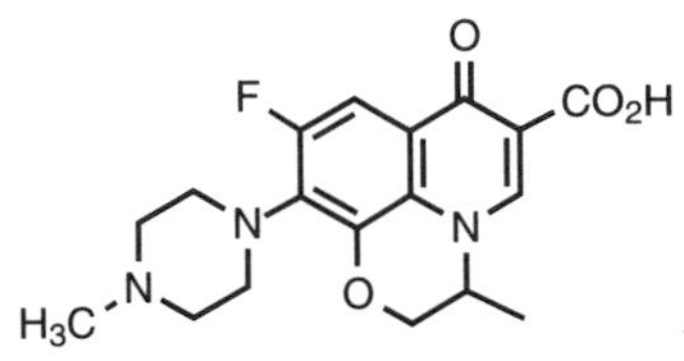

Ofloxacin

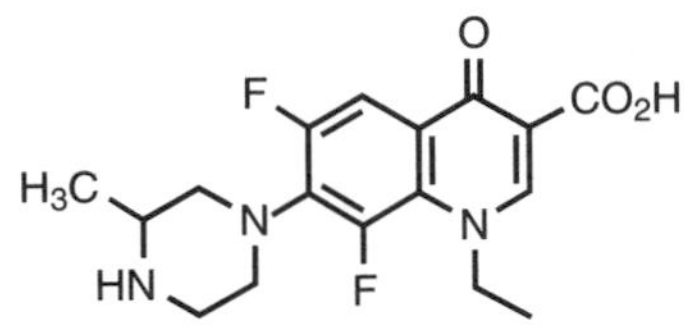

Lomefloxacin

consequence of their inhibition of DNA gyrase, a key bacterial enzyme that dictates the conformation of DNA so that it can be stored properly, unwound, replicated, repaired, transcribed on demand, and so forth. Inhibition of DNA gyrase makes a cell's DNA inaccessible and leads to cell death, particularly if other toxic effects are simultaneously present. Humans shape their DNA with a topoisomerase II, an analogous enzyme that, however, does not bind quinolones at normally achievable doses so the manufactured quinolones do not kill host cells.

Resistance to the quinolones is becoming more frequent and takes the form of lesser cellular uptake or lesser sensitivity of the DNA gyrase. Resistance by plasmid mediated mechanisms does not seem to take place.

The quinolones chelate polyvalent metal ions such as (Ca[II], Mg[II], Al[III], and Fe[II]) to form less water-soluble complexes and thereby lose considerable potency. Thus coadministration of certain antacids, hematinics, tonics, and consumption of dairy products soon after quinolone administration is contraindicated.

Among the toxicities associated with quinolones is a proconvulsant action, especially in epileptics, but this is mainly associated with the first-generation agents. Other CNS problems include hallucinations, insomnia, and visual disturbances. Some patients also experience diarrhea, vomiting, abdominal pain, and anorexia. The second generation (fluoroquinolone) drugs are generally much better tolerated. The quinolones are associated with erosion of the load-bearing joints of young animals. This does not seem to have been seen in humans but as a precaution these drugs are nonetheless not used casually before puberty nor in sexually-active females of child-bearing age. They are also potentially damaging in the first trimester of pregnancy because of a risk of severe metabolic acidosis and hemolytic anemia. Coadministration with theophylline potentiates the action of the latter and should be monitored closely.

Second-generation quinolones are more widely used than first. Whereas norfloxacin is mainly used for urinary tract infections (enterobacter, enterococcus, or Pseudomonas aeruginosa), the others, particularly ciprofloxacin, are also used for prostatitis, upper respiratory tract infections, bone infections, septicemia, staphylococcal and pseudomonal endocarditis, meningitis, sexually transmitted diseases (gonorrhea and chlamydial), chronic ear infections, and purulent osteoarthritis. Anaerobes, staphylococci, and pseudomonads must be watched carefully for any emergence of resistance. Lomefloxacin is used once daily for urinary tract and upper respiratory tract infections caused by susceptible microorganisms.

Miscellaneous Agents

Nitrofurans

Nitrofurantoin. Also called macrodantin, this is a widely used oral antibacterial substance available since World War II. It is used for prophylaxis or treatment of urinary tract infections when kidney function is not impaired, and it inhibits kidney stone growth. Nausea and vomiting are common side effects. They are avoided in part by slowing the rate of absorption of the drug through use of wax-coated large particles (macrodantin).

Furazolidone. This agent was also introduced during the 1940s and is not much used today. Its present indication is primarily for oral treatment of bacterial diarrhea. It is not effectively absorbed systemically when given orally. Its toxicity is manifested in nausea and vomiting.

Nitroimidazoles

Metronidazole. Initially introduced for the treatment of vaginal infections by amoeba, it is also useful for the treatment of trichomoniasis, giardiasis, and Gardnerella vaginalis vaginal infections. It has found increasing use of late in the parenteral treatment of anaerobic infections and for treatment of pseudomembranous entercolitis due to Clostridium difficile. C. difficile is an opportunistic pathogen that occasionally flourishes as a consequence of broad spectrum antibiotic therapy and its infections can be life-threatening.

Pentamidine Isethionate

This is a synthetic antiprotozoal agent with an unknown mode of action. The two basic arylamidine groups both

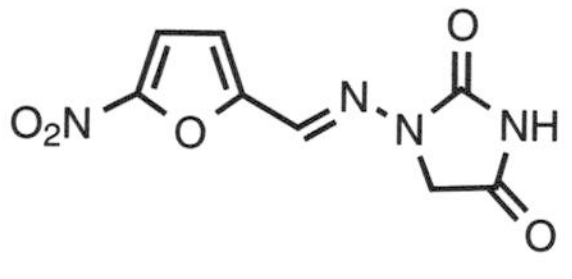

Nitrofurantoin (Macrodantin)

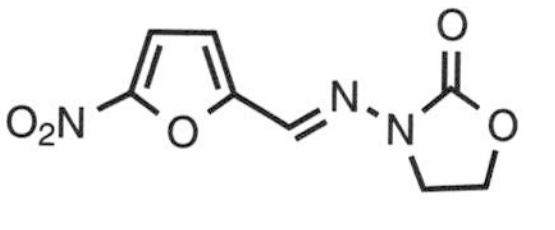

Furazolidone

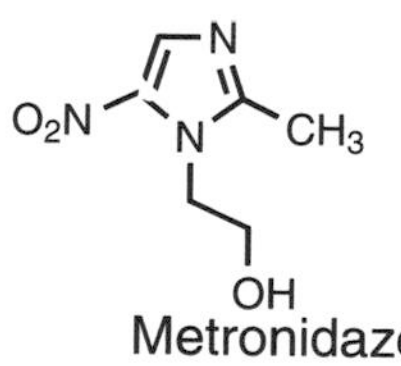

Metronidazole

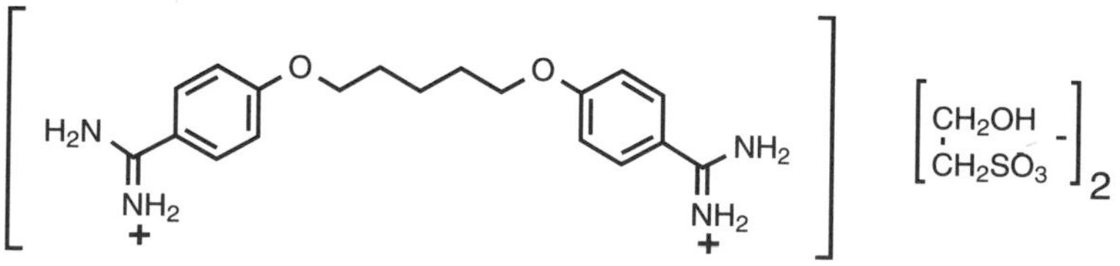

Pentamidine Isethionate

form salts with 2-hydroxyethanesulfonic acid (isethionic acid). It is particularly effective against trypanosomes and is used in the United States primarily in AIDS patients for the treatment of pneumonia caused by Pneumocystis carinii infection. It is used intramuscularly or in the form of oral nebulization. The injectable form is more efficacious but less convenient. It is used as a back up for trimethoprim-sulfamethoxazole. Hypotension, hyper- and hypoglycemia, renal failures, and cardiac arrhythmic side effects must be watched for.

Methenamine

A venerable drug taken for the disinfection of acidic urine, structurally it is a low polymer of ammonia and formaldehyde that reverts to its components under mildly acidic conditions. Formaldehyde is the active antimicrobial component.

$$\text{Methenamine} \xrightarrow{H_3O^+} 6\ H_2C{=}O \quad + \quad 4\ NH_3$$

Methenamine

ANTIBIOTICS: INHIBITORS OF BACTERIAL CELL WALL BIOSYNTHESIS

Bacterial Cell Wall

The bacterial cell wall differs dramatically in structure and function from the outer layers of mammalian cells and thus provides a number of potentially attractive targets for selective chemotherapy of bacterial infections. For one thing, the bacterial cell wall is constructed of enzymes that often have no direct counterpart in mammalian cell construction. Three of the main functions of the bacterial cell wall are: (1) to provide a semipermeable barrier interfacing with the environment through which only desirable substances may pass; (2) to provide a sufficiently strong barrier so that the bacterial cell is protected from changes in the osmotic pressure of its environment; and (3) to prevent digestion by host enzymes. The initial units of the cell wall are constructed within, but soon the growing and increasing complex structure must be extruded; further assembly takes place outside of the inner membrane. This circumstance makes the enzymes involved later in the process more vulnerable to inhibition. Whereas individual bacterial species differ in specific details, the following generalized picture of the process is sufficiently accurate to illustrate our process.

Gram-positive Bacteria

The cell wall of gram-positive bacteria, although complex enough, is rather simpler than that of gram-negatives. A schematic representation is shown in Figure 34–2. On the very outside is a set of characteristic carbohydrates and proteins that together make up the antigenic determinants that differ from species to species and that also cause adherence to particular target cells. Another group of proteins, the porins, have mainly hydrophobic amino acids on the outside by which circumstance, the assembly is embedded in the outer membrane. The inner layer of amino acids is mainly hydrophilic and forms a water-lined pore through which water-soluble substances can pass if their dimensions are suitable. The next significant barrier that the wall presents is the peptidoglycan layer. This is a spongy, gel-forming layer consisting of a series of alternating sugars (N-acetyl-glucosamine and N-acetylmuramic acid) linked (1,4)-β in a long chain (Fig. 34–3). To the free lactic acid carboxyl moieties of the N-acetylmuramic acid units is attached, through an amide linkage, a series of amino acids of which L-ala-D-glu-L-lys-D-ala is typical. One notes the D-stereochemistry of the glutamate and the terminal alanine. This feature is presumably important in protecting the peptidoglycan from hydrolysis by host peptidases, thereby enabling successful parasitism. The terminal D-alanyl unit is bonded to the lysyl unit of an adjacent tetrapeptide strand through a pentaglycyl unit. This last step in the biosynthesis is a transamidation wherein the terminal amino moiety on the last glycine of the A strand displaces the terminal D-ala unit on the nearby B strand. This step is catalyzed by a cell wall transamidase (one of the penicillin-binding proteins) that forms a transient covalent bond during the synthesis phase with a particular serine hydroxyl on the enzyme. Completion of the catalytic cycle regenerates the enzyme function. This process gives the resulting structure strength through adding a third dimension, much as would be achieved by glueing the pages of a book together. This provides the strong barrier needed against osmotic stress and accounts for the retention of the characteristic morphologic shape of bacteria (globes and rods, for example). This step is highly sensitive to inhibition by beta-lactam antibiotics. The peptidoglycan layer is transversed by teichoic and teichuronic acids. Beneath this layer is the lipoidal cytoplasmic cell membrane in which a number of important protein molecules that float in a lipid bilayer. Among these are the beta-lactam receptors, known as the penicillin binding-proteins. There are at least seven types of these. Their functions are not entirely understood but they are important in construction and repair of the cell wall. It is also clear that various beta-lactam antibiotics display different patterns of binding to these proteins. These proteins must alternate in a controlled and systematic way between their active and inert states in order that bacterial cells can grow and multiply. Selective interference by beta-lactam antibiotics with their functioning prevents normal growth and repair and creates serious problems for bacteria, particularly young cells needing to grow and mature cells needing to repair damage or to divide. The rapid bactericidal effect of penicillins on such cells can readily be imagined.

Binding of beta-lactam antibiotics to PBP-1 (transpeptidase) of E. coli leads to cell lysis; to PBP-2 (transpeptidase) leads to oval cells deficient in rigidity and to inhibition of cell division; to PBP-3 (transpeptidase) gives abnormally long, filamentous shapes by failure to produce a septum; and to PBP-4-6 (carboxypeptidases) leads to no lethal effects. Approximately 8% of a dose of benzyl penicillin binds to PCP-1; 0.7% to PCP-2; 2% to PBP-3;

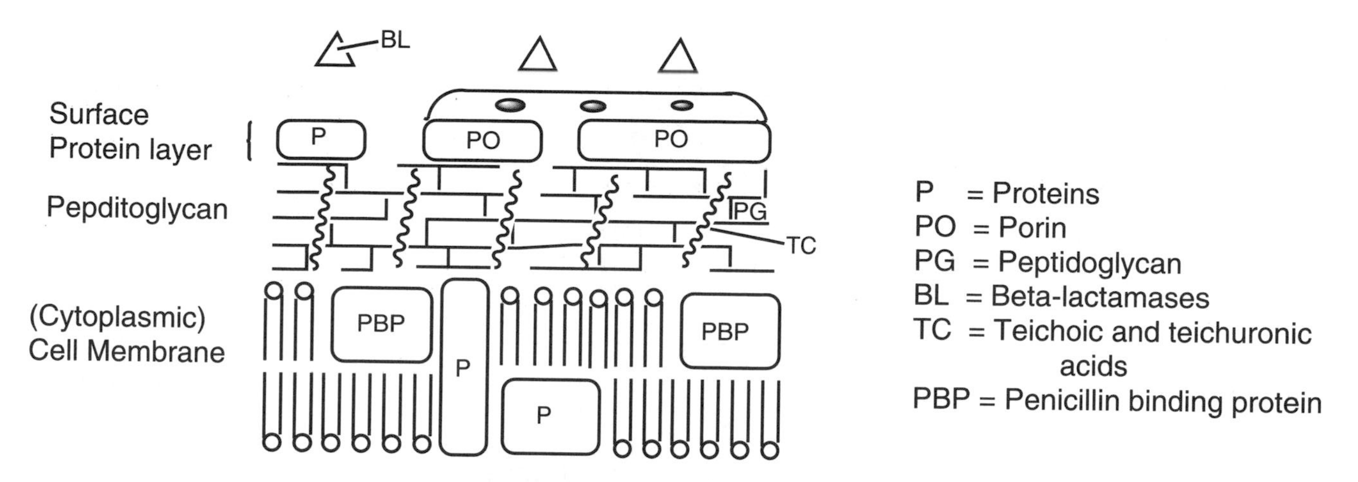

Fig. 34–2. Schematic of the gram-positive bacterial cell wall.

4% to PBP-4; 65% to PBP-5, and 21% to PBP-6. Thus, the majority of the dose bonds to PBPs whose function remains obscure. Binding to PBP-1 is lethal. Other beta-lactam antibiotics display different binding patterns. Amoxicillin and the cephalosporins bind more avidly to PBP-1; mecillinam and cefotaxime to PBP-2, and mezlocillin and cefuroxime to PBP-3. The beta-lactamases are secreted outside the gram-positive cell.

Gram-negative Bacteria

With the Gram-negatives, the cell wall is more complex and more lipoidal (Fig. 34–4). These cells usually contain an outer membrane and a periplasm. The outer layer itself has a complex layer of lipopolysaccharides that encode antigenic responses, cause septic shock, provide the serotype, and influence morphology. This exterior layer

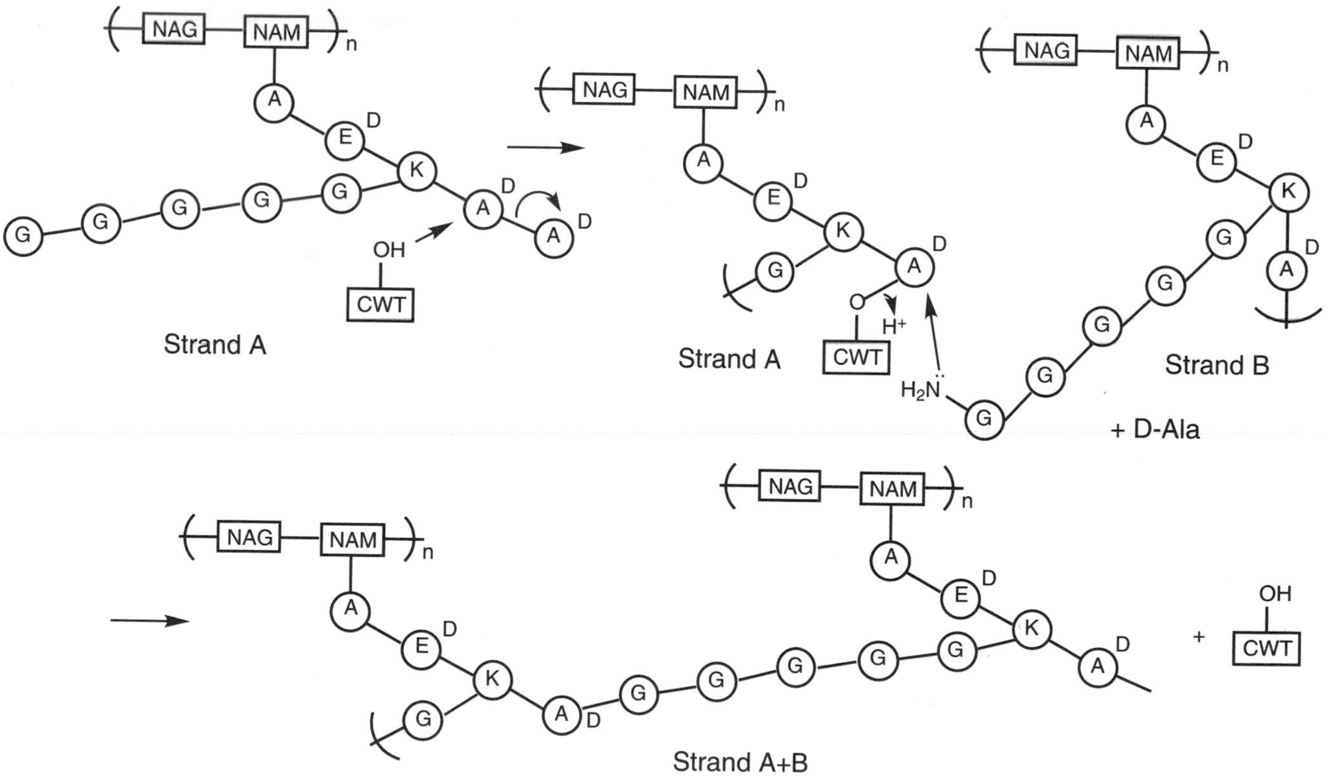

Strand A

Strand A

Strand B

+ D-Ala

Strand A+B

NAG = N-Acetylglucosamine; NAM = N-Acetylmuramic Acid, CWT = Cell Wall Transamidase (A Penicillin-binding Protein); A = Alanine; E = Glutamic Acid; K = Lysine and G = Glycine.

Fig. 34–3. Schematic of the cross-linking step in bacterial cell wall biosynthesis.

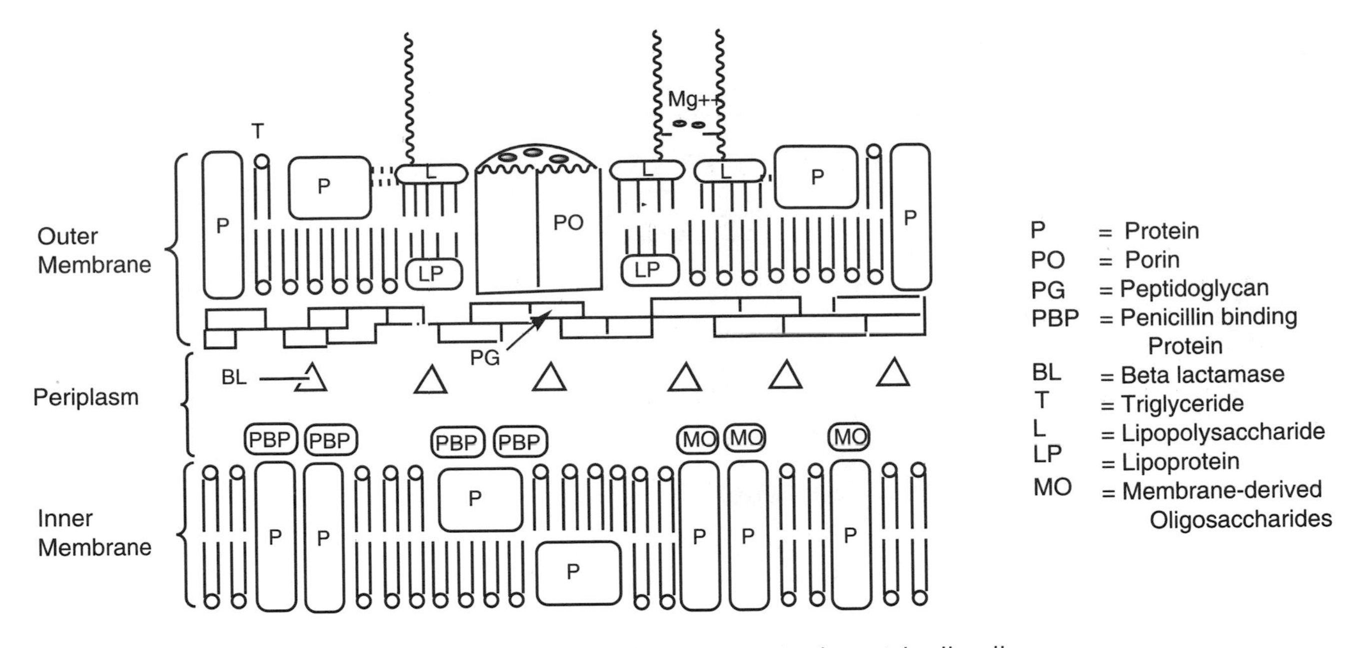

Fig. 34–4. Schematic of the gram-negative bacterial cell wall.

also contains a number of enzymes and exclusionary proteins. The porins penetrate at least to the end of this structural unit. Below this lies a less impressive, as compared to Gram-positive's, layer of peptidoglycan. Next comes a periplasmic space followed by a phospholipid-rich, cytoplasmic membrane in which floats a series of characteristic proteins with various functions. The beta-lactam receptors (PBPs) are found here. Penicillin hydrolyzing proteins, the beta-lactamases, lie in the periplasm. The inner membrane proteins are involved in transport, energy, and biosynthesis.

Beta-lactam Antibiotics

The name lactam is given to cyclic amides and is analogous to the name lactone given to cyclic esters. In an older nomenclature, the second carbon in an aliphatic carboxylic acid was designated alpha, the third beta, and so on. Thus, a beta-lactam is a cyclic amide with four atoms in its ring. The contemporary name for this ring system is azetidinone. This structural feature was very rare when it was found to be a feature of the structure of the penicillins so its name came to be a generic descriptor for the whole family. It is fortunate that this ring ultimately proved to be the main component of the pharmacophore so the term possesses medicinal as well as chemical significance.

The general outlines of the story of the discovery of the penicillins are widely known. In 1929, Alexander Fleming, a physician and a clinical microbiologist, was preserving a culture of a pathogen and the plate became contaminated with an airborne fungus, Penicillium notatum, which not only also grew but produced a clear zone of inhibition around its colony. Recognizing the potential significance of this antibiotic effect, he preserved the fungus and tried to identify its active constituent. The state of development of the art and his background and training were insufficient for the task at hand at that time. It was not until a decade later that another group of English chemists including Abraham, Chain, Florey, and Heatley succeeded in purifying the unstable antibiotic. Finally, on 12 February, 1941, following heroic efforts necessitated by wartime requirements and lack of suitable equipment and technology, enough material was available for clinical examination and the demonstration that penicillin actually worked in humans. Much new technology had to be developed before large scale use of penicillin could take place. The efforts of an international team solved, for example, the problems of large-scale sterile aerobic submerged fermentation, directed fermentation, strain improvement, and many other vexing problems. By 1943 penicillin was being produced in large quantities for the use of the armed forces. The penicillin field gradually expanded to include orally active penicillins, broad spectrum and enzymatically stable penicillins, and then the cephalosporins of three generations, the monobactams, carbapenems, beta-lactamase stable penicillins and cephalosporins, and beta-lactamase inhibitors were discovered. In 1993, about half of the money spent worldwide on antibiotics buys beta-lactams—over a hundred thousand of these compounds have been prepared by partial or total chemical synthesis.

Penicillins

The medicinal classifications, chemical structures, and generic names of the penicillins currently available are set forth in Table 34–1.

Preparation of Penicillins. The original fermentation-derived penicillins were produced by uncontrolled fermentation with the result that they were mixtures differing in the identity of the side chain moiety. When a suffi-

Table 34–1. Commercially Significant Penicillins and Related Molecules

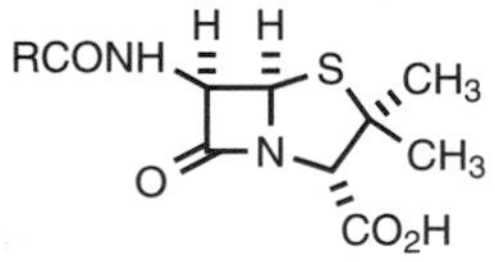

RCONH, H H, S, CH_3, CH_3, O, N, CO_2H

Fermentation-Derived Penicillins :

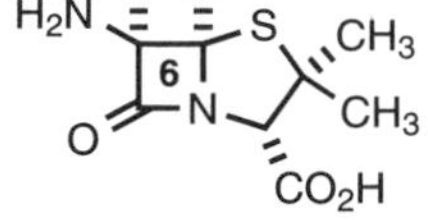

H_2N, H H, S, CH_3, 6, CH_3, O, N, CO_2H — 6-Aminopenicillanic Acid (6-APA)

R = CH_2 — Benzylpenicillin / Penicillin G

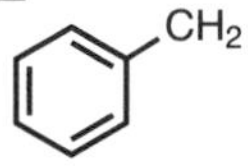

R = OCH_2 — Phenoxymethylpenicillin / Penicillin V

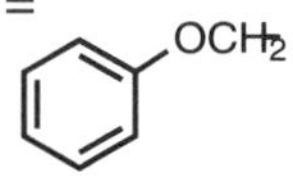

Penicillinase-Resistant Parenteral Penicillins :

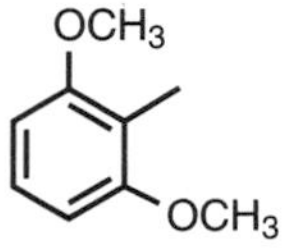

OCH_3, OCH_3 — Methicillin

Penicillinase-Resistant Oral Penicillins :

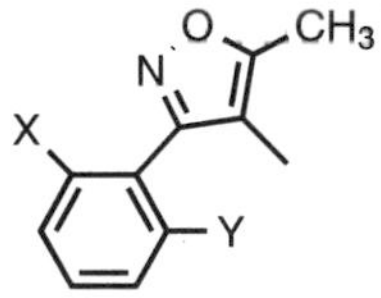

O, CH_3, N, X, Y

Oxacillin	X = Y = H
Cloxacillin	X = H, Y = Cl
Dicloxacillin	X = Y = Cl

Penicillinase-Sensitive, Broad-Spectrum, Parenteral Penicillins :

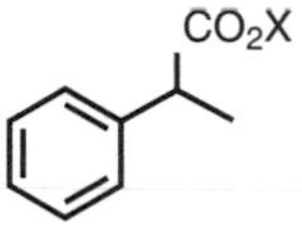

CO_2X Carbenicillin X = H
Carindacillin X = [indanyl]

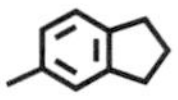

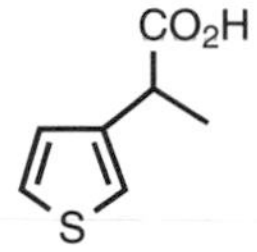

CO_2H, S — Ticarcillin

X–N, O, O, N, N, H, **D**
Azlocillin X = H
Mezlocillin X = CH_3SO_3

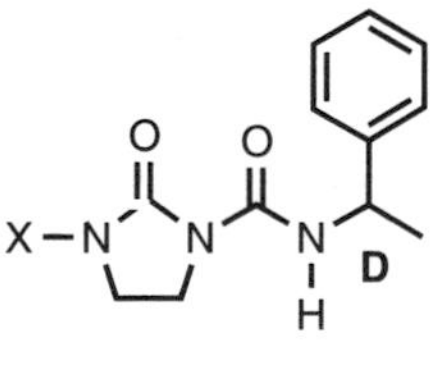

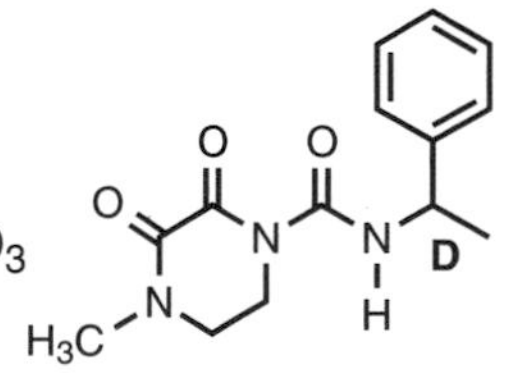

O, O, O, N, N, **D**, H, H_3C, N — Piperacillin

(continued)

Table 34–1. *(Continued)*

Penicillinase-Sensitive, Broad-Spectrum, Oral Penicillins

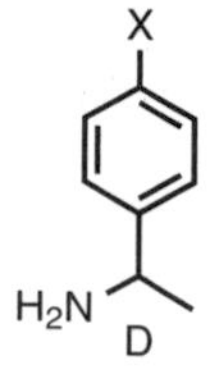

Ampicillin X = H
Amoxicillin X = OH

Miscellaneous

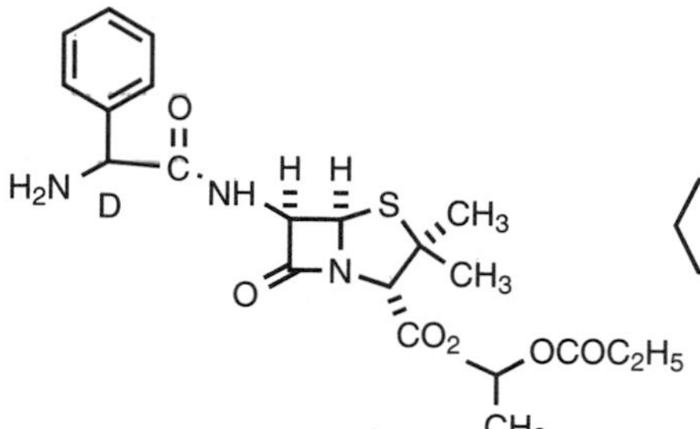

Bacampicillin

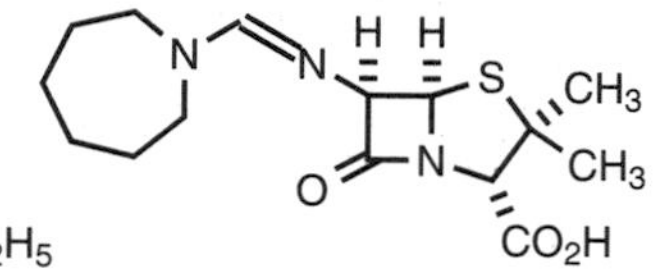

Mecillinam

Beta-Lactamase Inhibitors

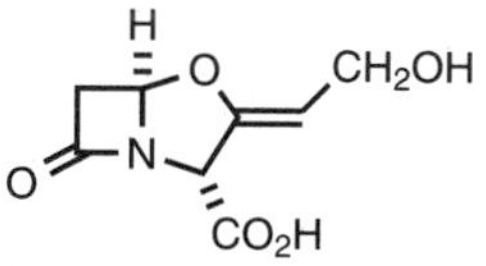

Clavulanic Acid

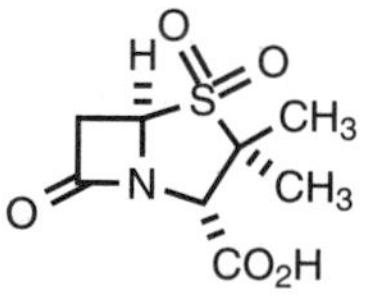

Sulbactam

cient supply of phenylacetic acid is present in the medium, this is preferentially incorporated into the molecule to produce benzylpenicillin (penicillin G in the old nomenclature). Use of phenoxyacetic acid instead leads to phenoxymethyl penicillin (penicillin V). More than two dozen different penicillins have been made in this way, but only these remain in clinical use. The bicyclic penicillin nucleus itself is prepared biosynthetically via a complex process from an acylated cysteinyl valyl peptide. The complete exclusion of side chain precursor acids from the medium produces the fundamental penicillin nucleus, 6-aminopenicillanic acid (6-APA), but in poor yield. With this key precursor isolated, limitations caused by enzyme specificities in biosynthesis could be overcome by use of partial chemical synthesis. A more practical modern process for making 6-APA employs a naturally occurring fungal enzyme that selectively hydrolyzes away the side chain of natural penicillins without cleaving the beta-lactam bond. More recently, ingeniously selective chemical processes have been devised for accomplishing this from biosynthetic penicillins. The operational chemical freedom of 6-APA has led to partial synthesis of many thousands of analogs. The sodium and potassium salts of penicillins are crystalline and water soluble; they are employed orally or parenterally. The procaine and benzathine salts of benzylpenicillin, on the other hand, are water insoluble. They are used for repository purposes following injection when long-term blood levels are required.

Nomenclature. In penicillins, as with most antibiotics, nomenclature is complex. The Chemical Abstracts system is definitive and unambiguous but too complex for ordi-

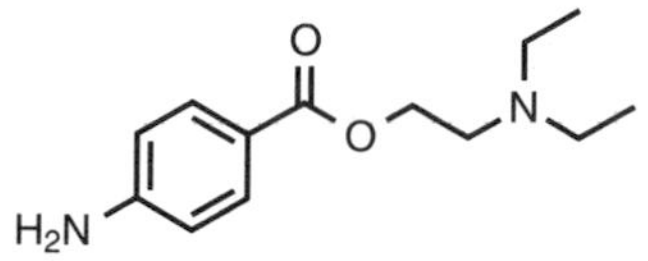

Procaine

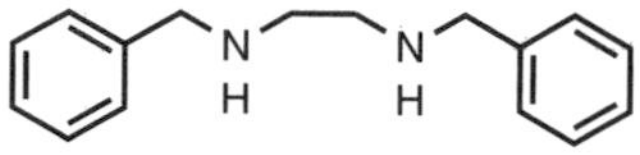

Benzathine

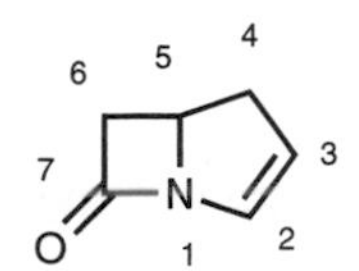

Penam
(4-Thia-1-azabicyclo-
[3.2.0]heptane)-7-one

Penem
4-Thia-1-azabicyclo-
[3.2.0]hept-2-ene)-7-one

Carbapenem
(1-Azabicyclo[3.2.0]-
hept-2-ene)-7-one

Cefem
(5-Thia-1-azabicyclo-
[4.2.0]oct-2-ene)-8-one

Monobactam
(1-Azacyclobutan-4-one)

Ring and numbering systems of clinically available beta-lactams.

nary use. For example, the chemical name for benzylpenicillin sodium is monosodium (2S,5R,6R)-3,3-dimethyl-7-oxo-6-(2-phenylacetamindo)-4-thia-1-azabicyclo[3.2.0] heptane-2-carboxylate! Confusingly, the United States Pharmacopia uses a different system, which results in the atoms' being numbered differently! The simplest system has stood the test of time and it involves taking the repeating radical, carbonyl-6-aminopenicillanic acid, and using the chemical trivial name for the radical which completes the structure. Thus, use of the names benzylpenicillin and phenoxymethylpenicillin makes practical sense. There are three asymmetric centers in the benzylpenicillin molecule as indicated in Table 34–1.

Clinically Relevant Degradation Reactions of Penicillins. The most unstable bond in the penicillin molecule is the highly strained and reactive beta-lactam amide bond. This bond cleaves only moderately slowly in water, but much more rapidly in alkaline solutions to produce penicilloic acid that readily decarboxylates as penilloic acid. Penicilloic acid has a negligible tendency to reclose to the corresponding penicillin so this reaction is essentially irreversible under physiologic conditions. Because the beta-lactam ring is an essential portion of the pharmacophore, this reaction deactivates the antibiotic. The enzyme beta-lactamase catalyzes this reaction also and is a principal cause of bacterial resistance in the clinic. Alcohols and amines bring about the same cleavage reaction but the products are the corresponding esters and amides. When proteins serve as the nucleophiles in this reaction, the antigenic conjugates that cause much penicillin allergy are produced. The small molecules that are not inherently antigenic but react with proteins to produce antigens in this manner are called haptens.

In acidic solutions, the reaction is more complex. Hydrolysis of the beta-lactam bond can be shown through kinetic analysis to involve participation of the side chain amide oxygen because the rate of this reaction differs widely depending on the nature of R. The main endproducts of the acidic degradation are penicillamine, penilloic acid, and penicilloaldehyde. The intermediate penicillenic acid is highly unstable and undergoes subsequent hydrolysis to the corresponding penicilloic acid. An alternate pathway involves sulfur ejection to a product that in turn fragments to liberate penicillamine (itself used clinically as a chelating agent) and penaldic acid. Penaldic acid decarboxylates to produce penicilloaldehyde. Several related fragmentations to a variety of other products take place, such as penillic acid. None of these products has bioactivity.

The installation of a side chain R group that is electron withdrawing decreases the electron density on the side chain carbonyl and protects these penicillins in part from acid degradation. This activity has clinical implications because these compounds survive passage through the stomach better and many agents can be given orally for systemic purposes.

These degradation reactions can also be retarded clinically by keeping the pH of solutions between 6.0 and 6.8 and by refrigerating them. Metal ions such as zinc and copper catalyze the degradations of penicillins so they should be kept from contact with their solutions. The container lids used today are routinely made of inert plastic to minimize such problems.

Protein Binding. The more lipophilic the side chain of a penicillin, the more serum protein bound is the antibiotic. Whereas this has some advantages in terms of protection from degradation, it does reduce the effective bacteriocidal concentration of the drug. Contrary to popular assumption, the degree of serum protein binding of the penicillins has comparatively little influence upon their half-lives. The penicillins are actively excreted into the urine via an active transport system for negatively charged ions and the rate of release from their bound form is sufficiently rapid that the controlling rate is the kidney secretion rate. Benzyl penicillin and methicillin lie in the range of 35 to 80% bound, ampicillin and amoxicillin are about 25 to 30% bound, carbenicillin is about 50% bound, and oxacillin and cloxacillin are over 90% bound.

Molecular Mode of Action. The generally accepted molecular mode of action of the beta-lactam antibiotics is a selective and irreversible inhibition of the enzymes processing the developing peptidoglycan layer. Just before crosslinking occurs, the peptide pendant from the lactate carboxyl of a muramic acid unit terminates in a D-ala-D-ala unit. This is cleaved between these two amino acids by hydrolysis catalyzed by a cell wall transamidase. This enzyme uses a serine hydroxyl group to attack the penultimate D-ala unit forming a covalent bond and the terminal D-ala which is released by this action diffuses away. The enzyme-protopeptidoglycan complex is attacked by the free amino end of a pentaglycyl unit of an adjacent strand regenerating the transpeptidases active site for further catalytic action and producing a new amide bond that glues two adjacent strands together.

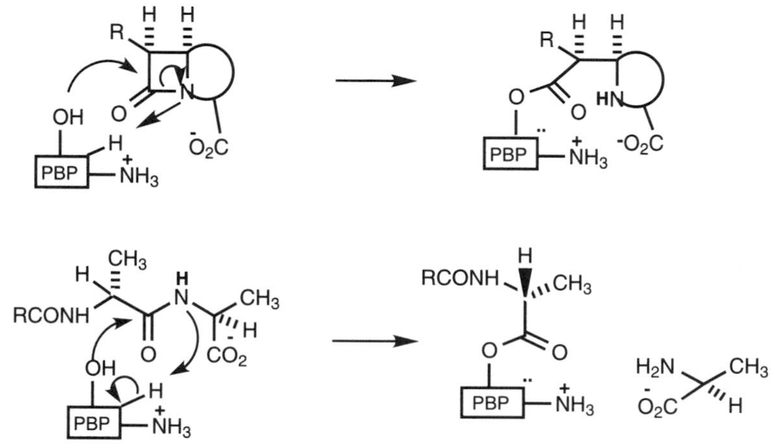

The three-dimensional geometry of the active site of the enzyme perfectly accommodates to the shape and separation of the amino acids of its substrate. Because the substrate has unnatural stereochemistry at the critical residues, this enzyme is not expected to attack host peptides or even other bacterial peptides composed of natural amino acids.

The present belief is that the penicillins and the other beta-lactam antibiotics resemble closely the geometry of acetylated D-ala-D-ala and that the enzyme mistakenly accepts it as its normal substrate. The highly strained beta-lactam ring is much more reactive than a normal amide, particularly when fused into the appropriate bicyclic system with a suitable heterocycle. The intermediate acyl-enzyme complex, however, is rather different structurally from the normal intermediate in that the hydrolysis does not break penicillins into two pieces as it does with its normal substrate. In the penicillins, a heterocyclic residue is still covalently bonded and cannot diffuse away as the natural terminal D-ala unit does. This presents a barrier to approach by the pentaglycyl unit and keeps the enzyme's active site from being regenerated and the cell wall precursors from being cross linked. The result is a defective cell wall and an inactivated enzyme. The relief of strain that is obtained on enzymatic beta-lactam bond cleavage is so pronounced there is virtually no tendency for the reaction to reverse. Water is also an insufficiently effective nucleophile and cannot hydrolyze the complex either. Thus, the cell wall transamidase is stoichiometrically inactivated.

More details of the putative drug-enzyme interaction will be discussed with the other classes of beta-lactams.

Another important class of enzymes exists whose action

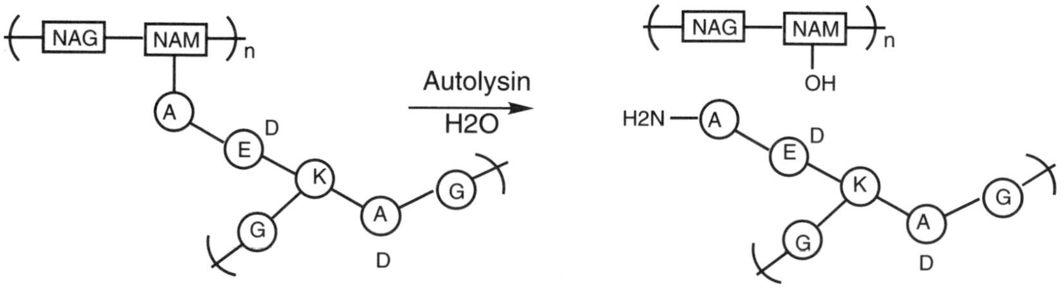

Autolysin H_2O

Peptidoglycan Segment

enhances the bactericidal action of beta-lactams. These are the autolysins. The autolysins cleave the N-acetylmuramic acid-peptide bond to L-ala. The result of this is that the whole amino acid side chain falls away with effects quite lethal to the bacterium.

Resistance. Resistance to beta-lactam antibiotics is unfortunately becoming increasingly common. It can be constituitive and involve decreased cellular uptake of drug, or involve lower binding affinity to the PBPs. This is particularly the case with methacillin-resistant Staphylococcus aureus (MRSA). Much more common, however, is the elaboration of a beta-lactamase. Beta-lactamases are enzymes (serine proteases) elaborated by microorganisms, which catalyze hydrolysis of the beta-lactam bond and inactivate β-lactam antibiotics before they can reach the PCPs. In this they resemble somewhat the cell wall transaminase but hydrolytic regeneration of the beta-lactamase's active site is dramatically more facile than is the case with cell wall transamidase, so that a comparatively smaller amount of enzyme can destroy a larger amount of drug. With gram-positive bacteria, the beta-lactamases are usually shed continuously into the medium and meet the drug outside the cell wall. With gram-negative bacteria, a more conservative course is found. Here the beta-lactamases are secreted into the periplasmic space so, although still distal to the PBPs, they do not readily escape into the medium and need not be resynthesized as often.

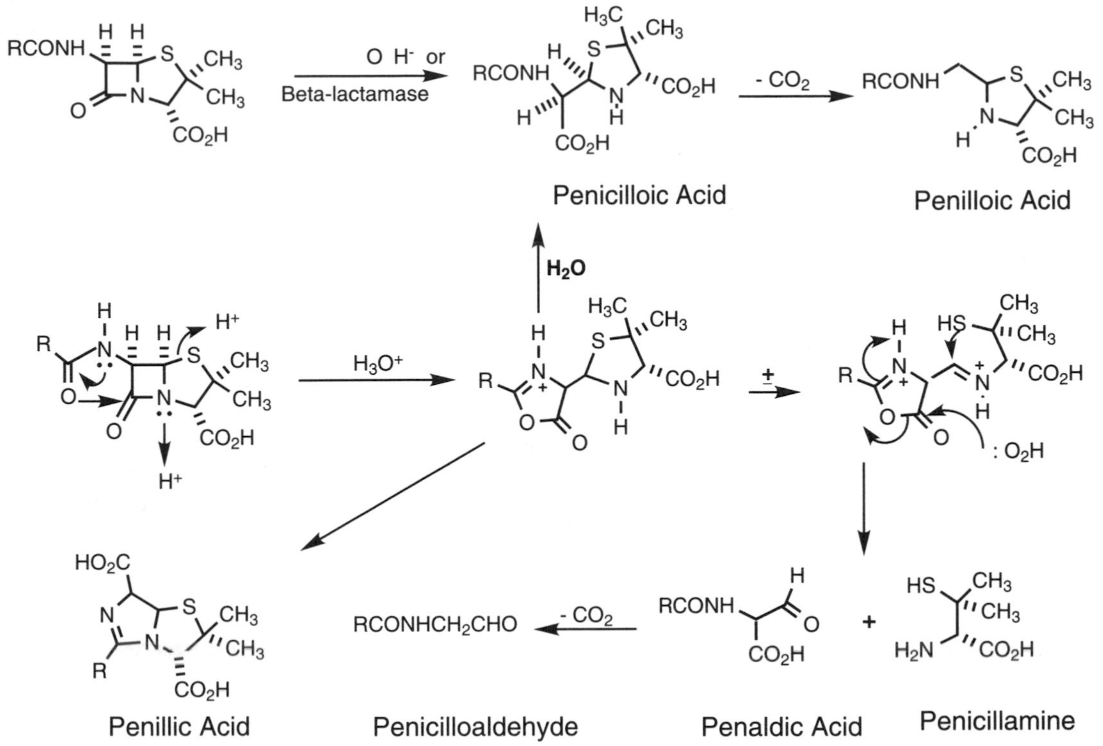

Penicilloic Acid

Penilloic Acid

Penillic Acid

Penicilloaldehyde

Penaldic Acid

Penicillamine

Elaboration of beta-lactamases is often R-factor mediated and, in some cases, is even induced by the presence of beta-lactam antibiotics. The normal function of beta-lactamases in the absence of antibiotic is not readily apparent.

Allergenicity. About 6 to 8% of the U.S. population is sensitive to beta-lactam antibiotics. Most commonly this is expressed as drug rash or itching and is of delayed onset but occasionally the reaction is immediate and profound. In the latter cases, cardiovascular collapse and shock can result in death. This is most common with injections. Sometimes penicillin allergy can be anticipated by taking a medication history and often patients likely to be allergic are those with a history of hypersensitivity to a wide variety of allergens, e.g., foods and pollens. A prior history of allergy to penicillins is a contraindicating factor to their use. Topical wheal and flare tests are available when there is doubt. When an allergic reaction develops, the drug must be discontinued and, because cross sensitivity is common, other beta-lactam drugs should generally be avoided. Considering all therapeutic categories, penicillins, especially the ones most commonly employed (benzylpenicillin and ampicillin/amoxycillin), are probably the drugs of all types most associated with allergy. Erythromycin and clindamycin are useful alternate choices in many cases of penicillin allergy.

Sometimes the patient has become sensitized without knowing it due to prior passive exposure through contaminated food stuffs or cross-contaminated medications. For some time it has been required that penicillins be manufactured in facilities separate from those used to prepare other drugs in order to prevent cross contamination and possible sensitization. Animals treated with penicillins are required to be drug free for a significant time before products prepared from them can be consumed. The number of pharmacists who unknowingly override these protective measures by failing to properly cleanse their pill counters between prescriptions is unknown.

Because the origin of the allergy is a haptenic reaction with host proteins and the responsible bond in the drug is the beta-lactam moiety, this side effect is caused by the pharmacophore of the drug and is unlikely to be overcome by molecular manipulation.

Antimicrobial Spectrum. With the exception of Neisseria gonorrhea, the useful antimicrobial spectrum of benzylpenicillin is primarily against gram-positive microorganisms.

Clinical Uses. Because of its cheapness, efficacy, and remarkable lack of toxicity (except for acutely allergic patients), benzylpenicillin remains a remarkably useful agent for treatment of diseases caused by susceptible microorganisms. It is the drug of choice for more infections than any other antibiotic. This justifies the screener's lament, "first is best"! As with most antibiotics, susceptibility tests must be performed as many formerly highly sensitive microorganisms are now comparatively resistant. Infections of the upper respiratory tract and the genitourinary tract are the particular province of benzylpenicillin. Infections caused by Group A beta-hemolytic streptococci (pharyngitis, scarlet fever, cellulitis, pelvic infections, and septicemia) are commonly responsive. Group B hemolytic streptococcus infections, especially of neonates (acute respiratory distress, pneumonia, meningitis, septic shock, and septicemia) usually respond. Pneumococcal pneumonia, H. influenza pneumonia of children, S. pneumoniae and S. pyogenes caused otitis media and sinusitis, meningicoccal meningitis and brain abscess, meningococcal and pneumococcal septicemia, streptococcal endocarditis (often by S. viridans), pelvic inflammatory disease (often caused by N. gonorrhoeae and S. pyogenes), uncomplicated gonorrhea (N. gonorrhoeae), syphilis (Treponoema pallidum), Lyme disease, gas gangrene (Cl. perfringens), and tetanus (Cl. tetani) are among the diseases that commonly respond to benzylpenicillin therapy, alone or sometimes with other drugs used in combination. Other, less common, bacterial diseases also respond.

Because of its cheapness, mild infections with susceptible microorganisms can be treated with comparatively large oral doses, although the most effective route of administration is parenteral because five times the blood level can be regularly achieved in this manner.

Deficiencies. The need to improve defects in benzylpenicillin use stimulated an intense research effort that persists to this day. Overcoming such negative features as comparative instability (particularly to acid), comparatively poor oral absorption, allergenicity, sensitivity to beta-lactamases, and relatively narrow antimicrobial spectrum, have been objectives of this work. Only antigenicity has failed to respond significantly to this effort.

Phenoxymethyl Penicillin (Penicillin V). Penicillin V is considerably more acid-stable than benzylpenicillin. This is thought to be due to the electronegative oxygen atom in the C-7 amide side chain inhibiting participation in beta-lactam bond hydrolysis. In any case, penicillin V was the first of the so-called oral penicillins giving higher and more prolonged blood levels than penicillin G itself. Its antimicrobial and clinical spectrum is roughly the same as that of benzylpenicillin, although it is somewhat less potent and is not, as a rule, used for acutely serious infections. Penicillin V has approximately the same sensitivity to beta-lactamases and allergenicity as penicillin G.

Penicillinase-resistant Parenteral Penicillins (Methicillin). Fortunately beta-lactamases are much less tolerant to the presence of steric hindrance near the side-chain amide bond than are the penicillin binding proteins. When the aromatic ring is attached directly to the side chain carbonyl and both ortho positions are substituted by methoxy groups, beta-lactamase stable methicillin results. It still retains clinically significant antibiotic activity. Movement of one of the methoxy groups to the para position or replacing one of them by a hydrogen results in an analog sensitive to beta-lactamases. Putting in a methylene between the aromatic ring and 6-APA likewise produces a beta-lactamase-sensitive agent. These findings provide strong support for the hypothesis that resistance to degradation is based on differential steric hinderance. Methicillin is significantly narrower in antimicrobial spectrum and less potent so it is restricted to clinical use primarily for parenteral use in infections due to beta-lactamase producing Staph. aureus. Nosocomial (treatment related) staphylococcal cellulitis is the primary indication. When the strain is sensitive to penicillin G or V, these are greatly to be preferred. Lately, an increasing number of infections are being found that are caused by methicillin-resistant Staph. aureus (MRSA). The mode of resistance in these cultures appears to be alterations in the PBPs so that they are less sensitive to inhibition by methicillin. Vancomycin is the current favorite for treatment of infections by MRSA. Methicillin is an efficient inducer of penicillinases so it should be restricted for use in infections that uniquely require it. Methicillin is also very sensitive to acid hydrolysis thus requiring it to be injected.

Penicillinase-resistant Oral Penicillins: Isoxazoyl Penicillins (Oxacillin, Cloxacillin, and Dicloxacillin). Using an isoxazoyl ring as a bioisosteric replacement for the benzene ring and a methyl on one flank and a substituted benzene ring on the other in place of the methoxyls of methicillin produces the isoxazoyl penicillins. These are

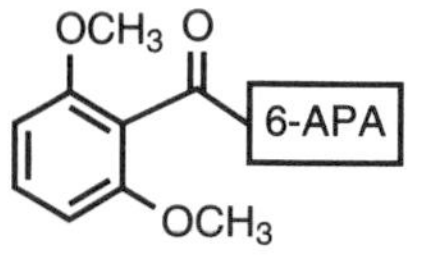

OCH_3 O 6-APA OCH_3

Beta-lactamase resistant

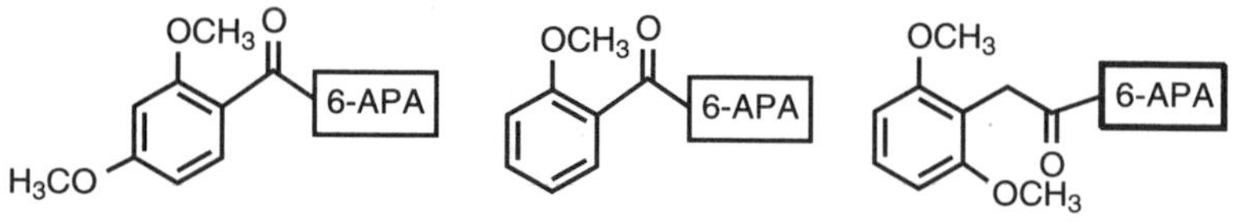

OCH_3 O 6-APA H_3CO | OCH_3 O 6-APA | OCH_3 O 6-APA OCH_3

Beta-lactamase sensitive

oxacillin, cloxacillin, and dicloxacillin. Chemically they differ from one another in the degree and nature of substitution in the benzene ring. Like methicillin these are generally less potent than benzyl penicillin against gram-positive microorganisms (generally staphylococci and streptococci) that do not produce a beta-lactamase but retain their potency against those that do. An added bonus exists in their being comparatively acid stable, they may thus be taken orally. Because they are highly serum protein bound (more than 90%), they are not good choices for treatment of septicemia. Microorganisms resistant against methicillin are generally also resistant to the isoxazoyl group of penicillins.

Penicillinase-sensitive, Broad Spectrum, Oral Penicillins: D-Phenylglycine Derived Penicillins. The important, first member of this group, ampicillin, is a benzyl penicillin analog in which one of the hydrogen atoms of the side chain phenylacetic acid moiety has been replaced with a primary amino acid group to produce a D-phenylglycine moiety. In addition to significant acid stability and, therefore, its successful oral use, the antimicrobial spectrum is shifted so that many common gram-negative pathogens are sensitive to ampicillin. The acid stability is generally believed caused by the electron-withdrawing character of a protonated primary amine group in the side chain reducing its participation in hydrolysis of the beta-lactam bond as well as to the comparative difficulty of bringing another positively charged species (H_3O^+) into the vicinity of the protonated amino group. Ampicillin is very widely prescribed by general practitioners. It unfortunately lacks stability towards beta-lactamases and resistance is an ever increasing phenomenon. To assist in dealing with this, several additives have been developed. Clavulanic acid, for example, is a mold product with only weak intrinsic antibacterial activity, but it is an excellent irreversible inhibitor of beta-lactamases. It is believed to acylate irreversibly the active site serine hydroxyl by mimicking the normal substrate. This leads to its classification as a mechanism-based inhibitor (or so-called suicide substrate). The precise chemistry is not well understood but when clavulanic acid is added to ampicillin preparations, the potency against beta-lactamase producing strains is markedly enhanced. Another such beta-lactamase sparing agent is sulbactum. Sulbactam is prepared by partial chemical synthesis from penicillins. The oxidation of the sulfur atom to a sulfone greatly enhances the potency of sulbactam.

Although comparatively well-absorbed, ampicillin's oral efficacy for systemic infections can be enhanced significantly through the preparation of prodrugs. In contrast to ampicillin itself, which is amphoteric, bacampicillin is a weak base, which is very well absorbed in the duodenum. Enzymic ester hydrolysis in the body liberates innocuous propionic acid and, next, acetaldehyde is spontaneously lost from this intermediate to produce ampicillin. The acetaldehyde is metabolized oxidatively by alcohol dehydrogenase to produce acetic acid, which joins the normal pool.

In addition to the usual mode of penicillin allergenicity, concentrated preparations of ampicillin can self condense to form high-molecular-weight aggregates through reaction of its primary amino group with the beta-lactam bond of another molecule. These aggregates are thought to be antigenic and to be responsible for ampicillin allergenicity—a form of hypersensitivity that differs in some details from the usual penicillin allergenicity. Ampicillin is the penicillin most commonly associated with drug-induced rash. Avoiding use of outdated preparations of ampicillin is a sometimes effective means of dealing with this potential problem.

Ampicillin is essentially equivalent to benzyl penicillin for pneumococcal, streptococcal, and meningococcal in-

β-lactamase — OH; OH; H+; O; N; CO_2H → β-lactamase — O; H H; O; N; H+ CO_2H; O...H...O →

n:; β-lactamase — O; O; HN; CO_2H; O...H...O → β-lactamase; n; O; O; H–N; CO_2H; O...H...O

Speculative mechanism for irreversible inactivation of β-lactams by clavulanic acid. An analogous mechanism can be drawn for sulbactams

O O; S; N; O; CO_2H + beta-lactamase → β-lactamase; n; O; O; H–N; SO_2H; CO_2H

fections and many strains of gram-negative Salmonella, Shigella, Proteus mirabilis and E. coli, as well as many strains of Haemophilus influenzae respond well to oral treatment with ampicillin. It is particularly widely used for outpatient therapy of uncomplicated community-acquired urinary tract infections.

Amoxicillin is a close analog of ampicillin in which a *para*-phenolic hydroxyl group has been introduced into the side chain amide moiety. This adjusts the isoelectric point of the drug to a more acidic value and perhaps enhances blood levels obtained with amoxicillin as compared to ampicillin. Better oral absorption leads to less disturbance of the normal gastrointestinal flora and, therefore, less drug-induced diarrhea. The antimicrobial spectrum and clinical uses of amoxicillin are approximately the same as those of ampicillin itself and it is presently one of the most popular drugs in North America. The addition of clavulanic acid serves to protect it to a considerable extent against beta-lactamases.

Penicillinase-sensitive, Broad Spectrum, Parenteral Penicillins. Azlocillin, mezlocillin and piperacillin are ampicillin derivatives in which the D-side chain amino group has been converted by chemical processes to a variety of ureas. They are collectively known as acylureidopenicillins. They preserve the useful anti-gram-positive activity of ampicillin but have higher anti-gram-negative potency. Some strains of Pseudomonas aeruginosa, for example, are sensitive to these agents. It is speculated that the added side chain moiety mimics a longer segment of the peptidoglycan chain than ampicillin. This would give more points of attachment to the penicillin-binding proteins and perhaps these features are responsible for their enhanced antibacterial properties. These agents are used parenterally with particular emphasis on gram-negatives, especially Klebsiella pneumoniae and the anaerobe, Bacteroides fragilis. Resistance due to beta-lactamases is a feature of their use so disk testing and incorporation of additional agents (such as an aminoglycoside) for the treatment of severe infections is advisable.

Carbenicillin and Ticarcillin. At the other extreme of polarity, carbenicillin is a benzyl penicillin analog in which one of the methylene hydrogens of the side chain has been substituted with a carboxylic acid moiety. The specific stereochemistry of this change is not very important because both diastereoisomers have unstable configurations and mutarotate with time to produce the same mixture of epimers. The introduction of the side chain carboxyl enhances anti-gram-negative activity. In fact, carbenicillin is intrinsically one of the broadest spectrum penicillins. The clinical use of carbenicillin is nonetheless restricted primarily to high-dose therapy of P. aeruginosa and Proteus vulgaris as well as some Enterobacter and Serratia infections. Carbenicillin is an order of magnitude less potent than the acylureidopenicillins. The drug is susceptible to beta-lactamases and is acid unstable and so must be given by injection. To enhance the likelihood of clinical success with carbenicillin, it is often administered in conjunction with an aminoglycoside antibiotic. When used in this manner, the antibiotics are chemically incompatible so they should not be administered in the same solution.

Being a malonic acid hemiamide with a carbonyl (amide) moiety beta to the carboxyl group, carbenicillin can decarboxylate readily to produce benzyl penicillin. Although still an antibiotic, this degradation product has no activity against the organisms for which carbenicillin is indicated, so this is still fairly scored as a degradation. In addition, the large doses employed, i.e., multigrams per day, result in ingestion of a significant amount of sodium ion, which could be a consideration with heart patients. Many of these problems are avoided with the oral use of the prodrug indanyl ester carcindacillin. Unfortunately, the potency of this preparation does not allow it to be used as a full substitute for carbenicillin but it is instead primarily used for oral treatment of urinary tract infections. Ticarcillin is a sulfur bioisotere of carbenicillin. This agent is more potent against pseudomonads, especially when laced with sulbactam.

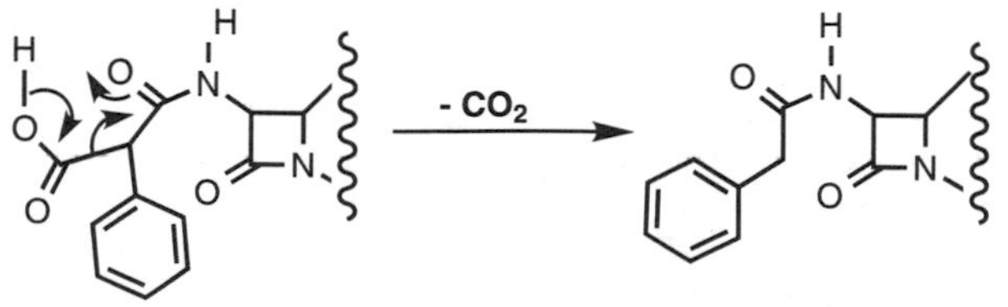

Miscellaneous Agents (Mecillinam). Mecillinam is a penicillin analog whose structure differs substantially from the others. It does not have a side chain amide linkage at all but rather an imide. This results in an agent with a different binding pattern to the PBPs. With its range of potency against gram-negatives better than against gram-positives, it is fading in popularity.

Cephalosporins

In contrast to the discovery of the penicillins in which the first agent had such remarkable properties that it entered clinical use with comparatively little change, the cephalosporins are remarkable for the level of persistence required after their initial discovery to yield economic returns. The original Cephalosporium acremonium culture was discovered in a sewage outfall off the Sardinian coast by Brotsu. In England, Abraham and Newton pursued it because one of the constituents had the useful property of being active against penicillin-resistant cultures due to its stability to beta-lactamases. Cephalosporin C is not potent enough to be a useful antibiotic but removal, through chemical means, of the natural side chain produced 7-aminocephalosporanic acid (7-ACA) which, like 6-aminopenicillanic acid, could be fitted with unnatural side chains by chemists. Many of the compounds produced in this way are remarkably useful antibiotics. They differ from one another in antimicrobial spectrum, beta-lactamase stability, absorption from the gastrointestinal tract, metabolism, stability, and side effects as detailed here. Exploitation of sulfenic acid chemistry by Morin at Lilly resulted in the conversion of penicillins to cephalosporins, including 3-desacetoxy-7-aminocephalosporanic acid (7-ADCA). This process is practical because the penicillin fermentation is much more efficient than the cephalosporin C fermentation.

Unfortunately, the chemistry involved is too complex to be covered in the space available here. Intensive investigation of the chemistry of 7-ACA and 7-ADCA has resulted in the subsequent preparation of many thousands of analogs from these two starting materials.

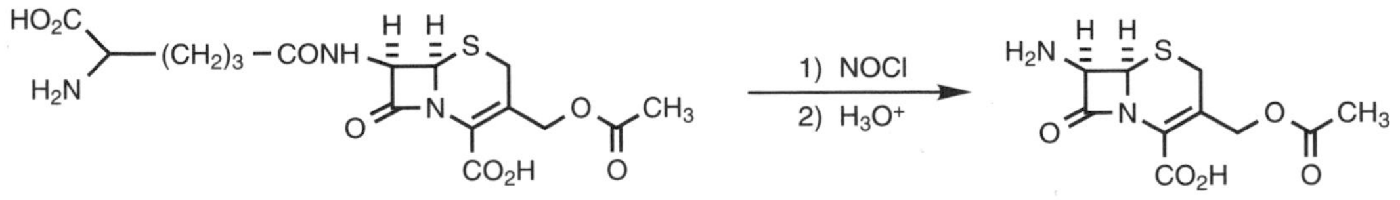

Cephalosporin C → 1) NOCl 2) H_3O^+ → 7-Aminocephalosporanic Acid (7-ACA)

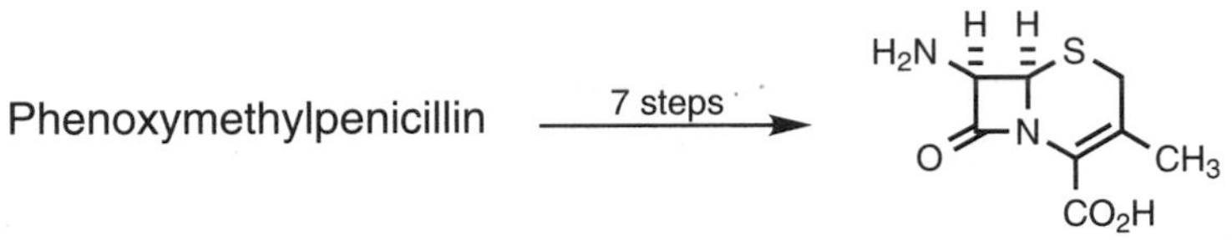

Phenoxymethylpenicillin → 7 steps → 7-Amino-3-deacetoxycephalosporanic Acid (7-ADCA)

The cephalosporins have their beta-lactam ring annealed to a 6-membered dihydrothiazine ring in contrast to the penicillins wherein the beta-lactam ring is fused to a 5-membered thiazolidine ring. As a consequence the cephalosporins should be less strained and less reactive/potent. Much of the reactivity loss is, however, made up by possession of an olefinic linkage at C-2,3 and an acetoxy group at C-3. When the beta-lactam ring is opened by hydrolysis, the acetoxy group at C-3 can be ejected carrying away the developing negative charge. This greatly reduces the energy required for the process. Thus, the facility with which the beta-lactam bond of the cephalosporins

Base or other nucleophile → + CH_3CO_2H

is broken is modulated both by the nature of the C-7 substituent (analogous to the penicillins) as well as the nature of the C-3 substituent and its ability to carry away a negative charge. Considerable support for this hypothesis comes from the finding that isomerization of the olefinic linkage to C-3,4 leads to great losses in antibiotic activity. In practice, most cephalosporins are comparatively unstable in aqueous solutions, the pharmacist is often directed to keep injectable preparations frozen before use. Being carboxylic acids, they form water-soluble sodium salts, whereas the free acids are comparatively water insoluble. In many cases where the free acids are supplied, the injectable forms contain sodium bicarbonate to facilitate solution.

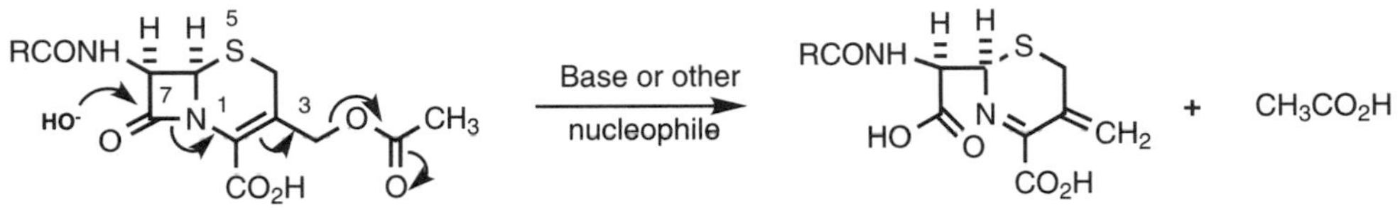

Molecular Mode of Action. The cephalosporins are believed to act in a manner analogous to that of the penicillins by binding to the penicillin-binding proteins followed by cell lysis. The full details of the manner in which bacterial cells are killed are obscure as yet. Cephalosporins are bactericidal in clinical terms.

Resistance. Analogous to the penicillins, susceptible cephalosporins are hydrolyzed by beta-lactamases before they reach the penicillin-binding proteins. Many beta-lactamases are known. Some are more efficient at hydrolysis of penicillins, some of cephalosporins and some are indiscriminate. Certain beta-lactamases are constituitive (chromosomally encoded) in certain strains of gram-negative bacteria (Citrobacter, Enterobacter, Pseudomonas, and Serratia) and are normally repressed. These are induced (or derepressed) by certain beta-lactam antibiotics, e.g., imipenem, cefotetam and cefoxitin. As with the penicillins, specific examples will be seen below wherein resistance to beta-lactamase hydrolysis is conveyed by strategic steric bulk near the side-chain amide linkage.

Allergenicity. Allergenicity is less commonly experienced and is less severe with cephalosporins than with penicillins. Cephalosporins are frequently administered to patients who have had a mild or delayed penicillin reaction, however cross allergenicity is comparatively common and this should be done with caution for patients with a history of allergies. Patients who have had a rapid and severe reaction to penicillins should not be treated with cephalosporins, either.

Nomenclature and Classification. Most cephalosporins have generic names beginning with prefixes ceph- or cef-. This is convenient for classification but makes discriminating among individual members a true memory test. The cephalosporins are classified by a trivial nomenclature system loosely derived from the chronology of their introduction but more closely related to their antimicrobial spectrum. The first-generation cephalosporins are primarily active in vitro against gram-positive cocci (penicillinase positive and negative Staph. aureus and S. epidermidis), group A beta-hemolytic streptococci (Strep. pyogenes), group B streptococci (Strep. agalactiae) and Strep. pneumoniae. They are not effective against methicillin-resistant Staph. aureus. They are not significantly active against gram-negatives although some strains of E. coli, K. pneumoniae, P. mirabilis and Shigella sp. may be sensitive. The second-generation cephalosporins generally retain the anti-gram-positive activity of the first but include H. influenzae as well and add to this better anti-gram-negative activity so that some strains of Acinetobacter, Citrobacter, Enterobacter, E. coli, Klebsiella, Neisseria, Proteus, Providencia, and Serratia are also sensitive. Cefotetan and cefoxitin have some antianaerobic activity as well. The third-generation cephalosporins are less active against staphylococci than the first-generation agents, but are much more active against gram-negatives than either the first or the second. They are frequently particularly useful against nosocomial multidrug-resistant hospital-acquired strains. One adds also Morganella, Bacteroides fragilis and Pseudomonas aeruginosa to the list of species that are often sensitive. Unfortunately, the third-generation agents are more expensive.

Clinical Uses. The incidence of cephalosporin resistance is such that it is usually preferable to do disk testing before instituting therapy. Infections of the upper and lower respiratory tract, skin and related soft tissue, urinary tract, bones and joints, as well as septicemias and endocarditis and intra-abdominal and bile tract infections caused by susceptible gram-positives are usually responsive to cephalosporins. When a gram-positive agent is involved, a first-generation agent is preferable. When the pathogen is gram-negative and the infection is serious, parenteral use of a third-generation agent is preferred. For treatment of Pseudomonas infections, a combination with an aminoglycoside antibiotic is recommended. For pelvic inflammatory disease (PID), the number one cause of sterility in sexually active young women, a combination with doxycycline is preferred because the infections are often mixed and frequently include Chlamydia trachomitis, anaerobes, and other microorganisms, which are not cephalosporin sensitive, along with Neisseria gonorrheae, with penicillinase-producing N. gonorrheae (PPNG) or without beta-lactamase, which are sensitive.

Toxicity. Aside from mild or severe allergic reaction, the most commonly experienced cephalosporin toxicities are mild and temporary nausea, vomiting, and diarrhea associated with disturbance of the normal flora. Rarely, a life-threatening pseudomembranous colitis diarrhea associated with the opportunistic and toxin-producing anaerobic pathogen, Clostridium difficile, can be experienced. Rare blood dyscrasias, which can even include aplastic anemia, are also seen. Certain structural types are associated with prolonged bleeding times and an antabuse-like acute alcohol intolerance.

Clinically Relevant Degradation Reactions. The principal chemical instability of the cephalosporins is associated with beta-lactam bond hydrolysis. The role of the C-7 and C-3 side chains in these reactions was discussed previously. Ejection of the C-3 substituent following beta-lactam-bond cleavage is usually drawn for convenience as though this were an unbroken (concerted) process. Evidence on this point being equivocal, ejection of the C-3 side chain may at certain times and with specific cephalosporins involve a discrete intermediate with the beta-lactam bond broken, but the C-3 substituent not yet ejected. The natural group at C-3 is an acetoxy and this is an excellent so-called leaving group.

Some cephalosporins contain nonejectable C-3 substituents and do not profit from this ejection. With certain other cephalosporins, a tetrazolethiomethyl group is ejected from C-3. This moiety is believed to be responsible in part for clotting difficulties and acute alcohol intolerance in certain patients. The role of the C-7 side chain in all of these processes is clearly important, but active participation of the amide moiety in a manner analogous to the penicillins is rarely specifically invoked. The same considerations that modulate the chemical stability of cephalosporins are also involved in dictating beta-lactamase sensitivity, potency, and allergenicity as well.

Metabolism. Those cephalosporins that have an acetyl side chain are subject to enzymatic hydrolysis in the body. The result is molecules with an OH at C-3. This is a rather poor leaving group so this change is considerably deactivating with respect to breakage of the beta-lactam bond. This would be bad enough were this all, however, the particular geometry of this part of the molecule leads to facile lactonization with the carboxyl group attached to C-2. Whereas in principle this should result in reattachment of a different good leaving group, the result is, instead, inactivation of the drugs involved. The penicillin binding proteins have an absolute requirement for a free carboxyl group to mimic that of the terminal carboxyl of the D-ala-D-ala moiety in their normal substrate. Lactonization removes this docking function and removes the affinity of the enzymes and their normal inhibitors. Thus, the inactivation by this metabolic sequence is readily understood.

Currently Available Cephalosporins. The cephalosporin antibiotics in present use in the United States are illustrated in Table 34–2.

RCONH, S, N, O, CO_2H, O, CH_3 — Active → Serum Esterase → RCONH, S, N, O, OH, CO_2H — Less Active → $- H_2O$ (spontaneous) → RCONH, S, N, O, O, O — Inactive

Table 34–2. Commercially Significant Cephalosporins

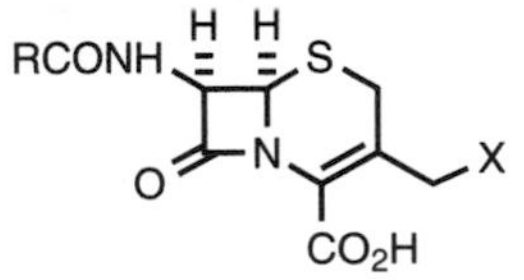

First Generation

RCONH, H H, S, O, N, X, CO_2H

Parenteral Agents	R	X	Salt
Cephalothin	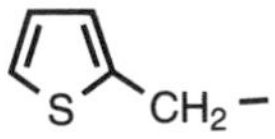 S, CH_2-	$OCOCH_3$	Na
Cephapirin	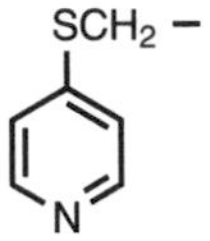 SCH_2-, N	$OCOCH_3$	Na
Cefazolin	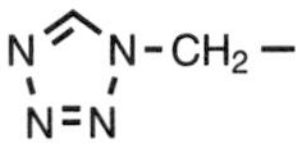 N, N–CH_2–, N=N	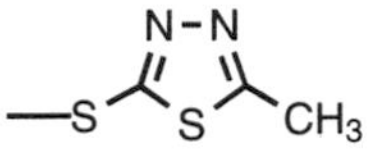 N–N, –S, S, CH_3	Na
Oral Agents			
Cephalexin	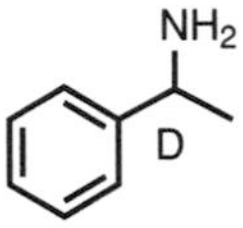 NH_2, D	H	HCl
Cefadroxil	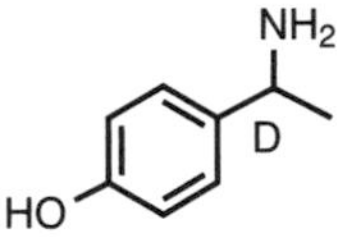 NH_2, D, HO	H	-
Oral and Parenteral Agents			
Cephradine	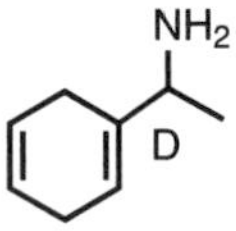 NH_2, D	H	-

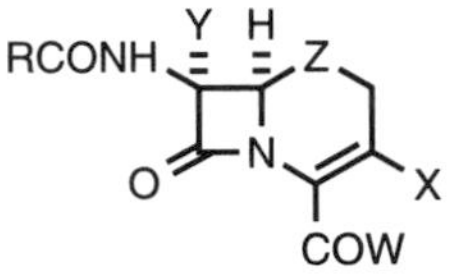

RCONH, Y H, Z, O, N, X, COW

Parenteral Agents	R	W	X	Y	Z	Salt
Cefamandole nofate	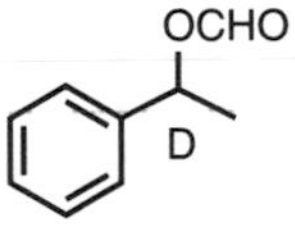 OCHO, D	OH	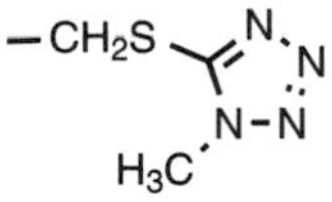 $-CH_2S$, N, N, N–N, H_3C	H	S	-
Cefonicid	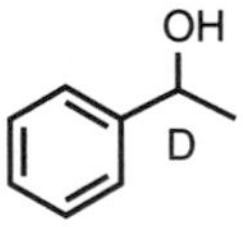 OH, D	OH	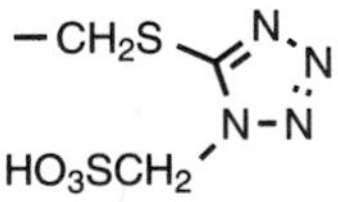 $-CH_2S$, N, N, N–N, HO_3SCH_2	H	S	diNa
Cefuroxime	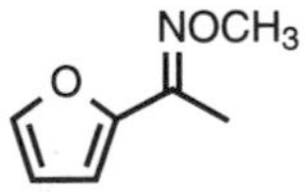 $NOCH_3$, O	OH	$-CH_2OCONH_2$	H	S	Na
Cefoxitin	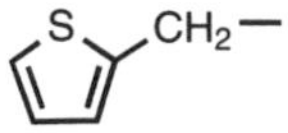 S, CH_2-	OH	$-CH_2OCONH_2$	OCH_3	S	Na

(continued)

Table 34–2. *(Continued)*

	R	W	X	Y	Z	Salt
Ceforanide	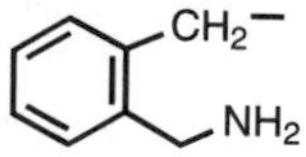 2-(aminomethyl)phenyl–CH_2– (CH_2—, NH_2)	OH	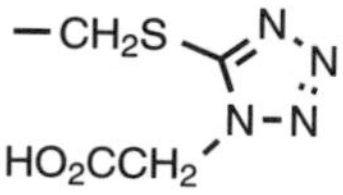 —CH_2S–tetrazolyl (N, N, N–N) with HO_2CCH_2	H	S	-
Cefotetan	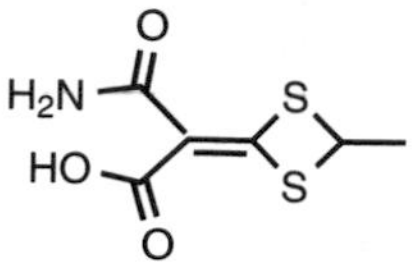 H_2N(C=O)–C(=1,3-dithietan-2-ylidene)–C(=O)HO (S, S)	OH	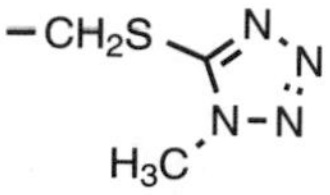 —CH_2S–tetrazolyl (N, N, N–N) with H_3C	OCH_3	S	diNa
Oral Agents						
Cefuroxime Axetil	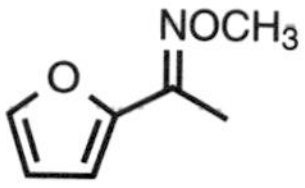 2-furyl–C(=$NOCH_3$)–	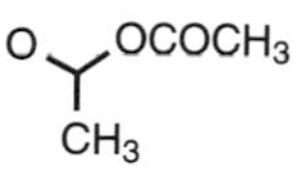 O–CH(CH_3)–$OCOCH_3$	— CH_2OCONH_2	H	S	-
Cefaclor	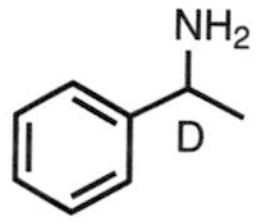 phenyl–CH(NH_2)– (D)	OH	Cl	H	S	-
Loracarbef	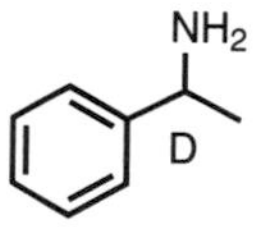 phenyl–CH(NH_2)– (D)	OH	Cl	H	CH_2	-
Cefprozil	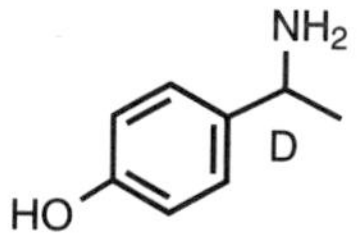 HO–phenyl–CH(NH_2)– (D)	OH	HC=$CHCH_3$	H	S	-

RCONH, Y, H, Z, O, N, X, COW

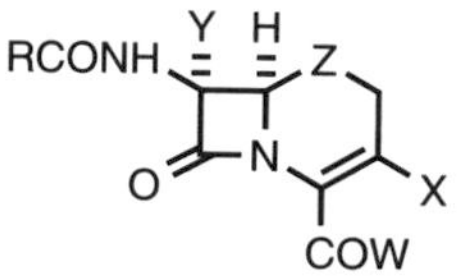

Third Generation Parenteral Agents	R	W	X	Y	Z	Salt
Cefotaxime	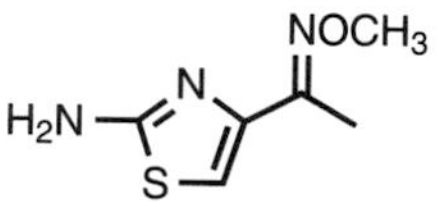 H_2N–thiazolyl (N, S)–C(=$NOCH_3$)–	OH	CH_2OCOCH_3	H	S	Na
Ceftizoxime	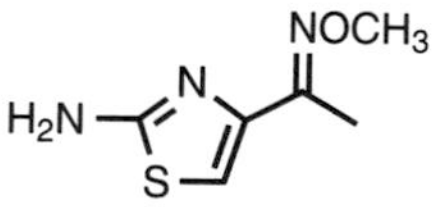 H_2N–thiazolyl (N, S)–C(=$NOCH_3$)–	OH	H	H	S	Na
Ceftriaxone	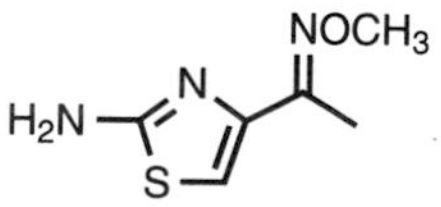 H_2N–thiazolyl (N, S)–C(=$NOCH_3$)–	OH	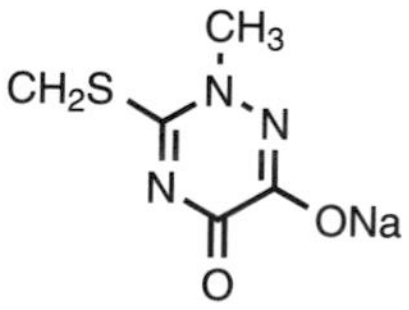 CH_2S–triazinyl (N–CH_3, N, N, O, ONa)	H	S	(di)Na
Ceftazidime	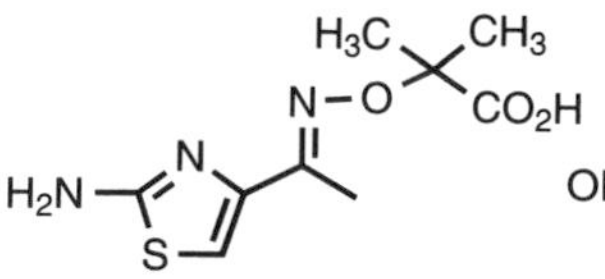 H_2N–thiazolyl (N, S)–C(=N–O–C(H_3C)(CH_3)CO_2H)–	OH	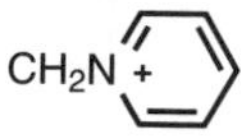 CH_2N^+(pyridinium)	H	S	H or Na

Table 34–2. *(Continued)*

Cefoperazone	OH; O O; C_2H_5-N NCONH D	OH	—CH_2S N N N–N H_3C	H	S	Na
Moxalactam	OH; HO_2C D	OH	—CH_2S N N N–N H_3C	OCH_3	O	diNa
Oral Agents						
Cefixime	$NOCH_2CO_2H$; H_2N N S	OH	$HC=CH_2$	H	S	-
Cefpodoxime proxetil	$NOCH_3$; H_2N N S	CH_3 CH_3; O O O; CH_3 O	CH_2OCH_3	H	S	-

Specific Agents—First Generation.

Parenteral Agents. Cephalothin has been on the market longest of the members of this group still in use. The C-7 side chain is the thiophene bioisostere of the benzyl group of penicillin G. Cephalothin is comparatively resistant to staphylococcal beta-lactamase and is a popular substitute for methicillin and the isoxazoyl subgroup of penicillins. It is not orally active. Following injection, it is excreted primarily in the urine, partly by glomerular filtration and partly by tubular secretion. It is sensitive to many beta-lactamases and also to host deacetylation, nonetheless it finds significant use in the parenteral treatment of infections due to susceptible bacteria. It is comparatively painful on intramuscular injection. Cephapirin has a pyridylthiomethylene side chain at C-7. It is closely related biologically to cephalothin. Cefazolin has the acetyl side chain at C-3 replaced by a thio-linked thiadiazole ring. Not only is this an activating good leaving group, but this moiety is also not subject to the inactivating host hydrolysis reaction that characterizes the precedent acetyl-containing examples. At C-7 it possesses a tetrazoylmethylene unit. Cefazolin is usefully less irritating on injection than its cohort in this generation of drugs.

Oral Agents. Use of the ampicillin side chain conveys oral activity to cephalexin. Whereas it no longer has an activating side chain at C-3, and as a consequence is somewhat less potent, it does not undergo metabolic deactivation and thus maintains its potency. Somewhat puzzling is that the use of the ampicillin side chain in the cephalosporins does not result in a comparable shift in antimicrobial spectrum. Cefadroxil has an amoxicillin-like side chain at C-7 so its oral activity is not surprising. There are some indications that cefadroxil has some immunostimulant properties mediated through T-cell activation and that this is of material assistance to patients in fighting infections. The prolonged biologic half-life of cefadroxil allows for one-a-day dosage. In cephradine an interesting drug design device has been used. The aromatic ring in the ampicillin side chain has been partially hydrogenated by a Birch reduction, such that the resulting molecule is still planar and pi-electron-excessive, but has no conjugated olefinic linkages. Cephradine has the useful characteristic that it can be used both orally and parenterally, so that parenteral therapy can be started in an institutional setting and then, the patient can be sent home with the oral form, avoiding the risk of having to establish a different antibiotic. This is consistent with the present economics requiring sending patients home earlier than some physicians prefer.

Specific Agents—Second-Generation Group.

Parenteral Agents. Cefamandole nafate has a C-7 side chain formate ester derived from D-mandelic acid. The formate ester is cleaved rapidly in the host to release more active cefamandole. The esterification apparently also overcomes instability of cefamandole when it is stored in dry form. This agent has increased activity against Haemophilus influenzae and some gram-negative bacilli as compared with the first-generation cephalosporins. Ejection of the 5-thio-1-methyl-1*H*-tetrazole moiety from C-3 is associated with prothrombin deficiency and bleeding problems as well as with an antabuse-like acute alcohol intolerance. Cefonicid has an unesterified D-mandelic acid moiety at C-7 and a different 5-thio-1*H*-tetrazole group at C-3. The extra acidic group in this side chain leads to this molecule's being sold as a disodium salt. Cefonicid has a longer half-life than the other members of its group but achieves this at the price of somewhat lower potency against gram-positives and aerobes. Cefuroxime has a *syn*-oriented methoxime moiety as part of its C-7 side chain. This conveys considerable resistance to beta-lactamase attack. This is believed to result from the steric demands of this group. This hypothesis is supported by the finding that the anti-analog is attacked by beta-lactamases. The carbamoyl moiety at C-3 is somewhat intermediate in metabolic stability between the classic acetyl moieties and the thiotetrazoles. Cefuroxime penetrates comparatively well into cerebral spinal fluid and is used in cases of H. influenzae meningitis. In the form of its axetil ester prodrug, cefuroxime axetil is more lipophilic and gives satisfactory blood levels on oral administration. The ester bond is cleaved metabolically and the resulting intermediate form loses acetaldehyde spontaneously to produce cefuroxime. Cefoxitin contains the same C-7 side chain as cephalothin and the same C-3 side chain as cefuroxime. The most novel chemical feature of cefoxitin is the possession of an alpha-oriented methoxyl group in place of the normal H atom at C-7. This increased steric bulk conveys very significant stability against beta-lactamases. The inspiration for these functional groups was provided by the discovery of the naturally occurring cephamycin C derived from fermentation of Streptomyces lactamdurans. Ingenious chemical transformations now enable synthetic introduction of such a methoxy group into cephalosporins lacking this feature. Cefoxitin has useful activity against gonorrhea and against some anaerobic infections than its congeners. On the negative side, cefoxitin has the capacity to induce certain broad spectrum beta-lactamases. Ceforanide has an *ortho*-aminomethylene moiety attached to its benzylpenicillin-like side chain at C-7 and this modification conveys significant protection from beta-lactamases. Cefotetan is clearly cephamycin C-inspired but has an unusual sulfur containing C-7 side chain amide. Possession of two carboxyl groups leads to its being marketed as a disodium salt. The C-3 side chain suggests caution in monitoring prothrombin levels and bleeding times when using this agent. Like cefoxitin, cefotetan has better activity against anaerobes than the rest of this group.

Oral Agents. Cefaclor differs from cephalexin primarily in the bioisosteric replacement of methyl by chlorine at C-3. Loracarbef is a synthetic C-5 carba analog of cefaclor. The smaller carbon atom (as compared with sulfur's), would be expected to make loracarbef more potent and this seems to be the case. It is more stable chemically, however, and this is not so readily explained. Cefprozil is a newly introduced oral agent, which has an amoxicillin side chain at C-7, but at C-3 there is a 1-propenyl group conjugated with the double bond in the six-membered ring.

Specific Agents—Third-Generation Group.

Parenteral Agents. Cefotaxime, like cefuroxime, has a *syn*-methoxime moiety at C-7 although this is connected to an aminothiazole ring. Like the other third-generation cephalosporins, it has excellent anti-gram-negative activity and is useful institutionally. It has a metabolically vulnerable acetoxy group attached to C-3. In ceftizoxime the whole C-3 side chain has been omitted to prevent deactivation by hydrolytic metabolism. Ceftriaxone has the same C-7 side chain moiety as cefotaxime and ceftizoxime, but the C-3 side chain consists of a metabolically stable and activating thiotriazinindione in place of the normal acetyl group. The C-3 side chain is sufficiently acid that at normal pH it forms a sodium salt so the article of commerce is a disodium salt. It is useful, given intra-arterially, in the treatment of some meningitis infections caused by gram-negatives. In ceftazidime the oxime moiety is rather more complex, containing two methyl groups and a carboxylic acid. The C-3 side chain has been replaced by a charged pyridinium moiety. The latter is quite water soluble and highly activates the beta-lactam bond towards cleavage. Cefoperazone has a C-7 side chain reminiscent of piperacillin's and also possesses the C-3 side chain that is often associated with the bleeding and alcohol intolerance problems among patients taking cephalosporins. Its useful activity against pseudomonads partly compensates for this. Moxalactam has a profoundly complex molecular structure. It has a C-7 side chain somewhat reminiscent of carbenicillin's, a C-3 side chain like that of cefoperazone and a C-7 methoxy like the cephamycins. More importantly, the S atom has been replaced by a bioisosteric oxygen. Moxalactam is one of the comparatively few antibiotics prepared by total chemical synthesis, so might be more properly discussed with the antimicrobials were it not so obviously inspired by the cephalosporins. Moxalactam's potency is usually higher than that of its sulfur-containing analogs. Despite its powerful antimicrobial properties, its popularity is lessened because of its tendency to cause bleeding problems in some patients.

Oral Agents. There are comparatively few orally active third-generation agents. This group is currently represented by cefixime and cefpodoxime proxetil. In cefixime, in addition to the *cis*-oximinoether at C-7, the C-3 side chain is a vinyl group analogous to the propenyl group of cefprozil. Cefixime has excellent anti-gram-positive activity, except against staphylococci, and its anti-gram-negative activity is intermediate between that of the second-generation and third-generation agents described previously. Cefpodoxime proxetil is a newly introduced prodrug. It is cleaved enzymically to isopropanol, carbon dioxide, acetaldehyde, and to cefpodoxime in the gut wall. It has better anti-Staph. aureus activity than cefixime and is used against pharyngitis, urinary tract infections, upper and lower respiratory tract infections, otitis media, skin and soft tissue infections, and gonorrhea.

Because the cephalosporin field is being very actively pursued, the student can expect continual developments into the foreseeable future.

Carbapenems

Thienamycin was isolated from Streptomyces cattleya. Because of its extremely intense and broad-spectrum antimicrobial activity and its ability to inactivate beta-lactamases, it combines in one molecule the functional features of the best of the beta-lactam antibiotics as well as the beta-lactamase inhibitors. It differs structurally in several important respects from the penicillins and cephalosporins. The sulfur atom is not part of the five-membered ring but rather has been replaced there by a methylene. It has been moved exocyclically to C-3 where it starts a functionalized side chain. This bioisosteric change dramatically increases the reactivity of the beta-lactam ring as does the endocyclic olefinic linkage. Both make thienamycin unstable and this has caused great difficulties in the original isolation. The terminal amino group in the side chain attached to C-3 is nucleophilic and attacks the beta-lactam bond of nearby molecules, so that the drug became less stable as it was purified and became more concentrated. Ultimately this problem was overcome by changing the amino group to a less nucleophilic N-forminoyl moiety by a semisynthetic process to produce imipenem. At C-6 there is a 2-hydroxyethyl group attached with alpha-stereochemistry. Thus the absolute stereochemistry of the molecule is 5R,6S,8S. With these striking differences from the penicillins and cephalosporins, it is not surprising that thienamycin binds differently to the penicillin-binding proteins (especially strongly to PBP2) but it is gratifying that the result is very potent broad spectrum activity. Thienamycin and imipenem penetrate very well through porins and are very stable, even inhibitory, to many beta-lactamases. Imipenem is not, however, orally active. Unfortunately for its use in urinary tract infections, renal dehydropeptidase-1 hydrolyzes imipenem and deactivates it. An inhibitor for this enzyme, cilastatin, is coadministered with imipenem to protect it. Inhibition of human dehydropeptidase does not seem to have deleterious consequences, making this combination highly efficacious.

The combination of imipenem and cilastatin is about 25% serum-protein bound. On injection it penetrates well into most tissues, but not cerebrospinal fluid, and it is subsequently excreted in the urine. It is broader in its spectrum than any other antibiotic presently available in the United States. This very potent combination is especially useful for treatment of serious infections by aerobic gram-negative bacilli, anaerobes, and Staph. aureus. It is used clinically for severe infections of adults of the gut, gastrointestinal tract, bones, skin, and endocardia. With allergic reaction as its main risk factor, imipenem also has the unfortunate property of being a good beta-lactamase inducer. Because of these features, imipenem-cilastatin is rarely a drug of first choice, but is reserved for use in special circumstances.

Monobactams

Aztreonam. Fermentation of unusual microorganisms led to the discovery of certain monocyclic beta-lactam antibiotics, called monobactams. None of these molecules has proven to be important in its own right but the group served as the inspiration for the synthesis of aztreonam. Aztreonam is a totally synthetic parenteral antibiotic whose antimicrobial spectrum is devoted almost exclusively to gram-negative microorganisms, and it even is capable of inactivating some beta-lactamases. Its molecular mode of action is closely similar to that of the penicillins, cephalosporins, and carbapenems, the action being characterized by especially strong affinity for PBP3 and producing filamentous cells as a consequence. Whereas the principal side chain closely resembles that of ceftazidime, the sulfamic acid moiety was unprecedented. Remembering the comparatively large size of sulfur atoms, this assembly may sufficiently spatially resemble the corresponding C-2 carboxyl group of the precedent beta-lactam antibiotics to confuse the penicillin binding protons. The strongly electron-withdrawing character of the sulfamic acid group probably also makes the beta-lactam bond more vulnerable to hydrolysis. In any case, a fused ring is not essential for antibiotic activity. The alpha-oriented methyl group at C-2 is associated with the stability of aztreonam towards beta-lactamases. The protein binding is moderate (about 50%) and the drug is hardly changed by metabolism. Aztreonam is given by injection

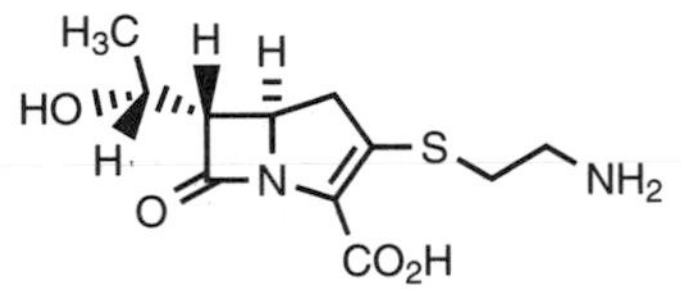

Thienamycin

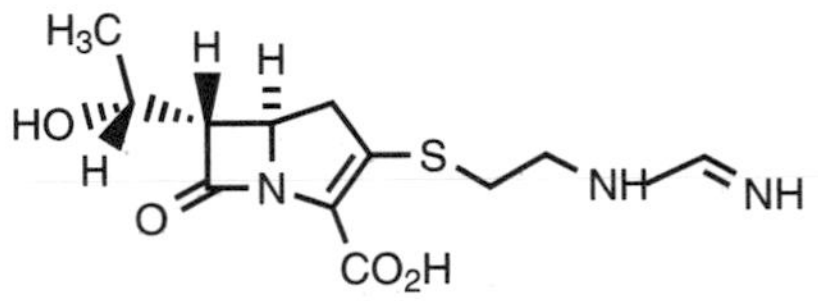

Imipenem

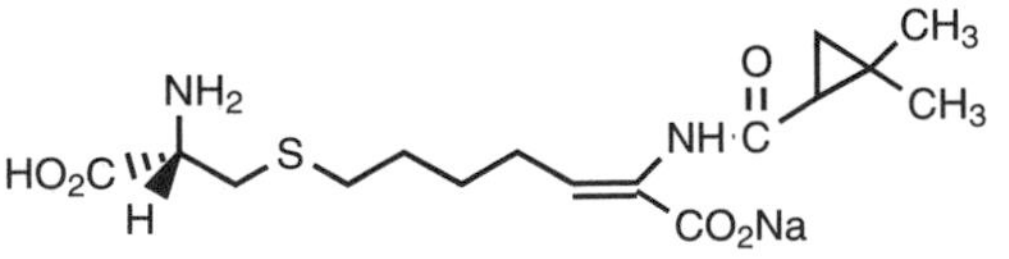

Cilastatin Sodium

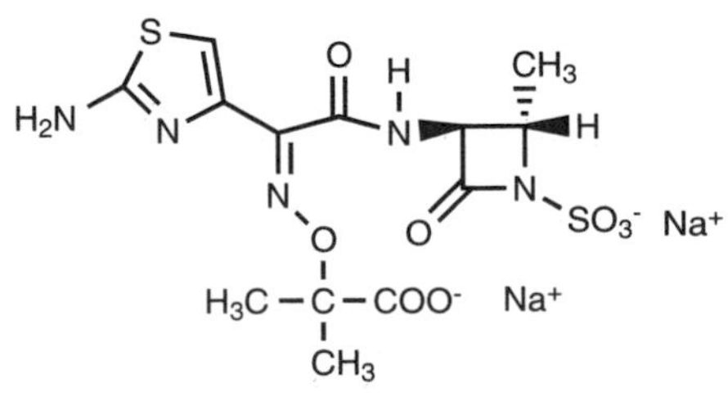

Aztreonam disodium

and is largely excreted in the urine. The primary clinical use of aztreonam is against severe infections caused by gram-negative microorganisms, especially those acquired in the hospital. These are mainly urinary tract, upper respiratory tract, bone, cartilage, abdominal, obstetric, and gynecologic infections and septicemias. The drug is well tolerated and side effects are infrequent. Interestingly, allergy would not be unexpected, but cross allergenicity with penicillins and cephalosporins has not often been reported.

ANTIBIOTICS: INHIBITORS OF PROTEIN BIOSYNTHESIS

Some antibiotic families exert lethal effects on bacteria by inhibiting ribosomally mediated protein biosynthesis. At first glimpse this may seem anomalous, because eukaryotic organisms also construct their essential proteins on ribosomal organelles and the sequence of biochemical steps in both biologic classes is closely analogous to that of prokaryotic microorganisms. At a molecular level, however, the apparent anomaly resolves itself. In E. coli, for example, the 70S ribosomal particle is composed not only of RNA but also of 55 different structural and functional proteins (21 on the 30S and 34 on the 50S subparticle). These proteins differ in structure from those in the eukaryotic 80S ribosome. The binding sites for the important antibiotics lie on the proteins of the bacterial 70S ribosome. The aminoglycosides, for example, bind to a site on the 30S subparticle associated with the presence of a specific protein, whereas the macrolides, lincosaminides, and chloramphenicol bind to a different site on the 50S subparticle. There is evidence that the tetracyclines bind to both subparticles. At normal doses, these antibiotics neither bind to, nor interfere with, the function of 80S ribosomal particles. The basis for the selective toxicity of these antibiotics is then apparent. Interference with bacterial protein biosynthesis prevents repair, cellular growth, and reproduction and the effect, in clinically achievable doses, is bacteriostatic or bactericidal.

Aminoglycosides and Aminocyclitols

The aminoglycoside/aminocyclitol class of antibiotics contains a pharmacophoric 1,3-diaminoinositol derivative: streptamine, 2-deoxystreptamine, or spectinamine. Some of the alcoholic functions of these are substituted through glycosidic bonds with characteristic aminosugars to form pseudo-oligosaccharides. The chemistry, spectrum, potency, toxicity, and pharmacokinetics of these agents are a function of the specific identity of the diaminoinositol unit and the arrangement of the attachments.

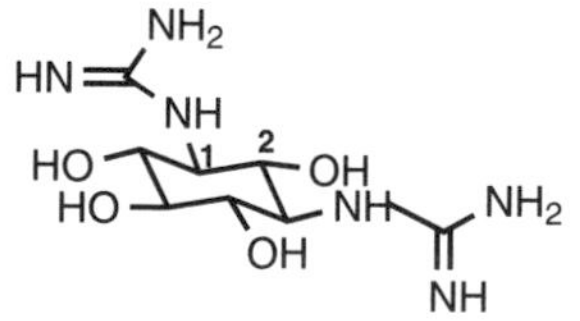

Streptamine 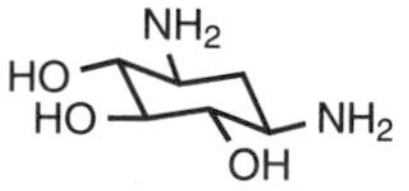 2-Deoxystreptamine 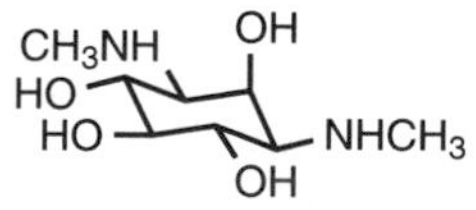 Spectinamine

The various aminoglycoside antibiotics are freely water-soluble at all achievable pHs, are basic and form acid addition salts, are not absorbed in significant amounts from the gastrointestinal tract, and are excreted in active form in fairly high concentrations in the urine. When the kidneys are not functioning efficiently, the concentrations injected must be reduced to prevent accumulation to toxic levels. When given orally, their action is primarily confined to the gastrointestinal tract. They are more commonly given intramuscularly or by perfusion. These agents have intrinsically broad antimicrobial spectra but their toxicity potential limits their clinical use to severe infections by gram-negative bacteria. These toxicities involve ototoxicity to functions mediated by the eighth cranial nerve, such as hearing loss and vertigo. Their use can also lead to kidney tubular necrosis producing decreases in glomerular function. These toxic effects are related to blood levels and are apparently mediated by the special affinity of these aminoglycosides to kidney cells and to the sensory cells of the inner ear. The effects may have a delayed onset, making them all the more treacherous as the patient can be injured significantly before symptoms appear. Less common is a curare-like neuromuscular blockade believed to be caused by competitive inhibition of calcium-ion-dependent acetylcholine release at the neuromuscular junction. This side effect can exaggerate the muscle weakness of patients with myasthenia gravis and Parkinson's disease. In current practice, all of these toxic phenomena are well known, therefore creatinine function is determined and the dose decreased accordingly so that these side effects are less common and less severe than previously. The aminoglycoside antibiotics are widely distributed (mainly in extracellular fluids) and have low levels of protein binding. They are bactericidal due to a combination of toxic effects. At less than toxic doses, they find to the protein portion of the 30S ribosomal subparticle leading to mistranslation of RNA templates and the consequent insertion of wrong amino acids and formation of so-called nonsense proteins. The most relevant of these unnatural proteins are involved in upsetting bacterial membrane function. Their presence destroys the semipermeability of the membrane and this damage cannot be repaired without de novo programmed protein biosynthesis. Among the substances admitted by the damaged membrane are large additional quantities of aminoglycoside. At these increased concentrations, protein biosynthesis ceases altogether. These combined effects are devastating to the target bacterial cells. Given their highly polar properties, the student may wonder how these agents can enter bacterial cells at all. Aminoglycosides apparently

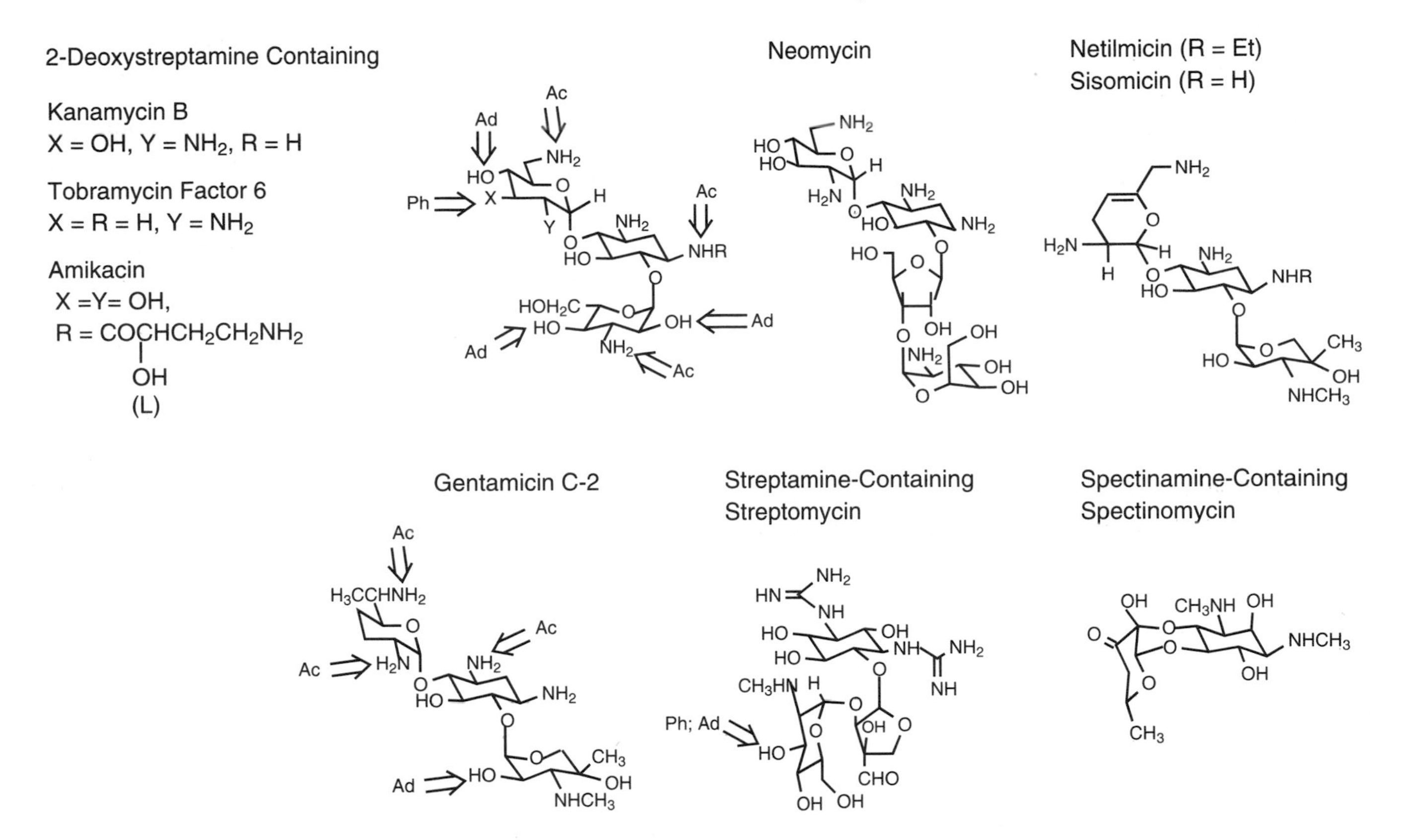

Fig. 34–5. The commercially important aminoglycoside antibiotics. Some points of inactivating attack by specific R-factor-mediated-enzymes are indicated by the following symbols. ⇒ Ad = adenylation; ⇒ Ac = acetylation; and ⇒ Ph = phosphorylation.

bind initially to external lipopolysaccharides and diffuse into the cells in small amounts through the porins. The uptake process is inhibited by Ca(II) and Mg(II) ions. These ions are, then, partially incompatible therapeutically.

Bacterial resistance to aminoglycoside antibiotics in the clinic is most commonly caused by bacterial elaboration of R-factor mediated enzymes that *N*-acetylate, *O*-phosphorylate and *O*-adenylate specific functional groups, preventing subsequent ribosomal binding. In some cases, chemical deletion of the functional groups transformed by these enzymes leaves a molecule that is still antibiotic but no longer a substrate, thus substances with intrinsically broader spectrum use can be produced semisynthetically in this way. In some other cases, novel functional groups can be attached to remote functionality, which converts these antibiotics to poorer substrates for these R-factor mediated enzymes and this expands their spectra, as is discussed later in this chapter.

These antibiotics have similar clinical targets to those of the quinolones and are decreasing in popularity as the quinolone use increases. Those aminoglycoside antibiotics in present clinical use are illustrated in Figure 34–5.

Kanamycin

Kanamycin is a mixture of at least three components (A, B, and C), isolated from Streptomyces kanamyceticus.

Enzymes APH (3')-I;II

Kanamycin A (Active)

Kanamycin A 3'-phosphate (inactive)

In addition to typical aminoglycoside antibiotic properties, it, along with gentamicin, neomycin, and paromomycin, is among the most chemically stable of the common antibiotics. These substances can be heated without loss for astonishing periods in acid or alkaline solutions and can even withstand autoclaving temperatures. Kanamycin is, however, unstable to R-factor enzymes, being *O*-phosphorylated on the C-3′ hydroxyl by enzymes APH(3′)-I and APH(3′)-II and *N*-acetylated on the C-6′ amino group, among others. These transformation products are antibiotically inactive. Kanamycin is used parenterally against some gram-negative bacteria, but Pseudomonas aeruginosa and anaerobes are usually resistant. Although it can also be used in combination with other agents against mycobacteria, its popularity is fading. Injections of kanamycin are painful, enough to require use of a local anesthetic.

Amikacin. This is made semisynthetically from kanamycin. Interestingly, the L-hydroxyaminobuteroyl amide (HABA) moiety attached to N-3 inhibits adenylation and phosphorylation in the distant amino sugar ring (at C-2′ and C-3′) even though the HABA substituent is not where the enzymatic reaction takes place. This effect is attributed to decreased binding to the R-factor mediated enzymes. With this change, potency and spectrum are strongly enhanced and amikacin is used competitively with gentamicin for the treatment of sensitive strains of Mycobacterium tuberculosis, Yersinia tularensis, and severe Pseudomonas aeruginosa infections resistant to other agents.

Tobramycin. One (factor 6) of a mixture is produced by fermentation of Streptomyces tenebrarius. Lacking the C-3′ hydroxy group, it is not a substrate for APH(3′)-I and -II and thus has an intrinsically broader spectrum than kanamycin. It is a substrate, however, for adenylation at C-2″ by ANT(2″), acetylation at C-3 by AAC(3)-I and -II and at C-2′ by AAC(2′). It is widely used parenterally for difficult infections, especially those by gentamicin-resistant Pseudomonas aeruginosa. It is believed by some clinicians to be less toxic than gentamicin.

Gentamicin

This is a mixture of several antibiotic components produced by fermentation of Micromonospora purpurea and other related soil microorganisms (hence it is spelled with an "i" instead of a "y"). Gentamicins C-1, C-2, and C-1a are the most prominent aminoglycoside antibiotics still in use. Gentamicin was, for example, one of the first antibiotics to have significant activity against Pseudomonas aeruginosa infections. This water-loving opportunistic pathogen is frequently encountered in burns, pneumonias, and urinary tract infections. It is highly virulent. As noted above, some of the functional groups that serve as targets for R-factor mediated enzymes are missing in the structure of gentamicins so their antibacterial spectrum is enhanced. It is, however, inactivated through C-2″ adenylation by enzyme ANT(2″) and acetylation at C-6′ by AAC(6′), at C-1 by AAC(3)-I and -II, and at C-2′ by AAC(2′). It is often combined with other anti-infective agents and an interesting incompatibility has been uncovered. With certain beta-lactam antibiotics, the two drugs react with each other so that N-acylation on C-1 of gentamicin by the beta-lactam antibiotic takes place, thus inactivating both antibiotics. The two agents should not, therefore, be mixed in the same solution and should be administered into different tissue compartments (usually one in each arm) to prevent this.

Gentamicin is used for urinary tract infections, burns, some pneumonias, and bone and joint infections caused by susceptible gram-negatives. It is often used to prevent fouling of soft contact lenses. It is also used in polymer matrices in orthopedic surgery to prevent sealed in sepsis. It is given topically, sometimes in special dressings, to burn patients.

Netilmicin

This has an N-ethyl substituent introduced semisynthetically at C-3 onto sisomicin, a now archaic antibiotic produced by fermentation of Micromonospora inyoensis and related soil microorganisms. Netilmicin is similar in its clinical properties to gentamicin and tobramycin, although its antimicrobial spectrum is broader against many R-factor carrying strains. Netilmicin and sisomicin are chemically unusual for this antibiotic class because of the unsaturation in the upper left sugar ring.

Kanamycin and two otherwise archaic members of the aminoglycoside antibiotic group, neomycin and paromomycin, find some oral use for the suppression of gut flora. Paromomycin is also used for the oral treatment of amoebic dysentery. Amoeba are persistent pathogens causing chronic diarrhea and are acquired most frequently by travelers from food supplies contaminated with human waste. Suppression of gut flora is otherwise mostly employed prophylactically before gut surgery to decrease the likelihood of post-surgical peritonitis. Neomycin is also used as an external ointment. Neomycin is considered too toxic by today's standards for parenteral use.

Gentamicin C-2a (active) + Beta-lactam Antibiotic (active) → (inactive)

Streptomycin

With a modified pharmacophore in that the diaminoinositol unit, streptamine, streptomycin has an axial hydroxyl group at C-2 and two highly basic guanido groups at C-1 and C-3 in place of the primary amine moieties of 2-deoxystreptamine. Streptomycin is produced by fermentation of Streptomyces griseus and several related soil microoganisms. It was introduced in 1943 primarily for the treatment of tuberculosis. The other aminoglycoside antibiotics are not nearly as useful against tuberculosis as is streptomycin. Tuberculosis was controlled only by public health measures before the advent of streptomycin. Control of this ancient scourge was greeted with such enthusiasm that Waksman, the discoverer of streptomycin, received a Nobel Prize in 1952. It is possible that the unusual pharmacophore of streptomycin accounts in large measure for its unusual antibacterial spectrum. Another feature, the alpha-hydroxy aldehyde function, is a center of instability such that streptomycin cannot be sterilized by autoclaving, so that streptomycin sulfate solutions that need sterilization are made by ultrafiltration. Streptomycin is rarely used today as a single agent. Other primary drugs for tuberculosis are isonicotinic acid hydrazide, rifampin, and para-aminosalicylic acid. M. tuberculosis is a slow-growing microorganism, prone to develop resistance during treatment. Treatment is long and recovery is slow. Alternation of the primary drugs during the course of therapy is employed to retard resistance emergence and combinations of drugs are employed for the same purpose. The resistance takes the now familiar course of *N*-acetylation, *O*-phosphorylation and *O*-adenylation of specific functional groups in the streptomycin molecule. Streptomycin is not very useful against the M. avium complex and the other unusual mycobacteria which have become more common pathogens among immunosuppressed people, including AIDS patients. Streptomycin is useful also against bubonic plague and tularemia. Candida albicans, an opportunistic yeast, is not inhibited by streptomycin and overgrowth leading to thrush and vaginal candidiasis can be a consequence of its long-term use because of its concomitant disturbance of the normal flora.

Spectinomycin

Another unusual aminoglycoside antibiotic, spectinomycin is produced by fermentation of Streptomyces spectabilis and differs substantially in its clinical properties from the others. The diaminoinositol unit (spectinamine) contains two mono-*N*-methyl groups and the hydroxyl between them has a stereochemistry opposite to that of streptomycin. The glycosidically attached sugar is also unusual in that it contains three consecutive carbonyl groups, either overt or masked, and is fused by two adjacent linkages to spectinamine. Spectinomycin is bacteriostatic as normally employed. It is almost exclusively used in a single bolus injection intramuscularly against Neisseria gonorrheae, especially penicillinase-producing strains (PPNG), in cases of urogenital or oral gonorrhea and does not apparently produce any serious oto- or nephrotoxicity when used in this way. It is particularly useful for the treatment of patients allergic to penicillin and patients not likely to comply well with a medication scheme. It would likely be more widely used except that syphilis does not respond to it. It causes significant miscoding following ribosomal binding but does not cause much inhibition of overall programmed protein biosynthesis.

Macrolide Antibiotics

The term macrolide is derived from its characteristic large (14-membered) lactone (cyclic ester) ring. The clinically important members of this antibiotic family (Fig. 34–6) have two or more characteristic sugars attached to this ring. One of these sugars usually carries a substituted amino group so their overall chemical character is weakly basic. They are not very water-soluble as free bases but salt formation with certain acids (glucoheptonic and lactobiono-S-lactone, for example) increases water solubility and it decreases with others (laurylsulfate, for sample). Macrolide antibiotics with 16-membered rings are popular outside the United States but one example, tylosin, finds extensive agricultural use in the United States. The 14-membered ring macrolides are biosynthesized from propionic acid units so that every second carbon of erythromycin, for example, bears a methyl group and the rest of the carbons, with one exception, are oxygen-bearing. Two carbons bear so-called extra oxygen atoms not present in a propionic acid unit and the two are glycosylated. The macrolides are often used for the treatment of upper and lower respiratory tract infections primarily caused by gram-positive microorganisms and also find some use for certain sexually transmitted diseases, such as gonorrhea and PID, which is caused by mixed infections involving cell-wall free organisms such as Chlamydia trachomitis. Thus the macrolides have a comparatively narrow antimi-

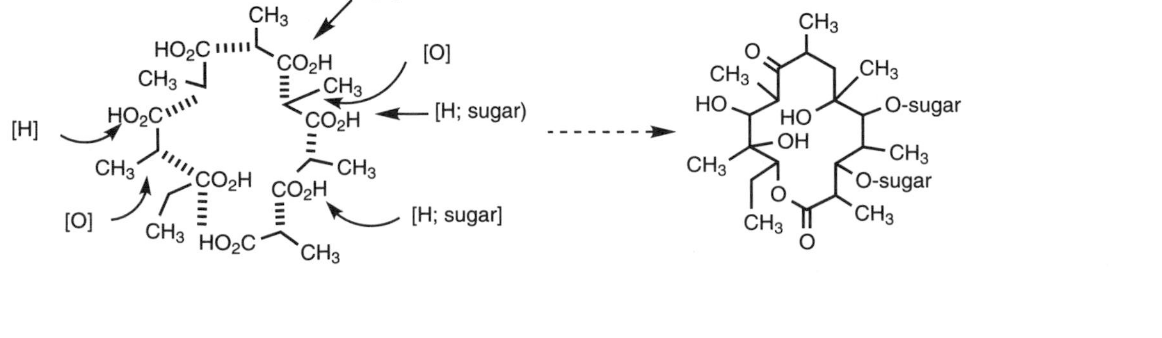

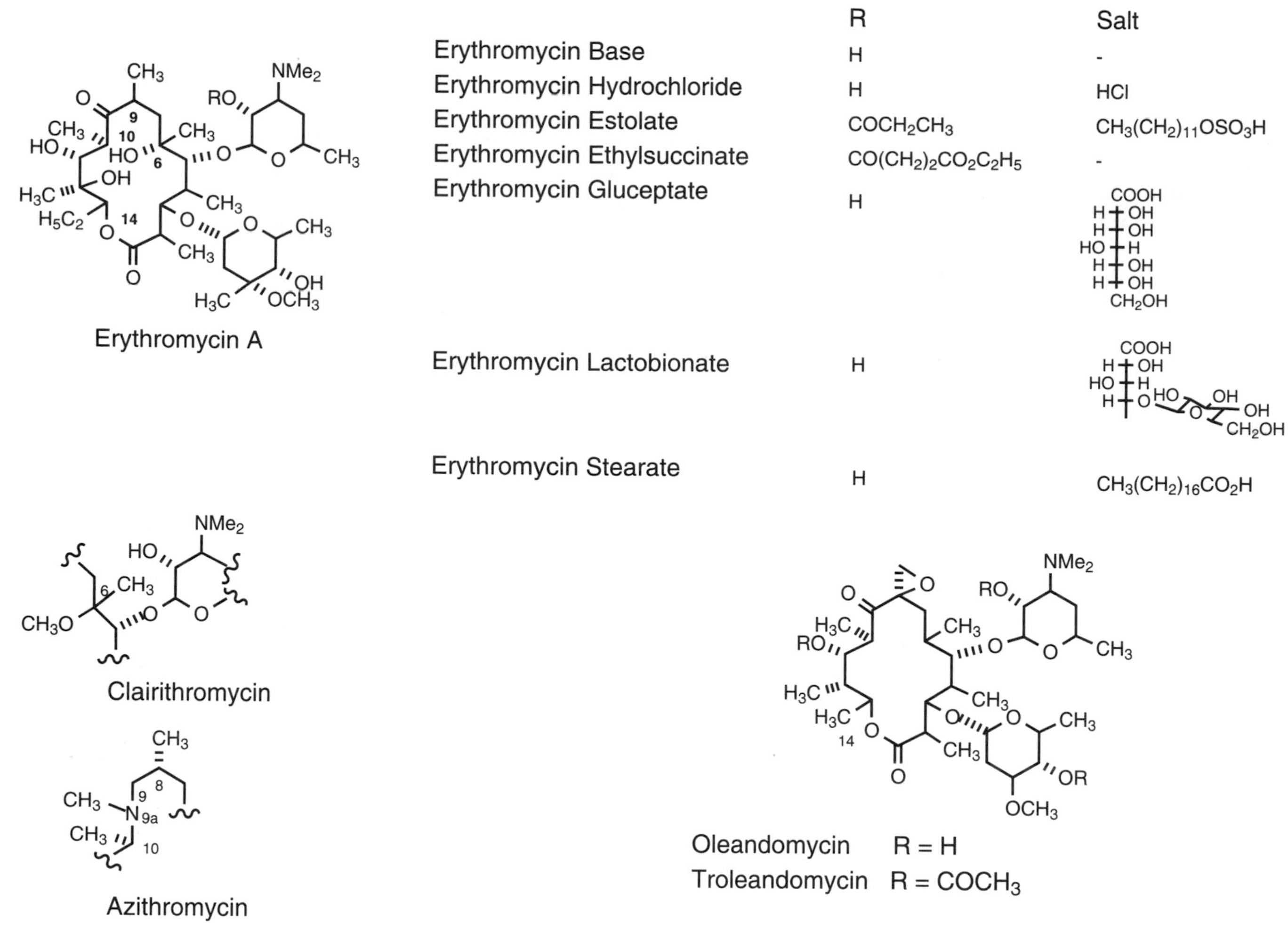

	R	Salt
Erythromycin Base	H	-
Erythromycin Hydrochloride	H	HCl
Erythromycin Estolate	$COCH_2CH_3$	$CH_3(CH_2)_{11}OSO_3H$
Erythromycin Ethylsuccinate	$CO(CH_2)_2CO_2C_2H_5$	-
Erythromycin Gluceptate	H	
Erythromycin Lactobionate	H	
Erythromycin Stearate	H	$CH_3(CH_2)_{16}CO_2H$

Erythromycin A

Clairithromycin

Azithromycin

Oleandomycin R = H
Troleandomycin R = $COCH_3$

Fig. 34–6. The commercially important 14-membered macrolide antibiotics.

crobial spectrum, reminiscent of the medium spectrum penicillins, but the organisms involved include many of the more commonly encountered community-acquired agents and the macrolides are remarkably free of serious toxicity to the host. The macrolides inhibit bacteria by interfering with programmed ribosomal protein biosynthesis by inhibiting translocation of aminoacyl t-RNA following binding to the 50S subparticle. They are bacteriostatic in the clinic in achievable concentrations. The macrolides are primarily used orally for mild systemic infections of the respiratory tract, liver, kidneys, prostate, and milk gland even though absorption is somewhat irregular, especially when taken with food. Some derivatives are propropulsive through stimulation of gastrin production. The resulting hyperperistalsis causes uncomfortable gastrointestinal cramps in some patients. Developed bacterial resistance is primarily caused by bacteria possessing R-factor enzymes which, interestingly, methylate a guanine residue on their own ribosomal RNA making them somewhat less efficient at protein biosynthesis but comparatively poor binders of macrolides. Some bacterial strains, however, appear to be resistant due to the operation of an active efflux process in which the drug is expelled from the cell at the cost of energy. Intrinsic resistance of gram-negatives is primarily caused by lack of penetration as the isolated ribosomes from these organisms are often susceptible. The macrolides of the erythromycin class are chemically unstable in acid due to rapid internal cyclic ketal formation leading to inactivity. This reaction is believed to be clinically important. Many macrolides have an unpleasant taste, which is partially overcome, as is at the same time the acid instability and the gut cramps, by using water-insoluble dosage forms, often combined with enteric coatings. The free base, the hydrochloride, and the stearate salts are examples.

Among the bacteria often susceptible to erythromycin are gram-positive cocci (Streptococci and Staphylococci), gram-positive bacilli (Listeria, Corynebacterium, and Bacillus anthracis), gram-negative cocci (Neisseria), various gram-negative species (Legionella, Campylobacter and Haemophilus), anaerobes (Propionibacterium, Clostridium, Bacteroides fragilis) and several other microorganisms (Mycoplasma, Ureaplasma, Chlamydia, Rickettsia, Treponema).

Drug-drug interactions with macrolides are comparatively common and usually involve competition for liver

Acid-catalyzed ketal formation

metabolism. Such drugs as ergotamine, theophylline, carbamazepine, bromocryptine, warfarin, digoxin, oral contraceptives, and methylprednisone can be involved. The result of this is a longer half-life and enhanced potential toxicity by increasing the effective dose over time. The main product of liver metabolism of erythromycin is N-demethylated.

Of the two most popular erythromycin prodrugs, erythromycin estolate is a C-2″-propionyl ester, N-laurylsulfate salt. Administration of erythromycin estolate produces higher blood levels following metabolic regeneration of erythromycin. In a small number of cases, a severe, dose-related, cholestatic jaundice occurs in which the bile becomes granular in the bile duct, impeding flow so that the bile salts back up into the circulation. This seems to be partly allergic and partly dose-related. If the drug causes hepatocyte damage, perhaps this releases antigenic proteins that promote further damage. When cholestatic jaundice occurs, the drug must be replaced by another, nonmacrolide antibiotic such as one of the penicillins or clindamycin. It is postulated that the propionyl ester group is transferred to a tissue component which is antigenic although the evidence for this is not compelling.

Erythromycin ethyl succinate (EES) is a mixed double ester in which one carboxyl of succinic acid esterifies the C-2″ hydroxyl of erythromycin and the other ethanol. This prodrug is frequently used in an oral suspension for pediatric use. Some cholestatic jaundice is beginning to be associated with the use of EES.

Recently, two new chemical entities have been introduced into the macrolide class. Clarithromycin differs from erythromycin in that the C-6 hydroxy group has been converted semisynthetically to a methyl ether. The C-6 hydroxy group is essential to the process initiated by protons in which an antibiotically inactive internal cyclic ketal is formed. This ketal, or one of the products of its subsequent degradation, is also associated with gastrointestinal cramping. Conversion of the molecule to its methyl ether not only gives better blood levels through chemical stabilization and better absorption but also appears associated with less gastric upset. An extensive saturable first-pass liver metabolism of clarithromycin leads to formation of its C-14 hydroxy analog, which has even greater antimicrobial potency, especially against Haemophilus influenzae. The enhanced lipophilicity of clarithromycin also allows for lower and less frequent dosage for mild infections. Azithromycin has been formed by semisynthetic conversion to a ring expanded analog in which an N-methyl group has been inserted between carbons 9 and 10 of erythromycin. Not only is azithromycin more stable to acid degradation than erythromycin, but it also has a considerably longer half life, attributed to greater and longer tissue penetration, allowing once-a-day dosage. This is convenient for patients who comply poorly. The drug should be taken on an empty stomach. Azithromycin has greater anti gram-negative activity than either erythromycin or clarithromycin. Both these new macrolides are quite similar in usage to erythromycin itself and are cross resistant with it, but their future impact on medical practice is not yet determined.

Troleandomycin is prepared semisynthetically by chemical transformation of oleandomycin, itself produced by fermentation of Streptomyces antibioticus. Troleandomycin is a prodrug requiring metabolic conversion back to oleandomycin in vivo. With a bacteriostatic spectrum similar to that of erythromycin it is significantly less active and frequently cross resistant with it. It is not common in current use.

Lincosamides

The lincosamides (lincomycin and clindamycin) contain an unusual thiomethyl amino-octoside (*O*-thiolincosamide) linked by an amide bond to an n-propyl substituted N-methylpyrrolidylcarboxylic acid (*N*-methyl-n-propyl-trans-hygric acid). Lincosamides are weakly basic and form clinically useful hydrochloric acid salts. They are chemically distinct from the macrolide antibiotics but possess many pharmacologic similarities to them. Lincomycin is a natural product isolated from fermentations of Streptomyces lincolnensis var. lincolnensis. It serves as the starting material for semisynthesis of clindamycin by an SN-2 reaction that inverts the R stereochemistry of the C-7 hydroxyl to a C-7 S chloride. Clindamycin is more bioactive and lipophilic than lincomycin, and is thus better absorbed following oral administration. The lincosamides bind to 50S ribosomal subparticles at a site partly overlapping with the macrolide site so they are

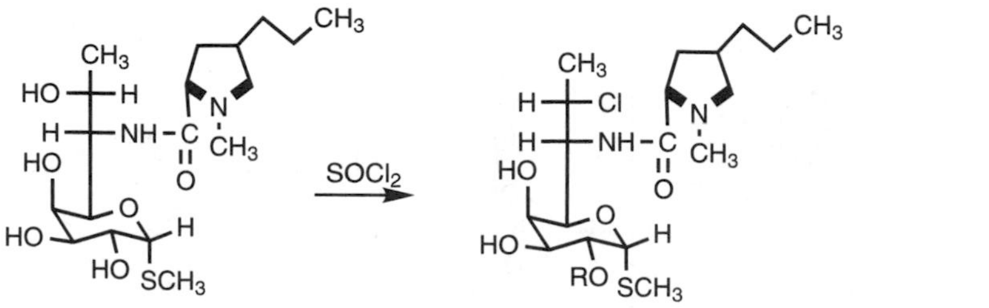

Lincomycin $\xrightarrow{SOCl_2}$ Clindamycin, R = H
Clindamycin Phosphate, R = PO_3H

mutually cross resistant with macrolides and work through essentially the same molecular mechanism of action. Clindamycin is significantly less painful than erythromycin when injected as a C-2 phosphate prodrug. Clindamycin has a clinical spectrum rather like the macrolides although it distributes better into bones. Clindamycin works well for gram-positive infections, especially in patients allergic to beta-lactams, and also has generally better activity against anaerobes. Unfortunately, however, the lincosamides are associated increasingly with drug-related diarrheas. The most severe of these is a pseudomembranous colitis caused by release of a glycoproteinaceous endotoxin produced by lysis of Clostridium difficile, an opportunistic anaerobe. Its overgrowth results from suppression of the normal flora, whose presence otherwise preserves a healthier ecologic balance. The popularity of clindamycin in the clinic has decreased even though pseudomembranous colitis is now also associated with several other broad-spectrum antibiotics. A less common side effect is exudative erythema multiforme (Stevens-Johnson syndrome). Clindamycin has excellent activity against Propionibacterium acnes when applied topically to comidones and because it is white, it can be cosmetically tinted to match flesh tones better than the yellow tetracyclines. Clindamycin and lincomycin undergo extensive liver metabolism resulting primarily in N-demethylation. The N-desmethyl analog is antibiotically active.

Tetracyclines

Chemistry

The tetracycline family is widely used in office practice. These antibiotics are characterized by a highly functionalized, partially-reduced-naphthacene (four linearly fused six-membered rings) ring system from which both the family name and numbering system are derived. They are amphoteric substances with 3 pK values revealed by titration (2.8 to 3.4, 7.2 to 7.8 and 9.1 to 9.7) and have an isoelectric point at about pH 5. The basic function is the C-4-alpha-dimethylamino moiety. Commercially available tetracyclines are comparatively water-soluble hydrochloride salts (Fig. 34–7). The conjugated phenolic enone system extending from C-10 to C-12 is associated with the pK_a at about 7.5, whereas the conjugated trione system extending from C-1 to C-3 in ring A is nearly as acidic as is acetic acid. Students who remember the principle of vinylogy will visualize this readily. These resonating systems can be drawn in several essentially equivalent ways. The formulae normally given are those settled upon by popular convention.

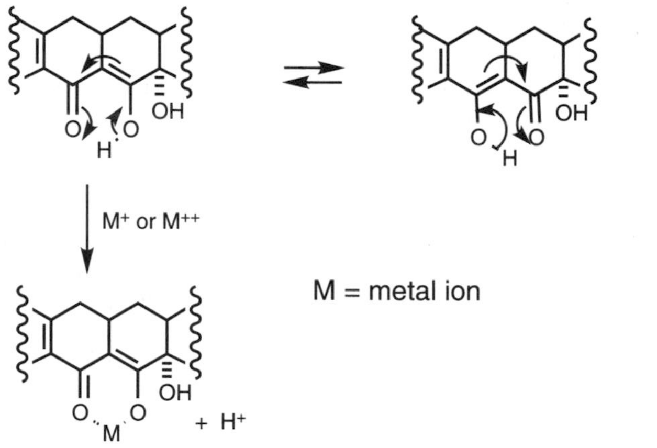

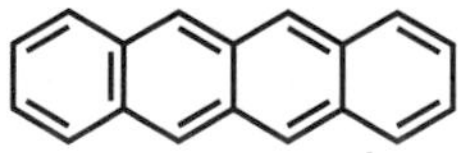

M^+ or M^{++}

M = metal ion

Naphthacene

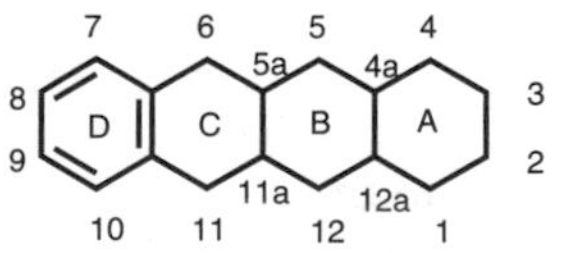

Tetracycline
(1,2,3,4,4a,5,5a,6,6a,11,11a,12,
12a-dodecahydronaphthacene)

Numbering and naming system of the tetracyclines.

Tetracycline	R = H, X = OH, Y = CH_3
Demeclocycline	R = Cl, X = OH, Y = H
Minocycline	R = $N(CH_3)_2$, X = Y = H
Sancycline	R = X = Y = H

Oxytetracycline	X = OH, Y = CH_3
Methacycline	X = Y = CH_2
Doxycycline	X = H, Y = CH_3

Fig. 34–7. The clinically useful tetracycline antibiotics.

Chelation and its Consequences. Chelation is an important feature of the chemical and clinical properties of the tetracyclines. The acidic functions of the tetracyclines are capable of forming salts through chelation with metal ions. The salts of polyvalent metal ions, such as Fe(II), Al(III), Ca(II), Mg(II), are all quite insoluble at neutral pHs. This insolubility is not only inconvenient for the preparation of solutions, but also interferes with blood levels on oral administration. Consequently, the tetracyclines are incompatible with coadministered multivalent ion-rich antacids and with hematinics, and concomitant consumption of dairy products rich in calcium ion is also contraindicated. Further, the bones, of which the teeth are the most visible, are calcium-rich structures at nearly neutral pHs and so accumulate tetracyclines in proportion to the amount and duration of therapy when bones and teeth are being formed. As the tetracyclines are yellow, this leads to a progressive and essentially permanent discoloration in which, in advanced cases, the teeth are even brown. The intensification of discoloration with time is said to be a photochemical process. This is cosmetically unattractive but does not seem to be deleterious except in extreme cases where so much antibiotic is taken up that the structure of bone is mechanically weakened. To avoid this, tetracyclines are not normally given to children once they are forming their permanent set of teeth (ages 6 to 12). Tetracyclines are painful upon intramuscular injection. This has been attributed in part to formation of insoluble calcium complexes. To deal with this, the injectable formulations contain EDTA and are buffered at comparatively acidic pH levels where chelation is less pronounced and water solubility is higher. When concomitant oral therapy with tetracyclines and incompatible metal ions must be done, the ions should be given 1 hour before or 2 hours after the tetracyclines.

Routes of Administration. Tetracyclines can also cause thrombophlebitis upon intravenous injection. The preferred route of administration is oral whereupon they are reasonably well absorbed in the absence of multivalent metal ion rich gut contents.

Stereochemistry and Epimerization. The alpha-stereo-orientation of the C-4 dimethylamino-moiety of the tetra-

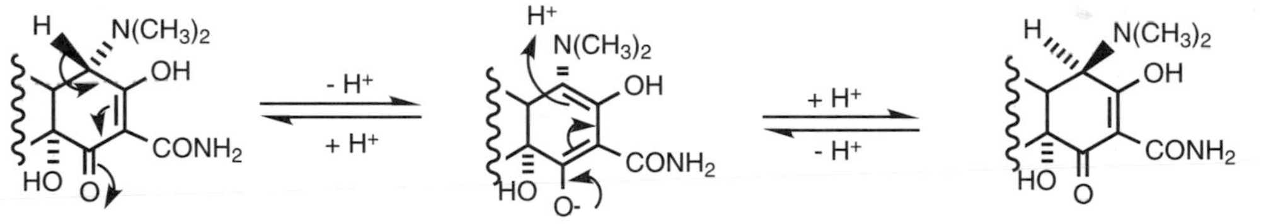

Tetracycline (active) 4-Epitetracycline (inactive)

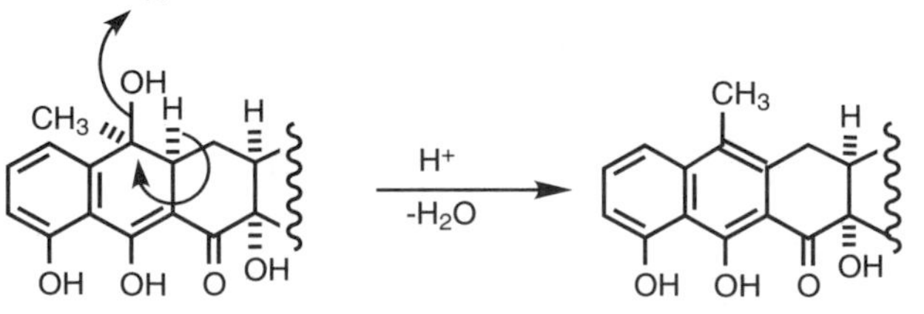

Tetracycline (Active) Anhydrotetracycline (Inactive)

cyclines is essential for their bioactivity. The presence of the tricarbonyl system of ring A allows enolization involving loss of the C-4 hydrogen. Reprotonation can take place from either the top or bottom of the molecule. Reprotonation from the top of the enol regenerates tetracycline. Reprotonation from the bottom, however, produces inactive 4-epi-tetracycline. At equilibrium the mixture consists of nearly equal amounts of the two diasteromers. Thus, old tetracycline preparations can lose approximately half their potency in this way. The custom is thus to overfill the capsules by about 15% during manufacture to allow for a longer shelf-life at or near labeled potency. The epimerization process is most rapid at about pH 4 and is relatively slower in the solid state.

Dehydration. Most of the natural tetracyclines have a tertiary and benzylic hydroxyl group at C-6. This function has the ideal geometry for acid catalyzed dehydration involving the C-5a α-oriented hydrogen (antiperiplanar trans) and the resulting product is a naphthalene derivative, so it has energetic reasons for the reaction proceeding in that direction. C-5a,6-anhydrotetracycline is much deeper in color than tetracycline and is biologically inactive. Discolored old tetracyclines are suspect and should be discarded. Not only can inactive 4-epitetracyclines dehydrate to produce 4-epianhydrotetracyclines, but anhydrotetracycline can epimerize to produce the same product. This degradation product is toxic to the kidneys and produces a Fanconi-like syndrome which, in extreme cases, has been fatal. Commercial samples of tetracyclines are closely monitored for the presence of 4-epianhydrotetracycline and injuries from this cause are now, fortunately, rare. Those tetracyclines, such as minocycline and doxycycline, which have no C-6-hydroxyl groups cannot undergo dehydration and so are completely free of this toxicity.

Tetracycline ⇌ 4-Epitetracycline
Tetracycline → (- H_2O) → Anhydrotetracycline
4-Epitetracycline → (- H_2O) → 4-Epianhydrotetracycline
Anhydrotetracycline ⇌ 4-Epianhydrotetracycline

Cleavage in Base. Another untoward degradation reaction involving a C-6-hydroxyl group is cleavage of the C-ring in alkaline solutions at or above pH 8.5. The lactonic product, an isotetracycline, is inactive. The clinical impact of this degradation under normal conditions is uncertain.

CH_3 H H OH O O OH —OH^-→ H_3C H H O O O OH

Isotetracycline

Photoxicity. Certain tetracyclines, most notably those with a C-7-chlorine, absorb light in the visible region leading to free radical generation and potentially cause severe erythremia to sensitive patients on exposure to strong sunlight. Patients should be advised to be cautious about such exposure for at least their first few doses to avoid potentially severe sunburn. This effect is comparatively rare with most currently popular tetracyclines.

Molecular Model of Action. The tetracyclines of clinical importance interfere with protein biosynthesis at the ribosomal level leading to bacteriostasis. Tetracyclines bind to the 30S subparticle with the possible cooperation of a 50S site by a process that remains imprecisely understood despite intensive study. The more lipophilic tetracyclines, typified by minocycline, are also capable of disrupting membrane function and have bactericidal properties. Tetracyclines enter bacterial cells passively by porin routes through the outer membrane, perhaps assisted by the formation of highly lipophilic calcium and magnesium ion chelates. Deeper passage, however, through the inner cytoplasmic membrane is an energy requiring active process suggesting that the tetracyclines are mistaken by bacteria as food.

Resistance. Resistance seems mainly to result from an unusual ribosomal protection process involving elaboration of bacterial proteins TET(M), TET(O), and TET(Q). These proteins associate with the ribosome, thus allowing protein biosynthesis to proceed even in the presence of bound tetracyclines, although exactly how this works is not understood. Another important resistance mechanism involves R-factor-mediated, energy-requiring, active efflux of magnesium-chelated tetracyclines from cells in exchange for protons. This is

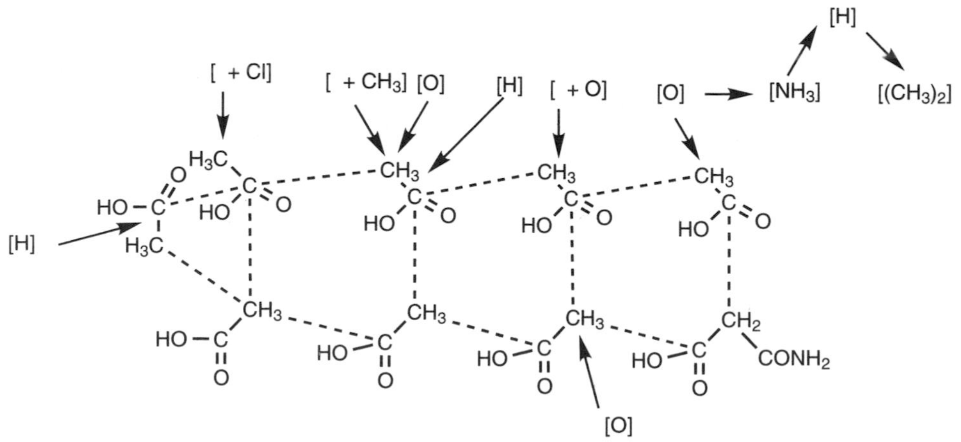

Fig. 34–8. Tetracycline biosynthesis.

particularly prominent in gram-negative cells. Certain other microbes, such as Mycoplasma and Neisseria seem to have modified enzymes which either accumulate fewer tetracyclines or those tetracyclines have defective passage through the porins. Because resistance is now widespread, these once extremely popular antibiotics are falling into comparative disuse. The tetracyclines distinguish imperfectly between the bacterial 70S ribosomes and the mammalian 80S ribosomes so, in high doses, or in special situations, i.e., intravenous use in pregnancy, these drugs demonstrate a significant antianabolic effect. This can lead to severe liver and kidney damage so tetracyclines are not administered in these situations. Diuretics can enhance tetracycline-associated azotemia so they are contraindicated as are inducers of metabolism such as hydantoins, carbamazepine, and barbiturates.

Biosynthesis. The biosynthesis of the tetracyclines (Fig. 34–8) proceeds by a complex sequence of transformations involving condensation of malonamoyl coenzyme A and eight acetate units in a process quite analogous to fatty acid biosynthesis, except that most of the carbonyl groups remain in the molecule and self condense in a controlled way to produce the partially reduced naphthacene nucleus. A sequence of reductions, oxidations, methylations, aminations, and dehydrations completes the biosynthesis.

Antimicrobial Spectra

The tetracyclines possess wide bacteriostatic activity. Because of resistance, and the comparative frequency of troublesome side effects, they are rarely the drugs of first choice. Nonetheless they remain popular for community health office use against susceptible microbes. The differences between the antimicrobial spectra of various tetracyclines are not large. They are popular for low dose oral therapy for acne, first-course community-acquired urinary tract infections (largely due to E. coli), brucellosis, borrelosis, upper respiratory tract infections, ophthalmic infections, sexually transmitted diseases, rickettsial infections, mycoplasmal pneumonia, prophylaxis for malaria, prevention of traveler's diarrhea, cholera, and many other less common problems. The tetracyclines are also used in tonnage quantities for agricultural use often in the form of feed supplements where it can be demonstrated that animals reach market weight more quickly and economically with their use.

Individual Agents

Tetracycline is produced by fermentation of Streptomyces aureofaciens, or by catalytic reduction of chlortetracycline. It is classical, typical, generic, and comparatively cheap. The blood levels achieved upon oral administration are often irregular. Demeclocycline lacks the C-6-methyl of tetracycline and is produced by fermentation of a genetically altered strain of S. aureofaciens. It is, being a secondary alcohol, more chemically stable than tetracycline and has served as the source of numerous semisynthetic analogs as well as enjoying its own clinical vogue. One important antibiotic produced from demeclocycline is minocycline. The C-7-chloro and the C-6-hydroxy groups can both be removed by catalytic reduction to produce sancycline, the chemically simplest tetracycline with characteristic tetracycline properties. Sancycline has not, however, become an important antibiotic. Nitration of sancycline is possible because of its great chemical stability and is followed by separation of the C-7 and the useless C-9 nitro analogs produced in the reaction. Reductive amination of the C-7 nitro analog with formaldehyde produces minocycline. Minocycline is much more lipophilic than its precursors, gives excellent blood levels following oral administration, and can be given once a day. It is less dependent on active uptake mechanisms and has a somewhat broader antimicrobial spectrum. It is also, apparently, less painful upon intramuscular or intravenous injection. It has, however, vestibular toxicities (e.g., vertigo, ataxia, and nausea) not shared by other tetracyclines. It is particularly broad spectrum with emphasis on important gram-positive pathogens, both staphylococci and streptococci, and so has become a very popular drug in North America. Oxytetracycline is also one of the classic tetracyclines being produced by fermentation of Streptomyces rimosus and other soil microorganisms. The most hydrophilic tetracycline on the market, it has largely now been replaced by its semisynthetic descendants. Methacycline is produced from oxytetracycline by an ingenious process involving blocking of C-11a by a halogen atom, dehydrating on an exocyclic course, and thereby carefully removing the blocking halogen. Although used infrequently in North America, it continues to serve as the chemical precursor, through further transformations including catalytic reduction processes, of doxycycline, one of the most important of the current tetracyclines. The direct reduction process has the capability of producing two isomers, so that the isomer with the same stereochemistry as the naturally occurring tetracyclines (alpha-oriented methyl) is significantly more potent. Doxycycline is well absorbed on oral administration, has a half-life permitting once a day dosage for mild infections, and is excreted partly in the feces and partly in the urine. Because it causes fewer gastrointestinal disturbances and cannot participate in degradation processes involving a C-6-hydroxyl group, it is the tetracycline of choice for many physicians.

Reserve or Special-Purpose Antibiotics

This group of antibiotics consists of a miscellaneous collection of structural types whose toxicities or narrow range of applicability give them a more specialized place in antimicrobial chemotherapy than those covered to this point. They are reserved for special purposes.

Chloramphenicol. Chloramphenicol was originally produced by fermentation of Streptomyces venezuelae but its comparatively simple chemical structure soon resulted in several efficient total syntheses. With two asymmetric centers it is one of four diastereomers only one of which (1R,2R) is active. As total synthesis produces a mixture of all four, the unwanted isomers must be removed before use. Chloramphenicol is a neutral substance only moderately soluble in water as both nitrogen atoms are non-basic under physiologic conditions (one is an amide and the other a nitro moiety). It was the first broad-spectrum oral antibiotic used in the United States (1947) and was once very popular. Severe potential blood dyscrasia has greatly decreased its use in North America, although its cheapness and efficacy makes it still very popular in much of the rest of the world. When given orally, it is rapidly and completely absorbed but has a fairly short half life. It is mainly excreted in the urine in the form of its metabolites, which are a C-3 glucuronide, and to a lesser extent, its deamidation product and the product of dehalogenation and reduction. These are all inactive. The aromatic nitro group is also reduced metabolically and this product can also undergo amide hydrolysis. The reduction of the nitro group, however, does not take place efficiently in humans per se but rather primarily occurs in the gut by the action of the normal flora. Chloramphenicol is also available for parenteral use. It is about 60% serum protein bound and diffuses well into tissues, especially into inflamed cerebrospinal fluid and is, therefore, of special value in meningitis. It also penetrates well into lymph and mesenteric ganglions rationalizing its particular value in typhoid and paratyphoid fever. Chloramphenicol is bacteriostatic by virtue of inhibition of protein biosynthesis in both the host and to a lesser extent in the bacterial ribosomes. Resistance is mediated by several R-factor enzymes that catalyze acetylation of the secondary and, to some extent, the primary hydroxyl groups in the aliphatic side chain. These products no longer bind to the ribosomes. Escherichia coli is frequently resistant due to chloramphenicol's lack of intercellular accumulation.

Its toxicities prevent chloramphenicol from being more widely used. Blood dyscrasias are seen in patients predisposed to them. The more serious form is a pancytopenia of the blood that is fatal in about 70% of cases and is believed caused by one of the reduction products of the aromatic nitro group. This side effect is known as aplastic anemia, and has even occurred following use of the drug as an ophthalmic ointment. There seems to be a genetic predisposition towards this in a very small percentage of the general population. This devastating side effect is estimated to occur once in every 25,000 to 40,000 courses of therapy. Less severe, but much more common, is a reversible inhibition of hematopoiesis, seen in aged patients or in those with renal insufficiency. If cell counts are taken, this can be controlled because it is dose-related and marrow function will recover if the drug is withdrawn.

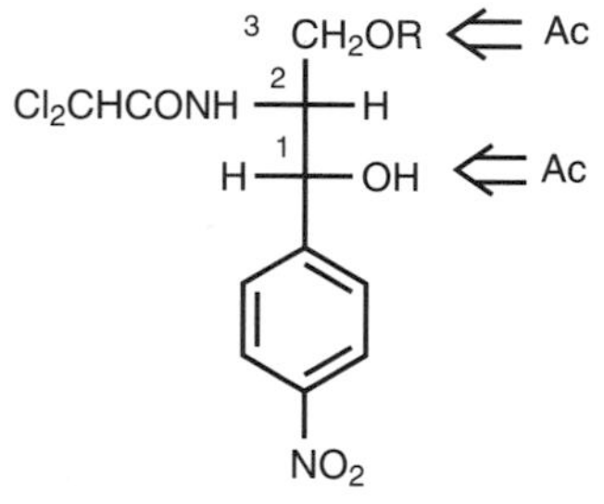

The so-called gray syndrome, a form of cardiovascular collapse, is encountered when chloramphenicol is given carelessly in the first 48 hours of life when liver glucuronidation is undeveloped and successive doses will lead to

3 CH_2OR ⇐ Ac
$Cl_2CHCONH$ — 2 — H
H — 1 — OH ⇐ Ac
NO_2

Chloramphenicol	R = H
Chloramphenicol hemisuccinate	$R = COCH_2CH_2CO_2H$
Chloramphenicol palmitate	$R = CH_3(CH_2)_{12}CO$

rapid accumulation of the drug due to impaired excretion. A dose-related profound anemia accompanied by an ashen gray pallor is seen, as are vomiting, loss of appetite, and cyanosis. Deaths have resulted, often involving cardiovascular collapse.

Two prodrug forms of chloramphenicol are available. The drug is intensively bitter. This can be masked for use as a pediatric oral suspension by use of the C-3 palmitate, which is cleared in the duodenum to liberate the drug. Its poor water solubility is largely overcome by conversion to the C-3 hemisuccinoyl ester, which forms a water-soluble sodium salt. This is cleaved in the body to produce active chloramphenicol. Because cleavage in muscles is too slow, this prodrug is used intravenously rather than intramuscularly.

Chloramphenicol potentiates the activity of some other drugs by inducing liver metabolism. Such agents include anticoagulant coumarins, sulfonamides, oral hypoglycemics, and phenytoin.

Despite potentially serious limitations, chloramphenicol is an excellent drug when used carefully. Its special value is in typhoid and paratyphoid fevers, Haemophilus infections (especially meningitis and epiglottitis), pneumococcal and meningococcal meningitis in beta-lactam allergic patients, anaerobic infections (especially by Bacteroides), rickettsial infections, and so on.

Safer antibiotics should be used whenever possible.

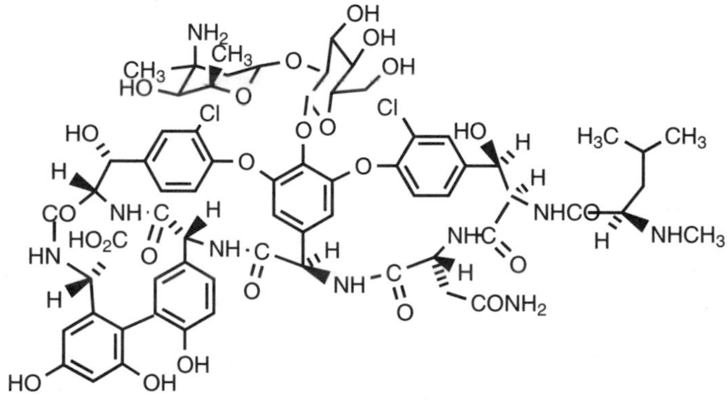

Vancomycin. Vancomycin is produced by fermentation of Amycolatopsis orientalis and is the only one of about 200 known glycopeptide antibiotics that is in clinical use. It has been available for about 35 years, but its popularity has increased significantly only in the last decade. Its useful spectrum is restricted to gram-positive pathogens with particular utility against multiple resistant coagulase negative staphylococci and methicillin-resistant Staphylococcus aureus causing septicemias, endocarditis, skin and soft tissue infections, and infections associated with venous catheters. It is used orally against Clostridium difficile where its use can be life-saving. It is usually bactericidal. It does not give useful blood levels following oral administration and is very irritating on intravenous injection. To prevent thrombophlebitis it is slowly instilled rather than pushed. It is only very recently, despite decades of intensive use, that some vancomycin-resistant bacteria are emerging. The mechanism of resistance appears to be alteration of the D-ala-D-ala units on the peptidoglycan precursors to D-ala-D-lactate. Vancomycin has little affinity for the new depsipeptide linkage. Chemically, vancomycin has a glycosylated hexapeptide chain rich in unusual amino acids, many of which contain aromatic rings that are halogenated and cross linked by aryl ethers bonds into a rigid molecular framework. The binding site for its target is a peptide-lined cleft having high affinity for acetyl-D-ala-D-ala and related peptides. Thus it is a specific peptide receptor and attacks bacterial cell wall biosynthesis at the same step as the beta-lactams but with a different mechanism. By covering the substrate for cell wall transamidase, it prevents cross linking resulting in osmotically defective cell walls. Manufactured vancomycin is a mixture of related substances of which vancomycin B predominates. In addition to the danger of thrombophlebitis accompanying rapid administration, higher doses of vancomycin can cause nephrotoxicity and auditory nerve damage. A significant drug rash (the so-called red man syndrome) can occur, mediated by histamine release.

Vancomycin

Cyclic Peptide Group. The usual physiologically significant peptides are linear. Several bacterial species, however, produce antibiotic mixtures of cyclic peptides, some of the constituent amino acids of which have D absolute stereochemistry. These cyclic substances often have a pendant fatty acid chain as well. One of the consequences of this unusual architecture is that these agents are not readily metabolized. These drugs are usually water soluble and are highly lethal to susceptible bacteria as they attach themselves to the bacterial membranes and interfere with their semipermeability so that essential metabolites leak out and undesirable substances pass in. Unfortunately they are also highly toxic in humans, so their use is reserved for serious situations with few alternatives. Bacteria are rarely able to develop significant resistance to this group of antibiotics. They are generally unstable so solutions should be protected from heat, light, and extremes of pH. Bacitracin is a mixture of similar peptides produced by fermentation of the bacterium Bacillus subtilis. The A component predominates. Bacitracin got its name from the genus of the producing organism and the family name of the first patient to be treated with it, who

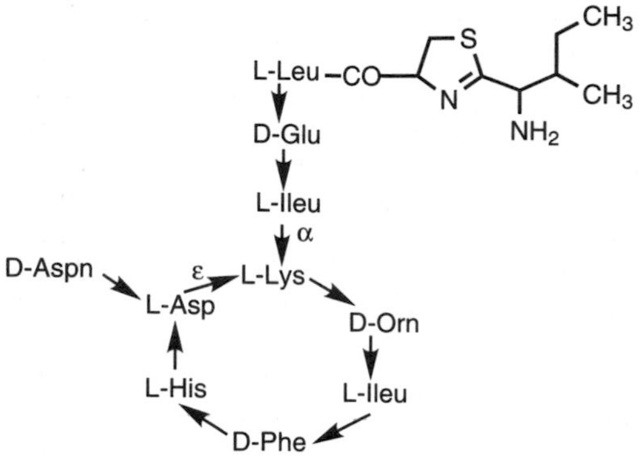

Bacitracin A

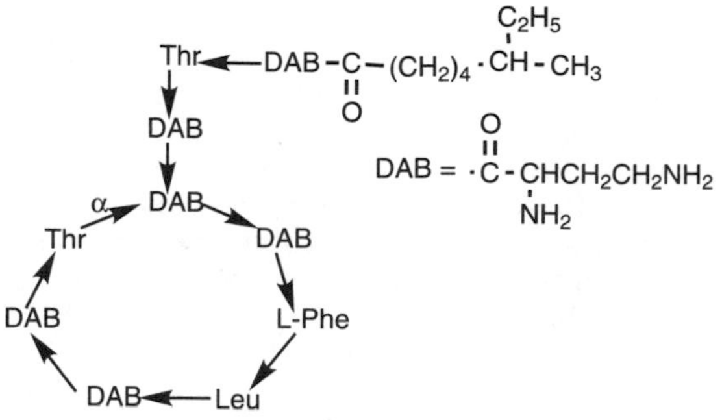

Polymyxin B

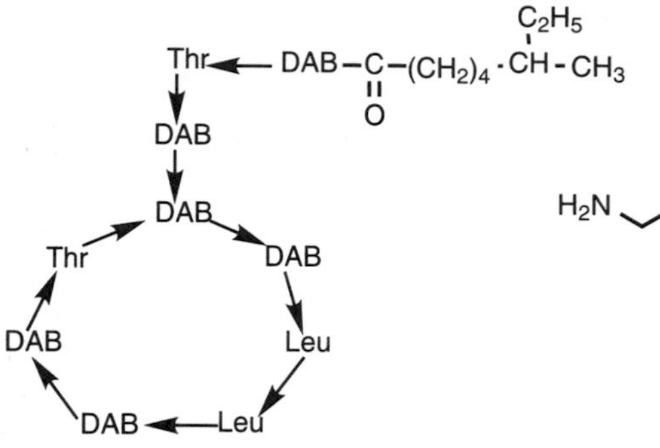

Colistin A

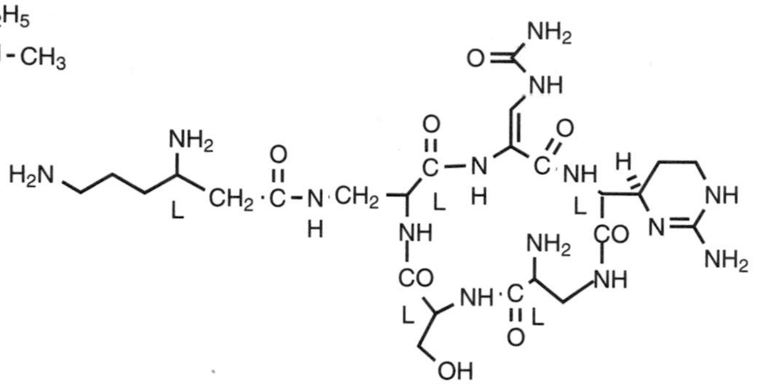

Capreomycin 1A

was a little girl, named Tracy. It is predominantly active against gram-positive microorganisms and is used intramuscularly against staphylococci resistant to other agents. It is also used orally for enteropathogenic diarrhea and, especially, against Clostridium difficile. It can also be used for preoperative bowel sanitization. It is rather neuro- and nephrotoxic so is employed with caution. Zinc(II) ion enhances the activity of bacitracin. Its mode of action is to inhibit both peptidoglycan biosynthesis and membrane function. Polymyxin B sulfate is produced by fermentation of Bacillus polymyxa. It is separated from a mixture of related cyclic peptides and is primarily active against gram-negative microorganisms. It apparently binds to phosphate groups in bacterial cytoplasmic membranes and disrupts their integrity. It is used intramuscularly or intravenously to treat serious urinary tract infections, meningitis, and septicemia primarily caused by Pseudomonas aeruginosa but some other gram-negatives will also respond. It is also used orally to treat enteropathogenic E. coli and Shigella sp. diarrheas. Irrigation of the urinary bladder with solutions of polymyxin B sulfate is also employed by some to reduce the incidence of infections subsequent to installation of indwelling catheters. When given parenterally, the drug is quite neuro- and nephrotoxic, so it is employed only after other drugs have failed. Colistin sulfate is a cyclic polypeptide drug produced by Bacillus polymyxa var. colistinus. The A component is the primary article of commerce. It is bactericidal primarily to gram-negative bacteria following destruction of the integrity of their cytoplasmic membranes. Resistance development is rare. It is used sparingly against severe gram-negative caused infections because of its nephro- and neurotoxicity. It is also used orally against diarrhea caused by E. coli and Shigella sp. Capreomycin is somewhat related structurally. It is produced by fermentation of Streptomyces capreolus as a mixture of about four components. It is bacteriostatic against certain mycobacterial strains including M. tuberculosis, M. bovis, M. kansasii, and M. avium by an unknown mechanism. Resistance develops, often during the course of therapy, also by an unknown mechanism. The drug must be employed with caution because of its oto- and nephrotoxicity.

Novobiocin is a coumarmycin- (coumarin-containing) family antibiotic produced by fermentation of Streptomyces niveus. It inhibits the function of DNA gyrase by binding to a different subunit than the fluoroquinolones with

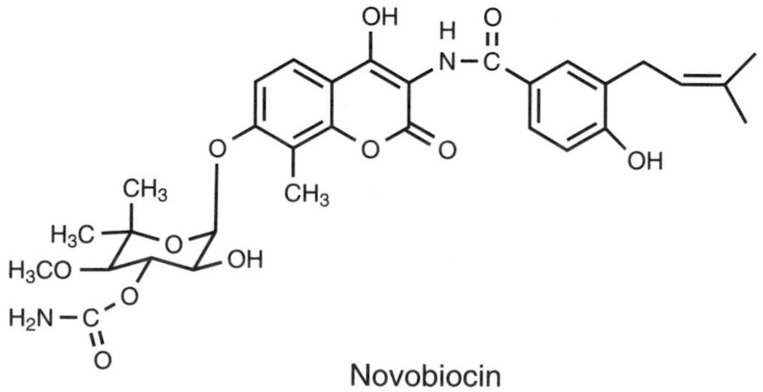

Novobiocin

which it is synergistic. Novobiocin binding interferes with ATP metabolism, which otherwise provides the energy needed for the conformational work of the enzyme. It is orally active and has a fairly broad range of activity against gram-positives and some gram-negatives but common side effects (rash, blood dyscrasias, and liver damage) and resistance emergence have decreased its utilization significantly. The 4-hydroxy coumarin moiety is vinylogously equivalent to a carboxy group so novobiocin readily forms a sodium salt.

ANTIMYCOBACTERIAL DRUGS

Next is a heterogeneous group of agents that, like streptomycin and capreomycin, are used in the treatment of tuberculosis. Some are antibiotics and others antimicrobials. Tuberculosis was one of the most lethal in infections, killing millions until the advent of antibiotics. Many famous people (Frédéric Chopin, Percy Shelley, Friedrich Schiller, George Orwell, Vivien Leigh, John Keats, Anton Chekhov, among others) fell victim to it before suitable therapy was found. Casual behavior and the emergence of AIDS patients as a particularly susceptible reservoir of people susceptible to the disease has changed all that. Tuberculosis and nontubercular infections by a variety of less common mycobacteria are rapidly gaining ground and appear to be a dramatically accelerating public health threat. Treatment of tuberculosis is long and recovery from it is slow. Further, the organism becomes resistant to antimicrobials during the course of therapy, in part because patients with the disease comply poorly with the physician's instructions. This is partially due to the unpleasant characteristics of many of the antimycobacterial drugs. Aggressive drug therapy, prophylactic treatment of patient contacts (the modern equivalent of quarantine), the rotation of drugs, and the use of combination therapy are among present day treatments. Para-aminosalicylic acid (PAS) was discovered as a synthetic antagonist of aspirin. (See Chapter 33 for structures of antimycobacterial agents.) Aspirin seems to stimulate the growth of mycobacteria prompting the search resulting in PAS. PAS is bacteriostatic and works by the same general mechanism as the sulfonamides. Its clinical spectrum is essentially restricted to tuberculosis. It causes significant gastric distress, which limits patient acceptance. Cycloserine is an antibiotic produced by fermentation of Streptomyces orchidaceus, Sm. garyphalus or by synthesis. It is a rigid cyclic analog of D-serine, which interferes with the formation and use of D-alanine from L-alanine in bacterial cell wall biosynthesis. It is effective as a component of oral combinations against M. tuberculosis and M. bovis. Its most significant side effects involve the central nervous system. Ethambutol is a synthetic agent whose bacteriostatic activity against M. tuberculosis, M. bovis, and M. marinum is mediated through an unknown mechanism. Its most serious side effect is optic neuritis. Isoniazide, INH, a synthetic agent, is one of the principal antimycobacterial agents. Its mode of action is not known but its resistance mechanism appears to involve lack of cellular uptake. Peripheral neuritis is its most common side effect, although a fatal hepatitis can also occur. It is often given in fixed ratio combinations with rifampin. Ethionamide is a synthetic structural analog of isonicotinic acid hydrazide possessing bacteriostatic or bactericidal activity, depending on the dose, against M. tuberculosis, M. bovis, and M. kansasii. It is also useful against leprosy. Gastrointestinal upset is its principal deleterious side effect. Pyrazinamide is a synthetic antitubercular whose mode of action is not known. It undergoes metabolic hydrolysis to pyrazinoic acid which is itself very active. Rifampin is an ansamacrolide antibiotic produced semisynthetically from rifamycin SV, a natural antibiotic derived by fermentation of Streptomyces mediterranei. Having a hydroquinone moiety in its structure, rifampin is easily oxidized to reddish products. It is excreted primarily in bile so that feces are tinted orange to red, but is also excreted in the urine, sweat, sputum, and tears all of which can also turn reddish. This is quite disturbing unless the patient is forewarned. Rifampin inhibits DNA-dependent RNA-polymerase and is a fairly broad spectrum agent against bacteria. In addition to many gram-positives and gram-negatives, rifampin also is active against many mycobacteria including M. tuberculosis, bovis, marinum, kansasii, fortuitum, avium, intercellulare, and leprae. Resistance is unfortunately fairly common and involves modification of the target enzyme. Rifampin is normally used in combination with other agents and is used prophylactically for those in contact with infected individuals. It is also used in the treatment of bacterial meningitis. Clofazimine is a synthetic phenazine dye, active orally against many mycobacteria (M. tuberculosis, leprae, marinum, avium-intercellulare, bovis, chelonae, fortuitum, kansasii, scrofulaceum, simiae, and ulcerans). Its mode of action is unknown although it is known to have a particular affinity to mycobacterial DNA. Gastrointestinal upset is experienced as well as discoloring red spots of skin and conjunctiva. The compound itself is a red phenazine dye.

SUGGESTED READINGS

A. Albert, *Selective Toxicity, 6th ed.,* New York, Chapman and Hall (Wiley), 1979.

Association Française les Enseignants de Chemie Thérapeutique, *Medicaments Antibiotiques, Vol. 2,* Paris, Tec & Doc Lavoisier, 1992.

K. Bartmann, *Antitubercular Drugs, Handbook of Experimental Pharmacology, Vol. 84,* New York, Springer, 1988.

M. Burnet and D. O. White, *Natural History of Infectious Disease, 4th ed.,* New York, Cambridge University Press, 1975.

P. de Kruif, *Microbe Hunters,* New York, Pocket Books, 1965 (first published 1926).

A. L. Demain and N. A. Solomon, Eds., *Antibiotics Containing the Beta-lactam Structure, Vols. 1 and 2, Handbook of Experimental Pharmacology, Vol. 67,* New York, Springer, 1983.

E. F. Gale, et al., Eds., *The Molecular Basis of Antibiotic Action, 2nd ed.,* New York, Wiley, 1981.

J. J. Hlavka and J. H. Boothe, *The Tetracyclines, Handbook of Experimental Pharmacology, Vol. 78,* New York, Springer, 1985.

A. Kucers and N. McK. Bennet, *The Use of Antibiotics, 4th ed.,* Philadelphia, Lippincott, 1987.

G. Lukacs and M. Ohno, Eds., *Recent Progress in the Chemical Synthesis of Antibiotics,* New York, Springer, 1990.

G. L. Mandell, et al., *Principles and Practice of Infectious Disease, 2nd ed.,* New York, Wiley, 1985.

The Medical Letter on Drugs and Therapeutics, *Handbook of Antimicrobial Therapy,* New Rochelle, N.Y., The Medical Letter, 1993.

L. A. Mitscher, *The Chemistry of Tetracycline Antibiotics,* New York, Marcel Dekker, 1978.

L. A. Mitscher, et al., *Antibiotic and Antimicrobial Drugs,* in D. F. Smith, Ed., *Handbook of Stereoisomers: Therapeutic Drugs,* Boca Raton, FL, CRC Press, 1989.

R. B. Morin and M. Gorman Eds., *Chemistry and Biology of Beta-lactam Antibiotics, Volumes 1–3,* New York, Academic Press, 1982.

S. Omura, Ed., *Macrolide Antibiotics,* New York, Academic Press, 1984.

D. Perlman, Ed., *Structure Activity Relationships among the Semisynthetic Antibiotics,* New York, Academic Press, 1977.

M. Plempel and H. Otten, *Walter/Heilmeyers's Antibiotika Fibel: Antibiotika und Chemotherapie, 5th Ed.,* Stuttgart, Georg Thieme, 1982.

W. B. Pratt, *Fundamentals of Chemotherapy,* New York, Oxford University Press, 1973.

T. Rosebury, *Microbes and Morals,* New York, Ballantine Books, 1976.

F. Ryan, *The Forgotten Plague: How the Battle Against Tuberculosis was Won—and Lost,* Boston, Little, Brown, 1993.

J. C. Sheehan, *The Enchanted Ring: The Untold Story of Penicillin,* Cambridge, MA, MIT Press, 1982.

G. W. Stewart, *The Penicillin Group of Drugs,* Amsterdam, Elsevier, 1965.

J. Sutcliffe and N. H. Georgopapadakou, Eds., *Emerging Targets in Antibacterial and Antifungal Chemotherapy,* New York, Chapman and Hall, 1992.

H. Umezawa and I. R. Hooper, Eds., *Aminoglycoside Antibiotics, Handbook of Experimental Pharmacology, Vol. 62,* New York, Springer, 1982.

M. Verderame, Ed., *Handbook of Chemotherapeutic Agents, Vols. 1 and 2,* Boca Raton, FL, CRC Press, 1986.

A. Whelton and H. C. Neu, Eds., *The Aminoglycosides,* New York, Marcel Dekker, 1982.

J. S. Wolfson and D. C. Hooper, Eds., *Quinolone Antimicrobial Agents, 2nd Edition.,* Washington, D.C., American Society for Microbiology, 1993.

H. Zinsser, *Rats, Lice and History,* Boston, Little, Brown, 1950.

Chapter 35

ANTIFUNGAL AGENTS

Eugene D. Weinberg

PATHOGENICITY OF FUNGUS

Fungi are plant-like, nonphotosynthetic eukaryotes growing either in colonies of single cells (yeasts) or in filamentous multicellular aggregates (molds). Most fungi live as saprophytes in soil or on dead plant material and are important in the mineralization of organic matter. Unfortunately, some species are parasites of terrestrial plants and can cause serious crop damage. A smaller number produce disease in humans and animals. Mycotic illnesses in humans are divided into three groups: contagious skin and hair infections; noncontagious soilborne or airborne systemic infections; and noncontagious foodborne toxemias. The responsible organisms and methods of prevention and treatment differ with each group.

The organisms that cause contagious skin and hair infections are specialized saprophytes that digest keratin in soil as well as in skin. A new host usually becomes infected by contact with a host or object previously contaminated by the previous host rather than with soil. The lesions caused by these filamentous dermatophytes generally are superficial and small. Nevertheless, in the United States, approximately 10% of the population is infected; in tropical areas of the world, the dermatophytoses pose an even greater medical problem.

Yeasts and molds are fungi that cause noncontagious soilborne or airborne infections. In some cases, individual organisms can shift between the yeast and mold (mycelial) phases according to the environment and nutritional milieu. With most infections, e.g., histoplasmosis, the yeast form is the pathogen, whereas the mold form is found in the soil environment. In candidiasis, however, both the yeast and mycelial forms can be present in invaded tissues.

Most fungi that cause systemic infections are soil saprophytes that accidentally invade human hosts through skin inoculations or inhalation of soil debris. Exceptions are strains of *Candida,* which are part of the normal flora of the human gastrointestinal (GI) tract and vagina. Candidal cells can cause minor lesions of the skin and mucous membranes and, in immunocompromised individuals, serious systemic disease.

A high percentage of humans in certain geographic areas are infected by various soilborne or airborne fungi, but few healthy persons either develop serious diseases or die. Individuals who do become seriously ill often have been weakened by immunosuppressive treatment for other clinical conditions or have a lowered cellular immune response due, for example, to pregnancy,[1] congenital thymic defects, or acquired immune deficiency syndrome (AIDS).[2–4] Fungal infection has become a major cause of death in patients with cancer and transplant recipients.[5] Clinical manifestations of diseases caused by soilborne or airborne fungi include chronic granulomas with necrosis and abscess formation. The principal mycotic infectious diseases of humans are listed in Table 35–1.

The molds that cause noncontagious, food-borne toxemias grow saprophytically and produce toxins on harvested crops stored in damp environments. Clinical symptoms in humans after ingestion or through other contact with the low molecular mass toxins include blood dyscrasias, hepatic cirrhosis and carcinoma, hallucinations, and dermatitis. Control is directed toward suppression of fungal growth in foods and includes prevention of mechanical damage during harvesting as well as reduction of the moisture content of susceptible crops during storage.

Currently, there are no clinically available immunizing agents nor useful antisera for either mycotoxicoses or mycotic infections. Moreover, because of the widespread prevalence and the soilborne or airborne transmission of the systemic fungal pathogens, sanitary methods are insufficient to eradicate the diseases caused by these microorganisms. Inasmuch as the eukaryotic anatomic and metabolic nature of fungal cells is similar to that of animal cells, it has been difficult to devise therapeutic strategies specifically against the pathogens and yet nontoxic to the host. As indicated in this chapter, some progress has been made. Nevertheless, the search for new and improved antifungal agents continues.

DEMONSTRATION OF ANTIFUNGAL ACTIVITY[6,7]

The in vitro and in vivo methods used for detection of antifungal potency are similar to those used in antibacterial screening. As with bacteria, it is easy to discover several synthetic and natural compounds that, in small quantity, can retard or prevent growth of fungi in culture media. Many of these substances, however, are disappointing in vivo. Some have insufficiently selective toxicity; others are unable to reach the site of infection, and still others metabolize too rapidly. Moreover, few pathogenic fungi grow in culture in exactly the same biologic form in which they multiply in vivo. Some compounds, e.g., fluconazole, may be quite useful in therapy but show little in vitro activity. Animal models are instrumental in distinguishing between fungicidal and fungistatic com-

Table 35–1. Important Mycotic Infectious Diseases in Humans

Disease	Etiologic Agents	Principal Tissues Affected	Potentially Useful Anti-Infectives
Contagious, nonsystemic infections			Undecylenic acid, tolnaftate, naftifine, terbinafine, griseofulvin, haloprogin, clotrimazole, miconazole, itraconazole, econazole, oxiconazole, sulconazole, ciclopirox
Dermatophytoses (ringworm/tinea)	Epidermophyton, Microsporum, and Trichophyton spp.	Skin, hair, nails	
Noncontagious, systemic infections			
Aspergillosis	Aspergillus spp.	External ear, lungs, eye, brain	Amphotericin B + 5-fluorocytosine, itraconazole
Blastomycosis	Blastomyces dermatitidis	Lungs, skin, bone, testes	Amphotericin B, itraconazole
Candidiasis	Candida spp.	Respiratory, gastrointestinal, and urogenital tracts; skin	Amphotericin B, nystatin, miconazole, clotrimazole, ketoconazole, fluconazole, econazole, butoconazole, terconazole, ciclopirox
Chromomycosis	Cladosporium and Phialophora spp.	Skin	Itraconazole
Coccidioidomycosis	Coccidioides immitis	Lungs, skin, joints, meninges	Amphotericin B, ketoconazole, fluconazole, itraconazole
Cryptococcosis	Cryptococcus neoformans	Lungs, meninges	Amphotericin B $\pm$ 5-fluorocytosine, fluconazole
Histoplasmosis	Histoplasma capsulatum	Lungs, spleen, liver, adrenals, lymph nodes	Amphotericin B, itraconazole
Mucormycosis	Absidia, Mucor, and Rhizopus spp.	Nasal mucosa, lungs, blood vessels, brain	Amphotericin B
Paracoccidioidomycosis	Paracoccidioides brasiliensis	Skin, nasal mucosa, lungs, liver, adrenals, lymph nodes	Ketoconazole, itraconazole
Pneumocystosis	Pneumocystis carinii	Lungs	Trimethoprim + sulphamethoxazole, pentamidine
Pseudallescheriasis	Pseudallescheria boydii	External ear, lungs, eye	Amphotericin B, miconazole
Sporotrichosis	Sporothrix schenkii	Skin, joints, lungs	Potassium iodide, amphotericin B, itraconazole

* Molecular methods of taxonomy indicate that Pneumocystis spp. are fungi rather than protozoa. The drugs generally employed in prophylaxis and therapy, however, are anti-protozoan; they are described in Chapters 32 and 34.

Table 35–2. Mechanisms of Action of Antifungal Agents

Class	Selected Examples	Mechanism of Action
Thiocarbamate	tolnaftate	Suppression of fungal squalene epoxidase to result in accumulation of squalene and decreased ergosterol
Allylamine	naftifine, terbinifine	
Benzofuran cyclohexene	griseofulvin	Binding to fungal RNA to cause malformation of spindle and cytoplasmic microtubules
Polyene	amphotericin B, nystatin	Binding to ergosterol in fungal cell membranes to result in membrane disorganization
Pyrimidine	5-fluorocytosine	Deamination by fungal cells to 5-fluorouracil which is incorporated into RNA in place of uracil or is converted to 5-fluorouracil-2′-deoxyuridylic acid which inhibits thymidine synthetase
Azole	clotrimazole, miconazole, ketoconazole, fluconazole, itraconazole, terconazole	Inhibition of cytochrome P-450 that catalyzes 14α-demethylation of lanosterol to ergosterol, accumulation of 14-methylated sterols cause permeability disruption

pounds and between a drug's efficacy in normal and immunosuppressed hosts. Mechanisms of action of commonly used antifungal compounds are summarized in Table 35–2.

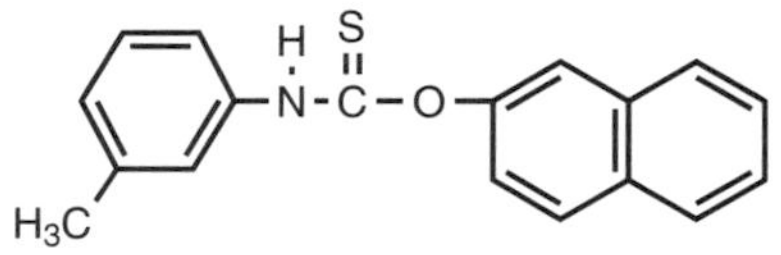

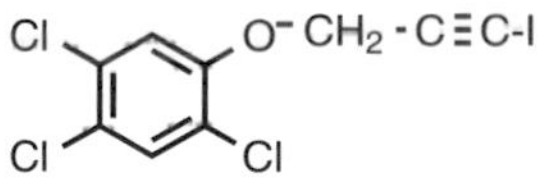

Haloprogin

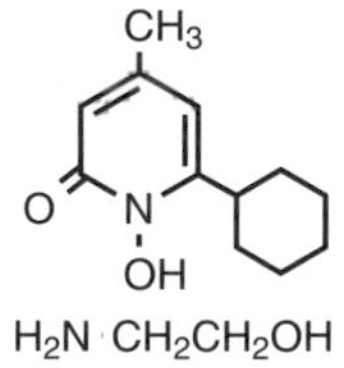

Ciclopirox olamine

$CH_2{=}CH{-}(CH_2)_8{-}COOH$

Undecylenic acid

TOPICAL ANTIFUNGAL AGENTS

Many synthetic compounds, generally incorporated in tinctures or ointments, have been developed since 1900 for topical use in dermatophytoses. Among those available early in this century are quinolines such as 5-chloro-7-iodo-8-quinolinol, salicylates such as N-phenylsalicylanilide, aniline dyes such as methylrosaniline chloride, and organic acids such as propionic, capryl, and 10-undecylenic acids and their salts. In recent years, many additional antidermatophytic drugs have been developed, as listed in Table 35–3.

Tolnaftate

Tolnaftate (0-2-naphthyl m,N-dimethylthiocarbanilate) has a high potency against dermatophytes, both in vitro and in vivo, but, because of poor permeability, is inactivate against most other fungi.[2] Like other thiocarbamates, tolnaftate inhibits squalene epoxidase resulting in increased squalene and concomitantly decreased ergosterol synthesis by the susceptible fungi. Activity is retained if the tolyl group is replaced by an α-naphthyl-β-methyl substituent. Other naphthalene isomer combinations are inert. Activity is lost if the methyl moiety in tolnaftate is replaced by a halogen, carboxylic acid, or nitro group. Potency is retained if hydrogen, hydroxy, or methoxy substituents are used. The carbamate methyl group is essential for activity.

Tolnaftate

Allylamines

Fungal squalene epoxidase is also inhibited by allylamines such as naftifine and terbinafine.[8] Topical ointments of these synthetic drugs are useful in dermatophytic fungal infections. Inasmuch as terbinafine has little affinity for mammalian squalene epoxidase, it is well tolerated by most patients when administered orally. The drug rapidly accumulates in sebum, maintains high concentrations in the stratum corneum and hair, and diffuses into nail plate. In a variety of dermatophytic infections, its efficacy has compared favorably with that of griseofulvin.

Griseofulvin

One of the few natural products useful against dermatophytes is griseofulvin. First obtained from Penicil-

Table 35–3. Antifungal Drugs

Chemical Class	Generic Name	Trade Name	Dosage Form and Dosage
Thiocarbamate	Tolnaftate	Tinactin (Schering)	Topical, 1%
		Aftate (Plough)	
Organic Acid	Undecylenate (salts)	Desenex (Zinc) (Pharmacraft)	Topical, 2–20%
		Cruex (Pharmacraft) (Calcium)	
Allylamine	Naftifine	Naftin (Herbert)	Topical, 1%
	Terbinafine	Lamisil (Sandoz)	Topical, 1%
Benzofuran	Griseofulvin	Fulvicin (Schering); Grifulvin (Ortho, Derm); Grisactin (Wyeth-Ayerst)	Tablet, microsize 500–1000 mg; ultramicrosize 330–375 mg
Polyene	Amphotericin B	Fungizone (Squibb)	IV, 0.25–1.0 mg/kg
	Nystatin	Mycostatin (Squibb)	Topical, 100,000 units
		Nilstat (Lederle)	Oral suspension, 100,000–1 million units/ml
		Nystex (Savage)	
	Natamycin	Natacyn (Alcon)	Topical
Pyrimidine	5-Fluorocytosine	Ancobon (Roche)	Capsule, 50–150 mg/kg/day
Azole Imidazole	Butoconazole	Femstat (Syntex)	Topical, 2%
	Clotrimazole	Gyne-Lotrimin (Schering)	Vaginal tablet, 500 mg
		Lotrimin (Schering)	Topical, 1%
		Mycelex (Miles Pharm.)	Topical, 1%
	Econazole	Spectazole (Ortho)	Topical, 1%
	Ketoconazole	Nizoral (Janssen)	Tablet, 200–400 mg/day
		Nizoral Cream (Janssen)	Topical, 2%
	Miconazole	Micatin (Advanced Care)	Topical, 2%
		Monistat IV (Janssen)	IV, 600 mg–3.6 g/day
	Oxiconazole	Oxistat (Glaxo Derm.)	Topical, 1%
	Sulconazole	Exelderm (ICI)	Topical, 1%
		Sulcosyn (Syntex)	Topical, 1%
	Tioconazole	Vagistat-1 (Fujisawa-SmithKline)	Topical, 6.5%
	Cloconazole	Clinical stage	
	Fenticonazole	Fentigyn (Ciba-Geigy)	
Tiazole	Fluconazole	Diflucan (Roerig)	Tablet, 100–200 mg/day
	Itraconazole	Sporanox (Janssen)	Tablet, 100–400 mg/day
	Terconazole	Terazol (Ortho)	Topical, 0.4–0.8%

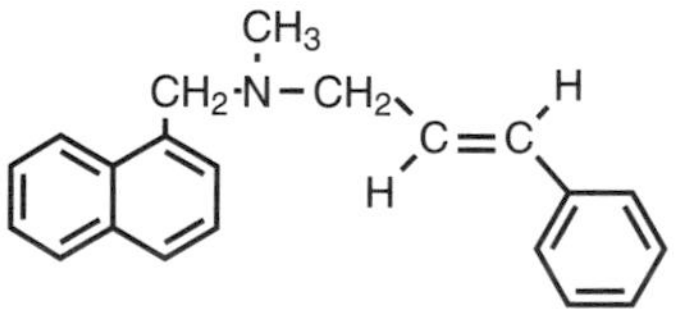

Naftifine hydrochloride

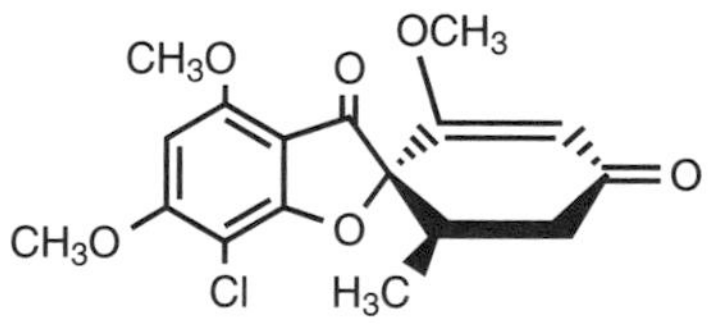

Griseofulvin

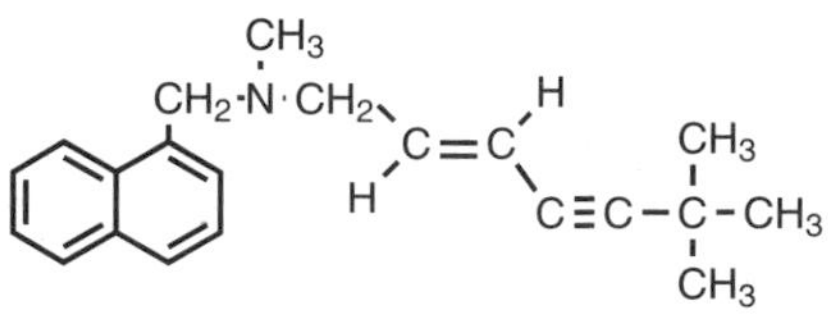

Terbinafine hydrochloride

lium griseofulvin in 1939, the antibiotic has been used against plant fungal pathogens since 1951 and against dermatophytes since 1958. Of the four possible stereoisomers, only the natural product is active. The chlorine atom can be replaced, with retention of potency, by fluorine but not by bromine or hydrogen. Replacement of the methoxy substituent on the cyclohexene with either

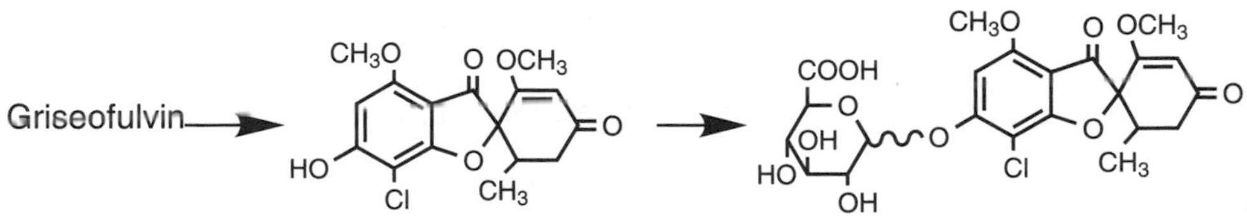

Fig. 35–1. Metabolism of griseofulvin.

propoxy or butoxy functions increases activity, but larger groups at this position decrease activity as does the replacement of the methoxy with hydrogen. Possibly this reflects a balance of oil-water distribution ratio. Replacement of the methoxy with an amino group or substitution of a ketone for the enol-methoxy (1,3-diketocyclohexane) eliminates biologic potency.

Occasionally, griseofulvin is used topically, but it is more effective when administered orally. Because of poor absorption, only a small portion of the drug becomes biologically available. To enhance absorption, micronized formulations are employed and the drug is taken with high-fat meals. A small but sufficient amount of the drug apparently diffuses from the blood to the cutaneous area of fungal multiplication so that hyphal penetration is slowed. Because the drug is fungistatic rather than fungicidal, a simultaneous outward thrust of keratinized host cells is required to cause death of the pathogens by blocking their access to nutrients.

The primary site of action of griseofulvin involves binding to RNA.[9] Manifestations include inhibition of hyphal cell-wall synthesis and malformation of spindle and cytoplasmic microtubules. Large doses in animals are teratogenic and tumorigenic. Side effects in humans with therapeutic doses consist of gastrointestinal, allergic, and neurologic reactions in occasional patients, and, more frequently, moderate increases in fecal protoporphyrin. The safety of the drug in pregnancy has not been established. Griseofulvin induces hepatic microsomal enzymes that decrease the activity of warfarin-type oral anticoagulants: patients receiving these drugs concomitantly may require dosage adjustment of the anticoagulant during and after griseofulvin therapy.

The major route of metabolism for griseofulvin is demethylation of the 6-methoxy, giving rise to 6-desmethylgriseofulvin. This material is found in the urine as its glucuronide derivative as shown in Figure 35–1. Neither metabolite possesses antifungal activity.

The antidermatophytic azoles listed in Tables 35–1 and 35–2 are applied directly to the infected area. Topical azoles are relatively free of side effects other than occasional irritation. The mechanism of action of the azoles and their utility in systemic fungal infections is described later in this chapter.

SYSTEMIC ANTIFUNGAL AGENTS COMPOUNDS USEFUL IN SYSTEMIC MYCOSES

Polyenes[5,9]

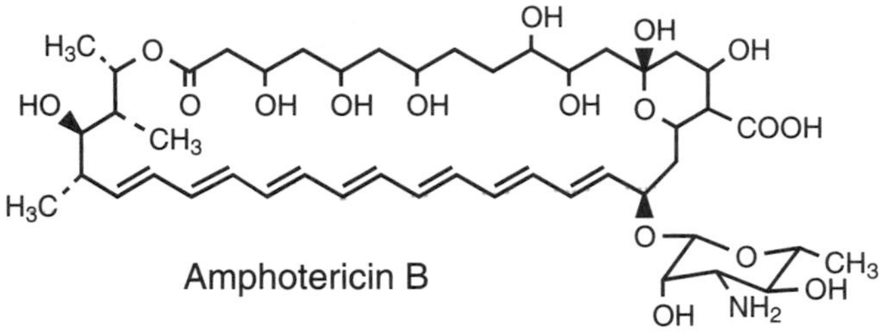

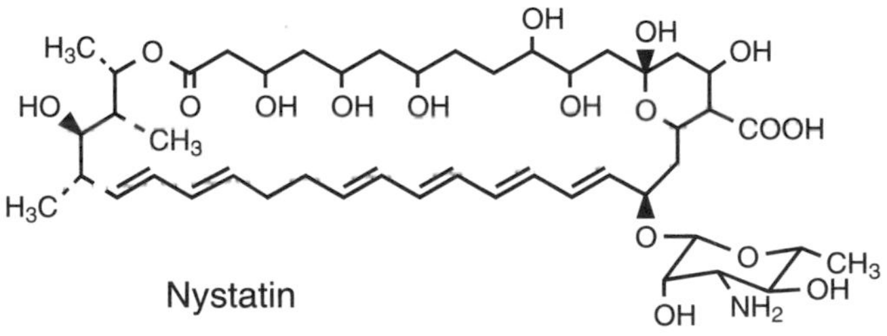

Approximately 60 polyene antibiotics, all produced by actinomycetes, have been described since 1951. Each contains a macrocyclic lactone ring with a lipophilic portion. The rings possess a chromophore of four, five, six, or seven conjugated double bonds and, in many cases, with linkage to an amino sugar. The polyenes are poorly soluble in water and common organic solvents but are reasonably soluble in polar organic solvents such as dimethyl formamide, dimethylsulfoxide, and pyridine.

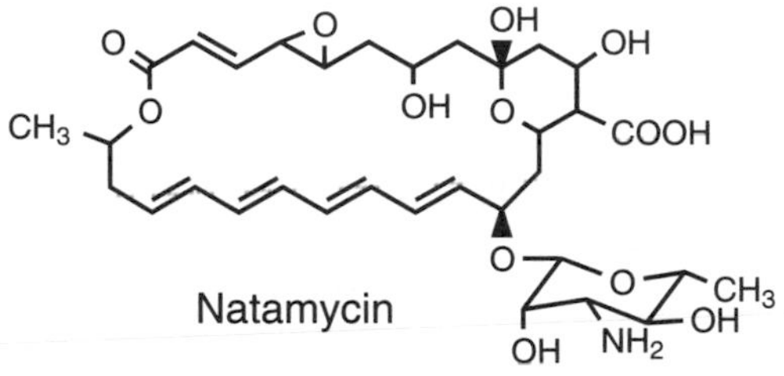

The heptaenes (seven conjugated double bonds) are approximately ten times more active in vitro than the pentaenes or tetraenes. Moreover, the heptaenes do less damage to host-cell membranes. Examples of heptaenes are amphotericin B, candicidin, hamycin, and trichomycin. Conjugated hexaenes include endomycin, pentaenes are represented by filipin, and tetraenes include nystatin and pimaricin.

Polyenes act against sensitive fungal, algal, protozoan, and metazoan cells by combining with membrane sterols (ergosterol in fungi) with subsequent alteration in permeability and loss of essential organic and inorganic cell constituents. Except for Mycoplasma, bacteria lack cell membrane sterols and therefore are not affected by the polyene antibiotics. Free sterols decrease the antifungal action by competing with cell membrane sterols for the drug molecules.

In mammalian cell membranes, the principal sterol is cholesterol. Amphotericin B, the most commonly used polyene, binds approximately ten times more tightly to vesicles containing ergosterol rather than to those containing cholesterol. For many years the intravenously administered colloid preparation of the drug, however, formulated with deoxycholate and phosphate buffer, has been associated with many side effects. The most serious are hypokalemia and distal tubular acidosis. Nephrotoxicity has been lessened to some extent by lipid-complexed formulations. The drug is incorporated into sonicated ampholiposomes composed of egg yolk, phosphatidylcholine, cholesterol, and stearylamine. The ampholiposomes are administered intravenously. These formulations are fairly well tolerated and doses as high as 3 to 5 mg/kg have been given without significant side effects.[10,11]

Neither amphotericin B nor other polyenes are usually employed in therapy of dermatophytoses because of the availability of effective and comparatively safer drugs as listed in the previous section. Nystatin is recommended for treatment of oral thrush or esophagitis caused by Candida albicans in AIDS patients. Likewise, amphotericin alone or in combination with 5-fluorocytosine is used for meningitis caused by Cryptococcus neoformans in AIDS patients. Although amphotericin B has a moderately broad antifungal spectrum, and development of resistance is rarely noted, not all fungal pathogens are susceptible. Moreover, the antibiotic may not readily penetrate into all infected tissues. For example, even when administered intrathecally, the drug cannot reach isolated areas in patients who have obstructive hydrocephalus.[12]

5-Fluorocytosine[5,9,11]

This fluorinated pyrimidine was first developed as a potential antileukemic agent. The drug is well absorbed

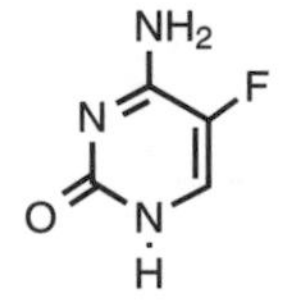

Flucytosine

through the gastrointestinal tract, has excellent penetration into the cerebrospinal fluid (CSF) and is of low toxicity. Unfortunately, its spectrum of antifungal activity is limited to susceptible strains of Candida, Cryptococcus, Torulopsis (Candida) glabrata, Cladosporium, and Aspergillus. When used alone in infections caused by initially susceptible strains, marked increase in resistance (MIC > 16 μg/ml) may occur during treatment. Accordingly, in such serious conditions as yeast-induced meningitis, 5-fluorocytosine, if used, generally is administered concurrently with amphotericin B. Thus, resistant clones can be killed by the polyene. Moreover, the latter drug may alter permeability barriers of the yeast cells to permit enhanced uptake of 5-fluorocytosine.

Susceptible yeast cells deaminate assimilated 5-fluorocytosine to 5-fluorouracil. The latter is converted to 5-fluoro-2′-deoxyuridylic acid, an inhibitor of thymidylate synthetase, thus suppressing DNA synthesis (Fig. 35–2). Alternatively, 5-fluorouracil can be incorporated into RNA to result in inhibition of protein synthesis. Fortunately, human cells have little or no cytosine deaminase activity, approximately 90% of the ingested dose is excreted unaltered in the urine. If renal excretion is inhibited, as might occur with concurrent amphotericin B therapy, plasma levels of 5-fluorocytosine can become excessive and cause hemolytic side effects. Accordingly, plasma levels should be monitored and maintained at 35 to 75 μg/ml by adjusting dosage. The antifungal action of 5-fluorocytosine is antagonized by purine and pyrimidine bases. Thus, in vitro assays for plasma levels and for strain susceptibility should be performed in culture media of defined chemical composition.

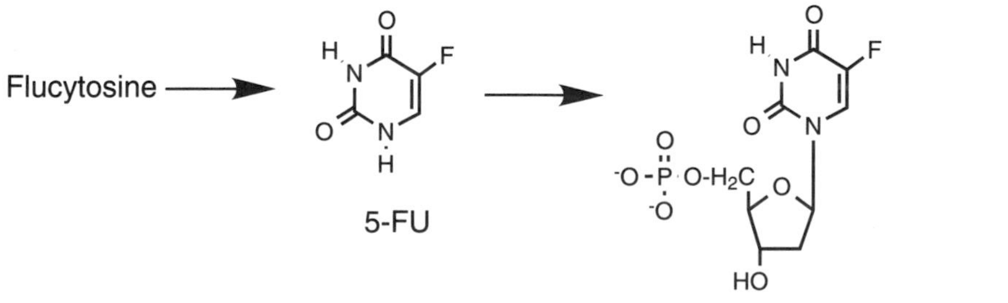

Flucytosine → 5-FU → 5-fluoro-2'-deoxyuridylate

Fig. 35–2. Metabolism of flucytosine.

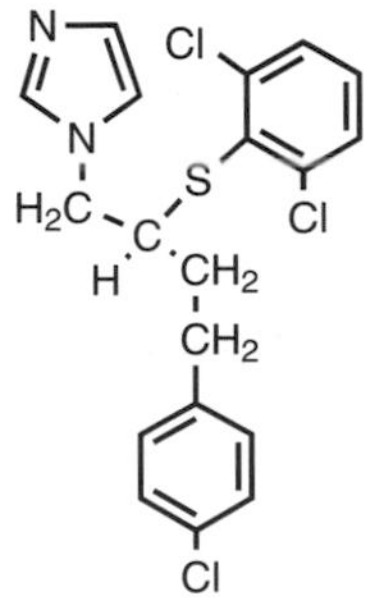

Butoconazole nitrate

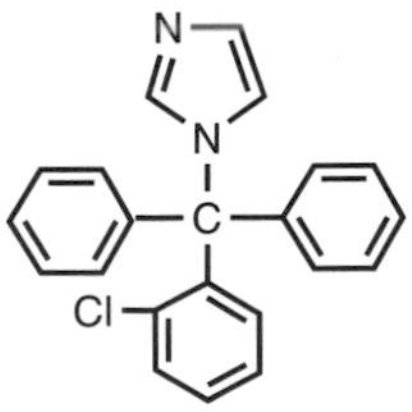

Clotrimazole

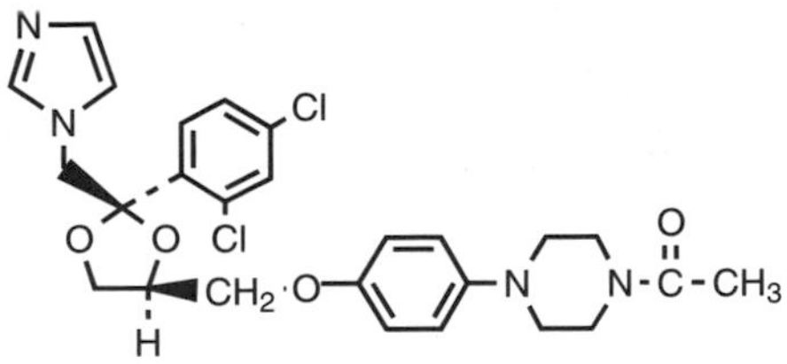

Ketoconazole

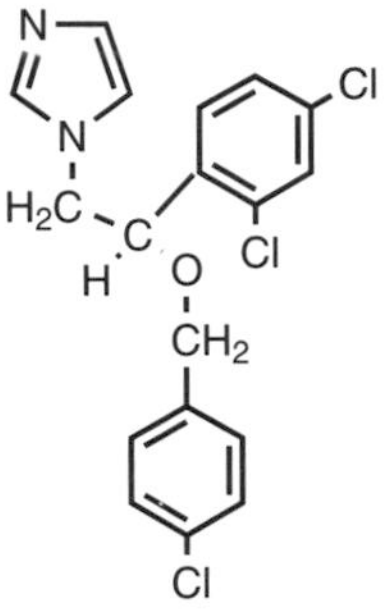

Econazole nitrate

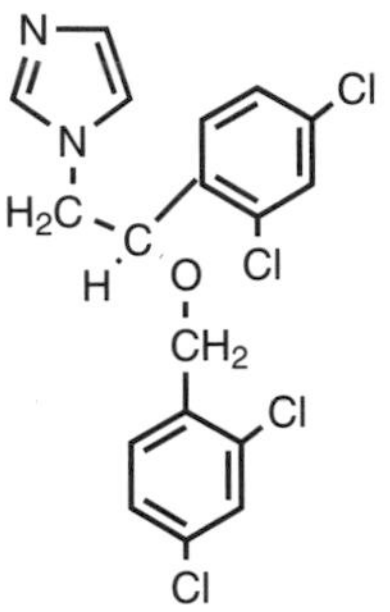

Miconazole

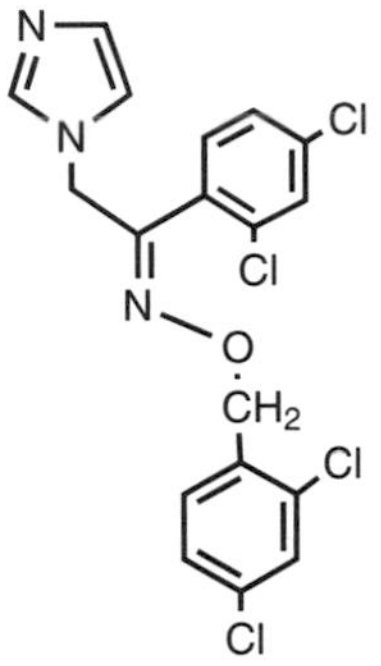

Oxiconazole nitrate

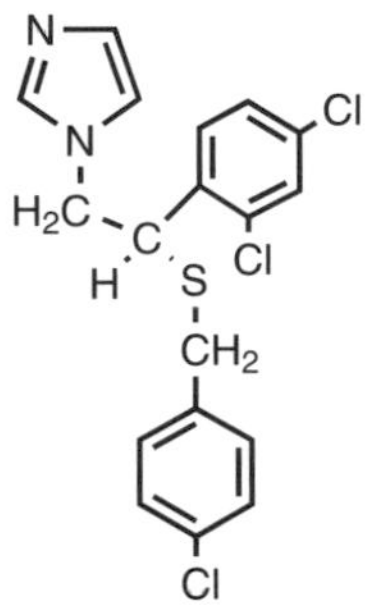

Sulconazole nitrate

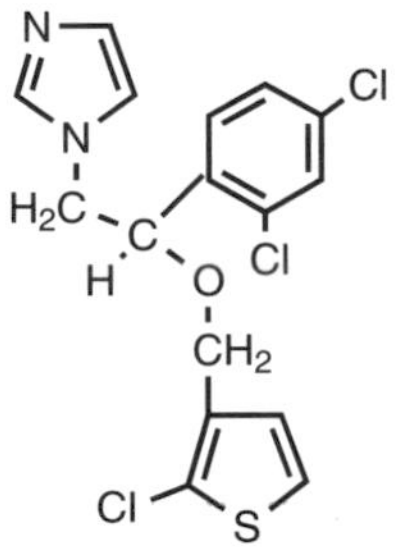

Tioconazole

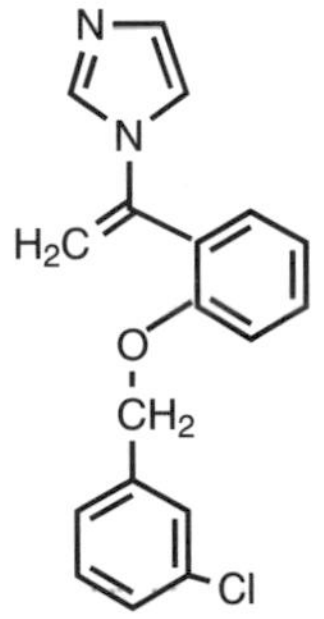

Cloconazole hydrochloride

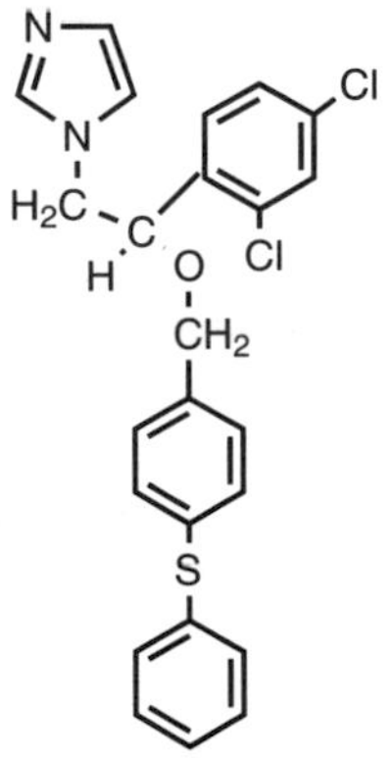

Fenticonazole nitrate

Azoles[2,9,11,13,14]

Beginning in the late 1960s, an extensive series of azole compounds have been synthesized and tested for antifungal activity. A few are used topically for the treatment of cutaneous dermatophytic and yeast infections. Several azoles are available for oral administration in patients with systemic mycoses. One of the drugs, fluconazole, also can be administered intravenously.

The antifungal azoles have a five-membered azole ring with the N-1 atom linked to other aromatic rings via an aliphatic carbon atom or atoms. Clotrimazole, miconazole, ketoconazole, butoconazole, econazole, cloconazole, fenticonazole, oxiconazole, sulconmazole, and tioconazole are imidazoles (two nitrogen atoms in a five-membered ring). Fluconazole, itraconazole, and terconazole are triazoles (three nitrogen atoms in a five-membered ring).

The antifungal azoles inhibit a cytochrome P-450, which catalyzes the 14 α-demethylation of lanosterol to ergosterol. The N-3 atom of the azole ring binds to the ferric iron atom in the heme prosthetic group to prevent the activation of oxygen for insertion into lanosterol. The

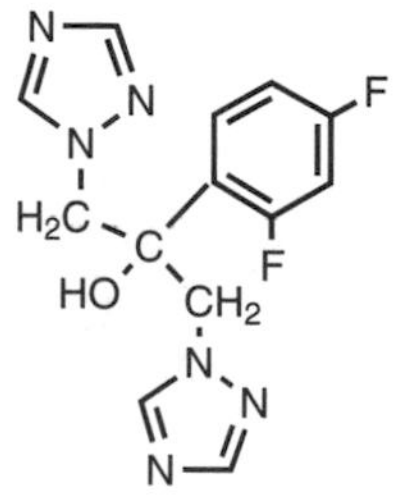

Fluconazole

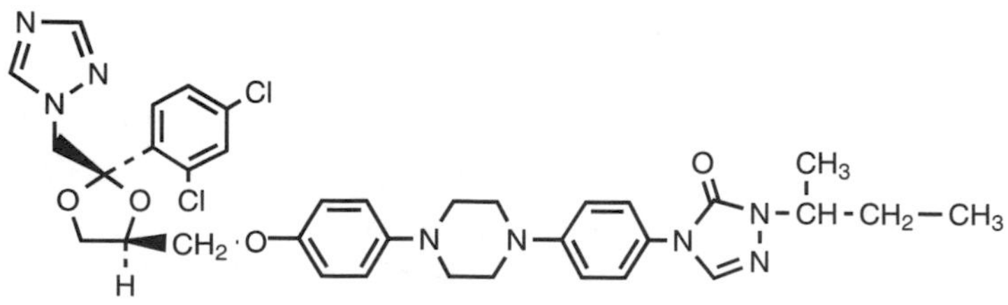

Itraconazole

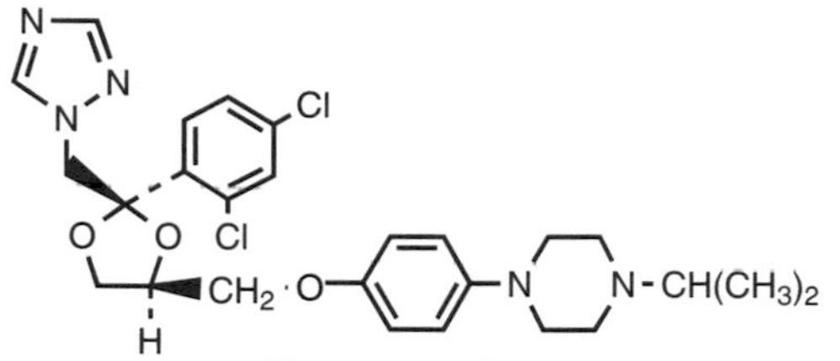

Terconazole

potency of specific azoles is determined not only by the strength of binding to the heme iron but also by the affinity of the N-1 substituent for the apoprotein of the cytochrome. Inhibition of the 14 α-demethylase results in accumulation of 14-methylated sterols that induce permeability changes, membrane leakiness, uncoordinated synthesis of chitin, cell disorganization, and fungal death.

The azole drugs have a relatively broad spectrum of antifungal activity but differences occur among individual compounds. For example, itraconazole has very strong activity against Aspergillus and Torulopsis. Although the azoles generally inhibit Candida albicans, ketoconazole and fluconazole are inactive, respectively, against C. tropicalis and C. krusei.[14]

These drugs differ also in pharmacokinetic properties. Orally administered clotrimazole induces liver microsomal enzymes that metabolize the drug so rapidly that it is ineffective against systemic fungi; thus it can be used only topically. Miconazole, in contrast, is not rapidly metabolized and has been effective when given systemically. Unfortunately, however, it caused multiple side effects mainly attributable to the polyethoxylated castor oil required for colloidal stabilization. Accordingly, miconazole, is now also reserved for topical use. Ketoconazole is slightly water soluble and, when converted to the hydrochloride by gastric acidity, is about 75% absorbed on oral administration. Absorption is blocked by antacids and H_2-receptor antagonists. Penetration of the absorbed drug into the CSF is poor.

Abnormal, transient elevation of liver function tests develops in 5 to 10% of patients on ketoconazole; occasionally, serious hepatitis occurs. The drug can block such host enzymes as 14α-demethylase, 11β-hydroxylase, and $C_{17,20}$-desmolases. Thus adrenal and gonadal steroidogenesis, especially of androgens, can be inhibited. Some loss of male libido and sexual potency, as well as development of gynecomastia, has been reported. Accordingly, ketoconazole has been used clinically to reduce steroidogenesis in such conditions as female acne and hirsutism, precocious puberty, prostatic and male breast carcinoma, and Cushing's syndrome.[15]

Fluconazole has a relatively low molecular weight (305), is water soluble, weakly protein bound with high bioavailability, penetrates well into CSF, and is excreted largely unchanged in the urine. Gastric acidity is not required for absorption. In the presence of either uninflamed or inflamed meninges, CSF concentrations of fluconazole are 60 to 80% of simultaneous plasma concentrations. In contrast, the ratio of CSF to plasma concentrations of miconazole, ketoconazole, and itraconazole is less than 10%. Thus fluconazole is a useful drug in patients with meningitis caused by susceptible fungal strains.

Fluconazole and itraconazole apparently have increased selectivity, as compared with ketoconazole, for fungal over human cytochrome P-450. Thus far, serious side effects have not been reported in human trials of the two triazole drugs. Plasma concentrations of itraconazole after oral administration show marked variation among patients. Except for the central nervous system, however, drug levels are substantially higher in tissue than in plasma. As with the polyenes and the imidazoles, the clinical development of strains resistant to the triazoles is rare.

The potential for drug-drug interactions is a concern in the systemic use of the antifungal azoles.[14,16] The plasma concentrations of the azoles may show substantial decreases if patients concurrently are treated with rifampin, isoniazid, phenytoin, or carbamazepine. The azoles, on the other hand, can retard metabolism of such drugs as cyclosporin, sulfonylureas, and phenytoin to result in untoward side effects.

NOVEL APPROACH TO FUNGAL CHEMOTHERAPY

In normal hosts, the rate of spontaneous cure, or at least remission, of fungal disease is high—particularly if the infections are acute, host defenses improve, or a predisposing foreign body is removed. On the other hand, in immunosuppressed hosts, the drugs now available for long-term therapy often are unable to eradicate the infection. Accordingly, novel approaches to fungal chemotherapy should be contemplated. One such approach might be to assist the infected host to withhold growth-essential iron from the fungi, but not from the host.

Poorly soluble iron chelators such as the hydroxypyridone, rilopirox, are being developed for topical use in dermatophytoses.[17] Iron chelators with high specific affinity for the metal also are now becoming available for oral administration.[18] Adjunctive therapy with such drugs might especially be indicated in cases of diabetic ketoacidosis. In such patients, the lowered pH of their plasma can suppress the natural ability of transferrin to withhold iron from fungal agents of mucormycosis.[4] Care must be taken, however, to avoid use of iron chelators that might enhance fungal acquisition of the metal. Note that deferoxamine, an actinomycete siderophore employed in patients with iron or aluminum overload, can stimulate growth of many strains of pathogenic fungi.[3]

REFERENCES

1. E. D. Weinberg, *Microb. Pathog., 3,* 393(1987).
2. H. V. Bossche and P. Marichel, *Am. J. Obstet. Gynecol., 165,* 1193(1991).
3. A. M. Sugar, *Clin. Infect. Dis., 14,* S126(1992).
4. S. E. Vartivarian, *Clin. Infect. Dis., 14,* S30(1992).
5. G. P. Bodey, *Ann. NY Acad. Sci., 544,* 431(1988).
6. R. S. Hare and D. Loebenberg, *Am. Soc. Microbiol. News, 54,* 240(1988).
7. T. J. Walsh, *Am. Soc. Microbiol. News, 54,* 240(1988).
8. J. A. Balfour and D. Faulds, *Drugs, 43,* 259(1992).
9. M. R. McGinnis and M. G. Rinaldi, in *Antifungal drugs,* in *Antibiotics in Laboratory Medicine,* 3rd ed., V. Lorian, Ed., Baltimore, Williams & Wilkins, 1991.
10. F. Meunier, et al., *Ann. NY Acad. Sci., 544,* 598(1988).
11. J. Cohen, *Lancet, 337,* 1577(1991).
12. J. R. Graybill, *Clin. Infect. Dis., 14,* S170(1992).
13. M. S. Saag and W. E. Dismukes, *Antimicrob. Agents Chemother., 32,* 1(1988).
14. G. P. Bodey, *Clin. Infect. Dis., 14,* S161(1992).
15. S. Venturoli, et al., *J. Clin. Endocrinol. Metab., 71,* 335(1990).
16. R. M. Tucker, et al., *Clin. Infect. Dis., 14,* 165(1992).
17. R. Kruse, et al., *Pharmacology, 43,* 247(1991).
18. E. D. Weinberg, *Drug Metab. Rev., 22,* 531(1990).

Chapter 36

ANTISEPTICS AND DISINFECTANTS

Thomas L. Lemke

GENERAL PRINCIPLES

Antimicrobial agents meant for topical application to living tissue are defined as antiseptics. Chemical substances which function as antimicrobial agents when applied to inanimate objects are referred to as disinfectants. Because antiseptics or disinfectants are nonspecific drugs and do not distinguish between pathogenic cells and human tissue, the only true difference between the two groups is in how they are used. Substances used as antiseptics might function well as a disinfectant at higher concentrations and the reverse. Disinfectants are usually expected to have cidal action on the microorganism, whereas an antiseptic may be either static or cidal, but this property is a concentration and time-of-contact-dependent variable. Antiseptics are regulated by the Food and Drug Administration (FDA) whereas disinfectants are regulated by the Environmental Protection Agency (EPA). Due to chemical similarity, several agents used primarily as preservatives in pharmaceutical preparations will be discussed in this chapter. Preservatives are chemicals which prevent the growth of microorganisms, and may come in contact with living tissue depending on the use of the particular product in which they appear. A common feature of a preservative is that it is used in low concentration.

In 1972 the FDA began an extensive review of the active ingredients in over-the-counter (OTC) drug products to ensure that the chemicals in them were both safe and effective. Because antiseptics are the responsibility of the FDA, the chemicals discussed in this chapter have been part of this review. The nature of the process and the depth of the review has extended the final decision of the safety and effectiveness so that many of the chemicals are still under consideration. For a discussion of the FDA review process, the reader is referred to the *Handbook of Nonprescription Drugs.*[1] The FDA will ultimately classify all chemical substances in OTC products into one of three categories: Category I—generally recognized as safe and effective for the claimed therapeutic indication; Category II—not generally recognized as safe and effective or unacceptable indications; Category III—insufficient data available to permit final classification. Throughout this chapter I will refer to the classification category of individual chemicals based on the most recently published results. Many of the antiseptics and disinfectants have been used for 100 or more years but rigorous biologic tests to substantiate their effectiveness may not yet have been carried out.

CHEMICAL CLASSES OF ANTISEPTICS AND DISINFECTANTS

The antiseptics and disinfectants are organized by chemical class in this discussion. Excluded from this presentation are the first-aid antibiotics, which the FDA has chosen to distinguish from antiseptics. Antibiotics and topical antibiotics cannot be referred to as topical antiseptic agents.[2] The topical antibiotics are discussed elsewhere in this text. The chemical classes of antiseptics and disinfectants include: phenols, alcohols, surface-active agents, dyes, heavy metals, halogen containing substances, and miscellaneous agents.

Phenols

Phenol and a wide variety of substituted phenols are used either as antiseptics, disinfectants, or preservatives. Historically, phenol was used as an antimicrobial agent by Lister as early as 1867.

Mechanism of Action

The phenols are nonspecific agents thought to disrupt a cell membrane by altering cell membrane/wall permeability and thereby reducing the surface tension between the aqueous media and the lipid membrane/wall through a balance of the hydrophilic-lipophilic physical properties of the drug. This mechanism is supported by the correlation between biologic activity and lipid solubility. The highly hydrophilic phenols (e.g., resorcinol) show low antimicrobial activity whereas the more lipophilic agents (e.g., hexylresorcinol) have increased biologic activity. A measure of biologic activity for phenols and other antimicrobials is the phenol coefficient (P.C.) which is defined as:

$$\text{P.C.} = \frac{\text{Dilution of disinfectant necessary to kill}}{\text{Dilution of phenol necessary to kill}}$$

This relationship to phenol is temperature and microorganism dependent, and when available, gives a general idea of the effectiveness of the antimicrobial agent.

Structure-Activity-Relationship

As indicated already, the biologic activity is related to the physical-chemical properties of the substituted phenol. An increase in the lipophilic character of the

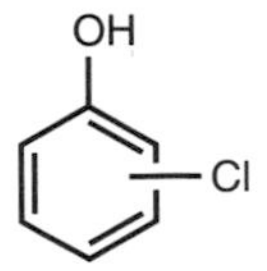

Monochlorphenol 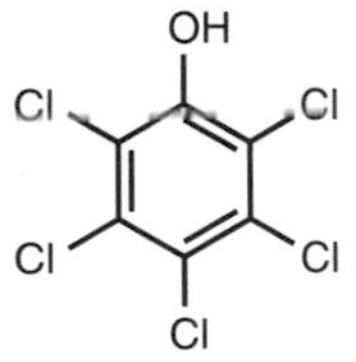 Pentachlorophenol

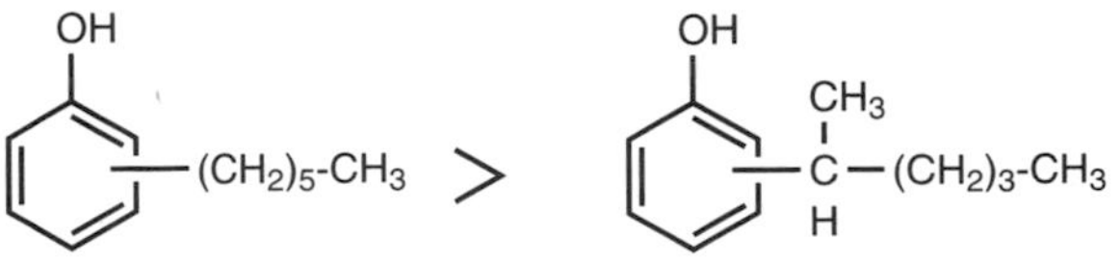

Branched alkanes

molecule results in a corresponding increase in its antimicrobial activity. Substitution of a hydrogen with a halogen increases activity and decreases water solubility. For example, monochlorophenol has a P.C. of approximately 4, whereas pentachlorophenol has a P.C. of approximately 50.

Pentachlorophenol is reported to act as an insecticide and wood preservative, but is water insoluble. Loss of water solubility reduces a compound's utility as a disinfectant because an aqueous medium is the usual vehicle for such products. Alkyl substitution also increases biologic activity with C_5 to C_6 chain giving maximum activity. Additionally, a straight chain alkyl group is usually better than the analogous branched alkane. The nitro group substituent results in a moderate increase in activity, e.g., picric acid has a P.C. of about 5.9. The alkoxy substitution also increases biologic activity, as it increases with the size of the alkyl portion.

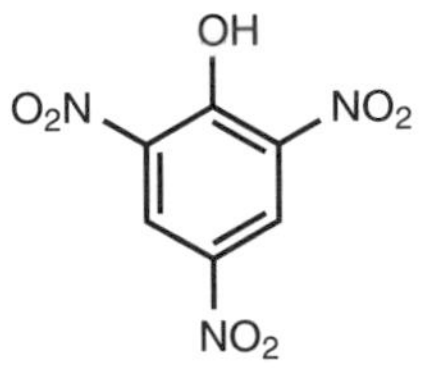

Picric acid

Specific Products

Phenol (Chloraseptic, [Procter & Gamble]). Phenol (water solubility of 1 g in 15 ml water) and sodium phen-

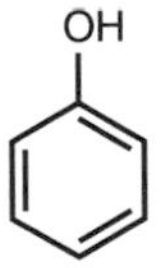

Phenol

olate are claimed to have moderate antimicrobial activity. The FDA panel on first-aid products has classified phenol as a safe and effective agent when used in a concentration of 0.5 to 1.5% (Category I).[3] Phenol and sodium phenolate appear in mouthwashes in a concentration of 1.5% or less.[4] Phenol is listed as being caustic to host tissue, which indicates its nonspecific nature and, additionally, exerts topical anesthetic action masking its destructive action on host tissue.

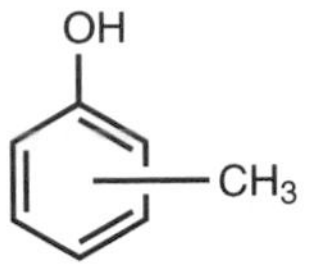

Cresol

Cresols (o, m, and ρ-methylphenol). Methylphenol, a mixture of the three isomeric compounds, has a water solubility of 1 g/50 ml of water, a P.C. of about 2.5, and is used as a disinfectant at 2% concentration.

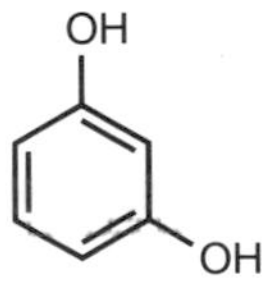

Resorcinol

Resorcinol. Resorcinol is significantly more hydrophilic than phenol with 1 g dissolving in 1 ml water. The molecule is classified as a mild antibacterial, antipruritic, and keratolytic with a P.C. of approximately 0.4.[5]

Hexylresorcinol (Sucrets [Beecham], S.T. 37 [Scot-Tussin]). Hexylresorcinol, water solubility of 1 g/2000 ml water, is reported to show an activity 50 to 100 times

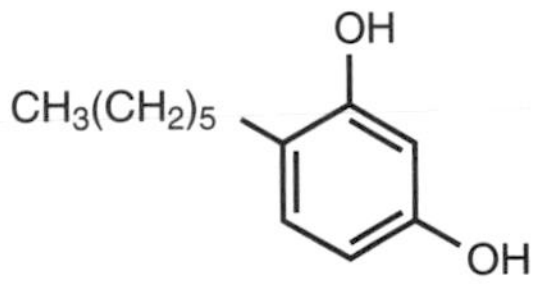

Hexylresorcinol

that of phenol against a variety of organisms. It is claimed to significantly lower the surface tension of a lipid-water interface (*S*urface *T*ension 37 dynes/cm, water has a surface tension of 71.2 dynes/cm), which is in agreement with the expected mechanism of action, water solubility, and claimed biologic activity. The FDA panel on first aid products has found hexylresorcinol to be safe and effective at a concentration of 0.1 to 0.2%.[3]

***ρ*-Chloro-m-xylenol (Chloroxylenol, Unguentine [Metholatum], Vaseline First-Aid [Chesebrough-Pond's]).**

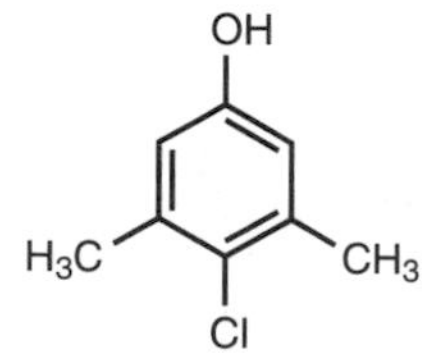

ρ–Chloro-m-xylenol

Chloroxylenol is found in a variety of topical antiseptic and sunburn products. This molecule demonstrates the cumulative effect that lipophilic substituents can have on the physical-chemical properties as well as biologic properties of phenols. The molecule's solubility is 1 g/3000 ml water (20°) with a P.C. of approximately 60. The FDA panel has placed chloroxylenol into Category III indicating insufficient evidence to prove safety or effectiveness.

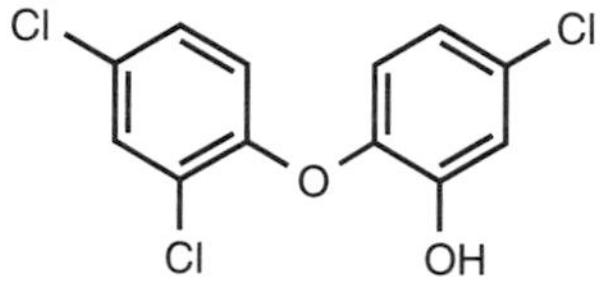

Triclosan

Triclosan. This is a bacteriostatic agent found in popular deodorant soaps, as well as deodorants and antiperspirants. It has served as a replacement for the once popular hexachlorophene after its removal from OTC products in the 1970s. This water-insoluble phenol has mild antibacterial activity, although evidence to substantiate this claim is not currently available.

Hexachlorophene (pHisoHex [Sanofi-Winthrop]). Hexachlorophene, first synthesized and patented in the early 1940s, became the standard for antimicrobial drugs

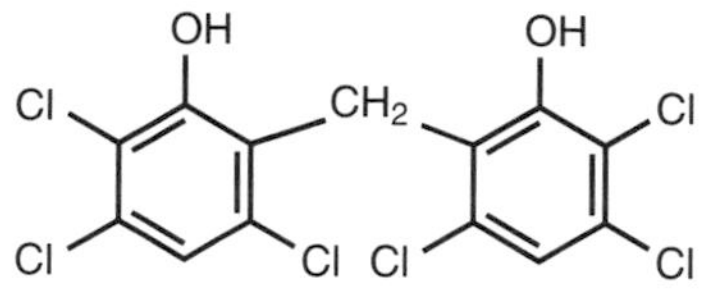

Hexachlorophene

in the 1960s. This highly lipophilic, water insoluble, phenol (P.C. about 125) has been used in nearly every OTC drug product from scalp preparations to foot creams. It was used in a 3% concentration to cleanse the skin and was especially popular for total body bathing of newborn infants, because it was an extremely effective prophylaxis against nursery epidemics of staphylococcal skin infections. In late 1971 and early 1972, data began to accumulate indicating that the drug was readily absorbed into the bloodstream through intact skin and was then distributed to the brain.[6] In August 1972, 40 French infants died, with hexachlorophene being implicated in these deaths. This led to its immediate ban from infant products, the eventual removal from all OTC products, and finally, the prescription-only status of this product. Animal and human toxicity was also reported at that time[7] and follow-up studies suggest vacuolar encephalopathy in both infants and experimental animals as well as major and minor birth defects.[8,9]

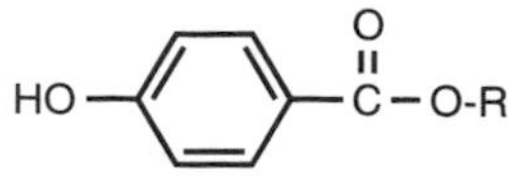

Alkyl ρ–hydroxybenzoates

Alkyl ρ-hydroxybenzoates (Alkyl Parabens). Certain alkyl parabens (methyl, ethyl, propyl and butyl) are commonly used in foods and some pharmaceuticals as preservatives. The usual concentration is 0.1 to 0.3%. Water solubility decreases proportionally as the alkyl chain is lengthened (R = CH_3, 1 g/400 ml; R = C_2H_5, 1 g/6500 ml). These chemicals are reported to cause contact dermatitis in a high proportion of the population, i.e., about 5%.

Alcohols

Alcohols are probably the first chemicals which come to mind when the general public chooses an antiseptic or disinfectant. The monohydric alcohols commonly used as antiseptics or disinfectants are ethyl alcohol (Alcohol, USP) and isopropyl alcohol. Propylene glycol and glycerin are two polyhydric alcohols with limited utility and the monohydric alcohol benzyl alcohol is occasionally encountered as a preservative.

Mechanism of Action

The alcohols are thought to produce their antimicrobial action by denaturation of cell wall protein resulting in disruption of the cell wall which in turn, will destroy the unicellular organism. In a similar fashion, alcohol can inactivate viruses by "dissolving" the capsid of the virus. By disruption of the outer surface of the virus, the virus is unable to attach itself to the host cell surface and infection of the host cell does not occur.[10]

Structure-Activity-Relationship

In general, primary alcohols are superior to the secondary alcohols, which are better than tertiary alcohols. As the molecular weight of the alcohol increases, the biologic activity is also expected to increase. Polyhydric alcohols are usually less active than monohydric analogs, suggesting the importance of a hydrophilic-lipophilic balance. Although many alcohols have been evaluated for their antiseptic/disinfectant utility, only ethyl alcohol and isopropyl alcohol have been assigned to Category I, e.g., safe and effective, by the FDA panel on topical antimicrobial agents.[3,11]

Specific Products

Ethanol (Ethyl alcohol, Alcohol, USP). Although ethanol has been found to be safe and effective as a topi-

$CH_3—CH_2-OH$

Ethanol

cal antimicrobial agent, it is best used only for minor cuts and scraps. Its use on open, extensive wounds is not recommended because the alcohol may do more harm than good on such wounds. The denaturation of the protein is a process requiring water, and therefore absolute alcohol is less effective as an antimicrobial than diluted aqueous alcohol solutions. Ethanol's germicidal action occurs with concentrations of 60 to 95%. It should also be recognized that the time of contact is important and the rapid evaporation of ethanol may reduce its effectiveness. Mouthwashes represent a class of pharmaceuticals containing ethanol in 14 to 54% concentration. The claim of antiseptic activity is not substantiated by clinical studies and the time variable would be expected to play a significant role. Additionally, a question has been raised about the value of destroying the commensal organisms found in the oral cavity. Ethanol is less effective against spores than the vegetative forms of microorganisms, reducing its usefulness in hospitals as a disinfectant for the sterilization of surgical instruments. Finally, in 30 to 70% concentration, ethanol is effective against a variety of viruses.[10]

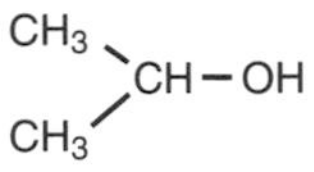

$(CH_3)_2CH-OH$

2-Propanol

2-Propanol (Isopropyl alcohol). The properties, both physical-chemical and biologic, of isopropyl alcohol are very similar to those of ethanol. Isopropyl alcohol is recognized as a safe and effective agent in concentrations of 50 to 91.3% for use as a topical antimicrobial agent. The compound also is active against viruses in concentrations of 30 to 95%.

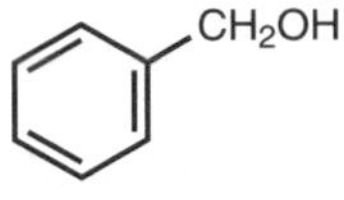

$C_6H_5-CH_2OH$

Benzyl alcohol

Benzyl alcohol. At present, benzyl alcohol is not considered to be safe and effective as a topical antimicrobial agent by the FDA panel considering such agents.[3] It is a slow-acting antibacterial and for this reason could not be recommended, but is found in pharmaceuticals, such as multiple dose parenterals, as a preservative. Preservatives are not expected to exhibit the same actions as topical antimicrobial agents.

Chlorobutanol (1,1,1-Trichloro-2-methyl-2-propanol). Chlorobutanol is a preservative in several classes of pharmaceutical products, but is not a topical antimicrobial agent.

$CH_3-C(CH_3)(OH)-CCl_3$

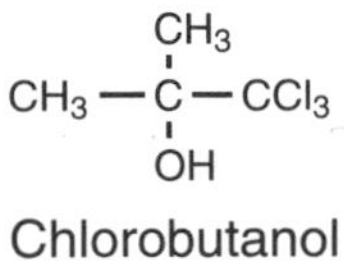

Chlorobutanol

Surface-Active Agents (SAA, Surfactant)

The term surface-active agents describes a group of organic chemicals meant to modify the characteristics of an interphase. The general characteristics for a SAA include that the molecule possess hydrophilic (polar) and lipophilic (nonpolar) portions. When considering antiseptics and disinfectants, the interphase in question is the cell wall/water interphase and thus lipid/water characteristics of the drug are important. The specific chemical classes considered as SAA are the quaternary amines and the nonionic surface-active agents.

Mechanism of Action

The primary activity of the SAA, whether quaternary amines or nonionic, is to disrupt the cell membrane. This occurs because of the ability of a SAA to dissolve partially in the lipid cell membrane and partially in the aqueous surroundings. The result of this physical-chemical property is a lowering of the surface tension leading to disruptive changes in the osmotic balance. The ultimate effect is lysis of the cell producing the cidal action.

Quaternary Amines (Cationic SAA)

In the case of the quaternary amines, the cationic center is the polar portion of the molecule whereas the hydrocarbon substituents comprise the lipophilic portions. A balance between the hydrophilic and lipophilic components is essential for optimal cidal activity. With chemicals composed of aliphatic lipophilic portions, it should be possible to find a specific chain length with maximum biologic activity, however, this maximum biologic activity would be expected to differ by organism. Because of the physical-chemical nature of cationic SAA, there are advantages and disadvantages to these molecules. The advantages include their water solubility, that they are nonstaining, and noncorrosive. They also manifest low toxicity. The disadvantages include inactivation at low pH, inactivity against spores, and that quaternary amines are adsorbed to glass and certain plastics such as polyvinyl chloride. Once adsorbed, the molecule loses its antimicrobial action.

Specific Products

Benzalkonium Chloride. This mixture of chemicals contains alkane chains of varying lengths, between C_8 and

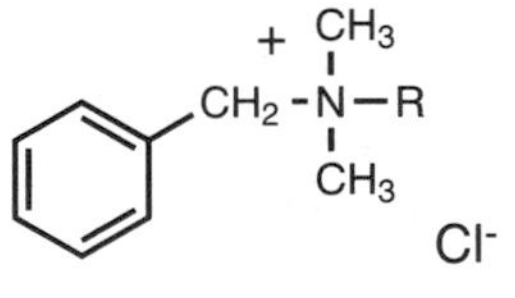

$C_6H_5-CH_2-N^+(CH_3)_2-R \quad Cl^-$

$R = C_8H_{17}$ to $C_{18}H_{37}$ (Mixture)

C_{18}. As an aqueous solution, the mixture is alkaline to litmus and foams strongly. The substance appears as an antiseptic in several topical antiseptics and burn medications. The mixture is used in concentrations of 1:750 to 1:10,000. An additional utility for benzalkonium chloride

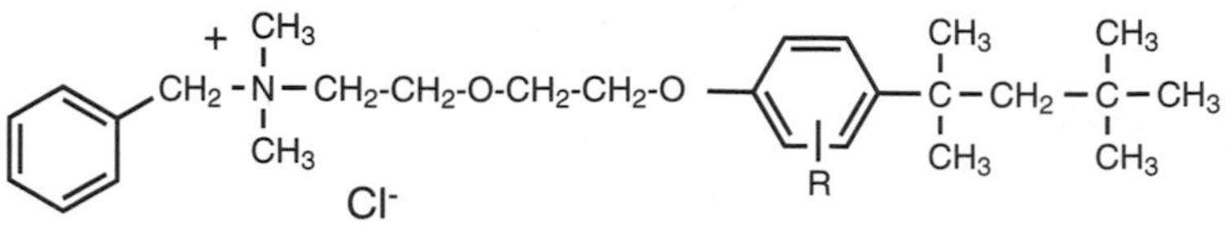

Benzethonium chloride R = H
Methylbenethonium chloride R = CH_3

is as a preservative and as such, is found in several contact lens' cleaning solutions.

Benzethonium Chloride (R = H, 18), Methylbenzethonium Chloride (R = CH_3, 19). The properties of benzethonium chloride and methylbenzethonium chloride are very similar to those of benzalkonium chloride. A 1% aqueous solution of either of these compounds has a pH between 4.8 and 5.5. The compounds are incompatible with soaps and detergents. These SAA are popular components of several diaper-rash products, although the therapeutic value against diaper rash is unproven.[12] The FDA panel on first-aid products has judged these agents to be safe and effective in the context of first aid products.[3]

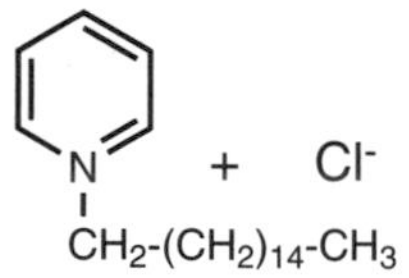

Cetylpyridinium chloride

Cetylpyridinium Chloride. This compound is structurally different from the traditional quaternary amines in that the quaternary head is part of an aromatic ring. The cetyl group (C_{16}) was found to give optimal biologic activity in this pyridinium derivative,[13] and a 0.1% solution significantly lowers surface tension from 71.2 dynes/cm (distilled water) to 40.8 dynes/cm. An aqueous solution of cetylpyridinium chloride produces an acidic pH of 6.0 to 7.0. Cetylpyridinium chloride is commonly found in mouthwashes and lozenges, and may also be found in other pharmaceuticals as a preservative. The antiseptic value of cetylpyridinium chloride is unproven and therefore its use in this context is questionable.

Nonionic Surface-Active Agents

Nonionic surface-active agents are neutral compounds made up of a hydrophilic polyethoxyethanol and the lipophilic alkyl-substituted aryl group. The balance of hydrophilic to lipophilic portion is based on the size of the polymer (x) and the size of the alkane substituent (R). Nonionic SAA are water soluble, and as neutral compounds, are compatible with both cationic and anionic compounds, and will not influence the pH of an aqueous medium.

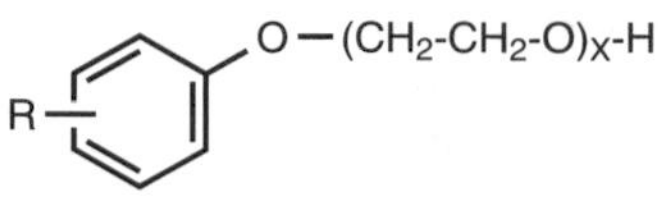

Specific Compounds

Nonoxynol-9 and Octoxynol-9. The nonionic SAA nonoxynol and octoxynol-9 act on a cell in the manner expected for a surface-active agent. By reduction of the surface tension of the lipid layer of the cell surface, these drugs lower surface tension, resulting in a disruption of the cell membrane. What distinguishes nonoxynol-9 and octoxynol-9 from other antiseptics is that the cells in question are not those of a potential pathogen but instead are spermatozoa with the result that these chemicals can be used as birth-control contraceptives. Their cell disruption results in the immobilization of the spermatozoa. Nonoxynol-9 is the most commonly used spermicide and is usually present in products in a 2 to 12% concentration.[14]

Dyes

Organic dyes have been used for their antimicrobial action since the early 1900s. The two major classes of dye are the triphenylmethane dyes (e.g., gentian violet and fuchsine) and the acridine dyes, 3,6-diaminoacridine and 9-aminoacridine. These agents were once popular antimicrobial agents, but today find very limited use, probably because they stain, which reduces patient acceptance.

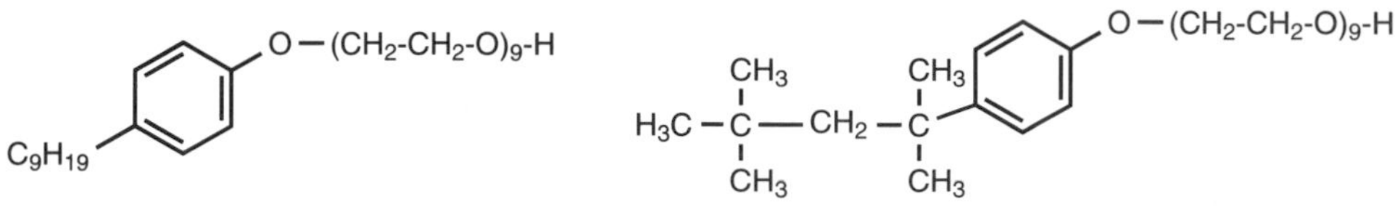

Nonoxynol

Octoxynol

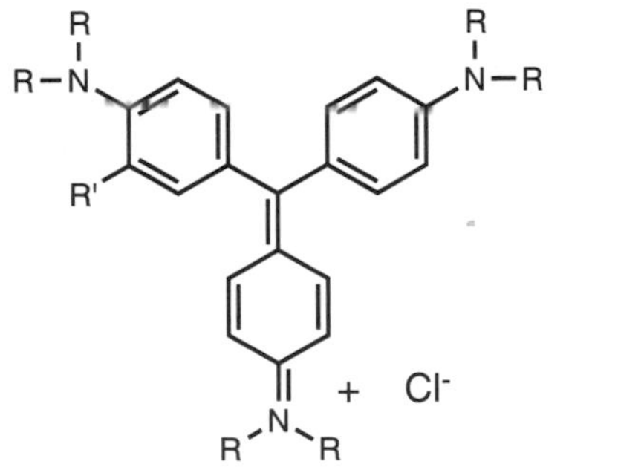

Gentian Violet (R = CH_3, R' = H)
Fuchsine (R = H, R' = CH_3)

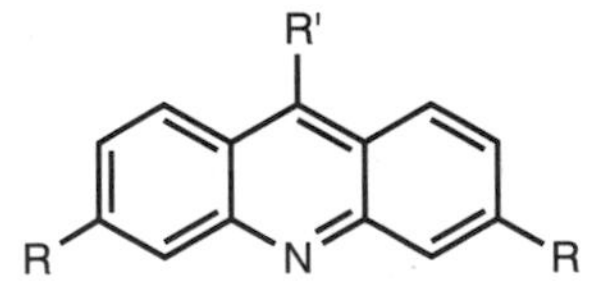

3,6-Diaminoacridine (R = NH_2, R' = H)
9-Aminoacridine (R = H, R' = NH_2)

Mechanism of Action

Triphenylmethane dyes are thought to act by forming either a nonionized complex or nearly irreversible complex with acidic cellular components. This may involve the carboxyl group of proteins or the phosphate of nucleic acids. These complexes disrupt the normal activity of the biopolymer adversely affecting the microorganism.

Acridine dyes may function in a similar manner, but there are also indications that these molecules may interact with DNA through intercalation, thus disrupting DNA replication.[15] The organic dyes have found utility against gram-positive bacteria and fungi. A FDA panel on topical antifungal agents found insufficient evidence to support claims that the triphenylmethane dye, fushsin or carbolfuchsin (Castellani's Paint) were effective antifungal agents.[16]

Heavy Metals

A variety of inorganic mercuric salts and organic mercury (II) compounds as well as inorganic silver salts have been used as antiseptics, disinfectants, and preservatives for many generations. Although many of these heavy metal antiseptics are household names, most are now recognized as neither safe nor effective.

Mercury Salts

The previously used inorganic salts of mercury have included mercuric chloride, mercurous chloride, mercuric oxide, and ammoniated mercury. Still used in various capacities are the organic mercury compounds: merbrom, nitromersol, thimerosal, and phenylmercuric acetate. A factor which has undoubtedly reduced the usefulness of these agents has been that they are only considered to possess bacteriostatic activity. This action can be best understood by considering the mechanism of action of all mercury compounds.

Mechanism of Action

Mercury has a strong affinity for sulfur and it has been suggested that mercury compounds react reversibly with sulfhydryl groups present in proteins and enzymes:

$$\text{Protein-SH} + \text{R-Hg-X} \rightleftharpoons \text{Protein-S-Hg-R}$$

This reversible binding to proteins in microorganisms results in at-best static action. Because there are many more effective cidal agents available, use of mercuric compounds has decreased.

Specific Products

Thimerosal. This water soluble, colorless compound is used primarily as a preservative. It has been placed by

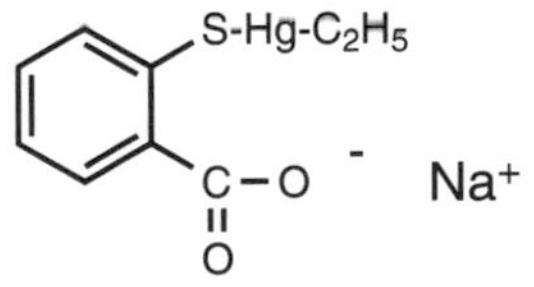

Thimerosal

the FDA panel on first-aid products in Category II as being neither safe nor effective. In addition, there is a sizable population of contact lens users who have become sensitized to thimerosal and experience contact dermatitis caused by thimerosal's presence as a preservative in certain contact lens' solutions. Products free of thimerosal should be recommended for such patients.

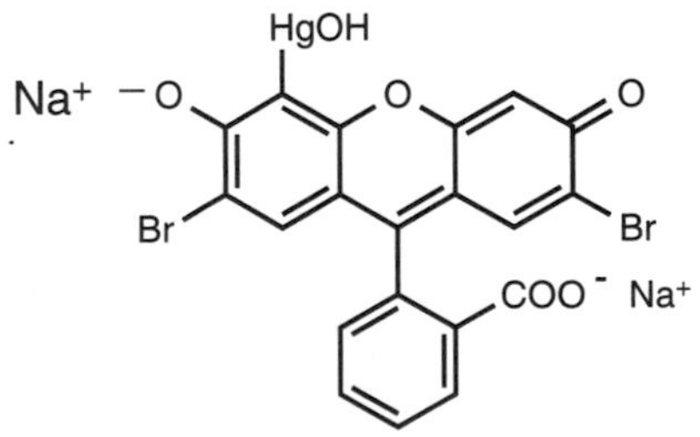

Merbromin

Merbromin. This is a water-soluble, red dye which can be found as a tincture in a 2% concentration as a topical antiseptic. Evidence supporting merbromin as being safe and effective has not been presented and as a result its continued use as a topical antiseptic appears unjustified.

Silver Salts

As with mercurial compounds, certain silver salts have a history of use as antiseptics and disinfectants dating back to Credé's first using silver nitrate as a prophylactic agent against gonorrheal opthalmia neonatorum in 1901. Silver picrate, colloidal silver protein, colloidal silver chloride, and silver nitrate-silver chloride (Toughened Silver Nitrate) have been available but most are no longer popular. The physical-chemical properties of silver salts do create some problems. Silver salts are easily reduced to a black metallic silver precipitate. Metallic silver is without biologic activity. This reaction is catalyzed by light, and thus silver salts must be protected from sunlight. The silver ion reacts with many anions (i.e., chloride) leading to precipitation reactions and many chemical incompatibilities. A precipitation reaction though also accounts for the mechanism of action of silver.

Mechanism of Action

The biologic activity of silver salts is derived from the free silver ion and its reactivity with a variety of anions: sulfhydryl, carboxyl, phosphate, amino groups, and chloride ion. Antiseptic action may result from the reaction of the silver ion with the carboxyl group on a key microbial protein. Precipitation of the protein in turn inactivates it and may produce a cidal effect. With human protein, this same reaction accounts for the astringent action of silver salts. Finally, the precipitation of silver protein or silver chloride may produce a protective layer over exposed tissue.

$$Ag^+ + \text{Protein-COOH} \rightleftharpoons Ag^+ \; ^-\text{OOC-Protein}$$

Silver Nitrate ($AgNO_3$). A 1% topical solution of silver nitrate is used to prevent gonococcal neonatal conjunctivitis in newborns. One drop in each eye is recommended, but in excess quantity, the solution can be an irritant. If dosed incorrectly, the irritation can lead to permanent damage and loss of vision. Erythromycin and tetracycline ointments are recommended by the CDC as replacements for silver nitrate.

Halogen Containing Drugs

Two inorganic halogen molecules have found uses as antiseptic-disinfectants since the early 1800s. Iodine (I_2) and hypochlorite ion (OCl^-), in a variety of forms, continue to be available for routine use.

Iodine

Iodine has been cited as "one of the best, if not the best, of the germicides".[17] Whereas this statement appears in a book published in 1985, most authoritative reference books neglect iodine completely. Iodine in its various transport vehicles has a long history of use, but potential toxicities, damage to tissue, difficulties with demonstrable activity, and lack of patient acceptance have raised questions about its suitability as an antiseptic. Recently, the FDA panel evaluating first-aid antiseptics has judged 2% iodine in aqueous and hydroalcoholic solutions as well as 5 to 10% concentration of povidone iodine as Category I agents, e.g., safe and effective. Evidence

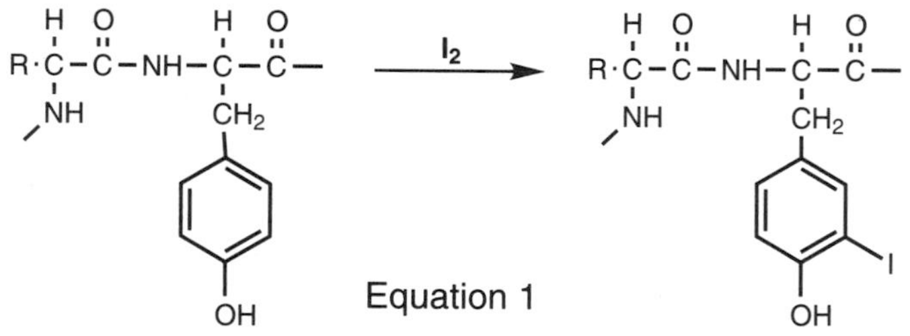

Equation 1

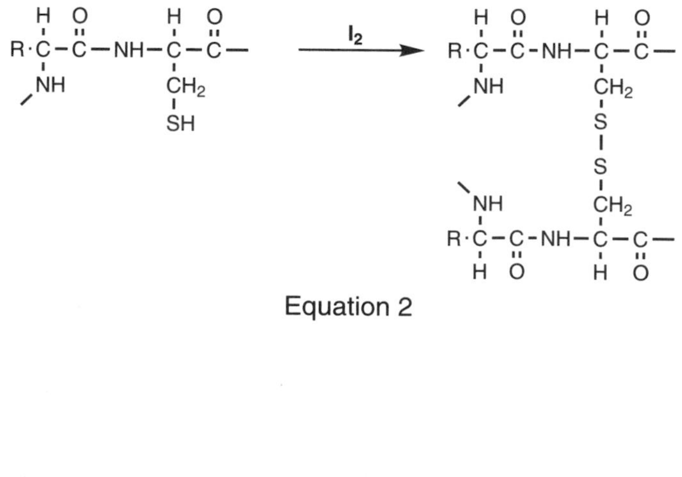

Equation 2

suggests that elemental iodine is active against bacteria, fungi, protozoans, and viruses.[3,10,18]

Mechanism of Action

Iodine is thought to attack protein in the pathogenic cell resulting in a detrimental change in the structure of the protein, and ultimately destroying the cell. The chemical changes suggested include iodinization or oxidation of specific amino acids in the protein. Tyrosine may be iodinated (equation 1) whereas the thio group of cysteine may be oxidized (equation 2). These destructive reactions are nonspecific, and the protein attacked may be that of the pathogen or of the host.

Specific Products

Tincture of Iodine. This common mixture consists of 2% iodine, and 2.4% sodium iodide in 47% alcohol/water. This preparation has good penetration into damaged tissue and dries rapidly. Its disadvantages include irritation of the wound, a stinging sensation, staining of the tissue, and the potential for toxicity. These have suggested caution in its use.

Aqueous Iodine. Aqueous iodine solution consists of 2% iodine, 2.4% sodium iodide in purified water. Iodine is insoluble in water, but in the presence of sodium iodide, a NaI_3 complex is formed, which is water soluble. The aqueous iodine solution is neither damaging to the wound nor does it cause a stinging sensation. Patients may find the product unacceptable because of the staining of the tissue, potential toxicity, and the slow-drying of the preparation caused by the solvent water.

Povidone-Iodine (Betadine [Purdue Frederick], Pharmadine [Sherwood Pharmaceuticals]). Povidone-iodine, is an example of an iodophor. An iodophor consists of a combination of iodine and an iodine-water solubilizing agent. In the case of povidone-iodine, polyvinylpyrrolidone is a water-soluble polymer that forms a water-soluble complex with iodine (equation 3).[19] The active ingredient remains iodine (I_2), but the iodine is strongly bound to polyvinylpyrrolidone, such that the mixture is nonstaining and has a low potential for toxicity. The absence of alcohol in the preparation reduces tissue damage and removes the stinging sensation upon application. Data to support the germicidal action of povidone-iodine has been presented allowing providone-iodine to be classified as a safe and effective agent when used in antimicrobial soap, as health-care personnel's handwash, as a surgical handscrub, as a skin preparation preoperatively, and for skin wound cleansing and protection.

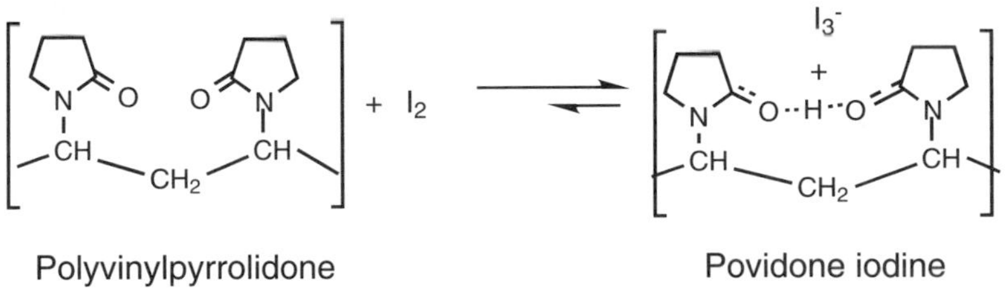

Polyvinylpyrrolidone + I_2 $\rightleftharpoons$ Povidone iodine (I_3^-)

Equation 3

Chlorine

Chlorine in the form of the hypochlorite ion is utilized as an effective disinfectant. Chlorine gas as well as several more stable forms of chlorine serve as a source of the hypochlorite.

Mechanism of Action

Several mechanisms have been proposed to account for the antimicrobial action of chlorine. Equation 4 outlines a mechanism that involves the chlorination of a protein. The result of chlorination is an altered microbial protein that would be expected to destroy the organism. Similar to iodine, chlorine is a nonspecific antimicrobial which cannot distinguish between host or pathogen protein.

Specific Products

Sodium Hypochlorite (liquid bleaches). Sodium hypochlorite (NaOCl) is a relatively strong base which was originally introduced as Modified Dakin's Solution, a 0.5 to 1% solution, during World War I, as an antiseptic.

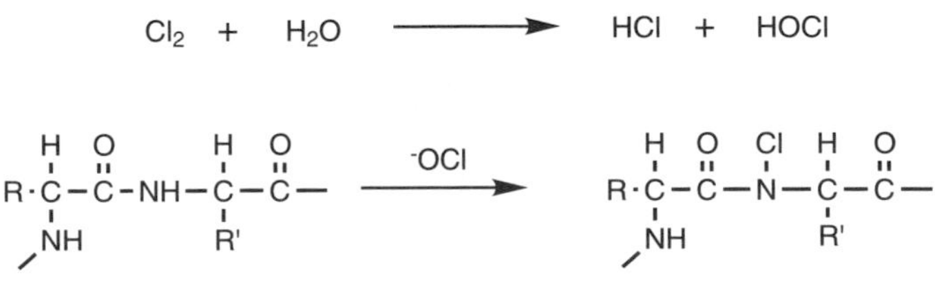

$$Cl_2 + H_2O \longrightarrow HCl + HOCl$$

$$R\text{-}CH(NH\text{-})\text{-}C(=O)\text{-}NH\text{-}CH(R')\text{-}C(=O)\text{-} \xrightarrow{^-OCl} R\text{-}CH(NH\text{-})\text{-}C(=O)\text{-}N(Cl)\text{-}CH(R')\text{-}C(=O)\text{-}$$

Equation 4

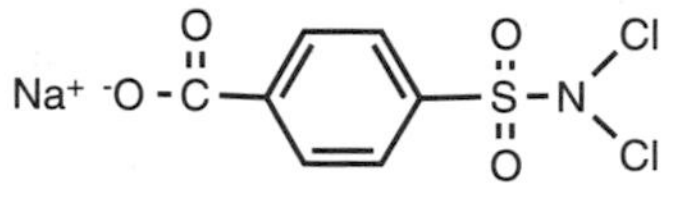

Halazone

Today, sodium hypochlorite in 5.25% concentration is a common ingredient in household bleach products and serves also as an effective disinfectant.

Halazone (Sodium N,N-dichloro-ρ-carboxybenzene sulfonamide). Halazone is a water soluble compound used as a water purifier. A chemical characteristic of this material is its instability towards water, including moisture in the air. In the presence of water, halazone decomposes into hypochlorous acid, the active ingredient of halazone (equation 5). Halazone tablets must be protected from atmospheric moisture to prevent decomposition of the product before use.

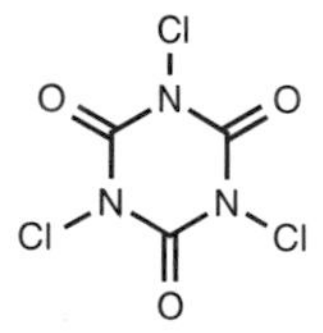

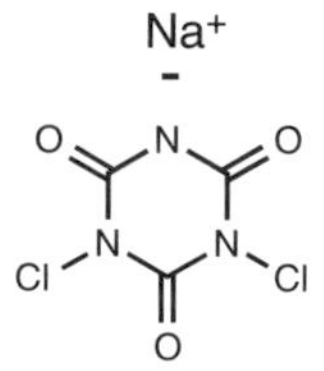

Equation 5

Trichloro-s-triazinetrione

Sodium dichloro-s-triazinetrione

Trichloro-s-triazinetrione (trichloroisocyanuric acid) and sodium dichloro-s-triazinetrione. Water-insoluble trichloro-s-triazinetrione and water-soluble sodium dichloro-s-triazinetrione are sources of hypochlorous acid, and are used as disinfectants. These products decompose in water, releasing either 3 mol or 2 mol, respectively, of hypochlorous acid. They are commonly found in commercial products used to purify swimming pools.

Chlorazodin. This water-insoluble source of 2 mol of hypochlorous acid, produced when chlorazodin is added to water.

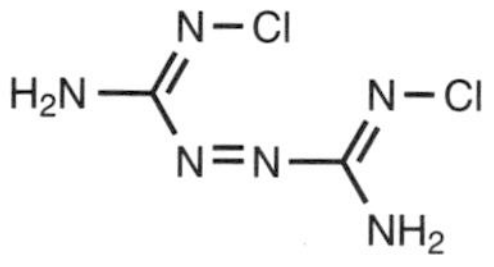

Chlorazodin

Miscellaneous Antiseptics and Disinfectants

Chlorhexidine Gluconate (Hibiclens [Stuart], Hibistat [Stuart]). Chlorhexidine gluconate is an antimicrobial antiseptic used as a skin cleaner. As such, the product is used as a surgical scrub, a health-care personnel handwash, and as a skin wound cleaner. The chemical has a high degree of substantivity but a low percutaneous absorption potential. Its mechanism of action may be due to strong binding of the drug to components of the bacterial cell wall.

Hydrogen Peroxide (H_2O_2). A weak acid, when hydrogen peroxide is used as a 3% concentration (10 volume) it is recognized as a topical antimicrobial agent. The hydrogen peroxide may prove beneficial as a skin wound cleaner because of its mechanical dislodging of oxygen. Hydrogen peroxide in the presence of a common tissue enzyme, catalase, is broken down to oxygen and water.

$$2\,H_2O_2 \xrightarrow{\text{Catalase}} 2\,H_2O + O_2\,(g)$$

The effervescent action of oxygen dislodges foreign material and cleanses the wound. In addition, hydrogen peroxide is an oxidizing agent and may oxidize chemicals in the microorganism to produce antimicrobial action. Hydrogen peroxide is a reactive substance which may undergo an oxidation-reduction reaction or precipitation reaction in the presence of metals. Light and heat also catalyze the destruction of hydrogen peroxide, resulting in the formation of water and oxygen.

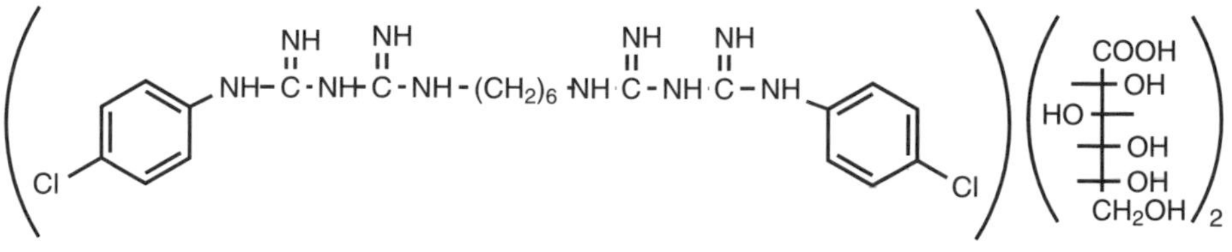

Chlorhexidine gluconate

REFERENCES

1. W. E. Gilbertson, *The FDA's OTC Drug Review,* In *Handbook of Nonprescription Drugs,* 9th ed., Washington, D.C., American Pharmaceutical Association, 1990.
2. Federal Register, *47,* 29986(July 9, 1982).
3. Federal Register, *56,* 33644(July 22, 1991).
4. K. A. Baker, *Oral Health Products,* in *Handboook of Nonprescription Drugs,* 9th ed., Washington, D.C., American Pharmaceutical Association, 1990.
5. *The Merck Index,* 11th ed., Rahway, N.J., Merck & Co., 1989.
6. FDA Drug Bulletin, December 1971 and February 1972, Washington, D.C., U.S. Department of Health, Education, and Welfare.
7. J. D. Lockhart, *Pediatrics, 50,* 229(1972).
8. R. M. Shuman, et al., *Arch. Neurol., 32,* 320(1975).
9. R. M. Shuman, et al., *Pediatrics, 56,* 689(1974).
10. H. N. Prince, *Part & Microb. Control, 2,* 54(1983).
11. Federal Register, *47,* 22324(1982).
12. G. H. Smith, *Diaper Rash and Prickly Heat Products,* in *Handbook of Nonprescription Drugs,* 10th ed., Washington, D.C., American Pharmaceutical Association, 1993.
13. C. L. Huyck, *Am. J. Pharm., 116,* 50(1944).
14. W. S. Pray, *U.S. Pharmacist,* 17(1992).
15. D. M. Neville and D. R. Davies, *J. Mol. Biol., 17,* 57(1966).
16. Federal Register, *47,* 12480(1982).
17. C. A. Discher, et al., *Modern Inorganic Pharmaceutical Chemistry,* 2nd ed., Prospect Heights, IL, Waveland Press, 1985.
18. A. J. Salle and B. W. Catlin, *J. Am. Pharm. Assoc., 36,* 129(1947).
19. H. Schenck, et al., *J. Pharm. Sci., 68,* 1505(1979).

Chapter 37

CANCER CHEMOTHERAPY

William O. Foye and Sisir K. Sengupta

NATURE OF CANCER

Occurrence

The most common cause of death in the United States in 1986 was heart diseases, accounting for 36.4% of all deaths. Cancer was second as a cause of death, accounting for 22.3%. This percentage had risen from 16.8% in 1967. As the U.S. population has aged, cancer death rates have increased. The cancer death rate per 100,000 population in 1930 was 143; in 1950, it was 157; and in 1986, it was 171.[1] (These data were statistically adjusted for age to make appropriate comparisons.) Lung cancer was the major cause for this increase, and in 1986, the deaths from lung cancer were approaching 50% of all reported deaths from cancer. Estimated deaths from various sites of cancer for 1990 are shown in Table 37–1.

Second only to cardiovascular ailments in incidence, cancer is feared more than any other disease. Most multicellular organisms can be afflicted by cancer, and cancerous lesions have been found in dinosaur bones. Only in the twentieth century, however, has there been much concern over the disease. Progress in the cure of the former major causes of death has inevitably led to a rise in the incidence of cancer. Cancer of the lung, alimentary tract, and breast account for more than 50% of all malignancies in the United States.

The outlook for survival rates from cancer shows improvement. Those who are alive 5 years after diagnosis were 1 in 5 in the 1930s, 1 in 4 in the 1940s, and 1 in 3 in the 1960s. More recently, 4 in 10 patients survive 5 years. In terms of normal life expectancy, 50% of cancer patients are now alive after 5 years. Poor Americans, regardless of race, have a 5-year survival rate that is 10 to 15% lower than the average.[1]

Terminology

Neoplasms

Various types of cancer have common properties that are definable. The medical term for "cancer" or "tumor" is neoplasm, which means "a relatively autonomous growth of tissue." Tumor is a general term indicating any abnormal mass or growth of tissue, not necessarily life-threatening. A "cancerous tumor" is a malignant neoplasm with potential danger. Anticancer drugs are also known as antineoplastic drugs.

The critical difference between benign and malignant neoplasms is that benign tumors do not metastasize, whereas malignant tumors (or cancers) do. A metastasis is a secondary growth originating from the primary tumor and growing elsewhere in the body (Fig. 37–1).

Because antineoplastic agents are often effective against only certain types of neoplasms, it is important to specify which types. Unfortunately, no system of nomenclature for tumors is accepted universally. Some tumors are named after the individual who first described the condition, such as Ewing's tumor of bone, Paget's disease, and Hodgkin's disease. Some are named after the tissue of origin, such as papillary, cystic, or follicular tumors. The suffix -oma literally means tumor, and words with this suffix refer to neoplasms. Exceptions are the terms granuloma, which is a growth of inflammatory tissue, and hematoma, a mass of blood within a tissue but outside the blood vessels.

Benign tumors are named with a prefix that refers to the tissue from which they arose and the suffix -oma. A benign tumor of fibrous tissue is called a fibroma; of cartilage, a chondroma; and of glandular tissue, an adenoma. Lymphoma is an exception, being a tumor of lymph tissue that may be malignant and dangerous.

Sarcoma

Beside indicating the tissue or origin, names for cancers are divided into two general categories according to their embryologic origin. In the early embryo of a multi-

Table 37–1. Estimated Deaths from Cancer Sites (1990)

Site	Estimated Deaths		
	Male	Female	Total
All sites	270,000	240,000	510,000
Oral	5,575	2,775	8,350
Colon-rectum	64,600	58,300	122,900
Respiratory system	95,900	51,200	147,100
Skin	5,700	3,100	8,800
Breast	300	44,000	44,300
Uterus		10,000	10,000
Genital (male)	30,600		30,600
Urinary organs	12,600	7,400	20,000
Leukemia	9,800	8,300	18,100
Other blood and lymph tissues	14,900	13,800	28,700

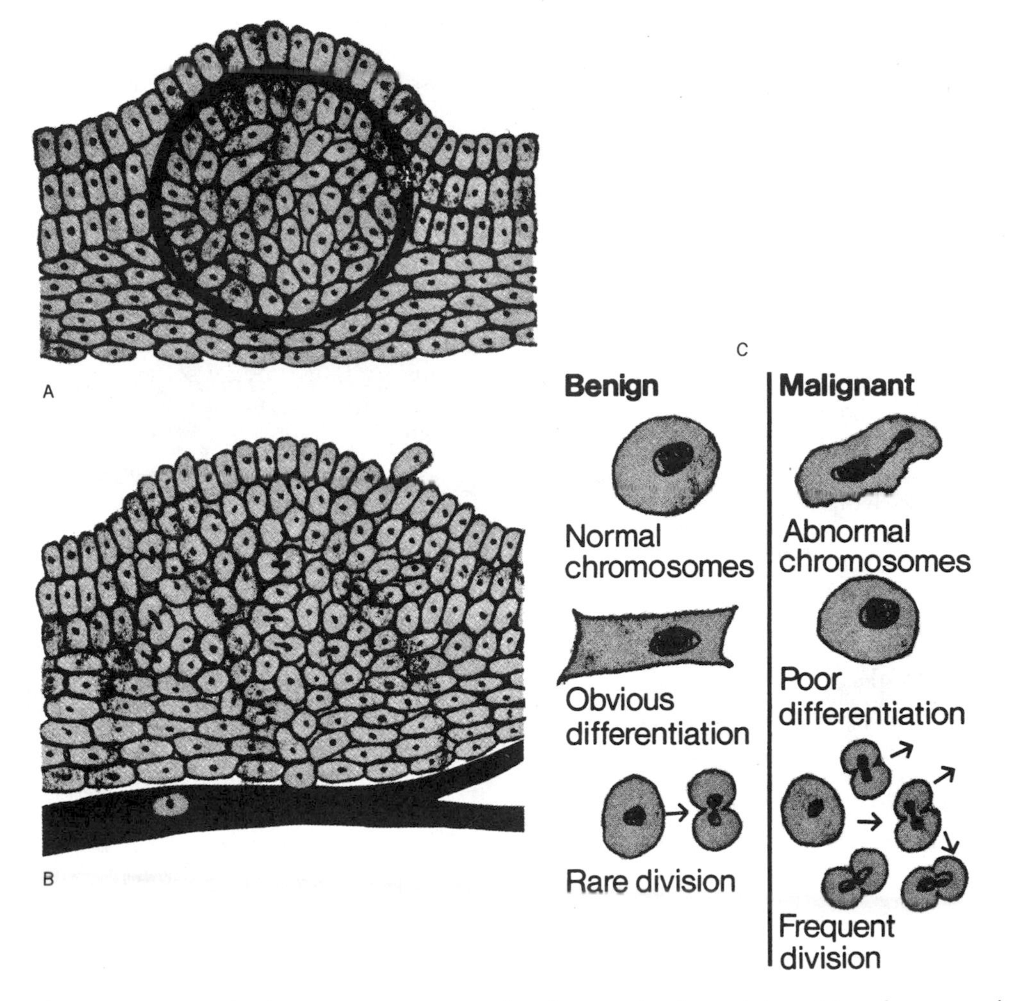

Fig. 37–1. Tumor characteristics. *A,* Benign tumors are encapsulated, grow slowly, and do not invade surrounding tissue, that is, metastasize. *B,* Malignant tumors grow rapidly, are nonencapsulated, invade surrounding tissue, and metastasize. *C,* Other characteristics. (From H. C. Pitot, *Chemistry, 50,* 11(1977), copyright by the American Chemical Society.)

cellular organism before organs begin to form, cells arrange themselves in three layers—ectodermal, mesodermal, and endodermal. Mesodermal cells form bone, muscle, cartilage, and related tissues. A cancer that arises from mesodermal tissue is called a sarcoma.

Carcinoma

Ectodermal cells form skin, its appendages, and nerve tissue. Endodermal cells form the intestinal system and its associated organs. A cancer that arises from ecto- or endodermal cells is called a carcinoma. A highly malignant tumor with the appearance of both a carcinoma and a sarcoma is termed a carcinosarcoma to indicate derivation from two embryonic layers. A tumor derived from all three embryonic layers, a teratoma, may be either benign or malignant.

By combining this nomenclature with more specific identification of the tissue from which the cancer arose, one may describe an adenocarcinoma of the stomach, pancreas, or breast. Adeno means glandular, and the organs named are endodermal. One may have a chondrosarcoma of cartilage, a fibrosarcoma from fibrous tissue, or an osteogenic sarcoma derived from bone.

Blastoma

Unfortunately, all cancer nomenclature does not fit neatly into this pattern. The suffix -blastoma is used to indicate certain types of tumors that have a primitive appearance resembling embryonic structures. Examples are the neuroblastoma of nerve tissue and the myoblastoma of muscle tissue.

Cancers of Blood

A cancer of the blood involving abnormal increase of leukocytes is called leukemia. In a normal person, the white blood cell count is about $7500/mm^3$. With leukemia, the number may increase to 10^5 to $10^6/mm^3$. Polycythemia is a slowly progressive disease, characterized by an increased number of erythrocytes and an increase in total blood volume.

Metastases

Cancer cells can detach themselves from the parent neoplasm and move to another location (metastasize) by several pathways. When a tumor reaches a critical size, it may shed cells into the circulatory system (see Fig. 37–1), but less than 0.01% of the cells in a tumor give rise to metastatic growth.[2] Another route for metastatic cells is the lymphatic system, which accounts for the frequency of growths in lymph glands near the site of the primary tumor. Improper surgical techniques in removing a tumor may deposit cancer cells in the surgical field. Surgery may also spread cancer cells by way of both blood and lymphatic systems.

In any case, cells from a primary tumor invade surrounding tissue, penetrating blood and lymph vessels. Single or multiple tumor-cell clumps, called emboli, are released into blood and lymph circulation. Circulating emboli are trapped in small blood vessels of other organs. Tumor cells penetrate the walls of these vessels into adjacent tissue, where they multiply. A new tumor grows and the process of metastasis may begin once again.[3]

Mutual Recognition of Cells

Cancer or cancer-like cells growing in cell cultures act differently from normal cells. Normal cells grow until an even layer of cells covers the culture dish, one cell deep, and then growth stops. Cancer cells grow beyond this point, piling up on one another. Cancer cells grow while floating freely, do not need to anchor themselves, and continue to divide even in culture dishes with surfaces coated with an adhesion-preventing substance such as agar. Normal cells must be attached to and spread out on the surface of a culture dish to grow. Such unrestrained growth and metastasis might occur if cancer cells do not "recognize" surrounding cells and stick to their surfaces. If lack of cell-to-cell adhesion is basic to cancer, the membrane forming the cell surface must play a key role, because all communication among cells has to go to or through this membrane.

Cell membrane structure, oversimplified, consists of a double layer of lipids, surfactant-like molecules having one end polar and water soluble and the other end nonpolar and water insoluble because it is composed of hydrocarbon chains. The polar ends of each side of the bilayer are toward the outside of the membrane, and the nonpolar ends are oriented toward the interior of the membrane. Several types of molecules are partially embedded in the lipid bilayer or are thicker than the bilayer, extending all the way through it. Examples of such molecules are glycolipids, proteins, and glycoproteins, all with mobility, at least part of the time.

One type of glycoprotein (about 250,000 m.w.) present in membranes is called the large external transformation-sensitive (LETS) protein. This protein is found in relatively large amounts on normal cells that have stopped growing and reached a resting state in culture. Actively growing or cancer-like cells contain relatively less or no LETS protein. It is possible that each exchange of sugar molecules by the LETS protein allows cells to recognize one another and triggers a response to stop dividing.[4]

Most membrane proteins are not glycoproteins. One protein in cancer cell membranes is a proteolytic enzyme or protease that can break down other proteins. Treating normal cells with proteases induces many changes similar to those seen in cancer-like cells. When normal cells touching each other and in a resting state are treated with protease, they can resume division, pile up on one another, lose their LETS protein, and undergo other structural changes.

Although cell surface changes may not be directly responsible for the onset or spread of cancer and might be secondary effects, a deeper understanding of membrane changes might provide some explanation of the action of presently known antineoplastic agents or the basis for design of future ones.

Cell Growth Cycles

A cell cycle is the period from the birth of a new cell to the time that the new cell divides. A new cell must double all of its parts before it can produce two new cells that can repeat the cycle themselves. DNA duplication is perhaps the most important doubling event because the genetic material of DNA must be copied exactly, and each daughter cell must receive a complete version. In a cell's nucleus, DNA replication occurs during only one specific part of the cell cycle, called S for synthesis. Between division and S is a period called G_1 (meaning gap 1) in which cells grow, but do not make DNA. In G_1, many molecules such as enzymes are synthesized. Another period called G_2 occurs between S and the period called M (mitosis), during which the two DNA copies separate. Although each of these four main periods of the cell cycle (G_1, S, G_2, M) is unique, all in proper order are necessary for new cell production. Normal and cancer cells behave differently.

Cells in tissue culture are in either a growing state, in which they advance through the cell cycle, or a resting state, in which they continue to use energy, but do not synthesize DNA or divide. This resting state, called G_0, is distinct from the other steps in the cycle of an active cell. When normal cells stop growing, they usually pass into a similar resting state, regardless of what stops their growth. Cells usually leave the cycle from the G_1 state to rest, but not the S or G_2 states. Resting cells most closely resemble G_1 cells in containing the same amount of DNA, which suggests that the G_1 state has a special, built-in mechanism that decides whether cells should proceed through the cycle or switch off into the G_0 state. Cells can return from the G_0 resting state back to G_1 if supplied with whatever they lack, such as hormones, or relieved of whatever might have been in oversupply, possibly LETS protein. Some sensing mechanism must inform the cells that conditions are right to resume progress through the cycle.

This mechanism does not work properly for cancer cells. Under conditions that send normal cells into the G_0 state, some cancer cells continue dividing. Yet in tissue culture, at least, cancer cells too may be made to stop growing by removing nutrients. Unlike normal cells, however, they stop not only in the G_0 resting state, but also anywhere in the cycle. If kept in this abnormal resting state, they do not later resume the cycle, but die. So some cancer cells are unusually dependent on favorable growth conditions. The unusual type of cell fragility might be useful in the design of new antineoplastic drugs. Many of these drugs are effective against cancer cells during only part of the cycle. Some drugs block DNA synthesis; therefore, these drugs act only during the S phase. Drugs blocking mitosis act only during the M state. Yet resting cancer cells are not affected by either of these types of anticancer agents, because they are not in the S or M state. Therefore, if a large fraction of cancer cells is resting, the cells are temporarily immune to the drugs. Worse still, some normal cells that happen to be in the S or M state will be killed. This outcome can create a major problem for cells that are supposed to replenish themselves at a relatively rapid rate, such as bone marrow cells, the source of erythrocytes, and intestinal cells, which are continuously sloughed off.

Mechanisms of Tumor Formation

Mutation

Mutation is the loss, substitution, and/or rearrangement of DNA in a cell. The link between the mutation theory and cancer was proposed by Boveri early in this century, is favored by many, yet is still controversial. In several cases, it was shown that tumors require no changes in genetic information and can develop without changes in cell genes (DNA).

Mouse teratocarcinoma illustrates how cancer cells may develop and perpetuate in mammals despite their having unaltered genes. The mouse teratocarcinoma is a highly malignant tumor made up of cells from all three embryonic germ layers. These tumors contain from 8 to 14 differentiated cell types such as bone, muscle, cartilage, epithelium, nerve, tooth buds, and hair follicles. These differentiated cells are mixed randomly with undifferentiated cells, called embryonal carcinoma cells, which are the malignant component of the tumor. When implanted in the abdomen of adult mice, the undifferentiated carcinoma cells produced tumors that contained all 8 to 14 differentiated cell types and rapidly killed the host. When they were implanted in early mouse embryos before normal cell differentiation had occurred, however, biochemically marked carcinoma cells, which could be identified as they multiplied in the developing mice, differentiated into most and perhaps all of the mouse's specialized cell types. These types included fully functional liver, kidney, thymus gland, and sperm cells, none of which are found in the teratocarcinoma itself. Most of the adult mice containing both normal and carcinoma-derived cells showed no malignancy. Somehow, single cancer cells differentiated with loss, or at least suppression, of malignant properties. The embryonal carcinoma cell of the mouse thus retains all of the genetic information necessary to reconstitute a normal mouse, apparently with no mutation of the DNA in the genes.

Addition of New Genetic Material

A second mechanism of tumor formation, which now appears well established, involves addition and integration of new viral genetic material into a cell's genes as a result of infection by tumor-producing viruses. A virus is known that causes cancer in chickens, and other viruses cause cancer in many other animals, including hamsters, mice, and monkeys.

Changed Gene Expression

The third mechanism, unlike the first two, involves no change in the integrity of the cell's genetic information, but rather involves a persistent change in the way a cell expresses or uses that information. Called epigenetic, this mechanism is similar to what occurs when plants or animals develop from single fertilized cells into different special cell types, such as nerve, liver, kidney, and skin cells. This process is called differentiation. If tumor production is epigenetic, the tumor state is potentially reversible and opens certain types of approaches to curative drugs. Tumor as well as normal cells differentiate according to which particular assortment of genes is active and which are inactive.

A cell's genetic information is not fixed, but is used according to conditions. For example, the Lucké adenocarcinoma of the frog is a highly malignant kidney tumor caused by a virus. Nuclei from this type of cancer cell were taken and implanted in frog eggs from which the nuclei were removed. Frog eggs that contained a cancer nucleus developed into normal, healthy tadpoles containing all the proper, specialized cell types, tissues, and organs. Therefore, cytoplasmic non-nuclear, and nonchromosomal factors in the normal frog eggs reprogram the cancer nuclei, just as they program noncancerous nuclei during development. The nuclei of Lucké adenocarcinoma cells, therefore, are not irreversibly altered by the virus.

In the second example, mouse neuroblastoma cells in culture can mature into stable, nondividing, specialized nerve cells when treated with the chemical dimethylsulfoxide.

External Causes of Cancer

Viruses

In 1911, Peyton Rous found a virus that causes cancer in chickens. As mentioned previously, it is also known that viruses cause cancer in other animals, including hamsters, mice, and monkeys. Since then, DNA and RNA viruses were found that cause cancer in humans. Among these oncogenic viruses is Epstein-Barr virus, which has an especially strong association with African Burkitt's lymphoma and nasopharyngeal cancer. Another DNA virus, hepatitis B virus, has been shown to cause hepatocellular carcinoma, mainly in Oriental populations. Herpes simplex virus type 2 (HSV-2) and human papilloma virus (HPV), two DNA viruses, have been closely associated with human cancer of the uterine cervix, but proof that they cause cervical cancer has not been established.[4]

One RNA virus, human T-lymphotropic virus type I (HTLV-1) has been proven to cause a form of leukemia, adult T cell leukemia-lymphoma (HTLL).[4] Both HTLV-1 and human immunodeficiency virus (HIV) have immunosuppressive properties from their effects on lymphocytes. Similar immunosuppressive oncogenic RNA viruses have been found in animals.

The isolation of several papovaviruses from human patients suggests they may be oncogenic; the same question has been raised with a human polyoma subgroup of papovaviruses. Recombinant DNA technology has enabled isolation of a number of virus genes that in cell culture are able to transform cells from a normal to a malignant phenotype. Development of vaccines based on oncogenic viruses has been difficult because of the long latent period of most cancers as well as the safety of tumor virus vaccines.[4]

Chemicals

A vast number of relatively simple substances cause some form of cancer. Long-term exposure to inhalation of asbestos particles produces a rare type of tumor. The shape and size of the asbestos fibers were found responsible, not their chemical composition. Compounds of beryllium, cadmium, chromium, nickel, and lead cause cancer in test animals, and it is thought that these, as well as arsenic, also cause human cancer.

Cigarette smoking is a major cause of cancers of the lung, larynx, oral cavity, and esophagus, and is at least a contributory factor in the development of cancers of the bladder, pancreas, and kidney. Cigar and pipe smokers have the same risk as cigarette smokers of developing cancers of the larynx, oral cavity, and esophagus.[5]

Skin cancers suffered by workers exposed to shale oil and coal tar are caused by numerous polycyclic aromatic hydrocarbons. One readily identified is benzo[a]pyrene,

Benzo[a]pyrene

Pyrene

but pyrene, which has a closely related structure, is noncarcinogenic. Bladder tumors in long-term workers in the dye-stuff industry were caused by aromatic amines, 2-naphthylamine in particular. A candidate for pesticide

Naphthylamine

2-Acetylaminofluorene

use, 2-acetylaminofluorene, was revealed to be a potent carcinogen. It was never used as a pesticide, but it has served researchers as a test carcinogen for over 30 years. Certain amino azo dyes, dialkylnitrosoamines, and alkylating agents proved carcinogenic in test animals.[6] Aflatoxins formed by the fungus Aspergillus flavus, besides having an acute toxic effect, are carcinogenic to livers of mice, rats, ducks, rainbow trout, and monkeys at much lower doses than any synthetic chemical. Many other naturally produced chemicals are carcinogenic, including safrole from sassafras root, cycasin from cycad nuts, pyrrolizidine alkaloids from the golden ragwort, and β-asarone from certain Calamus roots. Some of the medications, drugs, and other substances implicated as carcinogens in humans are listed in Table 37–2.

Aflatoxin B_1

Aflatoxin G_1

This great diversity of structures suggests different mechanisms of activity. Many theories of how certain carcinogens act on the proteins or DNA or normal cells are

Table 37–2. Drugs and Other Substances with Carcinogenic Potential*

Substance	Associated Neoplasms
Chemotherapeutic agents	
Cyclophosphamide	Leukemia, bladder cancer
Melphalan	Leukemia
Busulfan	Leukemia
Nitrosoureas	Leukemia
Chlorambucil	Leukemia
Nitrogen mustard	Leukemia
Procarbazine	Leukemia
Etoposide	Leukemia
Radioisotopes	
Radium	Osteogenic sarcoma
Thorium dioxide	Liver tumors
Radioactive iodine	Thyroid cancer
Hormones	
Prenatal diethylstilbestrol	Vaginal adenocarcinoma, seminoma
Androgenic steroids	Liver tumors
Estrogenic compounds	Endometrial carcinoma
Others	
Immunosuppressive therapy	Lymphoma
Aromatic amines	Bladder cancer
Phenacetin	Bladder and renal pelvic cancer
Inorganic arsenicals	Skin cancer
Phenytoin	Lymphoma
Chloramphenicol	Leukemia
Phenylbutazone	Leukemia
Coal tar ointments	Skin cancer
Intramuscular iron	Sarcoma at injection site

* From C.-H. Pui and W.M. Crist, in *American Cancer Society Textbook of Clinical Oncology,* A.I. Holleb, D.J. Fink, and G.P. Murphy, Eds., Atlanta, American Cancer Society, Inc., 1991, p. 456.

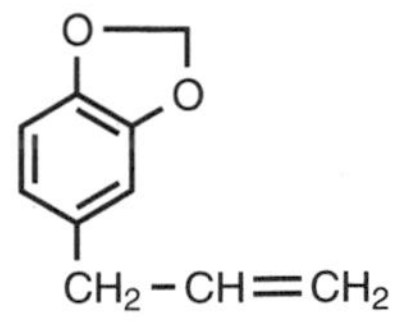

Safrole

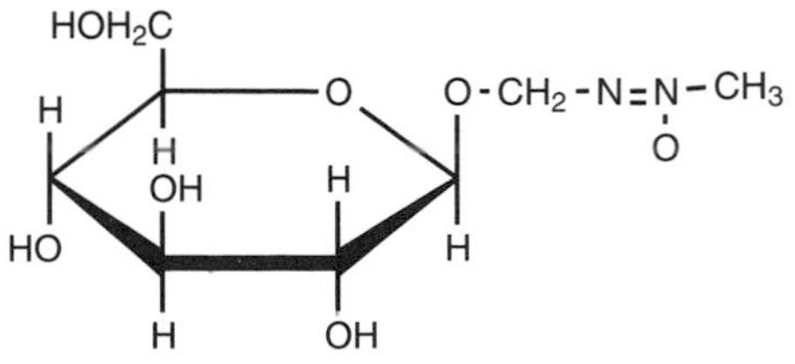

Cycasin

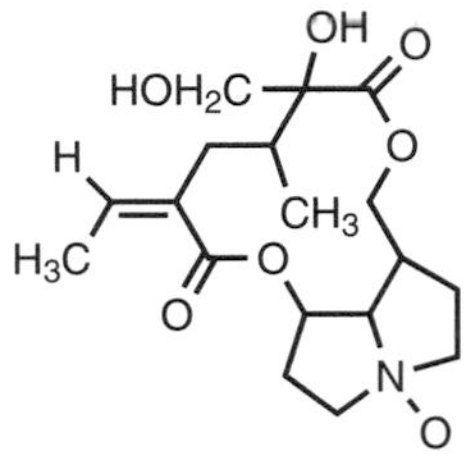

Isatidine (a pyrrolizidine alkaloid)

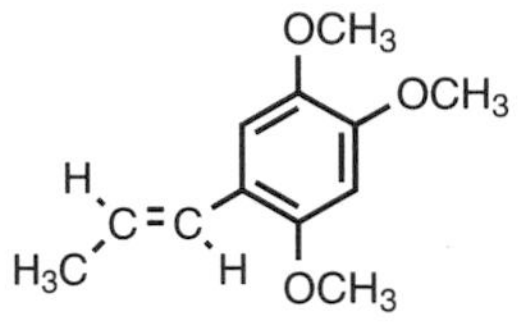

β-Asarone

disturbingly close to the theories rationalizing the action of antineoplastic drugs on cancer cells.

One baffling characteristic of chemical carcinogenesis in animals is latency—the occurrence of an extended period of time between the initial application of a carcinogen and the appearance of a tumor. Periods of at least 30 years have been observed between exposure of humans to carcinogens and the resultant formation of tumors.[6]

Radiation

Naturally occurring radioactive elements, such as uranium, radium, and thorium, are carcinogenic because of the ionizing radiation they produce. Miners in European uranium ore regions and in the Colorado Plateau of the United States suffer high cancer rates.[7] Women who painted watch dials with radium-containing paint absorbed the radium by pointing the paint brush with the lips and developed bone tumors. Bone tumors in test animals have been produced with x-irradiation or by feeding strontium-90. Atomic bomb explosions in Japan caused a large number of cases of human leukemia in addition to other types of cancers. Radiation can also increase metastasis of existing tumors.[3]

Survival Rates

Survival rates should not be used as a sole or primary measure of progress in cancer control, because factors unrelated to the efficiency of treatment play an important role in the determination of those rates and their trends. Survival rates give the probability of a person's remaining alive for a specified period after being diagnosed as having cancer; cure is usually defined as survival for at least 5 years. The rates are often expressed as the percentage of patients still alive at some specified time after diagnosis. Relative survival rates allow for the probability of dying from other causes. Real progress is being made, but optimism must be tempered by the knowledge that earlier diagnosis, itself subject to intense research,[8] may be increasing the time span between "onset" of cancer and death, rather than postponing death by efficacious treatment.

The 5-year survival rate for cancer of the lung is still only 13%. Incidence rates show a decrease in lung cancer in men (to 81.9 per 100,000 in 1986), but an increase in women (to 36.4 per 100,000).[1]

TREATMENT OF CANCER

Surgery

Even an extremely large tumor can be removed surgically with lasting benefit if it has not metastasized. Conversely, a small tumor that has dispersed even a few cells to other organs such as the lungs, liver, or brain cannot be treated successfully by removing the primary tumor alone. Disseminated forms of cancer such as leukemia cannot be attacked surgically. Some or all of the other forms of treatment discussed in this section are usually used in conjunction with surgery, sometimes as a precaution, but mostly as a recognized necessity for improving chances of long-term cure.

Photoradiation Therapy

Tumors can be localized and destroyed by making use of the selectivity and fluorescence of hematoporphyrin derivatives (HPD). To locate tumors, HPD is injected into the cancer patient. After a 24- to 72-hour interval, fluorescence can be observed in tumor tissues by use of probes with fiberoptics. Porphyrin also accumulates in certain normal organs of the host, including liver, kidney, spleen, and skin, the latter accounting for transient skin photosensitization. To destroy tumors, light from a tunable dye laser of 620 to 640-nm wavelength is directed onto the tumors, using fiberoptic extensions when necessary. Normal tissues, which have minimal absorbancy of 620 to

640 nm of light, are minimally damaged.[9] Therapeutic aspects are discussed in Chapter 38.

Radiation Therapy

Radiation therapy is superior to surgery when it effectively destroys a tumor and at the same time causes only minimal damage to the surrounding normal tissue. For example, in treating localized cancer of the larynx, surgery would require complete removal of the larynx, whereas radiation therapy allows it to be left intact, and patient survival rates are comparable to those undergoing surgery.

Combining surgery and radiotherapy often improves local control of a tumor. Radiotherapy can destroy microscopic cancer cells that may remain in the lymph nodes and surrounding tissue following surgery. Moreover, radiotherapy performed preoperatively can reduce large tumors and decrease local recurrence, and may decrease the chance of metastasis.

Radiation therapy is considered the treatment of choice for many cancers. It may preserve the function of an organ or minimize the mutilating effect of a surgical procedure. Gamma rays from radioisotopes, such as cobalt-60, and x-rays generated by instruments are the types of radiation used most often. Narrow beams are aimed at the tumor from different angles to minimize damage to surrounding tissues, because these electromagnetic waves are absorbed only partially by all types of living tissues. Experimental work seems to show that the use of densely ionizing radiations, such as beams of particles including protons and negative pi mesons, allows better concentration of the radiation damage in the tumor, with less damage to surrounding tissue. These particles have a short, definite range in which they are nearly completely absorbed.

Because all forms of ionizing radiation cause cancer under certain conditions, radiation therapy will always be subject to severe limitations. The use of compounds called radiation sensitizers, which increase the damaging effects of radiation to cells without an increase in the radiation dose, has given favorable results in some tumors not successfully treated by radiation or surgery alone. Examples of radiation sensitizers that have been used clinically[10] include halogenated pyrimidines such as 5-fluorouracil, dactinomycin, oxygen-mimetic nitromidazoles such as misonidazole, and razoxane, a bis(2,6-diketopiperazine) derivative.[11] Depletion of intracellular glutathione, such as L-buthionine sulfoximine, also increases radiosensitivity in hypoxic cells.[12]

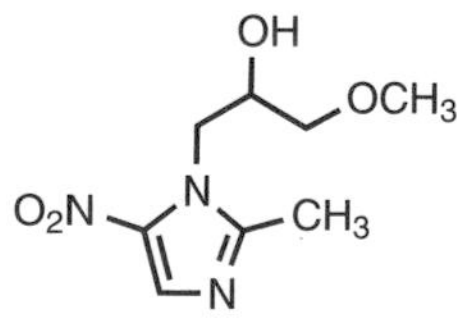

Misonidazole

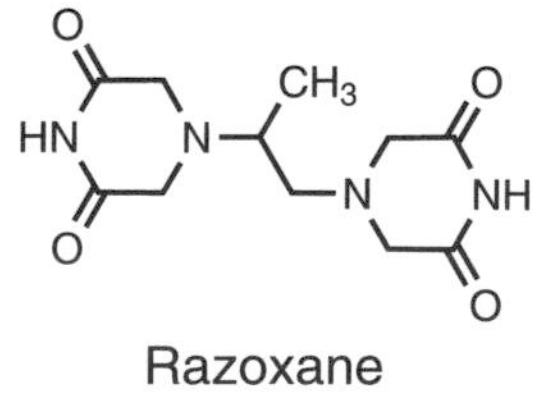

Razoxane

Immunotherapy

In follow-up treatment, the body's own defenses may be stimulated by immunotherapy to destroy the last few cancer cells remaining after surgery, radiotherapy, or chemotherapy.[15] This delays, sometimes for long periods, the spread or reappearance of a cancer.

Active nonspecific immunotherapy has been one kind of immune treatment studied. Such therapy is called nonspecific because it does not use antigens that are unique to the patient's own tumor. BCG (bacillus Calmette Guérin), the nonspecific agent most widely used, is a weakened derivative of the bacterium (Mycobacterium bovis) that causes tuberculosis in cattle. BCG is considered an active, nonspecific immunotherapeutic agent because it actively boosts general immune responses and stimulates macrophages (traveling cells able to neutralize invasive cells). BCG immunotherapy, begun after chemotherapy, produces remissions of greater duration than chemotherapy alone in some forms of childhood leukemia, adult leukemia, Hodgkin's disease, and head, neck, breast, skin, and colon cancer. Used before chemotherapy, BCG does not increase the remission rate.

The finding that various cytokines and other biologic response modifiers can augment immunologic reactivity against cancer has raised expectations that immune modulation may be successful in clinical cases. α-Interferon has been licensed by the U.S. Food and Drug Administration for treatment of hairy cell leukemia and Kaposi's sarcoma. Interleukin-2 has also shown strong metastatic effects in some experimental tumors. Immunotherapy for cancer is still at an early stage, however, and most of the promising strategies need to be developed in considerably greater detail.[13]

Chemotherapy

Limitations

In contrast to surgery and radiotherapy, chemotherapy is not so much limited by metastasis as by the total mass of the tumor(s). Although anticancer drugs permeate the body, acting on clumps of cells that may have lodged in other organs, the drugs have great difficulty in destroying all cells in a large tumor. A single course of therapy pushed to the limit of the patient's tolerance may destroy 99 or 99.9% of the tumor cells. A large tumor weighing 100 g and containing 10^{11} cells, after a 99.9% kill, would still leave 10^8 cells, often too much for the patient's immune response to control. The same treatment against a small clump of cells, 10^2 or 10^3, would leave so few cells that they might not be a threat.

The requirements for successful antineoplastic drugs in chemotherapy are not yet fully realized because the differences between cancer and normal cells are small according to present knowledge. Useful drugs without side effects do not yet exist. Rapidly dividing normal body cells, such as hair follicles, cells lining the gastrointestinal tract, and bone marrow cells involved in the immune defense system, are also destroyed by present antineoplastic drugs. This results in the common side effects of present-day chemotherapy—nausea, hair loss, increased susceptibility to infection, and many others in a discouraging list. At least the body recovers from most of these effects when the treatment is discontinued. Today's antineoplastic agents are generally palliative, not curative, with some exceptions.

Achievements

Cancer chemotherapy has been under intensive development for the past 40 years, resulting in cures of certain types of disseminated cancers that previously were fatal. Overt metastatic cancer should no longer be considered irrevocably fatal. In 12 of the more than 100 forms of clinical cancer, the use of present drugs results in longer life expectancy for a majority of patients. In advanced Hodgkin's disease, only about 5% of untreated patients reach normal life expectancy. Multiagent chemotherapy of advanced Hodgkin's disease produces a complete remission and 5-year disease-free survival in most patients. The MOPP regimen of nitrogen mustard (mechlorethamine), vincristine (Oncovin), procarbazine, and prednisone, first used by De Vita et al., gave a high complete response rate, with 66% of patients remaining disease-free more than 10 years.[14]

Mode of Action

Antitumor drugs are better at killing cells during DNA synthesis (S phase) and active division, that is, more active against cycling than noncycling cells. When a tumor is young, most of its cells are making DNA. This is defined as a large growth fraction. In this state, tumors are destroyed by drugs because the majority of their cells are making DNA and dividing. As it ages, the tumor's growth fraction decreases, and its drug sensitivity is reduced. The curable tumors are usually discovered while young, and their growth fraction is 30 to 100% of the total cells.

Only two fast-growing tumors, Burkitt's lymphoma and choriocarcinoma, have a large enough drug-susceptible fraction of cells to permit cure by a single agent when demonstrable metastases are present. Burkitt's lymphoma, most common in African children, appears in the jaw, rapidly spreads to many areas of the body, and eventually invades the central nervous system. Virtually 100% of its cells are in active mitosis, so that size doubles during 24 hours, and death occurs in 2 to 3 months. Because practically all cells in Burkitt's lymphoma are synthesizing DNA at the same time, it can be cured with a single dose of cyclophosphamide in 60% of patients. Most failures underscore an important concept in cancer chemotherapy, the existence of "pharmacologic sanctuaries" such as metastases in the central nervous system where otherwise highly active drugs cannot reach tumor cells. These exist because of the "blood-brain barrier," a barrier to polar compounds, which include most antineoplastic agents.

Gestational trophoblastic disease (GTD) and gestational choriocarcinoma originate in the trophoblastic elements of the placenta and grow rapidly. The cure rate was 20% with surgery and radiation therapy, now largely abandoned for this condition. With chemotherapy, over 75% of low-risk patients are curable by several courses of either dactinomycin or methotrexate, when 4 to 5 days of drug therapy are followed by a rest period of 2 to 3 weeks to allow repopulation of granulocytes, platelets, and orogastrointestinal epithelia. Oral etoposide has been administered to GTD patients with good results.

Antitumor agents are believed to act in many ways. They can react with the nuclei of cells as well as with the cell membrane, and in some cases with other cell organelles. They can act at all phases of the cell cycle by inhibiting several cellular processes, e.g., inhibiting the growth of the components in a cell such as DNA, RNA and protein. All these can go through a multitude of mechanisms, namely, interfering with the mode of action of enzymes such as folic acid reductive and thymidylate synthetic enzymes (see methotrexate and 5-fluorouracil). They act by interfering with or disrupting DNA-dependent enzymes such as DNA or RNA polymerases essential for replication and transcription of the cellular DNA. A large number of chemotherapeutic agents play a major role through these processes for their biologic activity, acting as inhibitors of growth and mammalian tumors or pathologic bacteria and viruses.

Combination Chemotherapy

Certain tumors with intermediate growth fractions (30 to 70%) are curable by drugs; however, combinations of drugs are necessary. Cells not active in DNA synthesis switch into active synthesis after treatment with a drug has killed the dividing cells. The tumor growth is slowed, but a single drug does not eradicate all cells because the reservoir repopulates the tumor soon after the course is completed. For example, a leukemic cell that is impervious to 6-mercaptopurine might still succumb to methotrexate, doxorubicin, or cyclophosphamide, because the chances of a given cell's development of resistance to all the drugs in a combination are small.

In addition to killing cells during DNA synthesis or mitosis, most antineoplastic drugs have a variable secondary killing capacity for cells in other stages of the cell cycle, designated G_1, G_2, or the proposed true latent phase, G_0. A combination of drugs may eradicate cells in these stages of the cell cycle, but it must be given more intensively and for relatively long periods. For example, in acute lymphocytic leukemia, single drugs or drugs given sequentially cure fewer than 1% of patients; but a combination of four drugs given intensively for 3 years or more in a regimen spanning several phases increases the cure rate to 25%. In the remission induction phase, vincristine and prednisone rapidly produce complete remission in about 90% of patients. Remission maintenance requires other drugs, however, most commonly methotrexate and 6-mercaptopurine. Other drugs are often added, and other variations of this regimen have been developed.

Perhaps the most gratifying success of combination chemotherapy has been the great improvement in survival rate of children with non-Hodgkin's lymphoma (a heterogeneous group of malignant lymphoproliferative disorders). For the one third of patients with localized disease, several regimens have produced cure rates of 90%. A recent trial of 6 weeks of therapy with prednisone, vincristine, cyclophosphamide, and doxorubicin followed by 6 months of maintenance therapy with 6-mercaptopurine and methotrexate, without radiation, gave prolonged disease-free survival in over 85% of cases. For high-risk forms of childhood lymphoblastic leukemia, 2-year disease-free survival is in the range of 60 to 75% (1985) with similar therapy.

Results have not been as satisfactory in leukemias of adults. Remissions in 50 to 70% of these patients can be achieved, but 75% of patients achieving remission will eventually relapse.

Adjuvant Chemotherapy

Nonresponsive tumors are usually diagnosed when they are old and have growth fractions below 8%. They require the combination of systemic chemotherapy with surgery and radiation, which are effective means of removing the old portions of the tumor. Chemotherapy should kill the two categories of tumor cells often left behind following local removal—the microscopic nests of cells in the tissue planes adjacent to the primary tumor left outside the surgical margin and clinically inapparent distant metastases. Both categories of cells are in the infancy of their cycle and are highly susceptible to drugs given after surgery.

In patients with osteogenic sarcoma, a bone cancer of young adults frequently occurring in the extremities, complete removal of the tumor by amputation of the affected limb successfully cures only about 20% of the patients. Within 6 to 18 months after surgery, tumor growth appears in the lungs. If intensive chemotherapy is given immediately after surgery, the small deposits of cells in the lungs are destroyed to give cure rates of 70 to 80%. This combined treatment method has improved the cure rate in Ewing's sarcoma, rhabdomyosarcoma, retinoblastoma, mycosis fungoides, and from a 50% cure rate to a 90% cure rate with Wilms' tumor.

Adjuvant chemotherapy, also called combined modality treatment, is also applied to the high-risk cancers such as cancer of the breast that has spread to the axillary lymph nodes under the arm and cancer of the large bowel that has spread to the lymph nodes in the bowel wall. One group of 211 patients with evidence of cancer that had spread to lymph nodes had 22% treatment failure in the 108 patients not receiving chemotherapy, and 10% failure in the 103 given melphalan.[15] In another study, a portion of a total of 386 patients with breast cancer and axillary node involvement were treated, after radical mastectomy, with a 2-week course each month of cyclophosphamide, methotrexate, and 5-fluorouracil; they had a 5% recurrence versus a 24% recurrence in the patients not given chemotherapy. A later report on the same group after 3 years, however, showed a recurrence rate of 26%, versus a 46% recurrence rate in the untreated group of patients.[15]

ANTINEOPLASTIC DRUGS

Antimetabolites

Antimetabolites inhibit a metabolic pathway essential for the survival or reproduction of cancer cells through

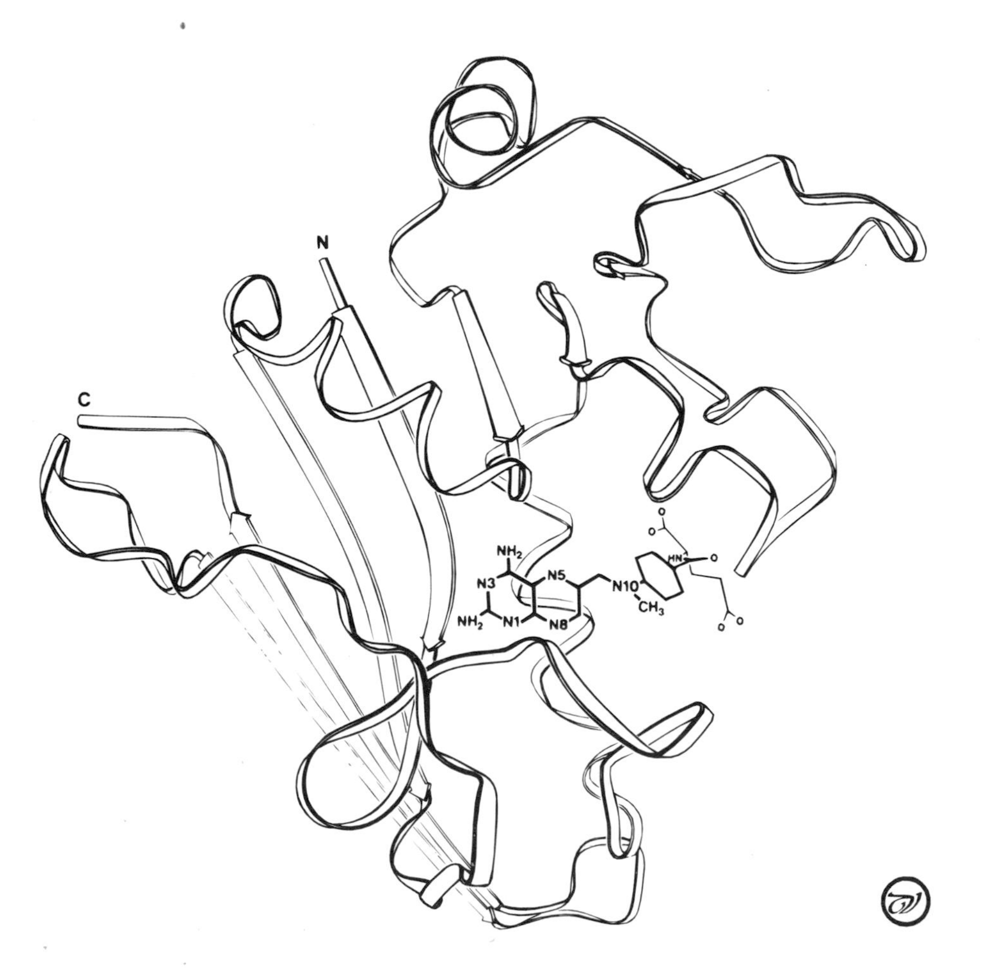

Fig. 37–2. Representation of the backbone chain folding of E. coli dihydrofolate reductase containing a bound methotrextae molecule. The drawing is derived from a computer-generated plot of all atoms in the drug and all α-carbon atoms of the enzyme. Strands of the central pleated sheet are shown as heavy arrows. (From D. A. Matthews, et al., *Science, 297,* 452(1977). Copyright by the American Association for the Advancement of Science.[17])

inhibition of folate, purine, pyrimidine, and pyrimidine nucleoside pathways required for DNA synthesis. No metabolic pathway unique for cancer cells has been found, yet antimetabolites can be used to kill tumor cells without killing the host, because of differences in cell growth fractions described previously. The antimetabolites have a narrower spectrum of use than the other classes of anticancer drugs.

Methotrexate

Previously known as amethopterin, this synthetic[16] acts as an antifolate by binding almost irreversibly to the enzyme dihydrofolate reductase and preventing the formation of the coenzyme tetrahydrofolic acid, essential for DNA synthesis and for replication of animal cells. The x-ray structure of the binary complex of dihydrofolate reductase from Escherichia coli with methotrexate has been determined[17] (Fig. 37–2). The methotrexate molecule is bound in a cavity that is 15 Å deep and cuts across one whole face of the enzyme. At least 13 amino acid residues (not shown) are involved in the binding. The drug is held in an open conformation. The pyrimidine end is buried in a primarily hydrophobic pocket. An explanation for the binding of methotrexate, 10^3-fold stronger than that of dihydrofolic acid, may be an indirect effect of the substitution of amino for hydroxyl on C_4. The substitution is known to increase the basicity of the pteridine ring system by 10^3, and it is thought that protonation of the ring occurs most readily at N_1. A fortuitously located carboxylate side chain of an aspartic acid residue in the enzyme interacts with N_1, and is thought responsible for the stronger binding of methotrexate. A direct consequence of this binding is that methotrexate inhibits not only the enzyme dihydrofolate reductase, but also thymidylate synthetase; these processes are vital for the metabolism of nucleic acid in all cells.

Physicians have learned to use methotrexate safely, despite its impressive potential toxicities, to the great benefit of patients with choriocarcinoma, the acute leukemias, osteosarcoma, and head, neck, and breast cancer. Methotrexate is readily absorbed from the gastrointestinal tract and distributed to the liver and kidneys; however, it is virtually excluded from the central nervous system. It is not significantlly metabolized, and excretion is mainly through the kidneys.

The recommended therapeutic dosages depend on normal kidney function, and in the presence of diminished clearance, plasma concentrations of toxic levels may result from typical dosages. Interactions of methotrexate with other drugs may change the concentration in either direction; concomitant use of aspirin raises the concentration of methotrexate by inhibition of tubular secretion. Vincristine increases the cellular uptake of methotrexate, and this effect has been used in the treatment of osteosarcoma.

Methotrexate can be given by any route, including the intrathecal (directly into spinal canal). Toxicity is a function of the duration of administration, and a prolonged 24-hour infusion is more toxic than a push injection of the same amount. Dosages are 2.5 to 10 mg/day, but larger doses, up to 1 g/day, can be given intravenously when "leucovorin rescue" is used. Leucovorin supplies the cells with tetrahydrofolate, and, when started after the methotrexate and continued for 72 hours, diminishes the toxicity to the host without abolishing the antitumor effect. It has been recommended to carefully monitor plasma levels of methotrexate to detect patients at high risk of toxicity, so that leucovorin rescue can be applied appropriately.[18]

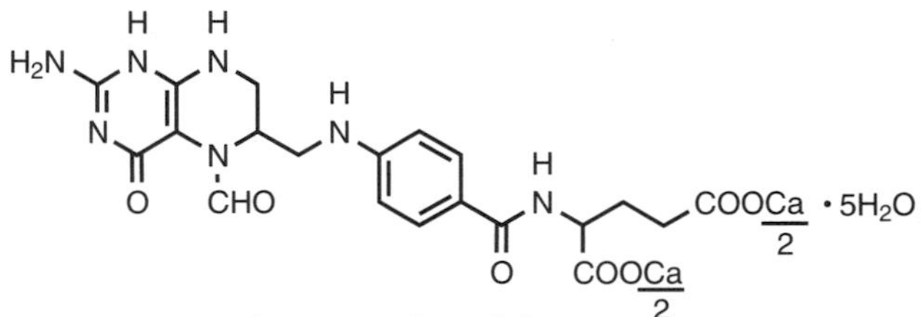

Leucovorin calcium

6-Mercaptopurine

Synthetic 6-mercaptopurine (6MP)[19] is an antagonist to purines, which are essential constituents of DNA. It serves as a specific replacement for hypoxanthine, a naturally occurring intermediate in DNA synthesis. 6-Mercap-

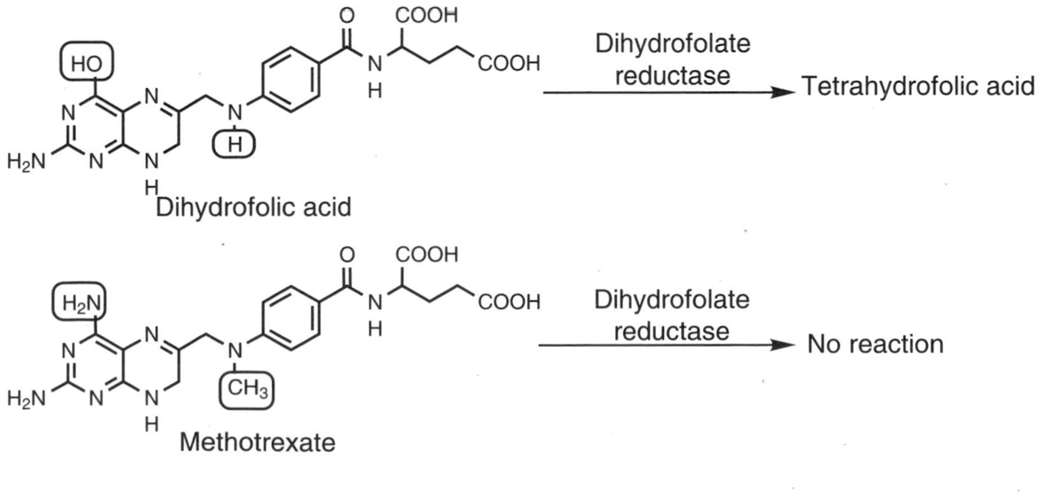

Dihydrofolic acid —Dihydrofolate reductase→ Tetrahydrofolic acid

Methotrexate —Dihydrofolate reductase→ No reaction

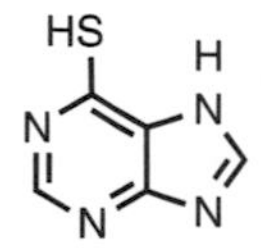

6-Mercaptopurine

topurine is first converted to the ribonucleotide and interferes with a number of metabolic pathways, most importantly with the first step in purine biosynthesis.

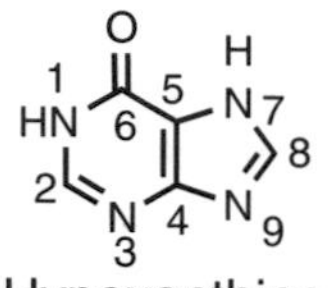

Hypoxanthine

6-Mercaptopurine and 6-thioguanine (6TG) are used almost exclusively for treatment of the leukemias. Absorption is incomplete after oral administration. There is rapid metabolic degradation of 6MP by xanthine oxidase to 6-thiouricacid; therefore, dosage must be reduced if a

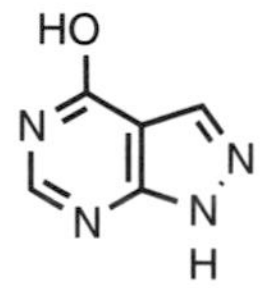

Allopurinol

patient is concomitantly receiving allopurinol, a xanthine oxidase inhibitor. Toxicity consists of bone marrow depression and orogastrointestinal damage similar to that seen with methotrexate.

Both 6MP and 6TG are inactive until metabolized to their respective monophosphate ribonucleotides, which result from action of the enzyme hypoxanthine-guanine phosphoribosyl transferase (HGPRTase). The monophosphates can inhibit de novo purine synthesis, i.e., the formation of adenylic and guanylic acids from inosinic acid. Further phosphorylation gives the triphosphate nucleotides. In the case of 6TG, conversion to the deoxyribose nucleotide allows the drug to be incorporated into DNA, the extent of which correlates with the production of strand breaks and cellular toxicity.

6-Thioguanine

The action of 6TG is similar to that of 6MP,[20] but this compound differs from 6MP in that it is actually incorporated into the DNA of normal and cancerous cells. Its

6-Thioguanine

Guanine

sole use is in the treatment of acute myelocytic leukemia in combination with cytarabine. It can be given orally or intravenously. Its pharmacology and toxicity are similar to those of 6MP, except that it requires no dose reduction in the presence of allopurinol.

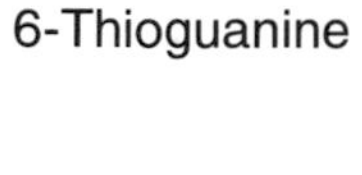

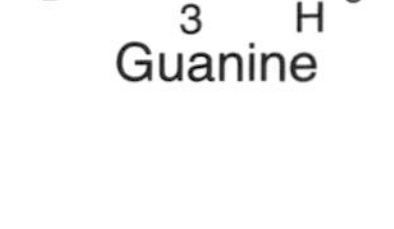

Cytarabine

Cytarabine (ARA-C), a pyrimidine nucleoside antagonist (synthesis[21]), is also known as cytosine arabinoside.

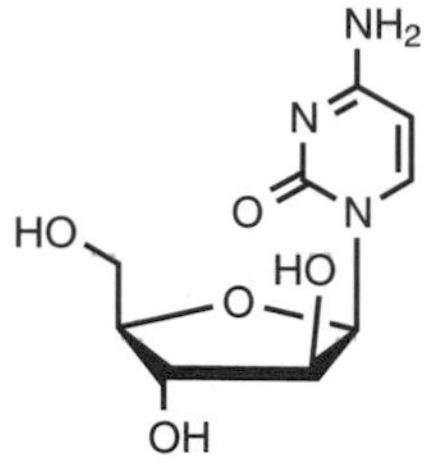

Cytarabine

Cytarabine must be converted into the nucleotide triphosphate ARA-CTP to be active and, as such, inhibits the binding of deoxycytidine triphosphate to DNA polymerase; DNA synthesis is thus arrested. ARA-CTP may also be incorporated into DNA, and cause defective ligation or incomplete synthesis of DNA fragments.

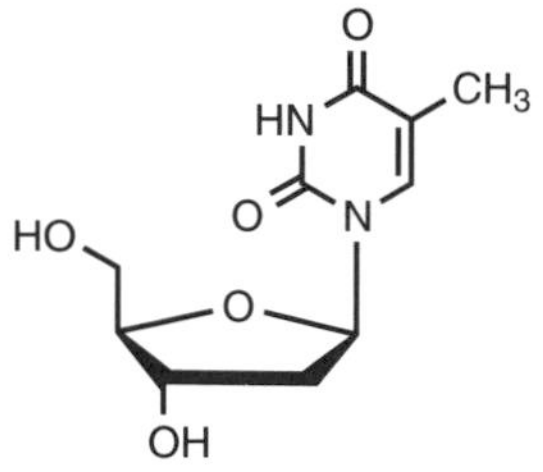

Thymidine

Like another adenosine analog, ARA-A, which has an adenosine ring instead of a cytosine, ARA-C is known to act as a triphosphate derivative in inhibiting DNA polymerase. These compounds have antiviral activity. ARA-A is also deaminated by adenosine deaminase, which can be reversed in the presence of another nucleoside, 2′-deoxycorfomycin or dihydrouracil. The combination of ARA-C and ARA-A with a deaminase inhibitor is likely to have selective effects only against tumors with high enzyme levels. It must be given intravenously and can safely be used intrathecally. In the body, it is rapidly deaminated by a kinase to an inert moiety, arabinosyl uridine, which is excreted in the urine. Because of this rapid degradation, ARA-C must be given by continuous infusion.

Its major use is in acute nonlymphocytic leukemia. Multiple agents have been used with varied success to combat the high incidence of relapse, but remission rates have been low (about 20%). It is a bone marrow depressant, causes lesions of orogastrointestinal epithelia, and occasionally gives rise to hepatic and renal toxicity.

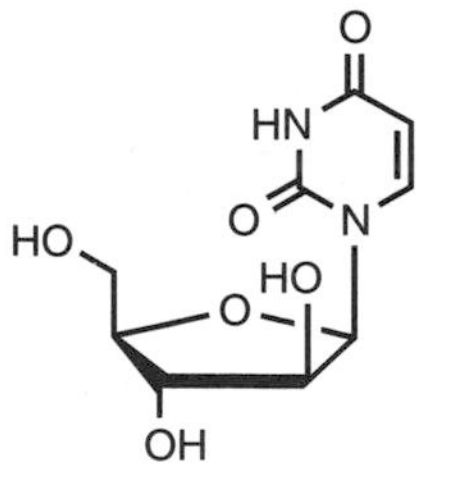

Arabinosyl uridine

5-Fluorouracil

In 5-fluorouracil (5-FU), a fluorine atom is substituted for hydrogen in the 5 position of the pyrimidine of uracil

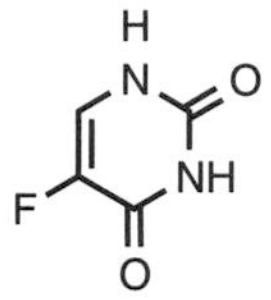

5-Fluorouracil 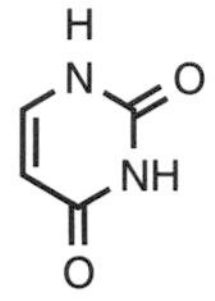   Uracil

(synthesis[22]). This carbon-fluorine bond is extremely stable and precludes addition of a methyl group in the 5 position, preventing the formation of thymidine. 5-FU must be phosphorylated to the nucleotide to be active and, as such, inhibits thymidylate synthetase, a key enzyme in the biosynthesis of DNA. It has at least two biochemical actions that may account for its cytotoxicity. It is converted first to the monophosphates 5-FUMP and 5-FdMP; the latter binds tightly to thymidylate synthetase and inhibits the eventual synthesis of DNA. On the other hand, 5-FUMP, after conversion to 5-FUTP, is incorporated into RNA and inhibits RNA processing of mRNA and rRNA, and may cause errors in base pairing during RNA transcription.

This pyrimidine antagonist is used for colorectal cancer and has modest activity in pancreatic and other gastrointestinal tumors. 5-FU has also been used intra-arterially for treatment of hepatic metastases from colorectal cancer, but it has considerable toxicity. Its effectiveness has been enhanced with concomitant treatment with leucovorin. When FdUMP binds to thymidylate synthetase, it forms a ternary complex with N^5,N^{10}-methylenetetrahydrofolate. High levels of leucovorin produce optimal conditions for formation of the complex. Clinical trials show the combination to be more effective than 5-FU alone in advanced-stage colon cancer.

Although some 5-FU is absorbed after oral administration, it should be used intravenously. The half-life of 5-FU in the blood is 15 to 30 minutes. It is rapidly catabolized; many products appear in the urine, and many of the carbon atoms are disposed of through respiratory CO_2. It crosses the blood-brain barrier, but when given intrathecally or into a carotid artery, it forms fluorocitrate, a metabolite that causes cerebellar ataxia. 5-FU is given on an intermittent schedule, either as a 4- to 5-day course at about 3- to 6-week intervals, or once weekly after an initial course. It can have devastating bone marrow and gastrointestinal toxicity, but these effects can be obviated by intermittent schedules and careful monitoring. Hair loss, skin rashes, or cerebellar dysfunction are noted occasionally.

Dacarbazine

Dacarbazine (DIC, DTIC), an antimetabolite (by synthesis[23]) of 5-aminoimidazole-4-carboxamide, has been shown to inhibit DNA synthesis, possibly by blocking the formation of inosinic acid.

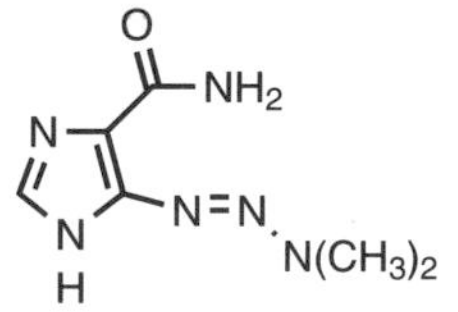

Dacarbazine 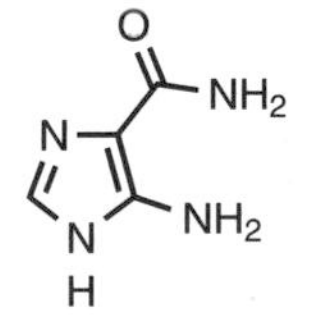   5-Aminoimidazole-4-carboxamide

It is used in malignant melanoma, but response rates average only 15 to 20%. Combinations of drugs have not been better than DTIC alone, but recent trials with cis-

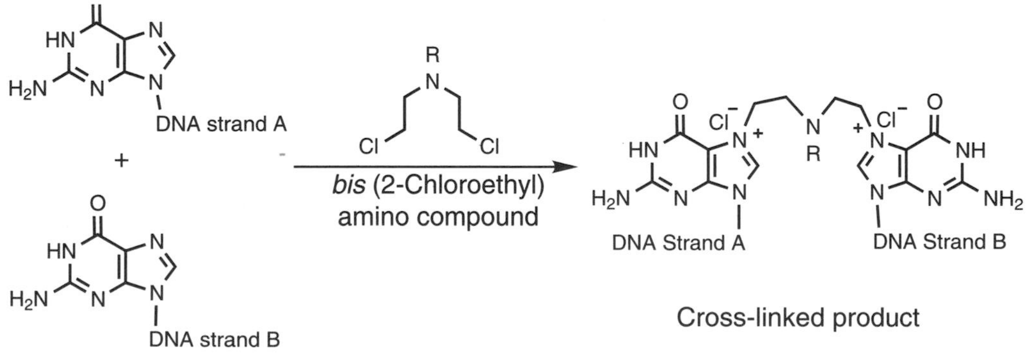

Guanine residues in DNA → *bis* (2-Chloroethyl) amino compound → Cross-linked product

platin have shown high response rates. It is also used in combination therapy for several soft-tissue sarcomas. Dacarbazine is used intravenously and has a half-life of 30 minutes. It is metabolized, and the products appear in the urine. Toxicity includes bone marrow depression, gastrointestinal erosions, and vomiting. It occasionally produces an influenza-like syndrome.

Deoxyribonucleic Acid (DNA)

DNA stores genetic information and is known to be the carrier of hereditary characteristics from one generation to the next. In the cell, it is combined with protein and is located in the chromatin or the chromosomes that are in the nucleus of cells. When a cell divides, the resulting daughter cells contain the genetic information, and all the characteristics of the parent cell unit are transferred and maintained in daughter cells until induced by evolution or mutation to do otherwise. The cell carries the information necessary for its exact duplication in its DNA in the precise sequence of purine and pyrimidine bases along its sugar phosphodiester backbone. DNA is actually a double-stranded helix of two individual molecular chains. The helices are held together by hydrogen bonding between pairs of bases in opposite positions in the two chains. The molecular geometry is such that the adenine forms a strong reciprocal bond with thymine (in AT) and guanine to cytosine (in CG) (Fig. 37–3). The two interwoven chains have exactly complementary structures. DNA is known to assume three different conformations: A, B, and Z.

In the three-dimensional conformation of DNA, two right-handed polynucleotide chains are coiled in helical fashion around the same axis, forming a double helix. The two chains are antiparallel, i.e., their 3′, 5′-internucleotide phosphodiester bridges run in opposite directions. In the B form of DNA, the sugar phosphate chains are on the outside of the helix coils, and purine and pyrimidine base pairs are stacked in the inner core of the helix with their planes parallel to each other with a center-to-center distance of 0.34-nm units from each other. Exactly 10 nucleotide residues are in each complete turn of the double helix, and therefore each helix pitch is 34 nm. In the Watson-Crick double-helix model, B DNA has one shallow (minor) and one deep (major) groove. The relatively hydrophobic bases form the inner core of the double helix and are shielded from water, while the polar sugar residues and the negatively charged phosphate groups are located on the periphery, exposed to water.

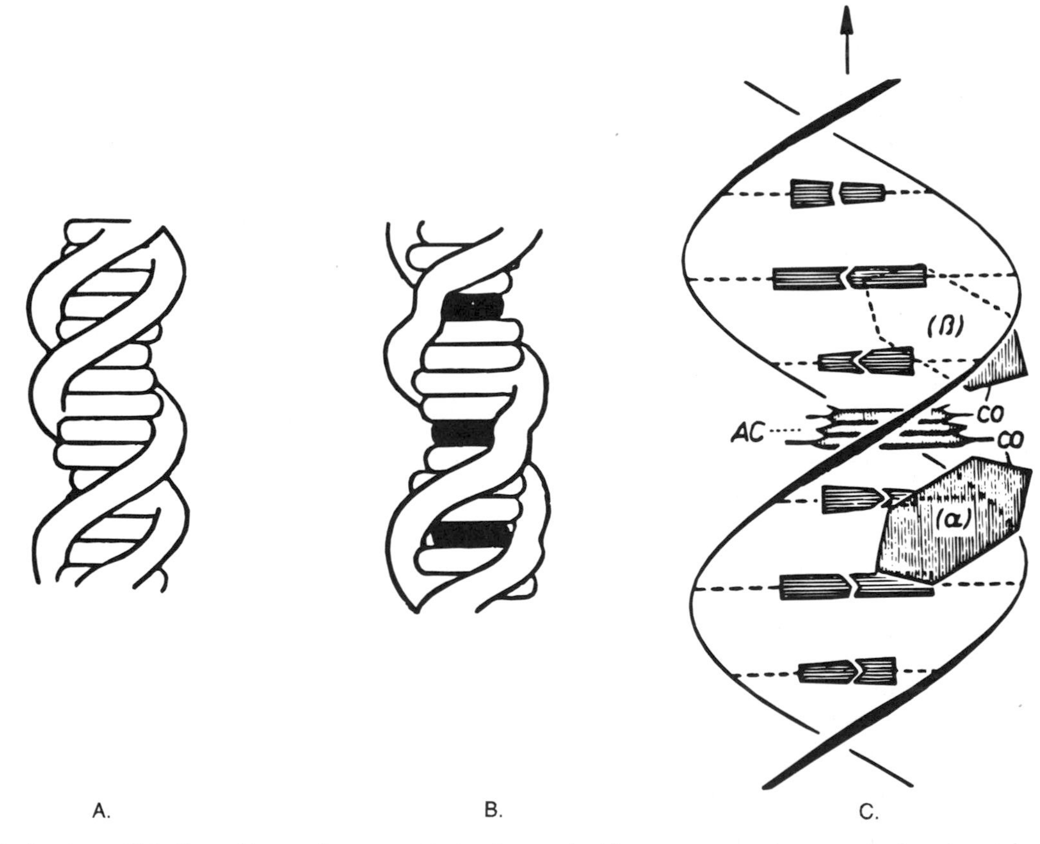

Fig. 37–3. Mechanism of binding of intercalators to DNA. The stacked base pairs are shown as hollow bands in B-DNA *(A)* and the sugar phosphate backbone is drawn as coils wound around the stacked base pairs. *B* illustrates an intercalated drug-DNA complex; the dark bands are intercalators. Deformation and lengthening of the DNA-backbone takes place due to intercalation. *C* is a representation of actinomycin D intercalation into DNA; the base pairs are shaded with horizontal lines and the actinomycin is shaded with vertical lines. The tricyclic chromophore is designated AC, and the cyclic peptides are α and β. The DNA helical axis is ↑ with the 3′–5′ phosphate linkage in the backbone.

The double helix is stabilized not only by hydrogen bonding of complementary pairs but also by electronic interactions between the stacked bases, as well as by some other hydrophobic interactions.

In addition to the B form, DNA can exist in the other helical A and Z forms. In the classic A conformation, it is believed that there are 11 nucleotide pairs per turn and the helix pitch is 28 nm, and the bases are tilted 20° from perpendicular to the helix axis (in the B form, bases are nearly perpendicular to the helix axis) but maintain the right-handed coil. Generally, highly A-T-rich segments of DNA tend to form the A conformation of DNA, and highly G-C-rich portions of DNA are presumed to take the Z form of DNA, especially in relatively high salt concentrations. The Z-DNA are believed to be in the left-handed helical conformation. The sugar phosphate groups are believed to be embedded in the inner core, thus exposing the hydrophobic base pairs on the outside. In this conformation, the bases appear to be more vulnerable to attack from outside agents, especially by alkylating agents, i.e., the agents that form chemical and covalent bonds with the electron-rich segments of DNA; these bonds are not dissociated.

The DNA in solution is constantly in a dynamic state; the chains elongate and compress in a series of waves alternately breaking and re-forming the hydrogen bonds. The phenomenon gives rise to the process known as "breathing." Breathing of DNA makes the core of DNA accessible for reaction by drug molecules. In this process, relatively flat and large molecules can be inserted in the space between adjacent base pairs. Antitumor actinomycins are known to bind to DNA by this process.

Figure 37–3 shows two simplified models of B DNA. In this model, the stacked base pairs (hollow bands) are wrapped around by the right-handed coils (hollow coils) of sugar phosphates. Intercalation is a process of binding to DNA that takes place by physical binding forces only. During intercalation, the intercalators (dark bands) are inserted between and parallel to the adjacent base pairs. One of the results of intercalation is that the length of the DNA helix is increased; this results in a change in the hydrodynamic properties, namely, the intrinsic viscosity and sedimentation coefficient of DNA. The process also introduces some distortion in the DNA helix, with the result that the DNA conformation is altered, and this change can be monitored by examining the circular dichroism and electrical dichroism of the complex. In some cases, when the intercalator itself has defined dichroic properties, as in actinomycin D, the intercalation can be examined by monitoring change in the circular dichroism of both, that is, the interactor as well as DNA. In Figure 37–3, the DNA intercalation of the antitumor drug actinomycin D is shown. Actinomycin D has a tricyclic chromophore that is symmetrically substituted with two identical peptide rings, α and β, which give the molecule dichroic properties. In the unbound state, the peptides are hydrogen bonded by interannular bonds. In the bound form, these bonds are broken and the peptides separate and bind to the narrow groove in DNA. Simultaneously, the tricyclic ring chromophore (AC) is inserted between the neighboring G-C base pairs in DNA, which causes change in the circular dichroism of both actinomycin and DNA.

Some agents may act by altogether different processes, e.g., by binding preferentially to the minor groove of DNA by some kind of covalent but labile bond formation; the antibiotic anthramycin is known to interact with DNA by this mechanism.

The molecule of DNA can acquire several well-defined conformations and orientations, and thus its many segments can assume numerous topologic shapes. This variability is especially true for DNAs that are present in the cell nuclei, and therefore can offer a number of discriminating sites for interaction by many DNA-reactive agents. Therefore, in spite of the apparent similarity in the mode of their reaction to DNA, e.g., covalent bond formation or intercalation, the drugs can elicit different biologic effects.

Alkylating Agents

Alkylating agents are reactive compounds that act on DNA, RNA, and certain enzymes. Their original use stems from the observation during World War I that individuals heavily gassed with mustard gas, *bis*(2-chloroethyl)sulfide, suffered damage to bone marrow and lymphoid tissues.

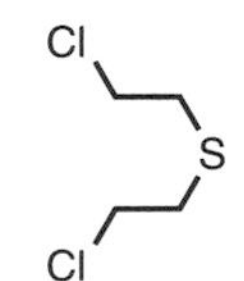

bis (2-Chloroethyl)sulfide

Studies on animals during World War II demonstrated that heavy exposure to the nitrogen mustards, *bis*(2-chloroethyl)amino compounds, destroyed lymphoid tissues. It was decided to use these chemicals cautiously to treat patients with cancers of the lymphoid tissues, such as lymphosarcoma and Hodgkin's disease. The drugs were successful in shrinking tumor masses, but also damaged normal bone marrow. Eventually, the damage during extended treatment made it impossible to continue therapy. Many derivatives of the nitrogen mustards have been synthesized with various improvements.[15] These agents are thought to react with the 7 position of guanine in each of the double strands of DNA, causing cross-linking, which interferes with separation of the strands and prevents mitosis.

Some members of this group of drugs are absorbed after oral administration, but many require intravenous use. Their active moieties disappear rapidly from the blood, usually within 2 to 15 minutes. They are distributed to all tissues except those of the central nervous system. All are toxic to bone marrow, cause immunodepression, and are carcinogenic[6] and mutagenic.

Mechlorethamine

The original nitrogen mustard mechlorethamine (HN2) has largely been supplanted by new congeners. It is, like mustard gas, a strong vesicant, cannot be taken by mouth, and was usually injected into the tubing of a

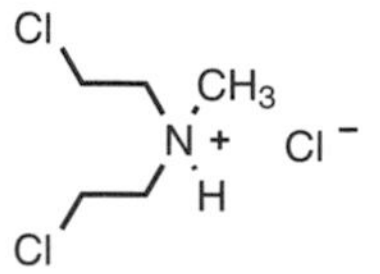

Mechlorethamine hydrochloride

running infusion, which was repeated at monthly intervals. The maximum effect on granulocytes, lymphocytes, and reticulocytes is at 10 days; recovery is complete at 21 to 28 days. Nausea and vomiting occur within minutes to an hour after injection in most patients, and prophylactic sedation and administration of antiemetics are advised. The principal use of mechlorethamine is in combination chemotherapy of Hodgkin's disease and the non-Hodgkin's lymphomas. It has wide activity, but more recent agents are safer and easier to use. A major disadvantage of this and other alkylating agents is a mutagenic, and ultimately carcinogenic, effect on bone marrow stem cells, culminating in a form of acute myelogenous leukemia. Mechlorethamine has had veterinary use, including treatment of lymphosarcoma and mast cell sarcoma in dogs and leukosis in chickens.

Thiotepa

An early drug, also called Thio-TEPA or triethylenethiophosphoramide (synthesis[24]), has effects similar to those of mechlorethamine. In bladder cancer, for which it is now largely used, thiotepa is instilled in the bladder once weekly for 4 weeks.

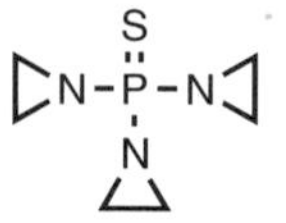

Thiotepa

Chlorambucil

This nitrogen mustard derivative (by synthesis[25]) is used chiefly in chronic lymphocytic leukemia, usually orally because of its favorable aqueous solubility as the sodium salt and rapid conversion to the free drug, which is easily absorbed by the gastrointestinal tract in daily doses. It is also used against non-Hodgkin's lymphomas. Side effects are anorexia, nausea, and vomiting. It has veterinary use in leukemias and, to a lesser degree, in solid tumors.

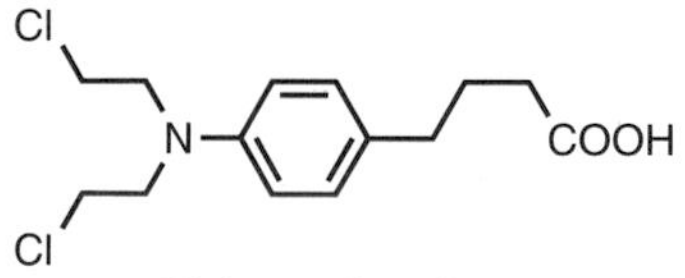

Chlorambucil

Melphalan

A phenylalanine mustard (by synthesis[26]), melphalan (L-PAM), is effective in multiple myeloma and has had a role in the treatment of breast and ovarian cancers. It is better absorbed orally because of its polar function, in 4- to 6-week courses, but it can be given intravenously. Oral dosage is subject to erratic gastrointestinal absorption.

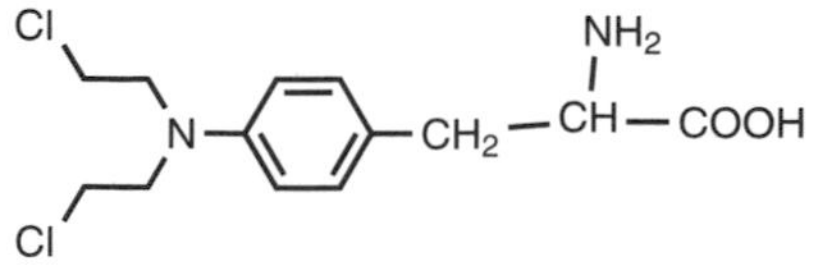

Melphalan

Carmustine (BCNU) and Lomustine (CCNU)

These unusual drugs (by synthesis[27]), are nitrosoureas with a broad spectrum of activity. They are classic alkylating agents, but they also inhibit DNA repair by isocyanate formation. They are distinguished by a pair of unusual properties. Toxicity to the bone marrow does not reach its maximum until 6 to 8 weeks after administration, which places certain limitations on the dose schedule. They are highly lipid soluble and easily transported across the blood-brain barrier, making them useful in tumors involving the central nervous system, such as glioblastoma multiforme. BCNU is given only intravenously as a 15-minute infusion in a single dose. CCNU can only be given orally.

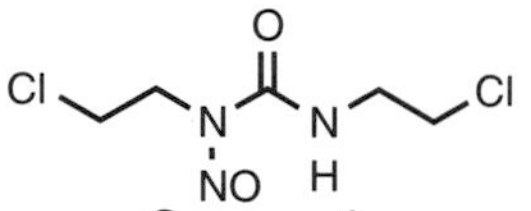

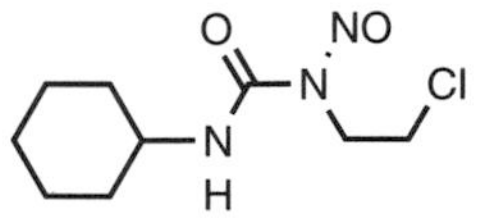

Carmustine Lomustine

The chloroethylnitrosoureas have limited clinical utility because of their tendency to cause more prolonged myelosuppression than other alkylators. The analog streptozocin, a glycosylated nitrosurea, has a spectrum of activity similar to that of the established agents, but it is associated with less bone marrow toxicity.

Chemical decomposition of these agents in aqueous solution yields two reactive intermediates, a chloroethyl diazohydroxide and an isocyanate group (Fig. 37–4). The former decomposes further to yield a relatively reactive chloroethyl carbonium ion that alkylates DNA, whereas the chloroethyl group reacts with another function, in the same manner as the classic mustards, producing cross links.

Cyclophosphamide

This most widely used alkylating agent[28] was synthesized[29] in the belief that it would be inactive in the body until its ring structure was broken down by an enzyme more common in cancer cells than in normal cells. It was thought that cyclophosphamide would be inert until it penetrated the cancer cell, where it would be converted

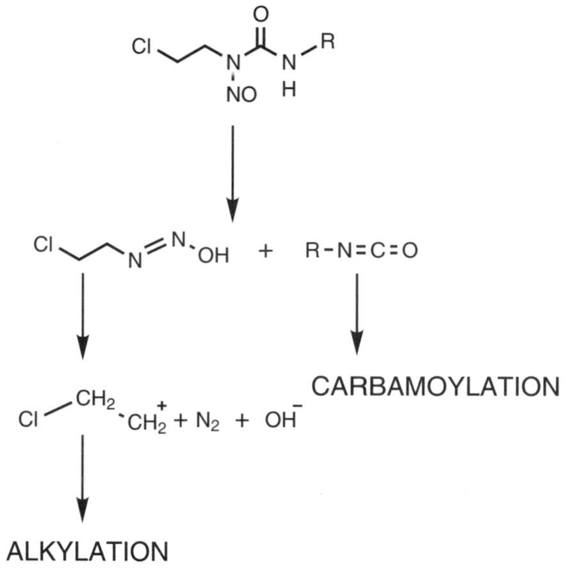

ALKYLATION

CARBAMOYLATION

Fig. 37–4. Decomposition of chloroethylnitrosoureas to form chloroethyl carbonium ion and a carbamoylating isocyanate group.

to the active derivative, thus damaging the cancer cell. This supposition did not turn out to be correct, however.

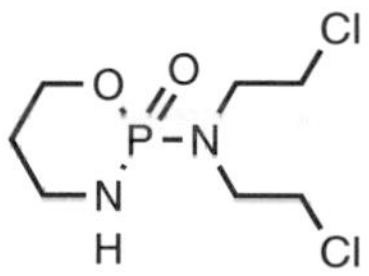

Cyclophosphamide

In the body, cyclophosphamide is converted to the active compound mainly in the liver rather than in the tumor.[15] Because it must be enzymatically oxidized with cleavage of the N—P bond to give an active metabolite, it is not active in vitro. It is converted to 4-hydroxycyclophosphamide and aldophosphamide. Aldophosphamide is chemically unstable, undergoing conversion to acrolein (CH_2═CH—CHO) and phosphoramide mustard.

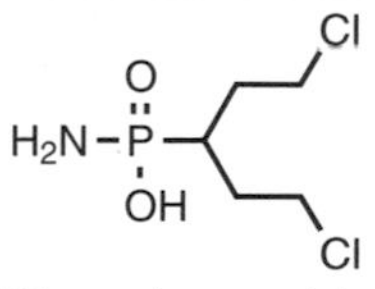

Phosphoramide mustard

Despite an incorrect premise concerning its action, cyclophosphamide is superior to other alkylating agents. It is frequently used to treat lymphosarcomas and Hodgkin's disease, as well as breast, ovarian, and lung cancer. Phenobarbital accelerates and proadifen slows the rate of active metabolite production. Cyclophosphamide may be given by mouth, even though the absorption is incomplete; however, it is often used intravenously to ensure maximum effectiveness. It may cause hair loss, sterility, testicular atrophy, ovarian fibrosis, and suppression of menstruation. Hemorrhagic cystitis occurs unless it is coadministered with N-acetylcysteine or 2-mercaptoethanesulfonate. Both are thiols that neutralize acrolein, the causative factor. The drug is teratogenic, and large doses cause myocardial necrosis.

Busulfan

Not a nitrogen mustard, this methanesulfonate ester (by synthesis[30]) has its greatest effect on granulocytes, and is used in chronic myelogenous leukemia. It is given orally daily, with titration of the dose to the granulocyte count. Overdosage can lead to severe irreversible granulocytopenia, and the dosage is cut or halted, depending on the white blood cell count. Side effects include skin pigmentation, ovarian suppression leading to cessation of menstruation, and (rarely) pulmonary fibrosis. Prolonged administration may produce a syndrome that mimics adrenal insufficiency.

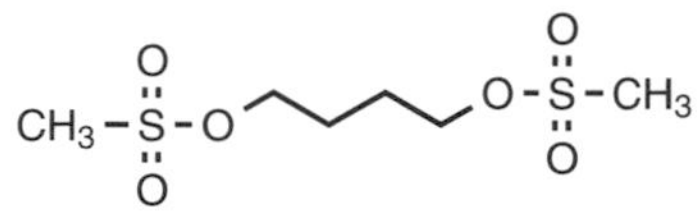

Busulfan

Mitomycin C

This natural product has been isolated[31] from Streptomyces verticillatus as well as from other sources (synthetic studies[32]). Unlike most other antibiotics, it is activated in vivo to a bifunctional or trifunctional DNA-acting agent. After activation, it binds preferentially to the guanine and cytosine moieties of DNA, leading to cross-linking of DNA, thus inhibiting DNA synthesis and function. Mitomycin C has been used mainly for bladder cancer, and in combination with cisplatin and a vinca alkaloid for lung cancer. A controlled study of this combination gave a survival rate at 3 years of 34%. The short duration of response (1 to 3 months) and the toxicity causing myelosuppression has limited the use of mitomycin C as a single agent.

Mitomycin C contains three groups that can damage cells: the quinone that can participate in free radical reactions generating superoxides, and aziridinyl and urethane functions that can take part in DNA alkylation (Fig. 37–5). Activation of mitomycin is known to proceed by means of reduction of the quinone and loss of a methoxy

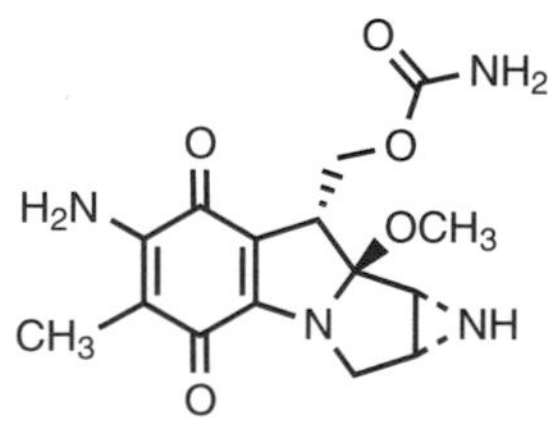

Mitomycin C

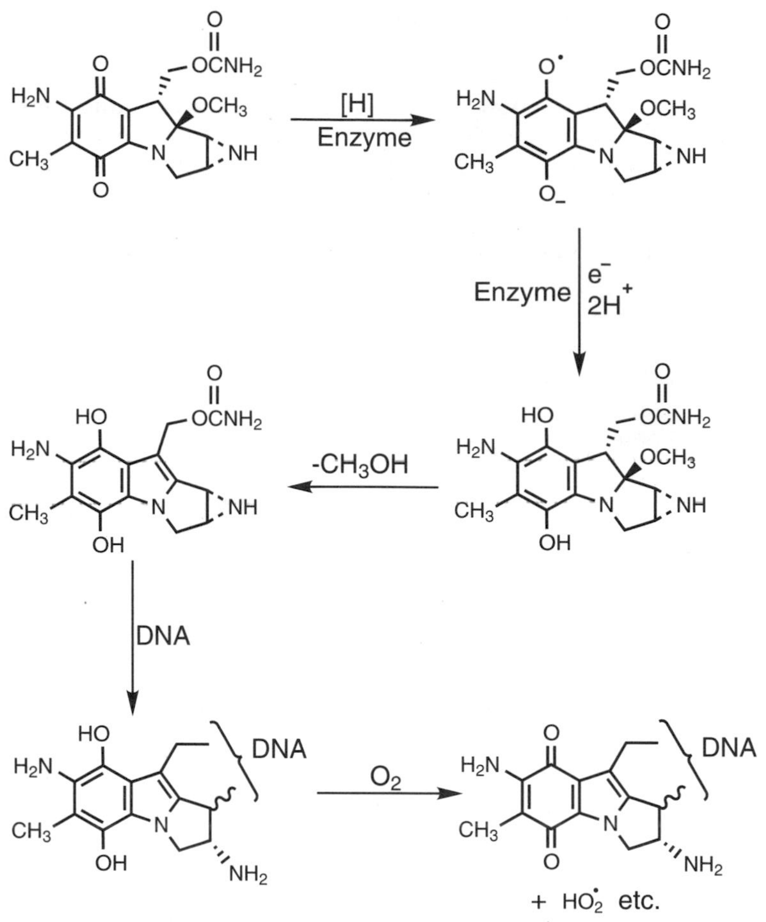

Fig. 37–5. Reaction scheme for bioreductive activation of mitomycin C and the subsequent interstrand cross-linking of DNA.

group. This process is followed by alkylation of DNA with the formation of interstrand cross links, resulting in inhibition of DNA synthesis, DNA fragmentation, and cell death.

Effect of Alkylating Agents on DNA

The alkylator mechlorethamine possesses the characteristics of forming a positively charged carbonium ion in aqueous solution, by way of an unstable aziridine immonium ion. The highly reactive carbonium ion species $R—CH_2—CH_2^+$ reacts with nucleophilic (electron-rich) sites on nucleic acids and proteins. The primary cytotoxic and mutagenic effects of these alkylating agents are probably the result of interaction with DNA. The favored site on DNA is at the N^7 position of guanine, although other nucleophilic positions on guanine, adenine, cytosine, and even the sugar phosphate groups are reported to be liable for such attack. It is not clear which of these sites of attack is most crucial in producing the end result, e.g., cytotoxicity or mutation. Alkylation of the N^7 position of guanine has a lesser effect on misreading of DNA and may result primarily in depurination (loss of the purine base through scission of the purine-sugar bond); alkylation of the N^3 of cytidine or the O^6 of guanine could conceivably interfere with precise base pairing in DNA. The proximal effects of base alkylation include misreading of the DNA codon and single-strand breakage of the DNA chain, and the distal effect is mutation and/or cell death. Cross-linking of DNA occurs when bifunctional alkylation agents with two potential alkylating functions react with DNA. The establishment of cross-stranded covalent binding correlates closely with the degree of exposure to mustards and to nitrosourea.

Alkylating agents as a class cause cytotoxic effects throughout the cell cycle, but have pronounced activity against rapidly dividing cells. DNA alkylation and consequent single-strand breakage occur primarily through enzymatic processes of repair; the alkylated base is excised by an endonuclease enzyme and the gap is joined by a ligating enzyme. The repair process that acts at sites of base alkylation or in regions that lack purine bases have less time to repair these damages in the rapidly dividing cell, because of the short span of the G_1 phase in the cell cycle. Therefore, the cell with damaged DNA is forced to enter the vulnerable DNA-synthetic S phase of the cycle without repair, making it prone to mutation or cell death.

Although the alkylating agents as a class share a common molecular mechanism of action and can promote cytotoxic, mutagenic, and carcinogenic activities, they differ substantially in their pharmacokinetic properties, lipid solubility, properties of transport across the blood-brain barrier, and cell membranes. They may even differ in chemical reactivity and in their sites of action in DNA.

Thus, the nitrosoureas, cyclophosphamide, and melphalan, in spite of belonging to the same class of agents, react differently in the treatment of lymphomas and multiple myelomas. Therefore, comprehensive knowledge of the individual agents is necessary to understand their unique properties and for the optimal usage of these agents in the treatment of cancer.

cis-Dichlorodiammineplatinum (II) (cisplatin, Platinol, DDP)

This coordination compound has broad application in human cancer chemotherapy. It seems to act with equal vigor against cells that are actively synthesizing nucleic acids (S phase) and against cells in mitosis (M phase). The platinum complex binds in a bidentate fashion to all the bases in DNA, but the preferred site of binding is the N^{-7} position of guanine. Because it is bifunctional (having two leaving C1 groups), cisplatin can form interstrand DNA cross-links, the number of which correlates with the drug's cytotoxicity. On intravenous injection, 90% of the drug binds rapidly to protein, and it is cleared from the plasma slowly. Free drug is cleared rapidly inside the kidneys.

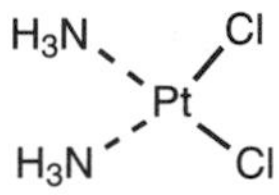

H_3N Cl Pt H_3N Cl

cis-Dichlorodiammineplatinum (II)

Cisplatin is the most active single agent against nonseminomatous testicular cancer, and combined with vinblastine and bleomycin (PVB), it is usually curative. It is also the most active single agent against ovarian cancer. Other applications include treatment of squamous (head and neck) and transitional cell (bladder) carcinomas, and treatment of small-cell lung cancer.

Cis-platinum is the only heavy metal compound in common use as a cancer chemotherapeutic agent. It has a divalent platinum bound to two potential leaving groups, the chloride ions; in *trans* position to the chlorides are two NH_3 groups bound irreversibly and in firm coordinate covalent bonds.

Only the *cis*-dichloro structure is an active antitumor agent. In vitro studies have revealed that only *cis*-DDP can form *interstrand* cross-links between adjacent guanine residues in DNA; the *trans*-DDP isomer lacks cytotoxicity. *Cis*-DDP undergoes a slow displacement by water, and the process is believed to accelerate in a low chloride concentration inside cells generating a positively charged hydrated activated complex that can react with a nucleophilic site of DNA (RNA or protein). Selective inhibition of DNA synthesis generally occurs rapidly, and correlation of DNA excision repair phenomenon with cytotoxicity has been observed in some cases.

The major toxic effect associated with cisplatin is renal damage; it is a direct tubular toxin. Rapid clearance of the drug, by use of saline or mannitol diuresis, minimizes the damage.

DNA-Intercalating Agents

Anthracyclines

Anthracyclines represent a major class of antineoplastic drugs, and their importance has increased since the discovery of daunorubicin (formerly daunomycin) and doxorubicin (formerly adriamycin). Doxorubicin is probably the most important anticancer drug available because of its relatively broad spectrum of activity, and daunorubicin is an important agent in the treatment of acute lymphocytic and myelocytic leukemia. Doxorubicin, on the other hand, has a significant role in the treatment of solid tumors such as carcinoma of the breast, lung, thyroid, and ovary, as well as soft tissue sarcomas.

These antibiotics are a part of a large group of highly colored Streptomyces products known as rhodomycins. In general, these compounds have a planar anthraquinone nucleus attached to an amino sugar. They exhibit a range of biologic activities, and are known to chelate divalent cations such as calcium and ferrous ions by virtue of the quinone and phenolic functions. These chelates show improved activity with less cardiac toxicity; although ferrous iron can react with O_2 to produce superoxide, $O_2^{\bullet}$, as well as H_2O_2 and $OH^{\bullet}$. because of the quinone-hydroquinone functionalities (characteristic of anthraquinone), they participate in biologic oxidation-reduction reactions. More significantly, because of the size and planar ring nature of the anthraquinones, they can also intercalcate between the base pairs in a DNA double helix. In addition, doxorubicin has been reported to react directly with cell membranes at low concentrations, which results in modification of membrane function and related cytotoxicity.

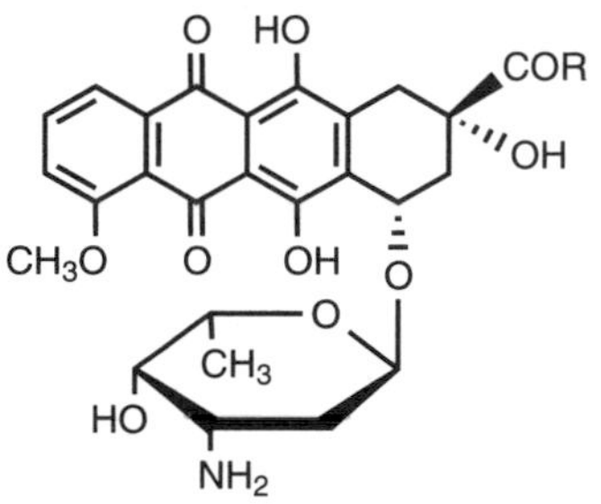

O HO COR OH CH_3O O OH O O CH_3 HO NH_2

Doxorubicin: R = CH_2OH
Daunorubicin: R = CH_3

When these compounds intercalate with DNA, the planar ring structure is inserted approximately perpendicularly to the long axis of the DNA double helix. The amino sugar appears to confer added stability to the binding through its interaction with the sugar phosphate backbone of DNA. Most evidence suggests that this DNA binding is necessary for inhibition of nucleic acid synthesis in tumor cells and for cytotoxic and antitumor activities. A correlation exists between DNA binding affinity and relative tumor inhibitory activity of the analogs of these compounds, with some notable exceptions.

Both of these agents can cause single-stranded DNA breaks and impair DNA repair. One theory suggests that intercalation leads to changes in the topography of the chromatin DNA that trigger "nicking" of the DNA by some repair enzyme. Another theory proposes that the anthracyclines are activated to free radicals that promote generation of highly active superoxides, and these mediate DNA damage. In fact, the existence of a nuclear membrane P-450 reductase with capacity to convert doxorubicin to a free radical has been reported. Results of in vitro studies confirm that doxorubicin radicals participate in oxygen-mediated single-strand breakage. It has been proposed that the cardiac toxicity of these agents is attributable to a free radical and that the tumor response is secondary to DNA binding. In support of this hypothesis, it was shown that radical scavenger agents such as tocopherol reduced cardiac toxicity in animal systems without affecting tumor response. Clinical trials with the radical scavengers vitamin E and N-acetylcysteine were unsuccessful, however. The cardiac tissues also appear to lack catalase, an enzyme that converts hydrogen peroxide to water and oxygen, suggesting that these anthracyclines are devoid of any mechanisms of disposing of this toxic metabolite. Damage to heart muscle is a distinct problem with use of anthracyclines, and increases with increasing cumulative doses of the drug.

Mitoxantrone and Bisantrene

Two anthracene derivatives that have been in phase II trials in cancer patients in recent years are mitoxantrone (dihydroxyanthacenedione, DHAQ) and bisantrene (bisguanylhydrazone of anthracene-9, 10-carboxaldehyde). These compounds, synthesized at Lederle Laboratories of American Cyanamid Co., show promising activity against several experimental rodent tumors, namely, P388 leukemia and B16 melanoma in mice. In phase I investigations with patients suffering from terminal cancer, these new agents showed low toxicity, and especially low cardiac toxicity, in contrast to the severe cardiac toxicity demonstrated by the anthracyclines. Although their clinical usefulness is not yet established, it is believed that the cytotoxicity of these agents is not due to the generation of reactive oxygen species.

Mitoxantrone

Bisantrene

Mitoxantrone, like other anthracyclines, is found to bind with a specificity for G-C-rich DNA, and bisantrene binds with little or no base-pair specificity. Again, like the anthracyclines, they appear to cleave DNA.

Antibiotics

Dactinomycin (Actinomycin D)

Isolated from Streptomyces parvulus, dactinomycin[33] (AMD) is the most active of a series of cyclic pentapeptides.

Dactinomycin

This complex structure[34] has been synthesized.[35] First discovered in the 1940s as a powerful bacteriostatic and cytostatic agent, it was tested in various infectious diseases as an antibiotic, but proved too toxic to the host for this use. Ten years later, it was tried as an anticancer agent with great success, particularly against Wilms' tumor (a kidney tumor of children). It is also a main agent in use for gestational trophoblastic disease, which refers to all neoplastic disorders arising from the human placenta. Against Wilms' tumor, it is used in combination with vincristine, cyclophosphamide, and doxorubicin (VACA). Good evidence exists that the actinomycins bind strongly, but reversibly, to DNA, interfering with synthesis of RNA and, consequently, with protein synthesis. Dactinomycin (AMD) binds with DNA by intercalation—insertion between base pairs as in a sandwich, and perpendicular to the main axis of the helix, as are the base pairs. Because of its flat rigid aromatic structure, the oxazine portion of dactinomycin can bind noncovalently between two successive bases in DNA, elongating the DNA. This process is considered a point mutation.

The chemical structure of the antibiotic is composed of a tricyclic, phenoxazin-3-one chromophore and two identical pentapeptide lactone groups, α and β, attached to the chromophore. The agent has been in use for treatment of human cancer for over 30 years, and is one of the most thoroughly studied agents in cancer chemotherapy. As mentioned previously, AMD binds to DNA by intercalation, with the phenoxazinone chromophore inserted nearly perpendicularly to the helical axis and preferably between the GC-base pairs. It is also known to prefer the pdG-dC sequence in DNA during this binding. It is the only agent known to show preference for this purine-pyrimidine DNA-base sequence. Its binding to DNA and nucleosides and oligonucleotides has been studied by physical techniques such as high-resolution NMR, x-ray crystallography, and circular and electrical dichroism. Hydrodynamic methods such as viscosity and sedimentation were performed using circular supercoiled DNA, and all confirmed the intercalative mode of binding to DNA. The peptide lactones bind to the minor groove in DNA by hydrogen bonding and hydrophobic interactions. Because of high cooperation between peptide and chromophore, the affinity of binding is strong, stronger than that known for anthracyclines (see Fig. 37–3). Once bound, the drug dissociates slowly, and thus blocks RNA polymerase from transcribing DNA.

The antibiotic is extremely potent in the inhibition of DNA-primed RNA synthesis, and inhibits preferentially preribosomal RNA and messenger RNA synthesis; transfer RNA synthesis is inhibited less effectively and so is DNA-dependent DNA synthesis. Actinomycin D appears to have little direct effect on the synthesis of protein in cells. Accordingly, the agent is used extensively in molecular biologic studies related to RNA metabolism.

Actinomycin D also causes single-strand breakage of DNA in chromosomes, like adriamycin and mitomycin. There are several possible reasons for this cleavage. Actinomycin D can be reduced via P-450 reductase, both nuclear and microsomal, to a radical intermediate, which can cause this damage. Alternatively, binding introduces strain in the helix, which induces the DNA topisomerase II to cause DNA strand breakage.

Actinomycin D is not metabolized either in tumors (in vitro) or in a tumor-bearing host. In humans, the unmetabolized drug remains tightly bound to the tissues, most probably in the nuclei of cells, for a prolonged period. The release of administered AMD from tissue pools and its excretion in the urine and bile is slow, which may be why the drug shows cumulative toxicity in humans. Also, the therapeutic and toxic dose levels of this drug are close, which makes its use in the clinic difficult. The dose-limiting toxicity of AMD is most commonly myelosuppression with intense gastrointestinal toxicity. Alopecia (hair loss) occurs and skin toxicity can be severe. Actinomycin D, like anthracyclines, is highly light sensitive and can give severe extravasation of skin during intravenous applications if sufficient drug comes in contact with skin epidermis. In conjunction with x-irradiation, it can lead to accelerated skin toxicity and lung and liver damage.

Bleomycin[36]

Bleomycin is the name given to a group of glycopeptides, with antitumor activity, isolated from Streptomyces verticillus[37] or produced by other means. The clinical preparation is a mixture of bleomycin A_2, A_2I, B_{1-4}, etc., with A_2 the predominant component. Bleomycin causes strand scission and fragmentation of DNA. It acts in the form of a cupric complex, inhibiting DNA ligase.

Bleomycin has been used in basal cell carcinoma and pericardial sclerotherapy, as well as in combination therapies, especially because it lacks bone marrow toxicity and immune suppression. It has modest activity in a variety of squamous cell cancers.

Bleomycin produces single- and double-strand breaks in DNA. The S-peptide binds to DNA, and the binding occurs preferentially to guanine bases in DNA. Fe^{++} ion, bound to imidazole and pyrimidine, undergoes oxidation to Fe^{+++}. Free radicals are generated on reaction with oxygen, superoxide, $O_2^{\bullet}$, and $OH^{\bullet}$, which attack phosphodiester bonds between the G-C or G-T sequence, leading to strand breaks.

Bleomycin is rapidly inactivated by all tissues except those of the skin and lung, where its specific toxicities lie. It causes erythema, pain, and hypertrophic changes in the skin in areas with a lot of keratin. Ulceration of these areas and pigmentation of the nails may occur. Pulmonary fibrosis, sometimes fatal, occurs in 5 to 15% of patients. Thus, bleomycin is usually not given to patients over 50 years of age or with pulmonary disease.

Mithramycin

An antibiotic produced by Streptomyces argillaceous and S. tanashiensis,[38] this complex glycoside binds to

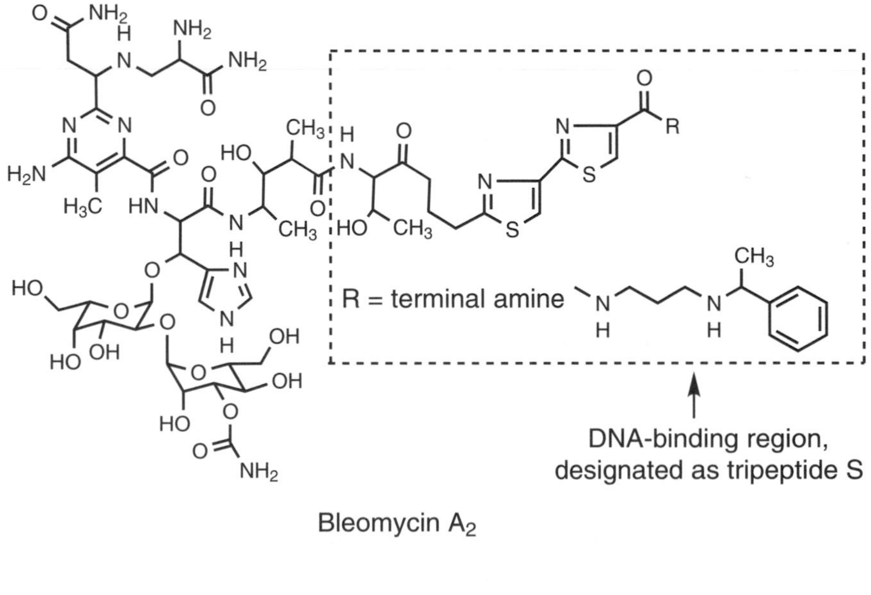

R = terminal amine

DNA-binding region, designated as tripeptide S

Bleomycin A_2

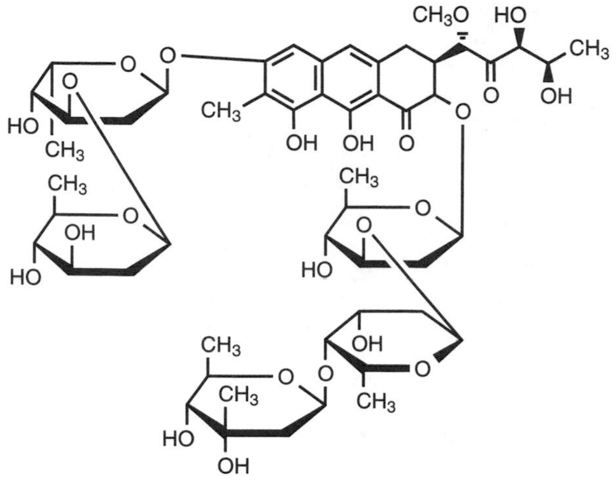

Mithramycin

DNA, but is more inhibitory to RNA. It has been studied for its usefulness in controlling hypercalcemia and Paget's disease of bone. Mithramycin has also led to improvements in patients with myeloid leukemia when used in conjunction with hydroxyurea. Only low doses of mithramycin are necessary to control hypercalcemia, and most patients do not develop side effects generally seen with high doses (thrombocytopenia, hypertension, liver function abnormalities, or nephrotoxicity).

Anthramycin

Anthramycin (AMC) is a representative of the pyrrolo [1,4]-benzodiazepine antibiotics derived from Streptomycetes. These antibiotics, in general, carry a potentially reactive carbinolamine group at neighboring C and N on the seven-membered diazepine ring.

Anthramycin binds in the narrow groove of DNA, preferentially with the guanine by a covalent binding with the 2-amino group. As a result of this adduct formation, the drug inhibits nucleic acid synthesis. Anthramycin com-

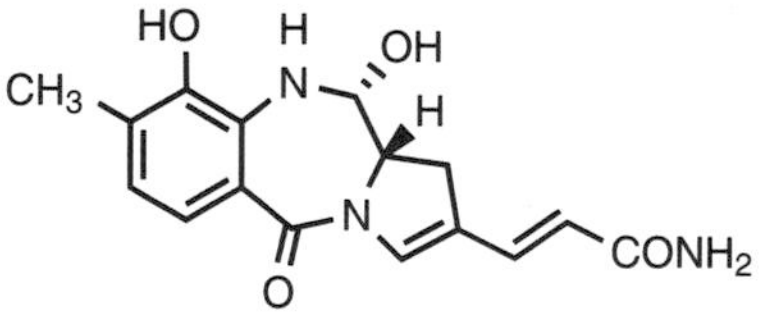

Anthramycin

petes with actinomycin D for the binding site because it shows preference to sites that are rich in the G-C base pair and also the minor groove in duplex DNA. This antibiotic has shown promising antitumor activity in several experimental tumors.

Antimitotic Agents

A family of alkaloids obtained from the periwinkle plant, Vinca rosea Linn.[39] have the complex structure shown.[40] Vincristine and vinblastine have well-established roles in the contemporary treatment of cancer. They appear to exert their major antitumor effect by binding to critical microtubular proteins within cells. Because these proteins are essential contractile proteins of the mitotic spindle of dividing cells, this binding leads to mitotic arrest. They also interfere with the synthesis of transfer RNA and of certain proteins.

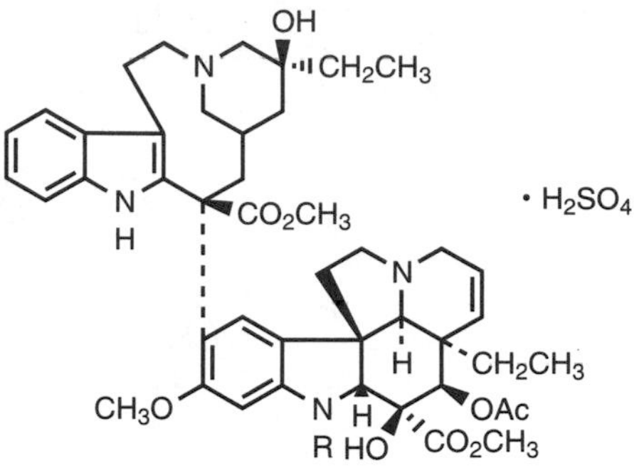

Vincristine: R = CHO
Vinblastine: R = CH_3

Vincristine is included in the highly effective combinations against acute lymphocytic leukemia, the lymphomas, breast cancer, sarcomas, and the various childhood neoplasms. Vinblastine has been important in combination regimens for testicular carcinoma. The latter is less popular in combination regimens because of its more marked myelosuppressive effect.

These compounds are not dependably absorbed after oral administration, so they are given intravenously. Their pharmacologic activity is poorly understood, but moderate amounts are excreted in the bile and smaller amounts in the urine. Metabolic degradation is extensive. Considerable vinblastine is absorbed into blood platelets. Vincristine's dose-limiting toxicity is its neurotoxicity; it does not cause myelosuppression with the usual doses. Vinblastine, however, is myelotoxic with neurotoxicity. Both drugs are extensively bound to proteins, producing rapid biphasic clearance from the plasma. They are eliminated mainly via the biliary tract.

Taxol

Taxol was isolated from the Pacific yew tree in the early 1960s, and its structure was determined by Wall and Wani in the late 1960s. Because of unfavorable testing results of the material available in the early 1970s, the compound was dropped from consideration by the National Cancer Institute. Later work showed the compound to have good activity against ovarian cancer and also activity against breast, lung, and other cancers. It has been approved by the Food and Drug Administration for use against ovarian cancer. Its only source was the bark of the yew tree, and taxol was a rare commodity until methods of synthesis were reported in 1994. A partial synthesis has been developed and the compound is marketed by Bristol-Myers Squibb. Some taxol derivatives are also showing promise.

Taxol is an antimitotic agent, acting against the spindle assembly of dividing cells, but in a different fashion from that of the Vinca alkaloids.

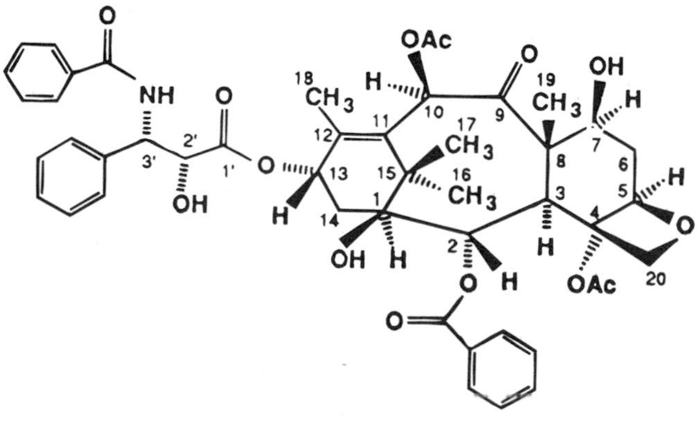

Taxol

Etoposide

Etoposide is a semisynthetic derivative of podophyllotoxin, a cytotoxic drug isolated from the root of the May apple plant. Like podophyllotoxin and the vinca alkaloids, etoposide causes metaphase arrest of the cell cycle. Unlike these other antimitotic compounds, etoposide does not inhibit microtubule assembly, apparently because of the sugar moiety absent in the other compounds. Etoposide can induce strand breaks in DNA, which is probably mediated by interaction with topoisomerase II.

Etoposide has a wide range of clinical uses. It is a component of curative combination chemotherapy for nonseminomatous testicular carcinoma and is included in combination regimens for aggressive forms of non-Hodgkin's lymphoma. It is also probably the most active single agent used in small-cell lung cancer.

Etoposide is usually given intravenously because its oral absorption is variable. The pharmacokinetics of the drug have not been clear, but it appears that 40 to 60% of a dose is excreted in the urine as unchanged drug and metabolites. The only peculiar toxicity is its tendency to cause transient hypotension if given by rapid intravenous injection.

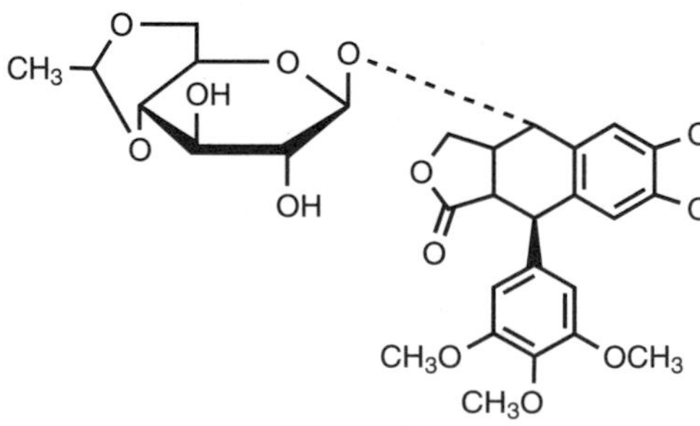

Etoposide

Other Compounds

Procarbazine

This hydrazine derivative (by synthesis[41]) is a component of the MOPP (mechlorethamine, vincristine [Oncovin], procarbazine, and prednisone) combination that is so effective in Hodgkin's disease. It has modest effects in meselothelioma. It must be converted into an azo derivative in vivo to become active against tumor cells. Alkylation of DNA or possible aberrant transmethylation may be the mode(s) of action.

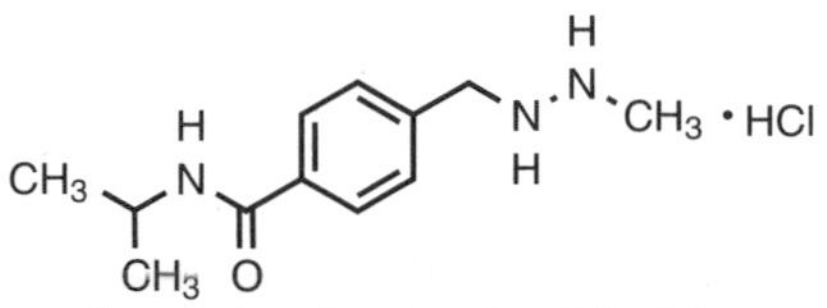

Procarbazine hydrochloride

Procarbazine is absorbed in the gastrointestinal tract, crosses the blood-brain barrier, has a plasma half-life of 10 minutes, is distributed to liver and kidney, and is extensively metabolized. Renal excretion is 70%, but 10 to 20% appears in respiratory CO_2. The drug is given by mouth, but can be given intravenously. In addition to the usual nausea, vomiting, and bone marrow depression, neurologic effects (somnolence, confusion, cerebellar ataxia) are related to the drug's ability to enter the central nervous system. Procarbazine is a weak monoamine oxidase inhibitor and is subject to precautions in concomitant use of other drugs and foods.

Hydroxyurea

This simple synthetic[42] interferes with DNA synthesis by inhibiting ribonucleoside reductase. Its main use is in rapidly (though temporarily) lowering the high leukocyte counts of chronic myelogenous leukemia in the blastic phase. Minor uses are against melanoma and as an adjuvant to radiotherapy for cervical cancer. Hydroxyurea can be given by mouth or intravenously, is extensively metabolized to urea, and crosses the blood-brain barrier. Its main toxicity is bone marrow depression.

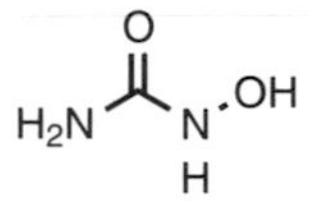

Hydroxyurea

Asparaginase

Isolated from Escherichia coli or other sources,[43] this enzyme is active in treating mouse and dog leukemia. It is thought that some cancer cells might be qualitatively different from normal cells in their ability to synthesize the amino acid asparagine. Normal cells do not require an outside source of the amino acid asparagine because they have an enzyme, asparagine synthetase, that can synthesize asparagine from aspartic acid and glutamine. Certain leukemic cells lack this enzyme and, as a result, depend completely on an outside source (blood) for their supply of asparagine. Thus, it is felt that administration of a large dose of asparaginase, which removes the amino group from asparagine, would destroy all the asparagine circulating in the blood plasma and the leukemic cells would be starved of this supposedly (for them) essential nutrient. Conversely, normal cells would continue to synthesize their own asparagine.

This therapy has produced complete remissions in about 50% of children with acute leukemia. Unfortunately, the leukemic cell eventually adapts to the altered situation by producing the enzyme asparagine synthetase, which makes asparagine. After a few months, the patient's leukemic cells have adequate asparagine and become resistant to asparagine therapy. Nevertheless, asparaginase is given intravenously to induce remissions in acute lymphatic leukemia. It is also used in sequence with ARA-C for cases of refractory or acute myelogenous leukemia.

Although marrow depression, hair loss, and effects on gastrointestinal mucosa are not seen with asparagine, it has many other serious toxicities in humans, especially on organs that synthesize large amounts of proteins, such as the liver and pancreas. Liver toxicity is moderate, but occasionally, patients develop fulminating acute pancreatitis. Allergic reactions occur in 5% of patients, and anaphylactic shock is sometimes observed.

Corticosteroids

Prednisone is the corticosteroid usually used. It is an essential and valuable component of curative combinations used against acute lymphocytic leukemia, Hodgkin's disease, and other lymphomas, particular when administered with cyclophosphamide and vincristine (CVP). It and dexamethasone are worthwhile in the management of metastatic and primary brain cancer, in part because of their rapid control of cerebral edema.

Interferons

Interferons are a class of small proteins produced and released by cells that have been invaded by a virus. Interferons signal noninfected cells to rearrange their internal machinery into a form that inhibits production of a virus.

A source of interferons has been the human white blood cell ingredient leukocyte, but recently, a DNA-recombinant technique has been applied to supplement this rare source of supply. Interferon, specifically IFN-α, has so far had a significant therapeutic impact only in the relatively rare hairy-cell leukemia.

Natural Killer Cells and Lymphokines

Lymphokines are glycoproteins produced in trace amounts by white blood cells; they regulate the body's natural immune responses. Certain lymphokines either suppress or promote the growth of cells known as B cells or T cells.

Generally, B cells are believed to produce antibodies that recognize foreign and invading cells, and T cells have the potential to promote production of cytotoxic or killer cells that attack cells foreign to the body, such as virus-infected cells, cancer cells, and even tissue grafts.

In vitro, natural killer cells have been shown to effect tumor rejection, but little evidence in vivo shows that this action occurs. Use of a T cell growth factor, interleukin 2 (IL-2), has produced regressions of renal cell carcinoma and melanoma, but the response rate is low. Because of high toxicity and a cumbersome technology for its production, this therapy has seen little use.

EXPERIMENTAL DRUGS

Test Methods

Most effective antineoplastic drugs have been found by pragmatic testing of compounds using rodents with transplanted tumors. The L-1210 mouse leukemia is the single most reliable system for detecting activity likely to benefit humans. One routine testing screen has been established with both L-1210 and another mouse leukemia, P388, because the latter responds to several natural products missed by L-1210.

Other tumors, including spontaneous as well as chemically and virally induced tumors, have been studied extensively. It seems that the tissue of origin and mode of transmission do not alter the value of a particular type of test animal tumor in predicting the usefulness of a drug in humans to any significant degree, because all tumor cells have in common a characteristic growth fraction, as discussed previously. Rapidly growing tumors, which have a large fraction of their cells initiating DNA replication at any given moment, are the ones chosen for screening compounds in test animals. Mouse tumors with small growth fractions are now known, however, and several of these, such as the B-16 melanoma and the Lewis lung tumor, have been investigated to see whether they might uncover agents uniquely active in small growth fraction tumors in humans.

The present screening program for anticancer agents at the National Cancer Institute, National Institutes of Health makes use of about 60 cell cultures for various tumor cell lines. Specificities for certain tumor cells may appear; if significant, the candidate agent may then be tested in appropriate animal tumors.

Over the years, mathematical formulas have been worked out for calculating the drug dose to be administered in the initial clinical studies in humans based on toxicity data in mice, and then in dogs and monkeys. It is said that in only 5% of cases do side effects appear that have not been detected first by careful work with dogs and monkeys. Such studies in the monkey and dog also indicate whether some laboratory test can be found that will give early warning of toxicity and whether a toxic effect can be reversed, as leucovorin reverses methotrexate.

Analog Searches

A cynical generalization about all drug development is that "the first compound discovered is always the most active in the series." This maxim is definitely not true of anticancer agents; the improved analogs are usually superior to the initial drug. This fact is one of the justifications for continued attempts to synthesize analogs and derivatives of active compounds, regardless of whether the activity was first found by random or rational search.

REFERENCES

1. A. I. Holleb, et al., Eds., *American Cancer Society Textbook of Clinical Oncology,* Atlanta, American Cancer Society, Inc., 1991. pp. 1–6.
2. M. Fishbein, Ed., *The New Illustrated Medical and Health Encyclopedia,* New York, H. S. Stuttman, 1969, p. 843.
3. I. J. Fidler and M. L. Kripke, *Chemistry, 50,* 1, 18(1977).
4. C. W. Heath, Jr., *Cancer prevention,* in *American Cancer Society Textbook of Clinical Oncology,* A. I. Holleb, et al., Eds., Atlanta, American Cancer Society, Inc., 1991, pp. 102–103.
5. U. S. Dept. of Health and Human Services, *The Health Consequences of Smoking; Cancer, a Report of the Surgeon General,* Washington, D.C., U.S. Government Printing Office, DHHS (PHS) 82-50179, (1982).
6. E. K. Weisburger, *Chemistry, 50,* 1, 42(1977).
7. J. E. Enstrom and D. F. Austin, *Science, 195,* 847(1977).
8. T. H. Maugh II, *Science, 197,* 543(1977).
9. D. Kessel and T. J. Dougherty, Eds., *Adv. Exp. Med. Biol., 160,* 1, 42(1983).
10. B. A. Teicher and E. A. Sotomayor, *Chemical radiation sensitizers and protectors,* in *Cancer Chemotherapeutic Agents,* W. O. Foye, Ed., Washington, D. C., American Chemical Society, 1994.
11. K. Hellmann, et al., Proceedings, 14th International Congress of Chemotherapy, Kyoto, 1985.
12. E. A. Bump, et al., *Science, 217,* 544(1982).
13. R. B. Herberman, *Principles of tumor immunology,* in *American Cancer Society Textbook of Clinical Oncology,* A. I. Holleb, et al., Eds., Atlanta, American Cancer Society, Inc., 1991, pp. 76–77.
14. V. T. DeVita, Jr., et al., *Malignant Lymphoma,* Baltimore, Williams and Wilkins, 1987, pp. 249–267.
15. J. H. Burchenal and J. R. Burchenal, *Chemistry, 50,* 6, 11(1977).
16. D. R. Seeger, et al., *J. Am. Chem. Soc., 71,* 1753(1949).
17. D. A. Matthews, et al., *Science, 197,* 452(1977).
18. R. G. Stoller, et al., *N. Engl. J. Med., 297,* 630(1977).
19. G. H. Hitchings and G. B. Elion, U.S. Patent 2,933,498(1960); A. G. Beaman and R. K. Robins, *J. Am. Chem. Soc., 83,* 4038(1961).
20. G. B. Elion and G. H. Hitchings, *J. Am. Chem. Soc., 77,* 1676(1955).
21. W. K. Roberts and C. A. Dekker, *J. Org. Chem., 32,* 816(1967).
22. D. H. R. Barton, et al., *J. Org. Chem., 37,* 329(1972).
23. Y. F. Shealy, et al., *J. Org. Chem., 27,* 2150(1962); K. Hano, et al., *Gann, 59,* 207(1968).
24. E. Kuh and D. R. Seeger, U.S. Patent 2,670,347 (1954); S. Saijo and M. Endo, Jap. Patent 218(1955).
25. J. L. Everett, et al., *J. Chem. Soc.,* 2386(1953); A. P. Phillips and J. W. Mentha, U.S. Patent 3,046,301 (1962).
26. F. Bergel and J. A. Stock, *J. Chem. Soc.,* 2409(1954); U.S. Patent 3,032,584 (1962).
27. G. S. McCaleb, et al., *J. Med. Chem.,* 6,669(1963).
28. D. L. Hill, *A Review of Cyclophosphamide,* Springfield, IL, Charles C Thomas, 1975.
29. H. Arnold, et al., *Nature, 181,* 931(1958); H. Arnold and F. Bourseaux, *Angew. Chem., 70,* 539(1958).
30. G. M. Timmins, U.S. Patent 2,917,432 (1959).
31. J. S. Webb, et al., *J. Am. Chem. Soc., 84,* 3185, 3187(1962).
32. G. J. Siuta, et al., *J. Org. Chem., 39,* 3739(1974).
33. E. Bullock and A. W. Johnson, *J. Chem. Soc.,* 3280(1957); H. Brockmann, et al., *Naturwissenschaften, 51,* 383, 435(1964).
34. H. Brockmann, *Ann. N. Y. Acad. Sci., 89,* 323(1960).
35. J. Meienhofer, *J. Am. Chem. Soc., 92,* 3771(1970).
36. S. K. Carter, et al., *Fundamental and Clinical Studies of Bleomycin,* Baltimore, University Park Press, 1976.
37. T. Takita, et al., *J. Antibiot. (Tokyo), 22,* 237(1969).
38. Y. A. Berlin, et al., *Nature, 218,* 193(1968).
39. G. H. Svoboda, *Lloydia, 24,* 173(1961); R. L. Noble, C. T. Beer, and J. H. Cutts, *Ann. N. Y. Acad. Sci., 76,* Art. 3, 882(1958).
40. N. Neuss, et al., *J. Am. Chem. Soc., 86,* 1440(1964); J. W. Moncrief and W. N. Lipscomb, *J. Am. Chem. Soc., 87,* 4963(1965).
41. Hoffman-La Roche & Co., Belg. Patent 618,638 (1962); Br. Patent 968,460 (1962).
42. A. Hantzsch, *Justus Liebig's Ann. Chem., 299,* 99(1898); R. J. Graham, U.S. Patent 2,705,727 (1955).
43. P. P. K. Ho, et al., *J. Biol. Chem., 245,* 3708(1970).

SUGGESTED READINGS

J. K. Barton, *Recognizing DNA Structure, Special Report, Chem. Eng. News, 66*(39), 30(1988).

F. F. Becker, Ed., *Cancer, A Comprehensive Treatise,* 2nd ed., Vol 5, *Chemotherapy,* New York, Plenum Press, 1982.

S. K. Carter, E. Glastein and R. B. Livingston, Eds., *Principles of Cancer Treatment,* New York, McGraw-Hill, 1982.

B. Chabner, *Pharmacologic Principles of Cancer Treatment,* Philadelphia, Saunders, 1982.

G. L. Chen and L. F. Liu, *DNA Topoisomerase as Therapeutic Targets in Cancer Chemotherapy,* in *Annual Reports in Medicinal Chemistry,* Vol. 21, D. M. Bailey, Editor-in Chief, New York, Academic Press, Inc., 1987.

V. T. DeVita, Jr., et al., Eds., *Important Advances in Oncology,* Philadelphia, J. B. Lippincott, 1987.

R. I. Glazer, Ed., *Developments in Cancer Chemotherapy,* Boca Raton, FL, CRC Press, 1984.

A. E. Gunn, Ed., *Cancer Rehabilitation,* New York, Raven Press, 1984.

C. M. Haskell, Ed., *Cancer Treatment,* 2nd ed., Philadelphia, Saunders, 1985.

A. I. Holleb, et al., Eds., *American Cancer Society Textbook of Clinical Oncology,* Atlanta, American Cancer Society, Inc., 1991.

M. Kirsch-Volders, Ed., *Mutagenicity, Carcinogenicity, and Teratogenicity of Industrial Pollutants,* New York, Plenum Press, 1984.

J. Lamb, et al., Eds., *T Cells,* New York, Wiley-Interscience, 1989.

S. H. Levitt and N. duV. Tapley, *Technological Basis of Radiation Therapy: Practical Clinical Applications,* Philadelphia, Lea & Febiger, 1984.

A. C. Sartorelli, et al., *Molecular Actions and Targets for Cancer Chemotherapeutic Agents,* New York, Academic Press, 1981.

M. G. Simic, et al., Eds., *Mechanism of DNA Damage and Repair,* New York, Plenum Press, 1986.

R. T. Skeel, *Manual of Cancer Chemotherapy,* Boston, Little, Brown & Co., 1982.

J. D. Watson, et al., *Molecular Biology of the Gene,* Vols. 1 and 2, Redwood City, CA, Benjamin/Cummings, 1987.

D. E. V. Wilman, Ed., *Chemistry of Antitumor Agents,* New York, Chapman and Hall, 1990.

Chapter 38

ANTIVIRAL AGENTS

Manohar L. Sethi

Viruses are microscopic organisms that can infect all living cells. They are parasitic and multiply at the expense of the host's metabolic system. Viruses may start their infectious cycle immediately on attack or remain dormant in the cellular site of the host for extended periods until an etiologic agent triggers them to reproduce. Once the effective particles become active, they may produce cytotoxic effects or cause numerous diseases in animals and humans. The major routes of transmission of viral infections in humans are through the respiratory, gastrointestinal, and genital tracts, and the skin, urine, blood, and placenta. Viral infections may occur through air, water, food, milk, or environmental sources. Whether the host survives the effects of the viral infection depends on the immune response of the host and also on the severity and type of infection. Immune response is obtained by the production of B lymphocytes derived from the bone marrow and T lymphocytes derived from the thymus with the help of macrophages. Specific immune response to viral diseases depends on antibodies formed by humoral (B cells), local (secretory IgA system), and cell-mediated (T cell) immunity. Discussion of immunizing biologics and virus-specific immunoglobulins that provide active and passive immunity is beyond the scope of this chapter. To understand the mechanisms of antiviral chemotherapy, however, some viral characteristics are reviewed, including viral replication and transformation of cells. Detailed information on the molecular biology of viruses is found in the suggested readings.

Viruses[1] are among the smallest microorganisms, varying in size from 0.02 to 0.40 μm. They are filterable through porcelain filters and can be seen and identified with the help of an electron microscope. Viruses consist of a nucleic acid core that contains either deoxyribonucleic acid (DNA) or ribonucleic acid (RNA), which constitutes the genetic material and provides a basis for classification of viruses. The nucleic acid core is surrounded by a protein coat known as a capsid. The entire structure is called the nucleocapsid. On the basis of the structural characteristics of the nucleocapsid, most viruses are divided into two groups on the basis of their symmetry or shape. One group of viruses may show helical symmetry and the other icosahedral (20-sided) symmetry of the nucleocapsid. The nucleocapsid may or may not be covered by another protein coat called an envelope. The envelope is composed of glycoproteins that are important virus antigens. The arrangement of coat proteins defines the overall shape of the viruses. Spheres, rods, filaments, bullets, rectangles, triangles, and elongated tubes are some of the shapes of viruses. The complete infectious virus particle is called a virion.

The protein coats of the nucleocapsid and envelope and the absence or the presence of the envelope play an important role in the initial stages of viral infection. Reactive sites on the capsid or envelope become attached to the receptor sites on the host cell. The penetration, uncoating, and release of the virions in the host cell depend on the structural coat proteins. This process influences the susceptibility of the virus to the actions of antiviral agents. The study of the structure of viral coat proteins and their properties is therefore important in the development of effective antiviral agents.

VIRAL DISEASES[2,3]

Viruses have long been recognized as the cause of a wide variety of infections in animals and humans. The disease type and symptoms depend on the group and species of viruses. Fundamentally, viruses are classified into two main groups, namely DNA- and RNA-containing viruses. Other methods of classification are based on traditional taxonomy, properties of the viruses (composition, structure, shapes, and relative sizes), induction and carriage of polymerases (DNA- or RNA-directed RNA), and presence or absence of a lipid-containing envelope.

Many diseases (Table 38–1) are produced by the different groups of viruses. Ocular viral diseases are primarily caused by the herpes group of viruses, namely herpes simplex virus (HSV), varicella-zoster virus (VZV), cytomegalovirus (CMV), and Epstein-Barr (EB) virus. Herpetic keratoconjunctivitis, a serious infection of the eye caused by the HSV, is the leading cause of corneal blindness in the United States. Herpes zoster (shingles) is a severe skin infection affecting mostly the elderly. It is caused by the VZV that also causes chickenpox. Herpes zoster is the reactivated form of the VZV infection. The virus enters the sensory nerve endings in the skin and remains dormant until reactivation. Varicella infection is also considered a minor form of smallpox (variola).

The herpes labialis (common cold sore) virus often lies dormant in early life, but may affect the mucous membrane, skin, eye, and genital tract in later life. Cytomegalovirus, HSV, and rubella virus produce chronic intrauterine and perinatal infections. Several viruses, such as EB virus, mumps, smallpox, CMV, HSV, viral hepatitis A and B, and togaviruses, are responsible for systemic and viral infections of immunosuppressed patients. Herpes sim-

Table 38–1. Viruses Infecting Animals and Humans

Virus Group	Species	Disease
DNA-Containing Viruses		
Adenovirus*	Many types	Respiratory tract and eye infections (keratitis)
Herpesvirus	Herpes simplex virus types 1 and 2	Encephalitis, eye infections (keratoconjunctivitis), skin diseases, genital infections
	Varicella-zoster (Varicella)	Chickenpox (children)
	Herpes zoster	Shingles (adults)
	Cytomegalovirus	Cytomegalovirus diseases
	Epstein-Barr virus	(Mononucleosis), hepatitis (animal and human), Marek's virus disease (avian leukemias), Epstein-Barr virus disease (infectious mononucleosis, Burkett's lymphoma, nasopharyngeal carcinoma)
Papovavirus†	Human wart virus	Animal and human warts
	Polyoma virus	Salivary gland (parotid) infection, progressive
	SV 40	Multifocal leukoencephalopathy (PML) in humans
Poxvirus	Variola	Smallpox (variola), cowpox (vaccinia), chickenpox,
	Vaccinia	infectious mononucleosis, eczema
RNA-Containing Viruses		
Arenavirus	Lymphocytic choriomeningitis (LCM) virus	Lymphocytic choriomeningitis (LCM), Lassa fever, hemorrhagic fever
	Lassa fever virus	
Orthomyxovirus	Influenza A, B, and C viruses	Influenza A, B, and C
Paramyxovirus	Parainfluenza virus	Mumps, measles, parainfluenza (pneumonia, bronchiolitis)
	Respiratory syncytial virus	
	Measles (rubeola)	
Picornavirus	Rhinoviruses	Respiratory diseases, gastrointestinal diseases, poliomyelitis, aseptic meningitis
	Enteroviruses (polio, coxsackie A, B, echovirus)	
Reovirus	Human reovirus (rotavirus)	Mild respiratory and gastrointestinal symptoms
Retrovirus (Oncornavirus)	Human T-cell lymphotropic viruses (type C viruses)	Leukemia, lymphoma and sarcoma in animals (birds, cats, rodents, cows, gibbon apes etc.), mouse mammary tumor, human breast cancer, human T-cell leukemia, nasopharyngeal carcinoma
	Human immunodeficiency virus type 1 (HIV-1, HTLV-III/LAV)	Acquired immunodeficiency syndrome (AIDS) and AIDS-related complex (ARC)
Rhabdovirus	Rabies virus	Rabies, encephalitis
Togavirus	Encephalitis virus	Rubella (German measles), meningoencephalitis,
	Rubella virus	encephalitis, hepatitis, yellow fever, and sand fly
	Arbovirus	fever

* Adenoviruses were originally isolated from humans (children and throats of patients with respiratory diseases).
† Papovaviruses are derived from the names of papilloma, polyoma, and vacuolating viruses.
Modified and adapted from G. J. Galasso, et al., Eds., *Antiviral Agents and Viral Diseases of Man,* New York, Raven Press, 1990, and W. L. Drew, Ed., *Viral Infections, A Clinical Approach,* Philadelphia, F. A. Davis, 1976.

plex virus types 1 and 2 are also involved in localized diseases of the skin.

Acute respiratory diseases are the most common manifestation of viral infections. Both DNA (adenovirus) and RNA viruses (influenza, parainfluenza, picornavirus, herpesvirus, and oncornavirus) are involved in respiratory diseases. Viruses are also associated with such diseases as viral rhinitis, pharyngitis, laryngitis, laryngotracheobronchitis, influenza, parainfluenza, and respiratory syncytial virus (RSV) pneumonia. Diseases of the nervous system include poliomyelitis, rabies, and meningoencephalitis associated with mumps, measles, vaccinia, and "slow" viral infections.

Viruses are also linked with various other diseases, such as rheumatoid arthritis, multiple sclerosis, diabetes mellitus, cancer of the cervix, certain heart diseases, hepatitis, and acquired immunodeficiency syndrome (AIDS). The common opportunistic infections in AIDS patients are attributable to bacteria, fungi, mycobacteria, and viruses. The prime candidate involved in AIDS, however, is the human T-cell lymphotropic virus HTLV-III, also known as human immunodeficiency virus type 1 (HIV-1). This infection is spread by sexual contact, blood transfusion, blood-derived products, and intravenous drug users or through intrauterine transmission.[4] AIDS has drawn much public attention because of its fatal nature and lack of a cure. A growing incidence of this disease has been reported in the United States and worldwide.

VIRAL REPLICATION AND TRANSFORMATION OF CELLS[5,6]

Because viruses do not multiply without living cells, they depend solely on the host cell to carry out their metabolic activities. The enzyme system of the host cell is used for the synthesis of DNA and virus replication. The viruses (DNA and RNA) may replicate and/or transform the cell simultaneously. When oncogenic or infectious viruses attack the host cell, they adsorb onto the cell surface by electrostatic interaction, penetrate it, and remove their viral coat, liberating the nucleic acid into the host cell. Viral nucleic acid is replicated within the host by enzymes resulting in synthesis of viral proteins. The assembled viral particles are released from the cell. In the case of DNA viruses (Fig. 38–1), the liberated DNA integrates into the host cell DNA. As a result, viral DNA becomes a permanent part of the host's genetic material. It may remain latent for years or duplicate into progeny viral DNA during cell division. Viral DNA transcribes into early and late mRNA, which is then translated into viral proteins under the direction of viral and host enzymes. In the case of oncogenic DNA viruses, synthesized viral proteins may act on the host cell to change the normal functions or morphologic characteristics. As a result, the infected cell behaves like a transformed or cancer cell. During the translation process, mRNA synthesizes the structural proteins of the viral capsid and envelope. The viral DNA in conjunction with the structural proteins assembles into progeny virions that escape from the cell.

Replication of oncogenic RNA viruses is conceivably different from that of the DNA viruses. In 1970, Temin and Mizutani[7] and Baltimore[8] discovered an enzyme from RNA viruses, reverse transcriptase, which reverses the usual information flow (DNA→RNA) in a cell. Consequently, DNA is produced on an RNA template (RNA→DNA). Reverse transcriptase (RNA-directed DNA polymerase) has been found in almost all oncogenic RNA viruses. In oncogenic transformation (Fig. 38–2), RNA viruses must first form a DNA copy that is integrated into host cell chromosomes. After an oncogenic RNA virus has invaded the host cell, viral DNA is formed on an RNA template with the help of reverse transcriptase. A double-stranded helix (RNA/DNA duplex) is formed in which one strand is the original RNA viral chromosome and the other is a new complementary DNA chain. The strand of viral RNA is then removed by another enzyme, leaving a single-stranded DNA molecule. Replication of single-stranded viral DNA with the host's enzyme results in the duplex viral DNA that is then integrated into the host's chromosomes. The integrated DNA copy of viral RNA into the host DNA is called a "provirus," which is transmitted to the daughter cell in the same way as other cellular genes. The provirus contains the same genetic information that was present in the viral RNA chromosome.

Subsequently, the provirus is transcribed into viral

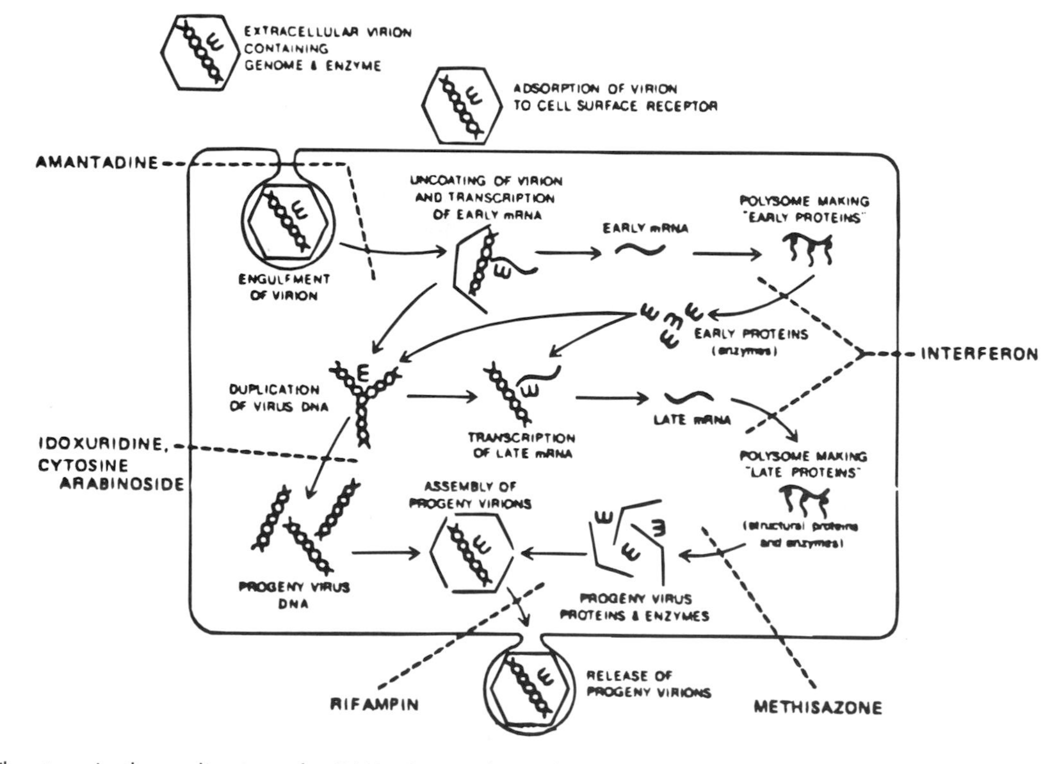

Fig. 38–1. The steps in the replication of a DNA virus and possible interruption of viral replication by various antiviral agents. Reprinted with permission from Drew, W. L., Ed., *Viral Infections: A Clinical Approach,* Philadelphia, F. A. Davis Company, 1976.

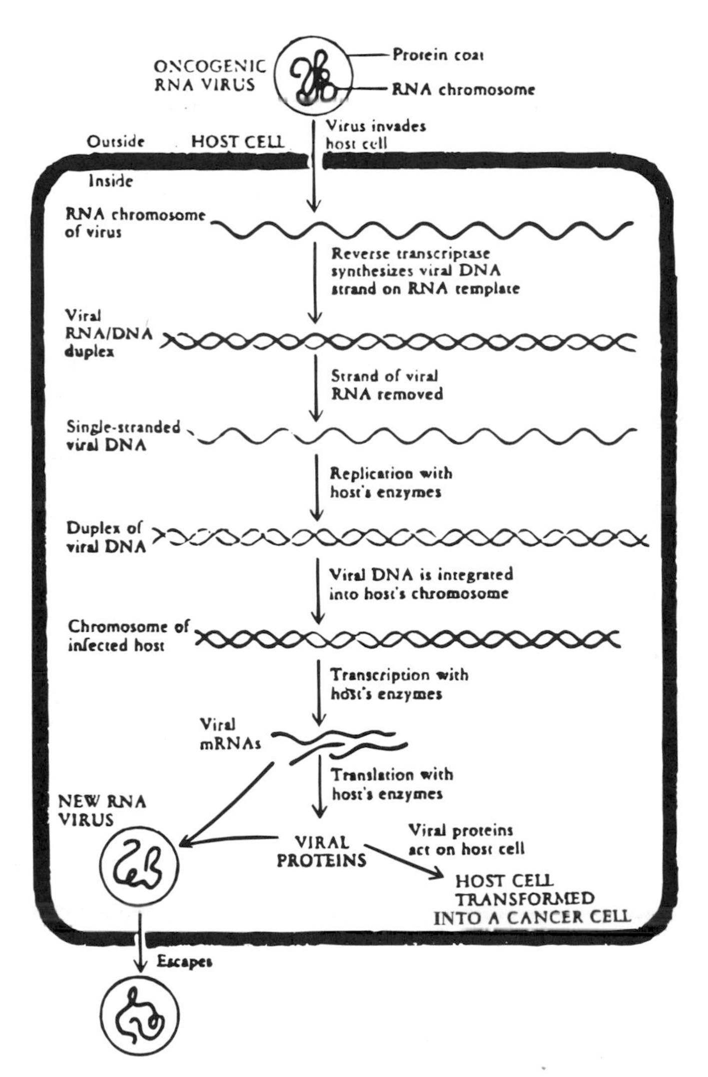

Fig. 38–2. Transformation of a host cell by oncogenic RNA virus. Reprinted with permission from Prescott Flexer, A. S., Ed, *Cancer; The Misguided Cell,* Sunderland, Massachusetts, Sinauer Associates, Inc., 1982.

mRNA, which is translated into viral proteins with the help of host enzymes. One or more of these viral proteins may alter cell behavior, resulting in the transformation of a cell into a cancer cell. Furthermore, viral mRNA synthesizes structural proteins for the viral capsid and envelope, which enclose viral nucleic acid. New RNA viral particles assembled by acquiring the outermost envelope layer escape from the host cell.

Infectious RNA viruses replicate in the cytoplasm via RNA polymerases. In some RNA viruses, for example poliomyelitis, the RNA of the infecting particles is capable of acting as mRNA. In other RNA viruses, such as influenza and measles, a complementary strand of mRNA is synthesized with the RNA of the infecting particles serving as a template. Translation of mRNA into viral proteins and the release of viral particles from the host cell are analogous to those described for oncogenic RNA viruses.

After the replication of DNA and RNA viruses, the release of viruses may occur after lysis of the host cell or by "budding" from the cell. The latter process causes less harm to the host cell. Released particles may infect adjacent cells immediately or be carried by tissue body fluids, lymph, or blood to distant cells, where the infectious cycle is repeated. This sequence is how a large number of infected cells are formed, leading to transformation of normal cells to cancer cells.

ANTIVIRAL AGENTS[9,10]

Because viruses are obligate intracellular parasites, their replication depends on the host's cellular processes. Ideally, a useful drug is considered most effective if it interferes with the viral replication without affecting normal cellular metabolic processes. Unfortunately, this objective has not been achieved with many antiviral compounds. Many of these drugs proved toxic to humans at therapeutic levels or had a limited spectrum of activity. This lack of success is one reason why antiviral drugs have not been developed as rapidly as antibacterial, antiprotozoal, or antifungal agents. Despite much research on the molecular biology of viruses and the complexity of the virus-host interaction, few antiviral agents have been licensed.

In most viral infections, immunity of the specific host cell is not well understood. Unfortunately, specific symptoms produced by viral infection may not appear until viral replication is complete or viral infection has already induced severe and sometimes irreversible changes in the infected cells. Because of this latent period, it may be difficult to determine the effectiveness of antiviral drugs. Administration of the antiviral agents may often be too late to inhibit a particular step of viral infection or to prevent host cells from performing abnormal functions. Perhaps because of these limitations and the life-long effects of viral vaccines, immunizing biologics rather than antiviral agents have been more successful in preventing viral infections.

Several antiviral agents that block specific enzymes essential for viral replication but not normal host cell metabolism have been discovered for the treatment of specific viral infections. This chapter is concerned mainly with antiviral agents that have been approved by the United States Food and Drug Administration (FDA) and are clinically effective in viral infections. Immunizing biologics and specific antineoplastic agents are not discussed in this chapter. Some important experimental antiviral agents are discussed under Investigational Antiviral Agents. The antiviral agents are classified and discussed according to their mode of action. These actions include inhibition of the stages of viral replication, interference with viral nucleic acid replication, and those affecting ribosomal translation. Possible interruptions of viral replication by these agents are shown in Figure 38–1.

Antibiotics are used primarily in the treatment of bacterial infections, but some also inhibit viral replication. For example, rifamycin derivatives (rifampicin) were reported to inhibit the growth of poxviruses and adenoviruses,[11] probably by blocking envelope formation and/or preventing release of infectious virions (see Fig. 38–1). Similarly, other antibiotics, such as bleomycin,[12] adriamycin,[13] and actinomycin D[14] inhibited RNA tumor viruses and DNA and RNA-directed polymerases in vitro. Because these antibiotics did not interfere with the transcription or translation of viral mRNA, and because large concentrations of the drug were required to inhibit the growth of viruses, such antibiotics are not used specifically for viral infections. Some inhibitors of reverse transcriptase activity from RNA tumor viruses are cytotoxic and have limited use in the chemotherapy of viral infections.[15] Antiviral and antitumor antibiotics are discussed in Chapter 37.

Agents Involving Inhibition of Stages of Viral Replication

Amantadine Hydrochloride, U.S.P. XXII

Amantadine, hydrochloride[16–18] (1-adamantanamine hydrochloride; Symmetrel; DuPont) is a symmetric tri-

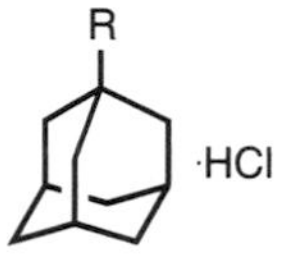

R

·HCl

Adamantane; R = H
Adamantine; R = NH_2
Rimantadine; R = $-CH(NH_2)CH_3$

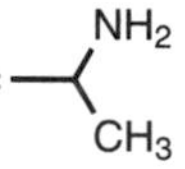

cyclic amine that inhibits penetration of virus particles into the host cell. It also inhibits the early stages of viral replication, blocking the uncoating of the viral genome and the transferring of nucleic acid into the host cell.

Amantadine hydrochloride is effective clinically in preventing and treating all A strains of influenza, particularly A2 strains of Asian influenza virus, and to a lesser extent, German measles (rubella) virus. It also shows in vitro activity against influenza B, parainfluenza, respiratory syncytial virus (RSV), and some RNA viruses (murine, Rous, and Esh sarcoma viruses). Many prototype influenza A viruses of different human subtypes (H1N1, H2N2, and H3N2) are also inhibited by amantadine hydrochloride in vitro and in animal model systems. If given within the first 48 hours, amantadine hydrochloride is effective in respiratory tract illness resulting from influenza A but not influenza B virus infection.

Amantadine hydrochloride is well absorbed orally and the usual dosage for oral administration is 100 mg twice daily. The FDA approved the amantadine hydrochloride capsule (100 mg) and syrup (10 mg/ml) for the treatment of HSV infection (keratoconjunctivitis). A 100-mg oral dose produced blood serum levels of 0.2 μg/ml within 1 to 8 hours. Maximum tissue concentration is reached in 48 hours when a 100-mg dose is given every 12 hours. Usually, no neurotoxicity is observed if the plasma level of amantadine is no more than 1.00 μg/ml.

Amantadine hydrochloride[18–20] crosses the blood-brain barrier and is excreted in saliva, nasal secretions, and breast milk. Approximately 90% of the drug is excreted unchanged by the kidney, primarily through tubular secretion. The half-life of the drug is 15 to 20 hours in patients with normal renal function; however, there are no reports of metabolic products of amantadine.

Generally, the drug has low toxicity at therapeutic levels but may cause severe central nervous system (CNS) symptoms such as nervousness, confusion, headache, drowsiness, insomnia, depression, and hallucinations. Convulsions and coma occur with high doses and in patients with cerebral arteriosclerosis and convulsive disorders. Chronic toxicity with amantadine hydrochloride may not be expected as this drug has been used continuously for Parkinson's disease for many years. Some serious reactions, however, include depression, orthostatic hypotension, psychosis, urinary retention, and congestive heart failure. Amantadine hydrochloride should be used with caution in patients who have a history of epilepsy and severe arteriosclerosis. Because amantadine hydrochloride does not appear to interfere with the immunogenicity of inactivated influenza A virus vaccine, patients may continue the use of amantadine hydrochloride for about 1 week after influenza A vaccination. A virus resistant to amantadine has been obtained in cell culture and from animals, but these reports are not confirmed in humans.

Rimantadine hydrochloride[19–21] (α-methyl-1-adamantanemethylamine hydrochloride; Flumadine, Hoffman-LaRoche), is an amantadine analog that appears more effective than amantadine hydrochloride against influenza A virus with fewer CNS side effects. Rimantadine hydrochloride appears to interfere with virus uncoating by inhibiting release of specific proteins. It may act by inhibiting reverse transcriptase or the synthesis of virus-specific RNA but does not inhibit virus adsorption or penetration. Rimantadine hydrochloride is not approved by the FDA in the United States, but it is used widely in Russia and Europe. The side effects are nightmares, hallucinations, and vomiting. Rimantadine hydrochloride is metabolized extensively before renal excretion. The half-life of rimantadine in adults ranges from 27 to 33 hours. Tromantadine (N-1-adamantyl-2-[2-(dimethylamino)ethoxy]) acetamide has been reported to have moderate activity against HSV types 1 and 2, herpes labialis, and herpes genitalis.

Interferon[22]

Interferon was first discovered by Isaacs and Lindenmann[23] in 1957. When they infected cells with viruses, viral interference was observed. Interferon was isolated and found to protect the cells from further infection. When interferon was given to other cells or animals, it displayed such biologic properties as inhibition of viral growth, cell multiplication, and immunomodulatory activities. The results led to the speculation that interferon may be a natural antiviral factor, possibly formed before antibody production, and may be involved in the normal mechanism of resistance against viral infection. Some investigators relate interferon to the polypeptide hormones and suggest that interferon functions in cell-to-cell communication by transmitting specific messages. Recently, antitumor and anticancer properties of interferon have evoked worldwide interest in the possible use of this agent in therapy for viral diseases, cancer, and immunodeficiency disorders.

Because viruses were found to induce release of interferon, the production or release of interferon in humans was attempted by the administration of other "inducers."[24] Different substances such as small molecules (substituted propanediamine) and large polymers (double-stranded polynucleotides) were used to induce interferons. Statolon, a natural double-stranded RNA produced in Penicillium stoloniferum culture, and a double-stranded complex of polyriboinosinic acid and polyribocytidylic acid (poly I:C) have been used as nonviral inducers for releasing preformed interferons. A modification of poly I:C stabilized with poly-L-lysine and carboxymethylcellulose (poly ICLC) has been used experimentally in humans. Clinically, it prevented coryza when used locally in the nose and conjunctival sacs. This substance was found to be a better interferon inducer than poly I:C. Another interferon inducer is ampligen, a polynucleotide derivative of poly I:C with spaced uridines. It has anti-HIV activity in vitro and is an immunomodulator.

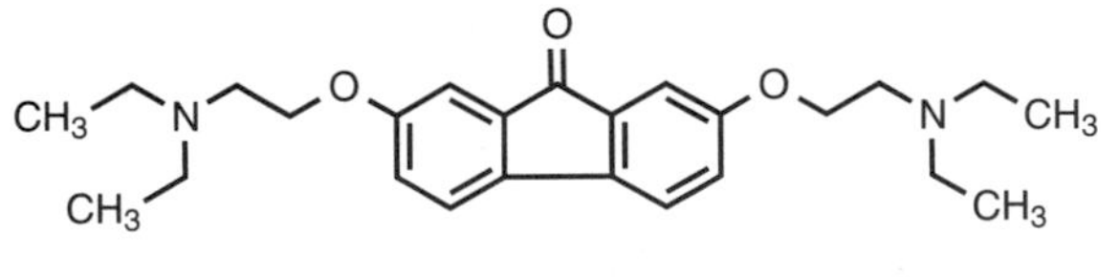

Tilorone

Other chemical inducers, such as pyran copolymers, tilorone, diethylaminoethyl dextran, and heparin, have also been used. Tilorone is an effective inducer of interferon in mice but it is relatively ineffective in humans. Initial use of interferon and its inducers instilled intranasally after rhinovirus exposure was successful in the prevention of respiratory diseases. The clinical success of interferon and its inducers has not yet been established, although they may play a significant role in cell-mediated immunity to viral infections and/or cancer. Disadvantages of interferon use include unacceptable side effects, such as fever, headache, myalgias, leukopenia, nausea, vomiting, diarrhea, hypotension, alopecia, anorexia, and weight loss.

Interferon consists of a mixture of small proteins with molecular weights ranging from 20,000 to 160,000. They are glycoproteins that exhibit species-specific antiviral activity. Broadly, human interferons are classified into three types[25]: alpha *(α)*, beta *(β)*, and gamma *(γ)*. The α-type is secreted by human leukocytes (white blood cells, non-T-lymphocytes) and the β-type by human fibroblasts. The γ-type of interferon is secreted by lymphoid cells (T lymphocytes), which either have been exposed to a presensitized antigen or have been stimulated to divide by mitogen. Gamma-interferon is also called "immune" interferon. Interferons are active in extremely low concentrations.

Interferon has been tested for use in chronic hepatitis B virus infection, herpetic keratitis, herpes genitalis, herpes zoster, varicella-zoster, chronic hepatitis, influenza, and common cold infections. Other uses of interferon are in the treatment of cancers, such as breast cancer, lung carcinoma, and multiple myeloma. Interferon has had some success when used as a prophylactic agent for CMV infection in renal transplant recipients. The scarcity of interferon and the difficulty in purifying it have limited clinical trials. Supplies have been augmented by recombinant DNA technology, which allows cloning of the interferon gene,[26] although the high cost still hinders clinical application. The FDA has approved recombinant interferon α-2a (Roferon A) and -2b (Intron A) for the treatment of hairy cell leukemia (a rare form of cancer), AIDS-related Kaposi's sarcoma, and genital warts (condyloma acuminatum). Subcutaneous injection of recombinant interferon α-2b has been approved for the treatment of chronic hepatitis C. Some foreign countries have approved α-interferon for the treatment of cancers such as multiple myeloma (cancer of plasma cells), malignant melanoma (skin cancer), and Kaposi's sarcoma (cancer associated with AIDS). B-, γ-interferons, and interleukin-2 may be commercial drugs of the future for the treatment of cancers and viral infections, including genital warts and the common cold.

Different mechanisms for the antiviral action of interferon have been proposed. Although α-interferon possesses broad-spectrum antiviral activity, it acts on virus-infected cells by binding to the specific cell surface receptors. It inhibits the transcription and translation of mRNA into viral nucleic acid and protein. Recent studies in cell-free systems have shown that the addition of adenosine triphosphate and double-stranded RNA to extracts of interferon-treated cells activates cellular RNA proteins and a cellular endonuclease. This activation causes the formation of translation inhibitory protein, which terminates production of viral enzyme, nucleic acid, and structural proteins.[27] Interferon may also act by blocking synthesis of a cleaving enzyme required for viral release.

Zidovudine[28–30]

Zidovudine (3′-azido-2′,3′-dideoxythymidine; azidothymidine; AZT; Retrovir, Burroughs Wellcome) is an analog of thymidine in which the azido group is substituted at the 3-carbon atom of the dideoxyribose moiety. It is active against RNA tumor viruses (retroviruses) that are the causative agents of AIDS and T-cell leukemia. Retroviruses, by virtue of reverse transcriptase, direct the synthesis of a provirus (DNA copy of a viral RNA genome). Proviral DNA integrates into the normal cell DNA, leading to the HIV infection. Zidovudine is converted to mono- di-, and triphosphates by the cellular thymidine kinase. These phosphates are then incorporated into proviral DNA, because zidovudine triphosphate is used as a substrate by reverse transcriptase. This process prevents normal 5′-3′-phosphodiester bonding, resulting in termination of DNA chain elongation owing to the presence of an azido group in zidovudine. The multiplication of HIV is halted by selective inhibition of viral DNA polymerase by zidovudine triphosphate at the required dose concentration. Zidovudine is a potent inhibitor of HIV-1 but also inhibits HIV-2 and EB virus. Zidovudine is used in

AIDS and AIDS-related complex (ARC) to control opportunistic infections by raising absolute CD4 (T4 helper/inducer) lymphocyte counts. A detailed discussion of zidovudine and other anti-AIDS agents is given in Chapter 39.

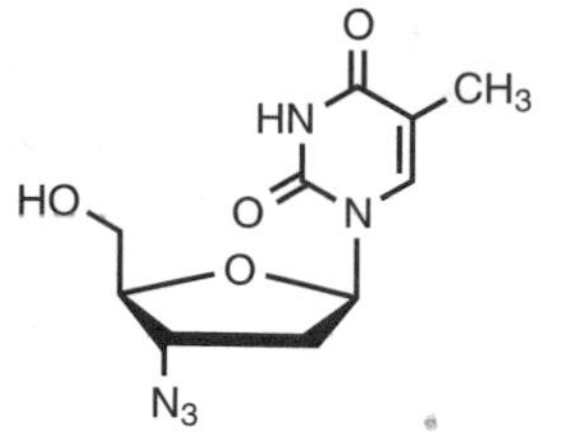

Zidovudine
(3'-Azido-2',3'-dideoxythymidine)

Zidovudine was first synthesized by Horwitz et al. in 1964, but its biologic activity was found by Ostertag et al. in 1974. In 1986, Yarchoan et al. demonstrated application of zidovudine in clinical trials of AIDS and related diseases. Zidovudine is recommended in the control of disease in asymptomatic patients in whom absolute CD4 (T4 helper/inducer) lymphocyte counts are less than $200/mm^3$. It prolongs the life of patients affected with Pneumocystis carinii pneumonia (PCP) and improves the condition of patients with advanced ARC by reducing the severity and frequency of opportunistic infections. Substantial benefits are obtained when the drug is given after the CD4 counts fall below $500/mm^3$. Therefore, zidovudine is used in early and advanced symptomatic treatment of AIDS or ARC. These patients have symptoms such as high fever, weight loss, lymphadenopathy, chronic diarrhea, myalgias, fatigue, and night sweats. The drug is toxic to the bone marrow and causes macrocytic anemia, neutropenia, and granulocytopenia. Other adverse reactions include headache, insomnia, nausea, vomiting, seizures, myalgias, and confusion.

Zidovudine is available in 100-mg capsules for oral administration. For asymptomatic adults, the initial recommended dosage is 1200 mg daily (200 mg every 4 hours) reducing to 600 mg daily (100 mg every 4 hours) for patients with advanced disease. The maintenance dose is 600 mg daily in symptomatic patients. Zidovudine is sensitive to heat and light because of its azide group and should be stored in colored bottles at 15°C to 25°C.

Zidovudine is well absorbed through the gastrointestinal tract. It concentrates in the body tissues and fluids, including cerebrospinal fluids. The bioavailability of drug was found to be approximately 60%. Its half-life is approximately 1 hour. Most of the drug is converted to its inactive glucuronide metabolite and is excreted unchanged through urine. Zidovudine also crosses the blood-brain barrier. Pentamidine (Pentam; Fujizawa), dapsone (Avlosulfon, Ayerst), amphotericin B (Fungizone; Apothecon), flucytosine (Ancobon, Roche), and doxorubicin (Adriamycin, Adria) may increase the toxic effects of zidovudine.

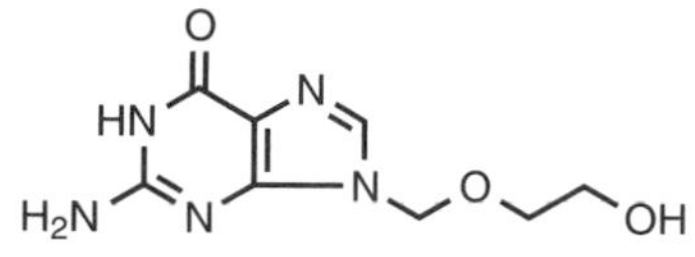

Acyclovir
(9-[(2-Hydroxyethoxy)methyl]guanine)

Agents Interfering with Viral Nucleic Acid Replication

Acyclovir[31]

Acyclovir (acycloguanosine, acyclo-G; Zovirax, Burroughs Wellcome) is a synthetic analog of deoxyguanosine in which the carbohydrate moiety is acyclic. Because of this difference in structure as compared to other antiviral compounds such as idoxuridine, vidarabine, and trifluridine, acyclovir possesses a unique mechanism of antiviral activity. Acyclovir is active against certain herpesvirus infections. These viruses induce virus-specific thymidine kinase and/or DNA polymerase; acyclovir inhibits these enzymes. Thus, acyclovir significantly reduces DNA synthesis in virus-infected cells without disturbing the active replication of uninfected cells. The mode of action consists of three consecutive mechanisms[32]: (1) Acyclovir is readily converted to active acyclovir monophosphate within cells by viral thymidine kinase. This phosphorylation reaction occurs faster by cells infected by herpesvirus than by normal cells because acyclovir is a poor substrate for the normal cell thymidine kinase. Acyclovir is further converted to di- and triphosphates by a normal cellular enzyme called guanosine monophosphate kinase. (2) Viral DNA polymerase is competitively inhibited by acyclovir triphosphate at lower concentrations than is cellular DNA polymerase. Acyclovir triphosphate is incorporated into the viral DNA chain during DNA synthesis. Because acyclovir triphosphate lacks the 3′-hydroxyl group of a cyclic sugar, it terminates further elongation of the DNA chain. (3) Preferential uptake of acyclovir by herpes-infected cells as compared to uninfected cells results in a higher concentration of acyclovir triphosphate, which leads to a useful drug to toxicity ratio of herpes-infected cells to normal cells.

The herpesviruses that are sensitive to acyclovir are HSV type 1, the common cause of labial herpes (cold sore), and type 2, the common source of genital herpes.[33] Varicella-zoster and some isolates of EB viruses are also affected by acyclovir. On the other hand, CMV is less sensitive to acyclovir, which has no activity against vaccinia virus, adenovirus, and parainfluenza infections.

An ointment containing 5% acyclovir has been used in a regimen of five times a day for up to 14 days for the treatment of herpetic keratitis and primary and recurrent infections of herpes genitalis; mild pain, transient burning, stinging, pruritus, rash, and vulvitis have been noted. The FDA has approved topical and intravenous acyclovir preparations for initial herpes genitalis and HSV types 1 and 2 infections in immunocompromised patients.[34] In these individuals, early use of acyclovir shortens the duration of viral shedding and lesion pain. Oral doses of 200 mg of acyclovir, taken five times a day for 5 to 10 days, have not proven successful because of the low bioavailability of current preparations. Oral doses of 800 mg of the drug given five times daily for 7 to 10 days have been approved, however, by the FDA for treatment of herpes zoster infection. This treatment shortens the duration of viral shedding in chickenpox and shingles. The intravenous injection of the drug (given 10 mg/kg three times daily for 10 to 12 days) has been approved for the treatment of herpes simplex encephalitis.[35] A soluble product, 6-deoxyacyclovir, is more useful because it is metabolized rapidly by xanthine oxidase to acyclovir. 6-Deoxyacyclovir is used in the treatment of varicella-zoster infection. Excessive and high doses of acyclovir have, however, caused viruses to develop resistance to the drug. This resistance results from reduction of virus-encoded thymidine kinase, which does not effectively activate the drug.

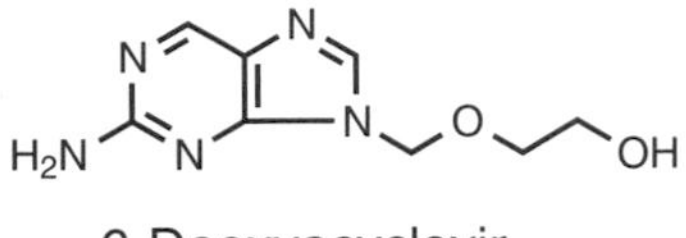

6-Deoxyacyclovir

Pharmacokinetic studies[36,37] show that acyclovir, after intravenous dose administration of 2.5 mg/kg, results in peak plasma concentrations of 3.4 to 6.8 mg/ml. The bioavailability of acyclovir is 15 to 30%. It is metabolized to 9-carboxymethoxymethylguanine, which is inactive. Plasma protein binding averages 15%, and approximately 70% of acyclovir is excreted unchanged in the urine by both glomerular filtration and tubular secretion. The half-life of the drug is approximately 3 hours in patients with normal renal function. In individuals with renal diseases, the half-life of the drug is prolonged. Therefore, acyclovir dosage adjustment is necessary for patients with renal impairment. Because of its low molecular weight and protein binding, acyclovir is easily dialyzed. Thus, a full dose of the drug should be given after hemodialysis. It should be infused slowly over at least 30 minutes to avoid acute transient and reversible renal failure. Acyclovir easily penetrates the lung, brain, muscle, spleen, uterus, vaginal mucosa, intestine, liver, and kidney. Acyclovir has relatively few side effects, except that intravenous injection causes reversible renal dysfunction and irritation, inflammation, and pain at the injection site. Infusion sites should therefore be inspected frequently and changed after every 72 hours. The drug is slightly toxic to bone marrow at higher doses. Less frequent side effects are nausea, vomiting, headache, skin rashes, hematuria, arthralgia, and insomnia.

Ganciclovir[38] (DHPG; Cytogene, Syntex) is an acyclic deoxyguanosine analog of acyclovir. It has greater activity than acyclovir against CMV and EB virus infection in immunocompromised patients. It is also active against HSV infection and in some mutants resistant to acyclovir. In AIDS patients, ganciclovir stopped progressive hemorragic retinitis and symptomatic pneumonitis related to CMV infection. Ganciclovir inhibits DNA polymerase. Its active form is ganciclovir triphosphate, which is an inhibitor of viral rather than cellular DNA polymerase. The phosphorylation of ganciclovir does not require a virus-specific thymidine kinase, the reason why the drug is active against CMV. The mechanism of action is similar to that of acyclovir; however, ganciclovir is more toxic to human cells than is acyclovir. Ganciclovir is absorbed as a prodrug and phosphorylated by infection-induced kinases of HSV and VZV infections. Common side effects are leukopenia, neutropenia, and thrombocytopenia. Ganciclovir with zidovudine causes severe hematologic toxicity. Ganciclovir is available only as an intravenous infusion because oral bioavailability is poor. It is given in doses of 5 mg/kg twice daily for 14 to 21 days. When ganciclovir is given by intravenous administration, concentrations of the drug in cerebrospinal fluid and in the brain vary from 25 to 70% of the plasma concentration. After minimal metabolism, ganciclovir is excreted in the urine. In adults with normal renal function, the serum half-life of the drug is approximately 3 hours. Ganciclovir has been approved by the FDA for the treatment of CMV retinitis in immunocompromised and AIDS patients.

Idoxuridine, U.S.P. XXII

Idoxuridine[39] (5-IUDR; Dendrid [Alcon], Herplex Liquifilm [Allergan], Stoxil [Smith Kline & French]) is a nucleoside containing a halogenated pyrimidine and is an analog of thymidine. It acts as an antiviral agent against DNA viruses by interfering with their replication given their similarity of structure. Idoxuridine is first phosphorylated by the host cell virus-encoded enzyme thymidine kinase to an active triphosphate form. The phosphorylated drug inhibits cellular DNA polymerase to a lesser extent than HSV DNA polymerase, which is necessary for the synthesis of viral DNA. The triphosphate form of the drug is then incorporated during viral nucleic acid synthesis by a false pairing system that replaces thymidine. On transcription, faulty viral proteins are formed, resulting in defective viral particles.[40]

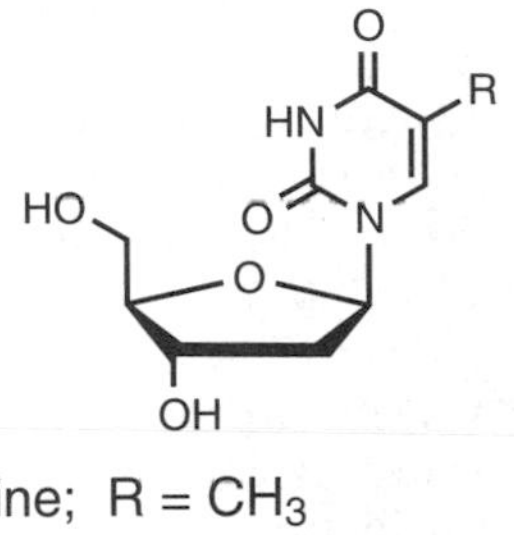

Thymidine; R = CH_3
5-Iodo-2'-deoxyuridine (Idoxuridine); R = I
5-Fluoro-2'-deoxyuridine; R = F
5-Bromo-2'-deoxyuridine; R = Br
2-Deoxy-5-(trifluoromethyl)uridine
(5-Trifluorothymidine); R = CF_3

Idoxuridine is available as ophthalmic drops (0.1%) and ointment (0.5%) for the treatment of HSV keratoconjunctivitis, the leading cause of blindness in the United States.[41] Because of its poor solubility, the drug is ineffective in labial or genital HSV or for cutaneous herpes zoster infection. Idoxuridine in dimethylsulfoxide (DMSO), however, has been used in mucocutaneous HSV infection of the mouth and nose. Because DMSO facilitates drug absorption and also has some therapeutic effect, a 40% solution of idoxuridine in DMSO is more effective than idoxuridine used without this vehicle. Idoxuridine, however, is approved by the FDA only for topical treatment of herpes simplex keratitis, and is more effective in epithelial than in stromal infections. It is less effective for recurrent herpes keratitis, probably because of the development of drug-resistant virus strains.

Adverse reactions of idoxuridine include such local reactions as pain, pruritus, edema, burning, and hypersensitivity. Systemic administration of idoxuridine by intravenous injection may be given in an emergency but leads to bone marrow toxicities such as leukopenia, thrombocytopenia, and anemia. It may also induce stomatitis, nausea, vomiting, abnormalities of liver functions, and alopecia. Idoxuridine has a plasma-half life of 30 minutes and is rapidly metabolized in the blood to idoxuracil and uracil.

Fluorodeoxyuridine has in vitro antiviral activity but is not used in clinical practice. 5-Bromo-2′-deoxyuridine is used in subacute sclerosing panencephalitis, a deadly virus-induced CNS disease. This agent appears to interfere with DNA synthesis in the same way as idoxuridine. The 5′-amino analog of idoxuridine (5-iodo-5′-amino-2′,5′-dideoxyuridine) is a better antiviral agent than idoxuridine and it is less toxic. It is metabolized in herpesvirus-infected cells only by thymidine kinase to di- and triphosphoramidates. These metabolites inhibit HSV-specific late RNA transcription, causing reduction of or less infective abnormal viral proteins.[42] 5-Bromo-2′-deoxyuridine has an action similar to that of other iodinated compounds.

Trifluorothymidine[43]

Trifluorothymidine (trifluridine, TFT, F3T; Viroptic, Burroughs Wellcome) is a fluorinated pyridine nucleoside structurally related to idoxuridine. It has been approved by the FDA and is a potent, specific inhibitor of replication of HSV type 1 in vitro. Its mechanism of action is similar to that of idoxuridine. Like other antiherpes drugs, it is first phosphorylated by thymidine kinase to mono-, di-, and triphosphate forms, which are then incorporated into viral DNA in place of thymidine to stop the formation of late virus mRNA and subsequent synthesis of the virion proteins. Trifluorothymidine, because of its greater solubility in water, is active against HSV types 1 and 2. It is also useful in treating infections caused by human CMV and VZV infections. The advantage of use of this agent over idoxuridine is its high topical efficacy in the cure of primary keratoconjunctivitis and recurrent epithelial keratitis.[44] It is also useful for difficult cases of herpetic iritis and established stromal keratitis.

Trifluorothymidine is available as a 1% ophthalmic solution, which is effective in dendritic ulcers. Generally, a 1% eye solution of trifluorothymidine is well tolerated. Cross-hypersensitivity and cross-toxicity between trifluorothymidine, idoxuridine, and vidarabine are rare. The most frequent side effects are temporary burning, stinging, localized edema, and bone marrow toxicity. It is less toxic but more expensive than idoxuridine. Trifluorothymidine, given intravenously, shows a plasma half-life of 18 minutes and is excreted in urine unchanged or as the inactive metabolite 5-carboxyuracil.

Vidarabine, U.S.P. XXII

Vidarabine[45] (adenine arabinoside, Ara-A, spongoadenosine; Vira-A, Parke-Davis) is an adenosine nucleoside obtained from cultures of Streptomyces antibioticus. Cellular enzymes convert ridarabine to mono-, di-, and triphosphate derivatives that interfere with viral nucleic acid replication, specifically inhibiting the early steps in DNA synthesis. This agent was used originally as an antineoplastic drug. Its antiviral effect is, in some cases, superior to that of idoxuridine or cytarabine.

This drug is used mainly in human HSV types 1 and 2 encephalitis, decreasing the mortality rate from 70 to 30%. Whitley and co-workers[46] reported that early vidarabine therapy is helpful in controlling complications of localized or disseminated herpes zoster in immunocompromised patients.

Vidarabine is also useful in neonatal herpes labialis or genitalis, vaccinia virus, adenovirus, RNA viruses, papovavirus, CMV, and smallpox virus infections. Given the efficacy of vidarabine in certain viral infections, the FDA approved a 3% ointment for the treatment of herpes simplex keratoconjunctivitis and recurrent epithelial keratitis, and a 2% intravenous injection for the treatment of herpes simplex encephalitis and herpes zoster infections. A topical ophthalmic preparation of vidarabine is useful in herpes simplex keratitis but shows little promise in herpes simplex labialis or genitalis. The monophosphate esters of vidarabine are more water soluble and can be used in smaller volumes and even intramuscularly. These esters are under clinical investigation for the treatment of hepatitis B, systemic and cutaneous herpes simplex,

Deamination

Vidarabine
9-β-D-Arabinofuranosyladenine (ara-A)

Arabinofuranosylhypoxanthine (ara-HX)

and herpes zoster virus infections in immunocompromised patients.

Vidarabine is deaminated rapidly by adenine deaminase, which is present in serum and red blood cells. The enzyme converts vidarabine to its principal metabolite, arabinosyl hypoxanthine (ara-HX), which has weak antiviral activity.[47] The half-life of vidarabine is approximately 1 hour, whereas ara-HX has a half-life of 3.5 hours. The drug is detected mostly in the kidney, liver, and spleen because 50% of it is recovered in the urine as ara-HX. Levels of vidarabine in cerebrospinal fluid are 50% of those in the plasma. Most side effects of vidarabine are gastrointestinal disturbances, such as anorexia, nausea, vomiting, and diarrhea. Central nervous system side effects include tremors, dizziness, pain syndromes, and seizures. Bone marrow suppression is reported at higher doses. Because vidarabine is reported to be mutagenic, carcinogenic, and teratogenic in animal studies, its use in pregnant women is to be avoided. Allopurinol and theophylline may interfere with the metabolism of vidarabine at higher doses because of the xanthine oxidase metabolism of vidarabine. Therefore, this agent should be avoided or given with caution to patients receiving these medications concurrently. Also, adjustment of the doses is necessary in patients with renal insufficiency.

Cytarabine, U.S.P. XXII

Cytarabine[48] (cytosine arabinoside, ara-C; Cytosar-U, Upjohn) is a pyrimidine nucleoside related to idoxuridine. It is used primarily as an anticancer rather than an antiviral agent. Cytarabine acts by blocking the utilization of deoxycytidine, thereby inhibiting the replication of viral DNA. The drug is first converted to mono-, di-, and triphosphates, which interfere with DNA synthesis by inhibiting both DNA polymerase and the reductase that promotes the conversion of cytidine diphosphate into its deoxy derivatives.

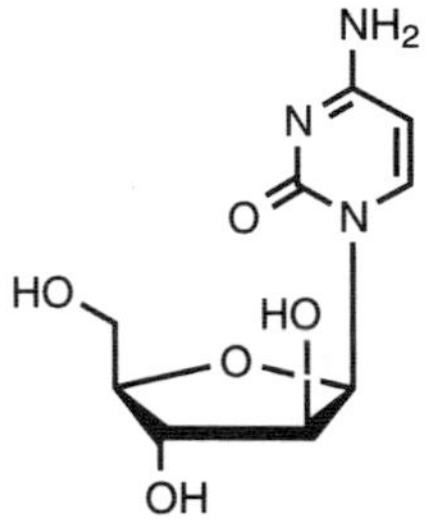

NH_2 N O N HO HO O OH

Cytarabine
1-β'-Arabinofuranosylcytosine (ara-C)

Cytarabine is used to treat Burkitt's lymphoma and both myeloid and lymphatic leukemias. Its antiviral use is in the treatment of herpes zoster (shingles) infection. It is also used to treat herpetic keratitis and viral infections resistant to idoxuridine. The drug is usually used topically, but it has been given by intravenous injection to individuals with serious herpes infection.[49] Cytarabine is deaminated rapidly in the body to an inactive compound, arabinosyluracil, which is excreted in the urine. The half-life of the drug in plasma is 3 to 5 hours. The toxic effects of cytarabine are chiefly on bone marrow, the gastrointestinal tract, and the kidney. The drug is not given in the early months of pregnancy because of its teratogenic and carcinogenic effects in animals.

Ribavirin[50,51]

Ribavirin (Virazole, ICN Pharmaceuticals), a purine nucleoside analog, is an investigational drug. It has broad-spectrum antiviral activity against both DNA and RNA viruses. It is phosphorylated by adenoxine kinase to the triphosphate resulting in inhibition of viral specific RNA polymerase, messenger RNA, and nucleic acid synthesis.

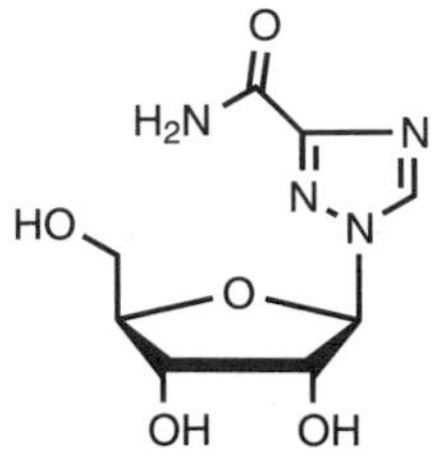

O H_2N N N N HO O OH OH

Ribavirin
(1-β-D-Ribofuranosyl-1,2,4-triazole-3-carboxamide)

Ribavirin is highly active against influenza A and B and the parainfluenza group of viruses, genital herpes, herpes zoster, measles, and acute hepatitis types A, B, and C. Aerosolized ribavirin has been approved by the FDA for the treatment of serious RSV infection, but it can cause cardiopulmonary and immunologic disorders in children. Ribavirin inhibits in vitro replication of HIV-1, which is involved in AIDS. Clinically, ribavirin was shown to delay the onset of full-blown AIDS in patients with early symptoms of HIV infection. Some viruses are less susceptible, for example, poliovirus, herpesviruses excluding varicella, vaccinia, mumps, reovirus, and rotavirus. A randomized double-blind study of aerosolized ribavirin treatment of infants with RSV infections indicated significant improvement in the severity of infection with a decrease in viral shedding.[52] Oral or intravenous forms of ribavirin are useful in the prevention and treatment of Lassa fever.[53] The clinical benefits of this agent are yet to be confirmed. Its few side effects are generally limited to gastrointestinal disturbances, such as nausea, vomiting, and diarrhea. The drug is contraindicated in asthma patients because of deterioration of pulmonary function. Viral strains susceptible to ribavirin have not been found to develop drug resistance, as is the case with other antiviral agents such as acyclovir, idoxuridine, and bromovinyldeoxyuridine (BVDU).

Agents Affecting Translation of Ribosomes

Methisazone and Others[54]

Methisazone (Marboran, Burroughs Wellcome) interferes with the translation of mRNA messages into protein synthesis on the cell ribosome. Ultimately, it produces a defect in protein incorporation into virus. Although viral

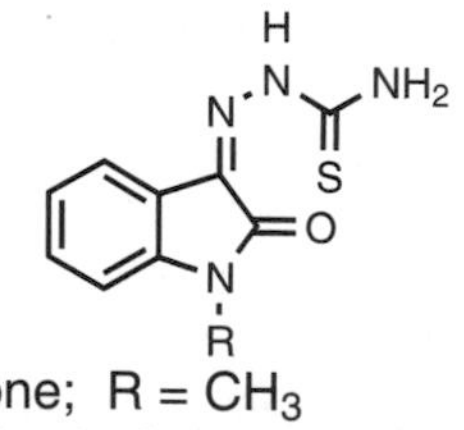

Methisazone; R = CH_3
(N-Methylisatin-β-thiosemicarbazone)
Isatin-β-thiosemicarbazone; R = H
1-Ethylisatin-β-thiosemicarbazone; R = C_2H_5

DNA increases and host cells are damaged, infectious virus is not produced.

Methisazone[55] is active against poxviruses, including variola and vaccinia. Some RNA viruses such as rhinoviruses, echoviruses, reoviruses, influenza, parainfluenza, and polio viruses are also inhibited. Therapeutically, methisazone is given in 1.5 to 3.0 g doses, twice daily by mouth. It has also been used as a prophylactic agent against smallpox. Historically, methisazone was one of the first antiviral compounds used in clinical practice. It is orally absorbed, with nausea and vomiting as the principal side effects. The drug is also used in vaccinia gangrenosa and disseminated vaccinia infections. This drug is not available in the United States, but it has been used in Europe for a long time. Isatin-β-thiosemicarbazone is active against variola and neurovaccinia. 1-Methyl- or ethyl-isatin-β-thiosemicarbazone is active against smallpox in mice and Rous sarcoma virus and vaccinia generalisata.

Investigational Antiviral Agents

Antiviral agents must be investigated because viral vaccines, although successful in preventing some viral diseases, are not effective as curative measures. Because viruses direct the host cell to synthesize enzymes and proteins for their own growth and multiplication, inhibition of such processes and development of antiviral drugs with selective toxicity to viruses is important. An effective antiviral agent should have broad-spectrum antiviral activity and completely inhibit viral replication. It should also be able to reach the target organ without interfering with the immune system of the patient. In addition, it should have minimal toxicity to the host cells and be effective against resistant mutant viruses.

Prospective antiviral agents that have been investigated represent a variety of chemical structures (Fig. 38–3). Among these agents are bromovinyldeoxyuridine (BVDU), AZT-related nucleosides, fluoroiodoaracytosine (FIAC), trisodium salt of phosphonoformic acid (foscarnet sodium), phosphonoacetic acid (PAA), levamisol hydrochloride, inosiplex (isoprinosine), different types of interferons, particularly α-interferon, 2-deoxy-D-glucose, arildone, oxychlorosene enviroxime, and some adenosine and guanosine analogs.

Among the 5-vinyl-2′-deoxyuridine analogs, BVDU[56] is one of the most potent and selective antiherpes viral nucleosides. Studies have shown this agent exerts an HSV type 2-preferential but not specific antiviral effect. It inhibits biosynthesis of HSV-1 glycoproteins and offers great promise for both topical and systemic treatment of HSV types 1 and 2 and VZV infections. In cell culture, it was active against these viruses in concentrations of 0.001 to 0.010 μg/ml but did not inhibit normal cell metabolism at concentrations of 30 to 100 μg/ml. The selective antiherpes action of BVDU depends primarily on inhibition of phosphorylation of HSV types 1 and 2 virus-induced thymidine kinase. Furthermore, differential sensitivity of HSV types 1 and 2 to BVDU can be used as a market test for differentiation of the two types in the clinical isolates. Oral administration of BVDU was found effective in the treatment of systemic herpesvirus and VZV infections, even in cancer patients. More recently, Jung-Cheng Lin and co-workers reported that BVDU was a potent inhibitor of EB virus replication in vitro.[57] The eye drops (0.1%) are effective in the treatment of herpes simplex keratitis and dendritic corneal ulcer. This agent appears to have no local or systemic toxicity and no mutagenic/teratogenic activity.

Fluoroiodoaracytosine (FIAC),[58] an analog of cytarabine, is a pyrimidine nucleoside as potent as BVDU. It is preferentially phosphorylated by HSV-specified thymidine kinase and is metabolized by a complex method. Fluoroiodoaracytosine and its metabolite fluoromethylarauridine are active in cell culture against HSV types 1 and 2, EBV, CMV, and VZV replications. They have greater antiviral activity than acyclovir against HSV type 1. These agents are being studied for their application as oral, topical, and injectable dosage forms.

Phosphonoformic acid (PFA)[59] and phosphonoacetic acid (PAA) have been investigated for their effects against human herpesviruses, including hepatitis B virus. They are also active against HIV. In the form of a cream, PFA is as effective as idoxuridine. It reduces lesions and pain and shortens the period of viral shedding, resulting in rapid healing. It also inhibits hepatitis B and CMV infections, but its activity against genital herpes and orolabial herpes infections is still under investigation. Major drawbacks in the use of PFA are its deposition in bones and skin irritation when applied topically. Also, herpesvirus mutants have been isolated from infected cell culture grown in the presence of PFA. The drug does not, however, undergo metabolic alterations.

Foscarnet sodium acts by selectively inhibiting viral DNA polymerases and reverse transcriptase. It is not phosphorylated into an active form by viral host cell enzymes. Therefore, it has the advantage of not requiring an activation step before attacking the target viral enzyme. It inhibits reverse transcriptase and is active against HIV. Foscarnet sodium (Foscavir, Astra) was approved by the FDA for the treatment of CMV retinitis in AIDS patients.[60] In combination with ganciclovir, the results have been promising, even in progressive disease with ganciclovir-resistant strains. Foscarnet sodium is also effective in the treatment of mucocutaneous diseases caused by acyclovir-resistant strains of HSV and VZV in AIDS patients. Foscarnet sodium is administered intravenously (60 mg/kg) three times a day for initial therapy and 90 to 120 mg/kg daily for maintenance therapy. The drug is neurotoxic and common adverse effects include anemia, nausea, vomiting, and seizures. Foscarnet sodium use poses the risk of severe hypocalcemia, especially with concurrent use of intravenous pentamidine.

Some immunomodulators are of interest in the treatment of viral infections. They include levamisol hydrochloride, inosiplex, and interferons. Levamisol hydrochloride is an immunotropic drug that is effective in acute genital herpes infection. The side effects of this drug include nervousness, nausea, vomiting, diarrhea, and skin rashes.[61] Inosiplex (isoprinosine) is a 1:3 complex of inosine and the 1-(dimethylamino)-2-propanol salt of 4-acetamidobenzoic acid. It is a weak inhibitor of HSV, adenovirus, rhinovirus, poliovirus, and influenza viruses in cell culture studies. Inosiplex is metabolized rapidly in humans. The half-life or the drug is 50 minutes after oral administration and 3 minutes after intravenous injection. Human use of inosiplex for viral infection and pre-AIDS patients is under investigation.[62] Its mode of action appears to inhibit both DNA and RNA viral replication. Clinically, inosiplex is active against infection caused by herpes, rhino, and influenza viruses. It is also reported to be useful in viral hepatitis and subacute sclerosing panencephalitis. In pre-AIDS and AIDS patients, it prolongs the progression of disease by stimulating host T cell-mediated immunity and by directly inhibiting viral replication. Interferon is administered by intramuscular or intravenous injection for the treatment of active hepatitis and localized herpes zoster. It delays CMV shedding as prophylaxis in renal transport recipients and also reduces HSV infections. Its side effects includes myalgia and gastrointestinal disturbances (nausea, vomiting, diarrhea, and anorexia).[63]

2-Deoxy-D-glucose[64] is active against both RNA and DNA enveloped viruses. It is useful in the treatment of herpes genitalis infection. Deoxyglucose replaces glucose in the nucleotide guanosine diphosphate glucose. Its analogs bind to a lipid carrier, dolichol pyrophosphate, thereby preventing glycosylation of proteins. The viruses produced are less effective because of their defective protein coat. Consequently, virions are unable to penetrate or liberate DNA into the host cell.

Arildone[65] is effective against both RNA and DNA vi-

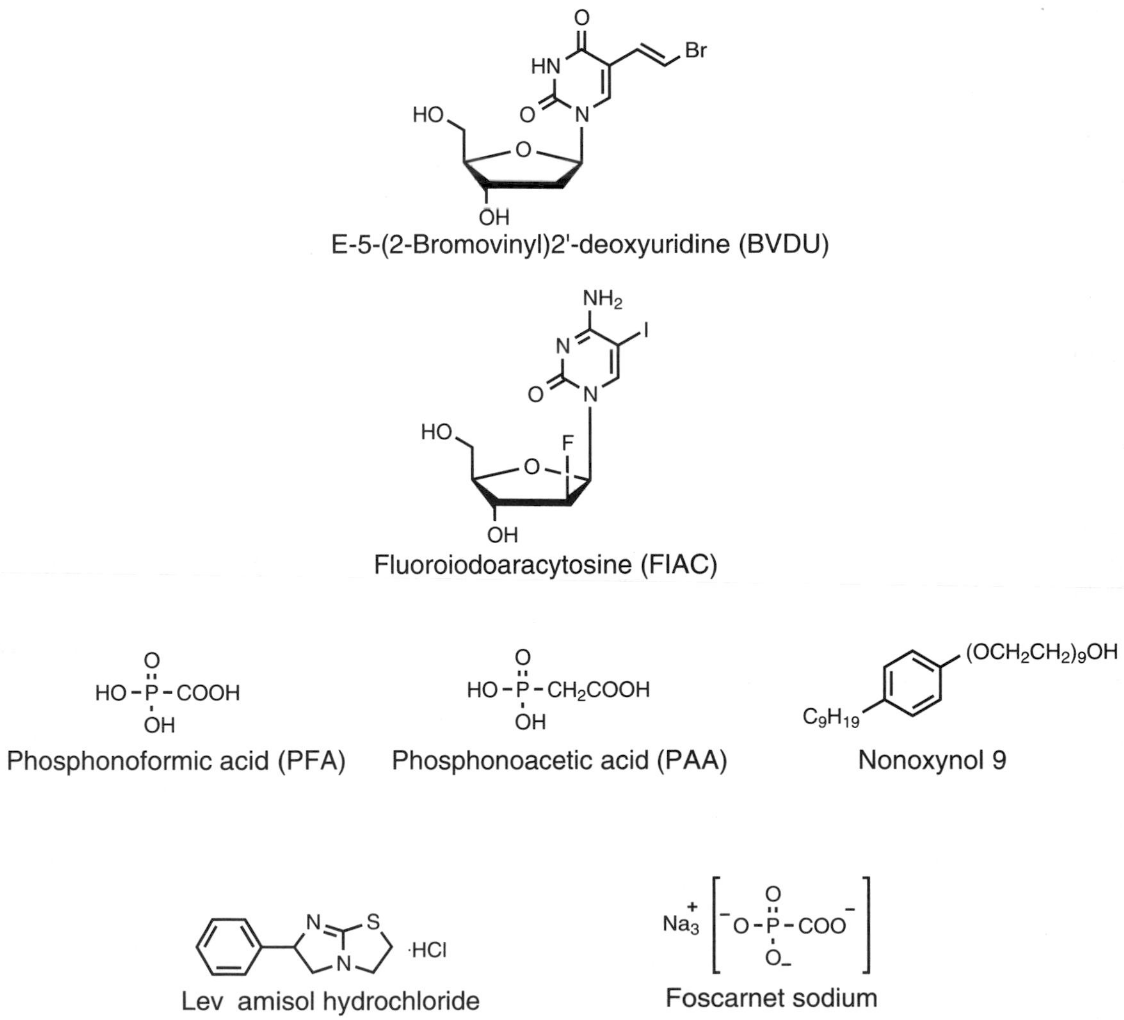

E-5-(2-Bromovinyl)2'-deoxyuridine (BVDU)

Fluoroiodoaracytosine (FIAC)

Phosphonoformic acid (PFA)

Phosphonoacetic acid (PAA)

Nonoxynol 9

Lev amisol hydrochloride

Foscarnet sodium

Fig. 38–3. Structural formulas of some investigational antiviral compounds.

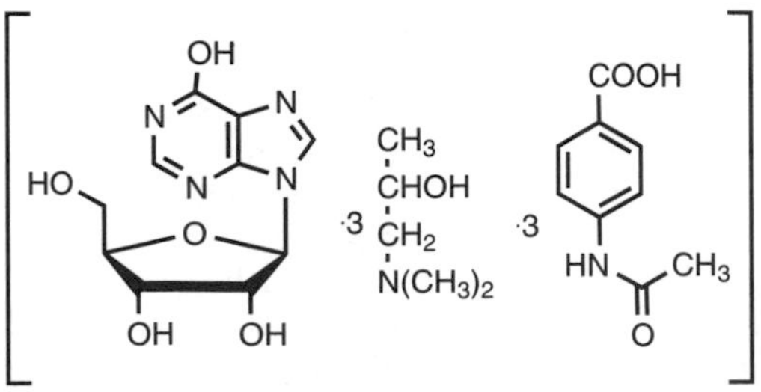

Inosiplex

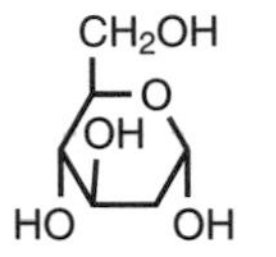

2-Deoxy-D-glucose

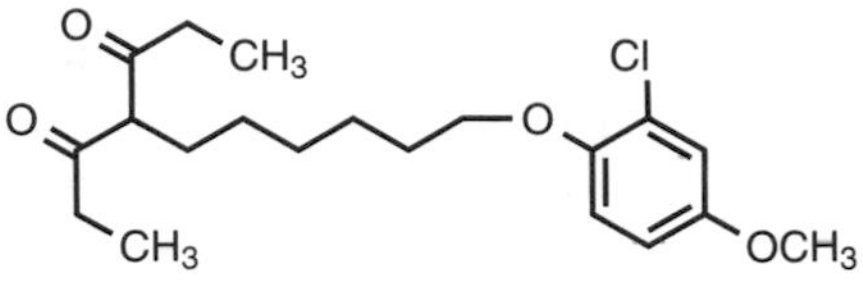

Arildone

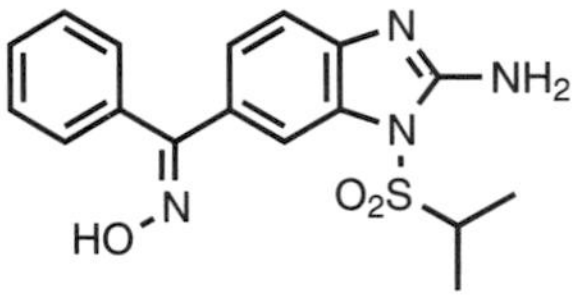

Enviroxime

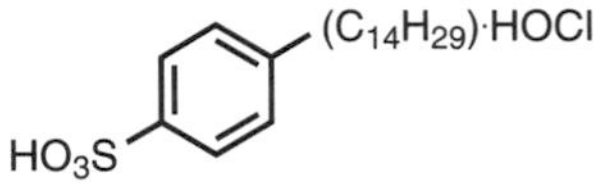

Oxychlorosene

Fig. 38–3. *(Continued).*

ruses, such as rhinovirus, herpesvirus, parainfluenza virus, and RSV. Arildone was found to be a potent inhibitor of CMV replication in cell culture. Oxychlorosene (clorpactin) is a mixture of straight and branched chain sodium tetradecylbenzenesufonates complexed with hypochlorous acid. This drug has been approved by the FDA as a topical antimicrobial agent for the treatment of localized infections. Its antiviral activity against HSV and genital and anogenital herpes infections is under clinical testing. This agent is promising in the treatment of human herpes simplex canker sores and poliovirus infections. Enviroxime is a benzimidazole derivative used against the common cold (rhinoviruses). It has antiviral effects against rhino-, coxsackie-, echo-, and polioviruses. It prevents viral uncoating and inhibits viral RNA polymerase from acting on late phase replication.[66] Orally, enviroxime has adverse effects that include nausea, vomiting, diarrhea, abdominal pain, and headache.

Some adenosine and guanosine analogs[67] (Fig. 38–4) are active against DNA and RNA viruses. Among the adenosine analogs is cyclaradine, the carbocyclic analog of cytarabine. The furanose ring oxygen of cytarabine is replaced by a methylene group. This agent is highly active against HSV-2 and is resistant to adenosine deaminase. Other adenosine analogs active against DNA and RNA viruses are 3-deazaadenosine, 3-deazaaristeromycin, neplanosin A, and *(s)*-9-(3-hydroxy-2-phosphonylmethoxypropyladenine (*(s)*HPMPA). The guanosine analogs are selenazofurin (2-β-D-ribofuranosylselenazole), 3-deazaguanosine, ganciclovir (9[2-hydroxy-1-(hydroxymethyl)ethoxy]methylguanine) (DHPG), and dihydroxybutylguanine (DHBG), which are promising against both DNA and RNA viral infections. Ganciclovir is related to acyclovir and is more soluble. Its mode of action is similar to that of acyclovir. It is clinically effective against human CMV infection in immunosuppressed patients. Studies revealed it inhibited replication of HSV types 1 and 2 and was active against herpesvirus encephalitis and vaginitis.[68] DHBG, a dihydroxybutyl analog of guanosine and *(s)*HPMPA, an acyclic adenosine analog, are antiherpetic agents that are under investigation. *(s)*HPMPA does not require thymidine kinase for activation. The active form, the diphosphoryl derivative, is probably phosphorylated intracellularly acting on viral DNA polymerase. Therefore, it has selective activity against broad-spectrum DNA viruses in vitro.[69] Tiazofurin (2-β-D-ribofuranosylthiazole) 4-carboxamide is a nucleoside structurally related to ribavirin. It is active against tomato spotted wilt virus even after the tomato plant has been infected.

Combination therapy[70,71] for viral infections is another approach being investigated. The synergistic antiviral effects of rimantadine with ribavirin and tiazofurin against influenza B virus and DHPG with foscarnet against HSV-1 and 2 are noteworthy. The synergistic action of either trifluorothymidine or acyclovir with leukocyte interferon has been used in the topical treatment of human herpetic keratitis.

During the past decade, combination antiretroviral

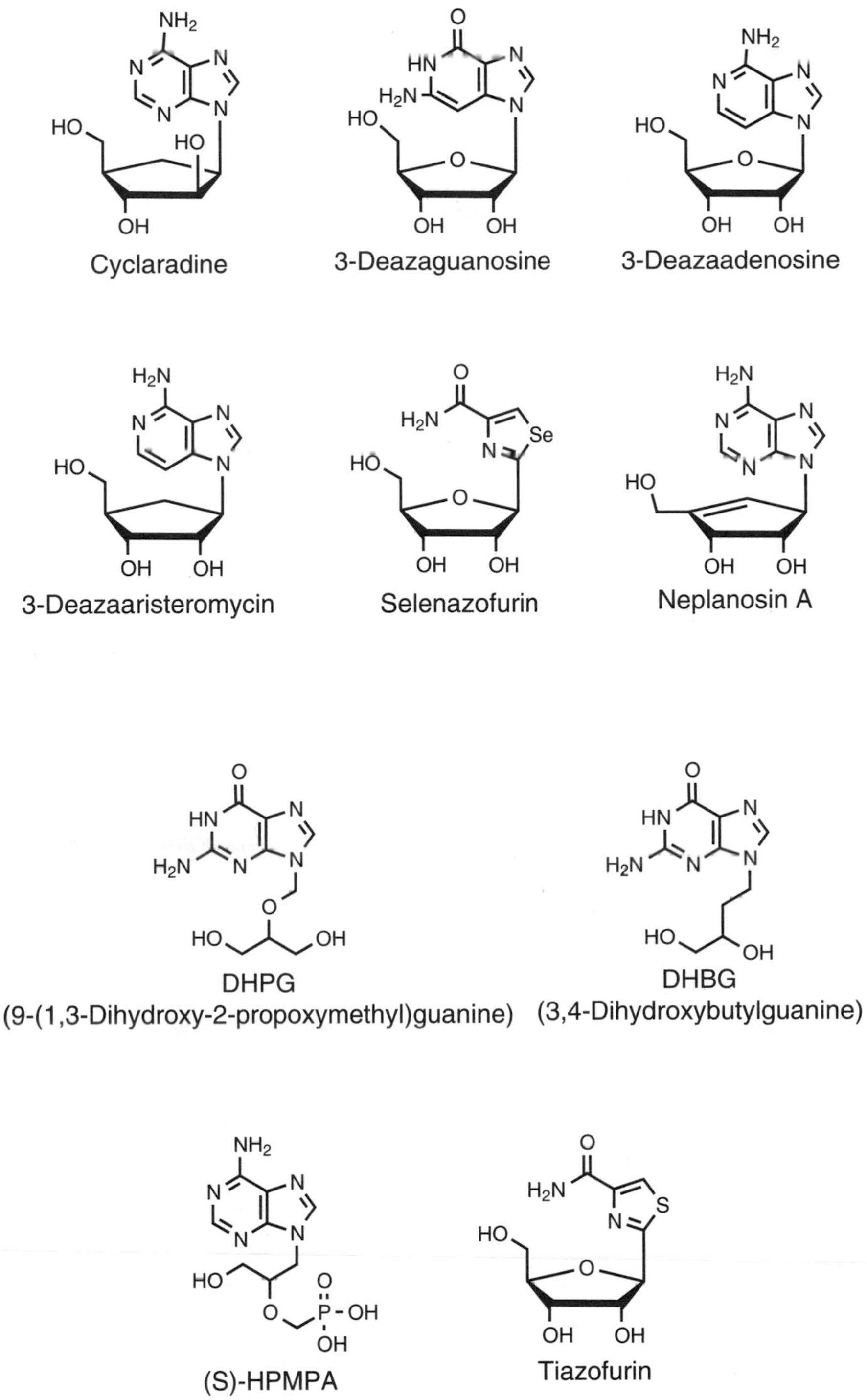

Cyclaradine 3-Deazaguanosine 3-Deazaadenosine

3-Deazaaristeromycin Selenazofurin Neplanosin A

DHPG (9-(1,3-Dihydroxy-2-propoxymethyl)guanine) DHBG (3,4-Dihydroxybutylguanine)

(S)-HPMPA Tiazofurin

Fig. 38–4. Structural formulas of some adenosine and guanosine analogs.

therapy for AIDS patients has made remarkable progress. AZT, the first approved drug for HIV-infected patients, produced bone marrow toxicity. To overcome toxic effects, combinations of AZT with foscarnet, DDC (2′,3′-dideoxycytidine) or DDI (2′,3′-dideoxyinosine) have been used. Such combination therapy indicated improved efficacy and decreased side effects as compared to either drug used alone. The combination of AZT with α-interferon has been used to treat patients with AIDS-related Kaposi's sarcoma.[72] The combination therapy delayed emergence of zidovudine-resistant HIV strains.

A combination of granulocyte-macrophage colony-stimulating factor with AZT and α-interferon has been successful in managing treatment-related cytopenia in HIV-infected patients. The advantages of combination therapy include therapeutic antiviral effect, decreased toxicity, and low incidence of drug-resistant infection. In recent years, emergence of drug resistance has been demonstrated in patients receiving single antiviral agent therapy. Resistance to amantadine, acyclovir, ribavirin, ganciclovir, AZT, and other antiviral agents is noteworthy.

REFERENCES

1. H. Fraenkel-Conrat, Ed., *The Chemistry and Biology of Viruses,* New York, Academic Press, 1969, pp. 45–109.
2. A. S. Evans, Ed., *Viral Infections of Humans, Epidemiology and Control,* 3rd ed., New York, Plenum Medical Book, 1991, pp. 77–806.
3. H. Rothschild, et al., Eds., *Human Diseases Caused by Viruses,* New York, Oxford University Press, 1978, pp. 61–258.
4. R. Calio and G. Nistico, Eds., *Antiviral Drugs, Basic and Therapeutic Aspects,* Rome-Milan, Pythagora Press, 1989, pp. 1–9.
5. W. L. Drew, Ed., *Viral Infections: A Clinical Approach,* Philadelphia, F. A. Davis, 1976, pp. 8–10.
6. G. J. Galasso, et al., Eds. *Antiviral Agents and Viral Diseases of Man,* New York, Raven Press, 1990, pp. 1–48; D. M. Prescott and A. S. Flexer, Eds., *Cancer, The Misguided Cell,* Sunderland, MA, Sinauer Associates, 1982, pp. 208–214.
7. H. H. Temin and S. Mizutani, *Nature, 226,* 1211(1970).
8. D. Baltimore, *Nature, 226,* 1209(1970).
9. J. R. Boyd, Editor-in-Chief, *Facts and Comparison,* St. Louis, J. B. Lippincott Company, 1983, p. 406.
10. R. G. Douglas, *Med. Clin. North. Am., 67,* 1163(1983); M. R. Keating, *Mayo Clin. Proc., 67,* 160(1992).
11. J. H. Subak-Sharpe, M. C. Timbury, and J. F. Williams, *Nature, 222,* 341(1969).
12. W. E. G. Müller, et al., *Biochem. Biophys. Res. Commun., 46,* 1167(1972).
13. W. E. G. Müller, et al., *Nature, 232,* 143(1971).
14. J. P. McDonnell, et al., *Nature, 228,* 433(1970).
15. V. S. Sethi and M. L. Sethi, *Biochem. Biophys. Res. Commun., 63,* 1070(1975); M. L. Sethi, *J. Nat. Prod., 42,* 187(1979).
16. N. Kato and H. J. Eggers, *Virology, 37,* 632(1969); H. J. Maasab and K. W. Cochran, *Science, 145,* 1443(1964).
17. W. C. Davis, et al., *Science, 144,* 862(1964); E. M. Neumayer, et al., *Proc. Soc. Exp. Biol. Med., 119,* 393(1965).
18. G. L. Mandell, et al., Eds., *Principles and Practice of Infectious Diseases,* New York, Churchill Livingstone, 1990, pp. 370–393; J. G. Tilles, *Annu. Rev. Pharmacol., 14,* 469(1974).
19. D. B. Burlington, et al., *Antimicrob. Agents Chemother., 21,* 794(1982); F. G. Hayden, et al., *Antimicrob. Agents Chemother., 23,* 458(1983).
20. F. G. Hayden, *J. Respir. Dis., 8*(Suppl. 11A), S45(1987).
21. W. L. Wingfield, et al., *N. Engl. J. Med., 28,* 579(1969); J. Mills and L. Corey, Eds., *Antiviral Chemotherapy: New Directions for Clinical Application and Research,* Vol. II, New York, Elsevier Science, 1989, pp. 117–142.
22. H. Monto, *Pharmacol. Rev., 34,* 119(1982); M. L. Francis, et al., *AIDS Res. Human Retroviruses, 8,* 199(1992).
23. A. Isaacs and J. Lindenmann, *Proc. R. Soc. London, B147,* 258(1957).
24. R. B. Pollard, *Drugs, 23,* 37(1982).
25. M. Strenli, et al., *Science, 209,* 1343(1980).
26. T. Taniguchi, et al., *Proc. Natl. Acad. Sci. USA, 77,* 5230(1980); D. M. Richards, et al., *Drugs, 26,* 378(1983).
27. C. Baglioni and P. A. Maroney, *J. Biol. Chem., 255,* 8390(1980).
28. M. A. Fischl, et al., *N. Engl. J. Med., 317,* 185(1987); P. A. Volberding, et al., *N. Engl. J. Med., 322,* 941(1990); R. Y. Yarchoan, et al., *N. Engl. J. Med., 321,* 726(1989).
29. M. A. Fischl, et al., *Ann. Intern Med., 112,* 727(1990); A. C. Collier, et al., *N. Engl. J. Med., 323,* 1015(1990); M. L. Sethi, in *Analytical Profiles of Drug Substances,* Vol. 20, K. Florey, Ed., San Diego, Academic Press, 1991, pp. 729–765.
30. M. Nasr, et al., *Antiviral Res., 14,* 125(1990); M. K. Sachs, *Arch. Intern. Med., 152,* 485(1992); M. A. Fischl, *in AIDS Clinical Reviews.* G. Volberding and M. A. Jacobson, Eds., New York, Marcel Dekker, 1991, pp. 197–214.
31. H. J. Field and I. Phillips, *J. Antimicrob. Chemother., 12* (Suppl. B), 1(1983); A. E. Nilsen, et al., *Lancet, 2,* 571(1982); L. J. Mayer, et al., *Am. J. Obstet. Gynecol., 158,* 586(1988).
32. G. B. Elion, et al., *Proc. Natl. Acad. Sci. USA, 74,* 5716(1977).
33. R. J. Whitley and C. A. Alford, *Hosp. Pract., 16,* 109(1981); J. J. Obrien and D. M. Campolirichards, *Drugs, 37,* 233(1989); H. C. Goodpasture, *Am. Fam. Physician, 43,* 197(1991).
34. C. D. Mitchell, et al., *Lancet, 1,* 1389(1981).
35. R. J. Whitley, et al., *N. Engl. J. Med., 314,* 144(1986); D. H. Shepp, et al., *Ann. Intern. Med., 103,* 368(1985).
36. D. Brigden, et al., *J. Antimicrob. Chemother., 7,* 399(1981); R. J. Whitley, et al., *Am. J. Med., 73,* 165(1982).
37. P. de Miranda, et al., *Clin. Pharmacol. Ther., 26,* 718(1979); P. de Miranda and M. R. Blum, *J. Antimicrob. Chemother., 12* (suppl. B), 29(1983).
38. C. V. Fletcher and H. H. Balfour, *Drug Intell. Clin. Pharm., 23,* 5(1989); D. Soucy, *Conn. Med., 55*(6), 345(1991).
39. H. E. Kaufman, *Proc. Soc. Exp. Biol. Med., 109,* 251(1962); J. A. Gold, et al., *Am. N. Y. Acad. Sci., 130,* 209(1965).
40. A. Farah, et al., Eds., *Handbook of Experimental Biology,* Vol. 38/2, Berlin, Springer, 1975, pp. 272–347; P. H. Fischer, et al., *Biochim. Biophys. Acta, 606,* 236(1980).
41. E. Maxwell, *Am. J. Ophthalmol., 56,* 571(1963); W. H. Prusoff, et al., *Pharmacol. Ther., 7,* 1(1979).
42. L. A. Babiuk, et al., *Antimicrob. Agents Chemother., 23,* 715(1983); H. Shiota, et al., Eds., *Herpesvirus: Clinical Pharmacological and Basic Aspects,* Amsterdam, Excerpta Medica International Congress Series 571, 1982, pp. 157–164.
43. C. Heidelberger and D. H. King, *Pharmacol. Ther., 6,* 427(1979).
44. A. A. Carmine, et al., *Drugs, 23,* 329(1982).
45. R. J. Whitley, et al., *N. Engl. J. Med., 297,* 289(1977).
46. R. J. Whitley, et al., *N. Engl. J. Med., 307,* 971(1982).
47. D. L. Chao and A. P. Kimball, *Cancer Res., 32,* 1721(1972).
48. R. L. Ward and J. G. Stevens, *J. Virol., 15,* 71(1975).
49. R. L. Nutter and F. Rapp, *Cancer Res., 33,* 166(1973).
50. R. W. Sidwell, et al., *Science, 177,* 705(1972); C. B. Hall, et al., *N. Engl. J. Med., 308,* 1443(1983); H. Fernandez, in *Ribavirin: A Broad Spectrum Antiviral Agent,* R. A. Smith and W. Kirkpatrick, Eds., New York, Academic Press, 1980, pp. 215–230.
51. C. B. Hall, *Am. J. Dis. Child, 140,* 331(1986); D. S. Stein, et al., *J. Infect. Dis., 141,* 548(1980); F. W. Moler, et al., *N. Engl. J. Med., 325,* 1884(1991).
52. C. F. Fox and W. S. Robinson, Eds., *Virus Research,* New York, Academic Press, 1973, pp. 415–436; L. H. Taber, et al., *Pediatrics, 72,* 613(1983).
53. J. B. McCormick, et al., *N. Engl. J. Med., 314,* 20(1986).
54. R. L. Thompson, et al., *J. Immunol., 70,* 229(1953).
55. L. A. do Valle, et al., *Lancet, 2,* 976(1965).
56. E. De Clercq, et al., *J. Infect. Dis., 143,* 846(1981); E. De Clercq, et al., *Br. Med. J., 281,* 1178(1980); P. C. Maudgal and E. De Clercq, *Curr. Eye Res., 10* (Suppl.), 193, (1991).
57. C. H. Stuart-Harris and J. S. Oxford, Eds., *Problems of Antiviral Therapy,* New York, Academic Press, 1983, pp. 295–315; J-C. Lin, et al., *Science, 221,* 578(1983).
58. A. J. Grant, et al., *Biochem. Pharmacol., 31,* 1103(1982); C. Lopez, *Antimicrob. Agents Chemother., 17,* 803(1980).
59. N. L. Schnipper, et al., *Appl. Environ. Microb., 26,* 264(1973); R. W. Honess and D. H. Watson, *J. Virol., 21,* 584(1977).

60. P. Chrisp and S. P. Clissold, *Drugs, 41,* 104(1991); M. A. Jacobson, et al., *Antimicrob. Agents Chemother., 33,* 736(1989).
61. G. G. Jackson, *J. Infect. Dis., 1335,* A 83(1976); M. Van Eygen, et al., *Lancet, 1,* 382(1976).
62. S. E. Reed, et al., *J. Infect. Dis., 1335,* 128(1976); R. H. Waldman and R. Granguly, *Am. N. Y. Acad. Sci., 284,* 153(1977); D. J. Morris, *J. Antimicrob. Chemother., 29,* 97(1992).
63. G. Emodi, et al., *J. Infect. Dis., 133,* A 199(1976); T. C. Merigan, Ed., *Antivirals with Clinical Potential,* Chicago, University Press, 1976, pp. 205–210; C. A. Schiffer, *Semin. Oncol., 18* (Suppl. 7), 1(1991).
64. G. Kaluza, et al., *J. Gen. Virol., 14,* 251(1972); S. Steiner, et al., *Cancer Res., 33,* 2402(1973).
65. G. D. Diana, et al., *J. Med. Chem., 20,* 750(1977); A. S. Tyms and J. D. Williamson, *Nature, 297,* 690(1982).
66. D. A. J., Tyrrell, et al., in *Problems of Antiviral Therapy,* C. H. Stuart-Harris and J. S. Oxford, Eds., London, Academic Press, 1983, pp. 265–276; R. Phillpotts et al., *Antimicrob. Agents Chemother., 23,* 671(1983); Y. Ninomiya, et al., *Virology, 134,* 269(1984).
67. R. K. Robins, *Chemical Engineering News, 64*(4), 28(1986); S. Shuto, et al., *J. Med. Chem., 35,* 324(1992).
68. Y-C Cheng, et al., *Proc. Natl. Acad. Sci. USA, 80,* 2767(1983); K. B. Frank, et al., *J. Biol. Chem., 259,* 1566(1984).
69. G. D. Diana, et al., *Ann. Rep. Med. Chem., 24,* 129(1989); U. Jehn, *Recent Results Cancer Res., 121,* 353(1991).
70. J. E. Groopman, et al., *Am. J. Med., 90 S,* 4A-18S(1991); V. A. Johnson and M. S. Hirsch, *Ann. N. Y. Acad. Sci., 616,* 318(1990).
71. E. W. J. De Koning, et al., *Arch. Ophthalmol., 101,* 1866(1983); H. J. Field, Ed., *Antiviral Agents: The Developments and Assessment of Antiviral Chemotherapy,* Vol. I, Florida, CRC Press Inc., 1988, pp. 1–18.
72. R. Mitsuyasu, *Br. J. Haematol., 79* (Suppl. 1), 69(1991).

SUGGESTED READINGS

L. H. Collier and J. S. Oxford, Eds., *Developments in Antiviral Therapy,* New York, Academic Press, 1980.
H. Fraenkel-Conrat, Ed., *Molecular Basis of Virology,* New York, Reinhold, 1968.
G. J. Galasso, et al., Eds., *Antiviral Agents and Viral Diseases of Man.,* 3rd ed., New York, Raven Press, 1990.
C. A. Knight, *Chemistry of Viruses,* 2nd ed., New York, Springer, 1975.
S. E. Luria, et al., Eds., *General Virology,* New York, John Wiley & Sons, 1978.
T. C. Merigan, *J. Infect. Dis., 133,* Suppl., 1–285(1976).
D. Shugar, Ed., *Viral Chemotherapy, International Encyclopedia of Pharmacology and Therapeutics,* Vol. II, New York, Pergamon Press, 1985.
R. W. Sidwell and J. T. Witkowski, *Antiviral Agents,* in *Burger's Medicinal Chemistry,* 4th ed., M. E. Wolff, New York, John Wiley & Sons, 1979, pp. 543–593.

Chapter 39

ANTI-AIDS AGENTS

Prem Mohan

Acquired immunodeficiency syndrome (AIDS) is a fatal pathogenic disease caused by a retrovirus.[1-3] The major modes of viral transmission have been through infected sexual partners and intravenous drug users. The epidemic in the United States is caused by human immunodeficiency virus type 1 (HIV-1), although scattered cases of infection by the type 2 virus (HIV-2) have also been reported. HIV-2 is endemic to western Africa and it is not unusual in certain African countries to find patients infected with both HIV-1 and HIV-2. AIDS is prevalent in almost every country on the globe, and it is estimated that 8 to 10 million people are infected worldwide, with approximately 1 to 1.5 million of these people in the United States. In the United States, the Centers for Disease Control reports a total of more than 240,000 reported AIDS cases as of the end of 1992, of which more than 158,000 individuals have died.[4] These alarming figures and the still uncurbed spread of the virus have posed a formidable challenge and mounted an unprecedented effort to search for a cure for the disease.

Because history documents the prevention or cure of viral diseases by using vaccines, it was first envisioned that the production of a vaccine would be a definitive approach to arrest the spread of AIDS. In fact, vaccine research is being hotly pursued in many laboratories. Research in this area has been hampered, however, by many factors. The major question is the selection of volunteers for phase III trials and the moral obligation to counsel control groups against risky behavior weighed by the need to obtain efficacy data.[5] Most importantly, the high mutation rate of the envelope glycoprotein has resulted in the identification of several isolates of HIV-1. Immunization against one isolate does not guarantee protection against another isolate. The lack of suitable animal models for HIV-1 infection and AIDS also poses a major problem. In addition, because HIV-1, unlike other viruses, is usually found internalized in host cells and is mostly transmitted in this manner, the vaccine approach to combat AIDS has been questioned.[6]

At the onset, the search for new drugs to fight AIDS can be debated on one crucial point. Because all patients with AIDS possess integrated proviral DNA in their genomes, the host has the blue print for guaranteed lifelong infection. Therefore, theoretically, a cure would only be possible if an agent specifically sought and excised the integrated proviral DNA fragment from every infected host cell.[7] If indeed this option is possible, the cure for AIDS may be at hand. On the other hand, a strong practical approach to circumvent this apparent problem is to design agents that would inhibit the various stages subsequent to activation of the viral genes. In other words, when new virions bud from infected cells and proceed to infect noninfected cells, several targets are attractive sites for therapeutic intervention. If future viral multiplication is halted, leading to the lessening of viral burden, the damaged immune system may be able to heal itself on its own or with the help of drugs that rejuvenate the immune system. It would be ideal to discover a drug that both annihilates the virus and reconstructs the injured immune system.

STRUCTURE AND LIFE CYCLE OF THE AIDS VIRUS

The AIDS virus is the pathogenic human virus studied most widely in medical history. Understanding the structure and life cycle of HIV-1 is crucial because it presents potential intervention sites for halting viral replication. Research has revealed that the genome of the virus is complex and contains at least nine genes. The functions of three of these genes, *gag* (coding for the core proteins p17 and p24), *pol* (coding for the enzymes reverse transcriptase, protease, and integrase) and *env* (coding for the envelope protein gp120) are known because they are similar to those present in many animal retroviruses.[8] The additional genes, however, *vif* (virion infectivity factor), *nef* (negative regulatory factor, now uncertain), *rev* (differential regulator), *vpu* (controls efficient virion budding), *vpr* (weak transcriptional activator), and *tat* (transactivator), working alone or in concert with other genes, endow the virus with several mechanisms for immune and drug evasion, resulting in its unprecedented pathogenic prowess.

The life cycle of the AIDS virus begins with its attachment to the host target cell. The high affinity receptor for this attachment on HIV-1 is the heavily glycosylated viral envelope protein gp 120. The corresponding receptor on the target cell is the CD4 receptor protein. It therefore follows that any cell that has the CD4 receptor can be a target for the virus, and indeed this is true for macrophage/monocyte cells. Glial cells, cells of the gut epithelium, and bone marrow progenitor cells, however, may also be infected[9] and may or may not involve the CD4 protein. The macrophage is less susceptible to the cytopathic effects of the virus and in essence serves as a reservoir for multiplication and harboring of HIV-1. In contrast, the targeted T4-helper/inducer cell is destroyed, leading to the gradual depletion of these cells and causing severe immune deficiency, opportunistic infections, and the onset of AIDS.[10]

After binding to the surface of the host cell, the outer membrane of the virus fuses with membrane of the host cell. At this point, the virion uncoats and unloads the genomic RNA and the enzyme reverse transcriptase (RT) into the cytoplasm of the target cell. At this juncture, RT performs three important functions. First, using the RNA (called a positive strand) as a template, it catalyzes an RNA-dependent DNA synthesis to produce a single negative strand of DNA (called the positive strand). Second, using the ribonuclease H (RNase H) section of RT, the enzyme systematically degrades the genomic negative strand of RNA. Third, using the newly synthesized positive DNA strand as a template, RT catalyzes a DNA-dependent DNA synthesis of a complementary copy of DNA (a negative strand). This newly formed DNA double helix, also called proviral DNA, is translocated into the nucleus in either a linear or circular form, where it is integrated into the host genome by the viral enzyme integrase.

With integration complete, the virus has accomplished an important requirement and is ready to produce more infectious particles and embark on its journey of immune destruction. More commonly, the virus remains latent for a few months or up to 8 to 10 years,[11] leaving the infected person asymptomatic during this entire period. The presence of host factors or gene products of viruses, like the Epstein-Barr virus, herpes simplex virus, cytomegalovirus, or HTLV-1, can activate the latent virus. This stimulation leads to expression of the viral genes and to the production of viral genomic RNA and messenger RNA, followed by the synthesis of viral proteins, movement to the surface, and viral budding. At this stage, the protease enzyme cleaves the viral polyproteins, and myristylation of gp17 takes place. The budding operation kills the host cell in the process and the newly formed virus particles seek out a target cell to once again begin its life cycle.[10]

POTENTIAL ANTI-HIV-1 AGENTS

To design a successful anti-HIV-1 agent, it is important to locate a target that is unique to the virus. Also, because the AIDS virus is found mostly inside cells, it is essential for the inhibitor to enter the host cell membrane to reach the virus. Another challenge for the design of such inhibitors is for the agent to be able to discriminate between a virally infected cell and a noninfected cell. Undoubtedly, this task is difficult, because the cellular membranes of virally infected cells differ only slightly from the membranes of noninfected cells. Further, at least in the case of the AIDS virus, the drug should be able to enter not only T cells, but also macrophages/monocytes, glial cells, and other host cells that are infected. Most important of all, if the agent enters noninfected healthy cells, it should be nontoxic, whereas once it enters the virally infected cell, it should obliterate the virus. These hurdles are but a few that hinder the development of potential anti-AIDS agents. Nevertheless, a variety of derivatives do inhibit the in vitro replication of HIV-1 and hold promise for the future development of anti-HIV-1 agents. So far, these compounds can be broadly divided into three major classes on the basis of enzymes or processes that they inhibit: (1) enzymes (reverse transcriptase, protease or glucosidase), (2) viral processes (syncytia formation, binding, or uncoating), or (3) the expression of genes or gene products.

Reverse Transcriptase Inhibitors

HIV-1 and HIV-2 Reverse Transcriptase Inhibitors

The anti-HIV-1 agents studied and researched most widely have been those that inhibit the retroviral reverse transcriptase (RT), an enzyme that is vital for replication of the AIDS virus. Because RT is not found in human cells and inhibitors for the RT found in animal viruses were available, a surge of activity to test many previously synthesized compounds against HIV-1 resulted.

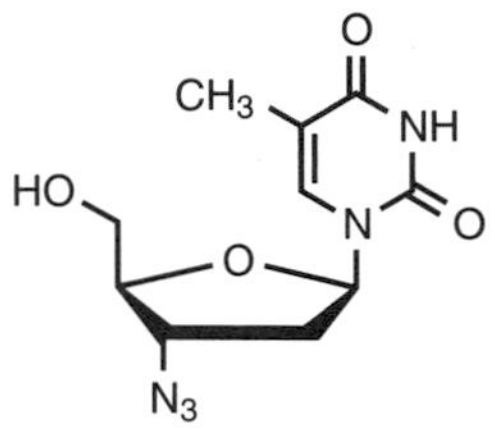

AZT

One of these compounds, 3′-azido-2′,3′-dideoxythymidine (AZT), first synthesized as an antitumor agent,[12] produced activity against a murine retrovirus, the Friend leukemia virus.[13] A decade later, AZT demonstrated in vitro inhibition of HIV-1,[14] was subsequently administered to patients,[15] and became the first drug to gain FDA approval for the treatment of AIDS and AIDS-related conditions (ARC).

This finding stimulated the evaluation of several other already synthesized or newly prepared nucleoside analogs, like 2′,3′-dideoxycytidine (DDC), 2′,3′-dideoxyadenosine (DDA), 2′,3′-dideoxyinosine (DDI), and 2′,3′-dideoxy-2′,3′-didehydrothymidine (D4T), with the consequent introduction of these agents into clinical trials. Since that time, DDI and DDC have also gained FDA approval. Other nucleosides undergoing active clinical development are 3′-deoxy-3′-fluorothymidine and 2′,3′-dideoxy-3′-thiacytidine.[16]

The mechanism of action of the above nucleoside compounds is the inhibition of RT. To achieve this inhibition, these agents must first undergo sequential 5′-phosphorylation by the relevant kinases to be converted to their active triphosphate species. For example, AZT undergoes phosphorylation by thymidine kinase, thymidylate kinase, and nucleoside diphosphate kinase. The nucleoside 5′-triphosphate inhibits RT by two modes: first, by acting as a competitive inhibitor of the normal nucleoside-5′-triphosphate for the enzyme; and second, although incorporated into the growing chain because of the presence of a 5′-hydroxy functionality, the lack of a necessary 3′-hydroxy moiety leads to the inability to form diester bonds between the 5′ and 3′ positions and hence causes chain termination.[17] On medicinal chemistry grounds,

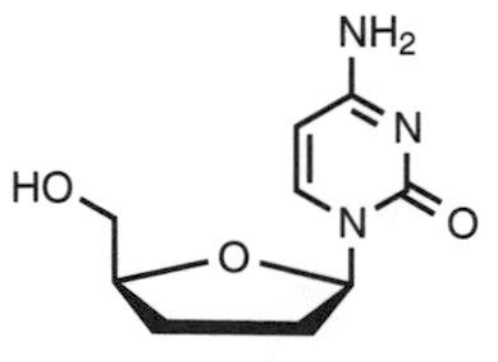

2',3'-Dideoxycytidine (DDC)

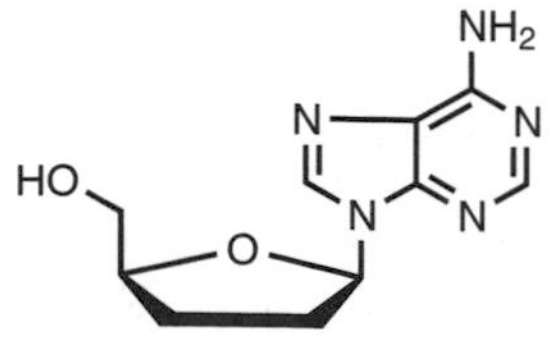

2',3'-Dideoxyadenosine (DDA)

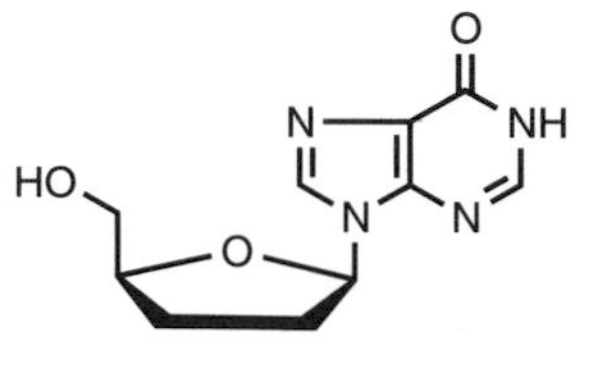

2',3'-Dideoxyinosine (DDI)

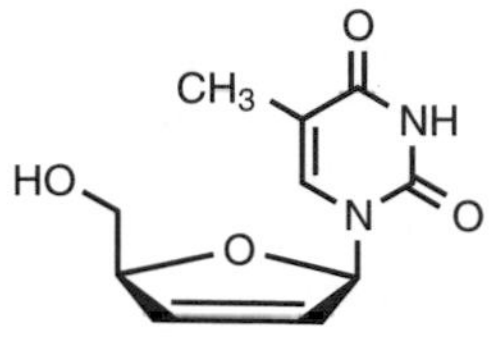

2',3'-Dideoxy-2',3'-didehydrothymidine (D4T)

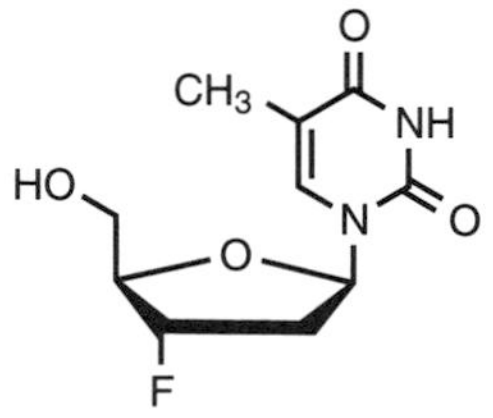

3'-Deoxy-3'-fluorothymidine

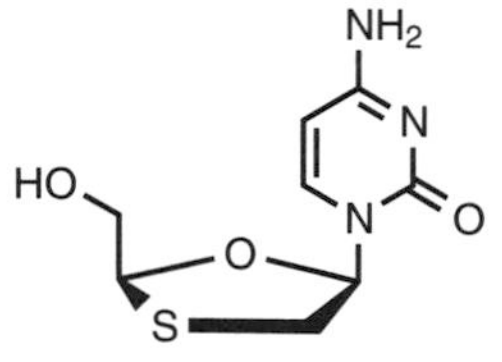

2',3'-Dideoxy-3'-thiacytidine

the role of the azido group is intriguing because of its uncommon occurrence on therapeutic agents. It has been suggested that in AZT, the azido group structure allows it to bind to the sites involved in both polynucleotide and mononucleotide 5′-phosphate binding.[18]

The clinical administration of AZT, DDI, DDC, and D4T reveals a toxicity profile that is different for each individual agent. Major problems with AZT have been bone marrow suppression and anemia, leading to patients requiring blood transfusions. Peripheral neuropathy has been associated with DDC, DDI, and D4T, and pancreatitis has been observed in patients receiving DDI.[19] In addition, it is disturbing that nucleoside derivatives have the ability to stimulate the emergence of HIV-1 strains that are resistant to these compounds. AZT-resistant mutants were the first to be discovered; amino acid mutations at positions 67, 70, and 215 were common to several isolates that were resistant to AZT.[20] Isolates that demonstrate decreased sensitivity to DDI therapy contain the Leu 74 → Val mutation,[21] whereas the Thr 69 → Asp substitution probably causes decreased susceptibility to DDC.[22]

The aforementioned properties of the nucleoside class of compounds have provided the impetus for further structural manipulation. Although it is possible to modify the nucleic acid base to produce newer analogs that are equally active or are more active than the nucleosides, a large amount of research has focused on modification of the sugar portion of the molecule. Systematic structure-activity relationship studies have revealed that AZT is the most potent of the 3′-azido compounds.[23] 3′-Fluoro nucleoside analogs are also active, with the thymidine derivative being the most potent.[24] Among the 2′,3′-didehydro analogs, D4T and 2′,3′-dideoxy-2′,3′-didehydrocytidine

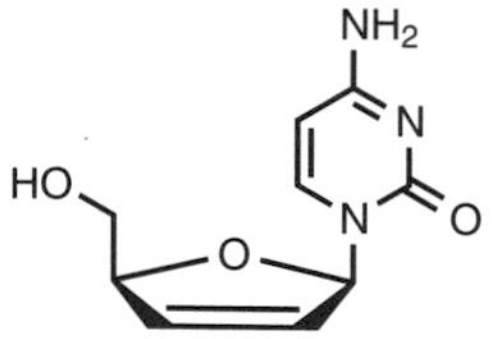

2',3'-Dideoxy-2',3'-didehydrocytidine (D4C)

(D4C) are the most active.[25,26] On the basis of these studies and more recent observations,[16] it can be said that general structural requirements for the nucleosides are (1) the presence of a 5′-hydroxy group, (2) an azido, hydrogen, or fluorine substitution at the 3′ position or a 2′,3′ double bond, (3) the presence of a heteroatom, preferably sulfur, in place of a methylene group in a 2′,3′-dideoxy sugar ring, and (4) suitable matching of the sugar functionality with an appropriate nucleic acid base.

Purine nucleosides (for example, DDA and DDI) possess short half-lives given their known degradation in an acidic environment. Although this problem can be remedied by preparing 2′-fluoro analogs, these modified analogs possess similar activity and increased toxicity as compared to the parent compounds.[27] 6-Halo derivatives of dideoxy purine nucleosides, like DDI, have shown increased central nervous system delivery as compared to DDI alone, because the 6-halo derivatives are substrates for adenosine deaminase (ADA), an enzyme that is more active in brain tissue than in plasma. In the brain, ADA converts 6-halo-DDI to DDI.[28]

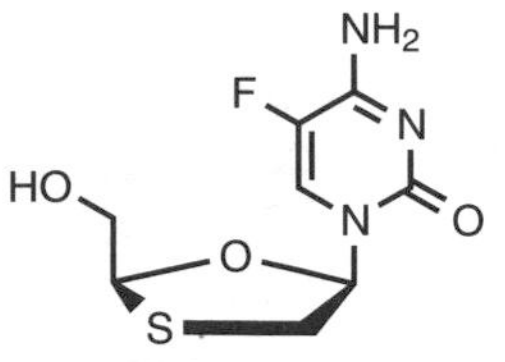

2′,3′-Dideoxy-5-fluoro-3′-thiacytidine

A newer nucleoside analog that is a promising candidate is 2′,3′-dideoxy-5-fluoro-3′-thiacytidine. This derivative demonstrates a bioavailability of 73% with oral administration, penetrates the blood-brain barrier, and shows resistance to deamination by cytidine-deoxycytidine deaminase for the L-(−)-isomer.[29] An additional modification in the preparation of nucleoside analogs is the opening of the furanose sugar ring to prepare acyclic nucleoside derivatives. The precedent for this approach is the use of acyclovir in the palliative treatment of genital herpes.[30]

Most notable analogs in this class of acyclic compounds are the phosphonylmethoxyalkylpurine and -pyrimidine compounds, a class that has broad-spectrum antiviral activity.[31] Among these analogs, the adenine derivatives 9-(2-phosphonylmethoxyethyl)adenine (PMEA) and 9-(2-phosphonylmethoxyethyl)-2,6-diaminopurine (PMEDAP) are more active than AZT in certain animal models.[32,33] In addition, PMEA and PMEDAP possess immunomodulating activity,[34] exhibit inhibitory activity against other viruses that infect AIDS patients, and demonstrate in vivo antiviral activity for several days after a single dose.[35] Addition of a fluoromethyl moiety to the side chain of PMEA to produce the class of 9-[(2RS)-3-fluoro-2-phosphonylmethoxypropyl] derivatives of purines produces greater selectivity in both in vitro and in vivo studies as compared to PMEA and PMEDAP.[36] The active species in these acyclic analogs is also the triphosphate species. Because these compounds have a terminal phosphonyl group, however, only two more in vivo phosphorylations are required to form the active compounds, which act as RT inhibitors. All the agents just mentioned interact with both HIV-1 and HIV-2 RT. The following discussion concerns agents that exhibit activity only against HIV-1 RT.

Specific HIV-1 Reverse Transcriptase Inhibitors

In the nucleoside class, spiro analogs belonging to the 2′,5′-*bis*-O-(tert-butyldimethylsilyl)-3′-spiro-5″-(4″-amino-1″,2″-oxathiole-2″, 2″-dioxide) (TSAO) class of purine and pyrimidine nucleosides have selective anti-HIV-1 activity and demonstrate potencies that range from 0.060 to 1.0 μM. In this group of compounds, structure-activity correlations for the sugar portion reveal that 3′-spiro-*xylo*-nucleosides are inactive. 3′-spiro-*ribo*-nucleosides having one or no silyl group at C-2′ or C-5′ were also inactive. The presence of silyl groups at both C-2′ and C-5′ is a requirement, however, and confers selectivity and high potency. The nucleic acid base is also an important determinant for activity. Therefore, the thymine derivative is more potent than the uracil analog. Introduction of a methyl group at the 3-position of the thymine analog produces a compound of similar potency but less toxicity than the thymine derivative. As a result, this compound exhibits an in vitro therapeutic index above 4000. Most interesting is the fact that these analogs are the first intact sugar ring

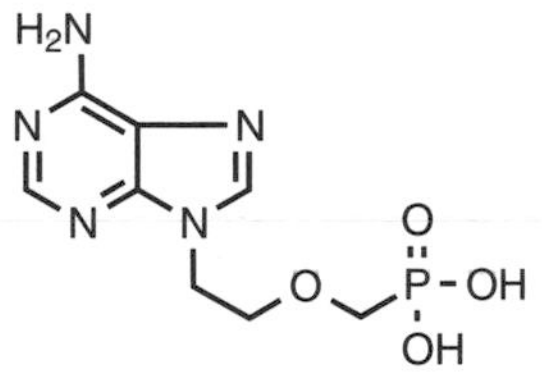

9-(2-Phosphonylmethoxyethyl)adenine (PMEA)

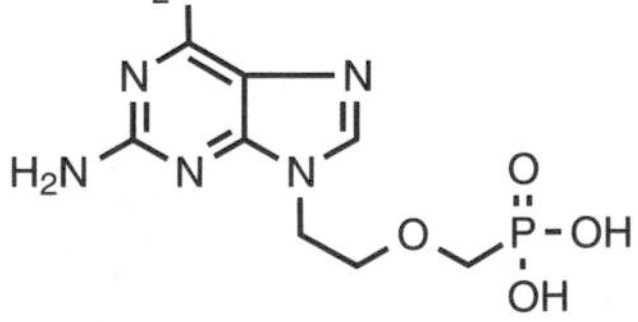

9-(2-Phosphonylmethoxyethyl)-2,6-diaminopurine (PMEDAP)

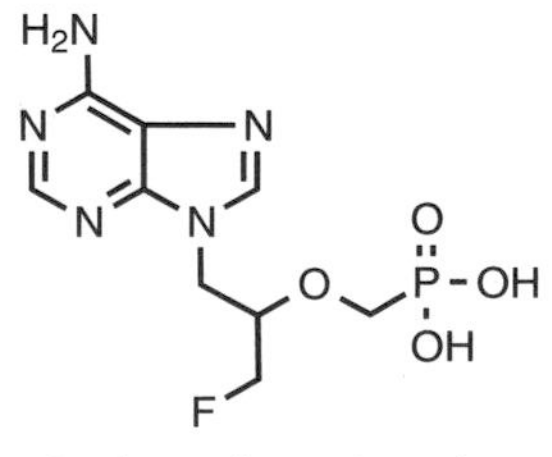

9-[(2RS)-3-Fluoro-2-phosphonylmethoxyproyl]adenine

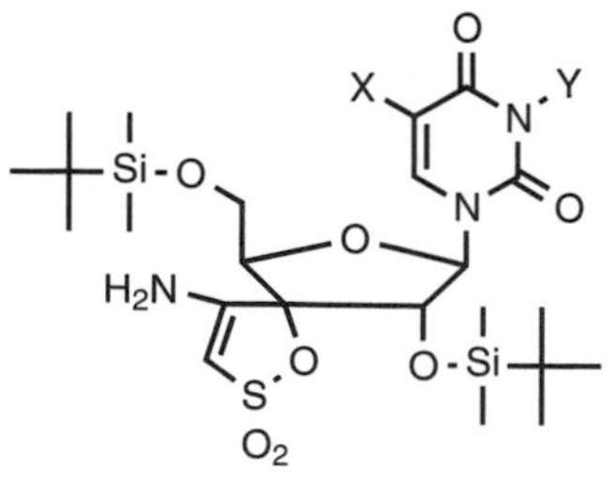

Thymine derivative $X = CH_3, Y = H$
Uracil analog $X = H, Y = H$
3-Methylthymine derivative $X = CH_3, Y = CH_3$

nucleoside derivatives that do not inhibit the replication of HIV-2 or Simian immunodeficiency virus (SIV).[37,38]

A systematic program that involved the design, synthesis, and antiviral evaluation of several 1-[(2-hydroxyethoxy) methyl]-6-(phenylthio)thymine (HEPT) derivatives has produced a new class of highly potent RT inhibitors. In this series, if the methyl at position 5 of the thymine nucleus (analog a, $EC_{50} = 6.5\ \mu M$) is replaced

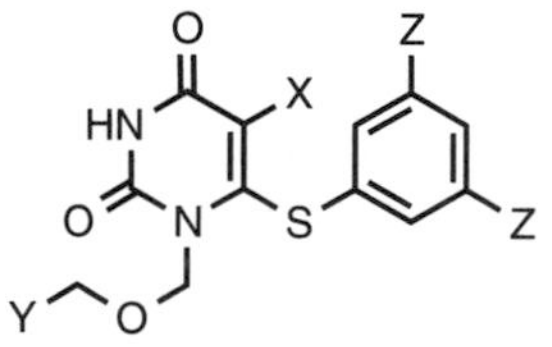

Analog a $X = CH_3, Y = CH_2OH, Z = H$
Analog b $X = C_2H_5, Y = CH_2OH, Z = H$
Analog c $X = C_2H_5, Y = CH_2OH, Z = CH_3$
Analog d $X = C_2H_5, Y = CH_3, Z = CH_3$
Analog e $X = C_2H_5, Y = Ph, Z = CH_3$

by an ethyl functionality (analog b, $EC_{50} = 0.12\ \mu M$), a distinct increase in potency results. Introduction of a dimethyl meta substitution on the phenyl ring (analog c, $EC_{50} = 0.016\ \mu M$) further increases activity. Replacement of the hydroxymethyl substituent in analog C by an ethoxymethyl moiety (analog d, $EC_{50} = 0.0062\ \mu M$) is also beneficial for increasing potency. In this series, further increase in activity is possible if the ethoxymethyl substituent is substituted by a benzyloxymethyl functionality (analog e, $EC_{50} = 0.0024\ \mu M$). The resulting derivative possesses an in vitro therapeutic index of greater than 8300.[39]

The search and anti-HIV analysis of compounds belonging to 600 different structural classes led to the discovery of the tetrahydroimidazo[4,5,1-jk][1,4]-benzodiazepin-2(1H)-one and -thione (TIBO) derivatives. These compounds display potencies that are similar to those of AZT and selectivities that are higher than those of AZT. Further, these derivatives are not active against HIV-2 or any other DNA or RNA viruses.[40] A parent member of this class acts at an allosteric site and inhibits the RNA-dependent DNA polymerization activity of RT.[41] A TIBO derivative containing a 9-chloro substitution has the ability to inhibit 13 different HIV-1 strains and demonstrates a median IC_{50} value of 0.15 μM. This compound also

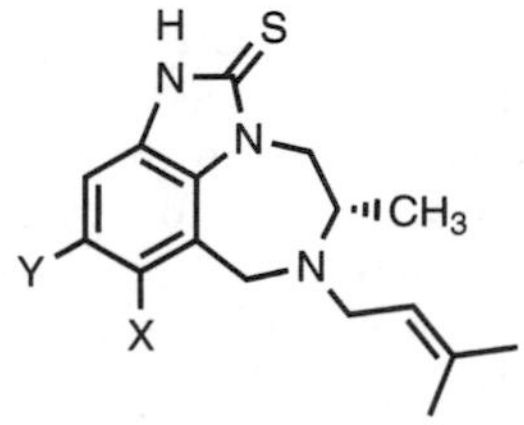

TIBO derivative $X = H, Y = H$
with 9-chlorosubstitution $X = H, Y = Cl$
8-halo-TIBO analogs $X = Cl, Y = H$
$X = F, Y = H$

exhibits an ID_{50} of 0.01 μM for inhibiting HIV-1 RT using a naturally occurring template.[42] 8-Halo-TIBO analogs have also shown high potencies in the nanomolar range and in vitro therapeutic indices of over 2000.[43] In this class of compounds, the cyclic urea moiety is needed for activity, and replacement of the urea oxygen with sulfur or selenium generates compounds with increased activity.[44]

In the diazepine class of compounds, nevirapine is also

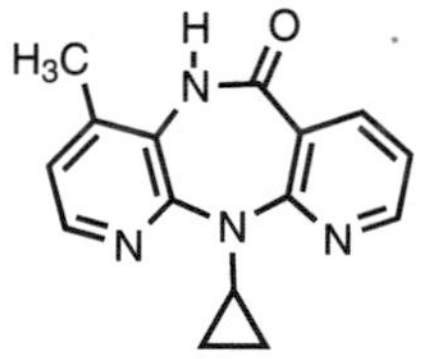

Nevirapine

a potent HIV-1 inhibitor, acting at the level of the RT.[45] Nevirapine is also active against an HIV-1 strain that is resistant to AZT.[46] Studies have revealed that Tyr-181 and Tyr-188 are required at the RT active site for susceptibility to both nevirapine and the TIBO analog containing a 9-chloro substitution. If these positions 181 and 188 on HIV-1 RT are replaced by the amino acids present in the corresponding positions in HIV-2, namely isoleucine or valine at 181 and leucine at 188, the newly constructed HIV-1 RT is resistant to nevirapine and the TIBO derivative just described. On the other hand,the inherent resistance of HIV-2 RT to these two compounds can be reversed if amino acids 176-190 from HIV-1 RT are introduced to the corresponding positions in HIV-2 RT.[47] The 3.5 Å x-ray crystal structure of nevirapine bound to the active site of RT has also been determined.[48]

For optimal activity in the nevirapine class of compounds, one of the two outer rings in the tricyclic nucleus must be a pyridine heterocycle. In general, the presence of two outer pyridine rings is preferred, and in this case, the position of the nitrogen in the pyridine rings is important. Maximum potency can be achieved in this subclass if the 4-position contains a small lipophilic group, like a methyl moiety, and the lactam nitrogen has a free hydrogen atom, as is observed in nevirapine.[49]

A general screening program led to the discovery of the lead compound belonging to the pyridinone class of

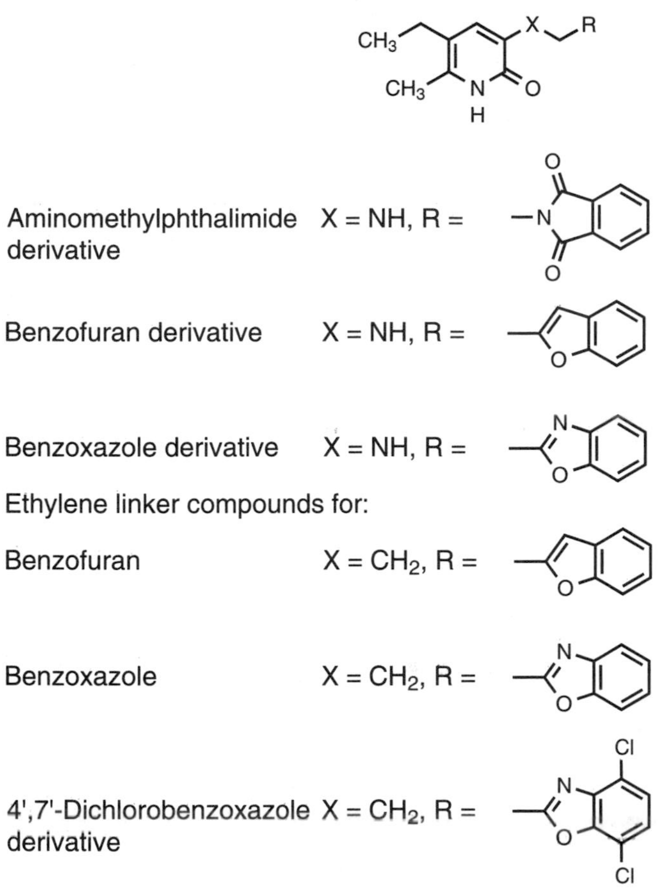

Aminomethylphthalimide derivative X = NH, R =

Benzofuran derivative X = NH, R =

Benzoxazole derivative X = NH, R =

Ethylene linker compounds for:

Benzofuran X = CH_2, R =

Benzoxazole X = CH_2, R =

4',7'-Dichlorobenzoxazole derivative X = CH_2, R =

HIV-1 RT inhibitors. Unfortunately, this aminomethylphthalimide derivative, although highly potent against HIV-1 RT (IC_{50} = 30 nM) had a half-life of approximately 2 hours under in vitro physiologic conditions.[50] To increase stability and not hamper the potency of this compound, analogs were prepared in which X could also be -CH_2- and the R group was replaced with a variety of heterocycles. Of these, the benzofuran (IC_{50} = 235 nM) and benzoxazole (IC_{50} = 235 nM) derivatives with the aminomethylene linker proved to be active in the RT assay. If however, X was -CH_2- instead of -NH- in these two compounds, these ethylene linker compounds for benzofuran (IC_{50} = 77 nM) and benzoxazole (IC_{50} = 22 nM) were more active. Introduction of substituents onto the benzoxazole moiety determined that the 4′ and 7′ positions were the best for enhancing potency. Thus, the 4′,7′-dichloro analog (IC_{50} = 9.6 nM) was the most potent analog in this series. It also had low oral bioavailability in animals. Further structure-activity relationship studies revealed that modifications in the pyridone ring generally led to less active compounds. Extensions in the ethylene and aminomethyl linkers to three atoms or shortening to one atom abrogated activity. Enforcement of *cis* or *trans* configurations by converting the ethylene linker to an olefin also reduced potency.[51] Metabolism studies with the ethylene linker compound of benzoxazole revealed the generation of several expected metabolites, with the major compounds being the 5α-hydroxyl ethyl and 6-hydroxymethyl metabolites. Whereas the former compound was significantly less active against RT than the parent compound, the latter was only fourfold

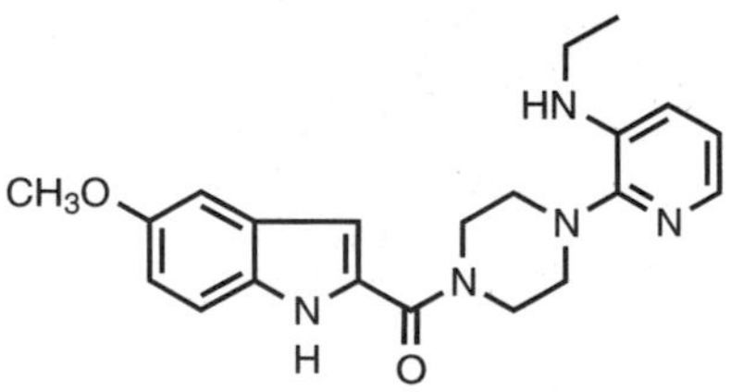

BHAP class of RT inhibitor

less active.[52] Another specific HIV-1 RT inhibitor belonging to the *bis*(heteroaryl)piperazine (BHAP) class was the culmination of a search involving 1500 compounds. This derivative demonstrated potencies equal to that of AZT and also showed activity in the HIV-1 infected SCID-hu mouse.[53] The α-anilinophenylacetamide (α-APA) deriva-

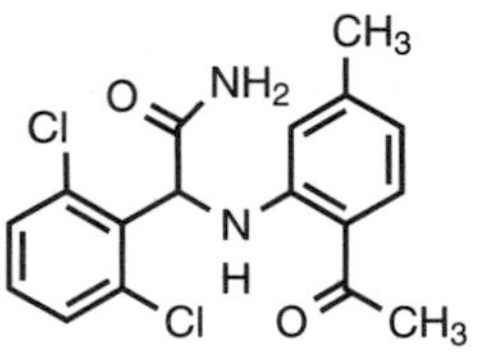

α-APA derivative

tives, as exemplified by the levo isomer, are also potent and selective anti-HIV-1 agents.[54]

As a whole, the RT inhibitors belonging to the HEPT, TIBO, nevirapine, pyridinone, BHAP, and α-APA classes are highly potent against HIV-1 but do not show any activity against HIV-2. It is remarkable that in spite of the differences in chemical structure, nevirapine, TIBO, and pyridinone compounds interact with the same site on RT.[50] A BHAP derivative and nevirapine have also been shown to share the same binding site.[55] A modulatory site capable of interacting with these different structural classes has been suggested as the target of these specific HIV-1 RT inhibitors.[56] In 1991, results of in vitro studies documented the occurrence of resistance against pyridinone, nevirapine, and TIBO derivatives.[57] HIV-1 mutants resistant to nevirapine were also cross-resistant to certain TIBO and HEPT derivatives, but they were sensitive to AZT. These mutants had cysteine substitution in place of tyrosine at position 181 of the RT.[58] One suggestion is that aromatic stacking of the amino acid side groups at residues 181 and 188 is required for inhibitory activity.[59]

The fact that resistance has been a problem for both nucleoside and non-nucleoside RT inhibitors, coupled with the inability of nucleosides like AZT, DDI, and DDC to cure AIDS in the clinic, brings into serious question the viability of the RT inhibitory mechanism as a target for anti-HIV-1 drug design. Because RT acts at the preintegration stage, nucleoside inhibitors like AZT may be valuable at the asymptomatic stage of HIV infection. Unfortunately, this premise has not gained much experimental support after several reports cited failures with AZT used in the postexposure prophylactic mode.[60–63] A small clinical trial designed to study the effects of AZT as an early treatment method also showed no clinical benefit.[64] These points, taken together with the evidence from a large, long-running European trial[65] supporting the inefficacy of AZT, reiterate that nucleoside RT inhibitors will not be singular curative agents and may only be beneficial in combination therapy regimens. A new in vitro study has shown, however, that AZT, DDI, and nevirapine can be used to cause mutations in HIV-1 that render it uninfectious.[66] The extrapolation of this approach in the clinical setting is eagerly awaited. In another vein, the search for other RT inhibitors can still be rationalized on the grounds of the possibility of discovering structures different from the nucleosides and the hope that they will bind in a dissimilar mode to produce a complete halt of viral replication. In this respect, the known x-ray[67] and solution secondary[68] structures of the RNase H domain of RT offer an additional target site for drug design.

Protease Inhibitors

The HIV-1 protease belongs to the family of aspartyl proteases and is an enzyme that elicits its action at a post-integration stage before or during viral budding. In this role, the retroviral protease cleaves the viral precursor polyprotein gp160 into proteins that make up the mature virion. It also seems that the protease may be needed in the early stages of the life cycle of the virus.[69] In addition, in vitro evidence shows that the HIV-1 protease can also attack host proteins like fibronectin, actin, troponin C, Alzheimer amyloid protein, and pro-interleukin-1β.[70–72] The prospect that the viral protease can exert its action outside the virion in vivo raises the possibility of additional cellular damage by the AIDS virus.

These facts, taken together with the known x-ray crystal structures of the HIV-1 protease,[73,74] have made the protease an alluring target for drug design.[75] Approaches to design protease inhibitors can be divided into three different classes. In the first class, agents are designed and synthesized to inhibit one of the many scissile bonds susceptible to the viral protease during its processing of polyproteins. In place of the scissile bond, the inhibitor possesses a nonhydrolyzable isostere that assumes tetrahedral geometry in the transition state. In one type of this inhibitor, agents are designed to inhibit the cleavage at tyrosine and proline or phenylalanine and proline. This group is appealing because mammalian endopeptidases are not expected to cleave these peptide linkages.[76] In another type of this inhibitor, design strategies are aimed at suppressing the cleavage between leucine and alanine. This approach suggests that antihypertensive agents that inhibit another aspartyl protease, renin, by preventing cleavage between leucine and valine may also be active.[76] Thus, these designed compounds incorporate the hydroxyethylene, dihydroxyethylene, hydroxyethylamine, statine, phosphinate, and reduced amide functions, and have resulted in numerous potent inhibitors.[77]

Structure-activity relationship studies have demonstrated that the M-Leu-Phe-NH_2 side chain of (2-phenylmethyl-3-hydroxy-4-amino-6-phenylhexanamide) (IC_{50} = 0.42 nM) could be replaced with 2-amino-3-hydroxyindan to produce Inhibitor A (IC_{50} = 0.42 nM) and still retain potency. It is important to note that the critical hydroxy functionality in both of these inhibitors has the S-configuration.[78] If on the other hand, the critical hydroxy functionality has an R-configuration, a different activity profile is obtained with varying substituents. This change, together with a slight modification of the indan substituent and replacement of the butyloxycarbonyl (Boc) group with a hydrogen atom as in Inhibitor B (IC_{50} = 63,000 Nm), has a dramatic effect on lowering activity. Reintroduction of the Boc group at the amino terminus and introduction of a *cis*-decahydroisoquinoline function as in Inhibitor C (IC_{50} > 3,000 nM) has an increasing influence on the potency. Substitution of the Boc group for a longer function containing an asparagine and quinoline moiety as in Inhibitor D (IC_{50} = 0.34 nM) has a substantial effect in improving potency. This truncation is also suitable with the hydroxyindan group at the amino terminus as in Inhibitor E (IC_{50} = 5.4 nM). Removal of the quinoline moiety to generate Inhibitor F (IC_{50} = 1100 nM), however, is detrimental for activity.[79]

The second approach to designing inhibitors has taken into consideration the homodimeric structure of the enzyme and the symmetry of the protease active site. Thus, at the catalytic site, two aspartyl residues result from the contribution of one residue from each half of the dimeric structure. This finding is not novel; this arrangement is similar to the active sites of other aspartyl proteases like rhizopuspepsin,[80] penicillopepsin[81] endothiapepsin,[82] and renin.[83] Using this fact, C_2 symmetric inhibitors were rationally designed, synthesized, and shown to possess high potencies.[84] The predicted symmetric mode of binding of these inhibitors was confirmed by the x-ray crystal structure of the inhibitor-enzyme complexes.[85,86] In a third computer-based approach, a program to create a negative image of the active site cavity based on the crystal structure of the protease were used. Using a search strategy, this image was matched with several thousand molecules from a crystallographic data base to come up with haloperidol as a nonpeptide-based protease inhibitor lead.[87] This strategy may also be used to discover drugs targeted at other active sites.[88]

In the design of effective protease inhibitors, certain hurdles need to be overcome during the course of chemical development. Because the normal function of humans requires the action of other aspartyl proteases like renin, pepsin, gastricsin, cathepsin D, and cathepsin E, it is important that the designed HIV-1 protease inhibitor does not suppress the function of these enzymes. In fact, it is comforting that this selectivity is attainable and has been demonstrated for inhibitor D.[89] Finally, because most potent protease inhibitors are peptides or have some degree of peptide character, there are concerns about absorption, bioavailability, distribution, and degradation by cellular proteases. One approach to this problem is the preparation of conformationally constrained β-turn peptidomimetic inhibitors, which can be designed

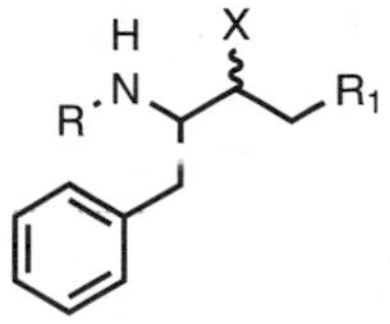

2-Phenylmethyl-3-hydroxy-4-amino-6-phenylhexanamide — R = Boc; X = OH; R_1 =

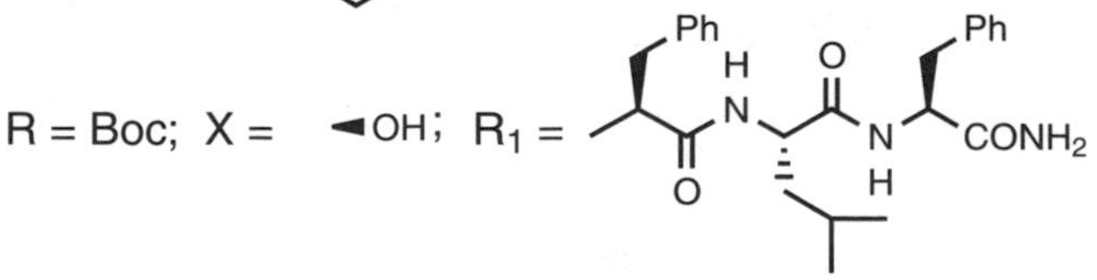

Inhibitor A — R = Boc; X = OH; R_1 =

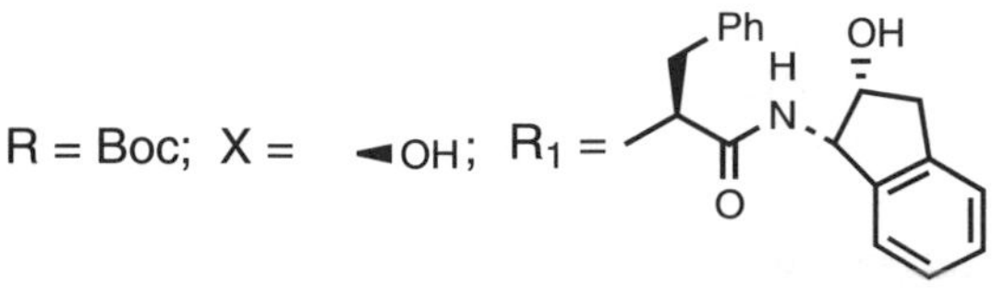

Inhibitor B — R = H; X = OH; R_1 =

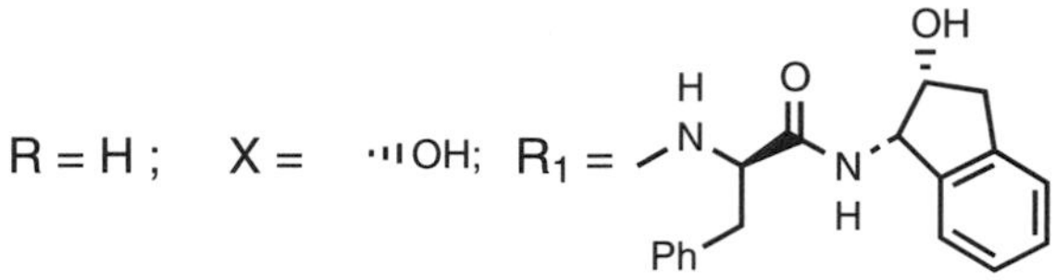

Inhibitor C — R = Boc; X = OH; R_1 =

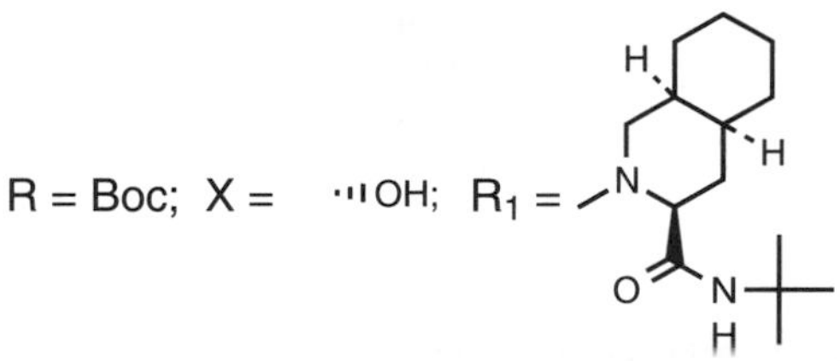

Inhibitor D

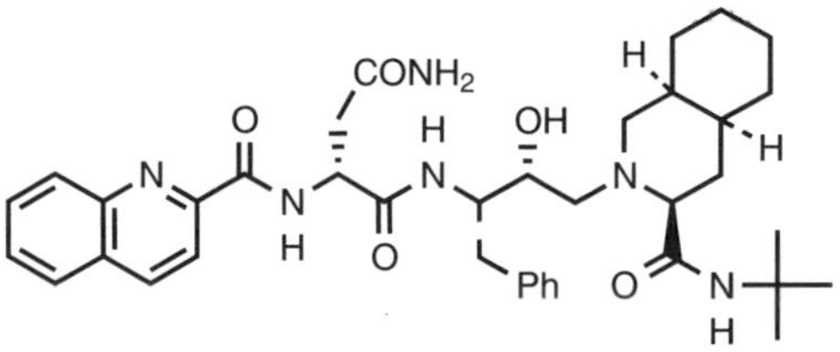

Inhibitor E

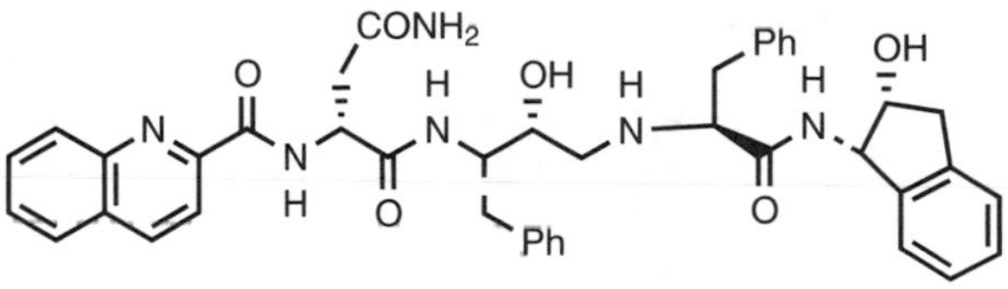

Inhibitor F

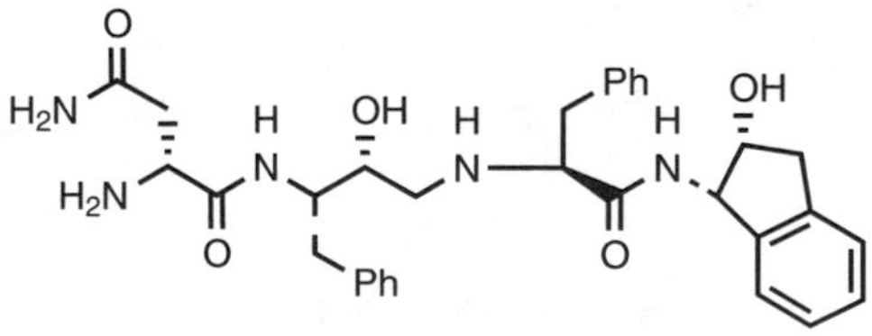

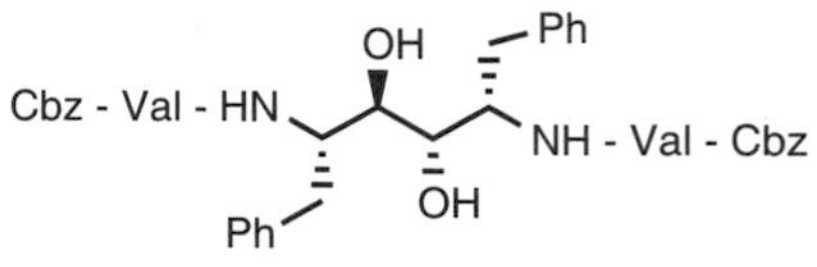

C_2 Symmetric protease inhibitor

to possess potencies at nanomolar concentrations.[90] In general, some of these apparent problems can be circumvented if an active derivative were discovered that had no resemblance to a peptide but was as highly potent as its peptide counterparts. This discovery will remain as a future challenge for the discovery of protease inhibitors.

Inhibitors of Gene Expression

Oligonucleotides

The antiviral potential of the antisense approach was first revealed in 1978, when an oligonucleotide was shown to inhibit replication of an animal RNA virus.[91] The therapeutic utility of this strategy and the pros and cons of its application have been reviewed.[92] In principle, an oligonucleotide is designed to be complimentary to a specific nucleotide sequence in a gene, a gene product, or a portion of mRNA. Once administered, the oligonucleotide is expected to seek the target site, hybridize with it, and thereby inhibit the normal function of the particular segment. As it turns out, the length of an antisense oligonucleotide needed to enforce successful hybridization is roughly the same length that is required to produce a unique sequence in the human genome. This fact assures selectivity and strengthens the argument in favor of the antisense approach.[93] The overall appeal of this approach is enhanced by documented methods that use vectors to transfer a gene that directs the synthesis of an antisense construct.[94] Oligonucleotides targeted at the *rev* gene have also shown anti-HIV-1 activity.[95,96]

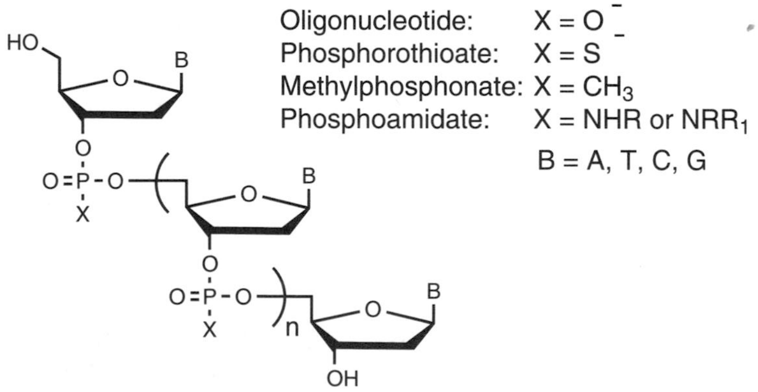

Oligonucleotide: X = O^-
Phosphorothioate: X = S^-
Methylphosphonate: X = CH_3
Phosphoamidate: X = NHR or NRR_1
B = A, T, C, G

Considering these agents as valid therapeutic entities led to some early and legitimate concerns. Because these agents are oligonucleotides, they should be expected to be susceptible to degradation by host cellular exonucleases. This problem can be overcome by replacing the negatively charged oxygen in these substrates with sulfur, methyl, or amino functions to produce phosphorothioates, methylphosphonates, or phosphoamidate derivatives.[97,99] During this structural modification, however, the phosphorus becomes asymmetric and the synthetic product is isolated as two isomers. Again, this problem can be surmounted by preparing dithiocompounds. Other possible modifications involve the sugar part of the molecule, although such altered derivatives may not be able to achieve effective hybridization.[100] Another strategy to increase stability and specificity is to replace the oligonucleotide backbone. Using computer modeling in the design process, a synthesized polyamide backbone derivative has shown the ability to recognize its target sequence with high specificity.[101]

Another question related to oligonucleotide development deals with the ability of these agents to cross cellular membranes to elicit their action. In this regard, cholesteryl[102] or poly-L-lysine[103] derivatives have been made to enhance penetration. Further development in this class must address the issues of susceptibility to endonuclease activity and the guarantee of in vivo hybridization.[100] Hybridization is governed by salt concentration and the amount of guanosine and cytosine residues.[104] Insufficient knowledge of the in vivo three-dimensional structure of most oligonucleotide targets coupled with the lack of complete understanding of the sense-antisense reaction[105] provide future challenges for the development of oligonucleotides. The identification of small molecule inhibitors of the regulatory genes or gene products has been a long sought-after goal. In this regard, a *tat* antagonist has been shown to exhibit anti-HIV-1 activity.[106]

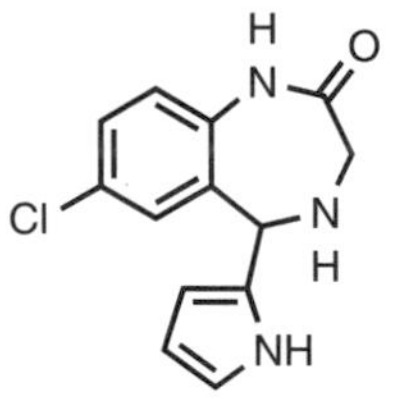

tat Antagonist

Glucosidase Inhibitors

Apart from reverse transcriptase and protease, the other enzymes that have received attention as a target for potential anti-HIV-1 agents are the glucosidase enzymes. In the processing stage before viral budding, the viral envelope protein gp120 is heavily N-glycosylated, and is followed by a process of sugar trimming by a group of glucosidase and mannosidase enzymes. Glucosidase I is responsible for trimming one glucose unit and glucosidase II cleaves off two glucose units. Further trimming is undertaken by the mannosidase enzymes. Trimming is vital for maturation and infectivity of the virus. Therefore, inhibitors that inactivate the glucosidase enzymes have the potential of arresting viral replication.

The first compound discovered as an inhibitor of this class was the natural product castanospermine, an alkaloid first isolated from Castanospermum australe. The activity of castanospermine against α- and β-D-glucosidases had been documented[107,108] and was found to be inhibitory to HIV-1 replication.[109,110] Because castanospermine was too toxic for clinical trials, synthetic studies revealed that an analog, N-butyldeoxynojirimycin (N-Bu-DNJ),[111] was more suitable.

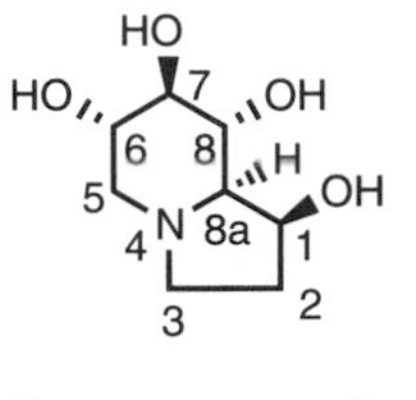

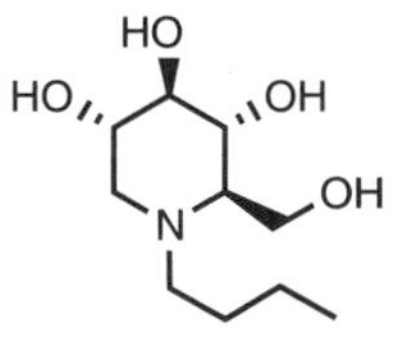

Castanospermine

N-Butyldeoxynojirimycin

Structurally, the amino sugar derivative N-Bu-DNJ can be considered as an open chain analog of castanospermine. The propensity of these agents to inhibit the glucosidases becomes apparent when their structures are studied in relation to glucose. The four aysmmetric centers at carbons 6,7,8, and 8a on the piperidine ring of castanospermine resemble the four asymmetric centers on carbons 2,3,4 and 5 on N-Bu-DNJ, which in turn parallel the orientations of carbons 2,3,4, and 5 on glucose.[112] Several research groups have synthesized structural variations of these agents,[113] among these, 6-O-butanoylcastanospermine is at least 50 times more active than N-Bu-DNJ.[114]

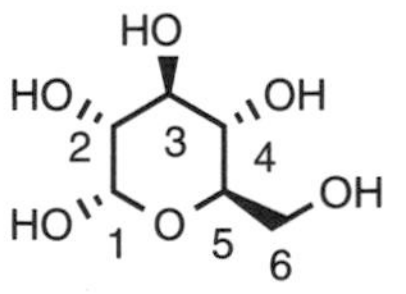

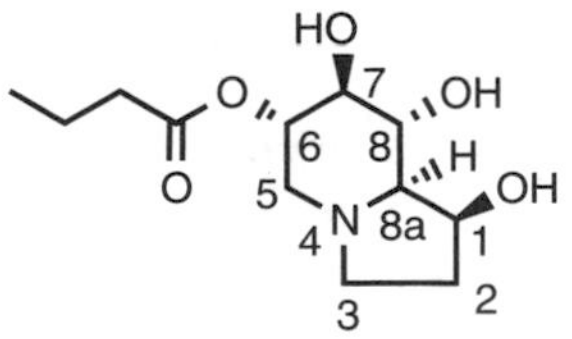

Glucose

6-O-Butanoylcastanospermine

Because both the virus and the host cell are glycosylated by the host glucosidase enzymes, concern has been raised regarding the selectivity of these agents. In addition, representatives of this class of compounds, including castanospermine, are inhibitors of mammalian intestinal disaccharidases.[115] Inhibition of the glucosidases can result in disruption of glycogen and glycolipid metabolism.[107,112] Nonetheless, a certain degree of selectivity for the viral process is observed and has been linked to the high proportion of carbohydrate in the gp120 precursor molecule, gp160.[116]

Inhibitors of Viral Binding

Soluble CD4 Derivatives

An important initial interaction in the initiation of pathogenesis of the AIDS virus is the high-affinity interaction between the viral glycoprotein, gp120 (120 refers to the approximate molecular weight in thousands) and the CD4 receptor (a 55,000-molecular weight protein found on the surface of, among other cells, the helper T lymphocyte, where it has a role in the immune response) of the target cell. This finding suggested that if the CD4 receptor could be administered in a soluble form, it could compete with the target cell CD4 protein for the viral antigen, gp120. Indeed, this premise was experimentally verified with the generation of several soluble CD4 derivatives.[117–121] These agents have also demonstrated the ability to inhibit infection of macrophages.[122]

Because soluble CD4 derivatives are proteinaceous, the short half-life of these agents was predicted and was found to be about 45 minutes.[123] This initial hurdle was efficiently overcome by hooking soluble CD4 to a portion of the immunoglobulin molecule to produce hybrid molecules called "immunoadhesions." This manipulation resulted in a 100-fold increase in half-life.[124] The observation that an immunoadhesion can cross the placenta of a non-human primate implies that these molecules may be used to prevent the passage of infection from mother to fetus.[125] Unfortunately, clinical trials with soluble CD4 have been disappointing.[126] Studies have also revealed that clinical isolates of HIV-1 can require as much as 2700-fold higher concentration of soluble CD4 for neutralization than is required for the laboratory strain of HIV-1,[127] in spite of these isolates having high affinity for the CD4 receptor.[128] Further investigations have shown that HIV-2 is less sensitive to CD4 than is HIV-1.[129,130]

A set-back in the discussion of this class of compounds

as a treatment method stems from the known ability of HIV-1 to enter cells that are devoid of the CD4 receptor, and suggests the role of other host cell receptors as targets for HIV-1.[131] In this context, galactosyl ceramide or its derivative, sulfatide,[132] a mannose-binding protein,[133] and other sites[134] have been proposed. Another legitimate concern involved the generation of antibodies to soluble CD4 preparations. Although it seems that this reaction does occur, it does not have any adverse effects.[135] From the drug design standpoint, it would be most desirable to have an agent that was small, nonimmunogenic, nonpeptidic, and therefore stable to biodegradation. A conformationally restricted β-turn mimetic has been shown to inhibit viral binding and to be relatively stable to proteolytic cleavage.[136]

Small Anionic Molecules

Anti-HIV-1 activity in the anionic class can be divided according to small and large molecules. In the small-molecule class, the first derivative administered to AIDS patients was the hexasulfonic acid suramin. This anionic compound has been used against trypanosomiasis for over 70 years and is also a potent inhibitor of the RT of RNA tumor viruses.[137] Suramin demonstrated activity against HIV-1 in vitro,[138] and was subsequently administered to patients.[139] Interest in suramin waned, however, after demonstrated lack of immunologic improvement and clinical efficacy.[140] In spite of this observation, several suramin analogs have shown anti-HIV-1 potential.[141]

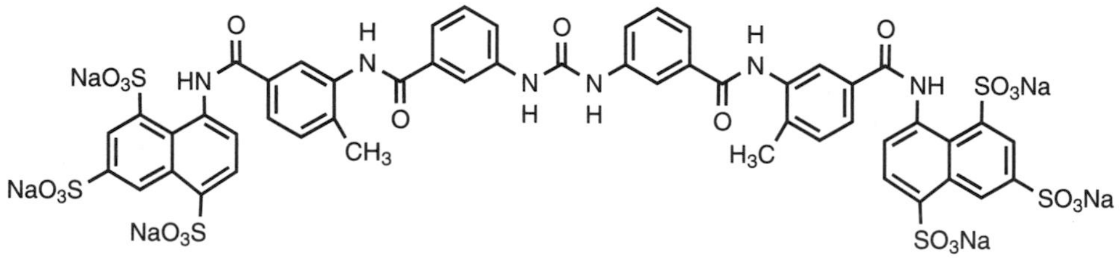

Suramin sodium

Related to suramin, the class of sulfonic acid azo dyes, for example, Evans blue, was also shown to be active against the AIDS virus.[142] The dye properties of these compounds and the fact that these agents can be reduced in vivo to produce carcinogenic amines,[143] however, precludes their use as singular agents. One way around this problem is to replace the carcinogen-generating fragment with a spacer that not only will give noncarcinogenic compounds, but also will abrogate the coloring property of the compound. In our laboratory, we have shown that activity in several symmetric as well as nonsymmetric compounds can be achieved at concentrations that were nontoxic to the host cells.[144–146] A potent representative of this class, is a naphthalenesulfonic acid derivative,[147] but lengthening of the spacer to 8 or 10 methylene units results in a decrease in potency.[147,148] Replacement of the hexamethylene chain with a biphenyl spacer results in an at least fivefold decrease in potency. Salient structure activity correlations in this class of naphthalenesulfonic acid derivatives are as follows.

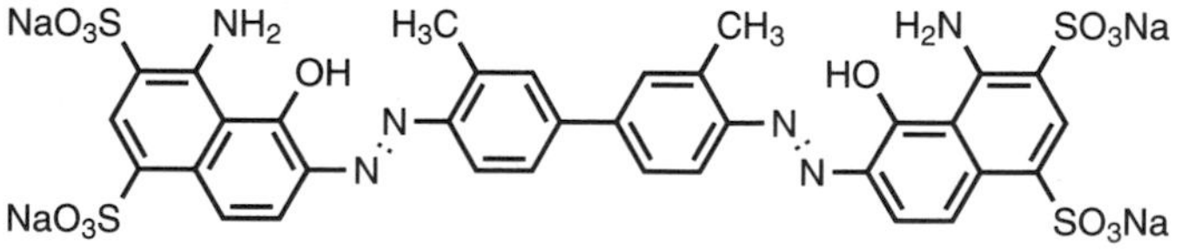

Evans blue

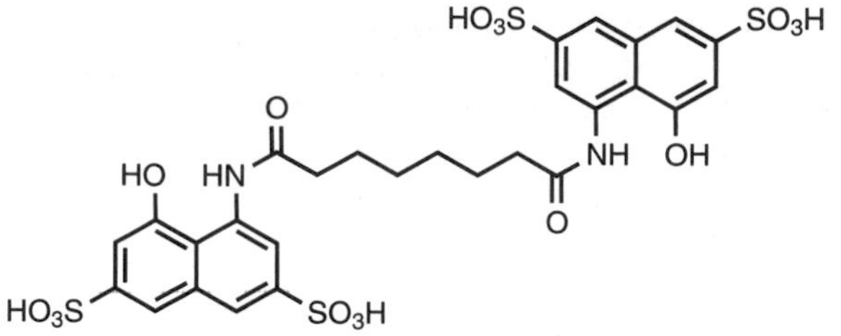

Symmetric bis-naphthalenedisulfonic acid derivative

(1) In the symmetric class of compounds, the bis-naphthalenedisulfonic acid compounds are more active than the naphthalenemono or disulfonic acid analogs. (2) The nature of the naphthalenedisufonic acid moiety, i.e., its substitution pattern, together with the nature of the other substituents on the naphthalene ring, are important determinants of activity. (3) A suitable naphthalenedisulfonic acid moiety needs to be coupled with a distinct spacer to optimize anti-HIV-1 activity. (4) In general, symmetric molecules possessing two naphthalenesulfonic acid units are more potent than nonsymmetric molecules in the whole virus assay.[149] (5) Against the retroviral RT, however, two classes of RT inhibitors are evident—the symmetric naphthalenedisulfonic acid compounds and

the nonsymmetric sulfonic acid derivatives containing a palmitoyl side chain.[145,148] Clearly, the symmetric deriva-

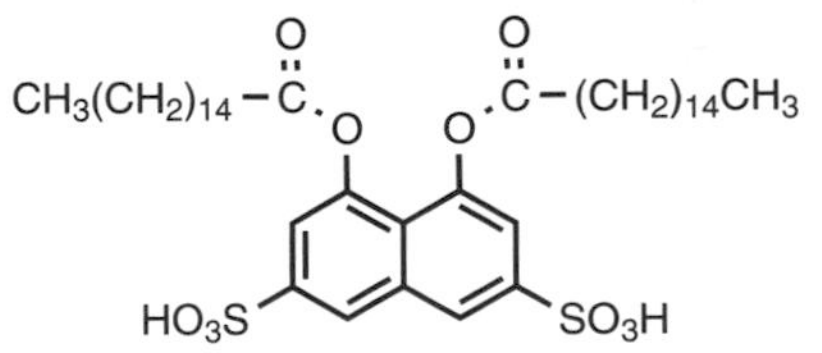

Symmetric disulfonic acid derivative

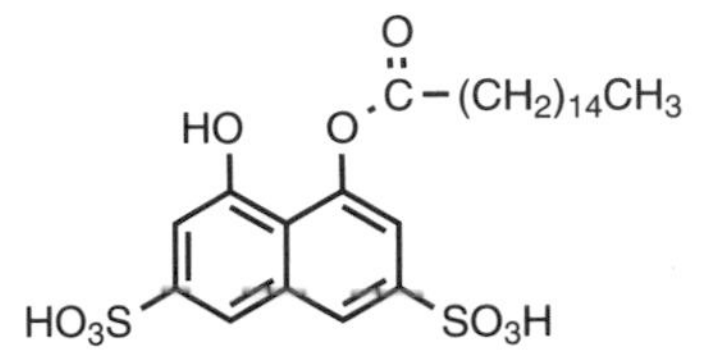

Nonsymmetric disulfonic acid derivative

tive is more active than that possessing only one palmitoyl side chain. (6) The active derivatives demonstrate antiviral activity against both HIV-1 and HIV-2. The naphthalenesulfonic acid compounds display their antiviral activity by inhibiting viral binding to the target cell by virtue of their selective interaction with the viral glycoprotein, gp120.[150] Additional credence to using naphthalenesulfonic acid derivatives stems from their known inhibition of human cytomegalovirus, a common opportunistic pathogen in AIDS patients.[161] Other small-molecule anionic compounds have included sulfonic acid derivatives and other phenolic carboxylic compounds.[152–158]

Polymeric Anionic Molecules

In the large-molecule class, the most notable and widely studied anionic derivative has been the sulfated polysaccharide dextran sulfate, established to exert its

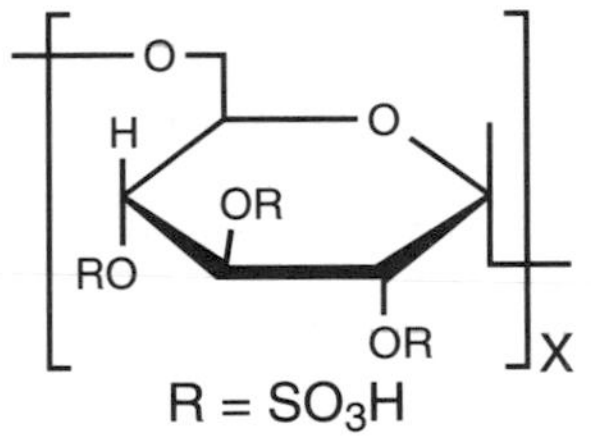

$R = SO_3H$

Dextran sulfate

anti-HIV-1 activity by inhibition of viral binding.[159] The advocation of dextran sulfate in therapy raised several concerns. At the forefront of the disputation was the anticoagulant potential of the derivative. This problem was circumvented, however, by showing that certain analogs can retain anti-HIV-1 activity but can be devoid of anticoagulant effects.[160] An important point raised against dextran sulfate was its in vivo instability, which can be predicted because of the polysaccharide nature of the molecule. Indeed, animal experimentation revealed, dextran sulfate cleaved to smaller units that were known to be inactive.[161] Therefore, it was not surprising that a clinical trial using parenteral dextran sulfate resulted in no clinical activity.[162] An analog of dextran sulfate, pentosan polysulfate, is known to be both desulfated and depolymerized in vivo.[163]

Our interest in the dextran sulfate problem was propelled by its documented in vivo instability. To remedy the depolymerization problem caused by cleavage, we considered replacing the polysaccharide backbone with an inert hydrocarbon chain, as was demonstrated for analogs of the sulfated polysaccharides.[164] To avoid desulfation, but still retain the necessary anionic character, we envisioned replacing the sulfate group with the sulfonic acid group. Support for this choice came from studies of aromatic sulfonic acid function in which metabolism experiments revealed little or no desulfonation.[165] Indeed, this working hypothesis revealed that four novel sulfonic acid polymers do demonstrate potent and selective activity against both HIV-1 and HIV-2. These agents inhibit viral binding and do so by interacting with the viral glycoprotein gp120. One active representative of this class is poly(vinylsulfonic

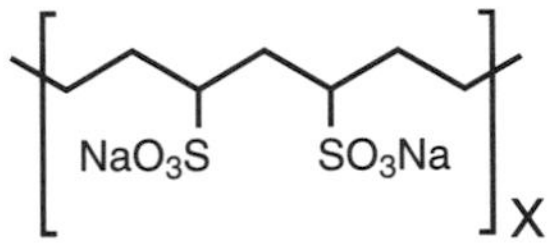

Poly(vinylsulfonic acid) (PVS)

acid) (PVS), which has a molecular weight of 2000.[166] By comparison, previously reported anti-HIV-1 anionic polymers have had molecular weights of at least 5000. The sulfonic acid polymers are also highly potent inhibitors of both HIV-1 and HIV-2 RT.[167] Another point that makes sulfonic acid polymers attractive is that a polymer of PVS was shown to interact with a site that interferes with all of the three major functions of the RT enzyme.[168]

Further consideration of these compounds leads to concern about their protein binding properties. This hurdle should not be major in that the evaluation of different analogs provides derivatives that have different binding properties. In fact, this principle has been demonstrated for two positional isomers in the class of sulfonic acid azo dyes, in which trypan blue is excreted more rapidly from the blood than its isomer, Evans blue.[169] An added benefit to the naphthalenesulfonic acid class of compounds is their established nonmutagenicity.[170] In the development of these agents, it must be decided whether RT or viral binding inhibition is the prefered strategy. If RT inhibition is desired, prodrugs will have to be prepared. Further considerations for improvement of these analogs have been discussed.[171]

Miscellaneous Compounds

The bicyclam class of compounds demonstrate potent and selective activity for HIV-1 and HIV-2 by interfering with the viral uncoating process. In this class of compounds, it is clear that a symmetric molecule is required for optimal activity. A representative of the bicyclam series is also active against an AZT-resistant strain of HIV-

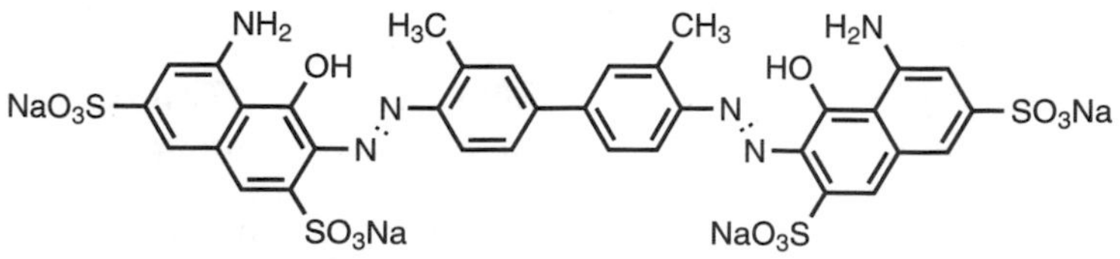

Trypan blue

1.[172] Bicyclam derivatives are being actively pursued for further development.

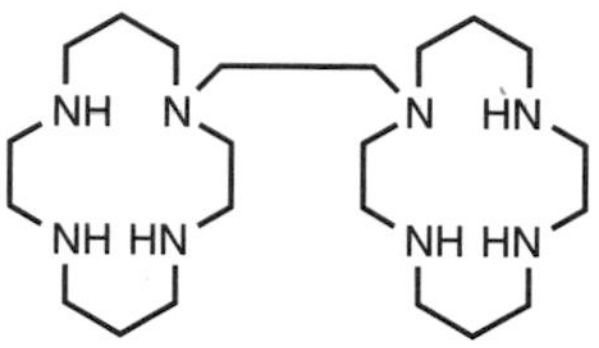

Bicyclam compound

CONCLUSIONS

A wide array of HIV-1 inhibitors are now available, demonstrating potencies as high as in the nanomolar range coupled with in vitro therapeutic indices that range from 1,000 to above 10,000. In addition, some compounds act at different sites of the viral life cycle. In spite of these encouraging in vitro data, the crucial determinant of the long-term success of a drug is its overall performance in the clinic. In this context, it seems more work remains to be done. The initial optimism offered by both the nucleoside and non-nucleoside RT inhibitors has translated into clinical failures.[173] Although it can be reasoned that toxicity may be controlled or annihilated by combination therapy, the emergence of mutant HIV-1 strains that are resistant to these compounds represents a formidable challenge for future efforts in drug design. Therefore, clinical evaluation of other classes of drugs possessing structures and mechanisms of action different from these RT inhibitors is pivotal. At this point, we recognize an urgent need to search for and evaluate new compounds with the hope that these agents will establish a new paradigm to halt or totally eradicate the AIDS virus.

Acknowledgements

The author gratefully acknowledges a Research Scholar Award from the Hans Vahlteich Research Foundation, and Mr. Man Fai Wong for diligent construction of the structures in this chapter.

REFERENCES

1. F. Barré-Sinoussi, et al., *Science, 220,* 868(1983).
2. R. C. Gallo, et al., *Science, 224,* 500(1984).
3. J. A. Levy, et al., *Science, 225,* 840(1984).
4. CDC HIV/AIDS Quarterly Report, Centers for Disease Control, Atlanta (1993).
5. W. C. Koff and D. F. Hoth, *Science, 241,* 426(1988).
6. A. Sabin,: Improbability of effective vaccination against human immunodeficiency virus because of its intracellular transmission and rectal portal of entry. *Proc. Natl. Acad. Sci.* USA, *89,* 8852(1992).
7. P. Mohan, *Pharm. Res., 9,* 703(1992).
8. B. R. Cullen, *Annu. Rev. Microbiol., 45,* 219(1991).
9. B. A. Castro, et al., *AIDS, 2*(Suppl 1), S17(1988).
10. W. C. Greene, *N. Engl. J. Med., 324,* 308(1991).
11. P. Bacchetti and A. R. Moss, *Nature, 338,* 251(1989).
12. J. P. Horowitz, et al. *J. Org. Chem., 29,* 2076(1964).
13. W. Ostertag, et al., *Proc. Natl. Acad. Sci. USA, 71,* 4980(1974).
14. H. Mitsuya, et al., *Proc. Natl. Acad. Sci. USA, 82,* 7096(1985).
15. R. Yarchoan, et al., *Lancet, 1,* 575(1986).
16. C. Flexner, *AIDS, 5,* 798(1992).
17. R. Yarchoan and S. Broader, *Pharmacol. Ther., 40,* 329(1989).
18. A. Camerman, et al. *Proc. Natl. Acad. Sci. USA, 84,* 8239(1987).
19. H. Mitsuya, et al., *Science, 249,* 1533(1990).
20. B. A. Larder, et al., *Science, 243,* 1731(1989).
21. M. H. St. Clair, et al., *Science, 253,* 1557(1991).
22. J. E. Fitzgibbon, et al., *Antimicrob. Agents Chemother., 36,* 153(1992).
23. C. K. Chu, et al., *Biochem. Pharmacol., 37,* 3543(1988).
24. J. Balzarini, et al., *Biochem. Pharmacol., 37,* 2847(1988).
25. M. Baba, et al., *Biochem. Biophys. Res. Commun., 142,* 128(1987).
26. J. Balzarini, et al., *Biochem. Biophys. Res. Commun., 140,* 735(1986).
27. V. E. Marquez, et al., *J. Med. Chem., 33,* 978(1990).
28. M. E. Morgan, et al., *Antimicrob. Agents Chemother., 36,* 2156(1992).
29. R. F. Schinazi, et al., *Antimicrob. Agents Chemother., 36,* 2432(1992).
30. J. E. Reardon and T. Spectro, *J. Biol. Chem., 264,* 7405(1989).
31. E. De Clercq, *Biochem. Pharmacol., 42,* 963(1991).
32. J. Balzarini, et al., *Proc. Natl. Acad. Sci. USA, 86,* 332(1989).
33. L. Naesens, et al., *Eur. J. Clin. Microbiol. Infect. Dis., 8,* 1043(1989).
34. V. D. Gobbo, et al., *Antiviral Res., 16,* 65(1991).
35. E. De Clercq, *J. Acquir. Immune Defic. Syndr., 4,* 207(1991).
36. J. Balzarini, et al., *Proc. Natl. Acad. Sci. USA, 88,* 4961(1991).
37. M.-J. Camarasa, et al., *J. Med. Chem., 35,* 2721(1992).
38. M. J. Perez-Perez, et al., *J. Med. Chem., 35,* 2988(1992).
39. M. Baba, et al., *Antiviral Res., 17,* 245(1992).
40. R. Pauwels, et al., *Nature, 343,* 470(1990).
41. Z. Debyser, et al., *J. Biol. Chem., 267,* 11769(1992).
42. E. L. White, et al., *Antiviral Res., 16,* 257(1991).
43. K. A. Parker and C. A. Coburn, *J. Org. Chem., 57,* 97(1992).
44. M. J. Kukla, et al., *J. Med. Chem., 34,* 3187(1991).
45. V. J. Merluzzi, et al., *Science, 250,* 1411(1990).
46. D. Richman, et al., *Antimicrob. Agents Chemother., 35,* 305(1991).

47. C.-K. Shih, et al., *Proc. Natl. Acad. Sci. USA, 88,* 9878(1991).
48. L. A. Kohlstaedt, et al., *Science, 256,* 1783(1992).
49. M. T. Skoog, et al., *Med. Res. Rev., 12,* 27(1992).
50. M. E. Goldman, et al., *Proc. Natl. Acad. Sci. USA, 88,* 6863(1991).
51. W. S. Saari, et al., *J. Med. Chem., 34,* 2922(1991).
52. S. K. Balani, et al., *Drug Metab. Dispos., 20,* 869(1992).
53. D. L. Romero, et al., *Proc. Natl. Acad. Sci. USA, 88,* 8806(1991).
54. R. Pauwels, et al., *Proc. Natl. Acad. Sci. USA, 90,* 1711(1993).
55. T. J. Dueweke, et al., *J. Biol. Chem., 267,* 27(1992).
56. P. M. Grob, et al., *AIDS Res. Hum. Retroviruses, 8,* 145(1992).
57. J. H. Nunberg, et al., *J. Virol., 65,* 4887(1991).
58. D. Richman, et al., *Proc. Natl. Acad. Sci. USA, 88,* 11241(1991).
59. V. V. Sardana, et al., *J. Biol. Chem., 267,* 17526(1992).
60. J. M. A. Lange, et al., *N. Engl. J. Med., 322,* 1375(1990).
61. D. F. M. Looke and D. I. Grove, *Lancet, 335,* 1280(1990).
62. E. Durand, et al., *N. Engl. J. Med., 324,* 1062(1991).
63. P. D. Jones, *Lancet, 338,* 884(1991).
64. B. Tindall, et al., *AIDS, 5,* 477(1991).
65. J.-P. Aboulker and A. M. Swart, *Lancet, 341,* 889(1993).
66. Y. K. Chow, et al., *Nature, 361,* 650(1993).
67. J. F. Davies II, et al., *Science, 252,* 88(1991).
68. R. Powers, et al., *J. Mol. Biol., 221,* 1081(1991).
69. C. Baboonian, et al., *Biochem. Biophys. Res. Commun., 179,* 17(1991).
70. A. G. Tomasselli, et al., *J. Biol. Chem., 22,* 14548(1991).
71. M. Oswald and K. von der Helm, *FEBS Lett., 292,* 298(1991).
72. L. D. Adams, et al., *AIDS Res. Hum. Retroviruses, 8,* 291(1992).
73. A. Wlodawer, et al., *Science, 245,* 616(1989).
74. R. Lapatto, et al., *Nature, 342,* 299(1989).
75. J. R. Huff, *J. Med. Chem., 34,* 2305(1991).
76. J. A. Martin, *Antiviral Res., 17,* 265(1992).
77. G. B. Dreyer, et al., *Proc. Natl. Acad. Sci. USA, 86,* 9752(1989).
78. D. Y. Young, et al., *J. Med. Chem., 35,* 1702(1992).
79. T. J. Tucker, et al., *J. Med. Chem., 35,* 2525(1992).
80. R. Bott, et al., *Biochemistry, 21,* 6956(1982).
81. M. N. G. James, et al., *Proc. Natl. Acad. Sci. USA, 79,* 6137(1982).
82. L. H. Pearl and T. Blundell, *FEBS Lett., 174,* 96(1984).
83. A. R. Sielecki, et al., *Science, 243,* 1346(1989).
84. D. J. Kempf, et al., *J. Med. Chem., 33,* 2687(1990).
85. J. Erickson, et al., *Science, 249,* 527(1990).
86. R. Bone, et al., *J. Am. Chem. Soc., 113,* 9382(1991).
87. R. L. DesJarlais, et al., *Proc. Natl. Acad. Sci.* USA, *87,* 6644(1990).
88. I. D. Kuntz, *Science, 257,* 1078(1992).
89. N. A. Roberts, et al., *Science, 248,* 358(1990).
90. M. Kahn, et al., *J. Med. Chem., 34,* 3395(1991).
91. P. C. Zamecnik and M. L. Stephenson, *Proc. Natl. Acad. Sci. USA, 75,* 280(1978).
92. E. Uhlmann and A. Peyman, *Chem. Rev., 90,* 543(1990).
93. J. S. Cohen, *Pharmacol. Ther., 52,* 211(1991).
94. S. Joshi, et al., *J. Virol., 65,* 5524(1991).
95. M. Matsukura, et al., *Proc. Natl. Acad. Sci., 86,* 4244(1989).
96. D. Kinchington, et al., *Antiviral Res., 17,* 53(1992).
97. C. A. Stein, et al., *AIDS Res. Hum. Retroviruses, 5,* 639(1989).
98. D. C. Montefiori, et al., *Proc. Natl. Acad. Sci. USA, 86,* 7191(1989).
99. P. S. Sarin, et al., *Proc. Natl. Acad. Sci. USA, 85,* 7448(1988).
100. J. S. Cohen, *Antiviral Res., 16,* 121(1991).
101. P. E. Nielsen, et al., *Science, 254,* 1497(1991).
102. R. L. Letsinger, et al., *Proc. Natl. Acad. Sci. USA, 86,* 6553(1989).
103. M. LeMaitre, et al., *Proc. Natl. Acad. Sci. USA, 84,* 648(1987).
104. M. K. Ghosh and J. S. Cohen, *Prog. Nucl. Acid Res. Mol. Biol., 42,* 79(1992).
105. G. Degols, *Antiviral Res., 17,* 279(1992).
106. M.-C. Hsu, et al., *Science, 254,* 1799(1991).
107. R. Saul, et al., *Proc. Natl. Acad. Sci. USA, 82,* 93(1985).
108. R. Saul, et al., *Arch. Biochem. Biophys., 230,* 668(1984).
109. B. D. Walker, et al., *Proc. Natl. Acad. Sci. USA, 84,* 8120(1987).
110. R. A. Gruters, et al., *Nature, 330,* 74(1987).
111. G. W. J. Fleet, et al., *FEBS Lett., 237,* 128(1988).
112. B. G. Winchester, et al., *Biochem. J., 269,* 227(1990).
113. A. Karpas, et al., *Proc. Natl. Acad. Sci. USA, 85,* 9229(1988).
114. D. L. Taylor, et al., *AIDS, 5,* 693(1991).
115. B. Winchester, *Biochem. Soc. Trans., 20,* 699(1992).
116. L. Ratner, *AIDS Res. Hum. Retroviruses, 2,* 165(1992).
117. R. E. Hussey, et al., *Nature, 331,* 78(1988).
118. D. H. Smith, et al., *Science, 238,* 1704(1987).
119. R. A. Fisher, et al., *Nature, 331,* 76(1988).
120. K. C. Deen, et al., *Nature, 331,* 82(1988).
121. A. Traunecker, et al., *Nature, 331,* 84(1988).
122. C. F. Perno, et al., *J. Exp. Med., 171,* 1043(1990).
123. S. Broder, et al., *Ann. Intern. Med., 113,* 604(1990).
124. D. J. Capon, et al., *Nature, 337,* 525(1989).
125. R. A. Byrn, et al., *Nature, 344,* 667(1990).
126. J. O. Kahn, et al., *Ann. Intern. Med., 112,* 254(1990).
127. E. S. Daar, et al., *Proc. Natl. Acad. Sci. USA, 87,* 6574(1990).
128. D. W. Brighty, et al., *Proc. Natl. Acad. Sci. USA, 88,* 7802(1991).
129. P. R. Clapham, et al., *Nature, 337,* 368(1989).
130. D. J. Looney, et al., *J Acquir. Immune Defic. Syndr., 3,* 649,1990.
131. M. Tateno, et al., *Proc. Natl. Acad. Sci. USA, 86,* 4287(1989).
132. J. M. Harouse, et al., *Science, 253,* 320(1991).
133. A. B. Ezelowitz, et al., *J. Exp. Med., 169,* 185(1989).
134. M. R. Koslowski, et al., *Brain Res., 553,* 300(1991).
135. C. Thiriart, et al., *AIDS, 2,* 345(1988).
136. S. Chen, et al., *Biochemistry, 89,* 5872(1992).
137. E. De Clercq, *Cancer Lett., 8,* 9(1979).
138. H. Mitsuya, et al., *Science, 226,* 172(1984).
139. S. Broder, et al., *Lancet, ii,* 627(1985).
140. L. D. Kaplan, et al., *Am. J. Med., 82,* 615(1987).
141. K. D. Jentsch, et al., *J. Gen. Virol., 68,* 2183(1987).
142. J. Balzarini, et al., *Int. J. Cancer, 37,* 451(1986).
143. M. C. Bowman, et al., *J. Anal. Toxicol., 7,* 55(1983).
144. P. Mohan, et al., *J. Med. Chem., 34,* 212(1991).
145. P. Mohan, et al., *Biochem. Pharmacol., 41,* 642(1991).
146. P. Mohan, et al., *Drug Des. Discovery, 8,* 69(1991).
147. P. Mohan, et al., *J. Med. Chem.,* Submitted.
148. G. T. Tan, et al., *J. Med. Chem., 35,* 4846(1992).
149. P. Mohan and M. Baba, *Drugs of the Future,* In press.
150. M. Baba, et al., *Antiviral Chem. Chemother.,* In press.
151. M. Baba, et al., *Antiviral Res., 20,* 223(1993).
152. J. Balzarini, et al., *Biochem. Biophys. Res. Commun., 136,* 64(1986).
153. R. F. Schinazi, et al., *Antiviral Res., 13,* 265(1990).
154. M. Baba, et al., *Biochem. Biophys. Res. Commun., 155,* 1404(1988).
155. A. D. Cardin, et al., *J. Biol. Chem., 266,* 13355(1991).
156. H. Suzuki, et al., *Agr. Biol. Chem., 53,* 3369(1989).
157. K. R. Gustafson, et al., *J. Natl. Cancer, Inst., 81,* 1254(1989).
158. M. Cushman, et al., *J. Med. Chem., 34,* 329(1991).
159. M. Baba, et al., *Proc. Natl. Acad. Sci. USA, 85,* 6132(1988).

160. M. Baba, et al., *J. Infect. Dis., 161,* 208(1990).
161. N. R. Hartman, et al., *AIDS Res. Hum. Retroviruses, 6,* 805(1990).
162. C. Flexner, et al., *Antimicrob. Agents Chemother., 35,* 2544(1991).
163. I. R. MacGregor, et al., *Thromb. Haemost., 51,* 321(1984).
164. M. Baba, et al., *Antimicrob. Agents Chemother., 34,* 134(1990).
165. W. R. Michael, *Toxicol. Appl. Pharmacol., 12,* 473(1968).
166. P. Mohan, et al., *Antiviral Res., 18,* 139(1992).
167. G. T. Tan, et al., *Biochim. Biophys. Acta, 1181:* 183, 1993.
168. I. W. Althaus, et al., *Experientia, 48,* 1127(1992).
169. M. I. Gregersen and R. A. Rawson, *Am. J. Physiol., 138,* 698(1943).
170. H. S. Freeman, et al., *Dyes and Pigments, 8,* 417(1987).
171. P. Mohan, *Drug Develop. Res., 29:* 1, 1993.
172. E. De Clercq, et al., *Proc. Natl. Acad. Sci. USA, 89,* 5286(1992).
173. J. Saunders, *Drug Des. Discovery, 8,* 255(1992).

Chapter 40

HORMONE ANTAGONISTS

V. Craig Jordan

Hormones, essential for successful reproduction, are also responsible for the development and growth of breast, prostate, and uterine cancers. It is therefore understandable that attention has been focused on the development of specific agents (antihormones) to control these cancers.

High doses of progestins or androgens are known to modulate and antagonize the proliferative action of estrogens. This knowledge is used in treating endometrial carcinoma with progestins and breast cancer with androgens. It is not clear, however, whether these steroids antagonize estrogen action through the estrogen receptor or modify it by an additional steroid receptor mechanism (physiologic antagonism).

Lerner and co-workers[1] described the first nonsteroidal antiestrogen, ethamoxytriphetol (MER25). This compound is an antiestrogen in all species tested and possesses no other hormonal or antihormonal properties. An interesting action of the compound is its ability to prevent the implantation of fertilized ova in laboratory

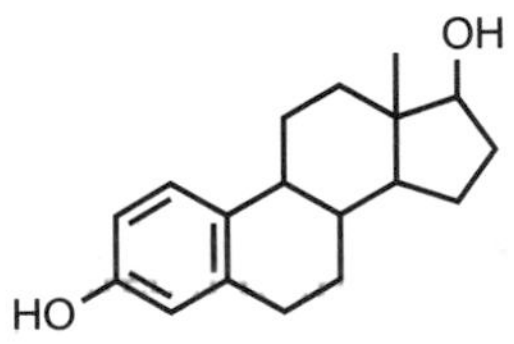

17β-Estradiol

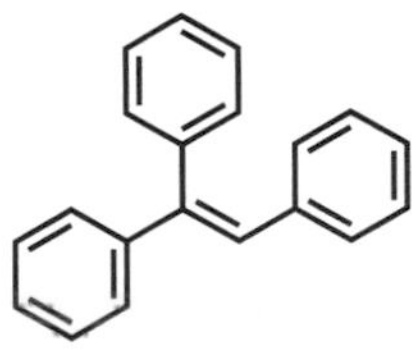

Triphenylethylene

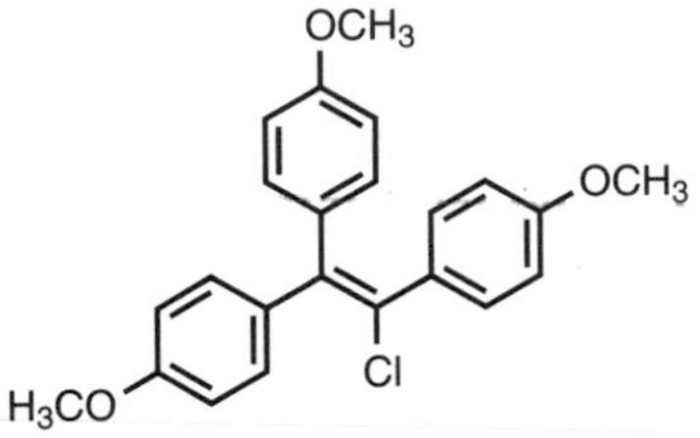

Chlorotrianisene

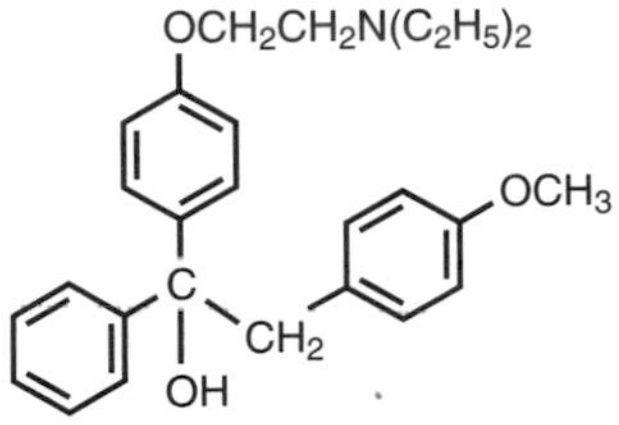

Ethamoxytriphetol

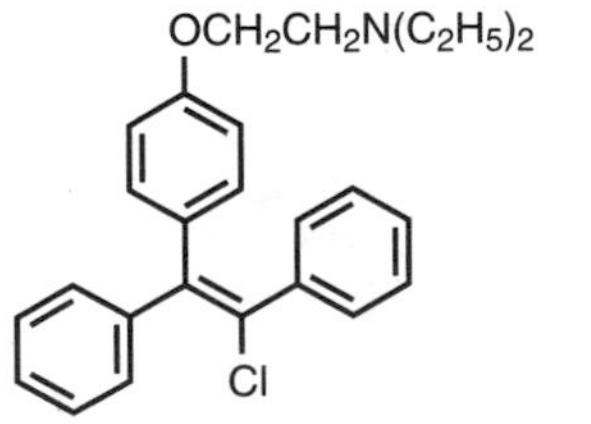

Clomiphene
(mixed E and Z isomers)

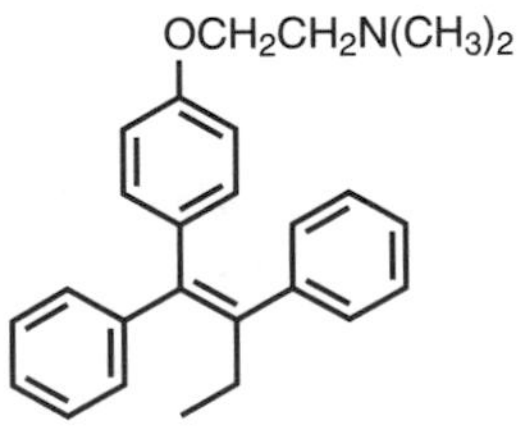

Tamoxifen
(Z isomer)

animals.[2] Unfortunately, clinical trials with MER25 showed that it had low potency, and the high doses required caused central nervous system (CNS) side effects.[3] A search for more potent compounds showed that MRL41 (later called clomiphene) has increased potency and is a postcoital contraceptive in laboratory animals. Clinical trials in humans, however, demonstrated that clomiphene induces ovulation and does not prevent implantation.[4] Clomiphene (Clomid), Marion Merrell Dow, Kansas City, MO) a mixture of E and Z geometric isomers, is used to induce ovulation in subfertile women.[5] The related compound tamoxifen (Nolvadex Zeneca, Wilmington, DE), the Z isomer of a substituted triphenylethylene, is also used in some countries to induce ovulation.[6]

Current interest in synthetic antifertility agents is focused on a search for specific antiprogestins. Progesterone is essential for the support and survival of the blastocyst following implantation in the endometrium. The administration of antiprogestins at the expected end of each menstrual cycle causes menstruation whether or not implantation has occurred.[7] The compound RU 38486 has entered clinical trials as the prototype antiprogestin. Unfortunately, the drug also has significant antiglucocorticoid activity, which may ultimately limit its usefulness.[8]

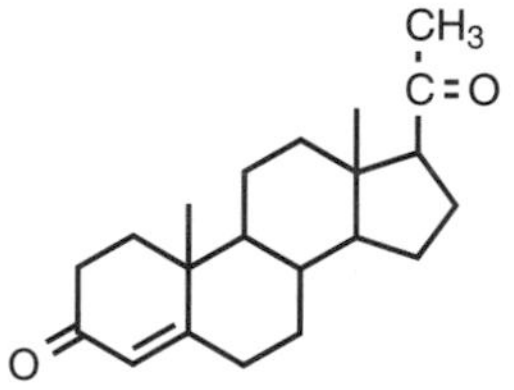

Progesterone

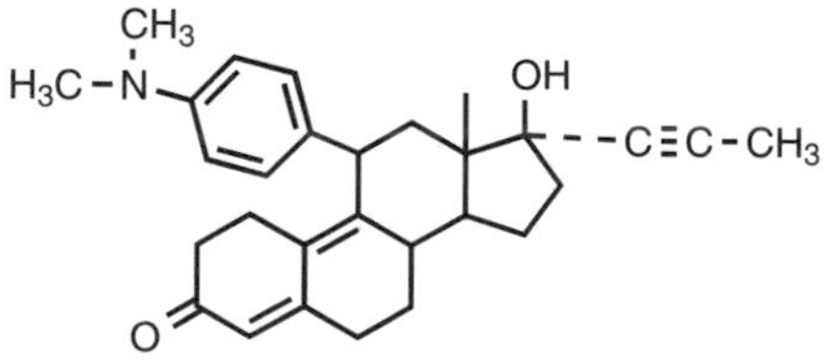

RU 38486

The most successful application of antihormones to date is in the treatment of hormone-dependent cancer, and this topic is considered in detail.

HORMONE-DEPENDENT BREAST CANCER

Breast cancer is one of the most common cancers in women; 1 of 11 women will have breast cancer in her lifetime.

In 1896, Beatson demonstrated that women with advanced breast cancer show improvement if their ovaries are removed.[9] It is now accepted that one in three premenopausal patients with breast cancer responds to oophorectomy. The reason for this empiric observation remained obscure until the early events involved in estrogen action were discovered.

Tritium-labeled hexestrol[10] and estradiol[11] bind to, and are retained by, the estrogen target tissues (uterus, vagina, pituitary gland) of laboratory animals. These findings led to the identification of an estrogen receptor protein in estrogen target tissues and the subsequent development of a subcellular estrogen receptor model by Jensen[12] and Gorski[13] and their co-workers. This model appeared to be consistent for all species and estrogen target tissues; however, it has been revised.[14–18] The simplified model proposes that the estrogen receptor is a nuclear protein, and therefore the steroid must diffuse into the nucleus to form a receptor complex and initiate estrogen action. The two models are compared in Figure 40–1.

Jensen reasoned that if the estrogen receptor is present in a tissue, estrogen must have a function in the cells. The concept was extrapolated to breast cancer to preselect patients who might respond to endocrine therapy. Different concentrations of estrogen receptors are present in breast cancers,[19,20] which can be explained by a heterogeneity of the tumor cell population. The more cells in the tumor that contain estrogen receptors, the higher the overall estrogen receptor content. Approximately 60% of estrogen receptor-positive (receptor-rich) patients are responsive to endocrine therapy, whereas only 10% of estrogen receptor-negative (receptor-poor) patients respond to endocrine therapy.[21,22]

These observations from controlled clinical studies have led to the determination of estrogen receptors in breast cancer before instituting therapy. The simultaneous determination of the progesterone receptor, an estrogen-stimulated protein, however, has increased the predictive nature of steroid hormone receptors. Of patients with estrogen and progesterone receptor-positive tumors, 80% will respond to endocrine manipulation. It is reasoned that the presence of the progesterone receptor indicates an active estrogen receptor system that is working efficiently and will be affected by estrogen deprivation.[23]

Nonsteroidal antiestrogens inhibit the binding of [^{3}H] estradiol to estrogen receptors,[24] and therefore their use has become a logical part of therapy for breast cancer. Clomiphene, nafoxidine, and tamoxifen have all been tested in clinical trial, but only tamoxifen is available for breast cancer treatment because of its low incidence of side effects.[25] Tamoxifen (Nolvadex) use has essentially replaced adrenalectomy as the treatment of choice in postmenopausal patients. Although response rates are similar with tamoxifen and high-dose estrogen (diethylstilbestrol) therapy or androgen (fluoxymesterone) therapy, tamoxifen has fewer side effects.

HORMONE-DEPENDENT PROSTATE CANCER

Dr. Charles Huggins was awarded the Nobel Prize for his pioneering research on endocrine control of prostatic carcinoma.[26] He introduced two important new concepts: that cancer is not necessarily an autonomous unre-

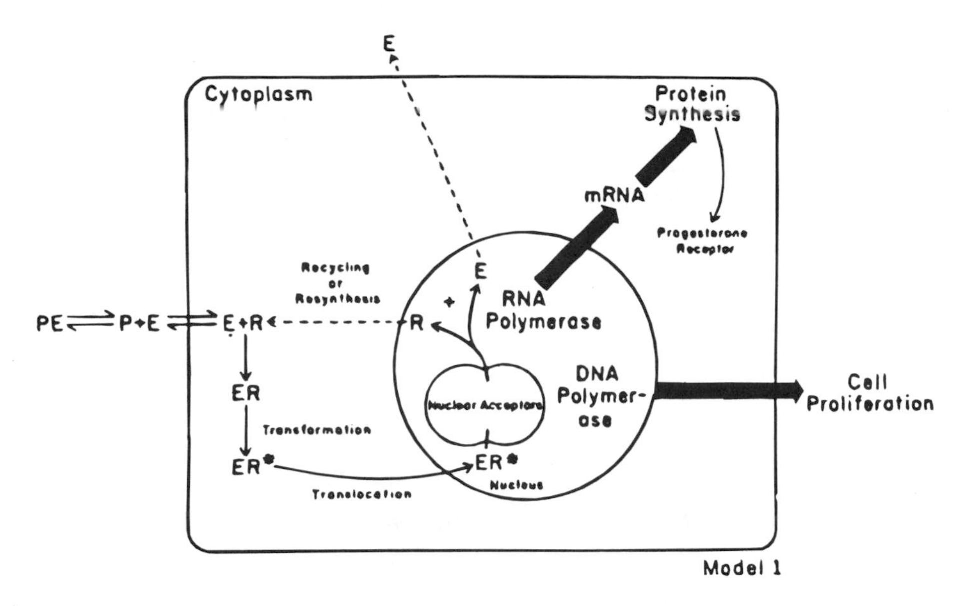

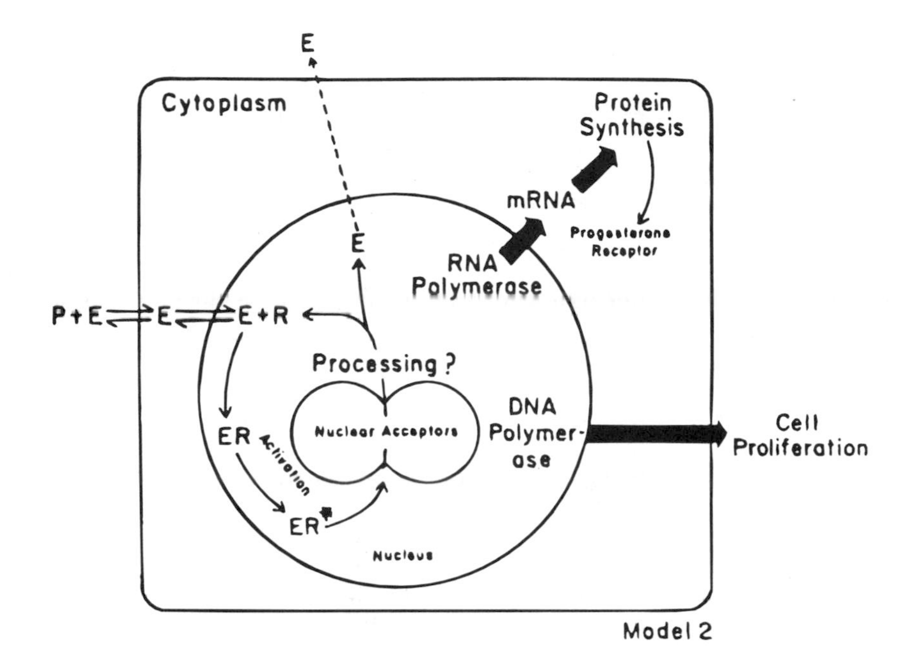

Fig. 40–1. Models for the subcellular mechanism of estrogen (E) action. E dissociates from plasma proteins (P) and diffuses into target tissue cells. E binds to estrogen receptors (R) in the cytoplasm (model 1) or the nucleus (model 2). The estrogen receptor complex (ER) is activated (*) (transformation) before initiating the events associated with estrogen action: activation of RNA polymerase and subsequent protein synthesis and activation of DNA polymerase and subsequent cell division. The fate of the nuclear ER complex is unknown.

lenting process, and that normal levels of hormones can sustain the growth of cancer of the prostate. As a result, orchiectomy has become a standard treatment for prostatic carcinoma because most prostatic tumors are androgen dependent.

The mechanism of action of androgens in the prostate differs from that of other steroid hormone systems because of an apparent metabolic activation of circulating androgens to a ligand with high binding affinity for the androgen receptor. Testosterone is converted to 5α-dihydrotestosterone by the enzyme 5α-reductase, located in the cytoplasm of prostate cells.[27,28]

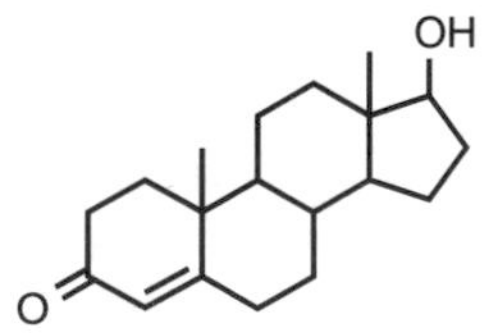

Testosterone

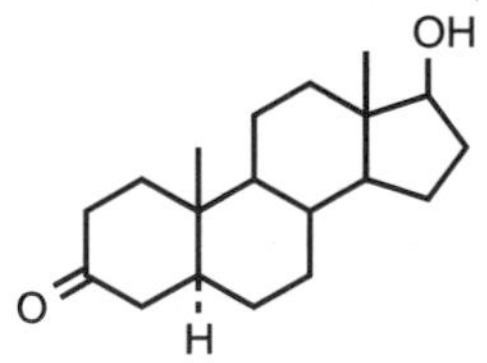

5α-Dihydrotestosterone

STRATEGIES FOR ANTIHORMONAL THERAPY

Breast Cancer

Hormone-dependent breast cancer can be treated in premenopausal women by removal of the ovaries. Ovarian estrogen production may also be inhibited with gonadotropin-releasing hormone (GnRH) agonists. Alternatively, the direct effects of estrogen on the growth of the breast tumor can be inhibited by administration of antiestrogens.

Most clinical experience with antiestrogens has been in postmenopausal patients with breast cancer. Although antiestrogen therapy might seem inappropriate in patients with inactive ovaries, significant estrogen production occurs by the peripheral aromatization of androstenedione (produced in the adrenals) to estrone. Adrenalectomy was an accepted treatment for hormone-dependent breast cancer in postmenopausal patients before the introduction of nonsteroidal antiestrogens. Similarly, high-dose estrogen (diethylstilbestrol), high-dose progestogen (megestrol, medroxyprogesterone acetate), or androgen (fluoxymesterone) therapies are used as "antiestrogen" treatments in postmenopausal patients, but the mechanism of tumor regression is unclear. An alternative approach for the treatment of breast cancer in postmenopausal patients is prevention of estrogen biosynthesis. Aminoglutethimide inhibits adrenal steroid biosynthesis and peripheral aromatase activity, although it also inhibits the synthesis of glucocorticoids, which results in a reflex rise in the secretion of adrenocorticotropic hormone (ACTH) by the pituitary. The large rise in ACTH overcomes the action of aminoglutethimide in the adrenal gland and steroid biosynthesis continues. The co-administration of aminoglutethimide and hydrocortisone avoids this problem. Another strategy is to develop drugs that are specific for the peripheral aromatization enzyme system. These agents can prevent estrogen biosynthesis without affecting adrenal physiology.

Each of the available types of hormone antagonists that can be used to treat hormone-dependent cancer and regulate other endocrine processes is considered in detail, with a brief description of its clinical applications.

Prostate Cancer

Disseminated androgen-dependent prostatic carcinoma can be controlled by removal of the primary source of androgens, the testes. Alternatively, androgen production in the testes can be reduced by the inhibition of luteinizing hormone (LH) release, which can be achieved by the administration of pharmacologic doses of estrogen (diethylstilbestrol). This therapy, however, is associated with numerous side effects. High-dose estrogen therapy can cause gynecomastia and life-threatening thromboembolic disorders. Release of LH can also be reduced by the continuous administration of luteinizing hormone-releasing hormone (LH-RH) agonists or antagonists. These agonists initially cause an increase in LH and testosterone, but their repeated administration paradoxically reduces LH levels, and eventually serum testosterone is reduced to castrate levels. The LH-RH antagonists block the binding of endogenous LH-RH to gonadotrophs in the pituitary gland and reduce serum LH levels without an initial increase in LH release.

The direct actions of androgens can be inhibited in patients with prostatic carcinoma by the administration of antiandrogens. Unfortunately, in noncastrated males, therapy can result in a reflex increase in testosterone secretion by the testes because the antiandrogen prevents the negative feedback produced in the hypothalamus by testosterone. This situation results in increased LH secretion by the pituitary gland, which stimulates Leydig cells in the testes to produce androgens. Increased production of androgen by the testis reverses the competitive blockade of androgen receptors by antiandrogen in the carcinoma.

INHIBITION OF STEROID ACTION

Nonsteroidal Antiestrogens

By definition, antiestrogens are compounds that block estrogen action. Antiestrogens inhibit estrogen-stimulated increases in uterine wet weight in immature or oophor-ectomized rats and prevent the vaginal cornification induced in oophor-ectomized rats by estrogen. Most compounds are structural derivatives of the estrogen triphenylethylene. Other than MER25, however, the nonsteroidal antiestrogens are all partial agonists in the rat.

Pharmacology and Mechanism of Action

The pharmacology of nonsteroidal antiestrogens is extremely complex and, as a result, it is not possible to define a unifying theory for their mechanism of action. Tamoxifen exhibits different actions in different species. The drug is an antiestrogen in the chick oviduct, a partial agonist/antagonist in the rat uterus, and a full estrogen in the mouse uterus. Although it is possible that differential metabolism causes the differences in pharmacology, no metabolic differences have been found. Antiestrogens inhibit the binding of [^{3}H]estradiol to estrogen target tissues, and at the subcellular level, they prevent the binding of [^{3}H]estradiol to estrogen receptors. Studies with [^{3}H]antiestrogens have demonstrated the direct binding to estrogen receptors. Differences in the intrinsic activity of the receptor complexes in different species and target tissues must be used to explain the different pharmacology. The potential mechanisms of action of antiestrogens have been reviewed.[29]

In studies with [^{3}H]antiestrogens investigators have identified a microsomal binding site with a high affinity for the triphenylethylene-type antiestrogens.[30] The binding site has been found in all tissues (unlike the estrogen receptor, which is found only in estrogen target tissues), although the highest concentration is found in the liver.[31] The function of the binding site is unknown, but the high binding affinity observed with antiestrogens can contribute to their long biologic half-life. The binding of the drug to high-affinity sites in the body prevents its metabolism and excretion. Several classes of drugs (tranquilizers, antihistamines, and antilipemics) bind to the "antiestrogen binding site," and therefore the name seems inappropriate. The compounds all have alkylamine side chains with a hydrophobic aromatic ring system.[32]

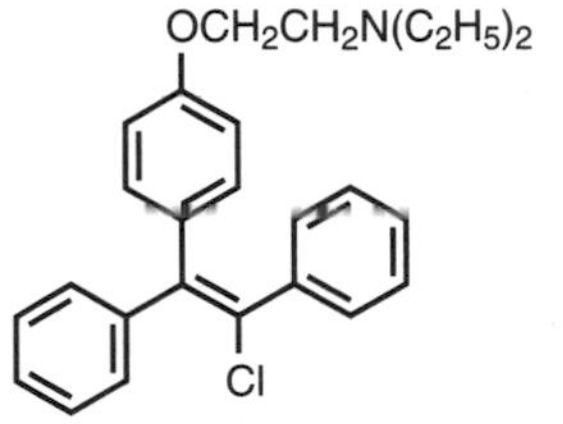

Enclomiphene
(*trans* isomer)

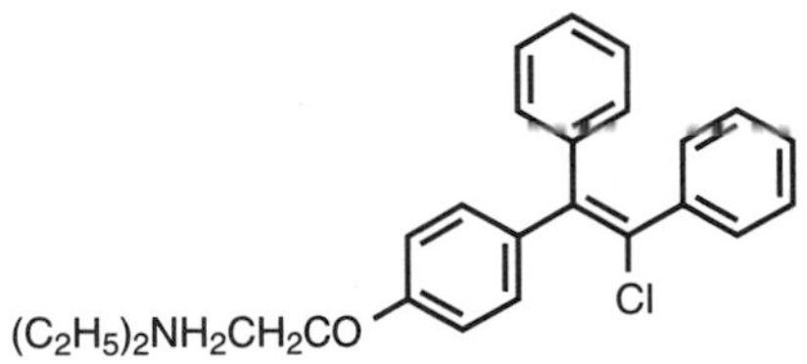

Zuclomiphene
(*cis* isomer)

Structure-Activity Relationships

A geometric requirement for substituted triphenylethylenes to inhibit estrogen action has been noted.[33] Clomiphene is a mixture of E and Z geometric isomers. The Z isomer *(cis)*(zuclomiphine) is an estrogen in rate uterine weight tests, whereas the E isomer *(trans)*(enclomiphene) is a partial agonist with antiestrogenic properties. Tamoxifen, however, the Z isomer *(trans),* is an antiestrogen in rat uterine tests, but the E isomer *(cis)* (ICI 47,699) is an estrogen.[34] Toremifene[35] is the *trans* isomer of the triphenylethylene that is an antiestrogen, but no information is available about its *cis* geometric isomer.

nylethylene type of structure by the introduction of a ketone bridging group that links the phenyl ring containing the pyrrolidinyl side chain with the rest of the molecule.[37] The compound is an antiestrogen in the rat, but removal of the basic side chain results in a compound that is neither estrogenic nor antiestrogenic.[38]

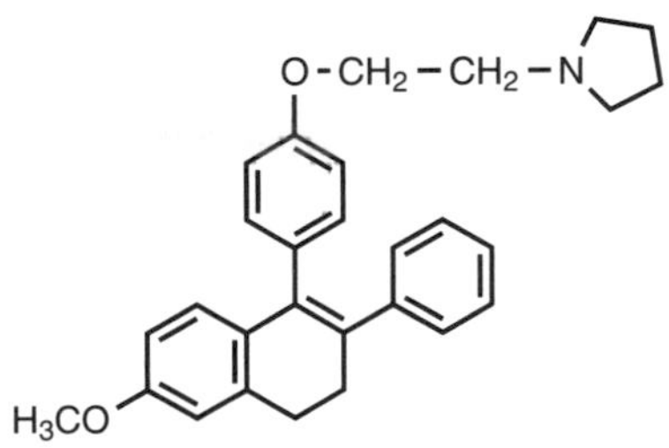

Most antiestrogens have a low binding affinity for the estrogen receptor. Tamoxifen has about 2% of the binding affinity for the estrogen receptor when compared with estradiol (100%). Nevertheless, introduction of a correctly positioned phenolic hydroxyl increases receptor binding affinity and antiestrogen potency. 4-Hydroxytamoxifen has a binding affinity for the estrogen receptor

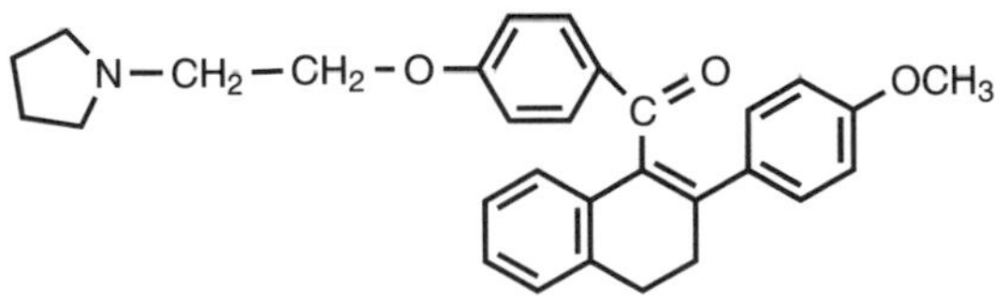

Nafoxidine (U-11,100A)

Trioxifene (LY 133314)

Several fixed-ring compounds have been tested to avoid potential problems of isomerization. Nafoxidine (U-11, 100A) is antiestrogenic; however, reduction in the number of carbons in the pyrrolidinyl side chain reduces antiestrogenic activity but increases estrogenic activity.[36] Trioxifine (LY 133314) deviates from the general triphe-

equivalent to that of estradiol.[39] This observation has led to a search for potent antiestrogens with reduced estrogenic activity for use as possible agents for breast cancer therapy. Keoxifene has a high affinity for the estrogen receptor and low estrogenic activity in rats and mice.[40,41] Unfortunately, antitumor activity in vivo is also low,[42] pos-

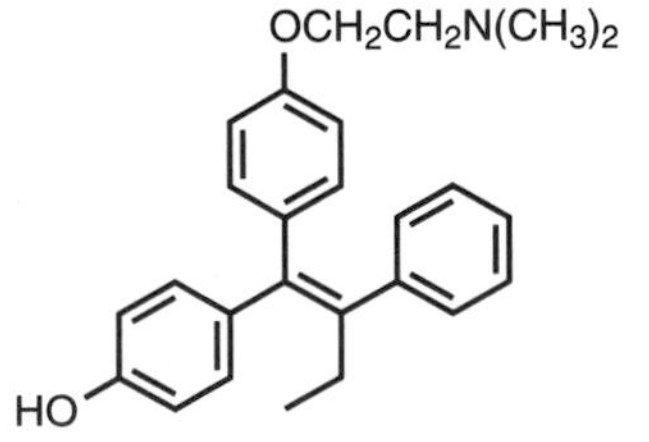

4-Hydroxytamoxifen

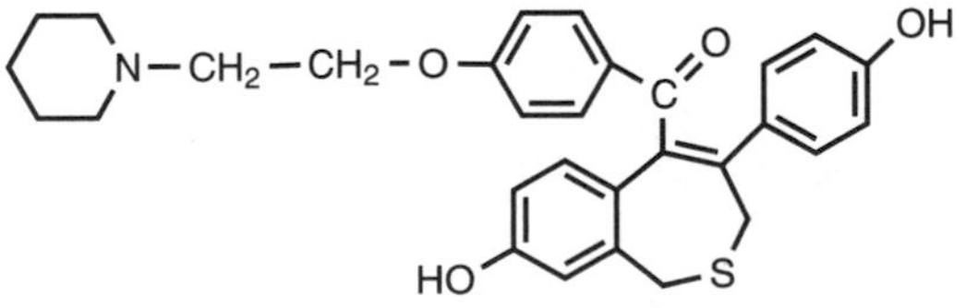

Keoxifene

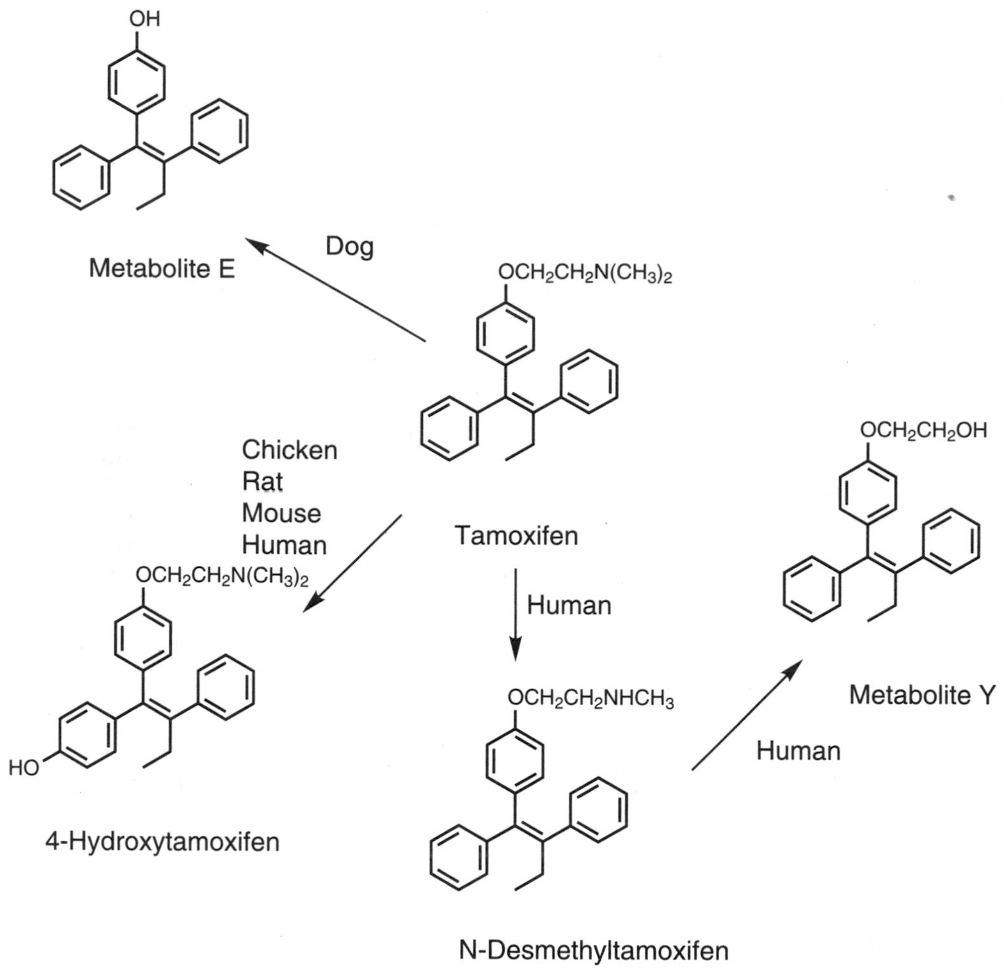

OH

Metabolite E

Dog

$OCH_2CH_2N(CH_3)_2$

Chicken
Rat
Mouse
Human

Tamoxifen

OCH_2CH_2OH

$OCH_2CH_2N(CH_3)_2$

Human

HO

4-Hydroxytamoxifen

$OCH_2CH_2NHCH_3$

Metabolite Y

Human

N-Desmethyltamoxifen

Fig. 40–2. Metabolism of tamoxifen in animals and humans.

sibly because compounds of this group are polar and have short biologic half-lives.[43]

Metabolism and Detection

Antiestrogens are metabolized extensively[44–46] (Fig. 40–2). Tamoxifen is hydroxylated to 4-hydroxytamoxifen, and nitromifene is demethylated to the phenol. Continuous administration of tamoxifen results in the formation of N-desmethyltamoxifen, the principal metabolite in humans.[47] N-Desmethyltamoxifen may be converted to Metabolite Y in humans.[48] All the metabolites that have been described in humans have antiestrogenic activity in rate uterine weight assays. Only Metabolite E, a metabolite of tamoxifen in dogs, is an estrogen. Tamoxifen and its metabolites (conjugates) are primarily excreted via the bile duct. Enterohepatic recirculation of hydrolyzed conjugated metabolites has been observed in rats.

Tamoxifen and its metabolites can be routinely identified and quantitated by thin-layer chromatography

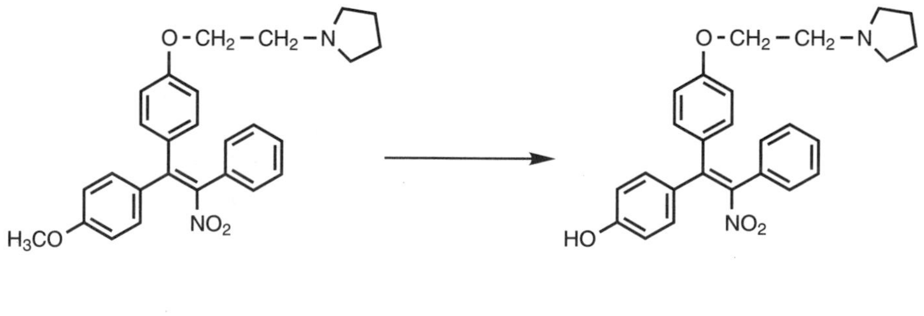

$O-CH_2-CH_2-N$

H_3CO

NO_2

Nitromifene

$O-CH_2-CH_2-N$

HO

NO_2

Phenolic metabolite with a high affinity for estrogen receptor

(TLC)[49] or high performance liquid chromatography (HPLC).[50,51] Detection of compounds using both TLC and HPLC involves the conversion of triphenylethylenes to fluorescent phenanthrenes by ultraviolet irradiation.

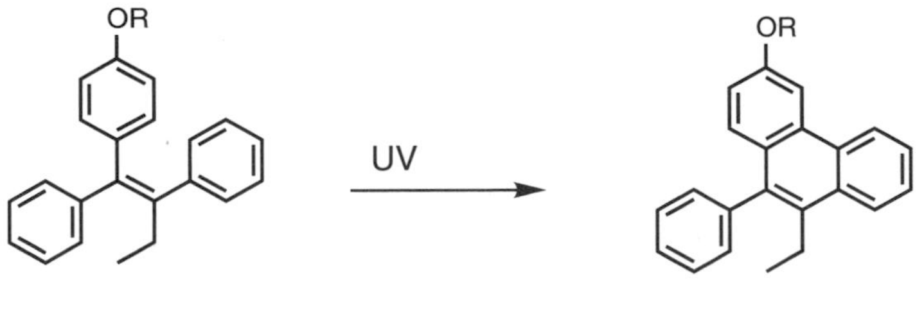

UV

Triphenylethylene Derivative
R = $CHECHEN(CH_3)_2$

Fluorescent Phenanthrene

Clinical Applications

Nafoxidine, clomiphene, and trioxifene have been tested in clinical trials of breast cancer therapy but are not available for clinical use. Tamoxifen is used in more than 70 countries around the world for the treatment of advanced (metastatic) breast cancer in women.[6] Patients with estrogen receptor-positive tumors have a higher response rate (50%) than those with estrogen receptor-negative tumors (13%). Recent clinical results have shown that the adjuvant use of tamoxifen therapy (i.e., treatment of patients after mastectomy) can delay the recurrence of the disease and improve survival.[52] Animal[53] and clinical studies[54] indicate that continuous treatment is the best strategy for adjuvant therapy. This is because tamoxifen causes a G1 blockade in the cell cycle of breast cancer cells, but this G1 blockade can be reactivated by estradiol. Tamoxifen is considered tumoristatic rather than tumoricidal. The drug is recommended as the adjuvant treatment of choice in postmenopausal women whose breast cancer was estrogen receptor-positive and who have microscopic tumors identified in axillary lymph nodes. Current overview analysis of all tamoxifen-containing clinical trials, however, has demonstrated an overall survival advantage for any women taking tamoxifen after a mastectomy.[55]

Side effects of tamoxifen use are minimal, nausea, vomiting, transient thrombocytopenia, and hot flashes (in premenopausal patients) are reported. Analysis of clinical trial data shows that only 2.5% of patients receiving tamoxifen therapy withdraw because of side effects.[6]

Tamoxifen therapy has been shown to be of benefit in male breast cancer and advanced endometrial cancer. Clomiphene is used to induce ovulation is subfertile women with an intact ovarian-hypothalamus-pituitary axis.[5] Tamoxifen is used in some countries to induce ovulation.[6]

Clinical Pharmacology

The recommended daily dosage of tamoxifen for breast cancer therapy is 10 or 20 mg twice daily. Tamoxifen accumulates to steady-state serum levels (100 to 200 ng/ml) within 3 to 5 weeks. The serum levels of the major metabolite, N-desmethyltamoxifen, are usually higher (150 to 300 ng/ml) than those of the parent drug. The calculated biologic half-lives of tamoxifen and N-desmethyltamoxifen are 7 and 14 days, respectively. It has been argued[56] that the administration of loading doses (100 mg twice daily) during the first week of therapy might achieve therapeutic serum levels sooner, but no clinical advantage has been noted.

Tamoxifen produces different endocrinologic effects in pre- and postmenopausal patients.[6] Estrogen-stimulated prolactin levels are reduced in premenopausal patients, whereas tamoxifen causes an increase in circulating estrogen levels. In postmenopausal patients, tamoxifen reduces prolactin levels but paradoxically causes a partial reduction in LH and follicle-stimulating hormone (FSH) levels. This partial agonist action is also seen with estrogen-like increases in sex hormone-binding globulin and estrogen-like alterations in vaginal cytologic studies.

Clomiphene is administered in five daily oral doses of 50 mg, starting on the fifth day of the menstrual cycle, to induce ovulation.[5] Clomiphene causes an increase in estrogen and gonadotropin secretion that induces ovulation.

Steroidal Antiestrogens

The nonsteroidal antiestrogens possess a significant amount of intrinsic estrogenic activity. Indeed, tamoxi-

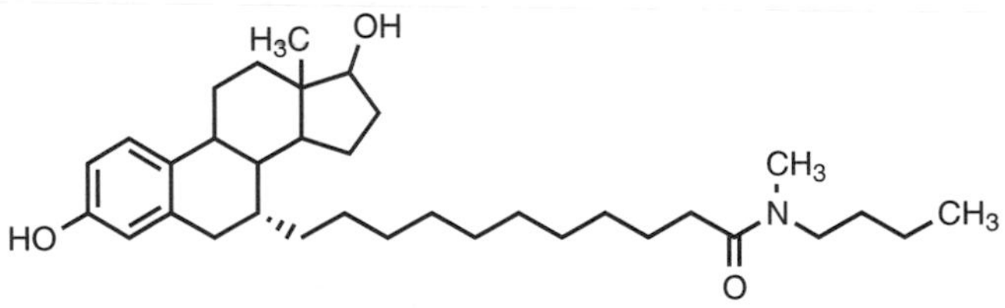

ICI 164,384

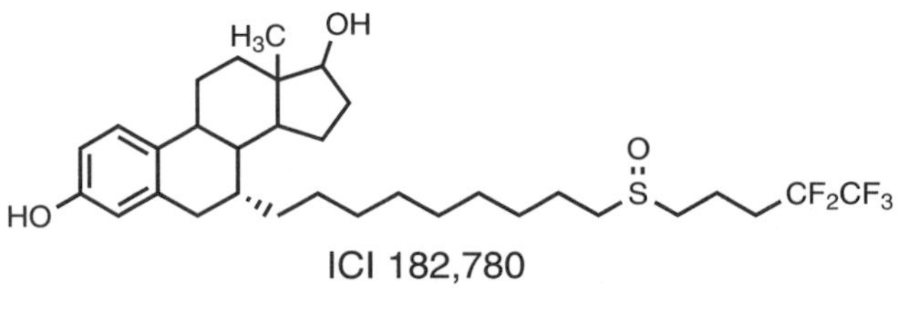

ICI 182,780

fen-stimulated breast[57] and endometrial cancer[58] have been studied in the laboratory. Pure antiestrogens that have no intrinsic estrogenic properties are being evaluated in laboratory and clinical studies. The lead compounds ICI164,384[59] and ICI182,780[60] are 7α substituted derivatives of estradiol. Bioavailability is poor and current clinical trials will evaluate depot injections.

Antiandrogens

The biologic activity of antiandrogens is determined in immature or castrated animals. The co-administration of test compounds and testosterone propionate can determine antiandrogen activity in assays such as chick comb, the costovertebral organ of hamsters, or weight changes in the prostate or seminal vesicles of mice or rats. In general, antiandrogens inhibit the binding of dihydrotestosterone to androgen receptors in target issues. Antiandrogens can be divided into steroidal and nonsteroidal compounds.

Steroidal Antiandrogens

Cyproterone acetate was the first antiandrogen discovered and used successfully in clinical studies.[61] The compound is a synthetic derivative of progesterone and possesses significant progestational activity. Several other progestational agents possess antiandrogenic properties (chlormadinone and megestrol), as do some 19 nortestosterone derivatives (TSAA 291).[62]

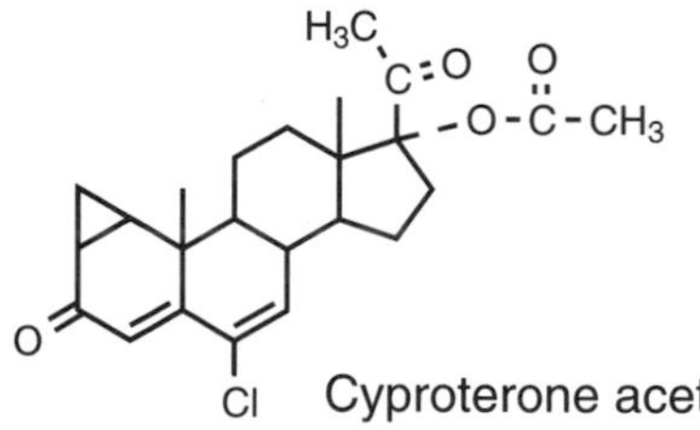

Cyproterone acetate

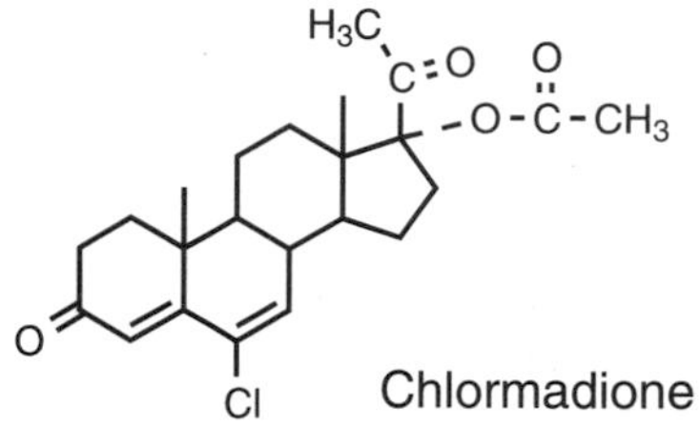

Chlormadione acetate

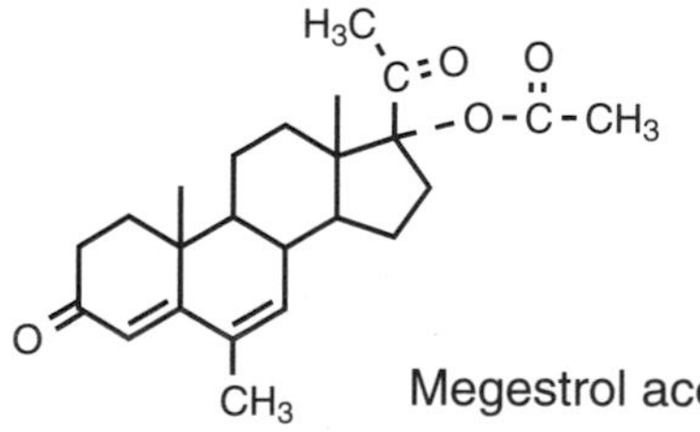

Megestrol acetate

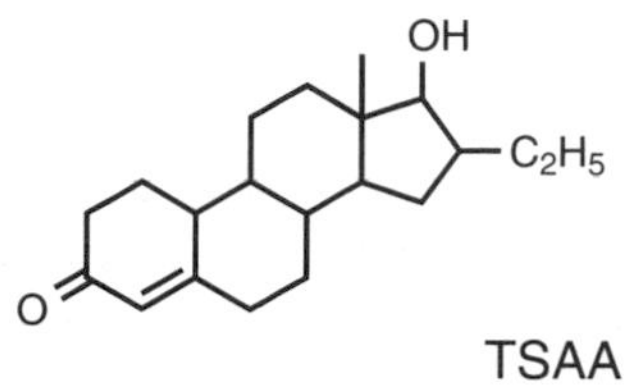

TSAA 291

Nonsteroidal Antiandrogens

Flutamide is the prototype of nonsteroidal antiandrogens.[63] It does not possess androgenic properties and is metabolically activated to the compound hydroxyflutamide.[64,65] This active metabolite must be used for studies in vitro. Other derivatives have been reported to have pure antiandrogenic properties (RU 23908),[66] and some have pure, peripherally selective androgenic activity (Casodex).[67] This latter property is seen as an advantage for clinical applications in treating prostate cancer, because pure antagonists cause a reflex rise in the secretion of gonadotropins from the pituitary gland.

Unrelated compounds have been found to possess antiandrogenic properties as unwanted side effects. The H2 antagonist cimetidine interacts with the androgen receptor and produces antiandrogenic actions in men during treatment for gastric ulcers.[68–70] Cimetidine antagonizes androgen-induced increases in seminal vesicle and ventral prostate weight in laboratory animals.[71,72] Etintidine, a potent H2 antagonist, has a reduced antiandrogenic activity and a lower affinity for the androgen receptor than cimetidine.[73,74]

Clinical Applications

Cyproterone acetate is used for the treatment of prostatic carcinoma in Europe.[75] Doses of 250 mg daily are effective. To avoid the uncertainty of patient compliance with oral medications, intramuscular injection of 300 mg weekly can be used. Side effects include edema (glucocorticoid-like), gynecomastia, and impotence. Flutamide has been studied in less detail than cyproterone acetate.[76,77] Oral daily doses of 0.75 or 1.5 g have been used in Europe to treat patients with metastatic prostate carcinomata. Side effects are minimal, but a rise in circulatory LH and testosterone has been noted in noncastrated men.[78]

Aldosterone Antagonists

Aldosterone is a potent mineralocorticoid that regulates electrolyte balance by promoting excretion of potassium and retention of sodium. Antimineralocorticoids are effective in treating essential hypertension and edematous diseases.

Spironolactone is an orally active aldosterone antagonist.[79–81] The drug also has antiandrogenic activity, which results in gynecomastia and impotence in men,[82,83] and weak progestational activity,[84] which produces menstrual irregularities in women.[85]

Current interest focuses on the synthesis of a specific aldosterone antagonist that does not interact with other steroid receptor systems.[86] The compound dihydrospirorenone has an increased antialdosterone activity. Spirorenone has a reduced affinity for the androgen receptor and an affinity for the progesterone receptor similar to

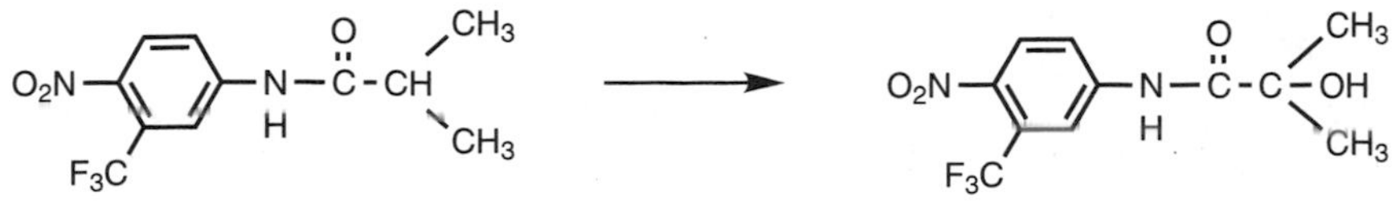

Flutamide → Hydroxyflutamide (active metabolite)

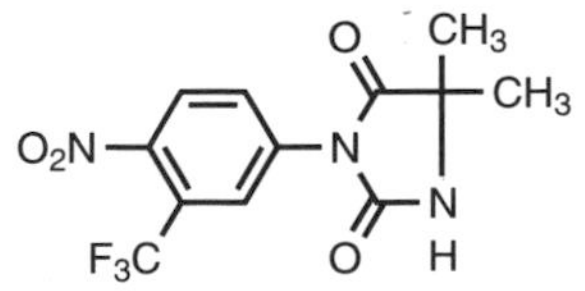

RU 23908

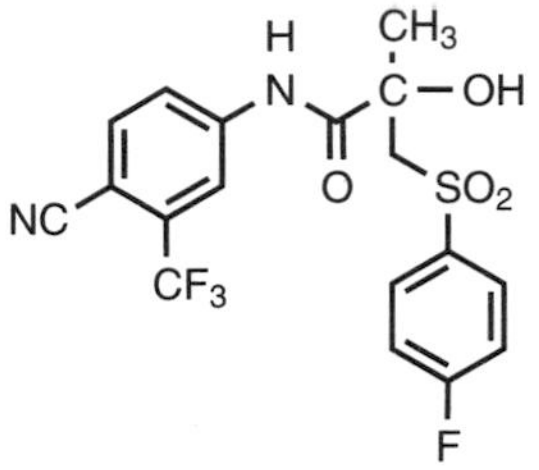

Casodex (ICI 176,334)

that of spironolactone, although it is converted into dihydrospirorenone by reduction of the 1,2 double bond in

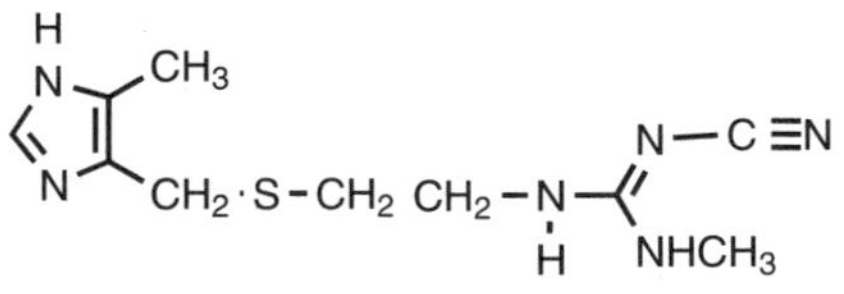

Cimetidine

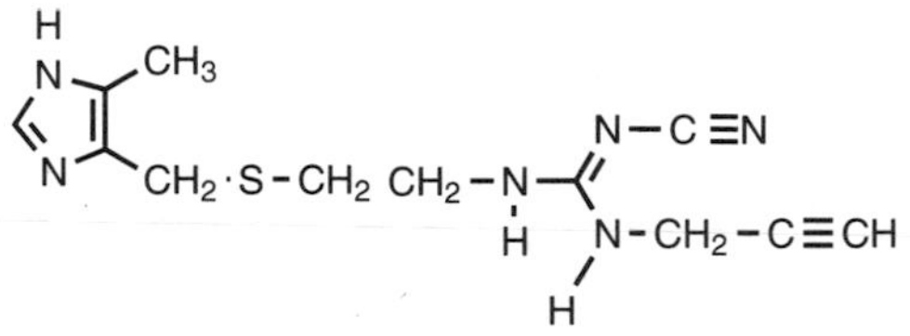

Etintidine

vivo. Introduction of a 1,2 methylene bridge (cf, cyproterone acetate) into dihydrospirorenone causes little effect on the aldosterone antagonist activity, when compared with spirorenone, but it also does not affect the interaction with androgen and progestin receptors. Antiandrogen or antiprogestin activities have not been determined.

The steroid 3-(9α-fluoro-17β-hydroxy-3-oxo-androst-4-en-17α-yl) propionic acid γ-lactone has negligible binding to androgen, progestin, estrogen, and glucocorticoid receptors, and some good antialdosterone activity in low doses. At higher doses, agonist activity predominates.[87]

It is clear that the search for specific steroidal agents to interact with mineralocorticoid, glucocorticoid, progestin, and androgen receptor systems is extremely complex because of the similarity of the binding ligands. In contrast, the aromatic A ring of estrogens appears to offer an advantage for developing specificity for its receptor system.

Antiprogestational Steroids

Interest in the development of a compound with antiprogestational activity is considerable because of its potential as an antifertility agent. Nevertheless, despite considerable research effort, no compound like the antiestrogens has been found that combines high potency with complete specificity of action.

RMI 12, 936 (17β-hydroxy-7α-methylandrost-5-en-3-one) is claimed to possess significant antiprogestational activity. The antifertility actions in laboratory animals are caused by accelerated egg transport.[88] The compound has weak estrogenic activity and binds to androgen receptors, but paradoxically does not compete with progesterone for binding to the progesterone receptor.[89] It has been suggested that RMI 12, 936 is converted to 7α-methyltestosterone by Δ^5-3-betasteroid isomerase.[90] This competitive action for a steroid biosynthetic enzyme system may contribute to antifertility effects. The compound R 2323 (13-ethyl-17-hydroxy-18,19-dinor-17α-pregna-4,9,11-triene-20-yne-3-one) has antiprogestational activity against exogeneous and endogenous progesterone.[91] The steroid has a mixture of activities; estogenic, progestational, androgenic, and antiestrogenic actions have been observed. R 2323 does compete with progesterone

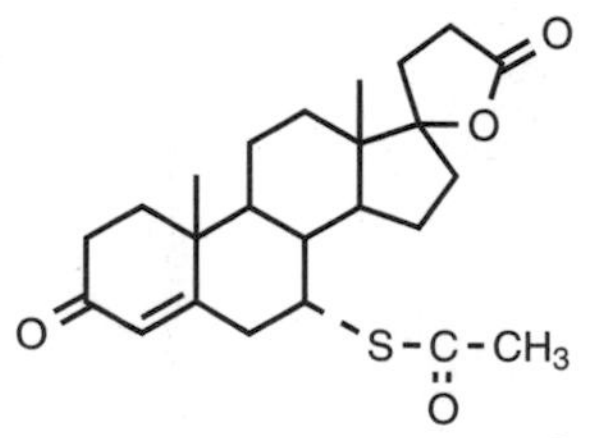

Spironolactone

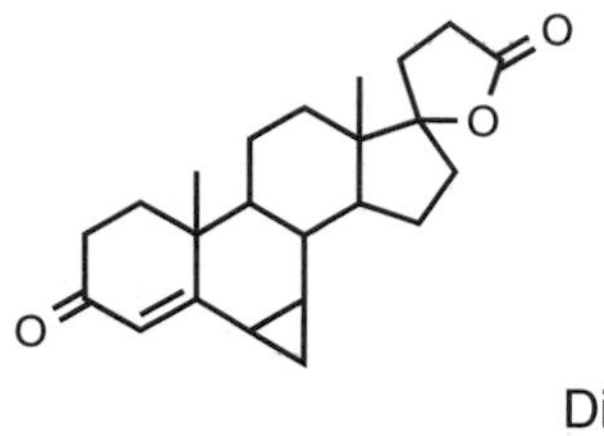

Dihydrospirorenone

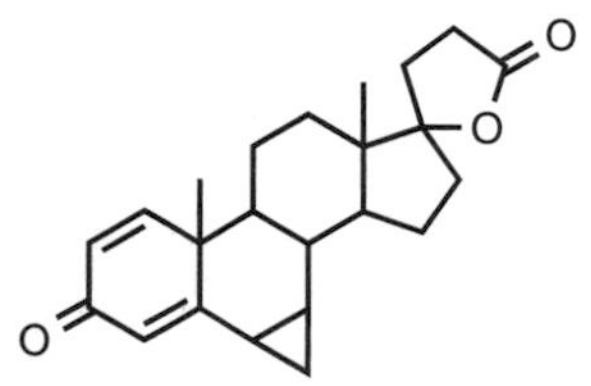

Spirorenone

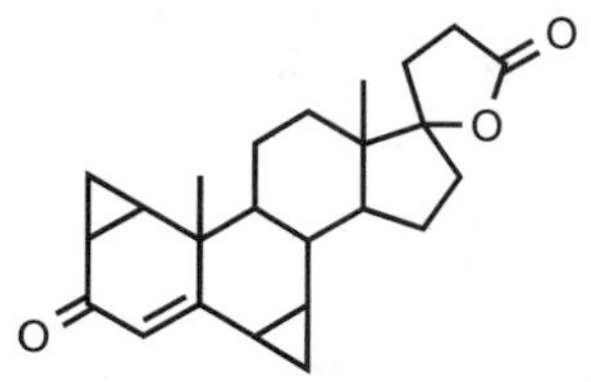

1,2-Methylene derivative
of dihydrospirorenone

for the progesterone receptor but also competes for androgen and aldosterone receptors.

The steroid RU 38486 has antifertility activity in laboratory animals and in women.[7] The compound binds to the progesterone receptor and shows antiprogestational activity. The affinity for the glucocorticoid receptor, however, is equivalent to that observed with dexamethasone, and antiglucocorticoid activity has also been reported.[8] The affinity for the androgen receptor is low.

Clinical Applications

In clinical trials, R 2323 was administered at a dosage of 2.5 mg or 5 mg weekly, but produced unacceptably high pregnancy rates compared with standard contraceptive producers.[92,93] Another approach has been to administer 50 mg daily on days 15 to 17 of the menstrual cycle to induce menses.[94] Few side effects are reported and

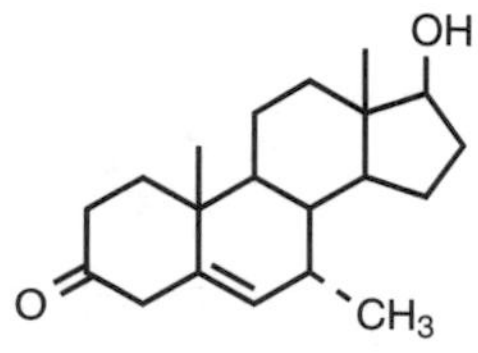

RMI 12,936

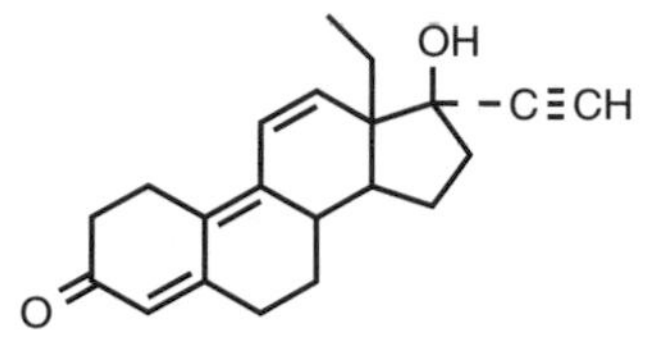

R 2323

good control of the menstrual cycle is achieved. The potent antiprogestin RU 38486 has been used in clinical trials as an abortifacient,[7] but the antiglucocorticoid activity may restrict its general use as a long-term therapy for breast cancer.

INHIBITION OF STEROID BIOSYNTHESIS

It can be argued that the most complete antagonist of hormone action is to have no hormone at all. This strategy is attempted with endocrine ablation. Removal of the ovaries or the testes causes a major decrease in sex steroids. The adrenals, however, produce androstenedione,

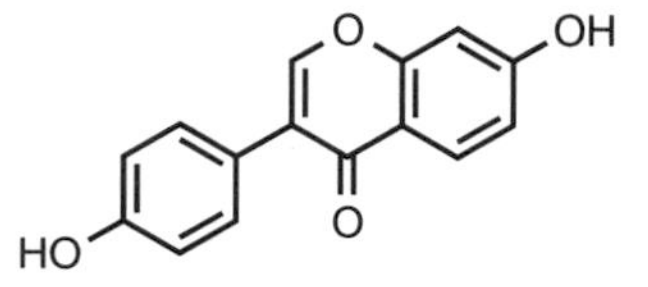

Diadzein

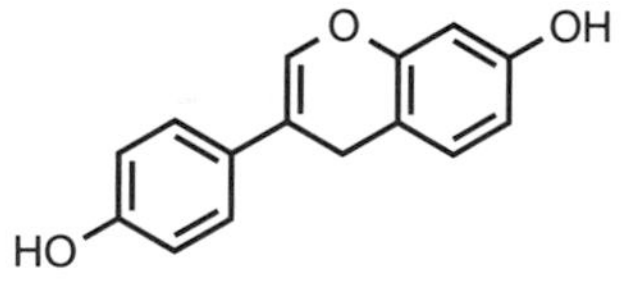

Equol

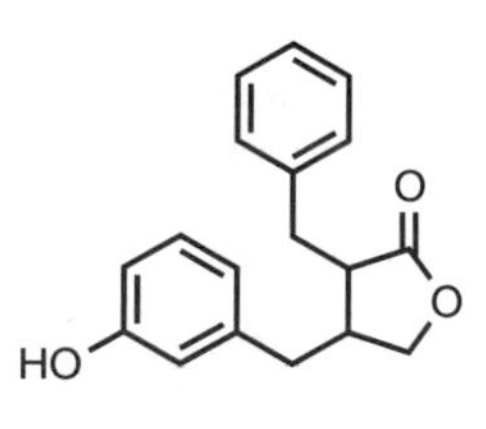

Enterolactone

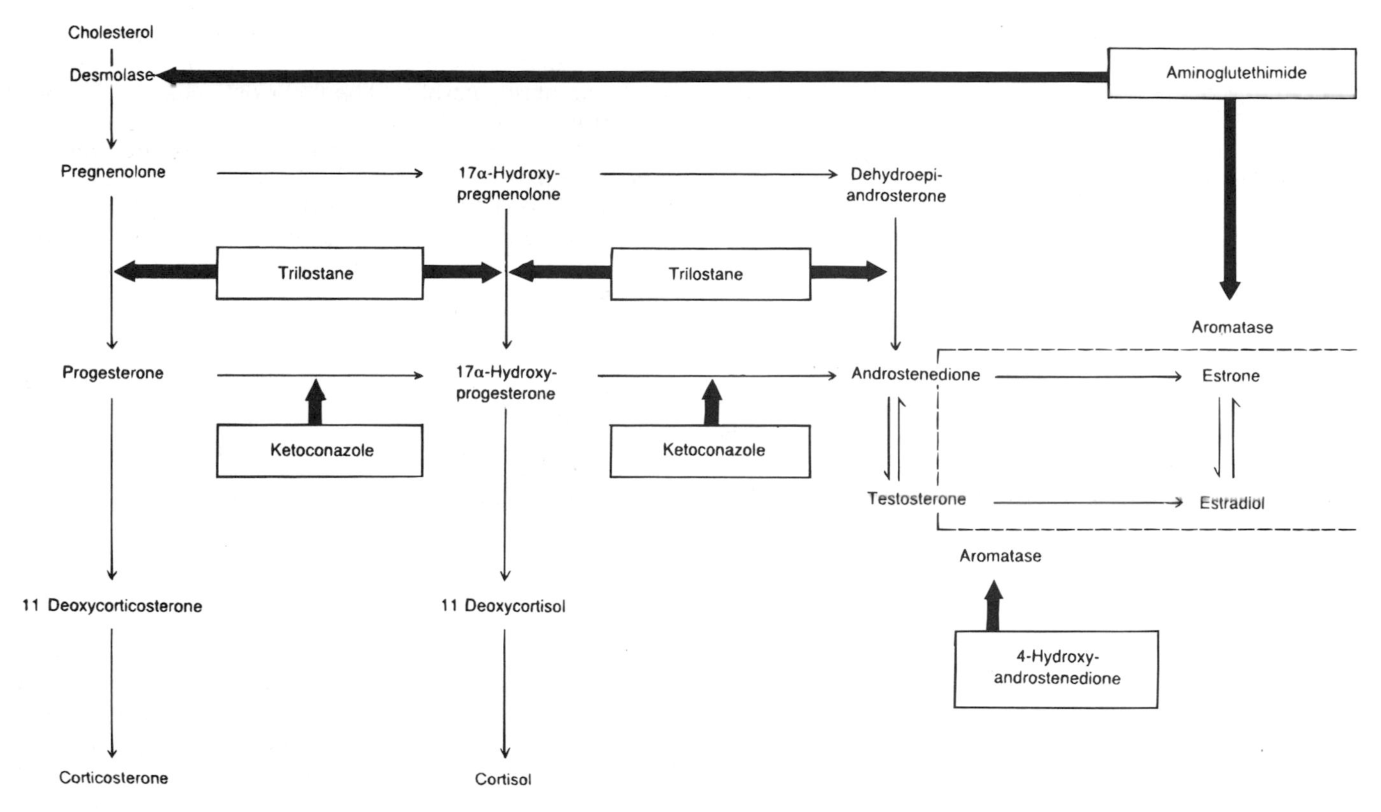

Fig. 40–3. Site of action of inhibitors of steroid biosynthesis.

and in women, this androgen can be converted to estrogens by aromatizing enzyme systems in peripheral sites.

Aminoglutethimide[95] inhibits two enzyme systems, the production of pregnenolone from cholesterol and the production of estrogens by peripheral aromatization. The drug is given with hydrocortisone to prevent ACTH from reversing the adrenal blockade. Recent interest had focused on the development of specific drugs to inhibit peripheral aromatizing enzyme systems. 4-Hydroxyandrostenedione is an example of this drug class that is in clinical trial.[96] It is argued that an increase in drug specificity may also increase patient acceptability and reduce side effects and toxicity. Nevertheless, the use of inhibitors of aromatization alone for the treatment of breast cancer may not be able to control hormone-dependent growth indefinitely. The biologic actions of phytoestrogens (plant estrogens) in the diet will not be controlled by inhibitors of steroid biosynthesis, so drugs that block estrogen receptors in the tumor must also be considered. Also of interest is the biologic impact of diadzein,[97] equol,[98,99] and enterolactone[100–102] as nonsteroidal phytoestrogens with the potential to stimulate the growth of breast cancer. It is possible that the phytoestrogens may also reverse antiestrogen therapy if the patient changes to a vegetarian diet.

The sites of action of various inhibitors of steroid biosynthesis are shown in Figure 40–3. Trilostane and ketoconazole have only moderate activity and, as yet, few applications.

Aminoglutethimide

Aminoglutethimide is the amino derivative of the hypnotic glutethimide. The drug was originally introduced

Glutethimide

Aminoglutethimide

as an anticonvulsant; however, low activity and an incidence of severe side effects caused its withdrawal from general use in 1966.

In 1967, aminoglutethimide was used to induce "medical adrenalectomy'' in patients with advanced breast cancer. The drug has been used extensively with a glucocorticoid for breast cancer therapy in postmenopausal women.[103] Dexamethasone was used originally to prevent the reflex rise in ACTH; however, aminoglutethimide induced metabolic tolerance to the synthetic glucocorticoid. Hydrocortisone is now used in combination with aminoglutethimide.

Mechanism of Action

The side-chain cleavage of cholesterol to form pregnenolone is achieved by a multienzyme system (desmolase) in mitochondria located in the adrenal glands, ovary, testis, and placenta. The terminal enzyme in the desmolase is a cytochrome P450. Aminoglutethimide inhibits the P450 in each of the steroid synthetic tissues.

The peripheral enzyme system that converts androstenedione and testosterone to estrone and estradiol is less well characterized. It is presumed that a P450 is associated with the aromatization steps.

In premenopausal women, aminoglutethimide rapidly lowers estrogen and progesterone levels in the luteal phase. In postmenopausal women who have undergone adrenalectomy and oophorectomy, aminoglutethimide prevents the aromatization of administered androstenedione.[95]

Structure-Activity Relationships

The amine is necessary for the antisteroidogenic activity of aminoglutethimide. Glutethimide has little or no activity to inhibit steroidogenesis.

The drug (Eliptin) used clinically is a mixture of two optically active isomers. The apparent Ki of D(+) aminoglutethimide is about two to five times less than the Ki of L(−) aminoglutethimide in inhibiting the side-chain cleavage of cholesterol and about 40 times less in inhibiting aromatase.[104] This situation translates into D(+) aminoglutethimide being 5 to 25 times more potent than the L(−) isomer in assays in vivo.[105]

Structure-activity relationship studies have been conducted with aminoglutethimide to avoid interaction with the desmolase system and increase selectivity for peripheral aromatase enzymes. The 4-pyridyl analog is a strong inhibitor of aromatase but is noninhibitory against desmolase.[106] The 2- and 3-pyridyl analogs are inactive against both enzyme systems. It is interesting to note that 1-amino-3-ethyl-3-phenylpiperidine-2,6-dione is a strong and selective inhibitor of desmolase,[107] but the 4-pyridyl derivative is only weakly active against desmolase and aromatase.

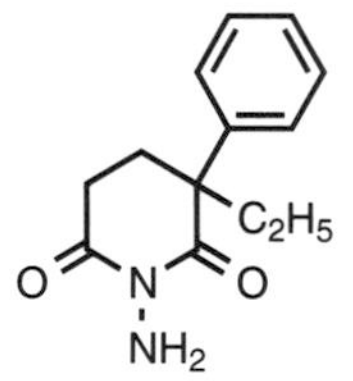

C_2H_5 O N O NH_2

Desmolase inhibitor

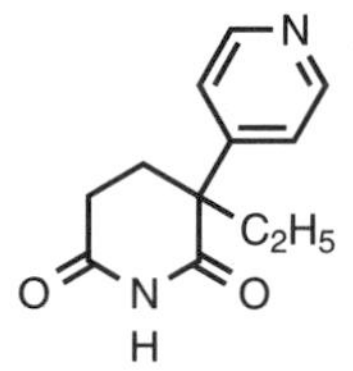

N C_2H_5 O N O H

Aromatase inhibitor

Clinical Applications

Aminoglutethimide does not inhibit ovarian aromatization completely in premenopausal women. Therefore, the drug is not recommended for treating breast cancer in these women. The combination of aminoglutethimide (1 g daily) and a glucocorticoid (hydrocortisone, 40 mg)[108] produces objective responses in approximately one third of unselected postmenopausal patients. The response rate is higher in estrogen receptor-positive patients.[103]

Objective responses are seen most frequently in skin and soft tissue diseases. Patients with bone metastases also achieve objective responses; however, patients with liver and visceral metastases rarely respond. This spectrum of responsiveness for endocrine therapies in breast cancer is well recognized. Aminoglutethimide can be used effectively as a second-line therapy after tamoxifen failure. The reason for the subsequent hormone responsiveness is unknown.

Toxicity[103]

Lethargy, the most frequent side effect (36%), is probably related to the close structural similarity to the parent compound glutethimide. Rashes occur in about 25% of patients. Other side effects include dizziness (15%) and nausea and vomiting (10%).

Steroidal Aromatase Inhibitors

Testololactone was first used in the 1960s for breast cancer therapy. The compound was tested because it is an androgen derivative; however, it is thought to have been effective because it acts as an inhibitor of aromatase.[109] 4-Hydroxyandrostenedione is the result of a systematic search for an inhibitor of the aromatization enzyme. Irreversible inhibition of the enzyme system converts either androstenedione or testosterone to estrone or estradiol, respectively. 4-Hydroxyandrostenedione has about 1% of the androgenic activity of testosterone.

Studies using the dimethylbenzanthracene-induced rat mammary carcinoma model have demonstrated the ability of 4-hydroxyandrostenedione to cause tumor regression and to reduce circulating levels of estrogen.[110] These findings have been used to support the development of 4-hydroxyandrostenedione for the treatment of breast cancer.

Structure-Activity Relationships

Studies in vitro with aromatase inhibitors are conducted using microsomal fractions from either human

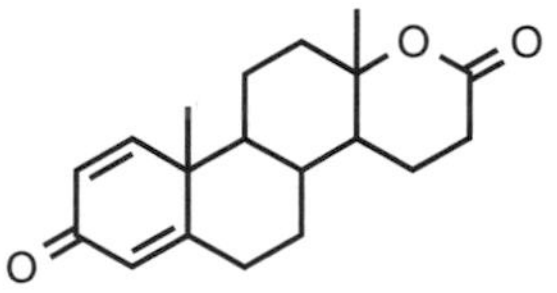

Δ1-Testololactone

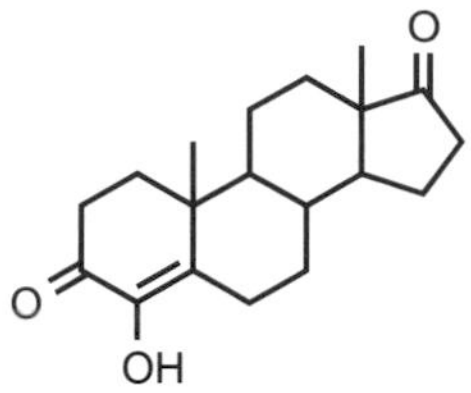

4-Hydroxyandrostenedione

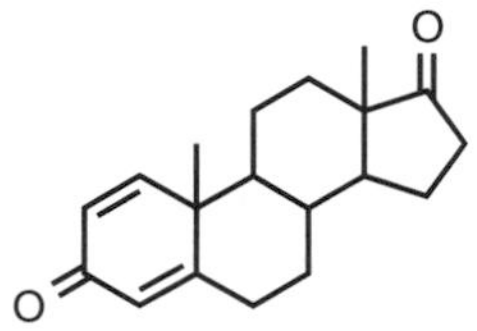

ATD

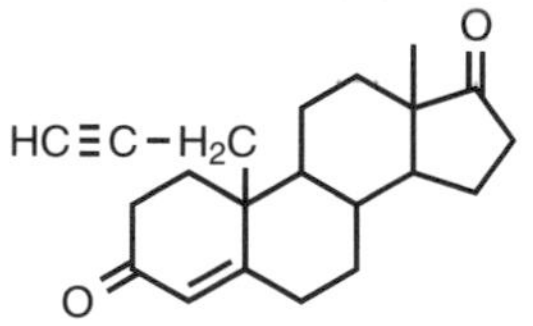

HC≡C–H_2C

MDL 18,962

placenta or pregnant mare's serum gonadotropin-primed rat ovaries.[111,112] The steroids 4-hydroxyandrostenedione, 4-acetoxyandrostenedione, and 1,4,6-androstatriene-3,17 dione (ATD) are all irreversible ("suicide") inhibitors of the aromatase enzyme system.[113]

The enzymatic mechanism that catalyzes estrogen formation is believed to involve two sequential hydroxylations of C-19 and then hydroxylation at the $C_2\beta$ position. The result is the loss of C_{19} as formic acid with elimination of 1β and 2β hydrogens, resulting in the aromatization of the A ring.[114,115]

The compound MDL 18,962 (10(2-propynyl) estra-4-ene-3,17 dione) is a highly potent suicide inhibitor of human placental aromatase.[116] The 2-propynyl group is essential for time-dependent enzyme inactivation because other derivatives without this particular group are inactive. The compound produces its irreversible effects with a specificity for aromatase; other P450 enzyme systems are unaffected.

Clinical Applications

4-Hydroxyandrostenedione is undergoing clinical evaluation in Europe for the treatment of breast cancer.[96] The drug is given by daily intramuscular injections (500 mg) and causes a reduction in serum estradiol in postmenopausal patients. Side effects are associated with the pain of daily injections. This route of administration and the rapid clearance of the steroid may limit the usefulness of the drug.

Nonsteroidal Aromatase Inhibitors

CGS 16949A, a new nonsteroidal inhibitor of aromatase that inhibits the growth of hormone-dependent

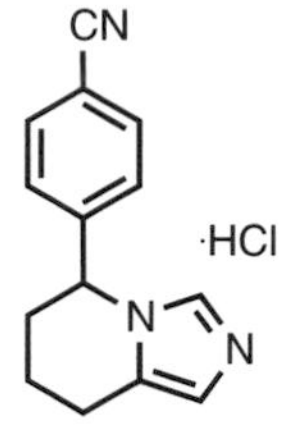

CN

·HCl

N

N

CGS 16949A

tumors in vivo,[117] is being evaluated as a treatment for breast cancer.[118,119] Serum estradiol is reduced at a dose of 2 mg twice daily, but daily doses of 16 mg cause a reduction in aldosterone production.[120]

INHIBITION OF GONADOTROPIN RELEASE

Luteinizing Hormone-Releasing Hormone (LH-RH)

It is well established that castration causes a rise in the pituitary secretion of LH and FSH, and it is equally well established that the administration of exogenous steroids (male: androgens, female: estrogens) alters gonadotrophin secretion. This negative feedback system, however, exists in the presence of a positive feedback system that operates at the correct time in the menstrual cycle to cause ovulation. A midcycle estrogen surge causes an LH surge that initiates ovulation by the prepared follicle in the ovary.

Luteinizing hormone release from the pituitary gland is initiated by LH-RH. The releasing hormone is produced in the hypothalamus and is secreted into the portal blood system to be carried to the adenohypophysis. The pulsetile release of LH-RH results in a pulsetile release of LH.

A single, large administration of LH-RH causes a large and prolonged release of LH. This effect has potential application for the induction of ovulation in subfertile women. Repeated administration of LH-RH, however, causes a reduction in LH release by a "down-regulation" of LH-RH receptors on the cells of the pituitary gland. This "antihormonal" action of LH-RH agonists has potential as a contraceptive method in women. Without an effective LH surge, ovulation cannot occur. Another application of "chemical gonadectomy" is in the treatment of either breast cancer in premenopausal women or prostatic carcinoma in men. The elucidation of the peptide

LH-RH	1 2 3 4 5 6 7 8 9 10 Pyro Glu—His—Trp—Ser—Tyr—Gly—Leu—Arg—Pro—Gly—NH_2
Superagonists	
HOE 766 (Buserelin)	1 2 3 4 5 6 7 8 9 Pyro Glu—His—Trp—Ser—Tyr—D Ser (But)—Leu—Ary—Pro NHET—
A 43818 (Leuprolide)	1 2 3 4 5 6 7 8 9 Pyro Glu—His—Trp—Ser—Tyr—D Leu—Leu—Ary—Pro NHET—
ICI 118630 (Zoladex)	1 2 3 4 5 6 7 8 9 10 Pyro Glu—His—Trp—Ser—Tyr—D Ser (But)—Leu—Arg—Pro Az—Gly NH_2
Antagonist	
ORG 30276	1 2 3 4 5 6 7 8 9 10 NAcD pClPhe—DpClPhe—D Trp—Ser—Tyr—D Arg—Leu—Arg—Pro—D Ala

Fig. 40–4. Amino acid sequence of luteinizing hormone-releasing hormone (LH-RH) and synthetic derivatives with agonist and antagonist actions.

chemistry of LH-RH by Andrew Schally, for which he was awarded the Nobel Prize,[121–124] has led to the investigation of more than 1000 synthetic analogs as potential agonists and antagonists.

Structure-Activity Relationships

LH-RH is a decapeptide (Fig. 40–4). Replacement of glycine in position 6 by D-tryptophan (D-Trp^6 LH-RH) produces a superactive LH-RH agonist that is 100 times more potent than natural LH-RH.[125] D-Leu^6-LH-RH ethylamide (Leuprolide, Lupron, Tap Pharmaceuticals, Deerfield, IL) and D-Ser $(But)^6$ AZ-Gly^{10}-LH-RH (ICI 118,630, Zoladex, Zeneca) are about 50 times more active than LH-RH.[126,127] The related compound (HOE 766) D-Ser (But) desglycine $NH_2{}^{10}$-LH-RH ethylamide (Buserelin, Hoechst-Roussell) is a long-acting compound that is 100 times more potent than LH-RH.

The first promising compound with LH-RH antagonist activity was D-Phe^2 LHRH.[128] The antagonist activity can be improved by combining 2-position modification with substitution of specific D-amino acids (Ala, Leu, Trp, Phe) in the 6 position. The elimination of the 10 terminal glycine to produce a C-terminal Pro^9 ethylamind (Fujino modification) to aid receptor binding and to produce resistance to enzymatic degradation actually decreases activity.[129] This result contrasts with the increases in potency produced by 6 and 10-position modifications of the superagonists.

The 3 position can be modified with Pro and D-Trp to increase antagonist activity. The triply substituted analogs D-Phe^2-Pro^3-D-Trp^6-LHRH and D-Phe^2-Pro^3-D-Phe^6-LHRH are fourfold more active than D-Phe^2-D-Ala^6-LHRH and have a long duration of action.[130] Position 1 has been modified to increase potency. Substitution in position 1 of D-Phe^2-D-$Trp^{3,6}$-LH-RH with either acetyl (Ac) dehydro-Pro, Ac-D-Ala, Ac-D-Phe, or D-pyroGlu produces potent long-acting antagonists.[131–134] The potency of the first effective antagonist, D-Phe^2-D-Ala^6-LHRH, has been improved 300 to 1000 times by the study of structure-activity relationships.

Animal Tumor Studies

Prolonged treatment of animals with transplanted Dunning R 3327H rat prostate tumors using the agonist D-Trp^6LHRH decreases tumor size. Serum LH, FSH, and testosterone levels are decreased, but progesterone levels are increased.[135]

The chronic administration of the antagonist Org 30276 causes a decrease in the growth rate of the Dunning R 3327H rat prostate tumor. Serum testosterone levels are decreased by 97%.[136] Studies with the LH-RH agonist ICI 118630 in the dimethylbenzanthracene-induced rat mammary carcinoma model demonstrate a decrease in tumor size and prevention of the appearance of new tumors.[137]

Clinical Applications

The LH-RH agonists were originally designed to induce ovulation in anovulatory patients. The paradoxic decrease in steroid synthesis during continuous therapy with LH-RH agonists has led to their consideration as contraceptive agents.

Leuprolide is used in the treatment of prostatic carcinoma. The recommended dose is 1 mg (0.2 ml), administered as a single daily subcutaneous injection. Clinical evaluation of leuprolide versus diethylstilbestrol has demonstrated therapeutic equivalence with fewer side effects. Leuprolide and diethylstilbestrol both cause a consistent decrease in circulating testosterone levels within the first 4 weeks of therapy.[138] Daily administration of ICI 118,630 decreases serum testosterone levels in men with prostatic carcinoma.[139,140] Clinical trials have been carried out to evaluate slow-release (depot) formulations of ICI 118,630.[141] The rice grain-sized depots, containing 3.5 mg of ICI 118,630, are implanted under the skin and replaced every 4 weeks.

Another successful approach to the treatment of prostate cancer is with the combination of Buserelin and the antiandrogen RU 23908. The therapy produces a decrease in LH with the LH-RH agonists, which decreases testosterone production in the testes, and the antiandrogen blocks the action of residual testosterone that is produced in the adrenals.[142] This use of antihormones has been successful and will probably become a standard therapeutic strategy for prostate cancer.

Side effects experienced with LH-RH agonists are dizziness, peripheral edema, and some nausea. About 50% of patients using leuprolide experience hot flashes. The initiation of therapy increases serum levels of testosterone. This change can be associated with a transient worsening of the signs and symptoms of the disease, but resolution occurs as therapy is continued and testosterone levels decline. Use of an LH-RH antagonist may have advantages during the early phase of the treatment of prostate cancer. As yet, no antagonists have had complete clinical evaluation.

Nonpeptide Inhibitors of Gonadotropin Release

Several relatively simple nonsteroidal compounds can inhibit the release of gonadotropins. The first compound, methallibure (ICI 33,828), had potential application as an antifertility agent,[143] but it also inhibited thyroid function.[144] A related compound, GP 48,989, inhibits the initiation and growth of dimethylbenzanthracene-induced rat mammary tumors[145] and has been found to decrease LH release in oophorectomized rats.[146] Because these drugs are orally active, they provide opportunities for treatment of prostatic carcinoma.

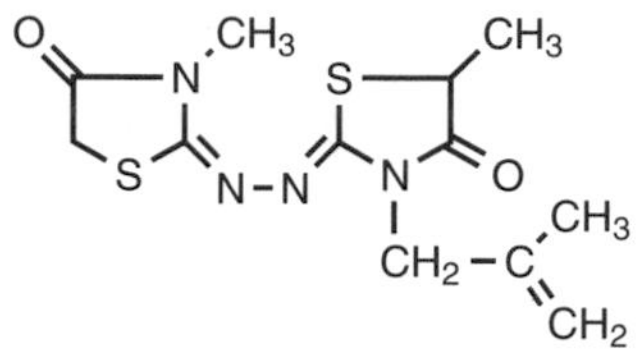

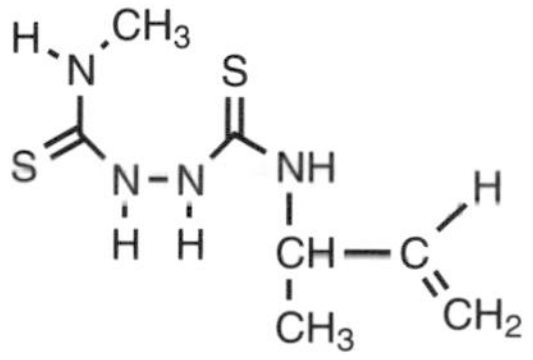

GP 48,989

Methallibure
(ICI 33,828)

CONCLUSION

The period from 1975 to 1992 saw the introduction of several hormone antagonists with broad clinical applications. These drugs are proving to be extremely valuable for the treatment of hormone-dependent cancer and also as tools to help understand hormonal control mechanisms. Considerable interest has been generated in the development of new and novel antihormones.

REFERENCES

1. L. J. Lerner, et al., *Endocrinology, 63,* 295(1958).
2. J. S. Segal and W. O. Nelson, *Proc. Exp. Biol. Med., 98,* 431(1958).
3. R. W. Kistner and O. W. Smith. *Fertil. Steril., 12,* 121(1961).
4. R. B. Greenblatt, et al., *Am. J. Obstet. Gynecol., 84,* 900(1962).
5. J. H. Clark and B. M. Markaverich, *Pharmacol. Ther., 15,* 467(1982).
6. B. J. A. Furr and V. C. Jordan, *Pharmacol. Ther., 25,* 127(1984).
7. W. Herrmann, et al., *C. R. Acad. Sci. [III], 294,* 933(1982).
8. M. N. Chobert, et al., *Biochem. Pharmacol., 32,* 3481(1983).
9. G. T. Beatson, *Lancet, ii,* 162(1896).
10. R. F. Glascock and W. G. Hoekstra, *Biochem. J., 72,* 673(1959).
11. E. V. Jensen and H. I. Jacobson, *Recent Prog. Horm. Res., 18,* 387(1962).
12. E. V. Jensen, et al., *Proc. Natl. Acad. Sci. USA, 59,* 632(1968).
13. J. Gorski, et al., *Recent Prog. Horm. Res., 24,* 45(1968).
14. D. M. Linkie and P. K. Siiteri, *J. Steroid Biochem., 9,* 1071(1978).
15. P. J. Sheridan, et al., *Nature, 282,* 579(1979).
16. W. J. King and G. L. Greene, *Nature, 307,* 745(1984).
17. W. V. Welshons, et al., *Nature, 307,* 747(1984).
18. V. C. Jordan, et al., *Endocrinology, 116,* 1845(1985).
19. P. Feherty, et al., *Br. J. Cancer, 25,* 697(1971).
20. S. G. Korenman and B. A. Dukes, *J. Clin. Endocrinol., 30,* 639(1970).
21. E. V. Jensen, et al., *NCI Monogr., 34,* 55(1971).
22. W. L. McGuire, et al., *Estrogen Receptors in Human Breast Cancer,* New York, Raven Press, 1975.
23. K. B. Horwitz, et al., *Science, 189,* 726(1975).
24. R. Hahnel, et al., *J. Steroid Biochem., 4,* 687(1973).
25. S. S. Legha and S. K. Carter, *Cancer Treat. Rev., 3,* 205(1976).
26. C. Huggins, et al., *Arch. Surg., 43,* 209(1941).
27. N. Bruchovsky and J. D. Wilson, *J. Biol. Chem., 243,* 2012(1968).
28. W. I. P. Mainwaring, *J. Endocrinol., 44,* 323(1969).
29. V. C. Jordan, *Pharmacol. Rev., 36,* 245(1984).
30. R. L. Sutherland, et al., *Nature, 288,* 273(1988).
31. K. Sudo, et al., *Endocrinology, 112,* 425(1983).
32. S. D. Lyman and V. C. Jordan, *Biochem. Pharmacol., 34,* 2795(1985).
33. M. J. K. Harper and A. L. Walpole, *Nature, 212,* 87(1966).
34. V. C. Jordan, et al., *Endocrinology, 108,* 1353(1981).
35. L. Kangas, et al., *Cancer Chemother. Pharmacol., 17,* 109(1986).
36. D. Lednicer, et al., *J. Med. Chem., 10,* 78(1967).
37. C. D. Jones, et al., *J. Med. Chem., 22,* 962(1969).
38. V. C. Jordan and B. Gosden, *Mol. Cell Endocrinol., 27,* 291(1982).
39. V. C. Jordan, et al., *J. Endocrinol., 75,* 305(1977).
40. L. J. Black, et al., *Life Sci., 32,* 1031(1983).
41. C. D. Jones, et al., *J. Med. Chem., 27,* 1057(1984).
42. J. A. Clemens, et al., *Life Sci., 32,* 2869(1983).
43. V. C. Jordan and B. Gosden, *Endocrinology, 1123,* 463(1983).
44. J. M. Fromson, et al., *Xenobiotica, 3,* 693(1973).
45. B. S. Katzenellenbogen, et al., *Recent Prog. Horm. Res., 35,* 259(1979).
46. V. C. Jordan, *Breast Cancer Res. Treat., 3,* 123(1982).
47. H. K. Adams, et al., *Biochem. Pharmacol., 27,* 145(1979).
48. V. C. Jordan, et al., *Cancer Res., 43,* 1466(1983).
49. H. K. Adam, et al., *J. Endocrinol., 84,* 35(1980).
50. Y. Golander and L. A. Sternson, *J. Chromatogr., 181,* 41(1980).
51. R. R. Brown, et al., *J. Chromatogr., 272,* 351(1983).
52. Nolvadex Adjuvant Trial Organisation, *Lancet 1,* 836(1985).

53. V. C. Jordan, *Breast Cancer Res. Treat., 3 (Suppl. 1),* 73(1983).
54. D. C. Tormey and V. C. Jordan, *Breast Cancer Res. Treat., 4,* 297(1984).
55. Early Breast Cancer Trials Organization, *Lancet i,* 1(1992).
56. C. Fabian, et al., *Cancer Treat. Rep., 64,* 765(1980).
57. M. M. Gottardis and V. C. Jordan, *Cancer Res., 48,* 5183(1988).
58. M. M. Gottardis, et al., *Cancer Res., 48,* 5183(1988).
59. A. E. Wakeling and J. Bowler, *J. Endocrinol., 112,* R7(1987).
60. A. E. Wakeling, et al., *Cancer Res., 51,* 3867(1991).
61. H. Hamada, et al., *Acta Endocrinol., 44,* 330(1963).
62. M. Masuoka, et al., *Acta Endocrinol., 92, suppl. 229,* 36(1979).
63. R. O. Neri, et al., *Endocrinology, 91,* 427(1972).
64. B. Katchen and S. Buxbaum, *J. Clin. Endocrinol. Metab., 41,* 373(1975).
65. R. Neri, et al., *Biochem. Soc. Trans., 7,* 565(1979).
66. J. P. Raynaud, et al., *J. Steroid Biochem. 11,* 93(1979).
67. B. J. A. Furr, *Hormone Res. 32, suppl. 1,* 69(1989).
68. D. H. Van Thiel, et al., *N. Engl. J. Med., 300,* 1012(1979).
69. N. R. Preden, et al., *Br. Med. J., 1,* 659(1979).
70. J. W. Funder and J. E. Mercer, *J Clin. Endocrinol. Metab., 48,* 189(1979).
71. S. J. Winters, et al., *Gastroenterology, 76,* 504(1979).
72. P. C. Sivelle, et al., *Biochem. Pharmacol., 31,* 677(1982).
73. R. L. Cavanagh, et al., *J. Pharmacol. Exp. Ther., 224,* 171(1983).
74. R. G. Foldesy, et al., *Proc. Soc. Exp. Biol. Med., 179,* 206(1985).
75. G. H. Jacobi, et al., *Br. J. Urol, 52,* 208(1980).
76. R. J. Irwin and G. J. Prout, *Surg. Forum, 24,* 536(1973).
77. P. C. Sogoni and W. F. Whitmore, *J. Urol, 122,* 640(1979).
78. G. R. Prout et al., *J. Urol. 113,* 834(1975).
79. J. A. Cella and C. M. Kagawa, *J. Am. Chem. Soc., 79,* 4808(1957).
80. J. A. Cella and R. C. Tweit, *J. Org. Chem., 24,* 1109(1959).
81. C. M. Kagawa, *Endocrinology, 67,* 125(1960).
82. R. Caminos-Tones, et al., *J. Clin. Endocrinol. Metab., 45,* 255(1977).
83. S. L. Steelman, et al., *Steroids, 14,* 449(1969).
84. H. P. Schane and G. O. Potts, *J. Clin. Endocrinol. Metab., 47,* 491(1978).
85. R. F. Spark and J. C. Melby, *Ann. Intern. Med., 69,* 685(1968).
86. K. Nickisch, et al., *J. Med. Chem., 28,* 546(1985).
87. S. Kamata, et al., *J. Med. Chem., 28,* 428(1985).
88. K. E. Kendle, *J. Reprod. Fertil., 43,* 505(1975).
89. L. P. Bullock, et al., *J. Reprod. Fertil., 52,* 365(1978).
90. K. E. Kendle, *J. Reprod, Fertil., 48,* 159(1976).
91. E. Sakiz and G. Azadian-Boulanger, in *Hormonal Steroids,* V. H. J. James and L. Martin, Eds., Amsterdam, Excerpta Medica, 1971, pp. 865–871.
92. E. Sakiz, et al., *Contraception, 14,* 275(1976).
93. G. Mora, et al., *Contraception, 10,* 145(1974).
94. G. Azadian-Boulanger, et al., *Am. J. Obstet. Gynecol., 125,* 1049(1976).
95. H. A. Salhanick, *Cancer Res., 42,* 3315s(1982).
96. R. C. Coombes, et al., *Lancet, ii,* 1237(1984).
97. C. Bannwart, et al., *Clin. Chim. Acta, 136,* 165(1984).
98. H. Adlercreutz, et al., *Lancet, ii,* 1295(1982).
99. K. D. R. Setchell, et al., *Am. J. Clin. Nutr., 40,* 560–579(1984).
100. S. R. Stitch, et al., *Nature, 287,* 738(1980).
101. K. D. R., Setchell, et al., *Nature, 287,* 740(1980).
102. V. C. Jordan, et al., *Environ. Health Perspec., 61,* 97(1985).
103. R. C. Stuart-Harris and I. E. Smith, *Cancer Treat. Rev., 11,* 189(1984).
104. V. I. Uzgiris, et al., *Biochemistry, 16,* 593(1977).
105. V. I. Uzgiris, et al., *Endocrinology, 101,* 89(1977).
106. A. B. Foster, et al., *J. Med. Chem., 28,* 200(1985).
107. A. B. Foster, et al., *J. Med. Chem., 26,* 50(1983).
108. R. J. Santen, et al., *J. Clin. Endocrinol. Metab., 45,* 469(1977).
109. R. M. Barone, et al., *J. Clin. Endocrinol., 49,* 672(1979).
110. A. M. K. Brodie, et al., *Endocrinology, 100,* 1684(1977).
111. W. C. Schwarzel, et al., *Endocrinology, 92,* 866(1973).
112. A. M. H. Brodie, et al., *J. Steroid Biochem., 7,* 798(1976).
113. D. F. Covey and W. F. Hood, *Endocrinology, 108,* 1597(1981).
114. J. Fishman and J. Goto, *J. Biol. Chem., 256,* 4466(1981).
115. E. A. Thompson and P. K. Siiteri, *J. Biol. Chem., 249,* 5373(1974).
116. J. O. Johnson, et al., *Endocrinology, 115,* 776(1984).
117. K. Schieweck, et al., *Cancer Res., 48,* 834(1988).
118. R. C. Stein, et al., *Cancer Res., 50,* 1381(1990).
119. P. E. Lonning, et al., *Br. J. Cancer, 63,* 789(1991).
120. L. M. DeMers, et al., *J. Clin. Endocrinol. Metab., 70,* 1162(1992).
121. A. V. Schally, et al., *Biochem. Biophys. Res. Commun., 43,* 393(1971).
122. A. V. Schally, et al., *J. Biol. Chem. 246,* 7230(1971).
123. H. Matsuo, et al., *Biochem. Biophys. Res. Commun., 43,* 1334(1971).
124. H. Matsuo, et al., *Biochem. Biophys. Res. Commun., 45,* 822(1971).
125. D. H. Coy, et al., *J. Med. Chem., 119,* 423(1976).
126. A. S. Dutta, et al., *J. Med. Chem., 21,* 1018(1978).
127. P. V. Maynard and R. I. Nicholson, *Br. J. Cancer, 39,* 274(1979).
128. R. W. Rees, et al., *J. Med. Chem., 17,* 1016(1974).
129. J. P. Yardley, et al., *J. Med. Chem., 18,* 1244(1975).
130. J. Humphries, et al., *J. Med. Chem., 21,* 120(1978).
131. J. Humphries, et al., *Biochem. Biophys. Res. Commun., 85,* 709(1978).
132. C. Y. Bowers, et al., *Endocrinology, 106,* 674(1980).
133. K. Channabasavaiah and J. Stewart, *Biochem. Biophys. Res. Commun., 86,* 1266(1979).
134. J. Rivier and W. Vale, *Life Sci., 23,* 869(1978).
135. T. W. Redding and A. V. Schally, *Proc. Natl. Acad. Sci. USA, 78,* 6509(1981).
136. A. V. Schally, et al., *Prostate, 4,* 545(1983).
137. R. I. Nicholson and P. V. Maynard, *Br. J. Cancer, 39,* 268(1979).
138. The Leuprolide Study Group, *N. Engl. J. Med., 311,* 1281(1984).
139. K. J. Waler, et al., *Lancet, ii,* 413(1983).
140. S. R. Ahmed, et al., *Lancet, ii,* 415(1983).
141. K. J. Walker, et al., *J. Endocrinol., 103,* R1–R4(1984).
142. F. Labrie, et al., *Prostate, 4,* 579(1983).
143. G. E. Paget, et al., *Nature, 192,* 119(1961).
144. M. I. Tollock, et al., *Nature, 199,* 288(1963).
145. K. H. Schmidt-Ruppin, et al., *Experientia, 29,* 823(1973).
146. V. C. Jordan, et al., *Eur. J. Cancer, 15,* 755(1979).

SUGGESTED READINGS

M. K. Agarwal, Ed., *Antihormones,* Elsevier, North Holland, Biomedical Press, 1979.

M. K. Agarwal, Ed., *Hormone Antagonists,* Berlin, Walter de-Gruyter, 1982.

B. J. A. Furr, Ed., *Clinics in Oncology,* Vol. 1, *Hormone Therapy,* Philadelphia, W. B. Saunders, 1982.

V. C. Jordan, Ed., *Estrogen/Antiestrogen Action and Breast Cancer Therapy,* Madison, WI, University of Wisconsin Press, 1986.

R. L. Sutherland and V. C. Jordan, Eds., *Non-Steroidal Antioestrogens,* Sydney, Academic, Press, 1981.

Chapter 41

PHOTOCHEMOTHERAPY

Francesco Dall'Acqua and Giulio Jori

Many diseases can be treated with photomedicine, the two important branches of which are phototherapy and photochemotherapy.[1,2] The terms phototherapy and photochemotherapy designate treatment methods in which light of an appropriate wavelength is used to induce a therapeutic response in the absence or in the presence of a photosensitizing drug, respectively. To produce its effect in either case, light must first be absorbed by a natural chromophore (phototherapy), or by a photosensitizing agent (administered topically or systemically) (photochemotherapy). After absorption of photons, the chromophore is excited and undergoes chemical reactions that induce biologic responses. At the doses used for photochemotherapy, radiation alone usually has little or no effect. Similarly, the drug alone is practically ineffective. The combined action of the photosensitizing drug and light provokes the therapeutic effect.[1]

Photosensitizing drugs used in photochemotherapy can act through various mechanisms, as illustrated in Scheme 1. The drug in the ground state (D_o), on absorption of light quanta in the UV-visible region, is promoted to an electronically excited short-lived (about 1 to 10 ns) state (singlet excited state, ^1D). The singlet excited state may decay to the ground state through nonradiative (internal conversion) or radiative (emission of light as fluorescence)[3] pathways. Alternatively, the singlet excited state can be converted to the triplet state through the so-called intersystem crossing ($^1D \xrightarrow{ISC} {^3D}$). The triplet state has a lifetime significantly longer than that of the singlet excited state (in the millisecond range for fluid systems) and has a greater intrinsic reactivity. The triplet state can also decay to the ground state through radiative (phosphorescence) and nonradiative pathways; however, the triplet excited state can originate different photochemical events by reacting with substrates, in particular, involving oxygen-dependent (type I or II) or oxygen-independent mechanisms (type III). Type I and type II are defined also as photodynamic pathways. Type I implies substrate photooxidation by radical species. Both activated oxygen species (superoxide anion and hydroxyl radical) and radical species form by electron transfer photoexcited furocoumarins. Type II involves generation of singlet oxygen by energy transfer. In the type III mechanism, the triplet photosensitizer may react directly with a substrate in a oxygen-independent process.[4] In the case of psoralens, structures carrying olefin double bonds can easily give C_4-cycloaddition with one of the two (or both) photoreactive sites of furocoumarins. Cycloadditions inducing biologically important consequences are those occurring between psoralens and pyrimidine bases of nucleic acids, membrane unsaturated fatty acids, or lecithins.[4-6] On the other hand, many dyes, such as methylene blue, eosin Y, and porphyrins, can transfer the excitation energy to a molecule of oxygen (which exists as a triplet in the ground state), generating singlet oxygen.[7] This is a reactive species that, in turn, can easily oxidize biologic substrates with toxic effects.[8]

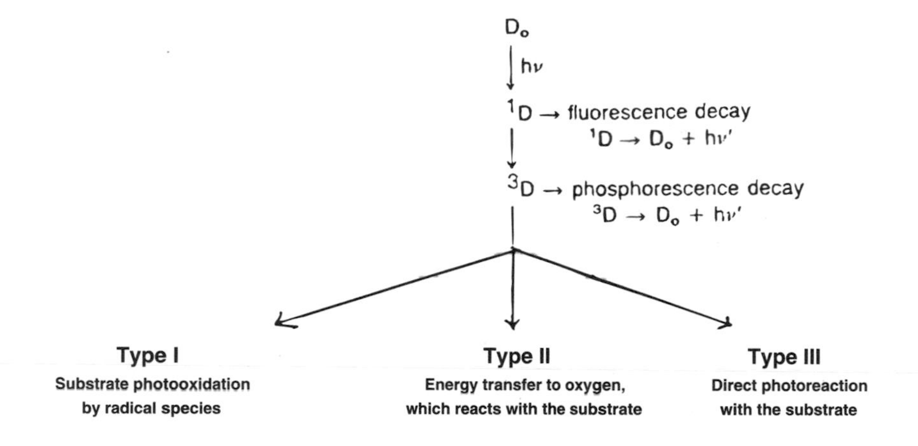

Scheme 1

Photochemotherapy is generally used to treat hyperproliferative skin diseases (psoriasis, mycosis fungoides) or cavitary tumors. In skin diseases, the drugs used are psoralens and the radiation used is in the interval of UV-A (320 to 400 nm). The therapy is generally called PUVA, derived from psoralen and UV-A. In cavitary tumors, the drugs used are hematoporphyrin and related compounds, and the radiation used is the red component (600 to 630 nm) of the visible spectrum: the treatment is defined as photodynamic therapy (PDT).[8] In PUVA treatment, lamps are generally used to irradiate large areas of the skin or the peripheral blood in an extracorporeal flow (photopheresis);[9] in PDT, laser sources coupled with optical fibers are also used to reach tumors located in the anatomic cavities. These two types of photochemotherapy are widely used. Other limited applications of photochemotherapy are in treatment of certain viral, bacterial, and fungal infections.[10,11] PUVA is used also in the preparation of antiviral vaccine[12] and in the photochemical decontamination system for the treatment of human blood products.[13]

Concerning the action of PUVA and PDT at the cellular level, psoralens realize their antiproliferaive activity through a type III mechanism (e.g., photoreactions with a macromolecule), whereas porphyrins exert a cytotoxic effect by damaging the cell membrane mainly through a type II (singlet oxygen-involving) mechanism.

PHOTOTHERAPY

Phototherapy exploits for therapeutic purposes the effect of nonionizing radiation (UV and visible light) on living tissue in various ways. The energy of photons in the UV and visible wavelength region can cause electronic excitation of specific chromophoric molecules, leading to specific chemical reactions. Thus the use of UV and visible radiation offers the possibility of photoactivating selected target molecules, thereby causing specific reactions that, in turn, bring about therapeutic effects.

Phototherapy of psoriasis[2] uses UV-B light, exploiting the antiproliferative effect of this radiation. Unfortunately, undesired side effects such as skin phototoxicity and risk of skin cancer have limited this therapeutic approach.

Phototherapy of neonatal jaundice[14] uses the ability of blue light (around 450 nm) to induce photoexcitation of the linear tetrapyrrolic pigment bilirubin. The accumulation of excess bilirubin (over 8 mg/dL of serum) in premature newborn babies is the origin of this disease, which can cause serious neurologic damage to infants. Blue light-excited bilirubin undergoes two types of photoprocesses (Scheme 2), both of which decrease its concentration in the body; phototherapy is usually performed until the bilirubin levels are reduced to normal values. The first photoprocess involves the conversion of bilirubin to more polar isomers (Fig. 41–1), which are readily excreted in the bile.

The second photoprocess (usually of minor importance) involves the formation of singlet oxygen by energy transfer from triplet bilirubin; singlet oxygen in turn attacks bilirubin, which is transformed into various water-soluble oxidation products; the latter are excreted in the urine. Although this kind of phototherapy induces some undesired side effects, it is now universally used for neonatal jaundice because these side effects are transient and alternative treatments, such as ex-sanguino transfusion, present greater risks.

PUVA Therapy

The term PUVA is derived from the therapeutic application of psoralen together with UV-A light. PUVA consists of oral administration or topical application of psoralen and subsequent long ultraviolet radiation (UV-A).

Photosensitization, and consequently antiproliferative effects, occur when psoralen reaches the skin, either by local absorption following topical application, or through the circulation after ingestion and absorption through the gastrointestinal tract. In systemic treatment, because serum level and skin photosensitivity peaks appear about 2 hours after oral administration (Fig. 41–2), UV-A exposures are given at this time.[1] Patients, however, should stay out of sunlight for the remainder of the day.

PUVA has been used mainly for the treatment of skin diseases characterized by hyperproliferative conditions, such as psoriasis and mycosis fungoides.[1] Other diseases have been treated with PUVA, among them vitiligo, atopic eczema, lichen planus, urticaria pigmentosa, polymorphous light eruption, and alopecia areata.[1]

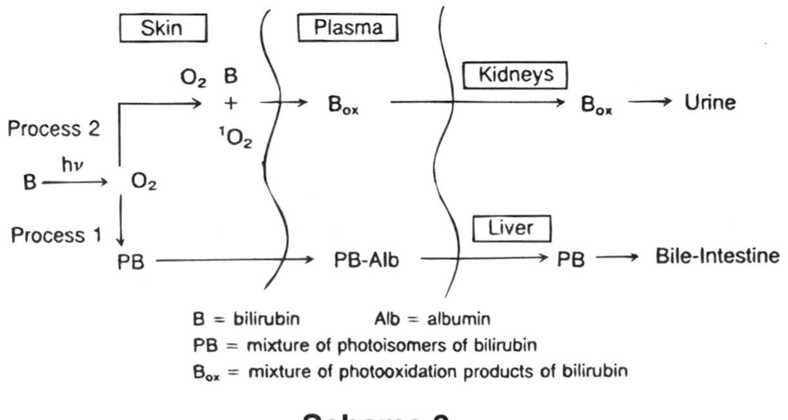

B = bilirubin Alb = albumin
PB = mixture of photoisomers of bilirubin
B_{ox} = mixture of photooxidation products of bilirubin

Scheme 2

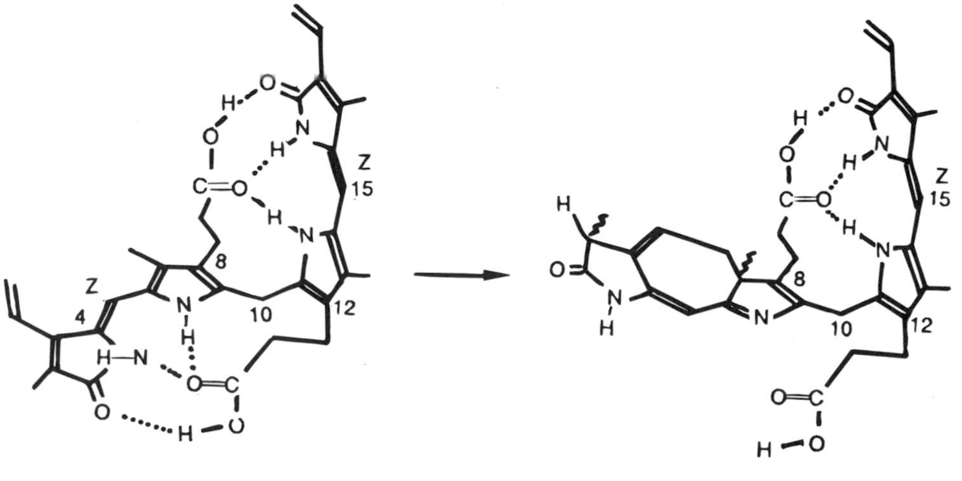

Fig. 41–1. Photochemical conversion of bilirubin (A) to one of its cyclic isomers (PB) (B), which is more rapidly eliminated in the bile.

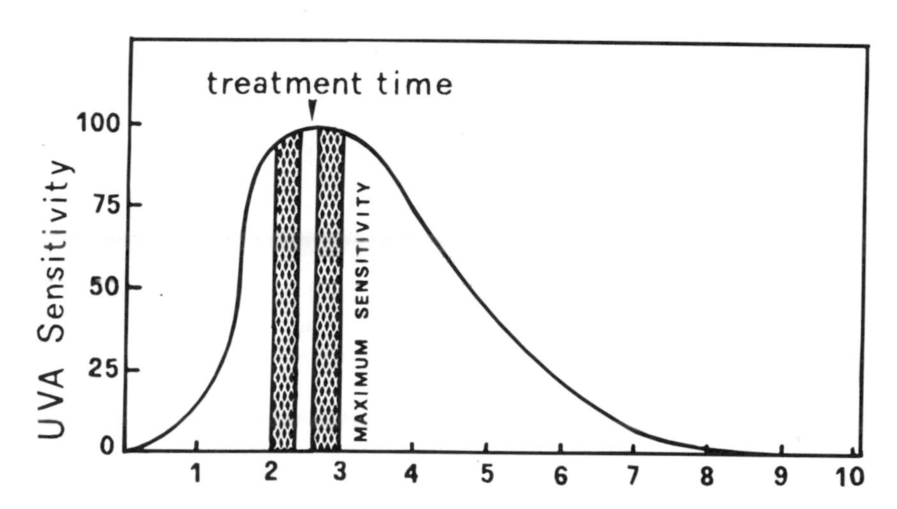

Fig. 41–2. After oral administration of 8-methoxypsoralen (8-MOP), patients become gradually reactive to UV-A and therefore to photochemotherapeutic treatment. The patients are maximally reactive 2 to 3 hours after ingestion of the drug, and during this period the irradiation is carried out.

A new application of PUVA therapy is represented by photophoresis. This approach involves the exposition of peripheral blood to photoactivated 8-MOP and is effective in the treatment of cutaneous T-cell lymphoma and other autoimmune disorders.[9]

Historic Aspects

The pharmacologic concept of photochemotherapy, i.e., the combined use of electromagnetic energy and a drug, although first reported in 1974,[15] is in fact thousands of years old, having been used in India and Egypt since 1200 and 2000 B.C., respectively. Photochemotherapy for the disfiguring disease vitiligo was practised in the ancient world by herbalists, who used boiled extracts of the seeds and plant Psoralea corylifolia in India and of the fruits of Ammi majus in Egypt; these plants contain psoralen, 8-methoxypsoralen (8-MOP) and 5-methoxypsoralen (5-MOP) (see Figs. 41–3 and 41–4).[16] These preparations were either applied to the skin or ingested as an infusion. Patients then exposed their skin to

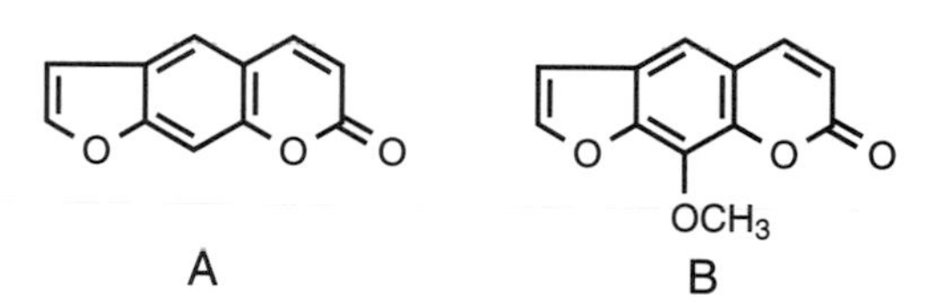

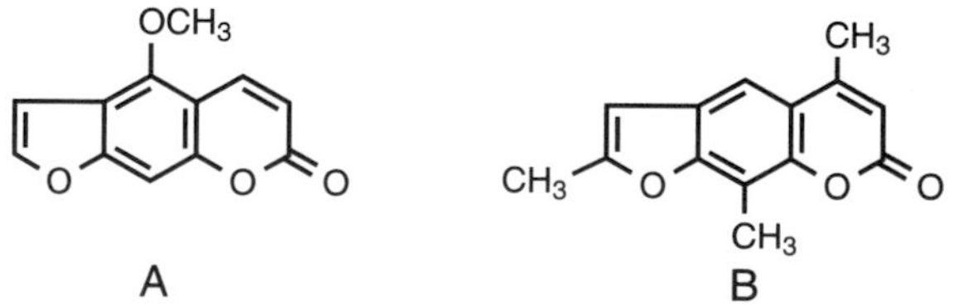

Fig. 41–3. Molecular structures of psoralen (A) and 8-methoxypsoralen (8-MOP) (B).

Fig. 41–4. Molecular structures of 5-methoxypsoralen (5-MOP) (A) and of 4,5′, 8-trimethylpsoralen (TMP) (B).

the intense Indian or Egyptian sunlight. In 1948, El Mofty, an Egyptian physician, practised this type of therapy, using the active component of Ammi majus (8-MOP) and 5-MOP, isolated in a chemically pure grade from the plant.[17] He was the first to use photochemotherapy for the treatment of vitiligo.

The rational discovery of photochemotherapy, however, was proposed by Parrish et al., Harvard University dermatologists, in 1974.[15] They first took into account that the antiproliferative effect of psoralen was attributable to its ability to photoreact with DNA, blocking its biologic functions and cell division. They considered that the light necessary to photoactivate the psoralen (UV-A) could interact only with skin layers (the stratum corneum, the epidermis, and partly the dermis) and decided therefore to treat skin diseases characterized by hyperproliferative conditions, such as psoriasis, with systemic psoralen and UV-A. They proposed the term photochemotherapy for this treatment.

Psoralens

Psoralens are heterocyclic aromatic compounds derived from the linear condensation of a coumarin nucleus (benzopyrone) with a furan ring. Although this fusion can occur in several ways (e.g., six linear and angular forms), only two forms, psoralens and angelicins and several derivatives of allopsoralen, have been found in the vegetable world.[18]

Natural furocoumarins have been isolated from four major plant families, the Umbelliferae (e.g., parsley, parsnip, celery, Ammi majus, Angelica archangelica); Rutaceae (e.g., bergamot fruit, lime, gas plant, cloves, common rue); Leguminosae (Psoralea corylifolia, Xanthoxylum flavum); and Moraceae (e.g., Ficus carica).[16]

Only linear psoralens such as psoralen (7H-furo-[3,2-g]1-benzopyran-2-one), 8-methoxypsoralen (8-MOP), 5-methoxypsoralen (5-MOP) (Fig. 41–4*A*), and the synthetic 4,5′,8-trimethylpsoralen (TMP) (Fig. 41–4*B*) are used in photochemotherapy.[1] These compounds are stable in the dark; however, under UV-A irradiation, they undergo modification yielding various photocompounds. They can form photodimers,[5] undergo fission of the furan ring by a singlet oxygen mechanism (discussed later),[19] and open the pyrone ring (Fig. 41–5). According to Potapenko,[20] another oxygen-dependent process involves the formation of photooxidized furocoumarin derivatives, which can react in the dark with several substrates (in particular membrane components), causing irreversible damage to the cell. The latter process is temperature dependent and may provoke important biologic consequences, such as hemolysis of red blood cells.

Psoralens have various and interesting photobiologic activities (Table 41–1); one of their widely known photosensitizing effects is their capacity to sensitize human or guinea pig skin.[18]

Application of even small amounts of the compound (0.5 to 5/μg/cm^2) and irradiation with solar or ultraviolet light is enough to cause erythema 10 to 12 hours after irradiation, followed in 2 to 3 days by dark pigmentation. This skin photosensitizing activity can be ascribed to phytophotodermatites, more frequent during the summer because of skin contact with plants containing psoralens and frequent exposure to sunlight.[21]

The antiproliferative activity is a beneficial effect, useful in skin diseases characterized by hyperproliferative conditions. Its mutagenic and photocarcinogenic properties are undesired side effects. Finally, it can affect the immune system. Effects on the Langerhans cells as well as on circulating T-lymphocytes and on other substrates have been observed.[22]

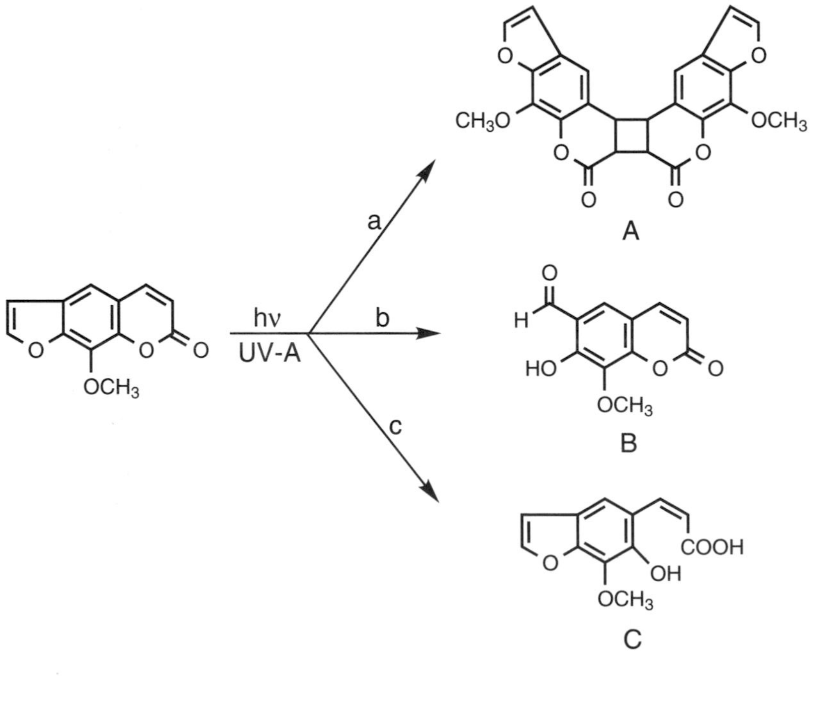

Fig. 41–5. Photodegradation of 8-MOP. A, Formation of 3,4-3,4-cyclodimer; B, formation of 7-hydroxy-8-methoxy-6-formalincoumarin by singlet oxygen; C, opening of the pyrone ring.

Table 41–1. Photobiologic Activities of Psoralens

Phototoxic effect on skin	Erythema on human or guinea-pig skin
	Induction of dark pigmentation
Antiproliferative effect	Inhibition of the tumor-transmitting capacity of Ehrlich ascites tumor cells
	Inhibition of growth or lethal effect (at higher doses) on bacterial or mammalian cells adapted in in vitro growth
	Inhibition of DNA, RNA and protein synthesis in E. coli and Ehrlich ascites tumor cells
	Inhibition of synthesis of epidermal DNA of mouse skin
	Inhibition of infectivity of DNA viruses
Mutagenic effect	Mutants in Sarcina lutea, Salmonella typhimurium, Escherichia coli, Chinese hamster cells grown in vitro, Sacchromyces cerevisiae, others.
Photocancerogenic activity	Photocancerogenic activity on mouse skin
	Inhibition of EGF cell binding
Other effects	Alteration of immune system

PUVA for Psoriasis

Psoriasis is a cutaneous disease of unknown cause, characterized by increased epidermal proliferation manifested clinically as red, raised, scaly plaques. It is a chronic disorder that may involve large portions of the skin and nails.[1] PUVA is effective in healing psoriasis and is used in the more serious and extended form of the disease.[1,15] The drugs used most often in PUVA are 8-MOP and 5-MOP; 4,5′,8-trimethylpsoralen, and psoralen have also found application. A mixture of 8-MOP and 5-MOP is used in Russia.

The drug is administered orally (0.6 mg/kg). Two hours after ingestion, the patient is irradiated in a suitable apparatus containing lamps emitting UV-A light. The UV-A dose is modulated according to the skin type of the patient. It is lower for patients who, when exposed to sunlight, burn and do not tan, and is increased for patients whose skin is less sensitive to UV light or sunlight.[1] In Finland, TMP is dissolved in a water bath, and after the bath, the patient is submitted to UV-A radiation.[23]

Another important therapeutic application of PUVA is the treatment of mycosis fungoides, an uncommon cutaneous T-cell lymphoma.[24] PUVA appears to be effective, especially in the early stage of the disease.

PUVA for Vitiligo

Vitiligo is an idiopathic hypomelanosis, in which a variety of factors including genetic, endocrine, metabolic, and autoimmune disorders may play a role. This disease is severely disfiguring especially in brown-skinned people, and the disfigurement may be ruinous psychologically and socially.[25] Several reports document the successful use of PUVA using topically applied psoralens.[25] PUVA is associated with some disadvantages, because topical psoralens are potent skin photosensitizers. Up to 70% of patients with vitiligo, however, improved when treated at least twice weekly for more than 12 months with oral TMP (0.6 to 0.8 mg/kg) 2 hours before sun exposure. Also, oral 8-MOP alone or in combination with TMP and psoralen itself is effective in most patients with vitiligo. PUVA increases melanin pigmentation. This action depends on the presence of functional melanocytes. It may involve both activation of melanocyte proliferation and transfer of melanocytes from hair follicles to the depigmented epidermis.[26] Results of studies in animals suggest that PUVA can also suppress degranulation of mast cells, which may explain the reported benefit of such treatment in cutaneous mast cell disease.[27]

KUVA for Vitiligo

A photochemotherapeutic treatment that uses khellin instead of psoralens is KUVA. Khellin, a naturally occurring furochromone, isolated from the seeds of Ammi visnaga, has been used as a coronary vasodilator and as a spasmolytic.[28] Khellin, the chemical structure of which closely resembles that of psoralens, is reported to be an efficient drug for the photochemotherapeutic treatment of vitiligo.[29] The treatment protocol consists of the oral administration of a dose of 100 mg of the drug to the patient who is irradiated with UV-A light 2½ hours after dosing. Irradiaton of UV-A ranges, in general, between 5 and 15 J/m^2 as delivered by a standard PUVA unit (e.g., Waldenmann PUVA, FRG).[30] In general, no skin phototoxicity test is performed, because previous tests either with oral or topical khellin have not revealed skin phototoxicity up to 100 J/cm^2. Treatments are given three times a week.[30]

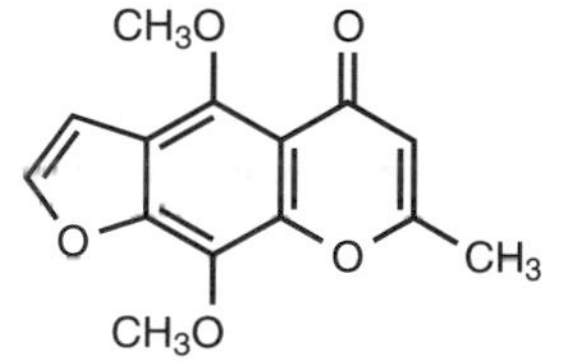

CH_3O O O O CH_3 CH_3O

Khellin

KUVA therapy is at least as effective as PUVA. After 12 months of continuous therapy, up to 70% of the skin surface originally involved with vitiligo may exhibit complete repigmentation. Usually khellin is well tolerated, although 30% of the patients report nausea or dizziness in the first 2 weeks of ingestion. Unrelated to this effect, transient and fully reversible elevation of liver transaminases may occur in about 30% of the patients. The mechanism of action of photochemotherapeutic effect on vitiligo is unknown.[30] Studies carried out with the aim of exploring the ability of KUVA to photoinduce lesions in DNA have shown a poor DNA-photobinding capacity, suggesting that a mechanism similar to that of psoralens is not involved. Also, the ability to generate activated oxygen species is low, indicating that a photodynamic mechanism is not involved.[28]

Photopheresis

Photopheresis is a process by which peripheral blood is exposed in an extracorporeal flow system to photoactivated 5-MOP and represents a new treatment for disorders caused by aberrant T lymphocytes. It is a standard therapy for advanced cutaneous T-cell lymphoma and shows promise in the treatment of two autoimmune disorders, pemphigus vulgaris and progressive systemic sclerosis (scleroderma).[9] Additional diseases for which clinical trials are in progress include multiple sclerosis, organ transplant rejection, rheumatoid arthritis, and AIDS. The mechanism of action appears to involve "vaccination" against the pathogenic T cells, in a clone-specific manner. Photoactivated 8-MOP initiates a cascade of immunologic events by forming covalent photoadducts with nuclear cell DNA and possibly with the cellular molecules.[9]

Mechanism of Action

Psoralens can exert their photosensitizing effect by either of the three mechanisms shown in scheme 1, even though the main biologic and therapeutic effects are generally ascribed to the type III mechanism, and in particular to photoreactions with DNA.[5,18] It has been shown that psoralens interact at the molecular level with DNA.[5,18] This interaction occurs in two successive steps, formation of a preliminary molecular complex in the ground state and subsequent covalent photoaddition of the psoralen to the macromolecule.[5]

Psoralens have a measurable affinity toward DNA in the dark also; in fact, they form a molecular complex, with the psoralens undergoing intercalation between two base pairs of the macromolecule (Fig. 41–6).[5,31,32] Considering the furocoumarins forming the molecular complex with DNA insert themselves inside the internal lipophilic part of the macromolecule, an influence on the lipophilicity of these ligands is to be expected. A good correlation has been observed between the partition coefficient in n-octanol-water and the association constants of the complexes between DNA and a series of methylated furocoumarins (methylangelicins).[33] The hydrophobicity of furocoumarins therefore affects the formation of the intercalated complex.[31] This complex is not strongly bound and can easily undergo dissociation.[5,18] The formation of such a complex does not have marked biologic consequences. The furocoumarin in the complex assumes a steric arrangement that favors successive photoaddition with a pyrimidine base (e.g., thymine) of DNA (Fig. 41–7).[31]

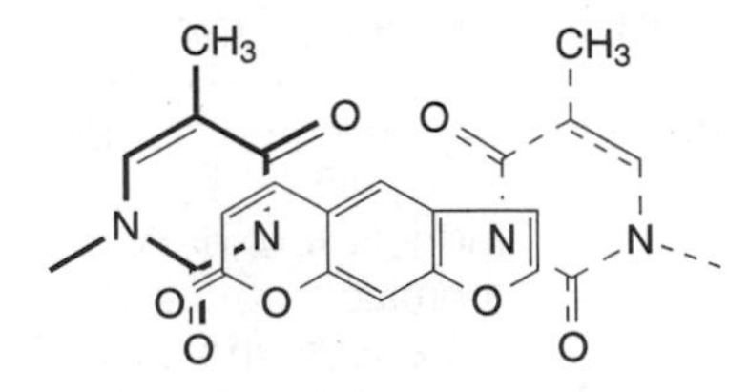

Fig. 41–7. Projection of psoralen between two base pairs of DNA (for simplicity, only the two thymine bases pertaining to the opposite strands are shown; the two complementary adenine bases are omitted).

Photocycloaddition

Psoralen has two photoreactive sites, the 3,4- and 4′,5′-double bonds. These double bonds can engage in reactions of C_4 cycloaddition with various substrates.[5,18,32]

Psoralen, when intercalated between two base pairs of DNA (see Fig. 41–7), assumes a position in which one of the two photoreactive double bonds is in the correct proximity to a 5,6 double bond of a pyrimidine base of DNA; when psoralen is irradiated with long UV light in such a position, it can give rise to different reactions of C_4-cycloaddition (Fig. 41–8).[5,31] It can engage its 3,4 double bond with the 5,6 double bond of the thymine, forming a 3,4 (1:1) C_4-monocycloadduct with thymine base.[34] Alternatively, it can engage its 4′,5′ double bond with the thymine base of the adjacent base pairs and form a 4′,5′-monocycloadduct (see Fig. 41–8).[5,18,32]

These adducts have been isolated from the products of hydrolysis of DNA irradiated in the presence of psoralens and characterized.[35] Nuclear magnetic resonance

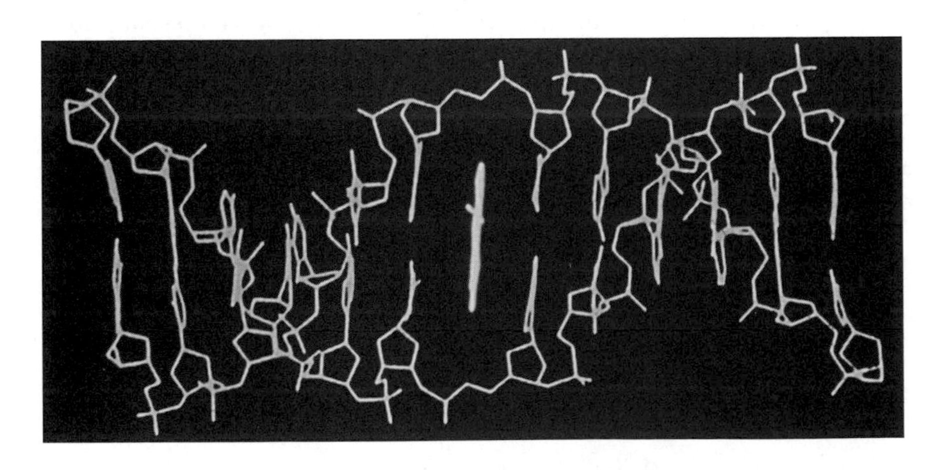

Fig. 41–6. Intercalation model of 8-MOP inside two base pairs of duplex DNA. The model was obtained by molecular modeling.

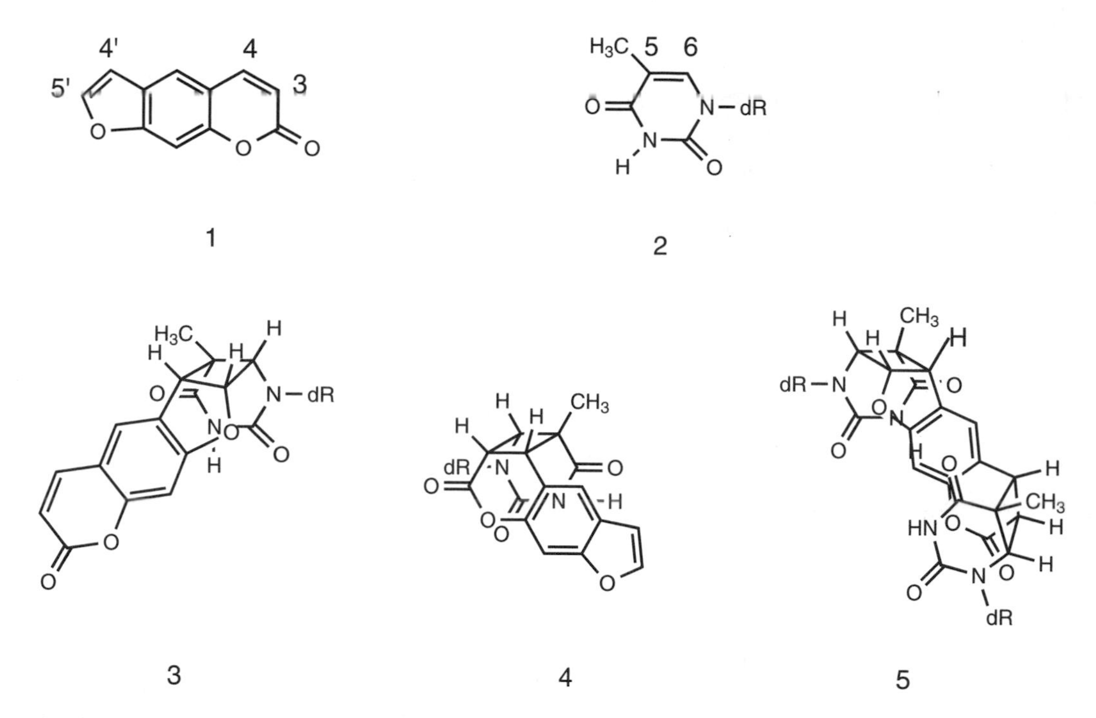

Fig. 41–8. Cycloadducts between psoralen (1) and thymidine (2): 4′,5′-monocycloadduct with thymidine (3), 3,4-monoadduct with thymidine (4), and diadducts with thymidine (5). All the compounds have a *cis*-syn configuration.

studies have also established the regiochemistry and stereochemistry of the 3,4 and 4′,5′ monocycloadducts formed in the photoreaction between 8-MOP, TMP, and DNA.[35] These photocompounds have a *cis*-syn structure. The 3,4 cycloadduct does not absorb above 320 nm, and because photoreaction and photosensitization experiments are carried out using UV-A light (320 to 400 nm, with preference for 365 nm), this adduct cannot absorb UV-A light and therefore cannot photoreact further.[18,36]

The 4′,5′-cycloadduct absorbs at 365 nm, can be excited, and therefore can engage the second photoreactive site in a second cycloaddition, linking a thymine base of the complementary strand of DNA to the first thymine base. In this way, an interstrand crosslinkage is formed, which covalently conjugates the two chains of the DNA molecule.[36] The capacity of the various furocoumarins to form interstrand cross-linkages is different. A correlation has been found between the rate constant of the formation of cross-linkages and the ability to induce skin erythema in a large series of psoralens.[33]

From a biologic point of view, however, cross-linkages provoke more pronounced biologic consequences. The repair of interstrand cross-linkages is less effective than the repair of the monofunctional adducts. This mechanism of action seems to be mainly responsible for the antiproliferative, mutagenic, and photocancerogenic effects.[37]

Another important C_4-photocycloaddition involves membrane constituent targets, such as unsaturated fatty acids[4] or lecithins.[6] By irradiation in vitro of a solution of a furocoumarin and an unsaturated fatty acid (UFA), generally methyl esters (e.g., oleic acid, linoleic acid, linolenic acid), a C_4-cycloaddition between the 3,4 double bond of the furocoumarin and one of the double bonds of UFA, takes place. The cycloadduct formed in the photoreaction has been isolated and characterized.[38] For example, in the photoreaction between TMP and oleic acid methyl ester (OAME), four isomeric products were isolated, instead of the eight possible. The two major products formed show a *cis* configuration, and the remaining two have a *trans* configuration (Fig. 41–9).[38] These adducts are formed also in vivo in the skin of rats treated with furocoumarins and UVA.[39] Analogous photocompounds are formed in the photoreaction between lecithins and 8-MOP.[6] According to Midden,[40] a furocoumarin-fatty acid adduct may inhibit a phospholipase and therefore prevent the activation of protein kinase C or other regulatory enzymes. Inhibition of protein kinase C may explain the antiproliferative effects of PUVA therapy.

Activated Form of Oxygen

As mentioned previously, furocoumarins may involve both oxygen-dependent (type I or II) and oxygen-independent (type III) mechanisms. The photodynamic mechanism involves both radical species formed by electron transfer from photoexcited furocoumarin (type I) with the formation of superoxide anion and hydroxyl radicals, or by energy transfer from the excited triplet furocoumarin to the triplet oxygen, thus generation singlet oxygen.[4,41]

It has been shown that 8-MOP, irradiated at 365 nm in solution, undergoes photodegradation. This photodegradation (Fig. 41–10) can be mediated by singlet oxy-

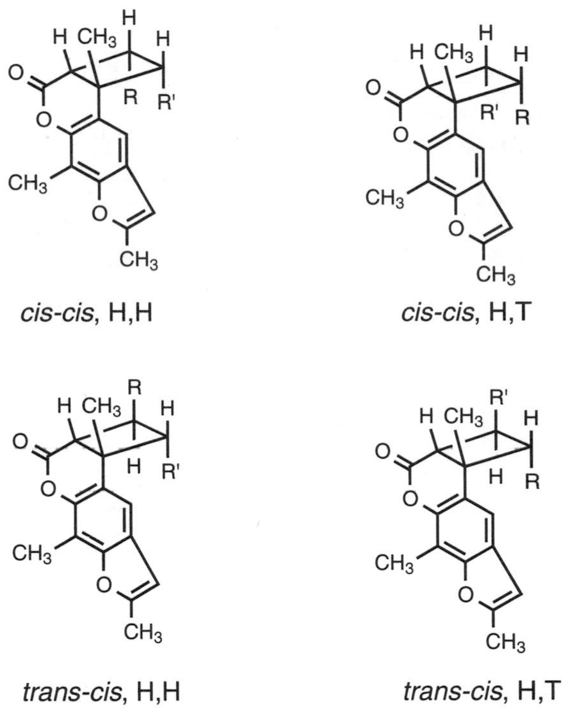

cis-cis, H,H

cis-cis, H,T

trans-cis, H,H

trans-cis, H,T

Fig. 41-9. Molecular structures of cycloadducts between TMP and OAME. H,H = head to head; H,T = head to tail; R = $(CH_2)_6COOCH_3$; R′, $(CH_2)_6CH_3$.

gen and lead to the formation of 6-formyl-7-hydroxy-8-methoxycoumarin.[19] This aldehyde, in the presence of proteins, can react with amino groups forming a Schiff base, linking a photomodified psoralen to a protein. Psoralens can also photoinactivate enzymes and ribosomes through a singlet oxygen mechanism.[5,18] A peroxidation of membrane lipids due to the singlet oxygen generation by the furocoumarins is thus possible. This mechanism may be responsible for the skin phototoxicity of psoralens.[21]

Another oxygen-dependent process involves the formation of photooxidized furocoumarin derivatives, which can react with several substrates (in particular membrane components) causing irreversible damage to the cell.[20] According to Potapenko,[20] these products may explain both toxic and therapeutic activity of furocoumarins.

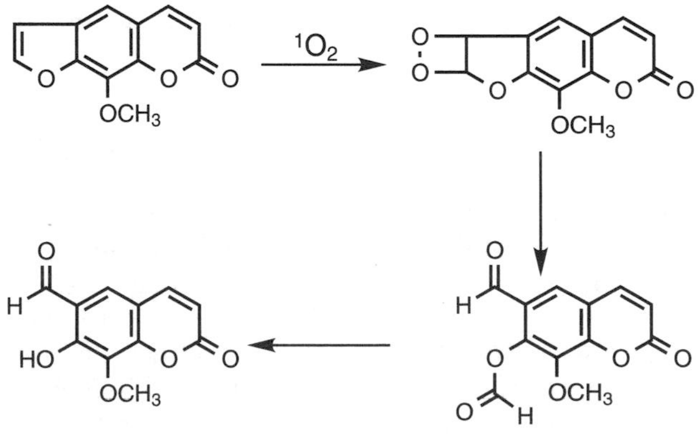

Fig. 41–10. Photooxidation pathway of 8-MOP through singlet oxygen.

Furocoumarin Receptors

Laskin et al.[42] identified specific, saturable, high-affinity binding sites for 8-MOP on HeLa cells. These authors detected specific binding of 8-MOP to four other cell lines and five mouse cell lines. In HeLa cells, binding is reversible and independent of the ability of the furocoumarin to form a molecular complex with DNA. Scatchard analysis indicates the presence of two classes of furocoumarin binding sites: high affinity and low affinity sites. The high-affinity binding sites become covalently modified by the psoralen molecule following UV-A irradiation. In particular, the psoralen receptor, with H^3-8-MOP, was visualized in the cytoplasmic and plasma membrane. The receptor has an apparent molecular weight of about 22,000 and was shown to be sensitive to protease but not to nuclease treatment. Covalent binding of the tritiated 8-MOP to the receptor protein is inhibited by excess of unlabeled 8-MOP, indicating the covalent furocoumarin-receptor binding is saturable.[43] Laskin and Laskin[44] proposed a model in which furocoumarin receptors are localized in the cell membrane close to the receptor of epidermal grow factor (EGF). Exposure to UV-A induces psoralen receptor activation, which initiates intracellular signals leading to biologic responses. These signals involve interaction with normal grow factor receptor toward grow factors. This activation modulates protein kinase activity, thus also modulating the cell's response to a growth factor stimulus.

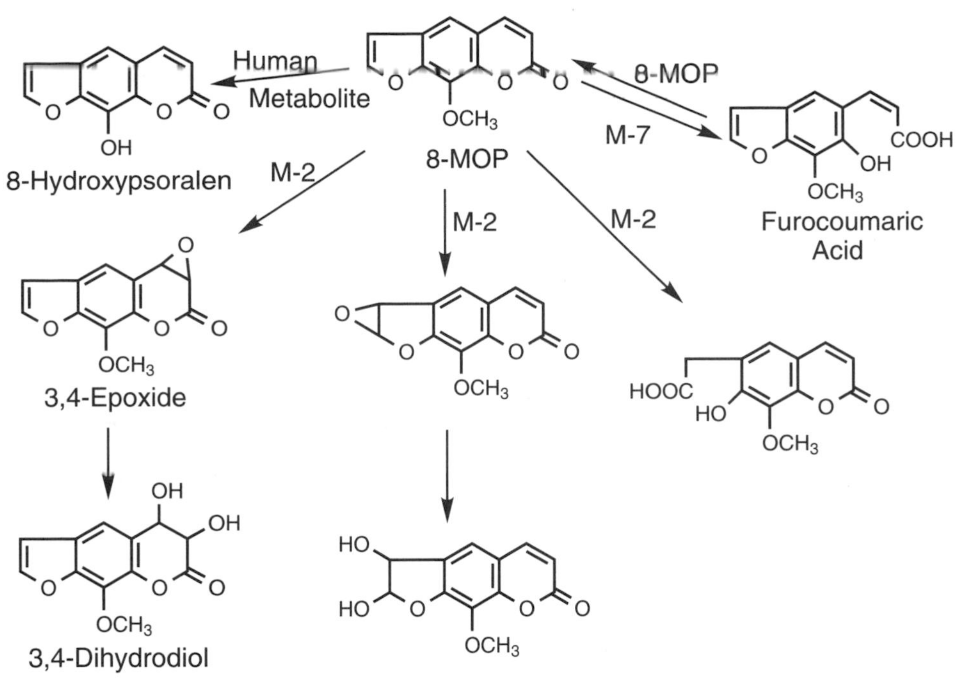

Fig. 41–11. Metabolism of 8-MOP in mice after oral administration (2 mg).

Metabolism of 8-MOP and TMP

The psoralens are absorbed rapidly after oral administration; photosensitivity is maximal 1 to 2 hours after ingestion of 8-MOP and 2 hours after TMP. The elimination half-lives of the drugs are about 2 hours, but the skin remains sensitive to light for 8 to 12 hours.[45] 8-MOP and TMP are extensively metabolized and can induce the hepatic cytochrome P450 monooxygenase system.[46] When administered topically, psoralens penetrate the epidermis, and substantial absorption can occur. If 1% methoxsalen ointment is applied to one half the body surface, plasma concentrations of drug are similar to those achieved after an oral dose of 0.5 mg/kg.[47]

8-MOP administered orally to mice is excreted in the urine as seven metabolites (four major, three minor). Chemical structures assigned to the four major metabolites are shown in Figure 41–11. 8-Hydroxypsoralen is formed in human metabolism of 8-MOP, but not in that of mice. Human volunteers receiving 0.6 to 1.2 mg/kg orally of 8-MOP revealed urinary excretion of four fluorescent metabolites, the most prominent of which were 8-hydroxypsoralen, 8-MOP, furocoumaric acid, and a glucuronate of 8-MOP. The metabolism of TMP involves oxidation of the 5′-CH_3 with formation of the intermediate 5′-OH-TMP and finally 4,8-dimethyl-5′-carboxypsoralen (major metabolite).[48]

Short- and Long-Term Side Effects of PUVA

Short-term reactions to PUVA therapy include nausea, which may be decreased by administration of the drug with milk, and pruritus.[45] Painful erythema and blistering can occur 36 to 48 hours after treatment, although this side effect is unlikely if the dose of UV-A is monitored carefully and the patient avoids exposure to sunlight or other sources of UV-A.[47]

Of greater concern are the long-term side effects such as the risk of skin cancer, cataract, and immune dysfunctions. It has been shown that psoralen, 8-MOP, and 5-MOP plus UV-A light can induce skin cancer in mice.[37] Some studies with 8-MOP in humans indicate a significant risk of developing basal cell and squamous cell carcinoma (but no melanoma). The risk is higher in patients whose skin type is sensitive to sunlight (easy to burn and difficult to tan),[49] and in those who previously have received ionizing radiations, arsenic coal tar or previous skin cancer. European studies, however, indicate a lower risk of skin cancer than that suggested by U.S. studies.[50]

Although psoralens diffuse out of the eye after 12 to 24 hours, concern is expressed that the photoadducts may accumulate in the lens;[51] animals given large doses of psoralens develop cataracts.[45] In one large study, however, no increase was reported in the incidence of symptomatic cataracts in patients treated with PUVA over a 5-year period.[52]

The effects of PUVA on the immune system are not yet definitely established. They may include both direct toxicity and long-term somatic alterations. The dermatologist should evaluate the risk-benefit ratio carefully for proper selection of patients to be treated with PUVA.

The undesirable side effects of psoralens used in photochemotherapy have led to a search for new photochemotherapeutic agents in which acute and chronic side effects of psoralens can be reduced or eliminated. The preparation and study of new monofunctional furocoumarins (Fig. 41–12) that photobind monofunctionally to DNA but lack the ability of psoralens to induce interstrand

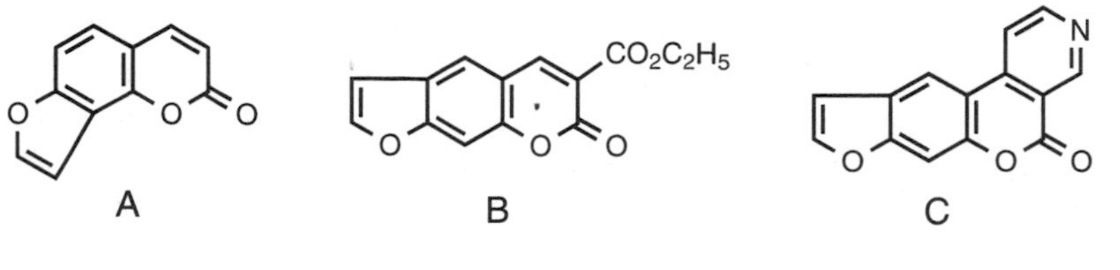

Fig. 41–12. Molecular structures of angelicin (A), 3-carbethoxypsoralen (3 CPs) (B), and pyrido(3,4-c)psoralen (C).

cross-linkages are promising. The repair of DNA photodamaged by monofunctional adducts is more effective than the repair of DNA containing interstrand cross-linkages.[53]

The new monofunctional drugs such as 3-CPs and pyridopsoralens, even though they form the molecular complex in the dark, cannot form interstrand cross-linkages in DNA because of steric hindrance of the 3-carbethoxy group[54] and the pyridine ring.[55] The angelicins that form a molecular complex with DNA, undergo intercalation between two base pairs of the macromolecule, but for geometric reasons cannot form interstrand cross-linkages.[53,56]

The new monofunctional drugs, such as 3-carbethoxypsoralen, pyridopsoralens, and methylangelicins, do not usually produce skin phototoxicity,[53,57] which makes their topical use easier because some long-term side effects connected with systemic use, such as the risk of cataract, are avoided. 3-CPs showed moderate activity in healing psoriasis by topical application, but low mutagenic activity, and complete inability to photoinduce tumors in mice.[54]

Among methylangelicins, 4,4′,6-trimethylangelicin (TMA) has been evaluated clinically for psoriasis, mycosis fungoides, and other skin diseases. TMA shows therapeutic activity with reduced side effects in comparison with psoralens (low mutagenicity, low risk of skin cancer, no skin phototoxicity).[58]

PHOTODYNAMIC THERAPY OF TUMORS

The property of dyes with a polycyclic hydrocarbon-type chemical structure to accumulate in greater amounts in tumor tissues than in normal tissues has been reported.[59,60] These dyes include acridines, xanthenes, psoralens, and porphyrins. The latter dyes, in particular hematoporphyrin (Hp) and some of its chemical derivatives (Hp D), have superior tumor-localizing properties, which are the basis of a novel phototherapeutic treatment of tumors by red light irradiation at predetermined times after systemic administration of the drug.[61] The chemical structures of Hp and HpD are shown in Figure 41–13. Moreover, HpD comprises some oligomeric material arising from intermolecular reaction between the secondary alcoholic functions in two different Hp units (with formation of an ether bond) or between one alcoholic and one carboxylic group in two Hp units (with formation of an ester bond).[62] Such oligomeric material appears to en-

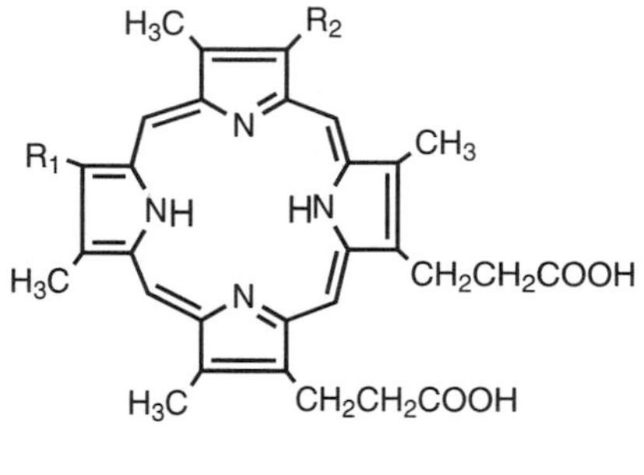

HpD is a mixture of Hp derivatives where:

R_1 =	R_2 =	
—CH(OH)–CH_3	—CH=CH_2	isomers of hydroxyethylvinyl-hematoporphyrin
—CH=CH_2	—CH(OH)–CH_3	
—CH($OCOCH_3$)–CH_3	—CH($OCOCH_3$)–CH_3	di-*O*-acetyl derivatives
—CH($OCOCH_3$)–CH_3	—CH(OH)–CH_3	mono-*O*-acetyl derivatives
—CH(OH)–CH_3	—CH($OCOCH_3$)–CH_3	

Fig. 41–13. Chemical structure of hematoporphyrin (Hp) and some of its chemical derivatives (HpD).

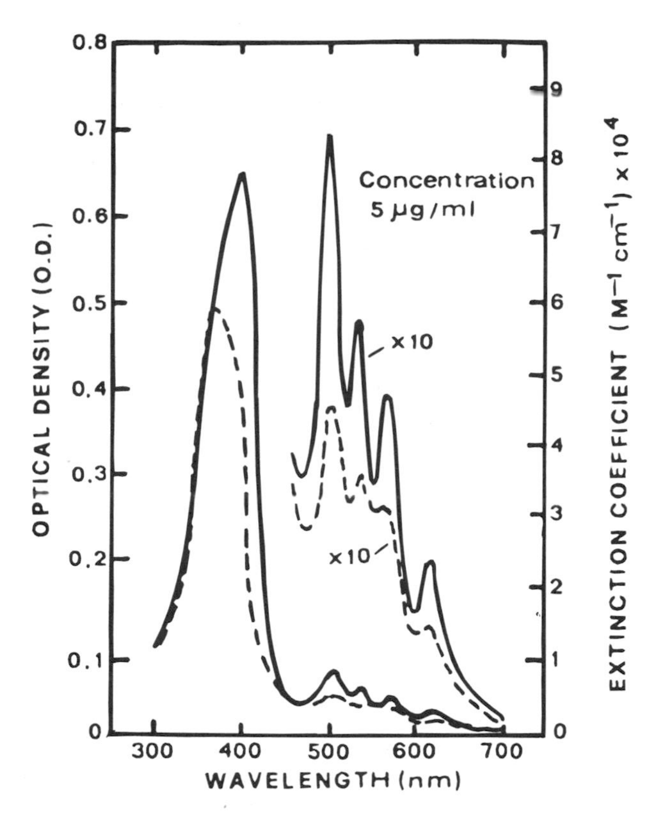

Fig. 41–14. Absorption spectra of hematoporphyrin derivative in saline with 10% serum. Solid line indicates HpD in saline with 10% serum; dotted line indicates HpD in saline.

hance the tumor affinity of HpD; therefore, a partially purified version of HpD, which is enhanced in Hp oligomers, has been obtained[63] and proposed for clinical applications under the commercial name of Photofrin II.

The properties of Hp and related porphyrins that make them suitable for this therapy are summarized in the following paragraphs.

Hp, injected intravenously into patients with tumors, is absorbed by neoplastic tissues in amounts the order of micrograms of drug per gram of tissue; of normal tissues, only the liver and spleen bind Hp in concentrations comparable with those accumulated by tumors.[61]

The clearance of Hp from tumor cells and tissues is slow, mainly because of the low activity of lymphatic drainage from tumors. Generally, at least 30% of the originally bound porphyrin is still present in tumor tissues 72 to 96 hours after administration. The clearance of Hp from normal tissues is essentially complete within 48 hours.[64]

Hp is not metabolized in the organism, and is eliminated in the feces with no structural modification. The disappearance of the porphyrin from the serum is biphasic: a major aliquot is eliminated with a half-life of about 12 hours, whereas a second fraction has a half-life of 8 to 10 days. The slowly decaying component is responsible for the general photosensitivity displayed by photodynamic therapy (PDT)-patients for several weeks after the phototreatment.[61]

Hp, like most other porphyrins, is a powerful photodynamic sensitizer. On photoexcitation by visible light, Hp generates singlet oxygen (type II photosensitization mechanism), which attacks several cell constituents. Because Hp is located mainly in the cytoplasmic membrane, the cell photodamage consists of the chemical modification of membrane proteins, sterols, and unsaturated lipids, with a consequent alteration of cell permeability and eventual cytolysis. No damage is usually seen at the nuclear level.[64]

Although Hp is maximally absorbing in the near-UV and blue spectral regions (Fig. 41–14), photoactivation of the tumor-localized porphyrin is usually achieved by irradiation with red light.[61] These light wavelengths (Fig. 41–15) have a particularly large penetrating power for human tissues,[65] thus allowing uniform irradiation of tissue volumes with a diameter of about 1 to 2 cm. Red light is not absorbed by the chromophores normally present in animal tissues (with the exception of melanin). As a consequence, no photoinduced damage is observed for normal tissues, thus ensuring a satisfactory degree of selectivity for tumor damage.

The chemical heterogeneity of Photofrin and the low molar absorptivity of this porphyrin in the wavelength region above 600 nm (Fig. 41–14) prompted the search

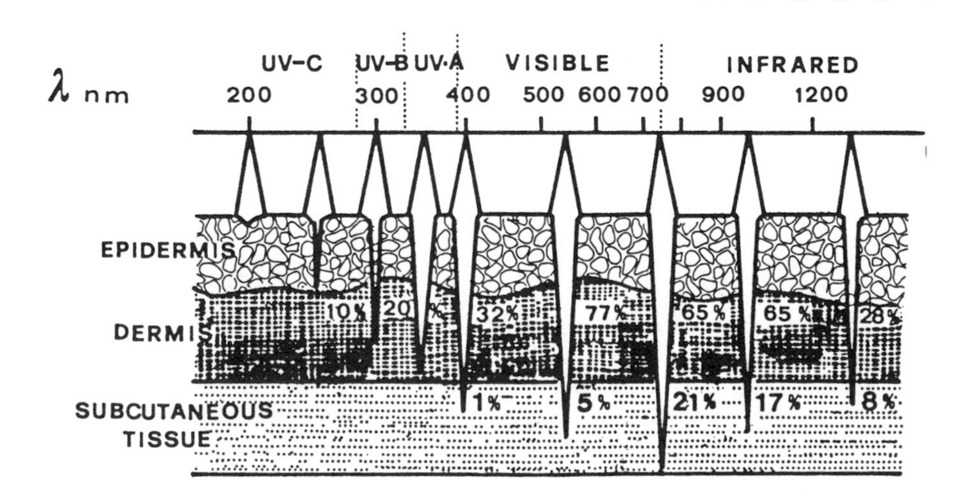

Fig. 41–15. Differential transmission of light with different wavelengths in various layers of the skin. Light used in PUVA can reach the dermis; light used in photodynamic therapy is more penetrating.

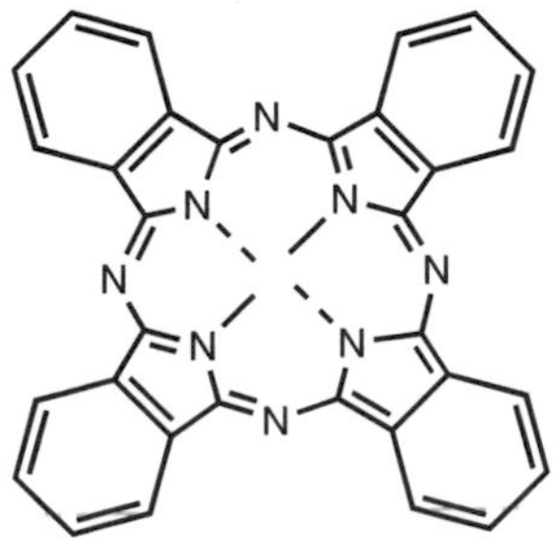

Fig. 41–16. Basic chemical structure of phthalocyanines showing the tetraazaisoindole chromophore. Phthalocyanines differ in the presence of metal ions coordinated with the pyrrole-type nitrogen atoms and/or peripheral substituents in the α or β position of the benzene rings. The linear condensation of one further benzene ring with each isoindole moiety originates naphthalocyanines.

for second-generation tumor photosensitizers, which should be chemically pure and exhibit intense absorbance in the wavelength interval between 700 and 850 nm, where the penetration of visible light into mammalian tissues becomes appreciable (Fig. 41–15), even when a significant degree of pigmentation is present (e.g., in melanotic melanoma).[66] Recently, the attention of investigators has been focused on tetraazaisoindole derivatives, such as phthalocyanines and naphthalocyanine (Fig. 41–16), whose molar absorptivity in the red spectral region is about two orders of magnitude larger than that typical of porphyrins. Both phthalo- and naphthalocyanines are excellent tumor localizers[67] and are the process of being proposed for the phase I/II clinical trials.

Clinical Applications

At present, the procedure for PDT[68] of tumors includes intravenous injection of a dose of Photofrin II ranging between 1 and 2.5 mg/kg body weight. Such a dose is far below the cytotoxic levels of this porphyrin (LD_{50} = 350 mg/kg), so that undesirable effects from Photofrin administration can be ruled out.

From 24 to 48 hours after Photofrin administration, the patient is irradiated with red light, using two main types of light sources. One consists of lamps with continuous emission (e.g., xenon arc-lamps, high efficiency halogen lamps) from which the 590 to 640 nm region is isolated by a set of optical filters. The red light thus obtained is focused on the patient either directly or through optical fibers and the light spot has a diameter of 0.5 to 1.5 cm. The other source is a laser light with monochromatic emission in the red spectral region. The most frequently used sources are gold vapor lasers (emission at 628 nm), or a combination of argon and dye lasers (emission at a wavelength within the 620 to 632 nm interval depending on the dye concentration). The laser emission can also be coupled with an optical fiber having a diameter of 0.4 to 1.0 mm. The continuously emitting lamp is usually preferred for the irradiation of tumors located at the skin level or in cavities accessible directly from the exterior (e.g., the oral or vaginal cavity). The laser source is essential in endoscopic or interstitial PDT, which involves the insertion of one or more fibers into the tumor to treat large or deep-sited neoplastic masses.

In all cases, the tumors are illuminated with a dose not higher than 200 mW/cm^2 to avoid the onset of thermal effects, which would be of nonspecific nature and overlap the photodynamic effect. The total delivered light dose again depends on the type and size of the tumor: thus, superficial tumors with a diameter below 1 cm can be cured by light doses as low as 80 to 100 J/cm^2. For larger pigmented tumors, light doses over 600 J/cm^2 must be delivered. Photoinduced tumor necrosis usually appears within 2 to 3 days after the phototreatment.

Mechanism

Porphyrins are typical type II photosensitizers,[69] and therefore it is likely that they exert their photosensitizing action mainly through generation of singlet oxygen. The attack of this activated species on unsaturated lipids and aromatic amino acid residues leads to the formation of endoperoxides, allylic hydroperoxides, or cross-linked products (Scheme 3). These chemical modifications are concentrated in the cytoplasmic and mitochondrial membrane, which is the preferential site of Hp-cell binding; as a result, the initially observed effects of PDT involve the rounding of the cell because of alteration of membrane permeability (Scheme 4). Prolonged irradiation causes lysis of the membrane and destruction of mitochondria.[70]

The morphologic changes at the cellular level are accompanied by simultaneous modifications of the vascular system, causing blockage of blood flow. The relative weight of the two types of photosensitized tissue damage depends on the actual dose of Photofrin administered to the patient. Low doses of Photofrin are completely bound by serum proteins, in particular by low-density lipoproteins (LDL), which deliver porphyrins inside the cell by receptor-mediated endocytosis. Therefore, in this case, direct cell damage is the principal effect of Photofrin photosensitization. In the presence of Photofrin doses exceeding the binding capacity of LDL or other lipoproteins, Photofrin exists in the serum as pseudomicellar aggregates, thereby inducing primary photoeffects at the level of blood vessels.[71] This type of photoprocess is characterized by the lack of any appreciable photodamage at the DNA level, which minimizes the risk of mutagenicity.

Present State

Several thousand patients affected by a large variety of neoplastic lesions have been treated by PDT with Photofrin, and the majority have benefited. Therefore, this therapy appears to be specific because it can be applied in the presence of tumors with different histologies or located in different sites.

Previous treatment of the tumor by other conventional techniques does not appear to affect the efficacy of PDT. Many patients receiving PDT had relapses after previous chemotherapy, radiation therapy, or surgical removal of the tumor. Recent reports indicate a positive synergism

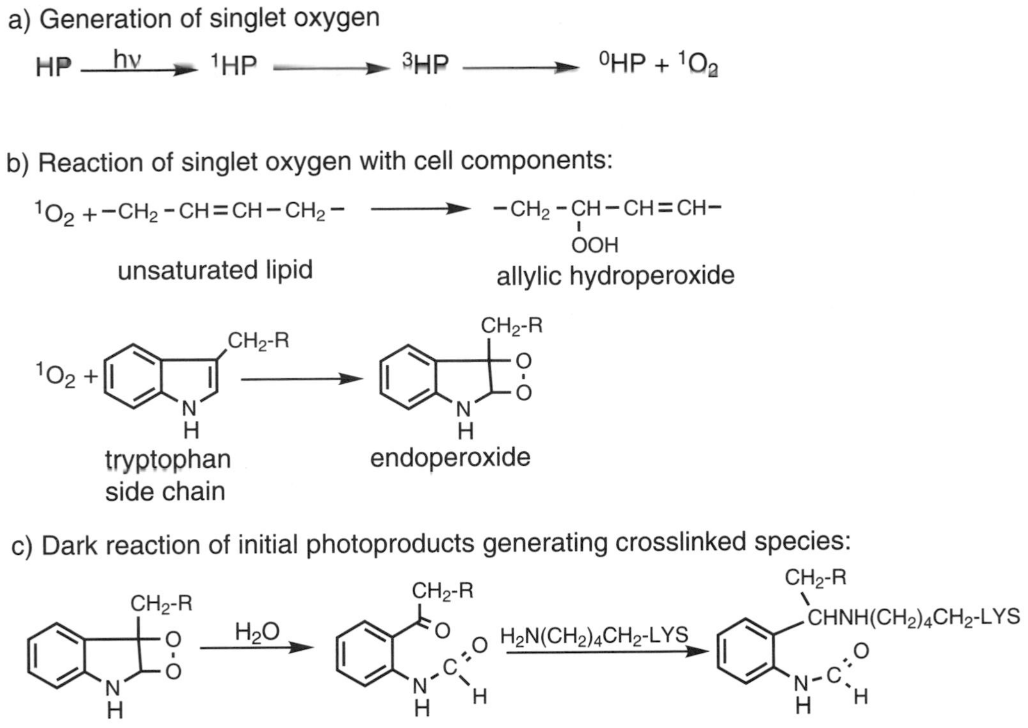

a) Generation of singlet oxygen

$$HP \xrightarrow{h\nu} {}^{1}HP \longrightarrow {}^{3}HP \longrightarrow {}^{0}HP + {}^{1}O_2$$

b) Reaction of singlet oxygen with cell components:

$${}^{1}O_2 + -CH_2-CH{=}CH-CH_2- \longrightarrow -CH_2-CH(OOH)-CH{=}CH-$$

unsaturated lipid → allylic hydroperoxide

${}^{1}O_2$ + tryptophan side chain (CH_2-R) → endoperoxide (CH_2-R)

c) Dark reaction of initial photoproducts generating crosslinked species:

endoperoxide (CH_2-R) $\xrightarrow{H_2O}$ N-formylkynurenine (CH_2-R) $\xrightarrow{H_2N(CH_2)_4CH_2\text{-LYS}}$ crosslinked product (CH_2-R, $CHNH(CH_2)_4CH_2$-LYS)

e.g., intermolecular crosslink between *N*-formylkynrenine (hydrolysis product of the endoperoxide formed from the amino acid tryptophan) and the side chain of the amino acid lysine

Scheme 3

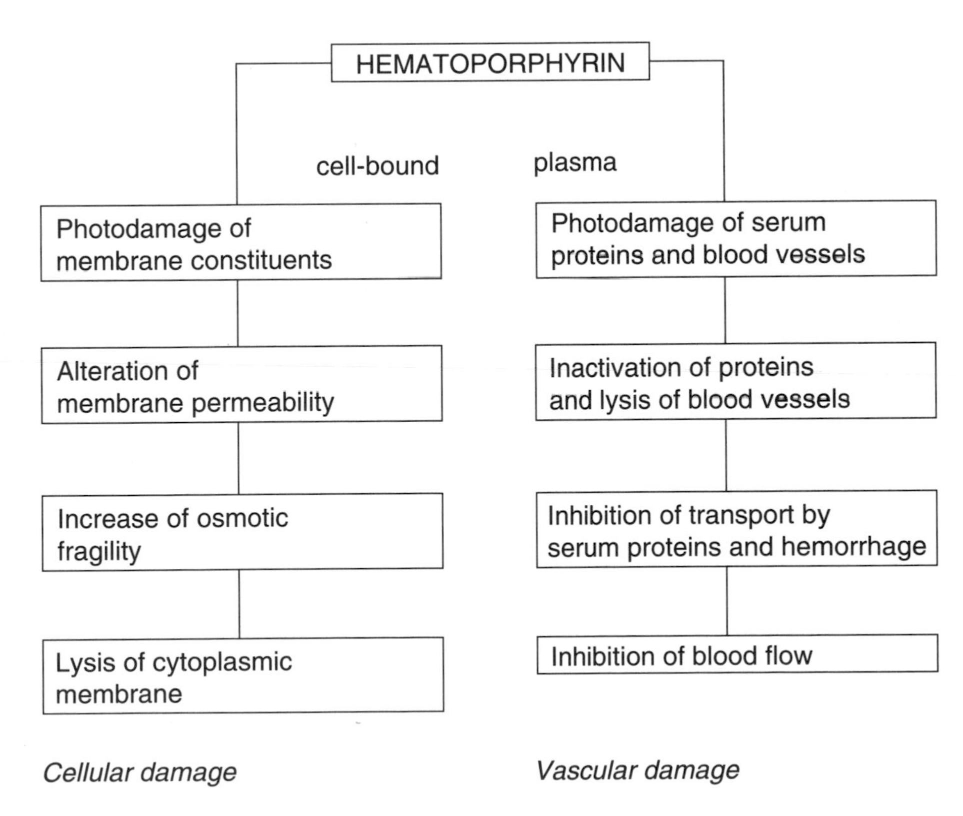

Scheme 4

between PDT and other treatments of tumors, such as surgery and hyperthermia.

Curative effects of PDT are most frequently observed in tumors of relatively small dimension (diameter below 2 cm). The investigations are especially advanced in the use of PDT in the treatment of early cancers of skin, bladder, vagina, mouth, and lungs.[72] Moreover, pilot studies are in progress to define the applicability of PDT to ocular[73] and brain tumors.[74] Neurosurgeons use PDT to sterilize the surgical cavity after removal of the tumor mass, hoping to destroy undetected metastatic infiltrations.

In its present formulation, PDT is less effective in advanced or larger tumors. PDT can, however, be used to reduce tumor masses or to slow tumor growth, thereby increasing the life span of the patient or obtaining palliative effects in patients with inoperable tumors. Particularly promising results are being obtained with PDT in the palliative treatment of obstructive cancers of the esophagus and bronchii.[75]

The further improvement of PDT depends on more detailed definition of treatment protocols for various types of tumors, in particular the doses of light and photosensitizer to be used in each specific situation. The fluorescence typically emitted by porphyrins at 630 to 690 nm can be used for early diagnosis of tumors; the sensitivity of this technique is being increased by the introduction of digital video imaging of the red emission from tumor-localized porphyrins.[76]

On the other hand, the preferential affinity displayed by porphyrins for cells with hyperproliferative activity suggests a possible extension of PDT for the treatment of diseases other than tumors. Both porphyrins and phthalocyanines have been shown in vivo and in vitro to accumulate by areas characterized by rapid mitosis, including some types of atheromatous plaques,[77] psoriatic areas,[78] metastatic abscesses,[79] and some viral infections.[80] In all of these cases, studies with cell cultures and laboratory animals appear to indicate a curative effect of PDT, although no definite conclusion can be drawn at present as to the feasibility of large-scale clinical use of the technique.

REFERENCES

1. J. A. Parrish, et al., *Photochemotherapy of Skin Diseases,* in *The Science of Photomedicine,* J. D. Regan and J. A. Parrish, Eds., New York, Plenum Press, 1982, p. 595.
2. J. A. Parrish, *Phototherapy of Psoriasis and Other Skin Diseases,* in *The Science of Photomedicine,* J. D. Regan and J. A. Parrish, Eds., New York, Plenum Press, 1982.
3. C. K. Smith, *Photosensitization,* in *The Science of Photobiology,* K. C. Smith, Ed., New York, Plenum Press, 1977, p. 87.
4. F. Dall'Acqua and P. Martelli, *J. Photochem. Photobiol., 8,* 235(1991).
5. F. Dall'Acqua and S. Caffieri, *Photomed. and Photobiol. (Japan), 10,* 1(1988).
6. S. Caffieri, et al., *C_4-Cycloaddition Reaction between Furocoumarins and Unsaturated Fatty Acids or Lecithins,* in *Frontiers of Photobiology,* A. Shima, et al. Eds., Amsterdam, Elsevier Science, 1993, p. 85.
7. D. Kessel and T. J. Dougherty, Eds., *Porphyrin Photosensitization,* New York, Plenum Press, 1983.
8. J. D. Spikes, *Ann. N.Y. Acad. Sci., 244,* 496(1975).
9. R. L. Edelson, *Yale J. Biol. Med., 62,* 565(1989).
10. M. Jaratt, et al., *Dye-light Phototherapy of Viral, Bacterial and Fungal Infections,* in *The Science of Photomedicine,* J. D. Regan and J. A. Parrish, Eds., New York, Plenum Press, 1982, p. 595.
11. J. L. Melnick and C. Wallis, *Photodynamic Inactivation of Herpes Virus,* in *The Science of Photomedicine,* J. D. Regan and J. A. Parrish, Eds., New York, Plenum Press, 1982, p. 545.
12. C. V. Hanson, *Inactivation of Viruses for Use as Vaccines and Immunodiagnostic Reagent,* in *Medical Virology II,* L. M. de la Maza and E. M. Peterson, Eds., New York, Elsevier, 1983, p. 45.
13. S. Isaacs, et al., *Photoinactivation of pathogens for medical applications,* Abstract S17-2, of the XI Intern. Congress on Photobiol., Kyoto, 1992, p. 168.
14. T. R. C. Sisson, and T. P. Vogl, *Phototherapy of Hyperbilirubinemia,* in *The Science of Photomedicine,* J. D. Regan and J. A. Parrish, Eds., New York, Plenum Press, 1982, p. 477.
15. J. A. Parrish, et al., *N. Engl. J. Med, 291,* 1207(1974).
16. M. A. Pathak, et al., *Photobiology and Photochemistry of Furocoumarins (Psoralens),* in *Sunlight and Man.,* M. A. Pathak et al., Eds., Tokyo, University of Tokyo Press, 1974, p. 335.
17. A. M. El Mofty, A Preliminary Clinical Report on the Treatment of Leukoderma with *Ammi majus Lin., J. Roy. Egyptian M.A., 31,* 651(1948).
18. F. Dall'Acqua, *Furocoumarin Photochemistry and Its Main Biological Implications,* in *Current Problems in Dermatology,* H. Hoenigsmann, Ed., Vienna, Karger, 1985.
19. M. K. Logani, et al., *Photochem. Photobiol., 35,* 565(1982).
20. A. Ya. Potapenko, *J. Photochem. Photobiol. B: Biol., 9,* 1(1991).
21. M. A. Pathak, *Phytophotodermatitis,* in *Sunlight and Man,* M. A. Pathak et al., Eds., Tokyo, University of Tokyo Press, 1974, p. 495.
22. R. Roelandts, *Arch. Dermatol., 129,* 662(1984).
23. M. Hannuksela and J. Karvonen, *Br. J. Dermatol., 99,* 703(1978).
24. H. H. Roenigk, Jr., *NCI Monogr., 66,* 179(1984).
25. D. B. Mosher, et al., *Vitiligo: Etiology, Pathogenesis, Diagnosis and Treatment,* in *Update Dermatology in General Medicine,* T. B. Fitzpatrick et al., Eds., New York, McGraw-Hill, 1983, p. 205.
26. J. P. Ortonne, et al., *Br. J. Dermatol., 101,* 1(1979).
27. K. Danno, et al., *J. Invest. Dermatol., 85,* 110(1985).
28. D. Vedaldi, et al., *Il Farmaco, 43,* 333(1988).
29. A. Abdel-Fattah, et al., *Dermatologica, 165,* 136(1982).
30. P. Morliere, et al., *J. Invest. Dermatol., 90,* 720(1988).
31. F. Dall'Acqua, *New Chemical Aspects of the Photoreaction between Psoralen and DNA,* in *Research in Photobiology,* A. Castellani, Ed., New York, Plenum Press, 1978, p. 245.
32. L. Musajo and G. Rodighiero, *Mode of Photosensitizing Action of Furocoumarins,* in *Photophysiology,* Vol. VII, A. G. Giese, Ed., New York, Academic Press, 1972, p. 115.
33. F. Dall'Acqua, et al., *QSAR on Furocoumarins, Agents for the Phototherapy of Psoriasis,* In *QSAR in Design of Bioactive Compounds,* M. Kuchar, Ed., Barcelona, J. R. Prous, 1984, p. 87.
34. L. Musajo, et al., *Rend. Accad. Naz. Lincei (Rome), 42,* 457(1967).
35. D. Kanne, et al., *J. Am. Chem. Soc., 104,* 9754(1982).
36. F. Dall'Acqua, et al., *Z. Naturforsch., 266b,* 56(1971).
37. F. Dall'Acqua and G. Rodighiero, *Bological and Medicinal Aspects of Furocoumarins* in *Primary Photo-Processes in Biology and Medicine,* R. V. Bensasson, et al., Eds., New York, Plenum Press, 1985, p. 277.
38. K. G. Specht, et al., *J. Org. Chem., 54,* 4125(1989).
39. S. Caffieri, et al., *Med. Biol. Environ., 17,* 796(1989).
40. R. W. Midden, *Chemical Mechanisms of the Bioeffects of Furocoumarins. The Role of the Reactions with Proteins, Lipids and*

other Cellular Components, in *Psoralen DNA Photobiology,* Vol. II, F. P. Gasparro, Ed., Boca Raton, CRC Press, 1988, p. 1.

41. L. I. Grossweiner, *NCI Monogr., 66,* 47(1984).
42. J. D. Laskin, et al., *Proc. Natl. Acad. Sci. USA, 82,* 6158(1985).
43. E. J. Yurkov and J. D. Laskin, *J. Biol. Chem., 262,* 8439(1987).
44. J. D. Laskin and D. L. Laskin, *Role of Psoralen Receptors in Cell Growth Regulation,* in *Psoralen DNA Photobiology,* Vol. II, F. P. Gasparro, Ed., Boca Raton, CRC Press, 1988, p. 135.
45. A. K. Gupta and T. F. Anderson, *J. Am. Acad. Dermatol. 17,* 703(1987).
46. D. R. Bickers and M. A. Pathak, *Psoralen Pharmacology: Studies on Metabolism and Enzyme Induction,* in *Photobiologic, Toxicologic and Pharmacologic Aspects of Psoralens,* M. A. Pathak and J. K. Dunnick, Eds., *NCI Monogr. 66,* 77(1984).
47. D. R. Bickers, et al., *Clinical Pharmacology of Skin Diseases,* New York, Churchill Livingstone, 1984.
48. M. A. Pathak, *NCI Monogr., 66,* 41(1984).
49. R. S. Stern, et al., *N. Engl. J. Med., 310,* 1156(1984).
50. T. Henseler and E. Christophers, *NCI Monogr., 66,* 217(1984).
51. S. Lerman, et al., *J. Invest. Dermatol., 74,* 197(1980).
52. N. H. Cox, et al., *Br. J. Dermatol., 116,* 145(1987).
53. G. Rodighiero, et al., *New Psoralen and Angelicin Derivatives,* in *Psoralen DNA Photobiology,* Vol. I, F. P. Gasparro, Ed., Boca Raton, CRC Press, 1988, p. 37.
54. L. Dubertret, et al., *Photophysical, Photochemical, Photobiological and Phototherapeutic Properties of 3-Carbethoxypsoralen,* in *Psoralen in Cosmetic and Dermatology,* G. Canhn, et al., Eds., Paris, Pergamon Press, 1981, p. 245.
55. L. Dubertret, et al., *Biochemie, 67,* 417(1985).
56. A. Guiotto, et al., *J. Med. Chem., 27,* 959(1984).
57. F. Dall'Acqua, et al., *Drugs of the Future, 10,* 307(1985).
58. F. Bordin, et al., *Pharmacol. Ther., 52,* 331(1991).
59. J. D. Spikes, *The Historical Development of Ideas on Applications of Photosensitized Reactions in the Health Sciences,* in *Primary Photoprocesses in Biology and Medicine,* R. V. Bensasson, et al., Eds., New York, Plenum Press, 1985, p. 209.
60. T. J. Dougherty, *JNCI, 51,* 1333(1974).
61. T. J. Dougherty, *Photoradiation Therapy—Clinical and Drug Advances,* in *Porphyrin Photosensitization,* by D. Kessel and T. J. Dougherty, Eds., New York, Plenum Press, 1983.
62. D. A. Bellnier, et al., *Photochem. Photobiol., 50,* 221(1989).
63. G. Bock and S. Harnett, Eds., *Photosensitizing Compounds: Their Chemistry, Biology and Clinical Use,* Chichester, J. Wiley & Sons, 1990, p. 146.
64. G. Jori, *Photochem. Photobiol., 52,* 439(1990).
65. J. Eichler, et al., *Radiat. Environ. Biophys., 14,* 239(1977).
66. B. W. Henderson and T. J. Dougherty, *Photodynamic Therapy,* New York, Marcel Dekker, Inc., 1992.
67. P. A. Firey, et al., *J. Am. Chem. Soc., 110,* 7626(1988).
68. T. J. Dougherty, et al., *Photoradiation Therapy of Human Tumors,* in *The Science of Photomedicine,* J. D. Regan and J. A. Parrish, Eds., New York, Plenum Press, 1982, p. 625.
69. G. Jori, *Molecular and Cellular Mechanisms in Photomedicine: Porphyrins in Cancer Treatment,* in *Primary Photoprocesses in Biology and Medicine,* R. V. Bensasson, et al., Eds., New York, Plenum Press, 1985, p. 381.
70. C. Zhou, *J. Photochem. Photobiol., B:Biol., 3,* 299(1989).
71. T. G. Truscott, *Photochemistry of Porphyrins and Bile Pigments in Homogeneous Solution,* in *Primary Photo-processes in Biology and Medicine,* R. V. Bensasson, et al., Eds., New York, Plenum Press, 1985, p. 309.
72. B. Krammer, et al., *J. Photochem. Photobiol., B:Biol., 17,* 109(1993).
73. D. Leupold and W. Freyer, *J. Photochem. Photobiol., B:Biol., 12,* 311(1992).
74. B. Roeder, et al., *Biophys. Chem., 35,* 303(1990).
75. G. D. Spikes and J. C. Bommer, *J. Photochem, Photobiol., B: Biol., 17,* 135(1993).
76. J. S. McCaughan, et al., *Arch Surg., 124,* 211(1989).
77. J. R. Spears, et al., *J. Clin. Invest., 71,* 395(1983).
78. W. Diezel, et al., *Dermatol. Monatsschr., 166,* 793(1980).
79. G. Jori and J. D. Spikes, *Photobiochemistry of Porphyrins,* in *Topics in Photomedicine,* K. C. Smith, Ed., New York, Plenum Press, 1984, p. 1833.
80. L. E. Schnipper, et al., *J. Clin. Invest., 63,* 632(1980).

Chapter 42

PESTICIDES

John L. Neumeyer and Raymond G. Booth

During the last 60 years, society has gained many benefits from the use of pesticides to prevent disease and to increase production of food and fibers. Our need for pesticides and other pest control methods will continue, but greater selectivity and specificity will be demanded, just as the need for less toxic and more effective drugs will be emphasized in the development of new pharmaceutical preparations.

Evidence has led to public concern about the unintentional effects of pesticides on various life processes within the environment and on human health. It is therefore becoming increasingly apparent that the benefits of using pesticides must, in every case, be considered in the context of present and potential risk.

Pesticide chemicals can be safe and are effective when used as recommended. They may be dangerous if directions are not followed. This chapter defines the extent to which insecticides in common use fit our current concepts of an ideal insecticide and form the basis for understanding the use, the chemistry, and the mode of action of these chemicals.

CLASSIFICATION OF PESTICIDES

These agents can be classified according to the plants or animals against which they are effective. Insects can be controlled by a variety of chemicals, which are further classified as insecticides, pheromones, hormones, antifeeding compounds, repellents, and chemosterilants. Rodents are controlled by rodenticides, fungi by fungicides, and plants by herbicides and plant growth regulators.

The discussion in this chapter is limited principally to insecticides and insect repellents.

PYRETHRUM

Use of pyrethrum powder as an insecticide originated in Asia about A.D. 1800. The substance occurs in plants belonging to the genus Chrysanthemum, family Compositae. The species with a significant content of pyrethrins, C. cinerariaefolium, is most widely used for commercial production. Flowers produced in Kenya contain an average of 1.3% pyrethrins. Pyrethrum extract is the principal export of Kenya, but pyrethrum is also produced in Japan and Ecuador.[1]

Pyrethrum extract is a mixture of pyrethrins, and it is becoming increasingly important as a household insecticide because of its low toxicity for mammals, lack of persistence in the environment, and remarkable property of instantaneously killing flying insects on contact.

The pyrethrins are most commonly used in an extract obtained from the dried group flowers. The extract is dissolved in an oil and used in a spray, generally in combination with a synergist. The principal uses of pyrethrins are in household aerosol sprays, for protection of food in warehouses, in dairy barns, and for control of fruitflies on harvested fruits and vegetables. Pyrethrum powder is also used in "mosquito coils," which are popular for outdoor use. These devices emit pyrethrin-containing fumes that affect mosquitoes and thereby inhibit their biting.

The toxicity of pyrethrum is shown in Table 42–1.

Chemistry

Pyrethrins are esters and therefore can be discussed in terms of the chrysanthemic or pyrethric acid moiety and the keto alcohol moiety, pyrethrolone and cinerolone. All pyrethrins and the more recently developed synthetic pyrethrins are esters or minor modifications of these four components.[1,3] This is evident from the structures shown in Table 42–2.

The search for replacements for the established residual insecticides used against mosquitoes, necessitated by the development of resistance to organochlorine compounds and concern about pollution of the environment,

Table 42–1. Toxicity of Pyrethrum and Synthetic Pyrethroids

Insecticide	Topical LD_{50} in Housefly (μg/g)	LD_{50} in Mosquito Larvae (ppm)	Oral LD_{50} in Rat (mg/kg)
Pyrethrins (natural)	15		580
Allethrin	8.5	0.14	770
Dimethrin	50	0.088	>10,000
Resmethrin	0.8	0.0081	1,400
Bioresmethrin	0.25		8,000
Phenothrin	2.6	0.0083	5,000
Permethrin	1.35	0.0039	600
Biopermethrin	0.85	0.0033	
Decamethrin	0.17	0.0001	125
Fenvalerate	4.0	0.0048	200

From R. L. Metcalf, in *Kirk-Othmer Encyclopedia of Chemical Technology,* Vol. 13, 3rd ed., New York, Wiley Intersciences, 1981, pp. 413–485.

Table 42–2. Structures of Active Ingredients of Pyrethrum and Synthetic Pyrethrins

	Acid Moiety	Alcohol Moiety
Active Constituents of Pyrethrum		
Pyrethrin I	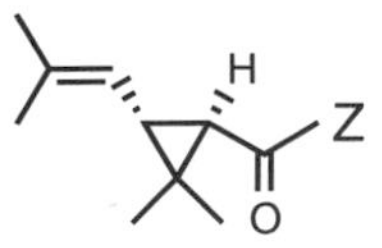	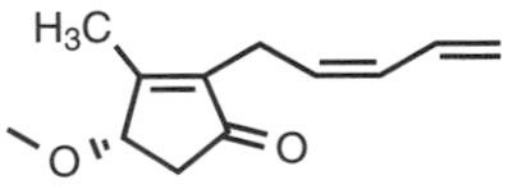
Cinerin I		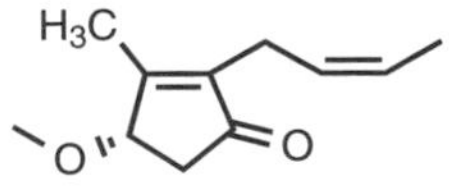
Pyrethrin II	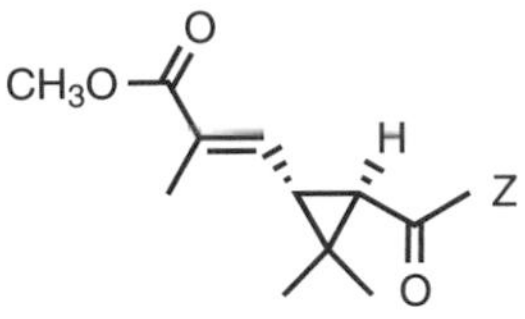	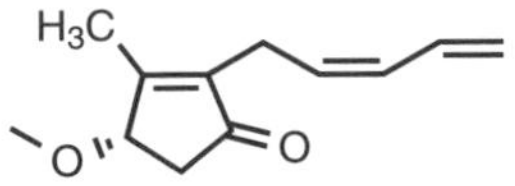
Cinerin II		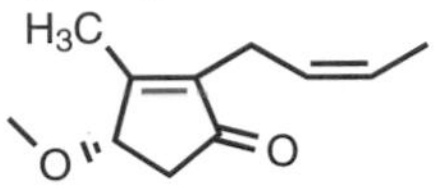
Jasmolin II		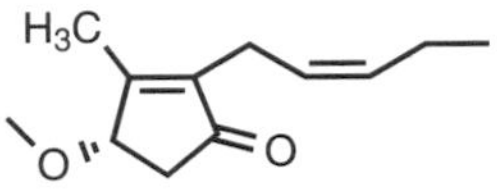
Synthetic Pyrethrins		
Allethrin ($R = CH_3$)		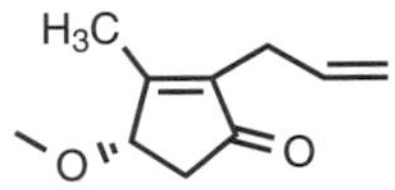
Dimethrin ($R = CH_3$)	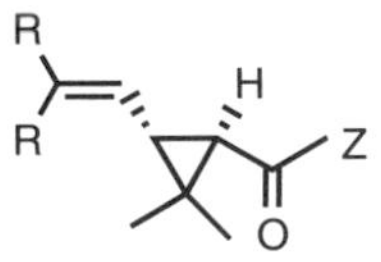	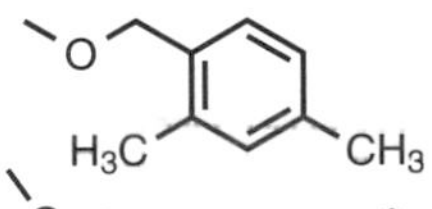
Resmethrin ($R = CH_3$)		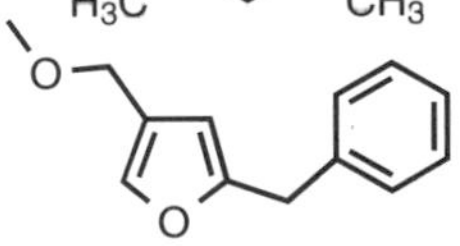
Permethrin ($R = Cl$)	(Mixture of *cis* and *trans* isomers)	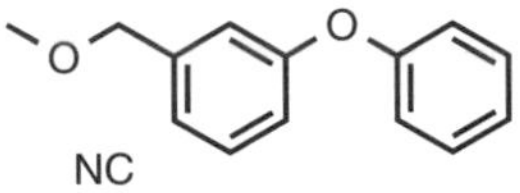
Decamethrin ($R = Br$)	*cis* isomer	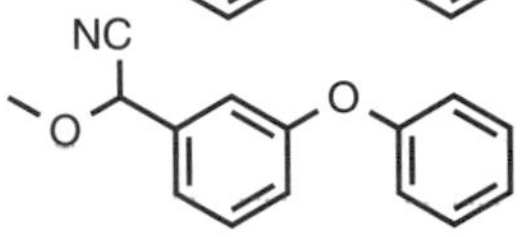
Fenvalerate	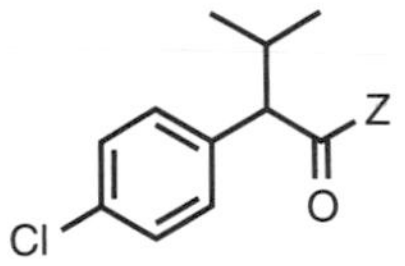	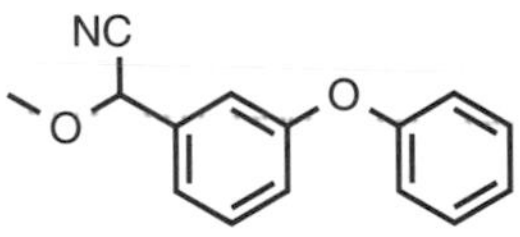

as focused attention on new synthetic pyrethroids. Although the natural pyrethrins are efficient insecticides with low toxicity to mammals, their use has been restricted because of high costs, limited availability, and rapid degradation by sunlight. Interest in developing effective synthetic pyrethrins was initiated during World War II and led to the synthesis of a number of esters of chrysanthemic acid, which culminated in the synthesis of allethrin.

Substantial structural changes have since been made, and highly effective synthetic pyrethroids have been marketed that possess substantially greater insecticidal activity. The structures of these new synthetic pyrethroids involve major departures from the natural pyrethrins in both the acid and alcohol moieties of the pyrethrins (see Table 42–2).

Pyrethroids from Chrysanthemic Acid

Substantial insecticidal activity is found in the substituted benzyl chrysanthemates of which 2-chloro-

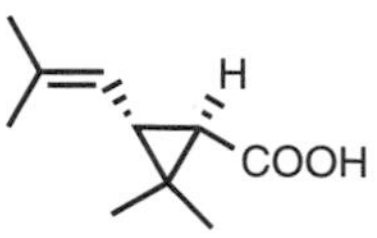

Chysanthemic acid 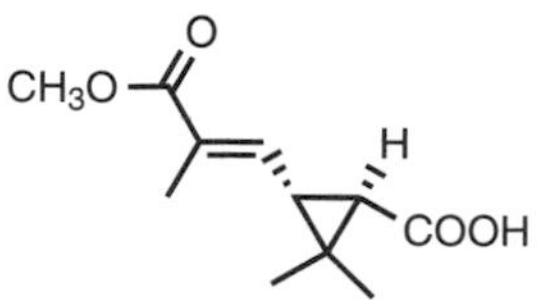   Pyrethric acid

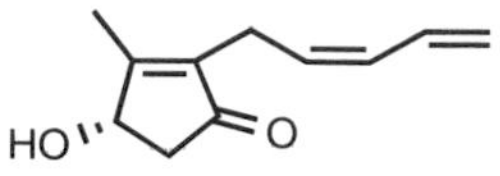

Pyrethrolone 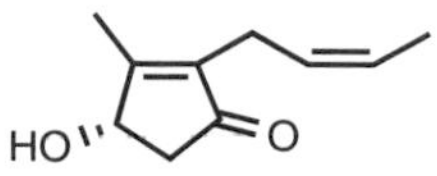   Cinerolone

3,4-methylenedioxybenzyl RS-*cis,trans*-chrysanthemate (barthrin) and 2,4-dimethylbenzyl RS-*cis,trans*-chrysanthemate (dimethrin) are the best known. Dimethrin has extremely low mammalian toxicity (oral LD_{50} in the rat, >10,000 mg/kg) and has been used as a mosquito larvicide that is safe for use in potable waters. 5-Benzyl-3-furylmethyl RS-*cis, trans*-chrysanthemate (resmethrin) is a pyrethroid of high insecticidal activity and low mammalian toxicity, but of short persistence. Activity in this compound is enhanced even more by esterification with R-*trans* chrysanthemic acid, which can be produced by microbial fermentation. Thus, bioresmethrin is about 2.2 times as toxic to the housefly and 0.18 times as toxic to the rat as resmethrin. A further structural modification that leads to increased stability to light and air is effected by 3-phenoxybenzyl RS-*cis,trans*-chrysanthemate (phenothrin).

Pyrethroids with Modified Acid Components

Recent developments in the synthesis and evaluation of synthetic pyrethroids have involved more substantial changes in the structure of both chrysanthemic acid and esterification with other alcohol moieties. Permethrin, the 3-phenoxybenzyl RS-*cis,trans*-3-(2,2-dichlorovinyl)-2,2-dimethylcyclopropanecarboxylate, has excellent insecticidal activity and enhanced persistence. The closely related 2-cyano-3-phenoxybenzyl RS-*cis*-3-(2,2-dibromovinyl)-2,2-dimethylcyclopropanecarboxylate (decamethrin) appears to be the most active synthetic pyrethroid yet synthesized.

Synergists

The efficacy of the pyrethroids can be greatly improved by using them in combination with other compounds that are virtually inactive when used alone. Synergists may or may not exhibit insecticidal activity, but when they are used with a toxicant, the combined effectiveness is significantly greater than the sum of the individual effects. (Pharmacologists tend to use the term potentiation for this phenomenon and synergism to mean a simple additive effect.)

Since the development of low-cost synergists for use with pyrethrum, these combinations have become popular. Pyrethrum synergists, such as piperonyl butoxide, are widely used to increase the effectiveness of the pyrethrins, and all are characterized by the methylenedioxyphenyl (MDP) group (Table 42–3). Sesame oil was first used as a synergist for pyrethrum and was shown to be effective because of its sesamin content.

A typical synergist-toxicant combination is pyrethrum and piperonyl butoxide. One micromilliliter of pyrethrum is required to kill the common housefly; piperonyl butoxide has no effect when used alone, but 1 μmL of a mixture of 1 part pyrethrum and 10 parts piperonyl butoxide causes death to the fly.

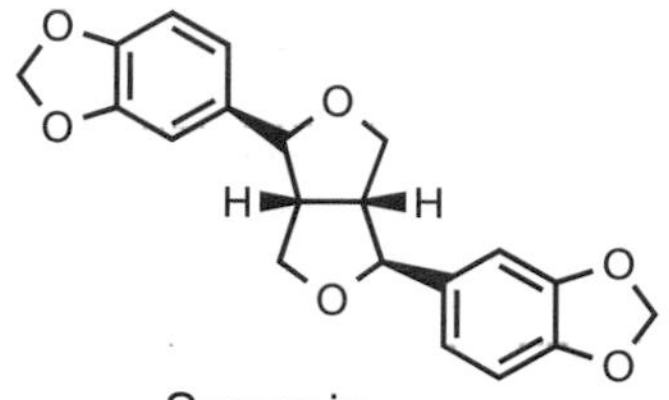

Sesamin

The synthetic pyrethrin allethrin has the same insecticidal potency as nonsynergized natural pyrethrins, but is less susceptible to synergism with MDP compounds; it is therefore less attractive for commercial use. Resmethrin and permethrin are unique in that they do not require a synergist for maximal insecticidal effectiveness.

Other classes of compounds besides the MDP group exhibit synergistic activity. Insecticidal synergists are believed to exert their effects by inhibiting the mixed function oxidase system of microsomes.[4] Such compounds therefore serve as alternative substrates, sparing the insecticide chemical from detoxification, or react with another site in the mixed function system, preventing oxidative detoxification. Interactions with other toxicants or drugs metabolized by such microsomal enzymes can occur in mammals treated with piperonyl butoxide or nonmethylenedioxyphenyl synergists[5] (Fig. 42–1). Synergists can therefore increase or decrease the toxicity of a chemical, depending on the shift in the balance of the competing activation or detoxification reactions caused by their presence.

Later in this chapter, we show that administration of piperonyl butoxide or a microsomal enzyme inhibitor such as 2-diethylamino-ethyl-2,2-diphenylvalerate (SKF 525 A) before administration of certain organophosphate insecticides decreases the toxicity of such insecticides by inhibiting the metabolic activation process.

Mechanism of Action

Pyrethrum and the synthetic pyrethroids are nerve membrane sodium channel toxins.[6–8] Pyrethroids slow the rate of inactivation of the sodium current elicted by membrane depolarization, thus prolonging the open time of the sodium channel. At lower doses, pyrethroids produce repetitive action potentials and neuron firing; at high concentration, they depolarize the nerve membrane completely and block excitability. The type II pyrethroids, which contain a cyano group at the α-position (eg., deca-

Table 42–3. Structures and Names of Some Pyrethrum Synergists in Commercial Use in the United States

$CH_2O(CH_2)_2O(CH_2)_2O(CH_2)_3CH_3$ — Piperonyl butoxide
$CH(O(CH_2)_2O(CH_2)_2O(CH_2)_3CH_3)_2$ — Tropital
MGK 264

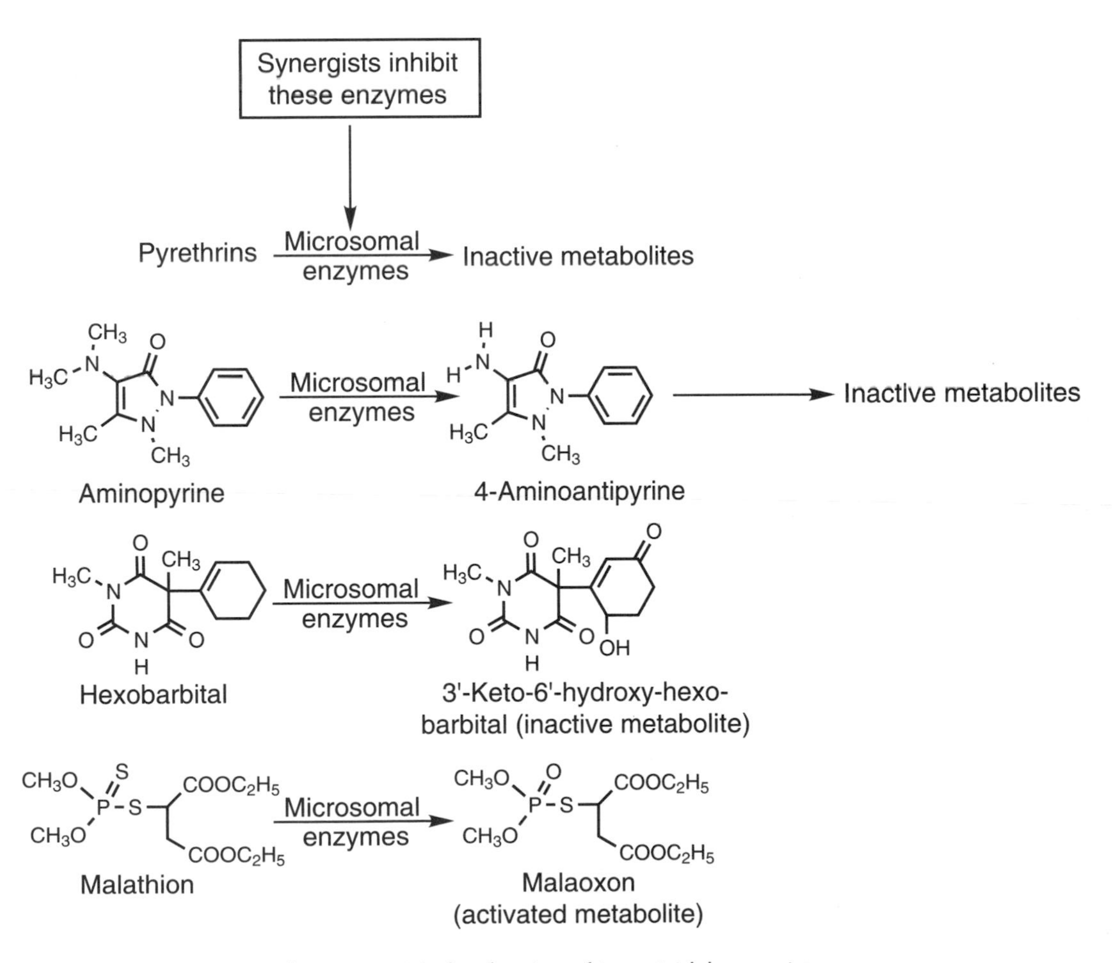

Fig. 42–1. Mode of action of insecticidal synergists.

methrin and fenvalerate, see Table 42–2), are generally more potent at producing membrane depolarization than noncyano-containing type I agents. This action is the basis for the parethesia caused by type II pyrethroids.[9]

The interaction of pyrethroids with the sodium channel complex is highly stereospecific,[10] dependent on the stereochemistry of the acid moiety: esters of 1R,3-*cis*- and 1R,3*trans*-cyclopropanecarboxylates are active (shown in Table 42–2). Although the corresponding 1S,3-*cis*- and 1S,3-*trans* isomers do not modify sodium channel function, they do bind to the same sites at the sodium channel as the R isomers and antagonize the effects of the active 1R forms. For type II pyrethroids, which contain a cyano group at the α-position, the αS isomer is more active as an insecticide than the αR isomer.

In mammals, the pyrethroid ester moiety is rapidly hydrolyzed in plasma and liver to inactive acid and alcohol metabolites; thus, mammals are relatively resistant to the toxic effects. The slower hydrolysis of the *cis* isomer by oxidases versus the *trans* isomer by esterases may account for some of the variations in reported mammalian toxicity of isomeric mixtures.

Other central nervous system mechanisms of action for pyrethroids have been proposed. For example, cyano-containing derivatives stimulate norepinephrine release from neurosynaptosomes,[11] an effect that may be related to their interaction with sodium or calcium channels.[12] Also, cyano-containing pyrethroids antagonize GABA receptor ligand binding and GABA-stimulated chloride flux, although the concentrations required for this action are higher than concentrations needed to enhance sodium flux.[9]

Selective Toxicity

Although pyrethroids, as well as chlorinated hydrocarbon pesticides such as DDT (see subsequent discussion) affect the sodium channels of both mammals and insects to a similar degree, these compounds show relatively selective toxicity to insects. The mechanism of this selective neurotoxicity may be related to body temperature differences of insects and mammals. Lower temperature increases the potency of pyrethroids and chlorinated hydrocarbons to affect sodium channel function. Thus, although sodium channels in mammals and insects may show similar intrinsic sensitivity to pyrethroids and chlorinated hydrocarbons, differences in body temperature (ca. 10°C) apparently make insect nerve membranes more sensitive to the insecticides.[9] Furthermore, because of body temperature differences, insects may metabolize insecticides more slowly (and differently) than mammals, allowing insecticides to reach their target site before metabolic inactivation occurs.

ROTENONE

This useful botanical insecticide, the major chemical constituent of derris (Derris elliptica and D. chinensis) and cube roots (Lonchocarpus species), has been used around the world for many years, first by primitive people as a fish poison to aid in hunting, and since 1848 as an insecticide.

The crude material is used as dusts at concentrations of 0.75 to 1.0% to control pests, such as Mexican bean beetles, cabbage worms, leaf hoppers, and other insects that attack vegetables. Rotenone is also used for controlling insect parasites, such as cattle grubs, lice, fleas, and ticks on pets and livestock. It is not recommended for louse control on humans because it often produces skin irritations. It should not be administered internally.

Rotenone

Rotenone is the most active insecticidal principle of derris, but the root also contains other rotenoids. Commercial derris and cube roots usually contain 5% rotenone. It decomposes on exposure to heat and light, limiting its insecticidal effectiveness to a few days. It has relatively low toxicity for warm-blooded vertebrates and is highly toxic to some insects.

Mechanism of Action

The mechanism of selective toxicity to insects, fish, and certain mammals and low toxicity to most mammals remains largely a mystery. Rotenone inhibits oxidation of pyruvate (but not succinate) by rat liver mitochondria by blockade of glutamic dehydrogenase (oxidation of nico-

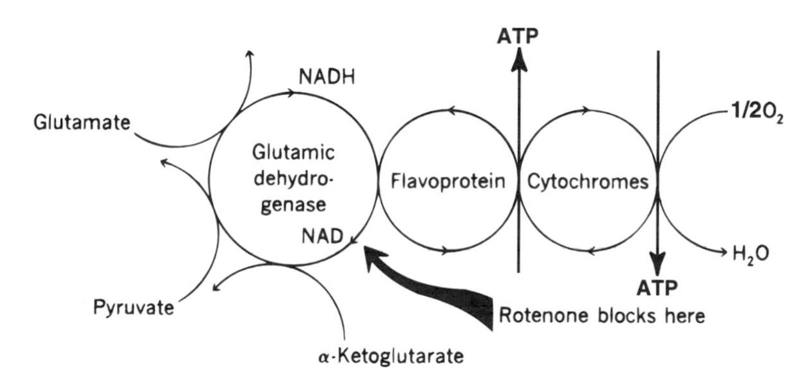

Fig. 42–2. Site of rotenone activity in blocking oxygen utilization.

tinamide adenine dinucleotide, $NADH_2 \rightarrow NAD$). Because oxidative phosphorylation shown in Figure 42–2 is the primary source of adenosine triphosphate (ATP) for the body, its disruption by blockade of oxygen utilization cuts off the body's energy source.[13] As expected, rotenone is synergized by piperonyl butoxide. Its metabolism in insects and mammals has been investigated.[14]

NICOTINOIDS

Nicotine has been of interest to pharmacologists because of its paralyzing effect on autonomic ganglia and its acetylcholine type of action. Use of nicotine as an insecticide dates back to 1746, when an infusion of tobacco leaves was recommended for plum curculio.

Nicotine, the major insecticidally active alkaloid in tobacco leaves, occurs in concentrations as high as 14%, depending on the species.

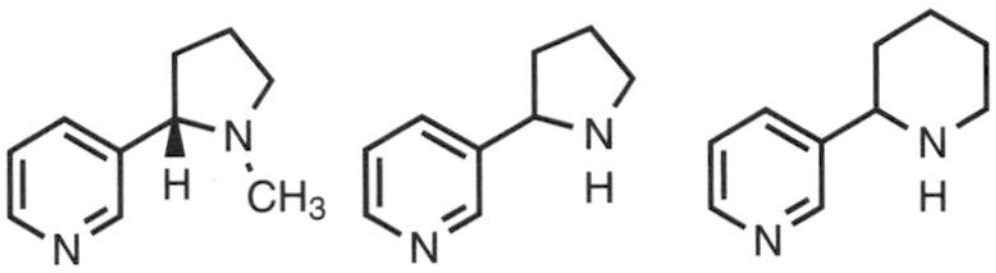

Nicotine Nornicotine Anabasine

Nornicotine and anabasine are two closely related alkaloids, also found in tobacco, that contribute to its high insecticidal activity.[15]

Nicotine is now commercially obtained from tobacco by steam distillation or solvent extraction. As an insecticide, it is available principally in two forms: as an aqueous solution containing 40% nicotine and as the sulfate salt. It is known commercially as Black Leaf 40.

Nicotine is used as 40% solutions in sprays or as dusts to control aphids (plant lice) and similar insects. Because of its high toxicity to mammals and the introduction of new and more selective insecticides, nicotine has lost much of its popularity in recent years.

Nicotine is a rapidly acting poison that acts primarily at the autonomic ganglia, where it exerts its effect first as a stimulant and then as a depressant, resulting in paralysis and functional failure of vital organs. Nicotine is intensely toxic to mammals following inhalation or dermal application because it is rapidly absorbed from the skin. The base tends to be more rapidly absorbed than the salts. For humans, the median lethal dose (MLD) is approximately 60 mg, and 4 mg may produce grave symptoms. The oral LD_{50} for rats is reported to be 30 mg/kg for the salt or the base.[2]

Nicotine is made up of a pyridine ring containing nitrogen that, because of the aromatic character of the ring, is only weakly basic (pK_a approximately 3). The second ring is a saturated five-membered pyrrolidine ring, which has a moderate basicity of about pK_a 8. Consequently, at the physiologic pH of 7.4, about 80% of the compound is protonated at the pyrrolidine ring.

Symptoms of nicotine poisoning resemble those caused by cholinesterase inhibitors: salivation, vomiting, muscular weakness, fibrillation, and acute poisoning. Because nicotine is rapidly metabolized in the body, death may be averted if supportive therapy (artificial respiration with oxygen) is instituted and the patient can be maintained during the interval necessary for detoxification and excretion of the poison.

CHLORINATED HYDROCARBONS

DDT

Since the discovery of its insecticidal effects in 1939 by Paul Muller, DDT (chlorophenothane) has been phenomenally successful and for many years was considered

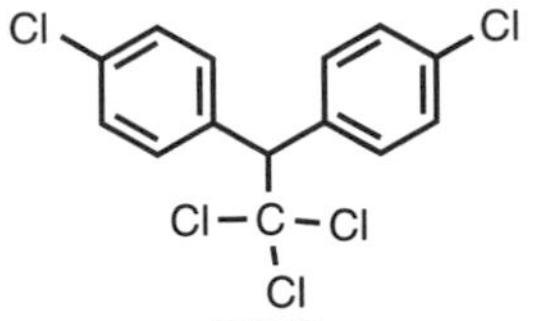

DDT

the ideal insecticide. It was cheap, nontoxic, persistent, and had a spectrum of activity that included many insects. It has, however, fallen into disfavor because of its potentially harmful effects on wildlife, the steady increase in insect resistance, and its accumulation in plants and animals. With the availability of more effective, less persistent insecticides that represent less of an ecologic hazard, there is little justification for the use of DDT for agricultural purposes or as a pediculicide in humans in the United States.[16] Both DDT and lindane are still widely used to mothproof clothes.

Although several hundred thousand tons of DDT have been used annually in agriculture, forestry, and public health for typhus, malaria, and yellow fever mosquito control, the only confirmed cases of human poisoning by DDT have been the result of massive accidental or suicidal ingestion.

The low vapor pressure of DDT is the cause of its remarkable persistence, killing insects for months and years on treated surfaces. It is unusually nonpolar, making it extremely oil soluble and water insoluble, which contributes to its accumulation in the food chain. It is chemically stable and insensitive to sunlight, which is beneficial for its insecticidal effects but undesirable from an ecologic point of view. Commercially available DDT contains approximately 80% of the p,p′-isomer and about 20% of

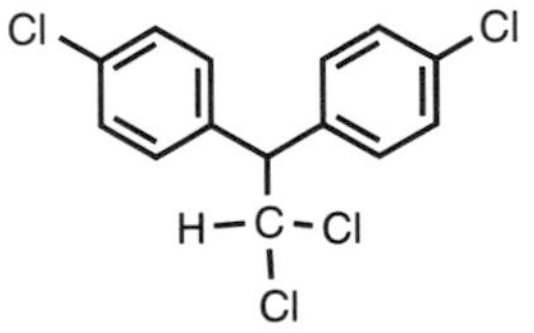

DDD

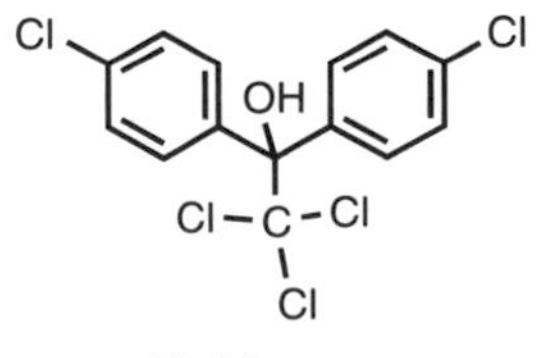

Kelthane

Table 42–4. Toxicity of DDT to Insects and Mammals

Organism	Route	LD_{50} (mg/kg)
Rat	Topical	3000
	Oral	400
	Intravenous	50
Rabbit	Topical	300–2820
	Oral	300
	Intravenous	50
Japanese beetle	Topical	93
American cockroach	Topical	10
Housefly	Topical	8–21

From R. D. O'Brien, Insecticides—Action and Metabolism, New York, Academic Press, 1967.

the o,p′-isomer. Some of the toxic effects of DDT have been attributed to the presence of the o,p′-isomer.

The compound 2,4′-dichlorodiphenyldichloroethane (o,p′-DDD), a minor metabolite of o,p′-DDT, is used in the chemotherapy of neoplasms of the adrenal cortex. Some DDT analogs, including kelthane, methoxychlor, DDD, and perthane, have been synthesized in attempts to remove the undesirable characteristics. Several have useful insecticidal properties—particularly methoxychlor, which is less toxic and more biodegradable. It is more expensive, however, which has contributed to its relative obscurity until recently.

Toxicity

The effects of one or a few massive doses of DDT administered by the oral, dermal, or respiratory route have been studied in at least 50 persons. Results show excellent agreement with reports of accidents involving DDT and indicate that 1.0 mg/kg is a threshold dose, causing moderate illness in some individuals and no symptoms in others.[17] As can be seen from Table 42–4, toxicity of DDT varies greatly, depending on the species and the route of administration.

Mechanism of Action

The mechanism of toxicity of chlorinated hydrocarbon pesticides is similar to that proposed for pyrethroids. DDT alters Na^+, K^+, and perhaps Ca^{2+} ion flow through nerve cell membranes. The mechanism may involve DDT-mediated inhibition of Na^+, K^+-, or Mg^{2+}-ATPase,[9,12,19] enzymes known to be involved in ion transport in the central nervous system.

The primary biochemical lesion in the nerve axon seems to result from the peculiar stereochemistry of the molecule, which is absorbed into a critical interface of the lipoprotein (e.g., the sodium gate), or from specific inhibition of ATPase.

The involvement of other ion transport mechanisms for DDT's neurotoxicity has been suggested by Matsumura and Ghiasuddin,[20] who showed that a Ca^{2+} ATPase formed in lobster nerve is highly sensitive to inhibition by DDT and other agents with neuroactivity similar to that of DDT. Chlorinated hydrocarbons also affect nerve cell phosphokinase activity and phosphorylation of sodium channels, and these activities may contribute to their neurotoxic effects.[19]

Metabolism

DDT is slowly metabolized by mammals into DDE (1,1-dichloro-2,2-bis(p-chlorophenyl) ethylene), DDA (bis(p-chlorophenyl)acetic acid), and DDD (1,1-dichloro-2,2-bis(p-chlorophenyl) ethane).[21] In insects, the best known metabolite is DDE, although DDA and dichlorbenzophenone are also found. The major cause of DDT resistance in houseflies is the buildup of DDT-dehydrochlorinase in these insects. Kelthane, an insecticide that is active against mites, is the metabolite produced in the domestic fruitfly and other species.[18]

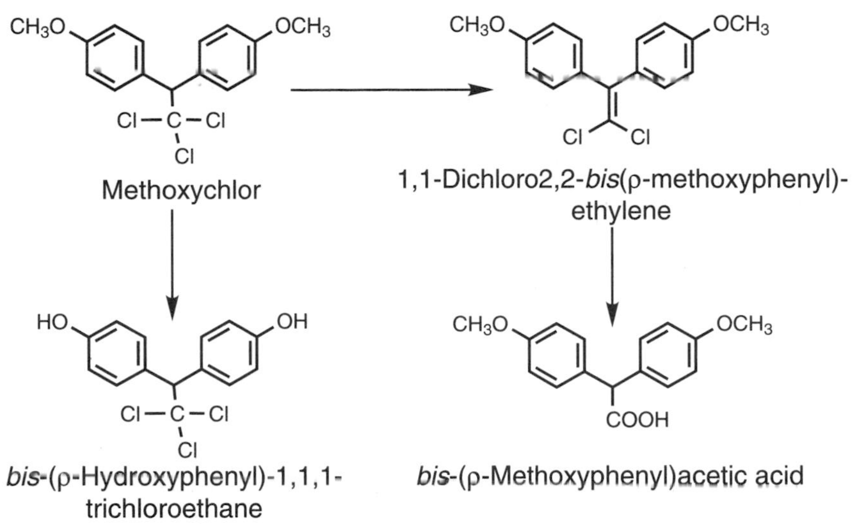

Methoxychlor → 1,1-Dichloro2,2-*bis*(ρ-methoxyphenyl)-ethylene

Methoxychlor → *bis*-(ρ-Hydroxyphenyl)-1,1,1-trichloroethane

1,1-Dichloro2,2-*bis*(ρ-methoxyphenyl)-ethylene → *bis*-(ρ-Methoxyphenyl)acetic acid

In the liver, DDT stimulates synthesis of many drug-metabolizing enzymes. Among the compounds metabolized more rapidly in the presence of DDT are hexobarbital, aminopyrine, aniline, p-nitrobenzoic acid, polycyclic hydrocarbons, and DDT.[22]

Kelthane and DDD are used principally as agricultural insecticides and have considerably less toxicity than DDT. (DDD has an oral LD_{50} of 3400 mg/kg).

Methoxychlor

Methoxychlor has a more limited spectrum of toxicity to insects than does DDT, but it has the advantage of being less toxic to mammals. (The oral LD_{50} for rats is 6000 mg/kg compared to 400 mg/kg for DDT.) Also, it is not readily stored in animal fat.

Since discontinuation of the use of DDT, methoxychlor is used more widely for controlling household pests, such as flies and moths; livestock pests, such as flies and lice; Mexican bean beetles; and a variety of insects attacking fruits, vegetables, forage crops, and for mosquito control.

Methoxychlor has a favorable rate of metabolism and biodegradability in animals.[22] In warm-blooded animals, it is detoxified by demethylation, which appears to eliminate the chronic effects seen with other chlorinated hydrocarbons. The two oxygen atoms facilitate an attack on the methoxychlor molecule, probably in the liver, to form phenolic derivatives, which can be conjugated and eliminated.

Perthane

This compound can also be considered to be closely related to DDT, but it may be more specifically considered an analog of DDD in which the para chlorine substituents on the phenyl ring are replaced by ethyl groups.

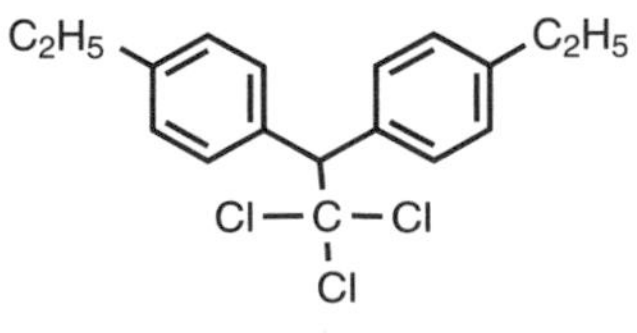

Perthane

This structural modification no doubt enables mammals to metabolize it more readily and results in greater toxicity to insects than other chlorinated hydrocarbon insecticides. The oral LD_{50} in the rat is 8 g/kg. Perthane is used as an insecticide on fruits and vegetables, in dairy and household formulations, and for mothproofing.

o,p′-DDD (Mitotane, Lysodren)

Although this compound has insecticidal properties and is a minor constituent of DDD (the p,p′-isomer), it is not used for its insecticidal activity. It is a minor metabolite of o,p′-DDT and is used in the treatment of adrenocortical neoplasms. Mitotane is available as tablets and is administered orally in doses of 8 to 10 g daily in the treatment of inoperable adrenocortical carcinoma.[23]

Lindane (γ-isomer of Hexachlorocyclohexane)

The term benzene hexachloride has been used incorrectly for this compound both in the chemical and pharmaceutical literature. The United States Pharmacopeia (U.S.P. XXIII) now uses lindane as the official name.

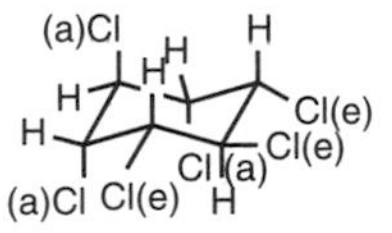

gamma isomer of hexachlorocyclohexane in chair conformation

conformation of chlorine substituent
α = Isomer: aaeeee
β = Isomer: eeeeee
γ = Isomer: aaaeee (lindane)
δ = Isomer: aeeeee
ε = Isomer: aeeaee
η = Isomer: aaeaee
θ = Isomer: aeaeee
ι = Isomer: aeaeae
a = axial; e = equatorial

This compound, the γ-isomer of hexachlorocyclohexane, is the most active of the eight possible isomers resulting from chlorination of benzene. Three of the chlorine substituents are in the equatorial conformation (in the plane of the ring), and three are in the axial conformation. Some of the possible isomers are enantiomorphs (optical isomers) rather than geometric isomers. The γ-isomer is 100 to 1000 times more toxic to insects than are any of the other isomers.

Lindane, the almost pure γ-isomer (99%), is more desirable because it does not have the musty odor and is less toxic to mammals than are the other isomers. It is active against a wide variety of insects, ticks, and mites and is generally effective in smaller doses than is DDT. Lindane is also widely used in household sprays, in dusts for animals, and for controlling pests on fruits and vegetables.

Both the lindane cream and the lotion are official in U.S.P. XXIII as pediculicides and scabicides. A shampoo is also available. Case reports of the central nervous system effects associated with both topical and accidental oral use of lindane have been reported.[24,25] It was known that percutaneous absorption can occur from its topical application in humans, but resulting neurotoxicity was not well established. Some of the effects were serious (convulsions), the result of the drug being applied too frequently or left on the skin for longer than the recommended time. Lindane has been used in vaporizers, but there is little justification for such use in the home because of the high mammalian toxicity.

Like other chlorinated hydrocarbons, the mode of action of lindane is presumably related to its affects on ion flow through nerve cell membranes by inhibition of ATPases. This drug, however, shows slightly greater inhibitory effects on Na^+, K^+-ATPase than on Mg^{2+}-ATPase, whereas DDT and chlordane (see the next section) behave in the opposite way.[120] This variation may explain some of the subtle differences in actions and toxicity of different chlorinated hydrocarbons.

Chlordane

This compound is the least toxic of the cyclodiene insecticides, which are products of a Diels-Alder reaction of hexachlorocyclopentadiene and a double-bonded compound followed by chlorination or further Diels-Alder reaction. Other compounds in this group include heptachlor, aldrin, and dieldrin, which have been banned from use in the United States as pesticides.

Chlordane was once used for controlling household pests such as cockroaches and ants; a variety of insects that live in soil; and lice, ticks, and other animal parasites, but in 1975 it was banned in the U.S. as a pesticide except for subsurface ground insertion for termite control. It has an acute oral toxicity of LD_{50} of 300 to 500 mg/kg in the mouse.

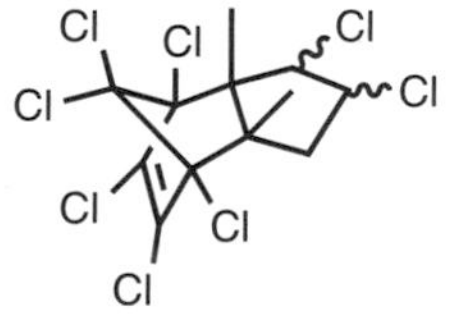

The mechanism of action of chlordane appears to be similar to that of DDT. Metabolism of cyclodiene insecticides involves epoxidation of an appropriate double bond. In the case of aldrin, epoxidation to dieldrin produces a faster acting toxicant but not necessarily a more toxic product.[18] The epoxide accumulates in the body fat, liver, and kidneys, from which it slowly disappears. In the case of chlordane, the metabolic products in mammals are principally hydrophilic metabolites and unchanged chlordane.

Chlordane

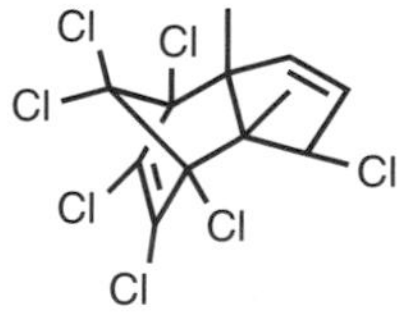

Heptachlor

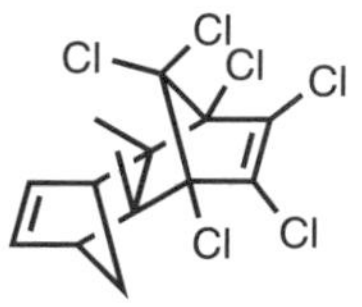

Aldrin

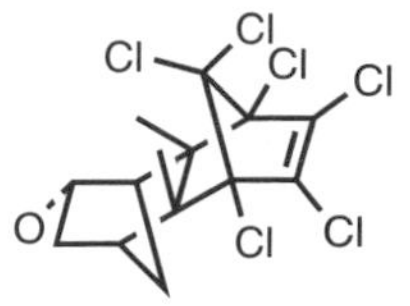

Dieldrin

ORGANOPHOSPHATES

The development of organophosphate insecticides started in 1932 with the observations of Lange and Kreuger in Germany that the vapors of diethyl phosphorofluoridate produced strong cholinergic effects in humans. Practical development of the compounds can be credited to Schrader, of the Bayer Company in Germany, who was interested in developing insecticidal substitutes for nicotine. His research resulted in the use of tetraethyl pyrophosphate as an insecticide in the late 1930s (Table 42–5). It is still used as an agricultural insecticide and in medicine for the treatment of glaucoma. Our technology has developed from tetraethyl pyrophosphate, which has an oral LD_{50} of 2.1 mg/kg, to the phosphate insecticide malathion, which has an LD_{50} of 1.5 g/kg in the rat.

During World War II, the Allies and the Germans devoted considerable effort to the preparation of organophosphates as cholinesterase inhibitors for use in warfare. Several such compounds, often referred to as nerve gases, were developed but fortunately were never used. Such applications stimulated research activity, however, and contributed greatly to our understanding of cholinesterase and its hydrolysis.

The development of this class of compounds is a good example of how molecular modification, based on a sound understanding of the chemistry and pharmacology of the drug, can lead to safer and more useful insecticides and therapeutic agents.

About 50 organophosphate insecticides are in agricultural, veterinary, and household use. They represent the largest and most versatile group of pesticides in use today. They have gained wide acceptance, largely because they can frequently replace persistent insecticides, such as the chlorinated hydrocarbons, are good substitutes for insecticides to which pests have developed a resistance, exhibit greater selective toxicity, and are rapidly metabolized and hydrolyzed.

The decreasing availability of chlorinated hydrocarbon insecticides and the consequent increased exposure to phosphate insecticides have caused a number of human fatalities and a variety of toxic effects. It is important that the more toxic compounds be handled with extreme caution and that medical and pharmaceutical practitioners

Table 42–5. Commonly Used Organophosphate Insecticides

Product	Chemical Name and Formula	Oral LD_{50} in the Rat (mg/kg)	Major Uses
Abate	O,O,O'O'-Tetramethyl-O,O'-thiodi-ρ-phenylene phosphorothioate	2030–2330	Mosquito and midge larvicide
Azinophosmethyl (Guthion)	O,O-Dimethyl-S-[4-oxo-1,2,3-benzotriazin-3(4H)-ylmethyl]phosphorodithioate	10–18	Broad-spectrum insecticide for cotton, tobacco, deciduous fruits, strawberries, beans, cole crops, potatoes, tomatoes, and other fruit, field and vegetable crops; ornamental plants, flowers, shrubs, and trees
Ciodrin	Dimethyl phosphate of α-methylbenzyl-3-hydroxy-*cis*-crotonate	125	Controls flies, lice, and ticks on dairy and beef cattle, swine, goats and sheep; also used on premises
Coumaphos (Co-Ral)	O,O-Diethyl-O-3-chloro-4-methyl-2-oxo-2H-1-benzopyran-7-yl phosphorothioate	90–110	Controls livestock ectoparasites, e.g., hornfiles, screw worms, lice, and ticks, and internal cattle grubs
Diazinon	O,O-Diethyl-O-(2-isopropyl-4-methyl-6-pyrimidyl)phosphorothioate	76–108	For controlling resistant soil insects, such as corn rootworms, wireworms, and cabbage maggots; also effective against many insect pests of fruits, vegetables, forage, and field crops, and ornamental plants; controls cockroaches and other household insects, nematodes, and insect pests of turf
Dichlorvos (DDVP, Vapona No-Pest Strip)	2,2-Dichlorovinyl dimethyl phosphate	56–80	Controls insects that are economically important in public health (human and livestock) and that attack stored products; effective against household pests
Dimethoate (Cygon)	O,O-Dimethyl-S-(N-methylcarbamoyl-methyl)phosphorodithioate	185–245	Used as a residual wall spray for controlling houseflies and insects on ornamental plants, vegetables, cotton, seed alfalfa, watermelons, bearing apples and pears, safflowers, and nonbearing citrus plants

(continued)

Table 42–5. *(Continued)*

Product	Chemical Name and Formula	Oral LD_{50} in the Rat (mg/kg)	Major Uses
Dursban	O,O-Diethyl-O-(3,5,6-trichloro-2-pyridyl) phosphorothioate 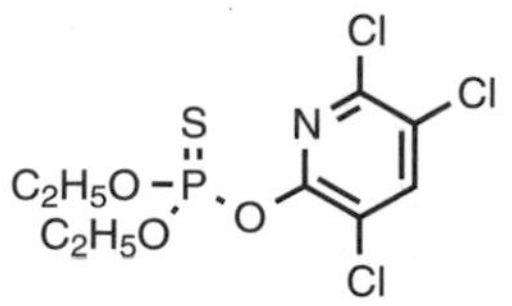	135	Broad-spectrum insecticide, particularly effective against mosquitoes, household pests, and other soil insects
Fenthion (Baytex, Queltox, Entex)	O,O-Dimethyl-O-[4-methylthio)-*m*-tolyl]-phosphorothioate 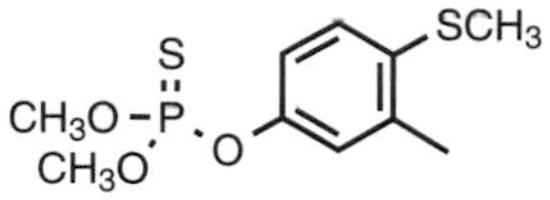	190–350	Controls mosquitoes and larvae, flies, cockroaches, ants, fleas, crickets, and wasps; Queltox formulated for bird control
Gardona	2-Chloro-1-(2,4,5-trichlorophenyl)vinyl dimethyl phosphate 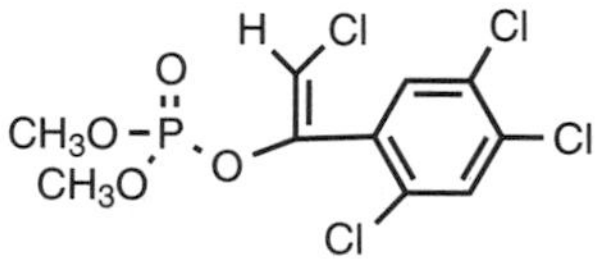	4000–5000	Controls corn earworms and fall armyworms, codling moths, gypsy moths, houseflies, and ectoparasites of livestock
Malathion (Cythion)	O,O-Dimethyl dithiophosphate of diethyl mescaptosuccinate 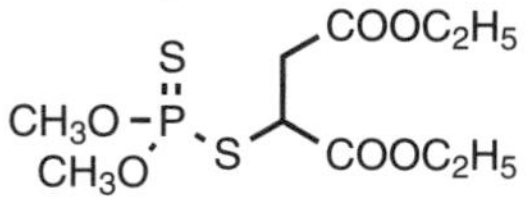	1509	Controls many insects, including mosquitoes, houseflies, spider mites, aphids, scales, and other sucking and chewing insects that attack fruits, vegetables, ornamental plants, animals, and stored products.
Naled (Dibrom)	1,2-Dibromo-2,2-dichloroethyl dimethyl-phosphate 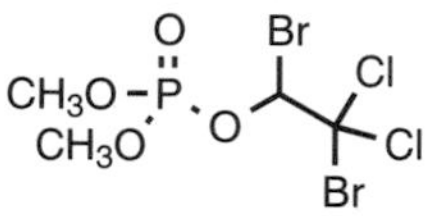	430	Contact insecticide and ascaricide with fumigant action; brief residual activity, used for vector control
Parathion	O,O-Diethyl-O-p-nitrophenyl phosphorothioate 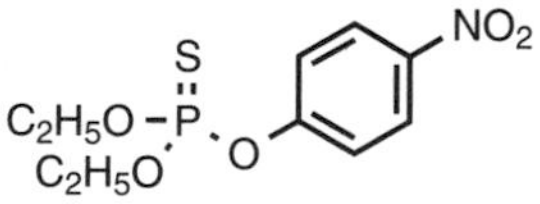	6–15	Broad-spectrum insecticide effective against aphids, mites, Lepidoptera, beetles, leaf-hoopers and thrips on fruits, vegetables, and forage crops; cotton insects, symphilids, rootworms, and other soil insects
Stirofos	2-Chloro-1-(2,4,5-trichlorophenyl)vinyl diethyl phosphate 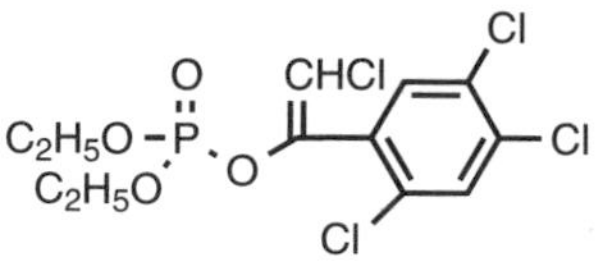	4,000–5,000	For protection against stored-product pests and vegetable and fruit insects

Table 42–5. *(Continued)*

Product	Chemical Name and Formula	Oral LD_{50} in the Rat (mg/kg)	Major Uses
TEPP	Tetraethyl pyrophosphate 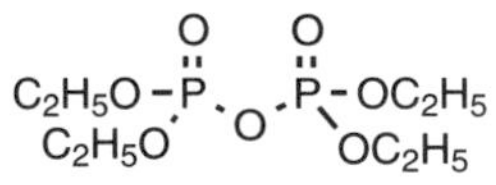	2.1	Contact insecticide effective against active stages of mites and other soft-bodied insects; used for treatment of glaucoma and myasthenia gravis
Trichlorfon (Dipterex, Neguvon)	O,O-Dimethyl-(2,2,2-trichloro-1-hydroxy-ethyl)phosphonate 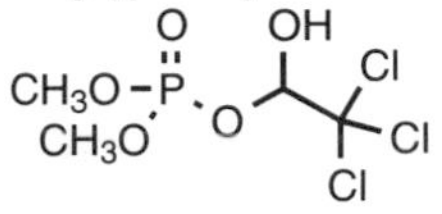	450–500	For control of flies, insects on crops, cattle and horse parasites

be aware of their mechanism of action so that they can initiate prompt and effective countermeasures.

Chemistry

The term organophosphate is used to designate all toxic organic compounds containing phosphorus. Compounds are named according to the substituents attached to the phosphorus atom (see below).

Most organophosphates can be thought of as esters of alcohols with a phosphorus acid or as anhydrides of a phosphorus acid with some other acid. Parathion, therefore, is an ester of thiophosphoric acid with two molecules of ethanol and the "acid" p-nitrophenol.

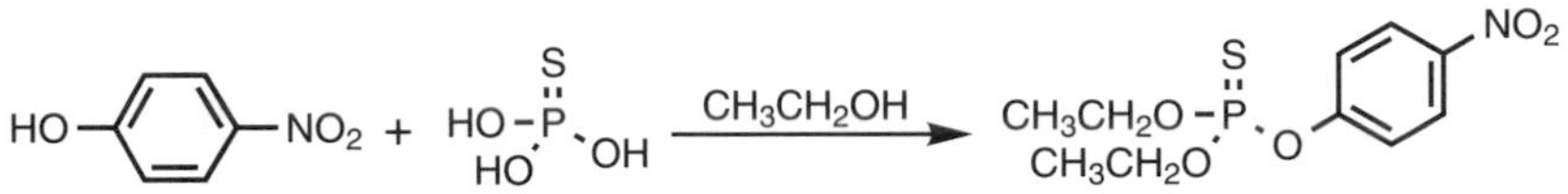

p-Nitrophenol + Thiophosphoric acid —(CH_3CH_2OH)→ Parathion

Phosphate · Phosphorothioate · Phosphorodithioate · Phosphorothiolate · Phosphoramidate · Phosphonate

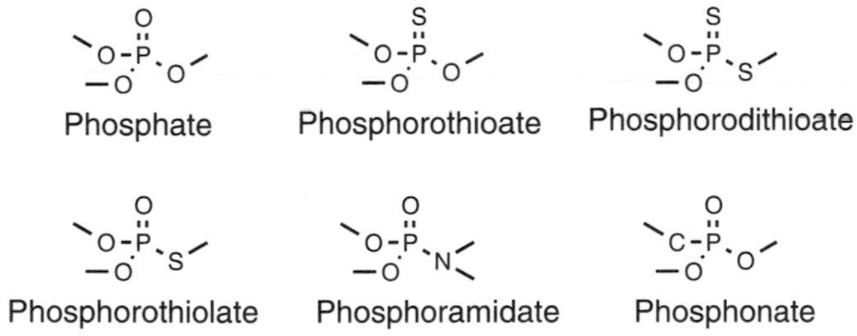

Nonenzymatic Hydrolysis

As esters, the organophosphates can be hydrolyzed at a rate directly related to the alkalinity of the base and the relative positive charge on the phosphorus atom. The mechanism involves nucleophilic attack on an electrophilic, or relatively positive phosphorus atom by a negative basic group, i.e., where the base is a hydroxide ion. The more positive the phosphorus atom, the more effective the attack. Consequently, the rate of hydrolysis is determined by the properties of the group attached to the phosphorus atom. When electrophilic (electron-withdrawing) groups are attached, the phosphorus atom is subjected to faster nucleophilic attack. When nucleophilic (electron-donating) groups are attached, they tend to make the phosphorus atom more negative.

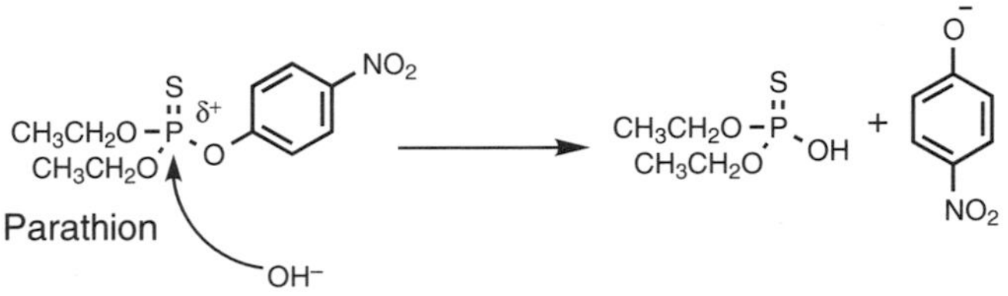

Parathion + OH^- → ...

The p-nitrophenyl group in paraoxon has an electrophilic inductive effect on the phosphorus atom, whereas the p-aminophenyl group in amino-paraoxon has a nucleophilic effect. Such chemical changes have a pro-

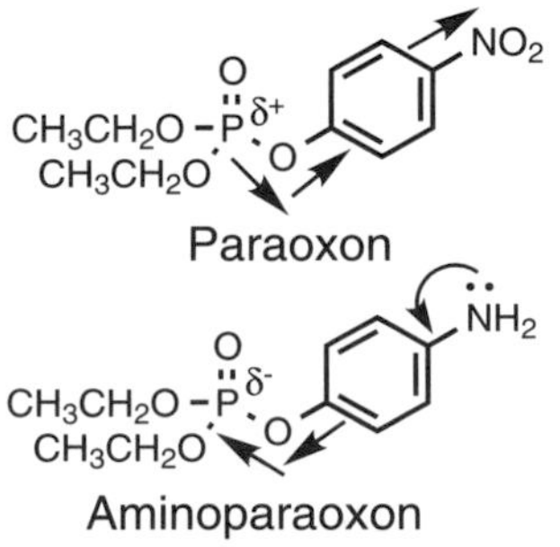

Paraoxon

Aminoparaoxon

nounced effect on the phosphorus atom and, consequently, the rate of hydrolysis, which is also directly related to its anticholinesterase activity and toxicity. Chemical or metabolic changes that modify the groups attached to the phosphorus atom alter the susceptibility of the organophosphates to hydrolysis. Because the oxygen atom in $>P(=O)—$ is more electrophilic than the sulfur in $>P(=S)—$, conversion of parathion to paraoxon makes it 30 times more susceptible to hydrolysis. An understanding of such mechanisms is not only important in appreciating the stability of organophosphates but also essential for an understanding of their interaction with cholinesterase.

Enzymatic Hydrolysis

The mode of action of the organophosphates in both insects and mammals is through irreversible inhibition of acetylcholinesterase in the cholinergic synapses. Interaction of acetylcholine (liberated by an impulse) with the postsynaptic receptor is therefore greatly potentiated.

Interaction Between Acetylcholinesterase and Its Substrates

Acetylcholine is the endogenous substrate of acetylcholinesterase, the enzyme present in great excess in the venous tissue of mammals and insects. Acetylcholinesterase, a member of a superfamily of serine hydrolases, hydrolyzes acetylcholine to produce acetate and choline. The three-dimensional structure of acetylcholinesterase in the electric organ of Torpedo californica electric fish has been determined by x-ray crystallography analysis.[26] The catalytic center of the enzyme contains two subsites, which lie near the bottom of a deep and narrow gorge that reaches halfway into the enzyme. The esteratic subsite is believed to resemble the catalytic subsites of other serine hydrolases. The esteratic site serine residue, with

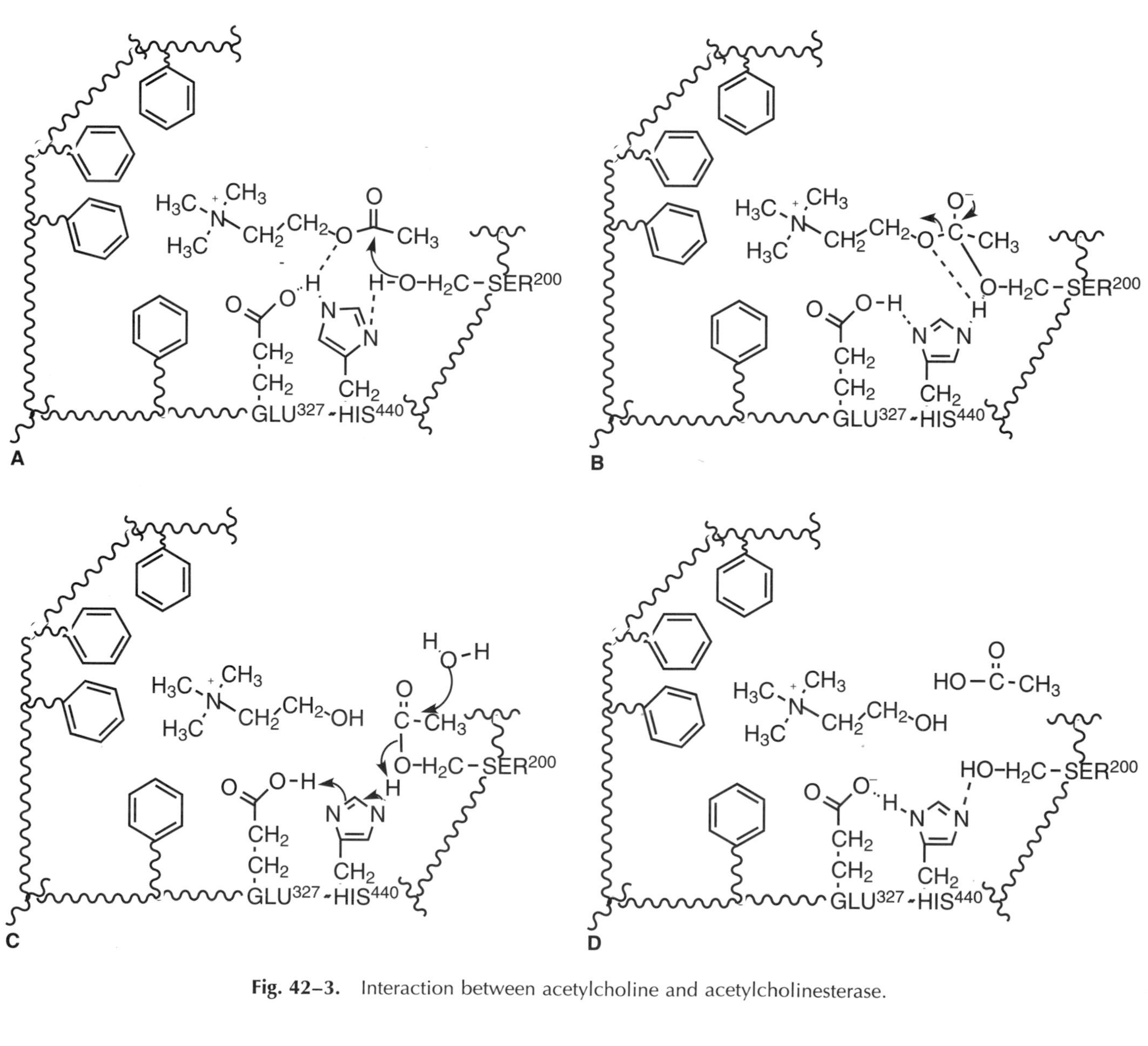

Fig. 42–3. Interaction between acetylcholine and acetylcholinesterase.

which organophosphates react, has been unequivocally established as Ser^{200}. The hydroxyl group Ser^{200} forms a hydrogen bond with the imidazole group of an adjacent histidine residue (His^{440}). In the absence of ligand, another hydrogen bond is formed between His^{440} and an unprotonated carboxylic acid group supplied by a glutamate residue (Glu^{327}). This catalytic triad forms a planar array at the active site, similar to that found in chymotrypsin and other serine proteases. Early proposals suggested an anionic subsite to bind the positively charged timethylammonium group of acetylcholine. Cohen et al., however, showed that uncharged analogs of acetylcholine also bind to this subsite.[27] The x-ray crystal structure of acetylcholinesterase shows that the "anionic" subsite is, in fact, uncharged and lipophilic, consisting mainly of aromatic residues. This misnamed "anionic" site apparently binds quaternary ammonium groups such as found in acetylcholine through interactions with the pi (π) electrons of some of the 14 aromatic residues that line the active site gorge. In addition to the two subsites of the catalytic center, acetylcholinesterase possesses one or more additional binding sites for quaternary ammonium groups.[28] These "peripheral anionic sites" are clearly distinct from the choline-binding pocket at the active site. A representation of the interaction between acetylcholine and acetylcholinesterase is shown in Figure 42–3*A*.

The catalytic mechanism of acetylcholinesterase resembles that of other serine proteases in which the serine hydroxyl group is rendered highly nucleophilic by an electron-proton charge-relay shuttle involving the imidazole group of His^{400} and the nearby unprotonated carboxylic acid group of Glu^{327}. At the esteratic subsite of acetylcholinesterase, the acetyl group of acetylcholine undergoes nucleophilic attack by the Ser^{200} hydroxyl group and a covalent bond is established, resulting in

$$H_3C{-}C(=O){-}O{-}CH_2CH_2{-}\overset{+}{N}(CH_3)_3 + H_2O \rightleftharpoons HO{-}CH_2CH_2{-}\overset{+}{N}(CH_3)_3 + AcO^- + H^+$$

Acetylcholine — Choline — Acetate

the formation of a tetrahedral intermediate between the enzyme and the substrate. This intermediate collapses to release choline and an acetylated enzyme complex that readily undergoes hydrolysis to form acetate and regenerated acetylcholinesterase. These catalytic events are depicted in Figure 42–3.

The reaction of organophosphates with cholinesterase parallels that involved in the hydrolysis of acetylcholine. The reaction of the Ser^{200} OH of acetylcholinesterase can be likened to the attack of an OH^- ion in the nonenzymatic hydrolysis of phosphate esters already discussed. Again the enzymatic hydrolysis can be illustrated with paraoxon: the nature of the inhibited enzyme, $(RO)_2{-}\overset{O}{\overset{\|}{P}}{-}O{-}$, is independent of the nature of the leaving group, $\left(^-O{-}C_6H_4{-}NO_2\right)$ for paraoxon. Many organophosphate cholinesterase inhibitors therefore yield the same phosphorylated enzyme. The ease of alkaline hydrolysis of the organophosphates parallels their potency against cholinesterase, which is directly related to the electrophilicity of the phosphorus atom.

Metabolism and Selective Toxicity of Organophosphates

Most of the common organophosphates used as insecticides are poor inhibitors of cholinesterase in vitro. When these same organophosphates are examined in vivo, however, they prove to be potent enzyme inhibitors because they are converted, or activated, in vivo to give compounds that are potent cholinesterase inhibitors.

The most common activation is the conversion of a phosphorothioate, or P=S compound, to a phosphate, or P=O compound. Parathion, therefore, is activated or oxidized to paraoxon, and malathion is converted to malaoxon. Metabolic activation can produce compounds that are more potent cholinesterase inhibitors by increasing the electrophilicity of the phosphorus atom. Similarly, metabolic conversions of phosphate esters can produce compounds in which the electrophilicity of the phosphorus atom is decreased, thereby resulting in less anticholinesterase potency. This process is illustrated by the partial hydrolysis of malathion by carboxyesterases to malathion acid, which occurs rapidly in mammals (see p. 922 for reaction sequence). Malathion acid is an extremely poor inhibitor of cholinesterase because the electrophilic character of the phosphorus atom is reduced by the closeness of the carboxyl group, resulting in inability of the compound to attack the nucleophilic center of the enzyme (the hydroxyl group on acetylcholinesterase).

$$(CH_3CH_2O)_2P(=O){-}O{-}C_6H_4{-}NO_2 + \text{Enzyme}{-}OH \longrightarrow (CH_3CH_2O)_2P(=O){-}O\text{Enzyme} + {}^-O{-}C_6H_4{-}NO_2$$

Paraoxon — Phosphorylated enzyme — leaving group

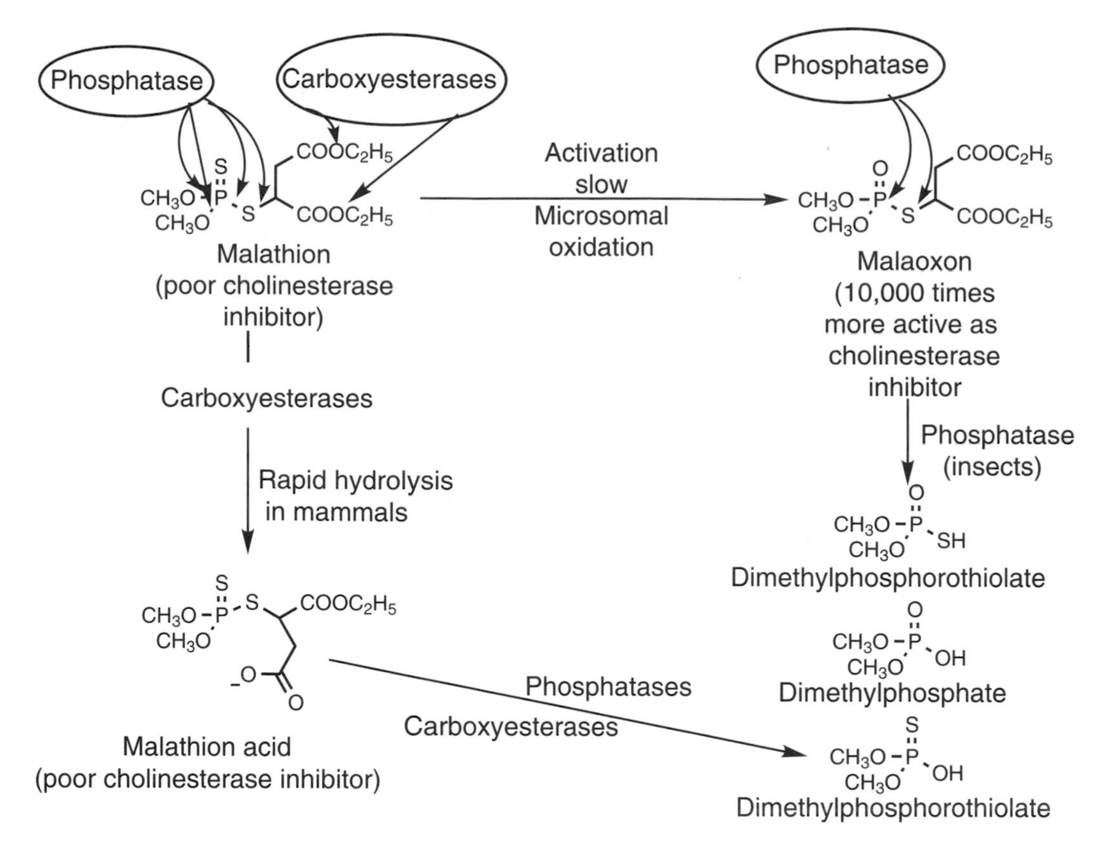

Fig. 42–4. Metabolism and activation of malathion in humans and insects. Carboxyesterases are present primarily in mammals and are responsible for hydrolysis at the sites indicated. Houseflies and mosquitoes that have developed a resistance to malathion have a high concentration of this enzyme. Blocking this enzyme in mammals results in greater toxicity of the compound. Phosphatases are present primarily in insects and are responsible for hydrolysis at the sites indicated.

Carboxyesterases, phosphatases, amidases, and microsomal oxidative enzymes degrade organophosphates and contribute to their selective toxicity. The metabolism and activation of malathion are shown in Figure 42–4. Differences in the rates of breakdown or of activation of compounds in different organisms are responsible for the differential toxicity.

Malathion is highly toxic to fish, and this difference in toxicity is thought to result from a decreased ability of fish to deactivate enzymatically the lethal metabolite malaoxon.[25] It follows that when two types of organisms, such as insects and mammals, metabolize a compound at different rates and activation is required, a greater differential toxicity results.

Carboxyesterase

Malathion → Malathion acid

The organophosphate insecticide malathion, for example, is toxic to most insects and has relatively low toxicity to mammals because of the different rates of activation, P═S → P═O, and of hydrolysis in insects and mammals (Table 42–6). In mammals, hydrolysis occurs primarily at the carboxyester group to give malathion acid, which is a poor cholinesterase inhibitor. In insects, however, oxidation to malaoxon, a potent cholinesterase inhibitor, followed by cleavage of the phosphate thioester bond, is the principal route of metabolism.

Anticholinesterase Intoxication[29]

In chronic or acute poisoning by anticholinesterase compounds such as the cholinesterase inhibitors (neostigmine, isoflurophate, and pyridostigmine), used clinically for the treatment of glaucoma or myasthenia gravis, cholinesterase levels in blood and tissue decrease to below normal pre-exposure levels. Such cholinesterase inhibition may produce a cholinergic crisis, characterized by increased activity of smooth muscle and secretory

Table 42–6. Toxicity of Malathion to Insects and Mammals

Organism	Route	LD_{50} (mg/kg)
Rat	Oral	1509
Guinea pig	Oral	570
Chicken	Oral	275
American cockroach	Topical	8.4
Housefly	Topical	30
Pea aphid	Topical	0.75

glands (nausea, vomiting, diarrhea, sweating, and increased salivary and bronchial secretions), bradycardia, central nervous system symptoms, muscular weakness, fasciculations, and finally coma and death. Symptoms may begin almost immediately after exposure to a direct inhibitor such as tetraethyl pyrophosphate (TEPP) or may be delayed several hours after exposure to an inhibitor such as parathion, which is metabolically activated. Administration of atropine combats the muscarinic parasympathetic effects of acetylcholine, but it has no effect on the skeletal neuromuscular blockade.

Pralidoxime chloride (2-PAM chloride) reactivates the inhibited enzyme at nicotinic as well as muscarinic sites and therefore relieves the skeletal neuromuscular block.

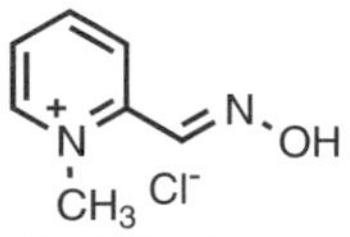

Pralidoxime chloride

Treatment of Poisoning

An effective treatment for intoxication with anticholinergic agents is as follows.[29]

Step 1. Remove clothing and any contamination. If victim is not breathing, institute artificial respiration.

Step 2. Administer atropine 1 mg (moderate poisoning) or 5 mg (severe poisoning) I.V. every 20 to 30 minutes until sweating and salivation disappear and slight flush and mydriasis appear.

Step 3. Institute resuscitation (mouth-to-mouth, mouth-to-nose, or mouth-to-oropharyngeal airway), if indicated.

Step 4. Give 1.0 g of pralidoxime I.V. over 20 to 30 minutes (moderate poisoning) or, if the patient shows no improvement (severe poisoning), start I.V. infusion at 0.5 g per hour.

The status of the patient should be monitored by repeated analysis of the degree of inhibition of serum cholinesterase and erythrocyte acetylcholinesterase activities, which are accurate indicators of the severity of poisoning by organophosphates or carbamates.

CARBAMATES

Esters of carbamic acid, $HOC(=O)NH_2$, have been known for many years to have pharmacologic effects associated with parasympathetic nerve transmission. Physostigmine, an alkaloid from the dried ripe seed of Physostigma venenosum, and neostigmine are two carbamates used principally for their effect on cholinesterase. The N-methyl and N,N-dimethylcarbamic esters of a variety of phenols and heteroxyclic enols have useful insecticidal properties and, like organophosphates, exert this toxic action on insects by inhibiting cholinesterase. The carbamates combine with the enzyme in the same manner as do the organophosphates, except that the carbamoylated enzyme reacts with water, more slowly than does the acylated enzyme. These compounds are therefore often referred to as reversible inhibitors of cholinesterase.

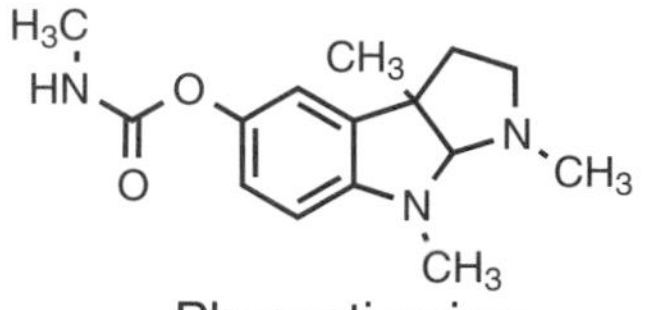

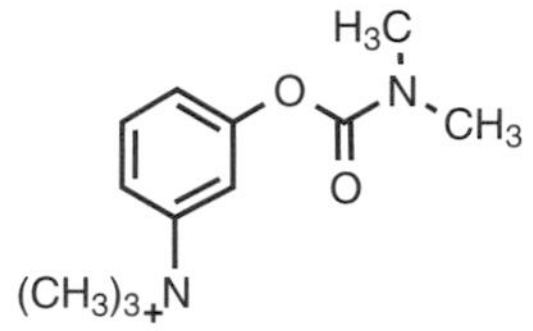

Physostigmine

Neostigmine

As shown previously, the organophosphates act at the esteratic site of cholinesterase to form a phosphorylated enzyme. With highly toxic compounds such as TEPP and diisopropyl phosphorofluoridate (isoflurophate), essentially no spontaneous hydrolytic reactivation of the enzyme occurs. Consequently, such compounds are referred to as nonreversible.

The carbamate insecticides also differ from the organophosphorus insecticides in that they do not require metabolic activation before exhibiting their enzyme-inhibiting properties. The carbamates are deactivated, however, by metabolism, and their insecticidal activity is therefore enhanced by synergists, such as piperonyl butoxide, which inhibit microsomal oxidation.

Even though the toxic carbamates are potent inhibitors of cholinesterase, no general correlation has been found between anticholinesterase activity and insecticidal activity. Some compounds that are good anticholinesterases were found to be nontoxic to houseflies, and others that had toxicity for houseflies were poor anticholinesterases.[30] Why is it that the medicinal carbamates have no insecticidal action? O'Brien[18] presents the point of view

Table 42–7. Commonly Used Carbamate Insecticides

Product	Chemical Name and Formula	Oral LD_{50} in the Rat (mg/kg)	Major Uses
Aldicarb (Temik)	2-Methyl-2-(methylthio)-propionaldehyde N-methylcarbamoyl oxime 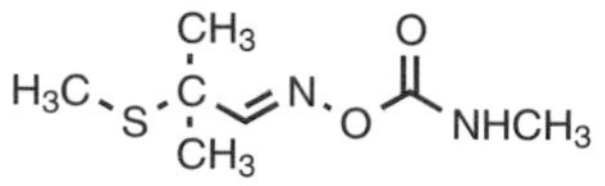	0.93	Broad-spectrum systemic insecticide used for soil and seed treatments
Carbaryl (Sevin)	1-Naphthyl-N-methylcarbamate 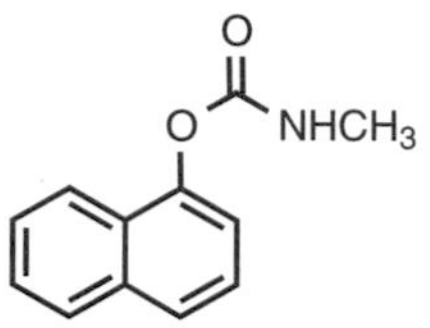	850	Control of insects on fruits, vegetables, forage, cotton and other economic crops, as well as poultry and pets; gypsy moth control; replacement for DDT
Carbofuran (Furadan)	2,3-Dihydro-2,2-dimethyl-7-benzofuranyl-methylcarbamate 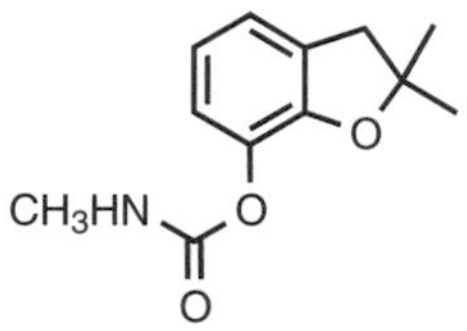	11	Systemic and contact insecticide and nematocide; broad-spectrum insecticide used on tobacco, rice, corn, peanuts, and sugar cane
Dimetilan (Snip Fly Bands)	2-Dimethylcarbamyl-3-methyl-5-pyrazolyl dimethylcarbamate 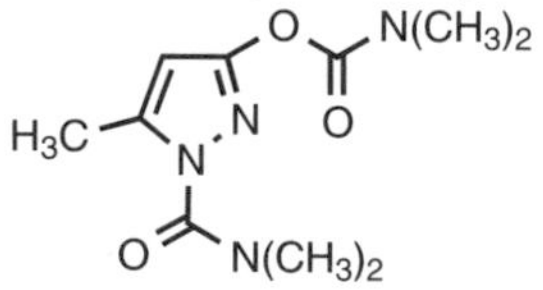	64	Impregnated in plastic fabric bands placed near ceilings in farm buildings for fly control
Propoxur (Baygon)	2-Isopropoxyphenyl-N-methylcarbamate 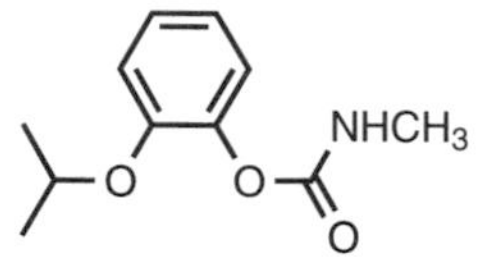	83	For mosquitoes, flies, ants, cockroaches, spiders, and other household pests
Zectran	4-Dimethylamino-3,5-xylenyl-N-methylcarbamate 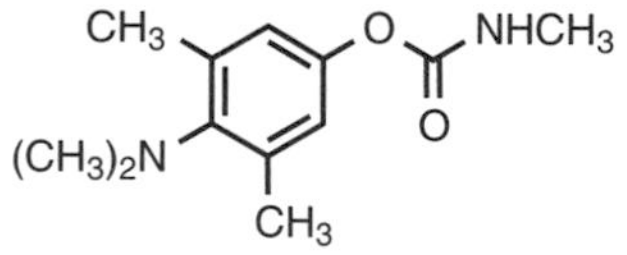	37	Broad-spectrum insecticide effective against lepidopteran larvae, primarily in the control of forest insects

that insects do not use cholinesterase in their neuromuscular junctions and that their vital cholinesterase is central and protected by a barrier system that hinders penetration of ionized molecules. All medicinal carbamates are ionized or ionizable and therefore have little effect on insects because they cannot penetrate the insect's central nervous system.

The carbamates have great variations in toxicity to insects. They are not regarded as broad-spectrum insecticides. In most cases, the oral toxicity of the insecticidal carbamates presents less of a hazard to mammals than does that of the organophosphates The structures, chemical names, toxicity, and uses of some widely used carbamates are shown in Table 42–7.

In case of poisoning by carbamate insecticides, only atropine should be administered, in addition to general supportive therapy. 2-PAM chloride is not helpful and *should not* be used for carbamate poisoning.

INSECT REPELLENTS

These compounds represent the oldest and, for many years, most widely used method to protect man from insects. When applied to the skin, repellents discourage insect attack and, as a result, offer protection from insect bites. These compounds can be categorized as chemical repellents,[31] as distinguished from purely physical methods of repelling insects.

Chemical Repellents

The first chemical repellent was undoubtedly discovered soon after man found that a fire (especially a smoky, slow-burning one) was a fairly efficient method of repelling insects. Early folklore indicates extensive use of plants, plant extracts, or other strong-smelling odoriferous and pungent compounds as insect repellents. Oil of citronella, turpentine, pennyroyal, cedarwood, eucalyptus, and wintergreen are still used in insect-repellent formulations.

Oil of citronella, which is extracted from Cymbopagon nardus (L), contains geraniol as its primary component, with lesser amounts of citronellol, citronella, borneol, and other terpenes. Citronellol and the corresponding aldehyde, citronellal, are considered the principal mosquito repellents in oil of citronella.[32] On the other hand, geraniol is an attractant for the oriental fruitfly.[33] Geraniol, citronellol, and citronellal are attractants for the Japanese beetle. Certain compounds that are attractants at low concentrations are repellents at higher concentrations.

Insect repellents were developed during World War II to keep mosquitoes and other annoying and disease-carrying pests from troops in tropical and semitropical regions. Thousands of compounds were tested by the U.S. Department of Agriculture, but only a few proved effective and safe enough to be used on the skin. An ideal repellent must be nontoxic, nonirritating, harmless to clothing, nonallergenic, have an inoffensive odor, and protect for several hours.

As a result of these studies during World War II, a repellent for flies, mosquitoes, chiggers, ticks, and gnats was developed for use on skin and clothing. It consisted of 60 parts dimethyl phthalate, 20 parts ethohexadiol, and 20 parts butyl mesityl oxide.

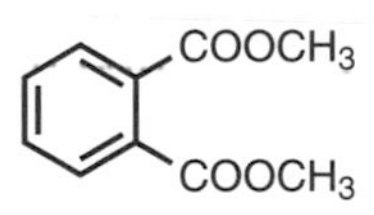

Dimethyl phthalate

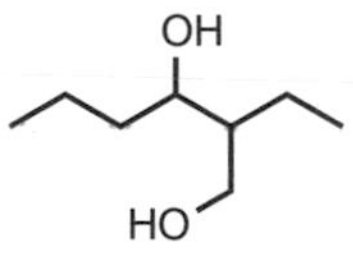

Ethohexadiol

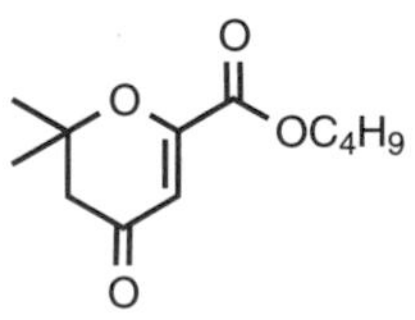

Butyl mesityl oxide

Several compounds have been marketed for use on humans, but the market is dominated by ethohexadiol and diethyl toluamide. The latter compound is the best singlet repellent available. It is effective against a wide range of insects, including mosquitoes, flies, chiggers, and biting flies.

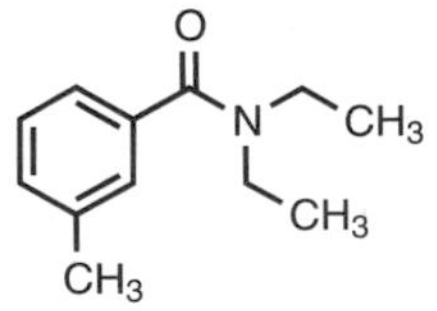

Diethyl toluamide

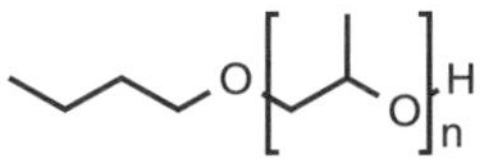

Butoxypolypropyleneglyciol
(Stabilene, Crag fly repellent)

Relatively little work has been carried out to expand our knowledge of the mechanism of attraction or in developing an oral or systemic repellent for use in humans. Differences in body chemistry may indeed be responsible for the relative unattractiveness of some individuals to some insects. Beroza and his group[33] reported that L(+)-lactic acid is the major component isolated from humans that attracts the female yellow fever mosquito Aedes aegypti (L). Good correlation was obtained between the attractiveness of an individual to mosquitoes and the quantity of lactic acid present in acetone washing of hands.

MOTHPROOFING COMPOUNDS

Clothes moths and carpet beetles cause as much as 500 million dollars damage annually to clothing, rugs, and upholstery. Therefore, the demand is great for insecticidal protection of these articles. Sodium fluosilicate (Na_2SiF_6), used at a concentration of 0.5 to 0.7% in water solution with 0.25 to 0.5% wetting agent, is fixed in wool and is an effective mothproofing agent. The use of chlorinated hydrocarbons for this purpose can hardly be justified. Perthane, which is closely related to DDT but less toxic, is also being sold as an aerosol mothproofing agent for household use. Paradichlorobenzene and naphthalene are the mothballs used as space fumigants to destroy fabric pests in confined storage spaces.

REFERENCES

1. J. E. Casida, Ed., *Pyrethrum, the Natural Insecticide,* New York, Academic Press, 1973.
2. R. L. Metcalf, in *Kirk-Othmer Encyclopedia of Chemical Technology,* Vol. 13, 3rd Ed., New York, Wiley Interscience, 1981, pp. 413–485.
3. M. Elliot, Ed., *Synthetic Pyrethroids,* A.C.S. Symposium Series 42, Washington, D.C., American Chemical Society, 1977.
4. J. E. Casida, *J. Agr. Food Chem., 18,* 753(1970).
5. D. M. Soderlund and J. E. Casida, in *Synthetic Pyrethroids,* A.C.S. Symposium Series 42, Washington, D.C., American Chemical Society, 1977, p. 162.
6. T. Narahashi, *Neurotoxicology, 6,* 3(1985).

7. H. P. M. Vijverberg and J. R. de Weille, *Neurotoxicology, 6,* 23(1985).
8. A. Lombet, et al., *Brain Res., 459,* 45(1988).
9. T. Narahashi, *Trends Pharmacol. Sci., 13,* 236(1992).
10. D. M. Soderlund, *Neurotoxicology, 6,* 35(1985).
11. M. W. Brooks and J. M. Clark, *Pestic. Biochem. Physiol., 28,* 127(1987).
12. D. E. Ray, in *Handbook of Pesticide Toxicology,* Vol. 2, W. J. Hayes, Jr. and E. R. Laws, Jr., Eds., San Diego, Academic Press, 1991.
13. N. Y. Kahn, in *Pesticide Chemistry: Human Welfare and the Environment,* J. Migomoto and P. C. Kearney, Eds., Oxford, Pergamon, 1983, pp. 115–121.
14. L. Crombie and R. Peace, *J. Chem. Soc. (Lond),* 5445(1961).
15. I. Yamamoto, *Adv. Pest Control Res., 6,* 231(1965).
16. U.S. Department of Health, Education and Welfare, *Report of the Secretary's Commission on Pesticides and Their Relationship to Environmental Health.* Washington, D.C., U.S. Government Printing Office, 1969.
17. W. J. Hayes, Jr., in *DDT, The Insecticide Dichlorodiphenyltrichlorothane and Its Significance,* Vol. 2, Paul Muller, Ed., Berlin, Birkhauser Verlag Busch, 1959.
18. R. D. O'Brien, *Insecticides—Action and Metabolism,* New York, Academic Press, 1967.
19. Y. Ishikawa, et al., *Biochem. Pharmacol., 38,* 2449(1989).
20. F. Matsumura and S. M. Ghiasuddin, in *Neurotoxicology of Insecticides and Pheromones,* T. Narahashi, Ed., New York, Plenum, 1979, pp. 245–257.
21. W. J. Hayes, *Annu. Rev. Pharmacol., 5,* 27(1965).
22. R. L. Metcalf, in *Scientific Aspects of Pest Control,* Washington, D.C., National Academy of Sciences, National Research Council, 1966.
23. P. Calabrisi and J. M. Clark, in *The Pharmacological Basis of Therapeutics,* 8th ed., A. G. Gilman, T. W. Rall, A. S. Nies, and P. Taylor, Eds., New York, Pergamon, 1990.
24. L. W. Cunningham, *Science, 125,* 1145(1957).
25. S. D. Murphy, *Toxicol. Appl. Pharmacol., 12,* 22(1969).
26. J. L. Sussman, et al., *Science, 253,* 872(1989).
27. S. G. Cohen, et al., *Biochim. Biophys. Acta, 997,* 167(1989).
28. P. Taylor and S. Lappi, *Biochemistry, 14,* 1989(1975).
29. D. J. Ecobichon, in *Casarett and Doull's Toxicology,* 4th ed., M. D. Amdour, et al., Eds., New York, Pergamon, 1991.
30. J. E. Casida, *J. Econ. Entomol., 53,* 1021(1960).
31. R. R. Painter, in *Pest Control,* W. W. Kilgore and R. L. Doutt, Eds., New York, Academic Press, 1967.
32. G. F. Shambaugh, et al., *Adv. Pest Control Res., 1,* 277(1957).
33. N. Green, et al., *Adv. Pest Control Res., 3,* 129(1960).

Chapter 43

AGENTS FOR ORGAN IMAGING

Raymond E. Counsell and Jamey P. Weichert

NONINVASIVE ORGAN IMAGING PROCEDURES

The clinician now has at his or her disposal many diagnostic procedures to help obtain information about the pathophysiologic status of internal organs without resorting to exploratory surgery. The most widely used methods for noninvasive imaging are scintigraphy (nuclear medicine scanning), radiography—including conventional x-ray and computed tomography (CT), ultrasonography (US), and magnetic resonance imaging (MRI). Diagnostic agents are important adjuncts for the first three of these procedures. For example, scintigraphy requires the administration of tracer doses of a radiopharmaceutical, whereas roentgen examinations and CT frequently require the administration of a contrast agent or radiopaque.

In many instances, more than one imaging method is used to answer a specific diagnostic question. Not only are scintigraphy and CT used to complement one another, but US and MRI are frequently called on to provide additional diagnostic information. Efforts to develop pharmaceuticals to enhance images obtained by US or MRI are in their early stages. Although we mention these developments, the focus of this chapter is on those agents currently used as adjuncts for scintigraphy and radiography.

RADIOPHARMACEUTICALS FOR SCINTIGRAPHY

Historical Background

In 1895, Roentgen discovered that a new penetrating radiation, called x-rays, could be produced by a discharge tube. Shortly thereafter, Becquerel observed that uranium salts spontaneously emitted a penetrating radiation similar to Roentgen's x-rays. Marie Curie, working in collaboration with Becquerel and her husband, Pierre, suggested the term "radioactivity" for this spontaneous form of radiation and set out to determine its presence in other minerals. In 1898, she succeeded in isolating from pitchblende (uranium oxide) trace amounts of two new elements, polonium and radium, which possessed 400 and 900 times, respectively, the radioactivity of uranium itself.

By 1900, the spontaneous radiations that Rutherford had named alpha (α) rays and beta (β) rays were known. That same year, Villard discovered the electromagnetic radiation of nuclear origin that is now known as gamma (γ) rays.

An understanding of nuclear phenomena and radiation developed in the years that followed, but the potential medical applications of radiation were recognized early. In 1903, Becquerel[1] stated in his Nobel Lecture that the emitted radiations were being explored for the treatment of cancer. The first clinical studies were not made, however, until Blumgart and Weiss[2] measured the "velocity of circulation" in 1924. They injected a radioactive bismuth solution ($^{214}_{83}Bi$) into one arm of a patient and then measured the appearance rate of radioactivity in the other. With these studies, the clinical use of radioactive tracers had its beginning.

The development of the cyclotron in the late 1930s and the nuclear reactor in the early 1940s was the next important phase and made available a variety of radionuclides for potential application in medicine. Indeed, modern-day nuclear medicine was ushered in with reactor-produced radioiodine and its use as the "radioactive cocktail" for diagnosis of thyroid dysfunction.

Other radioactive tracers as well as sensitive radiation-detecting instruments and cameras became available in the 1960s. This combination of events made it possible to study many organs of the body such as the liver, kidney, and lung. The obvious diagnostic value of these new noninvasive techniques served to establish nuclear medicine as a distinct medical specialty.

Broadly defined, a radiopharmaceutical is any pharmaceutical that contains a radionuclide. Accordingly, radiopharmaceuticals find many other medical applications in addition to organ imaging. These include use in innumerable in vitro assays (e.g., radioimmunoassay) and therapy, such as in the treatment of polycythemia vera with radioactive phosphorus. Despite the importance of these other uses of radiopharmaceuticals, this discussion focuses only on those agents that are currently approved by the Food and Drug Administration (FDA) for use in organ-imaging procedures.

Nature of Radioactivity

The atom is composed of a central nucleus surrounded by "orbiting" electrons. This nucleus contains neutrons and protons that are known collectively as nucleons. The number of these two types of nucleons characterizes a specific atom or "nuclide." Each nuclide is symbolized as $^{A}_{Z}X$, wherein A represents the total number of nucleons or the atomic mass, Z represents the number of protons or atomic number, and X is the symbol of the element. Thus, carbon with atomic mass 12 and atomic number 6 is expressed as $^{12}_{6}C$ (Fig. 43–1). Informally, the nuclides

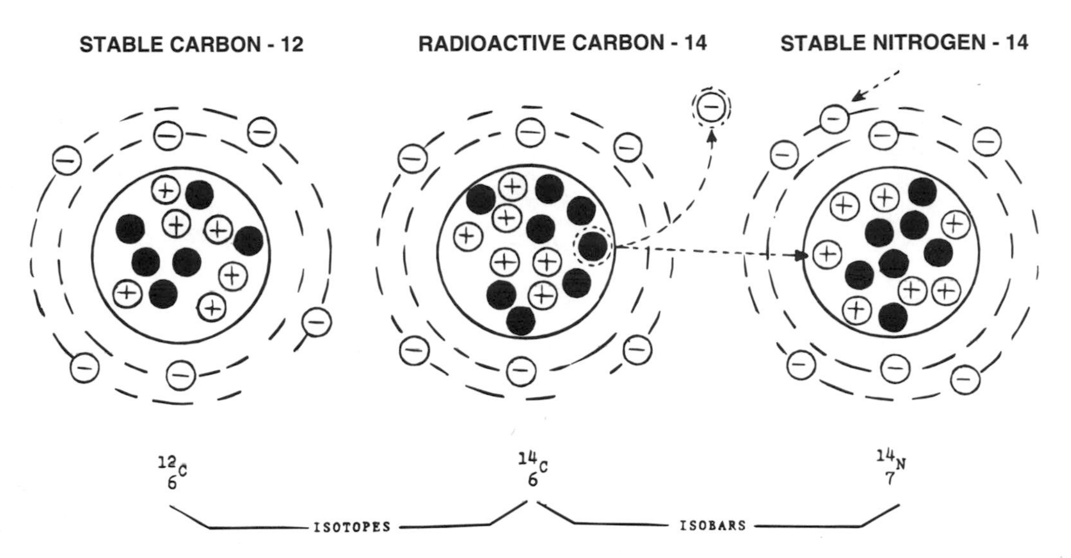

Fig. 43–1. Atomic structure distinctions between nuclides of carbon and nitrogen.

are often simply referred to by name and atomic mass, e.g., carbon-12.

Isotope, isotone, and isobar are terms used to express the relationship between two or more nuclides. For example, two or more different nuclides are isotopes if their proton numbers (i.e., atomic numbers) are equal; they are isotones if their neutron numbers (A–Z) are equal; or they are isobars if their nucleon numbers (i.e., atomic mass) are equal.

In an electrically neutral atom, the number of electrons always equals the number of protons. Thus, in carbon 12 there are 6 protons and 6 electrons. In the case of carbon-14 a radioisotope of carbon-12, six orbiting electrons remain. Because the chemical properties of an atom are determined by the electrons, the chemical or biochemical behavior of carbon-14 is, for all intents and purposes, the same as that of carbon-12. The higher mass number of carbon-14 reflects the two extra neutrons in the nucleus. The right side of Figure 43–1 depicts the radioactive decay of carbon-14 to stable nitrogen-14.

In radionuclides, the numbers of neutrons and protons in the nucleus are not equal because of an excess or deficiency of neutrons. This represents an unstable situation in which the atoms tend to change to more stable nuclei by a process known as radioactive decay.

The rate of decay of a specific radionuclide is often referred to as "activity" and is measured as disintegrations per second (dps or a Becquerel [Bq]) or disintegrations per minute (dpm). The standard unit of activity is the Curie, which is equivalent to 3.7×10^{10} dps or Bq. The time required for the activity of a given amount of radionuclide to decay to one half its initial value is called the physical half-life. This value is a constant that is characteristic for each radionuclide.

Figure 43–2 shows a portion of the "Trilinear Chart" of nuclides developed by Brucer.[3] Approximately 2250 different nuclides, including the 250 stable nuclides, are represented in this chart. The manner whereby these nuclides arise from the 103 known elements, as well as their interrelationships, is clearly depicted in this chart. Interestingly, all of the nuclides above bismuth (atomic number > 82) in the chart are radioactive. Unfortunately, none of these radionuclides with high atomic numbers are useful for radiopharmaceuticals.

Types of Radiation

When a radionuclide decays, energy is released in the form of radiation. This radiation may be emitted in the form of particles such as Rutherford's so-called α- or β-rays, or it may be nonparticulate (i.e., without mass) such as γ-rays. This radiation may then interact with matter to cause additional radiation in the form of electrons or x-rays. These latter types of radiations are distinguished from the former in that they are of extranuclear origin (secondary radiation) rather than the result of changes within the nucleus itself (primary radiation).

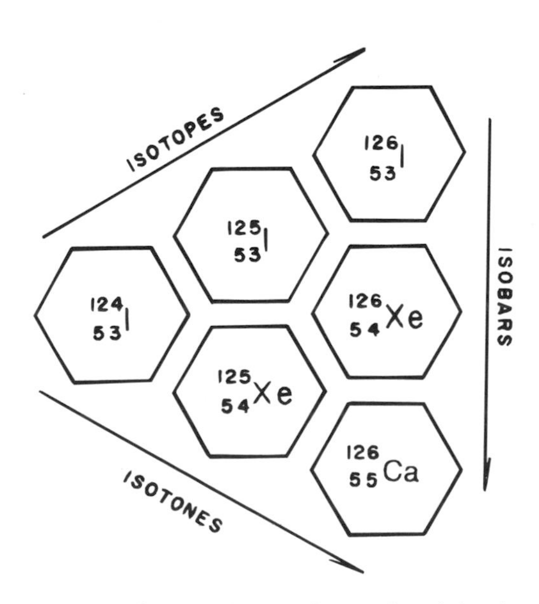

Fig. 43–2. Part of the "Trilinear Chart" of nuclides developed by Brucer.[3]

The α-particle is the nucleus of a helium atom (^4_2He), consisting of 2 neutrons and 2 protons. Because the orbiting electrons are absent, the protons confer a positive charge on this particle. Nuclides heavier than lead decay primarily by emitting α-radiation, which may be expressed as:

$$^A_ZX \longrightarrow {}^{(A-4)}_{(Z-2)}Y + {}^4_2He$$

For example, polonium decays to lead and helium:

$$^{210}_{84}Po \longrightarrow {}^{206}_{82}Pb + {}^4_2He$$

The β-particles may be either negatively (negatron) or positively (positron) charged. Their decay can be summarized as follows:

$$^A_ZX \longrightarrow {}_{(Z+1)}^{\ \ A}Y + \beta^- + \overline{\nu}$$
$$^A_ZZ \longrightarrow {}_{(Z-1)}^{\ \ A}Y + \beta^+ + \nu$$

in which ν and $\overline{\nu}$ represent the neutrino and antineutrino, respectively, and are unimportant for our purposes. The atomic number increases in negatron decay and decreases in positron decay. For example:

$$^{14}_6C \longrightarrow {}^{14}_7N + \beta^- + \overline{\nu}$$
$$^{11}_6C \longrightarrow {}^{11}_5B + \beta^+ + \nu$$

In neither case does the total number of nucleons change, but the number of protons, and thus the chemical identity of the daughter nuclide, is altered.

An important characteristic of the positron is that it readily interacts with any one of the many electrons present in matter. This interaction results in annihilation of the particles and usually gives rise to two equally energetic (511 keV) γ-rays that are emitted in opposite directions to one another. This directional character of annihilation γ-rays permits images to be achieved with a high degree of spatial resolution, a characteristic that forms the basis for positron-emission tomography (PET).

Some radionuclides decay by the emission of γ-rays alone, a process called isomeric transition (IT). The γ-ray, which is a high-energy photon or quantum of electromagnetic radiation, is the product of an energetic (metastable) nucleus. Its emission from the nucleus changes neither the atomic number nor the atomic mass and thus produces a daughter that is isomeric with the parent nuclide.

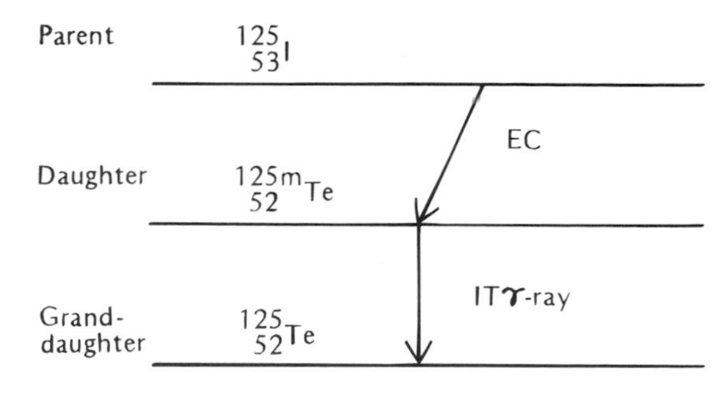

Fig. 43–4. Decay of iodine-125.

Often, however, IT is but one mode of decay in a nuclear reaction that proceeds through either particle emission or particle absorption. Hence, although negatron decay may be undesirable for organ imaging, that may not be a decisive factor determining the choice of radionuclide. For example, iodine-131 decays by emitting a negatron to become the metastable nucleus of xenon-131m (Fig. 43–3). The xenon-131m nucleus then emits a γ-ray by IT to become its stable isomer, xenon-131. Similarly, metastable technetium-99m decays by IT to the unstable technetium-99 daughter nucleus, which then decays by negatron emission. In other words, the nuclear reactions continue until a stable nucleus is formed.

Conversely, an absorption reaction that generates γ-rays is electron capture (EC). The orbital electron captured by the iodine-125 nucleus, for example, immediately combines with a proton to form a neutron (Fig. 43–4). The daughter, tellurium-125m, then decays by isomeric transition to stable tellurium-125. Some of the x-rays accompanying EC, such as those associated with the decay of iodine-125, are produced as the vacated electron orbital is refilled (electronic transition).

The application of these various modes of electronactive decay to diagnostic medicine depends on the ability of this radiation to penetrate matter (Fig. 43–5). Because α-particles are the largest particles emitted, they have the lowest penetrating power. Consequently, the massive α-particle cannot escape the body and dissipates its energy by producing tissue-damaging ions. Hence, radionuclides that emit α-particles have limited application in nuclear medicine.

Beta particles have the smaller mass of an electron and, thus, have appreciably more penetrating power than the α-particle. Occasionally, high-energy negatrons can be detected outside the body with sensitive instruments, but their short range usually makes negatrons more useful

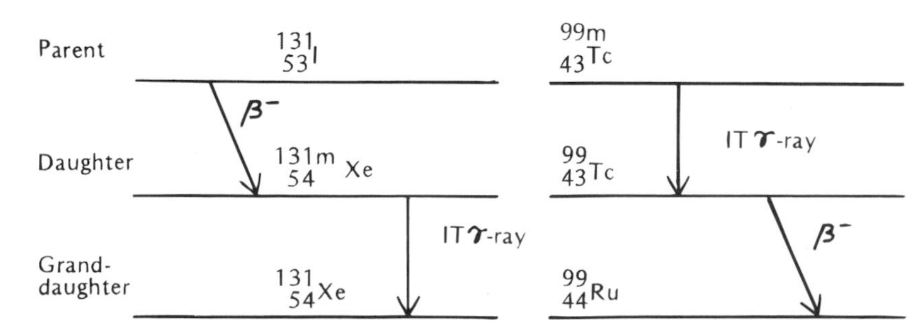

Fig. 43–3. Decay of iodine-131 and technetium-99.

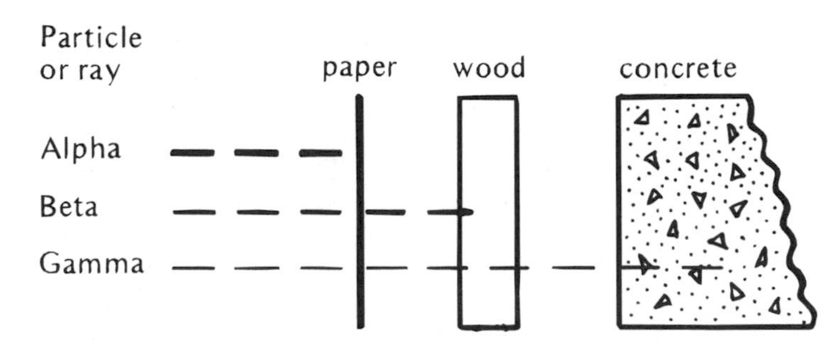

Fig. 43–5. Relative penetration of alpha, beta, and gamma radiation.

for therapy than for organ imaging. As previously noted, ^{32}P-sodium phosphate has been used in the treatment of polycythemia vera.

On the other hand, the annihilation γ-rays produced by positrons render positron-emitting radionuclides suitable for organ imaging. The massless γ-rays, being the most penetrating form of radiation, can be readily detected outside the body with appropriate instruments.

With present-day instrumentation, the best images are achieved with γ-rays in the energy range of 100 to 400 keV. Radiation of lower energy fails to reach the detectors with sufficient intensity, and that of higher energy causes secondary radiation that interferes with instrument performance. For the most part, γ-rays form the foundation of most diagnostic procedures in nuclear medicine, analogous to the function of x-rays in radiology.

Biologic Effects of Radiation

Radiopharmaceuticals differ from pharmaceuticals in that they are usually administered as tracer doses for diagnostic purposes rather than in pharmacologic doses to elicit a physiologic response. The toxicity of a radiopharmaceutical is therefore related to the amount of radioactivity received by the patient.

Nuclear radiation is often spoken of as ionizing radiation because it causes disruption of molecules with the release of highly reactive ions (ionizing interactions). These ions can undergo reaction with each other or with neighboring molecules. Because of the high water content of cells, ionizing reactions most frequently involve water molecules and lead to the production of free radicals and peroxides. These products chemically alter sulfhydryl groups of proteins and react with other cell components to bring about toxic effects. These toxic effects are, consequently, due to actions of the products of radiation (indirect action) rather than to actions of the radiation itself.

The other major biologic effect of radiation involves a direct action on important macromolecules, such as the nucleic acids. The mutagenic and carcinogenic capability of ionizing radiation has been ascribed to this direct effect on the structure of nuclear DNA.

Whichever is the mode of action, the result of radiation exposure is finally the dissipation of energy by transference to the environment. Any radiation that is so transferred within tissues is absorbed radiation, and the total energy of this transferred radiation per unit mass is called the radiation absorbed dose. The unit of radiation absorbed dose is called the rad and is equal to 100 ergs absorbed per gram of material.

Because the rad is not a measure of the biologic effects of ionizing radiation, a unit called the roentgen equivalent man (rem) was proposed and subsequently adopted. The rem is the amount of radiation that produces in humans the same biologic effects as the absorption of 1 rad of γ-rays or x-rays. Hence, where the absorbed dose is due to x-rays or γ-rays, the rem and rad are approximately equivalent.

The radiation dose associated with administration of a radiopharmaceutical depends on a number of factors: (1) type and energy of all radiations emitted; (2) amount of radionuclide deposited in the body (e.g., total body dose); (3) radionuclide distribution in the body (e.g., critical organ dose), (4) relative size and sensitivity of tissues to different types of radiation. For example, tissues most sensitive to ionizing radiation are rapidly differentiating cells, such as the fetus; the hematopoietic tissues, such as blood-forming bone marrow; and the germ cells of the reproductive organs. Examples of less sensitive tissues are skin, thyroid, and bone; (5) physical half-life ($t_{1/2}$) of the radionuclide; (6) the biologic half-life ($B_{1/2}$), or the time required for half the deposited radionuclide to be eliminated from the body by natural processes; and (7) the effective half-life ($E_{1/2}$), or the time required for the amount of radioactivity in vivo to fall to half its initial value. $E_{1/2}$ is related to $t_{1/2}$ and $B_{1/2}$ according to the expression: $E_{1/2} = (t_{1/2}B_{1/2})/(t_{1/2} + B_{1/2})$.

Consequently, the amount of absorbed energy, which is the appropriate basis on which to estimate radiation dose, involves theoretic and technical difficulties. In the years following the adoption of the rad in 1953, these difficulties received considerable attention. Then, in 1964, John Hidalgo, the president of the Society of Nuclear Medicine, initiated the Medical Internal Radiation Dose (MIRD) Committee, whose task was to improve the situation with regard to estimating the absorbed dose to patients, resulting from the diagnostic or therapeutic use of internally administered radiopharmaceuticals.[4] Beginning in 1968, the MIRD Committee provided nuclear medicine physicians with improved equations and reference data for estimating whole-body and organ doses of radiopharmaceuticals.

RADIONUCLIDES AND RADIOPHARMACEUTICALS FOR ORGAN IMAGING

In general, the ideal radionuclide for organ imaging would have the following characteristics: (1) decay by γ-

radiation alone and with an energy of about 200 keV; (2) have a physical half-life in the range of 6 to 12 hours (i.e., long enough for radiopharmaceutical preparation, but short enough to minimize the radiation absorbed dose); (3) be readily available; (4) be readily incorporated into carrier molecules (i.e., chemically versatile); and (5) form radiopharmaceuticals that are stable in vitro and in vivo. Unfortunately, no one radionuclide is currently available that fulfills all of these criteria. Indeed, only a few of the 2000 possible radionuclides are used today for organ-imaging purposes. The principal instrument for displaying images is the gamma camera. These images may be planar or tomographic. In the latter category, cameras are now available for use with radionuclides in which the emitted radiation is either a single photon (single photon emission computed tomograph, or SPECT) or a positron (positron emission tomography or PET). Table 43–1 lists the radionuclides and radiopharmaceuticals commonly used in these and more conventional nuclear medicine procedures.

Radiohalogens

Historically, radioiodine has a special place in nuclear medicine. Using iodine-128, Hertz, Roberts, and Evans[5] in 1938 clearly demonstrated for the first time the enhanced uptake of radioiodine by the hyperplastic thyroid gland. It was also the first publication concerning the use of radioiodine.[6] The short half-life of this nuclide ($t_{1/2}$, 25 minutes) severely limited its clinical utility, however, and it was not long before the longer-lived iodine-131 became available for clinical use in the diagnosis of thyroid function.

Iodine is unique among the physiologically important elements because of the large numbers of radioisotopic forms. Although 24 iodine isotopes are known, only iodine-123, iodine-125, and iodine-131 are used extensively in medicine. Iodine-125 is not generally used for organ imaging because the emitted photons are energetically too weak (28 and 35 keV). Nonetheless, the longer half-life of iodine-125 ($t_{1/2}$, 60 days) makes it useful for a variety of in vitro assays, such as radioimmunoassay procedures.

Iodine-123 more closely fulfills the criterion for an ideal radionuclide, but it is expensive. Improvements in production and purification, however, are bringing about cost reductions, and this nuclide is beginning to be used more frequently in nuclear medicine procedures, especially SPECT.

Iodine-131 is the most frequently used radionuclide of iodine in organ-imaging radiopharmaceuticals. Although it is readily available and economical, it has several disadvantages: (1) The γ-radiation of 359 keV is accompanied by high-energy β-particles that add significantly to the radiation dose. (2) The long half-life of 8.04 days also prolongs radiation exposure in tissues where it concentrates. (3) Compounds labeled with radioiodine tend to deiodinate rapidly in vivo.

Despite these limitations, one important feature of iodine-131 and of radioiodine in general is that it is readily incorporated into a variety of organic molecules either by direct iodination or by isotope exchange procedures. This distinctive feature enhances the interest in iodine-123 and other radiohalogens such as bromine-77 ($t_{1/2}$, 58 hours, 245 keV γ-photon). Chlorine does not have a radioisotope suitable for use in radiopharmaceuticals.

Fluorine-18 is the only other radiohalogen approved for radiopharmaceutical use. Its major limitation is its short half-life, which puts a severe constraint on the number of chemical manipulations that can be performed. As a positron-emitting nuclide, however, it is finding greater use in PET, especially when it is linked to neurotransmitter receptor antagonists and fluorodeoxyglucose (FDG).

Technetium-99m

The chemical properties of technetium are similar to those of its homologs manganese and rhenium. There has been interest in applying technetium-99m in nuclear medicine ever since Harper and co-workers[7] used it to visualize the thyroid and brain in 1964. While maintaining a low radiation dose to the patient, its monoenergetic 140 keV γ-ray is sufficiently energetic and of suitable half-life for deep organ imaging.

Technetium-99m is produced by negatron decay of molybdenum-99m, which has a 2.7-day half-life. The molybdenum is initially attached on an ion-exchange column (molybdenum-technetium generator). These generators, or "cows," are available from commercial suppliers. Radioactive decay changes the atomic number, thus altering both the radionuclide's chemistry and, hence, its attachment on the ion-exchange column. The technetium is then eluted from the column with saline as a clear, colorless solution of sodium pertechnetate.

Although technetium can assume valences from +7 to −1, it is most stable in water as either the pertechnetate anion (TcO_4^-), valence +7, or insoluble technetium oxide (TcO_2), valence +4. Chemically, the pertechnetate anion resembles the permanganate anion (MnO_4^-), whereas it is physically comparable to the iodide anion in size and charge distribution. Technetium can be reduced to other less stable oxidation states by such agents as ascorbic acid and ferrous or stannous salts. When chelated by appropriate ligands, these oxidation states are stabilized and can also exist in aqueous media.

Broadly speaking, then, most technetium-labeled organ-imaging agents in current clinical use can be classified as either pertechnetate salts, colloids, or chelates. No chemical problems are associated with using the aqueous pertechnetate form, once it has been prepared. The chemical structure and oxidation state of technetium, however, are difficult to identify in all but a few radiopharmaceuticals, and the chemistry and structure-activity relationships are inadequately understood.

Technetium sulfur colloid, for example, is thought to exist in the form of a heptasulfide (Tc_2S_7), which aggregates into particles. The exact sizes of these aggregates range in diameter from about 50 to 500 μm, depending on the method of preparation. Figure 43–6 shows the structure of a technetium-99m-labeled dimer of N-(2, 6-dimethylphenylcarbamoylmethyl) iminodiacetic acid

(HIDA), a chelate in which the organic moiety retains the physiologic capacity to enter the enterohepatic cycle while it remains bound together with the technetium.[8]

Other Radionuclides

Indium, gallium, and thallium are homologs of aluminum. Indium-113m is comparable to technetium with regard to the following properties: (1) conveniently available from a tin-indium generator; (2) a short physical half-life (1.73 hours); (3) decays primarily by single γ-ray (392 keV); (4) good radiopurity; and (5) high specific activity. Although the lower +1 and +2 valence states are known, its chemistry in the dominant +3 valence state resembles the chemistry for ferric ion.

Cyclotron-produced gallium-67, like indium, assumes +3, +2, and +1 valence states. Most gallium salts are hydrolyzed in aqueous media and are insoluble at physiologic pH. Therefore, the soluble gallium citrate salt is the agent used clinically. The principal energies of gallium-67 γ-rays are 94, 184, and 296 keV.

Thallium is chemically atypical, because it is stable in

Table 43–1. Radionuclides and Radiopharmaceuticals Used in Organ Imaging

Nuclide	$t_{1/2}$	Principal Imaging γ-ray, keV (mode of decay)	Target Organ/System	Radiopharmaceutical
Halogens				
^{18}F	1.83 hrs.	511 (β^+)	Skeleton	^{18}F-sodium fluoride
			Brain, tumors	^{18}F-fluorodeoxyglucose (FDG)
^{123}I	13.3 hrs.	159 (EC)	Thyroid	^{123}I-sodium iodide
			Brain	^{123}I-N-isopropyl-*p*-iodoamphetamine
				^{123}I-HIPDM*
			Cardiovascular	^{123}I-iodoheptodecanoic acid
				^{123}I-iodophenylpentadecanoic acid
^{131}I	8.04 days	364 (β^-)	Cisternography	^{131}I-HSA*
			Cardiovascular	^{131}I-HSA
			Thyroid	^{131}I-sodium iodide
			Hepatobiliary	^{131}I-rose bengal
			Kidney	^{131}I-*ortho*-iodohippurate
			Adrenal (cortex)	^{131}I-6β-iodomethyl-19-norcholesterol
			Adrenal (medulla)	^{131}I-MIBG*
Technetium				
^{99m}Tc	6.04 hrs.	140 (IT)	Brain	^{99m}Tc-sodium pertechnetate
				^{99m}Tc-hexamethylpropylamine oxime (HM-PAO)
			Lung	^{99m}Tc-MAA*^{99m}Tc-albumin microspheres
			Cardiovascular	^{99m}Tc-sodium pertechnetate
				^{99m}Tc-albumin
				^{99m}Tc-pyrophosphate
			Thyroid	^{99m}Tc-sodium pertechnetate
			Liver-spleen	^{99m}Tc-sulfur colloid
			Hepatobiliary	^{99m}Tc-Disofenin
				^{99m}Tc-Lidofenin
				^{99m}Tc-Mebrofenin
			Skeleton	^{99m}Tc-Etidronate
				^{99m}Tc-Medronate
				^{99m}Tc-Oxidronate
				^{99m}Tc-polyphosphate
			Kidney	^{99m}Tc-DTPA*
				^{99m}Tc-dimercaptosuccinate
				^{99m}Tc-gluceptate
				^{99m}Tc-mercaptoacetyltriglycine

Table 43–1. *(Continued)*

Nuclide	$t_{1/2}$	Principal Imaging γ-ray, keV (mode of decay)	Target Organ/System	Radiopharmaceutical
Other radionuclides				
^{11}C	20.40 mos.	511 (β^+)	Cardiovascular	^{11}C-palmitic acid
^{13}N	9.96 mos.	511 (β^+)	Cardiovascular	^{13}N-ammonia
^{15}O	2.04 mos.	511 (β^+)	Cardiovascular	^{15}O-water
^{67}Ga	79.2 hrs.	94 (EC)	Tumors and inflammatory lesions	^{67}Ga-gallium citrate
^{75}Se	120 days	265 (EC)	Pancreas	^{75}Se-selenomethionine
^{85}Sr	65 days	514 (EC)	Skeleton	^{85}Sr-strontium nitrate
^{87m}Sr	2.8 hrs.	388 (IT)	Skeleton	^{87m}Sr-strontium citrate
^{111}In	2.8 days	173, 247 (EC)	Cisternography	^{111}In-DTPA
			Cardiovascular	^{111}In-antimyosin
			Tumors	^{111}In-oxine-labeled leukocytes
				^{111}In-satumomab pendetide
^{113m}In	1.73 hrs.	393 (IT)	Brain	^{113m}In-DTPA
			Lung	^{113m}In-$Fe(OH)_3$ particles
			Cardiovascular	^{113m}In-transferrin
			Liver-spleen	^{113m}In-colloid
^{133}Xe	5.3 days	81 (β^-)	Lung	$^{133}Xenon$ gas
^{169}Yb	31.8 days	198 (EC)	Cisternography	^{169}Yb-DTPA
^{197}Hg	2.7 days	77 (EC) (and Au x-rays)	Brain tumors	^{197}Hg-chlormerodrin
			Kidney	^{197}Hg-chlormerodrin
^{198}Au	2.7 days	412 (β^-)	Liver-spleen	^{198}Au-colloid
^{201}Tl	73 hrs.	135, 167 (β^-) (and Hg x-rays)	Cardiovascular	^{201}Ti-thallium chloride
^{203}Hg	46.9 days	279 (β^-)	Brain tumors	^{203}Hg-chlormerodrin

* HIDPM, N,N,N[1]-trimethyl-N[1]-[2-hydroxy-3-methyl-5-iodobenzyl]-1,3-propanediamine; HSA, human serum albumin; MIBG, *m*-iodobenzylquanidine; MAA, macroaggregated albumin; DTPA, diethylenetriamine pentaacetic acid or pentetate.

neutral aqueous solutions as thallous ion (+1 valence). Biologically, it resembles potassium and other univalent alkaline metal cations. The principal organ-imaging γ-rays of cyclotron-produced thallium-201 are 135 and 167 keV γ-rays and 71 keV x-rays.

Within the family of sulfur, selenium, and tellurium, only selenium, as selenium-75, is found among currently used radiopharmaceuticals. Sulfur-35 is a pure β-emitter, and tellurium isotopes have been explored only recently. In addition to the chemical similarity of sulfur to selenium, the atomic radius (−2 valence) of selenium is close to that of sulfur. Accordingly, ^{75}Se-selenomethionine, an analog of methionine, behaves biochemically like the latter and is used to study the pancreas.

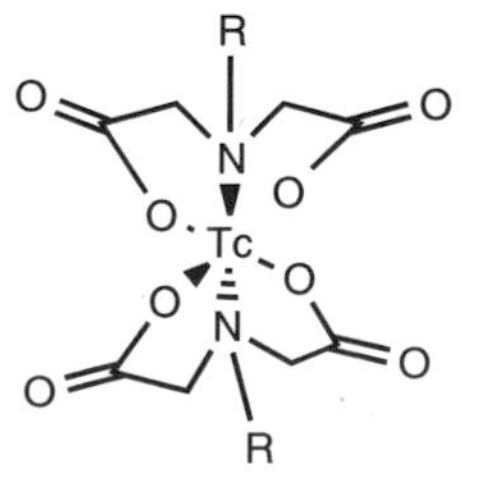

Fig. 43–6. Proposed structure of a ^{99m}Tc-labeled dimer of HIDA.

Radionuclides of strontium, being homologs of calcium, are used to obtain images of bone, where, in fact, it exchanges with calcium. The shorter physical half-life ($t_{1/2}$, 2.88 hours), and lower monoenergetic 388 keV γ-ray of strontium-87m result in a significantly lower radiation dose to the patient compared to that received from the 514 keV γ-rays and secondary radiation of strontium-85. Furthermore, strontium-87m is more conveniently available from an yttrium-strontium generator.

The divalent mercuric ion concentrates in liver and kidneys. Reactor-produced mercury-203 ($t_{1/2}$, 46.9 days) emits 279 keV γ-rays following negatron emission, which results in a significantly greater radiation dose, particularly to the kidney, than is received from cyclotron-produced mercury-197, ($t_{1/2}$, 2.7 days). The imaging radiations of mercury-197 are 77 keV γ-rays and 191 keV x-rays. ^{203}Hg- and ^{197}Hg-chlormerodrin (1-[3-(chloromercuri)-2-methoxypropyl]-urea), are the only mercury-labeled radiopharmaceuticals of consequence that are used in nuclear medicine. It binds to plasma proteins and was once a widely accepted agent for imaging brain tumors and assessing renal function.

Xenon-133 undergoes negatron decay with the emission of 81 keV γ-rays and 31 keV x-rays. It is one of the noble gases and is relatively insoluble and chemically unreactive. Radioactive gases, which may be used to evaluate regional prefusion and ventilation, include both soluble gases, such as ^{11}C-labeled carbon dioxide, and relatively insoluble gases, such as xenon. Reactor-produced xenon-133 is inexpensive and available with high specific activity.

Ytterbium is one of the heavier, more acidic lanthanons (with smaller atomic radii) that localizes primarily in bone. Ytterbium-169 emits γ-rays with a wide range of energies, but the 198 keV γ-ray is of interest for organ imaging. The disadvantages of ytterbium-169 relate to high radiation dose and its unknown chemistry. Nevertheless, complexed forms such as ^{169}Yb-DTPA(diethylenetriaminepentaacetic acid) have found use in cisternography.

Gold-198 salts are relatively insoluble, but readily form colloids in physiologic media or in vivo. The colloidal particles localize in organs of the reticuloendothelial system, where they are retained indefinitely. This event tends to result in a large radiation dose to the patient—a disadvantage to its use.

ORGAN IMAGING

Radiopharmaceuticals for organ imaging have undergone rapid development during the past three decades. They are now an important part of diagnostic medicine.

To be useful as an organ-imaging agent, a radiopharmaceutical must be able to localize selectively in a particular organ or tumor. Various physiologic and physical processes have been exploited to accomplish this task, such as active transport, compartmental storage, simple diffusion, phagocytosis, and capillary blockade. Usually, a 10-fold difference in uptake of radioactivity by the target organ versus surrounding nontarget areas is sufficient for imaging purposes.

The Brain

Before the advent of CT, the brain scan was the most frequently requested nuclear medicine procedure. Today it is largely used as an adjunct to CT and MRI.

IMP

HIPDM

Fig. 43–8. Chemical structures for IMP and HIPDM.

The ability of various agents to function as brain-imaging agents arises largely from the difference in permeability between a normal and an abnormal brain. Most intracranial lesions alter the blood-brain barrier so that various radiopharmaceuticals can penetrate inaccessible regions and localize in and around the lesion. The procedure, therefore, is nonspecific, because neoplasms, vascular accidents, or malformations and extracerebral hematomas all appear as areas of increased radioactivity on a scan (Fig. 43–7). The most commonly used agent for brain imaging is ^{99m}Tc-pertechnetate.

The ability of pertechnetate to localize in brain lesions appears to be proportional to the protein content of the lesion. Chelating pertechnetate with DTPA or gluceptate affords the radiopharmaceuticals that clear from the blood more rapidly than pertechnetate and thereby furnish higher lesion-to-brain concentrations more rapidly.

Other radiopharmaceuticals have become available that are sufficiently lipophilic to penetrate the blood-brain barrier. Two such agents are N-isopropyl-[^{123}I]-*p*-iodoamphetamine (IMP) and N,N,N^1-trimethyl-N^1-[2-hydroxy-3-methyl-5-iodobenzyl]-1,3-propanediamine (HIPDM) (Fig. 43–8). These radioiodinated amines are extracted by the brain and their distribution reflects local cerebral blood flow. The lipophilic ^{99m}Tc-labeled hexamethylpropylamine oxime (HM-PAO) is used widely for the same purpose.

An exciting development in recent years has been the application of positron emission tomography (PET) for

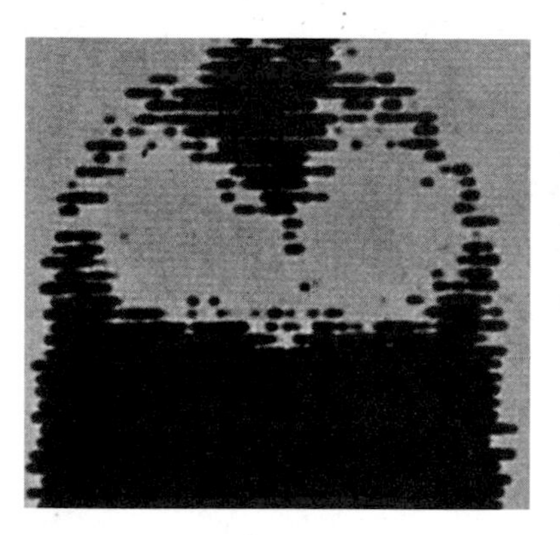

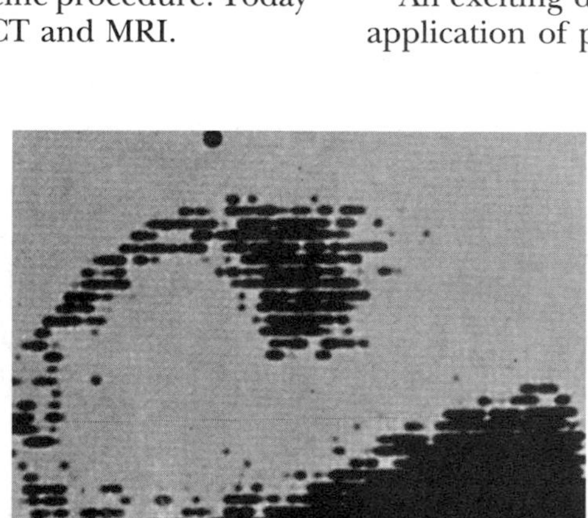

Fig. 43–7. Parasagittal meningioma seen on anterior and right lateral studies (courtesy of *Physicians' Desk Reference for Radiology and Nuclear Medicine*).

measuring various cerebral metabolic states. For example, ^{18}F-FDG is transported across the blood-brain barrier in the same way as glucose, the principal source of energy for the brain. Such a tracer therefore becomes important for characterizing disease states associated with altered cerebral glucose metabolism, such as stroke, epilepsy, and Huntington's disease. Other clinical studies include the use of ^{18}F- or ^{11}C-labeled neurotransmitter agonists and antagonists (e.g., ^{18}F-spiperone) for assessing neutotransmitter receptor levels in a variety of neurologic disorders (e.g., parkinsonism).

Cardiovascular System

Cardiovascular Blood Pool

Images of the heart are obtained by appropriate labeling of the blood pool, which is accomplished by intravenous administration of autologous red blood cells labeled with technetium-99m. Because the cardiac blood pool is the most vascular structure in the chest, it is readily visualized by this procedure.

Myocardial Scanning

Assessing of acute myocardial necrosis in the patient is an important and challenging problem because immediate and long-term management often depends critically on whether or not the patient is thought to have sustained irreversible loss of myocardial function. Ideally, the isotopic delineation of acute myocardial infarction should be performed with a radiopharmaceutical that is selectively taken up by irreversibly damaged myocardium. The presence of such an agent would provide positive evidence of necrosis, the "hot spot" scan. ^{99m}Tc-pyrophosphate has been found to complex with the elevated levels of calcium associated with ischemic and necrotic tissue, thereby providing infarct-avid imaging. One disadvantage of this agent is that maximum uptake in myocardial infarcts usually occurs 48 to 72 hours after the onset of symptoms. Alternatively, the leakage of intracellular cardiac myosin to extracellular sites that occurs with myocardial infarction now allows for imaging with radiolabeled myosin antibody in the form of indium-111 antimyosin.

Another technique for scanning acute myocardial infarction is one in which the radiopharmaceutical is accumulated in normal tissue, but is not taken up by the infarcted or ischemic zone, the "cold spot" scan. Long-chain fatty acids are an important energy source for the heart, and they are highly extracted from the blood by normal myocardium. Several radioiodinated fatty acid analogs such as ^{123}I-16-iodo-9-hexadecanoic acid and ^{123}I-p-iodoiphenylpentadecanoic acid are therefore being evaluated clinically for their ability to distinguish between normal and ischemic myocardium.

Radioactive potassium and its analogs, rubidium, cesium, and thallium, have been used for similar studies. Thallium-201, as a thallium chloride solution, is extensively used in the diagnosis of ischemic cardiac disease.

The Lung

Regional Perfusion

This procedure depends on the physical entrapment of the radiopharmaceutical in peripheral arterioles and capillary beds of the lung (Fig. 43–9). Radioactive particles that measure 10 to 50 μm in diameter are administered intravenously. Although physically entrapped, these particles are eventually eliminated from the lung by phagocytosis or other processes. Biodegradable albumin microspheres, having a greater uniformity in size and a technetium-99m label, e.g., ^{99m}Tc-MAA, are used widely for this purpose.

Regional Ventilation

Generally, pulmonary regions that are poorly perfused are also poorly ventilated. Thus, radioactive gases have been used to determine regions of decreased ventilation and perfusion. The most widely studied radiopharmaceutical gases are isotopes of the noble gases krypton and

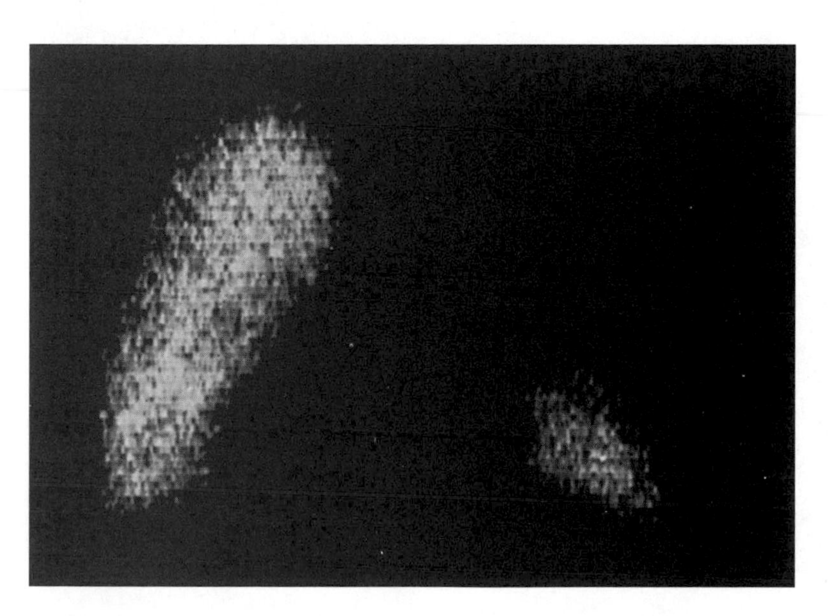

Fig. 43–9. Uptake of ^{99m}Tc microspheres by the lung indicating bronchiogenic carcinoma of the right lung.

xenon. Relative to the other elements, these gases are biochemically inert and are excreted rapidly by the pulmonary system. Xenon-127 and xenon-133 are the only gases currently approved for this purpose.

The Thyroid

Iodide anion is actively accumulated by the thyroid gland, where it serves as an important precursor in the biosynthesis of thyroxine. When radioactive isotopes of iodine became available in the 1940s, they were evaluated for their ability to diagnose thyroid dysfunction. Iodine-131 has received the widest clinical application, but iodine-125 and iodine-123 have been used as well. In fact, iodine-123 is the preferred isotope on the basis of physical half-life ($t_{1/2}$, 13.3 hours) and decay radiation (nearly monoenergetic 159 keV IT γ-ray), but this cyclotron-produced isotope is still extremely expensive.

Other radionuclides have been studied in an effort to decrease the high radiation dose associated with the use of iodine-131. Technetium-99m, as the pertechnetate anion, has been shown to accumulate in the thyroid, but, unlike iodine, it is not a precursor for thyroxine synthesis. Consequently, pertechnetate does not become organically bound within the organ. This shorter biologic half-life in the thyroid, along with the shorter physical half-life, makes ^{99m}Tc-sodium pertechnetate the preferred radiopharmaceutical for thyroid scanning (Fig. 43–10).

The Liver

Radiopharmaceuticals for liver imaging are based on the knowledge that the liver is composed of 60% polygonal cells and 40% Kupfer cells. The polygonal cells (hepatocytes) actively clear many dyes and various drugs and promptly excrete them through the biliary tracts into the gastrointestinal system.

It was on this basis that the dye rose bengal was labeled with iodine-131 and used for many years as a hepatobiliary imaging agent. The long physical half-life of iodine-131, however, limited the dose that could be safely administered to patients. This limitation led to the search for better agents that culminated in the introduction of ^{99m}Tc-lidofenin (^{99m}Tc-HIDA)[9] (see Fig. 43–6). This product and two closely related analogs (Table 43–2) represent the current agents of choice for hepatobiliary imaging. For example, ^{99m}Tc-mebrofen (^{99m}Tc-BrIDA) exhibited high specificity for the liver and only minimal excretion by the kidneys in animal studies,[10] and subsequent clinical trials confirmed the efficacy and safety of this agent as a hepatobiliary agent (Fig. 43–11).

The Kupfer cells of the reticuloendothelial system on the other hand, are able to phagocytize colloidal particles. This finding led to the development of ^{99m}Tc-sulfur colloid. This agent is taken up and retained by the Kupfer cells and provides excellent scans of the liver (Fig. 43–12). Abnormal tissues fail to concentrate this agent, and thus appear as "cold spots" on the scan. Approximately 80 to 85% of the colloidal particles localize in the liver, with most of the remainder taken up by the spleen and bone marrow.

Spleen and Bone Marrow

The spleen and bone marrow are part of the RES and thus can be studied with the same colloidal agents used for scanning the liver. Small colloids (less than 10 μm) are less efficiently extracted by the spleen and tend to accumulate preferentially in bone marrow. Because the spleen normally functions as a scavenger of red blood cells (RBC), thermally denatured RBC labeled with technetium-99m are used for imaging purposes. These radiolabeled cells are rapidly sequestered by the normally functioning spleen, whereas abnormal or pathologic areas appear on a scan as zones of decreased uptake. Splenomegaly (enlargement of the spleen) is associated with a variety of diseases and can readily be discerned with the aid of a spleen scan.

The Pancreas

Amino acids are actively accumulated by the pancreas at early periods. Unfortunately, the elements that make

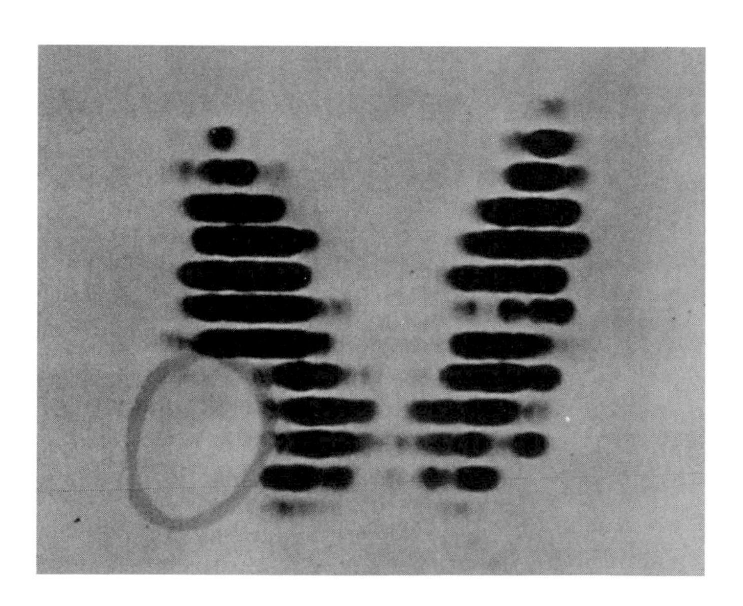

Fig. 43–10. Cyst in the low lateral portion of the right thyroid lobe presenting as a hypofunctioning nodule. The clinically palpable nodule is noted on the scan. (Courtesy of *Physicians' Desk Reference for Radiology and Nuclear Medicine.*)

Table 43–2. Chemical Structure of Ligands for ^{99m}Tc-HIDA and Analogs

2	3	4	5	6	Acronym
—CH_3	—H	—H	—H	—CH_3	HIDA
—C_2H_5	—H	—H	—H	—C_2H_5	DIDA
—$CH(CH_3)_2$	—H	—H	—H	—$CH(CH_3)_2$	DISIDA
—CH_3	—Br	—CH_3	—H	—CH_3	BrIDA
—H	—H	—$CH(CH_3)_2$	—H	—H	PIPIDA
—H	—H	—$(CH_2)_3CH_3$	—H	—H	BIDA

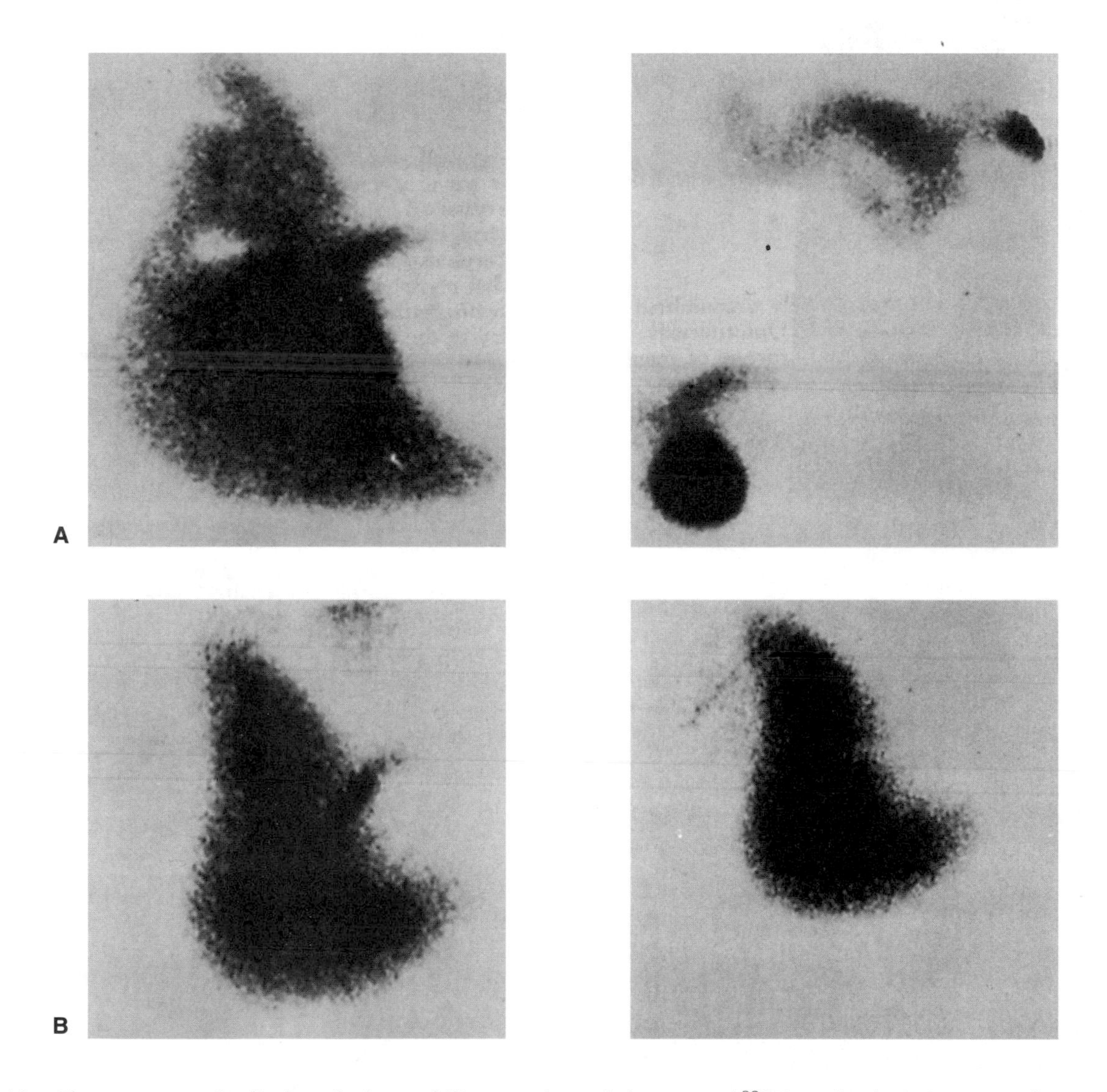

Fig. 43–11. The upper row (A) displays the hepatobiliary uptake and clearance of ^{99m}Tc-mebrofenin in a normal patient. The lower row (B) shows the scans obtained for a patient with liver obstruction. In both cases the scans were performed at 10 and at 45 minutes (courtesy of Squibb Diagnostics).

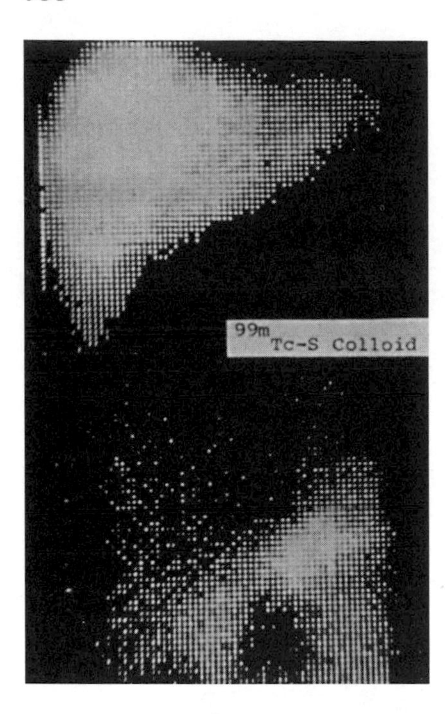

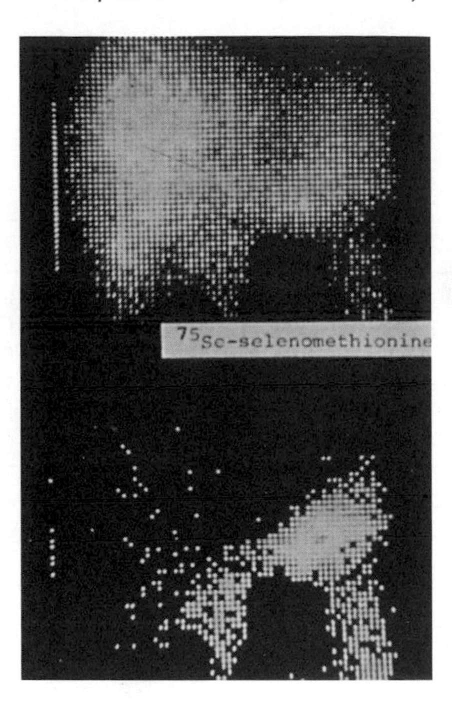

Fig. 43–12. Normal pancreas subtraction study performed with tape storage system. Jejunal activity also is present in the left side of the abdomen. Initial studies shown in top row (courtesy of *Physicians' Desk Reference for Radiology and Nuclear Medicine*).

up the structure of amino acids, namely C, H, N, O, and S, do not have a suitable γ-emitting nuclide for diagnostic purposes. On the other hand, replacement of the sulfur in methionine with selenium furnishes the amino acid analog, selenomethionine, which exhibits biologic properties similar to those of methionine.^{75}Se-selenomethionine is incorporated into tissues that are rapidly synthesizing proteins, such as the pancreas. Carcinoma and pancreatitis are both characterized by areas of diminished image intensity or as "cold spots." Pancreas scanning is a difficult procedure in nuclear medicine, not only because the normal pancreas can assume a variety of shapes, but also because the pancreas-to-liver (target-to-nontarget) ratio is usually less than 10. Images of the pancreas may be significantly enhanced by obtaining a liver scan with ^{99m}Tc-sulfur colloid and electronically "subtracting" this image from the one achieved with ^{75}Se-selenomethionine (see Fig. 43–12).

The Skeletal System

Calcium-45, strontium-85, and fluorine-18 are noted for their ability to localize in bone lesions. The mechanism of action of these nuclides appears to be related to an ion exchange at the surface of the hydroxyapatite crystal. If $[Ca_3(PO_4)_2]\cdot 3Ca(OH)_2$ is taken as representative of the hydroxyapatite crystal structure, the possibility for Sr^{+2} to exchange with Ca^{+2} and F^- to exchange with OH^- becomes apparent. Moreover, the knowledge that phosphate is an important component of the hydroxyapatite and is taken up by bony structures led to the development of other skeleton-imaging radiopharmaceuticals, such as ^{99m}Tc-polyphosphate and related chelates (Fig. 43–13). The technetium-labeled diphosphonate complexes shown in Table 43–3 are now used most commonly for bone scintigraphy.

The Kidney

The mercurial diuretic chlormerodrin, labeled with mercury-197 or mercury-203, was once used widely to obtain images of the kidneys. These agents have now been replaced by several chelate complexes of technetium-99m such as ^{99m}Tc-iron ascorbate, ^{99m}Tc-gluceptate, and ^{99m}Tc-dimercaptosuccinate. These complexes localize in the kidney like chlormerodrin, but are excreted more rapidly.

Renal imaging agents have also been developed by the appropriate labeling of drugs that are known to be cleared rapidly by the kidney. Sodium iodohippurate is used as a urographic contrast media because of this property. Labeling this agent with radioiodine afforded ^{131}I-*ortho*-iodohippuric acid, which has been used widely for measuring effective renal plasma flow and associated disorders. This agent has been replaced, however, with ^{99m}Tc-mercaptoacetyltriglycine (^{99m}Tc-MAG3).

Tumors

Gallium-67 citrate is used most widely for tumor imaging. This agent has been shown to concentrate in certain lymphomas and other soft tissues to such a degree that these lesions are detectable on whole body scintillation scanning.

Other tumors, such as lung cancer and melanoma, show an affinity for gallium-67, whereas still others such as colonic and uterine cancers are less responsive. Moreover, some benign inflammatory lesions, such as abscesses, also localize gallium-67. Although the mechanism of gallium-67 uptake is poorly understood at present, both tumor cells and inflammatory cells show an affinity for gallium agents. Although gallium is a group III element, its biologic properties more closely resemble ferric

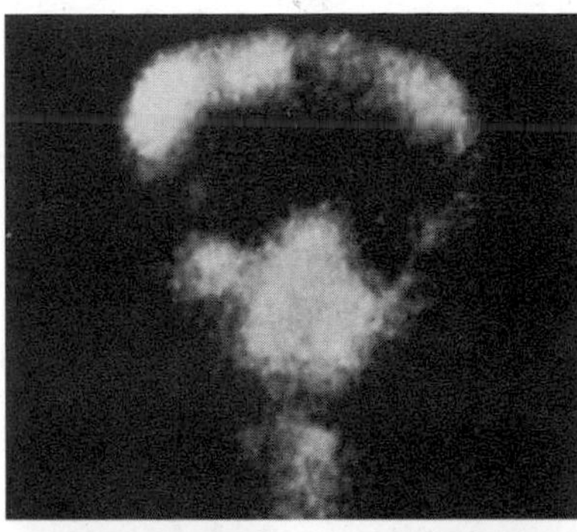

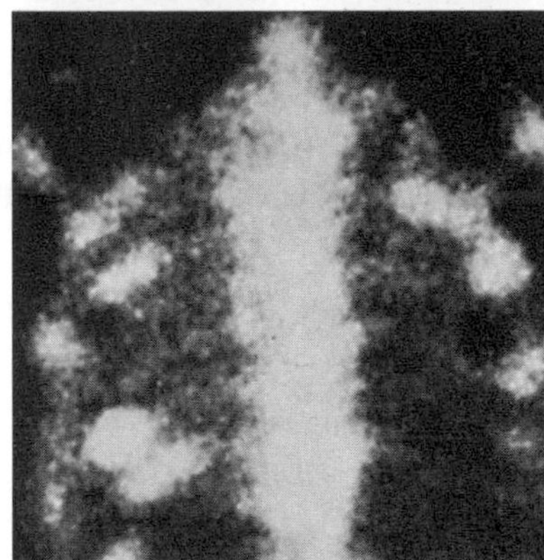

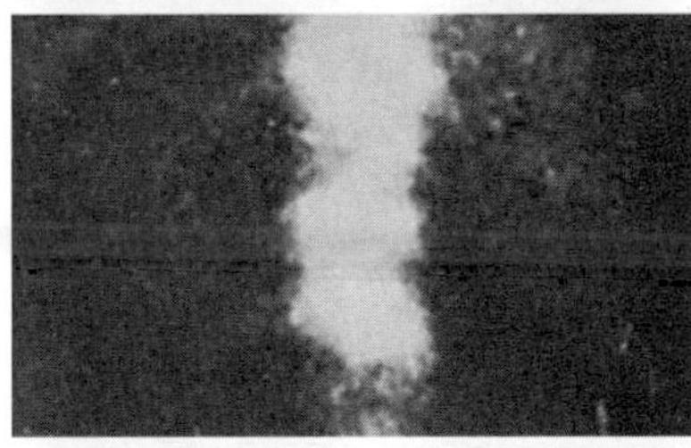

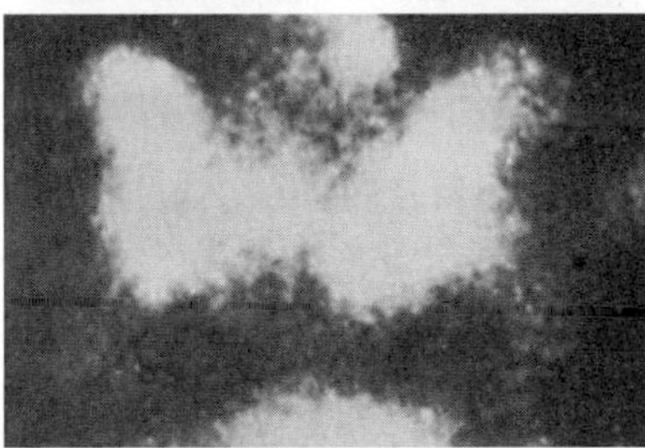

Fig. 43–13. Diffuse bony metastases from a breast carcinoma are noted in this posterior ^{99m}Tc-polyphosphate study. The skull, ribs, entire spine, and pelvis are involved (courtesy of *Physicians' Desk Reference for Radiology and Nuclear Medicine*).

ions, even to the extent of being taken into cells by transferrin, the iron transport system.

Two agents are available for the diagnosis of adrenal tumors. ^{131}I-*m*-Iodobenzylguanidine (MIBG), a radioiodinated analog of the adrenergic drug guanethedine, has been used to study the adrenal medulla and associated tumors, such as pheochromocytoma. On the other hand, tumors of the adrenal cortex have been visualized with radioiodinated cholesterol derivatives such as ^{131}I-6β-iodomethyl-19-norcholesterol (NP-59).

A more recent introduction has been the somatostatin analog ^{111}In-octreotide, which has shown promise in the visualization of primary and metastatic somatostatin receptor-rich tumors such as carcinoids, islet cell tumors of the pancreas, and paragangliomas.[11]

In addition to these approaches, numerous investigators have been examining the use of tumor-specific antibodies as carriers of radionuclides to tumors. Most tumor cells express specific antigens on their surface. Antibodies to such antigens can then be radiolabeled with radioiodine, technetium-99m, or indium-11 for tumor imaging. Although the concept is valid, progress has been slowed by the ability of the antigen to shed from the tumor into the systemic circulation and thereby reduce tumor specificity of the agent. In 1992, however, the FDA approved ^{111}In-satumomab pendetide. This product is derived from a murine monoclonal antibody to a glycoprotein expressed by a variety of adenocarcinomas. Its current clinical usage is focused on imaging of colorectal carcinoma and ovarian carcinoma.

Table 43–3. Diphosphonate Ligands for Technetium-99m

$$\mathrm{HO{-}\overset{\overset{\displaystyle O}{\|}}{\underset{\underset{\displaystyle HO}{|}}{P}}{-}\overset{\overset{\displaystyle R_1}{|}}{\underset{\underset{\displaystyle R_2}{|}}{C}}{-}\overset{\overset{\displaystyle O}{\|}}{\underset{\underset{\displaystyle OH}{|}}{P}}{-}OH}$$

Names	R_1	R_2
Methylene diphosphonate (MDP, Medronate)	—H	—H
Hydroxymethylene diphosphonate (HMPD), Oxidronate)	—H	—OH
Hydroxyethylidene diphosphonate (HEDP, Etidronate)	—CH_3	—OH

RADIOPAQUES FOR RADIOGRAPHY

In conventional roentgen examination and to a greater extent in CT, tissues are visualized according to their ability to attenuate a beam of x-rays (e.g., photons) before such rays strike a detector. The degree of attenuation varies according to the relative density of the tissue. Bone, for example, with an average density of 1.16 g/cm^3, absorbs a large percentage of the photons and thus appears light or white on the exposed film. The air-filled lung, on the other hand, is translucent relative to bone and thus appears darker on the exposed film. Using complex computer amplification techniques and mathematic analysis, current CT scanners are able to detect differences in tissue densities as small as 0.5%.[11]

Types of Radiopaques

Depending on their relative opacity to x-rays, agents can be categorized as either negative or positive contrast agents. Substances such as gases (air, oxygen, etc.) are translucent to x-rays and are known as negative contrast agents. Conversely, substances that increase the density of a tissue, thereby rendering it opaque to x-rays, are called positive contrast agents. Although the chemical structure and pharmacologic properties of radiopaques may vary widely, most possess the following desirable properties:

1. Maximal opacification to x-rays
2. Chemical stability
3. Low toxicity
4. High water solubility
5. Low viscosity
6. Minimal osmotic effect
7. Selective tissue uptake and excretion

In addition, agents administered orally should possess an optimal balance of lipo- and hydrophilicity to allow for sufficient intestinal absorption.

Historically, substances used as radiopaques have included heavy (high atomic number) metals (tantalum, lead, gold), heavy metal salts (bismuth nitrate, barium sulfate, iron oxide, thorium dioxide), and a multitude of iodine-containing compounds. Currently, virtually all contrast-enhanced radiographic procedures use barium sulfate, or any iodine-containing compound (refer to Fig. 43–14).

Inorganic Salts

Several inorganic salts were among the first compounds evaluated for use as radiographic contrast agents. Bismuth subnitrate was used for visualization of the gastrointestinal tract until it was replaced with barium sulfate, which, because of its low toxicity, effectiveness, and low cost, has remained the agent of choice for more than 70 years. Commercial preparations of barium sulfate are available, differing mainly in particle size, ionic charge, pH, and viscosity. This salt is administered orally for visu-

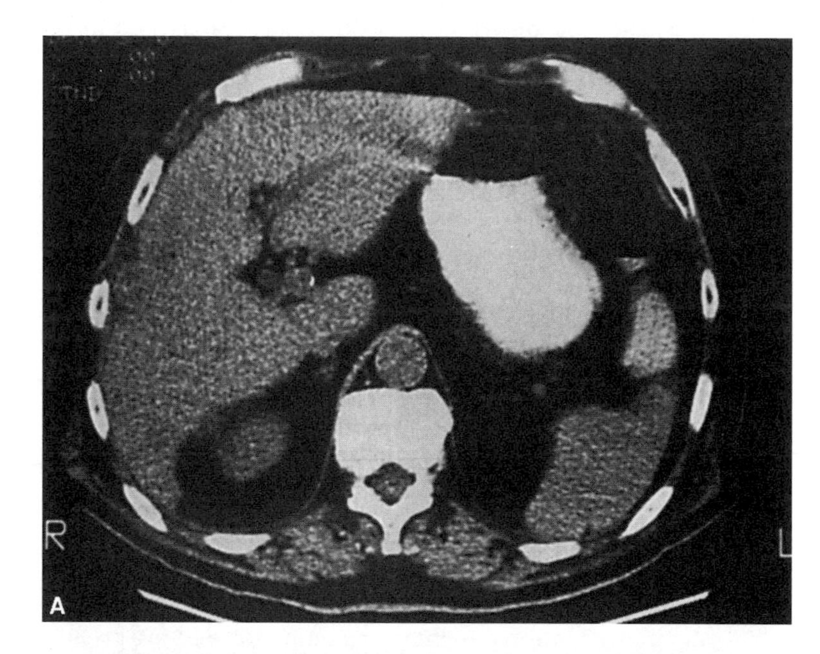

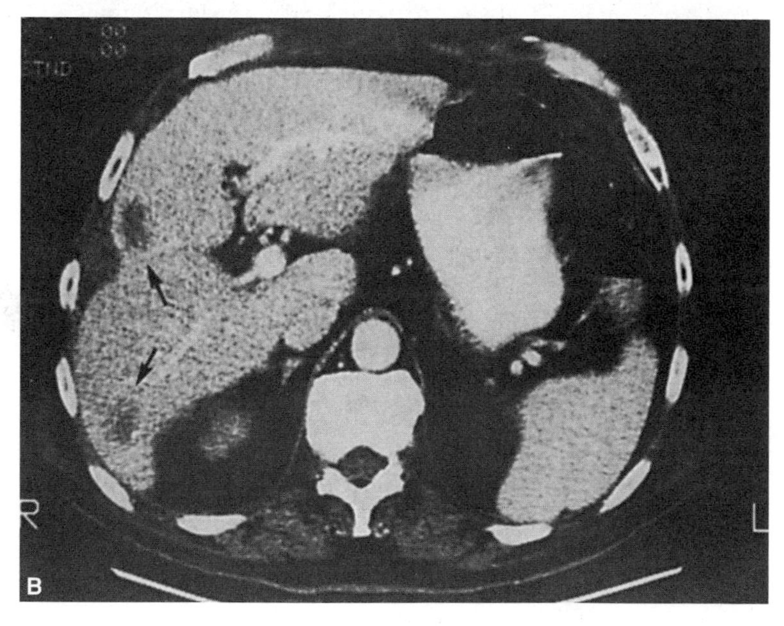

Fig. 43–14. Abdominal CT scans of a patient with liver metastases (arrows) from adenocarcinoma of the colon before *(A)* and after *(B)* administration of iodinated contrast media. (From M. B. Alpern et al., Radiology, *158,* 45(1986) with permission.)

alization of the esophagus and stomach or as a barium enema for examination of the colon.

Other inorganic salts such as tantalum oxide and thorium dioxide have been used for splenohepatography and angiography, respectively, but toxicity has limited their application to special cases.

Iodinated Compounds

Iodine, with an atomic number of 53 and a density of 4.94 g/cm^3, effectively absorbs x-rays at a variety of energy levels. Compounds containing ionically bound iodine, such as sodium iodide, were among the first evaluated as radiopaques in the early 1900s, but their use was discontinued owing to a considerable degree of toxicity. Conversely, compounds containing covalently bound iodine were less toxic, and as a result, subsequent research focused on these types of agents, which can be categorized as either water soluble or water insoluble.

Water-soluble Iodinated Compounds. Before the middle of the 1960s, most water-soluble contrast agents consisted of triiodinated benzoic acid salts that, after intravascular administration, dissociate into two particles—a triiodinated anion and an accompanying cation. Thus, for every three iodine atoms in solution, two ions were giving rise to an iodine-to-dissolved ion ratio of 3/2 or 1.5. By convention, these water-soluble agents, often referred to as ionic ratio-1.5 contrast agents, have been classified as triiodinated monomers (Table 43–4) and hexaiodinated dimers (Table 43–5). The former, represented by diatrizoate, iothalamate, metrizoate, ioxithalamate, and iodamide, are used primarily for urography, whereas the latter, represented by iodipamide, ioglycamide, iodoxamate, and iotroxamide, are used in cholangiography. Because of the necessity to administer these ionic ratio-1.5 agents in high concentrations, the plasma becomes severely hypertonic, frequently producing numerous osmotoxic effects, including vascular pain.

In hope of alleviating such osmotoxic effects, emphasis was placed on the development of agents with a higher iodine-to-ion ratio. Accordingly, two approaches were adopted. One of these, based on an ionic concept, afforded several hexaiodinated dimers similar to ioxaglate, shown in Table 43–5. The other approach, proposed by Almen in the late 1960s, resulted in the introduction of the first nonionic water-soluble contrast agents.[12] For example, metrizamide, which dissolves in water in a nondissociated form, affords three iodine atoms per molecule in solution, and is thus referred to as a nonionic ratio-3.0 contrast agent. Other such agents, including iopamidol and iohexol, have subsequently been introduced (see Table 43–4).

The ratio concept has been extended to a series of nonionic ratio-6.0 dimers as represented by iotrol (see Table 43–5). Although these agents are essentially isotonic with plasma, inherently high viscosities have thus far precluded their clinical utility.

Water-insoluble Iodinated Compounds. Water-insoluble contrast media are composed of ester derivatives of iodinated poppyseed oil, iodinated pyridones, and iopanoic acid analogs.

Iodized oils are prepared by iodination of the unsaturated fatty acid moieties present in vegetable oils. The resulting iodinated saturated fatty acids are subsequently converted to ethyl or glyceryl esters. Ethiodol and lipiodol, representing commercial ethyl and glyceryl ester preparations, respectively, contain from 35 to 45% iodine

Table 43–4. Monomeric Water-Soluble Contrast Agents

(General structure: 2,4,6-triiodobenzene ring bearing C(O)R, X, and Y substituents)

Nonproprietary Name	Ratio*	R	X	Y
Ionic				
Diatrizoate	1.5	—O(Na, Meg, Na/Meg)	—NHAc	—NHAc
Iothalamate	1.5	—O(Na, Meg, Na/Meg)	—NHAc	—$C(O)NHCH_3$
Metrizoate	1.5	—O(Na, Meg, Na/Meg)	—NHAc	—$N(CH_3)C(O)CH_3$
Ioxithamate	1.5	—O(Na, Mg, Meg)	—NHAc	—$C(O)NHCH_2CH_2OH$
Iodamide	1.5	—O(Na, Meg, Na/Meg)	—NHAc	—$CH_2NHC(O)CH_3$
Nonionic				
Metrizamide	3.0	(glucosamine ring: CH_2OH, O, OH, OH, HO, NH-)	—$N(CH_3)C(O)CH_3$	—NHAc
Iopamidol	3.0	—$NHCH(CH_2OH)_2$	—$NHC(O)CH(OH)CH_3$	—$C(O)NHCH(CH_2OH)_2$
Iohexol	3.0	—$NHCH(CH_2OH)_2$	—$N(C(O)CH_3)CH_2CH(OH)CH_2OH$	—$C(O)NHCH_2CH(OH)CH_2OH$

* Ratio, no. iodine atoms/no. of dissolved species; Meg, methylglucamine; Ac, $C(O)CH_3$.

Table 43–5. Dimeric Water-Soluble Contrast Agents

(General structure: two triiodinated benzene rings, the first bearing R and Y, the second bearing R_1 and Y_1, joined by X)

Nonproprietary Name	Ratio*	R	R_1	X	Y	Y_1
Iodipamide	1.5	-COOMeg	-COOMeg	$-NH-CO-(CH_2)_4-CO-NH-$	H	H
Ioglycamide	1.5	-COOMeg	-COOMeg	$-NH-CO-CH_2-O-CH_2-CO-NH-$	H	H
Iodoxamate	1.5	-COOMeg	-COOMeg	$-NH-CO-(CH_2CH_2O)_4-CH_2CH_2-CO-NH-$	H	H
Iotroxamide	1.5	-COOMeg	-COOMeg	$-NH-CO-(CH_2-O-CH_2)_3-CO-NH-$	H	H
Ioxaglate	3.0	-COO(Na/Meg)	$-N(CH_3)Ac$	$-NH-CO-CH_2-NH-CO-$	$-CO-NH-CH_2CH_2OH$	$-CO-NH-CH_3$
Iotrol	6.0	$-CO-NH-CH(CH_2OH)-CH(OH)-CH_2OH$	$-CO-NH-CH(CH_2OH)-CH(OH)-CH_2OH$	$-N(CH_3)-CO-CH_2-CO-N(CH_3)-$	$-CO-NH-CH(CH_2OH)-CH(OH)-CH_2OH$	$-CO-NH-CH(CH_2OH)-CH(OH)-CH_2OH$

* Ratio, no. iodine atoms/no. of dissolved species; Meg, methylglucamine; Ac, $C(O)CH_3$.

by weight and are emulsified with surfactants before administration in lymphography and intra-arterial hepatography. A disadvantage of iodized oils is their relative instability. Because of the weak nature of the aliphatic carbon-iodine bond, iodized oils tend to decompose after exposure to air or light.

Aromatic iodides, on the other hand, are more resistant to such photochemical and oxidative effects. Water-insoluble iodinated phenylalkanoic acids and esters that have found clinical utility include iophendylate for myelography, pyridone analogs for bronchography, and iopanoate analogs for oral cholecystography. These agents are discussed in subsequent sections.

Physicochemical Factors Associated with the Use of Radiopaques

Contrast agents administered intravascularly, such as urographic and angiographic agents, must exhibit a high degree of water solubility. Orally administered cholecystographic agents, on the other hand, must exhibit a balance of water and oil solubility to permit intestinal absorption, transport to the liver, and subsequent biliary excretion.

In general, water solubility is enhanced by incorporation of various polar functional groups including acetamido, carboxyl, carbamyl, or hydroxyalkyl into the structure of the agent. Radiopaque carboxylic acids, for example, usually consist of a benzoic acid derivative containing iodine atoms in the 2, 4, and 6 positions. The remaining 1, 3, and 5 positions contain polar substituents, which provide optimal solubility as well as minimal toxicity characteristics. Moreover, benzoic acid analogs are usually administered as either sodium, calcium, magnesium, or meglumine salts, or as mixtures containing both sodium and meglumine salts. In general, meglumine salts are better tolerated than the corresponding sodium salts. The high viscosity found to be associated with meglumine salts, however, has led to the development of meglumine and sodium salt mixtures that possess a more suitable combination of acceptable viscosity and diminished toxicity.

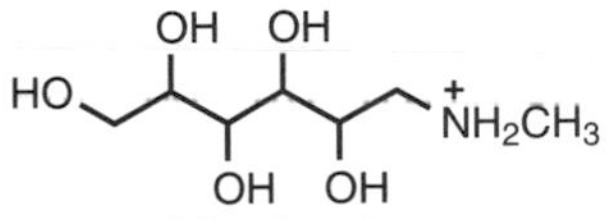

Meglumine

The osmolality of a solution depends on the number of different molecular or ionic species present. Ionic compounds (e.g., salts) dissociate in solution to form an anion and a cation, whereas nonionic molecules give rise to a single un-ionized species when dissolved in water. Intravascular administration of high concentrations of ionic compounds creates hyperosmolality in the vasculature, resulting in the rapid compensatory influx of water and electrolytes from surrounding tissues. As mentioned previously, the hyperosmolality induced by rapid administration of ionic contrast media often results in pain, decreased blood pressures, and vascular damage. To reduce or prevent such effects, synthetic efforts have focused on increasing the number of iodine atoms per molecule. This approach gave rise to several classes of ionic and nonionic monomers, dimers, and trimers, which have been shown to have reduced toxicity.

Another factor that influences the use of contrast agents is viscosity. The viscosity of a solution depends on the molecular structure, concentration, and charge of the dissolved species as well as the viscosity of the vehicle itself. Commercial water-soluble contrast preparations are typically from 2 to 18 times as viscous as water at 37°C.[13] The need for contrast preparations with suitable viscosity is obvious when considering the large volumes (50 to 150 ml) and rapid injection rates required in several common procedures, including urography. In addition to facilitating the rate of administration, low-viscosity preparations produce fewer adverse reactions.

Toxicity, metabolism, and excretion of iodinated contrast agents are influenced by their association with plasma proteins in the circulation. Agents that lack substituents at the 5-position of the benzene ring generally bind avidly to serum albumin and are cleared through the liver, whereas agents containing a substituent in the 5-position lack such binding and are generally excreted by the kidneys.

Moreover, fully substituted benzene analogs such as the water-soluble uro- and angiographic agents are generally excreted intact. Agents lacking a substituent in the 5-position, including the oral cholecystopaques, on the other hand, are first converted to glucuronide conjugates before biliary excretion. This relationship fails, however, in the case of dimeric cholangiographic agents such as iodipamide, ioglycamide, iodoxamate, and iotroxamide (see Table 43–5). These agents are excreted unchanged despite the absence of substituents in the 5-position of both benzene rings.

RADIOGRAPHIC PROCEDURES

The names of several common radiographic procedures that use contrast agents are included in Table 43–6. In the discussion that follows, we describe procedures in frequent use and then some less frequently used procedures.

Table 43–6. Uses of Contrast Agents in Radiographic Procedures

Radiographic Procedure	Tissue/Organ Visualized
Angiography	Blood vessels
Bronchography	Lungs
Cholangiography	Gallbladder, bile ducts
Cholecystography	Gallbladder
Esophagography	Esophagus
Hepatography	Liver
Hysterosalpingography	Uterus, fallopian tubes
Lymphography	Lymph nodes, vessels
Myelography	Spinal cord, subarachnoid space
Urography	Urinary tract

Urography

The use of radiopaque agents for visualization of the urinary tract is based on the rapid renal excretion of water-soluble iodinated benzoic acid salts as well as several dimeric ionic and monoionic analogs. Current agents in widespread use consist of salts of 2,4,6-triiodobenzoic acid, which differ from each other in the substituents in the 3- and 5-positions of the benzene nucleus. Examples of these monomeric salts include diatrizoate, iothalamate, metrizoate, ioxithalamate, and iodamide (see Table 43–4). They are administered as either sodium or meglumine salts or as a mixture of both, depending on the radiographic procedure. These agents are equally effective for urographic contrast and appear to exhibit the same degree of toxicity. In an attempt to suppress the hyperosmolar effects associated with these ionic agents, efforts have focused on the development of ionic agents with higher ratios of iodine per ion formed in solution, such as ioxaglate; ionic dimers, including iotrol; and also several nonionic compounds including metrizamide, iopamidol, and iohexol. Of these agents, the nonionics appear the most promising because they indeed elicit fewer hyperosmolarity-related adverse reactions than the ionic compounds.

Cholecystography

Agents useful for opacification of the gallbladder and bile ducts can be classified as either oral cholecystographic or intravenous cholangiographic agents. In 1951, Hoppe and Archer reported the synthesis and radiopacifying properties of several triiodophenylalkanoic acids.[14] One of these agents, iopanoic acid (Table 14–7), provided superior opacification and less toxicity than any previous agent. It not only became the most widely used oral cholecystographic agent for the next three decades, but also served as a model for many of the commercial agents that followed.

Currently, all oral cholecystopaques are analogs of 2,4,6-triiodinated alkylbenzoic acids. Although substituents in the 1- and 3-positions vary with each compound, all have enough hydro- and lipophilicity to allow intestinal absorption and hepatic excretion. Examples of current oral cholecystographic agents include iopanoic acid, iocetamic acid, tyropanoic acid, ipodic acid, and iopronic acid.

Once in circulation, these agents bind avidly to serum albumin, thus enhancing hepatic uptake as a result of decreased renal excretion of the protein-bound agent. In the liver, the oral contrast agent is converted into a water-soluble glucuronide conjugate, which subsequently is excreted in the bile and stored in the gallbladder.

Cholangiography

Contrast agents used in intravenous cholangiography for visualization of the gallbladder and bile ducts are all moderately strong dibasic triiodobenzoic acid dimers linked by a polymethylene-like chain of varying length. Examples of these agents are iodipamide, ioglycamide, iodoxamate, and iotroxamide (see Table 43–5). In contrast to their orally administered counterparts, intravenous biliary contrast agents are highly water soluble and completely ionized at physiologic pH. These agents usually are administered as meglumine salts to reduce adverse reactions.

Compared to the orally administered compounds, intravenous cholangiopaques are even more highly bound to plasma serum albumin while in transport. Such avid binding has, however, been related to an increase in toxicity as well as a reduction in hepatobiliary clearance. Moreover, biliary excretion of intravenous agents, unlike that of oral compounds, does not depend on prior glucuronide conjugation or other chemical modification.

Table 43–7. Oral Cholecystopaques

I, X, COOH, I, I, Y

Nonproprietary Name	X	Y
Iopanoic acid	CH_2CH_3	$—NH_2$
Iocetamic acid	N, CH_3OC, CH_3	$—NH_2$
Tyropanoic acid	CH_2CH_3	$—NHCOCH_2CH_2CH_3$
Ipodic acid		$—N{=}CHN(CH_3)_2$
Ipronic acid	O, O, CH_2CH_3	$—NHCOCH_3$

Angiography

Early attempts to visualize blood vessels with heavy metal salts, sodium iodide, and iodinated vegetables oils were unsuccessful because of the inherent toxicities associated with these agents. Several other di- and triiodinated compounds, including acetrizoate, were found to be less toxic and were used until the early 1950s. A series of ionic ratio-1.5 acetrizoate analogs including diatrizoate, iothalamate, and metrizoate soon followed (see Table 43–4). Acetrizoate was associated with certain toxic effects, but the others were better tolerated and have since found widespread use as angiographic agents.

Several late-generation nonionic compounds, including metrizamide, have been useful as angiographic agents. Moreover, because metrizamide is well tolerated and causes few side effects, it is considered the agent of choice for most cardiac angiography procedures.

Myelography

Both oil- and water-soluble contrast agents have been used to opacify the spinal cord and subarachnoid space. Oily agents such as lipiodol dominated myelography until iophendylate was introduced in 1944. Iophendylate is a mixture of ethyl iodophenylundecanoate isomers, and is immiscible with water and spinal fluid. Moreover, its slow absorption form the subarachnoid space necessitates its removal from the spinal canal by aspiration following the procedure. Overall, iophendylate has provided reliable and relatively safe evaluation of the spinal cord, and although it is still widely used today, several nonionic ratio-

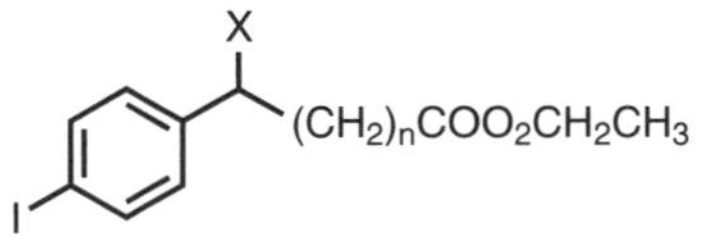

X
$(CH_2)_nCOO_2CH_2CH_3$
I

Iophendylate

X	n
H	9
CH_3	8

3.0 agents, including metrizamide, iopamidol, and iohexol, have gained popularity because of improved image detail and diminished toxicity.

Hysterosalpingography

Many myelographic agents have also been used for visualization of other body cavities and structures within, such as the uterus and fallopian tubes. Oil agents, lipiodol and ethiodol, were used extensively before the 1960s, but because of poor absorption and side effects including abdominal pain, irritation, foreign body reaction, and embolism, they were replaced by water-soluble agents. One commonly used preparation consists of a mixture of sodium acetrizoate and polyvinylpyrrolidone, a well-known plasma extender, whereas another consists of a 2:1 mixture of diatrizoate meglumine and iodipamide meglumine. Although the frequency of abdominal pain and hypersensitivity associated with both preparations actually increased relative to their oil predecessors, the severity was diminished. Overall, these water-soluble preparations are considered nontoxic and relatively safe.

Bronchography

Visualization of the bronchial tree was originally achieved with iodized oils, suspensions of iodized oils and inert powders, and solutions or suspensions of iodine compounds. Since the 1950s, however, several pyridone analogs have received the most attention. Preparations have included a mixture of iodopyracet and carboxymethylcellulose, aqueous or oily propylidone suspension, and an aqueous suspension of iopydol and iopydone. After intratracheal administration, these preparations coat the bronchial tree and alveolar spaces and are subsequently excreted from the lungs by coughing, mucociliary action, and expectoration, or by absorption. Clearance times can be months or years in certain disease states. Irritation, a common drawback of these agents, is generally severe enough to require complete anesthesia before administration of the agent. The use of bronchographic agents has declined rapidly in recent years.

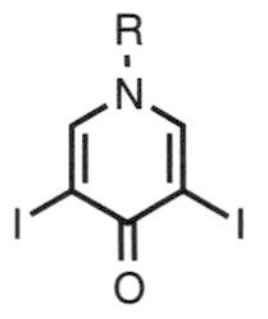

R
N
I I
O

Iopydone	R = H
Iopydol	R = $-CH_2CH(OH)CH_2OH$
Iodopyracet	R = $-CH_2COO^-$ $[NH_2(CH_2CH_2OH)_2]^+$
Propylidone	R = $-CH_2COOCH_2CH_2CH_3$

Lymphography

Visualization of the lymphatic system was initially achieved with water-soluble agents, but rapid diffusion from the lymph vessels precluded their use. After injection into a lymph duct, oil agents such as ethiodol and lipiodol traverse the vessel and concentrate in neighboring nodes, wherein they may remain for months until passing into the venous system through the thoracic duct. Although complications such as pulmonary and cerebral embolism resulting from lymphography have been reported, the incidence has been reduced by minimizing the volume of oil contrast agent administered.

Hepatography

Opacification of the liver is currently achieved with water-soluble urographic agents. Nonspecific hepatic uptake and rapid diffusion of the agent out of the vasculature into surrounding interstitial spaces are two common drawbacks of these agents. In an effort to develop more liver-specific agents, recent experimental approaches have been based on the known phagocytic action of the reticuloendothelial system. Accordingly, ethyl esters of water-soluble agents, metal colloids, ethiodol emulsions, and radiopaque liposomes have been evaluated as liver-specific agents; none are commercially available.

Adverse Reactions

The diverse and unpredictable nature of adverse reactions is considered a common disadvantage of all iodinated contrast agents. A useful classification of reactions, based on the severity and mode of treatment, includes those that are minor and require no therapy, those that are moderate and require treatment as needed, and finally those that are severe and life-threatening and require immediate intervention and treatment. Minor reactions include nausea, vomiting, flushing, "light-headedness," slight breathing difficulty, and mild urticaria. Among the moderate reactions are facial and laryngeal edema, urticaria, bronchial spasm, and transitory hypotension. Severe reactions include prolonged hypotension, circulatory collapse, angina, myocardial infarction, renal failure, ventricular fibrillation, coma, convulsions, and paralysis. For the most part, these adverse effects are associated with the hypertonicity of the ionic agents relative to the blood, calcium binding, and direct chemotoxicity of the agents themselves. Calcium binding, for example, is known to have a negative effect on myocardial contractility.[15]

The relative complication rate varies among the different procedures. Several large multicenter studies[16] have shown the nonfatal incidence of reactions to be about 2 to 3% for vascular procedures, 6% for intravenous urography, and about 10% for intravenous cholangiography. Overall, the total incidence for intravascular examinations has been estimated at from 5 to 8%. Mortality has been reported in only 1 to 50,000 procedures.

Although sex, age, and weight do not seem to influence the incidence of adverse reactions, a twofold increase in the incidence rate is noted in patients with a history of allergy.

Although well documented in several ongoing worldwide studies, the overall incidence of adverse reactions and toxicity of radiographic contrast agents is surprisingly low in light of the large multigram doses traditionally administered.

MAGNETIC RESONANCE IMAGING

Image contrast in conventional radiography is produced by a single parameter, namely the attenuation of x-rays. In magnetic resonance imaging (MRI), however, four parameters, including proton density, relaxation characteristics T1 and T2, and blood flow, influence image characteristics. Variation of these parameters determines the image contrast between normal and pathologic tissue. Despite many recent technologic advancements in MRI, many disease states and organ systems remain poorly characterized. It is in these areas that the application of contrast-enhancing agents have significantly extended the diagnostic potential of MRI.

Imaging agents currently used for MRI differ from their iodinated radiographic counterparts both chemically and in their basic mechanism of action. Unlike iodinated contrast agents that produce a direct contrast effect by absorbing x-rays, MR agents act indirectly by alternating the local magnetic environment of the surrounding observed nuclei.

In the development of MR contrast media, image contrast efforts to date have been devoted to affecting changes in the relaxation parameters T1 and T2. Accordingly, these efforts have centered on the use of paramagnetic materials to alter the T1 and/or T2 of tissues. In basic quantum mechanical terms, paramagnetic substances contain several orbitals with similar energy levels in which only one electron can be put in each orbital, all with parallel spins, resulting in a net electron spin. It is the generation of local magnetic fields created by these unpaired electrons that shorten T1 and T2 of neighboring protons and ultimately influence the image intensity of the tissue. Diamagnetic materials, on the other hand, contain electrons distributed in pairs into various orbitals, each with opposite spins. The two opposite spins cancel each other out, resulting in no net electronic spin and thus no change in the magnetic properties of neighboring protons.

Because of its seven unpaired electrons, gadolinium has evolved as the paramagnetic element of choice for production of MR contrast-enhancing agents. Early studies with gadolinium and other transition metals indicated that a high degree of toxicity was associated with the free ion form of the metal. It became necessary therefore to bind the metal ion to multidentate chelates to reduce toxicity.

Two types of paramagnetic agents are commercially available for clinical use as MR image-enhancing agents. The most common of these agents, gadolinium diethylenetetraminepentaacetic acid (gadopentetate, Fig.

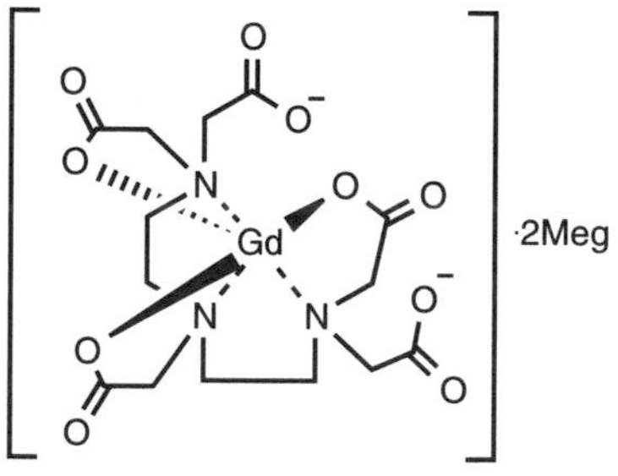

Gadopentetate ·2Meg

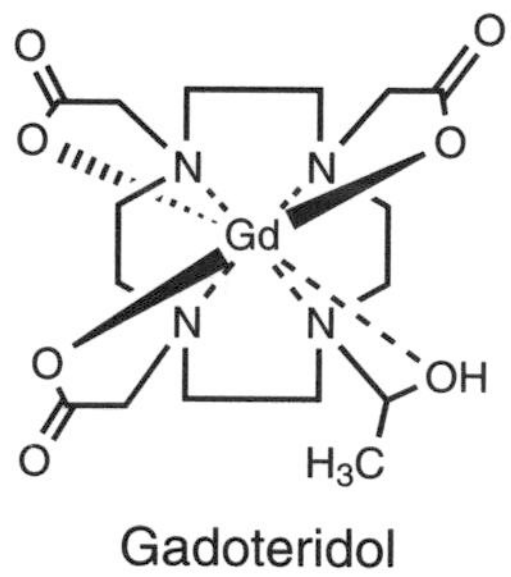

Gadoteridol

Fig. 43–15. Chemical structures for magnetic resonance image-enhancing agents.

43–15), possesses favorable in vivo stability and proton relaxivity characteristics and is also well tolerated by patients when administered as a dimeglumine salt. Based on previous experience with ionic CT contrast agents, efforts are underway to prepare nonionic versions of such MR agents. One of these agents, gadoteridol (Fig. 43–15), was approved recently for clinical use in the United States. It remains unclear, however, whether the newer nonionic versions offer any real clinical advantages. Apparently, the untoward effects associated with ionic CT contrast agents are infrequently encountered because of the smaller volumes of gadolinium-DTPA-dimeglumine administered relative to the CT agents.

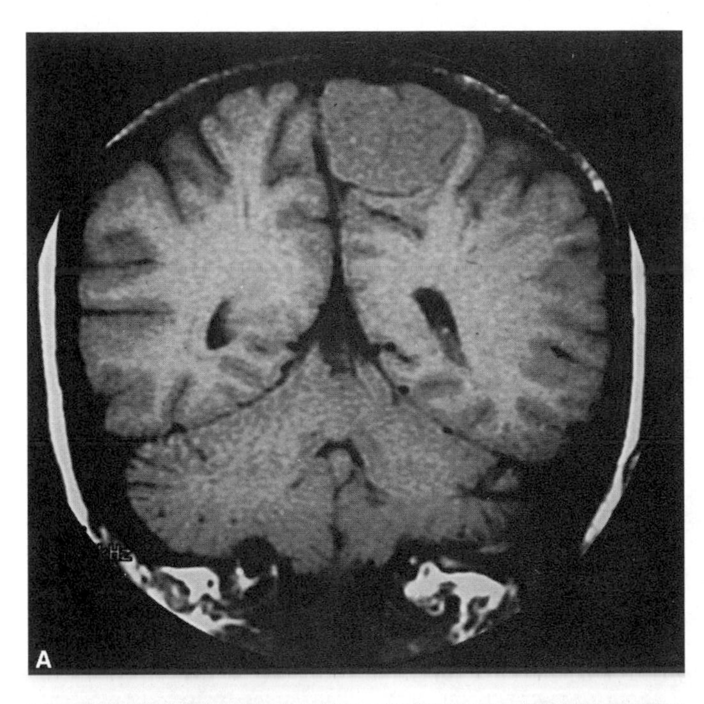

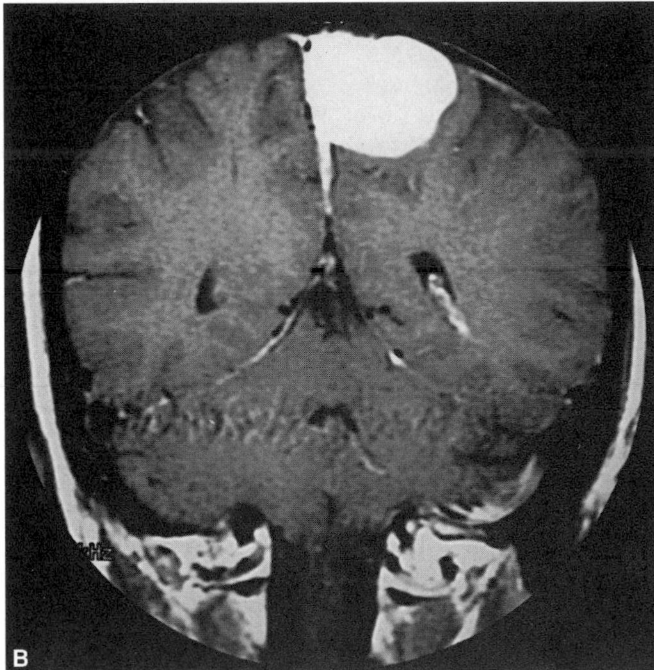

Fig. 43–16. Brain tumor detection with gadopentetate dimeglumine. Coronal T1-weighted MR head image before *(A)* and after *(B)* intravenous administration of gadopentetate dimeglumine demonstrates a 3-cm diameter enhancing area at the inner edge of the skull consistent with a meningioma. (Courtesy of J. Brunberg, Department of Radiology, University of Michigan.)

Once these agents are administered intravenously, they behave much like the urographic CT contrast agents: they are distributed extracellularly and rapidly undergo renal excretion, with a clearance half-life of about 20 minutes. Because of its nonspecific biodistribution, gadolinium-DTPA has been used for contrast enhancement in a variety of clinical situations. The agent has been used not only to detect or characterize many renal and myocardial conditions, but also to detect inflammatory and neoplastic musculoskeletal, gynecologic, liver, breast, and bronchial lesions. Perhaps the best results, however, have been obtained in the brain (Fig. 43–16). Gadolinium-DTPA does not cross the blood-brain barrier and therefore does not accumulate in normal brain tissue or in lesions that possess an intact blood-brain barrier. In those diseases associated with a breakdown in the blood-brain barrier, accumulation of the agent generally results at the pathogenic site and facilitates detection by MRI.

REFERENCES

1. A. H. Becquerel, *On radioactivity, a new property of matter,* in *Nobel Lectures, Including Presentation Speeches and Laureates' Biographies. Physics, 1901–1921,* New York, Elsevier, 1967, pp. 47–73.
2. H. L. Blumgart and S. Weiss, *J. Clin. Invest., 4,* 399(1927).
3. M. Brucer, *Trilinear Chart of the Nuclides,* St. Louis, Mallinckrodt/Nuclear, 1979.
4. E. M. Smith, *J. Nucl. Med. (Suppl.), 1,* 5(1968).
5. S. Hertz, et al., *Proc. Soc. Exp. Biol. Med., 38,* 510(1938).
6. R. D. Evans, *Med. Phys., 2,* 105(1975).
7. P. V. Harper, et al., *The use of technetium-99m as a clinical scanning agent for thyroid, liver and brain,* in *Medical Radioisotope Scanning,* Proc. Athens Symposium, Vol. 2, IAEA, Vienna, April, 1964.
8. M. D. Loberg and A. T. Fields, *Int. J. Appl. Radiat. Isot., 29,* 167(1978).
9. M. D. Loberg, et al., *J. Nucl. Med., 17,* 633(1976).
10. A. Nunn, et al., *J. Nucl. Med., 24,* 423(1983).
11. J. D. Meindl, *Science, 215,* 792(1982).
12. T. Almen, *Invest. Radiol. (Suppl.), 20(1),* S2(1984).
13. *Physicians' Desk Reference for Radiology and Nuclear Medicine,* 1977/78. Oradell, N.J., Medical Economics Co., 1977.
14. S. J. Archer, et al., *J. Am. Pharm. Assoc., 40,* 617(1951).
15. M. A. Bettman, *Am. J. Roentgenol., 139,* 787(1982).
16. W. H. Shehadi and G. Toniolo, *Radiology, 137,* 299(1980).

SOURCE MATERIAL AND SUGGESTED READING

G. Bydder, et al., Eds., *Contrast Media in MRI,* The Netherlands, Medicom Europe, 1990.

H. Katayama and R. Brasch, Eds., *New Dimensions of Contrast Media,* New York, Elsevier Science Publishing, 1991.

A. D. Nunn, Ed., *Radiopharmaceuticals,* New York, Marcel Dekker, 1992.

D. P. Swanson, et al., Eds., *Pharmaceuticals in Medical Imaging,* New York, MacMillan, 1990.

Appendix

pK_a VALUES FOR SOME DRUGS AND MISCELLANEOUS ORGANIC ACIDS AND BASES

pH VALUES FOR TISSUE FLUIDS

David A. Williams

Table A–1. pK_a Values for Some Drugs and Miscellaneous Organic Acids and Bases

Drug	pK_a Values		Reference
	HA	HB^+	
Acebutolol		9.2	1
Acenocoumarol	4.7		1
Acetaminophen	9.7		1
Acetanilide		0.5	1
Acetazolamide	7.4, 9.1		3
Acetohydroxamic acid	9.4		1
α-Acetylmethadol		8.6	1
Acetylsalicylic acid	3.5		1
Acyclovir	9.3	2.3	1
Adriamycin		8.2	1
Ajamaline		8.2	1
Albuterol	10.3	9.3	9
Alclofenac	4.3		1
Alfentanil		6.5	1
Allobarbital	7.8		1
Allopurinol	9.4		1,4
Alphaprodine		8.7	1
Alprenolol		9.7	1,5
Altretamine		10.3	1
Amantadine		9.0	1
Amdinocillin	3.4	8.9	1
Amiloride		8.7	1
Aminacrine		10.0	1
p-Aminobenzoic acid	4.9	2.5	1
Aminocaproic acid	4.4	10.8	1
Aminohippuric acid	3.8		1
Aminopterin	5.5		1
Aminopyrine		5.0	1
p-Aminosalicyclic acid	3.6	1.8	1
Aminothiadiazole		3.2	1
Amiodarone		6.6	1
Amitriptyline		9.4	15
Amobarbital	7.8		1
Amoxapine		7.6	1,4
Amoxicillin	2.4	9.6	12
Amphetamine		10.0	1
Amphotericin B	5.5	10.0	11
Ampicillin	2.5	7.2	16
Anileridine		3.7, 7.5	1
Antazoline		2.5, 10.1	1
Antipyrine		1.5	1
Apomorphine	8.9	7.0	1

Table A–1. *(Continued)*

Drug	pKa Values HA	HB$^+$	Reference
Aprobarbital	8.0		1
Ascorbic acid	4.2, 11.6		1
Atenolol		9.6	6
Atropine		9.8	1
Azatadine		9.3	1
Azathioprine	8.0		9
Azlocillin	2.8		1
Aztrenam	0.7, 2.9	3.9	1
Bacampicillin		6.8	1
Baclofen	5.4	9.5	1
Barbital	7.9		1
Bendroflumethiazide	8.5		5
Benzocaine		2.5	1
Benzphetamine		6.6	1
Benzquinamide		5.9	1
Benztropine		10.0	1
Betahistine		3.5, 9.8	1
Betaprodine		8.7	1
Bethanidine		10.6	1
Bromazepam	11.0	2.9	1
Bromocriptine[a]		9.8	1
Bromodiphenhydramine		4.9	7
Brompheniramine		3.6, 9.8	1
Brucine		8.2, 2.5	1
Bufuralol		8.9	1
Bumetanide	5.2, 10.0		1
Bunolol		9.3	1
Bupivacaine		8.1	1
Bupropion		7.0	8
Burimamide		7.5	1
Butabarbital	7.9		1
Butacaine		9.0	1
Butaclamol		7.2	1
Butamben		5.4	1
Butorphanol		8.6	1
Butylated hydroxytoluene	17.5		1
Butylparaben	8.5		3
Caffeine	>14.0	0.6	1
Camptothecin		10.8	1
Captopril	3.7, 9.8		4
Carbachol		4.8	1
Carbenicillin	2.7		1
Carbenoxolone	6.7, 7.1		1
Carbinoxamine		8.1	1
Carisoprodol		4.2	4
Carpindolol		8.8	1
Cefaclor	1.5	7.2	8
Cefamandole	2.7		9
Cefazolin	2.1		1
Cefoperazone	2.6		4
Cefotaxime	3.4		4
Cefoxitin	2.2		10
Ceftazidime	1.8, 2.7	4.1	11

(continued)

Table A–1. *(Continued)*

Drug	pK_a Values		Reference
	HA	HB^+	
Ceftizoxime	2.7	2.1	4
Ceftriaxone	3.2, 4.1	3.2	1
Cefuroxime			4
Cephacetrile			3
Cephalexin[b]	3.2		1
L-Cephaloglycin	4.6	7.1	1
Cephaloridine	3.4		1
Cephalothin	2.5		1
Cephapirin			4
Cephradine			4
Chenodiol	4.3		1,4
Chloral hydrate	10.0		16
Chlorambucil	5.8		4
Chlorcyclizine		2.1, 8.2	1
Chlordiazepoxide		4.8	1
Chlorhexidine		10.8	1
Chlorocresol	9.6		1
Chloroquine		8.1, 9.9	1
8-Chlorotheophylline	8.2		1
Chlorothiazide	6.8, 9.5		1
Chlorpheniramine		9.0	1
Chlorphentermine		9.6	1
Chlorpromazine		9.3	1
Chlorpropamide	4.9		1
Chlorprothixene		8.8, 7.6	16
Chlortetracycline[c]	3.3, 7.4	9.3	1
Chlorthalidone	9.4		1
Chlorzoxazone	8.3		1
Cimetidine		6.8	1
Cinchonine		4.3, 8.4	1
Ciprofloxacin	6.0	8.8	1
Clindamycin		7.5	9
Clofibrate	3.5		1
Clonazepam	10.5	1.5	1
Clonidine		8.3	1
Clopenthixol		6.7, 7.6	1
Clotrimazole		4.7	10
Cloxacillin	2.8		1
Clozapine		8.0	1
Cocaine		8.7	1
Codeine		8.2	1
Colchicine		1.9	1
Cromolyn	1.1, 1.9		1
Cyanocobalamin		3.4	9
Cyclacillin	2.7	7.5	4
Cyclazocine		9.4	1
Cyclizine		8.0, 2.5	1
Cyclobarbital	8.6		1
Cyclobenzaprine		8.5	1
Cyclopentamine		11.5	1
Cyclopentolate		7.9	1
Cycloserine		4.5, 7.4	1
Cyclothiazide[c]	9.1, 10.5		1
Cyproheptadine		8.9	4

Table A–1. *(Continued)*

Drug	pKa Values		Reference
	HA	HB^+	
Cytarabine		4.3	1
Dacarbazine		4.4	1
Dantrolene	7.5		1
Dapsone		1.3, 2.5	1
Daunorubicin		8.4	1,4
Debrisoquin		11.9	1
Dehydrocholic acid	5.12		1
Demeclocycline	3.3, 7.2	9.4	1
Demoxepam		4.5, 10.6	1
Deserpidine[d]		6.7	1
Desipramine		10.4	1
Dextroamphetamine		9.9	1
Dextrobrompheniramine		9.3	1
Dextrochlorpheniramine		9.2	1
Dextrofenfluramine		9.1	1
Dextroindoprofen	4.6		1
Dextromethorphan		8.3	1
Dextromoramide		7.0	1
Diacetylmorphine (heroin)		7.8	1
Diatrizoic acid	3.4		1
Diazepam		3.4	1
Diazoxide	8.5		1
Dibenzepin		8.3	8
Dibucaine		8.9	1
Dichlorphenamide	7.4, 8.6		1
Diclofenac	4.5		1
Dicloxacillin	2.8		1
Dicoumarol	4.4, 8.0		1
Dicyclomine		9.0	1
Diethazine		9.1	1
Diethylcarbamazepine		7.7	1
Diflunisal	3.0		1
Dihydroergocriptine		6.9	1
Dihydroergocristine		6.9	1
Dihydroergotamine		6.9	1
Dihydrostreptomycin		7.8	1
Dilevolol		9.5	1
Diltiazem		7.7	1
Dimethadione	6.1		1
Dimethisoquin		6.3	1
Dinoprost[e]	4.9		1
Dinoprostone	4.6		1
Diperodon		8.4	11
Diphenhydramine		9.1	1
Diphenoxylate		7.1	1
Diphenylpyraline		8.9	1
Dipipanone		8.5	1
Dipyridamole		6.4	1
Disopyramide	10.2	8.4	1
Dobutamine		9.5	1,4
Dopamine	10.6	8.9	1
Doxepin		9.0	1
Doxorubicin		8.2, 10.2	1

(continued)

Table A–1. *(Continued)*

Drug	pK$_a$ Values		Reference
	HA	HB^+	
Doxycycline	3.4, 7.7	9.5	1
Doxylamine		4.4, 9.2	1
Droperidol		7.6	1
Emetine		8.2, 7.4	1
Enalapril	3.0	5.5	1
Enalaprilat	2.3, 3.4	8.0	1
Ephedrine		9.6	1
Epinephrine	8.9	10.0	1
Ergometrine		7.3	1
Ergonovine		6.8	1
Ergotamine		6.4	1
Erythromycin		8.8	1
Estrone[f]	10.8		13
Ethacrynic acid	3.50		1
Ethambutol		6.3, 9.5	1
Ethoheptazine		8.5	1
Ethopropazine		9.6	1
Ethosuximide	9.5		1
Ethoxazolamide	8.1		1
Ethyl biscoumacetate	7.5		1
Ethylmorphine		8.2	1
Ethylnorepinephrine		8.4	1
Etidocaine		7.9	1
Etileprine	9.0	10.2	1
Etomidate		4.2	1
Eugenol	9.8		1
Fenclofenac	4.5		1
Fenfluramine		9.1	1
Fenoterol	10.0	8.6	1
Fenprofen	4.5		1
Fentanyl		8.4	14
Floxuridine	7.4		1
Flubiprofen	4.3		1
Flucloxacillin	2.7		1
Flucytosine	10.7	2.9	1
Flufenamic acid	3.9		10
Flumizole	10.7		1
Flunitrazepam		1.8	1
Fluorouracil	8.0, 13.0		1
Flupentixol		7.8	1
Fluphenazine enanthate		3.5, 8.2	1
Fluphenazine		3.9, 8.1	1
Flupromazine		9.2	1
Flurazepam	8.2	1.9	1
Furosemide	3.9		1
Fusidic acid	5.4		1
Gentamicin[b]		8.2	1
Glibenclamide	5.3		9
Glipizide	5.9		1
Glutethimide	9.2		1
Glyburide	5.3		1
Glycyclamine		5.5	1
Guanethidine		8.3, 11.9	1
Guanoxan		12.3	1

Table A–1. *(Continued)*

Drug Substance	pK_a Values		Reference
	HA	HB^+	
Haloperidol		8.3	1
Hexetidine		8.3	12
Hexobarbital	8.2		1
Hexylcaine		9.1	1
Hexylresorcinol	9.5		1
Hippuric acid	3.6		1
Histamine		5.9, 9.8	1
Homatropine		9.7	1
Hycanthone		3.4	1
Hydralazine		0.5, 7.1	1
Hydrochlorothiazide	7.0, 9.2		1
Hydrocodone		8.9	1
Hydrocortisone sodium succinate	5.1		1
Hydroflumethiazide	8.9, 10.7		1
Hydromorphone		8.2	1
Hydroquinone	10.0, 12.0		1
Hydroxyamphetamine		9.3	1
Hydroxyzine		2.0, 7.1	1
Hyoscyamine		9.7	1
Ibuprofen	5.2		1
Idoxuridine	8.3		1
Imipramine		9.5	1
Indapamide	8.8		5
Indomethacin	4.5		1
Indoprofen	5.8		1
Indoramin		7.7	1
Iocetamic acid[g]	4.1 or 4.3		4
Iodipamide	3.5		1
Iodoquinol	8.0		1
Iopanoic acid	4.8		4
Iprindole	8.2		1
Ipronidazole		2.7	1
Isocarboxazid		10.4	1
Isoniazid		2.0, 3.5, 10.8	1
Isoproterenol	10.1, 12.1	8.6	1
Isoxsuprine	9.8	8.0	1
Kanamycin		7.2	1
Ketamine		7.5	1,11
Ketobemidone		8.7	1
Ketoconazole		2.9, 6.5	1,4
Ketoprofen[h]	4.8		1,9
Labetalol	8.7	7.4	1
Leucovorin	3.1, 8.1, 10.4		1
Levallorphan tartrate	4.5	6.9	1
Levobunolol		9.2	1
Levodopa	2.3, 9.7, 13.4	8.7	1
Levomethorphan		8.3	1
Levomoramide		7.0	1
Levonordefrin	9.8	8.6	1
Levopropoxyphene		6.3	1
Levorphanol		9.2	1

(continued)

Table A–1. *(Continued)*

Drug Substance	pK$_a$ Values		Reference
	HA	HB$^+$	
Levothyroxine	2.2, 6.7	10.1	1
Lidocaine		7.8	1
Lincomycin		7.5	1
Liothyronine	8.4		1
Lisinopril	1.7, 3.3, 11.1	7.0	1
Loperamide		8.6	1
Lorazepam	11.5	1.3	1
Loxapine		6.6	1
Lysergide		7.5	1
Maprotiline		10.2	4
Mazindol		8.6	1
Mecamylamine		11.2	1
Mechlorethamine		6.4	1
Meclizine		3.1, 6.2	1
Meclofenamic acid	4.0		4
Medazepam		6.2	1
Mefenamic acid	4.2		1
Mepazine		9.3	1
Meperidine		8.7	1
Mephentermine		10.4	1
Mephobarbital	7.7		1
Mepindolol		8.9	1
Mepivacaine		7.7	1
Mercaptomerin	3.7, 5.1		1
Mercaptopurine	7.8	11.0	1
Mesalamine	2.7	5.8	1
Mesna	9.1		1
Metaproterenol	11.8	8.8	1
Metaraminol		8.6	1
Methacycline	3.5, 7.6	9.5	1
Methadone		8.3	1
Methamphetamine		10.0	1
Methapyrilene		3.7, 8.9	1
Methaqualone		2.5	1
Metharbital	8.2		1
Methazolamide	7.3		1
Methdilazine		7.5	1
Methenamine		4.8	4
Methicillin	3.0		1
Methohexital	8.3		1
Methotrexate	3.8, 4.8	5.6	1
Methotrimeprazine		9.2	1
Methoxamine		9.2	1
Methoxyphenamine		10.1	1
Methyclothiazide	9.4		1
Methyl nicotinate		3.1	1
Methyl paraben	8.4		1
Methyl salicylate	9.9		1
Methyldopa	2.3, 10.4, 12.6	9.2	1
Methylergonovine		6.6	1
Methylphenidate		8.8	1

Table A–1. *(Continued)*

Drug	pK$_a$ Values		Reference
	HA	HB$^+$	
Methylthiouracil	8.2		1
Methyprylon	12.0		1
Methysergide		6.62	1
Metoclopramide		0.6, 9.3	1
Metolazone	9.7		1
Metopon		8.1	1
Metoprolol		9.7	1
Metronidazole		2.6	4
Metyrosine	2.7, 10.1		1
Mexiletine		9.1	1
Mezlocillin	2.7		1
Miconazole		6.7	1
Midazolam		6.2	1
Minocycline	2.8, 5.0, 7.8	9.5	1
Minoxidil		4.6	4
Mitomycin		10.9	1
Molindone		6.9	1
Morphine	9.9	8.0	1
Moxalactam	2.5, 7.7, 10.2		4
Nabilone[b]	13.5		9
Nadolol		9.4	5
Nafcillin	2.7		1
Nalbuphine	10.0	8.7	4
Nalidixic acid	6.0		1
Nalorphine		7.8	1
Naloxone		7.9	1
Naphazoline		10.9	1
Naproxen	4.2		1
Natamycin	4.6	8.4	8
Neostigmine		12.0	1
Niacin	2.0	4.8	1
Nicotinamide		0.5, 3.4	1
Nicotine		3.1, 8.0	1
Nikethamide		3.5	1
Nitrazepam	10.8	3.2	1
Nitrofurantoin	7.1		1
Norcodeine		5.7	1
Nordefrin	9.8	8.5	1
Norepinephrine	9.8, 12.0	8.6	1,2
Norfenephrine		8.7	1
Normorphine		9.8	1
Nortriptyline		9.7	1
Noscapine		6.2	1
Novobiocin	4.3, 9.1		1
Nystatin[i]	8.9	5.1	11
Octopamine	9.5	8.9	1
Orphenadrine		8.4	1
Oxacillin	2.7		1
Oxazepam	11.6	1.8	1
Oxprenolol		9.5	5
Oxybutynin		7.0	4
Oxycodone		8.9	1

(continued)

Table A–1. *(Continued)*

Drug Substance	pK$_a$ Values		Reference
	HA	HB$^+$	
Oxymorphone	9.3	8.5	1
Oxyphenbutazone	4.7		1
Oxypurinol	7.7		1
Oxytetracycline[c]	3.3, 7.3	9.1	1
Pamaquine		1.3, 3.5, 10.0	1
Papaverine		6.4	1
Pargyline		6.9	1
Pemoline		10.5	1,2
Penbutolol[c]		9.3	1
Penicillamine	1.8, 10.5	7.9	1
Penicillin G	2.8		1
Penicillin V	2.7		1
Pentamidine		11.4	4
Pentazocine	10.0	8.5	2,9
Pentobarbital	8.1		1
Pentoxiphylline		0.3	1
Perphenazine		3.7, 7.8	1
Phenacetin		2.2	1
Phenazocine		8.5	1
Phencyclidine		8.5	2
Phendimetrazine		7.6	1
Phenethicillin	2.8		1
Phenformin		2.7, 11.8	1
Phenindamine		8.3	1
Phenindione	4.1		1
Pheniramine		4.2, 9.3	1
Phenmetrazine		8.5	1
Phenobarbital	7.4		1
Phenolphthalein	9.7		1
Phenolsulfonphthalein	8.1		1
Phenothiazine		2.5	1
Phenoxybenzamine		4.4	4
Phenoxypropazine		6.9	1
Phentermine		10.1	1
Phentolamine		7.7	1
Phenylbutazone	4.5		1
Phenylephrine	10.1	8.8	1
Phenylpropanolamine		9.4	1
Phenyltoloxamine		9.1	1
Phenyramidol		5.9	1
Phenytoin	8.3		1
Physostigmine		2.0, 8.2	1
Pilocarpine		1.6, 7.1	1
Pimozide		7.3, 8.6	1
Pindolol		8.8	1
Piperazine		5.6, 9.8	1
Pipradrol		9.7	1
Pirbuterol		3.0, 7.0, 10.3	1
Piroxicam	4.6		1
Pivampicillin		7.0	1
Polymyxin		8.9	1
Polythiazide	9.8		1
Practolol		9.5	1
Pralidoxime		7.9	1

Table A–1. *(Continued)*

Drug Substance	pKa Values		Reference
	HA	HB^+	
Pramoxine		6.2	1
Prazepam		2.9	1
Prazosin		6.5	1
Prenalterol	10.0	9.5	1
Prilocaine		7.9	1
Probenecid	3.4		1
Procainamide		9.2	1
Procaine		8.8	1
Procarbazine		6.8	1
Prochlorperazine		3.7, 8.1	1
Promazine		9.4	1
Promethazine		9.1	1
Proparacaine		3.2	11
Propiomazine		9.1	1
Propoxycaine		8.6	1
Propoxyphene		6.3	1
Propranolol		9.5	1
Propylhexedrine		10.4	1
Propylthiouracil	7.8		1
Pseudoephedrine		9.5	1
Pyrathiazine		8.9	1
Pyrazinamide		0.5	1
Pyridoxine	8.96	5.0	1
Pyrilamine		4.0, 8.9	1
Pyrimethamine		7.3	1
Pyrrobutamine		8.8	1
Quinacrine		8.2, 10.2	1
Quinethazone	9.3, 10.7		1
Quinidine		4.2, 7.9	1
Quinine		4.2, 8.5	1
Ranitidine		2.3, 8.2	4
Rescinnamine		6.4	4
Reserpine		6.6	1
Rifampin	1.7	7.9	1
Rimoterol	10.3	8.7	1
Ritodrine		9.0	1
Rolitetracycline	7.4		1
Rotoxamine		8.1	1
Saccharin	1.6		1
Salicylamide	8.2		3
Salicylic acid	3.0, 13.4		1
Salsalate	3.5, 9.8		1,2
Scopolamine		7.6	1
Secobarbital	7.9, 12.6		1
Serotonin	9.8	4.9, 9.1	1
Sotalol	8.5	9.8	1
Sparteine		4.8, 12.0	1
Spiperone		8.3, 9.1	1
Streptozocin		1.3	4
Strychnine		2.3, 8.0	1
Succinylsulfathiazole	4.5		1
Sufentanil		8.0	4
Sulfacetamide	5.4	1.8	1
Sulfadiazine	6.5	2.0	1

(continued)

Table A–1. *(Continued)*

Drug Substance	pK$_a$ Values		Reference
	HA	HB^+	
Sulfadimethoxine	6.7	2.0	1
Sulfaguanidine	12.1	2.8	1
Sulfamerazine	7.1	2.3	1
Sulfamethazine	7.4	2.4	1
Sulfamethizole	5.5	2.0	1
Sulfamethoxazole	5.6		1
Sulfaphenazole	6.5	1.9	1
Sulfapyridine	8.43	2.6	1
Sulfasalazine	2.4, 9.7, 11.8		1
Sulfathiazole	7.1	2.4	1
Sulfinpyrazone	2.8		1
Sulfisoxazole	5.0		1
Sulindac	4.5		1
Sulpiride		9.1	1
Sulthiame	10.0		1
p-Synephrine	10.2	9.3	1
Talbutal	7.8		1
Tamoxifen		8.9	4
Temazepam		1.6	1
Terbutaline	10.1, 11.2	8.8	1
Tetracaine		8.4	1
Tetracycline	3.3, 7.7	9.7	1
Tetrahydrocannabinol (THC)	10.6		2
Tetrahydrozoline		10.5	1
Thenyldiamine		3.9, 8.9	1
Theobromine	10.1	0.1	1
Theophylline	8.6	3.5	1
Thiabendazole		4.7	4
Thiamine		4.8, 9.0	1
Thiamylal	7.5		1
Thioguanine	8.2		3
Thiopental	7.5		1
Thiopropazate		3.2, 7.2	1
Thioridazine		9.5	1
Thiothixene		7.7, 7.9	1
Thiouracil	7.5		1
Thonzylamine		2.2, 9.0	1
L-Thyroxine	2.2, 6.7	10.1	1
Ticarcillin	2.6, 3.4		1
Ticrynafen	2.7		1
Timolol		8.8	1
Timoprazole		3.1, 8.8	1
Tiotidine		6.8	1
Tiprofenic acid	3.0		1
Tobramycin		6.7, 8.3, 9.9	2
Tocainide		7.5	1
Tolamolol		7.9	5
Tolazamide	3.1	5.7	1
Tolazoline		10.3	1
Tolbutamide	5.4		1
Tolmetin	3.5		1
Tramzoline		10.7	1
Tranylcypromine		8.2	1
Trazodone		6.7	1

Table A–1. *(Continued)*

Drug Substance	pKa Values		Reference
	HA	HB^+	
Triamterene		6.2	1
Trichlormethiazide	8.6		1
Trifluperazine		3.9, 8.1	1,3
Triflupromazine		9.2	1
Trimeprazine		9.0	1
Trimethobenzamide		8.3	1
Trimethoprim		6.6	3
Trimipramine		8.0	4
Tripelennamine		4.2, 8.7	1
Triprolidine		6.5	1
Troleandomycin		6.6	1
Tromethamine		8.1	1
Tropicamide		5.3	1
Tuaminoheptane		10.5	1
Tubocurarine		8.1, 9.1	1
Tyramine	10.9	9.3	1
Valproic acid	4.8		1,3
Verapamil		8.9	1
Vidarabine		3.5, 12.5	1
Viloxazine		8.1	1
Vinbarbital	8.0		1
Vinblastine		5.4, 7.4	1
Vincristine[j]		5.0, 7.4	1
Vindesine		6.0, 7.7	1
Warfarin	5.1		1
Xylometazoline		10.2	1
Zimeldine		3.8, 8.74	1

Miscellaneous Organic Acids and Bases

Substance	HA	HB^+
Acetic acid	4.8	
Allylamine		10.7
6-Aminopenicillanic acid	2.3	4.9
Ammonia		9.3
Aniline		4.6
Benzoic acid	4.2	
Benzyl alcohol	18.0	
Benzylamine		9.3
Butryic acid	4.8	
Carbonic acid	6.4, 10.4	
Citric acid	3.1, 4.8, 5.4	
Diethanolamine		8.9
Diethylamine		11.0
Dimethylamine		10.7
p-Dimethylaminobenzoic acid	5.1	
Ethanol	15.6	
Ethanolamine		9.5
Ethylamine		10.7
Ethylenediamine		7.2, 10.0
Fumaric acid	3.0, 4.4	

(continued)

Table A–1. *(Continued)*

Miscellaneous Organic Acids and Bases	pK_a Values HA	HB$^+$
Gluconic acid	3.6	
Glucuronic acid	3.2	
Guanidine		13.6
Imidazole		7.0
Isopropylamine		10.6
Lactic acid	3.9	
Maleic acid	1.9	
Mandelic acid	3.4	
Monochloroacetic acid	2.9	
N-propylamine		10.6
Nitromethane	11.0	
Phenol	9.9	
Phthalic acid	2.9	
Resorcinol	9.2, 11.3	
Sorbic acid	4.8	
Succinimide	9.6	
Tartaric acid	3.0, 4.4	
p-Toluidine		5.3
Trichloroacetic acid	0.9	
Triethanolamine		7.8
Triethylamine		10.7
Tropic acid	4.1	
Tropine		10.4
Uric acid	5.4	

[a] Determined in methyl cellosolve-water, 8:2 w/w mixture.
[b] Determined in 66% dimethylformamide.
[c] Determined in 25–30% ethanol.
[d] Determined in 40% methanol.
[e] Prostaglandin $F_{2\alpha}$.
[f] Spectrophotometric determination.
[g] The pK_a values are 4.1 for two and 4.25 for two of the four optical isomers.
[h] Determined in methanol-water, 3:1 mixture.
[i] Determined in dimethylformamide-water, 1:1 mixture.
[j] Determined in 33% dimethylformamide.

Table A-2. pH Values for Tissue Fluids

Fluid	pH
Aqueous humor	7.2
Blood, arterial	7.4
Blood, venous	7.4
Blood, maternal umbilical	7.3
Cerebrospinal fluid	7.4
Duodenum	5.5
Feces[a]	7.12 (4.6–8.8)
Ileum, distal	8.0
Intestine, microsurface	5.3
Lacrimal fluid (tears)	7.4
Milk, breast	7.0
Muscle, skeletal[b]	6.0
Nasal secretions	6.0
Prostatic fluid	6.5
Saliva	6.4
Semen	7.2
Stomach	1.0–3.5
Sweat	5.4
Urine	5.8 (5.5–7.0)
Vaginal secretions, premenopause	4.5
Vaginal secretions, postmenopause	7.0

[a] Value for normal soft, formed stools; hard stools tend to be more alkaline, whereas watery, unformed stools are acidic.
[b] Studies conducted intracellularly on the rat.

REFERENCES

1. C. Hansch, et al., *Comprehensive Medicinal Chemistry,* Vol. 6, New York, Pergamon Press, 1990.
2. A. Albert and E. P. Serjeant, *The Determination of Ionization Constants,* 3rd ed., New York, Chapman and Hall, 1984.
3. Merck Index, 11th ed., Rahway, N.J., Merck and Co., 1989.
4. G. K. McEvoy, Ed., *American Hospital Formulary Service: Drug Information 94,* Bethesda, American Society of Hospital Pharmacists, 1994.
5. P. H. Welling, et al., *J. Pharmacokinet. Biopharm., 12,* 263(1984).
6. K. Florey, Ed., *Analytical Profiles of Drug Substances,* Vol. 13, New York, Academic Press, 1984.
7. Ibid., Vol. 8, 1979.
8. Ibid., Vol. 9, 1980.
9. Ibid., Vol. 10, 1981.
10. Ibid., Vol. 11, 1982.
11. Ibid., Vol. 6, 1977.
12. Ibid., Vol. 7, 1978.
13. Ibid., Vol. 12, 1983.
14. Ibid., Vol. 5, 1976.
15. Ibid., Vol. 3, 1974.
16. Ibid., Vol. 2, 1973.

SUGGESTED READINGS

A. Albert and E. P. Serjeant, *The Determination of Ionization Constants,* 3rd ed., New York, Chapman and Hall, 1984.

A. Martin, *Physical Pharmacy,* 4th ed., Philadelphia, Lea & Febiger, 1993.

INDEX

Page numbers in *italics* refer to figures and chemical structures; those followed by the letter "t" refer to tables.

D

G

I